Structural Inorganic Chemistry

Structural Inorganic Chemistry

A. F. WELLS

FIFTH EDITION

CLARENDON PRESS · OXFORD 1984

Oxford University Press, Walton Street, Oxford OX2 6DP

London Glasgow New York Toronto
Delhi Bombay Calcutta Madras Karachi
Kuala Lumpur Singapore Hong Kong Tokyo
Nairobi Dar es Salaam Cape Town
Melbourne Wellington

and associate companies in Beirut Berlin Ibadan Mexico City

Published in the United States by
Oxford University Press, New York

Oxford is a trademark of Oxford University Press

© Oxford University Press 1975, 1984

First published   1945
Second edition   1950
Third edition   1962
Fourth edition   1975
Fifth edition   1984

British Library Cataloguing in Publication Data
Wells, A. F.
Structural inorganic chemistry. — 5th ed.
1. Chemistry, Inorganic
I. Title
546   QD151.2
ISBN 0-19-855370-6

Library of Congress Cataloging in Publication Data
Wells A. F. (Alexander Frank), 1912–
Structural inorganic chemistry.
Bibliography: p.
Includes index.
1. Chemical structure.   2. Stereochemistry.
3. Crystallography.   I. Title.
QD481.W44 1983   546'.252   82-18866
ISBN 0-19-855370-6 (Oxford University Press)

Printed in Great Britain by
The Thetford Press Ltd., Thetford, Norfolk

Preface

In the introduction to the first (1945) edition the author stated his conviction that the structural side of inorganic chemistry cannot be put on a sound basis until the knowledge gained from the study of the solid state has been incorporated into chemistry as an integral part of the subject, and that it is not sufficient merely to add information about the structures of solids to the descriptions of elements and compounds as usually presented in a systematic treatment of inorganic chemistry. Since the results of structural studies of crystals are described initially in crystallographic language, the first requirement is that these results be made available in a form intelligible to chemists. It was this challenge that first attracted the author, and it is hoped that this book will continue to provide teachers of chemistry with facts and ideas which can be incorporated into their teaching. However, while the incorporation of even the most meagre information about the structures of solids into the conventional teaching of chemistry is to be welcomed, a real understanding of the structures of crystals and of the relations between different structures is not possible without a knowledge of certain basic geometrical and topological facts and concepts. This essential background material includes the properties of polyhedra, the nature and symmetry of repeating patterns, and the ways in which spheres, of the same or different sizes, can be packed together. Because many students find difficulty in appreciating three-dimensional structures from two-dimensional illustrations (even stereoscopic photographs) the examination of, and preferably also the construction of, models should play a large part in the study of these subjects.

The general plan of the book is as follows. Part I deals with a number of general topics, including those mentioned above, and is intended as an introduction to the more detailed Part II, which forms the larger part of the book. In Part II the structural chemistry of the elements is described systematically, and the arrangement of material is based on the groups of the Periodic Table. The author believes that the numerous revisions and additions result in a reasonable picture of the subject at the end of the eighth decade of the twentieth century. It is evident that there are specialists who are better qualified than the author to write many of the chapters of a book which attempts to cover such a large field. Over the years, however, a number of good friends of the author have expressed the opinion that there is some virtue in having a book written by a single author, if only because this ensures a uniform style of writing. There may be readers who would like to see more thorough treatments of some topics which have become of great scientific or technological interest in recent years. For example, very careful and detailed studies have been made of the structures of certain groups of crystalline inorganic compounds with the object of correlating structure with physical properties. Also, there are certain types of crystal structure which are of crystallographic rather than chemical interest (for example, shear structures and 'infinitely adaptive' structures). The author can do little more than mention structures of these types in a book which is addressed to chemists and attempts a survey, necessarily sketchy in parts, of the structures of elements and compounds not only in the solid state but also in the gaseous state and, to a very limited extent, in the liquid state.

References. The present volume has never been intended as a work of reference, though it may serve as a useful starting-point when information is required on a particular topic. As an essential part of the educational process the advanced student should be encouraged to adopt a critical attitude towards the written word (including the present text); he must learn where to find the original literature and to begin to form his own judgment of the validity of conclusions drawn from experimental data. The numerous references to the scientific literature included in Part II generally refer to the latest work; these usually include references to earlier work. To save space the names of scientific journals have been abbreviated to the forms listed on pp. xxix–xxxi.

Indexes. There are two indexes. The arrangement of entries in the formula index is not entirely systematic, for there is no wholly satisfactory way of indexing inorganic compounds which retains chemically acceptable groupings of atoms. The formulae have been arranged so as to emphasize the feature most likely to be of interest to the chemist. The subject index is largely restricted to names of minerals and organic compounds and to topics which are not readily located in the list of contents.

Acknowledgements. During the writing of this book, which of necessity owes much to the work and ideas of other workers in this and related fields, I have had the benefit of helpful discussions with a number of colleagues, of whom I would particularly mention Dr B. L. Chamberland. I wish to thank Dr B. G. Bagley and the editor of *Nature* (London) for permission to use Fig. 4.3, Dr H.T. Evans and John Wiley and Sons for Figs. 11.5(c), 11.7, 11.10, 11.11. and 11.13(b), Dr. H.G. von Schnering for Fig. 19.4(e), and Drs G. T. Kokotailo and W. M. Meier for Fig. 23.27. It gives me great pleasure to acknowledge the debt that I owe to my wife for her support and encouragement over a period of many years.

Storrs, Connecticut   A.F.W.
1982

Contents

PART II

Abbreviations

The following abbreviations are used in references to Journals throughout this book.

AANL	Atti dell'Accademia nazional dei Lincei
AC	Acta crystallographica
AcM	Acta Metallurgica
ACSc	Acta Chemica Scandinavica
ACSi	Acta Chimica Sinica
AJC	Australian Journal of Chemistry
AJSR	Australian Journal of Scientific Research
AK	Arkiv för Kemi
AKMG	Arkiv för Kemi, Mineralogi och Geologi
AlC	Analytical Chemistry
AM	American Mineralogist
AnC(IE)	Angewandte Chemie (International Edition)
AP	Annalen der Physik
APURSS	Acta Physicochimica URSS
ARPC	Annual Review of Physical Chemistry
ASR	Applied Scientific Research
B	Berichte
BB	Berichte der Bunsengesellschaft für physikalische Chemie
BCSJ	Bulletin of the Chemical Society of Japan
BSCB	Bulletin des Sociétés chimiques Belges
BSCF	Bulletin de la Société chimique de France
BSFMC	Bulletin de la Société française de minéralogie et de cristallographie
C	Chimia (Switzerland)
CB	Chemische Berichte
CC	Chemical Communications (Journal of the Chemical Society, Chemical Communications)
CJC	Canadian Journal of Chemistry
CJP	Canadian Journal of Physics
CR	Comptes rendus hebdomadaires des Séances de l'Académie des Sciences (Paris)
CRURSS	Comptes rendus de l'Academie des Sciences de l'URSS
CSR	Chemical Society Reviews
DAN	Doklady Akademii Nauk SSSR
E	Experientia
FM	Fortschritte der Mineralogie
GCI	Gazzetta chimica italiana
HCA	Helvetica Chimica Acta
IC	Inorganic Chemistry
ICA	Inorganica Chimica Acta
IEC	Industrial and Engineering Chemistry
INCL	Inorganic and Nuclear Chemistry Letters
JACS	Journal of the American Chemical Society
JACeS	Journal of the American Ceramic Society

JACr	Journal of Applied Crystallography
JAP	Journal of Applied Physics
JCG	Journal of Crystal Growth
JCMS	Journal of Crystal and Molecular Structure
JCP	Journal of Chemical Physics
JCS	Journal of the Chemical Society (London)
JES	Journal of the Electrochemical Society
JINC	Journal of Inorganic and Nuclear Chemistry
JLCM	Journal of the Less-common Metals
JM	Journal of Metals
JMS	Journal of Molecular Spectroscopy
JMSt	Journal of Molecular Structure
JNM	Journal of Nuclear Materials
JOC	Journal of Organometallic Chemistry
JPC	Journal of Physical Chemistry
JPCS	Journal of the Physics and Chemistry of Solids
JPP	Journal de Physique (Paris)
JPSJ	Journal of the Physical Society of Japan
JSSC	Journal of Solid State Chemistry
K	Kristallografiya
KDV	Kongelige Danske Videnskabernes Selkab Matematisk-fysiske Meddeleser
MH	Monatshefte für Chemie und verwandte Teile anderer Wissenschaften
MJ	Mineralogical Journal of Japan
MM	Mineralogical Magazine (and Journal of the Mineralogical Society)
MMJ	Mineralogical Magazine (Japan)
MRB	Materials Research Bulletin
MSCE	Mémorial des Services chimiques de l'état (Paris)
N	Nature
NBS	Journal of Research of the National Bureau of Standards
NF	Naturforschung
NJB	Neues Jahrbuch für Mineralogie
NPS	Nature (Physical Sciences)
NW	Naturwissenschaften
PCS	Proceedings of the Chemical Society
PKNAW	Proceedings koninklijke nederlandse Akademic van Wetenschappen
PM	Philosophical Magazine
PNAS	Proceedings of the National Academy of Sciences of the U.S.A.
PR	Physical Review
PRL	Physical Review Letters
PRR	Philips Research Reports
PSS	Physica Status Solidi
QRCS	Quarterly Reviews of The Chemical Society
RCR	Revue de Chimie Minérale
RJIC	Russian Journal of Inorganic Chemistry
RMP	Reviews of Modern Physics
RPAC	Reviews of Pure and Applied Chemistry (Royal Australian Chemical Institute)
RS	Ricerca scientifica
RTC	Recueil des Travaux chimiques des Pays-Bas et de la Belgique
SA	Spectrochimica Acta
Sc	Science

SMPM	Schweizerische mineralogische und petrographische Mitteilungen
SPC	Soviet Physics: Crystallography
SR	Structure Reports
SSC	Solid State Communications
TAIME	Transactions of the American Institute of Mining and Metallurgical Engineers
TFS	Transactions of the Faraday Society
TKBM	Tidsskrift for Kjemi, Bergvesen og Metallurgi
ZaC	Zeitschrift für anorganische (und allgemeine) Chemie
ZE	Zeitschrift für Elektrochemie
ZFK	Zhurnal fizicheskoi Khimii
ZK	Zeitschrift für Kristallographie
ZN	Zeitschrift für Naturforschung
ZP	Zeitschrift für Physik
ZPC	Zeitschrift für physikalische Chemie
ZSK	Zhurnal strukturnoi Khimii

Part I

1

Introduction

In this introductory chapter we discuss in a general way a number of topics which are intended to indicate the scope of our subject and the reasons for the choice of topics which receive more detailed attention in subsequent chapters.

The number of elements known exceeds one hundred, so that if each one combined with each of the others to form a single binary compound there would be approximately five thousand such compounds. In fact not all elements combine with all the others, but on the other hand, some combine to form more than one compound. This is true of many pairs of metals, and other examples, chosen at random, include:

$$YB_2, \ YB_4, \ YB_6, \ YB_{12}, \text{and } YB_{66};$$
$$CrF_2, \ CrF_3, \ CrF_4, \ CrF_5, \ CrF_6, \text{ and } Cr_2F_5;$$
$$CrS, \ Cr_7S_8, \ Cr_5S_6, \ Cr_3S_4, \text{ and } Cr_2S_3.$$

The number of binary compounds alone is evidently considerable, and there is an indefinitely large number of compounds built of atoms of three or more elements. It seems logical to concentrate our attention first on the simplest compounds such as binary halides, chalconides, etc., for it would appear unlikely that we could understand the structures of more complex compounds unless the structures of the simpler ones are known and understood. However, it should be noted that simplicity of chemical formula may be deceptive, for the structures of many compounds with simple chemical formulae present considerable problems in bonding, and indeed the structures of some elements are incomprehensibly complex (for example, B and red P). On the other hand, there are compounds with complex formulae which have structures based on an essentially simple pattern, as are the numerous structures described in Chapter 3 which are based on the diamond net, one of the simplest 3-dimensional frameworks. We shall make a point of looking for the simple underlying structural themes in the belief that Nature prefers simplicity to complexity and also because structures are most easily understood if reduced to their simplest terms.

The importance of the solid state

Since we shall devote most of the first part of this book to matters directly concerned with the solid state it is appropriate to note a few general points, to some of which we return later in this chapter.

(i) Most of the elements (some 90 per cent) are solids at ordinary temperatures, and this is also true of the majority of inorganic compounds. It happens that most of the important reagents are liquids, gases, or solutions, but these constitute a small proportion of the more common inorganic compounds. Also, although it is true that chemical reactions are usually carried out in solution or in the vapour state, in most reactions the reactants or products, or both, are solids. Chemical reactions range from those between isolated atoms or discrete groups of atoms (molecules or complex ions), through those in which a solid is removed or produced, to processes such as the corrosion of metals where a solid product builds up on the surface of the solid reactant. In all cases where a crystalline material is formed or broken down, the process involves the lattice energy of the crystal. The familiar Born–Haber cycle for the reaction between solid sodium and gaseous chlorine to form solid NaCl provides a simple example of the interrelation of heats of dissocation, ionization energy and affinity, lattice energy, and heat of reaction.

(ii) Organic compounds (other than polymers) exist as finite molecules in all states of aggregation. This means, first, that the structural problem consists only in discovering the structure of the finite molecule, and second, that this could in principle be determined by studying its structure in the solid, liquid, or vapour state. Apart from possible geometrical changes such as rotation about single bonds and small dimensional changes due to temperature differences, the basic topology and geometry could be studied in any state of aggregation. Some inorganic compounds also exist as finite molecules in the solid, liquid, and gaseous states, for example, many simple molecules formed by non-metals (HCl, CO_2) and also some compounds of metals (SnI_4, $Cr(CO)_6$). Accurate information about the structures of simple molecules, both organic and inorganic, comes from spectroscopic and electron diffraction studies of the vapours, but these methods are not applicable to very complex molecules. Because crystalline solids are periodic structures they act as diffraction gratings for X-rays and neutrons, and in principle the structure of any molecule, however, complex, can be determined by diffraction studies of the solid.

In contrast to organic compounds and the minority of inorganic compounds mentioned above, the great majority of solid inorganic compounds have structures in which there is linking of atoms into systems which extend indefinitely in one, two, or three dimensions. Such structures are characteristic only of the solid state and must necessarily break down when the crystal is dissolved, melted, or vaporized. The study of crystal structures has therefore extended the scope of structural chemistry far beyond that of the finite groups of atoms to which classical stereochemistry was restricted to include all the periodic arrangements of atoms found in crystalline solids.

Because the great majority of inorganic compounds are compounds of one or more metals with non-metals, and because most of them are solids under ordinary conditions, the greater part of structural inorganic chemistry is concerned with the structures of solids. The only compounds of metals which have any structural chemistry, apart from that of the crystalline compound, are those molecules or ions that can be studied in solution or the molecules of compounds that can be melted or vaporized without decomposition. It is unlikely that very much accurate structural information will ever be obtained from liquids, whereas electron diffraction or spectroscopic studies can be made of molecules in the gas phase provided they are not too complex. It is important therefore to distinguish between solid compounds which can be vaporized without decomposition and those which can exist *only as solids*. By this we mean that their existence depends on types of bonding which are possible only in the solid state. Many simple halides and some oxides of metals have been studied as vapours, as also have some salts (for example, $LiBeF_3$, $NaAlF_4$, $LiBO_2$, $Cu(NO_3)_2$ and K_2SO_4). Since in most cases the (molecular) vapour species is not present in the crystal the information so obtained is complementary to that obtained by studying the solid. On the other hand, many simple compounds M_mX_n are unlikely to exist in the vapour state because the particular ratio of metal to non-metal atoms is only realizable in an infinite array of atoms between which certain types of bonding can operate. Crystalline Cs_2O consists of infinite layers, but nevertheless we can envisage molecules of Cs_2O in the vapour. However, oxides such as Cs_3O and Cs_7O depend for their existence on extended systems of metal-metal bonds which would not be possible in a finite molecule.

Large and important groups of compounds which can exist only in the crystalline state include complex halides and oxides, 'acid' and 'basic' salts, and hydrates. One particularly important result of the study of crystal structures has been the recognition that *non-stoichiometric compounds* are not the rarities they were once thought to be. A non-stoichiometric compound may be very broadly defined as a solid phase which is stable over a range of composition. This definition covers at one extreme all cases of 'isomorphous replacement' and all kinds of solid solution, the composition of which may cover the whole range from one pure component to the other. At the other extreme there are phosphors (luminescent ZnS or ZnS–Cu), which owe their properties to misplaced and/or impurity atoms which act as 'electron traps', and coloured halides (alkali or alkaline-earth) in which some of the halide-ion-sites are occupied by electrons (F-centres); these defects are present in very small concentration, often in the range 10^{-6}–10^{-4}. Of more interest to the inorganic chemist is the fact that many simple binary compounds exhibit ranges of composition, the range depending on the temperature and mode of preparation. The non-stoichiometry implies disorder in the structure and usually the presence of an element in more than one valence state, and can give rise to semiconductivity and catalytic activity. Examples of non-stoichiometric binary compounds include many oxides and sulphides, some hydrides, and interstitial solid solutions of C and N in metals. More complex examples include various complex oxides with layer and framework structures, such as the bronzes (p. 612). The existence of green and black NiO, with very different physical properties, the recent preparation for the first time of stoichiometric FeO, and the fact that Fe_6S_7 is not FeS containing excess S but FeS deficient in Fe (that is, $Fe_{1-x}S$) are matters of obvious importance to the inorganic chemist.

The compositions and properties and indeed the very existence of non-stoichiometric compounds can be understood only in terms of their structures. This is particularly evident in cases where the non-stoichiometry arises from the inclusion of foreign atoms or molecules in a crystalline structure. It can occur in crystals built of finite molecules or crystals containing large finite ions. For example, if $Pd_2Br_4[As(CH_3)_3]_2$ is crystallized from dioxane the crystals can retain non-stoichiometric amounts of the solvent in the tunnels between the molecules, and these molecules can be removed without disruption of the structure. In the mineral beryl (p. 1021) the large cyclic $(Si_6O_{18})^{12-}$ ions are stacked in columns, and helium can be occluded in the tunnels. Some crystals with layer structures can take up material between the layers. Examples include the lamellar compounds of graphite (p. 923) and of clay minerals (p. 1031). An unusual type of layer structure is that of $Ni(CN)_2 . NH_3$ which can take up molecules of H_2O, C_6H_6, $C_6H_5NH_2$, etc. between the layers (Fig. 1.9(a)). Structures of this kind are called 'clathrates', and examples are noted on p. 30.

(iii) The great wealth of information about atomic arrangement in crystals and in particular the detailed information about bond lengths and interbond angles provided by studies of crystal structures is the raw material for the theoretician interested in bonding and its relation to physical properties.

All elements and compounds can be solidified under appropriate conditions of temperature and pressure, and the properties and structures of solids show that we must recognize four extreme types of bonding:

(a)   the polar (ionic) bond in crystalline salts such as NaCl or CaF_2;

(b)   the covalent bond in non-ionizable molecules such as Cl_2, S_8, etc., which exist in both the crystalline elements and also in their vapours, and in crystals such as diamond in which the length of the C—C bond is the same as in molecules such as $H_3C–CH_3$;

(c)   the metallic bond in metals and intermetallic compounds (alloys), which is responsible for their characteristic optical and electrical properties; and

(d)   the much weaker van der Waals bond between chemically saturated molecules such as those just mentioned—witness the much larger distances between atoms of different molecules as compared with those within such molecules. In crystalline Cl_2 the bond length is 1.99 Å, but the shortest distance between Cl atoms of different molecules is 3.34 Å. The van der Waals bond is responsible for the cohesive forces in liquid or solid argon or chlorine and more generally between neutral molecules, chains, and layers in numerous crystals whose structures will be described later.

Although it is convenient and customary to recognize these four extreme types of bonding it should be realized that bonds of these 'pure' types—if indeed the term 'pure' has any clear physical or chemical meaning—are probably rather rare, particularly in the case of the first two types. Bonds of an essentially ionic type occur in salts formed from the most electropositive combined with the most electronegative elements and between, for example, the cations and the O atoms of the complex ion in oxy-salts such as $NaNO_3$. Covalent bonds occur in the non-metallic elements and in compounds containing non-metals which do not differ greatly in 'electronegativity' (see p. 289). However, it would seem that the great majority of bonds in inorganic compounds must be regarded as intermediate in character between these extreme types. For example, most bonds between metals and non-metals have some ionic and some covalent character, and at present there is no entirely satisfactory way of describing such bonds.

Evidently many crystals contain bonds of two or more quite distinct types. In molecular crystals consisting of non-polar molecules the bonds within the molecule may be essentially covalent (e.g. S_6 or S_8) or of some intermediate ionic–covalent nature (e.g. SiF_4), and those between the molecules are van der Waals bonds. In a crystal containing complex ions the bonds within the complex ion may approximate to covalent bonds while those between the complex ion and the cations (or anions) are essentially ionic in character, as in the case of $NaNO_3$ already quoted. In other crystals there are additional interactions between certain of the atoms which are not so obviously essential as in these cases to the cohesion of the crystal. An example is the metal–metal bonding in dioxides with the rutile structure, a structure which in many cases is stable in the absence of such bonding.

It is also necessary to recognize certain other types of interactions which, although weaker than ionic or covalent bonds, are important in determining or influencing the structures of particular groups of crystalline compounds — for example, hydrogen bonds (bridges) and charge-transfer bonds. Hydrogen bonds are of rather widespread occurrence and are discussed in more detail in later chapters.

(iv) It is perhaps unnecessary to emphasize here that there is in general no direct relation between the chemical formula of a solid and its structure. For example, only the first member of the series

HI	AuI	CuI	NaI	CsI	AX
1	2	4	6	8	(c.n. of A by X)*

consists of discrete molecules A–X under ordinary conditions. All the other compounds are solids at ordinary temperatures and consist of infinite arrays of A and X atoms in which the metal atoms are bonded to, respectively, two, four, six, and eight X atoms. Figure 1.1 shows some simple examples of systems with the composition AX. Two are finite groups, (a) the dimer and (b) the tetramer; the remainder are infinite arrangements, (c) 1-dimensional, (d) and (e) 2-dimensional, and (f) 3-dimensional. The number of ways of realizing a particular ratio of atoms may be large; Fig. 1.2 shows some systems with the composition AX_3.

Examples of all the systems shown in Figs 1.1 and 1.2 will be found in later chapters. In the upper part of Table 1.1 we list seven different ways in which an F:M ratio of 5:1 is attained in crystalline pentahalides; the list could be extended if anions $(MX_5)^{n-}$ are included, as will be seen from the structures of complex halides in Chapter 10. Conversely, we may consider how *different formulae* MX_n arise with the *same coordination number of* M. For tetrahedral and octahedral

*c.n. = coordination number

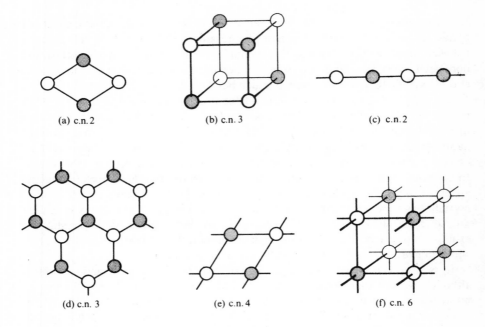

FIG. 1.1. Arrangements of equal numbers of atoms of two kinds.

coordination this problem is considered in some detail in Chapter 5; the examples of Table 1.1 may be of interest as examples of the less usual coordination number nine. In order to gain a real understanding of the meaning of the formulae of inorganic compounds it is evidently necessary to think in three rather than two dimensions and in terms of infinite as well as finite groups of atoms.

(v) The chemist is familiar with *isomerism* (p. 52), which refers to differences in the structures of *finite* molecules or complex ions having a particular chemical composition. If infinite arrangements of atoms are permitted, in addition to finite groups, the probability of alternative atomic arrangements is greatly increased, as is evident from Figs 1.1 and 1.2.

An element or compound is described as *polymorphic* if it forms two or more crystalline phases differing in atomic arrangement. (The earlier term *allotropy* is still used to refer to different 'forms' of *elements*, but except for the special case of O_2 and O_3 allotropes are simply polymorphs.) Polymorphism of both elements and compounds is the rule rather than the exception, and the structural chemistry of any element or compound includes the structures of all its polymorphs just as that of a molecule includes the structures of its isomers. The differences between the structures of polymorphs range from minor differences such as the change from fixed to random orientation (or complete rotation) of a molecule or complex ion in the high temperature form of a substance (for example crystalline HCl, salts

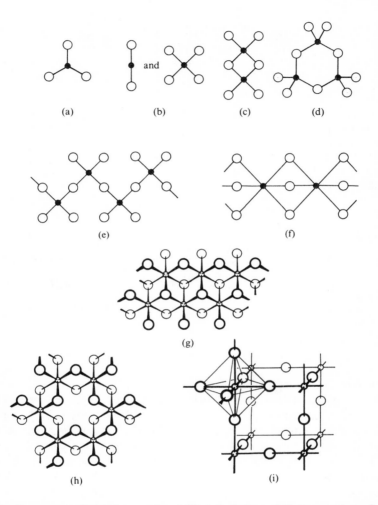

FIG. 1.2. Some ways of realizing a ratio of 3X:A in finite or infinite groupings of atoms: (a)–(d), finite groups AX_3, AX_2 and AX_4, A_2X_6 and A_3X_9; (e)–(g), infinite linear systems; (h) infinite two-dimensional system; (i) infinite three-dimensional complex.

containing NH_4^+, NO_3^-, CN^-, and other complex ions), or the α–β changes of the forms of SiO_2, to major differences involving reconstruction of the whole crystal (the polymorphs of C, P, SiO_2, etc.).

Originally the only variable in studies of polymorphism was the temperature, and substances are described as enantiotropic if the polymorphic change takes place at a definite transition temperature or monotropic if one form is stable at all temperatures under atmospheric pressure. Extensive work by Bridgman showed that many elements (and compounds such as ice) undergo structural changes when

TABLE 1.1
Structures of crystalline pentahalides

Halide	C.N. of M	Structural units in crystal
PBr_5	4	$(PBr_4)^+Br^-$
PCl_5	4 and 6	$(PCl_4)^+(PCl_6)^-$
$NbCl_5$	6	Nb_2Cl_{10}
MoF_5	6	Mo_4F_{20}
BiF_5	6	Chains $(BiF_5)_n$
$PaCl_5$	7	Chains $(PaCl_5)_n$
β-UF_5	8	(3D ionic structure)

Examples of tricapped trigonal prism coordination

AX_9 *sharing*	X : A	*Examples*
—	9	$[Nd(H_2O)_9](BrO_3)_3$, K_2ReH_9
2 edges	7	K_2PaF_7
2 faces	6	$[Sr(H_2O)_6]Cl_2$
2 edges, 4 vertices	5	$LiUF_5$
2 faces, 4 vertices	4	NH_4BiF_4, $NaNdF_4$
2 faces, 6 edges	3	UCl_3
2 faces, 12 edges	2	$PbCl_2$
(but see p. 273)		

subjected to pressure, the changes being detected as discontinuities in physical properties such as resistivity or compressibility. In some cases the high-pressure structure can be retained by quenching in liquid nitrogen and studied under atmospheric pressure by normal X-ray techniques. In recent years the study of high-pressure polymorphs has been greatly extended by the introduction of new apparatus (such as the *tetrahedral anvil*) which not only increase the range of pressures attainable but also permit the X-ray (and neutron) diffraction study of the phase while under pressure. Studies of halides and oxides, in addition to elements, have produced many new examples of polymorphism and some of these are described in later chapters.

We noted above that some high-pressure polymorphs do not revert to the normal form when the pressure is reduced. Many high-temperature phases do not revert to the low-temperature phase on cooling through the transition temperature, witness the many high-temperature polymorphs found as minerals. This non-reversibility of polymorphic changes is presumably due to the fact that the activation energies associated with processes involving a radical rearrangement of the atoms may be large, regardless of the difference between the lattice energies of the two polymorphs.

Members of families of closely related structures, the formation of which is dependent on the growth mechanism of the crystals, are termed *polytypes*. They are not normal polymorphs, and are formed only by compounds with certain types of structure. The best-known examples are SiC, CdI_2, ZnS, and certain complex oxides, notably ferrites, to which reference should be made for further details.

(vi) When atoms are bonded together to form either finite or infinite groupings complications can occur owing to the conflicting requirements of the various atoms due to their relative sizes or preferred interbond angles. It is well known that this problem arises in finite groups of atoms, as may be seen from scale models of molecules and complex ions. It is, however, less generally appreciated that geometrical and topological restrictions enter in much more subtle ways in 3D structures and may be directly relevant to problems which seem at first sight to be purely chemical in nature. As examples we may instance the relative stabilities of series of oxy-salts (for example, alkali-metal orthoborates and orthosilicates), the crystallization of salts from aqueous solution in the anhydrous state or as hydrates, and the behaviour of the nitrate ion as a bidentate or monodentate ligand. We return briefly to the subject later in this chapter and consider it in more detail in Chapter 7.

Structural formulae of inorganic compounds

Elemental analysis gives the relative numbers of atoms of different elements in a compound; it yields an 'empirical' formula. The simplest type of structural formula indicates how the atoms are linked together, and to this simple topological picture may be added information describing the geometry of the system. The nature of a structural formula depends on the extent of the linking of the atoms. If the compound consists of finite molecules it is necessary to know the molecular weight and then to determine the topology and geometry of the molecule:

$$HNO \longrightarrow H_2N_2O_2 \longrightarrow \underset{HO}{}N{=}N{\overset{OH}{}} \longrightarrow \text{bond lengths and interbond angles}$$

elemental analysis	molecular weight	infrared and Raman spectroscopy indicate *trans* configuration

If the atoms (in a solid) are linked to form a 1-, 2-, or 3-dimensional system the term molecular weight has no meaning, and the structural formula must describe some characteristic set of atoms which on repetition reproduces the arrangement found in the crystal. The repeat unit in an infinite 1-dimensional system is readily found by noting the points at which the pattern repeats itself:

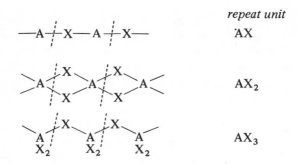

	repeat unit
—A—X—A—X—	AX
	AX_2
	AX_3

The complete description of the chain requires metrical information as in the case of a finite group. It should be noted that if the geometry of the chain is taken into account, that is, the actual spatial arrangement of the atoms in the crystal, then the (crystallographic) repeat unit may be larger than the simplest 'chemical' repeat unit. The crystallographic repeat unit is that set of atoms which reproduces the observed structure when repeated *in the same orientation*, that is, by simple translations in one, two, or three directions. The chemical repeat unit is not concerned with orientation. This distinction is illustrated in Fig. 1.3(a) for the HgO chain. The chemical repeat unit consists of one Hg and one O atom whereas if we have regard to the geometrical configuration of the (planar) chain we must recognize a repeat unit containing 2 Hg + 2 O atoms. The various forms of AX_3 chains formed from tetrahedral AX_4 groups sharing two vertices (X atoms) provide further examples (p. 1022); one is included in Fig. 1.3(b).

Similar considerations apply to structures extending in two or three dimensions. The repeat unit of a 2D pattern is a *unit cell* which by translation in the directions of two (non-parallel) axes reproduces the infinite pattern. One crystalline form of As_2O_3 is built of infinite layers of the kind shown in Fig. 1.4(a), the unit cell being indicated by the broken lines. The pattern arises from AsO_3 groups sharing their O atoms with three similar groups, or alternatively, the repeat unit is $As(O_{1/2})_3$. These units are oriented in two ways to form the infinite layer, with the result that the crystallographic repeat unit – which must reproduce the pattern merely by translations in two directions – contains two of these $As(O_{1/2})_3$ units, or As_2O_3.

The crystallographic repeat unit of a 3D pattern is a parallelepiped containing a representative collection of atoms which on repetition in the directions of its edges forms the (potentially infinite) crystal. As in the case of a 1- or 2-dimensional pattern this unit cell may, and usually does, contain more than one basic 'chemical' unit (corresponding to the simplest chemical formula).

The following remarks may be helpful at this point; they are amplified in later chapters. There is no unique unit cell in a crystal structure, but if there are symmetry elements certain conventions are adopted about the choice of axes (directions of the edges of the unit cell). For example, crystalline NaCl has cubic symmetry (see

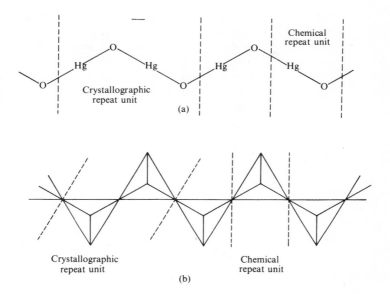

FIG. 1.3. Repeat units in chains.

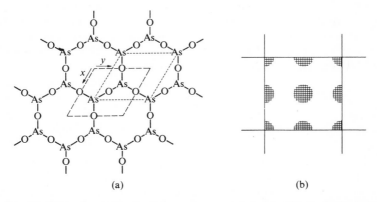

FIG. 1.4. (a) Alternative unit cells of layer structure of As_2O_3. (b) Projection of unit cell of a structure containing four atoms.

Chapter 2) and the structure is therefore referred to a cubic unit cell. This cell contains 4 NaCl, but the structure may be described in terms of cells containing 2 NaCl or 1 NaCl; these alternative unit cells for the NaCl structure are illustrated in Fig. 6.3 (p. 243). It is sometimes convenient to choose a different origin, that is, to translate the cell in the directions of one or more of the axes, and the origin is not necessarily taken at an atom in the structure. For example, the unit cell of the projection of Fig. 1.4(a) does not have an atom at the origin but it is a more convenient cell than the one indicated by the fine dotted lines because it gives the coordinates $\pm$ ($\frac{1}{3}$ $\frac{2}{3}$) rather than (00) and ($\frac{2}{3}$ $\frac{1}{3}$) for the two equivalent As atoms.

If there are atoms at the corners or on the edges or faces of a unit cell it may be difficult to reconcile the number of atoms shown in a diagram with the chemical formula — see, for example, the cells outlined by the broken lines in Fig. 1.4(a). It is only necessary to remember that the cell content includes all atoms whose centres lie within the cell and that atoms lying at the corner or on an edge or face count as follows:

    unit cell of 2D pattern:
        atom at corner belongs to four cells,
        atom on edge belongs to two cells;
    unit cell of 3D pattern:
        atom at corner belongs to eight cells,
        atom on edge belongs to four cells,
        atom in face belongs to two cells.

The cell content in each case could alternatively be shown by shading that portion of each atom which lies wholly within the cell (Fig. 1.4(b)).

Each atom shown in a *projection* repeats at a distance c above and below the plane of the paper, where c is the repeat distance in the structure along the direction of projection. Figure 1.5 represents the projection on its base of a cube containing an atom at its centre (body-centred cubic structure). The atom A has eight equidistant neighbours at the vertices of a cube, since the atoms at height 0 (i.e. in the plane of the paper) repeat at height 1 (in units of the distance c). Similarly the atom B has the same arrangement of eight nearest neighbours.

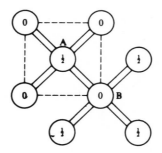

FIG. 1.5. Projection of body-centred structure showing 8-coordination of atoms.

In order to simplify an illustration of a structure it is common practice to show a set of nearest neighbours (coordination group) as a polyhedral group. Thus the projection of the rutile structure of one of the forms of TiO_2 may be shown as

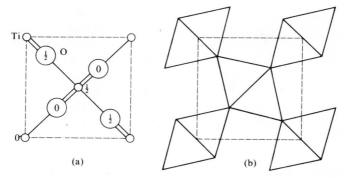

FIG. 1.6. Projections of the structure of rutile (TiO_2): (a) showing atoms and their heights; (b) showing the octahedral TiO_6 coordination groups.

either (a) or (b) in Fig. 1.6. In (a) the full lines indicate Ti–O bonds, and it may be deduced from the coordinates of the atoms that there is an octahedral coordination group of six O atoms around each Ti atom. In (b) the lines represent the edges of the octahedral coordination group. Since it is important that at least the two commonest coordination polyhedra should be recognized when viewed in a number of directions we illustrate several projections of the tetrahedron and octahedron at the beginning of Chapter 5.

Atoms arranged around 3-, 4-, or 6-fold helices project along the helical axis as triangle, square, or hexagon respectively. A pair of lines may be used to indicate that a number or atoms do not form a closed circuit but are arranged on a helix perpendicular to the plane of the paper (Fig. 1.7).

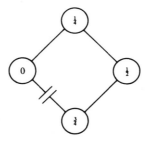

FIG. 1.7. Projection of 4-fold helix along its axis. The atom at height ¾ is connected to the atom vertically above the one shown at height 0.

It is perhaps unnecessary to stress that a formula should correspond as closely as possible to the structure of the compound, that is, to the molecule or other grouping present, as, for example, $Na_3B_3O_6$ for sodium metaborate, which contains cyclic $B_3O_6^{3-}$ ions. Compounds containing metal atoms in two oxidation states are of interest in this connection. If the oxidation numbers differ by unity the formula does not reduce to a simpler form (for example, Fe_3O_4, Cr_2F_5), but if they differ by two the formula appears to correspond to an intermediate oxidation state:

Empirical formula	*Structural formula*
$GaCl_2$	$Ga^I(Ga^{III}Cl_4)$
PdF_3	$Pd^{II}(Pd^{IV}F_6)$
$CsAuCl_3$	$Cs_2(Au^ICl_2)(Au^{III}Cl_4)$
$(NH_4)_2SbCl_6$	$(NH_4)_4(Sb^{III}Cl_6)(Sb^VCl_6)$

Studies of crystal structures have led to the revision of many chemical formulae by regrouping the atoms to correspond to the actual groups present in the crystal. This is particularly true of compounds originally formulated as hydrates; some examples follow.

Hydrate	*Structural formula*
$HCl.H_2O$	$(H_3O)^+Cl^-$
$NaBO_2.2H_2O$	$Na[B(OH)_4]$
$Na_2B_4O_7.10H_2O$	$Na_2[B_4O_5(OH)_4].8H_2O$
$FeCl_3.6H_2O$	$[FeCl_2(H_2O)_4]Cl.2H_2O$
$ZrOCl_2.8H_2O$	$[Zr_4(OH)_8(H_2O)_{16}]Cl_8.12H_2O$

The formulae of many inorganic compounds do not at first sight appear compatible with the normal valences of the atoms but are in fact readily interpretable in the light of the structure of the molecule or crystal. In organic chemistry we are familiar with the fact that the $H:C$ ratios in *saturated* hydrocarbons, in all of which carbon is tetravalent, range from the maximum value four in CH_4 to two in $(CH_2)_n$ owing to the presence of C–C bonds. Similarly the unexpected formula, P_4S_3, of one of the sulphides of phosphorus arises from the presence of P–P bonds; the formation by P(III) of three and by S(II) of two bonds would give the formula P_4S_6 if all bonds were P–S bonds.

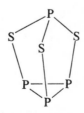

Bonding between atoms of the same element also occurs in many crystalline binary compounds and leads to formulae such as CdP_4, PdP_2, and PdS_2 which are not reconcilable with the normal oxidation numbers of Cd and Pd until their crystal structures are known. The structures of PdP_2 and PdS_2 are described shortly; for CdP_4 see p. 842.

Our final point relating to the structural formulae of solids is that in general crystallographers have not greatly concerned themselves with interpreting the structures of solids to chemists. As a result much of the structural chemistry of solids became segregated in yet another subdivision of chemistry (crystal chemistry, or more recently, solid-state chemistry), and many chemists still tend to make a mental distinction between the structures of solids and of the finite molecules and complex ions that can be studied in solution or in the gaseous state. The infinite layer structures of black and red phosphorus are manifestly only more complex examples of P forming three bonds as in the finite (tetrahedral) P_4 molecule of white phosphorus, but equally the chain structure of crystalline $PdCl_2$ is simply the end-member of the series starting with the $PdCl_4^{2-}$ and $Pd_2Cl_6^{2-}$ ions:

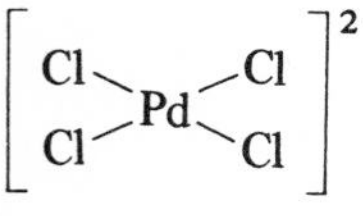

$$PdCl_4^{2-} \qquad Pd_2Cl_6^{2-} \qquad \text{crystalline } PdCl_2$$

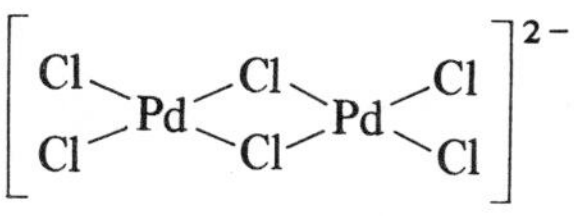

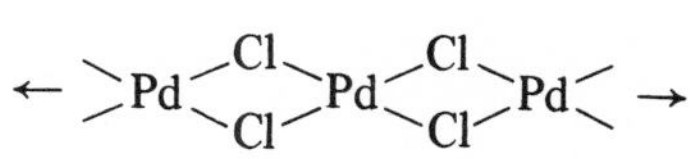

Diagrams purporting to show the origin of the electrons required for the various bonds are often given in elementary texts but only for (finite) molecules and complex ions — not for solids. It might help if 'Sidgwick-type' formulae were given for solids such as SnS (in which each atom forms three bonds), if only to show that the 'rules' which apply to finite systems also apply to some at least of the infinite arrays of atoms in crystals:

Many compounds of Pd(II) may be formulated in a consistent way so that the metal atom acquires a share in six additional electrons and forms planar dsp^2 bonds. Of the two simple possibilities (a) and (b) the former enables us to formulate the

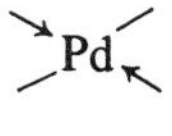

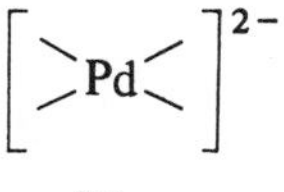

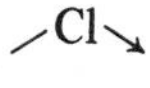

infinite chain of $PdCl_2$ and the $Pd_2Cl_6^{2-}$ ion (since a bridging Cl is represented as at (c)), while (b) represents the situation in $PdCl_4^{2-}$, though the actual state of the ion (e) is presumably intermediate between the 'ionic' picture (d) and the covalent one (f):

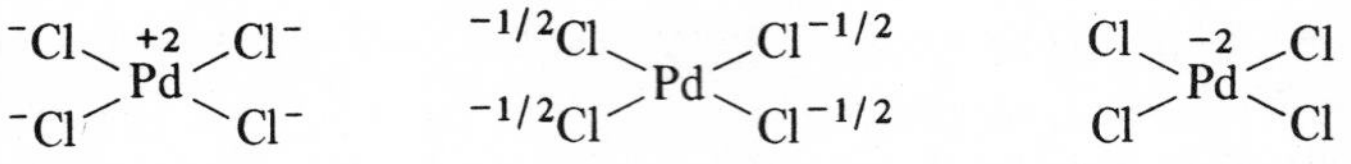

$$
\begin{array}{ccc}
\underset{^-Cl}{^-Cl}\!\!\searrow\!\!\overset{+2}{Pd}\!\!\swarrow\!\!\underset{Cl^-}{Cl^-} & \underset{^{-1/2}Cl}{^{-1/2}Cl}\!\!\searrow\!\!Pd\!\!\swarrow\!\!\underset{Cl^{-1/2}}{Cl^{-1/2}} & \underset{Cl}{Cl}\!\!\searrow\!\!\overset{-2}{Pd}\!\!\swarrow\!\!\underset{Cl}{Cl} \\
(d) & (e) & (f)
\end{array}
$$

In crystalline PdO (and similarly for PdS and PtS) O forms four tetrahedral bonds and the metal forms four coplanar bonds, and we have the bond pictures

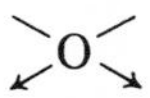

$$\searrow\!O\!\nwarrow \quad \text{and} \quad \searrow\!Pd\!\swarrow$$

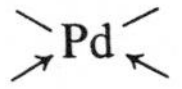

The compounds PdS_2 and PdP_2, which might not appear to be compounds of Pd(II), can be formulated in the following way. The disulphide consists of layers (Fig. 1.8) in which Pd forms four coplanar bonds to S atoms which are bonded in pairs by covalent bonds of length 2.13 Å. We therefore have the bond pictures

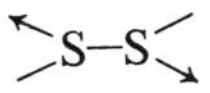

$$\searrow\!S\!-\!S\!\nwarrow \quad \text{and} \quad \searrow\!Pd\!\swarrow$$

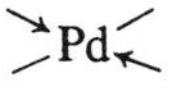

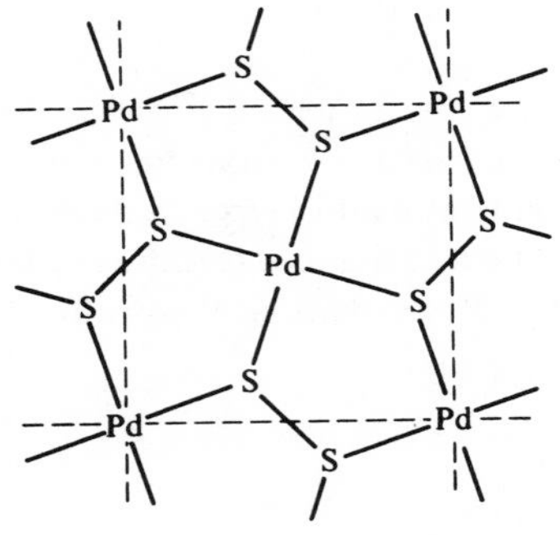

FIG. 1.8. Part of a layer in crystalline PdS_2.

Crystalline PdP_2 can be visualized as built from layers of the same general type, but each P atom forms a fourth bond to a P atom of an adjacent 'layer' so that there are continuous chains of P atoms. The structure is therefore not a layer structure like PdS_2 but a 3D framework in which the nearest neighbours are

$$\text{Pd}\!-\!4\,\text{P (coplanar)} \quad \text{and} \quad \text{P} \begin{cases} 2\,\text{P} \\ 2\,\text{Pd} \end{cases} \text{(tetrahedral)}$$

and the bond pictures are

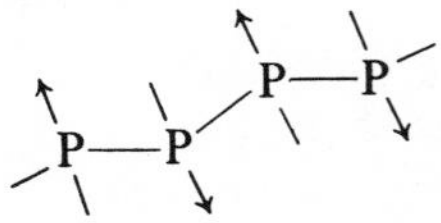

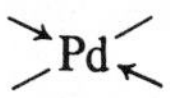

It is natural to enquire whether this somewhat naive treatment can be extended to related compounds. The compounds most closely related to PdS_2 are NiS_2 and PtS_2. All three compounds have different crystal structures and this, incidentally, is also true of the dichlorides. PtS_2 crystallizes with the CdI_2 structure in which Pt forms six octahedral bonds and S forms three pyramidal bonds, consistent with

$$-\overset{\downarrow}{\underset{\nearrow}{Pt}}\overset{}{\underset{\uparrow}{\diagdown}} \qquad \text{and} \qquad \diagup \overset{..}{\underset{\downarrow}{S}} \diagdown$$

The Pt atom thus acquires a share in eight additional electrons and forms octahedral d^2sp^3 bonds. NiS_2 crystallizes with the pyrites structure (p. 242) in which Ni is surrounded by six S atoms of S_2 groups, the Ni atoms and S_2 groups being arranged like the ions in the NaCl structure. For the pyrites structure of FeS_2, we write the bond pictures

$$\overset{\nwarrow}{\underset{\diagdown}{S}}\!\!-\!\!\overset{\diagup}{\underset{\diagdown}{S}}\!\!\overset{\rightarrow}{} \qquad \text{and hence} \qquad -\overset{\downarrow}{\underset{\nearrow\uparrow}{Fe}}\overset{\diagup}{}\!\!-$$

so that with a share in ten additional electrons Fe acquires the Kr configuration. There are also compounds containing As–S, As–As, or P–P groups, instead of S–S, which adopt the pyrites or closely related structures. These groups supply nine or eight electrons instead of the ten supplied by S_2. In fact we find the pyrites (or a similar) structure for

$$FeS_2 \qquad CoAsS \qquad NiP_2 \text{ (high pressure)}$$
$$PtAs_2$$

as expected, but we also find the same structure adopted by $FeAsS$ and $FeAs_2$. The latter present no special problem, there being respectively one or two *fewer* electrons to be accommodated in the d shell of the metal atom as compared with FeS_2. However, there are other compounds with the pyrites structure which have an *excess* of electrons, namely:

	excess electrons		
1	2	3	4
CoS_2	NiS_2	CuS_2 *	ZnS_2 *
NiAsS			

* high-pressure phases

which obviously do present a bonding problem. There is further discussion of the pyrites and related structures in Chapter 17.

Geometrical and topological limitations on the structures of molecules and crystals

Under this general heading we wish to draw attention to the importance of geometrical and topological factors which have a direct bearing not only on the details of molecular and crystal structures but also on the stability and indeed existence of some compounds. In spite of their relevance to structural problems, some of the factors we shall mention seem to have been completely ignored.

In any non-linear system of three atoms X—M—X (a), the distances M—X (bond length) and X—X (van der Waals contact) and the interbond angle X—M—X are necessarily related. In the regular tetrahedral molecule CCl_4 the C—Cl bond length (1.77 Å) implies a separation of only 2.9 Å between adjacent Cl atoms, a distance much less than the normal (van der Waals) distance between Cl atoms of different molecules (around 3.6 Å). If the X atoms are bonded to a second M atom, (b), there is formed a 4-ring, which we shall assume to be a planar parallelogram as it is in the examples we shall consider. The (non-bonded) distances M—M and X—X are related

$$\overset{\longleftarrow 2\cdot9\ \text{Å} \longrightarrow}{\underset{\text{(a)}}{\text{Cl}\diagdown_{C}\diagup\text{Cl}}} \qquad \underset{\text{(b)}}{X\diagdown_{M}^{M}\diagup X} \qquad \underset{\text{(c)}}{M-X\diagdown_{M}^{M}\diagup X-M}$$

to the bond length M—X and the angle X—M—X. In the lithium chloride dimer these distances are 2.5 Å, 3.6 Å, and 2.2 Å, and the angle Cl—Li—Cl is approximately 110°. The Li—Li separation, which is shorter than the interatomic distance in the crystalline metal (3.1 Å) can obviously not be discussed in isolation since it is one of a number of interrelated quantities. Rings of type (b) also occur in molecules such as Fe_2Cl_6 and Nb_2Cl_{10} formed by the sharing of two X atoms between two tetrahedral MX_4 or two octahedral MX_6 groups.

Now suppose that each X atom is bonded to additional M atoms, as is the case in many crystals in which the X and M atoms form a 3D system. The bond angles around X are still related to those of the M atom. An example of (c) is the structure of rutile (one of the polymorphs of TiO_2) in which every Ti is surrounded octahedrally by six O atoms and each O by three coplanar Ti atoms. Evidently if the TiO_6 octahedra are regular the angle O—Ti—O is 90°, and hence one of the O bond angles would also be 90°. It follows that there cannot be regular octahedral coordination of M *and* the most symmetrical environment of X (three bond angles of 120°) in a compound MX_2 with the rutile structure. In the high-temperature form of BeO there is tetrahedral coordination of Be by four O atoms and the BeO_4 tetrahedra are linked in pairs by sharing an edge, these pairs being further linked by sharing vertices to form a 3D framework. The structure is illustrated in Fig. 12.3 (p. 539); the important feature in the present context is the presence of 4-rings. Here also regular coordination (in this case tetrahedral), of both O and Be atoms is not possible, and the distance Be—Be across the ring is only 2.3 Å, the same as in metallic beryllium.

The structure of crystalline PtS (see Fig. 17.3, p. 756) is a 3D structure which may be visualized as built up from two sets of planar chains in which Pt(II) forms

four coplanar bonds. The two sets of chains lie in planes which are perpendicular to one another and each S atom is common to a chain of each set. It is clearly impossible to have both *regular* tetrahedral coordination of S and the most symmetrical arrangement of four coplanar bonds around Pt, for the sulphur bond angle α is the supplement of the Pt bond angle. The actual sulphur bond angles represent a compromise between the values $90°$ and $109\frac{1}{2}°$; they are two of $97\frac{1}{2}°$ and four of $115°$.

In these examples we have progressed from the simple 4-ring molecule of Li_2Cl_2 to examples of 4-rings of which the X atoms are involved in further bonds to M atoms, and the last three examples are all crystalline solids in which the M—X bonding extends throughout the whole crystal (3D complex). A somewhat similar problem arises in *finite* molecules (or complex ions) if there are connections between the X atoms attached to the central atom. The 'ideal' stereochemistry of a metal atom forming six bonds might be expected when it is bonded to six identical atoms in a finite group MX_6. If two or more of the X atoms form parts of a polydentate ligand (that is, they are themselves bonded together in some way) restrictions have been introduced which may alter the angles X—M—X and possibly also the lengths of M—X bonds. Many atypical stereochemistries of metal atoms have been produced in this way. For example, all four As atoms of the molecule (a) can bond to the same metal atom, as in the ion (b) where the tetradentate ligand leads to a bond arrangement unusual for divalent Pt. On the other hand, if there is sufficient flexibility in the ligand this may suffer deformation rather than distort the bond arrangement around the metal aton. In the tetramethyldipyrromethene derivative (c) it appears that the very short distances between the CH_3 groups in a model constructed with the usual coplanar bonds from Pd(II) might lead to a tetrahedral metal stereochemistry. In this case, however, the ring systems buckle in preference to distortion of the metal-bond arrangement.

(a)   (b)   (c)

The angles subtended by the A atoms at X atoms (commonly oxygen or halogen) shared between two coordination groups have long been of interest as indications of the nature of the A–X bonds. Simple systems of this type include the 'pyro'-ions, in which A is Si, P, or S, and X is an O atom. If we make the reasonable assumption that the X atoms of different tetrahedra should not approach more closely than they do within a tetrahedral group then a lower limit for the angle A–X–A can be calculated; the upper limit is $180°$. A similar calculation can be made for octahedral (or other) coordination groups meeting at a common X atom. Such simple geometrical considerations are obviously relevant to discussions of observed A–X–A bond angles, but even more interesting are the deductions that can be made concerning the sharing of edges between octahedra in complex oxide structures. These points are discussed in more detail in Chapter 5. It will also be evident that similar considerations limit the number of tetrahedral AX_4, octahedral AX_6, or other coordination groups AX_n that can meet at a point, that is, have a common vertex (X atom). These limits in turn have a bearing on the stability and indeed existence of crystalline compounds such as oxy-salts and nitrides, as we shall show in Chapter 7.

In contrast to these limitations on the structures of molecules and crystals which arise from metrical considerations there are others which may be described as topological in character. For example, the non-existence of compounds A_2X_3 (e.g. sesquioxides) with simple layer structures in which A is bonded to 6 X and X to 4 A is not a matter of crystal chemistry but of topology; it is concerned with the non-existence of the appropriate plane nets, as explained on p. 83. The non-existence of certain other structures for compounds A_mX_n, such as an AX_2 structure of $10:5$ coordination, may conceivably be due to the (topological) impossibility of constructing the appropriate 3D nets. On the other hand this problem may alternatively be regarded as a geometrical one, of the kind already mentioned, concerned with the numbers of coordination polyhedra of various kinds which can meet at a point, a matter which is further discussed on p. 192.

This interrelation of geometry (that is, metrical considerations) and topology (connectedness) seems to be a rather subtle one. It is well known that it is impossible to place five *equivalent* points on the surface of a sphere, if we exclude the trivial case when they form a pentagon around the equator, a fact obviously relevant to discussions of 5-coordination or the formation of five equivalent bonds. The most general (topological) proof of this theorem results from considering the linking of the points into a connected system of polygons (polyhedron) and showing that this cannot be done with the same number of connections to each point. Alternatively we may demonstrate that there is no *regular* solid with five vertices, when metrical factors are introduced. When we derive some of the possible 3D 4-connected nets in Chapter 3 simply as systems of connected points we find that the simplest (as judged by the number of points in the smallest repeat unit) is a system of 6-gons which in its most symmetrical configuration represents the structure of diamond. Although this net is derived as a 'topological entity', without reference to bond angles, it appears that it cannot be constructed with *any* arbitrary interbond angles, for example, with four coplanar bonds meeting at each point. Apparently geometrical limitations of a similar kind apply to other systems of connected points; this neglected field of 3D Euclidean geometry should repay study.

The complete structural chemistry of an element or compound

Having emphasized the importance of the solid state in inorganic chemistry we now examine what place this occupies in the complete structural chemistry of a substance — element or compound.

The complete structural chemistry of a substance could be summarized as in Chart 1.1. It includes not only the structures of the substance in the various states

CHART I.1
The complete structural chemistry of a substance

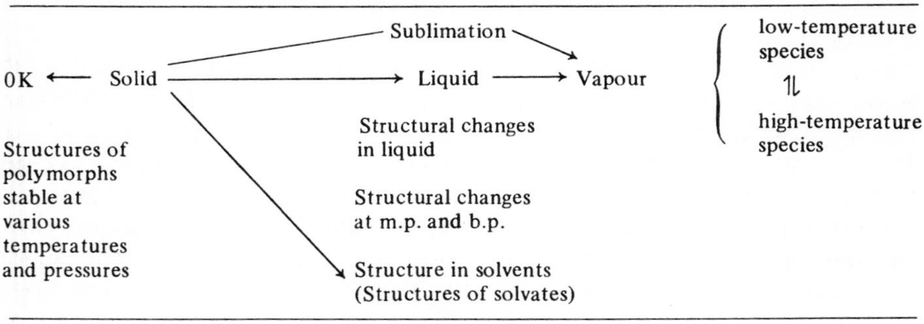

of aggregation but also the structural changes accompanying melting, vaporization of liquid or solid, or dissolution in a solvent, and those taking place *in* the solid, liquid, and vapour states. The complexity of the structural chemistry of an element or compound varies widely. At one extreme there are the noble gases which exist as discrete atoms in all states of aggregation. In these cases the only entries in the chart would be the arrangement of the atoms in the solid and the relatively small changes in structure of the simple atomic liquid with temperature. Next come gases such as H_2, N_2, O_2, and the halogens which persist as diatomic molecules from the solid through the liquid to the gaseous state and dissociate to single atoms only at higher temperatures. Sulphur, on the other hand, has an extremely complex structural chemistry in the elementary form which is summarized in Chapter 16. Unfortunately it is not possible to present anything like a complete picture of the structural chemistry of many compounds simply because the structural data are not available. Structural studies by particular investigators are usually confined to solids *or* gases (rarely liquids and rarely solids *and* gases); structural studies of a particular substance in more than one state are unusual. (We select $FeCl_3$ later to illustrate the structural chemistry of a relatively simple compound.) Here, therefore, we shall consider briefly the various entries in Chart 1.1 and illustrate a number of points with examples.

Structure in the solid state

The structural chemistry of the solid includes the structures of its various crystalline forms (if it is polymorphic), that is, a knowledge of its structures over the temperature range $0°K$ to its melting point and also under pressure. In recent years studies of substances subjected to high pressures have widened the scope of structural chemistry in two ways. First, new polymorphs of many elements and known compounds have been produced in which the atoms are more closely packed, the higher density being achieved often, though not always, by increasing the coordination numbers of the atoms. For example, the 4:4 coordinated ZnO structure transforms to the 6:6 coordinated NaCl structure at 100 kbar, but higher density is achieved in coesite (a high-pressure form of SiO_2) without increasing coordination numbers. Second, new compounds have been produced which, though they cannot be made under atmospheric pressure, can persist under ordinary conditions once they have been made; examples include PbS_2, CuS_2, ZnS_2, and CdS_2. Stoichiometric FeO, a compound normally deficient in iron, has been made by heating the usual $Fe_{0.95}O$ with metallic Fe at $770°C$ under a pressure of 36 kbar. It is not always appreciated that our 'normal' chemistry and ideas on bonding and the structures of molecules and crystals strictly refer only to 'atmospheric pressure chemistry'. It has been possible to vary the temperature over a considerable range for a long time, and examples of polymorphism have been largely restricted to those resulting from varying the temperature. The large pressure range now readily attainable (accompanied unavoidably by an increase in temperature) may well prove more productive of structural changes than variation of temperature alone. The phase-diagram of water (Chapter 15) illustrates well the extension of the structural chemistry of a simple compound beyond that of the two forms (hexagonal and cubic ice) stable under atmospheric pressure.

Structural changes on melting

The structural changes taking place on melting range from the mere separation against van der Waals forces of atoms (noble gases) or molecules (molecular crystal built of non-polar molecules) to the catastrophic breakdown of infinite assemblies of atoms in the case of crystals containing chains, layers, or 3D frameworks. There is likely to be little difference in the immediate environment of an atom of a close-packed metal at temperatures slightly below and above its melting point, though there is a sudden disappearance of long-range order. On the other hand there is a larger structural change when metallic bismuth melts. Instead of the usual decrease in density on melting, which may be illustrated by the behaviour of a metal such as lead, there is an *increase* of two and a half per cent due to the collapse of the rather 'open' structure of the solid to a more closely packed liquid:

	Bi (271 °C)	Pb (328 °C)
Density of solid	9.75	11.35 g/cm^3
Density of liquid	10.00	10.68

So little work has been done on the structures of liquids that it is not possible to give many examples of structural changes associated with melting. Crystals consisting of 3D complexes must break down into simpler units, while those built of molecules form a molecular melt. Simple ionic crystals such as NaCl melt to form a mixture of ions, but crystals such as SiO_2, which are built from SiO_4 groups sharing O atoms, do not break down completely. The tetrahedral groups remain linked together through the shared O atoms; accordingly the melt is highly viscous and readily forms a glass which is a supercooled liquid. Rearrangement of the tangled chains and rings of linked tetrahedra is a difficult process. The three crystalline forms of $ZnCl_2$ also consist of tetrahedral groups ($ZnCl_4$) linked to form layers or 3D frameworks, and this compound also forms a very viscous melt. The behaviour of $AlCl_3$ is intermediate between that of NaCl and $ZnCl_2$. The crystal consists of layers formed from octahedral $AlCl_6$ groups sharing three edges as shown diagrammatically for $FeCl_3$ on p. 27. These infinite layers break down on melting (or vaporization) to form dimeric molecules Al_2Cl_6, also of the same type as those in the vapour of $FeCl_3$. If crystalline $AlCl_3$ is heated the (ionic) conductivity, due to movement of Al^{3+} ions through the structure, increases as the melting point is approached and then drops suddenly to zero when the crystal collapses to a melt consisting of (non-conducting) Al_2Cl_6 molecules. The crystal structure of SbF_5 presumably consists of octahedral SbF_6 sharing a pair of vertices (F atoms), as in BiF_5, CrF_5, and other pentafluorides. The n.m.r. spectrum of the highly viscous liquid has been interpreted in terms of systems of SbF_6 groups sharing pairs of adjacent F atoms.

Structural changes in the liquid state

With certain important exceptions, notably elementary S and H_2O, changes in structure taking place between the melting and boiling points have not been very much studied. The changes with temperature of the radial distribution curves of simple atomic liquids show only the variation in average number of neighbours at various distances. On the other hand, the structural changes in liquid sulphur are much more complex and have been the subject of a great deal of study. The complete structural chemistry of elementary sulphur is briefly described in Chapter 16; it provides an example of the opening up of cyclic molecules (S_6 or S_8) into chains followed by polymerization to longer chains. A note on the structure of water is included in Chapter 15.

Structural changes on boiling or sublimation

Since we know so little about the structures of liquids we are virtually restricted here to a comparison of structure in the solid and vapour states. In the prestructural era knowledge of the structures of vapours was confined to the molecular weight and its variation with temperature and pressure. The considerable amount of information now available relating to interatomic distances and interbond angles in vapour molecules has been obtained by electron diffraction or from spectroscopic studies of various kinds. This information is largely restricted to comparatively simple molecules, not only because it is impossible to determine the large number of parameters required to define the geometry of a more complex molecule from the limited experimental data, but also because the geometry of many molecules becomes indeterminate if the molecules are flexible. (In addition, certain methods of determining molecular structure are subject to special limitations; for example, in general only molecules with a permanent dipole moment give microwave spectra.) Information about the molecular weights of vapour species of certain compounds can be obtained by mass spectrometry, though the method has not as yet been widely applied to inorganic compounds – see, for example, the alkali halides and MoO_3. A less direct source of information is the study of the deflection of simple molecules in a magnetic field, which shows, for example, that some halide molecules AX_2 are linear and others angular (see Chapter 9). An interesting development in infrared spectroscopy is the trapping of species present in vapour at high temperatures in a (solid) noble gas matrix at very low temperatures, making possible the study of molecules which are too unstable to be studied by the more usual techniques.

In general we shall indicate what is known about the structural chemistry of a substance, in all states of aggregation, when it is described in the systematic part of this book. However, the following note on ferric chloride is included at this point to emphasize that even a compound as simple as $FeCl_3$ may have a very interesting structural chemistry of its own. In the vapour at low temperatures the compound is in the form of Fe_2Cl_6 molecules. When these condense to form a crystal a radical rearrangement takes place and instead of a finite molecule, in which each iron atom is attached to four chlorine atoms, there is an infinite 2-dimensional layer in which every Fe is bonded to six Cl atoms. When this crystal dissolves in a non-polar solvent such as CS_2 the Fe_2Cl_6 molecules reform, but in polar solvents such as ether simple tetrahedral molecules $(C_2H_5)_2O \rightarrow FeCl_3$ are formed. We are accustomed to regard a process like the dissolution of a crystal as a simple reversible process, in the sense that removal of the solvent leads to the recovery of the original solute. This is usually true for non-polar solvents but not for dissolution in a highly polar solvent like water. When a crystal of ferric chloride dissolves in water separation into Fe^{3+} and Cl^- ions occurs, and many of the properties of an aqueous solution of $FeCl_3$ are the properties of these individual ions rather than of ferric chloride. Thus the solution precipitates $AgCl$ from a solution of $AgNO_3$ (property of Cl^- ion) and it oxidizes $SnCl_2$ to $SnCl_4$ or $TiCl_3$ to $TiCl_4$; these reactions are due to the transformation $Fe^{3+} + e \rightarrow Fe^{2+}$. In aqueous solution the ferric ion is hydrated and evaporation at ordinary temperatures leads to crystallization of the hexahydrate; the original anhydrous $FeCl_3$ cannot be recovered from the solution in this way. The 'hexahydrate' consists of octahedral $[FeCl_2(H_2O)_4]^+$ ions, Cl^- ions, and H_2O molecules which are not attached to metal atoms. We may set out the structural chemistry of $FeCl_3$ in the following way:

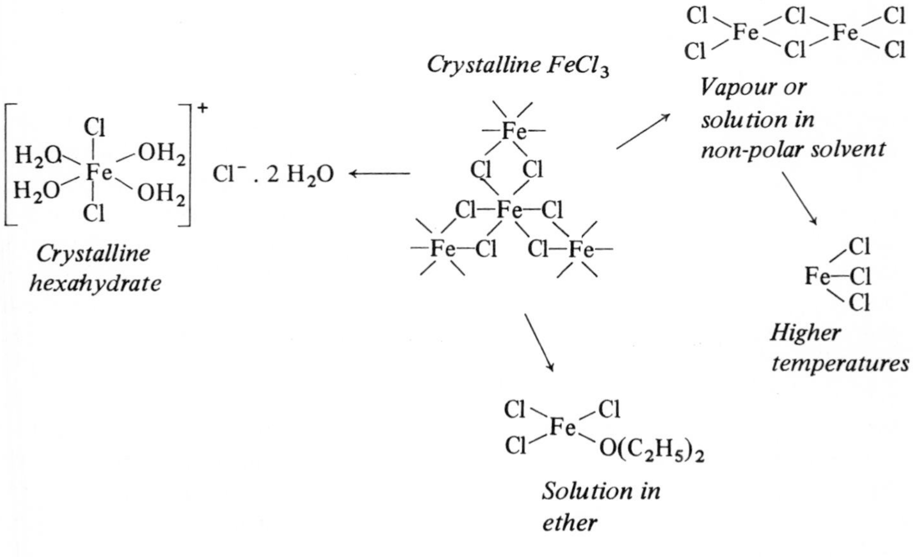

A classification of crystals

We have seen that in some crystals we may distinguish tightly-knit groups of atoms (*complexes*) within which the bonds are of a different kind from (and usually much shorter than) those between the complexes. The complexes may be finite or extend indefinitely in one, two, or three dimensions, and they are held together by ionic, van der Waals, or hydrogen bonds. The recognition of these types of complexes provides a basis for a broad geometrical classification of crystal structures, as shown in Chart 1.2. At first sight it might appear that the most obvious way of classifying structures would be to group them according to the types of bonds between the atoms, recognizing the four extreme types, ionic, covalent, metallic, and van der Waals. A broad division into ionic, covalent, metallic, and molecular crystals is

CHART 1.2

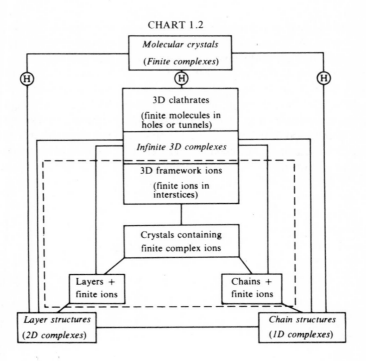

often made. However, bonds approximating to pure types (particularly ionic or covalent) are rare, and moreover in most crystals there are bonds of several different types. Numerous intermediate classes have to be recognized, and it is found that classifications based on bond types become complicated without being comprehensive. They also have the disadvantage that they over-emphasize the importance of 'pure' bond types, bonds of intermediate character being treated as departures from these extremes. It would seem preferable to be able to discuss the nature of the bonds in a particular crystal without having prejudged the issue by classifying a crystal as, for example, an ionic crystal.

We shall, of course, find structures that do not fall neatly into one of the main compartments of Chart 1.2. For example, in crystalline $HgBr_2$ and $CuCl_2$ the shortest bonds define finite molecules and chains respectively. These units are held together by weaker metal–halogen bonds to form layers, the bonds between which are the still weaker van der Waals bonds. Such structures find a place along one of the connecting lines of the chart; so also do structures in which molecules are linked by hydrogen bonds into chains, layers, or 3D complexes, at points marked H on the chart. It is inevitable that we refer in this section to many structures and topics which properly belong in later chapters; it may therefore be found profitable to refer back to this section at a later stage.

The four groups enclosed within the broken lines on the chart are closely related types of ionic structure, all of which involve two distinct kinds of charged unit. These are finite, 1-, 2-, or 3-dimensional ions in addition to discrete ions (usually monatomic). Examples of these four types of structure would include, for example, K_2SO_4, $KCuCl_3$, K_2CuF_4, and KB_5O_8, containing K^+ ions and respectively finite, chain, layer, and 3D complex ions. In each class of structure in the chart there may be departures from strictly regular structure such as rotation or random orientation of complex ions or molecules, absent or misplaced atoms, and so on, as in the various types of 'defect' structure. We shall now discuss the four main structural types somewhat briefly as they will be considered in more detail in subsequent chapters.

Crystals consisting of infinite 3-dimensional complexes

The simplest structures of this group are those of:

(a) the noble gases;
(b) metals and intermetallic compounds; and
(c) many simple ionic and covalent crystals;

that is, all crystals in which no less extensive type of grouping is recognizable. It is immaterial whether crystals containing small molecules such as H_2, N_2, and H_2S are included here or with molecular crystals. Essentially covalent crystals of this class include diamond (and the isostructural Si, BN, etc.) and related compounds such as the various forms of SiC. Some of the simplest structure types for binary compounds in which the infinite 3D complex extends throughout the whole crystal are set out in Table 1.2.

TABLE 1.2

Simple structures for compounds A_mX_n

Coordination number of A	Coordination number of X				
	2	3	4	6	8
4	SiO_2		ZnS		
6	ReO_3	TiO_2	Al_2O_3	NaCl	
8			CaF_2		CsCl
9		LaF_3			

In crystals of groups (a), (b) and (c) all the atoms together form the 3D complex. Closely related to these is another type of structure in which there is a 3D framework extending throughout the crystal but in addition discrete ions or molecules occupy interstices in the structure.

(d) Three-dimensional frameworks may be electrically neutral or they may be charged; in the latter case the charge on the framework must be balanced by ions of opposite charge accommodated in interstices. Crystals consisting of positively charged frameworks (3D cations) are rare. Examples, all containing B sub-group metals, which are described in later chapters, include: $(NHg_2)NO_3$; $(Ag_3S)NO_3$; $(Ag_7O_8)NO_3$;

$(Hg_3S_2)Cl_2$; and $(Sn_2F_3)Cl$. The water framework of $HPF_6 . 6H_2O$ may also be regarded as a 3D cation since the proton of the acid is associated with it; certain other hydrates to be mentioned shortly are examples of electrically neutral frameworks. Structures consisting of negatively charged frameworks enclosing cations are extremely numerous, particularly those built from tetrahedral or octahedral groups. In the zeolites (p. 1036), a group of aluminosilicates, there are rigid frameworks built from $(Si, Al)O_4$ tetrahedra, and through these frameworks there run tunnels which are accessible from the surfaces of the crystals. Foreign ions and molecules can therefore enter or leave the crystal without disturbing the structure. The term *zeolitic* is applied to crystals of this type. Some crystals, on the other hand, incorporate foreign molecules during growth in cavities from which they cannot escape until the crystal is dissolved or vaporized. These crystals are termed *clathrate* compounds. It would be more accurate to say that the framework of the 'host' forms around the included molecules, since the characteristic structure does not form in the absence of the 'guest' molecules. The clathrate structure is not stable unless a certain minimum proportion of the cavities is occupied; the structures of all the clathrate compounds of Fig. 1.9(a), (b), and (c) are different from those of the pure 'host' compounds. Clathrate compounds may have layer structures (for example, $Ni(CN)_2.NH_3.C_6H_6$, Fig. 1.9(a)) but more generally they have 3D structures, most of which are frameworks of hydrogen-bonded molecules. In the urea–hydrocarbon complexes (Fig. 1.9(b)) the urea molecules form a framework enclosing parallel tunnels around the linear hydrocarbon molecules, a much less dense packing of $CO(NH_2)_2$ molecules than in crystalline urea itself. The 'β-quinol' structure, Fig. 1.9(c), consists of two interpenetrating systems of hydrogen-bonded $HO.C_6H_4.OH$ molecules which enclose molecules of water or certain gases if these are present during crystal growth (see also p. 114). The ice-like hydrates have frameworks of hydrogen-bonded water molecules which in the simplest cases form around neutral atoms or molecules such as those of the noble gases, chlorine, etc. They are described in Chapters 15 and 3.

Layer structures

It is conventional to describe as layer structures not only those in which all the atoms are incorporated in well-defined layers but also those in which there are ions or molecules between the layers. Since the layers may be electrically neutral, bonded together by van der Waals or hydrogen bonds, or charged, and since they may be identically constituted or have different structures, we may elaborate our chart as shown in Fig. 1.10. Of the families of structures indicated in Fig. 1.10 by far the largest is (a_1), in which identical layers are held together by van der Waals bonds. The structure of the layer itself can range from the simplest possible planar network of atoms (graphite, white BN), through As_2O_3 (monoclinic), numerous layers formed from tetrahedral or octahedral coordination groups (halides AX_2 and AX_3), to the multiple layers in talc and related aluminosilicates. (The rarity of layer and chain structures among metal oxides and fluorides may be noted; these structures are more

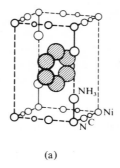

(a)

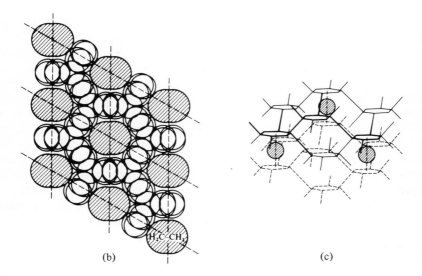

(b)   (c)

FIG. 1.9. Inclusion of foreign molecules in crystals of (a) $Ni(CN)_2NH_3 \cdot nC_6H_6$; (b) a urea–
hydrocarbon complex; (c) β-quinol.

commonly adopted by the other chalconides and halides.) In the classes (a_2) and
(c_2) identical layers are interleaved with molecules or ions, and there are many pairs
of structures in which the layers are of the same or closely similar geometrical types
but electrically neutral in (a_1) and charged in (c_2). Some examples are given in
Table 1.3. For example, the $(Si_2O_5)_n^{2n-}$ layer in a number of alkali-metal silicates
$M_2Si_2O_5$ is of the same general type (tetrahedral groups sharing three vertices)
as the neutral layer in one polymorph of P_2O_5, and there is the same relation
between the neutral layer of PbF_4 and the 2-dimensional ion in $K_2(NiF_4)$ or

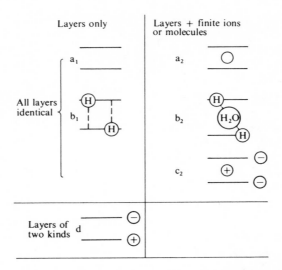

FIG. 1.10. Types of layer structure (diagrammatic).

between talc, built of neutral layers $Mg_3(OH)_2 Si_4 O_{10}$, and the mica phlogopite, $KMg_3(OH)_2 Si_3 AlO_{10}$, in which the charged layers are held together by the K^+ ions. An interesting example of an (a_2) structure is the clathrate compound in which benzene or water molecules are trapped between neutral layers of composition $Ni(CN)_2 . NH_3$ (Fig. 1.9(a)).

Examples of layer structures of types (b_1) and (b_2) are much less numerous. The former, in which the layers are held together by hydrogen bonds, include $Al(OH)_3$, $HCrO_2$, γ-FeO.OH and related compounds. Instead of direct hydrogen bonding between atoms of adjacent layers there is the possibility of hydrogen bonding through an intermediate molecule such as H_2O situated between the layers (class (b_2)). The structure of gypsum, $CaSO_4 . 2H_2O$ (p. 684) provides an example of a structure of this type.

TABLE 1.3

Related structures containing layers of the same general type

(a_1)	(a_2) *or* (c_2)
Graphite	Interlamellar compounds of graphite, e.g. C_8K, $C_{16}K$
	$CaSi_2$
P_2O_5	$Li_2(Si_2O_5)$
PbF_4	$K_2(NiF_4)$
Talc	Phlogopite ⎱ (micas)
Pyrophyllite	Muscovite ⎰

The last group, (d), of structures in which positively and negatively charged layers alternate, is also a small one and includes the chlorite minerals (p. 1032) and some hydroxyhalides such as $[Na_4Mg_2Cl_{12}]^{4-}[Mg_7Al_4(OH)_{22}]^{4+}$ (p. 262).

A property of certain layer structures which should be mentioned here is that of taking up 'foreign' atoms, ions, or molecules between the layers, with expansion of the structure only in a direction normal to the layers. It has long been known that this is a property of graphite (p. 735) and clay minerals (p. 1031), and in recent years interest has developed in such *intercalation* compounds of, for example, NbS_2 and TaS_2 (p. 758).

In this very brief survey we have not been concerned with the detailed structure of the layers, various aspects of which are discussed in more detail later, in particular the basic 2D nets and structures based on the simplest 3- and 4-connected plane nets (Chapter 3), and layers formed from tetrahedral and octahedral coordination groups (Chapter 5). The CdI_2 layer and more complex structures derived from this layer are further discussed in Chapter 6. We shall see that there are corrugated as well as plane layers, and also composite layers consisting of two interwoven layers (red P, $Ag[C(CN)_3]$); these are included in the chapters just mentioned.

Between the four major classes our classification allows for structures of intermediate types. We can envisage, for example, structures in which the distinction between the bonds (and distances) between atoms within the layers and between those in different layers is not so clear-cut as we have supposed here. Alternatively, there may be well-defined layers but different types or strengths of bonds between the units comprising the layers, so that the layer itself may be regarded as a rather loosely knit assembly of chains or finite molecules. We have already noted $CuCl_2$ and $HgBr_2$ as examples of structures of this kind, and other examples will be found in later chapters.

Chain structures

In principle the possible types of chain structure are similar to those for layer structures, and some are shown diagrammatically in Fig. 1.11, labelled in the same way. Examples are known of all the types of chain structure shown in the figure, the most numerous being those of class (c_2) which includes all oxy-salts containing infinite linear ions—the simplest of which are the linear meta-salts—and all complex halides containing chain anions. In some hydrated complex halides water molecules also are accommodated between infinite 1-dimensional anions, as in $K_2HgCl_4 . H_2O$ (Fig. 1.12).

A simple example of class (b_1) is the structure of $LiOH . H_2O$, in which there are hydrogen bonds between OH groups and H_2O molecules of neighbouring chains (see Fig. 15.23, p. 685), and the structure of borax illustrates the rare (d_1) type. Here the Na^+ ions and water molecules form a linear $[Na(H_2O)_4]_n^{n+}$ chain (consisting of octahedral $Na(H_2O)_6$ groups sharing two edges) and the anion is an infinite chain formed from hydrogen-bonded $[B_4O_5(OH)_4]^{2-}$ ions, as described on p. 1074. As in the case of the $(HCO_3)_n^{n-}$ chain in $NaHCO_3$ this borate anion is composed of

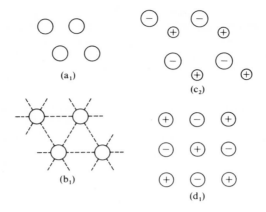

FIG. 1.11. Major classes of chain structure (diagrammatic) viewed along direction of chains.

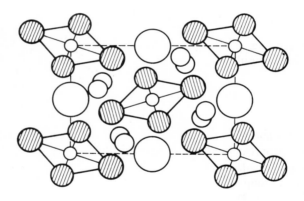

FIG. 1.12. The structure of $K_2HgCl_4 \cdot H_2O$ viewed along the direction of the infinite chains. The K^+ ions (pairs of overlapping circles) and H_2O molecules (large circles) lie, at various heights, between the chains.

sub-units which are hydrogen-bonded along the chain and is therefore not strictly a chain of the simplest type.

The simplest type of chain structure, (a_1), consists of identical chains whch are packed parallel to one another. Since the basic requirement for the formation of a chain is that the structural unit can bond to two others, the simplest possibility is a chain consisting of similar atoms as in plastic sulphur or elementary Se or Te:

TABLE 1.4

Structures containing chains of the same general type

Type of chain	Neutral molecule	Infinite ion
$-A-X-A-X-$ $\quad X \quad\quad X$	SeO_2	$(BO_2)_n^{n-}$ in CaB_2O_4 ⎫
$-A-X-A-X-$ (tetrahedral) $\quad X_2 \quad\quad X_2$	SO_3	$(SiO_3)_n^{2n-}$, $(PO_3)_n^{n-}$ ⎬ meta- ions in meta-salts ⎭
(planar)	AuF_3	$(CuCl_3)_n^{n-}$ in $CsCuCl_3$
$-A-X-A-X-$ $\quad X_4 \quad\quad X_4$	BiF_5	$(AlF_5)_n^{2n-}$ in Tl_2AlF_5
$\underset{X_2}{\overset{X}{\diagup}}A\underset{X\ X_2}{\diagdown}A\diagup$	NbI_4	$(HgCl_4)_n^{2n-}$ in $K_2HgCl_4 \cdot H_2O$
$\overset{X}{\underset{X}{\diagup}}A-X-A\diagdown$	ZrI_3, Cs_3O	$(NiCl_3)_n^{n-}$ in $CsNiCl_3$

An obvious elaboration is the chain consisting of alternate atoms of different elements, as in HgO or AuI:

$$\diagup Hg \overset{O}{\diagdown} Hg \underset{O}{\diagdown} Hg \overset{O}{\diagdown} Hg \diagdown$$

More complex 2-connected units are groups AX_3, AX_4, AX_6, etc., sharing one or more X atoms, where AX_3 is a planar group (e.g. $(BO_2)_n^{n-}$) or a pyramidal group (as in SeO_2), AX_4 a tetrahedral group sharing two vertices (SO_3) or two edges (SiS_2), AX_6 an octahedral group sharing two vertices (BiF_5), or two edges (NbI_4), or two faces (ZrI_3), and so on. These chains are described in more detail under tetrahedral and octahedral structures in Chapter 5. The structure of $ReCl_4$ provides an example of a more complex chain in which the repeat unit is a pair of face-sharing octahedra which are further linked by sharing a pair of vertices:

$$-\begin{bmatrix} Cl & Cl & Cl \\ Cl-Re-Cl-Re-Cl \\ Cl & Cl & Cl \end{bmatrix}\diagup$$

As in the case of layer structures the same general type of structure can serve for infinite molecules as for infinite linear ions (Table 1.4).

We do not propose to discuss here the geometry of chains, but we may note that whereas chains built of tetrahedral or octahedral groups sharing a pair of opposite edges or of octahedral groups sharing a pair of opposite faces are strictly linear, there are numerous configurations of chains built from tetrahedra sharing two vertices (p. 1022) and two configurations are found for AX_5 chains in which octahedral AX_6 groups share two vertices, the latter being *cis* or *trans* to one another. A rotation combined with a translation produces a helix, the simplest form of which is the plane zigzag chain (as already illustrated for HgO) generated by a 2-fold screw axis. In a second polymorphic form of HgO the chain, like that in 'metallic' Se, is generated by a 3-fold screw axis, while the infinite chain molecule in crystalline AuF_3 results from the repetition of a planar AuF_4 group around a 6-fold screw axis.

Crystals containing finite complexes

Finite complexes comprise all molecules and finite complex ions. We are therefore concerned here with two main classes of Chart 1.2, namely, molecular crystals and crystals containing finite complex ions. Molecular crystals include many compounds of non-metals, some metallic halides AX_n and a number of polymeric halides A_2X_6 and A_2X_{10}, metal carbonyls and related compounds, and many coordination compounds built of neutral molecules (e.g. $Co(NH_3)_3(NO_2)_3$). In the simplest type of molecular crystal identical non-polar molecules are held together by van der Waals bonds. The structures of these crystals represent the most efficient packing of units of a given shape held together by undirected forces. When the molecules are roughly spherical the same structure types may occur as in crystals described as 3-dimensional complexes, a group of atoms replacing the single atom. As the shape of the molecule deviates from spherical so the structures bear more resemblance to those of crystals containing 1- or 2-dimensional complexes. More complex types of molecular crystal structure arise when there are molecules of more than one kind, as in $AsI_3 . 3S_8$, and when there are hydrogen bonds between certain pairs of atoms of different molecules.

Finite complex ions include the numerous oxy-ions and complex halide ions, the aquo-ions in some hydrates (e.g. $Al(H_2O)_6^{3+}$), and all finite charged coordination complexes, in addition to the very simple ions such as CN^-, C_2^{2-}, O_2^-, O_2^{2-}, and many others. Crystals containing the smaller or more symmetrical complex ions often have structures similar to those of A_mX_n compounds; see, for example, the section on the NaCl structure in Chapter 6.

Relations between crystal structures

Our aim in the first part of this book is to set out the basic geometry and topology which is necessary for an understanding of the 3-dimensional systems of atoms which constitute molecules and crystals and to enable us to describe the more important structures in the simplest possible way. In view of the extraordinary variety of atomic arrangements found in crystals it is important to look for any principles which will simplify our task. We therefore note here some relations between crystal structures which will be found helpful.

(a) The same basic framework may be used for the structures or crystals with relatively complicated chemical formulae, as will be illustrated for the diamond net in Chapter 5.

(b) Structures may be related to a simple $A_m X_n$ structure in one of the following ways:

    (I) Substitution structures
        (i) regular (superstructures),
        (ii) random.

   (II) Subtraction and addition structures
        (i) regular,
        (ii) random.

 (III) Complex groups replacing A and/or X.

 (IV) Distorted variants
        (i) minor distortion,
        (ii) major distortion.
        (A minor distortion implies that the topology of the structure has not radically altered, further distortion leading to changes in coordination number.)

We shall illustrate these relations by dealing in some detail with some of the simpler structures in Chapter 6. It will also be shown how the CdI_2 layer (Fig. 6.14) is used in building the structures of a number of 'basic salts', some with layer and others with 3D structures.

(c) The structures of some oxides and complex oxides are built of blocks or slices of simpler structures. Examples described in later chapters include oxides of Ti and V with formulae $M_n O_{2n-1}$ built from blocks of rutile structure and oxides of Mo and W (for example $M_n O_{3n-1}$) related in a similar way to the structures of MoO_3 and WO_3 (shear structures). More complex examples include oxides $Sr_{n+1} Ti_n O_{3n+1}$ formed from slices of the perovskite (ABX_3) structure and the 'β-alumina' ($NaAl_{11}O_{17}$) and magnetoplumbite ($PbFe_{12}O_{19}$) structures in which slices of spinel-like ($AB_2 X_4$) structure are held together by Na^+ and Pb^{2+} ions respectively.

2

Symmetry

Symmetry elements

Symmetry, in one or other of its aspects, is of interest in the arts, mathematics, and the sciences. The chemist is concerned with the symmetry of electron density distributions in atoms and molecules and hence with the symmetry of the molecules themselves. We shall be interested here in certain purely geometrical aspects of symmetry, namely, the symmetry of finite objects such as polyhedra and of repeating patterns. Inasmuch as these objects and patterns represent the arrangements of atoms in molecules or crystals they are an expression of the symmetries of the valence electron distributions of the component atoms. In the restricted sense in which we shall use the term, symmetry is concerned with the relations between the various parts of a body. If there is a particular relation between its parts the object is said to possess certain *elements of symmetry*.

The simplest symmetry elements are the centre, plane, and axes of symmetry. A cube, for example, is symmetrical about its body-centre, that is, every point (xyz) on its surface is matched by a point $(\bar{x}\bar{y}\bar{z})$. It is said to possess a centre of symmetry or to be centrosymmetrical; a tetrahedron does not possess this type of symmetry. Reflection of one-half of an object across a plane of symmetry (regarded as a mirror, hence the alternative name mirror plane) reproduces the other half. It can easily be checked that a cube has no fewer than nine planes of symmetry. The presence of an n-fold axis of symmetry implies that the appearance of an object is the same after rotation through $360°/n$; a cube has six 2-fold, four 3-fold, and three 4-fold axes of symmetry. We postpone further discussion of the symmetry of finite solid bodies because we shall adopt a more general approach to the symmetry of repeating patterns which will eventually bring us back to a consideration of the symmetry of finite groups of points.

Repeating patterns, unit cells, and lattices

A repeating pattern is formed by the repetition of some unit at regular intervals along one, two, or three (non-parallel) axes which, with the repeat distances, define the *lattice* on which the pattern is based. For a 1-dimensional pattern the lattice is a line, for a 2-dimensional pattern it is a plane network, and for a 3-dimensional pattern a third axis is introduced which is not coplanar with the first two (Fig. 2.1). The parameters required to define the three types of lattice in their most general forms are

Lattice	Lattice translations *(repeat distances)*	Interaxial *angles*
1-dimensional	a	–
2-dimensional	a, b	γ
3-dimensional	a, b, c	γ, α, β

These parameters define the *unit cell* (repeat unit) of the pattern, which is accentuated in Fig. 2.1. Any point (or line) placed in one unit cell must occupy the same relative position in every unit cell, and therefore any pattern, whether 1-, 2-, or 3-dimensional, is completely described if the contents of one unit cell are specified.

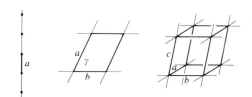

FIG. 2.1.   One-, two-, and three-dimensional lattices.

Clearly, every lattice point has the same environment. A further property is that if along any line in the lattice there are lattice points distance x apart then there must be points at this separation (and no other lattice points) along this line when produced indefinitely in either direction. (We refer to this property of a lattice when we describe the closest packing of spheres in Chapter 4.) Note that the lattice has no physical reality; it does not form part of the pattern.

One- and two-dimensional lattices; point groups

The strictly 1-dimensional pattern is of somewhat academic interest since only points lying along a line are permitted. The only symmetry element possible is an inversion point (symmetry centre) and accordingly the pattern is either a set of points repeating at regular intervals a, without symmetry, or pairs of points related by inversion points (Fig. 2.2(a)). Although the inversion points (small black dots) were inserted only at the points of the lattice (that is, a distance a apart) further inversion points appear midway between the lattice points. This phenomenon, the

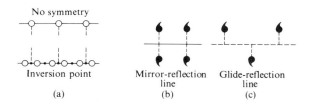

FIG. 2.2.   Symmetry elements in (a) one-dimensional, (b) and (c) two-dimensional patterns.

appearance of additional symmetry elements midway between those inserted at lattice points, is an important feature of 2- and 3-dimensional symmetry groups.

Two-dimensional patterns introduce two concepts that are of importance in the 3-dimensional patterns which are our main concern. First, the inversion (reflection) point is replaced by a reflection line (mirror line) (Fig. 2.2(b)) and two new types of symmetry element are possible, involving *translation* and *rotation*. The *glide-reflection line* combines in one operation reflection across a line with a translation of one-half the distance between lattice points (Fig. 2.2(c)). The translation is necessarily one-half because the point must repeat at intervals of the lattice translation. An *n*-fold *rotation* element produces sets of points related by rotation through 360°/*n*, as, for example, the vertices of a regular *n*-gon. (When considering plane patterns we should imagine ourselves to be 2-dimensional beings capable of movement only in a plane and unaware of a third dimension. The operation which produces a set of *n* points arranged symmetrically around itself in the plane is strictly a 'rotation point'. It is, however, easier for three-dimensional beings to visualize such a rotation point as the point of intersection with the plane of an *n*-fold symmetry axis normal to the plane.) When illustrating symmetry elements it is convenient to use an asymmetric shape such as a comma so that the diagram shows only the intended symmetry. For example, Fig. 2.3(a) illustrates only 4-fold rotational symmetry, whereas (b) possesses also reflection symmetry. We shall see shortly that it is important to distinguish between the simple 4-fold symmetry (class 4) and the combination of this symmetry with reflection symmetry (class 4*mm*). These symbols are explained later in this chapter.

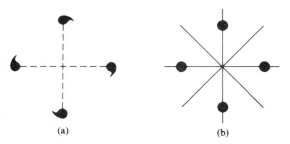

(a)   (b)

FIG. 2.3. Representation of symmetry elements (see text).

We now come to the second point concerning plane patterns. An isolated object (for example, a polygon) can possess any kind of rotational symmetry but there is an important limitation on the types of rotational symmetry that a plane repeating pattern *as a whole* may possess. The possession of *n*-fold rotational symmetry would imply a pattern of *n*-fold rotation axes normal to the plane (or strictly a pattern of *n*-fold rotation points in the plane) since the pattern is a repeating one. In Fig. 2.4 let there be an axis of *n*-fold rotation normal to the plane of the paper at *P*, and at *Q* one of the nearest other axes of *n*-fold rotation. The rotation through 2π/*n* about

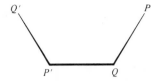

FIG. 2.4. Axial symmetry possible in plane patterns (see text).

Q transforms P into P' and the same kind of rotation about P' transforms Q into Q'. It may happen that P and Q' coincide, in which case $n = 6$. In all other cases PQ' must be equal to, or an integral multiple of, PQ (since Q was chosen as one of the nearest axes), i.e. $n \leqslant 4$. The permissible values of n are therefore 1, 2, 3, 4 and 6. Since a 3-dimensional lattice may be regarded as built of plane nets the same restriction on kinds of symmetry applies to the 3-dimensional lattices, and hence to the symmetry of crystals.

In Fig. 2.1 we illustrated the most general form of the 2-dimensional lattice, but it is clear that if a pattern has 3-, 4-, or 6-fold symmetry the lattice also will be more symmetrical than that of Fig. 2.1. It is found that there are five, and only five, 2-dimensional lattices consistent with the permissible symmetry of 2-dimensional patterns (Fig. 2.5).

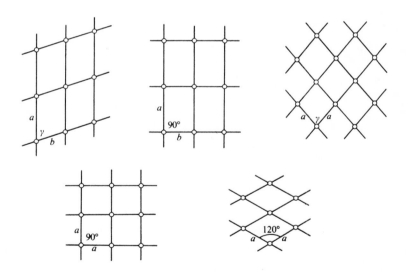

FIG. 2.5. The five plane lattices.

We now have to find which combinations of symmetry elements are consistent with the five fundamental plane lattices. Let us consider the square lattice. We may first have simply a 4-fold axis (□) at each point of the lattice (Fig. 2.6(a)). This automatically introduces a 4-fold axis at the centre of each square unit cell and

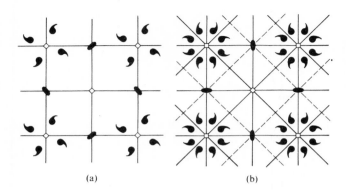

FIG. 2.6. 4-fold symmetry in plane patterns.

2-fold axes (●) at the mid-points of the sides. Any point placed in the plane is repeated in the manner shown as sets of four points (commas) arranged around each point of the lattice (unless it lies *on* the axis, when it will not be repeated). We could also make the pattern symmetrical across each side of the cell, as at (b), by making the edges of the cells mirror reflection lines. This combination of rotational and reflection symmetry turns a single point into a group of eight points around each lattice point. By finding all the permissible (different) combinations of symmetry with the various 2-dimensional lattices we arrive at a total of 17 *plane groups*, which form the bases of all 2-dimensional patterns.

Evidently different symmetries are associated with certain points in the unit cell. For example, in Fig. 2.6(a) the corners and centre of the unit cell have associated with them 4-fold symmetry, and the mid-points of the edges 2-fold symmetry. Similarly in (b) four reflection lines and 4-fold rotational symmetry are associated with the origin, but only two perpendicular reflection lines and 2-fold symmetry with the mid-points of the sides. The set or combination of symmetry elements associated with a point in a repeating pattern is called the *point group*. The symmetry elements all pass through the point and generate a set of symmetry-related (*equivalent*) points around it. There are ten different combinations (point groups) in two dimensions.

Three-dimensional lattices; space groups

As in the case of the 1- and 2-dimensional patterns we consider first the various possible lattices on which the patterns are based and then the possible combinations of symmetry elements which can be associated with the lattices. There are fourteen 3-dimensional lattices consistent with the types of rotational symmetry which a 3D repeating pattern may possess. These infinite 3D frameworks are the 14 Bravais lattices (Fig. 2.7 and Table 2.1). The repeat distances (unit translations) along the axes define the unit cell, and the full lines in Fig. 2.7 show one unit cell of each lattice.

In a simple or primitive (*P*) lattice there are lattice points only at the corners of

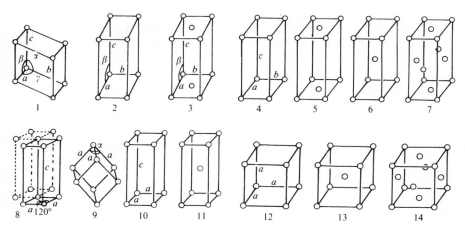

FIG. 2.7. The 14 Bravais lattices.

the unit cell. If there is a lattice point at the centre of the unit cell the lattice is described as body-centred (*I* lattice), and if there are lattice points at the centres of some or all of the faces (in addition to those at the corners) the lattice is described as *A*-, *B*-, or *C*-face-centred, or as an *F* lattice if it is centred on all faces. The unit cell of a centred lattice contains more than one lattice point (two for *I*, *A*, *B*, or *C*, and four for *F* lattices); a smaller cell containing only one lattice point could be found, but such cells correspond to lattices of lower symmetry already included in Fig. 2.7. It will be observed that Fig. 2.7 and Table 2.1 include only a limited number of centred lattices. There is, for example, no *B*-face-centred monoclinic lattice listed, because such a lattice may alternatively be described as a monoclinic *P* lattice with a different β angle. Similarly there is a *C*-face-centred orthorhombic lattice in Table 2.1 but no *C*-face-centred tetragonal lattice. The latter could equally well be described as a *P* lattice having *a* and *b* axes at 45° to those of the *C* lattice, whereas

TABLE 2.1

The 14 Bravais lattices

		Position in Fig. 2.7			Position in Fig. 2.7
Triclinic	*P*	1	Hexagonal	*P*	8
Monoclinic	*P*	2	Rhombohedral	*R*	9
	C (or *A*)	3	Tetragonal	*P*	10
Orthorhombic	*P*	4		*I*	11
	C (*B* or *A*)	5	Cubic	*P*	12
	I	6		*I*	13
	F	7		*F*	14

in the orthorhombic case the change to the smaller cell would lead to an interaxial angle not equal to 90°. It may easily be verified that a tetragonal F lattice reduces to the tetragonal I lattice. The only types of centred lattice consistent with cubic symmetry (see p. 50) are the body-centred and all-face-centred lattices.

The symmetry elements which can be inserted into 3-dimensional lattices are more numerous than those associated with 2-dimensional lattices. In addition to inversion (centre of symmetry), reflection (mirror plane), and simple rotational symmetry (axes of simple n-fold rotation, where $n = 1, 2, 3, 4,$ or 6), there are *axes of rotatory inversion* and two kinds of operation involving translation, namely, *glide planes* and *screw axes*. An axis of rotatory inversion, $\bar{n}$, combines the operations of rotation through 360°/n with *simultaneous* inversion through a centre of symmetry. For example, the axis $\bar{4}$ (normal to the plane of the paper) converts a point (xyz) into a set of four points, as shown in Fig. 2.8, in which points above and

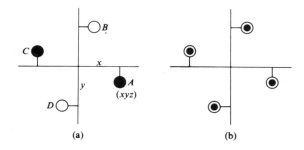

(a)   (b)

FIG. 2.8. Operation of a rotatory–inversion axis (see text).

below the plane of the paper are distinguished as full and open circles respectively. Rotated in a clockwise direction through 90° and then inverted A becomes B ($y\bar{x}\bar{z}$), B becomes C ($\bar{x}\bar{y}z$), and C becomes D ($\bar{y}x\bar{z}$). It should be emphasized that the two operations implied by an axis $\bar{n}$ are not separable, that is, $\bar{4}$ is not equivalent to a 4-fold rotation axis plus a centre of symmetry. This combination produces the set of eight points shown in Fig. 2.8(b) as compared with the four points produced by $\bar{4}$. It can easily be verified that $\bar{1}$ is a centre of symmetry, $\bar{2}$ is equivalent to a plane of symmetry (also written m), $\bar{3}$ to an ordinary 3-fold axis plus a centre of symmetry, and $\bar{6}$ to a 3-fold axis plus a plane of symmetry perpendicular to it (Fig. 2.9). The symbol $3/m$ is simply a convenient way of printing $\frac{3}{m}$ (3 over m) and indicates that the plane of symmetry is perpendicular to the 3-fold axis. The symbol $3m$ indicates that the plane is parallel to the symmetry axis. It is instructive to examine solid objects such as polyhedra and models of crystals which possess various types of symmetry, when it is found that $\bar{3}$ generates the six faces of a rhombohedron, $\bar{4}$ those of a tetrahedron, and $\bar{6}$ those of a trigonal bipyramid. When describing the symmetry of a polyhedron the symmetry element is operating on faces and not on points as in Figs. 2.8 and 2.9.

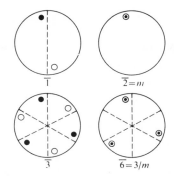

FIG. 2.9. The four kinds of rotatory–inversion axis.

The *glide plane* is the 3-dimensional analogue of the glide-reflection line of the 2-dimensional patterns. As its name implies, it combines in one operation a movement with a reflection. If we imagine a point A (Fig. 2.10) on one side of a mirror moved first to A' and then reflected through the plane of the mirror to B, then A is converted into B by the operation of the glide plane. The same operation performed on B would bring it to C, the translation being always a constant amount $\frac{1}{2}a$, where a is the unit translation of the lattice. As we noted earlier for the glide-reflection

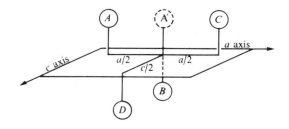

FIG. 2.10. Operation of a glide plane.

line the translation associated with a glide plane must be one-half of the lattice translation because the point must repeat at intervals equal to the lattice translation. A more complex type of glide plane transforms A into D, this involving translations of $\frac{1}{2}a + \frac{1}{2}c$, followed by reflection. If the unit translations a and c of the crystal lattice are not equivalent we clearly have three types of glide plane, with which are associated translations of $\frac{1}{2}a$, $\frac{1}{2}c$, and $\frac{1}{2}a + \frac{1}{2}c$ respectively; the corresponding symbols are a, c, and d (for diagonal).

The *screw axis* derives its name from its relation to the screw thread. Rotation about an axis combined with simultaneous translation parallel to the axis traces out a helix, which is left- or right-handed according to the sense of the rotation. Instead of a continuous line on the surface of a cylinder there could be a series of discrete points, one marked after each rotation through $360°/n$. After n points we arrive back at one corresponding to the first but translated by x, the pitch of the helix, which in a 3-dimensional pattern corresponds to a lattice translation. The symbol for a screw axis indicates the value of n (rotation through $360°/n$) and, as a subscript, the translation in units of x/n where x is the pitch. The translation associated with each rotation of an n-fold screw axis may have any value from x/n to $(n-1)x/n$, and therefore the possible types of screw axis in periodic 3D patterns are the following:

$$2_1; \quad 3_1 \text{ and } 3_2; \quad 4_1, 4_2, \text{ and } 4_3; \quad 6_1, 6_2, 6_3, 6_4, \text{ and } 6_5.$$

A convenient way of showing the sets of points generated by screw axes is to represent them by sets of figures giving the heights of the points above the plane of the paper in terms of x/n:

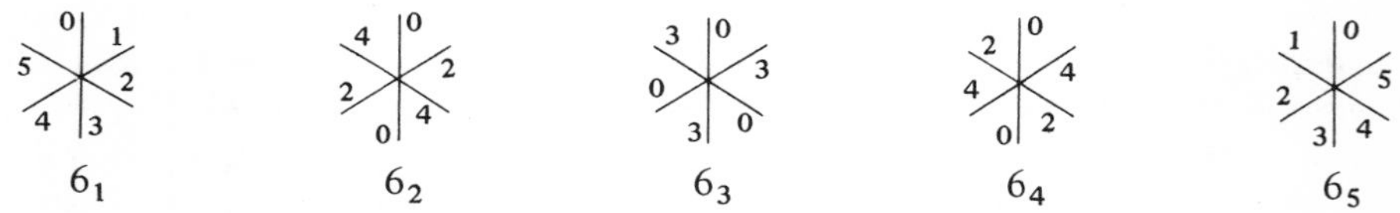

$$6_1 \qquad 6_2 \qquad 6_3 \qquad 6_4 \qquad 6_5$$

Starting in each case at the top of the diagram (height 0) and proceeding clockwise each point rises x/n for each rotation through $360°/n$. For 6_1 all the six points successively generated by the axis fall within one lattice translation normal to the plane of the paper, being at heights 0, 1/6, 2/6, 3/6, 4/6, and 5/6, but for the other 6-fold screw axes the points extend through 2, 3, 4, or 5 unit cells. For 6_4, for example, the heights of successive points are 0, 4, 8, 12, 16, and 20 units of $x/6$. However, since by definition each unit cell must contain the same arrangement of points we may deduct 6 or multiples of 6 from these values, giving 0, 4, 2, 0, 4, 2, a set of six points lying within one unit cell. Examination of the above diagrams shows that 6_1 and 6_5 are related as clockwise and anticlockwise helices, and similarly for 6_2 and 6_4.

The four types of symmetry element, axes of simple rotation or rotatory inversion, screw axes, and glide planes, have now to be inserted into the appropriate lattices. The only symmetry elements consistent with the first Bravais lattice of Fig. 2.7 (the triclinic lattice) are the axes 1 and $\bar{1}$, the former implying no symmetry and the latter a centre of symmetry. The highest rotational symmetry consistent with the lattices 2 and 3, which have two interaxial angles of 90° and one of β (hence the name monoclinic) is a 2-fold axis (2 or 2_1), the direction of which must correspond to the b axis. Alternatively, or in addition, one plane of symmetry is permissible, which must be perpendicular to the b axis. This may be a mirror plane (m or $\bar{2}$) or a glide plane. It is found that a total of 14 types of 3-dimensional symmetry (space-groups) can be associated with the two monoclinic lattices. It is interesting that the very considerable problem of determining the total number of space groups arising from all fourteen Bravais lattices was being studied independently during the same period (1885–1894) by Fedorov in Russia, Schoenflies in Germany, and Barlow in England. It was established that there are 230 such space groups.

Point groups; crystal systems

We saw earlier that different symmetries are associated with particular positions in a plane pattern, and that the number of point groups in 2-dimensional patterns is ten. For 3D patterns the number of point groups is 32. These point groups are the possible symmetries of finite groups of points arranged around particular positions in a lattice; they cannot therefore include symmetry elements involving translations, namely, screw axes or glide planes, though they may include axes of rotatory inversion. We have approached the subject of symmetry from the standpoint of repeating patterns and lattices, leading to the enumeration of the symmetries of 3D patterns (the 230 space groups) and incidentally of the combinations of symmetry elements that can be associated with particular points in a lattice (the 32 point groups). Historically the development of symmetry theory was very different.

The science of crystallography began, in the seventeenth century, with the study of the shapes of crystals. It was observed that there is considerable variation in the overall shape of crystals of a particular substance (or of crystals of one form if it is polymorphic), but that however much a crystal departed from the 'ideal' shape, such as a cube or regular prism, the angles between corresponding pairs of faces were constant on all crystals. For this reason the crystallographer always works with face-normals, and the symmetry of a crystal shape (and hence its point symmetry) refers to the symmetry of the set of face-normals and not to the actual (often distorted) shape of a particular crystal. Variations in the relative developments of crystal faces are described as variations in *crystal habit*; they are not characteristic of the internal structure of the crystal but are attributable to variations in external conditions, for example, nearness to other crystals or presence of impurities in the solution or melt. Some examples are shown in Fig. 2.11.

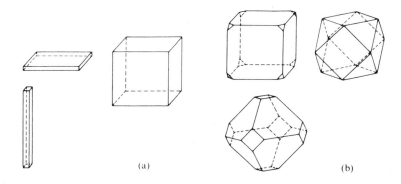

FIG. 2.11. Variation in habit of crystals. Different relative development of (a) cube faces; and (b) cube and octahedron faces.

Owing to the regular internal structure of a crystal the symmetry of its external shape (the 'ideal' face development as defined above) is subject to the same limitations as regards types of rotational symmetry that apply to any 3D repeating pattern. Also, the symmetry elements describing the shape of a crystal must all pass through a point and cannot involve translations, since they describe the arrangement of faces on a finite crystal. The 32 classes of crystal symmetry, which were derived as early as 1830, are therefore the 32 point groups to which we have already referred. They are grouped into seven crystal systems as shown in Table 2.2, which gives the Hermann–Mauguin symbols used by crystallographers and also the earlier Schoenflies symbols which are still favoured by spectroscopists.

TABLE 2.2

The thirty-two classes of crystal symmetry

	Hermann–Mauguin	Schoenflies		Hermann–Mauguin	Schoenflies
Triclinic	1	C_1	Trigonal	3	C_3
	$\bar{1}$	C_i, S_2		$\bar{3}$	C_{3i}, S_6
				32	D_3
Monoclinic	2	C_2		$3m$	C_{3v}
	m	C_s, C_{1h}		$\bar{3}m$	D_{3d}
	$2/m$	C_{2h}			
			Hexagonal	6	C_6
Orthorhombic	222	D_2, V		$\bar{6}$	C_{3h}
	$mm2$	C_{2v}		$6/m$	C_{6h}
	mmm	D_{2h}, V_h		622	D_6
				$6mm$	C_{6v}
Tetragonal	4	C_4		$\bar{6}m2$	D_{3h}
	$\bar{4}$	S_4		$6/mmm$	D_{6h}'
	$4/m$	C_{4h}			
	422	D_4	Cubic	23	T
	$4mm$	C_{4v}		$m3$	T_h
	$\bar{4}2m$	D_{2d}, V_d		432	O
	$4/mmm$	D_{4h}		$\bar{4}3m$	T_d
				$m3m$	O_h

The characteristic symmetries of the crystal systems and also the parameters required to define the unit cells are summarized in Table 2.3.

Comparison of Fig. 2.7 with Table 2.3 will show that whereas we have listed *trigonal* and hexagonal among the crystal systems, Fig. 2.7 includes *rhombohedral* and hexagonal lattices. The reason for this difference is the following. All crystals with a single axis of 3-fold or 6-fold symmetry can be referred to a hexagonal lattice (and unit cell), that is, all the points shown as open circles in Fig. 2.12 have the

TABLE 2.3

The crystal systems: unit cells and characteristic symmetry

System	Relations between edges and angles of unit cell	Lengths and angles to be specified	Characteristic symmetry
Triclinic	$a \neq b \neq c$ $\alpha \neq \beta \neq \gamma \neq 90°$	a, b, c α, β, γ	1-fold (identity or inversion) symmetry only
Monoclinic	$a \neq b \neq c$ $\alpha = \gamma = 90° \neq \beta$	a, b, c β	2-fold axis (2 or $\bar{2}$) in one direction only (y axis)
Orthorhombic	$a \neq b \neq c$ $\alpha = \beta = \gamma = 90°$	a, b, c	2-fold axes in three mutually perpendicular directions
Tetragonal	$a = b \neq c$ $\alpha = \beta = \gamma = 90°$	a, c	4-fold axis along z axis only
Trigonal† and Hexagonal	$a = b \neq c$ $\alpha = \beta = 90°$ $\gamma = 120°$	a, c	3-fold or 6-fold axis along z axis only
Cubic	$a = b = c$ $\alpha = \beta = \gamma = 90°$	a	Four 3-fold axes each inclined at 54°44′ to cell axes (i.e. parallel to body-diagonals of unit cell)

† Certain trigonal crystals may also be referred to rhombohedral axes, the unit cell being a rhombohedron defined by cell edge a and interaxial angle α ($\neq 90°$).

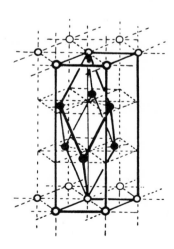

FIG. 2.12. Relation between hexagonal and rhombohedral unit cells.

same environment. Some trigonal crystals also have the special property that the points marked as black circles have the same environment as those at the corners of the hexagonal unit cell. These points are vertices of a rhombohedron (with one-third the volume of the hexagonal cell), and the structure of such a crystal may therefore be referred to rhombohedral axes and a rhombohedral unit cell.

Attention should perhaps be drawn to the characteristic symmetry of the cubic system which is not, as might be supposed, the 4-fold (or 2-fold) axes of symmetry or planes of symmetry but four 3-fold axes parallel to the body-diagonals of the cubic unit cell. This combination of inclined 3-fold axes introduces either three 2-fold or three 4-fold axes which are mutually perpendicular and parallel to the cubic axes. Further axes and planes of symmetry may be present but are not essential to cubic symmetry and do not occur in all the cubic point groups or space groups.

Equivalent positions in space groups

We have stated that the arrangement of faces on a crystal (or more accurately the arrangement of face-normals) can be described by its point symmetry. The six triangular faces of a regular hexagonal pyramid are related by a 6-fold axis, but the basal face is not converted into a group of six faces because its normal is coincident with the symmetry axis. Similarly, six of the vertices are related by the symmetry axis but the seventh is a unique point, the difference being that the six basal vertices lie *off* the axis whereas the seventh lies *on* the axis. Similar considerations apply to the more complex systems of symmetry elements constituting the 230 space groups. The number of equivalent points (positions) in a unit cell depends not only on the types of symmetry present but also on the location of a point (atom) relative to the symmetry elements. Figure 2.13 represents the projection of a unit cell of a crystal. Planes of symmetry perpendicular to the paper intersect the plane of the projection in full lines. The planes repeat at intervals of the lattice spacing (unit cell edge) because we are dealing with a 3D lattice, and it will be observed that the planes through the origin also give rise to additional planes of symmetry midway between

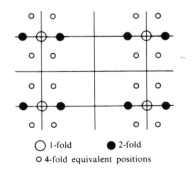

FIG. 2.13. Equivalent positions in a unit cell.

them. A point (atom) situated on the intersection of the symmetry planes is not operated upon by these symmetry elements; it is a 1-fold equivalent position. A point (solid circle) situated on one of the planes is operated on only by the second plane(s) − 2-fold position − while a point (small open circle) which lies on neither plane is operated on by both − 4-fold position. The latter is called the *general* position and the 2- and 1-fold positions are called *special* positions.

Two points should be emphasized. First, according to classical structure theory, all the equivalent positions of a given set should be occupied and moreover they should all be occupied by atoms of the same kind. In later chapters we shall note examples of crystals in which one or both of these criteria are not satisfied; an obvious case is a solid solution in which atoms of different elements occupy at random one or more sets of equivalent positions. (The occupation of *different* sets of equivalent positions by atoms of the *same* kind occurs frequently and may lead to quite different environments of chemically similar atoms. Examples include the numerous crystals in which there is both tetrahedral and octahedral coordination of atoms of the same element—in the same oxidation state—as noted in Chapter 5, and crystals in which there is both coplanar and tetrahedral coordination of Cu(II), p. 1120, or Ni(II), p. 1222). The second point for emphasis is if a molecule (or complex ion is situated at one of the *special* positions it should possess the point symmetry of that position. A molecule lying on a plane of symmetry must itself possess a plane of symmetry, and one having its centre at the intersection of two places of symmetry must itself possess two perpendicular planes of symmetry. If, therefore, it can be demonstrated that a molecule lies at such as position as, for example, would be the case if the unit cell of Fig. 2.13 contained only one molecule (a fact deducible from the density of the crystal), this would constitute a proof of the symmetry of the molecule. Such a conclusion is not, of course, valid if there is any question of random orientation or free rotation of the molecules. Moreover, there is another reason for caution in applying this type of argument to inorganic crystals.

Suppose, for example, that we wish to know the arrangement of X atoms around B in a complex halide A_2BX_4 which is thought to contain planar or tetrahedral BX_4^{2-} ions. We shall further suppose that the X-ray diffraction data show that the B atoms are situated at centres of symmetry. If the BX_4^{2-} ions are finite they cannot be tetrahedral, since a tetrahedral group is not centrosymmetrical. However, the BX_4^{2-} anion could be infinite in extent, and an infinite linear ion consisting of octahedral BX_6 groups sharing opposite edges is also consistent with the observed point symmetry.

One final point about equivalent positions may be noted. There is no restriction on the symmetry of a molecule which occupies a *general* position in a space group (for example, the 4-fold position of Fig. 2.13); it may possess 8-fold, icosahedral, or any type of symmetry or no symmetry at all.

Examples of 'anomalous' symmetry

Special interest attaches to molecules or crystals that possess symmetry which is lower or higher than might be expected. This statement calls for a brief explanation. The packing of equal spheres is discussed in Chapter 4 where we shall see that the majority of metals adopt certain highly symmetrical structures. Some metals, however, crystallize with less symmetrical variants of these structures, and these are obviously of interest since the lower symmetry presumably indicates some peculiarity in the bonding in such crystals. Some crystals containing only highly symmetrical ions (for example FeO) exhibit lower symmetry when cooled. A particularly interesting case is that of metallic tin, the *higher* temperature form of which has the *lower* symmetry. There are a number of compounds which do not adopt the most symmetrical form of a particular structure but a distorted variant of the structure. In some cases a reasonable explanation can be given for the lower symmetry as, for example, some kind of interaction between metal atoms in NbI_4 or VO_2 or the asymmetry of the d^9 configuration of Cu^{2+} leading to the distorted rutile structure of CuF_2. In other cases there is as yet no obvious explanation for the lower symmetry, for example, of crystalline PdS as compared with PtS.

Conversely there are many cases where the symmetry is *higher* than might be expected. Many crystals containing complex ions have, either at room temperature or higher temperatures, structures indicating a symmetry for the complex ion higher than that corresponding to its known (or expected) structure (for example, spherical symmetry for OH^- or C_2^{2-}). In such cases the higher symmetry results from random orientation or free rotation of the non-spherical group. We have already mentioned the random occupation of sets of equivalent positions by atoms of more than one kind in solid solutions; the random arrangement of vacancies in a 'defect' structure can also lead to retention of the symmetry of the ideal structure.

It may appear surprising that molecules or ions presumably possessing symmetry do not always exhibit this symmetry in crystals, that is, they occupy positions of lower point symmetry. If the molecular symmetry is of a non-crystallographic type, for example, the 5-fold symmetry of a planar ring or an icosahedral group, clearly this cannot be exhibited in the crystal. The highest symmetry such a group could show would be, for example, a plane or 2-fold axis. The croconate ion in $(NH_4)_2C_5O_5$ has exact 5-fold symmetry to within the accuracy of the structure determination, but the ions must pack in a way consistent with one of the 230 space groups. Similarly, even if the molecules possess symmetry of a crystallographic type (e.g. 4- or 6-fold axis), the basic requirement is that they pack efficiently, and this may not be possible if they are arranged with their symmetry axes parallel, as would be necessary in a structure with tetragonal or hexagonal symmetry.

Isomerism

Isomerism is a rather comprehensive term embracing several types of structural differences between molecules (ions) having the same chemical composition. It is therefore closely related to polymorphism, for both are concerned with differences between the spatial arrangements of a given set of atoms which in the one case form a finite group and in the other an infinite array (crystal). The structural differences between isomers range from those between *topological isomers*, which represent different ways of connecting together the same set of atoms and are usually regarded as different chemical compounds, to those between a pair of *optical isomers*, which are structurally and chemically identical in every way except that they are related as object and mirror image. Because optical activity can be exhibited by both molecules and crystals and because molecules or ions having this property do not necessarily exhibit any other type of isomerism, we shall discuss optical activity separately. Between these two extremes lie various types of *geometrical isomerism* or *stereoisomerism*. As the term implies, stereoisomerism includes all cases where a molecule (ion) can be obtained in two (or more) distinct forms which differ in the spatial arrangement of the constituent atoms but not in their topology. Thus in principle, one isomer could be converted into another simply by relative movements of certain of the atoms without breaking and remaking any of the σ bonds.

Between a pair of geometrical isomers there is an energy barrier which may be due to π-bonding (as in *cis–trans* isomerism of substituted ethylenes or the eclipsed and staggered forms of ferrocene) or to the fact that interconversion takes place through an intermediate with different geometry (for example, the rearrangement of a *cis* square-planar complex into the *trans* isomer via a tetrahedral intermediate). Geometrical isomers show differences in physical properties and in rare cases are recognized as different chemical compounds (for example, fumaric and maleic acids, the *trans* and *cis* forms of HOOC.CH=CH.COOH).

The subject of isomerism is closely concerned with symmetry. In the formulae (i)–(iv) X represents an atom (H or halogen) or group (e.g. CH_3) attached by a single bond. Molecules of type (i) are formed by O and S and molecules of types (ii) and (iv) by N and P. In (i) the two X atoms are not coplanar with the two O (S) atoms; the molecule is *enantiomorphic*, that is, it cannot be brought into coincidence with its mirror image; it exists in left- and right-handed forms.

| (i) | (ii) | and | (iii) | (iv) |

Description of the molecular geometry requires a knowledge of the dihedral angle XOO/OOX. On the other hand all atoms of (ii) are coplanar, and such molecules exist in *cis* and *trans* forms. The four atoms of (iii) are collinear, it has cylindrical symmetry, and there are no isomers. A number of molecules of type (iv) are formed by nitrogen and phosphorus and they provide examples of two of the three extreme

eclipsed (*cis*)   staggered (*trans*)   *gauche*

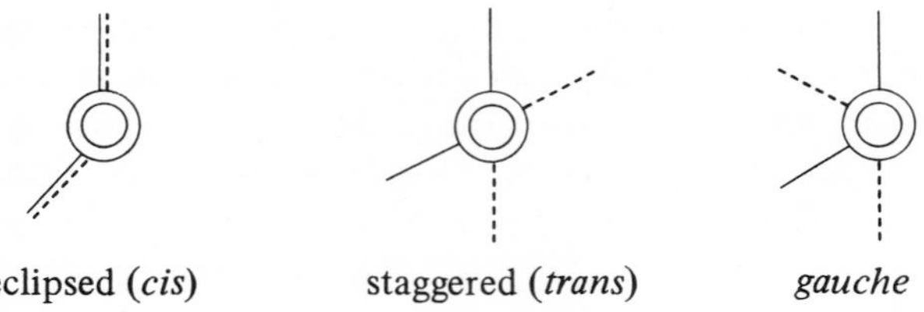

configurations possible for a molecule of this kind. The sketches show the molecule viewed along the N–N (P–P) axis. Examples include:

staggered: P_2H_4 (solid) P_2F_4 P_2Cl_4 P_2I_4
gauche: N_2H_4 P_2H_4 (gas)
staggered and *gauche*: N_2F_4

The dithionite ion ($S_2O_4^{2-}$) provides an example of a group of this kind with the eclipsed configuration. Examples of more complex types of isomerism, of hetero-polyacid ions, are described in Chapter 11.

The numbers and types of isomers of molecules such as Ma_2b_2 or Ma_2b_4 etc. depend on the spatial arrangement (symmetry) of the bonds around the central atom. Much of the 'classical' work on isomerism was expressly designed to distinguish between one or more plausible bond arrangements, such as the coplanar or tetrahedral arrangement of four bonds, or octahedral or trigonal prismatic arrangement of six bonds. Although determination of bond arrangements by enumeration of isomers and the resolution of optically active compounds has been superseded by direct structural studies, the older methods played an important part in the development of structural chemistry, and a few examples will be mentioned later.

Structural (topological) isomerism

As we have already remarked structural isomers are usually different chemical compounds, as in the case of NH_4NCO and $CO(NH_2)_2$, the earliest example of a pair of compounds to which the term 'isomers' was applied. This type of isomerism is extremely common in organic chemistry, simple examples being hydrocarbons such as normal and iso butane, *o*-, *m*- and *p*-substituted benzenes, and so on. The special case of two interlocked *n*-rings, isomeric with the 2*n*-ring, is mentioned in Chapter 3 as the simplest example of this kind of topological isomer. If the isomerism involves only movement of H atoms the term *tautomerism* is used, and this also is frequently encountered in organic chemistry, a simple example being the keto-enol tautomerism of acetylacetone:

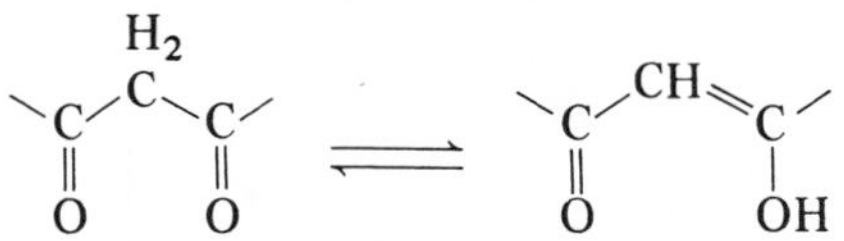

Thiocyanic acid (a), is tautomeric, existing essentially (95 per cent) in the iso form. Esters of both types, RSCN and SCNR, are known, and in some metal-coordination compounds the SCN ligand is attached to the metal through S and in others through N. On the other hand, the oxygen analogue cyanic acid exists entirely in the iso form and forms only one type of ester, RNCO. Both isomers (b) of S_2F_2

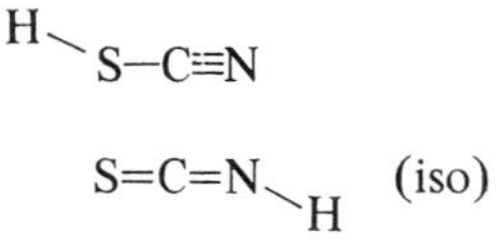

$$H\diagdown S{-}C{\equiv}N$$

$$S{=}C{=}N\diagdown H \quad \text{(iso)}$$

(a)

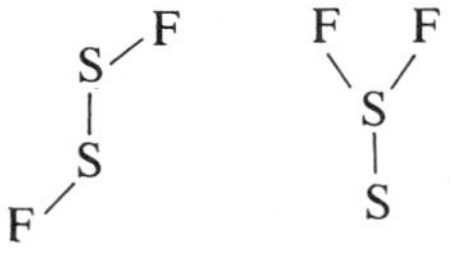

(b)

are formed when a mixture of dry AgF and S is heated in a glass vessel *in vacuo* to 120 °C. The 'symmetrical' isomer FSSF has a dihedral structure similar to that of H_2O_2 (and is enantiomorphic), while SSF_2 is pyramidal and very similar geometrically to OSF_2. Although nitrous acid could be formulated as at (c) and (d) there is no evidence for the existence of (c), the gas being a mixture of the *cis* and *trans* isomers (d).

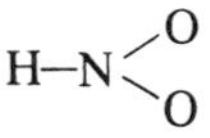

$$H{-}N\diagdown O^{O}$$

(c)

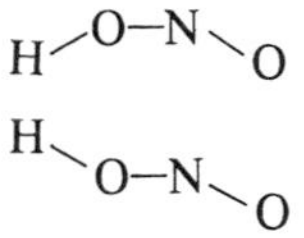

$$H\diagdown O{-}N\diagdown O \quad cis$$

$$H\diagdown O{-}N\diagdown O \quad trans$$

(d)

More complex examples of isomerism include that of carboranes, in which C atoms are introduced into polyhedral frameworks of B atoms, and of compounds such as $S_n(NH)_{8-n}$. These form cyclic molecules similar to the S_8 molecule of elementary sulphur but with some S atoms replaced by NH groups. Three of the four possible isomers of $S_6(NH)_2$ have been characterized, namely the 1,3, 1,4, and 1,5 isomers.

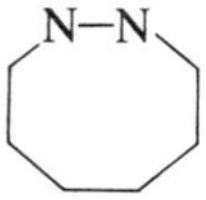

1,2 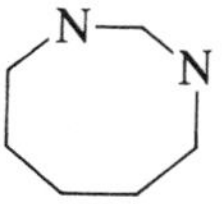   1,3 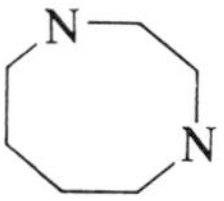   1,4 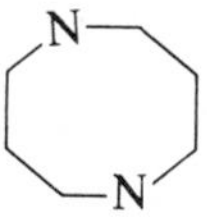   1,5

Coordination compounds provide numerous examples of structural isomerism, of which the following are a selection:

(i) 'Bond' isomerism

$[Co(NH_3)_5ONO]Cl_2$ and $[Co(NH_3)_5NO_2]Cl_2$.

(ii)   'Ionization' isomerism

   $[Pt(NH_3)_4Cl_2]Br_2$ and $[Pt(NH_3)_4Br_2]Cl_2$.

(iii)   'Coordination' isomerism

   $[Co(NH_3)_6] [Cr(CN)_6]$ and $[Co(CN)_6] [Cr(NH_3)_6]$

(iv)   'Polymerization' isomerism

   $[Pt(NH_3)_4] [PtCl_4]$ and $Pt(NH_3)_2Cl_2$.

Geometrical isomerism

Two parts of a molecule which are connected by a single bond may be free to rotate relative to one another, but isomerism arises only if there are two (or more) configurations separated by energy barriers sufficiently large to prevent inter-conversion (see p. 794). The $S_2O_6^{2-}$ ion is found in crystalline $K_2S_2O_6$ with two configurations, the *trans* (centrosymmetrical) and an almost eclipsed configuration. The oxalate ion, $C_2O_4^{2-}$, is planar in the sodium and potassium salts but non-planar in the ammonium salt. These differences are due to interactions with adjacent ions in the crystal and are characteristic of the crystalline state, and are not retained in solution.

Some simple examples of *cis–trans* isomerism have already been noted. More complex examples include those of cyclic molecules such as chair-shaped 6-rings, (a), where a distinction must be made between equatorial 'e' and axial 'a' bonds, and bridged molecules (b); the latter could also exhibit 'structural' isomerism, in the unsymmetrical isomer $a_2MX_2Mb_2$.

(a)   (b) *(trans)*

Much attention has been paid in the past to the isomerism of mononuclear molecules and ions because of the relation between numbers of isomers and the bond arrangement around the central atom. Whereas a tetrahedral molecule Ma_2b_2 exists in only one form a planar molecule or ion of this type can in principle exist in *cis* and *trans* forms, and many pairs of Pt compounds of this type have been prepared. Before it became possible to determine crystal structures there was considerable interest in the numbers of isomers of complexes such as Coa_4b_2 which would be expected to have two isomers if octahedral but three if trigonal prismatic (or coplanar). Two points may be noted in connection with the deduction of bond arrangements from the numbers of isomers. First, failure to obtain the expected number of isomers may be due to large differences in stability or to interconversion in solution followed by crystallization of the less soluble isomer. For example, whereas many compounds Pta_2b_2 are known in both *cis* and *trans* forms, the

$$a \diagdown \underset{b}{M} \diagup a \quad \text{and} \quad a \diagdown \underset{b}{M} \diagup b$$

corresponding Pd compounds are usually known only in the *trans* form (exceptions include $Pd(NH_3)_2Cl_2$ and $Pd(NH_3)_2(NO_2)_2$). In solutions of compounds $(R_3Sb)_2PdCl_2$ the two forms are in labile equilibrium with the *trans* form predominating, but the much less soluble *cis* isomer crystallizes out. Second, it is necessary to confirm that the various forms obtained are in fact geometrical isomers. For example, the isomers of the compound with the empirical composition $Pt(NH_3)_2Cl_2$ are structural isomers, one being a planar molecule $Pt(NH_3)_2Cl_2$ and the other the salt $[Pt(NH_3)_4][PtCl_4]$. A rather similar case, which for a time confused the stereochemistry of tellurium, concerned the existence of two forms of $Te(CH_3)_2I_2$, thought to be *cis* and *trans* isomers of the square planar complex. It has since been shown that one isomer is the salt $[Te(CH_3)_3]^+[TeCH_3I_4]^-$. A number of compounds of cobalt with the composition CoX_2R_2 exist in two forms, one consisting of finite tetrahedral complexes and the other of infinite chains in which the metal atom is octahedrally coordinated.

Optical activity

Optical activity, the property of rotating the plane of polarization of plane-polarized light, may be exhibited by matter in all states of aggregation. When an optically active compound is prepared there will normally be equal numbers of *d*- and *l*-molecules. They may generally be separated by forming a compound (e.g. a salt) with either the *d*- or *l*-form of a second active compound and utilizing differences between the physical properties of the resulting compounds $(d-A)(l-B)$ and $(l-A)(l-B)$ — a process called *resolution* — tough special methods (such as the hand-sorting of crystals) have been used in some cases. The resolution of tri-*o*-thymotide has been carried out[1] by forming a molecular compound with an *inactive* second component (benzene). This was possible because although the unsolvated material crystallizes from methanol as the racemate (see later) the molecular compound formed with benzene is optically active and large single crystals could be grown, each of which was either a *d*- or an *l*-crystal. The *d*- or *l*-forms of a compound exhibit optical activity in the crystalline, liquid, dissolved, and vapour states, though the activity may not persist indefinitely in the last three cases if interconversion of the two forms is possible; an inactive mixture of the two forms then results (*racemization*).

For a finite group of atoms the criterion for enantiomorphism is the absence of an axis of rotatory inversion. Such a group is described as *dissymmetric*, as opposed to *asymmetric*, which implies the absence of all symmetry. An axis $\bar{n}$ implies a centre of symmetry if n is odd; it introduces planes of symmetry if n is a multiple of 2 but not of 4; and if n is a multiple of 4 the system can be brought into coincidence with its mirror image. Of the simplest axes of these three types, $\bar{1}$ is synonymous with a centre of symmetry, and $\bar{2}$ with a plane of symmetry. Since axes of rotatory inversion $\overline{4n}$ are likely to occur very rarely in molecules, we may for practical purposes take as the criterion for enantiomorphism and for optical activity in a *finite* molecule or complex ion the absence of a *centre* or *plane of symmetry*.

One isomer of the tetramethyl-*spiro*-bipyrrolidinium cation provides an example of a molecule with $\bar{4}$ symmetry. Each methyl group can project either above or below the plane of the ring to which it is attached, and the two rings lie in perpendicular planes. There are accordingly four forms of this molecule, which may be shown diagrammatically as set out below. The molecules are viewed along the direction of the dotted line and the thick lines indicate the rings:

| *cis/cis* | *cis/trans* | *trans/trans* | *trans/trans* |
| (active) | (active) | (active) | (meso) |

All four forms have been prepared, the meso *trans/trans*-form being inactive[2], having $\bar{4}$ symmetry.

(Certain substituted cyclobutanes possess this unusual symmetry axis, e.g. that shown in I below. This molecule also, however, possesses diagonal planes of symmetry parallel to the $\bar{4}$ axis, so that its lack of enantiomorphism cannot be attributed to the presence of that axis. These planes of symmetry would disappear if groups C(XYZ) were substituted for the methyl groups, giving the molecule II(a). In the plan at II(b) only these C(XYZ) groups are shown to illustrate the absence of diagonal symmetry planes. This molecule would not be optically active since it is identical with its mirror image, though it contains no fewer than eight asymmetric carbon atoms.)

| I | II(a) | II(b) |

The relation between optical activity and enantiomorphism is not quite so simple for a crystal. Of the thirty-two crystal classes eleven are enantiomorphic:

$$1, 2, 3, 4, 6, 23, 222, 32, 42, 62, \text{ and } 432.$$

A particular crystal having the symmetry of one of these classes is either left- or right-handed, and if suitable faces happen to develop when the crystal grows hand-sorting may be possible. Optical activity is also theoretically possible in four of the non-enantiomorphic classes:

$$\overline{2} \, (= m), \, mm, \, \overline{4}, \text{ and } \overline{4}2m.$$

In these classes directions of both left- and right-handed rotation of the plane of polarization must exist in the same crystal. An earlier claim that optical activity is exhibited by a crystal in class m (mesityl oxide oxalic methyl ester) has since been disproved, but it has been experimentally verified in both $\overline{4}$ ($CdGa_2S_4$)[3] and $\overline{4}2m$ ($AgGaS_2$).[4]

In general the optical activity exhibited by a crystal will persist in other states of aggregation only if it is due to the dissymmetry of the finite molecule or complex ion. In this case it is also necessary that the energy of activation for racemization (change $d \rightleftharpoons l$) must exceed a certain value ($\sim 80 \, kJmol^{-1}$ at room temperature). The optical activity cannot, of course, be demonstrated unless there is a means of resolution (separation of d and l forms) or at least of altering the relative amounts of the two isomers. The optical activity of many crystals (e.g. quartz, cinnabar, $NaClO_3$) arises from the way in which the atoms are linked in the crystal; it is then a property of the crystalline material only.

Optical activity was first studied in compounds of carbon. In 1874 van't Hoff showed how optical activity could arise if the four bonds from a carbon atom were arranged tetrahedrally. Le Bel independently, and almost simultaneously, put forward somwhat similar views. A molecule $Cabcd$ in which a C atom is attached to four different atoms or groups should exist in two forms related as object and mirror image. The central C atom was described as an asymmetric C atom. A simple molecule of this sort is that of lactic acid, $CH(CH_3)(OH)COOH$, which exists in d- and l-forms. There is a third form of such a compound, the racemic form. A true racemate has a characteristic structure different from that of the active forms, there being equal numbers of d- and l-molecules in the unit cell of the crystal, which is optically inactive. We may note that inactive crystalline forms of optically active compounds are not necessarily racemates. For example, the inactive β-phenylglyceric acid (m.p. 141 °C) is not a racemate but the crystals are built of submicroscopic lamellae which are alternately d- and l-rotatory. Slow recrystallization yields single crystals of the d- and l-forms (m.p. 164 °C).[5]

A molecule containing two similar asymmetric C atoms, $cbaC–Cabc$, exists not only in d- and l-forms (which combine to form a racemate) but also in an inactive *meso* form. Although there are possible conformations of the mesotartaric acid molecule with $\bar{1}$ or $\bar{2}$ (m) symmetry, the molecules in the crystalline acid are in fact dissymmetric and arranged in pairs related by centres of symmetry:[6]

d		*l*			crystalline meso acid
	racemate				

A similar configuration is found for the mesotartrate ion in the K and Rb salts; there is a rotation through 19° from the staggered configuration in the dimethyl ester.[7] It is interesting that the molecule of the closely related meso-erythritol is centrosymmetric in the crystal.[8]

We now know that: (i) the presence of asymmetric carbon (or other) atoms does not necessarily lead to optical activity; and (ii) that optical activity may arise in the absence of asymmetric atoms. The criterion is the symmetry of the molecule as a whole. Some examples of (i) have already been given; another one is provided by dimethyl-diketopiperazine. This compound exists in *cis-* and *trans-*forms, each containing two asymmetric carbon atoms. The *trans-*form possesses a centre of symmetry and is therefore not optically active, while the *cis-*form is resolvable into optical antimers (*d-* and *l-*forms).

The numerous examples of (ii) include molecules such as

(in which the bonds shown as dotted and heavy lines lie in a plane perpendicular to and respectively behind and in front of the plane of the paper) and many molecules which cannot adopt the most symmetrical configurations for steric reasons. In some cases the optical activity is due to restricted rotation around a single C—C bond. The presence of groups projecting from the aromatic rings in the molecules (1)–(3) leads to enantiomorphic forms. The replacement of SO_3^- in (3) by the smaller COO^- group destroys the optical activity.

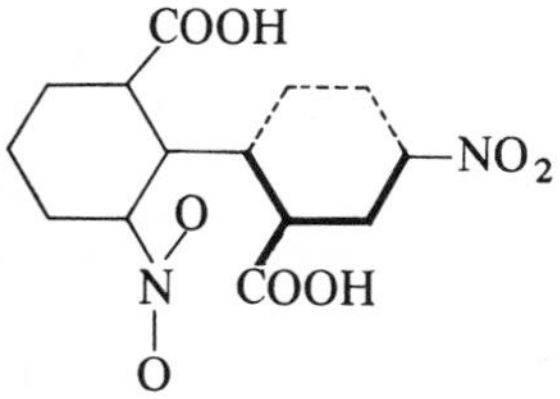

(1) 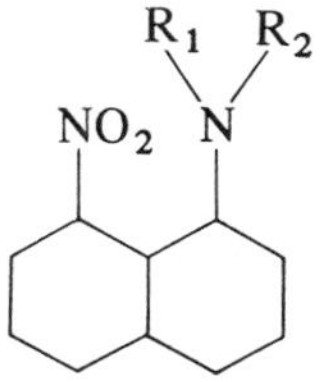 (2) 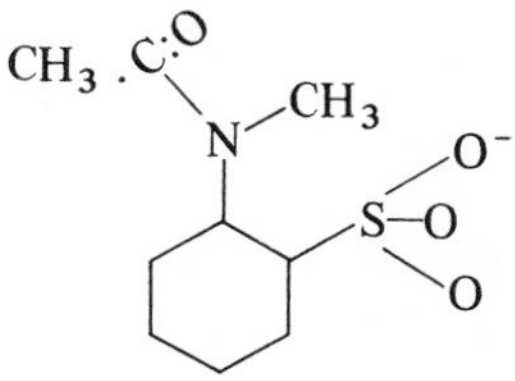 (3)

The molecules of many aromatic compounds would be highly symmetrical if planar, but if the planar configuration would imply impossibly short distances between certain atoms the molecules are forced into non-planar configurations which are often enantiomorphic. Examples of such 'overcrowded' molecules include:

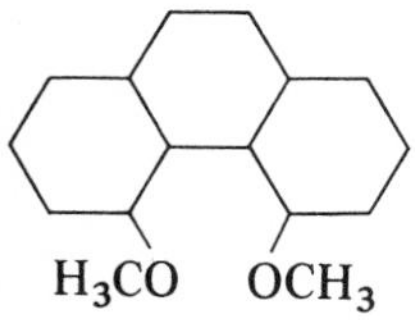

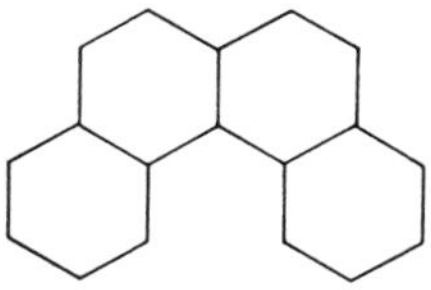

and

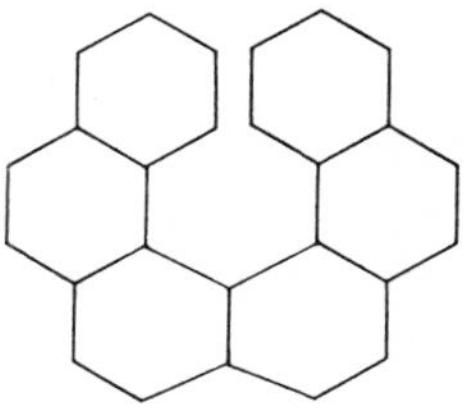

The compound on the right (hexahelicene) is also of interest because it was resolved by forming a crystalline molecular complex with a polynitro compound.[9]

We conclude these remarks on optical activity by giving a few examples of molecules or ions the resolution of which demonstrates the general arrangement of the bonds formed by the central atom.

Three pyramidal bonds

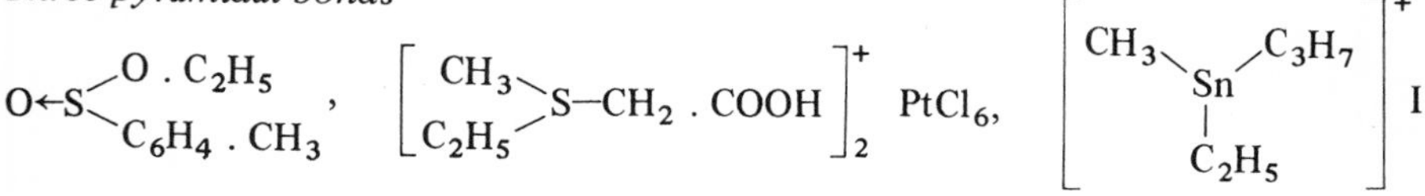

Four tetrahedral bonds

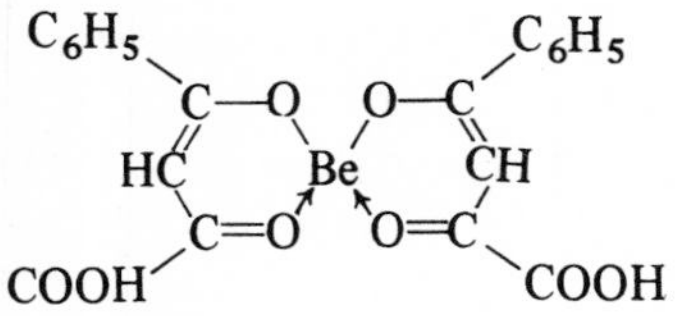

(also Cu^{II} and Zn)

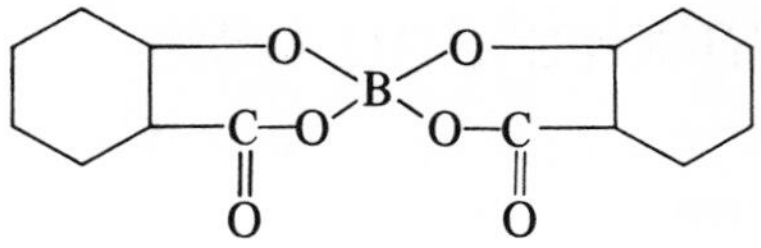

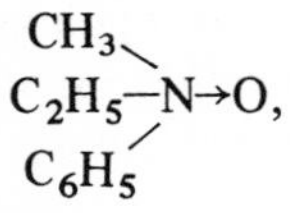

$$CH_3 \diagdown \atop C_2H_5 \!-\! N \!\rightarrow\! O, \atop C_6H_5 \diagup$$

(also P)

$$C_6H_5 . CH_2 \!-\! \underset{C_3H_7}{\overset{C_2H_5}{Si}} \!-\! CH_2 . C_6H_4SO_3H,$$

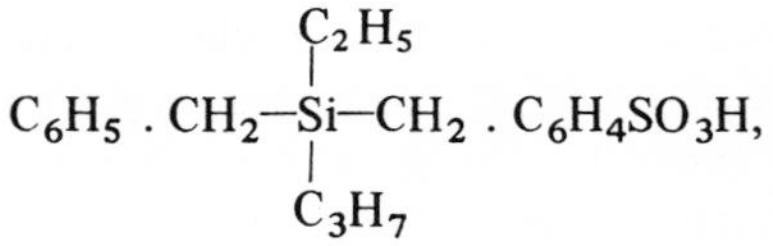

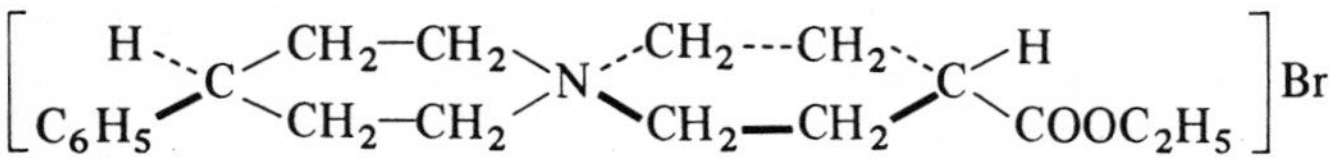

(1) JCS 1952 3747
(2) JACS 1955 **77** 4688
(3) AC 1969 **A25** 633
(4) AC 1968 **A24** 676
(5) ACSc 1950 **4** 1020

(6) AC 1967 **22** 522
(7) AC 1973 **B29** 1278
(8) PRS 1959A **250** 301
(9) JACS 1956 **78** 4765

3

Polyhedra and nets

Introduction

For a number of reasons it seems logical to postpone any discussion of structure and bonding until we have considered some basic geometry and topology:

(i) It is customary to discuss crystal and molecular structures in terms of other known structures. Ideally we should be able to set out the possible types of structure for a molecule or crystal composed of certain numbers of atoms with known bonding requirements so that the observed structures can be compared with all (geometrically and topologically) possible structures, up to some arbitrary limit of complexity. It is important to know why some apparently reasonable structures are never adopted by known molecules or crystals. Systematic studies of possible structure types have not been numerous; we shall survey here the simpler systems of connected points and the closest packing of equal spheres as essential parts of this basic geometry. A knowledge of the possible types of 3D nets throws at least some light on questions such as: why is diamond a system of rings of 6 carbon atoms, and why do certain crystalline forms of B_2O_3 and P_2O_5 consist of rings of 10 B (P) and 10 O atoms? We shall find that the structure of diamond represents the simplest 3D 4-connected net and that these oxide structures are examples of some of the simpler structures that can be formed by units (BO_3 triangles or PO_4 tetrahedra) that are to be joined to three other similar units.

(ii) We noted in Chapter 1 that many molecular and crystal structures represent a compromise between conflicting packing (or bonding) requirements of atoms of various kinds. Studies of the simpler types of structure for compounds A_mX_n in which X atoms are close-packed (Chapter 4) and of the linking together of tetrahedra and octahedra (Chapter 5) reveal interesting restrictions on bond angles which are obviously relevant to discussions of such simple structures as those of rutile and corundum.

(iii) Descriptions of the structures of ionic crystals usually start from a consideration of the environment of the individual ions (relation of coordination polyhedra to relative ionic sizes). However, the (energetically) preferred coordination polyhedron for an individual ion may be impossible (geometrically) in a 3D structure. For example, we find cubic coordination in two of the simplest ionic structures (CsCl and CaF_2) rather than antiprismatic or dodecahedral coordination. These two less symmetrical 8-coordination polyhedra are not possible in 3D AX or AX_2 structures but are important in more complex structures where the geometrical restrictions are less severe.

(iv) Crystal structures can be illustrated in various ways according to the feature

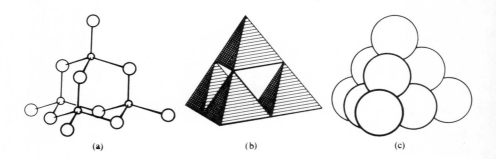

FIG. 3.1. Three representations of the P_4O_{10} molecule.

it is wished to emphasize. For example, the structure of the P_4O_{10} molecule may be shown as a 'ball and spoke' model (Fig. 3.1(a)), showing the atoms as spheres of arbitrary radii, to indicate the bond arrangement around the P and O atoms. Alternatively we may wish to emphasize the tetrahedral coordination of the P atoms, (b), or to focus attention on the group of ten close-packed O atoms, (c). There are similar types of representation of crystal structures, and it is important that the reader should be conversant with the types of illustration used by crystallographers.

(v) While some structures are associated more or less exclusively with bonding of a particular type (for example, the rutile structure of essentially ionic oxides and fluorides AX_2), other simple A_mX_n structures are not indicative of a particular type of bonding. For example, the NaCl structure is adopted not only by ionic oxides and halides but also by transition-metal nitrides and carbides and by some inter-metallic compounds. The physical properties of compounds such as PbS, PbSe, and PbTe (NaCl structure) suggest more complex interactions than the essentially ionic bonding in the alkali halides; the oxides RuO_2 (rutile structure) and ReO_3 are metallic conductors. The zinc-blende structure, with tetrahedral coordination of both kinds of atom, is suitable for both ionic and covalent bonding. It is therefore preferable to describe the simple A_mX_n structures as geometrical entities rather than typical ionic, covalent, or metallic structures.

For these reasons three chapters are devoted to essentially topological and geometrical topics which are basic to an understanding of crystal structures. The present chapter is concerned with the ways in which points can be joined together to form finite or infinite systems, and includes some account of polyhedra and connected systems extending indefinitely in one, two, or three dimensions. In Chapter 4 we consider the packing of spheres, in particular the closest packing of equal spheres. Chapter 5 deals with the two most important coordination polyhedra in inorganic chemistry, the tetrahedron and the octahedron, and attempts a systematic account of the types of structure that can be built from these units by sharing vertices, edges, and/or faces.

The basic systems of connected points

However complex the units that are to be joined together the problem may be reduced to the derivation of systems of points each of which is connected to some number (p) of others. In the simplest systems this number (the *connectedness*) is the same for all points:

$p = 1$: there is one solution only, a pair of connected points.

$p = 2$: the only possibilities are closed rings or an infinite chain.

$p \geqslant 3$: the possible systems now include finite groups (for example, polyhedra) and arrangements extending indefinitely in one, two, or three dimensions (*p*-connected nets).

It is not necessary to include singly-connected points ($p = 1$) in nets since they can play no part in extending the net. Also 2-connected points may be added along the links of any more highly connected net ($p > 2$) without altering the basic system of connected points. We therefore do not include either 1- or 2-connected points when deriving the basic nets though it may be necessary to add them to obtain the structures of actual compounds from the basic nets. For example, 2-connected points (representing —O— atoms) are added along the edges of the 3-connected tetrahedral group of four P atoms to form the P_4O_6 molecule, and additional singly-connected points (representing =O atoms) at the vertices to form P_4O_{10}.

In chain structures the repeating unit may be a single atom or a group of atoms:

Repeat unit

Se ∕Se∖ Se ∕Se∖ ∖ Se

PdCl₂: ∕Pd<Cl∖>Pd<Cl∖>Pd< → $PdCl_2$

SO₃: ∕O∖S∕O∖S∕O∖ → SO_3

but any simple chain may be reduced to the basic form —A—A— (or ⊃A⊃A⊃A etc.), each unit being connected to two others.

Still retaining the condition that all points have the same connectedness we find that there are certain limitations on the types of system that can be formed. For example, there are no polyhedra with all vertices 6-connected, and there is only one 6-connected plane net. These limitations are summarized in Table 3.1, which may be considered in two ways:

TABLE 3.1
Systems of p-connected points

p	Polyhedra	Infinite periodic nets	
		Plane nets	3D *nets*
3			
4			
5			
6		(one only)	
7			
.			
.			
.			
12			

(i) Together with the simple linear systems ($p = 2$) the *vertical* subdivisions correspond to the four main classes of crystal structure, namely, molecular crystals, chain, layer, and three-dimensional (macromolecular) structures. The division into polyhedra, plane, and 3D nets is also the logical one for the systematic derivation of these systems.

(ii) From the chemical standpoint, however, we are more interested in the *horizontal* sections of Table 3.1. If we wish to discuss the structures which are possible for a particular type of unit, for example, an atom or group which is to be bonded to three others, we have to select the systems having $p = 3$.

We shall first say a little about the three vertical subdivisions of Table 3.1, confining our attention for the most part to systems in which $p = 3$ or 4. The coordination numbers from 6 to 12 are more conveniently considered under sphere packings, and the very high coordination numbers (13–16 in many transition-metal alloys, 22 and 24 in other intermetallic compounds, 24 in CaB_6, etc.) will only be mentioned incidentally. We shall then give examples of polyhedral, cyclic, and linear systems and of structures based on 3- and 4-connected nets, dealing at some length with the three most important nets, namely, the plane hexagonal net, plane square net, and the diamond net.

The general topological relations already noted are well illustrated by the structures of the elements chlorine, sulphur, phosphorus, and silicon (Fig. 3.2). The number of bonds formed is $8-N$, where N is the ordinal number of the Periodic Group, that is, 4 for Si, 5 for P, etc.

$p = 1$: Cl–Cl molecule in all states of aggregation.

$p = 2$: cyclic S_6, S_8, and S_{12} in three of the crystalline forms and in their vapours, and infinite chains in plastic S.

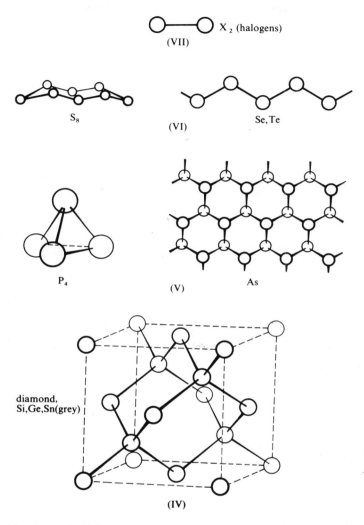

FIG. 3.2. The 8–N rule illustrated for certain elements of the IVth, Vth, VIth, and VIIth Periodic Groups.

$p = 3$: tetrahedral P_4 molecule, 3-connected (buckled) 6-gon layer in black phosphorus, and complex 3-connected layer in red P.

$p = 4$: diamond-like structure of elementary silicon.

The same principles determine the structures of the oxides and oxy-ions of these elements in their highest oxidation states, the structural units being tetrahedral oxy-ions MO_4^{p-}:

p		Oxy-ion				Oxide molecule
1	pyro	$Si_2O_7^{6-}$	$P_2O_7^{4-}$	$S_2O_7^{2-}$	Cl_2O_7	finite
2	meta	$(SiO_3)_n^{2n-}$	$(PO_3)_n^{n-}$	$(SO_3)_n$		cyclic or linear
3	infinite 2D	$(Si_2O_5)_n^{2n-}$	$(P_2O_5)_n$			polyhedral, 2D, or 3D
4	infinite 3D	$(SiO_2)_n$				infinite 3D

Polyhedra

Coordination polyhedra: polyhedral domains

Perhaps the most obvious connection of polyhedra with practical chemistry and crystallography is that crystals normally grow as convex polyhedra. The shapes of single crystals are subject to certain restrictions arising from the fact that only a limited number of types of axial symmetry are permissible in crystals, as explained in Chapter 2. We shall not be concerned here with the external shapes of crystals but with polyhedra which are of interest in relation to their internal structures and more generally to the structures of molecules and complex ions.

There are three obvious ways of describing a simple 3D structure of a compound A_mX_n. One is to describe it as a packing of spheres, usually of two different sizes, an approach developed in Chapter 4. Assuming that each atom (ion) is surrounded in the same way by its nearest neighbours of the other kind and that the numbers of such nearest neighbours are a and x respectively, then $ma = nx$, since these quantities are simply alternative ways of counting the number of A–X bonds in a large volume of the crystal. To simplify the discussion, let us consider a compound AX_n in which A is surrounded by 6 X atoms. If n is less than 6 some X atoms must be common to the coordination groups of a number of A atoms. The structure may therefore be described in terms of the way in which the AX_6 groups share X atoms. This approach is explored in some detail in Chapter 5 for tetrahedral AX_4 and octahedral AX_6 groups.

In the simplest and most symmetrical A_mX_n structures A has a well-defined coordination group consisting of a number of X atoms (often 4 or 6) at the same distance from A, but in some crystals the environment of an A atom is less regular and it is necessary to define what is meant by *coordination number* and *coordination polyhedron*. This can be done by describing the structure in terms of the polyhedral (Dirichlet) domains of the atoms. The Dirichlet domain is the polyhedron enclosed by planes drawn midway between the atom and each of its neighbours and perpendicular to the lines joining them. With each atom is associated its Dirichlet domain, and the whole structure may be represented as a space-filling array of polyhedra, of one or more kinds. The coordination number (c.n.) of an atom is then defined as the number of faces of the domain, and the coordination polyhedron as the polyhedron whose vertices are the mirror images of the centre of the domain in each of its faces; to each *face* of the domain there corresponds a *vertex* of the coordination polyhedron. This concept is not especially useful for simple structures in which an atom (ion) has a well-defined set of approximately equidistant nearest neighbours, but it does provide a more unambiguous definition of c.n. in crystals in which the environment of an atom is less regular. Also, it is the logical way of describing structures composed of atoms of comparable size, notably metals and intermetallic compounds, where it is often difficult to define c.n.'s in terms of distances to nearest neighbours. A simple example is the body-centred cubic structure, in which each atom has 8 nearest neigbours at a distance d and 6 more at a slightly greater distance $(1.15d)$ at the body-centres of the six adjacent unit cells. The domain of an atom in this structure is a truncated octahedron, with 8 hexagonal and 6 square faces corresponding to the two sets of neigbours; the coordination polyhedron is the rhombic dodecahedron, with 14 vertices. The domain concept also offers a definition of an interstice, and so is complementary to the sphere-packing description of certain structures, a point discussed on p. 178.

Space-filling arrangements of polyhedra, in which every polyhedron face is common to two polyhedra, are of interest in another connection. If we imagine atoms placed at the vertices of the polyhedra and bonded along the edges, either directly or through a 2-connected atom such as oxygen, then we have a framework with polyhedral cavities or tunnels. In some structures of this kind the cavities or tunnels are large enough to accommodate or allow the passage of ions or molecules. This topic is discussed briefly at the end of this chapter.

The regular solids

The Greeks had a considerable knowledge of polyhedra, but only during the past two hundred years or so has a systematic study been made of their properties, following the publication in 1758 of Euler's *Elementa doctrinae solidorum*. From Euler's relation between the numbers of vertices (N_0), edges (N_1), and faces (N_2) of a simple convex polyhedron

$$N_0 + N_2 = N_1 + 2$$

equations may be derived relating to special types of polyhedra. If 3 edges meet at every vertex (3-connected polyhedron) and f_n is the number of faces with n edges (or vertices):

$$3f_3 + 2f_4 + f_5 \pm 0f_6 - f_7 - 2f_8 - \ldots = 12 \tag{1}$$

from which it follows that if all the faces are of the same kind they must be 3-gons, 4-gons, or 5-gons, and that the numbers of such faces must be four, six, and twelve, respectively. The corresponding equations for 4- and 5-connected polyhedra are:

$$2f_3 \pm 0f_4 - 2f_5 - 4f_6 - \ldots = 16 \tag{2}$$

and

$$f_3 - 2f_4 - 5f_5 - 8f_6 - \ldots = 20, \tag{3}$$

of which the special solutions are: $f_3 = 8$ and $f_3 = 20$.

In the analogous equation for 6-connected polyhedra the coefficient of f_3 is zero and all the other coefficients have negative values. There is therefore no simple convex polyhedron having six edges meeting *at every vertex*. Polyhedra with more than six edges meeting at every vertex are impossible because all coefficients of f_n are negative.

The special solutions of equations (1)–(3), namely,

3-connected: $f_3 = 4$; $f_4 = 6$; $f_5 = 12$,
4-connected: $f_3 = 8$,
5-connected: $f_3 = 20$,

correspond to polyhedra having all vertices of the same kind and all faces of the same kind. This proof of the existence of only five such polyhedra is purely topological, the above equations making no reference to the regularity or otherwise of the faces.* The forms of these polyhedra when the faces are regular polygons are the five regular (Platonic) solids: tetrahedron, cube, pentagonal dodecahedron, octahedron, and icosahedron. The symbol (n, p) or (n^p) describes a polyhedron with p n-gon faces meeting at each vertex. The values of n and p (Table 3.2) show that there is a special (dual or reciprocal) relationship between certain pairs of these solids, the number of edges being the same but the numbers of faces and vertices are interchanged. Pairs related in this way are the cube and octahedron and the pentagonal dodecahedron and icosahedron; the dual of the tetrahedon is another tetrahedron (inverted).

TABLE 3.2
The regular (Platonic) solids

	n, p	Vertices (N_0)	Edges (N_1)	Faces (N_2)	Dihedral angle
Tetrahedron	3, 3	4	6	4	70° 32′
Octahedron	3, 4	6	12	8	109° 28′
Cube	4, 3	8		6	90°
Dodecahedron	5, 3	20	30	12	116° 34′
Icosahedron	3, 5	12		20	138° 12′

We noted earlier that the Dirichlet domain of an atom and its coordination polyhedron are related in the following way: each vertex of the latter corresponds to a face of the former. It should be noted that in spite of this relationship the Dirichlet domain and the coordination polyhedron are not necessarily duals. They are duals in the case of the face-centred cubic structure (cubic closest packing),

---

*In addition to the five *convex* regular solids there are four stellated bodies, produced by extending outwards the faces of the convex solids until they meet.

namely, rhombic dodecahedron and cuboctahedron, but they are not in, for example, the body-centred cubic structure, in which the domain is the truncated octahedron and the coordination polyhedron is the rhombic dodecahedron.

All the regular solids are encountered in structural chemistry, but those with triangular faces are of special importance (notably the tetrahedron and octahedron). A triangulated coordination polyhedron represents the most compact arrangement of a set of atoms around another, but we shall see later that there are other reasons for the widespread occurrence of tetrahedral and octahedral coordination. The cube is the coordination polyhedron is certain simple structures (of both ions in CsCl and of the Ca^{2+} ion in CaF_2), but other arrangements of eight neighbours, which are described later, are more usual in finite molecules and ions and also in complex ionic crystals. The tetrahedron, cube, and pentagonal dodecahedron are now all known as the carbon skeletons of molecules (see p. 913). The dodecahedron, which has 20 vertices, is not of interest as a coordination polyhedron, but certain space-filling combinations of dodecahedra and related polyhedra are directly related to the structures of a family of hydrates, as described later. Icosahedral coordination is referred to under sphere packing (Chapter 4), it is found in numerous alloy structures and as the arrangement of the 12 oxygen atoms bonded to the metal in, for example, the $Ce(NO_3)_6^{3-}$ ion and of the carbonyl groups in $Fe_3(CO)_{12}$ and $Co_4(CO)_{12}$. There are icosahedral B_{12} groups in elementary boron and certain borides, and the boron skeleton in many boranes is composed of icosahedra or portions of icosahedra.

Semi-regular polyhedra

The regular solids have all vertices equivalent (isogonal) and all faces of the same kind (isohedral). If we retain the first condition but allow regular polygonal faces of more than one kind we find a set of semi-regular polyhedra, which are listed in Table 3.3. The symbols show the types of *n*-gon faces meeting at each vertex, the index being the number of such faces. These solids comprise three groups. The first consists of 13 polyhedra (Archimedean solids), derivable from the regular solids by symmetrically shaving off their vertices (a process called truncation), of which only the first five are of importance in crystals. The other two groups are the prisms and antiprisms, both of which have, in their most regular forms, a pair of parallel regular *n*-gon faces at top and bottom and are completed by *n* square faces (regular prisms) or 2*n* equilateral triangular faces (antiprisms). The second prism, in its most symmetrical form, is the cube, and the first antiprism is the octahedron. The regular solids and some of the semi-regular solids are illustrated in Fig. 3.3.

The fact that the number of isogonal bodies is limited to the five regular solids and the semi-regular solids is of considerable importance in chemistry. There are many molecules and complex ions in which five atoms or groups surround a central atom. The fact that it is not possible to distribute five equivalent points uniformly over the surface of a sphere (apart from the trivial case when they form a regular pentagon) is obviously relevant to a discussion of the configuration of molecules or

TABLE 3.3
The Archimedean semi-regular solids, prisms, and antiprisms

	Symbol	Name	Number of		
			faces	vertices	edges
1	$3, 6^2$	Truncated tetrahedron	8	12	18
2	$3, 8^2$	Truncated cube	14	24	36
3	$4, 6^2$	Truncated octahedron	14	24	36
4	$3^2 4^2$	Cuboctahedron	14	12	24
5	$4, 6, 8$	Truncated cuboctahedron	26	48	72
6	$3, 4^3$	Rhombicuboctahedron	26	24	48
7	$3^4, 4$	Snub cube	38	24	60
8	$3, 10^2$	Truncated dodecahedron	32	60	90
9	$3^2, 5^2$	Icosidodecahedron	32	30	60
10	$5, 6^2$	Truncated icosahedron	32	60	90
11	$4, 6, 10$	Truncated icosidodecahedron	62	120	180
12	$3, 4, 5, 4$	Rhombicosidodecahedron	62	60	120
13	$3^4, 5$	Snub dodecahedron	92	60	150
14	$n, 4^2$	Regular prisms	$n + 2$	$2n$	$3n$
15	$n, 3^3$	Regular antiprisms	$2n + 2$	$2n$	$4n$

ions AX_5 or more generally of 5-coordination in crystals. Similar considerations apply to 7-, 9-, 10-, and 11-coordination.

Corresponding to the semi-regular polyhedra of Table 3.3 there is a set of polyhedra which are the duals of the Archimedean solids, prisms, and antiprisms. They are named after Catalan, who first described them all in 1865. Of these we need only note the rhombic dodecahedron, which is the reciprocal of the cuboctahedron, and the family of bipyramids which are related in a similar way to the prisms.

Polyhedra related to the pentagonal dodecahedron and icosahedron

In equation (1) for 3-connected polyhedra (p. 69) the coefficient of f_6 is zero, suggesting that polyhedra might be formed from simpler 3-connected polyhedra by adding any arbitrary number of 6-gon faces. Although such polyhedra would be consistent with equation (1) it does not follow that it is possible to construct them. The fact that a set of faces is consistent with one of the equations derived from Euler's relation does not necessarily mean that the corresponding convex polyhedron can be made. Three of the Archimedean solids are related in this way to three of the regular solids:

tetrahedron, $f_3 = 4$ truncated tetrahedron, $f_3 = 4$
$f_6 = 4$

cube, $f_4 = 6$ truncated octahedron, $f_4 = 6$
$f_6 = 8$

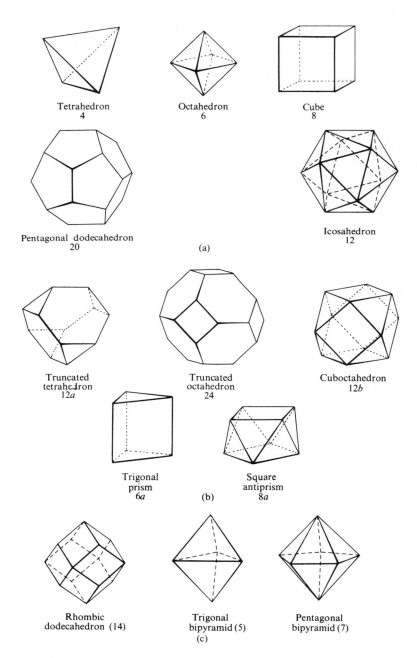

Tetrahedron
4

Octahedron
6

Cube
8

Pentagonal dodecahedron
20

Icosahedron
12

(a)

Truncated
tetrahedron
12*a*

Truncated
octahedron
24

Cuboctahedron
12*b*

Trigonal
prism
6*a*

Square
antiprism
8*a*

(b)

Rhombic
dodecahedron (14)

Trigonal
bipyramid (5)

Pentagonal
bipyramid (7)

(c)

FIG. 3.3. Polyhedra: (a) the regular solids; (b) some Archimedean semi-regular solids; (c) some Catalan semi-regular solids. (The numbers are the numbers of vertices.)

and   dodecahedron, $f_5 = 12$ truncated icosahedron, $f_5 = 12$
$$f_6 = 20$$

All the polyhedra intermediate between the dodecahedron and truncated icosahedron can be realized except $f_5 = 12$, $f_6 = 1$, and some in more than one form (different arrangements of the 5-gon and 6-gon faces). A number of these polyhedra are of interest in connection with the structures of clathrate hydrates (p. 660), because certain combinations of these solids with dodecahedra form space-filling assemblies in which four edges meet at every vertex. Two of these polyhedra, a tetrakaideca-hedron and a hexakaidecahedron, are illustrated in Fig. 3.4.

(a)   (b)

FIG. 3.4. (a) The tetrakaidecahedron: $f_5 = 12, f_6 = 2$; (b) the hexakaidecahedron: $f_5 = 12, f_6 = 4$.

The dual polyhedra are triangulated polyhedra with twelve 5-connected vertices and two or more 6-connected vertices. Together with the icosahedron they are found as coordination polyhedra in numerous transition-metal alloys.

Some less-regular polyhedra

Other equations may be derived from Euler's relation which are relevant to polyhedra with, for example, a specified number of vertices or faces. The former are required in discussions of the coordination polyhedra possible for a particular number of neighbours; the latter are of interest in space-filling by polyhedra. For 8-coordination we need polyhedra with eight vertices. These must satisfy the equation $\Sigma(n-2)f_n = 12$, that is,

$$f_3 + 2f_4 + 3f_5 + 4f_6 + 5f_7 = 12.$$

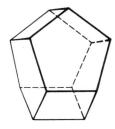

FIG. 3.5. The octahedron: $f_4 = f_5 = 4$.

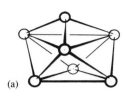 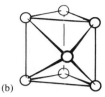

(a)   (b)

FIG. 3.6. 7-coordination polyhedra: (a) monocapped octahedron, (b) monocapped trigonal prism. The shaded circles represent H_2O molecules in $Ho(\phi COCHCO\phi)_3 . H_2O$ and $Yb(acac)_3 . H_2O$ respectively.

Solutions include

	N_1
f_3 = 12: triangulated dodecahedron or bisdisphenoid	18
f_3 = 8, f_4 = 2: square antiprism	16
f_3 = 4, f_4 = 4:	14
f_4 = 6: cube	12

The triangulated dodecahedron (Fig. 3.7(a)) is the third dodecahedron we have encountered, the others being the pentagonal and rhombic dodecahedra (with 5-gon and 4-gon faces respectively).

On the other hand we may be interested in polyhedra with eight faces (8-hedra or octahedra), which must satisfy the equation

$$v_3 + 2v_4 + 3v_5 + 4v_6 + 5v_7 = 12,$$

where v_p is the number of p-connected vertices.

The regular octahedron has $v_4 = 6$, and in this book the term octahedron will normally refer to this solid. Among numerous other 8-hedra are the truncated tetrahedron, hexagonal prism and the polyhedron $v_3 = 12$ which has $f_4 = f_5 = 4$ (Fig. 3.5) and is the reciprocal of the triangulated dodecahedron mentioned above. This 8-hedron has the property of packing with a particular kind of 17-hedron to fill space, and the vertices of this space-filling arrangement are the positions of the water molecules in the hydrate $(CH_3)_3CNH_2.9\frac{3}{4}H_2O$ (p. 662). Two 8-hedra of interest as 7-coordination polyhedra are the monocapped octahedron and the monocapped trigonal prism (Fig. 3.6). The prefix 'capped' means that there is an atom above the (approximate) centre of a face of the simpler polyhedron. In the case of the trigonal prism (or the antiprism of Table 3.4) the capped face is a square (or rectangular) face; for an octahedron it is necessarily a triangular face.

Table 3.4 summarizes the polyhedra most frequently found as the arrangements of nearest neighbours in finite molecules (or ions) and crystals. The polyhedra listed at the left of the Table have only triangular faces; those to the right have some triangular and some 4-gon faces. Polyhedra on the same horizontal line have the same number of vertices, and are related in the following way. Buckling of a 4-gon

TABLE 3.4
Polyhedral coordination groups

Vertices	Polyhedron	Faces			Polyhedron
		f_3	f_3	f_4	
4	Tetrahedron (4)	4			
5	Trigonal bipyramid (5)	6	4	1	Square pyramid
6	Octahedron (6)	8	2	3	Trigonal prism (*6a*)
7	Pentagonal bipyramid (7)	10			
	Monocapped octahedron	10	6	2	Monocapped trigonal prism
8	Dodecahedron	12	8	2	Square antiprism (*8a*)
		10		1	Bicapped trigonal prism
9	Tricapped trigonal prism	14	12	1	Monocapped antiprism
12	Icosahedron (12)	20	8	6	Cuboctahedron (12b)

Numbers in brackets refer to Fig. 3.3. The polyhedra with 7, 8, or 9 vertices other than the pentagonal bipyramid (Fig. 3.3) are illustrated in Figs. 3.6, 3.7, and 3.8.

face produces two triangular faces, so that the polyhedron on the right is converted into the one on the left. Such relationships are best appreciated by adding additional edges to 'outline' models of the former polyhedra to produce the triangulated polyhedra, and they are important when considering the geometry of the less symmetrical coordination groups. If there is appreciable departure from the most symmetrical form of one of the polyhedra of Table 3.4 (for example, trigonal bipyramid or square pyramid) the description of the coordination polyhedron may become somewhat arbitrary and of dubious value; a precise description of the geometry is then to be preferred.

Because of the outstanding importance of tetrahedral and octahedral coordination we devote the whole of Chapter 5 to systems built from these two polyhedra. Trigonal prism coordination is very rare in finite complexes (for example, the chelates $M(S_2C_2R_2)_3$ of Mo, W, Re, V, and Cr, p. 1191), and the Er chelate listed in Table 3.5) and not common in 3D structures other than NiAs, MoS_2, AlB_2 and related structures, and a few ionic crystals. Mono- or bicapped trigonal prism coordination is also not very common, and the following are examples in chemically related compounds:

monocapped trigonal prism:	EuO.OH	YO.OH	$Na_5Zr_2F_{13}$
bicapped:	$Eu(OH)_2.H_2O$		$N_2H_6(ZrF_6)$
(tricapped:	$Eu(OH)_3$	$Y(OH)_3$)	

We now comment briefly on 5-, 7-, 8-, and 9-coordination.

5-coordination

The most symmetrical (polyhedral) arrangement of 5 ligands is the trigonal bipyramidal one, which is found in a few ionic crystals, in some ions and molecules formed by non-transition elements (see the 40-electron group, p. 298), and in numerous transition-metal complexes of which examples are given in later chapters.

The closely related square pyramidal arrangement of 5 ligands is expected for non-transition elements having a 42-electron valence group which includes a lone pair (p. 298), and the configuration of some transition metal complexes is closer to square pyramidal than to trigonal bipyramidal; these are described in appropriate places.

7-coordination

Examples of the more well-defined coordination polyhedra include:

	Finite complexes	*Infinite systems*
Pentagonal bipyramid	K_3ZrF_7 K_3UF_7	$PaCl_5$
monocapped trigonal prism	K_2NbF_7 K_2TaF_7	
monocapped octahedron	chelates (Table 3.5)	A-La_2O_3

The earlier transition metals provide many examples of 7-coordinated complexes, particularly those containing halogens and bidentate ligands such as diarsine (a), though there is not 7-coordination in all complexes originally formulated in this

(a)   (b)   (c)

way. For example, $TaBr_5(diarsine)_2$ consists of dodecahedral 8- and octahedral 6-coordinated ions, $[TaBr_4(diarsine)_2]^+$ and $(TaBr_6)^-$. On the other hand $WOCl_4(diarsine)$ consists of pentagonal bipyramidal molecules, (b). We showed in Table 1.1. (p. 10) how different X:A ratios arise from sharing of X atoms in various ways between 9-coordination groups. Compounds of 5f elements, especially U, provide many examples of pentagonal bipyramidal coordination (see Table 28.4, p. 1256), and the $PaCl_5$ chain, (c), is also the form of the anion in $N_2H_5[InF_4(H_2O)]$. Other examples of this type of coordination include $Cr(O_2)_2(NH_3)_3$ and $Cr(O_2)_2O$ (dipyridyl), p. 504.

With bidentate ligands of the type R.CO.CH.CO.R yttrium and the smaller 4f ions form 7-coordinated complexes in which the seventh ligand is a water molecule. It is convenient to restrict the term monocapped octahedron to groups possessing exact or pseudo 3-fold symmetry with one ligand above one face of the octahedron, this face being somewhat enlarged. The complex $Ho(\phi CO.CH.CO\phi)_3.H_2O$ has 3-fold symmetry with a propeller-like arrangement of the three rings. The 6 O atoms are situated at the vertices of a distorted octahedron, and the H_2O molecule caps one face (Fig. 3.6(a)). With this structure compare that of $Yb(acac)_3.H_2O$, Fig. 3.6(b), in which the water molecule is not the capping ligand.

If we include metal–metal bonds there is 7-coordination of Nb in NbI_4 and of W in $K_3W_2Cl_9$, the seventh bond being perpendicular to an edge or face of the octahedral group of halogen atoms. The 7-coordination in $PbCl_2$ (p. 273) may be described as pseudo-9-coordination since there are ligands beyond the three rectangular faces of a trigonal prism, but the distances to two ligands are much greater than those to the other seven. Similarly the coordination of Ca^{2+} in $CaFe_2O_4$ is bicapped trigonal prismatic (8 + 1) rather than 9-coordination. The 7-coordination polyhedron of Zr^{4+} in monoclinic ZrO_2 (p. 542) may be described as related to either a capped trigonal prism or a capped octahedron. (See: AC 1970 **B26** 1129, and for a discussion of 7-coordination, CJC, 1963 **44** 1632.) For references see Table 3.5.

TABLE 3.5

Chelate molecules and ions containing acac *and related ligands*

C.n.	Coordination polyhedron	Complex	Reference
6	Trigonal prism	$Er[(CH_3)_3CCOCHCOC(CH_3)_3]_3$	AC 1971 **B27** 2335
7	Monocapped octahedron	$Ho(\phi COCHCO\phi)_3 . H_2O$	IC 1969 **8** 2680
	Monocapped octahedron	$Y(\phi COCHCOCH_3)_3 . H_2O$	IC 1968 **7** 1777
	Monocapped trigonal prism	$Yb(CH_3COCHCOCH_3)_3 . H_2O$	IC 1969 **8** 22
8	Dodecahedron	$[Y(CF_3COCHCOCF_3)_4]Cs$	IC 1968 **7** 1770
	Antiprism	$[Y(CH_3COCHCOCH_3)_3(H_2O)_2] . H_2O$	IC 1967 **6** 499

8-coordination

This is found in many molecules and crystals. Apart from the exceptional cubic 8-coordination in the body-centred cubic, CsCl, and CaF_2 structures (which are discussed in Chapter 7), the coordination polyhedron is usually either the Archimedean antiprism or the triangulated dodecahedron (bisdisphenoid). The bicapped trigonal prism is found in a few crystals, for example, $CaFe_2O_4$, $Sr(OH)_2 . H_2O$, $N_2H_6(ZrF_6)$, and NdP_5O_{14}. As implied by their listing in Table 3.4, these three polyhedra are closely related. The antiprism may be converted successively into the bicapped trigonal prism (by joining the vertices a and b in Fig. 3.7(b)) and then into the dodecahedron by adding a diagonal to the other square face. The dodecahedron of Fig. 3.7(a) is also called a *bisdisphenoid* since it consists of two interpenetrating disphenoids (tetrahedra), one elongated (A) and the other flattened (B). Examples of the two more usual 8-coordination groups include:

Dodecahedral	Antiprismatic
$Na_4[Zr(C_2O_4)_4] . 3H_2O$	$Zr(IO_3)_4$
K_2ZrF_6	$Na_3(TaF_8)$
$K_4[Mo(CN)_8] . 2H_2O$	$H_4[W(CN)_8] . 6H_2O$
$Ti(NO_3)_4$, $Sn(NO_3)_4$	$Ca(BF_4)_2$, $CaCO_3 . 6H_2O$
$TiCl_4$ (diarsine)$_2$	Rb^+ in $RbLiF_2$

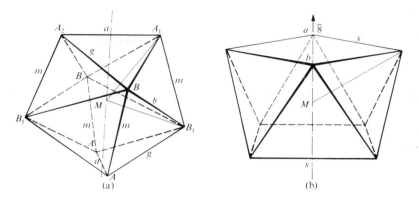

FIG. 3.7. 8-coordination polyhedra: (a) triangulated dodecahedron (bisdisphenoid); (b) antiprism.

Calculations of the ligand–ligand repulsion energies for 8-coordination polyhedra show that both of these polyhedra are more stable than the cube, for all values of n from 1 to 12 in a repulsion law of the type, force proportional to r^{-n}, and that the antiprism is marginally more stable than the dodecahedron. The very small difference in stability is shown by the adoption of the two configurations by chemically similar pairs of complexes such as those shown above and in Table 3.5.

It should be emphasized that there is an important difference between the triangulated dodecahedron and the antiprism. There is an 'ideal' antiprism, for this is one of the semi-regular solids, but there is no unique configuration of the dodecahedron. The maximum symmetry ($\bar{4}2m$) of this polyhedron permits unequal bond length M–A and M–B and also edges of four different lengths, a (two), b (four), m (four), and g (eight). Calculations must therefore be made for special configurations such as the 'hard-sphere' model ($a = m = g$). For this configuration the radius ratio (p. 315) is 0.668, and for the hard-sphere model of the antiprism ($l = s$ in Fig. 3.7(b)) the value is 0.645. We shall not give details of the geometries of these two polyhedra since it is now clear that there is a delicate balance between the factors determining the choice of coordination polyhedron and that the detailed geometry is dependent on the size and nature of the ligands and also on the interactions with more distant neighbours in the crystal. We refer the reader to discussions of 8-coordination: JCP 1950 **18** 746; IC 1963 **2** 235; JCS A 1967 345; IC 1968 **7** 1686; and to the sections on the complex fluorides of Zr (p. 471) and on octacyanide ions (p. 941).

9-coordination

Examples of tricapped-trigonal-prism coordination include the finite aquo complex in $Nd(BrO_3)_3 \cdot 9H_2O$ and the infinite linear one in $SrCl_2 \cdot 6H_2O$ (where these groups are stacked in columns in which they share basal faces), and numerous compounds of 4f and 5f elements, for example, YF_3 and compounds with the UCl_3 $(Y(OH)_3)$ structure, and the complex fluorides of Th and U (Chapter 28). Some examples of 9-coordination were listed in Table 1.1 (p. 10) to illustrate how different formulae MX_n can arise with the same c.n. of M. The less common monocapped antiprism occurs as the arrangement of Te around La atoms in $LaTe_2$ (Fe_2As, C38 structure); its relation to the tricapped trigonal prism may be seen by joining the vertices a and b in Fig. 3.8(a). The 'hard-sphere' model of the tricapped trigonal prism is shown in Fig. 3.8(b); the distance from the centre to the vertices is 0.86a, corresponding to the radius ratio, 0.732.

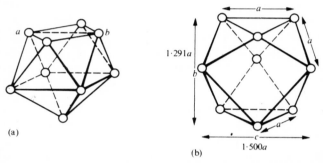

(a)

(b)

FIG. 3.8. 9-coordination polyhedra: (a) monocapped antiprism; (b) tricapped trigonal prism.

10- *and* 11-*coordination*

Very high coordination numbers are possible with bidentate ligands such as NO_3^- and CO_3^{2-} owing to the small O–O separation of around 2.15 Å compared with the normal separation (2.8 Å) in MO_n coordination groups. Because some edges of the coordination polyhedron are short and others of normal length the polyhedron is unlikely to be very symmetrical. The type of coordination is sometimes described more simply by regarding the bidentate group as one ligand. For example, in the molecule $Th(NO_3)_4(OP\phi_3)_2$,[1] the four NO_3 groups occupy the equatorial positions in an 'octahedral' coordination group. In the $[M(CO_3)_5]^{6-}$ ions formed by Ce and Th the ten O atoms are at the vertices of (two different) hexakaidecahedra which are intermediate between the triangulated dodecahedron and icosahedron:[2]

	Vertices	Faces	Edges
Triangulated dodecahedron	8	12	18
Hexakaidecahedron	10	16	24
Icosahedron	12	20	30

but these are irregular because of the five short edges belonging to the bidentate ligands. In the bidentate oxalato group the O–O distance (between O atoms bonded to different C atoms) is close to 2.8 Å, and in $K_4Th(C_2O_4)_4.4H_2O$ there are chains in which Th is bonded to *five* C_2O_4 groups, two of which are tetradentate:[3]

The coordination polyhedron of Th is a nearly-regular bicapped antiprism, a coordination polyhedron found in a less regular form in U(IV) acetate.[4] In $La_2(CO_3)_3.8H_2O$ there are complex layers consisting of La^{3+} and CO_3^{2-} ions and 6 of the 8 H_2O molecules.[5] Within a layer both kinds of non-equivalent La^{3+} ions are 10-coordinated. In each coordination group bidentate CO_3^{2-} ions occupy the positions B_1 in Fig. 3.7(a), producing two more edges parallel to the lower edge AA; the other 6 neighbours are H_2O molecules or O atoms of monodentate carbonate ions.

There is 11-coordination of Th^{4+} by 4 bidentate NO_3^- ions and 3 H_2O in $Th(NO_3)_4.5H_2O$[6] and of La^{3+} by 3 bidentate NO_3^- ions and 5 H_2O in $La(NO_3)_3.6H_2O$, which is $[La(NO_3)_3(H_2O)_5].H_2O$.[7]

In the hexanitrato ions formed by Ce and Th there is (icosahedral) 12-coordination (p. 825), and this coordination number is found in numerous close-packed structures (Chapter 4) of compounds of various kinds and in metals and intermetallic compounds, where higher coordination numbers are also found (Chapter 29).

(1) AC 1973 **B29** 2687
(2) AC 1975 **B31** 2607, 2612, 2615, 2620
(3) AC 1975 **B31** 1361
(4) AC 1964 **17** 758
(5) IC 1968 **7** 1340
(6) AC 1966 **18** 836, 842
(7) IC 1980 **19** 1207

Plane nets

Derivation of plane nets

The division of an infinite plane into polygons is obviously related to the enumeration of polyhedra, which can be represented as tessellations of polygons on a simple closed surface such as a sphere. In fact we find equations somewhat similar to those for polyhedra except that instead of the *number* f_n of *n*-gon faces we have ϕ_n as the *fraction* of the polygons which are *n*-gons, since we are now dealing with an infinite repeating pattern. These equations are:

3-connected: $3\phi_3 + 4\phi_4 + 5\phi_5 + 6\phi_6 + 7\phi_7 + 8\phi_8 + \ldots n\phi_n = 6,$ (4)

4-connected: $= 4,$ (5)

5-connected: $= 10/3,$ (6)

6-connected: $= 3.$

There are three special solutions corresponding to plane nets in which all the polygons have the same number of edges (and the same number of lines meet at every point), namely:

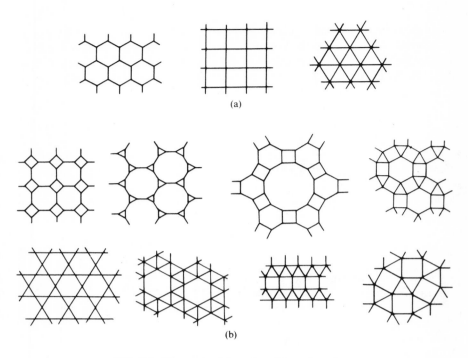

FIG. 3.9. Plane nets: (a) regular; (b) semi-regular.

$$\text{3-connected: } \phi_6 = 1,$$
$$\text{4-connected: } \phi_4 = 1,$$
and
$$\text{6-connected: } \phi_3 = 1.$$

They are illustrated in Fig. 3.9(a). The last is the only plane 6-connected net, and evidently plane nets with more than six lines meeting at every point are not possible. For 3-, 4-, and 5-connected nets there are other solutions corresponding to combinations of polygons of two or more kinds. For example, the next simplest solutions for 3-connected nets are: $\phi_5 = \phi_7 = \frac{1}{2}$, $\phi_4 = \phi_8 = \frac{1}{2}$, and $\phi_3 = \phi_9 = \frac{1}{2}$. The first of these is discussed later in connection with configurations of plane nets; the most symmetrical configuration of the second is illustrated in Fig. 3.9(b) as the first of the semi-regular plane nets. Examples of crystal structures based on these two nets are noted later (p. 109).

Corresponding to the five regular and thirteen semi-regular (Archimedean) solids, all of which have regular polygonal faces, there are three regular plane nets (the special solutions listed above) and eight semi-regular nets in which there are regular polygons of two or more kinds (Fig. 3.9(b)). The dual relations between plane nets are similar to those between pairs of polyhedra, for the nets (6, 3) and (3, 6) are related in this way while the dual of (4, 4) is the same net (compare the tetrahedron).

The duals of the eight semi-regular nets of Fig. 3.9(b) may be drawn by joining the mid-points of adjacent (edge-sharing) polygons; they represent divisions of the plane into congruent polygons – compare the relation of the Catalan to the Archimedean solids. We illustrate (Fig. 3.10) only two of these dual nets, those consisting of

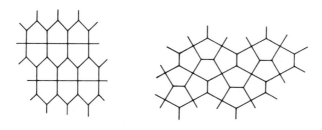

FIG. 3.10. Two plane nets consisting of pentagons.

congruent pentagons, which are included in Table 3.7 (p. 91). These two nets, which are the duals of the two at the bottom right-hand corner of Fig. 3.9(b), complete the series:

p	n-gons
3	6
3, 4	5
4	4
6	3

but, unlike the other three nets, cannot be realized with regular polygons.

In order to derive the general equation for nets containing both 3- and 4-connected points we must allow for variation in the proportions of the two kinds of points. If the ratio of 3- to 4-connected points is R it is readily shown that in a system of N points the number of links is $N(3R + 4)/2(R + 1)$. Using the same method as for deriving equations (4)–(6) it is found that $\Sigma n\phi_n = 2(3R + 4)/(R + 2)$. The value of $\Sigma n\phi_n$ ranges from 6, when $R = \infty$, to 4, when $R = 0$, and has the special value 5 if the ratio of 3- to 4-connected points is 2:1. The special solution of the equation $\Sigma n\phi_n = 5$ is $\phi_5 = 1$, corresponding to the 5-gon nets of Fig. 3.10. Although (3,4-connected plane nets are not of much interest in structural chemistry the 3D nets of this type form the bases of a number of crystal structures (p. 89).

Plane nets in which points of two kinds (p- and q-connected) alternate are of interest in connection with layer structures of compounds $A_m X_n$ in which the coordination numbers of both A and X are 3 or more. For example, the simple CdI_2 layer may be represented as the plane (3,6)-connected net. It has been shown (AC 1968 B24 50) that the only plane nets composed of alternate p- and q-connected points are those in which the values of p and q are 3 and 4, 3 and 5, or 3 and 6. The impossibility of constructing a plane net with alternate 4- and 6-connected points implies that a simple layer structure is not possible for a compound $A_2 X_3$ if A is to be 6-coordinated and X 4-coordinated. The nonexistence of an octahedral layer structure for a sesquioxide is, therefore, not a matter of crystal chemistry in the sense in which this term is normally understood but receives a very simple topological explanation.

Configuration of plane nets

Two points call for a little further amplification. The equations (4)–(6) are concerned only with the *proportions* of polygons of different kinds and not at all with the *arrangement* of the polygons relative to one another. Moreover, all nets have been illustrated as repeating patterns. For the special solutions $\phi_n = 1$ there is no question of different arrangements of the polygons since each net is entirely composed of polygons of only one kind. However, the *shape* of the hexagons in the net $\phi_6 = 1$ determines whether the net is a repeating pattern and if it is, the size of the unit cell. Figure 3.11 shows portions of periodic forms of this net; clearly a net in which the

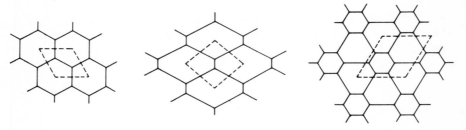

FIG. 3.11. Forms of the plane 6-gon net.

hexagons are distorted in a random way has no periodicity, or alternatively, it has an indefinitely large repeat unit (unit cell). The situation is more complex if the net contains polygons of more than one kind. Consider first the net $\phi_4 = \phi_8 = \frac{1}{2}$, which is one of the three solutions of equation (4) corresponding to nets consisting of equal numbers of polygons of only two kinds. If the 4-gons and 8-gons are regular there is a unique form of the net, that shown in Fig. 3.9(b). Let us drop the requirement that the polygons are regular but retain the same relative arrangement of 4-gons and 8-gons, that is, each 4-gon shares its edges with four 8-gons and each 8-gon shares alternate edges with 4-gons and 8-gons. If we make alternate 4-gons of different sizes (Fig. 3.12(a)) the content of the unit cell is doubled, to $Z = 8$. If we proceed a stage further, that is, we do not insist on the same relative arrangement of 4-gons and 8-gons but regard the net simply as a 3-connected system of equal numbers of the two kinds of polygon, then we find an indefinitely large number of nets (in which there is edge-sharing between 4-gons). An example is shown in Fig. 3.12(b).

The nets $\phi_5 = \phi_7 = \frac{1}{2}$ and $\phi_3 = \phi_9 = \frac{1}{2}$ do not have configurations with regular polygons of both kinds, but here again there is an indefinitely large number of ways of arranging equal numbers of polygons of the two types. Four of the simplest ways

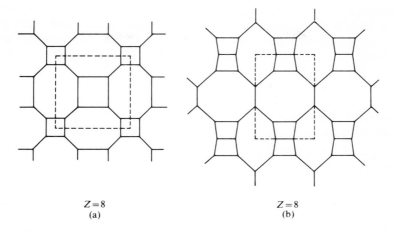

Z = 8
(a)

Z = 8
(b)

FIG. 3.12. (a) A less regular form of the plane net (4.8^2); (b) a less regular net consisting of equal numbers of 4- and 8-gons but having non-equivalent points of three kinds, 4.8^2, 4^28, and 8^3. Z is the number of points in the repeat unit (unit cell) enclosed by the broken lines.

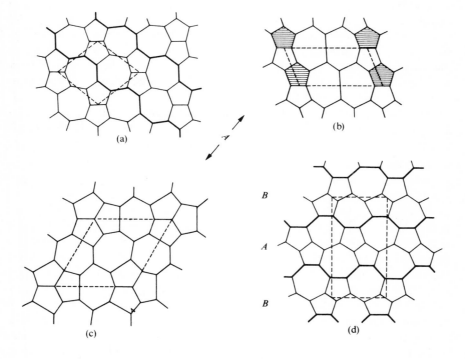

(a)

(b)

(c)

(d)

B

A

B

Fig. 3.13 Configurations of the plane net: $\phi_5 = \phi_7 = \frac{1}{2}$.

of arranging equal numbers of 5-gons and 7-gons are shown in Fig. 3.13. Two of these, (a) and (d), are closely related, being built of the same sub-units, the strip A and its mirror-image B. It is of some interest that the form of this net adopted by ScB_2C_2 is not the simplest configuration (Fig. 3.13(a)) with eight points in the repeat unit but that of Fig. 3.13(d) with $Z = 16$.

Three-dimensional nets

Derivation of 3D nets

No equations are known analogous to those for polyhedra and plane nets relating to the proportions of polygons (circuits) of different kinds. A different approach is therefore necessary if we wish to derive the basic 3D nets.

Any pattern that repeats regularly in one, two, or three dimensions consists of units that join together when repeated *in the same orientation*, that is, all units are identical and related only by translations. In order to form a 1-, 2-, or 3-dimensional pattern the unit must be capable of linking to two, four, or six others, because a one-dimensional pattern must repeat in both directions along a line, a two-dimensional pattern along two (non-parallel) lines, and a three-dimensional pattern along three (non-coplanar) lines (axes). The repeat unit may be a single point or a group of connected points, and it must have at least two, four, or six free links available for attachment to its neigbours. The requirement of a minimum of four free links for a 2D pattern may at first sight appear incompatible with the existence of 3-connected plane nets, but it can be seen (Fig. 3.11, p. 84) that even in the simplest of these nets ($\phi_6 = 1$) the repeat unit consists of a pair of connected points, this unit having the minimum number (4) of free links.

Evidently the simplest unit that can form a 3D pattern is a single point forming six links, but for 4- and 3-connected 3D nets the units must contain respectively two and four points, as shown in Fig. 3.14. The series is obviously completed by

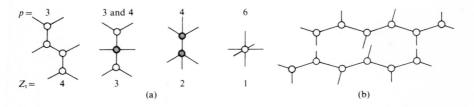

$p = $　3　　　3 and 4　　　4　　　6

$Z_t = $　4　　　3　　　2　　　1

(a)　　　　　　　　　　　　　　(b)

FIG. 3.14. Structural units for three-dimensional nets (see text).

the intermediate unit consisting of one 4- and two 3-connected points, which also has the necessary number (6) of free links. These values of Z, the number of points in the repeat unit, enable us to understand the nature of the simplest 3D nets. Similarly oriented units must be joined together through the free links, each one to six others. This implies that the six free links from each unit must form three pairs, one of each pair pointing in the opposite direction to the other. Identically oriented links repeat at intervals of $(Z+1)$ points, so that circuits of $2(Z+1)$ points are formed. We therefore expect to find the family of basic 3D nets listed in Table 3.6, where p is the number of links meeting at each point and n is the number of points in the smallest circuits. Note that the symbols for these nets are of the form (n, p); for example, $(10, 3)$ is a 3-connected net consisting of 10-gons.

TABLE 3.6
The basic 3-dimensional nets

Fig. 3.15	p	n	Z_t	Z_c	x	y
(a)	3	10	4	8	15	10
(b)	3, 4	8	3	6	$13\frac{1}{3}$†	8
(c)	4	6	2	8	12	6
(d)	6	4	1	1	12	4

† Weighted mean.

By analogy with the regular polyhedra and plane nets we might expect to find a set of regular 3D nets which have all their links equal in length and equivalent, all circuits (defined as the shortest paths including any two non-collinear links from any point) identical, and the most symmetrical arrangement of links around every point. Three of the nets of Table 3.6 satisfy all these criteria, namely, a 3-connected net, $(10, 3)$, the diamond net, $(6, 4)$, and the simple cubic framework (primitive cubic lattice), $(4, 6)$. These nets, all of which have cubic symmetry, are illustrated in Fig. 3.15(a), (c), and (d). There is a second 3-connected net $(10, 3)$, also with $Z = 4$ (Fig. 3.15(e)), and a second 4-connected net $(6, 4)$ with a less symmetrical (coplanar) arrangement of bonds from each point (Fig. 3.15(f)). It is convenient to refer to the nets (c), (e), and (f), as the diamond, $ThSi_2$, and NbO nets respectively because of their relation to the structures of these substances. The NbO net has cubic symmetry but the highest symmetry of the $ThSi_2$ net, and also of the net (b) is tetragonal.

A word of explanation is needed here concerning the values of Z_t and Z_c in Table 3.6. If the numbers of points in the unit cells of Fig. 3.15(a)–(d) are counted they will be found to be eight, six, eight, and one respectively. In all cases except (d), which is a primitive lattice, these numbers Z_c are multiples of the values of Z_t in the Table. The reason for this is that the nets are illustrated in Fig. 3.15 in their most symmetrical configurations, and it is conventional to describe a structure in terms of a unit cell the edges of which are related to the symmetry elements of the structure. Such a unit cell is *usually* larger than the smallest one that could be chosen without relation to the symmetry; it contains Z_c points (atoms). Thus the cubic unit cell of diamond (c) contains 8 atoms, but the structure may also be described in terms of a tetragonal cell containing four atoms or a rhombohedral cell containing

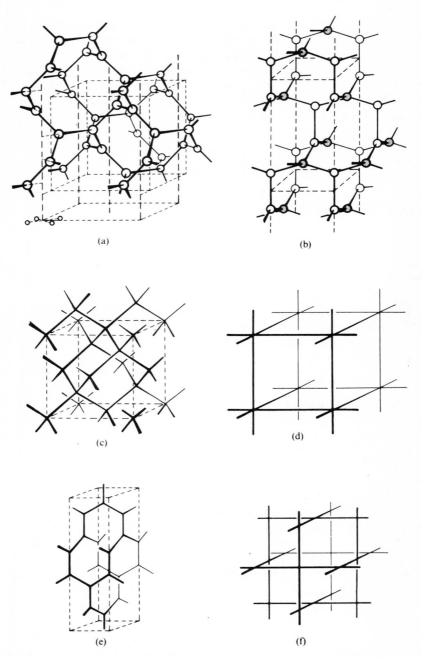

FIG. 3.15. Three-dimensional nets: (a)-(d) the four nets of Table 3.6; (e) ThSi$_2$ net; and
(f) NbO net.

two atoms. The (10,3) net of Fig. 3.30 (p. 114), on which the structure of B_2O_3 is based, is an example of a net for which the simplest topological unit has $Z_t = 6$ and this is also the value of Z_c for the most symmetrical (trigonal) form of the net.

No example appears to be known yet of a crystal structure based on the simplest 3D (3,4)-connected net (Fig. 3.15(b)), but two more complex nets of this general type do represent crystal structures. A particularly interesting family of (3,4)-connected nets includes those in which each 3-connected point is linked only to 4-connected points and vice versa. For such systems Z must be a multiple of 7, and the next two nets illustrated are of this type. Figure 3.16(a) represents the structure of Ge_3N_4, the open circles being N and the shaded circles Ge atoms. Essentially the same atomic arrangement is found in Be_2SiO_4 (phenacite), where 2 Be and 1 Si replace the Ge atoms in the nitride. This structure is suitable for atoms forming four tetrahedral bonds (Ge, Be, and Si) or three approximately coplanar bonds (N and O).

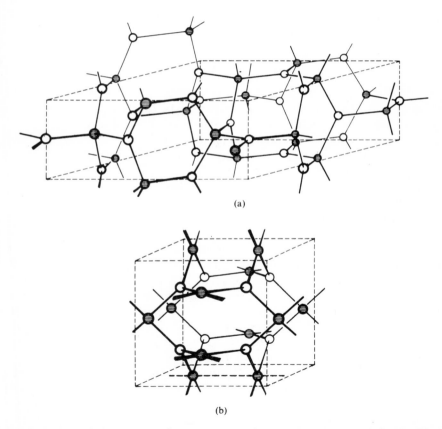

(a)

(b)

FIG. 3.16. Two (3,4)-connected 3D nets representing (a) the structure of Ge_3N_4; (b) the arrangement of Pt and O atoms in $Na_xPt_3O_4$ (shaded and open circles respectively).

The net of Fig. 3.16(b), on the other hand, is suited to a 4-connected atom forming four coplanar bonds (shaded circles), and represents the framework of Pt and O atoms in $Na_xPt_3O_4$, a compound formed by oxidizing Pt wire in the presence of sodium. The Na atoms are situated in the interstices and are omitted from Fig. 3.16(b).

Uniform nets. Certain nets have the property that the shortest path, starting from any point along any link and returning to that point along any other link, is a circuit of n points. Such nets may be called *uniform* nets. This condition is satisfied by the first three nets of Table 3.6, but it is not applicable in this simple form to nets with $p > 4$. However, it may be applied to 6-connected nets if we exclude circuits involving a pair of collinear bonds at any point. We may then include among uniform nets the $(4,6)$ net (primitive lattice) and also the pyrites structure as a net $(5,\frac{4}{6})$ containing 4-connected points (S atoms) and 6-connected points (Fe atoms) and consisting of 5-gon circuits. It is not yet known how many uniform nets there are, but uniform nets of all the following types have been derived and illustrated:

p	n	
3	12, 10, 9, 8, 7	
3, 4	9, 8, 7	
4		6
4, 6		5
6		4

The upper limit of n in 3-connected nets appears to be 12, but since no examples of this net are known it is not illustrated here or included in Table 3.7.

The author's work on 3D nets is summarized in *Three-dimensional nets and polyhedra*, Wiley-Interscience, New York, 1977 and in Monograph No. 8 of the American Crystallographic Association, 1979.

Further characterization of 3D nets

In addition to the two $(10,3)$ nets of Fig. 3.15(a) and (e) and the third $(10,3)$ net mentioned above there are also more complex 3-connected nets consisting of 10-gon circuits. These nets are not interconvertible without breaking and rejoining links, and therefore represent different ways of joining 3-connected points into 3D nets consisting of decagons. Evidently the symbol (n, p) is not adequate for distinguishing such nets which differ in *topological symmetry*. Unlike the crystallographic symmetry the topological symmetry does not involve reference to metrical properties of the nets, but only to the way in which the various polygonal circuits are related to one another. Two quantities that may be used as a measure of the topological symmetry of nets are x, the number of n-gons (here 10-gons) to which each *point* belongs, and y, the number of n-gons to which each *link* belongs. For a plane net x is equal to p, the connectedness, and y is always 2. In 3D nets x and y can attain quite high values, and for the very symmetrical nets of Table 3.6 y is equal to n. For these (regular) nets x and y are related: $x = py/2$, an expression which holds for the $(3,4)$-connected net if weighted mean values of x and p are used. In the second 3-connected net $(ThSi_2)$ there are two kinds of non-equivalent link, and the weighted mean value of y is $6\frac{2}{3}$, as compared with 10 in the regular $(10, 3)$ net. The decagon net of B_2O_3 forms the third member of the series of 3-connected 10-gon nets, as shown by the values of x:

Z_t	Maximum symmetry	x
4	Cubic	15
4	Tetragonal	10
6	Hexagonal	5

For the second 4-connected net $(6, 4)$, the NbO net, $y = 4$. The values of x and y are not of interest from the chemical standpoint, but it is recommended that they be determined, if only to ensure that the models are examined carefully and not merely assembled and dismantled.

The basic systems of connected points, polyhedra, plane and 3D nets are summarized in Table 3.7, which shows that the four plane and four 3D nets form series with $n = 6$, 5, 4, and 3, and with $n = 10$, 8, 6, and 4 respectively. All systems

TABLE·3.7

Relation between polyhedra, plane, and 3-dimensional nets

p \ n	3	3 and 4	4	6
3	Tetrahedron	Trigonal bipyramid	Octahedron	Plane (3, 6)
4	Cube	Rhombic dodecahedron	Plane (4, 4)	P lattice ①
5	Pentagonal dodecahedron	Plane (5, $\frac{3}{4}$)	(5, 4)	
6	Plane (6, 3)	(6, $\frac{3}{4}$)	Diamond ②	
7	(7, 3)	(7, $\frac{3}{4}$)	(The numbers enclosed in circles are values of Z_t.)	
8	(8, 3)	(8, $\frac{3}{4}$) ③		
9	(9, 3) Z_t	(9, $\frac{3}{4}$)		
10	Cubic (10, 3) ④			

on the same horizontal line are composed of n-gon circuits, and all those in a vertical column have the same value(s) of p. In the first column the 3-connected systems start with three of the regular solids, with $n = 3$, 4, and 5, and the series continues through the plane net ($n = 6$) into the 3D nets with $n = 7$, 8, 9, and 10. We shall not have occasion to refer to the nets with $n = 7$, 8, or 9, since no examples of crystal structures based on these nets are known. Two forms of $(5, \frac{3}{4})$ were illustrated in Fig. 3.10, and the 3D net $(8, \frac{3}{4})$ in Fig. 3.15(b); the latter is intermediate between the diamond, $(6, 4)$, and 3-connected $(10, 3)$ nets. There are also 3D $(3, 4)$-connected composed of 6-, 7-, and 9-gons which are not illustrated since they are not known to occur in crystals. The 4-connected system of 5-gons, $(5, 4)$, has been incorrectly described by the author in the first work cited on p. 90 as an infinite radiating net; it is in fact a finite set of 600 points, the polytope $\{5, 3, 3\}$ of Coxeter (H. S. M. Coxeter, *Polytopes*, Dover, New York, 1973).

The fifth regular solid, the icosahedron, and other 5-connected systems are omitted from Table 3.7 because there are no 5-connected plane or 3D nets composed of polygons of one kind only. For coordination numbers greater than six only 3D nets are possible. Two aspects of 3D nets which are important in structural chemistry should be mentioned here; they will be discussed in more detail later.

Nets with polyhedral cavities

In certain 3D nets there are well-defined polyhedral cavities, and the links of the net may alternatively be described as the edges of a space-filling assembly of polyhedra. At least four links must meet at every point of such a net, and the most important nets of this kind are, in fact, 4-connected nets. Space-filling arrangements of polyhedra leading to such nets are therefore described after we have dealt with the simpler 4-connected nets.

Interpenetrating nets

We have supposed that in any system all the points together form one connected net, that is, it is possible to travel along links from any point to any other. There are some very interesting structures in which this is not possible, namely, those consisting of two or more interpenetrating (interlocking) structures (Fig. 3.17).

The simplest of these is a pair of linked n-rings, isomeric with a single $2n$-ring. It is likely that some molecules of this type are formed in many ring-closure reactions in which large rings are formed. No examples of intertwined linear systems are known in the inorganic field, but an example of two 'interwoven' layers is found in crystalline silver tricyanomethanide, $Ag[C(CN)_3]$ (p. 105). Several examples of crystals built of two or more identical interpenetrating 3D nets will be mentioned in connection with the diamond structure (p. 127).

We shall now give examples of molecular and crystal structures based on 2-, 3-, and 4-connected systems. Although logically the cyclic and chain systems (corresponding to $p = 2$) should precede the polyhedral ones ($p \geqslant 3$) we shall deal with the latter first so that we proceed from finite to infinite groups of atoms. This

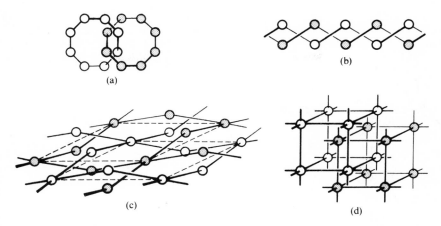

FIG. 3.17 Interpenetrating systems: (a) finite; (b)–(d) 1-, 2-, and 3-dimensional.

kind of treatment cuts right across the chemical classification of molecules and crystals and also is not concerned with details of structure, that is, it is topological rather than geometrical. It shows in a striking way how a small number of very simple patterns are utilized in the structures of a great variety of elements and compounds. The following are the main sub-divisions:

Polyhedral molecules and ions
Cyclic molecules and ions
Infinite linear molecules and ions
Structures based on 3-connected nets:
    the plane 6-gon net
    3D nets
Structures based on 4-connected nets:
    the plane 4-gon net
    the diamond net
       interpenetrating systems
    more complex nets
Polyhedral frameworks

Polyhedral molecules and ions

Structural studies have now been made of a number of polyhedral molecules and ions, and of these the largest class comprises the tetrahedral complexes.

Tetrahedral complexes

Most of these are of one of the four types shown in Fig. 3.18(a)–(d), the geometry of which is most easily appreciated if the tetrahedron is shown as four vertices of a

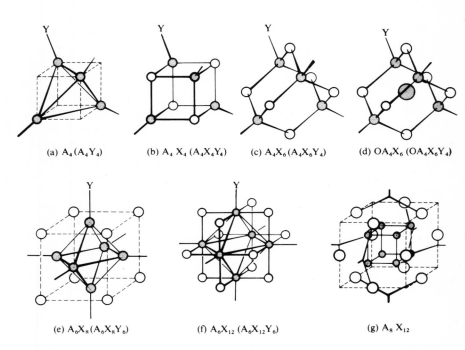

(a) $A_4 (A_4Y_4)$ (b) $A_4 X_4 (A_4X_4Y_4)$ (c) $A_4X_6 (A_4X_6Y_4)$ (d) $OA_4X_6 (OA_4X_6Y_4)$

(e) $A_6X_8 (A_6X_8Y_6)$ (f) $A_6X_{12} (A_6X_{12}Y_6)$ (g) $A_8 X_{12}$

FIG. 3.18. Polyhedral molecules and ions: (a)–(d) tetrahedral; (e) and (f) octahedral; (g) cubic.

cube. Disregarding the singly-attached ligands Y, there is first the simple tetrahedron A_4, (a). To this may be added either 4 X atoms situated above the centres of the faces, (b), or 6 X atoms bridging the edges, (c). These X atoms are, of course, wholly or largely responsible for holding the group of atoms together. In the most symmetrical form of (b) A and X form a cubic group, but some distortion is to be expected from this configuration, which would imply bond angles of 90° at A and X. In (c) the six X atoms delineate an octahedron; for the most familiar representation of the molecules of P_4O_6 and P_4O_{10} see Fig. 19.8 (p. 854). In class (d) there is an oxygen atom at the centre of the tetrahedron, and in several molecules of this type X is a bidentate chelate group.

To all of the basic types of molecule A_4, A_4X_4, A_4X_6, and OA_4X_6 there is the possibility of adding singly attached Y ligands to each A atom. There is usually one such atom, since A is already bonded to 3 X (in addition to any A—A interactions) and a fourth bond to a Y atom completes a tetrahedral group. Thus in $Cu_4I_4(AsEt_3)_4$ X = I and Y = $AsEt_3$, and in $[Fe_4S_4(S\phi)_4]^{2-}$ X = S and Y = Sϕ. In other complexes containing an A_4X_4 nucleus there is a fourth ligand attached to each A and X, giving $A_4X_4Y_8$, as in $Al_4N_4\phi_8$, or 3 Y atoms attached to each A atom completing an octahedral coordination group, as in $Pt_4I_4(CH_3)_{12}$. In the less symmetrical tetrahedral $Co_4(CO)_{12}$ molecule there are 3 CO bonded to the apical Co, 3 CO groups bridge the edges of the base, and two further CO are attached to each basal Co atom.

Octahedral molecules and ions

Compact groupings built around an octahedron of metal atoms are prominent in the halogen chemistry of the elements Nb, Ta, Mo, and W, where such groups are found as isolated ions or molecules and also as sub-units which are further linked through the Y atoms into layer and 3D structures. The two basic types of unit are shown (idealized) in Fig. 3.18(e) and (f). Few examples of the simple A_6X_8 unit are known, but a number of complexes of the general type $A_6X_8Y_6$ have been studied (Table 3.8). The A_6X_{12} unit of Fig. 3.18(f) represents the structure of the hexameric molecule in one polymorph of $PdCl_2$ and $PtCl_2$ and of the $Bi_6(OH)_{12}^{6+}$ ion in hydrolysed solutions of bismuth salts. Addition of a further six halogen atoms to A_6X_{12} gives the structure of the hexameric tungsten trichloride molecule, $W_6Cl_{12}Cl_6$ and the ion $Nb_6Cl_{18}^{4-}$. These complex halogen compounds are described in more detail in Chapter 9. The oxy-ions $M_6O_{19}^{8-}$ formed by V, Nb, and Ta could also be described as related to the A_6X_{18} complex, an additional O atom at the centre completing octahedral groups around the metal atoms. These and other complex oxy-ions of Groups VA and VIA metals are alternatively described as assemblies of MO_6 octahedra, as in Chapter 11.

TABLE 3.8
Polyhedral molecules and ions

	Fig. 3.18		
Tetrahedral	(a)	A_4	P_4, As_4
		A_4Y_4	B_4Cl_4
	(b)	$A_4X_4{}^a$	$Li_4(C_2H_5)_4$, $Tl_4(OAlk)_4$, $Pb_4(OH)_4^{4+}$
		$A_4X_4Y_4$	$Cu_4I_4(AsEt_3)_4$, $Fe_4S_4(S\phi)_4^{2-}$
		$A_4X_4Y_8$	$Al_4N_4\phi_8$
		$A_4X_4Y_{12}$	$Pt_4I_4(CH_3)_{12}$
	(c)	A_4X_6	P_4O_6, As_4O_6, Sb_4O_6, $N_4(CH_2)_6$, $(CH)_4S_6$,
			$(CH)_4(CH_2)_6$, $(SiR)_4S_6$
		$A_4X_6Y_4$	P_4O_{10}, $P_4O_6[Ni(CO)_3]_4$
	(d)	OA_4X_6	$OBe_4(ac)_6$, $OZn_4(ac)_6$, $OZn_4(S_2PR_2)_6$,
			$OZn_4(OSPR_2)_6$, $OCo_4[C(CH_3)_3COO]_6$
		$OA_4X_6Y_4{}^b$	$OMg_6Br_6(C_4H_{10}O)_4$, $OCu_4Cl_6(pyr)_4$, $(OCu_4Cl_{10})^{4-}$
Octahedral		A_6Y_6	$B_6H_6^{2-}$
	(e)	A_6X_8	$Sn_6O_4(OCH_3)_4$
		$A_6X_8Y_6$	$Mo_6Cl_{14}^{2-}$, $Mo_6Br_{12}(H_2O)_2$
	(f)	A_6X_{12}	Pd_6Cl_{12}, Pt_6Cl_{12}, $Bi_6(OH)_{12}^{6+}$
		$A_6X_{12}Y_6$	$W_6Cl_{12}Cl_6$, $Nb_6Cl_{18}^{4-}$, $Ta_6Cl_{12}Cl_2(H_2O)_4$
Cubic		A_8X_8	C_8H_8
	(g)	A_8X_{12}	$Cu_8[S_2C_2(CN)_2]_6$
		$A_8X_{12}Y_8$	$Si_8O_{12}\phi_8$

For references see other chapters and also: *a* N 1970 **228** 648; *b* IC 1969 8 1982.

The molecule of $Rh_6(CO)_{16}$ consists of an octahedral Rh_6 nucleus having 2 CO attached to each metal atom and four more Co situated above four faces of the octahedron, each of these bridging three Rh atoms; these four CO groups are arranged tetrahedrally.

Cubic molecules and ions

The simplest example is the molecule of 'cubane', C_8H_8. Figure 3.18(g) shows an A_8X_{12} complex closely related to the A_6X_{12} complex of Fig. 3.18(f). There is the same cuboctahedral arrangement of 12 X atoms but with a cube of A atoms instead of an octahedral group. This structure is adopted by the anion $Cu_8^IL_6^{4-}$ in salts with the large cations $[N(C_6H_5)(CH_3)_3]^+$ or $[As(C_6H_5)_4]^+$, L being the ligand shown below. The 12 X atoms are made up of six pairs of S atoms.

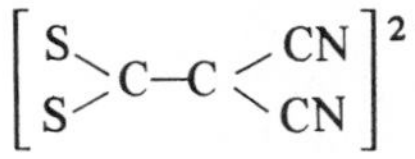

$$\begin{bmatrix} S \\ S \end{bmatrix} > C-C < \begin{matrix} CN \\ CN \end{matrix} \end{bmatrix}^{2-}$$

Miscellaneous polyhedral complexes

These include the B subgroup anions (p. 305), numerous compounds of boron (hydrides, hydride ions, and sub-halides), and siloxanes, for which reference should be made to later chapters.

Cyclic molecules and ions

In Table 3.9 are listed some examples of homocyclic and heterocyclic molecules and ions. Examples of rings containing atoms of more than two kinds are noted under the chemistry of N, P, and S. Four-membered rings are present in the numerous doubly-bridged ions and molecules in which the coordination groups of two A atoms share an edge or face and, of course, in the larger doubly-bridged rings such as (a) and (b).

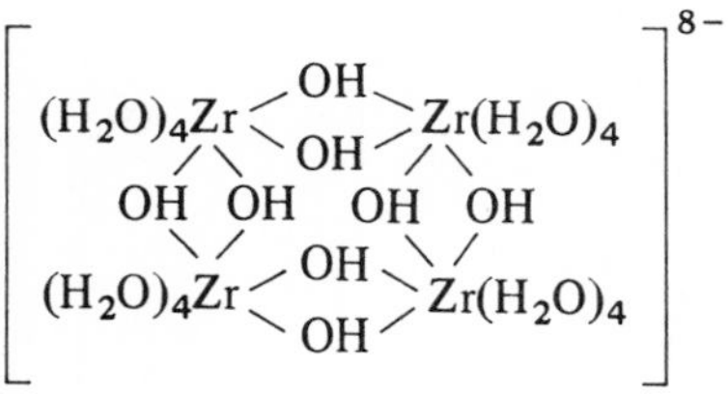

$$\left[\begin{matrix} (H_2O)_4Zr < \overset{OH}{\underset{OH}{}} > Zr(H_2O)_4 \\ OH \quad OH \quad OH \quad OH \\ (H_2O)_4Zr < \overset{OH}{\underset{OH}{}} > Zr(H_2O)_4 \end{matrix} \right]^{8-}$$

(a)

$$\begin{matrix} RS-Ni-SR \\ Ni-SR \quad RS-Ni \\ RS \quad SR \quad RS \quad SR \\ Ni-SR \quad SR-Ni \\ RS-Ni-SR \end{matrix}$$

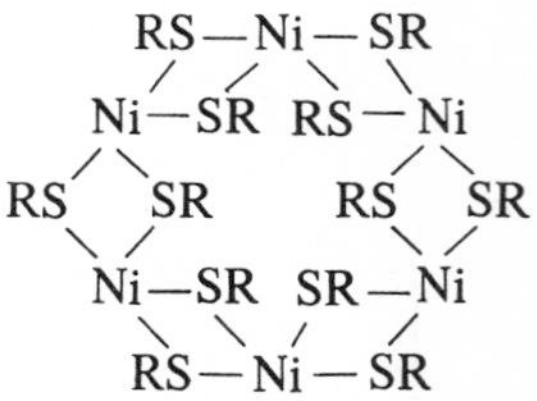

(b)

The larger rings most frequently encountered are 6- or 8-membered rings. Examples of planar 6-rings include $B_3N_3H_6$ and $N_3S_3^-$, but the configuration is most often chair-shaped, this being one of the two characteristic forms of a ring built of atoms

TABLE 3.9
Cyclic molecules and ions

Number of atoms in ring	Homocyclic	Heterocyclic A_nX_n with alternating A and X atoms
3	$Os_3(CO)_{12}$	—
4	$(PCF_3)_4$, $(AsCF_3)_4$, Se_4^{2+}	$(SSiCl_2)_2$, $(ClAlCl_2)_2$, $(ClNbCl_4)_2$
5	$(PCF_3)_5$, $(AsCH_3)_5$	—
6	$(AsC_6H_5)_6$, $(PC_6H_5)_6$, $(P_6O_{12})^{6-}$, S_6, $(Sn\phi_2)_6$	$(Si_3O_9)^{6-}$, $(P_3O_9)^{3-}$, S_3O_9, $B_3N_3H_6$, $(NSCl)_3$, $(PNCl_2)_3$, $Pd_3[S_3(C_2H_4)_2]_3$, $S_3N_3^-$
7	S_7, S_7O	—
8	S_8, S_8^{2+}, Se_8^{2+}	N_4S_4, $(Si_4O_{12})^{8-}$, $(P_4O_{12})^{4-}$, Se_4O_{12}, $(BNCl_2)_4$, $(PNCl_2)_4$, Mo_4F_{20}, $(OGaCH_3)_4$, $[(OH)Au(CH_3)_2]_4$, $S_4N_4^{2+}$, Sn_4F_8
10	S_{10}, Se_{10}^{2+}	$(PNCl_2)_5$, $S_5N_5^+$
12	S_{12}	$(Si_6O_{18})^{12-}$, $(P_6O_{18})^{6-}$. $[PN(NMe_2)_2]_6$
16		$[PN(OCH_3)_2]_8$
18	S_{18}	
20	S_{20}	

forming tetrahedral bonds. The 'twisted boat' has also been recognized as a distinct configuration in $Si_3S_3(CH_3)_3\phi_3$ and the dihedral (torsion) angles listed (ref. p. 996).

The stereochemistry of 8-rings is more complex, and this ring is much more flexible, different configurations being found in polymorphs of, for example, $N_4P_4Cl_8$ (chair and boat), in closely related salts (see the tetrametaphosphates, p. 862), and even in the same crystal in the case of $[(CH_3)_2SiNH]_4$. Models of four forms of the 8-membered ring are readily constructed from atoms forming tetrahedral bonds; the atoms are numbered consecutively around the ring.

Chair: 1,2,4,7 coplanar, 3,5,6,8 coplanar, in parallel planes. This centrosymmetric form is the analogue of the chair form of the 6-ring; highest axial symmetry, 2-fold.

Boat: 1,2,5,6 coplanar, 3,4,7,8 coplanar, in parallel planes; highest axial symmetry $\bar{4}$.

Saddle (cradle): 1,3,5,7 coplanar, 2,4,6,8 bisphenoidal. If the pairs of planes 1,2,3–5,6,7 and 3,4,5–7,8,1 are made parallel, the model represents the molecule of N_4S_4 with short separations 2–6 and 4–8. The molecules $N_4P_4F_4(CH_3)_4$ and $N_4P_4F_6(CH_3)_2$ have the maximum symmetry, $\bar{4}2m$.

Crown: 1,3,5,7 coplanar, 2,4,6,8 coplanar, in two parallel planes. Axial symmetry $\bar{8}$ (in crystalline $S_4N_4H_4$, 4mm).

Examples of molecules with these configurations include:

Chair	Boat	Saddle	Crown
$[(CH_3)_2Ga(OH)]_4$	$(NSF)_4$	N_4S_4	S_8
$[(CH_3)_2Si(NH)]_4$	$N_4P_4Cl_8$ (K form)	$N_4P_4F_4(CH_3)_4$	$S_4N_4H_4$

Infinite linear molecules and ions

Examples of the more important types of chain are given in Table 3.10. In the simplest chain, $-A-A-A-$, A may be a single atom as in plastic sulphur, or it may be a group of atoms, (a), as in the anion in $Ca[B_3O_4(OH)_3].H_2O$. In all the other chains of Table 3.10 the A atoms are linked through atoms X of a different element. In most of the examples the singly-attached X atoms are of the same kind as the bridging X atoms, but these may be ligands of two different kinds, as in oxyhalides, in a chain such as (b), and in the octahedral chains AX_2L_2 which are mentioned later.

$$-Sn-NC(CH_2)_3CN-Sn-$$
$$(Cl_4) \qquad (Cl_4)$$

(a)   (b)

The linear molecule of $Zn[S_2P(OEt)_2]_2$, (c), is an example of a chain in which a ligand is behaving both as a bidentate and as a bridging ligand — contrast the chain in $Zn[O_2P\phi(n-C_4H_9)]_2$ described later. Since the chains in class (ii) of the table consist of coordination groups AX_n, which share two X atoms, they may be described as 'vertex-sharing' chains.

(c)

Chains in class (iii) may similarly be described as 'edge-sharing' chains. This class includes the important chain consisting of octahedral groups sharing two edges. These edges may be opposite (NbI_4) or not $(TcCl_4)$, as described in Chapter 5, the

TABLE 3.10

Infinite linear molecules and ions

Type of chain	Formula	Molecules	Ions
(i) $-A-$	A	S (plastic), Se, Te	$[B_3O_4(OH)_3]_n^{2n-}$
(ii) $-A-X-$	AX	AuI, AuCN, HgO, $In(C_5H_5)$, $(\phi SeO_2)H$	$(HCO_3)_n^{n-}$, $(HSO_4)_n^{n-}$
$\begin{array}{c}-A-X-\\X\end{array}$	AX$_2$	SeO$_2$, Pb(C$_5$H$_5$)$_2$	$(BO_2)_n^{n-}$, $[Cu(CN)_2]_n^{n-}$, $(AsO_2)_n^{n-}$
$\begin{array}{c}-A-X-\\X_2\end{array}$	AX$_3$	SO$_3$, CrO$_3$, AuF$_3$, PNCl$_2$, SiOCl$_2$	$(SiO_3)_n^{2n-}$ etc., $(Cu^ICl_3)_n^{2n-}$ $(Cu^{II}Cl_3)_n^{n-}$
$\begin{array}{c}-A-X-\\X_4\end{array}$	AX$_5$	(*trans*): BiF$_5$, UF$_5$ (*cis*): CrF$_5$ etc., MoOF$_4$	$(AlF_5)_n^{2n-}$, $(PbF_5)_n^{n-}$ (in SrPbF$_6$)
(iii) 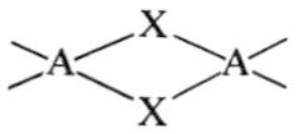	AX$_2$	(planar): PdCl$_2$, CuCl$_2$ (tetrahedral): BeCl$_2$, SiS$_2$	
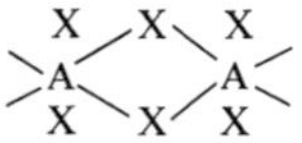	AX$_4$	NbI$_4$, TcCl$_4$	$(HgCl_4)_n^{2n-}$
	AX$_5$	PaCl$_5$	
Similarly	AX$_6$		$(ZrF_6)_n^{2n-}$ in K$_2$ZrF$_6$
	AX$_7$		$(PaF_7)_n^{2n-}$ in K$_2$PaF$_7$
(iv) 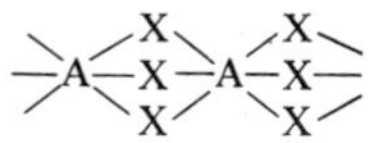	AX$_3$	ZrI$_3$ etc.	$(NiCl_3)_n^{n-}$ in CsNiCl$_3$
(v) 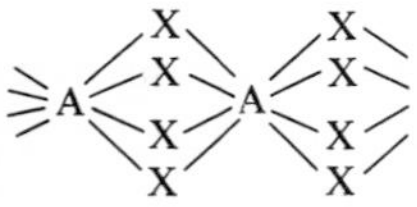	AX$_4$	U(ac)$_4$	
Hybrid chains		See text	
Multiple chains		Sb$_2$O$_3$, NbOCl$_3$	$(Si_4O_{11})_n^{6n-}$, $(CdCl_3)_n^{n-}$, $(Cu_2^ICl_3)_n^{n-}$

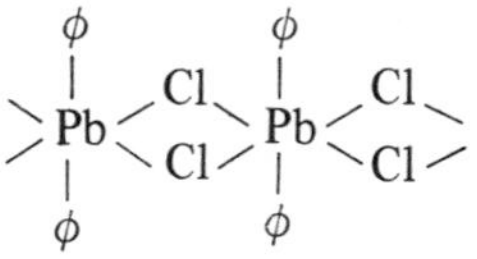

(d)

former type of chain (*trans*) being more usual. As in class (ii) the unshared ligands may be different from the bridging ones, giving the composition AX$_4$ or AX$_2$L$_2$ where L is a ligand capable of forming only one bond to A, as in the example (d). Other examples are listed in Table 25.4 (p. 1139).

More complex chains of this kind include the chain molecule $PaCl_5$ (pentagonal bipyramidal $PaCl_7$ groups sharing two equatorial edges), and the anions in K_2ZrF_6 (dodecahedral ZrF_8 groups sharing two edges), and K_2PaF_7 (tricapped trigonal prismatic groups PaF_9, sharing two edges).

In the much smaller classes, (iv) and (v), coordination groups share a pair of faces. For octahedral groups the formula is AX_3, as in ZrI_3 and $[Li(H_2O)_2]ClO_4$, and for tricapped trigonal prismatic groups it is AX_6, as in $[Sr(H_2O)_6]Cl_2$. In these chains triangular faces of coordination groups are shared; in class (v), $U(OOCCH_3)_4$, two opposite quadrilateral faces of antiprisms are shared, the bridges consisting of acetate groups, $-OC(CH_3)O-$.

The term 'hybrid chain' in Table 3.10 means a chain in which there are bridges of more than one kind. Examples include

$Zn[O_2P\phi(n\text{-}C_4H_9)]_2$:

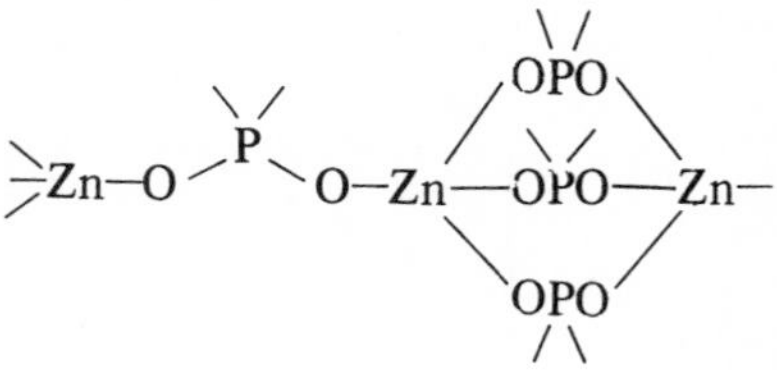

and $ReCl_4$:

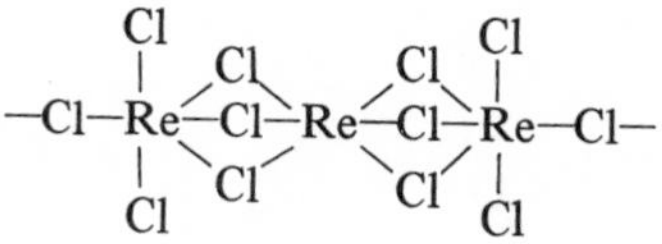

which are examples of unexpectedly complex structures for compounds with formulae of the simple basic types AX_2 and AX_4.

Under the heading 'multiple chains' in Table 3.10 we include all systems of bonded atoms which extend indefinitely in one dimension but have more complex backbones than the simple 2-connected one, $-A-A-A-$, of a simple chain. The latter is built of 2-connected points ($p = 2$), and we have seen that for $p \geqslant 3$ there are sufficient connections to form 2- and 3-dimensional systems (in addition to poly-hedra). However, if there are more than two connections to each point the system formed is not necessarily an infinite 2D or 3D one, and here we examine briefly some types of 1D systems for $p = 3$ or 4, confining our attention to those in which all points are equivalent. The simple 'ladder' of Fig. 3.19(a_1) represents a chain in orthorhombic Sb_2O_3 (O atom along each link), and it also represents the double chain ion $(Cu_2Cl_3)_n^{n-}$ formed from tetrahedral groups sharing three edges. Elaborations of this system (a_2) and (a_3), represent the chain molecule of $NbOCl_3$ (p. 221), in which octahedral NbO_2Cl_4 groups share one edge and two vertices, and the chain ion $(Si_4O_{11})_n^{6n-}$, in which some SiO_4 groups share three and others two vertices. The 4-connected ribbon (b_1) represents the chain anion in $NH_4(CdCl_3)$ formed from octahedral $CdCl_6$ groups sharing four edges.

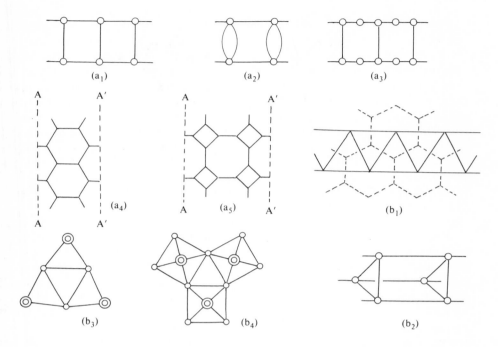

FIG. 3.19. Formation of multiple chains from 3- and 4-connected points (see text).

An indefinitely large number of tubular systems may be formed by wrapping strips of 2D nets around a cylinder and joining at lines such as AA and A'A' in (a_4) or (a_5). Examples of tubular 'chains' based on the simplest 3-connected planar net do not appear to be known, but the tubular 4.8^2 net corresponding to (a_5) is found in certain silicates (p. 1026). The simplest tubular system formed from the planar 4^4 net would be topologically similar to (a_2), but the next member of this family, (b_2), can be built from tetrahedra or octahedra sharing four vertices, as shown in the end-on views (b_3) and (b_4). An example of (b_4) is the $(CrF_4)_n^{n-}$ ion (p. 208).

Crystal structures based on 3-connected nets

Types of structural unit

The simplest units leading to structures of this kind are illustrated in Fig. 3.20. At (a) we have an atom forming three bonds which could be the structural unit in an element or in compounds in which all atoms are 3-connected. If atoms X are placed along the lines of any 3-connected net, as at (b), the formula is A_2X_3, and this arrangement is found in a number of oxides M_2O_3 and sulphides M_2S_3. At (c) we show a tetrahedral group AX_4 sharing three of its vertices with other similar groups.

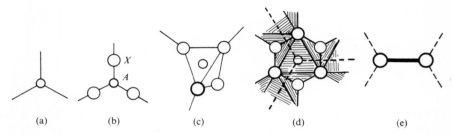

FIG. 3.20. Structural units forming 3-connected nets.

If this holds throughout the crystal the composition is A_2X_5. The most symmetrical way in which octahedral groups AX_6 can be joined to three others by sharing three edges is shown at (d). Since each X is common to two octahedra the formula is AX_3.

The four structural units (a)–(d) form the bases of the structures of numerous inorganic compounds in which the bonds are covalent or covalent–ionic. It is also convenient to represent diagrammatically, by means of 3-connected nets, the structures of certain crystals in which the structural units (which may be ions or molecules) form three hydrogen bonds. In the hydrates of some acids one H^+ is associated with H_2O to form H_3O^+ which is a unit of type (a). In certain hydroxy-compounds each OH group is involved in two hydrogen bonds, and accordingly a molecule of a dihydroxy-compound may be represented as at (e), it being necessary to indicate only the atoms involved in hydrogen bonding. The units of Fig. 3.20 may join together to form finite groups or arrangements extending indefinitely in one, two, or three dimensions. Finite (polyhedral) and infinite linear systems have already been mentioned, the only simple example of a chain being the double chain in the orthorhombic form of Sb_2O_3. The simplest plane 3-connected net is that consisting of hexagons. This is by far the most important plane 3-connected net in structural chemistry and will be dealt with in most detail. Examples will then be given of structures based on more complex plane nets before proceeding to 3D nets.

The plane hexagon net

Examples are known of structures based on this net incorporating all the five types of unit of Fig. 3.20. In its most symmetrical form this net is strictly planar and the hexagons are regular and of the same size. This form represents the structure of a layer of graphite or of B atoms in AlB_2. As the interbond angle decreases from 120° (in the plane regular hexagon) so the layer buckles.

It is well known that the bond angles (assumed to be equal) do not define the shape of an isolated 6-membered ring; the chair and boat forms are familiar to all chemists. It is less generally appreciated that the 6^3 net presents an analogous 'conformational' problem, which we may illustrate by considering nets in which all rings are chair-shaped and all bond angles have the value 109½°. There are alternative positions for the third bond from each atom, namely, e approximately in the mean

(a)

plane of the ring and p perpendicular to the mean plane of the ring (a). The numbers of possible combinations of e and p bonds for the ring as a whole are evidently the same as the numbers of substituted benzenes. Indicating the number of p positions by the superscript numeral, these numbers are:

$$\left.\begin{array}{l} p^1 \text{ and } p^5, \text{ one each,} \\ p^2 \text{ and } p^4, \text{ three each,} \\ p^3, \text{ three} \end{array}\right\} \text{ total 11.}$$

With the arrangements p^0 (or e^6) and p^6 there are 13 different ways of arranging the six extra-annular bonds. It can be shown (by trial) that if we require all the rings to be chair-shaped *with the same disposition of bonds from each*, then only three of these combinations lead to periodic 2D nets. The three permissible bond arrangements are p^0, p^1, and the centrosymmetric p^2 (a). Elevations of the three layers, built with tetrahedral bond angles, are shown in Fig. 3.21(a), the layers being viewed approximately in the direction of the arrow in the sketch (a). In Fig. 3.21(b) and (c) are shown the idealized forms of the layers p^0 and p^1 built with 90° bond angles to show their relation to the cubic lattice; in (b) the mean plane of the layer is perpendicular to a body-diagonal of the cell. No example is yet known of the p^1 layer; the p^0 layer is that of crystalline As, Sb, and Bi (and the low-temperature form of GeTe), and p^2 is the layer in black P and the isostructural GeS, SnS, GeSe, and SnSe. The actual configuration of the p^0 layer of As is close to (b), since the bond angle is 96°, but in the p^0 Si_2 layer in $CaSi_2$ the angles are close to the tetra-hedral value. The p^2 layer of black P has a configuration intermediate between those of Fig. 3.21(a) and (c), the mean bond angle being 100°.

If alternate atoms in the 6-gon layer are of different elements the composition is AB (Fig. 3.22), and we find hexagonal BN with plane layers and GeS (and SnS) with buckled layers similar to those in black P.

Orthoboric acid is a trihydroxy-compound, $B(OH)_3$, and in the crystalline state each molecule is hydrogen-bonded to neigbouring molecules by six $O-H---O$ bonds. These are in pairs to three adjacent molecules, an arrangement similar to that in the dimers of carboxylic acids:

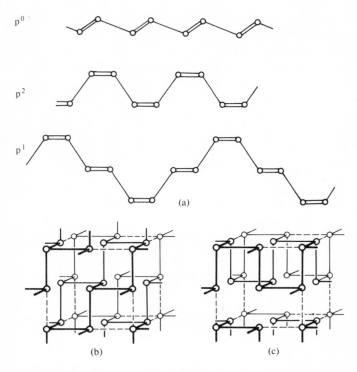

FIG. 3.21 (a) Elevations of 6^3 layers built with tetrahedral bond angles; (b) and (c) relation of p^0 and p^2 layers to the primitive cubic lattice.

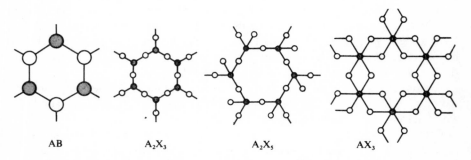

FIG. 3.22. The structures of binary compounds based on the plane 6-gon net.

and the molecules are arranged in layers based on the 6-gon net as shown in Fig. 24.11 (p. 1067).

In the crystalline hydrates of some acids a proton is transferred to the water molecule forming the H_3O^+ ion, which can form three hydrogen bonds. Accordingly, the structures of crystalline $HCl.H_2O$ and $HNO_3.H_2O$ consist of layers in which anions and H_3O^+ ions alternate in 6-rings, as in the AB layer of Fig. 3.22. In these two monohydrates the number of H atoms (3) is the number required for each unit to be hydrogen-bonded to three neigbouring ones. A comparison with the structures of $HClO_4.H_2O$ and $H_2SO_4.H_2O$ illustrates how the structures of these hydrates are determined by the number of H atoms available for hydrogen bonding rather than by the structure of the anion. The low-temperature form of $HClO_4.H_2O$ has a structure of exactly the same topological type as that of $HNO_3.H_2O$. As in $HNO_3.H_2O$ there are three hydrogen bonds connecting each ion to its neigbours, and one O of each ClO_4^- ion is not involved in hydrogen bonding. In the monohydrate of sulphuric acid, on the other hand, there are sufficient H atoms for an average of four hydrogen bonds from each structural unit (H_3O^+ or HSO_4^-), and we find three hydrogen bonds from each H_3O^+ and five from each HSO_4^-; the resulting structure is no longer based on the hexagon net. These structures are described in more detail in Chapter 15.

Trithiane, $S_3(CH_2)_3$, a chair-shaped molecule like cyclohexane, forms many metal complexes. Two silver compounds have layer structures based on the 6-gon net. In both structures Ag is tetrahedrally coordinated, the fourth ligand (X in Fig. 3.23(a)) being H_2O in Ag(trithiane)$ClO_4.H_2O$ and O of NO_3^- in Ag(trithiane)-$NO_3.H_2O$ (JCS A 1968 93). Silver tricyanmethanide, $Ag[C(CN)_3]$, has a very

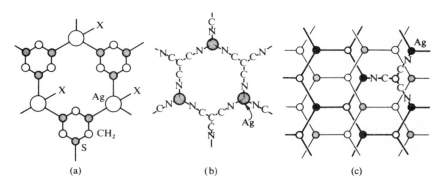

FIG. 3.23. Layers in the structures of (a) $Ag[S_3(CH_2)_3]NO_3.H_2O$; (b) $Ag[C(CN)_3]$; (c) two interwoven layers of type (b).

interesting structure. The ligand $C(CN)_3$ forms with Ag (here forming only three bonds) a layer of the same basic type as in the trithiane complex (Fig. 3.23(b)), but pairs of layers are interwoven, as at (c). This is one of the two examples of interwoven 2D nets at present known, the other being the much more complex multiple layer of Hittorf's P to which we refer later.

If A atoms at the points of a 6-gon net are joined through X atoms (Fig. 3.22) the result is a layer of composition A_2X_3. Crystalline As_2S_3 (the mineral orpiment) is built of layers of this kind (Fig. 3.24(a)), and both forms of claudetite (As_2O_3)

have very similar layer structures. (The other polymorph, arsenolite, is built of As_4O_6 molecules with the same type of structure as the P_4O_6 molecule.)

'Acid' hydrogen) salts provide many interesting examples of hydrogen-bonded systems which are particularly simple if the ratio of H atoms to oxy-anions is that required for a 3-, 4-, or 6-connected net:

Salt	H : *anion ratio*	*Type of net*
$NaH_3(SeO_3)_2$	3 : 2	3-connected
KH_2PO_4	2 : 1	4-connected
$(NH_4)_2H_3IO_6$	3 : 1	6-connected

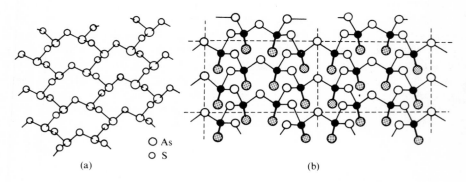

O As
O S

(a)   (b)

FIG. 3.24. Layers in (a) As_2O_3 (orpiment); (b) P_2O_5.

In $NaH_3(SeO_3)_2$ the SeO_3^{2-} ions are situated at the points of the plane 6-gon net and a H atom along each link, giving the required H:anion ratio, 3:2.

We note on p. 166 that a (close-packed) layer of composition AX_2 based on the 6^3 net can be built from tetrahedral AX_4 groups sharing one edge and two vertices. No example is yet known, but see p. 167 for the closely related $GaPS_4$ layer.

Layers formed by joining tetrahedral AX_4 groups through three of their vertices have the composition A_2X_5. Such a layer is electrically neutral if formed from PO_4 groups but the similar layer built from SiO_4 groups is a 2D ion $(Si_2O_5)_n^{2n-}$. Figure 3.24(b) shows a projection of the atoms in a layer of one of the polymorphs of P_2O_5. Although at first sight this layer appears somewhat complex, removal of the shaded circles (the unshared O atoms) leaves a system of P and O atoms very similar to that of As and S in As_2S_3 (Fig. 3.24(a)). The charged layer in $Li_2Si_2O_5$ and other silicates of this kind is of the same topological type but has a very buckled configuration, presumably adjusting itself to accommodate the cations between the layers.

The fourth structural unit, (d) of Fig. 3.20, is an octahedral group AX_6 sharing three edges. The mid-points of these edges are coplanar with the central A atom, so that the octahedra linked together in this way form a plane layer based on the 6-gon net. This layer is found in many compounds AX_3, the structures differing in the way in which the layers are superposed, that is, in the type of packing of the X atoms. In the BiI_3 structure there is hexagonal and in YCl_3 cubic closest packing of the halogen atoms, as described in Chapter 4; in $Al(OH)_3$ the more open packing of the layers is due to the formation of O–H---O bonds between OH groups of adjacent layers.

We describe in Chapter 5 the formation of composite layers formed by the sharing of the remaining vertices of a tetrahedral A_2X_5 layer with certain of the vertices of an octahedral layer AX_3, both layers being based on the plane 6-gon net. These complex layers are the structural units in two important classes of minerals, the clay minerals (including kaolin, talc, and the bentonites) and the micas. One of these layers is illustrated as an assembly of tetrahedra and octahedra in Fig. 5.44 (p. 233).

Two quite unexpected examples of structures based on the 6^3 net are those of UBr_4 and ThI_4, in which the metal atoms are 7- and 8-coordinated respectively. In UBr_4 pentagonal bipyramidal groups share three edges (Fig. 28.4, p. 1259) so that 6 of the 7 Br atoms are shared between two U atoms $(1 + 6(\frac{1}{2}) = 4)$. In ThI_4 each antiprismatic ThI_8 group shares one edge and two faces with the result that each I is bonded to two Th atoms: $8(\frac{1}{2}) = 4$. This structure is illustrated in Fig. 3.25.

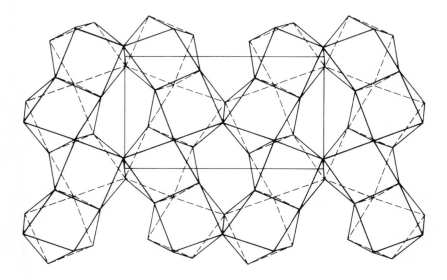

FIG. 3.25. Layer in crystalline ThI_4.

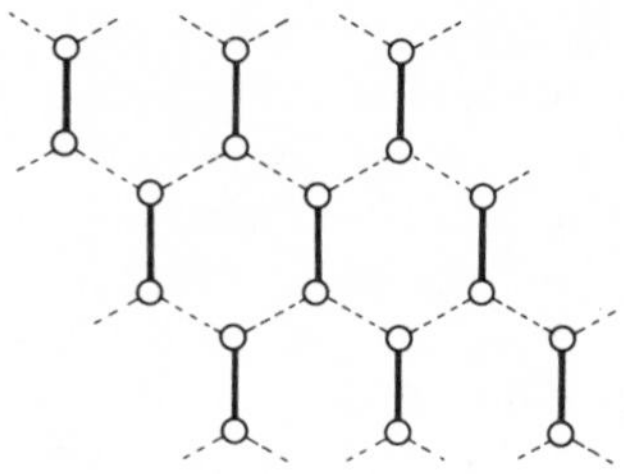

FIG. 3.26. Layer of molecules in crystalline γ-quinol (diagrammatic).

As an example of a unit of type (e) of Fig. 3.20 we illustrate diagrammatically the structure of one of the polymorphs of quinol, *p*-dihydroxybenzene. Each OH group can act as the donor and acceptor end of an O—H---O bond, and the simplest arrangement of molecules of this kind is the plane layer illustrated in Fig. 3.26. The structures we have described are summarized in Table 3.11.

TABLE 3.11
Layers based on the simple hexagon net

Layer type	Examples
A	C (graphite), As, Sb, Bi, P (black) $CaSi_2$, AlB_2 $B(OH)_3$
AB	BN, GeS, SnS $(H_3O)^+Cl^-$, $(H_3O)^+NO_3^-$, $(H_3O)^+ClO_4^-$ $Ag[S_3(CH_2)_3]ClO_4 . H_2O$, $Ag[S_3(CH_2)_3]NO_3 . H_2O$ $Ag[C(CN)_3]$ (two interwoven layers)
A_2X_3	As_2O_3, As_2S_3 $Na[H_3(SeO_3)_2]$, $(Te_2O_3)SO_4$
A_2X_5	P_2O_5 $Li_2(Si_2O_5)$, $Rb(Be_2F_5)$, $Pb_2(Ga_2S_5)$
AX_3	YCl_3, BiI_3, $Al(OH)_3$
AX_4	ThI_4

This kind of topological representation of crystal structures may be extended to more complex compounds if we focus our attention on the limited number of stronger bonds that hold the structure together. Nylon is formed by condensing hexamethylene diamine with adipic acid and forms long molecules which may be represented diagrammatically as in Fig. 3.27(a). The molecules are linked into layers by hydrogen bonds between the CO and NH groups of different chains, so that if we are interested primarily in the hydrogen bonding we may show only the CO and

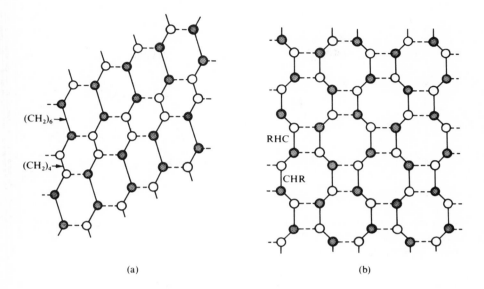

FIG. 3.27. Hydrogen bonding systems in layers of (a) nylon; (b) caprolactam.

NH groups (open and shaded circles respectively) and omit the CH_2 groups. The hydrogen bonds are shown as broken lines. Reduced to this simple form the structure appears as the 6-gon net. The type of 3-connected net depends on the sequence of pairs of CO and pairs of NH groups along each chain molecule. If these groups alternate (−CO−NH−CO−NH−) as in caprolactam, the natural way for the chains to hydrogen-bond together is that shown in Fig. 3.27(b), which is one of the next simplest groups of plane 3-connected nets; the 4.8^2 net.

Structures based on other plane 3-connected nets

Examples of crystal structures based on more complex plane 3-connected nets are less numerous than those based on the 6^3 net. Borocarbides MB_2C_2 are formed by Sc, Y, La, and the 4f elements (and also Ca); they consist of layers of composition B_2C_2 interleaved with metal atoms. The Sc compound is of special interest as the only example at present known of the layer consisting of equal numbers of 5-gons and 7-gons. A number of structures are based on the 4.8^2 net, illustrated as the first of the semi-regular nets in Fig. 3.9(b), p. 82. In LaB_2C_2 (AC 1980 B36 1540) the pattern of B and C atoms is that shown at (a). The same net forms the basis of the structures of $GaPS_4$ and $Pb_2Ga_2S_5$. The former consists of layers formed from tetrahedral groups (alternately GaS_4 and PS_4) sharing one edge and two vertices (see Fig. 4.19, p. 166). In the latter there are layers of GaS_4 tetrahedra sharing three vertices, with the Pb atoms between the layers in positions of 8-coordination (AC 1980 **B36** 1990). Silicate structures based on the 4.8^2 and more complex

(a)   (b)

3-connected nets are described in Chapter 23 and illustrated in Fig. 23.13, p. 1024. The layer in Sn_3F_5Br (p. 1183) is also based on the 4.8^2 net but with additional Sn atoms along the links common to the 8-gons, as shown at (b).

We mentioned earlier the structure of Hittorf's phosphorus as a second example of interwoven layers. In this structure each individual layer is itself a multiple layer with a complex structure which is described on p. 839.

Structures based on 3D 3-connected nets

There are very few examples as yet of simple inorganic compounds with structures based on 3-dimensional 3-connected nets. We might have expected to find examples among the crystalline compounds of boron, an element which forms three coplanar bonds in many simple molecules and ions. The structure of the normal form of B_2O_3 is in fact based on a simple 3-connected net; however, boron is 4-coordinated in many borates, and both triangular and tetrahedral coordination occur in many compounds. The complex crystalline forms of elementary boron are not simple covalent structures but electron-deficient systems of a quite special kind, the structures of which are briefly described in Chapter 24.

There are two 3-connected nets $(10, 3)$ with four points in the simplest unit cell, but if the nets are constructed with equal bonds and interbond angles of $120°$ they have eight points in their unit cells and cubic and tetragonal symmetry respectively. The cubic net (Fig. 3.28) is clearly the 3-connected analogue of the diamond net. It represents the arrangement of Si atoms in $SrSi_2$ (p. 990). Although the symmetry (space group $P\ 4_332$) is lower than that of the most symmetrical configuration (space group $I\ 4_13$) with exactly coplanar bonds, it retains cubic symmetry. This net also represents the structure of crystalline H_2O_2. If the molecules are represented as at (a) and O atoms are placed at the points of the net, then two-thirds of the links represent hydrogen bonds between the molecules. The net is not in the most 'open' configuration of Fig. 3.28(a), which is drawn with equal coplanar bonds from each point, but is in the most compact form consistent with normal van der Waals contacts between non-hydrogen-bonded O atoms and with O—H---O distances

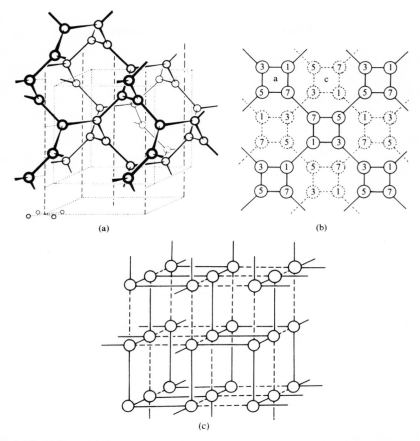

FIG. 3.28. (a) The cubic 3-connected 10-gon net (10,3)–a. (b) Projection of two interpenetrating nets. (c) Configuration of the net in $Hg_3S_2Cl_2$.

of 2.70 Å and HO–OH (intramolecular) equal to 1.47 Å. An interesting property of this net is that it is enantiomorphic; accordingly, crystalline H_2O_2 is optically active.

Examples of more complex 3-connected 3D nets formed by dihydroxy-compounds include the structures of α- and β-resorcinol (*m*-dihydroxybenzene) and of $OH . Si(CH_3)_2 . C_6H_4 . Si(CH_3)_2 . OH$.

The projection of the cubic $(10,3)$ net on a face of the cubic unit cell, the full circles and lines of Fig. 3.28(b), shows that the net is built of 4-fold helices which are all anticlockwise upwards. The figures indicate the heights of the points in terms of $c/8$, where c is the length of the cell edge. A second net can be accommodated in the same volume, and if the second net is a mirror-image of the first in no case is the distance between points of different nets as short as the distance between (connected) points within a given net. In the second net of Fig. 3.28(b) (dotted circles and lines) the helices are clockwise. This type of structure, which would be a 3D racemate, is not yet known, but in view of its similarity to the β-quinol structure described later, there is no reason why it should not be adopted by some suitable compound. An even more interesting structure is that of $[(Hg_2)_3O_2H]Cl_3$, p. 1158, which consists of *four* interpenetrating $(10,3)$ nets of this kind, the O atoms being situated at the points of a net and bonded through pairs of Hg atoms.

We showed in Fig. 3.21 the relation of the layers of As and black P to the simple cubic structure, from which they may be derived by removing one-half of the links. We may derive 3D 3-connected nets in a similar way, by removing different sets of bonds. Figure 3.28(c) shows the cubic $(10,3)$ net drawn in this way with interbond angles of 90°; this configuration of the net is close to the arrangement of S atoms in one form of $Hg_3S_2Cl_2$. The Hg atoms lie along the links of the net and the Cl ions are accommodated in the interstices of the Hg_3S_2 framework.

The second $(10,3)$ net is illustrated in Fig. 3.29(a) as the arrangement of Si atoms in α-$ThSi_2$, the large Th atoms being accommodated in the interstices of the framework. This net also represents the structure of the third crystalline form of P_2O_5, in which PO_4 tetrahedra are placed at all points of the net and joined by sharing three vertices (O atoms). As might be expected, this polymorph has the highest melting point of the three. It is interesting to find that a single compound, P_2O_5, has three crystalline forms which illustrate three of the four main types of crystal structure, namely, a finite (in this case polyhedral) group, a layer structure, and a 3D framework structure. A similar system of tetrahedra forms the framework in $La_2Be_2O_5$, in which the large La^{3+} ions occupy positions which are surrounded by irregular groups of 10 O atoms; contrast La–O, 2.42 Å with the mean Be–O, 1.64 Å, within the tetrahedra. As in the case of the cubic $(10,3)$ net in Fig. 3.28(c) there is a configuration of the $ThSi_2$ net with interbond angles equal to 90°. This configuration (Fig. 3.29(b)) is a close approximation to the positions of the Be atoms in $La_2Be_2O_5$. For a more complex example of this net see $Na_4Sn_3S_8$ (p. 1181).

For a rather complex example of a structure in which there are two interpenetrating $ThSi_2$ nets see neptunite (p. 1033).

A third 3-connected net, also consisting of 10-gons, forms the basis of the structure of the normal form of B_2O_3. This net has six points in the unit cell; it is illustrated in Fig. 3.30. It is related in a simple way to the $ThSi_2$ net, for both can be constructed from sets of zig-zag chains which are linked to form the 3D net. In the tetragonal net these sets are related by rotations through 90° along the direction of the c axis, whereas in the trigonal net successive sets of chains are rotated

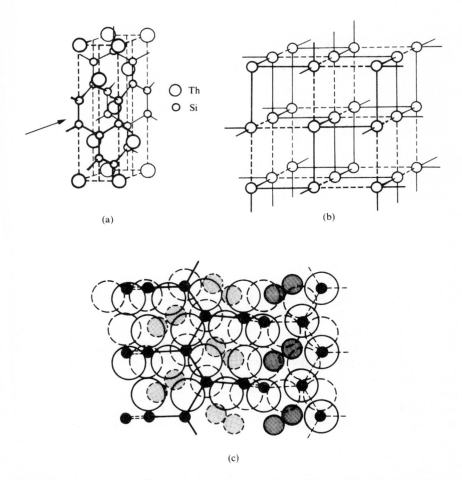

FIG. 3.29. (a) The tetragonal 3-connected 10-gon net (10,3)–b in α-ThSi$_2$. (b) Configuration of the net in La$_2$Be$_2$O$_5$. (c) Part of the same net in (Zn$_2$Cl$_5$)(H$_5$O$_2$).

through 120°.

There is an indefinitely large number of more complex nets. Those in which the same types of circuit meet at each point may be given symbols analogous to those of polyhedra (Table 3.3, p. 72) and plane nets. For example, the truncated octahedron is 4.6^2, the semi-regular net consisting of 4-gons and 8-gons is 4.8^2, and there are 3D nets 4.12^2 and 4.14^2. A net 4.12^2 is the basis of the structure of AuOCl (p. 1145); Au and O atoms are situated at alternate points of the net. In FeSO$_3$.2½H$_2$O (p. 675) Fe and S atoms occupy alternate points of a 4.14^2 net. In this somewhat unexpected example of a 3-connected net each sulphite ion is bonded to three different Fe atoms and each Fe is surrounded octahedrally by three O

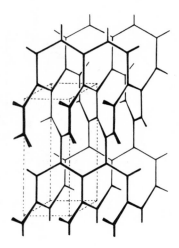

FIG. 3.30. The 10-gon net (10,3)–c which is the basis of the structure of B_2O_3.

atoms of different sulphite ions (and by three H_2O molecules). We illustrate in Fig. 3.31 the net 6.10^2. This net is readily visualized in relation to a cube (or, more generally, a rhombohedron) with one body-diagonal vertical, the cube being outlined by the broken lines. A plane hexagon is placed with its centre at each vertex and its plane perpendicular to the body-diagonal of the cube. Each hexagon is connected to six others by the slanting lines. There is a large unoccupied volume at the centre of the cube. In fact a second, identical net can interpenetrate the first, displaced by one-half the vertical body-diagonal of the cube, so that a ring A of the second net occupies the position A'. This system of two interpenetrating nets forms the basis of the structure of β-quinol, in which the long slanting lines in Fig. 3.31 represent the molecules of $C_6H_4(OH)_2$ and the circles the terminal OH groups which are hydrogen-bonded into hexagonal rings. This structure is not a true polymorph of quinol, as was originally supposed, for it forms only in the presence of foreign atoms such as argon or molecules such as SO_2 or CH_3OH. These act as 'spacers' between rings such as A and A'; the structure is illustrated in Fig. 1.9 (p. 31) as an example of a 'clathrate' compound.

A *single* net of the type of Fig. 3.31 forms the framework of crystals of $N_4(CH_2)_6 . 6H_2O$. The hexagons represent rings of six water molecules, and the $N_4(CH_2)_6$ molecules are suspended from three water molecules belonging to the eight hexagonal rings surrounding each large cavity, that is, they occupy the positions occupied by the rings of the second framework in β-quinol. This structure is illustrated in Fig. 15.8 (p. 665).

The obvious sequel to describing the topology of structures in terms of the basic 3-connected (or other) net is to enquire why a particular net is chosen from the indefinitely large number available. The clathrate β-quinol structure forms only in

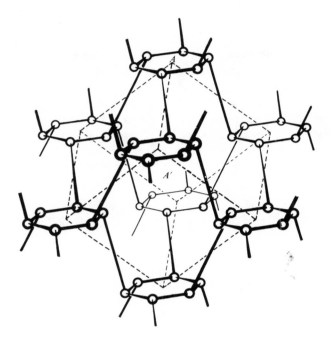

FIG. 3.31. 3D net of hydrogen-bonded molecules in β-quinol.

the presence of a certain minimum number of foreign atoms or molecules which must be of a size suitable for holding together the two frameworks. The choice of particular 3D framework in $N_4(CH_2)_6.6H_2O$ is obviously determined by the dimensions and hydrogen bonding requirements of the solute molecule. The growth of the crystal (which is synonymous with the formation of the hydrate) consists in the building of a suitable framework around the solute molecules which is compatible with the hydrogen bonding requirements of the H_2O molecules. Similarly, in the crystallization of $La_2Be_2O_5$ the Be_2O_5 framework must build up around the La^{3+} ions, this packing being the factor determining the choice of the particular net, since many 10-gon (or other) 3-connected nets could be built from tetrahedra sharing three vertices. The choice of the very simple 10-gon nets by H_2O_2, P_2O_5, and B_2O_3 is presumably also a matter of packing, that is, of forming a reasonably efficient packing of all the O atoms in the structure.

   Although we do not consider the closest packing of spheres until Chapter 4 it is convenient to illustrate at this point the synthesis of topological and packing requirements by a further example of a structure based on the $ThSi_2$ net of Fig. 3.15(e). From an acid aqueous solution of $ZnCl_2$ it is possible to crystallize a hydrate with the empirical composition $ZnCl_2.\frac{1}{2}HCl.H_2O$. The crystals consist of a framework of $ZnCl_4$ tetrahedra, of the composition $(Zn_2Cl_5)_n^{n-}$, topologically similar to P_2O_5 or $(Be_2O_5)_n^{6n-}$ in $La_2Be_2O_5$. The framework encloses pairs of hydrogen-bonded water molecules, $(H_5O_2)^+$, so that the structural formula is $(Zn_2Cl_5)^-(H_5O_2)^+$. The Cl atoms are arranged in a distorted hexagonal closest packing, the distortion arising from the need to incorporate the $(H_5O_2)^+$ ions. Figure 3.29(c) illustrates the structure as close-packed layers, and shows that the growth of the crystalline hydrate involves the construction of a suitable (Zn_2Cl_5) framework around the aquo-ions; it also shows how the $ThSi_2$ net can be achieved for a compound A_2X_5 at the same time as closest packing of the X atoms, the structure now being viewed along the direction of the arrow in Fig. 3.29(a).

Table 3.12 summarizes the examples we have described of structures based on 3D 3-connected nets.

TABLE 3.12

Structures based on 3D 3-connected nets

Net	*Fig.*	*Example*
Cubic $(10, 3)$–a	3.28	$SrSi_2$, H_2O_2, $Hg_3S_2Cl_2$, $CsBe_2F_5$, $(Sn_2F_3)Cl$, $[(Hg_2)_3O_2H]Cl_3$
Tetragonal $(10, 3)$–b	3.29	α-$ThSi_2$, h.p. $SrSi_2$, P_2O_5, $La_2Be_2O_5$, $(Zn_2Cl_5)(H_5O_2)$, $RbH_3(SeO_3)_2$
Trigonal $(10, 3)$–c	3.30	B_2O_3
More complex nets	3.31	$N_4(CH_2)_6 \cdot 6H_2O$, β-quinol, α- and β-resorcinol, $AuOCl$, $FeSO_3 \cdot 2\frac{1}{2}H_2O$, $HO \cdot Si(CH_3)_2C_6H_4Si(CH_3)_2OH$

Crystal structures based on 4-connected nets

Types of structural unit

Some units that can form 4-connected nets are shown in Fig. 3.32, and of these the simplest is a single atom capable of forming four bonds. There is no crystalline element in which the atoms form four coplanar bonds, but we have noted that the simplest 3D net in which each point is coplanar with its four neighbours (Fig. 3.15(f)) represents the structure of NbO. More complex nets in which some of the points have four coplanar and the remainder four tetrahedral neighbours represent the structures of PtS and PdP_2. The simplest 3D 4-connected net in which all points have four tetrahedral neighbours is the diamond net, with which we shall deal in more detail shortly. Other units we may expect to find in 4-connected systems include tetrahedral AX_4 groups sharing all vertices (X atoms), giving the composition AX_2, octahedral AX_6 groups sharing four vertices to form 2D or 3D structures of composition AX_4, and molecules such as H_2O and H_2SO_4 (or $O_2S(OH)_2$) which have sufficient H atoms to form four hydrogen bonds to adjacent molecules.

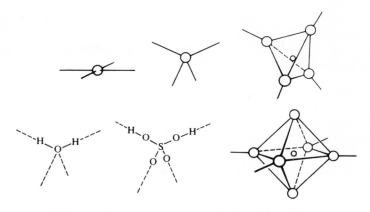

FIG. 3.32. Structural units forming 4-connected nets.

TABLE 3.13
Structures based on 4-connected nets

Plane 4-gon net

   A *layers*
      $SO_2(OH)_2$, $SeO_2(OH)_2$, $[B_3O_5(OH)]Ca$, $[B_6O_9(OH)_2]Sr \cdot 3 H_2O$

   AX *layers* (4 : 4)
      PbO, LiOH, $Pd(S_2)$

   AX_2 *layers* (4 : 2)
      HgI_2 (red), γ-$ZnCl_2$, $(ZnO_2)Sr$, $Zn(S_2COEt)_2$, $Cu(CN)(N_2H_4)$

   AX_2 *layers with additional ligands attached to* A
      AX_2Y: $[Ni(CN)_2 \cdot NH_3]C_6H_6$
      AX_2Y_2: $SnF_2(CH_3)_2$, $UO_2(OH)_2$
      AX_4: SnF_4, $(NiF_4)K_2$, $(AlF_4)Tl$

3-dimensional nets

	Diamond (see Table 3.14)	
	More complex nets	*Derived* 4 : 2 *structures*
	ZnS (wurtzite)	SiO_2 (tridymite), ice-I
	Ge (high pressure)	SiO_2 (keatite), ice-III
Tetrahedral	β-BeO	
	SiC polytypes	
		GeS_2 β-$ZnCl_2$
		Other forms of SiO_2 and H_2O
		Aluminosilicates

Planar   NbO

Planar and
  tetrahedral   PtS, PdO, PdS, CuO; PdP_2

               Polyhedral frameworks
Tetrahedral   { Aluminosilicates
                Clathrate hydrates

Table 3.13 includes examples of structures based on 2D and 3D 4-connected nets. Only the simplest of the plane 4-connected nets is of importance in crystals as the basis of layer structures; in its most symmetrical configuration it is one of the three regular plane nets.

Structures based on the plane 4-gon net

The simplest types of layer based on this net are shown in Fig. 3.33:

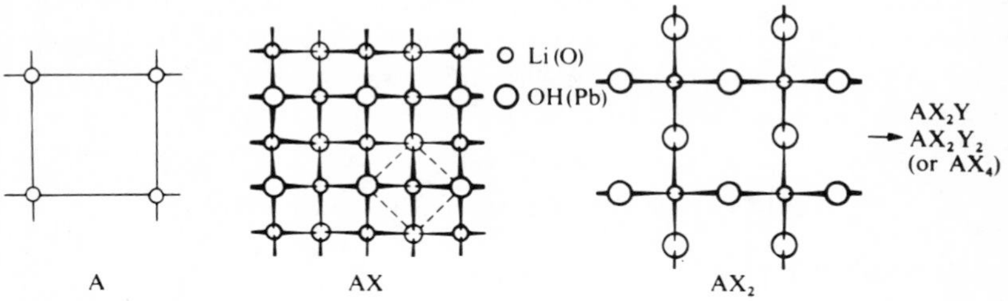

A: all units of the same kind
AX: alternate units A and X, both 4-connected
AX$_2$: A (4-connected) linked through X (2-connected) which may be a single atom or a more complex ligand. In an AX$_2$ layer all the X atoms are not necessarily of the same kind (for example, Cu(CN)(N$_2$H$_4$)). Further (singly-connected) ligands Y may be attached to A; they may be the same as X (K$_2$NiF$_4$) or they may be different (SnF$_2$(CH$_3$)$_2$).

FIG. 3.33. Layers based on the plane 4-gon net.

Layers of type A. Simple examples include the layer structures of crystalline H$_2$SO$_4$ and H$_2$SeO$_4$ (p. 374) in which each structural unit is hydrogen-bonded to four others; contrast the topologically similar H$_2$PO$_4^-$ ion which in KH$_2$PO$_4$ forms a 3D structure based on the diamond net, to which we refer later. Two borates provide examples of more complex units forming layers of this kind. In CaB$_3$O$_5$(OH) the units of Fig. 3.34(a) are linked together by sharing the O atoms distinguished as

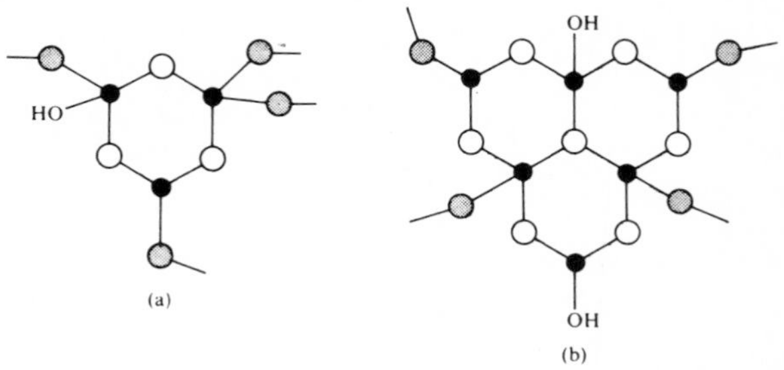

FIG. 3.34. Structural units in layers of (a) CaB$_3$O$_5$(OH); (b) SrB$_6$O$_9$(OH)$_2$.3H$_2$O.

shaded circles to form 2D ions; the layers are held together by the Ca^{2+} ions. In $SrB_6O_9(OH)_2 \cdot 3H_2O$ the primary structural unit of the anion is the tricyclic unit of Fig. 3.34(b), which is similarly joined to four others by sharing four O atoms.

Layers of type AX. Layers in which all A and X are coplanar are not known. Crystalline LiOH is built of the AX layers shown in Fig. 3.33. The small circles represent Li atoms in the plane of the paper and the larger circles OH groups lying in planes above (heavy) and below (light) that of the Li atoms. The layer is alternatively described as consisting of tetrahedral $Li(OH)_4$ groups each sharing four edges which are shown for one tetrahedron as broken lines. If Li is replaced by O and OH by Pb the layer is formed from tetrahedral OPb_4 groups, with Pb atoms. on both the outer surfaces of the layer forming four pyramidal bonds to O atoms. This remarkable layer is the structural unit in the red form of PbO; the same structure is adopted by one polymorph of SnO.

A portion of the layer structure of PdS_2 was illustrated in Fig. 1.8 (p. 18); this may be regarded as an AX layer built from the 4-connected units.

$$\mathord{>}Pd\mathord{<} \quad \text{and} \quad \mathord{>}S\!-\!S\mathord{<}$$

Layers of type AX_2. The simplest AX_2 layer based on the square net has A atoms at the points of the net joined through X atoms (2-connected) along the links of the net. The layer in which all A and X atoms are coplanar is not known. $Ni(CN)_2$ and $Pd(CN)_2$ might be expected to form layers of this type but their structures are not known. The clathrate compound $Ni(CN)_2 \cdot NH_3 \cdot C_6H_6$ is built of layers in which Ni atoms are joined through CN groups to form a square net, and NH_3 molecules complete octahedral coordination groups around *alternate* Ni atoms. Owing to the presence of the NH_3 molecules projecting from the layers there are cavities between the layers which can enclose molecules such as C_6H_6. The structure is illustrated in Fig. 1.9(a), p. 31.

If the rows of X atoms lie alternately above and below the plane of the A atoms, as in Fig. 3.33, the layer consists of tetrahedral AX_4 groups, sharing all their vertices. The layers in red HgI_2 (and the isostructural γ-$ZnCl_2$) and in $SrZnO_2$ are illustrated in this way in Fig. 5.6 on p. 000. Other examples of AX_2 layers include zinc ethyl xanthate and cuprous and argentous salts of nitrile complexes $M(nitrile)_2X$ $(X = NO_3, ClO_4)$, in which the bridging ligands are (a) and (b), and $CuCN(N_2H_4)$,

$$\begin{array}{c}\diagdown S \diagdown \\ \diagup S \diagup\end{array}\!\!C\!-\!OC_2H_5$$

(a)

$$-NC\!-\!(CH_2)_n\!-\!CN-$$

(b)

$$\begin{array}{ccc}
\;\mid\; & & \;\mid\; \\
-Cu\!-\!N\!-\!N\!-\!Cu- \\
\;\mid\; & & \;\mid\; \\
C & & N \\
\;\mid\; & & \;\mid\; \\
N & & C \\
\;\mid\; & & \;\mid\; \\
-Cu\!-\!N\!-\!N\!-\!Cu- \\
\;\mid\; & & \;\mid\;
\end{array}$$

(c)

(c), in which there are two different kinds of ligand X. The nitrile complexes, (b), provide examples of all three types of infinite complex AX_2:

	$n = 2$	$n = 3$	$n = 4$
$Cu(nitrile)_2NO_3$	chain	layer	3D net

We refer later to the adiponitrile complex ($n = 4$).

Layers of type AX_4. The plane layer consisting of octahedral groups sharing their four equatorial vertices occurs in a number of crystals. If formed from AX_6 groups the composition is AX_4, and this is the form of the infinite 2D molecules in crystalline SnF_4, PbF_4, and NbF_4, and of the 2D anions in $TlAlF_4$ and in the K_2NiF_4 structure. The latter structure is adopted by numerous complex fluorides and oxides and is illustrated in Fig. 5.16 (p. 209). If the unshared octahedron vertices are occupied by ligands Y different from the shared X atoms the composition is AX_2Y_2, as in the infinite 2D molecules of $SnF_2(CH_3)_2$ and $U(OH)_2O_2$. In $U(OH)_2O_2$ the two short U–O bonds characteristic of the uranyl ion are approximately perpendicular to the plane of the layer, the metal atoms being linked through the equatorial OH groups.

An interesting elaboration of the octahedral layer is found in the 'basic' salt $CuHg(OH)_2(NO_3)_2(H_2O)_2$. Distorted octahedral groups around Cu and Hg alternate in one direction, in which they share a pair of opposite edges, and are joined by sharing vertices (O atoms of NO_3^- ions) to form layers. This is one of the rare examples of

the octahedral AX_3 layer of Fig. 5.26 (p. 220). The layers are held together by O·H···O bonds between unshared O atoms of NO_3^- ions and H_2O molecules.

Structures based on the diamond net

The diamond net (Fig. 3.35(a)) is well suited to showing how a very simple structural theme may serve as the basis of the structures of a variety of compounds of increasing complexity. We saw on p. 87 that the simplest three-dimensional framework in which every point is joined to four others is a system of puckered hexagons. Any atom or group which can form bonds to four similar atoms or groups can be linked up in this way provided that the interbond angles permit a tetrahedral rather than coplanar arrangement of nearest neighbours.

The simplest structural unit which can form four tetrahedral bonds is an atom of the fourth Periodic Group, and accordingly we find this net ·as the structure of

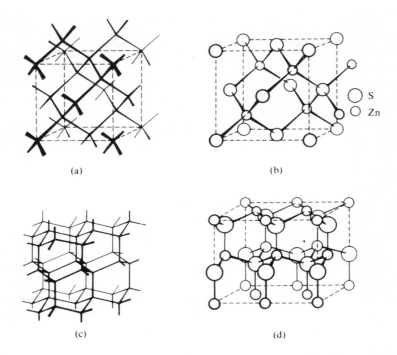

FIG. 3.35. 4-connected nets: (a) diamond; (b) zinc-blende; (c) hexagonal diamond; (d) wurtzite.

diamond, of the forms of silicon and germanium stable at atmospheric pressure, and of grey tin. The diamond-like structure of grey tin is stable at temperatures below 13.2 °C, above which transformation into white tin takes place. The structure of white tin, the ordinary form of this metal, is related in an interesting way to that of the grey form. In Fig. 3.36 the grey tin structure is shown referred to a tetragonal

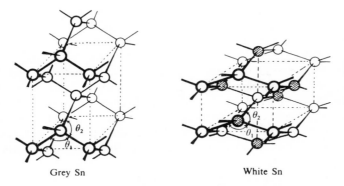

Grey Sn   White Sn

FIG. 3.36. The structures of grey and white tin.

unit cell (a axes at $45°$ to the axes of the conventional cubic unit cell). Compression along the c axis gives the white tin structure. In the diamond structure the six interbond angles at any atom are equal ($109\frac{1}{2}°$). In white tin two of these angles, distinguished as θ_1 in Fig. 3.36, are enlarged to $149\frac{1}{2}°$ and the other four angles θ_2 are reduced to $94°$. In this process the nearest neighbours of an atom have changed from (4 at 2.80 Å and 12 at 4.59 Å) to (4 at 3.02 Å, 2 at 3.18 Å and 4 at 3.77 Å) so that if we include the next two neighbours at 3.18 Å there are now six neighbours in white tin at approximately the same distance forming a distorted octahedral group. This structural change is remarkable, not only for the 26 per cent increase in density (from 5.75 to 7.31 g/cm^{-3}) but also because white tin is the *high*-temperature form. A high-temperature polymorph is usually less dense than the form stable at lower temperatures. This anomalous behaviour is due to the fact that an electronic change takes place, the grey form consisting of Sn(IV) atoms and the white form Sn(II) atoms. (If white and grey tin are dissolved in hydrochloric acid the salts crystallizing from the solutions are respectively $SnCl_2 . 2H_2O$ and $SnCl_4 . 5H_2O$.) A more normal behaviour is that of Ge or InSb, which crystallize with the diamond structure under atmospheric pressure but can be converted into forms with the white tin structure under high pressure.

The total of eight electrons for the four bonds from each atom in the diamond structure need not be provided as two sets of four (as in diamond itself, elementary silicon, or SiC), but may be derived from pairs of atoms from Groups III and V, II and VI, or I and VII. Binary compounds with the same atomic arrangement as in the diamond structure therefore include not only SiC (carborundum) but also one form of BN and BP and also ZnS (sphalerite, zinc-blende), BeS, CdS, HgS (which also has another form, cinnabar, with a quite different structure described on p. 1166), cuprous halides, and AgI. The process of replacing atoms of one kind by equal numbers of atoms of two kinds (Zn and S), Fig. 3.35(b), goes a stage further in CuFeS$_2$ the mineral chalcopyrite, or copper pyrites). One-half of the Zn atoms in zinc-blende are replaced by Cu and the remainder by Fe in a regular manner. Because of the regular replacement the repeat unit of the pattern is now doubled in one direction, since there must be atoms of the same kind at each corner of the unit cell. Further replacement of one-half of the Fe atoms by Sn gives Cu$_2$FeSnS$_4$.

Replacement of three-quarters of the metal atoms in ZnS by Cu and one-quarter by Sb gives Cu$_3$SbS$_4$. Other compounds have structures related in less simple ways to the zinc-blende structure. For example, in Zn$_3$AsI$_3$ only three-quarters of the Zn positions are occupied, and 3 I + 1 As occupy the S positions at random. If only two-thirds of the metal positions of the ZnS structure are occupied the formula becomes $X_{8/3}Y_4$, or X_2Y_3, and this is the structure of one of the (disordered) high-temperature forms of Ga$_2$S$_3$. An alternative description of this sesquisulphide structure is that the metal atoms occupy at random two-thirds of the tetrahedral holes in a cubic close-packed assembly of S atoms. A further possibility is that some of the S positions of the ZnS structure are vacant. If one-quarter are unoccupied, as in Cu$_3$SbS$_3$ and Cu$_3$AsS$_3$, each Sb or As has three pyramidal instead of four tetra-

TABLE 3.14

Structures based on the diamond net

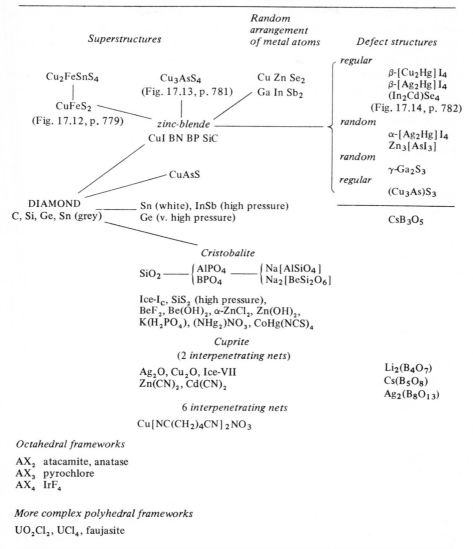

		Random arrangement of metal atoms	
Superstructures			*Defect structures*

Cu$_2$FeSnS$_4$ Cu$_3$AsS$_4$ Cu Zn Se$_2$
    | (Fig. 17.13, p. 781)   Ga In Sb$_2$
CuFeS$_2$ |
(Fig. 17.12, p. 779)   *zinc-blende*
              CuI BN BP SiC

                — CuAsS

DIAMOND ———— Sn (white), InSb (high pressure)
C, Si, Ge, Sn (grey)   Ge (v. high pressure)

              Cristobalite

SiO$_2$ —— { AlPO$_4$ —— { Na[AlSiO$_4$]
         { BPO$_4$ { Na$_2$[BeSi$_2$O$_6$]

Ice-I$_c$, SiS$_2$ (high pressure),
BeF$_2$, Be(OH)$_2$, α-ZnCl$_2$, Zn(OH)$_2$,
K(H$_2$PO$_4$), (NHg$_2$)NO$_3$, CoHg(NCS)$_4$

              Cuprite
         (*2 interpenetrating nets*)
Ag$_2$O, Cu$_2$O, Ice-VII
Zn(CN)$_2$, Cd(CN)$_2$

         6 interpenetrating nets
Cu[NC(CH$_2$)$_4$CN]$_2$NO$_3$

regular
β-[Cu$_2$Hg]I$_4$
β-[Ag$_2$Hg]I$_4$
(In$_2$Cd)Se$_4$
(Fig. 17.14, p. 782)
random
α-[Ag$_2$Hg]I$_4$
Zn$_3$[AsI$_3$]
random
γ-Ga$_2$S$_3$
regular
(Cu$_3$As)S$_3$

CsB$_3$O$_5$

Li$_2$(B$_4$O$_7$)
Cs(B$_5$O$_8$)
Ag$_2$(B$_8$O$_{13}$)

Octahedral frameworks

AX$_2$ atacamite, anatase
AX$_3$ pyrochlore
AX$_4$ IrF$_4$

More complex polyhedral frameworks

UO$_2$Cl$_2$, UCl$_4$, faujasite

hedral S neighbours. These structures are described in more detail in Chapter 17 where a number of them are illustrated, as indicated in Table 3.14.

The structures of CuFeS$_2$ and Cu$_3$SbS$_4$ are derived from the zinc-blende structure simply by replacing one-half or one-quarter of the metal atoms by atoms of another kind. It is also possible to use the same basic structure for a compound consisting of *equal* numbers of atoms of *three* kinds, as in CuAsS. The unit cell of diamond (or ZnS) contains 8 atoms, but we now require a cell containing $3n$ atoms. The relation of the unit cell of CuAsS to the diamond cell is shown in Fig. 3.37, where it is seen that the dimensions of the CuAsS cell are $a' = 3a/\sqrt{2}$ and $b = a/\sqrt{2}$, and the third dimension (perpendicular to the plane of the projection) is the same as that of the diamond structure (a). The cell of CuAsS therefore contains twelve atoms, four each of Cu, As, and S.

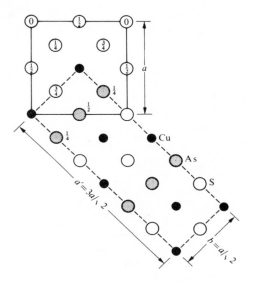

FIG. 3.37. Projection of the structure of CuAsS showing relation to the diamond structure (upper left).

AX$_2$ structures. The next way in which we may elaborate on the same basic pattern is to place atoms of a second kind along the links of the diamond net. The resulting AX$_2$ structure may alternatively be described as built of tetrahedral AX$_4$ groups sharing all their vertices (X atoms) with other similar groups. In order to simplify the diagram (Fig. 3.38(a)) the X atoms are shown lying on and at the mid-points of the links. This very symmetrical (cubic) structure, with Si–O–Si angle equal to 180°, was originally assigned to the high-temperature form of the mineral cristobalite, one of the polymorphs of silica (p. 1004). It is now known that the structural chemistry of cristobalite is complex and that the Si–O–Si angle is 147°. In fact no compound AX$_2$ has the structure of Fig. 3.38(a) with A–X–A bond angles of 180°, though the NHg$_2^+$ framework in Hg$_2$NOH.2H$_2$O has the 'anti-cristobalite' structure with collinear Hg bonds. It is nevertheless convenient to refer to any structure of the general topological type of Fig. 3.38(a), that is, consisting of vertex-sharing AX$_4$

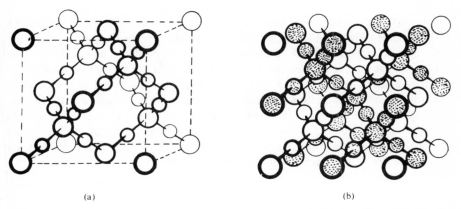

(a)　　　　　　　　　　　　　　　　(b)

FIG. 3.38. (a) Structure of cristobalite (idealized); (b) two interpenetrating frameworks of type (a) forming the structure of Cu_2O.

tetrahedra and based on the diamond net, as the cristobalite structure. The less symmetrical variants, with smaller A—X—A bond angles, arise by rotations of the tetrahedra about various axes resulting in denser packing of the anions, as in the high-pressure polymorphs of SiS_2 and GeS_2 and in α-$ZnCl_2$ (cubic closest packing of the Cl atoms). These forms of the vertex-sharing tetrahedral AX_2 structure (AC 1976 **B32** 2923) may be compared with those of the vertex-sharing octahedral AX_3 structure, which range from the ReO_3 structure (—O—, 180°) to the hexagonal closest-packed RhF_3 structure (—F—, 132°) — see pp. 208 and 265.

We noted that BP crystallizes with the zinc-blende structure. Replacement of alternate Si atoms in the cristobalite structure by B and P atoms gives the structure of BPO_4; in fact, BPO_4 crystallizes with all the structures of the ambient pressure polymorphs of SiO_2.

Because of the relatively open structures of cristobalite and tridymite (see below), both of which occur as minerals, these compounds usually — if not invariably — contain appreciable concentrations of foreign ions, particularly of the alkali and alkaline-earth metals. The holes in the structures of the high-temperature forms are sufficiently large to accommodate these ions without much distortion of the structure, but their inclusion is possible only if part of the Si is replaced by some other tetrahedrally coordinated atom such as Al or Be. The framework then acquires a negative charge which is balanced by those of the included ions. So we find minerals such as carnegeite, $NaAlSiO_4$, with a cristobalite-like structure, and kalsilite, $KAlSiO_4$, with a structure closely related to that of tridymite. In $Na_2BeSi_2O_6$ one-third of the Si in the cristobalite structure is replaced by Be as compared with the replacement of one-half of the Si by Al in $NaAlSiO_4$.

In the silica structures tetrahedral SiO_4 groups are linked together through their common oxygen atoms. We noted earlier (Fig. 3.32) that the same topological possibilities are presented by the H_2O molecule and by a molecule or ion of the type $O_2M(OH)_2$. There are a number of forms of crystalline H_2O (ice). In addition to those stable only under higher pressures, which are described on p. 653, there are two forms of ice stable under atmospheric pressure. Ordinary ice (ice-I_h) has the tridymite structure, but at temperatures around –130 °C water crystallizes with the cristobalite structure. The tridymite structure is related to the net of Fig. 3.35(c), which represents the structure of hexagonal diamond, in the same way that cristobalite is related to the cubic diamond net of Fig. 3.35(a). The two AX structures related to the nets of Fig. 3.35(a) and (c), with alternate A and X atoms, are the zinc-blende and wurtzite structures which are shown at (b) and (d). Both H_2O and SiO_2 crystallize with the cubic and hexagonal AX_2 structures:

Net	Fig. 3.35	ZnS	SiO_2	H_2O
Cubic diamond	(a)	zinc-blende	cristobalite	ice-I_c
Hexagonal diamond	(c)	wurtzite	tridymite	ice-I_h

We shall point out later other analogies between hydrate structures and silicates.

We mentioned earlier that crystalline H_2SO_4 has a layer structure, an arrangement of hydrogen-bonded molecules based on the simplest plane 4-connected net. In contrast to this layer structure the anion $O_2P(OH)_2^-$ in KH_2PO_4 is a 3D framework of hydrogen-bonded tetrahedral groups arranged at the points of the diamond net. A simplified projection of this anion is shown in Fig. 3.39, where the dotted lines

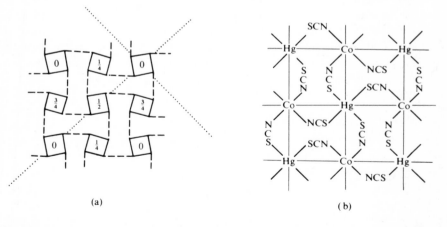

(a)   (b)

FIG. 3.39. Projections of the diamond-like structures of (a) the $PO_2(OH)_2^-$ anion in KH_2PO_4; (b) $CoHg(SCN)_4$.

indicate the directions of the cubic axes of the diamond structure. The K^+ ions, which are accommodated in the interstices of the framework, are omitted from Fig. 3.39(a). The structure of $CoHg(SCN)_4$ provides another example of a structure based on the diamond net (Fig. 3.39(b)). There is tetrahedral coordination of all the metal atoms, Co having 4 N and Hg 4 S nearest neighbours.

Diamond-like frameworks built from octahedra and other polyhedra. In the cristobalite structure tetrahedral AX_4 groups at the nodes of the diamond net are joined to four others through vertices. Diamond-like structures may be formed from octahedral AX_6 groups in various ways. Octahedral AX_6 groups sharing four vertices form a distorted framework of this type in IrF_4, the two *cis* vertices of each group being unshared. Similar groups sharing all vertices and arranged around octahedral cavities at the points of the diamond net form the AX_3 framework of the pyrochlore structure (Fig. 7.4(b)), while groups of four edge-sharing octahedra enclosing tetrahedral cavities at the points of the diamond net form the AX_2 (atacamite) structure of Fig. 7.4(a). The anatase structure may be described as an AX_2 structure formed from octahedral AX_6 groups sharing four edges (p. 216). Structures built from octahedral coordination groups are dealt with in more detail in Chapter 5.

Pentagonal bipyramidal UO_3Cl_4 groups sharing two edges (4 Cl) and two vertices (O atoms) form the diamond-like structure of UO_2Cl_2 (p. 1258), and dodecahedral UCl_8 groups share four edges in UCl_4. A more complex example is the structure of the zeolite faujasite, which is formed from truncated octahedra joined through hexagonal prisms on four of the eight hexagonal faces; it is illustrated in Fig. 23.26 (p. 1040).

Structures based on systems of interpenetrating diamond nets. In the structures we have been describing it is possible to trace a path from any atom in the crystal to any other along bonds of the structure, that is, along C–C bonds in diamond, Si–O–Si bonds in cristobalite, etc. In Cu_2O (the mineral cuprite) each Cu atom forms two collinear bonds and each O atom four tetrahedral bonds, and these atoms are linked together in exactly the same way as the O and Si atoms respectively in cristobalite. However, the distance O–Cu–O (3.7 Å) is appreciably greater than Si–O–Si (3.1 Å). As a result the Cu_2O framework has such a low density, or alternatively there is so much unoccupied space, that there is room for a second identical framework within the volume occupied by a single framework. The crystal consists of two identical interpenetrating frameworks which are not connected by any primary (Cu–O) bonds (Fig. 3.38(b)). Since ice-I_c has the cristobalite structure it is not surprising that one of the high-pressure forms of ice has the cuprite structure (see p. 654).

An even greater separation of the 4-connected points of the diamond net results if instead of a single atom X we use a longer connecting unit, that is, a group or molecule which can link at both ends to metal atoms. The CN group acts as a ligand of this kind in many cyanides of the less electropositive metals. The four atoms M–C–N–M are collinear. The simplest possibility is the linear chain in AgCN and AuCN:

$$-Au-CN-Au-CN-Au-CN-$$

The square plane net, for a metal forming four coplanar bonds, is not known in a cyanide $M(CN)_2$, though it is the basis of the structure of the clathrate compound

$Ni(CN)_2.NH_3.C_6H_6$, as already noted. In Prussian blue and related compounds CN groups link together Fe or other transition-metal atoms arranged at the points of a simple cubic lattice (Fig. 22.5). More directly related to our present topic is the crystal structure of $Zn(CN)_2$ and the isostructural $Cd(CN)_2$. Both compounds crystallize with the cuprite structure, the metal atoms forming tetrahedral bonds (like O in Cu_2O) and $-C-N-$ replacing $-Cu-$.

A longer ligand of a similar type which can also link two metal atoms is an organic dicyanide, $NC-(CH_2)_n-CN$; one or more CH_2 groups are interposed between two CN groups which bond to the metal atoms through their terminal nitrogen atoms. These dicyanides function as neutral coordinating molecules in salts of the type $[Cu(NC.R.CN)_2]NO_3$. In the glutaronitrile compound, $[Cu\{NC(CH_2)_3CN\}_2]NO_3$, the metal atoms are arranged at the points of a square plane net, but the next member of the series, the adiponitrile compound, $[Cu\{NC(CH_2)_4CN\}_2]NO_3$, has a quite remarkable structure. If metal atoms are placed at the points of the diamond net and joined through adiponitrile molecules the distance between the metal atoms

is approximately 8.8 Å, and the density of a single framework is so low that there is room for no fewer than six identical interpenetrating frameworks of composition $Cu(adiponitrile)_2$ in the same volume. The frameworks are, of course, positively charged, and the NO_3^- ions occupy interstices in the structure. It is of interest to see how this unique structure arises.

Figure 3.40(a) shows the simple diamond net, again referred to the tetragonal

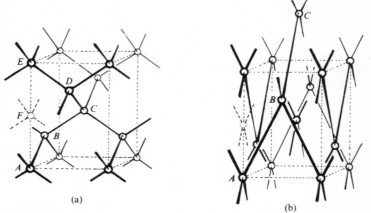

FIG. 3.40. (a) Single diamond net. (b) System of three interpenetrating diamond-like nets.

unit cell as in Fig. 3.36 (grey tin). The point B has coordinates of the type $(0, \frac{1}{2}, \frac{1}{4})$ and D is of the type $(\frac{1}{2}, 0, \frac{3}{4})$, while C is at the body-centre of the unit cell. It is clear that we can travel along lines of the net from A to B, to C, to D, to E, and in fact to any other point of the three-dimensional network of points. (The point F is the point at which a second net would start if we have two interpenetrating nets as in Cu_2O.) If, however, instead of joining the points in this way we take exactly the same set of points and join them as shown in Fig. 3.40(b), with a vertical component for each link of three-quarters of the height of the cell instead of one-quarter, then we find that the resulting system of lines and points has a remarkable property. The point A is no longer connected to the next point vertically above it (like the point E in (a)), and if in Fig. 3.40(b) we travel from $A \rightarrow B \rightarrow C$ etc. we do not arrive at a point vertically above A *which belongs to the same net as A* until we have travelled vertically the height of *three* unit cells. In other words, the points and lines of Fig. 3.40(b) form not one but three identical and interpenetrating frameworks. If, in addition, we add the points corresponding to the dotted circle (as F in (a)) and join these together in the same way, we should have a system of *six* identical interpenetrating nets. It should also be considered remarkable that this elaborate geometrical system can be so easily formed. It is necessary only to dissolve silver nitrate in warm adiponitrile, add copper powder (when metallic silver is deposited), and cool the solution, when the salt $[Cu\{NC(CH_2)_4CN\}_2]NO_3$ crystallizes out.

Further examples of structures based on two interpenetrating diamond nets are the anions in the borates $Li_2B_4O_7$, CsB_5O_8, and $Ag_2B_8O_{13}$ (Table 3.14) which are described in Chapter 24.

In the structures we have just been describing the interpenetrating nets are identical. We may also envisage structures consisting of two (or more) interpenetrating nets which have different structures. In the pyrochlore structure (p. 258) adopted by certain complex oxides $A_2B_2O_7$, we may distinguish a 3D framework of octahedral BO_6 groups each of which shares its vertices with six others, giving the composition BO_3 (or B_2O_6). This framework (Fig. 7.4) is stable without the seventh O atom (as in $KSbO_3$) and may be constructed from octahedra arranged tetrahedrally around the points of the diamond net or alternatively from octahedra placed along each link of that net. (Another way of arranging groups of four octahedra tetrahedrally around the points of the diamond net is to make edge-sharing groups, and this gives the atacamite structure, also illustrated in Fig. 7.4.) The structure of $Hg_2Nb_2O_7$ may be described as a Nb_2O_6 framework of this kind through which penetrates a cuprite-like framework of composition Hg_2O with the same structure as one of the two interpenetrating nets of Cu_2O. The Hg(II) atoms have as nearest neighbours 2 O atoms of the Hg_2O framework and at a rather greater distance 6 O of the Nb_2O_6 framework.

Structures based on more complex 4-connected nets

Three classes of more complex 3D 4-connected nets are included in Table 3.13, namely: (i) nets in which the arrangement of bonds from each point is tetrahedral, these nets being suitable for the structures of the same types of elements and compounds as is the diamond net; (ii) nets in which the links from some points are tetrahedral and from other coplanar, these nets being suitable only for compounds A_mB_n; and (iii) nets in which there are polyhedral cavities. Inasmuch as most nets of class (iii) are tetrahedral nets these constitute a sub-group of class (i), but because of their special characteristics it is convenient to describe them separately. The sole example of a net in which all atoms form four coplanar bonds (NbO) has been illustrated in Fig. 3.15(f).

(i) *More complex tetrahedral nets.*   All other 3D 4-connected nets are more complex than the diamond net, which is the only one with the minimum number (2) of points in the (topological) unit cell. The closely related net of hexagonal diamond has already been mentioned. Like diamond it is an array of hexagons, but in its most symmetrical configuration it consists of both chair-shaped and boat-shaped rings, in contrast to those in diamond which are all of the former type. The positions of alternate points in these two nets (that is, the positions of the S (or Zn) atoms in zinc-blende and wurtzite) are related in the same way as cubic and hexagonal closest packing (Chapter 4), and accordingly there is an indefinite number of closely related structures which correspond to the more complex sequences of close-packed layers. Many of these structures have been found in crystals of SiC (see the polytypes of carborundum, p. 986). The structure of high-temperature BeO (p. 538) is closely related to the wurtzite structure of the low-temperature form.

There are many AX_2 structures derived from more complex tetrahedral nets by placing an X atom along each link. They include those of quartz and other forms of SiO_2, the unexpectedly complex GeS_2, many aluminosilicates (some of which are mentioned later in class (iii)), and the high-pressure forms of Si, Ge, and of ice.

Among the high-pressure forms of elementary Si there is one with a body-centred cubic structure which is related to the diamond structure in an interesting way. The diamond net is shown in Fig. 3.41(a) viewed along the direction of a C–C bond,

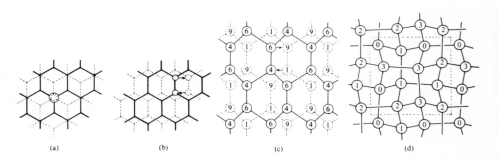

FIG. 3.41. (a) Projection of diamond structure along C–C bond. (b) and (c) Structure of a high-pressure form of Si. (d) Structure of a high-pressure form of Ge.

the 6-gon layers (distinguished as full and broken lines) being buckled. Alternate atoms of a 'layer' are joined to atoms of the layers above and below by bonds perpendicular to the plane of the paper, as indicated by the pair of circles. Suppose now that the latter bonds are broken and that alternate layers are translated relative to the neighbouring ones, as in (b), and that connections between the layers are remade as shown by the arrows. An arrow indicates a bond rising to an atom of a layer above the layer from which it originates. The fourth bond from each atom is no longer normal to the plane of the paper, and the structure (which is that of a high-pressure form of elementary silicon) is therefore more dense than the diamond-like structure (a). The numerals in (c) are the heights of the atoms in units of $c/10$ where c is the repeat distance normal to the plane of the projection. Atoms such as 1, 4, 6, and 9 form a helical array, the bond 9 . . . 1 bringing the helix to a point at a height c above the original point 1.

Figure 3.41(d) shows the projection of the structure of a high-pressure form of Ge, the numerals here indicating the heights of the atoms above the plane of the paper in units of $c/4$. Note that the circuits 0-1-2-3 projecting as 4-gons represent 4-fold helices, and that the smallest circuits in this 3D net are 5-gons and 7-gons, in contrast to the 6-gon circuits in (c). The net of Fig. 3.41(d) is also the basis of the structures of the silica polymorph keatite and of ice-III.

The structures of the high-pressure forms of ice are described in Chapter 15, but it is worth noting here how some of the more dense structures arise. Polymorphs with different 3D H_2O frameworks include the following:

	I_h	I_c	II	III	V	VI	VII
Density	0.92	1.17	1.16	1.23	1.31	1.50	g/cm^3

In contrast to ice-III, which has a structure based on the same net as keatite, ice-II and ice-V have structures not found as silica polymorphs, though that of ice-II has features in common with the structure of tridymite. All these three forms achieve their higher density by distortion of the tetrahedral arrangement of nearest neighbours and/or by the formation of more compact ring systems; there are 4-gon rings in ice-V and 5-gon rings in ice-III. As a result the next nearest neighbours are appreciably closer than in ice-I (4.5 Å). For example, the next nearest neighbours of a H_2O molecule in ice-V are at 3.28 Å and 3.64 Å. The even more dense forms VI and VII have structures consisting of two interpenetrating 4-connected frameworks, within each of which the H_2O molecules are hydrogen-bonded; there are no such bonds between molecules belonging to different frameworks. The arrangement of H_2O molecules in each net of ice-VI is similar to that of SiO_4 tetrahedra in the fibrous zeolite edingtonite (p. 1037), while ice-VII consists of two frameworks of the cristobalite type which interpenetrate to form a pseudo-body-centred cubic structure (Fig. 15.2, p. 655). Each O atom is equidistant from *eight* others, but it is hydrogen-bonded to only *four*, each of the two interpenetrating nets being similar to the single net of cubic ice-I.

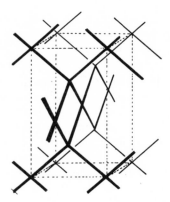

FIG. 3.42. The 4-connected net representing the structure of PtS, in which Pt forms 4 coplanar and S forms 4 tetrahedral bonds.

(ii) *Nets with planar and tetrahedral coordination.* Two simple compounds provide examples of nets of this type. In PtS (Fig. 3.42), equal numbers of atoms form coplanar and tetrahedral bonds; less symmetrical variants of this structure are adopted by PdS and CuO. Just as the diamond net may be dissected into buckled 6-gons (3-connected) layers, so the PdP_2 structure may be visualized in terms of puckered 5-gon (3,4-connected) layers (Fig. 3.43). The 4-connected points represent

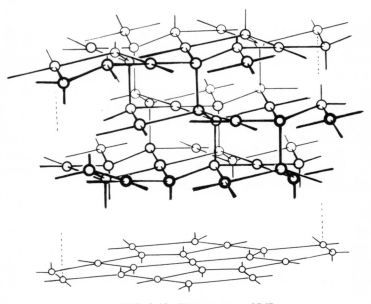

FIG. 3.43. The structure of PdP_2.

metal atoms forming four coplanar bonds and the 3-connected points represent the P atoms. The latter atoms of different layers are connected together to form the 3D structure, the interlayer bonds being directed to one or other side of the layer as indicated by the short vertical lines in the lower part of Fig. 3.43. In this structure each Pd is bonded to 4 P and each P to 2 Pd and 2 P, from which it follows that continuous chains of linked P atoms can be distinguished in the 3D structure.

(iii) *Nets with polyhedral cavities or tunnels.* For the purposes of deriving nets and illustrating the basic topology of structures it is convenient to base the classification on the connectedness (*p*) of the points. There are, however, other general features of nets which are important in relation to both the chemical and physical properties of structures based on them. In any repeating pattern the number of points is necessarily the same in each unit cell, and on this scale there is a uniform distribution of points throughout space. However, the distribution of points *within* the unit cell may be such that there is an obvious concentration of points (and links) around points, around lines, or in the neighbourhood of planes. If there is a marked concentration of atoms around certain lines or sets of planes, with comparatively few links between the lines or planes, the crystal may be expected to imitate the properties of a chain or layer structure. In the fibrous zeolites the chains of (Si, Al)O_4 tetrahedra are cross-linked by relatively few bonds, with the result that although the aluminosilicate framework is in fact a 3D one the crystals have a marked fibrillar cleavage.

In Chapter 1 we suggested a very simple classification of crystal structures based on recognition of the four main types of complex: finite (F), chain (C), layer (L), or 3D, hybrid systems including crystals containing complexes of more than one type (for example, chain ions and finite ions). If we arrange the four basic structure types at the vertices of a tetrahedron there are six families of intermediate structures, (1)–(6), corresponding to the six edges of the tetrahedron. Structures may be regarded as being intermediate between the four main types in one of two ways:

(a) Weaker (longer) bonds link the less extensive into the more extensive grouping. For example, finite molecules (ions) could be linked by hydrogen bonds into chains (class (1)), layers (class (2)), or 3D complexes (class (4)), or chains could be linked into layers (class (3)). This is the obvious chemical interpretation of the intermediate groups and could be represented on a diagram:

(b) A purely topological interpretation would be based not on differences in bond type (length) but simply on the relative numbers of bonds within and between the primary structural units. Thus in the intermediate group (1) relatively few

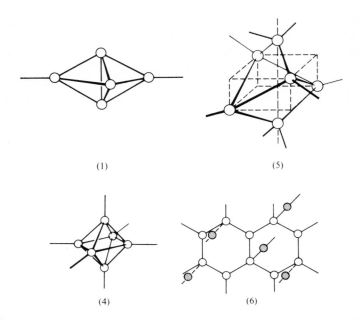

(1)   (5)

(4)   (6)

FIG. 3.44. Structures with uneven distributions of atoms in space (see text).

bonds link the finite groups into chains, as in the hypothetical 4-connected chain of Fig. 3.44(a). This is a chain structure but one with a very uneven distribution of points along the chain, as in the $ReCl_4$ or $[B_3O_4(OH)_3]_n^{2n-}$ chains mentioned earlier. An example of (2) is the 2D anion in $Sr[B_6O_9(OH)_2]$ and of (3) the layer in Hittorf's P in which most of the P–P bonds link the atoms into tightly knit chains which are further linked by a very small number of bonds into the double layers illustrated in Fig. 19.2 (p. 840). Both (2) and (3) are layer structures but with obvious concentrations of atoms in finite groups and chains respectively. Similarly, (4), (5), and (6) are special types of 3D complexes. The intermediate classes correspond to structures of types different from those in (a). Examples are shown in Fig. 3.44: (4) is the octahedral B_6 group which is linked to six other similar groups at the points of a simple cubic lattice to form the boron framework in CaB_6: (5) is a portion of the infinite chain (Si atoms only indicated) in a fibrous zeolite, each chain being linked laterally to four others; and (6) is a portion of a layer of Si_2N_2O. This interesting compound is made by heating a mixture of Si and SiO_2 to $1450\,^{\circ}C$ in a stream of argon containing nitrogen, and is intermediate between SiO_2 and Si_3N_4 not only as regards its composition but also structurally:

	SiO_2	Si_2N_2O	Si_3N_4
Coordination numbers	4:2	4:3:2	4:3

All the bonds from N and three-quarters of those from Si lie within the layers, which are held together only by the bonds through the 2-coordinated O atoms.

Instead of focussing attention on the regions of high density of points (atoms) we may consider the voids between them. These range from isolated cavities to complex 3D systems of tunnels. Crystals possessing these features are of interest as forming 'clathrate' compounds or exhibiting ion-exchange or molecular sieve properties. Three simple arrangements of tunnels found in crystals are shown in Fig. 3.45. In (a) and (b) the tunnels do not intersect, whereas in (c) they intersect at the

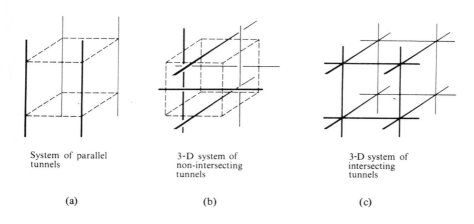

System of parallel tunnels

3-D system of non-intersecting tunnels

3-D system of intersecting tunnels

(a)   (b)   (c)

FIG. 3.45 Types of tunnel systems in crystals. The full lines represent the axes of the tunnels.

points of a primitive cubic lattice. An example of (a) is illustrated in Fig. 1.9(b) (p. 31), the tunnels resulting from the formation around hydrocarbon chains of a 3D net in which the hydrogen-bonded urea molecules form the walls of the tunnels. The actual tunnel system in (c) can be imagined to be formed by inflating the links of a 3D net so that they become tubes, the outer surfaces of which are covered by tessellations of bonded atoms; certain zeolite structures may be regarded in this way.

We shall conclude our remarks on nets by describing briefly two types of structure which are of particular interest in structural chemistry, namely, those in which there are well-defined polyhedral cavities or infinite tunnels. The B_6 groups in CaB_6 are preferably not described as polyhedral cavities because they are too small to accommodate another atom, but the large holes surrounded by 24 B atoms at all the vertices of a truncated cube accommodate the Ca atoms. We shall suppose all the links in a net to be approximately equal in length, since the general nature of a net can obviously be changed if large alterations are permitted in the relative lengths of certain sets of bonds.

Space-filling arrangements of polyhedra

The division of space into polyhedral compartments is analogous to the division of a plane surface into polygons. One aspect of this subject was studied in 1904 by von Fedorov,[1] namely, the filling of space by identical polyhedra *all having the same orientation*. He noted that this is possible with five types of polyhedron, the most symmetrical forms of which are the cube, hexagonal prism, rhombic dodecahedron, elongated dodecahedron, and truncated octahedron (Fig. 3.46). The third

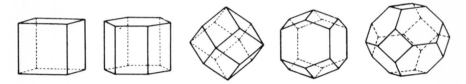

FIG. 3.46. The five space-filling polyhedra of Fedorov.

of von Fedorov's space-fillings is related to the closest packing of equal spheres, for the domain of a sphere in cubic closest packing is the rhombic dodecahedron; on uniform compression c.c.p. spheres are converted into rhombic dodecahedra. The space-filling by truncated octahedra (t.o.) corresponds to the body-centred cubic packing of equal spheres. in which the t.o. is the polyhedral domain.

Another aspect of this subject was studied by Andreini[2] (1907), who considered the filling of space by regular and Archimedean semi-regular solids, either alone or in combination, but without von Fedorov's restriction on orientation. This restriction is not important in the present context, for evidently there may be polyhedra of the same kind in different orientations within the unit cell. Since a number of polyhedra meet around a common edge it is a matter of finding which combinations of dihedral angles add up to 2π. For example, neither regular tetrahedra nor regular octahedra alone fill space but a combination in the proportion of two tetrahedra to one octahedron does do so. (These polyhedra are the domains of the ions in the CaF_2 structure, which may therefore be represented in this way as a space-filling assembly of these two polyhedra.)

The polyhedra in the space-fillings of Andreini do not include the regular or semi-regular polyhedra with 5-fold symmetry. It is not possible to fill space with regular dodecahedra or icosahedra or the Archimedean solids derived from them (or with combinations of these polyhedra) owing to the unsuitable values of the dihedral angles. However, there are space-filling assemblies of polyhedra including (irregular) pentagonal dodecahedra and polyhedra of the family $f_5 = 12, f_6 \geqslant 2$ which were mentioned on p. 74. In addition to these special families of space-filling polyhedra there are an indefinite number of ways of filling space with less regular polyhedra (of one or more kinds); an example is the packing of 8-hedra and 17-hedra noted on p. 75 as the basis of a hydrate structure.

In any space-filling arrangement of polyhedra at least four edges meet at each point (vertex). This number is not necessarily the same for all points; for example, in the space-filling by rhombic dodecahedra four edges meet at some points and eight at others. A special class of space-fillings comprises those in which the *same* number of edges (e.g. 4, 5, or 6) meet at each point, the edges forming a 3D 4-, 5-, or 6-connected net. The boron framework in CaB_6 (Fig. 3.47(c)) may be regarded as a space-filling by octahedra and truncated cubes in which 5 edges meet at every point. A more complex 5-connected net is that corresponding to the space-filling by truncated tetrahedra, truncated octahedra, and cuboctahedra (Fig. 3.47(d)), which represents the boron framework in UB_{12}. Here the uranium atoms are located in the largest (truncated octahedral) holes. Some space-filling arrangements of polyhedra are listed in Table 3.15.

TABLE 3.15

Space-filling arrangements of polyhedra

Fig. 3.47	p	Polyhedra
(a)	4	Truncated octahedra
(b)	4	Cube, truncated octahedron, truncated cuboctahedron
(c)	5	Octahedron, truncated cube
(d)	5	Truncated tetrahedron, truncated octahedron, cuboctahedron
(e)	6	Cube, cuboctahedron, rhombicuboctahedron
(f)	8	Octahedron, cuboctahedron

The space-fillings of greatest interest in structural chemistry are those in which four edges meet at each point. These structures are suitable for H_2O forming four hydrogen bonds and the topologically similar $(Si, Al)O_4$ tetrahedron linked to four others through its vertices; they belong to two families which we now consider in more detail.

(1) ZK 1904 **38** 321.
(2) Mem. Soc. Ital. Sci. 1907 (3) **14** 75.

Space-fillings of regular and Archimedean solids

The simplest representative of this group is von Fedorov's space-filling by truncated octahedra (t.o.) illustrated in Fig. 3.47(a). It represents the framework of Si (Al) atoms in ultramarine, $Na_8Al_6Si_6O_{24} \cdot S_2$, in which Na^+ and S_2^{2-} ions occupy the t.o. interstices. At higher temperatures it is possible for ions to move from one cavity to another through the 6-rings of the framework. Accordingly, if the mineral sodalite (which has the ultramarine structure with 2 Cl^- replacing S_2^{2-}) is fused with sodium sulphate it is converted into $Na_8Al_6Si_6O_{24} \cdot SO_4$, which also occurs as a mineral. The structure of the oxy-metaborate $Zn_4B_6O_{13}$ (or $OZn_4(BO_2)_6$) is also based on this framework. Tetrahedral BO_4 groups at the vertices of a t.o. framework form a 3D anion and in the interstices are situated tetrahedral OZn_4 groups. The same frame-

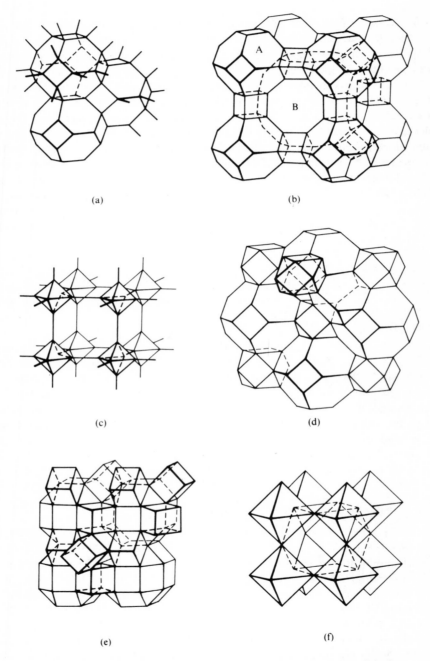

FIG. 3.47. Space-filling arrangements of regular and semi-regular polyhedra.

work also represents the arrangement of water molecules in $HPF_6 . 6H_2O$, the PF_6^- ions occupying the cavities, and in a modified form in $N(CH_3)_4OH . 5H_2O$. Since the ratio of vertices to cavities is $6 : 1$. the framework in the last compound is made up of $(5H_2O + OH^-)$.

The space-filling of Fig. 3.47(b) results from inserting cubes between the square faces of the t.o.s in the Fedorov packing, thereby producing much larger holes (truncated cuboctahedra, t.c.o.). The two kinds of larger polyhedron are distinguished as A and B in the figure. This framework represents the structure of the synthetic zeolite A, placing $Si(Al)O_4$ tetrahedra at all the points of the net. It is shown in Fig. 3.47(b) as a space-filling array of polyhedra of three kinds. Alternatively we could accentuate the framework formed by the t.o.s and cubes, the space left being a 3D system of intersecting tunnels formed from t.c.o.s sharing their 8-gon faces. It is the tunnel system that is the important feature as regards the behaviour of the crystal as a molecular sieve, the large tunnels (with free diameter approximately 3.8 Å) allowing the passage of gas molecules. These aluminosilicate structures are more fully described in Chapter 23 p.1036), where we also illustrate the structure of the $(Si, Al)O_2$ framework in the mineral faujasite, $NaCa_{0.5}(Al_2Si_5O_{14}) . 10H_2O$. This is a diamond-like arrangement of t.o.s each of which is joined to four others through hexagonal prisms; this is not a space-filling arrangement of polyhedra.

Space-fillings of dodecahedra and related polyhedra

Our second family of polyhedral space-fillings comprises those formed from pentagonal dodecahedra in combination with one or more kinds of polyhedra of the type $f_5 = 12$, $f_6 \geqslant 2$. They represent the structures of hydrates of substances ranging from non-polar molecules of gases, such as chlorine and methane, and liquids such as chloroform, to amines and substituted ammonium and sulphonium salts. These hydrates may be described as expanded ice-like structures, since they are not stable at temperatures much above $0\,°C$ and they consist of polyhedral frameworks of hydrogen-bonded water molecules built around the 'guest' molecules or ions. The latter almost invariably occupy the larger polyhedral cavities, from which they can escape only if the crystal is destroyed by dissolution or vaporization. The volumes of the cavities in these clathrate hydrates are: dodecahedra, $170\,Å^3$, 14-hedra, $220\,Å^3$, and 16-hedra, $240\,Å^3$, and the van der Waals diameter of the largest non-polar molecule forming a hydrate of this type is approximately $8\,Å$ (n-propyl bromide).

The compositions of these hydrates indicate large numbers of molecules of water of crystallization. The ratio of the number of water molecules to 'guest' molecules is equal to that of the number of points (vertices) to the number of larger cavities in the framework, and is not necessarily integral. For example, in $CHCl_3 . 17H_2O$ the molecules of chloroform occupy all the large (16-hedral) sites, and since there are 8 such sites in a unit cell containing $136\,H_2O$ the ratio of H_2O to $CHCl_3$ molecules is $136/8 = 17$. On the other hand, the unit cell of $(C_2H_5)_2NH . 8\frac{2}{3}H_2O$ contains 104 H_2O and there are 12 large (18-hedral) cavities ($f_5 = 12$, $f_6 = 6$), in addition to some less regular ones. The ratio of water molecules to amine molecules is therefore $104/12$ or $8\frac{2}{3}$.

In hydrated salts such as $R_4NF.mH_2O$ and $R_3SF.nH_2O$ the positively charged N^+ or S^+ together with F^- ions replace some of the H_2O molecules in the framework. In such hydrates the bulky organic groups are accommodated in the appropriate number of cavities adjacent to the N or S atom to which they are bonded, that is, in the positions occupied by the guest molecules in the simpler hydrates. These structures, and also the structures of the hydrates mentioned in the previous section, are described in some detail in Chapter 15.

4

Sphere packings

Periodic packings of equal spheres

Any arrangement of spheres in which each makes at least three contacts with other spheres may be described as a sphere packing. For example, equal spheres may be placed at the points of 3- or 4-connected nets. The densities of the resulting packings are low, density being defined as the fraction of the total space occupied by the spheres. If spheres are placed at the points of the diamond net and are in contact at the mid-points of the links the density is only 0.3401. Moreover, arrangements with small numbers of neighbours (low c.n.s) are stable only if there are directed bonds between the spheres, and structures of this kind are therefore described under Nets. We consider here only those packings in which each sphere is in contact with six or more neighbours (Table 4.1).

TABLE 4.1

Densities of periodic packings of equal spheres

Coordination number	Name	Density
6	Simple cubic	0.5236
8	Simple hexagonal	0.6046
8	Body-centred cubic	0.6802
10	Body-centred tetragonal	0.6981
11	Tetragonal close-packing	0.7187
12	Closest packings	0.7405

Simple cubic packing

For 6-coordination the most symmetrical packing in three dimensions arises by placing spheres at the points of the simple cubic lattice; slightly more than one-half of the space is occupied by the spheres. Each sphere is in contact with six others situated at the vertices of an octahedron, the contacts being along the edges of the unit cube (Fig. 4.1(a)). There is no contact between spheres along face-diagonals or body-diagonals.

The low c.n. and low density of this structure make it unsuitable for most metals; this structure has been assigned to α-Po. Mercury does, however, crystallize with a closely related structure which can be derived from the simple cubic packing by extension along one body-diagonal so that the cube becomes a rhombohedron (interaxial angle 70½° instead of 90°). The atoms retain 6-coordination.

The structures of crystalline As (Sb and Bi) and of black P illustrate two different

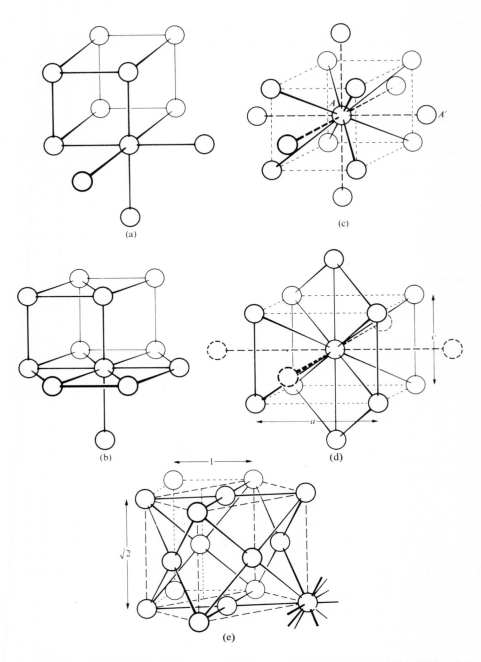

FIG. 4.1   Types of sphere packing: (a) simple cubic;   (b) simple hexagonal;   (c) body-centred cubic;   (d) body-centred tetragonal (c.n. 10);   (e) cubic closest packing.

ways of distorting the simple cubic packing so that each atom has three nearest and three more distant neighbours. Both are layer structures which may alternatively be described as forms of the plane hexagon net. They are described and illustrated in the section on the plane hexagon net (p. 102).

The body-centred cubic packing

The simple hexagonal sphere packing (c.n. 8), Fig. 4.1(b), is not of great importance as a crystal structure; it is mentioned again later. A more dense arrangement with the same c.n. is the body-centred cubic packing illustrated in Fig. 4.1(c). Spheres are placed at the body-centre and corners of the cubic unit cell, and are in contact only along the body-diagonals. This structure, with density 0.6802, is adopted by a number of metals (Table 29.3, p. 1281). Although each sphere is in contact with only eight others it has six next-nearest neighbours at a distance only 15 per cent greater than that to the eight nearest neighbours. For the atom A these six neighbours are those at the centres of neighbouring unit cells. If coordination number is defined in terms of the polyhedral domain (p. 68) of an atom it has the value 14 for this structure, the domain being the truncated octahedron.

The relatively small difference between the distances to the groups of eight and six neighbours suggests that structures with c.n.s intermediate between eight and fourteen should be possible. The six more distant neighbours are arranged octahedrally By compressing the structure in the vertical direction we may bring two of these neighbours to the same distance as the eight at the corners of the cell. Alternatively, by elongating the structure we may bring the four equatorial neighbours to this distance. If the height of the (tetragonal) cell is c and the edge of the square base is a, there are the two cases:

$$a^2/2 + c^2/4 = c^2, \text{ or } c:a = \sqrt{2}/\sqrt{3} \text{ for 10-coordination (Fig. 4.1(d))},$$

or $\qquad a^2/2 + c^2/4 = a^2, \text{ or } c:a = \sqrt{2} \text{ for 12-coordination (Fig. 4.1(e))}.$

Only one metal (Pa) is known to crystallize with the 10-coordinated structure under atmospheric pressure; it is also the structure of $MoSi_2$ (p. 988). For Pa the axial ratio $c:a$ (0.825) is very close to the ideal value (0.816) and accordingly each atom has ten very nearly equidistant neighbours. Further compression of the b.c.c. packing brings the two neighbours along the c axis closer (2 + 8 coordination), as in the high-pressure form of Hg, with $c:a$ equal to 0.707. The packing of Fig. 4.1(d) has a density (0.6981) somewhat higher than the b.c.c. packing, but the packing with 12-coordination and density 0.7405, is of outstanding importance. Because the height of the cell ($\sqrt{2}a$) is equal to the diagonal of the square base there is an alternative unit cell (Fig. 4.1(e)) which is a cube with spheres at the corners and at the mid-points of all the faces, hence the name face-centred cubic (f.c.c.) structure. This is one of the forms of *closest* packing of equal spheres.

It is instructive to view a model of b.c.c. sphere packing along a direction parallel to a face-diagonal of the unit cell (Fig. 4.2(a)). The open circles represent spheres lying in the plane of the paper, and the shaded circles spheres in parallel planes

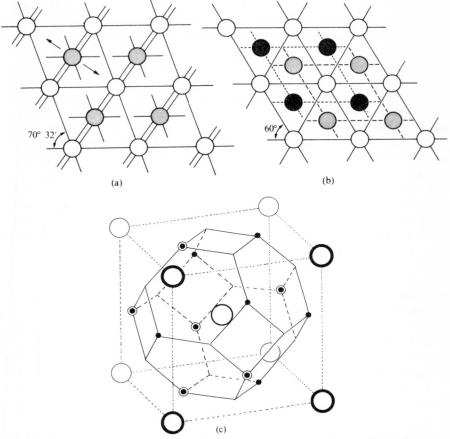

(a)   (b)

(c)

FIG. 4.2. (a) Body-centred cubic packing viewed along face-diagonal. (b) Projection of face-centred cubic packing. (c) The small black circles indicate the positions of one-half of the interstices, and the small open circles the positions of one-quarter of the interstices, those occupied in Ni_3S_2.

above and below that of the paper. The lines connect spheres which are in contact. A small alteration in the structure of the layer and small shifts of the layers relative to one another convert (a) into (b), which is a projection of the (cubic) closest packing (f.c.c.). This close relationship between b.c.c. and f.c.c. sphere packings is of interest in view of the fact that both are important as the structures of many metals. Also of interest is the fact that although b.c.c. packing is less dense than f.c.c. (or other closest packings) and therefore has more free space, the structures of interstitial hydrides, carbides, and nitrides are, almost without exception, based on the f.c.c. and not on b.c.c. packing of the metal atoms (but see V and Nb hydrides, p. 350).

We noted on p. 68 that the description of a structure as a space-filling assembly of polyhedral (Dirichlet) domains offers a way of defining an interstice in a structure. From the definition of a Dirichlet domain it follows (JSSC 1970 **1** 237) that a vertex of a domain is equidistant from at least four non-coplanar atoms and is therefore the centre of a polyhedral interstice. In the b.c.c. structure the interstices therefore correspond to the vertices of a truncated octahedron (Fig. 4.2(c)). They are *distorted* tetrahedral holes, in contrast to the regular tetrahedral and octahedral interstices in close-packed structures (p. 151), and there are $6N$ such interstices capable of accommodating an atom of radius 0.291 in an infinite packing of N spheres of unit radius. However, these interstices are arranged in compact groups of four, and because of their proximity not more than one-half can be occupied simultaneously, giving the formula AX_3, X being the interstitial atom. This is the situation in OCr_3, and possibly in β-W (p. 1283). If only one-quarter of the interstices are occupied the formula is A_2X_3. This is the structure of S_2Ni_3 (Ni_3S_2), which is a very slightly distorted b.c.c. packing of S with Ni atoms in the interstices indicated in Fig. 4.2(c) (AC 1980 **B36** 1179). In this structure Ni has four S neighbours at 2.27 Å and also 2 Ni at 2.51 Å (mean values — compare Ni–Ni in the metal (2.49 Å).

Tetragonal close packing

It has recently been noted (MRB 1981 **16** 339) that there is a periodic 3D packing of equal spheres in which the spheres are symmetrically equivalent, each has eleven equidistant neighbours, and the density of the packing is only 3 per cent less than that of closest packing (Table 4.1). The coordination polyhedron, which approximates to a tricapped tetragonal prism, is formed from groups of five and six spheres, each group being situated at certain of the vertices of (different) cuboctahedra. This is the packing of the anions in form (i) of the rutile structure (p. 248), with $c:a = 2-\sqrt{2}$ and $x = \frac{1}{2}(2-\sqrt{2})$; it approximates closely to hexagonal closest packing (p. 168). We note on p. 164 that although the number of tetrahedral interstices in a hexagonal closest packing of N spheres is $2N$ the filling of more than N of these results in the sharing of faces between tetrahedral coordination groups. In tetragonal close packing it is possible to fill $5N/4$ tetrahedral interstices without face-sharing, accounting for the formation of, for example, Li_4GeO_4, with this packing of the anions.

The closest packing of equal spheres

Models of closest sphere packings are usually built from closest packed layers, in which each sphere is in contact with six others. Such layers may be superposed in various ways which we examine shortly. The closest packing is achieved if each sphere touches three others in each adjacent layer, making a total of twelve contacts. However, it is not justifiable to assume that the closest possible periodic 3D packing will necessarily be formed from the most densely packed layers, for in such layers each sphere has a special (coplanar) arrangement of six neighbours. We now look at the matter in a different way.

Three spheres are obviously most closely packed in a triangular arrangement, and a fourth sphere will make the maximum number of contacts (three) if it completes a tetrahedral group. We might then expect to obtain the densest packing by continuing to place spheres above the centres of triangular arrangements of three others. The lines joining the centres of spheres in such an array would outline a system of regular tetrahedra, each having faces in common with four other tetrahedra, that is, we should have a space-filling arrangement of regular tetrahedra. However, it is not possible to pack together regular tetrahedra to fill space because the dihedral angle of a regular tetrahedron ($70°32'$) is not an exact submultiple of $360°$.

Icosahedral sphere packings. Alternatively, suppose that we continue placing spheres around a central one, all the spheres having the same radius. The maximum number that can be placed in contact with the first sphere is 12. There is in fact more than sufficient room for 12 equal spheres in contact with a central sphere of the same radius but not sufficient for a thirteenth. There is therefore an infinite number of ways of arranging the twelve spheres, of which the *most symmetrical* is to place them at the vertices of a regular icosahedron – the only regular solid with 12 vertices. The length of an edge of a regular icosahedron is some five per cent greater than the distance from centre to vertex, so that a sphere of the outer shell of twelve makes contact only with the central sphere. (Conversely, if each sphere of an icosahedral group of twelve, all touching a central sphere, is in contact with its five neighbours, then the central sphere must have a radius of 0.902 if that of the outer ones is unity.) The icosahedral arrangement of nearest neighbours around every sphere does not lead to a periodic 3-dimensional array of spheres, but it is of sufficient interest to justify a brief description. Some other sphere packings which are not periodic in three dimensions but, like the icosahedral packing, can extend to fill space, are of interest from the structural standpoint. A number of these non-crystallographic sphere packings have been described, including an infinite packing with density (0.7236) approaching that of closest packing and having a unique axis of 5-fold symmetry (Fig. 4.3).[1] A closely related packing with even higher density (0.7341) has also been described.[2]

First we note an interesting relation between the icosahedron and the cuboctahedron (Fig. 4.4(a)). If 24 rigid rods are jointed together at the 12 vertices of a cuboctahedron, rotation of the groups forming the triangular faces about their normals transforms the cuboctahedron into a regular icosahedron. In the course of the change each joint moves in towards the centre, so that the distance from centre to vertex contracts by some five per cent. If we imagine spheres of unit radius situated at the vertices and centre of the cuboctahedron the same transformation can be carried out, but now there is no radial contraction and in the icosahedral configuration the outer spheres are no longer in contact. If a close-packed arrangement is built starting with a cuboctahedral group of 12 spheres around a central sphere it will be found that the second shell of spheres packed over the first contains 42 spheres – in general $10n^2 + 2$ for the nth layer. A second shell packed around an

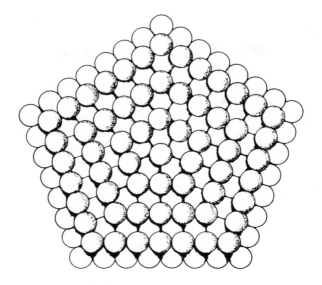

FIG. 4.3. A sphere-packing with 5-fold symmetry.

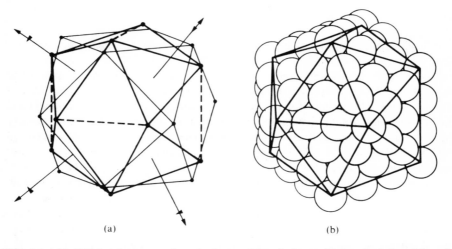

(a)   (b)

FIG. 4.4. (a) Relation between cuboctahedron and icosahedron. (b) Icosahedral packing of equal spheres, showing the third layer ($n = 3$). (AC 1962 **15** 916.)

icosahedral group also contains 42 spheres, the spheres being in contact along the 5-fold axes, and further layers of $10n^2 + 2$ spheres may be added as for the cuboctahedron. The whole assembly is not periodic in three dimensions but is a radiating structure, having a unique centre (Fig. 4.4(b)). It is less dense than the packing based on the cuboctahedron, the limiting density being 0.6882,[3] intermediate between that of body-centred cubic packing (0.6802) and the 'cuboctahedral' packing (0.7405), and it can be converted into the latter by the mechanism of Fig. 4.4(a).

Since the *most symmetrical* arrangement of 12 neighbours (the icosahedral coordination group) does not lead to the densest possible 3D packing of spheres we have to enquire which of the infinite number of arrangements of twelve neighbours lead to more dense packings and what is the maximum density that can be attained in an infinite sphere packing. It was shown by Barlow in 1883 that two coordination groups, alone or in combination, lead to infinite sphere packings which all have the same density (0.7405). These two coordination groups are the cuboctahedron and the related figure (the 'twinned cuboctahedron') obtained by reflecting one-half of a cuboctahedron cut parallel to a triangular face across the plane of section (Fig. 4.5). These are the arrangements of nearest neighbours in sphere packings formed

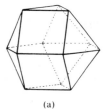
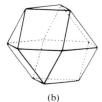

(a)   (b)

FIG. 4.5. Coordination polyhedra in (a) hexagonal and (b) cubic closest-packing.

by stacking in the closest possible way the close-packed layers mentioned at the beginning of this section. It is an interesting fact that it has yet to be proved that the density 0.7405 could not be exceeded in an infinite sphere packing of some (unknown) kind, though Minkowski showed that the packing based on the cubocta-hedron (cubic closest packing) is the densest *lattice packing* of equal spheres. (A lattice packing has the following property. If on any straight line there are two spheres a distance a apart then there are spheres at all points along the line, extended in both directions, at this separation. Note that hexagonal closest packing (p. 154) is *not* a lattice packing — cubic closest packing is unique as a *closest lattice packing*.) The difficulty arises from the fact that whereas the closest packing of spheres in a plane is a unique arrangement, there is an infinite number of ways of placing twelve equal spheres in contact with a single sphere of the same radius.

(1)   N 1965 **208** 674; JCG 1970 6 323   (3)   AC 1962 **15** 916
(2)   N 1970 **225** 1040

Sphere packings based on closest-packed layers

A sphere in a closest-packed layer is in contact with six others (Fig. 4.6(a)). When such layers are stacked parallel to one another the number of additional contacts

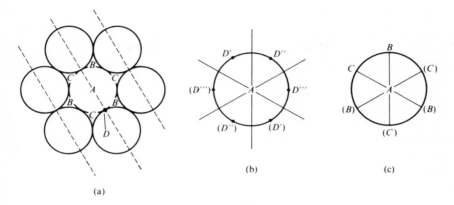

(a)

(b)

(c)

FIG. 4.6. Superposition of layers of c.p. spheres (see text).

(on each side) will be 1, 2, or 3 if the centres of spheres in the adjacent layers fall above or below points such as *A*, *D*, or *B* (or *C*) respectively. We shall refer to such layers as *A, D, B,* or *C* layers. There are three *D* positions relative to a given sphere, *D'*, *D''*, and *D'''*, the diametrically opposite point in each case is the position of another sphere in the same layer and is shown in parentheses in Fig. 4.6(b). If the same (vector) relationship is maintained between successive pairs of layers it is immaterial whether it is *AD'*, *AD''*, or *AD'''*, but if these translations are combined in a stacking sequence different structures result. The sequences *AD'AD'* ..., *AD''AD''* ..., and *AD'''AD'''* ... correspond to the same structure, but *AD'AD''* ... etc. represent different structures. There are only two different positions of the type *B* or *C* because all three *B* (or *C*) positions in Fig. 4.6(c) are positions of spheres in a given layer. The positions *B* and *C* are equivalent as regards the relation between two adjacent layers, so that the sequences *ABAB* ... and *ACAC* ... are the same structure, but *ABAC* ... and other sequences involving mixtures of *B* and *C* layers represent different structures.

The total number of contacts (coordination number) can therefore have any value from eight to twelve, that is, $6 + 1 + 1$ to $6 + 3 + 3$, depending on the stacking sequence. Of these types of sphere packing only those with coordination numbers of 8, 10, and 12 are found in crystals.

c.n. 8   The layer sequence is *A A A* .... This packing, in which the layers fall vertically above one another, is the *simple hexagonal* (lattice) *packing*.

c.n. 10   This can arise in two ways: (a) by a *D*-type contact with both adjacent layers $(6 + 2 + 2)$ or (b) by an *A*-type contact on one side and a *B*- (or *C*-) type contact on the other $(6 + 1 + 3)$.

(a) Layer sequence *A D A D* .... There is an infinite number of structures because there is a choice between positions *D'*, *D''*, and *D'''* at each layer junction. For a 10-coordinated structure of this kind, with somewhat distorted close-packed layers, see γ-Pu (p. 1285) in which there is (4 + 4 + 2)-coordination. In the special case where all pairs of successive layers are related by a translation of the same kind, for example AD', there is also closest packing of the spheres in a second set of planes *perpendicular* to the original c.p. layers. These planes intersect the plane of the paper in the broken lines of Fig. 4.6(a). This is the body-centred tetragonal packing described on p. 143 and listed in Table 4.1.

(b) Layer sequence $\begin{smallmatrix}B\\C\end{smallmatrix} A \, A \begin{smallmatrix}B\\C\end{smallmatrix}$. Again there is an infinite number of possible structures because we must distinguish between B and C positions. For example, the simplest sequence is $AA\,BB\,AA\,BB$ ... but any pair of B layers could be changed to a pair of C layers, and the sequence $AA\,BB\,AA\,CC$ ... represents a different structure.

c.n. 12   The layer sequence may be any combination of A, B, and C provided that no two adjacent layers are of the same type. The two simplest possibilities are therefore $AB\,AB$ ... and $ABC\,ABC$ .... These closest-packed structures are considered in more detail later.

Of these four main classes of layer sequence only those corresponding to c.n. 10(a) and 12 are found as the structures of crystals *in which the only atoms are those of the c.p. layers*. Metals do not crytallize with the simple hexagonal packing $AAAA$ ... because a small relative displacement of the layers gives the more closely packed 10- or 12-coordinated structures. Examples of 10-coordinated structures are confined to a few disilicides and to Pa, h.p. Hg, and Pu, but many metals form one or other of the most closely packed sequences. In general we should not expect to find adjacent layers directly superposed unless there is some special reason for this less efficient packing. In fact it occurs only if there are either atoms in the (trigonal prism) holes between such layers or hydrogen atoms directly along the $A \ldots A$ etc. contacts between the layers (see later). The reason for the

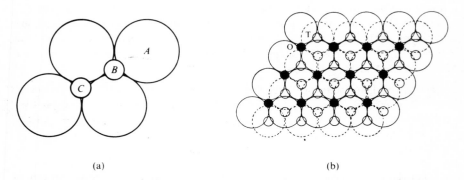

(a)   (b)

FIG. 4.7. (a) Trigonal prismatic holes between two directly superposed A layers. (b) Tetrahedral holes (small open circles) and octahedral holes (small black circles) between superposed A and B layers.

great importance of the most closely packed structures is that in many halides, oxides, and sulphides the anions are appreciably larger than the metal atoms (ions) and are arranged in one of the types of closest packing. The smaller metal ions occupy the interstices between the c.p. anions. In another large group of compounds, the 'interstitial' borides, carbides, and nitrides, the non-metal atoms occupy interstices between c.p. metal atoms.

Interstices between close-packed layers

In the plane of a c.p. layer there are small holes surrounded by triangular groups of c.p. atoms, and between the layers there are holes surrounded by polyhedral groups of c.p. atoms. These holes may be described as *a*, *b*, or *c* if they fall above positions *A*, *B*, or *C* (Fig. 4.7(a)). The types of hole are:

————— *A* —————	————— *A* —————
b or *c* (trigonal prismatic)	*c* (octahedral)
————— *A* —————	*a* or *b* (tetrahedral)
	————— *B* —————

and similarly for the *a* (octahedral) and *b* or *c* (tetrahedral) holes between *B* and *C* layers, and *b* (octahedral) and *a* or *c* (tetrahedral) holes between *A* and *C* layers. We

Sphere packing	Type of hole	Number	Maximum radius of interstitial sphere
Simple hexagonal	Trigonal prismatic	2 *N*	0.528
Closest packing	Tetrahedral	2 *N*	0.225
	Octahedral	*N*	0.414

give above the maximum radii of the spheres which can be accommodated in the interstices in packings of spheres of unit radius and also the numbers of these interstices in an assembly of *N* spheres. There are twice as many trigonal prismatic holes between a pair of *AA* layers as there are octahedral holes between a pair of *AB* layers.

Structures with some pairs of adjacent layers of type *A*

If all layers are directly superposed (for example, all of type *A*), all interstices are of type *b* or *c* surrounded by a trigonal prism of the larger spheres. Trigonal prismatic coordination is not to be expected in essentially ionic crystals, but may occur in other crystals for a number of special reasons: (i) an atom may have a preference for this arrangement of bonds; (ii) there may be some kind of interaction between the six atoms forming the trigonal prism coordination group which stabilizes this coordination group relative to the more usual octahedral one; or (iii) an interaction between the interstitial atoms themselves, in which case the coordination around these atoms may be incidental to satisfying other bonding or packing requirements.

This is probably the case in AlB_2. It is obviously not easy (or perhaps possible) to distinguish between (i) and (ii). In the two structures shown as diagrammatic elevations in Fig. 4.8(a) and (b) there are metal–metal bonds between the atoms in the layers; NbS crystallizes with both these structures. For the alternative, more usual, description of the NiAs structure see p. 753.

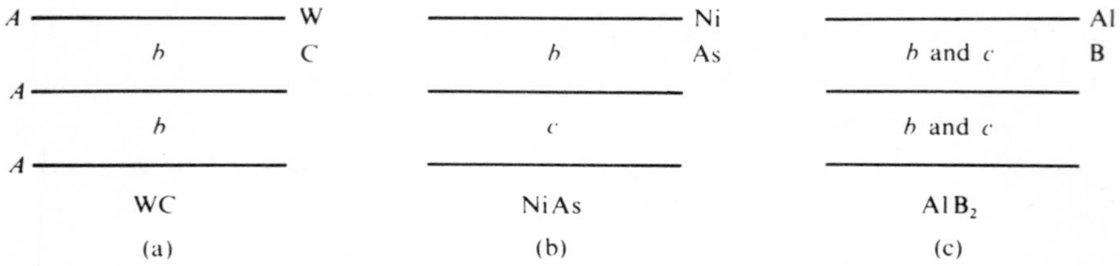

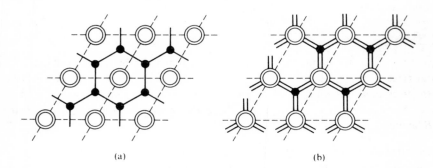

FIG. 4.8. Diagrammatic elevations of structures with trigonal prism coordination.

Between a given pair of A layers all interstices b and c will not be occupied simultaneously because of their close proximity, unless the interstitial atoms are themselves bonded. This occurs in AlB_2 (Fig. 4.8(c)). The boron atoms form plane 6-gon nets, occupying all the b and c positions between directly superposed layers of Al atoms, as shown in a projection of the structure (Fig. 4.9(a)). In hexagonal MoS_2 metal atoms occupy one-half of the trigonal prismatic holes between alternate pairs of layers of S atoms (Fig. 4.9(b)).

FIG. 4.9. Projections of atoms in trigonal prism holes (small circles) between two directly superposed c.p. layers (large circles): (a) AlB_2, showing linking of B atoms into plane 6-gon nets; (b) layer of the structure of hexagonal MoS_2.

Some further structures in which there are pairs of adjacent AA layers are shown diagrammatically as elevations in Fig. 4.10, where the type of coordination of the interlayer atoms is indicated. The simplest (layer) structure of the type $AbA\,BaB$, with trigonal prismatic coordination of all metal atoms, is that of the ordinary form of MoS_2 to which we have just referred. Because there are alternative b and c

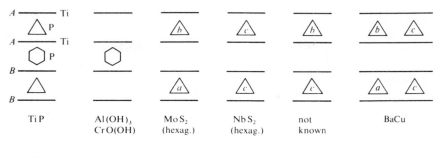

| TiP | Al(OH)₃ CrO(OH) | MoS₂ (hexag.) | NbS₂ (hexag.) | not known | BaCu |

△ indicates trigonal prism hole between layers

⬡ indicates octahedral hole between layers

FIG. 4.10. Diagrammatic elevations of structures with trigonal prismatic and/or octahedral coordination.

positions between two *A* layers and *a* and *c* positions between two *B* layers there are two other possible structures with 4-layer repeat units (*AABB*) namely, *AcA BcB* and *AbA BcB*. One is the structure of one form (hexagonal) of NbS_2; the anti-NbS_2 structure is adopted by Hf_2S. No example of the other structure is known. It may interest the reader to check that there are only two structures with a repeat unit consisting of three pairs of layers, namely,

$\qquad\qquad$ *AbA BcB CaC* (rhombohedral forms of NbS_2 and MoS_2)

and $\qquad\qquad$ *AbA BcB CbC* (no example known),

but no fewer than ten structures with a repeat unit consisting of four pairs of layers, namely, *AA BB AA CC* (7 structures) and *AA BB AA BB* (3 structures). The three *AA BB* structures and the two *AA BB CC* structures are illustrated as elevations in Fig. 4.11.

There is an indefinite number of more complex layer sequences in which some pairs of adjacent layers are of the same kind (for example, *A A*), and others different (*A B*, etc.). An example is the structure of ThI_2, in which some Th atoms occupy trigonal prismatic and others octahedral holes in the following layer sequence (8-layer repeat):

$\qquad$ . . . *A $\quad$ A C $\quad$ B A $\quad$ A B $\quad$ C A $\quad$ A* . . .
$\qquad\qquad$ t.p. $\qquad$ o $\qquad$ t.p. $\qquad$ o $\qquad$ t.p.

Such structures may be described as consisting of interleaved portions of CdI_2 and MoS_2 structures.

So far we have been concerned with the type of coordination of the (metal) atoms inserted between the c.p. layers. The way in which the double layers *AA* etc. are superposed also, of course, determines the coordination *around the c.p. atom*. This is not of special interest in MX_2 compounds since the contacts to one side are only of the van der Waals type. For example, in hexagonal MoS_2, with the structure

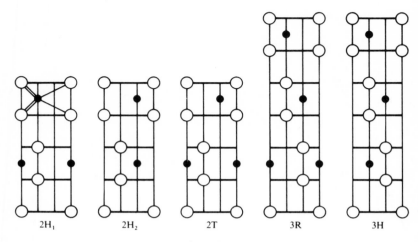

FIG. 4.11.   Projections (along *a* axis) of the three *AA BB* and two *AA BB CC* structures for compounds MX$_2$ in which the coordination of M is trigonal prismatic.

AbA BaB . . . each S has a trigonal prismatic arrangement of nearest neighbours (3 Mo + 3 S), but the S—S contacts are not of much chemical interest. The situation is, however, different in the anti-MX$_2$ structures of this type, such as the anti-2H$_2$ structure (Fig. 4.11) of Hf$_2$S. Here the nearest neighbours of Hf are 3 Hf + 3 S, Hf being bonded to 3 Hf by metal–metal bonds. The coordination groups *of the c.p. atoms* in the three simplest structures of Fig. 4.11 are:

2 H$_1$	all trigonal prismatic
2 H$_2$	all octahedral
2 T	one-half trigonal prismatic
	one-half octahedral.

Hexagonal and cubic closest packing of equal spheres

We now consider the sphere packings in which close-packed plane layers are stacked in the closest possible way. If we label the positions of the spheres in one layer as *A* (Fig. 4.12) then an exactly similar layer can be placed above the first so that the

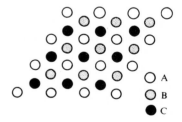

FIG. 4.12.   Three successive layers of cubic closest packing.

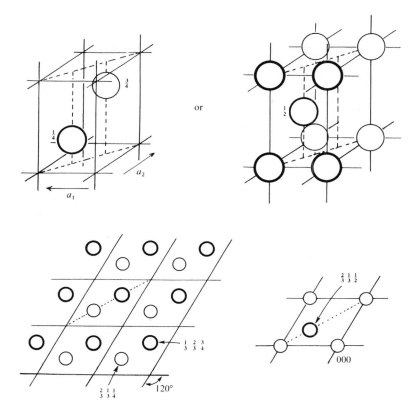

FIG. 4.13. Hexagonal close-packing of equal spheres: alternative unit cells and their projections.

centres of the spheres in the upper layer are vertically above the positions *B*. It is obviously immaterial whether we choose the positions *B* or the similar positions *C*, as may be seen by inverting Fig. 4.12. When the third layer is placed above the second (*B*) layer there are alternatives: the centres of the spheres may lie above either the *C* or *A* positions. The two simplest sequences of layers are evidently *ABAB* ... and *ABCABC* ..., but there is also an indefinite number of more complex sequences. In all such sphere packings the volume occupied per sphere is $5.66\,a^3$ for spheres of radius a. Alternatively, the mean density is 0.7405 if that of the spheres is unity and that of the space between the spheres is zero.

The sequence *ABAB* ... is referable to a hexagonal unit cell (Fig. 4.13) and is called hexagonal closest packing (h.c.p.). In this arrangement, also illustrated in Fig. 4.14(a), the twelve neighbours of an atom are situated at the vertices of the coordination polyhedron of Fig. 4.5(a) ('twinned cuboctahedron').

The sequence *ABCABC* ... possesses cubic symmetry, that is, 3-fold axes of symmetry in four directions parallel to the body-diagonals of a cube, and is therefore described as cubic closest packing (c.c.p.). A unit cell is shown in Fig. 4.1(e) (p. 142) and the packing is also illustrated in Fig. 4.14(b) and (c). The close-packed layers seen in plan in Fig. 4.12 are perpendicular to any body-diagonal of the cube. Since the atoms in c.c.p. are situated at the corners and mid-points of the faces of the cubic unit cell the alternative name 'face-centred cubic' (f.c.c.) is also applied to this packing. All atoms have their twelve nearest neighbours at the vertices of a cuboctahedron (Fig. 4.5(b)).

More complex types of closest packing

In the sequences AB . . . and ABC . . . every spheres has the same arrangement of nearest neighbours. Also the arrangements of more distant neighbours are the same for all spheres in a given packing, h.c.p. or c.c.p. In more complex sequences this is not necessarily true, and it can be shown that the next family of closest packed structures, with two (and only two) types of non-equivalent sphere, comprises four sequences.

In a concise nomenclature for c.p. structures a layer is denoted by h if the two neighbouring layers are of the same type (i.e. both A, both B, or both C) or by c if they are of different types. Hexagonal close-packing is then denoted simply by h (i.e. hhh . . .) and cubic close-packing by c. The symbols for more complex sequences include both h and c. Now the spheres in h and c layers have different arrangements of nearest neighbours (the coordination polyhedra of Fig. 4.5(a) and (b)) so that any more complex sequence necessarily has two types of non-equivalent sphere differing in arrangement of *nearest* neighbours. If we are to have *only* two kinds of non-equivalent sphere we must ensure that all spheres in h layers have identical arrangements of *more distant* neighbours, and similarly for spheres in c layers, that is, the sequence of h and c layers on each side of an h (or c) layer must always be the same.

Starting with hc we may add c or h. If we add c, giving hcc, the central c layer is surrounded by h and c. The second c layer must be surrounded in the same way, and therefore the next symbol added must be h. The addition of the third symbol c thus decides the choice of the fourth (h). We may then add either h or c. In the former case we have $hcchh$ which fixes the environment of an h layer (fourth symbol) as c and h. This means that the next symbol added must be c, and since c must have h and c neighbours this in turn makes the next layer c. It is easily verified that the sequence becomes $hhcc$ . . . . In a similar way we find that if we add c as the fifth symbol, instead of h, we should arrive at the sequence hcc . . . . Going back now to the original hc point, addition of h does not determine what the fourth symbol should be, and we find two possible sequences, hc and hhc. It can be shown in this way that only four sequences are possible for two types of non-equivalent sphere:

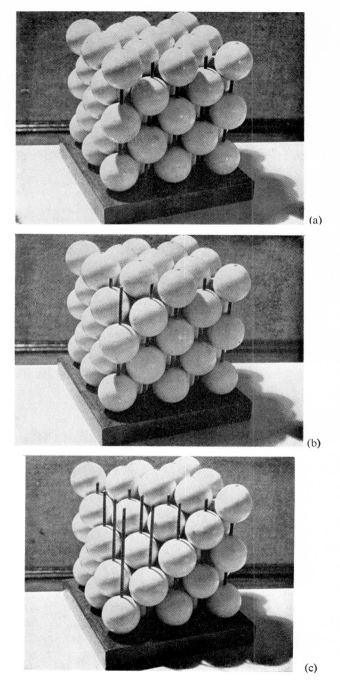

FIG. 4.14. (a) Hexagonal close-packing; (b) cubic close-packing; (c) cubic close-packing with atoms removed to show a layer of close-packed atoms. Arrangement of tetrahedral and octahedral interstices around an atom in (d) and (e) hexagonal close-packing; (f) and (g) cubic close-packing.

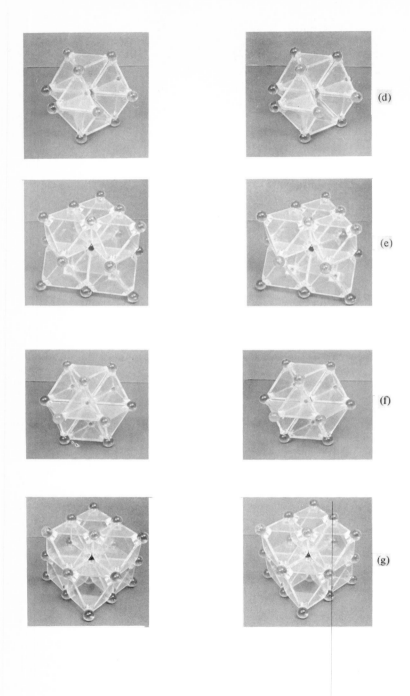

(d)

(e)

(f)

(g)

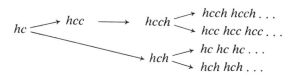

A symbol of the type *h, c, hc,* etc. obviously does not indicate the number of layers after which the sequence repeats, for this number is 2 for *h* (*AB* . . .), 3 for *c* (*ABC* . . .), and 4 for *hc* (*ABCB* . . .). The number of layers in the repeat unit of a particular stacking sequence may readily be found by deriving the latter from the symbol starting arbitrarily with *AB* . . . and continuing until the layer sequence repeats:

$$
\begin{array}{lll}
h \quad c \quad h \quad c \quad h \quad c & \text{or} & h \quad c \quad c \quad h \quad c \quad c \quad h \quad c \quad c \\
\text{(B)}\ A \quad B \quad C \quad B \quad A \quad B \ldots & & \text{(B)}\ A \quad B \quad C \quad A \quad C \quad B \quad A \quad B \ldots \\
\text{(4 layer repeat)} & & \text{(6 layer repeat)}
\end{array}
$$

TABLE 4.2

The closest packing of equal spheres

Number of kinds of non-equivalent sphere	Symbol			Number of layers in repeat unit
1	*h*	1 $\bar{1}$	2 H	2
	c	3	3 C	3
2	*hc*	2 $\bar{2}$	4 H	4
	hcc	3 $\bar{3}$	6 H	6
	chh	$(1\ \bar{2})_3$	9 R	9
	cchh	$(1\ \bar{3})_3$	12 R	12

The numerical symbols in Table 4.2 indicate both the layer sequence and the number of c.p. layers in the repeat unit in a direction normal to the layers. The change from $A \rightarrow B \rightarrow C$ is represented as a unit vector +1 and in the reverse direction as -1. A succession of *n* positive (or negative) units is abbreviated to *n* ($\bar{n}$), so that $2\bar{2}$ stands for $11\bar{1}\bar{1}$ . . . and starting from *A* gives the sequence *ABCBA* . . . . The symbols of the third kind show the number of layers in the repeat unit and the type of lattice (cubic, hexagonal, or rhombohedral).

The six c.p. sequences of Table 4.2 are those most frequently found in crystals, but the number of different sequences that repeat in 12 or fewer layers is very much larger if no account is taken of the number of kinds of non-equivalent sphere (Table 4.3). Only for 2, 3, 4, and 5 layers in the repeat unit is there a unique layer sequence.

The simplest examples of the closest packing of identical units are the structures of crystalline metals or noble gases and of crystals built of molecules of one kind

TABLE 4.3
Numbers of different sequences of c.p. spheres

Number of layers in repeat unit	Number of different sequences
2	1
3	1
4	1
5	1
6	2
7	3
8	6
9	7
10	16
11	21
12	43

only. In the latter the molecules must either be approximately spherical in shape or become effectively spherical by rotation or random orientation. Metals provide many example of hexagonal and cubic closest packing and a few examples of more complex sequences (for example, *hc*: La, Pr, Nd, Am; and *chh*: Sm). Close-packed molecular crystals include the noble gases, hydrogen, HCl, H_2S, and CH_4.

The great importance of closest packing in structural chemistry arises from the fact that the anions in many halides, oxides, and sulphides, are close-packed (or approximately so) with the metal atoms occupying the interstices between them. As already noted, the polyhedral interstices between c.p. atoms are of two kinds, tetrahedral and octahedral (T and O in Fig. 4.7(b)). The number of tetrahedral holes is equal to twice the number of c.p. spheres; the number of octahedral holes is equal to the number of c.p. spheres. By far the most important types of closest packing are the two simplest ones, namely, hexagonal and cubic, and a great variety of structures are derived by placing cations in various fractions of the tetrahedral and/or octahedral holes in such c.p. arrangements of anions. We shall consider these groups of structures in some detail shortly. As a matter of interest some examples of more complex layer sequences are given in Table 4.4. The sulphides of titanium have been included because they provide examples of a number of c.p. sequences (in addition to *h* in TiS and TiS_2); contrast the sulphides Cr_2S_3, Cr_5S_6, Cr_7S_8, and Cr_3S_4, which are all h.c.p. with different proportions and arrangements of metal atoms in octahedral holes.

Close-packed arrangements of atoms of two kinds

When discussing the crystal structure of KF (p. 316) we inquire why a 12-coordinated structure is not found, as this c.n. would be expected for spheres of the same radius. It is, however, evidently not possible to form a c.p. layer in which each ion is surrounded entirely by ions of the other kind, for if six X ions are placed around an

TABLE 4.4

Examples of more complex sequences of c.p. layers

Number of layers in repeat unit	Symbol	Close-packed X layers	Close-packed AX_3 layers	Reference
4	hc	La, Pr, Nd, Am		
		$HgBr_2$		
		HgI_2 (yellow)		IC 1967 **6** 396
		Ti_2S_3		
		Cd(OH)Cl		
			high-$BaMnO_3$	AC 1962 **15** 179
			$TiNi_3$	
5	hhccc		$Ba_5Ta_4O_{15}$	AC 1970 **B26** 102
6	hcc		VCo_3	
			$CsMnF_3$	JCP 1962 **37** 697
			high-$K_2(LiAl)F_6$	AC 1954 **7** 33
			$BaTiO_3$	
			$BaTi_5O_{11}$[1] (see text)	AC 1969 **B25** 1444
7	chcccch		$Ti(Pt_{0.89}Ni_{0.11})_3$	AC 1969 **B25** 996
8	ccch		$Sr_4Re_2SrO_{12}$	IC 1965 **4** 235
9	chh	Sm		
			$BaPb_3$	AC 1964 **17** 986
			high-YAl_3	AC 1967 **23** 729
		Mo_2S_3		JSSC 1970 **2** 188
			$BaRuO_3$	
			$CsCoF_3$	ZaC 1969 **369** 117
10	chhch	Ti_4S_5		
	cchhh		$BaFe_{12}O_{19}$[2]	ZK 1967 **125** 437
12	cchh	Fe_3S_4		AM 1957 **42** 309
		Ti_5S_8		RTC 1966 **85** 869
			$BaCrO_3$	IC 1969 **8** 286
			$Ba_4Re_2MgO_{12}$	IC 1965 **4** 235
			$Ba_6Ti_{17}O_{40}$[3]	AC 1970 **B26** 1645
14	hhhchhc		$Ba(Pb_{0.8}Tl_{0.2})_3$	AC 1970 **B26** 653

(1) Two thirds of layers BaO_7—remainder O_8

(2) Every fifth layer BaO_3—remainder O_4

(3) Equal numbers of BaO_{11} and Ba_2O_9 layers (O vacancy in latter).

A ion each X ion already has two X ions as nearest neighbours. This complication does not arise if all the c.p. atoms are anions, as in a hydroxyhalide. One of the forms of Cu(OH)Cl is built of c.p. layers of the type illustrated in Fig. 4.15(a), and many compounds containing atoms of two kinds which are similar in size are built of c.p. layers of composition AX_2 or AX_3. The AX_4 layer of Fig. 4.15(e) is the

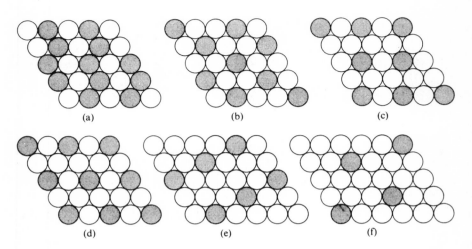

FIG. 4.15. Close-packed layers of spheres of two kinds: (a) AX; (b) AX_2; (c) and (d) AX_3;
(e) AX_4; (f) AX_6.

basis of the structure of $MoNi_4$, and the AX_6 layer of Fig. 4.15(f) represents the arrangement of K^+ and F^- ions in $KOsF_6$.

The only c.p. AX_2 layer in which A atoms are not adjacent is that of Fig. 4.15(b). This type of layer is found in WAl_5, where alternate layers have the compositions Al_3 and WAl_2, and in a number of silicides with closely related structures (p. 988).

The two simplest c.p. AX_3 layers in which A atoms are not adjacent are shown in Fig. 4.15(c) and (d). They form the bases of the structures of three groups of compounds.

(i) Intermetallic compounds AX_3, for example, $TiAl_3$ and $ZrAl_3$, which are superstructures of the c.c.p. structure of Al, being built of layers (c) and layers (c) and (d) respectively. Both $SnNi_3$ and $TiNi_3$ are built of (d) layers, the layer sequences being h and hc respectively.

(ii) Complex halides and oxides $A_xB_yX_{3x}$. The formation of numerous complex halides containing K and F or Cs and Cl in the ratio 1:3 is related to the fact that these pairs of ions of similar sizes can form c.p. AX_3 layers. Similarly there are many complex oxides containing a large ion such as Ba^{2+} which is incorporated in close-packed MO_3 layers between which smaller ions of a second metal occupy octahedral holes between six O atoms. The two AX_3 layers illustrated in Fig. 4.15 differ in an important respect. Layers (c) cannot be superposed so as to produce octahedral holes surrounded by 6 X atoms without at the same time bringing A atoms into contact, and therefore (c) layers are not found in complex halides or oxides. Later in this chapter we survey briefly the structures of these compounds which are based on c.p. layers of type (d) in Fig. 4.15.

(iii) Hydroxyhalides $M_2X(OH)_3$. In these compounds the c.p. layers are composed entirely of anions (halide and hydroxyl), so that whereas only (d) layers are permitted in (ii) no such restriction applies to hydroxyhalides. Their structures are described in more detail in Chapter 10.

We now derive the simpler structures for compounds M_mX_n in which the X atoms are close-packed and M atoms occupy various fractions of tetrahedral and/or octahedral holes. (For the systematic derivation of structures built of AX_n layers see: ZK 1967 **124** 104; AC 1968 **B24** 1477.)

Close-packed structures with atoms in tetrahedral interstices

At the outset we should emphasize two facts. The first is that atoms in the relative positions of closest packing are not necessarily close-packed, that is, in contact with twelve other atoms of their own kind. In general the coordinates of the atoms in a M_mX_n structure do not completely describe a structure by indicating which atoms (other than the A and X atoms) are in contact, a subject which is discussed in more detail with reference to the sodium chloride and fluorite structures in Chapter 6. The second is that in halides and chalconides M_mX_n in which there is octahedral coordination of the metal ions the c.p. description does provide a reasonably realistic description of the structure, but this is not so for the 'tetrahedral' structures. There are relatively few structures of compounds M_mX_n with M in tetrahedral coordination and close-packed X atoms (ions), for example, very few 3D MX_2 structures, but it is nevertheless convenient to describe the idealized structures in terms of c.p. X atoms with M atoms in certain fractions of the total number of tetrahedral interstices (Table 4.5). There are twice as many tetrahedral interstices as c.p. atoms and the formula is therefore M_2X if all the tetrahedral positions are occupied. In the c.c.p. M_2X structure (*antifluorite structure*) X is surrounded by 8 M atoms at the vertices of a cube (Fig. 4.16(a)). If only the fraction ¾, ½, or ¼ of the tetrahedral positions is occupied (in a regular manner) the coordination number of X falls to 6, 4, or 2 respectively. The corresponding coordination groups are derived by removing the appropriate number of M atoms from the cubic coordination group of Fig. 4.16(a) to give those shown at (b)–(f). These groups represent the arrangements of nearest neighbours of

  (a) O^{2-} in Li_2O (antifluorite structure) or Ca^{2+} in CaF_2 (fluorite);
  (b) and (c) two kinds of Mn^{3+} in O_3Mn_2;
  (d) Zn or S in ZnS (zinc-blende);
  (e) Pt in PtS; and
  (f) Pb in PbO.

This purely geometrical treatment regards Ca as the c.p. species in CaF_2, Mn in Mn_2O_3, Pt in PtS, and Pb in PbO. As noted on p. 252, the description of the fluorite structure as a c.c.p. assembly of Ca^{2+} ions with F^- ions in the tetrahedral interstices is not a realistic description of the structure since F^- is appreciably larger than Ca^{2+}. It is preferable to place in Table 4.5 the antifluorite structure of compounds such as the alkali-metal compounds M_2X (X = O, S, Se, Te). In the most

TABLE 4.5
Close-packed tetrahedral structures

Fraction of tetrahedral holes occupied	Sequence of c.p. layers		Formula	c.n.s of M and X
	AB...	ABC...		
All	*	Li_2O (antifluorite)	M_2X	4 : 8
$\frac{3}{4}$	*	O_3Bi_2 O_6Sb_4 O_3Mn_2	M_3X_2	4 : 6
$\frac{1}{2}$	Wurtzite β-BeO	Zinc-blende PtS PbO	MX	4 : 4
$\frac{3}{8}$	Al_2ZnS_4	Al_2CdS_4 Cu_2HgI_4	M_3X_4	
$\frac{1}{3}$	β-Ga_2S_3	γ-Ga_2S_3	M_2X_3	
$\frac{1}{4}$	γ-$ZnCl_2$	HgI_2 ZnI_2 SiS_2 OCu_2	MX_2	4 : 2
$\frac{1}{6}$	Al_2Br_6	In_2I_6	MX_3	
$\frac{1}{8}$	$SnBr_4$	SnI_4 OsO_4	MX_4	4 : 1

*See p. 164

favourable case, Li_2Te, the Te atoms are approximately close-packed, but in the oxides the O^{2-} ions are by no means close-packed; compare the O–O distances, 3.27, 3.92, and 4.55 Å in Li_2O, Na_2O, and K_2O with twice the radius of O^{2-}, approximately 2.8 Å. Clearly the description of the C-Mn_2O_3 structure in terms of c.p. Mn^{3+} ions is as unrealistic as that of CaF_2 in terms of c.p. Ca^{2+} ions; this structure is described on p. 545. We now note some other points relating to tetrahedral structures.

The first is that in the cubic CaF_2 (and Li_2O) and ZnS structures the atomic positions are exactly those of cubic closest packing and certain of the tetrahedral interstices. In PtS and PbO this is not so. In the 'ideal' PtS (PdO) structure, with c.c.p. metal atoms, S would be in regular tetrahedral holes but the four S atoms would form a plane rectangular group around Pt, with edges in the ratio $\sqrt{2}$: 1. If the structure is elongated, to become tetragonal with $c:a = \sqrt{2}$, the coordination group around the metal would become a square. The actual structure is a compromise, with $c:a = 1.24$. Similarly, the actual structure of PbO does not have c.c.p. Pb atoms but is tetragonal with $c:a = 1.25$. This represents a compromise between the cubic structure ($c:a = 1$) and a tetragonal structure ($c:a = \sqrt{2}$) with body-centred cubic packing of Pb atoms. The PbO and related structures are discussed in more detail on p. 269.

The geometrical relations between the structures derived from the fluorite structure by removing one-half or three-quarters of the tetrahedrally coordinated

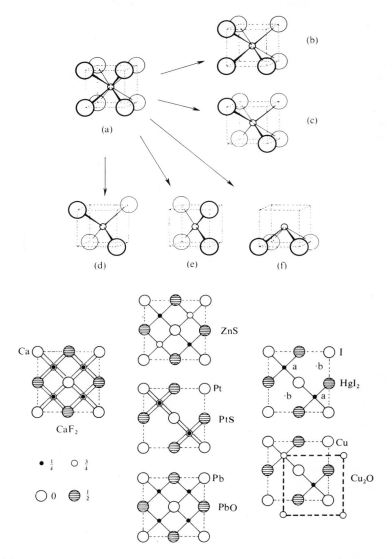

FIG. 4.16. (a)–(f) Coordination groups in the fluorite family of structures shown in projection at the right.

atoms are shown in Fig. 4.16, referred in each case to a cubic unit cell corresponding to that of fluorite. The simplest layer structure for a compound MX_2 of this family, having atoms at the positions *aa* in each cell, is not known, but a variation with doubled *c* axis and the positions *a* occupied in the lower half of the cell and *b* in the upper half is the structure of the red form of HgI_2. The other MX_2 structure of Fig. 4.16, the cuprite (Cu_2O) structure, has collinear bonds from the Cu atoms (large circles). This structure is usually referred to the unit cell indicated by the heavier broken lines. It is unique among structures of inorganic compounds MX_2 for it consists of two identical interpenetrating networks which are not connected by any primary Cu—O bonds.

The second point regarding Table 4.5 is that the structures marked by an asterisk are not formed. The hypothetical structure in which all tetrahedral holes are occupied in h.c.p. is illustrated in Fig. 4.17(a). Placing the c.p. atoms at $\pm(\frac{2}{3}\ \frac{1}{3}\ \frac{1}{4})$ the coordinates

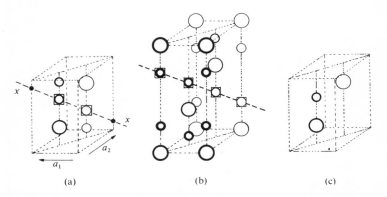

FIG. 4.17. Tetrahedral holes (smaller open circles) in close-packed structures (see text).

of the 'interstitial' atoms would be $\pm(\frac{2}{3}\ \frac{1}{3}\ \frac{5}{8})$ and $\pm(\frac{1}{3}\ \frac{2}{3}\ \frac{1}{8})$, that is, these atoms would occur in pairs separated by only $c/4$, where $c/2$ is the perpendicular distance between c.p. layers. In others words there would be pairs of *face-sharing* tetrahedra, as may be seen in Fig. 4.14(d). Such close proximity of interstitial atoms would also obtain in the corresponding $ABAC$ . . . structure, but not in the c.c.p. antifluorite structure, which is shown in Fig. 4.17(b) referred to a hexagonal unit cell. There would be the same close approach of interstitial atoms in a h.c.p. structure M_3X_2 in which three-quarters of the tetrahedral holes are occupied, but this is no longer so if only one-half of these holes are filled, as in the wurtzite structure. However, it is possible to have an $ABAC$ . . . structure for a compound M_3X_2 in which M atoms occupy three-quarters of the tetrahedral holes provided that no very close pairs ('face-sharing tetrahedral holes') are occupied simultaneously. Such a structure is adopted by high-Be_3N_2 (p. 833), in the unit cell of which two sets of tetrahedral sites are fully occupied and the third Be is disordered between two mutually exclusive tetrahedral sites.

Our third point concerns tetrahedral *layer* structures. Two of the structures of Fig. 4.16, those of LiOH (OPb) and HgI_2, are layer structures, that of LiOH being simply a slice of the anti-fluorite structure of Li_2O. These two structures may be derived from cubic closest packing by filling respectively ½ and ¼ of the tetrahedral interstices, but in both the planes of the layers are inclined to the c.p. layers. (See Fig. 5.6, p. 196) for an illustration of the HgI_2 layer.) The important octahedral AX_2 and AX_3 layer structures arise by filling all or two-thirds of the octahedral interstices between alternate pairs of c.p. layers; it might be expected that there would be layer structures in which some or all of the tetrahedral interstices are filled *between alternate pairs of c.p. layers*. Occupation of all tetrahedral sites between a pair of c.p. layers gives a layer of composition MX in whch each MX_4 tetrahedron shares three edges with three other tetrahedra and also three vertices with a second set of six tetrahedra (Fig. 4.18(a)). The structure preferred by LiOH

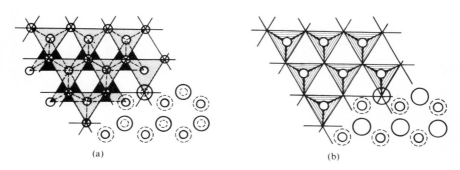

(a)   (b)

FIG. 4.18.   (a) Pair of c.p. X layers with M atoms in all tetrahedral holes between them (composition MX).   (b) Pair of c.p. X layers with M atoms in one-half of the tetrahedral holes (composition MX_2).

and OPb is the layer based on the square net in which each $Li(OH)_4$ or OPb_4 tetrahedron shares four edges (p. 119). Both layers are found as components of the $La_2O_3(Ce_2O_2S)$ and U_2N_2Sb structures respectively (p. 1271). Occupation in the most symmetrical way of one-half of the tetrahedral holes between a pair of c.p. layers gives a layer of composition MX_2. In this layer (Fig. 4.18(b)) each MX_4 tetrahedron shares vertices with three others and the fourth vertex is unshared, so that one-half of the X atoms have three equidistant M neighbours and the others only one M neighbour. While this may be the reason why halides and oxides do not have such a structure this type of non-equivalence of the c.p. atoms is inevitable in a molecular crystal such as Al_2Br_6. In this crystal pairs of adjacent tetrahedral holes indicated by the squares in Fig. 4.17(a) are occupied, leading to the formation of discrete Al_2Br_6 molecules consisting of two $AlBr_4$ groups sharing an edge. One-third of the c.p. atoms have 2 Al and two-thirds have 1 Al neighbour (Fig. 9.5, p. 418). A layer closely related to that of Fig. 4.18(b) is that of AlOCl (and GaOCl), but here there are not two c.p. layers. There is the same type of linking of the tetrahedra but equal numbers of the unshared vertices project on opposite sides of one layer of c.p. X atoms (see Fig. 10.16, p. 486).

There are, however, two other ways of filling one-half of the tetrahedral interstices between a pair of c.p. layers to form layers which can be described as the tetrahedral analogues of the CdI_2 layer. Each MX_4 tetrahedron shares one edge and two vertices, forming a structure of 4:2 coordination. Because each MX_4 group is joined to three others the layer is based on a 3-connected plane net. It appears that the only possible layers are based on the regular 6^3 net or on 4.8^2, one of the three semi-regular plane nets (Fig. 4.19). There are similar layers based on the other two semi-regular nets 3.12^2 and $4.6.12$, but in these only two-thirds and three-quarters respectively of the c.p. X positions are occupied. No example seems to be known of MX_2

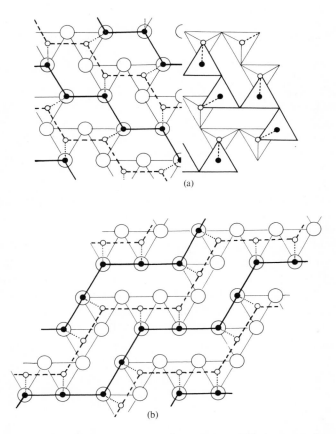

(a)

(b)

FIG. 4.19. Two AX_2 layers based on (a) the 6^3 and (b) the 4.8^2 planar nets formed by filling one half of the tetrahedral interstices between a pair of c.p. X layers. In (a) are shown both the topological representation (left) and two of the rings of six tetrahedra (right). The atoms of the lower c.p. layer only are shown (larger circles). The tetrahedral interstices at the two levels are distinguished as small open and black circles, and heavy full and broken lines in (a) (left) and (b) connect the A atoms at the upper and lower levels respectively.

structures of these types, but the structure of GaPS$_4$ (Fig. 4.19(b)) is an example of the 4.8^2 layer. It is interesting that one form of GeS$_2$ has a tetrahedral layer structure, but this is of a more complex sort in which one-half of the GeS$_4$ tetrahedra share all vertices and the remainder share two vertices and one edge; this layer is therefore based on a (3,4)-connected plane net. This is not a c.p. layer; it is illustrated on p. 197.

Finally, we may note that there is no h.c.p. analogue of the SiS$_2$ structure. Continued sharing of opposite edges of MX$_4$ tetrahedra to give a compound MX$_2$ is not possible in hexagonal closest packing because there are no tetrahedral sites at the points marked X in Fig. 4.17(a), but it can occur in cubic closest packing as may be seen by comparing Fig. 4.17(b) with (a). Occupation of one-quarter of the tetrahedral holes in a c.c.p. array of S atoms as at (b) gives chains formed from tetrahedra sharing opposite edges.

Close-packed structures with atoms in octahedral interstices

The most important structures of this kind are set out in Table 4.6. The relations between these structures may be illustrated in two ways: (a) we may consider the ways in which various proportions of the M atoms can be removed from the MX structure, in which all the octahedral holes are occupied, or (b) we may focus our attention on the patterns of sites occupied by the M atoms between the close-packed layers.

(a) The MX structures for hexagonal and cubic closest-packing are the NiAs and NaCl structures, in which the M atoms are arranged around X at the vertices of a trigonal prism or octahedron respectively. A feature of the NiAs structure is the close proximity of Ni atoms, suggesting interaction between the metal atoms in at least some compounds with this structure, which is not adopted by ionic compounds. In NaCl itself the Cl$^-$ ions are not in contact, since Na$^+$ is rather too large for an octahedral hole between c.p. Cl$^-$ ions. These ions are in contact in LiCl, compare

$$\text{Cl--Cl in NaCl} \quad 3.99 \text{ Å,}$$
$$\text{Cl--Cl in LiCl} \quad 3.63 \text{ Å.}$$

The c.c.p. structure would therefore more logically be called the LiCl structure, but since it has become generally known as the NaCl structure we keep that name here.

Removal of a proportion of the M atoms from these structures in a regular manner leaves X with fewer than six neighbours. For example, we may remove one-half of the Ni atoms from the NiAs structure so as to leave the close-packed (As) atoms with three neighbours arranged in one of the following ways (Fig. 4.20(a)):

(i) *abc*, a pyramidal arrangement with bond angles of 90°;

(ii) *abf*, a nearly planar arrangement with bond angles 90° and 132° (two);

(iii) *abd*, a pyramidal arrangement with bond angles 70°, 90°, and 132°.

Of these (iii) is not known; (i) corresponds to the CdI$_2$(*C*6) structure. Figure

TABLE 4.6
Close-packed octahedral structures

Fraction of octahedral holes occupied	Sequence of c.p. layers		Formula	c.n.s of M and X
	AB...	ABC...		
All	NiAs[a]	NaCl	MX	6 : 6
$\frac{2}{3}$	α-Al$_2$O$_3$ LiSbO$_3$	*	M$_2$X$_3$	6 : 4
$\frac{1}{2}$[b]	*Layer structures* CdI$_2$ CdCl$_2$ *Framework structures* CaCl$_2$ Rutile Atacamite NiWO$_4$ Anatase α-PbO$_2$ α-AlOOH		MX$_2$	6 : 3
$\frac{1}{3}$	*Chain structures* ZrI$_3$ – *Layer structures* BiI$_3$ YCl$_3$ *Framework structures* RhF$_3$ –		MX$_3$	6 : 2
$\frac{1}{4}$	IrF$_4$ (framework) α-NbI$_4$ (chain) NbF$_4$ (layer)		MX$_4$	6 : $\frac{2}{1}$
$\frac{1}{5}$	Nb$_2$Cl$_{10}$ Ru$_4$F$_{20}$ } (molecular) U$_2$Cl$_{10}$ Mo$_4$F$_{20}$ } (molecular) UF$_5$ (chain)		MX$_5$	6 : $\frac{2}{1}$
$\frac{1}{6}$[c]	α-WCl$_6$ –		MX$_6$	6 : 1

[a] For structures intermediate between the NiAs and CdI$_2$ (C6) structures see sulphides of chromium (p. 768) and the NiAs structure (p. 753).
[b] For other structures in which one-half of the octahedral holes are occupied in a c.p. XY$_3$ assembly see hydroxyhalides M$_2$X(OH)$_3$ and in particular Table 10.15 (p. 490).
[c] For other structures in which $\frac{1}{4}$, $\frac{1}{6}$, or $\frac{1}{8}$ of the octahedral holes are occupied in a c.p. AX$_3$ assembly see 'Structures built from c.p. AX$_3$ layers' and in particular Table 4.10 (p. 185).

4.21(a) shows an elevation of the NiAs structure (viewed along an *a* axis) in which the small circles represent rows of Ni atoms perpendicular to the plane of the paper. Removal of those shown as dotted circles gives an MX$_2$ structure in which X has three nearly coplanar M neighbours. This is the CaCl$_2$ structure. A slight adjustment of the positions of the c.p. atoms gives the rutile structure (Fig. 4.21(b)), in which X has three coplanar neighbours. It is seen here viewed along the tetragonal *c* axis, that is, parallel to the chains of octahedra which are emphasized in (c).

Similarly, removal of one-half of the M atoms from the NaCl structure leaves X

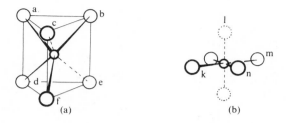

FIG. 4.20. Trigonal prismatic and octahedral coordination groups.

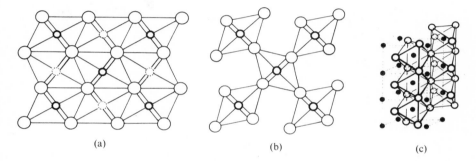

FIG. 4.21. Relation between the NiAs and rutile structures (see text).

with three neighbours at three of the vertices of an octahedron, one of the combinations *lmn* or *klm* in Fig. 4.20(b). The two simplest ways of leaving X with the pyramidal arrangement of neigbours (*lmn*) are to remove alternate layers of X atoms parallel to the plane (111) giving the $CdCl_2$ (layer) structure (Fig. 4.22(a)) or to remove alternate rows of X atoms as in Fig. 4.22(b). It is an interesting fact that no dihalide crystallizes wih this structure although the environments of M (octa-hedral) and X (pyramidal) are the same as in the $CdCl_2$ structure. Figure 4.22(b) does represent, however, the idealized structure of atacamite, $Cu_2(OH)_3Cl$, three-quarters of the X positions being occupied by OH groups.

The coplanar arrangement of neighbours *klm* in Fig. 4.20(b) arises by removing one-half of the Na atoms from the NaCl structure in the manner showin in Fig. 4.22(c); this necessitates doubling the *c* axis of the unit cell. The configuration *klm* of nearest neighbours is not suitable for an essentially ionic crystal, but by a slight adjustment of the atomic positions the coordination group around X becomes an approximately equilateral triangle. The hypothetical structure would have $c = 2a$ and $z = ¼$; in anatase, one of the polymorphs of TiO_2, $c:a = 2.51$ and $z_0 = 0.21$.

(b) Alternatively, we may consider the ways in which various proportions of the

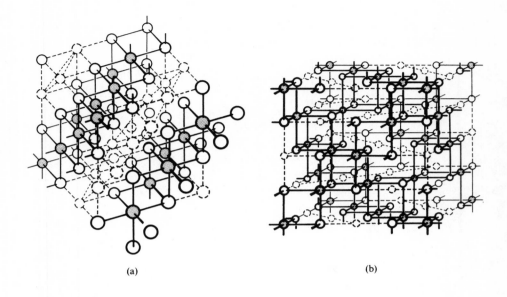

(a)   (b)

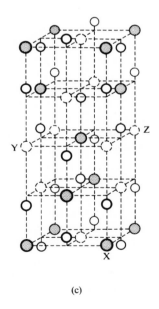

(c)

FIG. 4.22. Structures geometrically related to the NaCl structure: (a) $CdCl_2$; (b) atacamite, $Cu_2(OH)_3Cl$; (c) anatase, TiO_2. Metal ions removed from the NaCl structure are shown as dotted circles.

octahedral holes may be occupied in each type of closest packing. There are four variables which, however, are not all independent, as noted under (iv), p. 172:

(i) the sequence of c.p. layers;

(ii) the degree of occupancy of sites between successive pairs of layers;

(iii) there are different *patterns* of sites for a given fraction occupied; and

(iv) there is still the possibility of translating the sets of M atoms between successive pairs of layers relative to one another.

(i) The sequence of c.p. layers has already been discussed.

(ii) A particular composition implies the occupation of some fraction of the *total* number of octahedral sites, for example, one-half for MX_2. This may be achieved by filling all sites between *alternate* pairs of c.p. layers, one-half of the sites between each pair of layers, or alternately one-third and two-thirds or one-quarter and three-quarters of the sites between pairs of successive layers, and, of course, in more complex ways.

These possibilities are shown diagrammatically in Fig. 4.23 together with some simple arrangements for compositions M_2X_3 and MX_3.

Two of the structures of Fig. 4.23 are *simple* layer structures, that is, metal atoms occur only between alternate pairs of c.p. X layers. Note that a simple layer structure is not possible for a c.p. octahedral M_xX_y structure if $x/y > \frac{1}{8}$ (as for M_3X_4 or M_2X_3) because all the octahedral holes between alternate pairs of c.p. layers are filled when x/y reaches the value ½. More generally it can be shown that a periodic 2-dimensional system of composition M_2X_3 is not possible if M is to be 4-, and X 6-coordinated; a simple octahedral M_2X_3 layer is a special example of the more general topological limitation. Bi_2Se_3 provides an example of a more complex layer structure.

(iii) Figure 4.24 shows various ways of filling certain fractions of the octahedral sites between a pair of c.p. layers. Sites (a) and (a') are occupied between successive pairs of layers in $CaCl_2$ and in rutile (distorted h.c.p.); (b) and (b') in α-PbO_2 (h.c.p.) and anatase (distorted c.c.p.); (c) and (c') in ξ-Nb_2C (h.c.p.); (d) and (d') in $Cu_2(OH)_3Cl$ (c.c.p.) (atacamite); and (e) and (e') in ϵ-Fe_2N (h.c.p.).

Occupation of sites (d) or (e) between *alternate* pairs of layers gives layer structures:

(d) Nb_3Cl_8;

(e) BiI_3 (h.c.p.), OTi_3 (h.c.p.), YCl_3 (c.c.p.).

Sites of type (d') are occupied between *all* pairs of layers in IrF_4; sites (e) in α-Al_2O_3; sites (e') in a number of MX_3 structures which are discussed shortly; and sites (f) in $LiSbO_3$.

(iv) Evidently a knowledge of the c.p. sequence and the 2-dimensional site pattern, that is, one of the patterns of small black circles in Fig. 4.24, is not sufficient to describe a structure, for the relative translations of the sets of metal atoms are not defined in relation to some fixed frame of reference. For example, an indefinitely large number of structures can be built from c.p. layers with the sites (e') occupied between each successive pair of layers. Three simple structures result if the same

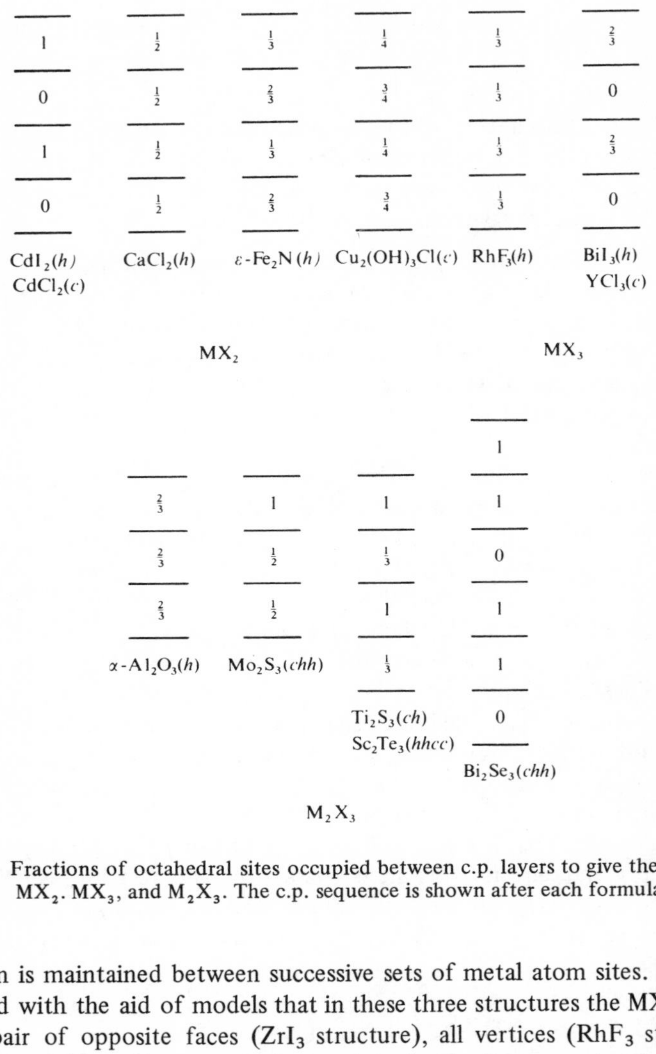

$$
\begin{array}{cccccc}
\underline{\;1\;} & \underline{\;\frac{1}{2}\;} & \underline{\;\frac{1}{3}\;} & \underline{\;\frac{1}{4}\;} & \underline{\;\frac{1}{3}\;} & \underline{\;\frac{2}{3}\;} \\[4pt]
0 & \frac{1}{2} & \frac{2}{3} & \frac{3}{4} & \frac{1}{3} & 0 \\[4pt]
\underline{\;1\;} & \underline{\;\frac{1}{2}\;} & \underline{\;\frac{1}{3}\;} & \underline{\;\frac{1}{4}\;} & \underline{\;\frac{1}{3}\;} & \underline{\;\frac{2}{3}\;} \\[4pt]
0 & \frac{1}{2} & \frac{2}{3} & \frac{3}{4} & \frac{1}{3} & 0 \\
\end{array}
$$

$CdI_2(h)$	$CaCl_2(h)$	$\varepsilon\text{-}Fe_2N(h)$	$Cu_2(OH)_3Cl(c)$	$RhF_3(h)$	$BiI_3(h)$
$CdCl_2(c)$					$YCl_3(c)$

$$MX_2 \qquad\qquad\qquad\qquad MX_3$$

Lower diagram (M_2X_3):

$$
\begin{array}{cccc}
 & & & \underline{\;1\;} \\
\underline{\;\frac{2}{3}\;} & \underline{\;1\;} & \underline{\;1\;} & \underline{\;1\;} \\
\underline{\;\frac{2}{3}\;} & \underline{\;\frac{1}{2}\;} & \underline{\;\frac{1}{3}\;} & \underline{\;0\;} \\
\underline{\;\frac{2}{3}\;} & \underline{\;\frac{1}{2}\;} & \underline{\;1\;} & \underline{\;1\;} \\
 & & \underline{\;\frac{1}{3}\;} & \underline{\;1\;} \\
 & & & \underline{\;0\;} \\
\end{array}
$$

$\alpha\text{-}Al_2O_3(h)$	$Mo_2S_3(chh)$	$Ti_2S_3(ch)$	$Bi_2Se_3(chh)$
		$Sc_2Te_3(hhcc)$	

$$M_2X_3$$

FIG. 4.23. Fractions of octahedral sites occupied between c.p. layers to give the compositions MX_2, MX_3, and M_2X_3. The c.p. sequence is shown after each formula.

translation is maintained between successive sets of metal atom sites. It can easily be verified with the aid of models that in these three structures the MX_6 octahedra share a pair of opposite faces (ZrI_3 structure), all vertices (RhF_3 structure), or two opposite edges and two opposite vertices. In the first two structures there are h.c.p. X atoms and in the third, c.c.p. X atoms, illustrating the point that the variables (i) and (iv) are not independent. In this connection it is of interest to note that there is no c.c.p. structure of the same topological type as the RuF_3 structure; in the ReO_3 structure the O atoms occupy only three-quarters of the positions of cubic closest packing.

For a more complete description of a c.p. structure we may give an elevation of the kind shown in Fig. 4.25, the structure being viewed in the direction of the arrow in Fig. 4.24(e). Note that the plane of the diagram is not perpendicular to

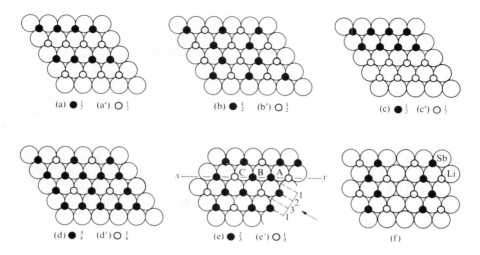

FIG. 4.24. Patterns of octahedral sites between pairs of c.p. layers (see text).

the arrow but is the plane intersecting the paper in the line xy in Fig. 4.24(e). In the corundum (α-Al$_2$O$_3$) structure the pattern of sites occupied by Al atoms is that of the small black circles of Fig. 4.24(e). If the positions occupied between a particular pair of layers are 1 and 2 (and all equivalent ones, that is all the black circles) then the positions occupied between successive pairs of (h.c.p.) layers are 2 and 3, 3 and 1, and so on. The elevation shown in Fig. 4.25 shows both the layer sequence and the pattern of sites occupied by the metal atoms. See also Fig. 6.19 (p. 267) for FeTiO$_3$ and LiNbO$_3$, which are superstructures of the corundum structure.

In Chapter 5 we show how certain structures may be constructed from octahedral coordination groups sharing vertices, edges, or faces (or combinations of these), and it is instructive to make models relating this way of describing structures to the

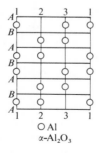

FIG. 4.25. Elevation of h.c.p. structure.

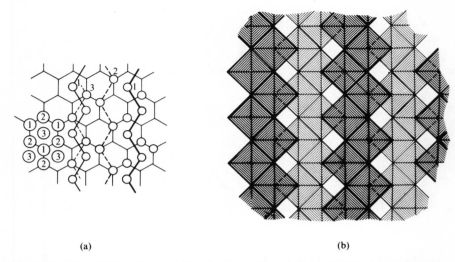

(a)   (b)

FIG. 4.26. The structure of anatase shown (a) as a c.p. structure; (b) as an assembly of edge-sharing octahedra. In (a) the numbers at the left indicate the positions of O atoms in the lowest (1) and succeeding layers, (2), (3), and the numbers against the Ti atoms (smaller circles at right) correspond to the c.p. layers on which they rest.

description in terms of c.p. anions. For this reason we show in Fig. 4.26 both these types of representation of the anatase structure which should be compared with Fig. 4.22(c), p. 170. Note that Fig. 4.26(a) is a projection in a direction normal to the c.p. layers, which are parallel to a plane such as XYZ in Fig. 4.22(c).

Whether vertices, edges, or faces are shared in the final structure depends not only on the patterns of sites occupied between pairs of layers but also on the relation between the sets of sites occupied between successive pairs of layers. For example, occupation of the sites of Fig. 4.24(e') implies no sharing of X atoms between the octahedra formed around the e' sites *in one layer*, but vertex-sharing occurs in the 3D structure (RhF_3) which results from the repetition of this pattern of site-filling. On the other hand, the occupation of more than one-third of the octahedral sites between a given pair of c.p. layers already implies edge-sharing between octahedra of one layer. Elevations of the type of Fig. 4.25 show at a glance whether octahedral coordination groups are sharing vertices, edges, or faces with one another, as indicated in Fig. 4.27. Note that because of the 3-fold symmetry (a) and (b) in Fig. 4.27 imply sharing of *three* vertices or edges.

Some related MX_2, $MM'X_4$, and $M_2M'X_6$ structures

Any structure A_mX_n in which A atoms occupy some fraction of the octahedral holes in a c.p. assembly of X atoms is potentially a structure for more complex compounds, as we have already noted for $FeTiO_3$ and $LiNbO_3$, where atoms of two kinds occupy the Al sites in the corundum structure (Fig. 4.25). There are many

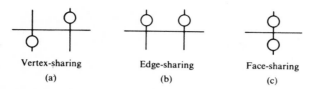

Vertex-sharing   Edge-sharing   Face-sharing
(a)   (b)   (c)

FIG. 4.27 The sharing of octahedron vertices, edges, or faces, as indicated in the elevation of Fig. 4.25.

complex halides and oxides containing cations of two or more kinds which have similar radii and can therefore occupy, for example, octahedral sites in a c.p. structure (Table 4.7). If the sites occupied by M and M' in MM'X_4 or M_2M'X_6 are

TABLE 4.7

Relations between some h.c.p. structures

Pattern of octahedral sites (Fig. 4.24) occupied	(a) (a')	(b) (b')	(e) (e')	—
Fractions of sites occupied between successive pairs of c.p. layers	$\frac{1}{2}$	$\frac{1}{2}$	$\frac{2}{3}$ $\frac{1}{3}$	$\frac{4}{9}$ $\frac{5}{9}$
MX_2 $MM'X_4$ $M_2M'X_6$	Rutile MgUO$_4$ Trirutile, ZnSb$_2$O$_6$	α-PbO$_2$ NiWO$_4$ Niobite, (Fe, Mn)Nb$_2$O$_6$	ϵ-Fe$_2$N — Li$_2$ZrF$_6$ Li$_2$NbOF$_5$	— Na$_2$SiF$_6$ NiU$_2$O$_6$

those occupied by M in MX_2 the more complex structures may be described as superstructures of the simpler MX_2 structure. For example, MgUO$_4$ and ZnSb$_2$O$_6$ are related in this way to the rutile structure. The 'trirutile' structure of ZnSb$_2$O$_6$ and other complex oxides is illustrated in Fig. 4.28(a), which is related in an obvious way to Fig. 4.24(a). A number of superstructures of the α-PbO$_2$ structure are known. In NiWO$_4$ the octahedral sites between alternate pairs of c.p. layers in the α-PbO$_2$ structure are occupied by Ni and W atoms, and in the columbite structure there is a more complex type of replacement:

$$Pb;Pb;Pb;Pb, \qquad Ni;W;Ni;W. \qquad \frac{Fe}{Mn};Nb;Nb;\frac{Fe}{Mn};Nb;Nb.$$

Three superstructures of α-PbO$_2$ are illustrated in Fig. 4.24(b)–(d). It is possible to illustrate a h.c.p. structure in this way only if the metal atom arrangement repeats after each two layers. In NaNbO$_2$F$_2$, (c), the repeat unit consists of four layers. We show the structures of two successive layers; in the next two layers the positions of Na and Nb are interchanged. Other ways of filling one-half of the octahedral sites are illustrated in Fig. 4.28(e)–(g), as found in α-NaTiF$_4$, Li$_2$ZrF$_6$, and Na$_2$SiF$_6$ respectively.

Close-packed structures with atoms in tetrahedral and octahedral interstices

We have noted that all tetrahedral holes will not be occupied in hexagonal close-packing because they are too close together. If all octahedral and all tetrahedral holes are occupied in a cubic close-packed assembly the atomic positions are those of Fig. 9.7 (p. 422), where the small circles would be the c.p. atoms and the large open and shaded circles would represent the atoms in tetrahedral and octahedral holes respectively. It is the structure of BiLi$_3$ and other intermetallic phases. Now although one-third of the Li atoms are in octahedral holes and two-thirds in tetrahedral holes in a c.c.p. of Bi atoms and consequently have respectively six octahedral or four tetrahedral Bi neighbours, these are not the nearest (or the only equidistant) neighbours. Each Li has in fact a cubic arrangement of eight nearest neighbours:

$$\text{Li}_\text{I} \left\{ \begin{array}{l} 8 \text{ Li}_\text{II} \text{ (cubic) at } \sqrt{3}a/4 \\ 6 \text{ Bi (octahedral) at } a/2 \end{array} \right. \qquad \text{Li}_\text{II} \left\{ \begin{array}{l} 4 \text{ Bi (tetrahedral)} \\ 4 \text{ Li}_\text{I} \text{ (tetrahedral)} \end{array} \right\} \text{ at } \sqrt{3}a/4$$

This is therefore an 8-coordinated structure and it is grouped with the NaTl and related structures in Chapter 29 (p. 1301).

It is only in structures in which small proportions of the two types of hole are occupied that the nearest neighbours of an atom in a tetrahedral (octahedral) hole are *only* the nearest 4 (6) c.p. atoms. This is true, for example, in the spinel and olivine structures (Table 4.8). The Co$_9$S$_8$ structure is closely related to the spinel structure, which has 32 c.p. O atoms in the cubic unit cell. In Co$_9$S$_8$ there are 32 c.p. S atoms with 4 Co in octahedral and 32 Co in tetrahedral holes; compare Co$_3$S$_4$ with a slightly distorted spinel structure, also with 32 S in the cubic unit cell, but 16 Co in octahedral and 8 Co in tetrahedral holes. The spinel structure is

TABLE 4.8

Tetrahedral and octahedral coordination in close-packed structures

Fractions of holes occupied		c.c.p.	h.c.p.
Tetrahedral	*Octahedral*		
$\frac{1}{8}$	$\frac{1}{8}$		Al$_2$CoCl$_8$
$\frac{1}{8}$	$\frac{1}{6}$	Cr$_5$O$_{12}$	
$\frac{1}{8}$	$\frac{1}{2}$	MgAl$_2$O$_4$	Mg$_2$SiO$_4$
		(spinel)	(olivine)
$\frac{1}{8}$	$\frac{1}{4}$	CrVO$_4$	CrPS$_4$
$\frac{1}{2}$	$\frac{1}{8}$	Co$_9$S$_8$	
$\frac{1}{2}$	$\frac{1}{2}$	β-Ga$_2$O$_3$	

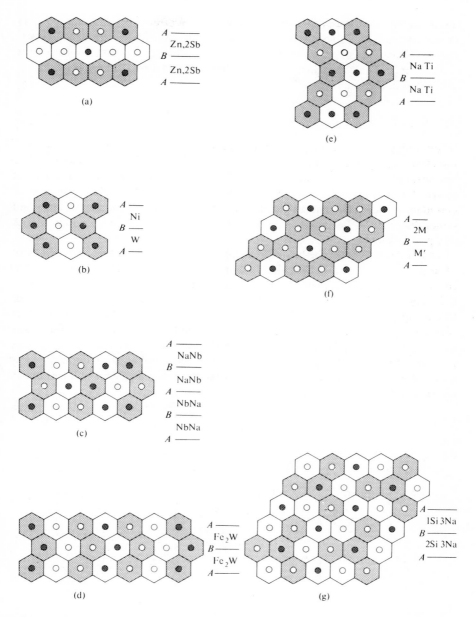

FIG. 4.28. The h.c.p. structures of (a) $ZnSb_2O_6$; (b) $NiWO_4$; (c) $NaNbO_2F_2$; (d) Fe_2WO_6; (e) α-$NaTiF_4$; (f) Li_2ZrF_6; (g) Na_2SiF_6. The open and filled circles represent M and M' atoms in octahedral holes in a h.c.p. assembly of X atoms. Clear and shaded octahedra contain cations at heights 0 and $c/2$. The sketches at the right show the types of atoms between successive pairs of c.p. layers.

discussed in detail on p. 593; for the olivine structure see p. 592. An expanded version of part of Table 4.8 will be found on p. 765, where the structures of a number of metallic sulphides are described, and in Chapter 5 we list some structures (not necessarily close-packed) in which there is tetrahedral and octahedral co-ordination of the same element.

An alternative representation of close-packed structures

In the foregoing treatment of structures emphasis is placed on the coordination of the atoms occupying the interstices in the c.p. assemblies. Except for certain of the 'tetrahedral' structures of compounds M_2X and M_3X_2 of Table 4.5 these are metal atoms, the c.p. assembly being that of the anions. As we point out elsewhere the determining factor in some structures appears to be the environment of the anion rather than that of the cation, and this is emphasized in an alternative representation of c.p. structures.

In our description of the b.c.c. structure on p. 145 we noted that the description of structures in terms of Dirichlet domains implies that interstices are situated at the vertices of such domains. The Dirichlet domains for h.c.p. and c.c.p. are the polyhedra of Fig. 4.29. They provide an alternative way of representing relatively simple c.p. structures (particularly of binary compounds). The (8) vertices at which three edges meet are the tetrahedral interstices, and those (6) at which four edges meet are the octahedral interstices. Table 4.9 shows the octahedral positions occupied in some simple structures; c.p. structures in which tetrahedral or tetrahedral and octahedral sites are occupied may be represented in a similar way. (For examples see JSSC 1970 1 279.)

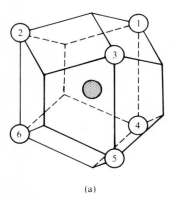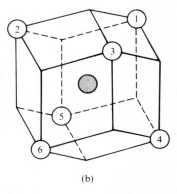

(a)   (b)

FIG. 4.29. The polyhedral domains for (a) hexagonal and (b) cubic closest packing, showing the six positions of octahedral coordination around a c.p. sphere. The remaining vertices of the domains are tetrahedral sites.

TABLE 4.9

Representation of c.p. structures by polyhedral domains

	c.n. of c.p. X	Vertices occupied	Structure MX$_n$
h.c.p. structures (Fig. 4.29 (a))	6	1 2 3 4 5 6	NiAs
	4	2 3 4 5	α-Al$_2$O$_3$
	3	2 4 5	CaCl$_2$
		4 5 6	CdI$_2$
	2	2 5	RhF$_3$
		2 6	ZrI$_3$
		5 6	BiI$_3$
c.c.p. structures (Fig. 4.29 (b))	6	1 2 3 4 5 6	NaCl
	4	2 4 5 6	Sc$_2$S$_3$
	3	4 5 6	CdCl$_2$
		1 4 6	TiO$_2$ (anatase)
	2	4 6	YCl$_3$
		1 6	ReO$_3$

Structures built from close-packed AX$_3$ layers

We noted earlier that certain pairs of ions of similar size can together form c.p. layers AX$_3$ and that of the two possible AX$_3$ layers with non-adjacent A atoms only one is found in complex halides and oxides A$_x$B$_y$X$_{3x}$. This layer (Fig. 4.15(d)) can be stacked in closest packing to form octahedral X$_6$ holes between the layers without bringing A ions into contact. Figure 4.30 shows two such layers and the positions

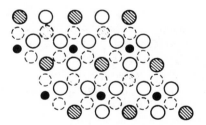

FIG. 4.30. Two close-packed AX$_3$ layers (full and dotted circles) showing the positions, mid-way between the layers, for metal atoms (small black circles) within octahedra of X atoms.

of octahedral coordination for the B atoms between the layers. The number of these X$_6$ holes is equal to the number of A atoms in the structure. All of these positions or a proportion of them may be occupied in a complex halide or oxide by cations B carrying a suitable charge to give an electrically neutral crystal A$_x$B$_y$X$_{3x}$, where A and X are, for example, K$^+$ and F$^-$, Cs$^+$ and Cl$^-$, Ba^{2+} and O^{2-}, etc. The formula depends on the proportion of B positions occupied:

all occupied	ABX_3
two-thirds	$A_3B_2X_9$
one-half	A_2BX_6

These fractions are, of course, respectively $\frac{1}{4}$, $\frac{1}{6}$, and $\frac{1}{8}$ of the total number of octahedral holes if we disregard the difference between the two kinds of c.p. atom A and X. In A_2BX_6 there are discrete octahedral BX_6 groups; in $A_3B_2X_9$ and ABX_3 the X:B ratios show that X atoms must be shared between BX_6 groups. A feature of the latter structures, which will be evident from Fig. 4.31, is that only *vertices* or *faces* of octahedral BX_6 groups are shared; this is in marked contrast to the edge-sharing found in many oxides and halides. This absence of edge-sharing is simply a consequence of the structure of the AX_3 layers, as may be seen by studying models. When a third layer is placed on a pair of layers there are two possible orientations of this layer, one leading to vertex-sharing and the other to face-sharing between BX_6 octahedra.

All the c.p. layer sequences are possible for AX_3 layers, and it is found that those most frequently adopted are the simplest ones having only one or two kinds of non-equivalent sphere, namely, those sequences which repeat after every 2, 3, 4, 6, 9, or 12 layers. Examples of structures based on c.p. AX_3 layers are included in the more general Table 4.4. (p. 159). Many compounds crystallize with more than one type of closest packing, for example, $BaMnO_3$ with the 2, 4, and 9 layer structures. In some cases this phenomenon is probably more accurately described as polytypism rather than polymorphism, as in the case of the very numerous polytypes of SiC and ZnS. For example, crystals of $BaCrO_3$, which is only formed under a pressure exceeding 3000 atmospheres, have been shown to have 4-, 6-, 9-, 14-, and 27-layer sequences of c.p. BaO_3 layers.

If all the octahedral X_6 holes are occupied by B ions there is only one possible structure (ABX_3) corresponding to each c.p. sequence, but if fewer are occupied there are various possible arrangements of the B ions in the available holes. This is exactly comparable to c.p. structures for binary compounds, there being one structure (NiAs, NaCl, etc.) for each type of closest-packing if all the octahedral holes are occupied but alternative arrangements of the same number of cations if only a fraction of the holes are filled.

We shall indicate here only structures with 2, 3, 4, and 6 layer sequences of c.p. layers, and for compounds $A_3B_2X_9$ and A_2BX_6 we shall illustrate first only the simplest of the structures for each c.p. sequence. The possible structures may be derived in the following way. By analogy with the nomenclature for c.p. layers of identical spheres we call layers *A, B,* and *C,* these positions referring to the larger cell of the AX_3 layer (Fig. 4.31). All A atoms of an *A* layer fall vertically above points such as *A,* those of a *B* layer over *B* and similarly for a *C* layer. A translation of *d* converts an *A* into a *B* layer, *B* into *C,* and *C* into *A.* In any sequence of these layers, stacked to form octahedral X_6 holes between the layers, the positions for

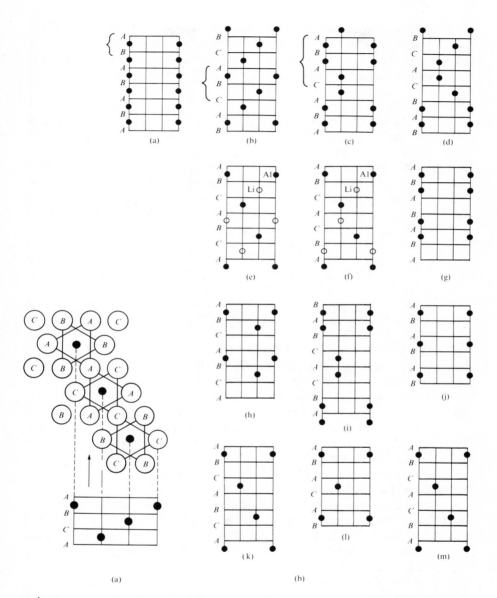

FIG. 4.31. At the left the plan shows the X atoms only of the c.p. AX_3 layers in the relative positions A, B and C, and the sites between pairs of layers AB, BC, and CA for B metal atoms (small black circles) surrounded octahedrally by six X atoms. The elevation shows the layers (horizontal lines) viewed in the direction of the arrow. (a)–(m), diagrammatic elevations of c.p. structures $A_xB_yX_{3x}$, in which there is octahedral coordination of B by 6 X atoms; (a)–(d) ABX_3; (e) and (f) low- and high-temperature forms of K_2LiAlF_6; (g)–(i) $A_3B_2X_9$; (j)–(m) A_2BX_6.

the B atoms in a compound ABX_3 also lie above the positions *A, B,* or *C.* Between *A* and *B* layers they lie above *C,* between *B* and *C* layers above *A,* and between *A* and *C* layers above *B.* The positions of the B atoms may therefore be shown in elevations of the same kind as those of binary c.p. structures given earlier.

ABX_3 *structures*

In these structures all the octahedral X_6 holes are occupied by B atoms. Figure 4.31(a)–(d) shows the structures based on the layer sequences *h, c, hc,* and *hcc.*

In the simple structure (a), with hexagonal closest packing of $A + 3X$, the octahedral BX_6 groups are stacked in columns sharing a pair of opposite faces. This is the structure of $CsNiCl_3$, $BaNiO_3$, and $LiI.3H_2O$ (or $ILi(H_2O)_3$). A variant of this structure is adopted by $CsCuCl_3$, in crystals of which the hexagonal closest-packing is slightly distorted (by a small translation of the layers relative to one another) so as to give Cu only four (coplanar) nearest neighbours and two more at a considerably greater distance completing a distorted octahedral group. The Cu atoms are not vertically above one another, as implied by the elevation of Fig. 4.31(a) but are displaced a little off the axis of the chain of octahedra, actually in a helical array around a 6-fold screw axis (Fig. 4.32).

The structure (b), with cubic close-packed $A + 3X$ atoms, is the perovskite structure. This is illustrated in Fig. 4.33 with an A atom at the origin showing part of a c.p. AX_3 layer. It is more easily visualized as a 3D system of vertex-sharing BX_6 octahedra having B atoms at the corners of the cubic unit cell and X atoms midway along the edges, that is, as the ReO_3 structure of Fig. 5.18(a) (p. 210) with the A atom added at the body-centre of the cell. The ideal (cubic) perovskite structure (or a variant with lower symmetry) is adopted by many fluorides, ABF_3 and complex oxides ABO_3; it is discussed in more detail in Chapter 13.

The structure (c) is adopted by the high-temperature form of $BaMnO_3$. The MnO_6 octahedra are grouped in pairs with a face in common; the pairs are linked together by sharing vertices.

The 6-layer structure (d), with single and also face-sharing pairs of BX_6 octahedra, is that of hexagonal $BaTiO_3$ and $CsMnF_3$.

The basic structures (a)–(d) can be utilized by compounds with more complex formulae in a number of ways. The complex halide with empirical formula $CsAuCl_3$ is actually $Cs_2Au^IAu^{III}Cl_6$. It has a very distorted perovskite structure in which there are linear $(Cl-Au^I-Cl)^-$ and square planar $(Au^{III}Cl_4)^-$ ions; it is described in Chapter 10. The fluoride K_2LiAlF_6 has two enantiotropic forms. In the low-temperature form the Li^+ and Al^{3+} ions occupy in a regular manner the B positions of the perovskite structure, (b); this superstructure is the cryolite structure. The high-temperature form is a superstructure of the simple 6-layer structure (d). Elevations of these two superstructures are shown in Fig. 4.31(e) and (f). A much more complex variant of the 6-layer structure may be mentioned here though it is not built exclusively of c.p. AX_3 layers. In $BaTi_5O_{11}$ the layers are of two kinds, some consisting entirely of oxygen (O_8 in the unit cell) and others in which one-eighth of

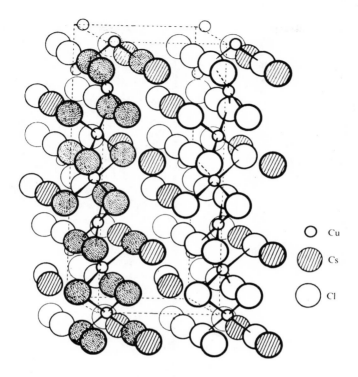

FIG. 4.32. The crystal structure of $CsCuCl_3$.

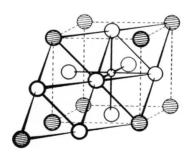

FIG. 4.33. The perovskite structure of $RbCaF_3$ showing a Ca^{2+} ion (small circle) surrounded by six F^- ions and a layer of close-packed Rb^+ and F^- ions (large shaded and open circles respectively).

the O atoms are replaced by Ba (giving the composition BaO_7). There is room for 4 Ti between a BaO_7 and an O_8 layer but only for 2 octahedrally coordinated Ti atoms between BaO_7 layers, so that the structure may be represented diagrammatically

$$(BaO_7 \quad BaO_7 \quad O_8 \quad BaO_7 \quad BaO_7 \quad O_8 \quad BaO_7 \qquad Ba_4 Ti_{20} O_{44}$$
$$2\,Ti \qquad 4\,Ti \quad 4\,Ti \quad 2\,Ti \qquad 4\,Ti \quad 4\,Ti \quad 2\,Ti \qquad =4\,(BaTi_5O_u)$$

References to some of the structures mentioned are included in the summarizing Table 4.4. (p. 159); others are included in later chapters.

$A_3B_2X_9$ *structures*

In these structures B atoms occupy two-thirds of the X_6 holes. Elevations of structures with not more than six layers in the repeat unit are shown in Fig. 4.31(g)–(i). The structures (g) and (i) contain complex ions B_2X_9 consisting of two BX_6 octahedra sharing a face. The simplest structure of this type is the h.c.p. structure (g), of $Cs_3Fe_2F_9$ and $Cs_3Ga_2F_9$, which results from removing every third B atom from the structure (a); removal of alternate B atoms gives the K_2GeF_6 structure containing discrete BX_6 ions, $Cs_3Tl_2Cl_9$ has a closely related structure (with hexagonal closest packing of $A + 3\,X$) but a more uniform spatial distribution of the $Tl_2Cl_9^{3-}$ ions. In Fig. 4.34 the broken lines indicate the unit cell to which the structures of Fig. 4.31

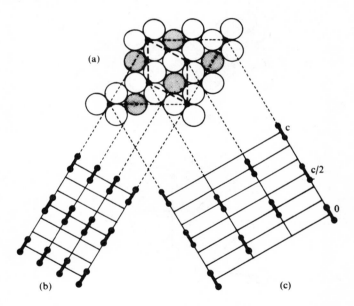

FIG. 4.34. The crystal structure of $Cs_3Tl_2Cl_9$ (see text).

can be referred. In structure (g) the $Tl_2Cl_9^{3-}$ ions would lie on lines (perpendicular to the plane of the paper) through all the small black circles and would be at the same heights on each line, as in the elevation (b). The more uniform distribution shown in the elevation (c) is that found in $Cs_3Tl_2Cl_9$. The same structure is adopted by $Ba_3W_2O_9$. Structure (i) is that of $K_3W_2Cl_9$. The structure (h) consists of

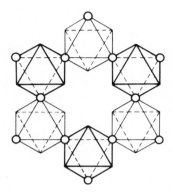

FIG. 4.35. Octahedral A_2X_9 layer.

corrugated layers, shown in plan in Fig. 4.35, formed from octahedra sharing three vertices with three other octahedra. The only example at present known of this structure is $Cs_3Bi_2Br_9$.

In addition to these types of $A_3B_2X_9$ structure, which contain finite or infinite 2D B_2X_9 ions, there are structures of a third kind in which the B atoms are displaced from the centres of their BX_6 octahedra to form discrete pyramidal molecules BX_3.

TABLE 4.10

Structures based on close-packed AX_3 layers

Layer sequence	Symbol	Fraction of octahedral X_6 holes occupied by B atoms		
		All ABX_3	$\frac{2}{3}$ $A_3B_2X_9$	$\frac{1}{2}$ A_2BX_6
AB ...	h	CsNiCl$_3$ BaNiO$_3$ CsCuCl$_3$† (a)‡	Cs$_3$Fe$_2$F$_9$ (g)	K$_2$GeF$_6$ Cs$_2$PuCl$_6$ (j)
ABC ...	c	Oxides ABO$_3$ RbCaF$_3$ CsAuCl$_3$† (b) (e)	Cs$_3$Bi$_2$Br$_9$ (h)	K$_2$PtCl$_6$ (k)
$ABAC$ ...	hc	High-BaMnO$_3$ (c)	§	K$_2$MnF$_6$ (l)
$ABCACB$ ...	hcc	BaTiO$_3$ (hexag.) CsMnF$_3$ (d) (f)	K$_3$W$_2$Cl$_9$ (i)	— (m)

† Distorted variants of ideal c.p. structure.
‡ The inset letters refer to Fig. 4.31.
§ There is no $A_3B_2X_9$ structure with $ABAC$ ... packing.

Examples include $Cs_3As_2Cl_9$, $Cs_3Fe_2Cl_9$, $Cs_3Sb_2Cl_9$, and $Cs_3Bi_2Cl_9$. References to these structures are given on p. 466.

A_2BX_6 *structures*

These arise by filling one-half of the available X_6 holes in the c.p. stacking of AX_3 layers. Elevations of the structures with *h, hc,* and *hcc* layer sequences are shown in Fig. 4.31(j)–(m). In all these structures there are discrete BX_6 ions. Numerous examples of the first three structures, the trigonal K_2GeF_6 (Cs_2PuCl_6), the cubic K_2PtCl_6, and the hexagonal K_2MnF_6 structures, are given in Chapter 10, where the first two structures are illustrated. The structures of complex halides and oxides ABX_3, $A_3B_2X_9$, and A_2BX_6 are summarized in Table 4.10.

This account of these structures has been based on the mode of packing of the c.p. layers. For the alternative description in terms of the way in which the octahedral BX_6 groups are joined together by sharing X atoms, see Chapter 5. The structures based on cubic closest-packing are normally illustrated and described in terms of the cubic unit cell. We noted earlier that the cryolite structure is a superstructure of perovskite; the relation between the perovskite, cryolite, and K_2PtCl_6 structures is described in Chapter 10 (p. 460).

5

Tetrahedral and octahedral structures

Structures as assemblies of coordination polyhedra

Illustrations of crystal structures in which X atoms are bonded to more than one X atom may be simplified by showing how the coordination groups AX_n are joined together by sharing X atoms. The two most important coordination polyhedra are the tetrahedron and octahedron, and since these will appear in various orientations in illustrations of structures it is important that they should be recognized when viewed in different directions. Figure 5.1 shows projections of a tetrahedron and an

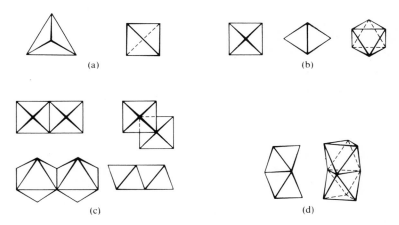

FIG. 5.1.   (a) Tetrahedron viewed along an axis of 3 or $\overline{4}$ symmetry. (b) Octahedron viewed along a 4-, 2-, or 3-fold symmetry axis. (c) Various projections of a pair of octahedra sharing an edge. (d) Projection and perspective view of a pair of octahedra sharing a face.

octahedron and also projections of a pair of octahedra sharing one edge or one face. In this chapter we shall describe the more important structures that may be built from these two polyhedra by sharing vertices, edges, or faces, or combinations of these polyhedral elements. With regard to the sharing of X atoms between AX_n coordination groups we shall adopt two conventions. The first is that if an edge is shared its two vertices are not counted as shared vertices, and similarly the three edges and three vertices of a shared face are not counted as shared edges or vertices. The second facilitates the description and classification of polyhedral structures. Consider the sharing of edges between tetrahedral groups. If two 'opposite' edges are shared as in Fig. 5.2(a), where a tetrahedron is viewed along an S_4 ($\overline{4}$) axis, each tetrahedron is joined to two others by edge-sharing and each X atom is common to

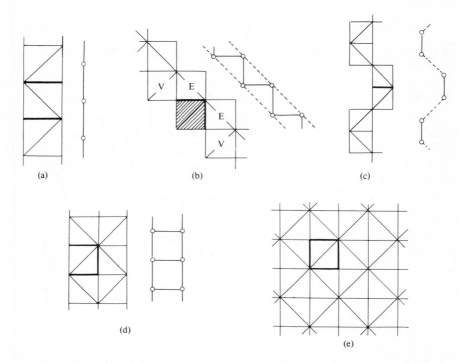

FIG. 5.2. The sharing of edges and vertices between tetrahedral coordination groups (see text). In (a)–(d) the topological diagrams at the right show how the tetrahedra are connected by edge- or vertex-sharing (full and broken lines respectively). The shared edges of one tetrahedron are shown as heavy lines.

only two tetrahedra. If, however, the two shared edges have a common vertex, as in the chain (b), a tetrahedron such as the one shaded shares X atoms not only with two others (E) through edge-sharing but also with two (V) by vertex-sharing, with the result that three X atoms of each tetrahedron are common to three tetrahedra (that is bonded to three A atoms). This vertex-sharing is incidental to the edge-sharing and does not affect the topology, which in both (a) and (b) is represented by a (2-connected) chain of A atoms. Alternatively, we could say that the edge-sharing corresponds to the shortest A–A distances and the vertex-sharing to the longer distances shown by the broken lines in (b). Similar considerations apply to structures in which more than two edges are shared, and also, of course, to systems of connected octahedra. We shall find it convenient to refer to structures such as (b) as edge-sharing structures, in spite of the sharing of vertices. On the other hand, if the sharing of vertices is *not* incidental to the sharing of edges as, for example, in (c), we shall refer to vertex *and* edge-sharing. This distinction is made in Table 5.2 for tetrahedral structures and in Table 5.4 for octahedral structures.

Our main concern in this chapter is the description of known structures. These are most easily visualized as resulting from the sharing of vertices, edges, and/or faces between the tetrahedral or octahedral groups, and this is the natural approach to the building of models, the study of which is essential to a real understanding of structures. However, it is of considerable interest to know which structures of a particular type are geometrically possible, even though examples may not be known, for it would be desirable to know why certain structures are not found. For a general survey of the indefinitely large number of possible tetrahedral or octahedral structures a more satisfactory approach is based on the coordination numbers of the X atoms. Consider the simplest structures, in which all AX_4 (or AX_6) groups are topologically equivalent, that is, they share X atoms in the same way. Tetrahedral structures may be derived systematically by starting with solutions of the equations $\Sigma v_x = 4$ and $\Sigma v_x/x = n/m$, where v_x is the number of X atoms of each AX_4 group which are common to x such groups in a compound of composition $A_m X_n$, that is, x is the coordination number of X. For example, for tetrahedral AX_2 structures the solutions are[1]

v_1	v_2	v_3	v_4
	4		
1		3	
1	1		2

We then examine how each of these solutions can be realized by vertex-, edge-, or face-sharing. For the class $v_2 = 4$, for example, the 2-coordination of each X atom can result from the sharing of (a) each vertex with a different tetrahedron; (b) one edge and two vertices; or (c) two edges which have no common vertex. The other solutions are analysed in a similar way. For octahedral structures the corresponding equations are $\Sigma v_x = 6$ and $\Sigma v_x/x = n/m$.

In the following pages the treatment will be essentially topological, that is, we shall be primarily concerned with the way in which the coordination polyhedra are connected together rather than with the detailed geometry of the systems. However, interatomic distances and interbond angles are of great interest to the structural chemist though they are usually discussed without reference to certain basic geometrical limitations which we shall examine first.

In descriptions of the structures of ionic crystals it is usual to point out that the sharing of edges and more particularly of faces of coordination polyhedra AX_n implies repulsions between the A atoms which lead, for example, to shortening of shared edges and to the virtual absence of face-sharing in essentially ionic structures. There are also, however, purely geometrical limitations on the interbond angles A–X–A at shared X atoms even if only vertices are shared, and these are clearly relevant to discussions of such angles in crystalline trifluorides with the ReO_3 or RhF_3 structures or in the cycle tetramers M_4F_{20} of certain pentafluorides.

(1) AC 1983 B **39**, 39

Limitations on bond angles at shared X atoms

It is convenient to consider first the angle A—X—A at an X atom shared between two regular tetrahedra or octahedra. This atom may be a shared vertex or it may belong to a shared edge or face, so that there are six cases to consider. If tetrahedra or octahedra share a face the system is invariant and the angle A—X—A has the value F in Fig. 5.3(a) or (b). For edge- or vertex-sharing there is in each case a *maximum*

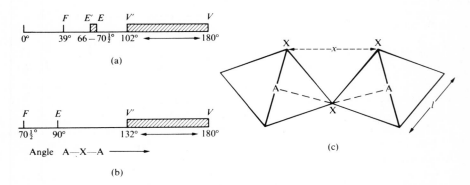

(a)

(b)

(c)

FIG. 5.3. Angles A—X—A for (a) tetrahedra AX_4 or (b) octahedra AX_6 sharing a face (F), an edge (E), or a vertex (V). (c) Restriction on distance between X atoms of different polyhedra (see text). The values given for certain of these angles are approximate. More precise values of those which are not directly derivable from the geometry of the polyhedra are: $E' = 65°58'$ $(\cos^{-1} 11/27)$; $V'_{tetr.} = 102°16'$ $[\cos^{-1} (-17/81)]$; and $V'_{oct.} = 131°48'$ $(2 \sin^{-1}\sqrt{(5/6)})$.

value of the angle A—X—A (points E and V for edge- and vertex-sharing respectively) corresponding to the fully extended systems with maximum separation of the A atoms (centres of polyhedra). However, there is freedom to rotate about a shared edge or vertex which reduces this distance and also reduces the distance (x) between X atoms of different polyhedra (Fig. 5.3(c)). The *minimum* value of the angle A—X—A therefore depends on the lower limit placed on x. If we suppose that x will not be less than the X—X distance in AX_4 (or AX_6), that is, the edge-length l, we calculate the lower limits shown in Fig. 5.3 as E' and V'. In the tetrahedral case the range EE' is small and we shall neglect it.

From the A—X—A angles we may calculate the distance between the centres of the polyhedra (A—A) for face-, edge-, or vertex-sharing. These distances may be expressed in terms of either the edge-length X—X or the distance A—X from centre to vertex (bond length); both sets of values are given in Table 5.1. They illustrate the increasingly close approach of A atoms as edges or faces are shared; sharing of faces by tetrahedral coordination polyhedra does not occur, and sharing of faces by octahedral coordination groups is confined to a relatively small number of structures.

The angular range of A—X—A in Fig. 5.3(b) corresponds closely to the observed —F— bond angles in trifluorides of transition metals with ReO_3- or RhF_3-type

TABLE 5.1

The distance (A—A) between centres of regular AX_4 *or* AX_6 *groups sharing X atoms*

	In terms of X—X			In terms of A—X		
	Vertex†	Edge†	Face	Vertex†	Edge†	Face
Tetrahedron	1·22	0·71	0·41	2·00	1·16	0·67
Octahedron	1·41	1·00	0·82	2·00	1·41	1·16

† Maximum value

structures and in cyclic M_4F_{20} molecules of transition-metal pentafluorides. In the MnF_4 layer of $BaMnF_4$ values of the angle Mn—F—Mn close to both the extreme values of Fig. 5.3(b), namely, 139° and 173°, are observed for the two kinds of non-equivalent F atom (Fig. 5.15(d), p. 207). Of more interest is the gap between 90° and 132°, from which the following conclusions may be drawn:

(i) In the rutile structure there cannot be both regular octahedral coordination of A and planar equilateral coordination of X, for the latter would require the angle A—X—A to be 120°.

(ii) In the corundum structure (α-Al_2O_3) there cannot be both regular octahedral coordination of Al and also regular tetrahedral coordination of O; the value 109½° occurs in the gap between E and V' (Fig. 5.3(b)). On this point see also p. 266.

(iii) The fact that the minimum value of A—X—A for vertex-sharing is greater than 120° implies that if three (or more) octahedra meet at a point at least one *edge* (or *face*) must be shared. We shall see that edge-sharing is a feature of many octahedral structures and arises in structures of compounds AX_n if $n \leqslant 2$ for this purely geoemetrical reason; it *may*, of course, occur if $n > 2$.

It should be emphasized that the validity of conclusion (iii) rests on our assumptions (a) that the distance of closest approach of X atoms of different vertex-sharing octahedra (x in Fig. 5.3(c)) may not be less than the distance X—X within an octahedron, and (b) that the octahedra are regular. Structures in which three octahedra share a common vertex but no edges or faces contravene (a) and/or (b).

(a) If x is less than the edge length X—X it is, of course, possible for three octahedra to meet at a point with only a vertex in common. This situation arises in the trimeric chromium oxyacetate ion in $[OCr_3(CH_3COO)_6(H_2O)_3]Cl.6H_2O^{(1)}$, in which the acetato groups bridge six pairs of vertices (Fig. 5.4).* The shortest O—O distances within the octahedral CrO_6 groups lie in the range 2.63–2.90 Å, but O—O in the acetato group is only 2.24 Å. In the ion $[OFe_3(SO_4)_6(H_2O)_3]^{5-\,(2)}$, which has a similar structure, the bridges are formed by six SO_4^{2-} ions. A somewhat similar problem arises in certain structures in which octahedral coordination groups share *faces*, for which see p. 229.

*In this chapter references to the literature are given only for compounds which are not described in other chapters.

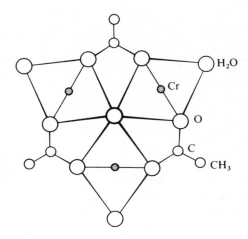

FIG. 5.4. The ion $[OCr_3(OOC.CH_3)_6(H_2O)_3]^+$.

(b) It might have been expected that there would be a simple 3D AX_2 structure in which AX_6 octahedra share only vertices and each vertex is common to three octahedra — contrast the rutile structure, in which each AX_6 shares *edges* and *vertices*. Such a structure is not possible unless there is appreciable distortion of the octahedra and/or close contacts between vertices of different octahedra; in this connection see the discussion of the structures of AgF_2, PdS_2, and HgO_2 on p. 275.

(1)   AC 1970 **B26** 673   (2)   AC 1980 **B36** 1278

The maximum number of polyhedra with a common vertex

The maximum number of *regular* tetrahedra that can meet at a point is eight, and the maximum number of *regular* octahedra that can meet at a point is six, assuming in each case that the distance between X atoms of different polyhedra is not less than the edge-length of the polyhedron. (If the sharing of edges or faces is not permitted, the numbers are 4 and 2.) The numbers 8 and 6 are, of course, the numbers of tetrahedral and octahedral interstices surrounding a sphere in a closest packing of equal spheres. Accordingly there are two arrangements of six regular octahedra AX_6 (plus eight regular tetrahedra AX_4) meeting at a common vertex, corresponding to h.c.p. and c.c.p. X atoms (Fig. 4.14(d) and (e)). In cubic closest packing the eight tetrahedral holes are separated by the six octahedral holes, but in hexagonal closest packing six of the tetrahedral holes form three face-sharing pairs. These three larger 'double tetrahedral' holes offer the possibility of inserting three more octahedra with the central X atom as common vertex, and hence the possibility of having up to nine (suitably distorted) octahedra meeting at a common vertex. Of these nine octahedra six would share three faces and two edges and three would share four faces with other octahedra of the group of nine octahedra.

These considerations are relevant to the existence of structures A_2X_3 of $9:6$ coordination and structures A_3X_4 of $8:6$ coordination, in which respectively nine and eight XA_6 groups would meet at each A atom (assuming equivalence of all A and of all X atoms). No structure of the former type is known, but Th_3P_4 is of the latter type. Evidently regular octahedral coordination of X is not possible, and moreover there must be considerable face-sharing between the XA_6 coordination groups. Accordingly we find that in Th_3P_4 one-half of the (distorted) octahedral PTh_6 groups share two faces (and three edges) and the others share three faces (and one edge) with other octahedra meeting at the common Th atom. It should be emphasized that the numbers of shared edges and faces given here are those shared with other octahedra belonging to the group of six, eight, or nine octahedra meeting at the common vertex; the numbers of edges and faces shared by each octahedron in the actual crystal structure will, of course, be greater. Some other consequences of these theorems are discussed in later chapters. They have a bearing on the stability (and therefore existence) of certain oxy-salts (p. 327), nitrides (p. 277), and cation-rich oxides (p. 329).

We now survey structures built from tetrahedra and octahedra. A survey of the kind attempted in this chapter not only emphasizes the great number and variety of structures arising from such simple building units as tetrahedra and octahedra but also draws attention to:

(a) the unexpected complexity of the structures of certain compounds for which there are geometrically simpler alternatives. For example, the unique layer structure of MoO_3 may be contrasted with the topologically simpler ReO_3 structure;

(b) the lack of examples of compounds crystallizing with certain relatively simple structures, which emphasizes the fact that the *immediate* environments of atoms are not the only factors determining their structures — note the non-existence of halides AX_2 with the atacamite structure;

(c) the large number of geometrically possible structures for a compound of given formula type (for example, AX_4), knowledge of which is necessary for a satisfactory discussion of any particular *observed* structure;

(d) the importance of the coordination group around the *anion*.

We may perhaps remind the reader at this point of the general types of connected system which are possible when any unit (atom or coordination group) is joined to a number (p) of others:

$$p = 1 \quad \text{dimer only;}$$
$$p = 2 \quad \text{closed ring or infinite chain;}$$
$$p \geqslant 3 \quad \text{finite (polyhedral), 1-, 2-, or 3D systems.}$$

In Chapter 3 we observed that although 3- or 4-connected points can form 3D structures they may also form systems extending only in one or two dimensions. Some of the less obvious 1D systems were illustrated in Fig. 3.19, p. 101.

Tetrahedral structures

The most numerous and most important tetrahedral structures are those in which only vertices are shared. The sharing of faces of tetrahedral groups AX_4 would result in very close approach of the A atoms (to 0.67 AX or 0.41 XX, where XX is the length of the tetrahedron edge) and a very small A–X–A angle ($38°56'$ for regular tetrahedra) and for these reasons need not be considered.

Vertex-sharing structures

When a vertex is shared between two tetrahedra the maximum value of the A–A separation is 2 AX, for collinear A–X–A bonds, but a considerable range of A–X–A angles ($102°16'$ - $180°$) is possible, consistent with the distance between X atoms of different tetrahedra being not less than the edge-length of the tetrahedron. Observed angles in oxy-compounds are mostly in the range 130–150°, though collinear O bonds apparently occur in ZrP_2O_7 and $Sc_2Si_2O_7$.

In most structures each X atom is common to only two tetrahedra, and if all the tetrahedra are topologically equivalent (that is, share the same number of vertices in the same way) the formulae are A_2X_7, AX_3, A_2X_5, and AX_2 according to whether 1, 2, 3, or 4 vertices are shared (Table 5.2). The first group includes Cl_2O_7 and the pyro-ions $S_2O_7^{2-}$ etc. and the second group contains cyclic and infinite linear molecules $(AX_3)_n$ and the meta-ions of the same types. The AX_3 chain is illustrated in Fig. 5.5(a). If three vertices of each tetrahedron are shared the possible structures include finite polyhedral groups such as P_4O_{10} (Fig. 3.1, p. 64) and infinite systems of which the simplest is the double chain of Fig. 5.5(b). A cylindrical chain formed from a strip of the plane 4.8^2 net wrapped around a cylinder (see p. 101) is found in $CuNa_2(Si_4O_{10})$, p. 1026. Examples of 2D and 3D systems are included in Chapter 3 under 3-connected nets. Tetrahedra sharing four vertices could also form cylindrical chains, of which an example is shown in Fig. 5.5(c), in addition to layers, double

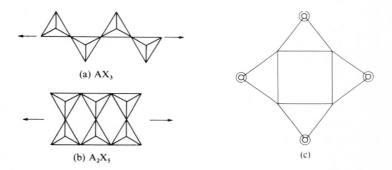

(a) AX_3

(b) A_2X_5

(c)

FIG. 5.5. Portions of infinite chains in which all tetrahedra share (a) two, (b) three, or (c) four vertices. In (c) the chain is viewed along its length.

TABLE 5.2

Tetrahedral structures

	Formula	Type of complex	Examples
Number of shared vertices	*Vertex-sharing* *Vertices common to two tetrahedra*		
1	A_2X_7	Finite molecule or pyro-ion	$Cl_2O_7 . S_2O_7^{2-}$, etc.
2	$(AX_3)_n$	Cyclic molecule or meta-ion, infinite chain	S_3O_9, $Se_4O_{12} . (PNCl_2)_n$ $(P_4O_{12})^{4-}$, $(Si_3O_9)^{6-}$, $(SO_3)_n$, $(PO_3)_n^{n-}$
3	$(A_2X_5)_n$	Finite polyhedral, Double chain, Cylindrical chain Layer 3D structure	P_4O_{10} $Al(AlSiO_5)$ $CuNa_2(Si_4O_{10})$ P_2O_5, $Li_2(Si_2O_5)$, P_2O_5, $La_2(Be_2O_5)$
4	$(AX_2)_n$	Layer, Double layer, 3D structure	HgI_2 (red) $CaSi_2Al_2O_8$ (hexag.) SiO_2 structures, GeS_2
	Vertices common to three tetrahedra		
3	$(AX_2)_n$	Infinite layer	AlOCl, GaOCl
Number of shared edges	*Edge-sharing* *Edges common to two tetrahedra*		
1	A_2X_6	Finite dimer	Al_2Cl_6, Fe_2Cl_6
2	$(AX_2)_n$	Infinite chain	$BeCl_2$, SiS_2, $Be(CH_3)_2$
3	$(A_2X_3)_n$	Infinite double chain	$Cs(Cu_2Cl_3)$
4	$(AX)_n$	Infinite layer	LiOH, PbO
5	$(A_4X_3)_n$	Infinite double layer	KCu_4S_3
6	$(A_2X)_n$	3D structure	Li_2O, F_2Ca
	Vertex- and edge-sharing		
	$(AX_2)_n$	Layer	$GaPS_4$
	$(AX)_n$	3D structure	β-BeO

layers, and 3D structures. Some layer and framework structures are described in Chapter 3 under plane and 3D 4-connected nets. Two configurations of the simple AX_2 layer formed from tetrahedra sharing four vertices are shown in Fig. 5.6. In $SrZnO_2$ (Fig. 5.6(b)) the layer is the anion, and the Sr^{2+} ions are accommodated between the layers.

A multiple unit ('super-tetrahedron') consisting of four tetrahedra, each sharing three vertices, has the composition A_4X_{10}. This represents the idealized structure of

P_4O_{10} and similar molecules and of the $Si_4S_{10}^{4-}$ and $Ge_4S_{10}^{4-}$ ions (p. 1173), and is also the structural sub-unit in several other structures. Replacement of the simple tetrahedra in Fig. 5.6(a) by super-tetrahedra gives a layer of composition A_4X_8

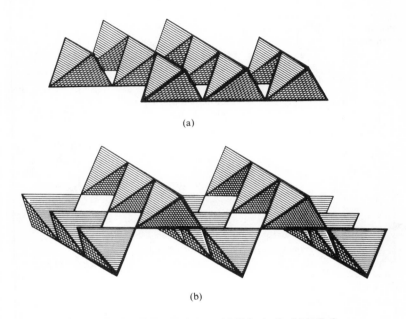

(a)

(b)

FIG. 5.6. Tetrahedral AX_2 layers: (a) HgI_2 (red); (b) $SrZnO_2$.

(AX_2), each of the four outermost X atoms being common to two A_4X_{10} groups. One form of HgI_2 consists of layers of this kind. Linking of the A_4X_{10} groups in this way can also give a 3D framework (compare the cristobalite structure), as in ZnI_2 (p. 412). If super-tetrahedra Be_4O_{10} are joined together in the same way as the ZnS_4 tetrahedra in sphalerite each of the four outermost O atoms becomes common to four such groups and the composition of the 3D framework is Be_4O_7. The O atoms occupy 7/8 of the positions of cubic closest packing, and if Te atoms occupy certain of the octahedral interstices this is the structure of Be_4TeO_7 (p. 737).

For SiO_2 structures and aluminosilicates see Chapter 23.

The examples of Table 5.2 are restricted to systems in which all the tetrahedra are topologically equivalent. In finite linear systems consisting of more than two tetrahedra the terminal AX_4 groups share only one vertex but the intermediate tetrahedra share two vertices. In such hybrid systems the X : A ratio lies between 3½ and 3, as in the $P_3O_{10}^{5-}$ and $S_3O_{10}^{2-}$ ions. Similarly there are chains in which some tetrahedra share two and others three vertices, as in the double-chain ions with X : A ratios between 3 and 2½ found in some silicates (Fig. 5.7(a) and (b)) and in the

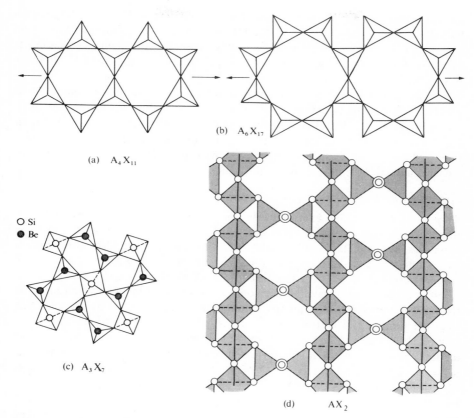

FIG. 5.7. (a) and (b) Portions of infinite chains in which some tetrahedra share two and others three vertices. (c) Portion of A_3X_7 (melilite) layer in which some tetrahedra share three and other four vertices. (d) Portion of a layer in h.t. GeS_4 in which equal numbers of GeS_4 tetrahedra share four vertices or two vertices and one edge. A double circle indicates a shared edge normal to the plane of the paper (diagrammatic).

$(P_5O_{14})_n^{3n-}$ chain (p. 863). The X:A ratio falls to values between 2½ and 2 if some of the tetrahedra share three and the remainder all their vertices, as in the layer of Fig. 5.7(c). This interesting layer is found in a number of compounds isostructural with melilite. $Ca_2MgSi_2O_7$ ($Ca_2SiAl_2O_7$, $Ca_2BeSi_2O_7$, $Y_2SiBe_2O_7$). In the A_3X_7 layer the tetrahedrally coordinated atom is Si, Al, Be, or Mg, and the larger Ca or Y ions are in positions of 8-coordination (distorted antiprism). This layer is based on one of the (3,4)-connected pentagon nets mentioned on p. 83. For a more complex A_3X_7 layer see $Na_2Si_3O_7$ (p. 1026).

Because SiO_4 tetrahedra can share any number of vertices from none, in the orthosilicate ion $(SiO_4)^{4-}$ to four, in SiO_2, the extensive oxygen chemistry of silicon provides examples of all types of tetrahedral structures. Their variety is considerably increased owing to the fact that Al may replace Si in some of the tetrahedra, leading to the aluminosilicates which account for most of the rock-forming minerals and soils. The only element rivalling silicon in this respect is germanium (the eka-silicon of Mendeleef) which has been shown to form oxy-salts analogous to all the families of silicates (or aluminosilicates).

In the structures we have considered above a shared vertex is common to two tetrahedra only. One structure in which each shared vertex is common to three tetrahedra is that of AlOCl (and the isostructural GaOCl); the layer is illustrated in Fig. 10.16 (p. 486). If we describe the ZnS structures as built from ZnS_4 tetrahedra then each vertex (S atom) is common to four such groups, but the description as 4-connected nets is simpler, as noted in Chapter 3.

Edge-sharing structures

We noted on p. 188 that except for the special case of the sharing of two edges to form a linear chain there are also vertices shared with other tetrahedra in these structures, but that this is incidental to the edge-sharing. When edges of tetrahedral AX_4 groups are shared the maximum A–A separation is 1.16 AX (or 0.71 XX), and a small variation in A–X–A is possible ($66°$–$70\frac{1}{2}°$) if we assume that X atoms of different tetrahedra may approach only as closely as within a tetrahedron. Edge-sharing between tetrahedral groups is not found in the more ionic crystals other than the fluorite and antifluorite structures, which may be described in terms of edge-sharing FCa_4 and LiO_4 groups. Here the edge-sharing is an unavoidable feature of the geometry of the 4 : 8 coordinated structure. Examples of edge-sharing tetrahedral structures are included in Table 5.2. They include the dimeric molecules A_2X_6 of certain halides (other than fluorides) in the vapour state and in some crystals, the corresponding infinite chain formed by sharing of opposite edges of each tetrahedron ($BeCl_2$ etc.), the double chain in a number of complex halides ($CsCu_2Cl_3$), and the layers in crystalline LiOH and PbO; the latter are described in Chapter 3 under plane 4-connected nets and also on p. 162 as examples of close-packed tetrahedral layers.

The family of edge-sharing tetrahedral structures in which from one to six edges are shared may all be regarded as portions of, or formed from, the layer shown diagrammatically in Fig. 5.2(e) in which each tetrahedron shares four edges. Any pair of tetrahedra represents the A_2X_6 molecule (e.g. of Al_2Cl_6), a strip of the layer is the AX_2 chain (Fig. 5.2(a)), and a strip two tetrahedra in width is the A_2X_3 double chain (d), the topological representation of which is the (3-connected) ladder. In the layer (e) each tetrahedron shares four edges, and an infinite stacking of such layers in which the remaining two edges are shared is the 3D antifluorite (A_2X) structure in which six edges are shared. There is an intermediate structure consisting of a pair of layers in which five edges of each tetrahedron are shared. This double layer has been found in $KCu_4S_3{}^{[1]}$, a compound with high metallic conductivity ('2D metal'), the double layers being held together by K atoms (ions). In the A_4X_3 layer the inner X atoms are bonded to 8 A and the outer ones to 4 A, and there are two of each in each AX_4 tetrahedron. There is another double layer which corresponds to Fig. 5.2(b), the layer being normal to the plane of the paper. In this layer each tetrahedron shares four edges; the inner X atoms are bonded to 6 A and the outer ones to 2 A atoms. Since there are three of the first type in each tetrahedron the composition is AX. No example appears to be known of this double layer, which would be more suitable for a composition $A_2X'X''$; compare $A_4X'X_2''$ for the A_4X_3 layer.

(1)   IC 1980 **19** 1945

Vertex- and edge-sharing structures

The sharing of vertices *and* edges of tetrahedral groups in a given structure is rare. We noted on p. 167 the structure of $GaPS_4$ as consisting of pairs of c.p. layers of S atoms between which Ga and P together occupy one-half of the tetrahedral interstices. Each GaS_4 or PS_4 group shares two vertices and one edge with a tetrahedron of the other kind. An example of a 3D structure is that of the high-temperature form of BeO, in which pairs of edge-sharing tetrahedra are further linked into a 3D structure by sharing the four remaining vertices. In these two structures all the tetrahedra are topologically equivalent, and the structures are therefore included in Table 5.2. In $Ba_7Fe_6S_{14}$[1] there are chains formed from tetrahedra of two kinds, one-third sharing a pair of opposite edges and the remainder one vertex and one edge. The high-

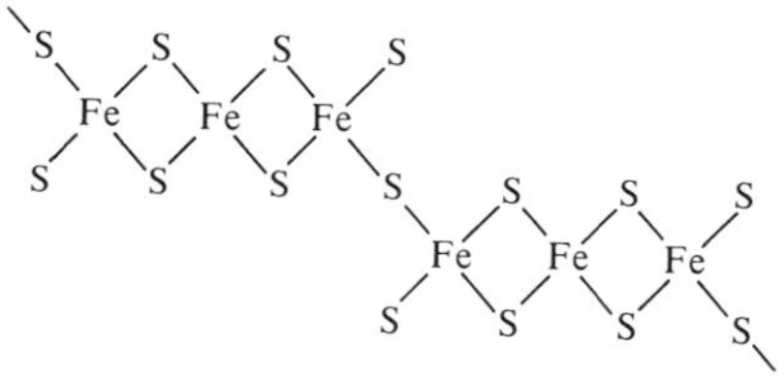

temperature form of GeS_2 consists of layers in which equal numbers of tetrahedra share four vertices or two vertices and one edge; it is illustrated in Fig. 5.7(d).

(1)   IC 1971 **10** 340

Octahedral structures

The greater part of this survey of octahedral structures will be devoted to systems which extend indefinitely in one, two, or three dimensions. However, in contrast to the limited number of types of finite complex built from tetrahedral groups the number of finite molecules and complex ions formed from octahedral units is sufficient to justify a separate note on this family of complexes. In our survey of infinite structures we shall deal systematically with structures in which there is sharing of vertices, edges, faces, or combinations of these elements. We shall not make these subdivisions for finite structures, which will simply be listed in order of increasing numbers of octahedra involved.

Some finite groups of octahedra

Some of the simpler octahedral complexes are illustrated in Fig. 5.8 and some examples are included in Table 5.3. We comment here on only a few of these complexes. An interesting example of a complex of type (a) is the cation in $\{Ca_2[(CH_3)_3AsO]_9\}^{4+}(ClO_4)_4{}^{(1)}$. The nucleus consists of two CaO_6 groups sharing a face, each O atom belonging to a $(CH_3)_3$ AsO molecule. Examples of (b) are numerous. with double halogen, double OH bridges in, for example, $[Al_2(OH)_2(H_2O)_8]^{4+}$ and cobaltammines (p. 1220), double O bridges in $[I_2O_8(OH)_2]^{4-}$ and $[Te_2O_6(OH)_4]^{4-}$, and S and/or halogen bridges in transition metal complexes such as those of Pd and Pt (p. 1236).

The A_3X_{12} and A_4X_{16} complexes represent the structures of the trimeric Ni and

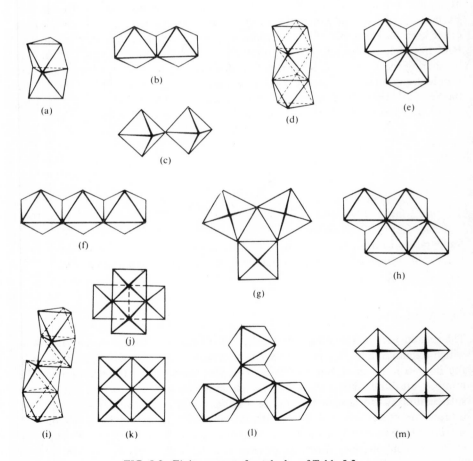

FIG. 5.8. Finite groups of octahedra of Table 5.3.

TABLE 5.3

Finite groups of octahedra

Fig. 5.8	*Formula*	*Examples*
(a)	A_2X_9	$Fe_2(CO)_9$, $W_2Cl_9^{3-}$, $Tl_2Cl_9^{3-}$, $I_2O_9^{4-}$, $Fe_2F_9^{3-}$
(b)	A_2X_{10}	Nb_2Cl_{10}, Mo_2Cl_{10}, U_2Cl_{10}
(c)	A_2X_{11}	$(Nb_2F_{11})^-$, $[(NH_3)_5Co . NH_2 . Co(NH_3)_5]^{5+}$
(d)	A_3X_{12}	$[Ni(acac)_2]_3$, Co_3L_6 (see text)
(e)	A_3X_{13}	$Te_3Cl_{13}^-$
(f)	A_3X_{14}	
(g)	A_3X_{15}	Nb_3F_{15}, Ta_3F_{15}, Sb_3F_{15} (vapour)
(h)	A_4X_{16}	$[Ti(OC_2H_5)_4]_4$, $(TeI_4)_4$. $[Al_4(OH)_{16}]Ba_2$
(i)	A_4X_{16}	$[Co(acac)_2]_4$
(j)	A_4X_{16}	Te_4Cl_{16}, $[Pt(CH_3)_3Cl]_4$
(k)	A_4X_{17}	
(l)	A_4X_{18}	$[Co_4(OH)_6(NH_3)_{12}]^{6+}$
(m)	A_4X_{20}	Mo_4F_{20}, $W_4O_4F_{16}$

More complex groups

Number of octahedra	*Examples*	*Number of octahedra*	*Examples*
6	$Nb_6O_{19}^{8-}$	12	$PW_{12}O_{40}^{3-}$, $H_4Co_2Mo_{10}O_{38}^{6-}$
7	$Mo_7O_{24}^{6-}$	13	$MnNb_{12}O_{38}^{12-}$, $CeMo_{12}O_{42}^{8-}$
8	$Mo_8O_{26}^{4-}$	18	$P_2W_{18}O_{62}^{6-}$
9	$Mn_9O_{32}^{6-}$		
10	$V_{10}O_{28}^{6-}$		

tetrameric Co acetylacetonates (Fig. 5.9); the coordination groups around the metal atoms consist of O atoms of $CH_3CO.CH.CO.CH_3$ ligands. In the Co_3L_6 complex of Table 5.3[2] L represents the ligand $(C_2H_5O)_2PO.CH.CO.CH_3$ shown here. The

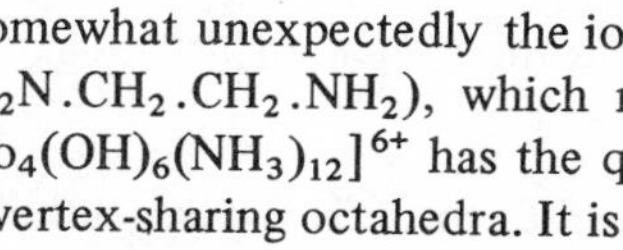

$$\left[\begin{matrix} C_2H_5 . O \diagdown & & CH & & \\ & P & & C & CH_3 \\ C_2H_5 . O \diagup & | & & | & \\ & O & & O & \end{matrix} \right]$$

A_4X_{18} complex of Table 5.3 consists of an octahedral $Co(OH)_6$ group sharing three edges with $Co(OH)_2(NH_3)_4$ groups. This ion is enantiomorphic, and is of special interest as the first purely inorganic coordination complex to be resolved into its optical antimers. Somewhat unexpectedly the ion $[Cr_4(OH)_6en_6]^{6+}$ (where en is ethylene diamine, $H_2N.CH_2.CH_2.NH_2$), which might have had the same structure as the Co ion $[Co_4(OH)_6(NH_3)_{12}]^{6+}$ has the quite different structure of Fig. 5.10[3] with edge- and vertex-sharing octahedra. It is interesting that the simple A_4X_{16} unit of Fig. 5.8(j) is not known as a finite oxy-ion in solution or as a molecule. It has cubic $(\bar{4}3m, T_d)$ symmetry and arises by adding a fourth octahedron below the centre of the unit (e). Each octahedron shares three edges, and the octahedra

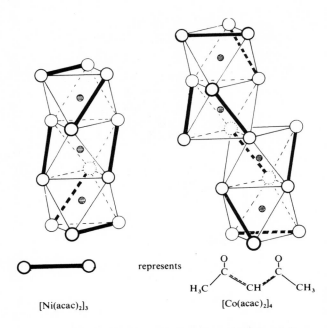

represents

$[Ni(acac)_2]_3$ $[Co(acac)_2]_4$

FIG. 5.9. The molecules $[Ni(acac)_2]_3$ and $[Co(acac)_2]_4$.

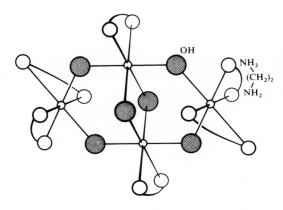

FIG. 5.10. The ion $[Cr_4(OH)_6(en)_6]^{6+}$.

enclose a tetrahedral hole at the centre. This unit does, however, occur in a crystalline tungstate noted on p. 516.

Two 3-octahedra units of type (e) may be joined by sharing one more edge (the broken line in Fig. 5.11) to form the centrosymmetrical 6-octahedra unit which

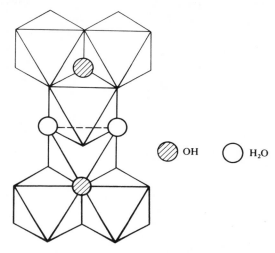

FIG. 5.11. The molecule $Ni_6(CF_3COCHCOCH_3)_{10}(OH)_2(H_2O)_2$.

would have, in the simplest case, the composition A_6X_{24}. The vertices shown in Fig. 5.11 as shaded circles are common to three octahedra. These are OH groups, the open circles represent H_2O molecules, and the remaining twenty vertices are occupied by ten bidentate ligands, $CF_3.CO.CH.CO.CH_3$ (L) in the complex $Ni_6L_{10}(OH)_2(H_2O)_2$[4] — an interesting example of a comparatively complicated formula arising from an essentially simple system of six edge-sharing octahedra.

Of special interest is the 'super-octahedron' A_6X_{19} which is illustrated in Fig. 11.3(a), p. 515, as the structure of the ions $M_6O_{19}^{n-}$, in which M is Nb, Mo, or W. Super-octahedra joined in pairs by Mn atoms form the $MnNb_{12}O_{38}^{12-}$ ion, and they are found as sub-units in the structures of $Cu_{7-x}V_6O_{19-x}$ and $Sn_{10}W_{16}O_{48}$.

Certain elements, notably V, Nb, Ta, Mo, and W, form complex oxy-ions built from larger numbers of octahedral coordination groups. Examples are included in Table 5.3. In the heteropolyacid ions such as $PW_{12}O_{40}^{3-}$ and $P_2W_{18}O_{62}^{6-}$ P atoms occupy the tetrahedral holes at the centres of the complexes. These more complex oxy-ions are described in Chapter 11.

(1) AC 1977 **B33** 931
(2) IC 1968 7 18
(3) JACS 1969 **91** 193
(4) IC 1969 8 1304

Infinite systems of linked octahedra

Of the indefinitely large number of structures that could be built from octahedra sharing vertices, edges, and/or faces, a large number are already known, and a complete account of octahedral structures would cover much of the structural chemistry of halides and chalconides. The number of octahedral structures is large because (a) an octahedron has six vertices, eight faces, and twelve edges, various numbers of each of which can be shared, and (b) there are numerous ways of selecting, for example, a particular small number of edges from the twelve available. Figure 5.12 shows four ways of selecting four edges. Moreover, it is not necessary

(a)   (b)   (c)   (d)

FIG. 5.12. Four ways of selecting four edges of an octahedron.

that all the octahedra in a structure are topologically equivalent, that is, share the same numbers and arrangements of vertices, edges, and/or faces. It may be assumed that all octahedra are topologically equivalent in the simpler systems we shall describe unless the contrary is stated. (Some comparatively simple octahedral complexes containing non-equivalent octahedra include the $[Na(H_2O)_4]_n^{n+}$ chain in borax, in which alternate $Na(H_2O)_6$ octahedra share a pair of opposite or non-opposite edges, and the Al_3F_{14} layer in $Na_5Al_3F_{14}$, in which some AlF_6 octahedra share four and the remainder two vertices.)

It is important to distinguish between the topological equivalence of the octahedra and the equivalence or otherwise of the X atoms (vertices). In the group of four octahedra of Fig. 5.13(a) there are two kinds of non-equivalent octahedron: A

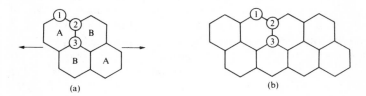

(a)   (b)

FIG. 5.13. Topological equivalence of octahedra (see text).

(sharing two edges), and B (sharing three edges). There are three kinds of non-equivalent X atom, belonging to 1, 2, and 3 octahedra respectively. If this group is extended indefinitely in the directions of the arrows to form the 'double chain' of Fig. 5.13(b) all the octahedra become equivalent but there are still three kinds of non-equivalent X atom belonging, as before, to 1, 2, or 3 octahedra. In some of the simplest structures (for example, rutile, ReO_3) all the X atoms are in fact equivalent, but in MoO_3, for example, there are three kinds of non-equivalent oxygen atoms.

We may distinguish as a special set those structures in which all the octahedra are equivalent and each X atom belongs to the same number of octahedra, two in the

AX_3 and three in the AX_2 structures. The simplest structures of this type are those in which each octahedron shares

(i)	6 vertices with 6 other octahedra	AX_3
(ii)	the 3 edges of Fig. 5.14(a)	AX_3
(iii)	the 6 edges of Fig. 5.14(b) or the 6 edges of Fig. 5.14(c)	AX_2
(iv)	2 opposite faces	AX_3

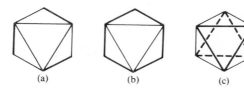

(a)   (b)   (c)

FIG. 5.14. Selection of (a) three, (b) and (c) six edges of an octahedron.

We shall see later that (i) and (ii) correspond to families of related structures while (iii) and (iv) produce a single structure in each case. In addition to these structures in which octahedra share only vertices, edges, *or* faces, there are structures in which vertices *and* edges are shared in which all the octahedra and all the X atoms are equivalent. The rutile structure is a simple example — others are described later.

Octahedral structures may be classified according to the numbers and arrangements of shared vertices, edges, and/or faces as in Table 5.4, which includes only infinite structures. We shall relate our treatment to the main classes of Table 5.4, dealing first with structures in which only vertices or only edges are shared and then proceed to the more complex structures.

Vertex-sharing structures

We noted earlier that the number of *regular* octahedra that can share a common vertex *without sharing edges or faces* is limited to two, assuming that the distance between any pair of non-bonded X atoms of different AX_6 groups is not less than the edge-length, X–X, taken to be the minimum van der Waals distance. We therefore have to deal here only with structures in which each shared vertex is common to two octahedra. For such structures there is a simple relation between the formula and the number of shared vertices (X atoms):

Number of shared vertices	1	2	3	4	5	6
Formula	A_2X_{11}	AX_5	A_2X_9	AX_4	A_2X_7	AX_3

The finite A_2X_{11} group is illustrated in Fig. 5.8(c). Structures in which three or five vertices of every octahedron are shared are few in number. The A_2X_9 layer has been illustrated in Fig. 4.35, p. 185. It will be convenient to describe vertex-sharing A_2X_7 structures after we have dealt with the much more numerous structures in which *even* numbers of vertices are shared.

TABLE 5.4
Infinite structures built from octahedral AX_6 groups

Vertex-sharing

2	AX_5 chains: *cis*: VF_5, CrF_5 *trans*: BiF_5, $(CrF_5)^{2-}$, α-UF_5	3	A_2X_9 layer: $(Bi_2Br_9)Cs_3$

4	AX_4 chain: $(CrF_4)Cs$
	layers:
	    cis: $(ZnF_4)Ba$
	    trans: SnF_4, $(NiF_4)K_2$
	framework: IrF_4
		5	A_2X_7 layer:
			$(Ti_2O_7)Sr_3$

6	AX_3 frameworks: ReO_3, $Sc(OH)_3$, RhF_3, etc., perovskite, W bronzes, pyrochlore

Vertex- and edge-sharing

AX_3, A_3X_8, A_2X_5 layers (V oxyhydroxides)
AX_2 frameworks
   Rutile structure

$\left.\begin{array}{l}\alpha\text{-AlO.OH}\\ Eu_3O_4\\ CaTi_2O_4\\ \alpha\text{-MnO}_2\\ \hline BeY_2O_4\end{array}\right\}$ (Table 5.5)

AX_3 layer: MoO_3
AX_3 framework: $CaTa_2O_6$

Edge-sharing

2	AX_4 chains: $TcCl_4$, NbI_4, ZrI_4
3	AX_3 layer: YCl_3, BiI_3
4	AX_3 double chain: NH_4CdCl_3 AX_3 layer: NH_4HgCl_3 A_3X_8 layer: Nb_3Cl_8
6	AX_2 layer: CdI_2, $CdCl_2$ AX_2 double layer: $MOCl$, γ-$MO.OH$ AX_2 framework: $Cu_2(OH)_3Cl$

Face-sharing

2	AX_3 chain: ZrI_3, $BaNiO_3$, $CsNiCl_3$

Vertex- and face-sharing

ABO_3 structures: hexagonal $BaTiO_3$; high-$BaMnO_3$, $BaRuO_3$ (Table 5.6)

Edge- and face-sharing

Nb_3S_4

Vertex-, edge-, and face-sharing

α-Al_2O_3 (corundum)
γ-$Cd(OH)_2$

Two vertices of an octahedron are either adjacent (*cis*) or opposite (*trans*). Sharing of two *cis* vertices by each octahedron leads to cyclic molecules (ions) or zigzag chains (Fig. 5.15(a) and (b)). Sharing of *trans* vertices could lead to rings of eight or more octahedra (since the minimum value of the angle A—X—A is 132° for vertex-sharing octahedra) but such rings are not known. The simpler possibility (A—X—A = 180°) is the formation of linear chains (Fig. 5.15(c)).

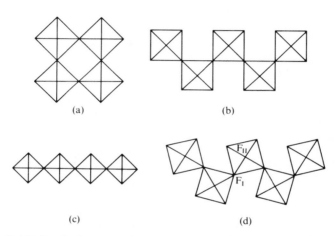

FIG. 5.15. Octahedra sharing *cis* vertices to form (a) cyclic tetramer; (b) the *cis* chain. (c) Octahedra sharing *trans* vertices to form the *trans* (ReO_3) chain. (b) and (c) also represent end-on views (elevations) of the *cis* and *trans* layers. The actual configuration of the *cis* layer in $BaMnF_4$ is shown at (d), where F_I and F_{II} are the two non-equivalent F atoms referred to on p. 191. The bond angle $M-F_I-M$ is 139°; the bonds from F_{II} are approximately perpendicular to the paper ($M-F_{II}-M$, 173°).

The cyclic tetramers in crystals of a number of metal pentafluorides M_4F_{20} are of two kinds, with collinear or non-linear M—F—M bonds:

F bond angle 180°: M = Nb, Ta, Mo, W
132°: M = Ru, Os, Rh, Ir, Pt.

Other pentafluorides form one or other of the two kinds of chain shown in Fig. 5.15:

cis chain: VF_5, CrF_5, TcF_5, ReF_5, $MoOF_4$; $K_2(VO_2F_3)$
trans chain: BiF_5, α-UF_5; $WOCl_4$; $Ca(CrF_5)$, $Tl_2(AlF_5)$.

The factors determining the choice of cyclic tetramer or of one of the two kinds of chain are not understood, and this is true also of the more subtle difference (in —F— bond angle) between the two kinds of tetrameric molecule, a difference which is similar to that between fluorides MF_3 to be noted shortly. The difference between $MoOF_4$ (chain) and WOF_4 (cyclic tetramer) is one of many examples of structural differences between compounds of these two elements (compare the structures of MoO_3 and WO_3, later). These two types (*cis* and *trans*) of AX_5 chain are also found in oxyhalides and in complex halides. More complex types of chain with the same composition can be built by attaching additional octahedra (through shared vertices) to those of a simple AX_5 chain. An example of such a 'ramified' chain is found, together with simple *trans* chains, in $BaFeF_5$, and is illustrated in Fig. 10.2 (p. 454).

The most important systems in which octahedral AX_6 groups each share four vertices are AX_4 layers, but before describing these we should note that the sharing of four vertices can lead to 3D or 1D structures. An example of a 3D structure is that of IrF_4, based on a distorted diamond net, while in $(CrF_4)Cs$ there is a cylindrical chain anion which projects along the chain axis as Fig. 5.8(g). In both these structures the *un*shared vertices are adjacent (*cis*). The next member of the family of cylindrical chains (see p. 101) would be that projecting as Fig. 5.8(m). Corresponding to the two chains formed by sharing two *cis* or *trans* vertices there are layers formed by sharing four vertices, the *un*shared vertices being *cis* or *trans*. If each square in Fig. 5.15(b) and (c) represents a chain of vertex-sharing octahedra perpendicular to the plane of the paper these diagrams are also elevations of the *cis* and *trans* layers. Figure 5.15(d) shows the elevation of the *cis* layer which is the form of the anion in the isostructural salts $BaMF_4$ (M = Mg, Mn, Co, Ni, Zn) and in (triclinic) $BiNbO_4$. In the *trans* layer there is sharing of the four equatorial vertices of each octahedron. This AX_4 layer is alternatively derived by placing AX_6 octahedra at the points of the plane 4-gon net with X atoms at the mid-points of the links. Examples of neutral molecules with this structure include SnF_4, PbF_4, $Sn(CH_3)_2F_2$, and $UO_2(OH)_2$. This layer also represents the structure of the 2D anion in $TlAlF_4$ and in the K_2NiF_4 structure (Fig. 5.16) which is adopted by numerous complex fluorides and oxides (see Chapters 10 and 13). Distorted variants of this structure are adopted by K_2CuF_4 and $(NH_4)_2CuCl_4$, in which two of the equatorial bonds (broken lines in Fig. 5.16(c)) are longer than the other four Cu—X bonds. A fourth structure containing the *trans* layer is that of isostructural salts Ba_2MF_6 (M = Co, Ni, Cu, Zn) which contain additional separate F^- ions, that is, the structural formula is $Ba_2(MF_4)F_2$ (Fig. 5.17).

In the AX_4 layer of K_2NiF_4 the two bonds from X are collinear. There is a characteristic buckled configuration of this layer which is suitable for X atoms forming non-collinear bonds. It is shown in plan and elevation in Fig. 5.16(d) and (e), and represents the layer in $CaCl_2 . 2H_2O$ (p. 687), in which the H_2O molecules are situated at the unshared vertices of the $CaCl_4(H_2O)_2$ coordination groups.

The limit of vertex-sharing is reached when each vertex is shared with another octahedron, giving 3D structures of composition AX_3. This is the first of the very symmetrical octahedral structures listed on p. 205. Since each A atom is connected to six others (through X atoms) the A atoms lie at the points of a 3D 6-connected net, and there is therefore a family of such structures of which the simplest corresponds, in its most symmetrical configuration, to the primitive cubic lattice. A unit cell of the structure is illustrated in Fig. 5.18. A model built of rigid octahedra but with flexible joints at all vertices can adopt an indefinitely large number of configurations. The most symmetrical of these has cubic symmetry and represents the structure of crystalline ReO_3; WO_3 has this structure only at very high temperatures and adopts

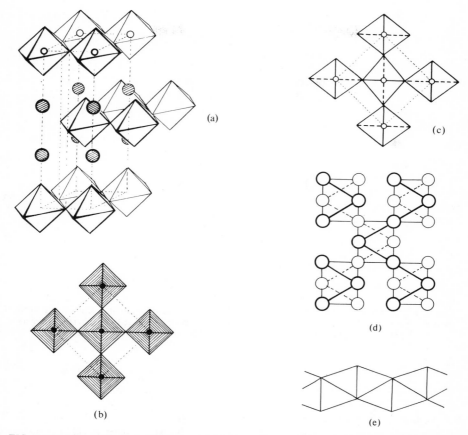

FIG. 5.16. The octahedral AX_4 layer. (a) The K_2NiF_4 structure; (b) plan of layer in K_2NiF_4; (c) plan of layer in $(NH_4)_2CuCl_4$; (d) and (e) plan and elevation of layer in $CaCl_2 \cdot 2H_2O$.

less symmetrical variants of the structure at lower temperatures.

In the ReO_3 structure the oxygen atoms occupy three-quarters of the positions of cubic closest packing, the position at the body-centre of the cube being unoccupied. (Occupation of this position by a large ion B comparable in size with O^{2-}, F^-, or Cl^- gives the perovskite structure for compounds ABX_3, in which the B and X ions together form the c.c.p. assembly.) For the fully-extended configuration of Fig. 5.18(a) the A—X—A angle is 180° but variants of the structure with smaller angles are also found. The most compact is that in which the X atoms are in the positions of hexagonal closest packing. This structure is adopted by a number of transition-metal trifluorides (see Table 9.16) (p. 435).

Just as α-$Zn(OH)_2$ and $Be(OH)_2$ crystallize with the simplest 3D framework structure possible for a compound AX_2 with 4:2 coordination (the cristobalite

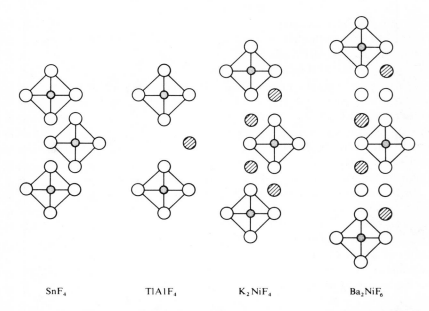

SnF₄   TlAlF₄   K₂NiF₄   Ba₂NiF₆

FIG. 5.17. Structures containing the *trans* AX₄ layer formed from octahedral AX₆ groups sharing their four equatorial vertices (diagrammatic elevations showing one octahedron of each layer). The larger shaded circles represent cations situated between the layers.

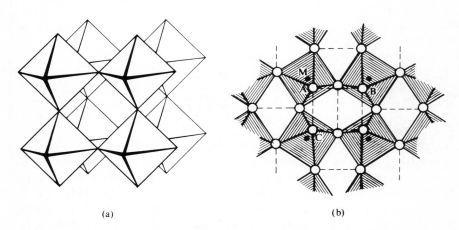

(a)   (b)

FIG. 5.18. The crystal structures of (a) ReO₃ and (b) Sc(OH)₃.

structure), distorted so as to bring together hydrogen-bonded OH groups of different coordination groups, so Sc(OH)₃ and In(OH)₃ have the simplest 3D framework structure of the AX₃ type (the ReO₃ structure), distorted so as to permit hydrogen bonding between OH groups of different M(OH)₆ octahedra. The nature of the distortion can be seen from Fig. 5.18(b). Instead of lying on the straight lines joining metal atoms the OH groups lie off these lines, each being hydrogen-bonded to two others. The OH group A is bonded to the metal atom M and to a similar atom vertically above M and hydrogen-bonded to the OH groups B and C.

From the topological standpoint, the ReO_3 structure is the simplest 3D framework structure for a compound AX_3 built of octahedral AX_6 groups, for it is based on the simplest 3D 6-connected net. More complicated structures of the same general type are known, that is, structures in which every octahedron is joined to six others through their vertices. The tungsten bronzes have structures of this kind which are described in Chapter 13. A feature of the tungsten bronze structures is that there are tunnels parallel to the 4- or 6-fold axes, that is, in one direction only.

There is another framework structure built of octahedral groups each of which shares its vertices with six others. In the basic BX_3 framework of the pyrochlore structure octahedra which share all vertices are grouped tetrahedrally around the points of the diamond net. In this framework there are rather large holes, the centres of which are also arranged like the carbon atoms in diamond, and they can accommodate two larger cations and also one additional anion X for each $(BX_3)_2$ of the framework. The rigid octahedral framework is stable without either the large cations A or the additional X atoms, and it therefore serves as the basis of the structures of compounds of several types. If all the A and X positions are occupied the formula is $A_2B_2X_7$ (as in oxides such as $Hg_2Nb_2O_7$), but the positions for cations A may be only half occupied ($BiTa_2O_6F$) or they may be unoccupied as in $Al_2(OH,F)_6(H_2O)_{\frac{3}{4}}$ where, in addition, there is incomplete occupancy of the seventh X position. The pyrochlore structure is illustrated in Fig. 7.3 (facing p. 318) and the structure is further discussed in Chapters 6 and 13.

A structure in which each octahedron shares five vertices is a member of a family of structures related to the ReO_3 structure. Figure 5.19 shows a projection of this structure along a 4-fold axis, each square representing a chain of octahedra sharing opposite vertices, in addition to the four equatorial vertices shown in the projection. Slices of the structure of type (a) have the general formula A_nX_{3n+2}; the first member is a single AX_5 chain and the others are layers, A_2X_8 ($= AX_4$), A_3X_{11}, A_4X_{14} ($= A_2X_7$), and so on. Examples include: ($n = 2$) $BaZnF_4$ and $NaNbO_2F_2$; ($n = 4$) $Ca_2Nb_2O_7$; ($n = 5$) $Ca_4NaNb_5O_{17}$. Slices of type (b) have the general formula A_nX_{3n+1}, that is, AX_4, A_2X_7, and so on. Examples include: ($n = 1$) K_2NiF_4 and Sr_2TiO_4, ($n = 2$) $Sr_3Ti_2O_7$, ($n = 3$) $Sr_4Ti_3O_{10}$. Only the first two members of each family have topologically equivalent octahedra, namely, the AX_5 chain, the two AX_4 layers, and the A_2X_7 layer of type (b). In this layer each octahedron shares five vertices, whereas in the A_2X_7 layer (a) equal numbers of octahedra share four or six vertices; compare the 3D weberite framework (p. 906) in which MgF_6 octahedra share six and AlF_6 octahedra four vertices.

Edge-sharing structures

We noted earlier that an indefinitely large number of structures could be built from

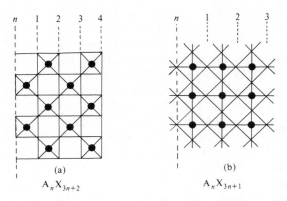

FIG. 5.19. Two families of layers which are slices of the ReO_3 structure.

octahedra which share only edges because not only can the number of shared edges range from two up to a maximum of twelve, but there is the additional complication that there is more than one way of selecting a particular number of edges. It is not necessary to consider the sharing of more than six edges, for this is observed only in the NaCl structure, which may be represented as octahedral $NaCl_6$ or $ClNa_6$ coordination groups sharing all twelve edges.·

There are four different ways of selecting two edges of an octahedron, namely: (i) *cis* edges (with a common vertex) inclined at 60°; (ii) *cis* edges inclined at 90°; (iii) 'skew' edges; and (iv) *trans* (opposite) edges. Of these (i) gives the finite group of three octahedra in Fig. 5.8(e) and (ii) gives the finite group of four octahedra of Fig. 5.8(k), or a zigzag chain (compare the elevation of the MoO_3 layer in Fig. 5.30(a), p. 222; this is the chain in white $MoO_3.H_2O$ (p. 519). Examples of (iii) and (iv) are shown in Fig. 5.20, where each octahedron is shown lying on a face. Two of the simplest structures formed by the sharing of skew edges are shown in Fig. 5.20(a), namely, the ring of six octahedra in $TeMo_6O_{24}^{6-}$ and the planar zigzag chain of $TcCl_4$ (and other tetrahalides), of the TeO_4^{2-} ion in the Na and Ca salts (p. 735), and of the $(CuCl_3.H_2O)^-$ ion in $LiCuCl_3.2H_2O$. In this crystal one-half of the molecules of water of crystallization are incorporated in the chain and the remainder, together with the Li^+ ions, are accommodated between the chains. In addition to these simple structures there is an indefinitely large number of more complex rings and skew chains; examples include the helical (4_1) chain in Li_2TeO_4 (p. 735) and the chains in crystalline HfI_4 and ZrI_4. Some simple examples were noted in *Models in structural inorganic chemistry* (Clarendon Press, Oxford (1970), pp. 98 and 148); a more systematic study has since been made (AC 1981 **B37** 532).

In the more complex chain which represents the cation-water complex in borax, $[Na(H_2O)_4]_2[B_4O_5(OH)_4]$, alternate octahedra share a pair of opposite *or* skew edges. In this crystal the chain apparently adopts this configuration in order to pack

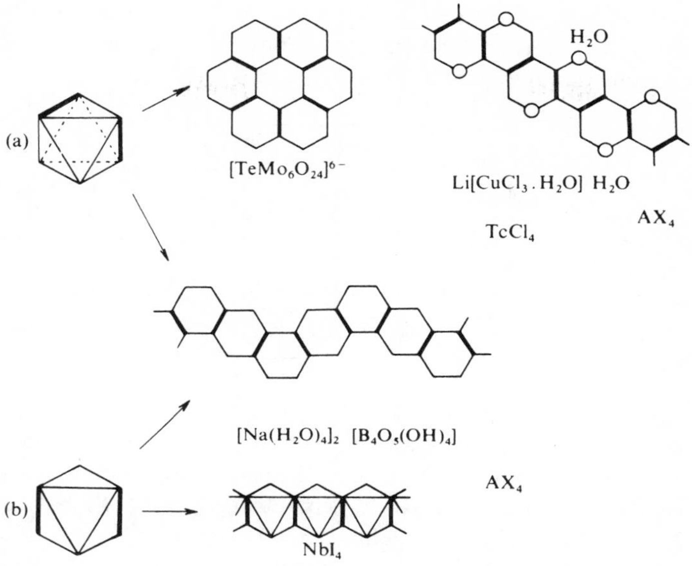

FIG. 5.20. Octahedra sharing (a) 'skew' edges and (b) 'trans' edges.

satisfactorily with the hydrogen-bonded chains of anions, but it is less obvious why a similar chain is found in $Na_2SO_4.10H_2O$. In this hydrate only eight of the ten molecules of water of crystallization are associated with the cations to form a chain with the same ratio $4\,H_2O:Na$ as in borax.

Sharing of two opposite edges of each octahedron leads to the infinite AX_4 chain of Fig. 5.20(b) which represents the structure of the infinite molecules in crystalline NbI_4 and of the infinite anion in, for example, $K_2HgCl_4.H_2O$. In this AX_4 chain only 4 X atoms of each octahedral AX_6 group are acting as links between the A atoms. The other two are attached to one A atom only and may be ligands of a second kind, not necessarily capable of bridging two metal atoms (formula AX_2L_2) or they may be atoms forming part of a bidentate ligand, when the formula is AX_2B. This simple octahedral chain thus serves for compounds of three types:

AX_4: NbI_4

AX_2L_2: $PbCl_2(C_6H_5)_2$, $CoCl_2.2H_2O$

AX_2B: $CuCl_2.C_2N_3H_3$

For $CuCl_2.C_2N_3H_3$ and other examples see p. 1139.

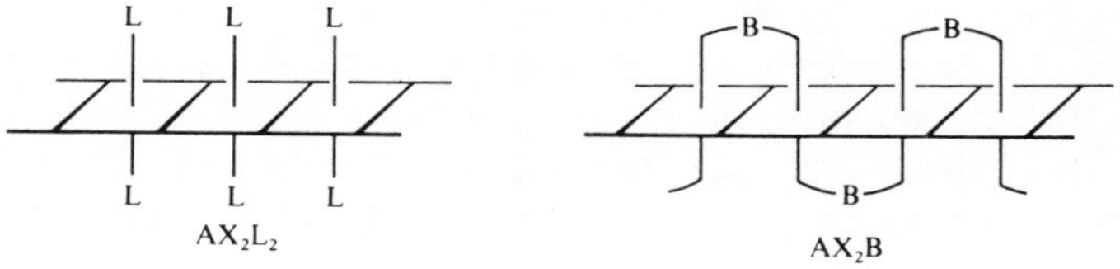

There is also a family of structures containing rutile-like chains which are held together by ions of a second kind, the chain ions being arranged to give suitable coordination groups around these cations:

$M'_n MX_4$	c.n. of M'	Reference
Sr_2PbO_4, Ca_2PbO_4		NW 1965 **52** 492
Ca_2SnO_4, Cd_2SnO_4	6	NW 1967 **54** 17
Na_2MnCl_4		AC 1971 **B27** 1672
High-pressure Mn_2GeO_4		AC 1968 **B24** 740
Na_2CuF_4, Na_2CrF_4	7	ZaC 1965 **336** 200
Ca_2IrO_4	6, 7, 9	ZaC 1966 **347** 282

The same (orthorhombic) structure, Fig. 5.21, has been described several times for a number of compounds. It is interesting for the trigonal prismatic coordination of the M' atoms between the chains. The monoclinic Na_2CuF_4 structure is a version of the structure distorted to give $Cu(II)$ $(4 + 2)$-coordination; Na^+ has 7 F neighbours in the range 2.27–2.66 Å. In the hexagonal Ca_2IrO_4 structure there is a somewhat different arrangement of the chains, but all three structures are basically of the same type. In $K_2HgCl_4 . H_2O$ there are H_2O molecules in addition to K^+ ions between the rutile-like $(HgCl_4)_2^{2n-}$ chains. Introduction of additional S atoms at the mid-points of the cell edges in Fig. 5.21 gives the structure of La_2SnS_5, or $La_2(SnS_4)S$,

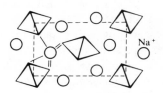

FIG. 5.21. Projection of the structure of Na_2MnCl_4 along the direction of the chain ions.

in which La^{3+} has $(7 + 2)$-coordination (tricapped trigonal prism). Introduction of two additional S atoms between the rutile chains and also a third cation produces the structure of Sm_3InS_6, or $Sm_3(InS_4)S_2$, in which Sm atoms are 7- and 8-coordinated.

When resting on one set of octahedral faces the *trans* AX_4 chain has the appearance of Fig. 5.22(a), and when viewed along its length it appears as shown on the r.h.s. of the figure. Two such chains may be joined laterally to form the double chain of Fig. 5.22(b) with the composition AX_3, in which each octahedron shares four edges. This is the form of the anion in NH_4CdCl_3 and $KCuCl_3$. It is convenient to refer to the chains of Fig. 5.22(a) and (b) as the rutile and 'double rutile' chains. The end-on views will be used later in representations of more complex structures

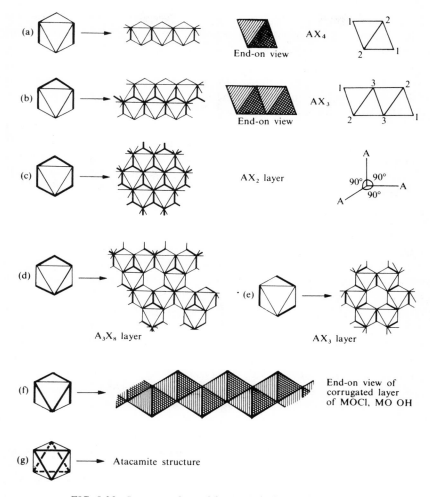

FIG. 5.22. Structures formed from octahedra sharing edges.

which result when these chains form a corrugated layer by sharing additional edges or 3D frameworks by sharing the projecting vertices.

We showed in Fig. 5.12 (p. 204) four ways of selecting four edges of an octahedron. The four edges of Fig. 5.12(a) are shared in the double chain of Fig. 5.22(b), and those of Fig. 5.12(b) in the A_3X_8 layer of Fig. 5.22(d). The sharing of four equatorial edges, Fig. 5.12(c), leads to a layer of composition AX_3 in which 4 X atoms of each AX_6 are common to four octahedra and two are unshared. This layer is found in $NH_4(HgCl_3)$, though the octahedral coordination group is so distorted that the alternative description in terms of $HgCl_2$ molecules is to be preferred. The sharing of the four edges of Fig. 5.12(d) is found in anatase (TiO_2). However, in this structure an octahedron also has *vertices* in common with four other octahedra, and this is therefore not a simple edge-sharing structure. Nevertheless it is convenient to emphasize the sharing of these edges for they are arranged tetrahedrally, a fact more easily appreciated from a model. The sharing of this special set of edges leads to a 3D diamond-like structure, just as the sharing of four vertices in IrF_4 leads to a diamond-like vertex-sharing structure.

Continued lateral linking of simple AX_4 chains leads to the infinite layer in which the six edges of Fig. 5.22(c) of each octahedron are shared and each X atom is common to three octahedra. The composition of the layer is AX_2; this is the layer of the CdI_2 and $CdCl_2$ structures, for which see also Chapters 4 and 6. We have included in Fig. 5.22 a further choice of four edges giving the A_3X_8 layer found in crystalline Nb_3I_8.

We have seen that linear and zigzag chains result from the sharing of different pairs of edges. Similarly plane and corrugated layers arise from the sharing of different sets of six edges. Whereas the sharing of the very symmetrical arrangement of six edges of Fig. 5.22(c) gives a plane layer, the sharing of the edges of Fig. 5.22(f) leads to a corrugated layer. This layer is the basis of the structures of a number of oxychlorides MOCl and oxyhydroxides MO(OH). The structures of pairs of compounds such as FeOCl and γ-FeO(OH) (the mineral lepidocrocite) differ in the way in which the layers are packed together, there being hydrogen-bonding between O atoms of different layers in the latter compound. It is interesting (and unexplained) that unlike the other 3d-metal dihydroxides with the simple CdI_2 layer structure $Cu(OH)_2$ apparently crystallizes with the corrugated layer structure more characteristic of compounds MOCl and MO(OH). The corrugated layer of lepidocrocite is found also in the compound $Rb_xMn_xTi_{2-x}O_4$ ($0.60 < x < 0.80$), where replacement of some Ti^{4+} in the octahedra by Mn^{3+} gives charged layers which are held together by the Rb^+ ions (Fig. 5.23). The same kind of layer is thus found held together by van der Waals bonds in FeOCl, by O–H–O bonds in γ-FeO(OH), and by Rb^+ ions in the complex oxide.

The sharing of the six edges (g) of Fig. 5.22 with six other octahedra leads to an interesting 3D framework structure which is related to a number of cubic structures. It will be observed that the shared edges (g) are the six unshared ones of (c); sharing of all twelve gives the NaCl structure. Accordingly there are two very simple structures derivable from the NaCl structure by removing one-half of the metal ions. Removal of alternate *layers* of cations as in Fig. 4.22(a) (p. 170) gives the $CdCl_2$ (layer) structure, in which the octahedra share edges (c). On the other hand, if alternate *rows* of cations are removed as indicated by the dotted circles of Fig. 4.22(b) we obtain a structure in which each octahedron shares the edges (g) with its neighbours. This structure is illustrated as a system of octahedra in Fig. 7.3. Although the immediate environments of the A and of the X ions are exactly the same in both the structures of Fig. 4.22(a) and (b) it is an interesting fact that no dihalide crystallizes with the structure (b), in contrast to the considerable number that

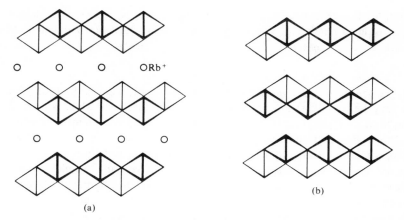

FIG. 5.23. Elevations of the structures of (a) $Rb_xMn_xTi_{2-x}O_4$; (b) γ-FeO.OH.

crystallize with the $CdCl_2$ structure. A distorted form of this structure is, however, adopted by the mineral atacamite, one of the polymorphs of $Cu_2(OH)_3\,Cl$, a second polymorph having a CdI_2 type of layer structure. This 3D octahedral framework is the BX_2 framework of the spinel (AB_2X_4) structure, as shown in Fig. 7.3(a) facing p. 318. It is also related to the diamond structure, for the octahedra are in groups of four around (empty) tetrahedral cavities at the nodes of the diamond net. However, the topological representation is the 6-connected net of Fig. 29.16(d), p. 1308, which is the 3D net formed by joining each A to its six nearest A neighbours across the six shared edges.

We have emphasized here the structures arising from the sharing of two, four, or six edges since they illustrate the progression from simple and double chains to layers and 3D structures. One of the most important edge-sharing structures is the AX_3 layer of Fig. 5.22(e), in which each octahedron shares three edges. This is the structural unit in $Al(OH)_3$, many trichlorides, and some tribromides and triiodides. We may also note here that there is an infinite family of variants of the double chain of Fig. 5.22(b). The simplest of these is shown at (a), in which one-half of the octahedra share three and the remainder five edges. This is the form of the $Cd_2Cl_5(H_2O)$ chain in $Cd_4NiCl_{10}.10H_2O$ (p. 687), in which octahedral coordination

H_2O

(a)

groups $CdCl_5(H_2O)$ and $CdCl_6$ share three and five edges respectively. There is an indefinitely large number of variants of this chain in which octahedra share three, four, and five edges.

Vertex- and edge-sharing structures

It is convenient to describe more complex octahedral structures as built of sub-units which may be, for example, finite groups or infinite chains which are linked by sharing additional vertices and/or edges to form the layer or 3D framework structure. While this device is often helpful in illustrating or constructing a model of a more complex structure it should be remembered that the sub-units have only been distinguished for this purpose and do not necessarily have any chemical or physical significance. We shall encounter structures built from edge-sharing and face-sharing pairs of octahedra, and in particular many structures may be described as built from chains, either the edge-sharing (rutile) AX_4 chain of Fig. 5.22(a) or the vertex-sharing (ReO$_3$) AX_5 chain of Fig. 5.15(c). Structures are often viewed along the direction of such a chain, and it is therefore important to note how the X atoms are shared in an isolated chain or multiple chain.

In the isolated AX_4 (rutile) chain the X atoms are of two kinds, as indicated by the numerals at the top right-hand corner of Fig. 5.22. Some belong to one octahedron only while those in shared edges belong to two octahedra. Similarly in the double (AX_3) chain of Fig. 5.22(b) X atoms belong to 1, 2, or 3 octahedra. We shall differentiate these different types of X atom as X_1, X_2, and X_3 respectively. There are obviously many ways in which such chains could be linked to form more complex structures by sharing X atoms. The most important group of these structures arises by sharing X_1 atoms of one chain with X_2 atoms of other chains.

The simplest structure of this kind is the rutile structure, shown in plan in Fig. 5.24. Because an X_1 atom of one chain is an X_2 atom of another chain each X is common to three octahedra, and the formula is accordingly AX_2. The rutile structure is described in more detail in Chapter 6. Corresponding to the rutile structure, which is built from single chains, there are 3D frameworks derived from the double

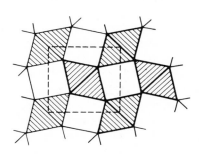

FIG. 5.24. The rutile structure.

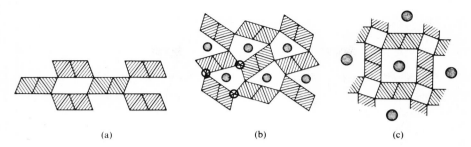

(a)   (b)   (c)

FIG. 5.25. Structures built from double rutile chains.

chain of Fig. 5.22(b). Three of these are illustrated in Fig. 5.25. In all the structures of Fig. 5.25 each vertex is common to three octahedra; the formula is therefore AX_2. Figure 5.25(a) represents the structure of α-AlO(OH), the mineral diaspore. In the structures (b) and (c) the more open packing of the chains leaves room for additional atoms (cations), and these charged frameworks (or 3D ions) form around suitable ions indicated by the circles in (b) and (c). Examples of compounds with these structures are given in Table 5.5.

TABLE 5.5

Structures built from double octahedral chains

Nature of basic single chain	Type of structure built from double chain			
	Chain	*Layer*	*3D framework*	
	AX_3	AX_2	AX_2	
Edge-sharing AX_4	$NH_4(CdCl_3)$	γ-AlO . OH	α-AlO OH	Eu_3O_4 $CaFe_2O_4$	α-MnO_2 (hollandite)
	AX_4	AX_3	AX_3	
Vertex-sharing AX_5	$NbOCl_3$	MoO_3	$CaTa_2O_6$	—	—

There is another family of possible structures in which the shared vertices are either X_1 atoms of both chains (which may be single or multiple) or X_2 atoms of both chains. The simplest structure of this kind is the AX_3 layer of Fig. 5.26(a) consisting of rutile chains, perpendicular to the plane of the paper, joined by sharing X_1 vertices. The analogous layer formed in the same way from double chains is shown at (c), and the intermediate possibility (from single and double chains) at (b). There are clearly two kinds of non-equivalent octahedra in (b). Vanadium oxyhydroxides (p. 569) provide examples of such layers. In Chapter 3 we gave the (layer) structure of the compound $CuHg(OH)_2(NO_3)_2(H_2O)_2$ as a more complex

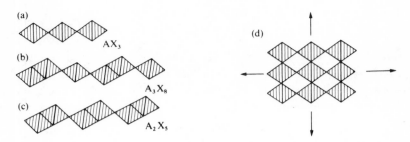

FIG. 5.26. Layers built from single and double octahedral chains which are perpendicular to the plane of the paper: (a) AX_3; (b) A_3X_8; (c) A_2X_5; (d) projection of layer (a).

example of the plane 4-gon net. This is in fact the AX_3 layer of Fig. 5.26(a), a point more easily appreciated if this layer is shown in projection (on the plane of the layer) as in (d).

Three-dimensional framework structures, analogous to those of Fig. 5.25, are also possible. An example is the structure of $CaTi_2O_4$ (Fig. 5.27(a)) which should be compared with Fig. 5.25(b). Further sharing of edges between the double chains gives the structure of Fig. 5.27(b), adopted by $Na_xFe_xTi_{2-x}O_4$ $(0.90 > x > 0.75)$.

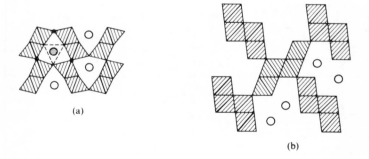

FIG. 5.27. Projections of the structures of (a) $CaTi_2O_4$; (b) $Na_xFe_xTi_{2-x}O_4$.

Yet another way of utilizing the double rutile chain is exemplified by the structure of $NaTi_2Al_5O_{12}$, where a framework built from double and single chains accommodates Al in tetrahedral and Na in octahedral holes (AC 1967 **23** 754).

Structures built from other types of multiple chain can also be visualized. The family of minerals related to MnO_2 provides examples of several structures of this general type, and the complex oxide BeY_2O_4 is built of quadruple rutile chains. These structures are described in Chapters 12 and 13 where further examples of compounds with the above structures are given. An interesting structure of the same general type as Fig. 5.25(b) is that of $Na_{3-x}Ru_4O_9$ (AC 1974 **B30** 1459). Alternate rows of double chains in the projection are replaced by alternating single and triple chains.

All the chains are parallel in the structures we have described as built from rutile-like chains. In BaU_2O_7 such chains run in two perpendicular directions to form a 3D framework (Fig. 5.28). Alternate O atoms of shared edges in one chain are also

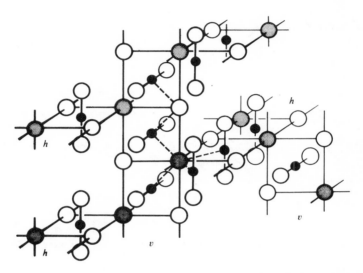

FIG. 5.28. The U_2O_7 framework in BaU_2O_7. The axes of the chains v and h lie respectively in and perpendicular to the plane of the paper. The UO_2 groups of chains v are therefore perpendicular to the paper and those of chains h parallel to the plane of the paper. The shaded circles represent O atoms bonded to 4 U and the open circles O atoms bonded to 2 U.

the corresponding atoms of perpendicular chains, so that in each UO_6 octahedron two O atoms are bonded to 1 U, two to 2 U, and two to 4 U. The ratio of O:U atoms in the framework is therefore 7:2 $[2(1) + 2(\frac{1}{2}) + 2(\frac{1}{4}) = 3\frac{1}{2}]$.

We now come to structures built from the AX_5 chain composed of octahedral AX_6 groups sharing a pair of opposite vertices. This chain is conveniently represented by its end-on view, as also is the double-chain formed from two single chains by edge-sharing (Fig. 5.29). This double chain is the infinite 'molecule' in crystalline $NbOCl_3$, the shared vertices being the O atoms; compare the dimeric Nb_2Cl_{10} molecule in the crystalline pentachloride, formed from two octahedra sharing one edge. Further *edge*-sharing between these double chains gives the corrugated layers seen end-on in Fig. 5.30(a) and (b). In each case three X atoms in each octahedron are bonded to three A atoms, two to 2 A and one to 1 A, so that the composition is AX_3; $(3 \times \frac{1}{3}) + (2 \times \frac{1}{2}) + 1 = 3$. Figure 5.30(a) represents a layer of crystalline MoO_3; the complexity of this structure may be compared with the simplicity of the ReO_3

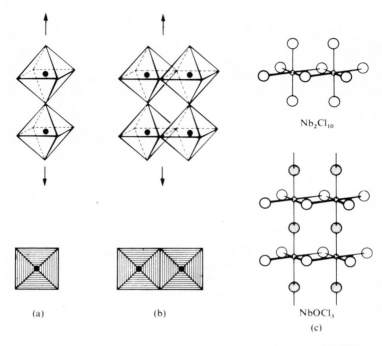

FIG. 5.29.   (a) ReO_3 (AX_5) chain; (b) double ReO_3 (AX_4) chain; and (c) the Nb_2Cl_{10} molecule and $NbOCl_3$ chain molecule.

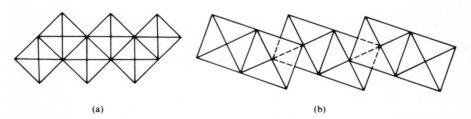

FIG. 5.30.   Layers of composition AX_3 built from 'double ReO_3' chains: (a) MoO_3: (b) layer in $Th(Ti_2O_6)$.

structure, in which every oxygen atom is bonded to two metal atoms. The layer of Fig. 5.30(b) is not known as a neutral AX_3 layer but it represents the arrangement of TiO_6 octahedra in $ThTi_2O_6$ (and the isostructural compounds UTi_2O_6, CdV_2O_6, and $NaVMoO_6$).

The double chains of Fig. 5.29(b) can be further linked to similar chains by vertex-sharing to form a family of 3D structures analogous to those of Fig. 5.25. In the frameworks of Fig. 5.31 each X atom is bonded to two A atoms; the formula is

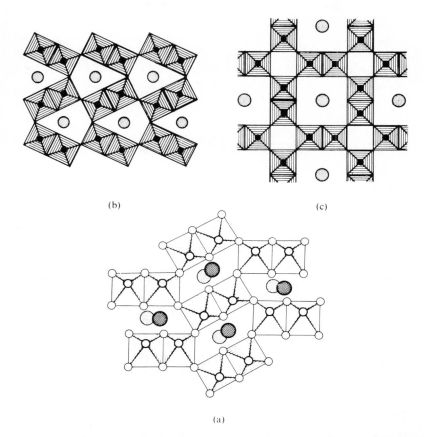

(b)

(c)

(a)

FIG. 5.31. Three frameworks of composition AX_3 built from double ReO_3 chains. (a) is a projection of the structure of $CaTa_2O_6$.

therefore AX_3. Examples of (b) and (c) are not known, but (a) represents the crystal structure of $CaTa_2O_6$ if the larger circles are the Ca^{2+} ions occupying the interstices in the framework.

We should not expect to find structures based on combinations of vertex-sharing AX_5 and edge-sharing AX_4 chains built from octahedra of the same type for the purely geometrical reason that the repeat distances along these two chains are not the same; they are $2(AX)$ and $\sqrt{2}(AX)$, respectively, if AX is the distance from centre to vertex of the octahedral AX_6 groups. However, such structures are possible if the octahedra are of different sizes, as is the case if, for example, the atoms A are of different elements or of one element in different oxidation states. Chromous fluoride, CrF_2, has the rutile structure (distorted to give Cr^{II} a coordination group of four closer and two more distant neighbours), while CrF_3 has a ReO_3-type structure with only vertex-sharing between octahedral CrF_6 groups. Owing to the fact that the Cr^{III}–F bonds (1.89 Å) are shorter than the Cr^{II}–F bonds (four of 1.98 Å and two of 2.57 Å), the repeat distance is the same along the two types of chain. They can therefore fit together as shown diagrammatically in Fig. 5.32 to give the fluoride Cr_2F_5.

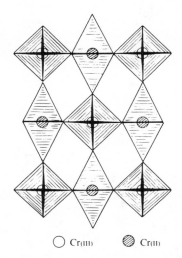

$\bigcirc$ Cr$_{(III)}$ $\oslash$ Cr$_{(II)}$

FIG. 5.32. Projection of the structure of Cr_2F_5 showing the two kinds of octahedral chain linked by vertex-sharing into a 3D framework.

Structures in which vertices and edges of coordination groups are shared include many which are more complex than those listed in Table 5.4. A relatively simple example is the AX_3 framework formed from edge-sharing pairs of octahedra which are further linked to form the 3D framework of Fig. 5.33. This framework is the basis of the structure of one form of $KSbO_3$ and of $KBiO_3$, where K^+ ions occupy the interstices. If only one-sixth of the positions occupied by K^+ in $KBiO_3$ are occupied by OLa_4 groups (consisting of an O atom surrounded by a tetrahedron of La^{3+} ions) in a framework built of ReO_6 octahedra the formula becomes $(OLa_4)Re_6O_{18}$ or $La_4Re_6O_{19}$. In this crystal there is interaction between the metal atoms within the edge-sharing pairs of octahedra (Re–Re = 2.42 Å) so that there is a physical basis for recognizing these sub-units in the structure.

The structures of two molybdenum bronzes differ from the tungsten bronzes in much the same way as does MoO_3 from WO_3; we have noted that there is edge-sharing as well as vertex-sharing in MoO_3 in contrast to the sharing of vertices only in WO_3. Similarly there is only vertex-sharing of WO_6 octahedra in the tungsten bronze structures, but two molybdenum bronzes with very similar compositions have layer structures which are built from edge-sharing sub-units containing respectively six and ten octahedra. These sub-units can be seen in Fig. 5.34, which shows how they are

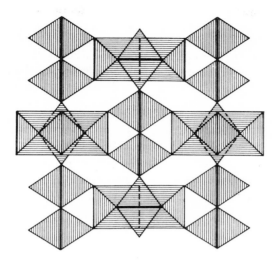

FIG. 5.33. Framework built from pairs of edge-sharing octahedra further linked by vertex-sharing.

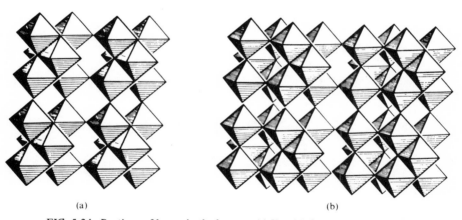

(a)   (b)

FIG. 5.34. Portions of layers in the bronzes (a) $K_{0.33}MoO_3$ and (b) $K_{0.30}MoO_3$.

further linked into layers by sharing eight vertices with other similar units. Both layers have the composition MoO_3 and they are held together by the potassium ions.

We comment elsewhere on the inadequacy of current bonding theory to account for the complexity of some binary systems in which solid phases appear with unexpected formulae and/or properties — for example, the oxides of caesium and the nitrides of calcium. Certain transition metals, notably Ti, V, Nb, Mo, and W, have a surprisingly complex oxide chemistry, and although it may be difficult to appreciate all the features of their structures from diagrams, these compounds are sufficiently important to justify mention here.

The structures in question are built from slices or blocks of the simpler rutile or ReO_3 structures which are displaced relative to one another to form structures with formulae that correspond in some cases to normal oxides (for example, Nb_2O_5) and in others to oxides with complex formulae implying non-integral (mean) oxidation numbers of the metal atoms. When the rutile structure is sheared along certain (regularly spaced) planes, sharing of *faces* of TiO_6 coordination groups occurs, giving a family of related structures with composition Ti_nO_{2n-1}. All members of this family have been prepared and characterized, in the titanium oxides for $n = 4$–10 inclusive, and in the vanadium oxides for $n = 4$–8. Their compositions are Ti_4O_7, Ti_5O_9, etc. Shearing the ReO_3 structure, in which there is only vertex-sharing, leads to sharing of *edges* of octahedral coordination groups and to homologous series of structures with formulae such as W_nO_{3n-2}. Since the ReO_3-type structures are more easily illustrated than those derived from the rutile structure we shall describe examples of the former. The structures will be shown as projections along the direction of the AX_5 (ReO_3) chains so that each square represents an infinite chain of vertex-sharing octahedra perpendicular to the plane of the paper. First we have to note that ReO_3 chains may be joined by edge-sharing in two essentially different ways, as shown in Fig. 5.35. In (a) the 'equatorial' edge (parallel to the plane of the

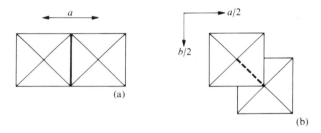

FIG. 5.35. Two ways of joining ReO_3 chains (perpendicular to paper) by edge-sharing.

paper) is shared, that is, the two chains are related by a translation a; in (b) the shared edge is inclined to the plane of the paper, one chain being displaced relative to the other by $a/2 + b/2 + c/\sqrt{2}$. Because there is now a translation in the direction of the length of the chain (i.e. perpendicular to the paper), we may refer to the chains in case (b) as being at different levels. Blocks of ReO_3 structure may be joined by sharing edges at the perimeters of the blocks in either or both of these two ways. As examples of these two types of edge-sharing we shall describe the structures of the following oxides:

Type of edge-sharing: (a) Mo_8O_{23}
(a) and (b) V_6O_{13},
(b) $WNb_{12}O_{33}$ and high-Nb_2O_5

Further examples are given in Chapters 12 and 13.

The first type of edge-sharing, (a), corresponds to shearing the ReO_3 structure so that the chains shown as black squares in Fig. 5.36(a) are displaced as in (b), and if

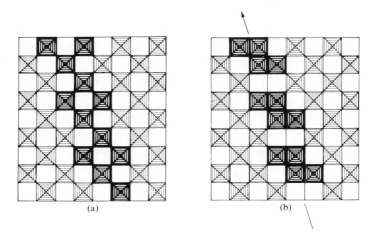

(a)   (b)

FIG. 5.36. Formation of shear structure (diagrammatic).

this shearing occurs at regular intervals the new structure is composed of infinite slabs of ReO_3 structure cemented together along the 'shear-planes' (arrow in Fig. 5.36(b)) by edge-sharing. The composition of the resulting oxide depends on the number of octahedra in the edge-sharing groups at the junctions and on the distance apart of the shear-planes. The examples of Fig. 5.37 show the structure of the oxide Mo_8O_{23} in the series M_nO_{3n-1} and of a hypothetical $Mo_{11}O_{31}$ in the series M_nO_{3n-2} (as a simpler example than the known oxide $W_{20}O_{58}$).

The structure of V_6O_{13} illustrates both the types of edge-sharing of Fig. 5.35. Blocks of ReO_3 structure consisting of six chains (3×2) are joined by sharing

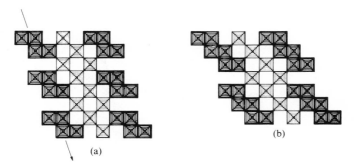

(a)   (b)

FIG. 5.37 Examples of shear structures: (a) Mo_8O_{23} in the M_nO_{3n-1} series; (b) hypothetical oxide $Mo_{11}O_{31}$ in M_nO_{3n-2} series.

equatorial edges and then the second type of edge-sharing results in the 3D structure of Fig. 5.38.

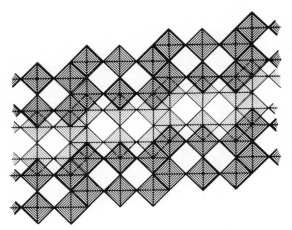

FIG. 5.38. The crystal structure of V_6O_{13}.

The octahedra drawn with heavier lines are displaced both in the plane of the paper and also in a perpendicular direction with respect to those drawn with light lines. Each block extends indefinitely normal to the plane of the paper, but as noted earlier it is convenient to refer to the blocks as being at different levels. The formula depends on the size of the block, and for blocks of $(3 \times n)$ octahedra at each level the general formula is $M_{3n}O_{8n-3}$:

n	Examples
2	V_6O_{13}
3	$TiNb_2O_7$ (M_9O_{21})
4	$Ti_2Nb_{10}O_{29}$ ($M_{12}O_{29}$)

In Fig. 5.39 we illustrate the structure of $WNb_{12}O_{33}$ as an example of a structure in which there is edge-sharing of the second type only, that is, there is no edge-sharing between blocks at the same level. This particular structure is built of ReO_3 blocks consisting of (3×4) octahedra. This type of structure has the peculiarity that there are small numbers of *tetrahedral* holes at the points indicated by the black circles. The high-temperature form of Nb_2O_5 has a more complex structure of the same general type (see p. 611). There is one such tetrahedral hole for every 27 octahedra, so that the structural formula is $NbNb_{27}O_{70}$. The existence of structures of this kind is very relevant to discussions of bonding in transition-metal oxides. For example, $TiNb_{24}O_{62}$ is a normal valence compound, with one tetrahedral hole (occupied by Ti) to every 24 octahedral holes (occupied by Nb). However, the same structure is adopted by a niobium oxide, but the formula, $Nb_{25}O_{62}$, implies a fractional oxidation number or, presumably, one delocalized electron, $Nb_{25}^V O_{62}$ (1e).

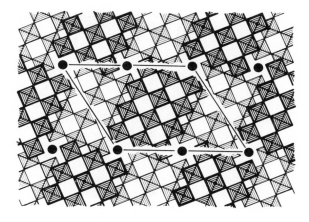

FIG. 5.39. The crystal structure of $WNb_{12}O_{33}$.

Face-sharing structures

Only one structure is possible if each octahedron shares two opposite faces. An infinite chain of composition AX_3 is then formed. A number of crystalline trihalides consist of infinite molecules of this type, and chain ions of the same kind exist in $BaNiO_3$ and $CsNiCl_3$. In the ZrI_3 structure the metal atoms occupy one-third of the octahedral holes in a close-packed assembly of halogen atoms to form infinite chains perpendicular to the planes of c.p. halogen atoms:

$$\underset{\diagdown I \diagup}{\overset{\diagup I \diagdown}{-I=}}Zr\underset{I}{\overset{I}{-}I=}Zr\underset{I}{\overset{I}{-}}I=Zr\underset{I}{\overset{I}{-}}I=$$

In $BaNiO_3$ there is close packing of Ba + 3 O, and the Ni atoms occupy octahedral holes between 6 O atoms to form an arrangement similar to that of the Zr and I atoms in ZrI_3.

There are a few structures in which there is sharing of octahedral faces other than opposite ones, and these are of considerable interest from the geometrical standpoint. There are three ways of selecting two faces of an octahedron or, stated in a different way, three configurations of a group of three face-sharing octahedra; all have the composition M_3X_{12} (Fig. 5.40). The faces shared by the central octahedron are (a) two opposite faces, (b) two faces with a common vertex, and (c) two faces with a common edge. The configuration (a), which is another view of the group shown in Fig. 5.8(d), p. 200, is found as a sub-unit in $Cs_4Mg_3F_{10}$ (Fig. 5.42) and, for example, in certain basic iron phosphates.[1] Examples of (b) do not come to mind, but (c) presents a problem similar to that of three regular octahedra sharing a common vertex, a subject discussed on p. 191. In (c), where the octahedra are viewed in the direction of the common edge, the distance aa' would be only 0.47 of

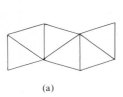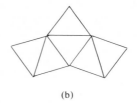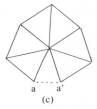

(a)   (b)   (c)

FIG. 5.40. The three configurations of a group of three face-sharing octahedra.

the edge length. This difficulty can be overcome by distorting the octahedra so as to close the gap, as in the O_3Cs_{11} group (Fig. 12.1, p. 535) or Nb_3S_4 (Fig. 17.8(a), p. 771), or by bridging aa' by an anion as in Fig. 5.4 (p. 192). In the mineral seamanite, $Mn_3(OH)_2[B(OH)_4]PO_4$[2] sub-units of type (c) share edges to form chains (Fig. 5.41) and within each sub-unit the gap aa' is bridged by a tetrahedral

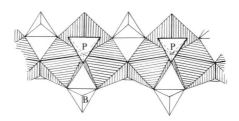

FIG. 5.41. Portion of a chain in seamanite.

$B(OH)_4$ group (O–O 2.35 Å). In $Fe_2(PO_4)Cl$[3] there are chains of octahedral coordination groups (around Fe) consisting of face-sharing octahedra of both types (a) and (c).

(1) AM 1970 **55** 135   (3) AC 1976 **B32** 2427
(2) AM 1971 **56** 1527

Vertex- and face-sharing structures

Although the sharing of faces of octahedral coordination groups in 3D (as opposed to chain) structures is not common it does occur in some close-packed structures. When considering structures built of c.p. AX_3 layers, between which smaller B ions occupy positions of octahedral coordination between 6 X atoms, we saw that only vertices and/or faces of BX_6 octahedra can be shared. The sharing of octahedral edges, which is a feature of so many oxide structures, cannot occur in $A_xB_yX_{3x}$ structures for purely geometrical reasons. The sharing of octahedron faces leads first to pairs (or larger groups) of octahedra which are then linked by vertex-sharing to form a family of structures which are related to the ReO_3 and perovskite structures.

A linear group of face-sharing octahedra is topologically similar to a single octahedron because it has three vertices at each end through which it can be linked, like a single octahedron, to six other groups or to single octahedra. The simplest of these structures are listed in Table 5.6 and also in Chapter 4 as examples of some of the

TABLE 5.6

Close-packed ABX_3 *structures in which octahedral coordination groups share faces and vertices*

Face-sharing groups	Structure	Other examples
Pairs + single All pairs Groups of 3 Infinite chains	Hexagonal $BaTiO_3$ High-$BaMnO_3$ $BaRuO_3$ High-$BaNiO_3$ Low-$BaMnO_3$	$CsMnF_3$, $RbNiF_3$ Low-$BaNiO_3$ $CsCoF_3$ $CsNiCl_3$, $CsCuCl_3$ $BaTiS_3$, $BaVS_3$, $BaTaS_3$

For further examples of complex oxides with structures of these types see Table 4.4 (p. 159) and Table 13.5 (p. 581).

more complex sequences of c.p. layers, for in all of these structures the transition-metal atoms occupy octahedral holes between six X atoms of the c.p. AX_3 layers.

An interesting structure of a quite different type is that of $Cs_4Mg_3F_{10}$, in which groups of three face-sharing MgF_6 octahedra are linked by *four* of their terminal vertices to form layers (Fig. 5.42), between which Cs^+ ions occupy positions of 10- and 11-coordination. Inasmuch as all the bonds in this crystal are presumably essentially ionic in character, the description of the structure in terms of the

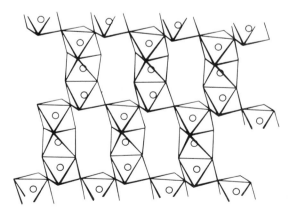

FIG. 5.42. Linking of MgF_6 octahedra into layers $(Mg_3F_{10})_n^{4n-}$ in $Cs_4Mg_3F_{10}$.

coordination groups around the smaller cations is not intended to imply that this is a layer structure. Isostructural crystals include the corresponding Co, Ni, and Zn compounds, and it is interesting that the groups of three face-sharing CoF_6 octahedra in $Cs_4Co_3F_{10}$ are also present in the structure of $CsCoF_3$ (Table 5.6).

We dealt separately at the beginning of this chapter with finite groups of octahedra (molecules and complex ions) but it seems justifiable to draw attention here to the anion $CeMo_{12}O_{42}^{8-}$ in view of its relation to some of the structures we have been describing. The four vertices of a pair of face-sharing octahedra which are marked A and B in Fig. 11.13(a) (p. 529) correspond to four vertices of an icosahedron. Six pairs of octahedra may therefore be joined together by sharing these vertices to form an icosahedral group of twelve octahedra. The Ce^{4+} ion occupies the position of 12-coordination at the centre of the complex ion.

Edge- and face-sharing structures

An example of this rare phenomenon is provided by the remarkable structure of Nb_3S_4. Three rutile chains can coalesce into a triple chain by face-sharing provided that the octahedra are modified to make the dihedral angle α equal to 120°. The shared faces in Fig. 5.43(a) are perpendicular to the plane of the paper and have a

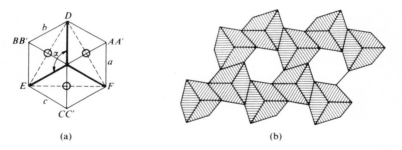

(a)   (b)

FIG. 5.43. The crystal structure of Nb_3S_4: (a) projection (idealized) of multiple chain formed from three rutile chains sharing the faces projecting as heavy lines; (b) projection of the Nb_3S_4 structure.

common edge (also perpendicular to the paper) at the centre of the diagram. These triple chains may now join together by sharing all the edges projecting as *a*, *b*, and *c* to form the 3D structure shown in projection in Fig. 5.43(b). If built from MX_6 octahedra the structure has the composition M_3X_4. A model, which can be made from 'D-stix', shows that the pairs of vertices AA' etc. and D, E, and F outline a tricapped trigonal prism, so that the triple column of Fig. 5.43(a) may also be regarded as a column of tricapped trigonal prisms sharing basal faces (assuming all M and X atoms within the column to be removed). We then see a general similarity to the structure of UCl_3, which is built of columns of face-sharing tricapped trigonal prisms joined by edge-sharing into a structure of the same general form as Fig.

5.43(b). This will be evident from a comparison of Fig. 9.8 (p. 422) with Fig. 5.43(b). We have described the Nb_3S_4 structure here in terms of the NbS_6 coordination groups, that is, in terms of Nb–S bonds, but it should be noted that metal–metal bonding plays an important part in this structure (p. 770) as in a number of the other structures we have described.

Vertex-, edge-, and face-sharing structures

Structures of this type include the corundum (α-Al_2O_3) and γ-$Cd(OH)_2$ structures. In the corundum structure there are pairs of octahedra with a common face, and in addition there is sharing of edges and vertices of coordination groups around the Al^{3+} ions. The complexity of this structure for a compound A_2X_3 appears to arise from the need to achieve a close packing of the anions and at the same time a reasonably symmetrical environment for these ions—see also the remarks on p. 266. In the structure of γ-$Cd(OH)_2$ (Fig. 5.44) pairs of rutile chains are joined through common faces and the double chains then share vertices.

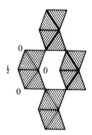

FIG. 5.44. The structure of γ-$Cd(OH)_2$.

Other structures in this class include the high-pressure form of Rh_2O_3 (p. 545) and $K_2Zr_2O_5$.[1]

(1)   JSSC 1970 **1** 478

Structures built from tetrahedra and octahedra

Structures which can be represented as assemblies of tetrahedral and octahedral coordination groups are very numerous. For example, they include all oxy-salts in which the cations occupy positions of octahedral coordination and the anion is a discrete BO_4 ion (as in Na_2SO_4 or $MgSO_4$) or a more complex ion built from such units (as in pyro- or meta- salts). In many such structures the oxygen atoms are close-packed, so that these structures, like those of many complex oxides, may be described as c.p. assemblies in which various proportions of tetrahedral and octahedral holes are occupied (Table 4.8, p. 176). Alternatively the structures may be described in terms of the way in which the tetrahedra and octahedra are joined together by sharing vertices, edges, or faces. It is not proposed to attempt an elaborate classification of this type but it is of interest to note a few simple examples of one class if only to show the relation between certain structures which are mentioned in later chapters.

We have already noted examples of structures in which tetrahedra *or* octahedra share various numbers of vertices. Structures in which tetrahedra *and* octahedra share only vertices form an intermediate group. Some of the simplest of this large family of structures are set out in Table 5.7; the structures are described in later chapters. The table shows the number of octahedra (o) with which each tetrahedron (t) shares vertices, the relative numbers of tetrahedra and octahedra, the contributions to the chemical formula made by each (a shared vertex counting as $\frac{1}{2}$ X and an unshared vertex as X), the formula, and examples of compounds with the structure.

Structures of very simple topological types arise if each tetrahedron shares p vertices with octahedra and each octahedron shares the same number of vertices with tetrahedra, for the structure is then based on a p-connected net. For $p = 3$ the simplest nets are the 2D 6-gon net and the 3D 10-gon nets; for $p = 4$ they are the 2D 4-gon net and the 3D diamond net. The formulae in the two cases would be ABX_7

TABLE 5.7

Structures in which tetrahedral AX_4 and octahedral BX_6 groups share only vertices

			Formula			*Example*
All vertices of t and o shared	$\left.\begin{array}{l} t-4\ o \\ o-6\ t \end{array}\right\}$ t_3o_2		$(AX_2)_3(BX_3)_2 = A_3B_2X_{12}$			Garnet framework (see text)
	$\left.\begin{array}{l} t-4\ o \\ o-4\ t \\ \ \ -2\ o \end{array}\right\}$ to		AX_2	BX_3	ABX_5	$NbOPO_4$ §
	$\left.\begin{array}{l} t-2\ o \\ \ \ -2\ t \\ o-2\ t \\ \ \ -4\ o \end{array}\right\}$ to		AX_2	BX_3	ABX_5	$Ca_2[Fe_2O_5]$ †
Each t and each o sharing 4 vertices	$\left.\begin{array}{l} t-4\ o \\ o-4\ t \end{array}\right\}$ to		AX_2	BX_4	ABX_6	$Na[PWO_6]$ §
	$\left.\begin{array}{l} t-2\ t \\ \ \ -2\ o \\ o-4\ t \end{array}\right\}$ t_2o		$(AX_2)_2\ BX_4$		A_2BX_8	P_2MoO_8 §
t sharing 2 vertices o sharing 4 vertices	$\left.\begin{array}{l} t-2\ o \\ o-2\ t \\ \ \ -2\ o \end{array}\right\}$ to		AX_3	BX_4	ABX_7	Re_2O_7 ‡ $Na_2[Mo_2O_7]$ ‡
	$\left.\begin{array}{l} t-2\ o \\ o-4\ t \end{array}\right\}$ t_2o		$(AX_3)_2\ BX_4$		A_2BX_{10}	$K_2[Mo_3O_{10}]$ ‡ $H_5[As_3O_{10}]$ ‡

† Square brackets enclose that part of the formula which corresponds to the t-o complex.
‡ Same element in t and o (A = B).
§ For these structures see p. 619.

and ABX_6 respectively. An example of an ABX_6 structure of this kind, but based on a more complex 3D net, is the PWO_6 framework in $NaPWO_6$ (p. 623); examples of the simpler ABX_6 layer do not appear to be known. Fig. 13.17(a), p. 621, would serve as an illustration of this layer, for this figure is the projection of the 3D ABX_5 structure which results from sharing of the remaining vertices of the octahedra of ABX_6 layers. The same illustration represents the structure of the closely related layer in $Na_2(SiTiO_5)$ (AC 1978 **B34** 905) if the squares represent square pyramidal TiO_5 groups whose vertices point alternately to opposite sides of the layer. In Na_2SiTiO_5 the layers are held together by the Na^+ ions, but if the latter are removed and the layers are brought closer together so as to form (vertex-sharing) octahedral groups around the B atoms, the formula of the 3D structure is ABX_5; this is the type of structure of $PNbO_5$ and related compounds. Fig. 24.18(b), p. 1079, shows that the structure of bandylite, $BCu(OH)_4Cl$, is of the same general type, though the octahedral coordination groups are distorted in different ways in the two structures (compare the bond lengths listed on pp. 622 and 1079). Note that on p. 620 the $PNbO_5$ structure is described in terms of chains of octahedra, perpendicular to the projection of Fig. 13.17(a), joined through tetrahedra; here we have emphasized the layers of tetrahedra and octahedra.

In the garnet structure the $A_3B_2X_{12}$ system of linked tetrahedra and octahedra is a charged 3D framework which accommodates larger ions C, (in positions of 8-coordination) in the interstices. The general formula is $C_3A_3B_2X_{12}$ or $C_3B_2(AX_4)_3$ if we wish to distinguish the tetrahedral groups in, for example, an orthosilicate such as $Ca_3Al_2(SiO_4)_3$. In some garnets the same element occupies the positions of tetrahedral and octahedral coordination, when the formula reduces to $Y_3Al_5O_{12}$, for example. In all the last four examples of Table 5.7 there is both tetrahedral and octahedral coordination of one element in the same structure, and as a matter of interest we have collected together in Table 5.8 a more general set of examples of this phenomenon, not restricted to vertex-sharing structures. It may also be of interest to set out the elements which exhibit both these coordination numbers when *in the same oxidation state*, in relation to the Periodic Table:

						Mg	Al		
		Mn^{2+}	Fe^{3+}	Co^{2+}	Ni^{2+}	Zn	Ga	Ge	As
Nb^{5+}	Mo^{6+}					Cd			
		Re^{7+}							

Referring again to the last four examples of Table 5,7 it may be noted that Re_2O_7 has a layer structure (p. 549), while the other three are chain structures. In the simple chain of Fig. 5.45(a), the topological repeat unit consists of one tetrahedron and one octahedron. This is the form of the anion in $Na_2Mo_2O_7$. The more complex chain of Fig. 5.45(b) is found in $K_2Mo_3O_{10}$, and a closely-related chain (Fig. 20.7) is the structural unit in $H_5As_3O_{10}$, an intermediate dehydration product of $As_2O_5 \cdot 7H_2O$. (For $K_2Mo_3O_{10}$ see also p. 516.)

TABLE 5.8
*Structures in which an element exhibits both tetrahedral and
octahedral coordination*

Structure	Element
γ-Fe_2O_3	Fe
β-Ga_2O_3, θ-Al_2O_3	Ga, Al
$X^{II}X_2^{III}O_4$ (regular spinel)	Mn, Co
$Y(XY)O_4$ (inverse spinel)	Fe in Fe_3O_4 and $Fe(MgFe)O_4$, Al in $Al(NiAl)O_4$, Co in $Co(SnCo)O_4$, Zn in $Zn(SnZn)O_4$
$Na_2Mo_2O_7$, $K_2Mo_3O_{10}$	Mo
$H_5As_3O_{10}$, As_2O_5	As
Re_2O_7	Re
$M_2^{II}Mo_3O_8$	Mg, Mn, Fe, Co, Ni, Zn, Cd
$Zn_5(OH)_6(CO_3)_2$, $Zn_5(OH)_8Cl_2 . H_2O$	Zn
$(Mg_2Al)(OH)_4SiAlO_5$, $KAl_2(OH)_2Si_3AlO_{10}$	Al
$Na_4Ge_9O_{20}$, $K_3HGe_7O_{16}$	Ge
$Y_3Fe_2(FeO_4)_3$, $Y_3Al_5O_{12}$ (garnets)	Fe, Al
$Nb_{25}O_{62}$, high-Nb_2O_5	Nb
Cr_5O_{12}, KCr_3O_8	Cr(VI and III)

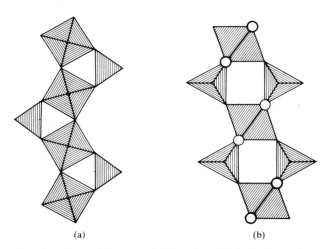

(a)   (b)

FIG. 5.45. Chains formed from tetrahedra and octahedra sharing only vertices.

Of the structures listed in Table 5.8 the sesquioxide, spinel, and other complex oxide structures are described in Chapters 12 and 13. The structures of the two 'basic' zinc salts $Zn_5(OH)_6(CO_3)_2$ and $Zn_5(OH)_8Cl_2 . H_2O$ are also described in more detail on p. 264. The composite layers with general formulae $Al_2(OH)_4Si_2O_5$ (or $Mg_3(OH)_4Si_2O_5$) (Fig. 5.46) and $Al_2(OH)_2Si_4O_{10}$ (or $Mg_3(OH)_2Si_4O_{10}$) which

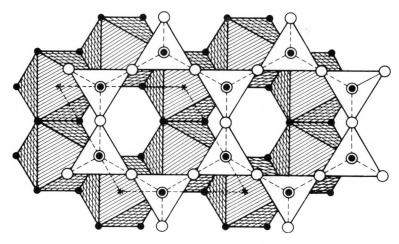

FIG. 5.46. The composite kaolin layer.

form the bases of the structures of many clay minerals and micas are described in Chapter 23. Germanates provide examples of crystals containing Ge in positions of tetrahedral and octahedral coordination. A particularly elegant example, the 3D $Ge_7O_{16}^{4+}$ framework ion in $K_3HGe_7O_{16}.4H_2O$, is shown in Fig. 5.47. We have referred

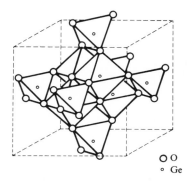

○ O
∘ Ge

FIG. 5.47. Framework of linked GeO_4 and GeO_6 groups in $K_3HGe_7O_{16}.4H_2O$.

earlier to oxides such as $Nb_{25}O_{62}$ and the high-temperature form of Nb_2O_5 in which a very small fraction of the metal atoms occupy tetrahedral holes in an essentially octahedral structure.

6

Some simple AX$_n$ structures

We described in Chapter 3 a number of A$_m$X$_n$ structures in which the A atoms form three or four bonds. We now consider four simple 3D structures in which A has six or eight nearest neighbours and, as in the case of the diamond structure, we shall show how these structures can be adapted to suit the bonding requirements of various kinds of A and X atoms. The structures are:

AX   Sodium chloride (6:6); Caesium chloride (8:8).
AX$_2$ Rutile (6:3); Fluorite (8:4).

In this chapter the emphasis is on the geometry of these pairs of AX and AX$_2$ structures. The reasons for the particular types of coordination (octahedral in NaCl and rutile, cubic in CsCl and CaF$_2$) and for the changes in structure with change in relative ionic sizes in halides and oxides are discussed in Chapter 7. Then follow sections on:

The CdI$_2$ and related structures;
The ReO$_3$ and related structures.

Next we consider some groups of structures:

The PbO and PH$_4$I structures;
The LiNiO$_2$, NaHF$_2$, and CsICl$_2$ structures;
The CrB, TlI (yellow), and related structures;
The PbCl$_2$ structure;
The PdS$_2$, AgF$_2$, and β-HgO$_2$ structures;

which are related in the following way. All structures in a particular group have the same symmetry (space group) and the atoms occupy the same sets of equivalent positions, but owing to the different values of one or more variable parameters, the similar analytical descriptions refer to atomic arrangements with quite different geometry (and/or topology). The chapter concludes with notes on the relations between the structures of some nitrides and oxy-compounds and on superstructures and other related structures.

The sodium chloride structure

In this structure (Fig. 6.1(a)) the A and X atoms alternate in a simple cubic sphere packing. it is the only AX structure in which there is regular octahedral coordination of both atoms, and it is notable for a number of reons. It is the structure of more than two hundred compounds of all chemical types ranging from the essentially ionic halides and hydrides of the alkali metals and the monoxides and monosulphides

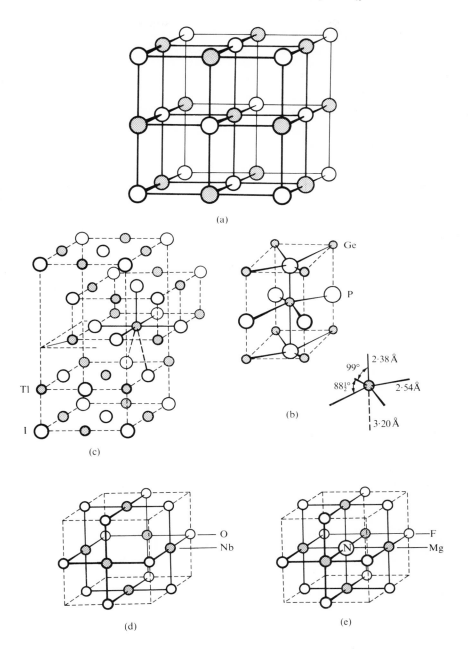

FIG. 6.1. The sodium chloride and related structures: (a) NaCl; (b) tetragonal GeP; (c) yellow
TlI; (d) NbO; (e) Mg$_3$NF$_3$.

of Mg and the alkaline-earths, through ionic-covalent compounds such as transition-metal monoxides to the semimetallic compounds of B subgroup metals such as PbTe, InSb, and SnAs, and the interstitial carbides and nitrides (Table 6.1). Since

TABLE 6.1

Compounds AX *with the cubic* NaCl *structure*

Alkali halides and hydrides, AgF, AgCl, AgBr

Monoxides of
Mg								
Ca	Ti	V	–	Mn	Fe	Co	Ni	
Sr	Zr	Nb						Cd
Ba	Hf							

Eu; Th, Pa, U, Np, Pu, Am

Monosulphides of
Mg	
Ca	Mn
Sr	
Ba	Pb

Ce, Sm, Eu; Th, U, Pu

Interstitial carbides and nitrides

	TiC,	VC,	ZrC,	UC
ScN,	LaN,	TiN,	ZrN,	UN

Phosphides etc.
InP, InAs, SnP (high pressure), SnAs

---

the *positions* of A *or* of X in the NaCl structure are those of cubic closest packing there are clearly two extreme configurations of this structure:

(i) A atoms close-packed, X in octahedral interstices: $r_A \geqslant r_X$,

(ii) X atoms close-packed, A in octahedral interstices: $r_X \geqslant r_A$,

and an indefinitely large number of intermediate configurations. In all cases there is regular octahedral coordination of A by 6 X and of X by 6 A. In (i) A would be in contact also with 12 A, as in TiC, TiN, ZrC, and ZrN; for example, in the Ti compounds Ti–Ti is close to 3 Å as in the metal. The other extreme case, (ii), is to be expected for very small cations and very large anions; for example, LiI in the alkali halides and MgTe in the group II chalconides. In fact the NaCl structure is found for most compounds of these two families although the ratio $r_A : r_X$ covers a very wide range, including compounds such as KF and BaO where the ions are of approximately the same size; see the discussion of radius ratio in Chapter 7. The relevant distances are those between X atoms and for the alkali fluorides these are:

	Li–F	Na–F	K–F	Rb–F	Cs–F	
F–F	2.84	3.26	3.77	3.98	4.25	Å

which should be compared with 2.7 Å, twice the radius of F^-, the value to be expected for close-packed anions.

The position is similar for the monoxides of Mg and the alkaline-earths. Comparison with 2.8 Å (twice the radius of O^{2-}) shows that only in MgO are the anions approximately close-packed:

	Mg—O	Ca—O	Sr—O	Ba—O	
O—O	2.98	3.40	3.65	3.90	Å

When subjected to high pressure certain phosphides and arsenides of metals of Groups IIIB and IVB adopt either the cubic NaCl structure (InP, InAs) or a tetragonal variant of this structure (GeP, GeAs) shown in Fig. 6.1(b) in which there are bonds of three different lengths and effectively 5-coordination. The compounds of Ge and Sn, with a total of nine valence electrons, are metallic conductors. This property is not a characteristic only of the distorted NaCl structure, for SnP forms both the cubic and tetragonal structures, and both polymorphs exhibit metallic conduction. (IC 1970 **9** 335; JSSC 1970 **1** 143.)

Unique distorted variants of the NaCl structure are adopted by the Group IIIB monohalides owing to the presence of the lone pair of electrons. The structure of the yellow form of TlI (Fig. 6.1(c)) is formed from slices of the NaCl structure which are displaced relative to one another with the result that the metal ion has five nearest neighbours at five of the vertices of an octahedron and then two pairs of nearly equidistant next-nearest neighbours; for these halides see p. 411. The low-temperature form of NaOH has the yellow TlI structure.

We have included two other structures in Fig. 6.1 in which atoms occupy certain of the positions of the NaCl structure, namely, those of NbO (p. 539) and Mg_3NF_3 (p. 475).

The full cubic symmetry ($m3m$) of the NaCl structure is retained in solid solutions such as (Mg,Ni)O or high-temperature forms of complex oxides in which there is statistical distribution of cations of two or more kinds (for example, h.t. $LiFeO_2$, Na_2SnO_3, Li_3TaO_4). There may also be a statistical distribution of vacancies in *defect* structures such as non-stoichiometric $Fe_{1-x}O$ (p. 551). At lower temperatures there may be rearrangement of the atoms to an ordered *superstructure* or of the vacancies in a defect structure (see TiO, p. 562). Full cubic symmetry may also be retained in crystals containing complex ions either if the ions have full cubic symmetry, as in $[Co(NH_3)_6](TlCl_6)$, or if there is rotation or random orientation of less symmetrical groups, as in the high-temperature forms of alkaline-earth carbides, KSH, or KCN. The S_2 group may, however, be arranged along the 3-fold axes of a lower class of cubic symmetry, as in the cubic pyrites (FeS_2) structure of Fig. 6.2. Here the 3-fold axes are directed along the body-diagonals of the octants of the cell, and do not intersect. This structure is adopted by numerous chalconides of 3d metals (Mn–Cu), in some cases only under pressure (see Chapter 17), and by the high-pressure form of SiP_2. The pyrites structure and the CaC_2 structure to be described shortly are also suitable for peroxides of the alkaline-earths (containing the O_2^{2-} ion) and for the superoxides of the alkali metals, which contain the O_2^- ion (Table 6.2). At higher temperatures the structures of some of these compounds

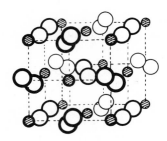

FIG. 6.2. The structure of FeS$_2$ (pyrites).

TABLE 6.2

Crystal structures of peroxides and superoxides

Pyrites structure	CaC$_2$ structure
MgO$_2$, ZnO$_2$, CdO$_2$ NaO$_2$	CaO$_2$, SrO$_2$, BaO$_2$ KO$_2$, RbO$_2$, CsO$_2$

(NaO$_2$, KO$_2$) change to the disordered pyrites structure, which is also the structure of the high-temperature form of KHF$_2$. In this structure there is random orientation of the (F—H—F)$^-$ ions along the directions of the four body-diagonals of the cube.

The structure of PdS$_2$ results from elongation of the pyrites structure in one direction so that Pd has only four (coplanar) nearest S neighbours as compared with the six octahedral neighbours of Fe in FeS$_2$; it is preferably regarded as a layer structure, and was so described in Chapter 1: for another way of describing the PdS$_2$ and pyrites structures see p. 275.

The two kinds of distortion of a cubic structure which preserve the highest axial symmetry are extension or compression parallel to a cube-edge or body-diagonal, leading to tetragonal or rhombohedral structures respectively. The structure of the tetragonal form of CaC$_2$ is an example of a NaCl-like structure in which the linear C$_2^{2-}$ ions are aligned parallel to one of the cubic axes (Fig. 22.6, p. 948), while the structures of certain monoalkyl substituted ammonium halides (for example, NH$_3$CH$_3$I and NH$_3$C$_4$H$_9$I) illustrate a much more extreme extension of the structure along a 4-fold axis. In these halides there is presumably random orientation or rotation of the paraffin chains. Parallel alignment of the CN$^-$ ions in the low-temperature form of KCN (and the isostructural NaCN) leads to orthorhombic symmetry (Fig. 22.1, p. 939).

It is usually easy to see the relation of the unit cell of a tetragonal NaCl-like structure to the cubic unit cell of the NaCl structure, as in the case of CaC$_2$ (Fig. 22.6, p. 948). The relation of rhombohedral NaCl-like structures to the cubic NaCl structure is best appreciated from models. The cubic NaCl structure itself may be

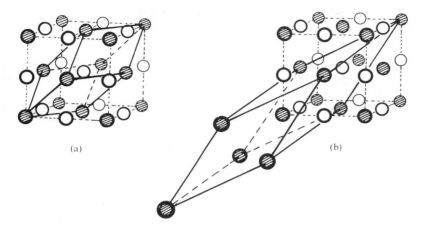

(a)   (b)

FIG. 6.3. Alternative rhombohedral unit cells in the NaCl structure.

referred to rhombohedral unit cells containing one or two formula-weights, as compared with four for the normal cubic cell, as shown in Fig. 6.3(a) and (b):

Unit cell	Edge	Angle α	$Z^\dagger$
Cubic	a	$90°$	4
Rhombohedral (a)	$a/\sqrt{2}$	$60°$	1
Rhombohedral (b)	$\sqrt{3}a/\sqrt{2}$	$33°\ 34'$	2

$^\dagger Z$ is the number of formula-units (NaCl) in the unit cell.

The smallest rhombohedral distortion of the NaCl structure is found in crystals such as the low-temperature forms of FeO and SnTe, the high-temperature polymorphs of which have the normal NaCl structure. The rhombohedral structure of low-KSH is referable to a cell of the type of Fig. 6.3(a) with $\alpha = 68°$ instead of the ideal value, $60°$. Other rhombohedral NaCl-like structures include those of LiNiO$_2$ and NaCrS$_2$, which are superstructures of NaCl, and those of NaHF$_2$, NaN$_3$, and CaCN$_2$. These groups of structures are further discussed on p. 270. The planar anions in NaNO$_3$ and CaCO$_3$ possess 3-fold symmetry and in the calcite structure of these and other isostructural salts they are oriented with their planes perpendicular to one of the 3-fold axes of the original cubic structure; the cations and the centres of the anions are arranged in the same way as the Na$^+$ and Cl$^-$ ions in NaCl. The structure of non-stoichiometric Sc$_2$S$_3$ is referable to the rhombohedral cell of Fig. 6.3(b) with α equal to the value $33°34'$, with one Sc in the position (000) but only 0.37 Sc in the position ($\frac{1}{2}\frac{1}{2}\frac{1}{2}$).

In the SnS (GeS) structure the distortion of the octahedral coordination group is such as to bring three of the six original neighbours much closer than the other three, so that the structure may alternatively be described as consisting of very buckled 6-gon nets in which alternate atoms are Sn and S. The structure has been described in this way in Chapter 3.

Sodium chloride superstructures

These are structures in which A and/or X atoms of more than one kind are arranged in a regular way in the Na and Cl positions of the NaCl structure. The simplest type is obviously a compound ABX_2 (or A_2XY), of which complex oxides and sulphides provide many examples; more complex examples include Li_2TiO_3 and Li_4UO_5. We consider here the structures possible for a compound ABX_2 in which each X atom has 3 A and 3 B nearest neighbours. There are two ways of arranging 3 A and 3 B in an octahedral arrangement around X, (a) or (b), and five superstructures in which the environments of X are:

(1): all (a): the tetragonal $InLiO_2$ structure (Fig. 6.4)

(2)⎫
     ⎬ all (b): rhombohedral $NaFeO_2$, p. 270
(3)⎭   cubic ($Fd3m$), h.p. $LiVO_2$

(4)⎫   monoclinic, $NaErO_2$
     ⎬ 50%A, 50%B:
(5)⎭   tetragonal β-$FeLiO_2$

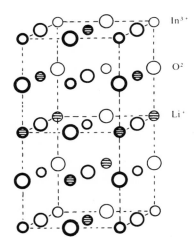

FIG. 6.4. The structure of $LiInO_2$.

Few examples seem to be known of the structures (3) and (5), but the lanthanide oxides $NaLnO_2$ (p. 577) and sulphides $MLnS_2$ (M = Li, Na, K) provide many examples of (1), (2), and (4). We have noted the relationship of SnS to the NaCl structure. There is similar $(3+3)$-coordination of Sn(II) in $BaSnS_2$ (AC 1973 **B29**

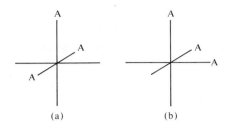

(a)   (b)

1480) and $BaSn_2S_3$ (AC 1980 **B36** 2940), which are distorted superstructures of the NaCl type.

Structures related to the NaCl structure are summarized in Chart 6.1.

CHART 6.1

The NaCl *and related structures*

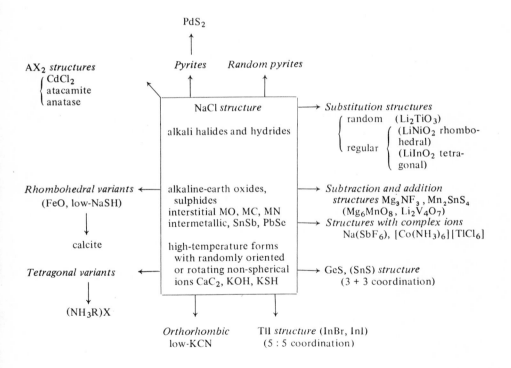

The caesium chloride structure

In this AX structure (Fig. 6.5) each atom (ion) has eight equidistant nearest neighbours arranged at the vertices of a cubic coordination group. Compared with

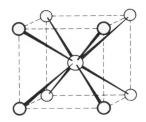

FIG. 6.5. The CsCl structure.

the NaCl structure this is an unimportant structure. It is adopted by some inter-metallic compounds, for example the ordered forms of β-brass (CuZn) and phases such as FeAl, TlSb, LiHg, LiTl, and MgTl, but not by any 'interstitial' compounds MC, MN, or MO. Only three of the alkali halides (and TlCl, TlBr, and TlI) have the CsCl structure at ordinary temperature and pressure, though some K and Rb halides adopt this structure under pressure; on the other hand those halides which do adopt the CsCl structure can be induced to crystallize with the NaCl structure on a suitable substrate.

Of the alkali hydroxides, hydrosulphides, and cyanides only CsSH and CsCN have CsCl-like structures; the structure of CsOH is not known. Only one form of CsSH is known, and it has the cubic CsCl structure in which SH⁻ is behaving as a spherically symmetrical ion with the same radius as Br⁻. The high-temperature form of CsCN has the cubic CsCl structure; the low-temperature form has the less symmetrical structure noted later. The bifluorides behave rather differently, the K and Rb salts showing a preference for CsCl-like structures. Ammonium salts differ from the corresponding alkali metal salts if there is the possibility of hydrogen bonding between cation and anion. For example, the structures of NH_4HF_2 and KHF_2 (low-temperature form) are similar in that both have CsCl-like arrangements of the ions (Fig. 8.6, p. 366) but owing to the different relative orientations of the $(F–H–F)^-$ ions in the two crystals NH_4^+ has only four nearest F^- neighbours while in KHF_2 K^+ has eight equidistant F^- neighbours. This breakdown of the coordination group of eight into two sets of four is due to the formation of N–H–F bonds; in NH_4CN the cation makes contact with only four anions, as described later. The random pyrites structure of the high-temperature form of KHF_2 has been noted under the sodium chloride structure.

As in the case of the NaCl structure, the two simplest types of distortion are those leading to rhombohedral or tetragonal structures. An example of the former is the low-temperature polymorph of CsCN, in which (with 1 CsCN in the unit cell) all the CN^- ions are oriented parallel to the triad axis (Fig. 22.2, p. 940). The edge of the tetragonal unit cell of the structure of NH_4CN (which is not known to exhibit polymorphism in the temperature range –80° to +35°C) is doubled in one direction because the CN^- ions have two different orientations (Fig. 22.3(b), p. 940). Although each NH_4^+ ion is surrounded by 8 CN^- ions, owing to the orientations of these ions there are 4 at a distance of 3.02 Å and 4 more distant (at 3.56 Å). Other compounds with CsCl-like packing of ions include KSbF$_6$, AgNbF$_6$, and isostructural compounds, [Be(H$_2$O)$_4$]SO$_4$, and [Ni(H$_2$O)$_6$]SnCl$_6$.

The rutile structure

This tetragonal structure (Fig. 6.6(a)) is named after one of the polymorphs of TiO$_2$; it is also referred to as the cassiterite (SnO$_2$) structure. The coordinates of the atoms are:

$$\text{Ti: } (000), (\tfrac{1}{2}\tfrac{1}{2}\tfrac{1}{2}).$$
$$\text{O: } \pm(xx0), (\tfrac{1}{2}+x, \tfrac{1}{2}-x, \tfrac{1}{2}).$$

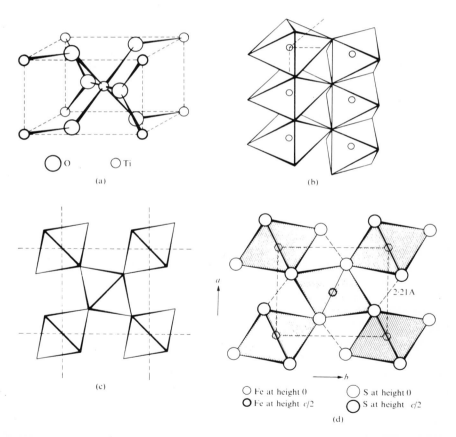

FIG. 6.6. The rutile structure: (a) unit cell; (b) parts of two columns of octahedral TiO$_6$ coordination groups; (c) projection of structure on base of unit cell; (d) corresponding projection of marcasite structure.

(In the *anti-rutile* structure the positions of metal and non-metal atoms are interchanged, as in Ti_2N.)

The structure consists of chains of TiO_6 octahedra, in which each octahedron shares a pair of opposite edges (Fig. 6.6(b)), which are further linked by sharing vertices to form a 3D structure of 6:3 coordination as shown in the projection of Fig. 6.6(c). With the above coordinates each O has three coplanar neighbours (2 at the distance d and 1 at e), Ti has six octahedral neighbours (4 at the distance d and 2 at e), and all Ti–Ti distances (between centres of octahedra along a chain) are equal. Although there are two independent variables, $c:a$ and x, in the structure as described, this number reduces to one if all the Ti–O distances are made equal ($d = e$), when there is the following relation between x and c/a: $8x = 2 + (c/a)^2$. The special cases would be: (i) regular octahedral coordination of Ti ($c/a = 0.58$, $x = 0.29$); and (ii) equilateral triangular coordination of O ($c/a = 0.817$, $x = 0.33$). We saw in Chapter 5 that both of these highly symmetrical arrangements of nearest neighbours are not possible simultaneously, for (i) implies bond angles at O of $90°$ and $135°$ (two). In dioxides and difluorides with the tetragonal rutile structure there is usually very little difference between the two M–O or M–F distances, the structure being much closer to case (i) than to (ii), with x close to 0.30. Typical data are:

	$c : a$	M–O *or* M–F 4 of	2 of
TiO_2	0·645	1·94 Å	1·99 Å
ZnF_2	0·665	2·03 Å	2·04 Å
NiF_2	0·663	1·98 Å	2·04 Å

The normal rutile structure is restricted to the difluorides of Mg, Zn and some 3d metals, a number of dioxides, some oxyfluorides (FeOF, TiOF, VOF), and MgH_2; it is *not* adopted by disulphides, by other dihalides, or by intermetallic phases, In general, compounds MF_2 and MO_2 containing larger ions adopt the fluorite structure, and since in this sense the rutile and fluorite structures are complementary we group together in Table 6.3 compounds that crystallize with one or other of these structures. Some interesting modifications of the rutile structure show how it can be adapted to suit the special bonding requirements of certain atoms (ions).

In a number of transition metal dioxides (of V, Nb, Mo, W. Tc, and Re) the distances between successive metal atoms in a chain alternate, so that there are metal-metal interactions between the close pairs, as detailed on p. 541. In the structures of CrF_2 (and the isostructural CuF_2) and $CrCl_2$ there are quite different types of distortion of the rutile structure. In both structures there is (Jahn–Teller) distortion of the octahedral MX_6 group resulting in 4 shorter and 2 longer bonds. It can be seen from Fig. 6.7(a) that in CrF_2 the metal atom at the centre of the cell is connected through F atoms to four others by stronger bonds (full lines), these five

TABLE 6.3

Dioxides and difluorides with the rutile or fluorite structures

Dioxides					Rutile structure					
								Si[b]		
Ti	V[a]	Cr	Mn					Ge		
Zr	Nb	Mo	Tc	Ru	Rh[b]			Sn		
Hf	Ta	W	Re	Os	Ir	Pt[b]		Pb	Po	

				Fluorite structure		
Ce	Pr		Tb			
Th	Pa	U	Np	Pu	Am	Cm

[a] Structure modified by metal–metal bonding.
[b] When crystallized under pressure.

Difluorides						Rutile structure		
Mg								
Ca	Cr[c]	Mn	Fe	Co	Ni	Cu[c]	Zn	
Sr							Cd	
Ba		*Fluorite structure*					Hg	Pb

[c] Distorted (see text).

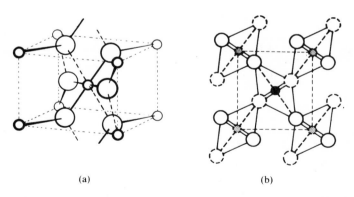

(a)   (b)

FIG. 6.7. (a) Distorted rutile structure of CuF_2. (b) Projection of the structure of $CrCl_2$.

atoms defining a plane passing through the centre of the cell. In the limit this type of distortion would lead to a layer structure, though the differences between the two M—F bond lengths in these fluorides is far too small to justify such a description. In

CrCl$_2$ (Fig. 6.7(b)) there is a different arrangement of the short and long bonds; this type of distortion would lead in the limit to chains in which the metal atoms form four coplanar bonds as in PdCl$_2$ (p. 414).

The nature of the distortion from ideal hexagonal closest packing of the packing of the X atoms in the tetragonal rutile structure can be seen by comparing Fig. 4.21 (a) and (b), p. 169. In the h.c.p. structure X would lie slightly out of the plane of its 3 M neighbours (interbond angles 90° (one) and 132° (two)) as compared with 90° (one) and 135° (two) in the rutile structure. The structure of CaCl$_2$ approximates closely to the h.c.p. structure and so also do the structures of CdF(OH) and InO(OH). In the tetragonal rutile structure built from regular octahedra there is maximum separation of X atoms of adjacent chains ($\sqrt{3}/\sqrt{2}$ or 1.22 times the length of the octahedron edge). Rotation of adjacent chains, in opposite senses, through approximately 10° gives the h.c.p. structure with reduced distances between X atoms of adjacent chains, as in the hydrogen-bonded InO(OH) which is shown in projection in Fig. 14.11 (p. 639). In marcasite (FeS$_2$) the greater relative rotations of the chains result from the presence of S–S bonds (2.21 Å) as shown in Fig. 6.6(d). Examples of compounds with the marcasite and the closely related löllingite and arsenopyrite structures are given in Chapter 17.

We have seen in the previous chapter that of the two simple octahedral MX$_2$ structures with 6:3 coordination, namely, one in which only vertices are shared and one in which the three octahedra meeting at each X atom share only one edge, the former is not to be expected for regular octahedral coordination; only modified forms of the structure are found, as described on p. 275. It is interesting to compare these with the modified forms of the edge-sharing rutile structure:

MX$_2$ *structure*	*Regular octahedra*	*Distorted octahedra* (4 + 2) (2 + 4)		X–X *bonds*
Edge-sharing	MO$_2$, MF$_2$ CaCl$_2$ InO(OH), CdF(OH)	CrF$_2$, CuF$_2$ CrCl$_2$	HgF(OH)	FeS$_2$ (pyrites)
Vertex-sharing	—	AgF$_2$	β-HgO$_2$	FeS$_2$ (marcasite)

Compounds ABX$_4$, A$_2$BX$_6$, etc. with rutile-like structures

The cation sites in the rutile structure may be occupied by cations of two or more different kinds, either at random (random rutile structure) or in a regular manner (superstructure). In the former case the structure is referable to the normal rutile unit cell, but the unit cell of a superstructure is usually, though not necessarily, larger than that of the statistical structure. Phases with disordered structures arise if the charges on the cations are the same or not too different. If the charges are the same the composition may be variable, as in solid solutions (Mn,Cr)O$_2$, but it is necessarily fixed if the charges are different (CrNbO$_4$, FeSbO$_4$, AlSbO$_4$, etc).

Ordered arrangements of cations arise when either
  (a) the ionic charges are very different, or
  (b) special bonding requirements of particular atoms have to be satisfied.

(a) A number of complex fluorides and oxides A_2BX_6 (X = F or O) crystallize with the *trirutile structure* (Fig. 6.8). Fluorides $A^{2+}B_2^{2+}F_6$ (e.g. FeMg$_2$F$_6$) adopt

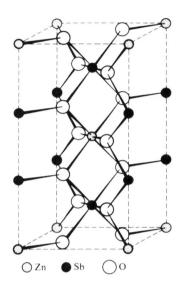

$\bigcirc$ Zn $\bullet$ Sb $\bigcirc$ O

FIG. 6.8. The trirutile structure of ZnSb$_2$O$_6$.

the disordered rutile structure but Li$_2^+$Ti^{4+}F$_6$ and one form of Li$_2$GeF$_6$ have the trirutile structure. There are three possible types of fluoride ABCF$_6$, namely, A$^+$B$^+$C^{4+}F$_6$, A^{1+}B^{2+}C^{3+}F$_6$, and A^{2+}B^{2+}C^{2+}F$_6$. The random rutile structure is adopted if all the ions are M^{2+} (as in FeCoNiF$_6$), but the trirutile structure is adopted by compounds of the first two classes. The arrangement of the different kinds of ion in the trirutile structure is such as to give maximum separation of the M^{4+} ions in the first group of compounds (e.g. Li$_2$TiF$_6$) or of the M$^+$ ions in the second group (e.g. LiCuFeF$_6$), that is, these ions occupy the Zn positions in Fig. 6.8. (ZN 1967 **22b** 1218; JSSC 1969 **1** 100). Oxides with the trirutile structure include MgSb$_2$O$_6$, MgTa$_2$O$_6$, Cr$_2$WO$_6$, LiNbWO$_6$, and VTa$_2$O$_6$ (JSSC 1970 **2** 295).

(b) In MgUO$_4$ the U atom has, as in many uranyl compounds, two close neighbours (at approximately 1.9 Å) and four more distant O neighbours (at 2.2 Å). Within one rutile chain all metal atoms are U atoms. The environment of Mg^{2+} is similar to that of U, 2 O at 2.0 Å and 4 O at 2.2 Å, this evidently being a secondary result of the U coordination. In CuUO$_4$, on the other hand, Cu and U atoms alternate along each rutile chain, and here the rutile structure is modified to give U(2 + 4)-

coordination (at the mean distances 1.9 Å and 2.2 Å close to those in $MgUO_4$) but Cu has (4 + 2)-coordination (at 1.96 Å and 2.59 Å).

Homologous series of oxides structurally related to the rutile structure. Vanadium and titanium form series of oxides M_nO_{2n-1} which have structures related to the rutile structure in the following way. Slabs of rutile-like structure, *n* octahedra thick but extending indefinitely in two dimensions, are connected across 'shear planes'. These structures can be derived from the rutile structure by removing the O atoms in certain planes and then displacing the rutile slabs so as to restore the octahedral coordination of the metal atoms. This operation results in the sharing of faces between certain pairs of octahedra (in addition to the vertex- and edge-sharing in the rutile blocks) as in the corundum structure (p. 173) — compare the shear structures of Mo and W oxides derived in a similar way from the vertex-sharing ReO_3 structure to give structures in which some octahedron edges are shared. References to the Ti oxides ($4 \leqslant n \leqslant 9$) and the V oxides ($3 \leqslant n \leqslant 8$) are given in Chapter 12.

The fluorite (AX_2) and antifluorite (A_2X) structures

In the fluorite structure (Fig. 6.9) for a compound AX_2 the A atoms (ions) are surrounded by 8 X at the vertices of a cube and X by 4 A at the vertices of a regular

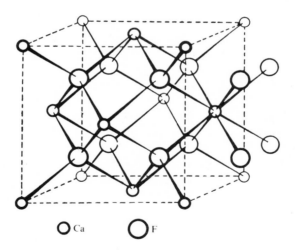

O Ca   O F

FIG. 6.9. The fluorite structure.

tetrahedron. It is the structure of a number of difluorides and dioxides (Table 6.3) and also of some disilicides (see later), intermetallic compounds (for example, $GeMg_2$, $SnMg_2$, $PtAl_2$ etc), and transition-metal dihydrides.

The *positions* of the Ca^{2+} ions in the fluorite structure are those of cubic closest packing and the *positions* of the F^- ions correspond to all the tetrahedral interstices.

There are two extreme configurations of the structure:

(i) A atoms (ions) in contact (c.c.p.), X in tetrahedral interstices: $r_X \ll r_A$;

(ii) X atoms in contact (simple cubic packing), A in cubical holes: $r_X > r_A$;

and an indefinite number of intermediate configurations such as BaF$_2$, in which $r_X \approx r_A$. Assuming 'hard' spheres these structures differ as regards the contacts between atoms in the following way:

		(i)		Intermediate structures	(ii)	
Contacts	A	12A	8X	8X	8X	
	X		4A	4A	4A	6X
Examples		YH$_2$		BaF$_2$, SrF$_2$	CaF$_2$	

In YH$_2$ a Y atom has, in addition to its 8 H neighbours, 12 Y neighbours at the same distance (3.65 Å) as in metallic yttrium. In CaF$_2$ the distance (2.70 Å) between F$^-$ and its 6 nearest F$^-$ neighbours is equal to twice its ionic radius, and the structure is therefore an example of the configuration (ii). The shortest distance between Ca^{2+} ions is 3.8 Å, which may be compared with twice the radius of Ca^{2+} (2.0 Å).

In the *antifluorite* (A$_2$X) structure the positions of cations and anions are interchanged, and again there are two extreme configurations:

(iii) X atoms close-packed, A in tetrahedral holes;

(iv) A atoms in contact, X atoms in cubical holes.

The contacts are as shown above, interchanging A and X throughout. Examples are:

	(iii)	Intermediate structures	(iv)
	Li$_2$Te	K$_2$O	Be$_2$C

The nearest approach to the c.p. structure (iii) is found for the combination of the smallest alkali metal and the largest chalcogen. In Li$_2$Te each Te atom has 12 Te neighbours at 4.6 Å, a distance close to twice the radius of Te^{2+} (4.4 Å). In the alkali metal oxides M$_2$O the O^{2-} ions are not most closely packed; compare O–O in Li$_2$O, 3.3 Å, and in Na$_3$O, 3.9 Å, with twice the radius of O^{2-} (2.8 Å). Their structures are intermediate between the configurations (iii) and (iv). An example of the configuration (iv) is Be$_2$C, in which Be has 6 Be neighbours at 2.2 Å, which is the interatomic distance in the metal. The antifluorite structure is adopted by all the oxides M$_2$O and sulphides M$_2$S of Li, Na, K. and Rb; for Cs$_2$O and Cs$_2$S see pp. 537 and 750 respectively. Note that in K$_2$O and Rb$_2$O the cations are comparable in size with O^{2-}, yet 4-coordination of the cations persists. We return to this subject in our discussion of ionic structures in Chapter 7. Other compounds with this structure include Mg$_2$Si, Mg$_2$Ge, and similar compounds.

A considerable number of fluorides, oxides, and oxyfluorides have structures which are related more or less closely to the fluorite structure. The relation is closest for compounds with cation : anion ratio equal to 2:1 with random arrangement of cations or anions, these structures having the normal cubic unit cell; examples include the high-temperature forms of $NaYF_4$ and K_2UF_6 and the oxyfluorides AcOF and HoOF. As in the case of the NaCl structure there are tetragonal and rhombohedral superstructures, which are described in Chapter 10. These are the structures of a number of oxyfluorides:

> Tetragonal: YOF, LaOF, PuOF;
> Rhombohedral: YOF, LaOF, SmOF, etc.

The fluorite superstructure of γ-Na_2UF_6 is illustrated in Fig. 28.3 (p. 1258). There is slightly distorted cubic coordination of M^{5+} in Na_3UF_8 (and the isostructural Na_3PuF_8). Owing to the arrangement of cations in this structure (Fig. 6.10) the unit cell is doubled in one direction and is a b.c. tetragonal cell.

The ternary fluorides $SrCrF_4$ (and isostructural $CaCrF_4$, $CaCuF_4$, and $SrCuF_4$) and $SrCuF_6$ have structures which are related less closely to fluorite. The cell dimensions correspond to cells of the fluorite type doubled and tripled respectively in one direction, and the cation positions are close to those of the fluorite structure of, for example, SrF_2, as shown in Fig. 6.10. There is, however, considerable movement of the F^- ions from the 'ideal' positions to give suitable environments for the transition-metal ions. The Sr^{2+} ions retain cubic 8-coordination but Cr^{2+} (Cu^{2+}) ions are surrounded by an elongated tetrahedron of anions (two angles of $94°$, four of $118°$). In a series of oxides MTe_3O_8 (M = Ti, Zr, Hf, Sn) the cation positions are close to those of the fluorite structure, as shown in the sub-cell of Fig. 6.10, but the actual unit cell has dimensions twice as large in each direction and the O atoms are displaced to give very distorted octahedral coordination of both M and Te (for example Te–4O, $2.04 Å$, Te–2 O, $2.67 Å$).

An interesting distortion of the fluorite structure to give weak metal–metal interactions is reminiscent of the distorted rutile structures of certain dioxides. Whereas $CoSi_2$ and $NiSi_2$ crystallize with the cubic fluorite structure, this structure is distorted in β-$FeSi_2$ to give Fe two Fe neighbours, the metal atoms being arranged in squares of side $2.97 Å$ (compare $2.52 Å$ in the 12-coordinated metal). The metal atom also has 8 Si neighbours at the vertices of a deformed cube. This distortion of the fluorite structure represents a partial transition towards the $CuAl_2$ structure (p. 1313), in which the coordination group of Cu is a square antiprism and there are chains of bonded Cu atoms (AC 1971 **B27** 1209).

Addition of anions to the fluorite structure: the Fe_3Al structure

In a number of metal–non-metal systems the cubic fluorite-like phase is stable over a range of composition. For the hydrides of the earlier 4f metals (La–Nd) this phase is stable from about $MH_{1.9}$ to compositions approaching MH_3. (The range of stability of the 'dihydride' phase of the later 4f metals is more limited, and separate from that of the 'trihydride', which has a different (LaF_3) structure.) The only sites available for the additional anions are at the mid-points of the edges and at the

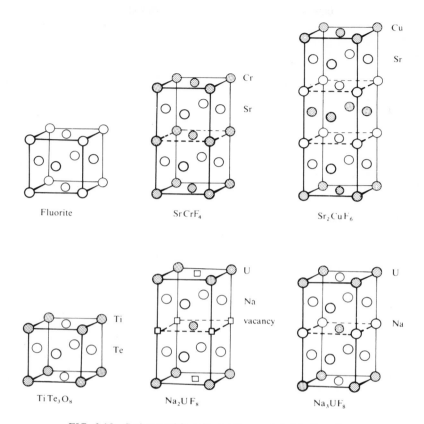

FIG. 6.10. Cation positions in structures related to fluorite.

body-centre of the fluorite cell (the larger shaded circles in Fig. 6.11). The resulting AX$_3$ structure is sometimes described as the BiF$_3$ structure because it was once thought (erroneously) to be the structure of one form of that compound (see p. 422). It is preferably termed the Fe$_3$Al (or Li$_3$Bi) structure since it is the structure of the ordered form of a number of intermetallic phases.

In the Fe$_3$Al structure there is cubic 8-coordination of all the atoms, and writing the formula AX$_2$X' the nearest neighbours are:

$$A:8\ X,\ X:4\ A,\ and\ X':8\ X.$$
$$4\ X'$$

The environment of the (additional) X' atoms is therefore a group of 8 X at 0.433a (a = cell edge), and the nearest A neighbours are an octahedral group of 6 A at 0.5a. Although satisfactory in an intermetallic compound this is an unlikely environment for an anion, and although fluorite-like phases certainly exist for the compounds

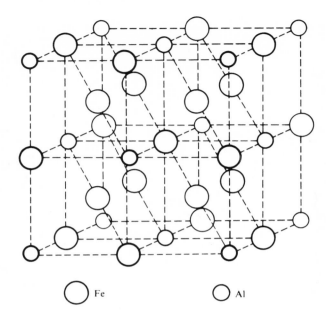

FIG. 6.11. The Fe$_3$Al structure.

MH$_{2+x}$ and UO$_{2+x}$ it seems likely that some rearrangement of the anions occurs in these structures. It seems improbable that any phases, other than intermetallic compounds form the ideal structure of Fig. 6.11 with its full complement of X' atoms. For further details of compounds with this type of structure see the discussion of the 4f hydrides (p. 348), Bi$_2$OF$_4$ (p. 422), U oxides UO$_{2+x}$ (p. 1263), and also Table 29.7 (p. 1302).

Defect fluorite structures

Fluorite structures deficient in cations seem to be rare; an example is Na$_2$UF$_8$ (earlier described as Na$_3$UF$_9$) in which there are discrete cubic UF$_8$ groups. The arrangement of the Na$^+$ ions is the same as that of two-thirds of those ions in Na$_3$UF$_8$, as shown in Fig. 6.10. The compound Li$_9$N$_2$Cl$_3$ has a defect antifluorite structure, the 10% Li$^+$ vacancies leading to ionic conductivity. On the other hand, there are a number of phases (mostly oxides) with structures derivable from fluorite by removing a fraction of the *anions*. These structures are summarized in Table 6.4.

The C-M$_2$O$_3$ structure (p. 545) may be derived from fluorite by removing ¼ of the anions and slightly rearranging the remainder. All the M atoms are octahedrally coordinated, the coordination being rather more regular for one-quarter of these atoms. Solid solutions UO$_{2+x}$-Y$_2$O$_3$ form defect f.c.c. structures with a degree of anion deficiency depending on the U content and on the partial pressure of oxygen. When the O deficiency becomes too large rearrangement takes place to form a

TABLE 6.4
Structures related to the fluorite structure

Excess cations	Cation defective
Fe$_3$Al structure	Na$_2$UF$_8$[1]
(MH$_3$)	Li$_9$N$_2$Cl$_3$
	(antifluorite)

FLUORITE
STRUCTURE

Random fluorite	Anion defect[6]	Anion defective
High-NaYF$_4$	$\frac{1}{14}$	Zr$_5$Sc$_2$O$_{13}$, Zr$_7$O$_{11}$N$_2$[5]
High-K$_2$UF$_6$	$\frac{1}{12}$	Pr$_6$O$_{11}$
AcOF		
Superstructures	$\frac{1}{8}$	Na$_3$UF$_7$
LaOF etc.		(pyrochlore A$_2$B$_2$X$_7$)
Na$_2$UF$_6$	$\frac{1}{7}$	M$_7$O$_{12}$(Ce, Pr, Tb)
Na$_3$UF$_8$[2]		Zr$_3$Sc$_4$O$_{12}$, Zr$_7$O$_8$N$_4$[5]
SrCrF$_4$[3]	$\frac{1}{4}$	C–M$_2$O$_3$ structure
Sr$_2$CuF$_6$[3]		
Bi$_2$UO$_6$[3a]		
TiTe$_3$O$_8$[4]		

(1) IC 1966 **5** 130
(2) JCS 1969 **A** 1161
(3) JSSC 1970 **2** 262
(3a) AC 1975 **B31** 127
(4) AC 1971 **B27** 602
(5) AC 1968 **B24** 1183; for other compounds see later chapters
(6) For a systematic treatment of the homologous series M$_n$O$_{2n-2}$ see: ZK 1969 **128** 55

rhombohedral phase M^{VI}M$_6^{III}$O$_{12}$ (M = Mo, W, U) in which M^{VI} is 6- and M^{III} 7-coordinated. This phase can also be formed by oxides M$_4^{III}$M$_3^{IV}$O$_{12}$ (e.g. Sc$_4$Zr$_3$O$_{12}$) and also as an oxide M$_7$O$_{12}$ by elements which can form ions M^{3+} and M^{4+} (e.g. Ce, Pr, Tb) (p. 607). Intermediate between this phase and a dioxide is the phase Zr$_5$Sc$_2$O$_{13}$, corresponding to removal of only $\frac{1}{14}$ of the O atoms, instead of $\frac{1}{7}$, from a fluorite structure. In these oxygen-deficient phases O retains its tetrahedral coordination, while the coordination of the M ions falls:

Oxygen deficiency:	0	$\frac{1}{14}$	$\frac{1}{7}$	$\frac{1}{4}$
Composition:	MO$_2$	M$_7$O$_{13}$	M$_7$O$_{12}$	M$_2$O$_3$
Coordination of M:	8	6, 7, 8	6, 7	6

Other defect fluorite structures with intermediate degrees of anion deficiency are noted elsewhere: Pr$_6$O$_{11}$ (p. 543) and Na$_3$UF$_7$ (p. 1260).

The pyrochlore structure

This structure (Fig. 7.3, between pp. 318 and 319) forms a link between the defect fluorite structures we have been describing and a topic discussed at the end of this chapter. With its complete complement of atoms this (cubic) structure contains eight $A_2B_2X_6X'$ in the unit cell, and there is only one variable parameter, the x parameter of the 48 X atoms in the position $(x, \frac{1}{8}, \frac{1}{8})$ etc. The structure may be described in two ways, according to the value of x:

(i) $x = 0.375$; coordination of A cubic and of B a very deformed octahedron (flattened along a 3-fold axis). This is a defect superstructure of fluorite, $A_2B_2X_7\square$ ($\square$ = vacancy).

(ii) $x = 0.3125$; coordination of A a very elongated cube (i.e. a puckered hexagon + 2), and coordination of B, a regular octahedron. This describes a framework of regular octahedra (each sharing vertices with six others) based on the diamond net, having large holes which contain the X' and 2 A atoms, which themselves form a cuprite-like net A_2X' interpenetrating the octahedral framework, as noted in Chapter 3. Because of the rigidity of the octahedral B_2X_6 framework this structure can tolerate the absence of some of the other atoms as illustrated for complex oxides in Chapter 13.

The coordination groups of the A atoms for (i) and (ii) are illustrated in Fig. 6.12. In fact no examples of this structure are known with x as high as 0.375, the

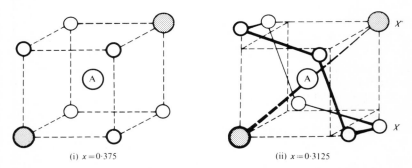

(i) $x = 0.375$ (ii) $x = 0.3125$

FIG. 6.12. Coordination of A atoms in the pyrochlore structure, $A_2B_2X_6X'$.

maximum observed value being 0.355, but x is less than 0.3125 in some cases ($Cd_2Nb_2O_7$, $x = 0.305$), when the BX_6 octahedron is elongated. The relation of the limiting (unknown) structure (i) to fluorite is interesting because certain compounds show a transition from the pyrochlore structure to a defect fluorite structure.

The CdI_2 and related structures

The simple ($C6$) CdI_2 structure (also referred to as the brucite, $Mg(OH)_2$, structure) is illustrated in Fig. 6.13 as a 'ball-and-spoke' model. It has been described in

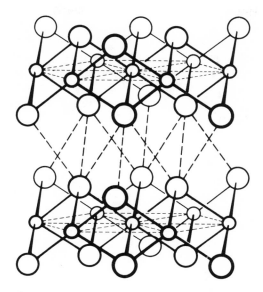

FIG. 6.13. Portions of two layers of the CdI$_2$ structure. The small circles represent metal atoms.

previous chapters in two other ways, as a hexagonal closest packing of I$^-$ ions in which Cd^{2+} ions occupy all the octahedral holes between alternate pairs of c.p. layers (that is one-half of all the octahedral holes), and as built from layers of octahedral CdI$_6$ coordination groups each sharing an edge with each of six adjacent groups. These three ways of illustrating one layer of the structure are shown in Fig. 6.14. From the c.p. description of the structure it follows that there is an indefinite number of closely related structures, all built of the I—Cd—I layers of Fig. 6.14, but differing in the c.p. layer sequences. The three simplest structures of the family are often designated by their Strukturbericht symbols:

	c.p. sequence	
$C\,6$	h $(AB\ldots)$	2 H
$C\,19$	c $(ABC\ldots)$	3 C
$C\,27$	hc $(ABAC\ldots)$	4 H

CdI$_2$ itself, mixed crystals CdBrI, PbI$_2$, etc. crystallize with more than one of these structures and also with structures with much more complex layer sequences; more than 80 polytypes of CdI$_2$ have been characterized. These are not of special interest here since their formation is obviously connected with the growth mechanism and they are all built of the same basic layer. We wish to show here how this very simple layer is utilized in the structures of a variety of compounds, and we shall not be particularly concerned with the mode of stacking of the layers. We shall describe five types of structure, starting with those in which the layer is of the simplest

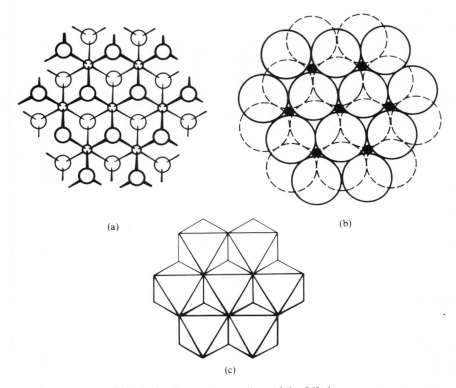

(a)

(b)

(c)

FIG. 6.14. Three representations of the CdI_2 layer.

possible type, namely, in compounds MX_2 or M_2X in which all M atoms or all X atoms are of the same kind.

(i) MX_2 *and* M_2X *structures*

The numerous compounds with these simple layer structures are described in other chapters; they include:

C 6 structure: many dibromides and diiodides, dihydroxides, and disulphides,
C 19 structure: many dichlorides,
C 27 structure: β-TaS_2, some dihalides.

The following compounds crystallize with onr or other of the anti-CdI_2 or anti-$CdCl_2$ structures, i.e. structures in which non-metal atoms occupy octahedral holes between layers of metal atoms:

Ag_2F, Ag_2O (under pressure), Cs_2O, Ti_2O, Ca_2N, Ti_2S, and W_2C.

In the rhombohedral form of $CrO.OH$ (and the isostructural $CoO.OH$) CdI_2-type layers are directly superposed and held together by short $O—H—O$ bonds.

The salts $Zr(HPO_4)_2.H_2O$ (IC 1969 8 431) and the γ form of $Zr(SO_4)_2.H_2O$ (AC 1970 **B26** 1125) have closely related structures based on $C\,6$ layers in which the 3-connected unit is a tetrahedral oxy-ion bridging three metal ions. In $Zr(HPO_4)_2.H_2O$ Zr^{4+} is 6-coordinated by O atoms of $O_3P(OH)^{2-}$ ions which lie alternately above and below the plane of the metal ions (Fig. 6.15). The P—OH

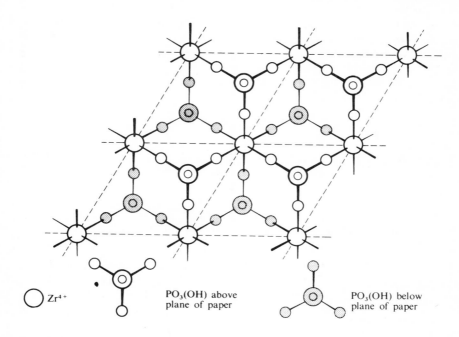

FIG. 6.15. Layer of the structure of $Zr(HPO_4)_2.H_2O$. The linking of anions and cations is similar in $\gamma\text{-}Zr(SO_4)_2.H_2O$.

bonds are perpendicular to this plane, and the H_2O molecules occupy cavities between the layers. The sulphate has a similar system of Zr^{4+} and bridging oxy-ions, but here the H_2O molecule is bonded to Zr^{4+}, which is thus 7-coordinated.

(ii) *Anions or cations of more than one kind in each* MX_2 *layer*

Hydroxyhalides $M(OH)Cl$, $M_2(OH)_3Cl$, and $M(OH)_xCl_{2-x}$ are of a number of different kinds. In some there is random arrangement of OH and Cl(Br), in others a regular arrangement of OH and Cl in each c.p. layer ($Co(OH)Cl$, $Cu_2(OH)_3Cl$), and in $Cd(OH)Cl$ alternate c.p. layers consist entirely of OH^- or Cl^- ions.

In complex hydroxides such as $K_2Sn(OH)_6$ two-thirds of the Mg positions of the $Mg(OH)_2$ layer are replaced by K and the remainder by Sn; the structure contains discrete $Sn(OH)_6^{2-}$ ions.

(iii) *Replacement of cations to form charged layers*

If part of the Mg^{2+} in a (neutral) $MgCl_2$ layer is replaced by Na^+ the layer becomes negatively charged; conversely, if part of the Mg^{2+} in a $Mg(OH)_2$ layer is replaced by Al^{3+} the layer becomes positively charged. Layers of these two types can together form a structure consisting of alternate negatively and positively charged layers, as shown at (a). The layers extend throughout the crystal, the chemical formula simply representing the composition of the repeating unit of the layers. Alternatively, the charged layers may be interleaved with ions of opposite charge (together with water, if necessary, to fill any remaining space), as at (b) and (c):

$[Na_4Mg_2Cl_{12}]_n^{4n-}$	$[Mg_6Fe_2^{III}(OH)_{16}]_n^{2n+}$	$[Ca_4Al_2(OH)_{12}]_n^{2n+}$
	$CO_3^{2-}(H_2O)_4$	$SO_4^{2-}(H_2O)_6$
$[Mg_7Al_4(OH)_{22}]_n^{4n+}$	$[Mg_6Fe_2^{III}(OH)_{16}]_n^{2n+}$	$[Ca_4Al_2(OH)_{12}]_n^{2n+}$
(a)	(b)	(c)

Note how much more informative are the 'structural' formulae of these compounds:

	Analytical formula	*Structural formula*
(a)	$4NaCl.4MgCl_2.5Mg(OH)_2.4Al(OH)_3$	$[Na_4Mg_2Cl_{12}][Mg_7Al_4(OH)_{22}]$
(b)	$MgCO_3.5Mg(OH)_2.2Fe(OH)_3.4H_2O$	$[Mg_6Fe_2(OH)_{16}]CO_3.4H_2O$
(c)	$3CaO.Al_2O_3.CaSO_4.12H_2O$	$[Ca_4Al_2(OH)_{12}]SO_4.6H_2O$

(For details see N 1967 **215** 622; ZK 1968 **126** 7; AC 1968 **B24** 972.)

Hydrogen bonding between water molecules and oxy-anions presumably contributes to the stability of these structures.

(iv) *Replacement of some OH in $M(OH)_2$ layer by O atoms of oxy-ions*

If one O atom of a nitrate ion replaces one OH^- in a $M(OH)_2$ layer the layer remains neutral, the remaining atoms of the NO_3^- ion extending outwards from the surface of the layer. The hydroxynitrate $Cu_2(OH)_3NO_3$ consists of neutral layers of this kind, O of NO_3^- replacing one-quarter of the OH^- ions in the layer, as shown diagrammatically in Fig. 6.16(a). There is hydrogen bonding between O of NO_3^- and OH of the adjacent layer. If O atoms of sulphate ions replace one-quarter of the OH^- ions in a $C6$ layer of composition $Cu(OH)_2$ the composite layer is charged, and it is necessary to interpolate cations; H_2O molecules occupy the remaining space (Fig. 6.16(b)). The structure of $CaCu_4(OH)_6(SO_4)_2.3H_2O$ is of this type, in which the relative numbers of Cu, OH, and oxy-ions are the same as in $Cu_2(OH)_3NO_3$. (In the mineral serpierite (AC 1968 **B24** 1214) there is also replacement of about one-third of the Cu by Zn.)

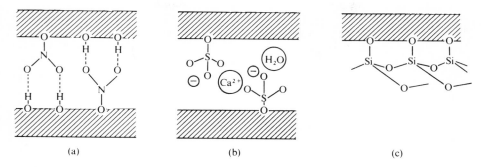

FIG. 6.16. Layers related to the CdI$_2$ layer: (a) Cu$_2$(OH)NO$_3$; (b) [Cu$_4$(OH)$_6$(SO$_4$)$_2$]$_n^{2n-}$; (c) Mg$_3$(OH)$_4$Si$_2$O$_5$.

A more complex arrangement is found in certain silicates and aluminosilicates. If one-third of the OH groups on one side of a Mg(OH)$_2$ layer are replaced by O of SiO$_4$ groups the other three O atoms of each SiO$_4$ can be shared with other SiO$_4$ groups (Fig. 6.16(c)). The formula of this composite layer, which is uncharged, is Mg$_3$(OH)$_4$Si$_2$O$_5$, as explained in Chapter 23. If there is a layer of linked SiO$_4$ groups on both sides of the Mg(OH)$_2$ layer the composition is Mg$_3$(OH)$_2$Si$_4$O$_{10}$. Crystals of chrysotile and talc consist of such neutral layers. Replacement of some of the Si by Al gives negatively charged layers which are the structural units in the micas, for example, KMg$_3$(OH)$_2$[Si$_3$AlO$_{10}$]. Corresponding to each of the Mg-containing layers there is an Al-containing layer in which the Mg^{2+} ions in the Mg(OH)$_2$ layer are replaced by two-thirds their number of Al^{3+} ions, as in KAl$_2$(OH)$_2$[Si$_3$AlO$_{10}$]. In the chlorite minerals negatively charged mica-like layers are interleaved with positively charged brucite layers in which some Mg has been replaced by Al, as in [Mg$_3$(OH)$_2$Si$_3$AlO$_{10}$]$^-$[Mg$_2$Al(OH)$_6$]$^+$; compare the layers in (iii) (a)–(c).

(v) *Attachment of additional metal atoms to the surface of a layer*

Since the X atoms of a CdX$_2$ layer are close-packed there are two kinds of site, on either side of such a layer, for additional metal ions, above an (occupied) octahedral interstice or above an empty tetrahedral site (Fig. 6.14(b)). From an energetic standpoint neither type of site is suitable for an added M ion, whether tetrahedrally or octahedrally coordinated (Fig. 6.17(a)), because of the proximity to either one or three cations of the layer, However, in the first case such close approach of metal ions can be avoided by removing the metal ion within the layer, the black circle of Fig. 6.17(a). The mineral chalcophanite, ZnMn$_3$O$_7$.3H$_2$O, consists of layers of the C6 type built of Mn^{4+} and O^{2-} ions from which one-seventh of the metal ions have been removed, giving the composition Mn$_3$O$_7$ instead of MnO$_2$. Attached to each side of this layer, directly above and below the unoccupied Mn^{4+} positions, are Zn^{2+} ions, and between the (uncharged) layers are layers of water molecules. Three of these complete the octahedral coordination group around Zn^{2+} (Fig. 6.17(b)).

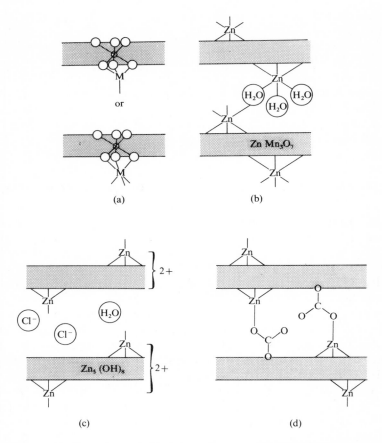

FIG. 6.17. (a) Attachment of metal ion to CdI$_2$-like layer. Diagrammatic representations of the structures of (b) ZnMn$_3$O$_7$.3H$_2$O; (c) Zn$_5$(OH)$_8$Cl$_2$.H$_2$O; (d) Zn$_5$(OH)$_6$(CO$_3$)$_2$.

In Zn$_5$(OH)$_8$Cl$_2$.H$_2$O the basic structural unit is a charged layer. One-quarter of the Zn^{2+} ions are absent from a Zn(OH)$_2$ layer, and for every one removed 2 Zn^{2+} are added beyond the surfaces of the layer, the composition being therefore Zn$_5$(OH)$_8^{2+}$. The charge is balanced by Cl$^-$ ions between the layers, with water molecules filling the remaining space (Fig. 6.17(c)). There is tetrahedral coordination of the Zn^{2+} ions on the outer surfaces of the layers, in contrast to the octahedral coordination within the layers. For another example of a layer structure of the same general type see Zn$_5$(OH)$_8$(NO$_3$)$_2$.2H$_2$O (p. 650).

Our last example is hydrozincite, Zn$_5$(OH)$_6$(CO$_3$)$_2$, a corrosion product of zinc which is also found accompanying zinc ores that have been subjected to weathering. Its structure arises from a combination of (iv) and (v). As in the previous example one-quarter of the Zn^{2+} ions are absent from a Zn(OH)$_2$ layer and replaced by twice their number of similar ions, equally distributed on both sides of the layer. In addition there is replacement of one-quarter of the OH$^-$ ions by O atoms of CO$_3^{2-}$ ions, so that the composition has changed from Zn$_4$(OH)$_8$ to Zn$_2$[Zn$_3$(OH)$_6$](CO$_3$)$_2$. A second O atom of each CO$_3$ is bonded to the 'added' Zn atoms, which are tetrahedrally coordinated as in the hydroxychloride. This is shown diagrammatically in Fig. 6.17(d). Although the structure as a whole is not a layer structure, for there is bonding between Zn and O atoms throughout, nevertheless the c.p. layers of O atoms around the octahedrally coordinated Zn atoms remain a prominent feature of the structure and may well act as a template in the growth of the crystal. It is not always appreciated that in the case of a compound of this kind, which exists only as a solid and can be formed on the surface of another solid (metallic Zn, ZnO, etc.), crystal growth is synonymous with the actual formation of a chemical compound.

The ReO$_3$ and related structures

The cubic ReO$_3$ structure has been described in Chapter 5 as the simplest 3D structure formed from vertex-sharing octahedral groups; the distorted variant adopted by Sc(OH)$_3$ was also illustrated. The cubic structure or a less symmetrical variant is adopted by NbF$_3$ and several trioxides and oxyfluorides (Table 6.5), but

TABLE 6.5
The ReO$_3$ and related structures

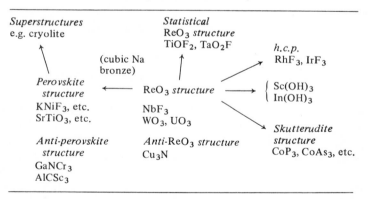

it is not possible for an ionic trinitride, since it would require cations M^{9+}; however, the anti-ReO$_3$ structure is found for Cu$_3$N. Similarly, the perovskite structure, derived from ReO$_3$ by adding a 12-coordinated ion at the body-centre of the unit cell of Fig. 13.3 (p. 584), is formed by complex fluorides and oxides (often with symmetry lower than cubic) and the anti-perovskite structure by a number of ternary nitrides and carbides, in which N or C occupy the positions of octahedral coordination. Alternatively, these nitrides and carbides may be described as c.c.p.

metal systems in which N or C atoms occupy one-quarter of the octahedral interstices, a more obvious description if all the metal atoms are similar (as in Fe_4N, Mn_4N, and Ni_4N). Since the O atoms in ReO_3 occupy three-quarters of the positions of cubic closest packing, rearrangement to a more dense structure is possible, giving in the limit the h.c.p. structure of RhF_3 and certain other trifluorides (Chapter 9).

The relation between the ReO_3 and RhF_3 structures can easily be seen from a model consisting of two octahedra sharing a vertex. In Fig. 6.18(a) the vertices

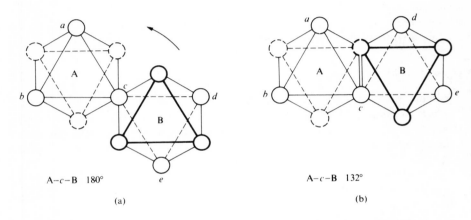

A–c–B 180°   A–c–B 132°

(a)   (b)

FIG. 6.18. Relation between vertex-sharing octahedra in (a) ReO_3; (b) RhF_3.

a, *b*, *c*, *d*, and *e* are coplanar, and the octahedra are related as in ReO_3, the angle AcB being 180°. Keeping these vertices coplanar, anticlockwise rotation through 60° of the upper octahedron B about the vertex *c* brings it to the position shown in (b), with the angle AcB equal to 132° and the X atoms in positions of hexagonal closest packing. This is the relation between adjacent octahedra in the h.c.p. RhF_3 structure, and it may be noted that the arrangement of RhF_6 octahedra is essentially the same as that of the LiO_6 *or* the NbO_6 octahedra in $LiNbO_3$, a superstructure of corundum. This point is illustrated by the diagrammatic elevations of Fig. 6.19, where (a), (b), and (c) are (approximately) h.c.p. structures (M–O–M in the range 120°–140°) and (d) represents the c.c.p. perovskite structure, with a similar system of vertex-sharing octahedra but M–O–M equal to 180°. (The representation of c.p. structures by elevations of the type shown in Fig. 6.19 is described in Chapter 4; see Figs. 4.25 and 4.27 and accompanying text.) It is interesting to note that the corundum structure (and its superstructures) illustrate the limitations on the values of A–X–A angles for octahedra set out in Fig. 5.3, since in this structure octahedral MO_6 groups share vertices, edges, and one face:[1]

	Octahedral element shared		
	Face	*Edge*	*Vertex*
Observed in α-Al_2O_3	85°	94° (two)	120° and 132° (two)
'Ideal' values for regular octahedra (p. 190)	$70\frac{1}{2}$°	90° (two)	132°–180° (three)

Since these are the bond angles at an O atom they also show the considerable departures from the regular tetrahedral coordination of that atom.

Although trinitrides MN_3 of transition metals do not occur, the analogous compounds with the heavier elements of Group VB are well known. They include

$$CoP_3 \quad NiP_3 \quad CoSb_3 \quad CoAs_3$$
$$RhP_3 \quad PdP_3 \quad RhSb_3 \quad RhAs_3$$
$$IrSb_3 \quad IrAs_3$$

which all crystallize with the skutterudite structure, named after the mineral $CoAs_3$ with that name. A point of special interest concerning this structure is that it contains well defined As_4 groups.

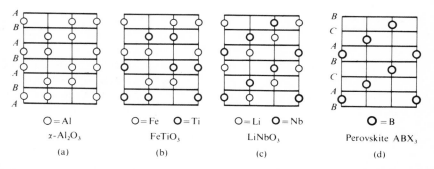

$O = Al$	$O = Fe$ $O = Ti$	$O = Li$ $O = Nb$	$O = B$
α-Al_2O_3	$FeTiO_3$	$LiNbO_3$	Perovskite ABX_3
(a)	(b)	(c)	(d)

FIG. 6.19. Elevations of close-packed structures containing 3D systems of linked octahedra: (a) corundum; (b) $FeTiO_3$; (c) $LiNbO_3$; (d) perovskite.

The skutterudite structure is related in a rather simple way to the ReO_3 structure. The non-metal atoms in the ReO_3 structure situated on four parallel edges of the unit cell are displaced into the cell to form a square group, as shown for two adjacent cells in Fig. 6.20(a). From the directions in which the various sets of atoms are moved it follows that one-quarter of the original ReO_3 cells will not contain an X_4 group. Figure 6.20(b) shows a (cubic) unit cell of the $CoAs_3$ structure, which has dimensions corresponding to twice those of the ReO_3 structure in each direction; the lower front right and upper top left octants are empty. (The unit cell contains 8 Co and 24 As atoms, the latter forming 6 As_4 groups.) Each As atom has a nearly regular tetrahedral arrangement of 2 Co + 2 As neighbours, and Co has a slightly

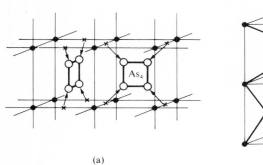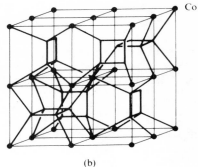

(a)   (b)

FIG. 6.20. The skutterudite ($CoAs_3$) structure: (a) relation to the ReO_3 structure; (b) unit cell. In (b) only sufficient Co–As bonds are drawn to show that there is a square group of As atoms in only six of the eight octants of the cubic unit cell. The complete 6-coordination group of Co is shown only for the atom at the body-centre of the cell.

distorted octahedral coordination group of 6 As neighbours. The planar As_4 groups are not quite square, the lengths of the sides of the rectangle being 2.46 Å and 2.57 Å.[2] The reason for the considerable reorientation of the octahedra which converts the ReO_3 to the $CoAs_3$ structure is that it makes possible the formation by each As atom of two bonds of the same length as those in the As_4 molecule. If $CoAs_3$ had the ReO_3 structure (with the same Co–As distance, 2.33 Å) each As would have 8 equidistant As neighbours at about 3.3 Å. (The relatively minor distortions from perfectly square As_4 groups and exactly regular octahedra are due to the fact that in the highly symmetrical cubic $CoAs_3$ structure it is geometrically impossible to satisfy both of these requirements simultaneously, and the actual structure represents a compromise.) In the phosphides with this structure the difference between the two P–P bond lengths is smaller (for example, 2.23 Å and 2.31 Å), the shorter bonds corresponding to normal single P–P bonds.[3]

We have noted that the perovskite structure is formed from the ReO_3 structure by filling the large cuboctahedral holes, of which there is one for every ReO_6 octahedron. In $CoAs_3$ these holes have become icosahedral and there is only one for every 4 Co (two in the cell of Fig. 6.20(b), which contains 8 Co.) The 'filled' $CoAs_3$ structure therefore has the formula AB_4X_{12}; it is the structure of compounds $LnFe_4P_{12}$, where Ln is La, Ce, Pr, or Nd.[4]

The fact that a compound such as CoP_3 contains cyclic P_4 groups suggests that it might be of interest to try to formulate it in the way that was done for FeS_2, PdS_2, and PdP_2 in Chapter 1. There we saw that the S_2 groups bonded to six metal atoms in the pyrites structure can be regarded as a source of ten electrons, so that Fe in FeS_2 acquires the Kr configuration. Each P in a cyclic P_4 group requires and additional electron, which it could acquire either as an ionic charge, (a), or by forming a normal covalent bond, (b):

(a)	(b)

We should then formulate CoP$_3$ either as $(Co^{3+})_4(P_4^{4-})_3$ or as a covalent compound Co$_4$(P$_4$)$_3$; in the latter case each metal atom would acquire a share in nine electrons, since each P$_4$ group is a source of 12 electrons. The Co atom in CoP$_3$ would thus attain the Kr configuration like Fe in FeS$_2$. Just as CoS$_2$ with the pyrites structure has an excess of one electron per metal atom, so this would be true of Ni in NiP$_3$, accounting for the metallic conduction and Pauli paramagnetism of the latter.

(1)   JSSC 1973 **6** 469   (3)   AK 1968 **30** 103
(2)   AC 1971 **B27** 2288   (4)   AC 1977 **B33** 3401

Structures with similar analytical descriptions

We include here a note on a subject which sometimes presents difficulty when encountered in the crystallographic literature, namely, the fact that the same analytical description may apply to crystal structures which are quite different as regards their geometry and topology. A simple example is a structure referable to a rhombohedral unit cell with M at (000) and X at $(\frac{1}{2}\frac{1}{2}\frac{1}{2})$. This describes the CsCl structure (8-coordination of M and X) if $\alpha = 90°$ but the NaCl structure (6-coordination of M and X) if $\alpha = 60°$. This complication arises if there is (at least) one variable parameter, which may be either one affecting the shape of the unit cell (e.g. the angle α in a rhombohedral cell or the axial ratio of a hexagonal or tetragonal cell) or one defining the position of an atom in the unit cell. The following are further examples.

The PbO *and* PH$_4$I *structures*

A number of compounds MX have tetragonal structures in which two atoms of one kind occupy the positions (000) and $(\frac{1}{2}\frac{1}{2}0)$ and two atoms of a second kind the positions $(\frac{1}{2}0u)$ and $(0\frac{1}{2}\bar{u})$. The nature of the structure depends on the values of the two variable parameters, u and the axial ratio $c{:}a$. For $c{:}a = \sqrt{2}$ and $u = \frac{1}{4}$, the atoms in $(\frac{1}{2}0u)$ and $(0\frac{1}{2}\bar{u})$ are cubic close-packed with the atoms at (000) and $(\frac{1}{2}\frac{1}{2}0)$ in tetrahedral holes. For $c{:}a = 1$ and $u = \frac{1}{4}$ the packing of the former atoms would be body-centred cubic. The structures of PbO and LiOH are intermediate between these two extremes; they are layer structures with only Pb–Pb (OH–OH) contacts between the layers (Fig. 6.21(a)). If $c{:}a = \frac{1}{2}\sqrt{2}$ and $u = \frac{1}{2}$ the structure becomes the CsCl structure, in which each kind of ion is surrounded by eight of the other kind at the vertices of a cube; the structure of PH$_4$I (Fig. 6.21(b)) approximates to the CsCl structure. Some compounds with these structures are listed in Table 6.6;

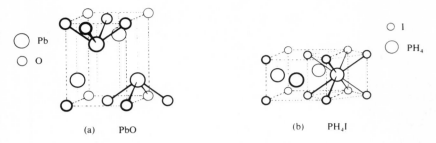

FIG. 6.21. The PbO and PH$_4$I structures.

TABLE 6.6

Compounds with the PbO and PH$_4$I (B 10) structures

	$c:a$	u	Atoms in	
			$(000), (\frac{1}{2}\frac{1}{2}0)$	$(\frac{1}{2}0u), (0\frac{1}{2}\bar{u})$
PbO	1·414	0·25		
	1·26	0·24	O	Pb
LiOH	1·22	0·20	Li	OH
	1·00	0·25		
InBi	0·95	0·38	In	Bi
N(CH$_3$)$_4$Cl, Br, I	0·71–0·72	0·37–0·39	N	Cl, Br, I
PH$_4$I	0·73	0·40	P	I
Ideal CsCl structure	0·707	0·50		
NH$_4$SH	0·667	0·34	N	SH

they fall into groups, with the exception of InBi ($c:a = 0.95$), with $c:a$ close to 1.25 or 0.70.

The LiNiO$_2$, NaHF$_2$, and CsICl$_2$ structures

Each of these three rhombohedral structures is described as having M at (000), X at $(\frac{1}{2}\frac{1}{2}\frac{1}{2})$, and two Y atoms at $\pm(uuu)$ situated along the body-diagonal of the cell. If $u = \frac{1}{4}$ and $\alpha = 33°34'$ this corresponds to the atomic positions of Fig. 6.3(b), one-half of the Na ions having been replaced by M, the remainder by X, and the Cl ions by Y. This is the type of structure adopted by a number of complex oxides and sulphides MXO$_2$ and MXS$_2$ which are superstructures of the NaCl structure (group (a) in Table 6.7). However, the same analytical description applies to two other quite different structures, the NaHF$_2$ and CsICl$_2$ structures, (b) and (c) of Table 6.7. All the compounds with the NaHF$_2$ structure have α in the range 30–40° and u about 0.40 (or 0.10 if M and X are interchanged), while for CsICl$_2$ $\alpha = 70°$ and $u = 0.31$.

If these structures are projected along the trigonal axis of the rhombohedron, that is, along the direction of the arrow in Fig. 6.22(a), all atoms fall on points of

TABLE 6.7

Data for some compounds MXY$_2$

Compound	α	u
(a) FeNaO$_2$ NiLiO$_2$ CrNaS$_2$ CrNaSe$_2$	*c.* 30°	*c.* 0·25
(b) CrCuO$_2$ FeCuO$_2$ NaHF$_2$ NaN$_3$ CaCN$_2$	*c.* 30° *c.* 40°	*c.* 0·40
(c) CsICl$_2$	70°	0·31

the type *A, B,* or *C* of (b), and these are the positions of closest packing. The sequence of layers depends on the parameter *u*, and discrete groups XY$_2$ appear only if the X atoms and the pairs of nearest Y atoms in adjacent layers are of the same type, that is, all *A*, all *B*, or all *C*. The elevations shown at (c), (d), and (e) show how different are the layer sequences along the vertical axis of Fig. 6.22(a).

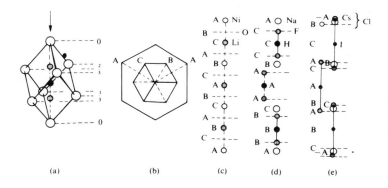

FIG. 6.22. The relation between the structures of LiNiO$_2$, NaHF$_2$, and CsICl$_2$ (see text). In (c)-(e) the horizontal dotted lines represent layers of oxygen or halogen atoms.

In (c) the sequence of O layers is that of cubic close packing, and all the metal ions are in positions of octahedral coordination between the layers. In (d) and (e) linear HF$_2^-$ and ICl$_2^-$ ions can be distinguished, but whereas in NaHF$_2$ a Na$^+$ ion has six nearest F$^-$ neighbours at the vertices of a distorted octahedron, in CsICl$_2$ a Cs$^+$ ion has six nearest Cl$^-$ neighbours nearly coplanar with it and two more not much further away (at 3.85 Å as compared with 3.66 Å). Comparing the compounds FeCuO$_2$ and FeNaO$_2$, both with $\alpha \approx 30°$, we see that the different values of *u* (0.39

and 0.25 respectively) lead to entirely different layer sequences. In $FeCuO_2$ Cu has only two nearest oxygen neighbours, whereas in $FeNaO_2$ there is regular octahedral coordination of both Na and Fe.

On the basis of both interatomic distances and physical properties it would seem justifiable to subdivide the compounds with structure (b) of Table 6.7 (Fig. 6.22(d)) into two classes. There are numerous oxides with this structure, sometimes called the *delafossite* structure after the mineral $CuFeO_2$:

$Cu^IFe^{III}O_2$	$AgFeO_2$	$AgRhO_2$	$PdCoO_2$	$PtCoO_2$
Co	Co	In	Cr	
Cr	Cr	Tl	Rh	
Al	Al			
Ga	Ga			

Although normally a structure for oxides $M^IM^{III}O_2$ it is also adopted by a few oxides containing Pd or Pt, but whereas $CuFeO_2$ and $AgFeO_2$ are semiconductors the Pd and Pt compounds are metallic conductors. The conductivity is very anisotropic, being much greater perpendicular to the vertical axis of Fig. 6.22(a) (parallel to the layers of Pd(Pt) atoms), in which plane it is only slightly inferior to that of metallic copper. In this plane Pd(Pt) has six coplanar metal neighbours at 2.83 Å, a distance very similar to that in the metal itself; contrast Cu–Cu in $CuFeO_2$ (3.04 Å) with Cu–Cu in the metal, 2.56 Å. Counting these metal–metal contacts the coordination is 6 + 2 (hexagonal bipyramidal). For reference see p. 577.

The CrB, *yellow* TlI (*B* 33), *and related structures*

All the structures of Table 6.8 are described by the space group *Cmcm* with atoms in 4(c), $(0y\frac{1}{4})$, $(0\bar{y}\frac{3}{4})$, $(\frac{1}{2},\frac{1}{2}+y,\frac{1}{4})$, and $(\frac{1}{2},\frac{1}{2}-y,\frac{3}{4})$. They should not be described as the same structure because there are four variables (two axial ratios, y_A, and y_X), and the nature of the coordination group around the X atoms is very different in the

TABLE 6.8
The CrB, ThPt, *and* TlI (*B* 33) *structures*

AX	a	b	c	y_A	y_X	X–2 X in chain	A–X (shortest)
CrB	2·97 Å	7·86 Å	2·93 Å	0·15	0·44	1·74 Å	2·19 Å
CaSi	4·59	10·79	3·91	0·14	0·43	2·47	3·11
ThPt	3·90	11·09	4·45	0·14	0·41	2·99	2·99
NaOH (rh.)	3·40	11·38	3·40	0·16	0·37	3·49	2·30
TlI	4·57	12·92	5·24	0·11	0·37	4·32	3·36

For a more complete list see: AC 1965 19 214 (which includes older data for NaOH and KOH).

various crystals. In CrB there is a trigonal prism of 6 Cr around B, but B also has 2 B neighbours at a distance clearly indicating B—B bonds; in CrB and CaSi these —B—B— and —Si—Si— chains are an important feature of the structure. In ThPt and a number of other isostructural intermetallic phases this distinction between the X—X and A—X distances has gone, and the nearest neighbours of Pt in ThPt are 2 Pt and 1 Th at 2.99 Å, 4 Th at 3.01 Å, 2 Th at 3.21 Å, etc. In yellow TlI there is no question of I—I chains (the covalent I—I bond length is 2.76 Å), and the structure is described in terms of (5+2)-coordination of Tl by I and of I by Tl. In (rhombic) NaOH, with the TlI structure, the shortest (interlayer) OH—OH distance (3.49 Å) and the positions of the H atoms rule out the idea of perferential bonding (in this case hydrogen bonding) between the OH$^-$ ions. The relative values of the X—X and A—X distances given in the last two columns of Table 6.8 show that the same analytical description covers three quite different structures.

The PbCl$_2$ structure

We include a note on this structure to emphasize not only how little information about a structure is conveyed by its analytical description but how one structure type is utilized by compounds in which the bonds are of very different kinds — a point already noted in connection with the NaCl structure. The orthorhombic unit cell of the PbCl$_2$ structure contains three groups of four atoms, all in 4-fold positions ($x\frac{1}{4}z$) etc. (space group *Pnma*). There are therefore eight variables, two axial ratios, three x, and three z parameters. More than one hundred compounds AX$_2$ or AXY are known to adopt a structure of this kind, and they may be divided into four groups which correspond to different values of the axial ratios (Table 6.9).

TABLE 6.9

Compounds with the PbCl$_2$ structure

Class	$a : c$	$(a+c)/b$	Examples
A	0·90 to 0·80	4·0 to 3·3	BaX$_2$, PbX$_2$, Pb(OH)Cl, EuCl$_2$, SmCl$_2$, CaH$_2$, etc., YbH$_2$ Mg$_2$Pb, ThS$_2$, US$_2$, Co$_2$P Ternary silicides, phosphides, etc.
B	0·75	4·3	TiP$_2$, ZrAs$_2$
C	0·74, 0·66	3·1, 3·3	Rh$_2$Si, Pd$_2$Al
D	0·55	5·3	Re$_2$P

For further examples and references see: AC 1968 **B24** 930.

The majority of the compounds fall into Class A, which includes the salt-like dihalides and dihydrides but also some sulphides, phosphides, silicides, and borides, and the intermetallic compounds Mg$_2$Pb and Ca$_2$Pb. Examples of compounds in the smaller Classes B and C, and the sole representative of Class D are given in the Table.

Although this is potentially a structure of $9:\frac{4}{5}$ coordination, with tricapped trigonal prismatic coordination of the large Pb^{2+} ion, there are never nine equidistant X neighbours. For example, in both PbF_2 and $PbCl_2$ the coordination group of the cation consists of seven close and two more distant anion neighbours, though in both structures there is appreciable spread of the interatomic distances:

PbF$_2$:   7 F at 2.41–2.69 Å (mean 2.55 Å); 2 at 3.03 Å.
PbCl$_2$; 7 Cl at 2.80–3.09 Å (mean 2.98 Å); 2 at 3.70 Å.

The two more distant neighbours are at two of the vertices of the trigonal prism, as shown in Fig. 6.23. This (7+2)-coordination is a feature also of the structures of

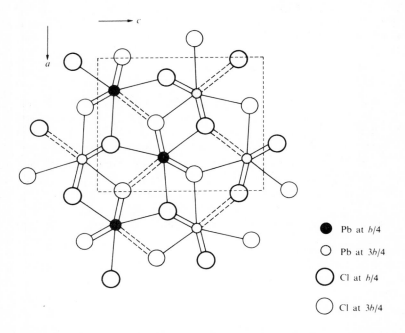

FIG. 6.23. Projection of the crystal structure of PbCl$_2$ on (010). Atoms at heights $b/4$ and $3b/4$ above the plane of the paper are distinguished as heavy and light circles. The broken lines indicate bonds from Pb^{2+} to its two more distant neighbours.

BaCl$_2$, BaBr$_2$, and BaI$_2$, but in mixed halides PbX'X" the A–X distances do not fall into well-defined groups of 7+2 (for example, in PbBrI there is only one outstandingly large bond length). In the numerous ternary phases with the 'anti-PbCl$_2$' structure (e.g. TiNiSi, MoCoB, TiNiP, etc.) the $\frac{4}{5}:9$ coordination is still recognizable but there are more neighbours within fairly close range, a general feature of intermetallic phases. For example, the following figures show the total numbers of neighbours up to $1.15(r_A + r_B)$, where r_A and r_B are the radii for 12-coordination (p. 1288):

	B	Mo	Co	Total
Mo	5	4	6	15
Co	4	6	2	12
B	–	5	4	9

The PdS$_2$, AgF$_2$, and β-HgO$_2$ structures and their relation to the pyrites structure

The structures of compounds AX$_2$ with octahedral coordination of A are of interest in connection with the 'theorem' stated on p. 191, that if three regular octahedral groups AX$_6$ have a common vertex at least one edge (or face) must be shared if reasonable distances are to be maintained between X atoms of different AX$_6$ groups. We saw that there is one vertex common to three octahedral groups in OCr$_3$(OOC.CH$_3$)$_6$.3H$_2$O, this being possible because certain pairs of O atoms of different octahedral groups belong to bridging acetate groups, with O–O much shorter (2.24 Å) than the normal van der Waals distance. It is worth while to examine a 3D structure in which three octahedral AX$_6$ groups meet at each vertex, without edge-sharing.

Consider the buckled layer of Fig. 6.24(a) built of square planar coordination groups AX$_4$. This represents a layer of the structures of PdS$_2$. If layers (a) and the (translated) layers (b) of Fig. 6.24 alternate in a direction normal to the plane of the paper, an octahedral group ABCDEF around a metal atom can be completed by atoms E and F of layers (b) situated $c/2$ above and below the layer (a). In the resulting 3D structure each X atom is a vertex common to three octahedra (which share no edges). The structure with regular octahedra and normal van der Waals distances between X atoms of different octahedra is not possible for the reason stated above, but versions of this structure are adopted by PdS$_2$, AgF$_2$, and β-HgO$_2$. The geometrical difficulty is overcome in one or both of two ways, namely, (i) distortion of the AX$_6$ coordination groups, or (ii) bonding between pairs of X atoms (such as E and F) in each 'layer'. In AgF$_2$ there is moderate distortion of the octahedral groups (Ag–4 F, 2.07 Å and Ag–2 F, 2.58 Å) and normal F–F distances between atoms of different coordination groups (shortest F–F, 2.61 Å) – case (i). Figure 6.24(c) shows one 'layer' of the structure of AgF$_2$. In PdS$_2$ the S atoms are linked in pairs (S–S, 2.13 Å) and also the octahedral PdS$_6$ groups are so elongated that the structure is a layer structure (Pd–4 S, 2.30 Å; Pd–2S, 3.28 Å). In β-HgO$_2$ the O atoms are bonded in pairs (O–O, 1.5 Å) and here the octahedra are compressed (Hg–2 O, 2.06 Å; Hg–4 O, 2.67 Å), so that instead of layers there are chains –Hg–O–O–Hg– in the direction of the c axis (normal to the plane of the layers in Fig. 6.24). In these structures, therefore, both factors (i) and (ii) operate. We thus have three closely related structures which are all derived from a hypothetical AX$_2$ structure built of regular octahedra which is not realizable in this 'ideal' form. All three structures have the same space group (*Pbca*) and the same equivalent positions are occupied, namely, (000) etc. for M and (xyz) etc. for X; there are rather similar

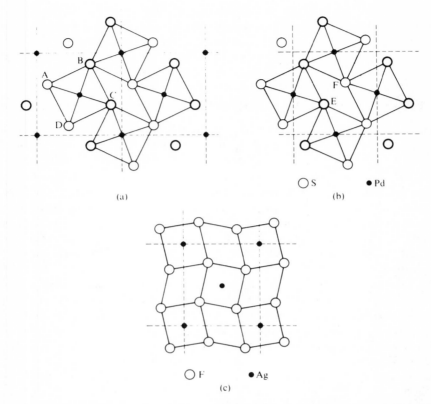

FIG. 6.24. (a) and (b) Successive layers of the PdS_2 structure. (c) One layer of the structure of AgF_2.

TABLE 6.10

The PdS_2, AgF_2, *and* β-HgO_2 *structures*

	a	b	c	x	y	z	*Reference*
PdS_2	5·46 Å	5·54 Å	7·53 Å	0·11	0·11	0·43	AC 1957 **10** 329
AgF_2	5·07	5·53	5·81	0·18	0·19	0·37	JPCS 1971 **32** 543
β-HgO_2	6·08	6·01	4·80	0·08	0·06	0·40	AK 1959 **13** 515

parameters for X but very different c dimensions of the unit cell (Table 6.10).

We have discussed these structures at this point as examples of structures with similar analytical descriptions. We noted earlier that the PdS_2 structure can be described as a pyrites structure which has been elongated in one direction to such an extent that it has become a layer structure. The pyrites structure may be described as a 3D assembly of FeS_6 octahedra, each vertex of which is common to three octahedra. Each vertex is also close to one vertex of another FeS_6 octahedron, these close contacts corresponding to the S–S groups and giving S its fourth (tetrahedral) neighbour. The pyrites structure is therefore one of a group of structures of the same topological type which represent compromises necessitated by the fact that a 3D AX_2 structure in which AX_6 groups share only vertices cannot be built with regular octahedra *and* normal van der Waals distances between X atoms of different octahedra:

Octahedral AX_2 structures in which each vertex is common to three octahedra and only vertices are shared

Regular octahedra and close X–X contacts	Distorted octahedra (4+2)-coordination: no close X–X contacts	Distorted octahedra (2+4)-coordination: close X–X contacts
Pyrites structure	AgF_2 *structure*	β-HgO_2 *structure*

$\downarrow$

PdS_2 *structure*

Highly distorted octahedra, giving planar 4-coordination and close X–X contacts

Relations between the structures of some nitrides and oxy-compounds

The close relation between structure and composition and simple geometrical considerations is nicely illustrated by some binary and ternary nitrides. In many of these compounds there is tetrahedral coordination of the smaller metal ions; in particular, there are many ternary nitrides containing tetrahedrally coordinated Li^+ ions. From the simple theorem that not more than eight regular tetrahedra can meet at a point (p. 192) it follows, for example, that there cannot be tetrahedral coordination of Li by N in Li_3N, a point of some interest in view of the unexpected (and unique) structure of this compound and the apparent non-existence of other alkali nitrides of type M_3N. It also follows that the limit of substitution of Be by Li in Be_3N_2 is reached at the composition BeLiN; for example, there cannot be a compound $BeLi_4N_2$ with tetrahedrally coordinated metal atoms.

These points are more readily appreciated if we formulate nitrides with 4 N atoms, for if all the metal atoms are 4-coordinated and all N atoms have the same coordination number this c.n. is equal to the total number of metal atoms in the formula. This number (n_t) is the number of MN_4 tetrahedra meeting at each N atom, and it must not exceed eight:

n_t	Compound	Structure type
3	Ge_3N_4	Phenacite (Be_2SiO_4)
4	Al_4N_4 (AlN)	Zinc-blende or wurtzite
5	($Al_2Be_3N_4$)	—
6	Ca_6N_4 (Ca_3N_2)	Anti-Mn_2O_3
7	($Be_5Li_2N_4$)	—
8	$Be_4Li_4N_4$ (BeLiN)	Anti-CaF_2

Examples of structures in which five or seven tetrahedra meet at each point do not appear to be known, but the structures with n_t = 3, 4, 6, and 8 are well known. We see that replacement of Be by Li in Be_6N_4 must stop at $Be_4Li_4N_4$ (BeLiN) since further substitution would imply that more than eight tetrahedra must meet at each N atom. The end-member of the series would be $Li_{12}N_4$ (Li_3N) in which, assuming tetrahedral coordination of Li, every N would be common to 12 LiN_4 tetrahedra.

More generally we can derive the limits of substitution of a metal M^{m+} by tetrahedrally coordinated Li^+ in nitrides in the following way. Writing the formula of the ternary nitride $(Li_xM^m)_3N_y$, we have

$$x+m=y \quad \text{and} \quad 3(x+1)/y \geqslant 2.$$

The first equation corresponds to the balancing of charges while the second states that the c.n. of N must not exceed eight, assuming tetrahedral coordination of all metal atoms. It follows that x cannot exceed $2m-3$, that is, the limiting values of Li:M in ternary nitrides are:

$$m = 2 \quad 3 \quad 4 \quad 5 \quad 6$$
$$\text{Li:M} \quad \ \ 1 \quad 3 \quad 5 \quad 7 \quad 9$$

corresponding to compounds such as

$$LiMgN, Li_3AlN_2, Li_5SiN_3, Li_7VN_4, \text{ and } Li_9CrN_5,$$

all of which have been prepared and shown to crystallize with the antifluorite structure. There is, of course, no limit to the replacement of more highly-charged ions by ions carrying charges $\geqslant 2$ since the end-members are the binary nitrides Ca_3N_2, AlN, etc. The foregoing restrictions, which are purely geometrical in origin, apply only to substitution by Li^+ or other singly-charged ions.

The feature common to the structures of Ge_3N_4, AlN, Ca_3N_2 and the numerous ternary nitrides with the antifluorite structure is the tetrahedral coordination of the metal atoms, the c.n. of N being 3, 4, 6, and 8 respectively. It would seem that in these structures the determining factor is the type of coordination around the metal ions.

A table similar to that given for nitrides can be drawn up for oxides, but here there is no geometrical limitation of the proportion of Li^+ (assumed to be tetrahedrally coordinated) since the limiting case (Li_2O) corresponds to eight tetrahedral LiO_4 groups meeting at each O atom. A point of interest here is the fact that certain simple compounds require 'awkward' numbers of tetrahedra meeting at a point, notably five in Li_4SiO_4 and seven in Li_6BeO_4:

n_t		n_t	
3	Be_2SiO_4, $LiAlSiO_4$	6	Li_5AlO_4
4	$BeLi_2SiO_4$, Li_3PO_4	7	Li_6BeO_4
5	Li_4SiO_4	8	Li_2O

Restrictions on composition would arise in oxides in which there is octahedral coordination of all metal atoms, since not more than six regular octahedra may meet at a point, though of course this number may be increased if the octahedra are sufficiently distorted; see the note on the Th_3P_4 structure in Chapter 5. Octahedral structures may be listed in the same way as the tetrahedral structures:

n_o		Known or probable structure type
2	ReO_3	ReO_3
3	Al_2ReO_6	Rutile?
4	Mg_3ReO_6	Corundum, etc.
5	$Na_2Mg_2ReO_6$, Mg_4TiO_6	?
6	Na_4MgReO_6	NaCl?

Structures with $n_o = 6$ (e.g. Na_6ReO_6) would be impossible for *regular octahedral* coordination of all metal atoms, while that of Mg_4TiO_6 would involve 5-coordination of O (five octahedra meeting at a point).

It is perhaps worthwhile to list the simpler structures in order of increasing numbers of tetrahedra or octahedra meeting at each X atom (Table 6.11).

TABLE 6.11

Structures with tetrahedral and octahedral coordination

n	Tetrahedral structures		Octahedral structures	
2	AX_2	SiO_2 structures	AX_3	ReO_3
3	A_3X_4	Ge_3N_4 (phenacite)	AX_2	Rutile
4	AX	ZnS structures	A_2X_3	Corundum
6	A_3X_2	Anti-M_2O_3	AX	NaCl
8	A_2X	Anti-CaF_2	A_4X_3	See p. 192

Superstructures and other related structures

In a substitutional solid solution AA' there is random arrangement of A and A' atoms in equivalent positions in the crystal structure. If on suitable heat treatment the random solid solution rearranges into a structure in which the A and A′ atoms occupy the same set of positions but in a regular way, the structure is described as a *superstructure*. We use the term in this book to describe relations such as

$$AX \rightarrow (AA')X_2 \quad \text{or} \quad A_2X_3 \rightarrow (AA')X_3$$

regardless of whether or not the superstructure is formed from a random solid solution or whether such a solid solution exists. In the superstructure the positions occupied by A and A′ are, of course, no longer equivalent. Corresponding to the relation between the parent structure and superstructure:

Parent structure	*Superstructure*
All of a set of equivalent positions occupied by A atoms	The same set of positions occupied in a regular way by atoms of two or more kinds A and A′

there are pairs of structures related in the following way:

'Normal structure'	*'Degenerate structure'*
A and A′ occupy different sets of equivalent positions	Both sets of equivalent positions occupied by A atoms

As regards the formulae the relation is similar to that between the parent structure and superstructure except that we have

$$(AA')X_2 - A_2X_2 \text{ instead of } A_2X_2 - (AA')X_2$$

Some pairs of structures related in this way are listed in Table 6.12. Structures in column (1) are normal structures and examples are given only of compounds isostructural with a compound in column (2) or (3). The structures (2) are also to be expected inasmuch as ions of the same metal carrying different charges have different sizes and bonding requirements, and from the structural standpoint are similar to ions of different elements. The structures (3), however, may be described as 'degenerate' since chemically indistinguishable atoms occupy positions with different environments. To the chemist the gradation from (a) to (c) is of some interest. In (3)(a) there is a marked difference between the environments of atoms of the same element in the same oxidation state; in (3)(b) the difference is smaller and in (3)(c) smaller still. In (3)(c) there is, to a first approximation, no difference between the immediate environments of the chemically similar atoms although they occupy crystallographically non-equivalent positions. In BeY_2O_4 both sets of non-equivalent Y atoms occupy positions of 6-coordination in an octahedral framework which is built from quadruple strips of rutile-like chains (p. 601). Any such strip composed of three or more single chains contains (topologically or crystallographically) non-equivalent octahedra:

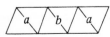   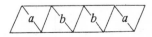

TABLE 6.12

Some related crystal structures

Difference between the various sets of equivalent positions	Different sets of equivalent positions occupied by:		
	(1) Atoms of *different* metals	(2) Atoms of *same* metal in *different* oxidation states	(3) Atoms of *same* metal in *same* oxidation state
(a) Different c.n.s	GeV$_3$, NiV$_3$ Ba$_2$CaWO$_6$ K$_2$NaAlF$_6$ Ca$_3$Al$_2$Si$_3$O$_{12}$ Na$_3$KAl$_4$Si$_4$O$_{16}$ CaFe$_2$O$_4$ ZnSb$_2$O$_4$	Eu$_3$O$_4$ Pb$_3$O$_4$	β-W Ba$_3$WO$_6$ K$_3$AlF$_6$ Y$_3$Al$_2$Al$_3$O$_{12}$ Na$_4$Al$_4$Si$_4$O$_{16}$
(b) Same c.n.s but different arrangements of nearest neighbours	Cd$_2$Mn$_3$O$_8$ ScTiO$_3$	Mn$_5$O$_8$	*C*-Mn$_2$O$_3$
(c) Same c.n.s and essentially the same coordination group	Fe$_2$TiO$_5$ ULa$_6$O$_{12}$ BFeCoO$_4$	Ti$_3$O$_5$ Tb$_7$O$_{12}$	BeY$_2$O$_4$

In the limit the process (a)$\rightarrow$(b)$\rightarrow$(c) in Table 6.12 would result in identical environments for all atoms of a given chemical species, that is, each would occupy its own set of equivalent positions. A structure in column (1) would then be a superstructure of the structure in column (3). Since any one of the actual structures in column (1) is a 'normal' structure, derived superstructures and statistical solid solutions are in principle possible. The following possibilities for oxides with the B(FeCo)O$_4$ and (BeY$_2$O$_4$) type of structure illustrate the relations between the types of structure we have been discussing:

Superstructure 　　　　*'Parent structure'* 　　　　*'Degenerate structure'*

$$M_2^{II}\left|\begin{array}{c}\text{Fe}\\ \hline \text{Ni}\\ \hline \text{Ti}_2\end{array}\right|O_8 \qquad M^{II}\left|\begin{array}{c}\text{Fe}^{II}\\ \hline \text{Ti}^{IV}\end{array}\right|O_4 \qquad M^{II}\left|\begin{array}{c}Y^{III}\\ \hline Y^{III}\end{array}\right|O_4$$

Statistical solid solution

$$M^{II}\left|\begin{array}{c}\text{Fe, Mg}\\ \hline \text{Ti}\end{array}\right|O_4$$

Symbols of metal atoms on different lines indicate sets of atoms in different equivalent positions. The above examples, except MY_2O_4, are purely hypothetical; it would be interesting to know the arrangement of metal ions in $B_2Mg_3TiO_8$ and related minerals, which form a related series containing B^{III} in place of M^{II}.

7

Bonds in molecules and crystals

Introduction

In conventional treatments of the 'chemical' bond it is usual for chemists to restrict themselves to the bonds in finite molecules and ions and for crystallographers and 'solid state chemists' to concern themselves primarily with the bonding in crystals. Moreover, effort is understandably concentrated on certain groups of compounds which are of special interest from the theoretical standpoint or on crystals which have physical properties leading to technological applications. As a result it becomes difficult for the student of structural chemistry to obtain a perspective view of this subject. The usual treatment of ionic and covalent bonds, with some reference to metallic, van der Waals, and other interactions, provides a very inadequate picture of bonding in many large groups of inorganic compounds. We shall therefore attempt to present a more general survey of the problem, the intention being to emphasize the complexity of the subject rather than to present an over-simplified approach which ignores many interesting facts. Clearly, a balanced review of bonding is not possible in the space available, and moreover would presuppose a knowledge of at least the essential structural features of (ideally) all finite groups of atoms and of all crystals. We shall therefore select a very limited number of topics and trust that the more detailed information provided in later chapters will provide further food for thought.

A glance at the Periodic Table will show the difficulty of making useful generalizations about bonding in many inorganic compounds. In Table 7.1 the full lines enclose elements which form ions stable in aqueous solution. At the extreme right the anions include only the halide ions; O^{2-} combines with H_2O to form (2) OH^-, and H^+ combines with H_2O to form H_3O^+. However, O^{2-} exists in many crystalline oxides, S^{2-}, etc., N^{3-} (and possibly also P^{3-}) in the appropriate compounds of the

TABLE 7.1

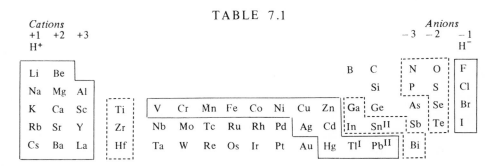

most electropositive metals. Next there are a few elements which do not form either cations or anions (B, C, Si: but note the formation of the finite C_2^{2-} ion in a number of carbides and the infinite 2D Si_n^{n-} ion in $CaSi_2$), a group which for most practical purposes also includes the neighbouring elements Ge, Sn(IV), As, and Sb. The Si—O bond is usually regarded as intermediate between covalent and ionic; Ge would be grouped with Si in view of the extraordinarily similar crystal chemistry of these two elements, though Ge—O bonds are probably closer to ionic than to covalent bonds.

The metals which form cations M^{n+} include the most electropositive elements at the left of the Table and a group of 3d metals in their lower oxidation states (usually 2 or 3), together with some of the earlier B subgroup metals, Tl(I) and Pb(II); note that Pb^{4+} is known only in crystalline oxides and oxy-compounds. In the lower centre of the Table is a group of metals which have no important aqueous ionic chemistry and probably not much tendency to form ions in the crystalline state — though the bonds in certain oxy- and fluoro-compounds may well have appreciable ionic character (for example, dioxides such as MoO_2, lower fluorides such as MoF_3, etc.).

The Group IVA metals have no important aqueous ionic chemistry but form ions in crystalline oxy-compounds. Bismuth can perhaps be grouped with Ti, Zr, and Hf; it has a strong tendency to form hydroxy-complexes, but salts such as $Bi(NO_3)_3 . 5H_2O$, $M_3^{II}[Bi(NO_3)_6]_2 . 24H_2O$, and $Bi_2(SO_4)_3$ presumably contain Bi^{3+} ions. It seems reasonable to put Ga(III), In(III), Tl(III), and Sn(II) in this group; the complex structural chemistry of these elements is outlined in Chapter 26.

The bonding in compounds of the non-metals, one with another, may be regarded as essentially covalent, but it is evident from the shortage of anions that the only large classes of essentially ionic binary compounds are those of the metals with O or F, sulphides etc. of the most electropositive elements (the I, II, and IIIA subgroups), the remaining monohalides of these metals and of Ag and Tl, and halides MX_2 or MX_3 of other metals enclosed within the full lines of Table 7.1. Clearly there remain large groups of compounds, even binary ones, which should be included in any comprehensive survey of bonding in inorganic compounds, and it must be admitted that a satisfactory and generally acceptable description of the bonding in many of these groups is not available. In the compounds of these metals with the more electronegative non-metals the bonding is probably intermediate between ionic and covalent, but as we proceed down the B subgroups the more metallic nature of the 'semi-metals' suggests bonding in their compounds of a kind intermediate between covalent and metallic. Even the structures of the crystalline B subgroup elements themselves present problems in bonding.

In Group IV there is a change from the essentially covalent 4-coordinated structure of diamond, Si, Ge, and grey Sn(IV) through white Sn(II) to Pb, with a c.p. structure characteristic of many metals. Group V begins with the normal molecular structure of N_2 and white P (P_4), but phosphorus also has the deeply coloured black and red forms, both with layer structures in which P is 3-coordinated. The structure of red P is unique and inexplicably complex. Then follow As, Sb, and Bi, with structures that can be described either as simple cubic structures, distorted to give (3+3)-coordination, or as layer structures in which the distinction between the two sets of neighbours becomes less as the metallic character increases:

	Distances to nearest neighbours	
	3 at	3 at
As	2.51 Å	3.15 Å
Sb	2.91	3.36
Bi	3.10	3.47

In Group VI crystalline oxygen exists as O_2 molecules, and sulphur also forms normal molecular crystals containing S_6 or other cyclic S_n molecules. However, this element also forms a fibrous polymorph built of chains with a unique configuration quite different from those in Se (metallic form) and Te. As regards its polymorphism Se occupies a position intermediate between S and Te. It has three red forms, all built of Se_8 molecules structurally similar to the S_8 molecule, and also a 'metallic' form built of helical chains. The only form of Te is isostructural with the metallic form of Se. In all these crystalline structures S, Se, and Te form the expected two bonds. However, the interatomic distances in Table 7.2 show that while rhombic S is adequately described as consisting of covalent S_8 molecules held together by van der Waals bonds this is an over-simplification in the cases of Se and Te. For example, there are contacts between atoms of different Se_8 molecules in β-Se which are as short as the shortest intermolecular contacts in rhombic S, and the shortest distances between atoms of different chains in Te are very little greater than the corresponding distances in metallic Se. A somewhat similar phenomenon is observed in Group VII. In crystalline Cl_2, Br_2, and I_2 the molecules are arranged in layers, the shortest intermolecular distances within the layers being appreciably less than those between the layers (Table 7.2); the latter are close to the expected van der Waals distances.

We have introduced the data of Table 7.2 at this point to emphasize that even the bonds in some crystalline elements cannot be described in a simple way. It appears that many bonds are intermediate in character between the four 'extreme' types, and of these the most important in inorganic chemistry are those which are intermediate in some way between ionic and covalent bonds.

A further complication is the presence of bonds of more than one kind in the same crystal or, less usually, in the same molecule. This is inevitable in any crystal containing complex ions, such as $NaNO_3$, the bonds between Na^+ and O atoms of NO_3^- ions being different in character from the N—O bonds within the NO_3^- ion; in some crystals bonds of three or four different kinds are recognizable. This complication also arises in much simpler crystals, for example, RuO_2, where there is electronic interaction leading to metallic conduction in addition to the metal–oxygen

TABLE 7.2
Interatomic distances in crystalline S, Se, *and* Te

	Intramolecular	*Shortest intermolecular*	*Pauling van der Waals distance* $(2r_{S^{2-}}$ *etc.*$)$
S_8 (rhombic)	2.04 Å	3.37 Å	3.6 Å
Se_8 (β)	2.34	3.35	⎱
Se (metallic)	2.37	3.44	⎰ 3.9
Te	2.84	3.50	4.4

Interatomic distances in the crystalline halogens

	Intramolecular	*Intermolecular*		$2r_{X^-}$
		In layer	*Between layers*	
Cl_2	1.98 Å	3.32 Å	3.74 Å	3.6 Å
Br_2	2.27	3.31	3.99	3.9
I_2	2.72	3.50	4.27	4.3

bonds which are the only kind of bonds we need to recognize in a crystal such as TiO_2 which has a rather similar crystal structure. In numerous other cases we shall find it convenient to indicate that one (or more) of the valence electrons appears to be behaving differently from the others, when we shall use formulae such as $Cs_3^+O^{2-}(e)$ or $Th^{4+}I_2^-(e)_2$.

TABLE 7.3
Types of bonds

No overlap of charge clouds	Valence electrons localized on particular atoms	*Van der Waals bond* charge-transfer bonds hydrogen bonds *Ionic bond* bonds between polarized ions Ionic–covalent bonds
Overlap of charge clouds	Shared electrons localized in particular bonds Partial delocalization of some valence electrons Conjugated systems	*Covalent bond*
Increasing delocalization of some or all of the valence electrons	($-C=C-C=C-$ etc.) 'resonating' systems (CO_3^{2-}, C_6H_6, $P_3N_3X_6$) finite metal clusters crystalline semi-metals 'electron-excess' solids	⎱ Localized and delocalized ⎰ bonds in the same system
	Complete delocalization of some proportion of valence electrons	*Metallic bond*

We may set out the various types of bonding as in Table 7.3, making our first broad subdivision according to whether there is or is not sharing of electrons between the atoms. In systems without appreciable overlapping of electron density we have interactions ranging from those between ions, through ion : dipole, dipole : dipole, ion : induced-dipole, dipole : induced-dipole, to induced-dipole : induced-dipole (van der Waals) bonds. In systems where electrons are shared between atoms we have the various type of covalent bond, with various degrees of delocalization of some or all of the bonding electrons, leading in the limit to the *metallic bond*. The latter is responsible for the electronic properties of metals and intermetallic compounds which are characteristic, in varying degrees, of the metallic state. It will be appreciated that, on the one hand, metallic conductivity is not exclusively a property of metals; it can occur in structures where the metal–metal interactions are not the strongest bonds, as in SmO, RuO_2, and ReO_3, and even in compounds which contain no metal atoms (graphite, $(SN)_x$). On the other hand, covalent bonds may be formed by atoms ranging from the non-metals to typical metals. There would seem to be no good reason for regarding the bonds between metal atoms in many molecules (p. 304) as differing in any essential respect from other covalent bonds. Indeed, Pauling showed that for discussing certain features of the structures of metals it is feasible to extend the valence-bond theory to the metallic state, as noted in Chapter 29. Certain *electron-deficient* molecules and crystals have, in common with metals, the feature that the atoms have more available orbitals than valence electrons; it may prove useful also to recognize *electron-excess* systems, which have more valence electrons than are required for the primary bonding scheme. This general term would include a number of groups of compounds of quite different types, for example, 'sub-compounds' such as Cs_3O, Ca_2N, and ThI_2, 'inert-pair' ions and molecules (p. 293), and transition-metal compounds such as those to the right of the vertical lines:

$$WO_3 \quad | \quad ReO_3$$
$$FeS_2 \quad | \quad CoS_2 \quad NiS_2$$
$$CoP_3 \quad | \quad NiP_3.$$

In such compounds there may be localized interactions between small numbers of metal atoms (as in MoO_2, NbI_4) or delocalization of a limited number of electrons leading to metallic conduction. We shall use the term 'm–m bond' in a later section to include all the various types of interaction between metal atoms which are implied by the examples of Chart 7.1 on p. 306.

We now discuss some aspects of the covalent bond, metal–metal bonds, the van der Waals bond, and the ionic bond.

The lengths of covalent bonds

From the early days of structural chemistry there has been considerable interest in discussing bond lengths in terms of radii assigned to the elements, and it has become customary to do this in terms of three sets of radii, applicable to metallic, ionic, and covalent crystals. Distances between non-bonded atoms have been compared with sums of 'van der Waals radii', assumed to be close to ionic radii. The earliest 'covalent radii' for non-metals were taken as one-half of the M–M distances in molecules or crystals in which M forms $8-N$ bonds (N being the number of the Periodic Group), that is, from molecules such as F_2, HO–OH, H_2N–NH_2, P_4, S_8, and the crystalline elements of Group IV with the diamond structure. This accounts for H and the sixteen elements in the block C–Sn–F–I. The origin of the covalent radii of metals was quite different owing to the lack of data from molecules containing M–M bonds. 'Tetrahedral radii' were derived from the lengths of bonds M–X in compounds MX with the ZnS structures, 'octahedral radii' were derived from crystals with the pyrites and related structures, assuming additivity of radii and using the $8-N$ radii assigned to the non-metals. As Pauling remarked at the time, it is unlikely that a bond such as Zn–S is a covalent bond in the same sense as C–C or S–S, and it is obviously difficult to justify the later use of such radii in discussions of the ionic character of other bonds formed by Zn. It could be added that it is also not obvious why the S–O bond length in SO_4^{2-}, in which S forms four tetrahedral bonds and O one bond, should be compared with the sum of r_S (from S_8, in which S forms two bonds) and r_O (from HO–OH, in which O forms two bonds).

The lengths of homonuclear bonds M–M are in general equal to or less than the standard single bond length in the molecules or crystals noted above. Exceptions include N–N in N_2O_3 (1.86 Å) and in N_2O_4, for which two determinations give 1.64 Å and 1.75 Å, both much longer than the bond in N_2H_4 (1.47 Å), S–S in $S_2O_5^{2-}$ (2.17 Å) and $S_2O_4^{2-}$ (2.39 Å), which are to be compared with the bond in S_8 (2.06 Å), and the bonds in I_3^- and other polyiodide ions which are discussed as a group in Chapter 9. Shorter bonds are regarded as having multiple-bond character. In some cases there are obvious standards for M=M and M≡M, as for

$$\begin{array}{cccc} H_3C\text{–}CH_3 & H_2C\text{=}CH_2 & \text{and} & HC\text{≡}CH \\ 1.54\,\text{Å} & 1.35\,\text{Å} & & 1.20\,\text{Å,} \end{array}$$

and the numerous bonds of intermediate length are assigned non-integral bond orders. In other cases (for example, N=N) there has been less general agreement as to the precise values, owing to the absence of data from, or the non-existence of, suitable molecules or ions.

The situation with regard to heteronuclear bonds (M–X) is different. Shortening due to π-bonding is to be expected in many bonds involving O, S, N, P, etc. and is presumably the major reason for variations in the length of a particular bond such as S–O. This is consistent with the values of the stretching frequencies of the bonds:

Molecule	S–O *stretching frequency*	Length
	(cm^{-1})	
SO	1124	1.49 Å
Cl_2SO	1239	1.45
SO_2	1256	1.43
F_2SO	1312	1.41

However, the longest measured S–O bond has a length of 1.65 Å (excluding those in $S_3O_{10}^{2-}$ and $S_5O_{16}^{2-}$, for which high accuracy was not claimed), as compared with 1.77 Å, the sum of the covalent radii. In other cases (for example $SiCl_4$) where there is no reason for supposing appreciable amounts of π-bonding (though it cannot be excluded) bonds are much shorter (Si–Cl, 2.00 Å) than the sum of the covalent radii (2.16 Å). In many cases the discrepancies between the observed lengths of (presumably) single bonds and radius sums appeared to be greater the greater the difference between the electronegativities of the atoms concerned, and it is now generally assumed that 'ionic character' of bonds reduces their lengths as compared with those of hypothetical covalent bonds. The introduction of electronegativity coefficients is thus seen to be a consequence of assuming additivity of radii. It is not proposed to discuss here the derivation of electronegativity coefficients, for which there is no firm theoretical foundation, but since an equation due to Schomaker and Stevenson has been widely used by those interested in relating bond lengths to

TABLE 7.4
Normal covalent radii and electronegativity coefficients

Electronegativity coefficients						Normal covalent radii				
H	C	N	O	F		H	C	N	O	F
2·1	2·5	3·0	3·5	4·0		0·37	0·77	0·74	0·74	0·72 Å
	Si	P	S	Cl			Si	P	S	Cl
	1·8	2·1	2·5	3·0			1·17	1·10	1·04	0·99
	Ge	As	Se	Br			Ge	As	Se	Br
	1·7	2·0	2·4	2·8			1·22	1·21	1·17	1·14
	Sn	Sb	Te	I			Sn	Sb	Te	I
	1·7	1·8	2·1	2·4			1·40	1·41	1·37	1·33

sums of covalent radii, we give in Table 7.4 some electronegativity coefficients and covalent radii. The empirical equation has the form:

$$r_{AB} = r_A + r_B - 0.09\,(x_A \sim x_B),$$

though a smaller numerical coefficient (0.06) has been suggested by some authors. This equation certainly removes some of the largest discrepancies, which arise for bonds involving the elements N, O, and F, but the following figures show that it does not account even qualitatively for the differences between observed and estimated bond lengths for series of bonds such as C–Cl, Si–Cl, Ge–Cl, and Sn–Cl:

Bond	$r_{obs.}$	$r_M + r_{Cl}$	Correction required	S–S correction
C–Cl	1·76 Å	1·76 Å	0·00 Å	0·045 Å
Si–Cl	2·00	2·16	0·16	0·11
Ge–Cl	2·08	2·21	0·13	0·12
Sn–Cl	2·30	2·39	0·09	0·12

Moreover, the electronegativity correction is by no means sufficient for bonds such as those in the molecules SiF_4 and PF_3:

	Si—F	P—F
$r_{obs.}$	1·56 Å	1·54 Å
$r_M + r_F$	1·89	1·82
r_{corr} (S.-S.)	1·69	1·65

In this connection it is interesting to note that the difference between pairs of bond lengths M—F and M—Cl is approximately equal in many cases to the difference between the *ionic* radii (0.45 Å) rather than to the difference between the covalent radii (0.27 Å) of F and Cl. This is to be expected for ionic crystals and molecules (for example, gaseous alkali-halide molecules) but it is also true for the following molecules:

	BX_3	CX_4	SiX_4	PX_3	SX_2
$r_{M-Cl} - r_{M-F}$	0.45 Å	0.44 Å	0.44 Å	0.50 Å	0.41 Å

though the difference becomes increasingly smaller for Br (0.38 Å), Cl (0.35 Å), O (0.28 Å), and F (0.21 Å), being finally equal to, or smaller than, the difference between the covalent radii.

It has long been recognized that an attractive alternative to the use of three sets of radii (metallic, ionic, and covalent) would be the adoption of one set of radii applicable to bonds in all types of molecules and crystals. This would obviate the need to prejudge the bond type. Such a set of radii was suggested by Bragg in 1920, and the idea has been revived by Slater (1964).[1] These radii agree closely in most cases with the calculated radii of maximum radial charge density of the largest shells in the atoms. The lengths of covalent or metallic bonds should therefore be equal to the sums of the radii since these bonds result from the overlapping of charge of the outer shells, and this overlap is a maximum when the maximum charge densities of the outer shells of the two atoms coincide. The radius of an ion with noble-gas configuration is approximately 0.85 Å greater or less than the atomic radius, depending on whether an anion or cation is formed. The length of an ionic bond is therefore not expected to differ appreciably from that of a covalent bond between the same pair of atoms. These radii have been shown to give the interatomic distances in some 1200 molecules and crystals with an average deviation of 0.12 Å. They were rounded off to 0.05 Å since for more precise discussions allowance would have to be made for coordination number and special factors such as crystal field effects. The agreement between observed bond lengths and radius sums is admittedly poor in some cases (bonds involving Ag, Tl, and the elements from Hg to Po), but some of these elements also present difficulty when more elaborate treatments are used. We do not give the Slater radii here because for metals they approximate to 'metallic radii' (Table 29.5, p. 1288) and for non-metals to Pauling's 'covalent radii' except for certain first-row elements, notably N (0.65 Å), O (0.60 Å), and F (0.50 Å). We have already noted that many M—F bond lengths suggest a radius for F much smaller than one-half the bond length in F_2 (0.72 Å); they agree much more closely with the Slater radius of 0.50 Å. Since bond length difficulties are most acute for certain bonds involving F, it should be noted that this element is abnormal in many ways, as shown by the data in Table 9.1 (p. 385). We have noted several examples of anomalously long bonds (N—N, S—S, I—I). If the bond in the F_2 molecule is also of this type (and possibly to some extent those in HO—OH and H_2N—NH_2) it may be necessary to re-examine the basis of the discussion of bond lengths and of the ionic–covalent character of bonds in terms of electronegativity coefficients.

(1) JCP 1964 **41** 3199

The shapes of simple molecules and ions of non-transition elements

The spatial arrangement of bonds in most molecules and ions AX_n formed by non-transition elements (and by transition elements in the states d^0, d^5, and d^{10}), where X represents a halogen, O, OH, NH_2, or CH_3, may be deduced from the total number of valence electrons in the system. If this number (V) is a multiple of eight the bond arrangement is one of the following highly symmetrical ones:

$$V = 16: \text{2 collinear bonds}$$
$$24: \text{3 coplanar bonds}$$
$$32: \text{4 tetrahedral bonds}$$
$$40: \text{5 trigonal bipyramidal bonds}$$
$$48: \text{6 octahedral bonds}$$
$$56: \text{7 pentagonal bipyramidal bonds}$$

For intermediate values of V the configuration is found by expressing V in the form $8n + 2m$ (or $8n + 2m + 1$ if V is odd). The arrangement of the n ligands and m unshared electron pairs then corresponds to one of the symmetrical arrangements listed above, for example:

V	n	m	$n + m$	Bond arrangement
32	4	0		Tetrahedral
26	3	1	4	Pyramidal
20	2	2		Angular

Compounds of non-transition elements containing odd numbers of electrons are few in number, but they can be included in the present scheme since an odd electron, like an electon pair, occupies an orbital. Thus a 17-electron system has the same angular shape as an 18-electron one, as described later.

It will be appreciated that the term $8n$ implies completion of the octets of the ligands X (usually O or halogen) rather than that of A, which seems reasonable since X is usually more electronegative than A; witness the 16-electron, 18-electron, and 24-electron systems in which A has an incomplete octet of valence electrons, and the existence of O–S–O and S–S–O but non-existence of S–O–S and S–O–O. To each ligand there corresponds one electron pair in the valence group of A. If the ligand is a halogen, OH, NH_2, or CH_3, one electron for each bond is provided by X and one by A, but in the case of O two electrons are required from A (or from A and the ionic charge). The negative formal charge on O is reduced by use of some of its electron density to strengthen the A–O bond, which is invariably close to a double bond. The use of other electron pairs on O in this way does not affect the stereochemistry appreciably, and it is not necessary to distinguish $=O$ from $-X$ in what follows. For example, the 24-electron systems include not only the boron trihalides, but also the carbonyl and nitryl halides, the oxy-ions BO_3^{3-}, CO_3^{2-}, and NO_3^-, and the neutral SO_3 molecule, all of which are planar triangular molecules or ions. An *equilateral* triangular structure is, of course, only to be expected if all the ligands are of the same kind; deviations from bond angles of 120° occur in less symmetrical molecules such as $COCl_2$ or O_2NF.

In this section we shall not in general distinguish multiple from single bonds in the structural formulae of simple molecules and ions.

In the examples given earlier the stereochemistry follows directly from the value of V (20, 26, or 32), since there is only one possible arrangement of two or three bonds derivable from a regular tetrahedral arrangement of four pairs of electrons. However, for some of the larger numbers of electron pairs there are several ways of arranging a smaller number of bonds. For example, there are two ways of arranging two lone pairs at two of the vertices of an octahedron (Fig. 7.1). Thus, although the

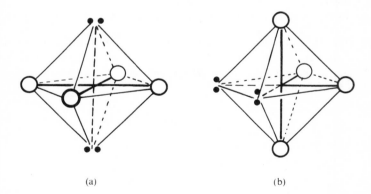

(a)   (b)

FIG. 7.1. The alternative ways of arranging two lone pairs at two of the vertices of an octahedron.

planar structure of the ICl_4^- ion is consistent with the octahedral disposition of the electron pairs so also would be the structure (b) of Fig. 7.1. Similarly, the irregular tetrahedral shape of $TeCl_4$, the T-shape of ClF_3, and the linear configuration of ICl_2^- are not the only arrangements derivable from the trigonal bipyramid for one, two, or three lone pairs. The highly symmetrical arrangements found for various numbers of *shared* electron pairs are, as might be expected, the same as the arrangements of a number of similar ions around a particular ion; moreover, the general validity of the $8n + 2m$ 'rule' suggests that electron pairs, whether shared or unshared, tend to arrange themselves as far apart as possible. However, in order to account for the arrangement of ligands in cases where there is a choice of structures (as in ICl_4^-) and for the finer details of the structures of less symmetrical molecules such as $TeCl_4$ or ClF_3 it is necessary to elaborate the very simple treatment. A refinement is to suppose that the repulsions between the electron pairs in a valence shell decrease in the order

$$\text{lone pair:lone pair} \underset{\delta_1}{>} \text{lone pair:bond pair} \underset{\delta_2}{>} \text{bond pair:bond pair}$$

as is to be expected since lone pairs are closer to the nucleus than bonding pairs. The structure (a) of Fig. 7.1 then clearly has the minimum lone pair:lone pair repulsion and is to be preferred if $\delta_1 > \delta_2$. In $TeCl_4$, with only one lone pair, the repulsions between the lone pair and the bond pairs favour the structure in which the lone pair occupies an equatorial rather than an axial position. Similar arguments can be applied to ClF_3 and to other molecules.

Special interest attaches to the arrangement of bonds formed by atoms with only one pair of non-bonding electrons. This is the situation in molecules MX_{N-2} formed by elements in oxidation states 2 units less than the number (N) of the Periodic group; they range from SiF_2 to XeF_6. They have the structures expected for 18-, 26-, 34-, 42-electron groups, but the 50-electron molecules and ions show abnormal behaviour as noted on p. 299. There is also one pair of electrons left, presumably in the valence shell, if elements of the earlier B sub-groups form ions $M^{(N-2)+}$.

N	III	IV	V	VI	VII	VIII
		MX_2	MX_3	MX_4	MX_5	MX_6
		(Si)	P	S	Cl	
	Ga	Ge	As	Se	Br	
	In	Sn	Sb	Te	I	Xe
	Tl	Pb	Bi			
	M^+	M^{2+}	M^{3+}			

Sidgwick gave the name 'inert pair' to this pair of electrons. However, the presence of this lone pair is very evident from the arrangement of nearest neighbours in many crystalline compounds formed with the most electronegative elements F and O, when it is described as 'stereochemically active', In some structures, on the other hand, the environment of atoms of the B sub-group atom is highly symmetrical, as in the octahedral 50-electron molecules and ions (stereochemically inactive pair). The structural chemistry of these elements is discussed in some detail in later chapters.

We shall now review the ions and molecules having $V \geqslant 16$ and then comment briefly on those with 10-14 electrons and those with 17 or 19 electrons.

Linear 16-electron molecules and ions

This group includes linear molecules and ions of Ag and Au (for example, $Ag(NH_3)_2^+$, $Au(NH_3)_2^+$, $H_3N.AuCl$, $AuCl_2^-$), mercuric halides, and a group of molecules and ions containing C, N, and O. (Note the absence of a linear BO_2^- ion; for metaborates see p. 1068). In the latter (Table 7.5) all the bonds are very short compared with single bonds and the overall length of the molecule or ion is close to 2.3 Å except for CN_2^{2-}:

TABLE 7.5

Bond lengths in linear molecules and ions

	Bond lengths		Lengths of single bonds	
O—C—O	1·16 Å	1·16 Å	C—C	1·54 Å
N—N—O	1·13	1·19	C—N	1·47
(O—N—O)$^+$	1·15	1·15	C—O	1·43
(N—C—O)$^-$	1·21	1·13	N—N	1·46
(N—N—N)$^-$	1·15	1·15	N—O	1·41
(N—C—N)$^{2-}$	1·22	1·22	O—O	1·47

If we wish to distribute the 16 electrons in the system A—B—C so that octets are maintained around all three atoms and all the bonds contain even numbers of electrons the possibilities are:

$$\begin{array}{ccccc} A{=}B{=}C & & & A{-}B{\equiv}C \\ 4 \quad 4 \quad 4 \quad 4 & \text{and} & \left\{ \begin{array}{cccc} 6 & 2 & 6 & 2 \\ & A{\equiv}B{-}C \\ 2 & 6 & 2 & 6 \end{array} \right. \end{array}$$

The bond lengths show that all the above molecules and ions approximate to the symmetrical form $A{=}B{=}C$, though the extreme shortness of most of the bonds suggests that there is further interaction of the electron systems in these compact linear molecules.

The cyanogen halides NCX provide examples of the alternative structure $A{\equiv}B{-}C$. All the molecules NCF, NCCl, NCBr, and NCI have been shown to be linear, with N—C, 1.16 Å; the C—X bonds are uniformly about 0.14 Å shorter than the corresponding bonds in the carbon tetrahalides.

The gaseous molecules of the alkaline-earth dihalides are also 16-electron molecules, but are presumably ionic rather than covalent molecules; some are linear, as expected, others apparently non-linear (p. 441).

Triangular arrangement of 3 electron pairs

18 electrons	24 electrons

The 18-*electron group.* It is convenient to arrange the examples according to Period Groups; note that there are no compounds of C and that the only halogen species AX_2 are those of elements of Group IV. (A molecule such as HFC=O containing H does not count as an 18-electron but as a 24-electron molecule; if it is desired to include H it must be counted as contributing seven electrons like a halogen). The ClO_2^+ exists in ClO_2 (AsF_6), formed from ClO_2F and AsF_5, according to i.r. evidence.

Group IV	*Group* V	*Group* VI	*Group* VII
$F-\overset{Si}{}-F$ 101°	$O-\overset{N}{}-O^-$ 115°	$O-\overset{O}{}-O$ 117°	$O-\overset{Cl}{}-O^+$
also	$O-\overset{N}{}-F$ 110°	$N\equiv\overset{S}{}-F$ 117°	
GeF_2 (94°)	also	also	
TiF_2 (130°)	NOCl (116°)	SSO (118°)	
SnX_2	NOBr (117°)	OSO (120°)	
PbX_2	NO(OH) (111°)		

Note the remarkable similarity in interbond angles in all the compounds of N, O, and S despite great differences in multiplicity of bonds.

The 24-*electron group.* Although potentially a large group examples are not very numerous. In fact one reason for setting these ions and molecules out in this way, with oxy-compounds at the top and halogen compounds at the bottom, is to draw attention to the 'missing' compounds. For example, only one of the *ionic* species containing halogen or halogen and O ligands appears to be known. The $(NOF_2)^+$

BO_3^{3-}	CO_3^{2-}	NO_3^-	SO_3
—	—	O_2NX	—
—	OCX_2	$(NOF_2)^+$	—
BX_3	—	—	—

cation exists in salts of the type $(NOF_2)BF_4$ and $(NOF_2)AsF_6$ formed from NOF_3 and the appropriate halides. The i.r. spectrum shows that the ion is structurally similar to the planar isoelectronic COF_2 molecule.

The three oxy-ions all have the form of equilateral triangles, as also does SO_3. Here we note the absence of planar SiO_3^{2-} and PO_3^- ions, metasilicate and metaphosphate ions being based on tetrahedral SiO_4 and PO_4 groups respectively; compare the preference for a higher c.n. shown in metaborates (see above). The neutral molecules, lying along the diagonal of the chart, range from SO_3 to BF_3, and include the nitryl halides (O_2NF and O_2NCl) with the angle O—N—O around $130°$ and the carbonyl halides, COF_2 (F—C—F, $112°$), $COCl_2$ (Cl—C—Cl, $111°$); all are planar molecules. No complete vertical family in the chart is known with O or halogen ligands, but with the isoelectronic NH_2 instead of halogen we have:

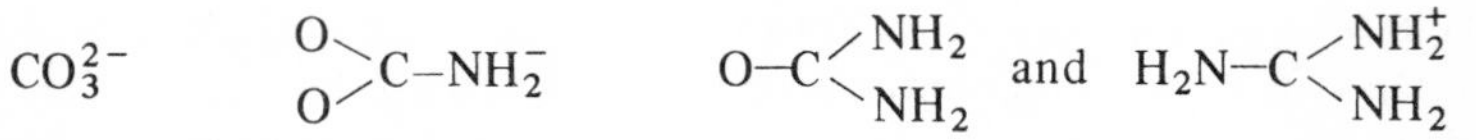

$$CO_3^{2-} \qquad \underset{O}{\overset{O}{>}}C-NH_2^- \qquad O-C\underset{NH_2}{\overset{NH_2}{<}} \text{ and } H_2N-C\underset{NH_2}{\overset{NH_2^+}{<}}$$

carbonate ion   carbamate ion   carbamide   guanidinium ion,

all of which have been shown to be planar.

Tetrahedral arrangement of 4 electron pairs

20 electrons   26 electrons   32 electrons

The 20-electron group. This small group includes a number of ions and molecules formed by elements of Groups V, VI, and VII:

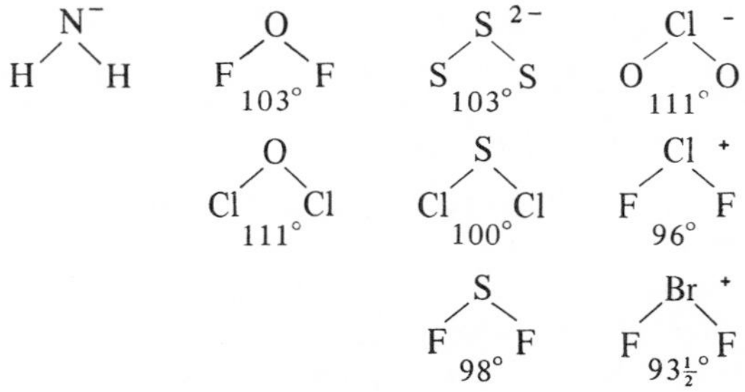

(Note the absence of SO_2^{2-}, $SOCl^-$, and $OClF$, for example.) SF_2 is an unstable species produced and studied by passing SF_6 through a radio frequency discharge, reacting the products with COS downstream from the discharge zone, and pumping them through a microwave cell. The ClF_2^+ ion occurs in compounds such as $ClF_2 \cdot AsF_6$ and the BrF_2^+ ion in $BrF_2 \cdot SbF_6$. In the latter the F—Br—F angle is $93½°$, but there are also two weaker Br—F bonds coplanar with the other two, suggesting a structure intermediate between BrF_2^+ and BrF_4^-. The FCl_2^+ ion is unsymmetrical, $(Cl-Cl-F)^+$. For halogen and interhalogen cations see p. 392.

The 26-electron group. Pyramidal AO_3 complexes include the SO_3^{2-}, SeO_3^{2-}, ClO_3^-, BrO_3^-, and IO_3^- ions and the XeO_3 molecule; note the absence of the pyramidal PO_3^{3-}

(and NO_3^{3-}) ion. Less symmetrical pyramidal molecules include the thionyl and chloryl halides; iodyl fluoride, IO_2F, a white crystalline solid, which is stable in dry air, is probably structurally different from the very reactive ClO_2F:

$$
\begin{array}{lll}
SO_3^{2-} & ClO_3^- & XeO_3 \\
SO_2F^- & ClO_2F & \\
SOCl_2 & ClOF_2^+ &
\end{array}
$$

$$SnCl_3^- \quad NF_3, PX_3 \quad SX_3^+(?)$$

It is possible that the ion SCl_3^+ may exist in SCl_4, and far i.r. studies of $SeCl_4$, $SeBr_4$, $TeBr_4$, and TeI_4, have suggested the formulation of these halides as $(SeX_3)X$ and $(TeX_3)X$. The pyramidal structure of $(SeF_3)^+$ has been established in $(SeF_3)(NbF_6)$ and $(SeF_3)(Nb_2F_{11})$, p. 739; for the structure of crystalline $TeCl_4$ see p. 709. The pyramidal structure of the isoelectronic ion $[(CH_3)_3Te]^+$ has been established in crystalline $[(CH_3)_3Te](CH_3TeI_4)$. We may include in this group the $SnCl_3^-$ ion in $K_2(SnCl_3)Cl.H_2O$, formerly thought to be $K_2SnCl_4.H_2O$.

The 32-electron group. This very large group ranges from the tetrahedral oxy-ions of Si, P, S, and Cl—such ions are not formed by the first-row elements C, N, and F —to the halogeno-ions MX_4 formed by numerous non-transition elements, and also includes many intermediate oxy-halogen ions and molecules. Compounds of P are given below because they form a more complete series than the N analogues. NF_4^+ and NOF_3 are known, but the highest oxyfluoride is NO_2F, though NS_3F has been prepared:

$$
\begin{array}{llllll}
B(OH)_4^- & SiO_4^{4-} & PO_4^{3-} & SO_4^{2-} & ClO_4^- & XeO_4 \\
& & PO_3X^{2-} & SO_3X^- & ClO_3F & \\
& & PO_2X_2^- & SO_2X_2 & & \\
& & POF_3 & & & \\
& BF_4^- & SiX_4 & PX_4^+ & & \\
BeCl_4^{2-} & AlCl_4^- & & & &
\end{array}
$$

Note the complete vertical column of P complexes and the complete diagonal series of neutral molecules, from SiX_4 to XeO_4.

Trigonal bipyramidal arrangement of 5 electron pairs

| 22 electrons | 28 electrons | 34 electrons | 40 electrons |

Particular interest attaches to the 22, 28, and 34 electron systems because of the non-equivalence of the axial and equatorial positions in trigonal bipyramidal coordination, there being no spatial arrangement of five equivalent bonds around an atom other than the coplanar (pentagonal) one.

The 22-electron group. Examples are here confined to linear ions formed by the heavier halogens ($ClBr_2^-$, $BrCl_2^-$, numerous ions IX_2^- and $(IX'X'')^-$, Br_3^-, and I_3^-) and the KrF_2 and XeF_2 molecules.

The 28-electron group. The T-shaped configuration has been established for ClF_3 and BrF_3; ions such as SCl_3^- would presumably have the same type of structure.

The 34-electron group. The expected structure derived from a trigonal bipyramid has been confirmed for all the molecules and ions set out below:

$$IO_2F_2^- \quad XeO_2F_2$$
$$IOF_3$$
$$SF_4 \quad BrF_4^+$$

and also for SeF_4 and molecules such as $Se(C_6H_5)_2Br_2$ and $Te(CH_3)_2Cl_2$.

The 40-electron group. The only complexes in this group with the full symmetry of the trigonal bipyramid are Group V pentahalides and the $SnCl_5^-$ ion. In SOF_4 the O atom occupies an equatorial position and the structure is necessarily somewhat less symmetrical. It is interesting that the $InCl_5^{2-}$ ion, isoelectronic with $SnCl_5^-$, has the form of a tetragonal pyramid in which the metal atom is 0.6 Å above the base; the IO_5^{3-} ion has a similar structure. The relation of this configuration to the trigonal bipyramid is discussed shortly. The $SnCl_3(CH_3)_2^-$ ion is trigonal bipyramidal like $SnCl_5^-$.

Octahedral arrangement of six electron pairs

$$>A< \qquad >A< \qquad >A<$$

36 electrons   42 electrons   48 electrons

The 36-electron group. The square planar configuration has been established for the ions BrF_4^- and ICl_4^- and for the XeF_4 molecule.

The 42-electron group. The arrangement of five ligands at five of the vertices of an octahedron has been demonstrated by structural studies of the following molecules and ions:

$$\begin{array}{ccc} & ClOF_4^- & XeOF_4 \\ SbF_5^{2-} \quad TeF_5^- & IF_5 & XeF_5^+ \\ SbCl_5^{2-} & & \end{array}$$

and also for IOF_4^-, ClF_5 and BrF_5. It may be noted that in no case of tetragonal pyramidal coordination is the central atom A located *in* the base of the pyramid. This atom is either situated about 0.5 Å above the base, as in the 40 electron $InCl_5^{2-}$ ion with five shared electron pairs, and in a number of transition-metal complexes when the bond arrangement is an alternative to — and closely related to — the trigonal bipyramidal configuration, or it is situated some 0.3–0.4 Å *below* the base of the pyramid. This is the case in all the pentafluoro-ions, BrF_5, and $SbCl_5^{2-}$, and also in TeF_4, which is built from TeF_5 pyramids sharing 2 F atoms (but on this point see p. 709). For details and references see Table 7.6.

TABLE 7.6

Square pyramidal ions and molecules of non-transition elements

Ion or molecule	M—X (apical)	M—X (basal)	X_a–M–X_b	Reference
$(TeF_5)^-$	1.85 Å	1.96 Å	79°	IC 1970 **9** 2100
$(TeF_4)_n$	1.80	2.03†	82°	JCS A 1968 2977
$(XeF_5)^+$	1.79	1.85	79°	IC 1973 **12** 1717
$(SbF_5)^{2-}$	2.00	2.04	83°	IC 1970 **9** 2100
$(SbCl_5)^{2-}$	2.36	2.62	85°	ACSc 1955 **9** 122
BrF_5	1.68	1.81	84°	JCP 1957 **27** 982
$IF_5(XeF_2)$	1.88	1.88	81°	IC 1970 **9** 2264

† Mean of lengths 1·90–2·26 A (unsymmetrical bridge, see p. 709).

The 48-*electron group.*   Octahedral molecules and ions of this group are more numerous, and include:

$$\text{Te(OH)}_6 \quad \text{IO(OH)}_5$$
$$\text{TeO}_6^{6-} \quad \text{IO}_6^{5-} \qquad \text{XeO}_6^{4-}$$
$$\text{IOF}_5$$
$$\text{AlF}_6^{3-} \quad \text{SiF}_6^{2-} \quad \text{PF}_6^- \quad \text{SF}_6 \quad \text{IF}_6^+$$

and other ions of Te noted in Chapter 16.

The arrangement of 7 and 9 electron pairs

It is convenient to deal first with the 56 electron group since the only molecule AX_n in which there is a valence group consisting of 7 shared electron pairs is the IF_7 molecule. Its structure has caused a great deal of discussion, but it would seem that this molecule probably has the expected pentagonal bipyramidal configuration.

The 50-electron group comprises the following molecules and ions which arise by adding X^- to the 42-electron systems noted earlier:

$$Sb^{III}X_6^{3-} \qquad Te^{IV}X_6^{2-} \qquad I^VF_6^- \qquad and \qquad Xe^{VI}F_6.$$

The oxidation number of A is two less than the 'group valence', so that there is a lone pair of electrons in addition to the six bonding pairs. The structure of the IF_6^- ion is not yet known. Careful studies of the crystal structures of $(NH_4)_4(Sb^{III}Br_6)$ (Sb^VBr_6) and of $(NH_4)_2TeCl_6$ and K_2TeBr_6 show that the ions under discussion form undistorted octahedra in spite of the presence of the seventh electron pair. Thus the latter does not occupy a bond position but is a 'stereochemically inert' pair.

No definite conclusion has yet been reached about the configuration of the XeF_6

molecule in the vapour state, except that it appears to be neither regular octahedral nor regular pentagonal bipyramidal. The structures of the crystalline forms are complex. The unique cyclic polymers can be described as built from tetragonal pyramidal XeF_5^+ groups and F^- ions, but there is no obvious simple description of the bonding.

Pauling commented many years ago on the abnormally long M–X bonds in $Se^{IV}X_6^{2-}$ and $Te^{IV}X_6^{2-}$ ions, comparing them with the sums of the tetrahedral covalent radii, which for Se and Te correspond to M(II). A more direct comparison could be made of the M–X bond lengths in the following pairs:

M–X			M–X
2·80 Å	$SbBr_6^{3-}$	$SbBr_6^-$	2.56 Å
?	TeF_6^{2-}	TeF_6	1.82
?	IF_6^-	IF_6^+	?

Unfortunately this comparison can be made at present only for Sb–X, for the only known hexahalide of Te is TeF_6 and the only accurate data are for the $TeCl_6^{2-}$ and $TeBr_6^{2-}$ ions; no data are available for IF_6^- or IF_6^+. It appears that although stereo-chemically inert the pair of 2s electrons increases the bond length, behaving as a spherically symmetrical shell resulting in an increase in the size of M(IV) as compared with M(VI).

The ion XeF_8^{2-} presents a somewhat similar problem, in that there are nine electron pairs, eight shared and one unshared, yet the shape is not far removed from a square antiprism. This is the sole representative of the 66-electron group.

The 10–14 electron groups

If we arrange from 10 to 14 electrons in a diatomic molecule (or ion) so as to maintain an octet of valence electrons in each atom the only possibilities are the symmetrical arrangements:

10	11	12	13	14
2 : 6 : 2	3 : 5 : 3	4 : 4 : 4	5 : 3 : 5	6 : 2 : 6

As more electrons are added to the system the number in the bond falls from six to two and the bond weakens and lengthens:

	10	11	12	13	14	
	(N_2)	O_2^+	O_2	O_2^-	O_2^{2-}	
Length	1.10	1.12	1.21	1.33	1.48	Å
Dissociation energy	941.4		493.7		154.8	kJ mol^{-1}
Stretching force constant	2290		1180		383	Nm^{-1}

(Since data are not available for the species O_2^{2+} the N_2 molecule is included as the representative of the 10-electron class.) The bond orders according to m.o. bond theory are 3, 2.5, 2, 1.5, and 1 respectively. The fact that the bond length increases on adding successive electrons to O_2^{2+} is due to the rearrangement of the electrons and corresponds to the change in the isoelectronic series:

10 electrons	12 electrons	14 electrons
$HC{\equiv}CH$	$H_2C{=}CH_2$	$H_3C{-}CH_3$
1.20 Å	1.35 Å	1.54 Å

Although this simple treatment accounts qualitatively for the bond lengths, it does not account for all of the facts. Not only are O_2^+ and O_2^- paramagnetic, as is to be expected for odd-electron ions, but so also is the O_2 molecule. This difficulty is overcome in m.o. theory by assigning separate (unpaired) electrons to two 'antibonding' orbitals.

Examples of the even-numbered members of this group include some important molecules and ions. The 10-electron group includes N_2, the high-temperature molecules PN and P_2, and the series:

C_2^{2-}	CN^-	CO	NO^+
1.20	1.17	1.13	1.06 Å.

The 12-electron group includes NO^-, O_2, SO, and S_2, and the 14-electron group, the molecules of the halogens and interhalogen compounds XX′, and halogen oxy-ions XO^-. There are few examples of the 11- and 13-electron groups:

11 electrons: NO, O_2^+; 13 electrons: O_2^-, Br_2^+.

Note that the simple arithmetic approach adopted here predicts facts such as the collinear arrangement of bonds $M{-}C{\equiv}O$ or $M{-}(N{=}O)^+$ for 10-electron ligands bonded to metal atoms as contrasted with the non-linear arrangement for a 12-electron ligand, $M{-}(N{=}O)^-$.

Odd electron systems AX_2 and their dimers $X_2A{-}AX_2$

The 17-electron and 19-electron systems are of special interest because they might be expected to dimerize to the 34- and 38-electron molecules and so get rid of the unpaired electrons.

† in Xe matrix (JCP 1969 **51** 4710)   ‡ in gas (PRS 1967 **A 298** 145)

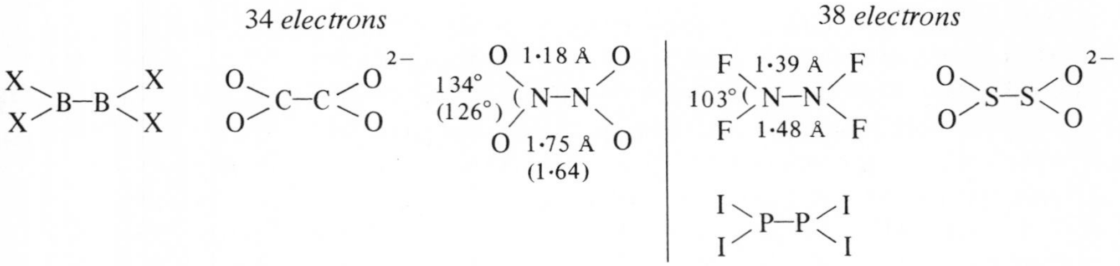

34 electrons   *38 electrons*

The effect on the stereochemistry of a system of adding first an odd electron and then a lone pair is nicely illustrated by the series: NO_2^+ (linear), NO_2, and NO_2^- (both angular), and it is noteworthy that the 17-electron system NO_2 is intermediate as regards both bond length and interbond angle between the 16- and 18-electron systems:

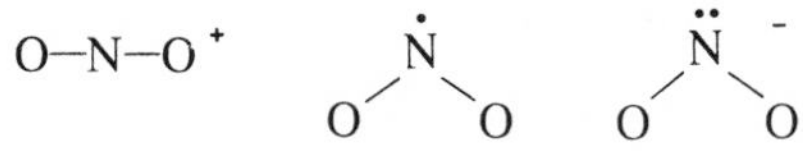

Angle O–N–O	$180°$	$134°$	$115°$
N–O	$1·15$ Å	$1·20$ Å	$1·24$ Å

The detailed structure of ClO_2^+ would be of great interest because, together with ClO_2 and ClO_2^- it forms another series including an odd-electron molecule (as also would SO_2^+, SO_2, and SO_2^-):

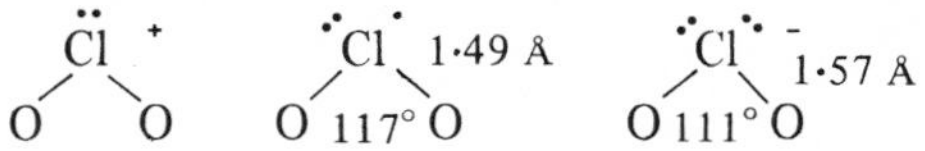

If eight electrons are assigned to each terminal O or X atom there are sufficient electrons for a single bond in the 34-electron systems while in the 38-electron systems there is a lone pair on each of the A atoms:

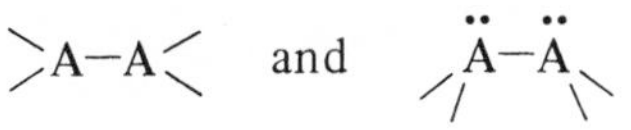

In the former there are therefore three coplanar bonds around A and in the latter three pyramidal bonds. In $C_2O_2^{2-}$ and B_2X_4 the central bonds are found to have the lengths expected for single bonds, but in N_2O_4 the N–N bond is abnormally long. Note the exactly similar structures of monomer and one-half of the dimer in the case of NO_2 and N_2O_4 and of NF_2 and N_2F_4. An interesting point about these systems is that N_2O_4 and B_2F_4 are planar (for B_2Cl_4 see p. 1059) while $C_2O_4^{2-}$ is planar in some salts and non-planar in others (see p. 919). The oxalic acid molecule is planar, with abnormally short C=O (1.19 Å) and C–OH (1.29 Å) bonds. For a discussion of the bonding in molecules A_2X_4 formed by B, N, and P see: JACS 1969 **91** 1922 and IC 1969 **8** 2086.

Unlike NO_2 ClO_2 shows no tendency to dimerize, whereas the isoelectronic SO_2^- can only be isolated as $S_2O_4^{2-}$ (in dithionites), though there is some evidence for an equilibrium between monomer and dimer in solution. Only one of the 19-electron halogen species (NF_2) is known; the dimeric species include N_2F_4, P_2F_4, P_2Cl_4, and P_2I_4. The free radical NF_2 is quite stable and can exist indefinitely in the gas phase in equilibrium with N_2F_4. It is formed by reacing NF_3 with metallic Cu at 400 °C and dimerizes on cooling. There is an interesting reaction between NF_2 and NO, another odd-electron molecule, to form $ON.NF_2$, a deep purple compound which at room temperature and atmospheric pressure rapidly dissociates to a colourless mixture of NO and NF_2.

The halogen compounds P_2X_4 apparently all exist in the centro-symmetrical *trans* form (in contrast to the *gauche* configuration of P_2H_4); the dithionite ion is notable for its eclipsed configuration and for the extraordinary length of the S—S bond (2.39 Å, as compared with 2.06 Å for the normal single bond in the various forms of elementary S).

The van der Waals bond

Under appropriate conditions of temperature and pressure it is possible to liquefy and solidify all the elements, including the noble gases, and all compounds, including those consisting of non-polar molecules such as CH_4, CCl_4, etc. The existence of a universal attraction between all atoms and molecules led van der Waals to include a term a/V^2 in his equation of state. For molecules with a permanent dipole moment (μ) Keesom calculated the mean interaction energy

$$E_K = -\frac{2\mu^4}{3r^6kT}$$

at a distance r, and to this expression Debye added the energy resulting from the interaction between the permanent dipole and the moments it induces in neighbouring molecules:

$$E_D = -\frac{2\alpha\mu^2}{r^6}$$

where α is the polarizability. The first expression, requiring the van der Waals factor a to be inversely proportional to the absolute temperature, is not consistent with observation, and moreover the sum of E_K and E_D is much too low (see Table 7.7). Neither expresson, of course, accounts for the interaction between non-polar molecules, a type of interaction first calculated by London and hence called London, or dispersion, energy. This (wave-mechanical) calculation gives

$$E_L = -\frac{3\,I\alpha^2}{4\,r^6}$$

TABLE 7.7
Lattice energy kJ mol^{-1}

	$\mu(D)$	$\alpha\,(\times10^{-24}\ cm^3)$	E_K	E_D	E_L	*Total*
Ar	0.00	1.63	0.000	0.000	8.49	8.49
CO	0.12	1.99	0.000	0.008	8.74	8.74
HI	0.38	5.40	0.025	0.113	25.86	25.98
HBr	0.78	3.58	0.686	0.502	21.92	23.10
HCl	1.03	2.63	3.31	1.004	16.82	21.13
NH$_3$	1.50	2.21	13.31	1.26	14.73	29.58
H$_2$O	1.84	1.48	36.36	1.92	9.00	47.28

for similar particles with ionization potential I, or

$$-\ \frac{3\,I_1 I_2 \alpha_1 \alpha_2}{2\,r^6(I_1 + I_2)}$$

for dissimilar particles.

The relative magnitudes of these three types of interaction can be seen from Table 7.7 for a few simple cases. For a non-polar molecule the London energy is necessarily the only contribution to the lattice energy; even for polar molecules such as NH$_3$ and H$_2$O for which E_K is appreciable, E_L forms an important part of the lattice energy. Note that these lattice energies are between one and two orders of magnitude smaller than for ionic crystals; for example, those of the 'permanent' gases are of the order of one per cent of the lattice energy of NaCl (approximately 753 kJ mol^{-1}). As a contribution to the lattice energy of salts the London interaction can be as large as 20 per cent of the total in cases where the ions are highly polarizable (for example, TlI).

Metal–metal bonding

The discussion of metal–metal (m–m) bonding by chemists is usually wholly or largely confined to finite molecules and complex ions. Here we shall attempt to place this aspect of the subject in perspective as part of a much more general phenomenon.

As noted on p. 287 we shall use the term m–m bond to include all types of interaction between metal atoms ranging from normal covalent bonds, through the various types of interaction in compounds of transition metals, to the *metallic* bonds in certain sub-halides and sub-chalconides.

The formation of a chemical compound implies bonding between atoms of different elements, and for simplicity we may start by considering binary compounds. In some binary compounds A$_m$X$_n$ all the primary bonds (between nearest neighbours) are between A and X atoms (as in simple ionic crystals) but this is by no means generally true. Bonds between atoms of the same element (A–A or X–X) are to be expected in a compound containing a preponderance of atoms of one kind, whether it consists of molecules (for example, S_7NH) or is a crystalline compound such as a phosphorus-rich phosphide (CdP_4), a silicon-rich silicide ($CsSi_8$), a boron-rich boride (BeB_{12}) or at the other extreme a metal-rich compound such as $Li_{15}Si_4$, Cs_3O, or Ta_6S. Of special interest are compounds in which the greater part of the bonding is between A and X atoms, but where there is also some interaction between A and A or X and X atoms. Disregarding the other bonds formed by A and X atoms (which may form part of a 3D arrangement in a crystal) we may show this diagrammatically:

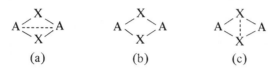

An example of (a) is the structure of NiAs, in which Ni has 6 As neighbours but is also bonded to 2 Ni; (b) represents a 'normal' binary compound with only A—X bonds, while an example of (c) is a boride such as FeB in which B is surrounded by 6 Fe but is also bonded to 2 B atoms.

In Chart 7.1 are set out the main types of structure (molecular and crystal) in which there is m–m bonding, and we now comment briefly on the various classes.

Class I. Molecules (ions) containing directly bonded metal atoms without bridging ligands

The simplest molecules of this kind are the diatomic molecules in the vapours of a number of metals (of Groups I–III, Co and Ni). The bond dissociation energies vary widely, from around 170 to 210 kJ mol^{-1} for Co_2, Ni_2, Cu_2, Ag_2 and Au_2 (very similar to those of Cl_2, Br_2, and I_2) and somewhat smaller values for the alkali metals (from 109 in Li_2 to 42 kJ mol^{-1} in Cs), to very much smaller values in Mg_2 (29 kJ mol^{-1}) and in Zn_2, Cd_2, and Hg_2 (around 4 to 8 kJ mol^{-1}). The next-simplest systems containing m–m bonds are those formed by a few B subgroup elements, and of these the most important are the mercurous compounds which contain the grouping X—Hg—Hg—X; compounds of Group IVB elements include Ge_2H_6, Sn_2H_6, and $(CH_3)_3Pb$—$Pb(CH_3)_3$.

Certain of the most metallic B subgroup elements form polynuclear *cations*, the structures of which are described in later chapters; Hg cations, p. 1156, Bi cations, pp. 877 and 888. More is known at the present time of the structures of *anions* formed by B subgroup elements. Groups of such atoms are recognizable in many alloys with the most electropositive metals, alkali and alkaline-earth. Examples include:

square Hg_4 in Na_3Hg_2, Bi_4 in $Ca_{11}Bi_{10}$;
tetrahedral Tl_4 in Na_2Tl, Ge_4, Sn_4, and Pb_4 in KGe, KSn, and NaPb;
planar cyclic Ge_5 in $Li_{11}Ge_6$;
polyhedral As_7 in Ba_3As_{14}.

Many types of metal anion exist in liquid ammonia solutions of alloys of Pb, Sb,

CHART 7.1

Types of structure with metal–metal bonding

Class I
  m–m bonds
  no bridging atoms

finite: Cu_2, etc.
  $X-Hg-Hg-X$, $Mn_2(CO)_{10}$, $Cl_4Re{\equiv}ReCl_4$
  Hg_3^{2+}, Hg_4^{2+}
  Sn_5^{2-}, Sb_7^{3-}, Sn_9^{4-}, Bi_9^{5+}

chains: Hg_∞ in Hg_3AsF_6
  $K_2[Pt(CN)_4]X_{0.3}\cdot3H_2O$

Class Ia
  m–m systems linked into more extensive system
  through X atoms (no further m–m bonding)

m–m grouping
  finite: See text
  chain: Gd_2Cl_3, Ta_2S, Ta_6S
  layer: ZrCl

Class IIa
Units of class II further linked through M—X—M
bridges to more complex systems, e.g.

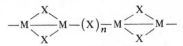

chains: NbI_4, $ReCl_4$, ReI_3, W_6Br_{16}
layers: $ReCl_3$, $MoCl_2$, WCl_2, Nb_6Cl_{14}, Ta_6I_{14}
3D: Nb_6I_{11}, Nb_6F_{15}, Ta_6Cl_{15}, MoO_2, WO_2, etc.

I–II hybrids, e.g.
$Co_4(CO)_{12}$

Class II
  m–m bonds and bridging:

$$M\overset{X}{\underset{X}{\diagup\!\!\!\diagdown}}M \quad \text{or} \quad M{-}X{-}M$$

  (NiAs)   (ReO₃)

finite: metal cluster compounds of Cu, Ag, Nb,
  Ta, Mo, W
  carbonyls, $Fe_2(CO)_9$, $Os_3(CO)_{12}$,
  $Re_4(CO)_{16}^{2-}$
  carboxylates (Cr, Mo, Re, Cu)

1D: AgCNO, AgNCO

2D: $PtCoO_2$

3D: ReO_3, RuO_2

Class III
Finite units, chains and layers bonded together by
m–m bonds: 'anti' chain and layer structures.

finite: Rb_9O_2, Rb_6O, Cs_7O

chains: Cs_3O

layers: Ag_2F, PbO, Ti_2O, Ti_3O

and Bi with alkali metals, but the highly coloured products isolated from such
solutions are amorphous salts containing $Na(NH_3)_n^+$ cations which lose NH_3 before
they can be recrystallized, and revert to the intermetallic compounds. However,
complexing with the ligand *crypt*, (a), prevents the delocalization of electrons and
re-formation of the alloy, as was first shown by the isolation of $[Na(crypt)]^+Na^{-}$.[1]
Subsequently, numerous polyatomic anions have been isolated in crystalline
compounds by the action of this ligand, in ethylene diamine solution, on inter-

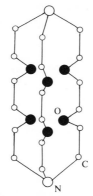

(a) The ligand 'crypt-222' $N(C_2H_4-O-C_2H_4-O-C_2H_4)_3N$.

TABLE 7.8
Polyatomic anions of B subgroup elements

Anion	Shape	M–M (Å)	Reference
Te_3^{2-}	Angular	2.70	IC 1977 **16** 632
Bi_4^{2-}	Square planar	2.94	IC 1977 **16** 2482
Sn_5^{2-}	Trigonal bipyramid	2.89 (×6) 3.10 (×3)	IC 1977 **16** 903
Pb_5^{2-}		3.00 (×6) 3.23 (×3)	JACS 1977 **99** 3313
Sb_7^{3-}	Fig. 19.10(c)	2.69–2.88	JACS 1976 **98** 7234
Ge_9^{2-} Ge_9^{4-}	See text	2.53–2.80	JACS **1977 99** 7163
Sn_9^{4-}	Monocapped antiprism	2.93–3.31	JACS 1977 **99** 3313

metallic compounds such as K_5Bi_4 and $NaSb_3$, which give $[K(crypt)]_2Bi_4$ and $[Na(crypt)]_3Sb_7$ respectively. Some examples are listed in Table 7.8.

The bonds in these anions are obviously of the same type as those in ions and molecules formed by the lighter elements of the B subgroups. For example, the Sb_7^{3-} anion has the same structure as the P_7^{3-} ion in Ba_3P_{14} or the As_7^{3-} ion in Ba_3As_{14} and the molecules P_4S_3, P_4Se_3, and As_4Se_3 (Fig. 19.10(c), p. 868). A point of special interest is that some of these *anions* are isostructural with isoelectronic *cations* of (other) B subgroup elements. For example, the angular Te_3^{2-} (bond angle 113°) has a structure similar to that of I_3^+ (angle 97°), and Bi_4^{2-} is square planar, like Te_4^{2+}. On the other hand, Sn_9^{4-} has a quite different structure from that of Bi_9^{5+}, which is isoelectronic with Pb_9^{4-}, expected to have the same structure as the Sn anion. The deep-red $[K(crypt)]_6Ge_9^{2-}Ge_9^{4-}$ contains two kinds of Ge_9 anion, one very similar to the monocapped antiprismatic Sn_9^{4-} and the other less symmetrical, with a configuration closer to that of the alternative 9-coordination polyhedron illustrated in Fig. 3.8(b), p. 80.

Infinite linear systems of bonded metal atoms are of particular interest for they give rise to metallic conduction. In Hg_3AsF_6 (p. 1156) the Hg chains are parallel to the tetragonal *a* axes, resulting in 'two-dimensional' conduction. Metal–metal bonding of a different kind exists in the 'one-dimensional' metal $[K_2Pt(CN)_4]Br_{0.3}.3H_2O$ (p. 1238). The $Pt(CN)_4^{2-}$ ions are stacked in columns with their planes parallel and along the chain there is overlap of d_z^2 orbitals. In the (hypothetical) absence of Br^- ions the delocalized d_z^2 band would be full, but the Br^- ions remove an average of 0.3 electrons from each complex so that the band is only 5/6 filled, leading to metallic conduction along the chain direction.

Molecules or ions containing pairs of singly-bonded metal atoms without bridging ligands are formed by various transition metals in low oxidation states when π-bonding ligands are attached to the metal atoms as in $[Ni_2(CN)_6]^{4-}$, $[Pd_2(NC.CH_3)_6]^{2+}$, $[Pt_2Cl_4(CO)_2]^{2-}$, $Mn_2(CO)_{10}$, $Re_2(CO)_{10}$, and $Tc_2(CO)_{10}$, and $[Co_2(NC.CH_3)_{10}]^{4-}$. In these complexes the m–m bonds are rather long (2.9–3.0 Å), possibly due to repulsions between the ligands on different metal atoms. Intermediate between compounds of this type and the Group IVB molecules noted above are numerous molecules containing a number of directly bonded metal atoms, in which some are B subgroup metals (Cu, Au, Hg, In, Tl, Ge, Sn) and others are transition metals (Mo, W, Mn, Fe, Co, Ir, Pt). Examples include:

$$(C_6H_5)_3Sn-Mn(CO)_5 \qquad\qquad (C_5H_5)Mo(CO)_3$$

$$(C_6H_5)_3Sn-Fe(CO)_2C_5H_5 \qquad C_5H_5(CO)_2Fe-Sn-Fe(CO)_2C_5H_5$$

$$[Cl_3Sn-Pt-SnCl_3][N(CH_3)_4]_2 \qquad\qquad (C_5H_5)Mo(CO)_3$$
$$\hspace{2.2cm}|$$
$$\hspace{2.2cm}Cl_2$$

In a small group of ions and molecules there is multiple bonding between the metal atoms of a type possible only for transition metals. They include the ions $Re_2Cl_8^{2-}$, $Tc_2Cl_8^{3-}$, and $Mo_2Cl_8^{4-}$, in which the lengths of the m–m bonds are respectively 2.24, 2.13, and 2.14 Å. These bonds are described as quadruple bonds $(\sigma\pi^2\delta)$, the δ component accounting for the eclipsed configuration of the ions and also of the similarly constituted molecule $Re_2Cl_6(PEt_3)_2$.

Class Ia. We can envisage systems of bonded metal atoms which are then cross-linked by ligands X, there being no metal–metal interactions between the different sub-units. The cyclic molecule (a) may be described in this way as consisting of two sub-units Mg–Mo–Mg joined through pairs of bridging Br atoms.[2] More complex examples of compounds in this class include Gd_2Cl_3, Ta_2S, Ta_6S, and ZrCl.

(1)   JACS 1974 **96** 7203   (2)   AC 1975 **B31** 852

(a)

Class II. Molecules (ions) containing directly bonded metal atoms and bridging ligands

This large class includes all compounds in which there is direct interaction between pairs of metal atoms which are also bridged by other ligands. These ligands may be halogen atoms, NH_2, CO, SR, PR_2, carboxylate groups, or other chelate organic ligands. Examples of finite systems with halogen bridges include those $M_2Cl_9^{3-}$ ions formed from a pair of face-sharing octahedra in which the M—M distance indicates interaction across the shared face, and the numerous metal cluster compounds formed by Nb, Ta, Mo, W, and Re, and also by Ag and Cu. For the polynuclear carbonyls in which some of the CO ligands act as bridges between metal atoms, and for compounds such as $[(CO)_3Fe(NH_2)]_2$ and $[(C_5H_5)Co(P\phi_3)]_2$ see Chapter 22. Two carboxylate bridges are present in compounds such as $Re_2(O_2CC_6H_5)_2I_4$ and four in $Mo_2(O_2CCH_3)_4$, molecules such as $Re_2(O_2CR)_4X_2$, and the monohydrates of Cr(II) and Cu(II) acetates.

Of special interest are molecules or ions containing very short m–m bonds,

The ligand DHMP
in W_2 (DHMP)$_4$

TABLE 7.9
*Some molecules containing metal-metal bonds**

(a)   (b)   (c)   (d)

Molecule	M—M (Å)	Reference
(a) $Re_2(O_2CC_6H_5)_2I_4$	2.20	IC 1969 8 1299
(b) $Cu_2(O_2CCH_3)_4(H_2O)_2$		
$Cu_2(O_2CCH_3)_4(pyr)_2$	see p. 897	
$Cr_2(O_2CCH_3)_4(H_2O)_2$	2.36	
$Rh_2(O_2CCH_3)_4(H_2O)_2$	2.39	AC 1971 **B27** 1664
$Mo_2(O_2CCH_3)_4$	2.09	AC 1974 **B34** 2768
$[Mo_2(SO_4)_4]^{3-}$	2.17	IC 1979 18 1159
$W_2(DHMP)_4$	2.16	IC 1979 18 1152
$Pd_2(S_2CCH_3)_4$	2.75	IC 1979 18 2258
$Pt_2(S_2CCH_3)_4$	2.77	IC 1980 19 3632
$Re_2(O_2CC_6H_5)_4Cl_2$	2.24	IC 1968 7 1570
(c) $Re_2OCl_3(O_2CC_2H_5)_2(P\phi_3)_2$	2.51	IC 1969 8 950
(d) $Re_2OCl_5(O_2CC_6H_5)(P\phi_3)_2$	2.52	IC 1968 7 1784

(e)   (f)

	M—M (Å)	Reference
(e) $(CO)_3Fe(CO)_3Fe(CO)_3$	2.46	JCS 1939 286
$(CO)_3Fe[Ge(CH_3)_2]_3Fe(CO)_3$	2.75	IC 1969 8 1424
(f) $(CO)_3Co(CO)_2Co(CO)_3$	2.52	AC 1964 17 732
$[Cl_3Mo(Cl_2)MoCl_3]Rb_3$	2.38	IC 1969 8 1060
$(CO)_3Fe(SC_2H_5)_2Fe(CO)_3$	2.54	IC 1969 8 2709

* For other examples see Chapters 9 and 22.

of which a number are included in Table 7.9. The simplest complexes are of the type M_2L_4, class (b). In the ion $[Mo_2(SO_4)_4]^{3-}$ there are four bridging SO_4 groups, and the length of the Mo—Mo bonds in $K_3[Mo_2(SO_4)_4].3.5H_2O$ and in $K_4[Mo_2(SO_4)_4]Cl.4H_2O$ corresponds to bond order 3.5. Quadruple bonds occur in molecules M_2L_4 in which L is, for example, 2,4-dimethyl 6-hydroxypyrimidine (DMHP in Table 7.9); their lengths are Cr—Cr, 1.91 Å; Mo—Mo, 2.07 Å; and W—W, 2.16 Å.

The structures of silver cyanate and fulminate are examples of simple chain structures in which there are m-m interactions along the chains:

$$
\begin{array}{cc}
\text{OCN} \diagdown & \text{ONC} \diagdown \\
\text{Ag} \diagdown & \text{Ag} \diagdown \\
\text{NCO} & \text{CNO} \\
\text{Ag} \diagup & \text{Ag} \diagup \\
\text{OCN} \diagdown & \text{ONC} \diagdown
\end{array}
$$

In this class come oxides such as ReO_3 and RuO_2 in which there is not only the 'normal' bonding of the metal atoms through O atoms but also some kind of less direct interaction ('super-exchange') involving the oxygen orbitals.

Class IIa. This class includes all structes in which units of class II are further linked, but without m–m interactions, into more extensive systems. Examples include the simple edge-sharing chain of NbI_4, with alternate short and long Nb–Nb distances, the chain molecules in crystalline $ReCl_4$ built from Re_2Cl_9 units, and the still more complex chains in ReI_3 (Re_3I_9 units) and in W_6Br_{16}. In the extraordinary structure of the latter compound $(W_6Br_8)Br_4$ clusters are linked into infinite chains by linear Br_4 groups. Infinite 2D structures in this class include $ReCl_3$, built from triangular Re_3Cl_9 units similar to those in Re_3I_9 but here further linked into layers rather than chains, and a number of halides built from the clusters of the two main types, M_6X_8 and M_6X_{12}. These units can also be connected into 3D systems, as in the following examples:

Type of metal cluster	2D structures	3D structures
M_6X_8	$MoCl_2$, WCl_2	Nb_6I_{11}
M_6X_{12}	Nb_6Cl_{14}, Ta_6I_{14}	Nb_6F_{15}, Ta_6Cl_{15}

These halide structures are described in more detail in Chapter 9. We may also include here those dioxides with distorted rutile structures in which there are alternate short and long m–m distances within each 'chain' (e.g. 2.5 Å and 3.1 Å in MoO_2 and WO_2), since the discrete pairs of close metal atoms are linked through oxygen bridges into a 3D structure.

Class III. Crystals containing finite, 1-, or 2-dimensional complexes bonded through metal–metal bonds

In Classes I and II we recognized units within which there is m–m bonding, with additional bonding through bridging X atoms in Class II. In Classes Ia and IIa these units are further bonded, through X atoms, without further m–m bonding. It is necessary to recognize a further class of structures in which there is extensive m–m bonding *between* finite, 1-, or 2-dimensional sub-units; there may also be metal–metal bonding *within* the units, but this is not necessary. This class includes halides and chalconides (usually metal-rich 'sub-compounds') which have structures geometrically similar to those of normal halides or chalconides but with positions of metal and non-metal interchanged. Just as we find polarized ionic structures with only van der Waals bonds between the halogen atoms on the outer surfaces of the chains or layers, so we find the 'anti' chain and layer structures of the same geometrical types in which the interactions between the chains (layers) are bonds between the metal atoms on their outer surfaces. The metal-rich oxides of Rb and Cs (p. 534) provide examples of structures built entirely of finite units held together by m–m bonds (Rb_9O_2) and of structures which might be described as consisting of finite units inserted into the metal, there being metal atoms additional to those in the sub-units, as in Rb_6O ($Rb_9O_2 . Rb_3$) and Cs_7O ($Cs_{11}O_3 . Cs_{10}$).

The ionic bond

The ionic bond is the bond between charged atoms or groups of atoms (complex ions) and is the only one of the four main types of bond that can be satisfactorily described in classical (non-wave-mechanical) terms. Monatomic ions formed by elements of the earlier A subgroups and ions such as N^{3-}, O^{2-}, etc., F^- etc., have noble gas configurations, but many transition-metal ions and ions containing two s electrons (such as Tl^+ and Pb^{2+}) have less symmetrical structures. We shall not be concerned here with the numerous less stable ionic species formed in the gaseous phase.

The simplest systems containing ionic bonds are the gaseous molecules of alkali halides and oxides, the structures of which are noted in Chapter 9 and 12. The importance of the ionic bond lies in the fact that it is responsible for the existence at ordinary temperatures, as stable solids, of numerous metallic oxides and halides (both simple and complex), of some sulphides and nitrides, and also of the very numerous crystalline compounds containing complex ions, particularly oxy-ions, which may be finite (CO_3^{2-}, NO_3^-, SO_4^{2-}, etc.) or infinite in one, two, or three dimensions.

The adequacy of a purely electrostatic picture of simple ionic crystals A_mX_n is demonstrated by the agreement between the values of the lattice energy resulting from direct calculation, from the Born–Haber cycle, and in a few cases from direct measurement.

Ionic radii

To calculate the electrostatic contribution to the lattice energy of the NaCl (or any other structure) it is necessary to know only the relative positions of the atoms and the distances between them, both of which are directly determined from the diffraction data. No knowledge of the relative sizes of the ions is required. If ions can be regarded as approximately incompressible spheres of various sizes having spherically symmetrical charge distributions we might expect to be able to relate the way in which they pack together to their relative sizes, that is, to relate the coordination numbers in different structures to the ionic radii. Since only *sums* of radii are measurable it is necessary to fix one ionic radius; that of O^{2-} is usually chosen.

Except for the (approximate) values for the ions P^{3-}, As^{3-}, NH_4^+, and SH^-, the ionic radii of Table 7.10 were derived from a critical review of interatomic distances in essentially-ionic crystals, and are based on the value 1.40 Å for O^{2-} in 6-coordination (AC 1976 **A32** 751). An alternative set of radii (JPCS 1964 **25** 31, 45) derived specifically for the alkali halides and extended on the basis of 1.19 Å for F^- has radii of cations 0.14 Å greater and those of anions 0.14 Å smaller than the values of Table 7.10. Obviously the increase in r_A and decrease in r_X values has a

TABLE 7.10
Ionic radii (Å)

-3	-2	-1	+1	+2	+3	+4			
			$Li^{(a)}$ 0.76	$Be^{(b)}$ 0.45					
N 1.46	$O^{(c)}$ 1.40	$F^{(d)}$ 1.33	$Na^{(g)}$ 1.02	Mg 0.72	Al 0.54		Ag^+ 1.15		
$\begin{bmatrix} P \\ 1.90 \end{bmatrix}$	S 1.84	Cl 1.81	K 1.38	Ca 1.00	Sc 0.75	Ti 0.61	Zn^{2+} 0.74		
$\begin{bmatrix} As \\ 2.20 \end{bmatrix}$	Se 1.98	Br 1.96	Rb 1.52	Sr 1.18	Y 0.90	Zr 0.72	Cd^{2+} 0.95		
	Te 2.21	I 2.20	Cs 1.67	$Ba^{(f)}$ 1.35	$La^{(e)}$ 1.03		Hg^{2+} 1.02	Tl^+ 1.50	Pb^{2+} 1.19
$\begin{bmatrix} NH_4^+ \\ 1.50 \end{bmatrix}$	OH^- 1.37	$\begin{bmatrix} SH^- \\ 2.00 \end{bmatrix}$							

	Ti^{2+}	V^{2+}	Cr^{2+}	Mn^{2+}	Fe^{2+}	Co^{2+}	Ni^{2+}	Cu^{2+}
Low spin			0.73	0.67	0.61	0.65		
High spin	0.86	0.79	0.80	0.83	0.78	0.75	0.69	0.73

	Ti^{3+}	V^{3+}	Cr^{3+}	Mn^{3+}	Fe^{3+}	Co^{3+}	Ni^{3+}	(Ga^{3+})
Low spin				0.58	0.55	0.55	0.56	
High spin	0.67	0.64	0.62	0.65	0.65	0.61	0.60	0.62

(a) c.n. 4, 0.59
(b) c.n. 4, 0.27
(c) c.n. 2, 1.35, c.n. 8, 1.42
(d) Note that M−F is at least 0.1 Å *greater* than M−O in ScOF and YOF (p. 315)
(e) c.n. 8, 1.16, c.n. 9, 1.22
(f) c.n. 8, 1.42, c.n. 12, 1.61 Å
(g) A radius close to that of I^- has been suggested for Na^- in $[Na(crypt)]^+Na^-$ (p. 306).

considerable effect on radius ratios (p. 315), and for this reason we shall prefer to discuss many aspects of the structures of ionic crystals in terms of interatomic distances, which are measured quantities, rather than radius ratios. (It may be of interest to some readers to note that the interionic distances in halides of the Groups IA and IIA metals can be reproduced very accurately using a set of 'soft sphere' radii (JCS 1978 **D** 1631) which for M^+ and M^{2+} in 6-coordination are the same as the metallic radii for 12-coordination. The interatomic distance d is derived from these radii by the relation: $d^{5/3} = M^{5/3} + X^{5/3}$.) As regards the derivation of ionic radii it should be noted that there are relatively few structures, even of fluorides and oxides, in which the ions have symmetrical arrangements of *equidistant* nearest neighbours. For metals which do not form binary compounds with structures of this type it is necessary to take mean interionic distances from structures such as the C–Mn_2O_3 or LaF_3 structures, in which there are groups of 'nearest neighbours' at somewhat different distances, or from complex fluorides or oxides in which there are highly symmetrical coordination groups. In some structures the departures from regular coordination groups (for example, around Ti^{4+} or V^{5+}) may indicate partial covalent character of the bonds, so that the significance of 'ionic radii' derived from such structures is doubtful. Allowance has always been made for the c.n. of the cation; it seems reasonable (see Table 7.10) also to allow for the c.n. of the anion and probably also to use different radii for low- and high-spin states of the 3d metal ions.

We shall not discuss here the differences between the various sets of ionic radii that have been proposed, but merely note the following general points.

(a) A positive ion is appreciably smaller than the neutral atom of the same element owing to the excess of nuclear charge over that of the orbital electrons, whereas the radius of an anion is much larger than the covalent radius:

Li	1.5 Å		Cl^-	1.8 Å
Li^+	0.7	compare	Cl	1.0

(b) In a series of isoelectronic ions the radii decrease rapidly with increasing positive charge, but there is no comparable increase in size with increasing negative charge (see the values in a horizontal row of Table 7.10).

(c) *Most* metal ions are smaller than anions; a few pairs of cations and anions of comparable size lead to the special families of c.p. structures noted above, and we discuss later some structures containing the exceptionally large Cs^+, Tl^+, and Pb^{2+} ions. The very small Li^+ and Be^{2+} ions are typically found in tetrahedral coordination in halide and oxide structures, Al^{3+} in both 4- and 6-coordination, the numerous ions M^{2+} and M^{3+} with radii in the range 0.7–1.0 Å are usually octahedrally coordinated in such structures, while the largest ions are found in positions of higher coordination (up to 12-coordination in complex structures).

(d) In contrast to the steady increase of radii down a Periodic family the radii of the 4f M^{3+} ions show a steady decrease with increasing atomic number. For example, the M–O distances (for 6-coordination) decrease from La^{3+}–O, 2.44 Å, to Lu^{3+}–O, 2.23 Å. As a result of this 'lanthanide contraction' certain pairs of elements in the same Periodic Group have practically identical ionic (and atomic) radii, for example, Zr and Hf, Nb and Ta; the remarkable similarity in chemical behaviour of such pairs of elements is well known. An effect analogous to the lanthanide contraction is observed also in the 5f ions.

(e) Note the irregular variation of the radii of 3d ions M^{2+} and M^{3+} with the numbers of e_g and t_{2g} electrons (p. 323).

Radius ratio and shape of coordination group

Suppose that we have three X ions surrounding an A ion. The condition for stability is that each X ion is in contact with A, so that the limiting case arises when the X ions are also in contact with one another. The following relation exists between r_A and r_X, the radii of A and X respectively: $r_A/r_X = 0.155$. (If the radius of X is a we have $r_A = \frac{2}{3}\sqrt{3}a - a = 0.155a$.) If the radius ratio $r_A:r_X$ falls below this value, then the X ions can no longer all touch the central A ion and this arrangement becomes unstable. If r_A increases the X ions are no longer in contact with one another, and when $r_A:r_X$ reaches the value 0.225 it is possible to accommodate 4 X around A at the vertices of a regular tetrahedron. In general, for a symmetrical polyhedron, the critical minimum value of $r_A:r_X$ is equal to the distance from the centre of the polyhedron to a vertex less one-half of the edge length. The radius ratio ranges for certain highly symmetrical coordination groups are:

$r_A:r_X$	0.155 ——— 0.225 ——————— 0.414 —————— 0.732
c.n.	3 4 6 8
	equilateral triangle regular tetrahedron regular octahedron cube

If the packing of X ions around A is to be the densest possible we should expect the coordination polyhedra to be those having equilateral triangular faces and appropriate radius ratio. For example, for 8-coordination we should not expect the cube but the triangulated dodecahedron or even the intermediate polyhedron, the square antiprism, which for a finite AX_8 group is a more stable arrangement than the cube. Similarly, the icosahedron would be expected for 12-coordination in preference to the cuboctahedron, which is the coordination polyhedron found in a number of complex ionic crystals. For this reason we include a number of triangulated polyhedra in Table 7.11. For the geometry of 7- and 9-coordination polyhedra the reader is referred to Chapter 3, where examples of 7-, 8-, and 9-coordination are given. The radius ratio is not meaningful for 5-coordination, either trigonal bipyramidal or tetragonal pyramidal, since for the former the minimum radius ratio

TABLE 7.11

Radius ratios

Coordination number	Minimum radius ratio	Coordination polyhedron
4	0.225	Tetrahedron
6	0.414	Octahedron
	0.528	Trigonal prism
7	0.592	Capped octahedron
8	0.645	Square antiprism
	0.668	Dodecahedron (bisdisphenoid)
	0.732	Cube
9	0.732	Tricapped trigonal prism
12	0.902	Icosahedron
	1.000	Cuboctahedron

would be the same as for equilateral triangular coordination (0.155), that is, not a value intermediate between those for tetrahedral (0.225) and octahedral (0.414) coordination, while for tetragonal pyramidal coordination the radius ratio would be the same as for octahedral coordination.

The correspondence between the relative sizes of ions and their c.n.s is generally satisfactory for complex fluorides and oxides and for crystals containing complex fluoro- or oxy-ions. It seems reasonable to suppose that in crystals containing complex ions there is a greater probability of arranging the cations in positions of suitable coordination than is the case in very simple structures. The very simple AX and AX_2 structures, however, present a number of problems. Of the alkali halides with radius ratio greater than 0.732 only CsCl, CsBr, and CsI normally adopt the 8-coordinated CsCl structure. We find the same persistence of the NaCl structure in monoxides, where the CsCl structure might have been expected for SrO, BaO, and PbO; in fact there is no monoxide with the CsCl structure. The c.n. six is not exceeded although the pairs of ions K^+ and F^-, Ba^{2+} and O^{2-}, have practically the same radius.

On p. 240 we noted that all configurations of the NaCl structure are geometrically possible between two extremes, (i) with c.p. A ions and (ii) with c.p. X ions. The CsCl structure is similarly adaptable, the limiting cases having either A or X ions in contact. The radius-ratio ranges for the two extreme configurations of these structures are:

	0.41	0.73	(1.00)	1.37	2.4
NaCl structure	(ii) ←			→	(i)
CsCl structure		(ii) ←	→	(i)	

Over a very wide range of radius ratios either structure is geometrically possible, and the lattice energy evidently favours the NaCl structure for all the alkali halides at ordinary temperatures except CsCl, CsBr, and CsI.

The reason for the cubical coordination, rather than the preferable square antiprismatic or dodecahedral, is a purely geometrical one. The two less symmetrical types of coordination are found in complex ionic crystals, in most 8-coordinated molecules and ions, and in AX_4 structures of 8:4 coordination (ThI_4, ZrF_4, and $ZrCl_4$). Cubical coordination occurs in CaF_2, because it is not possible to construct a 3D AX_2 structure with antiprismatic or dodecahedral coordination of A and tetrahedral coordination of X. In AX_4 structures the geometrical restrictions are far less severe than in an AX_2 structure because an X atom is common to only two 8-coordination polyhedra as opposed to four in an AX_2 structure. This argument applies with still more force to an AX structure, in which each X must be common to eight AX_8 coordination groups. The c.n. 12, to be expected when the radius ratio is close to unity, is not observed in AX (or, in fact, in any A_mX_n) structures, though 12-coordination of alkali and alkaline-earth ions occurs in various complex halides and oxides. This is also a geometrical matter, which is discussed later.

In contrast to the monohalides and monoxides the structures of Group II difluorides show the expected increase in c.n. with increasing cation radius:

c.n. of A	4	6	8
r_A	BeF$_2$ 0.27	MgF$_2$ 0.72	CaF$_2$–BaF$_2$ 1.00–1.35 Å
Structure	silica-like	rutile	fluorite

though, as already noted, the c.n. does not rise above 8. Metal dioxides show a clear-cut division into structures of 6- or 8-coordination, the smaller ions M^{4+} (of Ti, V, Cr, and Mn) forming the rutile structure and the larger ions (of Hf, Po, 4f, and 5f elements) the fluorite structure. We observed in Chapter 6 that the fluorite structure is geometrically possible for a range of radius ratios between two extreme configurations, as in the case of the NaCl and CsCl structures, and this is also true of the antifluorite structure. In this structure, which is adopted by all the alkali oxides M$_2$O except Cs$_2$O (which has the anti-CdCl$_2$ layer structure) the 4-coordination of Na$^+$, K$^+$, and Rb$^+$ by O^{2-} is obviously not consistent with the relative sizes of these ions, but higher coordination of M$^+$ in M$_2$O would require structures (presumably geometrically impossible) of, for example, 6:12 or 8:16 coordination.

In fluorides MF$_2$ the clear-cut change from 6-coordination in the rutile structure to 8-coordination in the fluorite structure is in marked contrast to the persistence of 6-coordination in the fluorides MF. We have seen that there are two limiting configurations of both the NaCl and CsCl structures and that both tolerate wide ranges of radius ratio. The fluorite structure also is possible for a wide range of radius ratios, the extreme structures having (i) A atoms in contact or (ii) X atoms in contact. In the case of the rutile structure only (ii) is realizable, in which there is approximately (hexagonal) closest packing of X with A in octahedral interstices. There is no structure of type (i) suitable for an ionic crystal because the A neighbours of A are 2 close and 8 more distant ones, the ratio of the two A–A distances being approximately 1.4:1. This structure, seen in elevation in Fig. 7.2, could collapse to the 10-coordinated sphere packing of Fig. 4.1(d) with pairs of X atoms in adjacent triangular interstices.

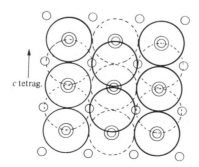

FIG. 7.2. Elevation of the structure (i) related to the rutile structure (see text).

Environment of anions

The introduction of the radius ratio concept has led to a tendency to underestimate the importance of the coordination of the anion. In a number of oxide structures the attainment of a satisfactory environment of the anion seems to be at least as important as the coordination around the cation. It is usual to describe the structures of metal oxides in terms of the coordination polyhedra of the *cations*. This is convenient because their c.n.s are usually greater than those of the anions, and a structure is more easily visualized in terms of the larger coordination polyhedra. However, this preoccupation with the coordination of the metal ions may lead us to underestimate the importance of the arrangement of ions around the *anions*, information which is regrettably omitted from many otherwise excellent descriptions of crystal structures. The formation of the unexpectedly complex corundum structure (with vertex-, edge-, and face-sharing of the octahedral AlO_6 groups) is presumably associated with the difficulty of building a c.p. M_2X_3 structure with octahedral coordination of M and an environment of the 4-coordinated X atom which approximates at all closely to a regular tetrahedral one. Similarly the two kinds of distorted octahedral coordination of Mn^{3+} ions in cubic Mn_2O_3 are probably incidental to attaining satisfactory 4-coordination around O^{2-} ions. The choice of tetrahedral holes occupied by the A atoms in a spinel AB_2O_4 (one-eighth of the total number) is that which gives a tetrahedral arrangement of 3 B + A around O^{2-}.

The outstanding feature of the AX_2 and AX_3 octahedral layer structures is the very unsymmetrical environment of the anions, which have their cation neighbours (3 in AX_2 and 2 in AX_3) lying to one side. This unsymmetrical environment is not, however, peculiar to layer structures; in Table 7.12 we set out the simplest structures for compounds AX_2 and AX_3 in which A is octahedrally coordinated by 6 X atoms (ions). We group together the tetragonal rutile and the $CaCl_2$ structures, in both of which the coordination of X is approximately planar, as opposed to the pyramidal coordination (with bond angles of 90° for regular octahedra) in the edge-sharing AX_2 structures. The simplest edge-sharing structures are of two types for both AX_2 and AX_3, namely, 3D structures (of which examples are shown in Fig. 7.3) and layer structures; for AX_3 there is also a chain structure. No examples are known of simple halides with either of the two types of 3D structure marked (*) and (**) (Table 7.12), though the idealized structure of one polymorph of $Cu_2(OH)_3Cl$ (atacamite) has a structure of this kind. There are on the other hand many compounds with the layer structures. The point we wish to emphasize is that the unsymmetrical environment of X is not a feature only of layer (and chain) structures; it arises from the sharing of *edges* or *faces* of the coordination groups, and is found also in the $PbCl_2$ and UCl_3 structures, which are 3D structures for higher c.n.s of A. It is not clear why the 3D structures marked (*) and (**) are not adopted by any halides AX_2 or AX_3, but it seems reasonable to suppose that the layer structures arise owing to the high polarizability of the larger halide ions. Layer structures are also adopted by many di- and tri-hydroxides, the polarizability of OH^- being intermediate between those of F^- and Cl^-. The positive charges of the anions will

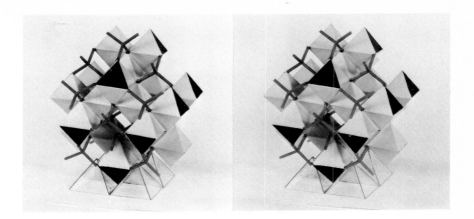

FIG. 7.3(a). The 3D octahedral framework, of composition AX_2, is derived from the NaCl structure by removing alternate rows of metal ions as shown in Fig. 4.22(b). Although each X forms three pyramidal bonds, as in the $CdCl_2$ or CdI_2 layers, this structure is not adopted by any compound AX_2, but it represents the idealized structure of one form of $Cu_2(OH)_3Cl$, the mineral atacamite. Additional atoms (B) in positions of tetrahedral coordination give the spinel structure for compounds A_2BX_4. The positions of only a limited number of B atoms are indicated.

FIG. 7.3(b). The BX_6 octahedra grouped tetrahedrally around the points of the diamond net form the vertex-sharing BX_3 (B_2X_6) framework of the pyrochlore structure for compounds $A_2B_2X_6(X)$; the seventh X atom does not belong to the octahedral framework. In $Hg_2Nb_2O_7$ a framework of the cuprite (Cu_2O) type is formed by the seventh O atom and the Hg atoms, O forming tetrahedral bonds and Hg two collinear bonds.

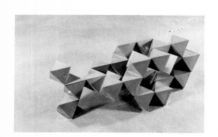

FIG. 7.3(c) and (d). Two of the (unknown) AX_3 structures in which octahedral AX_6 groups share three edges as in the $CrCl_3$ and BiI_3 layers, the A atoms being situated at the points of 3D 3-connected nets instead of the plane hexagon net as in the layer structures. In the structure (c), based on the cubic (10,3) net of Fig. 3.28, the X atoms occupy three-quarters of the positions of cubic closest packing – compare the ReO_3 structure. The spheres indicate the positions of the missing c.p. atoms. In the structure (d), based on the (10,3) net of Fig. 3.29, the X atoms occupy all the positions of cubic closest packing.

TABLE 7.12

The simplest structures for octahedral coordination of A *in compounds* AX₂ *and* AX₃

Type of structure AX₂	Octahedra sharing	Coordination of X		Octahedra sharing	Type of structure AX₃
3D: rutile CaCl₂	2 edges and 6 vertices	Triangular Pyramidal	Linear Non-linear	6 vertices 6 vertices	3D: ReO₃ RhF₃
3D: atacamite* Layer (CdI₂, CdCl₂)	6 edges 6 edges	Pyramidal Pyramidal	Non-linear Non-linear	3 edges 3 edges	3D** Layer (BiI₃, YCl₃)
			Non-linear	2 faces	Chain (ZrI₃)

* No examples known for simple halides AX₂ (Fig. 7.3(a)).

** Structures based on 3D 3-connected nets; no examples known (Fig. 7.3(c) and (d)).

be localized close to the cations, with the result that the outer surfaces of the layers are electrically neutral (Fig. 7.4(a)). In the AX₃ chain structure of Table 7.12 the chains of face-sharing octahedra are perpendicular to the planes of c.p. X atoms (Fig. 7.4(b)). The large van der Waals contribution to the lattice energy arising from the close packing of the X atoms, giving the maximum number of X–X contacts between the layers or chains, explains why there is little difference in stability between the layer and chain structures for halides AX₃, a number of which crystallize with both types of structure (TiCl₃, ZrCl₃, RuCl₃).

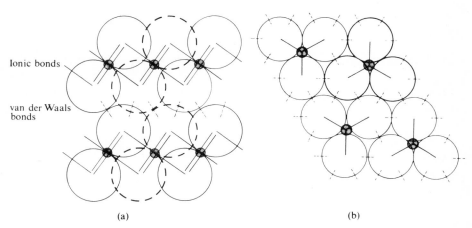

Ionic bonds

van der Waals bonds

(a)   (b)

FIG. 7.4. (a) Diagrammatic elevation of AX₃ layer structure (layer perpendicular to plane of paper); (b) AX₃ chain structure viewed along direction of chains.

Limitations on coordination numbers

We have commented on the absence of structures of ionic compounds $A_m X_n$ with coordination numbers of A greater than eight or nine. If we derive 2-dimensional nets in which A has some number (p) of X atoms and X has some number (q) of A atoms as nearest neighbours we find that the only possible (p,q)-connected nets (p and $q \geqslant 3$) in which p- and q-connected points alternate are the (3,4), (3,5), and (3,6) nets. Since there are upper limits to the values of p and q in plane nets, it is reasonable to assume that the same is true of 3D nets. We are not aware that this problem has been studied. However, the existence or otherwise of crystal structures with high c.n.s is not simply a matter of topology (connectedness); we must take account of metrical factors.

In a structure ${}^{a}A_{m}{}^{x}X_{n}$, where a and x are the c.n.s of A and X by X and A, the number a of coordination polyhedra XA_{x} meet at an A atom and x coordination polyhedra AX_{a} meet at an X atom (assuming that all atoms of each kind have the same environment). We saw in Chapter 5 that, allowing for the size of the X atoms, not more than six regular octahedral groups AX_{6} can meet at a point, that is, have a common vertex (X atom), though this number can be increased to eight (in $Th_{3}P_{4}$) and possibly nine if the octahedra are suitably distorted (p. 192). Thus we can construct a framework of composition AX_{2} in which each A is connected to 12 X and each X to 6 A — the AlB_{2} structure — but this structure cannot be built with octahedral coordination of B. It can be built with trigonal prism coordination of B because twelve trigonal prisms can meet at a point, though this brings certain of the B atoms very close together: B–B, 1.73 Å, compare Al–B, 2.37 Å, Al–Al, 3.01 Å and 3.26 Å. As c.n.s increase and more coordination polyhedra meet at a given point, not only does this limit the types of coordination polyhedra (six regular octahedra, eight cubes or tetrahedra, or twelve trigonal prisms can meet at a point), but also the coordination polyhedra have to share more edges and then faces, with the result that distances between *like* atoms decrease. The AlB_{2} structure is an extreme case, the very short B–B distances (covalent bonds) are those between the centres of BAl_{6} trigonal prism groups across shared rectangular faces. Although this particular 12:6 structure exists, not all 12:6 structures are possible, as may be seen by studying one special set of structures.

The $AuCu_{3}$ structure is one of 12:4 coordination in which Au has 12 equidistant Cu neighbours and Cu has 4 Au and 8 Cu neighbours.* This is one of a family of cubic close-packed $A_{m}X_{n}$ structures in which we shall assume that A has 12 X atoms as nearest neighbours. The neighbours of an X atom in these hypothetical structures would be $12m/n$ A atoms plus sufficient X atoms to complete the 12-coordination group:

	(AX)	A_3X_4	A_2X_3	AX_2	AX_3	AX_4	AX_6	(AX_{12})
Coordination group of X	(12 A)	9 A 3 X	8 A 4 X	6 A 6 X	4 A 8 X	3 A 9 X	2 A 10 X	(A) (11 X)

*For another (12,4)-connected net see Fig. 23.19 (p. 1034).

We may eliminate the first two structures immediately since X is obviously in contact with 4 X in the coordination shell around an A atom (cuboctahedron), but we can go further than this and show that all the structures to the left of the central vertical line are impossible. In cubic close packing the coordination polyhedron of an atom is a cuboctahedral group, and in the above structures the coordination group of X is made up of certain numbers of A and X atoms. Since A is to be surrounded entirely by X atoms we cannot permit A atoms to occupy *adjacent* vertices of the cuboctahedral group around an X atom. The problem is therefore to find the *maximum* number of vertices of a cuboctahedron that may be occupied by A atoms without allowing A atoms to occupy adjacent vertices. It is readily shown that this number is 4, and that there are two possible arrangements, shown in Fig. 7.5(a) and (b). (As a matter of interest we include in the figure a solution for

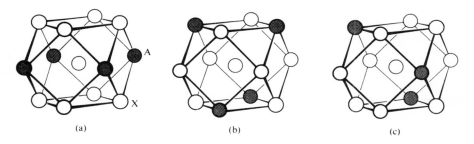

(a)   (b)   (c)

FIG. 7.5. Coordination of X atoms in close-packed AX_3 and AX_4 structures (see text).

$3 A + 9 X$, a coplanar arrangement of three A atoms which would be relevant to an AX_4 structure.) The arrangement (a) is that found in the $AuCu_3$ structure. Evidently c.p. structures for compounds such as KF (12:12) and BaF_2 (12:6) with 12-coordination of the cations are impossible. The $AuCu_3$ structure, with square planar coordination of Cu by 4 Au, is not found for the few trihalides in which M^{3+} ions are comparable in size with the halide ion (for example, the trifluorides of La^{3+} and the larger 4f and 5f ions) presumably because the observed structures have fewer X–X contacts and higher lattice energies; the nitrides Rb_3N and Cs_3N do not appear to be known. This structure is not adopted by any trioxides because there are no ions M^{6+} comparable in size with O^{2-}. It is, of course, not advisable to draw conclusions about the stability or otherwise of structures from the relative numbers of A–X, X–X, and A–A contacts. There are anion–anion contacts in many ionic structures, for example, BeO (O–O, 2.70 Å), LiCl (Cl–Cl, 3.62 Å), and CaF_2, where F has 4 Ca neighbours but is also in contact with 6 F (F–F, 2.73 Å).

Ligand field theory

Whereas in many ionic crystals there is a highly symmetrical arrangement of nearest neighbours around each type of ion, there are numerous structures with less symmetrical coordination groups. In the structures of some simple compounds, for example, La_2O_3 or monoclinic ZrO_2, the c.n. of the cation implies a much less symmetrical arrangement of the (seven) neighbours than is possible for c.n.s such as four or six. In other cases, the coordination group of an ion is not far removed from a highly symmetrical one, and the lower symmetry has a purely geometrical explanation as, for example, in the rutile or corundum structures, though other factors such as partial covalent character cannot in all cases be excluded. We may also mention here the very small deviations from the cubic NaCl structure shown by MnO, FeO, CoO, and NiO below their transition temperatures, which are connected with magnetic ordering processes, and the distortions from cubic symmetry of $BaTiO_3$, $KNbO_3$, and other perovskite-type compounds, which are associated with ferroelectric effects. These are noted in Chapters 12 and 13. A further type of distorted coordination group is characteristic of certain transition-metal ions such as Cr^{2+}, Cu^{2+}, and Mn^{3+}

It is to be expected that the behaviour of an ion when closely surrounded by others of opposite charge will depend not only on its size and polarizability but also on its outer electronic structure, and it is necessary to distinguish between ions with (a) closed shells of 8 or 18 electrons, or (b) incomplete outer shells with two s electrons (In^+, Tl^+, Pb^{2+}), or (c) transition-metal ions with incomplete d shells. The theoretical study of the interactions of transition-metal ions with the surrounding ions throws considerable light on the unsymmetrical coordination groups of certain of these ions.

The electrostatic or crystal field theory was originally developed by Bethe, Van Vleck, and others during the period 1929–35 to account for the magnetic properties of compounds of transition and rare-earth metals in which there are non-bonding d or f electrons. An alternative, molecular orbital, approach was also suggested as early as 1935 by Van Vleck. After a period of comparative neglect both theories have been widely applied since 1950 to the interpretation of the spectroscopic, thermodynamic, and stereochemical properties of finite complexes of transition metals and also to certain aspects of the structures of their crystalline compounds. The electrostatic and m.o. theories represent two approaches to the problem of dealing with the effect of the non-bonding d or f electrons on the behaviour of the transition (including 4f and 5f) elements, and they may be regarded as aspects of a general ligand field theory. In contrast to valence-bond theory, which is concerned primarily with the bonding electrons, ligand field theory considers the effect of the electric field due to the surrounding ligands on the energy levels of all the electrons in the outer shell of the central atom. Since an adequate account of ligand field theory could not be given in the space available and since we are not concerned in this book with such specialized subdivisions of physical chemistry as the optical spectra of transition-metal complexes, we shall simply note here some points relevant to the geometry of the structures of molecules (and complex ions) and crystals.

(a) *Preference for tetrahedral or octahedral coordination.* When a transition-metal ion is surrounded by a regular tetrahedral or octahedral group of ions (or dipoles) the five d orbitals no longer have the same energy but are split into two groups, a doublet e_g and a triplet t_{2g}. Ligands may be arranged in order of their capacity to cause d-orbital splitting (the spectrochemical series):

$$I^- < Br^- < Cl^- < F^- < OH^- < C_2O_4^{2-} \approx H_2O < -NCS < pyr \approx NH_3 < en < NO_2^- < CN^-.$$

The difference between the mean energies of the two groups e_g and t_{2g} (Δ) increases from left to right. If Δ is large (strong ligand field) as many electrons as possible occupy the orbitals of lower energy, which in an octahedral environment are the t_{2g} and in a tetrahedral environment are the e_g orbitals, while if Δ is small they are distributed so as to give the maximum number of parallel spins. This is shown in Table 7.13, from which low-spin (strong ligand field) tetrahedral complexes are

TABLE 7.13

Electronic structures of transition-metal ions

Number of d electrons	Octahedral environment				Tetrahedral environment		
	Weak ligand field		*Strong ligand field*		*Weak field*		
	t_{2g}	e_g	t_{2g}	e_g	e_g	t_{2g}	
1	↑	—	↑	—	↑	—	
2	↑ ↑	—	↑ ↑	—	↑ ↑	—	
3	↑ ↑ ↑	—	↑ ↑ ↑	—	↑ ↑	↑	b
4	↑ ↑ ↑	↑	a	⇅ ↑ ↑	—	↑ ↑	↑ ↑	c
5	↑ ↑ ↑	↑ ↑	⇅ ⇅ ↑	—	↑ ↑	↑ ↑ ↑	
6	⇅ ↑ ↑	↑ ↑	⇅ ⇅ ⇅	—	⇅ ↑	↑ ↑ ↑	
7	⇅ ⇅ ↑	↑ ↑	⇅ ⇅ ⇅	↑	⇅ ⇅	↑ ↑ ↑	
8	⇅ ⇅ ⇅	↑ ↑	⇅ ⇅ ⇅	↑ ↑	⇅ ⇅	⇅ ↑ ↑	b
9	⇅ ⇅ ⇅	⇅ ↑	a	⇅ ⇅ ⇅	⇅ ↑	⇅ ⇅	⇅ ⇅ ↑	c

a, b: Large tetragonal distortion (usually $c:a > 1$, but see p. 888).
c: Large tetragonal distortion ($c:a < 1$).

omitted since none is known. Accordingly cyanido- and nitrito-complexes are of the low-spin type but aquo- or halogen complexes are of the high-spin type. The main factors leading to spin-pairing appear to be high electronegativity of the metal (i.e. high atomic number and valence) and low electronegativity of the ligand, which should have a readily polarizable lone pair and be able to form d_π–p_π or d_π–d_π bonds by overlap of vacant p_π or d_π orbitals with filled t_{2g} orbitals of the metal.

In a regular octahedral field the energy of the e_g orbitals is higher (by Δ_{oct}) than that of the t_{2g} orbitals, and a simple electrostatic treatment gives for an ion with the configuration $(t_{2g})^m(e_g)^n$ the crystal field stabilization energy as $\Delta_{oct}(4m-6n)/10$. From the configurations for weak field (high-spin) of Table 7.13 it is readily found that the maximum value of this stabilization energy occurs for the following ions: d^3 (V^{2+}, Cr^{3+}, Mn^{4+}) and d^8 (Ni^{2+}). In a regular tetrahedral field the relative energies of the two sets of orbitals are interchanged, and the corresponding crystal field stabilization for a configuration $(e_g)^p(t_{2g})^q$ is $\Delta_{tetr}(6p-4q)/10$, which is a maximum for d^2 (Ti^{2+}, V^{3+}) and d^7 (Co^{2+}). These figures indicate that essentially ionic complexes of Co(II) will tend to be tetrahedral and those of Ni(II) octahedral. In the normal spinel structure for complex oxides AB_2O_4 the A atoms occupy tetrahedral and the B atoms octahedral sites, but in some ('inverse') spinels the ions arrange themselves differently in the two kinds of site. These stabilization energies are obviously relevant to a discussion of the types of site occupied by various transition-metal ions in this structure (see Chapter 13).

(b) *Distorted coordination groups.* When there is unsymmetrical occupancy of the subgroups of orbitals, more particularly of the subgroup of higher energy, a more stable configuration results from distorting the regular octahedral or tetrahedral coordination group (Jahn-Teller distortion). From Table 7.13 it follows that for weak-field octahedral complexes this effect should be most pronounced for d^4 (Cr^{2+}, Mn^{3+}) and d^9 (Cu^{2+}), The result is a tetragonal distortion of the octahedron, usually an extension ($c:a > 1$) corresponding to the lengthening of the two bonds on either side of the equatorial plane – (4+2)-coordination. Examples include the distorted rutile structures of CuF_2 and CrF_2 (compare the regular octahedral coordination of Cr^{3+} in CrF_3) and the (4+2)-coordination found in many cupric compounds. The structural chemistry of Cu(II) (and of isostructural Cr(II) compounds) is described in some detail in Chapter 25. Fewer data are available for Mn(III), but both types of octahedral distortion have been observed, in addition to regular octahedral coordination in Mn(acac)$_3$, a special case with three bidentate ligands attached to the metal atom:

MnF_3†	2 F at 1·79 Å		2 F at 2·09 Å	AC 1957 **10** 345
	2 F 1·91			
$K_2MnF_5 \cdot H_2O$ (p. 454)	4 F 1·83		2 F 2·07	JCS A 1971 2653
$K_2MnF_3(SO_4)$ (p. 721)	2 F 1·82		2 F 2·04	JCS A 1971 3074
			2 O 2·01	
Mn(acac)$_3$	6 O at 1·89 Å			IC 1968 7 1994

† From X-ray powder data; it may be significant that the mean of the four shorter bond lengths is close to the shorter Mn—F found in the other compounds.

In a tetrahedral environment the t_{2g} orbitals are those of higher energy, and large Jahn-Teller distortions are expected for d^3, d^4, d^8, and d^9 configurations. The distortion could take the form of an elongation or a flattening of the tetrahedral coordination group, as indicated in Table 7.13. The flattened tetrahedral $CuCl_4^{2-}$ ion is described in Chapter 25, but it is interesting to note that there is no distortion of the $NiCl_4^{2-}$ ion in $[(C_6H_5)_3CH_3As]_2NiCl_4$ (p. 1226). There are numerous crystal structures which are distorted variants of more symmetrical structures, the distortion being of the characteristic Jahn-Teller type. They include the following, the structures being distorted forms of those shown in parenthesis: CrF_2 and CuF_2 (rutile); $CuCl_2.2H_2O$ ($CoCl_2.2H_2O$); $CuCl_2$ (CdI_2); MnF_3 (VF_3), and γ-$MnO.OH$ (γ-$AlO.OH$). These are structures in which there is octahedral coordination of the metal ions. Examples of distorted tetrahedral coordination leading to lower symmetry include the spinels $CuCr_2O_4$ and $NiCr_2O_4$, containing respectively Cu^{2+} (d^9) and Ni^{2+} (d^8).

A further question arises if the six ligands in a distorted octahedral group are not identical: which will be at the normal distances from the metal ion and which at the greater distances in (4+2)-coordination? For example, in both $CuCl_2.2H_2O$ and $K_2CuCl_4.2H_2O$ we find

$$\text{Cu} \begin{cases} 2\,\text{Cl} & \text{at } 2.3\,\text{Å} \\ 2\,\text{H}_2\text{O} & \text{at } 2.0\,\text{Å} \end{cases} \quad \text{and} \quad 2\,\text{Cl at } 2.95\,\text{Å}$$

rather than 4 Cl or 4 H$_2$O as nearest neighbours. Similar problems arise in the structures of hydrated transition-metal halides (e.g. the structure of $FeCl_3.6H_2O$ is $[FeCl_2(H_2O)_4]Cl.2H_2O$), as noted in Chapter 15. It does not appear that satisfactory answers can yet be given to these more subtle structural questions.

The structures of complex ionic crystals

The term 'complex ionic crystal' is applied to solid phases of two kinds. In $MgAl_2O_4$ or $CaMgF_3$ the bonds between all pairs of neighbouring atoms are essentially ionic in character, so that such crystals are to be regarded as 3D assemblies of ions. The anions are O^{2-}, F^-, or less commonly, S^{2-} or Cl^-. The structures of many of these complex ('mixed') oxides or halides are closely related to those of simple oxides or halides, being derived from the simpler A_mX_n structure by regular or random replacement of A by ions of different metals (see, for example, Table 13.1, p. 576), though there are also structures characteristic of complex oxides or halides; these are described in Chapters 10 and 13. In a second large class of crystals we can distinguish tightly-knit groups of atoms within which there is some degree of electron-sharing, the whole group carrying a charge which is distributed over its peripheral atoms. Such *complex ions* may be finite or they may extend indefinitely in one, two, or three dimensions. The structures of many mononuclear complex ions are included in our earlier discussion of simple molecules and ions; the structures of polynuclear complex ions are described under the chemistry of the appropriate elements.

For simplicity we shall discuss complex oxides and complex oxy-salts, but the same principles apply to complex fluorides and ionic oxyfluorides. A complex oxide is an assembly of O^{2-} ions and cations of various kinds which have radii ranging from about one-half to values rather larger than the radius of O^{2-}. It is usual to mention in the present context some generalizations concerning the structures of complex ionic crystals which are often referred to as Pauling's 'rules'. The first relates the c.n. of M^{n+} to the radius ratio $r_M : r_O$. The general increase of c.n. with increasing radius ratio is too well known to call for further discussion here. We have seen that for *simple* ionic crystals the relation between c.n. and ionic size is complicated by factors such as the non-existence of alternative structures and the polarizability of ions, but in crystals containing complex ions there is a reasonable correlation between the c.n.s of cations and their sizes. For example, a number of salts MXO_3 crystallize with one or other (or both) of two structures, the calcite structure, in which M is surrounded by 6 O atoms, and the aragonite structure, in which M is 9-coordinated. (Aragonite is one of three high-pressure phases of $CaCO_3$, the form stable up to 15 kbar at ordinary temperatures (AC 1975 **B31** 343)). The choice of structure is determined by the size of the cation:

Calcite structure: $LiNO_3$, $NaNO_3$; $MgCO_3$, $CaCO_3$ $FeCO_3$; $InBO_3$, YBO_3
Aragonite structure: KNO_3 $CaCO_3$,$SrCO_3$; $LaBO_3$

(For salts such as $RbNO_3$ and $CsNO_3$ containing still larger cations neither of these structures is suitable.)

The second 'rule' states that as far as possible charges are neutralized locally, a principle which is put in a more precise form by defining the *electrostatic bond strength* (e.b.s.) in the following way. If a cation with charge $+ze$ is surrounded by n anions the strength of the bonds from the cation to its anion neighbours is z/n. For the Mg—O bond in a coordination group MgO_6 the e.b.s. is $\frac{1}{3}$, for Al—O in AlO_6 it is $\frac{1}{2}$, and so on. If O^{2-} forms part of several coordination polyhedra the sum of the e.b.s.s of the bonds meeting at O^{2-} would then be equal to 2, the numerical value of the charge on the anion. If a cation is surrounded by various numbers of non-equivalent anions at different distances it is necessary to assign different strengths to the bonds. In monoclinic ZrO_2 Zr^{4+} is 7-coordinated, by 3 O_I at 2.07 Å and 4 O_{II} at 2.21 Å, O_I and O_{II} having c.n.s three and four respectively. Assigning bond strengths of $\frac{2}{3}$ and $\frac{1}{2}$ to the shorter and longer bonds, there is exact charge balance at all the ions.

In many essentially ionic crystals coordination polyhedra share vertices and/or edges and, less frequently, faces. Pauling's third rule states that the presence of shared edges and especially of shared faces of coordination polyhedra decreases the stability of a structure since the cations are thereby brought closer together, and this effect is large for cations of high charge and small coordination number. A corollary to this rule states that in a crystal containing cations of different kinds those with large charge and small c.n. tend not to share polyhedron elements with one another, that is, they tend to be as far apart as possible in the structure.

These generalizations originated in the empirical 'rules' developed during the early studies of (largely) mineral structures, particularly silicates. They are not relevant to the structures of *simple* ionic crystals, in which sharing of edges (and sometimes faces) of coordination groups is necessary for purely geometrical reasons. For example, the NaCl structure is an assembly of octahedral $NaCl_6$ (or $ClNa_6$) groups each sharing all twelve edges, and the CaF_2 structure is an assembly of FCa_4 tetrahedra each sharing all six edges. In the rutile structure each TiO_6 group shares two edges, and in $\alpha\text{-}Al_2O_3$ there is sharing of vertices, edges, and faces of AlO_6 coordination groups. We have shown in Chapter 5 that the sharing of edges (and faces) of octahedral coordination groups in these simple structures has a simple geometrical explanation. It is, however, a feature of many more complex structures built from octahedral coordination groups, and in fact a characteristic feature of many complex oxides of transition metals such as V, Nb, Mo, and W is the widespread occurrence of rather compact groups of edge-sharing octahedra which are then linked, often by vertex-sharing, into 2- or 3-dimensional arrays, as described in Chapters 5 and 13. Even in some simple oxides such as MoO_3 there is considerable edge-sharing which is not present in the chemically similar WO_3. Since these compounds contain transition metals in high oxidation states it could be argued that this feature of their structures is an indication of partial covalent character of the bonds, and certainly there is often considerable distortion of the octahedral coordination. (It would be more accurate to say 'considerable range of M—O bond lengths', since the octahedra of O atoms may be almost exactly regular but with M displaced from their centres.)

We shall now show that a consideration of both the size factor and the principle of the local balancing of charges leads to some interesting conclusions about salts containing complex ions.

The structures and stabilities of anhydrous oxy-salts $M_m(XO_n)_p$

Much attention has been devoted over the years to that part of the electron density of complex ions which is concerned wih the bonding within the ions, since this determines the detailed geometry of the ion. Here we shall focus our attention on that part of the charge which resides on the periphery of the ion, since this has a direct bearing on the possibility, or otherwise, of building crystals of salts containing these ions and therefore on such purely chemical matters as the stability (or existence) of anhydrous ortho-salts (as opposed to hydrated salts or 'hydrogen' salts) or the behaviour of the NO_3^- and other ions as bidentate ligands. In so doing we shall observe the importance of purely geometrical factors which, somewhat surprisingly, are never mentioned in connection with Pauling's 'rules' for complex ionic crystals.

Consider a crystal built of cations and oxy-ions XO_n, and let us assume for simplicity that there is symmetrical distribution of the anionic charge over all the O atoms in XO_n so that the charge on each is -1 in SiO_4^{4-}, $-\frac{3}{4}$ in PO_4^{3-}, $-\frac{1}{2}$ in SO_4^{2-}, $-\frac{1}{4}$ in ClO_4^-, and similarly for ions XO_3. These O atoms form the coordination groups around the cations, which must be arranged so that charges are neutralized locally. For example, in the calcite structure of $NaNO_3$, $CaCO_3$, and $InBO_3$ there is octahedral coordination of the cations and therefore electrostatic bond strengths of $+\frac{1}{6}$, $+\frac{1}{3}$, and $+\frac{1}{2}$ respectively for the M—O bonds. These are balanced by arranging that each O belongs to two MO_6 coordination groups:

In the following series of salts the charge on O increases as shown:

	$NaClO_4$	Na_2SO_4	Na_3PO_4	Na_4SiO_4
Charge on each O atom	$-\frac{1}{4}$	$-\frac{1}{2}$	$-\frac{3}{4}$	-1

If there is octahedral coordination of Na^+ the e.b.s. of a Na—O bond is $\frac{1}{6}$ and therefore in Na_4SiO_4 each O must belong to one tetrahedral (SiO_4) and to six octahedral (NaO_6) groups, or in other words, one tetrahedron and six octahedra must meet at the common vertex (O atom). There are limits to the number of polyhedra that may meet at a point, without bringing vertices of different polyhedra closer than the edge-length, as was pointed out for tetrahedra and octahedra in Chapter 5. No systematic study appears to have been made of the permissible combinations of polyhedra of different kinds which can share a common vertex (subject to the condition noted above), but it is reasonable to suppose that in general the numbers will decrease with increasing size of the polyhedra. In the present case the factors which increase this purely geometrical difficulty are those which increase the size and/or number of the polyhedral coordination groups which (in addition to the oxy-ion itself) must meet at each O atom. The problem is therefore most acute when

the charge on the anion is large (strictly, charge on O of anion);
the charge on the cation is small; and
the c.n. of the cation is large (large cation).

We have taken here as our example a tetrahedral oxy-ion; the same problem arises, of course, for other oxy-ions XO_3, XO_6, etc. The figures in the self-explanatory Table 7.14 are simply the numbers of M—O bonds required to balance the charge on O of the oxy-ion; non-integral values, which would be mean values for non-equivalent O atoms, are omitted. It seems likely that structures corresponding to entries below and to the right of the stepped lines are geometrically impossible. If we wish to apply

TABLE 7.14

Number of coordination groups MO_x to which O must belong in alkali-metal salts (in addition to its XO_3, XO_4, or XO_6 ion)

c.n. of M^+	$-\frac{1}{3}$ NO_3^-	$-\frac{2}{3}$ CO_3^{2-}	-1 BO_3^{3-}	$-\frac{1}{4}$ ClO_4^-	$-\frac{1}{2}$ SO_4^{2-}	$-\frac{3}{4}$ PO_4^{3-}	-1 SiO_4^{4-}	$-\frac{5}{6}$ IO_6^{5-}	-1 TeO_6^{6-}
4			4	1	2	3	4		4
6	2	4	6		3		6	5	6
8			8	2	4	6	8		8
12	4	8	12	3	6	9	12	10	12

The column group above is headed *Charge on O atom of oxy-ion*.

this information to particular compounds we have to assume reasonable coordination numbers for the cations. For example, assuming tetrahedral coordination of Li^+ we see that none of the lithium salts presents any problem, nor do sodium salts (assuming 6-coordination) except Na_6TeO_6. Turning to the vertical columns, evidently the alkali nitrates and perchlorates present no problems, but this is not true of salts containing some of the more highly charged anions and larger cations. It will be appreciated that although the Table predicts the non-existence of, for example, orthoborates and orthosilicates of the larger alkali metals with their normal oxygen coordination numbers (in the range 8–12) it is possible that particular compounds will exist with abnormally low coordination numbers of the cations, for example, as a high-temperature phase. Lowering of the c.n. compensates for the increasing shortage of O atoms which occurs in a series of compounds such as MNO_3, M_2CO_3, and M_3BO_3. It is, of course, immaterial whether we regard a compound such as M_6TeO_6 as a salt containing TeO_6^{6-} ions or as a complex oxide $M_6^+Te^{6+}O_6^{2-}$; there is a charge of –1 on each O to be compensated by M–O bonds.

Systematic studies of cation-rich oxides $M_nM'O_6$ (M = alkali metal) have already produced interesting results. The compounds include Li_5ReO_6, Li_6TeO_6, Li_7SbO_6, and Li_8SnO_6. As regards accommodating the cations in (approximately regular) tetrahedral and/or octahedral interstices it is advantageous to have close-packed O atoms, when there are 12 tetrahedral and 6 octahedral positions per formula-weight. One of the latter is occupied by M′, and therefore there is no structural problem for values of $n \leqslant 5$. In fact Li_5ReO_6 has a c.p. structure rather similar to that of α-$NaFeO_2$. All cations occupy octahedral interstices, Li replacing Na and a mixture of cations replacing Fe between alternate pairs of c.p. layers:

$$\begin{array}{ccc} Na & Fe & O_2 \\ Li & (Li_{\frac{2}{3}}Re_{\frac{1}{3}}) & O_2 = Li_5ReO_6. \end{array}$$

None of the Li compounds presents any difficulty because Li^+ can go into the tetrahedral holes, and this apparently happens in all these structures (which are all close-packed):

	Tetrahedral	Octahedral	Reference
Li_5ReO_6	–	Li_5Re	ZN 1968 **23b** 1603
Li_6TeO_6	Li_6	Te	ZN 1969 **24b** 647
Li_7SbO_6	Li_6	LiSb	ZN 1969 **24b** 252
Li_8SnO_6	Li_6	Li_2Sn	ZaC 1969 **368** 248

A number of Na compounds are known, but they are few in number compared with the Li compounds: Na_5ReO_6 (also I, Tc, Os), Na_6AmO_6 (also Te, W, Np, Pu), Na_7BiO_6, and Na_8PbO_6 (also Pt). For example, 11 compounds Li_8MO_6 are known, 2 Na compounds, but none containing K, Rb, or Cs. For the larger alkali metals there are insufficient octahedral holes if $n > 5$, and for the Na compounds the possibilities are therefore (a) tetrahedral coordination or (b) very distorted octahedral coordination in a non-close-packed structure (compare Th_3P_4, p. 193, in which eight *distorted* 6-coordination polyhedra meet at a point). Tetrahedral coordination of all the alkali metals except Cs is found in the oxides M_2O (antifluorite structure), but the O^{2-} ions are by no means close-packed, even in Li_2O, as may be seen from the O–O distances:

Li_2O	Na_2O	K_2O	Rb_2O
3.27	3.92	4.55	4.76 Å

It will be interesting to see how the geometrical difficulties are overcome in these Na-rich oxides.

An alternative to the formation of the normal salt is to form 'hydrogen' salts. For example, assuming K^+ to be 8-coordinated, every O in K_3PO_4 would have to belong to 6 KO_8 coordination groups in addition to its own PO_4^{3-} ion. In KH_2PO_4, on the other hand, there is a charge of only $-\frac{1}{4}$ on each O, and moreover the H atoms can link the $PO_4H_2^-$ groups into a 3D network by hydrogen bonds. There is no difficulty in arranging that each O also has two K^+ neighbours situated in the interstices of the framework.

An entirely different problem arises if we *reduce* the charge on O of the oxy-ion, for example, by changing from $CaCO_3$ to $Ca(NO_3)_2$, and/or *increase* that on M, as in the series: $M(NO_3)_2$, $M(NO_3)_3$, and $M(NO_3)_4$. In calcite and aragonite, two polymorphs of $CaCO_3$, Ca^{2+} is 6- and 9-coordinated respectively. Octahedral coordination of M^{2+} in $M(NO_3)_2$ implies that the crystal consists of MO_6 groups linked through N atoms of NO_3^- ions, each O of which is bonded to only one M, as at (a). Increasing the c.n. to 12 gives (b), in which the charge on O is balanced in the same way as in $NaNO_3$ (p. 327). This is the situation in (cubic) $Pb(NO_3)_2$, where Pb^{2+} is surrounded by 6 NO_3^- and a nitrate ion is related to its three nearest Pb^{2+} neighbours as shown at (c). This structure is adaptable to ions of various sizes, for by rotation of the anions in their own planes the coordination of M can be changed to 6 + 6:

	$Cd(NO_3)_2$ (l.t.)	$Ca(NO_3)_2$	$Pb(NO_3)_2$
Coordination of M	6 O 2.34	6 O 2.50	12 O 2.81 Å
	6 O 3.12	6 O 2.93	

(a)   (b)   (c)

Now consider the extreme case of a compound such as $Ti(NO_3)_4$ containing a small ion Ti^{4+} which is normally coordinated octahedrally by 6 O. The Ti–O bond strength would be $\frac{2}{3}$, twice the charge on O of NO_3^-, so that it is impossible to achieve a charge balance. In order to reduce the e.b.s. of the Ti–O bond to $\frac{1}{3}$ the c.n. of Ti^{4+} would have to be increased to the impossibly high value 12. Clearly this difficulty is most pronounced when

the charge on the anion is small;
the charge on the cation is large; and
the c.n. of the cation is small.

To be more precise we may say that the charge on each O of the oxy-ion must be equal to or greater than the charge on the cation divided by its c.n. (the e.b.s. of the M–O bond). The values of this quantity (Table 7.15) show that this condition

TABLE 7.15

Electrostatic bond strengths of bonds M—O

c.n. of M	M^+	M^{2+}	M^{3+}	M^{4+}
4	$\frac{1}{4}$	$\frac{1}{2}$		1
6	$\frac{1}{6}$	$\frac{1}{3}$	$\frac{1}{2}$	$\frac{2}{3}$
8	$\frac{1}{8}$	$\frac{1}{4}$		$\frac{1}{2}$
9			$\frac{1}{3}$	
12	$\frac{1}{12}$	$\frac{1}{6}$	$\frac{1}{4}$	$\frac{1}{3}$

Limit for normal ionic $M(XO_4)_x$ $\longrightarrow$
$M(XO_3)_x$ $\longrightarrow$

cannot be satisfied for a number of anhydrous salts containing ions XO_3^- and XO_4^- assuming that the ionic charge is distributed equally over the O atoms, that is, $-\frac{1}{3}$ or $-\frac{1}{4}$ on each respectively. It is probable that the only normal ionic nitrates that can be crystallized from aqueous solution in the anhydrous form are those of the alkali metals, the alkaline-earths, Cd, and Pb. (The structures of the anhydrous nitrates of Be, Mg, Zn, and Hg do not appear to be known.) A number of anhydrous chlorates, bromates, iodates, and perchlorates are known (for example, $Ca(ClO_3)_2$, $Ba(BrO_3)_2$, $Ca(IO_3)_2$, $Ba(IO_3)_2$, and $M(ClO_4)_2$) (M = Cd, Ca, Sr, Ba). These are likely to be normal ionic salts, but again there is little information about their crystal structures. In the case of the corresponding salts containing ions M^{3+} or M^{4+} very few, if any, can be crystallized anhydrous from aqueous solution. Most of these compounds are highly hydrated, for example, $Al(ClO_4)_3.6$, 9, and $15H_2O$, $Ga(ClO_4)_3.6$ and $9H_2O$, and $Cr(NO_3)_3.9H_2O$, though in a few instances the anhydrous salts have been described ($Al(ClO_4)_3$, $In(IO_3)_3$). Anhydrous salts $M(XO_3)_4$ crystallizable from (acid) aqueous solution include $Ce(IO_3)_4$ and $Zr(IO_3)_4$, to which we refer shortly.

Clearly this charge-balance difficulty is overcome by the formation of hydrates when the salt is crystallized from aqueous solution. In a hydrate in which the cation is surrounded by a complete shell of water molecules the charge on the central ion M^{n+} is spread over the surface of the hydrated ion, and the aquo-complexes can be hydrogen-bonded to the oxy-ions. The structural problem is now entirely different, since there are no bonds between cations and O atoms of oxy-ions, and it is now a question of packing large $[M(H_2O)_x]^{n+}$ groups and anions as in salts such as $Nd(BrO_3)_3 . 9H_2O$ and $Th(NO_3)_4 . 12H_2O$. There is, however, an alternative structure for an anhydrous compound $M(NO_3)_3$, $M(NO_3)_4$, or the corresponding perchlorates, etc., which calls for the rearrangement of the charge distribution on the XO_3^- or XO_4^- ion from

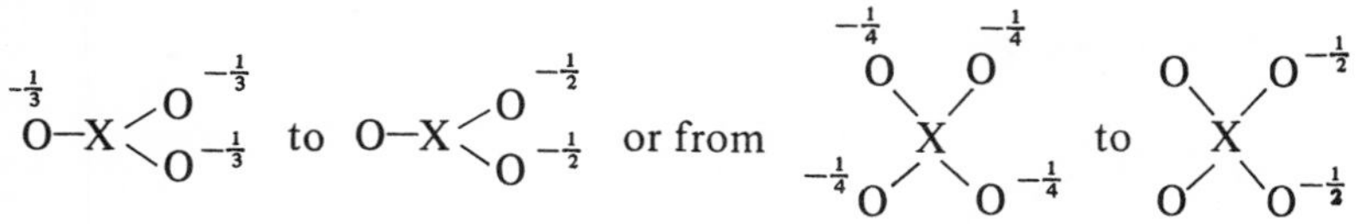

$$\overset{-\frac{1}{3}}{O}-X\overset{\diagup O^{-\frac{1}{3}}}{\underset{\diagdown O^{-\frac{1}{3}}}{}} \quad \text{to} \quad O-X\overset{\diagup O^{-\frac{1}{2}}}{\underset{\diagdown O^{-\frac{1}{2}}}{}} \quad \text{or from} \quad \overset{-\frac{1}{4}\;O\;\;\;O^{-\frac{1}{4}}}{\underset{-\frac{1}{4}\;O\;\;\;O^{-\frac{1}{4}}}{X}} \quad \text{to} \quad \overset{O\;\;\;O^{-\frac{1}{2}}}{\underset{O\;\;\;O^{-\frac{1}{2}}}{X}}$$

Only two of the O atoms of, for example, a NO_3^- ion are then coordinated to cations, and the ion can function either as a bidentate ligand, (a), or as a bridging ligand, in which case there are three possibilities (b), (c), and (d), corresponding to the linking of vertices, edges, or faces of different cation coordination groups by anions.

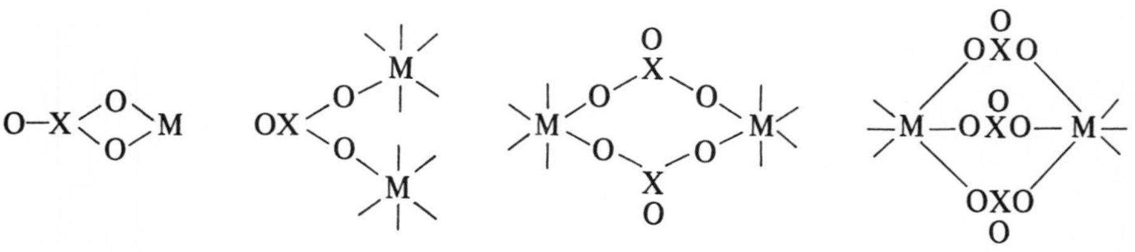

(a)   (b)   (c)   (d)

The use of new nitrating agents has led to the preparation of many anhydrous nitrates (nitrato compounds) which cannot be obtained from aqueous solution or by dehydrating salts. They include:

$M(NO_3)_2$: M = Be, Mg, Zn, Hg, Mn, Co, Ni, Cu, Pd
$M(NO_3)_3$: M = Al, Sc, Y, La, Cr, Fe, Au, In, Bi
$M(NO_3)_4$: M = Ti, Zr, Th, Sn.

Many of these compounds can be sublimed (under low pressure or *in vacuo*) and are clearly molecular, having bidentate NO_3 groups, as has been established for the vapour of $Cu(NO_3)_2$ and crystalline $Ti(NO_3)_4$ and $Sn(NO_3)_4$. Bidentate NO_3 groups are also present in the ion $[Co(NO_3)_4]^{2-}$, which has a structure very similar to that of the $Ti(NO_3)_4$ molecule, and also in $[Th(NO_3)_6]^{2-}$, $[Ce(NO_3)_6]^{2-}$, and $[Ce(NO_3)_6]^{3-}$ These compounds are discussed in Chapter 18.

This behaviour as a bidentate or bridging ligand should not be peculiar to NO_3^-; it is to be expected for ClO_3^-, BrO_3^-, IO_3^-, ClO_4^-, IO_4^-, and also SO_4^{2-}. The iodate ion exhibits all of the behaviours (b), (c), and (d) noted earlier. In all the crystalline salts $Ce(IO_3)_4$ (with which $Pu(IO_3)_4$ is isostructural), $Ce(IO_3)_4.H_2O$, and $Zr(IO_3)_4$ only two of the O atoms of each IO_3^- are used to coordinate the metal ions, and in all three compounds IO_3^- behaves as a bridging ligand. In anhydrous $Ce(IO_3)_4$ columns of CeO_8 coordination groups, intermediate in shape between cubic and square antiprismatic, are formed by bridging IO_3 groups as shown diagrammatically in Fig. 7.6(a). In $Ce(IO_3)_4.H_2O$ each of the eight O atoms around Ce^{4+} belongs to

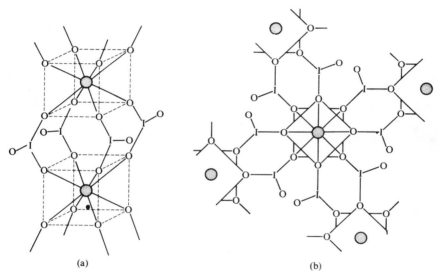

FIG. 7.6. Behaviour of the IO_3^- ligand in (a) $Ce(IO_3)_4$; (b) $Zr(IO_3)_4$ (diagrammatic).

a different IO_3^- ion, and the eight IO_3^- ions connect a particular Ce^{4+} to eight others. (The H_2O molecules are not involved in the coordination groups of the metal ions, but are situated in tunnels entirely surrounded by O atoms of iodate ions.) In $Zr(IO_3)_4$ also there is antiprismatic coordination of the metal, but here Zr^{4+} is surrounded by 8 IO_3^- ions which bridge in pairs as shown diagrammatically in Fig. 7.6(b). The pairs of O atoms involved in each bridge correspond to alternate slanting edges of the antiprism coordination group, and the structure consists of layers of Zr^{4+} ions linked together in this way by IO_3^- ions.

We have seen that ions XO_3^- cannot balance the e.b.s. of an 8-coordinated ion M^{4+} without rearranging the anionic charge and behaving as either bidentate or bridging ligands. For an ion such as SO_4^{2-} the possible types of behaviour are more numerous and are completely illustrated by the structures of $Zr(SO_4)_2$ and its hydrates. This compound has an unexpectedly complex structural chemistry, for in addition to three forms of the anhydrous salt there are numerous hydrates, and these are by no means a normal series of hydrates. It is not possible to prepare the higher hydrates in succession by simply increasing the water-vapour pressure over the anhydrous salt; for example, the 5- and 7-hydrates have lower saturated solution vapour pressures than the 4-hydrate and therefore cannot be made by vapour-phase hydration of the latter. The relations between some of the phases we shall describe are set out in Table 7.16.

The normal charge distribution of the sulphate ion, (a), could be rearranged to (b) or (c):

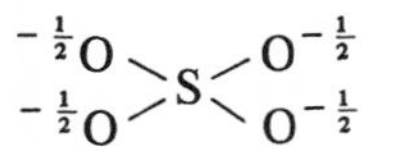

$$-\tfrac{1}{2}O\diagdown_{S}\diagup^{O^{-\tfrac{1}{2}}}_{O^{-\tfrac{1}{2}}} \quad -\tfrac{2}{3}O\diagdown_{S}\diagup^{O^{-\tfrac{2}{3}}}_{O^{-\tfrac{2}{3}}} \quad O\diagdown_{S}\diagup^{O^{-1}}_{O^{-1}}$$

$$\text{(a)} \qquad\qquad \text{(b)} \qquad\qquad \text{(c)}$$

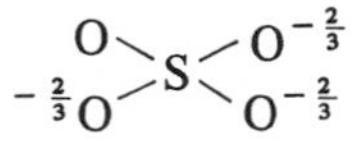

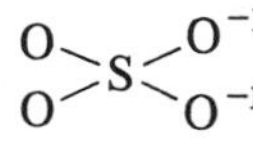

For a salt $M(SO_4)_2$ charge balance can be achieved with the normal structure (a), since the e.b.s. of M^{4+}–O for 8-coordination is $\tfrac{1}{2}$. However, if M^{4+} is coordinated by a smaller number of O atoms of sulphate ions, either because it has a smaller c.n. (for example, 7) or because its coordination group is partly composed of H_2O or other neutral molecules which do not neutralize any of the cationic charge, then rearrangement to (b) or (c) is necessary:

$$\begin{array}{lccc}
\text{c.n. of } M^{4+} \text{ by O of } SO_4^{2-}: & 8 & 6 & 4 \\
\text{form of anion:} & \text{(a)} & \text{(b)} & \text{(c)}
\end{array}$$

An intermediate c.n. such as 7 would require a mixture of (a) and (b).

The simpler structural possibilities implied by (a)–(c) are

(a) a_1 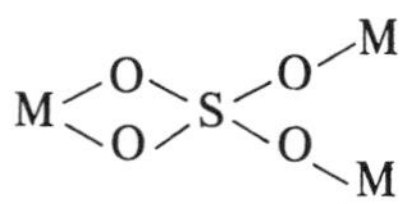 a_2 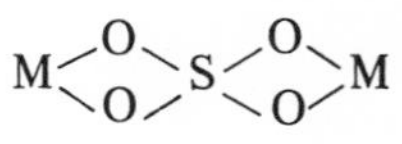 a_3

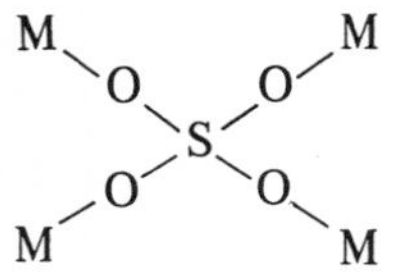

(b) b_1 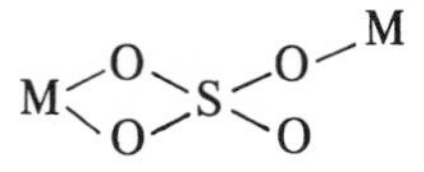 b_2

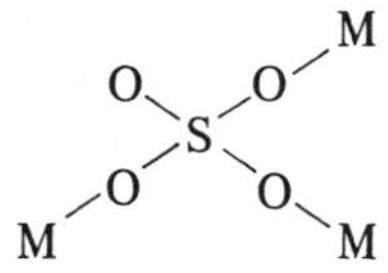

(c) c_1 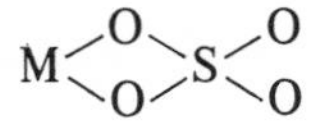 c_2

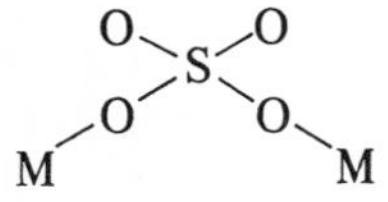

TABLE 7.16

Relations between the anhydrous and hydrated sulphates of zirconium

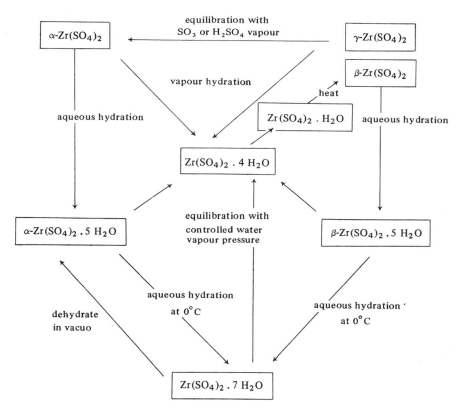

For further details see JSSC 1970 1 497.

In α-$Zr(SO_4)_2$ the coordination group around Zr^{4+} is necessarily composed entirely of O atoms of sulphate ions. The metal is 7-coordinated, and equal numbers of SO_4 groups are of types a_1 and b_1, that is, bonded to 4 or 3 M^{4+} ions (Fig. 7.7). The behaviour of the SO_4^{2-} ion in this salt is thus very similar to that of O^{2-} in the $7:\frac{3}{4}$ coordinated structure of monoclinic ZrO_2.

The monohydrate provides examples of (b). In both polymorphs the cation is 7-coordinated by 6 O + 1 H_2O. In the γ structure all SO_4 groups are of type b_1; each bridges 3 cations forming a very simple layer which was illustrated in Fig. 6.15 (p. 261) to show its relation to the CdI_2 layer. The α form also has a layer structure, but here there are SO_4 groups of both types b_1 and b_2. With increasing hydration transition to (c) occurs, and in fact the coordination group of Zr^{4+} in the 4-hydrate,

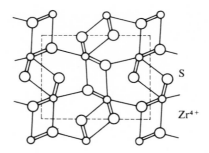

FIG. 7.7. The structure of α-Zr$(SO_4)_2$. The O atoms (one to each Zr–S line) are omitted.

both forms of the 5-hydrate, and the 7-hydrate is in all cases the same, namely, $4 O + 4 H_2O$. The tetrahydrate has a very simple layer structure based on the 4-gon net, all SO_4 groups being of type c_1, as shown diagrammatically in Fig. 7.8. Both forms of the pentahydrate and the heptahydrate are built of bridged molecules

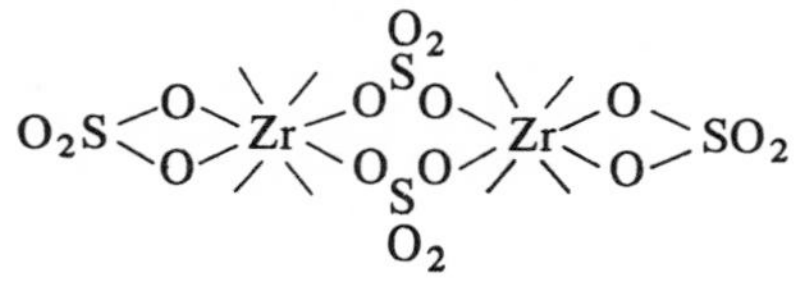

$$O_2S \underset{O}{\overset{O}{<}} \underset{\diagup}{\overset{\diagdown}{>}} Zr \underset{O}{\overset{O}{<}} \underset{S}{\overset{O_2}{\diagup}} O \underset{O_2}{\overset{O}{>}} \underset{\diagup}{\overset{\diagdown}{>}} Zr \underset{O}{\overset{O}{<}} \diagup SO_2$$

in which the SO_4 groups are of the two types c_1 and c_2. Dodecahedral 8-coordination groups around Zr^{4+} are completed by $4 H_2O$ and the remaining water molecules are not associated with the metal ions. The structural formulae are therefore $[Zr_2(SO_4)_4(H_2O)_8] . 2H_2O$ and $[Zr_2(SO_4)_4(H_2O)_8] . 6H_2O$. (For further details see: AC 1959 **12** 719; AC 1969 **B25** 1558, 1566, 1572; AC 1970 **B26** 1125, 1131, 1140.)

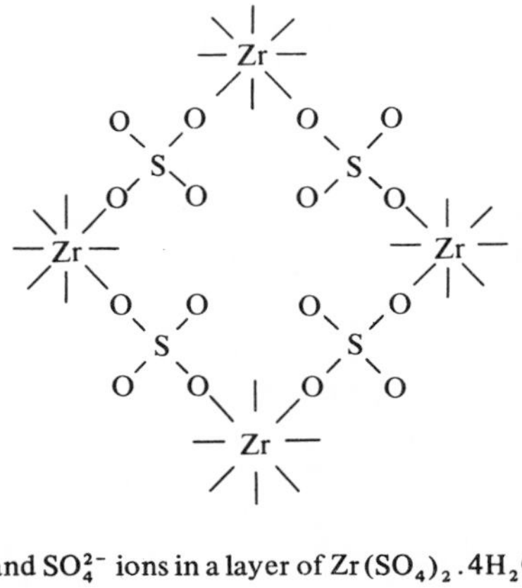

FIG. 7.8. Linking of Zr^{4+} and SO_4^{2-} ions in a layer of Zr$(SO_4)_2 . 4H_2O$ (water molecules omitted).

Summarizing, we see that problems of charge balance lead to geometrical difficulties in two distinct classes of salt:

	Charge on O of anion	*Cation size*	*Cation charge*
(a)	high	large	low
(b)	low	small	high

We have discussed (a) as it applies to the extreme cases Cs_3BO_3, Cs_4SiO_4, and Cs_6TeO_6, but the problem exists, though in a less acute form, in pyro-salts such as $M_6Si_2O_7$ and $M_4P_2O_7$. As in the case of K_3PO_4 and KH_2PO_4 it can be overcome by the formation of salts such as $Na_2H_2P_2O_7$. At the other extreme, case (b), the difficulties arise not only with ions XO_3^- and XO_4^-, as already discussed, but also with ions such as $S_2O_7^{2-}$ if combined with cations carrying high charges. Thus the normal ionic disulphate and dichromate of Ti^{4+} or even the salts of the larger Th^{4+} ion would not be possible with the distribution of the anionic charge over all (or even six) of the O atoms. For a family of pyro-ions we encounter difficulties at both ends of the series,

$$Cs_6(Si_2O_7) \quad ----- \quad Th(S_2O_7)_2$$
$$\text{case (a)} \qquad\qquad\qquad \text{case (b)}$$

just as we do for the simpler XO_3^{n-} ions

$$Cs_3BO_3 \quad ----- \quad Ti(NO_3)_4.$$

Part II

THE PERIODIC TABLE OF THE ELEMENTS

IA	IIA	IIIA	IVA	VA	VIA	VIIA	VIII	VIII	VIII	IB	IIB	IIIB	IVB	VB	VIB	VIIB	0
H 1																	He 2
Li 3	Be 4											B 5	C 6	N 7	O 8	F 9	Ne 10
Na 11	Mg 12											Al 13	Si 14	P 15	S 16	Cl 17	Ar 18
K 19	Ca 20	Sc 21	Ti 22	V 23	Cr 24	Mn 25	Fe 26	Co 27	Ni 28	Cu 29	Zn 30	Ga 31	Ge 32	As 33	Se 34	Br 35	Kr 36
Rb 37	Sr 38	Y 39	Zr 40	Nb 41	Mo 42	Tc 43	Ru 44	Rh 45	Pd 46	Ag 47	Cd 48	In 49	Sn 50	Sb 51	Te 52	I 53	Xe 54
Cs 55	Ba 56	La† 57	Hf 72	Ta 73	W 74	Re 75	Os 76	Ir 77	Pt 78	Au 79	Hg 80	Tl 81	Pb 82	Bi 83	Po 84	At 85	Rn 86
Fr 87	Ra 88	Ac‡ 89															

Group designations: A SUB-GROUPS, GROUP VIII TRIADS, B SUB-GROUPS.
SHORT PERIODS (Groups I, II) — LONG PERIODS.

† And the 4f series (lanthanides):

Ce 58	Pr 59	Nd 60	Pm 61	Sm 62	Eu 63	Gd 64	Tb 65	Dy 66	Ho 67	Er 68	Tm 69	Yb 70	Lu 71

‡ And the 5f series (actinides):

Th 90	Pa 91	U 92	Np 93	Pu 94	Am 95	Cm 96	Bk 97	Cf 98	Es 99	Fm 100	Md 101	No 102	Lw 103

8

Hydrogen: the noble gases

HYDROGEN

Introductory

The hydrogen atom has the simplest electronic structure of all the elements, having only one valence electron and only one orbital available for bond formation. Nevertheless hydrogen combines, in one way or another, with most of the elements, other than the noble gases, and plays an extremely important role in chemistry. The three simplest ways in which H can function are the following:

(a) By loss of the electron the H^+ ion (proton) is formed. An acid may be defined as a source of protons:

$$A \rightleftharpoons H^+ + B$$
$$\text{acid} \qquad \text{base}$$

According to this definition NH_4^+ is regarded as an acid ($NH_4^+ \rightleftharpoons NH_3 + H^+$) and water as both acid and base, since

$$H_2O \rightleftharpoons H^+ + OH^- \qquad \text{and} \qquad H^+ + H_2O \rightleftharpoons H_3O^+$$
$$\text{acid} \qquad \text{base} \qquad\qquad \text{base} \quad \text{acid}$$

assuming the existence of hydrated rather than simple protons in aqueous solution.

(b) By acquiring a second electron the H^- ion is formed. This ion cannot exist in aqueous solution because it immediately combines with H^+, but it is found in a number of crystalline hydrides.

(c) Hydrogen can form one normal covalent (electron-pair) bond. This occurs in the molecular hydrides of the non-metals and metalloids and in ions such as OH^- and NH_4^+, in organic compounds, and in a limited number of hydrido compounds of transition metals.

In addition there are many structures, both molecular and crystalline, in which a H atom is bonded in some way to two (or more) atoms.

(d) If H is bonded to a very electronegative atom the dipole $A^- - H^+$ results in an attraction to a second electronegative atom, which may be in the same or a different molecule,

$$\gtrsim A{-}H{\cdots}A \lesssim$$

forming a hydrogen bond (or bridge). This type of bond is of great importance in the structural chemistry of many groups of compounds, notably acids and acid salts, hydroxy-compounds, water, and hydrates. Since a comprehensive treatment of the hydrogen bond would evidently cover a considerable part of structural chemistry we shall discuss in this chapter only acids and acid salts.

(e) In certain electron-deficient compounds (of Li, Be, B, and Al) H atoms form bridges between pairs of atoms as, for example, in B_2H_6, and metal–H–metal bridges (both linear and non-linear) occur in certain carbonyl hydride ions such as $[(CO)_5CrHCr(CO)_5]^-$. Boron hydrides are described in Chapter 24 and carbonyl hydrides in Chapter 22; see also the discussion of bonds in Chapter 7.

In many crystalline hydrides H is bonded to larger numbers of metal atoms. This is readily understandable for the ionic hydrides (for example, LiH) where H^- has an environment similar to that of F^- in LiF, but the nature of the bonding is less clear in some transition-metal hydrides, in particular the non-stoichiometric interstitial hydrides.

We shall discuss in the present chapter the following aspects of the structural chemistry of hydrogen:

hydrides and hydrido complexes of transition metals;
the hydrogen bond; and
acids and acid (hydrogen) salts.

Hydrides

Binary hydrides are known of most elements other than the noble gases, the notable exceptions being Mn, Fe, Co, and the following members of the second and third long series:

Mo	Tc	Ru	Rh		
W	Re	Os	Ir	Pt	(Au)

These compounds are of three main types:
   (i)   the molecular hydrides;
  (ii) the salt-like hydrides of the more electropositive elements;
 (iii) the interstitial hydrides of the transition metals.

(i) *Molecular hydrides*

Hydrides MH_{8-N} are formed by the non-metals of the two short Periods (other than B which forms the exceptional hydrides described in Chapter 24) and by the B subgroup elements:

MH_4	MH_3	MH_2	MH
Tetrahedral	Pyramidal	Angular	
C	N	O	F
Si	P	S	Cl
Ge	As	Se	Br
Sn	Sb	Te	I
Pb	Bi	Po	

Structural studies have been made of all except PbH_4, BiH_3, and PoH_2, and details are given in other chapters; the stability of the heavy metal compounds is very low. These hydrides exist in the same molecular form in all states of aggregation, and except for those of the most electronegative elements there are only weak van der

Waals forces acting between the molecules in the crystals. We refer to crystalline NH_3, OH_2, and FH in our discussion of the hydrogen bond. Many short-lived hydride species are known to the spectroscopist, and the structures of some radicals have recently been studied in matrices at low temperatures (for example, CH_3 (planar), SiH_3 and GeH_3 (pyramidal)). Many of the non-metals form more complex hydrides M_xH_y in addition to the simple molecules noted above; the more important of these are included in later chapters under the chemistry of the appropriate non-metal.

(ii) *Salt-like hydrides*

The compounds to the left of the line in the following table are salt-like and contain H^- ions:

LiH[1]			AlH_3[5]	
NaH[2]	MgH_2[3]			
KH	CaH_2[4]			
RbH	SrH_2		EuH_2	YbH_2[6]
CsH	BaH_2			

They are colourless compounds which can be made by heating the metal in hydrogen (under moderate pressure in the case of MgH_2). They are more dense than the parent metals, the difference being greatest for the alkali metals (25–45 per cent) and less for the alkaline-earths (5–10 per cent). Solid LiH is as good an ionic conductor as LiCl (and 10^3 times better than LiF), and electrolysis of molten LiH, at a temperature just above its melting point with steel electrodes, yields Li at the cathode, hydrogen being evolved at the anode. These ionic hydrides have much higher melting points than the molecular hydrides of the previous section, for example, LiH, 691 °C, NaH, 700–800 °C with decomposition; the remaining alkali hydrides dissociate before melting. They all react readily with water, evolving hydrogen and forming a solution of the hydroxide. There are considerable structural resemblances between these compounds (in which the effective radius of H^- ranges from 1.3–1.5 Å) and fluorides (radius of F^-, 1.35 Å).

The alkali-metal hydrides have the NaCl structure, though the positions of the H atoms have been confirmed only in LiH (X-ray diffraction) and NaH (neutron diffraction); (Li–H, 2.04 Å, Na–H, 2.44 Å). The rutile structure of MgH_2 has been established by neutron diffraction of MgD_2; Mg–H, 1.95 Å, shortest H–H, 2.49 Å).

The alkaline-earth hydrides all have a $PbCl_2$ type of structure in which the metal atoms are arranged approximately in hexagonal closest packing. Of the two sets of non-equivalent H atoms one occupies tetrahedral holes while the other H atoms have (3+2)-coordination:

$$\text{Ca:} \begin{array}{l} 7\,\text{H } 2.32\,\text{Å} \\ 2\,\text{H } 2.85\,\text{Å} \end{array} \qquad H_I: \; 4\,\text{Ca } 2.32\,\text{Å} \qquad H_{II}: \begin{array}{l} 3\,\text{Ca } 2.32\,\text{Å} \\ 2\,\text{Ca } 2.85\,\text{Å} \end{array}$$

Note the tight packing of the H atoms; H_I has 8 H neighbours at 2.50–2.94 Å, and H_{II} has 10 H neighbours at 2.65–3.21 Å. The shortest metal–metal distances in these hydrides are *less* than in the metals (for example, Ca–Ca, 3.60 Å in CaH_2 as compared with 3.93 Å in the metal). The dihydrides of Eu and Yb, studied as the deuterides, have the same structure, and their formation from the metal is accompanied by a 13 per cent decrease in volume.

In the preparation of AlH_3 by the action of excess $AlCl_3$ on LiH in ether it has not proved possible to obtain pure AlH_3 (free from ether), but this hydride has been made by bombarding extremely pure Al with H ions in a number of crystalline forms, one of which is isostructural with AlF_3 (p. 418). The metal atoms occupy one-third of the octahedral interstices in an approximately hexagonal closest packing of H atoms, the structure being an assembly of vertex-sharing octahedral AlH_6 groups. The angle Al–H–Al is 141°, close to the ideal value (132°) for ideal closest packing. The structure suggests that this is an ionic hydride. The shortest distance between Al atoms is 3.24 Å, and the distances Al–6 H, 1.72 Å, and H–H, 2.42 Å, are very similar to the corresponding distances in AlF_3, namely, Al–6 F, 1.79 Å, and F–F, 2.53 Å. There are amine derivatives of AlH_3 which presumably contain H covalently bonded to Al, though the H atoms have not been located. Crystalline $AlH_3.2N(CH_3)_3$[7] consists of molecules in which the atoms N–Al–N are collinear (Al–N, 2.18 Å) with 3 H most probably completing a trigonal bipyramidal coordination group around Al. This would also appear to be true in $AlH_3[(CH_3)_2NCH_2CH_2N(CH_3)_2]$, the diamine molecules linking the AlH_3 groups into infinite chains.[8]

By the action of $LiAlH_4$ on the halide (or in the case of Be on the dialkyls) a number of other hydrides have been prepared, including the dihydrides of Be, Zn, Cd, and Hg, trihydrides of Ga, In, and Tl, and CuH. Some of these are extremely unstable, those of Cd and Hg decomposing at temperatures below 0 °C. Pure crystalline BeH_2 has not been prepared, but a crystalline amine complex has been made for which the bridged structure (a) has been suggested;[9] Be–H

(a)

(b)

bridges have also been postulated in the salt $Na_2[R_4Be_2H_2]$.[10] The etherate $[Na_2(C_4H_{10}O)_2][Be_2(C_2H_5)_4H_2]$[11] consists of H-bridged ions $Be_2H_2Et_4^{2-}$, isoelectronic with $B_2H_2Et_4$, linked through $NaOEt_2^+$ ions. The dotted lines between the metal atoms in (b) are perpendicular to the chain axis and are related by rotations of 90°, with the result that H is bonded, in a distorted tetrahedral arrangement, to 2 Na and 2 Be atoms. There is still some doubt about the existence of GaH_3, but in crystalline $GaH_3 \cdot N(CH_3)_3$ the bond length Ga—N has been determined as approximately 2.0 Å in a presumably tetrahedral molecule.[12]

The red-brown CuH is amorphous when prepared from CuI and $LiAlH_4$ in organic solvents but it has been obtained as a water soluble material with the wurtzite structure by reduction of Cu^{2+} by aqueous hypophosphorous acid. (Cu—H, 1.73 Å, Cu—Cu, 2.89 Å, compare 2.56 Å in the metal.)[13] The colour of this compound, which forms red solutions in organic solvents, is difficult to understand;[14] the n.d. study showed contamination with metallic Cu and Cu_2O.

Although we have included these B subgroup hydrides with the salt-like compounds it is possible that the bonding is at least partially covalent in some or all of these compounds; an obvious suggestion for BeH_2 is a hydrogen-bridged structure like that of $Be(CH_3)_2$.

We may mention here a B subgroup hydride which does not fall into any of our classes (i)–(iii), namely $PbH_{0.19}$. This is formed by the action of atomic hydrogen at 0 °C on an evaporated lead film, and apparently possesses considerable stability.[15]

(1)	ZPC 1935 **28B** 478	(9)	IC 1969 8 976
(2)	PR 1948 73 842	(10)	JCS 1965 692
(3)	AC 1963 **16** 352	(11)	AC 1981 **B37** 68
(4)	AC 1962 **15** 92	(12)	IC 1963 2 1298
(5)	IC 1969 8 18	(13)	AC 1955 8 118
(6)	AC 1956 9 452	(14)	JACS 1968 90 5769
(7)	IC 1963 2 508	(15)	PCS 1964 173
(8)	AC 1964 17 1573		

(iii) *Transition-metal hydrides*

The known compounds are listed in Table 8.1, which shows that a considerable number of these metals have not yet been shown to form binary hydrides. The criterion is that the formation of a hydrogen-containing phase should be accompanied by a definite structural change, since many of these metals when finely divided adsorb large volumes of hydrogen; it is clearly difficult to distinguish such systems from non-stoichiometric hydrides by purely chemical means. Characteristic properties of transition-metal hydrides include metallic appearance, metallic conductivity or semi-conductivity, variable composition in many cases, and inter-atomic distances appreciably larger than in the parent metals. The expansion which accompanies their formation is in marked contrast to the contraction in the case of the salt-like hydrides (including EuH_2 and YbH_2). Incidentally it is interesting that the number of H atoms per cubic centimetre in a number of metal hydrides is greater than in solid H_2 or in water. This fact, combined with the high thermal stability of some of these compounds, makes them of interest as neutron-shielding materials for nuclear reactors and also possibly for energy storage (p. 352).

Although these compounds were originally described as 'interstitial' hydrides, implying that they were formed by entry of H atoms into interstices (usually

TABLE 8.1
Binary hydrides of transition metals

	TiH	$VH^{(2)}$	$CrH^{(3)}$	–	–	–	$NiH_{0.6}{}^{(4)}$
ScH_2	$TiH_2{}^{(1)}$	VH_2	CrH_2				
		$NbH^{(14)}$	–	–	–	–	$PdH^{(5)}$
YH_2	ZrH_2	NbH_2					
YH_3							
		TaH	–	–	–	–	–
LaH_2	$HfH_2{}^{(1)}$						
LaH_3							

4f *metals*

    La Ce Pr Nd   Sm . Gd—Tm . Lu (Y)
      $MH_{1.9}$–MH_3 MH_2 fluorite structure
        f.c.c.$^{(6)}$ MH_3 hexagonal LaF_3 structure$^{(7)}$
                     (For EuH_2, YbH_2, and $YbH_{2.55}{}^{(8)}$ see text)

5f *metals*

	Ac	Th	Pa	U		$Np^{(13)}$	Pu	Am
MH_2	f.c.c.	f.c.t.$^{(9)}$					f.c.c. ⟶	
MH_3			β-UH_3	α-$UH_3{}^{(11)}$		⟵	LaF_3 ⟶	
				β-$UH_3{}^{(12)}$				
		$Th_4H_{15}{}^{(10)}$						

(1) AC 1956 **9** 607
(2) IC 1970 **9** 1678
(3) PSS 1963 **3** K249
(4) JPP 1964 **25** 460
(5) JPCS 1963 **24** 1141
(6) JPC 1955 **59** 1226
(7) JPP 1964 **25** 454

(8) IC 1966 **5** 1736
(9) AC 1962 **15** 287
(10) AC 1953 **6** 393
(11) JACS 1954 **76** 297
(12) JACS 1951 **73** 4172
(13) JPC 1965 **69** 1641
(14) SPC 1970 **14** 522

tetrahedral) in the metal structure, it is now known that the arrangement of metal atoms in the hydride is *usually* different from that in the parent metal. The hydride has a definite structure and in this respect is not different from other compounds of the metal:

$$Cr \text{ (b.c.c.)} \rightarrow CrH \text{ (h.c.p.)} \rightarrow CrH_2 \text{ (c.c.p.)}$$
$$Ti, Zr, \text{ and } Hf \text{ (h.c.p.)} \rightarrow MH_2 \text{ (c.c.p.)}.$$

The hydrides of V, Nb, and Ta probably come closest to the idea of an interstitial compound and are discussed later. In other cases where the arrangement of metal atoms in the hydride is the same as in (one form of) the metal there may be a discontinuous increase in lattice parameter when the hydride is formed (Pd) or there may be an intermediate hydride with a different metal arrangement. For example, the h.c.p. 4f metals Gd–Tm form hexagonal trihydrides, but the intermediate dihydrides have the fluorite structure with c.c.p. metal atoms. Moreover, the volume of the MH_3 phase is some 15–25 per cent greater than that of the metal. The hydride $YbH_{2.55}$ has, like Yb, a f.c.c. arrangement of metal atoms but this phase is only made under pressure and has a *smaller* cell dimension (5.19 Å) than the metal (5.49 Å) and there is an intermediate hydride YbH_2 with the quite different (CaH_2) structure already noted.

The 4f *and* 5f *hydrides.* All the 4f metals take up hydrogen at ordinary or slightly elevated temperatures to form hydrides of approximate composition $MH_{1.9}$ which, with the exception of EuH_2 and YbH_2 (CaH_2 structure), all have fluorite-type structures. In contrast to EuH_2 and YbH_2 the formation of these cubic 'dihydrides' is accompanied by expansion of the structure. Europium forms only EuH_2 and Yb forms only YbH_2 at atmospheric pressure; under a higher hydrogen pressure it forms $YbH_{2.55}$ (probably an ionic compound containing Yb^{2+} and Yb^{3+}), and a metastable cubic YbH_2 has been made by heating $YbH_{2.55}$ or YbH_2. The dihydrides of the other 4f elements react further with hydrogen at atmospheric pressure to form 'trihydrides', but the lanthanides now fall into two groups. The lighter (larger) 4f metals form continuous f.c.c. solid solutions from the approximate composition $MH_{1.9}$ to a composition approaching MH_3. Having occupied the tetrahedral holes (fluorite structure) the H atoms then occupy octahedral holes at random, as has been shown by n.d. for CeH_2 and $CeH_{2.7}$; the H atoms are slightly displaced from the ideal positions. Onwards from Sm (and Y) a new h.c.p. phase appears before the composition MH_3 is reached, and as in the dihydrides there is a range of composition over which the phase is stable. There is a gap between the stability ranges of the dihydride and trihydride which is small for Sm but larger for the other metals. Some typical figures for the H:M ratios at room temperature are:

| | Composition limits (H:M) *at room temperature* | |
	Cubic dihydride	*Hexagonal trihydride*
Sm	1.93–2.55	2.59–3.0
Ho	1.95–2.24	2.64–3.0
Er	1.95–2.31	2.82–3.0

The hexagonal MH_3 phase has the (revised) LaF_3 structure (p. 420) which may be regarded as derived from an expanded h.c.p. metal structure. A neutron diffraction study of HoD_3 (with which the other phases are isostructural) shows that H (D) occupies all the tetrahedral and octahedral holes, but because of the close proximity of the pairs of tetrahedral holes there is some displacement of the H atoms from the ideal positions and this in turn necessitates some displacement of the H atoms from the octahedral holes. In the resulting structure H atoms have 3 nearest metal atom neighbours and the metal atom has 9 (+2) H neighbours:

$$\text{Ho:}\quad \begin{array}{ll} 9\,\text{H} & 2.10\text{–}2.29\,\text{Å} \\ (2\,\text{H} & 2.48\,\text{Å}) \end{array} \qquad \text{H:}\quad 3\,\text{Ho} \begin{array}{ll} \text{at} & 2.10\,\text{Å} \\ \text{or} & 2.17\,\text{Å} \\ \text{or} & 2.24\text{–}2.29\,\text{Å} \end{array}$$

The typical 4f hydrides are pyrophoric and graphitic or metallic in appearance. The resistivity of the dihydrides, which may be formulated $M^{3+}(H^-)_2$ (e), is lower than that of the pure metal, but increases as more hydrogen is absorbed. For example, at 80 °K there is a 10^6-fold increase in resistivity when $LaH_{1.98}$ is converted into $LaH_{2.92}$ and a 10^4-fold increase in the Ce hydrides. It seems likely that the bonding in the dihydrides (other than EuH_2 and YbH_2, which are non-conductors) is a combination of ionic and metallic bonding, and that the addition of further hydrogen leads to the formation of H^- ions by removal of electrons from the conduction band, producing an essentially ionic trihydride.

The 5f hydrides are summarized in Table 8.1. In contrast to the other elements Th forms a dihydride with a distorted (f.c. tetragonal) fluorite-like structure, resembling in this respect Ti, Zr, and Hf, but the structure becomes cubic if a little oxygen is present. Th also forms Th_4H_{15}, in which there are two types of non-equivalent H atom, and the shortest Th—Th distance (3.87 Å) is appreciably greater than in the metal (3.59 Å):

$$Th: \begin{array}{ll} 9\,H & 2.29\,\text{Å} \\ 3\,H & 2.46\,\text{Å} \end{array} \qquad \begin{array}{lll} H_I: & 3\,Th & 2.29\,\text{Å} \\ H_{II}: & 4\,Th & 2.46\,\text{Å} \end{array}$$

compare Th—8 H at 2.41 Å in ThH_2.

Uranium forms only one hydride, UH_3, which is dimorphic and metallic in character. The metal atoms in β-UH_3 are in the positions of the β-W structure (p. 1283), a structure not adopted by metallic U. The U—U distances are much greater than those in α- or γ-uranium, even the shortest (U—2 U, 3.32 Å) indicating only very weak metal–metal bonds. (Compare U—8 U in γ-U, 2.97 Å, and the shortest bonds in α-U, 2.76 Å.) The H atoms have been shown by neutron diffraction to occupy very large holes in which they are surrounded, approximately tetrahedrally, by four U at 2.32 Å. Since each U atom has twelve H neighbours and the compound is metallic rather than salt-like it has been suggested that the atoms are held together by some kind of delocalized covalent bonds. In α-UH_3 the metal atoms occupy the positions of the shaded circles in Fig. 29.4 (p. 1283) and the H atoms the open circles. Here again the U—U bonds are extremely weak (U—8 U, 3.59 Å, and H is surrounded tetrahedrally by 4 U at 2.32 Å as in the β form. The trihydrides of Np, Pu, and Am are isostructural with the hexagonal 4f trihydrides.

Hydrides of the 3d, 4d, *and* 5d *metals*. We now comment briefly on the remaining hydrides of Table 8.1. All the (h.c.p.) elements Ti, Zr, and Hf form dihydrides with fluorite-type structures which are cubic above their transition points and have lower (tetragonal) symmetry at ordinary temperatures. Both phases have ranges of composition, which depend on the temperature, and the following figures (for room temperature) show that for Ti the cubic phase includes the composition TiH:

Cubic phase	Tetragonal phase
$TiH–TiH_2$	
$ZrH_{1.50}–ZrH_{1.61}$	$ZrH_{1.73}–ZrH_{2.00}$
$HfH_{1.7}–HfH_{1.8}$	$HfH_{1.86}–HfH_{2.00}$

In addition to these phases there are solid solutions, with small hydrogen concentrations, in the h.c.p. metal, and also solid solutions in the high-temperature (b.c.c.) forms; in the latter the concentration of hydrogen may be considerable, for example, up to $ZrH_{1.5}$. The fluorite structure of TiD_2 has been confirmed by neutron diffraction. It should be noted that hydrides previously formulated Zr_2H and Zr_4H are not distinct compounds but correspond to arbitrarily selected points on the phase diagram. The fact that the arrangement of the metal atoms in the tetragonal Zr and Hf hydrides is less symmetrical (for the composition MH_2) than in the defect structures emphasizes the important part played by the metal–hydrogen bonds in these structures. The following data for the Hf hydrides show that the formation of these hydrides is by no means a question of H atoms (or ions) simply occupying interstices in a c.p. metal structure:

α-Hf (hexagonal)	$HfH_{1.7}$ (cubic)	HfH_2 (tetragonal)
Hf−6 Hf 3·13 Å	Hf−12 Hf 3·31 Å	Hf−8 Hf 3·27 Å
−6 Hf 3·20	Hf−H 2·03	−4 Hf 3·46
		Hf−8 H 2·04
		H−4 Hf

The metals V, Nb, and Ta have b.c.c. structures. The α solid solution of H in this structure extends to $VH_{0.05}$, $NbH_{0.1}$ and $TaH_{0.2}$. The next distinct phase is the β hydride, a non-stoichiometric phase which is a (tetragonal) distorted version of the b.c.c. solid solution and is stable over a wide range of composition (for example, $VH_{0.45}$–$VH_{0.9}$). The tetragonal b.c. β-NbH becomes cubic at about 200 °C. A neutron diffraction study shows complete ordering of the H atoms in this phase when cooled below room temperature. In this region the phase diagram of these elements appears to be similar to that of Pd (Fig. 8.1), so that above a certain (consolute) temperature there is continuous absorption of hydrogen up to approximately the stage MH without radical rearrangement of the metal structure. This behaviour is presumably associated with the fact that in the b.c.c. structure there are six (distorted) tetrahedral sites per atom as compared with two (regular) tetrahedral sites per atom in a c.p. structure (see p. 145). There are accordingly many possible types of disordered structure and superstructure. Special treatment (for example, high H_2 pressure) is required to make the dihydrides of V and Nb, which have the usual f.c.c. structure, and the maximum hydrogen content of a Ta hydride is reached in $TaH_{0.9}$.

Two hydrides of Cr have been made (electrolytically), CrH with an anti-NiAs type of structure (Cr−6 H, 1.91 Å, Cr−Cr, 2.71 Å) and CrH_2 with apparently the fluorite (f.c.c.) structure.

Palladium absorbs large volumes of hydrogen. The phase diagram (Fig. 8.1) shows that at room temperature there is a small solubility (α solid solution) in the metal, then a two-phase region followed by the β hydride ($PdH_{0.56}$). The cell dimensions of the two phases are: α, 3.890 Å ($PdH_{0.03}$), and β, 4.018 Å ($PdH_{0.56}$).

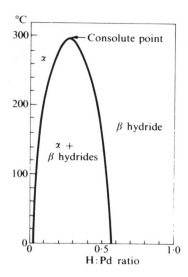

FIG. 8.1. The palladium–hydrogen phase diagram.

The maximum hydrogen content of the β phase corresponds to the formula $PdH_{0.83}$, at $-78\,°C$. Above $300\,°C$ only one phase is found up to hydrogen pressures of 1000 atmospheres. Nickel hydride has been prepared only as a thin film on a Ni surface by electrolysis, with a maximum H:Ni ratio of approximately 0.6. Neutron diffraction studies of both Pd and Ni hydrides indicate a f.c.c. structure (defect NaCl structure).

Ternary hydrides

The structures of these compounds (Table 8.2) suggest that they include both salt-like compounds and also compounds more akin to the transition-metal compounds

TABLE 8.2
Ternary metal hydrides

$LiBeH_3$[1]	$EuLiH_3$	$LiAlH_4$[3a]	$LiGaH_4$	$NiZrH_3$[5]
$NaBeH_3$	$SrLiH_3$	$NaAlH_4$[3b]	$LiInH_4$	$AlTh_2H_4$[6]
	$BaLiH_3$[2]	(also K, Cs)		
Li_2BeH_4		Li_3AlH_6		Mg_2NiH_4[7]
Na_2BeH_4		Na_3AlH_6[4]		Li_4RhH_4[8]
				Sr_2IrH_4[8]
		$Mg(AlH_4)_2$		
		$Ca(AlH_4)_2$		

(1) JCS A 1968 628. (2) JCP 1968 **48** 4660. (3a) IC 1967 **6** 669. (3b) AC 1979 **B35** 1454. (4) IC 1966 **5** 1615. (5) JPP 1964 **25** 451. (6) AC 1961 **14** 223. (7) IC 1968 **7** 2254. (8) IC 1969 **8** 1010.

of the last section. Methods of preparation include the action of hydrogen on a mixture of the metals (for example, $BaLiH_3$ from Ba + Li at 700 °C) or on an alloy (for example, Mg_2NiH_4 from Mg_2Ni and H_2 under pressure at 325 °C). The first product of the action of H_2 under pressure on a mixture of Na and Al in toluene at 165 °C is Na_3AlH_6; further reaction gives $NaAlH_4$.

A n.d. study has confirmed the perovskite structure of $BaLiH_3$ (and presumably therefore the structures of the isostructural Sr and Eu compounds). The M–H distances are appreciably larger in this structure than in MH_2:

	Eu–H	Sr–H	Ba–H
12-coordination	2.68 Å	2.71 Å	2.84 Å
7-coordination	2.45	2.49	2.67

and the Li–H distance is rather larger (2.01 Å) in $BaLiH_3$ than in $SrLiH_3$ (1.92 Å) or $EuLiH_3$ (1.90 Å); compare 2.04 Å in LiH.

In $LiAlH_4$ there is nearly regular tetrahedral coordination of Al (Al–H, 1.55 Å) and rather irregular 5-coordination of Li (4 H at 1.88–2.00 Å, 1 H at 2.16 Å). The four H atoms of an AlH_4 group are of two kinds, three being 2-coordinated (to Al and Li) and the fourth 3-coordinated (to Al + 2 Li). A similar nearly regular tetrahedral AlH_4 group exists in $NaAlH_4$ (Al–H, 1.53 Å) with Na in unusual triangular dodecahedral 8-coordination (Na–H, 2.50 Å). These structures appear to contain AlH_4^- ions; the structure of Na_3AlH_6 is said to be of the cryolite type.

Hydrogen is readily absorbed by $AlTh_2$ ($CuAl_2$ structure, p. 1313) forming ultimately $AlTh_2H_4$, in which H atoms occupy all of the tetrahedral holes between 4 Th atoms in the alloy structure. The Th–H distance (approximately 2.4 Å) is similar to that in ThH_2. The structure of $NiZrD_3$ is more complex, for here some of the D atoms occupy tetrahedral holes and some are in positions of 5-coordination:

$$D \begin{cases} \text{Zr} & 1.96\,\text{Å} \\ 2\,\text{Zr} & 2.18 \\ \text{Ni} & 1.77 \end{cases} \quad \text{and} \quad D \begin{cases} \text{Zr} & 1.95\,\text{Å} \\ 2\,\text{Zr} & 2.38 \\ 2\,\text{Ni} & 1.78 \end{cases}$$

To be useful as a means of storing and transporting energy a hydride must be a reasonably stable compound; it is also necessary that the hydrogen is recoverable. There has been some interest in ternary hydrides as potential energy storage materials, particularly those containing an element such as Mg or Ti capable of forming a stable hydride and a second metal (for example, Fe, Cu, etc.) which forms hydrides with difficulty or not at all. One such hydride is $CuTiH_{0.9}$,[1] formed from γ-CuTi, which has a structure closely related to the fluorite structure of TiH_2. Pairs of c.p. Cu and Ti layers alternate, and H atoms occupy tetrahedral sites only between the Ti layers.

(1) AC 1978 **B34** 2059

Hydrido complexes of transition metals

The d-type transition metals constitute the large block of elements lying between the electropositive metals which form ionic hydrides and the B subgroup and non-metallic elements which form covalent molecular hydrides. In addition to forming interstitial hydrides these transition metals also form covalent molecules MH_xL_y in which H atoms are directly bonded to the metal. Molecules of this general type are formally similar to substituted hydrides such as PHF_2, GeH_2Cl_2, etc. of non-metals and B subgroup elements, but a characteristic feature of the transition-metal compounds is that the ligands L must be of a particular kind, namely, those which cause electron-pairing in the d orbitals of the metal and are present in sufficient number to fill all the non-bonding d orbitals. They include CO, cyclopentadienyl, and tertiary phosphines and arsines; some examples are given in Table 8.3. The

TABLE 8.3

Some hydrido compounds of transition metals

HNb_6I_{11}	$[Cr_2H(CO)_{10}]^-$ $CrH(C_5H_5)(CO)_3$ $MoH_2(C_5H_5)_2$[3]	$MnH(CO)_5$ $Mn_2H(CO)_8(P\phi_2)$[1] $[TcH_9]^{2-}$[4]	$FeH_2(PP)_2$ $RuHCl(PP)_2$	$CoH(PP)_2$ $CoH(PF_3)_4$[2] $RhH(CO)(P\phi_3)_3$[5]		
$TaH_3(C_5H_5)_2$	$WH_2(C_5H_5)_2$ $[W_2H_2(CO)_8]^{2-}$[10]	$[ReH_9]^{2-}$[6] ReH_7P_2	$OsHBr(CO)(P\phi_3)_3$[7] OsH_4P_3	$RhHCl_2P_3$ $IrHCl_2P_3$	$PtHBrP_2$[8] $PtHClP_2$[9]	

(P stands for a tertiary phosphine, PR_3, and PP for $R_2P.CH_2.CH_2.PR_2$ or a cyclohexane ring with two PR_2 groups)

(1) JACS 1967 **89** 4323. (2) IC 1970 **9** 2403. (3) IC 1966 **5** 500. (4) IC 1964 **3** 567. (5) AC 1965 **18** 511. (6) IC 1964 **3** 558. (7) PCS 1962 333. (8) AC 1960 **13** 246. (9) IC 1965 **4** 773. (10) CC 1973 691.

formation by Re and Tc of the remarkable ions $(MH_9)^{2-}$ shows that H may bond to certain transition metals in the absence of such ligands. The majority of the elements of Table 8.3 form carbonyl hydrides, and these compounds provide many examples of metal clusters in which H atoms bridge two or three metal atoms. Hydrogen is also found in interstitial positions in certain metal clusters, HNb_6I_{11} (p. 436) and p. 966. Metal carbonyls are discussed in Chapter 22; here we note examples of molecules in which H acts as a direct link between two metal atoms:

$[(CO)_5Cr-H-Cr(CO)_5]^-$ (linear bridge, M–H–M, 3.4 Å), (p. 966),

$HMn_3(CO)_{10}(BH_3)_2$ (non-linear bridge, Mn–H, 1.65 Å), (p. 1092),

$(CO)_4MnH(P\phi_2)Mn(CO)_4$ (non-linear bridge, Mn–H, 1.87 Å),

$[W_2H_2(CO)_8]^{2-}$ (double hydrogen bridge).

In the remaining examples of this section H is bonded to one metal atom only, as a normal covalently bound ligand, with M–H, 1.6–1.7 Å.

Molecules or ions in which the H atoms have been definitely located by n.d. or X-ray diffraction include the following:

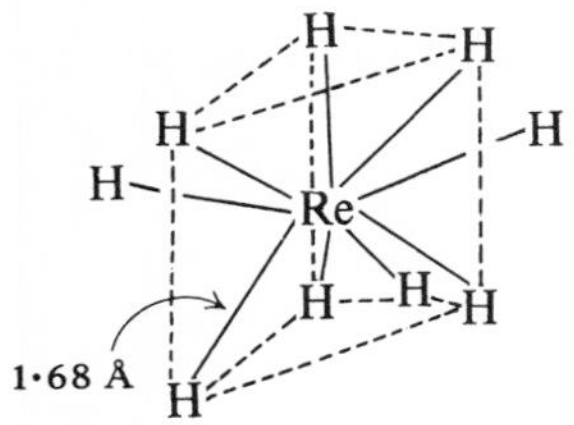

$[ReH_9]^{2-}$

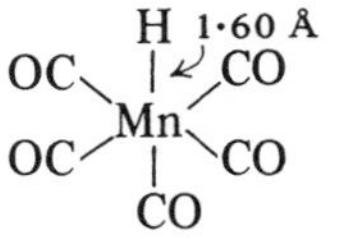

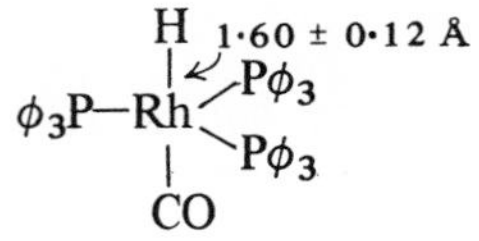

Less direct evidence for the positions of H atoms comes from X-ray studies of complexes in which the location of the other ligands strongly suggests the position of the H atom(s). In crystalline $PtHBr(PEt_3)_2$ the P and Br atoms were found to be

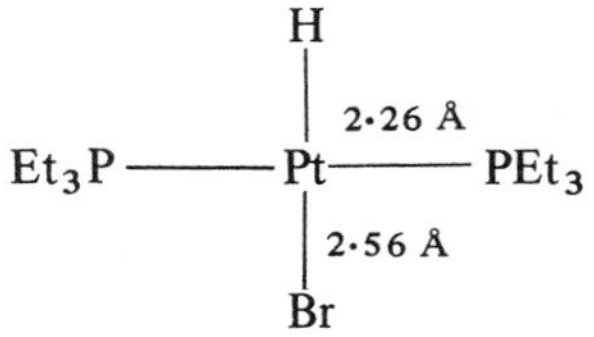

coplanar with the Pt atom, with P–Pt–Br angles of 94°, leaving little doubt that the H is directly bonded to the Pt in the *trans* position to the Br atom. The fact that the Pt–Br bond is rather longer than expected (sum of covalent radii, 2.43 Å) may be associated with the high chemical lability of this atom.

In an incomplete X-ray study of the diamagnetic $Os^{II}HBr(CO)(P\phi_3)_3$ the five heavier ligands were found to occupy five of the six octahedral positions, H presumably occupying the sixth position. In some cases the presence of H directly bonded to the metal is confirmed by the presence in the i.r. spectrum of an absorption known to be due to a M–H bond, as in the triply-bridged Ir ion shown below:[1]

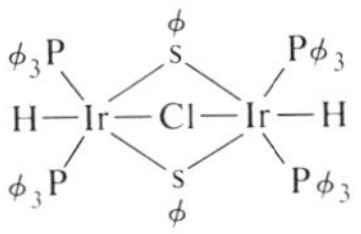

Still less direct evidence comes from spectroscopic[2] and dipole moment studies.[3] Configurations deduced in these ways include the following:

(P = tertiary phosphine)

(1)   AC 1976 **B32** 1513   (3)   PCS 1962 318
(2)   JACS 1966 88 4100

The hydrogen bond

It has been recognized for a long time that the properties of certain pure liquids and of some solutions indicate an unusually strong interaction between molecules of solvent, between solvent and solute, or between the solute molecules themselves. The first type of interaction, between the molecules in a pure liquid, led to the distinction between 'associated' and non-associated solvents. It may be illustrated by the properties of NH_3, H_2O, and HF, but many organic solvents, particularly those containing OH, COOH, or NH_2 groups, show somewhat similar abnormalities. If we compare such properties as melting point, boiling point, and heat of evaporation of a series of hydrides such as H_2Te, H_2Se, and H_2O, we find that these properties form a fairly regular sequence until we reach the last member of the series, H_2O. The hydrides of Groups VB and VIIB show similar behaviour, NH_3 and HF being 'abnormal' and HCl slightly so. The hydrides of Group IVB, including CH_4, form, on the other hand, a regular sequence; see Fig. 8.2. Instead of a melting point in the region of $-100\,^{\circ}C$, as might be expected from extrapolation of the melting points of H_2Te, H_2Se, and H_2S, we find that H_2O melts at $0\,^{\circ}C$. From the fact that the boiling points and heats of vaporization show the same type of abnormality as the melting points we deduce that many hydrogen bonds must exist in liquid NH_3, H_2O, and HF up to their boiling points. In the case of HF they persist in the vapour state.

Specific interaction between particular atoms of the solvent and solute molecules explains, for example, the much greater solubilities of aniline ($C_6H_5NH_2$) and phenol (C_6H_5OH) in water than in nitrobenzene in spite of the much larger dipole moment of the latter (4.19 D) as compared with water 1.85 D).

Association of solute molecules occurs when a substance like acetic acid, CH_3COOH, is dissolved in a non-associated solvent such as benzene, and is evident from determinations of molecular weight by the cryoscopic or ebullioscopic methods.

From the chemical nature of the molecules it is clear that these interactions are connected with the presence of hydrogen, usually as part of OH, COOH, NH_2, or other polar groups. If the hydrogen is replaced by, for example, alkyl groups, there is no association, showing that the H plays an essential part when it is attached to the electronegative N, O, or F atoms. Some abnormalities are also shown by compounds containing, in addition to H, the less electronegative S or Cl, but these are much less pronounced; we include some data on these weaker interactions in

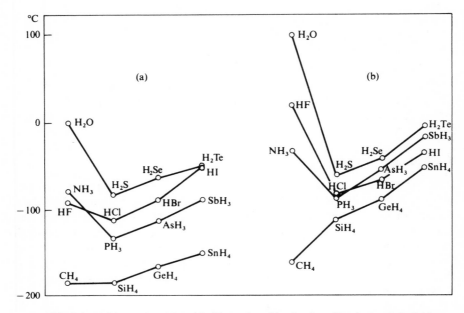

FIG. 8.2. Melting-points (a) and boiling-points (b) of series of isoelectronic hydrides.

Table 8.4. We shall confine our attention here largely to compounds containing N, O, and F. These intermolecular interactions, which we may write

$$\text{N---H---O,} \qquad \text{O---H---O,} \qquad \text{or} \qquad \text{F---H---F,}$$

are referred to as hydrogen bonds or hydrogen bridges; they may also be formed between atoms within a molecule or complex ion (intramolecular hydrogen bonds).

The properties of hydrogen bonds

The fact that strong hydrogen bonds are formed only between the most electronegative elements suggests that they are essentially electrostatic in character. The atoms concerned have lone pairs of electrons which not only influence the directional properties of bonding orbitals but also contribute substantially to the dipole moments of the molecules. Assuming various types of hybridization the orbital dipole moments of lone pairs on atoms such as N, O, and F can be calculated. These dipoles can interact with the $H^{\delta+}$ of FH, OH, NH_2, etc. to form hydrogen bonds which should be directed in accordance with the type of hybridization of the lone pair and bonding orbitals. This directional feature of hydrogen bonds is evident in ice, where each O atom has four tetrahedral neighbours, in HF, $H_2F_3^-$, and $H_4F_5^-$, and in most hydrates. There are, however, some crystals in which neither of the lone pairs of O is directed along the hydrogen bonds as, for example, in $K_2C_2O_4 . H_2O^{[1]}$

TABLE 8.4
Lengths of hydrogen bonds

Bond	Length (Å)	Compound	Reference
F---H---F	2.27	$NaHF_2$ ⎫	
	2.45	KH_4F_5 ⎬	See text
	2.49	HF ⎭	
	2.26	p-toluidine (HF_2)	JACS 1973 **95** 5780
O—H---O	2.40–2.56	Type (i) ⎫ See text	AC 1978 **B34** 2074
	2.40–2.63	Type (ii) ⎭	AC 1978 **B34** 436
O—H---O	2.7–2.9	Ice, hydrates,	See text
		hydroxy compounds	
O—H---F	2.56, 2.87	Hydrated metal fluorides	AC 1978 **B34** 355
	(mean 2.68)		
	2.50	$Te(OH)_6 . NaF$	AC 1976 **B32** 1025
O—H---Cl	2.95–3.10	HCl hydrates	See p. 692
	3.2–3.4	Hydrates of metal chlorides	AC 1977 **B33** 1608
O—H---Br	3.4–3.6	$NaBr.2H_2O$	AC 1979 **B35** 1679
	3.2–3.4	$HBr.4H_2O$	See p. 693
O—H---N	2.68	$N_2H_4.4CH_3OH$	ACSc 1967 **21** 2669
	2.79	$N_2H_4.H_2O$	AC 1964 **17** 1523
O—H---S	3.1–3.4	Hydrated thiosulphates	AC 1975 **B31** 135
N—H---O	2.86	$(NH_4)_2H_3IO_6$	AC 1980 **B36** 1028
	2.79	Cyanuric acid	AC 1971 **B27** 134, 146
N—H---F	2.6–2.96	$NH_4F, (N_2H_6)SiF_6$	AC 1980 **B36** 1917
N—H---Cl	3.00, 3.11	$(CH_3)_3NHCl, (CH_3)_2NH_2Cl$	AC 1968 **B24** 554, 549
	3.20	$(NH_3OH)Cl$	AC 1967 **22** 928
N—H---I	3.46	$[(CH_3)_3NH]I$	AC 1970 **B26** 1334
N—H---N	2.94–3.15	NH_4N_3, dicyandiamide	See Chapter 21
	3.35	NH_3	See text
N—H---S	3.23, 3.29	$N_2H_5(HS)$	AC 1975 **B31** 2355
(S—H---S)?	3.94	H_2S (cryst.)	N 1969 **224** 905
(C—H--O)?	2.92 +	Organic compounds	JCS 1963 1105

It may perhaps be considered doubtful whether some of the much weaker interactions should be described as hydrogen bonds or as van der Waals or weak ionic bonds. However, if the H atom is directed towards the other atom of the 'bond' many such interations are described in the literature as hydrogen bonds, and rather than discuss such cases in detail we give some references above. The H atoms have not been located in any of the 'C–H–O' bonds. In a number of cases we give a reference to a single compound since reference is often made to earlier work on related compounds.

where the two K^+ neighbours are nearly coplanar with the O—H bonds of the water molecule.

(1) JCP 1964 **41** 3616

Bond energies and lengths

The energies of hydrogen bonds range from $110 \, kJ \, mol^{-1}$ (or possibly higher values)[(1)] for the strongest (in the $F–H–F^-$ ion) through values around $30 \, kJ \, mol^{-1}$ for O—H–O to still smaller values for the weaker hydrogen bonds. There is also a considerable range of lengths for each particular type of bond X–H–X, and data for the various hydrogen bonds are summarized in Table 8.4.

(1)   IC 1963 **2** 996

Position of the H atom

The positions of the H atoms in hydrogen bonds in many crystals have been determined by X-ray and/or neutron diffraction, and indirectly from the proton-proton separations in some hydrates by p.m.r. It should be emphasized that the positions of H atoms as determined by X-ray diffraction are different from those resulting from n.d. studies, the former showing the H atom apparently closer to the heavier atom to which it is bonded. Typically the difference for an O–H distance is about 0.15 Å (n.d., 1.0 Å; X-ray, 0.85 Å). The effect (which is smaller for heavier atoms) appears to arise from the asphericity of the electron distribution due to chemical bonding. This affects the X-ray scattering factor of the atom (which depends on the orbital electrons) but not the neutron scattering which, for a diamagnetic atom, is purely nuclear. The methods for refining crystal structures involve the use of calculated atomic scattering factors, so that if a spherical electron distribution around an atom is assumed and the position of an atomic nucleus is determined as the centre of gravity of its electron cloud, the 'X-ray position' may be different from that determined by neutron diffraction.[1]

In many hydrogen bonds, particularly the stronger ones, the H atom lies on or near the straight line joining the bonded atoms, as in the F–H–F⁻ ion, many short O–H–O, O–H–F, and N–H–O bonds, but angles X–H–Y of 160–170° or less are observed in some structures (Fig. 8.3). Obviously these angles could be discussed

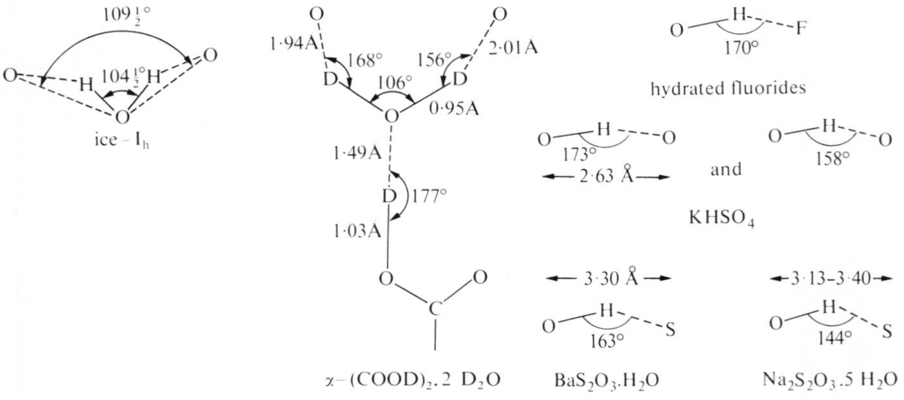

FIG. 8.3. Details of hydrogen bonds.

only in relation to the packing of the ions or molecules in each individual crystal structure. The position of the H atom in linear F–H–F and O–H–O bonds has been much studied, with the following results.

F–H–F *bonds*. All physical evidence (n.d., i.r., and n.m.r.) supports the view that the short F–H–F bond (2.27 Å) in the bifluoride ion is symmetrical in salts such as $NaHF_2$, but a n.d. study of *p*-toluidine bifluoride shows that the HF_2^- ion is linear but unsymmetrical, with a single minimum:

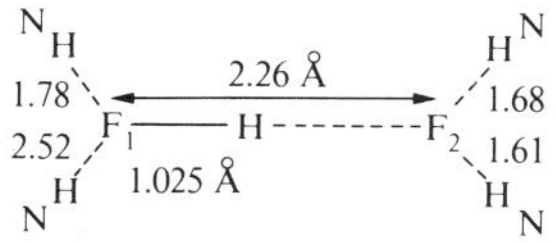

This exceptional asymmetry is attributed to the environment of the ion in the crystal, the atom F_2 being much more strongly hydrogen-bonded to its N neighbours than is F_1. The much longer hydrogen bonds (2.49 and 2.45 Å) in crystalline HF and KH_4F_5 are almost certainly not symmetrical; the nature of the intermediate bond (2.33 Å) in KH_2F_3 is not known.

O–H–O *bonds*. It appeared at one time that there might be a real difference between the structures of the long O–H–O bonds, of length 2.7 Å or more, in ice, hydrates, and hydroxy-compounds, and the shorter ones of length 2.55 Å or less in acids, and acid salts. In some crystals hydrogen bonds of two different lengths occur, as in $Na_3H(CO_3)_2 . 2H_2O$, where the 'acid salt' bond between pairs of CO_3^{2-} ions has a length of 2.53 Å while the bonds between water molecules and CO_3^{2-} ions have a mean length of 2.74 Å, and $(COOH)_2 . 2H_2O$, where the length of the short bonds between OH of $(COOH)_2$ and OH_2 is 2.53 Å as compared with 2.84–2.90 Å for the bonds between H_2O and CO groups of $(COOH)_2$ molecules (Fig. 8.3(c)).

All long O–H–O bonds in which the H atoms have been located have been shown to be unsymmetrical, as is to be expected if the H belongs to –OH or H_2O; similarly for N–H–O bonds formed by, for example, $-NH_2$ or NH_4^+. The situation in acids and hydrogen ('acid') oxy-salts is less simple. First we should note that the term 'symmetrical hydrogen bond' is used is two different senses, to mean either (i) that there is a symmetry centre or 2-fold axis at the centre of the bond, or (ii) that the potential curve for the H atom has a single minimum at the mid-point of the bond. Case (i) could imply (ii) *or* that there is disorder of the H atom and an equal probability of finding it at either of two sites equidistant from the centre of any bond. Short O–H–O bonds that are crystallographically symmetrical range in length from around 2.4–2.6 Å, and it appears that those with O–O less than about 2.5 Å have a single potential minimum. On the other hand, in the centrosymmetrical bonds of length 2.64 Å in $(NH_4)_2H_3IO_6$ the H atoms are dynamically disordered between two positions related by the centre of symmetry and about 0.7 Å apart. Short O–H–O bonds which are not astride a symmetry element also range in length from 2.4–2.6 Å but may be symmetrical (H at mid-point), as in KH (chloromaleate), (2.40 Å), nearly symmetrical in $KH_5(PO_4)_2$, (2.42 Å with two minima 0.08 Å apart), or grossly unsymmetrical as in tetragonal KH_2PO_4 (2.49 Å), (a), $(COOD)_2 \cdot 2D_2O$ (2.52 Å), (b), or $DCrO_2$ (2.55 Å), (c); for $HCrO_2$ see p. 642.

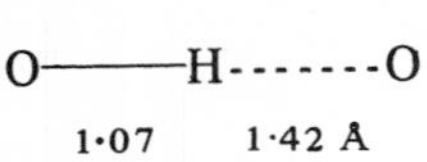

O————H-------O $\underset{}{\overset{}{>}}C—C\overset{O————D-------OD_2}{\underset{1.03 \quad\quad 1.49\ A}{<}}$ O————D-------O

1.07   1.42 A   0.96   1.59 A

(a) 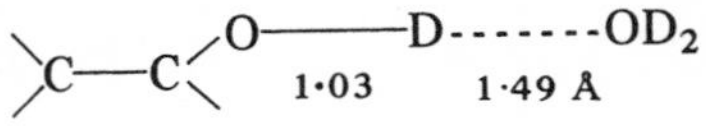   (b) 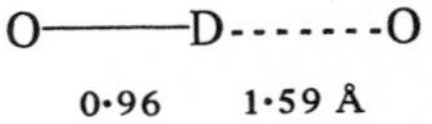   (c)

Neutron diffraction studies are required to locate the H atoms in the very short O---H---O bonds in molecules such as Ni ethyl methyl glyoxime[2a], (d). or the anion in $K_3[Pt\{(SO_3)_2H\}Cl_2]$[2b], (e). A recent review of strong hydrogen bonds is available.[2c]

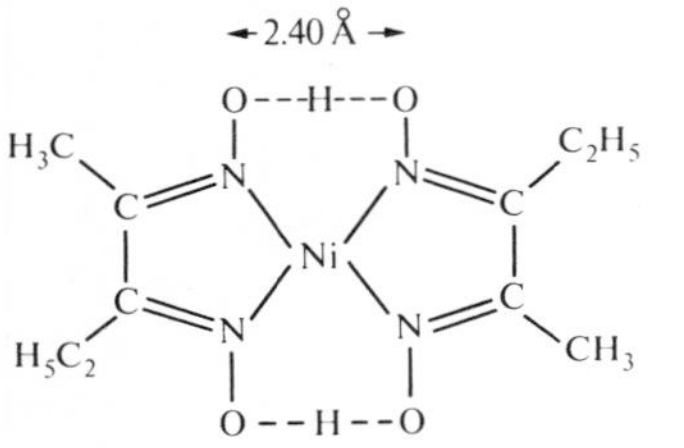

←2.40 Å→

(d)

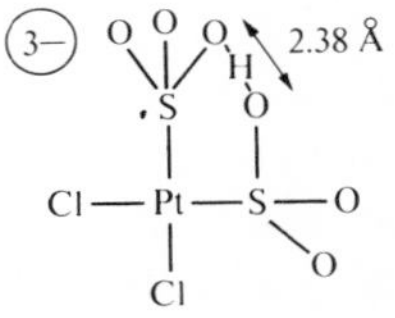

(e)

Less direct evidence for the position of the H atom also comes from proton magnetic resonance studies and from values of residual entropy. The positions of the protons in the H_2O molecules in gypsum have been determined indirectly from the fine structure of the n.m.r. lines; they are found to lie at a distance of 0.98 Å from the O atom along an O–H–O bond.[3] For n.m.r. studies of a number of hydrates see p. 690; the method has also been used to locate the H atoms in $Mg(OH)_2$ (p. 632). Measurements of residual entropy confirm the existence of two distinct locations for the proton in hydrogen bonds in ice (p. 655), salts of the type of KH_2PO_4, and $Na_2SO_4 \cdot 10H_2O$ (which has residual entropy about two-tenths that of ice).

(1)   AC 1969 **B25** 2451   (2c) CSR 1980 9 91
(2a) AC 1972 **B28** 2318   (3)   JCP 1948 **16** 327
(2b) AC 1980 **B36** 2545

The hydrogen bond in crystals

Hydrogen bonds play an important part in determining the structures of crystalline compounds containing N, O, or F in addition to H. We deal here with the hydrides of these elements, and with certain fluorides, oxy-acids, and acid salts. Ice and water, together with hydrates, are considered in Chapter 15, and hydroxy-acids are grouped with hydroxides in Chapter 14.

We noted in Chapter 3 that the interest, from the geometrical standpoint, of hydrogen-bonded structures lies in the variety of ways in which a group of atoms of limited extent (that is, a finite molecule or ion) can be linked by hydrogen bonds into a more extensive system, in the limit a three-dimensional network. These hydrogen-bonded systems may be:

finite groups: $[H(CO_3)_2]^{3-}$ (in $Na_3H(CO_3)_2 . 2H_2O$), dimers of carboxylic acids,
infinite chains: HCO_3^- and HSO_4^- ions,
infinite layers: $B(OH)_3$, $B_3O_3(OH)_3$, $N_2H_6F_2$, H_2SO_4,
3-dimensional nets: ice, H_2O_2, $Te(OH)_6$, $H_2PO_4^-$ ion in KH_2PO_4.

Crystalline organic compounds containing OH, COOH, CO.NH, NH_2, and other polar groups containing H atoms provide many further examples of all the above types of structure.

Hydrides

As already mentioned, the hydrides of N, O, and F alone show evidence of strong hydrogen bonds. The simplest structures would arise if every atom was joined to all its nearest neighbours by such bonds and if every H atom was utilized in forming hydrogen bonds. Then in the solid hydride AH_n every A atom would be surrounded by $2n$ A neighbours, that is, the coordination number of A by A atoms would be 6 for NH_3, 4 for H_2O, and 2 for HF. The structure of crystalline NH_3 may be described as a distorted form of cubic closest packing. Instead of 12 equidistant nearest neighbours each N atom has 6 neighbours at 3.35 Å and 6 more at the much greater distance 3.88 Å, indicating that a N atom forms six hydrogen bonds. The precise D positions in ND_3 have been determined by n.d.[1] With this ordered structure compare the structure of ordinary ice (ice-I_h). Each O is surrounded tetrahedrally by 4 others as expected, but there is a statistical distribution of H atoms between the equivalent positions at a distance of 1 Å from O along each hydrogen bond.

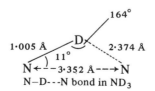

N–D---N bond in ND_3

The fact that there is only one H atom for each F limits the coordination numbers of both H and F in solid HF to 2, and the only possibilities are therefore infinite chains or closed rings. A neutron diffraction study of solid DF at 85 K[2] showed that the crystal consists of infinite planar zigzag chains:

There is no disorder in the polar chains. Many studies of HF vapour have been made,[3] indicating mixtures of monomers and various linear and cyclic polymers, with mean angle F–F–F 104° and F–H–F 2.53 Å. However, molecular beam deflection and mass spectrometric experiments show that the polymerization is continuous and presumably dependent on pressure and temperature. The behaviour of HCl and HBr in the crystalline state is complex, and careful n.d. studies have yielded the following results.[4]

Solid DCl undergoes a first-order phase transition at about 105 K (98.44° for HCl). At higher temperatures the structure is f.c.c. and there is disorder of a special kind. At any given time each molecule is in one of twelve orientations, so that it is hydrogen-bonded to one of its nearest neighbours which itself is hydrogen-bonded to one of its twelve nearest neighbours, and so on. Each hydrogen bond is only temporary (being broken when the molecule reorients) so that the 'instantaneous' structure of cubic DCl (HCl) is a mixture of shortlived hydrogen-bonded polymers of various lengths and shapes. At temperatures below the transition point the structure is orthorhombic. Each D is directed towards a neighbouring Cl atom (Fig. 8.4(a)) and the structure therefore consists of chains of hydrogen-bonded molecules (D–Cl, 1.28 Å; Cl–D–Cl, 3.69 Å; Cl–Cl–Cl, 93½°). Crystalline DBr shows two second-order (λ) phase transitions, at 93.5° and 120.3°K. At temperatures well below the lower λ region (for example, 74 K) the structure is the same ordered orthorhombic structure as for DCl (Fig. 8.4(a)), in which the bond lengths and angle are: D–Br, 1.38 Å, Br–D–Br, 3.91 Å, Br–Br–Br, 91°. As the temperature 93 K is approached there is a continuous change to the partially disordered structure of Fig. 8.4(b) in which there are two equilibrium orientations for each molecule. In contrast to the sudden loss of order in DCl this change in DBr is spread over some 20 K. At the higher λ transition there is presumably a gradual change from this disordered orthorhombic structure to a structure with the 12-fold orientational disorder of high-DCl. The behaviour of HCl is even more complex, this compound exhibiting three λ-type transitions. Details of the behaviour of HI are not yet available. (The presence of cyclic dimers in HCl condensed in solid Xe has been deduced from the i.r. spectrum, but their structure has not been determined.)[5]

(1)   JCP 1961 **35** 1730   (2)   AC 1975 **B31** 1998
(3)   JCP 1969 **50** 3611; ZN 1972 **27a** 983; JCP 1972 **56** 2442
(4)   N 1967 **213** 171; N 1967 **215** 1265; N 1968 **217** 541
(5)   JCP 1967 47 5303

Normal fluorides

Ammonium fluoride, NH_4F, crystallizes with a structure different from those of the other ammonium (and alkali) halides. The chloride, bromide, and iodide, have the

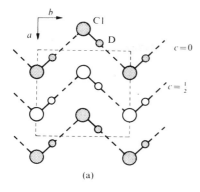

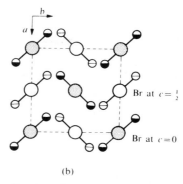

⬤   alternative positions of D atoms in
⊖   disordered molecules at $c=0$ or $\frac{1}{2}$

FIG. 8.4. (a) The structure of crystalline DCl at 77.4K. (b) The structure of crystalline DBr at 107 K. Small circles represent D and large circles Cl or Br.

CsCl structure at temperatures below 184.3, 137.8, and $-17.6\,°C$ respectively, and the NaCl structure at temperatures above these transition points, but NH_4F crystallizes with the wurtzite structure, in which each N atom forms N–H–F bonds of length $2.71\,Å^{(1)}$ to its four neighbours arranged tetrahedrally around it. This is essentially the same structure as that of ordinary ice.

Hydrazinium difluoride, $N_2H_6F_2$, provides a very interesting example of a crystal in which the structural units are linked by hydrogen bonds (N–H–F) into layers.[2] The structural units are $N_2H_6^{2+}$ and F^- ions. In the $N_2H_6^{2+}$ ion the N–N bond length is $1.42\,Å$ and the N–N–H bond angles $110°$. Viewed along its axis the ion has the configuration:

$$\begin{array}{c}\text{H}\\[2pt]\text{H}\diagdown \;\; \vdots \;\;\diagup\text{H}\\ \bigcirc\!\!\!\!\bigcirc\\ \text{H}\diagup \;\; | \;\;\diagdown\text{H}\\ \text{H}\end{array}$$

The position of the H atom along the hydrogen bond has been studied by p.m.r.:[3]

$$\text{N} \underline{\qquad} \text{--H--------F}$$
$$\quad 1.08\,Å \qquad 1.54\,Å$$

Each hydrazinium ion forms six octahedral N–H–F bonds and every F^- ion forms three such bonds (to different $N_2H_6^{2+}$ ions) arranged pyramidally. This is exactly the same bond arrangement as in the $CdCl_2$ or CdI_2 structures. In $N_2H_6F_2$ the halogen atoms are not close-packed as in $CdCl_2$ or CdI_2. The layers have a very open structure (F–F = 4.43 Å) owing to the directed N–H–F bonds (2.62 Å, compare 2.68 Å in NH$_4$F and 2.76 Å in NH$_4$HF$_2$). The mode of packing of the layers on one another is geometrically similar to that in the dihalides with layer structures. In these a halogen atom of one layer rests on three of the layer below, so that throughout the structure the halogen atoms are close-packed. In N$_2$H$_6$F$_2$ the layers are superposed in a similar way, but owing to the large separation of F atoms in a layer, a F atom of one layer (shown as a large open circle in Fig. 8.5) drops between

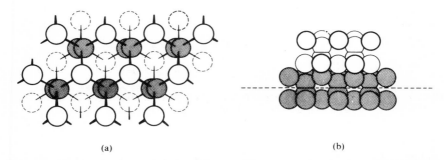

(a)   (b)

FIG. 8.5. The structure of hydrazinium fluoride N$_2$H$_6$F$_2$ (after Kronberg and Harker): (a) Plan of one layer. The lower N atoms of each N$_2$H$_6^{2+}$ ion (shaded) are slightly displaced. The open circles, full and dotted lines, represent F$^-$ ions respectively above and below the plane of the paper. All lines are N–H–F bonds. (b) Elevation, showing the packing of the layers. The broken line is the trace of the plane of the paper in (a).

the three F atoms of the next layer and comes in contact with a N atom (and/or its three associated H atoms). This interlayer F–N distance is 2.80 Å, and the F–F, 3.38 Å. It is estimated that the charge on the N$_2$H$_6^{2+}$ ion is about equally distributed over all the atoms, giving a charge of $+\frac{1}{4}$ on each. The interlayer bonds are therefore weak ionic bonds between F$^-$ ions and N (and/or H) atoms of adjacent layers. The weakness of these bonds accounts for the very good cleavage parallel to the layers.

Just as NH$_4$F and NH$_4$Cl have different structures owing to the fact that F but not Cl can form strong hydrogen bonds, so the corresponding chloride, N$_2$H$_6$Cl$_2$, has a different type of structure from N$_2$H$_6$F$_2$. It forms a somewhat deformed fluorite structure in which each N$_2$H$_6^{2+}$ ion is surrounded by 8 Cl$^-$ and each Cl$^-$ by 4 N$_2$H$_6^{2+}$ ions.[4]

(1)   AC 1970 **B26** 1635   (3)   TFS 1954 **50** 560
(2)   JCP 1942 **10** 309   (4)   JCP 1947 **15** 115

Bifluorides (acid fluorides), MHF$_2$

The crystal structures of the bifluorides MHF$_2$ of the alkali metals, NH$_4$, and Tl(i) are closely related to the NaCl and CsCl structures, while the structure of NH$_4$HF$_2$

differs in an interesting way from that of the ordinary form of KHF_2 owing to the formation of N–H–F bonds in the ammonium salt.

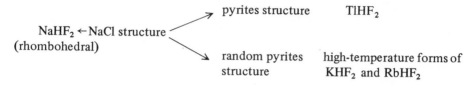

All (F–H–F)⁻
ions parallel

(F–H–F)⁻ *ions in 4 orientations parallel to the body-diagonals of the cube*

pyrites structure $TlHF_2$

$NaHF_2$ ←NaCl structure
(rhombohedral)

random pyrites   high-temperature forms of
structure KHF_2 and $RbHF_2$

The crystal structures of bifluorides MHF_2 are as follows:

M = Li   rhombohedral $NaHF_2$ structure[a]
Na   rhombohedral (p. 245)[b]
K   low temperature, tetragonal superstructure of CsCl type[c]
    high (above 196 °C), cubic (random pyrites)
Rb   low, tetragonal KHF_2 structure[d]
    high, cubic (random pyrites)
Cs   low, tetragonal[d]
    high, cubic (random CsCl, F–H–F along cube edges)
NH_4 Fig. 8.6(a)[e]
Tl(I)   pyrites structure[f] (no transition to lower symmetry).

(a) AC 1962 **15** 286	(d) JACS 1956 **78** 4256
(b) JCP 1963 **39** 2677; (n.d.)	(e) AC 1960 **13** 113
(c) JCP 1964 **40** 402 (n.d.)	(f) ZaC 1930 **191** 36

The structure of the ordinary (tetragonal) form of KHF_2 is illustrated in Fig. 8.6(b). Potassium and ammonium salts are often isostructural, the K^+ and NH_4^+ ions having very similar sizes. The structures of KHF_2 and NH_4HF_2 are both superstructures of the CsCl type, but the HF_2^- ions are oriented differently in the two structures (Fig. 8.6). In KHF_2 each K^+ ion is surrounded by eight equidistant F^- neighbours, but in NH_4HF_2 the NH_4^+ ion has only four nearest F^- neighbours (at 2.80 Å) and four at greater distances (3.02 and 3.40 Å). The breakdown of the coordination group of eight into two sets of four is to be attributed to the formation of N–H–F bonds.

The F–H–F distance, 2.27 Å, may be compared with H–F, 0.92 Å, in the HF molecule.

Other fluorides MH_nF_{n+1}

In addition to KHF_2 at least four other 'acid' fluorides of potassium have been prepared, namely, KH_2F_3, KH_3F_4, KH_4F_5, and $K_2H_5F_7$. An X-ray study of KH_2F_3[1]

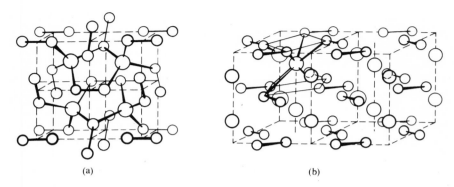

FIG. 8.6.   The crystal structures of (a) NH_4HF_2 and (b) KHF_2. Hydrogen atoms are not shown. The small circles represent fluorine atoms, joined in pairs by F–H–F bonds.

shows that the anion has the structure (a) with F–H–F slightly larger than in $(F-H-F)^-$ and a bond angle of 135° as compared with 120° in solid HF. In KH_4F_5

there are $(H_4F_5)^-$ ions consisting of a tetrahedral group (b) of four F atoms surrounding a central one, with F–H–F, 2.45 Å.[2]

(1)   AC 1963 **16** 58   (2)   JSSC 1970 **1** 386

Other ions $(X-H-X)^-$

We have already noted in our discussion of the hydrogen bond the formation of ions of this general type in salts $KH(RCOO)_2$, and we shall meet the ion $[H(CO_3)_2]^-$ in the later section on the structures of acid salts. With the exception of the stable bifluoride ion, which we have discussed, ions of this type, where X is Cl, Br, I, NCS, or NO_3, are stable only in the presence of large cations as in the salt $[P(C_6H_5)_4](NO_3 . H . NO_3)$.[1] Salts such as $[N(CH_3)_4]HCl_2$, $[N(C_2H_5)_4]HBr_2$, and $[N(C_4H_9)_4]HI_2$ are stable at room temperature, but all tend to lose HX rather readily and some are very unstable towards moisture and oxygen. The salt $CsHCl_2$ has been prepared but it is stable only at very low temperatures or under high HCl pressure, and salts prepared at room temperature from saturated aqueous CsCl and HCl are not $CsHCl_2$ as originally thought. One is probably $CsCl . \frac{3}{4}(H_3OCl)$,[2] and another is $CsCl . \frac{1}{3}(H_3O^+ . HCl_2^-)$. An X-ray study of the latter,[3] and of the

$$\begin{array}{c} \text{2·92 Å} \quad \text{O} \\ Cl_{110°}^{\cdots} \overset{|}{\underset{Cl}{\diagdown}} \cdots Cl \end{array}$$

isostructural bromide, shows pyramidal hydronium ions and linear ions $(Cl–H–Cl)^-$ or $(Br–H–Br)^-$, of overall length 3.14 Å and 3.35 Å respectively, but a n.d. study is required to locate the H atoms. Infrared and Raman studies of salts of ions such as $(Cl–H–Cl)^-$ suggest that their structures are very sensitive to their environment, and a final statement cannot yet be made on the structures of these ions.[4]

(1) JCS A 1966 1185
(2) IC 1963 2 657
(3) IC 1968 7 594
(4) JPC 1966 70 11, 20, 543

Acids and acid salts

Acids

For a discussion of their structures it is convenient to group the inorganic acids into the following classes:

(a) Acids H_mX_n. The most important of these (hydrogen halides and H_2S) are gases at ordinary temperatures and others are liquids (HN_3, H_2S_n, H_2O_2). The structures of the halogen acids HX and also H_2S have already been noted; the structures of other acids in this class are described under the chemistry of the appropriate element X.

(b) Polynuclear oxy-acids containing two or more directly bonded X atoms are few in number. They include $H_2N_2O_2$, $H_4P_2O_6$ (both crystalline solids), and the polythionic acids $H_2S_nO_6$ ($n = 3$–6), which are known only in aqueous solution. There are other oxy-*ions* of this type (e.g. $S_2O_3^{2-}$ and $S_2O_5^{2-}$) but we restrict our examples here to cases in which the free acid can be isolated.

(c) Per-acids. These contain the system O–O and are described after hydrogen peroxide in Chapter 11. (The description of certain ortho-acids as per-acids is regrettable but firmly established, for example, perchloric acid, $HClO_4$, and permanganic acid, $HMnO_4$.)

(d) Oxy-acids H_mXO_n. This large group contains all the simple oxy-acids. The molecule in the vapour or in the crystalline acid consists of a single X atom bonded to a number of O atoms, to one or more of which H atoms are bonded. In aqueous solution and in crystalline salts some or all of the H atoms are removed, leaving the oxy-ion. When discussing the structures of the crystalline acids it is convenient to distinguish as a subgroup the hydroxy-acids $X(OH)_n$ and to describe them under hydroxides (Chapter 14); examples include the crystalline H_3BO_3 and H_6TeO_6 – note that H_3PO_3 is not of this type. Inasmuch as all the H atoms of an oxy-acid H_mXO_n (other than certain acids of P where some H are directly bonded to P) are associated with O atoms to form OH groups (i.e. $XO_{n-m}(OH)_m$) these compounds are formally analogous to the oxyhydroxides of metals described in Chapter 14. The structural difference lies in the fact that the non-metal compounds form finite molecules in contrast to the infinite systems of metal atoms linked through M–O–M bonds in the metal oxyhydroxides.

Some oxy-acids are liquids at ordinary temperatures (HNO_3, H_2SO_4, $HClO_4$), many are solids (H_3PO_2, H_3PO_3, H_3PO_4, H_2SeO_3, H_2SeO_4, HIO_3, HIO_4, and H_5IO_6), and a further set of less stable acids are known only in aqueous solution (H_2CO_3, HNO_2, H_2SO_3, HFO, HClO, HBrO, HIO, $HClO_2$, $HClO_3$, $HBrO_3$). Replacement of OH in certain oxy-acids gives substituted acids such as HSO_3F and HSO_3NH_2.

(e) Pyro- and meta-acids. Complex oxy-ions formed from two or more XO_n groups sharing O atoms are numerous. Ions formed in this way from two ortho-ions are termed pyro-ions, and a few more complex ions of the same general type are known (for example, $P_3O_{10}^{5-}$, $S_3O_{10}^{2-}$, $P_4O_{13}^{6-}$). The sharing of two O atoms of every XO_n group leads to either cyclic or infinite linear meta-ions. Among the relatively few acids of this class which can be obtained as pure crystalline solids are metaboric acid, HBO_2 (three polymorphs) and $H_2S_2O_7$; HPO_3 and $H_4P_2O_7$ tend to form glasses. Of this same general type are the much more complex iso- and hetero-polyacids formed by metals such as Mo and W; they are described in Chapter 11.

Some of the inorganic acids, for example, sulphuric acid, were among the compounds known to the earliest chemists, but comparatively little is known even today of the structures of the inorganic acids as a class. This is largely due to the facts that many are liquids at ordinary temperatures or are too unstable to be isolated. However, the extension of X-ray crystallographic studies to lower temperatures has made possible the determination of the structures of a number of acids and their hydrates which are liquid at room temperature (for example, sulphuric acid, nitric acid and its hydrates). The spectroscopy of molecules trapped in inert matrices at low temperatures offers a means of studying the structures of acids which are too unstable to isolate under ordinary conditions (for example, HNO_2, p. 816). We deal in this chapter only with crystalline anhydrous acids; the hydrates of acids are discussed with other hydrates in Chapter 15. The structures of acids such as H_2S, HN_3, HNO_3, HNCS, and HNCO, which have been studied in the vapour state, are described in other chapters.

Acid salts

These are crystalline compounds intermediate in composition between the acid and a normal salt in which there is a system of hydrogen-bonded anions, these ions being either simple (F^-) or complex (XO_n). Apart from salts containing ions $(X–H–X)^-$ and the more complex ions formed by F which have already been discussed, anhydrous acid salts are formed only by oxy-acids. In a hydrated acid salt water molecules are incorporated *between* the hydrogen-bonded anion complexes, and their presence does not, as far as is known, affect the basic principle underlying the structures of these salts, namely, that the H atoms are associated exclusively with the oxy-ions to form the complex acid-salt anion. Since a crystalline oxy-acid also consists of an array of hydrogen-bonded oxy-ions the structural possibilities are exactly the same for the acid as for an acid-salt anion $H_x(XO_n)_y$

having the same $H:XO_n$ *ratio.* (In the acid salt the cations are accommodated between the hydrogen-bonded anions.)

The structures of acid salts and crystalline oxy-acids

The connectedness of the system of hydrogen-bonded anions in a crystalline oxy-acid or an acid salt is clearly *twice* the $H:XO_n$ ratio since each hydrogen bond connects two O atoms of different anions. Therefore the same topological possibilities are presented by pairs of compounds such as a substituted phosphoric acid $H(PO_4R_2)$ and an acid salt $KHSO_4$ or by sulphamic acid, which behaves as SO_3NH_3, and $(NH_4)_2H_3IO_6$. For the first pair of compounds there is a ratio of one H to one anion, so that only cyclic systems or infinite chains are possible. For the second pair the ratio is $3:1$, and with six hydrogen bonds from each anion to its neighbours the possible structures for the hydrogen-bonded system are 2- and 3-dimensional 6-connected nets.

The possible types of hydrogen-bonded anion complex are set out in Table 8.5, arranged in order of increasing $H:XO_n$ ratio.

(a) $H:XO_n$ *ratio* $1:2$. The only possibility is the linking of the XO_n ions in pairs as in $Na_3H(SO_4)_2$ and sodium sesquicarbonate, $Na_3H(CO_3)_2 . 2H_2O$ (Fig. 8.7).

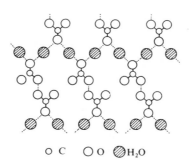

o C   ○ O   ⊘ H₂O

FIG. 8.7. Layer (diagrammatic) of the structure of $Na_3H(CO_3)_2 . 2H_2O$ showing CO_3^{2-} ions joined in pairs by short O–H–O bonds and to water molecules by long O–H–O bonds. The Na^+ ions are omitted.

In $Na_3H(SO_4)_2$ short unsymmetrical O–H–O bonds (O–H–O, 2.43 Å; O–H, 1.03 Å) link the SO_4 groups in pairs. In the sesquicarbonate the hydrogen-bond length within the pairs was determined as 2.53 Å; longer bonds (2.75 Å) link the pairs to water molecules. The pairs of hydrogen-bonded anions in salts such as potassium hydrogen phenylacetate have already been mentioned.

(b) $H:XO_n$ *ratio* $2:3$. This leads to a hybrid system in which all the XO_n ions are not equivalent, unlike all the other cases in Table 8.5. It is included because an

TABLE 8.5

Hydrogen-bonded anion complexes in acid salts and crystalline oxy-acids

	H : XO_n ratio	Number of O—H—O bonds from each anion	Type of $H_x(XO_n)_y$ complex	Examples — Acid salts	Examples — Acids
(a)	1 : 2	1	Dimer	$Na_3H(SO_4)_2$ [1] $Na_3H(CO_3)_2 \cdot 2H_2O$ [2]	
(b)	2 : 3	1 and 2	Trimer	$NH_4H_2(NO_3)_3$ [3]	
(c)	1 : 1	2	Cyclic dimer	$KHSO_4$, [4a] $KHCO_3$, [4b] $SnHPO_4$ [5]	$H[PO_2(OCH_2C_6H_5)_2]$ [12]
			1-dimensional	$NaHCO_3$ [6a] NH_4HSO_4 [6b]	α-HIO_3 [13] $H[SeO_2(C_6H_5)]$ [14] $(CH_3)_2PO \cdot OH$ [15] $HPO_2(OC_6H_4Cl)_2$ [16]
(d)	3 : 2	3	1-dimensional 2-dimensional 3-dimensional	$LiH_3(SeO_3)_2$ [7] $NaH_3(SeO_3)_2$ [8a] $RbH_3(SeO_3)_2$ [8b] $NH_4H_3(SeO_3)_2$ [8c]	
(e)	2 : 1	4	1-dimensional 2-dimensional	$(NH_4)_2H_2P_2O_6$ [9]	(β-oxalic acid) H_2SeO_3 [17] H_2SO_4 [18] H_2SeO_4 [19] (α-oxalic acid)
			3-dimensional	KH_2PO_4 [10]	
(f)	3 : 1	6	2-dimensional 3-dimensional	$(NH_4)_2H_3IO_6$ [11]	H_3PO_4 [20] $^-O_3S \cdot NH_3^+$ [21]

(1) AC 1979 **B35** 525. (2) AC 1956 9 82 (n.d.); AC 1962 **15** 1310 (p.m.r.). (3) AC 1950 **3** 305. (4a) AC 1975 **B31** 302. (4b) AC 1974 **B30** 1155. (5) AC 1976 **B32** 3309. (6a) AC 1965 **18** 818. (6b) AC 1971 **B27** 272. (7) JSSC 1972 **4** 255. (8a) AC 1977 **B33** 2108. (8b) JSSC 1973 **6** 254. (8c) AC 1974 **B30** 2497. (9) AC 1964 **17** 1352. (10) PRS 1955 **A230** 359. (11) AC 1980 **B36** 1028. (12) AC 1956 9 327. (13) AC 1949 **2** 128. (14) AC 1954 7 833. (15) AC 1967 **22** 678. (16) AC 1964 **17** 1097. (17) JCS 1949 1282. (18) AC 1965 **18** 827. (19) JCS 1951 968. (20) ACSc 1955 **9** 1557. (21) AC 1960 **13** 320.

example is found in $NH_4H_2(NO_3)_3$, an acid salt in which the nitrate ions are joined into linear groups of three by hydrogen bonds.

(c) H : XO_n *ratio* 1 : 1. Here the XO_n ions may be linked into cyclic systems or infinite chains. There are interesting differences between chemically similar salts. There are only (centrosymmetrical) dimers in $KHCO_3$, only chains in $NaHCO_3$ and NH_4HSO_4, but both chains and dimers in $KHSO_4$, (i). There is disorder in the O—H(D)—O bonds in $KHCO_3$ (2.59 Å) and $KDCO_3$ (2.61 Å), studied by neutron diffraction, but in $KHSO_4$ the S—O bond lengths indicate that H is attached to certain O atoms (details shown at (ii)). Dimers are also formed in $SnHPO_4$.

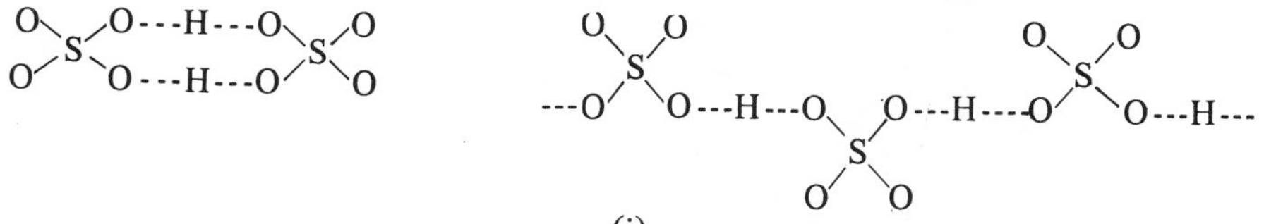

(i)

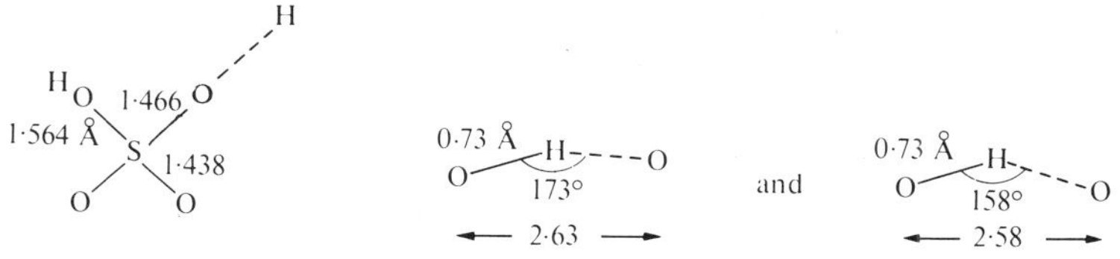

(ii)

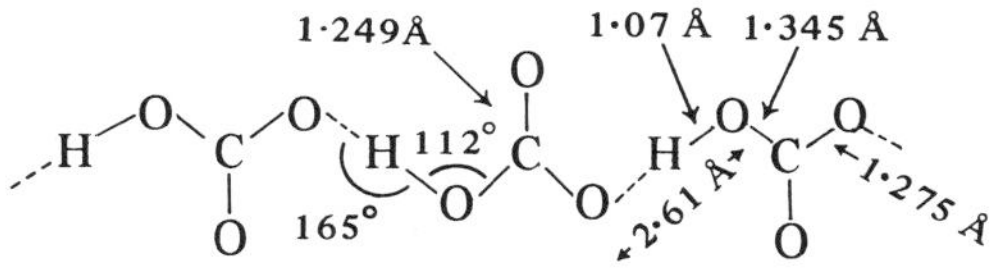

(iii)

In the bicarbonate ion in $NaHCO_3$ there are infinite chains of composition HCO_3^- which are held together laterally by the Na^+ ions. The H atoms have been located; the O—H—O bonds are unsymmetrical, showing that the linear anion is an assembly of HCO_3^- ions as shown at (iii).

Acids also provide examples of hydrogen bonding schemes of the same kinds. In α-HIO_3 the scheme is essentially similar to that in $NaHCO_3$. The distances between O atoms of different IO_3^- ions were originally interpreted in terms of 'bifurcated' hydrogen bonds. However, $I^{(V)}$ here exhibits the (3+3)-coordination characteristic of many lone-pair elements, and the three weaker bonds are shown (for one I atom) as dotted lines in Fig. 8.8. If we disregard those short 'intermolecular' O—O distances which are edges of the distorted octahedral IO_6 coordination groups, then each IO_3 group is joined to its neighbours by only two O—H—O bonds forming chains, as in the bicarbonate ion. A neutron diffraction study confirms that the O—H—O bonds (length 2.68 Å) are normal straight hydrogen bonds.

Disubstituted phosphoric acids, $HPO_2(OR)_2$, provide examples of both dimers and chains exactly comparable with the two types of HSO_4^- ion, and infinite chains also occur in crystalline benzene seleninic acid. The dimer in dibenzylphosphoric acid, $HPO_2(OCH_2C_6H_5)_2$, is shown at (iv) and the chain in $HSeO_2(C_6H_5)$ at (v). Very similar chains are found in dimethylphosphinic acid, $(CH_3)_2PO(OH)$ and in $HPO_2(OC_6H_4Cl)_2$ (see Table 8.5). This class also includes carboxylic acids, in

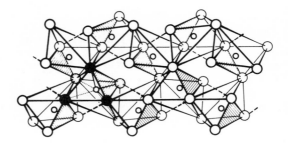

FIG. 8.8. The crystal structure of α-iodic acid shown as a system of linked IO_6 octahedra, or IO_3 groups (shaded) joined by normal hydrogen bonds (broken lines). The smaller circles represent I atoms.

which the ratio of H atoms to R.COO is necessarily unity. The dimers present in the vapours and crystals of these compounds are of the same general type as the $(HSO_4^-)_2$ dimer shown at (i).

$$\overset{\longleftarrow 2\cdot494\ \text{Å}\longrightarrow}{\begin{array}{ccc} RO\diagdown & O\text{-----}H\text{------}O \diagdown & OR \\ \diagup P & & P \diagup \\ RO\diagup & O\text{------}H\text{-----}O & OR \end{array}}$$

(iv)

(v)

(d) H:XO_n *ratio* 3:2. There are now three hydrogen bonds from each oxy-ion, making possible all kinds of 3-connected systems. Our examples are the structures of the alkali metal hydrogen selenites, $MH_3(SeO_3)_2$, in which there has been considerable interest in recent years because some of these salts exhibit ferro-electricity. This is a property of many complex oxides (e.g. non-cubic perovskites) and hydrogen-bonded structures such as tartrates (Rochelle salt) and dihydrogen phosphates (KH_2PO_4, $NH_4H_2PO_4$). In hydrogen-bonded ferroelectrics the interest lies in the positions and behaviour of the H atoms, but here we wish to comment

only on the 'heavy atom' topology of the structures. Each SeO_3 group is bonded to others by hydrogen bonds, and the alkali salts provide examples of three kinds of 3-connected systems. In $LiH_3(SeO_3)_2$ there are chains, (vi), in which each SeO_3 is bonded to *two* others, by a single and a double bridge:

$$
\text{---H---O---Se}\Big\langle{}^{\text{O---H---O}}_{\text{O---H---O}}\Big\rangle\text{Se---O---H---O---Se}\Big\langle{}^{\text{O---H---O}}_{\text{O---H---O}}\Big\rangle\text{Se---O---}
$$

(vi)

— ⬤〈 〉⬤ — ⬤〈 〉⬤ —

(vii)

The chain may be represented by the topological diagram (vii). In $NaH_3(SeO_3)_2$ each SeO_3 is hydrogen-bonded to *three* others to form the simplest 3-connected planar net (6^3). In the K, Rb, and NH_4 salts there is a 3-connected 3D framework $H_3(SeO_3)_2$, which in the Rb and NH_4 salts is a distorted form of the tetragonal (10,3)-b net (p. 113). The lengths of the Se—O bonds show that these compounds may be formulated $M^+(H_2SeO_3)(HSeO_3)^-$, the Se—O and Se—OH bond lengths being close to 1.66 and 1.75 Å respectively.

(e) $H:XO_n$ *ratio* 2:1. This ratio implies that four hydrogen bonds link each oxy-ion to its neighbours. We therefore have the possibility of all types of hydrogen-bonded complex, from finite to 3-dimensional, as in class (d). Acid salts provide examples of 1- and 3-dimensional complexes, but layer structures do not appear to be known in this class. In $(NH_4)_2H_2P_2O_6$ short O—H—O bonds (2.53 Å) link the anions into chains (Fig. 8.9) which are then further linked into a 3D framework

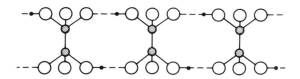

FIG. 8.9. Hydrogen bonded chain $(H_2P_2O_6)_n^{2n-}$ in $(NH_4)_2H_2P_2O_6$.

by N—H—O bonds. In Chapter 3, in our discussion of the diamond structure, we noted that the anion in KH_2PO_4 is a 3D array of hydrogen-bonded $PO_2(OH)_2^-$ ions (Fig. 8.10).

Inorganic acids provide a number of examples of layer structures. The layers in H_2SeO_3, H_2SeO_4, and H_2SO_4 are all of the same topological type, being based on the simplest 4-connected plane net, but whereas in acids H_2XO_4 (Fig. 8.11(a)) there

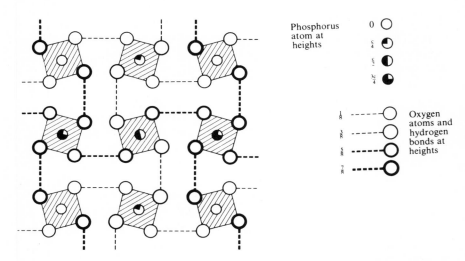

FIG. 8.10. Part of the structure of KH_2PO_4 projected on to the basal plane (K^+ ions omitted). The broken lines represent hydrogen bonds which link the PO_4 tetrahedra into an infinite 3-dimensional network.

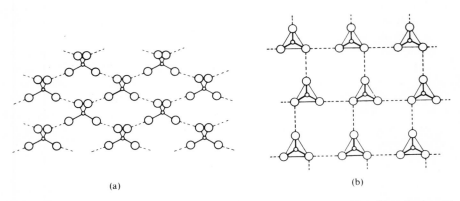

(a)   (b)

FIG. 8.11. Hydrogen bonding schemes in layers of (a) crystalline H_2SO_4; (b) orthorhombic H_2SeO_3 (diagrammatic). Broken lines indicate O–H–O bonds.

is one hydrogen bond from each O atom, in H_2SeO_3 (with a smaller number of O atoms) there are two hydrogen bonds from one O and one from each of the others (Fig. 8.11(b)). This illustrates the point that while the $H:XO_n$ ratio in these crystals determines the general type of structure it is necessary to take account of the nature of the XO_n group if the structure is to be discussed in more detail.

Oxalic acid $(COOH)_2$, has two polymorphs, one of which has a layer structure and the other a chain structure:

α-oxalic acid

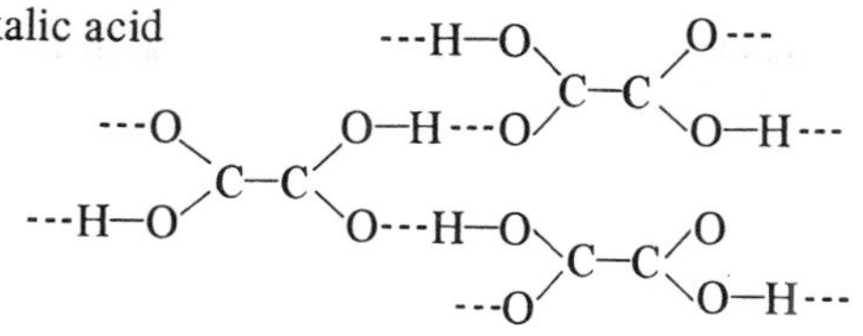

β-oxalic acid

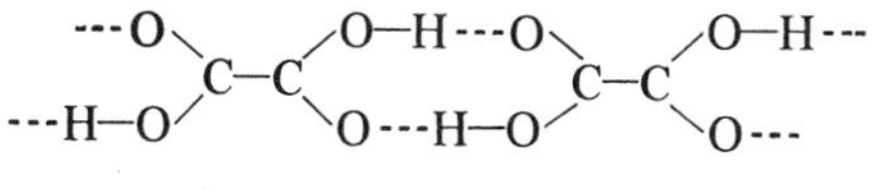

(f) H:XO$_n$ *ratio* 3:1. Crystalline H_3PO_4 has a layer structure. Within each layer each PO(OH)$_3$ molecule is linked to six others by hydrogen bonds which are of two lengths. Those between O and OH are short (2.53 Å) while those between OH groups of different molecules are longer (2.84 Å), as shown in Fig. 8.12. The salt

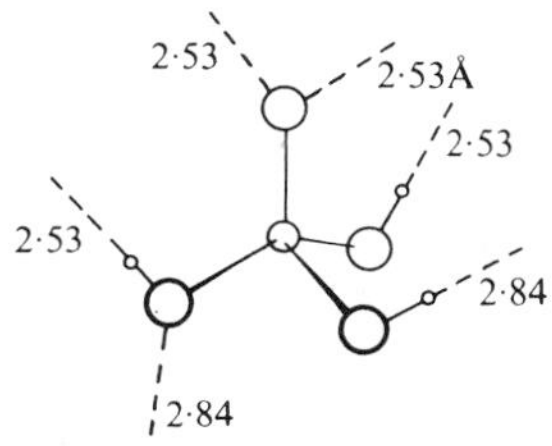

FIG. 8.12. The hydrogen bonds in crystalline H_3PO_4.

$(NH_4)_2H_3IO_6$ consists of a 3D framework in which each (regular octahedral) IO$_6$ groups is hydrogen-bonded to 6 others by disordered O–H–O bonds (2.64 Å). Each O also forms a O---H–N bond (2.85 Å) to a NH$_4^+$ ion, one H of which is not involved in hydrogen-bonding. The 3 hydrogen bonds from each NH$_4^+$ link the IO$_6$ groups into layers perpendicular to the 3-fold axis of the rhombohedral structure, and the O–H–O bonds operate between the 'layers'. Sulphamic acid, which apparently behaves as a zwitterion, $^-O_3S.NH_3^+$, in the crystal, is an example of an acid in which each unit is hydrogen-bonded to six others.

As already noted, the structures of hydroxy-acids, which include B(OH)$_3$ and Te(OH)$_6$, are described in Chapter 14 with hydroxides, to which they are more closely related structurally, the molecules being linked by long O–H–O bonds (two from each OH) as in many organic hydroxy-compounds. One form of metaboric acid, consisting of cyclic $B_3O_3(OH)_3$ molecules, is also of this type, for the three O atoms in the ring are not involved in hydrogen bonding.

We have confined our attention here to the structures of crystalline oxy-acids and 'acid salts', but other types of compound present similar possibilities as regards hydrogen bond systems. Two of the examples of Table 8.5 illustrate 3D systems of hydrogen-bonded ions, forming the simplest 4- and 6-connected nets. The 'hydrous oxides' of Sn and Pb extend the series with an example of a 3D system based on the 8-connected body-centred net:

	3D net
$(H_2PO_4)_n^{n-}$	Diamond
$(H_3IO_6)_n^{2n-}$	Simple cubic
$Sn_6O_4(OH)_4$	Body-centred net

COMPOUNDS OF THE NOBLE GASES

The existence of dimeric noble gas ions was first demonstrated in 1936 in the mass spectrometer, and there have been numerous later studies; they are formed by collisions involving an excited atom and a neutral atom $(X^* + X \rightarrow X_2^+ + e)$. Other unstable species have been shown to exist in discharge tubes, for example, ArO, KrO, XeO (band spectrum), $XeCl_4^-$ in decay of ^{129}I to ^{129}Xe (Mössbauer effect), and XeF in a γ-irradiated crystal of XeF_4 (e.s.r.). The existence of van der Waals complexes ArFH and ArClH has been demonstrated by mass spectrometry of beams made by expansion of Ar:HX mixtures through an ultrasonic nozzle, and the distances Ar–X in the angular complexes (Ar–X–H, 44°) have been determined as 3.54 and 4.01 Å respectively. Stable compounds are formed only with the most electronegative elements, F and O, and those of Xe are the most numerous. The first was prepared in 1962, the orange-yellow $XePtF_6$ formed directly from Xe and gaseous PtF_6 at room temperature. Krypton compounds include the crystalline KrF_2 and a number of adducts containing KrF^+ and $Kr_2F_3^+$ ions which are noted on p. 382. The preparation of KrF_4 has been described, but this compound does not seem to have been well characterized and its structure is not known.

We shall set out here the structures of the compounds of Xe without enlarging on their stereochemistry. The close analogy with the structural chemistry of iodine will be evident, for the bond arrangements are for the most part consistent with the simple view of the stereochemistry of non-transition elements set out in Chapter 7. It will be equally evident that the difficulties encountered with certain valence groups in connection with the stereochemistries of Sb, Te, and I are also encountered here.

Fluorides of xenon

These are prepared by direct combination of the elements under various conditions. They include the simple fluorides XeF_2, XeF_4, and XeF_6, which are all colourless crystalline solids with m.p. 140, 114, and 46 °C respectively. There is no reliable evidence for an octafluoride, but there is a molecular compound $XeF_2.XeF_4$. The fluorides form numerous adducts with other halides, for example,

	$XeF_2.AsF_5$, $2XeF_2.AsF_5$	
$XeF_2.MF_5$,	$XeF_2.2MF_5$, and $XeF_6.MF_5$	(M = Pt, Ir, etc.)

which form because the pentafluorides can abstract one F atom from the Xe fluoride to form cations as shown in the structural formulae:

$$(XeF)^+(M_2F_{11})^-, (Xe_2F_3)^+(AsF_6)^-, (XeF_5)^+(PtF_6)^-.$$

(On the formulation of the $(XeF)^+$ ion, see later.) Complex halides formed by XeF_6 include $RbXeF_7$, Rb_2XeF_8, and the corresponding Cs salts.

The structures of the XeF_2 and XeF_4 molecules are similar to those of the corresponding halogen ions ICl_2^- and ICl_4^-, namely, linear and square planar. In $XeF_2.XeF_4$ there are equal numbers of molecules of the two fluorides with structures the same as in the individual compounds. The same linear configuration for XeF_2 is found in $XeF_2.IF_5$, the shape of the IF_5 molecule being square pyramidal with I–F, 1.88 Å and bond angle to the axial I atom, 81°.

In contrast to XeF_2 and XeF_4 the hexafluoride has presented great difficulty. The latest electron diffraction data are not consistent with a regular octahedral structure or with the pentagonal bipyramidal model to be expected for a 14-electron valence group. It has been suggested that an irregular octahedral model with large amplitude of the bending vibrations and mean length of the (non-equivalent) bonds 1.89 Å is consistent with the data.

There are four crystalline forms of XeF_6, those formed at lower temperatures having increasingly complex structures, none of which contains discrete XeF_6 molecules:

IV	(-180°)	III	-25°	II	+10°C	I
cubic		monoclinic		orthorhombic		monoclinic
Z 144		64		16		8

All the forms I, II, and III consist of tetramers. The change from I to II involves a gross rearrangement of one-half of the tetramers, while the change from II to III consists only in the ordering of the right- and left-handed forms of the tetramer. In the more complex cubic form IV there are two kinds of polymeric unit, both formed from F^- ions and square pyramidal XeF_5^- ions which are linked through the F^- ions by weaker Xe–F bonds into cyclic tetramers and hexamers, these numbering respectively 24 and 8 per unit cell. In the tetramer (Fig. 8.13(a)) each F^- forms an unsymmetrical bridge between two Xe atoms, and therefore each Xe has two additional neighbours, but in the hexamer (Fig. 8.13(b)) each F^- bridges (symmetrically) three Xe atoms. The environments of Xe are therefore as shown in Fig. 8.13(c) and (d):

Tetramer		*Hexamer*	
	5 F 1.84 Å		1 F 1.76 Å
			4 F 1.92
Xe $\{$	F 2.23	Xe $\{$	
	F 2.60		3 F 2.56

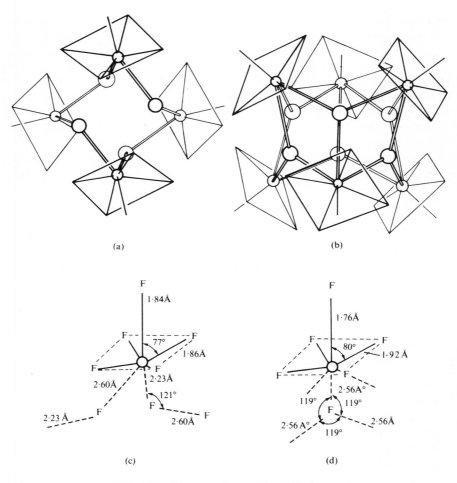

FIG. 8.13. Structure of crystalline XeF$_6$ (see text).

Fluoro-cations

In a number of adducts of XeF$_2$ Xe forms two collinear bonds and one F forms part of the coordination group of the metal atom. The terminal Xe–F bond is appreciably shorter than in XeF$_2$ (2.00 Å) and the bridging bond longer:

	Xe–F (terminal)	Xe–F (bridging)
F–Xe–F–RuF$_5$	1.87 Å	2.18 Å
F–Xe–F–WOF$_4$	1.89	2.04
F–Xe–F–Sb$_2$F$_{10}$	1.84	2.35

In F–Xe–OSO$_2$F also the terminal Xe–F bond is short (1.94 Å), but less so than in the above compounds, and the Xe–O bond is abnormally long (compare the other values in Table 8.6). The stuctures of these adducts suggest a formulation intermediate between a discrete molecule, for example, RuF$_6$XeF and an ionic compound (RuF$_6$)$^-$(XeF)$^+$.

The ion XeF$_3^+$ has a planar T shape like ClF$_3$, derived from a trigonal bipyramidal arrangement of 5 electron pairs. The bond lengths are 1.84 Å (equatorial) and 1.91 Å (apical), but there are additional weaker bonds of lengths 2.49 and 2.71 Å to F atoms of the anions (SbF$_6^-$ or Sb$_2$F$_{11}^-$).

The compound with the empirical formula 2XeF$_2$.AsF$_5$ contains planar ions (Fig. 8.14(a)) similar in shape to (H$_2$F$_3$)$^-$, in which the bond angle is 135°, and

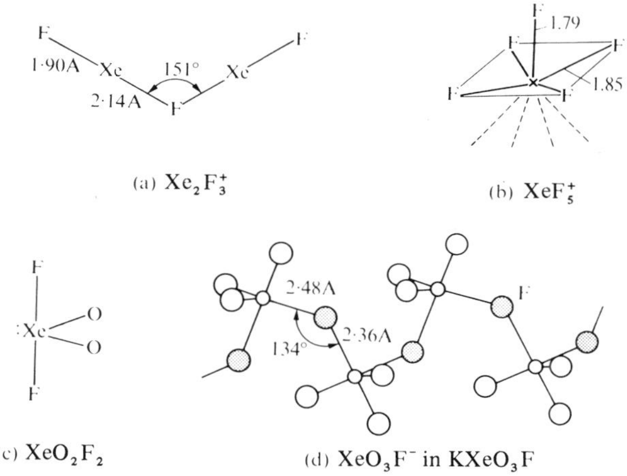

(a) Xe$_2$F$_3^+$ (b) XeF$_5^+$

(c) XeO$_2$F$_2$ (d) XeO$_3$F$^-$ in KXeO$_3$F

FIG. 8.14. Structures of compounds of xenon: (a) Xe$_2$F$_3^+$; (b) XeF$_5^+$; (c) XeO$_2$F$_2$; (d) XeO$_3$F$^-$.

I$_5^-$, bond angle 95°. The compound XeF$_4$.(SbF$_5$)$_2$ contains T-shaped ions XeF$_3^+$ similar in shape to ClF$_3$ and BrF$_3$. The salt (XeF$_5$)(PtF$_6$) is made by the action of fluorine under pressure at 200 °C on a mixture of Xe and PtF$_5$. It is an assembly of octahedral PtF$_6^-$ ions (mean Pt–F, 1.89 Å) and square pyramidal XeF$_5^+$ ions (Fig. 8.14(b)) in which the Xe atom is situated slightly below the base of the pyramid and the axial bond is a little shorter than the equatorial ones; the structure is very similar to that of IF$_5$ and TeF$_5^-$. Xe has four more distant F neighbours (broken lines) at 2.52, 2.65, and 2.95 Å (two); very similar distances are found in (XeF$_5$)(RuF$_6$). In a compound which has been formulated as (Xe$_2$F$_{11}$)$^+$(AuF$_6$)$^-$ two such square pyramids are joined through a F bridge and also associated with one AuF$_6$ group so that Xe forms 5 bonds of length 1.84 Å, one of 2.23, one of 2.64, and two of 3.40 Å. It is perhaps arguable whether the ion Xe$_2$F$_{11}^+$ should be recognized. Note that the above formulation as a compound of Xe(VI) implies that Au is in the oxidation state V (see p. 1096).

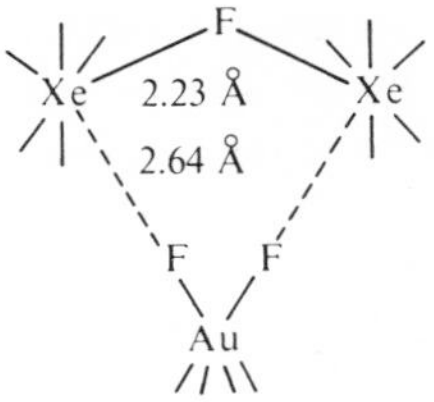

Xenon anions

Compounds containing two fluoro-anions have been described. The reaction of XeF_6 with an alkali fluoride (other than LiF) gives salts containing the XeF_7^- ion which decompose on heating to $XeF_6 + M_2XeF_8$. The anion XeF_8^{2-} has been studied in $(NO)_2^+(XeF_8)^{2-}$, formed from 2 NOF and XeF_6. It has the form of a very slightly deformed square antiprism; the mean values of the two sets of Xe—F distances to the F atoms at the corners of the two square faces are: Xe—4F, 1.96 Å; Xe—4F, 2.08 Å.

The linear anion in the Rb and Cs salts $M_9(XeO_3Cl_2)_4Cl$ is of special interest as the first example of a compound containing Xe—Cl bonds which is stable at ordinary temperatures. In the ion $(XeO_3Cl_2)_n^{2n-}$ there are nearly linear bridges between

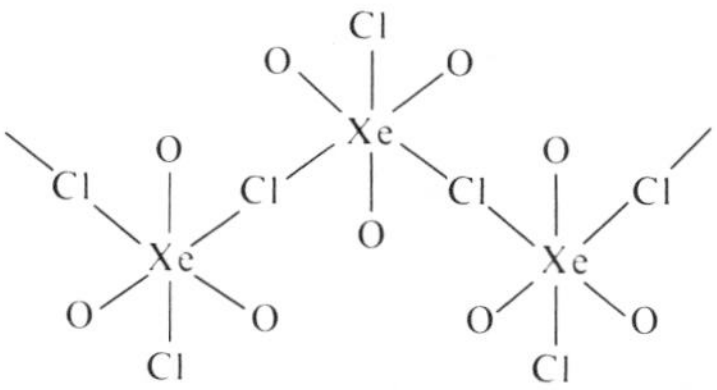

distorted octahedral groups XeO_3Cl_3 in which the bond lengths are: Xe—O, 1.77; Xe—Cl, 2.96 Å.

Oxides and oxy-ions

The trioxide forms colourless crystals which are stable in dry air but deliquesce in moist air and are highly explosive. The crystal structure is similar to that of α-HIO_3 (XeO_3 is isoelectronic with IO_3^-) except, of course, that there is no hydrogen bonding:

	M—O	O—M—O (mean)
XeO_3	1.76 Å	103°
IO_3^-	1.82	97°

The tetroxide is made by the action of H_2SO_4 on the perxenate, Na_4XeO_6. In the vapour state it forms a regular tetrahedral molecule, the Xe–O bond length (1.74 Å) being intermediate between the values found in XeO_3 (1.76 Å) and $XeOF_4$ (1.70 Å).

Xenates (containing Xe^{VI}) and perxenates (containing Xe^{VIII}) have been prepared. For example, $CsHXeO_4$ has been made from XeO_3 and CsOH in the presence of F^- ion, and hydrated salts containing the XeO_6^{4-} ion from XeF_6 and alkali hydroxides. The perxenate ion, of special interest because of the high oxidation state of Xe, forms a regular octahedron (Xe–O, 1.86 Å) in $Na_4XeO_6 \cdot 6H_2O$ and $8H_2O$ and in $K_4XeO_6 \cdot 9H_2O$.

Two other compounds containing Xe–O bonds may coveniently be included here. $Xe(OSeF_5)_2$ is the first compound to be prepared which contains Xe(II) bonded symmetrically to 2 O atoms

and $Xe(OTeF_5)_4$ is the only compound at present known in which Xe(IV) is bonded exclusively to O atoms.

Oxyfluorides and $KXeO_3F$

Three oxyfluorides are known. $XeOF_4$, a colourless liquid prepared from XeO_3 and XeF_6, shows a resemblance to halogen fluorides in forming compounds such as $CsF \cdot XeOF_4$ and $XeOF_4 \cdot 2SbF_5$. A microwave study indicates a square pyramidal configuration with Xe approximately in the basal plane, that is, essentially the same structure as XeF_5^-, one F being replaced by O.

The action of XeO_3 on $XeOF_4$ produces XeO_2F_2. Its molecular structure resembles that of $IO_2F_2^-$, that is, a trigonal bipyramid with a lone pair occupying one equatorial bond position (or the linear XeF_2 molecule plus two equatorial O atoms (Fig. 8.14(c)); bond angles are: O–Xe–O, 105.7°; F–Xe–F, 174.7°. Two weak additional equatorial bonds (Xe–O, 2.81 Å) from each Xe lead to a pronounced pseudo-octahedral layer of the SnF_4 type.

A very unstable oxyfluoride of Xe(VIII), XeO_3F_2, has been studied spectroscopically (matrix isolated) and assigned D_{3h} symmetry.

Closely related to these oxyfluorides is the salt $K(XeO_3F)$, which is prepared directly from CsF and XeO_3 in solution. In crsytals of this compound there are pyramidal XeO_3 units with the same structure as the molecule of XeO_3 itself:

	Xe–O	O–Xe–O
KXeO$_3$F	1.77 Å	100°
XeO$_3$	1.76	103°
	I–O	O–I–O
(compare IO$_3^-$	1.82	97°)

These XeO$_3$ units are joined through bridging F atoms to form infinite chains of composition XeO$_3$F (Fig. 8.14(d)). The Xe–F bonds are of two lengths, 2.36 and 2.48 Å, appreciably longer than in XeF$_4$ or XeF$_6$ (1.95–2.0 Å) but much shorter than would be expected for ionic or van der Waals contacts (of the order of 3.5 Å). The five bonds from Xe, three normal Xe–O bonds and two weak Xe–F bonds, are directed towards five of the vertices of an octahedron.

Compounds of krypton

Spectroscopic (i.r. and Raman) and electron diffraction studies have established the linear structure of the KrF$_2$ molecule. This fluoride forms adducts in which both the KrF$^+$ and Kr$_2$F$_3^+$ ions have been characterized by ^{19}F n.m.r. and Raman spectroscopy. The former include KrF.AsF$_6$ and KrF.Sb$_2$F$_{11}$ and the latter Kr$_2$F$_3$.AsF$_6$ and Kr$_2$F$_3$.SbF$_6$.

Structural details of the noble gas compounds and references to the literature are given in Table 8.6.

TABLE 8.6
Structures of noble gas compounds

	Configuration	Xe–F	Xe–O	*Reference*
XeF$_2$	Linear	1.98 Å		JCP 1969 **51** 2355 (i.r.)
		2.00		JACS 1963 **85** 241 (cryst.)
XeF$_4$	Square planar	1.95		NGC p. 238 (e.d.)*
		1.93		Sc 1963 **139** 1208 (n.d.)
XeF$_6$	Distorted octahedral (vapour)	1.89 (mean)		JCP 1968 **48** 2460, 2466
	Crystal (see text)			JACS 1974 **96** 43
XeF$_2$.XeF$_4$	See text:	XeF$_2$ 2.01		AC 1965 **18** 11
		XeF$_4$ 1.96		
		2.02		
XeF$_2$.IF$_5$		2.02		IC 1970 **9** 2264
FXeFWOF$_4$	Linear Xe bonds	1.89		AC 1975 **B31** 906
		2.04		
FXeFRuF$_5$	Linear Xe bonds	1.87		IC 1973 **12** 1717
		2.18		

	Configuration	Xe–F	Xe–O	Reference
XeF$_3^+$	Planar T	1.84 1.91		IC 1974 **13** 1690
Xe$_2$F$_3^+$	Linear Xe bonds	1.90 2.14		IC 1974 **13** 780
XeF$_5^+$	Square pyramidal	1.79 1.85		IC 1973 **12** 1717
Xe$_2$F$_{11}^+$	See text			IC 1974 **13** 775
XeF$_8^{2-}$	Square antiprism	1.96 2.08		Sc 1971 **173** 1238
Xe(OSeF$_5$)$_2$	Linear Xe bonds		2.12	IC 1976 **15** 2718
Xe(OTeF$_5$)$_4$				AnC(IE) 1978 **17** 356
FXeOSO$_2$F	Linear Xe bonds	1.94	2.16	IC 1972 **11** 1124
XeOF$_4$	Square pyramidal	1.90	1.70	JMS 1968 **26** 410
XeO$_2$F$_2$	Trigonal pyramidal	1.90	1.71	JCP 1973 **59** 453 (n.d.)
XeO$_3$F$_2$	See text			JCP 1971 **55** 1505
XeO$_3$F$^-$	See text	2.36 2.48	1.77	IC 1969 **8** 326
XeO$_3$	Pyramidal		1.76	JACS 1963 **85** 817
XeO$_4$	Tetrahedral		1.74	JCP 1970 **52** 812 (e.d.)
XeO$_6^{4-}$	Octahedral		1.86	IC 1964 **3** 1412, 1417; JACS 1964 **86** 3569
XeO$_3$Cl$_2^{2-}$	See text			JACS 1977 **99** 8202
XeCl$_2$	Linear			IC 1967 **6** 1758 (i.r.)
KrF$_2$	Linear	(Kr–F, 1.88)		JACS 1968 **90** 5690 (i.r.); JACS 1967 **89** 6466 (e.d.)
KrF$^+$, Kr$_2$F$_3^+$	See text			IC 1976 **15** 22
ArHCl, ArHF	See text			JCP 1973 **59** 2273; JCP 1974 **60** 3208

*NGC: Noble Gas Compounds, H. H. Hyman, Ed., University of Chicago Press, 1963. For references to earlier work see also: Sc 1964 **145** 773. For discussions of the bonding in noble gas compounds see JCS 1964 1442, and IC 1972 **11** 1124.

9

The halogens–Simple halides

Introduction

In every Periodic family the availability of only four orbitals differentiates the first member from the later ones which can make use of d in addition to s and p orbitals. We therefore find important differences between the first and later elements. In addition there are marked differences between iodine on the one hand and bromine and chlorine on the other; there is indeed little chemical similarity between the extreme members of this family, fluorine and iodine.

The first outstanding characteristic of fluorine is its extraordinary chemical reactivity. It combines directly with all non-metals except N and the noble gases (other than Kr and Xe) and with all metals. Moreover the product is (or includes) the highest fluoride of the element. For example, $AgCl \rightarrow AgF_2$, $CoCl_2 \rightarrow CoF_3$, and the platinum metals yield their highest known fluorides, RuF_5, OsF_6, IrF_6, and PtF_6. Many elements, particularly non-metals, exhibit higher oxidation states in their fluorides than in their other halides -- compare AsF_5, SF_6, IF_7, and OsF_6 (earlier described as OsF_8) with the highest chlorides, $AsCl_3$, SCl_4, ICl_3, and $OsCl_4$.

Fluorine seldom forms more than one essentially covalent bond -- see, for example, the interhalogen compounds and polyhalides -- though two collinear bonds (with appreciable ionic character) are formed in crystalline compounds MF_3, MF_5, etc. (see later) and in the ion $(C_2H_5)_3Al-F-Al(C_2H_5)_3^-$, which has been studied in the K salt.[1] Chlorine and bromine form up to four bonds to oxygen and exceed this number only in combination with fluorine (ClF_5, BrF_5, and BrF_6^-), while iodine is 6-covalent in some oxy-compounds and also forms a heptafluoride. An ion X_3^- is not formed by fluorine, and is most stable in the case of I_3^-. Another feature of fluorine is its small affinity for oxygen, a characteristic also of bromine and incidentally of Se, the element preceding Br in the Periodic Table.

Fluorine is the most electronegative of all the elements, so that the bonds it forms with most other elements have considerable ionic character. With the exception of the alkali halides most crystalline fluorides have structures different from those of the other halides of the same metal. A number of difluorides and dioxides have the same crystal structure, whereas the corresponding dichlorides, dibromides, and diiodides have in many cases structures similar to those of disulphides, diselenides, and ditellurides. The extreme electronegativity of fluorine enables it to form much stronger hydrogen bonds than any other element, resulting in the abnormal properties of HF as compared with the other acids HX, the much greater stability of the ion $(F-H-F)^-$ than of other ions HX_2^-, and structural differences between hydrated fluorides and other halides. The structures of the ions HX_2^-, $H_2F_3^-$, and $H_4F_5^-$ are described in Chapter 8.

The anomalously low dissociation energy of the F—F bond has usually been attributed to the repulsions of the unshared electrons of the two valence shells (Table 9.1); the extrapolated value on a $1/R$ plot, R being the bond length in the

TABLE 9.1
Some properties of the halogens

	Ionic radius	X—X in X_2	H—X in HX	Dissociation energy X—X	Ionization potential	Electron affinity
	(Å)	(Å)	(Å)	(kJ mol^{-1})	(eV)	(eV)
F	1·36	1·43	0·92	154·8	17·42	3·448
Cl	1·81	1·99	1·28	242·7	12·96	3·613
Br	1·95	2·28	1·41	192·5	11·81	3·363
I	2·16	2·66	1·60	150·6	10·45	3·063

X—X molecule, gives a value some $226\,kJ\,mol^{-1}$ higher. It has been pointed out[2] that there are similar anomalies in quantities involving only one F atom. For example, the plot of electron affinity against ionization potential is linear for I, Br, and Cl and would extrapolate to a value $1.14\,eV$ ($108.8\,kJ\,mol^{-1}$) greater than the observed value. There are similar anomalies in the dissociaton energies of covalent molecules such as HX or CH_3X or the gaseous alkali-halide molecules in which the bonds are ionic bonds. These facts suggest that an abnormally large repulsive force is experienced by an electron entering the outer shell *wherever it comes from*, being the same for an electron entering to form an ion as for one which is being shared in the formation of a covalent bond. When compared with twice the *atomic* radius (defined as the average radial distance from the nucleus of the outermost p electrons) it is seen that the bond in the F_2 molecule is abnormally long:

	F	Cl	Br	I
2 (atomic radius):	1.14	1.94	2.24	2.64 Å
X—X in X_2:	1.43	1.99	2.28	2.66 Å

We note later some cations containing exclusively halogen atoms. Iodine also forms ions $(R—I—R)^+$ in which the groups R are usually aromatic radicals (Ar). The halides (Ar—I—Ar)X are not very stable, but the hydroxides are strong bases, proving dissociation to $(Ar—I—Ar)^+$ ions. The iodonium compounds are the analogues of ammonium and sulphonium compounds:

$$(NR_4)X \qquad (SR_3)X \qquad (IR_2)X.$$

The expected angular shape of $(IR_2)^+$ has been demonstrated in crystalline $(C_6H_5)_2I^+(ICl_2)^-$, the angle between the iodine bonds in the cation being $94°$ and I—C $2.02\,Å$.[3] Ions $(R—I—R)^+$ in which R is bonded to I through a donor atom such as N, are linear like $(ICl_2)^-$, since 5 orbitals are used, three by lone pairs. Examples include $[(pyridine)_2I]^+I_7^-$ [4] and $[(HMT)_2I]^+I_3^-$ [5], where HMT stands for $(CH_2)_6N_4$. There is a similar trigonal bipyramidal arrangement of five electron pairs in $C_6H_5ICl_2$ and iodobenzene diacetate, (a), in which two lone pairs occupy equatorial positions.[6]

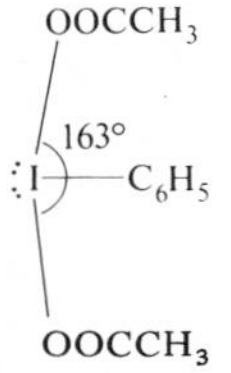

(a)

Other compounds peculiar to iodine include the iodoso and iodyl compounds of general formulae $Ar-I=O$ and $Ar-IO_2$ which undergo interesting reactions such as

$$C_6H_5IO + C_6H_5IO_2 + Ag^+ \rightarrow (C_6H_5)_2I^+ + AgIO_3.$$

Iodine forms a number of so-called salts, IPO_4, $I(CH_2Cl.COO)_3$ and others of unknown constitution which are sometimes regarded as containing I^{3+} ions and quoted to illustrate the 'basic' properties of iodine. It is, in fact, not certain that they are 'normal' salts, for the nitrate (prepared by the action of 100 per cent HNO_3 on iodine) is $IO(NO_3)$ and not $I(NO_3)_3$.[7] (Contrast the covalent nitrate of fluorine, p. 827). Both Br and I form nitrato *anions* which are stable in combination with large cations, as in the $N(CH_3)_4^+$ salts of $[Br(NO_3)_2]^-$, $[I(NO_3)_2]^-$, and $[I(NO_3)_4]^-$.[8] On the other hand iodosyl sulphate and selenate are both anhydrous compounds with the formulae $(IO)_2SO_4$ and $(IO)_2SeO_4$. In $I_2O_2SO_4$[9] helical chains consisting of alternate I and O atoms are arranged in layers and held together by SO_4 groups, each of which links two chains (Fig. 9.1(a)). The bond arrangement

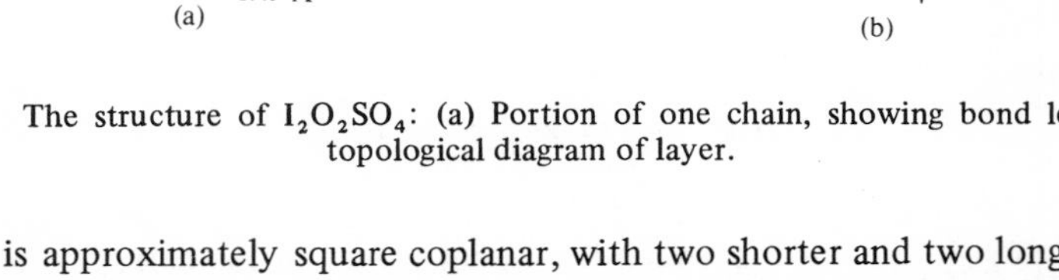

FIG. 9.1.   The structure of $I_2O_2SO_4$: (a) Portion of one chain, showing bond lengths; (b) topological diagram of layer.

around I is approximately square coplanar, with two shorter and two longer bonds; compare the somewhat similar situation in ICl_3 (p. 390). In the topological diagram (b) the O atoms, one along each link, have been omitted.

Nothing is yet known of the structural chemistry of astatine. The isotope At^{211}

is produced by bombardment of Bi by α particles. The formation of compounds has been studied in a time-of-flight mass spectrometer, those observed being HAt, $AtCH_3$, AtCl, AtBr, and AtI. There was no evidence for the formation of At_2 molecules.[10]

(1) AC 1963 **16** 185
(2) JACS 1969 **91** 6235
(3) AC 1960 **13** 1140
(4) ACSc 1961 **15** 407
(5) AC 1975 **B31** 1505
(6) AC 1977 **B33** 1620
(7) CR 1954 **238** 1229
(8) IC 1966 **5** 2124
(9) ACSc 1974 **A28** 71
(10) IC 1966 **5** 766

The stereochemistry of chlorine, bromine, and iodine

The stereochemistry of these elements is concerned with valence groups of 8, 10, 12, and 14 electrons, except for the few molecules such as ClO_2 which contain odd numbers of electrons. The shapes of all molecules and ions which are known at the present time are consistent with the following arrangements of 4, 5, 6, and 7 pairs of valence electrons: tetrahedral, trigonal bipyramidal, octahedral, and pentagonal bipyramidal, assuming that lone pairs occupy bond positions and that the gross stereochemistry is not affected by the π-bonding present in oxy-ions and molecules. Examples are given in Table 9.2. The configuration of XF_6^- ions presents the same problem as that of SeX_6^{2-} and similar ions.

TABLE 9.2

The stereochemistry of chlorine, bromine, and iodine

No. of σ-electron pairs	Type of hybrid	No. of lone pairs	Bond arrangement	Examples
4	sp^3	0	Tetrahedral	ClO_4^-, IO_4^-
		1	Trigonal pyramidal	ClO_3^-, BrO_3^-, HIO_3
		2	Angular	ClO_2^-, BrF_2^+, IR_2^+
5	$sp^3d_{z^2}$	0	Trigonal bipyramidal	IO_5^{3-}
		1	Distorted tetrahedral	$IO_2F_2^-$
		2	T-shape	ClF_3, BrF_3, $R \cdot ICl_2$
		3	Collinear	I_3^-, ICl_2^-, $BrICl^-$
6	$sp^3d_\gamma^2$	0	Octahedral	IO_6^{5-}
		1	Square pyramidal	BrF_5
		2	Square planar	BrF_4^-, ICl_4^-
7	$sp^3d_\gamma^2d_\epsilon$	0	Pentagonal bipyramidal	IF_7
		1	Distorted octahedral	IF_6^-?

The structures of the elements

Having seven electrons in the valence shell the halogens can acquire a noble gas structure by forming one electron-pair bond. Accordingly they form diatomic molecules in all states of aggregation. (The detailed structures of the two crystalline forms of F_2 are described on p. 1276; neither structure is similar to that of crystalline Cl_2, Br_2, and I_2.) A remarkable feature is that the molecules are arranged in layers, within which the shortest intermolecular distance is considerably less than that between the layers (Table 9.3). In Fig. 9.2 (below) we compare the packing of the

TABLE 9.3

Interatomic distances in crystalline halogens

	X—X (Å)	X—X (intermolecular) (Å)		Reference
		In layer	Between layers	
F	–	(See p. 1010)	–	α JSSC 1970 **2** 225
				β AC 1964 **17** 777
Cl	1·98	3·32 3·82	3·74	AC 1965 **18** 568
Br	2·27	3·31 3·79	3·99	AC 1965 **18** 568
I	2·72	3·50 3·97	4·27	AC 1967 **23** 90

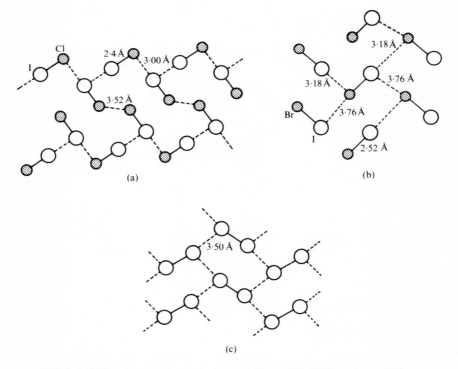

FIG. 9.2. Molecular arrangement in (a) crystalline α-ICl; (b) IBr; (c) crystalline I_2.

molecules in layers of crystalline I_2, ICl, and IBr. This additional weak bonding is very evident also in the structures of polyiodides, which are described later in this chapter. It invites comparison with the 'charge-transfer' bonds in the numerous molecular addition compounds formed by halogens, and halogen halides, in which these bonds are collinear with the X—X molecule and much shorter than the sums of the appropriate van der Waals radii:

$$\underset{(CH_3)_2}{\overset{\diagdown}{\underset{C}{\diagup}}}O\text{---}Br\text{---}Br\text{---}O\overset{C(CH_3)_2}{\underset{2\cdot82\text{ Å}}{\diagup}} \qquad\qquad \underset{2\cdot33\text{ Å}}{H_3CCN\text{---}Br\text{---}Br\text{---}NCCH_3}\overset{2\cdot84\text{ Å}\qquad(1)}{}$$

$$2\cdot28\text{ Å}$$

$$I\text{---}I\underset{2\cdot76\text{ Å}}{\overset{2\cdot91\text{ Å}}{\text{--------}}}Se\underset{H_2C\text{---}CH_2}{\overset{H_2C\diagdown CH_2}{}}\overset{(2)}{}$$

(1) AC 1968 **B24** 713
(2) AC 1964 **17** 712. See also: QRCS 1962 **16** 1

Interhalogen compounds

The compounds stable at ordinary temperatures are of the general form XX'_n where $n = 1, 3, 5$, or 7; none containing more than two different halogens is known. (Molecules with the composition Br_2Cl_2 are apparently formed when a mixture of the elements with a noble gas is passed through a microwave discharge and condensed at 20 K.[1] Their i.r. spectra suggest T-shaped molecules like ClF_3 (see below). Interhalogen compounds are made by direct combination of the elements, except IF_7, which is prepared from IF_5. Of the six possible compounds XX' all have been prepared except IF, for the existence of which there is only spectroscopic evidence. It has not yet been possible to prepare pure BrF, which readily breaks down into $BrF_3 + Br_2$; phase-rule studies indicate that BrF may exist in the liquid state, but it does not exist in the solid state. Of the higher compounds ClF_3, BrF_3, and ICl_3, ClF_5, BrF_5 and IF_5, and IF_7 are known. As far as is known fluorine forms no compounds of the kind FX_3, FX_5, or FX_7, and iodine alone exhibits a covalency of seven. Of the penta- and hepta-compounds only fluorides have so far been prepared.

(1) IC 1968 7 1695

The structures of interhalogen molecules

Bond lengths in gaseous molecules XX' are listed in Table 9.4. In crystalline α-ICl[1] the molecules (mean I—Cl, 2.40 Å) form zigzag chains with rather strong interaction between I and I (3.08 Å) and between I and Cl (3.00 Å) of different molecules (Fig. 9.1(a)). The angles between the I—Cl bond and these weak additional bonds are

TABLE 9.4

Interatomic distances in molecules XX′

F—Cl	1·628 Å	PR 1949 **76** 1723
F—Br	1·759	PR 1950 **77** 420
Cl—Br	2·138	PR 1950 **79** 1007
Cl—I	2·303	PR 1948 **72** 1268

102° at Cl and 94° at I. Similar bond lengths and angles are found in the second (β) form of solid ICl.[2] Crystalline IBr has a structure very similar to that of I_2, and here also there are short I—Br contacts in the layers (Fig. 9.1(b)).[3] The shortest X—X distances between layers are: I—Br, 4.18 Å, I—I, Br—Br, 4.27 Å. This tendency to form additional weaker bonds is also exhibited in polyhalides (see later).

Molecules XX′$_3$. ClF_3 has been studied both by m.w. and also in the crystalline state.[4] The (planar) molecule has the T-shape shown at (a), below, the corresponding results from the X-ray study being 1.716 and 1.621 Å and 86°59′. A m.w. study[5] shows that molecules of BrF_3 have the same shape as ClF_3, with Br—F, 1.810 and 1.721 Å and F—Br—F, 86°13′. The structure of ICl_3 in the solid state is entirely different,[6] for the crystal is built of planar I_2Cl_6 molecules (b) in which the terminal I—Cl bonds are of about the same length as in ICl, ICl_2^-, and ICl_4^-, but the central bonds are considerably longer. There is no obvious theoretical explanation of the form of this molecule.

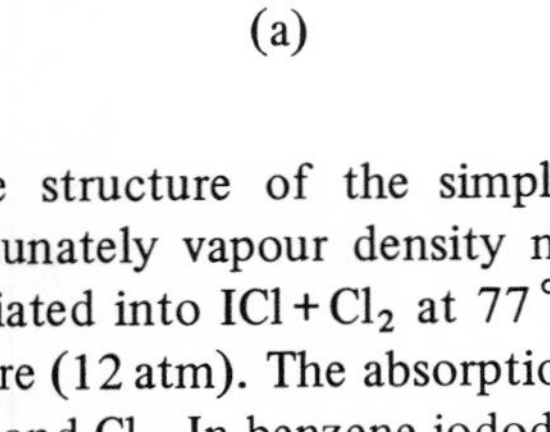

(a) 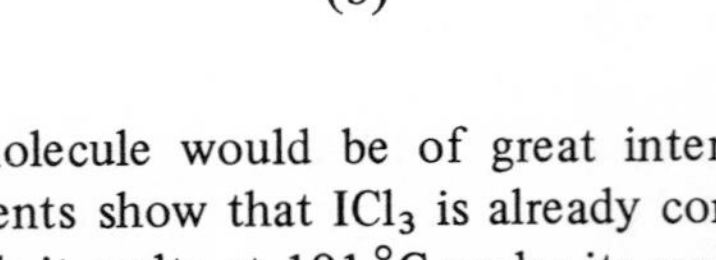   (b)

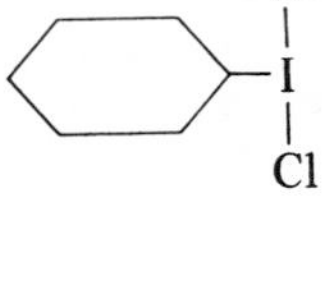

The structure of the simple ICl_3 molecule would be of great interest, but unfortunately vapour density measurements show that ICl_3 is already completely dissociated into $ICl + Cl_2$ at 77 °C, though it melts at 101 °C under its own vapour pressure (12 atm). The absorption spectrum of its solution in CCl_4 is the sum of those of ICl and Cl_2. In benzene iododichloride[7] the linear—ICl_2 system is approximately perpendicular to the plane of the benzene ring. Here the I—C bond occupies one of the equatorial bond positions. (In this molecule I—Cl is apparently rather longer (2.45 Å) than in ICl_2^-.)

Although compounds such as $IAlCl_6$ and $ISbCl_8$, formed by combining $AlCl_3$ or $SbCl_5$ with ICl_3, may be formulated as salts containing the $(ICl_2)^+$ ion, their structures are less simple.[8] The structure of $ISbCl_8$ consists of (angular) ICl_2 and octahedral $SbCl_6$ groups, but there is weak bonding between the groups which links them into chains, each I forming two weak bonds (of length about 2.9 Å) in addition to those (2.31 Å) within the ICl_2^+ ion, (c). Together with the, probably real,

(c)

lengthening of two of the Sb—Cl bonds, this suggests that the state of the compound is intermediate between

$$(ICl_2)^+(SbCl_6)^- \text{ and } (ICl_4)^-(SbCl_4)^+$$

the first form predominating. Similarly the aluminium compound is in a state intermediate between $(ICl_2)^+(AlCl_4)^-$ and $(ICl_4)^-(AlCl_2)^+$. These structures are of particular interest because the bond lengths in ICl_3 are consistent with the formulation:

Molecules XX'_5. The square pyramidal configuration expected for molecules of this type is confirmed by an electron-diffraction study of ClF_5[9a] (Cl—F, 1.57 Å), and by several X-ray studies. In crystalline BrF_5 (at $-120\,°C$) the Br atom lies just below the plane of the four F atoms[9b], and the apical bond (1.68 Å) is shorter than the other four (1.75–1.82 Å). The IF_5 molecule has been studied in crystalline IF_5 (at 193 K)[9c] and in $XeF_2.IF_5$ (p. 377). The shape is similar to that of the isoelectronic XeF_5^+ ion (p. 379).

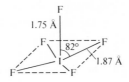

Molecules XX′$_7$. The pentagonal bipyramidal configuration suggested by i.r. and Raman data for IF$_7$ is consistent with the e.d. data[10] for a molecule having all I–F bond lengths approximately equal (1.825 ± 0.015 Å) but with the five equatorial F atoms not quite coplanar. The interpretation of the X-ray diffraction data, obtained from the orthorhombic form stable below –120°C, has presented much difficulty. The most definite statement in the latest review of the X-ray evidence is that it is not possible to demonstrate that the molecular symmetry is different from D_{5h}.[11]

(1)   AC 1956 **9** 274
(2)   AC 1962 **15** 360
(3)   AC 1968 **B24** 429, 1702
(4)   JCP 1953 **21**, 609, 602
(5)   JCP 1957 **27** 223
(6)   AC 1954 **7** 417
(7)   AC 1953 **6** 88

(8)   AC 1959 **12** 859
(9a) DAN 1978 **241** 360
(9b) JCP 1957 **27** 982
(9c) IC 1974 **13** 1071
(10) JCP 1960 **33** 182
(11) AC 1965 **18** 1018

Halogen and interhalogen cations

The discovery of these cations has resulted from the study of reactions in non-aqueous solvents. These ions are strongly electrophilic and may be prepared by the oxidation of the halogens by $S_2O_6F_2$, AsF_5, or SbF_5 in highly acidic solvents such as HSO_3F or the 'super-acid' HSO_3F-SbF_5-$3SO_3$, and isolated in combination with anions of extremely strong acids such as SO_3F^-, AsF_6^-, $Sb_2F_{11}^-$, or $Sb_3F_{16}^-$. The structures of some of these ions have been deduced from spectroscopic data; in a few cases the crystal structures of salts have been determined.

   The existence of the following ions has been established:

Cl_2^+	Br_2^+	I_2^+	FCl_2^+	ClF_2^+	BrF_2^+	IF_2^+
Cl_3^+	Br_3^+	I_3^+		ClF_4^+	BrF_4^+	IF_4^+
		I_4^{2+}	ICl_2^+	ClF_6^+	BrF_6^+	IF_6^+
	Br_5^+	I_5^+				

Ions X$_2^+$. Cl_2^+ is not a stable species either in solution or in the solid state, but its structure has been deduced from its electronic spectrum in the gaseous state. The ions Br_2^+ and I_2^+ are formed by oxidizing the elements by $S_2O_6F_2$ in 'super-acid' or HSO_3F respectively. The crystal structure of the paramagnetic scarlet salt $Br_2(Sb_3F_{16})$,[1a] which has $\mu = 1.6$ BM, shows that it contains Br_2^+ ions and anions consisting of three octahedral SbF_6 groups, the central one of which shares *trans* F atoms, (a). The I_2^+ cation ($\mu = 2.0$ BM), which is responsible for the blue colour of

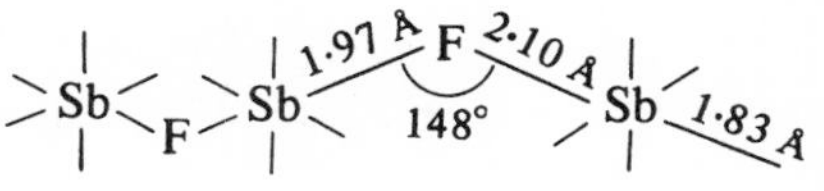

(a)

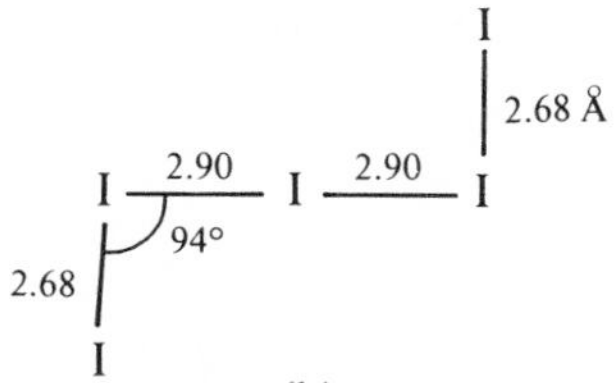

(b) 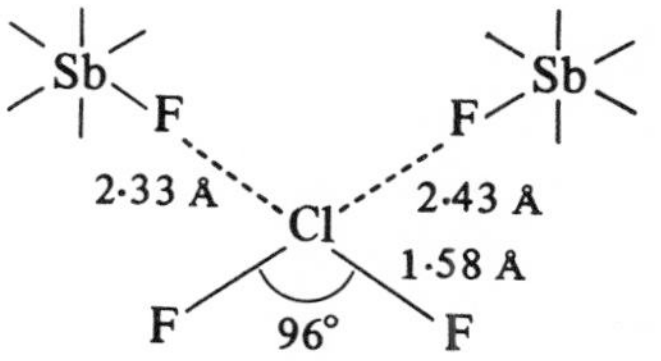   (c)

solutions of iodine in HSO_3F, has been studied in the dark blue $I_2(Sb_2F_{11})$,[1b] prepared from I_2 and SbF_5 in liquid SO_2. The bond lengths in the X_2^+ ions (Cl_2^+, 1.89 Å; Br_2^+, 2.15 Å; and I_2^+, 2.56 Å) are all about 0.1 Å less than the lengths in the X_2 molecules.

Ions X_3^+. The Raman spectrum of $Cl_3(AsF_6)$, isolated at $-78\,°C$, suggests an angular structure like that of the isoelectronic SCl_2.[1c] Crystalline $Br_3(AsF_6)$ has been prepared from $O_2(AsF_6)$ and Br_2, and the I_3^+ ion can be prepared, like I_2^+, by oxidizing the element in HSO_3F by the stoichiometric amount of $S_2O_6F_2$. The interbond angle in the angular I_3^+ ion was determined as $97.1°$ in an n.m.r. study of $(I_3)(AlCl_4)$.[1d]

Ions I_4^{2+} *and* I_5^+. The red, diamagnetic, I_4^{2+} ion is formed on cooling a solution containing I_2^+ ions in HSO_3F. The structure of the I_5^+ ion has been determined in crystalline $(I_5)(SbF_6)$.[1e] It has the planar structure (b), and these I_5^+ units are joined into groups of three by weaker I–I bonds (3.42 Å) similar in length to the shorter secondary bonds in the layer of crystalline iodine.

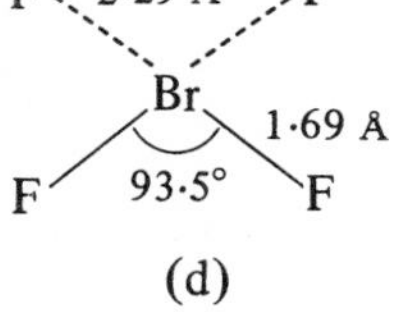

(d)

Ions $XX_2'^+$ (20 valence electrons) are expected to be angular. We have already noted (p. 391) that the structures of the adducts of ICl_3 with $SbCl_5$ and $AlCl_3$ may be formulated as containing ICl_2^+ ions but that there is considerable interaction between anions and cations in these crystals, giving an approximately square arrangement of 4 Cl around I. There is a similar type of interaction between ClF_2^+ and BrF_2^+ and the F atoms of the anions in $(ClF_2)(SbF_6)$[2] and $(BrF_2)(SbF_6)$,[3] as shown at (c) and (d). Raman spectra of $AsF_5.2ClF$ and $BF_3.2ClF$ show that these compounds contain unsymmetrical $(Cl—Cl—F)^+$ cations.[1c]

Ions XF_4^+. The ion BrF_4^+ in $(BrF_4)(Sb_2F_{11})$[4] has the expected SF_4 type of structure, a trigonal bipyramid with the lone pair occupying one equatorial bond position. The mean length of the four short Br—F bonds is 1.81 Å; two weak bonds in the equatorial plane complete distorted octahedral coordination, (e). The asymmetry of

(e)

the bridge in the $(Sb_2F_{11})^-$ ion suggests that this ion is not far removed from $(SbF_6)^-(SbF_5)$, the mean length of the six Sb—F bonds in SbF_6^- being 1.90 Å and of the five bonds in SbF_5 (i.e. excluding the bridge bond), 1.75 Å.

The IF_4^+ ion in $(IF_4)(Sb_2F_{11})$[5] has a similar shape, with equatorial and apical I—F bond lengths 1.77 and 1.84 Å and the corresponding interbond angles 92° and 160°; compare the similar angles in TeF_4 (89° and 161°) and GeF_2 (92° and 163°). The more distant neighbours of I in this crystal are, however, quite differently arranged from those of Br shown at (e), there being 2 F at 2.53 Å and 2 F at 2.88 Å.

Ions XF_6^+. Precise structural information is still required for these ions. The vibrational spectrum of ClF_6^+ has been studied in $(ClF_6)(PtF_6)$[6] and X-ray powder photography shows that $(BrF_6)(AsF_6)$[7] and $(IF_6)(AsF_6)$[8] have structures of the NaCl-type, presumably containing octahedral cations and anions.

(1a) JCS A 1971 2318   (3) JCS A 1969 1467
(1b) CJC 1974 **52** 2048   (4) IC 1972 **11** 608
(1c) IC 1970 9 811   (5) JCS D 1975 2174
(1d) IC 1975 14 428   (6) IC 1973 **12** 1580
(1e) CJC 1979 57 968   (7) IC 1975 14 694
(2) JCS A 1970 2697   (8) JCP 1970 **52** 1960

Polyhalides

These are salts containing *anions* of the types set out below or, in the case of poly-iodides, more complex anions which are described separately:

Cl_3^-	Br_3^-	I_3^-			
	BrF_2^-				
ClF_2^-	$BrCl_2^-$	ICl_2^-	$IBrF^-$		
$ClBr_2^-$	BrI_2^-	IBr_2^-	$IClBr^-$		
		I_2Br^-			
ClF_4^-	BrF_4^-	IF_4^-	ICl_3F^-	ICl_4^-	
	BrF_6^-	IF_6^-			

They may be prepared by direct combination of metal (or other) halide with inter-halogen compound (or halogen in the case of $CsIF_4$, $CsBrF_4$, $CsClF_4$, $CsBr_3$, and CsI_3) either in solution in a suitable solvent or by the action of the vapour of the interhalogen compound on the dry halide. $KBrF_4$ is prepared by dissolving KF in BrF_3 and distilling off the excess solvent; KIF_4 can be crystallized from a cold solution of KI in IF_5 and KIF_6 from the hot solution. Very unstable compounds must be prepared at low temperatures, for example, $NO(ClF_2)$ from NOF and ClF at $-78\,^{\circ}C$. In general the only crystalline salts stable at ordinary temperatures are those of the larger alkali metals and large organic cations. For example, the only salts MI_3, $MICl_2$, and $MICl_4$ are those of NH_4, Rb, and Cs and of ions such as $NH(CH_3)_3^+$ and $(C_6H_5N_2)^+$. The stability increases with increasing size of cation in a series such as $KICl_2$, $RbICl_2$, and $CsICl_2$, only the last salt being stable at room temperature. No anhydrous acids are known corresponding to polyhalides, but $HICl_4.4H_2O$ has been described.

The structures of polyhalide anions

Ions $(X–X–X)^-$. Structural data for these (linear) ions (Table 9.5) emphasize two important features of these ions. First, the ions Br_3^-, ICl_2^-, and I_3^- are found to be symmetrical in some salts and unsymmetrical in others. Second, the overall length of I_3^- ions, which have been studied in most detail, is approximately 0.5 Å greater than twice the single bond length, and varies by only about 0.1 Å for both symmetrical and unsymmetrical ions; compare the difference in I–I bond lengths, which ranges from zero to as much as 0.32 Å. The structural differences between I_3^- ions are attributed to the different environments of the terminal atoms in the crystals, but the effect is not simple. It is clearly not simply a question of the size of the cation. In $(\phi_4 As)I_3$ the anions are widely separated (shortest I---I between different anions 5.20 Å); the salt is pale yellow and contains symmetrical I_3^- ions. In the alkali salts, including $KI_3.H_2O$, there is a packing of the anions rather like that of I_2 molecules in crystalline iodine with contacts around 4 Å. Although Cs^+ is a large cation the I_3^- ion in CsI_3 is unsymmetrical, whereas in the black lustrous crystals of $KI_3.H_2O$ it is symmetrical to within 0.005 Å. For a survey of I_3^- ions see the reference to $KI_3.H_2O$.

The bond lengths in the $(I–I–Br)^-$ ion, where I–I is *shorter* than I–Br, suggest that the ion is tending towards

TABLE 9.5
Structures of ions $(X^1-X^2-X^3)^-$

Ion	Salt	X^1-X^2 (Å)	X^2-X^3 (Å)	Excess over single bond length (Å)		Reference
F–Cl–F	$KClF_2$	(i.r. indicates linear)				IC 1967 6 1159
Br–Br–Br	$(C_6H_5N_2)Br_3$	2.54	2.54	0.26	0.26	ACSc 1962 16 1882
	$[N(CH_3)_3H]_2Br(Br_3)$	2.54	2.54	0.26	0.26	PKNAW 1958 B61 345
	$(C_6H_7NH)_2(SbBr_6)Br_3$	2.54	2.54	0.26	0.26	IC 1968 7 2124
	$CsBr_3$	2.44	2.70	0.16	0.42	AC 1969 B25 1073
	$(PBr_4)Br_3$	2.39	2.91	0.11	0.63	AC 1967 23 467
Cl–I–Cl	$(PCl_4)ICl_2$					JACS 1952 74 6151
	$[N(CH_3)_4]ICl_2$					AC 1964 17 1336
	$KICl_2$	2.55	2.55	0.23	0.23	AC 1973 B29 2104
	$CsICl_2$					AC 1973 B29 2613
	$C_4H_{12}N_2(ICl_2)_2$	2.47	2.69	0.15	0.37	ACSc 1958 12 668
Cl–I–Br	$NH_4(ClIBr)$	2.91	2.51	0.59	0.04	AC 1967 22 812
Br–I–Br	$KIBr_2.H_2O$	2.71	2.71	0.24	0.24	AC 1973 B29 2556
	$CsIBr_2$	2.62	2.78	0.15	0.31	CC 1969 1374
I–I–Br	CsI_2Br	2.78	2.91	0.12	0.43	AC 1966 20 330
I–I–I	$KI_3.H_2O$	2.93	2.93	0.27	0.27	AC 1980 B36 2869
	$(As\phi_4)I_3$	2.92	2.92	0.26	0.26	AC 1972 B28 1331
	NH_4I_3	2.79	3.11	0.13	0.45	AC 1970 B26 904
	CsI_3	2.84	3.04	0.18	0.38	AC 1972 B28 1331

$$I\text{——}I \quad \cdots \quad Br^-$$
$$2.78 \quad\quad 2.91 \text{ Å}$$

and similarly the increasing asymmetry of the Br_3^- ion in a number of salts (see Table 9.5) suggests that the system is approaching the state $Br\text{--}Br\cdots Br^-$.

Ions $(XX'_4)^-$. Infrared studies[1] of the ClF_4^- ion in salts of NO^+, Rb^+, and Cs^+, indicate a planar configuration. X-ray studies confirm the square planar shape of the BrF_4^- ion in $KBrF_4$[2] (Br–F, 1.89 Å) and of the ICl_4^- ion in $KICl_4.H_2O$[3] and $HICl_4.H_2O$[4] (I–Cl, 2.53 Å).

(1) IC 1966 5 473
(2) JCS A 1969 1936

(3) AC 1963 16 243
(4) JACS 1972 94 1130

Polyiodide ions

These are ions I_n^- ($n = 3, 5, 7,$ or 9) or I_n^{2-} ($n = 4, 8,$ or 16) which form salts with large cations. As already noted, the tri-iodides consist of cations and I_3^- ions, but in the higher polyiodides the anion assembly is made up of a number of sub-units (I^-, I_3^-, and I_2 molecules) which are held together rather loosely to form the more complex I_n system. It is a feature of all polyiodide ions that some or all of the I–I distances are intermediate between the covalent I–I bond length (2.72 Å in I_2) and the shortest *inter*molecular separation, 3.50 Å in crystalline iodine.

In $[Cu(NH_3)_4]I_4$[1] I atoms of linear I_4^{2-} ions complete (4+2)-coordination of $Cu(\text{II})$ to form zigzag chains:

The bond lengths in the I_4^{2-} ion are 3.34, 2.80, and 3.34 Å and the next shortest I–I distance is 4.65 Å, indicating the formulation $(I---I---I---I)^{2-}$. Linear I_4 groups have also been reported in Tl_6PbI_{10}[2], but here all I–I distances lie in the range 3.11–3.18 Å, and it is less obvious that this compound should be formulated as a salt containing I_4^{2-} ions.

In $N(CH_3)_4I_5$[3] there are approximately square nets of I atoms (Fig. 9.3(a)) and the I–I distances suggest V-shaped I_5^- ions consisting of an I^- ion very weakly bonded to two polarized I_2 molecules. The arms of the V are linear to within 6°, and the V is planar to within about 0.1 Å. This ion is obviously not related structurally to the ICl_4^- ion.

In $[N(C_2H_5)_4]I_7$[4] the anion assembly consists of I_2 molecules and apparently linear I_3^- ions (see above) in the proportion of two I_2 to one I_3^-. Between these units the shortest contacts are I–I, 3.435 Å and within them, I–I, 2.735 Å and 2.904 Å respectively.

The violet-coloured salt with the empirical formula CsI_4 is diamagnetic; the I_4^- ion would be paramagnetic, having an odd number of valence electrons. An X-ray study of the salt[5] shows that it contains rather extraordinary groupings of eight coplanar I atoms as shown in Fig. 9.3(b). In this I_8^{2-} ion the terminal I_3 groups are very similar to the I_3^- ion in tri-iodides, so that the ion may be described as an assembly of two I_3^- ions and a polarized I_2 molecule. The less compact, but otherwise very similar, configuration of Fig. 9.3(c) is found for this ion in $[(CH_2)_6N_4CH_3]_2I_8$.[6]

In $N(CH_3)_4I_9$ there is an even more complex arrangement.[7] Five-ninths of the I atoms form rather densely packed layers (Fig. 9.3(d)) between which there are I_2 molecules (with I–I = 2.67 Å, close to the value in the I_2 molecule); the cations lie between the layers in the holes between the I_2 molecules. The other I–I distances are around 2.9, 3.2, and 3.5–3.7 Å, and the way in which the structure is described depends on the interpretation of these distances. The 2.9 and 3.2 Å bonds link I atoms into V-shaped I_5^- ions, but here it is a terminal atom which is the I^- ion, in contrast to the I_5^- ion in $N(CH_3)_4I_5$. The next shortest distances in a layer are those of length 3.49 Å, which would correspond to linking of the I_5^- ions in pairs to form

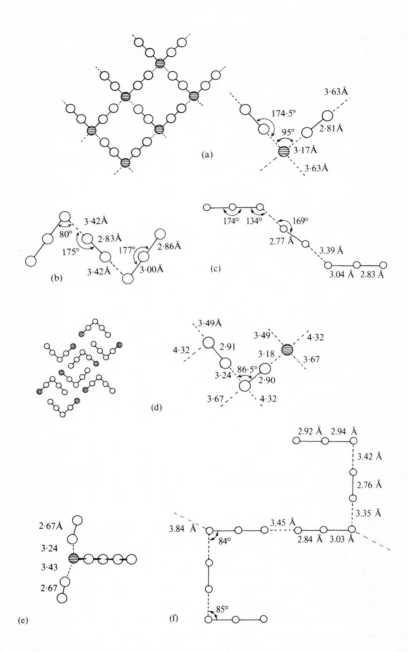

FIG. 9.3. Structures of the higher polyiodide ions: (a) I_5^- in $N(CH_3)_4I_5$; (b) I_8^{2-} in Cs_2I_8; (c) I_8^{2-} in $[(CH_2)_6 N_4CH_3]_2I_8$; (d) the I_5^- layer in $N(CH_3)_4I_9$; (e) environment of I^-(shaded circle) in $N(CH_3)_4I_9$; (f) I_{16}^{4-} in (theobromine)$_2$H$_2$I$_8$.

closed rings. The linking of the I^- ion in a layer to I_2 molecules between layers is shown in Fig. 9.3(e).

In the dark-blue crystals of $(theobromine)_2H_2I_8{}^{(8)}$ the I atoms are arranged in centrosymmetrical groups of 16 (Fig. 9.3(f)), the shortest distance between I atoms of different groups being 3.84 Å (dotted lines). This I_{16}^{4-} ion is not so well defined as the simpler polyiodide ions, for 3.84 Å is appreciably less than the distance between layers in crystalline I_2 (4.27 Å) and presumably indicates an interaction somewhat stronger than simple van der Waals bonding.

We see that in iodine itself and in polyiodides there is a tendency to form dense planes with I–I distances up to about 3.5 to 3.6 Å, with much larger (van der Waals) distances of about 4.3 Å between planes. In the polyiodides there is interaction between I^- ions and the easily polarizable I_2 molecules, giving I_3^-, I_5^-, I_8^{2-}, and I_9^- systems. In the higher members there is progressively less distortion of the I_2 molecule, which has the normal I–I bond length in the ennea-iodide. The angles between the weak bonds formed by I^- ions are approximately 90°, suggesting the use of p orbitals.

(1) HCA 1975 58 2604   (5) AC 1954 7 487
(2) ZaC 1977 432 5   (6) IC 1979 18 2416
(3) AC 1957 10 596   (7) AC 1955 8 814
(4) AC 1958 11 733   (8) CC 1975 677

Hydrogen halides and ions HX_2^-, $H_2X_3^-$, etc.

The interatomic distances, determined spectroscopically, in the hydrogen halide molecules are:

H–F	0.917 Å	H–Br	1.410 Å
H–Cl	1.275	H–I	1.600

The physical properties of the acids HF, HCl, HBr, and HI do not form regular sequences, HF differing from the others in much the same way as H_2O from H_2S, H_2Se, and H_2Te. The abnormal properties of HF are discussed in Chapter 8 with other aspects of hydrogen bonding, as also are the structures of the crystalline hydrogen halides.

Ions $(X–H–X)^-$ are formed by F, Cl, and Br, but in contrast to the stable and well-known bifluorides of the alkali metals the ions HCl_2^- and HBr_2^- are stable only in a few salts of very large cations. All attempts to prepare salts containing HCl_2^- ions from aqueous solution have given hydrated products (for example, $(C_6H_5)_4As.HCl_2.2H_2O$) which probably do not contain these ions (see below), but anhydrous $(C_6H_5)_4As.HCl_2$ can be prepared from $(C_6H_5)_4AsCl$ and anhydrous $HCl.{}^{(1)}$ The existence of HCl_2^- ions has been disproved in a number of salts once thought to contain these ions. For example, the compound $CH_3CN.2HCl$ is not $(CH_3CHN)^+(HCl_2)^-$ but an imine hydrochloride,$^{(2)}$ $[CH_3C(Cl)=NH_2]^+Cl^-$, and in $(Cl_2en_2Co)Cl.HCl.2H_2O$ the odd proton is associated with the two H_2O molecules in $H_5O_2^+$ units and not with the two Cl atoms in a HCl_2^- ion.$^{(3)}$ The structures of

$Cs_3(Cl_3.H_3O)HCl_2$ and the Br compound, which apparently do contain HX_2^- ions,[4] are described in Chapter 8.

Only F forms more complex ions $H_2X_3^-$, $H_4X_5^-$, (see Chapter 8). For the hydrates of HCl see p. 692.

(1) IC 1968 7 1921
(2) IC 1968 7 2577

(3) JINC 1961 **19** 208
(4) IC 1968 7 594

Metal hydride halides

The alkaline-earth salts MHX, where M = Ca, Sr, or Ba, and X=Cl, Br, or I, have been prepared by melting the hydride MH_2 with the dihalide MX_2 or by heating $M + MX_2$ in a hydrogen atmosphere at 900 °C.[1] The compound CaHCl was formerly thought to be the 'sub-halide' CaCl. All these compounds have the PbFCl structure (p. 486), but the H atoms were not directly located. Compounds described as MgHX (X = Cl, Br, I) have been shown to be physical mixtures of MgH_2 with MgX_2.[2]

Hydride-fluorides CaH_xF_{2-x}[3] have a statistical fluorite structure; they are of interest as ionic conductors.

(1) ZaC 1956 **283** 58; *ibid.* 1956 **288** 148, 156
(2) IC 1970 9 317

(3) JSSC 1979 **30** 183

Oxy-compounds of the halogens

Oxides

We list in Table 9.6 only those compounds of which structural studies have been made giving either quantitative data (X-ray, e.d., m.w.) or in the case of Raman and i.r. spectroscopic studies the probable shape of a molecule by analogy with molecules of known configuration.

Electron diffraction and spectroscopic studies give the configurations (a) and (b)

TABLE 9.6
Oxides, oxyfluorides, and oxyfluoride ions of the halogens

F_2O[1]	Cl_2O[3]	I_2O_5[7]	$ClOF_3$[8]	BrO_3F[11]	IOF_3[12]	IOF_4^-[16]
F_2O_2[2]	ClO_2[4]		ClO_2F_3[9]		IOF_5[13]	$IO_2F_2^-$[17]
	Cl_2O_6[5]		ClO_3F[10]		$(IO_2F_3)_2$[14]	$IO_2F_4^-$[18]
	Cl_2O_7[6]				$(IOF_3.IO_2F_3)_4$[15]	$ClO_2F_2^+$[19]

(1) JPC 1953 **57** 699. (2) See p. 502. (3) JCS A 1968 657. (4) JCS A 1970 46. (5) CR 1971 **272C** 1495. (6) TFS 1965 **61** 1821. (7) ACSc 1970 **24** 1912. (8) IC 1972 **11** 2196. (9) IC 1973 **12** 2245. (10) JCS A 1970 872. (11) IC 1970 9 622. (12) J. Fluorine Chem. 1974 **4** 173. (13) IC 1976 **15** 3009. (14 CC 1977 519. (15) JCS D 1980 481. (16) AC 1972 **B28** 979. (17) JCP 1976 **64** 3254. (18) JINC Suppl. 1976 91. (19) IC 1973 **12** 1356.

for the similarly constituted F_2O and Cl_2O, and a microwave study gives the structure (c) for ClO_2. This oxide is of special interest since the molecule contains an odd number of electrons; its structure is very similar to that of SO_2. The oxide Cl_2O_4 is described as chlorine perchlorate, $ClOClO_3$. A spectroscopic study of solid

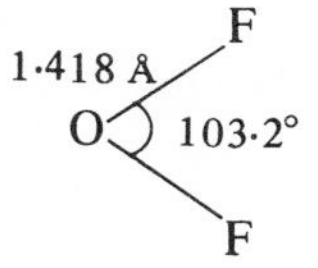

(a) 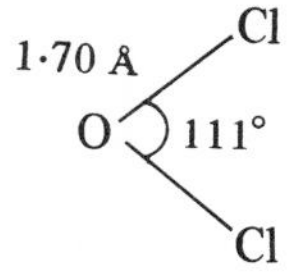   (b) 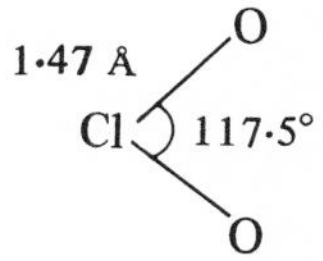   (c)

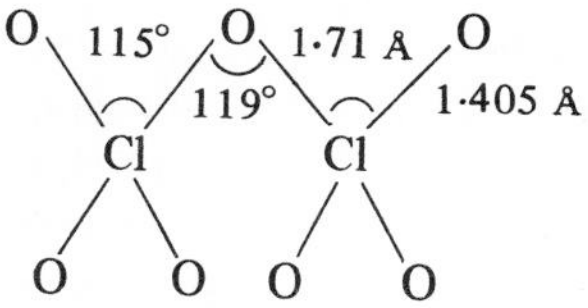

(d)

Cl_2O_6 indicated the ionic structure $(ClO_2)^+(ClO_4)^-$; the structure of the Cl_2O_6 molecule is not known.

The configuration of Cl_2O_7 is approximately eclipsed, (d).

Iodine pentoxide is the anhydride of iodic acid and is a white solid stable up to $300\,°C$. The crystal consists of molecules, (e), formed from two pyramidal IO_3 groups sharing one O atom, but as in other oxy-compounds of I there are secondary bonds of intermediate lengths, the stronger of these (2.23–2.72 Å) involving terminal O atoms. Two other oxides, both yellow powders, have been assigned the formulae I_2O_4 and I_4O_9, but the existence of the latter is doubtful.[1] The diamagnetism of the dioxide rules out the existence of IO_2 molecules, but the existence of discrete I_2O_4 molecules has not yet been proved. Mass spectrometric studies of iodine oxides give no evidence for I_2O_7 molecules.

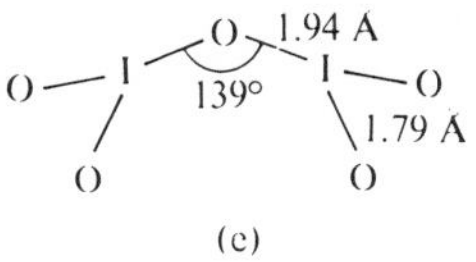

(e)

(1)   ACSc 1968 **22** 3309

Oxyfluorides and oxyfluoride ions

The structures of most of these molecules and ions present no special points of interest since they are consistent with the $8n + 2m$ 'rule' (p. 291). The 26-electron molecules XOF_2 (X = Cl, Br, I) and the ion $ClOF_2^+$ are presumably pyramidal, like

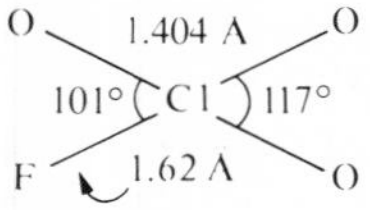

(f) 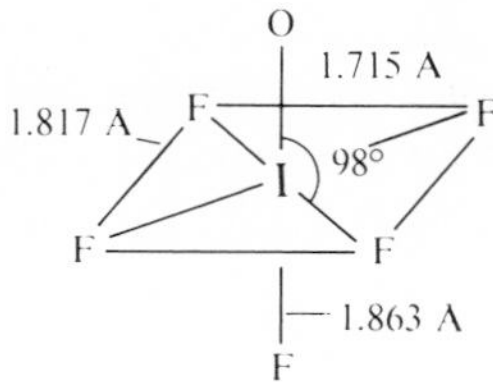   (g) 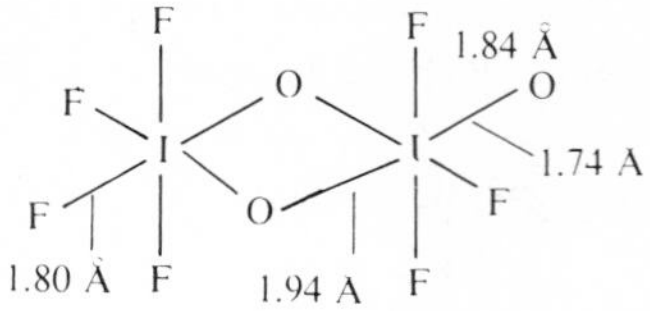   (h)

SF_2O, and the 32-electron molecules such as ClO_3F, (f) are tetrahedral. Details of the BrO_3F molecule are: Br–O, 1.582 Å; Br–F, 1.708 Å; O–Br–O, 114.9°; F–Br–O, 103.3°. The vibrational spectrum of $(ClO_2F_2)^+(PtF_6)^-$ closely resembles that of SO_2F_2. The 34-electron complexes XOF_3 (formed by Cl and I) have structures similar to SF_4 with the O atom occupying an equatorial bond position and the ion $IO_2F_2^-$ has a similar structure with two O atoms and the lone pair in equatorial positions; the structure of this ion is described on p. 407. The third equatorial position is occupied by a third F atom in the 40-electron ClO_2F_3 molecule. The 42-electron group is represented by the square pyramidal IOF_4^- ion which is described in detail on p. 407. Six octahedral bonds are formed in the 48-electron molecule IOF_5, (g), and in the ion $IO_2F_4^-$, which exists in *cis* and *trans* isomeric forms.

In contrast to the simple molecular form of ClO_2F_3 noted above IO_2F_3 forms dimers, (h), in the crystalline state, I showing its tendency to form octahedral bonds in its highest oxidation state.

There is octahedral coordination of one-half of the I atoms in $(IOF_3.IO_2F_3)_2$, the remainder forming three stronger and two weaker bonds. The crystal consists of cyclic 'molecules', (i). The I–O–I bridges in the 8-membered ring are unsymmetrical,

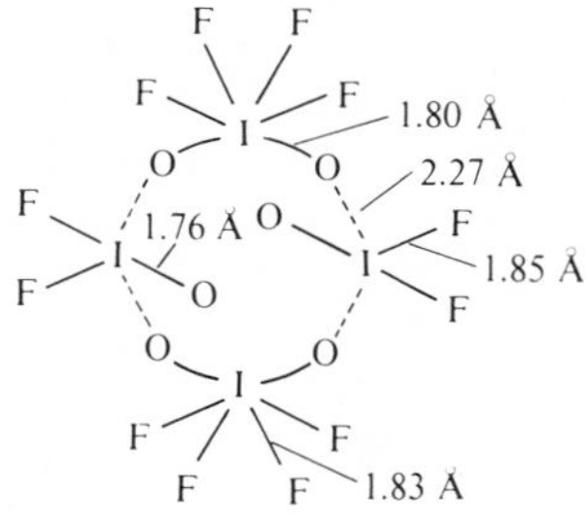

(i)

suggesting the alternative formulation $(IOF_2)^+(IO_2F_4)^-$. The ions are held together by weaker I–O bonds of length 2.27 Å; compare 1.76 and 1.80 Å within the ions.

Oxy-cations

Salts containing the ions ClO_2^{+},[1] BrO_2^{+},[2] and IO_2^{+}[3] have been prepared, the anion being AsF_6^-. Infra-red studies suggest that the compound prepared from IO_2F

and AsF$_5$, is an ionic solid. Iodyl fluorosulphate has been made as a hygroscopic powder stable up to 100 °C,[4] but the structures of the XO$_2^+$ ions are not yet known.

(1) IC 1969 8 2489
(2) AnC 1976 88 189

(3) IC 1965 4 257
(4) IC 1964 3 1799

Oxy-acids and oxy-anions

In the stability of their oxy-acids there are great differences between the halogens. The existence of only one oxy-acid of fluorine has yet been firmly established, and the only types of oxy-ion formed by all three halogens Cl, Br, and I are XO$^-$, XO$_3^-$, and XO$_4^-$ (Table 9.7). Structural studies have been made of HOCl, HOBr, α-HIO$_3$, HClO$_4$.H$_2$O, H$_5$IO$_6$, and HI$_3$O$_8$.

TABLE 9.7
Structures of oxy-ions XO$_n$ of the halogens

XO$_2^-$ angular	XO$_3^-$ pyramidal	XO$_4^-$ tetrahedral	XO$_5^{3-}$ square pyramidal	XO$_6^{5-}$ octahedral
ClO$_2^-$	ClO$_3^-$	ClO$_4^-$	–	–
(BrO$_2^-$)	BrO$_3^-$	BrO$_4^-$	–	–
	IO$_3^-$	IO$_4^-$	IO$_5^{3-}$	IO$_6^{5-}$

Acids HOX. Until recently neither the acid HOF nor its salts were known. Following its earlier detection in low-temperature matrices HOF has now been prepared in milligram quantities by passing a stream of F$_2$ at low pressure over water at 0°C. The product is separated in a trap at –183 °C.

The stability of all the acids HOX is low. Apart from HOF they are known only in solution, though HOCl and HOBr have been prepared by photolysis of Ar–HX–O$_3$ mixtures at 4 K and studied in the Ar matrix by i.r. spectroscopy.[1a] HOF has a sufficiently long half-life (5–60 min) to permit a m.w. study of the gas at room temperature.[1b] The structures found (with O–H, 0.96 Å) are

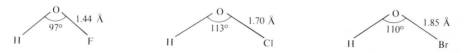

Salts, and therefore the structures of the XO$^-$ ions, are not known.

Acids HXO$_2$ *and their salts.* The appears to be no conclusive evidence that the acids HXO$_2$ exist if X is F, Br, or I, and the only salts of one of these acids that have been prepared are LiBrO$_2$ and NaBrO$_2$.[2] LiBrO$_2$ is apparently formed by the dry reaction between LiBr and LiBrO$_3$, and NaBrO$_2$ can be crystallized from a solution containing NaOH and Br$_2$. Pure chlorous acid, HClO$_2$, has not been isolated, but treatment of the barium salt with sulphuric acid gives a colourless solution which even at $0\,°C$ soon turns yellow owing to the formation of ClO_2 (and Cl_2).

The structure of the chlorite ion has been studied in a number of crystalline salts (Na, NH_4, Ag, Zn), giving values[3] close to $1.56\,Å$ for Cl—O and $108°$ for O—Cl—O, though the values $1.59\,Å$ and $111.4°$ were found in $Zn(ClO_2)_2 . 2H_2O$.[4] It should perhaps be noted that correction for thermal vibration can increase a bond length by as much as $0.02\,Å$.

Acids HXO_3 *and their salts.* The acids $HClO_3$, $HBrO_3$, and HIO_3 do not form a regular sequence, as is seen from the heats of formation: 100, 52, and $234\,kJ\,mol^{-1}$ respectively. The iodine acid is more stable than the chlorine and bromine compounds, for aqueous solutions of $HClO_3$ and $HBrO_3$ can be concentrated (*in vacuo*) only up to about 50 per cent molar, whereas an aqueous solution of HIO_3 may be concentrated until it crystallizes. Moreover, whereas in dilute aqueous solution iodic acid behaves as a moderately strong monobasic acid, the freezing points and conductivities of concentrated solutions suggest the presence of more complex ions or possibly basic dissociation to $IO_2^+ + OH^-$.[5]

Crystalline α-HIO_3[6] consists of pyramidal molecules (a) linked by hydrogen bonds (see Chapter 8). The mean interbond angle O—I—O is $98°$; the next nearest O neighbours, completing a very distorted octahedron around the I atom, are at 2.45, 2.70, and $2.95\,Å$.

(a)   (b)

The crystal structure of the so-called anhydro-iodic acid, HI_3O_8, formed by the dehydration of HIO_3 at $110\,°C$, shows that it is an addition compound of HIO_3 and I_2O_5 with hydrogen bonds between OH of HIO_3 and O of I_2O_5.[7a] Bond lengths in the pyramidal HIO_3 molecule are similar to those in crystalline HIO_3, and those in the non-planar I_2O_5 molecule are as shown at (b). This molecule consists of two IO_3 pyramids with a common O atom.

X-ray studies of crystalline salts demonstrate the pyramidal shape of the XO_3^- ions (Table 9.8). In crystalline iodates I has usually 3 more neighbours (at $2.6-3.2\,Å$) completing a distorted ('lone-pair') octahedral group, though in a few cases a less symmetrical arrangement of 2 or 4 next-nearest neighbours is found; for a review see the reference in Table 9.8. (Earlier structures assigned to one of the three forms of $LiIO_3$ and to $TlIO_3$ in which I had 6 equidistant neighbours were incorrect.)

TABLE 9.8
Structures of XO_3^- *ions*

	X–O (Å)	*Angle O–X–O*	*References*
ClO_3^-	1.49	107°	$\}$ AC 1977 **B33** 2698, 3601
BrO_3^-	1.65	104°	
IO_3^-	1.81	100°	AC 1980 **B36** 2130

Bonds I–OH are appreciably longer than the value of Table 9.8, as in HIO_3, (a) above. Three different bond lengths were found in $KH(IO_3)_2$,[7b] namely, I–O, 1.78 Å, I–OH(donor), 1.93 Å, and I–OH(acceptor), 1.87 Å.

Acids HXO_4 and their salts. In the formation of acids HXO_4 the halogens show still less resemblance to one another. Chlorine forms $HClO_4$ and perchlorates. Perbromates were unknown until 1968, when $KBrO_4$ was prepared (by oxidation of the bromate by F_2 in alkaline solution). Periodic acid is normally obtained (by evaporation of its solution) as H_5IO_6; it has been stated that HIO_4 results from heating H_5IO_6 at 100 °C *in vacuo*, and that an intermediate product $H_4I_2O_9$ is formed at 80 °C. A further peculiarity of the iodine acid is that on further heating HIO_4 does not form the anhydride (I_2O_7 apparently does not exist) but loses oxygen to form HIO_3. Whereas perchlorates are all of the type $M(ClO_4)_n$, periodates of various kinds can be obtained from a solution by altering the temperature of crystallization and the hydrogen ion concentration. For example, a solution of $Na_2H_3IO_6$ is formed by oxidizing $NaIO_3$ by chlorine in alkaline solution, and from this solution a number of silver periodates may be obtained (after adding $AgNO_3$), including $AgIO_4$, Ag_3IO_5, Ag_2HIO_5, Ag_5IO_6 and $Ag_2H_3IO_6$. It should be noted, however, that the formulae of compounds that analyse as hydrates must be interpreted with caution (see below).

The structure of the $HClO_4$ molecule in the vapour state has been studied by electron diffraction;[8a] it is very similar to that of ClO_3F (p. 402): Cl–O, 1.41 Å; Cl–OH, 1.64 Å; O–Cl–O, 113° or 117°.

The ions XO_4^- are tetrahedral with bond lengths similar to those in XO_3^-, namely, Cl–O, 1.43 Å,[8b] Br–O, 1.61 Å,[9] and I–O, 1.78 Å.[10a] In $NaIO_4$ the anion is a slightly compressed tetrahedron, having two bond angles of 114° and four of 107°.

Intermediate between the ions IO_4^- and IO_6^{5-} formed by I^{VII} is the ion IO_5^{3-} which has been studied in K_3IO_5.[10b] It has the form of a square pyramid, with I–O, 1.77 Å and an angle of 95° between the apical and equatorial I–O bonds; see the discussion of the 40-electron group, p. 298. The ions ReO_5^{3-}, OsO_5^{3-}, and TeO_5^{3-} have similar configurations.

Periodates containing octahedrally coordinated iodine. Of the halogens only I exhibits octahedral coordination by oxygen. Crystalline H_5IO_6 consists of nearly regular octahedral $IO(OH)_5$ molecules which are linked by O–H–O bonds (ten from each molecule) into a 3D array. The bond lengths are I–O, 1.78 Å, I–OH, 1.89 Å, and O–H, 0.96 Å.[11] The ion $IO_3(OH)_3^{2-}$ is present in salts such as $Cd[IO_3(OH)_3].3H_2O,$[12] and $[Mg(H_2O)_6][IO_3(OH)_3].$[13] In this ion (a), the bond lengths were determined as I–O, 1.86 Å, and I–OH, 1.95 Å. For $(NH_4)_2H_3IO_6$ see p. 375. It seems likely that many 'hydrated periodates' have been incorrectly formulated (compare borates, Chapter 24). For example, the Cd salt just mentioned was formulated as $Cd_2H_2I_2O_{10}.8H_2O$, apparently analogous to $K_4H_2I_2O_{10}.8H_2O$. In fact this K salt has also been formulated as $K_2HIO_5.4H_2O$ and $K_4I_2O_9.9H_2O$; it is obtained from a solution of KIO_4 in concentrated KOH. It actually contains ions, (b), formed from two octahedral $IO_5(OH)$ groups sharing an edge. The H atoms were not located, but their positions were deduced from the fact that all the bonds marked a have the length 2.00 Å while the length of the other (terminal) I–O bonds is 1.81 Å.[14a] Dehydration of $K_4H_2I_2O_{10}.8H_2O$, or crystallization of the solution above 78 °C gives $K_4I_2O_9$, which contains ions (c) consisting of two IO_6 octahedra sharing a face.[14b]

The simple octahedral IO_6^{5-} ion presumably occurs in anhydrous salts M_5IO_6. Like TeO_6^{6-} this ion has the property of stabilizing high oxidation states of metals (for example, Ni^{IV}, Cu^{III}), as in $KNiIO_6,$[15] in which K, Ni, and I all occupy octahedral positions in an approximately hexagonal closest packing of O atoms. (I–O, approx. 1.85 Å).

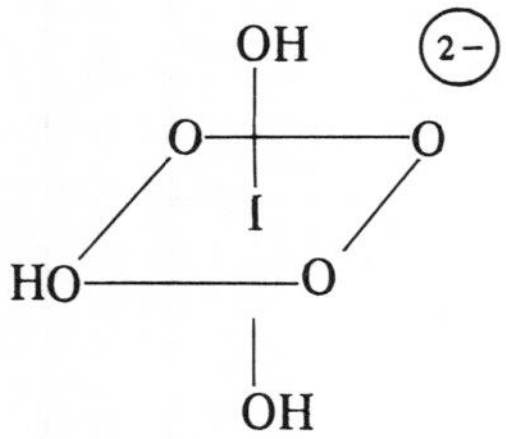

(a) 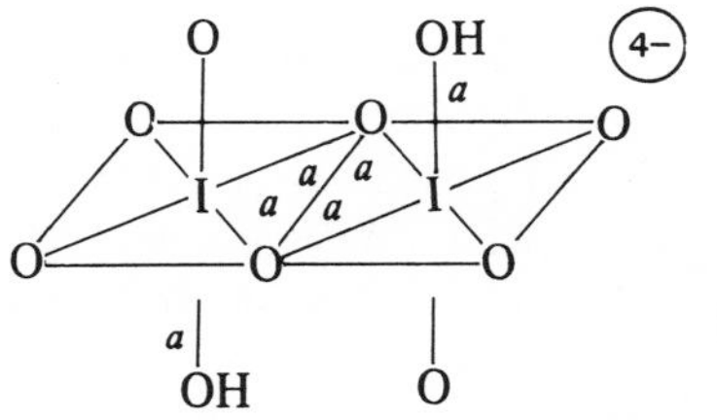   (b) 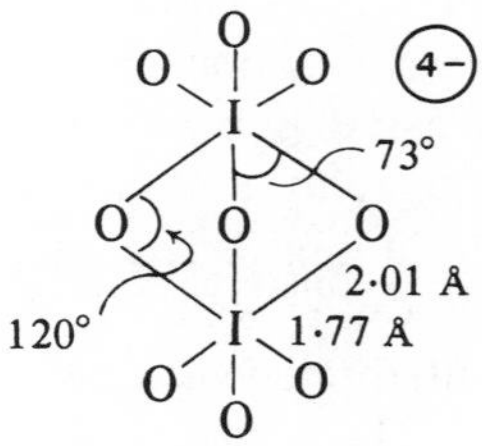   (c)

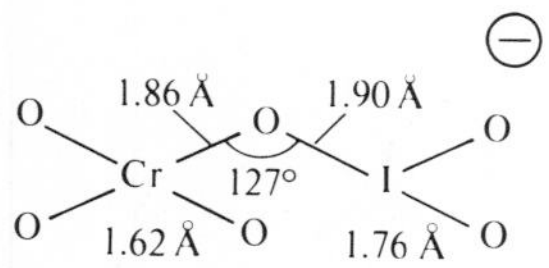

(d) 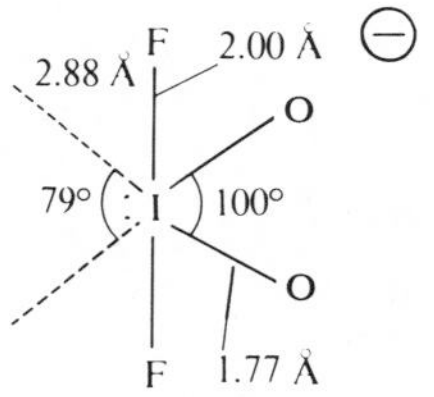   (e) 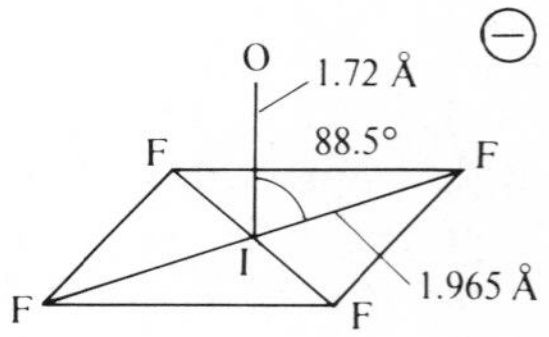   (f)

Three ions formed by I^V illustrate the stereochemistry of an atom with a lone pair which is bonded to 3, 4, or 5 ligands respectively (valence groups 2,6, 2,8, and

2,*10*). The arrangements of the 4, 5, and 6 electron pairs are respectively tetrahedral, trigonal bipyramidal, and octahedral.

The salt $KCrIO_6$, prepared from $K_2Cr_2O_7$ and HIO_3 in aqueous solution, contains ions (d) consisting of a tetrahedral CrO_4 group sharing one O atom with a pyramidal IO_3 group.[16] Potassium fluoroiodate, KIO_2F_2, prepared by the action of HF on KIO_3, contains ions with the structure (e). The four bonds are disposed towards four of the vertices of a trigonal bipyramid, one equatorial bond position being occupied by the lone pair. The plane of the I and O atoms is normal to the F—I—F axis, and the O—I—O bond angle is contracted to 100°.[17] In the anion in $CsIOF_4$, (f),[18] the ligands are situated at five of the vertices of an octahedron, the I atom lying slightly *below* the base of the square pyramid (contrast IO_5^{3-}, p. 405). The isoelectronic $XeOF_4$ molecule has a structure of the same kind but with F—Xe—O angle 91.8° (ref. p. 383).

(1a)	JACS 1967 **89** 6006	(10a)	AC 1970 **B26** 1782
(1b)	JCP 1972 **56** 1	(10b)	ZaC 1975 **411** 41
(2)	CR 1965 **260** 3974; ZaC 1970 **372** 127	(11)	AC 1966 **20** 765
(3)	AC 1976 **B32** 610	(12)	AC 1970 **B26** 1069
(4)	AC 1979 **B35** 2670	(13)	AC 1970 **B26** 1075
(5)	QRCS 1954 **8** 123	(14a)	AC 1965 **19** 629
(6)	AC 1949 **2** 128	(14b)	ZaC 1968 **362** 301
(7a)	AC 1966 **20** 769	(15)	ACSc 1966 **20** 2886
(7b)	AC 1977 **B33** 2795	(16)	ACSc 1967 **21** 2781
(8a)	JCS A 1970 1613	(17)	JCP 1976 **64** 3254
(8b)	AC 1977 **B33** 2918	(18)	AC 1972 **B28** 979
(9)	IC 1969 **8** 1190		

Halides of metals

The majority of metallic halides are solids at ordinary temperatures and relatively few consist of finite molecules in the crystalline state. Since these compounds melt and vaporize to finite molecules or ions a comprehensive review would call for a knowledge of their structures in the solid, liquid, and vapour states. Much of the available information relates to the solid state, but we shall indicate briefly what is known about these compounds in other states of aggregation.

Metal halides form a very large group of compounds for the following reasons.

(a) There are some 80 metals and four halogens (excluding astatine, about which very little is as yet known, see p. 386), and moreover most transition metals and many B subgroup metals form more than one compound with a given halogen. For example, Cr forms all five fluorides from CrF_2 to CrF_6 inclusive (and also Cr_2F_5, see below) and CrF has been identified as a vapour species formed from CrF_2 and Cr in a Knudsen cell. The iodides of Nb include Nb_6I_{11}, Nb_3I_8, NbI_3, NbI_4, and NbI_5. A particular halide may have more than one crystalline form; polymorphism is fairly common in halides.

(b) Apart from *solid solutions* of two or more halides and polyhalides (in which the halogen atoms are associated together in a polyhalide ion, as described earlier in this chapter) there are metal halides containing more than one halogen. Little is yet known of the structures of these compounds, but there are probably a number of types. Some are ionic (for example, SrClF, BaClF, and BaBrF),[1] others apparently molecular ($TiClF_3$, $TiCl_2F_2$, $WClF_5$, etc.). Compounds formed by Group V metals exist, like those of P, at least in some cases,[2] both as covalent liquids or low-melting molecular crystals and as ionic crystals. For example, liquid $SbCl_2F_3$ and crystalline $(SbCl_4)(SbF_6)$, molecular $NbCl_4F$ and $(NbCl_4)F$. Mixed halides such as $UClF_3$ may have structures related to those of the corresponding fluorides.

(c) In addition to normal halides MX_n there are halides in which:

(i) A metal exists in two definite oxidation states with quite different environments forming a stoichiometric compound with a simple formula. Examples include $Ga^I Ga^{III} Cl_4$ ($GaCl_2$), $Pd^{II} Pd^{IV} F_6$ (PdF_3), and Cr_2F_5[3] (p. 224).

(ii) A metal exists in two oxidation states in a phase exhibiting a range of composition. The coloured halides intermediate in composition between SmF_2 and SmF_3[4] (cubic $SmF_{2.00-2.14}$, tetragonal $SmF_{2.35}$, and rhombohedral $SmF_{2.41-2.46}$) result from addition of F^- ions to the fluorite structure of SmF_2 accompanied by replacement of some Sm^{2+} by Sm^{3+}.

(iii) There is metal–metal bonding. The mean oxidation state is abnormally low, and may be integral (BiBr, p. 877), ScCl, ZrCl, ZrI_2) or non-integral (Ag_2F, Nb_3Cl_8). We describe the structures of some 'sub-halides' containing metal clusters in two later sections on pp. 432 and 437.

(1) JCP 1968 **49** 2766   (3) AC 1964 **17** 823
(2) ZaC 1964 **329** 172   (4) IC 1970 **9** 1102

The structures of crystalline halides MX_n

We may make two generalizations about crystalline metal halides. First, fluorides differ in structure from the other halides of a given metal except in the case of molecular halides (for example, SbF_3 and $SbCl_3$ both crystallize as discrete molecules) and those of the alkali metals, all the halides of which are essentially ionic crystals. In many cases the fluoride of a metal has a 3D structure whereas the chloride, bromide, and iodide form crystals consisting of layer, or sometimes chain, complexes. (For exceptions, particularly fluorides MF_3–MF_6, see Table 9.9.) Second, many fluorides and oxides of similar formula-type are isostructural, while chlorides, bromides, and iodides often have the same types of structure as sulphides, selenides, and tellurides. The following examples illustrate these points:

$$
\left.
\begin{array}{l}
FeF_2 \\
PdF_2 \\
SnO_2
\end{array}
\right\}
\begin{array}{l}
\text{rutile} \\
\text{structure}
\end{array}
\qquad
\begin{array}{ll}
FeCl_2 & \text{layer structure} \\
PdCl_2 & \text{chain structure (and hexamer)} \\
SnS_2 & \text{layer structure}
\end{array}
$$

Metal halides provide examples of all the four main types of crystal structure: 3D complexes, layer, chain, and molecular structures, as shown in Table 9.9. The great

TABLE 9.9
Structures of crystalline metal halides

	C.n. of M	3D complex	Layer		Chain		Molecular	
MX	4	Zn blende, wurtzite	5+2	TlI (yellow)	2	AuI		
	6	NaCl						
	8	CsCl						
MX$_2$	4	Silica-like structures	4	HgI$_2$ (red)	3	SnCl$_2$	2	HgCl$_2$
	4	ZnI$_2$	4	HgI$_2$ (orange)	3	GeF$_2$		
	6	Rutile	6	$\begin{cases} CdI_2 \\ CdCl_2 \end{cases}$	4	BeCl$_2$	4	Pt$_6$Cl$_{12}$
	6	CaCl$_2$			4	PdCl$_2$	4	Pd$_6$Cl$_{12}$
	7	SrI$_2$, EuI$_2$	6	ThI$_2$				
	7+2	PbCl$_2$						
	8	Fluorite						
MX$_3$	6	ReO$_3$ and related structures (Table 9.16)	6	$\begin{cases} YCl_3 \\ BiI_3 \end{cases}$	4	AuF$_3$	3	SbF$_3$
					6	ZrI$_3$	4	Al$_2$Cl$_6$
	7+2	LaF$_3$						
	8+1	YF$_3$	8+1	PuBr$_3$			4	Au$_2$Cl$_6$
	9	UCl$_3$						
MX$_4$	6	IrF$_4$	6	PbF$_4$	5	TeF$_4$	4	SnBr$_4$
	8	ZrF$_4$	7	UBr$_4$	6	α-NbI$_4$	4	SnI$_4$
					6	ZrI$_4$		
					6	HfI$_4$		
	8	UCl$_4$	8	ThI$_4$	6	TcCl$_4$		
					6	ReCl$_4$		
MX$_5$	8	β-UF$_5$			6	BiF$_5$	5	SbCl$_5$
					6	CrF$_5$	6	Nb$_2$Cl$_{10}$
					7	PaCl$_5$	6	Mo$_4$F$_{20}$
MX$_6$					6	IrF$_6$		

majority of halides MX, MX$_2$, and MX$_3$ adopt structures shown to the left of the heavy line in the table, and most monohalides and most fluorides MF$_2$ and MF$_3$ crystallize with one of the following highly symmetrical structures suited to (though not exclusive to) essentially ionic crystals:

	Structure	Environment of M	Environment of X
MX	NaCl	Octahedral	Octahedral
	CsCl	Cubic	Cubic
MX$_2$	Rutile	Octahedral	Triangular
	Fluorite	Cubic	Tetrahedral
MX$_3$	ReO$_3$	Octahedral	Linear

These and most of the other structures of Table 9.9 have been described and illustrated in Chapters 3-6. We shall be more concerned here with structures that have not previously been described and we shall deal in turn with the six horizontal groups of halides of Table 9.9; a note will be included on the only two known metal heptahalides.

Monohalides

Alkali halides. At ordinary temperature all the alkali halides crystallize with the NaCl structure except CsCl, CsBr, and CsI, which have the CsCl structure. The latter is more dense (for a given halide) than the NaCl structure and is adopted under pressure by the Na, K, and Rb salts, but no structural changes have been induced in Li salts under pressure.[1] The reverse change, from CsCl to NaCl structure, takes place on heating CsCl to 469 °C, but neither CsBr nor CsI undergoes this transformation, at least up to temperatures within a few degrees of their melting points.[2] However, all the halides which normally crystallize with the CsCl structure can be grown with the NaCl structure from the vapour on suitable substrates (NaCl, KBr).[3] The interionic distances are listed in Table 9.10. (At 90 K LiI changes to a h.c.p. structure.)[4]

TABLE 9.10

Interionic distances in Cs and Tl halides

	NaCl *structure*	CsCl *structure*
CsCl	3·47 Å	3·56 Å
CsBr	3·62	3·72
CsI	3·83	3·95
TlCl	3·15	3·32
TlBr	3·29	3·44
TlI	3·47	3·64

Cuprous halides.[5] The cuprous halides (other than CuF, which has not been prepared pure at ordinary temperatures) crystallize with the zinc-blende structure. At the temperatures 435 °C, 405 °C, and 390 °C respectively CuCl, CuBr, and CuI transform to the wurtzite structure, and the last two exihibit a further transition at higher temperatures.

Argentous halides[5] AgF, AgCl, and AgBr crystallize with the NaCl structure and AgI adopts this structure under pressure. The iodide crystallizes with both the zinc-blende and wurtzite structures, the former being apparently metastable. Silver iodide transforms at 145.8 °C to a high-temperature form which is notable for its high ionic conductivity (1.3 ohm^{-1} cm^{-1} at 146 °C). In this form the arrangement of iodine atoms is b.c. cubic, that is, each I has only 8 I neighbours as opposed to 12 in the low-temperature polymorphs. The X-ray measurements, the high conductivity, and self-diffusion show that the silver ions move freely between positions of 2-, 3-, and 4-fold coordination between the easily deformed iodide ions. There are at least three high-pressure forms, one with the NaCl structure (Ag–I, 3.04 Å; compare 2.81 Å in the ZnS-type structures).[6]

Aurous halides. Structural information is available only for AuI, which consists of chain molecules with the structure[7]

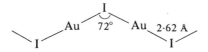

Subgroup IIIB *monohalides.* Gallium monohalides other than GaI (p. 1149) are known only as vapour species, and little seems to be known about InF. The structures of other monohalides of this group are summarized in Table 9.11. InCl has a yellow

TABLE 9.11

Structures of crystalline monohalides

	F	Cl	Br	I
In	–	N*/T[(1)]	T[(2)]	T[(3)]
Tl	?/N*[(4)]	C	C	T/C

N*: distorted NaCl structure   C: CsCl structure   T: TlI structure
(1) AC 1978 **B34** 3333.   (2) AJSR 1950 **3A** 581.   (3) AC 1955 8 847.   (4) JCS D 1974 1907

form stable below 390 K and a red high-temperature form. Red InCl, InBr, InI, and yellow TlI have the (B33) layer structure of Fig. 6.1 (p. 239). In TlI the nearest neighbours of Tl are: one I at 3.36 Å, four at 3.49 Å, and two at 3.87 Å, and two Tl at 3.83 Å. The two structures N^x in Table 9.11 are entirely different distortions of the NaCl structure. The yellow (α)-InCl contains four types of non-equivalent In^+ (and of Cl^-) ions. A detailed analysis[8] of the $InCl_6$ (and $ClIn_6$) groups shows that they represent two of the three main types of distortion of octahedral groups containing $(ns)^2$ ions. These have the lone pair directed along (a) a 4-fold, (b) a 2-fold, or (c) a 3-fold axis. The first is unlikely because the lone pair points towards a ligand, but is stabilized by removing the latter and replacing it by other ligands to provide more space, as in yellow TlI. In type (b) distortion the lone pair is directed towards the mid-point of an edge of the octahedron, which is elongated; in type (c) distortion the octahedron has one large face. In α-InCl the distortions are of types (b) and (c), the (mean) In-Cl distances being:

In_I :   2.93 (two), 3.26 (two), and 3.48 Å (two); type (b).

In_{II} :
In_{III}: } 2.91 (three), 3.50 Å (three); type (c).
In_{IV}:

There are two forms of TlF. There seems to be some doubt about the structure of tetragonal TlF, but orthorhombic TlF has a very distorted NaCl structure consisting of double layers. There are two kinds of Tl^+ ion, both having very distorted octahedral coordination with 4 neighbours closer than the other two: Tl_I 2.54, 2.62, 2.79 (two), 3.25, 3.50; Tl_{II} 2.25, 2.52, 2.67 (two), 3.07, and 3.90 Å. The four neighbours all lie to one side of the Tl^+ ion, and the longest bonds correspond to splitting the NaCl structure into double layers, within which Tl^+ has four close and one more distant neighbour. The layer is rather similar to that in yellow PbO. Under high pressure TlCl, TlBr, and TlI become metallic conductors; there is a very large (33 per cent) reduction in volume of TlI under a pressure of 160 kbar.[9]

Other monohalides include HfCl and ZrCl in Group IV and the solitary OsI in Group VIII.

(1) JCP 1968 48 5123   (6) JCP 1968 48 2446
(2) JACS 1955 77 2734   (7) ZK 1959 112 80
(3) AC 1951 4 487   (8) JSSC 1980 34 301
(4) ZP 1956 143 591   (9) JCP 1965 43 1381
(5) AC 1964 17 1341; JPC 1964 68 1111

Dihalides

It is convenient to divide these compounds into five groups.

(i) *Tetrahedral structures.* This group includes the silica-like structures of BeF_2 (low- and high-cristobalite and quartz structures) and the chain structure of one form of $BeCl_2$, which is trimorphic. The unique chain structure is that of the form

$$\text{>Be}\underset{Cl}{\overset{Cl}{<}}\text{>Be}\underset{Cl}{\overset{Cl}{<}}\text{>Be}<$$

stable over the temperature range 403–425 °C (m.p.), but this structure is formed by quenching the melt and is also formed on sublimation.[1] The structures of $BeBr_2$ and BeI_2 are not known.

The crystal chemistry of $ZnCl_2$ is more complex than at first appeared to be the case. Three polymorphs were originally described,[2a] two with 3D vertex-sharing structures and the third with the layer structure of red HgI_2 (p. 196). All are based on approximately closest packings of Cl and have the following layer sequences: α, ABC ...; β, ABAC ... (*hc*); and γ, ABC .... It now appears[2b] that the structure is very sensitive to the presence of water. If water is rigorously excluded during crystallization an orthorhombic h.c.p. (AB ...) structure is formed, and this changes into one of the other structures on exposure to the atmosphere. A possible explanation is that the presence of OH^- ions facilitates the rearrangement of the layers, the h.c.p. structure being the only one stable for pure anhydrous $ZnCl_2$. All four structures consist of tetrahedral $ZnCl_4$ groups each sharing vertices with 4 others.

Zinc iodide has a unique structure.[3] One quarter of the tetrahedral interstices in a c.c.p. array of I atoms are occupied in groups of four to form 'super-tetrahedra' (p. 196) Zn_4I_{10}. These groups are joined by sharing 4 vertices to form a 3D structure – compare the layer structure of orange HgI_2, which is also built of similar groups (p. 1161).

(1) JPC 1965 **69** 3839   (2b) IC 1978 **17** 3294
(2a) ZK 1961 .115 373   (3) AC 1978 **B34** 3160

(ii) *Octahedral structures* (Mg *and the* 3d *metals*). The structures containing ions of these metals, with radii close to 0.7 Å, are summarized in Table 9.12. The two

TABLE 9.12

Crystal structures of dihalides

	Mg	Ti	V	Cr	Mn	Fe	Co	Ni	Cu	Zn	
F_2	R		R	R*	R^a	R	R	R	R*	R	
Cl_2	C	I	I	R**	C	C	C	C	I*		tetrahedral structures
Br_2	I	I	I	I*	I	I	C/I	I	I*		
I_2	I	I	I	I*	I	I	I	C?			

R = rutile. R* and R** = distorted rutile structures. C = $CdCl_2$. I = CdI_2 structure. I* = distorted CdI_2 structure.
^{a} For detailed studies of 3d difluorides (rutile structure) see: JACS 1954 **76** 5279; AC 1958 **11** 488.

modified forms of the rutile structure, R* and R**, are quite different in nature, as described in Chapter 6. In CrF_2 (and the isostructural CuF_2) there is (4+2)-coordination of the metal, the four stronger bonds delineating layers based on the simple 4-gon plane net. In $CrCl_2$, on the other hand, the four stronger bonds define chains of the same kind as in $PdCl_2$ (and $CuCl_2$). Details of the Cu compounds are given in Chapter 25; for the Cr compounds details are given in Table 9.13.

(iii) *Dihalides of second and third series transition metals.* These elements present a very different picture from the 3d metals (Table 9.14). The following points are noteworthy: the almost complete absence of difluorides, the absence of dihalides of Tc, Rh, Hf, Ta, and Ir, the formation by Hf, Nb, and Ta of halides with non-integral oxidation numbers of the metal, and the presence of 'metal clusters' in compounds of Nb, Ta, Mo, W, Pd, and Pt. We deal with this last group of halides in a later section.

The only difluoride in this group, PdF_2, has the normal rutile structure. $PdCl_2$ is

TABLE 9.13

Coordination of Cr *in dihalides*

	4 X *at*	2 X *at*	*Reference*
CrF_2	2·00 Å	2·43 Å	PCS 1957 232
$CrCl_2$	2·39	2·91	AC 1961 **14** 927; HCA 1961 **44** 1049
$CrBr_2$	2·54	3·00	AC 1962 **15** 672
CrI_2	2·74	3·24	AC 1962 **15** 460; AC 1973 **B29** 1560

TABLE 9.14

Dihalides etc. of elements of 2nd and 3rd transition series

			Tc		Rh	
—	NbF$_{2.5}$	—		—		PdF$_2$
ZrCl$_2$	NbCl$_{2.33}$	MoCl$_2$	none known	RuCl$_2$	none known	PdCl$_2$
—	NbBr$_2$	MoBr$_2$		—		PdBr$_2$
ZrI$_2$	NbI$_{1.83}$	MoI$_2$		—		PdI$_2$
					Ir	
—	—	—	—	—		—
HfCl$_{2.5}$	TaCl$_{2.5}$	WCl$_2$	—	OsCl$_2$	none known	PtCl$_2$
—	TaBr$_{2.33}$	WBr$_2$	—	OsBr$_2$		PtBr$_2$
—	—	WI$_2$	ReI$_2$	OsI$_2$		PtI$_2$

apparently trimorphic,[1] one form (stable over the intermediate temperature range) having the simple chain structure of Fig. 9.4.[2] A second form[3] consists of hexameric molecules Pd$_6$Cl$_{12}$ (isostructural with Pt$_6$Cl$_{12}$ and Pt$_6$Br$_{12}$)[4] with the structure described on p. 436. Crystalline PdBr$_2$ is built of chains of the same general

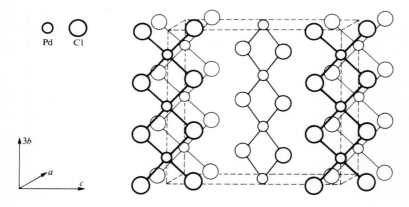

FIG. 9.4. The crystal structure of α-PdCl$_2$.

type as in α-PdCl$_2$ but here they are puckered, and apparently two of the Pd–Br bonds are shorter than the other two (2.34 and 2.57 Å).[5] The structure of PdI$_2$ is not recorded. It is interesting that the dichlorides of Ni, Pd, and Pt provide examples of three different structures:

		C.n. of M
NiCl$_2$:	CdCl$_2$ layer structure	6
PdCl$_2$:	chain structure and hexamer	4
PtCl$_2$:	hexamer Pt$_6$Cl$_{12}$	4

The chain structure of Fig. 9.4 has also been assigned to a form of $PtCl_2$, but on this point see ref. (4).

(1) JPC 1965 **69** 3669
(2) ZK 1938 **100** 189
(3) AnC 1967 **79** 244

(4) ZaC 1965 **337** 120
(5) ZaC 1966 **348** 162

(iv) *Dihalides of the alkaline-earths*, Cd, Pb, 4f *and* 5f *elements*. We include here Cd, which represents a case intermediate between our classes (iii) and (iv) and Pb, which clearly belongs in this class.

The alkaline-earth halides are surprisingly complex from the structural standpoint, the twelve compounds exhibiting at least six different structures. The fluorides adopt the simple fluorite structure (and the $PbCl_2$ structure under pressure), but the complexity of the other halide structures arises because the metal ions are too large to form the polarized AX_2 layer structures, and instead form structures with rather irregular 7- or 8-coordination. The ions Eu^{2+}, Sm^{2+}, and Pb^{2+} are similar in size to Sr^{2+} and show a behaviour very similar to that of the alkaline-earths (Table 9.15). The di-iodides of Tm and Yb crystallize with the CdI_2 structure, like CaI_2

TABLE 9.15

Structures of some crystalline dihalides

	Ca	Sr	Ba	Eu	Sm	Pb[10]	Cd	Am[11]
F	F/P	F/P[1]	F/P[1]	F	F	F/P[6]	F	
Cl	C[4]	C/F[3]	F/P[2]	P	P	P[3]	L	P
Br	C[5]	S_b[8]	P[2]	S_b[1]	S_b[1]	P	L	S_b
I	L	S_i[9]	P[2]	E[7]/S_i			L	L	E

L = $CdCl_2$ or CdI_2 layer structure. R = rutile structure. C = $CaCl_2$ structure. P = $PbCl_2$ structure. E = EuI_2 (monoclinic) structure. S_b = $SrBr_2$ structure. S_i = SrI_2 structure. F = CaF_2 structure.

(1) MRB 1970 **5** 759. (2) JPC 1963 **67** 2132. (3) JPC 1963 **67** 2863. (4) AC 1965 **19** 1027. (5) JINC 1963 **25** 1295. (6) AC 1967 **22** 744. (7) AC 1969 **B25** 1104. (8) IC 1971 **10** 1458. (9) ZaC 1969 **368** 62. (10) AC 1969 **B25** 769 (mixed halides of Pb). (11) JINC 1972 **34** 3427; JINC 1973 **35** 483.

and PbI_2. The unique structure of the golden-yellow ThI_2, which exhibits metallic conduction, is described in Chapter 4 as a 'hybrid' intermediate between the CdI_2 and MoS_2 structures. There is both octahedral and trigonal prismatic coordination of Th atoms, the bond length Th–I (3.20, 3.22 Å) being the same as the mean value in ThI_4. The compound is presumably $Th^{IV}(e)_2I_2$, two electrons being delocalized within the layers.[1] Two black chlorides $DyCl_2$ and $DyCl_{2.08}$, have been described, the former being isostructural with $YbCl_2$, but the structures do not appear to be known with certainty.[2] Halides of Nd(II) include $NdCl_{2.3}$, $NdCl_{2.2}$, $NdCl_2$ ($PbCl_2$ structure), and $NdI_{1.95}$ ($SrBr_2$ structure).[3]

In addition to the rutile and fluorite structures there are some 3D structures in Table 9.15 in which the environments of the ions are less symmetrical. The $CaCl_2$ structure (C) is closer to the ideal h.c.p. AX_2 structure (p. 168) than is the rutile structure; there is slightly pyramidal coordination of Cl^- (Br^-) suggesting a tendency towards the CdI_2 layer structure with its much more pronounced pyramidal coordination of the anions.

The $PbCl_2$ structure to which we refer here is that of salt-like compounds (dihalides, Pb(OH)Cl, CaH_2). There is considerable variation in axial ratios and atomic coordinates for compounds with this structure, and consequent changes in coordination number, so that in effect there are a number of different structures which should not all be described as the same '$PbCl_2$ structure'; this point is discussed in Chapter 6. The general nature of the structure is most easily visualized if the coordination group around Pb^{2+} is regarded as a tricapped trigonal prism (Fig. 6.23, p. 274), but this coordination group is very irregular. There are always two rather distant neighbours, and the structure is more accurately described as one of $7:\frac{3}{4}$ coordination. In the accurately determined structure of the low-temperature (or high-pressure) polymorph of PbF_2 with this structure the interionic distances are:

Pb–F: 2.41, 2.45 (two), 2.53, 2.64, 2.69 (two), and 3.03 Å (two).

The mean of the first seven distances (2.55 Å) is very close to the value (2.57 Å) for 8-coordinated Pb^{2+} in the polymorph with the fluorite structure. One-half of the X^- ions have their three close cation neighbours all lying to one side. This is true for all the anions in the UCl_3 structure (Fig. 9.8, p. 422), so that although these are both 3D structures some or all of the anions have the unsymmetrical environment of anions which is a feature of the layer, chain, and molecular structures.

The Eu dihalides are particularly interesting, for all four have different structures. The iodide is dimorphic; one form (orthorhombic) is isostructural with SrI_2, and the other has monoclinic symmetry. These are both structures of $7:\frac{3}{4}$ coordination, the 7-coordination being similar to that of Zr in baddeleyite. The chief difference between monoclinic EuI_2 and SrI_2 lies in the bond angles at the 3-coordinated I^- ions; these are two of 100° and one of 140° and one of 101° and two of 127° respectively (compare ZrO_2, 104°, 109°, and 146°). The three dihalides $AmCl_2$, $AmBr_2$, and AmI_2 are isostructural with the corresponding Eu dihalides.

In the inexplicably complex $SrBr_2$ structure there are two kinds of Sr^{2+} ion with different arrangements of 8 neighbours. One type of ion has 8 Br^- at the vertices of an antiprism (3.14 Å), while the other has a very irregular arrangement of nearest neighbours: 6 at a mean distance of 3.14 Å, 1 at 3.29 Å, and 1 at 3.59 Å.

(v) B *subgroup dihalides.* The structures of these compounds, which are usually characteristic of one compound or a small number of compounds, are described in Chapters 25 and 26; they include compounds of Cu, Ag, Hg, Ga, Ge, and Sn.

(1)   IC 1968 7 2257   (3)   IC 1964 3 993
(2)   IC 1966 5 938

Trihalides

Here it is sufficient to recognize three groups:

    (i) octahedral structures (Al, Sc, transition metals, In, Tl);

    (ii) structures of higher coordination (chiefly 4f and 5f metals);

    (iii) structures peculiar to small numbers of compounds.

(i) *Octahedral* MX_3 *structures.* Corresponding to the rutile structure for fluorides and the h.c.p. (CdI_2) and c.c.p. ($CdCl_2$) layer structures for other dihalides we have the 3D ReO_3 structure (and variants) and the BiI_3 and YCl_3 layer structures; the last two are the structures of the low- and high-temperature forms respectively of $CrCl_3$ and $CrBr_3$. In addition, the ZrI_3 chain structure (with face-sharing octahedra) is adopted by a number of chlorides, bromides, and iodides.

Trifluorides. In Chapter 5 we surveyed the structures in which octahedral groups are joined by sharing only vertices, and we noted that the structure in which all vertices are shared can have an indefinite number of configurations. Two special cases may be recognized, one in which the X atoms are arranged in hexagonal closest packing, and the other in which these atoms occupy three-quarters of the positions of cubic closest packing. A number of trifluorides adopt the h.c.p. structure, but in most of these compounds the packing of the F atoms is of an intermediate kind (Table 9.16). Moreover it appeared that no trifluoride adopts the cubic ReO_3

TABLE 9.16
Crystal structures of trifluorides

	Angle M—F—M
H.c.p. structure	
PdF_3, RhF_3, IrF_3 (also $Pt(O, F)_3$)	132°
Intermediate packings	
RuF_3, MoF_3[a]	Values around
GaF_3, TiF_3, VF_3, CrF_3, FeF_3, CoF_3	150°
MnF_3[b]	
ScF_3,[c] InF_3[c] (AlF_3?)	
Cubic ReO_3 structure	
NbF_3[d]	180°

[a] Rhombohedral VF_3 type but close to h.c.p. structure.
[b] Unique distorted (monoclinic) VF_3 type due to Jahn–Teller effect.
[c] Close to cubic ReO_3 structure.
[d] CR 1971 **C273** 1093.
For details and references see: Structure and Bonding, 1967 **3** 1.

Al								
Sc	Ti	V	Cr	Mn	Fe	Co		Ga
	Nb	Mo			Ru	Rh	Pd	In
					Ir			
					h.c.p.			

structure, compounds with this structure being either non-stoichiometric (mixed valence) compounds such as $Nb(O,F)_3$ or $Mo(O,F)_3$ or compounds such as Nb^VO_2F. Recently, however, it has been claimed that pure stoichiometric NbF_3 with the ReO_3 structure is obtained by heating $3\ NbF_5 + 2\ Nb$ at $750°C$ under a pressure of 3.5 kbar. The precise nature of the anion packing is of interest as it is related to the M–F–M bond angle. There is an interesting connection, which is not understood, between the structures of these compounds and the position of M in the Periodic Table. Confirmation is desirable of the structure of AlF_3, in which the neighbours of Al are apparently 3 F at 1.70 and 3 F at 1.89 Å. The distorted octahedral coordination in the complex monoclinic structure of MnF_3 (Mn–F bond lengths, 1.79, 1.91, and 2.09 Å) is referred to on p. 324, and may be contrasted with the regular octahedral coordination in VF_3. It is interesting that in both MoF_3 and VF_3 the bond length M–F is equal to 1.95 Å yet the compounds are not isostructural. The compound with the analytical formula PdF_3 is $Pd^{II}Pd^{IV}F_6$ with Pd^{II}–F, 2.17 Å and Pd^{IV}–F, 1.90 Å.[1] The structure is of the same type as those of two families of complex halides, $Pd^{II}M^{IV}F_6$ (M = Ge, Sn, Pt) and $M^{II}Pd^{IV}F_6$ (M = Mg, Ca, Zn, Cd),[2] and is the $LiSbF_6$ structure described on p. 457. PtF_3 is isostructural with PdF_3.[3]

(1)   CR 1976 **C282** 1069   (3)   MRB 1976 **11** 689
(2)   ZaC 1968 **359** 160

Other trihalides. Most of the other halides of metals of this group adopt one or more of the following structures (Table 9.17):

BiI_3:   h.c.p. layer structure (low-$CrCl_3$)
YCl_3:   c.c.p. layer structure (high-$CrCl_3$)
ZrI_3:   chain structure

In the ZrI_3 structure the M–M distance is comparable with that in the metal because the MX_6 octahedra share opposite faces, but there is not necessarily metal–metal bonding. The magnetic moment of $ZrCl_3$ (0.4 BM) indicates considerable overlap of

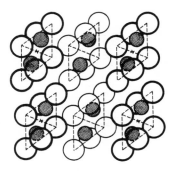

FIG. 9.5.   Projection [on (010)] of the crystal structure of $AlBr_3$ showing how molecules Al_2Br_6 arise by placing Al atoms in pairs of adjacent tetrahedral holes in a close-packed array of halogen atoms. The crosses mark the corners of the monoclinic unit cell.

TABLE 9.17
Crystal structures of trihalides

AlCl$_3$ (L)					
AlBr$_3$ (D)					
AlI$_3$ (D)					
ScCl$_3$ (L)	TiCl$_3$ (L/C)	VCl$_3$ (L)	CrCl$_3$ (L)[1]	FeCl$_3$ (L)	GaCl$_3$ (D)
	TiBr$_3$ (L/C)		CrBr$_3$ (L)[1]	FeBr$_3$ (L)	
	TiI$_3$ (C)	VI$_3$ (L)			GaI$_3$ (D)
YCl$_3$ (L)[2]	ZrCl$_3$ (L/C)			RuCl$_3$[3] (L/C)	InCl$_3$ (L)
	ZrBr$_3$ (C)		MoBr$_3$ (C)		
YI$_3$ (L)	ZrI$_3$ (C)				InI$_3$ (D)
	HfI$_3$ (C)		WCl$_3$ ReCl$_3$		TlCl$_3$ (L)
			see pp. 434, 432		

For the trihalides of As, Sb, and Bi see p. 884.
L = BiI$_3$ or YCl$_3$ layer structure; C = ZrI$_3$ chain structure; D = M$_2$X$_6$.

(1) JCP 1964 **40** 1958. (2) JPC 1954 **58** 940. (3) JCS A 1967 1038.
For various compounds of this group see also IC 1964 **3** 1236; IC 1966 **5** 281; IC 1969 **8** 1994.

metal orbitals, but the moment of TiCl$_3$ (1.3 BM) is not much less than the value expected for one unpaired electron (1.75 BM). Similarly, in CsCuCl$_3$ and CsNiCl$_3$ where the anions have this chain structure, the moments are normal for one and two unpaired electrons.[1]

We have included in Table 9.17 AlCl$_3$ and some B subgroup trihalides, but we should note the following points. In contrast to AlCl$_3$ AlBr$_3$ has a molecular structure. The Br atoms are close-packed and the Al atoms occupy pairs of adjacent *tetrahedral* holes (contrast the octahedral coordination in AlF$_3$ and AlCl$_3$), and the crystal is built of Al$_2$Br$_6$ molecules (Fig. 9.5). The same type of molecule is

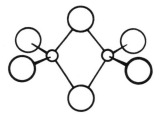

FIG. 9.6. The structure of the molecules Al$_2$Cl$_6$, Al$_2$Br$_6$, and Al$_2$I$_6$ according to Palmer and Elliot.

found in crystalline GaCl$_3$[2] and in InI$_3$, with which AlI$_3$ and GaI$_3$ are probably isostructural.[3] In InI$_3$ the bond lengths are: In–I$_t$, 2.64 Å, In–I$_b$, 2.84 Å; angles similar to those in GaCl$_3$. Dimeric molecules of the same kind exist in the vapours of both these halides and those of AlCl$_3$, AlBr$_3$, and AlI$_3$ (Fig. 9.6).

$$Cl \diagdown \underset{Cl \diagup}{Ga} \rangle 94° \diagdown \underset{Cl \diagup}{Ga} \rangle 123° \diagdown \underset{Cl}{Cl}$$

2·29(·09) Å   2·06(·03) Å

(1) IC 1966 5 277   (3) IC 1964 3 63
(2) JCS 1965 1816

(ii) *Structures of higher coordination*. The octahedral structures described in (i) are not possible for larger ions, and this is nicely illustrated by the Group IIIA trihalides, which have a structural chemistry comparable as regards its complexity with that of the alkaline-earth dihalides (Table 9.18). The nine compounds studied

TABLE 9.18
Structures of Group IIIA *trihalides*

	Sc	Y	La
F	ReO_3	YF_3[1]	LaF_3[2]
Cl	YCl_3	YCl_3	UCl_3
Br	–	–	UCl_3
I	–	BiI_3	$PuBr_3$

(1) JACS 1953 75 2453. (2) AC 1976 **B32** 94 (n.d.)

provide examples of no fewer than seven different structures, three layer structures (below the heavy line in the table) and four 3D structures.

The LaF_3 structure is also called the tysonite structure, after the mineral of that name which is a mixed fluoride, $(Ce, La, \ldots) F_3$. The structure earlier assigned to LaF_3 has been revised, and the description depends on the interpretation of the various La–F distances. The neighbours of La comprise: 7 F at 2.42–2.48 Å, 2 F at 2.64 Å, and a further 2 F at 3.00 Å. There are three kinds of non-equivalent F ion. If all the above F ions are counted as nearest neighbours the metal ions are 11-coordinated, two-thirds of F are 4-coordinated, and the remainder 3-coordinated. This 11-coordination group is a distorted trigonal prism capped on all faces. If only the 9 (or 7+2) F neighbours are included in the coordination group of the cation the structure is described as one of 9:3 coordination. There is no simple description of the cation coordination polyhedron, and the 3-coordination of the three kinds of fluoride ion ranges from nearly coplanar to definitely pyramidal. The structure is adopted by a number of 4f and 5f trifluorides (Table 9.19) and trihydrides (p. 348), some complex fluorides (e.g. $CaThF_6$, $SrUF_6$), and by $BiO_{0.1}F_{2.8}$ (p. 482). The structure of the high-temperature form of LaF_3 and the later 4f trifluorides is not known.

TABLE 9.19
Crystal structures of 4f and 5f trihalides

4f trihalides

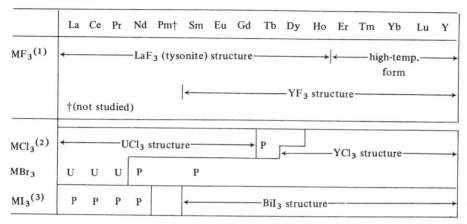

	La	Ce	Pr	Nd	Pm†	Sm	Eu	Gd	Tb	Dy	Ho	Er	Tm	Yb	Lu	Y
$MF_3^{(1)}$	←————LaF$_3$ (tysonite) structure————→	←——high-temp.——→
															form	
					←————————————YF$_3$ structure————————————→
	†(not studied)															
$MCl_3^{(2)}$	←————UCl$_3$ structure————→	P	←————YCl$_3$ structure————→
MBr$_3$	U	U	U	P		P										
$MI_3^{(3)}$	P	P	P	P		←————————BiI$_3$ structure———————— →

5f trihalides

	Ac	Th	Pa	U	Np	Pu	Am	Cm	Bk	Cf	Es
$MF_3^{(4)}$	L			L	L	L	L	L$^{(5)}$		L/Y$^{(5c)}$	
MCl$_3$	U			U	U	U	U	U$^{(5b)}$	U	U/P$^{(5a)}$	U$^{(5)}$
MBr$_3$	U			U	P	P$^{(5)}$	P$^{(5)}$	P$^{(5d)}$		A$^{(5d)}$	
MI$_3$				P$^{(7)}$	P	P$^{(3)}$	B$^{(3)}$	B$^{(5)}$			

A = AlCl$_3$ structure; B = BiI$_3$; L = LaF$_3$; P = PuBr$_3$; U = UCl$_3$;$^{(6)}$ Y = YF$_3$.
(1) IC 1966 **5** 1937. (2) IC 1964 **3** 185; JCP 1968 **49** 3007. (3) IC 1964 **3** 1137. (4) AC 1949
2 388. (5) IC 1965 **4** 985; INCL 1969 **5** 307. (5a) JINC 1973 **35** 1171. (5b) JINC 1973 **35**
1525. (5c) JINC 1973 **35** 3481. (5d) JINC 1975 **37** 743. (6) A refinement of the (UCl$_3$)
structure of GdCl$_3$ gives Gd–6 Cl, 2.82 Å, and Gd–3 Cl, 2.91 Å. (AC 1967 **23** 1112); and for
AmCl$_3$, Am–6 Cl, 2.874 Å, Am–3 Cl, 2.915 Å (AC 1970 **B26** 1885). A n.d. refinement of UCl$_3$
confirms the earlier structure: U–6 Cl, 2.931 Å, U–3 Cl, 2.938 Å, (AC 1974 **B30** 2803). (7)
AC 1975 **B31** 880 (n.d.): U–2 I, 3.165, U–4 I, 3.24, and U–2 I, 3.46 Å.

In the YF$_3$ structure there is also a distorted 9-coordination of M^{3+} (tricapped
trigonal prism), eight at approximately 2.3 Å and the ninth at 2.6 Å, so that the
coordination may be described as (8+1). This structure is adopted by the 4f
trifluorides SmF$_3$–LuF$_3$ and also by TlF$_3$ and β-BiF$_3$.$^{(1)}$

The cubic 'yttrium trifluoride' referred to in earlier literature is apparently
NaY$_3$F$_{10}$; other similar compounds include NH$_4$Ho$_3$F$_{10}$ and the Er and Tm
compounds. The structure of this type of compound is illustrated in Fig. 9.7. The
content of one unit cell comprises Na, 3 Y, and 10 F arranged at random in the

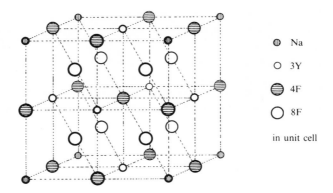

FIG. 9.7. The crystal structure of NaY_3F_{10}.

twelve positions shown by the larger circles.[2] A very similar cubic phase was also thought to be a cubic form of BiF_3, having a cell edge very similar in length to that of NaY_3F_{10}. It was assigned the structure of Fig. 9.7 with all metal ion positions occupied by Bi^{3+} and all twelve F^- positions occupied. (This would correspond to the fluorite structure with additional F^- ions—those shown as shaded circles at the body-centre and at the mid-points of the edges of the cubic unit cell.) This cubic material is actually Bi_2OF_4,[3] presumably with $Bi_4(O,F)_{10}$ in the unit cell; compare $(Na + 3 Y)F_{10}$. See also p. 897 for other oxyfluorides of bismuth.

The UCl_3 structure, in which the coordination group of the metal ion is a tricapped trigonal prism, is adopted by numerous 4f and 5f trihalides (Table 9.19) and also by the trihydroxides of La, Pr, Nd, Er, Sm, Gd, and Dy. Comparison of the plans of the UCl_3 structure (Fig. 9.8) and the $PbCl_2$ structure (Fig. 6.23, p. 274)

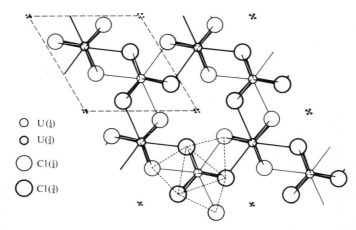

FIG. 9.8. Projection of the crystal structure of UCl_3 showing the tricapped trigonal prismatic coordination of U (see text).

shows that these two structures are related in much the same way as are the YCl_3 and $CdCl_2$ structures, though the latter are layer structures while UCl_3 and $PbCl_2$ are 3D arrangements of ions. We have remarked in Chapter 7 that the unsymmetrical arrangement of M ions to one side of Cl in these structures is similar to that in the layer structures.

In the UCl_3 structure the metal atom has nine approximately equidistant halogen neighbours (see ref. 6, Table 9.19). In the $PuBr_3$ structure, a projection of which is shown in Fig. 9.9, there is only 8-coordination of the metal atoms. This is a rather

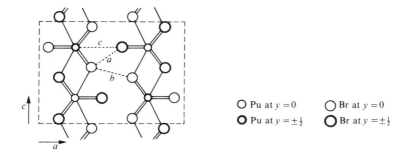

O Pu at $y = 0$ O Br at $y = 0$
O Pu at $y = \pm\frac{1}{2}$ O Br at $y = \pm\frac{1}{2}$

FIG. 9.9. The (layer) structure of $PuBr_3$. The planes of the layers are normal to that of the paper.

surprising structure in the sense that the halogen atoms are nearly in the positions required for 9-coordination of the metal atoms. The ninth Pu–Br bond would be that marked c, which would complete a coordination group of the same kind as in UCl_3. This distance is, however, 4.03 Å, as compared with 3.08 Å for the eight near neighbours. The layers are apparently prevented from approaching closer by the Br–Br contacts a and b which are unusually short for non-bonded Br atoms (3.81 and 3.65 Å respectively).

(1)   ACSc 1955 9 1206, 1209   (3)   ARPC 1952 3 369
(2)   JACS 1953 75 2453

(iii) *A number of structures peculiar to certain B subgroup halides* are described in other chapters: AuX_3 (Chapter 25), AsX_3, etc. (Chapter 20).

We noted earlier that PdF_3 is $(Pd^{II}F_6)(Pd^{IV}F_6)$, in which both types of Pd atom are in octahedral coordination. The structure of $PtBr_3$ is entirely different.[1] The Pt(II) atom forms Pt_6X_{12} hexamers as in $PtCl_2$ and $PtBr_2$, in which Pt has 4 coplanar Br at 2.45 Å (and Pt–Pt, 3.49 Å). The Pt(IV) atoms form the skew edge-sharing octahedral AX_4 chain (*cis* Br unshared) as in α-PtI_4, with Pt–Br, 2.44–2.57 Å. There are weak Pt–Br bonds (2.93 Å) between Pt(II) and Br atoms of the Pt(IV) chains.

(1)   AnCIE 1969 8 672

Tetrahalides

More than thirty metals form a tetrahalide with at least one of the halogens, but in relatively few cases are all four tetrahalides of a given metal known. All four tetrahalides are known of Ge, Sn, Ti, Zr, Hf, Nb, Ta, Mo, W, Th, and U, and at the other extreme MF_4 is the only tetrahalide known of Mn, Ru, Rh, Pd, Ir, Ce, Pr, and Tb. The crystal structures of more than one-half of the seventy or so known tetrahalides are known, and they include structures of all types, 3D, layer, chain, and molecular. Most of these structures have been described in previous chapters.

Tetrahedral molecules pack differently in $SnBr_4$[1] and SnI_4[2] with respectively h.c.p. and c.c.p. halogen atoms; compounds with the latter structure include GeI_4, $SnCl_4$, $TiBr_4$, TiI_4, and $ZrBr_4$; a second form of $TiBr_4$ is isostructural with $SnBr_4$, and ZrI_4 (see below) also crystallizes with the SnI_4 structure.

Octahedral MX_4 structures include examples of three of the four main types, namely, chain, layer, and 3D structures. In the 'trans' chain each octahedron shares a pair of opposite edges (α-NbI_4[3a], $NbCl_4$, $TaCl_4$, TaI_4, α-$MoCl_4$, WCl_4, $OsCl_4$, α-$ReCl_4$[3b]). In α-NbI_4 pairs of Nb atoms are alternately closer together and further apart (Fig. 9.10(a)), and the interaction between pairs of Nb atoms is sufficient to

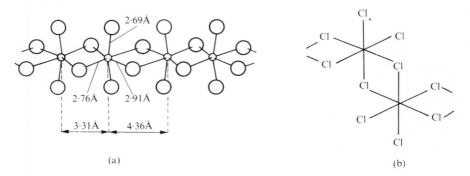

FIG. 9.10. Edge-sharing octahedral chains in (a) NbI_4, (b) $TcCl_4$.

destroy the paramagnetism expected for Nb(IV); the bridge bonds are appreciably longer than the terminal ones. In a skew chain two non-opposite edges of each octahedron are shared, and as noted on p. 212 there is an indefinite number of configurations of such a chain. In the simplest the metal atoms form a planar zigzag ($TcCl_4$,[4a] $ZrCl_4$,[4b] $PtCl_4$, PtI_4,[4c] UI_4[4d]); there are no m–m bonds in this chain (Fig. 9.10(b)); the shortest Tc–Tc distance in $TcCl_4$ is 3.62 Å. More complex configurations of the skew chain are found in HfI_4[5a] and ZrI_4[5b], with respectively 4 and 6 octahedra in the repeat unit of the chain as compared with 2 in $TcCl_4$.

The second form of $ReCl_4$ (β) has a quite different chain structure[6] in which pairs of face-sharing octahedra form an infinite chain by sharing one vertex at each end, as shown at (a).

(a)

In SnF_4 octahedral SnF_6 groups form a layer by sharing four equatorial F atoms as in the K_2NiF_4 structure (p. 208); NbF_4 and PbF_4 have the same structure.[7]

The first 3D octahedral structure for a tetrahalide has recently been assigned to IrF_4.[8a] In this structure each octahedral IrF_6 group shares four vertices, each with one other IrF_6 group, a pair of *cis* vertices being unshared. The structure is closely related to the rutile structure, from which it is derived by removing alternate metal atoms from each edge-sharing chain. This relationship may be compared with that between the NaCl and atacamite (AX_2) structures (Fig. 4.22, p. 170). Alternatively the IrF_4 structure may be described as related to the diamond net, since each Ir is bonded to four others through shared F atoms. The tetrafluorides of Rh, Pd, and Pt have the same structure.

It has been suggested, though not confirmed, that the structure of VF_4 is related to that of VOF_3.[8c] In the latter, edge-sharing pairs of octahedral coordination groups (here VOF_5) are further linked by sharing four of the remaining eight vertices to form layers as shown at (b). The structures of a number of 3d tetrahalides would

(b)

be of special interest because of the presence of excess d electrons, and in this connection we may note that the VCl_4 molecule was assigned a regular tetrahedral structure as the result of an early electron diffraction study of the vapour.[8d]

Other octahedral structures include that suggested for high-$MoCl_4$[9] that is, random occupation of three-quarters of the metal positions in the BiI_3 structure. There would be isolated $MoCl_6$ octahedra and portions of edge-sharing chains.

For a rare example of 7-coordination in a MX_4 structure see UBr_4 (p. 1258).

Of three structures with 8:2 coordination, two are examples of antiprism coordination and one of dodecahedral coordination. The remarkable ThI_4 structure[10] consists of layers formed from square antiprisms ThI_8 which share two triangular faces and one edge; it was illustrated in Chapter 3 as an example of a layer based on the plane 6-gon net. In contrast to $ZrCl_4$ and $ZrBr_4$, ZrF_4[11] has a typically ionic structure of 8:2 coordination. Both types of non-equivalent Zr^{4+} ion are surrounded by 8 F forming a slightly distorted square antiprism which shares vertices with eight others. This structure is confined to tetrafluorides of the larger M^{4+} ions (Hf, Ce, Pr,

Tb, and 5f elements Th–Bk). The tetrachlorides of Th, Pa, U, and Np (and also ThBr$_4$ and PaBr$_4$) adopt a different 8:2 structure. In the UCl$_4$ structure a metal atom has eight dodecahedral neighbours, and each coordination group shares one edge with each of four others in helical arrays around 4_1 axes, forming a 3D structure of 8:2 coordination (Fig. 9.11). In ThCl$_4$,[12] which has this structure, the

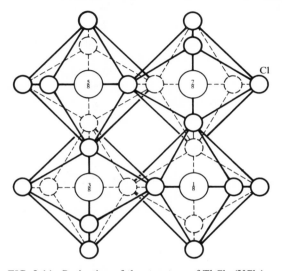

FIG. 9.11. Projection of the structure of ThCl$_4$ (UCl$_4$).

M–Cl distances are four of 2.72 Å and four of 2.90 Å. The configuration of the dodecahedral group is close to the value calculated for minimal ligand repulsion (see p. 79). A very similar coordination group is found in the isostructural PaBr$_4$[13] (Pa–Br, 2.83 Å and 3.01 Å), and in UCl$_4$[14] (U–4 Cl, 2.64, U–4 Cl, 2.87 Å).

For the unique structure of TeF$_4$, in which square pyramidal TeF$_5$ groups form infinite chains by sharing two F atoms, we refer the reader to Chapter 16 (Fig. 16.4).

The structures of crystalline tetrahalides provide a good illustration of the ways in which the same X:M ratio can be attained by sharing of increasing numbers of atoms between MX$_n$ coordination groups as n increases:

Coordination of M	C.n.	Elements of coordination polyhedra shared	Examples
Tetrahedral	4	None	SnI$_4$
Square pyramidal	5	2 vertices	TeF$_4$
Octahedral	6	4 vertices	SnF$_4$
		2 edges	TcCl$_4$, NbI$_4$
		1 face + 1 vertex	ReCl$_4$
Pentagonal bipyramidal	7	3 edges	UBr$_4$
Antiprismatic	8	8 vertices	ZrF$_4$
Dodecahedral	8	4 edges	ThCl$_4$
Antiprismatic	8	2 faces + 1 edge	ThI$_4$

(1)   AC 1963 **16** 446
(2)   AC 1955 **8** 343
(3a) AC 1962 **15** 903
(3b) IC 1968 **7** 2602
(4a) IC 1966 **5** 1197
(4b) ZaC 1970 **378** 263
(4c) ZaC 1969 **369** 154
(4d) IC 1980 **19** 672
(5a) JLCM 1980 **76** 7
(5b) AC 1979 **B35** 274
(6)   JACS 1967 **89** 2759

(7)   IC 1965 **4** 182
(8a) CR 1974 **278C** 1501
(8b) IC 1978 **17** 748
(8c) CC 1970 1475
(8d) JACS 1945 **67** 2019
(9)   ZaC 1967 **353** 281
(10) IC 1964 **3** 639
(11) AC 1964 **17** 555
(12) AC 1969 **B25** 2362
(13) JCS A 1971 908
(14) AC 1973 **B29** 1942

Pentahalides

Apart from the trigonal bipyramidal $SbCl_5$ molecule, which has the same structure in the crystalline as in the vapour state, we have to deal here with 6-, 7-, and 8-coordinated structures.

The octahedral structures, described in Chapter 5, include the dimeric and tetrameric molecules of types (a) and (b) (Fig. 9.12) and the '*cis*' and '*trans*' chains.

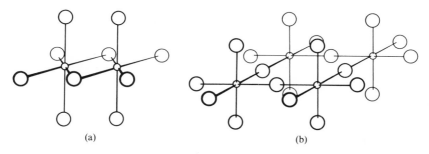

(a)   (b)

FIG. 9.12. Pentahalide molecules: (a) M_2Cl_{10}, (b) M_4F_{20}.

The dimers include chlorides, a bromide, and two iodides, while the cyclic tetramers and chains are fluorides. As in the case of other halides the molecules (a) or (b) can pack to give various types of closest packing of the halogen atoms. For both (a) and (b) the h.c.p. and c.c.p. structures are known, and for the dimers also the more complex *hc* packing (Table 9.20). In the dimeric molecules the metal atoms are displaced from the centres of the octahedra (by 0.2 Å in U_2Cl_{10}), and the bridge bonds are longer than the terminal ones (for example, 2.69 and 2.44 Å in U_2Cl_{10}).

The special interest of the two M_4X_{20} structures lies in the fact that the —F— bond angles are quite different, 180° (as shown at (b)) and 132° in the c.c.p. and h.c.p. structures respectively. (The cubic closest packing is distorted to give monoclinic symmetry in the former structures.) This is reminiscent of the trifluoride structures described earlier, as also is the choice of structure by the various pentafluorides (Table 9.20).

TABLE 9.20

Structures of crystalline pentahalides

Dimers M_2X_{10}

	Type of closest packing of X			*Reference*
	h	*c*	*hc*	
M_2Cl_{10}	Nb Ta Mo W			JCS A 1967 1825, 2017
		U		AC 1967 **22** 300
			Re	AC 1968 **B24** 874;
			Os	IC 1979 **18** 3081
M_2Br_{10}		β-Pa$_2$Br$_{10}$		AC 1969 **B25** 178
M_2I_{10}	Nb Ta			AC 1979 **B35** 2502

Tetrameric and linear pentafluorides

V[1]	Cr[1]				
Nb	Mo	Tc[1]	Ru	Rh[12]	
Ta	W[5]	Re[1]	Os[10]	Ir Pt	
					trans chain[6]
M_4F_{20}		*cis* chain	M_4F_{20}		—F— 180°
—F— 180°		—F— 150°	—F— 132°		BiF$_5$
c.c.p.			h.c.p.		α-UF$_5$
	also	also			pentagonal bipyramidal
	WOF$_4$[2]	MoOF$_4$[3]			chain
	NbCl$_4$F[8]	ReOF$_4$[4]			PaCl$_5$[7]
	TaCl$_4$F[9]	TcOF$_4$[11]			

(1) JCS A 1969 1651. (2) JCS A 1968 2074. (3) JCS A 1968 2503. (4) JCS A 1968 2511. (5) JCS A 1969 909. (6) JACS 1959 **81** 6375. (7) AC 1967 **22** 85. (8) ZaC 1968 **362** 13. (9) ZaC 1966 **346** 272. (10) JCS A 1971 2789. (11) CC 1967 462. Also green trimeric form with —F—, 161° (JCS A 1970 2521). (12) IC 1973 **12** 2640.

The two forms of octahedral MX_5 chain arise by sharing of adjacent or opposite vertices of octahedral MX_6 groups (*cis* and *trans* chains); here again it would be interesting to know what determines the choice of chain.

In the PaCl$_5$ structure (which is unique to that compound) pentagonal bipyramidal groups share two edges to form infinite chains (Fig. 9.13). Bond lengths are: Pa–Cl (bridge), 2.73 Å, Pa–Cl (terminal), 2.44 Å.

There is 8-coordination of U in β-UF$_5$ (p. 1255).

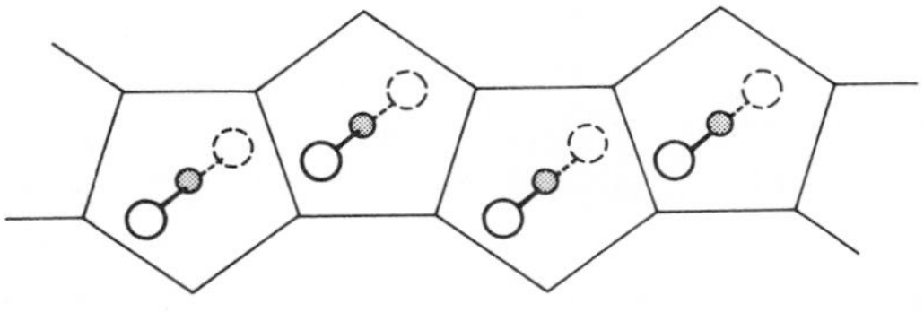

FIG. 9.13. Chain in crystalline PaCl$_5$.

It will be observed that apart from $SbCl_5$ 5-coordination is avoided in all the pentahalide structures described, as is also the case in crystalline PCl_5 and PBr_5.

Hexahalides

Although infinite complexes (for example, chains or layers built from 7-, 8-, or 9-coordination groups) are in principle possible, only octahedral molecular structures are as yet known for crystalline hexahalides. X-ray–powder–photographic examination of numerous hexafluorides of second- and third-row transition metals shows that they have isostructural low-temperature (orthorhombic) and high-temperature (disordered) b.c.c. polymorphs.[1] A more complete study has been made of the structurally similar low-temperature form of $Os^{VII}OF_5$,[2] which consists of nearly regular octahedral molecules. Like the halides M_2X_{10} the hexahalides adopt different kinds of closest packing of the halogen atoms; WCl_6 in fact has two h.c.p. forms differing in selection of octahedral holes occupied by the metal atoms:

$$UCl_6,^{(3)} \; WCl_6,^{(4)} \; MoCl_6$$
$$UF_6 \text{ (and } OsOF_5\text{), } MoF_6 \text{ (orthorhombic)}^{(5)}$$

A neutron diffraction study of UF_6[6] shows a significant departure from regular octahedral coordination, 5F at distances close to 1.91 Å and 1F at 2.28 Å, attributed to the unsymmetrical environment of the molecules in the structure.

(1) IC 1966 **5** 2187
(2) JCS A 1968 543
(3) AC 1974 **B30** 1481

(4) AC 1974 **B30** 1216
(5) AC 1975 **B31** 398
(6) AC 1973 **B29** 7

Heptahalides

The only known metal heptahalides are ReF_7 and OsF_7, the latter being stable only at low temperatures.[1] The only other heptafluoride known is IF_7. The evidence for the pentagonal symmetry of ReF_7 (and IF_7) has been summarized.[2]

No higher halides are known; the compound earlier described as OsF_8 proved to be OsF_6.

(1) CB 1966 **99** 2652

(2) JCP 1968 **49** 1803

Polynuclear complexes containing metal–metal bonds

Binuclear halide complexes formed by Mo, Tc, *and* Re

A number of elements form ions $M_2Cl_9^{n-}$ consisting of two octahedral groups sharing a face (p. 465), and similar units are joined by sharing one Cl at each end to form the infinite chain molecules in β-$ReCl_4$ (p. 424). Within the sub-units Re–Re is 2.73 Å, indicating a (single) metal–metal bond. The ions $Re_2Cl_9^{3-}$ and $Mo_2Cl_9^{3-}$ have a similar structure. The ion in the diamagnetic salt $M_3(Mo_2Cl_8H)^{(1)}$ was originally formulated

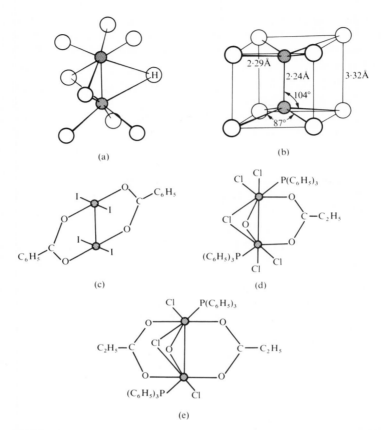

FIG. 9.14.   Binuclear halide ions and related molecules: (a) $Mo_2Cl_8H^{3-}$, (b) $Re_2Cl_8^{2-}$, (c) $Re_2I_4(OOC.C_6H_5)_2$, (d) $Re_2OCl_5(OOC.C_2H_5)(P\phi_3)_2$, (e) $Re_2OCl_3(OOC.C_2H_5)_2(P\phi_3)_2$.

$(Mo_2Cl_8)^{3-}$ with a structure like $(Mo_2Cl_9)^{3-}$ from which one bridging Cl has been removed (Fig. 9.14(a)). The replacement of one bridging halogen by H markedly decreases the length of the Mo-Mo bond:

	Mo–Mo		Mo–Mo
$Mo_2Cl_9^{3-}$	2.665 Å	$Mo_2Br_9^{3-}$	2.82 Å
$Mo_2Cl_8H^{3-}$	2.38	$Mo_2Br_8H^{3-}$	2.44

The $Mo_2Cl_8^{4-}$ ion has an entirely different structure, Fig. 9.14(b), which is also that of the $Tc_2Cl_8^{3-}$ and $Re_2Cl_8^{2-}$ ions. These ions have an extremely short M–M bond and an eclipsed configuration, the eight halogen atoms being at the vertices of an almost perfect cube. According to a m.o. treatment the metal–metal bond is regarded as quadruple ($\sigma\pi^2\delta$), the δ component accounting for the eclipsed configuration.

Substitution reactions may be performed on the Re_2X_8 nucleus, the general shape of which is retained in molecules such as $Re_2Cl_6(PEt_3)_2$, in which two (*trans*) X atoms are replaced by phosphine molecules, and $Re_2I_4(benzoate)_2$ (Fig. 9.14(c)); in both of these molecules Re–Re is close to 2.2 Å. In molecules such as (d) and (e), where there are bridging O and Cl atoms (and six ligands attached to each metal atom), the length of the metal–metal bonds is close to 2.5 Å. Note the range of Re–Re

TABLE 9.21

Metal–metal bond lengths in halide complexes

Ion or molecule	M–M (Å)	Reference
$ReCl_4$ (crystalline)	2·73	JACS 1967 89 2759
$(Re_2Cl_9)^-$	2·71	P. F. Stokely, Ph.D. Thesis M.I.T. 1969
$Re_2OCl_5(O_2CC_2H_5)(P\phi_3)_2$	2·52	IC 1968 7 1784
$Re_2OCl_3(O_2CC_2H_5)_2(P\phi_3)_2$	2·51	IC 1969 8 950
$Re_2I_4(O_2CC_6H_5)_2$	2·20	IC 1969 8 1299
$Re_2Cl_6(PEt_3)_2$	2·22	IC 1968 7 2135
$(Re_2Cl_8)^{2-}$	2·24	IC 1965 4 330, 334
$(Tc_2Cl_8)(NH_4)_3 . 2 H_2O$	2·13	IC 1970 9 789
$(Mo_2Cl_8)K_4 . 2 H_2O$	2·13	IC 1969 8 7
$(Mo_2Cl_8)Cl(NH_4)_5 . H_2O$	2·15	IC 1970 9 346
$(Mo_2Cl_8H)Cs_3$	2·38	IC 1969 8 1060
Re_3I_9	2·44, 2·51	IC 1968 7 1563
Re_3Cl_9	2·49	IC 1964 3 1402
$Re_3Cl_9(P\phi Et_2)_3$	2·49	IC 1964 3 1094
$(Re_3Cl_{11})(As\phi_4)_2$	2·44, 2·48	IC 1966 5 1758
$(Re_3Br_{11})Cs_2$	2·43, 2·49	IC 1966 5 1763
$(Re_3Cl_{12})Cs_3$	2·48	IC 1963 2 1166
$(Re_4Br_{15})(QnH)_2$	2·47	IC 1965 4 59

bond lengths, 2.2, 2.5, and 2.7 Å in these binuclear complexes. (For references see Table 9.21.)

(1)   IC 1976 **15** 522

Trinuclear halide complexes of Re

Rhenium forms a variety of ions and molecules based on a triangular unit of three Re atoms, for example, $Re_3Cl_9(pyr)_3$, $Re_3Cl_3(NCS)_8^{2-}$ and $Re_3Cl_3(NCS)_9^{3-}$, $Re_3X_{10}^-$, $Re_3X_{11}^{2-}$, and $Re_3X_{12}^{3-}$, (X = Cl, Br), the ions being isolated as salts of large cations such as Cs^+, $As(C_6H_5)_4^+$, etc. (Salts of the type $[N(C_2H_5)_4]_2Re_4Br_{15}$ do not contain Re_4 complexes but $ReBr_6^{2-}$ ions and Re_3Br_9 units.) The $Re_3Cl_{12}^{3-}$ ion, Fig. 9.15(a), is the anion in $CsReCl_4$. The Re–Re bond length, 2.48 Å, is much shorter than that in the metal (2.75 Å) and corresponds to bond order 2. The Re–Cl_a (2.36 Å) and Re–Cl_b

(2.39 Å) bonds are normal single bonds (compare 2.37 Å in $K_4ReOCl_{10}.H_2O$), and the length of Re–Cl_c (2.52 Å) has been attributed to overcrowding. The same Re_3 nucleus exists in $Re_3Cl_9(P\phi Et_2)_3$ and many other complexes, and the $Re_3Cl_{11}^{2-}$ ion has a structure very similar to $Re_3Cl_{12}^{3-}$ but devoid of one equatorial Cl atom.

The trihalides $ReCl_3$ and $ReBr_3$ have been shown by mass spectrometry to vaporize as Re_3X_9 molecules. Crystalline $ReCl_3$ contains Re_3Cl_9 molecules like the ions of Fig. 9.15(a) without the equatorial Cl_c atoms. These molecules are joined

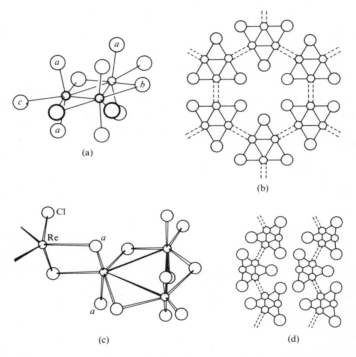

FIG. 9.15. Trinuclear halide complexes of Re: (a) $(Re_3Cl_{12})^{3-}$, (b) crystalline $ReCl_3$, (c) bridge in $ReCl_3$, (d) crystalline Re_3I_9.

through weaker Re–Cl bridges (2.66 Å) into layers based on the plane 6-gon net (Fig. 9.15(b)). In each bridge one Cl_a atom of one molecule occupies the Cl_c position in the adjacent molecule (Fig. 9.15(c)). In ReI_3 the Re_3X_9 molecules are of the same kind, but instead of being linked into layers by the bridges they are linked into chains (Fig. 9.15(d)).

Halide complexes of Nb, Ta, Mo, W, Pd, *and* Pt *containing metal 'clusters'*

The compounds to be discussed here are formed by certain metals of the second and third transition series,

Nb   Mo   Pd
Ta   W   Pt.

A limited number of B subgroup metals also form halides in which there is metal-metal bonding, notably Hg and Bi (p. 877); for mercurous compounds see Chapter 26, and for $Cu_4I_4(AsEt_3)_4$ see Chapter 25.

A feature of the chemistry of Nb and Ta is the formation of many halogen compounds in which the metal exhibits a non-integral oxidation number (Table 9.22). The halides are made by methods such as the following: Nb_3I_8 by thermal

TABLE 9.22

Some halides and halide complexes of Nb

Oxidation number of Nb							
1·83	2	2·33	2·50	2·67	3	4	5
		$(Nb_6Cl_{12})Cl_2$	Nb_6F_{15}		NbF_3	NbF_4	NbF_5
		$K_4(Nb_6Cl_{12})Cl_6$	$CsNb_4Cl_{11}$	Nb_3Cl_8	$NbCl_3$	$NbCl_4$	$NbCl_5$
		$(Nb_6Cl_{12})^{2+}$	$(Nb_6Cl_{12})^{3+}$	$(Nb_6Cl_{12})^{4+}$	$Cs_3Nb_2Cl_9$		
	$NbBr_2$						
Nb_6I_{11}				Nb_3I_8	NbI_3	NbI_4	NbI_5

decomposition of the higher iodides, Nb_6I_{11} by heating Nb_3I_8 with the metal, $CsNb_4Cl_{11}$ by a vapour transport reaction from a mixture of CsCl, Nb_3Cl_8, and Nb metal, and $K_4Nb_6Cl_{18}$ by heating together KCl and Nb_6Cl_{14}. A different set of compounds is obtained from solution, namely, those containing Nb_6X_{12} groups, which can be obtained in several different oxidation states. The $Nb_6Cl_{12}^{2+}$ ion may be regarded as the structural unit from which $(Nb_6Cl_{12})Cl_2$ and its octahydrate are formed. Oxidation of this ion in HCl–alcohol solution by air, followed by addition of NEt_4Cl, gives $(NEt_4)_3(Nb_6Cl_{12})^{3+}Cl_6$ ($\mu = 1.65$ BM), while oxidation by Cl_2 yields $(NEt_4)_2(Nb_6Cl_{12})^{4+}Cl_6$, with approximately zero moment.[1] Similarly, the $(Ta_6Cl_{12})^{n+}$ ion can be reversibly oxidized and reduced ($n = 2, 3, 4$).[2]

In all of the compounds of Table 9.22 except NbF_4 and the pentahalides there is some interaction between metal atoms. This ranges from bonding between alternate pairs of Nb atoms in the NbX_4 chain structure, through the formation of Nb_3 groups in the halides Nb_3X_8 and groups of four Nb atoms in $CsNb_4Cl_{11}$, to the formation of octahedral M_6 groups in all the compounds of Table 9.23.

Halides Nb_3X_8 form a layer structure[3] in which Nb atoms occupy $\frac{3}{4}$ of the octahedral holes between alternate layers of halogen atoms (Fig. 5.22(d), p. 215). The interaction between the Nb atoms is such that triangular Nb_3 groups can be distinguished which may be shown in idealized form as in Fig. 9.16(a), it being understood that these groups share all the outer Cl atoms with adjacent groups as

TABLE 9.23
Compounds containing M_6X_8 or M_6X_{12} 'clusters'

Type of complex		Reference
	M_6X_8	
Finite		
$(Mo_6Cl_8)Cl_6(NH_4)_2 . H_2O$		AK 1949 **1** 353
$(Mo_6Cl_8)(OH)_4(H_2O)_2 .12\,H_2O$		
1-dimensional		
$(W_6Br_8)Br_4 . Br_4$	W_6Br_{16}	ZaC 1968 **357** 289
2-dimensional		
$(Mo_6Cl_8)Cl_4$	$MoCl_2, Br_2, I_2$	ZaC 1967 **353** 281
	WCl_2, Br_2, I_2	
3-dimensional		
$(Nb_6I_8)I_3$	Nb_6I_{11}	ZaC 1967 **355** 295
$(Nb_6I_8)I_3H$	$Nb_6I_{11}H$	ZaC 1976 **355** 311
	$(Nb_6I_{11}H)Cs$	IC 1980 **19** 1241
	M_6X_{12}	
Finite		
Pd_6Cl_{12}	$PdCl_2$	AnC 1967 **79** 244
Pt_6Cl_{12}	$PtCl_2$	ZaC 1965 **337** 120
$(W_6Cl_{12})Cl_6$	WCl_3	AnC 1967 **79** 650
$(Ta_6Cl_{12})Cl_2(H_2O)_4 . 3\,H_2O$	$Ta_6Cl_{14} . 7\,H_2O$	IC 1966 **5** 1491
$(Nb_6Cl_{12}Cl_6)^{4-}$	$K_4Nb_6Cl_{18}$	ZaC 1968 **361** 235
$(Nb_6Cl_{12}Cl_6)^{2-}$	$[N(CH_3)_4]_2Nb_6Cl_{18}$	IC 1970 **9** 1347
$(Ta_6Cl_{12}Cl_6)^{2-}$	$H_2Ta_6Cl_{18} .6\,H_2O$	IC 1971 **10** 1460
2-dimensional		
$(Ta_6I_{12})I_2$	Ta_6I_{14}	JLCM 1965 **8** 388
$(Nb_6Cl_{12})Cl_2$	Nb_6Cl_{14}	ZaC 1965 **339** 155
3-dimensional		
Zr_6I_{12}	ZrI_2	JACS 1978 **100** 652
$(Nb_6F_{12})F_3$	Nb_6F_{15}	JLCM 1965 **9** 95
$(Ta_6Cl_{12})Cl_3$	$Ta_6Cl_{15}, Ta_6Br_{15}, Zr_6Cl_{15}$	ZaC 1968 **361** 259

indicated. (Of the 13 Cl atoms shown in the Figure, 6 are common to two and 3 to three such groups: $4 + 6(\frac{1}{2}) + 3(\frac{1}{3}) = 8$.)

In $CsNb_4Cl_{11}$ groups of 4 octahedra may be distinguished on the basis of the Nb—Nb distances, namely, $a = 2.95$, $b = 2.84$, $c = 3.56$, and $d = 3.95\,\text{Å}$ (Fig. 9.16(b)); the group of four octahedra shares edges and faces as shown.

We now come to the compounds of Table 9.23, the structures of which are based on one of two units, both containing a central octahedral group of metal atoms. These halides include not only the compounds of Nb and Ta in which the metal has non-integral oxidation numbers but also dihalides of Mo, W, Pd, and Pt and WCl_3. The units are illustrated in idealized form in Fig. 9.17. In (a) M atoms are at the centres of the faces of a cube and X atoms at the vertices, giving a finite group of composition M_6X_8. This group was first recognized as the basic structural

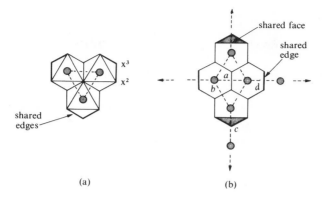

(a)   (b)

FIG. 9.16. (a) the M_3X_8 unit in Nb_3Cl_8; (b) the M_4X_{11} unit in $CsNb_4Cl_{11}$. In (a) the atoms X^2 and X^3 are common to two or three M_3X_8 groups respectively. In (b) all X atoms on shared edges or faces are common to two M_4X_{11} groups.

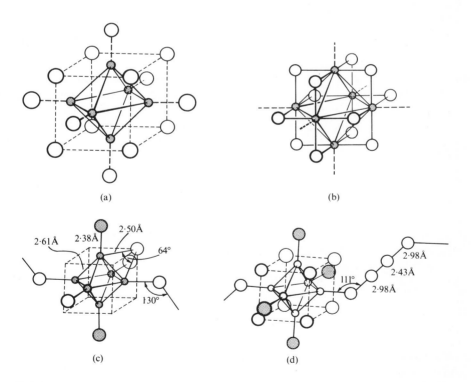

(a)   (b)

(c)   (d)

FIG. 9.17. Metal cluster complexes: (a) M_6X_8 (or $M_6X_8.X_6$), (b) M_6X_{12}, (c) crystalline $MoCl_2$ (details), (d) W_6Br_{16}.

unit in a number of chloro complexes of Mo. The yellow 'dichloride' is soluble in alcohol and from the solution alcoholic $AgNO_3$ solution precipitates only one-third of the chlorine. A saturated solution of the dichloride in aqueous HCl produces crystals of a 'chloro acid', $H_2(Mo_6Cl_{14}).8H_2O$, from which salts such as $(NH_4)_2(Mo_6Cl_{14}).H_2O$ can be prepared. Controlled hydrolysis of the solution of the chloro acid gives products in which part of the Cl is replaced by OH, H_2O or both. Attachment of 6 Cl to the M_6X_8 group of Fig. 9.17(a), as shown by the X atoms, gives the ion $(Mo_6Cl_8.Cl_6)^{2-}$ in the chloro acid, while attachment of 4 Cl and 2 H_2O gives a neutral group $[Mo_6Cl_8.Cl_4(H_2O)_2]6H_2O$ in the 'octahydrate' of the dichloride.

If the four equatorial X atoms of Fig. 9.17(a) are shared with other similar groups the result is a layer of composition $(Mo_6Cl_8)Cl_4$, and this is the structure of $MoCl_2$, WCl_2, etc. (Table 9.23). Details are shown in Fig. 9.17(c). Sharing of all six X atoms with other units gives a 3D framework with the composition M_6X_{11} which represents the structure of crystalline Nb_6I_{11}. At temperatures above 300 °C the paramagnetic Nb_6I_{11} absorbs hydrogen at atmospheric pressure to form a hydride with the limiting composition $Nb_6I_{11}H$, which is diamagnetic at low temperatures. A n.d. study shows the H atom at the centre of the Nb_6I_{11} group. This Nb_6I_{11} framework exists not only as the neutral iodide but also as an ion enclosing, for example, Cs^+ ions in its interstices. Moreover, like Nb_6I_{11} this salt can be hydrogenated to $Cs(Nb_6I_{11}H)$, which has almost identical cell dimensions.

The action of liquid Br_2 on WBr_2 (W_6Br_{12}) gives successively W_6Br_{14}, W_6Br_{16}, and W_6Br_{18}. In W_6Br_{16} ($W_6Br_8.Br_4$) groups are linked through linear Br_4 groups (Fig. 9.17(d)) to form infinite chain molecules.

The grouping of Fig. 9.17(b) has the composition M_6X_{12}, and represents the structure of the molecules in one of the crystalline forms of $PdCl_2$ and $PtCl_2$ and in their vapours. In these molecules there is square planar coordination of M. In ZrI_2 M_6X_{12} groups of this type are packed so that the coordination group around Zr is completed by a fifth I at the end of a dotted line in Fig. 9.17(b); this I atom is necessarily a bridging I atom of another Zr_6I_{12} cluster. One-half of the I atoms therefore bridge 2 Zr and the remainder 3 Zr atoms. Addition of a further X atom to each M gives M_6X_{18}, which is the structure of the molecule in tungsten trichloride. The $(Nb_6X_{12})^{n+}$ and $(Ta_6X_{12})^{n+}$ ions can be oxidized from $n = 2$ to $n = 3$ and 4, the corresponding $(M_6X_{18})^{m-}$ ions having $m = 4$, 3, and 2. Salts containing each of these ions have been made, and some of their structures have been determined (see Table 9.23):

$(Nb_6Cl_{18})^{4-}$:   Nb–Nb, 2.92 Å
$(Nb_6Cl_{18})^{2-}$:   Nb–Nb, 3.02 Å (see Table 9.23)
$(Ta_6Cl_{18})^{2-}$:   Ta–Ta, 2.96 Å.

As in the molybdenum complexes derived from M_6X_8 units some of the additional six ligands may be, for example, H_2O molecules, and this is so in $Ta_6Cl_{14}.7H_2O$, in which the structural unit is $[Ta_6Cl_{12}(Cl_2)(H_2O)_4]$. The basic structure of this unit has been deduced earlier from X-ray diffraction data obtained from a concentrated alcoholic solution of the compound.

If units M_6X_{12} are joined into layers through four additional (equatorial) X atoms the composition is M_6X_{14}, as in Nb_6Cl_{14} and Ta_6I_{14}. Finally, linking of such units to six others through additional X atoms gives a 3D framework with the composition M_6X_{15}, and this is the type of structure adopted by Nb_6F_{15}, Ta_6Cl_{15}, Ta_6Br_{15} and Zr_6Cl_{15}.

It will be observed that the atoms of the M_6X_{12} group of Fig. 9.17(b) also represent a unit cell of the structure of NbO, which contains the same octahedral Nb_6 groups as in the compounds we have been discussing. However, the system of Nb—Nb bonds in NbO is continuous throughout the crystal, accounting for the metallic lustre and conductivity of this oxide.

Other polyhedral metal groupings include (presumably) the trigonal bipyramidal Pt_3Sn_2 nucleus in the ion $(Pt_3Sn_8Cl_{20})^{4-}$ in $[N(CH_3)_4]_4Pt_3Sn_8Cl_{20}$, a compound produced by the interaction of $PtCl_2$ and $SnCl_2$ in acetone in the presence of $[N(CH_3)_4]^+$ ions.[4] There is i.r. evidence for the structure of the $(Pt_3Sn_8Cl_{20})^{4-}$ ion, and two other complexes containing $SnCl_3$ ligands have been studied. The detailed structure of the anion in $(\phi_3PCH_3)_3[Pt(SnCl_3)_5]$, (a), could not be deter-

mined owing to disorder in the crystal;[5] this is not a polyhedral cluster compound. A trigonal bipyramidal nucleus exists in $(C_8H_{12})_3Pt_3(SnCl_3)_2$,[6] (b), in which the following bond lengths were determined: Pt—Pt, 2.58 Å, Pt—Sn, 2.80 Å, and Sn—Cl, 2.39 Å. See also the molecules $Pt_4(OH)_4(CH_3)_{12}$ etc. described in Chapter 27.

(1) JACS 1967 89 159
(2) IC 1968 7 631, 636
(3) JLCM 1966 11 31; JACS 1966 88 1082
(4) IC 1966 5 109
(5) JACS 1965 87 658
(6) CC 1968 512

Sub-halides of elements of Groups IIIA and IVA

The study of high-temperature reactions between halides and metals has produced many sub-halides with structures based, like those of Table 9.23, on one or other of the clusters M_6X_8 or M_6X_{12}. The first compound of this kind to be characterized was Gd_2Cl_3, but the list now includes compounds with Cl:M ratios (mean oxidation numbers) down to 1.00. We have seen in the previous section that there are complexes with X:M ratios ranging from 3.00 to 2.00 based on octahedral clusters; it appears that at some point, as the X:M ratio decreases below 2.0 the M_6 octahedra coalesce to more extensive metal–metal systems by sharing edges. The infinite (single) chain of Fig. 9.18(a) is bonded through Cl atoms to form the 3D structure of Gd_2Cl_3 as shown at (b). Further condensation of the M_6 octahedra to form double chains occurs in Er_6I_7 and Tb_6Br_7 and finally double layers of metal atoms are found in ZrCl and ZrBr. Intermediate compositions arise by the addition of isolated M atoms in octahedral interstices between Cl atoms, as in Sc_5Cl_8, (c), and Sc_7Cl_{10}, (d). These M atoms do not form m–m bonds, but form chains of octahedral MCl_6 groups sharing opposite edges. These chains are shown with dotted outlines in Fig. 9.18(c) and (d) and the M_6 octahedra are shaded. (Compare the end-on views of single and double octahedral chains in Fig. 5.22(a) and (b), p. 215). In ZrCl there are double layers of M atoms enclosed between layers of Cl atoms, as seen end-on in Fig. 9.18(e). For detailed descriptions of these structures it is necessary to consider the positions of the X atoms relative to the M_6 octahedra, that is, to distinguish between X atoms which cap the octahedron faces and those which bridge octahedron edges. This is the significance of the entries (M_6X_{12} type) or (M_6X_8 type) in Table 9.24.

TABLE 9.24

Structures of sub-halides

X:M	Structure	Other examples	Type of m–m complex	Reference
1.71	Sc_7Cl_{12}	La_7I_{12} (Pr, Tb)	Discrete M_6X_{12} + M	JACS 1978 **100** 652
1.60	Sc_5Cl_8	Gd_5Br_8 (Tb)	Single chains (M_6X_{12} type) + M	JACS 1978 **100** 5039
1.50	Gd_2Cl_3	Y_2Cl_3, Y_2Br_3	Single chains (M_6X_8 type)	ZaC 1979 **456** 207
1.43	Sc_7Cl_{10}	–	Double chains (M_6X_8 type) + M	IC 1977 **16** 1109
1.17	Er_6I_7	Tb_6Br_7	Double chains (M_6X_8 type)	ZN 1980 **B35** 626
1.00	ZrCl ZrBr	GdCl, TbCl, YBr HfCl, ScCl, YCl	} Double layer } (M_6X_8 type)	IC 1976 **15** 1820 IC 1977 **16** 2029

References to many compounds with these structures are included in IC 1980 **19** 2128, which describes the structures of YCl, YBr, Y_2Cl_3, and Y_2Br_3. For Er_4I_5 and Er_7I_{10} see JLCM 1980 **76** 41.

   The structures of ZrCl and ZrBr are of interest in connection with the closest packing of spheres. Figure 9.18(f) shows the projection of one double layer of ZrCl, viewed in the direction of the arrow in (e). Disregarding the difference between Zr and Cl the packing of the four layers within one double layer is seen to be cubic

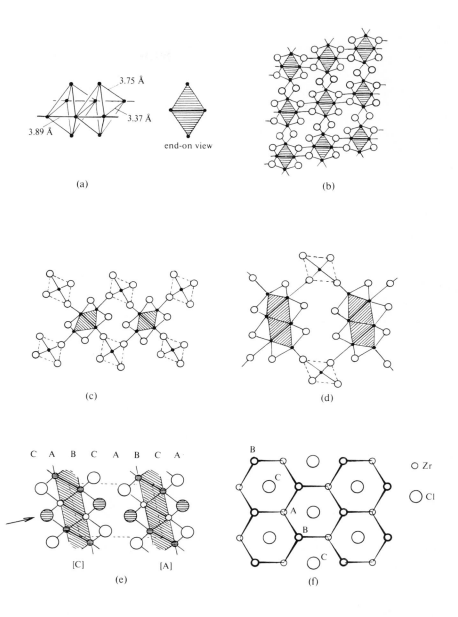

FIG. 9.18. Structures of sub-halides: (a) Chain of edge-sharing M_6 octahedra in Gd_2Cl_3, (b)-(e) projections of the structures of Gd_2Cl_3, Sc_5Cl_8, Sc_7Cl_{10}, and ZrCl, (f) the atoms of one double layer of ZrCl projected on to a plane perpendicular to the arrow in (e). In (e) open circles represent atoms in the plane (010) of the paper, and shaded circles atoms at $y = \frac{1}{2}$. In (f) light and heavy small circles represent the metal atoms at the two levels within one double layer.

closest packing, the positions being labelled according to the normal convention for c.p. structures (Fig. 4.7(a), p. 150). The structure of ZrCl may then be represented

```
Cl  Zr  Zr  Cl      Cl  Zr  Zr  Cl      Cl  Zr  Zr  Cl
A   B   C   A        B   C   A   B        C   A   B   C
        [A]                 [B]                 [C]
```

and we may label the double layer according to the positions of the Cl atoms on its surfaces as [A], [B], or [C]. There are clearly different possible sequences of double layers, which in the two closely related structures of ZrCl and ZrBr are

$$[A] \ [B] \ [C] \quad \ldots \quad ZrCl$$
$$[A] \ [C] \ [B] \quad \ldots \quad ZrBr$$

and polytypes are possible. Such structures differ only in the environment, trigonal antiprismatic (octahedral) or trigonal prismatic, of the X atoms on the outer surfaces of the double layers.

Two general points may be noted here. The first is that, as might be expected, quite different types of structure are found for the monohalides of B subgroup elements such as Bi (p. 877) and Te (p. 707). The second is that energy-level calculations suggest that the bonding between the chains in, for example, Gd_2Cl_3, may have considerable ionic character which in the limit would be expressed in a formulation of the type $[(M^{3+})_2(e^-)_3](Cl^-)_3$. This would be the opposite of the situation in the alkali suboxides, where the bonding within the oxygen–metal clusters is apparently ionic and the bonding between the clusters metallic in character.

Metal halides in the fused and vapour states

A complete picture of the structural chemistry of a compound would require a knowledge of its structure in the solid, liquid, and gaseous states. The amount of information obtainable about the structure of a liquid halide is very limited, and few X-ray studies have been made. (Examples include SnI_4,[1] $InCl_3$,[2] and CdI_2.[3]) We are therefore obliged to make direct comparisons of crystal and vapour. Strictly, these comparisons relate only to the process of sublimation, and if the compound is polymorphic the relevant crystal structure is that of the polymorph stable at the temperature of sublimation.

Physical properties are sometimes indicative of structural changes. Crystalline $AlCl_3$ has a layer structure, and the electrical conductivity of the solid increases rapidly as the melting point is approached, at which temperature it falls suddenly to nearly zero. At the melting point there is an unusually large decrease in density (about 45 per cent) as the ionic crystal, in which Al is 6-coordinated, changes to a melt consisting of Al_2Cl_6 molecules.[4] These molecules are the predominant species in the vapour at temperatures below about 400 °C, and consist of two tetrahedral $AlCl_4$ groups with a common edge. The viscosity of aqueous $ZnCl_2$ solutions rises sharply at high concentrations, and the viscosity of the molten salt is much higher[5] than that of normal unassociated dihalides such as $MgCl_2$ or $CdCl_2$, being intermediate between the value for such compounds and that of BeF_2. Like $ZnCl_2$ BeF_2 crystallizes as a system of tetrahedral MX_4 groups linked through vertices (silica-like structure), and the viscosity of the molten BeF_2 at its melting point[6] is of the same order of magnitude as the viscosities of GeO_2 and B_2O_3, that is, about 10^8 times that of 'normal' dihalides. The X-ray scattering curve of glassy BeF_2 at room temperature has been interpreted in terms of a random network of the silics type formed from BeF_4 groups. Similarly the high viscosity of SbF_5 (at room temperature) suggests a chain structure formed from SbF_6 octahedra sharing two vertices.

The structures of many halide molecules have been studied in the vapour state, generally by electron diffraction. Configurations found for molecules MX_n are:

MX_2: linear; Zn, Cd, and Hg dihalides ⎫
        angular; Sn and Pb dihalides   ⎬ (Table 9.25)
MX_3: planar; Al trihalides ⎭
        pyramidal; As, Sb, and Bi trihalides (Chapter 20)
MX_4: tetrahedral; Groups IV A and B tetrahalides (Table 21.1)
MX_5: trigonal bipyramidal; Nb, Ta, As, and Sb pentahalides (Tables 9.25 and 20.3)
MX_6: octahedral; MoF_6, WCl_6, TeF_6, and Table 9.25.

The conventional methods of studying vapour species (e.d. and spectroscopic) have been developed in a number of ways. Examples are the microwave spectroscopy of molecules formed by the direct vaporization of solids (Ag–Cl, 2.28 Å, Ag–Br, 2.39 Å[7] or of unstable species prepared in special ways. For example, the reaction of aluminium halides with metal gives monohalides (Al–F, 1.6544 Å, Al–Cl, 2.1298 Å[8]). Other methods include i.r. studies of molecules in the vapour or trapped in matrices,[9] giving estimates of bond angles in molecules such as GeF_2 ($94° \pm 4°$) and TiF_2 ($130° \pm 5°$), and dipole moment studies.[10] The deflection of molecular beams by inhomogeneous electric fields shows that some MX_2 molecules have permanent electric dipole moments and are therefore presumably non-linear.[11] The division of the Group II halides into two groups is somewhat unexpected and has been confirmed in a number of cases by i.r. studies of the halide in solid Kr at 20 K.

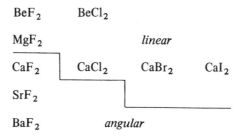

Estimates of the bond angles include CaF_2 ($140°$), SrF_2 ($108°$), and BaF_2 ($100°$). Other molecules whose shapes, but not detailed structures, have been determined include $FeCl_2$, $CoCl_2$, and $NiCl_2$, shown to be linear by matrix i.r.,[12] and SmF_2, EuF_2, and YbF_2, shown to be angular by molecular beam deflection.[13]

The relation between structure in the crystalline and vapour states is simplest in the case of a molecular crystal which vaporizes to molecules of the same kind as those in the crystal, a process which merely involves the separation of molecules against the van der Waals forces with very little change in the internal structure of the molecule. The molecule may be mononuclear (SnI_4, WCl_6) or, rarely, polymeric (Al_2Br_6). In all other cases there is breakdown of more extensive metal–halogen systems, either directly to MX_n molecules, or in some instances to polymeric species which dissociate to monomers at higher temperatures. The vaporization is accompanied by a reduction in the coordination number of the metal, which is shown in brackets in the following examples:

Type of crystal structure	Halide	Vapour species
3D complex	LiCl (6)	Li_2Cl_2 (2) → LiCl (1)
	CuCl (4)	Cu_3Cl_3 (2) [→CuCl (1)]
Layer	CdI_2 (6)	CdI_2 (2)
	SbI_3 (3 + 3)	SbI_3 (3)
	$AlCl_3$ (6)	Al_2Cl_6 (4) [→$AlCl_3$ (3)]
Molecular	Nb_2Cl_{10} (6)	$NbCl_5$ (5)

The structures of the vapour molecules of some halides are closely related to the structures of the crystals (Fig. 9.19). Mass spectrometric analysis of the ions produced by electron impact shows the presence of dimers, trimers, and in some cases tetramers, in the vapours of alkali halides. In particular the lithium halide vapours

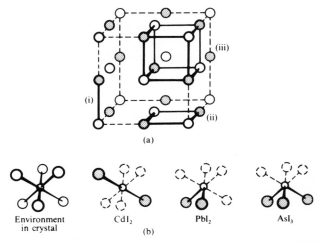

FIG. 9.19. Relation between crystal structure and configuration of molecule in the vapour state: (a) molecules AX, A_2X_2. and A_4X_4 derivable from the NaCl structure, (b) molecules AX_2 and AX_3 derived from halides with layer structures.

contain more dimers than monomers. The structures of three of the LiX dimers have been determined by electron diffraction. They form planar rhombus-shaped molecules as expected from a theoretical treatment assuming essentially Coulombic interaction between the oppositely charged ions. The bond lengths M–X have values intermediate between those in the monomer and in the crystal (Table 9.25). There are also some dimers in TlBr vapour.[14]

Planar structures (a) have been suggested for the dimers in the vapours of $FeCl_2$, $CoCl_2$, and $NiCl_2$,[12] and the non-planar structure (b) for the dimer of GeF_2,[15] similar to that of Se_2O_4.

At temperatures above 435 °C CuCl has a disordered wurtzite structure, and a mass spectrometric study showed that Cu_3Cl_3 molecules are the predominant species in the saturated vapour at 450 °C. An early e.d. study[16] showed that the probable configuration of the trimer is an essentially planar 6-ring, but the detailed structure was difficult to establish with certainlty. A later mass spectrometric study of the vapour in equilibrium with solid CuCl (280–430 °C) indicated comparable concentrations of Cu_3Cl_3 and Cu_4Cl_4 molecules and a smaller concentration of Cu_5Cl_5 molecules.[17] The vapours of $ReCl_3$ and $ReBr_3$ in the temperature range 240–350 °C consist almost exclusively of trimers. An e.d. study of Re_3Br_9 molecules in the vapour[18] indicated a structure similar to that of Re_3I_9 in the crystalline triiodide. Electron diffraction studies show that cyclic tetramers consisting of vertex-sharing octahedral groups exist in the low-temperature vapours of NbF_5 and TaF_5; they break up into trigonal bipyramidal molecules at higher temperatures (Table 9.25).

In contrast to the examples of Fig. 9.19 there is a more radical rearrangement of nearest neighbours in changes such as the following:

crystalline $AlCl_3$ (octahedral coordination) → Al_2Cl_6 dimer (tetrahedral coordination), and
crystalline $NbCl_5$ (octahedral Nb_2Cl_{10} dimers) → $NbCl_5$ (trigonal bipyramidal molecules).

Table 9.25 includes some interatomic distances not given in other chapters.

(1) JACS 1952 **74** 1763
(2) JACS 1952 **74** 1760
(3) JACS 1953 **75** 471
(4) JACS 1951 **73** 3151
(5) JCP 1960 **33** 366
(6) JCP 1960 **32** 1150
(7) JCP 1966 **44** 391
(8) JCP 1965 **42** 1013
(9) JCP 1965 **42** 902
(10) JCP 1964 **40** 3471
(11) JCP 1963 **39** 2023; JACS 1964 **86** 4544
(12) JCP 1968 **49** 4379
(13) JCP 1972 **56** 5392
(14) JPC 1964 **68** 3835
(15) IC 1968 **7** 608
(16) JACS 1957 **61** 358
(17) JCP 1971 **55** 4566
(18) ZSK 1971 **12** 315

TABLE 9.25 *Interatomic distances (Å) in some halide molecules*

	Li–X (A) Monomer	Li–X (A) Dimer	Li–X (A) Crystal	Angle X–M–X in dimer	Reference
Cl	2.02	2.23	2.57	108 (4)°	JCP 1960 **33** 685
		2.17		116°	
Br	2.17	2.35	2.75	110 (4)° ⎫	ZPC 1960 **213** 111
I	2.39	2.54	3.03	116 (4)° ⎭	

	Vapour (m.w.)[a]	crystal
LiF	1.5639	2.009
LiCl	2.021	2.566
LiBr	2.1704	2.747
LiI	2.3919	3.025
NaF	1.926	2.307
NaCl	2.3606	2.814
NaBr	2.5020	2.981
NaI	2.7115	3.231
KF	2.171	2.664
KCl	2.6666	3.139
KBr	2.8207	3.293
KI	3.0478	3.526
RbF	2.266	2.815
RbCl	2.7868	3.285
RbBr	2.9448	3.434
RbI	3.1769	3.663
CsF	2.3453	3.005
CsCl	2.9062	3.560†
CsBr	3.0720	3.713†
CsI	3.3150	3.950†

† CsCl structure.

Hg_2Cl_2[h]: Hg–Cl, 2·23

Linear MX_2 *molecules*[b]

	F	Cl	Br	I
Zn	1.81	2.05	2.21	2.38
Cd	1.97	2.21	2.37	2.55
Hg	–	2.25	2.44	2.61

Planar molecules MX_3

	M–X
AlF_3[k, l]	1.63
$AlCl_3$	2.06
AlI_3	2.42
$FeCl_3$[e]	2.14

Non-linear molecules MX_2[c]

	Cl	Br	I
Sn	2.42	2.55	2.73
Pb	2.46	2.60	2.79

CuCl[n]	2.051
CuBr	2.173
CuI	2.338

GaF[f]	1.775
GaCl	2.20
GaBr	2.35
GaI	2.57
InF	1.985
TlF	2.084
TlCl[g]	2.55
TlBr	2.68
TlI	2.87

$$M_2X_6^{(d)}: \quad {X_t \atop X}{>}M{<}{X_b \atop X}{>}M{<}{X \atop X}$$

	M–X_t	M–X_b
Al_2Cl_6[d]	2.07	2.25
Al_2Br_6[d]	2.22	2.41
Al_2I_6[d]	2.45	2.63
Ga_2Cl_6[d]	2.10	2.30
Ga_2Br_6[d]	2.25	2.45
Fe_2Cl_6[e]	2.11	2.28

Regular octahedral molecules MF_6[i]

	M–F		M–F
W	1.833	U	1.996
Os	1.831	Np	1.981
Ir	1.830	Pu	1.971

Trigonal bipyramidal molecules MX_5

	M–X[m]		M–X[j]		M–X[j]
NbF_5	1.88 A	$NbCl_5$	2.29 A	$TaCl_5$	2.30 A
TaF_5	1.86	$NbBr_5$	2.46	$TaBr_5$	2.45

(a) JCP 1964 **40** 156. (b) ZFK 1956 **30** 155, 951; BCSJ 1973 **46** 410. (c) TFS 1941 **37** 406.
(d) Q. Shen (1973), Dissertation, Oregon State University, Corvallis, Oregon. (e) Teplofiz Vys. Temp. 1964 **2** 705. (f) JCP 1966 **45** 263. (g) AP 1936 **26** 1. (h) AP 1940 57 21. (i) JCP 1968 **48** 4001. (j) TFS 1940 **36** 668. (k) K 1959 **4** 186. (l) ZSK 1967 8 391. (m) Vest. Moskov. Univ. Ser. Khim. 1968, No. 5, 7. (n) JCP 1975 **62** 1040, 4796; 1975 **63** 2724.
*Earlier results (JACS 1942 **64** 2514) for Ga_2Br_6 and In_2X_6 gave only mean values of bridging and terminal bond lengths.

10

Complex, oxy-, and hydroxy-halides

Complex halides

Complex halides are solid phases containing two or more kinds of metal ion (or other cation) and usually one kind of halogen atom. The simplest salts of this type contain only two kinds of cation and a common anion, and were formerly called 'double halides'. As generally understood the term complex halide does not include solid solutions such as (Na,K)Cl, the composition of which varies over a range dependent on the relative sizes of the two cations, but it does include stoichiometric compounds with random arrangements of two or more kinds of cation (see later). There are also compounds containing a metal and more than one halogen, for example, fluorohalides such as CaFCl with the PbFCl structure (p. 487) and PbClBr[1] etc. with the $PbCl_2$ structure (p. 274) and polyhalides such as $CsICl_2$, which are included in Chapter 9. Many complex halides are anhydrous, others are hydrated to various degrees. Except in one or two special cases we shall exclude hydrated salts from the present discussion; their structures are described in Chapter 15.

We shall confine our attention here to the structures of crystalline complex halides. Apparently the vapours of many complex halides contain molecules ($NaBeF_3$, Na_2BeF_4, $NaAlF_4$, Na_2AlF_5, and so on), and many e.d. studies have been made of the high-temperature vapours, mostly by Russian workers. We shall not discuss the various models suggested for such molecules because the interpretation of the e.d. data is complicated by the large amplitude of the thermal vibrations and also because it is not possible to do justice to this work in a brief summary. A review of much of this work is available in *The Molecular Geometries of Coordination Compounds in the Vapour Phase* by M. Hargittai and I. Hargittai, 1977, Elsevier, Amsterdam (English edition).

The empirical formulae of complex halides are of all degrees of complexity, as may be seen from the following selection: $CsAgI_2$, $CsHgCl_3$, $TlAlF_4$, Cs_3CoCl_5, Na_3AlF_6, $Na_5Zr_2F_{13}$, $Rb_2Fe_5F_{17}$, and $Na_7Zr_6F_{31}$. As is generally true throughout the chemistry of solids, similarity in formula-type does not imply that compounds have similar structures. For example, $KMgF_3$, NH_4CdCl_3, and $CsAuCl_3$ have quite different structures. Moreover, none of these salts ABX_3 contains a finite complex ion BX_3; in fact, $CsAuCl_3$ contains ions of two kinds and is preferably written $Cs_2(AuCl_2)(AuCl_4)$.

Most of the compounds we shall describe have formulae of the type A_mBX_n, though we shall mention some with more complex formulae. In particular there are a number with binuclear complex ions, for example, B_2X_{11}, B_2X_{10}, and B_2X_9, formed from two octahedral groups with a common vertex, edge, or face; the more complex 'metal cluster' ions have been described in the last chapter. Adopting the convention that B is the more highly-charged cation (or that the oxidation state of B is higher than that of A) it follows that most complex halides must contain ions A^+ (alkali metals, Ag^+, Tl^+, NH_4^+, etc.) or A^{2+} (alkaline-earths, etc.). For values of n up to 6 the possible types of complex halide are:

$$A^I B^I X_2 \qquad A^I B^{II} X_3 \qquad A^I B^{III} X_4 \qquad A^I B^{IV} X_5 \qquad A^I B^V X_6$$
$$A_2^I B^{II} X_4 \qquad A_2^I B^{III} X_5 \qquad A_2^I B^{IV} X_6$$
$$(A^{II} B^{II} X_4) \qquad A^{II} B^{III} X_5 \qquad A^{II} B^{IV} X_6$$
$$A_3^I B^{III} X_6$$
$$(A^{III} B^{III} X_6)$$

The great majority of known complex halides are fluorides; in particular, those with $n \geqslant 7$ are almost exclusively fluorides, of elements of Groups IV–VI, Ce, and 5f metals.

A simple possibility for a complex halide is that it adopts a structure of a halide (or oxide) $A_m X_n$ with A and B replacing, either statistically or regularly, the positions occupied by atoms of one kind in the binary halide (oxide); these form our class (a) in Table 10.1. These are ionic crystals, and the basic requirement is that the ions A and B are of similar size and carry charges appropriate to the structure, as in $Na^+ Y^{3+} F_4$ or $K_2^+ U^{4+} F_6$ with structures of the fluorite type. Quite different structures are formed if the ions are of very different sizes as, for example, in $CsBe_2F_5$ (p. 449) or $CsMnF_3$-II (p. 582). In both of these crystals there is a charged 3D framework built of vertex-sharing polyhedra containing the small ion (BeF_4 tetrahedra and MnF_6 octahedra respectively) in which the Cs^+ ions occupy very large cavities and are surrounded by 12 F^- ions.

Other complex halides are conveniently classified according to the type of grouping of the B and X atoms in the crystal. These atoms may form a finite group, in the simplest case a mononuclear group BX_n, or the B atoms may be linked by sharing X atoms to form 1- or 2-dimensional complex ions, giving the classes (d), (c), and (b) in Table 10.1. We have added subgroups for structures containing additional X^- ions (not coordinated to B atoms). It might seem logical to recognize also 3D systems of linked B and X atoms. For example, the B and X atoms in the perovskite (ABX_3) structure form a framework of vertex-sharing octahedra (ReO_3 structure) which extends throughout the crystal, with the 12-coordinated A ions in the interstices. However, there is little advantage to be gained from distinguishing the B–X complexes in the essentially ionic fluorides of class (a), particularly those with statistical structures (see later), because this would involve allocating these structures to classes (b), (c), or (d), and it is preferable to keep them together as a group to emphasize their relation to the simpler structures. For example, we may distinguish discrete BeF_4^{2-} ions in Li_2BeF_4, but since both Li^+ and Be^{2+} are tetrahedrally coordinated the structure is equally well described as one of the type A_3X_4. Similarly, Li_2TiF_6 (superstructure of rutile, AX_2 type) could be placed in class (d) since it contains TiF_6^{2-} ions. On the other hand, if the coordination number of B (or A) is greater than the ratio of X to B (or A) atoms, then the coordination

TABLE 10.1
Structures of complex halides

		A_mBX_3	A_mBX_4	A_mBX_5	A_mBX_6	A_mBX_7
(a)	*Statistical AX_n structures or superstructures*					
	Fluorite		α-NaYF$_4$		α-K$_2$UF$_6$	
	Trirutile				Li$_2$TiF$_6$	
	LaF$_3$				BaThF$_6$	
	ReO$_3$				CaPbF$_6$	
	VF$_3$				LiSbF$_6$	
	Ge$_3$N$_4$ (phenacite)		Li$_2$BeF$_4$			
	⎡3D *complexes*					
	⎣ Perovskite	ABX$_3$⎤				
(b)	*Layer structures*		K$_2$NiF$_4$			
			TlAlF$_4$			
			BaMnF$_4$			
	(layers + X$^-$ ions)				Ba(CrF$_4$)F$_2$	
(c)	*Chain structures*	K$_2$CuCl$_3$	K$_2$HgCl$_4$. H$_2$O		(NH$_4$)$_2$CeF$_6$	
		CsBeF$_3$		Tl$_2$AlF$_5$	K$_2$ZrF$_6$	
		CsNiCl$_3$		(NH$_4$)$_2$MnF$_5$	RbPaF$_6$	K$_2$PaF$_7$
		NH$_4$CdCl$_3$				
	(chains + X$^-$ ions)				Sr(PbF$_5$)F	
	Structures containing finite complex ions			For the ions		
(d)	*Mononuclear*		K$_2$PtCl$_4$	CuCl$_5^{3-}$,	ABX$_6$	K$_3$ZrF$_7$
				InCl$_5^{2-}$,	A$_2$BX$_6$	K$_2$NbF$_7$
				SnCl$_5^-$.	A$_3$BX$_6$	
				see p.	A$_4$BX$_6$	K$_2$TaF$_7$
	(+ X$^-$ ions)			Cs$_3$(CoCl$_4$)Cl		(NH$_4$)$_3$(SiF$_6$)
(e)	*Polynuclear*	} See text				
(f)	*Complex ions of two kinds*					

groups around B (A) must share X atoms. This is the situation in the (statistical) fluorite structure of K_2UF_6. The F:U ratio shows that there must be sharing of F atoms between UF_8 coordination groups, but the extent of the U–F complex varies in a random way throughout the structure, so that allocation to one of the classes (b)–(d) is not possible.

The formation of an ionic compound A_mBX_n implies the packing of A, B, and X ions to give the maximum lattice energy, and the sharing of X atoms between the coordination groups around B (or A) ions is presumably incidental to the attainment of suitable coordination numbers of A and B. So we find

	c.n.s of	
	A	B
$CaPbF_6$: ReO_3 superstructure (discrete PbF_6^{2-} ions)	6	6
$SrPbF_6$: linear $(PbF_5)_n^{n-}$ ions and F^- ions	10	6
$BaPbF_6$: $BaSiF_6$ structure (discrete PbF_6^{2-} ions)	12	6

The structure of $SrPbF_6$ illustrates a further point, namely, that the fact that the F:B ratio is equal to the c.n. of B does not necessarily mean that discrete BF_6 groups are present, although this is true in many ABX_6 and A_2BX_6 structures. Another example is the structure of Ba_2ZnF_6, which contains ZnF_4 layers and $2 F^-$ ions.

The similarity in size and electronegativity of F^- and O^{2-} gives rise to many structural resemblances between binary fluorides and oxides, whereas the structures of halides containing the heavier halogens are more often similar to those of sulphides. This difference between F and the other halogens is still evident in complex halides but is less marked than in simple halides. There are many examples not only of complex fluorides but also of complex chlorides with the same structures as complex oxides (Table 10.2).

TABLE 10.2
Isostructural complex chlorides and complex oxides

Complex oxide structure	*Complex chloride*	*Reference*
Ilmenite	$NaMnCl_3$	AC 1973 **B29** 1224
Perovskite	$KMnCl_3$	JCP 1966 **45** 4652
$BaTiO_3$ (hexagonal)	$RbMnCl_3$	AC 1977 **B33** 256
Sr_2TiO_4 (K_2NiF_4)	K_2MgCl_4	} AC 1977 **B33** 188
Al_2BeO_4	Na_2ZnCl_4	
β-K_2SO_4	Cs_2CoCl_4	K 1956 1 291
Sr_2PbO_4	Na_2MnCl_4	AC 1971 **B27** 1672
Spinel (inverse)	Li_2MCl_4 (M = Mg, Mn, Fe, Cd)	AC 1975 **B31** 2549
Ba_3MO_5 (M = Si, Ge, Ti, V, etc.)	Cs_3CoCl_5	AC 1980 **B36** 2893
$Sr_3Ti_2O_7$	$Rb_3Mn_2Cl_7$	AC 1978 **B34** 2617
Sr_4MO_6 (M = Pt, Ir, Rh)	K_4CdCl_6	AC 1959 **12** 519
Mg_6MnO_8	Na_6MCl_8 (M = Mg, Mn, Fe, Cd)	AC 1975 **B31** 770

It should perhaps be emphasized that although there are undoubtedly differences between the ionic–covalent character of A–X and B–X bonds in many complex halides, particularly chlorides, bromides, and iodides of transition and B subgroup metals, the recognition of chains, layers, etc. in complex fluorides is largely for convenience in describing and classifying the structures; it does not necessarily imply any large difference in character between the A–X and B–X bonds. For example, in $NaKThF_6$,[2] there is 6-coordination of Na^+ (the smallest cation) and 9-coordination of K^+ and Th^{4+}. The coordination groups around K and Th have F atoms in common but those around Na do not, but it is not desirable to single out the NaF_6 groups as complex ions simply because they do not share F atoms.

The stability of a crystalline complex halide depends on the nature (size, shape, and charge) of all the ions present, and in particular the stability of a finite complex ion BX_n depends on the size of the ion A. Many new complex ions have been isolated in combination with large cations. For example, the salts $[N(C_2H_5)_4]_2CeBr_6$ and $[N(C_2H_5)_4]_2CeI_6$ can be prepared from the solid chloro salt and the anhydrous acid HX or in acetonitrile solution. Ions prepared in combination with $[(C_6H_5)_3PH]^+$ include PrI_6^{3-} and other hexaiodo 4f ions, FeI_4^-, AuI_4^-, etc. Hexachloro ions MCl_6^{3-} of the 3d metals are not known in aqueous solution but are stable in salts with large cations such as $Co(NH_3)_6^{3+}$. Many complex chlorides, bromides, and iodides which cannot be prepared from aqueous solution have been made in recent years under anhydrous conditions, i.e. using non-aqueous solvents or crystallizing from the melt.

The number of complex halides of which structural studies have been made is already very large, but many groups have yet to be studied in detail. For example, compounds formed by the alkali and alkaline-earth halides include the following:[3]

$KCaCl_3$	K_2SrCl_4	K_2BaCl_4
	KSr_2Cl_5	

$NaCa_2Br_5$	$LiSr_2Br_5$	K_2BaBr_4
$KCaBr_3$	K_2SrBr_4	
	KSr_2Br_5	

KSr_2Cl_5 and KSr_2Br_5 are members of a group of complex halides which are iso-structural with the corresponding Pb(II) compounds (p. 1188).

There are numerous complex fluorides of alkali metals and metals of Groups I–IV.[4] The structures of $RbBe_2F_5$[5] and $CsBe_2F_5$[6] were mentioned in Chapter 3 as examples of the plane 6-gon net and the cubic (10,3) net. Each BeF_4 tetrahedron shares 3 vertices to form respectively a layer or a 3D framework of composition Be_2F_5.

We shall deal first with the more important groups of complex halides according to formula type, that is, with the vertical columns of Table 10.1 down to class (d), including the small group of halides ABX_2, then separately with classes (e) and (f), and finally with selected groups of complex halides of particular elements.

(1)   AC 1969 **B25** 796
(2)   AC 1970 **B26** 1185
(3)   IC 1965 **4** 1510; ACSc 1966 **20** 255
(4)   IC 1962 **1** 220 (130 refs.)
(5)   AC 1972 **B28** 1159
(6)   AC 1972 **B28** 2115

Halides ABX_2

These are necessarily few in number since only singly-charged cations are involved. Many pairs of alkali halides form solid solutions (in some cases over the complete range of composition) and only the extreme members form compounds, $RbLiF_2$ and $CsLiF_2$. In these fluorides tetrahedral LiF_4 groups share one edge and two vertices to form layers, between which the larger alkali-metal ion occupies large holes (distorted 8-fold antiprism coordination). In $RbLiF_2$ the eight Rb$-$F distances range from 2.78 to 3.16 Å (mean 2.95 Å), but in $CsLiF_2$ two of the eight neighbours are appreciably more distant than the other six (Cs$-$6 F, 2.96-3.15, mean 3.07 Å, Cs$-$2 F, 3.50, 3.53 Å).[1]

(1)   IC 1965 4 1510

Halides A_mBX_3

These form a very large class of complex halides, of which the structures of at least 60 have been determined. Two kinds of anion are found, chains and 3D frameworks. The simplest chain is that built from tetrahedral BX_4 groups sharing two vertices, (a). It is found in relatively few compounds, for example, $CsBeF_3$,[1] $KHgBr_3.H_2O$[2] and K_2CuCl_3, Cs_2AgCl_3, and Cs_2AgI_3, for which see Chapter 26. The chains are arranged in the various structures so as to provide suitable environments for the large A^+ ions, and in $KHgBr_3.H_2O$ also to accommodate the water molecules between the chains. With the structure of this monohydrate contrast the structure of α-$KZnBr_3.2H_2O$, which contains discrete tetrahedral groups $[ZnBr_3(H_2O)]^-$ and should therefore be formulated $K[ZnBr_3(H_2O)].H_2O$.[3] The chain (b) formed

(a)   (b)   (c)

from octahedral BX_6 groups sharing opposite faces is the form of the anion in $BaNiO_3$ to which we refer shortly. In the double chain (c) each octahedral BX_6 group shares four edges. Like (a) it is found in only a few structures, for example, NH_4CdCl_3 and $RbCdCl_3$, and in a modified form in $KCuCl_3$ (p. 1138).

In the great majority of complex halides ABX_3 there is octahedral coordination of B. All the elements Mg, V$-$Cu, Zn, Cd, Hg, and Pb form one or more compounds with Ag, Na$-$Cs, or large cations such as $[N(CH_3)_4]^+$, though not all possible compounds are formed by each combination of A and B. For example, in the Cs$-$Zn$-$F system the only compound formed is $Cs_4Zn_3F_{10}$ (though $RbZnF_3$ does exist), and in the Cs$-$Mg$-$F system only $Cs_4Mg_3F_{10}$ is found (but not $CsMgF_3$), while $CsMgCl_3$, $CsMgBr_3$, and $CsMgI_3$ are all known.

The structural chemistry of many of these compounds, particularly the halides containing the large cations K^+, Rb^+, and Cs^+, is very similar to that of the oxides ABO_3 based on c.p. AO_3 layers which is described in some detail in Chapter 13. The structures most commonly found are the perovskite (3C) and 2H ($BaNiO_3$)

structures,[4] but others are 4H (CsCuBr$_3$), 6H (CsFeF$_3$, CsMnF$_3$, RbMnCl$_3$, CsCdCl$_3$), and 9R (CsCoF$_3$,[5] CsMnCl$_3$). The 'ideal' structure, with maximum symmetry, is found in some cases, for example, KNiF$_3$, KMgCl$_3$, and CsPbCl$_3$ above 320 K[6], but in others distorted variants are found, as in oxides. For example, NaMgF$_3$ (and the Co, Ni, and Zn compounds) have the GdFeO$_3$ variant of the perovskite structure (p. 584), while NaMnF$_3$, NaCuF$_3$, and NaCrF$_3$ have super-structures of the perovskite type. The distortion in KCuF$_3$ gives Cu^{2+} the coordination group: 2F at 1.89, 2F at 1.96, and 2F at 2.25 Å.[7] There are several distortions of the 2H structure. The c.p. structure of CsNiCl$_3$ is modified for KNiCl$_3$[8] to give the smaller K$^+$ 9- instead of 12-coordination, and there are distortions of different kinds in compounds which contain the Jahn–Teller ions Cu^{2+} and Cr^{2+}. In one variant of the structure Cr^{2+} has (3+3)-coordination,[9] and both (4+2)- and (2+4)-coordination apparently occur in various polymorphs of RbCrCl$_3$, CsCrCl$_3$, and the corresponding iodides.[10] The distorted form of the CsNiCl$_3$ structure adopted by CsCuCl$_3$ is illustrated in Fig. 4.32 (p. 183).

The formulae of salts such as A$_2$(Pt$_2$Br$_6$), p. 464, and Cs$_2$(AuCl$_2$)(AuCl$_4$), p. 466, containing respectively a bridged binuclear anion and two different kinds of anion, should not be reduced to the simpler form ABX$_3$.

(1) AC 1968 **B24** 807
(2) AC 1969 **B25** 647
(3) AC 1968 **B24** 1339
(4) AC 1973 **B29** 1330
(5) ZaC 1969 **369** 117

(6) AC 1980 **B36** 1023
(7) ZaC 1963 **320** 150
(8) AC 1980 **B36** 28
(9) JCP 1972 **57** 3771; AC 1973 **B29** 1529
(10) AC 1978 **B34** 1973; JSSC 1980 **34** 65

Halides A$_m$BX$_4$

The major groups of compounds are ABX$_4$ (A^{I}B^{III}X$_4$ and A^{II}B^{II}X$_4$) and A$_2^I$B^{II}X$_4$. The structures range from typically ionic structures of fluorides containing large ions with high coordination numbers (for example, the 9-coordination of both types of cation in one polymorph of KCeF$_4$, p. 1261) through octahedral chain and layer structures (described in Chapter 5) to numerous compounds ABX$_4$ and A$_2$BX$_4$ containing discrete BX$_4$ ions. For compounds, ABX$_4$ containing the larger cations structures of the fluorite type are possible. This could be either a random fluorite structure, as adopted by the second form of KCeF$_4$, or a superstructure. An interesting distorted superstructure of fluorite is found for some fluorides ABF$_4$ in which A is Ca or Sr and B is Cr^{II} or Cu^{II}, where the transition-metal ions have a distorted tetra-hedral arrangement of nearest neighbours instead of the normal cubic 8-coordination group. This structure is described under the fluorite structure in Chapter 6.

There are two octahedral layers (see p. 208), the planar ('*trans*') layer formed from octahedral BX$_6$ groups sharing all equatorial vertices and the puckered ('*cis*') layer in which the two unshared vertices are *cis* to one another. Examples include:

'*trans*' layer $\begin{cases} \text{K}_2\text{NiF}_4 \text{ structure: K}_2\text{MF}_4 \text{ (M = Mg, Zn, Co, Cu (distorted))} \\ \text{TlAlF}_4 \text{ structure: } \beta\text{-RbFeF}_4^{(1)} \end{cases}$

'*cis*' layer   BaMnF$_4$ structure[2a]: SrNiF$_4$, BaMF$_4$ (M = Mn, Fe, Co, Ni, Cu, Zn)

In K$_2$NiF$_4$ (Fig. 5.16, p. 209) and TlAlF$_4$ (Fig. 10.10, p. 469), BF$_4$ layers alternate with layers of A ions. In [NH$_3$(CH$_2$)$_4$NH$_3$]MnCl$_4$[2b] the linear cations carry a positive charge at each end and are oriented perpendicular to the layers, with the terminal NH$_3^+$ groups tucked into cavities of adjacent layers.

Corresponding to the two BX$_4$ layers there are chains in which the *trans* or *cis* vertices of each octahedral BX$_6$ groups are unshared. We describe the first as the edge-sharing 'rutile' chain since there are pairs of adjacent shared vertices. This chain is found in K$_2$HgCl$_4$.H$_2$O (p. 34) and with distorted (4+2)-coordination in Na$_2$CrF$_4$ and Na$_2$CuF$_4$.[2c] There is also a family of cylindrical BX$_4$ chains in which two *cis* vertices of each BX$_6$ are unshared (p. 208), of which the simplest example has been found in CsCrF$_4$.[2d]

The majority of discrete ions BX$_4$ are tetrahedral and many complex halides are isostructural with oxy-salts of similar formula type:

NH$_4$BF$_4$ and RbBF$_4$ isostructural with BaSO$_4$,
Cs$_2$CoCl$_4$ and Cs$_2$ZnCl$_4$ isostructural with K$_2$SO$_4$.

The phenacite structure (p. 89) of Li$_2$BeF$_4$[3] may be described either as containing tetrahedral BeF$_4^{2-}$ ions or, since both Li$^+$ and Be^{2+} are tetrahedrally coordinated, as a 3D ionic structure. The Li and Be atoms are at the nodes of a (3,4)-connected net. In (N$_2$H$_5$)LiBeF$_4$[3a] these atoms are at the nodes of a 3D 4-connected net, with N$_2$H$_5^+$ ions in the channels.

A number of chloro-, bromo-, and iodo-ions which cannot be prepared from aqueous solution in metallic salts can be stabilized by very large cations, for example, (NEt$_4$)$_2$MBr$_4$ (M = Mn, Fe) and [(C$_6$H$_5$)$_3$As(CH$_3$)]$_2$FeI$_4$,[4] and the structures of a number of MCl$_4^{n-}$ ions have been studied in salts such as [(C$_6$H$_5$)$_3$As(CH$_3$)]$_2$NiCl$_4$[5] and [(C$_2$H$_5$)$_4$N]InCl$_4$.[6]

In Cs$_2$CuCl$_4$ and Cs$_2$CuBr$_4$ the anions have a flattened tetrahedral shape (see Chapter 25), and in K$_2$PtCl$_4$ and the isostructural Pd compound there are planar BX$_4^{2-}$ ions (Fig. 10.1). The structure of Pd(NH$_3$)$_4$Cl$_2$.H$_2$O is similar, Pd(NH$_3$)$_4^{2+}$ replacing PtCl$_4^{2-}$ and Cl$^-$ replacing K$^+$, with H$_2$O molecules at the points indicated by small black circles. On this point see also p. 1232.

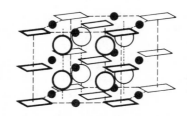

FIG. 10.1. The crystal structure of K$_2$PtCl$_4$.

There is 9-coordination (t.c. t.p.) of the 4f metal ions in $NaNdF_4$ (and the isostructural Ce and La compounds), the structure being related to, but different from, the β_2-Na_2ThF_6 structure (p. 1261) originally assigned to these fluorides.[7] The unique $Re_3Cl_{12}^{3-}$ anion in $Cs_3Re_3Cl_{12}$ ($CsReCl_4$) is described on p. 431.

(1) BSFMC 1969 **92** 335
(2a) JCP 1969 **51** 4928
(2b) AC 1980 **B36** 1355
(2c) ZaC 1965 **336** 200
(2d) ZaC 1978 **442** 151
(3) AC 1966 **20** 135

(3a) AC 1973 **B29** 2625
(4) IC 1966 **5** 1498, 1510
(5) JACS 1962 **84** 167
(6) AC 1969 **B25** 603
(7) IC 1965 **4** 881

Halides A_mBX_5

We have seen that there are very few crystalline pentahalides which consist of molecules MX_5. Similarly, 5-coordination is avoided in many complex halides containing $X:B$ in the ratio $5:1$. The following types of structure are found:

	C.n. of B	*Examples*
(a) 3D ionic structure	9	$LiUF_5$ (p. 1261)
(b) Octahedral BX_5 chain	6	Tl_2MF_5 (M = Al, Fe, Ga, Cr)[1], $(NH_4)_2MnF_5$[2],
		$CaCrF_5$[3], $BaFeF_5$[4], Rb_2CrF_5[5], $SrFeF_5$[6]
(c) Finite BX_5 ion	5	(t.b.) $[Cr(NH_3)_6]MCl_5$, M = Cu[7], Hg[8],
		$[Co(NH_3)_6]CdCl_5$[9], $(C_{28}H_{21}Cl)SnCl_5$[10],
		$(FeCl_5)^{2-}$, see p. 1205.
		(t.p.) $(NEt_4)_2InCl_5$[11], $(NH_4)_2SbF_5$[12]
		$(NH_4)_2SbCl_5$[13], $(bipyH_2)MnCl_5$[13a]
(d) BX_4^{2-} and X^- ions	4	$Cs_3(MCl_4)Cl$, M = Mn,[14] Ni, Co,[15] Hg[16]
		$[Cr(NH_3)_6](ZnCl_4)Cl$[17]

(t.b. = trigonal bipyramidal; t.p. = tetragonal pyramidal.)

We saw in Chapter 5 that there are two configurations of the simple octahedral AX_5 chain, the *trans* and *cis* chains, and that pentafluorides are known with both types of structure. Both types of chain are known also as the structures of anions BX_5. The choice of chain, *trans* or *cis*, and the configuration of the latter depend on the nature of the cation; $CaCrF_5$ contains the *trans* chain but Rb_2CrF_5 the *cis* chain. Note also that there are different conformations of the *cis* chain, coplanar in Rb_2CrF_5 but in $SrFeF_5$ a helical conformation. The isostructural $BaFeF_5$ and $SrAlF_5$ are of special interest since they contain equal numbers of chains of two kinds, the linear *trans* chain and a 'ramified' chain in which two additional octahedral BX_6 groups are attached to each member of the simple chain, as shown in Fig. 10.2. The composition of the chain remains MX_5. This complication is presumably due to the need to accommodate the larger Sr^{2+} and Ba^{2+} ions in positions of irregular 8- and 9-coordination between the chains—contrast the 7-coordination of Ca^{2+} in $CaCrF_5$.

Examples of the two configurations of finite ions, (c), are at present confined to B subgroup elements. The known trigonal bipyramidal ions are all stabilized by

 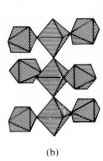

(a)   (b)

FIG. 10.2. The crystal structure of $BaFeF_5$ showing (a) the two types of $(FeF_5)_n^{2n-}$ chain; and (b) one ramified chain.

large cations, but large cations do not necessarily stabilize BX_5 ions, as is seen from the structures of $[Cr(NH_3)_6](ZnCl_4)Cl$ and $[Cr(NH_3)_6](CoCl_4)Cl$.[17]

There is a considerable number of monohydrated salts $A_2(MX_5.H_2O)$ which crystallize with the K_2PtCl_6 structure (p. 458) or distorted variants of that structure. The anion is here an octahedral group $(MX_5.H_2O)$ and A is one of the larger cations:

$(NH_4)_2(VF_5.H_2O)$ $(NH_4)_2(FeCl_5.H_2O)$
Rb_2 $Rb_2(CrF_5.H_2O)$ $(NH_4)_2(InCl_5.H_2O)$
Tl_2 Tl_2

However, all monohydrates of this type do not contain discrete $(MX_5.H_2O)$ groups. In $K_2AlF_5.H_2O$ and $K_2MnF_5.H_2O$[18] there are infinite MX_5 chains formed from octahedral MX_6 groups sharing a pair of opposite vertices; the H_2O molecules and the cations lie between the chains.

(1)	JSSC 1970 **2** 269	(11)	IC 1969 8 14
(2)	JCP 1969 **50** 1066	(12)	IC 1972 **11** 2322
(3)	MRB 1971 6 561	(13)	JCS A 1971 298
(4)	AC 1971 **B27** 2345	(13a)	CC 1971 803
(5)	AC 1974 **B30** 2688	(14)	AC 1976 **B32** 631
(6)	JSSC 1973 8 206	(15)	AC 1980 **B36** 2893
(7)	IC 1968 7 1111	(16)	AC 1976 **B32** 2905
(8)	JCS D 1975 2591	(17)	AC 1976 **B32** 2907
(9)	JCS A 1971 3638	(18)	JCS A 1971 2653
(10)	JACS 1964 **86** 6733		

Halides A_mBX_6

In some complex fluorides A_mBF_6 the A and B ions occupy the cation positions in simple ionic structures AX_n, either randomly or regularly (superstructure):

AX_n structure	C.N. of A and B	Random	Superstructure
Rutile	6	Mg_2FeF_6	Li_2TiF_6
Fluorite	8	$\alpha\text{-}K_2UF_6$	$\gamma\text{-}Na_2UF_6$
ReO_3	6	–	$CaPbF_6$
LaF_3	9	$BaThF_6$	–

We have already noted that there are discrete BX_6 groups in superstructures such as the trirutile structure of Li_2TiF_6 and the ReO_3 superstructure of $CaPbF_6$. In the latter structure the small open and black circles of Fig. 10.7 represent Ca^{2+} and Pb^{4+} ions respectively and the large open circles F^- ions. Such structures are therefore alternatively placed in class (d) of Table 10.1 as structures containing finite BX_6 ions. Similarly, the fluorite superstructure of $\gamma\text{-}Na_2UF_6$ contains UF_8 coordination groups which share two opposite edges to form chains, as may be seen from Fig. 28.5 (p. 1260).

The sharing of F atoms between BF_n coordination groups also occurs in a number of other complex fluorides (Table 10.3); it must occur for large B ions with c.n.>6.

TABLE 10.3

Structures of complex fluorides $A_m BF_6$

Compound	C.N. of B	Coordination group of B	Reference
1-dimensional BX_n complex			
K_2UF_6	9	t.c. trigonal prism sharing two faces	AC 1948 **1** 265
$RbPaF_6$	8 ⎫	dodecahedron sharing two edges	AC 1968 **B24** 1675
K_2ZrF_6	8 ⎭		AC 1956 **9** 929
$(NH_4)_2CeF_6$	8	antiprism sharing two edges	IC 1969 **8** 33
$\gamma\text{-}Na_2ZrF_6$	7	irregular sharing two vertices	AC 1969 **B25** 2164
$Sr(PbF_5)F$	6	octahedron sharing two vertices	ZaC 1957 **293** 251
2-dimensional BX_n complex			
$Ba_2MnF_4(F_2)$	6	octahedron sharing four vertices	ZaC 1967 **353** 13

The two closely related K_2UF_6 structures are described in Chapter 28. Dodecahedral coordination groups share two edges in $RbPaF_6$ and in K_2ZrF_6 (Fig. 10.3(a)), while antiprism groups share the edges accentuated in Fig. 10.3(b) in $(NH_4)_2CeF_6$. It is convenient to include here another chain structure which illustrates a different way of attaining a F:B ratio of $6:1$. In $SrPbF_6$ there are chains of composition PbF_5 formed from octahedral PbF_6 groups sharing opposite vertices, one-sixth of the fluorine being present as separate F^- ions (Fig. 10.3(c)). The structural formula is therefore $Sr(PbF_5)F$. Two-dimensional BX_n complexes are rare; they occur in salts Ba_2MF_6 (M = Cr, Mn, Fe, Co, Ni, Zn) which contain layers MF_4 of the same kind as

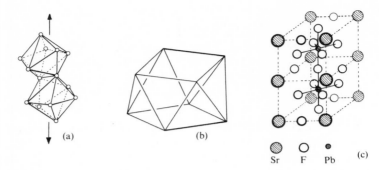

FIG. 10.3. (a) Portion of the infinite chain ion $(ZrF_6)_n^{2n-}$ in K_2ZrF_6; (b) edges of CeF_8 antiprismatic coordination groups shared in $(NH_4)_2CeF_6$; (c) the structure of $Sr(PbF_5)F$.

in K_2NiF_4 formed from octahedral MF_6 groups sharing all equatorial vertices, that is, $Ba_2(MF_4)F_2$.

Our group (d), comprising compounds containing discrete BX_6 groups, is an extremely large one. Not only are there compounds ABX_6, A_2BX_6, A_3BX_6, and A_4BX_6, but the first class includes $A^IB^VX_6$ and $A^{II}B^{IV}X_6$; at least one hundred fluorides $A^IB^VF_6$ have been prepared. The octahedral BX_6 ions are large, approximately spherical ions, and the crystal structures of these salts are determined in a general way by the relative sizes and charges of the A and BX_6 ions rather than by the chemical nature of the elements A and B. For example, $BaSiF_6$ has the same type of structure as $TlSbF_6$, while one form of K_2SiF_6 is isostructural with K_2PtCl_6, K_2SeBr_6, and numerous other compounds, as noted later.

The structures of many of these compounds may be described in two ways. Considering the A and BX_6 ions as the structural units we may describe the structures as derived from simple structures AX_n by replacing A or X by BX_6 ions. For example, in many compounds ABX_6 the A and BX_6 ions are arranged in the same way as the ions in the NaCl or CsCl structures. In the K_2PtCl_6 structure (Fig. 10.5) the positions of the K^+ and $PtCl_6^{2-}$ ions are those of the F^- and Ca^{2+} ions in the CaF_2 structure, hence the name 'anti-fluorite' structure. Alternatively, if the A and X atoms are close-packed, as in $KOsF_6$ and in many halides A_mBX_{3m}, we may describe the structures as close-packed assemblies of A and X atoms (ions) in which B atoms occupy certain of the octahedral holes between groups of 6 X atoms. This aspect of these structures has been emphasized in Chapter 4, and shows, for example, the relation between three A_2BX_6 structures based on close-packed $(A + 3X)$ layers:

Structure	Sequence of c.p. AX_3 layers
Cs_2PuCl_6 (K_2GeF_6)	*h*
K_2PtCl_6	*c*
K_2MnF_6	*hc*

ABX$_6$ *structures*. Most of the compounds $A^I B^V X_6$ and $A^{II} B^{IV} X_6$ are fluorides. Compounds containing the heavier halogens are much less stable or in many cases unknown, though some ions BCl_6^{n-} form salts of this type with large cations as, for example, $[Co(NH_3)_6]TlCl_6$ and $[Ni(H_2O)_6]SnCl_6$. All fluorides ABF_6 crystallize with structures of the NaCl or CsCl types, various modifications of the ideal cubic structures being necessary to give the A ions suitable numbers of X neighbours. Five structures have been recognized for fluorides and a sixth, a tetragonal NaCl-type structure, is adopted by $Na[Sb(OH)_6]$.

NaCl type

(a) LiSbF$_6$: A rhombohedral structure alternatively described as a super-structure of VF$_3$ with 6-coordination of both A and B.

(b) and (c): The NaSbF$_6$ and CsPF$_6$ structures are both cubic, with 6- and 12-coordination of A respectively. There still seems to be some confusion in the literature regarding these two structures.

CsCl type

(d) KSbF$_6$: Tetragonal, 8-coordination of A (dodecahedral).

(e) KOsF$_6$: Rhombohedral—the BaSiF$_6$ structure, which is also adopted by SrPtF$_6$, BaPrF$_6$, BaTiF$_6$, and CsUF$_6$.

Table 10.4 shows that the type of structure of compounds $A^I B^V F_6$ is largely determined by the size of the A ion, the c.n. of which is 6 in (a) and (b), 8+4 in (d), and 12 in (c) and (e).

The KOsF$_6$ structure may alternatively be described as a slightly distorted cubic

TABLE 10.4

Crystal structures of fluorides $A^I B^V F_6$

B	Li	Na	Ag	K	Tl	Rb	Cs
				A			
P		b	b/c	c/e	c	c	c
As			b/c				
V, Ru, Ir, Os	a					e	
Re, Mo, W, Sb, Nb, Ta		b		d			

(A few of the Ag and Tl compounds have not been studied.)
(a) LiSbF$_6$ structure: AC 1962 **15** 1098
(b) and (c) NaSbF$_6$ and CsPF$_6$ structures: AC 1956 **9** 539
(d) KSbF$_6$ structure: AC 1976 **B32** 2916
(e) KOsF$_6$ structure: JINC 1956 **2** 79; BaSiF$_6$: JACS 1940 **62** 3126
For a review see: JCS 1963 4408.

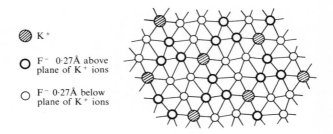

K⁺

F^- 0·27Å above plane of K^+ ions

F^- 0·27Å below plane of K^+ ions

FIG. 10.4. The slightly puckered KF_6 layer in $KOsF_6$.

closest packing of nearly plane layers of composition KF_6 (Fig. 10.4) between which Os atoms occupy octahedral interstices.

A_2BX_6 *structures.* We have already noted examples of halides A_2BX_6 with statistical AX_2 structures or with superstructures of AX_2 structures. Complex halides of this group necessarily contain alkali-metal ions, NH_4^+, Tl^+, or Ag^+. The majority contain cations which are comparable in size with one or more of the halide ions, and these form structures based on c.p. AX_3 layers. Such structures are not possible for the smaller Na^+ or Li^+ ions, which can be accommodated, like the B ions, in octahedral holes in a c.p. assembly of X ions. Accordingly we find a number of structures characteristic of compounds Li_2BX_6 and Na_2BX_6. These are closely related to (sometimes superstructures of) AX_2 structures, since they represent ways of filling (by $A_2 + B$) one-half of the octahedral holes in a closest packing of X ions (X_6). They are described in Chapter 4 and include three closely related structures:

Trirutile structure:	high-Li_2SnF_6, low-Li_2GeF_6: JSSC 1971 **3** 525
Na_2SiF_6 structure:	Na_2MF_6 (M = Si, Ge, Mn, Cr, Ti, Pd, Rh, Ru, Pt, Ir, Os): AC 1964 **17** 1408; JCS 1965 1559
Li_2ZrF_6 structure:	Li_2NbOF_5: ACSc 1969 **23** 2949

Several of these compounds are polymorphic; for example, the low–high transitions of Li_2SnF_6 (Li_2ZrF_6 to trirutile) and Li_2GeF_6 (trirutile to Na_2SiF_6).

We now describe the three c.p. structures (K_2PtCl_6, K_2GeF_6, and K_2MnF_6) in which $A + 3X$ together form the c.p. assembly.

The K_2PtCl_6 structure. In this structure (Fig. 10.5) the K^+ and $PtCl_6^{2-}$ ions occupy respectively the F^- and Ca^{2+} positions of the fluorite structure. In contrast to the K_2GeF_6 and K_2MnF_6 structures, which are largely confined to fluorides, this structure is more generally adopted by hexafluorides, hexachlorides, hexabromides, and a few hexaiodides (e.g. Rb_2SnI_6, Cs_2TeI_6) of the larger alkali metals, NH_4, and Tl. Some hexabromides and hexaiodides have less symmetrical variants of the structure. For example, K_2SnCl_6[1a] has the K_2PtCl_6 structure[1b] but K_2SnBr_6,[2] K_2TeBr_6, and K_2TeI_6 have distorted modifications of the structure in which the size of the cavity

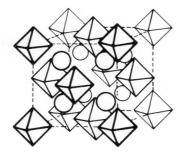

FIG. 10.5. The crystal structure of K_2PtCl_6.

for the A ion has been reduced by rotations of the BX_6 groups; a cavity of size suitable for K^+ in a hexachloride would be too large in the cubic structure of a hexabromide or hexaiodide.

Other compounds with this structure include hydrated compounds such as $(NH_4)_2(FeCl_5.H_2O)$, $K_2TcCl_5(OH)$,[3] $K_2(PtF_3Cl_3)$,[4] and $Cs_2Co^{IV}F_6$. The salt $(NH_4)_2SbBr_6$ is a superstructure of K_2PtCl_6 containing $SbBr_6^{3-}$ and $SbBr_6^-$ ions (p. 883). The relation of the K_2PtCl_6 to the cryolite and perovskite structures is discussed later.

The K_2GeF_6 (Cs_2PuCl_6) structure. This structure is confined to fluorides except for a few compounds such as Cs_2PuCl_6 and the trigonal forms of Cs_2ThCl_6 and Cs_2UCl_6. It is illustrated in Fig. 10.6. In the 'ideal' structure an A ion would have 12 equidistant X neighbours, but with the possible exception of Cs_2PuCl_6 the neighbours of A are either 9 approximately equidistant X atoms with 3 at a rather

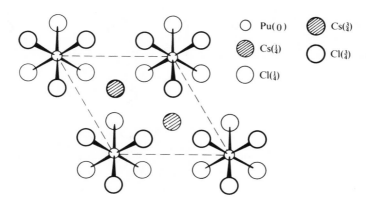

FIG. 10.6. The crystal structure of Cs_2PuCl_6 (or K_2GeF_6). Figures in brackets indicate the heights of atoms above the plane of the paper in terms of c (6.03 Å).

greater distance or groups of $3+6+3$ neighbours. Typical figures are:

$$\text{In } K_2GeF_6 \colon K \begin{cases} 9 \text{ F at } 2.85 \text{ Å} \\ 3 \text{ F at } 3.01 \text{ Å} \end{cases}; \quad \text{in } K_2TiF_6 \colon K \begin{cases} 3 \text{ F at } 2.75 \text{ Å} \\ 6 \text{ F at } 2.87 \text{ Å} \\ 3 \text{ F at } 3.08 \text{ Å} \end{cases}$$

In contrast, the neigbours of the A ions in the K_2PtCl_6 structure are always 12 equidistant X atoms, whether the close packing is perfect or whether there is some departure from perfect close packing. In the latter case the coordination group is no longer a regular cuboctahedron.

The K_2MnF_6 *structure.* This structure is adopted by numerous hexafluorides, many of which are polymorphic (Table 10.5).[5] Of special interest are the red salts

TABLE 10.5

Crystal structures of complex fluorides A_2BX_6

	K_2	Rb_2	Cs_2	$(NH_4)_2$	Tl_2
SiF_6	T C	C	C	T C	C
GeF_6	T H	T H	C	T	–
TiF_6	T H C	T H C	T C	T	T C
MnF_6	T H C	H C	C	–	–
$CrF_6{}^{(a)}$	H C	H C	C	–	–
$ReF_6{}^{(b)}$	T	T	T	–	–
$NiF_6{}^{(a)}$	C	C	C	–	–
PdF_6	H	H C	C	–	–
PtF_6	T	T	T	$C^{(c)}$	–

T = K_2GeF_6, H = K_2MnF_6, C = K_2PtCl_6 structure.
(*a*) ZaC 1956 **286** 136. (*b*) JCS 1956 1291. (*c*) AC 1980 **B36** 1921.

$M_2Ni^{IV}F_6$ (Ni–F, 1.70 Å) and the rose-coloured $M_2Cr^{IV}F_6$ (Cr–F, 1.72 Å). Since the feature common to all the three structures, K_2PtCl_6, K_2GeF_6, and K_2MnF_6, is the close packing of the A and X ions, there is no simple connection between the relative sizes of the A and BX_6 ions as in the ABX_6 structures of Table 10.4.

(1a) AC 1976 **B32** 2671
(1b) AC 1973 **B29** 1369
(2) AC 1979 **B35** 144

(3) JCS A 1967 1423
(4) JCS A 1966 1244
(5) JCS 1958 611

The cryolite family of structures for compounds A_3BX_6 *or* $A_2(B'B'')X_6$. The crystal structure adopted by certain salts A_3BX_6 is very simply related to that of K_2PtCl_6. In $(NH_4)_3AlF_6$ two-thirds of the NH_4^+ ions and the AlF_6^{3-} ions occupy the positions of the K^+ and $PtCl_6^{2-}$ ions in Fig. 10.5, and in addition there are NH_4^+ ions at the mid-points of the edges and the body-centre of the cubic unit cell. In this structure, as in K_2PtCl_6, the position of X along the cube edge is variable ($u00$, etc.) with $u \approx \frac{1}{4}$. It is convenient to describe this structure as the cryolite structure although the mineral cryolite itself (Na_3AlF_6)[1] does not crystallize with the most symmetrical form of the structure (compare the distorted form of the 'perovskite' structure adopted by the mineral of that name).

From the structural standpoint it is preferable to write the formulae of compounds of the cryolite family in the form $A_2(B'B'')X_6$ rather than A_3BX_6 because in this structure one-third of the A atoms in the formula A_3BX_6 are, like the B atoms, in octahedral holes in a cubic close-packed A_2X_6 (AX_3) assembly; the other two-thirds, corresponding to the K^+ ions in the K_2PtCl_6 structure, are surrounded by twelve equidistant X ions. Accordingly the cryolite structure is very closely related also to the perovskite structure. In Fig. 10.7 the A and X atoms are arranged in a c.c.p.

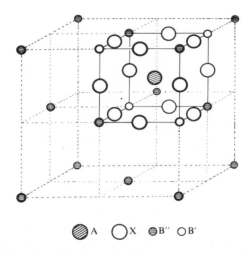

FIG. 10.7. Relation between perovskite, K_2PtCl_6, and cryolite structures.

arrangement of composition AX_3. Insertion of B'' atoms as shown gives the $A_2B''X_6$ (K_2PtCl_6) structure, which requires a unit cell (dotted lines) with eight times the volume of the small cube. Insertion of the B' atoms at the points marked gives the cryolite structure, $A_2(B'B'')X_6$, and if $B'=B''$ the structure is the perovskite structure (ABX_3), referable to the small unit cell. In other words, the very symmetrical structure of Fig. 10.7 (with $u=\frac{1}{4}$) is a superstructure of perovskite — compare the elevations of Fig. 4.31(b) and (e), p. 181.

As in the case of the perovskite structure there are variants of the cryolite structure with lower symmetry. In discussing these structures it is important to note that for full cubic symmetry (space group $Fm3m$) the X atoms must lie along the cube edges, their positions being determined by the value of the variable parameter u. Cubic symmetry may be retained, though of a lower class (space group $Pa3$) if the X atoms are not constrained to lie on the cell edges, and accordingly there is another (less symmetrical) structure, the *elpasolite* structure of compounds such as K_2NaAlF_6,[2] also with cubic symmetry. These structures are distinguished by the letters C and E in Table 10.6. There are also modified cryolite structures with only tetragonal or monoclinic symmetry.[3]

Most of the compounds A_3BX_6 with cubic symmetry at room temperature appear to be ammonium salts (e.g. $(NH_4)_3AlF_6$, $(NH_4)_3FeF_6$, and $(NH_4)_3MoO_3F_3$): K_3MoF_6 is an exception.[4] Generally departures from cubic symmetry are observed, but at higher temperatures the structure becomes cubic, as shown in Table 10.6.

TABLE 10.6
Compounds with cryolite-type structures

	$-180°$	$20°$	$300°$	$550°C$
K_2NaAlF_6	Cubic (E)	Cubic (E)	–	–
$(NH_4)_3AlF_6$ $(NH_4)_3FeF_6$	Tetragonal	Cubic (C)	–	–
K_3AlF_6	–	Tetragonal	Cubic (C)	–
Na_3AlF_6	–	Monoclinic	–	Cubic (C)

The behaviour of Li_3AlF_6, of which five polymorphs have been described, is apparently much more complex.[5] There is very irregular coordination of three types of non-equivalent Li^+ ion in the room-temperature form.

A similar behaviour is shown by the isostructural oxides R_3WO_6 and $R'_2R''WO_6$ and the corresponding Mo compounds.[6] For example, Ba_2CaWO_6 and Ba_2CaMoO_6 are cubic at ordinary temperatures, while Ba_2SrWO_6 is non-cubic but becomes cubic at 500°C. The simpler compounds Ca_3WO_6, Sr_3WO_6, and Ba_3WO_6 also crystallize at room temperature with distorted forms of the cubic structure.

The cryolite structure is also adopted by more than twenty chlorides $Cs_2NaM^{III}Cl_6$ in which M is In, Tl, Sb, Bi, Fe, Ti, Sc, Y, La, and 4f and 5f elements,[7] and by compounds with complex anions or cations. Examples include $I_3[Co(NH_3)_6]$ and numerous hexanitrites $M'_3[M''(NO_2)_6]$, in which $M' = NH_4$, K, Rb, Cs, or Tl, and $M'' = $ Co, Rh, or Ir. Salts containing two types of M' atoms are also known, for example, $K_2Ca[Ni(NO_2)_6]$; for $K_2Pb[Cu(NO_2)_6]$ see p. 1134. Oxyfluorides provide examples not only of the monoclinic and tetragonal forms of the cryolite structure, but also of the (cubic) elpasolite structure (see Table 10.11, p. 481).

(1) ACSc 1965 **19** 261
(2) JINC 1961 **21** 253
(3) AC 1953 **6** 49
(4) IC 1969 **8** 2694
(5) AC 1968 **B24** 225
(6) AC 1951 **4** 503
(7) IC 1970 **9** 1771

Halides A_4BX_6. These include $(NH_4)_4CdCl_6$[1] and the isostructural salts K_4MnCl_6, $K_3NaFeCl_6$, and K_4PbF_6, which contain octahedral ions $(BX_6)^{4-}$. The salt K_4ZnCl_6 is described in Groth's *Chemische Kristallographie*,[2] but its structure is not known.

If this salt contains $ZnCl_6^{4-}$ ions it would be the only known example of 6-coordination of Zn by Cl, but it could conceivably contain tetrahedral $ZnCl_4^{2-}$ (or trigonal bipyramidal $ZnCl_5^{3-}$) ions and additional Cl^- ions. There are numerous examples of octahedral coordination of Zn by F.

(1) ZaC 1956 **284** 10   (2) Vol. 1, p. 319 (1906)

Halides $A_m BX_7$

Complex fluorides of this type are formed by various metals of the IVA, VA, and VIA subgroups, by 5f elements, and by Tb in the 4f group. Cs_3TbF_7 is the first fluoro complex of Tb(IV) to be isolated.[1] These compounds provide examples of the following types of structure:

Finite BX_7 ion: pentagonal bipyramid—K_3UF_7, K_3ZrF_7, $(NH_4)_3ZrF_7$
                        capped trigonal prism—K_2TaF_7, Rb_2UF_7, Rb_2PuF_7[2]
Linear $(BX_7)_n$ ion: K_2PaF_7[3] (p. 1263)
Finite BX_6 ion + X^-: $(NH_4)_3(SiF_6)F$.

For further details of the IVA and VA compounds see p. 470, and for compounds of Th and U see Chapter 28. The structures of compounds such as $RbMoF_7$ and $RbWF_7$ are not yet known.

The chain ion in K_2PaF_7 (and the isostructural Rb, Cs, and NH_4 salts) is formed from tricapped trigonal prisms PaF_9 sharing two edges, and was mentioned in Chapter 1 as an interesting way of attaining an X : M ratio of 7 : 1 from a 9-coordination group.

(1) ZaC 1961 **312** 277   (3) JCS A 1967 1429, 1979
(2) JACS 1965 **87** 5803

Halides $A_m BX_8$

A considerable number of fluorides of this kind are formed by transition and 5f metals, for example, K_2MoF_8 and K_2WF_8[1] (structures not known), Na_3TaF_8, and compounds of Th, U, etc. (For $NaHTiF_8$ see p. 473.)

An ionic 3D structure with 8-coordination of both ions has been proposed for Na_2UF_8[2] (a compound earlier thought to be Na_3UF_9), and isolated UF_8^{4-} ions (distorted antiprisms) occur in $(NH_4)_4UF_8$ and the isostructural compounds of Ce, Pa, Np, Pu, and Am.[3] (There is square antiprism coordination also in Na_3TaF_8, for which see p. 473.) The corresponding Th compound, $(NH_4)_4ThF_8$, has a quite different structure.[4] There is a chain of composition ThF_7, rather similar to that in K_2PaF_7, formed from tricapped trigonal prism ThF_9 groups sharing two edges. The eighth F is a separate F^- ion, so that this compound, like $SrPbF_6$, has chain ions and separate F^- ions:

$SrPbF_6$: $(PbF_5)_n^{n-}$ chain ions + F^-,
$(NH_4)_4ThF_8$: $(ThF_7)_n^{3n-}$ chain ions + F^-;

compare the following compounds containing finite complex ions and additional X^- ions:

	Cs_3CoCl_5	$(NH_4)_3SiF_7$	$(NH_4)_5Mo_2Cl_9 \cdot H_2O$
Ions	$(CoCl_4)^{2-}$	$(SiF_6)^{2-}$	$(Mo_2Cl_8)^{4-}$
	Cl^-	F^-	Cl^-

In Na_3UF_8 (and the isostructural Pa and Np compounds) there is almost perfect cubic 8-coordination of the 5f ion.[5]

Complex chlorides of Mn(II) provide a number of examples of isostructural complex halides and complex oxides (Table 10.2, p. 448). An example is Na_6MnCl_8, which has the same structure as Mg_6MnO_8 (p. 606); another is Na_2MnCl_4, which is isostructural with Sr_2PbO_4. A third compound in this family is $Na_2Mn_3Cl_8$,[6] which is of interest as a partially closest-packed sequence of Cl layers with Na^+ in trigonal prismatic sites and Mn^{2+} in three-quarters of the octahedral sites:

$$A \quad\quad B \quad\quad B \quad\quad C \quad\quad C \quad\quad A \ldots$$
$$\text{oct.} \quad \text{t.p.} \quad \text{oct.} \quad \text{t.p.} \quad \text{oct.}$$

We have now dealt with the vertical subdivisions of Table 10.1. It remains to comment on the classes (e) and (f) and to add short sections on the complex halides of certain metals.

(1) JCS 1956 1242; 1958 2170
(2) IC 1966 **5** 130
(3) AC 1970 **B26** 38
(4) AC 1969 **B25** 1958
(5) JCS A 1969 1161
(6) AC 1975 **B31** 770

Complex halides containing finite polynuclear complex ions

Ions in which metal–metal interactions play a prominent part are included with 'metal cluster' compounds (p. 432); they are formed notably by Nb, Ta, Mo, W, Tc, Re, and Pt. The three simplest types of ion to be discussed here are those in which there are single, double, or triple halogen bridges, corresponding to the sharing of a vertex, edge, or face between the coordination groups of M atoms:

$$\text{M—X—M} \quad\quad\quad \text{M} \overset{X}{\underset{X}{<>}} \text{M} \quad \text{or} \quad \text{M} \overset{X}{\underset{X}{<-X->}} \text{M}$$

Single halogen bridges link planar $AgCl_3$ groups in the $Ag_2Cl_5^{3-}$ ion (p. 1146), pyramidal MX_3 groups in $NaSn_2F_5$ (p. 1184), tetrahedral MX_4 groups in $Te_4(Al_2Cl_7)_2$ (p. 703), $NH_4(Al_2Br_7)$,[1] and $K(Ga_2Cl_7)$,[2] octahedral MX_6 groups in $(SeF_3)(Nb_2F_{11})$ (p. 739), and trigonal prismatic MX_6 groups in $Na_5(Zr_2F_{13})$ (p. 472).

Double halogen bridges link planar MX_4 groups in $Pt_2Br_6^{2-}$,[3] tetrahedral MX_4 groups in $Cd_2Cl_6^{2-}$,[3a] and $Hg_2I_6^{2-}$,[3b] and octahedral MX_6 groups in $Ti_2Cl_{10}^{2-}$[4] and $Bi_2Br_{10}^{4-}$.[4a, b]

Triple halogen bridges occur in numerous enneahalides $A_3B_2X_9$ in which A^+ is a large cation and X is usually Cl, Br, or I. Numerous salts of this kind are formed by Cs and the structures are simple because Cs^+ is comparable in size with the halide ions and can form c.p. structures from CsX_3 layers between which M^{III} atoms occupy octahedral interstices (p. 184). There are three types of structure of salts $Cs_3M_2X_9$. In one each octahedral MX_6 group shares 3 vertices of one face to form the 2D anion illustrated in Fig. 4.35 (p. 185). The only known example is the (c.c.p.) structure of $Cs_3Bi_2Br_9$.[5] In the second type of structure the M atoms occupy pairs of face-sharing octahedral interstices to form ions $M_2X_9^{3-}$ with the structure shown in Fig. 10.8:

$$Cs_3M_2F_9: \quad M = Fe, Ga \text{ (see p. 469)}$$
$$Cs_3M_2Cl_9: \quad M = Ti, V, Cr, Mo, Tl, Rh$$
$$Cs_3M_2Br_9: \quad M = Cr, Mo$$
$$Cs_3M_2I_9: \quad M = Sb, Bi$$

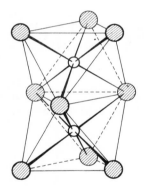

FIG. 10.8. The $Tl_2Cl_9^{3-}$ ion.

The structure of the $Ti_2Cl_9^-$ ion has been determined in $(PCl_4)(Ti_2Cl_9)$,[4] in which the bond lengths (terminal, 2.23, and bridging, 2.49 Å) are very similar to those in the $Ti_2Cl_{10}^{2-}$ ion (above). In the anion in $K_3W_2Cl_9$,[6] which has the same general type of structure as the above-mentioned Cs salts, there is bonding between the metal atoms (W–W, 2.44 Å) and consequent reduction of the magnetic moment, but in the other salts listed there is no metal–metal interaction (for example, Cr–Cr, 3.1 Å) and normal magnetic properties). In this respect the ions $Mo_2Cl_9^{3-}$ and $Mo_2Br_9^{3-}$ (Mo–Mo, 2.66 and 2.82 Å) seem to occupy an intermediate position.[7]

There are compounds $Cs_3M_2Cl_9$ of a third type which do not contain enneahalide ions, either finite or 2D. The M atoms are displaced from the centres of their MX_6 octahedra to form discrete pyramidal MCl_3 molecules embedded among Cs^+ and Cl^- ions. Examples include $Cs_3As_2Cl_9$[8] (As–3Cl, 2.25 Å, As–3Cl, 2.75 Å), both forms of $Cs_3Sb_2Cl_9$, and $Cs_3Bi_2Cl_9$.[9] The compounds $Cs_3Bi_2X_9$ form an interesting series, for the chloride contains $BiCl_3$ molecules, the bromide has the c.c.p. layer structure noted above, while the iodide contains finite $Bi_2I_9^{3-}$ ions.[9a] However, it should be observed that the nature of the enneahalide complex depends on the type of cation, for $(\phi_4P)_3Bi_2Br_9$[10] contains finite $Bi_2Br_9^{3-}$ ions. Ions $Sb_2Br_9^{3-}$ and separate Br^- ions occur in the salt $(C_5H_5NH)_5(Sb_2Br_9)Br_2$.[11] Compounds $A_mB_2X_9$ which do *not* contain enneahalide ions include the hydrate of $(NH_4)_5Mo_2Cl_9$, which is $(NH_4)_5(Mo_2Cl_8)Cl.H_2O$ (p. 431), and KU_2F_9, which has a typical ionic structure in which U^{4+} is 9-coordinated (tricapped trigonal prism) and K^+ is 8-coordinated (distorted cube).[12]

Closely related to the enneahalide ions are molecules such as $Ru_2Cl_5[P(C_4H_9)_3]_4$, (ref. 13) (a), and $Ru_2Cl_4[P(C_6H_5)(C_2H_5)_2]_5$.[14] As in $Cr_2Cl_9^{3-}$ there is no metal-

$$\xleftarrow{\hspace{1em}} 3{\cdot}115\ \text{A} \longrightarrow$$

$$
\begin{array}{c}
\text{P} \quad\ \ \text{Cl} \quad\ \ \text{P}\\
\diagdown\ \diagup\diagdown\ \diagup\\
\text{P—Ru—Cl—Ru—Cl}\\
\diagup\ \diagdown\ \text{\o}\ \diagup\ \diagdown\\
\text{Cl} \quad\ \ \text{Cl} \quad\ \ \text{P}
\end{array}
$$

$$[P = P(n\text{-}C_4H_9)_3]$$

(a)

metal bonding, as is shown by the M–M distances or by the values of the angle ϕ. This is $70\frac{1}{2}°$ for regular octahedra sharing a face, $77°$ in $Cr_2Cl_9^{3-}$, $79\frac{1}{2}°$ in (a), but only $58°$ in $W_2Cl_9^{3-}$.

(1)	AC 1975 **B31** 2177	(7)	IC 1971 **10** 1453
(2)	AC 1976 **B32** 247	(8)	JCP 1935 **3** 117
(3)	AC 1964 **17** 587	(9)	AC 1974 **B30** 1088
(3a)	AC 1980 **B36** 2616	(9a)	ACSc 1968 **22** 2943
(3b)	JCMS 1972 **2** 183	(10)	AC 1977 **B33** 2686
(4)	IC 1971 **10** 122	(11)	JCS A 1970 1359
(4a,b)	AC 1980 **B36** 2745, 2748	(12)	AC 1969 **B25** 1919
(5)	AC 1977 **B33** 2961	(13)	JCS A 1968 1981
(6)	AC 1958 **11** 689	(14)	JCS A 1968 2108

Complex halides containing complex anions of more than one kind

There are numerous complex halides in which both cation and anion are complex but little is known of compounds containing two kinds of complex anion. Examples include $Na_3(HF_2)(TiF_6)$, p. 473, and the deeply-coloured salts $(NH_4)_4(Sb^V Br_6)(Sb^{III}Br_6)$, p. 883, and $Cs_2(Au^ICl_2)(Au^{III}Cl_4)$.[1] The structure of the last compound is closely related to the perovskite structure, the structure being distorted so that instead of six octahedral neighbours the two kinds of Au atom have two collinear or four coplanar nearest neighbours. The linear $(Cl–Au^I–Cl)^-$ and the square planar $(Au^{III}Cl_4)^-$ ions can be clearly seen in Fig. 10.9(b). For comparison a portion of the ideal cubic perovskite structure is shown in Fig. 10.9(a), the

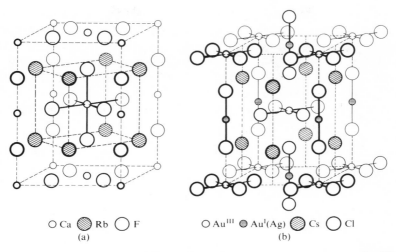

OCa ◕Rb ○F   ○Au^III ⊖Au^I(Ag) ◕Cs ○Cl
(a)   (b)

FIG. 10.9. The structures of (a) $RbCaF_3$, perovskite structure; and (b) $Cs_2Au^IAu^{III}Cl_6$ in similar orientations.

inscribed cube being the unit cell. For an alternative unit cell see Fig. 13.3(a) (p. 584). For the structure of $(NH_4)_6(AuCl_4)_3 Ag_2Cl_5$ see p. 1146.

(1) JACS 1938 **60** 1846

Miscellaneous complex halides

Apart from the special groups described in the present chapter other complex halides are described in other chapters: Sb, Chapter 20, B subgroup elements, Chapters 25 and 26, U and 5f elements, Chapter 28.

Complex chlorides $CsMCl_3$, Cs_2MCl_4, and Cs_3MCl_5 formed by 3d metals[1]

The eight elements from Ti to Cu inclusive all form ions M^{2+} and in these ions the number of d electrons increases from 2 to 9. All of these elements form one or more of the following types of complex chloride, $CsMCl_3$, Cs_2MCl_4, or Cs_3MCl_5:

$CsMCl_3$: formed by all these elements. The structure ($CsNiCl_3$) consists of c.p. Cs^+ and Cl^- ions with M^{2+} in *octahedral* coordination, which is distorted from regular octahedral coordination in the isostructural Cr and Cu compounds.

Cs_2MCl_4: found in all systems investigated (CsCl–$CrCl_2$ not studied) except V and Ni. The only compound found in the CsCl–VCl_2 system is $CsVCl_3$, while in the CsCl–$NiCl_2$ system Cs_2NiCl_4 does not exist. The salts Cs_2MCl_4 contain *tetrahedral* MCl_4^{2-} ions.

Cs_3MCl_5: known to be formed by Mn, Co, Cu (*not* by V) and by Ni, but Cs_3NiCl_5 is stable only at high temperatures. It can be prepared from the molten salts and quenched, but on slow cooling it breaks down into $CsCl + CsNiCl_3$. The salts are isostructural with Cs_3CoCl_5[2] and contain tetrahedral MCl_4^{2-} and Cl^- ions. (The ion $NiCl_4^{2-}$ can exist at ordinary temperatures in conjunction with very large cations, as in the salt $[(C_6H_5)_3CH_3As]_2NiCl_4$, in which it has a regular tetrahedral structure (Ni–Cl, 2.27 Å).)[3]

These findings agree well with the expectations from crystal field theory. A strong preference for octahedral coordination should be shown by V^{2+} (d^3) and Ni^{2+} (d^8). In the $CsMCl_3$ compounds the large distortion of the octahedral coordination group expected for the d^4 and d^9 configurations is found in the isostructural $CsCuCl_3$ and $CsCrCl_3$ ((4+2)-coordination). A non-regular tetrahedral coordination is found for Cu^{2+} in Cs_2CuCl_4, but Cs_2CrCl_4 has not been prepared.

As regards these chloro compounds Zn^{2+} (d^{10}) falls in line with the 3d ions, forming Cs_2ZnCl_4 and Cs_3ZnCl_5 with tetrahedral coordination of Zn but not $CsZnCl_3$. The preference for tetrahedral coordination by Cl is very marked, being found in all three forms of $ZnCl_2$ and even in $Na_2ZnCl_4 . 3H_2O$.[4] However, Zn^{2+} is 6-coordinated by F^- in ZnF_2 (rutile structure) and also in $KZnF_3$ which, like the corresponding compounds of Mn, Fe, Co, and Ni, has the ideal perovskite structure. $KCrF_3$ and $KCuF_3$, as expected, have a distorted variant of that structure. The structural chemistry of complex halides of Cu(II) is described in Chapter 25.

(1) For earlier refs. see JPC 1962 **66** 65   (3) IC 1966 **5** 1498
(2) AC 1964 **17** 506   (4) ZK 1960 **114** 66

Complex fluorides of Al[1] and Fe[2]

These compounds provide a good example of a family of structures based on a simple principle, namely, the sharing of vertices between octahedral AlF_6 (FeF_6) coordination groups to give various F : Al ratios. Most of the compounds discussed below are formed by K, Rb, and Tl (sometimes also NH_4), indicated by M, but certain of the structures are peculiar to the smaller Na^+ ion. An F : Al ratio of 6 : 1 implies discrete AlF_6 groups, as in the cryolite structure of compounds such as Na_3AlF_6 and in the more complex structure of cryolithionite, $Na_3Li_3Al_2F_{12}$. This mineral has the garnet structure, so that its formula may alternatively be written $Al_2Na_3(LiF_4)_3$. This illustrates the point that the description of these complex fluorides in terms of AlF_6 or FeF_6 octahedra is arbitrary in the sense that the Al–F bonds are not appreciably different in nature from the other (M–F) bonds in the structures. Sharing of 2 vertices of each MF_6 octahedron gives the MF_5 chain (*cis* or *trans*) and the sharing of 4 vertices the *cis* or *trans* layers. The latter is illustrated in Fig. 10.10(a). In the intermediate layer, of composition M_3F_{14}, found in chiolite, $Na_5Al_3F_{14}$, one-third of the octahedra share 4 equatorial vertices and the remainder

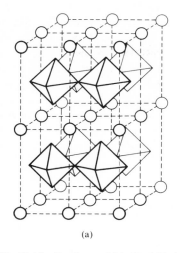
(a)

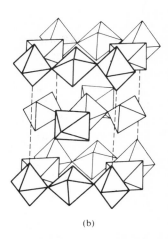
(b)

FIG. 10.10. Portions of (a) the AlF_4 layers in $TlAlF_4$; and (b) the Al_3F_{14} layers in $Na_5Al_3F_{14}$.

2 *trans* vertices. The puckered (*cis*) layer, in which the *un*shared vertices are adjacent, is found in $NaFeF_4$.

The sharing of odd numbers, 1, 3, or 5, of vertices of every MF_6 octahedron would give respectively the finite ion M_2F_{11}, the M_2F_9 layer of Fig. 4.35 (p. 185), or the M_2F_7 double layer which is a slice of the ReO_3 structure (p. 211). Only the last of these types of complex appears to be known in these families of compounds, as the 2D anion in $K_3Fe_2F_7$.[3] The structure of $Cs_3Fe_2F_9$[4] is of interest in this connection for it contains face-sharing pairs of FeF_6 octahedra, not the vertex-sharing layer that might have been expected.

An interesting structure has been suggested[2] for compounds $M_2Fe_5F_{17}$ based on the dimensional analogy with the oxide $MoW_{11}O_{36}$. In Fig. 10.11(a) each square

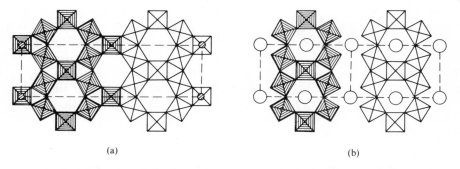
(a)

(b)

FIG. 10.11. Relation of octahedral layer M_5X_{17} (b) to 3D MX_3 structure (a).

represents a chain of vertex-sharing octahedra perpendicular to the plane of the paper (ReO_3-type chain), the small black circles representing Mo atoms. Removal of certain of the metal atoms and of the O atoms between them would leave layers of composition $W_{10}O_{34}$, suggesting the structure Fig. 10.11(b) for $M_2Fe_5F_{17}$ in which the complex layers are perpendicular to the plane of the paper. In this layer one-fifth of the octahedra share 6 and the remainder 5 vertices. The simplest 3D framework formed from octahedral groups sharing all their vertices represents the structures of AlF_3 or FeF_3, and a more complex framework is found in a hydrated hydroxyfluoride of Al with the pyrochlore structure (p. 258). These structures are summarized in Table 10.7.

TABLE 10.7
Structures of complex fluorides

Number of $Al(Fe)F_6$ *vertices shared*	*Type of complex*		*Examples*
0	Finite	M_3FeF_6	Na_3AlF_6 (cryolite)
			$Na_3Li_3Al_2F_{12}$ (cryolithionite)
2	Chain: *cis*	K_2FeF_5	
	trans		Tl_2AlF_5
2 and 4	Layer	$Na_5Fe_3F_{14}$	$Na_5Al_3F_{14}$ (chiolite)
4	'*cis*' layer	$NaFeF_4$	
	'*trans*' layer	$MFeF_4$	$MAlF_4$
5	Double layer	$K_3Fe^{II}_2F_7$	–
(5 and 6	Multiple layer	$M_2Fe_5F_{17}$)	
6	3D framework	FeF_3	AlF_3

(1)   ZaC 1937 **235** 139; 1938 **238** 201; 1938 **239** 301   (3)   BSCF 1965 **12** 3489
(2)   JSSC 1970 **2** 269   (4)   JLCM 1971 **25** 257

Some complex fluorides of group IV A and V A elements

In Group VA there is an interesting gradation in properties from V to Ta. In complex salts V prefers bonds to O rather than F—note the extensive oxygen chemistry of V and the formation of numerous oxyhalides by V and Nb; Ta forms only TaO_2F. The most stable salts obtained by the addition of metallic fluorides to a solution of Nb_2O_5 in HF are those containing the ion $NbOF_5^{2-}$. This ion is much more stable than simple NbF_n ions. For example, K_2NbF_7 is prepared from a solution of $K_2NbOF_5.H_2O$ in HF but is easily hydrolysed back to that salt, while Rb_2NbF_6 is made by repeated crystallization of Rb_2NbOF_5 from HF. Finally, tantalum prefers to form complex fluorides of the type $R_mTaF_n.xH_2O$ rather than oxyfluorides, and in these n rises to 8. The V and Nb analogues of salts like Na_3TaF_8 are not known.

The following scheme illustrates, for the potassium–niobium fluorides, how the composition of the product varies with the relative concentrations of the component halides and of HF:

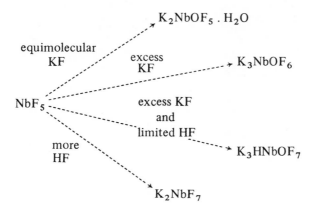

Complex fluorides and oxyfluorides provide many examples of isostructural compounds as, for example, compounds in the same horizontal row in the following table:

IV	V	VI
$K_2TiF_6.H_2O$	$K_2NbOF_5.H_2O$	$K_2WO_2F_4.H_2O$
$(NH_4)_3TiF_7$	$\begin{cases}(NH_4)_3NbOF_6\\(NH_4)_3TaOF_6\end{cases}$	
K_3ZrF_7	K_3NbOF_6	
$\left.\begin{array}{l}K_3HSnF_8\\K_3HPbF_8\end{array}\right\}$	K_3HNbOF_7	

The special interest of these compounds lies in the shape of the coordination group around the metal atom in the complex ion.

In complex fluorides Zr^{4+} is found in 6-, 7-, and 8-coordination. The 6-coordination is octahedral, the 7-coordination approximates to either pentagonal bipyramidal or monocapped trigonal prismatic, and the 8-coordination to either dodecahedral, square antiprismatic, or bicapped trigonal prismatic. We noted in Chapter 3 that certain pairs of polyhedra are closely related, and this is true of, for example, the square (tetragonal) antiprism and the bicapped trigonal prism (Fig. 10.12(a)). Unless the coordination in a particular crystal is very close to one of the 'ideal' polyhedra there may be doubt about the best way of describing the coordination type.

Octahedral coordination of Zr^{4+} is not common; one example, $CuZrF_6.4H_2O$, is mentioned later. The type of coordination obviously adjusts to the nature of the cations and to the geometrical requirements imposed by a particular F:Zr ratio. The inadvisability of trying to relate ionic sizes to c.n. and shape of coordination polyhedron is nicely illustrated by the structure of $Rb_5Zr_4F_{21}$. In this crystal there

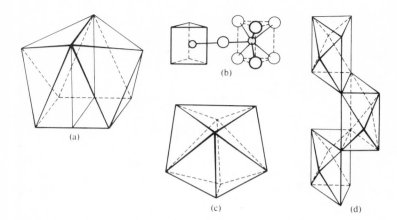

FIG. 10.12. Complex fluorides of Zr: (a) relation between bicapped trigonal prism and tetra-gonal antiprism;   (b) 7-coordination in $Na_5Zr_2F_{13}$;   (c) 8-coordination group in Li_6BeZrF_{12}; (d) chain of bicapped trigonal prisms in $(N_2H_6)ZrF_6$.

are four kinds of coordination of Zr^{4+} ions: 8, irregular antiprism; 7, irregular antiprism less 1 vertex; 7, pentagonal bipyramid; and 6, octahedron.

The simplest example of 7-coordination is the pentagonal bipyramidal anion in $(NH_4)_3ZrF_7$. Two such groups share an edge in the $Zr_2F_{12}^{4-}$ ion in $K_2Cu(Zr_2F_{12}).6H_2O$ and low-$BaZrF_6$. In $Na_5Zr_2F_{13}$ two 7-coordination groups, described as monocapped trigonal prisms, share one vertex (Fig. 10.12(b)).

The structure of the anion in K_2ZrF_6, formed from dodecahedral groups sharing two edges, has already been illustrated in Fig. 10.3 (p. 456); there are similar chains in the $(RbPaF_6)$ structure of high-$BaZrF_6$, $PbZrF_6$, etc. Well-defined dodecahedral coordination of Zr^{4+} is found in Li_6BeZrF_{12}, Fig. 10.12(c), in which crystal there is rather irregular octahedral coordination of Li^+, tetrahedral coordination of Be^{2+}, and discrete ZrF_8 groups. A second type of ZrF_6 chain occurs in $(N_2H_6)ZrF_6$, Fig. 10.12(d), the coordination here being bicapped trigonal prismatic. The coordination in the discrete ZrF_8^{4-} ion in $Cu_2ZrF_8.12H_2O$ is close to square antiprismatic. Two such groups form edge-sharing pairs in $Cu_3Zr_2F_{14}.16H_2O$, and this type of coordination is also found in $Na_7Zr_6F_{31}$ and the isostructural compounds of Pa, U, and other 5f elements. Thirty of the F atoms together with the Zr atoms form a 3D framework built from ZrF_8 antiprisms. The antiprisms share vertices in groups of six around cuboctahedral 'holes', and these subgroups consisting of six antiprisms then form the 3D framework by sharing some of their edges. The composition of this framework is $(ZrF_5)_n$; the remaining F^- ion and the Na^+ ions occupy interstices in the framework. (Some analogy may be drawn between this structure and structures such as pyrochlore (p. 258) and $La_4Re_6O_{19}$ (p. 224).)

The (hydrated) Cu–Zr fluorides provide examples of 6-, 7-, and 8-coordination of Zr^{4+} and also of various coordination groups for Cu^{2+}. In $CuZrF_6.4H_2O$ there are chains of alternate Cu and Zr coordination groups, $Cu(H_2O)_4F_2$ and ZrF_6:

$$
\begin{array}{ccccccc}
 & \text{F} & & \text{OH}_2 & & \text{F} & \\
 & | & \diagup\text{F} & | & \diagup\text{OH}_2 & | & \diagup\text{F} \\
-\text{F}-\!\!&\!\!\text{Zr}\!\!&\!\!-\text{F}-\!\!&\!\!\text{Cu}\!\!&\!\!-\text{F}-\!\!&\!\!\text{Zr}\!\!&\!\!-\text{F}- \\
 & \text{F}\diagup & | & \text{OH}_2\diagup & | & \text{F}\diagup & | \\
 & & \text{F} & & \text{OH}_2 & & \text{F}
\end{array}
$$

In $Cu_2ZrF_8.12H_2O$ there are antiprismatic ZrF_8 and octahedral $Cu(H_2O)_6$ groups, while in $Cu_3Zr_2F_{14}.16H_2O$ there are pairs of edge-sharing ZrF_8 antiprisms and two kinds of Cu aquo complex, namely $Cu(H_2O)_6$ and bridged groups formed

$$
(H_2O)_4Cu\!\!\diagdown_{\!\!H_2O}^{\!\!H_2O}\!\!\diagup Cu(H_2O)_4
$$

from two octahedra sharing an edge. The structural formulae of these last two compounds are therefore $(ZrF_8)[Cu(H_2O)_6]_2$ and $(Zr_2F_{14})[Cu(H_2O)_6][Cu_2(H_2O)_{10}]$. There is 7-coordination of Zr in $K_2CuZr_2F_{12}.6H_2O$, in which there are pairs of pentagonal bipyramidal groups sharing an edge. Accordingly, in these four compounds with F:Zr ratios of 6, 8, 7, and 6, the c.n.s of Zr are 6, 8, 8, and 7, again illustrating the point that there is no simple relation between F:Zr ratio and c.n. of Zr; only the first two contain finite ZrF_n groups. This point is elaborated in the upper part of Table 10.8, which includes references to crystal structures. The lower parts of this Table illustrate some features of the structures of complex fluorides of elements other than Zr, for example, the achievement of different F:Ce ratios with the same c.n. of Ce^{4+} and the formation of various types of complex ion in a number of complex fluorides and oxyfluorides. In contrast to K_3SiF_7 and $(NH_4)_3SiF_7$, which contain octahedral SiF_6^{2-} and separate F^- ions, $(NH_4)_3ZrF_7$ contains pentagonal bipyramidal ZrF_7^{3-} ions (compare K_3UF_7 and $K_3UO_2F_5$, p.1262), while the NbF_7^{2-} and TaF_7^{2-} ions are monocapped trigonal prisms. Yet another possibility is realized in $(NH_4)_3CeF_7.H_2O$, where two dodecahedral coordination groups share an edge to form $(Ce_2F_{14})^{6-}$ ions. These compounds thus illustrate three ways of attaining a F:B ratio of 7:1, with c.n.s of 6, 7, and 8 for the B atoms. Conversely, in the three ammonium ceric fluorides listed in Table 10.8 there is 8-coordination of Ce with F:Ce ratios of 6:1, 7:1, and 8:1, while $(NH_4)_2CeF_6$ is our third example of a salt A_2BX_6 in which the anion is an infinite chain built of 8-coordination polyhedra, all of different kinds:

	Coordination polyhedron
K_2ZrF_6	Dodecahedron
$(N_2H_6)ZrF_6$	Bicapped trigonal prism
$(NH_4)_2CeF_6$	Antiprism.

In Na_3TaF_8 the anion has the shape of a square antiprism; on the other hand there is no BX_8 ion in Na_3HTiF_8 but equal numbers of HF_2^- and TiF_6^{2-} ions. Similarly, K_3HNbOF_7 contains HF_2^- and octahedral $NbOF_5^{2-}$ ions; the latter are the anions in $K_2NbOF_5.H_2O$.

Equally interesting are the complex fluorides of Th and U(IV), which are described in Chapter 28, and in which 9-coordination also is important.

TABLE 10.8

Coordination polyhedra in complex fluorides and oxyfluorides

F : Zr ratio	Formula	C.n. of Zr	Coordination polyhedron	Reference
5.17	$Na_7Zr_6F_{31}$	8	Antiprism	AC 1968 **B24** 230
5.25	$Rb_5Zr_4F_{21}$	6, 7, 8	See text	AC 1971 **B27** 1944
6	K_2ZrF_6	8	Dodecahedron	AC 1956 9 929
6	$(N_2H_6)ZrF_6$	8	Bicapped trigonal prism	AC 1971 **B27** 638
6	$\gamma\text{-}Na_2ZrF_6$	7	Irregular	AC 1969 **B25** 2164
6	$K_2CuZr_2F_{12}.6H_2O$	7	Pentagonal bipyramid	AC 1973 **B29** 1958
6	low-$BaZrF_6$	7	Pentagonal bipyramid	AC 1978 **B34** 1070
6	$CuZrF_6.4H_2O$	6*	Octahedron	AC 1973 **B29** 1955
6.5	$Na_5Zr_2F_{13}$	7	Monocapped t.p.	AC 1965 18 520
7	$Cu_3Zr_2F_{14}.16H_2O$	8	Antiprism	AC 1973 **B29** 1963
7	Na_3ZrF_7	7*	Irregular	AC 1948 1 265
7	$(NH_4)_3ZrF_7$	7*	Pentagonal bipyramid	AC 1970 **B26** 2136
8	$Cu_2ZrF_8.12H_2O$	8*	Antiprism	AC 1973 **B29** 1967
8	$Li_6BeF_4(ZrF_8)$	8*	Dodecahedron	JCP 1964 **41** 3478

* Denotes finite complex ion BX_n

F : Ce ratio	Formula	C.n. of Ce	Coordination polyhedron	Reference
6	$(NH_4)_2CeF_6$	8	Antiprism	IC 1969 8 33
7	$(NH_4)_3CeF_7.H_2O$	8	Dodecahedron	AC 1971 **B27** 1939
8	$(NH_4)_4CeF_8$	8*	Antiprism	AC 1970 **B26** 38

* Denotes finite complex ion BX_n

Empirical formula	Type of complex ion	Shape	Reference
K_3SiF_7	SiF_6^{2-}	Octahedral	JACS 1942 **64** 633
K_2NbF_7, K_2TaF_7	NbF_7^{2-}, TaF_7^{2-}	Monocapped t.p.	AC 1966 **20** 220
K_3NbOF_6	$NbOF_6^{3-}$	(Pentagonal bipyramid)[a]	JACS 1942 **64** 1139
$K_2NbOF_5.H_2O$	$NbOF_5^{2-}$	Octahedral	JACS 1941 **63** 11
K_3HNbOF_7	$\begin{cases} NbOF_5^{2-} \\ HF_2^- \end{cases}$	Octahedral / Linear	JACS 1941 **63** 11
Na_3TaF_8	TaF_8^{3-}	Antiprism	JACS 1954 **76** 3820
Na_3HTiF_8	$\begin{cases} HF_2^- \\ TiF_6^{2-} \end{cases}$	Linear / Octahedral	AC 1966 **20** 534

[a] Probably as in $(NH_4)_3ZrF_7$, but no recent study has been made.

Metal nitride halides and related compounds

'Nitride halides' have been prepared in various ways, for example, Mg_2NCl (also F, Br, I) by direct combination of metal nitride and halide, Mg_3NF_3 by the action of nitrogen on a mixture of metal and halide at 900 °C, and ZrNI by heating $ZrI_4 \cdot nNH_3$. References to some structural studies are included in Table 10.9. The compound $VNCl_4$ is apparently vanadium chlorimide trichloride and has been assigned a structure in which the molecules are loosely associated in pairs. The bond arrangement

TABLE 10.9
Metal nitride halides and related compounds

	Structure	Reference
Mg_3NF_3	Defect NaCl (p. 195)	AC 1969 **B25** 1009
Mg_2NCl (F)	–	ZaC 1968 **363** 191; JSSC 1970 **1** 306
ZrNCl(Br, I) TiNCl(Br, I)	FeOCl (N not located)	ZaC 1964 **327** 207
ThNF ThNCl(Br, I)	LaOF (rhombohedral) BiOCl	ZaC 1968 **363** 258
UNCl(Br, I)	BiOCl	ZaC 1969 **366** 43
[U(NH)Cl?]		ZaC 1966 **348** 50
LaN_xF_{3-x}	LaF_3	MRB 1971 **6** 57
$VCl_3(NCl)$	See text	ZaC 1968 **357** 325
Ca_2NX		CR 1971 **C272** 1657; NW 1971 **4** 219

around V is square pyramidal or distorted octahedral according to whether the weak bonds to Cl atoms of other molecules are included; the accuracy of the bond lengths is not clear. The V–N bond length is similar to that in other transition-metal complexes in which N and Cl are directly bonded to the metal (e.g. K_2OsNCl_5, Os–N, 1.61 Å).

The defect NaCl structure of Mg_3NF_3 is illustrated in Fig. 6.1(e) (p. 239); Mg has 4 F and 2 N (octahedral) neighbours (at 2.11 Å), N has 6 Mg (octahedral) and F has 4 Mg (coplanar) nearest neighbours. The tetragonal structure of Mg_2NF (X-ray powder data) is apparently an ordered NaCl-like structure which is extended in one direction so that the nearest neighbours of Mg are 3N and 2 F (at a mean distance of 2.13 Å) with the sixth (F) at 2.86 Å. The coordination of N is distorted octahedral, and F has 4 close Mg neighbours as in Mg_3NF_3. The structure of the high-pressure form of Mg_2NF is apparently much closer to a cubic NaCl structure, as is to be expected since Mg_2NF is isoelectronic with MgO.

Thiohalides

Structural studies have as yet been made only of certain groups of thiohalides, and it is therefore too early to attempt a systematic treatment of these compounds. The thiohalides of P are described in Chapter 19; here we shall note the structures of a number of metal thiohalides. Compounds MSX are numerous, being formed by La, Y, the 4f elements, and B subgroup elements such as In, Sb, and Bi. More complex structures are found for $Hg_3S_2Cl_2$ (p. 1164) and NbS_2Cl_2 (see below).

The structures of 4f thiohalides, many of which are polymorphic, are probably peculiar to these compounds and contain the metal in 6-, 7-, or 8-coordination. For example, SmSI consists of layers of composition SmX formed from SSm_4 tetrahedra sharing 3 edges (Fig. 28.9(a), p. 1272) interleaved with I atoms. In the resulting structure Sm is 7-coordinated, to 3S at 2.73, 1S at 2.84, and 3I at 3.29 Å.[1] There is also 7-coordination of the metal (by 4S and 3Br) in the isostructural thiobromides of Pr, Sm, Nd, Gd, and Tb,[2] which are structurally similar to EuI_2 (p. 416). In the more complex structure of β-YSF (and isostructural Ho, Er, Yb, and Lu compounds)[3] there are CdI_2-like layers of composition YS_2 containing one-half of the Y atoms, which have 6 octahedral S neighbours. The S atoms in these layers are bonded to the remaining Y atoms, which are arranged in planar hexagonal layers of composition YF_2. These Y atoms have hexagonal bipyramidal coordination, to 6F and 2S. All the thiohalides of In have been prepared, and from X-ray photographic evidence InSCl and InSBr have been assigned a statistical $CdCl_2$ structure, but further examination would be desirable if single crystals could be grown.[4]

Of the twelve possible compounds ABX where A is Sb or Bi, B is S or Se, and X is Cl, Br, or I, all except SbSCl and SbSeCl have been prepared, and of these all except BiSeCl are isostructural. The structure of SbSBr[5] consists of pleated chains $(SbS)_n^{n+}$ between which lie Br^- ions (Fig. 10.13). The chain is the simplest 3-connected

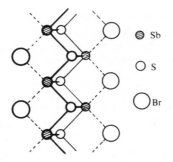

FIG. 10.13. The crystal structure of SbSBr.

linear system (see Fig. 3.19(a), p. 101), and it is closely related to the chain in Sb_2S_3 (p. 907). Within a chain, $Sb-S = 2.49$ Å (1) and 2.67 Å (2), Sb bond angles are 84° (2) and 96° (1), and S bond angles are all close to 96°. The nearest neighbours of a Br^- ion are two Sb (at 2.94 Å) and one S (at 3.46 Å). For $Hg_3S_2Cl_3$ see Chapter 26.

The compound NbS_2Cl_2 is not a simple thiohalide for it apparently contains S_2 groups, having a layer structure of the $AlCl_3$ type in which one Cl has been replaced by an S_2 group (S—S, 2.03 Å).[6] The nearest neighbours of Nb are 4 Cl at 2.61 Å and 4 S (of two S_2 groups) at 2.50 Å. There are Nb—Nb distances of 2.90 Å between pairs of metal atoms.

Molecular metal thiohalides are at present represented only by $WSCl_4$ and the similar, but not isostructural, $WSBr_4$.[7] Square pyramidal molecules are associated

in pairs, the bridging W—Cl distance (3.05 Å) being much larger than the mean W—Cl (2.28 Å) within the $WSCl_4$ molecule.

(1) AC 1973 **B29** 345
(2) AC 1973 **B29** 1532
(3) AC 1973 **B29** 1567
(4) ZaC 1962 **314** 303
(5) AC 1959 **12** 14
(6) ZaC 1966 **347** 231
(7) JCS A 1970 2815

Oxyhalides

Compounds containing one or more elements combined with oxygen and halogen atoms are of a number of quite different types, and as with complex halides the empirical formulae give no indication of the structures of compounds MO_xX_y. For example, calcium hypochlorite, CaO_2Cl_2, is a salt containing ClO^- ions—$Ca(OCl)_2$. Uranyl chloride, UO_2Cl_2 consists of UO_2^{2+} and Cl^- ions, that is, $(UO_2)Cl_2$, while sulphuryl chloride, SO_2Cl_2, is a molecular oxychloride. The following groups of compounds are included in other chapters:

salts of oxy-ions of the halogens (Chapter 9);

oxyhalides and oxyhalide ions of non-metals (e.g. POF_3, PO_2F^-, PO_3F^{2-}) are described under the appropriate element to which the O and X atoms are separately bonded;

complexes of $POCl_3$, $SeOCl_2$, etc. with halides are included in Chapters 19 and 16 respectively.

We include in this chapter a short section on oxyhalide ions of transition metals.

Oxyhalides of one kind or another are formed by one or more metals of most of the Periodic Groups except the alkali metals and possibly the elements of subgroup IB, though at present very few sets of compounds of the same formula-type are known containing a particular metal and all four halogens. An example is the series $NbOX_3$. We show in Table 10.10 some compounds of metals of groups VA–VIIA and VIII, emphasizing the predominance of fluorides and chorides and the very small number of iodides. Further study may, of course, alter this general picture. At least ten types of formula are found:

$$MOX, \quad MOX_2, \quad MOX_3, \quad MOX_4, \quad MOX_5,$$
$$MO_2X, \quad MO_2X_2, \quad MO_2X_3,$$
$$MO_3X, \quad MO_3X_2$$

and it is likely that, except in the case of molecular oxyhalides, structural differences will be found between crystalline oxyfluorides and other oxyhalides of a given metal.

Oxyhalide ions

A great variety of oxyhalide ions is formed by transition metals in their various oxidation states. The structures of some of the salts are similar to those of oxy-salts or complex halides. For example, the ion CrO_3F^- forms an almost regular tetrahedron in $KCrO_3F$, and because of the similar sizes of O and F there is random orientation of the anions in the crystal ($CaWO_4$-type structure). On the other hand, $KCrO_3Cl$ has a distorted form of that structure because of the difference between Cr–O (1.53 Å) and Cr–Cl (2.16 Å). $(NH_4)_3MoO_3F_3$ has the same structure as $(NH_4)_3AlF_6$, with octahedral $MoO_3F_3^{3-}$ ions, but whereas K_2OsCl_6 has the K_2PtCl_6 structure, the salt $K_2OsO_2Cl_4^{(1)}$ has a rather deformed version of that structure (Os–Cl, 2.38 Å and Os–O, 1.75 Å – compare 2.36 Å in K_2OsCl_6 and 1.66 Å in OsO_4). The salts $Cs_2M^VOCl_5$ (M = Nb, Cr, Mo, W) also have K_2PtCl_6-type structures,[2] and discrete octahedral ions exist in $Cs_2UO_2Cl_4^{(3)}$ (a), with the two short U–O bonds characteristic of uranyl compounds (Chapter 28).

Some oxyhalide ions are isolable only in combination with large cations. In $(C_2H_5)_4N[ReBr_4O(H_2O)]^{(4)}$ there is a square pyramidal arrangement of 4 Br and 1 O (with the metal atom 0.32 Å above the base of the pyramid), the sixth 'octahedral' bond to H_2O being very weak (b).

Other square pyramidal ions include the anions in $Ca_3(WO_5)Cl_2^{(5)}$ and $(As\phi_4)(ReOCl_4)$.[6] In the $(WO_5)^{4-}$ ion the axial W–O bond length is 1.725 Å and the others 1.90 Å; W is 0.55 Å above the square base, and the nearest atom in the sixth octahedral bond position is Cl^- at 3.40 Å. In $(ReOCl_4)^-$ Re is 0.41 Å above the base, Re–O, 1.63 Å, and Re–Cl, 2.34 Å; $(TcOCl_4)^-$ has a very similar structure.[6a]

Examples of more complex oxyhalide ions are shown at (c) and (d). The anion in $(NH_4)_3UO_2F_5^{(7)}$ has the pentagonal bipyramidal configuration (c). Details of the structures of O-bridged ions of type (d), namely, $M_2OCl_{10}^{4-}$ formed by W, Re, Ru, and Os, and $Re_2OCl_{10}^{3-}$, are included in Table 11.5 (p. 509).

Although not an oxyhalide ion, the linear anion (e) in the salt $K_3[Ru_2NCl_8(H_2O)_2]$ (ref. 8) may be mentioned here in view of its similarity to the ion (d).

In addition to finite oxyhalide ions there are infinite oxyhalide ions in some complex oxyhalides. Examples include the anion in $K_2(NbO_3F)$, which has the

TABLE 10.10

Some transition-metal oxyhalides

V		VI		VII		VIII
VOF VOCl		CrOCl	CrOF₃ Cl₃ CrO₂F₂ Cl₂ CrOF₄	MnO₃F MnOCl₃ MnO₃Cl MnO₂Cl₂ Mn₈O₁₀Cl₃		FeOF FeOCl
VOCl₂ Br₂	VOF₃ Cl₃ Br₃					
NbOCl₂ Br₂ I₂	NbOF₃ NbO₂F Cl₃ Br₃ I₃	MoOCl₂ MoO₂Cl	MoOCl₃ MoO₂F₂ MoOF₄ Br₃ Cl₂ Cl₄ Br₂	TcOCl₃ Br₃	TcOF₄	
	TaO₂F	WOCl₂	WOCl₃ WO₂Cl₂ WOF₄ Br₃ Br₂ Cl₄ I₂ Br₄	ReOCl₃ ReOF₄ ReOBr₃ ReOF₅ ReOCl₄ ReO₂F₃ ReO₃Cl ReO₃F		OsO₃F₂ OsOF₅ PtOF₄

(a)

(b)

(c)

(d)

(e)

K_2NiF_4 (layer) structure (see Table 10.11), and the *cis* octahedral AX_5 chain ion in $K_2(VO_2F_3)$[9] and in $(NH_4)_2(VF_4O)$,[10] F atoms forming the bridges.

(1)   AC 1961 **14** 1035
(2)   JCS 1964 4944
(3)   AC 1966 **20** 160
(4)   IC 1965 **4** 1621
(5)   AC 1974 **B30** 2587
(6)   AC 1977 **B33** 1248

(6a) IC 1979 **18** 3024
(7)   AC 1969 **B25** 67
(8)   JCS A 1971, 1792, 1795
(9)   AC 1971 **B27** 1270
(10) AC 1980 **B36** 1925

The structures of metal oxyhalides

We shall discuss these compounds in the following groups:

  (a) ionic oxyfluorides;
  (b) oxyhalides of transition metals in high oxidation states;
  (c) oxyhalides MOCl, MOBr, and MOI;
  (d) other oxyhalide structures.

(a) *Ionic oxyfluorides*

Examples are now known of oxyfluorides having the same structures as most of the simple oxides or fluorides MX_2 or MX_3 or structures characteristic of complex oxides or fluorides (Table 10.11). There is either random arrangement of O and F in the anion positions of such structures or, more rarely, regular arrangement (superstructure). We describe shortly the structures of the two forms of LaOF as examples of superstructures of the fluorite structure.

Yttrium forms a family of oxyfluorides $Y_nO_{n-1}F_{n+2}$ ($n = 4–8$) of which $Y_7O_6F_9$

TABLE 10.11

Crystal structures of oxyfluorides

Structure	Examples	Reference
Rutile	FeOF, TiOF, VOF (high press./temp.)	JSSC 1970 **2** 49
α-PbO$_2$	NaNbO$_2$F$_2$ (superstructure)	AC 1969 **B25** 847
ZrO$_2$ (baddeleyite)	ScOF	ACSc 1966 **20** 1082
Fluorite	AcOF, HoOF	AC 1970 **B26** 2129
		AC 1957 **10** 788
superstructure:	rhombohedral: YOF, LaOF, SmOF, etc.	JACS 1954 **76** 4734, 5237
	tetragonal: YOF, LaOF, PuOF	IC 1969 **8** 232
	distorted: TlOF	AC 1972 **B28** 3426
CaF$_2$-YF$_3$	Y$_n$O$_{n-1}$F$_{n+2}$	AC 1975 **B31** 1406
—	InOF	AC 1973 **B29** 627
PbCl$_2$	LaOF and 4f metals (h.p.)	MRB 1970 **5** 769
ReO$_3$	TiOF$_2$	AC 1955 **8** 25
	NbO$_2$F, TaO$_2$F	AC 1956 **9** 626
	WO$_{3-x}$F$_x$, MoO$_{3-x}$F$_x$	IC 1969 **8** 1764
MoO$_3$	Mo$_4$O$_{11}$F	
LaF$_3$	ThOF$_2$	AC 1949 **2** 388
	BiO$_{0.1}$F$_{2.8}$	ACSc 1955 **9** 1209
Perovskite	KNbO$_2$F, NaNbO$_2$F	ZaC 1964 **329** 211
	Tl^{I}Tl^{III}OF$_2$	MRB 1970 **5** 185
K$_2$NiF$_4$	K$_2$NbO$_3$F	JPC 1962 **66** 1318
	Sr$_2$FeO$_3$F	JPC 1963 **67** 1451
Spinel	Cu$_2$FeO$_3$F	JPCS 1963 **24** 759
	Fe$_3$O$_{4-x}$F$_x$	CR 1970 **270C** 2142
Cryolite	Na$_3$VO$_2$F$_4$ (monoclinic)	
	K$_3$TiOF$_5$ (tetragonal)	ZaC 1969 **369** 265
Elpasolite	K$_2$NaVO$_2$F$_4$	
Tetragonal bronze	KNb$_2$O$_5$F	ACSc 1965 **19** 1510
Pyrochlore	Cd$_2$Ti$_2$O$_5$F$_2$	CR 1969 **269C** 228
Garnet	Y$_3$Fe$_5$O$_{12-x}$F$_x$	MRB 1971 **6** 63
Magnetoplumbite	Y$_3$Al$_4$NiO$_{11}$F	JPCS 1963 **24** 759
	BaCoFe$_{11}$O$_{18}$F	

is an example. The unit cell, with dimensions a, na, a, consists of n fluorite-type sub-cells stacked above one another, two of which accommodate the excess anions. The unit cell contains two regions of almost undistorted CaF$_2$ and YF$_3$ structures, between which a gradual change in structure type occurs. There are also numerous oxyfluorides of transition metals which have structures similar to those of oxides of these metals and compositions very close to those of oxides. Examples noted in Chapter 13 include Nb$_3$O$_7$F and LiNb$_6$O$_{15}$F; more complex examples are Nb$_{17}$O$_{42}$F and Nb$_{31}$O$_{77}$F,[1] with structures related to that of one of the forms of Nb$_2$O$_5$. Members of the family M$_{3n+1}$X$_{8n-2}$, built from different sized blocks of the ReO$_3$ structure include

$$n = 9 \quad M_{28}O_{70} \text{ (Nb}_2O_5\text{)}$$
$$n = 10 \quad M_{31}X_{78} \text{ (Nb}_{31}O_{77}F\text{)}$$
$$n = 11 \quad M_{34}X_{86} \text{ (Nb}_{17}O_{42}F\text{)}.$$

In contrast to compounds of this type, which contain very little F, there are others, such as $Zr_7O_9F_{10}$,[2] with unique structures (in this case with 6- and 7-coordination of Zr^{4+}) more reminiscent of complex fluorides such as $Na_7Zr_6F_{31}$ (p. 472).

In some oxyfluorides the composition is variable within certain limits, in which case the formula given in Table 10.11 falls within the observed composition range. We noted earlier that the LaF_3 structure is apparently not stable for pure BiF_3 but becomes so if a small proportion of fluorine is replaced by oxygen (approximate formula, $BiO_{0.1}F_{2.8}$.)

Superstructures of fluorite. The cubic fluorite (CaF_2) structure is illustrated in Fig. 6.9 on p. 252. This structure may also be referred to a rhombohedral unit cell ($a = 7.02$ Å, $\alpha = 33\frac{1}{3}°$) containing two CaF_2 as shown in Fig. 10.14(a). The arrangement of the F^- and O^{2-} ions in the rhombohedral form of LaOF is such that the symmetry has dropped to rhombohedral, the ions being arranged with 3-fold symmetry around only one body-diagonal of the original cube (as at (b)). A recent investigation of rhombohedral YOF shows that the positions earlier assigned to O and F should be interchanged. The distances Y–O and Y–F are different, and it was originally assumed that the latter would be the shorter ones. The later results

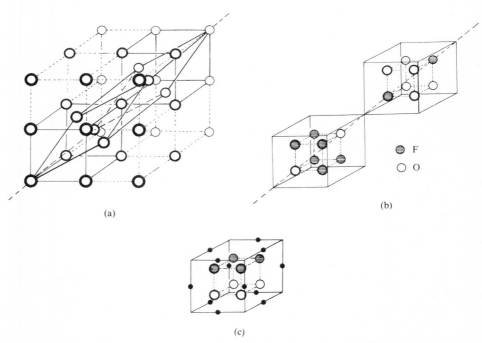

(a)

(b)

⊜ F

○ O

(c)

FIG. 10.14. (a) Alternative rhombohedral unit cell of the fluorite structure. Ca^{2+} ions lie at the centres of all faces of the smaller cubes, but only sufficient of these ions are shown to indicate the rhombohedral cell. (b) and (c) Structures of the two forms of LaOF (see text).

are: Y—O, 2.24 and 2.34 Å; Y—F, 2.41 and 2.47 Å. (In ScOF the Sc—F distances are approximately 0.1 Å greater than Sc—O.) In the tetragonal superstructure there is a quite different arrangement of the anions (Fig. 10.14(c)). This form is stable over a range of composition, expressed by the formula MO_nF_{3-2n} $(0.7 < n < 1.0)$. At the fluorine-rich limit the composition corresponds to the formula $MO_{0.7}F_{1.6}$, that is, there is an excess of F over the amount which can be accommodated in a fluorite-like structure. In this case it appears that the four O positions of Fig. 10.14(c) are occupied on the average by 2.8 O + 1.2 F and an excess of 1.2 F$^-$ ions occupy some of the interstices marked by small black circles.

(1)   ACSc 1965 **19** 1401   (2)   AC 1970 **B26** 830

(b) *Oxyhalides of transition metals in high oxidation states*

A number of the compounds of Table 10.10 have been studied in the vapour state. Details of the structures of a number of tetrahedral molecules MOX_3 and MO_2X_2 and of square pyramidal molecules MOX_4 are given in Table 10.12. It is likely that

TABLE 10.12
Structures of metal oxyhalide molecules

Molecule	M—O (Å)	M—X (Å)	Angles	Method	Reference
$VOCl_3$	1.57	2.14	O–V–Cl 108° Cl–V–Cl 111°	e.d.	JACS 1938 **60** 2360 ICA 1975 **13** 113
CrO_2Cl_2	1.57 (.03)	2.12 (.02)	O–Cr–O 105 (4)° O–Cr–Cl 109½ (3)° Cl–Cr–Cl 113 (3)°	e.d.	JACS 1938 **60** 2360
CrO_2F_2	1.58	1.74	O–Cr–O 102° F–Cr–F 119°	e.d.	CC 1973 633
MoO_2Cl_2	1.75 (.10)	2.28 (.03)	(O–Mo–O 109½°) O-Mo–Cl 108 (7)° Cl–Mo–Cl 113 (7)°	e.d.	H. A. Skinner, Thesis, Oxford 1941
MnO_3F	1.586 (.005)	1.724 (.005)	O–Mn–F 108° 27' (7')	m.w.	PR 1954 **96** 649
ReO_3F	1.692 (.003)	1.859 (.008)	O–Re–F 109° 31' (16')	m.w.	JCP 1959 **31** 633
ReO_3Cl	1.702 (.004)	2.229 (.004)	Cl–Re–O 109° 22' (7')	m.w.	
$MoOCl_4$	1.66	2.28	Cl–Mo–O 103°	e.d.	BCSJ 1975 **48** 666
$WOCl_4$	1.69	2.28	Cl–W–O 102°	e.d.	BCSJ 1975 **47** 1393

many of the compounds of Table 10.10 form molecular crystals as, for example, do $ReOCl_4$[1a] (square pyramidal) and both forms of $OsOF_5$.[1b] However, in other cases higher c.n.s of the metal atoms are achieved by forming finite polymeric molecules or 1, 2, or 3D systems, of which examples are set out in Table 10.13. For example, $NbOCl_3$ (like $VOCl_3$) is monomeric in the vapour state but has a polymeric chain structure (Fig. 10.15) in the crystal.[2] Octahedral NbO_2Cl_4 (*trans*) groups are linked into double chains by sharing an edge (2 Cl) and two opposite vertices (O atoms). There is a close relation between the structure of crystalline $NbOCl_3$

TABLE 10.13
Oxyhalides related to MX_4 and MX_5 structures

Halide structure		Oxyhalides	Reference
Tetramer	Nb_4F_{20}	$(WOF_4)_4$	JCS A 1968 2074;
Octahedral chain structures	$TcCl_4$	$ReOBr_3$, $TcOCl_3$ $MoOCl_3$ (monoclinic)	JCS A 1969 2415
	α-UF_5	$WOCl_4$, $WOBr_4$	ZaC 1966 344 157
	CrF_5	$MoOF_4$, $ReOF_4$, $TcOF_4$	See Table 9.20
Layer	SnF_4	WO_2Cl_2	ZaC 1968 363 58
3D	β-UF_5	UOF_4	AC 1974 **B30** 1701

and $NbCl_5$, for if the O atoms of Fig. 10.15 are replaced by Cl the portion of chain has the composition Nb_2Cl_{10} and represents the finite molecule in the crystalline pentachloride. Other compounds with the structure of Fig. 10.15 include: $MoOCl_3$ (tetragonal), $WOCl_3$, $WOBr_3$, $TcOBr_3$, $NbOBr_3$, and $MoOBr_3$.[3]

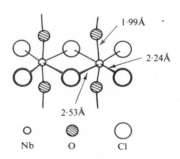

FIG. 10.15. Portion of the infinite chain molecule in crystalline $NbOCl_3$.

Extension of the double chain of $NbOCl_3$ gives a layer of composition MOX_2 which is apparently the structure of $NbOCl_2$, $NbOBr_2$, and $NbOI_2$.[4]

A number of oxyhalides have structures which are similar to those of halides, some of the halogen atoms being replaced by O atoms. Since many of the ionic 3D structures of Table 10.11 are adopted by both fluorides and oxides, many of the oxyfluorides listed therein could be included here. If, however, we restrict ourselves here to structures which are peculiar to halides (that is, are not adopted by oxides) the known examples are those of Table 10.13.

A number of oxychlorides and oxybromides MOX_3 have structures similar to $TcCl_4$ (octahedral '*cis*' chain structure), namely, $MoOCl_3$ (monoclinic), $TcOCl_3$, and $ReOBr_3$. In these compounds M is displaced from the centre of the octahedral coordination group towards the O atom. In $MoOCl_3$ (monoclinic), for example, the bridging atoms are Cl atoms and the bond opposite O is longer in the unsymmetrical bridge.

In $WOCl_4$ and $WOBr_4$ (α-UF_5 structure) the bridging atoms are O atoms, but these were not accurately located.

The structure assigned to WO_2Cl_2 consists of layers of octahedral WO_4Cl_2 groups which share four equatorial O atoms—compare the SnF_4 layer structure. The oxygen bridges are unsymmetrical, so that in both directions in the layer there are alternate short and long W—O bonds (mean lengths, 1.67 and 2.28 Å); in this connection see the structure of WO_3. The O bond angle is approximately 160° and W—Cl, 2.31 Å.

(1a) JCS D 1972 582   (3)   JCS A 1968 1061; AC 1970 **B26** 1161
(1b) JCS A 1968 536, 543   (4)   AnC 1964 76 833
(2)   AC 1959 **12** 21

(c) *Oxyhalides* MOCl, MOBr, *and* MOI

All the elements Al, Ga, La, Ti, V, and the 4f and 5f metals form an oxychloride MOCl and most of them also form MOBr and MOI. Antimony and bismuth also form more complex oxyhalides, which are described in Chapter 20, and Bi forms BiOF which belongs in this group. Other oxyfluorides MOF of the above elements were included in group (a), ionic oxyfluorides.

All the compounds of this group are structurally quite different from those in (a) or (b) and crystallize with one of three layer structures, in which there is respectively 4-, 6-, and 8(9)-coordination of M.

GaOCl. The structure of GaOCl[1] (and the isostructural AlOCl, AlOBr, and AlOI) consists of layers of tetrahedral GaO_3Cl groups (Ga—O, 1.91 Å, Ga—Cl, 2.22 Å) which are linked as shown in Fig. 10.16. Each O atom is common to three tetrahedra, the unshared vertices of which are occupied by the Cl atoms. Equal numbers of Cl atoms lie to one side or the other of the (puckered) layer of Al and O atoms.

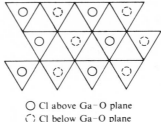

○ Cl above Ga−O plane
◌ Cl below Ga−O plane

FIG. 10.16. Layer of the structure of GaOCl (idealized).

FeOCl. The layers in this structure are of the same type as in γ-FeO.OH but they are packed together in a rather different way, as described on p. 641. The FeOCl structure is closely related to that of BiOCl as may be seen by comparing the elevation of FeOCl (Fig. 10.17) with that of BiOCl (Fig. 20.8(c), p. 898); the

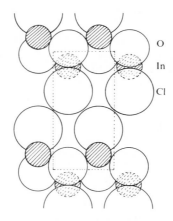

O

In

Cl

FIG. 10.17. Elevation of the crystal structure of InOCl or FeOCl.

essential difference is that in FeOCl M is 6-coordinated and in BiOCl 8-coordinated. Bond lengths in FeOCl[2] are shown at (a). Compounds isostructural with FeOCl include InOCl and InOBr,[3] InOI,[4] VOCl,[5] TiOCl,[6] and CrOCl.[7] In VOCl there is considerable distortion from regular octahedral coordination, (b).

BiOCl. The remaining oxyhalides in this group, which does *not* include SbOCl (for which see p. 896) are built of layers of a different kind and have the same type of structure as PbFCl and PbFBr. Each complex layer in the BiOCl structure consists of a central sheet of coplanar O atoms with a sheet of halogen atoms on each side, and the metal atoms between the Cl–O–Cl sheets. The plan of a layer is shown in Fig. 10.18 together with the sequence of atoms in a direction perpendicular to the

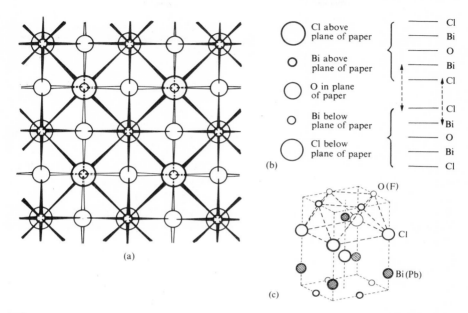

FIG. 10.18. (a) Atomic arrangement in a layer of the structure of BiOCl (or PbFCl). (b) Sequence of atoms in a direction normal to the layer (a). (c) Unit cell of the structure.

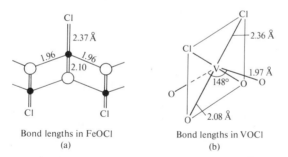

Bond lengths in FeOCl
(a)

Bond lengths in VOCl
(b)

layers. Within one of these complex layers a Bi atom is surrounded by 4 O and 4 Cl atoms at the vertices of a square antiprism. However, there is a Cl atom beyond the centre of the large square face of the antiprism, these additional Bi–Cl contacts being shown as broken lines in Fig. 10.18(b). The 8- or 9-coordination can be more easily seen in Fig. 10.18(c), which shows a (tetragonal) unit cell of the structure. The difference between the distance to the four nearest and to the fifth Cl(I) neighbour increases considerably going from BiOCl to BiOI or from LaOCl to LaOI, and there is also an increase, though much smaller, on going from BaFCl to BaFI. These structures range from the 9-coordinated structures of BaFCl and LaOCl to the 8-coordinated layer structures of LaOI and BiOI; BaFI, BiOCl, and also BiOF[8] (Bi–4F, 2.75; Bi–1F, 2.92 Å) occupy an intermediate position.

TABLE 10.14

Interatomic distances in oxyhalides and fluorohalides with the PbFCl structure

	BiOCl	BiOI	LaOCl	LaOI	BaFCl	BaFI
M–4X	3.07	3.36	3.18	3.48	3.29	3.58 A
M–X	3.49	4.88	3.14	4.79	3.20	3.84 A

The small alteration in the unit cell dimension a in the plane of the layer in the series LaOCl–LaOBr–LaOI (from 4.11–4.14 Å) compared with the large increase in the dimension c perpendicular to the plane of the layers (6.87–9.13 Å) reflects the rigidity of the M_2O_2 system consisting of a layer of M atoms on each side of an oxygen layer. This type of tetragonal metal–oxygen–metal layer is a characteristic feature of the structures of a number of compounds of Bi, La, and Pb (e.g. La_2MoO_6 and PbO, p. 557), in particular the complex oxyhalides and oxides of Bi which are described on pp. 897 and 893.

Compounds which crystallize with the PbFCl (BiOCl) structure include:

metal hydrogen halides MHX (p. 400);

fluorohalides MFX of Ca, Sr, and Ba;[9][10]

oxyhalides BiOX and MOCl, MOBr, and MOI of many 4f[11] and 5f elements (p. 1252);

oxysulphides of Zr, Th, U, and Np (p. 787); and

numerous semi-metallic ternary pnictides and chalconides such as ZrSiS and NbSiS in which the metals are not exhibiting their normal valences.

The structures of all these compounds have the same analytical description (see p. 269) but there is a wide range of bond types. In the first two groups the bonding is ionic or ionic–covalent and the structures range from a 3D 9-coordinated structure of an alkaline-earth fluorochloride or lanthanide oxychloride to the pronounced layer structures of the corresponding fluoroiodide or oxy-iodide. The structures of the fifth group of compounds form a distinct group which have been described as having a ternary version of the anti-Fe_2As structure.[12]

(1)   CR 1963 **256** 3477
(2)   AC 1970 **B26** 1058
(3)   ACSc 1956 **10** 1287
(4)   JCS A 1966 1004
(4a) ZaC 1978 **438** 203
(5)   AC 1975 **B31** 2521
(6)   ZaC 1958 **295** 268

(7)   ACSc 1975 **A28** 1171
(8)   ACSc 1964 **18** 1823
(9)   JSSC 1976 **17** 275
(10) AC 1977 **B33** 2790
(11) IC 1965 **4** 1637
(12) JSSC 1974 **9** 125

(d) *Other oxyhalide structures*

It is likely that many structures will be discovered which are peculiar to single oxyhalides or to small groups of compounds. For example there are compounds M_4OCl_6 and M_4OBr_6, made by heating together oxide and halide, suitable for the ions Ca^{2+}, Sr^{2+}, and Ba^{2+}. In Ba_4OCl_6 there is almost regular tetrahedral coordination of O by 4 Ba, and Ba is coordinated by either 7 Cl + O or by 9 Cl + O.[1]

We have noted in Table 10.10 the oxychlorides $MnOCl_3$, MnO_2Cl_2, and MnO_3Cl containing Mn(v), Mn(vɪ), and Mn(vɪɪ) and also $Mn_8O_{10}Cl_3$. This last compound is of a quite different kind, being an ionic crystal, $Mn^{2+}Mn_7^{3+}O_{10}Cl_3$.[2]

(1)   AC 1970 **B26** 16   (2)   AC 1977 **B33** 1031

Hydroxyhalides

The known hydroxyfluoride structures are similar to those of oxyhydroxides or oxides: $ZnF(OH)$,[1a] diaspore structure (p. 639), $CdF(OH)$,[1b] $CaCl_2$ structure (p. 250) with statistical distribution of F^- and OH^-, $HgF(OH)$,[1c] modified rutile structure (pp. 255 and 1159), and $InF_2(OH)$,[1d] distorted ReO_3 structure in which In^{3+} is surrounded octahedrally by 2 OH^- and 4 F^- (all at about 2.07 Å). Hydrated oxyfluorides of Al with the pyrochlore structure[2] (p. 258) have been prepared by adding ammonia to solutions of $Al_2(SO_4)_3$ containing varying amounts of AlF_3.

The largest groups of hydroxychlorides and hydroxybromides are the compounds $MX(OH)$ and $M_2X(OH)_3$ formed by Mg and the 3d metals Mn, Fe, Co, Ni, and Cu(ɪɪ). Their structures consist of c.p. layers of composition $X(OH)$ or $X(OH)_3$ as illustrated in Fig. 4.15 (p. 160) between which the metal atoms occupy respectively one-half or one-quarter of the octahedral interstices. These c.p. structures are not suitable for larger ions such as Pb^{2+} or Y^{3+}, and we deal later with compounds containing these ions.

Hydroxyhalides MX(OH)

Partial or complete X-ray studies have been made of a number of these compounds, for example, $CoBr(OH)$,[3] $CuCl(OH)$,[4] and β-$ZnCl(OH)$.[5] Various layer sequences are found comparable to the differences between the C 6, C 19, and C 27 structures of dihalides and dihydroxides. In $CuCl(OH)$ there is the usual distortion of the octahedral coordination group around $Cu(ɪɪ)-1$ Cl at 2.30, 3 OH at 2.01, and 2 Cl at 2.71 Å.

The structure of $CdCl(OH)$[6] differs from the above structures in that each layer contains all Cl or OH (Fig. 10.19) instead of equal numbers of Cl and OH in each c.p. layer.

Hydroxyhalides $M_2X(OH)_3$

The structures are of two general types, layers and 3D frameworks. In addition to the ordered structures described below (based on regular $X(OH)_3$ layers) the hydroxychlorides $M_2Cl(OH)_3$ of Mg, Mn, Fe, Co, and Ni form disordered phases of the C 6 (CdI_2) type in which the Cl and OH are randomly arranged.[7]

For hydroxyhalides $M_2Cl(OH)_3$ (M_2XY_3) the structural possibilities are essentially the same as for dihalides or dioxides MX_2 based on close-packed X atoms with M occupying octahedral holes; the relevant structure types have been

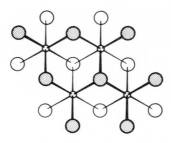

FIG. 10.19. Portion of one layer of Cd(OH)Cl. The OH groups (shaded) lie above, and the Cl atoms below, the plane of the metal atoms.

set out in Table 4.6 (p. 168). There are, however, two new factors to take into account: (i) There are two types of close-packed XY_3 layer, the rhombic and trigonal layers of Fig. 4.15(c) and (d), so that the number of structures possible for each kind of close-packing is doubled. (We shall suppose that all the layers in a given structure are of the same kind, as has been found up to the present time.) (ii) The number of possible structures is, on the other hand, reduced owing to the requirement that the X or Y atoms shall have their three M neighbours arranged pyramidally with X–M–X (Y–M–Y) bond angles of 90°. An asterisk in Table 10.15 indicates that structures satisfying this requirement are not possible.

The possible structures for hexagonal and cubic close packing are listed, with examples, in Table 10.15, which corresponds to the portion of the more general

TABLE 10.15

Crystal structures of hydroxyhalides $M_2X(OH)_3$

	Layer sequences	
	AB . . .	*ABC* . . .
Rhombic XY_3 layers (Fig. 4.15(c), p. 160)	*Layer structure* (CdI$_2$ $C6$ type) $Co_2Br(OH)_3$[a] $Cu_2Cl(OH)_3$[b] (botallackite) $Cu_2Br(OH)_3$[c] $Cu_2I(OH)_3$[c]	*Layer structure* (CdCl$_2$ type) —
	Framework structure *	*Framework structure* $Cu_2Cl(OH)_3$[d] (atacamite) β-Mg$_2$Cl(OH)$_3$[e]
Trigonal XY_3 layers (Fig. 4.15(d), p. 160)	*Layer structure* —	*Layer structure* —
	Framework structure *	*Framework structure* β-Co$_2$Cl(OH)$_3$[f]

(a) AC 1950 **3** 370. (b) MM 1957 **31** 237. (c) HCA 1961 **44** 2103. (d) AC 1949 **2** 175. (e) ASR 1954 **B3** 400. (f) AC 1953 **6** 359.

Table 4.6 (p. 168) for 'one-half octahedral holes occupied'. In the ideal c.p. structures the coordination group around a metal atom is an octahedral group (of $4\,OH + 2\,X$ or $5\,OH + X$) in which the M—O and M—X bonds have their normal lengths. As in the case of other cupric compounds there are distortions of the structures to give (4+2)-coordination (see Chapter 25). For a further note on $Cu_2Br(OH)_3$ and $Cu_2Cl(OH)_3$ and descriptions of the botallackite and atacamite structures see p. 1142.

Other hydroxyhalides

The hydroxychloride PbCl(OH) occurs as the mineral laurionite and its structure is very similar to that of $PbCl_2$ (p. 273) with (7+2)-coordination of Pb.

Other structures with higher coordination of the metal ions include those of the two forms of $YCl(OH)_2$,[8] in both of which there is 8-coordination of Y^{3+} (bicapped trigonal prism).

Hydrated hydroxychlorides

Two of these compounds present points of special interest. The action of a solution of $MgCl_2$ on MgO gives first a compound $5\,Mg(OH)_2 . MgCl_2 . 7H_2O$ (?) which slowly changes into $Mg_2(OH)_3Cl . 4H_2O$. The basis of the structure of this compound (and also apparently of a number of other hydrated hydroxy-salts of Mg) is the double chain of composition $Mg_2(OH)_3(H_2O)_3$ which has the same form as the double chain anion in $NH_4(CdCl_3)$. A portion of one of these chains is shown in Fig. 10.20(a), and (b) shows the structure viewed along the direction of the infinite chains, which

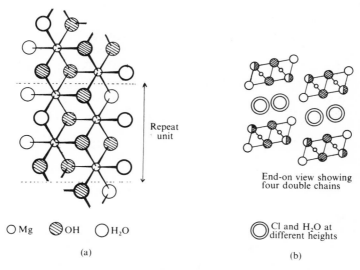

Repeat unit

○ Mg   ⊗ OH   ○ H_2O

(a)

End-on view showing four double chains

◎ Cl and H_2O at different heights

(b)

FIG. 10.20. The crystal structure of $Mg_2(OH)_3Cl . 4H_2O$.

are held together by Cl^- ions and H_2O molecules. The structural formula of this compound is accordingly $[Mg_2(OH)_3(H_2O)_3]Cl.H_2O$.[9]

The relationship of the layer structure of $Zn_5(OH)_8Cl_2.H_2O$ to the CdI_2 structure is described in Chapter 6. The structure is of interest because 3 Zn are in octahedral holes (Zn–O, 2.16 Å) and 2 Zn are in tetrahedral holes (Zn–3O, 2.02, Zn–Cl, 2.33 Å).[10]

(1a)	K 1978 **23** 951	(5)	ASc 1959 **13** 1049
(1b)	AC 1979 **B35** 2184	(6)	ZK 1934 **87** 110
(1c)	AC 1979 **B35** 949	(7)	HCA 1964 **47** 272
(1d)	ASc 1957 **11** 676	(8)	AC 1967 **22** 435
(2)	JACS 1948 **70** 105	(9)	AC 1953 **6** 40
(3)	HCA 1962 **45** 479	(10)	ZK 1962 **117** 238
(4)	HCA 1961 **44** 2095		

Amminohalides

The absorption of ammonia by many solid halides has been known for a long time. The products vary widely in composition and stability. For example, $CaCl_2.8NH_3$ and $CuCl_2.6NH_3$, prepared from the anhydrous halides and ammonia gas, readily lose ammonia, but many other amminohalides may be prepared from ammoniacal salt solutions and some possess considerable stability. The simplest compounds of this type have the general formula $MX_n.mNH_3$ (with sometimes water of crystallization), where m is the normal coordination number of M. They contain complex ions $M(NH_3)_m^{n+}$ and X^- ions. Typical ions of this sort are $Ag(NH_3)_2^+$, $Zn(NH_3)_4^{2+}$, and $Co(NH_3)_6^{3+}$. Many salts $[M(NH_3)_6]X_2$ and $[M(NH_3)_6]X_3$ have simple structures, for example, $[Co(NH_3)_6]Cl_2$ with the antifluorite structure of K_2PtCl_6, and $[Co(NH_3)_6]I_3$ with the $(NH_4)_3AlF_6$ structure.

The coordination group around M may, however, be made up of NH_3 and X as in $[Co(NH_3)_5Cl]Cl_2$ or $[Pt(NH_3)_3Cl_3]Cl$. Some of these Werner coordination compounds are described in Chapter 27. The compounds $M(NH_3)_xCl_x$ are non-electrolytes if $2x$ is the coordination number of M. Zinc, for example, forms $Zn(NH_3)_2Cl_2$ consisting of tetrahedral molecules (a) which may be compared with the ions $Zn(NH_3)_4^{2+}$, (b) in a salt such as $Zn(NH_3)_4Cl_2.H_2O$.

$$\begin{array}{cc} \text{Cl}\diagdown_{\diagup}\text{Cl} & \left[\text{NH}_3\diagdown_{\diagup}\text{NH}_3\right]^{2+} \\ \text{NH}_3^{\diagup}\text{Zn}^{\diagdown}\text{NH}_3 & \text{NH}_3^{\diagup}\text{Zn}^{\diagdown}\text{NH}_3 \\ \text{(a)} & \text{(b)} \end{array}$$

The molecule $Pd(NH_3)_2I_2$ is similar to (a) but planar instead of tetrahedral. Not all compounds with similar empirical formulae are of this type, for one form of $Pt(NH_3)_2Cl_2$ consists of planar ions $Pt(NH_3)_4^{2+}$ and $PtCl_4^{2-}$, while $Cd(NH_3)_2Cl_2$ is composed of infinite chains built from octahedral $Cd(NH_3)_2Cl_4$ groups sharing opposite edges:

$$\underset{\underset{NH_3}{|}}{\overset{\overset{NH_3}{|}}{>Cd}}\underset{Cl}{\overset{Cl}{<}}\underset{\underset{NH_3}{|}}{\overset{\overset{NH_3}{|}}{Cd}}\underset{Cl}{\overset{Cl}{<}}\underset{\underset{NH_3}{|}}{\overset{\overset{NH_3}{|}}{Cd}}<$$

This is one of a large family of compounds with octahedral chain structures which are listed in Table 25.4 (p. 1139).

For amminohalides of Cu, Ag, and Hg see Chapters 25 and 26, and for those of Group VIII metals see Chapter 27.

Ammines are compared with hydrates in Chapter 15.

11

Oxygen

After some introductory remarks on the stereochemistry of oxygen and some of the differences between oxygen and sulphur we deal with simple molecules and ions containing oxygen and then with the following topics:

peroxo- and superoxo-salts;
oxo molecules and ions; and
oxy-ions, isopoly, and heteropoly ions.

The stereochemistry of oxygen

With six electrons in the valence shell the simplest ways in which the octet can be completed are (a) the formation of the ion O^{2-}; (b) the acquisition of one electron and the formation of one covalent bond, as in the ion OH^-; and (c) the formation of two electron-pair bonds. The ions O^{2-} and OH^- are found in the oxides and hydroxides of metals. Although the O atom could, in principle, form a maximum of four covalent bonds, since there are four orbitals available, the formation of more than two *essentially covalent* bonds is rarely observed (see below). Assuming that the bond arrangement is determined by the number of σ bonds and lone pairs we may summarize the stereochemistry of oxygen as shown in Table 11.1. We have

TABLE 11.1
The stereochemistry of oxygen

No. of σ pairs	Type of hybrid	No. of lone pairs	Bond arrangement	Examples
2	sp	0	Collinear	$(Cl_5 RuORuCl_5)^{4-}$
3	sp^2	1	Angular	O_3
4	sp^3	0	Tetrahedral	$Be_4O(CH_3COO)_6$
		1	Pyramidal	H_3O^+
		2	Angular	H_2O, F_2O, H_2O_2

included one case where O forms two collinear bonds, the ion $Ru_2OCl_{10}^{4-}$ (p. 478), in which the bonds from the O atoms have considerable double-bond character; other examples will be found in the section on 'Oxo-salts'. The collinear bonds in $Sc_2Si_2O_7$ are mentioned later.

Differences between oxygen and sulphur

It is not proposed to discuss here the stereochemistry of S, which is much more complex than that of O because of the availability of d orbitals, but merely to summarize some points of difference between O and S. From O to Te the atoms increase in size and we may associate the change in behaviour of the outer valence electrons with the increased screening of the nuclear charge by the intervening completed shells of electrons. This shows itself in a number of ways:

(1) The decreasing stability of negative ions. The alkali oxides, sulphides, selenides, and tellurides all crystallize with typical ionic structures, showing that all the ions O^{2-}, S^{2-}, Se^{2-}, and Te^{2-} exist in solids. These compounds are all soluble in water, but the anhydrous compounds cannot be recovered from the solutions. If Na_2O is dissolved in water the solution on evaporation yields solid NaOH. A concentrated solution of Na_2S on evaporation yields the hydrate $Na_2S.9H_2O$, but in dilute solution the sulphur is almost entirely in the form of SH^- ions. This hydrosulphide ion is much less stable than the hydroxyl ion and, on warming, solutions of hydrosulphides evolve H_2S. Solutions of selenides and tellurides hydrolyse still more readily, liberating the hydrides, which are easily oxidized to the elements. In the case of oxygen, therefore, we have the combination of O^{2-} with H^+ giving the very stable OH^- ion. Sulphur behaves similarly, giving the less stable SH^- ion, but with Se^{2-} and Te^{2-} the process easily goes a stage farther to H_2Se and H_2Te. The increasing tendency for the divalent ion to form the hydride $(X^{2-} \rightarrow XH^- \rightarrow XH_2)$ is not due to the increasing stability of the hydrides, for this decreases rapidly from H_2O to H_2Te, the latter being strongly endothermic, but to the decreasing stability of the negative ions in aqueous solution.

These elements do not, with the exception of Te, form stable monatomic cations. The formation of the ion Te^{4+} (in TeO_2 and presumably in $Te(NO_3)_4$ and $Te(SO_4)_2$) is due to the inertness of the two s electrons (see Chapter 16 for details of Te compounds).

(2) The bonds formed by S with a particular element have less ionic character than those between O and the same element. Pauling assigns the following electronegativity coefficients:

O	3.5		F	4.0
S	2.5	compare	Cl	3.0

In many cases a dioxide has a simple ionic structure while the corresponding disulphide has a layer structure, and in general the same structure types are found for crystalline oxides as for fluorides of the same formula type, and similarly for sulphides and chlorides.

The dioxides of S, Se, Te, and Po, which are described later, range from molecular oxides to ionic crystals.

(3) The highest known fluoride of oxygen is OF_2, but S, Se, and Te all form hexafluorides. By the action of an electric discharge in O_2-F_2 mixtures at low temperatures O_4F_2 is produced, together with O_2F_2, but the only fluorides containing

directly bonded S, Se, or Te are S_2F_2, S_2F_{10}, and Te_2F_{10}. As in many other cases these three elements exhibit their highest valence in combination with fluorine; this is the only element with which S forms six bonds.

(4) Except in ozone, the alkali-metal ozonates, the unstable O_4F_2, and a number of organic peroxides and trioxides (e.g. $F_3C.OOO.CF_3$), the covalent linking of O to O does not proceed beyond O=O or —O—O—. Sulphur, on the other hand, presents a very different picture, a feature of its chemistry being the easy formation of chains of S atoms—in the element itself, in halides S_nX_2, polysulphides, and polythionates.

(5) Unlike sulphur, oxygen seldom forms more than two covalent bonds in simple molecules or ions. There are numerous ionic crystals in which O forms three bonds (rutile, phenacite, etc.). In salt hydrates where H_2O is bonded to a metal atom, (a), or in hydroxides where OH bridges two metal atoms, (b), the bonds to the metal atoms presumably have appreciable ionic character, and this is probably also true in AlOCl (p. 485) and $[Ti(OR)_4]_4$ (p. 1194). Three equivalent (covalent?)

(a)   (b)   (c)

bonds are formed in the hydronium ion, (c) found in the crystalline monohydrates of HCl and other acids, and a variety of methods indicate a rather flat pyramidal structure with O—H, 1.02 Å and angle H—O—H, 115-117°. The ion $O(HgCl)_3^+$, which is nearly planar (angle Hg—O—Hg, 118°), exists in crystalline Hg_3OCl_4 (p. 1164). Other examples of the formation of three nearly coplanar bonds are the borate ion in $SrB_6O_9(OH)_2.3H_2O$ (p. 1077), the cyclic molecules (d) and (e), and the structurally similar ions $[OCr_3(OOC.CH_3)_6(H_2O)_3]^+$, (f), and $[OFe_3(SO_4)_6(H_2O)_3]^{5-}$, (g),[1] in which pairs of acetate or sulphate groups are common to the octahedral coordination groups of pairs of metal atoms bonded to the central O atom. The molecule $Mn_{12}(CH_3COO)_{16}(H_2O)_4O_{12}$ is an interesting example of a complex molecule with S_4 ($\bar{4}$) symmetry.[2] There are Mn atoms with three types of environment, 16 bridging acetato groups (omitted from the topological diagram of Fig.

(d)   (e)   (f)

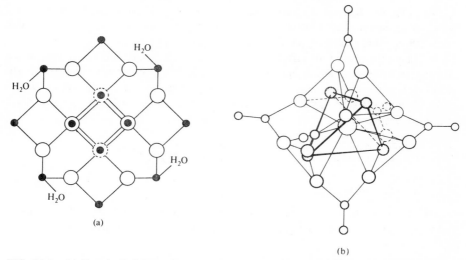

FIG. 11.1. (a) Topological diagram of the molecule $Mn_{12}(ac)_{16}(H_2O)_4O_{12}$. (b) The molecule of beryllium oxyacetate, $OBe_4(CH_3COO)_6$, drawn (not strictly to scale) to show the central oxygen atom (large shaded circle) surrounded tetrahedrally by four Be atoms (small shaded circles). An acetate group lies beyond each edge of the tetrahedron. Each Be atom has four O atoms arranged tetrahedrally as nearest neighbours, the central one and three of different acetate groups. (Small unshaded circles represent C atoms; H atoms are omitted.)

11.1(a)) and 12 μ_3 O atoms. The central nucleus is a cubane-like Mn_4O_4 group. The octahedral coordination groups around the metal atoms are made up as follows: Mn_1, 5 O + 1 b, Mn_2, 2 O + 4 b, and Mn_3, 2 O + 3 b + H_2O, where b represents an O atom of a bridging acetato group.

There are many molecules and crystals in which O forms four approximately regular tetrahedral bonds; some or all of the bonds are to metal atoms and presumably are fairly polar. Examples include monoxides such as ZnO and PdO (but note the 4 coplanar bonds in NbO, a crystal in which there is metal–metal bonding), tetrahedral molecules such as $OBe_4(CH_3COO)_6$ (Fig. 11.1(b)) and other similar molecules listed in Table 3.8, p. 95, $OZn_4B_6O_{12}$ (p. 1078), $OLa_4Re_6O_{18}$ (p. 224), and molecules containing cubane-like M_4O_4 groups such as $[TlOCH_3]_4$, p. 1172, methyl zinc methoxide, (h),[3] and the aminoethanolato complexes $Cu_4X_4L_4$,

p. 1126. A much more complex example is $Zr_{13}O_8(OCH_3)_{36}$.[4] A central ZrO_8 nucleus is surrounded by 12 Zr and 36 OCH_3 groups forming a complex with almost full cubic symmetry O_h ($m3m$) in which each O is bonded to 4 Zr atoms.

(6) The last important difference between O and S to be mentioned here is the formation by the former of relatively strong hydrogen bonds, a property of only the most electronegative elements. As a result many oxygen compounds have properties very different from those to be expected by analogy with the corresponding sulphur compounds.

(1)   AC 1980 **B36** 1278   (3)   AC 1980 **B36** 2046
(2)   AC 1980 **B36** 2042   (4)   AC 1977 **B33** 303

Simple molecules and ions

We deal here with the following topics: the molecules O_2 and O_3 and the ions O_2^+ and O_3^-; simple molecules OR_2; simple molecules and ions containing the system O–O, viz.

hydrogen peroxide, H_2O_2,
oxygen fluorides,
the peroxide, O_2^{2-}, and superoxide, O_2^-, ions.

The oxygen molecule and dioxygenyl ion[1]

The magnetic moment of the O_2 molecule indicates the presence of two unpaired electrons. The valence bond treatment formulates the molecule with one electron pair and two three-electron bonds. The m.o. method, on the other hand, accounts for the paramagnetism by showing that one state of the molecule ($^3\Sigma$), with the spins of two electrons parallel, is more stable than the other possible state ($^1\Sigma$) with opposed spins, and the molecule is formulated with a double bond (O=O, 1.211 Å). The lengths of selected O–O bonds are given in Table 11.2.

The O_2 molecule can be ionized by PtF_6 to the dioxygenyl ion, O_2^+, which exists in O_2PtF_6, a red crystalline solid formed directly from equimolar quantities of

TABLE 11.2
The lengths of O–O *bonds* (Å)

Molecule or ion	O–O	Reference
O_2^+	1.1227	GH 1950*
O_2	1.2107	PR 1953 **90** 537
O_3	1.278	JCP 1956 **24** 131
O_2^- (in α-KO$_2$)	1.28 (.02)	AC 1955 **8** 503
O_2^{2-} (in BaO$_2$)	1.49 (.04)	AC 1954 **7** 838
H_2O_2	{ 1.475	JCP 1962 **36** 1311
	{ 1.467	JCP 1965 **42** 3054

* G. Herzberg, Infra-red Spectra of Diatomic Molecules (2nd. ed.) Van Nostrand Co. N.Y.

oxygen gas and PrF_6 at $21\,°C$. This salt exists in·two crystalline forms both of which are structurally similar to salts such as $KPtF_6$ and $NO(OsF_6)$. The salts $O_2(BF_4)$ and $O_2(PF_6)$ have been prepared from O_2F_2 and BF_3 and PF_3 respectively. A careful n.d. study[2] of $O_2(PtF_6)$ has not provided a definite value for the O—O bond length but the stretching force constant of O_2^+ derived from the Raman spectra of $(O_2)PtF_6$ and $(O_2)SbF_6$ is the same as that derived from the electronic band spectrum and close to that of the isoelectronic NO molecule, indicating a bond length $1.12\,\text{Å}$.

Numerous transition-metal complexes have been prepared which contain an O_2 group of some kind as a ligand; peroxo- and superoxo-derivatives of metals are discussed shortly.

(1) IC 1969 **8** 828   (2) JCP 1966 **44** 1748

Ozone and ozonides

Microwave studies of O_3 give an angular configuration for the molecule of ozone, which has a dipole moment of $0.49\,D$.

Red ozonides MO_3 have been prepared of Na, K, Rb, and Cs by the action of ozone on the dry powdered hydroxide at low temperatures, the ozonide being extracted by liquid NH_3. Lithium forms $Li(NH_3)_4O_3$ which apparently decomposes when the NH_3 is removed, and NH_4O_3 also has not been prepared free from excess NH_3; $N(CH_3)_4O_3$ has, however, been prepared.[1] An X-ray powder photographic study of KO_3, based on a small number of reflections, gave O—O, $1.19\,\text{Å}$ and inter-bond angle, $100°$,[2] but a more detailed study of the O_3^- ion would be desirable.

(1) IC 1962 **1** 659; JACS 1962 **84** 34   (2) PNAS 1963 **49** 1

Peroxides, superoxides, and sesquioxides

All the alkali metals form peroxides M_2O_2, and peroxides MO_2 of Zn, Cd, Ca, Sr, and Ba are also known. Crystallographic studies of alkali-metal peroxides,[1] of ZnO_2 and CdO_2 (pyrites structure),[2] BaO_2, and $CaO_2.8H_2O$ show that the crystals contain O_2^{2-} ions in which the O—O bond length is $1.49\,\text{Å}$ (single bond).[3] For HgO_2 see pp. 275 and 1166. Heating an alkali-metal peroxide in oxygen under pressure at $500°C$ or the action of oxygen on the metal dissolved in liquid NH_3 at $-78°C$ produces the superoxide MO_2. Superoxides are formed by all the alkali metals, but the only superoxides prepared in a pure state and known to be stable at room temperature are those of Na and the heavier alkali metals and $[(CH_3)_4N]O_2$.[4] LiO_2 has been prepared in an inert gas matrix by oxidation of an atomic Li beam by a mixture of oxygen and argon at $15\,K$. The structure of the molecule is not known with certainty, but the shape is probably an isosceles triangle with the dimensions, O–O, 1.33 Å and Li–O, 1.77 Å.[5]

Superoxides contain O_2^- ions in which the bond length (1.28 Å) corresponds to bond order 1.5, as in ozone (Table 11.2). They are paramagnetic with moments close to 2 BM; compare the theoretical value (1.73 BM) for one unpaired electron. Sodium superoxide is trimorphic. At temperatures below –77 °C it crystallizes with the marcasite structure (p. 247), above that temperature with the pyrites structure (p. 241), and above –50 °C the structure becomes disordered owing to rotation of the anions. At ordinary temperatures KO_2, RbO_2, and CsO_2 are isostructural, crystallizing with the tetragonal CaC_2 structure (p. 948), but KO_2 also has a high-temperature cubic form with the NaCl (or disordered pyrites) structure.[6]

Sesquioxides of Rb and Cs have been prepared as dark-coloured paramagnetic powders (contrast the yellow M_2O_2 and MO_2). They have been assigned the anti-Th_3P_4 structure (p. 193), that is, they would be formulated $M_4(O_2)_3$.[7] The symmetry of this cubic structure implies equivalence of the three O_2 ions in $M_4(O_2^-)_2(O_2^{2-})$; it was not possible to determine the O–O bond length from the X-ray powder data. Further study of these compounds is desirable.

(1)   ZaC 1962 **314** 12
(2)   JACS 1959 **81** 3830
(3)   AC 1954 7 838
(4)   IC 1964 3 1798

(5)   JCP 1969 **50** 4288
(6)   AC 1952 5 851
(7)   ZaC 1939 **242** 201

Molecules OR$_2$

Molecules of this type are angular, the bond angle usually lying in the range 100–110°. For pure p bonds an angle of 90° would be expected, but partial use of the s orbital would lead to larger angles, and in the limiting cases 109½° for sp^3 or 120° for sp^2 bonds. When O is attached to two aromatic nuclei the angle is, in fact, close to 120°. In Table 11.3 we also include ethylene oxide, probably with 'bent' bonds as in

TABLE 11.3

Oxygen bond angles in simple molecules

Molecule	Oxygen bond angle	O–R (Å)	Method	Reference
$(CH_2)_2O$	61.6°	1.436	m.w.	JCP 1951 **19** 676
F_2O	103.3°	1.409	m.w.	JCP 1961 **35** 2240
H_2O	104.5°	0.97	i.r.	JCP 1965 **42** 1147
OCl_2	110.9°	1.700	m.w.	JCS A 1966 336
$O(CH_3)_2$	111.5°	1.416	m.w.	JCP 1959 **30** 1096
$p\text{-}C_6H_4(OCH_3)_2$	121°	1.36	X	AC 1950 3 279
$O(SiH_3)_2$	144°	1.634	e.d.	ASc 1963 **17** 2455
$O(GeH_3)_2$	126.5°	1.766	e.d.	JCS A 1970 315
$O(Ge\phi_3)_2$	135°	1.77	X	AC 1978 **B34** 119

cyclopropane. Much larger angles are found in $O(SiH_3)_2$, in pyro-ions and molecules X_2O_7, and also in meta-salts. Two collinear bonds are formed by O in the ReO_3 structure and in certain silicates (e.g. the $Si_2O_7^{6-}$ ion in $Sc_2Si_2O_7$) where the bonds presumably have considerable ionic character. (On the subject of apparently collinear O bonds in certain pyrophosphates see p. 860). There are also collinear O bonds in Tc_2O_7 and in metallic oxo-compounds containing M—O—M bridges; in these cases the bond lengths indicate appreciable double bond character.

Hydrogen peroxide

The configurations of molecules R_1—O—O—R_2 (and the analogous compounds of S, Se, and Te) are of particular interest because wave-mechanical calculations indicate that the bond O—R_1 will not lie in the plane of —O—O—R_2 but that owing to the strong repulsion of the unshared (p_π) electrons of the O atoms the molecule will have the configuration shown at (a). The stereochemistry of molecules O_2R_2, S_2R_2, etc., is discussed further on p. 727 with the S compounds, of which more examples

(a)

have been studied. For H_2O_2 the calculated values of θ and ϕ are close to $100°$ if sp hybridization of the lone pairs is assumed, but a later treatment assuming sp^3 hybridization predicts a dihedral angle of $120°$.

Many studies have been made of the structure of the H_2O_2 molecule giving values close to $1.47 Å$ for the O—O bond length, and O bond angle $103°$ (solid) or $98°$ (gas). However, the dihedral angle ranges from $90°$ in crystalline H_2O_2 to $180°$ in $Na_2C_2O_4 . H_2O_2$, where the H_2O_2 molecule is found to have a planar *trans* configuration ($\theta = 97°$). This large range of values for the dihedral angle indicates that the configuration of this molecule is very sensitive to its surroundings. Apparently the energy barrier corresponding to the *trans* configuration is only $3.8 kJ mol^{-1}$ above that of the equilibrium configuration (with $\phi = 111.5°$), as compared with $16.7 kJ$ mol^{-1} for the *cis* configuration. The value of ϕ would therefore easily be altered by hydrogen bonding, and also in $Na_2C_2O_4 . H_2O_2$ by the proximity of two Na^+ ions which complete a tetrahedral group around the O atom. Values of the dihedral angle found in organic peroxides also cover a wide range (Table 11.4). The value ($60°$) in the cyclic ion $[B_2(O_2)_2(OH)_4]^{2-}$ is a special case where the B—O—O—B system forms part of a 6-membered ring.

TABLE 11.4

Dihedral angles ROO/OOR

Compound	Dihedral angle	Reference
H_2O_2 (solid, n.d.)	90·2°	JCP 1965 **42** 3054
(gas, i.r.)	111·5°	JCP 1965 **42** 1931
	119·8°	JCP 1962 **36** 1311
$H_2O_2 . 2 H_2O$	129°	ACSc 1960 **14** 1325
$Na_2C_2O_4 . H_2O_2$	180°	ACSc 1964 **18** 1454
$Li_2C_2O_4 . H_2O_2$	180°	ACSc 1969 **23** 1871
$K_2C_2O_4 . H_2O_2$	101·6°	ACSc 1967 **21** 779
$Rb_2C_2O_4 . H_2O_2$	103·4°	ACSc 1967 **21** 779
O_2F_2	87·5°	JCS 1962 4585
O_2R_2 (organic)	Values from 81 to 146°	AC 1967 **22** 281
		AC 1968 **B24** 277
Theoretical	90–120°	CJP 1962 **40** 765
		JCP 1962 **36** 1311
		AC 1967 **22** 281

Oxygen fluorides

The (angular) molecule OF_2 has already been mentioned. Mixtures of oxygen fluorides have been prepared by passage of $O_2 + F_2$ through an electric discharge and more recently by radiolysis of liquid mixtures of these elements at low temperatures with 3-MeV bremsstrahlung. The existence of all compounds O_nF_2 (*n* from 2 to 6) has been claimed but the latest data support the existence of only O_2F, O_2F_2, and O_4F_2.[1] It is thought that O_3F_2 and probably also O_5F_2 and O_6F_2 are mixtures. An infrared study[2] of O_2F trapped in solid oxygen indicated a bent molecule with bond lengths similar to those in O_2F_2 (below). The molecule O_2F_2 is of the same geometrical type as H_2O_2, with dihedral angle 87.5° and bond angle 109.5° but with very abnormal bond lengths: O–O, 1.217 Å and O–F, 1.575 Å.[3] The extreme length of the latter (compare 1.40 Å in OF_2) and the shortness of O–O (the same as in O_2) suggest an analogy with the nitrosyl halides (q.v.).

The preparation of organic compounds containing the system –O–O–O– has been described[4] (e.g. $F_3COOOCF_3$) but their structures are not known.

(1) JACS 1969 **91** 4702   (3) JCS 1962 4585
(2) JCP 1966 **44** 3641   (4) JACS 1966 **88** 3288, 4316

Per-acids of non-metals

These include compounds of the following types:

(1) Permono-acids presumably containing systems H–O–O–A (e.g. H$_3$PO$_5$, H$_2$SO$_5$). Compounds described as 'perborates' and 'percarbonates' are prepared by the action of H$_2$O$_2$ on normal salts. The compound Na$_2$[B$_2$(O$_2$)$_2$(OH)$_4$].6H$_2$O,[1] originally formulated as NaBO$_3$.4H$_2$O or NaBO$_2$.H$_2$O$_2$.3H$_2$O, is in fact a peroxo-borate containing the cyclic ion of Fig. 11.2(a), with O–O, 1.47 Å as in H$_2$O$_2$. On the other hand, the important bleaching agent, sodium 'percarbonate', is a perhydrate, Na$_2$CO$_3$.1½H$_2$O$_2$,[2] consisting of layers of hydrogen-bonded CO$_3^{2-}$ ions and H$_2$O$_2$ molecules held together by Na$^+$ ions.

(2) Perdi-acids containing the grouping A–O–O–A include H$_2$C$_2$O$_6$, H$_4$P$_2$O$_8$, and H$_2$S$_2$O$_8$. The structure of the S$_2$O$_8^{2-}$ ion has been determined in the ammonium salt (Fig. 11.2(b)); the structures of other ions of this class have not been established.

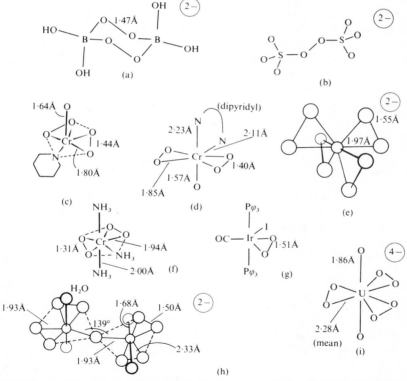

FIG. 11.2. Peroxo-ions and molecules: (a) [B$_2$(O$_2$)$_2$(OH)$_4$]$^{2-}$; (b) S$_2$O$_8^{2-}$; (c) (C$_5$H$_5$N)CrO$_5$; (d) (dipyridyl)CrO(O$_2$)$_2$; (e) [Mo(O$_2$)$_4$]$^{2-}$; (f) (NH$_3$)$_3$Cr(O$_2$)$_2$; (g) Ir(Pϕ$_3$)$_2$(CO)(O$_2$)I; (h) [W$_2$O$_3$(O$_2$)$_4$(H$_2$O)$_2$]$^{2-}$; (i) [UO$_2$(O$_2$)$_3$]$^{4-}$.

(1) AC 1978 **B34** 3551   (2) AC 1977 **B33** 3650

Peroxo- and superoxo-derivatives of metals

Both the peroxo- and superoxo-ions can function as ligands with both oxygen atoms bonded to the same metal atom or as bridges between two metal atoms:

$$M\underset{O}{\overset{O}{\lessgtr}} \quad \text{or} \quad M-O^{O-M}$$

Ions or molecules in which O_2 groups are coordinated to one metal atom ('side-on') are formed by many transition metals, including Ti, Nb, Ta, Cr, Mo, W, Co, Ni, Rh, Ir, Pt, and U. Complexes in which there is one O_2 ligand bonded to the metal include the square planar $Pt(O_2)(P\phi_3)_2$[1a] and $Ni(O_2)(\text{tert-butyl isocyanide})_2$[1b] and others mentioned later. There are also molecules and ions in which two or more O_2 groups are bonded to the metal atom. One or more of the ligands is usually a halogen, amine, phosphine, arsine, or CO.

The molecule $(C_5H_5N)CrO_5$ has the form of a distorted pentagonal pyramid (or trigonal pyramid if O_2 is counted as one ligand) as shown in Fig. 11.2(c), the two O atoms of each O_2 group being equidistant from the metal atom.[2] The blue form of $CrO(O_2)_2$ (dipyridyl) has a closely related structure (Fig. 11.2(d)) with ligand atoms at the vertices of a pentagonal bipyramid.[3] It has been suggested that the blue and violet perchromates are probably of this general type; for example, the explosive violet $KHCrO_6$ may be $K[Cr^{VI}O(O_2)_2OH]$. The red perchromates (and isostructural Nb and Ta compounds) such as K_3CrO_8 contain the ion $Cr^V(O_2)_4^{3-}$, in which

$$\underset{1\cdot85\text{ Å}}{\overset{1\cdot94\text{ Å}}{Cr \lessgtr}}\overset{O}{\underset{O}{\Big|}}\,1\cdot41\text{ Å}$$

four O_2^{2-} ions form a dodecahedral arrangement (as in $Mo(CN)_8^{4-}$) with apparently a slightly unsymmetrical orientation of the O_2 groups relative to the metal atom.[4] The paramagnetism of K_3CrO_8 indicates one unpaired electron, confirming its formulation as a peroxo-compound of Cr(v). The anion in the dark red $[Zn(NH_3)_4][Mo(O_2)_4]$[5] has a very similar structure (Fig. 11.2(e)) with all M–O, 1.97 Å, and O–O, 1.55 Å.

The action of H_2O_2 on $(NH_4)_2CrO_4$ in aqueous ammonia gives dark-reddish-brown crystals of $(NH_3)_3CrO_4$ which have a metallic lustre. They consist of pentagonal bipyramidal molecules (Fig. 11.2(f)) in which the four O atoms, one N, and Cr are coplanar (as also are Cr and the three N atoms).[6] As regards its stoichiometry this compound could be formulated as a Cr(ɪɪ) superoxide or a Cr(ɪv) peroxide. The magnetic moment (indicating 2 unpaired electrons) and its chemistry suggest the latter formulation, for in the former Cr(ɪɪ) would contribute 2 unpaired electrons and so also would the two O_2^- ions. The O–O bond is anomalously short, but the widely different lengths found for this bond in the tetraperoxo-ions (above) suggest that not all of the O–O bond lengths are very accurately known. On the basis of

O–O bond lengths the Ir iodo compound[7a] of Fig. 11.2(g) is described as a peroxo-compound of Ir(II) and the structurally similar chloro compound[7b] could be formally described as a superoxo-compound of Ir(II) (O–O, 1.30 Å), but there are then problems in accounting for its diamagnetism. In the ion $Rh(O_2)[P\phi(CH_3)_2]_4^{-}$[7c] the O–O bond length is 1.43 Å. Some compounds of this type (for example the Cl analogue of (g)) are of interest as reversible oxygen-carriers; in others the O_2 is irreversibly bonded, as in (g) and the cation $[Rh(O_2)As_4]^{+}$,[8] where As represents $As(CH_3)_2\phi$ (O–O, 1.46 Å). The arrangement of ligands (counting O_2 as one ligand) in these complexes is trigonal bipyramidal.

The colourless tetraperoxoditungstate (VI), $K_2W_2O_{11} \cdot 4H_2O$, contains the binuclear ion of Fig. 11.2(h).[9] Around each W atom the configuration of ligands is pentagonal bipyramidal, the bond lengths suggesting double bonds to the single O atoms (compare W–O in $CaWO_4$, 1.78 Å). The bonds to the water molecules are obviously weak. The diamagnetic $K_2Mo_2O_{11} \cdot 4H_2O$[10] has a very similar structure. Figure 11.2(i) shows the structure of the anion in $Na_4[UO_2(O_2)_3] \cdot 9H_2O$[11] in which three O_2 groups in the equatorial plane and the two uranyl O atoms form a trigonal (hexagonal) bipyramidal coordination group around U.

Niobium provides examples of peroxo-complexes in which respectively 2, 3, and 4 O_2^{2-} groups are attached to the metal atom: $(NH_4)_3[Nb(O_2)_2(C_2O_4)_2] \cdot H_2O$,[12] $K[Nb(O_2)_3C_{12}H_8N_2] \cdot 3H_2O$,[13] and $KMg[Nb(O_2)_4] \cdot 7H_2O$.[14] In all of these complexes there is a dodecahedral arrangement of atoms bonded to Nb, these being 8 O atoms or 6 O and 2 N in the phenanthroline complex. The $Nb(O_2)_4^{3-}$ ion is similar in structure to the Cr and Mo ions, and in all these compounds O–O is close to 1.50 Å in the peroxo-ligands. In the dioxalato complex the two O_2 ligands occupy *cis* positions.

Cobalt provides examples of both peroxo- and superoxo-bridges between metal atoms. There are pairs of ions having the same chemical composition but different charges, one series of salts being greenish-black in colour and paramagnetic with moments corresponding to one unpaired electron, while the other salts are reddish-brown and diamagnetic:

(a) $[(NH_3)_5Co-O_2-Co(NH_3)_5]^{5+}$[15]   (c) $[(NH_3)_5Co-O_2-Co(NH_3)_5]^{4+}$[17]

(b) $[(NH_3)_4Co\underset{O_2}{\overset{NH_2}{<>}}Co(NH_3)_4]^{4+}$[16]   (d) $[(NH_3)_4Co\underset{O_2}{\overset{NH_2}{<>}}Co(NH_3)_4]^{3+}$

green–paramagnetic   red–diamagnetic

The paramagnetic compounds contain superoxo-bridges and e.s.r. data indicate that the odd electron spends equal times on both Co atoms, which are both Co(III). In (a) the O–O bond length (1.30 Å) is typical of a superoxo-compound, and the system Co–O–O–Co is coplanar, and in (b) the central 5-ring is planar. In the red diamagnetic compounds the O–O bond is similar in length to that in H_2O_2 (1.47 Å).

Two configurations (c_1) and (c_2), have been found for the central portion of the ion (c). In the red diamagnetic disulphate[17a] there is a dihedral angle of $146°$, as shown at (c_1), but in the tetrathiocyanate[17b] the atoms Co—O—O—Co are coplanar, (c_2). This difference has been attributed to the difference between the hydrogen-bonding schemes in the two crystals; this would suggest that the environment in the crystal is more important in this respect than the number of electrons involved in the O—O bond, since the coplanar arrangement about the O—O bond is found in both the superoxo-bridge (a) and the peroxo-bridge (c_2). Note that in both these series of compounds the O_2 groups bridge by forming one bond from each O atom, as at (a), (b), (c_1) and (c_2) below.

A different type of bridge apparently occurs in a red diamagnetic salt which was originally regarded as an isomer of the green $[(en)_2Co(NH_2)O_2Co(en)_2](NO_3)_4 \cdot H_2O$. The latter contains a normal superoxo-bridge of type (b), but the red salt is apparently[18]

(1a) AC 1975 **B31** 2711
(1b) JACS 1969 **91** 2123
(2)   ACSc 1963 **17** 557
(3)   ACSc 1968 **22** 1439
(4)   ACSc 1963 **17** 1563
(5)   ACSc 1969 **23** 2755
(6)   JPC 1959 **63** 1279; JCS 1962 2136
(7a) IC 1967 **6** 2243
(7b) JACS 1965 **87** 2581
(7c) AC 1976 **B32** 1410
(8)   AC 1975 **B31** 2223

(9)   AC 1964 **17** 1127
(10) ACSc 1969 **22** 1076
(11) JCS A 1968 1588
(12) AC 1971 **B27** 1572
(13) AC 1971 **B27** 1582
(14) AC 1971 **B27** 1598
(15) AC 1981 **B37** 34
(16) IC 1969 **8** 291
(17a) IC 1968 **7** 725
(17b) AC 1974 **B30** 117
(18) JACS 1967 **89** 6364

Metal oxo-ions and molecules

Finite ions and neutral molecules containing O directly bonded to a metal atom are formed by many transition metals, including Nb, V, Cr, Mo, W, Re, and Os. Molecular oxyhalides and oxyhalide ions are included in Chapter 10, and vanadyl and uranyl compounds are described in more detail in Chapters 27 and 28. Examples of simple bond arrangements are the square pyramidal Os(VI) molecule (a) and (distorted) octahedral Nb compounds such as $NbOCl_3(NCCH_3)_2$, (b).

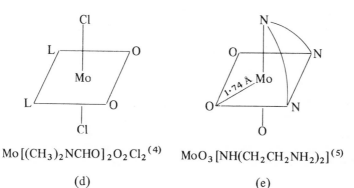

(a)[1]

(b)[2]

(c)[3]

Molybdenum provides many examples of oxo-complexes. The ion $[Mo^{IV}O_2(CN)_4]^{4-}$, (c), in $NaK_3[MoO_2(CN)_4] \cdot 6H_2O$ is octahedral and not, as earlier supposed, an 8-coordinated Mo^{IV} complex, $[Mo(OH)_4(CN)_4]^{4-}$. Examples of Mo^{VI} complexes with two or three O atoms bonded to the metal include $Mo[(CH_3)_2NCHO]_2O_2Cl_2$, (d), and $MoO_3 \cdot NH(CH_2CH_2NH_2)_2$, (e).

In (e) the MoO_3 portion of the molecule is very much like part of a tetrahedral MoO_4 group, having Mo–O, 1.74 Å, and O–Mo–O bond angles (106°) close to the tetrahedral value. The MoO_3 portions of the more complex ion (f) have the same bond lengths and interbond angles, as compared with Mo–O, 2.20 Å, and Mo–N, 2.40 Å, in the edta bridges.

$$Mo[(CH_3)_2NCHO]_2O_2Cl_2 \,^{(4)}$$

(d)

$$MoO_3[NH(CH_2CH_2NH_2)_2]^{(5)}$$

(e)

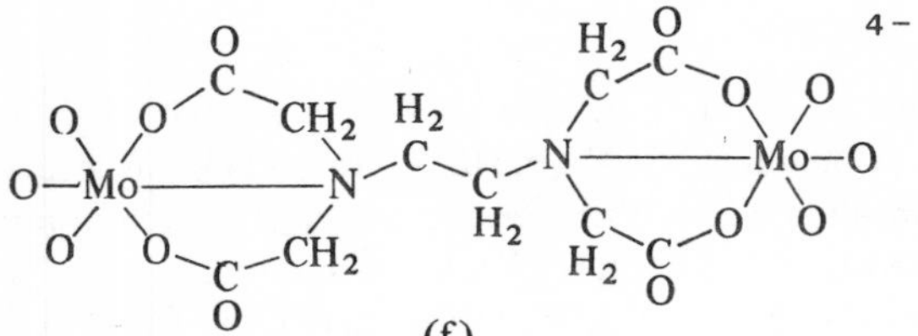

(f)

$[O_3Mo(edta)MoO_3]Na_4 . 8 H_2O$[6]

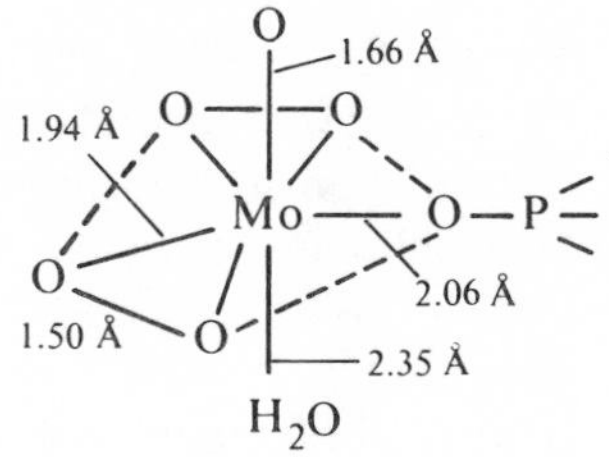

(g) $OMo(O_2)_2 L (H_2O)$ [7]

The peroxo molecule (g)[7] is of interest for a number of reasons. (The phosphine oxide is $OP[N(CH_3)_2]_3$.) There is a wide range of Mo—O bond lengths, a subject which has been comprehensively surveyed.[8] The bond arrangement is pentagonal bipyramidal (or trigonal bipyramidal if O_2 is counted as one ligand), and a very similar structure is found for the anion in $Cs[NbO(C_2O_4)_2(H_2O)_2].2H_2O$,[9] in which oxalato ligands replace the peroxo groups and the second coordinated H_2O molecule occupies the remaining equatorial position. The length of the axial Nb—O bond is 1.69 Å. If the phosphine oxide of (g) is replaced by O a bridged ion, $[(H_2O)(O_2)_2OMo—O—MoO(O_2)_2(H_2O)]^{2-}$ can be formed (Mo—O_b, 1.92 Å, Mo—O—Mo, 149°), in which the structure of each half is similar to (g). There is also an ion

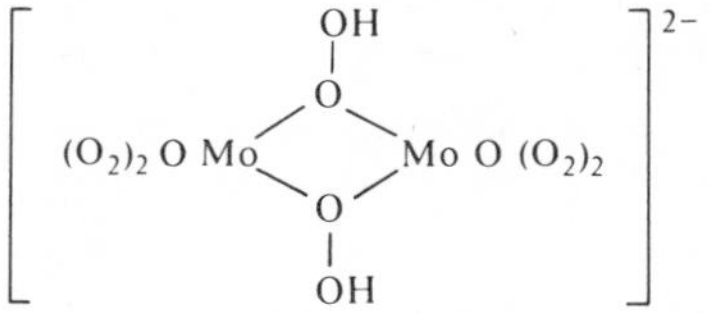

containing a double hydroperoxide bridge (O—OH, 1.46 Å).[10]

(1) AC 1975 **B31** 1814   (6) JACS 1969 **91** 301
(2) AC 1975 **B31** 1828   (7) AC 1972 **B28** 1278
(3) JACS 1968 **90** 3374   (8) AC 1975 **B31** 2294
(4) IC 1968 **7** 722   (9) AC 1973 **B29** 864
(5) IC 1964 **3** 397   (10) AC 1972 **B28** 1288

Metal oxo-compounds containing M—O—M *bridges*

We noted earlier that two collinear bonds to metal atoms are formed by O in some finite ions and molecules; some of the many examples are listed in Table 11.5. With the exception of the Al compound these are all compounds of transition metals. Details of some of the complexes which contain single M—O—M bridges are given at (a)–(c). The bonds in the transition metal complexes are regarded as having some multiple-bond character, but there is no obvious explanation of the anomalously

TABLE 11.5

Bridged metal-oxo-compounds

Compound	M–O (*bridge*)	Reference
Type M–O–M		
Al_2O (2-methyl 8-quinolinol)$_4$ (a)	1.68 Å	JACS 1970 **92** 91
$Cl_2(C_5H_5)Ti-O-Ti(C_5H_5)Cl_2$	1.78	JACS 1959 **81** 5510
$K_2[Ti_2O_5(C_7H_3O_4N)_2].5H_2O$ } p. 1194	1.83	IC 1970 **9** 2391
$Ti_2OCl_2(acac)_4.CHCl_3$	1.80	IC 1967 **6** 963
$[Fe_2O(tetraethylenepentamine)_2]I_4$	1.77	AC 1975 **B31** 1438
$Mo_2O_3(S_2COC_2H_5)_4$ (b)	1.86	JACS 1964 **86** 3024
$K_2[Mo_2O_5(C_2O_4)_2(H_2O)_2]$ (c)	1.88	IC 1964 **3** 1603
$Mo_2O_3[S_2P(OEt)_2]_4$	1.86	AC 1969 **B25** 2281
$Mn_2O(pyr)_2(phthalocyanin)_2$	1.71	IC 1967 **6** 1725
$(Cl_5Ru-O-RuCl_5)K_4.H_2O$	1.80	IC 1965 **4** 337
$(Cl_5Re-O-ReCl_5)K_4.H_2O$	1.86	AC 1962 **15** 851
$(Cl_5W-O-WCl_5)K_4$	1.87	AC 1975 **B31** 1783
$(Cl_5Os-O-OsCl_5)Cs_4$	1.78	ZaC 1973 **396** 664
$(Cl_5Re-O-ReCl_5)Cs_3$	1.83	AC 1976 **B32** 867
$[(NH_3)_5Cr-O-Cr(NH_3)_5]Cl_4.H_2O$	1.82	JACS 1971 **93** 1512
$O_3Re-O-ReO_3(H_2O)_2$	1.80, 2.10	AnC 1968 **80** 296, 291
$[Fe_2O(HEDTA)_2]^{2-}$	1.79	IC 1967 **6** 1825
$[ORe(CN)_4-O-Re(CN)_4O]^{4-}$	1.92	IC 1971 **10** 2785
$[(NH_3)_5Ru-O-Ru(en)_2-O-Ru(NH_3)_5]^{6+}$	1.87	IC 1971 **10** 1943
Type M$\genfrac{}{}{0pt}{}{\diagup O\diagdown}{\diagdown O\diagup}$M		
$Ba[Mo_2O_4(C_2O_4)_2].5H_2O$ (d)	1.91	IC 1965 **4** 1377
$[Mo_2O_4(cysteine)_2]_2Na_2.5H_2O$	1.93	AC 1969 **B25** 1857
$[Re_2O_2(C_2O_4)_4]K_4.3H_2O$	1.94	AC 1975 **B31** 1594

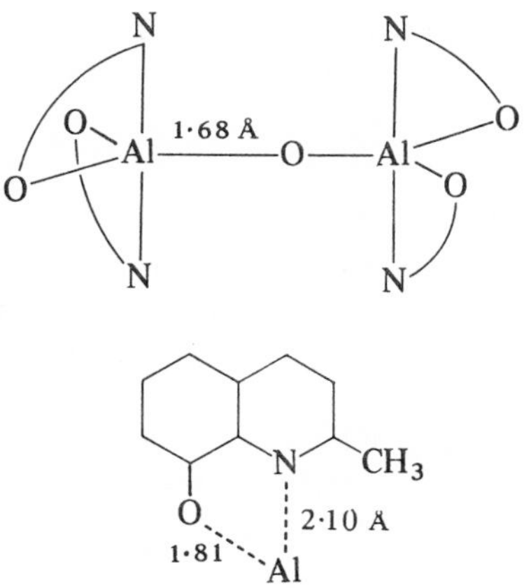

(a)

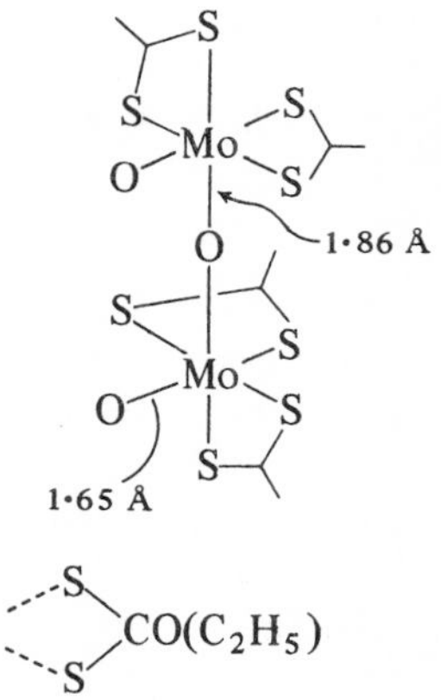

(b)

(c)   (d)

short and apparently collinear Al–O–Al bonds in (a); if the bond angle were 140° the Al–O bonds would be of normal length (1.79 Å). In a number of bridged Fe^{III} complexes the O bond angles range from 140° to 180°, but with Fe–O constant at about 1.75 Å; it is supposed that the O bond angle is very sensitive to the configuration of the complex (organic) ligands. It should be noted that terminal M–O bonds in complexes such as (b) and (c) are appreciably shorter than the collinear M–O–M bonds. This is also true in the ion $[O-Re(CN)_4-O-Re(CN)_4-O]^{4-}$ in which the terminal Re–O (1.70 Å) is much shorter than the bridging bonds (1.92 Å).

In addition to forming an ion of the type $(M_2OCl_{10})^{4-}$ Re also forms $(Re_2OCl_{10})^{3-}$, in which there is apparently one unpaired electron on the O atom which is responsible for the paramagnetism. There are also O-bridged cations as, for example, in $[Cr_2O(NH_3)_{10}]Cl_4 \cdot H_2O$; a more complex ion of this kind is that in 'ruthenium red'. This is a linear system of three octahedral groups, $[(NH_3)_5Ru-O-Ru(NH_3)_4-O-Ru(NH_3)_5]^{6+}$, in which the mean oxidation state of the metal is 10/3.

In double bridges M$\begin{smallmatrix}O\\O\end{smallmatrix}$M the O bonds are necessarily non-collinear. An example is the ion in $Ba[Mo_2O_4(C_2O_4)_2] \cdot 5H_2O$, (d), where the Mo–Mo bond length (2.54 Å) indicates sufficient interaction between the metal atoms to make the compound nearly diamagnetic (0.4 BM). Note the large range of Mo–O bond lengths in these compounds, ranging from the very short bonds (1.65 to 1.7 Å) to terminal O atoms, through values around 1.9 Å in the bridges to the long, presumably single, bonds of length 2.1 Å and above.

Oxy-ions

Oxygen accounts for some nine-tenths by volume of the earth's crust, and the greater part of inorganic chemistry is concerned with compounds which contain oxygen. In the mineral world pure oxides are rare. Compounds containing two or more elements in addition to oxygen may be roughly grouped into classes according to whether there is a small or large difference between the electronegativities of the elements. Since compounds $A_xX_yO_z$ containing two very electronegative elements A and X are not numerous, there are two main groups:

(i) A and X comparably electropositive:
(ii) A electropositve and X electronegative.

Compounds of the first group, *complex oxides*, may be regarded as assemblies of ions of two (or more) metals and O^{2-} ions. The numbers of oxygen ions surrounding the cations (their oxygen coordination numbers) are related to the sizes of the ions (Chapter 7). These coordination numbers are high (up to 12) for the largest ions, for example, Cs^+ and Ba^{2+}, and usually vary, within certain limits, in different structures. The crystal structures of complex oxides are described in Chapter 13.

Compounds of the second group are termed *oxy-salts* since they are formed by the combination of a basic oxide of the electropositive element A with the acidic oxide of the electronegative element X. The latter is usually a non-metal but may be a transition metal in a high oxidation state. Oxy-salts are assemblies of A ions and complex anions XO_n, the binding between the A ions and the oxygen atoms of the anions being essentially ionic in character. We should perhaps emphasize that we shall be dealing here exclusively with *anions*, for oxy-cations are few in number (for example, NO^+, NO_2^+, ClO_2^+) and are described in appropriate places. The complex ion is a charged group of atoms, which may be finite or extend indefinitely in one, two, or three dimensions, within which the bonds between the X and O atoms are essentially covalent in character. (Alternatively, to avoid reference to the nature of the X–O bonds we could adopt a topological definition. For example, the complex ion is that group of X and O atoms linked by X–O bonds, though to include the small number of oxy-ions in which there are X–X bonds (O_3S-S^{2-}, $O_3P-PO_3^{4-}$, etc.) or O–O bonds ($O_3S-O-O-SO_3^{2-}$, etc.) we should have to include also X–X and O–O bonds.) If the complex ion is an infinite grouping of atoms then breakdown of the crystal necessarily implies breakdown of the ion. If the complex ion is finite the same charged group of atoms XO_n *may* persist in solution or in the melt, but the stabilities of complex ions and of crystals containing them vary considerably. Salts of some of the very stable XO_n ions vaporize as molecules with simple formulae reminiscent of the structural formulae of a century ago. Their structures have been studied by e.d. of the vapours using nozzle temperatures ranging from 600–900 °C for $TlReO_4$, Tl_2SO_4, and Cs_2WO_4[1] to 1800–2000 °C for $BaWO_4$:[2]

LiNO₃   NaNO₃   K₂SO₄   Cs₂WO₄   TlReO₄   BaWO₄

Before describing the types of complex ion two points may be noted. The first is that there is not in all cases a clear-cut distinction between complex oxides and oxy-salts, particularly if the A–O bonds have appreciable covalent character. For example, a compound such as BPO_4 containing the elements B and P of rather similar electronegativity is structurally similar to silica, with presumably little difference between the B–O and P–O bonds as regards covalent character. The second is that there are crystalline compounds which may be called *oxide-oxy-salts* since they contain both discrete O^{2-} ions and oxy-ions. Examples referred to in other chapters include: Sr_3SiO_5, p. 1012, with a structure similar to that of Cs_3CoCl_5 (i.e. $Sr_3(SiO_4)O$); Fe_7SiO_{10}, p. 1017; Fe_3BO_6, p. 1080, isostructural with $Mg_3SiO_4(OH,F)_2$); $Bi_4O_3(BO_3)_2$, p. 1080, and $Cu_2O(SO_4)$, p. 1133. Oxide-phosphates whose structures have been determined include $Zr_2O(PO_4)_2$,[3] $Cu_5O_2(PO_4)_2$,[4] and $Pb_4O(PO_4)_2$ and $Pb_8O_5(PO_4)_2$.[5]

(1) ZSK 1973 **14** 548
(2) ZFK 1973 47 3030
(3) AC 1975 **B31** 1768
(4) AC 1977 **B33** 3465
(5) JSSC 1973 7 149

Types of oxy-ion

Oxy-ions range from the simplest mononuclear ions XO_n to 3D frameworks, and the following classes may be recognized.

(1) *Mononuclear ions.* These include the familiar ions of Table 11.6, the stereochemistry of which is discussed in Chapter 7 and in detail in other chapters. Although

TABLE 11.6
Shapes of oxy-anions

Non-linear				NO_2^-	ClO_2^-
Planar	BO_3^{3-}	CO_3^{2-}		NO_3^-	
Pyramidal				SO_3^{2-}	ClO_3^- BrO_3^- IO_3^-
Tetrahedral	SiO_4^{4-}	PO_4^{3-} AsO_4^{3-} VO_4^{3-}		SO_4^{2-}	ClO_4^- IO_4^- MnO_4^-
Tetragonal pyramidal					IO_5^{3-} ReO_5^{3-}
Octahedral				TeO_6^{6-}	IO_6^{5-}

the most important ions of these kinds are those formed by non-metals and by transition metals in high oxidation states, similar groups have been identified in, for example, K_2NiO_2 (linear $O-Ni-O^{2-}$ ions),[1] K_4PbO_3 (pyramidal PbO_3^{4-} ions, with bond angles close to 100° and Pb–O, 2.17 Å),[2] and Na_4FeO_3 (planar FeO_3^{4-} ions).[3]

(2) *Finite polynuclear ions.* There is a small number of polynuclear oxy-ions in which X or O atoms are directly bonded, as in the examples quoted above, but most polynuclear oxy-ions are formed from XO_n groups joined together by sharing O atoms. They are formed by many elements, notably by B, Si, P, Mo, and W, and to a much smaller extent by As, V, Nb, S, and Cr(VI). The possible types of complex

oxy-ion may be derived by considering how XO_3 (planar or pyramidal), XO_4 (tetrahedral), or XO_6 (octahedral) groups, for example, can link up by sharing vertices (O atoms) or edges; sharing of faces of polyhedral coordination groups in complex oxy-ions is rarely observed. From the ortho-ion XO_n we derive in this way:

$$O_{n-1}X{-}O{-}XO_{n-1}$$

pyro-ion

$$O_{n-2}X{\diagdown}O{\diagup}XO_{n-2} \text{ with } {\diagup}O{\diagdown}X{-}O_{n-2}$$

and other cyclic ions

$$\left[{\diagup}X{\diagdown}O_{n-2}{\diagdown}O{\diagup} \right]_{\infty}$$

meta-ions

where we have included the infinite chain ion, the alternative to a cyclic ion when 2 vertices are shared. Examples of X_2O_7 and cyclic $(XO_3)_n$ ions formed from tetrahedral XO_4 groups are numerous; the edge-sharing X_2O_6 group structurally similar to the Al_2Cl_6 molecule has been found in $K_6Fe_2O_6$[4] and $K_6Mn_2O_6$.[5] Octahedral ions X_2O_{11} formed from two octahedral groups sharing a vertex appear to be unknown, but two octahedral groups share an edge in the $Te_2O_{10}^{8-}$ ion and a face in $I_2O_9^{4-}$ and $W_2O_9^{6-}$.

Intermediate between the pyro-ion and the infinite linear meta-ion are the ions formed from three or more XO_n groups, examples of which now include:

X_3O_{10}: $P_3O_{10}^{5-}$, $As_3O_{10}^{5-}$, $S_3O_{10}^{2-}$, $Cr_3O_{10}^{2-}$, $PCr_2O_{10}^{3-}$, $AsCr_2O_{10}^{3-}$

X_4O_{13}: $S_4O_{13}^{2-}$, $Cr_4O_{13}^{2-}$

There are also ions $PCr_3O_{13}^{3-}$ and $PCr_4O_{16}^{3-}$ in which a central P atom is bonded (through O) to 3 (triangular) or 4 (tetrahedral) Cr atoms, that is, formed from 4 or 5 tetrahedral groups respectively. Details of such ions are described under the element X.

Two other classes of complex finite oxy-ion may be mentioned here. There is a small group of ions formed by the bonding of XO_n ions to a metal atom, either through X as in $[Co(NO_2)_6]^{3-}$ and $[Pd(SO_3)_4]^{6-}$, or through one or more of the O atoms, as in $[Co(NO_3)_4]^{2-}$ or $[Ce(NO_3)_6]^{2-}$. A more complex example is the ion $[UO_2(NO_3)_3]^-$, in which the coordination group of U is made up of two uranyl O atoms and six O atoms of the bidentate nitrato groups. A larger class, with which we deal shortly, comprises the numerous *isopoly* ions, composed of various numbers of octahedral groups (X = V, Nb, Ta, Mo, W) sharing vertices and edges, and the *heteropoly* ions, in which one or more atoms of a second element are incorporated in the complex oxy-ion.

(3) *Infinite polynuclear oxy-ions.* Numerous oxy-ions which extend indefinitely in 1, 2, or 3 dimensions are described in this book, in particular, borate ions (from planar BO_3 and/or tetrahedral BO_4 groups) in Chapter 24, silicate and aluminosilicate

ions (from tetrahedral $(Si, Al)O_4$ groups) in Chapter 23, and many structures built from tetrahedral or octahedral groups in Chapter 5. In this connection it is interesting to note that Fe(III), which forms the edge-sharing $Fe_2O_6^{6-}$ ion (a type not formed by Si) also forms a number of ions analogous to those of Si. These include the finite ions FeO_4^{5-} [6] and $Fe_2O_7^{8-}$,[7] the single chain in K_3FeO_3,[7a] the double

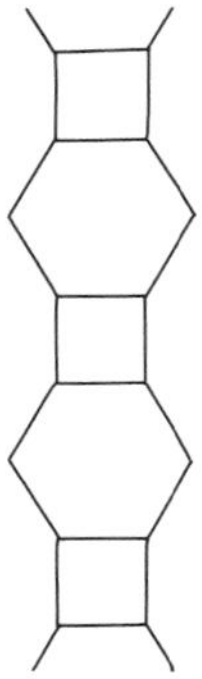

(a)

chain ion (a) in $Na_{14}Fe_6O_{16}$,[8] apparently similar to that in $Na_2Ca_3Si_6O_{16}$, and the anion in $Na_4Fe_2O_5$,[9] based on the planar 6^3 net.

(1) ZaC 1964 **324** 214
(2) ZaC 1978 **428** 105
(3) ZaC 1977 **437** 95
(4) ZaC 1974 **408** 151
(5) NW 1976 **63** 339

(6) NW 1975 **62** 138
(7) ZaC 1978 **438** 15
(7a) ZaC 1974 **408** 152
(8) AnCIE 1977 **16** 43
(9) NW 1977 **64** 271

Isopoly ions

Ions of V, Nb, and Ta

There is much experimental evidence for the formation of complex oxy-ions in acidified solutions of vanadates, niobates, and tantalates. We describe in Chapters 12 and 13 the structures of some crystalline vanadates; here we note first certain finite complex ions which exist both in solution and in crystalline salts. The ion of Fig. 11.3(a) has been shown to exist in $Na_7H(Nb_6O_{19}).15H_2O$.[1] Light scattering from an aqueous solution of $K_8(Ta_6O_{19}).16H_2O$ indicates that the anion species contains 6 Ta atoms and is presumably similar to the ion in the crystal.[2]

It seems that salts containing the corresponding V(v) ion cannot be crystallized from solution, though salts containing ions such as $VW_5O_{19}^{3-}$ and $V_2W_4O_{19}^{4-}$ have been prepared, and the structure of the guanidinium salt of the latter has been determined.[2a] Also, a super-octahedral V_6O_{19} group is found in α-copper vanadate, a non-stoichiometric compound of V(IV) with the composition $Cu_{7-x}V_6O_{19-x}$ $(x = 0.22)$.[2b]

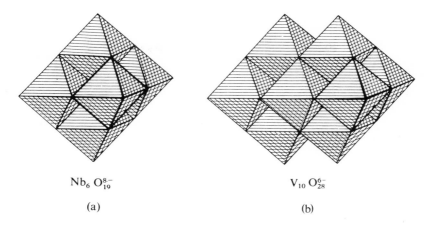

$$Nb_6O_{19}^{8-}$$

(a)

$$V_{10}O_{28}^{6-}$$

(b)

FIG. 11.3. The $Nb_6O_{19}^{8-}$ and $V_{10}O_{28}^{6-}$ ions.

The existence of a decavanadate ion, previously proved to exist in solution, has been demonstrated in crystalline salts, at least two of which occur as minerals, $Ca_3V_{10}O_{28}.27H_2O$ and $K_2Mg_2V_{10}O_{28}.16H_2O$. Salts of this type are formed by evaporation of vanadate solutions having pH 2-6. The $V_{10}O_{28}^{6-}$ ion (Fig. 11.3(b)), studied in the $Ca^{(3)}$ and $Zn^{(4)}$ salts, contains ten V atoms situated at the vertices of a pair of octahedra which share an edge. As regards their geometry both of the ions of Fig. 11.3 may be described as portions of NaCl structure, substituting Nb or Ta for Na and O for Cl. The analogous Nb ion has been studied in $[N(CH_3)_4]_6Nb_{10}O_{28}$.[5]

There is no finite complex ion in $Rb_4Nb_6O_{17}.3H_2O$.[6] The anion consists of layers formed from the A_6X_{22} sub-units of Fig. 11.4(b) sharing the 10 projecting vertices (O atoms), two each at top and bottom and three on each side. This is a rather unusual layer structure in that while some of the layers are held together only by (one half of the) Rb^+ ions, the remainder of these ions together with the water molecules are situated between the alternate pairs of layers.

(1)   AK 1953 **5** 247; ibid 1954 **7** 49
(2)   IC 1963 **2** 985
(2a) BCSJ 1975 **48** 889
(2b) JSSC 1973 **7** 17

(3)   AC 1966 **21** 397
(4)   IC 1966 **5** 967
(5)   AC 1977 **B33** 2137
(6)   JSSC 1980 **33** 83

Ions of Mo and W

Although the simpler ions are not strictly poly-ions we shall survey here all ions M_nO_x formed by Mo and W including the M_2O_7 ions. We note elsewhere the structural differences between the trioxides of S and Cr (tetrahedral coordination) and those of Mo and W (octahedral coordination, but different structures). There are rather similar differences between the structures of oxy-ions of these elements. Certain ortho-molybdates and tungstates resemble sulphates and chromates in that they contain tetrahedral MO_4 ions (p. 593), but otherwise there is little resemblance between the structures of sulphates and chromates on the one hand and those of molybdates and tungstates on the other. Polysulphates and polychromates are built of tetrahedral groups sharing vertices and are not known with $n > 4$. On the other hand polymolybdates and polytungstates are much more numerous and of much greater complexity. For example, all the following Cs compounds have been prepared from the melt:[1] $Cs_2Mo_2O_7$, $Cs_2Mo_3O_{10}$, $Cs_2Mo_4O_{13}$, $Cs_2Mo_5O_{16}$, and $Cs_2Mo_7O_{22}$. The poly-ions of Mo and W are almost exclusively built of MO_6 coordination groups, though there are exceptions. There is tetrahedral coordination of Mo in $MgMo_2O_7$[2] (anion similar to $Cr_2O_7^{2-}$) and tetrahedral coordination of one half of the M atoms in the isostructural $Na_2Mo_2O_7$[3] and $Na_2W_2O_7$[4] which contain chain anions (Fig. 5.43(a), p. 232). However, there are infinite chain ions formed from octahedral MO_6 groups in $Li_2W_2O_7$[5] and in $Ag_2Mo_2O_7$ and $Ag_2W_2O_7$,[6] two closely related vertex- and edge-sharing chains which are both different from the chain in $Li_2W_2O_7$.

It will be noted that polymolybdates and polytungstates of similar formula-type often have different structures (compare the trioxides), and that the structure of the anion is dependent on the nature of the cation; compare the quite different structures of $K_2Mo_3O_{10}$[7] and $K_2W_3O_{10}$.[8] The latter has a 3D framework structure whereas the anion in $K_2Mo_3O_{10}$ is an infinite 1D ion to which we refer later. This is illustrated in Fig. 5.43(b) as built from octahedra and tetrahedra, but should probably be described as containing 5- and 6-coordinated Mo. In the MoO_4 tetrahedron in $Na_2Mo_2O_7$ Mo–O ranges only from 1.71 to 1.79 Å, but in $K_2Mo_3O_{10}$ these groups are much less regular (Mo–O, 1.64–1.95 Å, with a fifth O at 2.08 Å), suggesting 5- rather than 4-coordination of Mo. In the MoO_6 groups in both these molybdates Mo–O ranges from around 1.70 to values 2.2–2.3 Å as in other polymolybdates and also in MoO_3. (The compound originally formulated $Rb_2W_3O_{10}$ is apparently $Rb_{22}W_{32}O_{107}$.[9])

At least four quite different structures have been found for compounds $M_2Mo_4O_{13}$ or $M_2W_4O_{13}$: in low-$Li_2Mo_4O_{13}$[10] an octahedral 3D framework related to V_6O_{13}, $K_2Mo_4O_{13}$,[11] a complex edge-sharing chain, $Tl_2Mo_4O_{13}$,[12] an octahedral layer structure, and $K_2W_4O_{13}$,[13] a vertex-sharing 3D framework reminiscent of the hexagonal bronze structure. Two rather compact groups of 4 octahedra with the composition A_4X_{16} are illustrated in Fig. 5.8 (p. 200). The unit of Fig. 5.8(h) is the anion in $Ag_8W_4O_{16}$.[14]

We noted in Chapter 5 that the very simple A_4X_{16} unit illustrated in Fig. 5.8(j) (p. 200) is not known as an ion in solution. However, a discrete W_4O_{16} unit with this configuration has been recognized in tungstates of the type $Li_{14}(WO_4)_3W_4O_{16}$·$4H_2O$[15] and the isostructural $(Li_{11}Fe)(WO_4)_3W_4O_{16}$,[16] in which it occurs together with tetrahedral WO_4^{2-} ions. There is, however, some disagreement as to whether one Li occupies the central tetrahedral cavity in the W_4O_{16} group or whether all the cations are situated between the $W_4O_{16}^{8-}$ and WO_4^{2-} ions in positions of octahedral coordination.

The ions $(Mo_6O_{19})^{2-}$ *and* $(W_6O_{19})^{2-}$. The anions in $[HN_3P_3(NMe_2)_6]_2Mo_6O_{19}$[17] and $(NBu_4)_2W_6O_{19}$[18a] have the structure of Fig. 11.3(a). Refinement of the structure of the W compound shows the usual range of W–O bond lengths (1.693, 1.924, and 2.325 Å). The 'super-octahedron', W_6O_{19}, is found as a sub-unit of the framework of $Sn_{10}W_{16}O_{48}$ formed from W_6O_{19} and W_2O_9 units.[18b]

The ion $Mo_7O_{24}^{6-}$. This anion (Fig. 11.4(a)) occurs in the salt $(NH_4)_6Mo_7O_{24}.4H_2O$, (ref. 19a), originally described as a paramolybdate. It is interesting that this ion does not have the coplanar structure of the $TeMo_6O_{24}^{6-}$ ion (see later) but is instead a more compact grouping of seven MoO_6 octahedra. There is a considerable range of Mo–O bond lengths, corresponding to the fact that there are oxygen atoms bonded to 1, 2, 3, and 4 metal atoms.

Sharing of O atoms between Mo_7O_{24} sub-units gives a layer of composition Mo_7O_{22} which is the 2D anion in $Tl_2Mo_7O_{22}$.[19b]

The ions $Mo_8O_{26}^{4-}$ *and* $H_2Mo_8O_{28}^{6-}$. X-ray studies[20a] of hydrates of $(NH_4)_4Mo_8O_{26}$ show that the octamolybdate ion has the structure shown in Fig. 11.4(c) and (d), in which six of the MoO_6 octahedra are arranged in the same way as in $Mo_7O_{24}^{4-}$, as shown in Fig. 11.4(b). This ion has now been designated the β-$Mo_8O_{26}^{4-}$ ion, for an isomeric ion, α-$Mo_8O_{26}^{4-}$ has been found in $[N(C_4H_9)_4]_4Mo_8O_{26}$.[20b] This consists of a ring of 6 octahedra like that in $TeMo_6O_{24}^{6-}$ (Fig. 11.8), but the central octahedral hole is unoccupied and bridged by tetrahedral MoO_4 groups, one above and one below the plane of the ring (Fig. 11.4(e)). Possible mechanisms for the change from one isomer to the other have been discussed.[20c]

Another finite octamolybdate ion has been found in the isopropylammonium salt $(C_3H_{10}N)_6[H_2Mo_8O_{28}].2H_2O$[21] (Fig. 11.4(f) and (g)). This complex consisting of 8 edge-sharing octahedral groups is of more than usual interest for it is the sub-unit in two other octamolybdates. Sharing of the two O atoms distinguished as circles in Fig. 11.4(g) leads to chains of the composition Mo_8O_{27} which are the form of the anion in $(NH_4)_6Mo_8O_{27}.4H_2O$.[22] This salt may be crystallized from a solution of either hexa- or hepta-molybdate, showing that various poly-ions can be inter-converted and are possibly in equilibrium in solution. The chain formed from the same sub-units by sharing the *edges* indicated in Fig. 11.4(g) has the composition Mo_8O_{26}, and is the structure of the anion in $K_2Mo_4O_{13}$ crystallized from the melt.

We noted earlier that the Nb_6O_{17} layer is built from the M_6X_{22} units of Fig. 11.4(b) by sharing 10 O atoms. A feature of many complex polymolybdates and polytungstates is that their structures contain sub-units consisting of *edge*-sharing groups of octahedra which are further joined into chains, layers, or 3D structures by sharing of O atoms which are usually vertices but sometimes edges of octahedral groups. Examples include:

Sub-unit

$[M_6X_{22}]^*$ → layer M_6X_{17} $(Rb_4Nb_6O_{17}.3H_2O)$

*Not known as a finite ion

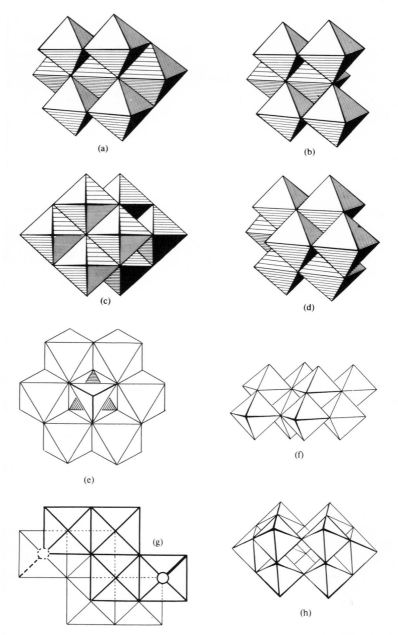

FIG. 11.4. (a) The $Mo_7O_{24}^{6-}$ ion. (c) and (d) Views of the β-$Mo_8O_{26}^{4-}$ ion. (b) The portion common to the two ions. (e) The α-$Mo_8O_{26}^{4-}$ ion. (f) and (g) Two views of the Mo_8O_{28} group. (h) The $W_{10}O_{32}^{4-}$ ion.

M_7X_{24} → layer M_7X_{22} $(Tl_2Mo_7O_{22})$
M_8X_{28} → chain M_8X_{27} by vertex-sharing $((NH_4)_6Mo_8O_{27}.4H_2O)^{(22)}$
chain M_8X_{26} by edge-sharing $(K_2Mo_4O_{13})$

The ion $W_{10}O_{32}^{4-}$. This ion (Fig. 11.4(h)), studied in the tributyl-ammonium salt,[23] is formed by removing one octahedron from each of two 'super-octahedral' W_6O_{19} groups and then joining the two W_5 units by sharing vertices, leaving an empty octahedral hole at the centre.

Molybdic and tungstic acids. If the (alkaline) solution of a normal alkali-metal tungstate is neutralized and crystallized by evaporation, a 'paratungstate' is obtained, which contains the $(H_2W_{12}O_{42})^{10-}$ ion. By boiling a solution of a paratungstate with yellow 'tungstic acid' (which results from acidification of a hot solution of a normal or a paratungstate) a 'metatungstate' results, and acidification of the solution gives the soluble metatungstic acid, which may be extracted with ether and forms large colourless crystals. The metatungstates contain the $(H_2W_{12}O_{40})^{6-}$ ion.

The final products of acidification of molybdate or tungstate solutions are molybdic and tungstic 'acids'. Yellow $MoO_3.2H_2O$ and $MoO_3.H_2O$ consist of vertex-sharing layers of composition $MoO_3(H_2O)$ (see p. 208) formed from $[MoO_5(H_2O)]$ coordination groups; in the dihydrate the second H_2O is situated between the layers.[24] The W compounds have the same structures. There is also a white $MoO_3.H_2O$ which has an edge-sharing chain structure,[25] the projection of which is the same as the elevation of the MoO_3 layer shown in Fig. 5.30(a), p. 222.

The paratungstate and metatungstate ions. The paratungstate ion has been studied in a number of salts including $(NH_4)_{10}(H_2W_{12}O_{42}).10H_2O$,[26] the 20- and 27-hydrates of the Na salt,[27] and $(NH_4)_6H_4[H_2W_{12}O_{42}].10H_2O$.[28] It is built from 12 WO_6 octahedra which are first arranged in four edge-sharing groups of three, two of type (a) and two of type (b), Fig. 11.5. These groups of three octahedra are then joined together by sharing vertices as indicated in the figure. In the resulting aggregate the 42 oxygen atoms are arranged approximately in positions of hexagonal

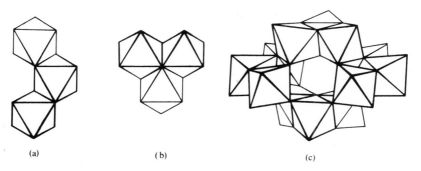

(a) (b) (c)

FIG. 11.5. (a) and (b) 3-octahedron sub-units from which the paratungstate ion (c) is built.

closest packing. The two non-replaceable H atoms have been shown to be associated with the O atoms lying on the central vertical axis. As in the polymolybdates there is a considerable range of metal–oxygen bond lengths, here approximately 1.8, 2.0, and 2.2 Å to O atoms bonded to 1, 2, and 3 W atoms respectively. It is interesting to compare with this ion the metatungstate ion, which is also built of 12 WO_6 octahedra but in this case arranged in four groups of type (b) to give an ion of composition $(H_2W_{12}O_{40})^{6-}$. This ion has the same basic structure as the heteropoly-acid ion $(PW_{12}O_{40})^{3-}$ which is described in more detail layer. Chemical reduction of metatungstates produces coloured ions, blue[29] and brown.[30] Study of the brown $Rb_4H_8(H_2W_{12}O_{40})$, in which 3 of the W atoms, probably in one W_3O_{13} cluster of the Keggin ion, have been reduced to W(IV), was complicated by disorder in the crystal.

(1)	AC 1975 **B31** 1293	(18a)	AC 1978 **B34** 1764
(2)	AC 1977 **B33** 3859	(18b)	AC 1980 **B36** 15
(3)	ACSc 1967 **21** 499	(19a)	JCS D 1975 505
(4)	AC 1975 **B31** 1200	(19b)	AC 1976 **B32** 1859
(5)	AC 1975 **B31** 1451	(20a)	JLCM 1977 **54** 283
(6)	JCS D 1976 1316	(20b)	JACS 1977 **99** 952
(7)	JCS A 1968 1398	(20c)	IC 1980 **19** 3866
(8)	AC 1976 **B32** 1522	(21)	AC 1978 **B34** 2728
(9)	AC 1977 **B33** 3345	(22)	AC 1974 **B30** 48
(10)	JSSC 1974 **9** 247	(23)	AC 1976 **B32** 740
(11)	JCS A 1971 2107	(24)	AC 1972 **B28** 2222
(12)	AC 1978 **B34** 3547	(25)	AC 1974 **B30** 1795
(13)	CC 1967 1126	(26)	AC 1971 **B27** 1393
(14)	MRB 1975 **10** 791	(27)	AC 1976 **B32** 1565
(15)	B 1966 **70** 598	(28)	AC 1979 **B35** 1675
(16)	K 1968 **13** 980	(29)	JINC 1976 **38** 807
(17)	IC 1973 **12** 2963	(30)	IC 1980 **19** 2933

Heteropoly ions

The most familiar heteropoly ions are the 12-phosphomolybdate and phospho-tungstate ions, $PMo_{12}O_{40}^{3-}$ and $PW_{12}O_{40}^{3-}$, which are readily formed in various simple reactions. Ammonium phosphomolybdate is precipitated from a solution containing a phosphate, ammonium molybdate, and nitric acid. Alkali phosphotungstates result from boiling a solution of the alkali phosphate with WO_3. The heteropoly acids are very soluble in water and ether and crystallize very well, often with large numbers of molecules of water of crystallization. The acids have the remarkable property of forming insoluble precipitates with many complex organic compounds such as dyes, albumen, and alkaloids. Both the acids and salts possess considerable stability towards acids, but are readily reduced by sulphur dioxide and other reducing agents. Like the acids, the salts of the smaller alkali metal ions are highly hydrated. On long standing the 12-ions transform into more complex ions such as $P_2W_{18}O_{62}^{6-}$, the structure of which is described later. The 12-ions consist of a globular cluster of vertex- and edge-sharing MoO_6 or WO_6 octahedra which enclose a central tetrahedral cavity occupied by the P atom. this central atom may also be one of a variety of elements, for example, B, Al, Si, Ge, As, Fe, Cu, or Co. Another ion in which the shell of octahedral groups totally encloses the hetero atom is the $CeMo_{12}$ ion, in which the central cavity is icosahedral.

In its widest sense the term heteropoly ion also includes at least two other groups of ions. There are ions in which there is less complete enclosure of the hetero atom which, however, is in all cases surrounded by a coordination group of O atoms belonging to the octahedral groups as in the 12- and 2 : 18-ions:

	PMo_9	Co_2Mn_{10}, $MnNb_{12}$, $TeMo_6$	CeW_{10}
coordination of hetero atom	tetrahedral	octahedral	antiprismatic

Although there are different types of coordination of the hetero atom in the $MnNb_{12}$ and CeW_{10} ions the structures of these two ions have much in common (Figs. 11.10 and 11.12). Finally there are ions in which the hetero atoms are not at all enclosed within a shell of octahedra; these include the P_2Mo_5, V_2Mo_6, and V_8Mo_4 ions, which are described later. (Other oxy-ions containing atoms of two kinds, in addition to O, include the $(PCr_2O_{10})^{3-}$ and $(PCr_4O_{16})^{3-}$ ions (p. 861), in which there is tetrahedral coordination of both P and Cr atoms.)

We shall describe the structures of the ions listed below, grouped according to the type of coordination of the hetero atom. We do not include here ions such as $VW_5O_{19}^{3-}$ and $V_2W_4O_{19}^{4-}$ (p. 514) in which atoms of two kinds occupy similar (octahedral) positions in the super-octahedral structure of $W_6O_{19}^{2-}$. Because of the close relation between the structures of the PMo_9, P_2Mo_{18}, and $H_2AsW_{18}O_{60}$ ions these three ions will be described together after the $PW_{12}O_{40}$ ion.

Coordination of hetero atom	Examples	
Trigonal pyramidal		$H_2AsW_{18}O_{60}^{7-}$
Tetrahedral	$PMo_9O_{34}H_6^{3-}$	$P_2Mo_{18}O_{62}^{6-}$
	$PW_{12}O_{40}^{3-}$	
Octahedral	$TeMo_6O_{24}^{6-}$	
	$MnMo_9O_{32}^{6-}$	$Co_2Mo_{10}O_{38}H_4^{6-}$
	$MnNb_{12}O_{38}^{12-}$	
Antiprismatic	$CeW_{10}O_{36}H_2^{6-}$	
Icosahedral	$CeMo_{12}O_{42}^{8-}$	
Tetrahedral		$P_2Mo_5O_{23}^{6-}$
		$V_2Mo_6O_{26}^{6-}$
Trigonal bipyramidal		$V_5Mo_8O_{40}^{7-}$
		$V_8Mo_4O_{36}^{8-}$

Tetrahedral coordination of hetero atom

The $PW_{12}O_{40}^{3-}$ *ion.* An early X-ray study of crystals assumed to be the pentahydrate of the acid correctly established the structure of the ion as shown in Fig. 11.6(b), often referred to as the Keggin structure.

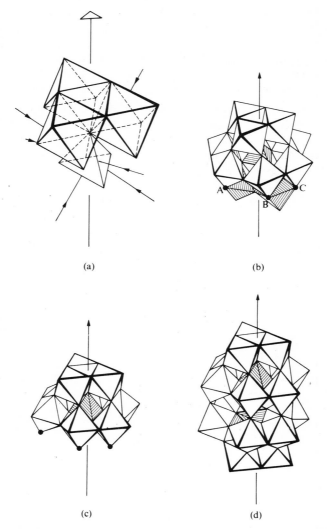

(a)

(b)

(c)

(d)

FIG. 11.6. (a) Arrangement of one group of three WO_6 octahedra relative to the central PO_4 tetrahedron in the $PW_{12}O_{40}^{3-}$ anion (b). (c) The 'half-unit'. (d) The $P_2W_{18}O_{62}^{6-}$ anion.

The $PW_{12}O_{40}^{3-}$ ion consists of a group of twelve WO_6 octahedra arranged around, and sharing O atoms with, a central PO_4 tetrahedron. The twelve WO_6 octahedra are arranged in four groups of three (of type (b) in Fig. 11.5) and within each group each octahedron shares two edges. These groups of three octahedra are then joined by sharing *vertices* to form the complete anion (Fig. 11.6(b)). Figure 11.6(a) shows the disposition of one group of three WO_6 octahedra relative to the central PO_4

tetrahedron. Each WO_6 octahedron shares two edges, one with each of the two neighbouring octahedra, and each O of the PO_4 tetrahedron belongs to three WO_6 octahedra. The formation of this ion from four similar sub-units, the groups of three edge-sharing octahedra, may be contrasted with the two types of sub-unit in the paratungstate ion described earlier.

More recently a detailed X-ray and n.d.-single-crystal study[1] showed that the 'pentahydrate' of the acid $H_3P_{12}O_{40}$ is actually a hexahydrate, containing $H_5O_2^+$ ions. This discovery leads to the more satisfactory formulation as $(H_5O_2^+)_3(PW_{12}O_{40})^{3-}$ and suggests also that the isostructural 'pentahydrates' of related acids are also hexahydrates, namely:

$$(H_5O_2)_3(HSiW_{12}O_{40})^{3-}$$
$$(H_5O_2)_3(H_2BW_{12}O_{40})^{3-}$$
$$(H_5O_2)_3(H_3H_2W_{12}O_{40})^{3-}.$$

There are respectively 1, 2, and 3 replaceable H atoms in the 12-silicotungstate, 12-borotungstate, and metatungstic acids; these are presumably bonded to non-bridging O atoms of the complex ion. Further, the Cs salt formulated as the dihydrate on the basis of the 'pentahydrate' structure becomes the anhydrous salt, Cs^+ replacing $(H_5O_2)^+$, not H_2O as previously thought. Reformulation of certain other hydrates is probably also necessary.

A recent addition to the list of Keggin type ions is the anion in $Na_4(GeMo_{12}O_{40})$.-$8H_2O$.[1a] Although ions containing Mo are of the same general type as the W compounds the coordination of Mo is much less regular. For example, in $H_3(PMo_{12}O_{40}) . \approx 30H_2O$[1b] there is square pyramidal coordination, the Mo atom being 0.41 Å above the centre of the base of the pyramid. A much more complex

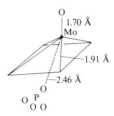

variant of the Keggin structure is found in the black $K_6\{V^V[V^{IV}V^VMo_{10}^{VI}]O_{40}\}^{6-}$.-$13H_2O$.[1c] The V^VO_4 tetrahedron at the centre is regular, and there is random arrangement of V^{IV}, V^V, and Mo^{VI} in the octahedral shell.

In the $PW_{12}O_{40}^{3-}$ ion the M atoms are arranged at the vertices of a cuboctahedron which has, however, only tetrahedral, as opposed to full cubic symmetry, for four of the triangular faces correspond to edge-sharing groups of three octahedra and the other four to vertex-sharing groups. Ions with the same structure include the meta-tungstate ion (p. 519), $[CoW_{12}O_{40}]^{5-}$,[2a] and the anion in α-$Ba_2[SiW_{12}O_{40}].16H_2O$.

If one of the groups of three octahedra of the $PW_{12}O_{40}^{3-}$ ion is rotated through

60° a structure is produced in which the 3-octahedron units are still joined only through vertices but the symmetry has dropped from cubic to trigonal. This 'isomer' is apparently the structure of the anion in β-$K_4[SiW_{12}O_{40}]$.$9H_2O$.[2b] If, in addition, a second group of three octahedra is rotated through 60° there is sharing of an edge between two of these sub-units; rotation of further sub-units leads to further edge-sharing:

	Shared between sub-units	Point symmetry	Examples
(i) ⎫	Vertices only	23	$(PW_{12}O_{40})^{3-}$, α-$(SiW_{12}O_{40})^{4-}$
(ii) ⎭		$3mm$	β-$(SiW_{12}O_{40})^{4-}$
(iii)	1 edge	$2mm$	
(iv)	3 edges	$3mm$	
(v)	6 edges	23	$[Al_{13}O_4(OH)_{24}(H_2O)_{12}]^{7+}$

The last of these five isomers of $M_{12}X_{40}$ has, like the first, cubic symmetry, and corresponds to a portion of the spinel structure. It represents the structure of complexes formed in partially hydrolysed solutions of aluminium salts. Its structure has been studied in $Na[Al_{13}O_4(OH)_{24}(H_2O)_{12}](SO_4)_4$.$13H_2O$[2c] and in the closely related $[Al_{13}O_4(OH)_{25}(H_2O)_{11}](SO_4)_3$.$16H_2O$[2d] and is illustrated in Fig. 11.7(a). The central tetrahedral cavity and the centres of the octahedra are all occupied by Al atoms. Figure 11.7(b) shows another view of the $[PW_{12}O_{40}]^{3-}$ ion with a 2-fold axis vertical instead of a 3-fold axis, as in Fig. 11.6(b).

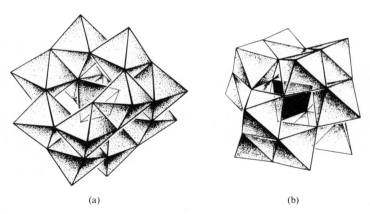

(a)　　　　　　　　　　　　　(b)

FIG. 11.7. (a) The c.c.p. $M_{12}X_{40}$ grouping of $[Al_{13}O_4(OH)_{24}(H_2O)_{12}]^{7+}$ and related ions. (b) The $PW_{12}O_{40}^{3-}$ ion.

The $PMo_9O_{34}H_6^{3-}$ *and* $P_2Mo_{18}O_{62}^{6-}$ *ions.*[3] These ions are related in a very simple way to the 12-ion. Removal of three octahedra from the base of the 12-ion leaves the 'half-unit' shown in Fig. 11.6(c). This is the structure of the anion in

$(PMo_9O_{34}H_6)Na_3 . \approx 7H_2O$. If two half-units are joined by sharing the exposed O atoms (three of which are shown as black dots in (c)) we obtain the $P_2Mo_{18}O_{62}^{6-}$ ion of Fig. 11.6(d). The upper half is related to the lower half by the plane of symmetry *ABC* in (b). This ion has been studied in $(P_2Mo_{18}O_{62})Na_4H_2 . \approx 20H_2O$, and the W analogue in $(P_2W_{18}O_{62})(NH_4)_6 . \approx 9H_2O$.

The $H_2AsW_{18}O_{60}^{7-}$ *ion.* If we remove the upper three, instead of the lower three, octahedra from the W_{12} unit of Fig. 11.6(b) and join two such units by sharing the exposed 6 O atoms, a different W_{18} complex is formed. The two possibilities may be summarized:

$$W_3O_{13} \begin{array}{c} \longrightarrow W_9O_{34} \\ \longrightarrow W_9O_{33} \end{array} \begin{array}{c} \rightarrow \quad W_{18}O_{62} \quad (a) \\ \rightarrow \quad W_{18}O_{60} \quad (b) \end{array}$$

(a) being realized in $PW_{18}O_{62}^{6-}$ as described above and (b) in $H_2AsW_{18}O_{60}^{7-}$.[3a] It might have been expected that isomers of both (a) and (b) would be possible, in which the two halves would be related either by a mirror plane or by a centre of symmetry. However, in (a) the equatorial ring of 6 O atoms is not a regular hexagon, for the octahedra to which they belong are alternately edge- and vertex-sharing. In the $W_{18}O_{62}$ complex, therefore, there is a horizontal plane of symmetry. Each half contains a tetrahedral cavity suitable for a hetero atom, giving the composition $P_2W_{12}O_{62}$. The form found for $H_2AsW_{18}O_{60}^{7-}$ is the centrosymmetrical one. This $W_{18}O_{60}$ cage does not contain the *two* tetrahedral cavities required for the normal hetero atoms but it does provide a suitable position of trigonal pyramidal coordination for *one* atom such as As(III) with a lone pair. Two such atoms could not be accommodated because their lone pairs would be directed towards one another along the central axis of the ion. One half of the cage therefore contains an As atom and the other half 2 protons.

Octahedral coordination of hetero-atom

The $TeMo_6O_{24}^{6-}$ *ion.* This ion (Fig. 11.8) exists in salts such as $(NH_4)_6TeMo_6O_{24}.7H_2O$ and has been studied in detail in $(NH_4)_6TeMo_6O_{24}.Te(OH)_6.7H_2O$.[4a] Crystals of this compound also contain $Te(OH)_6$ molecules (almost regular octahedra, Te–OH, 1.91 Å). The atomic arrangement in $TeMo_6O_{24}^{6-}$ is the same as in a portion of the CdI_2 layer but with some distortion from the idealized structure of Fig. 11.8 due

(a)

(b)

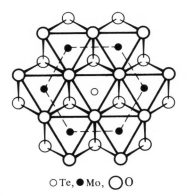

○ Te, ● Mo, ◯ O

FIG. 11.8. The $TeMo_6O_{24}^{6-}$ ion in $(NH_4)_6(TeMo_6O_{24}).7H_2O$, showing the mode of linking of the MoO_6 octahedra.

to the difference between the Mo–O and Te–O bond lengths, (a). Other ions with the same structure include $(Cr^{III}Mo_6O_{24}H_6)^{3-}$ [4b] and $(IMo_6O_{24})^{5-}$ [4c] (Cr–O, 1.975; I–O, 1.89 Å), with a central Cr or I atom. In the ion $[Co_4I_3O_{24}H_{12}]^{3-}$ the seven octahedra are occupied by 4 Co and 3 I as shown at (b).[4d]

The $MnMo_9O_{32}^{6-}$ *ion.*   The structure of a 9-ion has been determined in $(NH_4)_6MnMo_9O_{32}.8H_2O$.[5] In this ion, illustrated in Fig. 11.9, nine MoO_6 octahedra are grouped around, and completely enclose, the Mn^{4+} ion.

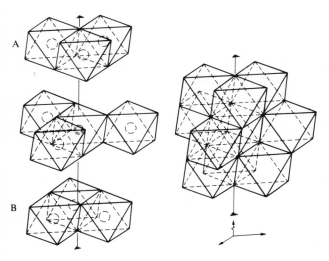

FIG. 11.9. The structure of the $MnMo_9O_{32}^{6-}$ anion with 'exploded' view at left. The single circles represent Mo and the double circle Mn.

The $MnNb_{12}O_{38}^{12-}$ *ion.*[6] This ion may be constructed from the same edge-sharing groups of three octahedra from which $PW_{12}O_{40}^{3-}$ is built and which are also sub-units in $MnMo_9O_{32}^{6-}$. We may refer to the unit as A or as B if it is upside-down (Fig. 11.9). If B is placed on A there is formed an octahedral group of six octahedra which is the structure of the $Nb_6O_{19}^{8-}$ ion of Fig. 11.3(a). Two of these units joined together by a Mn atom form the anion in $Na_{12}Mn^{IV}Nb_{12}O_{38}.H_2O$, as shown in Fig. 11.10.

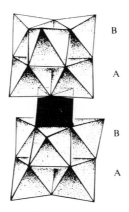

FIG. 11.10. The $[MnNb_{12}O_{38}]^{12-}$ anion.

The $H_4Co_2Mo_{10}O_{38}^{6-}$ *ion.*[7] A model of this ion, studied in $(NH_4)_6(H_4Co_2Mo_{10}O_{38})$.-$7H_2O$, is readily made from a chain of four edge-sharing octahedra by adding eight more octahedra, four in front and four behind, as shown in Fig. 11.11. Alternatively the ion may be constructed from two planar $CoMo_6$ units (similar to the $TeMo_6O_{24}^{6-}$ ion), from which one Mo is removed, and one unit is rotated relative to the other so that the Co atom of one unit occupies the position of a Mo atom in the other. The four H atoms were not located.

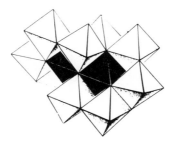

FIG. 11.11. The $[H_4Co_2Mo_{10}O_{38}]^{6-}$ anion. The black octahedra contain the Co atoms.

Antiprismatic coordination of hetero atom

The $CeW_{10}O_{36}H_2^{6-}$ *ion*. Removal of one octahedron from the super-octahedron of Fig. 11.3(a), which is the structure of the $W_6O_{19}^{2-}$ ion, leaves a W_5 portion which has a square array of 4 O atoms at one end. Two such sub-units, one rotated through 45° relative to the other, are joined through a Ce(IV) atom in square antiprismatic coordination in the anion of $Na_6[CeW_{10}O_{36}H_2].30H_2O,$[8] Fig. 11.12.

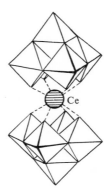

FIG. 11.12. The $(CeW_{10}O_{36}H_2)^{6-}$ anion.

The $[U(GeW_{11}O_{39})_2]^{12-}$ *ion*. Just as the removal of one W and its attached O atom from W_6O_{19} leaves W_5O_{18}, and two such units are joined to form $[M(W_5O_{18}H)_2]$, in which M is Ce or U, so the same process applied to the Keggin ion gives $[M(PW_{11}O_{39})_2]$. Here also the atom M is in antiprismatic 8-coordination and is a 4f or 5f element, and the central atom of each half-unit may be P, As, Si, Ge, B, etc. The structure of this ion has been established in $Cs_{12}[U(GeW_{11}O_{39})_2].13-14H_2O.$[9]

Icosahedral coordination of heteroatom

The $H_6CeMo_{12}O_{42}^{2-}$ *ion*.[10] Six pairs of face-sharing octahedra may be joined together by coalescing vertices of types *A* and *B* (Fig. 11.13(a)) of different pairs. The faces shaded in Fig. 11.13(a) then become twelve of the faces of an icosahedron and the hetero-atom occupies the nearly regular icosahedral hole at the centre of the group. The Ce—O distance (2.50Å) is the same as in ceric ammonium nitrate, and in fact the 12 O atoms of the six bidentate nitrate groups in the ion $Ce(NO_3)_6^{2-}$ are arranged in the same way as the inner 12 O atoms of Fig. 11.13(b). There is a considerable range of Mo—O bond lengths, 1.68, 1.98, and 2.28 Å for bonds to O atoms belonging to 1, 2, or 3 MoO_6 octahedra.

Heteropoly ions containing Mo *and* V (*or* P)

We describe now a number of new heteropoly ions built of various numbers of

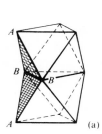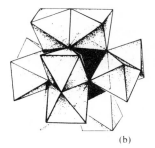

Fig. 11.13. The $[H_6CeMo_{12}O_{42}]^{2-}$ anion.

coordination polyhedra of two kinds which have for the most part much less regular structures than those we have already described. All contain Mo, and although we shall refer to MoO_6 octahedra it is to be understood that Mo is off-centre with its characteristic $(2+2+2)$-coordination (pairs of bonds of lengths around 1.7, 1.9, and 2.3 Å).

In the $P_2Mo_5O_{23}^{6-}$ ion,[11] Fig. 11.14(a), · five MoO_6 octahedra form a ring in which four of the junctions are edge-sharing and the fifth vertex-sharing. On each side of the ring a PO_4 tetrahedron shares 3 O with O atoms of the 5-membered ring. The $V_2Mo_6O_{26}^{6-}$ ion[12] consists of a ring of six MoO_6 octahedra capped on each side by VO_4 tetrahedra; there is an isostructural $Mo_8O_{26}^{4-}$ ion[13] (illustrated in Fig. 11.4(e)). The structure of the $V_8Mo_4O_{36}^{8-}$ ion[14] is not easily appreciated without a model. The basis is a ring of six octahedra, two MoO_6 and four VO_6, capped on each side by a pyramidal V group. Two additional MoO_6 octahedra and two VO_5 trigonal bipyramids complete a very compact grouping of 12 edge-sharing polyhedra.

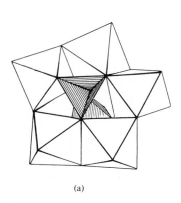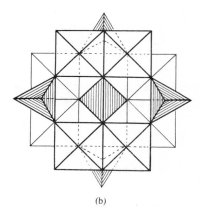

(a)

(b)

FIG. 11.14. The structures of heteropoly ions: (a) $P_2Mo_5O_{23}^{6-}$; (b) $V_5Mo_8O_{40}^{7-}$.

The description of the V coordination polyhedra is not clear-cut, and reference should be made to the original paper for details.

A much more symmetrical structure has been found for the anion in $K_7[V_5Mo_8O_{40}] . \approx 8H_2O$,[15] Fig. 11.14(b). Four pairs of MoO_6 octahedra form a puckered ring of 8 octahedra, and 4 VO_5 trigonal bipyramids each share two edges with edges of octahedra. There is a fifth V atom in a central VO_4 which is a fairly regular tetrahedron (V–O, 1.71 Å). In the VO_5 groups there are two short bonds (1.6 Å), two longer (1.85 Å), and one very long bond (2.7–2.8 Å), so that these groups could be described as distorted tetrahedral. As in all poly-ions the metal atoms are displaced within their O coordination groups in a direction away from the centre of the ion.

The possibility of the existence of other heteropoly ions of the same general type as that of Fig. 11.14(a) is suggested by the structure of the ion $[(CH_3)_2AsMo_4O_{15}H]^{2-}$, studied in its guanidinium salt,[16] which contains a ring built from only four MoO_6 octahedra. These share alternately an edge or face, as shown in the bond diagram (a), and two of the O atoms are bridged by the $(CH_3)_2As$ group.

(a)

There appears to be no limit to the complexity of heteropoly ions containing Mo (or W) and P, As, or V, and it seems appropriate to conclude this section with a note on the remarkable anion in the salt $(NH_4)_{23}[NH_4As_4W_{40}O_{140}Co_2(H_2O)_2] . nH_2O$.[17] We shall not attempt a detailed description of this ion, but some idea of its structure may perhaps be gathered from what follows. Removal of one group of 3 octahedra from the Keggin isomer $W_{12}O_{40}$ leaves a sub-unit W_9O_{33}, as noted in the description of the $H_2AsW_{18}O_{60}^{7-}$ ion on p. 525. The centre of such a unit may be occupied by As, with its lone pair directed towards the open end. If four of these AsW_9O_{33} units related by an S_4 axis are joined through additional WO_6 octahedra there is formed a cage-like structure of composition $As_4W_{40}O_{140}$ within which the 2 Co atoms are accommodated in tetrahedral interstices.

(1)   AC 1977 **B33** 1038
(1a) AC 1977 **B33** 3090
(1b) AC 1976 **B32** 1545
(1c) AC 1980 **B36** 1018
(2a) AC 1960 13 1139
(2b) CC 1973 648
(2c) ACSc 1960 14 771
(2d) AK 1963 20 324
(3)   AC 1976 **B32** 729
(3a) IC 1979 18 3010

(4a) AC 1974 **B30** 2095
(4b) IC 1970 9 2228
(4c) AC 1980 **B36** 661
(4d) AC 1980 **B36** 2530
(5)   ZK 1975 141 342
(6)   IC 1969 8 325
(7)   JACS 1969 91 6881
(8)   JCS D 1974 2021
(9)   AC 1980 **B36** 2012
(10) JACS 1968 90 3589

(11) AC 1977 **B33** 3083;
      AC 1979 **B35** 278
(12) AC 1979 **B35** 1995
(13) AnC 1976 88 385
(14) AC 1979 **B35** 1989
(15) AC 1980 **B36** 1530
(16) IC 1980 19 2531
(17) IC 1980 19 1746

12

Binary metal oxides

Introduction

We shall be largely concerned here with the structures of metal oxides in the crystalline state since nearly all these compounds are solids at ordinary temperatures. We shall mention a number of suboxides, but we shall exclude peroxides and superoxides (and ozonates), for these compounds, in which there are O—O bonds, are included in Chapter 11. Little is known of the structures of metal oxides in the liquid or vapour states, though several have been studied as vapours (Table 12.1).

TABLE 12.1

Metal oxides in the vapour state

Oxide	M—O (Å)	μ(D)	M—O—M	Method	Reference
Li_2O	1.59		180°	m.sp., i.r.	JCP 1963 **39** 2463
	1.55			m.sp., i.r.	JCP 1963 **39** 2299
LiO	1.62				
Li_2O_2	1.90		64°	m.sp., i.r.	JCP 1963 **39** 2463
Cs_2O			$\neq$180°	m.sp.	JCP 1967 **46** 605
SrO	1.92	8.91		m.sp.	JCP 1965 **43** 943
BaO	1.94	7.93		m.w.	JCP 1963 **38** 2705
MoO_3	See p. 572				
OsO_4	1.71			e.d.	ACSc 1966 **20** 385
RuO_4	1.705			e.d.	ACSc 1967 **21** 737

The structures of the oxides of the semi-metals and of the B subgroup elements are described in other chapters.

Metal oxides range from the 'suboxides' of Cs and of transition metals such as Ti and Cr, in which there is direct contact between metal atoms, and the ionic compounds of the earlier A subgroups and of transition metals in their lower oxidation states, to the covalent oxides of Cr(VI), Mn(VII), and Ru(VIII). In contrast to the basic oxides of the most electropositive metals there are the acidic CrO_3 and Mn_2O_7 which are the anhydrides of H_2CrO_4 and $HMnO_4$. As regards physical properties metal oxides range from high melting ionic compounds (HfO_2 melts at 2800 °C) to the volatile molecular RuO_4 (m.p. 25, b.p. 100 °C), and OsO_4 (m.p. 40, b.p. 101 °C), both of which are soluble in CCl_4 and H_2O, and Mn_2O_7 (m.p. 6 °C). The colours of metal oxides range over the whole spectrum, and their electrical properties show an equally wide variation, from insulators (Na_2O, CaO), through semiconductors (VO_2), to 'metallic' conductors (CrO_2, RuO_2). Binary oxides also include two of the rare examples of ferromagnetic compounds, EuO and CrO_2.

The crystal structures of metallic oxides include examples of all four main types, molecular, chain, layer, and 3D structures, though numerically the first three classes form a negligible fraction of the total number of oxides. The metals forming oxides with molecular, chain, or layer structures are distributed in an interesting way over the Periodic Table.

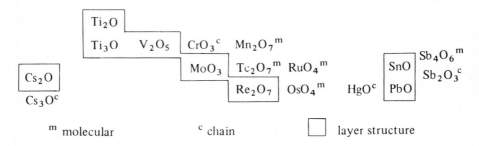

The following generalizations may be found useful, but it should be remembered that like all generalizations they are subject to many exceptions, some of which will be noted later.

(1) The majority of metal oxides have essentially ionic structures with high coordination number of the metal atom (often 6 or 8), the structures being in many cases similar to those of fluorides of the same formula type. However, the adoption of a simple structure characteristic of ionic compounds does not necessarily preclude some degree of covalent or metallic bonding — witness the NaCl structure not only of UO but also of UC and UN.

(2) The ionic radii (c.n.6) of metals other than Rb^+, Cs^+, and Tl^+ are smaller than that of O^{2-}. For example, the ionic radii of Al, Mg, and all the 3d metals lie in the range 0.5–0.8 Å. For this reason the oxygen ions in many oxides, both simple and complex, are close-packed or approximately so, with the smaller metal ions usually in octahedral holes.

(3) Comparison of Tables 12.2 and 17.2 shows that there is little resemblance between the structures of the corresponding oxides and sulphides of a particular metal except in the case of the ionic compounds of the most electropositive metals and the compounds of Be and Zn. In some cases oxides and sulphides of the same formula type do not exist or have very different stabilities. For example, PbO_2 is stable at atmospheric pressure but PbS_2 can only be made under higher pressures. In the case of iron the disulphide FeS_2 has the pyrites and marcasite structures, both containing S_2 groups, but FeO_2 is not known. A comparison of the highest known oxides of the Group VIII metals emphasizes the individuality of the elements and the difficulty of making generalizations:

$$\begin{array}{lll} Fe_2O_3 & Co_3O_4 & NiO \\ RuO_4 & Rh_2O_3 & PdO \\ OsO_4 & IrO_2 & PtO_2 \end{array}$$

(4) In contrast to earlier views on metal oxides, detailed diffraction studies and measurements of density and other properties have shown that many metal–oxygen systems are complex. In addition to oxides with simple formulae such as M_2O, MO, M_2O_3, MO_2, etc. containing an element in one oxidation state and oxides M_3O_4 containing $M(\text{II})$ and $M(\text{III})$ or $M\text{IV}$ there are numerous examples of more complex stoichiometric oxides containing a metal in two oxidation states, for example

$$U_3O_8 \qquad U_4O_9 \qquad Mn_5O_8 \qquad V_6O_{13} \qquad Tb_7O_{12}$$
$$Cr_5O_{12} \qquad Pr_6O_{11}.$$

Some metals form extensive series of oxides such as Ti_nO_{2n-1} or Mo_nO_{3n-1} with structures related to simple oxides MO_2 or MO_3. Many transition-metal oxides show departures from stoichiometry leading to semiconductivity and others have interesting magnetic and electrical properties which have been much studied in recent years. We shall illustrate some of these features of oxides by dealing in some detail with selected metal–oxygen systems and by noting peculiarities of certain oxides.

The structures of binary oxides

The more important structure types are set out in Table 12.2; others are mentioned later. Most of the simpler structures have already been described and illustrated in earlier chapters. In Table 12.2 the structures are arranged according to the type of complex in the crystal (3-, 2-, or 1-dimensional or finite) and for each structure the coordination number of the metal atom is given before that of oxygen. In a particular family of oxides the changes in structure type may be related in a general way to the change in bond type from the essentially ionic structures through the layer and chain structures to the essentially covalent molecular oxides. These changes may be illustrated by the structures of the dioxides of the elements of the fourth Periodic Group. After each compound we give the c.n.s of M and O and the structure type. Starting with the molecular CO_2 we pass through the silica structures with ionic-covalent bonds to the ionic structures of the later metal dioxides.

$$CO_2 \ (2:1) \ \text{Molecular}$$
$$SiO_2 \ (4:2) \ \text{various 3D networks}$$

TiO_2	$(6:3)$ rutile, etc.	GeO_2	$\begin{cases} (4:2) \\ (6:3) \end{cases}$	quartz / rutile
ZrO_2 $\Big\}$				
HfO_2 $\Big\}$	$(8:4)$ fluorite	SnO_2	$(6:3)$	rutile
ThO_2 $\Big\}$				
		PbO_2	$\begin{cases} (6:3) \\ (6:3) \end{cases}$	rutile / columbite

The structures of these compounds may be compared with those of the corresponding disulphides, which are set out in a similar way at the beginning of Chapter 17.

TABLE 12.2

The crystal structures of metal oxides

Type of structure	Formula type and coordination numbers of M and O		Name of structure	Examples
Infinite 3-dimensional complex	MO_3	6 : 2	ReO_3	WO_3
	MO_2	8 : 4	Fluorite	ZrO_2, HfO_2, CeO_2, ThO_2, UO_2, NpO_2, PuO_2, AmO_2, CmO_2, PoO_2, PrO_2, TbO_2
		6 : 3	Rutile	TiO_2, VO_2, NbO_2, TaO_2, CrO_2, MoO_2, WO_2, MnO_2, TcO_2, ReO_2, RuO_2, OsO_2, RhO_2, IrO_2, PtO_2, GeO_2, SnO_2, PbO_2, TeO_2
	M_2O_3	6 : 4	Corundum	Al_2O_3, Ti_2O_3, V_2O_3, Cr_2O_3, Fe_2O_3, Rh_2O_3, Ga_2O_3
		See text	$\left.\begin{array}{l} A\text{-}M_2O_3 \\ B\text{-}M_2O_3 \end{array}\right\}$	4f oxides
		6 : 4	$C\text{-}M_2O_3 \Big\}$	Mn_2O_3, Sc_2O_3, Y_2O_3, In_2O_3, Tl_2O_3
	MO	6 : 6	Sodium chloride	MgO, CaO, SrO, BaO, TiO, VO, MnO, FeO, CoO, NiO, CdO, EuO,
		4 : 4	Wurtzite	BeO, ZnO
	MO_2	4 : 2	Silica structures	GeO_2
	M_2O	2 : 4	Cuprite	Cu_2O, Ag_2O
		4 : 8	Antifluorite	Li_2O, Na_2O, K_2O, Rb_2O
Layer structures				MoO_3, As_2O_3, PbO, SnO, Re_2O_7
Chain structures				HgO, SeO_2, CrO_3, Sb_2O_3
Molecular structures				RuO_4, OsO_4, Tc_2O_7, Sb_4O_6

Note: this table does not distinguish between the most symmetrical form of a structure and distorted variants, superstructures, or defect structures; for more details the text should be consulted.

After noting some suboxides which are peculiar to Rb and Cs we shall describe the crystal structures of metal oxides in groups in the following order: M_3O, M_2O, MO, MO_2, MO_3, MO_4; M_2O_3, M_2O_5, M_2O_7, M_3O_4, and miscellaneous oxides M_xO_y. The remainder of the chapter is devoted to brief surveys of the oxides of certain metals which present points of special interest.

Suboxides of Rb *and* Cs

The elements Rb and Cs are notable for forming a number of suboxides which include Rb_6O and Rb_9O_2,[1] Cs_7O,[2] Cs_4O,[3] $Cs_{11}O_3$,[4] $Rb_7Cs_{11}O_3$,[5] and a metastable $Rb_{19}O_3$. We describe the structure of Cs_3O on p. 536. The two stable Rb oxides contain Rb_9O_2 groups consisting of two ORb_6 octahedra sharing one face (that is, an anti-M_2X_9 group). In Rb_6O layers of these groups alternate with layers of c.p. metal atoms, so that the structural formula is $(Rb_9O_2)Rb_3$. The oxide Rb_9O_2, which is copper-red with metallic lustre, consists of these Rb_9O_2 groups held together by Rb–Rb bonds of length 5.11 Å, rather longer than those in the element (4.85 Å) and much longer than the shortest Rb–Rb bonds within the groups (3.52 Å).

In contrast to the Rb compounds, all three Cs suboxides, the violet-blue $Cs_{11}O_3$, violet Cs_4O, and bronze Cs_7O, contain O_3Cs_{11} units (Fig. 12.1) formed from three

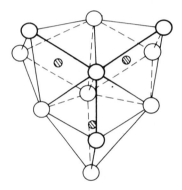

FIG. 12.1. The O_3Cs_{11} unit.

octahedral OCs_6 groups each sharing two adjacent faces to produce a trigonal group similar to the sub-unit in Nb_3S_4 (p. 770). The oxides Cs_4O and Cs_7O consist of these units together with additional Cs atoms, the structural formulae being therefore $(Cs_{11}O_3)Cs$ and $(Cs_{11}O_3)Cs_{10}$ respectively. The bond lengths are interpreted to indicate ionic bonds within the O_3Cs_{11} complexes (Cs–O, 2.7–3.0 Å) and metallic bonds between them. The Cs–Cs distances within the complex are the lengths of edges of OCs_6 octahedra and are therefore determined by Cs–O; the Cs–Cs distances outside the complexes are similar to those in metallic Cs; in Cs_7O these distances are respectively 3.76 Å and 5.27 Å. Of special interest is the oxide $Rb_7Cs_{11}O_3$, formed either from $Cs_{11}O_3$ and Rb or from a Rb-Cs alloy. It consists of $Cs_{11}O_3$ groups separated by layers of Rb atoms, and the structure is very similar to that of $Cs_{10}(Cs_{11}O_3)$.[5] Not only do the suboxides of Rb and Cs have quite different formulae and structures, but in $Rb_7Cs_{11}O_3$ there is separation of atoms of the chemically similar metals into positions with very different environments. The metallic appearance and conductivity of the Cs oxides suggest description as a Cs matrix in which O atoms (ions) occupy various proportions and arrangements of octahedral interstices.

(1)   ZaC 1977 **431** 5   (4)   ZaC 1977 **428** 187
(2)   ZaC 1976 **422** 208   (5)   IC 1978 **17** 875
(3)   ZaC 1976 **423** 203

Oxides M_3O

Two structures established for suboxides M_3O are 'anti' structures of halides MX_3. In Cs_3O,[1] which forms dark-green crystals with metallic lustre, there are columns of composition Cs_3O consisting of OCs_6 octahedra each sharing a pair of opposite faces (Fig. 12.2); this is the anti-ZrI_3 structure. The Cs–O bond length (2.89 Å) is

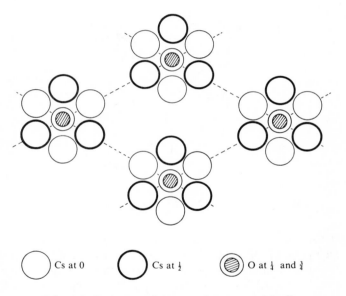

Cs at 0   Cs at $\frac{1}{2}$ O at $\frac{1}{4}$ and $\frac{3}{4}$

FIG. 12.2. Projection of the crystal structure of Cs_3O.

similar to that in Cs_2O. The Cs–Cs bonds between the chains (around 5.8 Å) are somewhat longer than in the element (5.26 Å), but in the distorted octahedra in the chains the lengths of the edges are 3.80 Å (six) and 4.34 Å (six). The former Cs–Cs distance (omitted from ref. 1) is similar to that in Cs_7O.

The oxide Ti_3O[2] has a structure closely related to the anti-BiI_3 layer structure which has been described in Chapter 4.

For Cr_3O, Mo_3O, and W_3O see p. 570.

(1)   JPC 1956 **60** 345   (2)   AC 1968 **B24** 211

Oxides M_2O

The structure adopted by the alkali-metal oxides M_2O and sulphides M_2S *other than those of* Cs is the anti-fluorite structure, with $4:8$ coordination. In this structure the alkali-metal ions occupy the F^- positions and the anions the Ca^{2+} positions of the CaF_2 structure. We have commented in Chapter 7 on the structures of Cs compounds. The unique structure of Cs_2S is described on p. 750. The structure of the orange-yellow Cs_2O is quite different from that of the other alkali-metal oxides and more closely related to Cs_3O. This oxide has the anti-$CdCl_2$ (layer) structure,[1] with Cs–Cs distances of 4.19 Å between layers and Cs–O distances (2.86 Å) within the layers, similar to those in Cs_3O. There is considerable polarization of the large Cs^+ ions.

Other oxides with closely related structures are Ti_2O, p. 562, (anti-CdI_2 structure) and Tl_2O, p. 1172 (anti-CdI_2 polytype).

The crystal structure of Cu_2O (and the isostructural Ag_2O) has been described in Chapter 3 (p. 127) together with other structures related to the diamond net.

Evidence for the existence of In_2O in the solid state is not satisfactory,[2] and the compound described as Sm_2O is almost certainly SmH_2.[3]

(1)   JPC 1956 **60** 338   (3)   IC 1968 7 660
(2)   ACSc 1966 **20** 1996

Oxides MO

The structures of metallic monoxides are set out in the self-explanatory Table 12.3.

TABLE 12.3
Crystal structures of monoxides

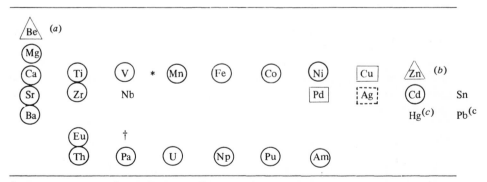

△ Wurtzite structure. ○ NaCl structure. □ PtS structure or variant.
(a) Also a high-temperature polymorph.
(b) NaCl structure at high pressure (4).
(c) Polymorphic.
* There is no reliable evidence for the existence of CrO.
† For h.p. lanthanide oxides see text.

The symbols mean that the structure of a particular compound is of the general type indicated and is not necessarily the ideal structure with maximum symmetry. An interesting point about the wurtzite structure is that unless $4a^2 = (12u-3)c^2$, where u is the z coordinate of O referred to Zn at (000), there is a small but real difference between one M—O distance and the other three, for example, in ZnO, 1.973 Å (3) and 1.992 Å (1).[1] For the ideal structure with regular tetrahedral bonds $u = 3/8$, and $c/a = 2\sqrt{2}/\sqrt{3} = 1.633$.

It seems probable that EuO is the only lanthanide monoxide which can be prepared pure under atmospheric pressure, but under higher pressure (10–80 kbar) and at temperatures between 500 and 1200 °C all the following oxides (NaCl structure) have been prepared: LaO, CeO, PrO, NdO,[2a] SmO,[2b] and YbO.[3] Both EuO and YbO are semi-conductors and are ionic oxides of the divalent metals. The other oxides are golden-yellow with metallic properties and contain the metal in the +3 oxidation state, $M^{3+}O^{2-}$(e), except SmO, the properties of which indicate the oxidation state +2.9.

Whereas the oxides of Mg, the alkaline-earths, and Cd have the normal NaCl structure those of certain transition metals depart from the ideal structure in the following ways.

(a) The structure has full (cubic) symmetry only at higher temperatures, transition to a less symmetrical variant taking place at lower temperatures, namely, FeO, NiO, MnO (rhombohedral), and CoO (tetragonal). For references see PR 1951 **82** 113.

(b) The oxide has a defect structure, that is, there is not 100 per cent occupancy of both anion and cation sites. At atmospheric pressure FeO is non-stoichiometric (see p. 551). Stoichiometric NiO is pale-green (like Ni^{2+} in aqueous solution) and is an insulator. By heating in O_2 or incorporating a little Li_2O it becomes grey-black and is then a semiconductor (due to the presence of Ni^{2+} and Ni^{3+} in the same crystal.[5])

The oxides TiO, NbO, and VO show appreciable ranges of composition, having either a superstructure or a defect structure at the composition MO. The oxides of Ti and V are discussed later in this chapter. The structure of NbO is described shortly.

The larger unit cells necessary to account for the neutron diffraction data from antiferromagnetic compounds such as MnO are due to the antiparallel alignment of the magnetic moments of the metal ions. They are not to be confused with the larger unit cells of many superstructures which are due to ordered as opposed to random arrangement of atoms of two or more kinds.

Table 12.3 shows that the majority of monoxides adopt the NaCl structure. Other structures are peculiar to one oxide or to a small number of oxides; they include the following.

BeO. In the high-temperature (β) form of BeO[6] pairs of BeO_4 tetrahedra share an edge and these pairs are then joined through vertices (Fig. 12.3). The arrangement of O atoms is essentially the same as in rutile (see Fig. 6.6) and since β-BeO is approximately h.c.p. it is closely related to the wurtzite structure of low-temperature BeO. One structure is converted into the other by moving one-half of the Be atoms into adjacent tetrahedral holes. Owing to the edge-sharing by BeO_4 tetrahedra in

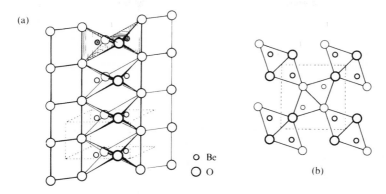

FIG. 12.3. The structure of β-BeO.

β-BeO there are necessarily short Be—Be distances (2.24 Å), the same as in metallic beryllium, but this does not imply metal–metal bonding. γ-LiAlO₂ has the same structure.

NbO. This oxide is unique in having the structure of Fig. 6.1(d) (p. 239)[7] in which both Nb and O form four coplanar bonds, occupying alternate positions in one of the simplest 3D 4-connected nets. Alternatively the structure is described as a defect NaCl structure, having 3 NbO in the unit cell with vacancies at (000) and ($\frac{1}{2}\frac{1}{2}\frac{1}{2}$). Note the existence of a 3D framework built from octahedral Nb₆ units (Nb—Nb, 2.98 Å) reminiscent of the halide complexes of Nb and Ta (p. 432). This structure has been confirmed by neutron diffraction.

PdO. In this structure also (which is that of PtS) the metal atoms form four coplanar bonds; the coordination of O is tetrahedral (Fig. 12.4). As noted in

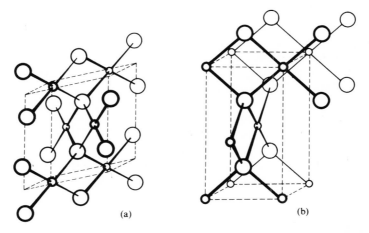

FIG. 12.4. The crystal structures of: (a) CuO (tenorite) and AgO; and (b) PdO and PtO.

Chapter 1 the bond angles represent a compromise between the ideal values of 90°
and $109\frac{1}{2}°$ due to purely geometrical factors. CuO has a less symmetrical variant of
the PtS structure, and it is interesting that whereas CuO has the less symmetrical
(monoclinic) structure and PtO apparently does not exist yet solid solutions
$Cu_xPt_{1-x}O$ may be prepared (under high pressure) which have the tetragonal PtS
structure $(0.865 > x > 0.645)$.[8] The structures of CuO and AgO are described in
more detail in Chapter 25.

For the structures of the other B subgroup monoxides see Chapters 25 and 26;
for PbO see also p. 557.

(1) AC 1969 **B25** 1233, 2254
(2a) JSSC 1981 **36** 261
(2b) IC 1980 **19** 2252
(3) JSSC 1979 **27** 29
(4) Sc 1962 **137** 993

(5) JPC 1961 **65** 2154
(6) AC 1965 **18** 393
(7) AC 1966 **21** 843
(8) JLCM 1969 **19** 209

Oxides MO_2

Many dioxides crystallize with one of two simple structures, the larger M^{4+} ions
being 8-coordinated in the fluorite structure and the smaller ions 6-coordinated in
the rutile structure. The term 'rutile-type' structure in Table 12.4 includes the most

TABLE 12.4
Crystal structures of dioxides

Ti*	V*	Cr	Mn*			rutile-type	Ge*	
Zr*	Nb*	Mo	Tc	Ru	Rh†	structure	Sn	
Hf*	Ta	W	Re*	Os	Ir	Pt†	Pb*	Po
	Ce	Pr			Tb		*fluorite*	
	Th	Pa	U	Np	Pu	Am	Cm *structure*	

*Polymorphic. † Amorphous when prepared under atm. press. but crystallizes under pressure.
For a survey (with many references) see: IC 1969 **8** 841.

symmetrical (tetragonal) form of this structure and the less symmetrical variants
referred to later—as indicated by the broken lines in the Table. Polymorphism is
common; for example, PbO_2 and ReO_2 also crystallize with the α-PbO_2 structure
(p. 175), and GeO_2 with the α-quartz structure, while TiO_2 and ZrO_2 are trimorphic
at atmospheric pressure. TiO_2 also adopts the α-PbO_2 structure under pressure, and
under shock-wave pressures greater than 150 kbar it converts to a still more dense
form which is not, however, like the α-PbO_2 structure, retained after release of
pressure.[1] PtO_2 is a high-pressure compound with two forms, α (hexagonal,
structure not known) and β with the $CaCl_2$ structure.[2] This is the only dioxide

known to have this structure, which is closer to the ideal h.c.p. structure than is the tetragonal rutile structure (see Chapter 4 and 6).

Although the coordination group around the metal ion in the tetragonal rutile structure approximates closely to a regular octahedron accurate determinations of the M—O distances show small differences. For example, in TiO_2,[3] Ti—4 O, 1.944 (0.004) and Ti—2 O, 1.988 (0.006) Å, and in RuO_2,[4] Ru—4 O, 1.917 (0.008) and Ru—2 O, 1.999 (0.008) Å. In the rutile (β) form of PbO_2, which is a major constituent of lead-acid batteries, the axial bonds (2.150 Å) are shorter than the equatorial bonds (2.169 Å).[5] Distortion of the octahedral coordination group is not confined to the rutile structure, for there are comparable differences in anatase,[6a] 1.934 and 1.980 Å, while in brookite[6b] there is a much less symmetrical environment of the cations, Ti—O ranging from 1.87–2.04 Å.

In the normal rutile structure each M is equidistant from two others in a chain of octahedra. The oxides enclosed within the broken lines in Table 12.4 have the tetragonal rutile structure as the high-temperature form (except ReO_2, the h.t. form of which has orthorhombic symmetry); below their transition temperatures VO_2, NbO_2, and ReO_2 have less symmetrical structures in which successive pairs of M atoms are alternately closer together and further apart. This less symmetrical (monoclinic) structure is the normal form of MoO_2, WO_2, and TcO_2, and the low-temperature form of VO_2 and ReO_2; the low-temperature form of NbO_2 has a complex (tetragonal) superstructure of the rutile type, also with alternating M—M distances in the chains. In Table 12.5 the high-temperature forms of the dimorphic dioxides are listed first. The close approach of pairs of M atoms in the distorted rutile structure can give rise to metallic conductivity and anomalously low paramagnetism, but the relations between structure, number of d electrons, and physical properties are complex. For example, RuO_2 has a high electrical conductivity but also a low paramagnetic susceptibility, while the low-temperature form of VO_2

TABLE 12.5
Metal–metal distances in some dioxides

Oxide		Number of d electrons in metal	M—M in chain (Å)		Reference
TiO_2		0		2.96	
VO_2	(tetragonal)	1		2.85	JPSJ 1967 **23** 1380
	(monoclinic)		2.62	3.17	ACSc 1970 **24** 420
NbO_2	(tetragonal)	1		(3.0)	AC 1979 **B35** 2836
	(superstructure)		2.71	3.30	
MoO_2		2	2.51	3.10	ACSc 1967 **21** 661
WO_2		2	2.48	3.10	AC 1979 **B35** 2199
TcO_2		3	2.5	3.1	} ACSc 1955 **9** 1378
ReO_2	(monoclinic)	3	2.5	3.1	
RuO_2				3.107	ACSc 1970 **24** 116
OsO_2				3.184	ACSc 1970 **24** 123

is a semi-conductor. If the very short M—M bonds of length close to 2.5 Å found in certain of these dioxides are multiple bonds it is necessary to recognize at least four types of structure in this family, namely, the undistorted structure (with or without M—M interactions) and distorted structures (with single or multiple M—M bonds). On this point see also p. 248.

The polymorphism of ZrO_2 presents several points of interest. The normal (monoclinic) form, the mineral baddeleyite, changes at around 1100°C to a tetragonal form and at around 2300°C to a cubic form (fluorite structure). The transition at 1100°C is of practical importance since it restricts the use of the pure oxide as a refractory material, thermal cycling through this temperature region causing cracking and disintegration. Solid solutions with the fluorite structure containing CaO or MgO are not subject to this change.[7] The tetragonal form cannot be quenched to room temperature, but tetragonal ZrO_2 can exist at room temperature if prepared by precipitation from aqueous solution or by calcining salts at a low temperature. It is apparently stabilized by the higher surface energy arising from the smaller particle size, which must not exceed 300 Å.[8] This tetragonal form[9] has a distorted fluorite structure. Instead of 8 O at 2.20 Å (the distance from Zr^{4+} to eight equidistant O^{2-} neighbours in the cubic high-temperature form) there are two sets of 4 (at 2.065 and 2.455 Å) forming flattened and elongated tetrahedral groups—a distorted cubic arrangement similar to that in $ZrSiO_4$. In monoclinic ZrO_2[10] Zr^{4+} is 7-coordinated and equal numbers of O^{2-} are 3- and 4-coordinated. The 3-coordinated ions (O_I) have a practically coplanar arrangement of 3 Zr neighbours (Zr—O, 2.07 Å) with interbond angles of 104°, 109°, and 143°, while the remainder (O_{II}) have 4 tetrahedral neighbours at a mean distance of 2.21 Å. All the interbond angles lie within the range 100–108° except one (134°). The 7-coordination group of Zr^{4+} is shown in idealized form in Fig. 12.5; the next

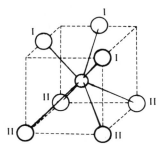

FIG. 12.5. The coordination of Zr^{4+} in monoclinic ZrO_2 (idealized).

nearest O^{2-} neighbour is at a distance of 3.58 Å, and accordingly the 7-coordination group is well defined.

The oxynitride TaON[11] is of interest in this connection. It has a low-temperature form with a complex hexagonal structure, but the high-temperature form has the baddeleyite structure with an ordered arrangement of (3-coordinated) O atoms in the positions I of Fig. 12.5 and the (4-coordinated) N atoms in the positions II.

The relation of the C-M_2O_3 structure to fluorite is described on p. 545. The structures of certain oxides of 4f and 5f elements are related in a more complex way to the structures of their dioxides (fluorite structure). For example, PrO_2 loses oxygen on heating to form Pr_6O_{11} with a fluorite-like structure from which one-twelfth of the anions have been removed at random.[12] Terbium forms, in addition to TbO_2, $TbO_{1.715}$, $TbO_{1.809}$, and $TbO_{1.823}$,[13] of which the first (Tb_7O_{12}) is noted on p. 607 in connection with the isostructural complex oxides $M'M''_6O_{12}$. PuO_2 loses oxygen to form substoichiometric oxides,[14] but in contrast UO_2 adds oxygen at high temperatures. The oxides of uranium are discussed in more detail on pp. 1263–5.

(1)	JCP 1969 **50** 319	(8)	JPC 1965 **69** 1238
(2)	JINC 1969 **31** 3803	(9)	AC 1962 **15** 1187
(3)	AC 1956 **9** 515	(10)	AC 1965 **18** 983
(4)	IC 1966 **5** 317	(11)	AC 1974 **B30** 809
(5)	AC 1980 **B36** 2394	(12)	JACS 1950 **72** 1386
(6a)	ZK 1972 **136** 273	(13)	JACS 1961 **83** 2219
(6b)	AC 1961 **14** 214	(14)	JINC 1965 **27** 541
(7)	PRS 1964 **279A** 395		

Oxides MO_3

These are few in number and include CrO_3, MoO_3, WO_3, ReO_3, TeO_3, and UO_3. The (tetrahedral) chain structure of CrO_3 is described under the structural chemistry of Cr(vi) in Chapter 27. One of the forms of TeO_3 is isostructural with FeF_3 (p. 717). The very simple ReO_3 structure is described on p. 210, and the structures of MoO_3 and WO_3 are described later in this chapter. For UO_3 see p. 1264.

Oxides MO_4

There are only two metal tetroxides, the volatile and low-melting RuO_4 and OsO_4. Both exist as regular tetrahedral molecules in the crystalline and vapour states:

	M–O (Å)	*Reference*
RuO_4	1.705 (vapour)	ACSc 1967 **21** 737
OsO_4	1.71 (vapour)	ACSc 1966 **20** 385
	1.74 (crystal)	AC 1965 **19** 157

Oxides M_2O_3

Of the structures listed in Table 12.2 only two are of importance for elements other than the 4f and 5f metals; these are the corundum (α-Al_2O_3) and the C-M_2O_3 structures. Many oxides M_2O_3 are polymorphic, including most (if not all) of the 4f and 5f oxides. References to recent work, particularly refinements of earlier structures, are included in Table 12.6.

TABLE 12.6
Crystal structures of sesquioxides

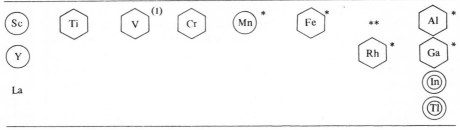

○ *C* structure.◎ *C* structure under atm. pressure; corundum structure under high pressure.

⬡ Corundum structure.

* Also other polymorphs.
(1) Changes to a semiconducting form at low temperatures.
** There is no evidence for the existence of Co_2O_3 or Ni_2O_3.

Crystal structures of 4f *and* 5f *sesquioxides*

La	Ce	Pr	Nd	Sm	Eu	Gd	Tb	Dy	Ho----Lu	U	Np	Pu	Am	Cm	Bk	Cf
		A										A	A	A	B/A	A
	?	?				B			under pressure							
?	?	?				C						C	C	C		

Corundum structure: AC 1969 **8** 1985; Rh_2O_3, AC 1970 **B26** 1876; In_2O_3, ACSc 1967 **21** 1046; Fe_2O_3, AM 1966 **51** 123; α-Ga_2O_3, JCP 1967 **46** 1862.
A-M_2O_3 structure: AC 1953 **6** 741; Pr_2O_3, IC 1963 **2** 791; Nd_2O_3, AC 1975 **B31** 2745; Bk_2O_3, Cf_2O_3, JINC 1973 **35** 4149.
B-M_2O_3 structure: Gd_2O_3, AC 1958 **11** 746; Sm_2O_3, JPC 1957 **61** 753; Tb_2O_3, ZaC 1968 **363** 145.
C-M_2O_3 structure: In_2O_3, AC 1966 **20** 723; Mn_2O_3, ACSc 1967 **21** 2871; Y_2O_3, AC 1969 **B25** 2140; Sc_2O_3, ZK 1967 **124** 136.
4f sesquioxides: IC 1965 **4** 426; IC 1966 **5** 754; IC 1969 **8** 165.
Other structures: β-Ga_2O_3, JCP 1960 **33** 676; Rh_2O_3 (high temp.), IC 1963 **2** 972; (high press.), JSSC 1970 **2** 134; Au_2O_3, AC 1979 **B35** 1435.

The corundum structure is an approximately h.c.p. array of O atoms in which Al^{3+} ions occupy two-thirds of the octahedral holes. In one sense the structure is surprisingly complex, for there is sharing of vertices, edges, and faces of AlO_6 coordination groups. Models of this and other c.p. structures suggest that this structure is adopted in preference to other geometrically possible ones because the arrangement of 4 Al^{3+} ions around O^{2-} approximates most closely to a regular tetrahedral one. The actual Al—O—Al bond angles in corundum, 85° (one), 94° (two), 120° (one), and 132° (two), may be compared with the 'ideal' values for vertex-, edge-, and face-sharing given on p. 190. We noted on p. 191 that for purely geometrical reasons a 3D M_2X_3 structure with regular octahedral coordination of M *and* regular tetrahedral coordination of X is not possible. Because faces are shared between pairs of AlO_6 octahedra in this structure there are two sets of M—X distances; a refinement of the structure of α-Fe_2O_3 gives Fe—3 O, 1.945 Å and Fe—3 O, 2.116 Å. (For β- and γ-Fe_2O_3 see p. 552.) Ti_2O_3 has the corundum structure at all temperatures but undergoes a change from semiconductor to metal in the temperature range 400–600 K. This is associated with a small *increase* in the Ti—Ti distance across shared octahedral faces (from 2.579 Å at 23 °C to 2.725 Å at 595 °C); the subtle structural changes are consistent with Goodenough's band-crossing model.[1]

The structures of three forms of Rh_2O_3 have been determined, the normal (corundum) form, a high-temperature orthorhombic form, and a high-pressure polymorph made at 1200 °C under 65 kbar. In this high-pressure form there are pairs of face-sharing octahedra (as in corundum) but a different selection of shared edges; the structure may be regarded as built from slices of the corundum structure.

The C-M_2O_3 structure is related to that of CaF_2, from which it may be derived by removing one-quarter of the anions (shown as dotted circles in Fig. 12.6) and then rearranging the atoms somewhat. The 6-coordinated M atoms are of two types. Instead of 8 neighbours at the vertices of a cube, two are missing. For one-quarter of the M atoms these two are at the ends of a body-diagonal, and for the remainder

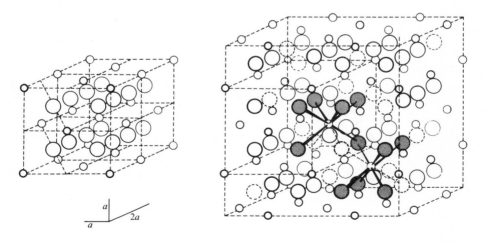

FIG. 12.6. The C-M_2O_3 structure of α-Mn_2O_3 and other sesquioxides, showing its relationship to the fluorite structure (left). The two kinds of coordination group of the metal ions are shown in the right-hand diagram.

at the ends of a face-diagonal. Both coordination groups may be described as distorted octahedra. The O atoms are 4-coordinated, and in this structure also it is probably the approximately regular tetrahedral coordination of O^{2-} which is responsible for the less regular coordination of the metal ions. In contrast to the corundum structure the C-M_2O_3 structure is an assembly of *edge*-sharing (distorted) octahedral coordination groups. In fluorite each cubic MX_8 coordination group shares an edge with each of 12 neighbouring MX_8 groups. Removal of 2 X at the ends of either a face-diagonal or body-diagonal leaves intact six of the original cube edges, so that in the C-M_2O_3 structure each octahedral coordination group shares six edges.

Not all compounds with the C-M_2O_3 structure have the most symmetrical (cubic) form of the structure, which is that of the mineral bixbyite, $(Fe,Mn)_2O_3$. In (cubic) In_2O_3 the first type of metal ion has 6 equidistant O neighbours (at 2.18 Å) and the other type has 2 O at each of the distances 2.13, 2.19, and 2.23 Å (mean 2.18 Å), and there is some distortion of the tetrahedral arrangement of neigbours around O^{2-} (four angles of 100° and two of 126°). There is much less regular coordination of M^{3+} in (synthetic) Mn_2O_3, which is not cubic. Each metal ion has 4 closer neighbours (at approximately 1.96 Å) and 2 more distant (at 2.06 and 2.25 Å for the two types of metal ion); compare the similar environment of Mn^{3+} (d^4) in $ZnMn_2O_4$, 4 O at 1.93 and 2 O at 2.28 Å.

The A-M_2O_3 and B-M_2O_3 structures are associated particularly with the 4f and 5f sesquioxides. When there is polymorphism the A, B, and C structures are characteristically those of high-, medium-, and low-temperature forms respectively. The structures of some 4f and 5f oxides are set out in Table 12.6. A striking feature of the A-M_2O_3 structure, which has been confirmed by X-ray, n.d., and e.d.,[2] is the unusual 7-coordination of the metal atoms (Fig. 12.7). In La_2O_3 the nearest oxygen neighbours of La^{3+} are 3 at 2.38, 1 at 2.45, and 3 at 2.72 Å.

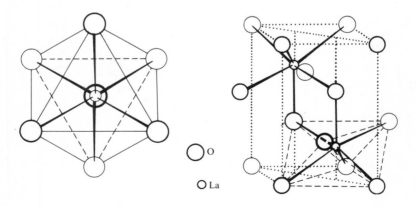

FIG. 12.7. The A-M_2O_3 structure of La_2O_3 and rare-earth sesquioxides. At the left is shown in projection the 7-coordination group around a M^{3+} ion, consisting of an octahedral group of O^{2-} ions with an additional O^{2-} ion (heaviest circle) above one of the octahedron faces.

In the rather complex B-M_2O_3 structure there are three kinds of non-equivalent metal ion, some with 6-coordination (octahedral) and the remainder with 7-coordination (monocapped trigonal prism). The seventh neighbour is appreciably more distant than the six at the vertices of the prism—at 2.73 Å (mean) as compared with 2.39 Å (in Sm_2O_3). On the basis of density differences the expected order of stability with increasing pressure would be

$$C \rightarrow \text{corundum} \rightarrow B$$

	C	corundum	B
c.n. of M	6	6	6 and 7

In fact, In_2O_3 and Tl_2O_3 (C structure) do change to the corundum structure under pressure, though the isostructural oxides of Mn, Sc, and Y, containing ions of comparable size, do not form the corundum structure but go straight to the B structure. The corundum structure is the high-pressure structure of Ga_2O_3 and also the structure at ordinary temperatures, but at high temperatures this oxide transforms to yet another sesquioxide structure. In this β-Ga_2O_3 structure one-half of the metal ions occupy tetrahedral and the remainder octahedral holes in an approximately cubic closest packing of O^{2-} ions. The same structure is adopted by θ-Al_2O_3; other polymorphs of this oxide are noted in a later section on the oxides of Al.

The structures of Sb_2O_3, Bi_2O_3, Pb_2O_3, and Au_2O_3 are described in other chapters.

(1)　PR 1960 117 1442　　　　　　　(2)　JSSC 1980 34 39

Oxides M_2O_5

We deal here with the Group VA pentoxides. The pentoxides of As, Sb, and Bi are described on p. 900, and Pa_2O_5 and $UO_{2.6}$ on pp. 1253 and 1263 respectively. A compound which may be Re_2O_5 has been described but not fully characterized.[1]

The structure of V_2O_5 is described under the oxygen chemistry of V (p. 565); there is a characteristic 5-coordination with one very short V—O bond. In marked contrast to V_2O_5 is the complexity of the crystal chemistry of Nb_2O_5 and Ta_2O_5. Many Roman and Greek letters have been used to distinguish the numerous polymorphs of Nb_2O_5,[2] in all of which there is sharing of vertices and edges of octahedral NbO_6 groups in various combinations. We refer the reader first to Chapter 5 (p. 226) and then to the discussion of complex oxides of Ti, V, Nb, etc. in Chapter 13 (p. 608) where we describe the formation of more complex structures from slices or blocks of ReO_3-like structure. The reason for the complexity of the crystal chemistry of Nb_2O_5 is that these sub-units can be joined together in an indefinitely large number of ways. Perhaps these 'forms' should be regarded as polytypes rather than polymorphs, since they result from a particular kind of growth mechanism and in some cases may be stabilized by foreign atoms. In this respect they have more in common with the polytypes of, for example, SiC or ZnS rather than with normal polymorphs. The sharing of edges of octahedral coordination groups reduces the O:M ratio below 3:1, and the same processes that give the exact ratio $5:2$ also give oxides such as $Nb_{12}O_{29}$, $Nb_{22}O_{54}$, and $Nb_{25}O_{62}$ and numerous oxides of Mo and W intermediate between MO_2 and MO_3. We give here only a few examples. The ReO_3 blocks can be joined as in Fig. 5.38 or as in Fig. 5.39; in the latter case tetrahedral holes are formed along the junction lines of the ReO_3 blocks. Examples of the first type include P-Nb_2O_5,[3a] M-Nb_2O_5,[3b] and N-Nb_2O_5.[3c] The structure of H-Nb_2O_5[3d] is described on p. 611. Figure 12.8(a) shows a projection of the relatively simple structure of P-Nb_2O_5. The structure of a h.p. high-temperature form, T-Nb_2O_5, has also been determined.[3e]

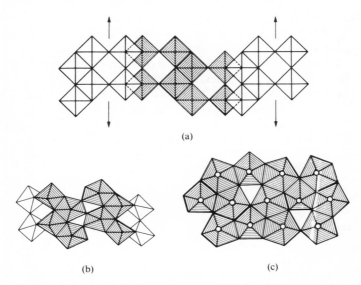

(a)

(b)   (c)

FIG. 12.8. The structures of crystalline oxides: (a) P-Nb_2O_5; (b) H-Ta_2O_5; (c) α-U_3O_8. The chains of vertex-sharing octahedra or pentagonal bipyramids are normal to the plane of the paper.

There are at least two structurally distinct forms of Ta_2O_5, with a transition point at $1360\,^{\circ}C$. As first prepared, by heating Ta in air or oxygen at $600\,^{\circ}C$, Ta_2O_5 is poorly crystalline, but the crystallinity is increased by heating at $1350\,^{\circ}C$ and the exact structure (as indicated by the superstructure lines) depends on the heat treatment. After extensive heat treatment the structure reaches an equilibrium state. (This is also true of a number of phases containing W up to the limiting composition $11\,Ta_2O_5 . 4\,WO_3$.) The structures of the low-temperature form, L-Ta_2O_5,[4] and of all these phases are of the same general type, consisting of chains built from octahedral or pentagonal bipyramidal groups sharing two opposite vertices, these chains being joined by vertex- or edge-sharing to give the 3D structure. The high-temperature form, H-Ta_2O_5, apparently undergoes various phase transitions on cooling, but the tetragonal structure can be stabilized by the addition of 2–4 mole per cent of Sc_2O_3. The structure[5] Fig. 12.8(b) is of the same general type as L-Ta_2O_5

and consists of blocks of octahedral and pentagonal bipyramidal coordination groups at two different 'levels' (see p. 228) so that there is both 6- and 7-coordination of the metal atoms. These Ta_2O_5 structures may be compared with the structure of α-U_3O_8 (Fig. 12.8(c)), in which all metal atoms are in chains of vertex-sharing pentagonal bipyramids, and with WNb_2O_8 and $NaNb_6O_{15}F$ (Fig. 13.14, p. 617) in which there is both distorted octahedral and pentagonal bipyramidal coordination of the transition-metal atoms. A third 'form' of Ta_2O_5, isostructural with one form of Nb_2O_5, may contain K as an essential constituent of the structure.[6]

(1) MRB 1973 8 627
(2) ZaC 1978 **440** 137
(3a) NW 1965 **52** 617
(3b) JSSC 1970 1 419
(3c) ZaC 1967 **351** 106

(3d) AC 1976 **B32** 764
(3e) AC 1975 **B31** 673
(4) AC 1971 **B27** 1037
(5) JSSC 1971 3 145
(6) JSSC 1970 2 24

Oxides M_2O_7

This small group consists of Mn_2O_7, which at ordinary temperatures is a deeply coloured oil, Tc_2O_7, and Re_2O_7. The last two are yellow, volatile, water-soluble solids. Mn_2O_7 is presumably molecular in all states of aggregation, and Tc_2O_7 consists of molecules in the crystalline state. In the centrosymmetrical molecule[1]

$$
\begin{array}{c}
1.67\,\text{Å} \\
\underset{O}{\overset{O}{\diagdown}}Tc\underset{1.84\,\text{Å}}{-}O\underset{\underset{180°}{\smile}}{-}Tc\overset{O}{\underset{O}{\diagup}}
\end{array}
$$

the Tc–O–Tc system is linear, the bridge bonds are similar in length to those in the $Re_2OCl_{10}^{4-}$ and $Ru_2OCl_{10}^{4-}$ ions, and the terminal bonds have approximately the same length as Re–O in ReO_3Cl, Re_2O_7 (1.70 Å), and $KReO_4$[1a] (1.72 Å).

Re_2O_7 vaporizes as molecules of the same type, but the crystal has an interesting layer structure[2] (Fig. 12.9(a)) in which equal numbers of metal atoms are in tetrahedral and octahedral coordination. (Contrast the chain ion in $Na_2Mo_2O_7$ which illustrates a third way of realizing the required O:M ratio of 7:2, also with equal numbers of MO_4 and MO_6 groups.) If chains of vertex-sharing octahedra at two different levels are linked by tetrahedra, rings consisting of two tetrahedra and two octahedra are formed which are perpendicular to the plane of the paper in Fig. 12.9(a). In each octahedral ReO_6 group three of the Re–O bonds are appreciably longer than the others. If two of these bonds are broken, as in Fig. 12.9(b), which shows one of the rings in crystalline Re_2O_7, molecules are produced, accounting for the volatility of the solid oxide. The structure of Re_2O_7 is also very closely related to that of the molecule $Re_2O_7(OH_2)_2$,[3] the first hydration product of the heptoxide (Fig. 12.9(c).)

(1) AnC 1969 **81** 328
(1a) AC 1975 **B31** 1764

(2) IC 1969 **8** 436
(3) AnC 1968 **80** 286

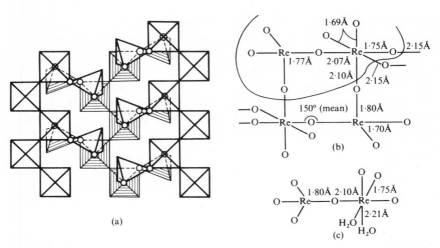

FIG. 12.9. (a) One layer of the crystal structure of Re_2O_7; the small circles represent O atoms shared between octahedral and tetrahedral groups, that is, the O atoms of the ring in (b). (b) Details of one ring of two tetrahedra and two octahedra. (c) The $Re_2O_7(H_2O)_2$ molecule.

Oxides M_3O_4

Relatively few elements form an oxide M_3O_4, which necessarily contains the element in two different oxidation states. Of the three possibilities $M_2^I M^{VI} O_4$, $M_2^{II} M^{IV} O_4$, and $M_2^{III} M^{II} O_4$, the first corresponds to an oxy-salt such as K_2WO_4, the second is realized in Pb_3O_4, and the third in Fe_3O_4, Co_3O_4, Mn_3O_4, and Eu_3O_4. The structure of Pb_3O_4 is described later in this chapter in the section on the oxides of lead and that of Eu_3O_4 on p. 601. The 3d oxides M_3O_4 have the spinel structure, which is described in Chapter 13 since it is adopted by many complex oxides XY_2O_4. For details of Co_3O_4 see p. 1209.

The structure of Fig. 12.10 was proposed many years ago for a compound with

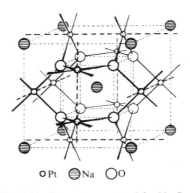

o Pt   ⊜ Na   ◯ O

FIG. 12.10. Structure proposed for $Na_xPt_3O_4$.

the limiting composition $NaPt_3O_4$.[1] Although it is doubtful if the full complement of Na atoms is ever present there are certainly phases $Na_xPt_3O_4$ ($0 < x < 1$), and Pd forms phases of the same type. Also it appears that the basic framework of Pt and O atoms can exist in the absence of Na atoms, when it is the oxide Pt_3O_4, though the phase $Na_xPd_3O_4$ cannot be prepared free from sodium.[2] In the Pt_3O_4 framework each O is bonded to 3 Pt (Pt–O, 1.97 Å) and each Pt forms 4 coplanar bonds to O atoms. Each metal atom also has 2 Pt atoms (at 2.79 Å) completing an elongated octahedral coordination group, as shown by the heavy broken lines in Fig. 12.10. The Na atoms occupy holes surrounded by 8 O at the vertices of a cube (Na–O, 2.46 Å).

(1)   JCP 1952 **20** 199   (2)   JLCM 1968 **16** 129

The oxides of iron

We have already referred to the structures of FeO and α-Fe_2O_3; we shall now see that the crystal chemistry of the oxides of iron is of some complexity. They include FeO, Fe_3O_4, and apparently four forms of Fe_2O_3.

FeO.   When prepared under atmospheric pressure this oxide is deficient in Fe, the 'wüstite' phase $Fe_{1-x}O$. Some Fe^{2+} ions are replaced by two-thirds their number of Fe^{3+} ions to maintain electrical neutrality. Stoichiometric FeO has been prepared at pressures greater than 36 kbar at 770 °C from $Fe_{0.95}O$ and metallic iron.[1] At about 1000°C the composition range is $Fe_{0.946}O$–$Fe_{0.875}O$, but at lower temperatures the composition limits converge. At 570 °C the composition is $Fe_{0.93}O$, and at still lower temperatures disproportionation to α-Fe + Fe_3O_4 occurs,[2] so that this oxide must be prepared at temperatures above 570 °C and quenched. At about 200 K a second-order change takes place over a range of temperature, accompanied by the development of antiferromagnetism and rhombohedral symmetry, but involving only a very minor structural change.[3] The length of the side of the unit cell varies, approximately linearly, with the composition, and typical values are: $a = 4.3010$ Å for 48.56 atomic per cent Fe and 4.2816 Å for 47.68 per cent Fe. It has been suggested that in quenched specimens of $Fe_{0.9}O$ there are periodically spaced clusters of vacant sites arranged at the vertices of tetrahedra surrounding interstitial (tetrahedrally coordinated) cations, the ratio of vacancies to interstitial cations being about 3.2:1.[4]

Fe_3O_4.   This oxide has the inverse spinel structure, $Fe^{3+}(Fe^{2+}Fe^{3+})O_4$, that is, one-third of the cations occupy tetrahedral interstices in a c.c.p. assembly of O^{2-} ions and equal numbers of Fe^{2+} and Fe^{3+} ions occupy octahedral interstices. The high electronic conductivity of Fe_3O_4 (magnetite) compared with that of, for example, Mn_3O_4 (distorted normal spinel structure) is attributed to the continuous interchange of electrons between Fe^{2+} and Fe^{3+} ions in the octahedral sites. (At 119K magnetite becomes antiferromagnetic, and the symmetry drops to rhombohedral or lower.[5])

α-Fe$_2$O$_3$. Corundum structure (h.c.p. O atoms).

β-Fe$_2$O$_3$. Apparently two different solid phases have been described as β-Fe$_2$O$_3$. One is a tetragonal material formed by the dehydration of β-FeOOH and the other, with the bixbyite (C-M$_2$O$_3$) structure prepared by hydrolysis of FeCl$_3$.6H$_2$O and later as a surface film by the decomposition of ferric trifluoroacetylacetonate.[6]

γ-Fe$_2$O$_3$. In this oxide there is a statistical distribution of $21\frac{1}{3}$ Fe^{3+} ions in the 8 tetrahedral and 16 octahedral sites of the unit cell of the spinel structure, which contains 32 c.c.p. O atoms.

The oxides FeO, Fe$_3$O$_4$, and γ-Fe$_2$O$_3$ are closely related, having the following numbers of cations in a cubic unit cell of edge approximately 8.5 Å containing 32 c.c.p. O atoms:

FeO: 32 Fe^{2+} in all octahedral interstices;
Fe$_3$O$_4$: 8 Fe^{3+} in tetrahedral sites, 8 Fe^{2+} + 8 Fe^{3+} in octahedral sites;
γ-Fe$_2$O$_3$: $21\frac{1}{3}$ Fe^{3+} statistically in the above 24 positions.

Accordingly γ-Fe$_2$O$_3$ and Fe$_3$O$_4$ are easily interconvertible. Careful oxidation of Fe$_3$O$_4$ yields γ-Fe$_2$O$_3$, which is converted back into Fe$_3$O$_4$ by heating *in vacuo* at 250 °C. The oxidation FeO→Fe$_3$O$_4$→γ-Fe$_2$O$_3$ may be visualized in the following way. The oxygen lattice of FeO is extended by the addition of oxygen, added as new layers of c.p. O atoms. Into these cations migrate, resulting in a continuous decrease in the concentration of Fe atoms in a given volume and an increase in the proportion of Fe^{3+} ions.

(1)   JCP 1967 **47** 4559
(2)   CR 1956 **242** 776
(3)   AC 1953 **6** 827
(4)   AC 1969 **B25** 275; AC 1975 **A31** 586; JSSC 1980 **33** 271
(5)   AC 1953 **6** 565; AC 1955 **8** 257
(6)   AC 1976 **B32** 667

The oxides of aluminium

The hydroxide, M(OH)$_3$, oxyhydroxide, MO(OH), and oxide, M$_2$O$_3$, of Al (and certain other trivalent elements) exist in α and γ forms. We give the mineral names of these Al compounds because they are often referred to by name and also because the American nomenclature is sometimes different from that used in England.

Al(OH)$_3$	α	bayerite	γ	gibbsite, hydrargillite
AlO(OH)	α	diaspore	γ	boehmite
Al$_2$O$_3$	α	corundum	γ	—

On heating diaspore dehydrates directly to α-Al$_2$O$_3$ but boehmite, both forms of Al(OH)$_3$, and compounds such as ammonium alum produce α-Al$_2$O$_3$ only after prolonged heating at high temperatures. At lower temperatures a variety of phases are formed which are collectively known as γ-Al$_2$O$_3$, and in various studies most of the Greek alphabet has been used in naming them. These phases represent various degrees of ordering of Al atoms in an essentially cubic closest packing of O atoms, described as a defect spinel structure (q.v.), since there are only $21\frac{1}{3}$ metal atoms arranged at random in the 16 octahedral and 8 tetrahedral positions of that structure. This γ-Al_2O_3 is important technically because of its great adsorptive power ('activated alumina') and its catalytic properties, and it is also used in the manufacture of synthetic sapphire in the Verneuil furnace. The degree of order and the pore structure depend on the starting material. For example, the arrangement of the O atoms in bayerite is h.c.p., and conversion to the c.c.p. spinel structure involves rebuilding the whole structure. In the case of boehmite, although the structure as a whole is not close-packed the O atoms within a layer are c.c.p., so that less rearrangement is required to form the γ-Al_2O_3 structure. The following[1] are typical of the schemes postulated as the result of X-ray or electron diffraction studies:

$$\text{boehmite} \xrightarrow{450°} \gamma \xrightarrow{750°} \delta \xrightarrow{1000°} \theta + \alpha \xrightarrow{1200°C} \alpha$$
$$\text{bayerite} \xrightarrow{230°} \eta \xrightarrow{850°} \theta \xrightarrow{1200°C} \alpha$$

The first phase formed in the dehydration of boehmite,[2] γ-Al_2O_3, has a defect spinel structure in which there is random arrangement of the metal ions. The δ phase is a tetragonal spinel superstructure with a tripled c-axis in which there are ordered vacancies in the octahedral sites. The θ phase is isostructural with β-Ga_2O_3 (p. 547). In all three forms the O atoms are c.c.p. or approximately so, and the structural changes therefore involve only movements of cations; compare the oxidation of FeO.

For β-alumina and related compounds see p. 598.

(1)   IEC 1950 **42** 1398; AC 1964 **17** 1312   (2)   JSSC 1980 **34** 315

The oxides of manganese

These comprise MnO, Mn_3O_4, Mn_2O_3, Mn_5O_8, MnO_2, and Mn_2O_7. The last is the anhydride of permanganic acid (a compound which is a volatile solid stable below $1°C$)[1] and forms an oil or dark green crystals. It breaks down irreversibly at temperatures above about $0°C$ with evolution of oxygen and formation of MnO_2; it is a powerful oxidizing agent, and obviously resembles Cl_2O_7 rather than metallic oxides.

The first three oxides, on the other hand, are closely similar to those of iron. The stability of the green MnO to oxygen depends on the temperature of preparation, low-temperature ignition of the carbonate giving very reactive material. It apparently oxidizes to the stage $MnO_{1.13}$ without forming new phases, and it has the NaCl structure, like FeO. There are two forms of the black Mn_2O_3, α having the C sesquioxide structure and γ being related to Mn_3O_4 in the same way as γ-Fe_2O_3 to Fe_3O_4 (p. 551). All oxides and oxyhydroxides of Mn on heating in air to about $1000°C$ form the purplish-red Mn_3O_4, which has a distorted spinel structure at ordinary temperatures but apparently becomes cubic at temperatures above

$1170\,^\circ\text{C}$.[2] Both γ-Mn_2O_3 and Mn_3O_4 show the same type of tetragonal distortion ($c:a = 1.16$).

The oxide Mn_5O_8 is $Mn_2^{II}Mn_3^{IV}O_8$ and is isostructural with $Cd_2Mn_3O_8$. The structure may be visualized as built from CdI_2-like layers from which one-quarter of the metal atoms are removed ($Mn_3^{IV}O_8$), Mn^{II} atoms then being added above and below the empty sites to give the composition M_5X_8 as in $Zn_5(OH)_8Cl_2 . H_2O$ (p. 264) or $Zn_5(OH)_8(NO_3)_2 . 2H_2O$ (p. 650). These Mn^{II} atoms join the 'layers' into a 3D structure. The coordination of Mn^{IV} is distorted octahedral (mean Mn^{IV}–O, 1.87 Å) and that of Mn^{II} is trigonal prismatic (mean Mn^{II}–O, 2.22 Å), (Fig. 12.11).

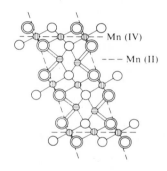

FIG. 12.11. Elevation of the structure of Mn_5O_8; the CdI_2-like layers are perpendicular to the plane of the paper.

There is considerable distortion of the latter coordination group (2.05, 2.17 (two), 2.28, and 2.31 Å (two)).[3]

The dioxide has long been known as the mineral pyrolusite with the simple tetragonal rutile structure, but many materials approximating in composition to MnO_2 have been prepared in the laboratory and recognized in the mineral world. They have been described as polymorphs of MnO_2, though their compositions are often appreciably different from pure MnO_2, for example, γ-MnO_2 (analysis corresponding to $MnO_{1.93}$) and α-MnO_2 (91.5 per cent MnO_2). The β form (pyrolusite) can be prepared pure, but not by hydrothermal methods. There is a large literature on the oxides of Mn, particularly the dioxide, which is important as a catalyst and as a component of dry batteries.[4] X-ray studies have now shown why the MnO_2 system is so complex and why different structures appear when the oxide is prepared from solutions containing ions of various kinds. Two main types of structure have been identified, 3D frameworks and layer structures.

(1)   JACS 1969 **91** 6200
(2)   JRNBS 1950 **45** 35
(3)   HCA 1967 **50** 2023

(4)   JACS 1950 **72** 856; RPAC 1951 **1** 203;
       ZaC 1961 **309** 1, 20, 121

Framework structures

These are built of single or multiple octahedral chains joined along their lengths
by sharing vertices, and they can conveniently be illustrated as projections along the
length of the chain. In this way Fig. 12.12(a) represents the rutile structure of

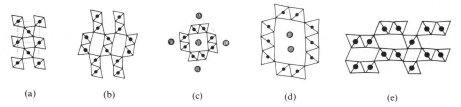

FIG. 12.12.   The structures of (a) β-MnO$_2$; (b) ramsdellite; (c) α-MnO$_2$ and hollandite; (b) psilomelane; (e) a 'γ-MnO$_2$' intermediate between (a) and (b).

β-MnO$_2$. The mineral ramsdellite and Glemser's γ-MnO$_2$ have the structure shown
at (b), built of double octahedral chains.[5] This is essentially the diaspore structure
(Fig. 14.12) except that the short O–O contacts of 2.65 Å due to hydrogen bonding
in AlO.OH (and MnO.OH) have lengthened to the much larger value 3.34 Å in
ramsdellite. The collective name γ-MnO$_2$ is given to a range of materials. The X-ray
diffraction patterns of the more poorly crystalline materials resemble that of
pyrolusite, but the patterns of more crystalline preparations show a stronger
resemblance to that of ramsdellite. It has been suggested[6] that these γ phases may
have structures intermediate between those of Fig. 12.12(a) and (b), being built of
slices of the β and γ structures, as shown at (e).

If some of the Mn^{4+} ions in a MnO$_2$ framework are replaced by Mn^{2+} or other
ions carrying smaller positive charges, then the framework is negatively charged
and can accommodate other positive ions in its interstices. Li$_2$Ti$_3$O$_7$ is an ionic
conductor in which disordered Li$^+$ ions occupy tetrahedral sites within the channels
of the ramsdellite structure, (b), and also some of the octahedral framework sites.
The structural formula $(Li_{1.72}\square_{2.28})(Ti_{3.43}Li_{0.57}O_8)$ has been suggested,[6a] $\square$
representing a vacant site. The structure of Fig. 12.12(c) corresponds to a group of
minerals, hollandite, cryptomelane, and coronadite, which with α-MnO$_2$ form an
isostructural series of the general formula A$_{2-y}$B$_{8-z}$X$_{16}$.[7] (A represents large ions
such as Ba^{2+}, Pb^{2+}, or K$^+$, B is Mn^{4+}, Fe^{3+}, or Mn^{2+}, and X is O^{2-} or OH$^-$.) Other
compounds of this large hollandite family include Ba(M^{IV},Al)O$_8$, where M^{IV} is Si,
Ti, or Sn, and Al may be replaced by Cr, Fe, or In.[7a] It is evident that in the more
open structures such as (c) and (d) some large ions are necessary to prevent collapse
of the framework, and accordingly the so-called α-MnO$_2$ can be prepared only in
the presence of a large ion such as K$^+$. Some of these compounds show cation
exchange like the zeolites. An even more open structure is found in psilomelane,[8]
(Ba,H$_2$O)$_2$Mn$_5$O$_{10}$, in which there are double and triple chains of octahedra and
larger tunnels accommodating Ba^{2+} ions and water molecules (Fig. 12.12(d)).

Dehydration is accompanied by structural change to hollandite, into which psilomelane is converted on heating to 550 °C.

(For a complex oxide containing Mn with a quite different type of structure see Mg_6MnO_8 (p. 606).)

An interesting new form of MnO_2 has been described as having the spinel framework of $LiMn_2O_4$ but virtually free from Li.[8a] The MnO_2 framework is therefore the 'atacamite' framework of Fig. 7.3(a), a framework hitherto unknown for a compound AX_2. In this connection see the structures of melanophlogite and silicalite (pp. 1008 and 1041).

(5)   ACSc 1949 **3** 163   (7a) AC 1978 **B34** 3554
(6)   AC 1959 **12** 341   (8) AC 1963 **6** 433
(6a) AC 1979 **B35** 798   (8a) JSSC 1981 **39** 142
(7)   AC 1950 **3** 146; AC 1951 **4** 469

Layer structures

There are many ill-defined hydrous manganese oxide minerals somewhat reminiscent of clay minerals, and probably like them having layer structures. A well-crystallized lithiophorite, $(Al_{0.68}Li_{0.32})Mn^{2+}_{0.17}Mn^{4+}_{0.82}O_2(OH)_2$, has a 2-layer structure in which both layers are of the $Mg(OH)_2$ type (p. 258) with ideal compositions MnO_2 and $(Li,Al)(OH)_2$.[9] The layers are held together by O—H—O bonds as in $Al(OH)_3$, though it may be deduced from the above formula that if all anions in the MnO_2 layer are O^{2-} and all are OH^- in the $(Li,Al)(OH)_2$ layer then the former carries a charge of –0.38 and the latter +0.38 per unit of formula. The detailed arrangement of the cations and the positions of the H atoms are not known with certainty.

A Mn-containing layer of essentially the same kind occurs in chalcophanite, $ZnMn_3O_7.3H_2O$,[10] but with one-seventh of the metal sites unoccupied so that the composition is Mn_3O_7 instead of MnO_2. Associated with each Mn layer and directly above and below the unoccupied Mn positions are the Zn^{2+} ions coordinated by three O atoms of the layer and on the other side by three H_2O, these six neighbours forming a somewhat distorted octahedron (Zn–3 O 1.95 (0.05) Å, Zn–3 H_2O 2.15 (0.05) Å). The layers are held together by O—H—O bonds between water molecules as shown in Fig. 12.13. The nature of the Zn—O bonds in this and other crystals is discussed on p. 1152.

(9) AC 1952 **5** 676   (10) AC 1955 **8** 165

The oxides of lead

The existence of three oxides of lead, PbO, Pb_3O_4, and PbO_2, has been recognized for a long time, but there has been far less general agreement concerning other oxides intermediate between PbO and PbO_2.[1] The black Pb_2O_3 is prepared either hydrothermally or by heating PbO at 600 °C under 1 kbar pressure of O_2. A brown-black $Pb_{12}O_{19}$ is described as formed by heating PbO or PbO_2 at 500 °C under

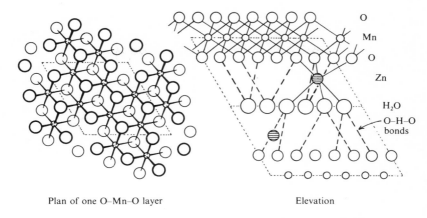

Plan of one O–Mn–O layer   Elevation

FIG. 12.13. The crystal structure of chalcophanite, $ZnMn_3O_7.3H_2O$.

175 bar pressure of O_2; the existence of a further oxide ($Pb_{12}O_{17}$?) is not fully confirmed.[2] Since the oxides PbO, Pb_3O_4, and PbO_2 are to some extent inter-convertible (by heating in air or in other ways) mixtures are often obtained. For example, it is stated that Pb_3O_4 results from heating either PbO or PbO_2 in air to 500 °C, whereas above 550 °C only PbO is stable. In practice, therefore, Pb_3O_4 and PbO_2 are not prepared thermally without further treatment if wanted pure in small quantities. Careful decomposition of $PbCO_3$ or $Pb(OH)_2$ gives PbO, oxidation of PbO followed by removal of unchanged PbO by acetic acid gives Pb_3O_4, while PbO_2 may be prepared by the action of nitric acid on the latter or by oxidation of an alkaline suspension of PbO by a hypochlorite. These oxides have striking colours —PbO yellow and red, Pb_3O_4 red, Pb_2O_3 black, PbO_2 maroon and black, showing, as in the case of PbS with its brilliant metallic lustre and semiconductivity, that the bonds are not of simple types.

(1)   JCS 1956 725   (2)   JACeS 1964 47 242

Lead monoxide, PbO

This oxide has two polymorphs, a red (tetragonal) form prepared by heating PbO_2 in air to 550 °C, and a yellow (orthorhombic) form made by heating PbO_2 to 650 °C or by rapidly cooling molten PbO. Tetragonal PbO has the layer structure of Fig. 12.14 in which the metal atom is bonded to 4 O atoms which are arranged in a square to one side of it, with the lone pair of electrons presumably occupying the apex of the tetragonal pyramid;[3] for this structure see also pp. 119 and 161. The structural chemistry of lead is summarized in Chapter 26. It has not proved possible to locate the O atoms in the yellow form of PbO by X-ray diffraction, but their positions have been determined by neutron diffraction.[4] The structure is built of

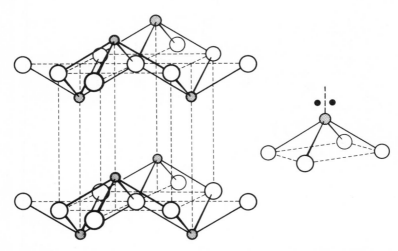

FIG. 12.14. The crystal structure of tetragonal PbO (and SnO). The small shaded circles represent metal atoms. The arrangement of bonds from a metal atom is shown at the right, where the two dots represent the 'inert pair' of electrons (see p. 1149).

layers which are recognizable as very distorted versions of that of Fig. 12.14, but instead of four equal Pb–O bonds of length 2.30 Å in the red form there are two of 2.20 Å and two of 2.49 Å. The shorter bonds delineate zigzag chains (O–Pb–O, 148°) which are bonded into layers by the longer Pb–O bonds. The shortest Pb–Pb distances in and between the layers are respectively 3.47 and 3.63 Å.

(3)   AC 1961 **14** 1304 (n.d.)   (4)   AC 1961 **14** 66, 80 (n.d.)

Red lead, Pb_3O_4

The structure of Pb_3O_4 is illustrated in Fig. 12.15. It consists of chains of $Pb^{IV}O_6$ octahedra sharing opposite edges, these chains being linked by the Pb^{II} atoms, each with a pyramidal arrangement of three O neighbours. The bond lengths and interbond angles are: Pb^{IV}–O, 6 at 2.14 Å, Pb^{II}–O, 2 at 2.18 Å and 1 at 2.13 Å; mean O–Pb–O bond angle 76° (as in PbO).[5]

(5)   ZaC 1965 **336** 104 (n.d.)

Lead sesquioxide, Pb_2O_3

In this (monoclinic) oxide the Pb^{II} atoms are situated between layers of distorted $Pb^{IV}O_6$ octahedra (Pb^{IV}–O, 2.08–2.28 Å, mean 2.18 Å). The Pb^{II} atoms are in positions of very irregular 6-coordination, these neighbours being arranged at six of the vertices of a distorted cube (those at the ends of one face-diagonal being absent). The six Pb^{II}–O distances fall into two sets, three short (2.31, 2.43, and 2.44 Å)

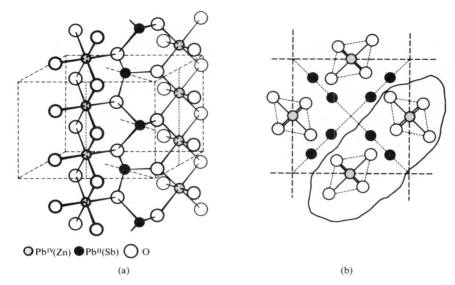

O Pb^{IV}(Zn) ● Pb^{II}(Sb) ○ O

(a)   (b)

FIG. 12.15. The crystal structure of Pb$_3$O$_4$ (and ZnSb$_2$O$_4$). (a) Portion of the structure outlined in the projection (b) showing the chains of Pb^{IV}O$_6$ octahedra joined by pyramidally coordinated Pb^{II} atoms.

and three longer (2.64, 2.91, and 3.00 Å), and there are Pb^{II}–Pb^{II} distances close to 3.5 Å, as in the metal.[6] These bond lengths are clearly difficult to reconcile with those in the other oxides and with the sum of the ionic radii (for 6-coordination), about 2.6 Å.

(6) AC 1970 **A26** 501 (X-ray and n.d.)

Lead dioxide, PbO$_2$

The usual maroon form of this oxide has the rutile structure, the mean distance of Pb to the six octahedral neighbours being 2.18 Å. At 300° under 40 kbar pressure this converts to a black orthorhombic polymorph, the structure of which has been described in the discussion of close-packed structures in Chapter 4.

Some complex oxides of Pb(IV)

Closely related to PbO$_2$ are some complex oxides in which Pb(IV) is octahedrally coordinated. In Sr$_2$PbO$_4$[7] there are rutile-like chains which are joined through 7-coordinated Sr^{2+} ions (instead of Pb^{II} in Pb$_3$O$_4$), while Ba$_2$PbO$_4$ has the K$_2$NiF$_4$ (layer) structure (p. 208) which accommodates the larger Ba^{2+} ions in positions of 9-coordination:

	Isostructural compounds
Ca_2PbO_4	Ca_2SnO_4
Sr_2PbO_4	Cd_2SnO_4
Ba_2PbO_4	Sr_2SnO_4
	Ba_2SnO_4

In contrast to the octahedral coordination of M(IV) in these compounds there is 5-coordination in the hygroscopic oxides $K_2M^{IV}O_3$ (M = Zr, Sn, Pb).[8] These contain the unusual MX_3 chain of Fig. 12.16 in which M has a tetragonal pyramidal

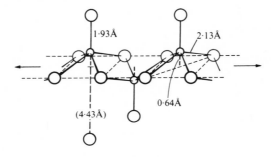

FIG. 12.16. MX_3 chain in the structure of K_2ZrO_3.

arrangement of 5 O neighbours (one closer than the other four), the M atom lying above the base of the pyramid. The detailed structure of the Pb compound was not determined; in K_2SnO_3 bond lengths are: Sn–O, 1.93 Å, Sn–2 O, 2.03 Å, and Sn–2 O, 2.21 Å. Note the numerous examples of isostructural compounds of Sn(IV) and Pb(IV).

(7)   ZaC 1969 **371** 225   (8)   JSSC 1970 **2** 410

The oxygen chemistry of some transition elements

For a discussion of certain aspects of their oxygen chemistry it is convenient to group together the six elements Ti, V, Nb, Mo, W, and Re. The following generalizations may be made. The diagonal relationship emphasized below is associated on the one hand with the larger size, more electropositive character, and reduced stability of lower oxidation states of Zr, Hf, and Ta as compared with the elements to their right, and on the other with the increasing resemblance in their highest oxidation states to the typical elements Si, P, S, and Cl as we go along the series Ti, V, Cr, and Mn. Thus Ti is typically ionic in its oxy-compounds, and while it can exist in lower oxidation states the ionic form Ti^{4+} in octahedral coordination is the preferred state of Ti in its oxy-compounds. In some of its oxy-compounds V^V shows a resemblance

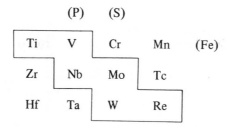

to P and the other 'tetrahedral' elements Si, Ge, and As (as in forming the tetrahedral VO_4^{3-} ion), while in others there is octahedral coordination by oxygen. However, in many crystals the coordination is more realistically described as trigonal bipyramidal or square pyramidal, the distinction between these two descriptions being somewhat arbitrary when there is appreciable distortion from the ideal arrangement.

As far as is known the oxygen chemistry of Cr^{VI} is based exclusively on tetrahedral coordination. Cr^{VI} is present in some, if not all, of the black oxides intermediate in composition between CrO_2 and CrO_3 (p. 1199) which are obviously metallic oxides more akin to those of Mo and W. Not much is known of the structural chemistry of oxy-compounds of Mn^{VII} apart from the fact that the permanganate ion MnO_4^- is tetrahedral. We have already referred to Mn_2O_7. Both Mo^{VI} and W^{VI} form tetrahedral oxy-ions but their complex oxy-chemistry is largely based on octahedral coordination, which appears even in the 'pyro'-salts $Na_2Mo_2O_7$ and $Na_2W_2O_7$ (p. 516). Some finite oxy-ions of V, Nb, Mo, and W are described in Chapter 11. Comparatively little is yet known of the complex oxy-chemistry of Re and still less in the case of Tc, while that of Ta has yet to be developed, but from what we know of these elements it would seem justifiable to group together Ti, V, Nb, Mo, W, and Re for a discussion of their complex oxy-chemistry.

The key to the structures of many of the complex oxy-compounds of these elements is the relation to the structures of TiO_2 (rutile) and ReO_3. VO_2 has normal and distorted forms of the rutile structure and NbO_2 has a complex superstructure of the rutile type. WO_3 crystallizes with distorted forms of the ReO_3 structure, but MoO_3 has a unique layer structure (p. 572). Octahedral MO_6 groups share only vertices in ReO_3, but by the sharing of various numbers and combinations of vertices and edges a great variety of more complex structures can be derived from portions of ReO_3 structure with formulae in the range MO_2-MO_3. Similarly, more complex structures may be built from portions of the rutile structure by the sharing of octahedral faces. In this chapter we shall devote sections to the following topics:

the oxides of Ti;
the oxygen chemistry of V; and
the oxides of Mo and W.

In Chapter 13 we discuss complex oxides of these elements, including the 'bronzes'.

The oxides of titanium

The elements Ti and V illustrate very well the extraordinary complexity of some metal–oxygen systems.

The h.c.p. metal Ti takes up oxygen to the stage $TiO_{0.50}$. and in this range three distinct phases have been recognized. First there is a random solid solution of O in Ti, and this is followed by preferential occupancy of certain positions leading to essentially ordered structures at compositions close to $TiO_{0.33}$ (anti-AX_3 layer structure) and $TiO_{0.50}$ (anti-CdI_2 structure). At higher oxygen content the hexagonal closest packing of the metal atoms breaks down. An oxide within the composition range $TiO_{0.68}$–$TiO_{0.75}$ results from oxidizing the metal under a low pressure of oxygen at temperatures below 900 °C. This oxide has the ϵ-TaN structure (Fig. 29.22, p. 1322) but is deficient in oxygen. The metal atoms are no longer close-packed but each Ti has two nearest Ti neighbours at 2.88 Å and twelve more at 3.22 Å forming a hexagonal prism. Each O has six Ti neighbours at the vertices of an elongated octahedron.

At the composition $TiO_{1.00}$ the phase has a defective NaCl structure. At high temperatures the structure has cubic symmetry with equal numbers of random vacancies in the cation and anion positions. At ordinary temperatures it has an ordered defective NaCl-type structure (Fig. 12.17) in which one-sixth of the sites

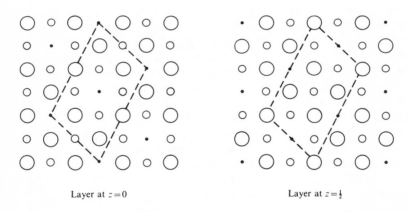

Layer at $z = 0$ Layer at $z = \frac{1}{2}$

FIG. 12.17. Successive layers in the structure of TiO (*defect* NaCl superstructure). The small black dots indicate vacant cation or anion sites (one-sixth of the total of each kind).

for each type of ion are unoccupied. The symmetry is monoclinic with the angle β very close to 90° (89.9°). At 800 °C the cubic NaCl-type structure has a narrow homogeniety range around $TiO_{1.18}$; at higher temperatures the range of composition increases, with a corresponding variation in the percentages of vacant sites. For example:

O : Ti *ratio*	*Percentage of sites occupied*	
	Ti	O
1.33	74	98
1.12	81	91
0.69	96	66

Interpolation gives 85 per cent of each type of site occupied at the composition TiO.

These phases are followed by Ti_2O_3, which has the corundum structure at ordinary temperatures and undergoes a structural change at 200 °C, Ti_3O_5, and no fewer than seven distinct phases in the range $TiO_{1.75}$–$TiO_{1.90}$. These form a series Ti_nO_{2n-1} (*n* from 4 to 10) comparable with the series of oxides formed by V, Mo, and W (see later). At temperatures below 100 °C Ti_3O_5 consists of a 3D array of TiO_6 octahedra sharing edges and vertices (Fig. 12.18). At 100 °C this oxide undergoes

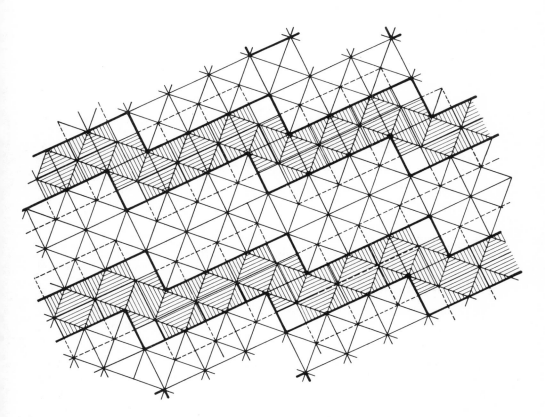

FIG. 12.18. Crystal structure of the low-temperature form of Ti_3O_5.

a rapid and reversible transformation to a form with a slightly distorted pseudo-brookite structure (p. 603), a structure which is apparently stabilized at lower temperatures by the presence of a little Fe. In the low-temperature Ti_3O_5 there are metal–metal distances of three kinds across shared octahedron edges, approximately 3.1, 2.8, and 2.6 Å (compare 2.93 Å in α-Ti). The oxides Ti_4O_7 and Ti_5O_9, of the Ti_nO_{2n-1} series, are built of slabs of rutile structure, infinite in two dimensions and respectively four and five octahedra thick, joined by sharing octahedron *faces* (compare corundum).

The highest oxide is the familiar TiO_2. The three forms stable at atmospheric pressure, rutile, anatase, and brookite, all occur as minerals. These three polymorphs, together with the high-pressure form with the α-PbO_2 structure, are all built of octahedral coordination groups, the Ti–O bond lengths in which have been noted (p. 541). A fifth polymorph, produced at pressures in excess of 150 kbar, is not stable at atmospheric pressure; it may contain Ti in a higher coordination state. Numerous complex oxides containing Ti are described in Chapter 13. The binary oxides are summarized in Table 12.7.

TABLE 12.7

The oxides of titanium

Oxide	Structure	Reference
Ti_3O	Anti-AX_3 (layer)*	AC 1968 **B24** 211
Ti_2O	Anti-CdI_2 (layer)	ACSc 1957 **11** 1641
δ-TiO_x	O-defective ϵ-TaN	ACSc 1959 **13** 415
TiO	NaCl superstructure	AC 1967 **23** 307
Ti_2O_3	Corundum	AC 1977 **B33** 1342
Ti_3O_5 ⎫		AC 1959 **12** 575
Ti_4O_7 ⎬	Members of homologous series Ti_nO_{2n-1} $(4 \leqslant n \leqslant 9)$	JSSC 1971 **3** 340
Ti_5O_9 ⎭		ACSc 1960 **14** 1161
TiO_2–I	Rutile, anatase, brookite	—
–II	α-PbO_2	JCP 1969 **50** 519

Ti–O system: ACSc 1962 **16** 1245; AK 1963 **21** 413; JCP 1967 **46** 2461, 2465.

* There is an indefinitely large number of c.p. layer structures AX_3 in which a particular set of octahedral sites is occupied between alternate pairs of c.p. layers. For the pattern of sites of Fig. 4.24(e) the two simplest layer sequences give the h.c.p. BiI_3 (low-$CrCl_3$) and the c.c.p. YCl_3 structures. In Ti_3O the arrangement of the (c.p.) Ti atoms is h.c.p. and the pattern of sites of O atoms is that of Fig. 4.24(e), but the relative translations (parallel to the c.p. layers) of the sets of O atoms are different from those of the metal atoms in the BiI_3 structure.

No lower oxides analogous to the homologous series Ti_nO_{2n-1} are formed by Zr or Hf, for the ions M^{4+} of these elements are too large to form the 6-coordinated rutile-type structures. In the next Periodic family there is also a marked difference between the oxide chemistry of the first member of the A sub-group (V) and the later members (Nb and Ta). Features of the Nb–O system are the unique NbO structure and the complex crystal chemistry of Nb_2O_5 and related oxides, to which we have already referred. In the Ta–O system[1] the following phases have been recognized: α, Ta metal with up to about 5 atomic per cent O; β, a deformed version of the body-centred cubic α-structure with a maximum O content corresponding to Ta_4O; γ, TaO with the NaCl structure; δ, TaO_2 with the rutile structure; and Ta_2O_5.

(1) ACSc 1954 **8** 240

The oxygen chemistry of vanadium

Lower oxides. Thermal and X-ray studies have shown that the vanadium–oxygen system shows marked similarities to the titanium–oxygen system up to the composition MO_2, though one notable difference is that both forms of V_3O_5 have structures different from those of the polymorphs of Ti_3O_5 (see below).

In addition to the α phase, a solid solution of O in b.c.c. V (up to $VO_{0.03}$ at $900\,^\circ C$) and the β phase ('V_4O'), $VO_{0.15}$–$VO_{0.25}$, consisting of a tetragonally deformed V lattice containing a small concentration of oxygen,[1] the oxides listed in Table 12.8 have been recognized.[2]

TABLE 12.8

The oxides of vanadium

Oxide	Structure		Reference
$VO_{0.53}$	—		ACSc 1960 **14** 465
VO	NaCl structure (homogeneous at $900\,^\circ C$ over range $VO_{0.80}$–$VO_{1.20}$)		Refs. 1 and 2 above
$VO_{1.27}$	B.c. tetragonal superstucture of NaCl type (V-deficient)		ACSc 1960 **14** 465
$VO_{1.50}$	Corundum structure		
$VO_{1.67}$ to $VO_{1.87}$	A series of six oxides forming a homologous series V_nO_{2n-1} (V_3O_5 to V_8O_{15})	V_3O_5 V_4O_7	AC 1980 **B36** 1332 AC 1972 **B28** 1404
VO_2	Monoclinic (MoO_2-type distorted rutile) ($>68\,^\circ C$) Tetragonal rutile structure		ACSc 1970 **24** 420 JPSJ 1967 **23** 1380
$VO_{2.17}$	V_6O_{13} ⎫		ACSc 1971 **25** 2675
$VO_{2.25}$	V_4O_9 ⎬ see text		ACSc 1970 **24** 3409
$VO_{2.33}$	V_3O_7 ⎮		AC 1974 **B30** 2644
$VO_{2.5}$	V_2O_5 ⎭		ZK 1961 **115** 110

Between V_2O_3 (corundum structure) and VO_2 (rutile structures) there is a family of oxides V_nO_{2n-1} ($3 < n < 8$) formed from slabs of rutile-like structure sharing *faces*. There is a similar family of 'sheared' rutile structures in the Ti oxides, but starting at Ti_4O_7. The sharing of octahedron faces found in both low- and high-V_3O_5 does not occur in the Ti_3O_5 structures. In these oxides V_nO_{2n-1}, as in V_2O_3

and VO_2, there is octahedral coordination of V but considerably distorted in some cases. For example, in V_4O_7 V—O ranges from 1.78 to 2.12 Å, and in monoclinic VO_2 from 1.76 to 2.06 Å. The structures of the oxides between VO_2 and V_2O_5 are more complex, and although some V atoms could be described as octahedrally coordinated the distortion from a regular octahedron is so great that the coordination is better described as 5-coordination.

The oxide V_6O_{13}, in which the mean oxidation state of the metal is 2.17, is thought to play an important part in the action of 'vanadium pentoxide' catalysts used in the oxidation of SO_2 to SO_3 and of naphthalene to phthalic anhydride. Its structure has been described on p. 227. There are equal numbers of three kinds of non-equivalent V atoms with nearest neighbours:

	V^1	V^2	V^3
	1.77 Å	1.66 Å	1.64 Å
	1.88 (two)	1.76	1.92 (two)
	1.96	1.90 (two)	1.93
	1.99	2.08	1.98
	2.06	2.28	2.26
Mean	1.92	1.93	1.94

It is clearly not possible to distinguish between the three kinds of V atom on the basis of V—O bond lengths. Similarly, in V_4O_9 all V atoms form one short bond (1.60–1.65 Å), four bonds in the range 1.87–2.02 Å and a sixth long bond. The lengths of the latter for the four non-equivalent V atoms are 2.23, 2.40, 2.50, and 3.00 Å, so that one-quarter of the V atoms are described as 5-coordinated. It does not seem possible to correlate the type of coordination with the oxidation state of individual atoms (unless the compound were $V_3^{IV}V^VO_8(OH)$).

The oxide V_3O_7 has a complex 3D structure formed by the sharing of vertices and edges between coordination polyhedra of three kinds. Of the 36 V atoms in the unit cell 12 have octahedral coordinaton (mean V—O, 1.94 Å, but one short, 1.62, and five around 2 Å); 16 have trigonal bipyramidal coordination with the same mean V—O; and 8 have square pyramidal coordination, mean V—O, 1.83 Å; compare the obvious formulation $V^{4+}V_2^{5+}O_7$. From the V—O distances in many V oxides the following approximate mean values have been derived, but it should be remembered that there is a very considerable range of bond lengths, particularly for 5- and 6-coordination. Even for tetrahedral V^{5+} the four bonds have the same length only in the simple VO_4^{3-} ion; there is an appreciable difference between V—O_t (1.69 Å) and V—O_b (1.80 Å) in, for example, $Ca_2V_2O_7 \cdot 2H_2O$.[3]

	6-coordination	5-coordination	4-coordination
V(III)	2.00 Å		
V(IV)	1.95	1.90 Å	
V(V)	1.90	1.83	1.72 Å

Vanadium pentoxide and vanadates. The structural chemistry of pentavalent V in V_2O_5 and vanadates is quite different from that of the lower oxides. Orthovanadates, containing tetrahedrally coordinated V^V are often isostructural with orthophosphates and orthoarsenates. For example, the dodecahydrates of Na_3PO_4, Na_3AsO_4, and Na_3VO_4 are isostructural, as also are the complex salts $Pb_5(PO_4)_3Cl$, $Pb_5(AsO_4)_3Cl$, and $Pb_5(VO_4)_3Cl$, while the rare-earth compounds $NdVO_4$, $SmVO_4$, $EuVO_4$, and YVO_4 all crystallize with the zircon ($ZrSiO_4$) structure like YPO_4 and $YAsO_4$. The pyrovanadate ZrV_2O_7 is isostructural with ZrP_2O_7.

Although V^V thus behaves like P^V in forming the ions VO_4^{3-} and $V_2O_7^{4-}$ it shows a much greater tendency than P to form condensed oxy-ions. Soluble ortho-, meta-, and pyro-vanadates are known, but the order of stability of these salts in aqueous solution is the reverse of that of the corresponding phosphates. Whereas meta- and pyro-phosphates are converted into orthophosphates on boiling in solution, that is, the complex ions break down into simple PO_4^{3-} ions, the orthovanadate ion VO_4^{3-} is rapidly converted into the pyro-ion $V_2O_7^{4-}$ in the cold, and on boiling further condensation into metavanadate ions $(VO_3)_n^{n-}$ takes place. The particular type of vanadate which is obtained from a solution depends on the temperature and also on the acidity of the solution. Acidification of a vanadate solution leads to the formation of deeply-coloured polyvanadates. A weakly acid solution of potassium vanadate contains the $V_{10}O_{28}^{6-}$ ion (p. 515) but on long standing or heating less soluble polyvanadates are precipitated, of which KV_3O_8[4] and $K_3V_5O_{14}$[5] are examples. Both these salts have layer structures, and there is apparently somewhat irregular coordination of V by 5 or 6 O atoms. In KV_3O_8 the metal atom has one close O neighbour (at 1.6 Å), four more at about 1.9 Å, these five forming a square pyramid, and a sixth O at a much greater distance (2.5 Å) completing a very distorted octahedron. This irregular coordination, which is characteristic of many complex oxy-compounds of vanadium, is in marked contrast to that of P, which retains tetrahedral coordination in all phosphates. There is, on the other hand, quite regular tetrahedral coordination of V^V in $Co_3V_2O_8$[5a] where V occupies tetrahedral and Co octahedral interstices in a c.c.p. array of O atoms. The similar coordination of V^V in $K_2V_3O_8$ is to be expected, since this compound is potassium vanadyl divanadate, $K_2(VO)V_2O_7$. The detailed environment of V in the colourless, hygroscopic $NaVO_3$ (diopside structure) has not been determined, but in both $KVO_3 . H_2O$[6] and V_2O_5[7] there is 5-coordination of V, though of rather different kinds.

In Fig. 12.19(a) we show a 'meta' chain formed of tetrahedral groups linked through two vertices. If now we place together two chains of type (a) each V has a fifth

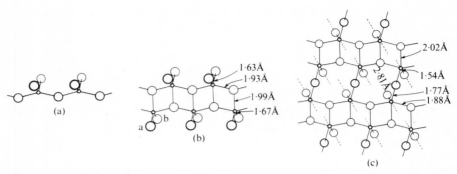

FIG. 12.19. (a) Metavanadate ion $(VO_3)_n^{n-}$; (b) double chain ion in $KVO_3.H_2O$; (c) layer of the structure of V_2O_5.

O neighbour in the other chain, as at (b). This double chain occurs in $KVO_3.H_2O$ with the bond lengths shown, the five O atoms around V being arranged at the vertices of a distorted trigonal bipyramid with V displaced towards two of the equatorial O atoms. (The K^+ ions and H_2O molecules are accommodated between the chains.) In $Sr(VO_3)_2.4H_2O^{(8)}$ there are chains similar to those in $KVO_3.H_2O$.

If these double chains are now joined through the atoms marked a to form layers we have the structure of V_2O_5, illustrated diagrammatically at (c). We may distinguish as O_1, O_2, and O_3 oxygen atoms attached to one, two, and three V atoms respectively. The bond lengths in $KVO_3.H_2O$ and V_2O_5 are as follows:

$$KVO_3.H_2O \qquad\qquad V_2O_5$$

$$V \begin{cases} 2\,O_1 & 1.63,\ 1.67\ \text{Å} \\ 2\,O_3 & 1.93 \\ 1\,O_3 & 1.99 \end{cases} \qquad V \begin{cases} 1\,O_1 & 1.59\ \text{Å} \\ 1\,O_2 & 1.78 \\ 2\,O_3 & 1.88 \\ 1\,O_3 & 2.02 \end{cases}$$

In $KVO_3.H_2O$ there are no further O neighbours which could be included in the coordination group, but in V_2O_5 V has a sixth neighbour in the adjacent layer at a distance of 2.8 Å, as indicated by the dotted lines in Fig. 12.19(c). The idealized structure of V_2O_5 is shown in Fig. 12.22(g) as built of octahedra to indicate its relation to the structure of MoO_3. In view of the V—O distances it is more realistic to describe the coordination as square pyramidal. If the structure of Fig. 12.22(g) is viewed in the direction of the arrow (top left) it appears as in Fig. 12.20(a), each

FIG. 12.20. The structure of V_2O_5 (see text).

square representing, as before, a ReO_3-type chain perpendicular to the paper, these chains now being those along the direction of the arrow. The longer (sixth) V–O bond in each 'octahedron' is shown as a broken line. Representing the coordination group of V as a square pyramid, that is, removing the sixth O from each octahedron, the structure is then represented as in Fig. 12.20(b). This characteristic coordination group is also found for all the V atoms in two vanadium bronzes, one of which has essentially the same structure as V_2O_5, while in some other bronzes there is both 5- and 6-coordination (for example, $Li_{1+x}V_3O_8$) or only 6-coordination ($Ag_{0.68}V_2O_5$) – see Chapter 13.

Vanadium oxyhydroxides. As noted in Chapter 5 some interesting layer structures arise by lateral vertex–vertex linking of rutile-type chains. A number of these are found as the structures of minerals, some of which are dark-coloured compounds containing V in more than one oxidation state. They are listed in Table 12.9 which

TABLE 12.9

Structures of vanadium oxyhydroxides

Figure 12.21	Formula	Mineral name	Oxidation state of V	Reference
(a)	$VO_2(OH)$	Paraduttonite	5 ⎱	AC 1958 **11** 56
	$VO(OH)_2$	Duttonite	4 ⎰	
(b)	$V_3O_4(OH)_4$	Doloresite	4	AM 1960 **45** 1144
	$V_3O_3(OH)_5$	Protodoloresite	$3\frac{2}{3}$	AM 1960 **45** 1144
(c)	$V_4O_4(OH)_6$	Häggite	$3\frac{1}{2}$	AM 1960 **45** 1144
	$VO(OH)$	Montroseite	3	AM 1953 **38** 1242

shows the apparent (mean) oxidation state of the metal and also includes the end-member of the series, $VO(OH)$, which is isostructural with goethite and diaspore. The layers in these structures are held together by O–H...O bonds as shown in Fig. 12.21. Comparison of Fig. 12.21(c) with the idealized structure of V_2O_5, Fig. 12.22(g), shows that häggite is closely related to V_2O_5, into which it is converted on heating. There is considerable distortion of the VO_6 octahedra in these minerals; for example in häggite, V has 1 O at 1.82 Å and 5 more at a mean distance of 2.00 Å, and in doloresite the short bond is 1.70 Å with again 5 more at 2.00 Å (mean).

(1) JM 1953 **5** 292
(2) ACSc 1954 **8** 221, 1599; ACSc 1960 **14** 465
(3) AC 1975 **B31** 2688
(4) IC 1966 **5** 1808
(5) ACSc 1959 **13** 377
(5a) AC 1973 **B29** 2304
(6) AC 1954 **7** 801
(7) ZK 1961 **115** 110

The oxides of molybdenum and tungsten

For an introduction to these and other oxides based on octahedral MO_6 coordination groups we refer the reader to the survey of such structures in Chapter 5.

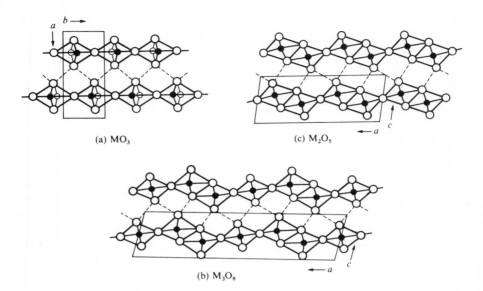

(a) MO$_3$

(c) M$_2$O$_5$

(b) M$_3$O$_8$

FIG. 12.21.　Structures of vanadium oxyhydroxides: (a) VO(OH)$_2$; (b) V$_3$O$_4$(OH)$_4$; (c) V$_4$O$_4$(OH)$_6$.

We shall discuss the oxides of Mo and W together, but it should be noted that there is only a general structural similarity between these compounds. For example, both trioxides are built of octahedral MO$_6$ groups and there are close relationships between certain of the intermediate complex oxides, but of the oxides of these two metals only the dioxides are isostructural and Mo does not, for example, form compounds analogous to the W bronzes at atmospheric pressure, though at higher pressures Mo does form bronzes with structures similar to those of the tungsten bronzes (p. 613).

In general the structural chemistry of Mo oxy-compounds is more complex than that of the analogous W compounds; the reason for this difference is not known.

Before proceeding to the higher oxides of these metals we should mention that both are said to form an oxide M$_3$O. The second form of tungsten metal described as 'β-tungsten' is produced by such methods as electrolysis of fused mixtures of WO$_3$ and alkali-metal phosphates or of fused alkali-metal tungstates at temperatures below 700 °C; above this temperature it turns irreversibly into α-W. It has been suggested[1] that β-W is in fact an oxide W$_3$O, and that the six W and two O atoms in the unit cell are arranged statistically in the eight positions (6-fold, open circles, and 2-fold, shaded circles) of Fig. 29.4 (p. 1283). (In Cr$_3$O[1a] the metal atoms are supposed to occupy the 6-fold and the O atoms the 2-fold positions.) On the other hand, it has been shown[2] that β-W can be prepared with less than 0.01 atoms of O per atom of W (by reducing WO$_{2.9}$ by hydrogen), though the presence of a small number of impurity atoms seems essential to the stability of the β-W structure. The

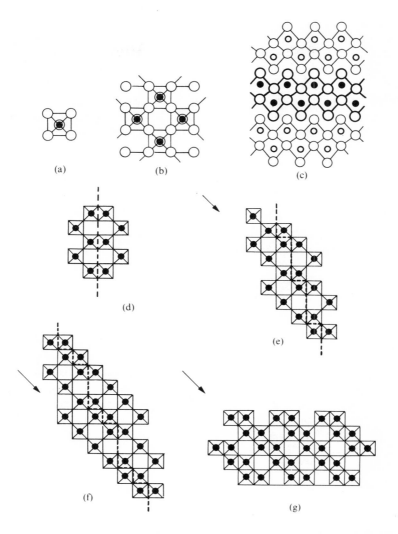

FIG. 12.22. (a) End-on view of infinite chain of MO_6 octahedra sharing opposite vertices; (b) ReO_3 structure viewed along direction of chains; (c)–(g) oxide structures in which chains of octahedra also share equatorial edges: (c) MoO_3 (layer) structure; (d) hypothetical structure for oxide M_nO_{3n}; (e) and (f) modes of linking slices of ReO_3 structure in oxides M_nO_{2n-1} and and M_nO_{3n-2}; (f) idealized structure of V_2O_5.

oxide Mo_3O apparently has a defective 'anti-BiF_3' structure (p. 422) with nine Mo arranged at random in nine of the twelve positions (000), etc., (4-fold), $(\frac{1}{4}\frac{1}{4}\frac{1}{4})$, etc., (8-fold), and three O in the 4-fold position $(\frac{1}{2}\frac{1}{2}\frac{1}{2})$, etc.[3] More recently doubt has been expressed about the existence of Mo_3O (or Cr_3O).[4]

Dioxides and trioxides. The (isostructural) dioxides have distorted rutile structures (p. 541), but the trioxides are not isostructural. MoO_3 has a unique layer structure (Fig. 12.22(c)).[5] Each octahedral MoO_6 group shares two adjacent edges with similar groups and, in a direction perpendicular to the plane of the paper in Fig. 12.22, the octahedra are linked through vertices. Three O atoms of each MoO_6 octahedron are therefore common to three octahedra, two are shared between two octahedra, and the sixth is unshared. (See also p. 222). The octahedral coordination in MoO_3 is highly distorted, and this characteristic distortion of $Mo^{VI}O_6$ groups is found also in other compounds (e.g. heteropolyacids, p. 520); for Mo–O bond lengths see p. 529.

WO_3 provides an extraordinarily complex example of polymorphism, for no fewer than eight phase transitions have been observed up to 900 °C. All polymorphs apparently represent distortions of the cubic ReO_3 structure:

	-40		17		330		740		900 °C
monoclinic (I)		triclinic[6]		monoclinic (II)[7]		orthorhombic[8]		tetragonal		cubic

The metal atom is displaced from the centre of the O_6 octahedron resulting in bond lengths ranging from 1.7–2.0 Å. A characteristic feature of the structures studied is the alternation of longer and shorter bonds along the axial directions.

A further (hexagonal) form of WO_3 has been prepared by, and only by, the dehydration of $WO_3 . \frac{1}{3} H_2O$.[9] This has the hexagonal W bronze framework – see Fig. 10.11(a), p. 469 and Fig. 13.12(d), p. 613. It is not a 'bronze', for it is a white powder, and there is no evidence for H_2O molecules or foreign ions in the tunnels. There is apparently very little distortion of the WO_6 octahedra in this structure.

The presence of polymeric species in the vapours of CrO_3, MoO_3, and WO_3 has been established by mass-spectrometric and e.d. studies. Molecules M_3O_9 (M = Cr, Mo, W) form chair-shaped rings like S_3O_9 in which the following bond lengths have been determined: Cr–O (mean), 1.68 Å; $Mo–O_t$, 1.67 Å and $Mo–O_b$, 1.89 Å; and W–O (mean), 1.77 Å.[10]

(1)	AC 1954 7 351	(6)	AC 1978 **B34** 1105
(1a)	ACSc 1954 8 221	(7)	AC 1969 **B25** 1420
(2)	ZaC 1957 **293** 241	(8)	AC 1977 **B33** 574
(3)	ACSc 1954 8 617	(9)	JSSC 1979 **29** 429
(4)	ACSc 1962 **16** 2458	(10)	JMSt 1971 8 31
(5)	AK 1963 **21** 357		

Intermediate oxides. By heating together the metal and trioxide *in vacuo* to temperatures up to 700 °C for varying times a number of intermediate oxides of these metals have been prepared.[1] They have been characterized by means of X-ray powder photographs and most of their structures have been determined from single-crystal data. These oxides, which are blue to violet in colour, include Mo_4O_{11},[2] Mo_5O_{14},[3] Mo_8O_{23},[4] Mo_9O_{26},[4] and $Mo_{17}O_{47}$,[5] $W_{18}O_{49}$[6] (in earlier literature W_2O_5 or W_4O_{11}), $W_{20}O_{58}$,[7] $W_{24}O_{70}$,[7a] $W_{25}O_{73}$,[7b] and $W_{40}O_{118}$.[8]

In the orthorhombic form of Mo_4O_{11} three-quarters of the Mo atoms are 6-coordinated and the remainder tetrahedrally coordinated, and the structure may be regarded as consisting of slices of a ReO_3-like structure connected by 4-coordinated Mo atoms. The oxides Mo_5O_{14}, $Mo_{17}O_{47}$, and $W_{18}O_{49}$ are of a different type, having both octahedral and pentagonal bipyramidal coordination of the metal atoms (see 'Bronzes', Chapter 13).

The oxides Mo_8O_{23} and Mo_9O_{26}, a number of mixed oxides, and $W_{20}O_{58}$ form a group based on a common structural principle. In the ReO_3 structure octahedra are joined through vertices only (Fig. 12.22(b)). If there is sharing of edges of octahedra at regular intervals there arise series of related structures which may be regarded as built of slices of ReO_3-like structure. The simplest possibility would be that shown in Fig. 12.22(d), with octahedra of pairs of chains sharing edges. The broken line marks the junction of the two portions of ReO_3 structure. Since no O atom is common to more than two octahedra the formula is still M_nO_{3n}. No example of this structure is known. More complex arrangements would involve octahedra belonging to groups of 4, 6, etc. ReO_3 chains sharing equatorial edges, as at (e) and (f), with formulae M_nO_{3n-1} and M_nO_{3n-2} respectively, the value of n depending on the repeat distance in the direction of the arrows in Fig. 12.22(e) and (f).[9]

Examples of binary oxides with these structures are Mo_8O_{23} and Mo_9O_{26} in the M_nO_{3n-1} series and $W_{20}O_{58}$, $W_{24}O_{70}$, $W_{25}O_{73}$, and $W_{40}O_{118}$ in the M_nO_{3n-2} series. A larger portion of the structure of Mo_8O_{23} is shown in Fig. 12.23. Higher members

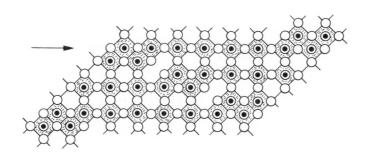

FIG. 12.23. The crystal structure of Mo_8O_{23}.

of the M_nO_{3n-1} series are represented by mixed (Mo, W) oxides, in which it appears that the proportion of W is of fundamental importance in determining the width of the ReO_3-like slices. The increasing value of n with increasing W content may be compared with the sharing of only vertices of WO_6 octahedra in WO_3 — contrast the complex structure of MoO_3.

Per cent W	Structure found[10]
0	Mo_8O_{23}, Mo_9O_{26}
14–24	$(Mo, W)_{10}O_{29}$, $(Mo, W)_{11}O_{32}$
32–44	$(Mo, W)_{12}O_{35}$
48	$(Mo, W)_{14}O_{41}$

In Fig. 12.22(d), (e), and (f) there are respectively 1, 2, and 3 *pairs* of adjacent MO_6 octahedra sharing edges; the limiting case would be the structure shown in (g), corresponding to the composition M_2O_5. This represents the idealized structure of V_2O_5 (p. 567).

The rather irregular coordination of V by five O may be described as square pyramidal or, since there is a sixth O at a much greater distance, as highly distorted octahedral. Although the sixth O is at a distance nearly twice as great as that to the nearest O, this interesting relation between the idealized structure of Fig. 12.22(g) and that of MoO_3 in Fig. 12.22(c) is apparently the reason for the considerable solubility of MoO_3 in V_2O_5. At 650 °C about 30 per cent of the V atoms can be substituted by Mo, and the structure of $(Mo_{0.3}V_{0.7})_2O_5$[11] is of the same general type as V_2O_5. MoO_3 does not, however, take up V_2O_5 in solid solution; this would require an excess of metal atoms in the structure.

Another series of molybdenum oxides, Mo_nO_{3n-m+1} ($n = 13$, $m = 2$; $n = 18$, $m = 3$; $n = 26$, $m = 4$), based on the MoO_3 structure, has been described.[12]

Two compounds isostructural with MoO_3 are mentioned in other chapters, namely, $Mo_4O_{11.2}F_{0.8}$ (listed in Table 10.11, p. 481 as $Mo_4O_{11}F$), and $Mo_2O_5(OH)$ (p. 618). The environments of Mo in these compounds are almost identical, $(1+4+1)$-coordination, but different from that in MoO_3, $(2+2+2)$-coordination:

	Distances to 6 nearest neighbours (Å)				
$Mo_4O_{11.2}F_{0.8}$	1.65		1.95 (four)	2.31	
$Mo_4O_{10}(OH)_2$	1.69		1.96 (four)	2.33	
MoO_3	1.67	1.73	1.95 (two)	2.25	2.33

The oxyfluoride $MoO_{2.4}F_{0.6}$ with higher F content is cubic (ReO_3 structure).

(1)   ACSc 1959 **13** 954; IC 1966 **5** 136
(2)   ACSc 1964 **18** 1571
(3)   AC 1975 **B31** 1666
(4)   ACSc 1948 **2** 501
(5)   ACSc 1963 **17** 1485
(6)   AK 1963 **21** 471
(7)   AK 1950 **1** 513
(7a) JSSC 1980 **35** 120
(7b) AC 1976 **B32** 2144
(8)   ACSc 1965 **19** 1514
(9)   AC 1953 **6** 495
(10) ACSc 1955 **9** 1382
(11) ACSc 1967 **21** 2495
(12) AK 1963 **21** 443

13

Complex oxides

Introduction

In Chapter 11 we distinguished between oxy-salts and complex oxides, but observed that there is no hard and fast dividing line between the two groups of compounds. As regards their crystal structures we may distinguish two main classes of complex oxide.

I. The positions of the atoms are the same (or essentially the same) as in a binary oxide.

(a) In most binary oxides in which all metal atoms are in the same oxidation state the environment of all the metal atoms is the same or approximately the same. (Exceptions include the minor difference between the two types of distorted octahedral coordination group in the C-M_2O_3 structure and the much larger difference in β-Ga_2O_3, with tetrahedral and octahedral coordination of the metal, and the B-M_2O_3 structure, with 6- and 7-coordination of the metal.) In the complex oxide with such a structure there may be random arrangement of atoms of two or more metals (*statistical structure*) or a regular arrangement (*superstructure*).

(b) If the binary oxide contains the metal in two oxidation states there may be appreciably different environments of the two kinds of metal ion as, for example, in Pb_3O_4 (3- and 6-coordination of Pb(II) and Pb(IV)) or Eu_3O_4 (6- or 8-coordination of Eu(III) and Eu(II)). Such structures are also possible for complex oxides, the structure usually being a regular one (like that of the binary oxide) rather than a statistical one. Some structures common to simple and complex oxides are listed in Table 13.1.

II. In some complex oxide structures the environments of the different kinds of metal ion are so different that the structure is not possible for a binary oxide. The size difference between the ions necessary for the stability of the structure may be too large (as in the perovskite and related structures) or the two (or more) oxidation states required for charge balance in the structure may not be possible for one metal. (The scheelite structure, for example, calls for equal numbers of 8- and 4-coordinated atoms in oxidation states totalling 8.) It should be noted that positions of different coordination number in complex oxide structures are not necessarily occupied by atoms of different metals. Just as one kind of ion occupies coordination groups of two kinds in certain exceptional binary oxides, as noted above, so we find the same phenomenon in some complex oxides. In the garnet structure (p. 605) there are positions of 4-, 6-, and 8-coordination for metal ions. In some garnets ions of the same kind occupy sites of 4- and 6-coordination and in others sites of 6- and 8-coordination. It is, however, unlikely that all three types of site would be occupied by ions of the same kind in a particular crystal. Structures of this second main

TABLE 13.1
Structures common to simple and complex oxides

Structure	Simple oxide	Complex oxide	
		statistical	*superstructure*
NaCl	MgO etc.	Li_2TiO_3, $LiFeO_2$	$LiNiO_2$, $LiInO_2$
Wurtzite	ZnO		$LiGaO_2$
β-BeO	β-BeO		$LiAlO_2$
Rutile	TiO_2 etc.	$CrTaO_4$, $CrNbO_4$	$ZnSb_2O_6$
Corundum	α-Al_2O_3		$FeTiO_3$
			$LiNbO_3$
C-M_2O_3	Mn_2O_3, etc.	$CaUO_3$	
Columbite ⎫	ReO_2 (high)		$FeNb_2O_6$
Wolframite ⎭	α-PbO_2		$NiWO_4$

	Oxide with same metal in two oxidation states	*Statistical*	*Regular structure*
	α-Sb_2O_4		$SbNbO_4$, $SbTaO_4$
	β-Sb_2O_4		$BiSbO_4$
Spinel	Fe_3O_4 etc.	Inverse spinels	$MgAl_2O_4$
			$LiAl_5O_8$ (low)
	Pb_3O_4		$ZnSb_2O_4$
$CaFe_2O_4$	Eu_3O_4		$SrEu_2O_4$
Pseudobrookite	Ti_3O_5		Fe_2TiO_5
	Mn_5O_8		$Cd_2Mn_3O_8$
	Tb_7O_{12}		$U^{VI}M^{III}_6O_{12}$
	$Nb_{12}O_{29}$	$Ti_2Nb_{10}O_{29}$	

group could be described as characteristically or exclusively *complex oxide structures*.

Many complex oxide structures can be described in such a way as to emphasize a particular feature of the structure and have been so described in previous chapters, in particular the following groups:

Structures containing chains, layers based on simple plane nets, or frameworks based on simple 3D nets (Chapter 3);

Structures in which the O atoms are close-packed and M atoms occupy tetrahedral and/or octahedral interstices. In a special group $A_nB_mO_{3n}$ the A and 3 O atoms together form the c.p. array (Chapter 4);

Structures, not necessarily close packed, built from tetrahedral and/or octahedral coordination groups (Chapter 5).

The grouping of structures in this way stresses the geometry or topology of the structure; we adopt here a more 'chemical' classification. We first deal with the simpler structures arranged in order of increasing complexity of chemical formula and then devote separate sections to certain selected groups of compounds.

One general point may be noted here. Owing to the similarity in size and electro-negativity of F^- and O^{2-} there are many resemblances between the structures of fluorides and oxides of similar formula-type, and also between those of oxyfluorides and oxides (Table 10.11, p. 481), but there are also many examples of isostructural complex chlorides and complex oxides (Table 10.2, p. 448).

Oxides ABO$_2$

Many oxides $M^I M^{III} O_2$ have structures closely related to those of oxides $M^{II} O$ (Table 13.2). The octahedral structures include the disordered NaCl structure and

TABLE 13.2
Structures of oxides $M^I M^{III} O_2$

Coordination of metal atoms	Structure	Examples
Both tetrahedral	Zinc-blende superstructure	h.p. LiBO$_2$ [1]
	Wurtzite superstructure	LiGaO$_2$ [2]
	β-BeO	γ-LiAlO$_2$ [3]
Both octahedral	Disordered NaCl	h.t. LiFeO$_2$, NaErO$_2$
	NaCl superstructures	
	Tetragonal	LiFeO$_2$, LiInO$_2$ [4]
		LiScO$_2$, LiEuO$_2$
	Rhombohedral	LiNiO$_2$, LiVO$_2$
		NaFeO$_2$, NaInO$_2$ [5]
	Cubic	h.p. LiVO$_2$
	Monoclinic	NaLnO$_2$ [6]
M^I_2 collinear	HNaF$_2$ (CuFeO$_2$)	CuCrO$_2$ [7]
M^{III}6 octahedral		δ-AgFeO$_2$ [8]

(1) JCP 1966 **44** 3348. (2) AC 1965 **18** 481. (3) AC 1965 **19** 396. (4) JCP 1966 **44** 3348 (review of oxides LiMO$_2$). (5) ZaC 1958 **295** 233. (6) JPCS 1972 **33** 1927; AC 1972 **B28** 722. (7) JACS 1955 **77** 896. (8) AC 1972 **B28** 1774

various superstructures described on p. 244. Of these the tetragonal superstructure is illustrated in Fig. 6.4, p. 244, and the relation between the rhombohedral superstructure and the NaHF$_2$ structure (which should be written HNaF$_2$ in the present context) has been described on p. 270. Oxides (NaLnO$_2$) containing lanthanide elements provide examples of three of the NaCl superstructures:

Lu	Yb	Tm	Er	Y	Ho	Dy	Tb	Gd	Eu	Sm	Nd	La
r^{3+} 0.82				0.89				0.94				1.06 Å

←—— Monoclinic ——→
←——Rhombohedral——→ ←——Tetragonal——→

The CuFeO$_2$ structure is adopted by numerous Ag and Cu compounds $M^I M^{III} O_2$ (e.g. M^{III} = Al, Cr, Co, Fe, Ga, Rh), and also by PdCoO$_2$, PdCrO$_2$, PdRhO$_2$, and

$PtCoO_2$.[1] It has been confirmed (Mössbauer) that $CuFeO_2$ contains Fe^{3+} (and therefore Cu^+), and apparently the Pd and Pt compounds contain the noble metal in the formal oxidation state +1. (There is some doubt about the stoichiometry of $PtCoO_2$; if it is metal-deficient it could be $Pt_{0.8}^{2+}Co_{0.8}^{3+}O_2$.) However, there is apparently metal–metal bonding in the Pd and Pt compounds. In the $HNaF_2$ structure for compounds ABO_2 there is linear 2-coordination of A by O, but A also has 6 A neighbours, hence the coordination of Pt and Pd in these compounds should probably be described as 2+6 (hexagonal bipyramidal), the six equatorial neighbours being metal atoms (Table 13.3). Delocalization of electrons in the hexagonal layers of

TABLE 13.3

Metal–metal distances in oxides ABO_2

Compound	A-6 A	Distance in metal	
$CuFeO_2$	3·04 Å	2·56 Å (Cu)	semiconductors
$AgFeO_2$		2·89 Å (Ag)	
$PtCoO_2$	2·83 Å	2·77 Å (Pt)	metallic conductors
$PdCoO_2$		2·75 Å (Pd)	

metal atoms is responsible for the metallic conduction, suggesting the formulation $Pt^{II}Co^{III}O_2(e)$.

Metaborates MBO_2 generally contain metaborate ions, but under high pressures $LiBO_2$ crystallizes as a zinc-blende superstructure.

Several structures of oxides $M^{II}M^{II}O_2$ have been determined in which ions of appreciably different size are accommodated. In $SrZnO_2$[2] layers of ZnO_4 tetrahedra (sharing all vertices) provide sites for Sr^{2+} ions in 7-coordination between the layers (see Fig. 5.6, p. 196 for the layer), while in $BaZnO_2$[3] the ZnO_4 tetrahedra form a quartz-like framework containing Ba^{2+} ions in positions of irregular 8-coordination. In contrast to $BaZnO_2$ (and the isostructural Co and Mn compounds) $BaCdO_2$[4] has a structure in which there is rather irregular (4+1)-coordination of Cd and 7-coordination of Ba. In $BaNiO_2$[5] there are perovskite-like layers from which one-third of the O atoms are missing. The layers are stacked in approximate h.c.p. with Ni atoms between the layers in what would have been positions of octahedral coordination, but because of the absence of two of the O neighbours Ni has only 4 (coplanar) neighbours.

(1) JSSC 1971 **10** 713, 719, 723
(2) ZaC 1961 **312** 87
(3) AC 1960 **13** 197; ZaC 1960 **305** 241

(4) ZaC 1962 **314** 145
(5) AC 1951 **4** 148

Oxides ABO_3

The two largest groups of oxides ABO_3 are (a) those containing ions A and B of approximately the same size and of a size suitable for octahedral coordination by oxygen (e.g. 3d ions M^{2+} and M^{3+}, Zn^{2+}. Mg^{2+}, In^{3+}, etc.), and (b) those containing a much larger ion which together with O^{2-} can form c.p. layers AO_3. Oxides of class (a) adopt a sesquioxide structure with either a random or ordered arrangement of A and B ions, or in some cases a structure peculiar to a small number of compounds (e.g. $LiSbO_3$). For examples of structures of class (a), with references, see Table 13.4. The ilmenite ($FeTiO_3$) and $LiNbO_3$ structures differ in the way shown in Fig. 6.19, p. 267, that is, the distribution of the two kinds of ion among a given set of octahedral sites. On the other hand a different selection of metal-ion sites is occupied in $LiSbO_3$, as shown in Fig. 4.24(f); $NaSbO_3$, and $NaBiO_3$ have the ilmenite structure. Typically, transition-metal oxides $M^{3+}M^{3+}O_3$ adopt a random corundum structure, andoxides $M^{2+}M^{4+}O_3$ the ilmenite superstructure, for example:

Corundum	FeV,	MnFe,	FeCr,	InFe	(under pressure)
	TiV		VCr	GaFe	
			NiCr		
Ilmenite	FeTi		$CoMn^{4+}$	CoV^{4+}	(distorted)
	CoTi		NiMn	NiV	
	NiTi		MgMn	MgV	

Structures that do not fall into the two classes (a) and (b) include those of $KSbO_3$, $LuMnO_3$, and $PbReO_3$ (and isostructural oxides). In the $KSbO_3$ structure described on p. 224 and illustrated in Fig. 5.33 on p. 225 the K^+ ions occupy cavities in a framework constructed from edge-sharing pairs of SbO_6 octahedra. At ordinary temperatures the arrangement of K^+ ions is an ordered one (space group Pn3), but if prepared from the ilmenite form of $KSbO_3$ under high pressure and temperature it forms a disordered structure (also cubic, space group Im3) with the

TABLE 13.4

Structures of oxides ABO_3

	C.n.s of A *and* B	Structure	Examples	Reference
(a)	6:6	C-M_2O_3 Corundum:	$ScTiO_3$, $ScVO_3$	IC 1967 6 521
		random superstructure	$A^{3+}B^{3+}O_3$	
		ilmenite	$A^{2+}B^{4+}O_3$	AC 1964 17 240; JSSC 1970 1 512
		$LiNbO_3$	$A^+B^{5+}O_3$	AC 1968 **A24** 583
		$LiSbO_3$		ACSc 1954 8 1021
			$BiSbO_3$	ACSc 1955 9 1219
(b)	12:6	c.p. structures	(see Table 13.5)	

same framework. Heat treatment at $1200\,^{\circ}$C under atmospheric pressure converts this to the ordered structure. All the compounds $MSbO_3$ (M = Li, Na, K, Rb, Tl, and Ag) form the disordered structure; the Li compound is an ionic conductor owing to the movement of these ions in the tunnels in the structure.[1] In $LuMnO_3$[2] there is 5- and 7-coordination of the metal ions.

In some systems the oxygen-deficient pyrochlore structure $A_2B_2O_{7-x}$ is stable at $x = 1$ (e.g. $PbReO_3$,[3] $BiYO_3$, $AgSbO_3$, $BiScO_3$, $PbTcO_3$). Pressure may convert it to the perovskite structure, as in the case of $BiYO_3$ and $BiScO_3$. This change involves a rearrangement of the vertex-sharing BO_6 octahedra to form a different 3D framework and an increase in the c.n. of A to 12. In other systems the phase stable at atmospheric pressure has a defect pyrochlore structure with $x < 1$. For example, heating the components in stoichiometric proportions gives $Pb_2Ru_2O_{7-x}$ which at $1400\,^{\circ}$C under a pressure exceeding 90 kbar converts to a perovskite-type $PbRuO_3$ with loss of oxygen.

The structures based on c.p. AO_3 layers have the composition ABO_3 if all the octahedral holes are occupied by B atoms. Together with some closely related structures, in which some of these holes are unoccupied, they form the subject-matter of the next section.

(1)   JSSC 1974 9 345   (3)   MRB 1969 4 191
(2)   AC 1975 B31 2770

Structures based on close-packed AO_3 layers

Certain cations comparable in size with O^{2-} form c.p. layers AO_3 which can be stacked in various c.p. sequences. Smaller cations can then occupy the octahedral holes between groups of six O^{2-} ions to form structures of the type $A_nB_mO_{3n}$. Some of these structures have been described in Chapter 4, and it was noted in Chapter 5 that these structures may alternatively be described in terms of the way in which the BO_6 octahedra are linked together. Only vertices and/or faces are shared, and the extent of face-sharing is indicated in the upper part of Table 13.5, which includes only structures ABO_3, that is, those in which all the octahedral B sites are occupied. The structure of hexagonal $BaTiO_3$ is compared with perovskite in Fig. 13.1, and in Fig. 13.2 we show sections through three structures to illustrate the relations between the BO_6 octahedra. We shall deal in detail only with the perovskite structure.

The structures most frequently found are those with 2-, 3-, 4-, 6-, and 9-layer repeat units, but the layer sequence can be changed by pressure, which also unavoidably involves a rise in temperature. It appears that the amount of 'hexagonal character' in a particular compound decreases with pressure and also with decreasing size of A in a series such as $BaMnO_3$, $SrMnO_3$, and $CaMnO_3$ (Table 13.6). The same basic ABO_3 structure types are found in compounds of two other kinds. (a) Some of the B sites are occupied by Li^+, when less usual stacking sequences are found, as in $Ba_4Nb_3LiO_{12}$ (8-layer) and $Ba_5W_3Li_2O_{15}$ (10-layer).[1] In the latter compound

TABLE 13.5

Structures with close-packed AO_3 *layers*

Number of layers	Type of closest packing	Face-sharing groups	Example	Reference
2	h	Chains	$BaNiO_3$	AC 1976 **B32** 2464
3	c	–	Perovskites	
4	hc	Pairs	High-$BaMnO_3$	AC 1962 **15** 179
6	hcc	Pairs	hex. $BaTiO_3$	AC 1948 **1** 330
9	chh	Groups of 3	$BaRuO_3$	IC 1965 **4** 306

Ratio of filled to vacant MO_6	Type of closest packing	Example	Reference
2:1	h	$Ba_3W_2\square O_9$	ZaC 1979 **455** 65
	hhc	$Ba_3Re_2\square O_9$	IC 1978 **17** 699
3:1	ccch	$Sr_4Re_2Sr\square O_{12}$	} IC 1965 **4** 235
	cchh	$Sr_4Re_2Mg\square O_{12}$	
4:1	hhccc	$Ba_5Ta_4\square O_{15}$	AC 1970 **B26** 102
5:1	hhcccc	$Ba_6B_3^{III}\square W_3O_{18}$	ZaC 1979 **448** 119

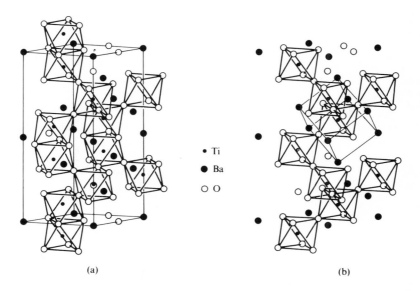

- Ti
- Ba
○ O

(a)   (b)

FIG. 13.1. The crystal structures of barium titanate, $BaTiO_3$: (a) hexagonal; (b) cubic.

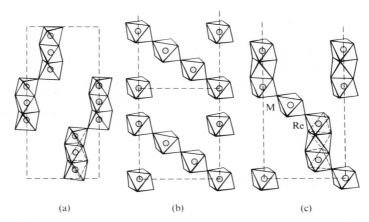

FIG. 13.2. Sections through structures built of AO_3 layers showing how the BO_6 octahedra are linked through vertex- or face-sharing: (a) the 9-layer structure of $BaRuO_3$; (b) the 5-layer structure of $Ba_5Ta_4O_{15}$, showing only the octahedra occupied by Ta atoms; and (c) the 12-layer structure of $A_4Re_2MO_{12}$. In all cases the A atoms are omitted, and in (c) only nine c.p. layers are shown.

TABLE 13.6

Effect of pressure on c.p. ABO_3 *and* ABX_3 *structures*

C.P. type	h	hhc	hc	cch	c
Symbol †	2 H	9 R	4 H	6 H	3 C
Groups of face-sharing octahedra	infinite chains	(3)	(2)	(2) + (1)	—

CsMnF$_3$-I ⟶ CsMnF$_3$-II

CsNiF$_3$-I ⟶ CsNiF$_3$-II ⟶ CsNiF$_3$-III
  5 kbar   50 kbar

BaRuO$_3$-I → BaRuO$_3$-II → BaRuO$_3$-III
            15 kbar   30 kbar

BaMnO$_3$-I ⟶ BaMnO$_3$-II → BaMnO$_3$-III
  30 kbar   90 kbar

SrMnO$_3$-I ⟶ SrMnO$_3$-II
        50 kbar

CaMnO$_3$

† H = hexagonal, R = rhombohedral, C = cubic.

For SrMnO$_3$ and BaMnO$_3$ see. JSSC 1969 **1** 103, 506; JSSC 1971 **3** 323.

the strings of vertex-sharing octahedra are occupied alternately by Li and W and the face-sharing pairs statistically by Li + W. In such crystals the cation distribution is such as to minimize the repulsions between the highly charged M^{5+} or M^{6+} ions. (b) Some of the B sites are vacant. The examples in the lower part of Table 13.5 are arranged in order of decreasing proportion of vacant octahedral sites. Occupation of two-thirds of the B sites leads to the simple formulae $Ba_3W_2O_9$ and $Ba_3Re_2O_9$ (a compound which is strictly $Ba_3Re_{2-x}O_9$). In $Ba_3W_2O_9$ there are pairs of face-sharing WO_6 octahedra, the structure being similar to that of $Cs_3Tl_2Cl_9$ (p. 184); the W atoms are displaced from the centres of their octahedra to give a W—W separation of 2.90 Å. In the 9-layer $Ba_3Re_2O_9$ structure, on the other hand, the ReO_6 octahedra are separated by empty O_6 octahedra. The elevation is similar to that of $BaRuO_3$ (Fig. 13.2(a)), the middle octahedron of each group of three being vacant in the ideal stoichiometric structure.

A notable feature of these oxides is the presence of highly charged ions such as Ni^{4+} or Fe^{4+} in adjacent face-sharing octahedra, in particular, in continuous columns of face-sharing octahedra in the $BaNiO_3$ structure. The sharing of faces is associated with short metal–metal separations (e.g. Ru–Ru, 2.55 Å in $BaRuO_3$) and often with abnormal magnetic and electrical properties. The structure of $BaNiO_3$ shows that face-sharing does not necessarily imply m–m bonding. Although Ni–Ni is only 2.41 Å, shorter than in the metal (2.49 Å), $BaNiO_3$ is not a metallic conductor, that is, there are no Ni–Ni bonds. In a purely ionic crystal we should expect strong repulsion between Ni^{4+} ions at this separation. However, the O—O distance in shared faces is only 2.47 Å, as compared with other distances, 3.16 Å in layers and 2.76 Å between layers, suggesting that the Ni^{4+} ions are shielded one from another along the c direction. (The Ni—O bond length is 1.86 Å.)

Many oxides ABO_3 have been prepared at different temperatures with different layer sequences, for example, $BaMnO_3$ (2-, 4-, 6-, 8-, 10-, and 15-layers)[2] and $BaCrO_3$ (4-, 6-, and 27-layers),[3] but of the '$BaMnO_3$' phases only the 2-layer structure corresponds to stoichiometric $BaMnO_3$. Those prepared at high temperatures show increasing departures from stoichiometry, and this is also true of $BaFeO_3$, the composition depending on the firing temperature and furnace atmosphere. These variables determine the proportion of Fe^{4+}; for example, $BaFeO_{2.5}$ contains all the iron in the 3+ state, while the phase with the 6-layer structure is $BaFeO_{2.67}$,[4] containing one-third of the iron in the 4+ state. The forms of compounds 'ABO_3' with different layer sequences are often referred to, somewhat loosely, as polytypes, probably for want of another term. It should be noted that the same term is applied to the forms of CdI_2, ZnS, and SiC with different layer sequences, but these polytypes are stoichiometric compounds.

(1)   AC 1974 **B30** 816
(2)   AC 1976 **B32** 1003
(3)   JSSC 1980 **34** 59
(4)   AC 1973 **B29** 1217

The perovskite structure

This structure is adopted by a few complex halides and sulphides and by many complex oxides. The latter are very numerous because the sum of the charges on A and B (+6) may be made of 1+5, 2+4, or 3+3, and also in more complex ways as in $Pb(B'_{1/2}B''_{1/2})O_3$, where $B' = Sc$ or Fe and $B'' = Nb$ or Ta, or $La(B'_{1/2}B''_{1/2})O_3$, where $B' = Ni$, Mg, etc. and $B'' = Ru(IV)$ or $Ir(IV)$. Also, many compounds ABO_3 are polymorphic with as many as four of five forms, some of which represent only small distortions of the most symmetrical form of the perovskite structure. The study of these compounds under higher pressures has produced many more examples of polymorphism (e.g. $InCrO_3$ and $TlFeO_3$). The perovskite structure of compounds such as $CaFeO_3$ (high pressure), $SrFeO_3$, and $BaFeO_3$, is of interest as an example of the stabilization of a high oxidation state (Fe^{IV}) in an oxide structure.

The 'ideal' perovskite sttructure, illustrated in Fig. 13.3(a), is cubic, with A surrounded by 12 O and B by 6 O. Comparatively few compounds have this ideal

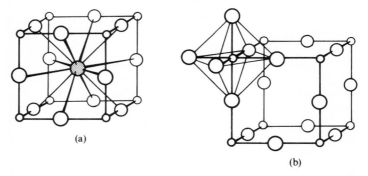

(a)

(b)

FIG. 13.3.   (a) The perovskite structure for compounds ABO_3 (or ABX_3). Large open circles represent O (or F) ions, the shaded circle A, and the small circles B ions.   (b) The crystal structure of ReO_3.

cubic structure, many (including the mineral perovskite, $CaTiO_3$,[4a] itself) having slightly distorted variants with lower symmetry. Some examples are listed in Table 13.7.

We have seen (p. 266) that the octahedra in the ideal ReO_3 structure (Fig. 13.3(b)) can be rotated relative to one another, producing in the limit a h.c.p. AX_3 structure. There is the same octahedral framework in the perovskite structure, and accordingly the octahedra in this structure may be tilted in various ways to give types of coordination of the A atoms other than the cuboctahedral 12-coordination of the ideal cubic structure. For example, in $GdFeO_3$[4b] the coordination of Gd is reduced to 8 (slightly distorted bicapped trigonal prism), while in certain oxides $AC_3B_4O_{12}$ (p. 588) the c.n. of A remains 12, but icosahedral, and that of C is reduced to 8. (In this connection see also Fig. 4.4(a), p. 147, and the description of

TABLE 13.7

Compounds with the perovskite type of structure

Ideal cubic structure	$SrTiO_3$, $SrZrO_3$, $SrHfO_3$ $SrSnO_3$, $SrFeO_3$ $BaZrO_3$, $BaHfO_3$, $BaSnO_3$ $BaCeO_3$ $EuTiO_3$, $LaMnO_3$

At least one form with distorted small cell ($a \approx 4$ Å):
     Cubic (C)
     Tetragonal (T)
     Orthorhombic (O)
     Rhombohedral (R)
	$BaTiO_3$ (C, T, O, R)
	 $KNbO_3$ (C, T, O, R)
	 $KTaO_3$ (C, ?)
	 $RbTaO_3$ (C, T)
	 $PbTiO_3$ (C, T)

Distorted multiple cells	$CaTiO_3$, $GdFeO_3$, $NaNbO_3$, $PbHfO_3$, $LaCrO_3$ low-$PbTiO_3$ low-$NaNbO_3$, high-$NaNbO_3$

the $CoAs_3$ structure, p. 267.) These variants of the perovskite structure have lower symmetry and are of special interest in connection with the dielectric and magnetic properties of these compounds. For example, many are ferroelectric, notably $BaTiO_3$, some are antiferroelectric, for example, $PbZrO_3$ and $NaNbO_3$ (see p. 587), and ferromagnetic ($LaCo_{0.2}Mn_{0.8}O_3$) and antiferromagnetic ($GdFeO_3$, $LaFeO_3$, etc.) compounds are known.

The complex oxide $BaTiO_3$ ('barium titanate') is remarkable in having five crystalline forms, of which three are ferroelectric. The structure of the high-temperature hexagonal form, stable from 1460 °C to the melting point (1612 °C), has already been described. The other forms are:

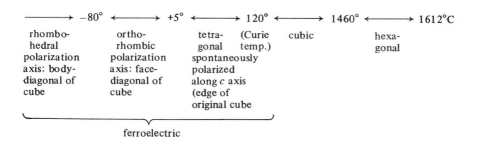

$KNbO_3$ shows the closest resemblance to $BaTiO_3$, having three transitions (to tetragonal, orthorhombic, and rhombohedral forms), while $NaNbO_3$ shows quite different dielectric behaviour, its transitions (at 350°, 520°, and 640 °C) being of a different kind from those in $KNbO_3$ and $BaTiO_3$.[5]

In attempts to understand the distortions from the cubic structure these oxides ABO_3 were first regarded as purely ionic crystals. From the geometry of the structure it follows that for the 'ideal' structure there is the following relation between the radii of the A, B, and O^{2-} ions:

$$r_A + r_O = \sqrt{2}(r_B + r_O).$$

Actually the cubic perovskite structure or slightly deformed variants of it are found for ions which do not obey this relation exactly, and this was expressed by introducing a 'tolerance factor':

$$r_A + r_O = t\sqrt{2}(r_B + r_O).$$

Provided that the ionic sizes are approximately right, the only other condition to be fulfilled is that the structure is electrically neutral, that is, that the sum of the charges on A and B is 6.

It then appeared that for all the compounds with the perovskite-type of structure the value of t lies between approximately 0.80 and 1.00—for lower values of t the ilmenite structure is found—and that for the ideal cubic structure t must be greater than 0.89. However, it is now evident that although in a given ABO_3 series (with either A or B constant) the use of a self-consistent set of radii does enable one to use the tolerance factor t to predict the approach to the ideal structure, it is not possible to compare two series of perovskites in which both A and B have been changed because the effective ionic radii are not constant in all the crystals. For example, in yttrium compounds B—O varies by as much as 0.09 Å, and in Fe compounds Fe—O varies by 0.05 Å. Moreover, it has been assumed in the above discussion that the ionic radii of Goldschmidt were used. If other 'ionic radii' (e.g. those of Pauling) are used, no such simple relation is found between t and the departures from cubic symmetry. Moreover, the properties of solid solutions suggest that the above picture is much over-simplified. For example, replacement of Ba^{2+} by a smaller ion such as Sr^{2+} or Pb^{2+} might be expected to have the same effect as replacing Ti^{4+} by the larger Zr^{4+} or Sn^{4+}. In fact, replacement of Ti by Zr or Sn lowers the Curie temperature of $BaTiO_3$ and replacement of Ba by Sr has the same effect, but the smaller Pb^{2+} has the opposite effect. Indeed, the situation is even more complex, for in the $(Ba,Pb)TiO_3$ system the upper transition temperature rises but the two lower transition temperatures fall as Pb replaces Ba. In pure $PbTiO_3$ the transition (at 490 °C) is associated with structural changes so extensive as to crack large crystals as they cool through the Curie temperature.

The determinations of the structures of the tetragonal forms of $BaTiO_3$[6] and $PbTiO_3$[7] provide good examples of the power of combined X-ray and neutron diffraction studies in cases where the X-ray method alone fails to give an unambiguous answer. The shifts of Ti and Ba (Pb) are conventionally shown relative to the O_6 octahedra around the original Ti position. In $BaTiO_3$ Ti shifts by 0.12 Å and Ba in the same direction by 0.06 Å (Fig. 13.4(a)). In tetragonal $PbTiO_3$ Ti is shifted by 0.30 Å with respect to the oxygen octahedra and Pb moves in the same direction by

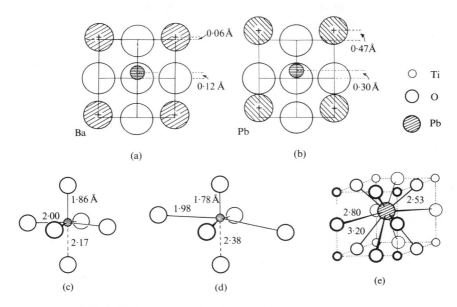

FIG. 13.4. Environments of ions in perovskite-type structures (see text).

0.47 Å (Fig. 13.4(b)). The Ti environments in these crystals are shown at (c) and (d). In both cases Ti is displaced from the centre of its octahedron giving one short Ti–O distance of 1.86 Å in $BaTiO_3$ and 1.78 Å in $PbTiO_3$ (compare the mean value 1.97 Å in rutile). As regards the Ba, the shift has a negligible effect on the twelve Ba–O distances, but in the Pb compound the large shift of Pb does have an appreciable effect on the Pb–O distances as shown at (e). These distances may be compared with 2.78 Å, the sum of the Goldschmidt ionic radii for 12-coordination. This difference is the most marked one between ferroelectric $BaTiO_3$ and $PbTiO_3$, and the environment of Pb, with four short Pb–O distances to one side, is reminiscent of that in tetragonal PbO. Whether the important factor in the transition to ferroelectric forms is the long-range forces of dipole interaction or an increase in the homopolar character of the Pb–O bonds is still a matter for discussion and is outside the scope of this book. Detailed studies have also been made of the ionic displacements in other perovskites, including $CaTiO_3$, $CdTiO_3$, $NaTaO_3$, $KNbO_3$,[8] and $NaNbO_3$.[9]

The fact that the structures of compounds ABO_3 are dependent not only on size factors but also on the nature of B has been demonstrated in many comparative studies. For example, while $AFeO_3$ (A = lanthanide) all have perovskite-type structures this is true for $AMn^{3+}O_3$ only if A is La or Ce–Dy. The compounds in which A = Ho–Lu or Y adopt a new hexagonal structure with 5- and 7-coordination of Mn and A respectively.[10] In the series $Ba_{1-x}Sr_xRuO_3$ the structure changes from the 9-layer to the 4-layer and then to the perovskite structure for $x = 0$, $\frac{1}{6}$, and $\frac{1}{3}$, but $Ba_{1-x}Sr_xIrO_3$ shows a more complex behaviour.[11]

Superstructures of perovskite. If B is progressively replaced by a second metal a large size difference tends to lead to superstructures rather than random arrangements of the two kinds of ion. The relation of the cryolite structure to perovskite has already been described in Chapter 10, where we noted oxides such as Ba_2CaWO_6 as having this type of superstructure. (A distorted form of the cryolite structure is adopted by $Ca_2'(Ca''U)O_6$,[12] possibly due to the tendency of U(VI) to form two stronger bonds; the Ca'' ions have a similar environment (compare Mg in $MgUO_4$).) Similarly in compounds $Ba_3M^{II}Ta_2O_9$ there is random arrangement of M^{II} and Ta in the octahedral positions when M^{II} is Fe, Co, Ni, Zn, or Ca, but $Ba_3SrTa_2O_9$ has a hexagonal (ordered) superstructure.[13]

In a number of oxides

$$MNb_3O_9 \quad (M = La, Ce, Pr, Nd); \text{ and}$$
$$MTa_3O_9 \quad (M = La, Ce, Pr, Nd, Sm, Gd, Dy, Ho, Y, Er)$$

there is an octahedral framework of the ReO_3-type but incomplete occupancy of the 12-coordinated sites (i.e. a *defect* perovskite structure).[14] The B sites in Fig. 13.5 are all vacant and there is two-thirds occupancy of the A sites.

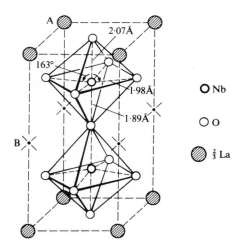

FIG. 13.5. Structure of $LaNb_3O_9$.

Some oxides $AC_3(B_4O_{12})$ have a structure related to perovskite in the following way. The cubic unit cell dimension is doubled and the A and C ions are ordered in the spaces in the framework of BO_6 octahedra. The latter are tilted so that the cuboctahedra of O atoms around A in the perovskite structure become regular

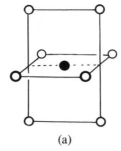

(a)

icosahedra while the C atoms have 8 nearest neighbours arranged at the corners of a square and rectangle, as shown at (a), with 4 much more distant neighbours, an arrangement suitable for Jahn–Teller ions. Examples include $CaCu_3Ge_4O_{12}$ (a h.p. phase with 6-coordinated Ge), $NaMn_3^{3+}Mn_4O_{12}$, and $CaCu_3Mn_4^{4+}O_{12}$.[15] In the first of these compounds the Cu–O bond lengths are: 1.96 Å (four), 2.68 Å (four), and 3.13 Å (four).

(4a) AC 1957 **10** 219
(4b) JCP 1956 **24** 1236
(5)   AC 1973 **B29** 2171
(6)   PR 1955 **100** 745
(7)   AC 1956 **9** 131; AC 1978 **B34** 1065
(8)   AC 1967 **22** 639
(9)   AC 1969 **B25** 851

(10) AC 1975 **B31** 2770
(11) IC 1966 **5** 335, 339
(12) AC 1966 **20** 508
(13) JACS 1961 **83** 2830
(14) AC 1967 **23** 740
(15) AC 1977 **B33** 3615

Oxides ABO_4

Numerous structures are found for compounds ABO_4, many of which are polymorphic at atmospheric pressure and also adopt different structures at higher pressures. These compounds range from those with silica-like structures, with tetrahedral coordination of A and B to oxysalts containing well-defined oxy-ions, $B\ddot{O}_4$, and high coordination number of A. Some of these structures are listed in Table 13.8, and we comment here only on a number of general points. (For uranates $CaUO_4$ etc. see Chapter 28.)

Some series of compounds illustrate the change in environment of A with its size, as in chromates of divalent metals:

Structure	C.n. of A	
$BaSO_4$	12	$PbCrO_4$, $BaCrO_4$, $SrCrO_4$
$ZrSiO_4$	8	$CaCrO_4$
$CrVO_4$	6	$ZnCrO_4$, $CdCrO_4$, $CuCrO_4$

The tetrahedral coordination of Mo(vi) in α-$MnMoO_4$ may be compared with its octahedral coordination in $CoMoO_4$, and the 6-coordination of Cr(iii) in $CrVO_4$ with the 4-coordination of Cr(vi) in the isostructural $CuCrO_4$. (The tetrahedral

TABLE 13.8

Crystal structures of compounds ABO_4

C.n. B	C.n. A	Structure	Examples	Reference
4	4	Silica-like structures	BPO_4, $BeSO_4$, $AlAsO_4$	
	6	$CrVO_4$		AC 1967 **22** 321
		$\alpha\text{-}MnMoO_4$		JCP 1965 **43** 2533
	8	Scheelite ($CaWO_4$)	$CaWO_4$	JCP 1964 **40** 501, 504
		Fergusonite	$M^{III}NbO_4$	AC 1963 **16** 888
				IC 1964 **3** 600
		Zircon ($ZrSiO_4$)	YVO_4	AC 1968 **B24** 292
		Anhydrite ($CaSO_4$)		
	12	Barytes ($BaSO_4$)		
6	6	Rutile superstructure	$MgUO_4$	AC 1954 **7** 788
			$AlWO_4$	JSSC 1975 **14** 144
		Rutile (statistical)	h.p. $CrVO_4$	AC 1962 **15** 1305
			$AlAsO_4$	AC 1964 **17** 1476
		Wolframite (statistical)	h.p. $FeVO_4$	AC 1964 **17** 1476
			$\left\{\begin{array}{l} FeWO_4 \\ MnWO_4 \\ NiWO_4 \end{array}\right.$	ZK 1967 **124** 192
		Wolframite		ZK 1967 **125** 120
				AC 1957 **10** 209
		Wolframite (distorted)	$CuWO_4$	AC 1970 **B26** 1020
		$CoMoO_4$		AC 1965 **19** 269
[4	8	Defect scheelite structure	$Eu_2(WO_4)_3$	JCP 1969 **50** 86]

coordination of V in $CrVO_4$ has been confirmed by examining the fine structure of the K absorption edges of Cr in $NiCrO_4$ (tetrahedral Cr(VI)) and $CrVO_4$ and of V in $CrVO_4$ and Na_3VO_4 (tetrahedral V(v)), the characteristic shape of the absorption curve showing that it is V that is tetrahedrally coordinated in $CrVO_4$.) Other compounds with the $CrVO_4$ structure include $InPO_4$, $TlPO_4$, and an unstable form of $CrPO_4$.

There are many examples of the change from tetrahedral to octahedral coordination of B under pressure. In the case of $AlAsO_4$ the c.n.s of both A and B increase from 4 to 6. In other cases the c.n. of B increases while that of A remains the same (e.g. $CrVO_4$ from 6:4 to 6:6 coordination) or is not greatly increased, as occurs with large ions such as Ba^{2+} or Pb^{2+} which are already in positions of high c.n. For example, in the change from $BaWO_4$-I to $BaWO_4$-II the c.n. of W changes from 4 to 6 but that of Ba only from 8 to 8 and 9,[1] and in h.p. $PbWO_4$-III, isostructural with h.p. $BaWO_4$, the c.n.s are 6 and 8.[2] In the form referred to as $PbWO_4$-II, which occurs as the mineral raspite,[3] there is octahedral coordination of W with the usual range of W–O distances (1.70–2.17 Å) and irregular 7-coordination of Pb.

There are many examples of high-pressure oxides having the same type of structure as ambient-pressure oxides containing the next element in the same vertical Periodic family (or subgroup), as for example h.p. ZnO and normal CdO (NaCl structure), or h.p. SiO_2 and normal GeO_2 (rutile structure). Examples among oxides ABO_4 include:

Wolframite structure	*Statistical wolframite*	*Rutile*
h.p. $MgMoO_4$, etc.	h.p. $FeVO_4$	h.p. $CrVO_4$
$MgWO_4$, etc.	$FeNbO_4$	$CrNbO_4$

There are many compounds ABO_4 with rutile-like structures, which are of two kinds. The statistical rutile structure is the normal form of many $M^{III}M^VO_4$ compounds, for example:

$CrTaO_4$	$CrNbO_4$	$AlSbO_4$	$RhVO_4$
Fe	Fe	Cr	
Rh	Rh	Fe	
V		Rh	
		Ga	

(The structure of $AlNbO_4$ (and the isostructural $GaNbO_4$) is mentioned later (p. 610).

We noted on p. 175 two examples of superstructures of rutile, one of which was $MgUO_4$; another is $AlWO_4$. In both of these structures the two kinds of metal atom are segregated in separate chains, but for different reasons. In $MgUO_4$ U(vI) forms 2 stronger U—O bonds, leading to considerable distortion of the UO_6 octahedra, and also (incidentally) of the MgO_6 octahedra. In the chains of WO_6 octahedra in $AlWO_4$ there is distortion due to m-m interactions across alternate shared edges leading to W—W distances of 2.613 and 3.108 Å along the chains. These interactions presumably stabilize the W(v) state. In (monoclinic) WO_2, where there is one more d electron available, the W—W distances are 2.48 and 3.1 Å.

(1) AC 1974 **B30** 2069   (3) AC 1977 **B33** 162
(2) AC 1976 **B32** 928

The wolframite structure

This structure takes its name from the mineral with composition $(Fe, Mn)WO_4$. The metal atoms each occupy one-quarter of the octahedral holes in a somewhat distorted hexagonal close packing of oxygen atoms. The pattern of sites occupied between the close-packed O layers is that of Fig. 4.24(b) (p. 173), in which the small black and open circles represent Ni and W atoms in octahedral sites between alternate pairs of c.p. layers. In $NiWO_4$ (isostructural with the Mg, Mn, Fe, Co, and Zn compounds) Ni has six equidistant O neighbours (at 2.08 (0.05) Å), but the octahedral group around W is considerably distorted, there being four O at 1.79 Å and two at 2.19 Å. As noted above this structure is also adopted by the high-pressure forms of the corresponding molybdates, and there are also oxides with the statistical variant of the structure. In the distorted wolframite structure of $CuWO_4$ Cu has 4 O at 1.98 Å and 2 O at 2.40 Å. As in all $W^{VI}O_6$ octahedra there is a range of W—O distances from 1.76–2.2 Å.

The scheelite and fergusonite structures

The scheelite structure is named after the mineral with the composition $CaWO_4$. Examples of compounds with this structure include:

$$M^IM^{VII}O_4: \quad NaIO_4, KIO_4, KRuO_4;$$
$$M^{II}M^{VI}O_4: \quad MMoO_4 \text{ and } MWO_4 \text{ (M = Ca, Sr, Ba, Pb)};$$
$$M^{III}M^VO_4: \quad MNbO_4 \text{ and } MTaO_4 \text{ (M = Y or 4f metal)}.$$

Detailed X-ray and n.d. studies of the (tetragonal) scheelite structure show very slight distortion of the WO_4 tetrahedra (four angles of 107.5° and two of 113.4°, W—O, 1.78 Å). Many compounds of the third group listed above have a distorted (monoclinic) variant of the structure (*fergusonite* structure). The oxides $YTi_{0.5}^{IV}Mo_{0.5}^{VI}O_4$ and $YTi_{0.5}W_{0.5}O_4$ show an interesting difference, the Mo compound having the tetragonal scheelite structure and the W compound the fergusonite structure. Such differences between compounds of Mo and W are numerous, and other examples are noted elsewhere.

A different deformed version of the scheelite structure is found for $KCrO_3Cl$.

Replacement of 3 Ca^{2+} in the scheelite structure by 2 M^{3+} gives a 'defect scheelite' structure, $M_{\frac{2}{3}}\square_{\frac{1}{3}}WO_4$, where $\square$ represents a vacancy. At least five different arrangements of the vacancies have been found in the structures of compounds such as $Bi_2(WO_4)_3$ and molybdates and tungstates of Sc, La, and 4f metals.[1]

(1)   AC 1973 **B29** 2074, 2433

Oxides AB_2O_4

This is a very large group of compounds, for it includes oxides and oxy-salts of three types: $A^{VI}B_2^IO_4$, $A^{IV}B_2^{II}O_4$, and $A^{II}B_2^{III}O_4$, sometimes referred to as 6:1, 4:2, and 2:3 compounds. Three structures of this type were described in our discussion of the closest packing of equal spheres, namely, those in which certain proportions of the tetrahedral and/or octahedral holes are occupied:

$\frac{3}{8}$ of tetrahedral holes: phenacite (Be_2SiO_4);
$\frac{1}{8}$ tetrahedral $\Big\}$. h.c.p. olivine (Mg_2SiO_4), chrysoberyl (Al_2BeO_4);
$\frac{1}{2}$ octahedral $\Big\}$. c.c.p. spinel (Al_2MgO_4).

Phenacite and olivine are orthosilicates, and since the present group includes all compounds $B_2(AO_4)$ containing tetrahedral AO_4 ions and ions B of various sizes it is a very large one and there are numerous structures, of which a selection is included in Table 13.9. For example, the alkali-metal tungstates have four different crystal structures:

TABLE 13.9
Crystal structures of compounds AB_2O_4

C.n. of B	4	6	8	9 and/or 10
C.n. of A				
4	Phenacite	Olivine Spinel	K_2WO_4	β-K_2SO_4
6			*	K_2NiF_4
8		$CaFe_2O_4$ $CaTi_2O_4$		

For a discussion of the crystal chemistry of compounds AB_2O_4 see, for example: IC 1968 **7** 1762; JSSC 1970 **1** 557.

* We could include here a number of structures in which rutile-like chains containing A are held together by B in positions of 6-coordination (Sr_2PbO_4, p. 214) or 6-, 7-, and 9-coordination (Ca_2IrO_4, ZaC 1966 **347** 282).

Li_2WO_4 (phenacite), Na_2WO_4 (modified spinel), K_2WO_4 (and the isostructural Rb_2WO_4), and Cs_2WO_4 (β-K_2SO_4 structure), with respectively 4-, 6-, 8-, and (9 and 10)-coordination of M^+ ions.

However, we are concerned here with complex oxides rather than oxy-salts containing well-defined oxy-ions, though the dividing line is somewhat arbitrary, as is illustrated by compounds such as Li_2BeF_4, Zn_2GeO_4, and $LiAlGeO_4$ with the phenacite structure and $LiNaBeF_4$, Cs_2BeF_4, and Ba_2GeO_4 with the olivine structure.

By far the most important of the structures of Table 13.9 is the spinel structure. A recent survey shows over 130 compounds with the cubic spinel or closely related structures; some 30 of these are sulphides but most of the remainder are oxides. After the spinel structure we shall deal briefly with two other structures in which there is 6-coordination of B, namely, those of $CaFe_2O_4$ and $CaTi_2O_4$, and then with the K_2NiF_4 structure.

There is considerable interest in the high-pressure forms of oxides AB_2O_4, for high-pressure forms of olivine $(Fe,Mg)_2SiO_4$, are believed to be important constituents of the earth's mantle. High pressure tends to convert a structure into one with higher c.n.s, that is, from top left to bottom right of Table 13.9. Examples include:

Ca_2GeO_4: olivine $\rightarrow K_2NiF_4$ structure (density increase 25 per cent)
Mn_2GeO_4: olivine $\rightarrow Sr_2PbO_4$ structure (density increase 18 per cent).

The normal and 'inverse' spinel structures

The spinel structure is illustrated in Fig. 7.4(a). It is most easily visualized as an octahedral framework of composition AX_2 (atacamite) which is derived from the

NaCl structure by removing alternate rows of metal ions (Fig. 4.22(b), p. 170). Additional metal ions may then be added in positions of tetrahedral coordination (Fig. 7.3(a), facing page 319). In the resulting structure each O^{2-} ion also has tetrahedral coordination, its nearest neighbours being three metal atoms of the octahedral framework and one tetrahedrally coordinated metal atom.

The crystallographic unit cell of the spinel structure contains 32 approximately cubic close-packed O atoms. (There are small departures from ideal closest packing which appear to be related to the charges on the B atoms, these atoms (ions) being in edge-sharing octahedral groups.[1a]) For every 32 c.p. O atoms there are 32 octahedral and 64 tetrahedral interstices but in the space group of spinel there are equivalent positions for 8 atoms in tetrahedral and 16 in octahedral coordination. It was therefore natural to place the A atoms in the former and the B atoms in the latter positions. The spinels MAl_2O_4 (where M is Mg, Fe, Co, Ni, Mn, or Zn) have this structure, but in certain other spinels the A and B atoms are arranged differently. In these the eight tetrahedral positions are occupied, not by the eight A atoms, but by one-half of the B atoms, the rest of which together with the A atoms are arranged at random in the 16 octahedral positions. These 'inverse' spinels are therefore conveniently formulated $B(AB)O_4$ to distinguish them from those of the first type, AB_2O_4. Examples of inverse spinels include $Fe(MgFe)O_4$ and $Zn(SnZn)O_4$.

The nature of a spinel is described by a parameter λ, the fraction of B atoms in tetrahedral holes; some authors refer to the degree of inversion y ($= 2\lambda$). For a normal spinel $\lambda = 0$, and for an inverse spinel $\lambda = \frac{1}{2}$. Intermediate values are found (e.g. $\frac{1}{3}$ in a random spinel), and λ is not necessarily constant for a given spinel but can in some cases be altered by appropriate heat treatment. For $NiMn_2O_4$ λ varies from 0.37 (quenched) to 0.47 (slow-cooled).[1b] Values of λ have been determined by X-ray and neutron diffraction, by measurements of saturation magnetization, and also by i.r. measurements. In favourable cases i.r. bands due to tetrahedral AO_4 groups can be identified showing, for example, that in $Li(CrGe)O_4$ Li occupies tetrahedral positions.[2]

If there is sufficient difference between the X-ray scattering powers of the atoms A and B it is possible to determine the distribution of these atoms by the usual methods of X-ray crystallography, but in spinel itself ($MgAl_2O_4$), for example, this is not possible. However, the scattering cross-section for neutrons of Mg is appreciably greater than that of Al, and this makes it possible to show that $MgAl_2O_4$ has the normal spinel structure.[3] Many 2:3 spinels have the normal structure, though some (including most 'ferrites') have the inverse structure, as for example $Ga(MgGa)O_4$ ($\lambda = 0.42$).[4] Both types of structure are found also for 4:2 spinels. Normal spinels include Fe_2GeO_4, Co_2GeO_4, Ni_2GeO_4, and Mg_2GeO_4[4a] (which transforms to the olivine structure at $810\,°C$); others have the inverse structure, for example, $Zn(ZnTi)O_4$ and $Fe(FeTi)O_4$ ($\lambda = 0.46$).[5]

The spinels of composition between $MgFe_2O_4$ and $MgAl_2O_4$, which have been studied magnetically and also by neutron diffraction,[6] are of interest in this connection. $MgFe_2O_4$ has an essentially inverse structure ($\lambda \approx 0.45$), that is, nine-tenths of the Mg^{2+} ions are in octahedral (B) sites. As Fe is replaced by Al the latter goes into B sites and forces Mg into tetrahedral (A) sites, so that there is a continuous transformation from the inverse structure of $MgFe_2O_4$ to the normal structure of $MgAl_2O_4$.

The structures of the spinels present two interesting problems. First, why do some compounds adopt the normal and others the inverse spinel structure? Second, there are some spinels which show distortions from cubic symmetry. This is part of the more general problem of minor distortions from more symmetrical structures which was mentioned in connection with ligand field theory in Chapter 7. We deal with these points in turn.

Calculations of the lattice energy on the simple electrostatic theory, without allowance for crystal field effects, indicate that while the inverse structure should be more stable for 4:2 spinels, the preferred structure for 2:3 spinels should be the normal structure. In fact a number of the latter have the inverse structure, as shown by the (approximate) values of λ in Table 13.10. The cation distribution in spinels

TABLE 13.10

Cation distribution in $2:3$ *spinels (values of* λ*)*

B^{3+} \ A^{2+}	Mg^{2+}	Mn^{2+}	Fe^{2+}	Co^{2+}	Ni^{2+}	Cu^{2+}	Zn^{2+}
Al^{3+}	0	0	0	0	0·38	–	0
Cr^{3+}	0	0	0	0	0	0	0
Fe^{3+}	0·45	0·1	0·5	0·5	0·5	0·5	0
Mn^{3+}	–	0	–	–	–	–	0
Co^{3+}	–	–	–	0	–	–	0

has been discussed in terms of crystal field theory.[7] Although values of Δ (see p. 323) appropriate to spinels are not known, estimates of these quantities may be made and from these the stabilization energies for octahedral and tetrahedral coordination obtained. The differences between these quantities, of which the former is the larger, give an indication of the preference of the ion for octahedral as opposed to tetrahedral coordination.

Excess octahedral stabilization energy (kJ mol^{-1})

Mn^{2+}	Fe^{2+}	Co^{2+}	Ni^{2+}	Cu^{2+}	Ti^{3+}	V^{3+}	Cr^{3+}	Mn^{3+}	Fe^{3+}
0	17	31	86	64	29	54	158	95	0

Thus while Cr^{3+} and Mn^{3+} occupy octahedral sites, most 'ferrites' are inverse spinels, Fe^{3+} having no stabilization energy for octahedral sites. The only normal, or approximately normal, ferrites are those of Zn^{2+} and Mn^{2+}, divalent ions which have no octahedral stabilization energy. The only inverse 'aluminate' is the Ni^{2+} spinel, and of the above ions Ni^{2+} has the greatest preference for octahedral coordination. Note that we are disregarding all other factors (covalent bonding and differences in normal lattice energies) when we take account only of the crystal field stabilization energies. For example, for $NiAl_2O_4$ a classical calculation of the lattice energy, as the sum of the electrostatic (Madelung) potential, the polarization energy, and the Born repulsion energy, shows that the normal structure would be some 105 kJ mol^{-1} more stable than the inverse one, so that the crystal field stabilization of octahedral Ni^{2+} almost compensates for this. The observed structure is in fact very close to the random one, for which $\lambda = 0.33$. The fact that Fe_3O_4 is an inverse spinel with cubic symmetry while Mn_3O_4 is a normal spinel with some tetragonal distortion (for which see p. 553) is explained by the much greater preference of Mn^{3+} than Fe^{3+} for octahedral coordination by oxygen, as shown by the stabilization energies quoted above. The simple crystal field theory is seen to be very helpful in accounting for the cation distributions in these 2:3 spinels.

The second problem concerns the departures from cubic symmetry. For example, $CuFe_2O_4$ is cubic at high temperatures (and at room temperature if quenched from temperatures above 760 °C),[8] but if cooled slowly it has tetragonal symmetry ($c:a = 1.06$ at room temperature). It is an inverse spinel with Fe^{3+} in the tetrahedral sites, and it is interesing that the FeO_4 tetrahedra are not distorted. The tetragonal symmetry appears to be due to the tetragonal distortion of the $Cu^{II}O_6$ octahedra (p. 324). $CuCr_2O_4$ is also tetragonal, but with $c:a = 0.91$.[9] Here the Cu atoms are in tetrahedral sites, but the bond angles of 103° and 123° show a tendency towards square coordination, or alternatively, the tetragonal distortion (type (c) of Table 7.13, p. 323) predicted by ligand field theory. The spinels of intermediate composition, $CuFe_{2-x}Cr_xO_4$, are even more interesting.[10] For $x = 0$ ($CuFe_2O_4$) there is a tetragonal distortion with $c:a = 1.06$, and for $x = 2$ ($CuCr_2O_4$) the structure is tetragonal with $c:a = 0.91$. Over the range $0.4 < x < 1.4$ the structure is cubic. Since for $x = 0$ the structure is almost completely inverse, and at $x = 2$ nearly normal, there is a continuous displacement of Fe by Cu in the tetrahedral sites. This is similar to the $MgFe_2O_4$–$MgAl_2O_4$ series but with the added complication that at one end of the series there is distortion due to Cu in octahedral sites and at the other due to Cu in tetrahedral sites, the elongation of the CuO_6 octahedra leading to $c:a > 1$ and the flattening of the tetrahedra to $c:a < 1$.

(Some sulphides with structures closely related to the spinel structure are noted on p. 764.)

(1a) AC 1974 **B30** 1872	(4) AC 1966 20 761	(7) JPCS 1957 **3** 318; PR 1955 **98** 391	
(1b) AC 1969 **B25** 2326	(4a) AC 1977 **B33** 2287	(8) AC 1956 9 1025	
(2) AC 1963 16 228	(5) AC 1965 18 859	(9) AC 1957 **10** 554	
(3) AC 1952 5 684	(6) AC 1953 6 57	(10) JPSJ 1956 **12** 1296	

Spinel superstructures

There are many spinel-like structures containing metal atoms of more than one kind. If these are present in suitable numbers they can arrange themselves in an orderly way; for example, the 16-fold position may be occupied by $8A + 8B$ or by $4A + 12B$. Large differences in ionic charge favour ordering of cations, and since in 'defect' structures holes also tend to order, superstructures with ordered vacancies are possible. Some examples of spinel superstructures are summarized in Table 13.11.

TABLE 13.11

Spinel superstructures

Compound	Occupancy of positions				Reference
	tetrahedral	*octahedral*			
$Zn(LiNb)O_4$	Zn	Nb Li		O_4	
$V^V(LiCu)O_4$	V	Li Cu		O_4	
Fe_3O_4 (<120°K)	Fe^{3+}	Fe^{2+} Fe^{3+}		O_4	AC 1955 8 257
	8	16		O_{32}	
Fe_5LiO_8 (Al_5LiO_8)	8 Fe	4 Li 12 Fe		O_{32}	JCP 1964 40 1988
γ-Fe_2O_3	8 Fe	$\frac{4}{3}Fe^{3+}\frac{8}{3}\square$ 12 Fe		O_{32}	N 1958 181 44
$(LiFe)Cr_4O_8$	4 Li 4 Fe	16 Cr		O_{32}	
In_2S_3	$\frac{16}{3}In$ $\frac{8}{3}\square$	16 In		S_{32}	
$Zn_2Ge_3O_8$	2 Zn	3 Ge $\square$		O_8	
$LiZn(LiGe_3)O_8$	Li, Zn	3 Ge Li		O_8	
$Co_3(VO_4)_2$ (low)	2 V	3 Co $\square$		O_8	J.-C. Joubert Thesis, Grenoble, 1965
$LiGaTiO_4$	2 Ga, Li	3 Ti, Ga, 2 Li		O_{12}	

$\square$ represents vacancy

A compound may be ordered at temperatures below a transition point and disordered at higher temperatures. For example, below 120 K there is ordering of Fe^{2+} and Fe^{3+} in octahedral sites in Fe_3O_4 but random arrangement at higher temperatures, leading to free exchange of electrons. The compounds $LiAl_5O_8$ and $LiFe_5O_8$ have ordered structures at temperatures below 1290° and 755 °C respectively and disordered structures at higher temperatures. In the low-temperature forms the 4 Li^+ and 12 M^{3+} ions are arranged in an ordered way in the 16 octahedral sites but in a random way at temperatures above the transition points.

'Ferrites' etc. with structures related to spinel

Certain oxides, notably those of the alkali metals, the alkaline-earths, and lead, form well-defined 'aluminates' and 'ferrites' with general formulae $M_2O.m(Al, Fe)_2O_3$ or $MO.n(Al, Fe)_2O_3$. Some of the simpler 'ferrites', for example, $LiFeO_2$, $NaFeO_2$, and $CuFeO_2$ (all with different structures) and $MgFe_2O_4$ and $CaFe_2O_4$ are included in other sections of this chapter. Examples of more complex compounds include:

$Na_2O.11Al_2O_3$, also formed by K
$K_2O.11Fe_2O_3$, also formed by Rb
$3CaO.16Al_2O_3$, also formed by Sr and Ba
$SrO.6Fe_2O_3$, also formed by Ba and Pb.

The first of these, $Na_2O.11Al_2O_3$, was originally thought to be a form of Al_2O_3, hence its name—β-alumina. X-ray investigation, however, showed it to have a definite structure closely related to that of spinel, no fewer than 50 of the 58 atoms in the unit cell being arranged in exactly the same way as in the spinel structure. The large Na or K ions are situated between slices of spinel structure and their presence is essential to the stability of the structure (Fig. 13.6(a)). They may be replaced by Rb^+, Ag^+ and other ions by heating with the appropriate molten salts, and Ag^+ may then be replaced by NO^+ (by heating in a $NOCl–AlCl_3$ melt) or by Ga^+.

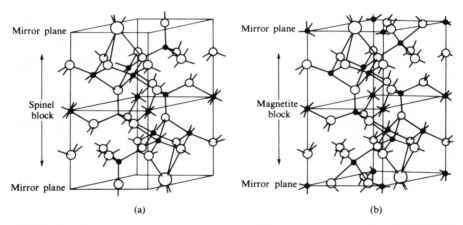

FIG. 13.6.   (a) The relation of the structure of β-alumina, $NaAl_{11}O_{17}$, to that of spinel (after Beevers and Ross). Large circles represent Na, small ones O, and black circles Al.   (b) The relation of the structure of magneto-plumbite, $PbFe_{12}O_{19}$, to that of magnetite Fe_3O_4 (after Adelsköld). Large circles represent Pb, small ones O, and black circles Fe.

Later work has shown that the Na content is somewhat higher than that corresponding to the formula $NaAl_{11}O_{17}$. The composition of a crystal used in a refinement of the crystal structure[11] was $Na_{2.58}Al_{21.8}O_{34}$, and it appears that the distribution of the Na^+ ions is much more complex than in the original structure. In a (hydrated) alkali-free β-Al_2O_3 the angle at the bridging O atom was found to be $158°$ rather than $180°$ and there are disordered H_2O molecules between the spinel slices.[12]

The Na^+ ions are highly mobile in β-Al_2O_3, between the spinel slices, and this compound has been intensively studied because of its potential use (in ceramic form) as a solid electrolyte in Na–S batteries. The slices of spinel-like structure consist of 4 c.p. O layers and they are joined through O layers in which the atoms are also in the *positions* of cubic closest packing but occupy only one quarter of those positions. These layers, which are 11.3 Å apart, contain the Na^+ ions and are the conducting layers. In β-alumina the planes of these layers are mirror planes, so that the sequence of layers perpendicular to the hexagonal axis is

$$C \ (ABCA) \ B \ (ACBA) \ C \ \ldots \ or \ C \ (\ ) \ B \ (\ ) \ C \ \ldots$$

where the parentheses enclose the 4 c.p. layers of the spinel slice and the intermediate symbols refer to the $\frac{1}{4}$-filled layers. In a second structure, β''-alumina, which has rhombohedral symmetry, the layer sequence is

$$C \ (ABCA) \ B \ (CABC) \ A \ (BCAB) \ C \ \ldots \ or \ C \ (\ ) \ B \ (\ ) \ A \ (\ ) \ C \ \ldots$$

The repeat distance along the c-axis is 33.85 Å as compared with 22.53 Å for β-Al_2O_3. The β'' structure is usually stabilized by some Mg and/or Li. Obviously the β and β'' structures are related as polytypes and may be given symbols 2H and 3R (p. 154), or if we wish to show the number of O layers in the basic sub-unit, 2H(5) and 3R(5). More complex structures in this family have been identified by electron microscopy (dynamic scattering from thick crystals) as 15R(5) and 30R(5) with repeat distances along the hexagonal axis of 169.5 Å and 339 Å respectively.[13] Other structures have been reported based on thicker spinel slices, for example, 2H(7) and 3R(7).

The structure of magnetoplumbite, $PbFe_{12}O_{19}$, is built of slices of magnetite (Fe_3O_4) structure alternating with layers containing Pb^{2+} ions. A detailed X-ray study has been made of the isostructural $SrAl_{12}O_{19}$.[14] In the magnetoplumbite structure (Fig. 13.6(b)) there is a sequence of 5 c.p. layers, four of O atoms (O_4 per cell) and one in which Pb^{2+} ions occupy one-quarter of the c.p. positions (PbO_3). The composition is therefore PbO_{19} plus the associated Fe atoms in tetrahedral and octahedral holes. The compounds containing Ba instead of Pb have been extensively studied in recent years, for they form a large family of ferrimagnetic compounds valuable in electronic devices. Early studies produced $BaFe_{12}O_{19}$, isostructural with magnetoplumbite, $BaFe_{15}O_{23}$[15] and $BaFe_{18}O_{27}$ with respectively five and six O_4 layers in the repeat unit. Already the hexagonal ferrites are more numerous than the polytypes of SiC, of which at least fifty are known, with c cell dimensions up to 990 Å.

These compounds fall into two distinct series, each formed by different stacking sequences based on three kinds of sub-unit formed from O_4 or BaO_3 layers:

—————O_4	—————O_4	—————O_4
—————O_4	—————O_3Ba	—————O_3Ba
2 layers	—————O_4	—————O_3Ba
S unit	—————O_4	—————O_4
$Me_2Fe_4O_8$	—————O_4	—————O_4
	5 layers	—————O_4
(Me is usually Zn, Ni, Co, or Fe)	M unit $BaFe_{12}O_{19}$	6 layers
		Y unit $Ba_2Me_2Fe_{12}O_{22}$

The blocks S, M, and Y stack in various ratios and permutations but as regards sequences the units show decided preferences; M is usually associated with Y (less commonly with S); Y not with S; and M, Y, and S not all together. There are therefore two main series: (i) M_nS (one S and n M blocks), giving in the limit, M ($BaFe_{12}O_{19}$); and (ii) M_pY_n, of which the limiting member is Y ($Ba_2Me_2Fe_{12}O_{22}$). Five of type (i) have been prepared, for example, M_6S is $Ba_6Me_2Fe_{70}O_{122}$ with $c = 223$ Å, but the much more extensive family (ii) includes nearly sixty members. For example, M_4Y_{33} has the sequence: $MY_6 MY_{10} MY_7 MY_{10}$, with composition $Ba_{70}Me_{66}Fe_{444}O_{802}$ and $c = 1577$ Å. It would not be feasible to elucidate these complex structures by X-ray diffraction alone, but fortunately the stacking sequence can be determined by electron microscopy of HNO_3-etched samples, using Pt-shadowing of carbon replicas. The etch patterns take the form of fine terracing along the sides of gently sloping etch pits, and knowing the absolute magnification the heights of the two types of steps can be determined from the replicas. For example, a M_2Y_7 specimen shows large and small steps corresponding to the units MY and MY_6 stacked in the sequence $MY MY_6 MY MY_6 \ldots$. The more complex four-step etch pattern of a M_4Y_{15} ferrite is interpreted as indicating the sequence $MY MY_2 \ MY_3 MY_9 \ldots (c \approx 793$ Å). (For references to the literature see ref. 16.)

The structure of $BaFe_{12}O_{19}{}^{[17]}$ is of interest as an example of a 10-layer c.p. sequence (*cchhh*):

$$B\ A\ B'\ A\ B\ C\ A\ C'\ A\ C\ldots$$

where B' and C' indicate BaO_3 layers. It is also remarkable for the three types of coordination of the Fe atoms:

$$\left. \begin{array}{ll} \text{octahedral} & 18\ \text{Fe} \\ \text{tetrahedral} & 4\ \text{Fe} \\ \text{trigonal bipyramidal} & 2\ \text{Fe} \end{array} \right\} \text{ for } 2(BaFe_{12}O_{19}).$$

(11) AC 1971 **B27** 1826
(12) AC 1977 **B33** 1596
(13) JSSC 1980 **33** 37
(14) AC 1975 **B31** 2940

(15) AC 1954 **7** 640
(16) Sc 1971 **172** 519
(17) ZK 1967 **125** 437

The $CaFe_2O_4$, $CaTi_2O_4$, and related structures

For larger A ions the spinel structure is not stable, and we find a number of structures built from 'double rutile' chains which form 3D frameworks enclosing the larger A ions in positions of 8-coordination. These structures have been described in Chapter 5.

No other compounds are yet known to have the $CaTi_2O_4$ structure, but there are numerous compounds with the $CaFe_2O_4$ structure, not only oxides $M^{II}M_2^{III}O_4$ (and some sulphides) but also more complex compounds in which ions M^{3+} and M^{4+} occupy at random the Fe^{3+} positions and Na^+ the Ca^{2+} positions:

$Eu^{II}Eu_2^{III}O_4$	$CaFe_2O_4$	$NaScTiO_4$	Eu	Y_2	
$SrEu_2O_4$	V_2	$NaScSnO_4$	Sm	Ho_2	S_4
	Cr_2	$NaAlGeO_4$	Pb	etc.	
	In_2	etc.	Sr		
	etc.		Ba		

(For references see: IC 1967 **6** 631; IC 1968 **7** 113, 1762; AC 1967 **23** 736).

In all cases the environment of Ca^{2+} (or the equivalent ion) consists of 8 O (S) neighbours, the ninth vertex of the tricapped trigonal prism coordination group being at a much greater distance.

$$\text{In } Eu_3O_4 \qquad Eu^{3+} - 6 \text{ O (octahedral) at } 2.34 \text{ Å}$$

$$Eu^{2+} \begin{cases} 6 \text{ O (prism) } 2.67 \text{ Å} \\ 2 \text{ O } 2.72 \text{ and } 2.96 \text{ Å} \\ 1 \text{ O } 3.99 \text{ Å} \end{cases}$$

Similarly in $NaScTiO_4$ Na^+ has 8 O neighbours at 2.43–2.65 Å and the ninth at 3.34 Å.

In the $CaFe_2O_4$ and $CaTi_2O_4$ structures a *larger* ion (Ca^{2+}) is accommodated between double octahedral chains which contain the *smaller* Fe^{2+} or Ti^{3+} ions. A very interesting structure is adopted by BeY_2O_4 in which the *smaller* Be^{2+} ion (Be–O, 1.55 Å, Y–O, 2.29 Å) is accommodated in positions of 3-coordination (coplanar) between quadruple chains. These chains are in contact not only at two terminal points but also at the mid-points of both sides, as may be seen from the projection of the structure (Fig. 13.7). (This unusual 3-coordination is also found for four of the seventeen Be^{2+} ions in $Ca_{12}Be_{17}O_{29}$.[1]) The 3D framework of Fig. 13.7(a) has the composition MO_2 (M_2O_4) and is found also in a number of oxyborates. In the synthetic $FeCoBO_4$ there is ordered arrangement of Fe and Co in the framework, and B replaces Be in Fig. 13.7(a). Since there are planar BO_3

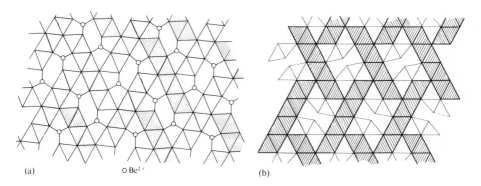

(a)   O Be^{2+} (b)

FIG. 13.7. (a) The structure of BeY_2O_4 projected along the direction of the quadruple rutile chains. (b) The structure of $TiIn_2O_5$.

groups in this structure the formula may be written $FeCo(BO_3)O$. In the mineral warwickite, $Mg_3TiB_2O_8$, Ti atoms occupy one-quarter of the metal positions in the framework, and Mg may be replaced by Fe. We may therefore write the formulae of these compounds

$$Be(Y_2O_4), \quad B(FeCoO_4), \quad \text{and} \quad B_2(Mg_3TiO_8),$$

and it would be interesting to know if the series can be completed by making compounds such as $Li(M^{III}M^{IV}O_4)$. (See also p. 280.)

There is essentially the same MX_2 framework in TiM_2O_5 $(M = La, Y, Gd, Sm, Eu)$, with Ti and a fifth O in the channels. Ti is 5-coordinated (trigonal bipyramidal) and shares two equatorial O atoms (not belonging to the framework) with other TiO_5 groups to form chains; the M atom of the framework acquires a seventh O neighbour. There are similar chains of 5-coordinated Ti atoms in $TiIn_2O_5$, the MO_2 framework of which is closely related (Fig. 13.7(b)) to that of BeY_2O_4. VIn_2O_5 has the same structure.

Framework	Compound	Composition of framework	Atom(s) in tunnels (c.n.)	Reference
Fig. 13.7(a)	BeY_2O_4	Y_2O_4	Be (3)	AC 1967 **22** 354
	$FeCoBO_4$	$FeCoO_4$	B (3)	ZK 1972 **135** 321
	$Mg_3TiB_2O_8$	Mg_3TiO_8	B_2 (3)	AC 1950 **3** 98, 473
	$TiLa_2O_5$	La_2O_4	Ti (5), O	AC 1968 **B24** 1327
Fig. 13.7(b)	$TiIn_2O_5$	In_2O_4	Ti (5), O	AC 1975 **B31** 1614

(1) AC 1966 **20** 295

The K_2NiF_4 structure

The structure of a number of complex oxides X_2YO_4 is closely related to the perovskite structure, which is that of the corresponding oxide XYO_3. A slice of the perovskite structure one unit cell thick has the composition X_2YO_4. If such slices are displaced relative to one another as shown in Fig. 13.8, there is formed a tetragonal structure with $c = 3a$, a being the length of the cell edge of the perovskite structure. In this structure the Y atoms have the same environment as in perovskite, namely six O arranged octahedrally, but the X atoms have an unusual arrangement of nine O instead of the original twelve neighbours. If the slices of the perovskite structure are n unit cells thick the composition of the crystal is $X_{n+1}Y_nO_{3n+1}$. In the complex Sr–Ti oxides the first three members of this series have been characterized, namely, Sr_2TiO_4, $Sr_3Ti_2O_7$, and $Sr_4Ti_3O_{10}$.[1] The X_2YO_4 structure was first assigned to K_2NiF_4. Examples of compounds with this structure are:

K_2MF_4: M = Mg, Zn, Co, Ni
M_2UO_4: M = K, Rb, Cs
Sr_2MO_4: (M = Ti, Sn, Mn, Mo, Ru, Ir, Rh)
Ba_2MO_4: (M = Sn, Pb)
$Sr_2FeO_3F, K_2NbO_3F, La_2NiO_4$.

For references see JPC 1963 **67** 1451.

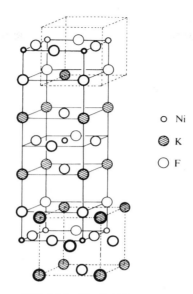

FIG. 13.8. The K_2NiF_4 structure.

○ Ni

◉ K

○ F

It is interesting to compare the isostructural Sr_2TiO_4, Ca_2MnO_4, and K_2NiF_4 with $SrTiO_3$, $CaMnO_3$, and $KNiF_3$ which all crystallize with the perovskite structure.

The distorted variants of this structure adopted by K_2CuF_4 and $(NH_4)_2CuCl_4$ are described in Chapter 5, where the structure is described as built of $(NiF_4^{2-})_n$ layers formed from octahedral coordination groups sharing their four equatorial vertices; the 2D ions are held together by the K^+ ions.

(1)   AC 1958 11 54

Further complex oxide structures

The pseudobrookite structure, A_2BO_5

In this structure[1] there is octahedral coordination of the B ions, but the coordination of A is unusual. It can be described either as a highly deformed octahedron or as a distorted tetrahedron, for in Fe_2TiO_5 Fe has four neighbours at a mean Fe–O distance of 1.92 Å and two more at 2.30 Å. Ti has six octahedral neighbours as in the forms of TiO_2, $FeTiO_3$, etc. Al_2TiO_5 also has the pseudo-brookite structure. For Ti_3O_5 see p. 563.

Oxides AB_2O_6

We noted in Chapter 4 a number of structures in which metal ions occupy one-half of the octahedral interstices in c.p. structures. Three h.c.p. structures of this kind are

trirutile structure: for which see p. 175 and 251;
columbite (niobite) structure (p. 175);[2]
Na_2SiF_6 structure (pp. 175 and 458).

Since both A and B ions occupy octahedral interstices in all of these structures they are suitable only for the smaller metal ions, as may be seen from the following examples: $ZnSb_2O_6$ (trirutile), $(Fe,Mn)Nb_2O_6$ (niobite), and NiU_2O_6[3] (Na_2SiF_6 structure).

The $CaTa_2O_6$ structure[4] has been illustrated with other octahedral structures in Chapter 5. As in the $CaFe_2O_4$ structure Ca^{2+} is 8-coordinated; the ninth, more distant, O would complete a tricapped trigonal prismatic coordination group.

(1) AC 1953 6 812	(3) ZaC 1968 358 226	
(2) AK 1963 21 407	(4) ACSc 1963 17 2548	

The pyrochlore structure, $A_2B_2O_7$

We have already mentioned this structure in connection with 3D nets in Chapter 3, and also as an example of a 3D framework of octahedra in Chapter 5. The geometry of the structure is further discussed in Chapter 6 (p. 258). We noted that this framework has the composition B_2O_6 (BO_3) and that it can exist without the A atoms or without the seventh O atom. In the latter case it serves as the structure of a number of oxides ABO_3 ('defect pyrochlore structure'). The structure is named after the mineral pyrochlore which has the approximate composition $(CaNa)Nb_2O_6F$, and since the seventh anion position may be occupied by OH^-, F^-, or even H_2O, we write the general formula $A_2B_2X_6'X''$ in Table 13.12. The largest group of compounds with this structure are oxides $A_2B_2O_7$ which include the following:

$M_2^{II}M_2^{V}O_7$: $Cd_2Nb_2O_7$, $Cd_2Re_2O_7$, $Ca_2Ta_2O_7$, $Hg_2Nb_2O_7$;
$M_2^{III}M_2^{IV}O_7$: $M_2Pt_2O_7$ (M = Sc, Y, In, Tl; made under high O_2 pressure).

The coordination group around A in $A_2B_2X_7$ consists of six X atoms of the B_2X_6 framework and two of the additional X atoms. In some of the oxide structures the

TABLE 13.12

Compounds with the pyrochlore structure $A_2B_2X_7$

Compound	A_2	B_2	X_6'	X''
$AgSbO_3$	Ag	Sb_2	O_6	–
Sb_3O_6OH	Sb^{III}	Sb_2^{V}	O_6	OH
$BiTa_2O_6F$	Bi^{III}	Ta_2	O_6	F
$Al(OH, F)_3 . \frac{3}{8} H_2O$	–	Al_2	$(OH, F)_6$	$(H_2O)_{\frac{3}{4}}$
$Hg_2Nb_2O_7$	Hg_2	Nb_2	O_6	O

Recent references to complex oxides with this structure include: IC 1965 4 1152; IC 1968 7 1649, 1704, 2553; IC 1969 8 1807.

8 O atoms are approximately equidistant from A, forming an approximately cubic coordination group, but in other cases there is an appreciable difference between A–6 O and A–2 O. It is then justifiable to describe the structure as an A_2O framework interpenetrating the B_2O_6 octahedral framework. $Hg_2Nb_2O_7$ has been described in this way in Chapter 3 as related to the diamond structure, Hg forming two collinear bonds to O (Hg–2 O, 2.26 Å) in a 3D framework of the same kind as one of the two equivalent nets in Cu_2O. (The other six O atoms are at a distance of 2.61 Å.) This aspect of the pyrochlore structure is emphasized in Fig. 7.3(b).

Layer structures $A_mB_2X_7$ related to perovskite

We describe on p. 211 two structures for compounds $A_mB_2X_7$ in which the layers are slices of the perovskite structure. There are two families of such structures, in which the layers have the compositions B_nX_{3n+1} and B_nX_{3n+2}; the composition B_2X_7 occurs in the first family for $n=2$ and in the second for $n=4$. In the B_2X_7 layers of the first type each BX_6 octahedron shares 5 vertices, while in those of the second kind equal numbers of BX_6 groups share 4 and 6 vertices. Examples include:

B_nX_{3n+1}: ($n=2$) $Sr_3Ti_2O_7$ AC 1958 **11** 54

B_nX_{3n+2}: ($n=4$) $Sr_2Ta_2O_7$ AC 1976 **B32** 2564

$Sr_2Nb_2O_7$ AC 1975 **B31** 1912

$La_2Ti_2O_7$ AC 1975 **B31** 2129

The garnet structure

The garnets have been known for a long time as a group of orthosilicate minerals $M_3^{II}M_2^{III}(SiO_4)_3$ in which M^{II} is Ca, Mg, or Fe, and M^{III} is Al, Cr, or Fe^{3+}. Since the total negative charge of -24 can be made up in a variety of ways the garnet structure can be utilized by a great number of complex oxides, of which some examples are given in Table 13.13. Interest in the optical properties of these compounds has led in recent years to many structural studies.

The backbone of the garnet structure is a 3D framework built of $M''O_6$ octahedra and $M'''O_4$ tetrahedra in which each octahedron is joined to six others through vertex-sharing tetrahedra. Each tetrahedron shares its vertices with four octahedra, so that the composition of the framework is $o_2t_3 = (M''O_3)_2(M'''O_2)_3 = M_2''M_3'''O_{12}$. Larger ions ($M'$) occupy positions of 8-coordination (dodecahedral) in the interstices of the framework, giving the final composition $M_3'M_2''M_3'''O_{12}$ or $M_3'M_2''(M'''O_4)_3$ if it is wished to emphasize the tetrahedral groups as in an orthosilicate. In some compounds the M'' and M''' positions are all occupied by atoms of the same element, when the formula becomes, for example, $Y_3Al_5O_{12}$. If, in addition, one-half of the M'' positions are occupied by M' atoms the formula becomes, for example, $Y_4Al_4O_{12}$ (or $YAlO_3$).

The ordered arrangement of two kinds of ion in the octahedral sites is also found in the tetragonal (pseudocubic) high-pressure forms of $CaGeO_3$ and $CdGeO_3$. The structures of the polymorphs of $CaGeO_3$ provide an excellent illustration of the

TABLE 13.13

Compounds with the garnet structure

	M′	M″	M‴		Reference
C.N.	8	6	4		
	Mg_3	Al_2	Si_3,	O_{12}	AC 1961 **14** 835; AM 1965 **50** 2023
	Ca_3	Al_2	Si_3		ZK 1966 **123** 81
	Y_3	Al_2	Al_3		
	$NaCa_2$	Mg_2	As_3		
	$NaCa_2$	Zn_2	V_3		
	Y_3	YAl	Al_3		
	$CaNa_2$	Ti_2	Ge_3		
	Ca_3	Te_2	Zn_3		IC 1969 **8** 1000
	Na_3	Te_2	Ga_3		IC 1969 **8** 1000
	Ca_3	$CaZr$	Ge_3		IC 1969 **8** 183
	Cd_3	$CdGe$	Ge_3		Sc 1969 **163** 386
	Ca_3	Al_2		$(OH)_{12}$	AC 1964 **17** 1329; JCP 1968 **48** 3037 (n.d.)

increase in cation coordination number with increasing pressure:

c.n.s of cations:	6, 4	(8 and 6)(6 and 4)	12, 6
	$CaGeO_3$-I	$CaGeO_3$-II	$CaGeO_3$-III
structure:	wollastonite	garnet-type	perovskite

The garnet framework can exist without the M′ ions, as in $Al_2(WO_4)_3$, or $M''_2M'''_3O_{12}$, when there is considerable distortion, with larger angles at the O atoms shared between M″ and M‴ (143–175°). The same framework is also found in the so-called 'hydrogarnets'. A neutron diffraction study of hydrogrossular, $Ca_3Al_2(OH)_{12}$, shows that tetrahedral $O_4H_4^{4-}$ groups take the place of SiO_4^{4-}, the centre of the group being the position of Si in the garnet structure. A slightly distorted tetrahedron of O atoms is surrounded by a tetrahedral group of H atoms. The coordination group of Ca is intermediate between cube and antiprism.

Miscellaneous complex oxides

The structure of Li_4UO_5[1] (with which Na_4UO_5 is isostructural) may be described as a NaCl structure in which U occupies $\frac{1}{5}$ and Li $\frac{4}{5}$ of the cation positions, or alternatively as containing chains of UO_6 octahedra sharing opposite vertices. The UO_6 groups are tetragonally distorted, having four shorter equatorial bonds (1.99 Å) and two longer bonds (2.32 Å); contrast the oxides UMO_{12} (below) with six equal U–O bonds and $BaUO_4$ with two shorter collinear bonds.

Mg_6MnO_8 and Cu_6PbO_8 have a slightly distorted NaCl structure[2] in which one-quarter of the cations M^{2+} have been replaced by half their number of M^{4+}

ions. Since the Mn^{4+} ion is rather smaller than Mg^{2+} (Mg–$6O$, $2.10\,\text{Å}$; Mn–$6O$, $1.93\,\text{Å}$) the cubic close packing of the oxygen ions is slightly distorted. From calculations of the electrostatic energies of defect structures it has been concluded that in general ionic crystals with vacancies will have ordered rather than random structures.

The structure of $Zn_2Mo_3^{IV}O_8$ presents several points of interest.[3] There is *ABCB* ... type of c.p. oxygen atoms. The Mo atoms occupy octahedral holes in groups of three, as shown in Fig. 5.22(d), p. 215. Owing to the metal–metal interactions (Mo–Mo, $2.52\,\text{Å}$) the compound is diamagnetic. One-half of the Zn atoms occupy tetrahedral and the remainder octahedral holes, the mean bond lengths being Zn–$4O$, $1.98\,\text{Å}$; Zn–$6O$, $2.10\,\text{Å}$.

Solid solutions UO_{2+x}:Y_2O_3[4] adopt a defect fluorite structure with a degree of anion deficiency depending on composition and partial pressure of oxygen. When the oxygen deficiency becomes too large rearrangement takes place to a rhombohedral structure $M'M_6''O_{12}$. This structure is derivable from fluorite by removing 2 O atoms along a 3-fold axis and ordering the vacancies to give a structure (called the δ-structure) with one cation site of 6-coordination (octahedral) and six sites of 7-coordination (at 7 vertices of a slightly distorted cube). This is the structure of Pr_7O_{12}[4a] (also Tb_7O_{12} and Ce_7O_{12}) and of compounds

$$UM_6O_{12} \qquad M = \text{La and all 4f metals from Pr to Lu,}$$
$$WM_6O_{12} \qquad M = \text{Y and all 4f metals from Ho to Lu, and}$$
$$MoM_6O_{12} \qquad M = \text{Y and all 4f metals from Er to Lu,}$$

and also of the complex oxides $Zr_3Sc_4O_{12}$ and $Zr_3Y_4O_{12}$. The oxides UM_6O_{12} are of interest as comparatively rare examples of U^{VI} with nearly regular octahedral coordination by 6 O; see also δ-UO_3, p. 1265. In an X-ray study of $Zr_3Y_4O_{12}$ it was not possible to determine the occupancy of the cation sites because of the similar scattering powers of the two kinds of cation, but the M–O distances suggest that the octahedral sites are occupied by Zr^{4+} and the six 7-coordination sites by 2 Zr^{4+} and 4 Y^{3+} arranged at random.[4b]

Of the numerous complex oxides formed by Re with La and the 4f elements we have noted the structure of $La_4Re_6O_{19}$[5] on p. 224 as containing OLa_4 groups in the cavities of a 3D framework formed from pairs of edge-sharing ReO_6 octahedra joined by sharing terminal vertices. Another oxide with an interesting structure has the empirical composition La_2ReO_5.[6] In a fluorite-like arrangement of O atoms La^{3+} ions occupy four-fifths of the cubic interstices, and the remainder are occupied by pairs of Re atoms with the Re–Re bond ($2.26\,\text{Å}$) parallel to a cube edge. The environment of the Re_2 group is similar to that in the $Re_2Cl_8^{2-}$ ion shown in Fig. 9.14(b) on p. 430. The formula $La_4Re_2O_{10}$ is therefore to be preferred.

The structures of a family of complex oxides with formulae $Bi_{12}GeO_{20}$, $Bi_{25}FeO_{40}$, and $Bi_{38}ZnO_{60}$ are described on p. 890.

In $Ca_2Fe_2O_5$ layers are formed from FeO_6 octahedra sharing 4 equatorial vertices, and these layers are joined through vertex-sharing tetrahedra FeO_4 to form a 3D framework of composition Fe_2O_5. This framework consists of equal numbers of tetrahedra and octahedra which share vertices:

$$t \begin{cases} 2\,o \\ 2\,t \end{cases} \quad \text{and} \quad o \begin{cases} 4\,o \\ 2\,t \end{cases} \text{(see p. 233)}$$

The Ca^{2+} ions occupy irregular interstices in the framework.[7]

(1)   JINC 1964 **20** 693
(2)   AC 1954 7 246
(3)   AC 1966 **21** 482
(4)   IC 1964 3 949; IC 1966 5 749
(4a) AC 1975 **B31** 971 (n.d.)
(4b) AC 1977 **B33** 281
(5)   AC 1968 **B24** 1466
(6)   AC 1976 **B32** 1485
(7)   AC 1970 **B26** 1469

Complex oxides containing Ti, V, Nb, Mo, or W

In Chapter 5 we derived some of the simpler structures that may be constructed by linking together octahedral AX_6 coordination groups by sharing vertices and edges and/or faces to give arrangements extending indefinitely in 1, 2, or 3 dimensions. Here we extend the treatment to include some rather more complicated structures. In Chapter 5 we used as building units first the AX_5 (ReO_3) chain formed from octahedra sharing two opposite vertices and the AX_4 (rutile) chain formed from octahedra sharing two opposite edges and then the corresponding double chains formed from two simple chains by edge-sharing. We also mentioned the 'shear' structures of certain Mo and W oxides which may be built from slices of the ReO_3 structure and some more complex structures built from blocks of that structure (V_6O_{13}, high-Nb_2O_5).

We saw that there are two ways in which blocks of ReO_3 structure can share edges to form more extended structures (Fig. 5.35(a) and (b)), that edge-sharing only of type (a) occurs in some Mo and W oxides, and only of type (b) in $WNb_{12}O_{33}$, but of both types (a) and (b) in one form of Nb_2O_5 and in the simple 'brannerite' structure of $NaVMoO_6$, and the metamict minerals $ThTi_2O_6$ and UTi_2O_6. The general types of structure that result are the following.

(a) Structures of binary or more complex oxides in which all the metal atoms are in octahedral coordination groups.

(b) 3-dimensional frameworks built around ions of Na, K, etc. in positions of higher coordination (>6), these ions being essential to the stability of the framework, which does not exist as the structure of an oxide of the transition element. Normal stoichiometric compounds of this kind include $Na_2Ti_6O_{13}$ and KTi_3NbO_9, with respectively 8- and 10-coordination of the alkali-metal ions. In certain frameworks with compositions MO_2 or MO_3, corresponding to a normal oxide, a valence change in a proportion of the M atoms leads to the formation of non-stoichiometric phases such as Na_xMO_2 or Na_xMO_3 which have very unusual physical and chemical properties. These *bronzes* are described in a later section.

(c) Layer structures of compounds such as $K_2Ti_2O_5$, $Na_2Ti_3O_7$, and $KTiNbO_5$, in which the c.n.s of the alkali-metal ions are respectively 8, 7 and 9, and 8, represent an alternative way of providing positions for larger cations.

(d) We shall also mention a further group of structures which are closely related to the bronzes in which there is bipyramidal coordination of certain metal atoms. This group includes the binary oxides Mo_5O_{14}, $Mo_{17}O_{47}$, and $W_{18}O_{49}$.

A very simple way of utilizing both types of edge-sharing by ReO_3 chains is to form first a double (or other multiple) chain and to put two of these together to form the more complex chain of Fig. 13.9(a) and (b). These chains may then join

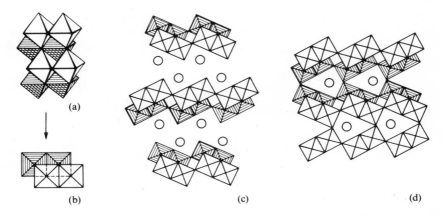

(a)

(b)   (c)   (d)

FIG. 13.9. (a) and (b) Multiple chain formed from four 'ReO_3' chains; (c) layer in $KTiNbO_5$; (d) 3D framework in KTi_3NbO_9.

by vertex-sharing into a layer, (c), or by further vertex-sharing to form a 3D structure, (d). Both types of structure can accommodate alkali or alkaline-earth ions, as in the examples shown. Starting with a triple chain we have the same possibilities (Table 13.14). More complex structures of the same general type can

TABLE 13.14
Structures built from multiple octahedral chains

Type of chain	Layer structure	3D structure	Reference
4-octahedra (Fig. 13.9)	M_2O_5	M_4O_9	
	$KTiNbO_5$	KTi_3NbO_9 $BaTi_4O_9$	AC 1964 **17** 623 JCP 1960 **32** 1515
6-octahedra	M_3O_7	M_6O_{13}	
	$Na_2Ti_3O_7$	$Na_2Ti_6O_{13}$	AC 1961 **14** 1245 AC 1962 **15** 194

be visualized with (2+3) or (3+4) octahedra in the primary 'chain', giving 3D structures with intermediate formulae M_5O_{11}, M_7O_{15}, etc. An example of the latter is $Na_2Ti_7O_{15}$.[1]

A second building principle is to take an infinite block of ReO_3 structure, join it to others to form either (i) isolated multiple blocks or (ii) an infinite layer, and then to join such units into 3D structures by edge-sharing of the second type to similar units at a different level. The structures are rather simpler in (ii) than in (i) and are therefore illustrated first. The smallest possible block of ReO_3 structure contains 4 ReO_3 chains and the 3D framework of $AlNbO_4$[2] is built of such sub-units. This framework is not known as the structure of a simple dioxide but it is found in Na_xTiO_2,[3] a bronze containing Ti^{3+} and Ti^{4+} in which there is partial occupancy by Na^+ of the positions indicated in Fig. 13.10. A number of oxide

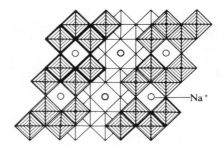

FIG. 13.10. The structure of Na_xTiO_2.

phases have structures of this type utilizing larger ReO_3 blocks, for example, V_6O_{13} ((3 × 2) blocks), illustrated in Fig. 5.38 (p. 228), $TiNb_2O_7$ ((3 × 3) blocks), $Nb_{12}O_{29}$ and $Ti_2Nb_{10}O_{29}$ ((3 × 4) blocks). All members of this family, in which the blocks of ReO_3 structure contain (3 × n) chains, can be represented by the general formula $M_{3n}O_{8n-3}$. The end-number is represented by Nb_3O_7F and V_2MoO_8 (Table 13.15).

In the structures we have just described the basic ReO_3 blocks are linked into

TABLE 13.15
The $M_{3n}O_{8n-3}$ family of oxide structures

n	Formula	Example	Reference
2	M_6O_{13}	V_6O_{13}	HCA 1948 **31** 8
3	M_9O_{21} (= M_3O_7)	$TiNb_2O_7$	AC 1961 **14** 660
4	$M_{12}O_{29}$	$Nb_{12}O_{29}$	ASc 1966 **20** 871
		$Ti_2Nb_{10}O_{29}$	AC 1961 **14** 664
∞	M_3O_8	V_2MoO_8	ASc 1966 **20** 1658
		Nb_3O_7F	ASc 1964 **18** 2233

infinite 'layers' by the first type of edge-sharing. There are also structures in which the blocks at both levels are joined in pairs, and here also there is the possibility of blocks of various sizes. Figure 13.11(a) shows the structure of $Nb_{25}O_{62}$[4] (and $TiNb_{24}O_{62}$) where the block size is (3×4). In an intermediate class of structure there is linking of the ReO_3 blocks of one family into infinite layers (those lightly shaded in Fig. 13.11(b)) while the blocks at the other level are discrete. This group of structures includes $Nb_{22}O_{54}$[5] and the high-temperature form of Nb_2O_5,[6] which differ in the sizes of the ReO_3 blocks at the two levels. In high-Nb_2O_5 (Fig. 13.11(b)) blocks of (3×5) octahedra are joined at one level to form infinite planar slabs, and these slabs are further linked by (3×4) blocks (heavy outlines). In both the structures of Fig. 13.11 there are tetrahedral holes, indicated by the black circles, some of which are occupied—in a regular way—by metal atoms. In high-

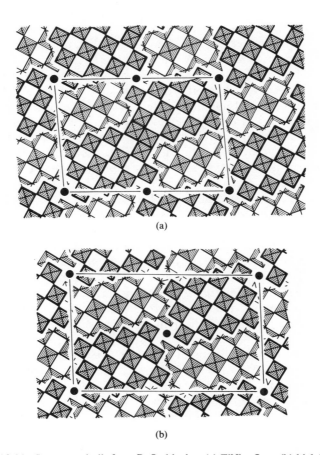

(a)

(b)

FIG. 13.11. Structures built from ReO_3 blocks; (a) $TiNb_{24}O_{62}$; (b) high-Nb_2O_5.

Nb_2O_5 27 Nb atoms in the unit cell are in positions of octahedral coordination and 1 Nb occupies a tetrahedral hole.

(1)   AC 1968 **B24** 392
(2)   ACSc **16** 421; AC 1965 **18** 874
(3)   AC 1962 **15** 201
(4)   AC 1965 **18** 724
(5)   ACSc 1965 **19** 1401
(6)   AC 1964 **17** 1545. For a survey of some of these compounds see HCA (Fasc. extra Alfred Werner) 1967 207.

Bronzes and related compounds

The name *bronze*, originally given to the W compounds by Wohler in 1824, is now applied to solid oxide phases with the following characteristic properties: intense colour (or black) and metallic lustre, metallic conductivity or semiconductivity, a range of composition, and resistance to attack by non-oxidizing acids. The bronzes Na_xWO_3, for example, have colours ranging from golden yellow ($x \approx 0.9$) through red ($x \approx 0.6$) to deep violet ($x \approx 0.3$). Most, but not all, bronzes consist of a host structure of composition MO_2 or MO_3 in which M is a (transition) metal capable of exhibiting a valence less than 4 or 6 respectively. If a proportion of the M(vi) atoms in MO_3 are converted to M(v) the requisite cations (alkali, alkaline-earth, La^{3+}, etc.) are incorporated into the structure to maintain electrical neutrality. The electrons liberated in this process are not, however, captured by individual metal ions of the host structure, but are distributed over the whole structure giving rise to the metallic or semi-metallic properties. From the geometry of the structures it would appear that the metallic conductivity of bronzes in not due to direct overlap of metal orbitals but to interactions through the oxygen atoms.

Bronzes have been prepared in which M is Ti, V, Nb, Ta, Mo, W, Re or Ru (see also $Na_xPt_3O_4$, p. 550). Typical methods of preparation include heating a mixture of Na_2WO_4, WO_3, and WO_2 *in vacuo* or reducing Na_2WO_4 by hydrogen or molten zinc (for Na—W bronzes), fusing V_2O_5 with alkali oxide, when oxygen is lost and V bronzes are formed, or reducing $Na_2Ti_3O_7$ by hydrogen at 950 °C, when blue-black crystals of Na_xTiO_2 are formed. The properties of bronzes and the reasons for their formation are by no means fully understood; in particular there is no clear connection between the appearance of the 'bronze' properties and the structure of the phase. For example, there are phases with both the tetragonal and hexagonal W bronze structures which are not bronzes (see later). Also, $Li_xTi_{4-x/4}O_8$ is a non-stoichiometric phase with the ramsdellite structure (p. 555) but it is colourless (i.e. it is not a bronze), whereas $K_{0.13}TiO_2$ has the closely related hollandite structure and is a bronze. Some bronzes have layer structures, examples of which are described later, but most bronzes have 3D framework structures, and the composition of the host structure corresponds to a normal oxide (as in Na_xTiO_2, $Na_xV_2O_5$, and Na_xWO_3) though the framework found in the bronze is not necessarily stable for the pure oxide. The hollandite structure of K_xTiO_2 is not stable for $x = 0$, and for all the isostructural K, Rb,, and Cs compounds x is approximately equal to 0.13. The cubic W bronze framework is, however, known not only as an oxide structure (ReO_3) and in distorted forms for WO_3 itself, but also with its full complement of Na^+ ions as the stoichiometric compound NaW^VO_3. This high-pressure phase (reference in Table 13.16) has the appearance (bronze colour) and metallic conductivity characteristic of bronzes. For a (colourless) form of WO_3 with the hexagonal W bronze framework see p. 572.

Numerous analogues of the alkali-metal bronzes which have been prepared include La_xTiO_3 and Sr_xNbO_3 (cubic),[1] In_xWO_3 (hexagonal),[2], Sn_xWO_3,[3] $Ba_{0.12}WO_3$ and $Pb_{0.35}WO_3$ (tetragonal),[4] Cu_xWO_3,[5] and 4f compounds $M_{0.1}WO_3$.[6] It is possible to replace some W by Ta in both the tetragonal and hexagonal bronze structures, as in $K_{0.5}(Ta_{0.5}W_{0.5})O_3$ (tetragonal) and $Rb_{0.3}(Ta_{0.3}W_{0.7})O_3$ (hexagonal).[7] These cream-white compounds are not bronzes but normal oxides. Bronzes containing Re have been made under high pressures.[8]

A Ru bronze, $Na_{3-x}Ru_4O_9$,[9] has a new type of 3D structure formed from single, double, and triple rutile chains which are linked together by sharing vertices.

(1) JACS 1955 77 6132, 6199
(2) AC 1978 **B34** 1433
(3) IC 1968 7 1646
(4) IC 1965 4 994
(5) JACS 1957 79 4048
(6) IC 1966 5 758
(7) JPC 1964 68 1253
(8) SSC 1969 7 299
(9) AC 1974 **B30** 1459

Tungsten bronzes

The approximate limits of stability of various bronzes are shown in Table 13.16, but somewhat different ranges have been found by different workers, and it is probable that the limits depend to some extent on the temperature of preparation.

The structures of these compounds provide a very elegant illustration of the formation of 3-dimensional networks MO_3 by the joining of MO_6 octahedra through all their vertices. The simplest and most symmetrical structure of this kind is the ReO_3 structure; addition of alkali-metal atoms at the centres of the unit cells gives the perovskite type of structure which is that of the cubic phases in Table 13.16 (Fig. 13.12(a) and (b)). Essentially the same basic framework occurs in the

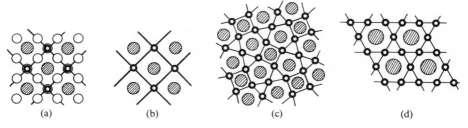

(a)　　　　　(b)　　　　　(c)　　　　　(d)

FIG. 13.12. Projections of the structures of tungsten bronzes: (a) perovskite structure of cubic bronzes; (b) the same showing only alkali-metal ions (shaded circles) and W atoms (small circles); (c) and (d) the W frameworks and alkali-metal ions in the tetragonal and hexagonal bronzes.

TABLE 13.16
Stability limits of tungsten bronzes

Li	Na	K	Rb	Cs	$\frac{x}{1 \cdot 00}$
		cubic†			
	cubic†				
		tetragonal			0·50
cubic					
	tetragonal-I				
		hexagonal	hexagonal	hexagonal	
tetragonal	tetragonal				
II	II				0·00

† The upper limits shown by the full heavy lines refer to compounds made at atmospheric pressure. Under high pressures the stability range of the cubic Na bronze includes stoichiometric $NaWO_3$, and K bronzes have been prepared with metal content up to $K_{0.9}$ (IC 1969 8 1183).

tetragonal-II Li and Na compounds, which have a slightly distorted perovskite-type structure.[1] Two quite different structures are found for the tetragonal K bronze ($x \approx 0.48$–0.57) and the hexagonal bronze formed by K, Rb, and Cs with $x \approx 0.3$.[2] (The tetragonal-I Na compound[3] has a structure closely related to the tetragonal K bronze.) Projections of these two structures are shown in Fig. 13.12(c) and (d), where it will be seen that the larger alkali-metal ions are situated in tunnels bounded by rings of five and six, instead of four, WO_6 octahedra. Note that these ions do not lie at the same level as the W atoms but $c/2$ above or below them. It will be appreciated that the upper limit of x in M_xWO_3 is determined by purely geometrical factors. The ratio of the number of holes for alkali-metal atoms to the number of W atoms falls from 1 in (b) to 0.6 in (c) and 0.33 in (d), so that from the compositions determined experimentally it would appear that nearly the full complements of these atoms can be introduced into the structures, and indeed are necessary for the stability of the tetragonal and hexagonal phases.

(1)   ACSc 1951 **5** 372, 670
(2)   ACSc 1953 **7** 315
(3)   AK 1949 **1** 269

The tetragonal bronze structure

We deal in more detail with this structure because it forms the basis of the structures of three groups of compounds, namely:

(i) the bronzes, the highly coloured, non-stoichiometric compounds to which we have just referred;

(ii) ferroelectric, usually colourless, compounds mostly with formulae of the type $M^{II}Nb_2O_6$ or $M^{II}Ta_2O_6$ but in some cases containing additional atoms, as in $K_6Li_4Nb_{10}O_{30}$; and

(iii) a large family of (stoichiometric) compounds with formulae such as $mNb_2O_5 . nWO_3$ in which metal atoms and additional O atoms occupy some of the pentagonal tunnels in the bronze structure. We shall also include here compounds such as $LiNb_6O_{15}F$ with structures of the same general type based on closely related octahedral frameworks.

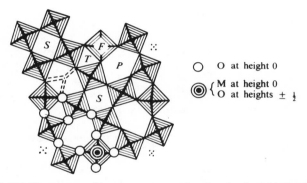

$\bigcirc$ O at height 0

$\circledcirc \left\{ \begin{array}{l} M \text{ at height } 0 \\ O \text{ at heights } \pm \frac{1}{2} \end{array} \right.$

FIG. 13.13. The tetragonal bronze structure, showing the three kinds of tunnel.

In the tetragonal bronze structure there are tunnels of three kinds (Fig. 13.13), of which only two (*S* and *P*) are occupied in the bronzes (class (i)). The environment of an atom in a tunnel depends on its height relative to the atoms in the framework. In compounds of classes (i) and (ii) the 'tunnel' atom is at height $\frac{1}{2}$, when the coordination groups of such atoms are:

T: tricapped trigonal prism (9-coordination), site suitable only for very small ion (Li^+);

S: distorted cuboctahedral (12-coordination);

P: in a regular pentagonal tunnel the coordination group would be a pentagonal prism with atoms beyond each vertical (rectangular) face. In fact the pentagonal tunnels have an elongated cross-section, and the coordination group is closer to a tricapped trigonal prism (9-coordination).

(*F* is the (distorted) octahedral site in the MO_3 framework.)

The examples in Table 13.17 show how certain of these positions are occupied in some of the ferroelectric compounds of class (ii), the numbers at the heads of the

TABLE 13.17

Compounds with structures related to the tetragonal bronze structure

Ferroelectrics (class (ii))

Cell content	4T	2S 4P	10F	30 O	Reference
	Li_4	K_6	Nb_{10}	O_{30}	JCG 1967 **1** 315, 318
		Ba_6	Ti_2Nb_8	O_{30}	AC 1968 **B24** 984
		$(Ba, Sr)_5$	Nb_{10}	O_{30}	JCP 1968 **48** 5048
		Pb_5	Nb_{10}	O_{30}	AC 1958 **11** 696

Superstructures (class (iii)) and structures based on other MO_3 *frameworks of the same general type (with T, S, and P tunnels).*

Formula type	Composition	Reference
M_3O_8	WNb_2O_8	NBS 1966 **70A** 281
	$NaNb_6O_{15}F$ (or OH)	ACSc 1965 **19** 2285
	$LiNb_6O_{15}F$	ACSc 1965 **19** 2274
	Ta_3O_7F	ACSc 1967 **21** 615
$M_{17}O_{47}$	$Mo_{17}O_{47}$	ACSc 1963 **17** 1485
	$Nb_8W_9O_{47}$	AC 1969 **B25** 2071
$M_{23}O_{63}$	$Nb_{12}W_{11}O_{63}$	AC 1968 **B24** 637
Miscellaneous	Mo_5O_{14}	AK 1963 **21** 427
	$W_{18}O_{49}$	AK 1963 **21** 471

On the subject of the nets in these compounds see: AC 1968 **B24** 50.

columns indicating the numbers of sites of each type in a unit cell containing 30 O atoms. In $Li_4K_6Nb_{10}O_{30}$ all the metal sites are occupied, including the T sites for very small ions. If the oxidation number of all the atoms in the framework sites is 5, only 5 M^{2+} ions are required for charge balance and these occupy statistically the $2S + 4P$ sites. The chemical formula reduces to a simple form if all these atoms are of the same element, as in $PbNb_2O_6$. If, however, some M^V is replaced (again statistically) by M^{IV} in the 10 F sites, the full complement of M^{II} atoms can be taken up, as in $Ba_6Ti_2Nb_8O_{30}$.

In class (iii) a new principle is introduced. A metal atom is placed in a P tunnel, assumed now to have approximately regular pentagonal cross-section, at the height 0 (see Fig. 13.13) so that it is surrounded by a ring of five O atoms at the same height. An additional O atom is introduced with each metal atom in the tunnel, the two kinds of atom alternating, giving M seven O neighbours at the vertices of a pentagonal bipyramid. The formula of the compound obviously depends on the proportion of P tunnels occupied in this way. For example, if one-half of the P tunnels in the bronze structure are occupied (by $2M + 2O$) the composition becomes $M_{10+2}O_{30+2}$ or M_3O_8. Occupation of one-third of the P tunnels (which implies a larger unit cell) would give the composition $M_{17}O_{47}$ ($M_{15+2}O_{45+2}$). There are numerous oxide phases with structures of this kind. Some (class (iii)) are superstructures of the tetragonal bronze structure (with multiple cells), while others are based on closely related but topologically different 3-dimensional MO_3

frameworks. The latter, of which there is an indefinitely large number, have different relative numbers of *T, S,* and *P* tunnels and/or different spatial arrangements of these tunnels (see reference in Table 13.17). It is interesting that compounds so closely related chemically as $LiNb_6O_{15}F$ and $NaNb_6O_{15}F$ have structures based on different 3D octahedral nets; both, like WNb_2O_8, are of the M_3O_8 type, but with additional alkali-metal atoms (not shown in Fig. 13.13).

These structures may alternatively be described as assemblies of composite units of the kind indicated in Fig. 13.14. Such a unit consists of a column of pentagonal

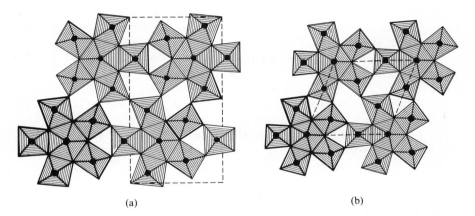

(a)   (b)

FIG. 13.14.   Unit cells of the structures of (a) $LiNb_6O_{15}F$; (b) $NaNb_6O_{15}F$. The alkali-metal ions are omitted.

bipyramidal coordination groups sharing edges with the surrounding columnds of octahedra; these multiple columns then form the 3D framework by vertex-sharing. In the series of stoichiometric oxides formed between Nb_2O_5 and WO_3 the structures are based on blocks of ReO_3 structure up to the composition $W_8Nb_{18}O_{69}$ but phases richer in W, starting at WNb_2O_8, have structures of the kind we have just described.

Molybdenum bronzes

The preparation under high pressure (65 kbar) of Mo bronzes with tungsten bronze structures has been described; for example, cubic Na_xMoO_3 and K_xMoO_3 ($x \approx 0.9$) and tetragonal K_xMoO_3 ($x \approx 0.5$).[1] The Mo bronzes prepared by the electrolysis of fused mixtures of alkali molybdates and MoO_3[2] have more complex structures. A bronze with the approximate composition $Na_{0.9}Mo_6O_{17}$ has a hexagonally distorted perovskite structure with an ordered arrangement of Na in one-sixth of the available sites and apparently a statistical distribution of 17 O atoms over 18 sites.[3] (The observed average structure apparently results from twinning of a monoclinic structure in which the O atoms are regularly arranged.)

Two potassium–molybdenum bronzes, a red $K_{0.33}MoO_3$[4] and a blue-black $K_{0.30}MoO_3$[5] have closely related layer structures. Edge-sharing groups of 6 and 10 octahedra respectively are further linked into layers by sharing vertices as indicated in Fig. 5.34 (p. 225), the composition of the layer being in each case MoO_3. The layers are held together by the K^+ ions which occupy positions of 8-coordination in the first compound and of 7- and (6+4)-coordination in the second. There is a closely related layer in $Cs_{0.25}MoO_3$,[6] in which the basic repeat unit is a different grouping of six octahedra. These units share the edges and vertices marked in Fig. 13.15 to form layers between which the larger Cs^+ ions can be accommodated.

Hydrogen molybdenum bronzes are formed by the reduction of MoO_3 by electrochemical or various chemical methods. They are coloured phases H_xMoO_3 which exhibit metallic conductivity. Three of these phases have homogeneity ranges (x values) as follows: blue orthorhombic (0.23–0.4); blue monoclinic (0.85–1.04), and red monoclinic (1.55–1.72); the green monoclinic phase is apparently stoichiometric ($x = 2$).[7] It appears that in the first phase the presence of the H atoms has little effect on the structure, which is essentially the same as that of MoO_3. It was first determined for a compound formulated as $Mo_2O_5(OH)$,[8] which corresponds to $x = 0.5$. A later study of $H_{0.36}MoO_3$ shows that the H atoms lie within the layers attached as OH groups to the bridging O atoms,[9] but further work is required to establish the positions of the H atoms in the phases with higher H contents.

(1) IC 1966 **5** 1559
(2) IC 1964 **3** 545
(3) AC 1966 **20** 59
(4) AC 1965 **19** 241; IC 1967 **6** 1682
(5) AC 1966 **20** 93
(6) JSSC 1970 **2** 16. For later work on lithium Mo bronzes see: JSSC 1970 **1** 327, and for Mo fluoro bronzes: JSSC 1970 **1** 332
(7) JSSC 1980 **34** 183; MRB 1978 **13** 311
(8) ACSc 1969 **23** 419
(9) ZaC 1977 **435** 247; JSSC 1979 **28** 185

Vanadium bronzes

These are made by methods such as fusing V_2O_5 with alkali-metal oxide or vanadate. Oxygen is lost and black semiconducting phases are formed which contain some alkali metal and V^{IV} atoms. Bronzes have been made containing various alkali metals, Cu, Ag, Pb, and so on, and in many systems there are several bronze phases with different composition ranges and structures.

The structure of $Li_{0.04}V_2O_5$ (Fig. 13.16(a)) is essentially that of V_2O_5 with a small number of Li^+ ions in positions of trigonal prism coordination. In LiV_2O_5 (Fig. 13.16(b)) also there is 5-coordination of the V atoms; both these compounds have layer structures.

In the next two compounds of Fig. 13.16 there is both 5- and 6-coordination of V. $Li_{1+x}V_3O_8$, (c), has a layer structure; in (d), which represents the structure of β-$Li_{0.30}V_2O_5$ and $Na_{0.15}V_2O_5$, the VO_5 and VO_6 coordination groups form a 3D framework. The sodium bronze $Na_xV_2O_5$ is stable over the range $0.15 < x < 0.33$, the upper limit corresponding to the formula NaV_6O_{15}. This corresponds to

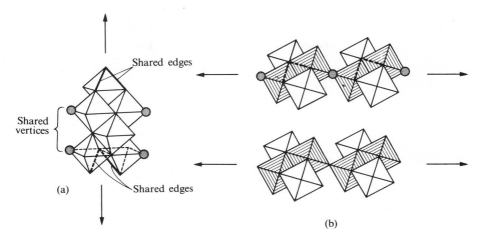

FIG. 13.15. The structure of the bronze $Cs_{0.25}MoO_3$: (a) the 6-octahedron unit which shares the edges and vertices indicated to form the layers (b), which are perpendicular to the plane of the paper.

occupation of only one-half of the available Na^+ sites; if they were all occupied in each tunnel Na^+ would have a close Na^+ neighbour in addition to 7 O atoms. The formula may therefore be written $Na_{2-y}V_6O_{15}$. A steel-blue $Pb_{0.20}V_2O_5$ with the structure of Fig. 13.16(d) has also been prepared.

In the $Ag_2O-V_2O_5$ system there is a bronze $Ag_{2-x}V_6O_{15}$ isostructural with the Na bronze of Fig. 13.16(d) and also $Ag_{0.68}V_2O_5$, (e), in which the V atoms may be described as octahedrally coordinated, though the V–O bond lengths range from 1.5–2.4 Å.

The structure of $Na_{0.2}TiO_2$ is included at (f) to show the close relation of its structure to that of the bronze (e). Further sharing of vertices of octahedra converts the layer structure (e) to the 3D framework (f) of the titanium bronze.

A number of Cs–V bronzes have been made, with either 3D framework structures ($Cs_{0.3}V_2O_5$, $Cs_{0.35}V_3O_7$) or layer structures (CsV_2O_5, $Cs_2V_5O_{13}$).[1]

In the systems $M_2O_3-VO_2-V_2O_5$ (M = Al, Cr, Fe) phases include one ($Al_{0.33}V_2O_5$) closely related to V_6O_{13}, from which it is derived by addition of Al and O atoms in the tunnels of that structure. There are also phases $M_zV_{2-z}O_4$ with a superstructure of the rutile type, having equal numbers of M^{III} and V^V replacing part of the V^{IV}, as in $Fe^{III}_{0.07}V^V_{0.07}V^{IV}_{1.86}O_4$). For references see Table 13.18.

(1)   AC 1977 **B33** 775, 780, 784, 789

Complex oxides built of octahedral AO_6 and tetrahedral BO_4 groups

Some of the simpler structures of this type were noted in the discussion in Chapter 5 of structures built from tetrahedra and octahedra. Here we describe some of the

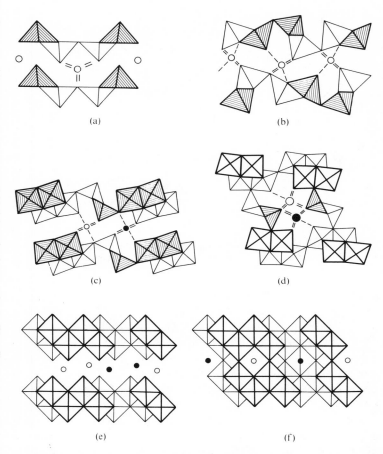

FIG. 13.16. The structures of vanadium and titanium bronzes: (a) $Li_{0.04}V_2O_5$, similar to V_2O_5 but with Li^+ ions (small circles) in trigonal prismatic coordination; (b) LiV_2O_5 (Li^+ in octahedral coordination); (c) $Li_{1+x}V_3O_8$ (octahedral coordination of Li^+); (d) $Na_{0.15}V_2O_5$ (7-coordinated Na^+); (e) $Ag_{0.68}V_2O_5$ (Ag^+ with 5 near neighbours); (f) $Na_{0.2}TiO_2$ (8-coordinated Na^+).

structures listed in Table 5.7 (p. 234) and also some more complex structures of the same general type. The composition depends on the relative numbers of octahedral and tetrahedral groups and on the way in which the vertices are shared. In the simplest case each tetrahedron shares all its vertices with octahedra and each octahedron shares four vertices with tetrahedra and two with other octahedra. The composition is then ABO_5.

Two simple structures of this kind may be regarded as built from ReO_3-type chains (in which each octahedron shares two opposite vertices) which are then cross-linked by BO_4 tetrahedra to form 3D structures. In $NbOPO_4$ (and the

TABLE 13.18
Coordination of V *in bronzes*

Fig. 13.16	Bronze	C.N. of V	V—O (A)	Reference
a	$Li_{0.04}V_2O_5$	5	1·6–2·45 (then 2·82)	BSCF 1965 1056
b	LiV_2O_5	5	1·61–1·98 (3·09)	AC 1971 **B27** 1476
c	$Li_{1+x}V_3O_8$	6	1·6–2·3 —	AC 1957 **10** 261
		5	1·6–2·1 (2·86)	
d	$Na_{0.15}V_2O_5$	6	1·6–2·3 —	AC 1955 **8** 695
		5	1·6–2·0 (2·68)	
e	$Ag_{0.68}V_2O_5$	6	1·5–2·4 —	ACSc 1965 **19** 1371
f	$Na_{0.2}TiO_2$			AC 1962 **15** 201;
				IC 1967 **6** 321
	$Pb_{0.20}V_2O_5$			BSMC 1969 **92** 17
	$Al_{0.33}V_2O_5$			BSCF 1967 227

For later work on $M_xV_2O_5$ phases see: JSSC 1970 **1** 339 and AC 1975 **B31** 1481, and for a new Ti bronze, $K_3Ti_8O_{17}$: JSSC 1970 **1** 319.

isostructural $VOPO_4$, $MoOPO_4$, $VOMoO_4$ and tetragonal $VOSO_4$) each BO_4 tetrahedron links four such chains (Fig. 13.17(a)) forming a structure which approximates to a c.c.p. assembly of O atoms in which Nb atoms occupy one-fifth of the octahedral holes and P atoms one-tenth of the tetrahedral holes (Fig. 13.17(b)). In a second form of $VOSO_4$ also each tetrahedral group (SO_4) shares its vertices with four octahedral groups (VO_6) but two of these belong to the same chain, so that the tetrahedral group links three ReO_3-type chains as compared with four in $NbOPO_4$.

Although it is convenient to describe these structures in terms of octahedral AO_6 and tetrahedral BO_4 groups it should be emphasized that while the BO_4 groups

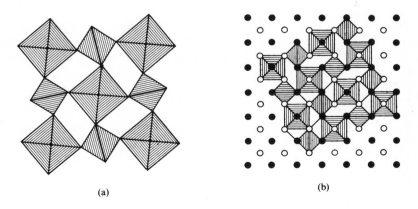

(a)   (b)

Fig. 13.17 Projections of the crystal structure of $NbOPO_4$ or $VOPO_4$ (see text).

are essentially regular tetrahdra the AO_6 groups are far from regular. The metal atom is displaced from the centre towards one vertex of the octahedron, and the arrangement of the five bonds may be described as tetragonal pyramidal. The distortion of the octahedral coordination group is of a kind found in many oxy-compounds of Ti^{IV}, V^{IV}, V^V, Nb^V, and Mo^V, with one very short and one very long bond and four bonds of intermediate length. Some examples are given in Table 13.19.

TABLE 13.19

Metal coordination groups in some oxy-compounds

Compound	Bond lengths in octahedron (Å)			Reference
	short (one)	(four)	long (one)	
$NbOPO_4$	1.78	1.97	2.32	ACSc 1966 **20** 72
Nb_3O_7F	1.86	1.98	2.20	ACSc 1964 **18** 2233
$MoOPO_4$	1.66	1.97	2.63	ACSc 1964 **18** 2217
$VOMoO_4$	1.68	1.97	2.59	ACSc 1966 **20** 722
$VOSO_4$ (β) (rh.)	1.59	2.03	2.28	ACSc 1965 **19** 1906
(α) (tetrag.)	1.63	2.04	2.47	JSSC 1970 **1** 394
$VOPO_4$	1.58	1.87	(2.87)	AC 1976 **B32** 2899
$Na_2(TiOSiO_4)$	1.70	1.99	–	AC 1978 **B34** 905

If the projection of Fig. 13.17(a) is regarded as a slice of the structure parallel to the plane of the paper, and if the very long bonds of each V coordination group are on the same side of the slice, then the structure has a very pronounced layer-like character, as in α-$VOPO_4$ (Table 13.19). The distortion of the 'octahedral' coordination is so great that the coordination should be described as tetragonal pyramidal.

A closely related structure is that of $K_2V_3O_8$. We refer to Fig. 5.7(c) on p. 197 which shows a tetrahedral layer of composition A_3X_7 (melilite layer). If the squares in that figure represent square pyramidal BO_5 groups instead of tetrahedral groups AX_4 and the triangles tetrahedral AO_4 groups then the composition would be A_2BO_8, or $BO(A_2O_7)$ if we separate the pairs of vertex-sharing tetrahedral AO_4 groups. This is the structure of the anion in $Ba_2[TiO(Si_2O_7)]$,[1] the mineral fresnoite, and of the similar anion in $K_2[V^{IV}O(V^V_2O_7)]$.[2] In the latter compound the coordination of V^{IV} is square pyramidal (V–O, 1.58 (one), and 1.98 Å (four)) and that of V^V tetrahedral:

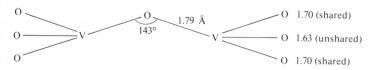

(The K^+ ions are situated between the layers in pentagonal antiprismatic 10-coordination.)

We now come to two related vanadates containing exclusively V^{IV} with the same coordination as in $K_2V_3O_8$. In Fig. 13.18 the short V–O bonds of the square pyramidal groups are directed upwards or downwards as indicated by the full or broken lines within their square outlines. The layers represent the structures of the anions in CaV_3O_7[(3)] and CaV_4O_9.[(4)]

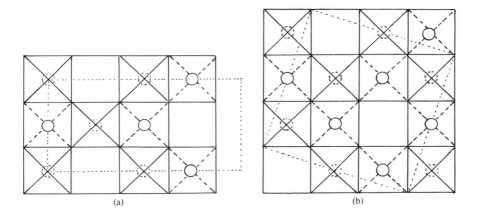

FIG. 13.18. Layers built of square pyramidal coordination groups in (a) CaV_3O_7 and (b) CaV_4O_9.

A number of oxy-compounds containing P and Mo (or W) with low Mo(W):P ratios have been prepared by methods such as the following: $Mo(OH)_3PO_4$ by heating a solution of MoO_3 in concentrated H_3PO_4 at 180 °C and then diluting with conc. HNO_3, and WOP_2O_7 from WO_3 and P_2O_5 on prolonged heating in an autoclave at 550 °C. These compounds are built from PO_4 tetrahedra and MoO_6 (WO_6) octahedra sharing vertices only, and a structural feature common to several of them is a chain consisting of alternate PO_4 and MoO_6 groups. In some compounds PO_4 shares vertices entirely with MoO_6 while in others there is linking of PO_4 groups in pairs or infinite chains, the further linking of these units through MoO_6 octahedra leading to infinite chains, layers, or 3D networks. The formulae assigned to these compounds in column 2 of Table 13.20 indicate the extent of sharing of O atoms between PO_4 groups, and the more elaborate structural formulae of column 3 show how many vertices of each PO_4 or MoO_6 group are shared and hence the nature of the fundamental net (column 4). The symbol ϕ represents an O atom shared between two coordination groups (PO_4 or MoO_6). We comment elsewhere on the surprisingly small number of pairs of isostructural Mo and W compounds. Here $Na(PMoO_6)$ and $Na(PWO_6)$ are isostructural, but the pairs P_2MoO_8 and P_2WO_8, $P_2Mo_2O_{11}$ and $P_2W_2O_{11}$ are not. We note now the structural principles in three of the compounds of Table 13.20.

TABLE 13.20

Some oxy-compounds built of PO_4 *and* MoO_6 (WO_6) *groups*

Compound		Structural formula	Nature of structure
$Mo(OH)_3PO_4$		$(PO\phi_3)[Mo\phi_3(OH)_3]$	3-connected double chain
P_2MoO_8	$MoO_2(PO_3)_2$	$(P\phi_4)_{2n}(MoO_2\phi_4)_n$	4-connected double layer
P_2WO_8	WOP_2O_7	$(P\phi_4 . PO\phi_3)(WO\phi_5)$	3,4,5-connected layer
$NaPWO_6$	$NaWO_2PO_4$	$Na[(P\phi_4)(WO_2\phi_4)]$	4-connected
$P_2W_2O_{11}$	$W_2O_3(PO_4)_2$	$(P\phi_4)(WO\phi_5)$	4,5-connected 3D net
$P_2Mo_2O_{11}$	$(MoO_2)_2P_2O_7$	$(P\phi_4 . P\phi_4)(MoO\phi_5)_2$	4,5-connected

References: AK 1962 **19** 51; ACSc 1964 **18** 2329

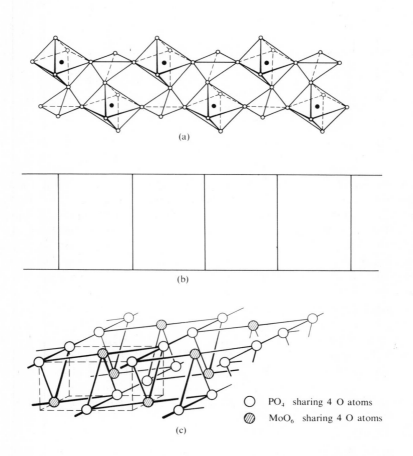

PO₄   sharing 4 O atoms

MoO₆   sharing 4 O atoms

FIG. 13.19. (a) and (b) The double chain $MoPO_4(OH)_3$. (c) The double layer P_2MoO_8.

The structural unit in $MoPO_4(OH)_3$ is the double chain of Fig. 13.19(a). The H atoms were not located but are presumably attached to the three unshared O atoms of each MoO_6 group and involved in hydrogen bonding between the chains. Since each PO_4 and MoO_6 shares 3 vertices the topology of the structure may be illustrated by the double chain of Fig. 13.19(b). In P_2MoO_8 chains of PO_4 tetrahedra sharing two O atoms are further linked by MoO_6 octahedra to form the double layer illustrated diagrammatically in Fig. 13.19(c). The simplest type of 3D network is found in $Na(PWO_6)$, in which every PO_4 shares all four vertices with MoO_6 groups and each of the latter shares four vertices with PO_4 groups. If we separate the two unshared O atoms of each MoO_6 the formula becomes $Na(PWO_4.O_2)$ – compare $NaAlSiO_4$. Basically the anion is a 4-connected net analogous to those found in aluminosilicates and in fact is very similar to those of the felspars (Chapter 23).

(1)   ZK 1969 **130** 438
(2)   AC 1975 **B31** 1794
(3)   AC 1973 **B29** 269
(4)   AC 1973 **B29** 1335

14

Metal hydroxides, oxyhydroxides, and hydroxy-salts

We shall be concerned in this chapter with the following groups of compounds:

Hydroxides, $M(OH)_n$ (with which we shall include the alkali-metal hydrosulphides MSH)
Complex hydroxides $M'_x M''_y (OH)_z$
Oxy-hydroxides MO(OH)
Hydroxy-salts (basic salts).

Metal hydroxides

Compounds $M(OH)_n$ range from the strongly basic compounds of the alkali and alkaline-earth metals through the so-called amphoteric hydroxides of Be, Zn, Al, etc. and the hydroxides of transition metals to the hydroxy-acids formed by non-metals ($B(OH)_3$) or semi-metals ($Te(OH)_6$). The latter are few in number and are included in other chapters.

Apart from the rather soluble alkali-metal (and Tl^I) hydroxides and the much less soluble alkaline-earth compounds, most metal hydroxides are more or less insoluble in water. Some of them are precipitated in a gelatinous form by NaOH and redissolve in excess alkali, for example, $Be(OH)_2$, $Al(OH)_3$, and $Zn(OH)_2$, while others such as $Cu(OH)_2$ and $Cr(OH)_3$ show the same initial behaviour but redeposit on standing. Such hydroxides, which 'dissolve' not only in acids but also in alkali have been termed amphoteric, the solubility in alkali being taken to indicate acidic properties. In fact the solubility is due in some cases to the formation of hydroxy-ions, which may be mononuclear, for example, $Zn(OH)_4^{2-}$,[1] or polynuclear. From a solution of $Al(OH)_3$ in KOH the potassium salt of the bridged ion (a) can be crystallized. (No difference was found between the lengths of the Al—O and Al—OH bonds (1.76 Å, mean),[2]) It is likely that hydroxy-ions of some kind exist in the solution. Certainly, from solutions containing Ba^{2+}, Al^{3+}, and OH^- ions salts such as $Ba_2[Al_2(OH)_{10}]$ and $Ba_2[Al_4(OH)_{16}]$ can be crystallized (p. 637). Hydroxy- or hydroxy-aquo complexes also exist in the 'basic' salts formed by the incomplete hydrolysis of

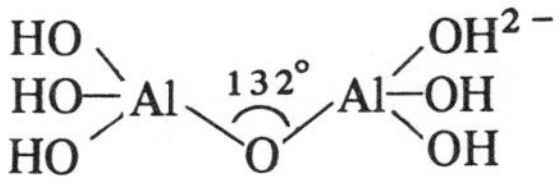

$$\begin{array}{c} HO \\ HO-Al \\ HO \end{array} \overset{132°}{\underset{O}{>}} Al \begin{array}{c} OH^{2-} \\ OH \\ OH \end{array}$$

(a)

normal salts (see later section), and some interesting polynuclear ions have been shown to exist in such solutions. The Raman and i.r. spectra[3] of, and also the X-ray scattering[4a] from, hydrolysed Bi perchlorate solutions show that the predominant species over a wide range of Bi and H ion concentrations is $Bi_6(OH)_{12}^{6+}$ (Fig. 14.1(a)). Salts containing this and related ions[4b] can be crystallized from such solutions. Similarly, X-ray scattering from hydrolysed Pb perchlorate solutions indicates at a OH:Pb ratio of 1.00 a complex based on a tetrahedron of Pb atoms, and at a OH:Pb ratio of 1.33 (the maximum attainable before precipitation occurs) a Pb_6 complex similar to that in the crystalline basic salt $Pb_6O(OH)_6(ClO_4)_4.H_2O$ —see Fig. 14.1(b).[5]

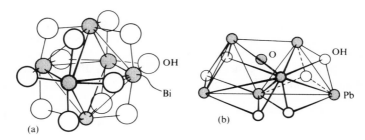

FIG. 14.1. Hydroxy-complexes: (a) $Bi_6(OH)_{12}^{6+}$; (b) $Pb_6O(OH)_6^{4+}$. In (b) there is an O atom at the centre of the central tetrahedron and OH groups above each face of the terminal tetrahedra.

In other cases the conductivities of 'redissolved' hydroxide solutions are equal to those of alkali at the same concentration, suggesting that the 'solubility' of the hydroxide is more probably due to the formation of a sol. For example, the freshly precipitated cream-coloured $Ge(OH)_2$ 'dissolves' in strongly basic solutions to form reddish-brown colloidal solutions.[6] Hydrolysis of metal-salt solutions often yields colloidal solutions of the hydroxides, as in the case of the trihydroxides of Fe and Cr and the tetrahydroxides of elements of the fourth Periodic Group such as Sn, Ti, Zr, and Th. It is unlikely that the hydroxides $M(OH)_4$ of the latter elements are present in such solutions, for the water content of gels $MO_2.xH_2O$ is very variable.

We must emphasize that our picture of the structural chemistry of hydroxides is very incomplete. Crystalline material is required for structural studies, preferably single crystals, and some hydroxides readily break down into the oxide, even on boiling in water (e.g. $Cu(OH)_2$, $Sn(OH)_2$, $Tl(OH)_3$). Many simple hydroxides are not known, for example, CuOH, AgOH, $Hg(OH)_2$, $Pb(OH)_4$. Though many of the higher hydroxides do not appear to be stable under ordinary conditions, nevertheless compounds such as $Ni(OH)_3$, $Mn(OH)_4$, $U(OH)_4$, $Ru(OH)_4$ etc. are described in the literature; confirmation of their existence and knowledge of their structures would be of great interest. Some may not be simple hydroxides; for example $Pt(OH)_4.2H_2O$ is $H_2[Pt(OH)_6]$. Conductometric titrations and molecular weight measurements show that 'aurates' and the gelatinous 'auric acid' prepared from them contain the

$Au(OH)_4^-$ ion.[7] The growth of crystals of $Pb(OH)_2$ in aged gels is described in the literature, but the most recent evidence suggests that this hydroxide does not exist.[8] The crystalline solid obtained by the slow hydrolysis of solutions of plumbous salts is $Pb_6O_4(OH)_4$ (see Fig. 14.15, p. 646, and p. 1186).

(1)   JCP 1965 **43** 2744

(2)   ACSc 1966 **20** 505

(3)   IC 1968 7 183

(4a) JCP 1959 **31** 1458

(4b) AC 1979 **B35** 448

(5)   IC 1969 8 856; ACSc 1972 **26** 3505

(6)   JINC 1964 **26** 2123

(7)   ZaC 1960 **304** 154, 164

(8)   IC 1970 9 401

The structures of hydroxides $M(OH)_n$

From the structural standpoint we are concerned almost exclusively with the solid compounds. Only the hydroxides of the most electropositive metals can even be melted without decomposition, and only the alkali hydroxides can be vaporized. Microwave studies of the vapours of KOH, RbOH, and CsOH and i.r. studies of the molecules trapped in argon matrices show that the (ionic) molecules are linear:

	M—O (A)	O—H (A)	μ(D)	*Reference*
KOH	2.18	–	–	JCP 1966 **44** 3131
RbOH	2.31	0.96	– ⎫	JCP 1969 **51** 2911;
CsOH	2.40	0.97	7.1 ⎬	JCP 1967 **46** 4768
(cf. CsF	2.35 (Cs-F)	–	7.87	PR 1965 **138A** 1303)

The structures of crystalline metal hydroxides and oxy-hydroxides MO(OH) are determined by the behaviour of the OH⁻ ion in close proximity to cations. We may envisage three types of behaviour: (a) as an ion with effective spherical symmetry; (b) with cylindrical symmetry; or (c) polarized so that tetrahedral charge distribution is developed. Attractions between the positive and negative regions of different OH⁻ ions in (c) leads to the formation of hydrogen bonds. The transition from (b) to (c) is to be expected with increasing charge and decreasing size of the cation. If

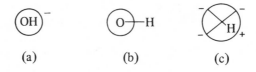

(a)   (b)   (c)

the polarizing power of a cation M^{n+} is proportional to ne/r^2 it can be readily understood why there is hydrogen bonding in, for example, $Al(OH)_3$ but not in $Ca(OH)_2$ or $La(OH)_3$, the relative values of the polarizing powers being

Ca^{2+}	La^{3+}	Al^{3+}
1.8	2.0	9.2

Case (a), implying random orientation or free rotation, is apparently realized in the high-temperature form of KOH (and RbOH?). The high-temperature form of NaOH was originally supposed to have the same (NaCl) structure but this is not confirmed by later work (see later). Apparently the SH⁻ ion behaves in this way in CsSH and in the high-temperature forms of NaSH, KSH, and RbSH, which is consistent with the fact that S forms much weaker hydrogen bonds than oxygen. Behaviour of type (a) is confined to (at most) one or two compounds, and on the basis of their structures we may recognize two main groups of crystalline hydroxides. The feature which distinguishes hydroxides of the first class is the absence of hydrogen bonding between the OH⁻ ions.

I. *Hydroxides* MOH, M(OH)$_2$, *and* M(OH)$_3$ *with no hydrogen bonding*

(a) *Random orientation or rotation of* OH⁻ *ions.* In the high-temperature form of KOH the OH⁻ ion is behaving as a spherical ion of effective radius 1.53 Å, intermediate between those of F⁻ and Cl⁻, and this compound has the NaCl structure. It is not known whether the cubic symmetry arises from random orientation of the anions or from free rotation.

Only the most electropositive metals, the alkali metals and alkaline-earths, form hydrosulphides. LiSH is particularly unstable, being very sensitive to hydrolysis and oxidation, and decomposes at temperatures above about 50 °C. It may be prepared in a pure state by the action of H$_2$S on lithium n-pentyl oxide in ether solution. Its structure is of the zinc-blende type and similar to that of LiNH$_2$, but there is considerable distortion of the tetrahedral coordination groups:

$$
\text{SH} \begin{cases} 2 & \text{Li} & 2.43\,\text{Å} \\ & \text{Li} & 2.46 \\ & \text{Li} & 2.67 \end{cases} \qquad \text{Li}_\text{I} - 4 \ \text{SH} \ 2.43\,\text{Å} \qquad \text{Li}_\text{II} \begin{cases} 2 \ \text{SH} & 2.46\,\text{Å} \\ 2 \ \text{SH} & 2.67 \end{cases}
$$

The high-temperature forms of NaSH, KSH, and RbSH have the NaCl structure; CsSH crystallizes with the CsCl structure. The fact that the unit cell dimension of CsSH is the same as that of CsBr shows that SH⁻ is behaving as a spherically symmetrical ion with the same radius as Br⁻. In contrast to the low-temperature forms of the hydroxides of Na, K, and Rb, which are described in (b), the hydrosulphides have a rhombohedral structure at room temperature. In this calcite-like structure the SH⁻ ion has lower symmetry than in the high-temperature forms and appears to behave like a planar group. It is probable that the proton rotates around the S atom in a plane forming a disc-shaped ion like a rotating CO$_3^{2-}$ ion.

(b) *Oriented* OH⁻ *ions.* The closest resemblance to simple ionic AX$_n$ structures is shown by the hydroxides of the larger alkali metals, alkaline-earths, and La^{3+}. The hydroxides of Li, Na, and the smaller M^{2+} ions form layer structures indicative of greater polarization of the OH⁻ ion.

The forms of KOH and the isostructural RbOH stable at ordinary temperatures have a distorted NaCl structure in which there are irregular coordination groups

(K—OH ranging from 2.69 to 3.15 Å). The shortest O—O distances (3.35 Å) are between OH groups coordinated to the same K^+ ion and delineate zigzag chains along which the H atoms lie. The weakness of these OH—O interactions is shown by the low value of the heat of transformation from the low- to the high-temperature form (6.3 kJ mol^{-1}; compare 21–25 kJ mol^{-1} for the O—H—O bond in ice). The structures of CsOH and TlOH are not known.

The larger Sr^{2+} and Ba^{2+} ions are too large for the CdI_2 structure which is adopted by $Ca(OH)_2$. $Sr(OH)_2$ has a structure of 7:(3,4)-coordination very similar to that of YO.OH which is a 3D system of edge-sharing monocapped trigonal prisms.[1] There are no hydrogen bonds (shortest O—H—O, 2.94 Å). A preliminary examination has been made of crystals of one form of $Ba(OH)_2$ but the complex structure has not been determined.[2] (For $Sr(OH)_2.H_2O$ see p. 685).

The trihydroxides of La, Y,[3] the rare-earths Pr, Nd, Sm, Gd, Tb, Dy, Er, and Yb. and also $Am(OH)_3$[4] crystallize with a typical ionic structure in which each metal ion is surrounded by 9 OH$^-$ ions and each OH$^-$ by 3 M^{3+} ions. This is the UCl_3 structure illustrated in Fig. 9.8 (p. 422) in which the coordination group around the metal ion is the tricapped trigonal prism. The positions of the D atoms in $La(OD)_3$ have been determined by n.d.[5a] In Fig. 14.2 all the atoms (including D) lie at heights $c/4$ or $3c/4$ above or below the plane of the paper (c = 3.86 Å). It follows that the D atom marked with an asterisk (at height $c/4$) lies approximately at the centre of a square group of four O atoms, two at $c/4$ below and two at $3c/4$ above the plane of the paper, with D—O (mean), 2.74 Å. In the OD$^-$ ion O—D is equal to 0.94 Å.

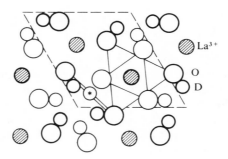

FIG. 14.2. The crystal structure of $La(OD)_3$. All atoms are at a height $c/4$ (light) or $3c/4$ (heavy), i.e. 0.96 Å above or below the plane of the paper.

Some references are given[5b] to recent work on individual compounds and to a review[5c] of $Ln(OH)_3$ structures.

A layer in the structure of LiOH is illustrated in Fig. 3.33 as an example of the simplest 4-connected plane layer. In a layer each Li$^+$ is surrounded tetrahedrally by 4 OH$^-$ and each OH$^-$ has 4 Li$^+$ neighbours all lying to one side. A n.d. study shows

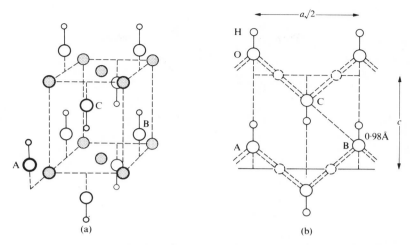

FIG. 14.3. The crystal structure of LiOH showing positions of protons: (a) unit cell; (b) section of structure parallel to (110) showing contacts between OH groups of adjacent layers.

that the O–H bonds are normal to the plane of the layer (O–H, 0.98 Å), and from the elevation of the structure in Fig. 14.3(b) it is evident that there is no hydrogen bonding between the layers.

The form of NaOH stable at ordinary temperatures has the TlI (yellow) structure. This is built of slices of NaCl structure in which Na$^+$ has five nearest neighbours at five of the vertices of an octahedron (Na–OH, 2.35 Å). The nearest neighbours of Na$^+$ in the direction of the 'missing' octahedral neighbour are 2 OH$^-$ at 3.70 Å, and the shortest interlayer OH–OH contacts are 3.49 Å. The H atoms have been located by n.d. at the positions indicated in Fig. 14.4, the system Na–O–H being collinear. The high-temperature form, which is stable over a very small temperature range (299.6–318.4 °C (m.p.)), is built of layers of the same kind. These apparently move relative to one another as the temperature rises (the monoclinic β angle varying) so that the structure tends towards the NaCl structure.

The third type of layer found in this group of compounds is the CdI$_2$ layer, this structure being adopted by the dihydroxides of Mg, Ca, Mn, Fe, Co, Ni, and Cd. (In addition to the simple CdI$_2$ structure some of these compounds have other structures with different layer sequences, as in the case of some dihalides.) Each OH$^-$ forms three bonds to M atoms in its own layer and is in contact with three OH$^-$ of the adjacent layer. The environments of OH$^-$ in LiOH and Mg(OH)$_2$ show that there are no hydrogen bonds in these structures:

$$
\begin{array}{ccc}
\begin{array}{c}
\text{Li Li}\\
\text{Li}\diagdown\;\diagup\;\diagup\text{Li}\\
\text{OH}\\
\text{OH}\diagup\;\diagdown\text{OH}\\
\text{OH OH}
\end{array}
&
\begin{array}{c}
\text{Mg}\\
\text{Mg}\diagdown\;|\;\diagup\text{Mg}\\
\text{OH}\\
\text{OH}\diagup\;|\;\diagdown\text{OH}\\
\text{OH}
\end{array}
\quad\text{compare}\quad
&
\begin{array}{c}
\text{Zn Zn}\\
\diagdown\;\diagup\\
\text{OH}\\
\diagup\;\diagdown\\
\text{OH OH}
\end{array}
\end{array}
$$

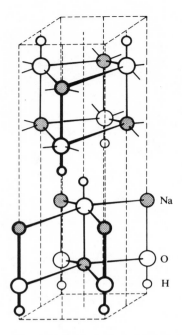

FIG. 14.4. The (yellow) TlI structure of low-temperature NaOH. Since the coordinates of the H atoms have not been determined O–H has been made equal to 1 Å.

As in LiOH the OH⁻ ions in these hydroxides are oriented with their H atoms on the outer surfaces of the layers, as shown by a n.m.r. study of $Mg(OH)_2$[6] and a n.d. study of $Ca(OH)_2$.[7] The elevation of the structure of $Ca(OH)_2$ (Fig. 14.5(b)) may be compared with that of LiOH in Fig. 14.3(b). It is interesting that a

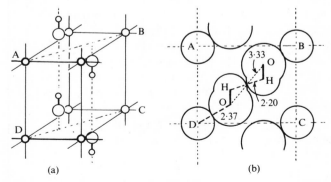

FIG. 14.5. The crystal structure of $Ca(OH)_2$ showing the positions of the protons: (a) unit cell; (b) section parallel to (110) plane. Ca^{2+} ions are at the corners of the cell and OH⁻ ions on the lines $\frac{1}{3}\frac{2}{3}z$ and $\frac{2}{3}\frac{1}{3}z$.

redetermination of the structure of $Mn(OH)_2$[8] gives Mn–OH equal to 2.21 Å, as compared with 2.22 Å for Mn–O in MnO, indicating that the OH^- radius of 1.53 Å is relevant only to structures such as high-KOH.

From X-ray powder-photographic data $Cu(OH)_2$ has been assigned a structure similar to that of γ-FeO.OH (which is described later), modified to give Cu^{II} a distorted octahedral arrangement of neighbours (4 OH at 1.94 and 2 OH at 2.63 Å).[9] This is a surprising structure for this compound since it has two types of non-equivalent O atom and is obviously more suited to an oxyhydroxide. Since the shortest distance between OH^- ions attached to different Cu atoms is 2.97 Å there is presumably no hydrogen bonding. The H positions in this crystal would be of great interest, but a detailed study would require larger single crystals than it has yet been possible to grow.

A number of dihydroxides are polymorphic. A new AX_2 structure has been proposed for γ-$Cd(OH)_2$,[10] consisting of octahedral coordination groups linked into double chains by *face*-sharing, these double chains sharing vertices to form a 3D framework of a new type (Fig. 14.6).

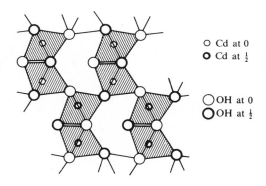

○ Cd at 0
◐ Cd at $\frac{1}{2}$

◯ OH at 0
◯ OH at $\frac{1}{2}$

FIG. 14.6. Projection along direction of double chains of the structure of γ-$Cd(OH)_2$.

References to the alkali-metal hydroxides and hydrosulphides are given in Table 14.1.

(1)   ZaC 1969 **368** 53
(2)   AC 1968 **B24** 1705
(3)   ACSc 1967 **21** 481
(4)   AC 1968 **B24** 979
(5a) JCP 1959 **31** 329
(5b) Nd(OH)$_3$, AC 1976 **B32** 2227; Tb(OH)$_3$, n.d., AC 1977 **B33** 3134; Ce(OH)$_3$, AC 1979 **B35** 2668
(5c) JINC 1977 **39** 65
(6)   JCP 1956 **25** 742
(7)   JCP 1957 **26** 563 (n.d.); AM 1962 **47** 1231 (n.m.r.)
(8)   ACSc 1965 **19** 1765
(9)   AC 1961 **14** 1041
(10) AC 1967 **22** 441

TABLE 14.1

Crystal structures of MOH and MSH

	LiOH	NaOH	KOH	RbOH	CsOH
Low-temperature form	Fig. 14.3 (a)	Orthorhombic[b], [e] (TlI structure) 299·6°C Monoclinic[e]	Monoclinic[d] 248° NaCl[c]	?	?
Transition temp. High-temperature form					?

	LiSH	NaSH	KSH	RbSH	CsSH
Low-temperature form Transition temperature High-temperature form	See text[f]	90°	Rhombohedral structure 160–170° NaCl structure[g]	?	CsCl structure

(a) ZK 1959 **112** 60 (n.d.); JCP 1962 **36** 2665 (i.r.). (b) JCP 1955 **23** 933. (c) ZaC 1939 **243** 138. (d) JCP 1960 **33** 1164. (e) ZK 1967 **125** 332. (f) ZaC 1954 **275** 79. (g) ZK 1934 88 97; ZaC 1939 **243** 86.

II. *Hydroxides* $M(OH)_2$ *and* $M(OH)_3$ *with hydrogen bonds*

Zinc hydroxide. A number of forms of this hydroxide have been described, one with the $C6$ (CdI_2) structure and others with more complex layer sequences, though some of these phases may be stable only in the presence of foreign ions.[1] This hydroxide may be crystallized by evaporating the solution of the ordinary precipitated gelatinous form in ammonia. The crystals of the ϵ form are built of $Zn(OH)_4$ tetrahedra sharing all vertices with other similar groups.[2] Representing the structure as basically a 3-dimensional network of Zn atoms each linked to four others through OH groups, the net is simply a distorted version of the diamond (or cristobalite) net (see p. 123). The distortion is of such a kind as to bring each OH near to two others attached to different Zn atoms, so that every OH is surrounded, tetrahedrally, by two Zn and two OH. This arrangement and the short OH–OH distances (2.77 and 2.86 Å) show that hydrogen bonds are formed in this structure. The β form of $Be(OH)_2$ has the same structure;[3] for α-$Be(OH)_2$ see reference (4).

One of the less stable forms of $Zn(OH)_2$, the so-called γ form, has an extraordinary structure[5] consisting of rings of three tetrahedral $Zn(OH)_4$ groups which are linked through their remaining vertices into infinite columns (Fig. 14.7). The upper vertices (1) of the tetrahedra of one ring are the lower vertices (2) of the ring above, so that all the vertices of the tetrahedra are shared within the column, which therefore has the composition $Zn(OH)_2$. Each column is linked to other similar ones only by hydrogen bonds (O–H–O, 2.80 Å; Zn–O, 1.96 Å).

Scandium and indium hydroxides. We noted on p. 209 that the simplest 3D AX_3 structure of 6:2 coordination is not normally found in its most symmetrical configuration, for this would imply collinear A–X–A bonds and an inefficient packing of the anions. The projection (Fig. 14.8) of the structure of $Sc(OH)_3$, or

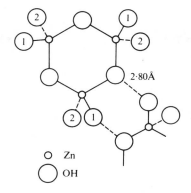

FIG. 14.7. The structure of γ-Zn(OH)$_2$ (see text).

In(OH)$_3$, shows that the OH groups lie off the lines joining the metal atoms and that each is hydrogen-bonded to two others. The OH group A is bonded to the metal atom M and to a similar atom vertically above M and hydrogen-bonded to the OH groups B and C. A neutron diffraction study of In(OH)$_3$[6] indicates statistical distribution of the H atoms, with O–H–O, 2.744 and 2.798 Å. Lu(OH)$_3$[6a] has the same (cubic) structure.

Aluminium hydroxide. A number of forms of Al(OH)$_3$ are recognized which are of two main types, α (bayerite), and γ (gibbsite), but there are minor differences within each type arising from different ways of superposing the layers. All forms are built of the same layer, the AX$_3$ layer shown in idealized form in Fig. 14.9. This may be described as a system of octahedral AX$_6$ coordination groups each sharing three

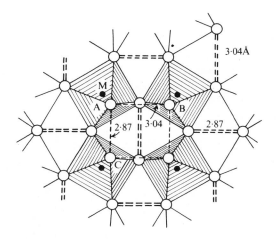

FIG. 14.8. The crystal structure of Sc(OH)$_3$.

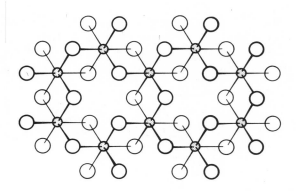

FIG. 14.9. Part of a layer of Al(OH)$_3$ (idealized). The heavy and light open circles represent OH groups above and below the plane of the Al atoms (shaded).

edges, or as a pair of approximately close-packed X layers with metal atoms in two-thirds of the octahedral interstices. The structures differ in the way in which the layers are superposed. In bayerite there is approximate hexagonal closest packing of the O atoms throughout the structure (as in Mg(OH)$_2$), but in hydrargillite (monoclinic gibbsite) the OH groups on the underside of one layer rest directly above the OH groups of the layer below, as shown in the elevation of Fig. 14.10.

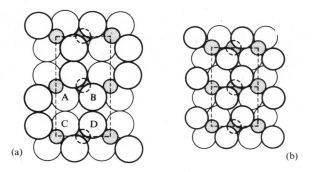

FIG. 14.10. The structures of (a) Al(OH)$_3$ and (b) Mg(OH)$_2$, viewed in a direction parallel to the layers, to illustrate the difference in packing of OH groups of different layers.

Nordstrandite appears to represent an intermediate case, with the layers superposed in nearly the same way as in hydrargillite; compare the interlayer spacings (d):

	d (Å)	*Reference*
Bayerite	4.72	ZK 1967 **125** 317
Nordstrandite	4.79	AC 1970 **B26** 649
Hydrargillite (gibbsite)	4.85	ZK 1974 **139** 129

The mode of superposition of the layer in hydrargillite suggested directed bonds between OH groups of adjacent layers rather than the non-directional forces operating in $Mg(OH)_2$ and similar crystals. There has been some discussion of the proton positions in gibbsite and bayerite, and calculations have been made of their positions in all three forms.[7]

(1)   HCA 1966 **49** 1971
(2)   ZaC 1964 **330** 170
(3)   ZaC 1950 **261** 94
(4)   JNM 1964 **11** 310

(5)   ACSc 1969 **23** 2016
(6)   ACSc 1967 **21** 1046
(6a) JINC 1980 **42** 223
(7)   AC 1976 **B32** 1719

Complex hydroxides

These compounds are more numerous than was once thought, for many have been formulated as hydrates of, for example, meta-salts (e.g. $BaSbO_3.3H_2O$ is in fact $NaSb(OH)_6$). Compounds $M'_x M''_y (OH)_z$ present the same general structural possibilitiees as complex halides. They range from compounds in which both metals are electropositive, when the crystal consists of an array of OH^- ions incorporating metal ions in holes of suitable sizes entirely surrounded by OH^- ions, to compounds in which one metal (or non-metal) is much less electropositive than the other and forms an essentially covalent hydroxy-ion. Examples of compounds of the first type include $Ca_3Al_2(OH)_{12}$ (which analyses as $3CaO.Al_2O_3.6H_2O$), in which Ca^{2+} and Al^{3+} ions are surrounded by 8 OH^- and 6 OH^- respectively, at 2.50 Å and 1.92 Å, $Ba_2Al_2(OH)_{10}$,[2] and $Ba_2Al_4(OH)_{16}$.[3] In the Ba compounds the more highly charged Al^{3+} ions associate in pairs or groups of four forming binuclear ions (a) and tetrameric ions of the type of Fig. 5.8(h), p. 200, respectively. In $[M(OH)_6]^{n-}$

ions formed by B subgroup or transition metal atoms the M–O bonds probably have appreciable covalent character, for example, in $Ge(OH)_6^{2-}$,[4] $Sb(OH)_6^-$ (p. 903), in $M_2[Pt(OH)_6]$, M = H,[5] Na, K,[6] and $Ba_2Cu(OH)_6$,[7] all with K_2PtCl_6 type structures, and $Ca[Pt(OH)_6]$ in which there is a NiAs-like packing of anions and cations.[8]

(1)   AC 1964 **17** 1329
(2)   AC 1970 **B26** 867
(3)   AC 1972 **B28** 519
(4)   ACSc 1969 **23** 1219

(5)   AC 1979 **B35** 3014
(6)   ZaC 1975 **414** 160
(7)   AC 1973 **B29** 1929
(8)   ZK 1975 **414** 169

Oxyhydroxides

Oxyhydroxides may be described as compounds of metals in which the coordination group around the metal is composed of O and OH ligands. The name is not applied to non-metal compounds such as $SO_2(OH)_2$, $PO(OH)_3$, or the ions derived from them, $SO_3(OH)^-$, and so on. Metals do not form such tetrahedral ions, though there is tetrahedral coordination of Al by O and OH in the $[(OH)_3Al-O-Al(OH)_3]^{2-}$ ion, as already noted on p. 626. The trigonal bipyramidal $[RuO_3(OH)_2]^{2-}$ [1a] and octahedral $[OsO_2(OH)_4]^{2-}$ [1b] ions are of some interest since both are anions in salts which were originally formulated as hydrates, namely, $BaRuO_4 . H_2O$ and $K_2OsO_4 . 2H_2O$. We have mentioned some examples of polynuclear oxyhydroxy ions formed by Bi and Pb and the neutral molecules $M_6O_4(OH)_4$ formed by Sn and Pb. However, most metal oxyhydroxides $MO_x(OH)_y$ consist of extended arrays of the component atoms (ions), having layer or 3D structures. Some of these compounds are described in other chapters — V oxyhydroxides, p. 569; H_xMoO_3, p. 618; $UO_2(OH)_2$, p. 1265.

The largest class of compounds of which the structures are known are of the type MO.OH, formed by Al, Sc, Y, V, Cr, Mn, Fe, Co, Ga, and In. A number of these compounds, like the trihydroxides and sesquioxides of Al and Fe, exist in α and γ forms. The so-called β-FeO.OH is not a pure oxyhydroxide; it has the α-MnO_2 structure, and is stable only if certain interstitial ions such as Cl^- are enclosed within the framework.[2a] The deep-brown, strongly magnetic δ-FeO.OH, made by rapid oxidation of $Fe(OH)_2$ in NaOH solution, has a very simple structure consisting of h.c.p. O(OH) with Fe^{3+} ions in certain of the interstices. The magnetic data agree better with a random distribution of the metal ions in all the octahedral sites than with partial occupation of some tetrahedral sites as was originally suggested.[2b] The structure of ϵ-FeO.OH is noted later.

As in the case of hydroxides we may distinguish between structures in which there is hydrogen bonding and those in which this does not occur; it is convenient to deal first with a structure of the latter type.

The YO.OH *structure*

All the 4f compounds MO.OH and YO.OH have a monoclinic structure in which M has 7 O(OH) neighbours arranged at the vertices of a monocapped trigonal prism (as in monoclinic Sm_2O_3).[3a] The interatomic distances suggest that the 3-coordinated O atoms belong to OH^- ions and the 4-coordinated ions are O^{2-}:

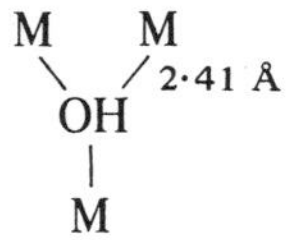

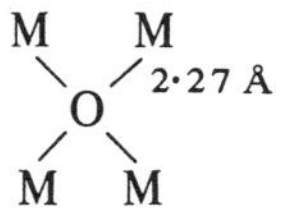

The mean O—OH distance, 3.01 Å, indicates that there is no hydrogen bonding, a conclusion confirmed by a n.d. study of YO.OD.[3b]

The InO.OH *structure*

This is a rutile-like structure modified by the formation of short unsymmetrical hydrogen bonds (2.54 Å) between certain pairs of atoms (Fig. 14.11).[4] β-CrO.OH (p. 643) and h.p. (ϵ) FeO.OH[5] also have this structure.

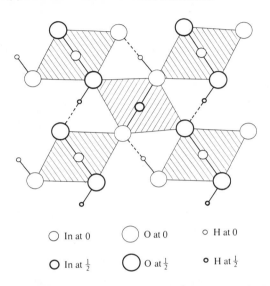

○ In at 0   ◯ O at 0   ∘ H at 0

◯ In at $\frac{1}{2}$ ◯ O at $\frac{1}{2}$ • H at $\frac{1}{2}$

FIG. 14.11. The distorted rutile-like structure of InO(OH) showing hydrogen bonds.

The α-MO.OH *structure*

Compounds with this structure include: α-AlO.OH (diaspore), α-FeO.OH (goethite), α-MnO.OH (groutite), α-ScO.OH,[6a] α-GaO.OH,[6b] and VO.OH (montroseite).

In our survey of octahedral structures in Chapter 5 we saw that double chains of the rutile type could be further linked by sharing vertices to form either 3-dimensional framework structures, of which the simplest is the diaspore structure, or a puckered layer as in lepidocrocite. These two structures are shown in perspective in Fig. 14.12. In Fig. 14.13(a) the double chains are seen end-on, as also are the c.p. layers of O atoms. The H atoms in diaspore have been located by n.d. and are indicated as small black circles attached to the shaded O atoms. The O—H—O bonds, of length 2.65 Å, as shown as broken lines, and the H atom lies slightly off the straight line joining a pair of O atoms:

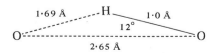

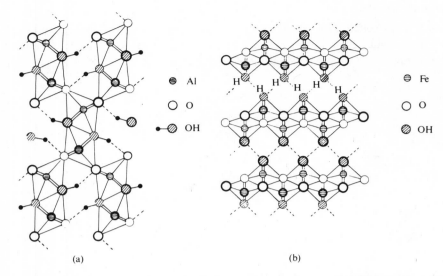

FIG. 14.13. Elevations of (a) the α-AlO(OH), diaspore, (b) the γ-FeO(OH), lepidocrocite, structures.

It will be seen that OH has 3 Al neighbours arranged pyramidally to one side, opposite to that of the H atom (Al–3 OH, 1.98 Å), while O is bonded to three more nearly coplanar Al neighbours (Al–3 O, 1.86 Å), the O–H–O bond being directed along the fourth tetrahedral bond direction. In the α-MO.OH structure each OH is hydrogen-bonded to an O atom; contrast the γ-MO.OH structure.

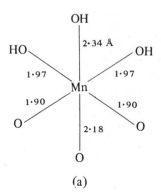

(a)

Bond lengths in the α structures are given in Table 14.2, which also includes γ-MnO.OH for comparison with α-MnO.OH. In groutite the length of the O–H–O bond is 2.63 Å. These Mn^{III} compounds are of special interest because of the

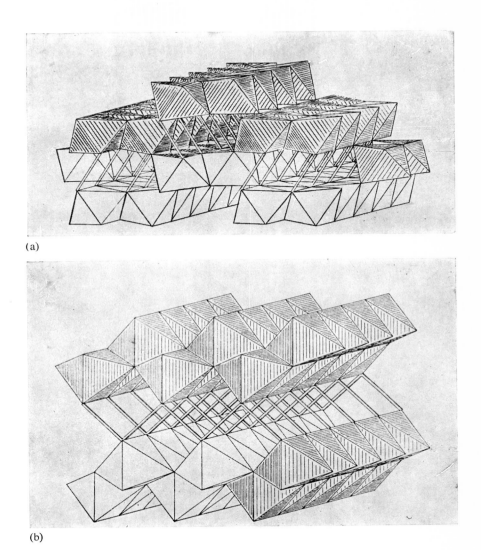

(a)

(b)

FIG. 14.12. The structures of (a) diaspore, AlO(OH), and (b) lepidocrocite, FeO(OH) (after Ewing). Oxygen atoms are to be imagined at the vertices of each octahedron and an Al (or Fe) atom at its centre. The double lines indicate O—H—O bonds.

TABLE 14.2

Bond lengths in MO.OH structures

	α-AlO.OH[a] (diaspore)	α-VO.OH[b] (montroseite)	α-FeO.OH[c] (goethite)	α-MnO.OH[d] (groutite)	γ-MnO.OH[e] (manganite)
M—O (one)	1.858 Å	1.94 Å	1.89 Å	2.18 Å	2.20 Å
(two)	1.851	1.96	2.02	1.90	1.87
M—OH (one)	1.980	2.10	2.05	2.34	2.33
(two)	1.975	2.10	2.12	1.97	1.97

(a) AC 1958 **11** 798. (b) AM 1955 **40** 861. (c) ZK 1941 **103** 73. (d) AC 1968 **B24** 1233. (e) ZK 1963 **118** 303.

expected Jahn–Teller distortion of the octahedral coordination group. This major distortion is superposed on the smaller difference between M—O and M—OH, as is seen by comparing MnO.OH with AlO.OH and VO.OH (Table 14.2). Details of the coordination group in groutite are shown at (a).

The γ-MO.OH structure

Compounds with this structure include: γ-AlO.OH (boehmite), γ-FeO.OH (lepidocrocite), γ-MnO.OH (manganite), and γ-ScO.OH.[6a]

Two X-ray studies of boehmite were based on the centrosymmetrical space group *Cmcm*, and gave 2.47 Å and 2.69 Å for the length of the O—H—O bond. A p.m.r. study[7] shows that this bond is not symmetrical; further study of this structure appears to be required.

The structure of lepidocrocite, γ-FeO.OH, is shown as a perspective drawing in Fig. 14.12(b). In the elevation of this structure (Fig. 14.13(b)) the broken circles indicate atoms $c/2$ (1.53 Å) above and below the plane of the full circles, which themselves repeat at c (3.06 Å) above and below their own plane. Each Fe atom is therefore surrounded by a distorted octahedral group of O atoms, and these groups are linked together to form corrugated layers. The H atoms have not been located in this structure, but from the environments of the O atoms it is possible to distinguish between O and OH. The oxygen atoms within the layers are nearly equidistant from 4 Fe atoms (2 Fe at 1.93 Å and 2 Fe at 2.13 Å), whereas the atoms on the surfaces of the layers are bonded to 2 Fe atoms only (at 2.05 Å). Also the distance between the O atoms of different layers is only 2.72 Å, so that there are clearly OH groups on the outer surface of each layer. In this structure there is hydrogen bonding only between the OH groups, each forming two hydrogen bonds.

The formation of O—H—O bonds between the OH groups of different layers accounts for the fact that the O atoms of adjacent layers are not packed together in the most compact way, though there is approximately cubic closest packing within a particular layer. As may be seen from Figs. 14.12 and 14.13, a hydroxyl group lies in the same plane as those of the next layer with which it is in contact, so that the H atoms must be arranged as indicated in Fig. 14.13(b). The oxychloride FeOCl is built of layers of exactly the same kind as in lepidocrocite but the layers pack together so that the Cl atoms on the outsides of adjacent layers are close-packed. This difference between the structures of FeOCl and γ-FeO.OH is quite comparable to that between Mg(OH)$_2$ and Al(OH)$_2$, being due to the formation of O—H—O bonds between the layers in the second of each pair of compounds.

The structures of the two forms of AlO.OH are of interest in connection with the dehydration of Al(OH)$_3$ and AlO.OH to give catalysts and absorbents. In diaspore, α-AlO.OH, the O atoms are arranged in hexagonal closest packing, and this compound dehydrates directly to α-Al$_2$O$_3$ (corundum) in which the O atoms are arranged in the same way. In boehmite, γ-AlO.OH, the structure as a whole is not close-packed but within a layer the O atoms are arranged in cubic closest packing. On dehydration boehmite does not go directly to γ-Al$_2$O$_3$ with the spinel type of structure, but instead there are a number of intermediate phases, and there is still not complete agreement as to the number and structures of these phases.[8] Among schemes suggested are the following:

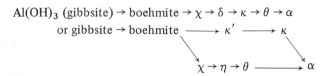

It seems likely that these intermediate phases, collectively known as forms of γ-alumina, represent different degrees of ordering of the Al atoms in a more or less perfect closest packing of O atoms (see also p. 552).

The α-CrO.OH (HCrO$_2$) structure

CrO.OH is normally obtained in a red or blue-grey (rhombohedral) α form. This is built of CdI$_2$-like layers which are superposed so that O atoms of one layer fall

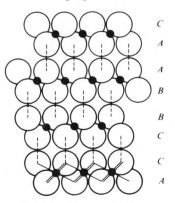

FIG. 14.14. Elevation of the crystal structure of HCrO$_2$ showing the way in which the layers are superposed. The broken lines represent O—H—O bonds.

directly above those of the layer below (Fig. 14.14).[9] The structure as a whole is therefore not close-packed. This mode of packing of the layers is due to the formation of O–H–O bonds, which are notable for their shortness (OH–O, 2.47 Å; OD–O, 2.55 Å). The n.d. and i.r. data show that the latter bonds are certainly unsymmetrical, but it has proved difficult to locate the H atom in the O–H–O bond. The latest n.d. results[9a] indicate $\frac{1}{2}$H atoms situated on each side of the centre of the bond at a distance 0.15 Å apart, giving the apparent O–H distance 1.16 Å (or 1.31 Å). This is perhaps a somewhat academic point since the potential barrier between the two minima is apparently very low. (Very similar results were obtained for β-CrO.OH,[9b] which has the InO.OH structure (p. 639).) The same structure is adopted by CoO.OH, formed by oxidizing β-Co(OH)$_2$.[10]

(1a)	AC 1976 **B32** 2413	(6a)	ACSc 1967 **21** 121
(1b)	ZSK 1962 **3** 685	(6b)	AC 1977 **B33** 3224
(2a)	MM 1960 **32** 545	(7)	JPC 1958 **62** 992
(2b)	JACeS 1968 **51** 594	(8)	JCS A 1967 2106
(3a)	ACSc 1966 **20** 896	(9)	AC 1963 **16** 1209
(3b)	ACSc 1966 **20** 2658	(9a)	JSSC 1977 **21** 325
(4)	ACSc 1970 **24** 1662 (n.d.)	(9b)	JSSC 1976 **19** 299
(5)	SSC 1975 **17** 1505	(10)	JCP 1969 **50** 1920

Hydroxy (basic) salts

The term 'basic salt' is applied to a variety of compounds intermediate between the normal salt and the hydroxide or oxide. This broad definition includes compounds such as Be and Zn oxyacetates (p. 497), oxy- and hydroxy-halides (Chapter 10), and hydroxy-oxysalts. We shall confine our attention here almost entirely to the last group of compounds. Coordination compounds with bridging OH groups, such as $[(NH_3)_4Co(OH)_2Co(NH_3)_4]Cl_4$, are mentioned under the element concerned.

Basic salts are numerous and of considerable interest and importance. They are formed by Be, Mg. Al, many of the A subgroup transition elements (e.g. Ti, Zr), 3d elements such as Fe, Co, and Ni, 4f and 5f elements (Ce, Th, U), and most of the B subgroup elements, particularly Cu(II), Zn, In, Sn, Pb, and Bi. Being formed by the action of oxygen and moisture on sulphide and other ores they form a large class of secondary minerals, some of which are also important as corrosion products of metals. The minerals brochantite, $Cu_4(OH)_6SO_4$, and atacamite, $Cu_2(OH)_3Cl$, form as patinas on copper exposed to town or seaside atmospheres, lepidocrocite, γ-FeO.OH, is formed during the rusting of iron, and hydrozincite, $Zn_5(OH)_6(CO_3)_2$, is the usual corrosion product of zinc in most air. White lead, $Pb_3(OH)_2(CO_3)_2$, is one of a considerable number of basic salts which have been used as pigments, while $Mg_2(OH)_3Cl.4H_2O$ is formed during the setting of Sorel's cement.

Basic salts may be prepared in various ways, which usually involve—directly or indirectly—hydrolysis of a normal salt. The hydrolysis may be carried out under conditions of controlled temperature, acidity, and metal-ion concentration, or indirectly by heating a hydrated salt. Many hydroxy-salts are formed by the latter process instead of the anhydrous normal salts, for example, $Cu_2(OH)_3NO_3$ by heating hydrated cupric nitrate. The precipitates formed when sodium carbonate solution is added to metallic salt solutions are often hydroxy-carbonates. Some metals do not form normal carbonates (e.g. Cu, see p. 1117); others such as Pb, Zn, Co, and Mg, form normal or hydroxy-salts according to the conditions of precipitation. Lead, for example, forms $2PbCO_3.Pb(OH)_2$ and $PbCO_3.Pb(OH)_2$, both of which occur as minerals, while Co forms $Co_4(OH)_6CO_3$ in addition to the normal carbonate.

Before describing the structures of a few hydroxy-salts we should mention that 'structural formulae' have been assigned to some of these compounds on the supposition that they are Werner coordination compounds. We may instance:

$$\left[Cu\left\{\begin{matrix}OH\\OH\end{matrix}{>}Cu\right\}_3\right]Cl_2 \quad \text{and} \quad \left[Cu\left\{\begin{matrix}OH\\OH\end{matrix}{>}Cu\right\}\right]CO_3$$

for atacamite, $Cu_2(OH)_3Cl$, and malachite, $Cu_2(OH)_2CO_3$, respectively. Such formulae are quite erroneous, for these compounds do not contain complex ions with a central copper atom. Malachite, for example, consists of an infinite array in three dimensions of Cu^{2+}, CO_3^{2-}, and OH^- ions. (For the structure of atacamite see p. 1142.) On the other hand, finite complexes containing M atoms bridged by OH groups do occur in some basic salts; examples will be given shortly.

The crystal structures of basic salts

The first point to note that is relevant to the structures of hydroxy-salts is that there is a wide range of OH : M ratios:

	$Ca_5(OH)(PO_4)_3$	$Cu_2(OH)PO_4$	$Cu(OH)IO_3$	$Zn_5(OH)_6CO_3$
OH : M ratio	$\frac{1}{5}$	$\frac{1}{2}$	1	$\frac{6}{5}$

	$Cu_2(OH)_3NO_3$	$Th(OH)_2SO_4$
OH : M ratio	$\frac{3}{2}$	2

The OH^- ions are associated with the metal ions, so that for small OH : M ratios the coordination group around M consists largely of O atoms of oxy-ions; with increasing OH : M ratio OH^- ions form a more important part of the coordination group of M. This is rather nicely illustrated by the hydrolysis of $Zr(SO_4)_2.4H_2O$ at 100 °C to $Zr_2(OH)_2(SO_4)_3.4H_2O$ and at 200 °C to $Zr(OH)_2SO_4$, the environment of Zr^{4+} being compared with that in (cubic) ZrO_2, the final product of heating these salts in air. (There is 7-coordination of the cations in the monoclinic form of ZrO_2.) The O atoms in the coordination groups around the metal ions in the following table are, of course, oxygen atoms of SO_4^{2-} ions.

	Coordination groups around Zr^{4+}		
$Zr(SO_4)_2 . 4 H_2O^{(a)}$	$Zr_2(OH)_2(SO_4)_3 . 4 H_2O^{(b)}$	$Zr(OH)_2SO_4^{(b)}$	ZrO_2
4 H_2O	2 H_2O	4 OH	8 O
4 O	2 OH	4 O	
	4 O		
Antiprism	Dodecahedron	Antiprism	Cube

(a) AC 1959 **12** 719   (b) IC 1966 **5** 284

In crystals such as those of hydroxyapatite, $Ca_5(OH)(PO_4)_3$, which is isostructural with $Ca_5F(PO_4)_3$, no hydroxy-metal complex can be distinguished, but with higher OH:M ratios there is generally sharing of OH between coordination groups of different M atoms so that a connected system of OH and M can be seen. Such a hydroxy-metal complex may be finite or infinite in 1, 2, or 3 dimensions, and in describing the structures of hydroxy-oxysalts it is convenient first to single out the M—OH complex and then to show how these units are assembled in the crystalline basic salt. The description of a structure in terms of the M—OH complex is largely a matter of convenience. For example, the M—OH system may extend only in one or two dimensions but the oxy-ions may then link these sub-units into a normal 3-D structure, that is, one which is not a chain or layer structure. Thus, while $Cu_2(OH)_3NO_3$ and $Zn_5(OH)_8Cl_2 . H_2O$ are layer structures, $Cu(OH)IO_3$, $Fe(OH)SO_4$, $Zn_2(OH)_2SO_4$, and $Zn_5(OH)_6(CO_3)_2$ are all 3D structures, though we shall see how they are constructed from 1- or 2-dimensional M—OH sub-units. It should be emphasized that there is no simple connection between the OH:M ratio and the possible types of M:OH complex. Assuming that every OH is shared between two M atoms, cyclic and linear M:OH complexes become possible for OH:M $\geqslant$ 1. An additional reason for distinguishing the M:OH complex is that some of the M:OH complexes in basic salts formed from dilute solution, with high OH:M ratios, are related in simple ways to the structures of the final hydrolysis products, namely, the hydroxide or oxide of the metal.

Finite hydroxy–metal complexes

One product of the hydrolysis of $Al_2(SO_4)_3$ by alkali is a salt with the empirical formula $Al_2O_3 . 2SO_3 . 11H_2O$ which contains bridged hydroxy-aquo complexes (a) and should therefore be formulated as $[Al_2(OH)_2(H_2O)_8](SO_4)_2 . 2H_2O$.[1] A more complex example is the dimeric $Th_2(OH)_2(NO_3)_6(H_2O)_6$ complex, (b),[2] formed by hydrolysis of the nitrate in solution. Three bidentate NO_3^- ions and 3 H_2O complete an 11-coordination group about each metal atom.

The compounds originally formulated $ZrOCl_2 . 8H_2O$ and $ZrOBr_2 . 8H_2O$ also contain hydroxy-complexes. The complicated behaviour of these compounds in aqueous solution suggests polymerization, and in fact the crystals contain square complexes (c) in which there is a distorted square antiprismatic arrangement of 8 O atoms around each Zr atom. The halide ions and remaining H_2O molecules lie between the complexes, and the structural formula of these compounds is therefore

$[Zr_4(OH)_8(H_2O)_{16}]X_8 . 12H_2O.$[3] We refer again to this type of complex later.

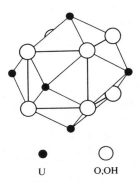

(a)   (b)

(c)

A polyhedral complex is found in $U_6O_4(OH)_4(SO_4)_6$[4] and the isostructural Ce compound. This complex (Fig. 14.15) consists of an octahedron of U atoms and an associated cubic group of 8 O atoms (the positions of the H atoms are not known) of the same general type as the $Mo_6Cl_8^{4+}$ ion (p. 434), but whereas in the Mo complex

FIG. 14.15. The finite $U_6O_4(OH)_4$ complex in $U_6O_4(OH)_4(SO_4)_6$.

the metal–metal distances are rather shorter than in the metal, here the U–U distances (3.85 Å) are as large as in UO_3, showing that the complex is held together by U–O bonds. The oxyhydroxides $M_6O_4(OH)_4$ of Sn and Pb form molecules of this type (p. 1186).

1-dimensional hydroxy–metal complexes

A type of chain found in a number of hydroxy-salts with OH : M = 1 consists of octahedral coordination groups sharing opposite edges which include the (OH) groups. The simplest possibility is that these chains are then joined through the oxy-ions to form a 3-dimensional structure, as illustrated by the plan and elevation of the structure of $Cu(OH)IO_3$[5] (Fig. 14.16). The other four atoms of the octahedral

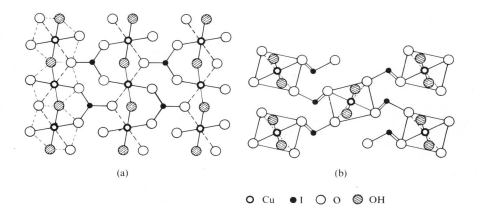

O Cu   ● I   O O   ⊘ OH

FIG. 14.16. The structure of $Cu(OH)IO_3$: (a) plan; (b) elevation, in which the octahedral chains are perpendicular to the plane of the paper.

coordination group of M in the chain are necessarily O atoms of oxy-ions. The structures of $Fe(OH)SO_4$[6] and $In(OH)SO_4$ are of the same general type. A further interesting possibility is realized in $Zn_2(OH)_2SO_4$.[7] Instead of the octahedral chains being cross-linked by finite oxy-ions as in $Cu(OH)IO_3$ they are linked by chains of a second type running in a direction perpendicular to that of the first set, as shown in Fig. 14.17. This structure is one of a number in which there is both tetrahedral and octahedral coordination of Zn^{2+}, a point referred to in Chapters 4 and 5.

A second kind of M–OH chain is related to the cyclic Zr complex already mentioned; it is a zigzag chain of M atoms joined by double hydroxy-bridges. The 4 OH all lie to one side of a particular M atom, and this chain is characteristic of larger metals such as Zr, Th, and U forming M^{4+} ions which are 8- (sometimes 7-) coordinated by oxygen. The chain can be seen in the plan (Fig. 14.18) of the structure of $Th(OH)_2SO_4$,[8] with which the Zr and U compounds are isostructural. The chains are bound together by the SO_4^{2-} ions, O atoms of which complete the antiprismatic coordination groups around the M atoms in the chains. In $Hf(OH)_2SO_4 . H_2O$[9] 4 OH and H_2O form a planar pentagonal group around Hf, and 2 O atoms of sulphate ions complete the pentagonal bipyramidal coordination group.

$$\begin{array}{c} \qquad\quad \text{H}_2\text{O} \qquad\qquad\quad \text{H}_2\text{O} \\ \diagdown\text{OH--OH} \quad \text{Hf} \quad \text{OH--OH} \quad \text{Hf} \quad \diagup\text{OH--} \\ \text{--OH} \quad ^{\text{Hf}} \quad \text{OH--OH} \quad ^{\text{Hf}} \quad \text{OH--OH}\diagdown \\ \qquad\quad \text{H}_2\text{O} \qquad\qquad\quad \text{H}_2\text{O} \end{array}$$

The dioxides of Ce, Zr, Th, and U all crystallize with structures of the fluorite type, and it is interesting to note that three of the complexes we have described may be regarded as portions of this structure. In the $[Zr_4(OH)_8(H_2O)_{16}]^{8+}$ ion the hydroxy-bridges are perpendicular to the plane of the Zr atoms, so that the arrangement of the 4 Zr and 8 OH is approximately that of Fig. 14.19(a), which is

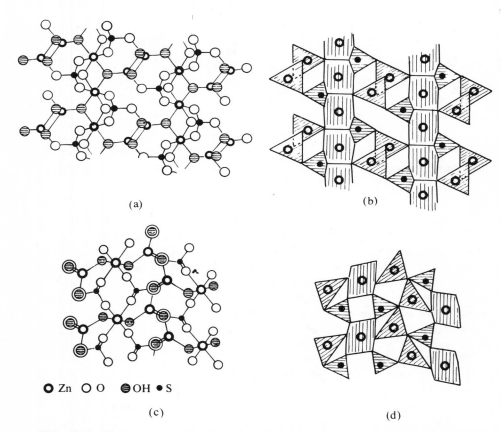

(a)

(b)

(c)

(d)

● Zn ○ O ⊜ OH • S

FIG. 14.17. The structure of $Zn_2(OH)_2SO_4$: (a) plan; (c) elevation; (b) and (d) diagrammatic representations of (a) and (c). In (a) and (b) the octahedral chains run parallel to the plane of the paper and the tetrahedral chains perpendicular to the paper; in (c) and (d) these directions are interchanged.

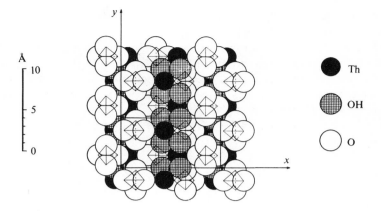

FIG. 14.18. The crystal structure of $Th(OH)_2SO_4$.

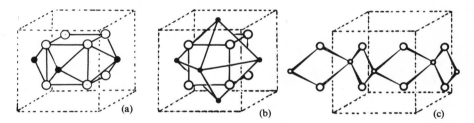

FIG. 14.19. The complexes (a) Zr_4O_8; (b) M_6O_8; and (c) the $(MO_2)_n$ chain shown as portions of the fluorite structure.

simply a portion of the fluorite structure. Similarly the $M_6O_4(OH)_4^{12+}$ ion is the portion (b), and the $[M(OH)_2]_n$ chain in $Th(OH)_2SO_4$ is the system of atoms (c) of the same structure.

2-dimensional hydroxy–metal complexes

The hydroxides of Ca, Mg, Mn, Fe, Co, and Ni crystallize with the CdI_2 (layer) structure. Although $Cu(OH)_2$ does not adopt this structure a number of basic cupric salts have structures closely related to it, for example, $Cu_2(OH)_3Br$ and the hydroxy-nitrate to be described shortly. Also, although Zn prefers tetrahedral coordination in $Zn(OH)_2$ itself, various hydroxy compounds with CdI_2-like structures appear to exist. In a number of basic salts Zn adopts a compromise, as in $Zn_2(OH)_2SO_4$, between the tetrahedral coordination in $Zn(OH)_2$ and the octahedral coordination in $ZnCO_3$ and other oxy-salts, by exhibiting both kinds of coordination in the same crystal. This results in more complicated structures, and for this reason we shall describe first the very simple relation between $Cu_2(OH)_3NO_3$ and the CdI_2-type structure of $CuCl_2$ and $Cu_2(OH)_3Br$.

Replacement by Br of one-quarter of the OH groups in a (hypothetical) $Cu(OH)_2$ layer of the CdI_2 type gives the layer of $Cu_2(OH)_3Br$, with suitable distortion to give Cu^{II} (4+2) instead of regular octahedral coordination. If instead of Br we insert one O atom of a NO_3^- ion, the plane of which is perpendicular to the layer, we have the (layer) structure of $Cu_2(OH)_3NO_3$,[10] illustrated in elevation in Fig. 14.20(a). The other O atoms of the NO_3^- ions are hydrogen-bonded to OH^- ions of the adjacent layer.

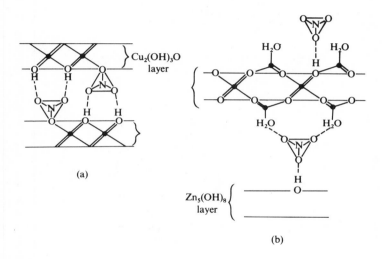

FIG. 14.20. Diagrammatic elevations of the structures of (a) $Cu_2(OH)_3NO_3$; (b) $Zn_5(OH)_8(NO_3)_2 . 2H_2O$.

In $Cu_2(OH)_3NO_3$ there is replacement of 1/4 of the OH positions in a $M(OH)_2$ layer by O atoms of NO_3^- ions. In $Zn_3(OH)_4(NO_3)_2$[11] there is replacement of 1/3 of the OH groups by O atoms of NO_3^- ions, and the structure is very similar to that of Fig. 14.20(a) with, of course, more NO_3^- ions between the layers. In both of these hydroxynitrates there is complete occupancy of the octahedral metal sites in the layer, the ratio of $(OH+O_{NO_3})$: M being in both structures 2 : 1. There are also structures in which some of the metal sites are not occupied.

We showed in Chapter 6 how the CdI_2 layer is used as the basic structural unit in a number of more complex structures, including

$Zn_5(OH)_8(NO_3)_2.2H_2O$	AC 1970 **B26** 860
$Zn_5(OH)_8Cl_2 .H_2O$	ZK 1968 **126** 417
$Zn_5(OH)_6(CO_3)_2$	AC 1964 **17** 1051

The first two structures are layer structures, the layer being derived from a hypothetical $Zn(OH)_2$ layer of the CdI_2 type. One-quarter of the (octahedrally coordinated) Zn atoms are removed and replaced by pairs of tetrahedrally coordinated Zn atoms, one on each side of the layer, giving a layer of composition $Zn_5(OH)_8$. In the hydroxynitrate a water molecule completes the tetrahedral coordination group of the added Zn atoms, and the NO_3^- ions are situated between the layers, being hydrogen-bonded to 2 H_2O of one layer and 1 OH of the adjacent layer (Fig. 14.20(b)). In $Zn_5(OH)_8Cl_2 . H_2O$ the fourth bond from the tetrahedrally coordinated Zn is to Cl, and the water molecules are situated between the layers. In the carbonate there is also replacement of one-quarter of the OH groups in the original $Zn(OH)_2$ layer by O atoms of CO_3^{2-} ions, so that the OH:Zn ratio becomes 6:5 instead of 8:5 as in the other salts. There is direct bonding between the tetrahedrally coordinated Zn atoms of one layer and the CO_3 groups of the adjacent layers, with the result that the structure is no longer a layer structure (see p. 264).

In the structures we have just described the basic layer is the octahedral CdI_2 $(Mg(OH)_2)$ layer suitable for ions of intermediate size (say, around 0.7 Å radius). A layer which can accommodate larger ions is found in $Pr(OH)_2NO_3$.[12] Figure 14.21 shows the end-on view of portions of two layers. Tricapped trigonal prismatic

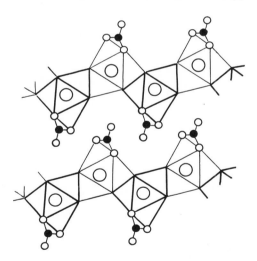

FIG. 14.21. Elevation of portions of two layers of the structure of $Pr(OH)_2NO_3$.

coordination groups consisting of 6 OH and 3 O atoms of NO_3 ions share their basal (triangular) faces to form chains normal to the plane of the paper. They also share edges, the result being a corrugated layer. Note that the 3 O atoms in each coordination group come from *two* bidentate NO_3^- ions, one O atom of each such ion being in the triangular base of the *tctp* and therefore common to two such metal coordination groups.

Some basic salts with layer structures are always obtained in a poorly crystalline state. For example, crystals of 'white lead', $Pb_3(OH)_2(CO_3)_2$, are too small and too disordered to give useful X-ray photographs, but from electron diffraction data it is concluded that there are probably $Pb(OH)_2$ layers interleaved with Pb^{2+} and CO_3^{2-} ions in a rather disordered structure.[13]

Three-dimensional hydroxy-metal frameworks are unlikely to occur in hydroxy-oxysalts because the number of O atoms belonging to OH groups is usually a relatively small proportion of the total number of O atoms, and therefore many M—O—M bridges involve O atoms of oxy-ions. On the other hand, in hydroxyhalides such as $M_2(OH)_3X$ the hydroxyl ions constitute three-quarters of the total number of anions, and in some of these compounds 3D hydroxy-metal frameworks do occur. (See the note on hydroxy-salts of Cu^{II} in Chapter 25 and the section on hydroxyhalides in Chapter 10.)

(1)   ACSc 1962 **16** 403
(2)   ACSc 1968 **22** 389
(3)   AC 1956 **9** 555
(4)   AK 1953 **5** 349
(5)   AC 1962 **15** 1105
(6)   ACSc 1962 **16** 1234
(7)   AC 1962 **15** 559
(8)   AK 1952 **4** 421
(9)   ACSc 1969 **23** 3541
(10)   HCA 1952 **35** 375
(11)   AC 1973 **B29** 1703
(12)   AC 1976 **B32** 2944
(13)   AC 1956 **9** 391

15

Water and hydrates

The structures of ice and water

The structure of the isolated water molecule in the vapour is accurately known from spectroscopic studies, but we have seen in Chapter 8 that compounds containing hydrogen are often abnormal owing to the formation of hydrogen bonds. Therefore it may not be assumed that the structural unit in the condensed phases water and ice has precisely this structure. It is convenient to discuss the structure of ice before that of water because we can obtain by diffraction methods much more information about the structure of a solid than about that of a liquid. In a liquid there is continual rearrangement of neighbours, and we can determine only the mean

$$H \diagdown \overset{104\frac{1}{2}°}{} \diagup H$$
$$O \diagup 0.96 \text{ Å}$$

environment, that is, the number and spatial arrangement of nearest neighbours of a molecule averaged over both space and time. The only information obtainable by X-ray diffraction from a liquid at a given temperature is the scattering curve, from which is derived the radial distribution curve, and the interpretation of the various maxima in terms of nearest, next nearest neighbours, and so on, is a matter of great difficulty.

Ice

At atmospheric pressure water normally crystallizes as ice-I_h, which has a hexagonal structure like that of tridymite. It may also be crystallized directly from the vapour, preferably *in vacuo*, as the cubic form ice-I_c with a cristobalite-like structure provided the temperature is carefully controlled (-120 to -140°C). (This cubic form is more conveniently prepared in quantity by warming the high-pressure forms from liquid nitrogen temperature.) Ice-I_c is metastable relative to ordinary ice-I_h at temperatures above 153 K. The existence of a second metastable crystalline form, ice-IV, has been firmly established for D_2O but less certainly for H_2O. A vitreous form of ice is formed by condensing the vapour at temperatures of -160°C or below.

In addition to I_h and I_c there are a number of crystalline polymorphs stable only under pressure, though the complete phase diagram is not yet established (Fig. 15.1). There are five distinct structures (differing in arrangement of O atoms) and there are low-temperature forms of two of them; the numbering is now unfortunately unsystematic:

II	III	V	VI	VII	
	↓			↓	
	IX			VIII	(low-temperature forms).

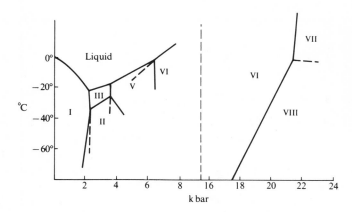

FIG. 15.1. Partial phase diagram for water.

The forms II-VII are produced by cooling liquid water under increasingly high pressures; cooling to –195 °C is necessary for II, which also results from decompressing V at low temperature and 2 kbar pressure. Ice-III converts to IX at temperatures below about –100 °C and VII to VIII below 0 °C. All the high-pressure forms II-VII can be kept and studied at atmospheric pressure if quenched to the temperature of liquid nitrogen.

The main points of interest of the structures of these polymorphs are: (i) the analogies with silica and silicate structures; (ii) the presence of two interpenetrating frameworks in the most dense forms VI and VII (VIII); and (iii) the ordering of the protons. Analogies with silica and silicate structures are noted in Table 15.1, namely ice-III with a keatite-like structure, ice-VI with two interprenetrating frameworks of the edingtonite type (p. 1037), and ice-VII (and VIII) with two interpenetrating cristobalite-like frameworks. In these structures, related to those of forms of silica or silicates, O atoms of H_2O molecules occupy the positions occupied by Si atoms in the silica structure or by Si or Al in the case of edingtonite. A projection of the Si atoms in keatite is similar to that of the Ge atoms in a high-pressure form of that element and has been illustrated in Fig. 3.41 (p. 130). The arrangement of the O atoms in ice-VII (or VIII) is body-centred cubic (Fig. 15.2). Each H_2O molecule has 8 equidistant neighbours but is hydrogen-bonded only to the 4 of its own framework. At atmospheric pressure (quenched in liquid N_2) the length of an O–H–O bond is 2.95 Å, and at –50 °C under a pressure of 25 kbar it is 2.86 Å.

The hydrogen-bonded frameworks in two of the ice polymorphs (II and V) are different from any of the 4-connected Si(Al)–O frameworks. The (rhombohedral) structure of ice-II has some similarity to the tridymite-like structure of ice-I, the greater density being achieved by the proximity of a fifth neighbour at 3.24 Å in addition to the four nearest neighbours at 2.80 Å – compare the distance of the next nearest neighbours (4.5 Å) in ice-I_h. The framework of ice-V also has no obvious

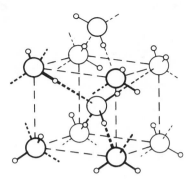

FIG. 15.2. The structure of ice-VIII.

silica or silicate analogue, in spite of the fact that the ratio of the densities of ice-v to ice-ɪ is the same as that of coesite (SiO_2) to tridymite. In ice-v H_2O has four nearest neighbours at a mean distance of 2.80 Å and then next nearest neighbours at 3.28 Å and 3.45 Å, but more striking is the very distorted tetrahedral arrangement of the nearest neighbours, the angles O–O–O ranging from 84° to 128°.

Proton positions in ice polymorphs.   In the early X-ray study of ice-I_h only the O atoms were located. Each is surrounded by a nearly regular tetrahedral arrangement of 4 O atoms, O–O being 2.752 Å for the bond parallel to the c axis and 2.765 Å for the other three, and the O–O–O angles are very close to 109½°. One H is to be placed somewhere between each pair of O atoms. The i.r. absorption frequencies indicate that the O–H distance in a H_2O molecule cannot have changed from 0.96 Å to 1.38 Å (one-half O–O in ice), therefore a H atom must be placed unsymmetrically along each O–O line. There are two possibilities: either H atoms occupy fixed positions (ordered protons) or there is random arrangement of H atoms (disordered protons) with the restriction that at any given time only two are close to a particular O atom, that is, there are normal H_2O molecules. It was suggested by Pauling that an 'average' structure, with $\frac{1}{2}$H at each of 4 N sites (each 0.96 Å from an O atom along an O–O line) within an array of N oxygen atoms would account for the observed value (0.82 e.u.) of the residual entropy of ice.

It has since been shown by a variety of physical techniques that in some of the high-pressure forms of ice the protons are ordered and in others disordered, in particular, the low-temperature forms ɪx and vɪɪɪ are the ordered forms of ice-ɪɪɪ and ice-vɪɪ respectively. The nature of the far infrared absorption bands indicates that the protons are ordered in ɪɪ and ɪx (sharp bands) but disordered in v (diffuse bands).[1] The dielectric properties of all the following forms of ice have been studied: I_c,[2] ɪɪ, ɪɪɪ, v, and vɪ,[3] vɪɪ and vɪɪɪ,[4] and the conclusions as to proton order/disorder are included in Table 15.1.

The second problem is the relation of the protons to the O–O lines. In ice-I_h the

TABLE 15.1
The polymorphs of ice

Polymorph	Density (g/cc)	Ordered or disordered protons	Analogous silica or silicate structure	Reference
I_h	0·92	D	Tridymite	AC 1957 **10** 70 (n.d.)
I_c	0·92	D	Cristobalite	JCP 1968 **49** 4365 (n.d.)
II	1·17	O	–	AC 1964 **17** 1437
				JCP 1968 **49** 4361 (n.d.)
III ⎫	1·16	D	Keatite	AC 1968 **B24** 1317
IX ⎭		O	Keatite	JCP 1968 **49** 2514 (n.d.)
[IV	–	–	–	JCP 1937 **5** 964]
V	1·23	D	–	AC 1967 **22** 706
VI	1·31	D	*Edingtonite	PNAS 1964 **52** 1433
VII ⎫	1·50	D	*Cristobalite	JCP 1965 **43** 3917
				NBS 1965 **69C** 275
VIII ⎭		O	*Cristobalite	JCP 1966 **45** 4360

* Structure consists of 2 interpenetrating frameworks.

angles O—O—O are close to 109½°, and n.d. shows that the H atoms lie slightly off these lines in positions consistent with an H—O—H angle close to 105°. The regular tetrahedral arrangement of the four O neighbours is due to the randomness of orientation of the molecules in the structure as a whole. The very large range of angles in the (proton-disordered) ice-v has already been noted. In the ordered II and IX phases also n.d. and p.m.r. studies[5] confirm that the H—O—H angle is close to 105°. However, the O—O—O angles are respectively 88° and 99°, and 87° and 99°; the protons therefore lie appreciably off the O—O lines.

(1) JCP 1968 **49** 775   (4) JCP 1966 **45** 3976
(2) JCP 1965 **43** 2376   (5) JCP 1968 **49** 4660
(3) JCP 1965 **43** 2384

Water

The fact that water is liquid at ordinary temperatures whereas all the hydrides CH_4, NH_3, HF, PH_3, SH_2, and HCl of elements near to oxygen in the Periodic Table are gases, indicates interaction of an exceptional kind between neighbouring molecules. That these interactions are definitely directed towards a small number of neighbours is shown by the low density of the liquid compared with the value $(1.84 \, \mathrm{g \, cm^{-3}})$ calculated for a close-packed liquid consisting of molecules of similar size, assuming a radius of 1.38 Å as in ice. When ice melts there are two opposing effects, the breakdown of the fully hydrogen-bonded system in the tridymite-like structure of ice to give a denser liquid, and thermal expansion operating in the opposite sense. The first process must take place over a range of temperature in order to explain the minimum in the volume/temperature curve. There has been considerable discussion of the concentrations and energies of hydrogen bonds in water at various temperatures. Raman spectroscopic studies and measurements of the viscosity indicate that the number of hydrogen bonds in water at 20 °C is about half that in ice. The idea that large numbers of hydrogen bonds are broken when ice melts or when water is heated is not generally accepted, and certainly does not appear probable if we accept Pauling's value of 19 kJ mol^{-1} for the hydrogen-bond energy. Alternative suggestions are the bending rather than breaking of these bonds or the breaking of weaker hydrogen bonds. Before describing actual structural models suggested for water it is relevant to mention two other properties of water, namely, the abnormal mobilities of H$^+$ and OH$^-$ ions and the fact that the dielectric constant of water is practically the same as that of ice for low frequencies.

The mobilities of the H$^+$ and OH$^-$ ions are respectively 32.5 and 17.8 × 10^{-4} cm s^{-1} for an applied field of 1 volt cm^{-1}, whereas the values for other ions are of the order of 6 × 10^{-4} cm s^{-1}. Calculations show that very little energy is required to remove a proton from one water molecule, to which it is attached as (H$_3$O)$^+$, to another, for the states H$_3$O, H$_2$O $\rightleftharpoons$ H$_2$O, H$_3$O have the same energy. If we assume that the H$^+$ ion acts as a bond between two molecules, then the process shown diagrammatically in Fig. 15.3(a) results in the movement of H$^+$ from A to B. The analogous mechanism for the effective movement of OH$^-$ ions through water is illustrated in Fig. 15.3(b). Whereas other ions have to move bodily through the water, the H$^+$ and OH$^-$ ions move by what Bernal termed a kind of relay race, small shifts of protons only being necessary.

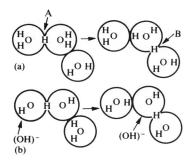

FIG. 15.3. The effective movement of (a) H$^+$ ion, and (b) OH$^-$ ion resulting from small shifts of protons (after Bernal and Fowler).

In order to account for the high dielectric constant of water it is necessary to suppose that there exist groups of molecules with a pseudocrystalline structure, that is, with sufficient orientation of the O—H—O bonds to give an appreciable electric moment. The upper limit of size of these 'clusters' has been estimated from studies of the infrared absorption bands in the 1.1–1.3 μm region as approximately 130 molecules at 0°, 90 at 20°, and 60 at 72 °C. In a liquid there is constant rearrangement of the molecules, and it is postulated that a given cluster persists only for a very short time, possibly of the order of 10^{-11} to 10^{-10} seconds.

The idea that water is in a general sense structurally similar to ice allowing, of course, for greater disorder in the liquid than the solid, is confirmed by the X-ray diffraction effects. Radial distribution curves have been derived from X-ray photographs taken at a number of temperatures from 1.5 to 83 °C and from the areas under the peaks the average numbers of neighbours at various distances can be deduced, at least in principle. The first peak on the curve suggests that at 1.5 °C there are on the average 4.4 neighbours at a mean distance of 2.90 Å; at 83° the corresponding figures are 4.9 at 3.05 Å. This first peak is succeeded by a curve which rises gradually to a poorly resolved maximum in the region 4.5–4.9 Å, indicating the presence of molecules at distances between those of the nearest and next nearest neighbours in ice (2.8 and 4.5 Å). Because of the small number of well-defined peaks the radial distribution curve has been interpreted in many ways.

Various arrangements of water molecules are possible if the only condition is that each is surrounded tetrahedrally by four neighbours, but the structure must be more dense than the tridymite-like structure of ice-I. Three suggestions have been made for the structure of the pseudocrystalline regions. The first is a quartz-like packing tending at high temperatures towards a more close-packed liquid, the density of quartz being $2.66 \, \mathrm{g \, cm^{-1}}$ compared with 2.30 for tridymite. The second is a structure resembling the water framework in the chlorine hydrate structure (p. 662) with non-hydrogen-bonded water molecules in all the polyhedral interstices, of which there are 8 in a unit cell containing 46 framework molecules. The third is a slightly expanded tridymite-like structure with interstitial molecules. It seems most likely that the higher density of water as compared with ice is due to the insertion of neighbours between the nearest and next nearest neighbours of an ice-like structure. In one model which has been refined by least squares to give a good fit with the observed radial distribution curve each framework H_2O has 4 framework molecules as nearest neighbours (1 at 2.77; 3 at 2.94 Å) and each interstitial molecule has 12 framework H_2O neighbours (3 at each of the distances 2.94, 3.30, 3.40, and 3.92 Å). In this model the ratio of framework to interstitial molecules is 4 : 1.

In view of the very extensive literature on the structure of water we give here only a few references to the more recent papers which include many references to earlier work.[1]

(1)   RTC 1962 **81** 904; JACS 1962 **84** 3965; JPC 1965 **69** 2145; JCP 1965 **42** 2563; JCP 1966 **45** 4719

Aqueous solutions

X-ray diffracton studies have been made of a number of aqueous solutions. It is possible to obtain some information about the nearest neighbours of a particular ion from the $\sigma(r)$ curve derived from concentrated solutions, though it is doubtful if the more detailed deductions made about the structures of solutions of HNO_3,

H_2SO_4, and NaOH represent unique interpretations of the experimental data.[1] In solutions containing complex ions interatomic distances corresponding to bonds within the ions can be identified, for example, peaks at 1.33 Å and 1.49 Å in the radial distribution curves of aqueous solutions of NH_4NO_3 and $HClO_4$ are attributable to the N–O bond and Cl–O bonds.[2] Deductions have been made about the environment of the cations in solutions of $EuCl_3$,[3] KOH,[4] $FeCl_3$,[5] $ZnCl_2$,[6] $ZnBr_2$,[7] and $CoCl_2$.[8] (For solutions of hydrolysed Pb and Bi salts containing hydroxy-complexes see p. 627.)

There is inevitably some doubt about the existence of octahedral as opposed to tetrahedral complexes in neutral solutions of $FeCl_3$ since the peak at 2.25 Å can be interpreted either in terms of tetrahedral coordination of Fe^{3+} by Cl^- (Fe–Cl, 2.25 Å) or as the mean of 4 Fe–O (2.07 Å) and 2 Fe–Cl (2.30 Å) in octahedral complexes of the type that exist in the crystalline hexahydrate (q.v.). On the other hand there seems to be no doubt about the existence of $FeCl_4^-$ in dilute solutions containing excess Cl^- ions or of Fe_2Cl_6 molecules in non-aqueous solvents of low dielectric constant such as methyl alchol.

Zinc chloride is extremely soluble in water even at room temperature, and the very viscous 27.5 molar solution approaches the molten salt in composition $(Zn_{0.20}Cl_{0.40}(H_2O)_{0.40})$. Its X-ray radial distribution curve is consistent with tetrahedral groups $ZnCl_3(H_2O)$, which must share, on the average, two Cl atoms with one another. In more dilute solutions ($5M$) groups $ZrCl_2(H_2O)_2$ predominate, and similar coordination groups are found in concentrated aqueous $ZnBr_2$. These change to $ZnBr_4^{2-}$ in the presence of sufficient excess Br^- ions. X-ray diffraction data from concentrated solutions of $CoCl_2$, on the other hand, have been interpreted in terms of octahedral $Co(H_2O)_6^{2+}$ groups (Co–O around 2.1 Å), but in methyl and ethyl alcohols there are apparently highly associated tetrahedral $CoCl_4$ groups (Co–Cl, approx. 2.3 Å).

(1) TKBM 1943 **3** 83; 1944 **4** 26, 5
(2) ACSc 1952 **6** 801
(3) JACS 1958 **80** 3576
(4) JCP 1958 **28** 464
(5) JCP 1969 **50** 4013
(6) IC 1962 **1** 941
(7) JCP 1965 **43** 2163
(8) JCP 1969 **50** 4313

Hydrates

One of the simplest ways of purifying a compound is to recrystallize it from a suitable solvent. The crystals separating from the solution may consist of the pure compound or they may contain 'solvent of crystallization'. For most salts water is a convenient solvent, and accordingly crystals containing water – hydrates – have been known from the earliest days of chemistry, and many inorganic compounds are normally obtained as hydrates. We have already noted in our survey of hydroxides that many compounds containing OH groups were originally formulated as hydrates, for example, $NaSb(OH)_6$ as $NaSbO_3.3H_2O$, $NaB(OH)_4$ as $NaBO_2.2H_2O$, and many others. The term hydrate should be used only for crystalline compounds containing

H_2O molecules or in the case of hydrated acids, ions such as H_3O^+, $H_5O_2^+$, etc. (see the discussion of these compounds later). Apart from inorganic and organic acids, bases, and salts, certain elements and non-polar compounds form hydrates, among them being chlorine and bromine, the noble gases, methane, and alkyl halides. Many of these hydrates contain approximately 6–8 H_2O or 17 H_2O for every molecule (or atom) of 'solute'. The hydrates of the gaseous substances are formed only under pressure and are extremely unstable, and the melting points are usually close to 0 °C. In these 'ice-like' hydrates there is a 3D framework built of H_2O molecules which encloses the solute molecules in tunnels or polyhedral cavities ('clathrate' hydrates). At the other extreme there are hydrates in which water molecules (and cations) are accommodated in tunnels or cavities within a rigid framework such as the aluminosilicate frameworks of the zeolites. Such a hydrate can be reversibly hydrated and dehydrated without collapse of the framework, in contrast to the clathrate hydrates in which the framework itself is composed of water molecules. Between these two extremes lie the hydrates of salts, acids, and hydroxides, in which H_2O molecules, anions, and cations are packed together to form a structure characteristic of the hydrate. Removal of some or all of the water normally results in collapse of the structure, which is usually unrelated to that of the anhydrous compound. The monohydrate of $Na_3P_3O_9$ is exceptional in this respect. In the structure of the anhydrous salt there is almost sufficient room for a water molecule, so that very little rearrangement of the bulky $P_3O_9^{3-}$ ions is necessary to accommodate it. For this reason the structures of $Na_3P_3O_9$ and $Na_3P_3O_9 . H_2O$ are very similar.

We shall be concerned in this chapter with the clathrate hydrates and the hydrates of salts, acids, and hydroxides. The structures of zeolites are described in Chapter 23, as also are the clay minerals, which can take up water between the layers. Examples of hydrated 'basic salts' and of hydrates of heteropoly acids and their salts are also discussed in other chapters.

Clathrate hydrates

In this type of hydrate the water molecules form a connected 3D system enclosing the solute atoms or molecules. We may list the latter roughly in order of inreasing interaction with the water framework: atoms of noble gases (Ar, Xe) and molecules such as Cl_2, CO_2, C_3H_8 and alkyl halides, compounds such as $N_4(CH_2)_6$ which are weakly hydrogen-bonded to the framework, and finally ionic compounds in which ions of one or more kinds are associated with or incorporated in the water framework. As regards their structures, the frameworks range from those in which there are well defined polyhedral cavities to 3- or (3+4)-connected nets in which there are no obvious voids of this kind. In the first group a H_2O molecule is situated at each point of a 4-connected net, the links of which (O—H—O bonds) form the edges of a space-filling arrangement of polyhedra. These 'ice-like' structures are not stable unless all (or most) of the larger voids are occupied by solute molecules, and the melting points are usually not far removed from 0 °C. In Table 15.2 we include

TABLE 15.2
Clathrate hydrates

	H_2O *molecules only in framework*	H_2O *and F, N, S in framework*	*Reference*
	Polyhedral frameworks		
Class (a)			
(i)	$HPF_6 \cdot 6 H_2O$		AC 1955 **8** 611
		$(CH_3)_4NOH \cdot 5 H_2O$	JCP 1966 **44** 2338
(ii)	$(CH_3)_3CNH_2 \cdot 9\frac{3}{4} H_2O$		JCP 1967 **47** 1229
Class (b)			
(i)	$Cl_2 \cdot 7\frac{1}{4} H_2O$		PNAS 1952 **36** 112
			JCS 1959 4131
	$6 \cdot 4 C_2H_4O \cdot 46 H_2O$		JCP 1965 **42** 2725
		$(n\text{-}C_4H_9)_3SF \cdot 20 H_2O$	JCP 1962 **37** 2231
(ii)	$CHCl_3 \cdot 17 H_2O$		JCP 1951 **19** 1425
	$7 \cdot 33 H_2S \cdot 8 C_4H_8O \cdot 136 H_2O$		JCP 1965 **42** 2732
(iii)		$(i\text{-}C_5H_{11})_4NF \cdot 38 H_2O$	JCP 1961 **35** 1863
(iv)	$Br_2 \cdot 8 \cdot 6 H_2O$		JCP 1963 **38** 2304
Class (c)	*Intermediate types*		
(i)		$(n\text{-}C_4H_9)_3SF \cdot 23 H_2O$	JCP 1964 **40** 2800
(ii)	$(C_2H_5)_2NH \cdot 8\frac{2}{3} H_2O$		JCP 1967 **47** 1222
(iii)	$4 (CH_3)_3N \cdot 41 H_2O$		JCP 1968 **48** 2990
Class (d)	*Non-polyhedral frameworks*		
(i)	$N_4(CH_2)_6 \cdot 6 H_2O$		JCP 1965 **43** 2799
(ii)		$(CH_3)_4NF \cdot 4 H_2O$	JCP 1967 **47** 414

some less regular structures to which we refer later. The polyhedral frameworks are divided into two classes, (a) and (b), the latter involving the pentagonal dodecahedron and related polyhedra with hexagonal faces to which reference was made in Chapter 3.

Polyhedral frameworks

Class (a). The analogy between the structural chemistry of H_2O and SiO_2 is illustrated by the structure of $HPF_6 \cdot 6H_2O$. In this crystal the H_2O molecules are situated at the vertices of Fedorov's space-filling by truncated octahedra, that is, at the same positions as the Si(Al) atoms in the framework of ultramarine (p. 1042). Each H_2O molecule in the framework (Fig. 15.4) is hydrogen-bonded to its four neighbours at a distance of 2.72 Å, and the PF_6^- ions occupy the interstices ($H_2O–F$, 2.74 Å; P–F, 1.73 Å).

If each link in the 'Fedorov' net is to be a hydrogen bond there are sufficient H atoms in $M \cdot 6H_2O$. In $HPF_6 \cdot 6H_2O$ there are 13 H, and H^+ is presumably associated in some way with the framework. An interesting distortion of this net occurs in $(CH_3)_4NOH \cdot 5H_2O$, where $OH^- + 5 H_2O$ are situated at the vertices of the truncated octahedra. Since there are only 11 H all the edges of the polyhedra cannot be

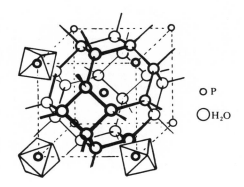

O P

O H$_2$O

FIG. 15.4. The crystal structure of HPF$_6$.6H$_2$O.

hydrogen bonds. The inclusion of the cations in the truncated octahedra expands them by extending certain of the edges to a length of 4.36 Å (Fig. 15.5).

Much less symmetrical polyhedral cavities are found in $(CH_3)_3C.NH_2.9\frac{3}{4}H_2O$. The framework represents a space-filling by 8-hedra ($f_4 = 4$, $f_5 = 4$) and 17-hedra ($f_3 = 3$, $f_5 = 9$, $f_6 = 2$, $f_7 = 3$). The unit cell contains 156 H$_2$O (16 formula units), and the amine molecules occupy the 16 large voids (17-hedra); the 8-hedra are not occupied.

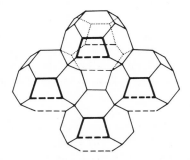

FIG. 15.5. The framework of hydrogen-bonded OH$^-$ ions and H$_2$O molecules in $(CH_3)_4N.OH.5H_2O$ showing distortion of the truncated octahedron in the 'Fedorov' net. The broken lines indicate O–O edges of length 4.36 A.

Class (b). In this class the (4-connected) networks are the edges of space-filling arrangements of pentagonal dodecahedra and one or more of the related polyhedra: $f_5 = 12, f_6 = 2, 3, 4$ (Table 15.3).

The hydrate of chlorine is of special interest as the solid phase originally thought to be solid chlorine but shown (in 1811) by Humphry Davy to contain water. It was later given the formula Cl$_2$.10H$_2$O by Michael Faraday. The unit cell of this

TABLE 15.3
The dodecahedral family of hydrate structures

Hydrate	Z	Vertices	Voids (n-hedra)				Total of large voids
			12	14	15	16	
$Cl_2 . 7\frac{1}{4} H_2O$	6	46	2	6	–	–	6
$CHCl_3 . 17 H_2O$	8	136	16	–	–	8	8
$(i\text{-}C_5H_{11})_4N^+F^- . 38 H_2O$	2	80	6	4	4	–	8
$Br_2 . 8\cdot6 H_2O$	20	172	10	16	4	–	20

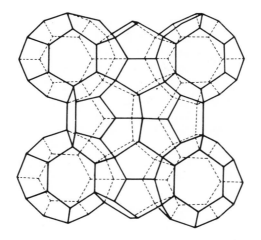

FIG. 15.6. The oxygen framework of the Type I gas hydrate structure. At the centre of the diagram are two of the six 14-hedra voids in the unit cell.

(cubic) structure ($a \approx 12$ Å) contains 46 H_2O which form a framework (Fig. 15.6) in which there are 2 dodecahedral voids and 6 rather larger ones (14-hedra). If all the voids are filled, as is probable for Ar, Xe, CH_4, and H_2S, this corresponds to $46/8 = 5\frac{3}{4}H_2O$ per atom (molecule) of solute. If only the larger holes are filled the formula of chlorine hydrate would be $Cl_2 .7\frac{2}{3}H_2O$. Earlier analyses suggested 8 H_2O, which would correspond to 2 H_2O in the smaller voids. However, more recent chemical analyses and density measurements indicate a formula close to $Cl_2 .7\frac{1}{4}H_2O$, suggesting partial ($\rlap{>}{\scriptstyle\sim}\, 20$ per cent) occupancy of the smaller voids by Cl_2 molecules.

The sulphonium fluoride $(n\text{-}C_4H_9)_3SF$ forms three hydrates, one of which (20 H_2O) has this structure, apparently with 2 S^+ statistically occupying framework sites, leaving 4 vacant sites, two of which may be occupied by 2 F^-. In hydrates of substituted sulphonium and ammonium salts the bulky alkyl groups occupy voids adjacent to the framework site occupied by S^+ or N^+. The host structure invariably has higher symmetry than is compatible with the arrangement of the guest ions in any one unit cell, and there is accordingly disorder of various kinds in these structures.

The second structure of Table 15.3, which is also cubic ($a \approx 17.2$ Å), is adopted by hydrates of liquids such as $CHCl_3$; a portion of the structure is illustrated in Fig. 15.7. The unit cell contains 136 H_2O and there are voids of two sizes, 16 smaller

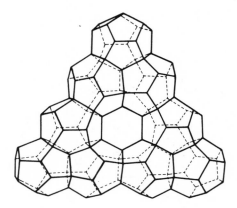

FIG. 15.7. Part of the framework of water molecules in a hydrate M.$17H_2O$ shown as a packing of pentagonal dodecahedra and hexakaidecahedra. At the centre part of one of the larger holes can be seen.

and 8 larger. Filling of only the larger voids gives a ratio of $136/8 = 17$ H_2O per molecule of $CHCl_3$, CH_3I, etc., but molecules of two quite different sizes can be accommodated, each in the holes of appropriate size, as in $CHCl_3.2H_2S.17H_2O$. In the H_2S-tetrahydrofuran hydrate listed in Table 15.2 there is rather less than half-occupancy of the smaller holes by molecules of H_2S.

In the third structure of Table 15.3 the 80 polyhedral vertices in a unit cell are occupied by 76 H_2O, 2 N^+, and 2 F^-. Alkyl groups project from N^+ into the four (tetrahedrally disposed) polyhedra meeting at that vertex (two 14- and two 15-hedra).

Class (c). In $(\dot{n}-C_4H_9)_3SF.23H_2O$ there are layers of pentagonal dodecahedra similar to those in the 38 hydrate structure just mentioned, and between them are large irregularly shaped cavities which accommodate the cations. The formation of this structure apparently represents an attempt to accommodate the $(n-C_4H_9)_3S^+$ cation in a hydrate with composition close to that of 20-hydrate which forms the cubic structure already described

Amines form numerous hydrates, with melting points ranging from $-35°$ to $+5\,°C$ and containing from $3\frac{1}{2}$ to 34 H_2O per molecule of amine. The structures include the cubic gas hydrate structure and some less regular ones. For example, in

$(C_2H_5)_2NH.8\frac{2}{3}H_2O$ layers of 18-hedra ($f_5 = 12$, $f_6 = 6$) are linked by additional water molecules to form less regular cages (12 in a cell containing 104 H_2O). The N atoms are not incorporated in the framework but are hydrogen-bonded to the H_2O molecules as in $N_4(CH_2)_6.6H_2O$ (see later).

Class (d). These hydrates are distinguished on the grounds that there are no well defined polyhedral cavities, the nets being 3- or (3+4)-connected. The 3-connected framework in $N_4(CH_2)_6.6H_2O$ (m.p. 13.5 °C) is that of Fig. 3.31 (p. 115), the same as one of the two identical interpenetrating frameworks in β-quinol clathrates. Since a framework of this kind built of H_2O molecules has only 9 links (O—H—O bonds) for every 6 H_2O there are 3 H atoms available to form hydrogen bonds to the guest molecules. The latter are suspended 'bat-like' in the cavities halfway between the 6-rings (Fig. 15.8), that is, they occupy the positions of the rings of the second network in the β-quinol structure. One-half of the H_2O molecules form three pyramidal O—H—O bonds and the remainder four tetrahedral

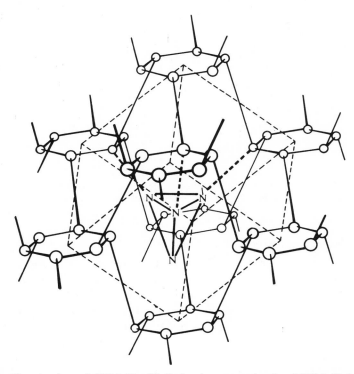

FIG. 15.8. The structure of $(CH_2)_6N_4.6H_2O$ showing one molecule of $(CH_2)_6N_4$ 'suspended' in one of the interstices. The $(CH_2)_6N_4$ molecule is represented diagrammatically as a tetrahedron of N atoms attached to the hydrogen-bonded water framework by N...H—O bonds (heavy broken lines).

hydrogen bonds (one to N). One-half of the protons are disordered (those in the 6-rings), the remainder are ordered—compare ice-I, in which all the protons are disordered, and ice-II, in which all protons are ordered.

In $(CH_3)_4NF.4H_2O$ the F^- ions and H_2O molecules form a hydrogen-bonded framework (Fig. 15.9) in which F^- has a 'flattened tetrahedral' arrangement of 4 H_2O neighbours (O–F–O, 156° (two) and $92\frac{1}{2}°$ (four)) and H_2O has three nearly coplanar neighbours (O–O, 2.73 Å, O–F, 2.63 Å). The cations occupy cavities between pairs of F^- ions.

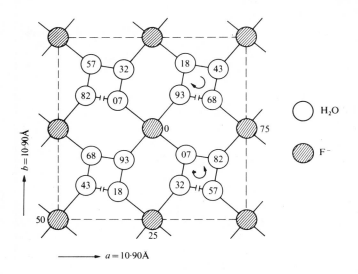

FIG. 15.9. The framework of F^- ions (4-connected) and H_2O molecules (3-connected) in $(CH_3)_4NF.4H_2O$. The heights of atoms above the plane of the paper are in units of $c/100$ ($c = 8.10$ Å).

Hydrates of oxy-salts, hydroxides, and halides

We exclude from the major part of our discussion the structures of hydrated complex salts, since the principles determining their structure are much less simple. For example, in hydrated complex halides $A_m(BX_n).pH_2O$, which we consider briefly later, the water is in some cases attached to B (e.g. in $(NH_4)_2(VF_5.H_2O)$) while in others (e.g. $K_2(MnF_5).H_2O$, p. 454) it is situated together with the A ions between BX_n complexes (here infinite octahedral chain ions). Similarly there are octahedral $(HgCl_4)_n^{2n-}$ chains in $K_2HgCl_4.H_2O$ between which lie the K^+ ions and the H_2O molecules.

Before reviewing the crystal structures of these compounds we note some general points. They form an extremely large group of compounds, ranging from highly hydrated salts such as $MgCl_2.12H_2O$ and $FeBr_2.9H_2O$ to monohydrates and even hemihydrates, for example, $Li_2SO_4.H_2O$ and $CaSO_4.\frac{1}{2}H_2O$. Moreover, a particular compound may form a series of stoichiometric hydrates. Many simple halides form three, four, or five different hydrates; $FeSO_4$ crystallizes with 1, 4, 5, 6 and 7 H_2O, and NaOH is notable for forming hydrates with 1, 2, $3\frac{1}{2}$, 4, 5, and 7 H_2O. The degree of hydration depends on the nature of both anion and cation. In some series of alkali-metal salts containing large anions such as SO_4^{2-} or $SnBr_6^{2-}$ the Li and Na salts are hydrated while those containing the larger ions of K, Rb, and Cs are anhydrous. The alkali-metal chlorides behave similarly, but the fluorides show the reverse effect, and the figures in Table 15.4 illustrate the difficulty of generalizing about the degree of hydration of series of salts.

TABLE 15.4
Hydrates of some alkali-metal salts

	Li	Na	K	Rb	Cs
MF	–	–	2, 4	$1\frac{1}{2}$	$\frac{2}{3}, 1\frac{1}{2}$
MCl	1, 2, 3, 5	2	–	–	–
M_2CO_3	–	1, 7, 10	2, 6	1, $1\frac{1}{2}$	$3\frac{1}{3}$
M_2SO_4	1	1, 7, 10	–	–	–

We have seen that the structures of the ice polymorphs and of the ice-like hydrates indicate that the H_2O molecule behaves as if there is a tetrahedral distribution of two positive and two negative regions of charge. The arrangement of nearest neighbours of water molecules in many crystalline hydrates is consistent with this tetrahedral character of the water molecule. In hydrated oxy-salts we commonly find a water molecule attached on the one side to two O atoms of oxy-ions and on the other to two ions M^+ or to one ion M^{2+} thus:

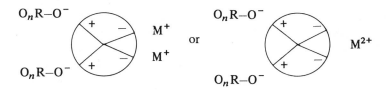

The neighbours of a water molecule may equally well be other water molecules suitably oriented so that oppositely charged regions are adjacent. In this way groups of water molecules may be held together as in $H_3PW_{12}O_{40}.29H_2O$, and water molecules in excess of those immediately surrounding the metal ions may be present, as in $NiSO_4.7H_2O$, which may be written $[Ni(H_2O)_6]H_2O.SO_4$. In our discussion of the structures of hydroxides we saw that in suitable environments, the OH group is polarized to the stage where an O—H---O bond is formed, and in hydrates we find an analogous effect. In hydrated oxy-salts and hydroxides the short distances (2.7–2.9 Å) between O atoms of water molecules and those of the oxy-ions are similar to those found in certain hydroxides and oxyhydroxides, and indicate the formation of hydrogen bonds. In hydrated fluorides there are O—H---F bonds of considerable strength, and we give examples later of O—H---Cl bonds. We deal separately with hydrated acids and acid salts in which protons are associated with some or all of the water molecules to form H_3O^+ or more complex groupings.

Some of the numerous possible environments of a water molecule in hydrates are shown in Fig. 15.10. It might seem logical to classify hydrates according to the

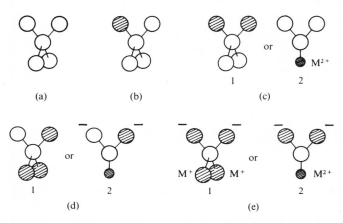

FIG. 15.10. Environment of water molecules in crystals. The larger shaded circles represent M^+, OH^-, F^-, or oxygen of oxy-ion.

way in which the water molecules are bonded together. If all the H_2O molecules in a particular hydrate have environments of one of the types shown in Fig. 15.10, with 4, 3, 2, 1, or zero H_2O molecules as nearest neighbours then the systems of linked H_2O molecules (aquo-complex) would be as follows: (a) and (b) all possible types up to and including 3D frameworks; (c) rings or chains; (d) pairs of H_2O molecules; and (e) no aquo-complex. For example, each H_2O in $Li_2SO_4.H_2O$ is of type (c), hydrogen-bonded to two others, so that chains of water molecules can be distinguished in the crystal, whereas in $KF.2H_2O$ each H_2O is surrounded tetrahedrally by 2 F^- and 2 K^+ ions, as at (e), that is, there is no aquo-complex. However, any classification of this kind would be impracticable because the water molecules in many hydrates do not all have the same kind of environment. There may be a major difference in environment, as when some of the H_2O molecules are bonded to the metal atoms and others are accommodated between the complexes as, for example, in $[CoCl_2(H_2O)_4].2H_2O$, or there may be differences between the environments of the various water molecules of one $M(H_2O)_n$ complex. Thus in $NiSO_4.7H_2O$ the seventh H_2O molecule is not in contact with a metal ion but in addition there is the further complication that the environments of the six H_2O molecules in a $Ni(H_2O)_6^{2+}$ group are not the same and are in fact of no fewer than four different types. Some have three approximately coplanar neighbours (Ni^{2+} and 2 O of SO_4^{2-} or 2 H_2O) and others four tetrahedral neighbours (Ni^{2+}, 2 O and H_2O or Ni^{2+}, O, and 2 H_2O). Much of the structural complexity of hydrates is due to this non-equivalence of water molecules which in turn is associated with the fact that so many arrangements of nearest neighbours are compatible with the tetrahedral charge distribution of the H_2O molecule as indicated in Fig. 15.10. It is perhaps unnecessary to mention that since there is an element of compromise in most crystal structures the environment of a water molecule is not always one of the arrangements of Fig. 15.10. For example, in $Na_2(H_2SiO_4).7H_2O$[1] certain of the water molecules have an unusual trigonal bipyramidal arrangement of five nearest neighbours, 3 Na and 2 H_2O, or 3 H_2O, O, and OH.

A detailed description of the bonding in hydrates evidently requires a knowledge of the positions of the H atoms. In the earlier X-ray studies it was not possible to locate these atoms, and it was assumed that they were responsible for certain unusually short O–O or O–X distances in the crystals. Later studies, particularly n.d. and n.m.r., have confirmed this and led to the precise location of the H atoms. It is now becoming possible to discuss not only the gross structures of the compounds, that is, the spatial arrangement of the heavier atoms, but also two aspects of the finer structure, namely, the positions of the H atoms in hydrogen bonds and the ordering of the protons. Reference to these topics will be made later.

Since in the hydrates under discussion the H_2O molecules tend to associate, albeit not exclusively, with the cations, it is convenient to show in any classification the nature of the coordination group around the cations. In a hydrated oxy-salt or halide this will be made up of O atoms of oxy-ions, halide ions, or O atoms of H_2O molecules (the H atoms of which will be directed away from M). If n is the coordination number of M in $M_xX_y.zH_2O$, where X represents the anion (Cl^-, SO_4^{2-}, O_2^{2-}, OH^-, etc.) we may list hydrates according to the value of z/xn (Table 15.5). Evidently, very simple structures may be expected if $z/xn = 1$ ($z = xn$), since there is in this case exactly the number of H_2O molecules necessary to form complete coordination groups $M(H_2O)_n$ around every M ion. However, the situation is more complicated than this because if H_2O molecules are common to two $M(H_2O)_n$ coordination groups there can be complete hydration of M with smaller values of z/xn. For example, for octahedral coordination of M ($n = 6$):

z/xn		Example
1	Discrete $M(H_2O)_6$ groups	$MgCl_2 . 6 H_2O$
$\frac{2}{3}$	$M(H_2O)_6$ sharing two edges	$KF . 4 H_2O$
$\frac{1}{2}$	$M(H_2O)_6$ sharing two faces	$LiClO_4 . 3 H_2O$

There is therefore no simple relation between the value of z/xn and the composition of the coordination group around M, as is clearly seen in Table 15.5.

TABLE 15.5

A classification of salt hydrates $M_xX_y.zH_2O$

		M fully hydrated		M incompletely hydrated	
	z/xn	*excess* H_2O I	II	*excess* H_2O III	IV
A	2 9/6 8/6 7/6 1	$MgCl_2.12H_2O$[1] $FeBr_2.9H_2O$ see text $NiSO_4.7H_2O$[2] $NaOH.7H_2O$[3]	$Sm(BrO_3)_3.9H_2O$[4] $Nd(BrO_3)_3.9H_2O$[5] $CaO_2.8H_2O$[6] $Sr(OH)_2.8H_2O$[7] $Zn(BrO_3)_2.6H_2O$[8] $CoI_2.6H_2O$[9] $MgCl_2.6H_2O$[10] $AlCl_3.6H_2O$[11] $CrCl_3.6H_2O$[12] $Mg(ClO_4)_2.6H_2O$[13] $BeSO_4.4H_2O$[14]	$CoCl_2.6H_2O$[15] $NiCl_2.6H_2O$[16] $FeCl_3.6H_2O$[17] $CrCl_3.6H_2O$[18]	
B	5/6 3/4	$Na_2SO_4.10H_2O$[19]	$Na_2CO_3.10H_2O$[20]	$CuSO_4.5H_2O$[26]	$CaCO_3.6H_2O$[29] $GdCl_3.6H_2O$[30]

	C	B	A
2/3	$SrCl_2.6H_2O$(21) $KF.4H_2O$(22)	$SnCl_2.2H_2O$(27)	$FeF_2.4H_2O$(31) $FeCl_2.4H_2O$(32) $MnCl_2.4H_2O$(32a)
7/12	$Na_2HAsO_4.7H_2O$(23) $NaOH.3\frac{1}{2}H_2O$(24) $LiClO_4.3H_2O$(25)		
1/2		$FeF_3.3H_2O$(28)	$CuSO_4.3H_2O$(33)
C 4/9		$CdSO_4.\frac{8}{3}H_2O$(34)	
1/3			$KF.2H_2O$(35), $NaBr.2H_2O$(36) $CoCl_2.2H_2O$(37) $NiCl_2.2H_2O$(37a) $BaCl_2.2H_2O$(38), $CaCl_2.2H_2O$(38a) $SrCl_2.2H_2O$(39)
1/4			$CaSO_4.2H_2O$(40) $LiOH.H_2O$(41) $CuSO_4.H_2O$(42) $Li_2SO_4.H_2O$(43)
1/6			$Sr(OH)_2.H_2O$(44)
1/8			$SrBr_2.H_2O$(45)
1/9			$Na_2CO_3.H_2O$(46)
1/12			

(1) AC 1966 20 875. (2) AC 1964 17 1167, 1361; AC 1969 B25 1784. (3) CR 1953 236 1579. (4) AC 1969 A25 621. (5) JACS 1939 61 1544. (6) AC 1951 4 67. (7) AC 1953 6 604. (8) ZK 1936 95 426. (9) ZSK 1963 4 63. (10) ZK 1934 87 345. (11) AC 1968 B24 954. (12) ZK 1934 87 446. (13) ZK 1935 91 480. (14) AC 1969 B25 304, 310. (15) JPSJ 1961 16 1574. (16) JCP 1969 50 4690. (17) JCP 1967 47 990. (18) AC 1966 21 280. (19) AC 1978 B34 3502. (20) AC 1969 B25 2656. (21) AC 1977 B33 2938. (22) JCP 1964 41 917. (23) AC 1970 B26 1574, 1584. (24) BSFMC 1958 81 287. (25) AC 1977 B33 3954. (26) PRS A 1962 266 95. (27) JCS 1961 3954. (28) AC 1964 17 1480. (29) IC 1970 9 480. (30) AC 1961 14 234. (31) AC 1960 13 953. (32) AC 1971 B27 2329. (32a) IC 1964 3 529; IC 1965 4 1840. (33) AC 1968 B24 508. (34) PRS A 1936 156 462. (35) AC 1951 4 181. (36) AC 1979 B35 1679. (37) AC 1963 16 1176. (37a) AC 1967 23 630. (38) AC 1978 B34 2290. (38a) AC 1977 B33 1608. (39) KDV 1943 20 Nr.5. (40) AC 1974 B30 921. (41) AC 1971 B27 1682. (42) CR 1966 B262 722. (43) JCP 1968 48 5561. (44) AC 1967 22 252. (45) JPC 1964 68 3259. (46) AC 1975 B31 890.
Miscellaneous: $NaOH.4H_2O$ JCP 1964 41 924.

For descriptive purposes it is convenient to make three horizontal subdivisions of the Table.

A. $(z/xn) \geqslant 1$

In these hydrates there is sufficient (or more than sufficient) water for complete hydration of the cations without sharing of H_2O molecules between $M(H_2O)_n$ coordination groups. There are no known exceptions to the rule that if $z/xn > 1$ (class AI) M is fully hydrated and the excess water is accommodated between the $M(H_2O)_n$ complexes or, alternatively, associated with the anions. A set of very simple structures is found if $z/xn = 1$ (class AII), but the structures of greatest interest are those of class AIII, at present represented by only two structures, both of halides of 3d metals. Although $z/xn = 1$ some of the coordination positions around M are occupied by Cl in preference to H_2O. There are no entries in Class AIV, for if $z/xn \geqslant 1$ either the cation is fully hydrated or there is excess water of crystallization.

B. $1 > (z/xn) \geqslant \frac{1}{2}$

Here there is sufficient water for complete hydration of M assuming that H_2O molecules can be shared between two (and only two) $M(H_2O)_n$ coordination groups. Structures of all four types I–IV are known, and those of greatest interest are perhaps those of types I and III where, as in AIII, structures could be envisaged in which all the water would be associated with M ions; instead, only part of the water hydrates the cations.

C. $(z/xn) < \frac{1}{2}$

There is insufficient water for complete hydration of M even allowing sharing of H_2O molecules between two $M(H_2O)_n$ coordination groups. Apart from one case the examples of Table 15.5 are all mono- or di-hydrates, and as might be expected are of type IV. Because of its charge distribution a water molecule is unlikely to have more than two cation neighbours, and accordingly there are no hydrates in classes CI or CII.

We have already remarked that in many hydrates the H_2O molecules are not all equivalent, often having very different environments. It is also found that in some hydrates there are two or more kinds of non-equivalent cation. This complication is less frequently encountered; an example is $Na_4P_2O_7 \cdot 10H_2O$,[2] in which one-half of the Na^+ ions have $6 H_2O$ while the remainder have $4 H_2O$ and O atoms of anions as nearest neighbours. We confine our examples in Table 15.5 and the following account to hydrates in which all the cations have similar arrangements of nearest neighbours, as is the case in most hydrates. This survey based on Table 15.5 is followed by short sections on hydrates of complex halides, and of calcium chloride, calcium oxy-salts, and vanadyl sulphate.

(1)   AC 1976 **B32** 705
(2)   AC 1957 **10** 428

Hydrates of Class A: $z/xn \geqslant 1$.

Type A1. This is apparently a very small group of compounds, of which few structures have been determined.

$MgCl_2 . 12H_2O$. This compound is of special interest as the most highly hydrated simple salt of which we know the structure. It is stable only at low temperatures, as shown by the transition points:

$$MgCl_2 . 12H_2O \xrightarrow{-16.4°} 8H_2O \xrightarrow{-3.4°} 6H_2O \xrightarrow{116.7°} 4H_2O \xrightarrow{181\,°C} 2H_2O$$

(The known hydrates of $MgBr_2$ and MgI_2 contain respectively 6 and 10, and 8 and 10 H_2O.) The structure consists of very regular $Mg(H_2O)_6$ octahedra and very distorted $Cl(H_2O)_6$ octahedra (Cl–O, 3.11–3.26 Å, but edges 3.86–5.55 Å). Each octahedral coordination group shares four vertices, $Mg(H_2O)_6$ with 4 $Cl(H_2O)_6$ and $Cl(H_2O)_6$ with 2 $Cl(H_2O)_6$ and 2 $Mg(H_2O)_6$, as shown diagrammatically in Fig. 15.11. The layers, of composition $MgCl_2(H_2O)_{12}$, are held together by O–H–O bonds between H_2O molecules. The H_2O molecules attached to Mg^{2+} have two other neighbours (H_2O or Cl^-); the others, one-half of the total, have four tetrahedral neighbours ($2\,Cl^- + 2\,H_2O$ or $1\,Cl^- + 3\,H_2O$). The structure provides a beautiful illustration of the behaviour of the water molecule in this type of hydrate.

($z/xn = 8/6$. We may include here two salts containing hexanitrato-ions, namely, $Mg[Th(NO_3)_6] . 8H_2O^{(3)}$ and $Mg_3[Ce(NO_3)_6]_2 . 24H_2O,^{(4)}$ in both of which Mg^{2+} is

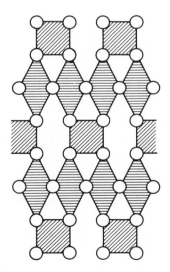

FIG. 15.11. Layer formed from vertex-sharing $[Mg(H_2O)_6]^{2+}$ and $[Cl(H_2O)_6]^-$ groups in $MgCl_2 . 12H_2O$ (diagrammatic). The squares represent $[Mg(H_2O)_6]^{2+}$ and the rhombuses $[Cl(H_2O)_6]^-$ groups. Two (unshared) vertices of each octahedron are not shown.

completely hydrated and one-quarter of the water of hydration is not attached to cations.)

$NiSO_4.7H_2O$. Reference has already been made to the fact that as regards their environment in the crystal there are five different kinds of H_2O molecule in this hydrate. From the chemical standpoint, however, we need only distinguish between those forming octahedral groups around Ni^{2+} ions and the seventh H_2O which is situated between three water molecules of $Ni(H_2O)_6$ groups and one O atom of a sulphate ion.

Type AII. In this, the simplest type of salt hydrate, all the water molecules are associated with the cations. The number of O atoms which can be accommodated around the ion M is determined by the radius ratio $r_M:r_O$, and since the bonds M–O are essentially electrostatic in nature the coordination polyhedra are those characteristic of ionic crystals.

Shape of $M(H_2O)_n$ *complex*

Tetrahedron:	$BeSO_4.4H_2O$
Octahedron:	$MgCl_2.6H_2O$; $CoI_2.6H_2O$; $AlCl_3.6H_2O$;
	$CrCl_3.6H_2O$; $Mg(ClO_4)_2.6H_2O$, etc.
Square antiprism:	$CaO_2.8H_2O$; $Sr(OH)_2.8H_2O$
Tricapped trigonal prism:	$Nd(BrO_3)_3.9H_2O$; $Sm(BrO_3)_3.9H_2O$

The environment of the water molecules in, for example, $BeSO_4.4H_2O$ is consistent with our description of the H_2O molecule; it corresponds to (e) 2 of Fig. 15.10.

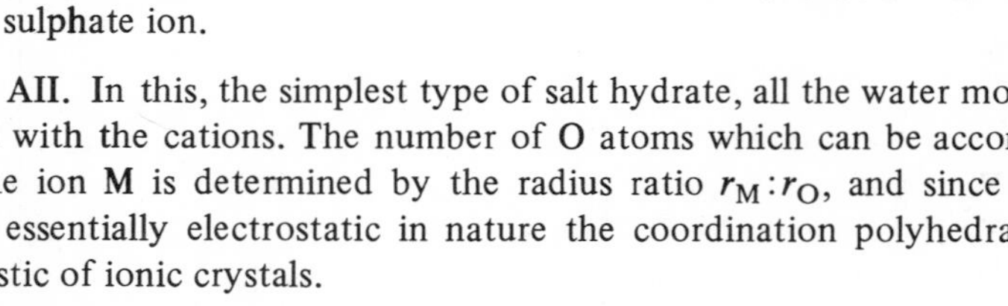

If the only factor determining the structure of $Mg(H_2O)_6Cl_2$ were the relative sizes of $Mg(H_2O)_6^{2+}$ and Cl^- this hydrate could have the fluorite structure, but this would mean that each H_2O would be in contact with 4 Cl^- ions. In fact, the corresponding ammine, $Mg(NH_3)_6Cl_2$, does crystallize with this structure, but the hexahydrate has a less symmetrical (monoclinic) structure in which each water molecule is in contact with only 2 Cl^- ions. Similarly, the ammine $Al(NH_3)_6Cl_3$ has the YF_3 structure, but $Al(H_2O)_6Cl_3$ has a much more complex rhombohedral structure in which every H_2O molecule is adjacent to only 2 Cl^- ions, giving it an environment very similar to that in $BeSO_4.4H_2O$ as shown above.

Type AIII. The two structures in this class, both of hexahydrates of 3d metal halides, form a striking contrast to two structures of AII. Both contain octahedral coordination groups $M(H_2O)_4Cl_2$ with the *trans* configuration. ($NiCl_2.6H_2O$ is similar to the cobalt compound.)

Coordination group of M		Coordination group of M		Structural formula
$MgCl_2 . 6 H_2O$	$6 H_2O$	$CoCl_2 . 6 H_2O$	$4 H_2O\ 2\ Cl$	$[CoCl_2(H_2O)_4] . 2 H_2O$
$AlCl_3 . 6 H_2O$	$6 H_2O$	$FeCl_3 . 6 H_2O$	$4 H_2O\ 2\ Cl$	$[FeCl_2(H_2O)_4] Cl . 2 H_2O$

In contrast to $CoCl_2 . 6H_2O$ the iodide is $Co(H_2O)_6I_2$ (type AII). For the trichlorides the stabilities of the two hydrate structures are presumably not very different, for the hexahydrate of $CrCl_3$ forms structures of both types. Early chemical evidence (for example, the proportion of the total chlorine precipitated by $AgNO_3$) indicated the following structures for the hydrates of chromic chloride:

blue hydrate: $[Cr(H_2O)_6]Cl_3$
green hydrates: $[CrCl(H_2O)_5]Cl_2 . H_2O$ and $[CrCl_2(H_2O)_4]Cl . 2H_2O$

X-ray studies have confirmed the first and third structures. With these hydrates compare $GdCl_3 . 6H_2O$ of type BIV (later).

(3)   AC 1964 18 698   (4)   JCP 1963 39 2881

Hydrates of Class B: $1 > z/xn \geqslant \frac{1}{2}$

   Type BI.
   $Na_2SO_4 . 10H_2O$. In two well-known decahydrates, those of Na_2SO_4 and Na_2CO_3, the 5:1 ratio of $H_2O:M$ is achieved in different ways, Na^+ being completely hydrated in both hydrates. In the sulphate there are infinite chains of octahedral $Na(H_2O)_6$ groups sharing two edges so that there is an excess of 2 H_2O to be accommodated between the chains: $[Na(H_2O)_4]_2SO_4 . 2H_2O$. It appears that the sharing of vertices between octahedral $M(H_2O)_6$ groups is extremely rare, possibly because the minimum value of the angle M—O—M would be approximately $130°$ (see p. 190). For example, the vertex-sharing $[M(H_2O_5]_n$ chain does not seem to have been found, though there is sharing of one H_2O between a pair of octahedral coordination groups $[Fe(H_2O)_3O_3]$ in $FeSO_3 . 2\frac{1}{2}H_2O;$[1] for the topology of this structure see p. 113.

   Type BII. This is an interesting group of hydrates, in all of which the cations are completely hydrated, in which different $H_2O:M$ ratios arise as the result of sharing different numbers of H_2O molecules of each coordination group, as illustrated by the following examples in which there is octahedral coordination of the cations:

Number of H_2O of each $M(H_2O)_n$ group shared	$H_2O : M$ ratio in hydrate	Examples
2	5	$Na_2CO_3 . 10 H_2O$
4	4	$KF . 4 H_2O$
5	$3\frac{1}{2}$	$Na_2HAsO_4 . 7 H_2O$, $NaOH . 3\frac{1}{2} H_2O$
6	3	$LiClO_4 . 3 H_2O$

$Na_2CO_3 . 10H_2O$. Here the $Na(H_2O)_6$ groups are associated in pairs to form units $Na_2(H_2O)_{10}$ of the same general type as the dimers of certain pentahalides. Contrast the structure of $Na_2SO_4 . 10H_2O$ above.

$KF . 4H_2O$. The $K(H_2O)_6$ coordination groups share two edges to give a $H_2O : M$ ratio of $4 : 1$.

$Na_2HAsO_4 . 7H_2O$. This hydrate contains a unique chain of composition $Na_2(H_2O)_7$ (Fig. 15.12) formed from face-sharing pairs of $Na(H_2O)_6$ octahedra

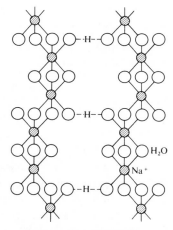

FIG. 15.12. Chains of composition $Na_2(H_2O)_7$ in $Na_2HAsO_4 . 7H_2O$.

which also share two terminal edges. The chains are cross-linked at intervals by $O-H-O$ bonds, and the $AsO_3(OH)^{2-}$ ions are accommodated between the chains. There is the same $H_2O : M$ ratio in $NaOH . 3\frac{1}{2}H_2O$, in which there is complete hydration of the cations as in the arsenate.

$LiClO_4 . 3H_2O$. The value $\frac{1}{2}$ for z/xn in this hydrate results from the formation of infinite chains by the sharing of opposite faces of octahedral $Li(H_2O)_6$ groups (Fig. 15.13). The four nearest neighbours of a water molecule (other than water molecules of the same chain, contacts with which are clearly not contacts between oppositely charged regions of H_2O molecules) are 2 Li^+ of the chain and 2 O atoms

FIG. 15.13. Plan of the structure of $LiClO_4 \cdot 3H_2O$. The heavy full and the broken lines indicate O–H–O bonds between H_2O molecules and oxygen atoms (of ClO_4^- ions) which are approximately coplanar. These bonds thus link together the columns of $Li(H_2O)_6$ octahedra, which share opposite faces, and the ClO_4^- ions. The small black circles represent Li^+ ions in planes midway between the successive groups of 3 H_2O.

of different ClO_4^- ions. These Li–H_2O–Li and O–H_2O–O bonds lie in perpendicular planes, so that the four neighbours of H_2O are arranged tetrahedrally. The structure of $LiClO_4 \cdot 3H_2O$ is closely related to that of a large group of isostructural hexa-hydrates $M(BF_4)_2 \cdot 6H_2O$ and $M(ClO_4)_2 \cdot 6H_2O$ in which M is Mg, Mn, Fe, Co, Ni, or Zn. In $LiClO_4 \cdot 3H_2O$ there is a Li^+ ion at the centre of each octahedral group of water molecules in the infinite chain of Fig. 15.13. If we remove these ions and place a Mg^{2+} ion in each alternate octahedron the crystal has the composition $Mg(ClO_4)_2 \cdot 6H_2O$, and there are discrete $Mg(H_2O)_6^{2+}$ groups (Type AII). Instead of 2 Li^+ neighbours, each H_2O has now only 1 Mg^{2+} neighbour; compare

$$Mg^{2+}-O\overset{\displaystyle .O}{\underset{\displaystyle H..}{\overset{H'}{<}}}_{O} \qquad \text{with} \qquad \overset{\displaystyle Li^+}{\underset{\displaystyle Li^+}{>}}O\overset{\displaystyle ..O}{\underset{\displaystyle H..}{\overset{102° \quad H' \quad 130°}{<}}}_{O}$$

$SrCl_2 \cdot 6H_2O$. Here there are columns of tricapped trigonal prisms, each sharing a pair of opposite faces (Fig. 15.14). This structure is adopted by the hexahydrates of all the following halides:

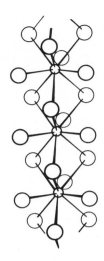

FIG. 15.14 Portion of the infinite 1-dimensional cation–water complex in crystalline $SrCl_2 \cdot 6H_2O$, in which each Sr^{2+} ion (small circle) is surrounded by nine water molecules.

$CaCl_2$	$SrCl_2$	
$CaBr_2$	$SrBr_2$	
CaI_2	SrI_2	BaI_2

Type BIII. Although there is sufficient water to hydrate M completely, not only is there incomplete hydration of the cations but there are H_2O molecules not attached to cations.

$CuSO_4 \cdot 5H_2O$. In this hydrate the metal ion is 6-coordinated, but although there are only 5 H_2O molecules for every Cu^{2+} ion the fifth is not attached to a cation. Instead, the coordination group around the cupric ion is composed of 4 H_2O and 2 O atoms of sulphate ions, and the fifth H_2O is held between water molecules attached to cations and O atoms of sulphate ions, as shown in Fig. 15.15.

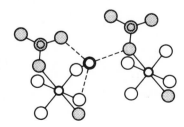

FIG. 15.15.   The environment of the fifth H_2O molecule (centre) in $CuSO_4 \cdot 5H_2O$. The octahedral coordination group around a copper atom is composed of four H_2O molecules and two oxygen atoms of SO_4^{2-} ions (shaded circles).

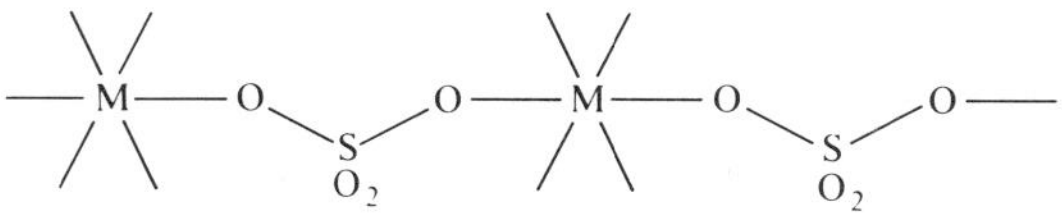
M
O
S
O_2
O
M
O
S
O_2
O

This structure may also be represented as chains of SO_4^{2-} ions and octahedral $M(H_2O)_4O_2$ coordination groups with the fifth H_2O molecule between the chains. Isostructural hydrates include those of $MgCrO_4$ and $MgSO_4$,[2] the latter being one of 8 hydrates (with 1, 5/4, 2, 3, 4, 5, 6, and 7 H_2O) of which only the 1-, 6-, and 7-hydrates are stable in contact with their solutions. $MgSO_4 . 5H_2O$ is of interest as one of the few cases of isostructural Mg and Cu salts; another example is the pair of salts $M(NH_4)_2(SO_4)_2 . 6H_2O$.

$SnCl_2 . 2H_2O$. This hydrate is of more interest in connection with the structural chemistry of divalent tin. Discrete pyramidal complexes $SnCl_2(H_2O)$ in which the

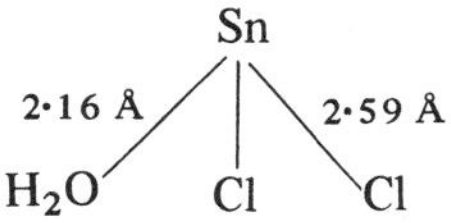
Sn
2·16 Å
2·59 Å
H_2O
Cl
Cl

mean interbond angle is 83°, are arranged in double layers which alternate with layers of water molecules.

$FeF_3 . 3H_2O$. Instead of other simpler structural possibilities, such as finite $FeF_3(H_2O)_3$ groups, one form of this hydrate consists of infinite chains (Fig. 15.16(f)) of composition $FeF_3(H_2O)_2$ between which the remaining water molecule is situated. In the chain there is random distribution of 2 F and 2 H_2O among

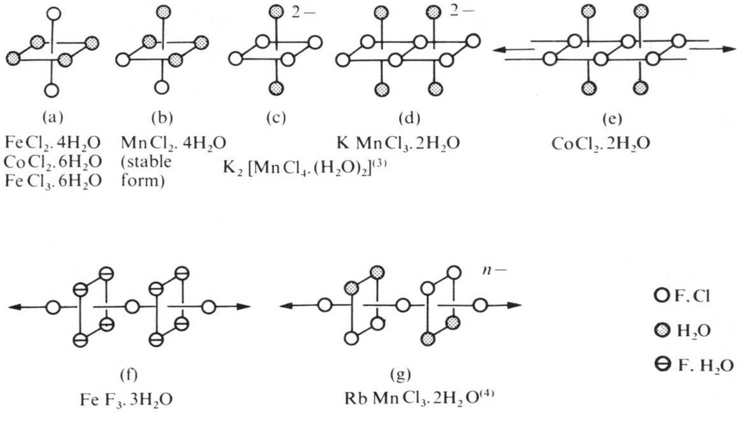
2−
2−
(a)
$FeCl_2.4H_2O$
$CoCl_2.6H_2O$
$FeCl_3.6H_2O$
(b)
$MnCl_2.4H_2O$
(stable form)
(c)
(d)
$K MnCl_3.2H_2O$
$K_2[MnCl_4.(H_2O)_2]$[3]
(e)
$CoCl_2.2H_2O$
n−
(f)
$FeF_3.3H_2O$
(g)
$RbMnCl_3.2H_2O$[4]
F, Cl
H_2O
F, H_2O

FIG. 15.16. Coordination groups of 3d ions in some hydrated halides and complex halides (see text).

the four equatorial positions in each octahedron. In the variant of this chain in RbMnCl$_3$.2H$_2$O there is, however, a regular arrangement of 2 Cl and 2 H$_2$O in these positions (Fig. 15.16(g)).

Type **BIV**.

CaCO$_3$.6H$_2$O. In addition to the well-known anhydrous forms CaCO$_3$ also crystallizes with 1 and 6 H$_2$O. In the remarkable structure of the hexahydrate there are isolated ion-pairs surrounded by an envelope of 18 water molecules. Of these, six complete the 8-coordination group around Ca^{2+} and the remainder are bonded to other cations. (Each water molecule is adjacent to (only) one Ca^{2+} ion.) This type of hydrate, in which ion pairs (resembling the classical picture of a CaCO$_3$

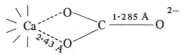

'molecule') are embedded in a mass of H$_2$O molecules, may be contrasted with MgCl$_2$.12H$_2$O in which separate Mg^{2+} and Cl$^-$ ions are completely surrounded by water molecules (some of which are shared between the coordination groups) and with the clathrate hydrates described earlier.

GdCl$_3$.6H$_2$O. Here also there is 8-coordination of the cations (by 6 H$_2$O and 2 Cl$^-$)−contrast [Al(H$_2$O)$_6$]Cl$_3$ and [Fe(H$_2$O)$_4$Cl$_2$]Cl.2H$_2$O. This type of structure is adopted by the hexahydrates of the trichlorides and tribromides of the smaller 4f metals and by AmCl$_3$.6H$_2$O and BkCl$_3$.6H$_2$O.[5] The larger M^{3+} ions of La, Ce, and Pr form MCl$_3$.7H$_2$O, the structure of which is not yet known.

FeF$_2$.4H$_2$O and FeCl$_2$.4H$_2$O. In both of these hydrates there are discrete octahedral molecules FeX$_2$(H$_2$O)$_4$. In the fluoride it was not possible to distinguish between F and H$_2$O; the chloride has the *trans* configuration. Although the water content of these hydrates is sufficient to form an infinite chain of octahedral Fe(H$_2$O)$_6$ groups sharing opposite edges, finite groups with 2 Cl attached to the cation are preferred.

MnCl$_2$.4H$_2$O. Two polymorphs of this hydrate have been known for a long time (both monoclinic), one of which is described as metastable at room temperature. This form has the same structure as FeCl$_2$.4H$_2$O, i.e. it consists of *trans* MnCl$_2$(H$_2$O)$_4$ molecules (Fig. 15.16(a)). Somewhat unexpectedly the stable form is also built of molecules with the same composition but with the *cis* configuration (Fig. 15.16(b)).

CuSO$_4$.3H$_2$O. Cupric sulphate forms hydrates with 1, 3, and 5 H$_2$O. In contrast to the pentahydrate, in which only 4 H$_2$O are associated with the cations, all the water is coordinated to Cu^{2+} in the trihydrate. The coordination group around the metal ion consists of 3 H$_2$O + 1 O (mean Cu−O, 1.94 Å) with two more distant O atoms of SO$_4^{2-}$ ions (at 2.42 Å) completing a distorted octahedral group.

(1)   AC 1980 **B36** 1184
(2)   AC 1972 **B28** 1448
(3)   ACSc 1968 **22** 647
(4)   ACSc 1967 **21** 889; ACSc 1968 **22** 641
(5)   IC 1971 **10** 147

Hydrates of Class C: $z/xn < \frac{1}{2}$

Type CIII.

$CdSO_4 . \frac{8}{3}H_2O$. Our only example of this type of hydrate has a rather complex structure owing to the unusual ratio of water to salt. There are two kinds of cadmium ion with slightly different environments, but both are octahedrally surrounded by two water molecules and four oxygen atoms of sulphate ions. There are four kinds of crystallographically non-equivalent water molecules and, of these, three-quarters are attached to a cation. The remaining water molecules have no contact with a metal ion, but have four neighbours, two other water molecules and two oxygen atoms of oxy-ions. It appears that the important point is the provision of three or four neighbours, which can be

$$Cd^{2+} \quad \begin{matrix} \bar{}H_2O_+^+ \\ \bar{}H_2O_+^+ \\ \bar{}H_2O_+^+ \end{matrix} \quad \text{or } Cd^{2+} \quad \begin{matrix} \bar{}H_2O_+^+ \\ \bar{}H_2O_+^+ \end{matrix} \quad \text{or} \quad \begin{matrix} O^- & H_2O_+ & O^- \\ & \bar{}H_2O_+^+ & \\ H_2O^+ & & O^- \end{matrix}$$

It is as well to remember that in the present state of our knowledge we are far from understanding why a hydrate with a formula so extraordinary as $CdSO_4 . \frac{8}{3}H_2O$ should form at all; its structure and therefore its chemical formula represent a compromise between the requirements of Cd^{2+}, SO_4^{2-}, and H_2O.

Type CIV. Our examples here are of mono- and dihydrates. We describe first the structures of some dihydrated halides.

$KF.2H_2O$. In this hydrate each $K^+(F^-)$ ion is surrounded by 4 H_2O and 2 $F^-(K^+)$ at the vertices of a slightly distorted octahedron. Each H_2O has 2 K^+ and 2 F^- ions as nearest neighbours arranged tetrahedrally (Fig. 15.17).

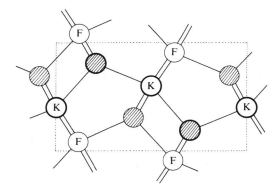

FIG. 15.17. Projection of the crystal structure of $KF.2H_2O$. Heavy circles represent atoms lying in the plane of the paper and light circles atoms lying in planes 2 Å above and below that of the paper. Shaded circles represent water molecules.

NaBr.$2H_2O$. There are similar coordination groups (i.e. *cis* $NaX_2(H_2O)_4$) in this crystal and also in the isostructural NaCN.$2H_2O$ and NaCl.$2H_2O$.[1] The (layer) structure is related to that of $MnCl_2$.$4H_2O$ in the following way. If an octahedral MX_3 layer of the $AlCl_3$ ($Al(OH)_3$) type is built of *cis* $MX_2(H_2O)_4$ groups its composition is $MX.2H_2O$, and this is the layer in NaBr.$2H_2O$, shown diagrammatically in Fig. 15.18. Removal of one-half of the cations (those shown as small open circles in Fig. 15.18) leaves a layer of composition MX_2.$4H_2O$, consisting of isolated $MX_2(H_2O)_4$ molecules. This represents the structure of the stable form of $MnCl_2$.$4H_2O$ to which we have already referred. The relation between these two hydrates should be compared with that between the structures of $LiClO_4$.$3H_2O$ and $Mg(ClO_4)_2$.$6H_2O$ noted above.

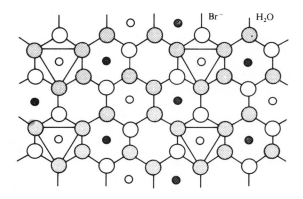

FIG. 15.18. Layer in NaBr.$2H_2O$.

CoCl$_2$.$2H_2O$. The dihydrates of a number of 3d dihalides are built of the edge-sharing octahedral MX_4 chains of Fig. 15.16(e); they include the dichlorides and dibromides of Mn, Fe, Co, and Ni. There is an interesting distortion of the chain in NiCl$_2$.$2H_2O$, which is not isostructural with the Co compound (Fig. 15.19). The planes of successive equatorial $NiCl_4$ groups are inclined to one another at a

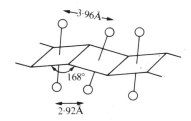

FIG. 15.19. Distorted octahedral chain in NiCl$_2$.$2H_2O$.

small angle and the distances between successive H_2O molecules along each side of the chain are alternately 2.92 Å and 3.96 Å.

A quite different type of distortion of the octahedron occurs in $CuCl_2 . 2H_2O$ where the nearest neighbours of Cu are: 2 H_2O at 1.93 Å, 2 Cl and 2.28 Å, and 2 Cl at 2.91 Å. In view of the very large difference between the Cu–Cl bond lengths the structure is preferably described as built of planar *trans* $CuCl_2(H_2O)_2$ molecules. A n.d. study showed that the whole molecule is planar, including the H atoms, and that the latter lie close to the lines joining O atoms of H_2O molecules to their two nearest Cl neighbours (Fig. 15.20.)

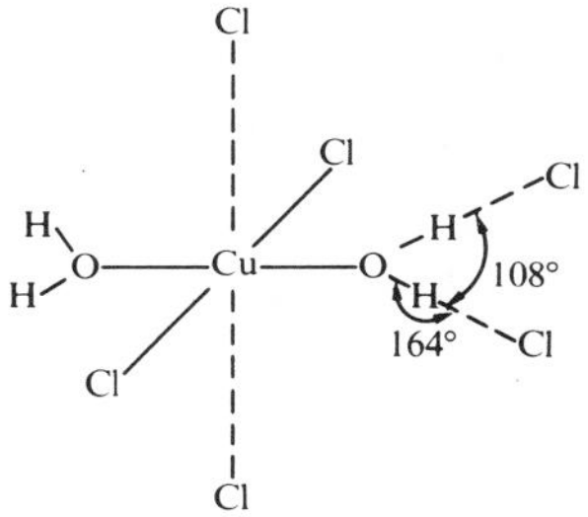

FIG. 15.20. The structure of $CuCl_2(H_2O)_2$.

$SrCl_2 . 2H_2O$ and $BaCl_2 . 2H_2O$. These hydrates have closely related layer structures in both of which the coördination group around the metal ion is composed of 4 Cl^- ions and 4 H_2O molecules. The layers are illustrated in Fig. 15.21.

$CaCl_2 . 2H_2O$. This hydrate also has a layer structure, but the smaller Ca^{2+} ion is octahedrally coordinated, by 4 Cl and 2 H_2O, in a puckered form of the vertex-sharing SnF_4 (K_2NiF_4) layer. The Cl ions are situated at the shared (equatorial) vertices of the octahedral coordination groups (see also p. 209).

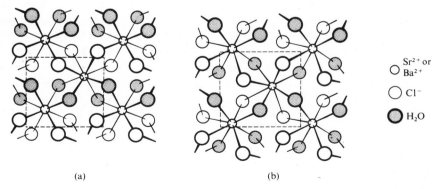

(a)   (b)

◯ Sr^{2+} or Ba^{2+}

◯ Cl^-

◉ H_2O

FIG. 15.21. Plans of layers in the crystal structures of (a) $SrCl_2 . 2H_2O$; and (b) $BaCl_2 . 2H_2O$. Atoms above or below the plane of the metal ions are shown as heavy or light circles respectively.

As an example of a dihydrate of an oxy-salt we describe the structure of $CaSO_4.2H_2O$.

$CaSO_4.2H_2O$. This hydrate, the mineral gypsum, has a rather complex layer structure (Fig. 15.22) in which the layers are bound together by hydrogen bonds of

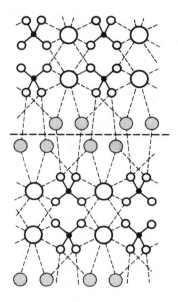

FIG. 15.22. Section through the crystal structure of gypsum perpendicular to the layers (diagrammatic). The layers are composed of SO_4^{2-} and Ca^{2+} ions (large open circles) with H_2O molecules on their outer surfaces (shaded). The heavy broken line indicates the cleavage, which breaks only O—H—O bonds.

lengths 2.816 and 2.896 Å between water molecules and O atoms of sulphate ions. Each H_2O molecule is bonded to a Ca^{2+} ion and to a sulphate O atom of one layer and to an O atom (of SO_4^{2-}) of an adjacent layer, so that its environment is similar to that of a water molecule in $BeSO_4.4H_2O$ or $Zn(BrO_3)_2.6H_2O$. These O—H—O bonds are the weakest in the structure, accounting for the excellent cleavage and the marked anisotropy of thermal expansion, which is far greater normal to the layers than in any direction in the plane of the layers. The location of the H atoms on the lines joining water O atoms to sulphate O atoms has been confirmed by n.m.r. and n.d. studies.

The structures of $CaHPO_4.2H_2O$ (brushite) and $CaHAsO_4.2H_2O$ (pharmacolite) are very similar to that of gypsum but with additional hydrogen bonding involving the OH groups of the anions.[2]

In monohydrates H_2O molecules form a decreasingly important part of the coordination group of the cations as the coordination number of cation increases:

	C.N. of M	*Coordination group*
LiOH . H$_2$O	4	2 H$_2$O 2 OH
Li$_2$SO$_4$. H$_2$O	4	$\begin{cases} 1\ H_2O\ 3\ O \\ \qquad 4\ O \end{cases}$
Na$_2$CO$_3$. H$_2$O	6	$\begin{cases} 2\ H_2O\ 4\ O \\ 1\ H_2O\ 5\ O \end{cases}$
SrBr$_2$. H$_2$O	9	2 H$_2$O 7 Br

(In two of these monohydrates there are two sets of cations with different environments.)

LiOH.H$_2$O. Each structural unit, Li$^+$, OH$^-$, and H$_2$O, has four nearest neighbours. The Li$^+$ ions are surrounded by 2 OH$^-$ and 2 H$_2$O and these tetrahedral groups share an edge and two vertices to form double chains (Fig. 15.23) which are

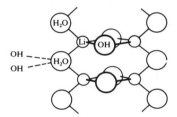

FIG. 15.23. Portion of one double chain in LiOH.H$_2$O (diagrammatic).

held together laterally by hydrogen bonds between OH$^-$ ions and H$_2$O molecules. Each water molecule also has four tetrahedral neighbours, 2 Li$^+$ and 2 OH$^-$ ions of different chains, a similar type of environment to that in LiClO$_4$.3H$_2$O.

Li$_2$SO$_4$.H$_2$O. This structure provides a good example of the tetrahedral arrangement of four nearest neighbours (2 H$_2$O, Li$^+$, and O of SO$_4^{2-}$) around a water molecule and of water molecules hydrogen-bonded into infinite chains. The orientation of the H–H vector in the unit cell agrees closely with the value derived from p.m.r. measurements. The O–H–O bonds from water molecules are of two kinds, those to sulphate O atoms (2.87 Å) and to H$_2$O molecules (2.94 Å).

Na$_2$CO$_3$.H$_2$O. The larger Na$^+$ ion is usually surrounded octahedrally by six neighbours in oxy-salts. In this hydrate the cations are of two types, one-half being surrounded by one water molecule and five O atoms and the remainder by two water molecules and four O atoms of carbonate ions. Here also the H$_2$O molecule has four tetrahedral neighbours, 2 Na$^+$ ions and 2 O atoms of CO$_3^{2-}$ ions.

SrBr$_2$.H$_2$O. It is interesting to note that Sr^{2+} is not 8-coordinated in this hydrate as in SrBr$_2$ but has an environment more resembling that in BaBr$_2$, two of the nine nearest neighbours being H$_2$O molecules.

Sr(OH)$_2$.H$_2$O.   The structure of this hydrate (and the isostructural

$Eu(OH_2 . H_2O)$ is of interest as an example of bicapped trigonal prismatic coordination (Sr–6 OH and 2 H_2O).

The classification adopted here is not, without undue elaboration, applicable to hydrates in which there are different degrees of hydration of some of the cations. For example, in $ZnCl_2 . 1\frac{1}{3}H_2O^{(3)}$ two-thirds of the Zn atoms are tetrahedrally coordinated by Cl in infinite chains of composition $(ZnCl_3)_n^{n-}$, and the remaining Zn atoms lie between these chains surrounded (octahedrally) by 4 H_2O and 2 Cl (of the chains), as shown in Fig. 15.24. Bond lengths found are: Zn–4 Cl, 2.28 Å; Zn–2 Cl, 2.60 Å and 4 H_2O, 2·03 Å.

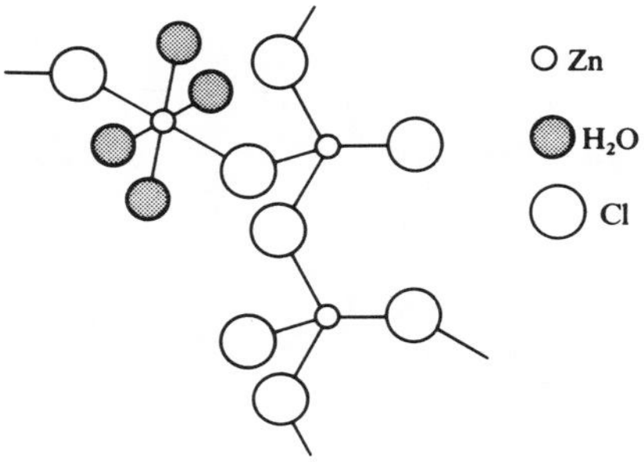

O	Zn
◉	H_2O
○	Cl

FIG. 15.24. Environments of the two kinds of Zn^{2+} ion in $ZnCl_2 . 1\frac{1}{3}H_2O$.

(1)   AC 1974 **B30** 2363   (3)   AC 1970 **B26** 1679
(2)   AC 1969 **B25** 1544

Hydrates of 3d *halides.*   The structures described above suggest the following generalizations.

(i) $M(H_2O)_n$ groups share edges or faces but not vertices; this is true generally in hydrates.

(ii) There is complete hydration of M in $CoI_2 . 6H_2O$ and in one form of $CrCl_3 . 6H_2O$, but in a number of cases where there is sufficient water for complete hydration of M this does not occur. Usually H_2O is displaced by X from the coordination group around M; sometimes both H_2O and X are partly coordinated to M and partly outside the cation coordination group, as in $[CrCl_2(H_2O)_4]Cl . 2H_2O$. In a 4f trihalide, Cl is displaced (from a larger coordination group) in preference to H_2O, namely, in $[GdCl_2(H_2O)_6]Cl$.

(iii) If sharing of X and/or H_2O is necessary X is shared in preference to H_2O, as in numerous $MX_2 . 2H_2O$.

(iv) Many of these compounds are polymorphic, and there is probably not very much difference between the stabilities of structures containing different types of cation coordination group, witness the three forms of $CrCl_3.6H_2O$ and the two forms of $MnCl_2 . 4H_2O$.

Hydrates of complex halides. These have been mentioned only incidentally. For example, it was noted on p. 454 that hydrated complex halides of the same formula type may have quite different structures, and that in $A_m(BX_n).pH_2O$ there is not necessarily any H_2O attached to B; compare the coordination of Mn by 6 F in $K_2MnF_5.H_2O$ with that of V by 5 F + H_2O in $(NH_4)_2(VF_5.H_2O)$. The anions in the dihydrates of $KMnCl_3$ and $RbMnCl_3$ and in $K_2MnCl_4.2H_2O$ were illustrated in Fig. 15.16, p. 679. The distribution of H_2O molecules between the two cations in some more highly hydrated complex halides proves to be of interest. The CsCl-like structure of $NiSnCl_6.6H_2O$ built of $(SnCl_6)^{2-}$ and $[Ni(H_2O)_6]^{2+}$ groups was mentioned on p. 247. The same cation is found in $Cd_2NiCl_6.12H_2O$, in which the anion is the octahedral AX_4 chain (a); the remaining H_2O molecules are not included

(a)   (b) H_2O

in the complexes. In $Cd_4NiCl_{10}.10H_2O$ there are multiple octahedral chains, (b), in which equal numbers of octahedra share 3 or 5 edges (compare the AX_3 chain of NH_4CdCl_3, in which all octahedra share 4 edges). These chains have the composition $Cd_2Cl_5(H_2O)$. In $Mg_2CaCl_6.12H_2O$ all the Cl is associated with Ca and the H_2O with Mg. The hydrate $Ca_2CdCl_6.12H_2O$ is related in an interesting way to the isostructural β-$CaCl_2.4H_2O$ (see below). In the latter one-third of the Ca^{2+} ions are octahedrally surrounded by 6 Cl^- and the remainder by 1 Cl^- + 7 H_2O. In the complex halide the 6-coordinated Ca^{2+} ions are replaced by Cd^{2+}. The structural formulae of these hydrates are accordingly:

$Cd_2NiCl_6.12H_2O$:	$[CdCl_3(H_2O)]_2[Ni(H_2O)_6].4H_2O$	AC 1980 **B36** 3088
$Cd_4NiCl_{10}.10H_2O$:	$[Cd_2Cl_5(H_2O)]_2[Ni(H_2O)_6].2H_2O$	AC 1980 **B36** 3090
$Mg_2CaCl_6.12H_2O$:	$(CaCl_6)[Mg(H_2O)_6]_2$	AC 1980 **B36** 2736
$Ca_2CdCl_6.12H_2O$:	$[CdCl_6 Ca_2 (H_2O)_{12}]_n$	AC 1978 **B34** 3341

Hydrates of $CaCl_2$. This salt forms hydrates with 1, 2, 4, and 6 H_2O, and three forms of the tetrahydrate are known. The dihydrate consists of octahedral AX_4 layers in which $CaCl_4(H_2O)_2$ groups share their four equatorial vertices (Cl atoms). The layer is a modified form of that found in $K_2(NiF_4)$ as described on p. 208. The unstable (γ) form of $CaCl_2.4H_2O$ consists of octahedral molecules $CaCl_2(H_2O)_4$ having the *trans* configuration, but the other two forms have much more interesting structures. The α form consists of dimers, (a), in which Ca is 7-coordinated (by 3 Cl and 4 H_2O), and the β form of unexpectedly complex chains in which there are two

(a)   (b)

kinds of non-equivalent Ca atoms (ions), (b). Pairs of Ca atoms connected through double H_2O bridges are joined into chains via octahedral $CaCl_6$ groups. The hexahydrate is isostructural with $SrCl_2.6H_2O$, and consists of linear aquo-cations built from tricapped trigonal prismatic $Ca(H_2O)_9$ groups sharing their basal faces (Fig. 15.14, p. 678).

These structures may be summarized as follows:

Hydrate	C.n. of Ca	Coordination group		Bridges	Reference
$CaCl_2.2H_2O$	6	4Cl	$2H_2O$	Single Cl	AC 1977 **B33** 1608
$CaCl_2.4H_2O$-γ	6	2Cl	$4H_2O$	None	AC 1980 **B36** 2757
-α	7	3Cl	$4H_2O$	Double Cl	AC 1979 **B35** 585
-β	$\begin{cases} 6 \\ 8 \end{cases}$	6Cl		Single Cl and	AC 1978 **B34** 900
		1Cl	$7H_2O$	Double H_2O	
$CaCl_2.6H_2O$	9	$9H_2O$		Triple H_2O	AC 1977 **B33** 2938

Hydrates of Ca *oxy-salts.* We have already noted the structures of several hydrates of calcium oxy-salts. There is complete hydration, by 8 H_2O, of the cations in $CaO_2.8H_2O$ and in $CaK(AsO_4).8H_2O^{(1)}$ (where the coordination groups around the cations necessarily have H_2O molecules in common), and 8-coordination in $CaCO_3.6H_2O$ (6 H_2O, 2 O) and $CaSO_4.2H_2O$ (2 H_2O, 6 O). Higher c.n.s are readily attained in nitrates because NO_3^- can act as a bidentate ligand with a distance of only about 2.15 Å between the O atoms, a point well illustrated by the structures of three hydrates of calcium nitrate. The dihydrate has a layer structure in which Ca^{2+} is 10-coordinated (2 H_2O and 8 O at the vertices of a bicapped Archimedean antiprism. The NO_3^- ions bridge the Ca^{2+} ions as shown at (a), which shows the elevation of a portion of the layer.[2] The trihydrate consists of tetramers, (b), in which Ca^{2+} is 9-coordinated and one half of the NO_3^- ions form bridges of the same kind as in the dihydrate.[3] The tetrahydrate consists of dimers, (c), also with 9-coordination of Ca^{2+} (tctp).[4] In all three structures there is only hydrogen-bonding between the sub-units, layers, tetramers, or dimers.

(1)   AC 1972 **B28** 3056   (3)   AC 1976 **B32** 235
(2)   AC 1974 **B30** 605   (4)   CR 1970 **271C** 1555

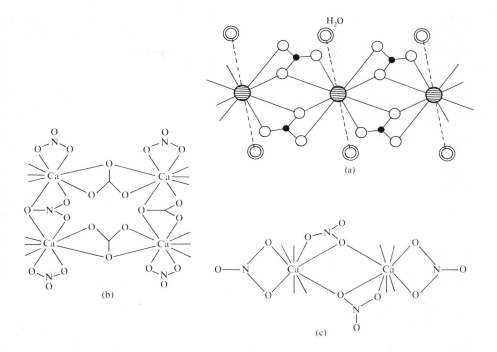

Hydrates of vanadyl sulphate. Vanadyl sulphate forms hydrates with 1 H_2O, 3 H_2O, 5 H_2O (three polymorphs), and 6 H_2O; the hydrates of $MgSO_4$ include, among others, those with 4, 6, and 7 H_2O. There is an interesting relation between the structures of pairs of these hydrates, for in both the Mg and VO salts there is octahedral coordination of the metal, but in the vanadyl salts one coordination position is occupied by the vanadyl O atom. The structure of a hydrate $MgSO_4 . mH_2O$ might therefore be similar to that of a hydrate $VOSO_4 .(m-1)H_2O$, as in fact is found in several cases. A curious point is that there is no tetrahydrate of $VOSO_4$, which might have been expected to be similar to $MgSO_4 . 5H_2O$, isostructural with $CuSO_4 . 5H_2O$ (p. 678). In the vanadyl compounds there is, of course, considerable distortion of the octahedral coordination, (a).

Crystalline $VOSO_4.3H_2O^{(1)}$ consists of dimeric units, (b), of the same kind as in $MgSO_4.4H_2O.^{(2)}$ The stable (monoclinic) form of $VOSO_4.5H_2O^{(3)}$ consists of units of type (c), the fifth H_2O being an isolated molecule of water of crystallization, $[VO(H_2O)_4SO_4].H_2O$, an arrangement not found in the $MgSO_4$ hydrates. However, one of the two less stable orthorhombic forms consists of units $[VO(H_2O)_5]^{2+}$ and discrete SO_4^{2-} ions;$^{(4)}$ compare the structure of hydrates $[M(H_2O)_6]SO_4$ formed by Fe, Ni, and Mg. In $VOSO_4.6H_2O^{(5)}$ also there are octahedral groups $[VO(H_2O)_5]^{2+}$, so that the structural formula is $[VO(H_2O)_5]SO_4.H_2O$ — compare $[Mg(H_2O)_6]SO_4.H_2O$ for $MgSO_4.7H_2O$.

(1)   AC 1980 **B36** 2873   (4)   AC 1980 **B36** 1757
(2)   AC 1962 **15** 815   (5)   AC 1980 **B36** 249
(3)   JACS 1968 **90** 3305

Hydrated acids and acid salts

Special interest attaches to the hydrates of acids and acid salts because of the possibility that some or all of the protons may be associated with the H_2O molecules to form ions H_3O^+ or more complex species. The p.m.r. spectra show groups of 3 H atoms at the corners of an equilateral triangle in the monohydrates of a number of acids (HNO_3, $HClO_4$, H_2SO_4, H_2PtCl_6). The existence of short O–H–O bonds (2.4–2.6 Å) between certain pairs of O atoms which are hydrogen-bonded to others by longer bonds (2.7–2.8 Å) is taken to indicate the formation of $H_5O_2^+$, $H_7O_3^+$, or more complex hydronium ions, though the interpretation may not always be clear-cut. For example, in $HAuCl_4.4H_2O$ there are O–H–O bonds of length 2.57 Å between certain pairs of O atoms and bonds of length 2.74 Å to surrounding H_2O molecules, but n.d. shows that the H atom is disordered in two positions along the shorter bonds 0.62 Å apart, suggesting $H_3O^+(H_2O)$ as an alternative description.

The association of one proton with one H_2O molecule to form H_3O^+ might be expected in hydrates with the $H^+:H_2O$ ratio equal to unity, as in $HX.H_2O$ or $H_2SO_4.2H_2O$. Similarly the formation of $H_5O_2^+$ might be expected if the $H^+:H_2O$ ratio is 1:2, as in $HX.2H_2O$ or $H_2SO_4.4H_2O$, and so on for other $H^+:H_2O$ ratios. Some hydrates may accordingly be described as containing the 'expected' hydronium ion, and in Table 15.6 we distinguish these from other hydrates with less simple structures; the latter are marked with an asterisk. Just as some H_2O molecules are not associated with the cations in certain salt hydrates, so there are hydrates of acids in which some of the H_2O molecules are not involved in the protonated complexes. The converse situation arises if there are more protons available than are sufficient to form hydronium complexes, as may be the case if the degree of hydration is very low, when not all the acid molecules will be ionized. These three cases may be represented

$$[A^-][H_mO_n^+]$$
$$[A^-][H_mO_n^+]\ [H_2O]$$
$$[AH]\quad [A^-][H_mO_n^+]$$

TABLE 15.6

Proton-water complexes in hydrates

Complex	$H^+:H_2O$ ratio	Hydrate	Reference
H_3O^+	2:1	$F_3CSO_3H.\frac{1}{2}H_2O$	AC 1975 **B31** 2208
	1:1	Monohydrates of HCl, HBr, HNO_3, $HClO_4$, F_3CSO_3H $H_2SO_4.2H_2O$	AC 1973 **B29** 1923
	2:1 {	$H_2SO_4.H_2O$, $H_2SeO_4.H_2O$	AC 1979 **B35** 2384
	2:5	$HClO_4.2\frac{1}{2}H_2O*$	AC 1971 **B27** 898
$H_5O_2^+$	1:2	Dihydrates of HCl, $HClO_4$, F_3CSO_3H $H_2SO_4.4H_2O$ $(ZnCl_2)_2.HCl.2H_2O$	AC 1975 **B31** 2202 AC 1972 **B28** 1692 AC 1978 **B34** 1330
	1:3	$HCl.3H_2O*$ $HAuCl_4.4H_2O*$	
$H_7O_3^+$	1:3	Trihydrates of HNO_3, $HClO_4$	AC 1975 **B31** 1489 AC 1972 **B28** 481
$H_9O_4^+$	1:4	Tetrahydrate of F_3CSO_3H $HBr.4H_2O**$	AC 1978 **B34** 2428
	1:6	$HCl.6H_2O*$	AC 1978 **B34** 2424
$H_{13}O_6^+$	1:6	$[(C_9H_{18})_3NH_2Cl]$ $Cl_2H.6H_2O$	Sc 1975 **190** 151
$H_{14}O_6^{2+}$	1:3	$HSbCl_6.3H_2O$	AC 1980 **B36** 2001

*Also H_2O molecules.   **Also H_2O and $H_7O_3^+$

where A^- is an anion of an acid and $H_mO_n^+$ is any hydronium complex. Structures of the third type include those of $F_3C.SO_3H.\frac{1}{2}H_2O$ and $H_2SO_4.H_2O$; the more numerous structures of the second class include:

$HClO_4.2\frac{1}{2}H_2O$	$(H_3O^+)_2$	$(H_2O)_3$	
$HCl.3H_2O$	$(H_5O_2^+)$	(H_2O)	
$HAuCl_4.4H_2O$	$(H_5O_2^+)$	$(H_2O)_2$	
$HBr.4H_2O$	$(H_7O_3^+)$	$(H_9O_4^+)$	(H_2O)
$HCl.6H_2O$	$(H_9O_6^+)$	$(H_2O)_2$	

We now comment briefly on the structures of certain hydrates.

The acids HX form hydrates with the following numbers of molecules of water of crystallization:

HF	$\frac{1}{4}$	$\frac{1}{2}$	1		
HCl			1	2	3
HBr			1	2	3	4
HI				2	3	4

The structures of the hydrates of HF do not appear to be known. The hydrates of HCl melt at progressively lower temperatures (-15.4, -17.4, and -24.9 °C) and the structures of all three are known. The monohydrate consists of corrugated 3-connected layers of the simplest possible type (Fig. 15.25(a)) in which each O or

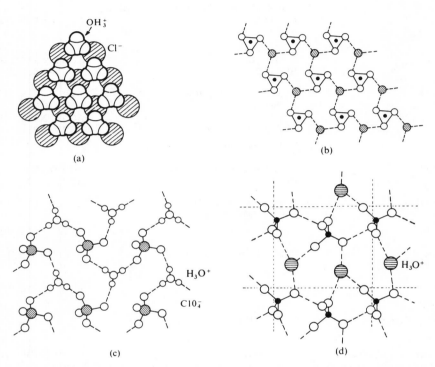

FIG. 15.25. Atomic arrangement in layers of the structures of (a) $HCl.H_2O$; (b) $HNO_3.H_2O$; (c) $HClO_4.H_2O$; (d) $H_2SO_4.H_2O$.

Cl atom has three pyramidal neighbours, so that the units are Cl^- and H_3O^+ ions, (i). The distance O–H---Cl is 2.95 Å and the interbond angle O–H–O is 117°. The angle subtended at O by two Cl^- is 110°, the H atoms lying about 0.1 Å off the O–H---Cl lines. (The shortest distance between O and Cl of adjacent layers is 3.43 Å.)

The dihydrate of HCl contains $H_5O_2^+$ units, (ii), in which the central O–H–O bond is extremely short, and the O–H---Cl bonds have a mean length of 3.07 Å (range 3.04–3.10 Å), appreciably longer than in $(H_3O)^+Cl^-$. The same units are found in the trihydrate, here hydrogen-bonded to 2 H_2O (O–H–O, 2.70 Å) and to 2 Cl^- (O–H–Cl, 3.03 Å) instead of to 4 Cl^- as in $(H_3O)^+Cl^-$. The third water molecule has a normal tetrahedral environment, being surrounded by 2 Cl^- (mean

O–H–Cl, 3.10 Å) and 2 H$_2$O (mean O–H–O, 2.70 Å). These three hydrates are therefore

$$(H_3O)^+Cl^-, \quad (H_5O_2)^+Cl^-, \quad and \quad (H_5O_2)^+Cl^-.H_2O$$

(i)   (ii)

(iii)   (iv)

The structure of HBr.4H$_2$O is even more complex, and corresponds to the structural formula $(H_9O_4)^+(H_7O_3)^+(Br^-)_2.H_2O$. The environment of the odd water molecule is tetrahedral (Br$^-$ and 3 H$_2$O), the shortest distance to neighbouring water molecules being 2.75 Å. The structures of the O$_3$ and O$_4$ units are shown at (iii) and (iv), the latter being pyramidal. The O–O distances within these units are all short compared with other O–H–O bonds in the structure (2.75 Å), though rather longer than in H$_5$O$_2^+$. On the grounds that the O–Br distance in the unit (iii) is rather shorter than other O–Br distances, the next shortest being 3.28 Å, this unit could alternatively be described as a pyramidal unit $(H_7O_3^+Br^-)$ somewhat similar to $(H_9O_4)^+$.

In $(H_3O)^+(NO_3)^-$ the number of H atoms is that required to form three hydrogen bonds from each ion to three neighbours, and the structure (Fig. 15.25(b)) is a further example of the simple hexagonal net. The trihydrate contains $(H_7O_3)^+$ groups, as expected for the H$^+$:H$_2$O ratio 1:3. In the form of HClO$_4$.H$_2$O stable at temperatures above –30 °C there is some rotational disorder of the H$_3$O$^+$ ions, but the low-temperature form has a structure of exactly the same topological type as that of HNO$_3$.H$_2$O. As in HNO$_3$.H$_2$O there are three hydrogen bonds connecting each ion to its neighbours, so that one O of each ClO$_4^-$ ion is not involved in hydrogen bonding (Fig. 15.25(c)). In the pyramidal H$_3$O$^+$ ion the angle H–O–H is 112° and the mean O–H–O bond length is 2.66 Å.

The system H$_2$SO$_4$–H$_2$O is complex, freezing-point curves indicating the existence of six hydrates, with 1, 2, 3, 4, 6.5, and 8 H$_2$O. In H$_2$SO$_4$.H$_2$O the number of H atoms is sufficient for four hydrogen bonds per unit of formula. The crystals consist of SO$_4$H$^-$ and H$_3$O$^+$ ions arranged in layers (Fig. 15.25(d)) and linked by hydrogen

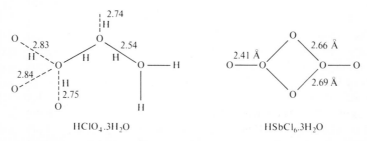

$$H_2SO_4 . H_2O$$

bonds, three from H_3O^+ and five from SO_4H^-. Since there are equal numbers of the two kinds of ion SO_4H^- is necessarily hydrogen-bonded to 3 H_3O^+ and 2 SO_4H^-; there are no hydrogen bonds between the layers. The three hydrogen bonds from H_3O^+ are disposed pyramidally (interbond angles, 101°, 106°, and 126°) and have lengths 2.54, 2.57, and 2.65 Å. Note that it is O_2 and not O_3 that forms two hydrogen bonds. In contrast to the monohydrate, which is $(HSO_4)^-(H_3O)^+$, the dihydrate is $(SO_4)^{2-}(H_3O)_2^+$. Each H_3O^+ is hydrogen-bonded to three different sulphate ions forming a 3D framework in which SO_4^{2-} forms six hydrogen bonds, two from two of the O atoms and one from each of the others (O—H–O, 2.52-2.59 Å; S–O, 1.474 Å). The tetrahydrate has the expected structure, $(H_5O_2)_2^+(SO_4)^{2-}$.

In $HClO_4 . 2H_2O$, with a ratio of one proton to two water molecules, we find $H_5O_2^+$ units as in $HCl . 2H_2O$. The O—H–O bond length in this unit is very similar to that in $HCl . 2H_2O$ (2.424 Å) but the geometry of this ion is somewhat variable. It has a staggered configuration in $HClO_4 . 2H_2O$, nearly eclipsed in $HCl . 3H_2O$, and an intermediate shape in $HCl . 2H_2O$. As already noted, $HClO_4 . 2\frac{1}{2}H_2O$ is $(H_3O)_2^+(H_2O)_3(ClO_4)_2^-$, but in the trihydrate there are sufficient protons to form $(H_7O_3)^+(ClO_4)^-$.

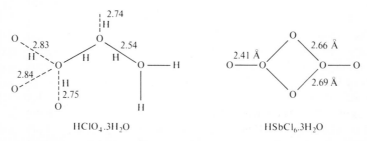

$HClO_4 . 3H_2O$ $HSbCl_6 . 3H_2O$

Trifluoromethanesulphonic acid forms five hydrates, with ½, 1, 2, 4, and 5 H_2O. The hemihydrate is of special interest as the lone example in Table 15.6 of a hydrate with the $H^+ : H_2O$ ratio equal to 2:1. One other structure calls for special note. In $HSbCl_6 . 3H_2O$ there are $H_{14}O_6^{2+}$ groups rather than the simpler $H_7O_3^+$ groups which might have been expected.

Very few structural studies have been made of hydrated acid salts. In $Na_3H(CO_3)_2 . 2H_2O$ the proton is associated with a pair of anions, $(CO_3—H–CO_3)$, and the water is present as normal H_2O molecules. On the other hand, in salts such as $(Coen_2Br_2)Br.HBr.2H_2O$ and $ZnCl_2.\frac{1}{2}HCl.H_2O$ in which there is the same ratio of $H^+ : H_2O$, namely, $1 : 2$, there are $H_5O_2^+$ ions similar to those in certain acid hydrates. As noted in Chapter 3, the addition of one Cl^- to two $ZnCl_2$ permits the formation of a 3D network of composition Zn_2Cl_5 built of $ZnCl_4$ tetrahedra sharing three vertices (compare one form of P_2O_5). This framework forms around, and encloses, the $H_5O_2^+$ ions, so that the structural formula is $(Zn_2Cl_5)^-(H_5O_2)^+$. The structure approximates to a hexagonal closest packing of Cl in which one-sixth of these atoms have been replaced by the rather more bulky $H_5O_2^+$ ions.

The location of H atoms of hydrogen bonds; residual entropy

We noted in Chapter 8 that in many hydrogen bonds the H atom lies to one side of the line joining the hydrogen-bonded atoms, and we have seen that this is also true in ice-I and hydrates such as $CuCl_2.2H_2O$; in the latter the angle O–H---Cl is $164°$. Very similar values are found for O–H---F in $FeSiF_6.6H_2O$[1] (which is isostructural with $Ni(H_2O)_6SnCl_6$ and consists of a CsCl-like packing of $Fe(H_2O)_6^{2+}$ and SiF_6^{2-} ions) and in $CuF_2.2H_2O$.[2] The angle subtended at the O atom of H_2O by the two O atoms to which it is hydrogen-bonded in oxy-salts and acids ranges from the value of $84°$ for $(COOH)_2.2H_2O$ to around $130°$ (possibly larger). In chrome alum, $[K(H_2O)_6][Cr(H_2O)_6](SO_4)_2$[3] this angle is less than $104\frac{1}{2}°$ for both types of H_2O molecule, while in $CuSO_4.5H_2O$[4] all but one of these angles are greater than H–O–H, the largest being $130°$ and the mean value $119°$. Probably the only conclusion to be drawn from these angles if that they show that the H_2O molecule retains its normal shape in hydrates and that hydrogen bonds are formed in directions as close to the O–H bond directions as are compatible with the packing requirements of the other atoms in the crystal.

Like ice-I certain hydrates possess residual entropy owing to the possibility of alternative arrangements of H atoms. The crystal structure of $Na_2SO_4.10H_2O$ has already been described as having 8 H_2O in octahedral chains $[Na(H_2O)_4]_n$ and 2 H_2O not attached to cations. For the H_2O molecules in the chains there is no alternative orientation, but the hydrogen bonds involving the remaining 2 H_2O are arranged in circuits such that there are two ways of arranging the H atoms in any given ring so as to place one on each O–H–O bond. If the choice of arrangement of the H atoms in different rings throughout the crystal is a random one this would account for the observed entropy of $2 \ln 2$ per mole.

It is not feasible to refer here to all the studies of the positions of H atoms in hydrates. A comparison of n.m.r. with n.d. results on twelve hydrates[5] showed that there is generally very close agreement between the two methods. Many references are given in a paper[6] summarizing n.m.r. determinations of proton positions in hydrates.

(1) AC 1962 **15** 353 (3) PRS A 1958 **246** 78 (5) JCP 1966 **45** 4643
(2) JCP 1962 **36** 50 (4) PRS A 1962 **266** 95 (6) AC 1968 **B24** 1131

Ammines and hydrates

Ammines are prepared by the action of ammonia on the salt, using either the gas or liquid and the anhydrous salt, or crystallizing the salt from ammoniacal solution. In preparing ammines of trivalent cobalt from cobaltous salts, oxidation of the cobalt to the trivalent state must accompany the action of ammonia, and this may be accomplished by passing air through the ammoniacal solution. No general statement can be made about the stability of ammines as compared with hydrates, for, as we shall see, the term ammine covers a very large number of compounds which differ greatly in constitution and stability. As extremes we may cite the very unstable ammines of lithium halides and the stable cobalt- and chrom-ammines.

A comparison of the empirical formulae of hydrates and ammines leads to the following conclusions:

(i) *Oxy-salts*. Many sulphates crystallize with odd numbers of molecules of water of crystallization, e.g. the vitriols $MSO_4 7H_2O$ (M = Fe, Co, Ni, Mn, Mg, etc.), $MnSO_4 . 5H_2O$ and $CuSO_4 . 5H_2O$, $MgSO_4 . H_2O$, and many others. Many of the salts which form heptahydrates also form hexahydrates, but the higher hydrate is usually obtained if the aqueous solution is evaporated at room temperature. However, ammines of these salts contain even numbers of NH_3 molecules, usually four or six, for example, $NiSO_4 . 6NH_3$; note particularly the ammine of cupric sulphate, $Cu(NH_3)_4SO_4 . H_2O$.

The nitrates of many divalent metals crystallize with 6 H_2O (Fe, Mn, Mg, etc.), but a few nitrates have odd numbers of water molecules, e.g. $Fe(NO_3)_3 . 9$ (and 6)-H_2O, $Cu(NO_3)_2 . 3H_2O$. Where the corresponding ammine is known it has an even number of NH_3, as in $Cu(NH_3)_4(NO_3)_2$. There are many other series of hexahydrated salts—perchlorates, sulphites, bromates, etc.—and although in such cases the ammines often contain 6 NH_3, the structures of ammine and hydrate are quite different—compare, for example, the structure of $[Co(H_2O)_6](ClO_4)_2$, described on p. 677, with the simple fluorite structure of $[Co(NH_3)_6](ClO_4)_2$.

The reason for the non-existence of ammines containing *large odd* numbers of NH_3 molecules is presumably that some of these molecules would have to be accommodated between the cation coordination groups and the anions (compare the environment of the fifth H_2O in $CuSO_4 . 5H_2O$ or the seventh H_2O in $NiSO_4 . 7H_2O$). The NH_3 molecule does not have the same hydrogen-bond-forming capability as H_2O, particularly to other NH_3 molecules.

(ii) *Halides*. In contrast to the oxy-salts, halides rarely crystallize with an odd number of water molecules. When they do, the number is usually not greater than the coordination number of the metal (see Table 15.5 and note, for example, $FeBr_2 . 9H_2O$). In a hydrated oxy-salt containing water in excess of that required to complete the coordination groups around the metal ions additional water molecules can be held between the $M(H_2O)_n$ groups and the oxy-ions by means of O—H—O bonds. It seems, however, that additional water molecules are very weakly bonded between $M(H_2O)_n$ groups and Cl^- ions, The electrostatic bonding is sufficiently strong in the system (a) in a crystal such as $[Mg(H_2O)_6]Cl_2$ but much weaker in (b), where H_2O is bonded between a hydrated metal ion and Cl^-. For example, the hydrates $MgCl_2.8H_2O$ and $MgCl_2.12H_2O$ are stable only at low temperatures

$$\overset{\diagdown}{\underset{\diagup}{-}}M-OH_2-Cl^-$$

(a)

$$\overset{\diagdown}{\underset{\diagup}{-}}M-OH_2-H_2O-Cl^-$$

(b)

(p. 673). The ammines of those halides which contain *odd* numbers of H_2O in their hydrates contain *even* numbers of NH_3 molecules, e.g. $ZnCl_2.H_2O$ but $ZnCl_2.2NH_3$. Although the ammines and hydrates of halides usually contain the same numbers of molecules of NH_3 and H_2O respectively,

$NiCl_2.6H_2O$,	$NiCl_2.6NH_3$	contrast	$NiSO_4$	$\begin{cases} 7\ H_2O\ \text{but}\ 6\ NH_3 \\ 6\ H_2O \end{cases}$
$CuBr_2.4H_2O$,	$CuBr_2.4NH_3$		$CuSO_4.5H_2O$	$4NH_3.H_2O$
$CuCl_2.2H_2O$,	$CuCl_2.2, 4,\ \text{or}\ 6NH_3,$		$Cu(NO_3)_2.3H_2O$	$4NH_3$
	($CuCl_2.4NH_3.2H_2O$ from solution)			

the crystal structures of the two sets of compounds are different, as has already been pointed out for the pairs $Al(NH_3)_6Cl_3$ and $Al(H_2O)_6Cl_3$, and $Mg(NH_3)_6Cl_2$ and $Mg(H_2O)_6Cl_2$.

From this brief survey we see that the ammines $MX_x.nNH_3$ with n greater than the coordination number of M are likely to be rare and that n is commonly 4 or 6, depending on the nature of the metal M. In these ammines the NH_3 molecules form a group around M, and these groups $M(NH_3)_n$ often pack with the anions into a simple crystal structure. The linking of the NH_3 to M in the ammines of the more electropositive elements such as Al and Mg is of the ion-dipole type. The stability of these compounds is of quite a different order from that of the very stable ammines of trivalent cobalt and chromium (see Chapter 27). Magnetic measurements show that the (much stronger) $M-NH_3$ bonds in the ammine complexes of Co^{III} (and also of Ir^{III}, Pd^{IV}, Pt^{IV}) would be described as 'covalent' or, in ligand-field language, 'strong-field' bonds. In the earlier Periodic Groups the more covalent ammines of the B subgroup metals are more stable than, and quite different in structure from, the ammines of the A subgroup metals. We may conclude, therefore, that the formal similarity between hydrates and ammines is limited to those ammines in which the $M-NH_3$ bonds are essentially electrostatic, and that even in these cases there are significant differences between the crystal structures of the two compounds $MX_x.nNH_3$ and $MX_x.nH_2O$. This latter point has already been dealt with; the former is best illustrated by a comparison of the ammines and hydrates of the salts of the metals of Groups I and II of the Periodic Table.

Of the alkali-metal chlorides only those of Li and Na form ammines, $MX.yNH_3$

($\nu_{max} = 6$), and these are very unstable. Similarly LiCl forms a pentahydrate, NaCl a dihydrate (stable only below $0\,^{\circ}$C), while KCl, RbCl, and CsCl crystallize without water of crystallization. The formation of ammines thus runs parallel with the formation of hydrates. As the size of the ion increases, from Li to Cs, the charge remaining constant, the ability to polarize H_2O or NH_3 molecules and so attach them by polar bonds decreases. In the B subgroup we find an altogether different relation between hydrates and ammines. Ammino-compounds are commonly formed by salts which form no hydrates and they resemble in many ways the cyanido- and other covalent complexes. The only halide of silver which forms a hydrate is the fluoride, but AgCl and the other halides form ammines. Again, the nitrate and sulphate are anhydrous but form ammines, $AgNO_3.2NH_3$ and $Ag_2SO_4.4NH_3$ respectively. In the latter, the structure of which has been determined, and presumably also in the former, there are linear ions $(NH_3-Ag-NH_3)^+$ exactly analogous to the ion $(NC-Ag-CN)^-$ in $KAg(CN)_2$. Gold also forms ammines, e.g. $[Au(NH_3)_4](NO_3)_3$, and several of the ammines of copper salts have already been mentioned.

In the second Periodic Group we find a similar difference between the stability and constitution of ammines in the A and B subgroups. Magnesium forms a number of ammines, but the tendency to form ammines falls off rapidly in the alkaline-earths. The compound $CaCl_2.8NH_3$, for example, readily loses ammonia. The degree of hydration of salts also decreases from Ca to Ba except for the octahydrates of the peroxides and hydroxides ($SrO_2.8H_2O$, $Ba(OH)_2.8H_2O$), which presumably owe their stability to hydrogen bond formation between the O_2^{2-} or OH^- ions and the water molecules, In the B subgroup, however, as in IB, the ammines and hydrates are structurally unrelated. We find

$$ZnCl_2.H_2O \qquad \text{but} \qquad ZnCl_2.2NH_3$$
$$\text{and}$$
$$CdCl_2.2H_2O \qquad \text{and} \qquad CdCl_2.2NH_3$$

but these two diamminodichlorides have quite different structures, in which the bonds $M-NH_3$ must possess appreciable covalent character. These structures have already been noted in the section on amminohalides in Chapter 10.

16

Sulphur, selenium, and tellurium

We describe in this chapter the structural chemistry of sulphur, selenium, and tellurium, excluding certain groups of compounds which are discussed elsewhere. Metal oxysulphides and sulphides (simple and complex) are described in Chapter 17, sulphides of non-metals are included with the structural chemistry of the appropriate element, and alkali-metal hydrosulphides are grouped with hydroxides in Chapter 14.

The stereochemistry of sulphur

The principal bond arrangements are set out in Table 16.1 which has the same general form as the corresponding Table 11.1 for oxygen. For the sake of simplicity we have assumed sp^3 hybridization in H_2O though the interbond angle (92°) is nearer to that expected for p bonds. For other molecules SR_2 see Table 16.2 (p. 705). The stereochemistry of S is more complex than that of O because S utilizes d in addition to s and p orbitals. Although covalencies of two and four are more usual than three the latter is found in SO_3, sulphoxides, $R_2S{=}O$, sulphonium salts, $(R_3S)X$, and in a number of oxy-ions. The multiple character of most S—O bonds does not complicate the stereochemistry because the interbond angles are determined by the number of σ bonds and lone pairs, as shown in Table 16.1.

Elementary sulphur, selenium, and tellurium

At high temperatures the vapours of all three elements consist of diatomic molecules (in which the bond lengths are: S═S, 1.89 Å; Se═Se, 2.19 Å; and Te═Te, 2.61 Å),

TABLE 16.1

The stereochemistry of sulphur

No. of σ pairs	Type of hybrid	No. of lone pairs	Bond arrangement	Examples
3	sp^2	0	Plane triangular	SO_3
		1	Angular	SO_2
4	sp^3	0	Tetrahedral	SO_2Cl_2, $O_2S(OH)_2$
		1	Trigonal pyramidal	$SOCl_2$, $S(CH_3)_3^+$
		2	Angular	SH_2, SCl_2, S_8
5	$sp^3d_{z^2}$	0	Trigonal bipyramidal	SOF_4
		1	'Tetrahedral'	SF_4
6	$sp^3d_\gamma^2$	0	Octahedral	SF_6, S_2F_{10}

but at lower temperatures and in solution (e.g. in CS_2) S and Se form S_8 and Se_8 molecules respectively. Owing to the insolubility of Te there is no evidence for the formation of a similar molecule by this element.

Sulphur

Sulphur is remarkable for the number of solid forms in which it can be obtained. These include at least four well-known 'normal' polymorphs stable under atmospheric pressure, numerous high-pressure forms[1] of which one is the fibrous form made from 'plastic' sulphur, so-called amorphous forms, characterized by small solubility in CS_2, and coloured forms produced by condensing the vapour on surfaces cooled to the temperature of liquid nitrogen. Some at least of the 'amorphous' sulphur preparations (e.g. milk of sulphur) give X-ray diffraction lines indicative of some degree of crystallinity, and it is perhaps preferable to use the term μ-sulphur rather than amorphous sulphur.[2] The term 'normal' polymorph used above refers to the forms now known to consist of S_6 or S_8 molecules. In recent years cyclic molecules S_n ($n = 7, 9, 10, 12, 18, 20$) have been prepared by special methods (see below).

In the orthorhombic form of sulphur stable at ordinary temperatures the unit of structure is the cyclic S_8 molecule (crown configuration)* in which the bond length is 2.06 Å, interbond angle 108°, and dihedral angle 98°.[3] At 95.4 °C orthorhombic α-S_8 changes into a monoclinic form which normally reverts fairly rapidly to rhombic S but can be kept at room temperature for as long as a month if it is pure and has been annealed at 100 °C. This form, β-S_8, consists of S_8 molecules with the same configuration as in rhombic S and is remarkable for the fact that two-thirds of the molecules are in fixed orientations but the remainder are randomly oriented.[4] This randomness leads to a residual entropy of $\frac{1}{3}R$ ln 2 per mole of S_8 molecules (0.057 e.u./g atom), which agrees reasonably well with the observed value (0.045 e.u./g atom). A second monoclinic form, Muthmann's γ-S_8 , also consists of crown-shaped S_8 molecules.[5a] Closely related to the S_8 molecules are cyclic molecules S_5Se_3, isostructural with γ-S_8,[5b] S_4Se_4[5c] with alternate S and Se atoms in the ring, and molecules $S_{8-n}(NH)_n$ (see p. 831), of which $S_6(NH)_2$ also is isostructural with γ-S_8.

The rhombohedral S prepared by Engel in 1891 by crystallizing from toluene rapidly breaks down into a mixture of amorphous and orthorhombic S. The crystals consist of S_6 molecules (chair configuration) with S–S, 2.06 Å, interbond angle 102°, and dihedral angle 75°.[6] A mass spectrometric study shows that Engel's sulphur vaporizes as S_6 molecules,[7] whereas the vapour of rhombic S at temperatures just above the boiling point consists predominantly of S_8 molecules with the same configuration as in the crystal.[8] (The vapour from FeS_2 at 850 °C consists of S_2 molecules, the same units as those in the crystal,[9] and mass spectrometric studies of solid solutions S–Se show peaks due to S_7Se, S_6Se_2, S_5Se_3, and possibly other molecules.[10])

By reactions such as those between polysulphanes and dichloropolysulphanes or between $(C_5H_5)_2TiS_5$ and $S_{2m}Cl_2$ many new crystalline forms of sulphur have been prepared which consist of cyclic S_n molecules ($n = 7*, 9*, 10*, 11, 12, 13, 18,$ 20) (Fig. 16.1), some of which (*) are not stable at room temperature. References are given to those whose structures have been determined.[11] In the highly

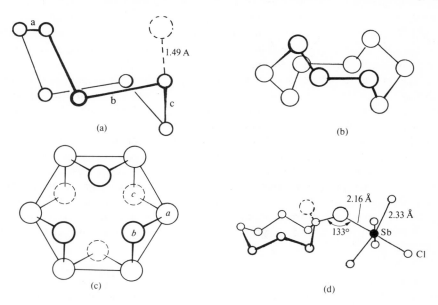

FIG. 16.1 (a) the S_7 molecule; dotted circle indicates the position of O in S_7O or I in S_7I^+; (b) the S_{10} molecule; (c) the cyclic S_{12} molecule; (d) the molecule $S_8O.SbCl_5$.

symmetrical S_{12} molecule (S–S, 2.05 Å, interbond angle, 106½°, dihedral angle 86°) six of the atoms of the ring, those of type *a* in Fig. 16.1(c), are coplanar, three (*b*) lie above and three (*c*) below the plane of the hexagon of *a* atoms. For the S_7 ring see also S_7O (p. 713) and S_7I^+ (p. 705). For S_{11} see ref. 11(e).

For cyclic and polyhedral S, Se, and Te cations see p. 703.

In sulphur vapour dissociation to S atoms is appreciable only at temperatures above 1200 °C, even at a pressure of 0.1 mm. If the vapour at 500 °C and 0.1–1.0 mm, consisting essentially of S_2 molecules, is condensed on a surface cooled to the temperature of liquid nitrogen, sulphur condenses as a purple solid which on warming reverts to a mixture of crystalline and amorphous sulphur. The paramagnetism of this form supports the view that it consists of S_2 molecules. The vapour from *liquid* sulphur at lower temperatures (200–400 °C) can be condensed to a green solid which is not a physical mixture of the purple and yellow forms; the vapour from *solid* sulphur condenses to yellow S.[12] In addition to the larger rings noted above the S_3, S_4, and S_5 molecules have been identified in the vapour and in liquid S.[13]

Fibrous or plastic sulphur results from quenching the liquid from temperatures above 300°C and stretching. It contains some γ-S_8, the monoclinic S_8 form noted above, which is extracted by dissolving in CS_2. The crystallinity of the residual S_ψ can be improved by heating the stretched fibres at $80\,^{\circ}C$ for 40 hours. An X-ray study of such fibres shows them to consist of an approximately hexagonal close-packed assembly of equal numbers of left- and right-handed helical molecules in which the S–S bond length is $2.07\,\text{Å}$, the interbond angle $106°$, and dihedral angle $95°$.[14] The helices are remarkable for containing 10 atoms in the repeat unit of 3 turns; contrast Se and Te (below). This fibrous sulphur is apparently identical with one of the high-pressure forms.

The properties of liquid S are complex, and although there has been much theoretical and practical work on this subject it seems that no existing theory accounts satisfactorily for all the properties.[15] On melting sulphur forms a highly mobile liquid (S_λ) consisting of cyclic S_8 molecules, but at $159\,^{\circ}C$ an extremely rapid and large increase in viscosity begins, which reaches a maximum at around $195\,^{\circ}C$ (S_μ), above which temperature the viscosity falls off. The specific heat also shows a sudden rise at $159\,^{\circ}C$. The typical λ-shaped curve is due to sudden polymerization, and estimates have been made from e.s.r. and static magnetic susceptibility of the average chain length, which ranges from 10^6 atoms at about $200\,^{\circ}C$ to 10^3 at $550\,^{\circ}C$.[16] The X-ray diffraction effects from liquid sulphur show that a S atom has an average of two nearest neighbours at approximately the same distance as in crystalline S.[17]

(1)   IC 1970 **9** 1973, 2478
(2)   JACS 1957 **79** 4566
(3)   AC 1965 **18** 562, 566; AC 1972 **B28** 3605
(4)   IC 1976 **15** 1999
(5a) AC 1974 **B30** 1396
(5b) AC 1978 **B34** 911
(5c) Ind. J. Chem. 1978 **16A** 335
(6)   JPC 1964 **68** 2363
(7)   JCP 1964 **40** 287
(8)   JACS 1944 **66** 818
(9)   JCP 1963 **39** 275
(10) JINC 1965 **27** 755
(11) (a) S_7: AnCIE 1977 **16** 715; (b) S_{10}: AnCIE 1978 **17** 57; (c) S_{12}: AnCIE 1966 **5** 965; (d) S_{18}, S_{20}: ZaC 1974 **405** 153 (e) S_{11}: CC 1982 1312
(12) JACS 1953 **75** 848, 6066
(13) For a review of elemental S see Chemical Reviews 1976 **76** 367
(14) JCP 1969 **51** 348
(15) JPC 1966 **70** 3528, 3531, 3534
(16) TFS 1963 **59** 559
(17) JCP 1959 **31** 1598

Selenium

The ordinary form of commerce is vitreous Se which, if heated and cooled slowly, is converted into the grey 'metallic' form, the stable polymorph. From CS_2 solution two metastable red polymorphs can be crystallized,[1, 2] and a third form from the reaction of CS_2 with dipiperidinotetraselenane, $Se_4(NC_5H_{10})_2$.[3] This is an unusual case of polymorphism for all three forms have the same space group and consist of cyclic Se_8 molecules of the same kind (Se–Se, $2.33\,\text{Å}$; interbond angle $106°$); they differ only in the mode of packing of the molecules. The metallic form consists of infinite helical chains (Se–2Se, $2.37\,\text{Å}$ and interbond angle $103°$).[4] The atomic radial distribution curve of vitreous Se has been determined by both X-ray and neutron diffraction, and shows that this form contains the same helical chains as the metallic form (peaks at 2.33, 3.7, and 5.0 Å).[5]

(1) AC 1972 **B29** 313   (4) IC 1967 **6** 1589
(2) AC 1953 **6** 71   (5) JCP 1967 **46** 586
(3) JCS D 1980 624

Tellurium

The normal form stable at atmospheric pressure is the hexagonal (metallic) form isostructural with metallic selenium, in which Te—Te is 2.835 Å and interbond angle, 103.2°.[1] Two high-pressure forms have been recognized, one stable at 40–70 kbar (structure not known) and one above 70 kbar. The latter is apparently isostructural with β-Po, having a structure resulting from the compression of a simple cubic structure along the [111] axis.[2] The nearest neighbours are

$$\text{Te} \begin{cases} 6 \text{ Te at } 3.00 \text{ Å} \\ 6 \text{ Te} \quad 3.72 \\ 2 \text{ Te} \quad 3.82 \end{cases} \text{compare Te} \begin{cases} 2 \text{ Te at } 2.84 \text{ Å} \\ 4 \text{ Te} \quad 3.50 \end{cases} \text{in hexagonal Te}$$

(1) AC 1967 **23** 670   (2) JCP 1965 **43** 1149

Cyclic and polyhedral S, Se, and Te cations

In contrast to the single cation O_2^+ formed by oxygen, S, Se, and Te form numerous cations. The ion S_5^+ has been identified as the paramagnetic species in the coloured solutions in oleum, and the structures of a number of ions S_n^{2+} have been determined in crystalline salts. These ions are highly electrophilic and can be produced only in very acidic solvents and isolated as stable solids in combination with anions of extremely strong acids such as SO_3F^-, AsF_6^-, $Sb_2F_{11}^-$, and $AlCl_4^-$. Methods of preparation include the oxidation of S by AsF_5 in solution in anhydrous HF or SO_2, which gives red $S_{19}(AsF_6)_2$ and blue $S_8(AsF_6)_2$; excess $S_2O_6F_2$ and S in liquid SO_2 form colourless $S_4(SO_3F)_2$. The Se and Te compounds may be prepared by somewhat similar methods and also in $Se(Te)Cl_4$–$AlCl_3$ melts containing $Se(Te)$–compare the Hg cations, p. 1156.

The structures of these ions have been determined in salts containing cations such as AsF_6^-, SbF_6^-, or in the case of Te_4^{2+}, $AlCl_4^-$ or $Al_2Cl_7^-$. The ions S_4^{2+},[1] Se_4^{2+},[2] Te_4^{2+},[3] and *trans*-$Te_2Se_2^{2+}$ are planar and almost exactly square with bond lengths slightly less than the single-bond values; compare the planar N_2S_2 ring in $N_2S_2(SbCl_5)_2$, p. 828. The 8-membered cyclic cations in $S_8(AsF_6)_2$[4] and $Se_8(AlCl_4)_2$[5] have very similar shapes, Fig. 16.2(a), intermediate between the crown configuration of S_8 (48 valence electrons) and that of N_4S_4 (44 valence electrons), p. 829. The bond lengths are normal but there is one short distance across the ring (broken line), S–S, 2.86, Se–Se, 2.83 Å, and two others which are also shorter than normal van der Waals contacts (dotted lines), S–S, 2.97, Se–Se, 3.33 Å.

Cations containing 10 atoms include Se_{10}^{2+}[6] and $Te_2Se_8^{2+}$.[7] Both ions have a similar structure, Fig. 16.2(b), alternatively described as a boat-shaped 6-membered

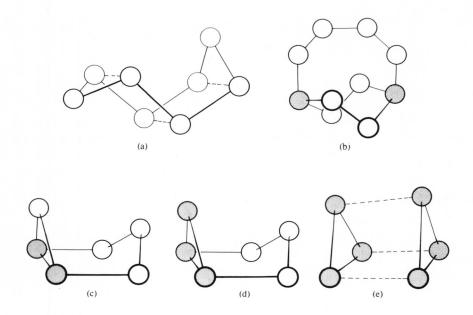

FIG. 16.2. The structures of cations: (a) S_8^{2+}; (b) Se_{10}^{2+} or $Te_2Se_8^{2+}$ (Te atoms shaded); (c) $Te_2Se_4^{2+}$; (d) $Te_3Se_3^{2+}$; (e) Te_6^{4+}.

ring bridged by a chain of 4 atoms or as formed from two fused 8-membered rings. The shaded atoms in (b) represent Te atoms in $Te_2Se_8^{2+}$. The most remarkable sulphur cation is that found in $S_{19}(AsF_6)_2$;[8] it consists of two 7-membered rings joined by a chain of 5 S atoms.

Cations S_6^{2+}, Se_6^{2+}, and Te_6^{2+} have not yet been prepared, but two 'hetero-cations', $Te_2Se_4^{2+}$ and $Te_3Se_3^{2+}$, have the structures (c) and (d), Fig. 16.2.[9] Closely related to these is the ion Te_6^{4+},[10] the first example of a polyhedral cation of Te, Fig. 16.2(e). The bonds in the triangular 'faces' are of the normal length found in these cations (around 2.67 Å), but those shown as broken lines are appreciably longer (3.12 Å).

(1) CSR 1979 **8** 315
(2) IC 1971 **10** 2319
(3) IC 1972 **11** 357
(4) IC 1971 **10** 2781
(5) IC 1971 **10** 2749

(6) IC 1980 **19** 1432
(7) IC 1976 **15** 765
(8) IC 1980 **19** 1423
(9) IC 1977 **16** 892
(10) IC 1979 **18** 3086

Molecules SR_2, SeR_2, and TeR_2 : cyclic molecules

Structural data for the (angular) molecules SR_2, SeR_2, and TeR_2 are listed in Table 16.2. Interbond angles close to $100°$ are also found in cyclic molecules based on strain-free 6-membered rings, which have the chair configuration; examples include 1,4-dithiane,[1] 1,3,5-trithiane,[2] S_5CH_2,[3] and $S_6(CH)_4$,[4] structurally similar to

TABLE 16.2

Structures of molecules SR$_2$, SeR$_2$, and TeR$_2$

Molecule	R—S—R	S—R	Method	Reference
SH$_2$	92°	1.35	m.w.	JMS 1975 **28** 237
SF$_2$	98°	1.59	m.w.	Sc 1969 **164** 950
SCl$_2$	100°	2.00	R	JCP 1955 **23** 972
S(CH$_3$)$_2$	99°	1.80	m.w.	JCP 1961 **35** 479
S(CF$_3$)$_2$	106°	1.83	e.d.	TFS 1954 **50** 452
S(CN)$_2$	98°	1.70	m.w., i.r.	JCP 1965 **43** 3423
S(SiH$_3$)$_2$	97°	2.14	e.d.	ACSc 1963 **17** 2264
S(GeH$_3$)$_2$	99°	2.21	e.d.	JCS A 1970 315
	R—Se—R	Se—R		
SeH$_2$	91°	1.46	m.w.	JMS 1962 **8** 300
Se(CH$_3$)$_2$	98°	1.98	e.d.	JACS 1955 **77** 2948
Se(SiH$_3$)$_2$	97°	2.27	e.d.	ACSc 1968 **22** 51
Se(SCN)$_2$	101°	2.21	X	JACS 1954 **76** 2649
Se(CF$_3$)$_2$	104°	1.96	e.d.	TFS 1954 **50** 452
Se[SeP(C$_2$H$_5$)$_2$Se]$_2$	104°	2.35	X	ACSc 1969 **23** 1398
	R—Te—R	Te—R		
TeBr$_2$	98°	2.51	e.d.	JACS 1947 **69** 2102
Te[S$_2$P(OCH$_3$)$_2$]$_2$	98°	2.44	X	ACSc 1966 **20** 24
Te[SeP(C$_2$H$_5$)$_2$S]$_2$	101°	2.50	X	ACSc 1969 **23** 1389

(CH$_2$)$_6$(CH)$_4$ and (CH$_2$)$_6$N$_4$ (p. 795). Smaller angles are, of course, found in cyclic molecules such as CH$_2$–CH$_2$ (66°)[5] and CH=CH–CH=CH (91°).[6]
$\underset{\text{S}}{\underline{\quad\quad}}$ $\underset{\text{S}}{\underline{\quad\quad\quad}}$

(1) AC 1955 **8** 91
(2) AC 1969 **B25** 1432
(3) ZaC 1976 **423** 103
(4) AK 1956 **9** 169
(5) JCP 1952 **19** 676
(6) JACS 1939 **61** 1769

The halides of sulphur, selenium, and tellurium

The compounds set out below appear to be well characterized. Some have been known for many years, while others have been prepared more recently by special methods, such as Se$_2$F$_2$ and SeF$_2$ which have been identified among the products of reaction of Se with much-diluted fluorine. Before describing their structures we note some points of special interest. The first is that they vary greatly in stability; moreover, some exist only in the crystalline state and others only in the vapour state. Little is known of iodides other than those of Te; no iodides of S are known which are stable at room temperature, though there is spectroscopic evidence for the existence of S$_2$I$_2$.[1a] We comment later on some sub-halides which are peculiar to Te. The novel cation (S$_7$I)$^+$, Fig. 16.1(a), has been studied in the salt (S$_7$I)(SbF$_6$),[1b] prepared from S, I, and SbF$_5$ in AsF$_3$ solution. The same reaction produces (S$_{14}$I$_3$)(SbF$_6$)$_3$.2AsF$_3$, in which the cation consists of two S$_7$I units joined through a linear —I— bridge.[1c] The ion $S_2I_4^{2+}$ has been studied in the salt $(S_2I_4)(AsF_6)_2$.[1d] It consists of two planar S_2I_2 units joined at the common S—S bond, the angle between the two planes being $90°$. Its structure may be contrasted with that of the isoelectronic P_2I_4 molecule (Fig. 19.5, p. 848).

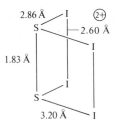

The chlorides S_nCl_2 up to S_5Cl_2, consisting of S_n chains terminated by Cl atoms, can be prepared from SCl_2 and H_2S_n ($n = 1, 2, 3$) and those up to S_8Cl_2 from S_2Cl_2 and H_2S_n. By the action of HBr the chlorides can be converted into S_nBr_2 (n from 2 to 8).

S_2F_2	S_2Cl_2	S_2Br_2	Se_2F_2	Se_2Cl_2	Se_2Br_2				
SF_2	SCl_2		SeF_2	$SeCl_2$	$SeBr_2$		$TeCl_2$	$TeBr_2$	
SF_4	SCl_4		SeF_4	$SeCl_4$	$SeBr_4$	TeF_4	$TeCl_4$	$TeBr_4$	TeI_4
SF_6			SeF_6			TeF_6			
S_2F_{10}	S_nCl_2	S_nBr_2				Te_2F_{10}			

Sub-halides of Te: Te_2Cl, Te_3Cl_2; Te_2Br; Te_2I, TeI (α and β).

SF_2 is an unstable species which was studied in a microwave cell but not isolated; its structure (Table 16.2) was determined from the moments of inertia of $^{32}SF_2$ and $^{34}SF_2$.

The structure of SCl_4 has been discussed for many years. This compound forms a yellow solid which melts at $-30°C$ to a red liquid which is a mixture of SCl_2 and Cl_2 and evolves the latter gas, and it is completely dissociated into SCl_2 and Cl_2 at ordinary temperatures. The dielectric constant of the liquid of composition SCl_4 rises from 3 to 6 on freezing, and it has been suggested that the solid is a salt, $(SCl_3)Cl$. Like SCl_4, $SeCl_4$ and $SeBr_4$ dissociate on vaporization to a mixture of dihalide and halogen; these compounds exist only in the solid state. On the other hand, the vapour density of $TeCl_4$ corresponds to the formula $TeCl_4$ up to $500°C$; this is the only one of these tetrahalides which can be studied in the vapour state.

As in the rest of its chemistry Te stands rather apart from S and Se as regards its halides. It is the only one of these elements that forms all four tetrahalides, the dihalides $TeCl_2$ and $TeBr_2$ are known only in the gaseous state, and Te forms well-defined sub-halides. In its sub-halides Te exhibits a variety of bond arrangements, being bonded to 2 Te only or to 2 Te and 1 or 2 halogen atoms. In Te_3Cl_2[2] there are helical chains of Te atoms with 2 Cl attached to every third atom; this Te atom has a 'trigonal bipyramidal' coordination, (a), like that in TeS_7Cl_2,[3] a molecule in which two Cl atoms are bonded to the Te atom in an 8-membered ring (a). In halides Te_2X[2] there are double chains, (b), in which Te is bonded alternately to 3 Te or to 2 Te + 2 X. The (metastable) β form of TeI,[2] (c), has a closely related structure; there is square planar and trigonal pyramidal coordination of alternate Te atoms. The more stable α-TeI consists of cyclic Te_4I_4 molecules which are linked into chains by weaker bonds shown at (d) as broken lines. These stoichiometric iodides replace the material previously described in the literature as Te_xI or Te_nI_n.

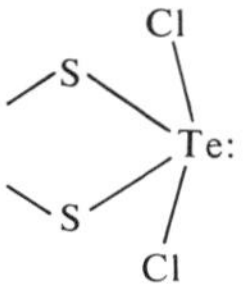

(a) 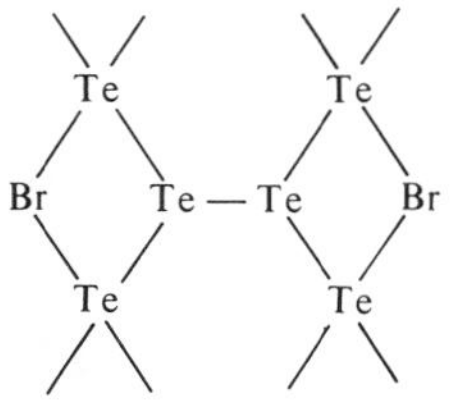   (b) 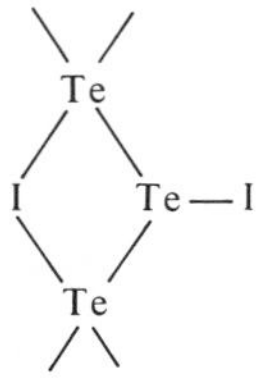   (c) 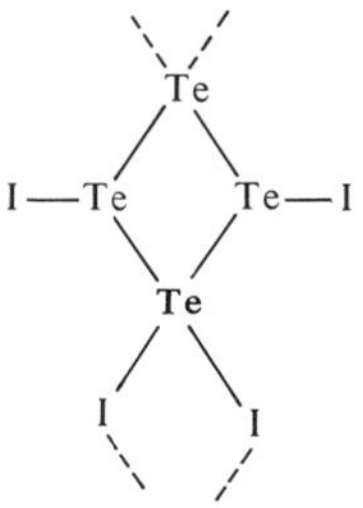   (d)

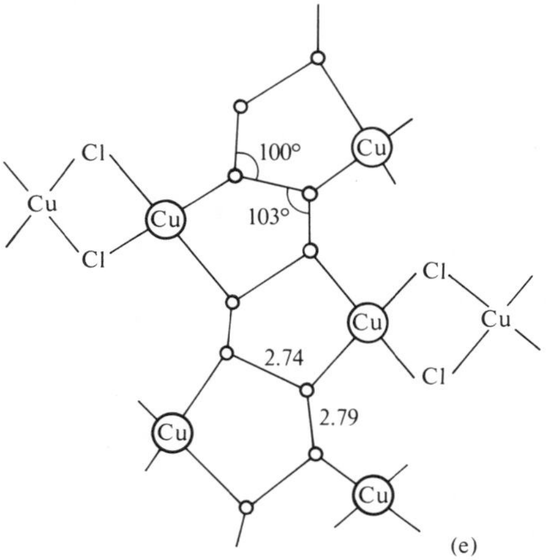

(e)

It is convenient to mention here another 'sub-halide' which also contains —Te—Te— chains, namely, $CuTe_2Cl$.[4] This compound has a layer structure formed from such chains which are connected by $Cu\!\!\begin{array}{c}Cl\\ Cl\end{array}\!\!Cu$ bridges as shown at (e). Each Cu atom is bonded to 2 Te and 2 Cl, and forms four tetrahedral bonds.

Halides SX_2 etc.

Structural details of the angular molecules SCl_2 and $TeBr_2$ are included in Table 16.2.

Halides S_2X_2 etc.

In addition to the isomer with the structure XS–SX (which is enantiomorphic) a topological isomer S–SX$_2$ is possible. S$_2$F$_2$ exhibits this type of isomerism. Both isomers are formed when a mixture of dry AgF and S is heated in a glass vessel *in vacuo* to 120°C. The isomer S=SF$_2$ is pyramidal[5] and very similar geometrically to OSF$_2$ (Fig. 16.3(a) and (b)); for OSF$_2$ see Table 16.3. The isomer FS–SF has the H$_2$O$_2$ type of structure with an unexpectedly short S–S bond (1.89 Å, compare 1.97 Å in S$_2$Cl$_2$ and 2.055 in S$_2$H$_2$[5a]) and S–F abnormally long (1.635 Å). For details of the structures of S$_2$F$_2$, S$_2$Cl$_2$, and S$_2$Br$_2$ see Table 16.6 (p. 728).

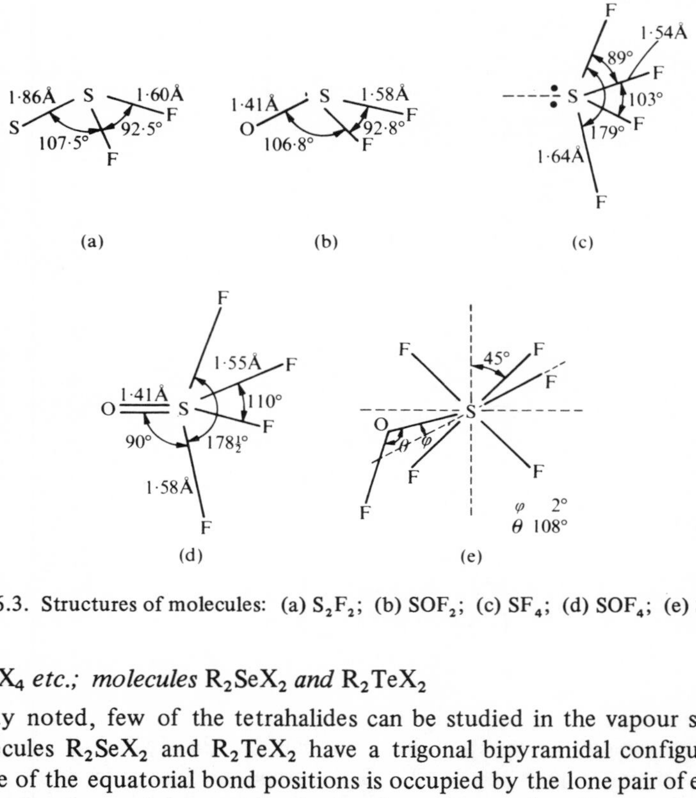

(a)   (b)   (c)

(d)   (e)

FIG. 16.3. Structures of molecules: (a) S$_2$F$_2$; (b) SOF$_2$; (c) SF$_4$; (d) SOF$_4$; (e) SF$_5$OF.

Halides SX$_4$ etc.; molecules R$_2$SeX$_2$ and R$_2$TeX$_2$

As already noted, few of the tetrahalides can be studied in the vapour state; SF$_4$ and molecules R$_2$SeX$_2$ and R$_2$TeX$_2$ have a trigonal bipyramidal configuration in which one of the equatorial bond positions is occupied by the lone pair of electrons. The stereochemistry of these molecules has been discussed in Chapter 6. A feature of these molecules is that the axial bonds are collinear to within a few degrees and longer than the equatorial ones in a molecule SX$_4$ (or longer than the covalent radius sum in a molecule R$_2$SeX$_2$). The lone pair of electrons in SF$_4$ is used to form the fifth bond in OSF$_4$ and these molecules have rather similar configurations (Fig. 16.3):

		SF_4[6]	OSF_4[7]
axial	S–F	1.643 Å	1.583 Å
equatorial	S–F	1.542	1.550

A m.w. study of SeF_4[8] shows that this molecule has the same general shape as SF_4 with a similar difference between the equatorial and axial bond lengths: Se–F, 1.68 Å (equatorial), 1.77 Å (axial); F–Se–F (equatorial) 100.55°, 169.20° (axial). For compounds with NbF_5 containing the ion SeF_3^+ see p. 739. It is not profitable to discuss the early e.d. results for $TeCl_4$ vapour[8a] since it was assumed that all four bonds were of equal length (2.33 Å); a molecular configuration of the SF_4 type would be consistent with the fact that $TeCl_4$ has a dipole moment (2.54 D) in benzene solution.

Other molecules with the same general shape as SF_4 include $Se(C_6H_5)_2Cl_2$ and $Se(C_6H_5)_2Br_2$, $Te(CH_3)_2Cl_2$ and $Te(C_6H_5)_2Br_2$. In all these molecules the halogen atoms occupy the axial positions; for details see p. 741.

Crystalline TeF_4[9] is not a molecular crystal but consists of chains of TeF_5 groups sharing a pair of *cis* F atoms (Fig. 16.4(a)). There are various ways of describing the coordination of Te^{IV}, as 3 F at the vertices of a trigonal pyramid (mean interbond angle 86½°), as 4 F (shaded) at the vertices of a trigonal bipyramid (lone pair equatorial), or as 5 F at the vertices of an octahedron (lone pair in sixth bond position). The Te atom is situated 0.3 Å below the base of the square pyramid in Fig. 16.4.

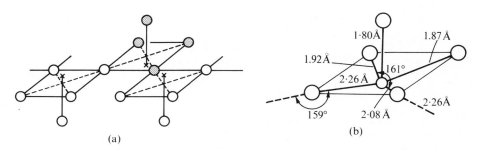

FIG. 16.4. The structure of crystalline TeF_4: (a) linear molecule; (b) bond lengths.

Crystalline $SeCl_4$, $TeCl_4$,[10] $SeBr_4$,[10a] and $TeBr_4$ all consist of tetramers of the same general type as $[Pt(CH_3)_3Cl]_4$ with the idealized structure of Fig. 5.8(j), p. 200. In $TeCl_4$ Te atoms are displaced outwards along 3-fold axes of the 'cube' so that the coordination of Te is distorted octahedral: Te–3Cl, 2.31 Å; Te–3Cl, 2.93 Å; Cl_t–Te–Cl_t, 95°; Cl_b–Te–Cl_b, 85°. (Alternatively the crystal may be described as distorted ccp Cl with Te atoms in one-quarter of the octahedral interstices but displaced from their centres.) In the limit this (3+3)-coordination would lead to separation into $TeCl_3^+$ and Cl^- ions, and successive removal of one or two $TeCl_3^+$ would lead to $Te_3Cl_{13}^-$ and $Te_2Cl_{10}^{2-}$ ions. The first of these has been

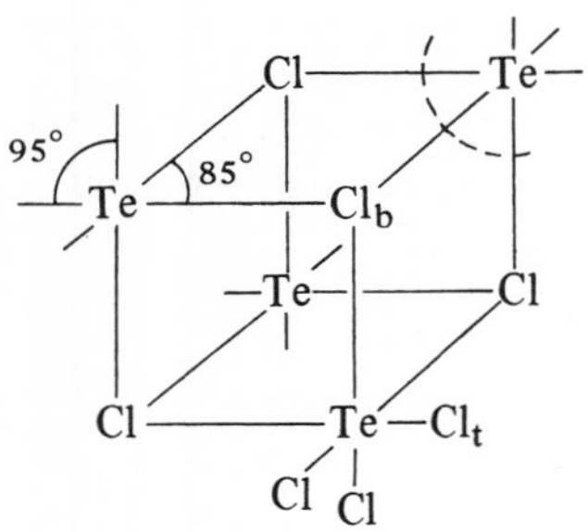

$$Cl-Te-Cl$$

Te—3 Cl_t 2·31 Å	95°
Te—3 Cl_b 2·93	85°

studied in $[(C_6H_5)_3C](Te_3Cl_{13})^{(10b)}$; it consists of the portion marked off by the broken lines in the diagram, with the idealized structure of Fig. 5.8(e). The structures of chloro complexes of Te^{IV} emphasize that the stereochemical effect of the lone pair is not a simple matter. In $TeCl_3^+$ and $TeCl_4$ there is (3+3)-coordination; in the *cis* octahedral chain of $TeCl_5^-$ there is a larger range of Te–Cl bond lengths than would be expected to arise simply from the sharing of two *cis* vertices, while the $TeCl_6^{2-}$ ion shows no distortion from a regular octahedron:

	Octahedral bond angles	*Neighbours of* Te	*References*
$(TeCl_3)^+(AlCl_4)^-$	95°	3 at 2.28 Å; 3 at 3.06 Å	ZaC 1971 **386** 257
$TeCl_4$		3 at 2.31; 3 at 2.93	IC 1971 **10** 2795
$(TeCl_5)^-(PCl_4)^+$ }	90 ± 5°	4 at 2.31–2.51; 2 at 2.68, 2.87	AC 1972 **B28** 1260
$(TeCl_6)^{2-}(NH_4)_2^+$	90°	6 at 2.51	ACSc 1966 **20** 165

A further point of interest is that the $(TeCl_5)_n^{n-}$ ion has a quite different structure from the isoelectronic (square pyramidal) $SbCl_5^{2-}$ ion (p. 886); on the other hand the TeF_4 chain is very similar to the $(SbF_4)_n^{n-}$ ion in $NaSbF_4$.

Crystalline $TeI_4^{(10c)}$ consists of tetrameric molecules of a quite different kind from those in $TeCl_4$, namely, octahedral edge-sharing groups like those of $[Ti(OR)_4]_4$ (Fig. 5.8(h), p. 200). The bond lengths are 2.77, 3.11, and 3.23 Å to I atoms bonded to one, two, or three Te (compare 2.93 Å in TeI_6^{2-}), a range of bond lengths comparable with those in the Ti compound; the lone pair is obviously having little effect on the stereochemistry.

Hexahalides

Electron diffraction or microwave studies have shown that the molecules SF_6,[11] SeF_6,[11] TeF_6,[12] and also $SF_5Cl^{(13)}$ and $SF_5Br^{(14)}$ have octahedral configurations —regular in the case of the hexafluorides:

S–F   1.56 Å   S–Cl   2.03 Å   S–Br   2.19 Å
Se–F   1.69
Te–F   1.82

Disulphur decafluoride, S_2F_{10}

The e.d. data for this halide are consistent with a model consisting of two $-SF_5$ groups joined in the staggered configuration so that the central S–S bond is collinear with two S–F bonds, with all interbond angles 90° (or 180°), S–F, 1.56 Å, and S–S, 2.21 Å.[15] The length of the S–S bond may be due to repulsion of the F atoms.

(1a) ZaC 1978 **443** 217	(8a) JACS 1940 **62** 1267
(1b) CC 1976 689	(9) JCS A 1968 2977
(1c) CC 1979 901	(10) IC 1971 **10** 2795
(1d) CC 1980 289	(10a) ZaC 1979 **451** 12
(2) ZaC 1976 **422** 17	(10b) ZN 1979 **34b** 900
(3) AC 1972 **B28** 3653	(10c) AC 1976 **B32** 1470
(4) AC 1976 **B32** 3084	(11) TFS 1963 **59** 1241
(5) JACS 1963 **85** 2028	(12) ACSc 1966 **20** 1535
(5a) JMS 1972 **41** 534	(13) IC 1976 **15** 3004
(6) JCP 1963 **39** 3172	(14) JCP 1963 **39** 596
(7) JCP 1969 **51** 2500	(15) TFS 1957 **53** 1545, 1557
(8) JMS 1968 **28** 454	

Oxyhalides of S, Se, and Te

A listing of these compounds emphasizes the large number of S compounds and the virtual absence of oxyiodides. Details of the pyramidal molecules of thionyl halides (SOX_2) and of the tetrahedral molecules of sulphuryl halides (SO_2X_2) are included in Tables 16.3 and 16.4. The structures of $SeOF_2$ and $SeOCl_2$ have been determined in a number of adducts (p. 740).

SOF_2	SOF_4	SO_2F_2	S_2OF_{10}	$SeOF_2$	Te_2OF_{10}
Cl_2	SOF_6	Cl_2	$S_2O_2F_{10}$	Cl_2	$Te_2O_2F_8$
Br_2	$SOClF_5$	FCl	$S_2O_5F_2$	Br_2	$TeF_n(OTeF_5)_{6-n}$
I_2	SO_3F_2	FBr	$S_2O_6F_2$	$SeOF_4$	
FCl			$S_3O_8F_2$	SeO_2F_2	
			$S_3O_8Cl_2$	Se_2OF_{10}	
				$Se_2O_2F_8$	

The structure of SOF_4[1] is compared with that of SF_4 in Fig. 16.3. Apart from the expansion of the equatorial F–S–F bond angle from 103° (mean of e.d. and m.w. values) to 110° the molecules are very similar in shape (S–O, 1.41 Å). The S–F bond lengths were compared earlier.

The constitution of SF_5.OF as *pentafluorosulphur hypofluorite* has been confirmed by an e.d. study (Fig. 16.3(e)).[1a] Bond lengths are: S–F, 1.53; S–O, 1.64; and O–F, 1.43 Å in the essentially octahedral molecule. The values of the angles are $\phi = 2°$, $\theta = 108°$. SO_3F_2 is presumably also a hypofluorite.

Disulphur decafluorodioxide, $S_2F_{10}O_2$, is a derivative of hydrogen peroxide, to which it is related by replacing H by $-SF_5$. The dihedral angle is similar to that in H_2O_2 (107°) and the angle S–O–O is 105°. Approximate bond lengths (e.d.) are: O–O, 1.46; S–O, 1.66; and S–F, 1.56 Å.[2]

Peroxodisulphuryl difluoride, $S_2O_6F_2$, is also apparently of this type. It is a colourless liquid prepared by the reaction of F_2 with SO_3 at a temperature above 100 °C; its reactions, i.r. and n.m.r. spectra suggest the formula $FO_2S-O-O-SO_2F$.[3]

The structures of the polysulphuryl fluorides $S_2O_5F_2$, (a), and $S_3O_8F_2$, (b), have been established by e.d. studies.[4] They are generally similar to those of the isoelectronic $S_2O_7^{2-}$ and $S_3O_{10}^{3-}$ ions. A third molecule in which there is octahedral

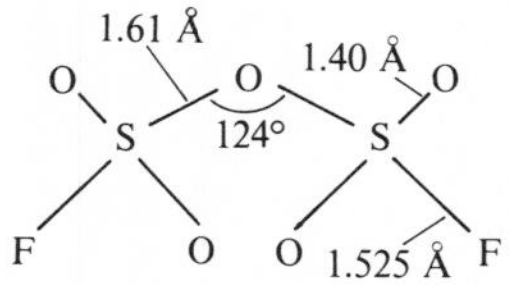

(a) 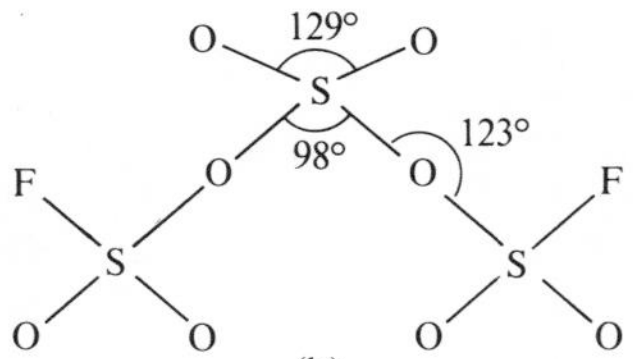   (b) 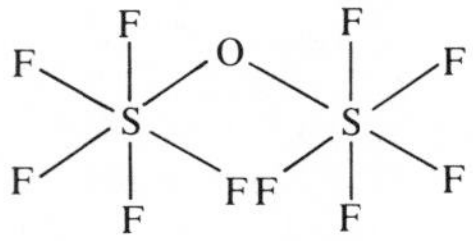   (c)

coordination of S is S_2OF_{10}, (c), which has an eclipsed structure with a short intra-molecular F–F distance (2.4 Å): S–F, 1.563; S–O, 1.586 Å; angle S–O–S, 142.5°.[5] The molecules Se_2OF_{10} and Te_2OF_{10} have similar structures.[5a]

There are planar 4-membered rings in the similarly constituted $Se_2O_2F_8$ and $Te_2O_2F_8$ molecules,[6] (d): Se–O, 1.78; Te–O, 1.92 Å; Se–O–Se, 97.5°; Te–O–Te, 99.5°. A number of compounds $F_nTe(OTeF_5)_{6-n}$ have been prepared, some in *cis*

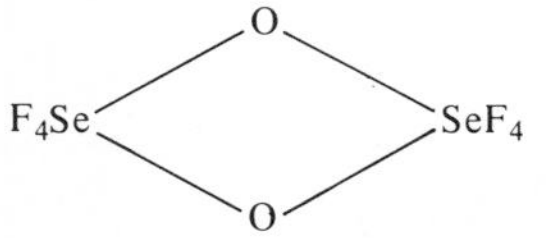

(d) 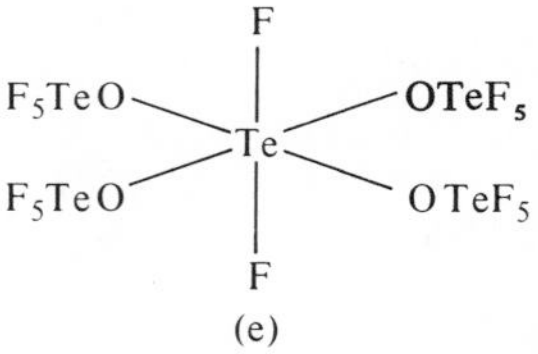   (e)

and *trans* forms, for example, (e).[7a] The structures of two forms of the end-member of the series, $Te(OTeF_5)_6$[7b] have been determined; U forms the similar molecule $U(OTeF_5)_6$.[8]

(1)   JCP 1969 **51** 2500
(1a) JACS 1959 **81** 5287
(2)   JACS 1954 **76** 859
(3)   JACS 1957 **79** 513
(4)   CJC 1973 **51** 2047
(5)   AnC 1978 **90** 66

(5a) IC 1978 **17** 1435
(6)   IC 1979 **18** 2226
(7a) IC 1977 **16** 2685
(7b) IC 1978 **17** 1926
(8)   IC 1976 **15** 2720

The oxides of S, Se, and Te

In addition to SO_2 and SO_3 a number of less stable oxides are known, including SO, SO_4, S_2O, and S_2O_2, and a group of cyclic oxides S_5O, S_6O, S_7O, S_7O_2, and S_8O. Of the last group S_5O has only been detected in solution, but the others have been prepared pure and well characterized. The S_6 and S_7 compounds have been prepared by the oxidation of the appropriate form of S by SF_3CO_3H, and S_8O by the reaction between $SOCl_2$ and H_2S_n (p. 731). The molecules S_7O[1a] and S_8O[1b] have one O atom attached to the ring, which has a configuration similar to that of S_7 or S_8 but with longer bonds (approximately 2.2 Å) to the S atom to which O is bonded. These longer bonds (*b* and *c* in Fig. 16.1(a)) are similar in length to the long bond *a* in the S_7 molecule itself (S–O, 1.49 Å). In S_8O the O atom occupies an axial position indicated by the dotted line in Fig. 16.1(d), which shows the structure of the molecule $S_8O.SbCl_5$.[1c] In this molecule the S_8O portion may be described as a geometric isomer of the free molecule; the S–O bond length has increased from 1.48 Å in the free molecule to 1.55 Å, and the length of the adjacent S–S bonds has decreased from 2.20 Å to 2.11 Å.

Disulphur monoxide, S_2O

This very unstable oxide is formed when a glow discharge is passed through a mixture of sulphur vapour and SO_2 and in other ways. A microwave study[2a] shows

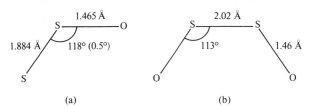

(a)   (b)

that the S–O bond is essentially a double bond as in SO_2, and the bond angle, (a), is very similar to that in SO_2.

Sulphur monoxide, SO

The term 'sulphur monoxide' was originally applied (erroneously) to an equi-moledcular mixture of S_2O and SO_2. The oxide SO is a short-lived radical produced, for example, by the combination of O atoms produced in an electric discharge with solid sulphur. By pumping the products rapidly through a wave-guide cell the micro-wave spectrum can be studied: S–O, 1.48 Å; $\mu = 1.55$ D (compare SO_2, 1.59 D).[2b]

Disulphur dioxide, S_2O_2

The S_2O_2 molecule, (b), has been shown to be planar in the gas phase (m.w.).[2c]

Sulphur dioxide, SO_2

The structure of this molecule has been studied in the crystalline and vapour states by spectroscopic and diffraction methods, most recently by i.r. in a solid Kr matrix.[3] All these studies give a bond angle close to 119.5° and the bond length 1.43 Å. The dipole moment in the gas phase is noted above.

Sulphur dioxide can behave as a ligand in transition-metal complexes in various ways: (i) the S atom forms a single bond M—S to the metal atom, the three bonds from S being arranged pyramidally as in $Ir(P\phi_3)_2Cl(CO)SO_2$;[4] (ii) the S atom uses its lone pair to form the σ bond to the metal which then forms a π-bond making M=S a multiple bond, as $[Ru^{II}(NH_3)_4SO_2Cl]Cl$;[5a] (iii) the SO_2 molecule bonds to a single metal atom through both S and O;[5b] or (iv) it bridges two metal atoms.[5c]

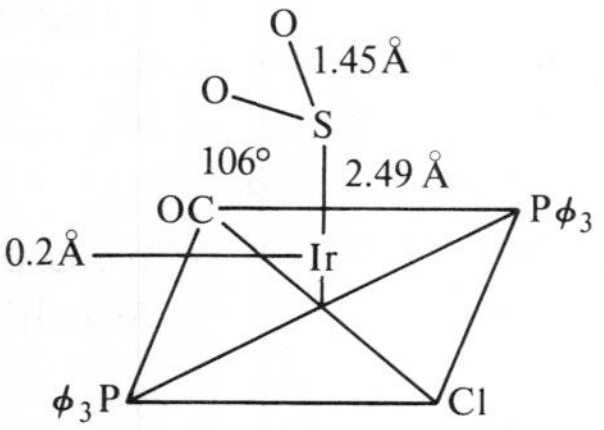

(i) 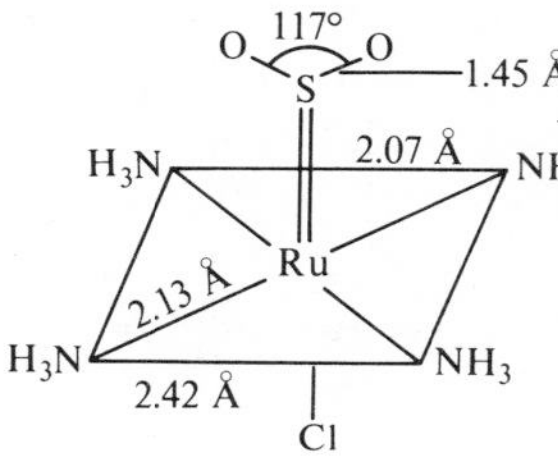   (ii) 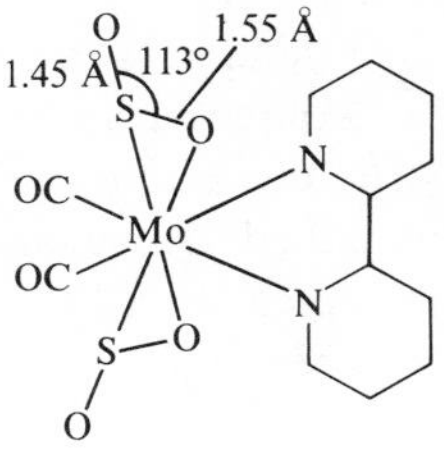   (iii)

In (iii) the structure of the SO_2 molecule is appreciably different from that of the free molecule, in contrast to cases (i) and (ii). The situation in (iv) is less clear-cut (S—O, 1.42-1.49 Å; O—S—O, 114-117°).

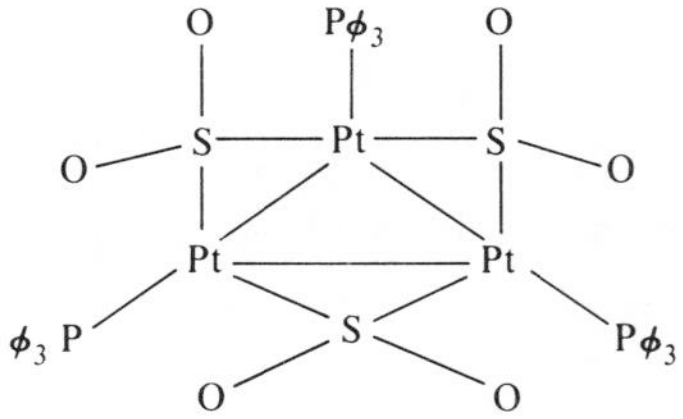

(iv)

Selenium dioxide, SeO_2

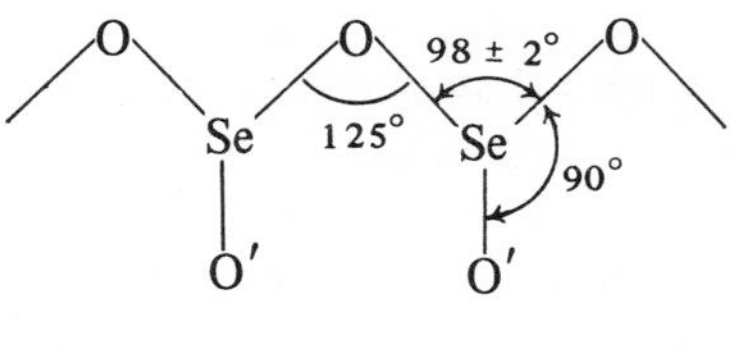

Whereas crystalline sulphur dioxide consists of discrete SO_2 molecules, crystalline SeO_2[6] is built of infinite chains of the type in which the (approximate) bond lengths were determined as Se–O, 1.78 (0.03) Å, and Se–O', 1,73 (0.08) Å. A m.w. study[7] of the SeO_2 molecule in the vapour gave Se–O, 1.607 Å and O–Se–O, 114°. Dimers also exist in the vapour of SeO_2, and the bridged structure shown was derived from an i.r. study (matrix isolation method).[8]

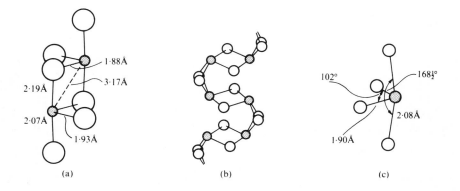

Tellurium dioxide, TeO_2

There are two crystalline forms, a yellow orthorhombic form (the mineral tellurite; the h.p. form[9a]) and a colourless tetragonal form (paratellurite). There is 4-coordination of Te in both forms, the nearest neighbours being arranged at four of the vertices of a trigonal bipyramid, suggesting considerable covalent character of the Te–O bonds. Tellurite has a layer structure[9b] in which TeO_4 groups form edge-sharing pairs (Fig. 16.5(a)) which form a layer, (b), by sharing the remaining

FIG. 16.5. Crystalline TeO_2: (a) and (b) tellurite; (c) paratellurite.

vertices. The short Te–Te distance, 3.17 Å (compare the shortest, 3.74 Å, in paratellurite) may be connected with its colour. In paratellurite[10] very similar TeO_4 groups share all vertices to form a 3D structure of 4:2 coordination in which the O bond angle is 140° (Fig. 16.5(c)). (There are Te_2O_6 groups very similar to those in tellurite in the mineral denningite, $(Mn, Ca, Zn)Te_2O_6$.)

These group VIB dioxides form an interesting series:

	C.n.	Type of structure
SO_2	2	molecular
SeO_2	3	chain
TeO_2	4	layer and 3D
PoO_2	8	3D (fluorite)

Sulphur trioxide, SO_3

The molecule SO_3 has zero dipole moment and e.d. shows it to be a symmetrical planar molecule (S—O, 1.43 Å, O—S—O, 120°).[11] Like P_2O_5 SO_3 has a number of polymorphs, and as in the case of P_2O_5 the forms differ appreciably in their stability towards moisture.

The orthorhombic modification consists of trimers (Fig. 16.6(a)). The bond length in the ring is 1.62 Å, and there is an unexpected (and unexplained)

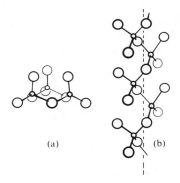

(a)   (b)

FIG. 16.6. (a) The cyclic S_3O_9 molecule in the orthorhombic form of sulphur trioxide; (b) the infinite chain molecule in the asbestos-like form.

difference between the lengths of the axial and equatorial bonds, 1.37 Å and 1.43 Å respectively.[12] The O bond angle is 121.5°.

In the asbestos-like form[13] the tetrahedral SO_4 groups are joined to form infinite chain molecules (Fig. 16.6(b)), with S—O in the chain 1.61 Å and to the unshared O, 1.41 Å. In addition to these two well-defined forms of known structure the existence of others has been postulated to account for the unusual physical properties of solid and liquid sulphur trioxide. The properties of the liquid are still incompletely understood and the system is obviously complex—compare sulphur itself, to which SO_3 is topologically similar in that both can form chain and ring polymers, and units being —S— and —O(SO_2)O—.

Other oxides of sulphur which have been described include SO_4, produced by the action of an electric discharge on a mixture of O_2 or O_3 with SO_2 or SO_3. The existence of S_2O_7 seems less certain.

Selenium trioxide, SeO_3

Pure SeO_3 may be prepared by heating K_2SeO_4 with SO_3, which gives a liquid mixture of SeO_3 and SO_3 from which the more volatile SO_3 may be distilled. Like SO_3 it is at least dimorphic. Although the cyclic modification forms mixed crystals

with S_3O_9 containing 25-100 molar per cent SeO_3 it consists[14] of tetrameric molecules with a configuration very similar to that of $(PNCl_2)_4$. Similar molecules predominate in the vapour at $120°C$.[14a]

A compound Se_2O_5 formed by the thermal decomposition of SeO_3 is presumably $Se^{IV}Se^{VI}O_5$ since it is decomposed by water to an equimolecular mixture of selenious and selenic acids.

Tellurium trioxide, TeO_3

The trioxide formed by dehydrating H_6TeO_6 with concentrated H_2SO_4 in an oxygen atmosphere is a relatively inert substance, being unattacked by cold water or dilute alkalis, though it can act as a strong oxidizing agent towards, for example, hot HCl. There is a second, even more inert, form which has been crystallized under pressure.[15a] The first form is isostructural with FeF_3, that is, it is a 3D structure formed from TeO_6 octahedra sharing all vertices.[15b]

The oxides Te_2O_5, Te_4O_9, *and the oxyhydroxide* $Te_2O_4(OH)_2$

These three compounds are conveniently described together because all contain Te^{IV} forming trigonal prism bonds (TeO_4E), E being the lone pair, and Te^{VI} forming six octahedral bonds. The most closely related from the structural standpoint are Te_2O_5 and $Te_2O_4(OH)_2$. In Te_2O_5[16] $Te^{VI}O_6$ octahedra form AX_4 layers by sharing four vertices (Te–O, $1.92 Å$) and the Te^{IV} atoms are situated between the layers bonded to two O atoms of the layers and to two additional O atoms which form chains ($Te^{IV}O$) as shown in Fig. 16.7(a). The result is a 3D framework of TeO_6 and TeO_4 groups. The structure, $(Te^{VI}O_4)Te^{IV}O$, has therefore considerable similarity to that of Sb_2O_4 (p. 899), $(Sb^VO_4)Sb^{III}$, but with an additional O atom for each Te^{IV} atom between the layers. In $Te_2O_4(OH)_2$[17] the $Te^{VI}O_6$ octahedra form AX_5 chains of composition $[TeO_3(OH)_2]_n^{2n-}$ which are linked through chains

$$-\overset{\displaystyle \diagdown\diagup}{Te}-O-Te-O-$$

in much the same way as in Te_2O_5. The result is a layer structure with the structural formula $[Te^{VI}O_3(OH)_2]Te^{IV}O$. Te_4O_9 also has a layer structure,[18] but there the layer contains both Te^{IV} and Te^{VI} (Fig. 16.7(b)) and has the composition

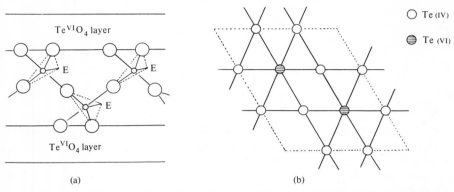

FIG. 16.7. The structures of (a) Te_2O_5 and (b) a layer of Te_4O_9 (diagrammatic) shown as a system of 4-connected Te(IV) and 6-connected Te(VI) atoms.

$Te_3^{IV}Te^{VI}O_9$. The coordination of Te^{IV} in all three structures is similar to that in the two forms of TeO_2 (Fig. 16.5(c)).

(1a)	AnCIE 1977 **16** 716	(9a)	PR 1975 **B12** 1899
(1b)	CB 1976 **109** 180	(9b)	ZK 1967 **124** 228
(1c)	CC 1980 180	(10)	ACSc 1968 **22** 977
(2a)	JMS 1959 **3** 405	(11)	JACS 1938 **60** 2360
(2b)	JCP 1964 **41** 1413	(12)	AC 1967 **22** 48
(2c)	JCP 1974 **60** 5005	(13)	AC 1954 **7** 764
(3)	JCP 1969 **50** 3399	(14)	AC 1965 **18** 795
(4)	IC 1966 **5** 405	(14a)	RTC 1965 **84** 74
(5a)	IC 1965 **4** 1157	(15a)	ZaC 1968 **362** 98
(5b)	IC 1980 **19** 3003	(15b)	CR 1968 **266** 277
(5c)	IC 1977 **16** 1052	(16)	AC 1973 **B29** 643
(6)	JACS 1937 **59** 789	(17)	AC 1973 **B29** 956
(7)	JMS 1970 **34** 370	(18)	AC 1975 **B31** 1255
(8)	JMSt 1971 **10** 173		

Oxy-ions and molecules formed by S, Se, and Te

Our chief concern will be with the compounds of sulphur since these are the most numerous and most important. We shall, however, include those species formed by Se and Te which are structurally similar to S compounds, for example, XO_3^{2-} and XO_4^{2-} ions. We deal in a later section with ions and molecules with stereochemistries peculiar to Se and Te such as complex oxy-ions of Te. First we consider pyramidal and tetrahedral molecules and ions, then ions formed from tetrahedral SO_4 groups by sharing O atoms, and finally the ions $S_2O_4^{2-}$ and $S_2O_5^{2-}$. Thionates are included in the next main section with compounds containing S_n chains. Per-ions are included in Chapter 11, the crystal structures of acids (H_2SeO_3, $C_6H_5SeO.OH$, H_2SO_4, SO_3NH_3) in Chapter 8, and $Te(OH)_6$ in Chapter 14.

Pyramidal ions and molecules

These include the following types:

$$\left[\begin{matrix} O \\ O \end{matrix} \!\!>\!\! S\!-\!O \right]^{2-} \quad \left[\begin{matrix} R \\ O \end{matrix} \!\!>\!\! S\!-\!O \right]^{-} \quad \begin{matrix} R \\ R \end{matrix} \!\!>\!\! S\!-\!O \quad \left[\begin{matrix} R \\ R \end{matrix} \!\!>\!\! S\!-\!R \right]^{+}$$

(a)   (b)   (c)   (d)

(a) The sulphite, selenite, and tellurite ions have all been shown to have the shape of flat pyramids with the structures given in Table 16.3.

As a ligand SO_3^{2-} bonds to transition metals through S in compounds such as $Coen_2(NCS)SO_3$,[1] $Pd(SO_3)(NH_3)_3$,[2] $Na_2[Pd(SO_3)_2(NH_3)_2].6H_2O$,[3] $Na_6[Pd(SO_3)_4].2H_2O$,[4] and $K_3[Pt\{(SO_3)_2H\}Cl_2]$ (p. 360), but it functions as a bridging ligand in $K_2[Pd(SO_3)_2].H_2O$[5] in which it is bonded through both S and O atoms.

(b) The ion SO_3H^- might have been expected to be of this type, that is, $(O_2S-OH)^-$. Even the existence of hydrogen sulphites ('bisulphites') has been in doubt until recently, but NH_4, Rb, and Cs salts $M(SO_3H)$ are now well known and spectroscopic studies indicate the probable structure $(H-SO_3)^-$, containing a S–H bond. The first X-ray study of a crystalline salt, $CsHSO_3$, shows that the ion has C_{3v} symmetry, and although the H atom was not located it seems likely that the ion has a tetrahedral structure HSO_3^- analogous to the HPO_3^{2-} ions. It is therefore included in Table 16.4(f).

Substituted sulphite ions include SO_2F^- (salts of the alkali metals being prepared by the direct action of SO_2 on MF), and $O_2S.CH_2OH^-$. Sodium hydroxy-methane-sulphinate (dihydrate) is used as a reducing agent in vat dyeing. It is made by reducing a mixture of $NaHSO_3$ and formaldehyde by zinc dust in alkaline solution. The ion has the structure:

$$\left[\begin{array}{c} \text{O} \\ \text{109}^\circ \quad \text{S} \quad \text{CH}_2 \quad \text{OH} \\ \text{O} \end{array} \right]^-$$

with labels: 1·51 Å, 101°, 1·40 Å, 1·83 Å, 109°, 101°

(c) The simplest members of this class are the thionyl halides SOX_2 and analogous Se compounds (Table 16.3): for adducts of $SeOF_2$ and $SeOCl_2$ see p. 740. The optical activity of unsymmetrical sulphoxides such as $OS(CH_3)(C_6H_4COOH)$ and structural studies of $(CH_3)_2SO$ and $(C_6H_5)_2SO$ have confirmed the pyramidal shape of these molecules.

(d) The non-planar configuration of sulphonium ions was deduced from the optical activity of unsymmetrical ions such as $[S(CH_3)(C_2H_5)(CH_2COOH)]^+$ and has been confirmed by X-ray studies of the salts $[S(CH_3)_3]I$ and $[S(CH_3)_2C_6H_5]ClO_4$ (Table 16.3).

(1) AC 1969 **B25** 946
(2) JCS A 1967 1194
(3) JCS A 1969 260

(4) AC 1981 **B37** 19
(5) AC 1979 **B35** 815

Tetrahedral ions and molecules

By the addition of one O atom to the four kinds of pyramidal ions and molecules of the previous section we have:

$$\left[\begin{array}{c} \text{O} \quad \text{O} \\ \text{O} \quad \text{S} \quad \text{O} \end{array} \right]^{2-} \qquad \left[\begin{array}{c} \text{R} \quad \text{O} \\ \text{O} \quad \text{S} \quad \text{O} \end{array} \right]^{-} \qquad \begin{array}{c} \text{R} \quad \text{O} \\ \text{R} \quad \text{S} \quad \text{O} \end{array} \qquad \left[\begin{array}{c} \text{R} \quad \text{R} \\ \text{R} \quad \text{S} \quad \text{O} \end{array} \right]^{+}$$

(e)   (f)   (g)   (h)

Replacement of one O by S in (e) and (f) gives respectively the thiosulphate ion, (i), and ions such as the methane thiosulphonate ion, (j):

$$\left[\begin{array}{c} \text{O} \quad \text{O} \\ \text{O} \quad \text{S} \quad \text{S} \end{array} \right]^{2-} \qquad \left[\begin{array}{c} \text{R} \quad \text{O} \\ \text{O} \quad \text{S} \quad \text{S} \end{array} \right]^{-}$$

(i)   (j)

(e) The regular tetrahedral shape of the sulphate ion has been demonstrated in many salts, and numerous selenates are isostructural with the sulphates. Te^{VI}, however, does not form a tetrahedral MO_4^{2-} ion; tellurates M_2TeO_4 and $MTeO_4$ contain anions built of octahedral TeO_6 groups (p. 735).

TABLE 16.3
Pyramidal molecules and ions

(a)

Compound	M—O	M—OH	O—M—O	Reference
$(SO_3)Na_2$	1.51 Å[a]		106°[a]	ACSc 1972 **26** 218
$(SO_3)Fe$	1.543		103°	AC 1980 **B36** 1297
$(SeO_3)Mg.6H_2O$	1.69		101°	AC 1966 **20** 563
$(SeO_2OH)Li$	1.66	1.79		AC 1979 **B35** 3011
$SeO(OH)_2$	1.64	1.74		ACSc 1971 **25** 1233
$(TeO_3)Na_2.5H_2O$	1.86		99.5°	AC 1979 **B35** 1337

(a) These bond lengths and angles are dependent on the environment of the O atom(s); S—O ranges from 1.45–1.55 Å and O—S—O from 114–102°.

(b)

Molecule or ion	S—O	S—C	O—S—O	Reference
$O_2S(CH_2OH)^-$	1.51 Å	1.83 Å	109°	JCS 1962 3400
$O_2S.C(NH_2)_2$	1.49	1.85	112°	AC 1962 **15** 675
$HO.OS(CH_3)$	1.50 (S—O)	1.79	108°	AC 1969 **B25** 350
	1.60 (S—OH)			

(c)

Molecule	R—S—O	R—S—R	S—O	S—R	Method	Reference
F_2SO	106°	92°	1.42 Å	1.58 Å	e.d.	JMSt 1973 **16** 69
Cl_2SO	106°	114°[a]	1.45	2.07	e.d.	JACS 1938 **60** 2360
Br_2SO	108°	96°	1.45[b]	2.27	e.d.	JACS 1940 **62** 2477
$(CH_3)_2SO$	107°	97°	1.53	1.80	X	AC 1966 **21** 12
$(C_6H_5)_2SO$	106°	97°	1.47	1.76	X	AC 1957 **10** 417
	R—Se—O	R—Se—R	Se—O	Se—R		
$SeOF_2$	105°	92°	1.58	1.73	m.w.	JMS 1968 **28** 461
$SeOCl_2$	106°	97°	1.61	2.20	e.d.	JMS 1976 **31** 261

(d)

Salt	S—C	C—S—C	Reference
$S(CH_3)_3I$	1.83 Å	103°	ZK 1959 **112** 401
$[S(CH_3)_2(C_6H_5)]ClO_4$	1.82	103°	AC 1964 **17** 465

(a) Angles less than 106° not tested. (b) Assumed.

The various ways in which SO_4^{2-} can behave as a simple bridging or as a chelate ligand are discussed on pp. 333 *et seq.* Simple examples include the finite ion $[(en)_2Co(NH_2)(SO_4)Co(en)_2]^{3+}$ (p. 1220) and the infinite chain ion in $K_2MnF_3SO_4$,[1]:

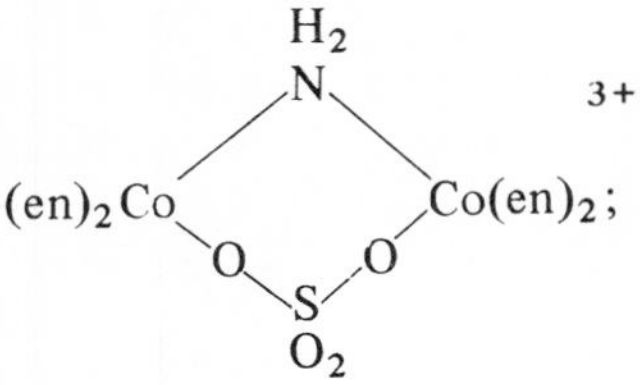

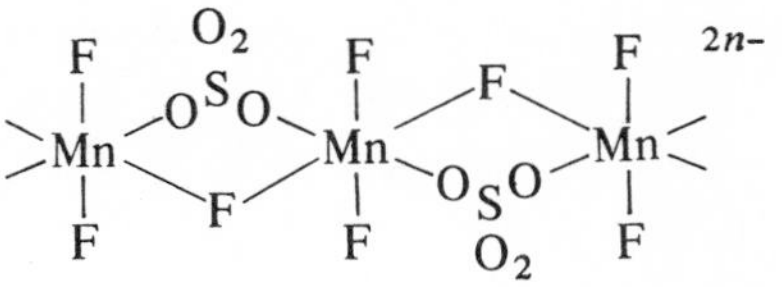

(f) We have already commented on the ion $(HSO_3)^-$. Sulphonate ions RSO_3^- have been studied in a number of salts. The very stable alkali fluorosulphonates, KSO_3F etc., are isostructural with the perchlorates. In the ethyl sulphate ion in $K(C_2H_5O.SO_3)$ and in the bisulphate ion in salts such as $NaHSO_4.H_2O$ and $KHSO_4$ the bond S–OR or S–OH is appreciably longer than the other three S–O bonds (Table 16.4). There are minor variations in these bond lengths depending on the environment of the O atoms:

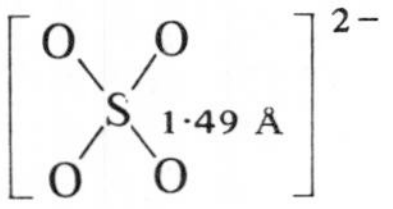

M_2SO_4

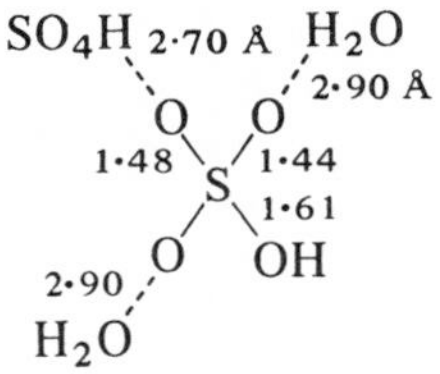

$NaHSO_4 . H_2O$

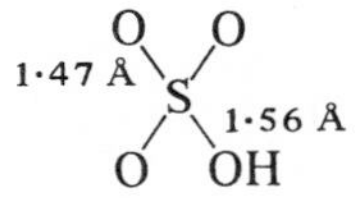

$KHSO_4$

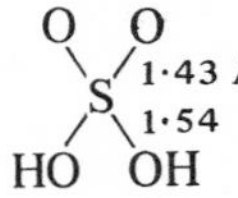

H_2SO_4

With the $(HSO_4)^-$ ion we may compare the $(HSeO_4)^-$ ion in $(H_3O)(HSeO_4)$, p. 691, in which the bond lengths are Se–O, 1.62 Å, and Se–OH, 1.71 Å. The structure of the aminosulphonate (sulphamate) ion, $SO_3NH_2^-$, has been accurately determined in the K salt, (I). Since the angles S–N–H and H–N–H are both 110° the N bonds are not coplanar, as suggested by an earlier X-ray study. We may include here two zwitterions. A neutron diffraction study of sulphamic acid, $^-O_3S.NH_3^+$, shows that the molecule has approximately the staggered configuration, (II). Of special interest are the N–H---O bonds (2.93–2.98 Å) in which the N–H bond makes angles of 7°, 13°, and 14° with the N–O vectors; there are three hydrogen bonds from each NH_3^+. Hydroxylamine-O-sulphonic acid has the structure (III):

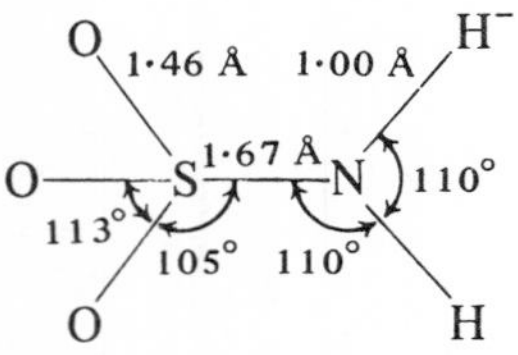

(I)

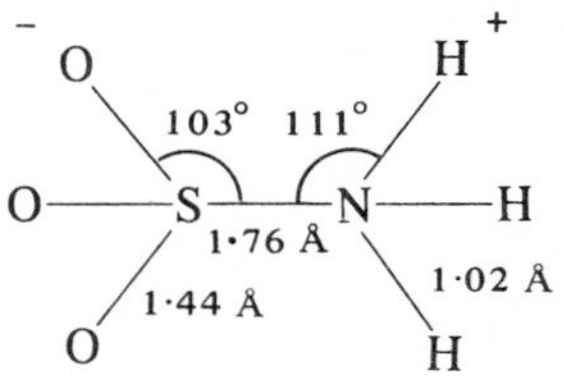

(II)

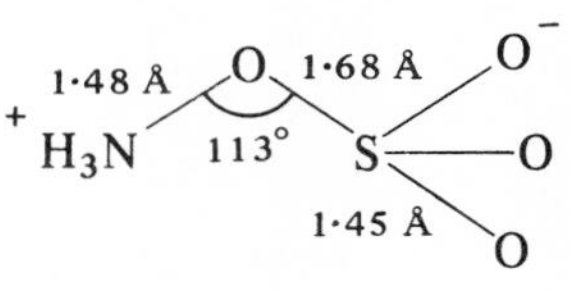

(III)

The structures of some hydroxylamine O- and N-sulphonates are described on p. 797.

(g) Structural studies have been made of the sulphuric acid molecule, of several sulphones, R_2SO_2, of sulphamide, $(NH_2)_2SO_2$, and of the sulphuryl halides SO_2F_2 and SO_2Cl_2. Microwave results for SO_2F_2 may be compared with those for SOF_2:

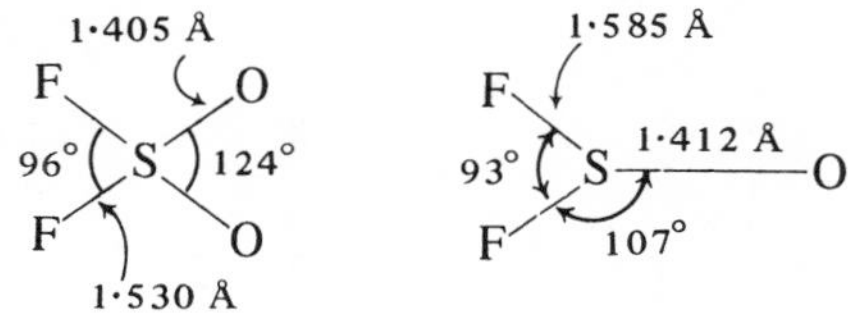

(h) The trimethyl oxosulphonium ion, $(CH_3)_3SO^+$, has been shown to be tetrahedral in the perchlorate and tetrafluoroborate, with C–S–O, 112.5°, and C–S–C, 106°.

(i) A warm solution of an alkali sulphite M_2SO_3 dissolves S to form the thiosulphate. In this reaction there is simple addition of S to the pyramidal sulphite ion forming the tetrahedral $S_2O_3^{2-}$ ion which has been studied in a number of salts (S–S, 2.01 Å). Thiosulphuric acid has been prepared by the direct combination of H_2S with SO_3 in ether at $-78\,°C$ and isolated as the ether complex, $H_2S_2O_3.2(C_2H_5)_2O$.

As a ligand the $S_2O_3^{2-}$ ion appears to behave in several ways, as a bridging ligand utilizing only one S atom in $Na_4[Cu(NH_3)_4][Cu(S_2O_3)_2]_2$,[2] or utilizing both S and O in $Zn(etu)_2S_2O_3$,[3] as a monodentate ligand bonded through a S atom, in $[Pd(en)_2][Pd(en)(S_2O_3)_2]$,[4] and as a bidentate ligand forming a weak bond from

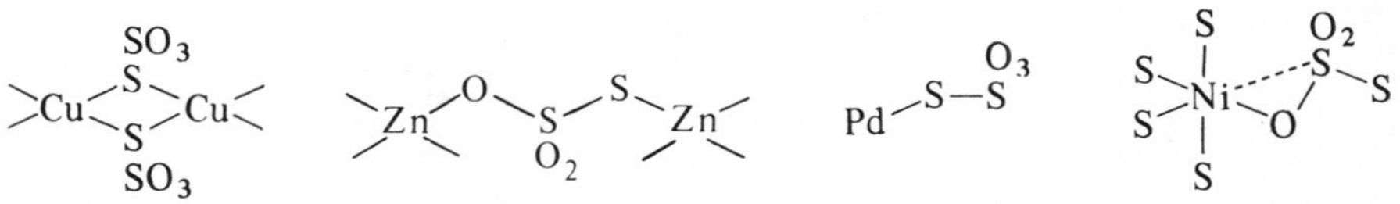

S, in $Ni(tu)_4S_2O_3.H_2O$.[5] The Ni–O bond has the normal length (*c.* 2.1 Å) but Ni–S (2.70 Å) is longer than normal (2.4–2.6 Å).

(j) An example of an ion of this type is the methane thiosulphonate ion, which has the structure shown in the monohydrated Na salt.

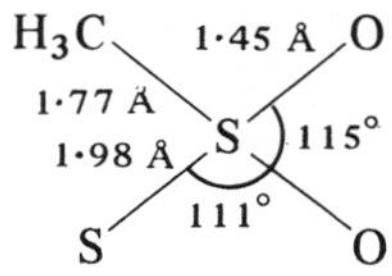

TABLE 16.4
Tetrahedral molecules and ions

(e)

Ion	M–O	Reference
$(SO_4)^{2-}$	1.49 Å	AC 1972 **B28** 2845
$(SeO_4)^{2-}$	1.65	AC 1977 **B33** 592

(f)

Salt or zwitterion	S–O	S–X	O–S–O	O–S–X	
$(SO_3H)^-Cs$	1.45 Å	–	113°	–	AC 1980 **B36** 2523
$(SO_3F)^-Li$	1.43	1.56 Å	113°	106°	AC 1978 **B34** 38
$(SO_3OH)^-Na.H_2O$	1.44, 1.48	1.61	113°	106°	AC 1965 **19** 426
$(SO_3OH)^-K$	1.47	1.56	113°	106°	AC 1964 **17** 682
$(SO_3OC_2H_5)^-K$	see text				AC 1958 **11** 680
$(SO_3NH_2)^-K$	1.46	1.67	113°	110°	AC 1967 **23** 578
$^-(O_3SONH_3)^+$	1.45	1.68	116°	102°	IC 1967 **6** 511
		(S–ONH₃)			
$^-(O_3SNH_3)^+$	1.44	1.77	116°	102°	AC 1976 **B32** 1563

(g)

Molecule	R–S–R	O–S–O	S–O	S–R	
F_2SO_2	97°	123°	1.40 Å	1.53 Å	JMSt 1978 **44** 187
Cl_2SO_2	111°	120°	1.43	1.99	JACS 1938 **60** 2360
$(NH_2)_2SO_2$	112°	119½°	1.39	1.60	AC 1956 **9** 628
$(OH)_2SO_2$	104°	119°	1.43	1.54	AC 1965 **18** 827
	R–Se–R	O–Se–O	Se–O	Se–R	
F_2SeO_2	94.1°	126.2°	1.575	1.685	JMSt 1978 **44** 187

(h)–(j)

Salt	S–O	S–S	S–C	
$[(CH_3)_3SO]^+ClO_4$	1.45 Å	–	1.78 Å	AC 1963 **16** 676, 883
$(S_2O_3)^{2-}Na_2.5H_2O$	1.47	2.02		AC 1978 **B34** 1975
$(CH_3S_2O_2)^-Na.H_2O$	1.45	1.98	1.77	ACSc 1964 **18** 619

Details of the structures of tetrahedral ions and molecules (e)–(j) are given in Table 16.4.

(1)	JCS A 1971 3074	(4)	AC 1970 **B26** 1698
(2)	AC 1966 **21** 605	(5)	AC 1969 **B25** 203
(3)	AC 1974 **B30** 2166		

The pyrosulphate and related ions

The three related ions

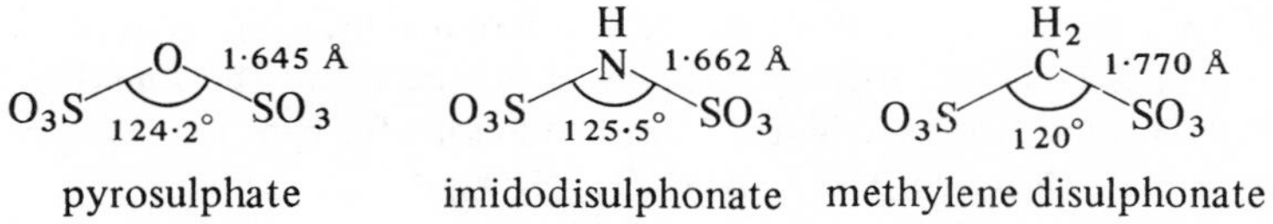

pyrosulphate   imidodisulphonate   methylene disulphonate

and also the ion $(O_3S-N-SO_3)^{3-}$ have very similar structures (Table 16.5). A neutron diffraction study of the ion $(O_3S-NH-SO_3)^{2-}$ shows that the H atom is statistically distributed between the two tetrahedral positions.

TABLE 16.5

The pyrosulphate and related ions (in K salts)

	S–O terminal	S–O (N, C) bridging	S–M–S	*Reference*
$(O_3S-O-SO_3)^{2-}$	1.437 Å	1.645 Å	124.2°	JCS 1960 5112
$(O_3S-NH-SO_3)^{2-}$	1.46	1.68	124°	AC 1979 **B35** 1308
$(O_3S-CH_2-SO_3)^{2-}$	1.461	1.770	120°	JCS 1962 3393
$(O_3S-N-SO_3)^{3-}$	1.48	1.62	121°	AC 1979 **B35** 1308

In $(NO_2)(HS_2O_7)$ and $Se_4(HS_2O_7)_2$ (q.v.) the anion consists of a chain of hydrogen-bonded $S_2O_7H^-$ ions:

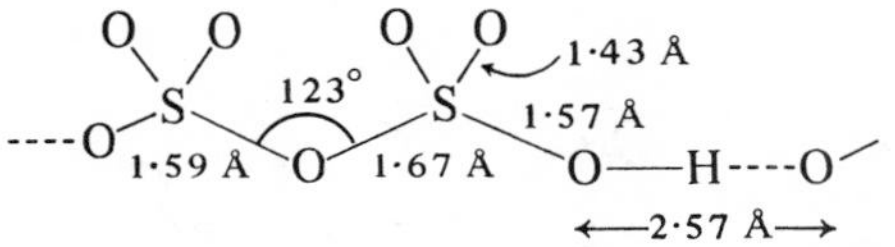

The trisulphate and pentasulphate ions, $S_3O_{10}^{2-}$ and $S_5O_{16}^{2-}$, have been studied in the K salts but accurate bond lengths were not determined in the former ion.[1,2] They are linear chains of three and five SO_4 groups respectively. The configuration found for $S_5O_{16}^{2-}$ has nearly planar terminal SO_3 groups:

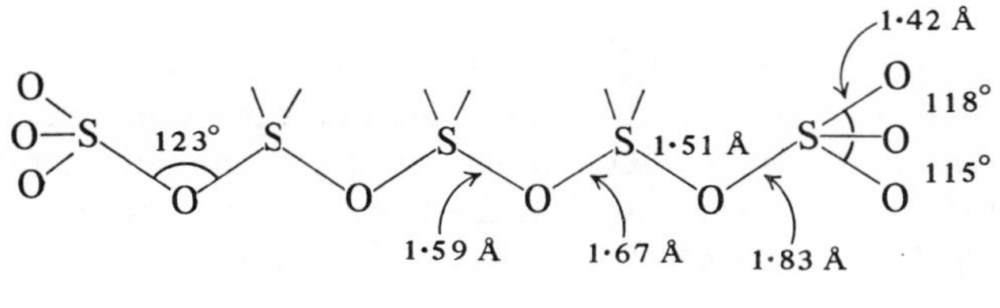

In addition to the apparent slight alternation in S–O bond lengths in the central portion there is an extremely long S–O bond (1.83 Å) to the terminal SO_3 groups, suggesting that the system is closer to a $S_3O_{10}^{2-}$ ion weakly bonded to two SO_3 molecules. It would be desirable to have an independent confirmation of the long

S—O bond in another salt of this type. A long terminal bridging bond (1.72 Å) was also found in $S_3O_{10}^{2-}$, but high accuracy was not claimed. The ions $HS_2O_7^-$ and $S_3O_{10}^{2-}$ exist, together with nitronium ions, in the salts formed from SO_3 and N_2O_5 or HNO_3, namely, $(NO_2)HS_2O_7$[3] and $(NO_2)_2S_3O_{10}$.[4]

In the molecule $Te(S_2O_7)_2$ the two S_2O_7 groups act as bidentate ligands, the lone pair occupying an equatorial position in a trigonal bipyramidal coordination group.[4a]

Dithionites

The $S_2O_4^{2-}$ ion exists in sodium dithionite, the reducing agent, 'hydrosulphite'. This salt is prepared from the Zn salt, which results from the reduction of SO_2 by

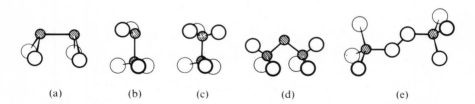

Zn dust in aqueous suspension. In solution it rapidly decomposes to disulphite and thiosulphate: $2S_2O_4^{2-} \rightarrow S_2O_5^{2-} + S_2O_3^{2-}$. The structure of this ion is rather extraordinary (Fig. 16.8(a)).[5] It has an eclipsed configuration with the planes of the two SO_2 nearly parallel (angle between normals 30°), and the S—S bond is very

FIG. 16.8. The structures of some complex sulphur oxy-ions: (a) the dithionite ion, $S_2O_4^{2-}$; (b) the metabisulphite ion, $S_2O_5^{2-}$; (c) the dithionate ion, $S_2O_6^{2-}$; (d) the trithionate ion, $S_3O_6^{2-}$; (e) the perdisulphate ion, $S_2O_8^{2-}$.

long, corresponding to a (Pauling) bond-order of only about 0.3. This very weak bond is consistent with the fact that almost instantaneous exchange of ^{35}S occurs between the dithionite ion and SO_2 in neutral or acid solution; none takes place between $S_3O_6^{2-}$ and SO_2. Note also that the isoelectronic ClO_2 does not dimerize appreciably. The length of this S—S bond has been confirmed in $ZnS_2O_4 \cdot$ pyridine.[6]

Disulphites ('metabisulphites'), diselenites, and ditellurites

If a solution of K_2SO_3 is saturated with SO_2 the salt $K_2S_2O_5$ may be crystallized from the solution. The disulphite ion has the configuration shown in Fig. 16.8(b);[7] addition of SO_2 to the pyramidal SO_3^{2-} ion has produced the unsymmetrical ion

$O_2S-SO_3^{2-}$ in which the two parts are held together by a rather weak S–S bond, (a). The structures of the analogous Se and Te compounds are quite different. In $(NH_4)_2Se_2O_5$[8] and $ZnSe_2O_5$[9] there are well-defined $Se_2O_5^{2-}$ ions with the expected O-bridged structure (b) and pyramidal SeO_3 groups (O–Se–O close to 100°). There are similar anions in $CuSe_2O_5$[10] where Se has one next nearest O neighbour at 2.83 Å (compare the next shortest Se–O in the Zn salt, 3.02 Å). There are quite similar Se_2O_5 groups in VSe_2O_6[11] (Se–O_b, 1.80 Å; Se–O_t, 1.69 Å) which

(a)   (b)   (c)   (d)

may be regarded as an assembly of VO^{2+} and $Se_2O_5^{2-}$ ions. In this structure all O atoms except the bridging O atoms of $Se_2O_5^{2-}$ are bonded to V(IV), which is off-centre in an octahedral coordination group (V–O_{axial}, 1.61 and 2.32 Å; V–$O_{equat.}$, 1.98 Å). In $(NH_4)_2Te_2O_5$[12] the shorter Te–O bonds delineate Te_2O_5 groups similar to the $Se_2O_5^{2-}$ ion, but each Te forms a fourth bond (of different lengths) which link the Te_2O_5 groups into chains, (c). The Te coordination here mey be described as (distorted) trigonal bipyramidal (TeX_4E). In $CuTe_2O_5$[13] also there are Te_2O_5 groups but the coordination of the two Te atoms is different, (d).

(1)	AC 1964 **17** 684	(7)	AC 1971 **B27** 517
(2)	AC 1969 **B25** 1696	(8)	AC 1980 **B36** 675
(3)	AC 1954 **7** 402	(9)	AC 1974 **B30** 2840
(4)	AC 1964 **17** 680	(10)	AC 1976 **B32** 2664
(4a)	AC 1981 **B37** 218	(11)	AC 1974 **B30** 2834
(5)	AC 1956 **9** 579	(12)	AC 1978 **B34** 2830
(6)	AC 1978 **B34** 1499	(13)	AC 1973 **B29** 963

The stereochemistry of molecules and ions containing S_n chains

The atoms of a chain S_3 are necessarily coplanar, but chains of four or more S atoms present interesting possibilities of isomerism because in a system $S_1S_2S_3S_4\ldots$ the dihedral angles $S_1S_2S_3/S_2S_3S_4$, etc., are in the region of 90°. We have already noted that in O_2H_2 the O–H bonds are not coplanar, and this applies also to S_2R_2 molecules and the similar compounds of Se and Te. To describe the structure of such a molecule it is necessary to specify not only the interbond angles and bond lengths but also the dihedral angle ϕ (see (a), p. 501). For S these dihedral angles range from 70° to 110°; for Se recorded values range from 75° to 120° and for Te from 72° to 110°). Some of these dihedral angles are listed in Table 16.6.

Since the stereochemistry of a molecule $S_{n-2}R_2$ is similar to that of a sulphur chain S_n the following remarks apply equally to three main groups of compounds:

TABLE 16.6

Structures of molecules S_2R_2, Se_2R_2, Te_2R_2,
and molecules containing chains of S (Se) *atoms*

Compound	Dihedral angle	S–S–R	S–S	S–R	Method	Reference
S_2H_2	$\approx 90°$	$(95°)$	2.05 Å	(1.33 Å)	e.d.	JACS 1938 **60** 2872
S_2F_2	88°	108°	1.89	1.64	m.w.	JACS 1964 **86** 3617
S_2Cl_2	83°	107°	1.97	2.07	e.d.	} BCSJ 1958 **31** 130
S_2Br_2	84°	105°	1.98	2.24	e.d.	
$S_2(CH_3)_2$	84°	104°	2.02	1.81	e.d.	TFS 1971 **67** 3216
$S_2(C_6H_5)_2$	96°	106°	2.02	1.79	X	AC 1975 **B31** 327
$S_2(C_6F_5)_2$	76.5°	101°	2.06	1.77	X	JOC 1976 **121** 333
$S_2(C_6H_5CH_2)_2$	92°	103°	2.02	1.85	X	AC 1969 **B25** 2497
$S_3(CN)_2$		See text			X	AC 1977 **B33** 2264
$S_3(CH_3)_2$	93°	104°	2.04	1.78	e.d.	JCP 1948 **16** 92
$S_3(CF_3)_2$	–	104°	2.03	1.85	e.d.	TFS 1954 **50** 452
$S_3(CCl_3)_2$	94°	106° (SSS)	2.03	1.90	X	ZK 1961 **116** 290
		102° (SSC)				
$S_3(C_2H_4I)_2$	84°	113° (SSS)	2.05	1.86	X	JACS 1950 **72** 2701
		98° (SSC)				
		Se–Se–R	Se–Se	Se–R		
$Se_2(CH_3)_2$	87.5°	99°	2.33	1.95	e.d.	JCP 1971 **55** 1071
$Se_2(C_6H_5)_2$	82°	106°	2.29	1.93	X	AC 1952 **5** 458
$Se_2(C_6F_5)_2$	75°	99°	2.32	1.91	X	JOC 1976 **121** 333
$Se_2[(C_6H_5)_2HC]_2$	82°	100°	2.39	1.97	X	AC 1969 **B25** 1090
$Se_2[S(C_2H_5)_2P]_2$	104.5°	106°	2.33	2.28	X	ACSc 1966 **20** 51
$Se_2(p\text{-}Cl.C_6H_4)_2$	74.5°	101°	2.33	1.93	X	AC 1957 **10** 201
$Se_2(CN)_2$		See text			X	ACSc 1954 **8** 1787
		R–Te–Te	Te–Te	Te–R		
$Te_2(C_6H_5)_2$	88.5°	99°	2.71	2.12	X	AC 1972 **B28** 2438
$Te(p\text{-}ClC_6H_4)_2$	72°	94°	2.70	2.13	X	AC 1957 **10** 201
$Te_2(p\text{-}CH_3C_6H_4)_2$	86°	101°	2.70	2.13	X	AC 1979 **B35** 1727

molecules $S_{n-2}R_2$, polysulphide ions S_n^{2-}, polythionate ions $S_nO_6^{2-}$ and closely related molecules such as $TeS_{n-1}O_4R_2$.

The stereochemistry of azo compounds, $RN=NR$, is similar to that of, for example, dihalogenoethylenes, the compounds existing in planar *cis* and *trans* forms:

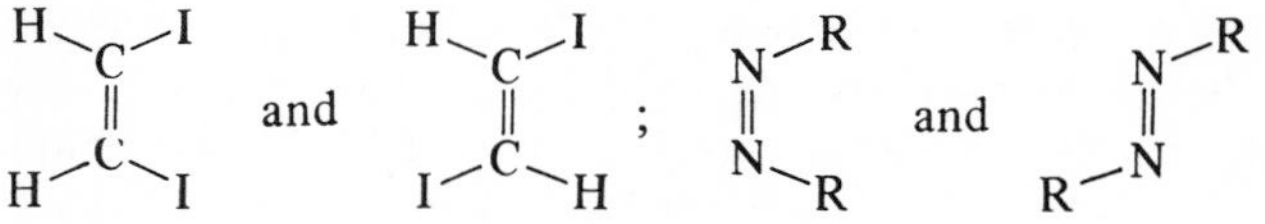

The two forms of a molecule, R_1SSR_2, on the other hand, differ in having R_2 on different sides of the plane SSR_1:

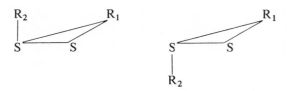

Although they appear to correspond to the *cis* and *trans* isomers of an azo compound they are in fact *d* and *l* enantiomorphs. Examples include S_2H_2, S_2F_2, and S_2Cl_2, in all of which the dihedral angle is close to 90°.

In order to illustrate the possible types of isomers of longer S_n chains it is convenient to idealize the dihedral angle to 90°. In Fig. 16.9(a) and (b) we show the *d* and *l* forms of RSSR or $-S_4-$. The possible structures of the S_5 chain are found by joining a fifth S atom to S_4 in (a) so that the dihedral angle of 90° is maintained. Three of the five positions for S_5 are coplanar with $S_2S_3S_4$ leaving the

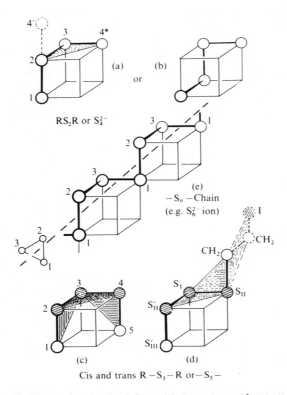

FIG. 16.9. Stereochemistry of molecules S_2R_2 and S_3R_2 and ions S_n^{2-}, idealized with interbond and dihedral angles equal to 90°. (a) is given for comparison with (c), (d), and (e), and (b) for comparison with the sketch of H_2O_2 on p. 501.

two possibilities shown in (c) and (d), which are described as the *cis* and *trans* forms of the $-S_5-$ chain (or RS_3R molecule). Of these the *trans* form is enantiomorphic. A similar procedure shows that there are three forms of the $-S_6-$ chain, all of which are enantiomorphic (Table 16.7). The *cis* form of S_5 and the *cis–cis* form of S_6 correspond to portions of the S_8 ring, while the *trans* form of S_5 and the *trans–trans* form of S_6 correspond to portions of the infinite helical chain of 'metallic' Se or Te (or fibrous S) – Fig. 16.9(e).

Molecules S_2R_2 and S_3R_2

Structural data and references for these compounds are summarized in Table 16.6 (p. 728). The halides S_2X_2 have already been mentioned (p. 708).

The two forms, *cis* and *trans*, of a molecule RS_3R are shown, in idealized form, in Fig. 16.9(c) and (d), the three shaded circles representing S atoms. Dicyano-trisulphane, $S_3(CN)_2$, has the *cis* configuration with S–S, 2.07; S–C, 1.70; C–N, 1.16 Å, and angles S–S–S, 105°; S–S–C, 99°; and S–C–N, 178°. The molecule $Se_3(CN)_2$ has a similar structure, with Se–Se, 2.33 Å and Se–Se–Se, 101°. The molecule 2:2′-diiododiethyltrisulphane, $S_3(C_2H_4I)_2$, has the configuration indicated in Fig. 16.9(d), where the shaded circles represent the S atoms and one of the $-CH_2-CH_2-I$ groups is shown at the right. The terminal portion $-S-CH_2-CH_2-I$ is planar and has the *trans* configuration, so that $S_{II}S_IS_{II}$, $S_IS_{II}C$, and $S_{II}CCI$ lie in three mutually perpendicular planes. The corresponding dihedral angles are close to 90° (82° and 85° respectively), and the S–S–S interbond angle is 113°.

Examples of S_n chains with various configurations are listed in Table 16.7.

TABLE 16.7
Configuration of S_n chains

	Forms	*Fig. 16.9*	*Examples*
S_4 or RSSR	*d* or *l*	(a), (b)	S_4^{2-}, $S_4O_6^{2-}$, $S_4O_6(CH_3)_2$
S_5 or RS_3R	*cis*	(c)	$S_5O_6^{2-}$ in Ba salts
	trans (d and l)	(d)	$S_3(C_2H_4I)_2$, $S_5O_6^{2-}$ in K salt
S_6 or RS_4R	*cis–cis (d and l)*		$S_6O_6^{2-}$ in $K_2Ba(S_6O_6)_2$
	cis–trans (d and l)		
	trans–trans (d and l)	(e)	S_6^{2-}, $S_6O_6^{2-}$ in $[Co(en)_2Cl_2]S_6O_6 \cdot H_2O$

Polysulphides

These are formed, with the normal sulphide, when an alkali or alkaline-earth carbonate is fused with S or, preferably, when S is digested with a solution of the sulphide or hydrosulphide. From the resulting deeply-coloured solutions various salts can be obtained containing ions from S_2^{2-} to S_6^{2-}. The alkali-metal compounds may also be prepared directly from the elements in liquid ammonia. The relative stabilities of the crystalline alkaline polysulphides vary from metal to metal. The following well-defined compounds have been prepared: for M_2S_n, $n = 2, 4$, and 5 for Na; 2, 3, 4, 5, and 6 for K; 2, 3, and 5 for Rb; and 2, 3, 5, and 6 for Cs.

By treating a solution of Cs_2S with S the salt $Cs_2S_5 \cdot H_2O$ is obtained. This can be dehydrated *in vacuo* to Cs_2S_5, but evaporation of an alcoholic solution yields Cs_2S_6, suggesting that a solution of Cs_2S_5 contains S_6^{2-} and also lower S_n^{2-} ions. In fact, it seems probable that all polysulphide solutions are equilibrium mixtures of various S_n^{2-} ions (and M^+ or M^{2+} ions). Such solutions are immediately decomposed by acids, but if a polysulphide solution is poured into aqueous HCl a yellow oil separates from which pure acids H_2S_n (n, 2-5) may be separated by fractional distillation as yellow liquids. H_2S_6, H_2S_7, and H_2S_8 have been prepared by the reaction: $2H_2S_2 + S_nCl_2 \rightarrow 2HCl + H_2S_{4+n}$.

The structures of all the ions from S_2^{2-} to S_6^{2-} have been studied in crystalline salts, usually of alkali or alkaline-earth metals. (For the Te_3^{2-} anion see p. 307). Unfortunately most of these salts contain a very heavy cation and the accuracy of the dimensions of the polysulphide ions is therefore difficult to estimate. References are given to some of the more recent studies. The ions S_4^{2-} and S_6^{2-} are shown, in idealized form, in Fig. 16.9. A number of determinations give terminal S–S close to 2.07 Å, but higher values (e.g. 2.16 Å) have been given for the intermediate bonds in the longer chains. Interbond angles range from 105° (K_2S_3) to 115° (BaS_3) and dihedral angles from around 70° ($BaS_4 \cdot H_2O$ and Tl_2S_5) to 106° (Cs_2S_6). Points of general interest include the following. K_2S_2 and one form of Na_2S_2 are isostructural with Na_2O_2. The salt Ba_2S_3 contains anions of two kinds, $Ba_2S^{2-}(S_2)^{2-}$, and rather more unexpectedly there are two kinds of S atom in LaS_2. One half form S_2 groups (S–S, 2.10 Å) while the remainder are discrete S^{2-} ions; the structural formula may therefore be written $La_2(S^{2-})_2(S_2)^{2-}$. In Cs_2S_6 the anion has the form of an unbranched chain and there are relatively short van der Waals distances (3.4 Å) between the ends of chains. These weak attractions between successive S_6^{2-} ions link them into infinite helices of S atoms very similar to the helical chains in metallic Se and Te.

The S_2 group is found not only as the S_2^{2-} ion in the ionic disulphides noted above but also in the numerous transition metal sulphides MS_2 with the pyrites and marcasite structures (pp. 758-60), and acting as a chelate ligand in the ions $[Mo_2(S_2)_6]^{2-,(1a)}$ (a), $[Mo_2S_2(S_2)_2O_2]^{2-,(1b)}$ (b), and $[Mo_3S(S_2)_6]^{2-,(1c)}$ (c).

By bonding to a metal atom through the two terminal S atoms the S_5^{2-} ion forms

(a)   (b)   (c)

chelate complexes $M(S_5)_2$ and $M(S_5)_3$. In $(NH_4)_2Pt(S_5)_3 \cdot 2H_2O^{(2)}$ the 6-membered rings so formed have the chain configuration, with Pt–S, 2.39 Å; S–S, 2.05 Å; S–S–S, 111°; and dihedral angles 72–83°.

References to polysulphides. Na_2S_2, K_2S_2: ZaC 1962 **314** 12. BaS_2: AC 1975 **B31** 2905. SrS_2: AC 1976 **B32** 3110. LaS_2: AC 1978 **B34** 403. K_2S_3: ZaC 1977 **432** 167. BaS_3, Ba_2S_3: IC 1975 **14** 129. Na_2S_4: AC 1973 **B29** 1463. $BaS_4 \cdot H_2O$: AC 1969 **B25** 2365. K_2S_5: JCS D 1976 1314. Tl_2S_5: AC 1975 **B31** 1675. Cs_2S_6: ACSc 1968 **22** 3029.

(1a) IC 1980 **19** 2835   (1c) B 1979 **112** 778
(1b) AC 1980 **B36** 2784   (2) AC 1969 **B25** 745

Thionates and molecules of the type $S_n(SO_2R)_2$

The lower thionate ions have the general formula $S_nO_6^{2-}$, where $n = 2, 3, 4, 5,$ or 6. The dithionate ion, $S_2O_6^{2-}$, is formed by the direct union of two SO_3^{2-} ions, for example by the electrolysis of neutral or alkaline solutions of a soluble sulphite or by passing SO_2 into a suspension of MnO_2, in water. This ion is not formed by the methods which produce the higher thionates, for these require S in addition to SO_3 groups, as may be seen from their formulae, $(O_3S-S_n-SO_3)^{2-}$, and they are formed whenever S and SO_3^{2-} ions are present together in aqueous solution. A mixture of all the ions from $S_3O_6^{2-}$ to $S_6O_6^{2-}$ is usually produced in such circumstances (compare the polysulphides), and, moreover, a solution of any one of the higher thionates is converted, though very slowly, into a mixture of them all. The alkali and alkaline-earth salts, however, are very stable in solution and interconversion of ions takes place only in the presence of hydrogen ions. Certain reactions tend to produce more of one particular $S_nO_6^{2-}$ ion, a fact which is utilized in their preparation. The controlled decomposition of a thiosulphate by acid gives largely $S_5O_6^{2-}$, while the oxidation of $S_2O_3^{2-}$ by H_2O_2 gives $S_3O_6^{2-}$ and by iodine (or electrolytically) yields the tetrathionate ion, $S_4O_6^{2-}$, e.g.

$$2S_2O_3^{2-} + I_2 \rightarrow S_4O_6^{2-} + 2I^-$$

Some of the relations between these ions are indicated in the following scheme:

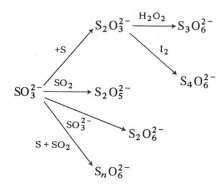

Thiosulphuric acid, $H_2S_2O_3$ may be prepared directly from H_2S and SO_3 in ether solution at $-78\,°C$. If H_2S_n is used instead of H_2S, acids $H(O_3S.S_nH)$ are produced which by the action of further SO_3 give polythionic acids $H_2(O_3S.S_n.SO_3)$. Oxidation by iodine gives acids $H_2(O_3S.S_{2n}.SO_3)$, by which means $H_2S_8O_6$, $H_2S_{10}O_6$, and $H_2S_{12}O_6$ have been prepared.

The structure of the dithionate ion has been studied in a number of salts including $Na_2S_2O_6.2H_2O$,[1] $K_2S_2O_6$,[2] $MgS_2O_6.6H_2O$,[3] and $BaS_2O_6.2H_2O$.[4] Figure 16.8(c) shows the centrosymmetrical configuration of the ion, but various configurations are found, sometimes in the same salt (for example, staggered and nearly eclipsed in $K_2S_2O_6$). A recent study gives S–S, $2.14\,Å$; S–O, $1.45\,Å$; and bond angles, S–S–O, $105°$; O–S–O, $114°$.

The structure of the trithionate ion is shown in Fig. 16.8(d), derived from the crystal structure of $K_2S_3O_6$[5] (S–S, $2.15\,Å$). In the selenotrithionate ion, $(SeS_2O_6)^{2-}$ the Se–S bond length is $2.26\,Å$ and the angle S–Se–S, $98°$,[5a] while in the closely related molecules $S(SO_2C_6H_5)_2$[6] and $Se(SO_2C_6H_5)_2$[7] the bond angles at the central S(Se) atom were determined as $106.5°$ and $105°$ respectively, and the bond lengths S–S and S–Se as 2.07 and $2.20\,Å$.

In the tetrathionate ion, studied in $BaS_4O_6.2H_2O$,[8a] $Na_2S_4O_6.2H_2O$,[8b] and $K_2S_4O_6$,[9] the S_4 chain has the configuration of Fig. 16.9(b), with dihedral angles close to $90°$. There appears to be a definite difference between the lengths of the central and terminal S–S bonds, (a), and this also seems to be the case in dimethane sulphonyl disulphide, (b),[10] with the same S_4 chain, though here it is doubtful if

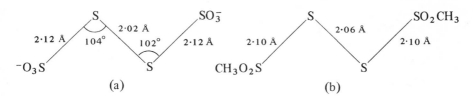

(a) (b)

the differences are larger than the probable experimental errors. The tetrathionate ion is dissymetric, and it is interesting that while crystals of the Na salt contain only d or l forms, the K and Ba salts contain both d and l forms of the ion.

Very different dihedral angles were found in the selenotetrathionate ion, $(SeS_3O_6)^{2-}$ ion in the K^+ and $[Co(en)_2Cl_2]^+$ salts, namely, $81°$ and $119°$,[10a] suggesting that the dihedral angle is very susceptible to changes in the environment of the ion in the crystal. On the other hand, very similar values, $88.5°$ and $92.6°$, both close to the 'expected' value ($90°$), were found in two polymorphs of $[Co(en)_2Cl_2]_2(Se_2S_2O_6).H_2O$.[10b]

Both configurations of the $S_5O_6^{2-}$ ion have been found, *cis* (Fig. 16.9(c)) in various polymorphs and hydrates of $Ba_2S_5O_6$, and *trans* (Fig. 16.9(d)) in $K_2S_5O_6.1\tfrac{1}{2}H_2O$,[11] and different configurations of substituted ions in hydrates of the same salt; for example, *cis* in $BaTeS_4O_6.2H_2O$[12] but *trans* in $BaTeS_4O_6.3H_2O$.[13] Dihedral angles appear to lie in the range 104–110° for the *cis* ions but smaller values are found in the *trans* ions, usually 80–90°, though larger values (88° and 99°) are recorded for $BaTeS_4O_6.3H_2O$. There appears to be a real difference between the lengths of the central and terminal S–S bonds in $S_5O_6^{2-}$ (2.03 and 2.12 Å). The two different configurations have also been found for the closely related molecules of the type

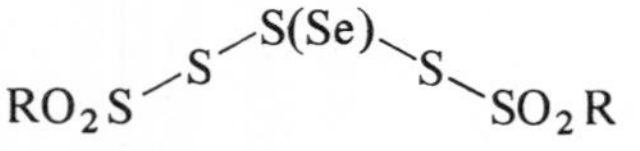

cis 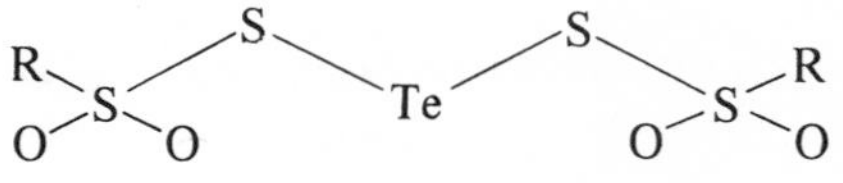   *trans*

where R is CH_3, C_6H_5 etc.

The hexathionates are the most complex salts of this family which have been isolated in the pure state. As already mentioned, there are three possible isomers of a $-S_6-$ chain (each capable of existing in d and l configurations), assuming that free rotation around the S–S bonds is not possible. In $K_2Ba(S_6O_6)_2$ the anion has the *cis-cis* configuration[14] which, like the *cis* configuration of the pentathionate ion, corresponds to a portion of the S_8 ring. In $[Co(en)_2Cl_2]_2S_6O_6.H_2O$, on the other hand, it has the extended *trans-trans* configuration.[15]

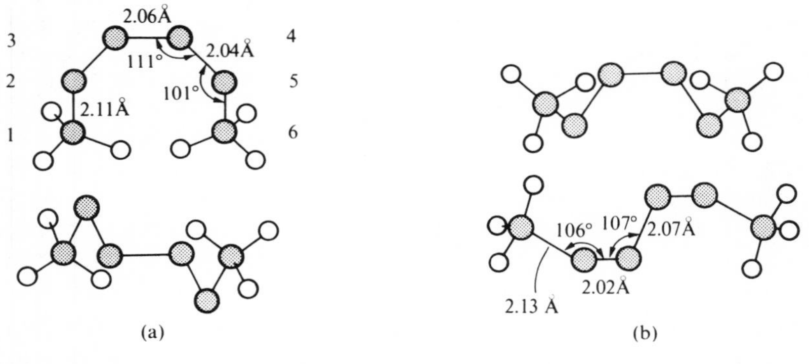

FIG. 16.10. Configurations of the $S_6O_6^{2-}$ ion in (a) $K_2Ba(S_6O_6)_2$; (b) $[Co(en)_2Cl_2]_2S_6O_6.H_2O$.

Two views of the $S_6O_6^{2-}$ ion in these two salts are shown in Fig. 16.10. There are two different dihedral angles in each configuration, the values being:

	$S_1S_2S_3/S_2S_3S_4$	$S_2S_3S_4/S_3S_4S_5$
$K_2Ba(S_6O_6)_2$	107° (mean)	89°
$[Co(en)_2Cl_2]_2S_6O_6.H_2O$	86°	71°

(1)   AC 1980 **B36** 223
(2)   AC 1956 **9** 897
(3)   AC 1975 **B31** 615

(4)   AC 1966 **21** 672
(5)   ZK 1934 **89** 529
(5a) IC 1980 **19** 1063

(6) JCS 1949 724
(7) ACSc 1954 8 42
(8a) ACSc 1954 8 459
(8b) ACSc 1964 18 662
(9) AC 1979 **B35** 1971
(10) ACSc 1953 7 1
(10a) IC 1980 **19** 1040

(10b) IC 1980 **19** 1044
(11) ACSc 1971 **25** 2580
(12) ACSc 1973 **27** 1695
(13) ACSc 1973 **27** 1653
(14) ACSc 1973 **27** 1684
(15) ACSc 1973 **27** 1705

The structural chemistry of selenium and tellurium

This is summarized in Table 16.8 where the horizontal divisions correspond to the formal oxidation states. The examples given have been established by structural studies unless marked by a query (?). (For example, the structures of the gaseous molecules of TeO_2, SeO_3, and TeO_3 are not known.) Configurations to the left of the heavy line are also known for sulphur compounds. Simple examples of some of these valence groups have already been given, for example, elementary Se and Te and molecules SeR_2 and Se_2R_2 (4,4), SeO_3^{2-} (2,6), SeO_4^{2-} (8), molecules SeR_2X_2 (2,8) and SeX_6 (12), and the Te analogues. We are concerned here particularly with stereochemistries exhibited by Se and/or Te and not by S, that is, those to the right of the heavy line, but we shall also note some additional examples of the above-mentioned valence groups found in compounds peculiar to Se or Te.

Se(vi) *and* Te(vi)

Valence group (12). Reference has already been made to the regular octahedral structures of the SeF_6 and TeF_6 molecules. We discuss here the oxygen chemistry of Te(vi), which is the only one of the elements S, Se, and Te found in octahedral coordination by 6 O atoms. The resemblance to iodine, the only halogen exhibiting 6-coordination by O, is marked; compare the isoelectronic pairs, $TeO(OH)_5^-$ and $IO(OH)_5$, or $[Te_2O_8(OH)_4]^{6-}$ and $[I_2O_8(OH)_4]^{4-}$. Note, however, the absence of the TeO_4^{2-} ion.

The structural chemistry of tellurates forms a beautiful illustration of the ways in which octahedral coordination groups can be joined together by sharing O atoms to form finite groups, chains, or 3D structures. The coordination group around Te consists of 6 O atoms, to some or all of which H is attached. Such groups may join either by sharing one O between each pair of octahedra (vertex-sharing structures) or by sharing pairs of adjacent O atoms (edge-sharing structures). The simplest infinite vertex-sharing systems are the *trans* and *cis* AX_5 chains, the AX_4 layer (SnF_4 etc., p. 208), and the 3D AX_3 structure (ReO_3). In the edge-sharing AX_4 chain the edges shared may be opposite (*trans*) or skew. There is only one form of the *trans* chain, the strictly linear chain in, for example, $K[TeO_3(OH)]$, but there is an indefinitely large number of configurations of the skew edge-sharing chain (p. 212); two of these have been found in tellurates. In Na_2TeO_4[1] (and also $CaTeO_4$ and $SrTeO_4$[2]) the chain is the planar zigzag chain of $TcCl_4$ (Fig. 5.20, p. 213), but in Li_2TeO_4[3] it has a helical configuration (4_1 axis), another of the

TABLE 16.8
The stereochemistry of Se and Te

Total number of σ pairs	3	4	5	6	7
Arrangement	Triangular	Tetrahedral	Trigonal bipyramidal	Octahedral	[Octahedral]
Valence group	(6)	(8)	(10)	(12)	(14)
Se(VI)	SeO_3?	SeO_4^{2-}, Se_4O_{12}		SeF_6	
Te(VI)	TeO_3 molecule?			TeO_3 (cryst.), TeF_6 See CHART 16.1	
Valence group	(2,4)	(2,6)	(2,8)	(2,10)	(2,12)
Se(IV)	SeO_2 molecule	SeO_2 (cryst.), SeO_3^{2-} $SeOF_2$, $SeOCl_2$ $Se(CH_3)_3^+$ $(SeF_3)(NbF_6)$ $SeOCl_2$ adducts	SeR_2X_2 $Br_2 . SeC_4H_8S$ $Cl_2SeC_4H_8SeCl_2$	$SeOCl_2(pyr)_2$	$SeCl_6^{2-}$?
Te(IV)	TeO_2 molecule?	TeO_3^{2-} TeS_3^{2-} $Te(CH_3)_3^+$	TeX_4(?) TeR_2X_2, TeO_2 $Te_2O_4 . HNO_3$ $Te_2(C_6H_4O_2)_2$	TeF_4 (cryst.) $(TeF_5)K$ $Te(CH_3)I_4^-$	$TeCl_6^{2-}$, $TeBr_6^{2-}$ $Te(tmtu)_2Cl_4$ TeI_4 (cryst.)
Valence group		(4,4)	(4,6)	(4,8)	
Se(II)		Se (element) SeR_2, Se_2R_2			
Te(II)		Te (element) $TeBr_2$	$Te\phi Cl(tu)$ (26) $[Te\phi(tu)_2]^+Cl^-$	$Te(tu)_4^{2+}$, $Te(tu)_2X_2$ $Te_2(tu)_6^{4+}$, $Te(etu)_2X_2$	

three simplest skew edge-sharing AX_4 chains. This salt has a distorted inverse spinel structure, $Li(LiTe)O_4$.

The structures of Li_6TeO_6, Li_4TeO_5, Li_2TeO_4, and TeO_3 illustrate the sharing of 0, 2, 4, and 6 O atoms of TeO_6 groups. There are discrete TeO_6^{6-} ions in Li_6TeO_6,[4] dimeric ions $Te_2O_{10}^{8-}$ in Li_4TeO_5,[5] chains in Li_2TeO_4,[5] and finally a 3D framework in TeO_3 (p. 717). Tellurate structures may be classified according to the coordination group of Te; Chart 16.1 includes only known structures. Four of the hydroxy-tellurate ions are illustrated in Fig. 16.11. The neutral systems include TeO_3,

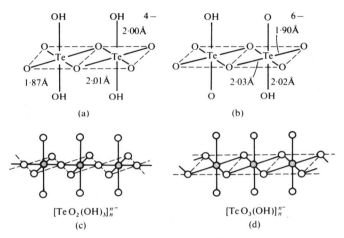

FIG. 16.11. Structures of Te(VI) compounds: (a) and (b) bridged binuclear ions; (c) and (d) infinite anions.

$TeO_2(OH)_2$, and $Te(OH)_6$. The second of these (H_2TeO_4) has the same type of AX_4 layer structure as $UO_2(OH)_2$ (p. 208). Obviously there are many other possible structures for tellurates and 'telluric acids', such as H_4TeO_5, a linear chain formed from $TeO_2(OH)_4$ groups. Bond lengths in these compounds include 1.90 Å in $Te(OH)_6$ (n.d.), 1.86 and 1.97 Å for terminal and bridging Te—O in $K[TeO_3(OH)]$, 1.87 and 1.96 Å for Te—O and Te—OH in $(NH_4)_2[TeO_2(OH)_4]$, and 1.92 Å in Ca_3TeO_6[5a] (monoclinic cryolite structure, p. 462).

The very elegant (cubic) structure of Be_4TeO_7[6] may be described as derived from the sphalerite structure by replacing Zn and S by TeO_6 and Be_4O groups respectively. Alternatively it may be described as a 3D framework built from tetrahedral BeO_4 groups which form 'super-tetrahedra' Be_4O_{10} (see p. 195), and it is described in this way together with other structures built from super-tetrahedra in Chapter 5.

(1) AC 1977 **B33** 2596
(2) AC 1979 **B35** 728
(3) JSSC 1977 **22** 113, 385
(4) ZN 1969 **B24** 647, 1656

(5) AC 1978 **B34** 3156
(5a) AC 1981 **B37** 220
(6) AC 1977 **B33** 3959
See also Chart 16.1

CHART 16.1

Structures of oxy-compounds of Te(VI)

Oxygen : tellurium ratio

6:1	5:1		4:1		3:1
Coordination group	Vertex-sharing chain	Edge-sharing dimer	Vertex-sharing layer	Edge-sharing chain	Vertex-sharing 3D framework
*TeO_6^{6-} $TeO_5(OH)^{5-}$ $TeO_4(OH)_2^{4-}$ $TeO_3(OH)_3^{3-}$ *$TeO_2(OH)_4^{2-}$ (a) *$TeO(OH)_5^-$ (b) *$Te(OH)_6$ (c)	$[TeO_2(OH)_3]_n^{n-}$ (d)	$Te_2O_{10}^{8-}$ $[Te_2O_8(OH)_2]^{6-}$ (e) $[Te_2O_6(OH)_4]^{4-}$ (f)	$TeO_2(OH)_2$ (h)	$(TeO_4)_n^{2n-}$ $[TeO_3(OH)]_n^{n-}$ (g)	TeO_3

*Known as discrete units

(a) AC 1979 **B35** 1684
(b) IC 1964 **3** 364
(c) AC 1980 **B36** 2565 (cubic); ACSc 1973 **27** 85 (monoclinic)
(d) ZaC 1965 **334** 225
(e) ACSc 1969 **23** 3062
(f) ACSc 1966 **20** 2138
(g) ACSc 1972 **26** 4107
(h) AC 1974 **B30** 1813

Se(IV) *and* Te(IV)

Valence group (2,6). There are numerous examples of Se(IV) forming three pyramidal bonds and (usually) three weaker bonds, the simplest being the ions SeO_3^{2-}, SeF_3^+, and $SeCl_3^+$.

The reaction of SeF_4 with NbF_5 produces $(SeF_3)(NbF_6)$[1] and $(SeF_3)(Nb_2F_{11})$.[2] Both salts contains pyramidal SeF_3^+ ions, (a), and Se also forms three weaker bonds to F atoms of the $(NbF_6)^-$ or $(Nb_2F_{11})^-$ ions, (b), so that in addition to its three

nearest (pyramidal) neighbours Se has three more distant F neighbours completing a very distorted octahedral coordination group. The large F–F distances between these three more distant neighbours (3.4–4.0 Å) may be due to the lone pair directed tetrahedrally as shown at (c)—compare the 7-coordination group in the A-La_2O_3 structure. In $(SeF_3)(NbF_6)$ Se–F was found to be 1.73 Å (and 2.35 Å) and the interbond angle 95°.

The compound $AlSeCl_7$[3] consists of tetrahedral $AlCl_4^-$ ions (Al–Cl, 2.13 Å) and pyramidal $SeCl_3^+$ ions (Se–Cl, 2.11 Å), and Se forms 3 weaker bonds of length 3.04 Å. In crystalline $Se(CH_3)_3I$[4] there are pyramidal cations but each is associated

rather closely with only one I^-. Although Se---I is only a weak bond (3.78 Å) this is less than the estimated van der Waals separation (4.15 Å) and in fact shorter than S---I (3.89 Å) in the sulphur analogue. Moreover this weak Se---I bond is almost collinear with one of the Se–C bonds, and the ion pair has been described as a charge-transfer complex. Similarly, the Se---I distance in the complex of I_2 with 1,4-diselenane is slightly less than the S---I distance in the S analogue. At the same time there is an increase in I–I from the value 2.66 Å in the free I_2 molecule to 2.79 Å in the S compound and to 2.87 Å in the Se compound. It appears that the stronger the bond from S or Se to I the weaker is the I–I bond:

	S (Se) − I	I−I	*Reference*
$C_4H_8S_2.2I_2$	2.87 Å	2.79 Å	AC 1960 13 727
$C_4H_8Se_2.2I_2$	2.83	2.87	AC 1961 14 940
$C_4H_8OSe.I_2$	2.76	2.96	IC 1966 5 522
$C_4H_8OSe.ICl$	2.63	2.73 (I−Cl)	IC 1968 7 365

We have noted in our description of the crystalline elements and of polyiodides that the formation of weak additional bonds intermediate in length between normal covalent bonds and van der Waals bonds is a feature of Se, Te, and I. Note the different molecular structures of $I_2Se(C_4H_8)SeI_2$ and $Cl_2Se(C_4H_8)SeCl_2$ (Fig. 16.12).

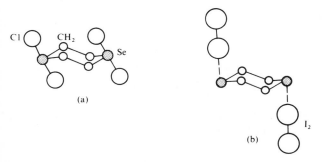

FIG. 16.12. Molecular structures of (a) $Cl_2SeC_4H_8SeCl_2$; (b) $I_2SeC_4H_8SeI_2$.

Details of the $SeOF_2$ and $SeOCl_2$ molecules are included in Table 16.3 (p. 721). The term adduct is used for compounds in which the $SeOX_2$ (or other molecule) retains essentially the same structure as that of the free molecule, in contrast to compounds such as $SeOCl_2.(pyridine)_2$—see later—in which there is rearrangement of the Se, O, and Cl atoms and formation of a (2,10) valence group. For example, in $SeOF_2.NbF_5$[5] the O atom forms part of an octahedral group around Nb (Nb−O, 2.13 Å, Nb−F, 1.85 Å) and the structure of the $SeOF_2$ portion of the molecule is very similar to that of the free molecule (which has a pyramidal shape). Similarly, in $SbCl_5 . SeOCl_2$ [6] and $SnCl_4 . 2SeOCl_2$ [7] the O atoms of $SeOCl_2$ complete the octahedral group around the metal atom. Weaker Se---Cl bonds usually complete a distorted octahedral coordination group around Se. For example, in $SnCl_4 . 2SeOCl_2$ these are Se—Cl, 3.01 Å (intramolecular) and Se—2 Cl, 3.34 and 3.38 Å to Cl in adjacent molecules. In 8-hydroxyquinolinium trichloro-oxyselenate, $C_9H_8NO^+(SeOCl_3)^-$, there are two additional bonds from each Se (2.98 Å) to Cl⁻ ions and together the

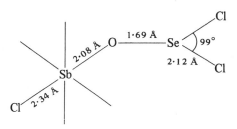

$SeOCl_2$ molecules and Cl⁻ ions form infinite 3-dimensional $SeOCl_3^-$ complexes.[8] The sixth Se---Cl contact is at 3.3 Å, still shorter than the van der Waals radius sum (3.8 Å). The compound $[N(CH_3)_4]Cl . 5SeOCl_2$ is an assembly of $N(CH_3)_4^+$ and Cl⁻ ions and $SeOCl_2$ molecules.[9]

Examples of Te(IV) forming 3 pyramidal bonds include $Te(CH_3)_3^+$ (see later), the TeO_3^{2-} ion (Table 16.3), and the TeS_3^{2-} ion. In $BaTeS_3$ [10] Te forms 3 pyramidal bonds (Te—S, 2.35 Å) and has 3 more distant S neighbours at 3.4 Å.

(1)	JCS A 1970 1891	(6)	ACSc 1967 **21** 1313
(2)	JCS A 1970 1491	(7)	AC 1960 **13** 656
(3)	AC 1971 **B27** 386	(8)	IC 1967 **6** 1204
(4)	AC 1966 **20** 610	(9)	ACSc 1967 **21** 1328
(5)	JCS A 1969 2858	(10)	AC 1976 **B32** 444

Valence group (2,8). This valence group occurs in molecules of the type SeX_2R_2 [1] and TeX_2R_2 [2] which have already been mentioned (p. 708), and it was noted that in these molecules the halogen atoms occupy the axial positions and the lone pair one equatorial bond position. Molecules studied include $(C_6H_5)_2SeCl_2$ and dibromide, $(C_6H_5)_2TeBr_2$ and substituted derivatives, and molecules such as $Cl_2 . Se(CH_2)_4Se . Cl_2$ [3] (Fig. 16.12(a)). In molecules $R_2Se(Te)X_2$ the equatorial angle C—Se—C is 106–110° and in the Te compounds 96–101°, and the X—Se(Te)—X angle is close to 180°. The (axial) Se—X and Te—X bonds are much longer than the sums of the Pauling covalent radii, while the Se—C and Te—C bond lengths are close to the radius sums (Se—C, 1.94 Å; Te—C, 2.14 Å):

	Observed	*Radius sum*
Te−Cl	2.51 Å	2.36 Å
Te−Br	2.68	2.51
Te−I	2.93	2.70

We noted earlier the structures of the two crystalline forms of TeO_2. The similarity of the bond lengths in Te (IV) catecholate[4] and of the four shorter Te−O bonds in

both forms of TeO_2, and in $Te_2O_4.HNO_3$[5] suggests that the same bonding orbitals are used in all these compounds.

	α-TeO_2	β-TeO_2	$Te_2O_4.HNO_3$	$Te(C_6H_4O_2)_2$
Te−O (axial)	(2) 2.08 Å	2.19 Å	2.17 Å	2.11 Å
		2.07	2.05	2.01
Te−O (equatorial)	(2) 1.90	1.93, 1.88	1.93, 1.89	(2) 1.98
O_{ax}−Te−O_{ax}	169°	169°	148°	154°
O_{eq}−Te−O_{eq}	102°	101°	100°	98°

The 'basic nitrate', $Te_2O_4.HNO_3$, crystallizes from a solution of Te in HNO_3. Its structure is built of puckered layers (Fig. 16.13) based on the planar 6^3 net. There are single and double O bridges between pairs of Te atoms which form four 'trigonal bipyramidal' bonds. The O atoms of the NO_3 groups are not bonded to Te, and the distinction between the formulations $Te_2O_4.HNO_3$ and $[Te_2O_3(OH)]^+NO_3^-$ rests on the positions of the H atoms. The bond lengths in the Te_2O_4 layer and the symmetrical nature of the NO_3 group (N−O, 1.25 (0.01) Å) suggest that H is attached to the atom shaded in Fig. 16.13, and that this is one of the rare examples of 2D cations.

$Te_2O_3.SO_4$[6] also has a layer structure based on the 6^3 net (compare monoclinic As_2O_3) which results from the three shortest bonds from each Te atom (1.89, 1.91, and 2.00 Å). Pairs of adjacent Te atoms are bridged by SO_4 groups as shown diagrammatically in Fig. 16.14(a) and (b); the length of these Te−O bonds is 2.26 Å. Since the only other bonds formed by Te are appreciably longer, (c), Te forms four bonds of lengths similar to those in other oxy compounds, but the

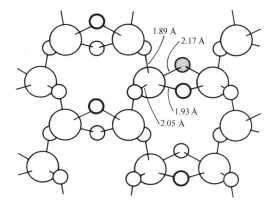

FIG. 16.13. Tellurium–oxygen layer in Te$_2$O$_4$.HNO$_3$ (HNO$_3$ omitted).

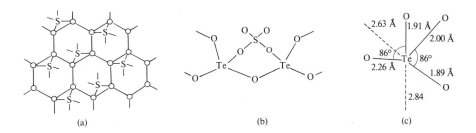

FIG. 16.14. (a) Projection of layer in Te$_2$O$_3$.SO$_4$ – the O atoms on each link are omitted; (b) bridging by bidentate SO$_4$ groups; (c) environment of Te(IV).

arrangement is a very distorted trigonal bipyramidal one. A very similar structure is adopted by Te$_2$O$_3$.HPO$_4$.[7]

(1) AC 1953 **6** 746
(2) AC 1962 **15** 887; IC 1970 **9** 797
(3) AC 1961 **14** 940
(4) ACSc 1967 **21** 1473
(5) MH 1980 **111** 789
(6) AC 1976 **B32** 2720, 3115
(7) ZK 1975 **141** 354

Valence group (2,10). Relatively few structural studies appear to have been made of compounds in which Se(IV) forms five bonds; the arrangement is square pyramidal, or octahedral MX$_5$E. In the finite molecule SeOCl$_2$.2C$_5$H$_5$N,[1] (a), the five ligands form a slightly distorted square pyramidal coordination group, the nearest neighbour in the direction of the sixth octahedral bond being a Cl atom of another molecule.

In the dimeric molecule $(CF_3SeCl_3)_2$,[2] (b), there is apparently a very large difference between the lengths of the terminal and bridging Se—Cl bonds.

The unique structure of TeF_4 has been described with the other halides (p. 709). Te(IV) apparently does not form TeF_6^{2-} but only TeF_5^-, which has been studied in $KTeF_5$.[3] This ion exists as discrete units with the same pyramidal configuration as in TeF_4 and a structure very similar to that of the isoelectronic XeF_5^+ ion in

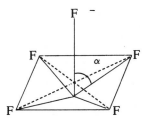

$(XeF_5)(PtF_6)$. The Te atom lies about 0.4 Å below the basal plane of the square pyramid.

Ion or molecule	M—X_{apical}	M—X_{basal}	α
TeF_5^-	1.85 Å	1.96 Å	79°
XeF_5^+	1.81	1.88	79°
SbF_5^{2-}	2.00	2.04	83°
BrF_5	1.68	1.81	84°

(a)   (b)   (c)

The (2,10) valence group is also found in the $Te(CH_3)I_4^-$ ion in $[Te(CH_3)_3]^+$ $[Te(CH_3)I_4]^-$[4], a compound originally thought to be a geometrical isomer of $Te(CH_3)_2I_2$. This salt, which reacts with KI to give $Te(CH_3)_3I$ and $K[Te(CH_3)I_4]$, consists of pyramidal $[Te(CH_3)_3]^+$ and square pyramidal $[Te(CH_3)I_4]^-$ ions. Te---I contacts (3.84–4.00 Å) complete the octahedral environment of both types of Te (IV) atom, (c) and (d).

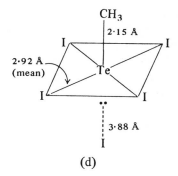

(d)

Although the 'typical' coordination of Te(IV) by oxygen may be described as that of Fig. 16.5(a) or (c), there are exceptions. We have noted certain compounds that contain both Te(IV) and Te(VI), namely, Te_2O_5, Te_4O_9, and $Te_2O_4(OH)_2$, in which Te has this type of coordination. In two complex oxides containing Te(IV) and Te(VI) there is square pyramidal coordination of Te(IV). In $BaTe_2O_6$[5] there are equal numbers of Te(IV) and Te(VI) atoms, with square pyramidal and octahedral coordination respectively. In $K_2Te^{IV}Te_3^{VI}O_{12}$[6] each $Te^{VI}O_6$ octahedron shares 4 equatorial vertices to form AX_4 layers based on the net 3^26^2 (see Fig. 13.12(d)), and these layers are joined through the 5-coordinated Te(IV) atoms. The bond lengths in these oxides are:

		$BaTe_2O_6$	$K_2Te_4O_{12}$
Te–O	(apical)	1.83	1.92 Å
	(equatorial)	2.13 (four)	2.02 (two)
			2.28 (two)

In contrast to the square pyramidal arrangement of bonds in TeF_4 and TeF_5^- there is (distorted) octahedral coordination of Te(IV) in the compound with the empirical formula $TePCl_9$. The anion has the form of an infinite chain in which $TeCl_6$ groups share *cis* vertices, with terminal and bridging Te–Cl bonds of lengths 2.4–2.5 Å and 2.7–2.9 Å respectively.[7] This is similar to the structure of the anion in $(piperidinium)_2BiBr_5$ (p. 888) but in marked contrast to the square pyramidal coordination of Sb in the isoelectronic $SbCl_5^{2-}$ ion. The cation in $TePCl_9$ is the tetrahedral PCl_4^+ ion; the structural formula is accordingly $(TeCl_5)(PCl_4)$.

(1) AC 1959 **12** 638
(2) AC 1977 **B33** 139
(3) IC 1970 **9** 2100
(4) JCS A 1967 2018

(5) AC 1979 **B35** 1439
(6) AC 1978 **B34** 1782
(7) AC 1972 **B28** 1260

Valence group (2,12). Octahedral Te(IV) complexes imply a valence group of 14 electrons. Careful X-ray studies of $(NH_4)_2TeCl_6$[1] and K_2TeBr_6[2, 3] show that the anions have a *regular* octahedral shape, and are therefore exceptions to the generalization that the arrangement of bonds formed by non-transition elements corresponds to the most symmetrical disposition of the total number of bonding and lone pairs of electrons. The electronic spectra suggest that the $5s^2$ electrons are partially

(e)

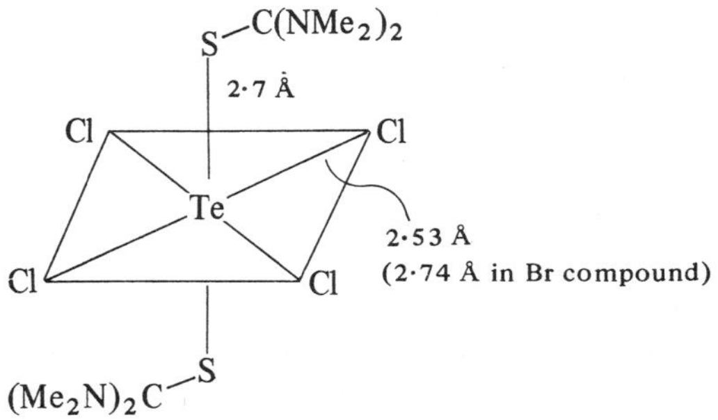

delocalized to the halide ligands. The same problem is presented by SbX_6^{3-} and IF_6^-. The Te—Cl bond length in $TeCl_6^{2-}$ is 2.54 Å, similar to that in molecules TeX_2R_2. In the octahedral molecule, $TeCl_4(tmtu)_2$[4] (where tmtu is tetramethyl thiourea), there is negligible angular distortion of the bonds (e).

(1)   ACSc 1966 **20** 165   (3)   JACS 1970 **92** 307
(2)   CJC 1964 **42** 2758   (4)   IC 1969 **8** 313

Se(II) *and* Te(II)

Valence groups (4,6) *and* (4,8). For Se(II) and Te(II) the valence groups of 8, 10, and 12 electrons include in each case two lone pairs. The non-linear bond arrangement arising from (4,4) has been encountered in elementary Se and Te, molecules SeR_2 and Se_2R_2 and their Te analogues. The 10-electron group (4,6) leads to a T-shaped molecule, as in ClF_3. Examples include the molecule (f) and the ion (g), in which S represents the sulphur atom of thiourea in $C_6H_5.Te(tu)Cl$[1] or $[C_6H_5.Te(tu)_2]Cl$.[2] In (f) and its Br analogue the Te—X bonds are remarkably long (but Te—C is equal

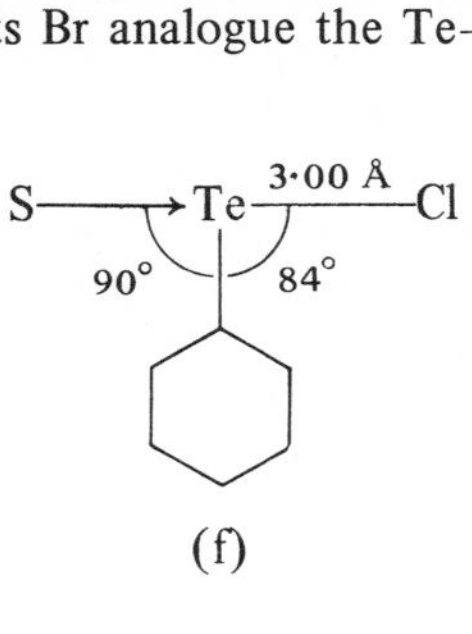

(f) 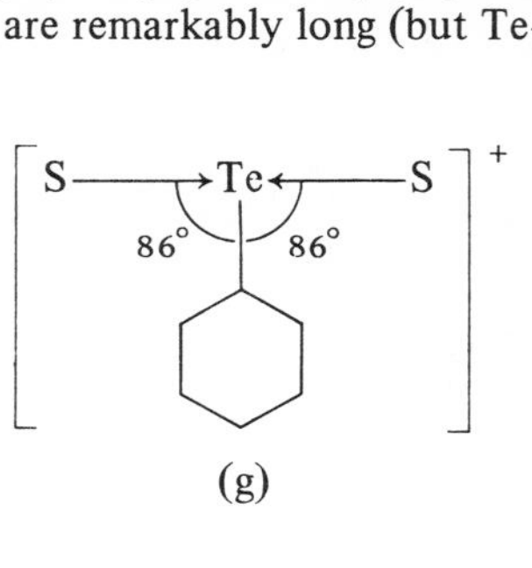   (g)

to the radius sum) and Te—S is 2.50 Å, while in (g) Te—C is again normal but Te—S is abnormally long (2.68 Å).

Four coplanar bonds from Te(II) are expected for the valence group (4,8), as in ICl_4^-, that is, the four equatorial bonds of an octahedral group. This bond arrangement occurs in molecules such as $Te(tu)_2X_2$,[3] in the cation in salts $[Te(tu)_4]X_2$,[4] and in the bridged ion in $[Te_2(tu)_6](ClO_4)_4$.[5] There appear to be interesting differences

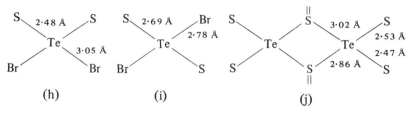

(h) (i) (j)

between the Te—S and Te—X bond lengths in the *cis* molecule (h) and those in the ethyl thiourea analogue $Te(etu)_2X_2$[6] which have the *trans* configuration, (i). The same long Te—S bond (2.68 Å) is found in the $[Te(tu)_4]^{2+}$ ion, as compared with values around 2.37 Å in compounds of 2-covalent Te(II) such as $Te(S_2O_2CH_3)_2$. Even longer Te—S bonds were found in the bridged $[Te_2(tu)_6]^{4+}$ ion, (j), and apparently also lower symmetry for which there would appear to be no obvious explanation.

For the sake of completeness we note here the structures of simple selenides and tellurides which are not included in other chapters:

antifluorite structure: Li_2Se, Na_2Se, K_2Se,
Li_2Te, Na_2Te, K_2Te

	Be	Mg	Ca	Sr	Ba	Zn	Cd	Hg
Se	Z	N	N	N	N	WZ	W	Z
Te	Z	W	N	N	N	WZ	Z	Z

N = NaCl structure: W = wurtzite structure: Z = zinc-blende (sphalerite) structure

(1) ACSc 1966 **20** 132
(2) ACSc 1966 **20** 123
(3) ACSc 1966 **20** 113
(4) ACSc 1965 **19** 2336
(5) ACSc 1965 **19** 2395
(6) ACSc 1965 **19** 2349

17

Metal sulphides and oxysulphides

The structures of binary metal sulphides

Introduction

All metal sulphides are solid at ordinary temperatures, and we are therefore concerned here only with their crystal structures. These compounds may be divided into three groups, the sulphides of

    (a) the more electropositive elements of the I, II, and III A subgroups;
    (b) transition metals (including Cu^{II});
    (c) elements of the II, III, IV, and V B subgroups.

The polysulphides of the more electropositive elements, which contain ions S_n^{2-}, are described in Chapter 16.

The crystal structures of some sulphides are set out in Table 17.1, and comparison with the corresponding Table 12.2 for oxides shows that with few exceptions (e.g. MnO and MnS) structural resemblances between sulphides and oxides are confined to the ionic compounds of group (a) and a few compounds of group (c) such as ZnO and ZnS, HgO and HgS. The B subgroup sulphides are for the most part covalent compounds in which the metal atom forms a small number of directed bonds, and while the oxide and sulphide of some elements may be of the same topological type (e.g. with 3:2 or 4:2 coordination) the compounds are not usually isostructural. For example, the normal (low-temperature) form of GeS_2 consists of a 3D framework of GeS_4 tetrahedra linked through all vertices (compare the silica-like structures of GeO_2), but the actual framework is peculiar to GeS_2 and is not found in any form of GeO_2 (or SiO_2).

A bond M–S is more covalent in character than the bond M–O, and accordingly while there is often a structural resemblance between oxides and fluorides, the sulphides tend to crystallize with the same type of structure as chlorides, bromides, or iodides of the same formula type – compare MgF_2, MnF_2, TiO_2, and SnO_2 with the ionic rutile structure with $MgBr_2$, MnI_2, TeS_2, and SnS_2 (CdI_2 or $CdCl_2$ layer structures). Layer structures are rare in oxides and fluorides but are commonly found in sulphides and the other halides. When discussing the change in type of structure going down a Periodic Group towards the more electropositive elements the dioxides of the elements of Group IV were taken as examples. The structures of the corresponding disulphides are set out below, and they may be compared with those of the dioxides shown on p. 533.

CS_2 (finite molecules)
SiS_2 (infinite chains in normal form)

TABLE 17.1
The crystal structures of metal sulphides

Type of structure	Coordination numbers of M and S	Name of structure	Examples
Infinite 3-dimensional complexes	4 : 8	Antifluorite	Li_2S, Na_2S, K_2S, Rb_2S
	6 : 6	Sodium chloride	MgS, CaS, SrS, BaS, MnS, PbS, LaS, CeS, PrS, NdS, SmS, EuS, TbS, HoS, ThS, US, PuS
	6 : 6	Nickel arsenide	FeS, CoS, NiS,[a] VS, TiS
	6 : 6[b]	Pyrites or marcasite	FeS_2, CoS_2, NiS_2, MnS_2, OsS_2, RuS_2
	4 : 4	Zinc-blende	BeS, ZnS, CdS,[c] HgS
		Wurtzite	ZnS, CdS, MnS
	4 : 4	Cooperite	PtS
Layer structures	6 : 3	Cadmium iodide (C 6)	TiS_2, ZrS_2, SnS_2, PtS_2, TaS_2, HfS_2
		Cadmium chloride (C 19)	TaS_2
	6 : 3	Molybdenum sulphide	MoS_2, WS_2
Chain structures			Sb_2S_3 Bi_2S_3, HgS

[a] Also the millerite structure (5 : 5 coordination)
[b] The coordination numbers here are those of Fe by S and of S_2 groups by Fe or other metal.
[c] Sodium chloride structure under high pressure (JSSC 1980 **33** 197, 203).

$\left.\begin{array}{l} TiS_2 \\ ZrS_2 \\ HfS_2 \end{array}\right\}$ (CdI_2 layer structure) GeS_2 (structures built of GeS_4 tetrahedra)
SnS_2 (CdI_2 structure)
PbS_2 (stable only under pressure; see p. 1179)

A number of important structure types are found in transition-metal sulphides which have no, or few, counterparts among oxide structures, notably the various layer structures, the NiAs structure, and the pyrites and marcasite structures. (Note, however, the peroxides ZnO_2 and CdO_2 with the pyrites structure, p. 242, and the superoxide NaO_2 with both the pyrites and marcasite structures.) Further, many sulphides, particularly of the transition metals, show resemblances to intermetallic compounds in their formulae, in which the metals do not exhibit their normal chemical valences, as in Co_9S_8, Pd_4S, TiS_3, and so on, their variable composition, and their physical properties—metallic lustre, reflectivity, and electronic conductivity. The crystal structures of many transition-metal sulphides show that in addition to M–S bonds there are metal–metal bonds as, for example, in monosulphides with the NiAs structure (see later), in the Mo 'cluster' compounds, and in many metal-rich sulphides such as Hf_2S, Ta_2S, Pd_4S, and Ta_6S. We shall describe the structures of these compounds later; a simple example is the structure of Ni_3S_2, which was noted in Chapter 4 as an example of an approximately b.c.c. packing of S atoms in which Ni atoms occupy some of the (distorted) tetrahedral interstices. The resemblance to intermetallic compounds is even more marked in the case of some selenides and tellurides; CoTe and $CoTe_2$ are described later to illustrate this point.

The sulphides and oxides of many metals of groups (b) and (c) do not have similar formulae; for example, there is no S analogue of Pb_3O_4 or O analogue of FeS_2. In cases where the oxide and sulphide with the same type of formula do exist they generally have quite different structures, as in the following pairs: CuO and CuS, Cu_2O and Cu_2S, NiO and NiS, PbO and PbS. Many metal–sulphur systems, like the metal–oxygen systems, are surprisingly complex, as we shall show shortly for the sulphides of In, Cr, Ti, V, Nb, and Ta. We shall devote short sections to certain groups of sulphides which have structural features or physical properties of special interest, namely, *intercalation compounds, cluster compounds,* and *'infinitely adaptive' structures.* First, we shall describe some of the simpler sulphides M_2S, MS, MS_2, M_2S_3, and M_3S_4.

Sulphides M_2S

The alkali-metal compounds other than Cs_2S have the antifluorite structure; Cs_2S has a unique structure.[1] The S^{2-} ions are in the *hcp* positions with Cs^+ ions in all the positions of octahedral and one half of the positions of tetrahedral coordination. The interatomic distances (S–S, 5.15; Cs–Cs, 4.00; and Cs–S, 3.4 Å) show, however, that the S^{2-} ions are by no means close-packed, and that metal–metal bonding is important (compare the Cs oxides, pp. 534–7). The structure of Tl_2S (a layer structure of the anti-CdI_2 type) is noted in Chapter 26. In Group IB there are Cu_2S and Ag_2S, both known as minerals; for Cu_2S see Chapter 25. There are three polymorphs of Ag_2S:

$$\text{monoclinic} \xrightarrow{\ 176\ °C\ } \text{b.c. cubic} \xrightarrow{\ 586\ °C\ } \text{f.c. cubic}$$

In the monoclinic form (acanthite) there are two kinds of non-equivalent Ag atoms with respectively 2 and 3 close S neighbours, but in the high-temperature forms, which are notably for their electrical conductivity, there is movement of Ag atoms between the interstices of the sulphur framework.[2] A structure of the cuprite type has been assigned to Au_2S.[3]

Transition-metal sulphides M_2S include those of Ti, Zr, and Hf. The physical properties of these compounds and the fact that Ti_2S[4] and Zr_2S[5] (and also the selenides) are isostructural with Ta_2P[6] show that these are not normal valence compounds but essentially metallic phases. In the complex structure of Ti_2S there are six kinds of non-equivalent Ti atom with from 3 to 5 close S neighbours and also various numbers of Ti neighbours at distances from 2.8 Å upwards. Each S has at least 7 close Ti neighbours arranged at the vertices of a trigonal prism with additional atoms capping some rectangular faces. Metal–metal bonding clearly plays an important part in this structure, which may be compared with the anti-CdI_2 structure of Ti_2O. (In the much more complex structure of Ti_8S_3[7] there are 16 sets of non-equivalent Ti atoms which have only 1–4 S neighbours, Ti atoms making the total c.n. 9–12. The coordination groups of Zr in the metal-rich $Zr_{21}S_8$ and Zr_9S_2[8] are also necessarily composed largely of metal atoms.) In contrast to Ti_2S and Zr_2S, Hf_2S[9] has the anti-$2H_2$ (hexagonal NbS_2) structure which has been described in Chapter 4. Within a layer S has 6 (trigonal prism) Hf neighbours, while Hf has an octahedral arrangement of nearest neighbours, 3 S (at 2.63 Å) in the layer and 3 Hf (at 3.06 Å) of the adjoining layer. This compound is diamagnetic and a metallic conductor, and if bond strengths are calculated from Pauling's equation (p. 1292), including the six weaker Hf–Hf contacts at 3.37 Å, the total is close to a valence of 4 for Hf:

Number of bonds	Bond	Length	Bond order	
3	Hf–S	2.63 A	0.56	
3	Hf–Hf	3.06	0.50	Total 4.08
6	Hf–Hf	3.37	0.15	

suggesting that all the $5d^2s^2$ electrons are used for Hf–S and Hf–Hf bonds.

(1) ZaC 1977 **429** 118
(2) JSSC 1970 **2** 309
(3) CR 1966 **B263** 1327
(4) AC 1967 **23** 77
(5) MRB 1967 **2** 1087

(6) ACSc 1966 **20** 2393
(7) AC 1974 **B30** 427
(8) AC 1972 **B28** 1399
(9) ZK 1966 **123** 133

Monosulphides

These compounds provide examples of all our classes (a), (b), and (c). MgS and the alkaline-earth compounds form ionic crystals, but the same (NaCl) structure is adopted by MnS (but no other 3d monosulphide) and by the 4f and 5f compounds. This illustrates the point that similarity in geometrical structure type does not imply similar bond character; witness the silver coloured PbS and the metallic gold colour of LaS. The d transition-metal monosulphides (class (b)) are considered shortly.

Monosulphides are formed by the following B subgroup metals (class (c)): Cu, Zn, Cd, and Hg, Ga, In, and Tl, Ge, Sn, and Pb. Details of structures peculiar to one sulphide are given in Chapters 25 and 26 (CuS, hexagonal HgS, GaS, InS, and TlS). Compounds with the zinc-blende and wurtzite structures are listed in Table 17.1.

The structures of the monochalconides of the Group IVB metals are set out in Table 17.2. Both the black P and As structures are layer structures in which M forms only three strong bonds, forming corrugated versions of the simple 6-gon net as explained in Chapter 3. Further details of the sulphides are given in Chapter 26.

Monosulphides of transition metals. Our present knowledge of the crystal structures

TABLE 17.2
Structures of Group IVB chalconides[1]

	S	Se	Te
Ge	P	P	As[(a)]
			N
Sn	P	P	N
Pb	N	N	N

[(a)] Below 400°C. (1) IC 1965 4 1363

P = black P structure.
As = As structure.
N = NaCl structure.

of these compounds is summarized in Table 17.3, concerning which we may note the following points.

(i) In contrast to the monosulphides with the NiAs structure, the monoxides of Ti, V, Fe, Co, and Ni crystallize with the NaCl structure.

TABLE 17.3
Structures of transition-metal monosulphides

Heavy type indicates structures peculiar to single compounds MS to which reference is made in the text.

† Also the millerite structure (5 : 5 coordination)
‡ An earlier study[1] indicated a tetragonal structure which was a distorted NaCl structure. This phase was not found in later studies[2],[3] which indicate a cubic structure, apparently a defect NaCl structure with ordered vacancies. The same cell (a = 10·25 Å) is found over the entire composition range $Zr_5S_8-Zr_9S_8$.

(1) AK 1954 7 371 (2) ZaC 1957 **292** 82 (3) AJC 1958 **11** 445

(ii) Note the special behaviour of Mn, with a half-filled 3d shell (d^5).

(iii) The structures of the d^4 (Cr^{II}) and d^9 (Cu^{II}) compounds are peculiar to these monosulphides. That of Cr is described shortly; that of CuS (covellite) is quite inexplicable as a compound of Cu^{II}, and in fact the Cu–S system is extremely complex; it is summarized in Chapter 25.

(iv) PdS and PtS have structures with planar 4-coordination of the metal atoms, that of PdS being rather less symmetrical than that of PtS, which is described later.

The nickel arsenide structure. The structure most frequently encountered is the NiAs structure (Fig. 17.1), which is also that of many phases MX in which M is a

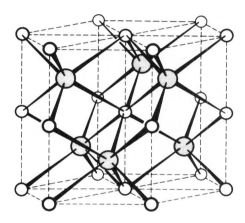

FIG. 17.1. The structure of NiAs (As atoms shaded). The Ni atom in the centre of the diagram is surrounded octahedrally by six As atoms and has also two near Ni neighbours situated vertically above and below.

transition metal and X comes from one of the later B subgroups (Sn, As, Sb, Bi, S, Se, Te). We noted this structure in Chapter 4 as having h.c.p. X atoms with M in all the octahedral interstices, X therefore having 6 M neighbours at the vertices of a trigonal prism. The immediate neighbours of a Ni atom are 6 As arranged octahedrally (at 2.43 Å), but since the $NiAs_6$ octahedra are stacked in columns in which each octahedron shares a pair of opposite faces with adjacent octahedra there are also 2 Ni atoms (at 2.52 Å) sufficiently close to be considered bonded to the first Ni atom. In the metallic phases with this structure (e.g. CoTe or CrSb) these 8 neighbours are in fact equidistant from the transition-metal atom. It seems likely that the metal–metal bonds are essential to the stability of the structure. Compounds with this structure have many of the properties characteristic of intermetallic phases, opacity and metallic lustre and conductivity, and from the chemical standpoint their most interesting feature is their variable composition. In some systems the phase MX with the ideal NiAs structure is stable only at high temperatures. There are several possibilities for the structure of the phase MX at lower temperatures. A less symmetrical variant of the NiAs structure may be formed, an entirely new structure may be more stable, or there may be disproportionation to $M + M_{1-x}X$.

Some phases $M_{1-x}S$ have a sequence of c.p. S layers different from that in MS and are therefore recognized as definite compounds, as in the case of the sulphides of Ti described later. Alternatively the packing of the S atoms remains the same as in MS and there are vacant metal sites. For the removal of M atoms from the NiAs structure there are two simple possibilities:

(a) random vacancies in (000) and $(00\frac{1}{2})$ — see Fig. 17.2;

(b) (000) fully occupied but $(00\frac{1}{2})$ only partly filled.

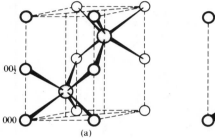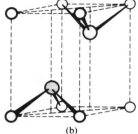

$00\frac{1}{2}$

000

(a)   (b)

FIG. 17.2.   Unit cells of the structures of (a) CoTe (NiAs structure), and (b) $CoTe_2$ (CdI_2 structure). Shaded circles represent Te atoms.

The further alternatives are then (i) random, or (ii) ordered vacancies in alternate metal layers.

All examples of (b) represent structures intermediate between the NiAs structure and the $C6$ (CdI_2) structure. The sulphides of Cr, which are described in a later section, provide examples of (b)(i) and (ii), and the Fe–S system illustrates (a) and (b)(ii).

The sulphides of iron include FeS_2 (pyrites and marcasite), Fe_2S_3, only recently prepared pure in an amorphous form from aqueous Fe^{3+} and Na_2S at 0 °C or below,[1] Fe_3S_4 (spinel structure), FeS (troilite), and a number of Fe-deficient phases called pyrrhotites. (A microcrystalline form of stoichiometric FeS has been described (mackinawite) and assigned a structure similar to that of LiOH.[2]) At temperatures above 138 °C stoichiometric FeS has the NiAs structure, but at lower temperatures there are small displacements of the atoms necessitating a larger unit cell, with a and c dimensions respectively $\sqrt{3}$ and 2 times those of the NiAs cell.[3] The structures of the pyrrhotites, which have compositions corresponding approximately to the formulae Fe_7S_8,[4] Fe_9S_{10}, $Fe_{10}S_{11}$, and $Fe_{11}S_{12}$,[5] provide many examples of superstructures of the NiAs type with various arrangements of the vacancies, and we give only a sample of the extensive literature.

In the Fe–S system the simple NiAs structure is stable over only a small range of composition, and if we include the defect structures the range is still only a few atomic per cent S. A number of metal–selenium and metal–tellurium systems show a different behaviour. In the Co–Te system the phase with the NiAs structure is homogeneous over the range 50-66.7 atomic per cent Te (using quenched samples), and at the latter limit the composition corresponds to the formula $CoTe_2$. Over this whole range the cell dimensions vary continuously, but change only by a very small amount:

	a	c
50 per cent Te	3.882 Å	5.367 Å
66.7 per cent Te	3.784	5.403

The explanation of this unusual phenomenon, a continuous change from CoTe to $CoTe_2$, is as follows. The compound $CoTe_2$ crystallizes at high temperatures with the CdI_2 structure which is retained in the quenched specimen. (On prolonged annealing it changes over to the marcasite structure.) This CdI_2 structure is very simply related to the NiAs structure of CoTe, as shown in Fig. 17.2. In both the extreme structures the Te atoms are arranged in hexagonal close packing. At the composition CoTe all the octahedral holes are occupied by Co. As the proportion of Co decreases some of these positions become vacant, and finally, at the composition $CoTe_2$, only one-half are occupied, and in a regular manner, forming the CdI_2 structure. It should be emphasized that the homogeneity range and structures of samples depend on the temperature of preparation and subsequent heat treatment. For samples annealed at $600\,^\circ C$ it is found that at the composition CoTe the product is a mixture of $Co + Co_{1-x}Te$ (defect NiAs structure). As the Te content increases the structure changes towards the $C6$ structure, but before the composition $CoTe_2$ is reached the structure changes to the marcasite structure.

The PtS *(cooperite) structure.* We have seen that FeS, CoS, and NiS crystallize with the NiAs structure, in which the metal atoms form six (or eight) bonds. Palladium and platinum, however, form four coplanar bonds in their monosulphides. The structure of PtS[6] is illustrated in Fig. 17.3. Each Pt atom forms four coplanar bonds and each S atom four tetrahedral bonds. The bond angles in this structure (and in the isostructural PtO and PdO) are not exactly 90° for Pt and $109\frac{1}{2}^\circ$ for S because they represent a compromise. The angle α in the planar chain has a value ($97\frac{1}{2}^\circ$) intermediate between that required for square Pt bonds (90°) and the tetrahedral value ($109\frac{1}{2}^\circ$). The Pt bond angles are accordingly two of $82\frac{1}{2}^\circ$ and two of $97\frac{1}{2}^\circ$, while those of S are two of $97\frac{1}{2}^\circ$ and four of 115°. In PdS[7] the coordination of the two kinds of atom is similar to that in PtS, but the structure is not quite so regular.

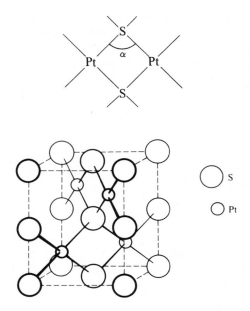

FIG. 17.3. The structure of PtS, showing the planar coordination of Pt by four S and the tetrahedral coordination of S by four Pt.

(1)   JACS 1978 **100** 2553
(2)   ZK 1968 **361** 94
(3)   Sci 1970 **167** 621
(4)   AC 1971 **B27** 1864; AC 1979 **B35** 722
(5)   AC 1975 **B31** 2759
(6)   MM 1932 **23** 188
(7)   ZK 1937 **96** 203

Disulphides

Disulphides are formed by the elements of Group IV and also by many transition elements. The structures of GeS_2 and SnS_2 are noted on pp. 1173 and 1179; PbS_2 has been made under pressure (p. 1179). La and the $4f$ metals form sulphides MS_x with $x = 1.7$–2.0 for La and the lighter lanthanides and 1.7–1.8 for the heavier $4f$ elements.[1] The structure of LaS_2 is noted on p. 731. BiS_2 has been made as a high-pressure phase.[2] Most of the disulphides of the transition metals have either a layer structure or the pyrites or marcasite structures (Table 17.4), in contrast to the essentially ionic rutile-type structures of the dioxides.

The layer structures are of two main types, with octahedral or trigonal prismatic coordination of the metal atoms. The simplest structure of the first kind is the CdI_2 ($C6$) structure of TiS_2, ZrS_2, HfS_2, TaS_2, and PtS_2. Polytypes are found, as in the case of the Cd halides; for example, TiS_2 has been shown to have the 4H, 8H, 10H, 12H, 12R, 24R, and 48R structures.[3] (For the nomenclature, see Table 4.2, p. 157.)

Two distorted forms of the $C6$ structure have been described in which there are short metal–metal distances indicating metal–metal bonds. In one, the structure of

TABLE 17.4
Crystal structures of disulphides

Ti	V	Cr	Mn	Fe	Co	Ni	Cu	Zn
C	–	–	P	P/M	P	P	$P^{(a)}$	$P^{(a)}$
Zr	Nb	Mo	Tc	Ru	Rh	Pd	Ag	Cd
C	$W^{(b)}$	$W^{(b)}$	–	P	P	PdS_2	–	$P^{(a)}$
Hf	Ta	W	Re	Os	Ir	Pt		
C	$C^{(b)}$	W	W	P	$P^{(c)}$	C		

C = CdI_2 (C 6) structure; W = MoS_2 structure (or polytype); P = pyrites structure;
M = marcasite structure.
[a] Synthesized under pressure (IC 1968 7 2208).
[b] Also more complex layer sequences.
[c] Cation vacancies at atmospheric pressure; two different pyrites-like structures under 60 kbar.
(IC 1968 7 389).

WTe_2 and the high-temperature form of $MoTe_2$,[3] the metal atoms are situated off-centre in the octahedra leading to corrugated layers and two short M–M distances (compare the 8-coordination in NiAs), for example:

$$WTe_2 \quad \begin{array}{ll} W-2W & 2.86\,\text{Å} \\ W-6Te & 2.71-2.82 \end{array}$$

In $ReSe_2$[5] the shifting of Re from the octahedron centres leads to three nearest Re neighbours at distances from 2.65–3.07 Å which may be compared with 2.75 Å in the metal.

The structures in which there is trigonal prism coordination of the metal atom contain pairs of adjacent S layers which are directly superposed (and therefore not close-packed), but the multiple S–M–S layers are then packed in the same way as simple layers in normal c.p. sequences. The simplest structures of this kind are illustrated in Fig. 4.11 (p. 154), in which the nomenclature is similar to that used for mica polytypes; the numeral refers to the number of composite (S–M–S) layers in the repeat unit. Examples include:

2 H_1 ($C7$): MoS_2 (molybdenite), WS_2, $MoSe_2$, WSe_2, low-$MoTe_2$;[4]
2 H_2:   hexagonal NbS_2;[6]
3 R: rhombohedral MoS_2,[7] NbS_2[8] TaS_2, WS_2, ReS_2.

The plan of a layer of the structures of Fig. 4.11 has been illustrated in Fig. 4.9(b) (p. 152). As noted in Chapter 4, polytypes of some chalconides have been characterized in which the sequence of c.p. layers leads to both octahedral and trigonal prism coordination of M (e.g. TaS_2 and $TaSe_2$[9]); the classification of such structures has been discussed.[10]

(1) IC 1970 9 1084
(2) IC 1964 3 1041
(3) AC 1976 **B32** 1302
(4) AC 1966 **20** 268
(5) ACSc 1965 **19** 79
(6) N 1960 **185** 376
(7) SMPM 1964 **44** 105
(8) AC 1974 **B30** 551
(9) ACSc 1967 **21** 513
(10) AC 1965 **18** 31

Intercalation complexes

The term intercalation complex is applied to solid phases formed by the insertion of 'foreign' ions, atoms, or molecules between layers in a crystalline material, the original structure being stable in the absence of the intercalated material. Compounds of this kind have been known for a long time; some are stoichiometric (e.g. those formed by alkali metals and graphite, p. 924), while others have wide ranges of composition (e.g. hydrated clay minerals). The compounds formed by disulphides such as NbS_2 and TaS_2 have received considerable attention in recent years on account of their physical properties. These two sulphides are metallic superconductors, and this property is also possessed by their intercalation complexes. As regards their geometry we may distinguish two classes of complex: those in which only single atoms (ions) are inserted between the layers, without altering appreciably the distance between them, and those in which the inter-layer distance is appreciably increased by the insertion of bulky foreign molecules (ions).

Between the S—M—S layers of NbS_2 or TaS_2 there are vacant octahedral (and tetrahedral) holes, and these can be occupied to various extents by 'foreign' atoms of, for example, Li, Ag, Cu, or any 3d metal, or by additional atoms of Nb (Ta). In the latter case the result is simply a sulphide intermediate in composition between MS_2 and MS, with M atoms in both octahedral and trigonal prismatic coordination. Systems M_xNbS_2 and M_xTaS_2 containing atoms of a second metal are of a number of kinds. For example, in compounds M_xNbS_2 where M is Cu or Ag the foreign atoms occupy tetrahedral sites in the $2H_1$ structure of NbS_2, while there are ordered arrangements of 3d atoms in octahedral sites in $M_{\frac{1}{3}}NbS_2$ and $M_{\frac{1}{4}}NbS_2$, both being superstructures of the $2H_2$ NbS_2 structure.[1]

Many compounds have been prepared in which the separation of the layers of the sulphide structure is increased, sometimes by very large amounts, by the incorporation of molecules and/or ions between the layers. In one group the inter-layer spacing of about 6 Å is increased to 9 or 12 Å by a single or double layer of hydrated Na^+ or K^+ ions. In these compounds, formed by cathodic or chemical reduction of MS_2 in electrolytes, electrons are taken up by the conduction band of the MS_2 layer, as in $K_{0.5}^+(H_2O)_{0.5}(TaS_2)^{-0.5}$.[2] Very much larger inter-layer distances are found in intercalation complexes containing an extraordinary variety of inorganic or organic molecules (ions) ranging from NH_3 and N_2H_4 to aliphatic and aromatic amines. Many, possibly all, of these complexes are stoichiometric compounds, for example, $NbS_2(C_6H_5NH_3^+)_{1.0}$ and $TaS_2(C_5H_5NH^+)(TaS_2)_{\frac{1}{2}}$.[3]

(1)   JSSC 1971 **3** 154   (3)   Sc 1970 **168** 568; Sc 1971 **174** 493
(2)   JSSC 1980 **34** 97, 253

The pyrites and marcasite structures. The pyrites and marcasite structures, named after the two forms of FeS_2, contain discrete S_2 groups in which the binding is homopolar, the S—S distances being 2.17 Å and 2.21 Å respectively. Geometrically the pyrites structure is closely related to that of NaCl, the centres of the S_2 groups and the Fe atoms occupying the positions of Na^+ and Cl^- in the NaCl structure. Every Fe atom lies at the centre of an octahedral group of six S atoms, and the coordination of S is tetrahedral (S + 3 Fe). The c.n.s are the same in the marcasite structure, which is derived from the rutile structure by rotating the chains of edge-sharing octahedra so that there are short S–S distances (2.21 Å) between S atoms of different chains, as shown in Fig. 6.5(d) (p. 246). The two structures are illustrated in Fig. 17.4 and are further described in Chapter 6, the pyrites structure on p. 242, and the marcasite structure on p. 250.

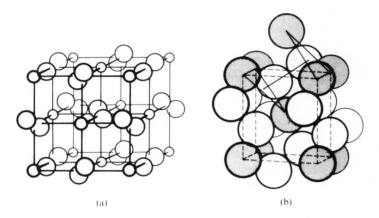

(a)   (b)

FIG. 17.4. The structures of the two forms of FeS_2: (a) pyrites, and (b) marcasite. In (a) the S–S distance has been reduced to accentuate the resemblance of this structure to that of NaCl. In (b) the shaded circles represent Fe atoms, six of which surround each S_2 group as is also the case in the pyrites structure.

There are interesting changes in the S–S and M–S distances in the 3d disulphides with the pyrites structure, differences which have been correlated with the numbers of e_g electrons.[1]

	S–S	M–S	Number of e_g electrons
MnS_2	2.086 Å	2.59 Å	2
FeS_2	2.171	2.259	0
CoS_2	2.124	2.315	1
NiS_2	2.065	2.396	2

Note the abnormal distances in MnS_2, attributed to the stability of the half-filled 3d shell. A cupric sulphide with the composition $CuS_{1.9}$ having the pyrites structure has been prepared from covellite (CuS) and S under high pressure at a temperature of 350 °C or above.[2]

Just as the NiAs structure is adopted by a number of arsenides, stibides, etc. in addition to sulphides, selenides, and tellurides, so the pyrites structure is adopted by compounds such as $PdAs_2$, $PdSb_2$ (but not PdP_2), PtP_2, $PtAs_2$, and $PtSb_2$. Also, instead of S_2 or As_2 groups we may have mixed groups such as AsS or SbS, and we find a number of compounds structurally related to pyrites and marcasite but with lower symmetry due to the replacement of the symmetrical S_2 or As_2 group by AsS, etc. So FeS_2 and $PtAs_2$ have the pyrites structure but NiSbS a less symmetrical structure related to it. Similarly, FeS_2 and $FeAs_2$ have the marcasite structure, but FeAsS and FeSbS have related structures of lower symmetry (arsenopyrite structure).

All three minerals, CoAsS, cobaltite, NiAsS, gersdorffite, and NiSbS, ullmannite, have structures which are obviously closely related to pyrites. There are three simple structural possibilities:

(i) Each S_2 group in pyrites has become As–S (Sb–S).
(ii) One-half of the S_2 groups have become As_2 (Sb_2).
(iii) There is random arrangement of S and As (Sb) in the S positions of pyrites.

Structure (i) was originally assigned to all three compounds, but as the result of later studies[3] the following structures have been proposed: CoAsS, (ii); NiAsS, (iii); and NiSbS, (i). It would seem that the greatest reliance may be placed on the later work on CoAsS, which shows that the apparently cubic structure is a polysynthetic twin of a monoclinic structure, and it is not impossible that other structures in this family are in fact superstructures of lower symmetry. As a result of the difference between Ni–Sb, 2.57 Å, and Ni–S, 2.34 Å, the symmetry of NiSbS has dropped to the enantiomorphic crystal class 23; the *absolute* structure has been determined.[4]

The marcasite structure is adopted by only one disulphide (FeS_2) but also by a number of other chalconides and pnictides. Their structures fall into two groups, with quite differently proportioned (orthorhombic) unit cells, and there are apparently no intermediate cases: the differences in bonding leading to the two types of marcasite structure are not yet understood.

	c/a	c/b			
Normal marcasite structure	0.74	0.62	FeS_2		
			$FeSe_2$	$CoSe_2$	β-$NiAs_2$
			$FeTe_2$	$CoTe_2$	$NiSb_2$
Compressed marcasite structure	0.55	0.48	$CrSb_2$	FeP_2	RuP_2
(löllingite structure)[5]				$FeAs_2$	$RuAs_2$
				$FeSb_2$	$RuSb_2$
			OsP_2	$OsAs_2$	$OsSb_2$

In marked contrast to Pt, which forms only PtS and PtS_2 (CdI_2 structure), Pd forms a variety of sulphides, selenides, and tellurides,[6] the sulphides including Pd_4S, Pd_3S, $Pd_{2.2}S$, PdS, and PdS_2. Some are high-temperature phases, for example, Pd_3S, which can be quenched but on slow cooling converts to $Pd_{2.2}S + Pd_4S$. Both Pd_3S and Pd_4S are alloy-like phases, with high c.n.s of Pd:

$$Pd_3S: \quad \begin{array}{l} Pd_I: \quad 2\,S + 10\,Pd \\ Pd_{II}: \quad 2\,S + 7\,Pd \end{array} \qquad Pd_4S: \quad Pd \left\{ \begin{array}{l} 2\,S \\ 10\,Pd \end{array} \right.$$

Whereas RhS_2 and $RhSe_2$ have the normal pyrites structure, PdS_2 and $PdSe_2$ have a very interesting variant of this structure[7] which results from elongating that structure in one direction so that Pd has four nearest and two more distant S (Se) neighbours instead of the octahedral group of six equidistant neighbours. Alternatively the structure can be described as a layer structure, the layer consisting of Pd atoms forming four coplanar bonds to S_2 (Se_2) groups, as shown in Fig. 17.5(a).

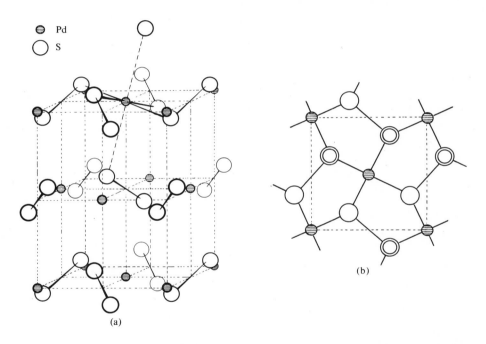

FIG. 17.5. (a) The crystal structure of PdS_2. The large open circles represent S atoms. (b) Single layer of PdPS.

With this structure compare that of CuF_2, with (4+2)-coordination, derived in a somewhat similar way from the 6-coordinated rutile structure (p. 247). The following interatomic distances were found:

$$\begin{array}{llll} \text{In } PdS_2 & \text{Pd--4S,} & 2.30\,\text{Å} & \\ & \text{Pd--2S,} & 3.28 & \text{S--S,} \quad 2.13\,\text{Å} \\ \text{In } PdSe_2 & \text{Pd--4Se,} & 2.44\,\text{Å} & \\ & \text{Pd--2Se,} & 3.25 & \text{Se--Se,} \quad 2.36\,\text{Å} \end{array}$$

The bonding within a layer of PdS_2 is more clearly seen in the projection of one layer shown in Fig. 1.8, p. 18. Layers of this kind are recognizable also in the structures of PdPS,[8] PdP_2, marcasite, and pyrites. In pyrites such layers are super-posed so that M forms strong bonds to S atoms of layers above and below, thus completing an octahedral coordination group; in PdS_2 the inter-layer bonds are weak. In PdPS one-half of the S atoms are replaced by P (Fig. 17.5(b)) and each of these atoms forms a (fourth) bond to a P atom of the same adjacent layer, forming in this way a double layer. If all the non-metal atoms are P atoms and if they form their fourth bonds to similar atoms in layers both above and below, then a 3D structure is formed in which there are chains of P atoms, as shown in Fig. 3.43, p. 132. These pentagonal layers may also be joined so as to form octahedral MX_6 groups sharing *edges*, with additional interactions between the metal atoms. this is the situation in the marcasite structure. An attempt is made in the accompanying diagrams to show these features of this group of related structures, the layers being approximately normal to the plane of the paper.

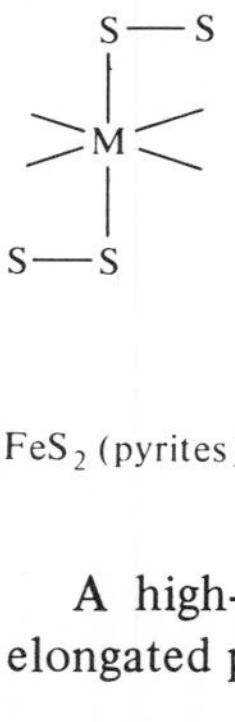

FeS₂ (pyrites)   PdS₂   PdPS   PdP₂   FeS₂ (marcasite)

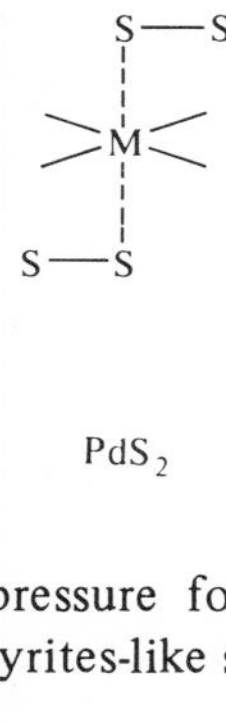

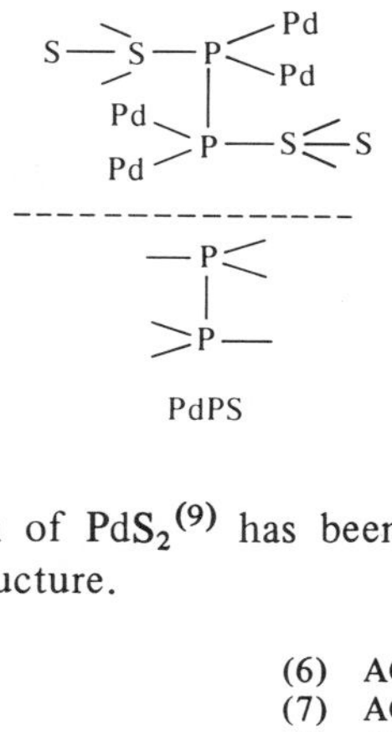

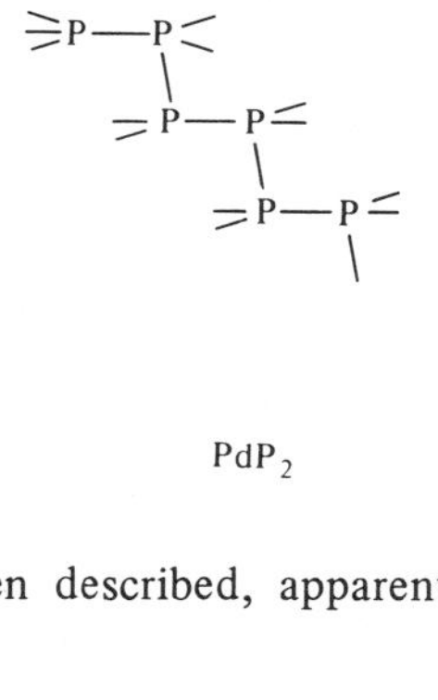

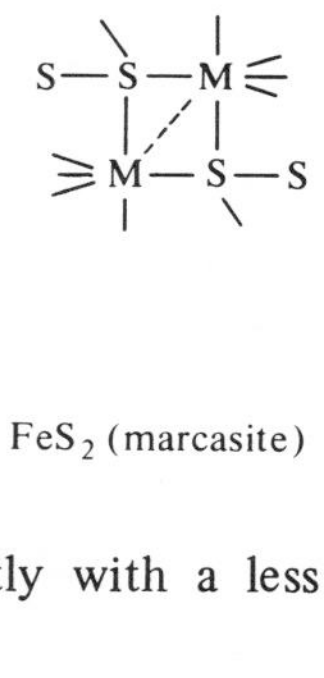

A high-pressure form of PdS_2[9] has been described, apparently with a less elongated pyrites-like structure.

(1)   JCP 1960 **33** 903
(2)   IC 1966 **5** 1296
(3)   AC 1957 **10** 764
(4)   MJ 1957 **2** 90
(5)   ACSc 1969 **23** 3043
(6)   ACSc 1968 **22** 819
(7)   AC 1957 **10** 329
(8)   AC 1974 **B30** 2565
(9)   IC 1969 **8** 1198

Sulphides M_2S_3 and M_3S_4

Most of the known sesquisulphide structures may be placed in one of three groups, corresponding to metal coordination numbers of 3, 4, and/or 6, or greater than 6 (Table 17.5).

Structures which do not fit into this simple classification include Sn_2S_3 and Rh_2S_3. In Sn_2S_3[1] double rutile chains (as in NH_4CdCl_3) of composition $Sn^{IV}S_3$ are connected through Sn^{II} atoms. In the octahedra Sn—S ranges from 2.50-2.61 Å (mean 2.56 Å) and Sn^{II} has 2 S at 2.64 Å and 1 S at 2.74 Å (mean 2.67 Å). The

TABLE 17.5

Crystal structures of sesquisulphides M_2S_3 *and related compounds*

Class (i)

Characteristic M_2X_3 structures with 3-coordination of M (but see text):

As_2S_3 (layer) structure

Sb_2S_3 (chain) structure: Bi_2S_3, Th_2S_3, U_2S_3, Np_2S_3

Class (ii)

Structures with close-packed S and 4- or 6-coordination of M

Coordination of M	h.c.p.	c.c.p.	More complex sequences
Tetrahedral	Random wurtzite (Al_2S_3, β-Ga_2S_3) Ordered defect wurtzite (α-Ga_2S_3)	Random zinc-blende (γ-Ga_2S_3)	
Tetrahedral and octahedral		α- and β-In_2S_3 (see Table 17.6)	
Octahedral	NiAs superstructure (Cr_2S_3) Corundum structure (Al_2S_3)	Ordered defect NaCl (Sc_2S_3)	Mo_2S_3, Bi_2Se_3 Bi_2Te_3 Sc_2Te_3

Other octahedral structures: Rh_2S_3

Class (iii)

Structures with higher coordination of M:

6 and 7: Ho_2S_3

7 and 8: Gd_2S_3

8: Ce_2S_3 (La_2S_3, Ac_2S_3, Pu_2S_3, Am_2S_3)

structure of Rh_2S_3 (and the isostructural Ir_2S_3)[2] consists of pairs of *face*-sharing octahedra which are linked into a 3D structure by further sharing of S atoms, so that a structure of 6:4 coordination (distorted octahedral and tetrahedral) results. This structure has obvious resemblances to the corundum structure. The shortest Rh—Rh distance (3.2 Å) shows that there are no metal–metal bonds.

Class (i): M_2S_3 *structures with 3-coordination of* M. It might be expected that some of the simplest structures for sulphides M_2S_3 would be found among the compounds of Group V elements. Strangely enough, phosphorus forms no sulphide P_2S_3 (or P_4S_6), though it forms four other sulphides (p. 867). As_2S_3 has a simple layer structure[3] (p. 907), but that of Sb_2S_3 is much more complex (p. 907). The structure of Sb_2S_3 is illustrated in Fig. 20.13; it has been confirmed by a later study of the isostructural Sb_2Se_3.[4]

Class (ii): *structures with close-packed* S. In these structures metal atoms occupy tetrahedral and/or octahedral holes in a c.p. assembly of S atoms. With the exception of one form of Al_2S_3 which crystallizes with the corundum structure these are not typically M_2X_3 structures but are defect structures, that is, MX structures (zinc-blende, wurtzite NiAs, or NaCl) from which one-third of the M atoms are missing, or spinel structures $M'M_2''X_4$ of the type $M_{2/3}'M_2''X_4$. In some structures the arrangement of the vacancies is random and in others regular.

The structures of a number of sulphides are related in the following way. A cubic block of the zinc-blende structure with edges equal to twice those of the unit cell (Fig. 3.35(b), p. 121) contains 32 c.c.p. S atoms. Removal of one-third of the metal atoms at random gives the structure of γ-Ga_2S_3.[5] (At temperatures above 550 °C β-Ga_2S_3 has a random wurtzite structure.) In this structure an average of $21\frac{1}{3}$ tetrahedral holes are occupied in each block of 32 S atoms. In Co_9S_8[6] the metal atoms occupy the same number of tetrahedral holes (32) as in ZnS—but a different selection of 32 holes—and in addition 4 octahedral holes. The neighbours of a Co atom in a tetrahedral hole are 1 S at 2.13 Å and 3 S at 2.21 Å but also 3 Co at 2.50 Å (the same as Co—Co in the metal). In $Rh_{17}S_{15}$,[7] the corresponding phase in the Rh—S system, the metal–metal bonds (2.59 Å) are even shorter than in the metal (2.69 Å).

If the appropriate sets of 8 tetrahedral and 16 octahedral holes are occupied in an assembly of 32 c.c.p. S atoms we have the spinel structure, and this is the structure of Co_3S_4, though in this case the cubic closest packing is somewhat distorted. (Zr_3S_4 is another example of a sulphide of this type.) The spinel-type structure of Co_3S_4 extends over the composition range $Co_{3.4}S_4$ to $Co_{2.06}S_4$ for solid phases prepared from melts, i.e. it includes the composition Co_2S_3, but Co_2S_3 prepared by heating together Co, S, and a flux, has a defect spinel structure, so that Co_2S_3 is related to Co_3S_4 in the same way as Fe_2O_3 (cubic) is to Fe_3O_4.[8]

Four forms of Al_2S_3 have been described, α and β with defect wurtzite-like structures, γ (corundum structure), and a high-pressure tetragonal polymorph with a defect spinel structure.[9]

The (unique) structure of Sc_2S_3[10] is referable to a cell containing 48 c.c.p. S atoms, which has dimensions $2a$, $\sqrt{2}a$, and $3\sqrt{2}a$, where a would be the cell dimension of a simple NaCl structure containing 4 c.c.p. anions. All Sc atoms occupy octahedral holes, so that the structure is a defect NaCl structure with ordered vacancies. Each S atom has 4 Nb neighbours at four of the vertices of an octahedron (*cis* vertices vacant) and there is very little disturbance of the original NaCl structure, for all the bond angles are 90° to within 1.2°. This is the structure of the yellow stoichiometric Sc_2S_3, which is a semiconductor. There is also a black non-stoichiometric Sc_2S_3 which is a metallic conductor $(Sc_{2+x}^{3+}(e)_{3x}S_3^{2-})$, also with a NaCl-type structure referable to the simple rhombohedral cell of Fig. 6.3(b). The cell content is presumably 1 Sc at (000), 0.37 Sc at $(\frac{1}{2}\frac{1}{2}\frac{1}{2})$, and 2 S at $(\frac{1}{4}\frac{1}{4}\frac{1}{4})$. This group of c.p. structures is summarized in Table 17.6.

It is interesting that although ScTe has, like ScS, the NaCl structure, the close packing of Te in Sc_2Te_3 is of the *cchh* (12-layer) type, with alternate layers of octahedral metal sites one-third occupied (statistically).[11] The same layer sequence

TABLE 17.6

Sulphides with cubic close-packed S atoms

	Structure	Interstices occupied	
		Octahedral	*Tetrahedral*
CeS	NaCl	All	–
Sc_2S_3	Defect NaCl	2/3	–
β-In_2S_3	Defect spinel (superstructure)	1/2	1/12
Co_3S_4	Spinel	1/2	1/8
Co_9S_8		1/8	1/2
γ-Ga_2S_3	Defect zinc-blende (random)	–	1/3
ZnS	{ Zinc-blende	–	} 1/2
	{ Wurtzite	–	

is found in Fe_3S_4, with every fourth layer of octahedral sites unoccupied. For Cr_3S_4 and Ti_3S_4 see pp. 768 and 772.

In contrast to Bi_2S_3, the corresponding selenide and telluride have structures in which Bi occupies octahedral holes in close-packed assemblies of Se or Te atoms. Interesting examples of some of the more complex types of close packing are found in Bi_2Se_3, Bi_2Te_2S, Bi_2Te_3 (the last two being the minerals tetradymite and tellurobismuthite respectively), and in Bi_3Se_4.

Representing the S, Se or Te layers by *A, B,* or *C* (p. 150), and the Bi atoms as *a, b,* or *c* in the octahedral positions between the layers we find the 9-layer sequence

$$chh: \quad A \quad BA \quad B \quad CB \quad C \quad AC \quad A \ldots$$
$$\qquad\quad c \qquad c \quad a \qquad a \quad b \qquad b$$

in Bi_2Se_3, Bi_2Te_2S, and Bi_2Te_3, and the 12-layer sequence

$$cchh: \quad C \quad B \quad AB \quad A \quad C \quad BC \quad B \quad A \quad CA \quad C \ldots$$
$$\qquad\quad a \quad c \qquad c \quad b \quad a \qquad a \quad c \quad b \qquad b \quad a$$

in Bi_3Se_4.

The *cch* sequence also occurs in Mo_2S_3,[12] but this is not a layer structure since the following fractions of octahedral sites are occupied between successive pairs of (approximately) c.p. layers:

$$c \quad h \quad h \quad c \quad h \quad h$$
$$1 \quad \tfrac{1}{2} \quad \tfrac{1}{2} \qquad \text{etc.}$$
$$M_I \quad M_{II} \quad M_{II}$$

There is appreciable distortion from the ideal c.p. structure (in which the metal atoms would be at the centres of the octahedral holes) owing to the formation of zigzag chains of metal–metal bonds in both the fully occupied and the half-occupied metal layers. These bonds are not much longer than in the metal. In the isostructural Nb_2Se_3 (and Ta_2Se_3)[13] there is metal–metal bonding only in the fully occupied layers, possibly because Nb and Ta each has one fewer d electrons than Mo:

	M—M in chains within metal layers	
	Mo_2S_3	Nb_2Se_3
M_I layer	2·85 Å	2·97 Å
M_{II} layer	2·87	3·13
Compare b.c. metal	2·73	2·86

The compounds we have been discussing illustrate three ways of attaining the composition M_2X_3 by occupying two-thirds of the octahedral holes in c.p. assemblies. The fractions of holes occupied between successive pairs of c.p. layers are:

									c.p. sequence
Bi_2Se_3	0	1	1	.	0	1	1		*chh*
Mo_2S_3	1	$\frac{1}{2}$	$\frac{1}{2}$	.	1	$\frac{1}{2}$	$\frac{1}{2}$		*chh*
Sc_2Te_3	1	$\frac{1}{3}$	.	1	$\frac{1}{3}$	.	1	$\frac{1}{3}$	*cchh*

Class (iii): *structures with higher coordination of* M. We now come to a group of structures adopted by sesquisulphides of Y, La, and the 4f and 5f elements in which the c.n. of some or all of the metal atoms exceeds 6.

The smallest Ln^{3+} ions, of Lu and Yb, form the 6-coordinated corundum structure, followed by the Ho_2S_3 structure (Tm–Dy), the Gd_2S_3 structure, and the Ce_2S_3 structure (Table 17.7).

The Ho_2S_3 structure is not a simple c.p. structure but has one-half of the metal atoms 6- and the remainder 7-coordinated, and two-thirds of the S 4-coordinated, and one-third 5-coordinated. The Gd_2S_3 structure also is complex, with equal numbers of metal atoms 7- and 8-coordinated (mono- and bi-capped trigonal prism),

TABLE 17.7
Crystal structures of 4f *sesquisulphides**

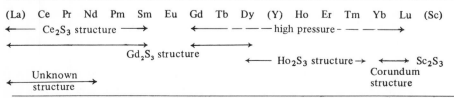

(La) Ce Pr Nd Pm Sm Eu Gd Tb Dy (Y) Ho Er Tm Yb Lu (Sc)

←——— Ce_2S_3 structure ———→   ←——— — — high pressure – — — —→

←————————————————————→
              Gd_2S_3 structure

←——— Ho_2S_3 structure →   ←——→ Sc_2S_3

←————————→   Corundum
   Unknown   structure
   structure

*For references see: IC 1969 8 2069

and all S atoms 5-coordinated (two-thirds square pyramidal and one-third trigonal bipyramidal).

In the Ce_2S_3 structure metal atoms occupy 8/9 of the metal positions in the Th_3P_4 structure, that is, $10\frac{2}{3}$ Ce are distributed over 12 positions in a cell containing 16 S atoms. The formula is therefore preferably written $Ce_{2.68}S_4$. The Th_3P_4 structure is one of 8:6 coordination, the coordination polyhedra being the triangulated dodecahedron and (distorted) octahedron. (Reference was made to this structure on p. 193.)

The study of the compounds of the larger lanthanides is complicated by the existence of sulphides M_3S_4 with the Th_3P_4 structure, formed by all the elements from La to Eu inclusive, and by the fact that this phase is readily formed from M_2S_3 by loss of S. Because of the relationship between the M_2S_3 and M_3S_4 structures a continuous change from one structure to the other is possible. There is, however, an interesting difference between the La and Ce sulphides on the one hand and those of Sm and Eu. The cell dimensions of the La and Ce phases remain nearly constant over the composition range $M_{2.68}S_4$ (M_2S_3) to M_3S_4, whereas in the Sm and Eu systems there is an increase in cell dimensions. An increase is to be expected if M^{3+} ions are changing into the larger M^{2+} ions. It therefore appears that in the La and Ce compounds the ions remain M^{3+}, the extra electron being delocalized, $(M^{3+})_3S_4(e)$, whereas the Sm and Eu compounds should be formulated $M^{2+}M_2^{3+}S_4$. This difference is consistent with the different relative stabilities of the di- and trivalent states of these elements and with the fact that Ce_3S_4 is a good metallic conductor whereas Sm_3S_4[14] is a semiconductor. Compare the difference between CeS and SmS. The former is metallic with a magnetic moment corresponding to Ce^{3+}, that is, it is $Ce^{3+}(e)S^{2-}$, where SmS is a semi-conductor with magnetic moment corresponding to $Sm^{2+}S^{2-}$. Other compounds with the Ce_2S_3 structure include Ac_2S_3, Pu_2S_3, and Am_2S_3.

(1)	AC 1967 **23** 471	(8)	PNAS 1955 **41** 199
(2)	AC 1967 **23** 832	(9)	JSSC 1970 **2** 6
(3)	MJ 1954 **1** 160	(10)	IC 1964 **3** 1220
(4)	AC 1957 **10** 99	(11)	IC 1965 **4** 1760
(5)	ZaC 1955 **279** 241; AC 1963 **16** 946	(12)	JSSC 1970 **2** 188
(6)	AC 1962 **15** 1195	(13)	AC 1968 **B24** 1102
(7)	AC 1962 **15** 1198	(14)	AC 1978 **B34** 2084

The sulphides of indium

These compounds present a number of points of interest, apart from the unusual formulae of two of them, namely, metal–metal bonding and the effect of the lone pair in the monosulphide. Structural studies have been made of In_5S_4,[1] InS,[2] In_6S_7,[3] and In_2S_3;[4] we comment on In_3S_4 later.

The structure of In_5S_4 may be briefly described as a 3D network of vertex-sharing $In(S_3In)$ tetrahedra consisting of a central In atom bonded to 1 In and 3 S. These tetrahedra come together in groups of four sharing a common vertex, which is an

In atom bonded to 4 other In atoms (at 2.76 Å). Metal atoms of this kind comprise 1/5 of the total number, the remainder being bonded to 1 In and 3 S; each S is bonded on 3 In. The structure of InS is described on p. 1150, where it is compared with that of GeS. In InS all In atoms have a distorted tetrahedral arrangement of 1 In and 3 S neighbours, like four-fifths of those in In_5S_4; the In—In distance (2.80 Å) is similar to that in In_5S_4. To a first approximation the structure of In_6S_7 may be described as a c.c.p. assembly of S atoms in which metal atoms occupy certain of the octahedral interstices, but there is considerable distortion because 2 of the 6 non-equivalent In atoms are associated in pairs (In—In, 2.74 Å). There has been much discussion of the structures of the phases with higher S contents, and particularly of the nature of the changes taking place at temperatures around 420° and 750 °C in material with the composition In_2S_3. In all phases $InS_{1.33}$–$InS_{1.50}$ there is cubic closest packing of the S atoms with increasing numbers of vacancies in the tetrahedral sites, but their arrangement appears to depend on the method and temperature of preparation. Forms of In_2S_3 described include a defect spinel structure like that of γ-Al_2O_3, a tetragonal spinel superstructure with $a_{tetr.} = a_{cubic}/\sqrt{2}$ and $c_{tetr.} = 3a_{cubic}$ (a_{cubic} being the edge of the cubic spinel cell containing 32 c.c.p. S atoms), comparable with δ-Al_2O_3, and a high-temperature form with a double-layer structure,[5] *CbAcB CbAcB* . . . in which all the In atoms occupy octahedral sites, as opposed to tetrahedral and octahedral sites in the other structures. The existence of In_3S_4 has been claimed as a definite phase stable between 370° and 840 °C, but this has not been confirmed by all later workers. It is possible that the limiting composition In_2S_3, with all the tetrahedral sites in the spinel structure occupied, may not be attained under all experimental conditions. On this point see, for example, ref. (4).

(1) AC 1980 **B36** 2220
(2) AC 1966 **20** 566
(3) AC 1967 **23** 111

(4) JSSC 1980 **34** 353
(5) PSS 1976 **36** 517

The sulphides of chromium

The Cr–S system is much more complex than it was originally thought to be. Between the compositions CrS and Cr_2S_3 there are three definite solid phases:

CrS	monoclinic	$\approx CrS_{0.97}$
Cr_7S_8	trigonal	$Cr_{0.88}S$–$Cr_{0.87}S$
Cr_5S_6	trigonal	$Cr_{0.85}S$
Cr_3S_4	monoclinic	$Cr_{0.79}S$–$Cr_{0.76}S$
Cr_2S_3	trigonal	$Cr_{0.69}S$
Cr_2S_3	rhombohedral	$Cr_{0.67}S$

In all the trigonal Cr sulphides Cr has six S neighbours at 2.42-2.46 Å but there are also Cr–Cr bonds of length approximately 2.80 Å, that is, there are ionic Cr–S bonds but also metal–metal bonds.

All these sulphides except $CrS^{(1)}$ have structures intermediate between the NiAs and CdI_2 ($C6$) structures. In Cr_7S_8 there are random vacancies in alternate metal layers of the NiAs structure, while the others have ordered vacancies in alternate metal layers. The proportions of *occupied* metal sites between c.p. layers are:

$$1 \tfrac{2}{3} \qquad 1 \tfrac{1}{2} \qquad 1 \tfrac{1}{3} \qquad \text{and} \quad 1 \ 0$$
$$M_5S_6 \qquad M_3S_4 \qquad M_2S_3 \qquad\qquad MS_2$$

The patterns of *vacant* metal sites in Cr_2S_3 (trigonal and rhombohedral forms) and Cr_5S_6 are shown in Fig. 17.6; Cr_3S_4 is also of this type. A further sulphide, Cr_5S_8, has been produced under pressure;[2] it also has a structure of the same general type, as also does the isostructural V_5S_8.[3]

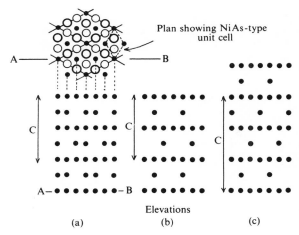

Plan showing NiAs-type unit cell

Elevations

(a) (b) (c)

FIG. 17.6. Ordered vacancies in the NiAs structure: (a) Cr_5S_6; (b) trigonal Cr_2S_3; (c) rhombohedral Cr_2S_3.

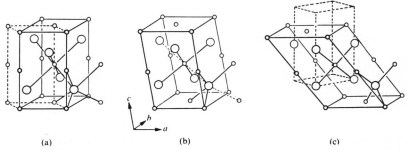

(a) (b) (c)

FIG. 17.7. The relation between the structures of (a) NiAs; (b) CrS; and (c) PtS. In (a) and (c) the broken lines indicate the conventional unit cells.

The structure of CrS is unique and intermediate between that of NiAs and PtS. The neighbours of Cr are four S at 2.45 Å (mean) and two much more distant (2.88 Å) —compare CrF_2 with a deformed rutile structure and also (4+2)-coordination. Although CrS is formally isotypic with PtS the cells are of very different shapes and CrS is best regarded as a new structure type. It is illustrated in Fig. 17.7.

(1)   AC 1957 **10** 620   (3)   CR 1964 **258** 5847
(2)   IC 1969 8 566

The sulphides of vanadium, niobium, and tantalum

The formulae and structures of the sulphides and a comparison with the oxides formed by these metals illustrate the general points noted at the beginning of this chapter. For example, at least nine crystalline Nb–S phases have been characterized, and none has its counterpart among niobium oxides. There are two forms of NbS_{1-x} (low temperature, NiAs superstructure, high temperature, MnP structure), two forms of $Nb_{1+x}S_2$, and two of NbS_2 (hexagonal and rhombohedral MoS_2 structures), in addition to the other sulphides noted:

$V_3S^{(1)}$	$Nb_{21}S_8{}^{(3)}$	$Ta_6S^{(5)}$
		$Ta_2S^{(6)}$
VS	NbS_{1-x} (2 forms)	
V_3S_4	$Nb_3S_4{}^{(4)}$	
$V_5S_8{}^{(2)}$	$Nb_{1+x}S_2$ (2 forms)	
	NbS_2 (2 forms)	TaS_2 (CdI_2, Cd(OH)Cl,
		$CdCl_2$ structures)
	NbS_3	TaS_3
VS_4		

Some of these compounds have complex structures which are not easily described, such as the two forms of V_3S and $Nb_{21}S_8$. The latter is a compound with metallic properties in which there are six kinds of Nb atoms with from 1 to 4 S neighbours.

The structure of Nb_3S_4 was illustrated in Fig. 5.43 as built of triple columns of face-sharing octahedral NbS_6 groups, each of which also shares four edges to form a 3D structure with a general geometrical similarity to the UCl_3 structure. The simplified projection of Fig. 17.8(a) shows that there are empty tunnels through the structure and that the S atoms are of two kinds. Those (S_1) on the surfaces of the tunnels have a very unsymmetrical arrangement of 4 Nb neighbours, while those (S_2) on the central axes of the columns have 6 (trigonal prism) Nb neighbours. The Nb atoms are displaced from the centres of their octahedral coordination groups so that Nb–Nb bonds are formed as zigzag chains perpendicular to the paper as indicated in projection by the broken lines in Fig. 17.8(a). The compound is a metallic conductor; Nb–Nb, 2.88 Å, compare 2.86 Å in the metal.

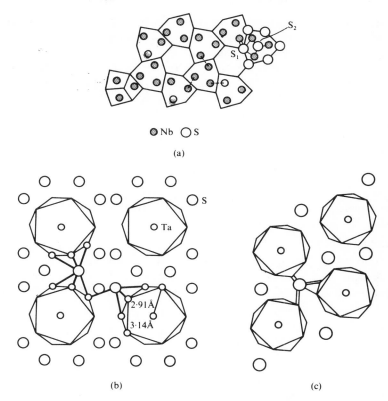

ⓞ Nb ◯ S

(a)

FIG. 17.8. Projections of the structures of (a) Nb_3S_4; (b) Ta_2S; (c) Ta_6S.

The structures of Ta_2S and Ta_6S are closely related. In both these sulphide structures (Fig. 17.8(b) and (c)) the metal atoms form columns consisting of body-centred pentagonal antiprisms sharing their basal (pentagonal) faces. (Since the Ta atoms at the centres of the antiprisms have an icosahedral arrangement of 12 nearest neighbours—10 forming the antiprism and 2 at the body-centres of adjacent anti-prisms—the columns could also be described as built of interpenetrating icosahedra.) These columns of metal atoms are held together by the S atoms, which in Ta_2S are of two kinds (with 4 or 6 neighbours) and in Ta_6S are all similar and have 7 Ta neighbours (monocapped trigonal prism). The Ta atoms at the centres of the columns are entirely surrounded by (12) Ta atoms; those on the periphery have 2 or 3 S neighbours (in Ta_2S) or 1 or 2 S neighbours (in Ta_6S), the remaining close neighbours being Ta atoms of the same column. There are no very short Ta–Ta contacts between the columns, but although the main metal–metal interactions are within the columns the interatomic distances indicate that there may be some interaction of this kind between the columns. Thus in Ta_2S the shortest Ta–Ta distances are those between atoms at the centres of the antiprisms (2.80 Å), as compared with 3.14 Å between those on the surfaces of the columns, but there are some comparable contacts (3.10 Å) between the columns.

It is noteworthy that although S.and P are similar in size, and Ti_2S and Zr_2S are isostructural with Ta_2P, yet Ta_2S has a unique structure with largely Ta–Ta interactions, that is very low Ta–S coordination numbers; compare the 3–5 close S neighbours of a metal atom in Ti_2S.

For a group of complex sulphides having a general structural resemblance to these 'sub-sulphides' see the section *Complex sulphides containing metal clusters* on p. 784.

(1)   AC 1959 **12** 1022
(2)   CR 1964 **358** 5847
(3)   AC 1968 **B24** 412

(4)   AC 1968 **B24** 1614
(5)   AC 1970 **B26** 125
(6)   AC 1969 **B25** 1736

The sulphides of titanium

Apart from the extreme compounds TiS_3 and Ti_2S the sulphides of Ti include TiS_2 and TiS, with the simple (h.c.p.) CdI_2 and NiAs structures, and a series of phases with compositions intermediate between those of the disulphide and monosulphide (Table 17.8). These phases have structures based on more complex c.p. sequences in

TABLE 17.8

The sulphides of titanium: TiS_2-TiS; $Li_xTi_{1.1}S_2$

Sulphide	S Layer sequence	Fractional site occupancy between successive S layers	Reference
TiS_2	h	$h_{1.0}h_0$	–
$Ti_5S_8(Ti_3S_5)$	$cchh$	$c_{1.0}c_{0.2}h_{1.0}h_{0.2}$	RTC 1966 **85** 869
"Ti_2S_3"	ch	$c_{1.0}h_{0.4}$	AC 1975 **B31** 2800
$Ti_{2.45}S_4$		$c_{1.0}h_{0.23}$	JSSC 1970 **2** 36
Ti_3S_4	$chhchch$	$c_{0.5}h_{1.0}h_{0.5}c_{0.95}h_{0.7}c_{0.7}h_{0.95}$	
Ti_4S_5	$chchh$	$c_{0.9}h_{0.9}c_{0.6}h_{1.0}h_{0.6}$	JSSC 1970 **1** 519
Ti_8S_9	chh	$c_{0.83}h_{1.0}h_{0.83}$	
TiS	h	$h_{1.0}h_{1.0}$	–
$Li_xTi_{1.1}S_2$			Sc 1972 **175** 884

which Ti atoms occupy octahedral interstices. Certain layers of metal atom sites are fully (or almost fully) occupied while others are only partially occupied (at random) — contrast the Cr sulphides Cr_7S_8, Cr_5S_6, and Cr_3S_4, in which the packing of the S atoms remains the same (h.c.p.).

Compounds $Li_xTi_{1.1}S_2$ prepared by melting together metallic Li and $Ti_{1.1}S_2$ (CdI_2 structure) have the *ch* packing for x between 0.1 and 0.3, with apparently random occupancy by Li and Ti atoms of octahedral sites in the partially filled layers. At higher Li concentrations $(0.5 < x < 1.0)$ a completely different (tetragonal) structure is adopted. These compounds, and also compounds Na_xMoS_2 and M_xZrS_2 and M_xHfS_2 (M = Na, K, Rb, or Cs) are notable for developing superconductivity at relatively high temperatures (10–13 K), whereas none of the binary compounds of Li, Ti, and S become superconducting above 1 K.

Complex sulphides and thio-salts

The number of solid phases now known which consist of sulphur combined with more than one kind of metal is very large. These phases comprise compounds synthesized in the laboratory and an extraordinary variety of sulphide minerals. A few simple examples have been noted in earlier chapters to illustrate certain types of structure—$Pb_2Ga_2S_5$ as an example of a structure based on the planar 4.8^2 net (p. 109), La_2SnS_5 and Sm_3InS_6 as structures containing rutile-like chains (p. 214), and Mn_2SnS_4 as a defect NaCl structure (p. 245). In the sulphide minerals, as in oxides, isomorphous replacement is widespread, leading to random non-stoichiometric compounds, but we may recognize three main types of compound. If for simplicity we describe the bonds A–S and B–S in a compound $A_xB_yS_z$ as essentially ionic or essentially covalent we might expect to find three combinations:

	A–S	B–S
(a)	Ionic	Ionic
(b)	Ionic	Covalent
(c)	Covalent	Covalent

In (a) and (c) there would be no great difference between the characters of the A–S and B–S bonds in a particular compound, while in (b) the B and S atoms form a covalent complex which may be finite or infinite in one, two, or three dimensions. By analogy with oxides we should describe (a) and (c) as complex sulphides and (b) as thio-salts. Compounds of type (c) are not found in oxy-compounds, and moreover the criterion for isomorphous replacement is different from that applicable to complex oxides because of the more ionic character of the bonding in the latter. In ionic compounds the possibility of isomorphous replacement depends largely on ionic radius, and the chemical properties of a particular ion are of minor importance. So we find the following ions replacing one another in oxide structures: Fe^{2+}, Mg^{2+}, Mn^{2+}, Zn^{2+}, in positions of octahedral coordination, while Na^+ more often replaces Ca^{2+} (which has approximately the same size) than K^+, to which it is more closely related chemically. In sulphides, on the other hand, the criterion is the formation of the same number of directed bonds, and we find atoms such as Cu, Fe, Mo, Sn, Ag, and Hg replacing Zn in zinc-blende and closely related structures.

Obviously this naive classification is too simple to accommodate all known compounds, and because of its basis it has the disadvantage of prejudging the bond type. An essentially geometrical classification based on known crystal structures would, however, be of the same general type. Class (a) includes structures like those of complex oxides (see Table 17.9), and the analogy with complex oxides extends

TABLE 17.9

Crystal structures of some complex sulphides and thio-salts

$A_xB_yS_z$	C.n.s of A *and* B	Type of thio-ion	Reference
$(NH_4)_2WS_4$	(9, 10) : 4	Finite	AC 1963 **16** 719
Tl_3VS_4	(4+4) : 4	Finite	AC 1964 **17** 757
$Na_3SbS_4 \cdot 9H_2O$	6 : 4	Finite	AC 1950 **3** 363
$KFeS_2$	8 : 4	Chain (edge-sharing)	ZaC 1968 **359** 225
$NH_4(CuMoS_4)$	12 : 4	Chain (edge-sharing)	IC 1970 **9** 1449
Ba_2MnS_3	7 : 4	Chain (vertex-sharing)	IC 1971 **10** 691
Ba_2ZnS_3	7 : 4	Double chain	ZaC 1961 **312** 99
KCu_4S_3	8 : 4	Double layer	ZaC 1952 **269** 141
$(NH_4)Cu_7S_4$	8 : 4	3D framework	AC 1957 **10** 549

		Structure	
$BaTiS_3$ ⎫			
$BaVS_3$ ⎬	12 : 6	$CsNiCl_3$ ($BaNiO_3$)	AC 1969 **B25** 781;
$BaTaS_3$ ⎭			IC 1969 **8** 2784
$BaSnS_3$	9 : 6	NH_4CdCl_3	MRB 1970 **5** 789
$EuHfS_3, EuZrS_3$ ⎫			
$CaHfS_3, CaZrS_3$ ⎬	8 : 6	$GdFeO_3$	AC 1980 **B36** 2223
$BaZrS_3, BaUS_3$ ⎭		(p. 584)	
$NaCrS_2, NaInS_2$	6 : 6	NaCl superstructure	JPCS 1968 **29** 977
$LiCrS_2$	6 : 6	NiAs superstructure	IC 1970 **9** 2581
$NiCr_2S_4$	6 : 6	NiAs superstructure	IC 1966 **5** 977
$NaBiS_2$	6 : 6	NaCl (statistical)	RTC 1944 **63** 32
$FeCr_2S_4, CuCr_2S_4$ ⎫	4 : 6	Spinel	AKMG 1943 **17B** No. 12
$ZnAl_2S_4$ ⎭			ZaC 1967 **23** 142

as far as the adoption by a number of complex sulphides of the same modified form of the perovskite structure as found for oxides such as $GdFeO_3$. Class (a) tends to merge into class (c) as bond character changes from ionic to covalent or covalent–metallic. In class (a) the ions A and B are usually those of the more electropositive elements of the earlier A subgroups or of certain B subgroup elements (e.g. In^{3+}, Bi^{3+}). In thio-salts A may be an alkali metal, Ag, Cu(I), NH_4, or Tl(I), and B a non-metal or metalloid (Si, As, Sb) or a transition metal in a high oxidation state (V^V, Mo^{VI}). In compounds of class (c) both metals are typically from the B subgroups (Cu, Ag, Hg, Sn, Pb, As, Sb, Bi) but include some transition elements such as Fe.

We have noted one difference between complex oxides and sulphides, namely, the compounds of class (c) have no counterpart among oxy-compounds. A second difference is that sulphides other than those of the most electropositive elements show more resemblance to metals than do oxides. Metal–metal bonding occurs only rarely in simple oxides whereas it is more evident in many transition-metal sulphides. In many complex sulphides of class (c), as indeed in simple sulphides such as those of Cu, it is not possible to interpret the atomic arrangements and bond lengths in terms of normal valence states of the metals, suggesting a partial transition to metallic bonding, as is also indicated by the physical properties of many of these compounds. In the limit the formation of localized metal–metal bonds leads to clusters, some examples of which are given on p. 784.

Thio-salts

A considerable number of thio-salts containing alkali metals have been prepared in one of two ways:

(i) The sulphides of some non-metals and of certain of the more electronegative metals dissolve in alkali sulphide solutions, and from the resulting solutions compounds may be crystallized or precipitated by the addition of alcohol. These compounds are usually very soluble, often highly hydrated, and often easily oxidized and hydrolysed. Thiosilicates and thiophosphates have been prepared, and other compounds of this kind include $(NH_4)_3VS_4$, $Na_6Ge_2S_7.9H_2O$, K_3SbS_3, $Na_3SbS_4.9H_2O$, thiomolybdates, M_2MoS_4, and thiotungstates, M_2WS_4. The existence of tetrahedral thio-ions has been established in $Na_3SbS_4.9H_2O$, and in $(NH_4)_2MoS_4$ and $(NH_4)_2WS_4$, the last two being isostructural with one form of K_2SO_4. The structures of salts containing thio-ions MS_4^{n-} are noted under the chemistry of the atom M.

(ii) Prolonged fusion of a transition metal or its sulphide with sulphur and an alkali-metal carbonate, followed by extraction with water, yields compounds such as $KFeS_2$, $NaCrS_2$, and KCu_4S_3. (This method can also be used to give complex sulphides such as $KBiS_2$.) In all these compounds, which are insoluble and highly coloured, the alkali metal is apparently present as ions M^+, but the thio-ions are of various types, chain, layer, or 3D frameworks.

The steel-blue fibrous crystals of $KFeS_2$ are built of infinite chain ions formed of FeS_4 tetrahedra sharing opposite edges, and between these chains lie the K^+ ions surrounded by 8 S atoms (Fig. 17.9). An interesting elaboration of this chain occurs in $NH_4(Cu^IMo^{VI}S_4)$, where Cu^I and Mo^{VI} alternate along the chain. In ammoniacal cuprous chloride solution crystals of $KFeS_2$ change into the brassy, metallic $CuFeS_2$ with the zinc-blende type of structure which is described shortly, in which no thio-ions can be distinguished. The action of H_2S on a mixture of BaO and ZnO at $800\,^{\circ}C$ gives Ba_2ZnS_3, also containing an infinite one-dimensional thio-ion, in this case the double chain of tetrahedra found in the isostructural K_2CuCl_3 (q.v.). Ba_2MnS_3, with single vertex-sharing chains is isostructural with K_2AgI_3.

Copper forms a number of complex thio-salts with alkali metals. Dark-blue crystals of KCu_4S_3 can be prepared by fusing the metal with alkali carbonate and sulphur and extracting with water. The crystals are good conductors of electricity. Their structure consists of double layers built of CuS_4 tetrahedra, the layers being interleaved with K^+ ions surrounded by 8 S at the vertices of a cube. A 3D thio-ion

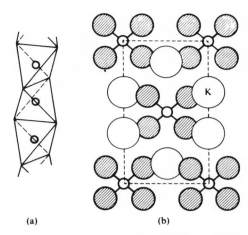

(a)   (b)

FIG. 17.9. (a) Arrangement of Fe atoms (small circles) in chains of FeS_4 tetrahedra in $KFeS_2$. (b) Projection of the structure of $KFeS_2$ along the direction of the chains.

is found in $NH_4Cu_7S_4$. Finely divided copper reacts slowly with ammonium sulphide solution in the absence of air to give, among other products, black, lustrous (tetragonal) crystals of $NH_4Cu_7S_4$. We are interested here only in the general nature of the structure (Fig. 17.10) which is a charged 3D framework of composition $Cu_7S_4^-$

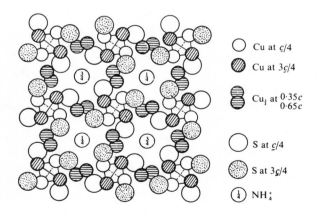

○ Cu at $c/4$

◨ Cu at $3c/4$

⊟ $Cu_{\frac{1}{2}}$ at $\begin{matrix} 0·35c \\ 0·65c \end{matrix}$

○ S at $c/4$

◉ S at $3c/4$

(¼) NH_4^+

FIG. 17.10. The crystal structure of $NH_4Cu_7S_4$.

built of columns cross-linked at intervals by Cu atoms arranged statistically in three-quarters of the positions indicated by the smallest circles. The framework encloses cubical holes between 8 S atoms which are occupied by NH_4^+ ions.

In all the above compounds there is tetrahedral coordination of the metal forming the thio-ion. We now give examples of thio-ions in which there is octahedral coordination of the metal. The (distorted) perovskite structure of $BaZrS_3$ is probably to be regarded as an ionic structure in which Ba^{2+} is 12- and Zr^{4+} 6-coordinated. (Inasmuch as the ZrS_6 octahedra are linked by vertex-sharing into a 3D framework the ZrS_3 complex could be distinguished as a 3D 'thio-ion'.) A number of compounds ABS_3 have h.c.p. structures built of AS_3 layers between which B atoms occupy columns of face-sharing octahedral holes (the $BaNiO_3$ or $CsNiCl_3$ structure), so forming infinite linear thio-ions. In $BaVS_3$ the V—V distance (2.81 Å) indicates metal–metal bonding consistent with the metallic conductivity, though $BaTaS_3$ (Ta—Ta, 2.87 Å) is only a semi-conductor. Double octahedral chain ions are found in $BaSnS_3$, $SrSnS_3$, and $PbSnS_3$ (NH_4CdCl_3 structure). Infinite 2D octahedral thio-ions occur in $NaCrS_2$ and other similar compounds. Crystals of $NaCrS_2$ are thin flakes which appear dark-red in transmitted light and greyish-green with metallic lustre in reflected light. In these crystals there are CrS_2^- layers of the same kind as in CdI_2 held together by the Na^+ ions (Fig. 17.11); both Cr and Na have 6 octahedral

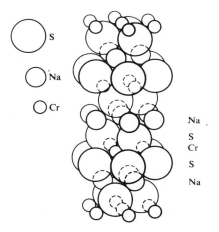

S

Na

Cr

Na

S

Cr

S

Na

FIG. 17.11. The crystal structure of $NaCrS_2$.

neighbours. Alternatively this structure may be described as a superstructure of NaCl; $NaInO_2$ and $NaInS_2$ are isostructural with $NaCrS_2$, as also are $KCrS_2$ and $RbCrS_2$. On the other hand, $LiCrS_2$ is a superstructure of NiAs, that is, Li and Cr alternate in columns of face-sharing octahedral coordination groups in a hexagonal closest packing of S atoms. Alternatively the structure may be described as CdI_2-type layers of composition CrS_2 interleaved by Li^+ ions. There are apparently no very strong Li—Cr bonds across the shared octahedron faces (Li—Cr, 3.01 Å), for the compound has a high resistivity. Just as $NaBiS_2$ (statistical NaCl structure) is related, in a structural sense, to PbS and other simple sulphides with the NaCl structure, so $FeCr_2S_4$ is isostructural with Co_3S_4 and $NiCr_2S_4$, and NiV_2S_4 with

Cr_3S_4, and many of the sulphides of class (c) are related to the simplest covalent sulphide ZnS.

Infinitely adaptive structures

We have described the structures of $KFeS_2$ and NH_4 $(CuMoS_4)$ in which the cations are situated between MS_2 chains formed of edge-sharing tetrahedral groups, the packing of the chains being such as to give K^+ 8- and NH_4^+ 12-coordination by S. These are normal structures, that is, the periodicity of the cations is the same as that of the atoms of the chains. Investigation of the Ba–Fe–S system has shown that, apart from $BaFe_2S_4$,[1] there is a series of compounds $Ba_{1+x}Fe_2S_4$ (or $Ba_p(Fe_2S_4)_q$) which have structures of an unusual type for values of x between 0.062 and 0.143 (or of p/q from 16/15 to 8/7). Examples are $Ba_9Fe_{16}S_{32}$ and $Ba_{10}Fe_{18}S_{36}$.[2] In these structures the spacing of the Ba ions between the (parallel) chains becomes 'out of step' with that of the S atoms along the chains, resulting in long repeat periods along the chain axis. For example, in $Ba_9Fe_{16}S_{32}$ the sub-cell for Ba is one-ninth of 44.41 Å, whereas that for the S atoms is one-eighth of 44.41 Å, requiring $c = 44.41$ Å; the tetragonal a parameter is 7.78 Å. The description 'infinitely adaptive' has been given to structures of this kind.

A somewhat similar phenomenon is observed in a series of compounds with compositions in the range $TB_{1.25}$–TB_2 where T is an element of Groups V–VIII and B is from Group IV; examples are $Mn_{15}Si_{26}$[3] and $Rh_{17}Ge_{22}$.[4] These compounds have tetragonal structures with a around 6 Å but c up to 300 Å. For example, the unit cell of $Mn_{15}Si_{26}$ has $c = 65.3$ Å and consists of 15 sub-cells stacked vertically above one another, each containing 4 Mn. These atoms are arranged in a helical manner with z coordinates increasing by $c/60$. The unit cell also contains 52 pairs of Si atoms whose z coordinates increase regularly by $c/52$. The structure as a whole therefore repeats after every 60 Mn and 104 Si atoms.

(1)   JSSC 1980 **32** 329   (3)   AC 1967 **23** 549
(2)   JCS D 1973 1107; AC 1975 **B31** 45   (4)   AC 1967 **22** 417

Sulphides structurally related to zinc-blende or wurtzite

This relationship is most direct for compounds such as α-$AgInS_2$ in which Ag and In atoms occupy at random the Zn positions in the wurtzite structure. Alternatively, there may be regular replacement of the Zn atoms in one of the two ZnS structures by atoms of two or more kinds. In both the random and regular structures the metal:sulphur ratio remains 1:1. We have seen that in some sulphides M_2S_3 and M_3S_4 only a proportion of the metal sites (tetrahedral or octahedral) in c.p. S assemblies are occupied. In the case of ZnS structures this implies occupancy of 2/3 or 3/4 of the Zn sites (that is, 1/3 or 3/8 of all the tetrahedral holes in the c.p. assembly). If instead of an incomplete set of metal atoms some of the S atoms are omitted, we have structures in which some of the metal atoms have 4 tetrahedral S neighbours and others 3 pyramidal S neighbours. Since the latter is a suitable bond arrangement for As or Sb we find compounds such as Cu_3AsS_3 with this kind of structure. A further possibility, the substitution of both Zn and S by other atoms, occurs in lautite, $CuAsS$,[1] in which the Zn and S positions are occupied (in a regular way) by equal numbers of Cu, As, and S. This structure is, however, preferably described as a substituted diamond structure, and it has been described in this way in Chapter 3. We may therefore recognize the following types of structure:

	M : S ratio	
(i)	1:	statistical or regular replacement of Zn in zinc-blende or wurtzite structures
(ii)	<1:	$[Al_2Cd]S_4$
(iii)	>1:	$[Cu_3As]S_3$

Examples of classes (i) and (ii) are set out in Table 17.10.

(i) The simplest example of a superstructure of zinc-blende is the structure of chalcopyrite, or copper pyrites, $CuFeS_2$, which arises by replacing Zn by equal numbers of Cu and Fe atoms in a regular way. As a result of this substitution the atoms at the corners of the original ZnS unit cell are not all of the same kind so that the repeat unit is doubled in one direction. The unit cell in $CuFeS_2$ is therefore twice as large as that of ZnS. We may proceed a stage further by replacing one-half of the Fe atoms in $CuFeS_2$ by Sn, and so arrive at the structure of stannite, Cu_2FeSnS_4. The structures of ZnS, $CuFeS_2$, and Cu_2FeSnS_4 are shown in Fig. 17.12. The nearest

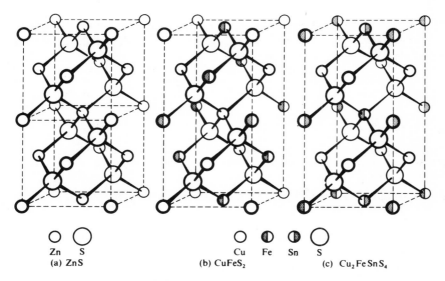

FIG. 17.12. The crystal structures of (a) ZnS, showing two unit cells. (b) $CuFeS_2$; and (c) Cu_2FeSnS_4.

TABLE 17.10

Structures related to the zinc-blende and wurtzite structures

Fraction of metal positions occupied	Zinc-blende			Wurtzite	
	Regular		*Random*	*Regular*	*Random*
All	Zinc-blende ZnS	BN, BP XY (X=Al, Ga, In Y=P, As, Sb)		AlN, GaN, InN	
	Chalcopyrite CuFeS$_2$	AgMX$_2$, CuMX$_2$ (M=Al, Ga, In X=S, Se, Te) ZnSnP$_2$ CdGeAs$_2$	CdZnSe$_2$ ZnSnAs$_2$ GaInSb$_2$	BeSiN$_2$ CuFe$_2$S$_3$	α-AgInS$_2$
	Stannite Cu$_2$FeSnS$_4$	Cu$_2$FeGeS$_4$ Cu$_2$CoGeS$_4$	–	Cu$_2$CdSiS$_4$ Cu$_2$CdGeS$_4$	–
		Cu$_3$AsS$_4$ (luzonite), Cu$_3$PS$_4$	–	Cu$_3$AsS$_4$ (enargite)	–
$\frac{3}{4}$(a)	*Regular*		*Random*	*Random*	
	Al$_2$CdS$_4$, β-Cu$_2$HgI$_4$ β-Ag$_2$HgI$_4$ In$_2$CdSe$_4$		α-Ag$_2$HgI$_4$ α-Cu$_2$HgI$_4$ Ga$_2$HgTe$_4$	β-Al$_2$ZnS$_4$	
$\frac{2}{3}$(a)	*Random*			*Random*	
	γ-Ga$_2$Se$_3$ γ-Ga$_2$S$_3$ γ-Ga$_2$Te$_3$ γ-In$_2$Te$_3$			Al$_2$Se$_3$ β-Ga$_2$S$_3$	

(a) These fractions correspond to three-eighths and one-third of the *total* number of tetrahedral holes in close-packed assemblies as listed in Table 4.5 (p. 162).

neighbours of an S atom in the three structures are:

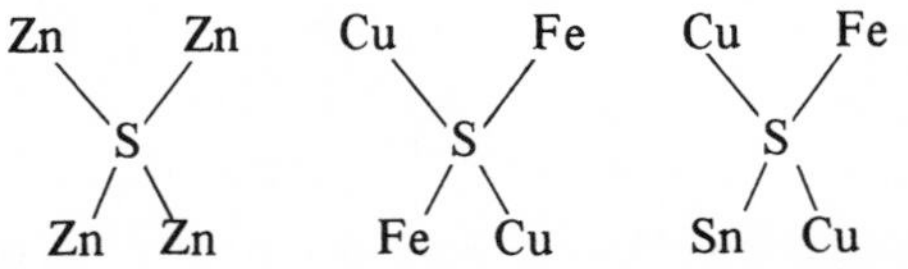

Zn   Zn   Cu   Fe   Cu   Fe

S   S   S

Zn   Zn   Fe   Cu   Sn   Cu

(ZnS)   (CuFeS$_2$)   (Cu$_2$FeSnS$_4$)

Examples of the analogous wurtzite superstructures include Cu_2CdSiS_4 and Cu_2CdGeS_4,[2a] in which the nearest neighbours of a S atom are 2 Cu, 1 Cd, and 1 Si (Ge).

It is worth noting that neither FeS nor SnS has the 4-coordinated zinc-blende structure, and also that CuS is apparently not a true sulphide of Cu^{II}. The valence of Cu (and other elements) in CuS, $CuFeS_2$, etc., is discussed on p. 1142. Examples of compounds[2b] with the $CuFeS_2$ structure are given in Table 17.10.

Some compounds have the ordered $CuFeS_2$ structure at ordinary temperatures and a statistical zinc-blende structure at higher temperatures, for example, $ZnSnAs_2$.[3]

Another example of a compound of this type is Cu_3AsS_4 (luzonite),[4a] the structure of which is of the zinc-blende type with three-quarters of the Zn atoms replaced by Cu and the remainder by As. This regular replacement of 4 Zn by 3 Cu + As leads to the larger unit cell of Fig. 17.13. Another form of Cu_3AsS_4 (enargite) is a superstructure of wurtzite.[4b]

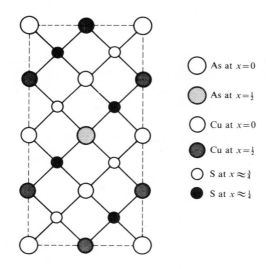

FIG. 17.13. Projection of the structure of luzonite, Cu_3AsS_4, along the *a* axis.

(ii) Three regular structures of this kind have been described, and although examples of *sulphides* with each of these structures are not known they are illustrated here (Fig. 17.14) because of their close relation to chalcopyrite and stannite; for examples see Table 17.10.

(iii) Examples of structures related to ZnS by omission of some S atoms include Cu_3AsS_3 and Cu_3SbS_3,[5] though minerals of this family usually contain iron and their formulae are more complex.

Although the M:S ratio is unity the ferromagnetic cubanite, $CuFe_2S_3$[6] is not

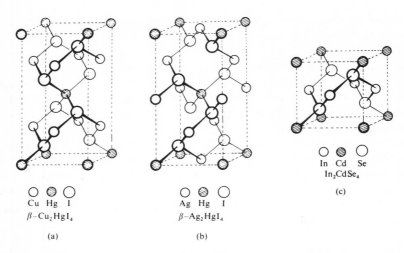

FIG. 17.14.   Structures related to the zinc-blende structure: (a) β-Cu_2HgI_4;   (b) β-Ag_2HgI_4;
(c) In_2CdSe_4.

of type (i), for its structure is related to wurtzite in a more complex way. It is built of slices of the wurtzite structure joined together in such a way that pairs of FeS_4 tetrahedra share edges (Fig. 17.15). The resulting Fe—Fe distances (2.81 Å) are rather long for metal–metal bonds, but presumably indicate appreciable interaction between the Fe atoms.

The structural relation of wolfsbergite, $CuSbS_2$[7] to wurtzite is less simple. This compound has a layer structure in which each Sb atom forms the usual three

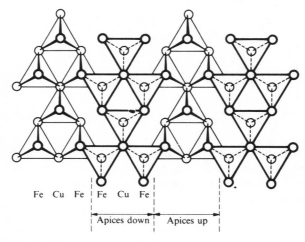

FIG. 17.15.   Portion of the structure of cubanite ($CuFe_2S_3$) showing slabs of the wurtzite structure with the tetrahedra pointing alternately up and down.

pyramidal bonds and Cu its four tetrahedral bonds to S atoms. In Fig. 17.16(a) is shown a plan of part of the wurtzite structure, in which plan Zn and S are superposed since they lie (at different heights) on the same lines perpendicular to the plane of the paper. If we take a vertical section through this structure between the dotted lines we see that a metal atom lying on these dotted lines loses one of its S neighbours, shown as a dotted circle, whereas a metal atom lying between the dotted lines retains all its four S neighbours. (It must be remembered that each metal atom is joined to one S atom lying vertically above or below it, so that the number of M—S bonds is one more than the number seen in the plan.) These latter atoms are Cu and the former Sb in the layers of $CuSbS_2$ which are viewed end-on in Fig. 17.17. Comparison of Figs. 17.16 and 17.17 shows the similarity in structure of a $CuSbS_2$

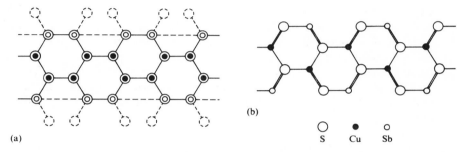

(a) (b)

◯ ● ○
S Cu Sb

FIG. 17.16. (a) Plan of a portion of the wurtzite structure. The metal atoms, which are all Zn atoms in wurtzite, are shown as small circles of two types to facilitate comparison with Fig. 17.17. (b) The same, with certain atoms removed, to show the relation to the layers in $CuSbS_2$.

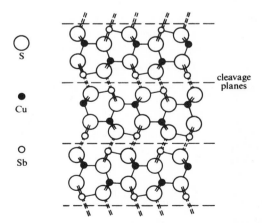

◯
S

●
Cu

○
Sb

cleavage planes

FIG. 17.17. Elevation of the structure of $CuSbS_2$, viewed parallel to the plane of the layers. The neighbours of an Sb atom in a layer are one S at 2.44 Å and two S at 2.57 Å. The broken lines between the layers indicate much weaker Sb—S bonds (Sb—S = 3.11 Å) which account for the very good cleavage parallel to the layers.

layer and the section of the wurtzite structure between the dotted lines (parallel to the crystallographic 1120 plane).

The number of complex sulphide minerals is very large, and although it is possible to regard many of them as derived from simple sulphide structures (e.g. ZnS, PbS) the relationship is not always very close. For example, the Ag and Sb positions in miargyrite, $AgSbS_2$,[8] are close to those of alternate Pb atoms in galena (PbS), but the S atoms are so far removed from the positions of the ideal PbS structure that there is no very close resemblance between the structures. A more important feature

$$\begin{array}{ccc} S & & S \\ | & & | \\ \diagdown S \diagup Sb \diagdown S \diagup Sb \diagdown S \diagup \end{array}$$

of $AgSbS_2$ is the pyramidal bonding of Sb to 3 S. In many complex sulphides Sb forms three such pyramidal bonds, and sometimes also two weaker ones, and these structures are best classified according to the nature of the Sb–S complex. This is the SbS_2 chain in berthierite, $FeSb_2S_4$,[9] and fragments (Sb_3S_7) of such chains in jamesonite, $FePb_4Sb_6S_{14}$.[10] In a very comprehensive classification of the structures of all known complex sulphide minerals[11] the general formula is written in the form $A'_q A''_r (B_m C_n) C_p$ where A' and A'' stand for metals with the following coordination numbers:

$$A' \begin{cases} 2 & Ag, Tl, Hg \\ 3 & Ag, Cu \\ 4 & Ag, Cu, Zn \end{cases} \qquad A'' \begin{cases} 6 & Pb, Fe, Co, Ni, Hg \\ 7 & Pb, Tl \\ 9 & Pb, \end{cases}$$

The atoms B (As, Sb, Bi) and C (S, Se, Te) form the system $B_m C_n$ built of pyramidal BC_3 and/or tetrahedral BC_4 groups, and C_p represents the S (Se, Te) which does not form part of the sulpho-salt complex. The basis of the classification is the nature of the $B_m C_n$ complex which, together with the coordination requirements of the A' and A'' atoms, determines the crystal structure.

(1)	AC 1965 **19** 543	(6)	AM 1955 **40** 213
(2a)	AC 1972 **B28** 1626	(7)	ZK 1933 **84** 177
(2b)	AC 1958 **11** 221	(8)	AC 1964 **17** 847
(3)	AC 1963 **16** 153	(9)	AM 1955 **40** 226
(4a)	ZK 1967 **124** 1	(10)	ZK 1957 **109** 161
(4b)	AC 1970 **B26** 1878	(11)	SMPM 1969 **49** 109 (180 references).
(5)	ZK 1964 **119** 437		

Complex sulphides containing metal clusters

Interactions between metal atoms which are also bridged by non-metal atoms occur in, for example, the clusters in halides of Nb, Ta, Mo, W, Pd, and Pt (p. 432), and in metal carbonyls, p. 964. There are metal clusters of various kinds in certain complex sulphides. Tetrahedral clusters have been found in sulphides such as $MMo_2Re_2S_8$ in which M is Zn, Fe, Co, or Ni[1] and 'cubane-like' clusters Mo_4S_4 may be recognized around vacant sites in ordered spinels $Al_{0.5}Mo_2S_4$ and $Ga_{0.5}Mo_2S_4$.[2] An interesting family of compounds containing clusters of Mo atoms has the general formula $M_m(Mo_{3n}S_{3n+2})$, where M is one of the elements Na, K, Rb, Cs, In, or Tl. The clusters are built from triangular sub-units (a) which are reminiscent of those in certain Ni and Pt carbonyls, but without the bridging CO groups. These sub-units

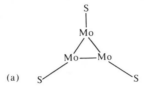

(a)

may be stacked, in staggered arrangement, to form either a discrete octahedral group or more complex groups of face-sharing octahedra. A cluster is completed by a S atom at each end, on the 3-fold axis, giving the formulae Mo_6S_8, Mo_9S_{11}, or $Mo_{12}S_{14}$ (Fig. 17.18). The atoms M are accommodated in cavities in the structures.

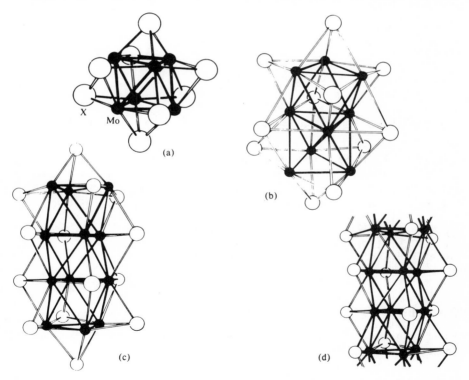

FIG. 17.18. Clusters built from MoS_6 (or $MoSe_6$) octahedra: (a) Mo_6X_8; (b) Mo_9X_{11}; (c) $Mo_{12}X_{14}$; (d) $(Mo_6X_6)_n$ (X = S, Se). (AC 1980 **B36** 1547)

Combinations of complexes of two kinds occur in $In_3Mo_{15}Se_{19}$ (Mo_6Se_8 and Mo_9Se_{11})[3] and in $Tl_2Mo_9S_{11}$, in which the clusters are not the simple M_9S_{11} complexes but a combination of Mo_6S_8 and $Mo_{12}S_{14}$, giving the structural formula $Tl_4(Mo_6S_8)(Mo_{12}S_{14})$.[4] The limiting structure is a column of face-sharing octahedra with its surrounding S atoms (composition Mo_6S_6).[5] In all these structures the M atoms are situated between the finite groups or between the chains in the compounds $M_2Mo_6S_6$. Figure 17.19 shows that the projection of the structure of $K_2Mo_6S_6$ or the

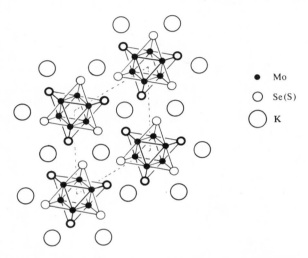

FIG. 17.19. Projection of the structure of $K_2Mo_6Se_6$ along the chain direction.

isostructural $K_2Mo_6Se_6$ along the chain axis is very similar to that of Ta_6S (Fig. 17.8(c)) except that the metal chains are surrounded more closely by S atoms and the atoms between the chains are K atoms. The Mo–Mo distances are 2.66 and 2.72 Å in and between the Mo_3 triangles, very close to the value in the metal itself (2.72 Å).

(1) JSSC 1976 **19** 305   (4) AC 1980 **B36** 1319
(2) JSSC 1975 **14** 203   (5) JSSC 1980 **35** 286
(3) AC 1979 **B35** 285   (6) AC 1980 **B36** 1545

Oxysulphides

These compounds appear to be formed by a relatively small number of elements. Those of non-metals are described in other chapters (COS, $P_4O_6S_4$). Three structures will be described here, though a few more complex compounds have also been studied (for example, $Ce_4O_4S_3$:[1]

the cubic ZrOS structure;[2]

the BiOCl (PbFCl) structure, of ThOS, PaOS, UOS, NpOS,[3] and the second form of ZrOS;[4]

the La_2O_2S structure,[5] also adopted by Ce_2O_2S and Pu_2O_2S.

ZrOS is prepared by passing H_2S over $Zr(SO_4)_2$ at red heat or over ZrO_2 at 1300 °C; it is dimorphic. In the cubic form the Zr^{4+} ion is 7-coordinated, its neighbours being 3 O^{2-} at 2.13 Å and 4 S^{2-} at 2.62 Å. The coordination group is a distorted octahedron consisting of 3 O + 3 S with the fourth S ion above the centre of the face containing the three oxygen ions. This coordination group is illustrated, in idealized form, in Fig. 3.6(a), p. 75.

The BiOCl structure has been described and illustrated in Chapter 10. In this structure there is 8-coordination of the metal atom, at the vertices of an antiprism, by 4 O + 4 Cl, but in the isostructural LaOCl a fifth Cl (beyond one square antiprism face) is at the same distance from the metal ion as the other four. In ThOS etc. the positions of the S atoms were not accurately determined, and these atoms were so placed as to give the metal atom 5 equidistant S neighbours, in addition to the 4 O neighbours. (Th—4 O, 2.40 Å, Th—5 S, 3.00 Å.)

The compound ZrSiS may be mentioned here since it was at one time mistaken for Zr_4S_3. It has a structure closely related to the PbFCl structure, the main difference being that in the oxysulphides with this structure O—O contacts correspond only to the van der Waals bonding whereas in ZrSiS there are Si—Si bonds (2.51 Å).[6]

The structure of La_2O_2S is closely related to that of the high-temperature form of La_2O_3 (the A—M_2O_3 structure of Fig. 12.7). One-third of the oxygen is replaced by sulphur, so that the metal ions have the same 7-coordination group as in cubic ZrOS, here made up of 4 O + 3 S. In La_2O_2S bond lengths are La—4O, 2.42 Å; La—3S, 3.04 Å. For further discussion of the La_2O_2S structure see p. 1271.

(1) AC 1978 **B34** 3564
(2) AC 1948 **1** 287
(3) AC 1949 **2** 291
(4) ACSc 1962 **16** 791
(5) AC 1973 **B29** 2647
(6) RTC 1964 **83** 776

18

Nitrogen

Introduction

Nitrogen stands apart from the other elements of Group V. It is the most electronegative element in the group, and from the structural standpoint two of its most important characteristics are the following. Only the four orbitals of the L shell are available for bond formation so that nitrogen forms a maximum of four (tetrahedral) bonds, as in NH_4^+, substituted ammonium ions, and amine oxides, $R_3N.O$. Only three bonds are formed in halides and oxy compounds, the bonds in many of the latter having multiple-bond character. Like the neighbouring members of the first short period, carbon and oxygen, nitrogen has a strong tendency to form multiple bonds. It is the only Group V element which exists as diatomic molecules at ordinary temperatures and the only one which remains in this form ($N{\equiv}N$) in the liquid and solid states in preference to polymerizing to single-bonded systems as in the case of phosphorus and arsenic. (For solid N_2 see Chapter 29.) There is an interesting difference between the relative strengths of single and multiple nitrogen–nitrogen, nitrogen–carbon, and carbon–carbon bonds:

C–C	347	C–N	305	N–N	163 kJ mol^{-1}
C=C	611	C=N	615	N=N	418
C≡C	837	C≡N	891	N≡N	946

For the above reasons nitrogen forms many compounds of types not formed by other elements of this group, and for this reason we deal separately with the stereochemistry of this element. For example, the only compounds of N and P which are structurally similar are the molecules in which the elements are 3-covalent and the phosphonium and ammonium ions. There are no nitrogen analogues of the phosphorus pentahalides, and there is little resemblance between the oxygen compounds of the two elements. Monatomic ions of nitrogen and phosphorus are known only in the solid state, in the salt-like nitrides and phosphides of the more electropositive elements. The multiple-bonded azide ion, N_3^-, is peculiar to nitrogen.

The following compounds of nitrogen are described in other chapters:

The stereochemistry of nitrogen

In all its compounds nitrogen has four pairs of electrons in its valence shell. According to the number of lone pairs there are the five possibilities exemplified by the series

$$\left[\begin{array}{c} H \\ H:\!\overset{\cdot\cdot}{N}:\!H \\ H \end{array}\right]^{+} \quad \begin{array}{c} H \\ H:\!\overset{\cdot\cdot}{N}: \\ H \end{array} \quad \left[\begin{array}{c} H \\ \overset{\cdot\cdot}{:}\!N:\!H \\ \overset{\cdot\cdot}{} \end{array}\right]^{-} \quad \left[\begin{array}{c} H \\ :\!\overset{\cdot\cdot}{N}:\! \\ \overset{\cdot\cdot}{} \end{array}\right]^{2-} \quad \text{and} \quad \left[:\!\overset{\cdot\cdot}{\underset{\cdot\cdot}{N}}:\right]^{3-}$$

The last three, the NH_2^-, NH^{2-}, and N^{3-} ions, are found in the salt-like amides, imides, and nitrides of the most electropositive metals, but with few exceptions the stereochemistry of nitrogen is based on N^+ with no lone pairs and N with one lone pair. These two states of the nitrogen atom correspond to the classical trivalent and 'pentavalent' states, now preferably regarded as oxidation rather than valence states.

The stereochemistry of N^+ is similar to that of C, with which it is isoelectronic:

(a) $\overset{\diagdown}{\underset{\diagup}{N^+}}\overset{\diagup}{\underset{\diagdown}{}}$ (b) $=\!N^+\!\!\overset{\diagup}{\underset{\diagdown}{}}$ and (c) $=\!N^+\!=$ or $-N^+\!\equiv$

  tetrahedral   planar   collinear

As in the case of carbon, the valence bond representations (b) and (c) do not generally represent the electron distributions around N in molecules in which this element is forming three coplanar or two collinear bonds. Such bonds often have intermediate bond orders which may be described in terms of resonance between a number of structures with different arrangements of single and multiple bonds or in terms of trigonal (sp^2) or digonal (sp) hybridization with varying amounts of π-bonding.

The central N atom of the N_3^- ion in azides approaches the state $=\!N^+\!=$ and diazonium salts[1] illustrate $-N^+\!\equiv$:

$$[N\!\!=\!\!=\!\!N^+\!\!=\!\!=\!\!N]^- \quad \text{compare}$$
$$\underset{1\cdot15 \quad 1\cdot15\,\text{Å}}{}$$

$$-N^+\!\equiv\!N$$
$$1\cdot385\ 1\cdot097\,\text{Å}$$

The bond length N–N in the diazonium ion is the same as in elementary nitrogen ($N\!\equiv\!N$); the C–N bond is shorter than a normal single bond owing to interaction with the aromatic ring.

Two collinear N bonds are formed in ions such as $[(H_2O)Cl_4Ru\text{–}N\text{–}RuCl_4\text{-}(H_2O)]^{3-}$ – compare $(Cl_5Ru\text{–}O\text{–}RuCl_5)^{4-}$, p. 478.

When one of the four orbitals is occupied by a lone pair the possibilities are:

(d) $:\!N\overset{\diagup}{\underset{\diagdown}{}}$ (e) $-N\!\!\overset{\diagup\diagup}{\underset{\cdot\cdot}{}}$ or (f) $:\!N\!\equiv$

  pyramidal   angular

The unshared pair of electrons in molecules such as NH_3 can be used to form a fourth bond, the coordinate link ($\rightarrow$) of Sidgwick, as in the metal ammines (see Werner coordination compounds, Chapter 27).

An interesting development in the chemistry of nitrogen is the demonstration that the N atoms of neutral N_2 molecules can utilize their lone pairs (like the isoelectronic CO) to bond to a transition metal. There is bonding either through one N or through both, when the N_2 molecule forms a bridge between two metal atoms. These are examples of case (f) above, and include the trigonal pyramidal molecule $CoHN_2(P\phi_3)_3$,[2] in which Co–N is appreciably shorter than in ammines (1.96 Å), the octahedral ion in $[Ru(NH_3)_5N_2]Cl_2$,[3] and the bridged anion (eclipsed

configuration) in $[(NH_3)_5RuN_2Ru(NH_3)_5](BF_4)_4$,[4] in which the Ru–N bond length in the linear bridge is markedly shorter than Ru–NH_3. The cation in the salt $[Ru(en)_2N_2N_3]PF_6$[5] is notable for having bonds from metal to N of three different kinds, to NH_2, N_2 molecules, and N_3^- ligands. For examples of transition metal complexes in which a single N atom is a ligand see p. 1190.

The stereochemistry of nitrogen is summarized in Table 18.1, in which the primary subdivision is made according to the number of orbitals used by bonding and lone pairs of electrons. The first main subdivision includes the simpler hydrogen and halogen compounds, and we shall deal with these first in the order of Table 18.1, except that amides and imides will be included with ammonia. The second and third sections include most of the oxy-compounds, and since it is preferable to deal with oxides and oxy-acids as groups we shall not keep rigidly to the subdivision

TABLE 18.1

The stereochemistry of nitrogen

Number of σ pairs (bonding and lone pairs)	Type of hybrid	Number of lone pairs	Bond arrangement	Examples
4	sp^3	0	Tetrahedral	NH_4^+, $(CH_3)_3N^+{-}O^-$
		1	Trigonal pyramidal	NH_3, NH_2OH, NF_3, NHF_2, NH_2F, N_2H_4, N_2F_4
		2	Angular	NH_2^-
3	sp^2	0	Trigonal planar	NO_2Cl, $NO_2(OH)$, NO_3^-, N_2O_4 $[ON(NO)SO_3]^-$, $[ON(SO_3)_2]^{2-}$
		1	Angular	$NOCl$, NOF, NO_2^-, N_2F_2
2	sp	0	Collinear	N_2O, NO_2^+, N_3H

according to bond arrangement. Certain other special groups, such as azides, sulphides, and nitrides, will also be described separately.

(1) ACSc 1963 **17** 1444
(2) IC 1969 **8** 2719
(3) AC 1968 **B24** 1289
(4) JACS 1969 **91** 6512
(5) IC 1970 **9** 2768

Nitrogen forming four tetrahedral bonds

The optical activity of substituted ammonium ions $(Nabcd)^+$ and of more complex ions and molecules (see Chapter 2) provided the earliest proofs of the essentially tetrahedral arrangement of four bonds from a nitrogen atom. Later, this was confirmed by structural studies of, for example, AlN (wurtzite structure) and BN (wurtzite and zinc-blende structures), of ammonium salts, amine oxides, and the series of molecules (a)–(c) in which N is bonded tetrahedrally to 3 C + 1 Al, 2 C + 2 Al, and 1 C + 3 Al respectively.

(a)[1]

(b)[2]

(c)[3]

Ammonium and related ions

It has not generally proved possible to establish the tetrahedral structure of the NH_4^+ ion in simple ammonium salts by X-ray studies, both because of the small scattering power of H for X-rays and because the ion is rotating or shows orientational disorder in the crystalline salts at ordinary temperatures. In NH_4F, however, rotation of the ions is prevented by the formation of N–H–F bonds, and the tetrahedral disposition of the four nearest F neighbours of a N atom indicates the directions of the hydrogen bonds. It is possible to locate H atoms by neutron diffraction, and detailed studies of several ammonium salts have been made. The N–H bond lengths in ammonium salts have also been determined by n.m.r. and several studies give values close to 1.03 Å.[4] The structures of substituted ammonium ions have been determined in $[N(CH_3)_4]_2SiF_6$, $[N(CH_3)_4]ICl_2$, etc.

The NF_4^+ ion exists in salts such as $(NF_4)(AsF_6)$ and $(NF_4)(SbF_6)$ prepared from NF_3, F_2, and MF_5 heated under pressure. $(NF_4)(AsF_6)$ is a stable, colourless, non-volatile, hygroscopic solid at 25 °C, presumably structurally similar to $(PCl_4)(PCl_6)$; n.m.r. shows that all four F atoms attached to N are equivalent.[5]

Diffraction studies of salts containing the hydroxylammonium ion give N–O, 1.40 Å in $Li(NH_3OH)SO_4$[6] and 1.42 Å in $NH_3OH.ClO_4$ (n.d.).[7]

Amine oxides. In the tetrahedral molecule of trimethylamine oxide, $((CH_3)_3NO)$, N–O = 1.40 Å.[8] Trifluoramine oxide, NF_3O, is a stable colourless gas prepared by the action of an electric discharge on a mixture of NF_3 and oxygen. Infrared[9] and n.m.r.[10] data are consistent with a nearly regular tetrahedral shape.

(1)   AC 1969 **B25** 377
(2)   JACS 1970 **92** 285
(3)   AC 1972 **B28** 1619
(4)   JCP 1954 **22** 643, 651
(5)   IC 1967 **6** 1156
(6)   AC 1973 **B29** 2628
(7)   AC 1974 **B30** 1167
(8)   AC 1964 **17** 102
(9)   IC 1968 **7** 2064
(10) JCP 1967 **46** 2904

Nitrogen forming three pyramidal bonds

Ammonia and related compounds

In general three single bonds from a nitrogen atom are directed towards the apices of a trigonal pyramid, as has been demonstrated by spectroscopic and electron diffraction studies of NH_3 and many other molecules NR_3; some results are summarized in Table 18.2. A pyramidal configuration is to be expected whether we

TABLE 18.2

Structural data for molecules NR_3, NR_2X, and NRX_2

Molecule	N—H (Å)	N—C (Å)	N—X (Å)	Bond angles	Method	Reference
NH_3	1.015			HNH 106.6°	e.d.	ACSc 1964 **18** 2077
NH_2CH_3	1.011	1.474		HNH 105.9°	m.w.	JCP 1957 **27** 343
				CNH 112.1°		
$NH(CH_3)_2$	1.022	1.466		CNC 111.6°	m.w.	JCP 1968 **48** 5058
				CNH 108.8°		
$N(CH_3)_3$		1.451		CNC 110.9°	m.w.	JCP 1969 **51** 1580
$N(CH_3)_2Cl$			1.75		e.d.	TFS 1944 **40** 164
NH_2Cl				HNCl 102°	i.r.	JACS 1952 **74** 6076
$NHCl_2$			(1.76)	ClNCl 106°		
NCl_3			1.75	ClNCl 106.8°	X	ZaC 1975 **413** 61
NF_3			1.371	FNF 102.2°	e.d.	JACS 1950 **72** 1182
					m.w.	PR 1950 **79** 513
NHF_2	1.026		1.400	FNF 103°	m.w.	JCP 1963 **38** 456
				HNF 100°		
$N(C_6H_5)_3$		1.42		CNC 116°		JCP 1959 **31** 477
		N—Si	Si—N—Si			
$N(SiH_3)_3$		1.74	120°		e.d.	JACS 1955 **77** 6491
$N(SiH_3)_2H$		1.725	128°		e.d. ⎫	JCS A 1969 1224
$N(SiH_3)_2CH_3$		—	126°		e.d. ⎬	
$Al[N\{Si(CH_3)_3\}_2]_3$		1.75	118°		X	JCS A 1969 2279
		N—Ge	Ge—N—Ge			
$N(GeH_3)_3$		1.84	120°		e.d.	JCS A 1970 2935

regard the orbitals as three p or three of four sp^3 orbitals, the unshared pair of electrons occupying one of the bond positions in the latter case. For pure p orbitals bond angles of 90° would be expected, and mutual repulsion of the H atoms then has to be assumed to account for the observed bond angles, which are close to the tetrahedral value. It seems likely that the bonds have some s character, though less than for sp^3 hybridization, so that the lone-pair electrons have some p character. Since this lone-pair orbital is directed it leads to an atomic dipole, and it is this which is responsible for most of the dipole moment (1.44 D) of NH_3. The importance of the atomic dipole is shown by the fact that the pyramidal NF_3 molecule is almost non-polar (0.2 D), due to cancellation of the moments due to the polar N–F bonds by the lone-pair moment. In contrast to NF_3, NHF_2 has an appreciable moment (1.93 D).

We should expect a molecule N*abc* containing three different atoms or groups to exhibit optical activity, since the molecule is enantiomorphic. However, no such molecule has been resolved in spite of many attempts, and this at one time cast doubt on the non-planar arrangement of three bonds from a neutral N atom. In the case of NH_3 itself it has been suggested that the potential energy curve of the molecule is of the form shown in Fig. 18.1(a), where the energy is plotted against the distance *r*

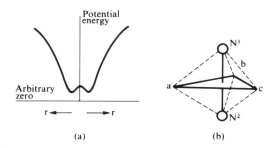

(a)   (b)

FIG. 18.1.   (a) Probable form of the potential energy curve for NH_3. (b) The two possible configurations for the molecule NH_3 or N*abc*.

of the N atom from the plane of the H atoms. There are two minima corresponding to the two possible positions of the N atom (on each side of the plane of the H atoms), but they are separated by an energy barrier of only about 25 kJ mol^{-1}. For the resolution of *d* and *l* modifications of a substance to be detectable at room temperature it has been calculated that the energy of activation for racemization (change of $d \rightleftharpoons l$) must be not less than 84 kJ mol^{-1}. The failure to resolve substituted ammonias may therefore be due to the fact that the molecule can easily turn inside out by movement of the N atom from position N^1 to N^2 (Fig. 18.1(b)). Calculations[1] of the rates of racemization of pyramidal molecules XY$_3$ with N. P. As, Sb, and S as central atoms, based on a potential energy function derived from known vibrational frequencies and molecular dimensions, suggest that optically active compounds of As, Sb, and S should be stable towards racemization at room temperature, and in the case of P possibly at low temperatures.

Three pyramidal bonds are formed by N in hexamethylenetetramine (HMT), $N_4(CH_2)_6$ (Fig. 18.2), which is a quite remarkable compound. Not only is it made, at room temperature, by a very simple reaction, the interaction of NH_3 with H_2CO, but it has an extraordinary crystal chemistry. The structure of the HMT molecule has been studied many times, in the crystalline state by X-ray and neutron diffraction

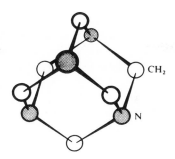

FIG. 18.2. The structure of the molecule of hexamethylene-tetramine, $N_4(CH_2)_6$. N atoms are shaded and H atoms omitted. This molecule, based on a tetrahedral group of N atoms, should be compared with those of $Be_4O(CH_3COO)_6$, P_4O_6, and P_4O_{10}, Figs. 11.1 and 19.8.

and in the vapour state by e.d. (C–N, 1.476 Å; C–H, 1.088 Å; NCN, 113.6°; CNC, 107.2°).[2] HMT forms a hexahydrate (p. 665) and crystalline compounds with numerous organic compounds and inorganic salts. An example of the latter is $CaCr_2O_7 \cdot (HMT)_2 \cdot 7H_2O$,[3] which is a hydrogen-bonded assembly of units of three kinds: $[Ca(H_2O)_7]^{2+}$ groups; $Cr_2O_7^{2-}$ ions; and HMT molecules. The Ca^{2+} ion is surrounded only by 7 H_2O molecules; there are no contacts between Ca^{2+} ions and O atoms of $Cr_2O_7^{2-}$ ions. The analogous Mg compound is a hexahydrate,[4] in which there is the more usual octahedral coordination of the metal ions. HMT forms two adducts with iodine, $HMT.I_2$ (orange), which changes slowly into the brown $[(HMT)_2I]I_3$ with the same composition, and the red-brown $HMT.2I_2$; the structures of all three have been determined. The adducts $HMT.I_2$[5] and $HMT.2I_2$[5] consist of molecules in which one or two I_2 molecules are bonded to N forming linear –N–I–I systems as shown at (a) and (b), the N–I bonds utilizing the lone pairs of N atoms. The molecule (b) rearranges to ions $[(HMT)_2I]^+$, (c), with collinear N–I bonds of the same length as in $NI_3 \cdot NH_3$ (p. 799), and I_3^- ions (slightly unsymmetrical, I–I, 2.90 and 2.93 Å).[6]

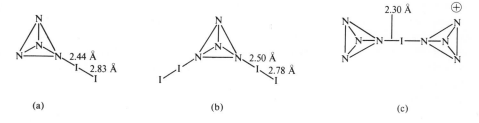

(a)   (b)   (c)

Ions and molecules in which d_π-p_π bonding leads to coplanarity of three bonds from N include a number of molecules containing the silyl group (in the lower part of Table 18.2). The molecule $Al[N\{Si(CH_3)_3\}_2]_3$[7] is also of interest as an example of Al forming three coplanar bonds. The $Si(CH_3)_3$ groups lie out of the AlN_3 plane, the dihedral angle $NAlN/AlNSi$ being $50°$. There is the same unusual arrangement of three bonds from the metal atom in $Fe[N(SiMe_3)_2]_3$, and P forms three coplanar bonds in the closely related $(Me_3Si)_2N-P-(NSiMe_3)_2$.[8]

(1) JACS 1954 **76** 2645   (5) AC 1975 **B31** 1589
(2) PRS 1963 **273A** 435, 455   (6) AC 1975 **B31** 1505
(3) AC 1975 **B31** 423   (7) JCS A 1969 2279
(4) AC 1974 **B30** 22   (8) AnCIE 1975 **14** 261

Amides. imides, and nitride hydrides. These colourless, crystalline salts are formed only by the more electropositive metals. Amides and imides contain the NH_2^- and NH^{2-} ions respectively. Nitride hydrides contain N^{3-} and H^- ions; ionic nitrides are described on p. 833.

Amides are formed by the action of NH_3 on the heated metal or in solution in liquid NH_3. A solution of $NaNH_2$ in liquid NH_3 is a good conductor of electricity, indicating ionization in this solvent. The structures of amides, both simple and complex, show strong resemblances to those of halides and hydroxides. For example, the high-temperature forms of K, Rb, and Cs amides have the NaCl structure but at ordinary temperatures these compounds have less symmetrical structures. For example, the Cs salt has a (tetragonal) deformed CsCl structure (like γ-NH_4Br). In $NaNH_2$ and $LiNH_2$ at ordinary temperatures there are distorted coordination groups around the NH_2^- ions. In $NaNH_2$[1] there is tetrahedral coordination of both anion and cation, but the arrangement of 4 Na^+ around N is a very distorted tetrahedron, showing that the NH_2^- ions are not rotating. $LiNH_2$ has a slightly distorted zinc-blende structure.[2] The tetrahedral coordination is somewhat irregular (NH_2^- has one Li^+ at 2.35 Å and three Li^+ at 2.15 Å), apparently due to the orientation of the NH_2^- dipoles. With $LiNH_2$ contrast the layer structure of LiOH, in which Li^+ ions occupy a different set of tetrahedral holes in the same arrangement of c.c.p. anions.

Two forms of $Ca(NH_2)_2$ and $Sr(NH_2)_2$ have been described, one having a structure similar to that of anatase[3] and the other a defect NaCl structure.[4] In both polymorphs the NH_2^- ions are apparently in fixed orientations, as in $Be(NH_2)_2$,[5] which has a 3D framework structure of the same general type as β-Be$(OH)_2$, p. 634. From spectroscopic data the N–H bond length in NH_2^- has been determined as 1.03 Å.[6] Structural studies have been made of complex amides such as $K[Be(NH_2)_3]$[7] and $Na[Al(NH_2)_4]$,[8] in which there is trigonal planar and tetrahedral coordination of Be and Al respectively (Be–N, 1.59 Å; Al–N, 1.84 Å), and $K_2Sn(NH_2)_6$[9] and $K_3La(NH_2)_6$,[10] in both of which there are octahedral $M(NH_2)_6$ ions.

The alkali and alkaline-earth imides have typical ionic structures. Li_2NH has the anti-CaF_2 structure,[11] like Li_2O, and the alkaline-earth imides crystallize with the NaCl structure, as do the corresponding oxides, the rotating NH^{2-} ion possessing spherical symmetry. The following radii have been deduced for the amide and imide ions: NH_2^-, 1.73 Å, and NH^{2-}, 2.00 Å; compare Cl^-, 1.81 Å, and SH^-, 2.00 Å.

Some nitride hydrides are structurally similar to nitride halides (p. 475) of similar formula type. For example, Ba_2NH has a NaCl-like structure with statistical distribution of N^{3-} and H^- ions in the anion positions (compare Mg_2NF), but Ca_2NH and Sr_2NH are more complex. The former, made by heating Ca_3N_2 with CaH_2, is apparently a NaCl superstructure with one quarter of the H^- ions statistically distributed in interstitial sites;[12] it is related to Ca_2NH_x ($x \approx 0.5$), previously described as Ca_2N.[13]

(1)	JPC 1956 **60** 821	(8)	AC 1973 **B29** 925
(2)	ZaC 1959 **299** 33	(9)	AC 1977 **B33** 1076
(3)	ZaC 1963 **324** 278	(10)	ZaC 1974 **410** 104
(4)	CR 1969 **268** 175	(11)	ZaC 1951 **266** 325
(5)	ZaC 1976 **427** 1	(12)	JSSC 1976 **17** 135
(6)	JPC 1957 **61** 384	(13)	IC 1968 **7** 1757
(7)	AC 1974 **B30** 2579		

Hydroxylamine

The H atoms were not located in crystalline NH_2OH (N–O, 1.47 Å[1]) but their probable positions were deduced from a reasonable hydrogen bonding scheme. For hydroxylammonium salts see p. 792.

(1)   AC 1958 **11** 511, 512

Sulphonate ions derived from NH_3 *and* NH_2OH

Progressive substitution of SO_3^- for H in NH_3 gives the series of ions

$$\underset{(a)}{\overset{H}{\underset{H}{\Large{\diagdown}}}\!\!N\!-\!SO_3^-,} \qquad \underset{(b)}{H\!-\!N\!\!\overset{SO_3^-}{\underset{SO_3^-}{\Large\diagup}}} \qquad \text{and} \qquad \underset{(c)}{{}^-O_3S\!-\!N\!\!\overset{SO_3^-}{\underset{SO_3^-}{\Large\diagup}}}$$

The action of $KHSO_3$ on KNO_2 gives the trisulphonate, $K_3N(SO_3)_3$, which may be hydrolysed successively to the imidodisulphonate, $K_2NH(SO_3)_2$, then to the amino-sulphonate (sulphamate), KNH_2SO_3, or to the hydroxylamine-*N*-sulphonate, $K[NH(OH)SO_3]$. The structures of (a) and of the zwitterion sulphamic acid,

$^+H_3N.SO_3^-$, have been described on p. 722, where we include another zwitterion, hydroxylamine *O*-sulphonic acid, the acid corresponding to the ion (f), below. The imidosulphonate ion (b) is the NH analogue of the $O_3S.O.SO_3^{2-}$ ion, and is included in Table 16.5 (p. 725) together with (b′).

Ions of the hydroxylamine family are more numerous because H atoms of either or both the NH_2 or OH groups may be replaced, giving *N*- or *O*-sulphonates and finally the trisulphonate ion (h):

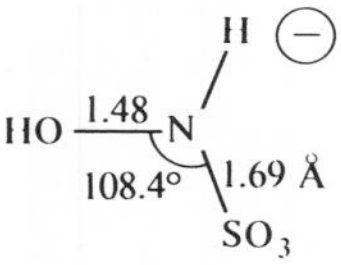
(d)[1]

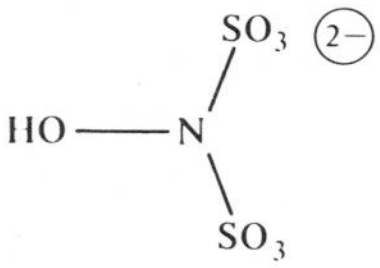
(e)

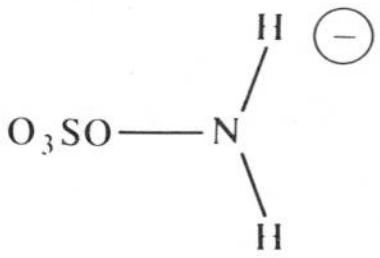
(f)

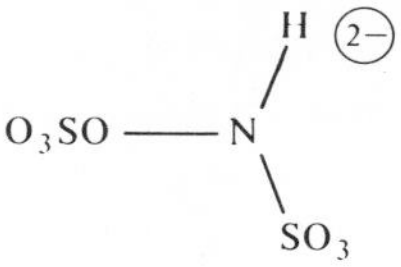
(g)

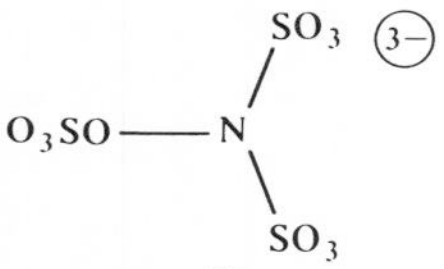
(h)[2]

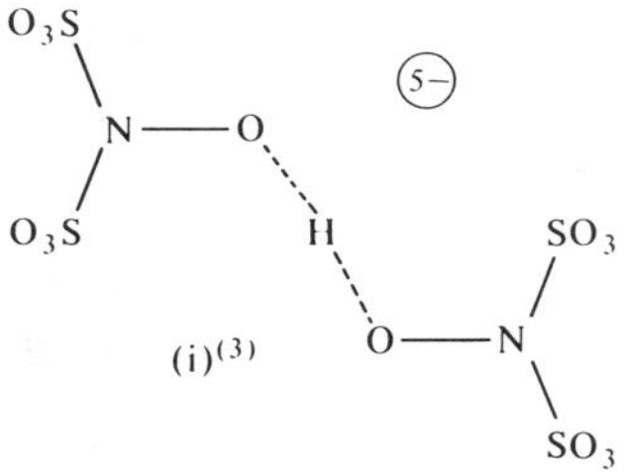
(i)[3]

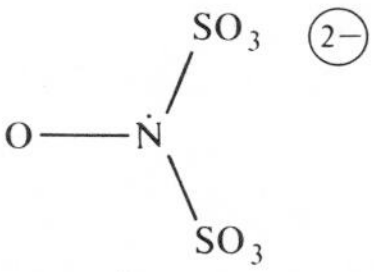
(j)[4]

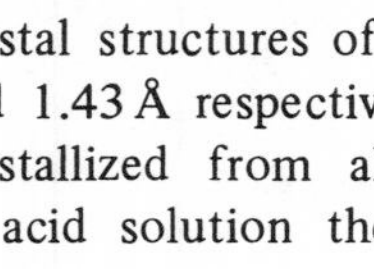
(e′)[5]

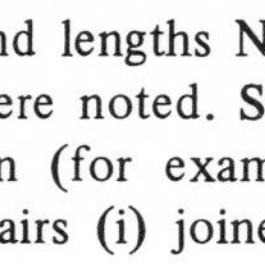
(b′)[6]

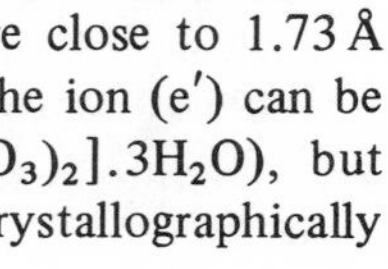
(k)[7]

We shall not describe the structures of these ions in detail, but give references to crystal structures of salts. The bond lengths N–S and N–O are close to 1.73 Å and 1.43 Å respectively except where noted. Salts containing the ion (e′) can be crystallized from alkaline solution (for example, $Na_3[ON(SO_3)_2].3H_2O$), but in acid solution the ions form pairs (i) joined through a (crystallographically

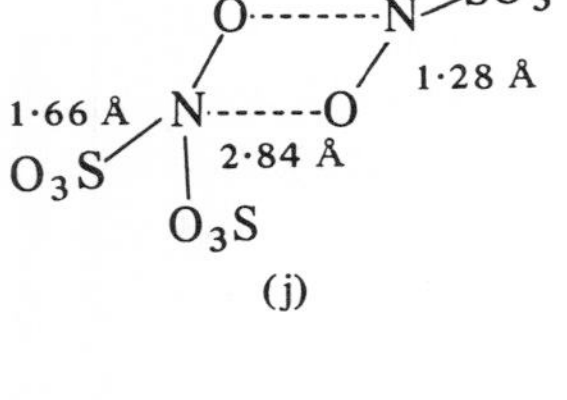
(j)

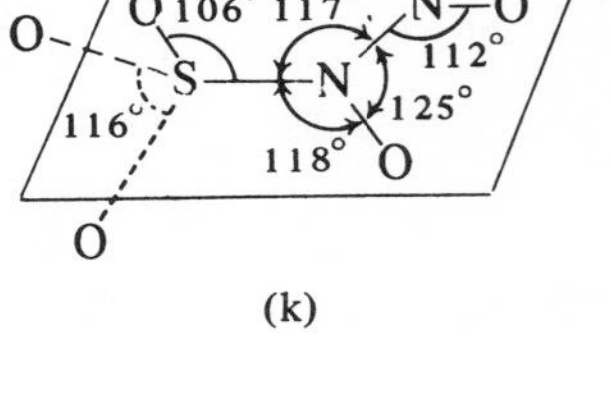
(k)

symmetrical) O---H---O bond of length 2.42 Å as in $Rb_5\{[ON(SO_3)_2]_2H\}.3H_2O$. In Frémy's salt, $K_2[ON(SO_3)_2]$ (j), the bonds from N are coplanar, as in the nitroso-hydroxylaminesulphonate ion (k), but N–S is appreciably shorter (1.66 Å). The anions are very weakly associated in pairs related by a centre of symmetry, reminiscent of the dimers in crystalline NO (p. 808). The K salt of (k) is made by absorbing NO in alkaline K_2SO_3 solution. The bonds drawn as full lines all lie in one plane, and the bond lengths indicate a single S–N bond (1.79 Å) but considerable π-bonding in the N–N bond (1.33 Å); N–O and S–O are 1.28 Å and 1.44 Å respectively.

It will be appreciated that the ions (j) and (k) contain odd numbers of electrons, leading to paramagnetism and colour. Frémy's salt is quite highly coloured, there being two crystalline forms, one orange and the other orange-brown. The reference is to the structure of the latter.

(1) AC 1966 **21** 819
(2) AC 1977 **B33** 1591
(3) IC 1974 **13** 2062
(4) JCS A 1968 3043
(5) IC 1975 **14** 2537
(6) AC 1979 **B35** 1308
(7) JCS 1951 1467

The trihalides of nitrogen

Two halides are well characterized, NF_3 and NCl_3. The former is quite stable when pure and has been studied in the gaseous state; the explosive NCl_3 has been studied in the crystalline state (at $-125\,°C$). Partially halogenated compounds which have been prepared include NH_2F, NHF_2, NH_2Cl, and $NHCl_2$. Pure NBr_3 does not appear to have been made, though NBr_3 (and also NH_2Br and $NHBr_2$) has been identified in dilute aqueous solution (resulting from bromination of NH_3 in buffered solutions) by u.v. absorption and chemical analysis.[1] There is some evidence for the existence of an ammine, $NBr_3.6NH_3$, stable at temperatures below $-70\,°C$. The so-called tri-iodide made from iodine and ammonia is $NI_3.NH_3$, not the simple halide. This ammine does not contain discrete NI_3 molecules but consists of chains of NI_4 tetrahedra sharing two vertices, with one NH_3 attached by a weaker bond to alternate unshared I atoms along the chain.[2] Similar chains are found in NI_3.pyridine and $CH_3NI_2.½$(pyridine)[3] and also in $(CH_3)_2NI$;[4] this characteristic N–I bond of

$NI_3.NH_3$

$(CH_3)_2NI$

length 2.3 Å is the same as that in $[(HMT)_2I]I_3$ (p. 795) and is appreciably greater than the sum of the covalent radii (2.07 Å).

For structural details of molecules NX_3, NHX_2, and NH_2X see Table 18.2.

(1) IC 1965 **4** 899   (3) ZaC 1974 **409** 228
(2) ZaC 1968 **357** 225   (4) AC 1977 **B33** 3209

Hydrates of ammonia

Ammonia forms two hydrates, $NH_3 . H_2O$ and $2NH_3 . H_2O$, and the crystal structure of the latter has been studied at 170 K.[1] The crystal apparently consists of a hydrogen-bonded framework of composition $NH_3 . H_2O$, in which NH_3 and H_2O each form one strong N–H–O bond (2.84 Å) and three weaker ones (3.13, 3.22, and 3.22 Å). The second NH_3, which may be rotating, is attached to this framework only by a single N–H–O bond (of length 2.84 Å). In $NH_3 . H_2O$ the H_2O molecules are linked by hydrogen bonds into chains which are then cross-connected by further hydrogen bonds to NH_3 molecules into a three-dimensional network.[2] There is no hydrogen bonding between one NH_3 molecule and another. As in $2NH_2 . H_2O$ the unshared pair of electrons on the N atom forms one strong N–H–O bond (2.8 Å) and three weaker ones (3.25 Å). It is interesting to compare with the hydrates the structure of the cubic form of solid NH_3.[3] This has a slightly distorted cubic close-packed structure in which each H atom has 6 N neighbours at 3.4 Å and 6 more at 3.9 Å, suggesting that three weak N–H–N bonds are formed by each lone pair of electrons.

(1) AC 1954 **7** 194   (3) AC 1959 **12** 832
(2) AC 1959 **12** 827

Ammines

All the alkali metals and alkaline-earths (but not Be) and also Eu and Yb dissolve in liquid NH_3. The alkali-metal solutions are blue, paramagnetic, and conduct electricity. Such solutions also dissolve other metals which are not themselves soluble in liquid NH_3, for example, the Na solution dissolves Pb, and in this way a number of inter-metallic compounds (Na_2Pb, NaPb) have been prepared. From certain of the solutions crystalline ammines can be obtained, including $Li(NH_3)_4$ and hexammines of Ca, Sr, Ba, Eu, and Yb. The hexammines are body-centred cubic packings of octahedral molecules $M(NH_3)_6$. The Eu and Yb compounds are golden-yellow and are metallic conductors (M–N $\approx$ 3.0 Å), the magnetic and electrical properties suggesting that two electrons occupy a conduction band.[1]

Ammonia also combines with many salts to form ammines in which NH_3 utilizes its lone pair of electrons to bond to the metal. The M–NH_3 bond is extremely weak in the case of the alkali metals, but stronger to the B subgroup and transition metals. In $NaCl.5\frac{1}{7}NH_3$, for example, Na^+ has 5 N neighbours at 2.48 Å and 1 N at 3.39 Å, and the structure is an approximately c.c.p. array of NH_3 molecules and Cl^- ions.[2]

In $NH_4I.4NH_3$ NH_4^+ is hydrogen-bonded to 4 tetrahedral neighbours (N—N, 2.96 Å) and I^- is surrounded by 12 NH_3.[3] In contrast to the relatively weak bonding in the ammines of the more electropositive metals NH_3 is covalently bonded to metal atoms in numerous molecules and complex ions formed by B subgroup and transition metals, the structures of which are described under the individual element.

(1)   JSSC 1969 **1** 10   (3)   ACSc 1960 **14** 1466
(2)   AC 1965 **18** 879

Molecules containing the system $\diagdown$N—N$\diagup$

The only known symmetrical molecules of the type a_2N—Na_2 with single N—N bonds are those of hydrazine, N_2H_4, and dinitrogen tetrafluoride, N_2F_4. In some molecules in which O atoms are attached to N as, for example, nitramine, (a), substituted nitramines, (b), and dimers of C-nitroso compounds (c), the coplanarity

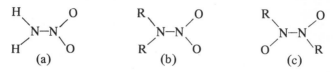

$$\qquad (a) \qquad\qquad\qquad (b) \qquad\qquad\qquad (c)$$

of the six atoms, the existence of *cis-trans* isomers of the nitroso compounds, and the N—N bond lengths show that the N—N bonds have some double-bond character. Note, however, the extremely *long* N—N bonds in N_2O_3(p. 809) and N_2O_4 (p. 810).

Hydrazine N_2H_4

Hydrazine exists in the form of simple N_2H_4 molecules in all states. Hydrogen bonding is responsible for the high b.p. (114 °C) and high viscosity of the liquid. Structurally the molecule has one feature in common with H_2O_2, for rotation of the NH_2 groups about the N—N axis is hindered by the lone pair of electrons—the free molecule has a configuration of the *gauche* type (p. 54). For structural details see Table 18.3.

The monohydrate, $N_2H_4.H_2O$, apparently has a structure of the NaCl type with disordered or rotating N_2H_4 and H_2O molecules.[1]

Hydrazine forms two series of salts, containing the $N_2H_5^+$ and $N_2H_6^{2+}$ ions. The structures of the halides are described in Chapter 8. From their formulae it is evident which type of ion they contain, but this is not true of all oxy-salts. For example, $N_2H_4.H_2SO_4$ might be $(N_2H_5)HSO_4$ or $(N_2H_6)SO_4$. The type of ion is deduced from the crystal structure (arrangement of hydrogen bonds) or from the p.m.r. spectrum. Of the sulphates, $N_2H_4.\frac{1}{2}$, 1, and $2H_2SO_4$, the first is $(N_2H_5)_2SO_4$[2] and the second, $(N_2H_6)SO_4$;[3] the third is presumably $(N_2H_6)(HSO_4)_2$. In $N_2H_4.H_3PO_4$ there is a 3D H_2PO_4 framework similar to that in KH_2PO_4, and the salt is accordingly $N_2H_5(H_2PO_4)$.[4] On the other hand, $N_2H_4.2H_3PO_4$ is $(N_2H_6)(H_2PO_4)_2$, in which the cation has the staggered configuration.[5]

The N—N bond lengths in the cations in $(N_2H_5)N_3$[5a] and $(N_2H_6)(TiF_6)$[5b] are not appreciably different from the value in N_2H_4 (1.45 Å), but the configuration of the $N_2H_5^+$ ion is variable. It has the staggered configuration in $(N_2H_5)H_2PO_4$ and $(N_2H_5)N_3$ but in $(N_2H_5)_2SO_4$ there are non-equivalent ions with approximately staggered and eclipsed configurations.

Hydrazine acts as a neutral bridging ligand in salts of the type $Zn(N_2H_4)_2X_2$ (X = Cl,[6] CH_3COO,[7] CNS,[8] etc.), double bridges linking the metal atoms into infinite chains.

(1)   AC 1962 **15** 803
(2)   ACSc 1965 **19** 1612
(3)   AC 1970 **B26** 536
(4)   ACSc 1965 **19** 1629
(5)   ACSc 1966 **20** 2483

(5a)   AC 1974 **B30** 2229
(5b)   AC 1980 **B36** 659
(6)   ZK 1962 **117** 241
(7)   GCI 1961 **91** 69
(8)   AC 1965 **18** 367

Dinitrogen tetrafluoride, N_2F_4

The molecule of this compound, which is prepared by passing NF_3 over various heated metals, has a hydrazine-like structure in which the angle of rotation from the eclipsed configuration is 70°. A Raman spectroscopic study of the liquid (–80 to –150 °C) indicated an equilibrium mixture of the *trans* and *gauche* isomers.[1] Electron diffraction data show that both conformations also exist in the vapour.[2] N_2F_4 dissociates appreciably to the free radical $\cdot NF_2$ at temperatures above 100 °C (compare $N_2O_4 \rightleftharpoons 2NO_2$), and this radical can exist indefinitely in the free state; its structure has been determined (Table 18.3). This behaviour of N_2F_4 is in marked contrast to that of N_2H_4, and is responsible for many reactions in which $-NF_2$ is inserted into molecules, as in the following examples:

$$N_2F_4 \rightleftharpoons \cdot NF_2 \quad + \quad \begin{cases} S_2F_{10} \rightarrow SF_5 . NF_2 \\ NO \rightarrow ON . NF_2 \\ Cl_2 \rightarrow Cl . NF_2 \end{cases}$$

(1)   JCP 1968 **48** 3216

(2)   JCP 1972 **56** 1691

Nitrogen forming two bonds $=N$

We confine our attention here to molecules or ions containing one or more N atoms each forming *one* multiple and *one* single bond. As regards bond character these are intermediate between the (rare) cases where N forms two angular single bonds as in NH_2^- and those in which it forms two collinear multiple bonds. It should be noted that two angular bonds are formed by N in many cyclic molecules and ions in

TABLE 18.3

Structural data for hydrazine and related molecules

Molecule	N–N (Å)	N–C(Si) (Å)	C–F (Å)	Angle of rotation from eclipsed	Method	Reference
N_2H_4	1.453			90–95°	i.r.	JCP 1959 **31** 843
	1.449				e.d.	BCSJ 1960 **33** 46
	1.46				X	JCP 1967 **47** 2104
$N_2(SiH_3)_4$	1.46	1.73		82.5°	e.d.	JCS A 1970 318
$N_2H_2(CH_3)_2$	1.45				e.d.	JACS 1948 **70** 2979
$N_2(CF_3)_4$	1.40		1.433	88°	e.d.	IC 1965 **4** 1346
		N–F	FNF			
N_2F_4	1.492	1.372	103.1°	64.2°	e.d.	JCP 1972 **56** 1691
·NF_2		1.363	102.5°		e.d.	IC 1967 **6** 304
			N–O			
$(CH_3)_2N.NO$	1.34	1.46	1.23 ⎫	C, N, O, all	e.d.	ACSc 1969 **23** 660
$(CH_3)_2N.NO_2$	1.38	1.46	1.22 ⎭	coplanar	e.d.	ACSc 1969 **23** 672

which both bonds have lengths corresponding approximately to double bonds; some finite cations also illustrate the difficulty of assigning simple bond formulae to systems in which there is π-bonding. In $N[S(CH_3)_2]_2^+$ the N bond angle is 111° and N–S, 1.64 Å;[1a] in $N(SCl)_2^+$ the bond angle is 149° and N–S, 1.535 Å[1b] (compare N–S, 1.72 Å, and N=S, 1.54 Å). In $[N(P\phi_3)_2]^+$ the N bond angle is 147° and N–P, 1.57 Å,[1c] close to the value for a double bond.

Molecules such as diazirine[2a] (cyclic diazomethane) and perfluorodiazirine[2b] are special cases like cyclopropane or ethylene oxide. If the lone-pair orbital has

diazirine   perfluorodiazirine

more s character (in the limiting case all s) then angles down to 90° could be expected. Apart from that in N_2H_2 (100°) bond angles lie in the range 110–120° in molecules in which the bonds are formally =N̑ but they are larger in HN=CO (128°), HN=CS (130°), and CH_3N=CS (142°). This bond arrangement attracted considerable attention in the pre-structural period for it occurs in numerous organic compounds. Early evidence for the non-linearity of one single and one double bond from N came from the resolution into its optical antimers of the oxime and from the stereoisomerism of unsymmetrical oximes (I).

Among the later demonstrations of this bond arrangement are the crystallographic studies of *cis* and *trans* azobenzene (II) and of dimethylglyoxime. The configuration of the latter molecule changes from the *trans* form (a) in which it exists in the crystalline dioxime to the *cis* form (b) when it forms the well known Ni, Pd, and Cu derivatives.

Di-imide (N_2H_2) *and difluorodiazine* (N_2F_2)

The simplest compound of the type RN=NR, di-imide, has been identified in the mass spectrometer among the solid products of decomposition of N_2H_4 by an electrodeless electrical discharge at 85 K. Both isomers have been identified (i.r.) in inert matrices at low temperatures (as also has imidogen, NH[3]), and the structure (a) has been assigned to the *cis* (planar) isomer.[4]

Difluorodiazine, N_2F_2, is a colourless gas stable at ordinary temperatures and is prepared by the action of an electric discharge on a stream of NF_3 in the presence of Hg vapour or, with less risk of explosion, by the action of KOH on N,N-difluorourea. The isomer F_2NN is not known, but the two isomers FNNF have been separated by gas chromatography, the *cis* isomer, (b), being the more reactive.[5] The structure of the *trans* isomer,[6] (c), which may be prepared in 45 per cent yield from N_2F_4 and $AlCl_3$, is similar to that of azomethane, (d).

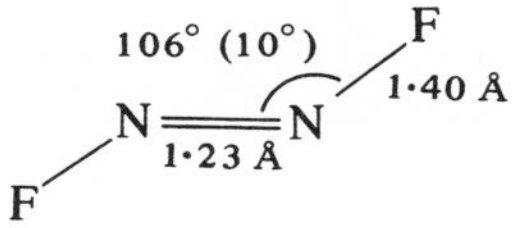

The crystalline compound of *cis*-N_2F_2 and AsF_5 is soluble (and stable) in anhydrous HF. It is isostructural with $(NO_2)AsF_6$ and is presumably $(N_2F)(AsF_6)$[7] containing the linear N_2F^+ ion which is isoelectronic with NO_2^+.

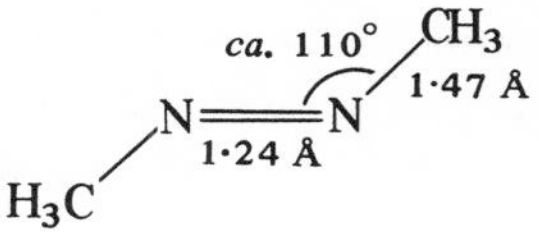

The centrosymmetric molecule of bis-trimethylsilyl diimine[8] is of interest as having unusually short N–N and long N–Si bonds.

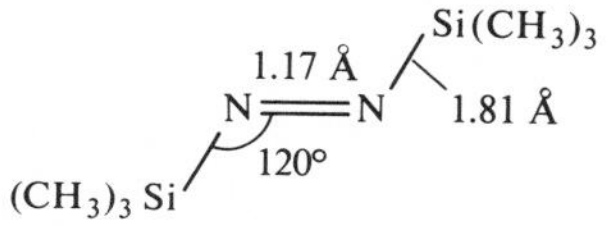

(1a) AC 1975 **B31** 893
(1b) AnCIE 1969 **8** 598
(1c) IC 1979 **18** 3024
(2a) JACS 1962 **84** 2651
(2b) JACS 1967 **89** 5527
(3) JCP 1965 **43** 507

(4) JCP 1964 **41** 1174
(5) JCP 1963 **39** 1030
(6) IC 1967 **6** 309
(7) JACS 1965 **87** 1889
(8) AC 1974 **B30** 1806

Compounds containing the system [N---N---N]

Apart from organic compounds such as the vicinal triazines, which do not concern us here, this group contains only hydrazoic acid and the azides.

Azides

Reference has already been made to the S_3^{2-} and the I_3^- ions, which may be formulated with single bonds, giving the central iodine atom in the latter ion a group of ten valence electrons. The azide ion (and radical) is of quite a different type, involving multiple bonds; no other element forms an ion of this kind. Hydrazoic acid, N_3H, can be prepared by the action of hydrazine (N_2H_4) on NCl_3 or HNO_2, but it is best obtained from its sodium salt, which results from a very interesting reaction. When nitrous oxide is passed into molten sodamide the following reaction takes place:

$$NH_2^- + N_2O \rightarrow N_3^- + H_2O.$$

When the vapour of hydrazoic acid is passed through a hot tube at low pressure a beautiful blue solid can be frozen out on a finger cooled by liquid air. On warming to $-125\,°C$ this is converted into NH_4N_3. The blue solid gives no X-ray diffraction pattern and is apparently glassy or amorphous. Although it was at first thought to be $(NH)_x$ the solid is substantially, if not entirely, NH_4N_3, and the colour is probably due to F-centres.[1]

Numerous structural studies of hydrazoic acid have been made. The three N atoms are collinear (compare the isoelectronic N_2O molecule, p. 808), and the molecule has the following structure:[2]

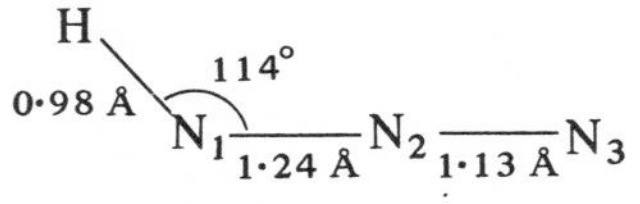

Calculations by the molecular orbital method of the orders of the bonds give 1.65 for N_1-N_2 and 2.64 for N_2-N_3.

The azides of the alkali and alkaline-earth metals are colourless crystalline salts which can almost be melted without decomposition taking place. X-ray studies have been made of several ionic azides (Li, Na,[3] K, Sr,[3] Ba[4]); they contain linear symmetrical ions with N–N close to 1.18 Å. (A number of azides MN_3 are isostructural with the difluorides MHF_2.) Other azides which have been prepared include $B(N_3)_3$, $Al(N_3)_3$, and $Ga(N_3)_3$ from the hydrides, $Be(N_3)_2$ and $Mg(N_3)_2$ from $M(CH_3)_2$ and HN_3 in ether solution, and $Sn(N_3)_4$ from NaN_3 and $SnCl_4$.

Azides of some of the B subgroup metals explode on detonation and are regarded as covalent compounds (e.g. $Pb(N_3)_2$ and AgN_3). In one of the four polymorphs of lead azide there are four non-equivalent N_3 groups, all approximately linear but with a small range of N–N bond lengths (1.16–1.20 Å, the mean in all cases being close to 1.180 Å).[5] In $Cu(N_3)_2$[6] there are two kinds of non-equivalent N_3 groups bonded in different ways to the Cu atoms. The differences between the results of two studies make it unprofitable to discuss the detailed structures of the N_3 groups. Any lowering of symmetry seems to be less than in the HN_3 molecule.

The azide group can function as a monodentate ligand in coordination compounds such as $[Co(N_3)_2en_2]NO_3$,[7a] $[Co(N_3)(NH_3)_5](N_3)_2$,[7b] and $Zn(N_3)_2(NH_3)_2$.[8] In the Co complexes the angle between the Co–N bond and the linear N_3 group is around 125°.

The azide group can act as a bridging ligand in one of two ways, (a) and (b). In (a)[9a] it is symmetrical, with bond lengths similar to those in ionic azides; in (b)[9b] it is unsymmetrical, the terminal N–N bond being shorter than the other, one N of which is bonded to one or two metal atoms.

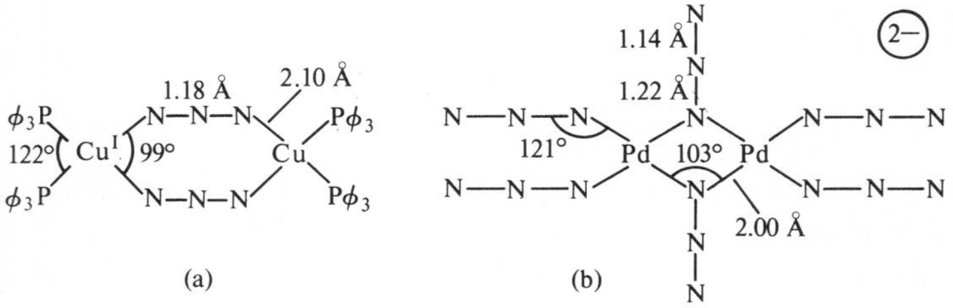

(a)   (b)

For the three structures (c), (d), (e):

(c): 1.27 Å, 1.13 Å, 115°, with NO₂ group
(d): 1.47 Å, 1.24 Å, 1.12 Å, 120°, H₃C
(e): 1.26, 1.12 Å, 1.355, 1.155, 114°

Covalent non-metal azides include the halogen azides $\dot{N}_3X$ (X = F, Cl, Br, I), silyl azide, H_3SiN_3, and organic azides. Cyanuric[10] and *p*-nitrophenyl[11] azides have been studied in the crystalline state, and methyl[12] and cyanogen[13] azides in the vapour state (e.d.). The structures of the last three are shown at (c), (d), and (e).

For comparison with the bond lengths in azides and other compounds standard bond lengths of N—N and C—N bonds are listed in Table 18.4.

TABLE 18.4

Observed lengths of N—N *and* C—N *bonds*

Bond	Length	Molecule
	(Å)	
N—N[a]	1.46	N_2H_4 and $(CH_3)_2N_2H_2$
N=N	1.23	N_2F_2 and $N_2(CH_3)_2$
N≡N	1.10	N_2
C—N[b]	1.47	CH_3NH_2, $N_4(CH_2)_6$, etc.
C_{ar}—N	≈1.40	C part of aromatic ring, e.g. $C_6H_5.NH.COCH_3$
C—N	1.33–1.35	Heterocyclic molecules (e.g. pyridine)
C=N	(1.28)	Accurate value not known for a non-resonating molecule
C≡N	1.16	HCN

(a) 1.86 Å in N_2O_3; 1.75 Å in N_2O_4; 1.53 Å in N_2F_4.
(b) 1.51 Å in amino acids and $(O_2N_2.CH_2N_2O_2)K_2$.

(1)	JCP 1959 **30** 349, **31** 564	(8)	ACSc 1971 **25** 1630
(2)	JCP 1964 **41** 999	(9a)	IC 1971 **10** 1289
(3)	AC 1968 **B24** 262	(9b)	JACS 1972 **94** 3377
(4)	AC 1969 **B25** 2638	(10)	PRS 1935A **150** 576
(5)	AC 1977 **B33** 3536	(11)	AC 1965 **19** 367
(6)	ACSc 1967 **21** 2647; AC 1968 **B24** 450	(12)	JPC 1960 **64** 756
(7a)	AC 1968 **B24** 1638	(13)	ACSc 1973 **27** 1531
(7b)	AC 1964 **17** 360		

The oxygen chemistry of nitrogen

Oxides

Five oxides of nitrogen have been known for a long time: N_2O, NO, N_2O_3, N_2O_4, and N_2O_5. A sixth, NO_3, is said to be formed in discharge tubes containing N_2O_4 and O_2 at low pressures. It has at most a short life, like others of this kind which have been described (PO_3, SO, SO_4) and radicals such as OH and CH_3, for the separate existence of which there is spectroscopic evidence. We shall confine our attention to the more stable compounds of ordinary chemical experience.

The trioxide and the pentoxide are the anhydrides of nitrous and nitric acids respectively, but N_2O is not related in this way to hyponitrous acid, $H_2N_2O_2$. Although N_2O is quite soluble in water it does not form the latter acid. Nitrous oxide cannot be directly oxidized to the other oxides although it is easily obtained by the reduction of HNO_3, HNO_2, or moist NO. The oxides NO, N_2O_3, and N_2O_4, on the other hand, are easily interconvertible and are all obtained by reducing nitric acid with As_2O_3, the product depending on the concentration of the acid. The trioxide, N_2O_3, probably exists only in the liquid and solid phases, which are blue in colour, for the vapour is almost completely dissociated into NO and NO_2, as may be seen from its brown colour. The tetroxide, N_2O_4, forms colourless crystals (at $-10\,^\circ C$), but its vapour at ordinary temperatures consists of a mixture of N_2O_4 and NO_2, and at about $150\,^\circ C$ the gas is entirely dissociated into NO_2. On further heating decomposition into NO and oxygen takes place, and this is complete at about $600\,^\circ C$. Since the pentoxide is easily decomposed to N_2O_4 there is the following relation between these oxides:

$$N_2O_5 \rightarrow N_2O_4 \rightleftharpoons \underbrace{NO_2 \rightleftharpoons NO + O}_{N_2O_3}$$

Nitrous oxide, N_2O. This molecule is linear and its electric dipole moment is close to zero ($0.17\,D$). Because of the similar scattering powers of N and O it is not possible to distinguish by electron diffraction between the alternatives NNO and NON, but the former is supported by spectroscopic data. X-ray diffraction data from crystalline N_2O[1] (which is isostructural with CO_2) are explicable only if the NNO molecules are randomly oriented. This disorder would account for the residual entropy of 1.14 e.u. which is close to the theoretical value ($2 \ln 2 = 1.377$ e.u.). The central N atom may be described as forming two sp bonds, the additional π-bonding giving total bond orders of approximately 2.5 ($N\equiv N$) and 1.5 ($N\text{---}O$) corresponding to the bond lengths 1.126 and $1.186\,\text{Å}$[2] and force constants 17.88 and 11.39 respectively.[3]

Nitric oxide, NO. Nitric oxide is one of the few simple molecules which contain an odd number of electrons. The N–O bond length is $1.1503\,\text{Å}$[4] and the dipole moment $0.16\,D$. The monomeric NO molecule in the gas phase is paramagnetic, but the condensed phases are diamagnetic. Evidence for polymerization comes not only from magnetic susceptibility measurements and from the infrared and Raman spectra of the liquid and solid[5] and of NO in a N_2 matrix at 15 K[6] but also from an examination of the structure of the crystalline solid.[7] The variation in degree of association with temperature gives the heat of dissociation of the dimer as 15.52 ± 0.62 kJ mol^{-1}. Unfortunately there is disorder in the crystal (resulting in the residual entropy of 1.5 e.u. per mole of dimer), and although the form of the dimer is probably

$$\begin{array}{ccc} \ddot{N} \text{--------} \ddot{O} & \\ \| & \| & 1 \cdot 12 \ (\cdot 02) \text{ Å} \\ \ddot{O} \text{--------} \ddot{N} & \\ & 2 \cdot 40 \text{ Å} \end{array}$$

it is not known whether the dimer has the structure

$$\text{structure} \quad \begin{array}{cc} N \cdots O \\ | \quad | \\ O \cdots N \end{array} \text{ or } \begin{array}{cc} N \cdots N \\ | \quad | \\ O \cdots O \end{array}$$

or whether the shape is strictly rectangular. The strength of the bonding between the two molecules of the dimer is certainly much less than that of a single bond.

Nitrogen dioxide, NO_2. This is also an odd-electron molecule, and some of its reactions resemble those of a free radical, for example, its dimerization, its power of removing hydrogen from saturated hydrocarbons, and its addition reactions with unsaturated and aromatic hydrocarbons. A microwave study[8] gives N–O, 1.197 Å and the O–N–O angle, 134° 15', in agreement with the results of earlier electron diffraction and infrared studies.

Dinitrogen trioxide, N_2O_3. Pure N_2O_3 cannot be isolated in the gaseous state since it exists in equilibrium with its dissociation products NO and NO_2. However, its m.w. spectrum has been studied[9] at -78 °C and 0.1 mm pressure and indicates the

$$\begin{array}{cccc} O & 105 \cdot 1° & 112 \cdot 7° & O \\ \diagdown & & & \diagup \ 1 \cdot 202 \text{ Å} \\ 1 \cdot 142 \text{ Å } N & \text{------} & N & \\ & 1 \cdot 864 \text{ Å} & & \diagdown \ 1 \cdot 217 \text{ Å} \\ & & & O \end{array}$$

planar structure shown. Some of the numerous suggestions as to the nature of the extraordinarily long N–N bond are discussed in ref. (9). The structure of crystalline N_2O_3 (at -115 °C) proved too complex to analyse.[10]

Dinitrogen tetroxide, N_2O_4. The chemistry of this oxide, the dimer of the dioxide, is most interesting, for the molecule can apparently dissociate in three different ways. Thermal dissociation takes place according to the equation

$$N_2O_4 \rightleftharpoons 2NO_2.$$

Reactions with covalent molecules are best explained by assuming

$$N_2O_4 \rightleftharpoons NO_2^+ + NO_2^-$$

(compare the ionization of HNO_3 in H_2SO_4), but in media of high dielectric constant there is ionization

$$N_2O_4 \rightleftharpoons NO^+ + NO_3^-.$$

A number of reactions of liquid N_2O_4 involving nitrosyl compounds and nitrates may be interpreted in terms of this type of ionization. It might appear that these two modes of ionization support the unsymmetrical formula (b) or the bridge formula (c) rather than one of type (a), in which the two NO_2 groups could be

(a)   (b)   (c)

either coplanar or inclined to one another. However, the weakness of this argument is shown by the fact that the isoelectronic oxalate ion, which is known to be symmetrical and of type (a), not only resembles N_2O_4 in its vibration frequencies but also in its chemical reactions:

with $N_2O_4 \rightarrow NO^+ + NO_3^- \xrightarrow{2H^+} NO^+ + H_2NO_3^+ \xrightarrow{H^+} NO^+ + NO_2^+ + H_3O^+$

compare $C_2O_4^{2-} \rightarrow CO + CO_3^{2-} \xrightarrow{2H^+} CO + H_2CO_3 \xrightarrow{H^+} CO + CO_2 + H_3O^+$.

The most probable structure of the planar N_2O_4 molecule in the vapour,[11] in the monoclinic form of the solid,[12] and in the crystalline complex with 1:4-dioxane[13] is that shown, but somewhat different dimensions were found in the cubic form of the solid (at $-40°C$), namely, N–N, 1.64 Å, and O–N–O, $126°$.[14] Note the extraordinary length of the N–N bond compared with the normal single bond

length (1.46 Å) in hydrazine, and compare with the long B–B bond in B_2Cl_4 which is also planar, Infrared studies have been made of N_2O_4 trapped in solid argon and oxygen matrices and evidence for various isomers has been given, for example, a twisted non-planar form of type (a) and the isomer (b).[15] The spectrum of this oxide made by the oxidation of NO in liquid ethane–propane mixtures at 80 K has been interpreted as that of $(NO)^+(NO_3)^-$.[16]

Dinitrogen pentoxide, N_2O_5. This oxide consists of N_2O_5 molecules in the gaseous state and also in solution in CCl_4, $CHCl_3$, and $POCl_3$. An early study of the gas by electron diffraction did not lead to a definite configuration of the molecule, and a more precise determination is desirable.[17]

When dissolved in H_2SO_4 or HNO_3 this oxide ionizes

$$N_2O_5 \rightleftharpoons NO_2^+ + NO_3^-$$

and from such solutions nitronium salts can be prepared (see later). The earlier recognition of NO_2^+ and NO_3^- frequencies in the Raman spectrum of crystalline

N_2O_5 is confirmed by the determination of the crystal structure. At temperatures down to that of liquid N_2 crystalline N_2O_5 consists of equal numbers of planar NO_3^- ions (N–O $\approx$ 1.24 Å) and linear NO_2^+ ions (N–O $\approx$ 1.15 Å).[18] (The i.r. spectrum of this oxide suggests that at the temperature of liquid He it is a molecular solid, possibly O_2NONO_2,[19] but this has not been confirmed by a structural study.)

The reaction of N_2O_5 with SO_3 is noted under nitronium compounds.

(1)	JPC 1961 **65** 1453	(11)	JCP 1972 **56** 4541
(2)	JCP 1954 **22** 275	(12)	ACSc 1963 **17** 2419
(3)	JCP 1950 **18** 694	(13)	ACSc 1965 **19** 120
(4)	JCP 1955 **23** 57	(14)	N 1949 **164** 915
(5)	JACS 1952 **74** 4696	(15)	IC 1963 **2** 747; JCP 1965 **42** 85
(6)	JCP 1969 **50** 3516	(16)	JCP 1965 **43** 136
(7)	AC 1953 **6** 760	(17)	JCP 1934 **2** 331
(8)	JCP 1956 **25** 1040	(18)	AC 1950 **3** 290
(9)	TFS 1969 **65** 1963	(19)	SA 1962 **18** 1641
(10)	AC 1953 **6** 781		

Nitrosyl compounds: NO^+ *ion*

The loss of an electron by NO should lead to a stable ion $[:N{\equiv}O:]^+$ since this contains a triple bond compared with a 2½-bond (double + 3-electron bonds) in NO. Hantzsch showed that nitrous acid does not exist as HNO_2 in acid solution but as NO^+ ions:

$$HNO_2 + H^+ \rightleftharpoons NO^+ + H_2O$$

and that the same NO^+ ions are present in solutions of 'nitrosylsulphuric acid', or nitrosyl hydrogen sulphate, $NO^+.HSO_4^-$. It has been shown by X-ray studies that certain nitrosyl compounds, $NO.ClO_4$, $NO.BF_4$, and $(NO)_2SnCl_6$ (originally written $SnCl_4.2NOCl$), are structurally similar to the corresponding ammonium salts, $NH_4^+ClO_4^-$, etc., and $NO(PtF_6)$ is isostructural with $O_2(PtF_6)$ and $KSbF_6$.[1] These salts contain the NO^+ ion, which is also found in the nitrato salts described on p. 824.

The nitrosyl halide molecules are non-linear (interbond angles, 110°, 113°, and 117° in NOF,[2] NOCl,[3] and NOBr[4] respectively) and the N–O bond length is close to 1.14 Å, as in NO. This bond length and the fact that the N–X bond lengths (1.52, 1.97, and 2.13 Å) are considerably greater than normal single bond lengths suggest that these molecules are not adequately represented by the simple structure (1) but that there may be resonance with the ionic structure (2):

(1)   (2)

Support for this polar structure comes from the dipole moments of pure NOCl and

NOBr (2.19 D and 1.90 D respectively) which are much greater than the values (around 0.3 D) estimated for structure (1).[5]

(1)   IC 1966 **5** 1217   (4)   JACS 1937 **59** 2629
(2)   JCP 1951 **19** 1071   (5)   RTC 1943 **62** 289
(3)   JCS 1961 1322

Nitroso compounds

Structural studies have been made of the free radical $[(CH_3)_3C]_2NO^{(1)}$ (N–O, 1.28 Å), of nitrosomethane, CH_3NO,[2] (CNO, 113° assuming N–O, 1.22 Å), and of F_3CNO,[3] in which N–O is 1.17 Å and angle CNO 121°. The C–N bond is abnormally long (1.55 Å) as in the nitro compounds, CF_3NO_2, etc. (p. 819). Many of these

$$O_2NCH_2CH_2 \diagdown \quad _{1\cdot 304\ \text{Å}} \diagup O$$
$$N\!-\!\!-\!\!-\!\!N$$
$$O \diagup \qquad \diagdown CH_2CH_2NO_2$$

compounds dimerize, the example shown being formed by the action of N_2O_3 on ethylene.[4] For the ions $[ON(NO)SO_3]^-$ and $[ON(SO_3)_2]^{2-}$, in both of which N forms three coplanar bonds, see p. 798.

(1)   ACSc 1966 **20** 2728   (3)   JPC 1965 **69** 3727
(2)   JCP 1968 **49** 591   (4)   JACS 1969 **91** 1371

Nitrosyl derivatives of metals

Numerous metallic compounds containing NO have been prepared, most of which are deeply coloured (red to black). Examples include the well known nitroprusside, $K_2[Fe(CN)_5NO]$, ammines such as $[Co(NH_3)_5NO]X_2$, and numerous other complex salts such as $K_3[Ni(S_2O_3)_2NO].2H_2O$, $K_2[Ni(CN)_3NO]$, and $K_2[RuCl_5NO]$. Many of these compounds can be prepared by the direct action of NO on the appropriate compound of the metal, e.g. nitrosyl halides (e.g. $Co(NO)_2Cl)^{(1)}$ from the halides and nitrosyl-carbonyls ($Fe(CO)_2(NO)_2$ and $Co(CO)_3NO$) from the carbonyls, while the action of NO on an ammoniacal solution of a cobaltous salt gives two series (red and black) of salts $[Co(NH_3)_5NO]X_2$. Apparently the red (more stable) form is a dimer, presumably a hyponitrite, $[(NH_3)_5Co.ON\!=\!NO.Co(NH_3)_5]X_4$, for equivalent conductivity measurements show that it is a 4:1 electrolyte.[2] The action of NO under pressure on $Fe_2(CO)_9$ gives $Fe(NO)_2(CO)_2$ or $Fe(NO)_4$ depending on the temperature. Solutions of ferrous salts react with NO in the presence of sulphides to give Roussin's black salts, $M[Fe_4S_3(NO)_7]$, which are converted by alkalis to less stable red salts $M_2[Fe_2S_2(NO)_2]$. If mercaptans are used instead of alkali sulphides esters of the red salt are obtained, for example, $Fe_2S_2(NO)_4(C_2H_5)_2$.

As a monodentate ligand in metal complexes NO behaves in one of two ways: (a) as a 12-electron unit (NO^-) when it forms a single bond to M with M–N–O around 125°, and (b) more usually, as a 10-electron unit (NO^+), when M–N is a multiple bond and M–N–O is approximately linear. The distinction between the

(a)   (b)

two classes (a) and (b) is perhaps an over-simplification, for a survey of all known structures of square pyramidal 5-coordinated Fe compounds[2a] shows that angles Fe—N—O ranging from 122° to 180° have been recorded. Certainly the smallest angles, 122° and 132° in Fe(salen)NO,[2b] are associated with the longest Fe—N bonds (around 1.8 Å), but there is disorder of the NO ligands in this crystal. On the other hand, angles as small as 155° (and Fe—N, 1.70 Å) have been found in a structure with ordered ligands, Fe(N_2S_2)NO,[2c] suggesting that crystal packing effects may also be relevant.

TABLE 18.5
Structures of metal nitrosyl compounds

	Angle	M—N	N—O	Reference
Type (a)	M—N—O			
[Co(en)$_2$Cl(NO)]ClO$_4$	124°	1.82 Å	1.11 Å	IC 1970 9 2760
[Ir(CO)Cl(Pφ$_3$)$_2$NO]BF$_4$	124°	1.97	1.16	JACS 1968 90 4486
[Ir(CO)I(Pφ$_3$)$_2$NO]BF$_4$.C$_6$H$_6$	125°	1.89	1.17	IC 1969 8 1282
Co(NO)[S$_2$CN(CH$_3$)$_2$]$_2$	(≈135°)	(1.7)	(1.1)	JCS 1962 668
Type (b)				
[Fe(NO)$_2$I]$_2$	161°	1.67	1.15	JACS 1969 91 1653
Ru(NO)$_2$(Pφ$_3$)$_2$	168°, 175°	1.72	1.22	AC 1975 B31 897
[Cr(CN)$_5$NO][Co(en)$_3$].2H$_2$O	176°	1.71	1.21	IC 1970 9 2397
Fe(NO)[S$_2$CN(CH$_3$)$_2$]$_2$	170°	1.72	1.10	JCS A 1970 1275
[Fe(CN)$_5$NO]$^{2-}$	176°	1.65	1.12	AC 1979 B35 2193
[Fe(CN)$_4$NO]$^{2-}$	177°	1.65	1.16	ZN 1977 B32 275
Cr(C$_5$H$_5$)Cl(NO)$_2$	170°	1.71	1.14	JCS A 1966 1095
Cr(C$_5$H$_5$)(NCO)(NO)$_2$	171°	1.72	1.16	JCS A 1970 605, 611
Mo(NO)(C$_5$H$_5$)$_3$	179°	1.75	1.21	JACS 1969 91 2528
Mn(CO)$_2$(Pφ$_3$)$_2$NO	178°	1.73	1.18	IC 1967 6 1575

Some of the compounds in Table 18.5 are also of interest in connection with the stereochemistry of the transition-metal atom, for example, the tetragonal pyramidal structures of [Ir(CO)I(Pφ$_3$)$_2$NO]$^+$ and the Fe and Co compounds M(NO)[S$_2$CN(CH$_3$)$_2$]$_2$, (c), and Mo(C$_5$H$_5$)$_3$(NO), (d). The Fe molecule of type (c) has nearly collinear M—N—O bonds, but the bonding is of type (a) in the Ir and probably also the Co compound, the structure of which is known less accurately. There are approximately collinear M—N—O bonds in the nitroprusside ion, [Fe(CN)$_5$NO]$^{2-}$, and the similar ions formed by Cr and Mn.

In the molecule (e) NO behaves both as a monodentate ligand of type (b) (angle Cr—N—O, 179°) and also as a bridging ligand. The Cr—Cr distance indicates a metal-

$$O$$
$$N$$
$$|$$

$(CH_3)_2NC$ $S=M=S$ $CN(CH_3)_2$
S —— S

(c)

$$O$$
$$|$$
$$N$$
$$|$$
$$Mo$$
$$|$$
$$(C_5H_5)_3$$

(d)

$$H_2$$

C_5H_5 N 1·94 A O
2·65 A N 1·20 A
Cr —————— Cr 1·65 A
O N 1·94 A N C_5H_5
1·12 A
O

(e)

metal bond.[3] An even more complex behaviour is exhibited in $(C_5H_5)_3Mn_3(NO)_4$.[4] Here NO is functioning in two ways, as a doubly and as a triply bridging ligand. In Fig. 18.3 the molecule is viewed in a direction perpendicular to the equilateral triangle of Mn atoms (Mn–Mn approx. 2.50 Å) and along the axis of the triply bridging NO, which lies beneath the plane of the paper.

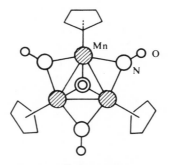

FIG. 18.3. The molecule $(C_5H_5)_3Mn_3(NO)_4$.

The crystal structure of one of the black Roussin salts, $CsFe_4S_3(NO)_7 . H_2O$, has been determined.[5] The nucleus of the ion consists of a tetrahedron of Fe atoms with S atoms above the centres of three of the faces (Fig. 18.4(a)). One iron atom (Fe_I) is linked to three S atoms and one NO group, and in addition forms weak bonds to the other three iron (Fe_{II}) atoms. The Fe_{II} atoms are bonded to two S and two NO and to the Fe_I atom. The length of the Fe_I–Fe_{II} bonds (2.70 Å) corresponds to about a 'half-bond', using Pauling's equation (p. 1292), but there are no bonds between Fe_{II} atoms $(Fe_{II}–Fe_{II}, 3.57 Å)$. The classical valence bond picture (a)

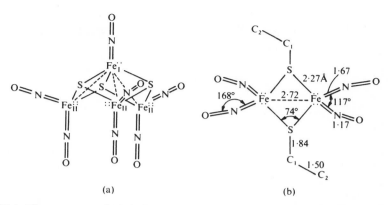

FIG. 18.4. The structures of (a) the $Fe_4S_3(NO)_7^+$ ion and (b) the $Fe_2S_2(NO)_4(C_2H_5)_2$ molecule.

gives the iron atoms an effective atomic number of 18 and is consistent with the diamagnetism of the compound, but it implies rather high formal charges on Fe_I (-2) and Fe_{II} (-3) suggesting resonance with structures which will reduce these charges. An alternative model having a four-centre molecular orbital linking the iron atoms has also been suggested. Other bond lengths in this molecule are: Fe–S, 2.23; Fe_I–N, 1.57; Fe_{II}–N, 1.67; and N–O, 1.20 Å.

The molecule of the red ester, $Fe_2S_2(NO)_4(C_2H_5)_2$, has the structure[6] shown in Fig. 18.4(b), in which Fe forms tetrahedral bonds to two S and two NO. The Fe–Fe distance (2.72 Å) is very similar to that in the black salt, and there must be sufficient interaction between these atoms to account for the diamagnetism of this compound. For nitrosyl carbonyls see p. 967.

(1) ACSc 1967 **21** 1183
(2) IC 1964 **3** 1038
(2a) IC 1980 **19** 3622
(2b) ICA 1979 **33** 119
(2c) IC 1977 **16** 3262

(3) AC 1970 **B26** 1899
(4) JACS 1967 **89** 3645
(5) AC 1958 **11** 594
(6) AC 1958 **11** 599

Nitryl halides and nitronium compounds

The existence of two nitryl halides, NO_2F and NO_2Cl, is well established; the existence of NO_2Br is less certain. The fluoride can conveniently be prepared by the action of fluorine on slightly warm dry sodium nitrite: $NaNO_2 + F_2 \rightarrow NO_2F + NaF$, and is a reactive gas which converts many metals to oxide and fluoride or oxyfluoride, and reacts with many non-metals (Br and Te excepted) to form nitronium salts. Examples of nitronium salts are $(NO_2)BF_4$, $(NO_2)_2SiF_6$, $(NO_2)PF_6$, and $(NO_2)IF_6$.

Nitryl chloride can be prepared from chlorosulphonic acid and anhydrous nitric acid. It combines with halides to form salts such as $(NO_2)SbCl_6$ which in an ionizing solvent (e.g. liquid SO_2) undergo reactions of the type:

$$(NO_2)(SbCl_6) + N(CH_3)_4 ClO_4 \rightarrow (NO_2)ClO_4 + N(CH_3)_4 SbCl_6.$$

Microwave data for $NO_2F^{(1)}$ and $NO_2Cl^{(2)}$ are consistent with planar molecules:

$$F \xrightarrow{\text{1·47 Å}} N \begin{array}{c} {}^O \\ {}_{1\cdot18} \searrow_O \end{array} 136° \qquad Cl \xrightarrow{\text{1·84 Å}} N \begin{array}{c} {}^O \\ {}_{1\cdot20} \searrow_O \end{array} 130°$$

Nitronium salts contain the linear NO_2^+ ion (N–O, 1.10 Å), studied in $(NO_2)(ClO_4)$.[3] We have already noted that N_2O_5 ionizes in H_2SO_4 to NO_2^+ and NO_3^- and that the crystalline oxide consists of equal numbers of these two ions. By mixing SO_3 and N_2O_5 in $POCl_3$ solution the compounds $N_2O_5 . 2SO_3$ and $N_2O_5 . 3SO_3$ can be obtained, and from a solution of SO_3 in HNO_3 the compounds $HNO_3 . 2SO_3$ and $HNO_3 . 3SO_3$ have been isolated. these compounds are in fact $(NO_2)_2S_2O_7$ and $(NO_2)_2S_3O_{10}$, $NO_2(HS_2O_7)$ and $NO_2(HS_3O_{10})$ respectively. X-ray studies show that crystals of $(NO_2)_2S_3O_{10}$[4] and $NO_2(HS_2O_7)$[5] consist of linear $(O–N–O)^+$ ions and $S_3O_{10}^{2-}$ or $HS_2O_7^-$ ions respectively. For the linear NS_2^+ ion see p. 827.

(1) JCS A 1968 1736
(2) JMS 1963 **11** 349
(3) AC 1960 **13** 855
(4) AC 1954 **7** 430
(5) AC 1954 **7** 402; IC 1971 **10** 2319

Oxyhalogen ions

Salts such as $(NOF_2)BF_4$ and $(NOF_2)AsF_6$ prepared from NOF_2 and the halide contain planar ions NOF_2^+ similar to the isoelectronic COF_2 (i.r. spectrum).[1] The $NOCl_2^+$ ion has been studied in $(NOCl_2)SbCl_6$, prepared by the interaction of NCl_3, $SbCl_5$, and $SOCl_2$, and shown to be planar with N–Cl, 1.61 and 1.72 Å, and N–O, 1.30 Å.[2]

(1)   IC 1969 **8** 1253
(2)   AnC 1977 **89** 569

Acids and oxy-ions

The existence of three oxy-acids of nitrogen is well established: nitrous, HNO_2; nitric, HNO_3; and hyponitrous, $H_2N_2O_2$. The preparation of other acids or their salts has been described. For example, Na_3NO_4 has been prepared from $NaNO_3$ and Na_2O at 300 °C as a compound very reactive toward H_2O and CO_2, and its Raman spectrum suggests a tetrahedral anion. We shall confine our remarks here chiefly to compounds of known structure.

Reference has already been made to the *cations* NO^+ and NO_2^+.

Nitrous acid. Only dilute solutions of nitrous acid have been prepared, and these rapidly decompose, evolving oxides of nitrogen. Two structures can be envisaged for the molecule of nitrous acid, the second of which could exist in *cis* and *trans* forms:

(a) (b)

This tautomerism is suggested by the fact that by the interaction of metal nitrites with alkyl halides two series of compounds are obtained, the nitrites—in which the NO_2 group is attached to carbon through an O atom—and the nitro compounds, in which there is a direct N—C link. From i.r. studies of nitrous acid gas it is concluded[1] that there is no evidence for the existence of form (a) and that the two isomers of (b) exist in comparable amounts, the *trans* isomer being rather more stable. The structure (c) has been assigned to the *trans* isomer.[2]

(c)

(1) JCP 1951 **19** 1599   (2) JACS 1966 **88** 5071

Metal nitrites and nitrito compounds. Until recently the only stable simple metal nitrites known were those of the alkali and alkaline-earth metals, Zn, Cd, Ag(I), and Hg(I). The alkali nitrites decompose at red heat, the others at progressively lower temperatures. The ionic nitrites of the more electropositive metals contain the NO_2^- ion which in $NaNO_2$[1] has the following configuration: ONO, 115.4°, and N—O, 1.236 Å. It may be of interest to summarize here the structural data for the series NO_2^+, NO_2, and NO_2^- with respectively 16, 17, and 18 valence electrons:

	NO_2^+	NO_2	NO_2^-
N—O	1.10	1.19	1.24 Å
ONO angle	180°	134°	115°

(Nothing is known of the structure of the NO_2^{2-} ion which may be present in sodium nitroxylate, Na_2NO_2,[2] formed by the reaction between $NaNO_2$ and Na metal in liquid ammonia as a bright-yellow solid. The paramagnetic susceptibility of this salt is far too small for $(Na^+)_2(NO_2)^{2-}$, and is interpreted[3] as due to about 20 per cent dissociation of the dimeric $N_2O_4^{4-}$ ion.)

The insoluble, more deeply coloured, nitrites of metals such as Hg are probably essentially covalent compounds. Anhydrous nitrites of Ni(II) and Co(II) have been made from NO_2 and a suitable compound of the metal, for example, $Ni(NO_2)_2$ from $Ni(CO)_4$ and NO_2 (each diluted with argon).[4] It is stable up to 260 °C.

Complex nitrites of transition metals (e.g. $Na_3[Co(NO_2)_6]$ and $K_2Pb[M(NO_2)_6]$ (M = Fe, Co, Ni, Cu)) are well known.

As a ligand directly bonded to a metal atom the NO_2^- ion can function in the following ways:

(i)   (ii)   (iii)   (iv)

(i) Bonded through N as in complex nitrito ions such as $[Co(NO_2)_6]^{3-}$ and $[Cu(NO_2)_6]^{4-}$ (p. 1134)—in $K_2Pb[Cu(NO_2)_6]$ the dimensions of NO_2 are the same as for the NO_2^- ion[5]—and in complexes such as $[Co(NH_3)_2(NO_2)_4]^-$ and $[Co(NH_3)_5NO_2]^{2+}$.[6]

(ii) Bonded through O, as in $Ni[(CH_3)_2NCH_2CH_2N(CH_3)_2]_2(ONO)_2$:[7a]

In $[Cr(NH_3)_5ONO]Cl_2$[7b] two equal N—O bond lengths (1.19 Å) were found, and O—N—O, 122°.

(iii) Bonded through both O atoms (bidentate), as in $K_3[Hg(NO_2)_4]NO_3$,[8a] where NO_2^- has the same structure as the free ion: N—O, 1.24 Å; O—N—O, 114½°. In $[Cu(bipyr)_2(ONO)]NO_3$[8b] there is nearly symmetrical bridging. Both Cu—O bonds are weak compared with the normal bond (≈ 2.0 Å), and their mean length is 2.29 Å. In $Cu(bipyr)(ONO)_2$,[9] on the other hand, one bond to each ONO^- is of normal length and the other very long, but the mean (2.24 Å) is close to that in the more symmetrical bridge.

(iv) Bridging (unsymmetrically) two M atoms, forming a planar bridge:

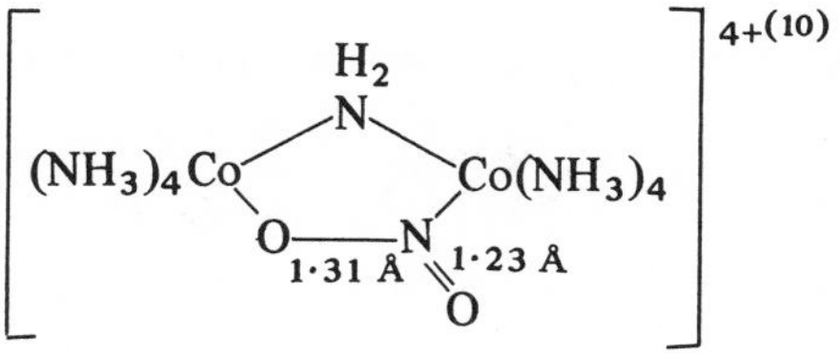

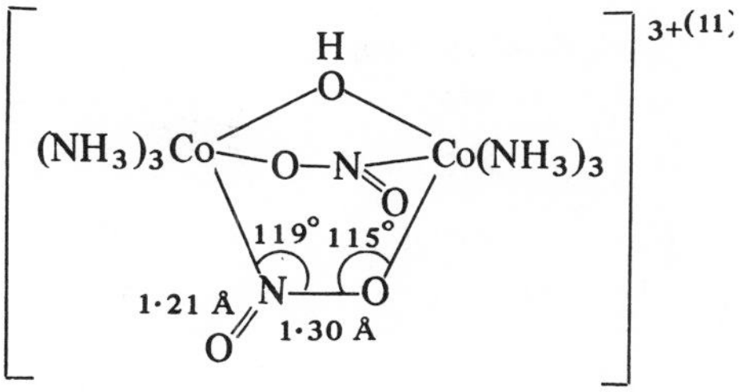

(1)   AC 1961 **14** 56
(2)   B 1928 **61** 189
(3)   ACSc 1958 **12** 578
(4)   IC 1963 **2** 228
(5)   IC 1966 **5** 514
(6)   AC 1968 **B24** 474
(7a) CC 1965 476

(7b) AC 1978 **B34** 2285
(8a) JCS D 1976 93
(8b) JCS A 1969 1248
(9)   JCS A 1969 2081
(10) IC 1970 **9** 1604
(11) AC 1970 **B26** 81

Organic nitro compounds. The NO_2 group is known to be symmetrical in organic nitro compounds, with N–O, 1.21 Å, and ONO 125–130° (nitromethane, CH_3NO_2,[1] and various crystalline nitro compounds). Rather larger angles (132° and 134°) have been found in CF_3NO_2 and CBr_3NO_2.[2] In these molecules the C–F, C–Br, and N–O bond lengths are normal, 1.33, 1.92, and 1.21 Å respectively, but C–N is unusually long (1.56 and 1.59 Å), and the system $-C-N\begin{smallmatrix}O\\O\end{smallmatrix}$ non-planar.

(1)   JCP 1956 **25** 42   (2)   JCP 1962 **36** 1696

Hyponitrous acid. In contrast to nitrous and nitric acids, hyponitrous acid crystallizes from ether as colourless crystals which easily decompose, explosively if heated. The detailed molecular structure of this acid has not been determined, but it is known that the molecular weights of the free acid and its esters correspond to the double formula, $H_2N_2O_2$, that it is decomposed by sulphuric acid to N_2O, and that it can be reduced to hydrazine, H_2N-NH_2. Infrared and Raman studies show conclusively that the hyponitrite ion has the *trans* configuration (a), but the N–N frequency suggests that the central bond has an order of rather less than two.[1]

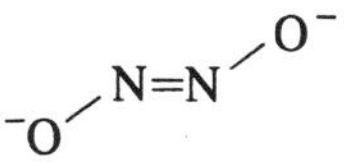

(a) 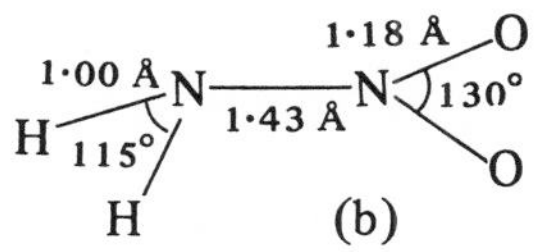   (b)

The isomer nitramide, $H_2N.NO_2$, which decomposes in alkaline solution to N_2O and which like $H_2N_2O_2$ may be reduced to hydrazine, has the structure (b), in which the angle between the NH_2 plane and the NNO_2 plane is 52°.[2]

(1)   JACS 1956 **78** 1820   (2)   JMS 1963 **11** 39

Oxyhyponitrite ion. The salt $Na_2N_2O_3$ (Angeli's salt) is prepared by the action of sodium methoxide on hydroxylamine and butyl nitrate at $0\,^{\circ}C$:

$$C_4H_9ONO_2 + H_2NOH + 2\,NaOCH_3 \rightarrow Na_2(ONNO_2) + C_4H_9OH + 2\,CH_3OH.$$

The structure of the (planar) $N_2O_3^{2-}$ ion has been determined in $Na_2N_3O_3 . H_2O$;[1] it is as follows:

(1)   IC 1973 **12** 286

Peroxynitrite ion. This ion, produced by oxidizing azide solutions by O_3 or NH_2Cl by O_2, apparently exists in methanol solution for weeks. Its presence has been demonstrated by its u.v. absorption spectrum, but no solid salts have been obtained. The structure $(O\!\!=\!\!N-O-O)^-$ has been suggested.[1]

(1)   JINC 1964 **26** 453; JCS A 1968 450; JCS A 1970 925

Nitric acid and nitrate ion. Pure nitric acid can be made by freezing dilute solutions (since the freezing point of HNO_3 is about $-40\,^{\circ}C$), but it is not stable. Dissociation into N_2O_5 and H_2O takes place, the former decomposes, and the acid becomes more dilute. If strong solutions of the acid are kept in the dark, however, decomposition of the N_2O_5 is retarded and the concentration of nitric acid remains constant. The anhydrous acid and also the crystalline mono- and tri-hydrates have been studied in the crystalline state and the structures of the hydrates in solution have also been studied.[1] In the vapour state HNO_3 exists as planar molecules with the structure (a).[2] Reliable data on the structure of the molecule in the crystalline anhydrous acid were not obtained owing to submicroscopic twinning of the crystals.[3] In the monohydrate[4] there are puckered layers of the kind shown in Fig. 15.25(b) (p. 692), where the shaded circles represent O atoms of water molecules. N.m.r.

(a)

studies indicate groups of three protons at the corners of equilateral triangles, showing that this hydrate is $(H_3O)^+(NO_3)^-$ and contains H_3O^+ ions, as do the monohydrates of certain other acids (p. 690). In $HNO_3 . 3H_2O$[5] short hydrogen bonds (2.48 and 2.63 Å) link H_3O^+ ions to 2 H_2O molecules to form $H_7O_3^+$ ions which in turn are bonded to other similar ions by hydrogen bonds of length 2.80 Å and to NO_3^- ions. The length of the N—O bonds (1.26 Å) indicates NO_3^- ions rather than HNO_3 molecules.

Simple ionic nitrates (see later) are quite different structurally from salts of the type $A_m(XO_3)_n$ formed by other Group V elements. They contain the planar equilateral NO_3^- ion, geometrically similar to the BO_3^{3-} and CO_3^{2-} ions, in contrast to the pyramidal SO_3^{2-} and ClO_3^- ions of the elements of the second short Period and the polymeric $(PO_3)_n^{n-}$ ions in metaphosphates. The N—O bond length in NO_3^- is close to 1.25 Å in a number of nitrates.[6] The value 1.27 Å was found in a n.d.

(b)

study of $Pb(NO_3)_2$,[7] 1.26 Å in $AgNO_3$,[8] and 1.256 Å in $HNO_3.H_2O$ and also in a n.d. study of the high-temperature form of $NaNO_3$,[9] in which there is complete disorder between alternative orientations of the NO_3^- ions, In the vapour state $LiNO_3$ and $NaNO_3$ exist as molecules with the structure (b) similar to that of HNO_3 (O—Li, 1.60; O—Na, 1.90 Å).[10]

In addition to normal nitrates salts $M[H(NO_3)_2]$ and $M[H_2(NO_3)_3]$ are known in which M is a large ion such as K^+, Rb^+, Cs^+, ammonium or arsonium ion. The $H(NO_3)_2^-$ ion consists of two NO_3^- ions linked by a short hydrogen bond, (c). The nitrate ions lie in perpendicular planes in $[Rh(C_5H_5N)_4Cl_2]H(NO_3)_2$[11] and $CsH(NO_3)_2$, but a coplanar configuration was found in $(As\phi_4)H(NO_3)_2$.[12] In $NH_4H_2(NO_3)_3$[13] the $H_2(NO_3)_3^-$ group consists of a central nitrate ion hydrogen-bonded to two HNO_3 molecules which lie in a plane perpendicular to that of the NO_3^- ion, (d).

(c)

(d)

(1) ACSc 1954 8 374
(2) JCP 1965 42 3106
(3) AC 1951 4 120
(4) AC 1975 B31 1486
(5) AC 1975 B31 1489
(6) QRCS 1971 25 289
(7) AC 1957 10 103

(8) JCS A 1971 2058
(9) AC 1972 B28 2700
(10) ZSK 1965 6 765
(11) AC 1979 B35 1099
(12) CC 1967 1211
(13) AC 1970 B26 1117

Metal nitrates and nitrato complexes. There are several groups of compounds in which the NO_3 group is directly bonded to one or more metal atoms by bonds which presumably have considerable covalent character, namely:

(i)   volatile metal nitrates, $M(NO_3)_n$;
(ii)   nitrato ions, $[M(NO_3)_n]^{m-}$;
(iii)   oxy-nitrato compounds, $MO_x(NO_3)_y$; and
(iv)   coordination complexes containing $-NO_3$ as a ligand.

Examples of compounds of classes (i)–(iii) are given in Table 18.6.

TABLE 18.6
Metal nitrates and nitrato complexes

Class (i)			
$Cu(NO_3)_2$[1]	$Al(NO_3)_3$	$Ti(NO_3)_4$[3]	
(also Be, Zn, Mn,	(also Cr, Fe, Au, In,	(also Sn,[4] Zr)	
Hg, Pd)	Co[2]		
Class (ii)			
			$[Al(NO_3)_5]Cs_2$[13]
			$[Ce(NO_3)_5](NO)_2$[14]
$[Au(NO_3)_4]K$[5]			$[Ce^{IV}(NO_3)_6].(NH_4)_2$[9]
		$[Al^{III}(NO_3)_4]NO$	$[Th^{IV}(NO_3)_6]Mg.8H_2O$[10]
		$[In^{III}(NO_3)_4](NEt_4)$	$[Ce^{III}(NO_3)_6]_2Mg_3.24H_2O$[11]
		$[Co^{II}(NO_3)_4](As\phi_4)_2$[6]	
		$[Mn^{II}(NO_3)_4](As\phi_4)_2$[7]	
		$[Fe^{III}(NO_3)_4](As\phi_4)$[8]	
Class (iii)			
$Fe^{III}O(NO_3)$	$V^VO_2NO_3$	$Re^{VII}O_3NO_3$	$Be_4O(NO_3)_6$
$Ti^{IV}O(NO_3)_2$	$Cr^{VI}O_2(NO_3)_2$		
$Nb^VO(NO_3)_3$	$UO_2(NO_3)_2$		
	$[UO_2(NO_3)_3]Rb$[12]		

(1) JACS 1963 **85** 3597. (2) CC 1968 871. (3) JCS A 1966 1496. (4) JCS A 1967 1949. (5) JCS A 1970 3092. (5a) JCS D 1976 989. (6) IC 1966 **5** 1208. (7) JCS A 1970 226. (8) CC 1971 554. (9) IC 1968 **7** 715. (10) AC 1965 **18** 698. (11) JCP 1963 **39** 2881. (12) AC 1965 **19** 205. (13) CC 1973 595 (14) CC 1978 580.

In these various compounds the NO_3 group behaves in one of the following three ways:

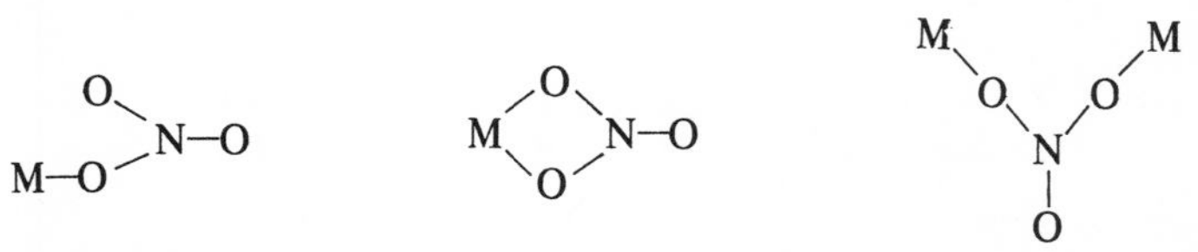

(a) monodentate   (b) bidentate   (c) bridging

In (b) and (c) it appears that the NO_3 group can bond symmetrically or unsymmetrically, that is, the two bonds to the metal atom(s) have the same or different lengths. This is discussed further on p. 1127 in connection with $Cu(NO_3)_2$ and related compounds. Owing to lack of sufficient structural information it is not yet possible to say whether all these possibilities are realized in each of the above four classes, but examples can certainly be given of the following:

Class (i)   (b)   (c)
      (ii)   (b)
     (iii)   (b)   (c)
     (iv)   (a)   (b)

We now deal briefly with the four groups of compounds listed above.

(i) *Anhydrous metal nitrates.* Anhydrous *ionic* nitrates include the familiar water-soluble compounds of the alkali and alkaline-earth metals and those of a few B subgroup metals (e.g. $AgNO_3$, $Cd(NO_3)_2$, $Pb(NO_3)_2$) and transition metals ($Co(NO_3)_2$, $Ni(NO_3)_2$). The nitrates of Mg and many 3d metals are normally obtained as hydrates, and while the hydrates are normal ionic crystals the anhydrous salts are in many cases surprisingly volatile and are less readily prepared. Heating of many hydrated nitrates produces hydroxy-salts or the oxide, but the volatile anhydrous nitrates of many transition metals and of the less electropositive metals (Be, Mg, Zn, Hg, Sn^{IV}, Ti^{IV}) may be prepared by the action of liquid N_2O_5 or of a mixture of N_2O_4 and ethyl acetate on the metal or a compound of the metal. For example, if metallic Cu is treated with N_2O_4 and dried ethyl acetate of room temperature for 4 hours, the addition of further N_2O_4 then precipitates $Cu(NO_3)_2 . N_2O_4$. On heating this is converted into anhydrous $Cu(NO_3)_2$, which may be purified by sublimation. Some of these anhydrous nitrates (e.g. $Sn(NO_3)_4$ and $Ti(NO_3)_4$) are powerful oxidizing agents and explode or inflame with organic compounds, whereas others (of Zn, Ni^{II}, Mn^{II}) are inert.

There are interesting differences in series such as Zn, Cd, and Hg nitrates. The Zn and Hg compounds are volatile, in contrast to $Cd(NO_3)_2$ which is a typical ionic nitrate (see p. 330).

Examples of volatile nitrates which have been studied in the crystalline state include $Cu(NO_3)_2$, which forms molecules of type (a) in the vapour—for the structure of the solid see p. 1128, $Co(NO_3)_3$, and $Ti(NO_3)_4$, which exist as molecules in the crystal. In all these molecules NO_3 is behaving as a bidentate ligand. The relative orientation of the two NO_3 groups in the vapour molecule $Cu(NO_3)_2$ was not determined. The molecule of $Co(NO_3)_3$ is essentially octahedral, allowing for the smaller angle subtended at the metal by the bidentate ligand; the bond angles not given in Fig. 18.5(a) are close to $100°$. The $Ti(NO_3)_4$ molecule is a flattened tetrahedral grouping of 4 NO_3 around Ti, the 8 O atoms bonded to the metal atom forming a flattened dodecahedral group (Fig. 18.5(b)). There is some lowering of symmetry of the NO_3 group, (b). The molecule of

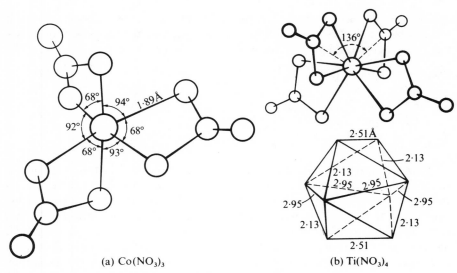

(a) Co(NO$_3$)$_3$ (b) Ti(NO$_3$)$_4$

FIG. 18.5. The molecules (a) Co(NO$_3$)$_3$; (b) Ti(NO$_3$)$_4$.

Sn(NO$_3$)$_4$ is similar. In all these molecules there is symmetrical bonding of the NO$_3$ group to the metal, that is, equal M—O bonds.

(ii) *Complex nitrato ions*. Salts containing ions [M(NO$_3$)$_x$]$^{n-}$ are known for x = 4, 5, and 6. Hydrated salts are prepared from aqueous solution and anhydrous salts by special methods such as reacting a metal chloride or complex chloride with liquid N$_2$O$_5$ or N$_2$O$_4$ in solvents such as CH$_3$CN or ethyl acetate. For example Cs$_2$[Sn(NO$_3$)$_6$] is prepared from Cs$_2$SnCl$_6$. Of special interest are compounds which analyse as 'N$_2$O$_4$ solvates', prepared from the metal chloride in a N$_2$O$_4$–ethyl acetate mixture. An example is Sc(NO$_3$)$_3$.2N$_2$O$_4$, which is actually [Sc(NO$_3$)$_5$]$^{2-}$(NO$^+$)$_2$.

In the [Au(NO$_3$)$_4$]$^-$ ion Au forms four coplanar bonds to monodentate NO$_3$ ligands (Fig. 18.6(a)). The four Au—O distances of 2.85 Å are too long to indicate appreciable interaction between the Au and O atoms. The structures of all the ions [Mn(NO$_3$)$_4$]$^{2-}$, [Fe(NO$_3$)$_4$]$^-$, and [Co(NO$_3$)$_4$]$^{2-}$ (studied in their tetraphenyl-arsonium salts) are of the same general type as the Ti(NO$_3$)$_4$ molecule (i.e. symmetry close to D_{2d}). However, in the first two ions there is symmetrical bidentate coordination of the NO$_3$ groups but in [Co(NO$_3$)$_4$]$^{2-}$ unsymmetrical coordination of NO$_3$ to the metal, though the two different Co—O bond lengths are apparently 2.03 and 2.36 Å for two NO$_3$ groups as compared with 2.11 and 2.54 Å for the other two. This difference in behaviour of the NO$_3$ ligand in these tetranitrato ions has been attributed to the lower symmetry of Co(II), d^7 (high spin) as compared with Ti(IV), d^0, Mn(II) and Fe(III), d^5 (high spin), and Sn(IV), d^{10}, but it should be noted that there is unsymmetrical bonding of the NO$_3$ ligands also in the [Zn(NO$_3$)$_4$]$^{2-}$ ion, where the longer bonds to Zn correspond to the shorter N—O bonds and vice versa.

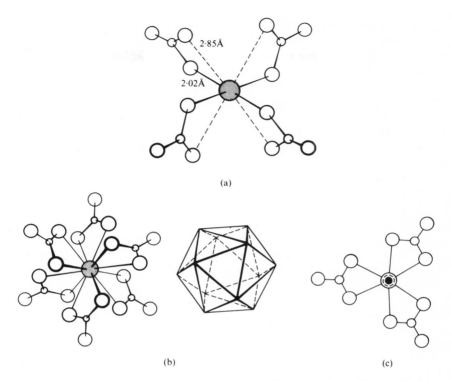

FIG. 18.6. The ions (a) $[Au(NO_3)_4]^-$; (b) $[Ce(NO_3)_6]^{3-}$; (c) $[UO_2(NO_3)_3]^-$.

The pentanitrato ions are notable for the different behaviours of the NO_3 groups. If all are bidentate the central metal atom is 10-coordinated, as in $[Ce(NO_3)_5]^{2-}$ and also in $[Y(NO_3)_5](NO)_2$. However, in $[Sc(NO_3)_5](NO)_2$ the smaller Sc^{3+} ion is 9-coordinated, by 4 bidentate and 1 monodentate NO_3. The still smaller Al^{3+} ion is 6-coordinated in $[Al(NO_3)_5]^{2-}$, by 1 bidentate and 4 monodentate ligands.

Several hexanitrato ions have been studied. There is nearly regular icosahedral co-ordination of the metal atom in $Mg[Th(NO_3)_6].8H_2O$ and $Mg_3[Ce(NO_3)_6]_2.24H_2O$. Since Mg^{2+} is completely hydrated in both these crystals they are preferably formulated $[Mg(H_2O)_6][Th(NO_3)_6].2H_2O$ and $[Mg(H_2O)_6]_3[Ce(NO_3)_6]_2.6H_2O$. For an ion of this type see Fig. 18.6(b). The anion in $(NH_4)_2[Ce(NO_3)_6]$ also has six bidentate NO_3 groups bonded to the metal atom, but here they are arranged with approximate T_h symmetry. There is marked asymmetry of the NO_3 group (N–O, 1.235 and 1.282 Å) but rather less than in $Ti(NO_3)_4$ (see above).

(iii) *Oxynitrato compounds.* Transition-metal compounds $MO_x(NO_3)_y$ are prepared by methods such as the action of liquid N_2O_5 on the metal oxide or hydrated nitrate to give addition compounds with N_2O_5 which on heating *in vacuo* give in some cases the anhydrous nitrate and in others an oxynitrato compound. The volatile $Be_4O(NO_3)_6$, formed by heating $Be(NO_3)_2$ *in vacuo*, presumably has a tetrahedral structure similar to that of $Be_4O(CH_3COO)_6$.

Compounds containing oxynitrato ions include $Rb[UO_2(NO_3)_3]$. In the anion in this salt (Fig. 18.6(c)) the axis of the linear UO_2 group is normal to the plane of the paper and three bidentate NO_3 groups complete the 8-coordination group around the metal atom.

(iv) *Coordination complexes containing NO_3 groups.* There are numerous molecules in which NO_3 behaves either as a monodentate or as a bidentate ligand, for example:

Monodentate:	$Cu(en)_2(ONO_2)_2$	AC 1964 **17** 1145
	$Cu(C_5H_5NO)_2(NO_3)_2$	AC 1969 **B25** 2046
Bidentate:	$Cu^I(P\phi_3)_2NO_3$	IC 1969 **8** 2750
	$Co(Me_3PO)_2(NO_3)_2$	JACS 1963 **85** 2402
	$Th(NO_3)_4(OP\phi_3)_2$	IC 1971 **10** 115
	$La(NO_3)_3(bipyridyl)_2$	JACS 1968 **90** 6548

In the *trans* octahedral molecule of $Cu(en)_2(NO_3)_2$, (a), Cu^{II} forms two long bonds to O atoms; in $Cu(P\phi_3)_2NO_3$ there is distorted tetrahedral coordination of Cu^I, (b). The structure of $Co(Me_3PO)_2(NO_3)_2$ may also be described as tetrahedral if the bidentate NO_3 is regarded as a single ligand, for the optical and magnetic properties of this molecule (and also the ion $Co(NO_3)_4^{2-}$) suggest that the metal atom is situated in a tetrahedral ligand field.

(a)   (b)   (c)

The molecules $Th(NO_3)_4(OP\phi_3)_2$ (and the isostructural Ce compound) and $La(NO_3)_3(bipyridyl)_2$ are interesting examples of 10-coordination. In the former, (c), the 'pseudo-octahedral' coordination group is composed of four bidentate NO_3 groups and two phosphoryl O atoms. (There is 11-coordination of Th in $Th_2(OH)_2(NO_3)_6(H_2O)_6$ and $Th(NO_3)_4.5H_2O$, in which the coordination groups around Th are $(OH)_2(H_2O)_3(NO_3)_3$ and $(H_2O)_3(NO_3)_4$ respectively.)

Covalent nitrates. Apart from organic nitrates covalent nitrates of non-metals are limited to those of H, F, and Cl. ($ClNO_3$ has been prepared from anhydrous HNO_3 and ClF as a liquid stable at $-40°C$ in glass or stainless steel vessels.[1]) Electron diffraction studies have been made of the explosive gas FNO_3[2] and of the (planar) methyl nitrate molecule.[3] A refinement of the crystal structure of pentaerythritol nitrate, $C(CH_2ONO_2)_4$,[4] shows that the nitrate group has the same structure, (d), as in nitric acid.

(d)

(1)   IC 1967 **6** 1938
(2)   JACS 1935 **57** 2693; ZE 1941 **47** 152
(3)   JCP 1961 **35** 191
(4)   AC 1963 **16** 698

The sulphides of nitrogen and related compounds

These compounds have little in common with the oxides of nitrogen, either as regards chemical properties or structure, though the S analogue of the linear nitronium ion NO_2^+ is known. The salt $(NS_2)(SbCl_6)$[1a] is formed by the reaction between S_7NH and $SbCl_5$ in liquid SO_2, and has been shown to contain linear $(S-N-S)^+$ ions (N–S, 1.46 Å). Only two sulphides, N_2S_4 and N_2S_5, have formulae similar to those of oxides. The structure of N_2S_5 is not known; for N_2S_4 see later.

There is an extraordinary number of cyclic compounds containing 4, 5, 6, 7, 8, and 10-membered rings, including planar ions or molecules of all these types. (We shall include here NSF and NSF_3 because of their close relation to $(NSF)_4$.) The simplest ring is that of N_2S_2, formed when N_4S_4 (see later) is sublimed through a

(a)

(b)

plug of silver wool at a pressure of 0.01 mm Hg at 300 °C and the products are cooled rapidly. The structure of this ring was first established in $N_2S_2(SbCl_5)_2$, (a).[1b] In the planar N_2S_2 ring the four bonds are equal in length (1.62 Å, as in N_4S_4). The Sb—N bonds are appreciably weaker than in $N_4S_4.SbCl_5$ (see later), and one $SbCl_5$ is easily removed, leaving $N_2S_2.SbCl_5$. There would appear to be a correlation between the Sb—N and Sb—Cl bond lengths in the adducts of $SbCl_5$:

	$N_2S_2.SbCl_5$	$CH_3CN.SbCl_5$	$N_4S_4.SbCl_5$
Sb—N	2.28	2.23	2.17 Å
Sb—Cl	2.31	2.36	2.39 Å

Polymeric $(NS)_x$ has attracted attention because of its physical properties. Polymerization of the colourless diamagnetic N_2S_2, which consists of square planar molecules (N—S, 1.655 Å), in the solid state at room temperature or slightly above produces a blue-black paramagnetic material which subsequently changes to diamagnetic golden-yellow crystals. These have electrical conductivity comparable with that of metals such as Hg at room temperature, and become superconducting at very low temperatures. They consist of aligned fibres, each of which is a single crystal of $(NS)_x$ built of almost planar chains in which N and S atoms alternate:[1c]

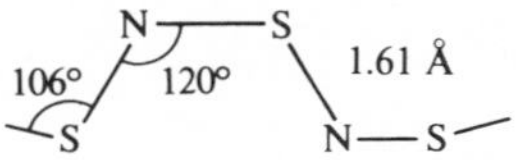

The 5-membered ring N_2S_3 is planar as the ion $N_2S_3^+$ in the salt $(N_2S_3)(AsF_6)$[2a] (S—S, 2.15; N—S, 1.58 Å) but puckered, (b), in $(N_2S_3Cl)Cl$,[2b] a salt formed by refluxing a suspension of NH_4Cl in S_2Cl_2 or from the interaction of N_4S_4 or $(NSCl)_3$ with S_2Cl_2.

The 6-membered ring, N_3S_3 is planar as the ion $S_3N_3^-$, formed by interacting, for example, the azide ion with N_4S_4 and isolated as salts of large cations such as Cs^+ or NR_4^+.[3a] It is chair-shaped in $(NSCl)_3$[3b] (Fig. 18.7(a)), formed by chlorination of N_4S_4, in the trisulphimide ion in $Ag_3(NSO_2)_3.3H_2O$,[3c] in the cyclic molecule $(CH_3NSO_2)_3$,[3d] and also in α-sulphanuric chloride, $N_3S_3O_3Cl_3$[3e] (Fig. 18.7(b)). All the N—S bonds are of equal length in these N_3S_3 rings, with values 1.61, 1.64, 1.67, and 1.57 Å respectively. The N_3S_3 ring has also been studied in the $N_3S_3O_4^-$

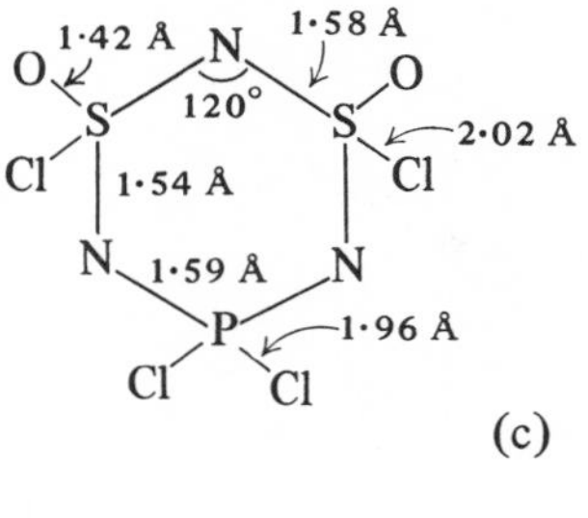

(c)

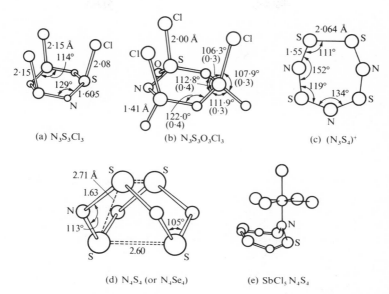

FIG. 18.7.   Cyclic nitrogen–sulphur systems: (a) $N_3S_3Cl_3$;   (b) $N_3S_3O_3Cl_3$;   (c) $(N_3S_4)^+$;   (d) N_4S_4;   (e) $N_4S_4 . SbCl_5$.

ion,[3f] in which there are 2 O atoms attached to each of two of the S atoms. In the related molecule $N_3S_2PO_2Cl_4$[4] containing one P atom in the ring, (c), all bond lengths are similar to those in $N_3S_3O_3Cl_3$, but here the ring is nearly planar (compare the planar $(PNCl_2)_3$ ring).

The 7-membered N_3S_4 ring has been studied as the planar cation in $(N_3S_4)(NO_3)$[5a] and $(N_3S_4)_2(SbCl_5)$,[5b] Fig. 18.7(c). The salt N_3S_4Cl is formed quantitatively when $N_2S_3Cl_2$ is refluxed with a mixture of S_2Cl_2 and CCl_4, and it is converted into $(N_3S_4)NO_3$ by dissolving in concentrated HNO_3. The S–S bond is apparently a single bond but the approximate equality of the S–N bond lengths and the well-defined u.v. absorption spectrum suggest a 10-electron π system.

The 8-membered N_4S_4 ring occurs both as a cation, $N_4S_4^{2+}$, and also in a very buckled form as the molecule N_4S_4. The ion has been studied in the salts $N_4S_4(SbCl_6)_2$ and $N_4S_4(SbF_6)(Sb_3F_{14})$[6a] and found to be approximately planar, but disorder in the orientation of the cations makes a discussion of the bonding difficult. The mean bond lengths are similar to those in $N_3S_4^+$ and $N_5S_5^+$. We return to N_4S_4 after concluding these notes on the nitrogen–sulphur ions.

The 9-membered ring is represented by the $N_5S_4^-$ ion[6b] and the 10-membered cyclic ion $N_5S_5^+$ ion has been studied in the salts $(N_5S_5)(AlCl_4)$,[7a] $(N_5S_5)(SnCl_5OPCl_3)$,[7b] and $(N_5S_5)(S_3N_3O_4)$.[3f] In the first of these the heart-shaped configuration of Fig. 18.8(a) was found, but in the other two salts the configuration of Fig. 18.8(b). The ring is almost planar, and is presumably a 14-electron π system.

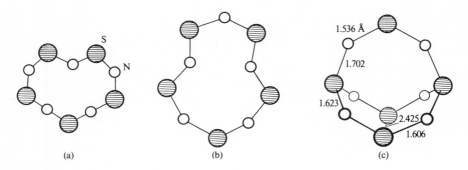

FIG. 18.8. (a) and (b) Configurations of the $N_5S_5^+$ ion; (c) the N_6S_5 molecule.

The parent compound from which many nitrogen–sulphur compounds can be derived is N_4S_4. It is readily prepared by the interaction of S with NH_3 in the absence of water or preferably by passing S_2Cl_2 vapour through hot pellets of NH_4Cl. It forms yellow crystals and has remarkable chemical properties. The N_4S_4 molecule, Fig. 18.7(d), has a tetrahedral arrangement of 4 S resulting from buckling the 8-membered ring so that there are two (non-bonded) S–S distances of 2.60 Å and four of 2.71 Å.[8a] These distances, intermediate between a single S–S (2.06 Å) and the van der Waals distance (3.3 Å) suggest some kind of weak covalent bonding. The molecule of N_4Se_4[8b] presents a similar problem.

In adducts of N_4S_4 with Lewis acids such as BF_3,[9] $SbCl_5$,[10] and AsF_5[10a] the latter are attached to one N only, and the configuration of the N_4S_4 portion changes (Fig. 18.7(e)), with the four S atoms coplanar as opposed to tetrahedral in the free molecule. In $N_4S_4(CuCl)$,[10b] however, N_4S_4 bridges two Cu atoms via the N atoms, and has the configuration of the free molecule (Fig. 18.7(d)).

Mild fluorination by AgF_2 in an inert solvent converts N_4S_4 into $(NSF)_4$.[11] In this puckered 8-membered ring (boat configuration) the F atoms are attached to S atoms, and there is a definite alternation of longer and shorter bonds, (d). More vigorous fluorination of N_4S_4 yields NSF, (e),[12] and NSF_3, (f),[13] in both of which F is bonded to S. Hydrolysis of NSF (or the action of NH_3 on $SOCl_2$) gives HNSO,[14] *cis*-thionyl imide, with the structure shown at (g).

NSF can act as a ligand bonded to a metal through the N atom, as in $[Co(NSF)_6](AsF_6)_2$.[14a] The structure of the ligand is similar to that of the free molecule.

We now come to the imides S_7NH, $S_6(NH)_2$, $S_5(NH)_3$, and $S_4(NH)_4$. The first three result from the action of NH_3 on S_2Cl_2 in dimethylformamide at low temperatures, and are separated chromatographically. $S_4(NH)_4$ is made by reducing a solution of N_4S_4 in benzene by alcoholic $SnCl_2$. Three of the four possible structural isomers of $S_6(NH)_2$, namely, $1:3$,[15] $1:4$,[16] and $1:5$,[17] have been characterized, and the crystal structures of several of these compounds have been studied. All are structurally similar to the S_8 molecule of rhombic sulphur, that is, they have the crown configuration. In $1:4$ $S_6(NH)_2$ both enantiomorphic forms of the molecule are present in the crystal. Neutron diffraction confirms that H is attached to N in $N_4S_4H_4$[18] (mean bond angles: SNS, 124°; NSN, 109°; S–N, 1.665 Å).

The N–S bond lengths in this family of compounds suggest the following standard values: S–N, 1.72 Å (compare 1.73 Å in O_3SNH_3) and S=N, 1.54 Å; lengths of formal triple bonds in NSF and NSF_3 are 1.45 Å and 1.42 Å.

If N_4S_4 is heated in solvents N_2S_5 is produced, while by the action of S in CS_2 solution under pressure N_4S_4 is converted into N_2S_4, a dark-red, rather unstable solid. Solutions of N_2S_4 in organic solvents are diamagnetic, showing that this sulphide is not dissociated, in contrast to N_2O_4. The detailed structure of the N_2S_4 molecule is not known, but a review of its spectra (i.r., Raman, u.v., and mass) indicates the structure (h).[19]

(h)

Thionitrosyl complexes of a number of metals have been prepared, for example, $Ni(N_2S_2H)_2$ and the similar Co and Fe compounds from N_4S_4 and the carbonyls, and $Pt(N_2S_2H)_2$ from the action of N_4S_4 on H_2PtCl_6 in solution in dimethylformamide. An X-ray study of the Ni compound indicates a planar molecule with the *cis* configuration (i).[20]

(i)

A recent addition to nitrogen–sulphur chemistry has been the preparation of N_6S_5, from Br_2 and $(Bu_4N)(N_5S_4)$ in methylene chloride.[21] The molecule, Fig. 18.8(c), is related to N_4S_4, the system $-N{=}S{=}N-$ bridging two S atoms.

(1a)　IC 1978 **17** 2975
(1b)　IC 1969 **8** 2426
(1c)　JACS 1975 **97** 6358
(2a)　CJC 1975 **53** 3147
(2b)　IC 1966 **5** 1767
(3a)　CC 1978 391
(3b)　AC 1966 **20** 192
(3c)　AC 1974 **B30** 2721
(3d)　AC 1974 **B30** 2724
(3e)　AC 1966 **20** 186
(3f)　CC 1975 735
(4)　 AC 1969 **B25** 651
(5a)　IC 1965 **4** 681
(5b)　ZaC 1972 **388** 158
(6a)　CC 1977 253
(6b)　AnC IE 1976 **15** 379
(7a)　ACSc 1972 **26** 1987
(7b)　JCS D 1976 928

(8a)　IC 1978 **17** 2336
(8b)　AC 1966 **21** 571
(9)　 IC 1967 **6** 1906
(10)　ZaC 1960 **303** 28
(10a)　AC 1980 **B36** 655
(10b)　AnC 1976 **88** 807
(11)　AC 1963 **16** 152
(12)　JACS 1963 **85** 1726
(13)　JACS 1962 **84** 334
(14)　JACS 1969 **91** 2437
(14a)　AC 1980 **B36** 141
(15)　AC 1971 **B27** 2480
(16)　AC 1969 **B25** 611
(17)　AC 1973 **B29** 915
(18)　AC 1977 **B33** 2309
(19)　JCS A 1971 136
(20)　AC 1978 **B34** 1999
(21)　CC 1978 642

Nitrides

Most elements combine with nitrogen, and the nitrides include some of the most stable compounds known. The following metals are *not* known to form nitrides: (Na?), K, Rb, Cs, Au, Ru, Rh, Pd, Os, Ir, and Pt. We can make a rough division into four groups, but it should be emphasized that there is no sharp dividing line, particularly in the case of the IIA–IVA subgroup metals. In some cases an element forms nitrides of quite different types, and behaviour as both an ionic compound and as a metallic conductor is a feature of a number of nitrides (for example, HfN, below). Thus Ca forms Ca_3N_2, one form of which (α) has the anti-Mn_2O_3 structure,[1] but there are also black and yellow forms of the same compound. Moreover Ca also forms Ca_2N, Ca_3N_4, and $Ca_{11}N_8$. Ca_2N forms lustrous green-black crystals with the anti-$CdCl_2$ (layer) structure[2] (Ca–N, 2.43 Å), in which there are Ca–Ca distances (3.23 and 3.64 Å) much shorter than in metallic Ca (3.88 and 3.95 Å). Nevertheless this is apparently an ionic compound, for the action of water produces $NH_3 + H_2$. The lustre and semiconductivity suggest formulation as $Ca_2^{2+}N^{3-}$ (e). There are even shorter Ca–Ca distances in $Ca_{11}N_8$ (3.11 Å),[3] a compound which is decomposed by water vapour. Zirconium forms not only the brown Zr_3N_4 but also blue Zr_xN

Li_3N Be_3N_2 Mg_3N_2 Ca_3N_2 etc.	Transition-metal nitrides Interstitial and more complex structures	BN AlN Si_3N_4 GaN Ge_3N_4 InN Hg_3N_2	Molecular nitrides of non-metals
ionic	*semi-metallic*	*covalent or ionic-covalent*	*covalent*

(defect NaCl structure)[4a] and the metallic yellow ZrN. Hafnium forms a nitride HfN which has a defect NaCl structure,[4b] the density indicating that one-eighth of the sites are unoccupied. When dissolved in HF all the nitrogen is converted into NH_3 (NH_4^+) and the compound is a metallic conductor ($Hf^{4+}N^{3-}(e)^?$).

In the first and third of our groups of nitrides the formulae *usually* correspond to the normal valences of the elements and this is also true of some transition-metal compounds (e.g. Sc, Y, and 4f nitrides MN, Zr_3N_4) and of some of the molecular nitrides (e.g. P_3N_5);[5] others, such as N_4S_4, have formulae and structures which are not understood. The molecular nitrides are few in number, and if their structures are known they are described under the structural chemistry of the appropriate element.

In addition to the binary nitrides described here, many ternary nitrides have been prepared, for example, Li_5TiN_3, Li_7NbN_4, and Li_9CrN_5, all with the fluorite type of structure (superstructures in most cases),[6] and alkaline-earth compounds with Re, Os, Mo, or W (e.g. $Sr_9Re_3N_{10}$, Ca_5MoN_5).[7]

Ionic nitrides

Nitrides of the alkali metals other than Li do not appear to be known in a pure state but Li_3N has been known for many years, and its structure was determined in 1935. It has become of interest because of its high ionic conductivity, both at ambient and high temperatures.[8] The compound is sensitive to moisture, but

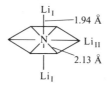

sintered material can be protected by coating the surface, and the large red crystals which can be pulled from the melt are stable in moist air owing to the formation of a surface layer (oxide, hydroxide, or carbonate). In the simple (and unique) structure there are Li ions of two kinds, Li_I and Li_{II}. The Li_{II} ions are in hexagonal layers of composition Li_2N with a N^{3-} ion at the centre of each ring, and between these layers are layers containing only Li_I ions. The resulting coordinations are: Li_I, 2 N collinear at 1.94 Å; Li_{II}, 3 N triangular at 2.13 Å; and N, 6 Li hexagonal bipyramidal, as shown. The unusual coordination of Li^+ is of course related to the very small size of this ion compared with N^{3-}, though we noted in Chapter 6 that a structure with tetrahedral coordination of Li^+ is geometrically impossible.

We have noted the ReO_3 structure of Cu_3N (and the perovskite structure of $GaNCr_3$) in Chapter 6.

The nitrides M_3N_2 (M = Be, Mg, Zn, Cd), and α-Ca_3N_2 have the anti-Mn_2O_3 structure, in which the metal ions have 4 tetrahedral neighbours and N has two types of distorted octahedral coordination groups. (Be_3N_2 also has a close-packed high-temperature form in which Be is tetrahedrally coordinated;[9] it is described on p. 164). The anti-Mn_2O_3 structure of α-Ca_3N_2 is remarkable not only for the tetrahedral coordination of Ca^{2+} but also because this leads to a very open packing of the N^{3-} ions. The Ca–N distance (2.46 Å) is about the value expected for 6-coordinated Ca^{2+}, but the N–N distances are very large. In this structure there are empty N_4 tetrahedra with all edges 4.4 Å, while the edges of a CaN_4 tetrahedron are 3.76 Å (four) and 4.4 Å (two). All these distances are large compared with twice the radius of N^{3-} (about 3 Å). Presumably the abnormally low c.n. of Ca^{2+} (and the resulting low density of the structure) is due either to the fact that there is no alternative structure of high coordination (say, 6:9 or 8:12) or to the fact that the structure is determined by the octahedral coordination of N^{3-}, to which the 4-coordination of Ca^{2+} is incidental.

The radius of the N^{3-} ion in these nitrides would appear to be close to 1.5 Å, larger than most metal ions. These compounds are the analogues of the oxides, and form transparent colourless crystals. They are formed by direct union of the metal with nitrogen, or in some cases by heating the amide. Compare, for example:

$$3\,Ba(NH_2)_2 \longrightarrow Ba_3N_2 + 4\,NH_3$$
$$\text{with} \qquad Ba(OH)_2 \longrightarrow BaO + H_2O.$$

The nitrides have high melting points, ranging from 2200 °C (Be_3N_2) to 900 °C (Ca_3N_2) and are hydrolysed by water to the hydroxides with liberation of ammonia.

The structures of compounds of the type M_3N_2 are summarized in Table 18.7, where A represents the anti-Mn_2O_3, B the Zn_3P_2, and C the anti-La_2O_3 structure.

TABLE 18.7

Crystal structures of compounds M_3N_2, etc.

	Be	Mg	Zn	Cd	Ca
N	A	A	A	A	A
P	A	A	B	B	
As		A	B*	B*	
Sb		C			
Bi		C			

* Neither of these compounds has the simple Zn_3P_2 structure but closely related structures (Zn_3As_2,[10] Cd_3As_2[11]).

The Zn_3P_2 structure is closely related to the anti-Mn_2O_3 structure. Zinc atoms are surrounded tetrahedrally by four P and P by six Zn at six of the vertices of a distorted cube.

Covalent nitrides

In contrast to the ionic nitrides of Zn or Cd, the chocolate-coloured explosive Hg_3N_2 made from HgI_2 and KNH_2 in liquid ammonia, is presumably a covalent compound. For the nitrides of B, Al, Ga, In, and Tl the geometrical possibilities are the same as for carbon, there being an average of 4 valence electrons per atom in MN.

Boron nitride has been known for a long time as a white powder of great chemical stability and high melting point with a graphite-like layer structure. It has also been prepared with the zinc-blende structure[12] by subjecting the ordinary form to a pressure of 60 kbar at 1400–1800 °C. This form, 'borazon', is very much more dense, 3.47 as compared with 2.25 g cm^{-3} for the ordinary form. In contrast, BP can be prepared by heating red P with B at 1100 °C under a pressure of only 2 atm; it also has the zinc-blende structure.[13] The same structure is adopted by BAs.[14a] The structures of the compounds of Group III elements with N, P, etc. are summarized below. While BN is presumably a covalent compound, the bonds in AlN, GaN, and InN probably have appreciable ionic character, and the compounds with the heavier Group V elements are essentially metallic compounds. (For InBi see p. 270.) The structure of $BeSiN_2$ is a superstructure of the wurtzite structure of AlN etc.[14b]

	B	Al	Ga	In
N	ZL	W	W	W
P	Z	Z	Z	Z
As	Z	Z	Z	Z
Sb	–	Z	Z	Z

L = BN layer structure; Z = zinc-blende; W = wurtzite

The structures of several nitrides, M_3N_4 are known. Both Si_3N_4 and Ge_3N_4 have two polymorphic forms.[15] The structures of both are closely related to that of Be_2SiO_4 (phenacite); M has 4 tetrahedral neighbours and N forms coplanar bonds to 3 M atoms. There is probably appreciable ionic character in the bonds in Ge_3N_4 and also in Th_3N_4. In Th_3N_4[16] there is closest packing of the metal atoms of the same kind as in metallic Sm (*chh*). One-half of the N atoms occupy tetrahedral and the other half octahedral holes in such a way that one-third of the metal atoms have 6 N and the remainder 7 N neighbours. The N atoms of both kinds are displaced from the centres of the holes in the direction of greatest N–N separation:

$$N_t \begin{cases} 3 \text{ Th} & 2.31 \text{ Å} \\ 1 \text{ Th} & 2.47 \end{cases} \qquad N_o \begin{cases} 3 \text{ Th} & 2.53 \text{ Å} \\ 3 \text{ Th} & 2.91 \end{cases}$$

(The compound originally described as Th_2N_3 is Th_2N_2O, in which N occupies tetrahedral and O octahedral holes in a h.c.p. arrangement of Th atoms; all Th atoms have 4 N + 3 O neighbours.) For an alternative description of this (Ce_2O_2S) structure see p. 1271.

Nitrides of transition metals

Some of these compounds have already been mentioned. They are extremely numerous, more than one nitride being formed by many transition metals. For example, five distinct phases (apart from the solid solution in the metal) have been recognized in the Nb–N system up to the composition NbN.[17] Of the great variety of structures adopted by these compounds we shall mention only some of the simpler ones.

In the so-called 'interstitial' nitrides the metal atoms are approximately, or in some cases exactly, close-packed (as in ScN, YN, TiN, ZrN, VN, and the rare-earth nitrides with the NaCl structure), but the arrangement of metal atoms in these compounds is generally *not* the same as in the pure metal (see Table 29.13, p. 1321). Since these interstitial nitrides have much in common with carbides, and to a smaller extent with borides, both as regards physical properties and structure, it is convenient to deal with all these compounds in Chapter 29.

The nitrides of Mn, Fe, Co, and Ni form a group of less stable compounds of greater complexity than those of the earlier Periodic Groups—compare the formulae of the nitrides of the metals of the first transition series:

$$
\begin{array}{llllllllll}
\text{ScN} & \text{Ti}_2\text{N} & \text{V}_3\text{N} & \text{Cr}_2\text{N} & \text{MnN (?)} & \text{Fe}_2\text{N} & \text{CoN} & \text{Ni}_3\text{N} & \text{Cu}_3\text{N} \\
 & \text{TiN} & \text{VN} & \text{CrN} & \text{Mn}_6\text{N}_5 & \text{Fe}_3\text{N} & \text{Co}_2\text{N} & \text{Ni}_4\text{N} \\
 & & & & \text{Mn}_3\text{N}_2 & \text{Fe}_4\text{N} & \text{Co}_3\text{N} \\
 & & & & \text{Mn}_2\text{N} & \text{Fe}_8\text{N} & \text{Co}_4\text{N} \\
 & & & & \text{Mn}_4\text{N} \\
 & & & & \text{Mn}_x\text{N (}\delta\text{)}
\end{array}
$$

Although the N atoms in these compounds are usually few in number compared with the metal atoms they seem to exert great influence, and it has been suggested[18] that the tendency of the N atoms to order themselves is great enough to cause rearrangement of the metal atoms in, for example, the change from ϵ-Fe$_2$N to ζ-Fe$_2$N.

Interstitial structures are not possible for the larger P and As atoms, and apart from a few cases such as LaP, PrP, and GeP with the NaCl structure there is usually little similarity between nitrides and phosphides (or arsenides). Compare, for

TABLE 18.8

Coordination numbers in crystalline nitrides

Nitride	Structure	C.N. of M	C.N. of N
Cu$_3$N	Anti-ReO$_3$	2	6
Ti$_2$N	Anti-rutile[20]	3	6
Co$_2$N	Anti-CdCl$_2$	3	6
Ca$_3$N$_2$	Anti-Mn$_2$O$_3$	4	6
ScN etc.	NaCl	6	6
Th$_3$N$_4$		6, 7	4, 6
Ge$_3$N$_4$	Phenacite	4	3
AlN	Wurtzite	4	4
Ca$_3$N$_2$	Anti-Mn$_2$O$_3$	4	6
AlLi$_3$N$_2$	Anti-CaF$_2$	4	8

example, the formulae of the nitrides and phosphides of Mo and W: Mo_2N, W_2N, MoN, and WN; but Mo_3P, W_3P; MoP, WP; MoP_2, and WP_2.[19]

A short section on the structures of metal phosphides is included in Chapter 19.

Table 18.8 summarizes the coordination numbers of metal and N atoms in a number of nitrides.

(1)   AC 1968 **B24** 494
(2)   IC 1968 7 1757
(3)   AC 1969 **B25** 199
(4a) ZaC 1964 **329** 136
(4b) ZaC 1967 **353** 329
(5)   JACS 1957 79 1765
(6)   ZaC 1961 **309** 276
(7)   IC 1970 9 1849
(8)   JSSC 1979 **29** 379
(9)   ZaC 1969 **369** 108
(10) AC 1956 9 685

(11) AC 1968 **B24** 1062
(12) JCP 1957 **26** 956
(13) N 1957 **179** 1075
(14a) AC 1958 **11** 310
(14b) ZaC 1967 **363** 225
(15) JACeS 1975 58 90; AC 1979 **B35** 800
(16) AC 1971 **B27** 243
(17) ZaC 1961 **309** 151
(18) AC 1952 5 404
(19) ACSc 1954 8 204
(20) ACSc 1962 **16** 1255

Phosphorus

The stereochemistry of phosphorus

Phosphorus, like nitrogen, has five valence electrons. These are in the third quantum shell, in which there are d orbitals in addition to s and p orbitals. Phosphorus forms up to six separate bonds with other atoms, but covalencies greater than four are *usually* exhibited only in combination with halogens (in the pentahalides and PX_6^- ions) and with groups such as phenyl, C_6H_5. (For an example of 5-coordination by O see p. 853). The d orbitals are used for σ-bonding if more than four bonds are formed, but they are also used for π-bonding in tetrahedral oxy-ions and molecules. A very simple summary of the stereochemistry of P may be given if it is assumed that the arrangement of bonds is determined by the number of σ electron pairs which are used for either σ bonds or lone pairs (Table 19.1). For PX_3 molecules this picture is valid if appreciable hybridization occurs. In PH_3 the bonds are likely

TABLE 19.1

The stereochemistry of phosphorus

No. of σ pairs	Type of hybrid	No. of lone pairs	Bond arrangement	Examples
4	sp^3	0	Tetrahedral	PH_4^+, PCl_4^+ $POCl_3$, PO_4^{3-}, $PO_2F_2^-$
		1	Trigonal pyramidal	PCl_3, $P(CN)_3$, $P(C_6H_5)_3$
5	$sp^3d_{z^2}$	0	Trigonal bipyramidal	PCl_5, PF_3Cl_2, $P(C_6H_5)_5$
6	$sp^3d_\gamma^2$	0	Octahedral	PCl_6^-

to be essentially p^3 bonds, and this may also be true in the trihalides, but the interpretation of bond angles around $100°$ is not simple. The bond angles in the P_3 rings in P_4 and P_4S_3 are to be treated as special cases. (For an example of P forming 3 coplanar bonds see p. 796.)

Elementary phosphorus

Phosphorus crystallizes in at least five polymorphic forms. The white form is metastable and is prepared by condensing the vapour. There are apparently two closely related modifications of white P, with a transition point at $-77°C$. The molecular weight in various solvents shows that white P exists in solution as P_4

molecules, presumably similar to those in the vapour of white P, the configuration of which was determined in an early e.d. study.[1] It is also virtually certain that both modifications of white P consists of regular tetrahedral P_4 molecules with the same structure as the vapour molecule (interbond angle 60°, bond length 2.21 Å), but the structure of neither form has yet been satisfactorily established.[2] In the P_2 molecule in the vapour of red P the P≡P bond length is 1.895 Å; in PN, P≡N is 1.49 Å.

Black P is formed under pressure at 200 °C, by crystallization from molten Bi, or by heating white P to 220-370 °C for 8 days with Hg on Cu as catalyst in the presence of a seed of black P. It is the most stable crystalline form of the element, as is shown by the densities (1.83, 2.31, and 2.69 g cm⁻³) and the heats of solution in Br_2(249, 178, and 160 kJ mol⁻¹) for white, red, and black P respectively.

Black P has a layer structure[3] in which each atom is bonded to three others. A layer of this structure is shown, in idealized form, in Fig. 19.1. In spite of its

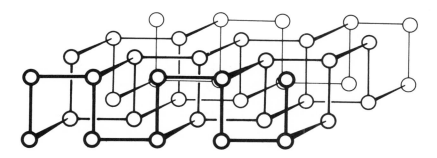

FIG. 19.1. The crystal structure of black phosphorus – portion of one layer (idealized).

appearance it is simply a puckered form of the simple hexagonal net, the interbond angles being two of 102° and one of 96½°. (See also p. 103 for the relation of this structure to the simple cubic lattice.) The P–P bond length is 2.23 Å.

Hittorf's, or monoclinic, P may be made by dissolving white P in thirty times its weight of molten lead, cooling slowly, and dissolving away the lead electrolytically. It has an extraordinarily complex structure.[4] Complex chains with 21 atoms in the repeat unit are linked into layers by bonds to similar chains which lie at right angles to the first set but in a parallel plane. Two such systems of cross-linked chains form a complex layer (Fig. 19.2) in which there are no P–P bonds between the atoms of one half of the layer (broken lines) and the other (full lines). The whole crystal is an assembly of composite layers of the type shown in Fig. 19.2(b) held together by van der Waals bonds. Within a layer the mean P–P bond length is 2.22 Å and the angle P–P–P, 101°.

Two high-pressure forms of P have been produced from black P – a rhombohedral form with the As structure at 50 kbar and a primitive cubic form at about 110 kbar.[5]

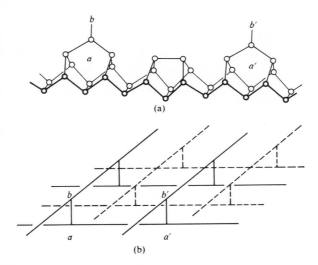

FIG. 19.2. The crystal structure of monoclinic phosphorus: (a) chain sub-unit;   (b) layer formed from cross-linked chains (diagrammatic).

The detailed structure of the former was not determined, but in the cubic (metallic) form P has 6 equidistant nearest neighbours at 2.38 Å.

Numerous other forms of P have been described, including the well-known red P which results from heating white P in an inert atmosphere or by the action of light or X-rays. The physical properties depend on the method of preparation. Commercial red P is amorphous and other preparations are poorly crystalline; their structures are not known.

It has been suggested that a brown form of P, formed by condensing the vapour of white P at 1000 °C or the vapour of red P at 350 °C on to a cold finger at -196 °C, may consist of P_2 molecules.[6] A vitreous form of P has been prepared by heating white P with Hg at 380 °C; at 450 °C black P is formed.[7]

(1) JCP 1935 **3** 699	(5) Sc 1963 **139** 1291
(2) N 1952 **170** 629	(6) JACS 1953 **75** 2003
(3) AC 1965 **19** 684	(7) IC 1963 **2** 22
(4) AC 1969 **B25** 125	

Phosphides of metals

Most elements combine with phosphorus, and many metals form more than one phosphide; for example, potassium forms K_3P, K_3P_2, K_5P_4, KP, K_4P_6 (two forms), and K_3P_7. Sufficient examples are given in Table 19.2 to show the extraordinary variety of formulae and types of structure. Certain of the less familiar structures which are described in other places include those of PdP_2 (p. 132), TiP (p. 153), and Th_3P_4 (pp. 193 and 1273). The relation of the $CoAs_3$ structure to that of ReO_3 is described on p. 267.

TABLE 19.2
Structures of metallic phosphides

Structure ·	System of linked P atoms	Examples
Zinc-blende		AlP, GaP, InP
NaCl		LaP, SmP, ThP, UP, ZrP, etc.
NiAs		VP
TiP	Isolated P atoms	HfP, β-ZrP
MnP		FeP, CoP, WP
Anti-CaF$_2$		Ir$_2$P, Rh$_2$P
Anti-PbCl$_2$		Co$_2$P, Ru$_2$P
Anti-Mn$_2$O$_3$		Be$_3$P$_2$, Mg$_3$P$_2$
Zn$_3$P$_2$		Cd$_3$P$_2$
Th$_3$P$_4$		Th$_3$P$_4$
Na$_3$As		Li$_3$P, Na$_3$P, K$_3$P
Pyrites or marcasite	P$_2$ groups	NiP$_2$ (h.p.),[1] FeP$_2$,[2]
		PtP$_2$,[2] OsP$_2$, RuP$_2$, CaP[3]
CoAs$_3$	P$_4$ rings	RhP$_3$, IrP$_3$, PdP$_3$[2]
	P$_6$ rings	Rb$_4$P$_6$[4]
	P$_7^{3-}$ ions	Li$_3$P$_7$, Ba$_3$P$_{14}$,[5]
	P$_{10}$ groups	Cu$_4$SnP$_{10}$[6]
	P$_{11}^{3-}$ ions	Na$_3$P$_{11}$[7]
	Chains — simple	LiP,[8] NaP (KP),[9] NiP$_2$, PdP$_2$,[10] CdP$_2$[11]
	— complex	BaP$_3$,[12] Au$_2$P$_3$,[13] RbP$_7$,[14] KP$_{15}$[15]
	Layers	CaP$_3$,[16] SrP$_3$,[17] TlP$_5$,[18] CdP$_4$ (MgP$_4$),[19]
		FeP$_4$,[20] RuP$_4$,[21] CrP$_4$,[22] MnP$_4$[23]
	3D nets	LiP$_5$,[24] LaP$_5$[25]

(1) IC 1968 7 998. (2) ACSc 1969 **23** 2677. (3) JLCM 1973 **30** 211. (4) AnC 1974 **86** 379. (5) NW 1973 **60** 429. (6) ZK 1980 **153** 339. (7) NW 1973 **60** 104. (8) NW 1967 **54** 225. (9) ZaC 1979 **456** 194. (10) AC 1963 **16** 1253. (11) AC 1970 **B26** 1883. (12) NW 1971 **58** 623. (13) AC 1979 **B35** 573. (14) NW 1972 **59** 78. (15) AnCIE 1967 **6** 356. (16) NW 1973 **60** 518. (17) NW 1973 **60** 429. (18) ACSc 1971 **25** 1327. (19) ZaC 1956 **285** 15. (20) AC 1978 **B34** 3196. (21) AC 1978 **B34** 2069. (22) AC 1972 **B28** 1893. (23) AC 1975 **B31** 574. (24) NW 1972 **59** 78. (25) NW 1975 **62** 180.

Phosphides range from compounds containing single P atoms (P^{3-} ions in compounds of the more electropositive elements) through those containing finite, one- or two-dimensional P_n complexes, to those in which all the P atoms form a (charged) 3D framework. The structures of ionic compounds such as Mg_3P_2 are consistent with a radius of about 1.9 Å for the P^{3-} ion. All phosphides containing systems of bonded P atoms contain some (or all) P atoms forming fewer than 3 P–P bonds; the bonding of every P to 3 others results in one of the structures of elementary P. The formation of one P–P bond by each P atom leads only to the P_2

group, found as the P_2^{4-} ion in CaP (Ca_2P_2) and also in the covalent diphosphides of transition metals with the pyrites or marcasite structures. The formation of 2 P–P bonds by each P atom results in rings or chains. The simplest chain

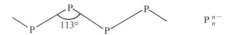

$$P_n^{n-}$$

is found, as a distorted 4-fold helix, in LiP and in NaP and the isostructural KP. In LiP there are equal numbers of chains of opposite chirality, and the structure is closely related to the NaCl structure, while in NaP all the chains are of the same chirality. The P–P bond length (2.23 Å) is close to the single-bond value, and this is true also in the chains in CdP_2 and in the covalent compounds such as NiP_2 and PdP_2. Rather longer P–P bonds have been found in some transition metal phosphides, including 2.4 Å (NiP), and 2.6-2.7 Å (CrP, MnP, FeP, CoP). Examples of cyclic P_4 and P_6 complexes are included in Table 19.2; we comment later on the cyclic P_6^{4-} ion.

It is convenient to describe as a group the polyphosphides of the more electropositive metals, but first we should mention the interesting adamantine-like P_{10} group in Cu_4SnP_{10}, a compound which has a defect ZnS type of structure, and also some features of the tetraphosphides listed in Table 19.2. In these (layer) structures each P is bonded to either 2 or 3 other P atoms, and the system of connected P atoms is derived from the planar 6^3 net by placing 4 2-connected P atoms on the links of each 6-gon to form a puckered net of 10-gons. There are many ways of doing this, five of which are realized in the MP_4 structures of Table 19.2. The layers of P atoms are held together by metal atoms in octahedral coordination, and these atoms complete the tetrahedral coordination of the P atoms, one half of which are bonded to 3 P + 1 M and the remainder to 2 P + 2 M. In CdP_4 (Fig. 19.3(a)) the puckered P_4 layer runs from left to right and is perpendicular to the plane of the paper. The P atoms are arranged around screw axes normal to the paper. The diagrammatic projection of the layer in Fig. 19.3(b) shows how the two kinds of

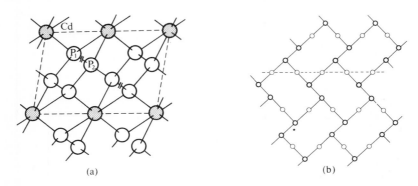

(a)   (b)

FIG. 19.3. Two projections of the structure of CdP_4 (see text).

P atoms are arranged in rings of ten, and the broken line represents one of the screw axes.

Numerous polyphosphides of the more electropositive metals have been prepared by the action of P vapour on metal–metal halide melts at temperatures ranging from 350 to 1100 °C. Not all of the types of compound set out in Table 19.3 are formed

TABLE 19.3

Formulae of polyphosphides

M^+	(alkali metals, Tl)	MP_n: n = (1), 5, 7, 11, 15;	M_4P_6, M_3P_7, M_3P_{11}
M^{2+}	(alkaline-earths, Mg, Cd, Eu(II))	MP_n: n = (1), 2, 3, 4, 7, 10;	M_4P_3, M_4P_5, M_3P_{14}
M^{3+}	(Al, La, 4f metals)	MP_n: n = 2, 3, 5, 7	

by a particular metal. These compounds range in colour from yellow (Li_3P_7) through red (LiP_5, LiP_7) to black (CaP_3, MgP_4). The stability of polyphosphides towards water varies considerably, and the products range from PH_3 and P_2H_4 through the polycyclic P_7H_3 and $P_{11}H_3$ to solid polymeric phosphanes. The polyphosphides contain systems of bonded P atoms, finite, 1-, 2-, and 3-dimensional, in which the P atoms are bonded to 2 or 3 other P atoms, and in most cases each 2-connected P atom contributes one negative charge.

The P_7^{3-} ion is geometrically similar to the P_4S_3 molecule (Fig. 19.10(c), p. 868), P^- replacing S; it occurs in the alkali metal compounds M_3P_7 and in Ba_3P_{14}. An infinite chain formed from these units, with one 2-connected P^- per unit, forms the anion in RbP_7 (Fig. 19.4(a)). The finite P_{11}^{3-} ion is found in Na_3P_{11} (Fig. 19.4(b)). Chains with obvious resemblances to those of Hittorf's P form the anion in KP_{15}, Fig. 19.4(c), together with P_7 chains in CsP_{11}, and also in TlP_5, Fig. 19.4(d). In the latter chain the repeat unit contains 9 P atoms, but in this compound the Tl(I) atoms function rather like the 2-connected P atoms, for one-half of them occupy positions indicated by the dotted circles, forming their two strongest bonds in the chains. The remainder, together with an equal number of P atoms, link the chains into layers, so that the composition becomes $TlP(TlP_9)$, or TlP_5. The metal atoms form, of course, numerous weaker bonds in addition to those emphasized in this description of the structure, and the P atoms are of three kinds, bonded to 2P, 3P, or 3P + 1Tl.

The structures of the anions in CaP_3, SrP_3, and BaP_3 represent three ways of removing one-quarter of the atoms from a corrugated 6^3 layer, that of black P (Fig. 19.4(e)). The actual geometry of the chain in BaP_3 (or Au_2P_3) is shown in Fig. 19.4(f). Another way of removing one-quarter of the atoms from a 6^3 net is realized in Rb_4P_6 (Fig. 19.4(g)). The net here is planar, like the B net in AlB_2; in this structure the charge on P_6^{4-} obviously does not correspond to the number of P atoms forming 2 P–P bonds. *Replacement* of one-quarter of the P atoms in a buckled (As-like) layer of P atoms by Sn atoms gives the layer of SnP_3, the projection of which is similar to (g), the dotted circles now representing Sn atoms. Halogen-

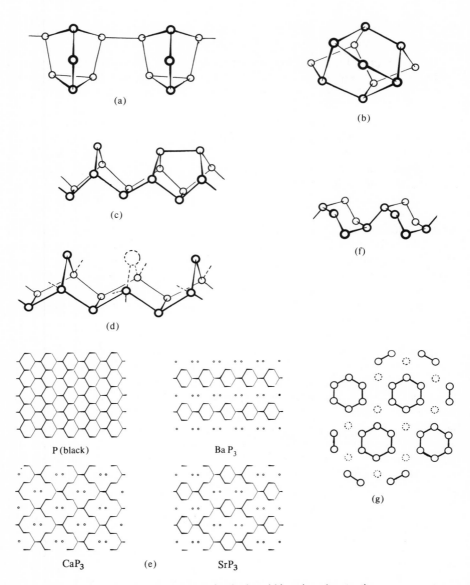

FIG. 19.4.   Structures of polyphosphide anions (see text).

containing compounds related structurally to the polyphosphides and to the polymorphs of P include Ba_2P_7Cl, which contains P_7^{3-} ions like those in Ba_3P_{14}, and $Au_7P_{10}I$,[1] in which some of the Au atoms are incorporated in the phosphorus layer. Examples of polyphosphides in which the anion is a 3D framework built of P atoms forming 2 or 3 bonds include LiP_5 and LaP_5.

(1)   AC 1979 **B35** 573

Polyphosphanes and related compounds

Just as the carbon skeletons in certain cyclic and polycyclic hydrocarbons are portions of the graphite or diamond structures so it is likely that there are many polyphosphanes related in this way to the P_n complexes in phosphides. For example, Rb_3P_7 and Na_3P_{11} hydrolyse rapidly to P_7H_3 and $P_{11}H_3$. For the simpler phosphanes and related compounds see p. 847.

We have already noted that the P_7^{3-} ion has a structure similar to that of P_4S_3. The same P_7 nucleus is found in compounds such as $(Me_3Si)_3P_7$,[1] where $(CH_3)_3Si-$ is bonded to the P atoms which are 2-connected in P_7^{3-}, and in the ion P_{16}^{2-}. This remarkable ion has been isolated in the salt $(\phi_4P)_2P_{16}$, which is prepared from Na_3P_7 and $(\phi_4P)Cl$ in tetrahydrofurane. It consists of two P_7 units linked through a pair of P atoms.[2]

$$P_{16}^{2-}$$

(1)   NW 1975 **62** 573   (2)   AnCIE 1981 **20** 594

The structures of simple molecules

Molecules PX_3

If p orbitals are used in simple molecules PX_3 interbond angles not much greater than 90° are expected. A value of 93.8° is found in PH_3 and angles around 100° in many pyramidal molecules (Table 19.4). It is of interest that PF_3 behaves like CO in forming complexes:

$Ni(PF_3)_4$	compare	$Ni(CO)_4$
$Pt(PF_3)_2Cl_2$		$Pt(CO)_2Cl_2$
$F_3P.BH_3$		$OC.BH_3$

```
    F                 1·207 A   H
      \                        /
  100°  \      1·836 A        /115°
   F ——— P ——————————— B ——— H
      /                        \
    F  1·538 A                   H
```

TABLE 19.4

Structural data for molecules PX₃, POX₃ and PSX₃, and PX₅

Molecules PX₃ etc.

Bond	Length (Å)	X—P—X	Method	Reference
P—H	1.4206 ⎫ 1.437 ⎭	93.8°	m.w. e.d.	⎫ JCP 1959 **31** 449
P—F	1.570	97.8°	e.d.	IC 1969 8 867
HFP—F	1.58	99°	m.w.	JACS 1968 **90** 1705
P—Cl	2.04	100.3°	m.w.	JCP 1962 **36** 589
P—Br	2.20	101.0°	e.d./m.w.	IC 1971 **10** 2584
	2.21	101°	X	AC 1979 **B35** 546
P—I	2.43	102°	e.d.	AC 1950 3 46
	2.46	102°	X	IC 1976 **15** 780
P—CH₃	1.843	98.9°	e.d.	⎫ JCP 1960 **32** 512, 832
H₂P—CH₃	1.858	—	e.d.	
P—C₆H₅	1.828	—	X	JCS 1964 3799
P—SiH₃	2.248	96.5°	e.d.	JCS A 1968 3002

Molecules POX₃ and PSX₃

	P—X	P—O (S)	X—P—X	Method	Reference
POF₃	1.524	1.436	101.3°		For references see:
POCl₃	1.993	1.449	103.3°	e.d.	IC 1971 **10** 344
PSF₃	1.53	1.87	100.3°		ICA 1976 **16** 29
PSCl₃	2.011	1.885	101.8°		
POCl₃ (cryst.)	1.98	1.46	(105°)	X	AC 1971 **B27** 1459
POBr₃ (cryst.)	2.14	1.44	(105½°)	X	AC 1969 **B25** 974

Molecules PX₅

		Method	Reference
P—F	1.577 (axial) 1.534 (equatorial)	e.d.	IC 1965 4 1775
P—Cl	2.14 (axial) 2.02 (equatorial)	e.d.	See: IC 1971 **10** 344

The d_π bonding postulated in the Ni and Pt compounds (p. 1244) cannot explain the shortness of the P—B bond in $F_3P.BH_3$ compared with that in BP (zinc-blende structure), 1.96 Å, or $H_3P.BH_3$, 1.93 Å (ref. p. 1064).

HCP *and the* PH₂⁻ *ion*

Many simple molecules containing P (e.g. the hydrides) are much less stable than the corresponding N compounds. This is also true of methinophosphine, HCP, the analogue of HCN. Prepared by passing PH_3 through a specially designed carbon arc, this reactive gas readily polymerizes; it is converted by HCl into $H_3C.PCl_2$ (H—C, 1.07 Å, C≡P, 1.54 Å).[1]

The PH_2^- ion, the effective radius of which has been estimated as 2.12 Å, occurs in KPH_2 and $RbPH_2$, both of which have distorted NaCl structures.[2]

(1)   JCP 1964 **40** 1170   (2)   AC 1962 **15** 420

Hydrides and molecules P_2X_4 and P_3X_5

In addition to PH_3 there are a number of lower hydrides.[1] P_2H_4 decomposes in the presence of water to a solid $P_{12}H_6$, but in the dry state to P_9H_4 or non-stoichiometric hydrides. P_3H_5 has been prepared by the photolysis of P_2H_4.[2] P_2H_4 has the *trans* configuration with P—P, 2.20 and P—H, 1.42 Å.[2a] The substituted diphosphines $P_2(CH_3)_4$ and $P_2(CF_3)_4$ have been prepared, and structural studies have been made of compounds $P_2R_4Y_2$ where R is an organic radical (CH_3, C_6H_5, etc.) and Y is S, BH_3, or $Fe(CO)_4$, for example;

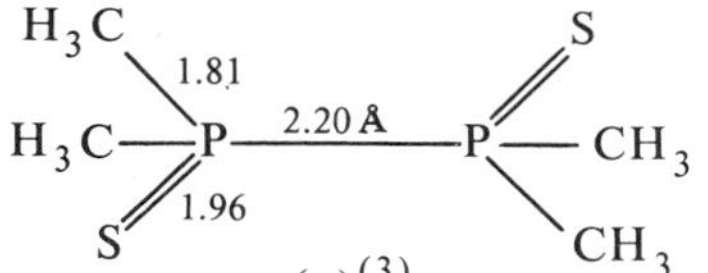

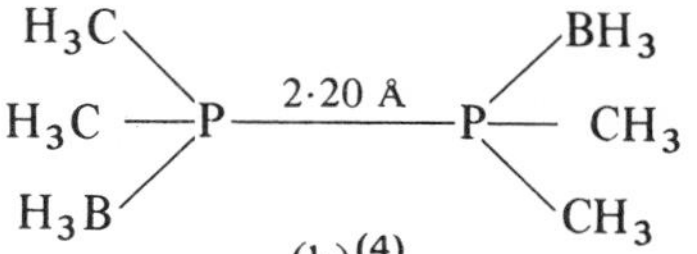

Bond lengths include: P—C, 1.83 Å; P—S, 1.94 Å; P—B, 1.95 Å; and P—P, 2.20 Å. In contrast to the symmetrical structure of $P_2S_2(CH_3)_4$ the analogous As compound has the unsymmetrical structure $(CH_3)_2S.As.S.As(CH_3)_2$ (p. 910). Infrared and ^{31}P n.m.r. suggest the symmetrical structure for $P_2I_4S_2$, which is made from the elements in CS_2 solution.[6]

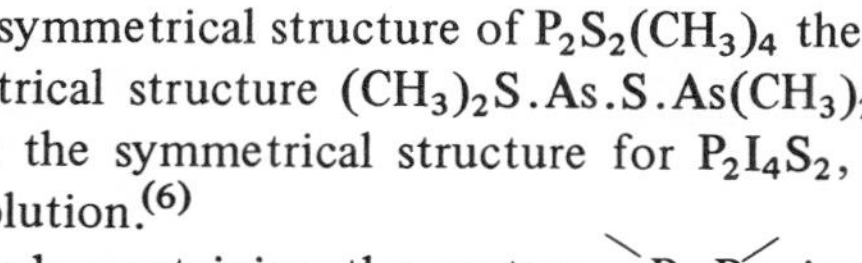

Other compounds containing the system ＞P—P＜ include the dihalides P_2X_4, hypophosphoric and diphosphorous acids and their ions (see later), the infinite chain in CuBr [$Et_2P—PEt_2$] (p. 1109), and salts containing the $P_2S_6^{4-}$ ion (p. 867).

Three dihalides are known, P_2F_4, P_2Cl_4, and P_2I_4. The fluoride is made from PF_2I and Hg, the chloride by subjecting a mixture of PCl_3 and H_2 to an electric discharge, and the iodide directly from the elements in CS_2 solution. The P_2I_4 molecule in the crystalline iodide is centrosymmetrical, with the structure shown in Fig. 19.5.[7] The bond lengths are: P—P, 2.21; P—I, 2.48 Å; and mean angle I—P—P,

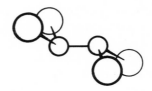

FIG. 19.5. The structure of the P_2I_4 molecule.

94°. An e.d. study of P_2F_4[8] gave the following details: P–P, 2.28; P–F, 1.587 Å; P–P–F, 95.4°; and F–P–F, 99.1°.

(1) JACS 1956 **78** 5726
(2) JACS 1968 **90** 6062
(2a) JACS 1974 **96** 2688
(3) AC 1971 **B27** 302
(4) AC 1968 **B24** 699

(5) JCS A 1968 622
(6) IC 1964 **3** 780
(7) JPC 1956 **60** 539. For further references to
     these compounds see: IC 1969 8 2086, 2797
(8) IC 1975 **14** 599

Phosphoryl and thiophosphoryl halides

Results of electron and X-ray diffraction studies of these molecules are listed in Table 19.4. All these molecules are tetrahedral, but since the X–P–X angles range from 100° to 108° the tetrahedra are not regular. The P–O bond length is close to 1.45 Å. In the Raman spectrum of solid $POBr_3$ there is an extra vibrational line indicating a lowering of the molecular symmetry, which is C_{3v} for the free molecule. This is attributed to weak charge-transfer bonds between O and Br atoms of different molecules, as indicated by the distance 3.08 Å in the infinite chains. All other intermolecular Br–O distances are greater than the sum of the van der Waals radii (3.35 Å). The intermolecular Cl---O distance in the chains in crystalline $POCl_3$ (3.05 Å) may indicate an interaction slightly stronger than van der Waals bonds. In

the thiophosphoryl halides the P–S bond length (1.87 Å) is, like the P–O bond length in the phosphoryl halides, close to that expected for a double bond. For other phosphorus thiohalides see p. 869.

$POCl_3$ and certain other molecules mentioned below react with some tetra- and penta-halides to form molecules in which an octahedral group around the metal atom has been completed by one or two O atoms of $POCl_3$ (or other) molecules.

(a) SbCl$_5$.POCl$_3$ (b) SnCl$_4$.2POCl$_3$ (c) (TiCl$_4$.POCl$_3$)$_2$

FIG. 19.6.   Molecular structures: (a) SbCl$_5$.POCl$_3$;   (b) SnCl$_4$.2POCl$_3$;   (c) (TiCl$_4$.POCl$_3$)$_2$
(O atoms shaded).

Figure 19.6 shows examples of three molecules of this general type:

(a) MCl$_5$.POCl$_3$ (M = Sb, Nb, Ta),[1] and SbCl$_5$.OP(CH$_3$)$_3$[1];
(b) MCl$_4$.2POCl$_3$ (M = Sn,[2] Ti[3]);
(c) (MCl$_4$.POCl$_3$)$_2$ (M = Ti[4]).

In these adducts there is an increase in the c.n. of the acceptor atom while the structure of the donor molecule is essentially unchanged. Note, however, the remarkable difference between the Sb–O bond lengths in the molecules SbCl$_5$.POCl$_3$ and SbCl$_5$.OP(CH$_3$)$_3$, 2.17 and 1.94 Å respectively (Sb–Cl, 2.34 Å). In the POCl$_3$ adducts the O bond angle lies in the range 140–152°. It is interesting that SnCl$_4$.2POCl$_3$ has, like SnCl$_4$.2SeOCl$_2$, the *cis* configuration, in contrast to molecules such as *trans*-GeCl$_4$(pyridine)$_2$.

The O-bridged molecule (a) has been identified by its ^{31}P n.m.r. spectrum.[5]

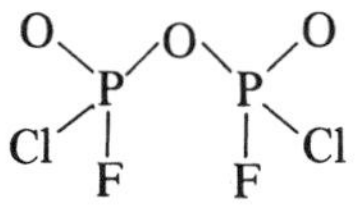

(a)

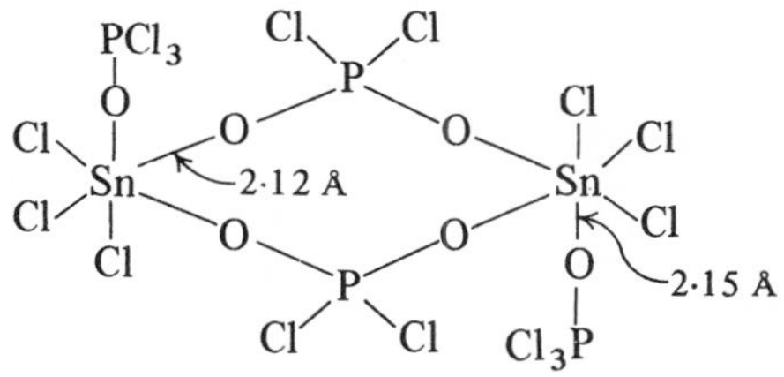

(b)

In the molecules of Fig. 19.6 POCl$_3$ is attached to a metal atom through the single O atom. The action of Cl$_2$O on a solution of SnCl$_4$ in POCl$_3$ gives a compound with the formula (SnO$_3$P$_2$Cl$_8$)$_2$. In addition to POCl$_3$-ligands on the metal atoms the latter are also bridged by tetrahedral O$_2$PCl$_2$ groups, (b), so that a (puckered) 8-membered ring is formed.[6] Bridges of the same kind occur in the compound Mn(PO$_2$Cl$_2$)$_2$(CH$_3$COOC$_2$H$_5$)$_2$[7] formed by the action on MnO of POCl$_3$ dissolved in ethyl acetate. The bridging PO$_2$Cl$_2$ groups lead to infinite chains (c) of octahedral coordination groups in which the $CH_3COOC_2H_5$ molecules occupy *cis* positions and successive 8-membered rings are in nearly perpendicular planes. There are chains of the same kind in $Mg(PO_2Cl_2)_2(POCl_3)_2$ with two $OPCl_3$ groups on each metal atom.[8]

(c)

(1) ACSc 1963 **17** 353	(5) IC 1964 **3** 280
(2) ACSc 1963 **17** 759	(6) AC 1969 **B25** 1720, 1726
(3) ACSc 1962 **16** 1806	(7) ACSc 1963 **17** 1971
(4) ACSc 1960 **14** 726	(8) ACSc 1970 **24** 59

Other tetrahedral molecules and ions

Further examples of tetrahedral molecules are shown at (a)[1] and (b).[2] Others which have been studied structurally include $F_3P.BH_3$,[3] $PS(OCH_3)(C_6H_5)_2$,[4a] $OP(NH_2)_3$,[4b] and many oxy- and thio-ions described later in this chapter.

(a)   (b)

The phosphorus analogue of the ammonium ion, the phosphonium ion, has a regular tetrahedral structure in the salts $(PH_4)X$. A n.d. study of PH_4I gives P–H, $1.414\,Å$ and shows that the P–H bonds are directed towards I^- ions. The anion has 8 H neighbours at the corners of a distorted cube, 4 at $2.87\,Å$ and 4 at $3.35\,Å$; the atoms P–H–I are practically collinear $(172°)$.[5] Ions PX_4^+ are included in the next section.

(1) IC 1968 **7** 2582	(4a) AC 1969 **B25** 617
(2) ACSc 1965 **19** 879	(4b) JCS A 1969 1804
(3) JCP 1967 **46** 357	(5) JCP 1964 **47** 1818

Phosphorus pentahalides, PX_4^+ *and* PX_6^- *ions*

The stability of the pentahalides decreases rapidly with increasing atomic weight of the halogen. The pentafluoride is stable up to high temperatures; PCl_5 is about half dissociated at $200\,°C$; PBr_5 is less stable, and PI_5 is not known. In the vapour state PF_5 and PCl_5 have been shown to exist as trigonal bipyramidal molecules, the stereochemistry of which has been discussed in Chapter 7. The axial bonds are longer than the equatorial ones as in other molecules of this type (see below). Ionization presumably takes place in nitrobenzene, for solutions of PCl_5 in this solvent have appreciable electrical conductivity, and crystalline PCl_5 is built of tetrahedral PCl_4^+ and octahedral PCl_6^- ions, which are packed together in much the same way as the ions in $CsCl$.[1] The PCl_4^+ ion also exists in the crystalline compounds PCl_6I,[2] $(PCl_4)FeCl_4$,[3] $(PCl_4)(Ti_2Cl_9)$, and $(PCl_4)_2(Ti_2Cl_{10})$; for the third compound see p. 465. PCl_6I, which is prepared by direct union of PCl_5 and ICl in CS_2 or CCl_4, ionizes in polar solvents and consists of PCl_4^+ and linear ICl_2^- ions. Crystalline $(PCl_4)FeCl_4$ is an aggregate of PCl_4^+ ions (P–Cl, 1.91 Å) and tetrahedral $FeCl_4^-$ ions (Fe–Cl, 2.19 Å). For $(PCl_4)(TeCl_5)$ see p. 710.

In contrast to the pentachloride PBr_5 crystallizes as $(PBr_4)Br$, the P–Br bond length in the PBr_4^+ ion being 2.15 Å.[4] The same cation occurs in PBr_7,[5] formed by the action of Br_2 on PBr_5, in which the anion is the linear (unsymmetrical) Br_3^- ion, and also presumably in compounds such as PBr_6I and PBr_5ICl.

Of metallic salts $M(PX_6)_n$ only hexafluorophosphates appear to be known, and of these the salts with large cations such as NH_4^+, K^+, Cs^+, Ba^{2+}, and $Co(NH_3)_6^{3+}$, are the most stable in the solid state. The first three have been shown to crystallize with a NaCl-like packing of M^+ and octahedral PF_6^- ions.[6] The crystal structure of $HPF_6.6H_2O$ is described on p. 661. In the PF_6^- ion in this crystal P–F was found to be 1.73 Å, but in $NaPF_6$[7] the length of this bond is apparently only 1.58 Å, and this is confirmed as the mean value in $(CH_3)_4N.PF_6$ (axial, 1.59, equatorial, 1.57 Å). Moreover, in $NaPF_6.H_2O$[9] the ion was found to be distorted, with four (equatorial) P–F bonds of length 1.58 Å (these F atoms have a Na^+ ion as nearest external neighbour) and two of 1.73 Å (these F atoms having a H_2O neighbour at 2.90 Å).

(1) JCS 1942 642
(2) JACS 1952 **74** 6151
(3) IC 1968 **7** 2150
(4) AC 1970 **B26** 443
(5) AC 1967 **23** 467

(6) ZaC 1951 **265** 229
(7) ZaC 1952 **268** 20
(8) AC 1980 **B36** 1523
(9) AC 1956 **9** 825

Molecules PR_5, $PR_{5-n}X_n$, *and mixed halides*

Molecules in which five independent ligands are attached to P have trigonal bipyramidal configurations, and in PR_5 and PX_5 the axial bonds are longer than the equatorial ones. For example, in $P(C_6H_5)_5$[1] the length of the axial bonds is 1.99 Å as compared with 1.85 Å for the equatorial bonds; the latter have the same length as in $P(C_6H_5)_3$. Assuming that the more significant steric interactions are those between groups at 90° to one another and that the magnitude of ligand repulsions increases in the order F–F < F–R < R–R it is expected that in molecules PF_nR_{5-n} the groups R will preferentially occupy equatorial positions. This has been confirmed (e.d.) for CH_3PF_4 and $(CH_3)_2PF_3$[2] and for CF_3PF_4.[3] The first two of these compounds have structures very similar to those of the corresponding molecules formed by S and Cl with respectively one and two lone pairs (Fig. 19.7). There is a

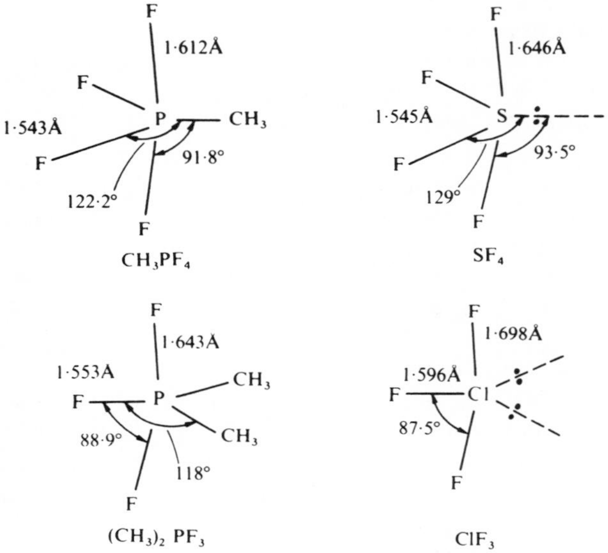

FIG. 19.7. Structures of CH_3PF_4, $(CH_3)_2PF_3$ and other molecules.

considerable and steady increase in P–F$_{equat.}$ and in P–F$_{axial}$ with increasing number of CH_3 groups in molecules $(CH_3)_nPF_{5-n}$ (n, 0–3).[3a] The values for the axial bonds range from 1.57 ($n = 0$) to 1.685 ($n = 3$) with the intermediate values for $n = 1$ and 2 shown in Fig. 19.7. An equatorial position is occupied by H in PHF_4,[4] but in many molecules of this general type there is apparently rapid interchange between axial and equatorial positions at ordinary temperatures. An extensive review of 5-coordination is available.[5] The molecule (a), with three isopropoxy groups and one phenanthrene quinone molecule bonded to P, is of special interest as an example of a trigonal bipyramidal arrangement of 5 O atoms around P. There is very little distortion of the bond angles from the ideal values (90° and 120°).[6] In molecules such as $(CH_3)P(C_6H_4O_2)_2$, (b),[7] containing two chelate ligands, on the other hand, the coordination tends towards square (or more strictly, rectangular) pyramidal. In this molecule all the angles $H_3C–P–O$ are close to 104°, while the diagonal O–P–O angle is 152.5°.

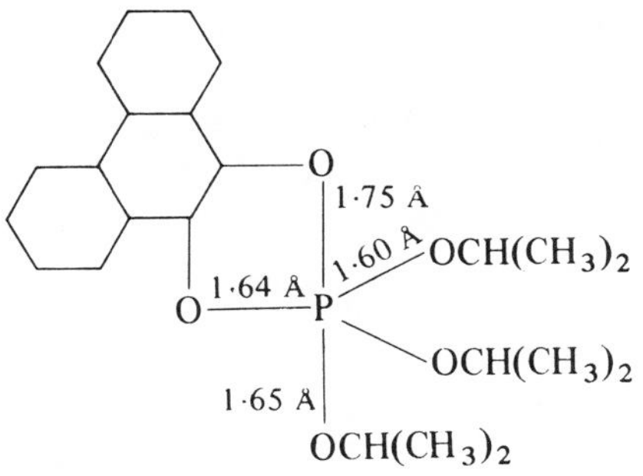

(a)

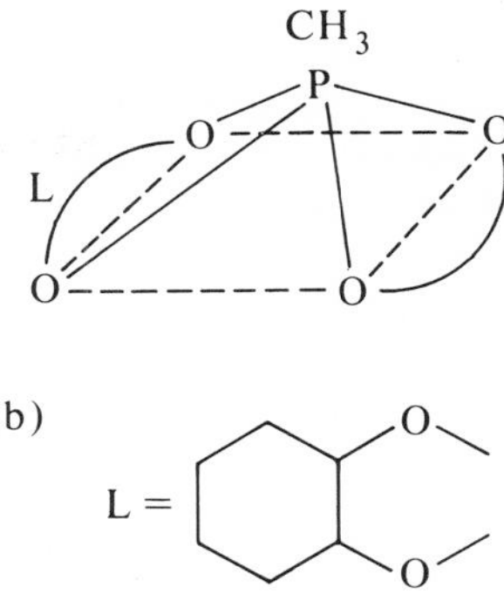

(b)

L =

All the mixed halides PCl_nF_{5-n} have been prepared. They have low-temperature forms in which they exist as trigonal bipyramidal molecules and these rearrange slowly at room temperature to form ionic crystals. For example, PCl_2F_3 is a gas at room temperature while the ionic form $(PCl_4)^+PF_6^-$ is a hygroscopic salt which sublimes, with some decomposition, at $135\,°C$. At temperatures above $70\,°C$ it changes into $PF_5 + PCl_4F$, and the latter exists both as a non-polar liquid and also as an ionic solid, $(PCl_4)F$.[8]

(1) JCS 1964 2206
(2) IC 1965 **4** 1777
(3) JCP 1968 **49** 1307
(3a) JMSt 1973 **15** 209
(4) JCP 1968 **48** 2118
(5) QRCS 1966 **20** 245
(6) JACS 1967 **89** 2267, 2272
(7) AC 1974 **B30** 935, 939
(8) ZaC 1957 **293** 147; IC 1965 **4** 738

The oxides and oxysulphide

In addition to the trioxide and pentoxide there are crystalline phases with compositions in the range $PO_{2.0}$–$PO_{2.25}$ which behave chemically as compounds containing P(III) and P(v). For example, they hydrolyse to mixtures of phosphorous and phosphoric acids. It is convenient to dicuss these phases after the trioxide and pentoxide.

Phosphorus trioxide

This oxide exists in the vapour state as tetrahedral molecules, P_4O_6, at moderate temperatures; the structure of crystalline P_2O_3 is not known. The results of e.d. studies of the vapour of this oxide and related molecules are given in Table 19.5, and the structure of the P_4O_6 molecule is illustrated in Fig. 19.8.

Phosphorus pentoxide

The structural chemistry of this oxide in the solid and liquid states is complex. In addition to a high-pressure form[1] and a glass there are three polymorphs stable at atmospheric pressure. As noted in Chapter 3 these three crystalline forms represent

TABLE 19.5

Observed interatomic distances and interbond angles

Molecule	M–O$_b$ (Å)	M–O$_t$ or M–S (Å)	Angle		Reference
			M–O–M	O–M–O	
P$_4$O$_6$	1.638	–	126°	100°	TFS 1969 **65** 1219
P$_4$O$_{10}$	1.604	1.429	124°	O$_b$PO$_b$ 102° O$_b$PO$_t$ 117°	TFS 1967 **63** 836
P$_4$O$_6$S$_4$	1.61	1.85	124°	O$_b$PO$_b$ 102° O$_b$PS 117°	JACS 1939 **61** 1130
As$_4$O$_6$	1.78	–	128°	99°	JACS 1944 **66** 818
Sb$_4$O$_6$	2.00	–	129°	98°	ZSK 1961 **2** 542

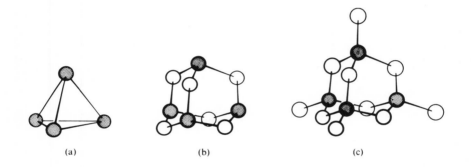

(a)　　　　　　　(b)　　　　　　　(c)

FIG. 19.8. The structures of the molecules (a) P$_4$; (b) P$_4$O$_6$; and (c) P$_4$O$_{10}$ in the vapour state (diagrammatic). Shaded circles represent P atoms.

different ways of linking tetrahedral PO$_4$ groups through *three* vertices to form (a) finite P$_4$O$_{10}$ molecules[2] of the same form as those in the vapour (Fig. 19.8(c)); (b) a layer structure[3] based on the simplest planar 3-connected net; and (c) a 3-dimensional framework[4] built of rings of ten PO$_4$ groups based on one of the two simplest 3D 3-connected networks. The accurate data for the layer structure (P–O, 1.56 Å; P–O′, 1.49 Å; and P–O–P 145°) may be compared with those for the P$_4$O$_{10}$ molecule given in Table 19.5. These crystal structures throw light on the physical and chemical properties of the three polymorphs. The metastable form, consisting of discrete P$_4$O$_{10}$ molecules, is the most volatile and hygroscopic; it is formed by condensation of the vapour. The rearrangement of the PO$_4$ tetrahedra required to form the infinite networks of the other two forms does not take place if the vapour is rapidly condensed; it occurs only if the molten oxide is maintained at a high temperature for a considerable time. These forms are much less volatile and less rapidly attacked by water because the extended systems of linked tetrahedra must be broken down when the crystal melts, vaporizes, or reacts with water.

The P_4O_{10} molecule has been studied in the vapour by e.d. with the results summarized in Table 19.5. The outer (unshared) O atoms are distinguished as O_t and the bridging O atoms as O_b. The (multiple) bond $P–O_t$ is the shortest recorded phosphorus–oxygen bond.

(1) JCP 1961 **35** 1271   (3) AC 1964 **17** 679
(2) AC 1964 **17** 677   (4) RTC 1941 **60** 413

$P(III)P(V)$ *oxides*

All three oxides intermediate between P_4O_6 and P_4O_{10}, namely, P_4O_7, P_4O_8, and P_4O_9, are known as stoichiometric crystalline solids, and their molecular structures have been determined.[1] Solid solutions consisting of molecules of more than one kind have also been studied, for example, those with compositions in the range $PO_{2.02}–PO_{2.25}$ (P_4O_8 and P_4O_9) or $PO_{2.0}–PO_{1.93}$ (statistical arrangement of P_4O_8 and P_4O_7 molecules).[2] The molecules P_4O_7, P_4O_8, and P_4O_9 are all structurally similar to P_4O_6 and P_4O_{10}, with the appropriate number of terminal O atoms, and the corresponding bond lengths and interbond angles are very similar. Bond lengths additional to those in Table 19.5 are those in the unsymmetrical bridges between $P(III)$ and $P(V)$ atoms, namely, $P(III)–O_b$, 1.59 Å; and $P(V)–O_b$, 1.67 Å.

(1) AC 1981 **B37** 222   (2) AC 1964 **17** 1593; AC 1966 **21** 34

Phosphorus oxysulphide

The oxysulphide $P_4O_6S_4$, formed by heating P_4O_6 with sulphur, forms a tetrahedral molecule similar to that of P_4O_{10}. The P–S bond length (1.85 Å) is similar to that in the thiophosphoryl halides.

Molecules of the same geometrical type as P_4O_{10}

It is convenient to mention here two molecules which are structurally similar to P_4O_{10}. The basal ring of P_4O_{10} can be replaced by a cyclohexane ring, as in the phosphoric ester and its S analogue[1] which is illustrated in Fig. 19.9. The action of

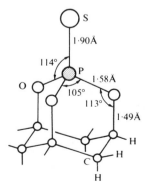

FIG. 19.9. The molecular structure of $C_6H_9O_3PS$.

excess $Ni(CO)_4$ on P_4O_6 yields $P_4O_6[Ni(CO)_3]_4$, in which a $Ni(CO)_3$ group is attached to each P.[2]

(1)   ACSc 1960 **14** 829   (2)   AC 1966 **21** 288

The oxy-acids of phosphorus and their salts

The number and stability of the oxy-acids of phosphorus are in marked contrast to those of nitrogen. The formulae and basicities of the acids are set out in Table 19.6.

TABLE 19.6
Oxy-acids of phosphorus

Acid	Formula	Basicity	Ion	Reference
Orthophosphorous	H_3PO_3	2	HPO_3^{2-}	AC 1969 **B25** 227
Hypophosphorous	H_3PO_2	1	$H_2PO_2^-$	AC 1979 **B35** 1041
Pyrophosphorous	$H_4P_2O_5$	2	$H_2P_2O_5^{2-}$	JACS 1957 79 2719
Diphosphorous	$H_4P_2O_5$	3	$HP_2O_5^{3-}$	JACS 1957 79 2719
	$H_5P_3O_8$	5	$P_3O_8^{5-}$	AC 1969 **B25** 1077
	$H_6P_6O_{12}$	6	$P_6O_{12}^{6-}$	ZaC 1960 **306** 30
Hypophosphoric	$H_4P_2O_6$	2	$H_2P_2O_6^{2-}$	AC 1964 17 1352
	$H_4P_2O_6 \cdot 2H_2O$	2	$H_2P_2O_6^{2-}$	AC 1971 **B27** 1520
Isohypophosphoric	$H_4P_2O_6$	3	$HP_2O_4^{3-}$	IC 1967 6 1137
Orthophosphoric	H_3PO_4	3	PO_4^{3-}	
Metaphosphoric	HPO_3	1	$(PO_3)_n^{n-}$	see text
Pyrophosphoric	$H_4P_2O_7$	4	$P_2O_7^{4-}$	

The so-called 'per-acids', H_3PO_5 and $H_4P_2O_8$, are omitted as nothing is known of their structures. The latter is presumably analogous to $H_2S_2O_8$. The structures, and indeed the existence, of meta- and pyro-phosphorous acids must still be regarded as doubtful. Crystalline anhydrous pyrophosphites are well known, for example, $Na_2(H_2P_2O_5)$, which is prepared by heating Na_2HPO_3. By analogy with ortho-phosphorous acid the pyro-acid would be

$$\left[\begin{matrix} O \\ H{>}P{-}O{-}P{<}H \\ O \qquad\quad O \end{matrix} \right] H_2$$

a formulation which is supported by the n.m.r. spectrum of $Na_2(H_2P_2O_5)$, which shows that the ion is symmetrical and that there is one H attached to each P atom.

Apart from the metaphosphoric acid, which is usually obtained as a glass, the other acids in Table 19.6 can be obtained crystalline at ordinary temperatures. Nevertheless, the crystal structures of only two of the anhydrous acids, H_3PO_3 and H_3PO_4, have been studied.

Orthophosphorous acid

A striking feature of the phosphorous acids is their abnormal basicities. We might have expected H_3PO_3 to form salts containing pyramidal PO_3^{3-} ions analogous to the pyramidal SO_3^{2-} and ClO_3^- ions, but the acid is in fact dibasic. Normal salts are of the type $Na_2(HPO_3)$ and $Ba(HPO_3)$, though acid salts such as $NaH(HPO_3)$ have also been prepared.

In crystalline phosphorous acid the P and three O atoms form a pyramidal group, and the fourth tetrahedral position is occupied by a H atom. Unlike the other two H atoms this third one does not take part in the hydrogen-bonding scheme. The H atoms were not located in the X-ray study, but their positions were inferred from the short intermolecular contacts O---H---O (around 2.60 Å); they have been confirmed by a n.d. study. The structure of the phosphorous acid molecule, $HPO(OH)_2$, is shown at (a).

Normal phosphites contain the ion $(HPO_3)^{2-}$, (b), which has been shown to have trigonal symmetry in $Mg(HPO_3).6H_2O^{(1)}$ with P–O, 1.51 Å and O–P–O angles of 110°. We give in Table 19.6 the reference to $Cu(HPO_3).2H_2O$, which includes references to work on H_3PO_3.

Hydrogen or 'acid' phosphites contain the ion $[HPO_2(OH)]^-$, (c), which has been studied in $LiH_2PO_3^{(2)}$ and $Cd(H_2PO_3)_2.H_2O.^{(3)}$

(1) AC 1956 9 991   (3) AC 1978 **B34** 32
(2) AC 1976 **B32** 412

Hypophosphorous acid

The structure of the monobasic H_3PO_2 (or $H(H_2PO_2)$) is not known, but several salts have been studied by X-ray or neutron diffraction, and the positions of the H atoms determined. The ion has the distorted tetrahedral structure (d). The reference given in Table 19.6 (for $NaH_2PO_2.\frac{4}{5}H_2O$) contains a review of earlier work.

We come now to a group of acids in which there are two or more P atoms directly bonded.

Hypophosphoric acid

Cryoscopic measurements show that the molecular weights of the ethyl ester and of the free acid correspond to the formulae $(C_2H_5)_4P_2O_6$ and $H_4P_2O_6$ respectively.

The magnetic evidence is conclusive on this point. The molecule H_2PO_3 would be paramagnetic since there would be an unpaired electron, but the Na and Ag salts are diamagnetic. An X-ray study of $(NH_4)_2H_2P_2O_6$ shows that the ion has the staggered configuration. The bond lengths P–OH and P–O are very similar to those in KH_2PO_4 (1.58 and 1.51 Å) and H_3PO_4 (1.57 and 1.52 Å); the ion has a similar structure in the 'dihydrate' of the acid, which is $(H_3O)_2^+(H_2P_2O_6)^{2-}$. For isohypophosphoric acid see later.

Diphosphorous acid

A salt described as a diphosphite, $Na_3HP_2O_5.12H_2O$, has been prepared by hydrolysing PBr_3 with ice-cold aqueous $NaHCO_3$. Its n.m.r. spectrum is consistent with the structural formula shown.

$H_5P_3O_8$ *and* $H_6P_6O_{12}$. The ion $P_3O_8^{5-}$ has been studied in $Na_5P_3O_8.14H_2O$. The P–P bond has the length of a normal single bond in this ion, in which P atoms are in the formal oxidation states III, IV, and III.

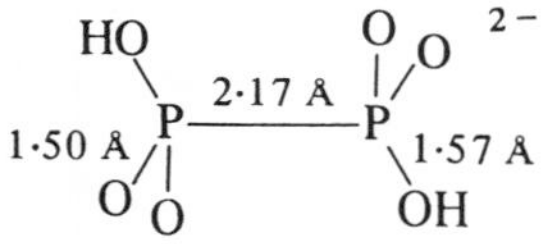

$$H_2P_2O_6^{2-}$$

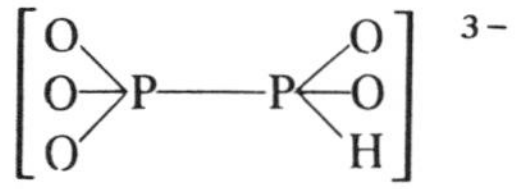

$$HP_2O_5^{3-}$$

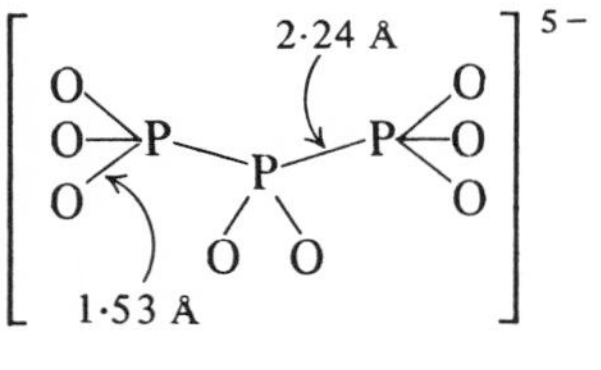

$$P_3O_8^{5-}$$

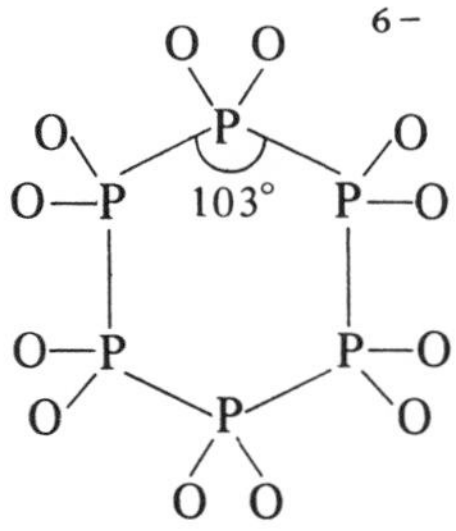

$$P_6O_{12}^{6-}$$

Salts of an acid $H_6P_6O_{12}$ have been prepared by treating red P suspended in alkali hydroxide solution with hypochlorite. An X-ray study of the Cs salt shows the ion to contain a chair-shaped ring of six directly bonded P atoms in which P–P is approximately 2.2 Å.

Isohypophosphoric acid

The hypophosphoric acid of an earlier paragraph results from the oxidation of P by moist air. The mild alkaline hydrolysis of PCl_3 gives a complex mixture of products which have been separated chromatographically. One component has been identified as isohypophosphoric acid, and the free acid has been prepared from PCl_3 and H_3PO_4. The trisodium salt, $Na_3(HP_2O_6).4H_2O$, results from heating a mixture of Na_2HPO_4 and NaH_2PO_3. The n.m.r. spectrum is consistent with the structures (a) and (b) for the ion and the acid.

(a)   (b)

Phosphoric acid and phosphates

We now come to the extensive oxygen chemistry of P(v) based on discrete PO_4^{3-} ions or on tetrahedral PO_4 groups sharing one or two O atoms; sharing of three O atoms by all PO_4 tetrahedra leads to the neutral oxide P_2O_5. Discrete PO_4^{3-} ions are found in the orthophosphates and systems of linked PO_4 groups in pyro-, poly-, and metaphosphates.

Orthophosphoric acid and orthophosphates

Discrete PO_4^{3-} ions exist in normal orthophosphates. In dihydrogen phosphates such as KH_2PO_4 and $(N_2H_5)H_2PO_4$ $PO_2(OH)_2$ units are hydrogen-bonded to form extended anions—see p. 373. Several X-ray and n.d. studies[1] give P—O close to 1.51 Å and P—OH 1.55-1.58 Å. The length of the hydrogen bonds is around 2.50 Å. In the (layer) structure of H_3PO_4[2] there are also hydrogen bonds of length 2.84 Å; this structure is described on p. 375. The P—O bond lengths found in H_3PO_4 and its hemihydrate[3] are P—O, 1.52 and 1.49 Å, and P—OH, 1.57 and 1.56 Å respectively.

The structures of a number of orthophosphates are similar to those of forms of silica or of silicates. For example, $AlPO_4$ crystallizes with all the three normal silica structures[4] and also undergoes a transition from a low- to a high-temperature form in each case; YPO_4 and $YAsO_4$ crystallize with the zircon ($ZrSiO_4$) structure.

(1) AC 1974 **B30** 1195   (3) AC 1974 **B30** 1470
(2) ASSc 1955 **9** 1557   (4) ZK 1967 **125** 134

Pyrophosphates–diphosphates

Pyrophosphates $M^{IV}P_2O_7$, $M_2^{II}P_2O_7$, and $M_4^{I}P_2O_7$ (hydrated) have been studied in some detail, particular interest being centred in the bond angle at the bridging O atom. Salts of divalent metals often have low- (α) and high-temperature (β) forms in which the configuration of the pyrophosphate ion varies with the size of M^{2+}. For small metal ions the configuration is approximately staggered, and for large ions nearly eclipsed.[1] There is considerable variation in the O bond angle, from

123° $(Ca_2P_2O_7.2H_2O)$,[2] 139° in one form of SiP_2O_7,[3] and 156° in α-$Mg_2P_2O_7$, to 180° in phosphates with the thortveitite structure. The apparent collinearity of the O bonds in the high-temperature forms of certain salts $M_2P_2O_7$ is now generally attributed either to positional disorder (random arrangement throughout the crystal of O_b atoms around but off the P–P line, P-----P) or to highly anisotropic motion of O_b; compare high-cristobalite. An early study of the cubic polymorph of ZrP_2O_7 (with which the Si, Ge, Sn, Pb, Ti, Hf, Ce, and U compounds are isostructural) indicated collinear bonds in the P_2O_7 ion. It has now been shown, by a careful study of the high-temperature form of SiP_2O_7[4] that the true unit cell has 27 times the volume of the cell to which the earlier structure was referred, and that the mean P–O–P angle is 150°, though there are still a few apparently collinear O bonds in the structure. There appears to be a correlation between the O bond angle and the difference in length between the terminal and bridging P–O bonds: P–O–P, 123°; P–O_b, 1.62; P–O_t, 1.52 Å; $(Ca_2P_2O_7.2H_2O)$; P–O–P, 156°; P–O_b, 1.58; P–O_t, 1.55 Å (α-$Mg_2P_2O_7$).

Closely related to the $P_2O_7^{4-}$ ion are the ions $O_3P.NH.PO_3^{4-}$ and $O_3P.CO.PO_3^{4-}$:

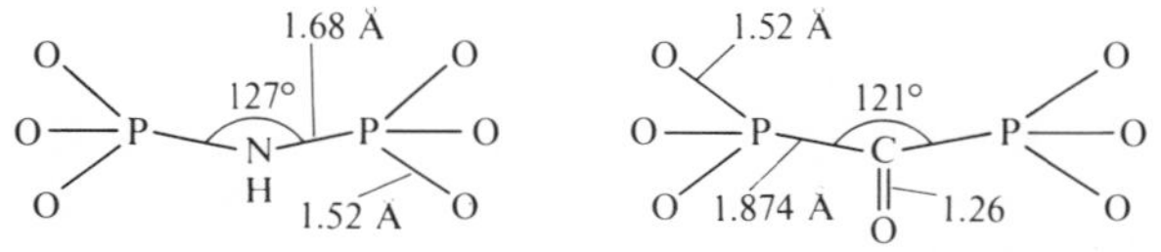

So closely similar are the structures of the O-bridged and NH-bridged ions that the decahydrates of the Na salts are isostructural.[5] The structure of the carbonyl-diphosphonate ion was determined in $Na_4(O_3P.CO.PO_3).2H_2O$.[6]

(1) JSSC 1970 **1** 120
(2) AC 1975 **B31** 1730
(3) AC 1970 **B26** 233

(4) JSSC 1973 7 69
(5) AC 1974 **B30** 522
(6) AC 1976 **B32** 488

Linear polyphosphates

These contain 'hybrid' ions intermediate between the pyrophosphate ion and the infinite linear metaphosphate ion, in which the terminal PO_4 groups share one O and the intermediate groups two O atoms. The normal sodium triphosphate, $Na_5P_3O_{10}$, is known anhydrous and as the hexahydrate; $Na_2H_3P_3O_{10}$ and $Na_3H_2P_3O_{10}$ have also been prepared. The triphosphate, which is used as a detergent in mixtures with soaps and sulphonates, can be prepared in various ways, for example, by heating together

$$2Na_2HPO_4 + NaH_2PO_4 \rightarrow Na_5P_3O_{10} + 2H_2O.$$

X-ray studies of the two crystalline forms of $Na_5P_3O_{10}$[1] show the structure of the ion to be though rather shorter bridging bonds (1.59 and 1.63 Å) were found in $Zn_5(P_3O_{10})_2 \cdot 17H_2O^{(1a)}$ and an oxygen bond angle of 129°.

The linear tetraphosphate ion results from alkaline hydrolysis of the cyclic metaphosphate, and the salts of large ions such as Pb^{2+}, Ba^{2+}, and $N(CH_3)_4^+$ can be crystallized from the acidified solution; they tend to remain in alkaline solution, forming viscous liquids. Material marketed as 'hexasodium tetraphosphate', $Na_6P_4O_{13}$, is not a simple chemical individual but is a mixture of $Na_5P_3O_{10}$ and $NaPO_3$. The acids $H_5P_3O_{10}$ and $H_6P_4O_{13}$ have been obtained by passing solutions of their tetramethylammonium salts through a cation exchange resin. The polyphosphates containing more than four P atoms are not phase-diagram entities,[2] and only Ca hexaphosphate can be prepared relatively easily. However, gram quantities of pure polyphosphates containing 4–8 P atoms have been prepared chromatographically. The basic starting material is a polyphosphoric acid with average chain length around five which results from heating 85 per cent aqueous H_3PO_4 in a gold dish at 400°C for 12 hours. For $(NH_4)_2SiP_4O_{13}$ see p. 984.

The salt $K_4(P_3O_9NH_2) \cdot H_2O$ contains an ion of the same type as $P_3O_{10}^{5-}$ in which one O has been replaced by NH_2.[3]

(1)   AC 1964 **17** 674
(1a) AC 1975 **B31** 2482

(2)   JACS 1967 **89** 2884; ZaC 1964 **330** 78
(3)   AC 1965 **19** 363

Poly ions containing P *and* Cr

Closely related to the diphosphate and linear polyphosphate ions are a number of ions containing both P and Cr in tetrahedral coordination, in which 1, 2, 3, or 4 of the O atoms of a PO_4 group are shared with tetrahedral CrO_4 groups:[1]

$(PCrO_7)^{3-}$ $(PCr_2O_{10})^{3-}$ $(PCr_3O_{13})^{3-}$ $(PCr_4O_{16})^{3-}$

Details of the structures of the first three ions will be found in ref. (1); the $PCr_4O_{16}^{3-}$ ion[2] is similar to the Si_5O_{16} group in the mineral zunyite (p. 1013).

(1)   JSSC 1980 **33** 325

(2)   JSSC 1980 **33** 439

Metaphosphates

In these compounds each PO_4 group shares two O atoms to form rings or chains of composition $(PO_3)_n^{n-}$ analogous to those in metasilicates. Metaphosphoric acid itself has not been obtained crystalline, but as a glass or syrup; it is obviously highly polymerized.

Some metaphosphates are rubbery or horn-like masses and some are mixtures of various metaphosphates. Thus the solid of empirical composition $NaPO_3$ obtained by heating $NaNH_4HPO_4$ or NaH_2PO_4 is not homogeneous, part being soluble and the remainder insoluble in water, but under controlled conditions numerous well defined crystalline salts can be prepared. Also, by rapidly cooling molten $NaPO_3$ to below $200\,°C$ a brittle, transparent, glassy form is obtained, called Graham's salt. A radial distribution function derived from X-ray data for a $NaPO_3$ glass shows that it consists largely of long chains of PO_4 tetrahedra linked by sharing two vertices and held together by O–Na–O bonds. However, these phosphate glasses also contain small quantities of cyclic metaphosphates. The penta- and hexa-metaphosphates have been isolated from Graham's salt, and the presence of the 7- and 8-ring ions proved by paper chromatography. Of these higher cyclic ions the 6-ring is most resistant to hydrolysis.

The two simplest water-soluble (cyclic) sodium metaphosphates, which are important water-softening agents, are $Na_3P_3O_9$, formed by heating NaH_2PO_4 to $550-600\,°C$, and $Na_4P_4O_{12}$, conveniently prepared by treating the soluble form of P_2O_5 with a cold suspension of $NaHCO_3$.

Trimetaphosphates contain cyclic $P_3O_9^{3-}$ ions with the chair configuration and the dimensions shown at (a).[1] An unusual feature of $Na_3P_3O_9.H_2O$[2] is that its crystal structure is almost identical with that of the anhydrous salt. In the structure

(a)   (b)

of the latter there is almost sufficient room for the H_2O molecule, and only a small expansion of the structure is necessary to accommodate it.

The tetrametaphosphate ring, (b), is flexible and adopts a configuration to suit its environment:[3]

$$2/m \quad \text{in} \quad (NH_4)_4P_4O_{12},$$
$$\bar{1} \quad \text{in} \quad Na_4P_4O_{12}.4H_2O,$$
$$\bar{4} \quad \text{in} \quad Tl_4P_4O_{12}.$$

The penta- and hexa-metaphosphate ions have been studied in $Na_4(NH_4)P_5O_{15}$.-$4H_2O^{(4)}$ and $Na_6P_6O_{18}.6H_2O.^{(5)}$ In the latter the six P atoms of the ring are coplanar, and bond lengths are $P-O_t$, 1.47 Å; $P-O_b$, 1.61 Å.

The insoluble metaphosphates contain infinite chain ions with the mean inter-bond angles and bond lengths shown:

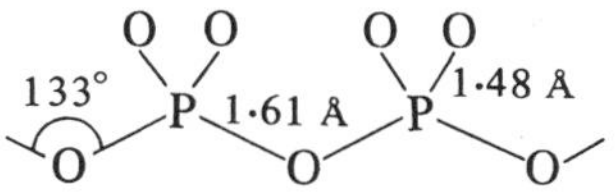

Some of these compound exist in more than one form with different configurations of the $(PO_3)_n^{n-}$ chain, as in the so-called Maddrell and Kurrol salts[6] (two forms of $NaPO_3$). The configurations of the chains in Li, Na, K, Rb, Ag, Pb, and Bi meta-phosphates are discussed on p. 1022. Hydrogen atoms are attached to some O atoms of the chains in salts such as $Na_3H(PO_3)_4^{(7)}$ and $BiH(PO_3)_4.^{(8)}$

Of special interest is the double chain ion in $BiP_5O_{14}^{(8)}$ and certain 4f phos-phates:[9]

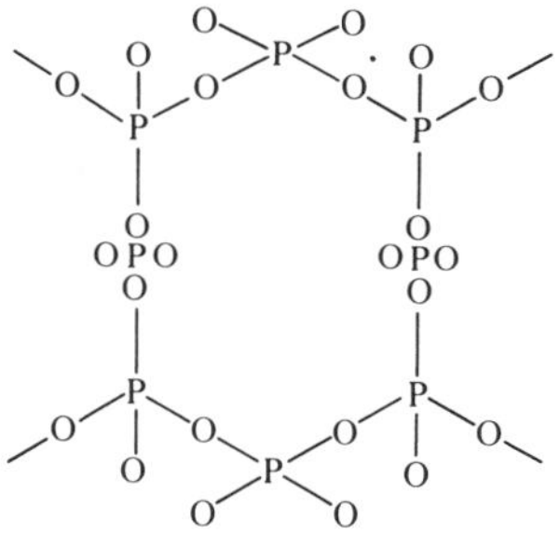

(1) AC 1975 **B31** 2264, 2680
(2) AC 1965 **18** 226
(3) AC 1974 **B30** 1979
(4) AC 1972 **B28** 732
(5) AC 1972 **B28** 2740; AC 1977 **B33** 2529

(6) AC 1968 **B24** 1621
(7) AC 1968 **B24** 992
(8) AC 1975 **B31** 2285
(9) AC 1974 **B30** 468, 1751

Mono- and di-fluorophosphoric acids

Intermediate between the neutral molecule POF_3 and the ion PO_4^{3-} are the ions $PO_2F_2^-$ and PO_3F^{2-}. Salts of both mono- and di-fluorophosphoric acids have been prepared, for example, the NH_4 and K salts and several others. Anhydrous mono-fluorophosphoric acid, which can be prepared in quantitative yield from anhydrous metaphosphoric acid and liquid anhydrous HF, is an oily liquid which sets to a glass at the temperature of solid CO_2 and shows a strong resemblance to H_2SO_4.

Both $BaPO_3F$ and KPO_2F_2 are structurally similar to $BaSO_4$, and $NaK_3(PO_3F)_2$ is isostructural with glaserite, $CaBa_3(SiO_4)_2$. There appears to be a regular trend in bond lengths and interbond angles in the series PO_4^{3-}, PO_3F^{2-}, $PO_2F_2^-$, and POF_3, as indicated by the following mean values:

		P–O (Å)	P–F (Å)	O–P–O	
	PO_4^{3-}	1.55	–	109.5°	
(a)	PO_3F^{2-}	1.51	1.59	114°	AC 1975 **B31** 1533
(b)	$PO_2F_2^-$	1.47	1.57	122°	AC 1975 **B31** 2506
	POF_3	1.44	1.52	–	IC 1971 **10** 344

The $PO_2F_2^-$ ion is appreciably distorted from regular tetrahedral shape and is extremely similar in structure to the isoelectronic SO_2F_2 molecule (c). The acid HPS_2F_2 and many of its salts have been prepared.

(a)   (b)   (c)

Phosphoramidates and amidothiophosphates

In $Na(H_3N.PO_3)$[1a] there are zwitterions $^+H_3N.PO_3^{2-}$, (d), analogous to the iso-electronic sulphamic acid, (e), $^+H_3N.SO_3^-$ (p. 722). The N–H–O bonds link the ions into a 3D framework, in the interstices of which are located the Na^+ ions. A later study[1b] of $K(H_3N.PO_3)$ gives P–N, 1.800 Å. Replacement of one O by S in $PO_3NH_2^{2-}$ gives the ion $[PO_2S(NH_2)]^{2-}$, in which the bond lengths shown at (f)

(d)   (e)

(f)

come from a study of the di-ammonium salt.[2] There are similar bond lengths in the diamidothiophosphate ion, studied in $NH_4[POS(NH_2)_2]$.[3]

(1a) AC 1964 **17** 671   (2) AC 1969 **B25** 1256
(1b) AC 1980 **B36** 2391   (3) ZaC 1968 **358** 282

Phosphorothioates

Sulphur can replace O in the PO_4^{3-} ion to give the whole range of phosphorothioate ions from PSO_3^{3-} to PS_4^{3-}. The sodium salts are all hydrated and generally hygroscopic. Thiophosphates have been made from $Na_2S-P_2S_5$ melts with S:P up to 3½:1 as glasses which are very unstable to water.

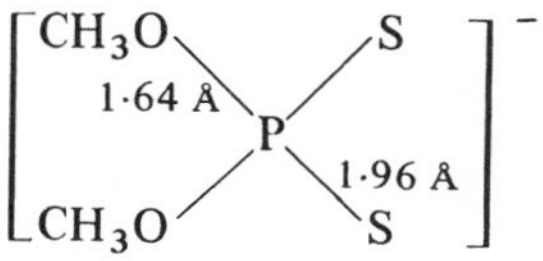

$$\begin{bmatrix} CH_3O & & S \\ _{1\cdot64\ \text{Å}} & P & \\ CH_3O & & {}_{1\cdot96\ \text{Å}}\ S \end{bmatrix}^-$$

(g)

In potassium O-O-dimethylphosphorodithioate the ion has the structure (g).[1]

(1) AC 1962 **15** 765

Other substituted phosphoric acids, etc.

Replacement of H in H_3PO_4 by a radical R gives substituted phosphoric acids $OP(OH)_2OR$ and $OP(OH)(OR)_2$ and finally the phosphoric ester, $OP(OR)_3$. The structure of dibenzylphosphoric acid, $HPO_2(OCH_2C_6H_5)_2$,[1] was described in Chapter 8; the molecules are linked in pairs by O—H—O bonds as in dimers of carboxylic acids. Examples of mono- and di-substituted ions are the phenylphosphate ion, (h), studied in the K salt,[2] and the diethylphosphate ion (i), in the Ba salt,[3] and of an ester, triphenylphosphate, (j).[4]

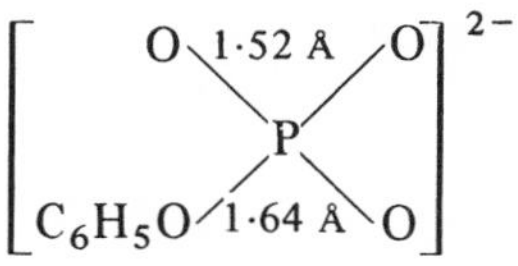

$$\begin{bmatrix} O & {}_{1\cdot52\ \text{Å}} & O \\ & P & \\ C_6H_5O & {}_{1\cdot64\ \text{Å}} & O \end{bmatrix}^{2-} \quad \begin{bmatrix} C_2H_5O & & O \\ & P & \\ C_2H_5O & & O \end{bmatrix}^{-} \quad \begin{matrix} O & & OC_6H_5 \\ & P & \\ C_6H_5O & & OC_6H_5 \end{matrix}$$

(h) 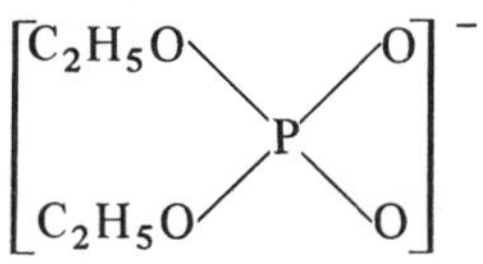   (i) 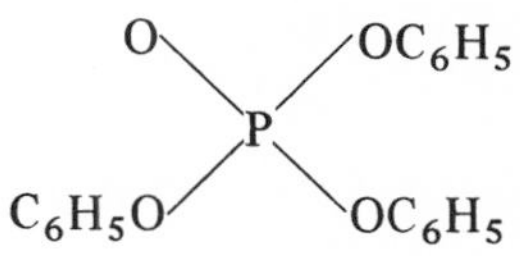   (j)

$$\begin{matrix} O & & OH \\ & P & \\ H_3C & & CH_3 \end{matrix}$$

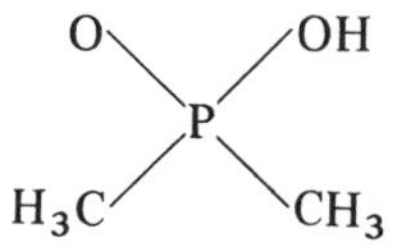

(k)

Replacement of OH in H_3PO_4 by R gives $RPO(OH)_2$ $R_2PO(OH)$, and finally the phosphine oxide, R_3PO. Structural studies have been made of dimethylphosphinic acid, (k),[5] (see also p. 371), and of its salts.[6] Some of the salts $M(R_2PO_2)_2$, of Be, Zn, and other divalent metals, have interesting properties. They range from high-melting crystalline salts to compounds which can be melted and drawn into fibres and others which are waxy solids. The linear chain in the Zn compound, with alternate single and triple $-OP(R_2)O-$ bridges, was mentioned in Chapter 3 as an interesting type of chain structure.

Since groups R_2PO_2 or R_2PS_2 can act either as bridging ligands or as bidentate ligands numerous types of polymeric system can be formed, of which the following are examples. The first two are dimers, the others infinite linear molecules.

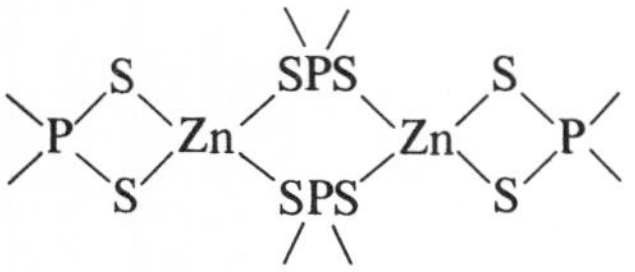

$\{Zn[PS_2(i\text{-}C_3H_7)_2]_2\}_2$ [7]

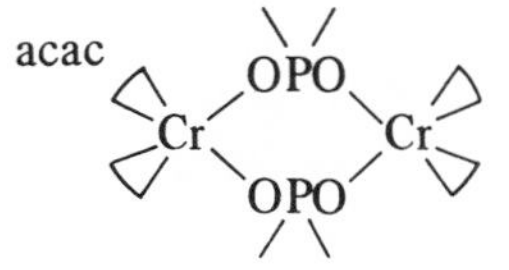

$[(acac)_2Cr \cdot PO_2\phi_2]_2$ [8]

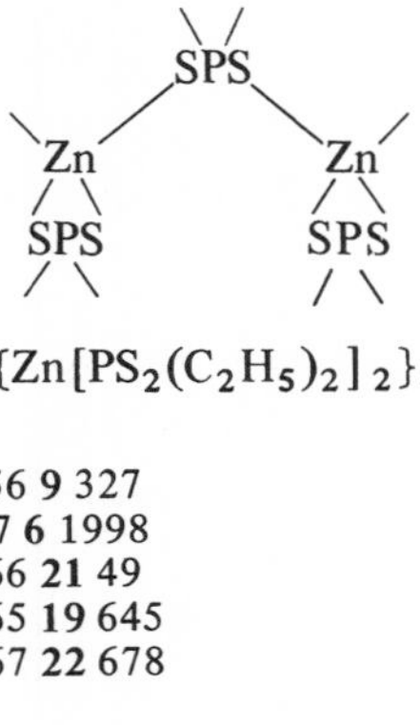

$\{Zn[PS_2(C_2H_5)_2]_2\}_n$ [9]

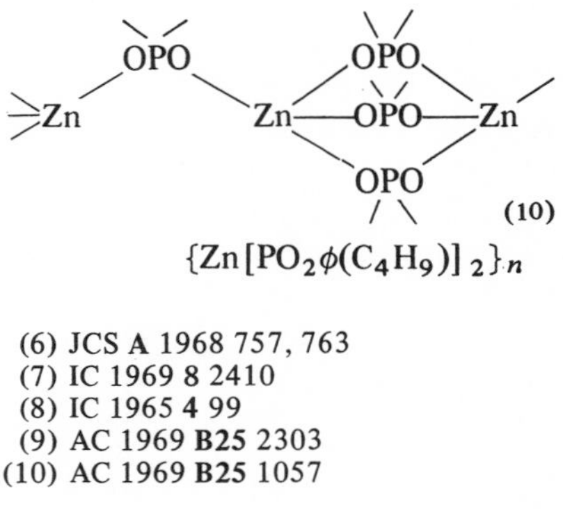

$\{Zn[PO_2\phi(C_4H_9)]_2\}_n$ (10)

(1) AC 1956 9 327
(2) IC 1967 6 1998
(3) AC 1966 21 49
(4) AC 1965 19 645
(5) AC 1967 22 678

(6) JCS A 1968 757, 763
(7) IC 1969 8 2410
(8) IC 1965 4 99
(9) AC 1969 B25 2303
(10) AC 1969 B25 1057

Thiophosphates and related compounds

The simplest thiophosphate ion, the tetrahedral PS_4^{3-} ion (P–S, 2.05 Å), has been studied in a number of compounds MPS_4:[1]

	Packing of S	Coordination of M
BPS_4		
$AlPS_4$	c.c.p.	tetrahedral
$InPS_4$		
$GaPS_4$	h.c.p.	tetrahedral
$CrPS_4$	h.c.p.	octahedral
$BiPS_4$	not c.c.p.	6- and 8-

$GaPS_4$ is mentioned on p. 167 as an example of a rare type of tetrahedral layer structure in which each PS_4 and GaS_4 group shares 1 edge and 2 vertices. Bismuth is too large to fit into interstices in a c.p. assembly of S atoms and has a much less symmetrical coordinaton than is found in the other compounds listed above.

More complex ions include $P_2S_6^{2-}$, (a), $P_2S_6^{4-}$, (b), $P_2S_7^{4-}$, (c), and $P_2S_8^{2-}$, (d). The $P_2S_6^{2-}$ ion, formed from two PS_4 groups sharing an edge, is bonded through tetrahedrally coordinated Ag^+ ions in $Ag_4P_2S_6$.[2] It is structurally similar to the

(a)   (b)   (c)   (d)

$P_2S_4(CH_3)_2$ molecule (p. 869). The $P_2S_6^{4-}$ ion is found in the Fe, Cd, and Sn salts[3] and both the $P_2S_6^{2-}$ and $P_2S_6^{4-}$ ions in $Zn_4(P_2S_6)_3$,[4] which should therefore be formulated $Zn_4(P_2S_6)_2^{2-}(P_2S_6)^{4-}$. The $P_2S_7^{4-}$ ion, the analogue of $P_2O_7^{4-}$, has been studied, like the $P_2S_6^{2-}$ ion, in its Ag salt.[5]

Pyridine reacts with P_4S_{10} to form $[(pyr)_2H]_2P_2S_8$.[6] which contains the cyclic ion (d), with a chair-shaped ring like that of S_6—compare $P_2S_6Br_2$ (p. 870).

(1)   AC 1977 **B33** 1399
(2)   AC 1978 **B34** 3561
(3)   MRB 1970 **5** 419
(4)   AC 1978 **B34** 384
(5)   AC 1977 **B33** 285
(6)   AC 1978 **B34** 1378

Phosphorus sulphides

The following sulphides of phosphorus, several of which are dimorphic, have been fully characterized by structural studies: P_4S_3; P_4S_4 (α); P_4S_5 (α and β); P_4S_7 (α and β); P_4S_9; and P_4S_{10}. Note the surprising absence of P_4S_6. At some compositions P–S melts can be quenched to brittle glasses (S:P ratio 3.5–3.0) or to viscous gums (S:P 1.25 and 1.00), but P_4S_3, P_4S_7, and P_4S_{10} crystallize very rapidly from melts. For S:P between 2 and 3 the melt viscosity shows a maximum at temperatures above 300°C like sulphur, suggesting that the incorporation of P into molten S leads first to branching and cross-linking and then to gradual breakdown of the polymeric S structures owing to the formation of small cage molecules.

All the phosphorus sulphides are yellow crystalline solids, all have molecular weights (determined either in CS_2 solution or in the vapour state) corresponding to the formulae given, and all are formed by direct union of the elements under various conditions. There are interesting relations between these compounds. For example, P_4S_3 readily combines with sulphur to give P_4S_5 or P_4S_7, and on heating, P_4S_5

forms P_4S_3 and P_4S_7. In spite of the fact that all the molecules are of the type P_4S_n structural studies show that there is no P_4 unit common to these molecules. The molecule of α-P_4S_4 is similar to that of As_4S_4 (Fig. 20.13(a)) except for the very long P–P bond; As—As in realgar has the normal length. A second form (β) of P_4S_4 has been isolated (CC 1976 809) for which n.m.r. and chemical evidence suggests the formula shown, but it was not possible to grow crystals suitable for a structural study. See also the third crystalline form of As_4S_4 (p. 908).

The molecule in β-P_4S_5 is similar to that of As_4S_5 (Fig. 20.13(b)).

The molecules of P_4S_3, α-P_4S_5, and P_4S_7 are illustrated in Fig. 19.10. In all three of these molecules there are bonds between P atoms. In the first two there are single bonds of about the same length as in white P; in P_4S_7, however, the only P–P bond is a weak one, of length 2.33 Å. In sulphur-deficient (β) P_4S_7 apparently some of the external S atoms are missing, and the unique P–P bond has a more normal length (2.26 Å). The P–S bond lengths are generally close to one of two values, 2.10 Å if S is forming two bonds, and 1.93 Å to an 'external' S attached to only one

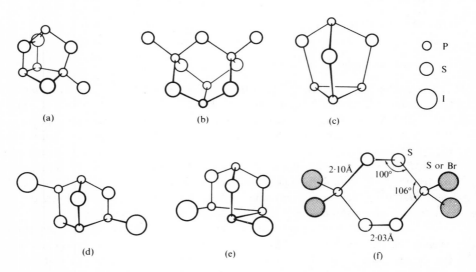

FIG. 19.10.   The molecular structures of (a) α-P_4S_5;   (b) P_4S_7;   (c) P_4S_3;   (d) α-$P_4S_3I_2$; (e) β-$P_4S_3I_2$;   (f) $P_2S_6Br_2$.

P atom. Similar bond lengths are found in P_4S_{10}, the structure of which is similar to that of the P_4O_{10} molecule (Fig. 19.8, p. 854). Although the S analogue of P_4O_6 is not known there is a sulphide P_4S_9 analogous to P_4O_9. Its structure is derived from that of P_4S_{10} by removing one of the terminal S atoms. It is therefore $P^{III}P_3^{V}S_9$. Interbond angles lie for the most part in the range 100–115° except those in the P_3 ring of P_4S_3, in the quadrilateral P_3S ring of P_4S_5 (mean angle, 87°), and an exterior angle of 125° in P_4S_5. References are included in Table 19.7.

TABLE 19.7

Phosphorus sulphides, thiohalides, and related molecules

Molecule	P—P (Å)	P—S (Å)	P=S (Å)	Reference
P_4S_3	2.235	2.090	–	AC 1957 **10** 574
α-P_4S_4	2.350	2.108	–	AC 1978 **B34** 1326
α-P_4S_5	2.25	2.11 (mean)	1.94	AC 1965 **19** 864
β-P_4S_5	2.295	2.12 (mean)	–	AC 1975 **B31** 2738
α-P_4S_7	2.33	2.10 (mean)	1.92	AC 1965 **19** 864
β-P_4S_7	2.36			AC 1965 **18** 221
P_4S_{9-x}				ZaC 1969 **366** 152
P_4S_9		2.11	1.93	AC 1969 **B25** 1229
P_4S_{10}		2.10	1.91	AC 1965 **19** 864
$P_2S_6Br_2$			1.98	IC 1965 **4** 186
α-$P_4S_3I_2$	2.20	2.10		AC 1959 **12** 455
β-$P_4S_3I_2$	2.22	2.12	(P–I, 2.49)	JCS A 1971 1100
$P_2S_4(CH_3)_2$		2.14	1.95	JCS 1964 4065
$P_2S_4(i$-$C_3H_7O)_4$		2.07	1.91	IC 1970 **9** 2269
P–S melts				JINC 1963 **25** 683
		P—Se	P—I	
$P_4Se_3I_2$	2.22	2.24	2.47	AC 1970 **B26** 2092

Closely related to the sulphides are the molecules (a) and (b) containing both single and double P–S bonds (see Table 19.7 for references).

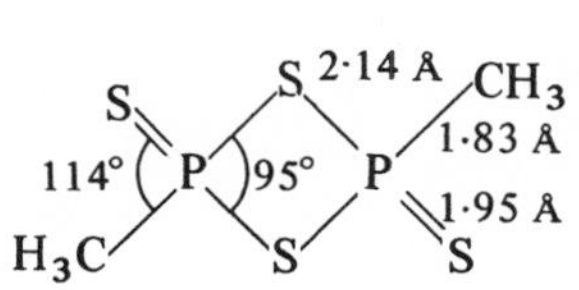

(a)

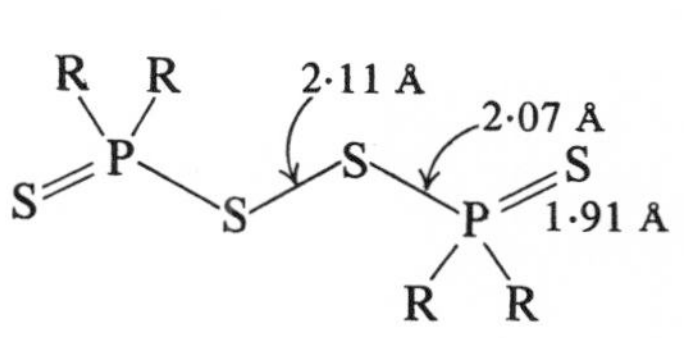

(b)

Phosphorus thiohalides

The known compounds are of four structural types:

(i) PSX_3 (X = F, Cl, Br): tetrahedral molecules $S{=}PX_3$ already summarized in Table 19.4 (p. 846).

(ii) $P_2S_2I_4$: structure not known but presumably a diphosphine (p. 847), $I_2SP{-}PSI_2$.

(iii) $P_2S_5Br_4$ and $P_2S_6Br_2$: the structure of $P_2S_5Br_4$ is not known. In crystalline $P_2S_6Br_2$ (formed by the action of Br_2 on P_4S_6) the molecule consists of a P_2S_4 ring, with the skew boat configuration, and to each P are attached two atoms which are either Br (P–Br, 2.07 Å) or S (P–S, 1.98 Å). There is apparently complete disorder as regards the choice of these atoms. The very flexible skew boat configuration is apparently the ring configuration which minimizes repulsions between non-bonded atoms consistent with S dihedral angles close to 100°. The dihedral angles for P are small (40° and 42°) – Fig. 19.10(f) and Table 19.7.

(iv) $P_4S_3I_2$. This compound exists in two forms. The α form is prepared in CS_2 solution from the elements or by treating the β form with excess iodine. The molecule (Fig. 19.10(d)) consists of one 6- and two 5-membered rings, but has different relative arrangements of the P and S atoms as compared with P_4S_3. This compound illustrates an interesting feature of the phosphorus–sulphur compounds, the facile rearrangement of the P and S atoms. The reaction of P_4S_3 with I_2 under mild conditions gives β-$P_4S_3I_2$. The structure of this form (Fig. 19.10(e)) is much more closely related to P_4S_3, one P–P bond having been broken and two P–I bonds formed. The $P_4Se_3I_2$ molecule has the same structure.

Cyclic phosphorus compounds

Compounds containing P_n rings

We have mentioned that connected systems of P atoms occur in some metal phosphides and that there are pairs of directly bonded P atoms in P_2H_4 and substituted diphosphines, in P_2Cl_4 and P_2I_4, in certain phosphite ions, and also in molecules such as $R_2SP{-}PSR_2$. The systems P–P and P–P–P form parts of the ring systems in phosphorus sulphides, while the P_4 molecule is built of P_3 rings. There are also molecules based on larger P_n rings; the acid $[PO(OH)]_6$ has already been mentioned.

(a)   (b)

The cyclic molecules (a) and (b), in which R is $-CF_3$, are prepared by the action of Hg on $F_3C.PI_2$. Both the P_4[1] and P_5[2] rings are non-planar, with P–P–P bond angles of 85° and 101° (mean) respectively and normal bond lengths P–P (close to 2.2 Å). Compounds originally thought to be the analogues of azo compounds, that is, RP=PR, and now known to be cyclic polymers, include $(CH_3P)_5$ and the phosphobenzenes $(PC_6H_5)_5$ and $(PC_6H_5)_6$ (three polymorphs). In the pentamer the P_5 ring, (c), is non-planar with dihedral angles ranging from $3°$ to $61°$ (mean $38°$),[3]

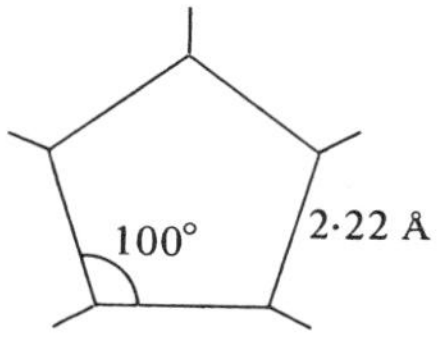

(c) 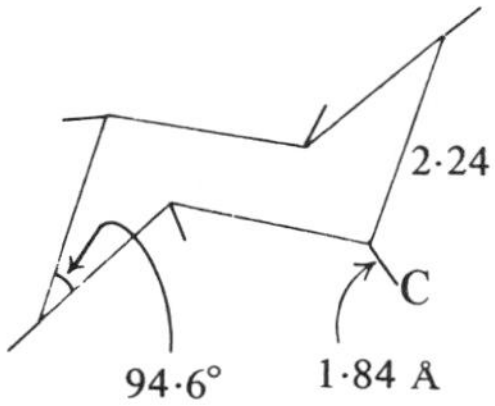　　　　　　　(d) 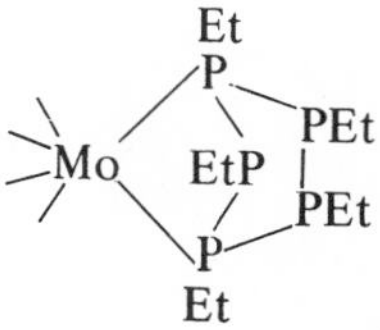　　　　　　　(e)

and in the hexamer the 6-ring, (d), has the chair configuration[4] with the six phenyl groups in equatorial positions and dihedral angles $85°$ (compare $69°$ in $(CsPO_2)_6$, p. 858, and $89°$ in $(AsC_6H_5)_6$). For a comparison of the structures of these cyclic molecules see reference.[5] The reaction between $(PC_2H_5)_4$, a molecule of type (a), and $Mo(CO)_6$ is of interest as converting a P_4 ring into a P_5 ring, the product being $(CO)_4Mo(PC_2H_5)_5$, (e).[6]

(1)　AC 1962 **15** 564
(2)　AC 1961 **14** 250; AC 1962 **15** 509
(3)　JCS 1964 6147
(4)　JCS 1965 4789
(5)　AC 1962 **15** 708
(6)　JCS A 1968 1221

Compounds containing systems of bonded N and P atoms

We have already noted several simple ions containing N–P bonds, namely $O_3P.NH_3^-$ and $O_2SP.NH_2^{2-}$ (p. 864) and $O_3P.NH.PO_3^{4-}$ (p. 860). The chlorine analogue of the latter, (a), is one of the many compounds which result from the reaction between PCl_5 and NH_4Cl under various conditions; the salt $(Cl_3PNPCl_3)(PCl_6)$[1] may also be made by reacting PCl_5 with S_7NH or S_4N_4. The mean P–Cl bond length (1.94 Å) is similar to that in PCl_4^+, as compared with 2.15 Å in PCl_6^-. The simplest cyclic

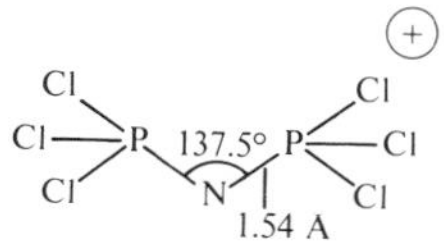

(a) 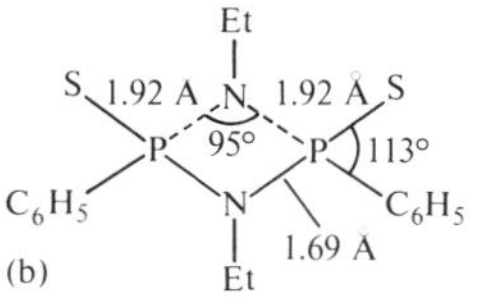　　　　　　　(b) 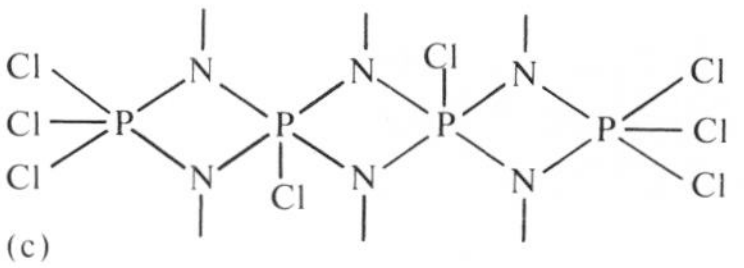　　　　　　　(c)

system is the 4-membered P_2N_2 ring, which has been studied in $(CH_3NPCl_3)_2$[2] (Fig. 19.11(a)), in both isomers (*cis* and *trans*) of the cyclodiphosphazane (b),[3] and also in the *trans* methyl analogue.[4] The $(CH_3NPCl_3)_2$ molecule can be regarded as built from two PCl_5 molecules, replacing one equatorial and one axial Cl by N and then joining to form a planar 4-membered ring. The corresponding angles and bond lengths are very similar to those in PCl_5.

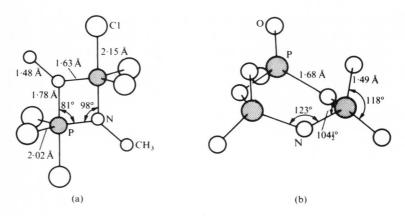

FIG. 19.11. The molecular structures of (a) $(CH_3NPCl_3)_2$; (b) $(HN.PO_2)_3^{3-}$.

The reaction between PCl_5 and CH_3NH_3Cl produces not only $(CH_3NPCl_3)_2$ but also $P_4(NCH_3)_6Cl_8$,[5] (c), in which there is trigonal bipyramidal coordination of P as in the simpler molecule of Fig. 19.11(a).

A number of larger $(PN)_n$ rings have been studied in the cyclophosphazenes, The phosphonitrile chlorides, $(PNCl_2)_n$, may be prepared by heating together PCl_5 and NH_4Cl, for example, in tetrachloroethane. The extract in light petroleum yields crystalline cyclic polymers $(PNCl_2)_n$, $n = 3$ to 8; the residue is an oil which contains polymers up to at least $n = 17$. The fraction insoluble in the petroleum is a viscous oil of composition $(PNCl_2).PCl_5$ probably consisting of linear polymers $Cl.(PNCl_2)_nPCl_4$. This material is converted by boiling in tetrachloroethane into the rubbery $(PNCl_2)_n$, presumably a linear polymer

$$\equiv N\diagdown \underset{Cl_2}{P}=N\diagdown \underset{Cl_2}{P}=N\diagdown$$

The fluorides have been prepared up to $(PNF_2)_{17}$. For example, both $(PNF_2)_3$ and $(PNF_2)_4$ are volatile solids stable up to $300\,^{\circ}C$, above which temperature they are converted into colourless liquid polymers.

The $(PN)_n$ rings have considerable stability. For example, $(PNCl_2)_4$ can be heated to $250\,^{\circ}C$ before further polymerization takes place, and depolymerization takes place only above $350\,^{\circ}C$. This chloride can be boiled with water, acid, or alkali without appreciable decomposition taking place, though slow hydrolysis occurs if it is shaken with water in ether solution. Even then, however, the ring system is not broken (see later). Aqueous NH_3 converts $(PNCl_2)_3$ into $P_3N_3Cl_4(NH_2)_2$ and liquid NH_3 gives $P_3N_3(NH_2)_6$.

X-ray studies have been made of crystalline compounds $(PNX_2)_3$ where X is F, Cl, Br, SCN, C_6H_5, etc. The P_3N_3 ring in $(PNF_2)_3$[6a] and $(PNCl_2)_3$[6b] is essentially planar with angles of $120°$ and six equal P—N bonds of length 1.56Å (compare 1.78Å expected for a single bond), and is clearly an aromatic system in which P uses its d orbitals. In all these compounds both X ligands are attached to P, and the lengths of P—F (1.51Å) and P—Cl (1.99Å) correspond to those of single bonds.

The 8-membered P_4N_4 ring shows considerable variation in shape in compounds $P_4N_4X_4$ and even more variation in the various isomers of $P_4N_4Cl_4(C_6H_5)_4$[7a]— compare the tetrametaphosphate ring (p. 862). It is planar in $(PNF_2)_4$[7b] with a rather shorter P—N bond (1.51Å) and angles P—N—P, $147°$, and N—P—N, $123°$, but in the stable form of $P_4N_4Cl_8$[8] the ring has the chair conformation and in the less stable (K) form[9] the ring is boat-shaped. The molecule $[NP(CH_3)_2]_4$ is also boat-shaped,[10] and the same conformation is retained in $[(NPMe_2)_4H]CuCl_3$.[11] By reacting $[NPMe_2]_4$ with certain transition-metal halides in methyl ethyl ketone compounds such as $[(NPMe_2)_4H]CuCl_3$ and $[(NPMe_2)_4H]_2CoCl_4$ are produced. In the former a $CuCl_3$ group is attached to one N of the ring and H to the opposite N atom, as shown at (a). The bond arrangement around Cu(II) is distorted square planar, the bond angles being $143°$, $134°$, and four of $97°$ (mean). In the Co compound[12] there is simply one H attached to the ring, (b), and the crystal contains discrete tetrahedral $(CoCl_4)^{2-}$ ions. An interesting feature of both structures is the formation of short N—H---Cl bonds (3.20Å). In the Co compound there are rings with both the boat and approximately the saddle conformations.

(a)   (b)

X-ray studies have also been made of molecules containing 10, 12, and 16-membered rings. Like the P_4N_4 ring the P_5N_5 ring varies in shape; it is planar in $P_5N_5Cl_5$,[13a] with no suggestion of alternating bond lengths (mean P—N, 1.52Å), but non-planar in $P_5N_5(CH_3)_{10}$[13b] P—N, 1.59Å; N—P—N, $116–120°$; P—N—P, $132–136°$) and $[P_5N_5(CH_3)_{10}H_2]^{2+}$.[13c] The 12-membered ring in $P_6N_6(NMe_2)_{12}$[14] is highly puckered with all bonds in the ring of length 1.56Å (*exo* P—N bonds, 1.67Å) and angles in the ring, $148°$ (at N) and $120°$ (at P). The 16-membered ring in $P_8N_8(OMe)_{16}$[15] consists of two approximately planar 6-segments in two parallel planes joined at a 'step'; the P—N bond length is the same as in the 12-membered ring.

Structural studies have been made of more complex bicyclic phosphazenes. In those of types (c) and (d)[16] the rings are nearly planar with only small differences between P–N bond lengths, all of which are close to 1.60 Å. In (c) the P–P bond is single (2.21 Å); in (d) the lengths of Ni–N and Ni–S are 1.94 and 2.23 Å respectively.

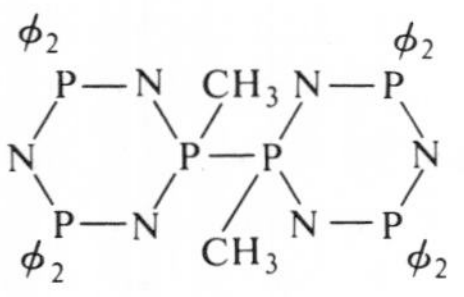

(c) 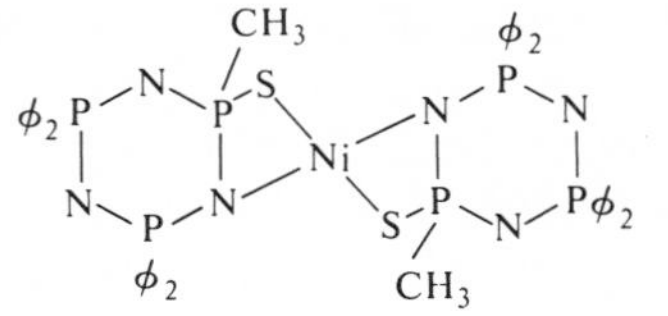 (d) 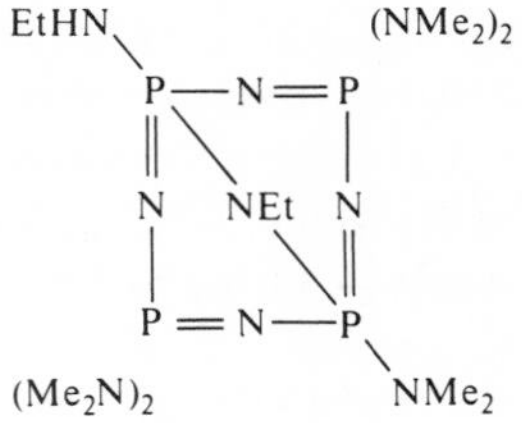 (e)

In (e)[17] opposite corners of a P_4N_4 ring are bridged by a N atom, forming two 6-membered rings. The cage-like molecule $P_2N_6(CH_3)_6$,[18] (f), is structurally similar to $N(CH_2CH_2)_3N$.

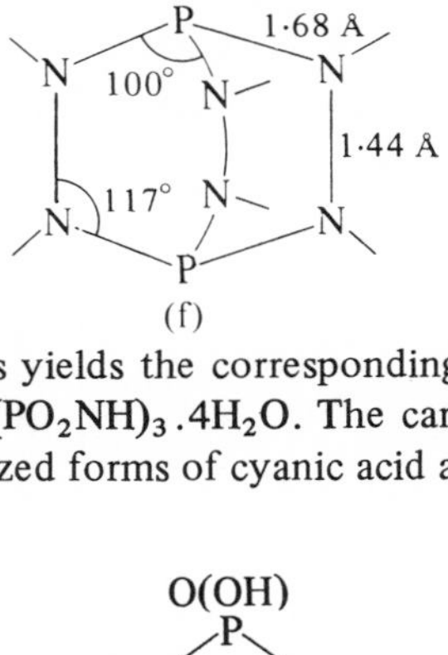

(f)

Hydrolysis of the halides yields the corresponding cyclic metaphosphimic acids or their salts, such as $Na_3(PO_2NH)_3.4H_2O$. The carbon and sulphur analogues of these acids are the polymerized forms of cyanic acid and sulphimide:

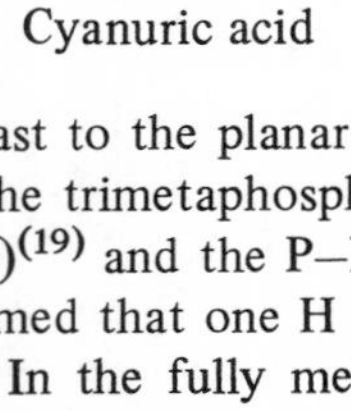

Cyanuric acid 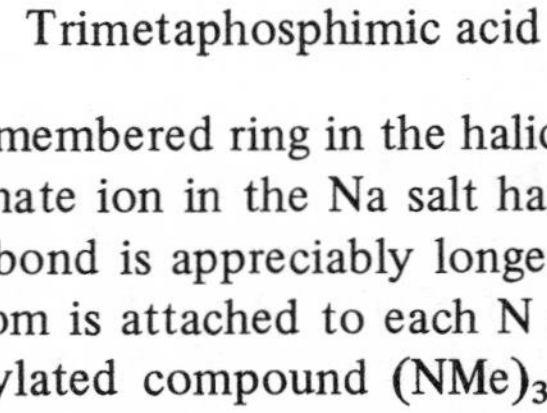 Trimetaphosphimic acid 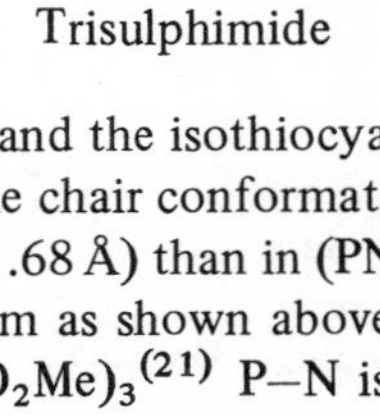 Trisulphimide

In contrast to the planar 6-membered ring in the halides and the isothiocyanate the ring of the trimetaphosphimate ion in the Na salt has the chair conformation (Fig. 19.11(b))[19] and the P–N bond is appreciably longer (1.68 Å) than in $(PNX_2)_4$. It is confirmed that one H atom is attached to each N atom as shown above for the acid.[20] In the fully methylated compound $(NMe)_3(PO_2Me)_3$[21] P–N is 1.66 Å, close to the value in the trimetaphosphimate ion, and the ring has a distorted boat shape. The tetrametaphosphimate ring shows considerable variation in shape like the 8-membered ring in $(PNX_2)_4$. It has the boat ($\bar{4}$) conformation in $(NH_4)_4[P_4N_4H_4O_8].2H_2O^{(22)}$ and in $(NMe)_4(PO_2Me)_4^{(23)}$ (P–N, 1.67; P=O, 1.47; and P–O, 1.58 Å), the chair conformation in the K salt (tetrahydrate)[24] and the saddle (cradle) conformation in the Cs salt (hexahydrate).[24] A study of the dihydrate of tetrametaphosphimic acid[25a] confirms the imino formulation as for trimetaphosphimic acid.

Heterocyclic systems containing N, P, and S atoms include the N_3PS_2 ring in $N_3PS_2O_2Cl_4$,[25b] (g); $N_3(PCl_2)(SOF)_2^{(26)}$ (S–F, 1.57 Å), in which the rings are nearly planar; and the deep-blue $N_3PS_2[NH(SiMe_3)]_2$,[27] (h). An example of the

N_3P_2S ring is shown at (i).[28] The anion in $K_6(P_{12}S_{12}N_{14})^{(29)}$ is an example of a complex cage structure.

(1)	AC 1980 **B36** 1014	(14)	AC 1968 **B24** 1423
(2)	ZaC 1966 **342** 240; ZaC 1966 **342** 240	(15)	JCS A 1968 2227
(3)	AC 1973 **B29** 1439	(16)	AC 1976 **B32** 3078
(4)	AC 1975 **B31** 2333	(17)	AC 1977 **B33** 443
(5)	ZaC 1967 **351** 152	(18)	IC 1971 **10** 2591
(6a)	JCS 1963 3211	(19)	AC 1965 **19** 596
(6b)	JCS A 1971 1450	(20)	ZaC 1976 **419** 139
(7a)	AC 1974 **B30** 2861	(21)	JCS A 1968 3026
(7b)	JCS 1961 4777	(22)	AC 1964 **19** 603
(8)	AC 1968 **B24** 707	(23)	AC 1977 **B33** 1367
(9)	AC 1962 **15** 539	(24)	AC 1971 **B27** 740; AC 1976 **B32** 435
(10)	JCS 1961 5471	(25a)	AC 1977 **B33** 605
(11)	JCS A 1970 455	(25b)	AC 1969 **B25** 651
(12)	JCS A 1970 460	(26)	AC 1974 **B30** 2795
(13a)	JCS A 1968 2317	(27)	AC 1977 **B33** 2272
(13b)	AC 1977 **B33** 295	(28)	AC 1976 **B32** 1192
(13c)	JCS D 1974 382	(29)	ZN 1976 **31b** 419

20

Arsenic, antimony, and bismuth

Elementary arsenic, antimony, and bismuth

The normal ('metallic') forms of As, Sb, and Bi are isostructural and have a layer structure (p. 67) in which each atom has three equidistant neighbours, the next set of three neighbours being at a greater distance. The distinction between these two sets of neighbours grows less going from As to Bi, as is seen from the following figures:

	3 at	3 at	M–M–M	Reference
As	2.52 Å	3.12 Å	96.65°	JACr 1969 2 30
Sb	2.91	3.36	95.6°	AC 1963 16 451
Bi	3.07	3.53	95.45°	AC 1962 15 865

The yellow (non-metallic) forms of As and Sb are metastable, and are prepared by condensing the vapour at very low temperatures. Arsenic also forms a polymorph isostructural with black P; it is prepared by heating amorphous As at 100–175 °C in the presence of Hg.[1] The yellow forms revert to the metallic forms on heating or on exposure to light; they probably consist of tetrameric molecules, but owing to their instability their structures have not been studied. The As_4 molecule in arsenic vapour has the same tetrahedral configuration as the P_4 molecule, with As–As, 2.44 Å. Mass spectrometric studies show that the molecules As_4, Sb_4, and Bi_4 and all combinations of these atoms exist in the vapours from liquid mixtures of the elements. No high-pressure forms of As have been recorded, but both Sb and Bi transform into simple cubic forms under pressure (M–M, 2.97 and 3.18 Å respectively) while at higher pressures Sb forms a hcp structure and Bi a bcc structure (Bi–8Bi, 3.29 Å).

(1) ZaC 1956 283 263

The structural chemistry of As, Sb, and Bi

The series P, As, Sb, and Bi show a gradation of properties from non-metal to metal. We shall be concerned here chiefly with the stereochemistry of As and Sb, for the Bi analogues of many of the simple molecules containing As and Sb are either much less stable or have not been prepared. Simple molecules containing multiple bonds are not formed by any of these elements. For example, $AsCF_3$ and $AsCH_3$ are not structurally similar to azomethane, $H_3C-N=N-CH_3$, but form cyclic molecules containing single As–As bonds. The former, isostructural with $(PCF_3)_4$, contains As_4 rings, non-planar, As bond angle 83.6°, dihedral angle 37°, and As–As, 2.45 Å.[1] Arsenomethane exists in two forms, yellow and red. Crystals of the yellow form consist of molecules $(AsCH_3)_5$ having the form of puckered 5-membered rings (mean As bond angle, 102°, As–As, 2.43 Å, and As–C, 1.95 Å).[2]

Arsenobenzene, on the other hand, consists of 6-membered rings $[As(C_6H_5)]_6$ which are chair-shaped, with As–As, 2.46 Å and As bond angle 91°.[3] Examples of rings containing alternate As and N atoms include $(\phi_2AsN)_3$,[4] $(\phi_2AsN)_4$,[5] and $As_4(NMe)_6$.[6] This last compound is prepared from $AsCl_3$ and CH_3NH_2, and is structurally similar to $N_4(CH_2)_6$. (For the $CoAs_3$ structure which contains As_4 rings, see p. 267). Among the few examples of molecules containing As–As bonds which are appreciably shorter than that in As_4 (2.44 Å) are the molecules (a) and (b)[7] which contain tetrahedral nuclei related to the As_4 molecule. An interesting difference between P and As is that only one pure arsenyl halide is known; this is

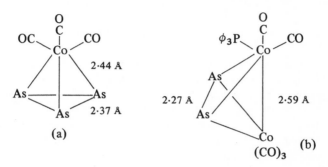

(a)   (b)

AsOF_3. All the elements As, Sb, and Bi form a hydride MH_3. In the preparation of AsH_3 and SbH_3 (by reduction of the trichlorides in aqueous HCl by sodium hydroborate) As_2H_4 and Sb_2H_4 are formed as secondary products, but no higher homologues have been observed.

In addition to compounds in which it exhibits the normal oxidation states III and V Bi also forms sub-halides and polynuclear cations and anions in which there are metal–metal bonds. In BiBr[8] and both crystalline forms of BiI[9] there are chains of the type shown in which the bonding of the Bi atoms is of two very

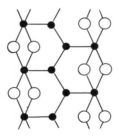

different kinds. One-half form pseudo-octahedral bonds to 4X+1Bi, while the remainder form three pyramidal bonds to 3 Bi atoms. The compound $Bi_{12}Cl_{14}$ is noted on p. 888; it contains $BiCl_5^{2-}$, $Bi_2Cl_8^{2-}$, and Bi_9^{5+} ions. From a solution of Bi in molten $NaAlCl_4$ the compounds Bi_4AlCl_4 and $Bi_5(AlCl_4)_3$ have been crystallized, possibly containing ions Bi_8^{2+} and Bi_5^{3+}.[10] For the Bi_4^{2-} *anion* see p. 307.

TABLE 20.1
The stereochemistry of As, Sb, *and* Bi

Total number of electron pairs	Bond type	Lone pairs	Bond arrangement	Proved for
4	sp^3	0	Tetrahedral	As^+
		1	Pyramidal	As, Sb, Bi
5	sp^3d	0	Trigonal bipyramidal	Sb
		1	See text	Sb
6	sp^3d^2	0	Octahedral	As^-, Sb^-, (Bi)
		1	Square pyramidal	Sb
7		1	Octahedral	Sb

The bond arrangements of Table 20.1 have been established by structural studies of simple ions or covalent molecules for the elements indicated. The presence of a lone pair implies M^{III}; otherwise M^V. The paucity of information about Bi is due to the more metallic character of this element, which does not form many of the simple covalent molecules formed by As and Sb. The octahedral bonds formed by Bi in, for example, the crystalline pentafluoride have considerable ionic character. For 7-coordination (sp^3d^3) a pentagonal bipyramidal bond arrangement would be expected. We noted in Chapter 7 that $SbBr_6^{3-}$ should form a distorted octahedron since there is a valence group of 14 electrons including one lone pair. In fact the ion shows very little distortion from a regular octahedron. (See p. 883).

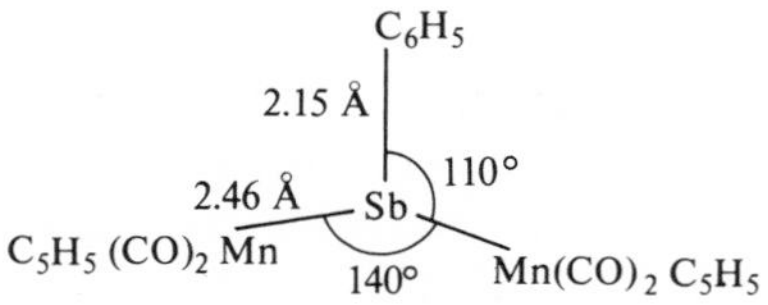

We have not included in Table 20.1 examples of 'abnormal' bond arrangements exhibited in combination with special ligands. An example is the planar molecule $(C_6H_5)Sb[Mn(CO)_2C_5H_5]_2$.[11]

(1)   AC 1971 **B27** 476
(2)   JACS 1957 **79** 859
(3)   AC 1961 **14** 369
(4)   IC 1973 **12** 2304
(5)   JCS D 1974 2527
(6)   ZaC 1967 **350** 9

(7)   JACS 1969 **91** 5631, 5633
(8)   ZaC 1978 **438** 53
(9)   ZaC 1978 **438** 37
(10) IC 1969 **7** 198
(11) AnCIE 1978 **17** 843

Molecules MX_3: *valence group* 2, 6

The trigonal pyramidal shape of many molecules MX_3 formed by As, Sb, and Bi has been demonstrated by e.d. or m.w. studies of the vapours or by X-ray studies of the solids (Table 20.2).

TABLE 20.2
Structural data for molecules MX₃

Molecule	M–X (Å)	X–M–X	Method	Reference
AsH₃	1.519	91.83°	m.w.	PR 1955 97 684
As(CH₃)₃	1.96	96°	m.w.	SA 1959 15 473
As(CF₃)₃	2.053	100°	e.d.	TFS 1954 50 463
AsF₃	1.706	96.2°	m.w.	} BCSJ 1966 39 71
AsCl₃	2.161	98.7°	m.w.	IC 1970 9 805
AsBr₃	2.33	99.7°	e.d.	JMSt 1976 35 67, 81
AsI₃	2.55	100.2°	e.d.	
AsCl₃	2.17	97.7°	X	CR 1978 287C 259
AsBr₃	2.36	97.7°	X	ZK 1967 124 375
AsI₃	2.59	99.7°	X	AC 1980 B36 914
As(SiH₃)₃	2.355	94°	e.d.	JCS A 1968 3006
As(CH₃)I₂	2.54	104°	X	AC 1963 16 922
AsBr(C₆H₅)₂	2.40	C–As–C 106°	X	JCS 1962 2567
		C–As–Br 95°		
As(CN)₃		90.5°	X }	AC 1966 20 777
As(CN)₂CH₃		94°	X }	
SbF₃	1.92	87.3°	X	JCS A 1970 2751
SbH₃	1.707	91.3°	m.w.	PR 1955 97 680
Sb(CF₃)₃	2.202	100.0°	e.d.	TFS 1954 50 463
SbCl₃	2.33	97.2°	e.d.	BCSJ 1973 46 404
	2.34 (one)	91° (one)	X	AC 1979 B35 3020
	2.37 (two)	96° (two)		
SbBr₃	2.49	98°	e.d.	BCSJ 1973 46 413
α (cryst.)	2.50	95°	X	JCS 1964 4162
β	2.49	95°	X	JCS 1962 2218
SbI₃	2.719	99.1°	e.d.	ACSc 1963 17 2573
	See text		X	ZK 1966 123 67
BiCl₃	2.50	84°, 94° (two)	X	AC 1971 B27 2298
BiCl₃	2.48	100°	e.d. }	TFS 1940 36 681
BiBr₃	2.63	100°	e.d. }	
Bi(CH₃)₃	2.27	96.7°	e.d.	JMSt 1973 17 429

For the structures of the crystalline halides see p. 884.

Tetrahedral ions MX₄⁺: valence group 8

There are no arsenic or antimony analogues of the simple ammonium or phos-phonium halides. The tetrahedral configuration of substituted arsonium ions has been demonstrated in crystalline $[As(CH_3)_4]Br^{(1)}$ and $[As(C_6H_5)_4]I_3$.

There are many compounds in which As^V and Sb^V form tetrahedral bonds (valence group 8) as, for example, GaAs, GaSb, InAs, and InSb which, like the corresponding phosphides, crystallize with the zinc-blende structure. Tetrahedral bonds are formed by As^V in the AsO_4^{3-} ion in orthoarsenates (and arsine oxides) but note the quite different behaviour of Sb^V in its oxy-compounds (see later).

(1) JCS 1963 4051

Ions MX$_4^-$: valence group 2, 8

The formation of 4 bonds by M^{III} would result in the valence group 2, 8, for which there are two closely related bond arrangements, (a) and (b). The structure of a

(a)   (b)   (c)

halide ion MX$_4^-$ is not yet known; finite ions (MX$_4$)$^-$ containing As, Sb, or Bi probably do not exist. Tetrachloroarsenites (for example, (CH$_3$)$_4$N.AsCl$_4$) have been prepared but they presumably do not contain finite (AsCl$_4$)$^-$ ions but chain ions analogous to (SbCl$_4$)$_n^{n-}$ etc. described on p. 887.

An example of this unusual valence group (2, 8) is found for Sb(III) in antimonyl tartrates. In the potassium salt, KSbC$_4$H$_4$O$_7$.½H$_2$O,[1] the bond arrangement around Sb appears to be close to (a), while in the ammonium salt[2] it is closer to (b). The ion is shown at (c). A similar stereochemistry is found for the anion in Rb[Bi(SCN)$_4$].[3]

(1)   AC 1965 **19** 197   (3)   AC 1976 **B32** 2319
(2)   DAN 1964 **155** 545

Pentahalides and molecules MX$_5$: valence group 10

The following pentahalides are known:

AsF$_5$ SbF$_5$ BiF$_5$
                   SbCl$_5$

The trigonal bipyramidal configuration of AsF$_5$ and SbCl$_5$ moledcules in the vapour state has been demonstrated by e.d., and the same stereochemistry has been confirmed for SbCl$_5$ in the crystalline state (Table 20.3). Other molecules shown to have this shape in the crystalline state include As(C$_6$H$_5$)$_5$[1] Sb(C$_6$H$_5$)$_4$OH,[2] Sb(C$_6$H$_5$)$_4$OCH$_3$ and Sb(C$_6$H$_5$)$_3$(OCH$_3$)$_2$,[3] Sb(CH$_3$)$_3$Cl$_2$,[4] and Bi(C$_6$H$_5$)$_3$Cl$_2$.[5] Oxygen-bridged molecules include (a),[6] the corresponding methyl compound,

(a)   (b)

TABLE 20.3
Structural data for pentahalides and adducts

Molecule		M–X (Å)		Reference
AsF$_5$		1.711 (axial)		IC 1970 9 805
		1.656 (equat.)		
SbF$_5$		Not determined (Raman)		JCP 1972 **56** 5829
SbCl$_5$	(cryst.)	2.34 (axial)		JACS 1959 **81** 811
		2.29 (equat.)		
	(liquid	–		JPC 1958 **62** 364
	(vapour)	2.43 (axial)		
		2.31 (equat.)		APL 1940 **14** 78
			Sb–O (Å)	
SbF$_5$.SO$_2$		1.85	2.13	JACS 1968 **90** 1358
SbCl$_5$.POCl$_3$		2.33	2.17	ACSc 1963 **17** 353
SbCl$_5$.OC(H).N(CH$_3$)$_2$		2.34	2.05	AC 1966 **20** 749
BiF$_5$ (cryst.)		1.90		ZaC 1971 **384** 111

which forms disordered crystals (Sb–O, 2.10 Å; O bond angle 126–9°),[7] the analogous Bi compounds, and the molecule (b),[8] which is notable for containing both 5- and 6-coordinated Sb(v) and also for the use of all three O atoms by the bridging CO$_3$ group.

In contrast to the analogous P and As compounds Sb(C$_6$H$_5$)$_5$[9] has a tetragonal pyramidal configuration with Sb approximately 0.5 Å above the base of the pyramid and an angle of 102° between the axial and equatorial bonds. At present this remains the only example of a Group V molecule of this kind (10-electron valence group) which is not trigonal bipyramidal.

BiF$_5$ has the α-UF$_5$ structure (p. 1254) in which there is octahedral coordination of the metal atom. An early e.d. study established the trigonal bipyramidal configuration of the molecules NbCl$_5$, NbBr$_5$, TaCl$_5$, and TaBr$_5$.[10]

(1)	JCS 1964 2206	(6)	AC 1973 **B29** 2221
(2)	JACS 1969 **91** 297	(7)	AC 1975 **B31** 1260
(3)	JACS 1968 **90** 1718	(8)	AC 1974 **B30** 103
(4)	ZK 1938 **99** 367	(9)	JACS 1968 **90** 6675
(5)	JCS A 1968 2539	(10)	TFS 1940 **36** 668

Formation of square pyramidal bonds by Sb^{III} *and* Bi^{III}: *valence group* 2,10

If the valence group of M in a molecule or ion MX$_5$ consists of six electron pairs the lone pair is expected to occupy the sixth octahedral position, giving a square pyramidal arrangement of bonds. Some complex halides of Sb^{III} are of interest in this connection (p. 884).

Formation of octahedral bonds by As(v), Sb(v), *and* Bi(v): *valence group* 12

The octahedral arrangement of six bonds from As(v) may be deduced from the optical activity of molecules such as the tricatechol derivative; it has been subsequently confirmed by an X-ray study.[1] The simplest examples of this coordination group are the MX_6^- ions:

AsF_6^- (As–F, 1.69 Å);[2] SbF_6^- (Sb–F, 1.88 Å);[3]   and $SbCl_6^-$ (Sb–Cl, 2.37 Å).[4][5]

Metallic salts containing the BiF_6^- ion do not appear to be known; the octahedral coordination of Bi in BiF_5 has already been noted. There is also octahedral coordination of Sb(v) in $(Sb_2F_{11})^-$ (see pp. 379 and 394) and in $(Sb_3F_{16})^-$ (p. 393), in oxyfluorides, adducts of $SbCl_5$ and in numerous oxy compounds described in later sections of this chapter. Here we note certain oxyfluorides and $SbCl_5$ adducts.

Both As and Sb form oxygen-bridged ions $M_2F_{10}O^{2-}$ and $M_3F_{12}O_3^{3-}$ formed from octahedral groups linked through bridging O atoms:

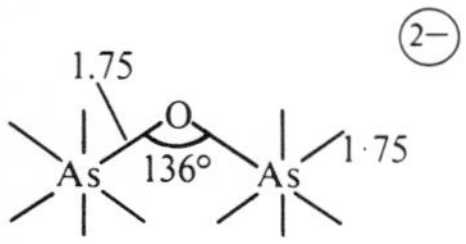

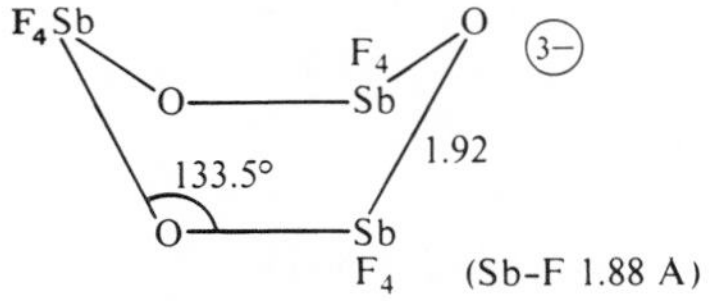

The $As_2F_{10}O^{2-}$ ion has been studied in the K and Rb salts,[6] $Sb_2F_{10}O^{2-}$ in the Rb salt,[7] and $Sb_3F_{12}O_3^{3-}$ in the Cs salt.[8] There is also an ion $As_2F_8O_2^{2-}$, in which the angle at the bridging O atom is only 96°.[9]

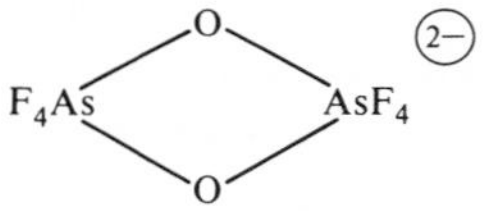

The valence group 10 in $SbCl_5$ is readily expanded to the octahedral group 12 not only in the $SbCl_6^-$ ion but also in numerous adducts in which an O or N atom occupies the sixth bond position. Crystalline $SbCl_5 . POCl_3$ consists of molecules

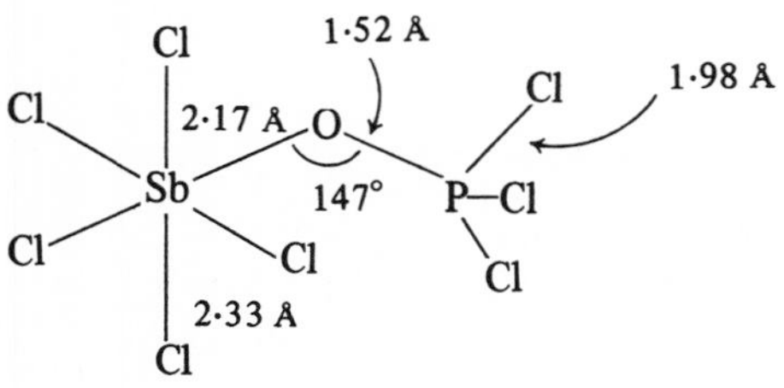

(a) 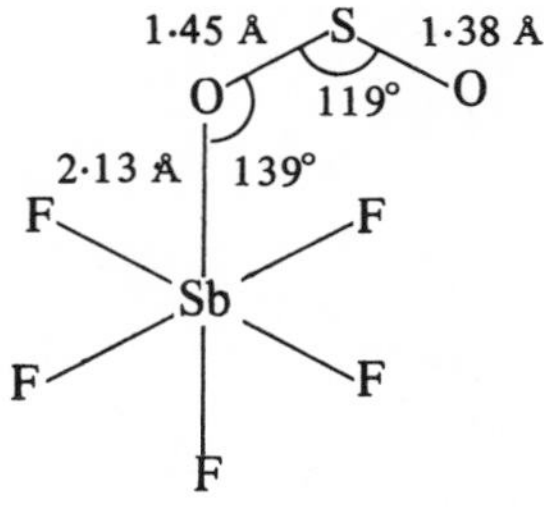   (b)

(a), and adducts are also formed with $PO(CH_3)_3$, $(CH_3)_2SO_2$, etc. (Table 20.3). The molecule $SbF_5 . SO_2$ is shown at (b). In $SbCl_5 . N_4S_4$ the sixth octahedral bond is formed to a N atom of the cyclic N_4S_4 molecule (Sb–N, 2.17 Å; Sb–Cl, 2.39 Å). In the compound with ICl_3, $ISbCl_8$ (p. 391) there are recognizable $SbCl_6^-$ ions, though the bond lengths suggest that the system may be intermediate between the states $(ICl_2)^+(SbCl_6)^-$ and $(ICl_4)^-(SbCl_4)^+$.

(1) AC 1972 **B28** 3446
(2) AC 1974 **B30** 250
(3) AC 1976 **B32** 2916
(4) AC 1972 **B28** 2673
(5) AC 1973 **B29** 2622

(6) AC 1974 **B30** 1722
(7) AC 1974 **B30** 2508
(8) AC 1974 **B30** 2465
(9) B 1974 **107** 1009

Formation of octahedral bonds by Sb(III): *valence group* 2, 12

We noted in Chapter 16 that not only does the $TeCl_6$ molecule, in which Te has the valence group 12, have the expected octahedral configuration but also the ion $TeCl_6^{2-}$ (valence group 2,12) has a regular octahedral configuration in salts such as K_2TeCl_6 (cubic K_2PtCl_6 structure). Ions $Sb^{III}X_6^{3-}$ are isoelectronic with the corresponding $Te^{IV}X_6^{2-}$ ions. The jet-black salt with the empirical composition $(NH_4)_2SbBr_6$ is in fact $(NH_4)_4(Sb^{III}Br_6)(Sb^VBr_6).$[1] Its structure is a superstructure of the K_2PtCl_6 type, the octahedral ions containing Sb^{III} and Sb^V alternating as shown in Fig. 20.1 so that the unit cell has twice the c dimension of the simple

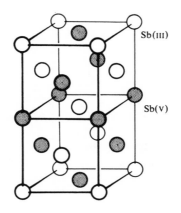

FIG. 20.1. Arrangement of Sb^{III} and Sb^V atoms in $(NH_4)_4(Sb^{III}Br_6)(Sb^VBr_6)$.

K_2PtCl_6 structure. In spite of the presence of the lone pair on Sb^{III} the $SbBr_6^{3-}$ octahedron is *un*distorted (Sb–Br, 2.795 Å); there is slight distortion of the $SbBr_6^-$ ions (Sb–Br, 2.564 Å). Salts $R_6Sb_4Br_{24}$[2] (R = pyridinium) also contain Sb^{III} and Sb^V, and should be formulated $(C_5H_5NH)_6(Sb^{III}Br_6)(Sb^VBr_6)_3$. The Sb–Br bond lengths are very similar to those just quoted.

(1) JACS 1966 88 616

(2) IC 1971 **10** 701

The crystalline trihalides of As, Sb, and Bi

The trigonal pyramidal shape of the molecules in the vapour state has already been noted; Table 20.2 also includes literature references to those crystalline halides in which there are well-defined MX_3 molecules, that is, all except BiF_3[1] and BiI_3.[2] The ionic (9-coordinated) structure of β-BiF_3 has been described on p. 421. There are two main types of structure of these trihalides, and we shall describe the simpler one first.

In the iodides MI_3 M occupies octahedral interstices in hcp assemblies of I atoms, but M is progressively further from the centre of the I_6 octahedron in the series BiI_3, SbI_3, and AsI_3:

$$\text{Bi–6I} \quad 3.07\,\text{Å}\,(90°); \quad \text{Sb} \begin{cases} 3I & 2.87\,\text{Å} \\ 3I & 3.32 \end{cases} (96°); \quad \text{As} \begin{cases} 3I & 2.59\,\text{Å}\,(100°) \\ 3I & 3.47 \end{cases}$$

(Bond angles in parentheses)

M–I in MI_3 molecule in vapour   ?   2.72   (99°)   2.55   (100°)

The structures of PI_3 and SbF_3 are of the same general type.

In the trichlorides and tribromides of Sb and Bi also there are well-defined molecules MX_3, but there are 5 next nearest neighbours at approximately the same distance which complete a bicapped trigonal prism coordination group:

	3X at d	2X at d_1	3X at d_2
$SbCl_3$	2.34	3.46	3.61 (one)
			3.74 (two)
$BiCl_3$	2.50	3.24	3.36 (mean)
PBr_3	2.21	3.93	3.88

However, there is a striking difference between the structures of $SbCl_3$ and $BiCl_3$. Not only is the ratio of the distance d_1 (or d_2) to d much greater in $SbCl_3$ than in $BiCl_3$ but the *actual* distances to the 5 next nearest neighbours are shorter in $BiCl_3$ than in $SbCl_3$, although Bi–Cl (2.50 Å) is greater than Sb–Cl (2.34 Å). This indicates a strong tendency towards 8-coordination, as observed in BiF_3. A much greater difference between the distances to the sets of 3 and 5 neighbours is seen in PBr_3 and a still greater one in $POBr_3$ (next 3 Br at 4.37 Å) which is structurally similar but with an O atom occupying the position of the lone pair in crystalline PBr_3.

(1)   ACSc 1955 **9** 1206, 1209   (2)   ZK 1966 **123** 67

Complex halides containing trivalent Sb or Bi

Structural studies have been made of complex fluorides of the following types formed with alkali metal or similar ions: $MSbF_4$, M_2SbF_5, MSb_2F_7, MSb_3F_{10}, and

MSb_4F_{13} and also of MSb_3F_{14}, which contains both $Sb(III)$ and $Sb(V)$. Salts M_3SbF_6 are apparently not known. K_2SbF_5 and $(NH_4)_2SbF_5$[1] contain SbF_5^{2-} ions with the same tetragonal pyramidal configuration as $SbCl_5^{2-}$ in $(NH_4)_2SbCl_5$ (see below). In $KSbF_4$[2] similar SbF_5 groups are joined by sharing two F atoms to form cyclic $Sb_4F_{16}^{4-}$ ions, while in $NaSbF_4$ [3] there are apparently infinite chains formed from SbF_5 groups linked in a similar way. In these complexes the Sb atom lies about $0.2-0.3$ Å *below* the square base of the pyramid, and this is also true of the 5-coordinated Sb atoms in Sb_2S_3 (see later) and certain complex sulphides (e.g. $FeSb_2S_4$). The type of Sb–F complex in salts MSb_2F_7 depends on the nature of the cation. In KSb_2F_7[4] there are distorted trigonal bipyramidal SbF_4^- ions, (a), and SbF_3 molecules (Sb–F, 1.94 Å) very similar to those in crystalline SbF_3. However, whereas in crystalline SbF_3 3 F at 2.61 Å complete a distorted octahedral group (or capped octahedron if the lone pair is included) in KSb_2F_7 there are only 2 additional (F_a) neighbours (at 2.41 Å and 2.57 Å) which, *together with the lone pair*, complete a distorted octahedron. The structure of $CsSb_2F_7$[5] is quite different. There are well defined $Sb_2F_7^-$ ions, (b), formed from two distorted trigonal bipyramidal SbF_4 groups sharing axial F atoms, with long bridge bonds. (The next shortest Sb–F distance is 2.77 Å). Note that the anion in KSb_2F_7 could be described as an infinite chain of $SbF_2(F_a)_2E$ groups sharing axial F atoms with pseudo-octahedral $SbF_3(F_a)_2E$ groups but with longer (unsymmetrical bridges of 2.41 Å and 2.57 Å).

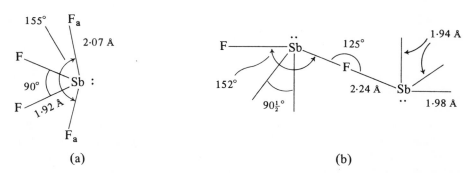

(a)   (b)

The salt $NaSb_3F_{10}$[6] has a layer structure formed from octahedral SbF_5E groups sharing three equatorial vertices, two between two octahedra and a third common to three octahedra, as shown diagrammatically in Fig. 20.2(a). The pseudo-octahedral coordination of $Sb(III)$ is very irregular (Fig. 20.2(b)), and there are unsymmetrical Sb–F–Sb bridges (1.98 and 2.60 Å). The nature of the complex ion in KSb_4F_{13}[7] is not very clear. In the $Sb_3F_{14}^-$ ion[7a] (c) there are unsymmetrical bridges (Sb(V)–F, 1.95 Å, Sb(III)–F, 2.30 Å) between octahedrally coordinated $Sb(V)$ and 4-coordinated $Sb(III)$ atoms – compare the structure of the $Sb_2F_7^-$ ion (b).

Another compound containing both $Sb(III)$ and $Sb(V)$ is the 'adduct' $SbF_3 . SbF_5$.[7b] It may be formulated $(Sb_2F_4)^{2+}(SbF_6^-)_2$ since it contains octahedral SbF_6^- ions, but the environment of one of the $Sb(III)$ atoms is unusual. That of Sb'

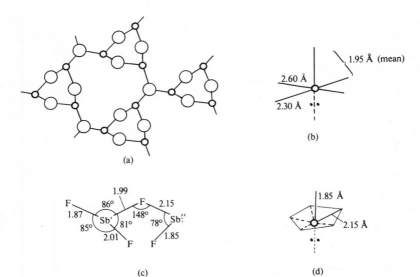

FIG. 20.2. (a) Topology of the 2D $Sb_3F_{10}^-$ ion. (b) Coordination of Sb(III). (c) The $Sb_2F_4^{2+}$ ion in Sb_4F_{16}. (d) Coordination of Sb(III).

is normal, SbF_3E (as in SbF_3), with four long contacts of 2.54–2.87 Å surrounding the lone pair (Fig. 20.2(c)). The Sb″ atom forms 2 short bonds and also 4 others (2.33–2.41 Å) approximately coplanar with one of the short bonds. The Sb″ atom is below the equatorial plane, suggesting a pentagonal bipyramidal arrangement with the lone pair axial (Fig. 20.2(d)).

Structural studies have been made of complex chlorides or bromides containing ions of the following types: $(SbX_4)^-$, $(SbX_5)^{2-}$, $(SbX_6)^{3-}$, and the enneahalide ion $(Sb_2X_9)^{3-}$. The anion in $(SbCl_4)(C_5H_5NH)$[8] is an infinite chain of very distorted octahedra sharing 'skew' edges (p. 212). All Cl–Sb–Cl angles are close to 90°, but the Sb–Cl bond lengths are: 2 of 2.38 Å; 2 of 2.64 Å; and 2 of 3.12 Å, the four latter being involved in the unsymmetrical bridges between the $SbCl_6$ groups, as shown in Fig. 20.3(a). In $(NH_4)_2SbCl_5$[9] the NH_4^+ and Cl^- ions form a distorted closest packing with vacant Cl^- sites in every third layer, and the Sb atoms occupy what would be octahedral holes in a normal closest packing. Owing to the absence of some Cl^- ions the nearest neighbours of Sb are 5 Cl arranged at five of the vertices of an octahedron, the bond lengths being: axial, 2.36 Å; equatorial 2.58 Å (two), and 2.69 Å (two); the lone pair presumably occupies the sixth bond position, Fig. 20.3(b).

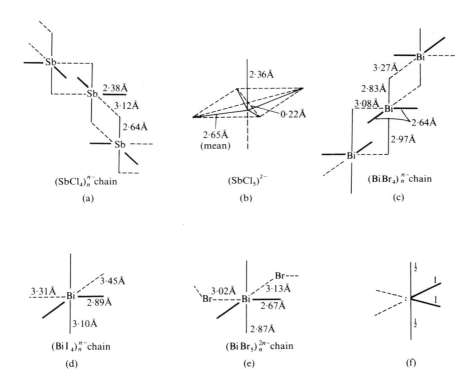

FIG. 20.3. Environments of Sb or Bi in complex halides (see text).

The $(SbBr_6)^{3-}$ ion has been studied in halides which also contain the $(SbBr_6)^-$ ion, as noted in an earlier section. The enneabromide ion occurs in the pyridinium salt $(C_5H_5NH)_5(Sb_2Br_9)Br_2$.[10] It consists of two octahedral $SbBr_6$ groups sharing a face, with Sb–Br: terminal, 2.63 Å; and bridging, 3.00 Å.

Complex fluorides of Bi are, like BiF_3, ionic compounds. For example, NH_4BiF_4[11] has an ionic structure in which the infinite $(BiF_4)_n^{n-}$ layer is built of BiF_9 coordination groups (Bi–F, 2.19–2.86 Å) of the same type as in orthorhombic BiF_3 (p. 421). When 6-coordinated by Cl, Br, or I Bi(III) forms 6 octahedral bonds. The octahedron is regular in isolated ions, as in the cubic $Cs_2NaBiCl_6$[12] and in Rb_3BiBr_6[12a] or $Rb_5I^-(I_3)^-(BiI_6)^{3-}.2H_2O$[13] but distorted to various extents when there is sharing of halogen atoms, as in $K_4(Bi_2Br_{10}).4H_2O$ (one edge shared)[14] or in the enneahalides. The structures of $Cs_3Bi_2Cl_9$, $Cs_3Bi_2Br_9$, $Cs_3Bi_2I_9$, and $(\phi_4P)_3Bi_2Br_9$ are described on p. 465.

The ions $(SbI_4)_n^{n-}$, $(BiBr_4)_n^{n-}$, and $(BiI_4)_n^{n-}$ in the 2-picolinium salts[15] consist of infinite 'skew' octahedral chains, Fig. 20.3(c) and (d), very similar in structure to the $(SbCl_4)^-$ chains shown at (a), with three pairs of bonds of similar length. In the bispiperidinium salt $(C_5H_{10}NH_2)_2BiBr_5$,[16] with which the Sb compound is isostructural, there are chains of octahedral MX_6 groups sharing *cis* vertices, and here also there is the same pattern of bond lengths, Fig. 20.3(e). If we regard the Bi–Br bond in the $BiBr_3$ molecule (2.63 Å) as a single bond then the two next shortest bonds in (c) and (e) correspond to bond order ½, using Pauling's equation, $d_n = d_1 - 0.6 \log n$. Similarly, if the six octahedral bonds in crystalline BiI_3 (3.07 Å) are regarded as having bond order ½ then the shortest bonds in (d) are single bonds. Thus in each case the total bond order is close to 3. It may be noted that the four stronger bonds could alternatively be regarded as distorted trigonal bipyramidal bonds, when the remaining two weak (equatorial) bonds would be in the plane of the lone pair as shown at (f), where the approximate bond orders are indicated.

An example of Bi forming tetragonal pyramidal bonds is provided by $Bi_{12}Cl_{14}$.[17] In this extraordinary compound, which is formed by dissolving the metal in the molten trichloride, there are groups of 9 Bi atoms (at the vertices of a tricapped trigonal prism), tetragonal pyramidal $BiCl_5^{2-}$ groups, and $Bi_2Cl_8^{2-}$ groups formed from two $BiCl_5^{2-}$ ions sharing two Cl atoms:

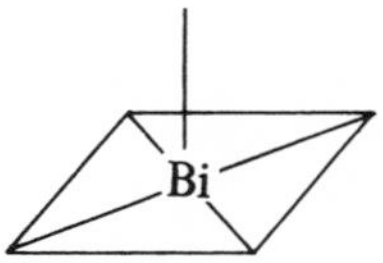

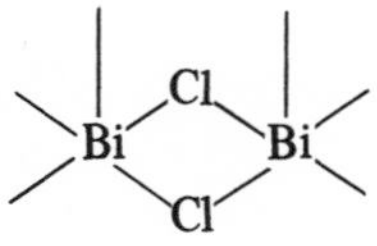

The structural formula of this halide may be written $(Bi_9^{5+})_2(BiCl_5^{2-})_4(Bi_2Cl_8^{2-})$ to indicate the relative numbers of the three kinds of structural unit. The existence of the Bi_9^{5+} ion has also been established in $Bi^+Bi_9^{5+}(HfCl_6)_3$.[18]

(1) IC 1972 **11** 2322
(2) AK 1952 **4** 175
(3) AK 1953 **6** 77
(4) IC 1971 **10** 1757
(5) IC 1971 **10** 2793
(6) AC 1975 **B31** 2322
(7) AK 1951 **3** 17
(7a) CC 1977 253
(7b) JCS D 1977 971
(8) JCS A 1970 1356
(9) JCS A 1971 298
(10) JCS A 1970 1359
(11) ACSc 1964 **18** 1554
(12) AC 1972 **B28** 653
(12a) AC 1978 **B34** 2288
(13) AC 1977 **B33** 1957
(14) AC 1977 **B33** 1954
(15) JPC 1967 **71** 3531
(16) JPC 1968 **72** 532
(17) IC 1963 **2** 979
(18) IC 1973 **12** 1134

The oxygen chemistry of trivalent As, Sb, and Bi

In this section we shall summarize what is known of the structures of the trioxides, complex oxides, and oxyhalides of these elements. In the trioxides and in those complex oxides of Sb^{III} which have been studied the stereochemistry of the Group VB atoms is simple; they form three pyramidal bonds like P^{III} in P_4O_6. The behaviour of Sb in its oxyhalides and Sb_3O_6OH is less simple, and moreover different from that of Bi. For this reason the oxyhalides of Sb and Bi are considered separately. Structurally related to the oxyhalides of bismuth are a number of complex oxides which will also be mentioned.

The trioxides of As, Sb, and Bi

Each of the trioxides exists in at least two polymorphic modifications (Table 20.4). Vaporization of As and Sb trioxides gives molecules As_4O_6 and Sb_4O_6 which break

TABLE 20.4

Polymorphic forms of As_2O_3, Sb_2O_3, *and* Bi_2O_3

	As_2O_3	Sb_2O_3	Bi_2O_3
Low-temperature form	Monoclinic (claudetite)	Orthorhombic (valentinite)	(α) Monoclinic
High-temperature form	Cubic (arsenolite)	Cubic (senarmontite)	(δ) Cubic (see text)
Transition temperature	110°C	606°C	729 °C (α-δ)

down to the simpler As_2O_3 and Sb_2O_3 molecules only at high temperatures. The structures of As_4O_6 and Sb_4O_6 in the vapour state have been shown to be similar to that of P_4O_6 (see Table 19.5, p. 854). The cubic forms of these oxides consist of molecules of the same kind. In cubic Sb_4O_6[1] Sb has 3 pyramidal neighbours at 1.98 Å (O–Sb–O, 96°) and no other nearer than 3.81 Å, indicating a simple description of the stereochemistry as SbO_3E, the lone pair occupying the fourth tetrahedral bond position; contrast the more complex situation in valentinite (below).

There are two forms, (I) and (II), both monoclinic, of low-temperature As_2O_3, the mineral claudetite.[2] Both have the simplest type of layer structure for a compound A_2X_3, namely, a 6^3 net of As atoms joined through O atoms. The layers are puckered (O–As–O, 95½°; As–O, 1.79 Å) and differ in the following way. Since each As atom could be either above or below the mean plane of the O atoms there are various possible configurations of the layer differing only in the arrangement of the As atoms of the two kinds. The two forms of monoclinic As_2O_3 differ in this way, and a third configuration of the layer is found in As_2S_3. In the accompanying sketch only As atoms are shown, and the filled circles represent As atoms above the plane of the O(S) layer.

The second form of Sb_2O_3, valentinite, has a different type of structure.[3] Instead of finite molecules Sb_4O_6 there are infinite double chains of the kind shown below, with Sb–O, 2.01 Å, O–Sb–O bond angles, 80°, 92°, and 98° and Sb–O–Sb angles, 116° and 131°. The description of the stereochemistry here is

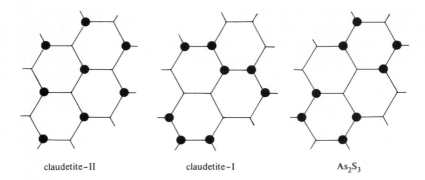

claudetite-II   claudetite-I   As₂S₃

not obvious, for in addition to its three (pyramidal) nearest neighbours Sb has one O in an adjacent chain at 2.52 Å and one in the same chain at 2.62 Å as next nearest neighbours.

The crystal chemistry of Bi_2O_3 is completely different from, and far more complex than, that of either As_2O_3 or Sb_2O_3 and there is a considerable literature on the subject. Some of the confusion regarding the number and structures of high-temperature forms is due to the fact that Bi_2O_3 is very easily contaminated. For example, fusion for a short time in a porcelain crucible gives a b.c. phase with the composition $SiBi_{12}O_{20}$ to which we refer shortly. The normal form of this oxide, α-Bi_2O_3, has monoclinic symmetry; it transforms at 729 °C to the cubic δ form, which is stable up to the melting point (824 °C). On cooling, however, two metastable phases can be obtained, the tetragonal β phase at 650 °C and b.c.c. γ at 639 °C. These are both metastable phases. A further complication is that both the β and δ phases have structures related to fluorite, there being ordered O vacancies in β but a complicated disorder of the O atoms in δ-Bi_2O_3. It is the statistical distribution of O vacancies and the movement of the O^{2-} ions which give rise to the high ionic conductivity of δ-Bi_2O_3.

The normal α form has a complex structure[4] containing two kinds of non-equivalent Bi atoms, both with distorted octahedral coordination, there being 5 neighbours at distances from 2.13-2.6 Å and either one more at 2.8 Å or two at 3.3 and 3.4 Å. The cubic γ form is related to a group of compounds with a common framework structure and almost identical cubic unit cell dimensions built from BiO_5E octahedra and having the basic composition $Bi_{24}O_{40}$. Within the cages of this structure there are two tetrahedral sites which must be occupied by ions with a total charge of +8 required to balance the charge on $(Bi_{24}O_{40})^{8-}$. This charge balance results most simply in the stoichiometric compounds $M^{IV}Bi_{12}O_{20}$ (M = Ge, Si),[5] but there are other compounds with this framework structure, probably including γ-Bi_2O_3:[6]

	Tetrahedral sites in cages
$Bi_{25}FeO_{40}$	Bi^{5+}, Fe^{3+}
$Bi_{38}ZnO_{60}$	$1\frac{1}{3} Bi^{5+}, \frac{2}{3} Zn^{2+}$
$(Bi_{25\frac{1}{3}}Zn_{\frac{2}{3}}O_{40})$	
γ-Bi_2O_3	Bi^{5+}, Bi^{3+}
$(Bi_{25}^{3+}Bi^{5+}O_{40})$	

In many Bi^{III} oxy-compounds the metal atom has 5 or 6 neighbours at distances from about 2.1 Å to 2.7 Å (mean close to 2.4 Å) and then a small number of additional neighbours. In some cases, for example, $GeBi_{12}O_{20}$, the coordination group consists of three very close O atoms, a, b, and c, (Fig. 20.4) at 2.08, 2.22,

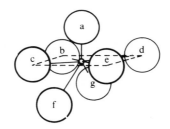

FIG. 20.4. Coordination of Bi in $GeBi_{12}O_{20}$.

and 2.23 Å, arranged pyramidally, then d and e at the same distance (2.63 Å) and finally f and g at 3.08 and 3.17 Å respectively. Regarded as a group of 5 neighbours the atoms a–e form a square pyramidal group, with mean Bi—O 2.36 Å as in many other B^{III} oxy-compounds (see Table 20.5).

(1)   AC 1975 **B31** 2016
(2)   MH 1975 **106** 755
(3)   AC 1974 **B30** 458
(4)   ZaC 1978 **444** 151, 167
(5)   JCP 1967 **47** 4034
(6)   JSSC 1975 **15** 1

Sulphates of Sb(III)

Like certain other B subgroup elements (see, for example, $Te_2O_3SO_4$ and $I_2O_4SO_4$) Sb forms a number of sulphates and 'basic' sulphates. Since these compounds contain SO_4 groups bridging Sb atoms (compare $Zr(SO_4)_2$ and its hydrates, p. 333) we may rewrite their formulae as follows:

TABLE 20.5
Coordination of Bi^{III} *in some oxy-compounds*

Compound	Total c.n.	Bi to 5 or 6 O atoms	Further O to	Mean Bi to 5 or 6 O	Reference
Bi$_4$Si$_3$O$_{12}$	9	2·15–2·62 Å	3·55 A	2·39 A	ZK 1966 **123** 73
Bi$_2$GeO$_5$	7	2·13–2·66	3·18	2·35	ACSc 1964 **18** 1555
Bi(OH)CrO$_4$ (monoclinic)	9	2·23–2·58	3·33	2·38 ⎫	ACSc 1964 **18** 1937
Bi(OH)CrO$_4$ (orthorhombic)	9	2·19–2·68	3·09	2·38 ⎬	
Bi$_2$O$_2$SO$_4$. H$_2$O	10	⎰ 2·22–2·57	3·45	2·38 ⎫	
		⎱ 2·10–2·59	3·35	2·32 ⎬	ACSc 1964 **18** 2375
Bi(OH)SeO$_4$. H$_2$O	9	2·19–2·64	3·16	2·40 ⎭	
Bi$_{12}$GeO$_{20}$	7	2·08–2·64	3·17	2·36	JCP 1967 **47** 4034

$$\text{Sb}_2\text{O}_3 . 3\text{SO}_3 : \text{Sb}_2(\text{SO}_4)_3 \qquad \text{AC 1976 } \textbf{B32} \text{ 2787}$$
$$\text{Sb}_2\text{O}_3 . 2\text{SO}_3 : \text{Sb}_2\text{O}(\text{SO}_4)_2 \qquad \text{AC 1975 } \textbf{B31} \text{ 2081}$$
$$3\text{Sb}_2\text{O}_3 . 2\text{SO}_3 : \text{Sb}_6\text{O}_7(\text{SO}_4)_2 \qquad \text{AC 1976 } \textbf{B32} \text{ 1771}$$

The first is the simplest from the topological, though not from the structural, standpoint, since each Sb is joined through bridging SO$_4$ groups to three others. The infinite linear molecules represent a considerable elaboration of the simple 3-connected ladder of Fig. 20.5(a). Along each link there is a bridging SO$_4$ group,

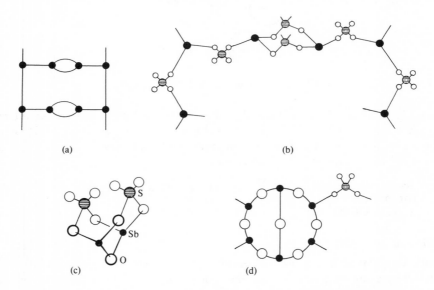

(a)   (b)

(c)   (d)

FIG. 20.5. (a) and (b), the structure of Sb$_2$(SO$_4$)$_3$; (c) the molecule Sb$_2$O(SO$_4$)$_2$; (d) Sb$_6$O$_7$(SO$_4$)$_2$ (see text).

as shown for a portion of the ladder at (b). In the other compounds there are bridging O atoms in addition to bridging SO_4 groups. The bicyclic molecule $Sb_2O(SO_4)_2$, (c), contains two chair-shaped rings, and it is interesting that $Sb_2(SO_4)_3$ does not adopt the analogous structure with three such rings instead of the much more complex structure (b). The structure of $Sb_6O_7(SO_4)_2$ is much more complex. There are cylindrical sub-units on the surface of which Sb and O atoms alternate to form a 3-connected net. One-third of the Sb atoms in this net, of composition $(Sb_6O_6)_n$, are cross-connected in pairs through additional O atoms within the cylinder, giving the composition $(Sb_6O_7)_n$, while two-thirds are connected to similar atoms of other cylinders through bridging SO_4 groups, as shown in the very diagrammatic sketch of Fig. 20.5(d). The result is a 3D structure of composition $(Sb_6O_7)_n(SO_4)_{2n}$, in which each Sb forms 4 bonds—3 in the range 2.0–2.2 Å and a fourth of length 2.33–2.38 Å.

Meta-arsenites

Incorrect formulae have been given to some of these compounds. For example, a salt sometimes formulated as Na_2HAsO_3 is actually the polymeta-arsenite $(NaAsO_2)_n$,

and an X-ray study shows that it consists of chains of pyramidal AsO_3 groups held together by Na^+ ions.[1]

(1)   AC 1958 11 742

Complex oxides of trivalent As and Sb

A number of complex oxides $M^{II}Sb_2O_4$ (M^{II} = Mg, Zn, Mn, Fe, Co, Ni) and also $NiAs_2O_4$ are isostructural with Pb_3O_4 (Fig. 12.15, p. 559); Sb replaces Pb^{II} and M^{II} replaces Pb^{IV}. Refinement of the structure of $ZnSb_2O_4$[1] shows small distortions of the pyramidal SbO_3 and octahedral ZnO_6 coordination groups. The mean bond lengths are: Sb–O, 1.97 Å, Zn–O, 2.11 Å, and O–Sb–O angles, 93.4° (two) and 96.4° (one). Ref. (1) includes references to recent work on the Mg and Fe compounds.

Some crystalline compounds formed by As_2O_3 with alkali and ammonium halides, for example, $NH_4Cl.As_2O_3.\frac{1}{2}H_2O$, $KCl.2As_2O_3$, etc., are closely related structurally to arsenious oxide. In the former,[2] there are hexagonal As_2O_3 sheets, with all O atoms directed to one side of the sheet, arranged in pairs with the O atoms turned toward each other and with water molecules enclosed between the layers. Between these pairs of As_2O_3 layers there are NH_4^+ and Cl^- ions.

(1)   AC 1982 B38 2022   (2)   AK 1955 8 245

Complex oxides of trivalent Bi

Bismuth forms with other metallic oxides an extraordinary variety of complex oxides, in some of which O can be replaced by F.

The systems $Ca(Sr, Ba, Cd, Pb)O-Bi_2O_3$. Studies have been made of some of the phases formed when Bi_2O_3 is fused with one of the oxides CaO,[1] SrO,[1] BaO,[1] CdO,[2] or PbO.[3] The phases studied show variation in composition over rather wide ranges. For example, in the system $BaO-Bi_2O_3$ there is a rhombohedral phase ($x = 0.10-0.22$) and a tetragonal phase ($x = 0.22-0.50$), where x is the atomic fraction of Ba in $Ba_{2x}Bi_{2-2x}O_{3-x}$. Structures have been proposed for a number of such phases, in which it is concluded that the metal content of the unit cell is constant but the oxygen content variable.

X-ray studies have also been made of a series of complex oxides with tetragonal or pseudotetragonal symmetry which all have the a dimension of the unit cell close to 3.8 Å, and are based on sequences of Bi_2O_2 layers of the same type as in Fig. 20.6, interleaved with portions of perovskite-like structure. They are summarized in Table 20.6, and illustrated in Fig. 20.6. The simplest member of the family is at

TABLE 20.6
Complex oxides and oxyhalides of bismuth

Number of perovskite layers	Formula	c	Reference
		(Å)	
1	Bi_2NbO_5F, Bi_2TaO_5F, $Bi_2TiO_4F_2$	~16·5	AK 1952 5 39
2	$PbBi_2Nb_2O_9$, Bi_3NbTiO_9	25·5	AK 1949 1 463
3	$Bi_4Ti_3O_{12}$	32·8	AK 1949 1 499
4	$BaBi_4Ti_4O_{15}$	41·7	AK 1950 2 519

present represented only by compounds in which part of the O is replaced by F. From the structural standpoint these compounds are closely related to the complex bismuth oxyhalides (p. 897) in which the square Bi_2O_2 layers are interleaved with halogen layers. A great variety of compounds can be prepared, for example, $(MBi)_6R_4O_{18}$, in which M can be Na, K, Ca, Sr, Ba, or Pb, and R is Ti, Nb, or Ta. Note that compounds of this family such as Bi_2TaO_5F are quite unrelated to those described on p. 904, for example, $BiTa_2O_6F$.

(1) AKMG 1943 **16A** No. 17   (3) ZK 1939 **101** 483
(2) ZPC 1941 **B49** 27

The oxyhalides of trivalent Sb

Structural data are available for two forms of one oxyfluoride. Four forms of SbOF have been prepared thermally from SbF_3 and Sb_2O_3. The L form consists of 'ladders' (Fig. 20.7),[1] while the M form has a layer structure.[2] In both forms the

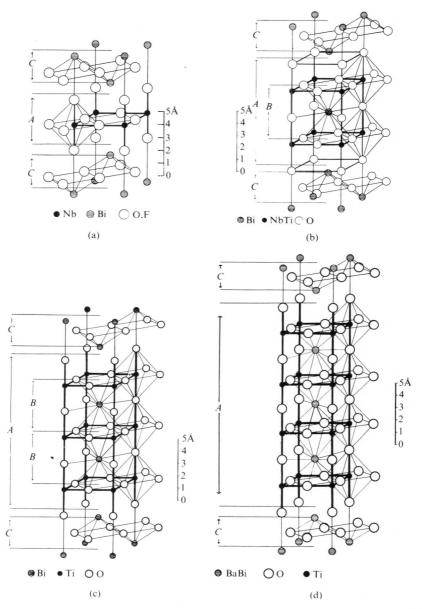

FIG. 20.6. The crystal structures of (a) Bi_2NbO_5F; (b) Bi_3NbTiO_9; (c) $Bi_4Ti_3O_{12}$; and (d) $BaBi_4Ti_4O_{15}$, showing one-half of each unit cell. In each diagram C represents the Bi_2O_2 layer (in (a), $Bi_2(O, F)_2$, in (d), $(Ba, Bi)_2O_2$). In (a) A denotes the layer of $Nb(O, F)_6$ octahedra and in (b), (c), and (d) the perovskite portions of the structure. The regions B in (b) and (c) mark off unit cells of the hypothetical perovskite structures $BaNb_{0.5}Ti_{0.5}O_3$ and $BaTiO_3$ respectively.

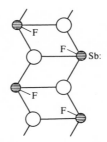

FIG. 20.7. Double chain (ladder) in L-SbOF (diagrammatic).

coordination group around Sb(III) consists of 3 O and 1 F at four of the vertices of a trigonal bipyramid, the lone pair occupying an equatorial bond position as in Sb_2O_4 (p. 899).

Other oxyhalides, mostly oxychlorides and oxybromides, result from the controlled hydrolysis of the trihalides, and are of interest for two main reasons. First, they are quite unrelated to the oxyhalides of bismuth. Although both antimony and bismuth form compounds MOX the structures of the antimony compounds are quite different from those of the compounds BiOX, which have been described on p. 486. The more complex oxyhalides of Sb have no analogues among Bi compounds. Second, a feature of the published structures of the antimony oxyhalides is the formation of extended Sb–O systems, generally layers interleaved with halogen ions. In these structures Sb is bonded to 3 or 4 O atoms in one of the arrangements (a) or (b) or some intermediate arrangement. It will be seen that in order to change from 4- to 3-coordination it is only necessary to raise the Sb atom above the equatorial plane.

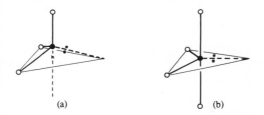

Structural studies have been made of the following compounds, listed in order of decreasing X:Sb ratio: SbOCl,[3] $Sb_4O_5Cl_2$[4] and $Sb_4O_5Br_2$, $[Sb_8O_8(OH)_4]$- -$[(OH)_{2-x}(H_2O)_{1+x}]Cl_{2+x}$,[5] and $Sb_8O_{10}(OH)_2X_2$[6] (X = Cl, Br, or I).

SbOCl is built of puckered sheets in which two-thirds of the Sb atoms are linked to 2 O + 1 Cl and the remainder to 4 O, the sheets $(Sb_6O_6Cl_4)_n^{2n+}$ being held together by Cl⁻ ions. (The distinction between the two types of Cl is based on the much greater Sb–Cl distance (2.9 Å) between Sb and interlayer Cl than Sb–Cl within a layer (2.3–2.5 Å).

$Sb_4O_5Cl_2$. Here also there are infinite layer ions, here of composition $(Sb_4O_5)_n^{2n+}$, in which one half of the Sb atoms are 3- and the others 4-coordinated, with Cl^- ions between the layers. The coordination of both kinds of Sb atom is of type (b), there being 3 closest neighbours at around 2 Å and a fourth at either 2.46 Å or 2.91 Å.

In the two more complex oxyhalides the characteristic structural feature is apparently a double chain, in which all the Sb atoms are 4-coordinated, which is joined through OH groups to form infinite layers $[Sb_8O_8(OH)_4]_n^{4n+}$ in the first compound and complex cylindrical chain ions $[Sb_4O_5(OH)]_n^{n+}$ in the second.

The following oxyiodides of Sb(III) are known: Sb_3O_4I, $Sb_8O_{11}I_2$, and Sb_5O_7I (low and high temperature forms, α and β). In α-Sb_5O_7I[7] the I atoms are situated between layers formed from 3-coordinated Sb atoms and O atoms, Sb forming either 2 or 3 weaker bonds to I.

(1) JSSC 1973 **6** 191
(2) ACSc 1972 **26** 3849
(3) AK 1953 **6** 89
(4) AC 1978 **B34** 2402
(5) AK 1955 **8** 257
(6) AK 1955 **8** 279
(7) AC 1975 **B31** 234

Oxyfluorides of Bi

We have seen that BiF_3 is structurally different from all other trihalides of As, Sb, and Bi. Similarly, Bi oxyfluorides, with the exception of BiOF, are not related structurally to the compounds formed by Bi (or Sb) with the other halogens. Oxidation of BiF_3 gives the following oxyfluorides, the structures of which are described elsewhere as indicated:

$BiO_{0.1}F_{2.8}$: tysonite (LaF_3) structure (p. 420): ACSc 1955 **9** 1206
$BiO_{0.5}F_2$ (Bi_2OF_4); cubic (p. 422); ARPC 1952 **3** 369
BiOF: BiOCl structure (p. 486): ACSc 1964 **18** 1823, 1851

Complex oxyhalides of Bi with Li, Na, Ca, Sr, Ba, Cd, and Pb

By fusing the bismuth oxyhalides with excess of one of the halides of the above metals, Sillén and co-workers have prepared a large number of complex oxyhalides in which the halogen is Cl, Br, or less usually I. They all form tetragonal leaflets with an a axis of about 4 Å, but c axes up to 50 Å are found. These compounds are closely related to the simple bismuth oxyhalides and like those compounds contain characteristic oxygen layers. The metal–oxygen sheet is shown in Fig. 20.8(a). The simplest possibility is that sheets of this type alternate with single, double, or triple halogen layers, forming the X_1, X_2, or X_3 structures of Fig. 20.8(b), (c), and (d). The simple bismuth oxyhalides are, of course, of Sillén's X_2 structure type. More complex structures arise if halogen layers of more than one kind occur in the same structure, as in $SrBi_3O_4Br_3$ (X_1X_2 structure). In the X_1 compounds, $M^{II}BiO_2X$ or $M^IBi_3O_4Cl_2$, and the X_1X_2 compounds $M^{II}Bi_3O_4Cl_3$, all the metal atoms are associated with the metal–oxygen layer (a), apparently arranged at random in the available metal positions, but in the X_3 structures there are also metal atoms between

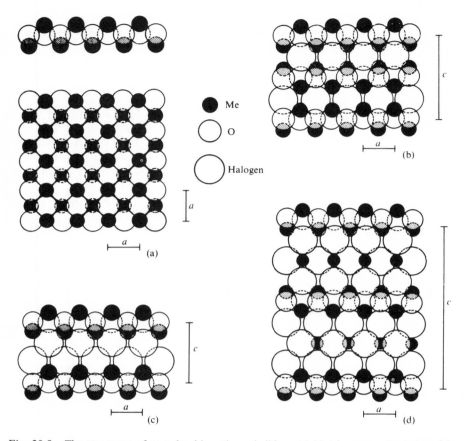

Fig. 20.8. The structures of complex bismuth oxyhalides. (a) Metal–oxygen sheet viewed in directions parallel and perpendicular to the sheet. (b), (c), (d) Elevations of structures built from (b) single halogen layers X_1, (c) double halogen layers X_2, and (d) triple halogen layers X_3 (Sillén).

the halogen layers (d). Not all these latter positions are occupied, however, so that structures containing X_3 layer are apparently always defect structures, as shown by the formulae of the cadmium bismuth oxyhalides in Table 20.7. A further structure type (O_1X_2) has been suggested for a barium bismuth oxyiodide in which the characteristic metal–oxygen layers are separated alternately by double halogen layers and oxygen layers.

Other compounds also related structurally to these oxyhalides include $Bi_{24}O_{31}Cl_{10}$ and the corresponding Br compound. The existence of Bi oxyhalides other than BiOX has been disputed, but according to Sillén the compounds $Bi_{24}O_{31}X_{10}$ (originally formulated $Bi_7O_9X_3$) do exist, and the chloride, for example, can be prepared by heating BiOCl to temperatures above 600 °C. A structure has been

TABLE 20.7

Structure types of bismuth oxyhalides[a]

Structure type	Examples
X_1	$BaBiO_2Cl$, $CdBiO_2Br$, $LiBi_3O_4Cl_2$
X_2	$BiOCl$, $LaOCl$
X_3	$Ca_{1.25}Bi_{1.5}O_2Cl_3$
X_1X_2	$SrBi_3O_4Cl_3$
X_2X_3	$Ca_{2-3x}Bi_{3+2x}O_4Cl_5$
$X_1X_1X_2$	$SrBi_2O_3Br_2$
$X_1X_2X_3$	$Cd_{2-3x}Bi_{5+2x}O_6Cl_7$

(a) ZaC 1942 **250** 173; ZK 1942 **104** 178; AKMG 1947 **25** 49.

proposed for these compounds. The various structure types proposed by Sillén for these oxyhalides are shown diagrammatically in Fig. 20.8, and some representative compounds are listed in Table 20.7. It should perhaps be remarked that certain features of these structures cannot be regarded as wholly satisfactory, though there seems little doubt about the general scheme according to which they are built. The general geometrical analogies with the layer-type hydroxyhalides will be obvious.

Oxides $M^{III}M^VO_4$

In addition to oxides $M_2^{III}O_3$ and $M_2^VO_5$ As and Sb form oxides $M^{III}M^VO_4$, and there are also isostructural compounds containing Nb and Ta:

	M^{III} *coordination*	M^V *coordination*	*Reference*
$As^{III}As^VO_4$	Trigonal pyramidal	Tetrahedral	AC 1980 **B36** 439
$Sb^{III}As^VO_4$		Tetrahedral	AC 1980 **B36** 1923
α-$Sb^{III}Sb^VO_4$	4-pyramidal		AC 1977 **B33** 1271
$Bi^{III}Sb^VO_4$	(Fig. 20.9(c))	Octahedral	AK 1952 **3** 153
$Sb^{III}Nb^VO_4$			CC 1965 611
β-$Sb^{III}Sb^VO_4$			PCS 1964 400

The hygroscopic $AsO_2(As^{III}As^VO_4)$ is prepared by heating cubic As_2O_3 in an autoclave under a pressure of O_2. It is formed of layers based on the 6-gon net (as in monoclinic As_2O_3, p. 889) with an additional O attached to alternate As atoms so that one half of these atoms form three pyramidal bonds (As–O, 1.81 Å, O–As–O, 90°) and the remainder four tetrahedral bonds (As–O, 1.72 Å, bridging, and 1.61 Å, terminal). This layer is intermediate between the M_2O_3 layer of As_2O_3 (all M 3-coordinated) and the M_2O_5 layer of one form of P_2O_5 (all M 4-coordinated).

In contrast to the trigonal pyramidal coordination of As^{III} in As_2O_4 the coordination of Sb^{III} in these compounds is the characteristic 'one-sided' pyramidal coordination illustrated in Fig. 20.9(c). The orthorhombic (α) form of Sb_2O_4 occurs as the mineral cervantite, isostructural with $SbNbO_4$ and $SbTaO_4$, and there

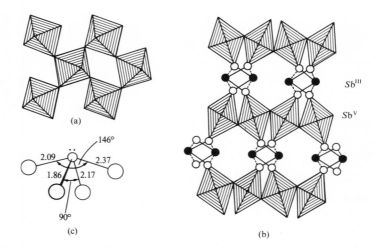

FIG. 20.9. (a) Layer of composition SbO_4 formed from Sb^VO_6 octahedra sharing equatorial vertices. (b) Showing Sb^{III} atoms situated between the SbO_4 layers, which are perpendicular to the plane of the paper. Dotted lines represent the longer Sb—O bonds. (c) The pyramidal coordination of Sb^{III} in α-Sb_2O_4.

is also a monoclinic β form made by heating the α form in air or oxygen. The structures of the two polymorphs are very similar, one being a 'sheared' version of the other. Both consist of corrugated layers (Fig. 20.9(a)) formed from Sb^VO_6 octahedra sharing all their equatorial vertices (as in the plane layer in K_2NiF_4) and the Sb(III) atoms lie between the layers in positions of rather irregular pyramidal 4-coordination. The structure of β-Sb_2O_4 is illustrated in Fig. 20.10(b). The environments of Sb(III) and Sb(V) are very similar in the two forms of Sb_2O_4, and the compounds $Sb^{III}Nb^VO_4$ and $Bi^{III}Sb^VO_4$ are isostructural with α-Sb_2O_4. A n.d. refinement of the structure of the latter gives the bond lengths and angles shown in Fig. 20.9(c).

The structure of $Sb^{III}As^VO_4$ differs from those of both As_2O_4 and Sb_2O_4 in having 4-coordination of Sb and As but of different kinds, 'one-sided' and tetrahedral. Together these atoms form layers based on the simplest planar 4-connected net (4^4), consisting of 8-membered M_4O_4 rings.

The oxygen chemistry of pentavalent arsenic and antimony

The oxy-compounds of pentavalent arsenic

The pentoxides of As, Sb, and Bi cannot be prepared, like P_2O_5, by direct oxidation of the element, for they decompose at high temperatures with loss of oxygen. As_2O_5 is prepared by dehydration of its 'hydrates':

$$
\begin{array}{ccccc}
& -30\,^\circ\text{C} & & 36\,^\circ\text{C} & & 170\,^\circ\text{C} \\
\text{As}_2\text{O}_5.7\text{H}_2\text{O} & \longrightarrow & \text{As}_2\text{O}_5.4\text{H}_2\text{O} & \longrightarrow & \text{As}_2\text{O}_5.\tfrac{5}{3}\text{H}_2\text{O} & \longrightarrow & \text{As}_2\text{O}_5 \\
(\text{H}_3\text{AsO}_4.2\text{H}_2\text{O}) & & (\text{H}_3\text{AsO}_4.\tfrac{1}{2}\text{H}_2\text{O}) & & (\text{H}_5\text{As}_3\text{O}_{10})
\end{array}
$$

The second of these compounds, $\text{H}_3\text{AsO}_4.\tfrac{1}{2}\text{H}_2\text{O}$, is a true hydrate of H_3AsO_4 for it consists of a hydrogen-bonded assembly of tetrahedral AsO(OH)_3 molecules and H_2O molecules.[1] The third compound, $\text{H}_5\text{As}_3\text{O}_{10}$, consists of infinite chains of octahedral and tetrahedral groups sharing vertices as shown in Fig. 20.10.[2] The

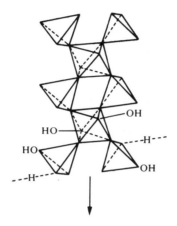

FIG. 20.10. The structure of $\text{H}_5\text{As}_3\text{O}_{10}$.

H atoms were not directly located, their probable positions being deduced from the O–O separations between the chains.

Crystalline As_2O_5 consists of a complex 3D framework of tetrahedral AsO_4 and octahedral AsO_6 groups (As–O, 1.68 and 1.82 Å respectively).[3]

It is apparently not possible to prepare the ortho-, pyro-, or meta-arsenic acids by dehydrating hydrates of H_3AsO_4, but salts of all three types are well known and structurally similar to the corresponding phosphates. In all these salts there is tetrahedral coordination of As(v), in AsO_4^{3-}, $\text{AsO}_3(\text{OH})^{2-}$, $\text{AsO}_2(\text{OH})_2^-$, and in the molecule AsO(OH)_3 noted above. YAsO_4 is isostructural with YPO_4 and ZrSiO_4, and KH_2AsO_4 with KH_2PO_4. The ion $\text{AsO}_3(\text{OH})^{2-}$ has been studied in a number of salts, including $\text{CaHAsO}_4.3\text{H}_2\text{O}$[4] and $\text{MgHAsO}_4.4\text{H}_2\text{O}$,[5] and the ion $\text{AsO}_2(\text{OH})_2^-$ in $\text{NaAsO}_2(\text{OH})_2$.[6] Bond lengths are close to: As–O, 1.67 Å; As–OH, 1.73 Å. Meta-arsenates contain linear or cyclic ions similar to those in phosphates and silicates; for example, the linear ion in NaAsO_3[7] and the cyclic $\text{As}_4\text{O}_{12}^{4-}$ ion in $\text{Tl}_4(\text{AsO}_3)_4$.[8] It is interesting that the anion in $\text{K}_5\text{As}_3\text{O}_{10}$[9] is not linear like $\text{P}_3\text{O}_{10}^{5-}$, in which the angle in the P---P---P skeleton is 173°, but is bent, with

As---As---As equal to 105°. This is also the case in the anion of $BaH(AsCr_2O_{10})$,[10] (a). Note also the $AsCr_3O_{13}^{3-}$ ion (b).[10a]

Tetrahedral coordination of As^V by oxygen has also been demonstrated in arsonic acids $R.AsO(OH)_2$,[11] and in cacodylic acid, $(CH_3)_2AsO(OH)$,[12] which forms hydrogen-bonded dimers in the crystal like those of carboxylic acids.

(a)   (b)   (c)

Octahedral coordination of As(v) by oxygen has already been noted in the catechol derivative and in $H_5As_3O_{10}$. We have mentioned that there is both tetrahedral and octahedral coordination of As(v) in As_2O_5 and $H_5As_3O_{10}$; this is also the case in the cyclic anion in $BaH_6As_4O_{14}$,[13] (c), in which there are discrete ions $As_4O_{14}^{8-}$. In $NaHAs_2O_6$[14] the anion consists of layers based on the simplest planar 4-connected net formed from the units (c) joined together by sharing the O atoms shown as shaded circles.

(1)	AC 1968 **B24** 987	(9)	AC 1980 **B36** 1634
(2)	AC 1966 **21** 808	(10)	AC 1979 **B35** 726
(3)	ZaC 1978 **441** 5	(10a)	AC 1978 **B34** 3350
(4)	AC 1973 **B29** 90	(11)	BCSJ 1962 **35** 1600
(5)	AC 1976 **B32** 1460	(12)	JCS 1965 4466
(6)	AC 1977 **B33** 2684	(13)	AC 1977 **B33** 3222
(7)	AC 1956 **9**, 87, 811	(14)	AC 1978 **B34** 3727
(8)	ZaC 1969 **367** 92		

The oxy-compounds of pentavalent antimony

The pentoxide is formed by dissolving Sb in HCl, precipitating with HNO_3, and heating at 780°C for 3 minutes. Its crystal structure is a complex 3D assembly of vertex- and edge-sharing SbO_6 octahedra[1] which is the same as found for one (B) of the many forms of Nb_2O_5 (q.v.). Bi_2O_5, prepared by oxidizing Bi_2O_3 in solution or by fusion with $KClO_3$, is a rather ill-defined compound, often containing alkali and/or water; it is not certain that pure Bi_2O_5 has been prepared.

The oxygen chemistry of pentavalent antimony was formerly very perplexing, and many compounds—some anhydrous, some hydrated—were described as ortho-, meta-, and pyro-antimonates and were assigned formulae analogous to those of phosphates. As a result of the study of the structures of many of these crystalline compounds it is found that their structural chemistry is very simple and quite different from that of phosphates, arsenates, and vanadates, being based not on tetrahedral but on octahedral coordination of Sb^V by oxygen. Many of the old formulae require revision. The compounds in question fall into two main groups:

(a) salts containing $Sb(OH)_6^-$ ions;
(b) complex oxides of the following (and possibly other) types, all based on octahedral SbO_6 coordination groups: $MSbO_3$, $M^{III}SbO_4$, $M^{II}Sb_2O_6$, $M^{II}Sb_2O_7$.

The free oxy-acids of antimony are even less stable than those of arsenic. Various hydrated forms of the trioxide and pentoxide have been described as antimonious and antimonic acids, but their structures are unknown. Finite oxy-ions SbO_4^{3-}, SbO_3^-, and $Sb_2O_7^{4-}$ do not exist. Compounds formerly described as ortho-, meta-, and pyro-antimonates all contain Sb^V octahedrally coordinated by oxygen, either as $Sb(OH)_6^-$ ions, in compounds which were once described as hydrated meta- or pyro-antimonates, or as SbO_6 groups in the compounds of type (b), which are best described as complex oxides, and not as antimonates, for reasons given later.

(a) *Salts containing* $Sb(OH)_6^-$ *ions*. Typical of the old formulae are $LiSbO_3.3H_2O$ and $Na_2H_2Sb_2O_7.5H_2O$ for hydrated meta- and pyro-antimonates. In fact no meta- or pyro-ions exist in these salts, which were formulated thus to bring them into line with the phosphorus compounds. In no case is there any structural similarity between the phosphorus and antimony compounds. A study of the crystal structure of sodium 'pyroantimonate', $Na_2H_2Sb_2O_7.5H_2O$, shows that its correct structural formula is $NaSb(OH)_6$, and the isostructural silver salt is $AgSb(OH)_6$.[2] In the pseudo-cubic sodium salt the Na and Sb atoms are arranged in the same way as the Na^+ and Cl^- ions in NaCl, and each Sb atom is the centre of an $Sb(OH)_6^-$ ion. By the action of 40 per cent HF, $Na[Sb(OH)_6]$ is converted into $Na(SbF_6)$ which has a quite similar structure. If a solution of sodium 'pyroantimonate' is added to one of a magnesium salt the precipitate has the empirical composition $Mg(SbO_3)_2.12H_2O$ and this was considered to be a hydrated meta-antimonate. Its formulation as $[Mg(H_2O)_6][Sb(OH)_6]_2$, resulting from an X-ray study of the crystals,[3] shows that it is a hydrated antimonate of exactly the same type as sodium pyroantimonate. The old and new formulae of such antimonates are summarized below.

Old formula	Correct structural formula
$Na_2H_2Sb_2O_7.5H_2O$	$Na[Sb(OH)_6]$
$Mg(SbO_3)_2.12H_2O$	$Mg[Sb(OH)_6]_2.6H_2O$ or $[Mg(H_2O)_6][Sb(OH)_6]_2$
(Co and Ni salts similar)	
$Cu(NH_3)_3(SbO_3)_2.9H_2O$	$[Cu(NH_3)_3(H_2O)_3][Sb(OH)_6]_2$
$LiSbO_3.3H_2O$	$Li[Sb(OH)_6]$

The partial hydrolysis of $NaSbF_6$ leads to the formation of a salt originally formulated as $NaF.SbOF_3.H_2O$. Its correct formula is $Na[SbF_4(OH)_2]$, showing that it is of the same type as $NaSbF_6$ and $NaSb(OH)_6$.

(1)   AC 1979 **B35** 539   (3)   RTC 1937 **56** 931
(2)   ZaC 1938 **238** 241

(b) *Complex oxides based on* SbO_6 *coordination groups.* In all these compounds antimony is octahedrally coordinated by oxygen, and in some cases the structures are similar to those of simple ionic crystals viz. $NaSbO_3$, ilmenite structure ($FeTiO_3$, p. 267). $FeSbO_4$, random rutile structure (p. 591), and $ZnSb_2O_6$, trirutile structure. It is therefore preferable to describe these compounds as complex oxides rather than as 'antimonates' of various types.

M^ISbO_3. The salt $NaSbO_3$ may be prepared by heating $NaSb(OH)_6$ in air or by fusing Sb_2O_3 with sodium carbonate—there are similar silver and lithium salts. The sodium salt has the ilmenite structure; the compound thought to be a polymorphic modification is actually $Na_2Sb_2O_5(OH)_2$ (see later). The structure of $LiSbO_3$[1] is related to that of $NaSbO_3$ in the way described in Chapter 4. In both structures M and Sb each occupy one-third of the octahedral holes in a h.c.p. assembly of O atoms, but the choice of holes occupied is different in the two structures (p. 171).

$KSbO_3$ is prepared by heating $KSb(OH)_6$, and the l.t. form has the ilmenite structure. At $1000\,°C$ it transforms to a cubic polymorph (space group $Im3$)[2a] and at higher pressures to a second cubic form with space group $Pn3$.[2b] Both of these cubic forms have essentially the same structure (Fig. 5.33, p. 225), which is built of edge-sharing pairs of octahedra, a structural feature of a number of these compounds. In $K_2Sb_4O_{11}$[3] also the 3D framework is built of pairs of octahedra, while in $K_3Sb_5O_{14}$[3] it is built of pairs together with single octahedra.

The compound originally formulated $CaSb_2O_5$ is actually $CaNaSb_2O_6OH$. The first product of heating $NaSb(OH)_6$ is $Na_2Sb_2O_5(OH)_2$, or possibly $NaSbO_3.xH_2O$, which finally decomposes to $NaSbO_3$.[4]

All the following compounds have very similar structures: $Na_2Sb_2O_5(OH)_2$; the minerals pyrochlore, $CaNaNb_2O_6F$, and romeite, $CaNaSb_2O_6(OH)$, $Sb_3O_6(OH)$, $BiTa_2O_6F$, and $AgSbO_3$. The compound $Sb_3O_6(OH)$ is of interest in connection with the structure of Sb_2O_4. When hydrated Sb_2O_5 is heated (at about $800\,°C$) it is not directly converted into Sb_2O_4, as was at one time supposed, but into $Sb_3O_6(OH)$. This latter compound is converted into the tetroxide only after prolonged heating at $900\,°C$, and $Sb_3O_6(OH)$ had been confused with the tetroxide, which is also formed by the oxidation of Sb_2O_3 in air.

The compound $BiTa_2O_6F$ is prepared by heating BiOF with Ta_2O_5. The structures of the above-mentioned compounds are related in the following way. In the unit cell of the pyrochlore structure (see also p. 258) there are eight formula weights, that is, $8[CaNaNb_2O_6F]$. The 8 Ca + 8 Na may be replaced by 8 Sb(III), the Nb by Sb(v), and the F by OH, giving $Sb^{III}Sb_2^VO_6(OH)$. Alternatively the 8 Ca + 8 Na may be replaced by 16 Ag and the 8 OH omitted, giving $AgSbO_3$. These relationships may be summarized as follows:

Pyrochlore	AgSbO$_3$	Sb$_3$O$_6$(OH)	BiTa$_2$O$_6$F
8 Ca + 8 Na	16 Ag	8 Sb(III)	8 Bi(III)
16 Nb	16 Sb(V)	16 Sb(V)	16 Ta
48 O	48 O	48 O	48 O
8 F	—	8 OH	8 F

$M^{II}Sb_2O_6$. Three types of structure have been found for these compounds. The small ions of Mg and certain 3d metals form the trirutile structure (p. 251), sometimes described as the tapiolite structure, after the mineral of that name, $FeTa_2O_6$. Compounds with this structure include:

MgSb$_2$O$_6$	FeSb$_2$O$_6$	MgTa$_2$O$_6$	FeTa$_2$O$_6$
Zn	Co		Co
	Ni		Ni

The somewhat larger Mn^{2+} ion prefers the niobite structure (p. 175). These structures are not possible for larger M^{2+} ions, and a hexagonal structure is adopted if the ionic radius of M is about 1 Å or more.[6] In this structure (Fig. 20.11) the M^{2+} and

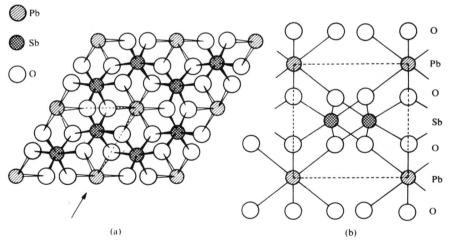

FIG. 20.11. The crystal structure of PbSb$_2$O$_6$; (a) plan, and (b) elevation (viewed in the direction of the arrow) on a slightly larger scale.

Sb^{5+} ions are segregated in layers so that ions ranging from Ca^{2+} to Ba^{2+} can be accommodated by pushing the Sb–O layers further apart.

$M^{III}SbO_4$. A number of compounds of this type have been shown to have a statistical rutile structure, that is, the ions M^{3+} and Sb^{5+} occupy at random the

Ti^{4+} positions in the rutile structure. They include $FeSbO_4$, $AlSbO_4$, $CrSbO_4$, $RhSbO_4$, and $GaSbO_4$.[7]

$M_2^{II}Sb_2O_7$. Two structures have been found for compounds of this type,[8] viz. the 'weberite' structure for $Ca_2Sb_2O_7$, $Sr_2Sb_2O_7$, and $Cd_2Sb_2O_7$, and the pyrochlore structure for $Pb_2Sb_2O_7$, $Ca_2Ta_2O_7$, $Cd_2Ta_2O_7$, and $Cd_2Nb_2O_7$.[9] These two closely related structures are adopted by a number of complex oxides and fluorides. In the weberite structure (of Na_2MgAlF_7)[10] we can distinguish a framework of linked MgF_6 and AlF_6 octahedra in which all vertices of each MgF_6 and four vertices of each AlF_6 octahedron are shared, so that the framework has the composition $(MgAlF_7)$ or more generally M_2X_7 when all atoms in the octahedra are of the same kind. (There is no essential difference in bond type between the Mg–F or Al–F and the Na–F bonds but the Na^+ ions lie in the interstices in positions of 8-, or strictly 4 + 4- or 2 + 4 + 2-coordination, and when comparing the two structures it is convenient to describe the Mg–Al–F frameworks.) In the pyrochlore framework all octahedral MX_6 groups share each X with another group, so that the composition of the framework is $MX_3(M_2X_6)$. There is room for a seventh X atom but this is not essential to the stability of the framework, and while compounds such as ralstonite, $Na_xMg_xAl_{2-x}(F,OH)_6 . H_2O$, $Ca_2Ta_2O_7$, etc. have this structure so also has $AgSbO_3$, already mentioned on p. 904. Since it is difficult to illustrate clearly complicated packings of octahedra we show in Fig. 20.12 only the Mg and Al atoms of these

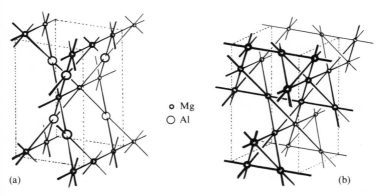

o Mg
O Al

FIG. 20.12. (a) The M_2X_7 framework in weberite, $Na_2(MgAlF_7)$; (b) the M_2X_6 (pyrochlore) framework in ralstonite, $Na_xMg_xAl_{2-x}(F,OH)_6H_2O$.

M_2X_7 and M_2X_6 frameworks. Each line joining two M atoms represents M–X–M, that is, the sharing of an octahedron vertex, though in general the X atoms do not lie actually along these lines. For the pyrochlore structure see also Chapters 6 and 13; the structure is illustrated in Fig. 7.4(c).

(1) ACSc 1954 **8** 1021	(4) CJC 1970 **48** 1323	(9) PR 1955 **98** 903
(2a) AK 1940 **14A** 1	(6) AKMG 1941 **15B** No. 3	(10) AKMG 1944 **18B** No. 10
(2b) JSSC 1973 **6** 493	(7) AKMG 1943 **17A** No. 15	
(3) AC 1974 **B30** 945	(8) AKMG 1944 **18A** No. 21	

The sulphides of arsenic, antimony, and bismuth

These elements form the following sulphides:

$$As_2S_3 \qquad As_2S_5 \qquad As_4S_3 \qquad As_4S_4 \qquad As_4S_5 \qquad As_4S_6$$
$$Sb_2S_3 \qquad Sb_2S_5$$
$$Bi_2S_3$$

The structures of the pentasulphides are not known; As_4S_6 is the dimeric form of As_2S_3 in the vapour, structurally similar to As_4O_6 and P_4O_6 (p. 854).

Crystalline As_2S_3 (orpiment)[1] has a layer structure (Fig. 20.13) of essentially the same kind as in the monoclinic forms of As_2O_3 (p. 889).

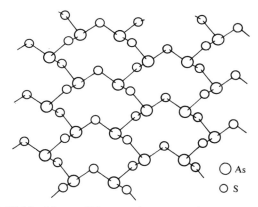

○ As
○ S

FIG. 20.13. A layer of the crystal structure of orpiment, As_2S_3.

A simple chain with the composition M_2S_3 for an element M forming three bonds is

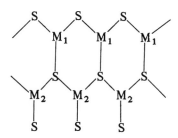

and chains of this type occur in the structure of Sb_2S_3 and of the isostructural Bi_2S_3. The interatomic distances indicate, however, that these chains are associated in pairs as shown diagrammatically in Fig. 20.14, for whereas the atoms M_1 (see above) have three S neighbours at 2.53 Å and no others nearer than 3.14 Å the atoms M_2 have S neighbours as follows: one at 2.46, two at 2.68, and two more at

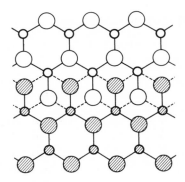

FIG. 20.14. The double chains in Sb_2S_3 (diagrammatic). The larger circles represent S atoms. All the atoms of one chain are shaded, and the weak secondary Sb–S bonds are shown as broken lines.

2.85 Å in addition to others at greater distances. The M_2–S bonds of length 2.85 Å are shown as dotted lines in Fig. 20.14. It is clear that no simple bond picture can be given for crystalline Sb_2S_3, but the bonds from the two kinds of non-equivalent Sb atoms can be given fractional bond numbers using Pauling's equation (p. 1292). This structure for Sb_2S_3 has been confirmed by a later study of the isostructural Sb_2Se_3.[2]

The mineral getchellite, $AsSbS_3$,[3] which is remarkable for its extraordinary softness, has a complex layer structure in which the metalloid atoms are disordered, with pyramidal coordination, and the S atoms 2-coordinated. Possibly the complications are due to the fact that As usually has 3 pyramidal S neighbours in its sulphur compounds whereas Sb is both 3- and 5-coordinated in Sb_2S_3.

In Bi_2S_3[4] Bi has 4 closest neighbours in a nearly planar rectangle and additional neighbours completing an irregular 7-coordination group. The coordination of Bi^{III} by S in Bi_2S_3 and other compounds such as $BiPS_4$[5] and $Bi_2In_4S_9$[6] is not easily described, and as in Sn^{II} compounds is often not the same for all the metal atoms in a particular structure. There are 3 or 4 S at around 2.7 Å and others at distances up to 3.5 Å completing very distorted 6-, 7-, or 8-coordination groups.

In the two forms of As_4S_3[7, 8] the molecule has the same configuration, which is that of the P_4S_3 molecule (Fig. 19.10(c), p. 868).

Realgar, As_4S_4, is a molecular crystal which sublimes readily *in vacuo* and changes into orpiment on exposure to light. At temperatures up to about 550°C the vapour density corresponds to the formula As_4S_4. At higher temperatures dissociation to As_2S_2 takes place, and is complete at 1000°C. The structure of the As_2S_2 molecule is apparently not known, but the $As_4\dot{S}_4$ molecule was shown to have the configuration of Fig. 20.15(a) by an early e.d. study of the vapour[9] and later by X-ray studies of crystalline realgar[1] and of a second crystalline form.[10] There is a third crystalline form of As_4S_4[11] with the isomeric molecular structure noted on p. 868 for P_4S_4. In this molecule the As atoms are of three kinds:

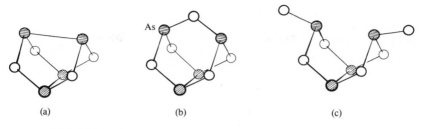

(a)   (b)   (c)

FIG. 20.15. The structures of (a) As_4S_4; (b) As_4S_5; and (c) $As_4S_6^{2-}$.

$$(\text{one}) \ As{<}_S^{2As}, \quad (\text{one}) \ As{-}3S, \quad \text{and} \ (\text{two}) \ As{<}_{2S}^{As}.$$

In realgar all four As atoms are bonded to 1As + 2S.

In crystalline As_4S_5 the molecule has the structure shown in Fig. 20.15(b),[12] which is similar to that of As_4S_4 but with an additional bridging S atom. The interposition of another S atom between the other pair of As atoms gives the As_4S_6 molecule.

Bond lengths in the As sulphides range from 2.15 Å to 2.20 Å for As–S (S terminal or bridging), 2.50-2.55 Å for As–As, and interbond angles are close to 100° except the As–As–As angles in As_4S_3 (60°).

(1) ZK 1972 **136** 48
(2) ZK 1972 **135** 308
(3) AC 1973 **B29** 2536
(4) Miner. petrogr. Mitt. 1970 **14** 55
(5) AC 1975 **B31** 2003
(6) AC 1972 **B28** 3128
(7) JCS A 1970 1800
(8) JCS D 1973 1737
(9) JACS 1944 **66** 818
(10) JCS D 1972 1347
(11) ZaC 1976 **419** 176
(12) JCS D 1973 1740

Thio ions of As *and* Sb; *cacodyl disulphide*

Thioarsenite ions include the pyramidal AsS_3^{3-} ion (As–S, 2.25 Å)[1] and the metadithioarsenite ion $(AsS_2)_n^{n-}$ formed from similar pyramidal groups (angles near 100°):[2]

$$\text{...As...As...} \quad 2.32 \text{ Å} \quad 2.17 \text{ Å}$$

Both ions were studied in their Na salts.

All the following ions are known, AsS_4^{3-},[3] AsS_3O^{3-}, $AsS_2O_2^{3-}$,[3a] and $AsSO_3^{3-}$;[4] in these tetrahedral ions bond lengths are As–O, 1.67 Å and As–S, 2.15 Å.

The $As_4S_6^{2-}$ ion has been studied in the piperidinium salt, $(C_5H_{12}N)_2(As_4S_6)$.[5]

It is related to As_4S_4 and As_4S_5 as shown in Fig. 20.15(c); the non-bonded As—As distance is 3.51 Å.

The molecule of 'cacodyl disulphide'[6] is of interest as containing three kinds of As—S bonds: As^{III}—S, As^V—S, and As^V=S. The coordination of As^{III} is pyramidal and that of As^V tetrahedral.

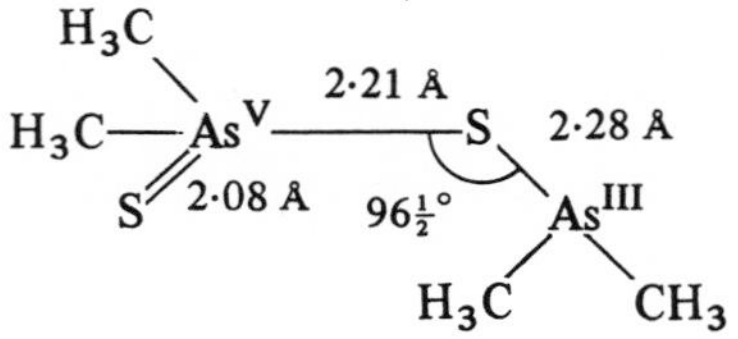

$$
\begin{array}{c}
H_3C \\
\diagdown \qquad \text{2·21 Å} \\
H_3C—As^V————S \quad \text{2·28 Å} \\
\diagup\diagup \; \text{2·08 Å} \quad 96\tfrac{1}{2}° \quad As^{III} \\
S \qquad\qquad\qquad\qquad \diagup \diagdown \\
\qquad\qquad\qquad H_3C \qquad CH_3
\end{array}
$$

For the SbS_4^{3-} ion see p. 774.

(1) AC 1976 **B32** 3175
(2) AC 1977 **B33** 908
(3) AC 1974 **B30** 2378
(3a) AC 1976 **B32** 2119

(4) AC 1976 **B32** 516
(5) JCS A 1971 3130
(6) JCS 1964 219

21

Carbon

Introduction

We shall be concerned here with only some of the simpler compounds of carbon, for the very extensive chemistry of this element forms the subject-matter of organic chemistry. Carbon is unique as regards the number and variety of its compounds based on skeletons of similar atoms directly linked together. There are also compounds based on mixed C–N, C–O, and C–N–O skeletons, including cyclic systems, to some of which we shall refer later. Two major classes of organic compounds are recognized: aliphatic compounds, based on the tetrahedral carbon atom, and aromatic compounds, based on the hexagonal C_6 ring, in which some of the C atoms may be replaced by N, etc. Corresponding to these two types of carbon skeleton in finite molecules are the two polymorphic forms of crystalline carbon, diamond–in which each C atom is linked by tetrahedral sp^3 bonds to its four neighbours, and graphite–in which each atom forms three coplanar sp^2 bonds leading to the arrangement of the atoms in sheets.

$$\diagdown \diagup C \diagdown \qquad \text{(diamond)–aliphatic compounds}$$

$$\diagdown C \diagup \qquad \text{(graphite)–aromatic compounds}$$

There are also 'mixed' compounds containing C_6 rings and, for example, aliphatic side chains.

In contrast to the chemistry of living organisms based on systems of linked C atoms there is the chemistry of silicates based on Si atoms linked via O atoms. We shall note some of the more important differences between C and Si at the beginning of Chapter 23. Here we note only one other feature of carbon chemistry which differentiates this element from silicon, namely, the presence of multiple bonds in many simple molecules, a characteristic also of nitrogen and oxygen.

The stereochemistry of carbon

The structural formulae of carbon compounds were originally based on the four bond pictures:

(a) $\diagdown \diagup C \diagdown$ (b) $\diagdown C \diagdown$ (c) $=C=$ and   (d) $\equiv C-$.

It is now known, however, that (b) is not an adequate representation of the behaviour of a carbon atom in many molecules in which it is attached to three other atoms.

Similarly (c) and (d) are unsatisfactory for many molecules containing a carbon atom forming two collinear bonds because there is resonance (π-bonding) leading to non-integral bond orders, when neither $=C=$ nor $\equiv C-$ accurately represents the electronic structure. Their lengths and physical properties indicate that many carbon–carbon bonds are intermediate between C–C, C=C, and C≡C. The stereochemistry of this element may accordingly be discussed under the following heads:

(a) *The tetrahedral carbon atom* – forming four sp^3 bonds.

(b) *3-covalent carbon* – forming sp^2 bonds. With the proviso that intermediate types of bond are to be expected, it will simplify our treatment if we recognize three cases, to which a number of molecules approximate fairly closely:

   (i) the simple system of one double and two single bonds already noted;

   (ii) one (single) and two equivalent bonds; and

   (iii) three equivalent bonds.

   (i) $=C\Big\langle$: $(H_3C)_2C=CH_2$, $H_2C=O$, $X_2C=O$,

   (ii) $-C\langle$: carboxylate ions, benzene, and cyclic C–N systems,

   (iii) $---C\langle$: carbonate, guanidinium, etc. ions, urea, graphite.

(c) *and* (d) *2-covalent carbon* (sp bonds). The compounds to be considered include CO_2, COS, and CS_2 with which the monoxide and suboxide are conveniently included), acetylene, cyanogen, cyanates, and thiocyanates. (There are also relatively unstable species containing 2-covalent carbon such as CF_2,[1] for which m.w. data give C–F, 1.30 Å and bond angle 105° (ground state) or 135° (excited state), and the CO_2^- ion. This ion, detected by its e.s.r. spectrum, is formed when CO_2 is adsorbed on u.v.-irradiated MgO and in γ-irradiated sodium formate. The ion persists for months if formed in a KBr matrix; the bond angle has been estimated as $127 \pm 8°$.[2])

The very simple scheme outlined above does not make allowance for special types of bond such as those in carboranes and metal alkyls in which C is bonded to more than four atoms; these are noted in other places.

Although we shall adopt the above rough classification according to the type of bonds formed by carbon, it will not be convenient to adhere strictly to this order. For example, cyanogen derivatives polymerize to cyclic molecules in which C forms three bonds; these cyclic C–N compounds are logically included after the simple cyanogen derivatives. The following groups of compounds are included in Chapter 22: metal cyanides, carbides, and carbonyls, and certain classes of organo-metallic compound.

(1) JCP 1966 **45** 1067, 1068   (2) JPC 1965 **69** 2182; JCP 1966 **44** 1913

The tetrahedral carbon atom

Diamond: saturated organic compounds

The earliest demonstration of the regular tetrahedral arrangement of carbon bonds was provided by the analysis of the crystal structure of diamond. In this crystal each C atom is linked to four equidistant neighbours and accordingly the linking shown in Fig. 21.1 extends throughout the whole crystal. The bond angles and the

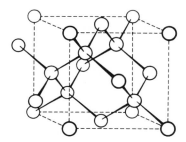

FIG. 21.1. The crystal structure of diamond.

C–C bond length (1.54 Å) are the same as in a simple molecule such as $C(CH_3)_4$ or in any saturated hydrocarbon chain. The hexagonal form of diamond, related to cubic diamond in the same way as wurtzite to zinc-blende, has been produced at high temperature and pressure.[1] The molecule of symmetrical tricyclo-decane ('adamantane'), $C_{10}H_{16}$, has the same configuration as $N_4(CH_2)_6$ (Fig. 18.2, p. 795), and the molecule may be regarded as a portion of the diamond structure which has been hydrogenated. Larger portions of the diamond network form the C skeletons of $C_{14}H_{20}$ ('diamantane') and $C_{22}H_{28}$ ('anti-tetramantane').[2] Polyhedral molecules of special interest include $C_4(n\text{-}C_4H_9)_4$, in which there is a central tetrahedron of C atoms, 'cubane', C_8H_8, with a cubic framework of C atoms (C–C, 1.552 Å),[3a] and methyl-substituted dodecahedranes. In 1,16-$C_{20}H_{18}(CH_3)_2$[3b] the C atoms are situated at the vertices of a nearly regular pentagonal dodecahedron (mean C–C, 1.546 Å). The octaphenyl derivative of cubane, $C_8(C_6H_5)_8$, is, however, a derivative of cyclooctatetraene (p. 917), the lengths of alternate bonds in the 8-membered ring being 1.343 Å and 1.493 Å.[4]

Many simple carbon compounds have been studied by diffraction or spectroscopic methods, and in all molecules Ca_4 the interbond angles have the regular tetrahedral value (109°28′) to within the experimental error. Some reliable values of bond lengths in simple molecules of carbon and other Group IV elements are listed in Table 21.1.

(1)   JCP 1967 **46** 3437
(2)   AC 1977 **B33** 2335
(3a) JACS 1964 **86** 3889
(3b) Sc 1981 **211** 575
(4)   JCS 1965 3136

TABLE 21.1
Bond lengths in molecules–Group IV *elements*

Bond	Bond length (Å)	Molecule	Reference
C–H	1.06	C_2H_2	PRS A 1956 **234** 306
	1.10	C_2H_4	JCP 1965 **42** 2683
	1.11	C_2H_6 ⎱	JCP 1968 **49** 4456
C–C	1.53–	C_2H_6 ⎰	
	1.56	C_2Cl_6	ACSc 1964 **18** 603
C=C	1.34	C_2H_4	JCP 1965 **42** 2683
C≡C	1.20_5	C_2H_2	PRS A 1956 **234** 306
C–N	See Table 18.4, p. 807		
C–O	See Table 21.5, p. 926		
C–S	See Table 21.6, p. 926		
C–F	1.32–	CF_4	ARPC 1954 **5** 395
	1.39	CH_3F, CCl_3F	JCP 1960 **33** 508
C–Cl	1.75_5–	CF_3Cl	JCP 1962 **36** 2808
	1.77	CCl_4	AC 1979 **B35** 1670
C–Br	1.91–	CF_3Br	TFS 1954 **50** 444
	1.94	CH_3Br, CBr_4	JCP 1952 **20** 1112
C–I	2.14	CF_3I, CH_3I	JCP 1958 **28** 1010
Si–H	1.48	SiH_4	JCP 1955 **23** 922
Si–C	1.83–	$H_3SiC≡CH$	JCP 1963 **39** 1181
	1.87	$(CH_3)_3SiH$	JCP 1960 **33** 907
Si–Si	2.32	Si_2F_6	JMSt 1976 **31** 237
	2.36	$Si[Si(CH_3)_3]_4$	IC 1970 **9** 2436
Si–O	1.64	$Si(OCH_3)_4$	JCP 1950 **18** 1414
	1.625	$Si(OOCCH_3)_4$	ZK 1975 **141** 97
Si–F	1.56–	$SiBrF_3$	JCP 1951 **19** 965
	1.59	SiH_3F	PR 1950 **78** 64
Si–Cl	1.98–	SiF_3Cl	JCP 1951 **19** 965
	2.05	SiH_3Cl	ACSc 1954 **8** 367
Si–Br	2.15–	SiF_3Br	JCP 1951 **19** 965
	2.21	SiH_3Br	PR 1949 **76** 1419
Si–I	2.39–	SiF_3I	JCP 1967 **47** 1314
	2.43	SiH_3I	N 1953 **171** 87
Si–N	1.72	$(CH_3)_3Si(NHCH_3)$	TFS 1962 **58** 1686
Ti–Cl	2.19	$TiCl_4$ ⎱	BCSJ 1956 **29** 95
Ti–Br	2.32	$TiBr_4$ ⎰	
Zr–Cl	2.33	$ZrCl_4$ ⎱	TFS 1941 **37** 393
Th–Cl	2.61	$ThCl_4$ ⎰	
Ge–H	1.52–	GeH_4	JCP 1954 **22** 1723
	1.54	$GeH_3F–GeH_3Br$	JCP 1965 **43** 333
Ge–C	1.90–	$H_3GeC≡CH$	JCP 1966 **44** 2602
	1.95	$GeH_2(CH_3)_2$	JCP 1969 **50** 3512
	1.957	$Ge(C_6H_5)_4$	AC 1972 **B26** 2889
Ge–Ge	2.41	Ge_2H_6, Ge_3H_8	JACS 1938 **60** 1605
Ge–Si	2.36	$H_3Ge.SiH_3$	JCP 1967 **46** 2007
Ge–F	1.69–	GeF_3Cl	PR 1951 **81** 819
	1.73_5	GeH_3F	JCP 1965 **43** 333

TABLE 21.1–*continued*

Bond	Bond length (Å)	Molecule	Reference
Ge—Cl	2.07– 2.15	GeF_3Cl GeH_3Cl	PR 1951 **81** 819
Ge—Br	2.31	GeH_3Br	JCP 1965 **43** 333
Ge—I	2.50	GeI_4	TFS 1941 **37** 393
Sn—H	1.70	SnH_4	JCP 1956 **25** 784
Sn—C	2.11–2.15	CH_3SnCl_3, $(CH_3)_2SnH_2$	AC 1974 **B30** 444
Sn—Cl	2.28	$SnCl_4$	BCSJ 1970 **43** 1933
	2.31–2.35	CH_3SnCl_3-$(CH_3)_3SnCl$	AC 1974 **B30** 444
Sn—Br	2.44–	$SnBr_4$	TFS 1941 **37** 393
	2.49	CH_3SnBr_3-$(CH_3)_3SnBr$	TFS 1944 **40** 164
Sn—I	2.64–	SnI_4	TFS 1941 **37** 393
	2.72	CH_3SnI_3-$(CH_3)_3SnI$	TFS 1944 **40** 164
Pb—C	2.20	$Pb(CH_3)_4$	JCP 1958 **28** 1007
Pb—Cl	2.43	$PbCl_4$	TFS 1941 **37** 393
Pb—Pb	2.88	$Pb_2(CH_3)_6$	TFS 1940 **36** 1209

Carbon fluorides (fluorocarbons)

The product of the decomposition of $(CF)_n$ (see later) or the direct fluorination of carbon is a mixture of fluorides of carbon from which the following have been isolated in quantity and studied: CF_4; C_2F_6; C_3F_8; C_4F_{10} (two isomers); C_2F_4; C_5F_{10}; C_6F_{12}; and C_7F_{14}. All are extremely inert chemically and are notable for the very low attractive forces between the molecules, approaching the atoms of the noble gases in this respect. There is a great difference between the boiling points of the fluorocarbons and the hydrocarbons, as shown in Fig. 21.2.

Mixed fluoro–chloro compounds have been prepared by fluorinating the appropriate chloro compounds; they include $CHClF_2$, CCl_2F_2, $CHCl_2F$, CCl_3F, and $CCl_2F.CCl_2F$. The unsaturated compound tetrafluoroethylene, C_2F_4, can be polymerized under pressure to a solid polymer which is, like the simple fluorocarbons, very inert chemically. Polytetrafluoroethylene, $(CF_2)_n$, consists of chain molecules $-CF_2-CF_2-$ but in contrast to the planar zigzag $(CH_2)_n$ chains in polyethylene they have a helical configuration. The solid carbon monofluoride $(CF)_n$ is closely related to graphite, the layers of which have been pushed apart by the introduction of F atoms (see later).

Carbon forming three bonds

On p. 912 we recognized three main groups, (i)-(iii), of compounds in which carbon forms three bonds. We shall now give examples of each group, and for comparison with observed bond lengths we give in Table 21.2 standard values for single, double, and triple bonds. It is possible to select examples illustrating a simple

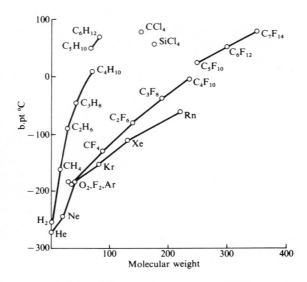

FIG. 21.2. Boiling-points of hydrocarbons, fluorocarbons, and noble gases.

TABLE 21.2
Observed bond lengths (Å)

	Single	Double	Triple
C–C	1.54	1.34	1.20
C–N	1.47	[1.28]	1.16
C–O	1.41	1.22	1.13

correlation between types of bond and interbond angles, in which the logical sequence would be:

(i)	(iii)	(ii)
125° 110° 125°	120° 120° 120°	115° 130° 115°
COCl$_2$	CO$_3^{2-}$	(R . COO)$^-$

but we shall see that there are no clear-cut dividing lines between these classes, particularly between (ii) and (iii); cyclic molecules such as C$_6$H$_6$ illustrate this point. As we proceed it will become evident that the bonds in many molecules cannot be described in any simple way; a good example is the molecule of ethyl carbamate, in which there are three different C–O bond lengths:[1]

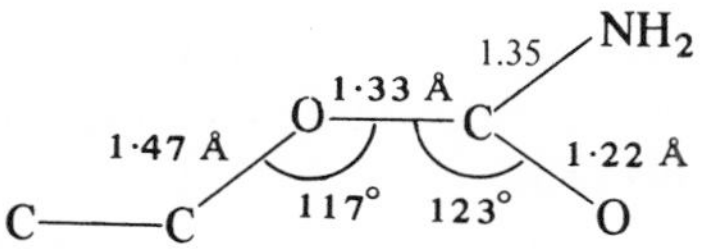

Bond arrangement $=C<$

For carbon forming one double and two single bonds the bond angles

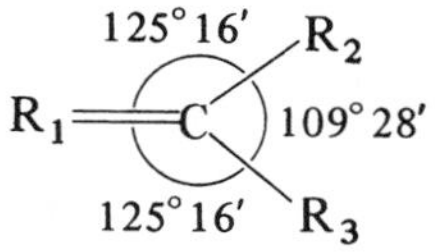

should be as shown and the lengths of the bonds $C=R_1$, $C-R_2$, and $C-R_3$ should be those expected for one double and two single bonds respectively. The most symmetrical molecules of this type include isobutene, $(CH_3)_2C=CH_2$, formaldehyde, $H_2C=O$, ketones, $R_2C=O$, carbonyl halides, $X_2C=O$, and their sulphur analogues, $X_2C=S$. Two studies of the planar $I_2C=CI_2$ molecule, by e.d. of the vapour[2a] and n.d. of the powder at 4 K[2b] give $C=C$, 1.363 and 1.295 Å, and $C-I$, 2.106 and 2.114 Å respectively. Less symmetrical molecules approximating to this bond arrangement include the boat-shaped molecule of cyclooctatetraene, (a),[3] the bond lengths in which indicate that there is comparatively little interaction between the bonds in the ring, formyl fluoride, (b),[4] and carboxylic acids.

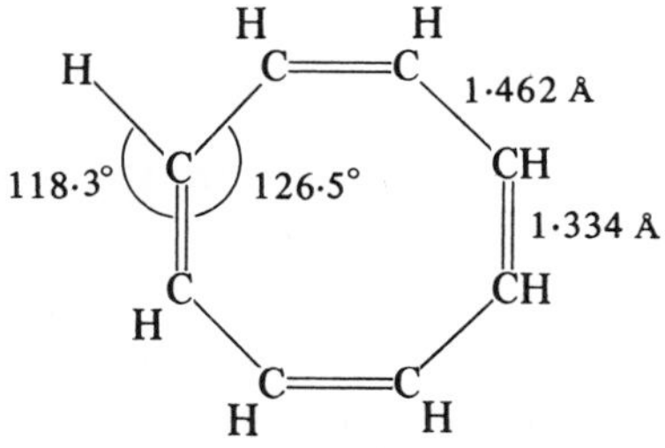

(a)

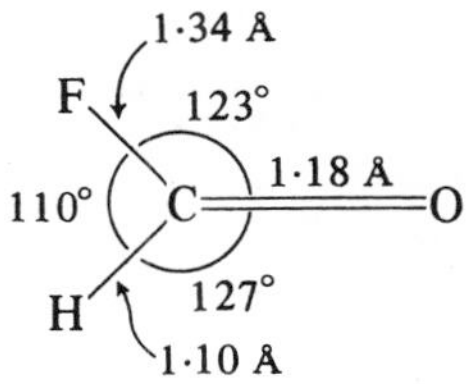

(b)

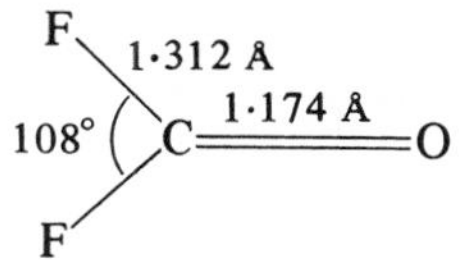

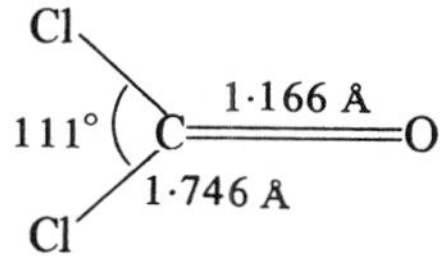

Carbonyl halides and thiocarbonyl halides. All the compounds COF_2,[5] $COCl_2$,[6] and $COBr_2$,[7] have been studied in the vapour state by e.d. and/or m.w., and $COCl_2$ (phosgene) has also been studied in the crystalline state.[8] The $C-X$ bond lengths are close to the values expected for single bonds, while the $C=O$ bonds are slightly shorter than 'normal' double bonds. There is no sign of disorder in crystalline phosgene, so that the residual entropy (1.63 e.u.) calculated from calorimetric and spectroscopic data is unexplained.

An early electron diffraction study of thiophosgene, $CSCl_2$, assumed C–S, 1.63 Å, which could profitably be checked. The dimer, produced by photolysis of $CSCl_2$, has the structure (c),[9] and the compounds with the empirical formulae $(COCl_2)_2$ and $(COCl_2)_3$ are (d) and (e).[10]

(c)   (d)   (e)

Carboxylic acids and related compounds. Acids such as formic and acetic dimerize in the vapour state. The structures of the monomer and dimer of formic acid are shown at (f) and (g)[11] as determined by e.d. studies of the vapour; the acid has

(f)   (g)

also been studied in the crystalline state.[12] The dimers of these acids are very nearly planar molecules;[13] it may be supposed that sp^2 hybridization of the C atom requires also sp^2 hybridization of the carbonyl O atoms. Consequently the two lone pairs of the latter are extended at 120° to the C=O bond and are coplanar with it and with the other C–O bond. The whole structure is therefore essentially planar. An electron diffraction study of thioacetic acid[14] gave the following result, values enclosed in brackets being assumed:

Many studies have been made of oxalic acid, which has two anhydrous poly-morphs[15] and of oxalates, $(NH_4)_2C_2O_4.H_2O$,[16] $K_2C_2O_4.H_2O$,[17] KHC_2O_4,[18] etc. Although the dihydrate of the acid is not an oxonium salt there is compara-tively little difference between the dimensions of the $(COOH)_2$ molecule in $(COOH)_2.2H_2O$,[19](h), and of the ion in alkali-metal salts, (i). Points of interest

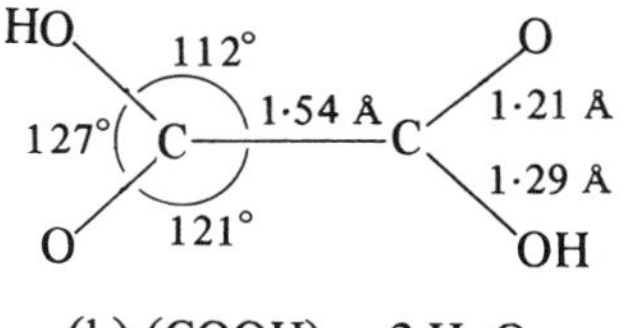

(h) $(COOH)_2 . 2 H_2O$ 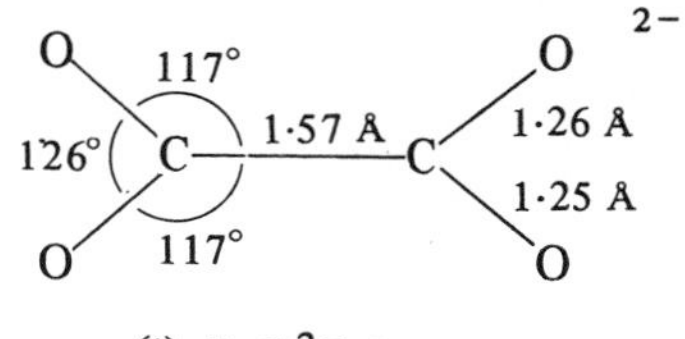   (i) $C_2O_4^{2-}$ ion

are the planarity of the oxalic acid molecule, in which the C–C bond is at least as long as a single bond, and the variation in conformation of the oxalate ion, which is planar in $K_2C_2O_4.H_2O$ but non-planar in KHC_2O_4 and $(NH_4)_2C_2O_4.H_2O$. In the last two crystals the angles between the planes of the COO groups are 13° and 27° respectively. The non-planarity is attributed to strong hydrogen bonding, which also affects slightly the C–O bond lengths in a number of oxalates. Evidently, although formic acid is close to $O=C\overset{H}{\underset{OH}{<}}$, with a marked difference between the C–O bond lengths, oxalic acid is closer to $-C\overset{<}{\underset{\cdots}{}}$, the structure expected for carboxylate ions. This is also true of oxamide[20] and dithio-oxamide[21] which are planar molecules in which the C–N bonds are appreciably shorter than single bonds:

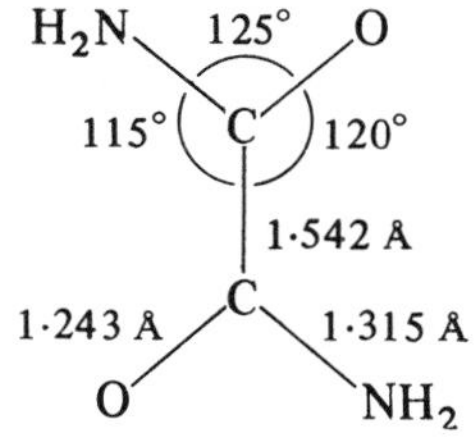

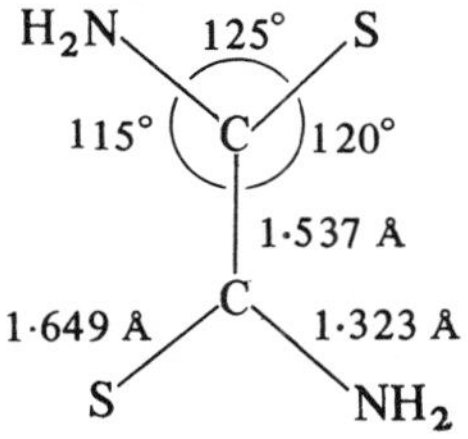

Two cyclic ions are conveniently included here. They are the square planar $C_4O_4^{2-}$ ion in the potassium salt of 3:4-diketocyclobutene diol, $K_2C_4O_4.H_2O$,[22] and the croconate ion, $C_5O_5^{2-}$, studied in $(NH_4)_2C_5O_5$.[23] The latter has almost perfect pentagonal symmetry. In both ions the bond lengths are C–C, 1.46 Å, C=O, 1.26 Å.

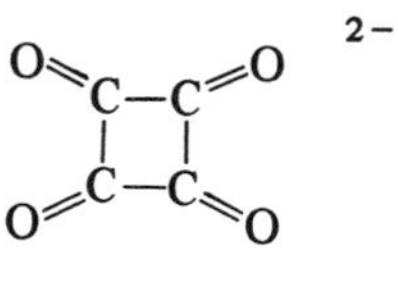

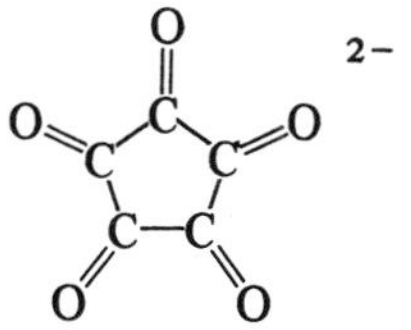

(1)	AC 1967 **23** 410	(12)	AC 1978 **B34** 2188
(2a)	ACSc 1967 **21** 2111	(13)	AC 1963 **16** 430
(2b)	AC 1977 **B33** 1765	(14)	JCP 1946 **14** 560
(3)	JCP 1958 **28** 512	(15)	AC 1974 **B30** 2240
(4)	JCP 1961 **34** 1847	(16)	AC 1965 **18** 410
(5)	JCP 1962 **37** 2995	(17)	AC 1969 **B25** 469
(6)	JCP 1953 **21** 1741	(18)	ACSc 1968 **22** 2953
(7)	JACS 1933 **55** 4126	(19)	AC 1969 **B25** 2423
(8)	AC 1952 **5** 833	(20)	AC 1954 **7** 588
(9)	ZaC 1969 **365** 199	(21)	JCS 1965 396
(10)	JCS 1957 618	(22)	JCP 1964 **40** 3563
(11)	ACSc 1969 **23** 2848	(23)	JACS 1964 **86** 3250

Bond arrangement $-C\overset{\nwarrow}{\underset{\searrow}{}}$

Carboxylate ions. Ions R–COO⁻ of carboxylic acids are symmetrical with two equal C–O bonds (1.26 Å) and O–C–O bond angle in the range 125–130°. Recent studies of the formate ion give C–O, 1.24 Å and the angle O–C–O, 126°.[1]

Carbamate ion. The following bond lengths have been determined in the $(H_2NCO_2)^-$ ion:[2] C–N, 1.36 Å, C–O, 1.28 Å.

Benzene. A number of studies give values for C–C close to 1.395 Å.

(1)   AC 1980 **B36** 1940
(2)   AC 1973 **B29** 2317

Bond arrangement $---C\overset{\nwarrow}{\underset{\searrow}{}}$

A number of symmetrical ions (and one free radical) have been shown to be planar, or approximately so, in the crystalline state. They are listed in Table 21.3, which also gives the other ion in the salt studied.

Carbonate ion. Redetermination of the C–O bond length in calcite and in $K_2CO_3 \cdot 3H_2O$ gives values close to 1.29 Å, though the lower value 1.25 Å was found in sodium sesquicarbonate (p. 369) which has been studied by both X-ray and neutron diffraction. In finite ions or molecules the CO_3 group behaves either as a monodentate ligand, as in $[Co(NH_3)_5CO_3]^+$, or as a bidentate ligand, in $[Co(NH_3)_4CO_3]^+$ and in ions $[M(CO_3)_5]^{6-}$ (M = Ce, Th) noted in Chapter 3 as examples of 10-coordination. The CO_3 group can also behave in both ways in $Sb_2CO_3\phi_8$, (a), p. 881, and $CuCO_3(NH_3)_2$, (b), p. 1127, and in yet another way in compounds $L_2Cu_2Cl_2CO_3$,[1] (c), in which L is, for example, $(CH_3)_2N(CH_2)_3N(CH_3)_2$.

(a)   (b)   (c)

TABLE 21.3
Planar ions CX_3^{2-}, CX_3^-, *and* CX_3^+

Ion	C–X	Reference
CO_3^{2-} : Ca^{2+} } $: K_2^+ . 3H_2O$ }	1.29 Å (C–O)	JCP 1967 **47** 3297
CS_3^{2-} : $K_2^+ . H_2O$	1.71 (C–S)	AC 1970 **B26** 877
$C(CN)_3^-$: NH_4^+	1.40 (C–C)	ASc 1967 **21** 1530
: K^+	1.39 (C–C)	AC 1971 **B27** 1835
: Ag^+	—	IC 1966 **5** 1193
$C(SO_2CH_3)_3^-$: NH_4^+	1.70 (C–S)	AC 1965 **19** 651
$C(p\text{-}C_6H_4NO_2)_3$	1.47 (C–C)	ASc 1967 **21** 2599
$C(NH_2)_3^+$: CO_3^-	1.34 (C–N)	AC 1974 **B30** 2191
$C(N_3)_3^+$: $SbCl_6^-$	1.34 (C–N)	AC 1970 **B26** 1671
$C(C_6H_5)_3^+$: ClO_4^-, BF_4^-	1.45 (C–C)	AC 1965 **18** 437
$C(C_6H_4NH_2)_3^+$: ClO_4^-	1.45 (C–C)	AC 1971 **B27** 1405

This molecule is also of interest as an example of a diamagnetic cupric complex, presumably a result of antiferromagnetic exchange across the nearly collinear Cu–O–Cu bonds; the bond arrangement around Cu is square pyramidal.

(1) IC 1979 **18** 2296

Triazidocarbonium ion. The dimensions found for this ion are shown at (d).

Tricyanomethanide ion. The very small deviations from planarity of this ion observed in certain salts have been attributed to crystal interactions. This ion can form interesting polymeric structures with metal ions. In the argentous compound the group acts as a 3-connected unit in layers based on the simplest 3-connected plane net, and these layers occur as interwoven pairs as described on p. 105.

Urea. This molecule, (e), is intermediate between CO_3^{2-} and $C(NH_2)_3^+$ (guanidinium ion). A number of studies have shown that the H atoms are coplanar with the C, N, and O atoms;[1] other molecules in which N forms three coplanar bonds include formamide, $H_2N.CHO$, and $N(SiH_3)_3$. Bond lengths in thiourea are C–N, 1.33 Å, and C–S, 1.71 Å.[2]

(1) AC 1969 **B25** 404   (2) JCS 1959 2551

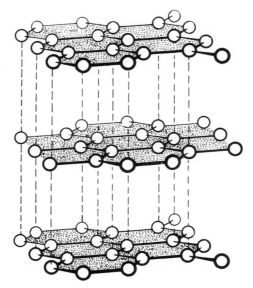

FIG. 21.3. The crystal structure of graphite.

Graphite. A portion of the crystal structure of graphite is illustrated in Fig. 21.3. The large separation of the layers (3.35 Å) compared with the C–C bond length of 1.42 Å in the layers indicates relatively feeble bonding between the atoms of different layers, which can therefore slide over one another, conferring on graphite its valuable lubricating properties. In the graphite structure (Fig. 21.3) the carbon atoms of alternate layers fall vertically above one another, and the structure can be described in terms of a hexagonal unit cell with $a = 2.456$ Å and $c = 6.696$ Å. The structure of many graphites is rather more complex.[1][2] In Fig. 21.4 the heavy and

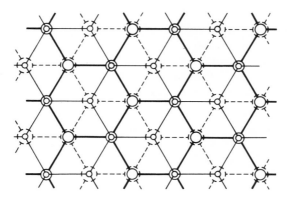

FIG. 21.4. Relative positions of atoms in successive layers of the two forms of graphite (see text).

light full lines represent alternate layers in the normal graphite structure, and it is seen that there is a third possible type of layer (broken lines) symmetrically related to the other two. A second graphite structure is therefore possible in which these three types of layer alternate, a structure with $a = 2.456$ Å and $c = \frac{3}{2}(6.696) = 10.044$ Å. Several natural and artificial graphites have been found to have partly this new structure and partly the normal structure. It is interesting to note that in the normal graphite structure the carbon atoms are not all crystallographically equivalent, whereas they are in the second form of Fig. 21.4. Also, this new form is closely related to diamond, from which it could be derived by making coplanar the atoms in the puckered layers parallel to (111) and lengthening the bonds between these layers. Graphite can be converted into diamond under a pressure greater than 125 kbar at about 3000 K.[3]

The great difference between the bonding within and between the layers of graphite is responsible for the formation of numerous compounds in which atoms or molecules are introduced between the layers. Some of these compounds will now be described.

Derivatives of graphite[4]

Charcoal is essentially graphitic in character but with a low degree of order, and because of its extensive internal surfaces and unsatisfied valences possesses the property of adsorbing gases and vapours to a remarkable extent and of catalysing reactions by bringing together the reacting gases. Crystalline graphite is far less reactive in this way since the layers of carbon atoms are complete and there is only a weak residual attraction between them. Nevertheless graphite takes up an extraordinary variety of elements and compounds between the layers. The graphite crystallites do not disintegrate but expand in the direction perpendicular to the layers.

Graphitic oxide. This name is applied to the pale, non-conducting product of the action of strong oxidizing agents such as nitric acid or potassium chlorate on graphite. The graphite swells in one direction, and measurements of the inter-layer distance in these compounds show that it increases from the value 3.35 Å to values between 6 and 11 Å, this expansion following the increasing oxygen content. The composition is not very definite but the limit corresponds approximately to the formula $C_4O(OH)$. The structure of the oxide is highly disordered, but e.d. results[5] on the dehydrated material (inter-layer spacing ≈ 6.2 Å) suggest that oxygen may be attached in two ways to the puckered graphite-like sheets:

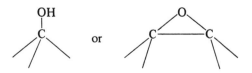

There is no regularity in the stacking of the layers.[6]

Graphitic 'salts'. If graphite is suspended in concentrated sulphuric acid to which a small quantity of oxidizing agent (HNO_3, $HClO_4$, etc.) is added, it swells and develops a steel-blue lustre and blue or purple colour in transmitted light. These products are stable only in the presence of the concentrated acid, being decomposed by water with regeneration of the graphite which, however, always contains some strongly held oxygen. In the 'bisulphate', the maximum separation of layers is approximately 8 Å. If the electrode potential of the graphite during oxidation or subsequent reduction in H_2SO_4 is plotted against composition, sharp breaks are observed in the curve corresponding to the ratios C_{96}^+, C_{48}^+, and C_{24}^+ to HSO_4^-.[7] Other graphitic 'salts', perchlorate, selenate, nitrate, etc., have been described.

Carbon monofluoride. When graphite, previously degassed, is heated in fluorine under atmospheric pressure the gas is taken up to form products ranging in colour from black, through grey, to snow-white. Careful temperature control ($627 \pm 3\,°C$) is necessary to produce this white fluoride, which has a limiting composition $CF_{1.12}$.[8] Apparently the C layers buckle and one F atom is attached to each (tetrahedrally bonded) C atom. The excess F over and above the composition CF results from the formation of CF_2 groups around the edges of the very small crystallites. Owing to the low crystallinity it has not been possible to determine in detail the structure of $CF_{1.12}$, which is thermally stable and is a good solid lubricant under extreme conditions (high temperature, high load, or oxidizing atmosphere). On heating to higher temperatures in fluorine graphite does not take up further fluorine, and the structure eventually decomposes into a mixture of various carbon fluorides together with soot—compare the further oxidation of 'graphitic acid' to mellitic acid, $C_6(COOH)_6$.

Compounds of graphite with alkali metals and bromine. On treatment with the molten metal or its vapour graphite forms the following compounds C_nM with the alkali metals:[9]

			n			
Li	6	12	18			
Na		only		64		
K						
Rb	8	10	24	36	48	60
Cs						

Absorption of bromine gives C_8Br and $C_{28}Br$,[10] and since it seems that C_8K and C_8Br are similar from the geometrical standpoint, having alternate layers of C and K (Br), these compounds are conveniently discussed together. The marked temperature-dependent diamagnetism of graphite, due to the large π-orbitals, is changed to a weak temperature-independent diamagnetism in C_8Br (comparable with that of metallic Sn or Zr), and a comparatively strong temperature-independent paramagnetism in C_8K (comparable with Ca), stronger than that of metallic K itself. Both C_8K and

C_8Br are better conductors of electricity than graphite. The interlamellar separation of 3.35 Å in graphite becomes 7.76 Å in C_8K and 7.05 Å in C_8Br. It would seem that if these structures were ionic in character there would be a larger separation of the carbon layers in the bromine compound (since the radius of the Br^- ion is 1.96 Å while that of K^+ is 1.38 Å), whereas the metallic radius of K is somewhat greater than the covalent radius of Br. It has therefore been suggested that there are 'metallic' bonds between K or Br and the graphite layers, the empty electron band in graphite being partly filled in the K compounds and the full band being partly emptied in the Br compound. The fact that Na does not form a similar compound, while Rb and Cs do, may be due to the fact that the ionization energy of Na is higher than that of K.

Graphite complexes with halides.   In addition to the well-known compounds with $FeCl_3$ and $AlCl_3$ complexes can be formed with many other halides, for example, $CuCl_2$, $CrCl_3$, 4f trihalides, $MoCl_5$, and also chromyl fluoride and chloride. (Crystalline BN also forms similar compounds with $AlCl_3$ and $FeCl_3$.) An electron diffraction study[11] has been made of the graphite–$FeCl_3$ system. The best characterized phase in the graphite–$MoCl_5$ system consists of layers of Mo_2Cl_{10} molecules between sets of four graphite layers.[12] Within a layer the Mo_2Cl_{10} molecules are close-packed. With AsF_5 graphite forms several intercalates, some of which are of interest in having electrical conductivity in directions parallel to the layers comparable with that of copper. Other intercalates include the blue C_8OsF_6 and compounds C_nSO_3F formed with $S_2O_6F_2$, with interplanar spacings of 7.86 and 11.3 Å respectively. Corresponding to the graphite compound $C_{12}SO_3F$ there is an intercalate $(BN)_4SO_3F$, with interplanar spacing close to 8.0 Å.[13,14]

(1) PRS A 1942 **181** 101
(2) ZK 1956 **107** 337
(3) JCP 1967 **46** 3437
(4) QRCS 1960 **14** 1
(5) AC 1963 **16** 531
(6) ZaC 1969 **369** 327
(7) PRS A 1960 **258** 329, 339
(8) JACS 1974 **96** 2628
(9) JCP 1968 **49** 434
(10) PRS A 1965 **283** 179
(11) AC 1956 **9** 421
(12) AC 1967 **23** 770
(13) CC 1977 389
(14) CC 1978 200

The oxides and sulphides of carbon

Carbon monoxide

In this molecule, which has a very small (possibly zero) dipole moment, the bond length is 1.131 Å. The length, force constant, and bond energy (Table 21.4) show that it is best represented C≡O.

Carbon dioxide and disulphide: carbonyl sulphide

The molecules OCO, SCS, and OCS are linear. Bond lengths are given in Tables 21.5 and 21.6. In carbon dioxide the carbon–oxygen bond is intermediate in length between a double and triple bond. The studies of CO_2 and CS_2 by high resolution

TABLE 21.4

Properties of carbon-oxygen bonds

Bond	Length	Force constant	Bond energy
	(Å)	N m^{-1}	(kJ mol^{-1})
C=O in H.CHO	1.209	0.0123	686
O=C=O	1.163	0.0155	803
C≡O	1.131	0.0186	1075

TABLE 21.5

Lengths of C—O bonds

Molecule	Bond length	Method	Reference
CO	1.131 Å	m.w.	PR 1950 **78** 140
COCl$_2$	1.166	m.w.	JCP 1953 **21** 1741
COSe	1.159	m.w.	PR 1949 **75** 827
CO$_2$	1.163	i.r.	JRNBS 1955 **55** 183
COS	1.164	m.w.	PR 1949 **75** 270
HNCO	1.184	X	AC 1955 **8** 646
H$_2$CO	1.209	e.d.	BCSJ 1969 **42** 2148
CH$_3$OH	1.427	m.w.	JCP 1955 **23** 1200
HCOO—CH$_3$	1.437	m.w.	JCP 1959 **30** 1529

TABLE 21.6

Lengths of C—S bonds

Molecule	Bond length	Method	Reference
HNCS	1.56 Å	m.w.	JCP 1953 **21** 1416
OCS	1.56	m.w.	PR 1949 **75** 270
TeCS	1.56	m.w.	PR 1954 **95** 385
SCS	1.55	i.r.	JCP 1958 **28** 682
(NH$_2$.CS)$_2$	1.65	X	JCS 1965 396
SC(NH$_2$)$_2$	1.71	X	JCS 1958 2551
CH$_3$SH	1.81	m.w.	PR 1953 **91** 464
(CF$_3$)$_2$S	1.83 ⎫	e.d.	TFS 1954 **50** 452
(CF$_3$)$_2$S$_3$	1.85 ⎭		

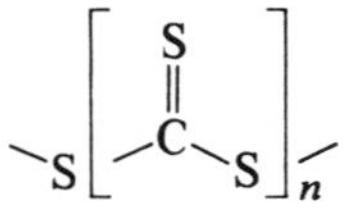

infrared spectroscopy provide an example of the use of this method for molecules which cannot be studied by the microwave method because they have no permanent dipole moment. The structures of the CO_2 and CS_2 molecules have also been studied in the crystalline state (C–O, 1.155 Å at 150 K,[1a] C–S, 1.56 Å[1b]). Under a pressure of 30 kbar CS_2 polymerizes to a black solid for which a chain structure has been suggested.[2]

Carbon suboxide

The third oxide of carbon, the so-called suboxide C_3O_2, is a gas at ordinary temperatures (it boils at +6 °C) and may be prepared by heating malonic acid or its diethyl ester with a large excess of phosphorus pentoxide at 300 °C. Its preparation from, and conversion into, compounds containing the –C–C–C– nucleus are consistent with its structure, O–C–C–C–O. The action of water regenerates malonic acid, ammonia yields malonamide, and HCl malonyl chloride:

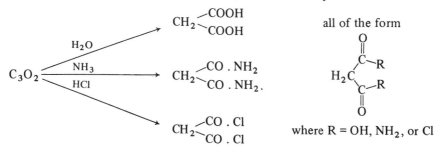

Carbon suboxide polymerizes readily at temperatures above –78 °C. An e.d. study indicates a linear molecule with C–C, 1.28 Å, and C–O, 1.16 Å,[3] but the possibility of small deviations from linearity could not be excluded.

The oxides C_2O[4] and CO_3 [5] have been identified in low-temperature matrices. The latter is an unstable species formed by the reaction of O atoms with CO_2; its i.r. spectrum has been studied in an inert matrix and the cyclic structure (a) suggested.

$$\begin{array}{c} O \\ | \\ O \end{array}\!\!>\!\!C\!=\!O \quad \text{(a)}$$

Two molecules with some structural resemblance to CO_2 and CS_2:

are remarkable for the non-linear bonds formed by C_1.[6,7] The C_1–C bond length corresponds to a triple bond and P–C to a double bond, the length of a single P–C bond being at least 1.85 Å.

(1a) AC 1980 **B36** 2750
(1b) JCP 1968 **48** 2974
(2) CJC 1960 **38** 2105

(3) JACS 1959 **81** 285
(4) JCP 1965 **43** 3734
(5) JCP 1966 **45** 4469

(6) JCS A 1966 1703
(7) JCS A 1967 1913

Acetylene and derivatives

A literature reference for acetylene is given in Table 21.1 (p. 914). All molecules R–C≡C–R are linear, and in all cases the length of the C≡C bond is close to the value in acetylene itself (1.205 Å). In compounds containing the system –C≡C–C≡C– the central bond is only slightly longer (1.37 Å) than the double bond in ethylene (1.34 Å). Also, the terminal bond in molecules X–C≡C–X is appreciably shorter than a normal single bond. For example, in the molecules $H_3C.C≡CX$ (X = Cl, Br, or I), H_3C–C is approximately 1.46 Å and C–X is 1.64,[1] 1.79,[2] and 1.99 Å[2] as compared with the single-bond lengths, C–C, 1.54; C–Cl, 1.77; C–Br, 1.94; and C–I, 2.14 Å.

Very short C–X bonds are also found in HCCF and HCCCl:[3]

$$\text{F} \underset{1.279}{\rule{2cm}{0.4pt}} \text{C} \underset{1.198}{\rule{2cm}{0.4pt}} \text{C} \underset{1.053 \text{ A}}{\rule{2cm}{0.4pt}} \text{H} \qquad \text{Cl} \underset{1.637}{\rule{2cm}{0.4pt}} \text{C} \underset{1.204}{\rule{2cm}{0.4pt}} \text{C} \underset{1.055 \text{ A}}{\rule{2cm}{0.4pt}} \text{H}$$

these C–X bond lengths being very similar to those in cyanogen halides.

(1) JCP 1955 **23** 2037   (3) TFS 1963 **59** 2661
(2) JCP 1952 **20** 735

Cyanogen and related compounds

Cyanogen

For the cyanogen molecule, which is linear, the simplest formulation is N≡C–C≡N. The length of the central bond, 1.38 Å, however, is the same as in diacetylene, HC≡C–C≡CH, showing that there is considerable π-bonding. The C≡N bond length is the same as in HCN. Bond lengths in cyanogen and related compounds are given in Table 21.7.

TABLE 21.7
Structural data for cyanogen and related compounds (Å)

	C≡N	C–C	C≡C	Reference
N≡C–C≡N	1.15	1.38	—	JCP 1965 **43** 3193
N≡C–C≡CH	1.157	1.382	1.203	JACS 1950 **72** 199
N≡C–C≡C–C≡N	1.14	1.37	1.19	AC 1953 **6** 350

Cyanogen and its derivatives are of special interest because of the ease with which they polymerize. When cyanogen is prepared by heating (to 450 °C) mercuric cyanide, preferably with the chloride, there is also obtained a quantity of a solid polymer, paracyanogen $(CN)_n$, which may also be made by heating oxamide in a sealed tube to 270 °C. The structure of this extremely resistant material is not known; some conjugated ring structures have been suggested. The structures of the trimeric (cyanuric) compounds are noted later.

HCN: *cyanides and isocyanides of non-metals*

Only one isomer of HCN is known, but organic derivatives are of two kinds, cyanides RCN and isocyanides RNC. The latter have been formulated R–N≡C, but in fact the N≡C bond length in CH_3NC is only slightly greater (1.167 Å) than in CH_3CN (1.158 Å), which is close to that in HCN itself (1.155 Å). The force constants of the C–N bonds in HCN, CH_3CN, and the cyanogen halides, XCN, are in all cases close to $1800 \, N \, m^{-1}$, and therefore cyanides are satisfactorily represented R–C≡N.

The structures in the vapour state of HCN and of a number of covalent cyanides and isocyanides containing the linear systems R(M)–C–N and R(M)–N–C respectively are summarized in Table 21.8. There are two crystalline forms of HCN,

TABLE 21.8
Structures of covalent cyanides and isocyanides

Molecule	C–N	Other data		Reference
HCN	1.155 Å	C–H	1.063 Å	TFS 1963 **59** 2661
CH_3CN	1.158	C–C	1.460	PR 1950 **79** 54
				JACS 1955 **77** 2944
CH_3NC	1.167	H_3C–N	1.426	RMP 1948 **20** 668
CF_3CN	1.15	C–C	1.475	JACS 1955 **77** 2944
		C–F	1.300	JCP 1952 **20** 591
	C–M–C	M–C	M–C≡N	
$S(CN)_2$	96°	1.73 Å	177°	AC 1966 **21** 970
$P(CN)_3$	93°	1.78	172°	AC 1964 **17** 1134
$As(CN)_3$	92°	1.90	171°	AC 1963 **16** 113
$C(CN)_4$	109½°	1.485	180°	AC 1974 **B30** 1818

the low-temperature orthorhombic form changing reversibly into a tetragonal polymorph at –102.8 °C. The structures of both forms are essentially the same, the molecules being arranged in endless chains:[1]

$$\text{---N≡C–H------N≡C–H---}$$
$$\leftarrow 3{\cdot}18 \, \text{Å} \rightarrow$$

The dipole moment of HCN is 2.6 D in benzene solution and 2.93 D in the vapour. The abnormally high dielectric constant of liquid HCN is attributed to polymerization, and there is i.r. evidence for dimerization in an inert matrix at low temperatures.[2]

HCN polymerizes to a dark-red solid which on recrystallization yields almost colourless crystals of $(HCN)_4$. This 'tetramer' is actually diaminomaleonitrile:[3]

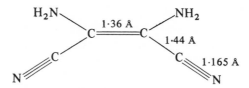

The approximately collinear configuration of M–C≡N has been confirmed in a number of molecular cyanides (Table 21.8). In some of these cyanides there are intermolecular M---N distances less than the sums of the van der Waals radii, and in $S(CN)_2$, for example, the two next nearest neighbours of S are coplanar with the intramolecular C atoms, suggesting that intermolecular interactions may be responsible for the small deviations of the M–C≡N systems from exact collinearity.

The structures of metal cyanides are described in the next chapter.

(1)   AC 1951 **4** 330   (3)   AC 1961 **14** 589
(2)   JCP 1968 **48** 1685

Cyanogen halides

All four cyanogen halides are known. The fluoride is a gas at ordinary temperatures and is prepared by pyrolysis of $(FCN)_3$ which in turn is made from $(ClCN)_3$ and NaF is tetramethylene sulphone. The chloride is a liquid boiling at 14.5 °C which readily polymerizes to cyanuric chloride, $Cl_3C_3N_3$, and is prepared by the action of chlorine on metal cyanides or HCN. The bromide and iodide, also prepared from halogens and alkali cyanides, form colourless crystals. All four halides have been studied in the vapour state and all except FCN also in the crystalline state (Table 21.9).

TABLE 21.9
Structures of cyanogen halides

	C–X	C–N	C–X in CX_4	X···N in crystal	Method	Reference
FCN	1.26 Å	1.16 Å	1.32 Å	–	m.w.	TFS 1963 **59** 2661
ClCN	1.63	1.16	1.77	–	m.w.	JCP 1965 **43** 2063
				3.01 Å	X	AC 1956 **9** 889
BrCN	1.79	1.16	1.94		m.w.	PR 1948 **74** 1113
				2.87	X	JCP 1955 **23** 779
ICN	2.00	(1.16)	2.14		m.w.	PR 1948 **74** 1113
				2.8	X	RTC 1939 **58** 448

The molecules are linear, and in the crystals are arranged in chains

$$----X–C≡N----X–C≡N----$$

There appears to be some sort of weak covalent bonding between X and N atoms of adjacent molecules, the effect being most marked in the iodide, where the intermolecular I---N distance is only 2.8 Å, far less than the sum of the van der Waals radii (see p. 285). The C–X bond lengths within the molecules are appreciably less than the single-bond values in the molecules CX_4 and are the same as in H_3C–C≡C–X, but C≡N is very close to the value in HCN.

Cyanamide, $H_2N.CN$

This colourless crystalline solid formed from NH_3 and ClCN readily polymerizes, for example, by heating to 150 °C, to the trimer melamine (see later). The bond lengths in the molecule are:[1] N–C, 1.15 Å, and C–NH_2, 1.31 Å. The group N–C–N is unsymmetrical also in (covalent) heavy metal derivatives such as $Pb(NCN)$[2] (bond lengths approximately 1.17 and 1.25 Å) presumably because it is more strongly bonded to the metal through one N. Derivatives of more electropositive metals are salts containing the linear symmetrical NCN^{2-} ion (studied in $CaNCN$);[3] compare N–C, 1.22 Å, with N–N, 1.17 Å, in the isoelectronic N_3^- ion. (Only approximate bond lengths were determined in $Sr(NCN)$.)[4]

Dicyandiamide, $NCNC(NH_2)_2$

The dimer of cyanamide is formed when calcium cyanamide is boiled with water. This also is a colourless crystalline compound, the molecule of which has the structure:[5]

Cyanuric compounds

The cyanogen halides polymerize on standing to the trimers, the cyanuric halides. Cyanamide is converted into the corresponding trimer, melamine, on heating to about 150 °C. Isocyanic acid, however, polymerizes far more readily. If urea is distilled, isocyanic acid is formed but polymerizes to cyanuric acid, $(NCOH)_3$, a crystalline solid, the vapour of which, on rapid cooling, yields isocyanic acid as a liquid which, above 0 °C, polymerizes explosively to cyamelide, a white porcelain-like solid. This latter material is converted into salts of cyanuric acid by boiling with alkalis. These reactions are summarised in Chart 21.1. Cyanuric derivatives may be interconverted. For example, if cyanuric bromide is warmed with water, cyanuric acid is formed. Cyanuric chloride is converted by the action of NaF in tetramethylene sulphone to the fluoride, $(FCN)_3$, pyrolysis of which provides pure FCN in good yield.

The structure of the cyanuric ring has been studied in several derivatives, for example, the triazide, triamide (melamine), and cyanuric acid (Fig. 21.5). The ring is a nearly regular hexagon with six equal C–N bonds of length 1.34 Å. In crystalline

CHART 21.1

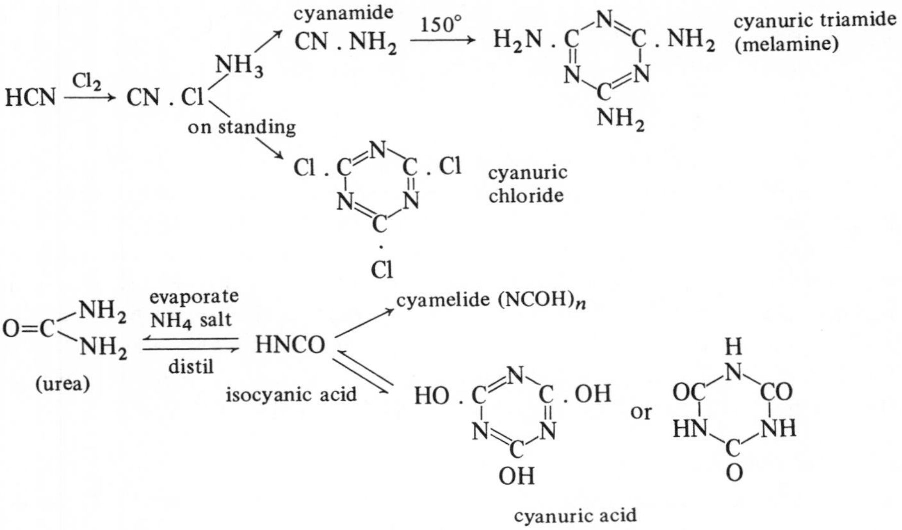

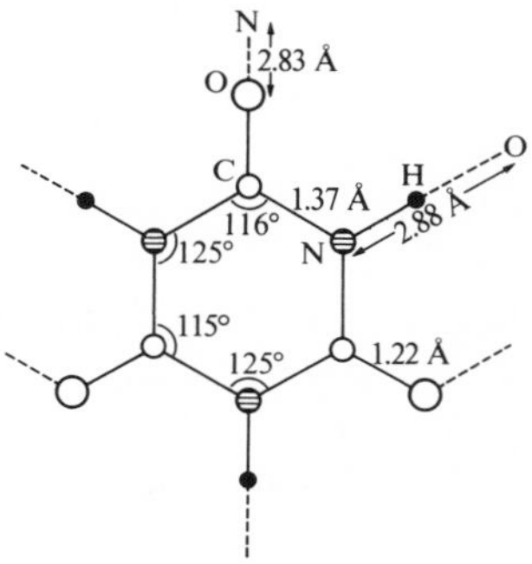

FIG. 21.5. The configuration of the molecule in crystalline cyanuric acid. Broken lines indicate N–H---O bonds to adjacent molecules.

cyanuric acid the molecules are arranged in layers,[6] six N–H–O bonds linking each molecule to four neighbours as shown in Fig. 21.6.

(1) SPC 1961 **6** 147
(2) AC 1964 **17** 1452
(3) ACSc 1962 **16** 2263
(4) ACSc 1966 **20** 1064
(5) JACS 1940 **62** 1258
(6) AC 1971 **B27** 134, 146

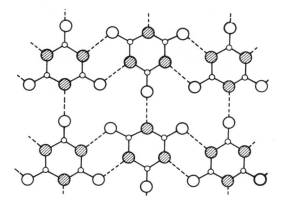

FIG. 21.6. Layer of hydrogen-bonded molecules in crystalline cyanuric acid.

Isocyanic acid and isocyanates

In the acids HNCO and HNCS there is the possibility of tautomerism:

$$H-O-C\equiv N \quad \text{or} \quad H-N=C=O \quad \text{(iso form)}$$
$$\quad\text{(S)} \qquad\qquad\qquad \text{(S)}$$

Two series of esters, normal and iso, might be expected but both forms of the acid would give the same ion $(OCN)^-$, isoelectronic with the azide ion. In fact, the O and S compounds show rather different behaviours:

O *compounds*	S *compounds*
Acid entirely in *iso* form, HNCO	Acid tautomeric (in CCl_4 solution the Raman spectrum indicates *iso* (HNCS); vapour at least 95 per cent *iso*)
Esters: one type only, *iso*, RN=C=O	Two series of esters: R–S–C≡N (normal) R–N=C=S (*iso*)

The suggestion that freshly prepared isocyanic acid may contain a trace of HO–C≡N is based on the presence of a very small amount of cyanate in the isobutyl ester prepared from the acid and diazobutane, but there are admittedly other explanations.[1]

The structure of the isocyanic acid molecule has been studied in the vapour[2] and crystalline[3] states. The H–N–C angle is close to 128° and the bond lengths are: N–H, 0.99 Å; N–C, 1.20 Å; and C–O, 1.18 Å. Values of the R–N–C bond angle in esters include 126° in an early e.d. study of $H_3C.NCO$[4] and $150 \pm 3°$ in $(CH_3)_3Si.NCO$.[5] The value 180° found in $Si(NCO)_4$ should perhaps be accepted with reserve.

Few studies have been made of ionic (iso)cyanates. The fact that the sodium salt is isostructural with the azide confirms the linear structure of the ion, but the bond lengths N–C, 1.21 Å, and C–O, 1.13 Å, were not claimed to be highly accurate.[6]

In AgNCO[7] the isocyanate groups are linked by Ag atoms which form two collinear bonds (Ag–N, 2.12 Å). (The silver-oxygen distances, 3.00 Å, are too long for normal bonds; compare 2.05 Å in Ag_2O.)

(There are two cyrstalline forms of the isomeric fulminate, AgCNO, one containing chains somewhat like those in the isocyanate and the other rings of six Ag atoms

linked through CNO groups.)[8] In the chains the Ag–Ag distance, 2.93 Å, is only slightly longer than in the metal (2.89 Å), and in the rings it is 2.83 Å.

(1) ACSc 1965 **19** 1768	(5) JACS 1966 **88** 416
(2) JCP 1950 **18** 990	(6) MSCE 1943 **30** 30
(3) AC 1955 **8** 646	(7) AC 1965 **18** 424
(4) JACS 1940 **62** 3236	(8) AC 1965 **19** 662

Isothiocyanic acid, thiocyanates, and isothiocyanates

The free acid, prepared by heating KSCN with $KHSO_4$, is in the iso form to at least 95 per cent in the vapour state, and the molecule has the structure (a).[1] A microwave study of methyl thiocyanate[2] gives the structure (b), values in brackets being

(a)   (b)   (c)

assumed. In crystalline $Se(SCN)_2$,[3] in which the system S–C–N was assumed to be linear, the configuration (c) was found, with dihedral angle SSeS/SeSC 79°, but the S–C and C–N bond lengths were not accurately determined.

In molecular isothiocyanates the angle R–N–C has been determined as 142° in an early m.w. study of $H_3C.NCS$,[4] 154° in $(CH_3)_3Si.NCS$,[5] but 180° in $H_3Si.NCS$[6] and $Si(NCS)_4$.[7]

(1)	JMS 1963 **10** 418	(5)	JACS 1966 **88** 416
(2)	JCP 1965 **43** 3583	(6)	TFS 1962 **58** 1284.
(3)	JACS 1954 **76** 2649	(7)	SA 1962 **18** 1529 (but see comments in ref. 5)
(4)	JACS 1949 **71** 927		

Metal thiocyanates and isothiocyanates

As in the case of metallic cyanides (Chapter 22) we may distinguish three main types:

(i) *ionic crystals containing the SCN⁻ ion;*

(ii) *compounds containing NCS bonded to one metal atom only, either through S (thiocyanate) or through N (isothiocyanate);* and

(iii) *compounds in which –NCS– acts as a bridge between two metal atoms.*

We shall give an example of a compound of type (ii) which is both a thiocyanate and isothiocyanate; $Ni(NH_3)_3(SCN)_2$ may represent a fourth class in which there are both bridging and singly attached NCS groups in the same crystal. In (i) and (iii) there is no distinction between thiocyanate and isothiocyanate.

(i) *Ionic thiocyanates.* The dimensions of the (linear) SCN⁻ ion in NaSCN and KSCN are close to the values: S–C, 1.65, and C–N, 1.17 Å.[1] (The approximate bond lengths in SeCN⁻ in KSeCN are:[2] Se–C, 1.83 Å; C–N, 1.12 Å.)

(ii) *Covalent thiocyanates and isothiocyanates.* Numerous ions and molecules containing the –SCN or –NCS ligands are formed by transition and B subgroup metals (e.g. Cd, Hg). It appears that in general metals of the first transition series except Cu(II) form isothiocyanates, –M–N–C–S, while those of the second half of the second series and those of the third series form thiocyanates, –M–S–C–N, or bridged compounds of class (iii). However, for some metals (e.g. Pd) the choice between M–NCS and M–SCN is affected by the nature of the other ligands attached to M. The evidence comes from i.r. spectroscopy, the stretching frequency of the S–C bond apparently making possible a decision between thiocyanate and isothiocyanate; for example, $Pd(P\phi_3)_2(NCS)_2$ but $Pd(Sb\phi_3)_2(SCN)_2$. The claim that both isomers of $Pd(As\phi_3)_2(SCN)_2$ have been isolated has not been confirmed by a structural study.[3] Nevertheless there is both a thiocyanato and an isothiocyanato ligand in the remarkable square planar Pd complex (a).[4]

X-ray crystallographic studies have been made of many compounds, of which the following are examples:[5]

Isothiocyanates   tetrahedral: $[Co(NCS)_4]^{2-}$
octahedral: $[Cr(NH_3)_2(NCS)_4]^-$, $Ni(NH_3)_4(NCS)_2$

Thiocyanates   planar: $[Pt(SCN)_4]^{2-}$
                      tetrahedral: $[Hg(SCN)_4]^{2-}$
                      octahedral: $[Rh(SCN)_6]^{3-}$, $Cu(en)_2(SCN)_2$.

The correlation of bond angle with bond lengths to be expected if the system approximates to $M-N^+{\equiv}C-S^-$ or $_M{\diagdown}N{=}C{=}S$ would not appear to be justified by the data at present available. Bond angles M–S–C in metal thiocyanato complexes are typically around 105°, as in (a) above and in $[(NH_3)_5CoSCN]^{2+}$.[6] The value (80°) in $Cu(en)_2(SCN)_2$[7a] is atypical because the bond Cu–S is extremely weak (3.27 Å), completing a (4+2)-coordination group. Bond angles M–N–C in isothiocyanato complexes tend to lie in the range 160–180° in, for example, $[Co(NCS)_4]^{2-}$ (166°), p. 1209, and $[Cr(NH_3)_2(NCS)_4]^-$ (180°), but 140° is recorded in $Ni(en)_2(NCS)_2$, (b), and angles of 145° and 179° in $[Fe(NCS)_6]^{3-}$;[7b] the very low values may be the result of packing problems. The bond lengths in both thiocyanato and isothiocyanato complexes appear to be close to the values, C–N, 1.16 Å, and C–S, 1.62 Å. Both –S–C–N and –N–C–S are linear to within the experimental errors.

(iii) *Thiocyanates containing bridging –S–C–N– groups.* A study of crystalline AgSCN showed that the crystal is built of infinite chains which are bent at the S atoms and apparently also at the Ag atoms (where the bond angle was determined

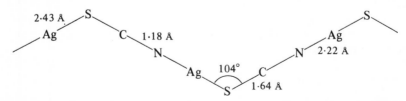

as 165°), though the positions of the lighter atoms are not very certain.[8] On the subject of the Ag bond angle see also p. 1104.

Bridging of metal atoms by —S—C—N— occurs in a number of compounds of Cd, Mg, Pb, Co, Ni, Cu, Pd, and Pt. An example of a finite molecule is the Pt complex (a),[9] while infinite linear molecules are formed in $Cd(etu)_2(SCN)_2$,[10] (b), and the isostructural Pb compound (etu = ethylenethiourea, $SC(NHCH_2)_2$). There are similar linear chains of type (b), H_2O replacing etu, in $Co(SCN)_2 . 3H_2O$;[11] the third H_2O molecule is not included in the chain. Mercury provides examples of 2D and 3D complexes (Chapter 26).

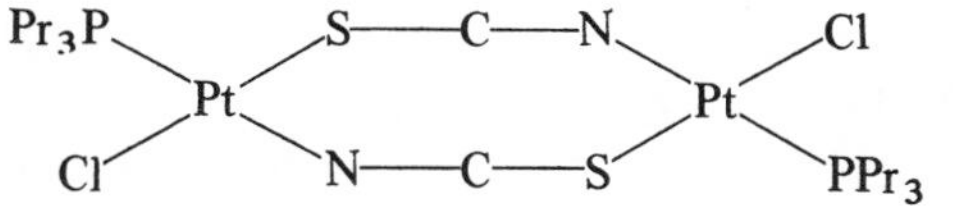

$(Pr = C_3H_7)$

(a)

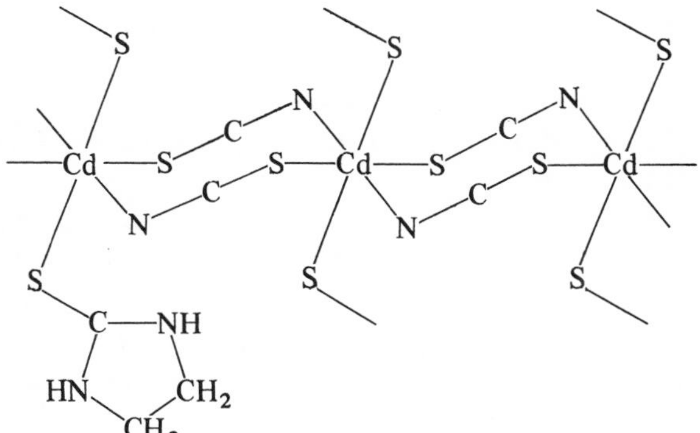

(b)

(1) AC 1977 **B33** 1534, 1542

(2) IC 1965 **4** 499

(3) IC 1966 **5** 1632

(4) IC 1970 **9** 2754

(5) For refs. see JCS 1961 4590; JACS 1967 89 6131; AC 1975 **B31** 172

(6) AC 1972 **B28** 1908

(7a) AC 1964 **17** 254

(7b) AC 1977 **B33** 2197

(8) AC 1957 **10** 29

(9) JCS 1961 1416 (see p. 1238)

(10) AC 1960 **13** 125

(11) AC 1976 **B32** 1526

22

Metal cyanides, carbides, carbonyls, and alkyls

In this chapter we deal first with compounds formed by metals containing the isoelectronic groups CN^-, C_2^{2-}, and CO. (HCN and non-metal cyanides, and also cyanates and thiocyanates are included in Chapter 21.) We then describe briefly compounds of metals with hydrocarbon radicals or molecules other than olefine compounds of Pd and Pt (Chapter 27) and certain compounds of Cu and Ag (Chapter 25).

Metal cyanides

We may recognize three classes of metal cyanides corresponding to the behaviour of the cyanide ion or group:

(a) *simple ionic cyanides, containing the ion* CN^-;

(b) *molecules or complex ions in which* —CN *is attached to one metal atom only (through the C atom); and*

(c) *compounds in which the* —CN— *group is bonded to metal atoms through both C and N, so performing the same bridging function as halogen atoms, oxygen, etc.*

In a few complex cyanides CN groups of types (b) and (c) are found, for example, $KCu(CN)_2$ and $K_2Cu(CN)_3 \cdot H_2O$. The structures of these salts are described in Chapter 25.

(a) *Simple ionic cyanides*

At ordinary temperatures NaCN, KCN, and RbCN crystallize with the NaCl structure, and CsCN and TlCN with the CsCl structure. These structures would be consistent with either free rotation of the CN^- ion, when it would behave as a sphere of radius about 1.9 Å (intermediate in size between Cl^- (1.81 Å) and Br^- (1.96 Å)), or with random orientations of CN^- ions along directions [111] consistent with cubic symmetry. A neutron diffraction study of KCN at room temperature indicates free rotation of the ion (C–N, 1.16 Å).[1]

All the above alkali cyanides exist in less symmetrical 'low-temperature' forms, in which the CN^- ion behaves as an ellipsoid of rotation, with long axis about 4.3 Å and maximum diameter about 3.6 Å. Their structures are closely related to the NaCl or CsCl structures of the high-temperature forms (Table 22.1). (There is also a metastable rhombohedral form of NaCN made by quenching the cubic form from 200–300 °C and also a second low-temperature form.[2] A second low-temperature form of KCN is stable only in the range −108 to −115 °C.[3])

TABLE 22.1
Crystal structures of the alkali cyanides

	Low-temp. form	*High-temp. form*	*Transition point*	*Reference*
NaCN	Orthorhombic (Fig. 22.1)	NaCl	+15·3°C	ZK 1938 **100** 201
KCN			−108°	RTC 1940 **59** 908
RbCN	Monoclinic?		Between −100° and −180°	SR 1942–4 **9** 138-142
CsCN	Rhombohedral (Fig. 22.2)	CsCl	−55°?	
NH₄CN	Tetragonal (Fig. 22.3(b))	?	Continuous change	RTC 1944 **63** 39
LiCN	Orthorhombic (Fig. 22.3(a))	−	No transition	RTC 1942 **61** 244

The structure of the (orthorhombic) low-temperature form of NaCN (and the isostructural KCN) is illustrated in Fig. 22.1; its close relation to the NaCl structure is evident. (The portion of structure shown is enclosed by (110) planes.) All the CN⁻ ions are parallel and there is 6-coordination of both anion and cation. Low-temperature RbCN also has a deformed NaCl structure, but the details of the structure are not known with certainty. The metastable rhombohedral form of NaCN is closely related structurally to the high-temperature form (for $\alpha = 53.2°$) and not, like low-temperature CsCN, to the CsCl structure.[4] The low-temperature form of CsCN is rhombohedral, having a slightly deformed CsCl structure, illustrated in Fig. 22.2, in which the CN⁻ ions lie along the trigonal axis. The coordination is 8-fold as in CsCl. Lithium cyanide differs from the other alkali cyanides in many ways. It has a low melting point (160°C, compare 550°C for NaCN and 620°C for KCN), a low density (1.025 g cm⁻³), and a loosely-packed crystal structure in which

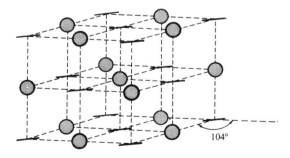

FIG. 22.1. The structure of the low-temperature form of NaCN. The CN⁻ ions are indicated by short straight lines.

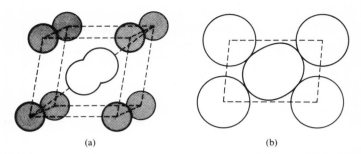

FIG. 22.2. The crystal structure of the low-temperature (rhombohedral) form of CsCN; (a) unit cell (Cs⁺ ions shaded); (b) section through the structure showing the packing of the Cs⁺ and CN⁻ ions.

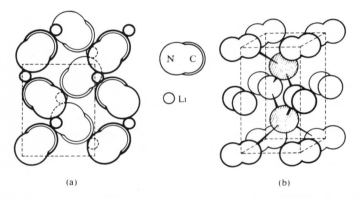

FIG. 22.3. The crystal structures of (a) LiCN, projected on (100); and (b) NH₄CN. In (a) the heavier circles represent atoms lying in a plane 1.86 Å above the others. In (b) the shaded circles represent NH₄⁺ ions.

each ion is only 4-coordinated (Fig. 22.3(a)). Also, it is possible to distinguish the C and N atoms in the CN⁻ ions, which is not possible in the other alkali cyanides. From an electron count in a Fourier projection it appears that the negative charge resides on the N atom, and that the tetrahedral coordination group around Li⁺ consists of one C and three N, while a CN⁻ ion is surrounded by four Li⁺, one near the C and three around the N. The change from 8→6→4-coordination in CsCN, NaCN, and LiCN is probably not a simple size effect, for the small Li⁺ ion appears to polarize the CN⁻ ion appreciably.

The structure of NH₄CN, which shows no polymorphism in the temperature range +35 to –80°C, also has some interesting features. The structure is very simply related to that of CsCl, the unit cell being doubled in one direction owing to the way in which the CN⁻ ions are oriented (Fig. 22.3(b)). Although each NH₄⁺ ion is surrounded by eight ellipsoidal CN⁻ ions, owing to the orientations of the latter there are strictly only four in contact with NH_4^+ (at 3.02 Å), the remaining four being farther away (3.56 Å).

(1) AC 1961 **14** 1018   (3) AC 1962 **15** 601
(2) JCP 1968 **48** 1018   (4) JCP 1949 **17** 1146

(b) *Covalent cyanides containing* —CN

It should be noted that in X-ray studies of the heavier metal cyanides it is not possible to distinguish between C and N atoms. However, from neutron diffraction studies of $K_3Co(CN)_6$[(1)] and $K_2Zn(CN)_4$[(2)] (Zn–C, 2.024 Å; C–N, 1.157 Å) it is concluded that the C atom is bonded to the metal, as has always been assumed.

Only one metal cyanide, $Hg(CN)_2$, is known to have a molecular structure of the type we are considering. The crystal is built of nearly linear molecules[(3)]

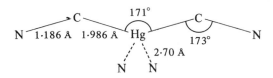

The two weaker bonds from Hg to N atoms in a plane perpendicular to the C–Hg–C plane may account for the departure of the C–Hg–C bond angle from 180°. In the ions to be discussed now the C–N bond is collinear with the metal–carbon bond; the system M–C–O in metal carbonyls is also linear.

Many heavy metal cyanides redissolve in the presence of excess CN^- ions to form complex ions $[M(CN)_n]^{m-}$. Since these ions are primarily of interest as illustrating the stereochemistry of the metal atoms, details and references are in most cases given in other chapters.

Bond arrangement	Examples
2 collinear	K[NC–Ag–CN] (also Au)
4 coplanar	Ba[Pd(CN)$_4$] . 4 H$_2$O (also Ni, Pt)
4 tetrahedral	K$_3$[Cu(CN)$_4$]
6 octahedral	K$_4$[Fe(CN)$_6$] . 3 H$_2$O
8 dodecahedral	K$_4$[Mo(CN)$_8$] . 2 H$_2$O (also W, Re)
8 antiprism	. Na$_3$[Mo(CN)$_8$] . 4 H$_2$O (also W)

The octahedral configuration of a number of $M^{III}(CN)_6^{3-}$ ions has been established by X-ray studies of, for example, the free acids $H_3Fe(CN)_6$ and $H_3Co(CN)_6$,[(4)] $K_3Co(CN)_6$, $[Co(NH_3)_6][Cr(CN)_6]$, and $[Co(NH_3)_5H_2O][Fe(CN)_6]$, the last two crystallizing with the $[Ni(H_2O)_6][SnCl_6]$ structure. The ferrocyanide ion has been studied in the free acid, $H_4Fe(CN)_6$ (Fe–C, 1.89 Å),[(5)] and $K_4Fe(CN)_6$. The structures of Prussian blue and related compounds are discussed shortly.

Octacyanide ions are formed by Mo(IV) and (V), W(IV) and (V), and Re(V) and (VI). The configurations most frequently found are the square antiprismatic

(D_{4d}, $\bar{8}m2$) and dodecahedral (D_{2d}, $\bar{4}m2$), but a bicapped trigonal prismatic configuration has been found for the anion in $Cs_3[Mo(CN)_8].6H_2O$.[6] However, it seems that few ions have shapes very close to these ideal shapes, as judged by the values of certain angles calculated for those configurations,[7] though the examples listed in Table 22.2 are close to the dodecahedral and antiprismatic configurations. Of greatest chemical interest are the facts that (i) the same ion, for example, $Mo(CN)_8^{3-}$, may have different configurations in different salts, and (ii) the same configuration may be adopted by ions such as $Mo(CN)_8^{3-}$ and $Mo(CN)_8^{4-}$ in spite of the presence of the extra electron in the latter. Evidently there is not a very large energy difference between the various configurations, and it appears that the geometry of these ions cannot be deduced from factors such as ligand–ligand repulsions, π-bonding, etc., but that lattice energy is the most important factor. An extreme example of the importance of lattice energy is the series of fluoro-tantalates, Na_3TaF_8, K_2TaF_7, and $CsTaF_6$, where the size of the cation actually determines the *composition* of compounds crystallizing from solution.

TABLE 22.2

Structures of octacyanide ions

Dodecahedral (D_{2d})	Antiprismatic (D_{4d})	Reference
$Mo(CN)_8^{3-}[(n\text{-}C_4H_9)_4N]_3$		IC 1970 **9** 356
	$Mo(CN)_8^{3-}Na_3.4H_2O$ $W(CN)_8^{3-}Na_3.4H_2O$ }	AC 1970 **B26** 684
$Mo(CN)_8^{4-}K_4.2H_2O$		JACS 1968 **90** 3177
	$W(CN)_8^{4-}H_4.6H_2O$	AC 1970 **B26** 1209
$Mo(CN)_8^{4-}(C_6H_6NO_2)_4$		AC 1980 **B36** 2025

The red salt with the empirical formula $K_2Ni(CN)_3$ is of interest since it is diamagnetic both in solution and in the solid state; Ni(I) would contain one unpaired electron. (Reduction yields the yellow salt $K_4Ni(CN)_4$ which presumably contains the tetrahedral $Ni^0(CN)_4^{4-}$ ion isoelectronic with the $Ni(CO)_4$ molecule.) The crystalline salt contains ions $[Ni_2(CN)_6]^{4-}$ in which the Ni(I) atoms are directly bonded (Ni–Ni, 2.32 Å), the two $Ni(CN)_3$ moieties lying in nearly perpendicular planes.[8] There are metal–metal bonds also in the structurally similar complexes containing Pd(I) and Pt(I), namely, $[Pd_2(NC.CH_3)_6]^{2+}$ [9] and $[Pt_2Cl_4(CO)_2]^{2-}$.[10]

(1)   AC 1959 **12** 674
(2)   AC 1966 **20** 910
(3)   ACSc 1958 **12** 1568
(4)   AC 1972 **B28** 2530
(5)   AC 1969 **B25** 1685

(6)   AC 1980 **B36** 1765
(7)   JACS 1974 **96** 1748
(8)   ZK 1972 **136** 122
(9)   IC 1976 **15** 535
(10) JCS D 1975 1516

(c) *Covalent cyanides containing* –CN–

In some metal cyanides the CN group acts as a bridge between pairs of metal atoms by forming bonds from both C and N, the atoms M–C–N–M being collinear or approximately so. The simplest finite molecule of this kind which has been studied is $(NH_3)_5Co.NC.Co(CN)_5$ (a), in its monohydrate.[1] A more complex example is

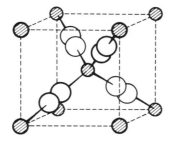

(a)

(b)

the cyclic molecule $[Au(n-C_3H_7)_2.CN]_4$ (b).[2] Examples of the other main types of structure include:

chain structures: AgCN and AuCN, $Ag_3CN(NO_3)_2$ (for which see Chapter 25);

layer structures: not known for simple cyanides, but found in $Ni(CN)_2.NH_3.-C_6H_6$ [3] (p. 32); and

3-dimensional frameworks: $Zn(CN)_2$ [4] and $Cd(CN)_2$ [5] with the anti-Cu_2O structure (p. 124), and Prussian blue and related compounds which are discussed in the next section. The anti-cuprite structure is shown in Fig. 22.4. As in Cu_2O the structure consists of two identical interpenetrating

FIG. 22.4. Structure proposed for $Cd(CN)_2$ and the isostructural $Zn(CN)_2$.

nets which are not cross-connected by any Cu—C or Cu—N bonds (see Chapter 3).

(1) IC 1971 **10** 1492
(2) PRS A 1939 **173** 147
(3) JCS 1958 3412
(4) APURSS 1945 **20** 247
(5) CRURSS 1941 **31** 350

Prussian blue and related compounds

When cyanides of certain metals, for example, Fe, Co, Mn, Cr, are redissolved by the addition of excess alkali cyanide solution, complex ions $M(CN)_6$ are formed.

The alkali and alkaline-earth salts of these complexes are fairly soluble in water and crystallize well. The ferrocyanides and chromocyanides (for example, $K_4Fe(CN)_6$ and $K_4Cr(CN)_6$) are pale-yellow in colour. The ferricyanides and manganicyanides, for example, $K_3Fe(CN)_6$ and $K_3Mn(CN)_6$, are deeper in colour and are made by oxidation of the $M(CN)_6^{4-}$ salts. Solutions of the manganese and cobalt $M(CN)_6^{4-}$ ions oxidize spontaneously in the air to the $M(CN)_6^{3-}$ salts. When solutions of these complex cyanides are added to solutions of salts of transition metals or Cu^{II}, insoluble compounds are precipitated in many cases. Thus addition of potassium ferrocyanide to ferric salt solutions gives a precipitate of Prussian blue, and cupric salts produce brown cupric ferrocyanide. These precipitates are usually flocculent and may form colloidal solutions on washing. When dried they are very stable towards acids, whereas alkali ferrocyanides yield the acid $H_4Fe(CN)_6$ on treatment with dilute HCl. Such compounds are obviously quite different in structure from the simple soluble ferrocyanides.

The iron compounds have been known for several centuries and valued as pigments. The name Prussian (Berlin) blue is given to the compound formed from a solution of a ferric salt and a ferrocyanide, while that obtained from a ferrous salt and a ferricyanide has been known as Turnbull's blue. It appears from Mössbauer studies that both of these compounds are (hydrated) $Fe_4^{III}[Fe^{II}(CN)_6]_3$, with high-spin Fe^{III} and low-spin Fe^{II} in the ratio $4:3$.[1] There is also a 'soluble Prussian blue' with the composition $KFe[Fe(CN)_6]$ which is also a ferrocyanide. Most, if not all, of the compounds we are discussing are hydrated; we consider the water of hydration later.

An early X-ray study[2] of Prussian blue and some related compounds showed that in ferric ferricyanide (Berlin green), $FeFe(CN)_6$, soluble Prussian blue, $KFeFe(CN)_6$, and the white insoluble $K_2FeFe(CN)_6$, there is the same arrangement of Fe atoms on a cubic face-centred lattice. In Fig. 22.5 ferrous atoms are distinguished as shaded and ferric as open circles. In (a) all the iron atoms are in the ferric state; in (b) one-half the atoms are Fe^{II} and the others Fe^{III}, and alkali atoms maintain

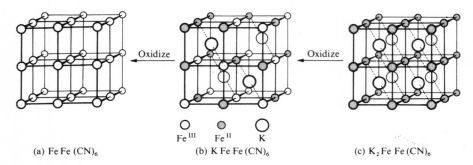

○	◉	○
Fe III	Fe II	K

(a) Fe Fe (CN)₆   (b) K Fe Fe (CN)₆   (c) K₂ Fe Fe (CN)₆

FIG. 22.5. The relationship between (a) Berlin green; (b) soluble Prussian blue; and (c) potassium ferrous ferrocyanide.

electrical neutrality. These are at the centres of alternate small cubes, and it was supposed that in hydrated compounds water molecules could also be accommodated in the interstices of the main framework. Lithium and caesium, forming very small and very large ions respectively, do not form compounds with this structure. In (c), with all the iron in the ferrous state, every small cube contains an alkali-metal atom. The CN groups were placed between the metal atoms along all the full lines of Fig. 22.5, so that each transition metal atom is at the centre of an octahedral group of 6 C or 6 N atoms. The whole system, of composition $M'M''(CN)_6$ thus forms a simple 6-connected 3D framework. Alkali-metal compounds with the Prussian blue structure include $KCuFe(CN)_6$, $KMnFe(CN)_6$, $KCoFe(CN)_6$, $KNiFe(CN)_6$, and $KFeRu(CN)_6$ (ruthenium purple).

Many ferro- and ferri-cyanides of multivalent metals also have cubic structures of the same general type. The unit cell of Fig. 22.5 contains 4 M', 4 M'', and 24 CN, where M' is the atom at the origin and face-centres and M'' at the body-centre and mid-point of each edge. For a complete 3D framework we therefore require a formula of the type $M'M''(CN)_6$, but in many of these complex cyanides, including Prussian blue itself, the ratio of M' : M'' is not 1 : 1. There are two possibilities for the structure of these compounds. If the metal–cyanide framework is to be maintained intact (requiring 24 CN per cell) there would be excess M' atoms in some of these structures, the group A in the left-hand column of Table 22.3.

TABLE 22.3

Prussian blue and related compounds

Number of atoms for 24 CN per cell		Example	Number of M'' for 4 M'	Number of CN per cell
M'	M''			
[C] 3	4	$Ti_3[Fe(CN)_6]_4$	$5\frac{1}{3}$	(32) [D]
4	4	$FeFe(CN)_6$	4	24
$5\frac{1}{3}$	4	$Fe_4[Fe(CN)_6]_3$	3	18
[A] 6	4	$Co_3[Co(CN)_6]_2$	$2\frac{2}{3}$	16 [B]
8	4	$Cu_2[Fe(CN)_6]$	2	12

Alternatively there could be a deficit of M'' atons (which implies fewer than 24 CN per cell) as at B in the right-hand column. Density measurements distinguish between A and B for $Co_3^{II}[Co^{III}(CN)_6]_2 \cdot 12H_2O^{(3)}$ (and for the isostructural Cd and $Mn^{(4)}$ cobalticyanides) and confirm B. The unit cell contains $1\frac{1}{3}$ formula weights, that is, $Co_4^{II}Co_{8/3}^{III}(CN)_{16} \cdot 16H_2O$, and therefore one-third of the Co^{III} positions are unoccupied and at the same time one-third of the CN positions are occupied by 8 H_2O molecules. The remaining 8 H_2O are situated at the centres of the octants of the cell (K^+ positions in Fig. 22.5(c)). The average environment of Co^{II} is $N_4(H_2O)_2$.

In Prussian blue, $Fe_4^{III}[Fe(CN)_6]_3.15H_2O$ (made by slowly diluting a solution of soluble Prussian blue in concentrated HCl), there is 1 formula weight per cell (i.e. 4 Fe^{III}), and replacement of one-quarter of the CN by H_2O, giving the mean environment of Fe^{III} as $N_{4.5}(H_2O)_{1.5}$.[5] Only three-quarters of the $Fe(CN)_6$ sites are occupied, and the cubic symmetry implies that there is a statistical distribution of vacant sites. A n.d. study[6] of Prussian blue in various states of hydration shows that there are two kinds of H_2O molecules; 6 H_2O at empty N sites and approximately 8 H_2O either at the centres of octants of the unit cell or connected to the first set by hydrogen bonds, that is, not randomly arranged. In a compound such as $Ti_3[Fe(CN)_6]_4$ the situation is the opposite of that in Prussian blue (see D in Table 22.3), and since the total number of CN per cell cannot exceed 24 the possibility C would presumably be realized if the same general type of structure is adopted, but this point has not yet been verified.

(1) JCP 1968 **48** 3597   (4) IC 1970 **9** 2224
(2) N 1936 **137** 577   (5) C 1969 **23** 194
(3) HCA 1968 **51** 2006   (6) IC 1980 **19** 956

Miscellaneous cyanide and isocyanide complexes

In addition to complex cyanides containing the simple CN group bonded to metal (through C) there are also metal complexes in which cyanide molecules R.CN or isocyanides R.NC are bonded to metal atoms. In the former case the bond to the metal is necessarily through N, as in $[Fe(NCH)_6](FeCl_4)_2$, a salt made by dissolving $FeCl_3$ in anhydrous HCN (Fe–Cl, 1.89 Å; Fe–N, 2.16 Å, angle Fe–N–C, 171°).[1] Isocyanides must bond to metal through C and numerous compounds of this type are formed by transition metals. In these isocyanide complexes CN.R behaves like CO—compare $Cr(CN.C_6H_5)_6$ with $Cr(CO)_6$, and $Fe(NO)_2(CNR)_2$ with $Fe(NO)_2(CO)_2$. Like CO isocyanides stabilize low oxidation states of metals, as in the diamagnetic $[Mn^I(CNR)_6]I$ and the yellow paramagnetic $[Co^I(CN.CH_3)_5]ClO_4$. The cation in this salt has a trigonal bipyramidal configuration with essentially linear Co–C–N–C and Co–C, 1.87 Å (compare 2.15 Å for a single bond).[2] Another example is the salt $[Fe(CN.CH_3)_6]Cl_2.3H_2O$, in which there is a regular octahedral arrangement of six $CN.CH_3$ molecules around Fe, with Fe–C, 1.85 Å, and C–N, 1.18 Å.[3] There are also mixed coordination groups in molecules such as the *cis* and *trans* isomers of $Fe(CN)_2(CN.CH_3)_4$.[4]

Organic dicyanides NC.R.CN can act as bridging ligands by coordinating to metal through both N atoms, and examples of structures of this kind were noted in Chapter 3.

For the nitroprusside ion, $[Fe(CN)_5NO]^{2-}$ see p. 812.

(1) JSSC 1970 **2** 421; AC 1975 **B31** 1838   (3) JCS 1945 799
(2) IC 1965 **4** 318   (4) JCS 1957 719

Metal carbides

Many metals form carbides M_xC_y which are prepared either by direct union of the elements, by heating the metal in the vapour of a suitable hydrocarbon, or by heating the oxide or other compound of the metal with carbon. Their chemical and physical properties suggest that they may be divided into four main groups:

(1) the salt-like carbides of metals of the earlier Periodic Groups;

(2) the carbides of 4f and 5f elements;

(3) the interstitial carbides of transition metals, particularly of the IVth, Vth, and VIth Periodic Groups; and

(4) a less well-defined group of carbides comprising those of metals of Groups VII and VIII and also some of those of elements of Group VI.

Although carbides MC_2 in groups (1) and (2) are similar structurally they have very different properties, so that it is not convenient to adopt a purely structural classification. Moreover the dividing lines in Table 22.4 are not entirely clear-cut. For example, Tb_2C and Ho_2C (like Y_2C) have the anti-$CdCl_2$ structure, and above $900\,^{\circ}C$ yttrium carbide has a range of composition (YC_x, $0.3 < x < 0.7$) and a defect NaCl structure.[1] These c.p. structures are characteristic of the interstitial carbides of class 3. On the other hand, UC has the NaCl structure like many interstitial carbides but unlike them belongs to the group of reactive heavy metal carbides (PuC, Pd_2C_3, CeC_2, UC_2, etc.) which have metallic properties but are readily hydrolysed. Also, whereas WC does not have a c.p. structure, WC_{1-x} has the NaCl structure with c.p. metal atoms.

TABLE 22.4

Examples of metal carbides

Be_2C	Al_4C_3	Ti_2C	V_2C	$Cr_{23}C_6$	Mn_4C	Fe_3C	Co_3C	Ni_3C
		TiC	V_4C_3	Cr_7C_3	$Mn_{23}C_6$	$Fe_{20}C_9$	Co_2C	
CaC_2	*Class 1*		V_6C_5	Cr_3C_2	Mn_3C			
SrC_2			V_8C_7		Mn_5C_2.			
BaC_2					Mn_7C_3	*Class 4*		
4f and 5f carbides				W_2C				
M_2C_3, MC_2	Y_2C			WC_{1-x}	WC	RuC	OsC	
ThC_2	Tb_2C	*Class 3*						
UC	Ho_2C							
Class 2								

This rough classification is illustrated by the examples of Table 22.4. Little appears to be known of the structures of carbides of B subgroup elements or of the carbides M_2C_2 formed by the alkali metals.[2]

Class 1. The carbides of the more electropositive elements have many of the properties associated with ionic crystals. They form colourless, transparent crystals which at ordinary temperatures do not conduct electricity. They are decomposed by water or dilute acids, and since the negative ions are unstable, hydrocarbons are evolved. According to the type of C_x^{n-} ions present in the crystal we may divide these carbides into two main groups:

(a) Those containing discrete C atoms (or C^{4-} ions).

(b) Those in which C_2^{2-} ions may be distinguished.

(There is no proof yet of more complex carbon anions in ionic carbides, but the fact that Mg_2C_3 on hydrolysis yields chiefly allylene, $CH_3-C\equiv CH$, has been taken to suggest the presence of C_3^{4-} ions in this crystal. It is interesting that MgC_2, which may be prepared by the action of acetylene on $Mg(C_2H_5)_2$ and yields acetylene on hydrolysis, is unstable at high temperatures and breaks down into Mg_2C_3 and carbon.)

Carbides of type (a) yield CH_4 on hydrolysis. Examples are Be_2C, with the anti-fluorite structure, and Al_4C_3. The structure of the latter is rather more complex and its details do not concern us here.[3] It is sufficient to note that each carbon atom is surrounded by Al atoms at distances from 1.90 to 2.22 Å, the shortest C–C distance being 3.16 Å. As in Be_2C therefore there are discrete C atoms, accounting for the hydrolysis to CH_4. The alkaline-earth carbides, type (b), crystallize at room temperature with the CaC_2 structure (Fig. 22.6). (There is some double about the

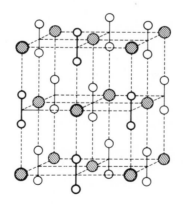

FIG. 22.6. The crystal structure of CaC_2.

structure of MgC_2.[4]) The CaC_2 structure is a NaCl-like arrangement of M^{2+} and C_2^{2-} ions, the symmetry having dropped from cubic to tetragonal owing to the parallel alignment of the anions. Carbides of type (b) yield acetylene on hydrolysis; compare the hydrolysis of BaO_2 (with the same structure) to H_2O_2.

Class 2. The 4f elements from La to Ho form carbides of three types[5]: M_3C, in which the C atoms occupy at random one-third of the octahedral holes in a NaCl-like structure, M_2C_3 with the Pu_2C_3 structure, and MC_2 with the CaC_2 structure. In the Pu_2C_3 structure[6] all the C atoms are present as C_2 groups; there are 6 C_2 groups in the unit cell, which contains 4 Pu_2C_3.

The structure of ThC_2 is very similar to the CaC_2 structure but of lower symmetry (monoclinic). The similarity to the NaCl structure is not readily seen from the conventional diagram showing a unit cell of the structure, but can be seen from Fig. 22.7 where the relation to the monoclinic axes is indicated.

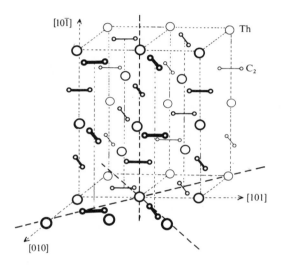

FIG. 22.7. The crystal structure of ThC_2. The heavy broken lines are edges of an orthogonal NaCl-type cell with dimensions 11.7, 11.7, and 10.3 A.

Although the 4f and 5f carbides MC_2 crystallize with the CaC_2 or ThC_2 structures these are not salt-like carbides like those of the alkaline-earths. For example, CaC_2 is an insulator but LaC_2, ThC_2, and UC_2 have metallic lustre and electrical conductivities similar to those of the metals. Also, whereas CaC_2 gives largely C_2H_2 on hydrolysis, ThC_2 gives almost exclusively CH_4, as also does UC at temperatures between 25 and 100 °C. The fact that there is no direct relation between the structures of heavy metal carbides and their hydrolysis products is clearly shown by the behaviour of LaC_2 and CeC_2.[7] At room temperature the product, though a complex mixture of hydrocarbons, consists predominantly of acetylene, but at 200 °C contains no acetylene.

The differences in physical and chemical properties between these carbides MC_2 and M_2C_3 appear to be related to the C–C bond lengths in the C_2^{2-} ions; a number of these have now been determined by neutron diffraction (Table 22.5). The value in CaC_2 is that expected for the $(C\equiv C)^{2-}$ ion, while those in the 4f and 5f carbides, approaching the value for a double bond, suggest that one or more electrons have

TABLE 22.5

Bond lengths in carbides

C—C	Crystal	Reference
1·19 Å	CaC_2	JCP 1961 **35** 1950
		ACSc 1962 **16** 1212
1·24	La_2C_3, Pr_2C_3, Tb_2C_3	JCP 1961 **35** 1960
1·276	Ce_2C_3	JCP 1967 **46** 4148
1·28–1·29	MC_2 (M = Y, La, Ce, etc.)	JCP 1967 **46** 1891
1·295	U_2C_3	AC 1959 **12** 159
1·34	UC_2 (cubic)	AC 1965 **18** 291
1·35	UC_2 (tetragonal)	JCP 1967 **47** 1188
1·30–1·35	ThC_2	AC 1968 **B24** 1121

entered the conduction band, leading to metallic conduction. (The cell dimensions of EuC_2 suggest that this compound is probably $Eu^{II}C_2$, but the C—C bond length is not known.[8])

Most, if not all, of the dicarbides of Classes 1 and 2 have one or more high-temperature forms in which the fixed alignment of the C_2^{2-} ions is relaxed. Compounds such as LaC_2 and UC_2, which at ordinary temperatures have the tetragonal CaC_2 structure, become cubic, the C_2^{2-} ions being randomly oriented along [111] axes (random pyrites, or high-KCN structure). On the other hand, ThC_2 changes first to a tetragonal structure in which there is random orientation of the anions in the basal (001) plane and at 1480 °C to a cubic form in which cations and anions are centred at the sites of the NaCl structure. As noted under KCN it is difficult to ascertain whether there is completely random orientation of the ions in a high-temperature cubic form, and if one ion is linear, orientation along one of the [111] directions is initially more likely.

Class 3. The interstitial carbides are refractory materials with certain of the characteristic properties of metals (lustre, metallic conductivity) and in addition extraordinary hardness and infusibility. They derive their name from the fact that the metal atoms are close-packed, the C atoms occupying octahedral interstices, but it should be noted that the arrangement of metal atoms is not always the same as in the metal itself. In such cases a carbide MC is not to be regarded as the limit of solid solution of carbon in the metal structure; on the contrary the presence of the C atoms has resulted in the rearrangement of the metal atoms to form the NaCl structure. Because the C atoms occupy octahedral holes in a c.p. metal structure there is a lower limit (around 1.3 Å) to the radius of the metal atom that can form interstitial carbides; metals with smaller radii form carbides with less simple structures. The radii (for c.n. 12) of a number of metals are set out in the accompanying table. The metals to the left of the vertical line, with the exception of Th (which forms ThC_2) form interstitial carbides; those to the right form carbides in which the metal atoms are not close-packed (Class 4).

Ti	1·47 Å	V	1·35 Å			Cr	1·29 Å
Zr	1·60	Nb	1·47	Mo	1·40 Å	Mn	1·37(a)
Hf ·	1·59	Ta	1·47	W	1·41	Fe	1·26
Th	1·80					Co	1·25
						Ni	1·25

(a) The crystal structures of the forms of Mn are complex (see p. 1282).

The main structural features of the transition-metal interstitial carbides are as follows. The maximum carbon content depends on the c.p. layer sequence. The two octahedral interstices on either side of an *h* layer are located directly above one another, and only one of these is ever occupied. This restriction gives the following limiting formulae:

Layer sequence	*h*	*hcc*	*hhc*	*hhcc*	*c*
Limiting formula	M_2C	M_3C_2	M_3C_2	M_4C_3	MC
Example	See below	Mo_3C_2	$(Ta_2V)C_2$	V_4C_3	See below

Examples of structures based on the two simplest c.p. sequences are numerous. The h.c.p. structures are examples of the various ways of filling one-half of the octahedral interstices, and the simpler patterns of sites have been illustrated in Chapter 4. Examples are given in Table 22.6; a review of these structures is available.[9] (The literature of these compounds is confusing. Many have different structures at different temperatures and the forms have been designated by Greek letters which

TABLE 22.6

Structures of interstitial carbides

Hexagonal c.p. metal atoms Octahedral sites occupied	Structure	Examples
One-half of all sites at random	—	β-V_2C, γ-Nb_2C
All sites between alternate pairs of layers	anti-CdI_2	α-Ta_2C, α-W_2C
One-half between each pair of c.p. layers	⌠ anti-$CaCl_2$ ⎨ ζ-Fe_2N ⌡ ζ-Nb_2C	Co_2C α-Mo_2C ζ-Nb_2C
One-third and two-thirds between successive pairs of c.p. layers	ϵ-FeN	ϵ-V_2C, ϵ-Nb_2C, ϵ-W_2C

Cubic c.p. metal atoms	Examples	Reference
All octahedral sites occupied (NaCl structure)	HfC, TiC, ZrC, VC, NbC, TaC, β-MoC_{1-x}	
Ordered defect NaCl structures	⌠ V_8C_7 ⎪ V_6C_5 ⎨ Sc_2C, Ti_2C, Zr_2C ⎪ anti-$CdCl_2$ (Ho_2C) ⌡ Y_2C, etc.)	AC 1970 **B26** 1882 PM 1968 **18** 177 AC 1970 **B26** 153

have not been used consistently by different authors.) The c.c.p. structures include not only the carbides MC with the NaCl structure but also a number in which there is an ordered arrangement of vacancies; a feature of certain of these structures is the helical arrangement of the vacancies (V_6C_5 and V_8C_7). As remarked earlier, the arrangement of C atoms in the carbides of the heavier metals cannot be regarded as firmly established unless neutron diffraction studies have been made, as in the examples given in Table 22.6.

Class 4. The carbides of this class do not possess the extreme properties of the interstitial carbides. For example, whereas titanium carbide is not attacked by water or HCl even at 600 °C, these carbides are decomposed by dilute acids (Fe_3C and Ni_3C) or even water (Mn_3C). Although the carbon is present as discrete C atoms the products include, in addition to hydrogen, complex mixtures of hydrocarbons.

We shall not attempt here to survey the structures of these carbides, which are often complex. At low carbon contents there are marked similarities to borides, but where there is more non-metal the similarities are much less marked, for C does not show the same tendency as B to form extended systems of non-metal–non-metal bonds. There are groups of isostructural carbides and borides (with different formulae types) in all of which the non-metal atom has very similar environments, often a trigonal prismatic arrangement with from one to three additional neighbours beyond the vertical prism faces. Both (6+1)- and (6+2)-coordination of C occur in Cr_3C_2,[10] and tricapped trigonal prismatic coordination in the following group of closely related structures:

Fe_3C (cementite),[11] Mn_3C, Co_3B, Pd_3P, Al_3Ni
Mn_5C_2 and Pd_5B_2
Cr_7C_3, Re_7B_3, Th_7Fe_3[12]

It will be noticed that some of these structures are adopted by metallic phases in addition to carbides and borides. The WC structure, illustrated in Fig. 4.8(a), p. 152, provides another example of a structure that is not peculiar to carbides, for it is adopted by MoP and WN in addition to WC, RuC, and OsC.[13]

In addition to the binary carbides noted above there are ternary carbides and also carbide-nitrides and oxide-carbides. The behaviour of metals in metal–carbon–nitrogen systems ranges from that of Al, which forms a 'homologous' series of compounds with definite compositions and structures intermediate between those of AlN and Al_4C_3 (namely, $Al_8C_3N_4$, $Al_7C_3N_3$, $Al_6C_3N_2$, and Al_5C_3N)[14] and that of Th, in which there is a complete series of solid solutions between ThC and ThN; there are also the compounds ThC_2 and Th_3N_4. ThCN is a carbide-nitride containing C_2 groups (C–C, 1.23 Å) and N atoms surrounded tetrahedrally by 4 Th. The metal atoms have 4 C and 4 N neighbours.[14a] The compound Yb_2OC is mentioned in Chapter 12 as originally mistaken for the monoxide. A number of the 4f elements form compounds of the type MO_2C_2[15] or M_4O_3C.[16] The conductivity of the silver-grey Nd_4O_3C, which has a NaCl-like structure, is similar to that of Nd metal; it hydrolyses largely to methane, and could be formulated $Nd_4^{3+}O_3^{2-}C^{4-}(e)_2$. The structure of Al_4O_4C has been studied.[17] It consists of a 3D framework formed from AlO_3C tetrahedra, an unusual feature of which is the association of certain of the tetrahedra in edge-sharing pairs, the shared edges being O---O edges.

(1) JCP 1969 **51** 3862, 3872
(2) For Li_2C_2 see AC 1962 **15** 1042
(3) AC 1963 **16** 559
(4) JACS 1943 **65** 602, 1482
(5) JACS 1958 **80** 4499
(6) AC 1952 **5** 17
(7) IC 1962 **1** 345, 683
(8) IC 1964 **3** 335
(9) AC 1970 **B26** 153
(10) ACSc 1969 **23** 1191
(11) AC 1965 **19** 463
(12) AC 1962 **15** 878
(13) AC 1961 **14** 200
(14) AC 1966 **20** 538
(14a) AC 1972 **B28** 1724
(15) IC 1966 **5** 1567
(16) JACS 1968 **90** 1715
(17) AC 1963 **16** 177

Metal carbonyls

Preparation and properties

The compounds of CO with the alkali metals and the alkaline-earths are quite different in structure from those of the transition metals. The compound $K_2(CO)_2$ formed by the action of CO on the metal dissolved in liquid NH_3 is a salt containing the linear acetylenediolate ion $(OC≡CO)^{2-}$, in which the bond lengths are C–C, 1.21 Å, and C–O, 1.28 Å.[1] This salt is also formed from CO and the molten metal at low temperatures; at higher temperatures (above 180 °C) the product is predominantly $C_6(OK)_6$.[2] The so-called carbonyls of the alkaline-earths made from the metals and CO in liquid NH_3 are mixtures of the metal acetylenediolates and methoxides with ammonium carbonate.[3]

Certain of the salts of the IB subgroup metals comhine with CO forming, for example, $CuCl(CO).2H_2O$, $Ag_2SO_4.CO$, and $AuCl(CO)$, of which the last is a comparatively stable volatile compound. The structures of these compounds are not known, and we shall confine our attention here to the carbonyls and related compounds of transition metals.

Transition metal compounds containing CO are numerous and of a number of different types. Some simple compounds of 3d metals are set out in the upper part of Table 22.7; there are also many compounds of the heavier Group VIII metals. Those containing only CO ligands range from the mononuclear carbonyls to molecules and ions containing large numbers of metal atoms which form a central metal cluster. There are molecules related to carbonyls $M_m(CO)_n$ by replacing some CO by other donor ligands (e.g. phosphines), by H or by halogens, and still more complex molecules or ions in which C, N, or S atoms are incorporated into the metal cluster. Literature references to the very numerous carbonyls which are not mentioned in the text are readily available in reviews which attempt to rationalize the shapes of metal clusters on the basis of molecular orbital theory[4] or to relate them to the arrangement of CO groups in the polyhedral shell.[5] The relevant polyhedra have been considered in detail for molecules and ions containing 12 CO and for 13-16 CO groups. A review is also available of hydrido compounds.[6]

TABLE 22.7

Carbonyls, carbonyl hydrides, and carbonyl halides

$V(CO)_6$	$Cr(CO)_6$ $Cr(CO)_5H_2$	$-$ $Mn(CO)_5H$ $Mn(CO)_5I$	$Fe(CO)_5$ $Fe(CO)_4H_2$ $Fe(CO)_nX_2$ $(n = 2, 4, 5)$	$-$ $Co(CO)_4H$ $Co(CO)X_2$	$Ni(CO)_4$
		$Mn_2(CO)_{10}$ $Mn_2(CO)_8I_2$	$Fe_2(CO)_9$	$Co_2(CO)_8$	$Ni_2(CO)_6H_2$

	Metal cluster			
3	Triangular		$Fe_3(CO)_{12}$ $Os_3(CO)_{12}$	$[Fe_3(CO)_{11}H]^-$ $[Re_3(CO)_{12}H_6]^-$
4	Rhombus		$[Re_4(CO)_{16}]^{2-}$	
	Tetrahedron		$Co_4(CO)_{12}$	$[Fe_4(CO)_{13}]^{2-}$
5	Trigonal bipyramid		$Os_5(CO)_{16}$ $[Ni_5(CO)_{12}]^{2-}$	$[Os_5(CO)_{15}I]^-$
	Tetragonal pyramid			$[Fe_5C(CO)_{15}]^-$
6	Octahedron		$Rh_6(CO)_{16}$ $[Co_6(CO)_{14}]^{4-}$	$Rh_6(CO)_{15}I$
	centred			$[Fe_6C(CO)_{16}]^{2-}$ $[Co_6H(CO)_{15}]^-$
	Trigonal prism (centred)			$[Rh_6C(CO)_{15}]^{2-}$ $[Co_6N(CO)_{15}]^-$
	Bicapped tetrahedron		$Os_6(CO)_{18}$	

We shall describe later the structures of the representative selection of compounds listed in Table 22.7 in order of increasing numbers of metal atoms.

The carbonyls are in general volatile compounds with an extensive chemistry which presents many problems as regards valence and stereochemistry. Some are reactive and form a variety of derivatives, as shown in Chart 22.1 for the iron compounds, while others are relatively inert, as for example, $Cr(CO)_6$ etc. and $Re_2(CO)_{10}$. This rhenium compound, although converted to the carbonyl halides by gaseous halogens, is stable to alkalis and to concentrated mineral acids. A few carbonyls may be prepared by the direct action of CO on the metal, either at atmospheric pressure $(Ni(CO)_4)$ or under pressure at elevated temperatures $(Fe(CO)_5, Co_4(CO)_{12})$. Others are prepared from halides or, in the case of Os and Re, from the highest oxide. The polynuclear carbonyls are prepared photosynthetically, by heating the simple carbonyls, or by other indirect methods.

The carbonyl hydrides are prepared either from the carbonyls, as shown in the chart for $Fe(CO)_4H_2$, by reduction with Na in liquid NH_3, for $Ni_2(CO)_6H_2$, or directly from the metal by the action of a mixture of CO and H_2. Cobalt in this way gives $Co(CO)_4H$.

$$\left[\begin{array}{c} OC \quad\quad CO \\ -Fe-Hg- \\ OC \quad\quad CO \end{array} \right]_n$$

(a)

The heavy metal derivatives of $H_2Fe(CO)_4$ and $HCo(CO)_4$, the so-called 'mixed metal carbonyls', show very considerable differences in properties. $HgFe(CO)_4$ is a stable yellow solid, insoluble in both polar and non-polar solvents; these properties suggest a polymeric structure for example, of type (a), whereas the derivatives of $HCo(CO)_4$, $M[Co(CO)_4]_2$, where M = Zn, Cd, Hg, Sn, or Pb, resemble the poly-nuclear carbonyls, being soluble in non-polar solvents, insoluble in water, and subliming without decomposition; they consist of simple linear molecules in the cystalline state (see later).

The formation of carbonyl halides does not coincide with that of carbonyls, as shown in the accompanying table, where full lines enclose the elements which form simple carbonyls or carbonyl ions and broken lines those which form carbonyl halides:

V	Cr	Mn	Fe	Co	Ni	Cu	Zn
Nb	Mo	Tc	Ru	Rh	Pd	Ag	Cd
Ta	W	Re	Os	Ir	Pt	Au	Hg

These compounds may be formed directly from the metal halides by the action of CO (for example, CO and anhydrous CoI_2 give $Co(CO)I_2$) or, in the case of iron indirectly from the carbonyls or carbonyl hydrides. The only carbonyl fluoride which has been well characterized has the empirical formula $Mo(CO)_2F_4$;[7] a dimeric structure with two bridging F atoms has been suggested. The compounds $Pt(CO)_2F_8$ and $Rh(CO)_2F_3$ have been described as the products of the action of CO on the tetrafluorides. Simple carbonyls of Pd do not seem to be known, but Pt forms a series of compounds $[Pt_3(CO)_6]_n^{2-}$ where n = 2, 3, 4, or 5 (see later). Also $PdCl_2$ and $PtCl_2$ combine with CO to form compounds with the empirical formulae $PdCl_2.CO$, $PtCl_2.CO$, and $PtCl_2(CO)_2$. The compound $PtCl_2(CO)_2$ is monomeric and a non-electrolyte, and the dipole moment (4.85 D) shows that it has the *cis* configuration,[8] but $PtCl_2.CO$, which is a solid melting at 195 °C is dimeric and non-polar and analogous to the compounds of $PtCl_2$ with arsines and phosphines, of the general type

$$\begin{array}{c} R \quad\quad Cl \quad\quad Cl \\ \quad Pt \quad\quad Pt \\ Cl \quad\quad Cl \quad\quad R \end{array}$$

All the carbonyl halides are decomposed by water, but their stability towards water increases from the chloride to the iodide.

CHART 22.1

Some reactions of iron carbonyls

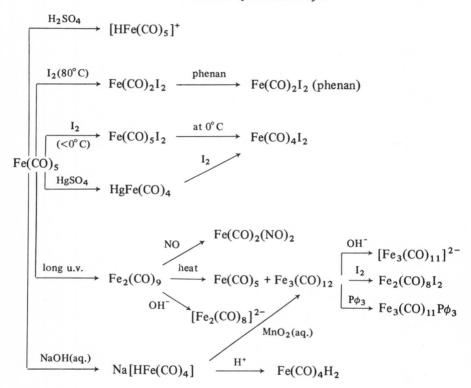

The nitrosyl carbonyls result from the action of NO on the polynuclear carbonyls of iron and cobalt. Like the carbonyls the nitrosyl carbonyls react with neutral molecules such as organic nitrogen compounds and with halogens, and in such reactions CO, and not NO, is replaced. Thus $Fe(CO)_2(NO)_2$ with iodine gives $Fe(NO)_2I$. No reaction has yet been detected between NO and the much more stable hexacarbonyls of Cr, Mo, and W. Whereas Mn, Fe, and Co form respectively $Mn(NO)_3CO$, $Fe(NO)_2(CO)_2$, and $Co(NO)(CO)_3$, Ni does not form a nitrosyl carbonyl of this type (see later) and reacts with NO only in the presence of water. A blue compound, $Ni(NO)OH$, is then formed which is soluble in water and possesses reducing properties. It is regarded as a compound of Ni(I) and is obviously quite different in type from the nitrosyl carbonyls of Co and Fe.

(1) HCA 1963 **46** 1121
(2) HCA 1964 **47** 1415
(3) HCA 1966 **49** 907
(4) JACS 1975 **95** 3802; JACS 1978 **100** 5305; JACS 1979 **101** 1979; *Adv. Inorg. Chem. Radiochem.* 1976 **18** 1; *Trans. Amer. Cryst. Assoc.* 1980 **16** 1, 17

(5) CC 1976 211; JCS D 1980 1743
(6) Acc. Chem. Res. 1979 **12** 176
(7) IC 1970 **9** 2611
(8) JACS 1954 **76** 4271

The structures of carbonyls and related compounds

The main features are as follows.

1. The CO molecule is bonded to the metal atom(s) through C; this has been proved for $Fe(CO)_5$ and $Cr(CO)_6$ and is assumed to be true in all other molecules.

2. The CO molecule behaves as a monodentate ligand, (a), as a bridge between two metal atoms, (b), or (less frequently) between three metal atoms, (c).

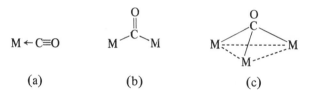

(a)   (b)   (c)

It is supposed that the single σ M–C bond in (a) above, formed by the lone pair of electrons of the ligand, is supplemented by π-bonding from the filled metal d orbitals to a vacant orbital of the ligand, as shown diagrammatically in Fig. 22.8. This 'back-bonding' reduces the negative formal charge on M implied by the formulation M←C≡O and accounts for the fact that CO and other π-bonding ligands stabilize low oxidation states of transition metals. In $Ni(CO)_4$ the formal oxidation state of Ni is zero, so that it is forming $4s4p^3$ (tetrahedral) bonds. This type of bonding, which also operates in the case of NO, isocyanides, substituted phosphines, sulphides, etc., is to be distinguished on structural grounds from that operative in

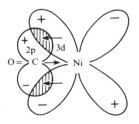

FIG. 22.8. Dative π-bond in $Ni(CO)_4$.

the pure π-complexes formed by $C_5H_5^-$, C_6H_6, etc. which are discussed later. The nodal planes of the π orbitals used to supplement the σ bonding in the carbonyls must include the axis of the σ bond, requiring that the system M—C—O must be linear. In π-complexes in which the bonding involves only the π orbitals of the ligand the metal atoms lie out of the molecular plane of the C_5H_5 or other ligand.

3. The formulae of most simple carbonyls and related compounds are consistent with the view that in many of these compounds the metal atom acquires a share in sufficient electrons to attain a noble gas configuration. Assuming that CO bonded as at (a) contributes 2 electrons, and NO 3 electrons, then in all the following compounds the metal would have the Kr configuration:

$Ni(CO)_4$

$Co(CO)_3NO$

$Fe(CO)_5$ $Fe(CO)_2(NO)_2$

$Mn(CO)(NO)_3$

$Cr(CO)_6$

An alternative statement of the 'noble gas rule' is that the total number of valence electrons is 18, required to fill the 9 orbitals, five d, one s, and three p orbitals. The simple electron counting is consistent with the formation of mononuclear carbonyls by the even-numbered elements, $Cr(CO)_6$, $Fe(CO)_5$, and $Ni(CO)_4$, and of binuclear carbonyls by Mn, Fe, and Co, assuming normal electron-pair bonds between the metal atoms. It is also consistent with the stereochemistries of molecules formed by replacing CO by other donor molecules as in $Fe(CO)_3(PR_3)_2$, trigonal bipyramidal like $Fe(CO)_5$, and $Mo(CO)_5PF_3$, octahedral like $Mo(CO)_6$, and with the structures of $Co(CO)_4SiH_3$ (trigonal bipyramidal) and $Mn(CO)_5CH_3$ (octahedral), the ligands providing 9 and 11 electrons for the last two elements, which do not form mononuclear carbonyls. It is not, however, consistent with the formation of $V(CO)_6$ or of certain 16-electron compounds of Pt and molecules such as $Ni(C_5H_5)_2$ (p. 973).

There has been considerable success in recent years in extending this idea to compounds containing metal clusters; those containing large clusters are of special interest as intermediate between normal chemical molecules and bulk metals. Calculations have been made of the energies of molecular orbitals for clusters of various types, the orbitals of lower energy being those suitable for ligand bonding or m–m bonds. The number of these Cluster Valence Molecular Orbitals (CVMOs) for certain types of cluster are: monomer, 9; dimer, 17; trimer, 24; tetrahedron, 30. trigonal bipyramid, 36; bicapped tetrahedron, 42; octahedron, 43; and trigonal prism, 45. The number of cluster valence electrons (CVE) is, of course, equal to twice the number of CVMOs, and for a given compound is equal to the number of valence electrons of the transition metal atom (and of the interstitial atoms, if any) plus 2 for each CO or similar ligand and the charge (if an anion). These CVE numbers for cluster molecules or ions correspond to the number 18 for the mononuclear carbonyls, and they are consistent with the stoichiometries of *most* known compounds, for example: 86 for $Co_6(CO)_{14}^{4-}$ and $Co_6H(CO)_{15}^-$ (octahedral), and 90 for $Co_6N(CO)_{15}^-$ (trigonal prism). Just as we find $Ni(CO)_4$, $Fe(CO)_5$, and $Cr(CO)_6$ consistent with the 18-electron 'rule' so we find the octahedral complex $Co_6(CO)_{16}$ with 86 electrons, but the corresponding octahedral Fe compound would require 19 CO, a number difficult to accommodate in an approximately spherical group. This difficulty, which is a shortage of valence electrons and therefore more acute the smaller the number of d electrons on the metal, can be alleviated by incorporating an interstitial atom and/or acquiring a negative charge as, for example, in $Fe_5C(CO)_{16}^{2-}$. The packing problem is also less acute for larger clusters, around which many more ligands can be accommodated, or by closer packing of the metal atoms which increases their c.n.s and hence decreases the number of CMVOs to be satisfied by the ligands. We return to these large clusters later.

Carbonyls $M(CO)_n$. These have the highly symmetrical structures listed in Table 22.8.

TABLE   22.8
Structures of carbonyls $M(CO)_n$

Molecule	Shape	M—C	C—O	Reference
$Ni(CO)_4$	Regular tetrahedral	1.83 Å	1.15 Å	AC 1952 5 795
$Fe(CO)_5$	Trigonal bipyramidal	1.806 (ax.)	1.145	AC 1969 **B25** 737
		1.833 (eq.)		ACSc 1969 **23** 2245
$Cr(CO)_6$		1.918	1.141	AC 1975 **B31** 2649
$Mo(CO)_6$	Regular octahedral	2.063	1.145	ACSc 1966 **20** 2711
$W(CO)_6$		2.058	1.148	

Closely related to $Fe(CO)_5$ are the trigonal bipyramidal molecules $Fe(CO)_4(PH\phi_2)$ (ref. 1a) and $Fe(CO)_3[P(OCH_3)_3]_2^{(1b)}$ (phosphine(s) in axial position(s)), and the π-complex $C_2H_4Fe(CO)_4,^{(2a)}$ in which C=C (1.46 Å) lies in the equatorial plane and the length of the Fe–C bond is 2.12 Å. The molecule $C_2F_4Fe(CO)_4^{(2b)}$ has a structure of the same general type, but the C–C and Fe–C bond lengths (1.53 and 1.99 Å) suggest description as a pseudo-octahedral complex.

Also related to $Ni(CO)_4$ and $Fe(CO)_5$ are the bridged molecules (a)$^{(3)}$ and (b),$^{(4)}$ in which the bond arrangements around Ni and Fe are the same as in the simple carbonyls. In (b) the axial Fe--C bond appears to be significantly shorter (1.71 Å) than the equatorial bonds (1.79 Å) and C–O longer (1.29 Å as compared with 1.16 Å). In many formulae in the following pages we shall indicate –CO simply by a stroke (–).

Carbonyls $M_2(CO)_n$. An early X-ray study of $Fe_2(CO)_9$[5] showed that the molecule has the structure of Fig. 22.9(a) with three bridging CO groups. The accuracy of location of the light atoms does not justify detailed comparison of the two different Fe–C and C–O bond lengths (see $Co_2(CO)_8$, later) but indicates appreciably lower bond orders in the bridging bonds than in the terminal ones. The Fe–Fe distance (2.46 Å), equal to twice the 'metallic' radius of Fe for 8-coordination, shows that there is a Fe–Fe bond, accounting for the diamagnetism and giving Fe the Kr closed-shell configuration.

There are many compounds related more or less closely to $Fe_2(CO)_9$ including (c) with a triple $(CH_3)_2Ge$ bridge,[6] compounds of type (d) with double bridges, where the bridging group is, for example, NH_2[7] or SC_2H_5 (see p. 970), and singly bridged compounds, (e), $(CO)_4Fe.R.Fe(CO)_4$, where R is $P(CH_3)_2$, $Ge(C_6H_5)_2$, etc. The Ge compounds are of interest as containing 3-membered $GeFe_2$ rings— compare the 5-membered ring in the compound noted on p. 976.

(c)   (d)   (e)

The $Co_2(CO)_8$ molecule,[8] Fig. 22.9(b), has essentially the same structure as $Fe_2(CO)_9$ with one of the bridging CO groups removed. By forming a metal–metal bond (2.52 Å) Co acquires the same closed-shell configuration as Fe in $Fe_2(CO)_9$.

In $Mn_2(CO)_{10}$[9] and $Re_2(CO)_{10}$[9a] (and the isostructural $Tc_2(CO)_{10}$) there are no bridging CO groups, the molecule consisting of two $-M(CO)_5$ joined by a metal–metal bond. Each metal atom forms six octahedral bonds (to 5 CO and M) and in the crystalline forms of these carbonyls the two octahedra are rotated through about 45° from the eclipsed position. (An e.d. study of $Re_2(CO)_{10}$ in the vapour state[10] indicated an eclipsed configuration.) The long M–M bonds (Mn–Mn, 2.92 Å; Re–Re, 3.02 Å; and Tc–Tc, 3.04 Å) are attributed to the large negative formal charges and to repulsions between the CO groups. Carbonyl ions $[M_2(CO)_{10}]^{2-}$ exist in the Na salts (M = Cr, Mo, W) prepared by reducing the hexacarbonyls with $NaBH_4$, and ions $[(CO)_5M-M'(CO)_5]^-$ have been isolated in $[N(C_2H_5)_4]^+$ salts (M = Mn, Re; M' = Cr, Mo, W).[10a] The red diamagnetic form of $[Co(CN.CH_3)_5]_2-$ $(ClO_4)_4$ contains dimeric ions of the same structural type as $Mn_2(CO)_{10}$, with Co–Co, 2.74 Å.[11]

Carbonyls $M_3(CO)_n$. The molecule of $Fe_3(CO)_{12}$ in the solid can be described as resulting from the replacement of one of the bridging CO groups in $Fe_2(CO)_9$ by

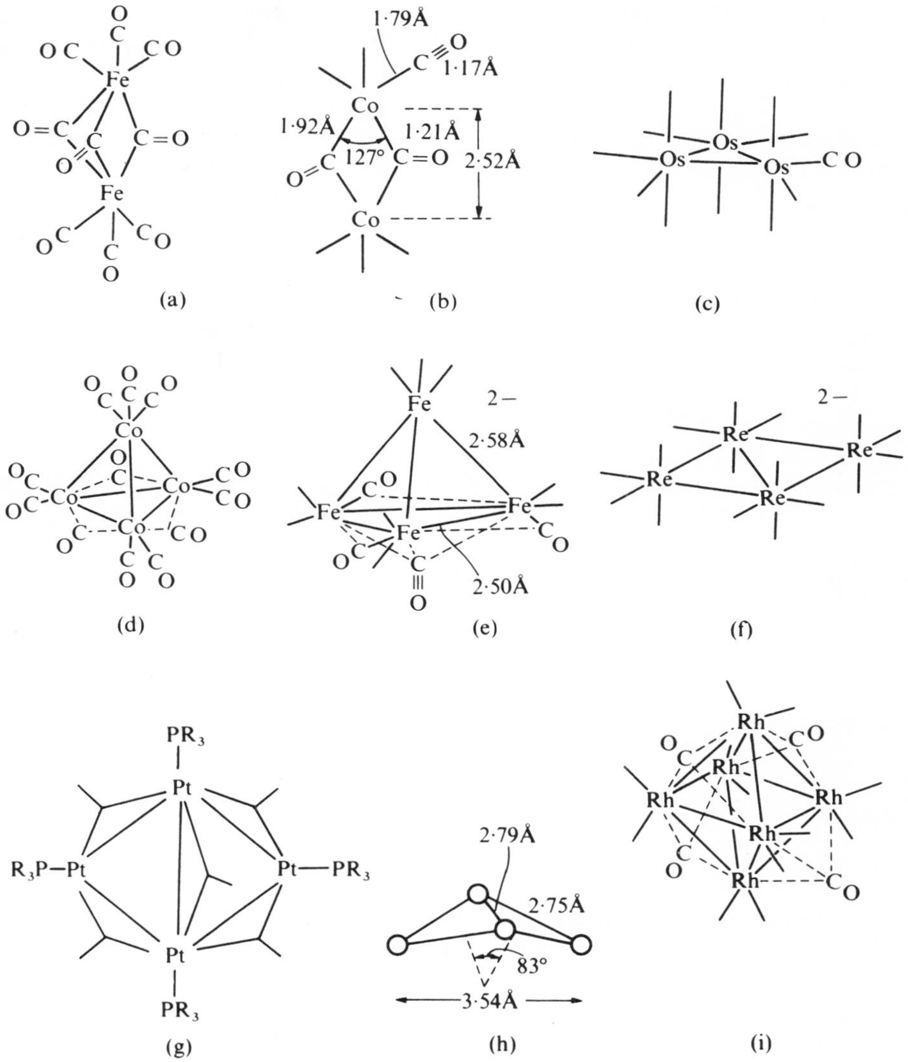

FIG. 22.9. The structures of metal carbonyls: (a) $Fe_2(CO)_9$; (b) $Co_2(CO)_8$; (c) $Os_3(CO)_{12}$; (d) $Co_4(CO)_{12}$; (e) $Fe_4(CO)_{13}]^{2-}$; (f) $[Re_4(CO)_{16}]^{2-}$; (g) and (h) $Pt_4(CO)_5[P(C_2H_4CN)_3]_4$; (i) $Rh_6(CO)_{16}$.

the unit $Fe(CO)_4$, the third Fe atom being equidistant from the other two.[12] The three Fe atoms are situated at the corner of an isosceles triangle (f), and the twelve CO are arranged at the vertices of an icosahedron (as is also the case in $Co_4(CO)_{12}$, below). The apparent incompatibility of the i.r. spectrum of $Fe_3(CO)_{12}$

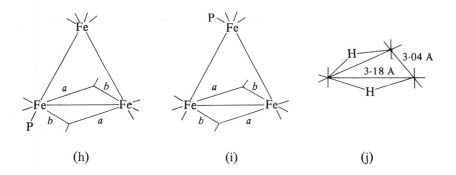

(f)   (g)

in solution with the model (f) may be due to rearrangement in solution, a view
supported by the fact that one isomer of $Rh_3(CO)_3(C_5H_5)_3$ has a structure similar
to (f) and the other has the symmetrical structure (g).[12a]

The anion in $[(C_2H_5)_3NH][Fe_3(CO)_{11}H]$[13] apparently has the same type of
structure as $Fe_3(CO)_{12}$, and in $Fe_3(CO)_{11}P\phi_3$ one terminal CO is replaced by the
phosphine. The crystal contains two structural isomers, (h) and (i), and the bridges
are slightly unsymmetrical,[14] with a and b approximately 2.04 Å and 1.87 Å (means).

(h)   (i)   (j)

In contrast to $Fe_3(CO)_{12}$ the molecules $Ru_3(CO)_{12}$[15] and $Os_3(CO)_{12}$[16] (originally formulated as enneacarbonyls) have the symmetrical structure of Fig. 22.9(c),
with 4 CO bonded to each metal atom. In the equilateral triangle of metal atoms
the bond lengths are: Ru–Ru, 2.85 Å, and Os–Os, 2.88 Å, and each metal atom
forms four normal and two 'bent' octahedral bonds—compare the similar 'bent'
bonds in $Co_2(CO)_8$, $Rh_2(CO)_4Cl_2$, etc.

The anion in $[Re_3(CO)_{12}H_2][As(C_6H_5)_4]$ has a somewhat similar structure,
(j).[17] The H atoms were not located but presumably bridge the two longer edges
of the Re_3 triangle. As in $Ru_3(CO)_{12}$ there are no bridging CO groups.

Carbonyls $M_4(CO)_n$. For a molecule $M_4(CO)_{12}$ the most symmerical shape is a
tetrahedron of metal atoms with 3 CO attached to each and a cuboctahedral arange-
ment of the 12 CO ligands, as in $Ir_4(CO)_{12}$.[18] Other tetrahedral complexes include
$Ru_4H_4(CO)_{12}$, which is mentioned later, and the anion $[Re_4(CO)_{12}H_6]^{2-}$.[19] In this

ion the H atoms are logically placed along or close to the edges of the tetrahedron. The mean Re—Re distance (3.16 Å) is close to the larger value in $[Re_3(CO)_{12}H_2]^-$ and appreciably larger than the value (around 3.0 Å) for non-bridged Re—Re distances in, for example, $Re_2(CO)_{10}$.

The tetrahedral molecules $Co_4(CO)_{12}$ [20a] and $Rh_4(CO)_{12}$ [20b] are less symmetrical (Fig. 22.9(d)), with three bridging CO groups. One metal atom forms octahedral bonds to three others and to 3 CO, while the bond arrangement around the basal metal atoms is approximately pentagonal bipyramidal. Owing to crystallographic complications (disorder and twinning) the bond lengths are not known accurately, but mean metal—metal distances are respectively 2.49 Å and 2.73 Å.

The carbonyl anion $[Fe_4(CO)_{13}]^{2-}$ has a tetrahedral structure (Fig. 22.9(e)) of a quite different kind.[21] An apical $Fe(CO)_3$ unit is bonded to the base of the tetrahedron only by Fe—Fe bonds, and a feature of the molecule is the triply bonded CO below the base. (There is very weak bonding between the C atom of each CO in the basal plane and a second metal atom.)

The ion $[Re_4(CO)_{16}]^{2-}$ (Fig. 22.9(f)) has an entirely different shape. The four metal atoms are coplanar, all Re—Re distances are close to 3.0 Å, and the arrangement of bonds around Re is octahedral.[22]

We noted in Chapter 3 some families of tetrahedral molecules, including molecules $M_4X_4Y_4$ and $M_4X_6Y_4$. The molecule $Fe_4(CO)_4(\pi\text{-}C_5H_5)_4$ is tetrahedral with a triply bridging CO on each face, while $Ni_4(CO)_6[P(C_2H_4CN)_3]_4$ has a bridging CO along each edge and a phosphine molecule attached to each vertex (Ni).[23] Removal of one of the edge-bridging CO molecules from a molecule of this kind leads to opening-up of the tetrahedral to a dihedral ('butterfly') structure, as in $Pt_4(CO)_5[P(CH_3)_2C_6H_5]_4$,[24] as shown diagrammatically in Fig. 22.9(g) and (h). The dihedral angle is 83° instead of 109½°, the value in the tetrahedron. Removal of one bridging CO and one metal—metal bond reduces the electron count to 16 for the 5-covalent metal atoms, leaving the two 7-coordinated atoms with a closed shell configuration.

Carbonyls $M_5(CO)_n$. A structural study of $Os_5(CO)_{16}$ [25] shows a trigonal bipyramidal nucleus with all CO terminal, three on each of four vertices and four on one of the equatorial vertices. $Os_5(CO)_{15}I$ [26] has essentially the same structure with one CO replaced by an I atom.

Carbonyls $M_6(CO)_n$. Three different structures have been found for M_6 clusters. The most symmetrical is that of $[Co_6(CO)_{14}]^{4-}$ [27] (and $[Co_4Ni_2(CO)_{14}]^{2-}$), with 6 terminal CO, one on each vertex, and 8 triply-bridging CO, one on each face. The 14 CO ligands are arranged at the vertices of an 'omni-capped cube', that is, corresponding approximately to the vertices of a rhombic dodecahedron. The black crystals of $Rh_6(CO)_{16}$ [28] are composed of octahedral molecules (Fig. 22.9(i)) in which there are two CO bonded to each Rh by normal Rh—C bonds and also triply-bridging CO above four of the faces of the octahedron. The Rh—Rh distance is 2.78 Å, and since the electron count for a Rh atom is 56 this compound is an exception to the 'noble gas rule'. Other octahedral clusters include those of $Rh_6(CO)_{15}I$ and $[Co_6(CO)_{15}]^{2-}$. An octahedral group of 6 metal atoms with a central C or H atom is found in $[Fe_6C(CO)_{16}]^{2-}$ [29] and the isoelectronic $[Co_6H(CO)_{15}]^-$ [30] (CVE count 86). In the isoelectronic anions $[Rh_6C(CO)_{15}]^{2-}$ [31] and $[Co_6N(CO)_{15}]^-$ (ref. 32) the metal atoms form a trigonal prismatic cluster, with an electron count of 96. A third type of metal cluster is found in $Os_6(CO)_{18}$,[33] a bicapped tetrahedron; three terminal CO are bonded to each Os atom.

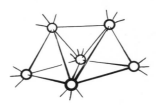

(1a)	JCS A 1969 1906	(16)	IC 1977 **16** 878
(1b)	AC 1974 **B30** 2798	(17)	JACS 1968 **90** 7135
(2a)	JOC 1970 **21** 401	(18)	IC 1978 17 3528
(2b)	AC 1973 **B29** 1499	(19)	JACS 1969 **91** 1021
(3)	JCS A 1967 1744	(20a)	IC 1976 15 380
(4)	JCS A 1968 622	(20b)	IC 1969 8 2384
(5)	JCS 1939 286	(21)	JACS 1966 **88** 4847
(6)	JACS 1968 **90** 3587	(22)	IC 1968 7 2606
(7)	JACS 1968 **90** 5422	(23)	JACS 1967 **89** 5366
(8)	AC 1964 **17** 732	(24)	JACS 1969 **91** 1574
(9)	AC 1963 **16** 419	(25)	AC 1977 **B33** 173
(9a)	IC 1965 4 1140	(26)	AC 1978 **B34** 3376
(10)	ZSK 1973 **14** 419	(27)	JOC 1969 **16** 461
(10a)	JACS 1967 89 539	(28)	JACS 1963 85 1202
(11)	PCS 1964 175	(29)	JCS D 1974 2410
(12)	JACS 1969 **91** 1351	(30)	ANCIE 1979 18 80
(12a)	JOC 1967 **10** P3, 331	(31)	JCS D 1973 651
(13)	IC 1965 4 1373	(32)	JACS 1979 **101** 7095
(14)	JACS 1968 **90** 5106	(33)	JACS 1973 **95** 3802
(15)	IC 1977 **16** 2655		

Carbonyls containing larger metal clusters

There are many carbonyls of the heavier elements containing larger clusters than those we have described. Examples include $Os_7(CO)_{21}$, $Os_8(CO)_{23}$, $Pt_9(CO)_{18}^{2-}$, and complexes containing 12–15 metal atoms of which we shall give a few examples. Those most closely related to the polyhedral molecules and ions already described contain compact metal groupings with obvious resemblances to portions of metal structures, with high coordination of the central atom. In Fig. 22.10 we show the metal clusters in some Rh complexes containing 13, 14, and 15 metal atoms.[1] Quite different from these globular clusters are those in families of carbonyls $[M_3(CO)_6]_n^{n-}$ formed by Ni and Pt.[2] These are formed from triangular sub-units of type (a) which are stacked above one another in the manner shown in Fig. 22.10(d). There are differences between the Ni and Pt complexes. For example, the anion

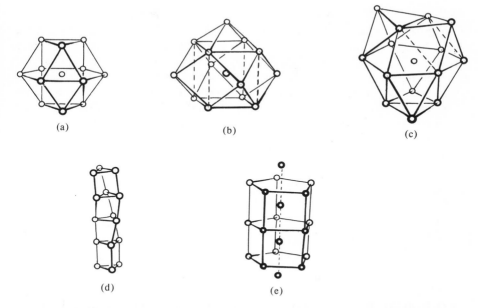

FIG. 22.10. The metal clusters in (a) $[Rh_{13}H_{5-n}(CO)_{24}]^{n-}$; (b) $[Rh_{14}(CO)_{25}]^{4-}$; (c) $[Rh_{15}(CO)_{27}]^{3-}$; (d) $[Pt_{15}(CO)_{30}]^{2-}$; (e) $[Pt_{19}(CO)_{22}]^{4-}$.

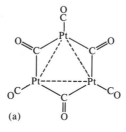

(a)

$[Ni_6(CO)_{12}]^{2-}$ has a distorted octahedral nucleus, with Ni–Ni distances of 2.38 Å in the triangular sub-units and 2.77 Å along the other six edges, whereas the Pt_6 nucleus is a distorted trigonal prism.

A quite different type of structure is found for the $[Pt_{19}(CO)_{22}]^{4-}$ anion, Fig. 22.10(e), which was isolated as the $(P\phi_4)^+$ salt.[3] The numerous terminal and bridging CO groups are omitted from Fig. 22.10, and not all of the metal–metal bonds are shown. Their lengths cover wide ranges even within a given ion; for details the reader is referred to the original papers.

(1) JACS 1978 **100** 7096
(2) JACS 1974 **96** 2614, 2616
(3) JACS 1979 **101** 6110

Carbonyl hydrides

Owing to their small scattering power for X-rays the positions of H atoms could not be determined in the earlier X-ray studies, nor could they be located in early e.d. studies of $Fe(CO)_4H_2$ and $Co(CO)_4H$. Direct location of H atoms by X-ray diffraction has become possible in some cases with improvements in technique, but in most X-ray studies the positions of H atoms have been deduced from the m–m distances. For example, the tetrahedral cluster in $Ru_4H_4(CO)_{12}$,[1] (a), has 4 longer edges, suggesting bridging H atoms, while in $Os_3H_2(CO)_{10}$[2] the fact that one edge

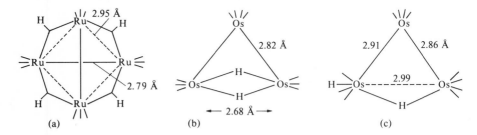

(a)   (b)   (c)

of the triangle is shorter than the other two, (b), is consistent with a double H bridge. In $Os_3H_2(CO)_{11}$[3] on the other hand there is one longer bond, indicating a single H bridge and one terminal H, (c). Neutron diffraction is making possible the direct location of H atoms and will eventually clear up many uncertainties in this field.

It appears that H can behave in a number of ways:

(i) As a normal ligand, occupying a coordination position as in K_2ReH_9, $HPtBr[P(C_2H_5)_3]_2$, etc., for which see Chapter 8, in $MnH(CO)_5$, where 5 CO and 1 H are situated at the vertices of an octahedron (Mn–H, 1.60 Å),[4] and in $[Fe_3H(CO)_{11}]^-$, and as a direct link between the metal atoms in $(CO)_5Cr-H-Cr(CO)_5^-$,[5] in which the Cr–Cr distance (3.41 Å) is too large for a normal Cr–Cr bond but 1.70 Å is a reasonable value for Cr–H (compare 1.68 Å in K_2ReH_9).

(ii) As a ligand bridging either 2 or 3 metal atoms in a cluster. We have given examples of single and double H bridges above; triply-bridging H atoms are found in $Ru_6H_2(CO)_{18}$,[6] above the two larger octahedron faces, and in $FeCo_3H(CO)_9$-$[P(OCH_3)_3]_3$,[7] where the H atom is situated below the Co_3 face of the tetrahedron.

(iii) As an interstitial atom enclosed within the metal cluster, as in $[Co_6H(CO)_{15}]^-$, noted earlier, and in $[Ni_{12}H(CO)_{21}]^{3-}$ and $[Ni_{12}H_2(CO)_{21}]^{2-}$.

(1) IC 1978 **17** 1271
(2) CC 1978 723
(3) IC 1977 **16** 878
(4) IC 1969 8 1928
(5) JACS 1966 **88** 366
(6) JACS 1971 **93** 5670
(7) CC 1975 684

Carbonyl halides

We have noted $Os_5(CO)_{15}I$ and $Rh_6(CO)_{15}I$ as having structures similar to those of the parent carbonyls. Structural studies have been made of several simpler compounds.

$Ru(CO)_4I_2$[1] forms octahedral molecules with the I atoms in *cis* positions (Ru–C, 2.01 Å; Ru–I, 2.72 Å), and a similar configuration has been assigned to $Fe(CO)_4I_2$ on the basis of i.r. evidence.[2] The structure of $[Rh(CO)_2Cl]_2$ is more complex.[3] This compound is dimeric in solution and also in the crystal, in which the dimers are bonded into infinite chains by Rh–Rh bonds as shown in Fig. 22.11(a). To account for the diamagnetism it has been suggested that bent metal–metal bonds are formed by overlap of d^2sp^3 orbitals at an angle of 56°.

$Mn_2(CO)_8Br_2$[4] forms a symmetrical bridged molecule (Fig. 22.11(b)) and molecules $M_2(CO)_8X_2$ are also formed by Tc, Re, Mo, and W, for example,

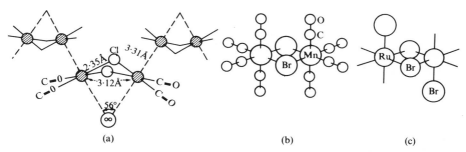

(a)   (b)   (c)

FIG. 22.11. The structures of (a) $[Rh(CO)_2Cl]_2$; (b) $[Mn(CO)_4Br]_2$; (c) $[RuBr_2(CO)_3]_2$.

$Mo_2(CO)_8I_2$.[5] The closely related molecule $Ru_2(CO)_6Br_4$[6] is shown in Fig. 22.11(c).

(1) AC 1962 **15** 946
(2) ZaC 1956 **287** 223
(3) JACS 1961 **83** 1761
(4) AC 1963 **16** 611
(5) AC 1976 **B32** 582
(6) AC 1968 **B24** 424

Nitrosyl carbonyls

Early e.d. studies[1] of $Fe(CO)_2(NO)_2$ and $Co(CO)_3NO$ indicated a tetrahedral arrangement of the four ligands with

$$Fe \underset{1\cdot84}{\text{——}} C \underset{1\cdot15}{\text{——}} O \quad \text{and} \quad Fe \underset{1\cdot77}{\text{——}} N \underset{1\cdot12\,\text{Å}}{\text{——}} O$$

and very similar bond lengths in the Co compound, though it is not likely that the bond lengths are of high accuracy. These studies did not prove the linearity of the –M–N–O system, but this has since been demonstrated by m.w. studies of molecules such as $(C_5H_5)NiNO$.[2] Manganese forms $Mn(NO)_3CO$ and $Mn(CO)_4NO$, and on irradiation the latter is converted into $Mn_2(CO)_7(NO)_2$ – compare the conversion of $Fe(CO)_5$ to $Fe_2(CO)_9$. The molecule $Mn(CO)_4NO$ has a trigonal bipyramidal structure, like the isoelectronic $Fe(CO)_5$, with NO at one equatorial position and linear M—N—O bonds.[3] Bond lengths are Mn—C, 1.89 Å (axial), and 1.85 Å (equatorial); Mn—N, 1.80 Å.

The Ru and Os compounds of the type $M_3(CO)_{10}(NO)_2$[4] are interesting for they possess double NO bridges. The two shorter edges of the isosceles triangle correspond to normal metal–metal bonds (2.87 Å) (a).

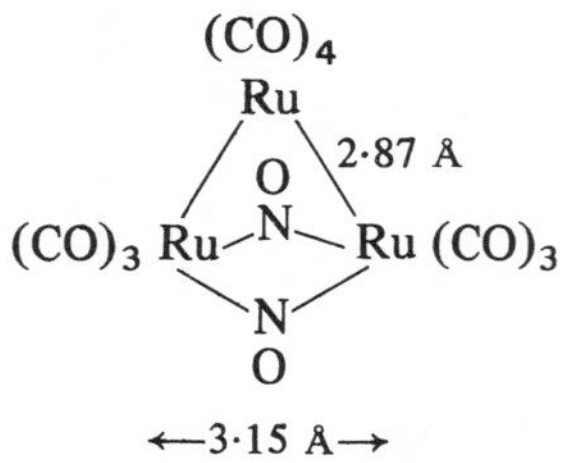

(a)

(1) TFS 1937 **33** 1233
(2) TFS 1970 **66** 557

(3) IC 1969 8 1288
(4) IC 1972 **11** 382

Mixed metal carbonyls

Many carbonyls have been prepared containing atoms of two or more different elements, some with only CO ligands and others with ligands of various types. The special interest lies in the metal–metal bonds and in the stereochemistry of the metal atoms. We give in Table 22.9 a few simple examples of typical molecules.

TABLE 22.9

Structures of mixed metal carbonyls

Molecule	M—M	Metal stereochemistry	Reference
$(CH_3)_3Sn-Mn(CO)_5$	2·674 Å	Sn tetr.; Mn oct.	JCS A 1968 696
$(CO)_5Mn-Fe(CO)_4-Mn(CO)_5$	2·815	Mn and Fe octahedral	AC 1967 **23** 1079
$(CO)_4Co-Zn-Co(CO)_4$	2·305	Co trig. bipyr.; Zn linear	JACS 1967 **89** 6362
$(CO)_4Co-Hg-Co(CO)_4$	2·50	Co trig. bipyr.; Hg linear	JCS A 1968 1005

Miscellaneous carbonyl derivatives of metals

Molecules containing CO and other ligands are very numerous, and range from simple molecules and ions in which all the ligands form a normal coordination group around the transition-metal atom to those in which there are more complex cyclic and polyhedral systems of metal atoms. Compounds containing unsaturated hydrocarbon ligands are described in the following pages, and we mention here only a few molecules which present points of special interest.

The compound $Fe_5(CO)_{15}C$ is produced in small quantity when $Fe_3(CO)_{12}$ is heated in petroleum ether with methylphenylacetylene. It has the tetragonal pyramidal structure shown in Fig. 22.12(a), in which C is located just below the base of the pyramid.[1] With the Fe—C bonds the Fe atom acquires the Kr configuration. Cobalt forms a number of compounds $Co_3(CO)_9C.R$, of which the methyl compound, Fig. 22.12(b), is remarkable in that an aliphatic C atom bridges

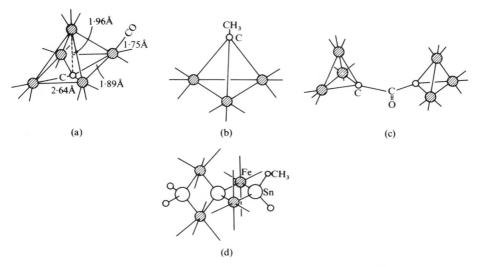

(a)

(b)

(c)

(d)

FIG. 22.12. The molecular structures of (a) $Fe_5(CO)_{15}C$; (b) $Co_3(CO)_9C(CH_3)$; (c) $[Co_3(CO)_9C]_2CO$; (d) $(CH_3)_4Sn_3Fe_4(CO)_{16}$.

three Co atoms. The compound is diamagnetic, Co is bonded octahedrally to 3 CO, 2 Co and C, and the length of the Co—Co bonds (2.47 Å) is similar to that in the metal (2.51 Å).[2] The molecule $[Co_3(CO)_9C]_2CO$, Fig. 22.12(c), is built of two tetrahedral $Co_3(CO)_9C$ units joined through a bridging CO.[3] There are other molecules and ions with skeletons similar to Fig. 22.12(b), with S replacing C and bridging H atoms. They include $Co_3SH_3(CO)_9$, $Os_3SH_2(CO)_9$, and $[Os_3SH(CO)_9]^{-}$,[3a] an ion studied in its tetramethylammonium salt. In the spiro molecule $(CH_3)_4Sn_3Fe_4(CO)_{16}$,[4] Fig. 22.12(d), the three Sn atoms are almost exactly collinear and are bridged by $Fe(CO)_4$ groups. There are two different Sn—Fe bond lengths, terminal 2.63 Å and central 2.75 Å. The bond arrangement around both types of Sn atom is tetrahedral: Fe—C, 1.75 Å; Sn—C, 2.22 Å.

There are many complex carbonyls containing S or $-SR$ ligands, some of which are structurally similar to molecules already described. For example, $[(C_2H_5S)Fe(CO)_3]_2$,[5] Fig. 22.13(a), is of type (d), p. 960, there being no bond between the S atoms (S–S, 2.93 Å), while in the closely related $[SFe(CO)_3]_2$,[6] Fig. 22.13(b), the S atoms are bonded (S–S, 2.01 Å). In $S_2Fe_3(CO)_9$ there is the more complex nucleus (c), which has been studied in the 1:1 molecular compound $[S_2Fe_2(CO)_6][S_2Fe_3(CO)_9]$.[7] The molecule $[CH_3SFe_2(CO)_6]_2S$,[8] (d), is of a different kind. There is a central S atom bonded tetrahedrally to 4 $Fe(CO_3$ units which are also bridged in pairs by SCH_3 ligands. It is convenient to include here also $(C_5H_5)_4Fe_4S_4$.[9] (e). The nucleus of the molecule is an elongated tetrahedron of Fe atoms (of which only two edges correspond to Fe–Fe bonds, the others being much longer (3.37 Å)), above the faces of which are arranged the S atoms, each bridging 3 Fe. The lengths of the Fe–S bonds in this and the other molecules of Fig. 22.13 are very close to that in pyrites (2.26 Å).

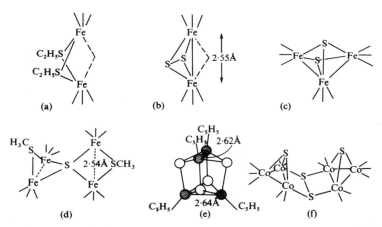

FIG. 22.13. The molecular structures of (a) $[(C_2H_5S)Fe(CO)_3]_2$; (b) $[SFe(CO)_3]_2$; (c) $S_2Fe_3(CO)_9$; (d) $[CH_3SFe_2(CO)_6]_2S$; (e) $(C_5H_5)_4Fe_4S_4$; (f) $[SCo_3(CO)_7]_2S_2$.

Of the numerous sulphur-containing Co carbonyls we give only a few examples. The molecule $SCo_3(CO)_9$[10] is of the same general type as $CH_3C.Co_3(CO)_9$, Fig. 22.12(b), and $[SCo_3(CO)_7]_2S_2$[11] consists of two identical tetrahedral units $SCo_3(CO)_7$ bridged by an S_2 group, as shown in Fig. 22.13(f). The following molecules[12] all contain bridging $-SC_2H_5$ groups: $Co_3(CO)_4(SC_2H_5)_5$, $Co_5(CO)_{10}$-$(SC_2H_5)_5$, and $SCo_6(CO)_{11}(SC_2H_5)_4$, the latter also containing a Co_3S unit resembling that in Fig. 22.13(f).

(1)	JACS 1962 **84** 4633	(5)	IC 1963 **2** 328	(10)	IC 1967 **6** 1229
(2)	JACS 1967 **89** 261	(6)	IC 1965 **4** 1	(11)	JACS 1967 **89** 3727
(3)	AC 1969 **B25** 107	(7)	IC 1965 **4** 493	(12)	JACS 1968 **90** 3960, 3969, 3977
(3a)	AC 1978 **B34** 3767	(8)	IC 1967 **6** 1236		
(4)	IC 1967 **6** 749	(9)	IC 1966 **5** 892		

Compounds of metals with hydrocarbons

Compounds of metals containing hydrocarbon ·radicals or molecules are numerous. The simplest contain only metal and hydrocarbon radical (metal alkyls and aryls) or unsaturated hydrocarbon molecules (for example, $Cr(C_6H_6)_2$), which in others there are also ligands such as halogens or CO. Because they are most closely related to the metal carbonyls we deal first with transition-metal compounds containing unsaturated hydrocarbons and CO ligands and later with the alkyls of non-transition metals. The formulae of many of the simpler transition-metal compounds are consistent with the view that the metal gains the following numbers of electrons from the ligands:

$$H_2C\text{---}CH\text{---}CH_2 \qquad \begin{array}{c} HC=CH \\ | \quad | \\ HC=CH \end{array}$$

3 4 5 6 7

and thereby reaches the electronic structure of a noble gas—compare the carbonyls and nitrosyl carbonyls. In all the following diamagnetic compounds of the 3d elements the metal has the Kr configuration:

A.N.

23	$V(CO)_4(C_5H_5)$		
24		$Cr(CO)_3(C_6H_6)$	$Cr(C_6H_6)_2$
25	$Mn(CO)_3(C_5H_5)$		
26		$Fe(CO)_3(C_4H_4)$ $Fe(C_5H_5)_2$	$Fe(CO)(C_3H_5)(C_5H_5)$
27	$Co(CO)_2(C_5H_5)$		
28		$Ni(C_5H_5)NO$	

while a molecule such as $Ni(C_5H_5)_2$ clearly has 2 electrons in excess of the noble gas number. The molecule of cyclooctatetraene, C_8H_8 (COT), functions as a 4-electron donor in $(CO)_3Fe(COT)$ but as two (cyclobutadiene) halves in $(CO)_3Fe(C_8H_8)$-$Fe(CO)_3$ (see later).

We have already noted the geometrical difference between the bond formed by a transition metal to CO, which is a true π bond (overlap of metal d orbitals with the π orbital of CO in the nodal plane) and the bond to an unsaturated hydrocarbon in which the metal orbital is perpendicular to the nodal plane of the π orbital. The term μ-bond has been suggested for this latter type of bond.

Acetylene complexes

The reactions of acetylene with metal carbonyls can be grouped into two main classes: (i) those in which no new C–C bonds are formed, and the acetylene is bonded to the metal atom(s) by μ-bonds, 'bent' σ bonds, or by a combination of μ and σ bonds, and (ii) those in which new C–C bonds are formed by combination of some of the CO with the acetylene to form a cyclic hydroxy-ligand or a lactone ring.

(i) Acetylenes replace two CO in $Co_2(CO)_8$, and a study of crystalline $Co_2(CO)_6$-(diphenylacetylene) shows that the molecule has the configuration of Fig. 22.14(a). The metal atoms form six bonds, arranged octahedrally, one to the other metal atom, two to the acetylene carbon atoms, and three to CO molecules; the C–C bond in the acetylene has lengthened to about 1.46 Å.[1] In $Co_4(CO)_{10}C_2H_5C.C.C_2H_5$ [2] the acetylene forms two σ bonds to two Co atoms and μ bonds to the other two metal atoms (Fig. 22.14(b)).

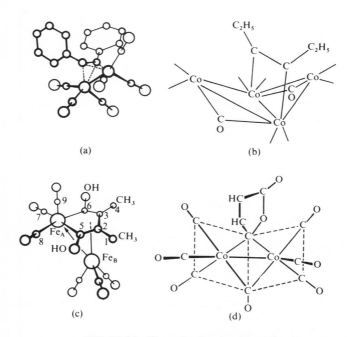

FIG. 22.14. The molecular structures of:
(a) (diphenylacetylene)$Co_2(CO)_6$; (b) $Co_4(CO)_{10}C_2H_5CCC_2H_5$;
(c) $(CO)_3Fe(C_6H_8O_2)Fe(CO)_3$; (d) $(CO)_3Co(CO)(C_4H_2O_2)Co(CO)_3$.

(ii) An interesting reaction is the displacing of two CO groups from two molecules of iron or cobalt carbonyl hydrides by acetylenes or other unsaturated molecules. A study of the crystal structure[3] of the but-2-yne complex from $Fe(CO)_4H_2$, which has the formula $(OC)_3Fe(C_6H_8O_2)Fe(CO)_3$, shows that the molecule has the structure shown in Fig. 22.14(c). All the atoms C^1 to C^8 are coplanar, and the group C^9O is perpendicular to this plane. The three CO groups of Fe_B are arranged with three-fold symmetry around an axis which is perpendicular to the plane of

$C^6C^3C^2C^5$. There is a covalent Fe—Fe bond, and the π-bonding between Fe_B and the ring of carbon atoms is apparently similar to that in cyclopentadienyl compounds.

The action of CO under pressure on $Co_2(CO)_6C_2H_2$ yields a compound with the empirical formula $Co_2(CO)_9C_2H_2$. In this molecule, Fig. 22.14(d), the two Co atoms are bridged by one CO and a C atom of a lactone ring.[4] Each Co is bonded to five C atoms in a square pyramidal configiration, and the two square pyramids are joined along a basal edge about which the molecule is folded until the Co—Co separation has the value 2.5 Å. The molecular structure is very similar to that of $Co_2(CO)_8$.

(1) JACS 1959 **81** 18
(2) JACS 1962 **84** 2451
(3) AC 1961 **14** 139
(4) PCS 1959 156

Cyclopentadienyl complexes

Cyclopentadiene, C_5H_6, is a colourless liquid which readily forms a monosodium derivative. This in turn reacts with anhydrous transition-metal halides to form derivatives $M(C_5H_5)_n$, some of which may be made direct from the hydrocarbon

$$
\begin{array}{c}
CH_2 \\
HC \quad CH \\
\| \quad\quad \| \\
HC—CH
\end{array}
$$

and the metal carbonyls at about 300 °C. Some of these compounds, including $Fe(C_5H_5)_2$ ('ferrocene') can be oxidized to cations. Structural studies have been made of compounds of the following types: $M(C_5H_5)_n$ (n = 1, 2, 3, 4), $M(C_5H_5)_2X$, $M(C_5H_5)_2X_2$, and numerous more complex molecules and ions. In most of the compounds to be described C_5H_5 acts as a π-donor, providing 5 electrons to a metal atom which is equidistant from all 5 C atoms. The formation of normal σ bonds by one or more C atoms (as, for example, in $C_5H_5SiH_3$) occurs in the molecules $Ti(C_5H_5)_3$ and $Ti(C_5H_5)_4$ and also in crystalline $Be(C_5H_5)_2$ and $In(C_5H_5)_3$.

The simplest type of metal derivatives, $M(C_5H_5)$, is formed by In and Tl,[1] the molecule (Fig. 22.15(a)) having C_{5v} symmetry. The most numerous derivatives are of the type $M(C_5H_5)_2$, formed by Be, Mg, all the 3d metals from V to Ni, Ru, Rh, Sn, and Pb. Of these the first to be discovered and the most studied is $Fe(C_5H_5)_2$ (ferrocene). In the vapour state all these molecules have the 'sandwich' structure of Fig. 22.15(b), either the staggered configuration (Mn, Sn, Pb, Be, Ni[1a]) or the eclipsed configuration (Fe, Ru,[1b] Cr). Whereas M is normally midway between the two rings, in $Be(C_5H_5)_2$[2] it oscillates between two positions 1.47 Å from the centres of the rings, the perpendicular distance between which is 3.375 Å. There are complications in the crystalline state. There are low- and high-temperature forms of solid ferrocene. In the former the configuration of the molecule is approximately eclipsed, but because of disorder in the crystal the apparent centrosymmetrical (staggered) configuration of the high-temperature form cannot be interpreted as the structure of each individual molecule.[3] The length of the Fe—C bond (2.06 Å)

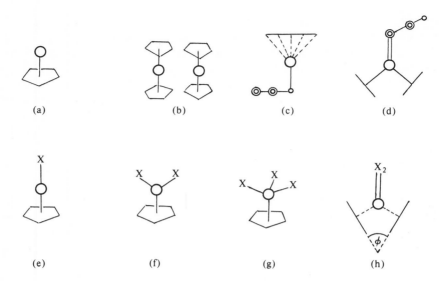

FIG. 22.15. Simple molecules containing C_5H_5 ligands.

is about the value expected for a single bond, in contrast to the mean value (1.84 Å) in $Fe(CO)_5$. Here also Fe has the Kr configuration. The ferricinium ion has been studied in $(C_5H_5)_2Fe(I_3)$,[4] but disorder in the crystal made it impossible to determine the detailed structure of the ion. The structure of the $Be(C_5H_5)_2$ molecule in the crystal, studied at $-120\,^\circ C$,[5] has a quite different structure from that of the vapour molecule. The metal atom is π-bonded to one ring (Be—C, 1.94 Å) but σ-bonded to one C atom of the other (Be—C, 1.81 Å), as shown in the end-on view of Fig. 22.15(c). In $Ti(C_5H_5)_3$[6] two of the cyclopentadiene rings are π-bonded to the metal atom, but the third is bonded through 2 C atoms, as shown diagrammatically in Fig. 22.15(d). Other simple molecules containing more than two C_5H_5 ligands include the tetrahedral molecules $Ti(C_5H_5)_4$[7] and $(C_5H_5)_3UCl$.[8] In $Ti(C_5H_5)_4$ Ti is π-bonded to two C_5H_5 and σ-bonded to the other two.

Molecules of types (e)–(h), Fig. 22.15, include:
(e) $CH_3Be(C_5H_5)$,[9] $ClBe(C_5H_5)$,[10] and $(NO)Ni(C_5H_5)$;[11]
(f) $(CH_3)_2Al(C_5H_5)$;[12]
(g) $(CO)_3Mn(C_5H_5)$;[13] and $Cl_3Ti(C_5H_5)$,[14]
(h) $X_2M(C_5H_5)_2$; dichloro compounds include those of Ti, Zr, Nb(IV), and Mo(IV).[15] In these molecules the angle ϕ is in the region of 50–60° and the angle X—M—X, 90–100°.

The structures described so far are those of discrete molecules or ions. In certain crystalline derivatives C_5H_5 acts as a bridging ligand. In crystalline $Pb(C_5H_5)_2$[16] there are chains, Fig. 22.16(a), in which the metal atoms are bonded through C_5H_5

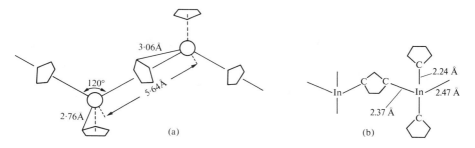

FIG. 22.16. The structures of (a) $Pb(C_5H_5)_2$; (b) $In(C_5H_5)_3$.

ligands and also one terminal C_5H_5 bonded to each Pb. The interbond angles at the Pb atoms are approximately 120°. In the chains in $In(C_5H_5)_3$[17] In forms σ bonds to C atoms of four C_5H_5 ligands in a nearly regular tetrahedral arrangement (b).

We conclude this section with some examples of binuclear molecules containing C_5H_5 ligands.

The interaction of $Fe(CO)_5$ with excess of C_5H_5 gives first the bridged compound (a) of Fig. 22.17, which decomposes at 200°C to ferrocene but can be

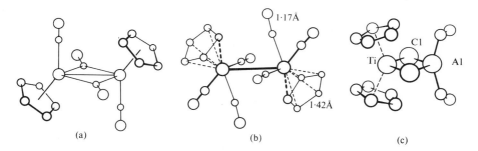

FIG. 22.17. The molecular structures of (a) $Fe_2(CO)_4(C_5H_5)_2$; (b) $Mo_2(CO)_6(C_5H_5)_2$; (c) $(C_5H_5)_2TiCl_2Al(C_2H_5)_2$.

isolated if the temperature is kept in the range 100–200°C. The bond length Fe–Fe (2.49 Å) is close to that in $Fe_2(CO)_9$, and the general resemblance to this molecule is evident.[18] An X-ray study of $Mo_2(CO)_6(C_5H_5)_2$[19] gives the structure (b) of Fig. 22.17. All the atoms drawn in heavily lie in one plane. The arrangement of bonds around a metal atom is such that if the Mo atom is imagined to be at the centre of a cube then a C_5H_5 ring lies centrally in one face and the other four bonds (to Mo and three C) are directed towards the vertices of the opposite cube face. In this molecule Mo has a completed outer shell of eighteen electrons and the compound is therefore diamagnetic. Bond lengths include: Mo–C, 2.34 Å; Mo–Mo, 3.22 Å; and the bond angle Mo–C–O is approximately 175°.

In the bridged molecule $(C_5H_5)_2TiCl_2Al(C_2H_5)_2$, Fig. 22.17(c), there is an approximately square bridge of edge 2.5 Å between the two metal atoms.[20] The bonds from both Ti and Al are arranged tetrahedrally, so that the general shape of the molecule is similar to that of Al_2X_6 molecules.

The molecule (a) in Fig. 22.18 is included here as an interesting example of a 5-membered ring of metal atoms of three different kinds.[21]

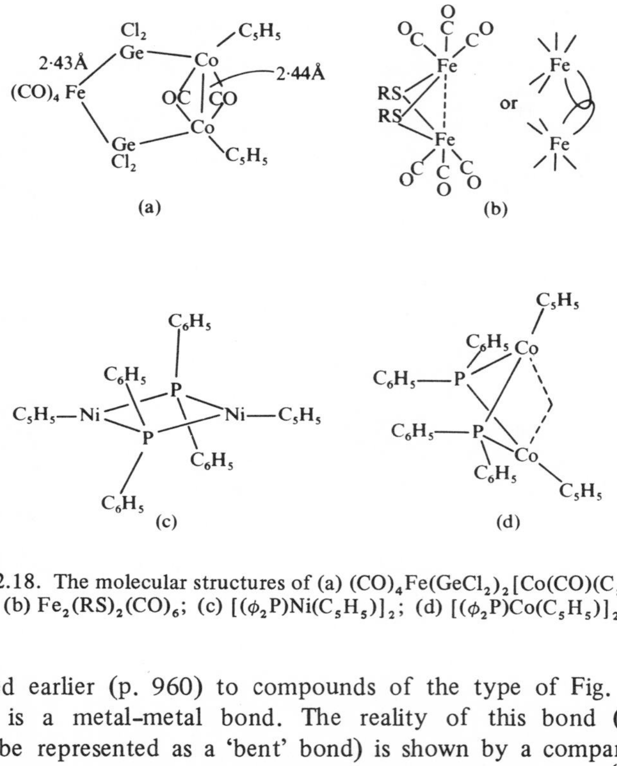

FIG. 22.18. The molecular structures of (a) $(CO)_4Fe(GeCl_2)_2[Co(CO)(C_5H_5)]_2$; (b) $Fe_2(RS)_2(CO)_6$; (c) $[(\phi_2P)Ni(C_5H_5)]_2$; (d) $[(\phi_2P)Co(C_5H_5)]_2$.

We referred earlier (p. 960) to compounds of the type of Fig. 22.18(b) in which there is a metal–metal bond. The reality of this bond (which may alternatively be represented as a 'bent' bond) is shown by a comparison of the structures of the dimeric $[(\phi_2P)Ni(C_5H_5)]_2$ and $[(\phi_2P)Co(C_5H_5)]_2$.[22] The Ni compound, (c), has a planar bridge system, but in the Co compound, (d), the two CoP_2 planes are inclined to one another owing to the formation of the Co–Co bond (Co–Co, 2.56 Å; compare Ni–Ni, 3.36 Å). An electron count, assuming that the two $P\phi_2$ ligands provide three, C_5H_5 five, and the metal–metal bond one electron to each metal atom, shows that in all these molecules a noble-gas configuration is attained: Fe, 26+9+1; Co, 27+8+1; and Ni, 28+8.

We have given examples of molecules in which two metal atoms are directly bonded or bridged by two ligands, with or without a m–m bond. Examples of molecules or ions in which the bridge consists of a single O atom include

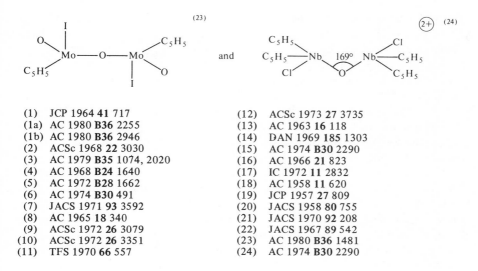

(1)	JCP 1964 **41** 717
(1a)	AC 1980 **B36** 2255
(1b)	AC 1980 **B36** 2946
(2)	ACSc 1968 **22** 3030
(3)	AC 1979 **B35** 1074, 2020
(4)	AC 1968 **B24** 1640
(5)	AC 1972 **B28** 1662
(6)	AC 1974 **B30** 491
(7)	JACS 1971 **93** 3592
(8)	AC 1965 **18** 340
(9)	ACSc 1972 **26** 3079
(10)	ACSc 1972 **26** 3351
(11)	TFS 1970 **66** 557

(12)	ACSc 1973 **27** 3735
(13)	AC 1963 **16** 118
(14)	DAN 1969 **185** 1303
(15)	AC 1974 **B30** 2290
(16)	AC 1966 **21** 823
(17)	IC 1972 **11** 2832
(18)	AC 1958 **11** 620
(19)	JCP 1957 **27** 809
(20)	JACS 1958 **80** 755
(21)	JACS 1970 **92** 208
(22)	JACS 1967 **89** 542
(23)	AC 1980 **B36** 1481
(24)	AC 1974 **B30** 2290

Complexes containing benzene or cyclooctatetraene

We give here only a few of the simpler complexes of these types. Closely related to the cyclopentadienyl compounds is the compound $Cr(C_6H_6)_2$, prepared by heating a mixture of $CrCl_3$, $AlCl_3$, Al powder, and benzene at 150 °C. This is a crystalline compound (m.p. 284 °C) in which molecules of the type shown in Fig. 22.19(a) are

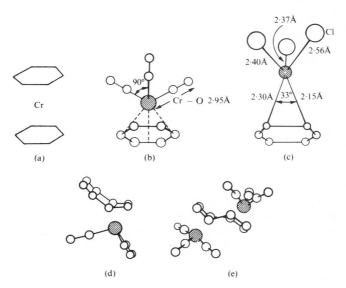

FIG. 22.19. The molecular structures of (a) $Cr(C_6H_6)_2$; (b) $(CO)_3Cr(C_6H_6)$; (c) $(C_6H_6)CuAlCl_4$; (d) $(CO)_3Fe(C_8H_8)$; (e) $(CO)_3Fe(C_8H_8)Fe(CO)_3$.

arranged in cubic closest packing. The Cr—C distance is 2.15 Å and the perpendicular distance between the rings is 3.23 Å.[1] The $Cr(C_6H_6)_2^+$ ion in $Cr(C_6H_6)_2I^{(1a)}$ has a similar structure, and Cr can also be sandwiched between the benzene rings of cyclophanes, as in $(3,3\text{-paracyclophane})^+I_3$. The $C_{18}H_{20}Cr^+$ ion consists of two slightly bent benzene rings bridged in the para positions by chains $-CH_2-CH_2-CH_2-$, with Cr^+ placed symmetrically between them.

It is interesting to compare with the molecule $(CO)_3Cr(C_6H_6)^{(2)}$ (Fig. 22.19(b)) the environment of Cu^+ in $C_6H_6 . CuAlCl_4$,[3] which is shown in Fig. 22.19(c). This compound contains tetrahedral $AlCl_4^-$ ions, and the neighbours of Cu are a C_6H_6 ring and 3 Cl atoms of $AlCl_4^-$ ions.

In crystalline $AgClO_4 . C_6H_6$[4] there are columns of C_6H_6 molecules and Ag^+ ions, two of the latter being associated with each benzene ring, and each Ag^+ is distant 2.50 and 2.63 Å from the two C atoms of the nearest C—C bond. The energy of the 'charge-transfer' bonds has been estimated at about 66 kJ/bond mole. In $AgNO_3 . C_8H_8$[5] also there are columns of Ag^+ ions and hydrocarbon molecules, between which the anions are situated, and as in the benzene compound each Ag^+ is closely associated with one bond of the organic ring, here with a double bond. These shortest Ag—C bonds are again close to 2.5 Å in length. There are two weaker bonds from Ag^+ to two other atoms of the ring (Ag—C, c. 2.8 Å). The even longer Ag—C bonds, of length around 3.2 Å, link together the $Ag^+ . C_8H_8$ units into columns.

In $AgNO_3 . C_8H_8$ the C_8H_8 ring has the boat configuration like the free molecule, but two other shapes of this ring are found,[6] namely, a dihedral shape in $(CO)_3Fe(C_8H_8)$ and a 'chair' shape in $(CO)_3Fe(C_8H_8)Fe(CO)_3$. In the tricarbonyl Fe is attached to part of the ring which behaves like butadiene (Fig. 22.19(d)), whereas in the second molecule both ends of the ring are behaving in this way, as shown at (e) in Fig. 22.19. There are two configurations of the hydrocarbon ring also in $Ti_2(C_8H_8)_3$. The molecule has a double sandwich structure, $C_8H_8-Ti-C_8H_8-$ $-Ti-C_8H_8$,[7] in which the outer rings are planar and the central one boat-shaped.

(1)   ACSc 1965 **19** 41
(1a) AC 1974 **B30** 838
(2)   JACS 1959 **81** 5510
(3)   JACS 1966 **88** 1877
(4)   JACS 1958 **80** 5075
(5)   JCP 1959 **63** 845
(6)   JCP 1962 **37** 2084
(7)   AC 1968 **B24** 58

Metal alkyls

Simple alkyls $M(alkyl)_n$ are formed by two groups of metals:

(i)   the alkali metals, Be, Mg, the alkaline-earths, and Al; and

(ii) the B subgroup metals; little appears to be known of alkyls of the intervening transition metals. (For compounds $[Pt(CH_3)_3X]_4$, one of which, the hydroxide, was earlier thought to be $Pt(CH_3)_4$, see p. 1242). It is convenient to deal first with group (ii) because with the possible exception of alkyls of Cu, Ag, and Au (of which the structures are not known) the B subgroup alkyls are normal covalent compounds like those of B, C, Si, etc. in which a C atom of the alkyl group forms a single bond to M.

Alkyls of B subgroup metals

Alkyls of IB metals include $CuCH_3$, $AgCH_3$, AgC_2H_5, and trimethyl gold. The last compound has been prepared at low temperatures in ether solution, in which it probably exists as $(CH_3)_3Au.O(C_2H_5)_2$. This etherate reacts with ethereal HCl to give dimethyl auric chloride, presumably a bridged molecule, $(CH_3)_2AuCl_2Au(CH_3)_2$.

The IIB alkyls are normal covalent compounds, the chemical reactivity of which decreases from that of the violently reactive $Zn(CH_3)_2$ through $Cd(CH_3)_2$ to the relatively inert $Hg(CH_3)_2$. The high resolution Raman spectra of these compounds show that the molecules are linear; the bond lengths are respectively 1.93, 2.11, and 2.09 Å.[1]

The alkyls of Ga, In, and Tl form planar molecules $Ga(CH_3)_3$ etc.—contrast Ga_2Cl_6, Ga_2H_6, and the dimeric methyl gallium hydrides - as shown by a spectroscopic study of $Ga(CH_3)_3$,[2] and an early e.d. study of $In(CH_3)_3$ in the vapour state,[3] and later by crystallographic studies of $In(CH_3)_3$[4] and $Tl(CH_3)_3$.[5] Since these are planar molecules a metal atom necessarily has neighbours in directions approximately normal to the plane of $M(CH_3)_3$, and these complete a distorted trigonal bipyramidal arrangement of nearest and next nearest neighbours. The intra- and shortest inter-molecular bond lengths are:

$In-C$, 2.16 Å (three), 3.11 Å (one), and 3.59 Å (one);
$Tl-C$, 2.29 Å (three), 3.16 Å (one), and 3.31 Å (one).

Alkyls of groups I and II metals and Al

The alkyls of Na, K, Rb, and Cs are colourless amorphous solids which are insoluble in all solvents except the liquid zinc dialkyls, with which they react to form compounds such as $NaZn(C_2H_5)_3$. The Li compounds are quite different, for they are soluble to varying extents in solvents such as benzene, in which they are polymeric, and both lithium methyl and ethyl exist as colourless crystalline solids at ordinary temperatures.

Group II provides the crystalline $Be(CH_3)_2$, the very reactive magnesium dialkyls, and compounds of the alkaline-earths such as the dimethyls of Ca, Sr, and Ba.[6]

A number of aluminium alkyls have been prepared: $Al(CH_3)_3$ by refluxing Al with CH_3I in a nitrogen atmosphere, and the ethyl, n- and iso-propyl compounds by heating Al with the appropriate mercury dialkyl at 110 °C for 30 hours. At ordinary temperatures they are water-white liquids (the trimethyl melts at 15 °C), very reactive and spontaneously inflammable in air. The trimethyl is dimeric in benzene solution and in the vapour state, the ethyl and n-propyl dimers show measurable dissociation, while the isopropyl compound is monomeric.[7] There is a striking similarity between $Al(CH_3)_3$ and $AlCl_3$. Both exist as dimers in the vapour state, and the energy of association is of the same order, viz. 84 and 121 kJ mol^{-1} respectively. Both combine with amines and ethers to give coordination compounds.

Since the structures of the compounds of the heavier alkali metals and of the alkaline-earths are not known our discussion is restricted to the alkyls of Li, Be, Mg, and Al. All are electron-deficient compounds in which some or all of the CH_3 groups form two or three bonds to metal atoms. These are described as 3- or 4-centre bonds formed by overlap of an sp^3 orbital of C with appropriate orbitals of the metal atoms.

Crystalline methyl lithium consists of a b.c.c. packing of molecules $Li_4(CH_3)_4$. The Li atoms are arranged at the vertices of a regular tetrahedron of edge 2.56 Å and a CH_3 group is situated above the centre of each face (Li–C, 2.28 Å). Since all Li and C atoms lie on body-diagonals of the unit cell Li has another CH_3 belonging to a neighbouring molecule as its next nearest neighbour (at 2.52 Å). The intra-molecular contacts are Li–3 Li (2.56 Å) and 3 CH_3 (2.28 Å).[8] In crystalline $Li_4(C_2H_5)_4$ the tetrahedral molecule is apparently less symmetrical, Li–Li ranging from 2.42 to 2.63 Å[9] and instead of three equidistant C at 2.28 Å Li has 3 C at 2.19, 2.25, and 2.47 Å, this last bond length being very close to the distance to the nearest C of an adjacent molecule (2.53 Å). These bond lengths are extremely difficult to understand, particularly in view of the strong bonding between the Li atoms—compare Li Li, 2.67 Å in Li_2 and 3.04 Å in the metal.

Dimethyl beryllium is a white solid which sublimes at 200 °C and reacts violently with air and water. It is isostructural with SiS_2, the crystal being built of infinite chains (a) in which Be forms tetrahedral bonds (Be–C, 1.93 Å) to four methyl

(a)

groups.[10] $Mg(CH_3)_2$ has the same structure (Table 22.10) and the same type of chain is found in $LiAl(C_2H_5)_4$ where the metal atoms along the chain are alternately Li and Al and CH_3 is replaced by C_2H_5 (Table 22.10).

In Group III we find a third type of structure for the metal alkyl, namely, the bridged dimer (b) of $Al_2(CH_3)_6$ which is structurally similar to the Al_2X_6 molecules. There are also less symmetrical molecules with the same type of structure, for example, $(CH_3)_5Al_2N(C_6H_5)_2$, where the bridge consists of one CH_3 (Al–C–Al,

(b)

(c)

79°) and the $N(C_6H_5)_2$ group (Al–N–Al, 86°) and the central 4-ring is not planar as in (b). The mixed alkyl $MgAl_2(CH_3)_8$, made by dissolving $Mg(CH_3)_2$ in $Al(CH_3)_3$, has the structure (c) intermediate between those of the simple Al and Mg methyls. Some details of the structures of these compounds are summarized in Table 22.10.

TABLE 22.10
Structures of metal alkyls

	M—M	M—C (bridge)	M—C—M	Reference
$[Mg(CH_3)_2]_\infty$	2.72 Å	2.24 Å	75°	JACS 1964 **86** 4825; JOC 1964 **2** 314
$[LiAl(C_2H_5)_4]_\infty$	2.71	2.02 (Al) 2.30 (Li)	77°	IC 1964 **3** 872
$Al_2(CH_3)_6$	2.60	2.14	75°	JACS 1967 **89** 3121
$MgAl_2(CH_3)_8$	2.70	2.10 (Al) 2.21 (Mg)	78°	JACS 1969 **91** 2538
$Al_2(CH_3)_5N(C_6H_5)_2$	2.72	2.14	79°	JACS 1969 **91** 2544

The fluoro-diethyl, $AlF(C_2H_5)_2$,[11] is a very viscous liquid soluble in benzene, in which it is tetrameric. The cyclic structure (d) has been confirmed for $[AlF(CH_3)_2]_4$,[12] and an e.d. study of $[AlCl(CH_3)_2]_2$[13] indicates the Cl-bridged structure (e).

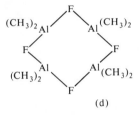

(d)

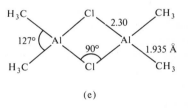

(e)

(1) CJP 1960 **38** 1516
(2) JCS 1964 3353
(3) JACS 1941 **63** 480
(4) JACS 1958 **80** 4141
(5) JCS A 1970 28
(6) JACS 1958 **80** 5324
(7) JACS 1946 **68** 2204

(8) JOC 1964 **2** 197
(9) AC 1963 **16** 681
(10) AC 1951 **4** 348
(11) IC 1966 **5** 503
(12) JOC 1973 **54** 77
(13) ACSc 1974 **A28** 45

23

Silicon

Introduction

Although the silicon atom has the same outer electronic structure as carbon its chemistry shows very little resemblance to that of carbon. It is true that elementary silicon has the same crystal structure as one of the forms of carbon (diamond) and that some of its simpler compounds have formulae like those of carbon compounds, but there is seldom much similarity in chemical or physical properties. Since it is more electropositive than carbon it forms compounds with many metals which have typical alloy structures (see the silicides, p. 987) and some of these have the same structures as the corresponding borides. In fact, silicon in many ways resembles boron more closely than carbon, though the formulae of the compounds are usually quite different. Some of these resemblances are mentioned at the beginning of the next chapter. Silicides have few features in common with carbides but many with borides, for example, the formation of extended networks of linked Si (B) atoms, though few silicides are actually isostructural with borides.

Silicon is compared with germanium in a short section on the latter element in Chapter 26. Here we shall note some points of difference from and resemblance to carbon. The outstanding feature of carbon chemistry is the direct bonding of carbon atoms. Simple molecules or ions containing Si–Si bonds are relatively few in number; they include the silanes, linear and cyclic, simple and substituted (p. 992), the ion $Si_2Te_6^{6-}$ with an ethane-like structure in $K_6Si_2Te_6$[1] (Si–Si, 2.40; Si–Te, 2.51 Å), and some oxyhalides. Many silicides contain systems of linked Si atoms, but apart from these cases the chemistry of complex compounds of silicon is largely based on linking of Si atoms through O atoms.

In its simple molecules and ions Si does not exhibit a covalency of less than four except (possibly) in the silyl ion (see later). Unlike carbon it does not form a small number of multiple bonds, as does carbon in CO, CO_2, $-CN$, etc. Only one oxide of Si is stable at ordinary temperatures, and this exists in a number of crystalline forms, in all of which—with the exception of stishovite (p. 1005)—there is a tetrahedral arrangement of four bonds from each Si atom, and in every case the Si and O atoms form an infinite 3D network. The lower oxide SiO has nothing in common with the gaseous CO. It is produced by heating SiO_2 with Si at temperatures above 1250 °C and disproportionates on slow cooling. It has been claimed that if it is quenched it can be obtained at room temperature, but apparently not in crystalline form.[2] The reaction $Si + SiCl_4 \rightarrow 2SiCl_2$ is said to take place at temperatures above 1100 °C.[3] Certainly SiF_4 reacts with Si at 1150 °C to form gaseous SiF_2, a species with a half-life at room temperature of 150 seconds (compare 1 second for CF_2 and a few hundredths of a second for CH_2); see also under silicon halides.[4] There is no stable sulphide CS, but material with the composition SiS can be sublimed from a mixture of $SiS_2 + Si$. At room temperature the coloured sublimates are not crystalline and have probably disproportionated to $SiS_2 + Si$.[5] The structures of crystalline SiS_2 bear no relation to that of the simple CS_2 molecule; the normal form of SiS_2 consists of infinite chains of SiS_4 tetrahedra sharing pairs of opposite edges. (There is also a high-pressure polymorph with a compressed cristobalite-like structure, M—S—M, $109\frac{1}{2}°$.[6])

Many of the differences between Si and C, like those between P and N, S and O, or Cl and F, are attributable to the fact that the elements of the second short Period can utilize d orbitals which are not available in the case of the first row elements. We find covalencies greater than 4 in compounds of Si, P, and S with the more electronegative elements. There are no carbon analogues of the fluorosilicates and there is no nitrogen analogue of the PF_6^- ion. Although the silyl compounds, containing $-SiH_3$, are formally analogous to methyl compounds, they behave in many ways very differently from their carbon analogues. The abnormal properties of many silyl compounds are probably connected with the powerful electron-accepting properties of $-SiH_3$. Although Si is less electronegative than C, $-SiH_3$ is a stronger electron-acceptor than $-CH_3$ because π bonding takes place between a p_π orbital on the atom attached to Si by a σ bond and a vacant d_π orbital of Si; there are no 3d orbitals on the C atom. (Usually π bonding is stronger the *more* electronegative is the donor atom.) This characteristic of the silyl group shows itself in, for example, the planarity of $N(SiH_3)_3$ (p. 795), and the weaker coordinating power of silyl compounds. For example, $N(SiH_3)_3$ forms no quaternary salts $[N(SiH_3)_4]X$ or $[N(SiH_3)_3H]X$, and the addition compound with BF_3 is much less stable than the compound with $N(CH_3)_3$.

The last point we have to mention is the great difference in stability towards oxygen and water between pairs of carbon and silicon compounds. In contrast to the stable CCl_4 and CS_2 the corresponding silicon compounds are decomposed by water to form 'silicic acid' (and, of course, HCl and H_2S respectively), a term applied to the colloidal silica which may be obtained hydrated to varying degrees. (The preparation of well defined di-, tri-, and tetra-silicic acids has been claimed[7] by the hydrolysis and disproportionation of ortho silicic esters.) Silicon forms a number of hydrides (silanes), and although these are stable towards pure water they are hydrolysed by dilute alkali, which has no action at all on the corresponding hydrocarbons. In view of the ease with which silicon compounds are converted into the oxide or other oxy-compounds, it is not surprising that silicon occurs in nature exclusively as oxy-compounds, and we shall devote much of this chapter to a review of the extensive chemistry of the silicates.

(1) AC 1978 **B34** 2390
(2) JPC 1959 **63** 1119; ZaC 1967 **355** 265
(3) ZaC 1953 **274** 250
(4) JACS 1965 87 2824
(5) JACS 1955 77 904
(6) Sc 1965 **149** 535
(7) ZaC 1954 **275** 176; **276** 33.

The stereochemistry of silicon

In the crystalline element and in the great majority of its compounds stable under ordinary conditions Si forms four tetrahedral bonds; the simplest examples include SiH_4, SiX_4, SiO_4^{4-}, and SiS_4^{4-}. The thiosilicate ion has been studied in Ba_2SiS_4,[1] which is isostructural with one form of K_2SO_4 (Si–S, 2.10 Å), and the 'super-tetrahedral' $Si_4S_{10}^{4-}$ ion is referred to on p. 1173. Silicon does not form salts with strong acids, but there is an acetate $Si(ac)_4$[1a] which presumably consists of tetrahedral molecules. The next most important coordination number is six, and there is apparently no tendency to exceed this number. Thus with acetylacetone Si forms $[Si(acac)_3]AuCl_4$ and $Si(acac)_3Cl.HCl$ (a compound which has been resolved into its optical antimers),[2] but it does not, like Ti and Zr, form $M(acac)_4$; K_3SiF_7, formed by heating K_2SiF_6 in dry air, is $K_3(SiF_6)F$.[2] The octahedral configuration has been established for finite complexes such as the SiF_6^{2-} ion—for example, K_2SiF_6 has the K_2PtCl_6 structure—, the ion (a) in its pyridinium salt,[3] and the neutral molecule (b),[4] and it is probable for the trisoxalato ions in salts

(a)   (b)

$(NR_4)_2[M^{IV}(C_2O_4)_3]$, (M = Si, Ge, Sn, Ti),[5] and derivatives $PcSiX_2$ of the phthalo-cyanin.[6a] Octahedral coordination by oxygen occurs in all forms of SiP_2O_7[6b] stable under atmospheric pressure, in $[Ca_3Si(OH)_6.12H_2O]SO_4.CO_3$[6c] and in a number of high-pressure oxide phases, including one form of SiO_2 itself (stishovite), synthetic $KAlSi_3O_8$, and a garnet.[7] There is also octahedral coordination of Si in the high-pressure forms of $SiAs_2$ and SiP_2. In the usual form of SiP_2 the coordination numbers of Si and P are 4 and 3, but in the high-pressure form (pyrites structure) they are 6 (octahedral) and 4 (3 Si + P, tetrahedral).[8a] The persistence of octahedral coordination in the *high-temperature* (cubic) form of SiP_2O_7 is noteworthy, as also is the fact that this structure is based on the same (6,4)-connected 3D net as the *high-pressure* form of SiP_2, from which it is derived (topologically) by inserting an O atom along each link (Si–P or P–P) of the SiP_2 (pyrites) structure. There is also 6-coordination of Si (and 4-coordination of P) in $(NH_4)_2SiP_4O_{13}$,[8b] in which the 2D anion is formed from P_4O_{13} sub-units linked through Si atoms which are in octahedral coordination.

There are several compounds in which Si is 5-coordinated. The most straight-forward case would be an ion SiX_5^- such as occurs in salts $(NR_4)(SiF_5)$,[9] in which R must be larger than CH_3, but their structures are not yet known. Structural studies have been made of various chelate species, for example, the anion (c) in its tetramethylammonium salt,[10] and more symmetrical molecules of which (d) is an example.[11] Note the very long Si—N bond compared with the bond length 1.74 Å for tetrahedral coordination. There is presumably trigonal bipyramidal coordination also in the cyclic molecule $[(CH_3)_2N.SiH_3]_5$,[12] (e), with 3 H equatorial and 2 N

(c)   (d)   (e)

axial, but the H atoms were not located. The Si—N bond has the same length as in the 6-coordinated molecule (b).

The action of metallic K on SiH_4 or Si_2H_6 in aprotic solvents such as 1:2-dimethoxyethane produces $KSiH_3$, which forms colourless crystals with the NaCl structure.[13] The H atoms were not located, but the SiH_3^- ion would contain 3-covalent Si. For the unstable SiF_2 and $SiCl_2$ species see p. 982.

Bond lengths in some simple molecules formed by Si are included in Table 21.1, p. 914.

(1)   AC 1974 **B30** 549
(1a) AC 1972 **B28** 321
(2)   JACS 1959 **81** 6372
(3)   JACS 1969 **91** 5756
(4)   AC 1969 **B25** 156
(5)   JCS A 1969 363
(6a) IC 1962 **1** 236; IC 1966 **5** 1979
(6b) AC 1979 **B35** 724
(6c) AC 1971 **B27** 594

(7)   AC 1970 **B26** 233
(8a) ACSc 1967 **21** 1374
(8b) AC 1976 **B32** 2957
(9)   IC 1969 **8** 450
(10) JACS 1968 **90** 6973
(11) JOC 1975 **101** 11; 1976 **121** 333;
      JCS D 1974 2051
(12) JACS 1967 **89** 5157
(13) JACS 1961 **83** 802

Elementary silicon and carborundum

We have mentioned in a previous chapter the two crystalline forms of carbon: diamond and graphite. The latter structure is unique among the elements, but a number of other elements of Group IV crystallize with the diamond structure; silicon, germanium, and tin (grey modification). In the B subgroup only lead has a typical metallic structure (cubic close-packed), whereas all the elements of the A subgroup crystallize with close-packed structures.

Under moderate pressure both Si and Ge transform to more dense (about 10 per cent) modifications with different 4-connected nets, both different from the diamond structure.[1] There is little change in bond lengths, but appreciable angular distortion—in Si to three angles each of 99° and 108° and in Ge to angles ranging from 88° to 135°. The Ge structure is particularly interesting since the same network is the basis of the structure of one of the high-pressure forms of SiO_2 (keatite), and also of ice III. These 4-connected nets have been described in Chapter 3.

High-pressure forms of both Si and Ge with the white Sn structure have been studied:

	White Sn structure	*Diamond structure*
Si	4 Si at 2.430; 2 Si at 2.585	4 Si at 2.35 Å
Ge	4 Ge at 2.533; 2 Ge at 2.692	4 Ge at 2.45 Å.

Elementary Si has also been obtained with the hexagonal diamond structure by heating the b.c.c. high-pressure form described above.

Carborundum, SiC crystallizes in a large number of forms. The cubic form (β-SiC) has the zinc-blende structure (p. 121), and in addition there are numerous crystalline modifications all with structures closely related to the zinc-blende and wurtzite structures, which are collectively referred to as α-SiC. The atomic positions in zinc-blende are the same as in diamond, alternate atoms being Zn and S. The structures of zinc-blende and wurtzite are not layer structures in the sense in which we use this term, but we may dissect these structures into layers of types (a) and (b) as shown in Fig. 23.1 (where alternate circles represent Si and C or Zn and S

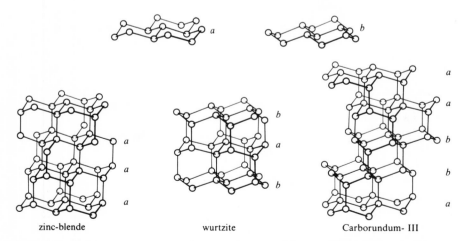

FIG. 23.1. The geometrical relationship between the structures of zinc-blende, wurtzite, and carborundum III.

atoms as the case may be). It will be seen that the zinc-blende structure is built up of layers of type (a) only, while in wurtzite (a) and (b) layers alternate. More complex sequences are found in the various forms of α-SiC, for example:

	Layer sequence
Zinc-blende	*a*
Wurtzite	*a b*
Carborundum III	*a a b b*
Carborundum I	*a a a b b*
Carborundum II	*a a a b b b*

The first three structures are illustrated in Fig. 23.1.

The carborundum structures set out above are the most abundant forms in the ordinary material and are relatively simple examples of a very large group of structures with increasingly complex layer sequences which in some cases repeat only after several hundred layers. The layer sequence has been determined in more than seventy of these structures, which are described as *polytypes* to distinguish them from normal polymorphs. Carborundum is apparently formed by a sublimation process in the electric furnace, and the explanation of these regular sequences must lie in the mode of growth of the crystals. It is possible that more than one mechanism may be operative, but growth by screw dislocations probably provides the best explanation of the long range order.[2]

(1)   AC 1964 **17** 752   (2)   AC 1969 **B25** 477; AC 1973 **B29** 1548

Silicides

Silicides form a large group of compounds the formulae of many of which are not explicable in terms of the normal valences of the elements. For example, the compounds of the alkali metals include $Li_{15}Si_4$ and Li_2Si, KSi and KSi_6, CsSi and $CsSi_8$. Bonding ranges from essentially metallic or ionic to covalent, and in many silicides there are bonds of more than one kind. We shall therefore adopt here only a very general geometrical classification, since a knowledge of the structure must necessarily precede any discussion of the nature of the bonding in a particular compound.

At one extreme lie silicides containing isolated Si atoms as, for example, Ca_2Si[1] with the anti-$PbCl_2$ structure, Mg_2Si, which has the antifluorite structure (like Mg_2Ge, Mg_2Sn, and Mg_2Pb), and a variety of compounds with typically metallic structures—V_3Si, Cr_3Si, and Mo_3Si (β-W structure), Cu_5Si (β-Mn structure), Mn_3Si (random b.c.c.), and Fe_3Si (Fe_3Al superstructure). The interpretation of interatomic distances in these structures is not simple (compare intermetallic compounds, Chapter 29); for example, in $TiSi_2$ (see later) Si has 2 Si and 2 Ti at 2.54 Å and 3 Ti and 3 Si at approximately 2.75 Å. In other silicides there are Si—Si distances close to that in elementary silicon (2.35 Å). A uniquely short Si—Si distance, 2.18 Å, has been found in IrSi$_3$.[2a] The metal atoms lie, in positions of 9-coordination, between plane 4-connected nets of Si atoms of the kind shown in Fig. 29.13(b). If we recognize distances up to 2.5 Å or so as indicating appreciable interaction between Si atoms we may distinguish systems of linked Si atoms of the following kinds:

Type of Si *complex*	*Examples*
Si$_2$ groups	U$_3$Si$_2$ (Si–Si, 2.30 Å)
Si$_4$ groups	MSi (M = Na, K, Rb, Cs)[2b]
Chains	USi (Si–Si, 2.36 Å)
	CaSi (Si–Si, 2.47 Å)
Plane hexagonal net	β-USi$_2$
Puckered hexagonal net	CaSi$_2$
3D framework	SrSi$_2$, α-USi$_2$ (ThSi$_2$)

It is interesting that the silicides of U provide examples of all the main types of silicide structure,[3] for in addition to those listed there is U$_3$Si which has a distorted version of the Cu$_3$Au structure (p. 1297), in which there are discrete Si atoms entirely surrounded by U atoms. They also show that some silicides adopt the same structures as certain borides which contain extended systems of linked B atoms, though because of the greater size of Si and its different electronic structure borides and silicides are not generally isostructural.

The FeB structure of USi and other 4f and 5f monosilicides[4] (and also TiSi and ZrSi) with infinite silicon chains, and the AlB$_2$ structure of β-USi$_2$ (and other 4f and 5f disilicides)[5] with plane hexagonal Si layers, are described and illustrated in the next chapter. Here, therefore, we shall first note the structures of some silicides with alloy-like structures containing discrete Si atoms and then describe the structure of CaSi$_2$, which contains puckered Si layers, and finally the structures of SrSi$_2$ and α-ThSi$_2$ as examples of 3D framework structures.

TiSi$_2$, CrSi$_2$, *and* MoSi$_2$. The structures of TiSi$_2$ (orthorhombic), the isostructural CrSi$_2$, VSi$_2$, NbSi$_2$, and TaSi$_2$ (hexagonal), and the isostructural MoSi$_2$, WSi$_2$, and ReSi$_2$ (tetragonal) are related in a very simple way, which is unexpected in view of their quite different symmetries. TiS$_2$ is built of close-packed layers of the type shown in Fig. 23.2(a), with the Si:Ti ratio 2:1. The layers are not, however, superposed in such a way as to make the structure as a whole close-packed (that is, to give each atom twelve equidistant nearest neighbours) but, instead, the Ti atoms of successive layers fall above the points *B, C,* and *D* respectively (Fig. 23.2(a)). Each atom has therefore ten nearest neighbours, Ti being approximately equidistant from 10 Si and each Si approximately equidistant from 5 Ti and 5 Si. Crystals of CrSi$_2$ contain the same type of close-packed layers, and, moreover, the layers are superposed in a similar way, giving each atom ten nearest neighbours (Cr to 10 Si, Si to 5 Cr + 5 Si). In this (hexagonal) structure the repeat unit comprises only three

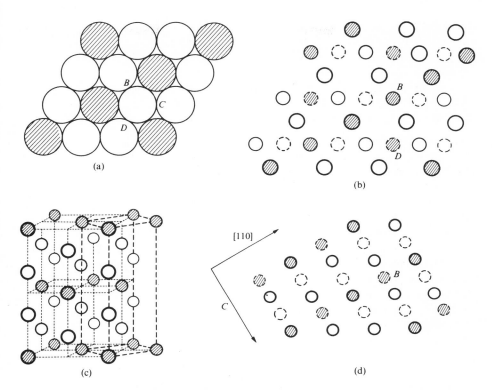

FIG. 23.2. (a) Close-packed MSi_2 layer in some disilicides. (b) Three successive layers in $CrSi_2$. (c) The crystal structure of $MoSi_2$. (d) Projection of the structure of $MoSi_2$ on (110) showing the relative positions of atoms in two adjacent layers (full and dotted circles). Metal atoms are represented by shaded circles.

layers, the Cr atoms of successive layers falling above the points B and D only (Fig. 23.2(b)). From the geometrical standpoint the structure of $MoSi_2$ is related to that of CaC_2, as may be seen by comparing Fig. 23.2(c) with Fig. 22.6, p. 948, though the Si atoms are not associated in pairs as are the C atoms in CaC_2. (Both structures are tetragonal, CaC_2 having $a = 3.89$, $c = 6.38$ Å, and the parameter of the C atom $(00z)$ $z_C = 0.406$; $MoSi_2$ has $a = 3.20$, $c = 7.86$ Å and $z_{Si} \approx 0.33$. The conventional body-centred unit cell is indicated by heavy broken lines in Fig. 23.2(c).) However, the projection of the structure on (110), Fig. 23.2(d), shows that $MoSi_2$ is very closely related to $TiSi_2$ and $CrSi_2$, In the (110) plane the atoms are arranged at the points of a hexagonal net of exactly the same type as in the latter structures, and here also successive layers are superposed in such a way that each type of atom has the same arrangement of ten nearest neighbours. In $MoSi_2$ atoms of *alternate* layers fall vertically above one another, adjacent layers being displaced with respect to one another, so that a Mo atom of an adjacent layer falls over the point *B* in Fig. 23.2(d). These three structures thus form an interesting series, all being built from the same type of close-packed layers superposed in essentially the same way, to give 10- instead of 12-coordination, but differing in the number of layers in the repeat unit, which is four layers in $TiSi_2$, three in $CrSi_2$, and two in $MoSi_2$. It is found that crystals with compositions intermediate between $TiSi_2$ and $MoSi_2$ adopt the intermediate (3-layer) $CrSi_2$ structure, and the same is true in the $TiSi_2$–$ReSi_2$ series.[6]

$CaSi_2$, $SrSi_2$, *and* $BaSi_2$. The structures of these compounds at normal pressure (NP structures) and under a pressure of 40 kbar at 1000 °C (HP structures) are set out below. (A second HP form of $BaSi_2$ has been made at 600–800 °C and 40 kbar.) The HP structure is retained for several months at NTP in the absence of air and moisture, but reverts to the NP structure on heating to 400 °C. These compounds illustrate all the main types of structure possible for an element forming three bonds, and emphasize the analogy between Si^- and the isoelectronic P atom. In particular, the systems of connected Si atoms in the three forms of $BaSi_2$ are the simplest 3-connected polyhedron, the simplest 3-connected plane net, and one of the two simplest 3-connected 3D nets (the cubic net of Fig. 3.28(a), p. 111). The same 3D net is found in $SrSi_2$, while the second of the two simplest 3D 3-connected nets (the tetragonal (10,3) net) is the arrangement of Si atoms in the α-$ThSi_2$ structure (see below) adopted by the HP forms of $CaSi_2$ and $SrSi_2$. The hexagonal layer is illustrated in Fig. 23.3, which actually represents the structure of $CaSi_2$

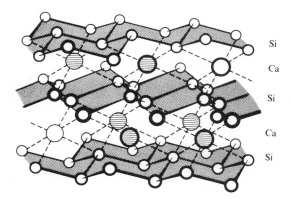

FIG. 23.3. The crystal structure of $CaSi_2$.

containing a little Sr. The layer is of the same type in pure $CaSi_2$ and somewhat less buckled in the form of $BaSi_2$ with a layer structure. Silicon bond angles range from 60° in the Si_4 group of $BaSi_2$ through 105° and 112° in the layer structures of $CaSi_2$ and $BaSi_2$ to values close to 120° in the 3D nets. There is little variation in Si–Si bond lengths (2.40–2.45 Å). Possible mechanisms for the polymorphic transformations have been discussed.[12]

Type of Si_n complex in disilicides

	NP structure	HP structure
$CaSi_2$	2D hexagon net[7]	
$SrSi_2$	Cubic 3D (10,3) net[9]	} Tetragonal 3D (10,3) net[8]
$BaSi_2$	Tetrahedral Si_4 group[7]	2D hexagon net[10]
		3D cubic net[11]

α-$ThSi_2$. Many 4f and 5f disilicides adopt the α-$ThSi_2$ structure (and/or a less symmetrical variant of this structure) and/or the AlB_2 structure.[5] In α-$ThSi_2$ the Si atoms form a very open, 3-coordinated, 3D network in which the Si–Si bond length is 2.39 Å, close to the value in elementary Si. This 3D network may be compared with the 3-coordinated boron layer in the AlB_2 structure (p. 1055); it is emphasized in Fig. 23.4. In the interstices are the Th atoms, each bonded to 12 Si

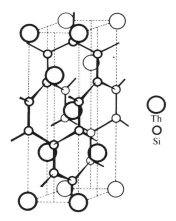

Th
Si

FIG. 23.4. The crystal structure of $ThSi_2$.

neighbours (at 3.16 Å). The next nearest neighbours of a Si atom are 6 Th at this distance. The close relation between the $ThSi_2$ and AlB_2 structures is emphasized by the facts that the disilicides of Th, U, and Pu crystallize with both structures, and the disilicides of the earlier 4f metals (La–Ho) crystallize with the $ThSi_2$ structure (in some cases distorted) while those of Er, Tm, Yb, and Lu adopt the AlB_2 structure.

(1) ZaC 1955 **280** 321
(2a) IC 1971 **10** 1934
(2b) ZaC 1964 **327** 260
(3) 1949 **2** 94
(4) AC 1966 **20** 572
(5) IC 1967 **6** 842
(6) AC 1964 **17** 450

(7) ZaC 1970 **372** 87
(8) JSSC 1977 **20** 173
(9) AC 1972 **B28** 2326
(10) AnC 1977 **89** 673
(11) AnC 1978 **90** 562
(12) JSSC 1978 **24** 199

Silanes

Hydrides from SiH_4 to Si_7H_{16} have been prepared by the action of 20 per cent HCl on magnesium silicide in an atmosphere of hydrogen, or preferably by the action of NH_4Br on the same compound in liquid NH_3. The mixture of gases evolved is cooled in liquid nitrogen and fractionated as in the case of the boron hydrides. All the above hydrides are obtained in this way in regularly decreasing amounts. The two isomers of Si_4H_{10} have been separated by gas chromatography and identified by their i.r. and n.m.r. spectra.[1] Mixed Si-Ge hydrides (e.g. Si_3GeH_{10} and Si_5GeH_{14}) have been prepared by hydrolysis of Mg-Ge-Si alloys and by pyrolysis of mixtures of silanes and germanes.[2] The thermal stability of the silanes decreases with increasing molecular weight and the higher members break down on heating into mixtures of the lower hydrides—compare the 'cracking' of hydrocarbons, which is a similar process. The silicon hydrides are not attacked by water but are hydrolysed by dilute alkali solutions, hydrogen being evolved and a silicate formed. Like the hydrocarbons they may be halogenated, and the hydrolysis of the products is described later.

A convenient general method of preparing Group IV hydrides, simple and substituted, is by the action of lithium aluminium hydride on the appropriate halide in ether solution:

$$4\,MR_yX_{4-y} + (4-y)LiAlH_4 \rightarrow 4\,MR_yH_{4-y} + (4-y)LiX + (4-y)AlX_3,$$

where R represents alkyl or aryl groups and X a halogen. In this way, $SiCl_4 \rightarrow SiH_4$, $Si_2Cl_6 \rightarrow Si_2H_6$, $Et_2SiCl_2 \rightarrow Et_2SiH_2$, etc. (The same method has been used to prepare GeH_4, SnH_4, and the three methyl stannanes.)

The silanes are very easily oxidized, Si_2H_6, for example, being spontaneously inflammable in air. In marked contrast to carbon, silicon appears to form no simple unsaturated hydrides corresponding to the olefines, acetylenes, and aromatic hydrocarbons. Solid unsaturated hydrides such as $(SiH_2)_n$, of unknown constitution, have been described, though in a study of the higher silanes no solid products were found at any stage. Structurally the silanes are presumably similar to the saturated hydrocarbons. Many substituted monosilanes have been prepared; for examples see Table 21.1 (p. 914), and also halogenated and alkylated disilanes such as $F_3Si.SiF_3$, Si_2Cl_6, and $Si_2(CH_3)_6$.

Various linear and cyclic fully methylated polysilanes have been made by catalytic rearrangement of $Si_6(CH_3)_{14}$, and the cyclic compounds $[Si(CH_3)_2]_n$ ($n = 5$–8) have been prepared by the action of Li on $Si(CH_3)_2Cl_2$ in tetrahydrofuran. The Si_4 rings are slightly folded across a diagonal in $Si_4(CH_3)_4(Bu_3C)_4$[3] and $Si_4(C_6H_5)_8$,[4] the dihedral angles being 37° and 13° respectively. The structure of the Si_5 ring has been determined in Si_5H_{10} (e.d.)[5] and in $Si_5(C_6H_5)_{10}$.[6] Details of the (non-planar) ring of Si_5H_{10} are: Si–Si, 2.342; Si–H, 1.496 Å; angle Si–Si–Si, 104.2°. In $Si_6(CH_3)_{12}$[7] the ring is chair-shaped (Si–Si, 2.34 Å). Bicyclic and cage polysilanes have been prepared by the action of Na on mixtures of $(CH_3)_2SiCl_2$ and

CH_3SiCl_3 in tetrahydrofuran. They include $Si_8(CH_3)_{14}$, $Si_9(CH_3)_{16}$, $Si_{10}(CH_3)_{16}$, $Si_{10}(CH_3)_{18}$, and $Si_{13}(CH_3)_{22}$. Photolysis of the first compound yields $Si_7(CH_3)_{12}$. Structures suggested for the Si_7 and Si_8 compounds are shown at (a) and (b). The

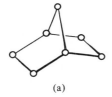

(a)   (b)   (c)

crystal structure of $Si_9(CH_3)_{16}$[8] has been determined and establishes the bicyclic structure (c) (Si–Si, 2.34 Å).

(1)   IC 1964 **3** 946
(2)   JCS 1964 1467; JCS A 1969 2937
(3)   Organomet. Chem. 1975 **91** 273
(4)   AC 1978 **B34** 883
(5)   ACSc 1976 **A30** 697
(6)   AC 1978 **B34** 3678
(7)   AC 1972 **B28** 1566
(8)   IC 1976 **15** 524

Silicon halides

The lower halides SiF_2 and $SiCl_2$ have already been mentioned. In SiF_2, Si–F = 1.591 Å and the angle F–Si–F = 101°;[1] this compound has also been studied in solid Ne and Ar matrices.[2] Condensation of gaseous SiF_2 at low temperatures yields a plastic polymer $(SiF_2)_n$ which burns spontaneously in moist air and on heating generates all the perfluorosilanes from SiF_4 to $Si_{14}F_{30}$. With BF_3 in a trap at the temperature of liquid nitrogen SiF_2 forms compounds such as Si_2BF_7, Si_3BF_9, etc. which have been identified mass-spectrometrically. These are presumably linear molecules $F_3Si–(SiF_2)_n–BF_2$.[3]

Electron diffraction and spectroscopic studies show that in the gaseous state the halides SiX_4 form tetrahedral molecules; the simple molecular structure of SiF_4 in the crystalline state has been confirmed by an X-ray study at –145 °C.[4]

Higher chlorosilanes have been prepared, some conveniently by methods such as the disproportionation of Si_2Cl_6 in the presence of trimethylamine, when $5\,Si_2Cl_6 \rightarrow 4\,SiCl_4 + Si_6Cl_{14}$.[5] The crystalline phase with the composition Si_6Cl_{16} has been shown to be a molecular compound, $Si_5Cl_{12} \cdot SiCl_4$[6] containing the tetrahedral molecules (d) (Si–Si, 2.34 Å; Si–Cl, 2.01 Å; and in $SiCl_4$, 2.04 Å).

$$
\begin{array}{c}
SiCl_3 \\
| \\
Si \\
Cl_3Si \diagup \;|\; \diagdown SiCl_3 \\
Si \\
Cl_3
\end{array}
$$

(d)

Controlled hydrolysis of these compounds is discussed later (p. 995); complete hydrolysis of these halides and of other chloro-silicon compounds gives oxy-compounds which are in many cases formally analogous to carbon compounds. The silicon compounds bear little resemblance, however, as regards physical or chemical properties to their carbon analogues, the final products of hydrolysis being highly polymerized solids. All chlorosilanes are readily hydrolysed, for example SiH_2Cl_2 gives $(H_2SiO)_n$, 'prosiloxane', which appears in many polymeric forms depending on the mode of hydrolysis. The final product of the hydrolysis of Si_2Cl_6 is a white powdery solid of empirical composition $(SiO_2H)_n$, called silico-oxalic acid. In spite of its name and formula this compound bears no chemical resemblance to oxalic acid, for it is insoluble in water and possesses no acidic properties. No salts or esters are known, and it dissolves in alkalis with evolution of hydrogen. It is obviously a complex polymer. The 'silico-mesoxalic acid', $(Si_3O_6H_4)_n$, formed by the action of moist air on Si_3Cl_8, has rather similar properties. Another polymerized product, also of unknown constitution, $(Si_2H_2O)_n$, is produced by the action of alcoholic hydrochloric acid on $CaSi_2$ (p. 990). It is spontaneously inflammable in air; it is a reducing agent, and is rapidly decomposed by hot water. It forms flakes of the same shape as the calcium silicide from which it was formed, which is reminiscent of the graphite derivatives described in Chapter 21.

(1) JCP 1967 **47** 5031
(2) JACS 1969 **91** 2536
(3) JACS 1965 **87** 2824, 3819

(4) AC 1954 **7** 597
(5) JINC 1964 **26** 421, 427, 435
(6) AC 1972 **B28** 1233

Oxyhalides and thiohalides

The oxyhalide $SiOCl_2$, the analogue of phosgene, is not known; it would imply the formation by silicon of a double bond to oxygen. The simplest oxychloride known is the liquid Si_2OCl_6, formed by passing a mixture of air and $SiCl_4$ vapour through a red-hot tube. An electron diffraction study of this compound gives the following data: Si–O. 1.64 (0.04); Si–Cl, 2.02 (0.03) Å; Cl–Si–Cl, 109 (2)°. The Si–O–Si angle was not determined but is probably close to 130°.[1]

The linear oxychlorides $Si_nO_{n-1}Cl_{2n+2}$ have been prepared by passing a mixture

$$Si_2OCl_6 \qquad Si_nO_{n-1}X_{2n+2} \qquad (SiOX_2)_n$$

of $O_2 + 2Cl_2$ over heated silicon and also by controlled hydrolysis of $SiCl_4$ in ether solution at $-78\,°C$ followed by fractional distillation. All up to and including $Si_7O_6Cl_{16}$ have been prepared and characterized.[2] They are colourless oily liquids, which are incombustible and miscible with solvents such as chloroform, carbon tetrachloride, and carbon disulphide. They are all hydrolysable, the lower ones most rapidly, and with alcohol they form esters, for example, $Si_6O_5(OC_2H_5)_{14}$. The cyclic compounds $(SiOCl_2)_n$, $n = 3$, 4 or 5, have been prepared.[3] The corresponding linear oxybromides are known up to $Si_6O_5Br_{14}$, and the cyclic $(SiOBr_2)_4$ has also been prepared.[4] Mixed fluoro-chloro compounds which have been prepared include $Si_2OF_3Cl_3$ and $Si_2OF_4Cl_2$.[5]

Whereas controlled hydrolysis of $SiCl_4$ at $-78\,°C$ gives the linear oxychlorides $Si_nO_{n-1}Cl_{2n+2}$, a similar treatment of Si_2Cl_6 gives another series $Si_{2n+2}O_nCl_{4n+6}$ containing pairs of linked Si atoms bonded into chains by O atoms:

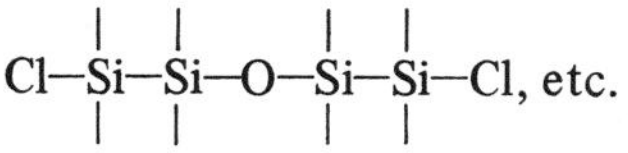

$$Cl-\overset{|}{\underset{|}{Si}}-\overset{|}{\underset{|}{Si}}-O-\overset{|}{\underset{|}{Si}}-\overset{|}{\underset{|}{Si}}-Cl, \text{ etc.}$$

In this way Si_4OCl_{10}, $Si_6O_2Cl_{14}$, and $Si_8O_3Cl_{18}$ have been obtained as colourless liquids.[6]

Thiochlorides that have been prepared include Si_2SCl_6 (I), the cyclic compound (II) and the crystalline $(SiSCl_2)_4$.

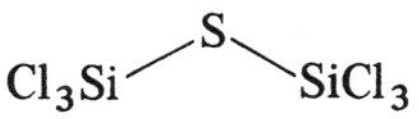

I

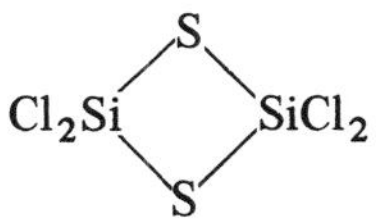

II

(1) JCP 1950 **18** 1414
(2) JACS 1941 **63** 2753
(3) JCS 1960 5088
(4) JACS 1937 **59** 261
(5) JACS 1945 **67** 1092
(6) JACS 1954 **76** 2091

Cyclic silthianes

Electron diffraction studies[1] have been made of tetramethyl cyclodisilthiane (a) and hexamethylcyclotrisilthiane (b):

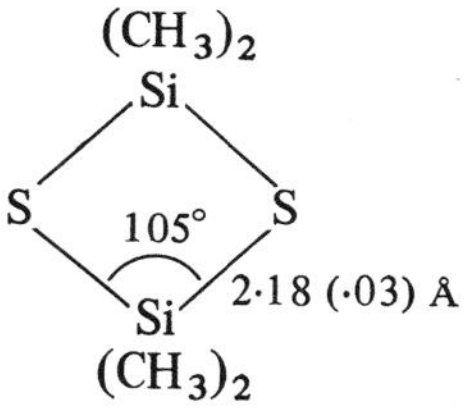

(a)

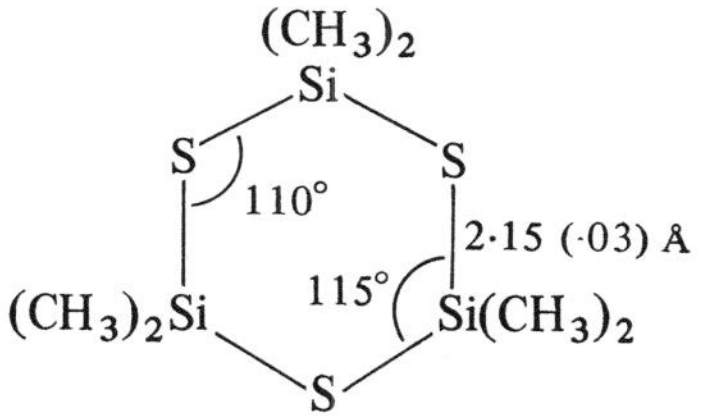

(b)

It was not possible to determine the configuration of the Si_3S_3 ring in the e.d. study, but in an X-ray study of $[Si(CH_3)(C_6H_5)]_3S_3$[2] the ring was found to have a 'twisted boat' configuration (see p. 97).

(1)   JACS 1955 **77** 4484   (2)   AC 1977 **B33** 1780

Some silicon–nitrogen compounds

We mention here a few compounds which present points of particular interest. The planarity of $N(SiH_3)_3$ has been noted; coplanar N bonds are also formed in the molecule (a),[1] in which all 4 N and 4 Si are coplanar, and in the cyclodisilazane (b).[2] More complex Si–N ring systems include the cyclotetrasilazane (c),[3] which is of interest because both chair and cradle forms of the molecule exist in the same crystal, and the fully methylated compound $[Si(CH_3)_2.N(CH_3)]_4$[4] which is

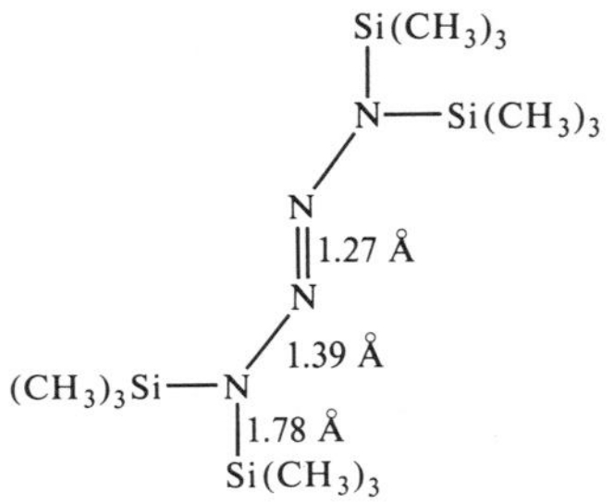

(a)

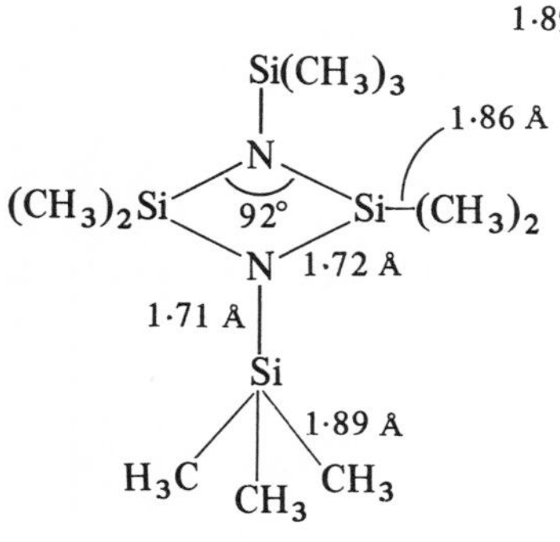

(b) 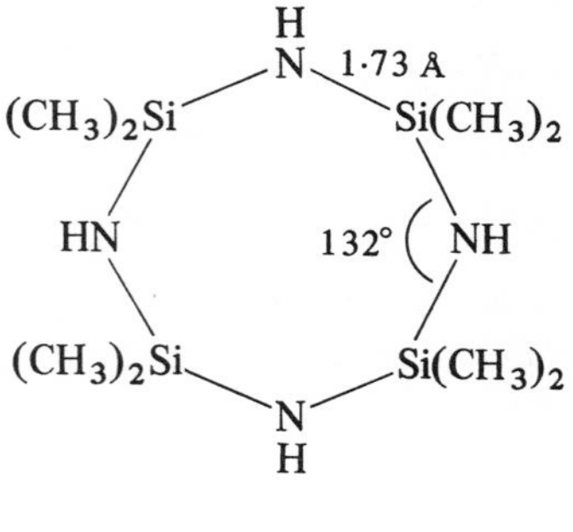   (c)

unusual from the crystallographic standpoint. There are 7 molecules in the rhombohedral unit cell, 6 ordered around $\bar{3}$ axes and 1 disordered.

The refractory Si_2N_2O,[5] made by heating to 1450 °C a mixture of $Si + SiO_2$ in a stream of argon containing 5 per cent N_2, is related structurally to both SiO_2 and

Si_3N_4.[6, 7] The latter has two crystalline forms, both with phenacite-like structures (p. 89) in which Si has 4 tetrahedral neighbours and N is bonded to 3 Si (Si–N, 1.74 Å). The oxynitride is built of SiN_3O tetrahedra which are linked to form a 3-dimensional framework. Puckered hexagonal nets consisting of equal numbers of Si and N atoms are linked together through O atoms which complete the tetra-hedra around the Si atoms. One half of the Si atoms are linked to the layer above and the remainder to the layer below. In the resulting structure every O atom is connected to 2 Si (as in SiO_2), and N and Si are respectively 3- and 4-connected, as in Si_3N_4 (Fig. 3.16(a)); see Figure 23.5. Ge_2N_2O[8] is isostructural with Si_2N_2O.

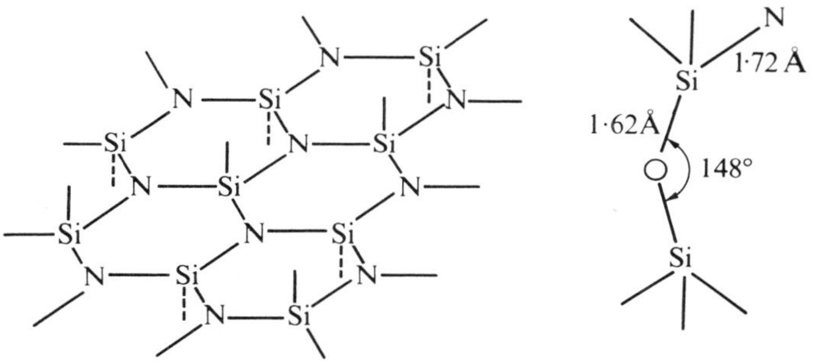

FIG. 23.5. The structure of Si_2N_2O (diagrammatic).

(1) AC 1975 **B31** 678
(2) JCS 1962 1721
(3) AC 1963 **16** 1015
(4) AC 1976 **B32** 648
(5) ACSc 1964 **18** 1879
(6) JACeS 1975 **58** 90
(7) AC 1980 **B36** 2891
(8) AC 1979 **B35** 141

Organo-silicon compounds and silicon polymers

Silicon forms a great variety of compounds containing carbon. In spite of the formal resemblance between these and purely organic compounds, there is little similarity in chemical properties except in the case of compounds such as the aliphatic quaternary silanes SiR_4. Here the resemblance is presumably due to the fact that the silicon atom is completely shielded by the organic groups. A few organo-silicon compounds are extremely stable. For example, $Si(C_6H_5)_4$ may be distilled unchanged in air at 425 °C, and although it is unwise to generalize about the stability of silicon compounds we may note two differences between the silicon and carbon compounds:

(1) Silicon has a much great affinity than carbon for oxygen. For example, $SiCl_4$ is violently hydrolysed by water, whereas CCl_4 is stable towards water and is only slowly hydrolysed by aqueous alkali. The silicon hydrides are hydrolysed to silicic acid, and disilane is spontaneously inflammable in air—contrast the stability of the corresponding hydrocarbons.

(2) No silicon compounds containing multiple bonds to silicon are known. The

hydrolysis of halogen compounds of silicon (discussed later) does not yield the silicon analogues of ketones and carboxylic acids by elimination of water from each molecule. Instead, water is eliminated from two or more molecules leading to polymerized products.

The following are some typical methods of synthèsizing organo-silicon compounds:

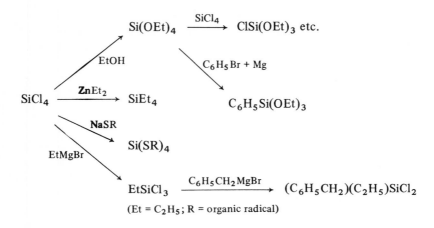

(Et = C_2H_5; R = organic radical)

By successive reactions with different Grignard reagents (RMgBr) compounds such as $(C_6H_5CH_2)(C_2H_5)(C_3H_7)Si(CH_2C_6H_4SO_3H)$ have been prepared. Here we shall be chiefly interested in the hydrolysis of substituted chlorosilanes.

Substituted chlorosilanes, silanols, and siloxanes

Chlorosilanes were formerly obtained by the action of Grignard compounds on $SiCl_4$ but it is also possible to prepare them directly from the alkyl or aryl halide and silicon. The vapour of the halide is passed over sintered pellets of silicon containing some 15 per cent of silver or copper as catalyst, the products being condensed and then fractionated. Most important technically are the methyl chlorosilanes, CH_3SiCl_3, $(CH_3)_2SiCl_2$, and $(CH_3)_3SiCl$. The following vinyl, allyl, and phenyl compounds are among those which have been prepared in this way:

$$C_2H_3SiCl_3 \quad (C_3H_5)HSiCl_2 \quad C_6H_5SiCl_3$$
$$(C_2H_3)_2SiCl_2 \quad C_3H_5SiCl_3 \quad (C_6H_5)_2SiCl_2$$
$$(C_3H_5)_2SiCl_2 \quad (C_6H_5)_3SiCl.$$

Intensive studies of the products of hydrolysis of substituted chloro-silanes and esters have led to an important new branch of chemistry. Many organo-silicon compounds were prepared by Kipping and his school in the early years of this century and it was observed that glue-like products often resulted from the hydrolysis of these compounds, as well as intractable solids of the kinds already mentioned. No study was made of these polymeric products which are now the basis of *silicone chemistry*.

By analogy with carbon we might expect mono-, di-, and tri-chloro silanes to hydrolyse to the silicon analogues of alcohols, $R_3Si.OH$, ketones $R_2Si{=}O$ (by loss of water from $R_2Si(OH)_2$), and carboxylic acids, $RSi{\overset{\textstyle O}{\underset{\textstyle OH}{\diagdown}}}$ (by loss of water from the trihydroxy compound). In fact compounds containing $Si{=}O$ are never formed, though the generic name *silicone* (by analogy with ketone) is applied to the various hydrolysis products of dichlorosilanes.

Silanols $R_3Si(OH)$ were obtained in 1911. These compounds tend to condense to the disiloxane

$$2\,R_3Si(OH) \rightarrow R_3Si{-}O{-}SiR_3$$

at rates depending on the nature of the group R. Dicyclohexylphenyl silanol was isolated by Kipping, and $(C_6H_5)_3Si.OH$ may be distilled under reduced pressure without forming the disiloxane, but the trialkyl silanols are not isolable. The term *disiloxane* is applied to all compounds $R_3Si{-}O{-}SiR_3$ and includes $H_3Si{-}O{-}SiH_3$, $(C_2H_5)_3Si{-}O{-}Si(C_2H_5)_3$, and oxyhalides, $X_3Si{-}O{-}SiX_3$. The first of these is made by the hydrolysis of SiH_3Cl by water at $30\,^\circ C$.

Silanediols $R_2Si(OH)_2$. In the early studies the only diols isolated were those in which R is an aromatic radical, but even these compounds condense to linear and cyclic polysiloxanes. All the compounds I, II, and III were isolated during the years

$$[\phi = C_6H_5]$$

1912–1914, the last two being the first cyclic silicon polymers to be characterized. The first dialkylsilanediol was prepared in 1946 by the hydrolysis, by weak alkali, of $(C_2H_5)_2SiCl_2$: it is a stable crystalline solid. Later, $(CH_3)_2Si(OH)_2$ was prepared by the hydrolysis of $(CH_3)_2Si(OCH_3)_2$ as a crystalline solid extremely sensitive to acid or alkali, which cause condensation to polysiloxanes, mostly the cyclic tri- and tetra-compounds. Other diols prepared include the n- and iso-propyl and butyl, and tert-butyl compounds, in accition to tetramethyldisiloxane-1,3-dio (IV). No silane-triol $RSi(OH)_3$ has been isolated.

$$
\begin{array}{ccc}
HO & & OH \\
H_3C\text{-}Si\text{-}O\text{-}Si\text{-}CH_3 \\
H_3C & & CH_3
\end{array}
$$

IV

The structures of silanols and siloxanes

Although the stereochemistry of silicon in these compounds presents no points of special interest, the bond arrangement being the normal tetrahedral one, the siloxanes include some interesting ring systems.

Silanols. Diethylsilanediol, $Si(C_2H_5)_2(OH)_2$, is a simple tetrahedral molecule:

O—Si—O, 110°; C—Si—C, 111°; Si—OH, 1.63 Å; and Si—C, 1.90 Å.[1]

For the crystal structure of $OHSi(CH_3)_2 . C_6H_4 . Si(CH_3)_2OH$[2] see p. 111.

Disiloxanes. The structures of a number of disiloxanes $R_3Si\text{-}O\text{-}SiR_3$ (R = H,[3,4] F, Cl, CH_3,[4] C_6H_5[5]) have been studied either in the vapour by electron diffraction and/or in the crystalline state. The O bond angle lies in the range 142–156° (and Si—O, 1.63 Å) except in crystalline $\phi_3Si\text{-}O\text{-}Si\phi_3$, where the crystal symmetry

indicates an apparent Si—O—Si angle of 180°. In the sulphur analogue of disiloxane, $H_3Si.S.SiH_3$, the S bond angle has the much smaller angle shown.

Cyclic polysiloxanes. Particular interest attaches to the configurations of these molecules, for the related carbon compounds paraldehyde and metaldehyde have puckered rings. The true silicon analogue of paraldehyde, tri-methylcyclotrisiloxane, has been prepared but its structure is not known.

A detailed electron diffraction study[7] of hexamethylcyclotrisiloxane established the planarity of the Si_3O_3 ring but indicated considerable thermal motion of the methyl groups. To maintain planarity in the Si_4O_4 ring the oxygen bond angle would have to increase to about 160°. An X-ray study of the low-temperature form of $(CH_3)_8Si_4O_4$[8a] showed that the Si_4O_4 ring is puckered, with bond angle Si—O—Si 142.5°. The conformation of this ring has been studied in other substituted derivatives, including $(CH_3)_4(C_6H_5)_4Si_4O_4$.[8b]

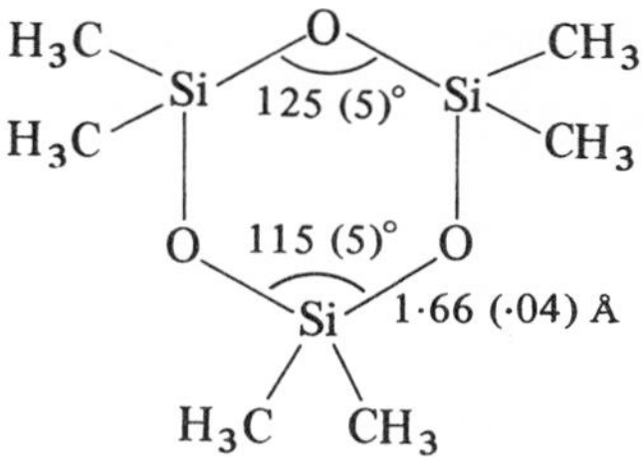

(Si–C = 1·88 (·04) Å)

'Spiro' compounds are also known, and the crystal structure of octamethylspiro (5.5) pentasiloxane has been determined.[9] The planar rings lie in two mutually perpendicular planes. There is evidence that the methyl groups 'librate' about the Si–O bonds as if the Si atoms to which they are attached were free to move in a 'ball-and-socket' joint. Data obtained include: Si–O, 1.64 ± 0.03 Å; Si–C, 1.88 ± 0.03 Å; Si bond angles tetreahedral; Si–O–Si angles 130°.

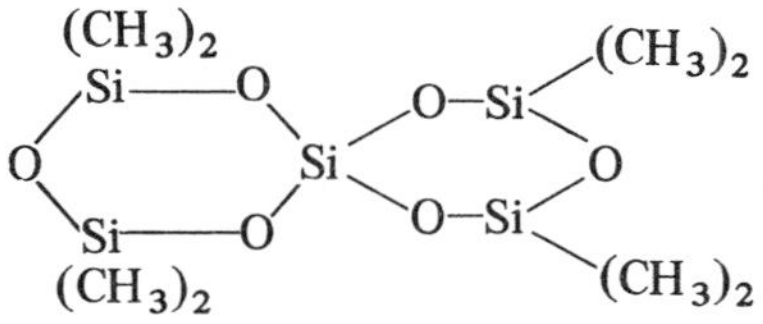

(1) JCP 1953 **21** 167
(2) JPC 1967 **71** 4298
(3) ACSc 1963 **17** 2455
(4) AC 1979 **B35** 2093
(5) AC 1978 **B34** 124

(6) JCP 1958 **29** 921
(7) JCP 1950 **18** 42
(8a) AC 1955 **8** 420
(8b) ACSc 1978 **B32** 171
(9) AC 1948 **1** 34

Silicone chemistry

The basis of silicone chemistry is the use of four types of unit. The monofunctional unit (*M*) results, for example, from the hydrolysis of $(CH_3)_3SiCl$ or $(CH_3)_3Si(OC_2H_5)$. These units alone can only give disiloxanes, and in a mixture (see later) an *M* unit ends a chain. (In these formulae O strictly represents half an oxygen atom, since on condensation O is shared between two Si atoms.)

The difunctional unit (*D*) is formed from $(CH_3)_2SiCl_2$ or $(CH_3)_2Si(OC_2H_5)_2$ and *D* units, alone, form cyclic polysiloxanes. The trifunctional unit (*T*) is formed

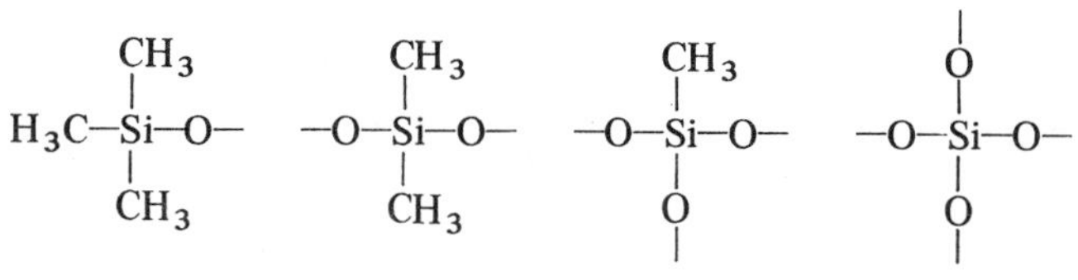

M   *D*   *T*   *Q*

from *mixtures* containing CH_3SiCl_3; complete hydrolysis of methyltrichlorosilane alone gives an infusible white powder (see later). The quaternary unit (Q) may be introduced by using $SiCl_4$. The polysiloxanes are the analogues of the silicon–oxygen ions found in silicates. The infinite chain polymer (a) is the analogue of the pyroxene chain ion (b) (see p. 1022) and the ring polymers are the analogues of the $Si_3O_9^{6-}$ and $Si_6O_{18}^{12-}$ ions found in the silicates benitoite and beryl (p. 1021).

(a)   (b)

compare

The next type of polymer would be the infinite 2-dimensional polymer containing Si atoms joined to three others via O atoms, the fourth bond being that to the organic group R (or halogen). The layer would have the composition $(Si_2O_3R_2)_n$ and would correspond to the petalite or apophyllite layers of Fig. 23.13, p. 1024. Polymers intermediate between infinite chains and layers would correspond to the amphibole chain, for example, having the composition $(Si_4O_5R_6)_n$. Although these 'pure' types, apart from the chains and rings, are not known, 2- and 3-dimensional linking by means of T and Q groups presumably occurs in the gels and resins. Polymers have also been made from silicon compounds containing unsaturated organic groups. The vinyl and allyl chlorosilanes already mentioned can be hydrolysed to alkylene polysiloxanes which can polymerize further through the unsaturated organic groups.

Technically useful silicones arise by co-hydrolysis of suitable mixtures of chlorosilanes or esters yielding mixtures of two or more of the basic M, D, T, or Q units. The product of hydrolysis of $(CH_3)_2SiCl_2$ is a colourless non-volatile oil, but this is not stable because the chains end in D groups and further condensation, to rings or longer chains, can take place. To stabilize the product it is necessary to introduce some M units, by cohydrolyzing $(CH_3)_3Si(OC_2H_5)$ and $(CH_3)_2Si(OC_2H_5)_2$ or $(CH_3)_2SiCl_2$. However, if the proportion of the latter is increased more cyclic compounds D_n are formed, and to obtain the desired mixtures of longer linear polysiloxanes (for oils and greases of various viscosities) a device known as *catalytic equilibration* is used. A liquid methylpolysiloxane mixture, the cyclic tetrasiloxane, or a mixture of $(CH_3)_3Si.O.Si(CH_3)_3$ and any source of D units is shaken with sulphuric acid for a few hours, when rearrangement takes place to a mixture of MD_xM:

$$M \qquad D_x \qquad M$$

These equilibrated silicone oils have valuable properties: they are very stable to heat and oxidation, they have good electrical properties, and are water-repellent. They find application as vacuum oils and greases, water-proofing agents, and paint media.

From these oils the compounds M_2D_5 to M_2D_9 have been obtained by fractional distillation under reduced pressure. All the following compounds have been prepared: linear M_2, M_2D to M_2D_9, cyclic D_3 to D_9. The last can be prepared by the hydrolysis of $(CH_3)_2SiCl_2$ or by the depolymerization of dimethylsiloxane high polymers. Rearrangement to low cyclic polymers takes place on heating the latter to 230–240 °C under reduced pressure with NaOH as catalyst, and these can be separated by fractional distillation under reduced pressure.

Various polycyclic methylpolysiloxane oligomers have been produced by thermal decomposition of branched-chain polymers; an example is the T_2D_3 molecule $Si_5O_6(CH_3)_8$ [1] with the structure (a). The molecule has approximately 3-fold

(a)   (b)

symmetry about the dotted line. The Si–O–Si bond angle is 146°; the CH_3 groups are omitted from the diagram.

Other types of polysiloxane obtainable by co-hydrolysis include TM_3 and QD_n. The co-hydrolysis product from $SiCl_4$ and $(CH_3)_2SiCl_2$ undergoes rearrangement in nitrogen at 400–600 °C to give spirosiloxanes such as $Si_5O_6(CH_3)_8$, QD_4, described on p. 1001. Combinations of appropriate proportions of Q and T give silicone resins which can be hardened by heat treatment. We noted earlier that total hydrolysis of a trifunctional silane $RSiX_3$ such as $CH_3Si(OR)_3$ or CH_3SiCl_3 gives an infusible solid ('*T*-gel') of unknown structure, but partial hydrolysis, for example of $CH_3Si(OC_2H_5)_3$ in benzene solution, gives solid mixtures from which well defined, high-melting solids have been obtained by fractional sublimation. Examples of poly-hedral silasesquioxanes whose structures have been determined include $\phi_8Si_8O_{12}$[2] which has the 'cubic' structure shown diagrammatically at (b), and $\phi_{12}Si_{12}O_{18}$,[3] in which the Si atoms are situated at the vertices of the polyhedron ($f_4 = f_5 = 4$) illustrated in Fig. 3.5, p. 74.

In siloxanes the silicon atoms are linked, as in complex silicates, through the oxygen atoms. It is also possible to link them through carbon atoms, and some molecules containing silicon–carbon chains have been prepared, for example

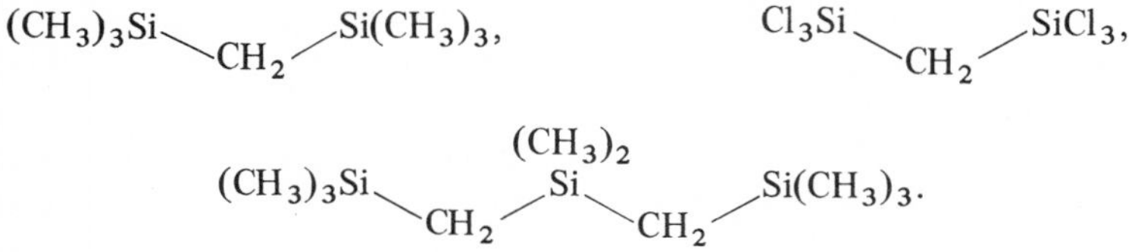

$$(CH_3)_3Si{\diagdown}_{CH_2}{\diagup}Si(CH_3)_3, \qquad\qquad Cl_3Si{\diagdown}_{CH_2}{\diagup}SiCl_3,$$

$$(CH_3)_3Si{\diagdown}_{CH_2}{\diagup}\overset{\displaystyle (CH_3)_2}{Si}{\diagdown}_{CH_2}{\diagup}Si(CH_3)_3.$$

The methyl compounds are prepared by reactions of the type

$$(CH_3)_3Si.CH_2.MgCl + (CH_3)_3SiCl \rightarrow (CH_3)_3Si.CH_2.Si(CH_3)_3$$

and the chloro compound by passing methylene chloride over silicon–copper (compare the preparation of the substituted chlorosilanes, p. 998).

The Si–O skeletons in mineral silicates can be recognized in the siloxanes prepared from them. The ground mineral is treated with a mixture of cold concentrrated HCl, isopropyl alcohol, and hexamethyldisiloxane. The siloxane layer is separated and analysed chromatographically. The trimethylsilyl derivatives correspond to the type of Si–O complex in the original silicate. For example, from olivine (an orthosilicate) 70 per cent of the Si is recovered as $Si[OSi(CH_3)_3]_4$, from hemi-morphite (pyrosilicate) 78 per cent as $(Me_3SiO)_3SiOSi(OSiMe_3)_3$ and $(Me_3SiO)_3$-$SiOSi(OH)(OSiMe_3)_2$; there is some cleavage of the original Si–O–Si bonds. Differences are found between the products from framework aluminosilicates depending on the sequence of Si and Al atoms in the framework.[4]

(1)   AC 1975 **B31** 1214   (3)   AC 1980 **B36** 3162
(2)   AC 1979 **B35** 2258   (4)   IC 1964 **3** 574

The crystalline forms of silica

At atmospheric pressure silica exists in three crystalline forms, which are stable in the temperature ranges indicated below:

$$\underset{Quartz}{\xrightarrow{\hspace{2cm}}} 870° \underset{Tridymite}{\xrightarrow{\hspace{2cm}}} 1470° \underset{Cristobalite}{\xrightarrow{\hspace{2cm}}} 1710°C \text{ (m.p.)}$$

Crystallization of the liquid is difficult, and it usually solidifies to a glass which has great value on account of its extremely small coefficient of thermal expansion and high softening point (in the region of 1500 °C). Silica in this glassy form is found in nature as the mineral obsidian. These three polymorphic forms of silica are not readily interconvertible as is shown by the fact that all three are found as minerals, though tridymite and cristobalite are rare in comparison with quartz. Furthermore, each of the three forms exists in a low- and high-temperature modification (α and β respectively) with the following transition points: α-β quartz, 573°; α-β tridymite, 110–180°; and α-β cristobalite, 218 ± 2 °C. The fact that in the last two cases these transitions can be studied at temperatures at which the particular polymorphic forms are metastable is a further indication of the difficulty of changing one of the three varieties of silica into another. The general structural relations between these crystalline forms of silica have been already noted in our discussion of nets in Chapter 4 and are as follows. The three main forms, quartz, tridymite, and cristobalite, are all built of SiO_4 tetrahedra linked together so that every oxygen atom is common to two tetrahedra (thus giving the composition SiO_2), but the arrangement of the linked tetrahedra is quite different in the three crystals. These structures are illustrated, as packings of tetrahedra, in Fig. 23.20, later in this chapter. The difference between, for example, tridymite and cristobalite is like that between wurtzite and zinc-blende, the two forms of ZnS. On the other hand, the α and β forms of one of the three varieties differ only in detail, there being, for example, slight rotations of the tetrahedra relative to one another without any alteration in the general way in which they are linked together. The change from quartz to tridymite thus involves the breaking of Si—O—Si bonds and the linking together of the tetrahedra in a different way, and accordingly is a very sluggish process, whereas the conversion of $\alpha \rightarrow \beta$ quartz merely involves a small distortion of the structure without any radical rearrangement and is an easy and reversible process. Quartz is optically active, in contrast to the other forms, and the fact that a crystal of left-handed α-quartz remains left-handed after conversion into the β form is further evidence of the relatively small change in structure at the transition point.

The high-temperature polymorphs, tridymite and cristobalite, differ from quartz in having much more open structures, as shown by their densities: quartz, 2.655; tridymite, 2.26 and cristobalite, 2.32 g cm^{-3}. There are also high-pressure forms of silica, coesite, of which two modifications are now recognized,[1] keatite,[2] which is the stable form in the pressure range between those for quartz and coesite, and stishovite, formed at very high pressures. The last form has the 6-coordinated rutile structure, (Si—4O, 1.76 Å; Si—2O, 1.81 Å)[3] while coesite and keatite are based on 4-connected frameworks in which the smallest rings are respectively 4- and 8-membered (coesite) and 5-, 7-, and 8-membered (keatite). For the basic keatite framework see p. 131.

In β-quartz, the structure of which is illustrated in Fig. 23.6, there may be distinguished helices of linked tetrahedra, and in a given crystal these are all either left- or right-handed, the crystal being laevo- or dextro-rotatory. Some idea of the

FIG. 23.6. Plan of the structure of β-quartz. Small black circles represent Si atoms. The oxygen atoms lie at different heights above the plane of the paper, those nearest the reader being drawn with heaviest lines. Each atom is repeated at a certain distance above (and below) the plane of the paper along the normal to that plane so that the Si_3O_3 rings in the plan represent helical chains.

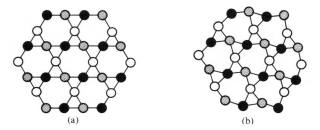

(a)   (b)

FIG. 23.7. The arrangement of the Si atoms (in plan) in (a) β-quartz and (b) α-quartz.

relation between the α and β forms of quartz may be obtained from Fig. 23.7, which shows, in plan, the arrangement of the silicon atoms in the two crystals. It will be seen that the α form is a distorted version of the β form, but that the general scheme according to which the tetrahedra are linked together is the same in the two forms.

In Fig. 23.8 the structures of β-tridymite and β-cristobalite are shown with the O atoms midway between, and on the straight lines joining, the pairs of Si atoms. In fact, the O atoms lie off the lines joining pairs of nearest Si atoms, giving Si–O, 1.61 Å and Si–O–Si close to 144°, values also found in α-quartz.[4] The (tetragonal) structure of α-cristobalite results from rotation of the tetrahedra of Fig. 23.8(b) to give Si–O–Si angles close to 147°. The cubic symmetry of β-cristobalite, which was the reason for placing the O atoms as shown, is a statistical symmetry, the crystals consisting of domains within any one of which the Si–O–Si angle is 147°.[5, 6]

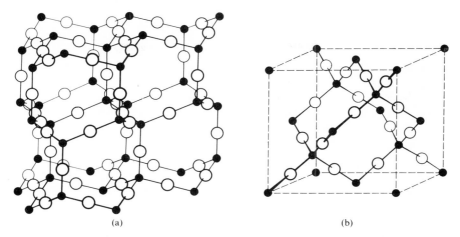

FIG. 23.8.   The idealized structures of (a) β-tridymite, and (b) β-cristobalite (see text). Small black circles represent Si atoms.

Tridymite presents a more complex picture. The less dense structures of tridymite and cristobalite are presumably maintained at high temperatures by the thermal movements of the atoms, and it is possible that they are stabilized at lower temperatures by the inclusion of foreign atoms.

Natural tridymite and cristobalite often, if not invariably, contain appreciable concentrations of foreign ions, particularly of the alkali and alkaline-earth metals,[7] suggesting that they may have been deposited as the stable phases at temperatures well below the respective transition points. The variable composition of these minerals is presumably the reason for wide ranges of temperatures recorded for the α–β transition points. The origin of the high-temperature tridymite studied by Gibbs is not known but it appears that the monoclinic form stable at room temperature undergoes some sort of structural change at approximately 107°C. Structural studies have been made of a number of (α) low-tridymites, including a synthetic monoclinic form[8, 9] and a triclinic form with a large pseudo-orthorhombic cell containing 320 SiO_2[10] found in volcanic rocks and containing 99 mol per cent SiO_2, 1 per cent Al, and traces of Na, K, and Ti. The Si–O bond length in these structures is close to 1.61 Å and the mean Si–O–Si angle close to 150°, though in both structures there are some much larger interbond angles (176° and 173° respectively). Above 180°C tridymite transforms to an orthorhombic structure in which the apparent Si–O bond length is 1.56 Å, but corrected for the very large thermal motion this becomes 1.61 Å.[11]

In addition to the polymorphs of SiO_2 already described there are two silica frameworks which are not polymorphs in the ordinary sense of the word. In clathrates such as urea-hydrocarbon, β-quinol, or gas hydrates, there are arrangements of 'host' molecules which form only if the 'guest' molecules (atoms) are present at the time of crystallization, and the framework is not sufficiently strongly bonded to withstand removal of all of the included material. However, a framework formed from Si and O atoms bonded as in the normal forms of silica is an extremely stable one from which the included atoms or molecules can be removed, leaving an open structure of composition SiO_2. The (black) mineral melanophlogite was discovered more than a century ago in the Sicilian sulphur deposits in limestone saturated with liquid bituminous matter. Studies during the past 20 years[12] have shown that the structure is similar to that of the chlorine hydrate illustrated in Fig. 15.6 on p. 663. After heating to $1050\,^{\circ}C$ this highly crystalline material has cubic symmetry (a = 13.4 Å) and at ordinary temperatures a tetragonal superstructure with doubled a dimensions. The superstructure is presumably due to the presence of the included molecules, since the heated material, of composition SiO_2, has cubic symmetry. The content of one-quarter of the tetragonal unit cell, that is, of the cubic 'pseudocell', is $46SiO_2.C_2H_{17}O_5$ (plus traces of S).

The formation in nature, presumably from a solution of SiO_2 (which is quite soluble in water at temperatures above $100\,^{\circ}C$) in the presence of hydrocarbon molecules, has been imitated in the hydrothermal preparation of 'silicalite' from a mixture of NR_4^+ ions (for example, $N(propyl)_4^+$), OH^- ions, and a reactive form of SiO_2 at temperatures between 100 and $200\,^{\circ}C$.[13] The product is a form of silica with a framework similar to that of the synthetic zeolites ZSM-5 and ZSM-11 (Fig. 23.29, p. 1042). Unlike normal silicate surfaces and zeolites this material is hydrophobic; it is crystalline, with cell dimensions very close to those of ZSM-5. A feature of the structure, as of that of melanophlogite, is the presence of numerous Si_5O_5 rings, which are not found in the normal forms of SiO_2 or in most silicate structures, though they are present in keatite and certain zeolites such as mordenite and epistilbite. The large tunnels, which confer molecular sieve and catalytic properties, are defined by $Si_{10}O_{10}$ rings, and are intermediate between those in, for example, zeolite A (Si_8O_8) and faujasite ($Si_{12}O_{12}$).

(1)   ZK 1977 **145** 108; ZK 1979 **149** 315   (8)   AC 1976 **B32** 2486
(2)   ZK 1959 **112** 409   (9)   AM 1976 **61** 971
(3)   AC 1971 **B27** 2133   (10)   AC 1978 **B34** 391
(4)   AC 1976 **B32** 2456   (11)   AC 1967 **23** 617
(5)   PM 1975 **31** 1391   (12)   Sc 1965 **148** 232; AM 1972 57 779
(6)   AC 1976 **B32** 2923   (13)   N 1978 **271** 512
(7)   AM 1954 **39** 600

Stuffed silica structures

This name is given to structures in which the mode of linking of the tetrahedra in one of the polymorphs of silica is retained, but some Si is replaced by Al and cations (usually alkali-metal ions) are accommodated in the interstices to maintain electrical neutrality. The holes in the high-quartz structure are only large enough for Li^+ (and possibly Be^{2+}), but those in the high-temperature forms of tridymite and cristobalite can accommodate the large alkali-metal ions (Table 23.1). We

TABLE 23.1
Stuffed silica structures

Silica polymorph	Stuffed silica structure	Reference
Quartz	α- and β-LiAlSiO$_4$ (eucryptite)	AC 1972 **B28** 2168, 2174
Tridymite	KAlSiO$_4$ (kalsilite)	AC 1948 **1** 42
	KNa$_3$Al$_4$Si$_4$O$_{16}$ (nepheline)	
	BaAl$_2$O$_4$	AC 1972 **B28** 1241
	K(LiBeF$_4$)	AC 1972 **B28** 1383
Cristobalite	NaAlSiO$_4$ (high-carnegieite)	DAN 1956 **108** 1077
	Na$_2$BeSi$_2$O$_6$	
	CsLiWO$_4$	AC 1980 **B36** 657
Keatite and quartz	LiAlSi$_2$O$_6$	AC 1971 **B27** 1132

note later that the normal form of LiAlSi$_2$O$_6$ (the mineral spodumene) contains metasilicate chains; it is one of the monoclinic pyroxenes. Heat treatment of glasses with this composition gives crystalline forms with stuffed β-quartz and keatite structures. Spodumene is therefore preferably formulated LiAl(Si$_2$O$_6$), having both Li$^+$ and Al^{3+} in positions of octahedral coordination between the infinite linear anions, while the stuffed silica polymorphs are Li(AlSi$_2$O$_6$), indicating that Al and Si are tetrahedrally coordinated in a silica-like framework.

Silicates

The earth's crust is composed almost entirely of silicates and silica, which constitute the bulk of all rocks and of soils, clays, and sands, the breakdown products of rocks. All inorganic building materials, ranging from natural rocks such as granite to artificial products such as bricks, cement, and mortar, are silicates, as also are ceramics and glasses. Metallic ores and other mineral deposits form an insignificant proportion of the mass of the earth's crust. Table 23.2, due to Goldschmidt, gives the average composition of the lithosphere; that of the hydrosphere is given for comparison. It is seen that more than nine-tenths of the volume of the earth's crust is occupied by oxygen. In many silicates the oxygen atoms are close-packed, or approximately so, and the more electropositive ions, which are nearly all smaller than oxygen, are found in the interstices between the close-packed oxygen atoms. We have already discussed the relation between radius ratio and coordination number in Chapter 7. Here we need only remind the reader of the usual oxygen coordination numbers of the ions commonly found in silicate minerals (less usual c.n.s in parentheses):

$$
\begin{array}{ll}
\text{Be}^{2+} & 4 \\
\text{Li}^+ & 4\ (6) \\
\text{Al}^{3+} & 4,\ 6 \\
\text{Mg}^{2+} & (4)\ 6 \\
\text{Fe}^{2+},\ \text{Ti}^{4+} & 6
\end{array}
$$

TABLE 23.2

Compositions of the lithosphere and hydrosphere

Lithosphere					Hydrosphere	
Element	% (weight)	% (atomic)	Radius (Å)[a]	% (volume)	Element	% (weight)
O	46.59	62.46	1.32	91.77	O	85.89
Si	27.72	21.01	0.39	0.80	H	10.80
Al	8.13	6.44	0.57	0.76	Cl	1.93
Fe	5.01	1.93	0.82	0.68	Na	1.07
Mg	2.09	1.84	0.78	0.56	Mg	0.13
Ca	3.63	1.93	1.06	1.48	S	0.09
Na	2.85	2.66	0.98	1.60	Ca	0.04
K	2.60	1.43	1.33	2.14	K	0.04
Ti	0.63	0.28	0.64	0.22		
All other elements are present in smaller amounts					All others less than 0.01%	

(a) These ionic radii differ slightly from those given in Chapter 7, since Goldschmidt based his values on O^{2-} 1.32 Å.

$$Na^+, Ca^{2+} \qquad\qquad 6\,(8)$$
$$K^+ \qquad\qquad 6-12$$

Since there are a number of ions of similar size which can occupy, for example, octahedral holes (i.e. coordination number 6), and since atoms of many kinds were present in the melts or solutions from which the minerals crystallized, isomorphous substitution is very common in the silicates, which rarely have the 'ideal' composition of a simple chemical compound. The most important and widespread substitution is that of Al for Si in tetrahedral coordination, so that most of the 'silicates' that occur in nature are in fact aluminosilicates. This substitution must be accompanied by the incorporation of cations to balance the charge on the silicon–oxygen anion. The fact that Al also occupies positions of octahedral coordination in aluminosilicates complicated the interpretation of the analytical formulae of minerals until the dual role of Al was discovered as the result of X-ray studies.

In almost all silicates which occur as minerals the anions are built of tetrahedral $(Si, Al)O_4$ groups. In hydrated silicates the water of crystallization is associated with the cations as normal H_2O molecules. There are, however, hydroxy-silicates containing $[SiO_3(OH)]^{3-}$ or $[SiO_2(OH)_2]^{2-}$ ions with which we deal briefly before proceeding to the main groups of silicate structures.

Hydroxy-silicates

In addition to silicates containing anions built of SiO_4 groups the alkali and alkaline-earth metals form hydroxy-silicates containing $[SiO_3(OH)]^{3-}$ or $[SiO_2(OH)_2]^{2-}$ ions. The sodium compounds can be crystallized from mixtures of Na_2O, SiO_2, and

H_2O kept in a sealed plastic bottle at room temperature. These compounds are unstable to CO_2, some are hygroscopic, and some of the more highly hydrated compounds lose water of crystallization to form lower hydrates. Sodium forms a series of hydrated compounds with the analytical formulae $Na_2SiO_3 . 5, 6, 8$, and $9H_2O$ which are in fact $Na_2SiO_2(OH)_2 . 4, 5, 7$, and $8H_2O$. All these compounds contain ions with the structure (a).[1] In the very hygroscopic $Na_3HSiO_4 . 2H_2O$[2]

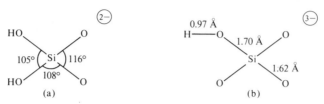

there are hydrogen-bonded chains of SiO_4H groups, (b), and in $Na_3HSiO_4 . 5H_2O$[3] the anions are associated in hydrogen-bonded pairs—compare the SO_4H^- anions in $KHSO_4$ (p. 370). Other hydroxy-silicates containing the $SiO_3(OH)^{3-}$ ion include $NaCa[SiO_3(OH)]$,[4] $Ca_3[SiO_3(OH)]_2 . 2H_2O$,[5] and $Ca_2[SiO_3(OH)]OH$.[6] Hydroxy-silicates apparently play an important part in the setting of Portland cement (p. 1017).

(1) AC 1976 **B32** 705
(2) AC 1979 **B35** 3024
(3) SPC 1973 **18** 173

(4) AC 1974 **B30** 864
(5) AC 1976 **B32** 475
(6) AC 1952 **5** 724

The classification of silicate structures

Only vertices of SiO_4 or $(Si,Al)O_4$ groups are shared in silicates. (The only recorded case of edge-sharing between SiO_4 groups is that postulated in fibrous silica;[1] it has not been observed in silicates.) The 'normal' Si–O bond length is close to $1.62\,Å$, but its value ranges from $1.54-1.70\,Å$ depending on the number and types of cations bonded to the O atoms. An empirical equation has been suggested:[2]

$$d(\text{Si–O}) = (1.44 + 0.09p)\,Å$$

where p is the total Pauling electrostatic bond strength of the cation–oxygen bonds. Replacement of Si by Al results in longer bonds in the tetrahedral groups, up to $1.72\,Å$ for AlO_4.

  An interesting feature of silica and silicate structures is the value of the Si–O–Si bond angle. This could have a wide range of values (up to $180°$, as observed in thortveitite, p. 1019, and in a number of other silicates and silica polymorphs[3]) but typical values are close to $145°$. (In coesite the majority of Si–O–Si bond angles lie in the range $137-150°$ but there is also collinear coordination of certain O atoms.) Since the O atoms are not always very closely packed whereas Si–Si distances in many silicates (and also in many other molecules in which they are bonded through one O or N atom) are usually close to $3.06\,Å$, this has led to the interesting suggestion that the detailed geometry of vertex-sharing SiO_4 groups is determined by Si–Si and Si–O contacts rather than by O–O contacts.[3, 4, 5]

Since there is only sharing of vertices between $(Si, Al)O_4$ tetrahedra it is convenient to make the first broad subdivision of silicate structures according to the numbers of shared vertices. Discrete $(SiO_4)^{4-}$ ions occur in orthosilicates. Higher $O:Si$ ratios than $4:1$ are possible if there are separate O^{2-} ions in addition to those in the SiO_4^{4-} ions. The structure of Sr_3SiO_5,[6] for example, is of this kind; it is very similar to that of Cs_3CoCl_5. Lower $O:Si$ ratios result if there is sharing of O atoms between the SiO_4 tetrahedra. The simplest structures arise if all the SiO_4 groups share the same number of vertices.

(1) ZaC 1954 **276** 95
(2) AM 1971 **56** 1573
(3) AC 1980 **B36** 2198
(4) AC 1977 **A33** 924
(5) AC 1978 **B34** 27
(6) AC 1965 **18** 453

(a) One atom of each SiO_4 is shared with another SiO_4 giving the pyrosilicate ion $Si_2O_7^{6-}$.

(b) Two O atoms of every SiO_4 are shared with other SiO_4 tetrahedra, resulting in closed rings or infinite chains, as in

and larger rings, and

These complexes are all of the type $(SiO_3)_n^{2n-}$.

(c) Three O atoms of each SiO_4 are shared with other SiO_4 tetrahedra to give systems of composition $(Si_2O_5)_n^{2n-}$. These may be finite or infinite in one, two, or three dimensions, as explained in Chapter 3 for 3-connected systems and in particular in Table 5.2 (p. 195) for tetrahedra sharing three vertices.

(d) Sharing of all vertices of each SiO_4 tetrahedron leads to infinite 3D frameworks, as in the various forms of silica, or to double layers. If some of the Si is replaced by Al the framework—always of composition $(Si, Al)O_2$—is negatively charged, and positive ions must be accommodated in the holes in the structure. We shall mention later the felspars and zeolites as the most important classes of aluminosilicates with 3D framework structures. An alternative way of linking $(Si, Al)O_4$ tetrahedra so that all vertices are shared is to form double layers, as found in the hexagonal forms of $CaAl_2Si_2O_8$ and $BaAl_2Si_2O_8$. Replacement of Si by Al to different extents is illustrated by the formulae of the following minerals:

		Si : Al *ratio*
The various forms of silica:	SiO_2	∞
Orthoclase	$K(AlSi_3O_8)$	3 : 1
Analcite	$Na(AlSi_2O_6).H_2O$	2 : 1
Natrolite	$Na_2(Al_2Si_3O_{10}).2H_2O$	3 : 2
Anorthite	$Ca(Al_2Si_2O_8)$	1 : 1

In the groups (a)–(d) above we have considered only those Si_xO_y complexes in which there are the same numbers of shared and unshared oxygen atoms around every Si atom, viz.

3 unshared, 1 shared	Si_2O_7
2 unshared, 2 shared	SiO_3
1 unshared, 3 shared	Si_2O_5
4 shared	SiO_2

There are other possible complexes in which the environment of every Si atom, in terms of shared and unshared O atoms, is not the same. This topological non-equivalence of SiO_4 tetrahedra could arise in complexes of the following kinds:

(i) polynuclear ions intermediate between $Si_2O_7^{6-}$ and $(SiO_3)_n^{2n-}$ chains;

(ii) ions formed by union of chain ions (multiple chains);

(iii) complex layers; or

(iv) frameworks in which SiO_4 tetrahedra share different numbers of vertices.

Polysilicate ions, (i), analogous to $P_3O_{10}^{5-}$ and $S_3O_{10}^{2-}$, are extremely rare; Si_3O_{10} groups can be distinguished in a rare arsenic vanadium silicate, ardennite[1a] and also, together with SiO_4^{4-} ions, in $Ho_4(Si_3O_{10})(SiO_4)$.[1b]

Groups of 5 SiO_4 tetrahedra consisting of a central one sharing a vertex with each of 4 others, (Si_5O_{16}), can be distinguished in the complex structure of zunyite, $Al_{13}F_2(OH, F)_{16}Si_5O_{20}Cl$.[1c] The structure can be built from such groups together with 1 AlO_4 group and an interesting complex built from octahedral coordination groups which as an isolated group would have the composition $Al_{12}O_{16}(OH)_{30}$. However, the Si_5O_{16} group is not yet known as a discrete unit in any simpler silicate structure, though the structurally similar $(PCr_4O_{16})^{3-}$ ion is known (p. 861). We might digress to mention that the $Al_{12}O_{16}(OH)_{30}$ group can be built, like the $PW_{12}O_{40}$ isomers (p. 524) from four edge-sharing groups of three octahedra, but with their central tetrahedral depressions facing inwards instead of outwards. Since each A_3X_{13} group then shares only 3 X instead of 6, the composition is $A_{12}X_{46}$ as compared with $A_{12}X_{40}$ for the Keggin isomers.

Multiple chains, (ii) can be formed by sharing vertices between tetrahedra of two or more simple chains placed side-by-side. The type of chain so formed depends on the relative orientations of successive tetrahedra in the simple chain, as illustrated in Fig. 23.12(d)–(f), p. 1022. For example, two simple chains (b) give first the double chain (e), but for *m* such chains joined laterally the composition is $Si_{2m}O_{5m+1}$. The limiting composition is Si_2O_5, for such chains are simply strips of the hexagonal Si_2O_5 layer. The value $m = 2$ corresponds to the Si_4O_{11} chain of Fig. 23.12(e). The chains with $m = 3$, 4, and 5 are the anions in $Ba_4Si_6O_{16}$, $Ba_5Si_8O_{21}$, and $Ba_6Si_{10}O_{26}$.[2a] Double and triple chains of these types, and a combination of the two, have also been identified in a number of minerals.[2b] As in the case of the pyroxenes and amphiboles, which contain single and double chains respectively (p. 1022), there are two series of compounds (with orthorhombic and monoclinic symmetry), and one cell dimension reflects the width of the multiple chain(s). For example, in all the orthorhombic compounds a and c are approximately 18 Å and 5.2 Å but the values of b are as follows:

Type of chain	b (Å)	
$m = 1$	9	pyroxene (enstatite, $MgSiO_3$)
2	18	amphibole (anthophyllite, $Mg_7Si_8O_{22}(OH)_2$)
3	27	$(Mg, Fe)_{10}Si_{12}O_{32}(OH)_4$
2 and 3	45	$(Mg, Fe)_{17}Si_{20}O_{54}(OH)_6$

Similar multiple chains have been identified by high resolution electron microscopy in jade.[2c]

We describe later an example of a more complex layer, (iii), in which two-thirds of the SiO_4 tetrahedra share 3 and the remainder 4 vertices, giving the composition Si_3O_7 (p. 1025).

Neptunite (p. 1033) provides an example of (iv).

We shall now describe some of the important silicate structures, dealing in turn with the following groups:

1. orthosilicates, containing discrete SiO_4^{4-} ions;
2. silicates with $Si_2O_7^{6-}$ ions;
3. silicates with cyclic $(SiO_3)_n^{2n-}$ ions;
4. silicates with chain structures—pyroxenes and amphiboles;
5. silicates with layer structures—clay minerals and micas;
6. silicates with framework structures—felspars and zeolites.

It should be remarked that although for convenience we adopt this classification in terms of the silicon–oxygen complex we are in fact classifying according to the (Al, Si)–O complex in many cases, particularly layer structures, where replacement of some Si by Al is common, and in framework structures, in which replacement of some Si by Al is essential to form a charged framework. For this reason it is more logical to recognize not only Al but other tetrahedrally coordinated ions as forming part of the complex. These include the ions of

$$\begin{matrix} \text{Li} & \text{Be} & \text{B} & & & \\ \text{(Na} & \text{Mg)} & \text{Al} & \text{Zn} & \text{Ga} & \text{Ge} \end{matrix}$$

For example, there is tetrahedral coordination of both Li and Si in Li_2SiO_3,[3] which is a superstructure of BeO or ZnO and may be formulated $(Li_2Si)O_3$. Similarly there is 4-coordination of both Be and Si in phenacite, Be_2SiO_4, which is more logically described as a 3D framework M_3O_4 than as an orthosilicate, as is shown by the fact that Ge_3N_4 has the same structure. Hemimorphite (p. 1019) contains Si_2O_7 groups, but tetrahedrally coordinated Zn and Si together form a 3D framework. Other examples of structures in which one of the above ions together with Si forms a 3D framework include $Rb_2[Be_2Si_2O_7]$ (p. 1020); beryl, $Al_2[Be_3Si_6O_{18}]$ (p. 1021); milarite (p. 1033); $Na_2[ZnSi_3O_8]$ (p. 1035); helvite and tugtupite (p. 1042); and $Pb(ZnSiO_4)$,[4] the mineral larsenite.

In the melilite family (p. 1019) there are layers in which XO_4 tetrahedra share 4 and YO_4 tetrahedra share 3 vertices, giving a layer of composition XY_2O_7, as in $Ca_2(Si_2MgO_7)$ or $Ca_2(Al_2SiO_7)$. Finally, there is the same type of X_2O_5 layer in gillespite, $BaFe(Si_4O_{10})$; datolite (p. 1078); $Ca(BSiO_4OH)$; herderite, $Ca(BePO_4F)$; and gadolinite, $FeY_2(Be_2Si_2O_{10})$.[5]

It should also be noted that the type of Si–O complex is not necessarily the most important feature of a crystalline silicate in certain contexts. In the transformation of a single crystal of one silicate into another of related structure (e.g. by heating) there is certainly rearrangement of the Si-O complex rather than disturbance of the cation–oxygen system.[6] Thus in the conversion of rhodonite (p. 1023) to β-$CaSiO_3$ there is movement of Si from the 5T chain to form a 3T chain which lies in a different direction from the original chain, This involves far less movement of the larger ions (particularly O^{2-}) than would a rearrangement of the 5T chain by twisting about its own axis. A similar case is the 'dehydration' of xonotlite to wollastonite; for these structures see p. 1023.

(1a) AC 1968 B24 845
(1b) NW 1972 59 35
(1c) AC 1960 13 15
(2a) ZK 1980 153 3
(2b) Sc 1977 198 359

(2c) CC 1977 910
(3) AC 1977 B33 901
(4) ZK 1967 124 115
(5) K 1959 4 324
(6) AC 1961 14 8, 18

Orthosilicates

These contain SiO_4^{4-} ions, the cations occupying interstices in which they are surrounded by a number of O^{2-} ions appropriate to their size:

	C.n. of M
Be_2SiO_2 (phenacite)	4
$(Mg, Fe)_2SiO_4$ (olivine)	6
$ZrSiO_4$ (zircon)	8

The phenacite structure has been mentioned in the discussion of (3,4)-connected nets in Chapter 3.

The garnets are a group of orthosilicates of the general formula $R_3^{II}R_2^{III}(SiO_4)_3$, where R^{II} is Ca^{2+}, Mg^{2+}, or Fe^{2+}, and R^{III} is Al^{3+}, Cr^{3+}, or Fe^{3+}, as in grossular, $Ca_3Al_2(SiO_4)_3$, uvarovite, $Ca_3Cr_2(SiO_4)_3$, and andradite, $Ca_3Fe_2(SiO_4)_3$. The garnet structure, in which R^{II} and R^{III} are respectively 8- and 6-coordinated by oxygen, has been described in Chapter 13. The synthetic garnet pyrope, a high-pressure phase of composition $Mg_3Al_2Si_3O_{12}$, is of interest as an example of 8-coordinated

Mg^{2+} (distorted cube, 4 O at 2.20, 4 O at 2.34 Å).[1]

The olivine structure may be described either as an assembly of SiO_4^{4-} and Mg^{2+} ions or as an array of approximately close-packed O atoms with Si in tetrahedral and Mg (or Fe) in octahedral holes. The idealized (h.c.p.) structure is illustrated, in plan, in Fig. 23.9(a). The plan of a layer of the $Mg(OH)_2$ structure is shown at (b). Owing to the similar packing of the oxygen atoms in these two compounds portions of the two structures fit together perfectly, giving rise to a series of minerals of general formula $m[Mg_2SiO_4].n[Mg(OH)_2]$ – the so-called chondrodite series. If we compare the compositions of the horizontal rows of atoms in the plans of Fig. 23.9(a) and (b) we see that they are of two kinds, x and y, the former containing Mg and O and the latter Mg, O, and Si. In brucite, $Mg(OH)_2$, the rows are all of the type x; in olivine x and y alternately. Other sequences are possible and those found in the chondrodite minerals are as follows:

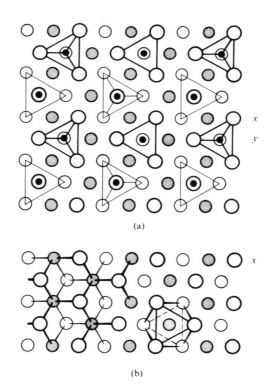

(a)

(b)

FIG. 23.9.   Plans of the structures of (a) Mg_2SiO_4, and (b) $Mg(OH)_2$. Small black circles represent Si, shaded circles Mg, and open circles O atoms. In (a) light and heavy lines are used to distinguish between SiO_4 tetrahedra at different heights. To the left in (b) the Mg–OH bonds are shown, and to the right an octahedral coordination group is outlined.

Repeat unit

xxy	Norbergite	$Mg_2SiO_4.Mg(OH)_2$
xy, xxy	Chondrodite	$2Mg_2SiO_4.Mg(OH)_2$
xy, xy, xxy	Humite	$3Mg_2SiO_4.Mg(OH)_2$
xy, xy, xy, xxy	Clinohumite	$4Mg_2SiO_4.Mg(OH)_2$

The compound Al_2SiO_5 exists in three polymorphic forms, cyanite, sillimanite, and andalusite. In each of the three forms of Al_2SiO_5 one-half of the Al atoms are 6-coordinated and the other half are in positions of 4-, 5-, and 6-coordination in sillimanite, andalusite, and cyanite respectively. The latter is accordingly the most compact of the three forms, the oxygen atoms being cubic close-packed. Just as the structures of the minerals of the chondrodite series are closely related to those of olivine and brucite, so we find a more complex mineral, staurolite—with the ideal composition $HFe_2Al_9Si_4O_{24}$—related in a rather similar way to cyanite. Two unit cells of the staurolite structure are shown, in plan, in Fig. 23.10, and the broken lines outline cyanite unit cells. It is seen that the cyanite structure is interrupted by the insertion of Fe and Al atoms.[2] Other compounds with chondrodite-type structures include $Mn_5(SiO_4)_2(OH)_2$[3a] and $Al_4Co(BO_4)_2O_2$ (h.p. form).[3b]

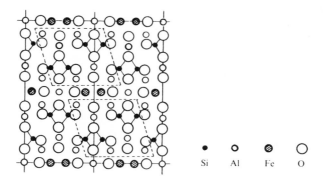

●	○	◔	◯
Si	Al	Fe	O

FIG. 23.10. Relation between the structures of staurolite, $HFe_2Al_9Si_4O_{24}$, and cyanite, Al_2SiO_5.

A much higher O:Si ratio is found in Fe_7SiO_{10}, extracted from the reheating furnace of a steel plant. In a c.c.p. assembly of O atoms Si occupies tetrahedral holes, 5 Fe^{2+} octahedral holes, and 2 Fe^{3+} is equally distributed between tetrahedral and octahedral holes.[4]

Portland cement. Portland cement may conveniently be mentioned here since orthosilicates are important constituents. It is formed by burning a mixture of finely ground limestone or chalk with clay or shale, using coal, oil, or gas fuel. The main constituents are CaO, 60–67 per cent, SiO_2, 17–25 per cent, Al_2O_3, 3–8 per cent, and some minor ingredients which may have an appreciable effect on its properties. The more important compounds formed are Ca_2SiO_4, Ca_3SiO_5,[5] $Ca_3Al_2O_6$, and

Ca_2AlFeO_5. Phase-rule studies show that tricalcium silicate is stable only between 1250 and 1900 °C, outside which range it decomposes into $CaO + Ca_2SiO_4$, though it will remain indefinitely in a metastable state when quenched to room temperature. Ca_2SiO_4 has four polymorphs, of which the two high-temperature forms (α' and α) are isostructural with β- and α-K_2SO_4 respectively. This polymorphism is important because the different forms do not all hydrate equally readily. The γ form (olivine structure)[6] is very inert towards water and therefore valueless as a component of cement, which owes its strength to the interlocking crystals of the hydration products. The β form, on the other hand, which is metastable at ordinary temperatures, sets to a hard mass when finely ground and mixed with water.[7]

Structural studies of Portland cement compounds and their hydration products are directed towards an understanding of the reactivity of the former towards water, which varies greatly; β-Ca_2SiO_4, Ca_3SiO_5, and $Ca_3Al_2O_6$ are very 'hydraulically active'. This activity also implies the existence of stable hydration products. These include hydrated silicates, 'aluminates', and more complex compounds containing SO_4^{2-} ions arising from the gypsum which is added to control the setting time. Hydroxy silicates (p. 1010) include afwillite, $Ca_3(SiO_3OH).2H_2O$,[8] and dicalcium silicate α-hydrate, $Ca_2(SiO_3OH)OH$,[9] both of which contain tetrahedral $SiO_3(OH)^{3-}$ ions in which OH is directly bonded to Si.

In $Ca_3Al_2O_6$[10] there are rings of 6 AlO_4 tetrahedra (Al–O, 1.75 Å) and in the (metastable) $Ca_5Al_6O_{14}$[11] there are layers consisting of rings of 5 AlO_4 tetrahedra (see the melilite structure (pp. 197 and 1019), but hydrated products are built from octahedral $Al(OH)_6$ groups. These are present as discrete groups in compounds such as $Ca_3Al_2(OH)_{12}$ (p. 637), as edge-sharing pairs in $Ba_2[Al_2(OH)_{10}]$ (p. 637), and as groups of four forming the complex of Fig. 5.8(h) in $Ba_2[Al_4(OH)_{16}]$.[12] It has been pointed out that these are not necessarily the most stable structures for the particular $M^{2+}:Al^{3+}$ ratios *in the solid state*, but that the ions found are those existing *in solution*, and these are necessarily finite groups. For example, there are less strong Al^{3+}–Al^{3+} repulsions in the chain found in two forms of $Ba[AlO(OH)_2]_2$:[13]

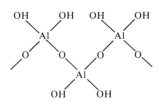

(1)   AM 1965 **50** 2023
(2)   AC 1958 **11** 862
(3a) ZK 1970 **132** 1
(3b) AC 1975 **B31** 2440
(4)   AC 1969 **B25** 1251
(5)   K 1975 **20** 721
(6)   AC 1971 **B27** 848

(7)   AC 1977 **B33** 1696
(8)   AC 1952 **5** 477
(9)   AC 1952 **5** 724
(10) AC 1975 **B31** 689, 1169
(11) AC 1978 **B34** 1422
(12) AC 1972 **B28** 519
(13) AC 1972 **B28** 760

Silicates containing $Si_2O_7^{6-}$ *ions*

We noted earlier that from the structural standpoint it is logical to group together all the tetrahedrally coordinated atoms in certain structures rather than single out only the Si_xO_y grouping as a particular type of silicate ion. The latter view regards as simple pyrosilicates only the rare minerals thortveitite, $Sc_2Si_2O_7$,[1] and barysilite, $MnPb_8(Si_2O_7)_3$,[2] and the 4f pyrosilicates, $M_2Si_2O_7$.[3] There has been considerable interest in pyrosilicates because of the variation in the Si–O–Si bond angle. The 4f pyrosilicates adopt one or more of four structures according to the size of M^{3+}:

(La) Ce . . . . . . Eu	Gd	Tb	Dy	(Y)	Ho	Er	Tm	Yb	Lu	Sc
c.n. of M ← 8 →				← 6 →			← 6 →			

← 7 → (thortveitite structure)

and these compounds provide examples of the whole range of Si–O–Si angles: 180° (Sc, Yb); 159° (Gd); and 133° (Nd). It is evident that the structure of the $Si_2O_7^{6-}$ ion is very dependent on its environment. There are also minerals which contain both SiO_4^{4-} and $Si_2O_7^{6-}$ ions:

Vesuvianite $Ca_{10}Al_4(Mg, Fe)_2(Si_2O_7)_2(SiO_4)_5(OH)_4$[4]
Epidote $Ca_2FeAl_2(SiO_4)(Si_2O_7)O(OH)$[5]

In the following crystals we may regard the Si_2O_7 groups as part of more extensive tetrahedral systems in which the larger cations are 8-coordinated (melilite, danburite):

Nature of tetrahedral complex

Melilite type	$Ca_2(MgSi_2O_7)$	A_3X_7 layer
Hemimorphite	$[Zn_4Si_2O_7(OH)_2]H_2O$	A_2X_3 3D framework
Danburite	$Ca(B_2Si_2O_8)$	AX_2 3D framework

Although some members of the melilite family are formally pyrosilicates (since they contain Si_2O_7 units) the basic feature of the structure is the A_3X_7 layer (p. 197). The tetrahedral positions may be occupied by various combinations of Si, Al, Be, Mg, and Zn, and the layer does not always have the composition (MSi_2O_7), as is evident from the formulae: $Ca_2(Al_2SiO_7)$, (gehlenite), $Y_2(Be_2SiO_7)$, and $Ca_{2.5}(Al_3O_7)$, p. 197.

Hemimorphite, an important zinc ore, is of interest as an example of the way in which X-ray studies lead to the assignment of correct structural formulae. It was formerly written $H_2Zn_2SiO_5$, but one-half of the H in this empirical formula is in the form of H_2O molecules which may be removed by moderate heating of the crystal, whereby its form and transparency are not destroyed. The removal of further H_2O, at temperatures above 500°C, results in destruction of the crystal. These facts are consistent with the crystal structure, which shows that the composition is

		Nature of tetrahedral complex
Melilite type	$Ca_2(MgSi_2O_7)$	A_3X_7 layer
Hemimorphite	$[Zn_4Si_2O_7(OH)_2]H_2O$	A_2X_3 3D framework
Danburite	$Ca(B_2Si_2O_8)$	AX_2 3D framework

$Zn_4(OH)_2Si_2O_7.H_2O$.[6] The structure may be regarded as built from $ZnO_3(OH)$ and SiO_4 tetrahedra: these share three vertices to form layers of the type shown, which are then joined through Si–O–Si or Zn–OH–Zn links to form a 3D framework. (The Si–O–Si angle is 150°.) There is a strong resemblance to the structure of Si_2N_2O, as may be seen by writing the formulae

$$Si_4 \quad Si_2 \quad N_6 \quad O \quad O_2$$
$$Zn_4 \quad Si_2 \quad O_6 \quad O \quad (OH)_2$$

and comparing with Fig. 23.5 (p. 997). The H_2O molecules are accommodated in the tunnels in the framework. We refer to danburite on p. 1035.

The structure of $Rb_2Be_2Si_2O_7$[7] is of interest as an example of planar 3-coordination of Be and also as an example of a framework incorporating both Be and Si. The Si_2O_7 groups (Si–O_b–Si, 171°) are joined through Be atoms, each of which is bonded to terminal O atoms of three different Si_2O_7 groups forming a 3D framework. Since each Be is bonded to 3 Si_2O_7 and each such group has 6 terminal O atoms, the composition of the framework is $Be_2Si_2O_7$. The Rb^+ ions are accommodated in large cavities in the structure, surrounded by 13 or 14 O atoms.

(1) AC 1962 **15** 491; SPC 1973 **17** 749
(2) AC 1966 **20** 357
(3) AC 1970 **B26** 484
(4) ZK 1931 **78** 422
(5) AC 1954 **7** 53
(6) ZK 1967 **124** 180
(7) AC 1977 **B33** 381

Silicates containing cyclic $(SiO_3)_n^{2n-}$ *ions*

The existence of the $Si_3O_9^{6-}$ and $Si_6O_{18}^{12-}$ ions in minerals has long been established. The former occurs in benitoite, $BaTiSi_3O_9$, and catapleite, $Na_2ZrSi_3O_9.2H_2O$, and the latter in beryl (emerald), $Be_3Al_2Si_6O_{18}$,[1] dioptase, $Cu_6Si_6O_{18}.6H_2O$[2] (originally formulated as an orthosilicate, CuH_2SiO_4), and tourmaline, (Na, Ca)-$(Li, Al)_3Al_6(OH)_4(BO_3)_3Si_6O_{18}$.[3] The structures of benitoite and beryl are similar in general type, as can be seen from the plans in Fig. 23.11. In each structure the cyclic ions are arranged in layers with their planes parallel, but the structures are not layer structures, as might appear from these plans, for the metal ions lie between the layers of anions and bind together the anions at different levels. Both the Ti^{4+} and Ba^{2+} ions in $BaTiSi_3O_9$ are 6-coordinated (the Ba–O distance being greater than the Ti–O), but in beryl the Be^{2+} ions are in tetrahedral and the Al^{3+} in octahedral holes. The low-temperature form of cordierite, $Al_3Mg_2(Si_5Al)O_{18}$, has a similar structure to beryl.

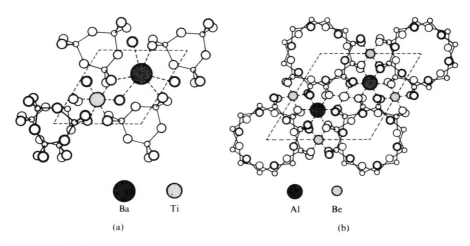

Ba   Ti   Al   Be

(a)   (b)

FIG. 23.11. Plans of the structures of (a) benitoite, $BaTiSi_3O_9$, and (b) beryl, $Be_3Al_2Si_6O_{18}$. Atoms belonging to complex ions lying at different heights above the plane of the paper are shown as heavy and light circles in order to show the nature of the coordination groups around the metal ions. Only one of the upper complex ions is shown in (a) and many oxygen atoms are omitted from (b) for the sake of clarity.

Since both Be and Si are tetrahedrally coordinated in beryl, and Be–O, 1.65 Å, is close to Si–O, 1.62 Å (contrast Al–O, 1.95 Å) it is reasonable to describe the structure as a 3D 4-connected net of composition $(Be_3Si_6O_{18})_n^{6n-}$, this formula showing that the framework is an AX_2 structure of 4:2 coordination. The Al^{3+} ion (0.61 Å) may be replaced in beryl by Cr^{3+} (0.70 Å) to th extent of 0.3 weight per cent to give emerald; contrast the replacement of Al in Al_2O_3 by Cr to form ruby.

The synthetic metasilicate $K_4(HSiO_3)_4$[4, 5] contains the silicon analogue of the cyclic $P_4O_{12}^{4-}$ ion. The $Si_4O_{12}^{8-}$ ion also exists in axinite, a complex borosilicate, but only the approximate structure is known;[6] the same ion may exist in the mineral baotite.[7] Recent additions to the crystal chemistry of metasilicates have been the discovery of the cyclic $Si_8O_{24}^{16-}$ ion in the mineral murite.[8] $Ba_{10}(Ca, Mn, Ti)_4Si_8O_{24}$-$(Cl, OH, O)_{12}.4H_2O$, and of $Si_9O_{27-y}(OH)_y$ in the complex mineral eudialyte.[9] Cyclic metasilicate ions are now known containing 3, 4, 6, 8, and 9 Si atoms.

(1) AC 1972 **B28** 1899
(2) AC 1955 8 425
(3) AC 1969 **B25** 1524
(4) AC 1964 **17** 1063
(5) AC 1965 **18** 574

(6) AC 1952 5 202
(7) K 1960 5 544
(8) Sc 1971 **173** 916
(9) Tschermaks Min. Petr. Mitt. 1971 **16** 105

Silicates containing chain ions

Single chains formed from tetrahedral MX_4 groups sharing two vertices adopt various configurations in crystals. Although there are different numbers of tetrahedra in the repeat unit, namely, 1, 2, and 3 for the chains (a), (b), and (c) of Fig. 23.12

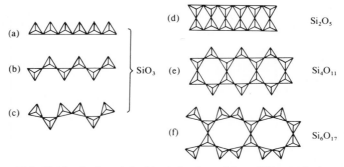

FIG. 23.12.   Single and double chains formed from SiO_4 tetrahedra.

(described as 1T, 2T, 3T, etc. chains) the composition is in all cases MX_3 since each tetrahedron shares two vertices. In the double chain (d) each tetrahedron shares three vertices, and in (e) and (f) there are different proportions of tetrahedra sharing 2 and 3 vertices. Examples of the various chains are summarized in Table 23.3. The most important single chain is (b) which occurs in synthetic metasilicates of Li and Na and in the pyroxene group of minerals, a group which includes enstatite, $MgSiO_3$; diopside, $CaMg(SiO_3)_2$; jadeite, $NaAl(SiO_3)_2$; and spodumene, $LiAl(SiO_3)_2$. In the various crystals the parallel chains pack so as to provide suitable environments for the cations—6-coordination of Mg^{2+}, 8-coordination of Ca^{2+} (in diopside).

Silicates $M_nSi_2O_5$ prefer 3-connected layer structures to the double chain (d) of Fig. 23.12—see next section. A folded form of this chain occurs in $Na_2Mg_2Si_6O_{15}$. However, one-half of the Mg atoms are in tetrahedral coordination (and the remainder in octahedral coordination) and if these are included a M_7O_{15} framework ($MgSi_6O_{15}$) would be recognized. In $(Na, K)Fe^{2+}(Fe^{3+}Si_6O_{15}).\frac{1}{2}H_2O$ Fe^{3+} replaces the tetrahedrally coordinated Mg in $Na_2Mg_2Si_6O_{15}$.

TABLE 23.3

Crystals containing single and multiple chain ions

Single chain $(MO_3)_n$		Multiple chain
1T	$CuGeO_3$, K_2CuCl_3	M_2O_5 (d) Al_2SiO_5 (sillimanite),
2T	Pyroxenes, Li_2SiO_3, Na_2SiO_3[1] $RbPO_3$,[2] $LiAsO_3$,[3] SO_3	$Na_2Mg_2Si_6O_{15}$,[18] (Na, K) $Fe_2Si_6O_{15}.\frac{1}{2}H_2O$[19]
3T	β-Wollastonite,[4] high-$NaPO_3$ (Maddrell salt),[5] $Ca_2NaHSi_3O_9$ (pectolite)[6]	M_4O_{11} (e) amphiboles M_6O_{17} (f) $Ca_6Si_6O_{17}(OH)_2$ (xonotlite)[20] P_5O_{14} (p. 863)
4T	Linear: Pb $(PO_3)_2$[7] Helical: $NaPO_3$ (Kurrol salt A),[8] $AgPO_3$[8]	Si_6O_{16}
5T	$CaMn_4Si_5O_{15}$[9] (rhodonite)	Si_8O_{21} } (p. 1014)
6T	$Bi_2P_6O_{18}$,[10] $Ca_2Sn_2Si_2O_{18}$[11]	$Si_{10}O_{26}$
7T	(Ca, Mg) (Mn, Fe)$_6Si_7O_{21}$[12] (pyroxmangite)	
8T	$Na_3H(PO_3)_4$[13]	
9T	$Fe_9Si_9O_{27}$[14]	
10T	$LiPO_3$,[15] $Ba_2K(PO_3)_5$[16]	
12T	$Pb_{12}Si_{12}O_{36}$[17]	

(1) AC 1967 **22** 37. (2) AC 1964 **17** 681. (3) AC 1956 **9** 87. (4) PNAS 1961 **47** 1884. (5) AC 1955 **8** 752. (6) ZK 1967 **125** 298. (7) AC 1964 **17** 1539. (8) AC 1961 **14** 844. (9) ZK 1963 **119** 98. (10) AC 1975 **B31** 2281. (11) MM 1963 **33** 615. (12) AC 1959 **12** 177. (13) AC 1968 **B24** 992. (14) Sc 1966 **154** 513. (15) AC 1976 **B32** 2960. (16) ZK 1975 **141** 403. (17) ZK 1968 **126** 98. (18) AC 1972 **B28** 3583. (19) Sc 1969 **166** 1399. (20) AC 1956 **9** 1002.

Sillimanite is strictly an orthosilicate, for it contains discrete SiO_4 groups. However, since one-half of the Al atoms are 4-coordinated it may alternatively be regarded as an aluminosilicate in which the Si atoms and these Al atoms form double chains of type (d). The octahedrally coordinated Al atoms lie between these chains, and the compound may be formulated $Al(AlSiO_5)$. In the mineral world the most important compounds containing double chains are the amphiboles, of which tremolite, $(OH)_2Ca_2Mg_5(Si_4O_{11})_2$, is a typical member. The term *asbestos* was originally reserved for the more fibrous amphiboles, for example, tremolites, actinolites,[1] and crocidolites.[2] The name is also, however, applied to fibrous varieties of serpentine which are not amphiboles but minerals more closely related to talc and the other clay minerals. In fact, some 50 per cent of commercial asbestos is chrysotile, for the structure of which see p. 1029. Synthetic amphiboles (NaMg, NaCo, NaNi) have been prepared hydrothermally as long, flexible, fibrous crystals.[3]

A different type of chain, also with the composition Si_4O_{11}, has been found in the mineral vlasovite, $Na_2Zr(Si_4O_{11})$.[4] This consists of rings of four tetrahedra and

like the amphibole chain has equal numbers of tetrahedra sharing two and three vertices. An example of the Si_6O_{17} chain is included in Table 23.3. Also included in the Table are some other meta-salts containing tetrahedral chain ions. Chains differ not only in the number of tetrahedra in the repeat unit along the chain (1T, 2T, etc.) but also in the relative orientations of the tetrahedra within a repeat unit. In

Table 23.3 we distinguish two types of 4T chain, one essentially linear (two pairs of tetrahedra) and the other helical.

(1)   AC 1955 **8** 301   (3)   IC 1964 **3** 1001
(2)   AC 1960 **13** 291   (4)   CRURSS 1961 **141** 958

Silicates with layer structures

Three main classes of layer structure may be recognized.

(i) There are single layers of composition Si_2O_5, or $(Si, Al)_2O_5$, held together by cations. The simplest layers of this kind are those in which each SiO_4 tetrahedron shares 3 vertices with 3 others, the structure being based on one of the plane 3-connected nets (Chapter 3). Silicate structures are known which are based on the (only) *regular* 3-connected plane net, (6^3), on two of the three *semi-regular* 3-connected nets, (4.8^2) and $(4.6.12)$, and on the more complex net $(6.8^2)(4.6.8)_2$, Fig. 23.13(a)–(d). It has been suggested, though not proved, that the mineral okenite is based on the net of Fig. 23.13(e). Examples of these nets include:

(a) (6^3) net: $Li_2Si_2O_5$ and the isostructural $H_2Si_2O_5$,[1] $Na_2Si_2O_5$,[2] $Ag_2Si_2O_5$,[3] $LiAlSi_4O_{10}$ (petalite)[4]

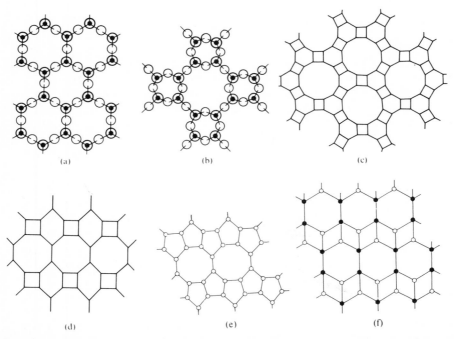

(a)   (b)   (c)

(d)   (e)   (f)

FIG. 23.13. (a)–(e) Layers of SiO_4 tetrahedra sharing 3 vertices. In (a) and (b) the small black circles represent Si atoms and the open circles O atoms. Oxygen atoms lying above Si atoms are drawn more heavily. The O atoms are omitted from the nets (c)–(f). (f) the Si_4O_9 layer built from equal numbers of SiO_4 tetrahedra sharing 3 or 4 vertices.

(b) (4.8²) net: BaFeSi₄O₁₀ (gillespite),[5] CaCuSi₄O₁₀ (Egyptian blue),[6]
Ca₄Si₈O₂₀.KF.8H₂O (apophyllite)[7]

(c) (4.6.12) net: $M^{II}_{16}Si_{12}O_{30}(OH, Cl)_{20}$ (manganpyrosmalite)[8]

(d) (6.8²)(4.6.8)₂ net: $K_{1.7}Na_{0.3}ZrSi_6O_{15}$ (dalyite)[9]

(e) (5.8²)(5²8)₂ net: $CaSi_2O_5.2H_2O$ (okenite).[10, 11]

An interesting 'topochemical' reaction is the conversion of a single crystal of
α-Na₂Si₂O₅ into one of Ag₂Si₂O₅ by heating in molten AgNO₃. Acids H₂Si₂O₅ have
been prepared which are structurally similar to two of the alkali disilicates, and
gillespite from which the metal ions have been leached by 5N HCl gives sufficient
X-ray diffraction to show that the layer structure is retained in the crystals (of
composition H₄Si₄O₁₀.½H₂O), which are colourless and have a density of 2.1 g cm⁻³
—compare 3.4 g cm⁻³ for the red crystals of gillespite.[12]

All the layers of Fig. 23.13 are *simple* layers in the sense that they can be
represented on a plane without any links overlapping or intersecting. More complex
layers can be envisaged which have some extension in a direction perpendicular to
the layer, as in the double layer described in (ii). An intermediate possibility is
realized in Na₂Si₃O₇.[13] We may represent the end-on view of an idealized metasilicate chain as in Fig. 23.14(a). In (b) each pair of vertex-sharing tetrahedra

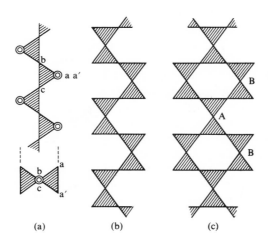

(a)   (b)   (c)

FIG. 23.14. (a) Metasilicate chain ion $(SiO_3)_n^{2n-}$ and end-on view. (b) Layer of composition
Si₂O₅ built of SiO₃ chains perpendicular to the plane of the paper. (c) Layer of composition
Si₃O₇ similarly represented.

represents a chain perpendicular to the plane of the paper, so that (b) represents a
buckled layer in which each SiO₄ group shares 3 vertices. This is approximately
the configuration of the Si₂O₅ layer in Li₂Si₂O₅. In (c) the tetrahedra in the B

chains share 3 vertices and those in the A chains all 4 vertices, these layers being members of the series Si_mO_{2m+1}:

$$SiO_3 \qquad Si_2O_5 \ \ Si_3O_7 \quad \ldots \quad SiO_2$$
$$\text{chain} \qquad \text{layers} \qquad \qquad \text{3D framework}$$

All the layers are in fact slices of the cristobalite structure, which is the end-member of the series, and the Si_3O_7 layer is the anion in $Na_2Si_3O_7$. This A_3X_7 layer, in which all circuits are rings of six tetrahedra, may be compared with the simple A_3X_7 layer in melilite and related compounds which is composed entirely of rings of five tetrahedra. We referred to this layer under pyrosilicates; it is illustrated in Fig. 5.7(c), p. 197. In both A_3X_7 layers two-thirds of the tetrahedra share 3 vertices and the remainder 4 vertices. A layer in which *equal* numbers of tetrahedra share 3 or 4 vertices is found in $K_2Si_4O_9$;[14] it is illustrated in Fig. 23.13(f).

It will be appreciated that the type of structure (finite, or infinite in 1, 2, or 3 dimensions) cannot in general be deduced from the O:Si ratio. Whereas the sharing of only 2 vertices can give only rings or chains (O:Si = 3), the sharing of 3 or 4 vertices by every SiO_4 group can result in complexes of all types, with O:Si equal to 2.5 or 2.0 respectively. (We saw in Chapter 5 that the sharing of 4 vertices can produce 1D (cylindrical), 2D or 3D structures.) The O:Si ratio has values between 2.5 and 2.0 if there is a mixture of 3- and 4-connected tetrahedra, and between 2.5 and 3.0 if there are 3- and 2-connected tetrahedra.

For the composition Si_2O_5 we have already noted the double chain (Fig. 23.12(d)) and various layers (Fig. 23.13(a)–(e)), there seem to be no silicates containing 3D Si_2O_5 frameworks. Another possibility is to wrap strips of plane nets around a cylinder. A simple vertical strip of the (4.8^2) net of Fig. 23.24(b), p. 1039, has the composition Si_4O_{11} (as in vlasovite, p. 1023), but strips two or more 4-membered rings wide can be wrapped around a cylinder; the composition is the same, $(Si_2O_5)_n$, as for the plane net since each tetrahedron shares 3 vertices. This type of cylindrical ion is found in $CuNa_2Si_4O_{10}$[15] and a number of minerals.

If some tetrahedra share 2 and others 3 or 4 vertices values of the O:Si ratio between 2 and 3 result. A simple example is the layer in the mineral zeophyllite;[16] the structure is based on the plane (6^3) net but there is a 2-connected SiO_4 group inserted along each link, giving the composition Si_5O_{14}. The above examples are summarized below.

(ii) If the (unshared) vertices of all the tetrahedra of a plane layer lie on the same side of the layer, two layers can be combined to form a double layer which has the composition $(Si, Al)O_2$, since all the vertices of each tetrahedron are now shared. There must be replacement of some of the Si by Al (or possibly Be) since otherwise the layer would be neutral, of composition SiO_2. The hexagonal forms of $CaAl_2Si_2O_8$[17] and $BaAl_2Si_2O_8$[18] contain double layers of this kind (Fig. 23.15) formed from two 6-gon layers of the type of Fig. 23.13(a).

The simplest plane 3-connected net is found as a single layer of composition

Formula	O : Si *ratio*	*Type of structure*	*Number of vertices shared*
$(Si, Al)O_2$	2.0	1-, 2-, or 3D (Chapter 5)	4
Si_4O_9	2.25	Layer of Fig. 23.13(f) $\}$	3 and 4
Si_3O_7	2.33	See text	
Si_2O_5	2.5	Double chain (Fig. 23.12(d)) Layers (Fig. 23.13(a)–(e)) Cylindrical ion (see text)	3
Si_4O_{11}	2.75	Single chain (vlasovite) Double chain (amphiboles)	3 and 2
Si_5O_{14}	2.8	Layer (zeophyllite)	
Si_6O_{17}	2.83	Double chain (xonotlite)	

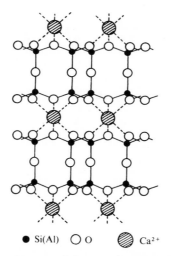

● Si(Al) ○ O ◍ Ca^{2+}

FIG. 23.15. Elevation of the structure of hexagonal $CaAl_2Si_2O_8$ showing the double layers interleaved with Ca^{2+} ions.

Si_2O_5 in (i) and forms the double layer in the structures just mentioned in (ii). There is a third way in which this net plays a part in silicate structures which is due to the approximate dimensional correspondence between a plane hexagonal Si_2O_5 layer and the $Al(OH)_3$ or $Mg(OH)_2$ layers.

(iii) The third, and by far the most important, family of layer structures is based on composite layers built of one or two silicon–oxygen layers combined with layers of hydroxyl groups bound to them by Mg or Al atoms. These composite layers are the units of structure, and we first consider their internal structure before seeing how they are packed together in crystals. Figure 23.16(a) shows a portion of a sheet of linked SiO_4 tetrahedra, each sharing three vertices, with all the unshared vertices pointing to the same side of the layer. Figure 23.16(b) shows the arrangement of the OH groups in a layer of the $Mg(OH)_2$ or $Al(OH)_3$ structures. The distance

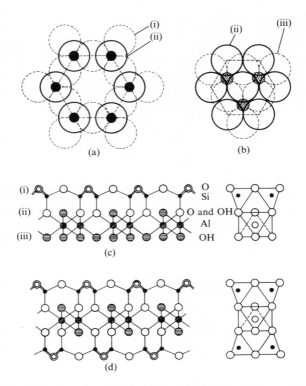

FIG. 23.16.   The formation of composite silicon–aluminium–oxygen or silicon–magnesium–
oxygen layers (see text).

between the upper O atoms in (a), drawn with heavy full lines, is approximately the
same as that between the corresponding OH groups in (b). It is therefore possible
by inverting (a) and placing it on top of (b) to form a composite layer having these
O atoms in common. The composite layer is shown in elevation in Fig. 23.16(c); it
has been illustrated in Fig. 5.46 (p. 237) as an example of a structure built from
tetrahedral and octahedral coordination groups. It is possible to repeat this process
of condensation on the other side of the (b) layer giving the more complex layer
shown at (d). In illustrating structures based on these layers we shall use the
diagrammatic forms shown to the right of (c) and (d) to represent layers of these
types extending indefinitely in two dimensions, a simplification suggested by
Pauling. Just as the $Mg(OH)_2$ and $Al(OH)_3$ layers have the same arrangement of OH
groups and differ, from the geometrical point of view, only in the number of
octahedral holes occupied by metal atoms, so in these composite layers there may
be either Mg or Al ions or, of course, other di- or tri-valent ions of appropriate size.
The ideal compositions of these layers in the extreme cases are:

(c) $Mg_3(OH)_4Si_2O_5$ or $Al_2(OH)_4Si_2O_5$
(d) $Mg_3(OH)_2Si_4O_{10}$ or $Al_2(OH)_2Si_4O_{10}$

according to whether the octahedral holes are occupied by Mg or Al. The structural chemistry of these layer minerals is further complicated by the fact that Si can be partly replaced by Al (in tetrahedral coordination) giving rise to charged layers. Such layers are either interleaved with positively charged alkali or alkaline-earth ions, as in the micas, by layers of hydrated ions in montmorillonite, or by positively charged $(Mg, Al)(OH)_2$ layers in the chlorites. We may distinguish five main structural types, remembering that in each of the five classes the composition may range from Mg_3 to Al_2 as already indicated.

Layers of type (c) *only.* The pure Mg layer occurs in chrysotile, $(OH)_4Mg_3Si_2O_5$, in which the structural unit is a kaolin-like layer of this composition, instead of $(OH)_4Al_2Si_2O_5$ as in Fig. 23.16. Since the dimensions of the brucite $(Mg(OH)_2)$ part of the composite layer do not exactly match those of the Si_2O_5 sheet, the layer curls up, the larger (brucite) portion being on the outside. The fibres are built of curled ribbons forming cylinders several thousand Å long with a dozen or so layers in their walls and overall diameters of several hundred Å;[19] the detailed structures of chrysotiles are still not known, nor is the nature of the material filling the tubes.

We referred earlier to the silylation of silicates to give trimethylsilyl derivatives corresponding to the Si–O–Si system in the original minerals. A development of this idea is to treat chrysotile with a mixture of HCl and $ClSi(CH_3)_3$ when the Mg and OH ions are stripped off the outside of the layer and replaced by $-OSi(CH_3)_3$, giving a gel which on treatment with water forms a fibrous mass of ribbons. When dry these curl up to form fibres similar in shape to those of the original chrysotile.[20]

The $(OH)_4Al_2Si_2O_5$ layer is the structural unit in the kaolin (china-clay) minerals (Fig. 23.17). The three minerals kaolinite, dickite,[21] and nacrite all have the composition $Al_2(OH)_4Si_2O_5$ and their structures differ only in the number of kaolin layers (1, 2, and 6 respectively) in the repeat unit. In halloysite,[22] the commonest of the kaolin minerals, there is an irregular sequence of layers; in amesite, with the ideal formula $(Mg_2Al)(SiAlO_5)(OH)_4$, the replacement of part of the Mg_3 by Al is balanced by substitution of Al for half of the Si.[23]

Layers of type (d) *only.* The two extremes are talc, $Mg_3(OH)_2Si_4O_{10}$, and pyrophyllite, $Al_2(OH)_2Si_4O_{10}$ (Fig. 23.17). As in the kaolins the layers are electrically neutral, and there are only feeble attractive forces between neighbouring layers. These minerals are therefore soft and cleave very easily, and talc, for example, finds applications as a lubricant (french chalk).

Charged layers of type (d) *interleaved with ions.* Replacement of one-quarter of the Si in talc and pyrophyllite layers gives negatively-charged layers which are interleaved with K^+ ions in the *micas* phlogopite and muscovite (Fig. 23.17):

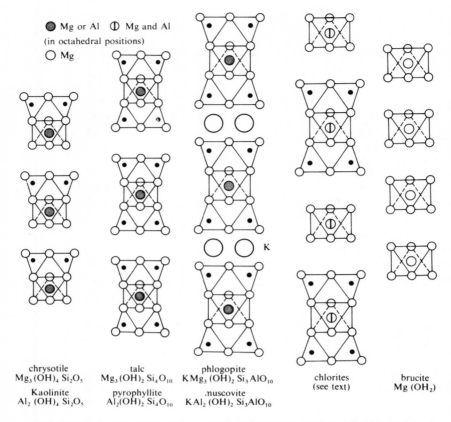

FIG. 23.17. The structures of some minerals with layer structures (diagrammatic).

$$KMg_3(OH)_2Si_3AlO_{10} \quad \text{phlogopite}[24]$$
$$KAl_2(OH)_2Si_3AlO_{10} \quad \text{muscovite}[25]$$

The potassium ions occupy large holes between twelve oxygen atoms so that the K–O electrostatic bond strength is only one-twelfth. These bonds are easily broken and the micas accordingly possess very perfect cleavage parallel to the layers. In the so-called brittle micas, typified by margarite, $CaAl_2(OH)_2Si_2Al_2O_{10}$, further substitution of Al for Si has taken place so that the negative charge on the layers is twice as great as in phlogopite and muscovite. The layers are therefore held together by divalent ions such as Ca^{2+}. The increased strength of the interlayer Ca–O bonds, with electrostatic bond strength now equal to one-sixth, results in greater hardness, as shown by the figures for the hardnesses on the Mohs scale:

	Hardness
Talc, pyrophyllite	1–2
Micas	2–3
Brittle micas	3½–5

Charged layers of type (d) *interleaved with hydrated ions*. In addition to the micas, which are anhydrous, there are some very important minerals, sometimes called 'hydrated micas', which are built of layers with smaller charges per unit area than in the micas, interleaved with layers of hydrated alkali or Mg^{2+} ions. Such feebly charged layers can arise by replacing part of the Al_2 in a pyrophyllite layer by Mg as in

$$[\underbrace{Mg_{0.33}Al_{1.67}}_{\text{total } 2.00}Si_4O_{10}(OH)_2]^{-0.33}Na_{0.33} \quad \text{(montmorillonite)}^{(26)}$$

total 2·00

or there can be substitution in a talc layer not only in the octahedral positions but also in the tetrahedral positions, as in vermiculite:

$$[(Mg_{2.36}Fe^{3+}_{0.48}Al_{0.16})(Si_{2.72}Al_{1.28})O_{10}(OH)_2]^{-0.64}[Mg_{0.32}(H_2O)_{4.32}]^{+0.64}$$

As might be expected, this mineral dehydrates to a talc-like structure. A section through the structure[27] is shown in Fig. 23.18, but only a proportion of the interlamellar Mg^{2+} and H_2O positions are occupied.

Clay minerals. It is customary to group under this general heading a number of groups of minerals with certain characteristic properties. They are soft, easily

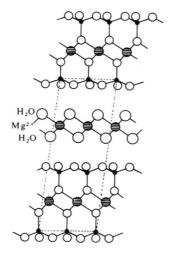

FIG. 23.18. Elevation of the structure of vermiculite.

hydrated, and exhibit cation exchange with simple ions and also with many organic ions. They include the kaolin group, the montmorillonite–beidellite group, and the illite clays, related to vermiculite. Apart from their fundamental importance as constituents of soils they find many industrial applications. Kaolins are used in pottery and ceramics, as rubber fillers, and for filling and coating paper. Montmorillonite (bentonite), an important constitutent of fuller's earth, finds many applications owing to its property of forming thick gelatinous suspensions at concentrations of only a few per cent. For example, sodium bentonite swells readily in water and is used as a binder for foundry sand and as a thickener in oil-well drilling muds, while the non-swelling calcium bentonite is converted by acids into adsorbents used in oil refining. Vermiculite has many uses, ranging from soil conditioner to a porous filler in light-weight plastics and concretes.

Charged layers of two kinds. In all these layer structures so far described all the layers in a given crystal are of the same kind. There are also more complicated structures with layers of two kinds, and the chlorite minerals are in this class. In kaolin, talc, and pyrophyllite the uncharged layers are held together by comparatively weak forces, and in the micas the negatively-charged layers are cemented together by positive ions. In the chlorites another possibility is realized, the alternation of negatively-charged layers of the mica type with positively-charged layers, that is, the structures contain infinite 2-dimensional ions of opposite charge. The mica-like layers have compositions ranging from $[Mg_3(AlSi_3O_{10})(OH)_2]^-$ to $[Mg_2Al(Al_2Si_2O_{10})(OH)_2]^-$. The positive layers result from the replacement by Al of one-third of the Mg in a brucite $(Mg(OH)_2)$ layer, giving the composition $Mg_2Al(OH)_6^+$. The sequence of layers in the chlorites is compared with that in $Mg(OH)_2$ and phlogopite in Fig. 23.17. Since there are many ways of stacking the layers in the chlorites, as in other layer structures, polytypes are numerous.[28]

(1)   ZK 1964 **120** 427
(2)   AC 1968 **B24** 13, 1077
(3)   AC 1961 **14** 537
(4)   AC 1961 **14** 399
(5)   AM 1943 **28** 372
(6)   AC 1959 **12** 733
(7)   AM 1971 **56** 1222, 1234, 1243
(8)   AC 1968 **B24** 690
(9)   ZK 1965 **121** 349
(10) DAN 1958 **121** 713
(11) MM 1956 **31** 5
(12) AM 1958 **43** 970
(13) N 1967 **214** 794
(14) AC 1974 **B30** 2206
(15) AC 1977 **B33** 1071
(16) AC 1972 **B28** 2726
(17) AC 1959 **12** 465
(18) MJ 1958 **2** 311
(19) AC 1956 **9** 855, 862, 865
(20) IC 1967 **6** 1693
(21) AC 1956 **9** 759
(22) AC 1975 **B31** 2851
(23) AC 1951 **4** 552
(24) AC 1966 **20** 638
(25) AC 1960 **13** 919
(26) TFS 1948 **44** 306, 349
(27) AM 1954 **39** 231
(28) AC 1969 **B25** 632

Silicates with framework structures

The possible types of 3D structure have been noted earlier in this chapter, namely:

(a) 3 vertices of each SiO_4 tetrahedron shared: composition $(Si_2O_5)_n^{2n-}$ or more highly charged if some Si is replaced by Al (or Be);

(b) 3 or 4 vertices shared;

(c) 4 vertices shared: composition SiO_2 or $(Si, Al)O_2$, the charge on the framework in the latter case depending on the extent to which Al replaces Si.

We have already described the forms of SiO_2 and noted some minerals with closely related structures in which some Si is replaced by Al or Be. In each class we could imagine additional tetrahedra inserted between the 3- or 4-connected ones and having the effect of simply extending the links in the basic 3-, (3,4)-, or 4-connected net. Little is known of structures of classes (a) or (b), but there is an interesting example of a framework based on tetrahedra sharing 3 vertices some of which are linked through 2-connected tetrahedra. In neptunite, $LiNa_2K(Fe, Mg, Mn)_2$-$(TiO)_2Si_8O_{22}$,[1] there are equal numbers of SiO_4 tetrahedra sharing 2 and 3 vertices. They form two interpenetrating 3-connected nets of the $ThSi_2$ type (Fig. 3.15(e), p. 88) in which there are two additional tetrahedra (sharing two vertices) inserted along one-third of the links of each net. Such nets have the same O:Si ratio as the amphibole chain: $OSi(O_{1/2})_2(Si_2O_7)_{1/2} \equiv Si_4O_{11}$. They account for all except two of the O atoms in the above formula; these are not bonded to Si but, together with the O atoms of the frameworks, complete octahedral coordination groups around the Ti atoms.

We shall be concerned here exclusively with frameworks in which tetrahedra share all four vertices. In all these structures some (often about one-half) of the tetrahedral positions are occupied by Al (rarely by Be), and positive ions are present to neutralize the negative charge of the $(Si, Al)O_2$ framework. These framework structures, and some of their physical properties, are more easily understood if they are subdivided according to whether or not there are obvious polyhedral cavities or tunnels in the structure. We shall deal here with three groups of minerals (and some related synthetic compounds), the felspars, zeolites, and ultramarines. The felspar structures are relatively compact, but in the other two groups there are polyhedral cavities or tunnels in which are accommodated water molecules (in the zeolites) or finite anions (Cl^-, SO_4^{2-}, S^{2-}, etc. in the ultramarines) in addition, of course, to the necessary numbers of cations. Between the felspars and the framework structures which may be described as space-filling arrangements of polyhedra lie structures of an intermediate nature. For example, milarite, $K_2Ca_4(Be_4Al_2Si_{24}O_{60}).H_2O$[2] contains tightly knit hexagonal prisms built from 12 tetrahedra which are linked through single tetrahedra into a 3D network (Fig. 23.19). The prisms have the composition $(Be_{0.10}Si_{0.90})_{12}O_{30}$, while the separate tetrahedra have the mean composition $(Be_{0.27}Al_{0.37}Si_{0.40})O_4$.

A simpler example of a compound with the milarite structure is $K_2Mg_5Si_{12}O_{30}$.[2a] The K^+ ions are in positions of 9- or 12-coordination and the Mg^{2+} ions are of two kinds, 2 Mg^{2+}, 6-coordinated, and 3 Mg^{2+}, 4-coordinated. The latter form the connections between the hexagonal prisms and together with the Si atoms form the AX_2 framework of composition $(Mg_3Si_{12}O_{30})_n^{6n-}$.

Felspars. These are the most important rock-forming minerals, comprising some

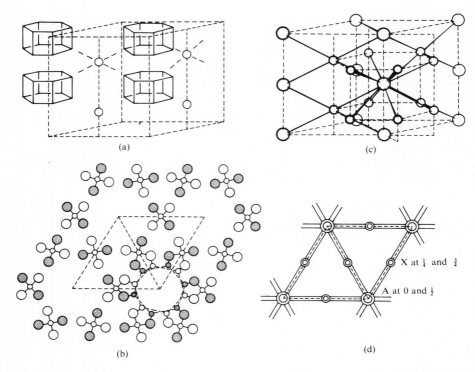

FIG. 23.19. The structure of milarite. In (a) the Si_{12} prisms have been shrunk to show how the Be atoms link them together to form the 3D framework. The actual environment of Be is a tetrahedral group of O atoms belonging to four different $Si_{12}O_{30}$ prisms, as shown in the projection (b). The shaded O atoms belong to the lower 6-rings of $Si_{12}O_{30}$ groups and the unshaded ones to the upper 6-rings of prisms below. If the Si_{12} prisms are shrunk to become 12-connected points the basic framework is the (12,4)-connected AX_3 net shown at (c) and in plan at (d).

two-thirds of the igneous rocks. Granite, for example, is composed of quartz, felspars, and micas. The felspars are subdivided into two groups according to the symmetry of their structures, typical members of the groups being:

(1) Orthoclase $KAlSi_3O_8$[3]
    Celsian $BaAl_2Si_2O_8$

(2) The plagioclase felspars: Albite $NaAlSi_3O_8$
                               Anorthite $CaAl_2Si_2O_8$.[4]

We have already noted that there are hexagonal forms of both $BaAl_2Si_2O_8$ and $CaAl_2Si_2O_8$ with double-layer structures. It will be noticed that in the first example in each group one-quarter of the tetrahedral positions are occupied by Al, requiring a monovalent ion to balance the charge on the $AlSi_3O_8$ framework. In celsian and

FIG. 23.20. Stereoscopic photographs showing the structures of the three forms of silica: (A) cristobalite; (B) quartz; and (C) tridymite, as systems of linked SiO$_4$ tetrahedra.

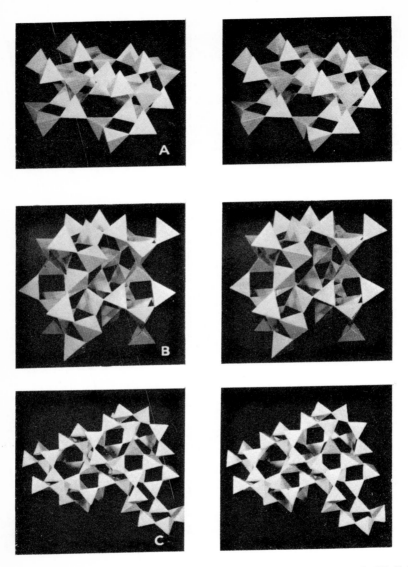

FIG. 23.21. Stereoscopic photographs showing the structures of (A) paracelsian, $Ba(Al_2Si_2O_8)$; (B) a fibrous zeolite, $Na(AlSi_2O_6).H_2O$; and (C) ultramarine, $Na_8(Al_6Si_6O_{24})S_2$, as systems of linked $(Al, Si)O_4$ tetrahedra.

anorthite further replacement of Si by Al necessitates the introduction of divalent ions. The plagioclase felspars, albite and anorthite, are of peculiar interest since they are practically isomorphous and many minerals of intermediate composition are known. The composition of labradorite, for example, ranges from AbAn to $AbAn_3$, where Ab = albite and An = anorthite. This isomorphous replacement of (K + Si) by (Ba + Al) or (Na + Si) by (Ca + Al) is characteristic of felspars and other framework silicates. The felspars of the first group contain the large K^+ and Ba^{2+} ions (radii 1.38 and 1.35 Å respectively), whereas the plagioclase felspars contain the smaller Na^+ and Ca^{2+} ions (radii 1.02 and 1.00 Å respectively). It may be noted here that only the comparatively large positive ions are found in felspars. The smaller ions of Fe, Cr, Mn, etc., so commonly found in chain and layer silicates do not occur in these minerals, presumably because the frameworks cannot close in around these smaller ions. The difference between the symmetries of the two groups of felspars is associated with this difference in size of the positive ions, for the framework is essentially the same in all the above felspars but contracts slightly in the second group around the smaller sodium and calcium ions.

The $(Si, Al)O_2$ frameworks of felspars may be regarded as built from layers of tetrahedra placed at the points of the plane 4:8 net of Fig. 23.13(b), p. 1024, the fourth vertex of each tetrahedron pointing either upwards or downwards out of the plane of the paper. These layers are then joined through the projecting vertices so that adjacent layers are related by planes of symmetry. Different 3-dimensional frameworks arise according to the relative arrangement of upward- and downward-pointing tetrahedra. The layers of Fig. 23.22(a) give the framework of paracelsian,[5] one of the forms of $Ba(Al_2Si_2O_8)$ (and the isostructural danburite, $Ca(B_2Si_2O_8)$), while that of the common felspars arises by joining up the more complex layers of Fig. 23.22(b). The idealized paracelsian framework is illustrated in Fig. 23.21(a). The compound $Ba(Al_2Si_2O_8)$ provides one of the best examples of polymorphism. Not only are there the two felspar structures of celsian and paracelsian with different 3-dimensional frameworks based on the nets of Fig. 23.22, but there is also the hexagonal form with the double-layer structure described on p. 1027). Moreover, the latter has low- and high-temperature modifications which differ in the precise configuration of the hexagonal layers, with different coordination numbers of the Ba^{2+} ions between the layers. The anion in $Na_2(ZnSi_3O_8)$[5a] has a similar framework to that of paracelsian, illustrating the incorporation of Zn in a tetrahedral framework along with Si.

It seems likely that the finer structural differences between felspars of similar types may be connected with the degree of ordering of Si and Al in the tetrahedra of the frameworks, for there is an appreciable difference between the distances Al–O and Si–O. Estimates have been made of the percentage of Si and Al occupying a particular set of tetrahedral sites in a number of structures, though there has been some discussion of the precise values of Si–O and Al–O to be adopted as standards.[6] For felspars values close to 1.605 Å for Si–O and 1.76 Å for Al–O are probably appropriate. The difference between the low- and high-temperature forms of albite,

 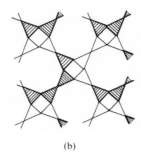

(a)   (b)

FIG. 23.22. Layers of tetrahedra which, by further linking through O atoms, form the framework structures of (a) paracelsian and (b) the common felspars. Open and shaded triangles represent tetrahedra the fourth vertices of which project either above or below the plane of the paper.

$NaAlSi_3O_8$, is apparently a question of the ordering of Si and Al in the tetrahedral sites. In the low-temperature form there is an ordered arrangement of Al in one-quarter of the tetrahedra and Si in the remaining sites, but there is a random arrangement in the high-temperature form $(Al(Si)-O, 1.64 Å)$.[7] A similar value was found for both of the two types of non-equivalent sites in sanidine, corresponding to a statistical arrangement of $\frac{1}{4}Al+\frac{3}{4}Si$ in each type of tetrahedron.[8]

We noted earlier the change of a synthetic sanidine $(KAlSi_3O_8)$ under high pressure (120 kbar, 900 °C) to the hollandite structure, in which Si is 6-coordinated.[9]

(1) AC 1966 **21** 200
(2) AC 1952 **5** 209
(2a) AC 1972 **B28** 267
(3) AC 1961 **14** 443
(4) AC 1962 **15** 1005, 1017
(5) AC 1953 **6** 613

(5a) AC 1977 **B33** 1333
(6) AC 1968 **B24** 355
(7) AC 1969 **B25** 1503
(8) AC 1949 **2** 280
(9) AC 1967 **23** 1093

Zeolites.   The zeolites, like the felspars, consist of $(Si, Al)_nO_{2n}$ frameworks, substitution of Al for some of the Si in SiO_2 giving the framework a negative charge which is balanced by positive ions in the cavities. The structures are much more open than those of the felspars, and a characteristic property of the zeolites is the ease with which they take up and lose water, which is loosely held in the structure. In addition to water, a variety of other substances can be absorbed, including gases like CO_2 and NH_3, alcohol, and even mercury. Also the positive ions may be interchanged for others merely by soaking the crystal in a solution of the appropriate salt. The 'permutites' used for water-softening are sodium-containing zeolites which take out the calcium from the hard water and replace it by sodium, thereby rendering it 'soft'. The permutite is regenerated by treatment with brine, when the reverse process—replacement of Ca by Na—takes place, making the permutite ready for further use. In this interchange of positive ions the actual number of ions in the crystal may be altered provided the total charge on them remains the same. Thus one-half of the Ca^{2+} ions in the zeolite thomsonite, $NaCa_2(Al_5Si_5O_{20}).6H_2O$, may be replaced by 2 Na^+, thereby increasing the total number of positive ions from three to four, there being sufficient unoccupied positions in the structure for the additional sodium ions.

In recent years many new zeolites have been synthesized and found to have other valuable properties in addition to the absorptive and ion-exchange properties already noted, notably their behaviour as molecular sieves and their catalytic properties. Nearly 40 different zeolite frameworks are described (and beautifully illustrated by stereoscopic pairs) in the *Atlas of Zeolite Structure Types* by W. M. Meier and D. H. Olson, published by the Structure Commission of the International Zeolite Association, 1978. Some of these synthetic compounds have the same framework structures as natural zeolites. others have more complex structures. For example, the closely related ZSM-5 and ZSM-11 are notable for containing many 5-membered rings and tunnels defined by 10-membered rings and for their very low Al:Si ratios (for example, $Na_3Al_3Si_{93}O_{192}. \approx 16H_2O$).[1][2] They are useful catalysts for the conversion of methanol to (water and) hydrocarbons suitable for motor fuels. The Al:Si ratio in most natural zeolites ranges from 1:1 to about 1:5; the very much lower ratios that are attainable in ZSM-5 and ZSM-11 suggest the possibility of pure SiO_2 frameworks of these types, and this has been realized in the case of the ZSM-5 structure (see p. 1041).

A characteristic feature of zeolites is the existence of tunnels (or systems of interconnected polyhedral cavities) through the structures, and we may therefore expect three main types in which the tunnels are parallel to (a) one line, giving the crystals a fibrous character; (b) two lines and arranged in planes, so that the crystals have a lamellar nature; or (c) three non-coplanar lines, such as cubic axes, when the crystals have no pronounced fibrous or lamellar structure. The most symmetrical frameworks have cubic symmetry, and in a number of zeolites the framework corresponds to the edges in a space-filling arrangement of polyhedra.

(a) The structure of a typical fibrous zeolite, edingtonite, $Ba(Al_2Si_3O_{10}).4H_2O$,[3] is illustrated in Fig. 23.23. The structure may be described in terms of a characteristic chain formed by the regular repetition of a group of five tetrahedra. These chains are linked to other similar chains through the projecting O atoms. These atoms, shown as shaded circles in Fig. 23.23(b), are at heights $3c/8$ and $5c/8$ (c being the repeat distance along the chain) above the Si atom which lies on the axis of the chain. Although the resulting structure contains a 3D framework there is a marked concentration of atoms in the chains, giving the crystal its fibrous character. The synthetic zeolite K-F[4] has the same framework. The structures of thomsonite, $NaCa_2(Al_5Si_5O_{20}).6H_2O$, and natrolite, $Na_2(Al_2Si_3O_{10}).2H_2O$,[5] are closely related to that of edingtonite.

The structure of a fibrous zeolite is illustrated in Fig. 23.21 (earlier in this chapter) as a packing of tetrahedra.

(b) Lamellar zeolites are frequently important constituents of sedimentary rocks. An example is phillipsite, $(K, Na)_5Si_{11}Al_5O_{32}.10H_2O$.[6] Each triangle in Fig.

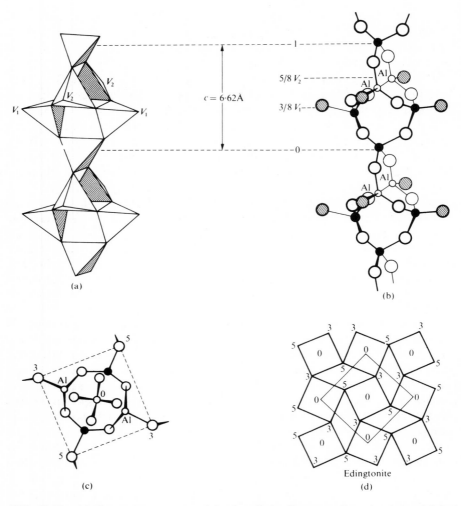

FIG. 23.23. (a) The common structural feature of the fibrous zeolites, the chain of linked tetrahedra. It is attached to neighbouring chains by the vertices V_1 and V_2. (b) The same chain, showing the silicon–oxygen arrangement. The vertices V_1 and V_2 are at heights 3/8 and 5/8 in the repeat of the pattern. (c) The chain viewed along its length. (d) Diagrammatic projection of the structure of edingtonite, with numbers 3, 5, to indicate the heights of attachment.

23.24(b) represents a pair of tetrahedra sharing a common vertex as shown at (a). These pairs are perpendicular to the plane of the paper in (b), so that one of the portions of (b) drawn with heavier lines represents a non-linear double chain of tetrahedra. In planes above and below that of the paper these double chains (light lines) are displaced, and the whole structure therefore consists of a 3D framework of tetrahedra each of which is linked to four others through its vertices. There are

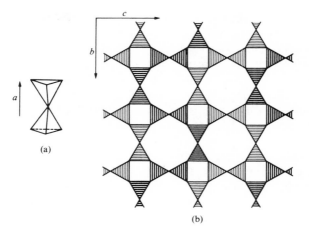

FIG. 23.24. The structure of a lamellar zeolite (phillipsite). The double tetrahedral unit shown at (a) projects in the plan (b) as a single (equilateral) triangle.

tunnels parallel to the a axis (perpendicular to the paper in (b)) of cross-sectional area $12\,\text{Å}^2$ and parallel to the b axis of approximately $9\,\text{Å}^2$ cross-sectional area, but none parallel to the c axis.

Not all zeolites fit neatly into any simple classification. In yugawaralite, $Ca_2Al_4Si_{12}O_{32}.8H_2O$, there is a 3D framework consisting of 4-, 5-, and 8-membered rings in which the widest tunnels form a 2D system (being parallel to the a and c axes) but there are also smaller channels parallel to the b axis.[7]

(c) We have remarked that the edges in certain space-filling arrangements of polyhedra represent the basic (4-connected) frameworks in some aluminosilicates. We shall refer to the simplest of these, the space-filling by truncated octahedra, in connection with the ultramarines. If the square faces of the truncated octahedra are separated by inserting cubes as shown in Fig. 23.25 we have another space-filling, by truncated octahedra (t.o.), cubes, and truncated cuboctahedra (t.c.o.). Examination

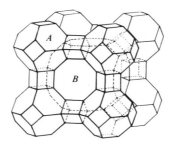

FIG. 23.25. Space-filling arrangement of truncated octahedra, cubes, and truncated cuboctahedra.

of Fig. 23.25 shows that this structure can be described in two other ways. The t.c.o.s form a continuous system in three dimensions, since each shares its octagonal faces with six other t.c.o.s; there are accordingly tunnels parallel to the three cubic axial directions through the larger cavities B. A third way of describing the same structure is illustrated in Fig. 23.27(a) later.

The framework of Fig. 23.25 is that of the synthetic zeolite 'Linde A', which can be prepared with different Si : Al ratios[8] as in $Na_{12}(Al_{12}Si_{12}O_{48}).27H_2O$ or $Na_9(Al_9Si_{15}O_{48}).27H_2O$. The ion-exchange properties of zeolites have long been known and utilized, but more recent commercial interest in these compounds has been due to their possible use as 'molecular sieves'. The type of gas molecule that can pass through a tunnel is determined by the diameter of the tunnel, so that if this has a critical value the zeolite can separate mixtures of gases by allowing the passage of only those with molecular diameters smaller than that of the tunnel. The cubes in Fig. 23.25 do not represent cavities in the actual aluminosilicate framework for each consists of a tightly knit group of eight tetrahedra, but water can enter both the small holes (A) and the larger holes (B), whereas O_2 can apparently enter only the larger (B) holes (compare the hydrate $CHCl_3.2H_2S.17H_2O$, p. 664).

In the space-filling arrangement of truncated octahedra each of the polyhedra is in contact with 14 others. If hexagonal prisms are placed in contact with alternate hexagonal faces, which are parallel to the faces of a tetrahedron, an open packing (not a space-filling) of these two kinds of polyhedron may be built (Fig. 23.26) which is equivalent to placing truncated octahedra at the points of the diamond net and joining through four of the eight hexagonal faces. This is the basic (Si, Al) framework in the natural zeolite, faujasite, $NaCa_{0.5}(Al_2Si_5O_{14}).10H_2O$.[9]

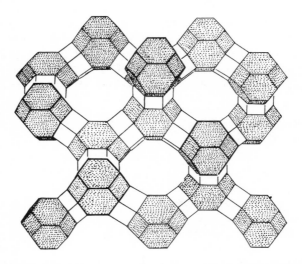

FIG. 23.26. The crystal structure of faujasite. The Si (Al) atoms are situated at the apices of the truncated octahedra, which are joined together to form a 3-dimensional framework.

Corresponding to these structures which result from joining t.o.s through cubes and hexagonal prisms there is another family derived from the t.c.o., each being joined to neighbouring ones through (a) 12 cubes (on square faces), (b) 8 hexagonal prisms (on hexagonal faces), or (c) 6 octagonal prisms (on 8-gon faces). These are illustrated in Fig. 23.27. The first, (a), is simply a third way of describing the 'Linde A' structure. The second (b), is the framework of the synthetic zeolite ZK-5, with composition approximating to $Na_{30}(Al_{30}Si_{66}O_{192}).98H_2O$.[10] There are two sets of (interprenetrating) 3D channel systems through the approximately planar 8-rings, with free diameter of about 3.8 Å, as compared with only one 3D system of channels in the 'Linde A' zeolite. The third, Fig. 23.27(c), is of interest because the polyhedra of Fig. 23.27(c) fill one-half of space. There are accordingly two sets of interpenetrating 3D systems of tunnels like those shown in Fig. 23.27(c). It is the framework in the synthetic zeolite RHO, $(Na, Cs)_{12}Al_{12}Si_{36}O_{96}.44H_2O$.[11]

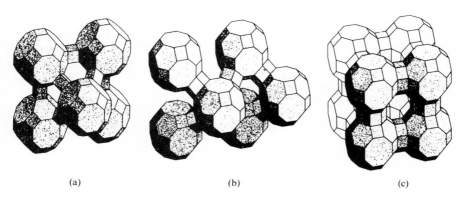

(a)   (b)   (c)

FIG. 23.27.   Aluminosilicate frameworks of zeolites: (a) Linde A;   (b) zeolite ZK 5;   (c) zeolite RHO.

Polyhedral cavities of less symmetrical types occur in the closely related minerals gmelinite and chabazite.[12] A Ca-chabazite has the approximate composition $Ca(Al_2Si_4)O_{12}.6H_2O$, but the Ca may be partly replaced by Na. These structures contain hexagonal prisms (as do millarite and the zeolite ZK-5) and larger cavities. Their structures may be built up by a process analogous to the closest packing of equal spheres, using hexagonal prisms as the units instead of spheres.[13] In Fig. 23.28(a) and (b) each hexagonal ring represents a hexagonal prism. The layer sequence AB... (h.c.p.) corresponds to the structure of gmelinite and the sequence ABC... to chabazite. The resulting polyhedral cavities are illustrated in Fig. 23.28(c) and (d).

We noted earlier that rings Si_5O_5 or $Si_{10}O_{10}$ are not common in silicates, but that they occur in certain zeolites, natural and synthetic, and in silicalite (p. 1008), while Si_5O_5 rings are prominent in the structure of melanophlogite. The family of framework structures comprising the synthetic zeolites ZSM-5 and ZSM-11, silicalite,

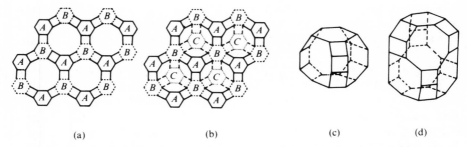

(a)   (b)   (c)   (d)

FIG. 23.28. The structures of gmelinite and chabazite (see text).

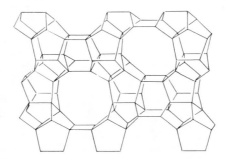

FIG. 23.29. Slice of a structure of the 'pentasil' family of zeolites.

and closely related natural zeolites, have been given the generic name 'pentasil'. Fig. 23.29 depicts a thin section through a structure of this type, viewed approximately along the direction of the tunnels enclosed by the 10-membered rings.[13a]

Ultramarines. The last group of framework silicates that we shall mention includes the materials called ultramarines, the coloured silicates which have been manufactured for use as pigments. The mineral lapis lazuli is of the same type, and since a number of colourless minerals such as sodalite are closely similar in structure, we shall for simplicity refer to all these silicates as ultramarines. Like the other framework silicates they are based on $(Si, Al)O_2$ frameworks with positive ions in the interstices, but a characteristic of the crystals of this group is that they also contain negative ions such as Cl^-, SO_4^{2-}, or S^{2-}. Like the felspars and in contrast to the zeolites the ultramarines are anhydrous. Formulae of representative members of the group are:

Ultramarine	$Na_8Al_6Si_6O_{24} . (S, Se, Te)$
Sodalite	$Na_8Al_6Si_6O_{24} . Cl_2$
Noselite	$Na_8Al_6Si_6O_{24} . SO_4$
Helvite	$(Mn, Fe)_8Be_6Si_6O_{24} . S_2$
Tugtupite	$Na_8Al_2Be_2Si_8O_{24}(S, Cl_2)$

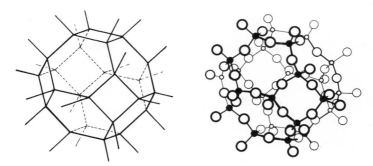

FIG. 23.30. The basket-like framework of linked SiO_4 tetrahedra which is the basis of the structure of the ultramarines. The silicon atoms are arranged at the vertices of the polyhedron shown on the left and the Si–O framework extends indefinitely in three dimensions.

All of these compounds contain essentially the same framework of linked tetrahedra (Figs. 23.21 and 23.30), the Si(Al, Be) atoms being situated at the vertices of Fedorov's space-filling array of truncated octahedra. The positive and negative ions are situated in the numerous cavities in the framework. A refinement of the structure of sodalite[14] gives Si–O, 1.628 Å, and Al–O, 1.728 Å, showing that the Si and Al atoms are arranged in an ordered way in alternate positions in the framework. There is a similar arrangement of the Be and Si atoms in helvite[15] where the higher charge on the framework is balanced by high-spin Mn^{2+} in tetrahedral coordination (Mn–3O, 2.08 Å, Mn–S, 2.43 Å) which may be replaced by Fe^{2+}. Tugtupite provides a further example of the replacement of Si by both Al and Be in a 3D framework. As in zeolites, exchange of interstitial ions for others is possible. For example, sodalite, which contains Cl^- ions, is converted into noselite by heating in fused sodium sulphate. Also, many ultramarines have been prepared containing ions such as Li^+, Tl^+, Ca^{2+}, and Ag^+ instead of Na^+ and with Se or Te replacing S. Considerable variations in colour, from nearly colourless to yellow, red, violet, and blue, result from these various replacements. It will be appreciated that the formulae assigned to the compounds in the above list are 'ideal' formulae and that considerable variation in composition is possible, subject always to the balancing of the total positive and negative charges.

(1)   N 1978 **272** 437
(2)   N 1978 **275** 119
(3)   AC 1976 **B32** 1623
(4)   AC 1976 **B32** 1714
(5)   ZK 1960 **113** 430
(6)   AC 1974 **B30** 2426
(7)   AC 1969 **B25** 1183
(8)   AC 1960 **13** 737; IC 1966 **5** 1537, 1539; ZK 1971 **133** 134
(9)   AM 1964 **49** 697
(10) ZK 1965 **121** 211
(11) Adv. Chem. Series 1973 **121** 106
(12) AC 1964 **17** 374
(13) N 1964 **203** 621
(13a) Chemical Society Special Publication, 1980 **33** p. 133
(14) AC 1967 **23** 434
(15) AC 1972 **B28** 114

24

Boron

Introduction

Boron has little in common with the other elements of Group III. Being the most electronegative element of the group it resembles in general the non-metals, particularly, silicon, far more than aluminium and the metals of the A and B subgroups. Boron is the only element of Group III which forms an extended series of hydrides. These bear no resemblance to the solid ionic hydrides of the alkali and alkaline-earth metals but are volatile compounds like the molecular hydrides of the non-metals of the later groups and form a remarkable group of electron-deficient compounds. Unlike aluminium, boron forms no stable normal oxy-salts (nitrate, sulphate, etc.). An acid sulphate is said to be produced by the action of SO_3 on boric acid, and tetraacetyl diborate, $(CH_3COO)_2BOB(CH_3COO)_2$, has been made (but no normal acetate). The salt $(CH_3)_4N[B(NO_3)_4]$ is formed as a white crystalline powder, stable at room temperature and insoluble in water, by the action of excess N_2O_4 at $-78\,^{\circ}C$ on solid $(CH_3)_4N(BCl_4)$. The compounds $B(ClO_4)_3$, $B(ClO_4)_2Cl$, and $B(ClO_4)Cl_2$ have all been prepared from BCl_3 and anhydrous $HClO_4$ at $-78\,^{\circ}C$ as low-melting salts which are sensitive to moisture and decompose on warming. On the other hand, boron forms numerous oxy-salts in which B is bonded to three or four O atoms, and it also forms the silica-like BPO_4 and $BAsO_4$.

The halides BX_3 are monomeric in the vapour, in contrast to the Al_2X_6 molecules in the vapours of $AlCl_3$, $AlBr_3$, and AlI_3, and these molecules persist in the crystalline halides (contrast $AlCl_3$). Boron also forms the halides B_2X_4, B_4X_4, B_8X_8, and B_9Cl_9; the last three are of a type peculiar to this element. The monofluoride, BF, can be prepared in 85 per cent yield by passing BF_3 over B at $2000\,^{\circ}C$; its life is even shorter than that of SiF_2. In Group III the extensive stereochemistry based on the formation of three coplanar bonds is characteristic of B alone; there are also many compounds in which four tetrahedral bonds are formed. For example, $B(CH_3)_3$ unites with NH_3 to form a stable solid $B(CH_3)_3 . NH_3$ which can be distilled without decomposition, and the halides form similar compounds. In this respect the aluminium halides behave similarly with organic oxy-compounds; $AlCl_3$ reacts with many ketones to give $R_2C{=}O.AlCl_3$, and also forms volatile esters like boron, for example, $Al(OC_2H_5)_3$.

The chemical, though not structural, resemblance of boron to silicon is marked. Its halides, like those of Si but unlike those of C, are easily hydrolysed, boric acid being formed. The acid HBF_4 is formed (together with boric acid) when BF_3 reacts with water, and $NaBF_4$ may be prepared from solutions of boric acid and $NaHF_2$. In the case of Si there is, of course, a change from SiF_4 to SiF_6^{2-}. Salts $M(BCl_4)$ can be prepared in anhydrous solvents and are stable only if M is a fairly large ion, for example, K^+, Rb^+, Cs^+, $N(CH_3)_4^+$, or PCl_4^+. The salts $N(C_2H_5)_4BBr_4$ and (pyridinium)-BI_4 have been prepared in the liquid hydrogen halides and isolated as very hygroscopic white solids. Boron rivals silicon in the number and complexity of its oxy-salts. In these compounds B forms three coplanar or four tetrahedral bonds to O; in many borates there is both planar and tetrahedral coordination in the same anion. Since various borate ions are in equilibrium in borate solutions it is not possible to predict the formula of the borate which will crystallize from a solution or be formed by double decomposition. Anhydrous borates with a particular composition are therefore made by fusion of a mixture of the metal oxide (or carbonate) with B_2O_3 in the appropriate proportions. Here again borates resemble silicates, for many borates made in this way supercool to glasses (like B_2O_3 itself). However, the formation of series of borates containing OH groups is not paralleled by silicon. We deal later with boric acids and borates.

Boron monoxide is prepared in a pure state by heating the element with the trioxide to 1350°C. At ordinary temperatures it forms an amorphous amber-coloured glass which reacts vigorously with water and alcohol. The molecular species in the vapour has been shown to be B_2O_2.[1] There is a second form of the monoxide which is a soft white hygroscopic solid (see Chart 24.1, p. 1058). Unstable boron–oxygen species include B_2O (p. 1061), BO_2[2] (formed by heating $ZnO+B_2O_3$ in a Knudsen cell), and the BO_2^- ion. The latter has been studied by introducing some KBH_4 into a pressed KBr disc and heating, when it is formed together with $B_3O_6^{3-}$ by oxidation due to included oxygen. The i.r. spectrum indicates a linear structure. The ion is not stable, its concentration decreasing appreciably after a few months.[3]

The sulphide B_2S_3 is formed by the decomposition *in vacuo* at 300°C of HBS_2 which in turn is produced by the action of H_2S on B at 700°C. Mass spectrometric study of the vapour of B_2S_3 shows the presence of many polymeric species.[4] Like B_2O_3 it usually forms a glass, but it was crystallized in 1962 and its structure has been determined.[5] It consists of layers formed from 4-membered B_2S_2 and 6-membered B_3S_3 rings similar to those in the finite $(BSSH)_2$ and $(BSSH)_3$ molecules. These rings are joined through S atoms to form $B_{16}S_{16}$ rings as shown in Fig. 24.1. Each B atom forms three coplanar bonds of mean length 1.81 Å. Notable features of the structure are the very short S–S distance across the B_2S_2 ring (2.88 Å as compared with 3.8 Å or more between the layers) and the compact form of the $B_{16}S_{16}$ rings which results in approximately hexagonal closest packing of the S atoms. This structure may be contrasted with the 3D framework structure of B_2O_3 built of $B_{10}O_{10}$ rings.

There is an extensive organo-boron chemistry. Among the simpler types of molecule are:

RB(OH)$_2$,
R$_2$B(OH)

$\begin{array}{c} R \\ R \end{array}$B—O—B$\begin{array}{c} R \\ R \end{array}$,

(OH)$_2$B—⬡—B(OH)$_2$, and

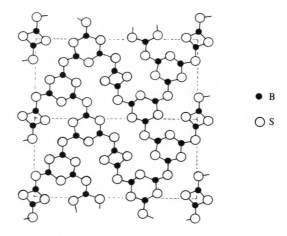

● B

○ S

FIG. 24.1. Structure of a layer of crystalline B_2S_3.

Compounds $RB(OH)_2$ polymerize to cyclic trimers only, in contrast to $R_2Si(OH)_2$ which give polymeric linear siloxanes. compare the formation of only one cyclic metaborate ion, $B_3O_6^{3-}$. The formation of 6-membered rings is a prominent feature of boron chemistry, as in boroxine (p. 1080) and the numerous substituted boroxines (boroxoles), (a); B_3N_3 rings (b); and B_3S_3 rings (c):

(a) R = H, F, OH CH₃, OCH₃   (b)   (c) X = Cl, Br, SH   (d)

The 6-membered B_3O_3 ring also occurs in many more complex borate anions. Cyclic boron–nitrogen molecules are considered later in this chapter. The cyclic molecule (c), X = SH, is the predominant species in the vapour of HBS_2 at temperatures below $100\,^{\circ}C^{(6)}$ and is the structural unit in crystalline $B_3S_3(SH)_3$.$^{(7)}$ The ring is planar, with the structure (d). There is a very similar ring in $B_3S_3Br_3$. The 5-membered B_2S_3 ring in $B_2S_3Cl_2$ has the structure (e).$^{(8)}$

(e)

(1) JCP 1956 **25** 498; JPC 1958 **62** 490   (5) AnC 1977 **89** 327
(2) CJP 1961 **39** 1738   (6) JACS 1966 **88** 2935
(3) IC 1964 **3** 158   (7) AC 1973 **B29** 2029
(4) JACS 1962 **84** 3598   (8) ACSc 1973 **27** 21

The stereochemistry of boron

The stereochemistry of boron is simple in many of its compounds with the halogens (but note B_nCl_n), oxygen, nitrogen, and phosphorus, three coplanar or four tetrahedral bonds being formed. The more complex stereochemistry of the element in electron-deficient systems (elementary B, some borides, and the boranes) is dealt with separately. The planar arrangement of three bonds from a B atom has been demonstrated in many molecules BX_3 and BR_3 (Table 24.1), in cyclic molecules of the types noted in the prevous section, in many oxy-ions (see later section), and in crystals such as the graphite-like form of BN (p. 1060) and AlB_2 (p. 1055).

TABLE 24.1

The structures of planar molecules BX_3 and BR_3

Molecule	B–X	Other data	Reference
BF_3	1.31 Å		JCP 1968 **48** 1571
BCl_3	1.74		BCSJ 1966 **39** 1134
BBr_3	1.89		BCSJ 1966 **39** 1146
BI_3	2.10		IC 1962 **1** 109 (crystal)
BI_3	2.12		BCSJ 1975 **47** 1974 (e.d., vapour)
HBF_2	1.31	B–H, 1.19 Å; F–B–F, 118°	JCP 1968 **48** 1
$B(CH_3)_3$	—	B–C, 1.58 Å	JCP 1965 **42** 3076
$B(CH_3)_2F$	1.29		JACS 1942 **64** 2686
$B(OCH_3)_3$	—	B–O, 1.38 Å	JACS 1941 **63** 1394

The non-coplanar arrangement of four bonds from B was established by the methods of classical stereochemistry. The borosalicylate ion, (a), would possess a plane of symmetry if the boron bonds were coplanar. The optical activity of this ion was demonstrated by resolution into optical antimers using strychnine as the active base. Although interpreted as indicating tetrahedral as opposed to coplanar bonds, this result only eliminates a strictly coplanar configuration of the complex (or other configuration possessing n̄ symmetry, as explained in Chapter 2). More

(a)   (b)

recently the optically active cation (b) was resolved by hand selection of crystals, and the detailed structure of the analogous bromo cation has been determined in the crystalline hexafluorophosphate.[1]

The simplest groups containing B forming four tetrahedral bonds are the ions BX_4^-. Some fluoroborates are isostructural with oxy-salts containing tetrahedral ions, for example $NaBF_4$ with $CaSO_4$, $CsBF_4$ and $TlBF_4$ with $BaSO_4$, and $(NO)BF_4$ with $(NO)ClO_4$. The B–F bond length is 1.41 Å in $NaBF_4$[2] and 1.406 Å in NH_4BF_4;[3] it is interesting that there is no hydrogen bonding in the ammonium salt, the shortest N–H–F distance being 2.93 Å (compare 2.69 Å in NH_4F). The ion BF_3OH^- occurs in $H(BF_3OH)$ and $(H_3O)(BF_3OH)$, originally respectively formulated as $BF_3.H_2O$ and $BF_3.2H_2O$.[4] and the tetrahedral ion BOF_3^{2-} (B–F, 1.43 Å, B–O, 1.435 Å) in salts such as $BaBOF_3$.[5] The $B(OH)_4^-$ ion and many examples of tetrahedral BO_4 coordination groups are mentioned under boron-oxygen compounds later in this chapter.

Simple tetrahedral molecules include $H_3B.N(CH_3)_3$, borine carbonyl, $H_3B.CO$,[6] (c), the numerous adducts of BF_3, and the remarkable molecule $B_4F_6.PF_3$,[7] (d). This compound is formed as a very reactive colourless solid when the high-temperature species BF is condensed on to a cold surface with PF_3. This molecule is of special

(c)

(d)

interest as containing one B atom bonded tetrahedrally to 1 P (B–P, 1.83 Å) and to 3 B (B–B, 1.68 Å), these B atoms forming three coplanar bonds (B–F, 1.31 Å; P–F, 1.51 Å). Other examples of tetrahedral complexes include the molecule (e)[8] and the anion in $K[B(SO_3Cl)_4]$, (f).[9]

(e)

(f)

Many molecules and crystals containing equal numbers of B and N (or P) atoms form structures similar to those of the corresponding compounds of Group IV elements. Examples include BN and BP with the diamond structure, the silica-like structures of BPO_4 already mentioned, and analogues of substituted cyclohexanes such as $[H_2B.N(CH_3)_2]_3$ with the same chair configuration as C_6H_{12}.

(1) IC 1971 **10** 667
(2) ZaC 1966 **344** 279
(3) AC 1971 **B27** 1102
(4) AC 1964 **17** 742
(5) AC 1957 **10** 199

(6) PR 1951 **78** 1482
(7) IC 1969 **8** 836
(8) AC 1972 **B28** 1961
(9) AC 1978 **B34** 1771

Elementary boron and related borides

Boron is notable for the structural complexity of its polymorphs, some of which are almost as hard as diamond. Decomposition of boranes or BI_3 in the range 800–1200 °C yields α rhombohedral boron (B_{12}); the subscript number is the number of atoms in the unit cell. This form is considered to be thermodynamically unstable relative to the β rhombohedral B_{105} which crystallizes from any fairly pure molten boron or by recrystallizing the element at any temperature above 1500 °C and is the one polymorph that is readily available. The reaction between BBr_3 and H_2 over a Ta filament at 1200–1400 °C gives α tetragonal B_{50}. Massive (glassy) boron and the thin filaments which have high tensile strength are apparently nearly 'amorphous', giving only two diffuse X-ray diffraction rings. All three known boron structures and also those of a number of boron-rich borides contain icosahedral B_{12} groups with, in some cases, additional B atoms which form an essential part of the framework.

In α B_{12} all the B atoms belong to icosahedral groups which are arranged in approximately cubic closest packing. This type of packing may be referred to a rhombohedral unit cell with interaxial angle close to 60°. If B_{12} icosahedra are placed at all the points of the rhombohedral lattice of Fig. 24.2(a) 6 B atoms of each

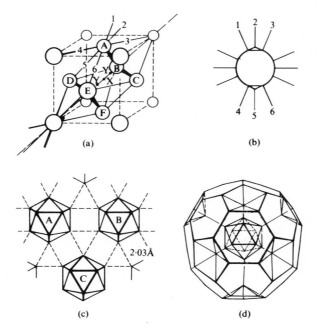

(a)

(b)

(c)

(d)

FIG. 24.2. The crystal structures of α-B_{12} and β-B_{105}. (a) and (b), packing of B_{12} units in α-B_{12}. (c) Section through the structure of α-B_{12} perpendicular to a 3-fold axis showing the 3-centre bonds. (d) the B_{84} unit of β-B_{105} showing the central icosahedron and six only of the 'half-icosahedra', one of which is attached to each B atom of the central B_{12} group.

(those of two opposite faces) can bond to a B atom of another icosahedron by a B—B bond of length 1.71 Å. These bonds lie along the lattice directions (cell edges) and result in a 3D framework in which each icosahedron is connected to six others; the directions of these bonds correspond ideally to certain of the 5-fold axes of the icosahedron (Fig. 24.2(b)). The remaining six B atoms of an icosahedron, which form a buckled hexagon around its equator, cannot bond in this way, the shortest contacts in the equatorial plane ABC (shaded) being 2.03 Å. The bonding in this plane is by means of '3-centre' bonds, as shown in (c).

This framework can also serve as the basis of the structures of certain borides. Groups of atoms YXY can be placed along the body-diagonal of the cell of Fig. 24.2(a), for at the centre of the cell there is an octahedral hole surrounded by the icosahedra ABCDEF. In B_4C these are linear C—C—C groups, and the 3-centre bonding in the equatorial plane is replaced by bonds between the six equatorial B atoms of the B_{12} groups and the terminal C atoms of the chains. One-half of the B atoms therefore form bonds to 6 B (as in α-B_{12}) and the remainder are bonded to 5 B + 1 C. The only congruently melting borocarbide has the composition $B_{13}C_2$, that is B_{12}.CBC, and it seems probable also that B_4C is $(B_{11}C)CBC$ rather than $(B_{12})C_3$: this field is still being actively studied. In AlC_4B_{40} there are apparently linear CBC and CBB chains and also non-linear C(Al)B chains. (Reference to these compounds are given in Table 24.2.)

There are many compounds with the same general type of structure as B_4C and rhombohedral B_{12} though the details of their structures are not in all cases certain. The C_3 or CBC chains may be replaced by S in $B_{12}S$, O—B—O in $B_{13}O$, or by 2 Si in

TABLE 24.2

Polymorphs of boron; icosahedral borides

Structure	Units	Reference
α-B_{12} (rhombohedral)	B_{12} (c.c.p.) only	AC 1959 **12** 503
B_4C	$B_{11}C$; chains CBC	AC 1975 **B31** 1797
$B_{13}C_2$	B_{12}; CBC	AC 1979 **B35** 1052
$B_{13}P_2$	B_{12}; (B,P)	AC 1974 **B30** 2549
$B_{12}As_{1.77}B_{0.23}$	See text	AC 1976 **B32** 972
$B_{2.89}Si$	$(B_{10.3}Si_{1.7})(Si_2)_3$	ACSc 1962 **16** 449
AlC_4B_{40}	B_{12}; chains CBC, CBB, CAlB	AC 1970 **B26** 315
$Al_{2.1}C_8B_{51}$	B_{12} (h.c.p.); CBC (CAlC) chains	AC 1969 **B25** 1223
NaB_{15}	B_{12}, 3 B, Na	JSSC 1970 **1** 150
$MgAlB_{14}$	B_{12}, 2 B, Mg, Al	AC 1970 **B26** 616
α-B_{50} (tetragonal	$(B_{12})_4$, 2 B	JACS 1958 **80** 4507
BeB_{12}	B_{12}, Be	ZaC 1960 **306** 266
β-B_{105} (rhombohedral)	$(B_{84})(B_{10}B.B_{10})$	AC 1977 **B33** 1951
YB_{66}		AC 1969 **B25** 237

$B_{2.89}Si$. Since there would appear to be room for only 2 Si replacing 3 C there is apparently some (statistical) replacement of B by Si in the icosahedra, giving a composition close to $B_{31}Si_{11}$ (i.e. $B_{31}Si_5 . Si_6$ or $B_{2.89}Si$). Earlier work on '$B_{13}P_2$' suggested 3-atom chains P—B—P in the spaces between the icosahedra. A later study indicates only pairs of atoms, B and P, statistically distributed in $B_{12}(P_{1.36}B_{0.64})$. Each such B(P) has a tetrahedral arrangement of bonds to 3 B_{12} icosahedra and to the other single atom. See also the reference to the As compound listed in Table 24.2.

There is a closely related (orthorhombic) phase $Al_{2.1}C_8B_{51}$ (variously described as AlB_{10}, AlB_{12}, or AlC_4B_{24}) in which there is distorted *hexagonal* closest packing of B_{12} icosahedra and linear CBC chains, in some of which Al replaces B. As in B_4C each B_{12} group is directly bonded to six others via B—B bonds (1.77 Å) so that six B atoms of each icosahedron form 6 B—B bonds and the remainder are bonded to 5 B + C. The bond lengths are similar to those in B_4C:

Bond	$Al_{2.1}C_8B_{51}$	B_4C
B—B (intra-icosahedral)	1·816 A	1·789 Å
B—B (inter-icosahedral)	1·774	1·718
$B_{icos.}$—C	1·623	1·604
B_{chain}—C	1·467	1·435

The close relation between these two structures is shown by the fact that the Al compound undergoes a topotactic transition by loss of Al at 2000 °C to form B_4C, a process involving a change from h.c.p. to c.c.p. B_{12} icosahedra.

In the isostructural NaB_{15} and $MgAlB_{14}$ the B_{12} icosahedra are packed in layers which are stacked vertically above one another, so that only two direct pentagonal pyramidal contacts between icosahedra result (perpendicular to the layers). In the layers each B_{12} is bonded to four others at the same distance (1.75 Å) but there is also bonding through Mg, Al, and additional B atoms. The interstitial atoms obviously play an essential part in stabilizing this mode of packing of the icosahedra.

A feature of the polymorphs of boron is the large proportion of atoms which form a sixth bond (pentagonal pyramidal coordination) directed outwards along the 5-fold axis of the icosahedron. The fractions are:

Rhombohedral B_{12} 50%
Rhombohedral B_{105} 80%
Tetragonal B_{50} 64%

If all the B atoms of each icosahedron were to form a sixth bond to a B atom of another icosahedron there would be formed a radiating icosahedral packing which is not periodic (Chapter 4). The structure of rhombohedral B_{105} may be visualized as built of units $(B_{84})(B_{10}BB_{10})$. The B_{84} unit consists of an icosahedron bonded to 12 half-icosahedra, one attached, like an umbrella, to each vertex (Fig. 24.2(d)). These units are placed at the points of the rhombohedral lattice of Fig. 24.2(a) so that six of the half-icosahedra of each B_{84} meet at the mid-points of the edges of the cell forming further icosahedra. The basic framework therefore consists of icosahedra at the points of the lattice of Fig. 24.2(a) each connected to six others situated at the mid-points of the cell edges. These icosahedra are directly bonded only to two others. The remaining six half-icosahedra of a B_{84} unit do not juxtapose to form icosahedra but together with the B_{10} units provide further bonding in the equatorial plane—compare the 3-centre bonds in B_{12} and the bonding through C atoms in B_4C. Along the diagonal of the cell, replacing the CBC chain in B_4C, are $B_{10}BB_{10}$ groups, there being a unique B atom at the body-centre of the cell.

It is convenient to mention here the structure of YB_{66} which is extremely complex, with 1584 B and 24 Y atoms in the unit cell, but has high (cubic) symmetry. The majority of the B atoms (1248) are contained in units consisting of 12 icosahedra (156 atoms), twelve around a central one, which are further linked by B—B bonds (1.62–1.82 Å) into a 3D framework. The grouping of twelve icosahedra around a central one represents the beginning of a radiating packing which, as already noted, would not be periodic in three dimensions. The packing of the B_{156} units leaves large channels in which the remaining B and the Y atoms are accommodated.

The structure of tetragonal B_{50} is comparatively simple. There are 50 atoms in the unit cell of which 48 form 4 nearly regular icosahedral groups which are linked together both directly and through the remaining B atoms. The latter form tetrahedral bonds, but the atoms of the icosahedra form six bonds, five to their neighbours in the B_{12} group and a sixth to an atom in another B_{12} or to an 'odd' B atom. Although all the icosahedral B atoms form 'pentagonal' pyramidal bonds, for 8 of the 12 B atoms there is considerable deviation ($20°$) of the external bond from the 5-fold axis of the icosahedron. The bond lengths in this structure are 1.60 Å for the 4-coordinated B, 1.68 Å for bonds between B_{12} groups and 1.81 Å within the B_{12} groups. Several borides have been prepared which have the same basic structure as tetragonal B_{50}, for example NiB_{25} and BeB_{12}. The metal atoms clearly occupy some of the numerous holes in the B_{50} structure.

Metal borides and borocarbides

We deal here with borides and borocarbides other than those containing icosahedral B_{12} groups. Like carbides and silicides they can be prepared by direct union of the elements at a sufficiently high temperature; alternatively some borides and silicides result from reduction of borates or silicates by excess metal. As in the case of silicides it is not possible to account for the formulae of borides in terms of the normal valences of the metals. Thus we have NaB_6, CaB_6, and YB_6; yttrium also forms YB_2, YB_4, YB_{12}, and YB_{66}, and chromium, for example, forms CrB, Cr_3B_4, CrB_2, and CrB_4.

We have described the structures of some boron-rich borides in which there are extensive 3D systems of B—B bonds. At the other extreme there are crystalline borides with low boron content in which there are isolated B atoms, that is, B atoms surrounded entirely by metal atoms as nearest neighbours. With increasing

TABLE 24.3

The crystal structures of metal borides

Nature of B complex	Structure	Examples	Reference
Isolated B atoms	Be_4B		ZaC 1962 **318** 304
	Fe_3C	Ni_3B, Co_3B	ACSc 1958 **12** 658
	$CuAl_2$	Mn_2B, Fe_2B, Co_2B, Ni_2B,	ACSc 1949 **3** 603;
		Mo_2B, Ta_2B	ACSc 1950 **4** 146
	CaF_2	Be_2B	DAN 1955 **101** 97
	Rh_2B		AC 1954 **7** 49
B_2 groups	$M'_2M''B_2$	$(Mo, Ti, Al)CrB_2$	AC 1958 **11** 607
Single chains	CrB	NbB, TaB, VB	AC 1965 **19** 214
	FeB	CoB, MnB, TiB, HfB	AC 1966 **20** 572
	MoB	WB	ACSc 1947 **1** 893
	Ni_4B_3		ACSc 1959 **13** 1193
	$Ru_{11}B_8$		ACSc 1960 **14** 2169
Double chains	Ta_3B_4	Cr_3B_4, Mn_3B_4, Nb_3B_4	ACSc 1961 **15** 1178
Layers	AlB_2	M = Ti, Zr, Nb, Ta, V, Cr, Mo, U,	AC 1953 **6** 870;
		Mg	ACSc 1959 **13** 1193
	ϵ—Mo—B	ϵ—W—B phase	ACSc 1947 **1** 893
3D frameworks	CrB_4		ACSc 1968 **22** 3103
	UB_4	M = Y, La, 4f, Th	AC 1974 **B30** 2933
	CaB_6	M = Sr, Ba, Y, La, 4f, Th, U	IC 1963 **2** 430;
			JSSC 1970 **2** 332
	UB_{12}	M = Y, Zr, 4f	AC 1965 **19** 1056

boron contents the B atoms link together to form first B_2 units, then chains, layers, or 3D frameworks extending throughout the whole crystal (Table 24.3).

Borides in which there are discrete B atoms include Be_4B and Ni_3B (and Co_3B) with the cementite (Fe_3C) structure. In this structure the metal atoms form slightly corrugated layers of the kind shown in Fig. 24.3(a). The B atoms lie between the layers with six nearest neighbours (B—Ni, 2.04 Å) at the apices of a trigonal prism. These metal layers are closely related to the layers (parallel to $10\bar{1}1$) in hexagonal

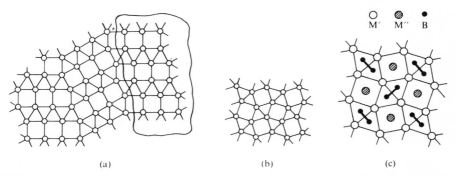

FIG. 24.3. (a) Layers of Ni atoms in Ni_3B. At the right is shown a portion of a layer parallel to the plane ($10\bar{1}1$) in hexagonal close-packing. (b) Layer of metal atoms in borides with the $CuAl_2$ structure. (c) The structure of borides $M'_2M''B_2$.

close-packing, as shown in the figure. Next we have Be_2B with the fluorite structure and the numerous borides M_2B with the $CuAl_2$ structure described and illustrated on p. 1313. In this structure the metal atoms form layers of the type of Fig. 24.3(b). Successive layers are rotated through 45° relative to one another and the B atoms are at the centres of antiprisms formed by two square groups in adjacent layers. In Fe_2B, for example, B has eight Fe neighbours at 2.18 Å, and the two nearest B atoms are at 2.12 Å; there are no B–B bonds.

Isolated pairs of B atoms occur in the structures of certain double borides $M_2'M''B_2$ in which M' is a large atom (Mo, Ti, Al, or a mixture of these) and M'' is a small atom (Cr, Fe, Ni). Here again the large atoms form the 5-connected net of Fig. 24.3(b) and the M'' and B atoms lie between the layers, M'' at the centres of tetragonal prisms and B at the centres of pairs of adjacent trigonal prisms (Fig. 24.3(c)). In the B_2 groups the B–B bond length is 1.73 Å.

The two polymorphs of Ni_4B_3 form a link between structures such as Ni_3B with isolated B atoms and those with chains of B atoms. In one form of Ni_4B_3 two-thirds of the B atoms form zigzag chains, the remainder being isolated B atoms, while in the other polymorph all the B atoms are in chains.

Next come three closely related structures (FeB, CrB, and MoB) in which there are zigzag chains of boron atoms. In FeB (and the isostructural CoB) each B atom is surrounded by 6 Fe at the vertices of a trigonal prism (Fe–B, 2.12–2.18 Å), but its nearest neighbours are two B atoms (at 1.77 Å) as shown in Fig. 24.4(a). The B atoms therefore form infinite chains.

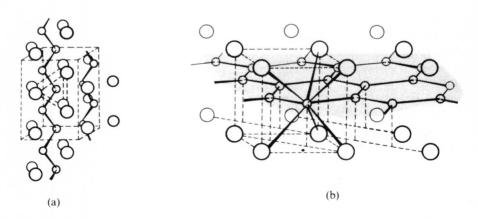

(a)   (b)

FIG. 24.4.   The crystal structures of (a) FeB, and (b) AlB_2. The smaller circles represent B atoms.

In borides of the type M_3B_4 the boron atoms form double chains in which the central bonds are apparently much stronger than those along the length of the chain (Fig. 24.5). The Ta_3B_4 structure is in this way intermediate between the chain and layer borides.

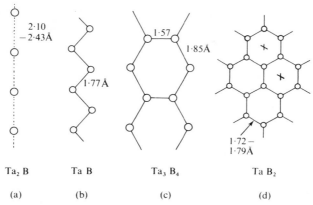

FIG. 24.5. Arrangements of B atoms in borides: (a) isolated B atoms; (b) single chain; (c) double chain; (d) hexagonal layer.

In AlB_2 the B atoms are arranged in layers with layers of Al interleaved between them. The structure of a B layer is the same as that of a layer in the graphite structure (p. 922). Each B is here equidistant from three other B atoms (at 1.73 Å), the next nearest neighbours being a set of six Al at the vertices of a trigonal prism (Fig. 24.4(b)). Most of the 'layer' borides have the AlB_2 structure, but in the Mo—B and W—B systems there are also phases with the ideal composition M_2B_5, apparently with additional B atoms at the points X of the hexagonal net of Fig. 24.5(d). The tantalum–boron system provides examples of four types of boride structure (Fig. 24.5); compare the silicon complexes in metallic silicides.

With higher boron–metal ratios some extremely interesting 3-dimensional boron frameworks are found. In CrB_4 the B atoms form a very simple 4-connected framework. Square groups of B atoms (B—B, 1.68 Å) at different heights are linked by somewhat longer bonds (1.91 Å) as shown in Fig. 24.6(a). Each B forms two bonds of each type to other B atoms, while Cr has 2 Cr and 12 B as nearest neighbours. In CaB_6 (and numerous isostructural compounds) the boron network consists of octahedral B_6 groups joined together as shown in Fig. 24.6(b). Alternatively this is a five-connected net the links of which are the edges of a space-filling array of octahedra and truncated cubes. There are large holes surrounded by 24 B atoms which accommodate the metal atoms (Ca–B, 3.05 Å). The B—B distances of about 1.7 Å are similar to those in the FeB and AlB_2 structures.

The UB_4 structure is of interest because of the geometrical relationships with the structures of borides MB_2 and MB_6. In the AlB_2 structure there are close-packed metal layers in which each M has six equidistant neighbours, and these layers are superposed directly above one another so that the B atoms in the holes between the layers are surrounded by six M at the vertices of a trigonal prism (Figs. 24.4(b) and 24.7(a)). In CaB_6 there is simple cubic packing of the M atoms (Fig. 24.7(c)) so

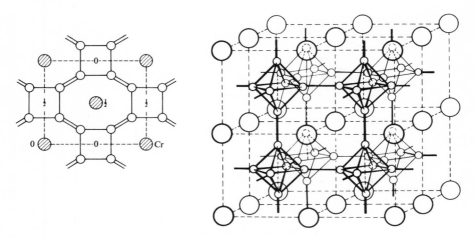

FIG. 24.6.   The crystal structures of (a) CrB_4 and (b) CaB_6.

that there are holes between groups of eight M atoms situated at the vertices of a cube. The CaB_6 structure can therefore be described as a simple cubic M lattice expanded by the introduction of the octahedral B_6 groups, though since these are joined to neighbouring B_6 groups there is a continuous 3-dimensional B framework. In UB_4 the metal atoms are in layers in which every M has five neighbours, and between the layers there are both 'trigonal prism' and 'cubic' holes, as may be seen from Fig. 24.7(b). In the holes of the first type there are single B atoms and in the latter B_6 groups, and since there are pairs of adjacent holes of the former type the actual B network is that shown in Fig. 24.7(d), with B_6 groups joined through pairs of B atoms. In this structure some B atoms are linked to three B, others to five B, and the B–B distances are in the range 1.7-1.8 Å. Although the B–B distances in the rigid boron framework structures are consistent with $r_B = 0.86$ Å, the metal-boron distances suggest larger effective radii—from 0.86 to 1.10 Å. Yet another way of describing the CaB_6 structure is as a CsCl-like packing of M atoms and B_6 groups. We noted in Chapter 22 that the bonding is not of the same type in all carbides MC_2; this is also true in borides MB_6. These have been prepared containing metals in various oxidation states, namely, 1 (KB_6), 2 (Ca, Sr, Yb), and >2 (La, Th). The electronic properties of KB_6 have not yet been studied, but compounds of the second group are semiconductors and those of the third are metallic conductors.

The boron framework in UB_{12} is illustrated in Fig. 3.47(d) (p. 138) as a space-filling arrangement of cuboctahedra, truncated octahedra, and truncated tetrahedra, the edges of which form a 5-connected 3D net. The U atoms are situated at the centres of the largest cavities in positions of 24-coordination—compare the 24-coordination, truncated cubic, of the metal atoms in the CaB_6 structure. Alternatively the structure may be described as a NaCl-like packing of B_{12} groups and U atoms. The cubocta-hedral B_{12} groups may be contrasted with the icosahedral groups in the polymorphs

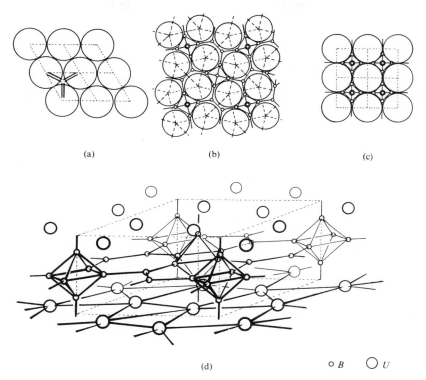

(a) (b) (c)

(d) ○ B ○ U

FIG. 24.7. The crystal structures of metallic borides: (a) close-packed metal layer in UB$_2$ (AlB$_2$) with B in trigonal prism holes between layers; (b) metal layer in UB$_4$ with positions of 6- and 8-coordination between layers; (c) metal layer in CaB$_6$ (ThB$_6$) structure; (d) the UB$_4$ structure.

of elementary B and related borides. The B–B bonds in the B$_{12}$ groups are appreciably shorter (1.68 Å) than those linking the groups (1.78 Å in ZrB$_{12}$; 1.80 Å in UB$_{12}$; 1.81 Å in YB$_{12}$).

Numerous borocarbides have been prepared, for example, those of Al already described and compounds such as Mo$_2$BC,[1] in which C is octahedrally coordinated by Mo and B has trigonal prismatic coordination and forms chains as in CrB. Of special interest are the borocarbides MB$_2$C$_2$ formed by Sc, La, and the 4f elements. The metal atoms are situated between plane 3-connected layers which in the La and 4f compounds are the 4:8 net but in ScB$_2$C$_2$ are the 5:7 net. In both LaB$_2$C$_2$ (p. 109) and ScB$_2$C$_2$ there are pairs of adjacent B and C atoms. The Sc atoms are in pentagonal prism holes between the layers, but there are also Sc–Sc distances very nearly the same as in h.c.p. Sc metal, so that Sc has (14+5)-coordination (C–C, 1.45: B–B, 1.59, B–C, 1.52-1.61 Å).[2]

(1) AC 1969 **B25** 698   (2) AC 1965 **19** 668

Lower halides and diboron compounds

Other compounds in which there are B—B bonds include the halides B_2X_4 (all four of which have been prepared), sub-halides B_nX_n, and diboron compounds. Salts containing the $B_2Cl_6^{2-}$ ion include $[(CH_3)_4N]_2B_2Cl_6$, prepared in liquid anhydrous HCl, and $(PCl_4)_2B_2Cl_6$. Methods of preparation and relations between some of the diboron compounds are indicated in Chart 24.1.

CHART 24.1
Preparation of diboron compounds †

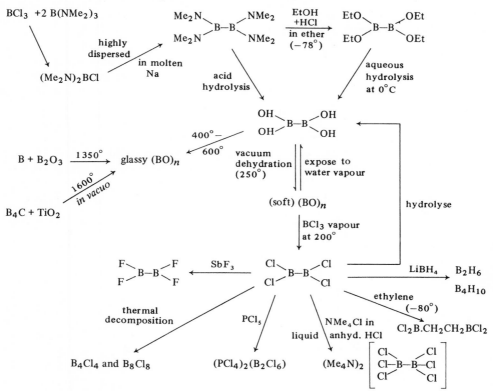

† See, for example: JACS 1954 **76** 5293; JACS 1960, **82**, 6242, 6245; JACS 1961 **83** 1766, 4750: PCS 1964 242.

Halides B_2X_4

The fluoride B_2F_4 is an explosive gas. The molecule has been shown to have the planar symmetrical structure (a) both in the crystal (studied at -120 °C)[1a] and in the vapour (e.d.).[1b] The bond lengths shown come from the latter study. The

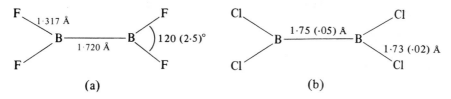

(a)   (b)

chloride B_2Cl_4 is formed by passing BCl_3 vapour through an electric discharge between mercury electrodes. It is a colourless liquid which decomposes at temperatures above $0\,°C$ into a mixture of the yellow crystalline B_4Cl_4, red B_8Cl_8, and the paramagnetic $B_{12}Cl_{11}$.[2] Among its derivatives are esters $B_2(OR)_4$ and the ethylene compound $Cl_2B.C_2H_4.BCl_2$ mentioned later. The compounds B_2Br_4 and B_2I_4 (a yellow crystalline compound) are formed by methods similar to that used for the chloride.

The crystal structure of B_2Cl_4 has been determined (at $-165\,°C$) and it is found that the molecule is planar with bond angles close to $120°$, (b).[3] The B–Cl bond is normal but B–B is long compared with that expected for a single bond (around $1.6\,Å$) – compare the long bond in N_2O_4 and the normal single bond in P_2I_4. In the vapour, however, the B_2Cl_4 molecule has the 'staggered' configuration, the planes of the two halves being approximately perpendicular (B–B, 1.70; B–Cl. 1.75 Å; Cl–B–Cl, $119°$).[4] The eclipsed configuration in the crystal may result from better packing of the molecules, for m.o. calculations show that the staggered form is more stable than the eclipsed by at least $4\ kJ\ mol^{-1}$.[5]

The molecule $Cl_2B.C_2H_4.BCl_2$ is also planar to within the limits of experimental error, though the bond lengths are only approximate:[6]

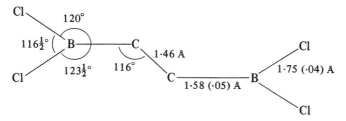

The B–C bond length may be compared with the (approximate) value $1.52\,(0.07)\,Å$ found in $C_6H_5.BCl_2$.[7]

Sub-halides B_nX_n

Unlike halides B_2X_4 these halides are electron-deficient. The structures of B_4Cl_4, B_8Cl_8, and B_9Cl_9 in the crystalline state have been determined. In B_4Cl_4[8] a nearly regular tetrahedron B_4 is surrounded by a tetrahedral group of 4 Cl, also nearly regular (Fig. 24.8(a)). The B–Cl bonds ($1.70\,Å$) are single, but the length of the B–B bonds (also $1.70\,Å$) corresponds to a bond order of about 2/3, assuming a boron radius of $0.8\,Å$ and using Pauling's equation (p. 1292). Assuming that one

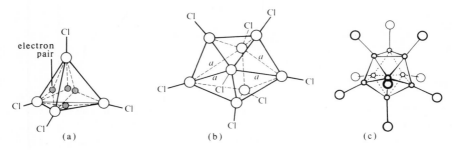

FIG. 24.8. The molecular structures of (a) B_4Cl_4; (b) B_8Cl_8; (c) B_9Cl_9.

electron of each B is used for a single B—Cl bond there remain 8 electrons (4 electron pairs) for the 6 B—B bonds, as shown diagrammatically in the figure. The molecule B_8Cl_8 also consists of a polyhedral boron nucleus to each B atom of which is bonded a Cl atom. The polyhedron (Fig. 24.8(b)) is closer to a dodecahedron (bis-disphenoid) than to a square antiprism, but there is a considerable range of B—B bond lengths. The four longer B—B bonds (a) range from 1.93 to 2.05 Å and the remainder from 1.68 to 1.84 Å. The mean B—Cl bond length is 1.74 Å.[9] Since there are 8 electron pairs available for B—B bonds in the polyhedron with 18 edges, of which four are appreciably longer than the other fourteen, no simple description of the bonding is possible In B_9Cl_9[10] the B atoms are at the vertices of a tricapped trigonal prism (Fig. 24.8(c)), and the bond lengths are similar to those in B_8Cl_8 (B—B, 1.73-1.81; B—Cl, 1.72-1.75 Å).

The colourless liquid B_2Br_4 decomposes thermally to sub-bromides B_nBr_n ($n = 7$, 8, 9, 10, 12), of which B_9Br_9 is the most stable. This compound can be sublimed, and large (red) crystals can be handled briefly in air.[11]

(1a)   JCP 1958 **28** 54
(1b)   JACS 1977 **99** 6484
(2)   IC 1963 **2** 405
(3)   JCP 1957 **27** 196
(4)   JCP 1969 **50** 4986
(5)   JCP 1965 **43** 503

(6)   AC 1956 **9** 668
(7)   JPC 1955 **59** 193
(8)   AC 1953 **6** 547
(9)   AC 1966 **20** 631
(10)   N 1970 **228** 659
(11)   JINC 1975 **37** 1593; IC 1980 **19** 3295

Boron–nitrogen compounds

Boron nitride

Boron nitride, BN, is made by the action of nitrogen or ammonia on boron at white heat and in other ways. In the crystalline state it is very inert chemically, though it can be decomposed by heating with acids, and is described as melting under pressure at about 3000 °C. The crystal structure of one form is very closely related to that of graphite (Fig. 21.3, p. 922) being built of hexagonal layers of the same kind, but in BN these are arranged so that atoms of one layer fall vertically above those of the layer below (Fig. 24.9). The B—N bond length is 1.446 Å.[1] In spite of the

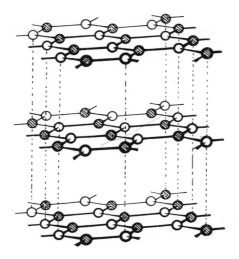

FIG. 24.9. The crystal structure of boron nitride, BN.

structural resemblance to graphite the physical properties of BN are very different from those of graphite. It is white, a very good insulator, and its diamagnetic susceptibility is very much smaller than that of graphite. Like graphite BN can be prepared with a turbostratic (unordered layer) structure, which can be converted to the ordered hexagonal structure by suitable heat treatment.[2] BN also crystallizes with the zinc-blende and wurtzite[3] structures. In cubic BN the length of the B—N (single) bond is 1.57 Å.[4]

Also isoelectronic with graphite is B_2O, prepared by reducing B_2O_3 with B or Li at high temperatures under high pressure. (In the 'tetrahedral anvil' pressures of 50–75 kbar and temperatures of 1200–1800 °C are reached.) The B and O atoms were not definitely located in the graphite-like structure of B_2O.[5]

(1)   AC 1952 **5** 356
(2)   JACS 1962 **84** 4619
(3)   JCP 1963 **38** 1144
(4)   JCP 1960 **32** 1569
(5)   IC 1965 **4** 1213

Boron–nitrogen analogues of carbon compounds

The simplest compounds of this kind are molecules

$$>N\!\!=\!\!\frac{1\cdot40\,\text{Å}}{}\!\!B<\qquad \text{and} \qquad >N\!\!-\!\!\frac{1\cdot60\,\text{Å}}{}\!\!B<$$

(a)   (b)

which correspond to substituted ethylenes and ethanes. In (a) both atoms form three coplanar bonds and the B—N bond length is 1.40±0.02 Å. This length is close to values in one group of cyclic molecules to which we refer shortly and corresponds to a double bond. In (b) both atoms form tetrahedral bonds and the B—N bond is a single bond of length close to 1.60 Å. Examples include:

	B—N	Reference
(a) $(CH_3)_2N—BCl_2$	1·38 A	IC 1970 9 2439
$(CH_3)_2N—B(CH_3)_2$	1·42	JCS A 1970 992
(b) $(CH_3)_3N—BX_3$	1·58–1·64	AC 1969 **B25** 2338
		IC 1971 **10** 200

(All four halides of type (b) have been studied, but since there are differences of as much as 0.05 Å between the B—N bond length found in crystal and vapour molecule detailed discussion of bond lengths is probably premature.)

Cyclic molecules with alternate B and N atoms include 'aromatic' 4-, 6-, and 8-membered rings and saturated molecules with 4- and 6-membered rings. In the former B—N is close to 1.42 Å and in the latter, 1.60 Å. Examples are given in Fig. 24.10. In the molecule (a) the 4-ring is planar and the three bonds from each N are coplanar, but the shortness of the *exo*cyclic B—N bond (1.44 Å, the same as in the ring) is not consistent with the fact that the planes NSi_2 are approximately perpendicular to that of the 4-ring (B—N, 1.45; N—Si, 1.75; Si—C, 1.87 Å).[1] Molecules of type (b) which have been studied include the symmetrical benzene-like molecule of borazine itself, $B_3N_3H_6$ (B—N, 1.44 Å),[2a] $B_3N_3Cl_6$[2b] (B—Cl, 1.76 Å; N—Cl, 1.73 Å), B-monoaminoborazine, $B_3N_4H_7$ (B—N_{ring}, 1.42; B—NH_2, 1.50 Å),[3] and B-trichloroborazine (B—N, 1.41; B—Cl, 1.75 Å).[4] (This compound is made by heating together BCl_3 and NH_4Cl at temperatures above 110 °C— compare $(PNCl_2)_n$.)

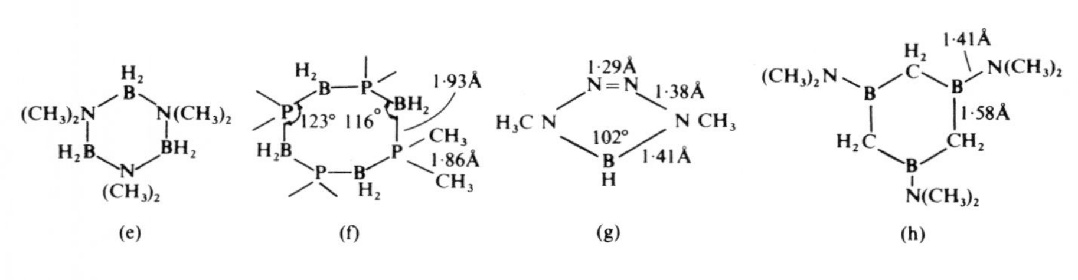

FIG. 24.10. Cyclic boron–nitrogen, boron–phosphorus, and boron–carbon molecules.

For references to numerous cyclic B—N compounds see reference (10). Compounds (RN.BX)₄ of type (c) have been prepared from BCl_3 and primary alkylamines and a number of derivatives have been made. In crystalline [(CH₃)₃C.NB.NCS]₄ both B and N form three coplanar bonds and there is a slight alternation in the B—N bond lengths in the boat-shaped ring:[5]

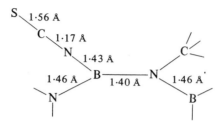

The ethylene-like molecules $R_2N.BX_2$ readily polymerize, and the dimers and trimers are the analogues of substituted cyclobutanes and cyclohexanes. Molecules of type (d) have planar 4-membered rings[6,7] with B—N close to 1.60 Å, and the molecule (e) has the chair configuration.[8] The phosphorus analogue of (e) has been shown to have a similar configuration with B—P, 1.94 Å,[9a] and the molecule [H₂B.P(CH₃)₂]₄ has the structure (f).[9b] (The As compounds [(CH₃)₂As.BH₂]ₙ (n = 3 and 4) have been prepared.)

Miscellaneous cyclic boron compounds include (g), with a 5-membered N₄B ring,[10] and (h), with a B₃C₃ ring and double bonds to the N(CH₃)₂ groups.[11]

(1)   AC 1969 **B25** 2342
(2a) IC 1969 8 1683
(2b) AC 1974 **B30** 731
(3)   JACS 1969 **91** 551
(4)   JACS 1952 **74** 1742
(5)   JCS 1965 6421
(6)   AC 1970 **B26** 1905

(7)   JCS A 1966 1392
(8)   AC 1961 **14** 273
(9a) AC 1955 **8** 199
(9b) JACS 1962 **84** 2457
(10) IC 1969 8 1677
(11) AC 1969 **B25** 2334

Boron–nitrogen compounds related to boranes

The simple molecule $H_3N.BH_3$ is mentioned later. Both $(CH_3)_3N.BH_3$ and the closely related aziridine borane[1] (a), have been studied.

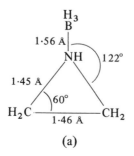

(a)

'Ammoniates' of boranes.   The ammonia addition compound of B_2H_6 has a molecular wieght in liquid ammonia corresponding to the formula $B_2H_6 \cdot 2NH_3$. An ionic structure $NH_4^+[BH_3 \cdot NH_2 \cdot BH_3]^-$ has been suggested, but the formulation as a borohydride, $[H_2B(NH_3)_2]^+(BH_4)^-$, is now preferred. This would appear to be consistent with its reaction with NH_4Cl:

$$[H_2B(NH_3)_2](BH_4) + NH_4Cl \rightarrow [H_2B(NH_3)_2]^+Cl^- + H_3N \cdot BH_3 + H_2$$

or

$$BH_4^- + NH_4^+ \rightarrow H_3N \cdot BH_3 + H_2$$

Crystalline (monomeric) $H_3N \cdot BH_3$ is a disordered crystal, and X-ray studies give B–N approximately $1.60\,Å.$[2] The other product, $[H_2B(NH_3)_2]Cl$ has a typical ionic crystal structure,[3] consisting of layers of $[H_2B(NH_3)_2]^+$ ions (b) interleaved with Cl^- ions.

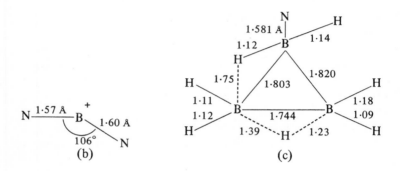

(b)   (c)

(High resolution n.m.r. shows that the phosphorus analogue of $H_3N \cdot BH_3$ is certainly (monomeric) $H_3P \cdot BH_3$ in the liquid state, and the i.r. and Raman spectra show that the same structure is maintained in the solid state.[4] The closely related $H_3B \cdot P(NH_2)_3$, prepared by the action of NH_3 on $H_3B \cdot PF_3$, consists of tetrahedral molecles in which the bond lengths are P–N, $1.65\,Å$, and P–B, $1.89\,Å$.)[5]

The action of sodium amalgam on diborane in ethyl ether gives a product of composition NaB_2H_6 which is actually a mixture of $NaBH_4$ and NaB_3H_8. The latter reacts with NH_4Cl in ether at $25\,°C$ to give $H_3N \cdot B_3H_7$. This is a white crystalline solid which has a disordered structure at temperatures above about $25\,°C$; the structure of the molecule has been studied in the low-temperature form.[6] The framework of the molecule (c) consists of a triangle of B atoms to one of which the N is attached. The B_3H_7 portion may be regarded as a distorted fragment of tetraborane resulting from symmetrical cleavage of the double bridge. (The estimated accuracy of the B–H bond lengths is $\pm0.05\,Å$ and of the other bond lengths $\pm0.005\,Å$.) For '$B_4H_{10} \cdot 2NH_3$' see under $B_3H_8^-$ ion.

Aminodiboranes.   These compounds are derived from B_2H_6 by replacing one H of the bridge by NH_2. Electron diffraction studies of $H_2N \cdot B_2H_5$ and $(CH_3)_2N \cdot B_2H_5$[7] give the following data:

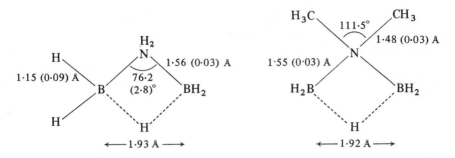

(In $H_2N.B_2H_5$ the following distances and angles were assumed: N—H, 1.02; B—H_{bridge}, 1.35 Å; angles HNH, 109½°; and HBH, 120°.) The B—N bonds are similar in length to those in the addition compounds of BF_3 described later, but the B—B distance is very much increased over the value (1.77 Å) in B_2H_6 owing to the interposition of the N atom.

(1) JCP 1967 **46** 357
(2) JACS 1956 **78** 502, 503
(3) JACS 1959 **81** 3551
(4) IC 1966 **5** 723
(5) AC 1960 **13** 535
(6) JACS 1959 **81** 3538
(7) JACS 1952 **74** 954

The oxygen chemistry of boron

Boron, like silicon, occurs in nature exclusively as oxy-compounds, particularly hydroxyborates of calcium and sodium. In borates based exclusively on BO_3 coordination groups there would be a simple relation between the O:B ratio and the number of O atoms shared by each BO_3 group, assuming these to be equivalent and each bonded to 2 B atoms:

O:B ratio		Number of O atoms shared
3	Orthoborates: discrete BO_3^{3-} ions	0
2½	Pyroborates: discrete $B_2O_5^{4-}$ ions	1
2	Metaborates: cyclic or chain ions	2
1½	Boron trioxide	3

Intermediate ratios (e.g. 1¾) would correspond to the sharing of different numbers of O atoms by different BO_3 groups, as in the hypothetical $(B_4O_7)_n^{2n-}$ chain analogous to the amphibole chain formed from SiO_4 groups. All of the above four possibilities are realized in compounds in which B is exclusively 3-coordinated, but there are two factors which complicate the oxygen chemistry of boron. First, there is tetrahedral coordination of B in many oxy-compounds, either exclusively or admixed with 3-coordination in the same compound. There is, therefore, no simple relation between O:B ratios and the structures of borates; we shall see later that a ratio such as 7:4 can be realized in a number of ways which, incidentally, do not include that mentioned above. Second, there are many hydroxyborates containing OH bonded to B as part of a 3- or 4-coordination group; this is in marked contrast to the rarity of hydroxysilicates. Examples are known of all 4-coordination groups from BO_4 through $BO_3(OH)$ and $BO_2(OH)_2$, both of which occur in $CaB_3O_4(OH)_3 . H_2O$, and $BO(OH)_3$ (in $Mg[B_2O(OH)_6]$), to $B(OH)_4$ in tetrahydroxyborates such as $NaB(OH)_4$. The recognition of this fact, as the result of structural studies, has led to the revision of many formulae as in the case of antimonates (p. 902), for example:

Colemanite, $Ca_2B_6O_{11} . 5H_2O$, is $CaB_3O_4(OH)_3 . H_2O$,
Bandylite, $CuCl_2 . CuB_2O_4 . 4H_2O$, is $CuClB(OH)_4$, and
Teepleite, $NaBO_2 . NaCl . 2H_2O$, is $Na_2ClB(OH)_4$.

The chemistry of borates is complex both in solution and in the melt. It is concluded from ^{11}B n.m.r. and other studies that $Na_3B_3O_6$ dissociates in solution to $B(OH)_4^-$ ions and that at low concentrations tetraborates dissociate completely into $B(OH)_3$ and $B(OH)_4^-$, but that in more concentrated solutions of borates various polyborate ions coexist in equilibrium with one another. From melts extensive series of borates are obtained, the product depending on the composition of the melt, for example:

	Li_3BO_3	$Li_4B_2O_5$	$Li_6B_4O_9$	$LiBO_2$	LiB_3O_5	$Li_2B_8O_{13}$	LiB_5O_8
O : B ratio	3	2.5	2.25	2	1.66	1.625	1.6

Little is yet known of the structures of metal-rich compounds of these types, but the structures of a number of anhydrous borates with O : B ratios between 1.75 and 1.55 are described later. For a perborate see p. 503.

Boron trioxide

This compound was known only in the vitreous state until as late as 1937, but it may be crystallized by dehydrating metaboric acid under carefully controlled conditions or by cooling the molten oxide under a pressure of 10–15 kbar. The normal form has a density of $2.56 g cm^{-3}$ and consists of a 3D network of BO_3 groups joined through their O atoms,[1] in which B–O = 1.38 Å. This net, an assembly of $B_{10}O_{10}$ rings, has been described in Chapter 3. Under a pressure of 35 kbar at 525 °C B_2O_3-II is formed (density, $3.11 g cm^{-3}$).[2] This much more dense poly-

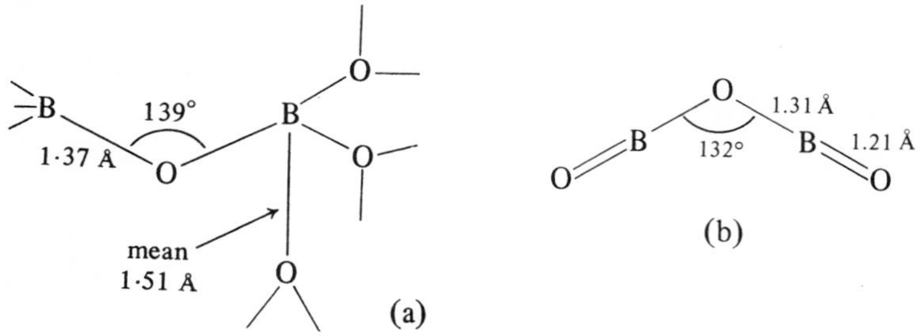

(a)

(b)

morph is built of irregular tetrahedra, three vertices of which are common to 3 BO_4 groups and one to 2 BO_4 groups; the structure may be described in terms of vertex-sharing pairs of tetrahedra (details shown at (a)). Mass spectrometric studies show that liquid B_2O_3 vaporizes predominantly as B_2O_3 molecules. The V-shape, (b), has been established by e.d. studies,[3] though there has been some doubt about the value of the angle B—O—B owing to the large amplitude of the thermal vibrations.

(1) AC 1971 **B27** 1662   (3) ZSK 1972 **13** 972
(2) AC 1968 **B24** 869

Orthoboric acid and orthoborates

In H_3BO_3 planar $B(OH)_3$ molecules (B—O, 1.36 Å) are linked into plane layers by O—H—O bonds of length 2.72 Å, the angle between which (at a given O atom) is 114° (Fig. 24.11). The H atoms were located by X-ray diffraction at positions about 1 Å from the O atoms to which they are bonded, a result confirmed by a n.d. study of D_3BO_3 (O—D, 0.97 Å along a linear O—D---O bond of length 2.71 Å).[1]

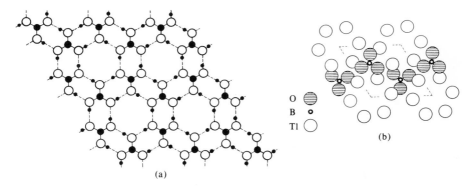

O
B
Tl

(a)   (b)

FIG. 24.11. (a) Portion of a layer of H_3BO_3. Broken lines indicate O—H---O bonds. (b) Projection of the structure of Tl_3BO_3.

Not very much is known about the structures of orthoborates of the alkali metals, and indeed not many appear to have been prepared. In α-Li_3BO_3 each Li^+ is surrounded by a very distorted tetrahedral group of 4 O atoms at approximately 2.0 Å, but two-thirds of them also have a fifth O neighbour at 2.5 Å.[2] Na_3BO_3 is formed from $Na_3B_3O_6$ at temperatures above 680 °C but has not been obtained pure, there being an equilibrium: $Na_3B_3O_6 \rightleftharpoons Na_3BO_3 + B_2O_3$. There is reason to suppose that some of the alkali orthoborates cannot exist (see p. 327). Known compounds $M_3(BO_3)_2$ include salts of Mg, Ca, Ba, Cd, and Co,[3] and there are complex borates such as $NaCaBO_3$ and $CaSn(BO_3)_2$, the latter being isostructural with dolomite, $CaMg(CO_3)_2$. From the structural standpoint the simplest orthoborates are $M^{III}BO_3$, which furnish examples of crystals isostructural with all three polymorphs of $CaCO_3$: h.p. $AlBO_3$,[4] $ScBO_3$; $InBO_3$, and $FeBO_3$ (calcite structure);

YBO$_3$ and LaBO$_3$ (aragonite structure); and SmBO$_3$ (vaterite structure). We refer to the structure of Tl$_3$BO$_3$ (Fig. 24.11(b)) on p. 1172. For other compounds containing discrete BO$_3^{3-}$ ions see 'oxide–borates' (p. 1080). Some recent references are given to orthoborates.[5]

(1) AC 1966 **20** 214
(2) AC 1971 **B27** 904
(3) ACSc 1949 **3** 660

(4) AC 1977 **B33** 3607
(5) AM 1961 **46** 1030; AC 1966 **20** 283;
      JCP 1969 **51** 3624

Pyroborates

Two rather different configurations of the pyroborate ion have been found, in the magnesium and cobalt salts. In Co$_2$B$_2$O$_5$,[1] (a), the planes of the BO$_3$ groups make angles of 7° with the plane of the central B–O–B system and are twisted in opposite directions (mean B–O approximately 1.30 Å). In Mg$_2$B$_2$O$_5$,[2] (b), the angle between

(a)   (b)

the planes of the BO$_3$ groups is found to be 22° 19′ and the mean B–O length is 1.36 Å. The oxygen bond angle in (b) is close to that found in metaborates, viz. 130° in CaB$_2$O$_4$, and 126½° in the B$_3$O$_6^{3-}$ and B$_5$O$_{10}^{5-}$ ions. No difference was found between the lengths of the central and terminal B–O bonds.

For the (OH)$_3$B–O–B(OH)$_3^{2-}$ ion see p. 1078.

(1) ACSc 1950 **4** 1054   (2) AC 1952 **5** 574

Metaboric acid and metaborates

The structure of the HBO$_2$ molecule in the vapour state has been determined by i.r. spectroscopy.[1] Similar structures have been found by e.d. for alkali and Tl meta-

borates in the vapour state,[2] with B–O and B=O as in B$_2$O$_3$ and M–O–B around 100–105° for the Na and K salts and 140° for CsBO$_2$ and TlBO$_2$.

Metaboric acid is known in three crystalline modifications, which provide a good example of monotropism (Fig. 24.12) and of the increase in density with change from 3- to 4-coordination of B (Table 24.4). The monoclinic form is readily prepared by the dehydration of H$_3$BO$_3$ in an open vessel at 140 °C; quenching of the molten material gives a glass which later recrystallizes as the orthorhombic form. If the melt is held at 175 °C the most stable (cubic) form is slowly precipitated, while complete dehydration at about 230 °C yields B$_2$O$_3$.

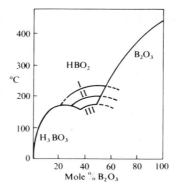

FIG. 24.12. Phase diagram for the B_2O_3–H_2O system.

TABLE 24.4

Crystalline forms of metaboric acid

	M.P.	Density	C.N. of B	Reference
Orthorhombic	176°C	1·784 g/cc	3	AC 1964 **17** 229
Monoclinic	201°	2·045	3 and 4	AC 1963 **16** 385
Cubic	236°	2·487	4	AC 1963 **16** 380

Orthorhombic metaboric acid is built of molecules $B_3O_3(OH)_3$ which are linked into layers by O–H–O bonds (Fig. 24.13). Monoclinic metaboric acid is apparently built of chains of composition $[B_3O_4OH(OH_2)]$, recognition of the OH group and the H_2O molecule resting on the location of the H atoms. The direct bonding of a water molecule to B is most unexpected. Cubic HBO_2 has a framework structure built from tetrahedral BO_4 groups with hydrogen bonds between certain pairs of O atoms.

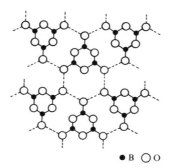

● B ○ O

FIG. 24.13. Arrangement of $B_3O_3(OH)_3$ molecules in a layer of one form of crystalline metaboric acid.

Metaborates are anhydrous compounds $M_x(BO_2)_y$; certain compounds originally formulated as hydrated metaborates contain $B(OH)_4^-$ ions; for example, $NaBO_2 \cdot 4H_2O$ is $NaB(OH)_4 \cdot 2H_2O$. The normal forms of these salts stable under atmospheric pressure contain either the cyclic $B_3O_6^{3-}$ ion (Fig. 24.14(a)), as in

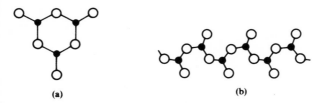

(a)   (b)

FIG. 24.14. Metaborate ions: (a) cyclic $(B_3O_6)^{3-}$; (b) infinite $(BO_2)_n^{n-}$ chain ion in CaB_2O_4 and $LiBO_2$–I, Small black circles represent B atoms.

$Na_3B_3O_6$, $K_3B_3O_6$, and $Ba_3(B_3O_6)_2$, or the infinite linear $(BO_2)_n^{n-}$ ion of Fig. 24.14(b), as in $LiBO_2$, CaB_2O_4, and SrB_2O_4. At higher pressure some of these compounds undergo changes to forms (designated by Roman numerals II, III, etc.) in which some or all of the B atoms become 4-coordinated (Table 24.5). For example, $LiBO_2$-II is a superstructure of the zinc-blende type, in which both Li and

TABLE 24.5
Crystalline forms of metaborates

	Pressure (kbar)	Density (g cm⁻³)	C.N. of B	C.N. of Ca	Reference
$Ca(BO_2)_2$-I	–	2·702	All 3	8	AC 1963 **16** 390
$Ca(BO_2)_2$-II	12–15	2·885	Half 3 Half 4	8	AC 1967 **23** 44
$Ca(BO_2)_2$-III	15–25	3·052	One-third 3 two-thirds 4	8 and 10	AC 1969 **B25** 955
$Ca(BO_2)_2$-IV	25–40	3·426	All 4	(9 + 3) and 12	AC 1969 **B25** 965

	Type of structure		
$Sr(BO_2)_2$-I			AC 1964 **17** 314
$Sr(BO_2)_2$-III	Isostructural with		AC 1969 **B25** 1001
$Sr(BO_2)_2$-IV	Ca compounds		
$Eu(BO_2)_2$ -I, III, IV	Isostructural with Ca compounds		IC 1980 **19** 983
$LiBO_2$-I	Infinite chain ion		AC 1964 **17** 749
$LiBO_2$-II	Zinc-blende superstructure		JCP 1966 **44** 3348
$Na_3B_3O_6$			AC 1963 **16** 594
$K_3B_3O_6$	Cyclic $B_3O_6^{3-}$ ions		AC 1970 **B26** 1189
$Ba(BO_2)_2$			AC 1966 **20** 819
$Cu(BO_2)_2$	3D tetrahedral framework		AC 1971 **B27** 677

B are surrounded tetrahedrally by 4 O, at 1.96 and 1.48 Å respectively. In the h.p. structures III and IV, which are common to the metaborates of Ca, Sr, and Eu, the boron–oxygen complexes are 3D frameworks.

A surprising difference is found between the structures of the $B_3O_6^{3-}$ ions in $Na_3B_3O_6$ and $K_3B_3O_6$, the structures of which have been carefully refined:

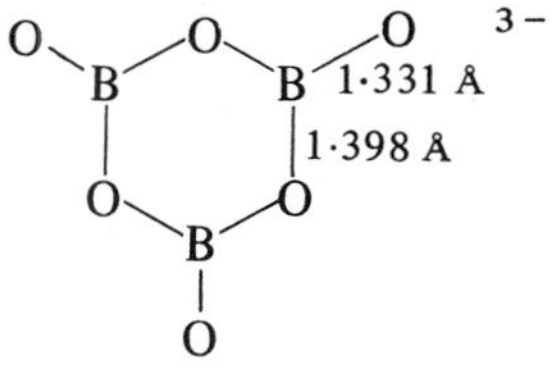

Since these compounds are isostructural these data are not consistent with Pauling's rules if the same relation between bond strength and length is assumed for both compounds.

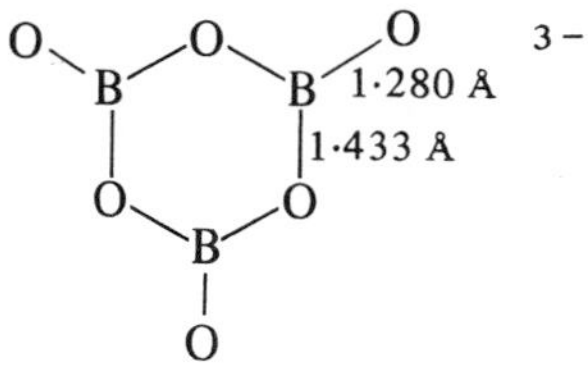

The structures of CaB_2O_4-IV and CuB_2O_4 are of special interest because the oxy-ion is a 3D framework formed from tetrahedral BO_4 groups sharing all vertices – compare silica structures. Since the framework contains planar B_3O_3 rings it may alternatively be described as built from rings of three tetrahedra similar to the S_3O_9 molecule or $Si_3O_9^{6-}$ ion which share the extra-annular O atoms with those of six similar rings. In CaB_2O_4-IV the framework forms around Ca^{2+} ions which are either 12- or (9+3)-coordinated:

$$Ca_I - 6\,O \quad 2.70\,Å \qquad Ca_{II} - 3\,O \quad 2.39\,Å$$
$$6\,O \quad 2.67 \qquad\qquad 3\,O \quad 2.48$$
$$3\,O \quad 2.60$$
$$3\,O \quad 3.14$$

In CuB_2O_4 the framework accommodates (or is cross-linked by) Cu^{2+} ions which have only 4-coplanar nearest neighbours:

$$Cu_I - 4\,O \quad 2.00\,Å \qquad Cu_{II} - 4\,O \quad 1.94\,Å$$
$$(O \quad 3.07)$$

In spite of the very different coordination groups around the cations in these crystals the volumes of the unit cells, one cubic and the other tetragonal, are almost equal (731 and 741 Å^3).

Tetrahedral coordination in metaborates is not confined to high-pressure polymorphs. An interesting new $(BO_2)_n^{n-}$ chain has been found in $AgBO_2$:[3]

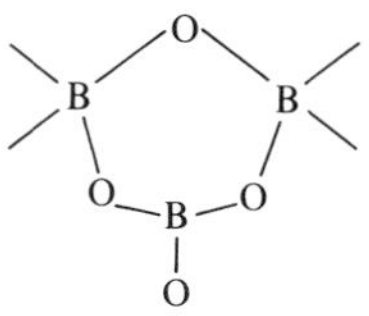

There are equal numbers of 3- and 4-coordinated B atoms, but in contrast to the simpler chain of Fig. 24.14(b) the BO_3 groups share only 2 O atoms. These structures raise a question of nomenclature.

The prefixes ortho, pyro, and meta applied to acids and salts of Si, P, and As refer to *tetrahedral* ions (or molecules in the case of the acids) which share respectively 0, 1, and 2 O atoms, that is, to compounds containing MO_4, M_2O_7, or $(MO_3)_n$ groups. As applied to borates they normally refer to acids and ions in which B is 3-coordinated, that is, ortho, BO_3^{3-}, pyro, $B_2O_5^{4-}$, and meta, $(BO_2)_n^{n-}$ (cyclic or linear). The tetrahedral B ions would be the (unknown)

$$\text{ortho, } BO_4^{5-}, \text{ pyro, } B_2O_7^{8-}, \text{ and meta, } (BO_3)_n^{3n-},$$

and in addition there would be oxy-ions in which all BO_4 groups share 3 or 4 O atoms, namely,

$$(B_2O_5)_n^{4n-}, \text{ layers or 3D frameworks,}$$
$$(BO_2)_n^{n-}, \text{ double layers or 3D frameworks.}$$

Sharing of two opposite edges of each BO_4 would give a chain $(BO_2)_n^{n-}$ structurally similar to the SiS_2 chain. Three of the ions of the tetrahedral family have formulae the same as those of the trigonal family. At present the only known type of borate oxy-ion in which all B atoms are tetrahedrally coordinated and all shared O atoms bonded to 2 B, as we have assumed above, is the $(BO_2)_n^{n-}$ framework ion in CaB_2O_4-IV and CuB_2O_4. These compounds are called metaborates, and indeed CaB_2O_4-IV is a high-pressure polymorph of the compound which in its normal form contains the metaborate ion formed from BO_3 groups. If we retain the term metaborate for all compounds $M(BO_2)_x$ it loses its earlier structural significance, for it includes not only the two extreme types of structure (all 3-coordinated B or all 4-coordinated B) but also the intermediate structures (e.g. CaB_2O_4-II and III) in which there is coordination of both types.

(1)   JCP 1960 **32** 488   (3)   NW 1980 **67** 606
(2)   ZSK 1972 **13** 972

Hydroxyborates and anhydrous polyborates

We discuss these compounds together because their anions either consist of or are built from simple cyclic units of the kind shown in Fig. 24.15. The number of possible anions is large because

(i) there are different basic ring systems, three of which, (a), (b), and (c), are illustrated;

(ii) the number of extra-annular O atoms can in principle increase until all the B atoms are 4-coordinated, as in the series (a_3)–(a_6);

(iii) the extra-annular O atoms can be O of OH groups or O atoms which could be either unshared or shared between two units. Some of the finite units of Fig. 24.15 in which all the extra-annular O atoms belong to OH groups exist in hydroxyborates (Table 24.6), but various numbers of these O atoms can be shared to form

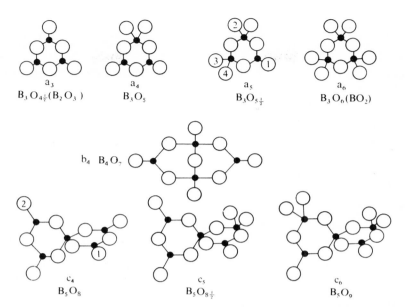

FIG. 24.15. Cyclic boron–oxygen systems in hydroxyborates and/or polyborates. The subscript is the number of extra-annular O atoms. The formula shows the compositions of the 3D anion formed if all of these are shared with other similar units.

ions extending indefinitely in one, two, or three dimensions. The *minimum* numbers of shared O atoms are obviously 2 for a chain and 3 for a layer or 3D system; the known layers are built from 4- or 5-connected units. If some OH groups remain the result is a hydroxy-anion, as found in many borates crystallized from aqueous solution; if all the extra-annular O atoms are shared the anion is of the type $B_x O_y$ characteristic of anhydrous polyborates prepared from the melt.

(iv) A particular borate anion may be built of units all of the same kind (for example, a_4) or it may be built from units of two or more kinds (for example, a_4 and c_4). We now amplify (ii)–(iv).

Of the fully 'hydroxylated' units of type (a), a_3 is the cyclic $B_3O_3(OH)_3$ molecule, a_4 is the anion in $Na_3[B_3O_3(OH)_4]$, and a_5 occurs in a series of hydrated calcium hydroxyborates[1] which includes the minerals meyerhofferite, $Ca[B_3O_3(OH)_5].H_2O$ and inyoite, the tetrahydrate. Finite ions may be envisaged which have some extra-cyclic O and some OH, for example. the intermediate members of the series:

$$a_3 \qquad B_3O_6^{3-} \qquad B_3O_5(OH)^{2-} \qquad B_3O_4(OH)_2^- \qquad B_3O_3(OH)_3.$$

The anion in $Na_3[B_3O_5(OH)_2]$[1a] is an example of an a_4 ion of this kind. Sharing of the two O atoms 1 and 2 of a_5 gives the infinite chain ion in $Ca[B_3O_4(OH)_3].H_2O$, colemanite (Fig. 24.16(a)), and sharing of all the atoms 1–4 gives the 2D ion in $Ca[B_3O_5(OH)]$, a layer based on the simplest plane 4-connected net. (The 3D

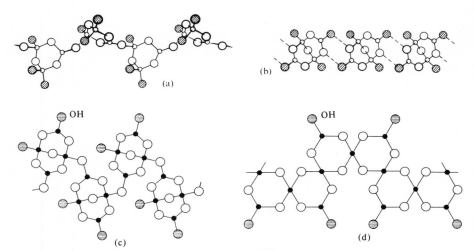

FIG. 24.16   (a) The infinite chain ion $[B_3O_4(OH)_3]_n^{2n-}$ in $CaB_3O_4(OH)_3.H_2O$;   (b) the system of hydrogen-bonded $[B_4O_5(OH)_4]^{2-}$ ions in borax;   (c) the anion in $Na_2[B_4O_6(OH)_2]$;   (d) the anion in $Na_2[B_4O_6(OH)_2].3H_2O$.

structure of CaB_2O_4-IV may be described as built of units a_6 sharing all extra-annular O atoms.)

The finite hydroxy-ion b_4 is the anion in $K_2[B_4O_5(OH)_4].2H_2O$ and also in borax,[2] a fact necessitating the revision of a familiar chemical formula, $Na_2B_4O_7.-10H_2O$, to $Na_2[B_4O_5(OH)_4].8H_2O$. Hydrogen bonds link the units into chains (Fig. 24.16(b)). The cation–water complex in borax was mentioned in Chapter 5 as an example of an infinite chain formed from octahedral $[Na(H_2O)_6]$ groups sharing two edges to give the $H_2O:Na$ ratio of 4:1.

The b_4 chain (c) forms the anion in $Na_2[B_4O_6(OH)_2]$,[2a] and it is interesting to find that the same composition is realized in an entirely different way in the trihydrate, $Na_2[B_4O_6(OH)_2].3H_2O$,[2b] (d), the mineral kernite.

The tetrahydroxy-ion c_4 occurs in $K[B_5O_6(OH)_4].2H_2O$,[3] and chains formed by sharing the O atoms 1 and 2 form the anion in the mineral larderellite, $NH_4[B_5O_7(OH)_2].H_2O$.[4] An intermediate possibility is realized in the mineral ammonioborite, $(NH_4)_3[B_{15}O_{20}(OH)_8].4H_2O$,[4a] in which the anion is a finite group formed from three c_4 units:

$$3-$$

The formation of 3D framework anions requires the sharing of *at least* 3 O by each sub-unit, but usually 4 or 5 are shared. The simplest structures arise from the units a_4, b_4, and c_4 which have their four extra-annular O atoms disposed at the vertices of (irregular) tetrahedra. These units can therefore link up to form 3D frameworks based on the diamond net, as noted in Chapter 3. There are a number of points of special interest. In CsB_3O_5[5] the a_4 units form one framework, but the more bulky b_4 and c_4 sub-units form two interpenetrating identical frameworks (in $Li_2B_4O_7$[6] and KB_5O_8[7] respectively). This is also true in $Ag_2B_8O_{13}$,[8] where there is the additional complication that each framework is composed of alternate units of two kinds (a_4 and c_4): $B_3O_5 + B_5O_8 = B_8O_{13}$. The units a_5 and c_5 have 5 extra-annular O atoms. In BaB_4O_7[9] alternate units of these types form a 3D 5-connected framework by sharing all these O atoms: $B_3O_{5\frac{1}{2}} + B_5O_{8\frac{1}{2}} = 2(B_4O_7)$. An even more complex system is the anion in CsB_9O_{14}[10] which consists of two interpenetrating 3D nets each built of two kinds of sub-unit. These are the a_3 and a_4 units of Fig. 24.15, which are present in the ratio $2a_3 : 1a_4$, so that the composition is $2(B_3O_{4\frac{1}{2}}) + B_3O_5 = B_9O_{14}$.

The c_5 unit is found in a number of structures with either two or four of the extra-annular O atoms shared, forming respectively a chain with composition $[B_5O_7(OH)_3]_n^{2n-}$ or a layer based on the planar 4-gon net with the composition $[B_5O_8(OH)]_n^{2n-}$. The former occurs in the mineral ezcurrite, $Na_2[B_5O_7(OH)_3].2H_2O$, and the latter in the minerals gowerite, veatchite, and nasinite, $Na_2[B_5O_8(OH)].2H_2O$[11] and in the isostructural (synthetic) potassium salt. Reference (11) includes literature references to the other hydroxyborates mentioned here.

The c_6 unit is found as the hexahydroxy ion in ulexite, $NaCaB_5O_6(OH)_6.5H_2O$, (ref. 12) as chains, $[B_5O_7(OH)_4]^{3-}$, in probertite, and as layers (sharing 4 O atoms), $[B_5O_8(OH)_2]^{3-}$ in heidornite and in $Na_3[B_5O_8(OH)_2].H_2O$.[12a] Table 24.6 summarizes the formulae of anions formed from the sub-units a_5, b_4, and c_4.

TABLE 24.6
Hydroxyborates and polyborates

Number of O atoms shared	Cyclic unit of Fig. 24.15		
	a_5	b_4	c_4
0	$[B_3O_3(OH)_5]Ca . H_2O$ $2 H_2O$ $4 H_2O$	$[B_4O_5(OH)_4]Na_2 . 8 H_2O$	$[B_5O_6(OH)_4]K . 2H_2O$
2	$[B_3O_4(OH)_3]Ca . H_2O$	$[B_4O_6(OH)_2]^{2-}$	$[B_5O_7(OH)_2]NH_4 . H_2O$
4	$[B_3O_5(OH)]Ca$†	$[B_4O_7]Li_2$ *	$[B_5O_8]K$*

† Layer structure. * Two interpenetrating 3D frameworks.

One of the outstanding features of polyborate chemistry is the extraordinary number of different ways in which relatively simple formulae result from combinations of the units of Fig. 24.15. We have noted the single 3D framework ion in CsB_3O_5 formed from units a_4. Two much more complex anions are found in the two forms of NaB_3O_5. In α-NaB_3O_5[13] there are two interpenetrating frameworks, each consisting of equal numbers of b_4 and c_4 units ($B_5O_8 + B_4O_7 = B_9O_{15}$). Layer structures are very rare in anhydrous polyborates, but in β-NaB_3O_5[14] there are double layers built from equal numbers of units of three kinds, tetrahedral BO_4, a_4, and c_4 ($BO_2 + B_3O_5 + B_5O_8 = B_9O_{15}$).

Borates containing B_4O_7 ions present an even greater variety of structures. We have described the two different framework ions in $Li_2B_4O_7$ and BaB_4O_7; two other possibilities are realized in $Na_2B_4O_7$ and $K_2B_4O_7$:

		Units	Reference
$Li_2B_4O_7$	⎫	b_4	(6)
MB_4O_7 (M = Cd, Mg, Fe, etc.)	⎬		AC 1966 **20** 132
BaB_4O_7		a_5 c_5	(9)
$Na_2B_4O_7$		a_4^* c_5	AC 1974 **B30** 578
$K_2B_4O_7$		BO_3 a_5 b_4	AC 1972 **B28** 3089
SrB_4O_7	⎫ isostructural	See text	AC 1966 **20** 274
EuB_4O_7	⎭		AC 1980 **B36** 2008

(The symbol a_4^* stands for a_4 sharing only 3 O atoms and therefore contributing $B_3O_{5\frac{1}{2}}$ to the formula, like a_5.)

In the first four types of structure all B atoms are 3- or 4-coordinated, and bridging O atoms are bonded to 2 B atoms. In SrB_4O_7 there is a 3D anion in which all B atoms are tetrahedrally coordinated and 2/7 of the O atoms are 3-coordinated. In $Eu_2(B_5O_9)Br$,[14a] and the isostructural Ca compound, the cations and Br^- ions are accommodated in tunnels in a 3D framework in which two-fifths of the B atoms are 3- and the remainder 4-coordinated. For a note on the chemical properties of Eu borates see p. 1250.

Even more complex anions built of the units of Fig. 24.15 are found. In $Pb_6B_{10}O_{21}$[15] there are discrete groups (a) formed from two b_4 units linked through two triangular BO_3 groups. The Pb–O bonds presumably have appreciable covalent character, for Pb has 3 or 4 O atoms (all to one side) at distances from

(a)

2.23–2.55 Å—compare the shortest bonds in PbO, 2.18 Å or the sum of the ionic radii, 2.6 Å. The anion in $K_5B_{19}O_{31}$[16] is a single 3D framework constructed from equal numbers of a_4, c_4, and BO_3 groups together with half that number of tetrahedral BO_4 groups: $(B_3O_5)_2 + (B_5O_8)_2 + B_2O_3 + BO_2 = B_{19}O_{31}$. Apparently the introduction of single 3- or 4-coordinated B atoms avoids the necessity of forming two interpenetrating frameworks as in some of the structures we have mentioned.

We conclude this section with examples of structures containing more complex sub-units. The tricyclic unit $B_6O_7(OH)_6^{2-}$ of Fig. 24.17(a) is planar, apart from the

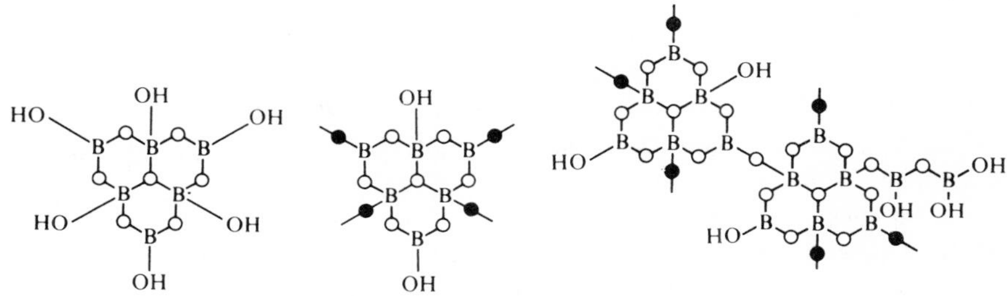

FIG. 24.17. Tricyclic boron–oxygen units in hydroxyborates (see text).

OH groups attached to the tetrahedral B atoms, and is of interest as containing a central O atom bonded to three B atoms. It is found as a discrete anion in $Mg_2[B_6O_7(OH)_6]_2 \cdot 9H_2O$.[17] Sharing of the O atoms shown as black circles in Fig. 24.17(b) results in a layer of composition $[B_6O_9(OH)_2]_n^{2n-}$ based on the simplest 4-connected plane net (Chapter 3); this is the anion in the mineral tunellite, $SrB_6O_9(OH)_2 \cdot 3H_2O$.[18] In $(Ca,Sr)_2B_{14}O_{20}(OH)_6 \cdot 5H_2O$[19] this tricyclic unit is found as the sub-unit in a much more complex layer. The multiple unit consists of two B_6 units to one of which is attached a B_2 side chain, and this B_{14} complex is linked into layers by sharing the six O atoms shown as black circles in Fig. 24.17(c). The polycyclic ion containing 12 B atoms shown at (b) is the anion in $Na_8[B_{12}O_{20}(OH)_4]$.[20]

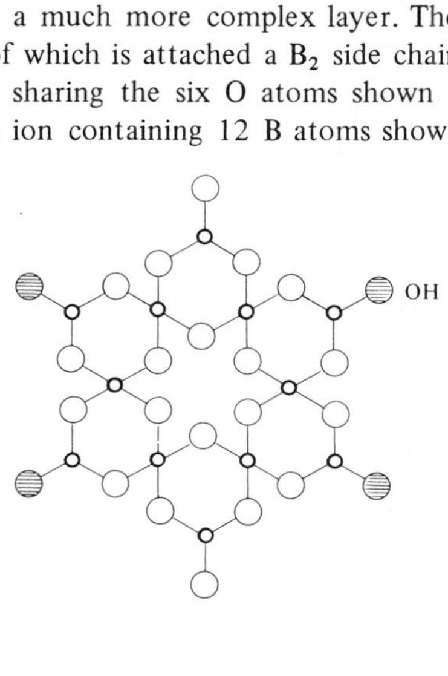

(b)

(1)	JINC 1964 **26** 73; AC 1974 **B30** 2194	(10)	AC 1967 **23** 427
(1a)	AC 1975 **B31** 1993	(11)	AC 1975 **B31** 2405
(2)	AC 1978 **B34** 3502	(12)	Sc 1964 **145** 1295
(2a)	AC 1978 **B34** 1080	(12a)	AC 1977 **B33** 3730
(2b)	AM 1973 **58** 21	(13)	AC 1974 **B30** 747
(3)	AC 1978 **B34** 45	(14)	AC 1972 **B28** 1571
(4)	AC 1969 **B25** 2264	(14a)	IC 1980 **19** 3807
(4a)	Sc 1971 **171** 377	(15)	AC 1973 **B29** 2242
(5)	AC 1974 **B30** 1178	(16)	AC 1974 **B30** 1827
(6)	AC 1968 **B24** 179	(17)	AANL 1969 **47** 1
(7)	AC 1972 **B28** 168	(18)	AM 1964 **49** 1549
(8)	AC 1965 **18** 77; AC 1969 **B25** 2153	(19)	AM 1970 **55** 1911
(9)	AC 1965 **19** 297	(20)	AC 1979 **B35** 2488

Other borate structures containing tetrahedrally coordinated boron

Coordination groups ranging from BO_4 to $B(OH)_4$ are found in some borates, and we shall note here a number of the more interesting structures.

The simplest structures containing BO_4 coordination groups are those of BPO_4 and $BAsO_4$, with silica-like structures. Further similarity to silicon is shown by the isomorphism of $TaBO_4$ and $ZrSiO_4$ and by the structure of $Zn_4B_6O_{13}$.[1] The B atoms in the 3D framework of this crystal are situated at the vertices of Fedorov's packing of truncated octahedra, the BO_4 tetrahedra being linked in the same way as the SiO_4 tetrahedra in, for example, sodalite, $Na_4Si_3Al_3O_{12}Cl$. One O atom does not form part of the B_6O_{12} framework, so that the compound may be formulated $Zn_4O(B_6O_{12})$.

Examples of crystals containing $BO_3(OH)$ coordination groups include the minerals datolite,[2] $CaBSiO_4(OH)$, and colemanite, $CaB_3O_4(OH)_3.H_2O$. The structure of the latter has been mentioned in the previous section. Datolite consists of apophyllite-like sheets of tetrahedra (p. 1025) held together by Ca^{2+} ions. The tetrahedral groups composing the sheets are alternately SiO_4 and BO_3OH, and each tetrahedron shares three vertices with tetrahedra of the other kind, the unshared vertices being O of SiO_4 and OH of BO_3OH groups. The composition of the layer is therefore $BO_{3/2}OH.SiO_{5/2}=BSiO_4OH$ (Fig. 24.18(a)).

The mineral pinnoite,[3] originally formulated $MgB_2O_4.3H_2O$, contains ions

$$\left[(OH)_3B \underset{124°}{\overset{O}{\diagup\diagdown}} B(OH)_3 \right]^{2-}$$

in which B forms tetrahedral bonds.

Tetrahedral $B(OH)_4^-$ ions are found in salts such as $LiB(OH)_4$[3a] (B–OH, 1.48 Å), $NaB(OH)_4.2H_2O$[4] and $Ba[B(OH)_4]_2.H_2O$[5] and in the minerals teepleite,[6] $Na_2ClB(OH)_4$, and bandylite,[7] $CuClB(OH)_4$. The latter is also of interest in connection with the stereochemistry of the cupric ion (p. 000). The structure may be dissected into puckered layers composed of tetrahedral $B(OH)_4$ groups connected

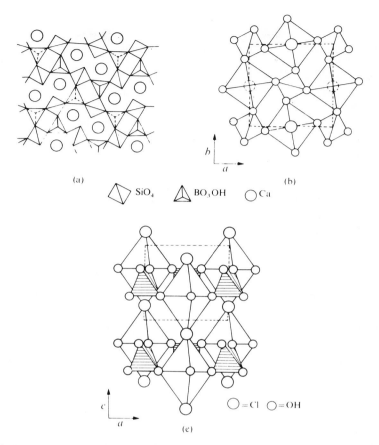

FIG. 24.18. (a) the $BSiO_4OH$ layer in datolite, $CaBSiO_4(OH)$; (b) mode of linking of distorted octahedral $[Cu(OH)_4Cl_2]$ and tetrahedral $[B(OH)_4]$ coordination groups in bandylite, $CuClB(OH)_4$; (c) elevation of the structure of bandylite showing the layers (b) held together by long Cu–Cl bonds.

by Cu^{II} atoms which are thereby surrounded by four coplanar OH groups at the corners of a square (Fig. 24.18(b)). The layers are held together by long Cu–Cl bonds between Cu atoms of adjacent layers and Cl atoms situated between them (Fig. 24.18(c)), the coordination group around Cu being a distorted octahedron (Cu–4OH, 1.98 Å; Cu–2Cl, 2.80 Å). The B–OH bond length is 1.42 Å.

The mineral hambergite,[8] $Be_2(BO_3)OH$, is a hydroxy-orthoborate containing planar BO_3^{3-} and OH^- ions, not tetrahedral $(BO_3OH)^{4-}$ ions. Similarly fluoborite,[9] $Mg_3(OH,F)_3BO_3$, consists of a c.p. assembly of O^{2-}, OH^-, and F^- ions with B in positions of triangular coordination and Mg in octahedral holes. In general, O:B ratios $>4:1$ do *not necessarily* imply tetrahedral coordination of B since all the O atoms are not necessarily bonded to B. For example, we may have an assembly of O atoms in which B atoms occupy some positions of 3-coordination and the metal atoms octahedral interstices, as in fluoborite. This is the case in the 'boroferrites' such as $Mg_2Fe_2^{III}B_2O_8$[10] and warwickite,[11] $Mg_3TiB_2O_8$ (p. 602). $Bi_4B_2O_9$[12] and $Co_4Fe_2^{III}B_2O_{10}$. In such oxide–borates there may be triangular BO_3 groups, as in $Bi_4B_2O_9$, which is $Bi_4O_3(BO_3)_2$, or tetrahedral BO_4 groups, as in Fe_3BO_6,[13] which is isostructural with norbergite, $Mg_3SiO_4(OH, F)_2$ (p. 1017).

The Mg^{2+} ions in the mineral boracite, $Mg_3B_7O_{13}Cl$, may be replaced by other ions M^{2+} (for example, Ni^{2+}) and also by Li^+, and these synthetic boracites have cubic high-temperature forms like the natural boracite. In the Li-boracite, $Li_4B_7O_{12}Cl$, part of the Li replaces the Mg nearly completely and is accommodated very loosely in large 'octahedral' holes. The remaining Li^+ ions occupy approximately one-third of another set of equivalent positions. Movement of the Li^+ ions through the crystal makes this boracite a good ionic conductor, a property further enhanced by partial replacement of Cl by Br, as in $Li_4B_7O_{12}Cl_{0.68}Br_{0.32}$.[14]

(1)	ZK 1961 **115** 460	(8)	AC 1963 **16** 1144
(2)	ZaC 1967 **125** 286; AC 1972 **B28** 326	(9)	AC 1950 **3** 208
(3)	AC 1967 **23** 500	(10)	AC 1950 **3** 473
(3a)	ZaC 1966 **342** 188	(11)	AC 1950 **3** 98
(4)	AC 1963 **16** 1233	(12)	AC 1972 **B28** 2007
(5)	AC 1969 **B25** 1811	(13)	AC 1975 **B31** 1662
(6)	RS 1951 **21** No. 7	(14)	AC 1977 **B33** 2767
(7)	AC 1951 **4** 204		

The lengths of B–O *bonds*

Observed lengths of B–O bonds range from 1.21 Å for B=O in the gaseous B_2O_3 molecule to around 1.55 Å. The mean value for triangular coordination is 1.365 Å and for tetrahedral coordination 1.475 Å, but there are considerable ranges of lengths for both types of coordination:

$$\underset{1{\cdot}20 \text{ Å}}{\text{B=O}} \qquad 1{\cdot}28 \xleftrightarrow{\underset{1{\cdot}365}{BO_3}} 1{\cdot}43 \xleftrightarrow{\underset{1{\cdot}475 \text{ Å}}{BO_4}} 1{\cdot}55 \text{ Å}$$

Cyclic $H_2B_2O_3$, *boroxine*, $H_3B_3O_3$, *and boranocarbonates*

Cyclic $H_2B_2O_3$ is formed as an intermediate in the oxidation of B_5H_9, B_4H_{10}, etc. as an unstable species with a half-life of only 2–3 days at room temperature. As the result of a m.w. study the molecule has been assigned the structure (a).[1] Boroxine, $H_3B_3O_3$, is prepared by the action of H_2 on a mixture of $B + B_2O_3$. Although it is a 'high-temperature' species it can be preserved for an hour or two at room temperature under a pressure of 1–2 torr in the presence of excess argon. It decomposes to $B_2O_3 + B_2H_6$ but can be oxidized to cyclic $H_2B_2O_3$. The structure (b) has been assigned to the molecule.[2]

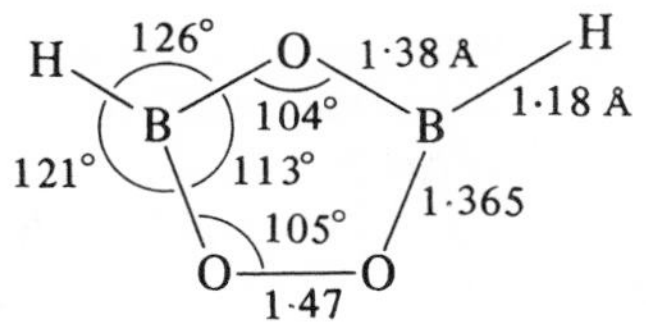

(a) 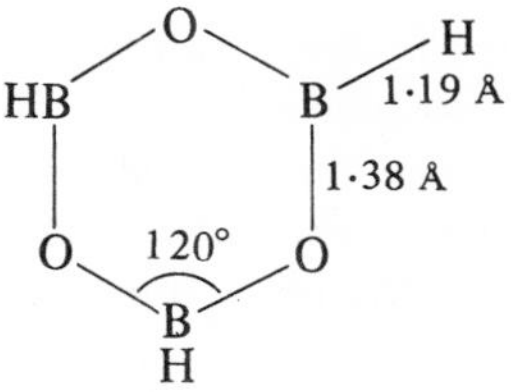   (b)

A BH_3-substituted carbonate ion has been made by reacting $H_3B.CO$ with KOH to give the boranocarbonate, $K_2(H_3B.CO_2)$.[3]

(1) JCP 1967 **47** 4186; AC 1969 **B25** 807   (3) IC 1967 **6** 817
(2) IC 1969 **8** 1689

Boranes and related compounds

Preparation and properties

The boron hydrides (boranes) were originally prepared by the action of 10 per cent HCl, or preferably 8N-phosphoric acid, on magnesium boride. The chief product of this method was B_4H_{10}, mixed with small quantities of B_5H_9, B_6H_{10}, and $B_{10}H_{14}$. This mixture was separated into its components by fractional distillation. B_2H_6 had to be obtained indirectly (with B_5H_9 and $B_{10}H_{14}$) by heating B_4H_{10} at $100\,°C$. It is interesting to note that the action of acid on Mg_2Si gives the silanes from SiH_4 to Si_6H_{14} in decreasing amounts. A later method of preparing boron hydrides was to pass the vapour of BCl_3 with hydrogen in a rapid stream at low pressure through an electric discharge between copper electrodes. The main boron-containing product of this reaction is B_2H_5Cl, which decomposes when kept at $0\,°C$ into B_2H_6 and BCl_3. These and other methods of preparing diborane were adequate while the compound was only of theoretical interest. Diborane is now a useful intermediate and reagent; for example, it converts metal alkyls to borohydrides, reduces aldehydes and ketones to alcohols, and decomposes to pure boron at high temperatures. Moreover, it appeared at one time to have a future as a rocket fuel, and efforts were therefore made to find better ways of preparing B_2H_6 on a large scale from readily accessible starting materials. Methyl borate can be prepared in over 90 per cent yield and in a very pure state by heating B_2O_3 with an excess of methanol and removing the methanol from the $CH_3OH-B(OCH_3)_3$ azeotrope so formed with LiCl. Methyl borate is then converted into $NaBH(OCH_3)_3$ by refluxing with NaH at $68\,°C$, and the following reaction gives a nearly theoretical yield of diborane:

$$8(C_2H_5)_2O.BF_3 + 6NaBH(OCH_3)_3 \rightarrow B_2H_6 + 6NaBF_4 + 8(C_2H_5)_2O + 6B(OCH_3)_3$$

Diborane may alternatively be made directly from B_2O_3 by heating the oxide with $Al + AlCl_3$ at $175\,°C$ under a pressure of 750 atm of H_2.

An outstanding feature of borane chemistry is the large number of reactions which result in the conversion of one or more boranes into others. These reactions make it possible to develop the whole of borane chemistry from one simple starting material, diborane. This point is emphasized in Chart 24.2, which shows only a small part of the very complex chemistry of boranes. Pyrolysis of B_2H_6 yields a number of the lower boranes including, for example, the thermally unstable B_5H_{11}, but this compound is preferably prepared by utilizing the equilibrium

$$B_2H_6 + 2B_4H_{10} \rightleftharpoons 2B_5H_{11} + 2H_2.$$

This borane can be converted catalytically into B_6H_{10}:

$$2B_5H_{11} \rightarrow B_6H_{10} + 2B_2H_6$$

CHART 24.2
Reactions of diborane

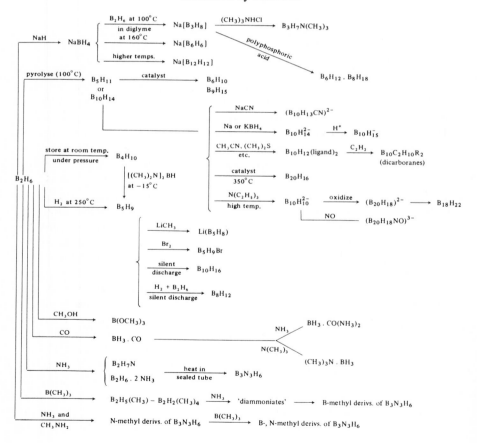

or reacted with B_4H_{10} to give B_5H_9:

$$B_5H_{11} + B_4H_{10} \rightarrow B_5H_9 + 2B_2H_6.$$

Recycling of the products of a particular reaction may give a satisfactory yield of a desired product (for example, $B_{10}H_{14}$ from the pyrolysis of B_2H_6), and the use of the silent electric discharge (with or without the addition of hydrogen) figures prominently in the preparation of boranes.

In some respects the hydrogen chemistry of boron resembles that of carbon and in others that of silicon. For example, boranes undergo many substitution reactions, H being replaced by halogen, CN, and organic ligands; derivatives of $B_{10}H_{14}$ include $B_{10}H_{13}I$ and $B_{10}H_{12}I_2$ and $B_{10}H_{12}L_2$, where L is R_2S, RCN, R_3N, etc. Mono-iodo-diborane reacts with sodium in exactly the same way as does ethyl iodide in the Wurtz reaction:

$$2B_2H_5I + 2Na \rightarrow B_4H_{10} + 2NaI$$

compare

$$2C_2H_5I + 2Na \rightarrow C_4H_{10} + 2NaI$$

In their reactions with halogens and hydrogen halides, however, the boranes behave quite differently from hydrocarbons. Diborane, for example, reacts with HCl to give a chloro derivative and hydrogen, and $B_{10}H_{14}$ —which from its formula would appear to be an unsaturated compound — forms with a halogen a substituted derivative and not an addition product as is the case with C_2H_4 and other unsaturated hydrocarbons. There are large differences in thermal stability between boranes such as B_5H_9 and the very unstable B_5H_{11}, but generally in their low stability and vigorous reaction with water the boranes resemble silanes rather than hydrocarbons.

In addition to the neutral boranes many borohydride ions have been prepared, ranging from BH_4^- to $B_{20}H_{18}^{2-}$ and including a remarkable series of polyhedral ions $B_nH_n^{2-}$ (n = 6 and 8 to 12). Unlike the boranes these ions have highly symmetrical structures and appear to be the boron analogues of the aromatic carbon compounds. There is extensive delocalization of a small number of electrons, in contrast to the localized 3-centre bonds in the neutral boranes, and the alkali-metal salts of $B_{10}H_{10}^{2-}$ and $B_{12}H_{12}^{2-}$ are extremely stable compounds. Borohydride ion chemistry is complicated by the fact that not only are there polyhedral ions $B_nH_n^{2-}$ but there are also (a) many substituted ions (for example, $B_{12}H_{11}OH^{2-}$), (b) ions with the same composition but different charge (for example, $B_8H_8^{2-}$ and the paramagnetic $B_8H_8^-$), and (c) ions with the same boron, or boron–carbon, skeleton but different numbers of H atoms and different charges (for example, $B_{11}H_{13}^{2-}$ and $B_{11}H_{14}^-$, which are reversibly interconvertible). Fully halogenated ions include $B_{12}Br_{12}^{2-}$ and $B_{10}Cl_{10}^{2-}$; the latter has been isolated in the free acid, $(H_3O)_2B_{10}Cl_{10} \cdot 5H_2O$. Furthermore, oxidation of $B_{10}H_{10}^{2-}$ gives the $B_{20}H_{18}^{2-}$ ion, which consists of two B_{10} units joined by two 3-centre bonds (contrast the borane $B_{20}H_{16}$). There are also ions in which the two 'halves' are linked by NO or a metal atom in place of the 3-centre bonds, for example $B_{20}H_{18}NO^{3-}$ and $Fe(\pi\text{-}B_9C_2H_{11})_2^{2-}$.

Isoelectric with ions $B_nH_n^{2-}$ are the dicarboranes, $B_nC_2H_{n+2}$, prepared from appropriate boranes and acetylene, which also have aromatic character. These also form composite π-bonded ions which form salts such as $[(C_2H_5)_4N]_2[Cu(C_2B_9H_{11})_2]$.

We have therefore three main groups of structures to summarize:

(a) the boranes and their derivatives, including the higher members formed from two simpler units joined at a common vertex ($B_{10}H_{16}$), a common edge ($B_{16}H_{20}$, $B_{18}H_{22}$), or a common face ($B_{20}H_{16}$). In Table 24.7 an asterisk distinguishes boranes that were prepared by Stock; all of these, and most of the others, fall into one of two families, B_nH_{n+4} and B_nH_{n+6}, the former being generally far more stable than the latter;

(b) borohydride ions and metal borohydrides, and

(c) the polyhedral ions $B_nH_n^{2-}$ and the isoelectronic carboranes $B_nC_2H_{n+2}$ and their derivatives, including the composite ions mentioned above.

Table 24.7 lists the boranes and some simple derivatives, with references to structural studies.

TABLE 24.7
Boranes and simple derivatives

B_nH_{n+4}	B_nH_{n+6}	Others	Reference
$B_2H_6^*$			JCP 1968 **49** 4456 (vapour)
			JCP 1965 **43** 1060 (crystal)
	$B_4H_{10}^*$		JACS 1953 **75** 4116 (vapour)
			JCP 1957 **27** 209 (crystal)
$B_5H_9^*$			JCP 1954 **22** 262 (vapour)
			AC 1952 **5** 260 (crystal)
	$B_5H_{11}^*$		JCP 1957 **27** 209
$B_6H_{10}^*$			JCP 1958 **28** 56
	B_6H_{12}		
B_8H_{12}			AC 1966 **20** 631
		B_8H_{16}	
		B_8H_{18}	
	B_9H_{15}		JCP 1961 **35** 1340
	i-B_9H_{15}		
$B_{10}H_{14}^*$			IC 1969 **8** 464 (n.d.)
	$B_{10}H_{16}$		JCP 1962 **37** 2872
		$B_{10}H_{18}$	
$B_{16}H_{20}$			IC 1970 **9** 1452
n-$B_{18}H_{22}$			AC 1966 **20** 631
i-$B_{18}H_{22}$			JCP 1963 **39** 2339
		$B_{20}H_{16}$	JCP 1964 **40** 866
$B_2H_2(CH_3)_4$			IC 1968 **7** 219
B_5H_8I			AC 1965 **19** 658
$B_5H_7(CH_3)_2$			IC 1966 **5** 1752
$B_{10}H_{13}I$			IC 1967 **6** 1281
$B_{10}H_{12}I_2$			JACS 1957 **79** 2726
$B_{10}H_{12}[S(CH_3)_2]_2$			AC 1962 **15** 410

The molecular structures of the boranes

Because of the number and complexity of the boranes and their derivatives we shall not attempt to describe all their structures in detail. Apart from B_2H_6 most of the boranes have boron skeletons which are usually described as icosahedral fragments. (They could equally well be related to the more recently discovered carboranes $B_nC_2H_{n+2}$ since the boron–carbon skeletons in these compounds are highly symmetrical triangulated polyhedra (Table 24.8) which include the icosahedron.) The great theoretical interest of the boranes stems from the fact that they are electron-deficient molecules, that is, there are not sufficient valence electrons to bond together all the atoms by normal electron-pair bonds. In these molecules the number of atomic orbitals (1 for each H and 4 for each B) is greater than the total number of valence electrons:

	Total number of atomic orbitals	Total number of valence electrons	Number of 3-centre bonds
B_2H_6	14	12	2
B_4H_{10}	26	22	4
B_5H_9	29	24	5
$B_{10}H_{14}$	54	44	10
$B_{18}H_{22}$	94	76	18

Structural studies show that in molecules of boranes some H atoms (on the periphery of the molecule) are attached to a single B (B–$H_{terminal}$, 1.2 Å) while others are situated between B atoms forming B–H–B bridges. These bridges are usually symmetrical (B–H_{bridge}, 1.34 Å); an exception is the slightly unsymmetrical bridge in $B_{10}H_{14}$, with B–H_b, 1.30 and 1.35 Å. The B atoms are at the vertices of triangulated polyhedra or fragments of such polyhedra, with B–B usually 1.70–1.84 Å but in rare cases smaller (1.60 Å for the basal B–B in B_6H_{10}) or larger (1.97 Å for two B–B bonds in $B_{10}H_{14}$). In general each B atom is bonded to one or more terminal H atoms, but in some boranes which consist of two 'halves' the B atoms involved in the junctions are bonded only to B atoms or to B atoms and bridging H atoms:

$$B_{10}H_{16}: \quad \text{two B bonded to} \quad \text{5 B only}$$
$$\text{i-}B_{18}H_{22}: \quad \text{one B} \quad \text{5 B + 2 } H_b$$
$$\text{one B} \quad \text{7 B only}$$
$$n\text{-}B_{18}H_{22}: \quad \text{two B} \quad \text{6 B + 1 } H_b.$$

In order to account for the molecular structures, retaining two-electron bonds and at the same time utilizing the excessive numbers of orbitals available it has been supposed that three or more atomic orbitals combine to form only one bonding orbital. The structures of the simpler boranes may be formulated with 3-centre bonds of three kinds (Fig. 24.19). In the central (closed) bond, (a), the three B

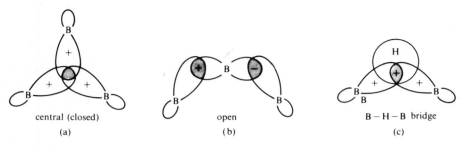

FIG. 24.19. Types of 3-centre bond in boranes.

atoms use hybrid orbitals and are situated at the corners of an equilateral triangle. In the open 3-centre bond, (b), they form an obtuse-angled triangle, and the intermediate B atom uses p orbitals, while in (c) the two B atoms are bridged by a H atom, and the orbitals used are 1s of H and hybrid orbitals of B. In some of the more complex systems, such as the polyhedral ions, complete delocalization of a number of electrons (implying the use of a larger number of orbitals) appears to be necessary to account for the 'aromatic' character—compare $C_5H_5^-$, C_6H_6, and $C_7H_7^+$.

Diborane, B_2H_6. The results of the latest electron diffraction (sector, microphotometer) study of this (diamagnetic) molecule are:

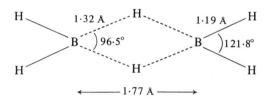

The plane of the central BH_2B system is perpendicular to those of the terminal BH_2 groups (Fig. 24.20(a)). An X-ray study of the crystalline β form (at 4.2 K) gives a similar B–B distance but shorter B–H bond lengths (B–H_t, 1.09 Å, and B–H_b, 1.24 Å) and angles H_tBH_t, 124°, and H_bBH_b, 90°. The absence of a direct B–B bond in diborane accounts for reactions such as

$$B_2H_6 + 2CO \rightarrow 2BH_3.CO$$
$$B_2H_6 + 2N(CH_3)_3 \rightarrow 2BH_3.N(CH_3)_3$$
and
$$B_2H_6 + 2NH_3 \rightarrow NH_4(H_3B.NH_2.BH_3).$$

Of the six H atoms in B_2H_6 only four can be replaced by CH_3 and in these methylated compounds there are never more than two CH_3 groups on a particular B atom. Also, $B(CH_3)_3$ is known, but not $BH(CH_3)_2$ or $BH_2(CH_3)$, which could obviously condense to $B_2H_2(CH_3)_4$ and $B_2H_4(CH_3)_2$ respectively. In the molecule $(CH_3)_2BH_2B(CH_3)_2$ the following distances were determined: B–H, 1.36 Å; B–C, 1.59 Å; and B–B, 1.84 Å.

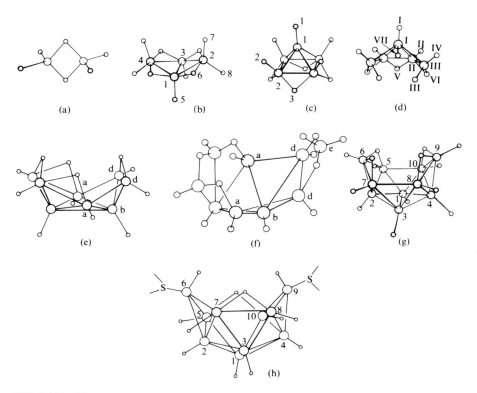

FIG. 24.20. The molecular structures of boranes and related compounds; (a) B_2H_6; (b) B_4H_{10}; (c) B_5H_9. (d) B_5H_{11}; (e) B_8H_{12}; (f) B_9H_{15}; (g) $B_{10}H_{14}$. (h) $B_{10}H_{12}[S(CH_3)_2]_2$.

Tetraborane (10), B_4H_{10}. The configuration of Fig. 24.20(b) has been established by e.d. and by a study of the crystal structure. The B atoms lie at the corners of two triangles hinged at the line B_1B_3 with dihedral angle $B_1B_3B_2/B_1B_3B_4 = 124\frac{1}{2}°$ and angle $B_2B_1B_4 = 98°$. This second angle would be $90°$ or $108°$ respectively if the group of four B atoms was a fragment of an octahedron or icosahedron.

Pentaborane (9), B_5H_9. The boron framework has the form of a tetragonal pyramid (Fig. 24.20(c))–compare the octahedral B_6 groups in CaB_6 and $B_6H_6^{2-}$ (see later). The mean B–B bond lengths are close to $1.80\,\text{Å}$ in the base (B_2B_2) and to $1.70\,\text{Å}$ for $B_1–B_2$. A very similar B_5 skeleton is found in B_5H_8I (I attached to apical B_1) and in $B_5H_7(CH_3)_2$ (CH_3 bonded to two adjacent basal, B_2, atoms).

Pentaborane (11), B_5H_{11}. In this hydride the boron skeleton (Fig. 24.20(d)) may be regarded as a fragment of the icosahedron-like arrangement in $B_{10}H_{14}$ or as related to the tetragonal pyramid of B_5H_9 by opening up one of the basal B–B bonds. The bond lengths are:

$$B_I-B_{II} = 1.72 \text{ Å (mean)}$$
$$B_{II}-B_{III} = 1.76 \quad \text{(mean)}$$
$$B_{II}-B_{II} = 1.77$$
$$B_I-B_{III} = 1.87$$

$$B-H_t = 1.10 \text{ Å (mean)}$$
$$B-H_b = 1.22$$
$$B_{III}-H_{VII} = 1.72 \quad \text{(mean)}$$

A feature of this molecule is the unique H_{VII} which is bonded to B_1 (1.09 Å) but is also at a distance of 1.72 Å from the two B_{III} atoms.

Hexaborane (10), B_6H_{10}. An X-ray study of B_6H_{10} showed that in this hydride the B atoms are arranged at the vertices of a pentagonal pyramid, so that here also the boron skeleton is a portion of an icosahedron. Bond lengths are shown in Fig. 24.21, the estimated standard deviations being about 0.05 Å for B–H and 0.01 Å for B–B. Note the unsymmetrical nature of the base of the pyramidal molecule, with one short B–B bond.

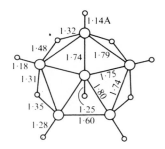

FIG. 24.21. Bond lengths in B_6H_{10}.

Octaborane (12), B_8H_{12}. This thermally unstable borane is produced in very small amounts by the action of a silent electric discharge on a mixture of B_5H_9, B_2H_6, and H_2. The molecule is closely related to B_9H_{15}, the doubly bridged BH_2 marked e in Fig. 24.20(f) being replaced by a bridging H atom. Certain of the B atoms in Fig. 24.20(e) and (f) are distinguished by letters to facilitate comparison of the molecular structures.

Enneaborane (15), B_9H_{15}. The much less regular skeleton of this hydride (Fig. 24.20(f)) can be derived from an icosahedron by removing three connected B atoms which do not form a triangular group and then opening out the structure around the 'hole' so formed.

Decaborane (14), $B_{10}H_{14}$. This (solid) hydride is one of the most stable boranes. The B atoms occupy ten of the twelve vertices of a distorted icosahedron, and ten of the H atoms project outwards approximately along the 5-fold axes of the icosahedron so that the outer surface of the molecule consists entirely of H atoms. We have already noted the longer B–B bonds (1.97 Å) in this molecule (they are the bonds B_7-B_8 and B_5-B_{10} in Fig. 24.20(g)) and the slight asymmetry of the hydrogen bridges. Two isomers of $B_{10}H_{13}I$ have similar structures to $B_{10}H_{14}$ and in $B_{10}H_{12}I_2$ the iodine atoms are attached to B_2 and B_4. In $B_{10}H_{12}[S(CH_3)_2]_2$, on the other hand, the substituents are bonded to B_6 and B_9 and there is rearrangement of the hydrogen bridges (Fig. 24.20(h)).

Decaborane (16), $B_{10}H_{16}$. This borane is produced directly from B_5H_9 (electric discharge) with loss of two H atoms. The molecule (Fig. 24.22(a)) consists of two B_5 units as in B_5H_9 joined by a single B–B bond.

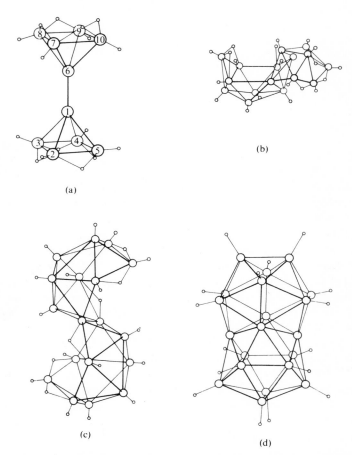

(a)

(b)

(c)

(d)

FIG. 24.22. Structures of boranes: (a) $B_{10}H_{16}$; (b) $B_{16}H_{20}$; (c) $B_{18}H_{22}$; (d) $B_{20}H_{16}$.

Boranes $B_{16}H_{20}$, $B_{18}H_{22}$, *and* $B_{20}H_{16}$. A number of higher boranes consist of two icosahedral fragments joined by sharing an edge ($B_{16}H_{20}$ and the two isomers of

$B_{18}H_{22}$) or of two icosahedra sharing two faces ($B_{20}H_{16}$). Their molecular structures are illustrated in Fig. 24.22(b)-(d).

Borohydride ions and carboranes

Metal derivatives of boranes and related compounds are of at least three types:

    (a)   salts containing ions, which range from BH_4^- and $B_3H_8^-$ to polyhedral ions;

    (b)   covalent molecules and ions in which BH_4 and B_3H_8 groups are bonded to metal atoms by hydrogen bridges; and

    (c) π-type sandwich structures formed by carboranes.

We shall not adhere strictly to this order.

The BH_4^- ion. Numerous borohydrides $M(BH_4)_n$ have been prepared by the action of metal halides on $NaBH_4$, obtained from $B(OCH_3)_3$ and NaH at 250 °C, and in other ways. In the series $NaBH_4$, $LiBH_4$, $Be(BH_4)_2$, and $Al(BH_4)_3$ there is a gradation in chemical and physical properties, for example, an increase in volatility and decreased stability towards air and water. The sodium compound is a crystalline solid stable *in vacuo* up to 400 °C, though very hygroscopic. At room temperature it has the NaCl structure[1] (like the K, Rb, and Cs salts), but below –83 °C there is a lowering of symmetry to body-centred tetragonal[2] (compare the ammonium halides). The lithium salt melts at 275 °C and apparently has the zinc-blende or wurtzite structure.[3] These salts presumably contain tetrahedral BH_4^- ions similar to the tetrahedral BH_4 groups in the covalent compounds to be described shortly.

$Al(BH_4)_3$ is a volatile liquid boiling at 44.5 °C and the borohydrides $M(BH_4)_4$ of U, Th, Hf, and Zr are the most volatile compounds known of these metals.

In the more covalent metal borohydrides BH_4 is bonded to the metal by hydrogen bridges, as shown for a number of molecules in Fig. 24.23. There have been many studies of the structure of $Be(BH_4)_2$ in the vapour state by e.d., i.r. (isolated in a noble gas matrix), and *ab initio* calculations. They have produced contradictory results and many structural models, and also the suggestion that the molecule may exist in more than one configuration in the vapour. The weight of evidence[4a] appears to favour the linear structure of Fig. 24.23(a) which was suggested in the earliest e.d. study, in which Be is surrounded octahedrally by 6 H. In crystalline $Be(BH_4)_2$[4b] there are helical arrays, (b), also with 6-coordination of Be. In the $Al(BH_4)_3$ molecule[5] there is trigonal prismatic coordination of 6 H around the metal atom, (c) and the distance Al–B is 2.14 Å. The structure of the $Zr(BH_4)_4$ molecule has been studied in the vapour by e.d.[6a] and also at a low temperature in the crystalline state.[6b] The BH_4 groups are arranged tetrahedrally around the

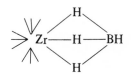

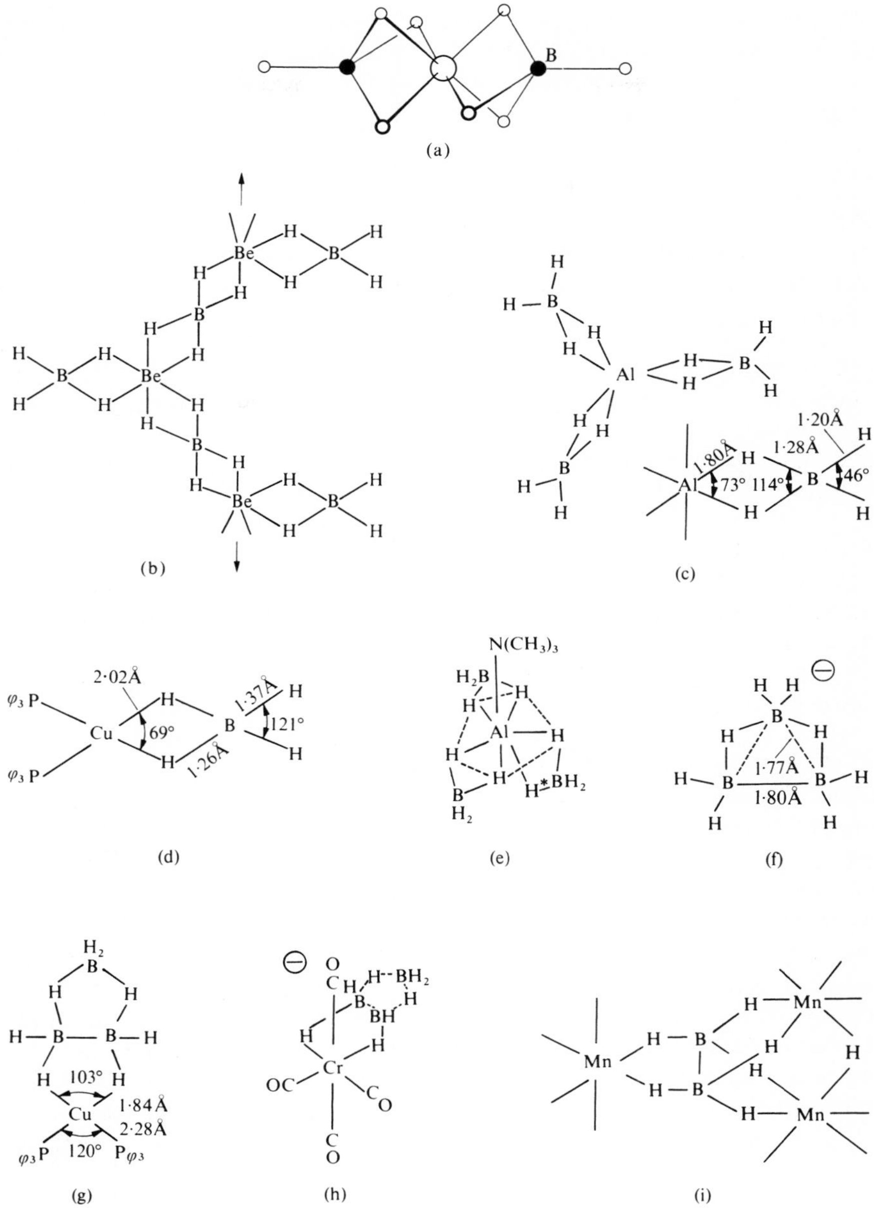

FIG. 24.23. Molecular structures of borohydrides and related compounds: (a) and (b), Be(BH₄)₂. (c) Al(BH₄)₃; (d) [(C₆H₅)₃P]₂Cu(BH₄); (e) Al(BH₄)₃.N(CH₃)₃; (f) (B₃H₈)⁻; (g) [(C₆H₅)₃P]₂Cu(B₃H₈); (h) [Cr(CO)₄B₃H₈]⁻; (i) HMn₃(CO)₁₀(BH₃)₂.

metal atom, with triple hydrogen bridges. Structural studies have also been made of $[(C_6H_5)_3P]_2 \cdot CuBH_4$,[7a] (d), and $Al(BH_4)_3 \cdot N(CH_3)_3$.[7b] It was not possible to locate the H atoms in the room temperature form of the latter, in which the gross structure of the molecule is tetrahedral (Al—N, 2.00 Å; Al—B, 2.19 Å), but in the low temperature form the three BH_4 groups are not equivalent. One is rotated so that one H (H*) occupies an apical position of a pentagonal bipyramidal group around Al, (e).

The $B_3H_8^-$ ion. The structure of this ion, Fig. 24.23(f), has been determined in $[H_2B(NH_3)_2](B_3H_8)$[8] a compound originally formulated $B_4H_{10} \cdot 2NH_3$ which is prepared from $B_4H_{10} + NH_3$ in ether at -78 °C.

Examples of covalent molecules in which B_3H_8 is attached to the metal by hydrogen bridges include $[(C_6H_5)_3P]_2 \cdot Cu(B_3H_8)$[9] and $[Cr(CO)_4B_3H_8]N(CH_3)_4$, (ref. 10), Fig. 24.23(g) and (h). All H atoms were located in an X-ray study of $HMn_3(CO)_{10}(BH_3)_2$, (i), the molecule of which contains not only metal—H—boron bridges but also Mn—H—Mn bridges.[11]

TABLE 24.8

Polyhedral borohydride ions and carboranes

Configuration	$B_nH_n^{2-}$	$B_{n-2}C_2H_n$ or derivative	Reference
Trigonal bipyramid			
		$B_3C_2H_5$	IC 1973 **12** 2108
Octahedron	$B_6H_6^{2-}$		IC 1965 4 917
		$2,3$-$B_4C_2H_6$	JCP 1970 **53** 1890
		$1,6$-$B_4C_2H_6$	IC 1973 **12** 2108
Pentagonal bipyramid			
		$B_5C_2H_7$	JCP 1965 **43** 2166
Dodecahedron	$B_8H_8^{2-}$		IC 1969 8 2771
		$B_6C_2H_6(CH_3)_2$	IC 1968 7 1070
Tricapped trigonal prism	$B_9H_9^{2-}$		IC 1968 7 2260
		$B_7C_2H_7(CH_3)_2$	IC 1968 7 1076
Bicapped square antiprism	$B_{10}H_{10}^{2-}$		JCP 1962 **37** 1779
11-skeleton	$B_{11}H_{11}^{2-}$		IC 1966 5 1955
		$B_9C_2H_9(CH_3)_2$	JACS 1966 **88** 4513
Icosahedron	$B_{12}H_{12}^{2-}$		JACS 1960 **82** 4427
		$B_{10}C_2H_{12}$	IC 1971 **10** 350
		$B_{10}C_2H_{10}Br_2$	IC 1967 6 874
Composite ions			
	$B_{20}H_{18}^{2-}$		IC 1971 **10** 151
	$B_{20}H_{18}^{2-}$ (photoisomer)		IC 1968 7 1085
	$B_{20}H_{18}NO^{3-}$		IC 1971 **10** 160
Icosahedral fragments			
	$B_{11}H_{13}^{2-}$		IC 1967 6 1199
		$B_7C_2H_{11}(CH_3)_2$	IC 1967 6 113
		$B_4C_2H_8$	IC 1964 3 1666

(1) JACS 1947 **69** 987
(2) JCP 1954 **22** 434
(3) JACS 1947 **69** 1231
(4a) JCP 1973 **59** 3777
(4b) IC 1972 **11** 820
(5) ACSc 1968 **22** 328
(6a) IC 1971 **10** 590

(6b) JCS D 1973 162
(7a) IC 1967 **6** 2223
(7b) IC 1968 **7** 1575
(8) JACS 1960 **82** 5758
(9) IC 1969 **8** 2755
(10) IC 1970 **9** 367
(11) JACS 1965 **87** 2753

From the structural standpoint the simplest polyhedral ions are the family $B_nH_n^{2-}$ which are known for $n = 6$ and 8 to 12 and are conveniently grouped with the isoelectronic carboranes $B_{n-2}C_2H_n$. These ions are usually isolated in alkali metal or substituted ammonium salts. The boron (or boron–carbon) skeletons are the highly symmetrical triangulated polyhedra listed in the upper part of Table 24.8, as shown by structural studies of ions, carboranes, or substituted carboranes for which references are given. One terminal H is attached to each vertex of the polyhedral B_n or $B_{n-2}C_2$ nucleus; there are no bridging H atoms.

In addition to ions $B_nH_n^{2-}$ numerous ions containing more H atoms have been prepared, and the structures of some have been studied. Such ions include: $B_5H_8^-$, $B_6H_9^-$, $B_9H_{14}^-$, $B_{10}H_{12}^{2-}$, $B_{10}H_{13}^-$, $B_{10}H_{14}^{2-}$, $B_{10}H_{15}^-$, and $B_{11}H_{13}^{2-}$. In ions of this type the excess of H atoms, over the number required for one terminal H on each B, are available for bridge formation, which results in rearrangement of the boron skeleton. The same situation arises in the carboranes. Thus $B_7C_2H_7(CH_3)_2$ is a substituted derivative of a borane of the $B_{n-2}C_2H_n$ family, and the B_7C_2 polyhedron is the tricapped trigonal prism in which C atoms cap two of the prism faces (Fig. 24.24(a)). On the other hand, $B_7C_2H_{11}(CH_3)_2$ is a derivative of $B_7C_2H_{13}$ and the skeleton is an 'opened out' icosahedral fragment with two H bridges and one H on each B and C atom (Fig. 24.24(b)). Similarly, $B_4C_2H_6(CH_3)_2$ has the form of a pentagonal pyramid (not an octahedron, like $B_6H_6^{2-}$), while $B_{11}H_{13}^{2-}$ is strictly an icosahedron less one vertex (two H bridges), Fig. 24.24(c), in contrast to $B_9C_2H_{11}$ which has the more symmetrical shape (d) formed by adding one B atom to the skeleton of the $B_{10}H_{14}$ type. Other carboranes with the pentagonal bipyramidal arrangement of a total of 6 B and/or C atoms include $B_2C_4H_6$,[1a] with 4 C atoms in the base, and B_5CH_9,[1b] in which there is one C atom in the base.

Isomerism is possible in carboranes. The *cis*- and *trans*- forms of $B_4C_2H_6$ are included in Table 24.8; the three isomers of $B_{10}C_2H_{12}$ are referred to as *o*-, *m*-, and *p*-carborane.

Composite ions include the two isomers of $B_{20}H_{18}^{2-}$, (e) and (f), and the $B_{20}H_{18}NO^{3-}$ ion, (g). The parent ion $B_{20}H_{18}^{2-}$, (e) is formed by oxidation of $B_{10}H_{10}^{2-}$ by ferric or ceric ion; action of u.v. light in acetonitrile solution gives the photoisomer (f). The ion (g) is formed by the action of NO on $B_{20}H_{18}^{2-}$.

Metal derivatives of carboranes. The last group of metal compounds to be mentioned here are those in which one or more carborane ions are π-bonded to a transition-metal atom to form either a composite ion or a neutral molecule.

Numerous molecules and ions of the type of Fig. 24.25(a) have been prepared

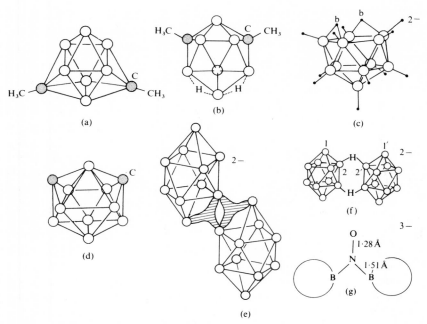

FIG. 24.24.   Molecular structures of (a) $B_7C_2H_7(CH_3)_2$;   (b) $B_7C_2H_{11}(CH_3)_2$;   (c) $B_{11}H_{13}^{2-}$;
(d) $B_9C_2H_{11}$;   (e) and (f) $B_{20}H_{18}^{2-}$;   (g) $(B_{20}H_{18}NO)^{3-}$

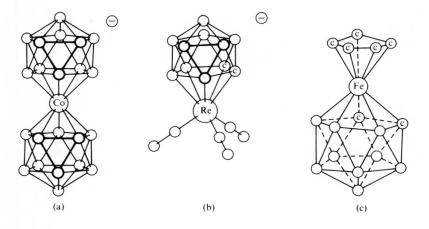

FIG. 24.25.   The molecular structures of (a) $[Co(B_9C_2H_{11})_2]^-$;   (b) $[B_9C_2H_{11}Re(CO)_3]^-$;
(c) $Fe(C_5H_5)(B_9C_2H_{11})$. H atoms are omitted.

containing various transition metals, for example, $Ni^{IV}(B_9C_2H_{11})_2$,[2a] the $Co(III)$[2b] and $Ni(II)$[2c] ions, and ions with substituents in place of some of the H atoms. Some of these complexes have the staggered configuration of Fig. 24.25(a), for example, $[Co(B_9C_2H_{11})_2]^-$, studied in the Cs salt, while others have a 'sheared' structure in which M is close to 3 B of each B_3C_2 ring. The difference may be associated with the number of d electrons of the transition metal atom. Ions with this type of structure include $[3,3'-M(1,2-B_9C_2H_{11})_2]^{n-}$, where $n = 2$ for $Cu(II)$ and $n = 1$ for $Cu(III)$ and $Au(III)$. In the $Au(III)$ ion, for example, the distances $Au-3B$ and $Au-2C$ are respectively 2.24 and 2.77 Å.[3]

'Ion-pairs' formed from one $B_9C_2H_{11}^{2-}$ unit and one metal ion are also known; an example is $(TlB_9C_2H_{11})^-$, studied in the $(P\phi_3CH_3)^+$ salt.[4] Complexes of the type of Fig. 24.25(b) include the anion in $Cs[B_9C_2H_{11}Re(CO)_3]$[5] and molecules such as $B_9C_2H_{11}ReH(P\phi_3)_2$, a catalyst for certain homogeneous hydrogenation reactions. The 'sandwich' molecule $Fe(\pi-C_5H_5)(\pi-B_9C_2H_{11})$[6] is illustrated in Fig. 24.25(c). In the anion in $Cs_2[(B_9C_2H_{11})Co(B_8C_2H_{10})Co(B_9C_2H_{11})].H_2O$[7] the metal atoms are bridged by the 10-atom icosahedral fragment and also bonded to 11-atom fragments.

(1a) CC 1973 928
(1b) IC 1971 **10** 1144
(2a) JACS 1970 **92** 1173
(2b) IC 1967 **6** 1911
(2c) JACS 1970 **92** 1187

(3) AC 1977 **B33** 3604
(4) AC 1978 **B34** 2373
(5) IC 1966 **5** 1189
(6) JACS 1965 **87** 3988
(7) IC 1969 **8** 2080

25

Copper, silver, and gold

Valence states

Each of these elements follows a transition metal (Ni, Pd, and Pt respectively) with a completed d shell. They might be expected to behave as non-transition metals and to form ions M^+ by loss of the single electron in the outermost shell or to use the s and p orbitals of that shell to form collinear sp or tetrahedral sp^3 bonds. In fact, both Cu and Ag form ions M^+ and bonds of both these types, but Au^+ is not known and Au(I) shows a marked preference for 2- as opposed to 4-coordination. Moreover, all these elements exhibit higher valences as a result either of losing one or more d electrons (e.g. Cu^{2+}) or of utilizing d orbitals of the penultimate shell in combination with the s and p orbitals of the valence shell. In these higher valence states these elements have some of the characteristic properties of transition metals, for example, the formation of coloured paramagnetic ions. This dual behaviour greatly complicates the structural chemistry of these metals. Since not very much is known of the structural chemistry of these elements in certain oxidation states we shall deal separately and in more detail with Cu(I), Ag(I), and Au(I), Cu(II), and Au(III), and include what is known about Cu(III), Cu(IV), Ag(II), Ag(III), and Au(V) in our preliminary survey.

In spite of the general similarity in the electronic structures of their atoms, Cu, Ag, and Au differ very considerably in their chemical behaviour. First, the valences exhibited in their common compounds are: Cu, 1 and 2; Ag, 1; and Au, 1 and 3. In addition, Cu and Ag may be oxidized to the states Cu(III) and Cu(IV) and Ag(II) and Ag(III) respectively, but no simple compounds of Au(II) are known. Most crystalline compounds apparently containing Au(II), for example, $CsAuCl_3$ and $(C_6H_5CH_2)_2S.AuBr_2$,[1] actually contain equal numbers of Au(I) and Au(III) atoms. The first fully characterized paramagnetic compound of Au(II) is $[Au^{II}(mnt)_2]$-$[(n-C_4H_9)_4N]_2$[2] (where mnt = maleonitrile dithiolate), which is stable in the absence of air but oxidizes rapidly in solution; according to an e.s.r. study the phthalocyanin is a derivative of Au(II).[3] The (unique) example of Au(V), $(Xe_2F_{11})(AuF_6)$ (p. 379) may be compared with the occurrence of Pt(V) in $(XeF_5)^+(PtF_6)^-$; the M—F bond lengths are very similar: Au—F, 1.86 Å; Pt—F, 1.89 Å. Second, the more stable ion of Cu is the (hydrated) Cu^{2+} ion, whereas that of silver is Ag^+. In contrast to copper and silver, there is no ionic chemistry of gold in aqueous solution, for the Au^+ and Au^{3+} ions do not exist in aqueous solutions of gold salts, at least in any appreciable concentration. The only water-soluble compounds of gold, aurous or auric, contain the metal in the form of a complex ion as, for example, in solutions of $K[Au(CN)_2]$ or $Na_3[Au(S_2O_3)_2].2H_2O$. Coordination compounds of Au are considerably more stable than the corresponding simple salts. For example, AuCl is readily decomposed by hot water, which does not affect $[Au(etu)_2]Cl.H_2O$. Similarly, aurous nitrate has not been made but $[Au(etu)_2]NO_3$ is a quite stable compound. Auric nitrate can be prepared under anhydrous conditions and complex auric nitrates such as $K[Au(NO_3)_4]$ are known.

As already noted, the more stable ion of copper in aqueous solution is the cupric ion. Cuprous oxy-salts such as Cu_2SO_4 are decomposed by water, $2\,Cu^+ \rightarrow Cu + Cu^{2+}$, and $CuNO_3$ and CuF are not known. Although anhydrous Cu_2SO_3 is not known, the pale-yellow $Cu_2SO_3.\tfrac{1}{2}H_2O$ can be prepared, and also NH_4CuSO_3 and $NaCuSO_3.-6H_2O$.[4] We refer later to salts containing both $Cu(I)$ and $Cu(II)$. The stable cuprous compounds are the insoluble ones, in which the bonds have appreciable covalent character (the halides, Cu_2O, Cu_2S), and the halides and the cyanide are actually more stable in the presence of water than the cupric compounds. Thus CuI_2 and $Cu(CN)_2$ decompose in solution to the cuprous compounds. The cuprous state is, however, stabilized by coordination, and derivatives such as $[Cu(etu)_4]NO_3$ and $[Cu(etu)_3]_2SO_4$ are much more stable than the simple oxy-salts (etu = ethylene-thiourea). In pyridine the equilibrium $2\,Cu^+ \rightleftharpoons Cu^{2+} + Cu$ is strongly in favour of Cu^+. In general the cupric salts of only the stronger acids are stable, $Cu(NO_3)_2$ and $CuSO_4$ for example; those of weaker acids are unstable, and only 'basic salts' are generally known, as in the case of the nitrite, sulphite, etc. However, if a coordinated ion such as $Cu(en)_2^{2+}$ is formed, then stable compounds result, for example, $[Cu(en)_2](NO_2)_2$, $[Cu(en)_2]SO_3$ etc. If methyl sulphide if added to a solution of a cupric salt, the cup*rous* compound is precipitated, while if ethylene diamine is added to a solution of cup*rous* chloride in KCl (in absence of air) the cup*ric* ion $Cu(en)_2^{2+}$ is formed with precipitation of Cu. From these facts it is clear that the relative stabilities of Cu^+ and Cu^{2+} cannot be discussed without reference to the environment of the ions, that is, the neighbouring atoms in the crystal, solvent molecules or coordinating ligands if complex ions are formed. For the reaction

$$2\,Cu^+ (g) \rightarrow Cu^{2+} (g) + Cu (s)$$

ΔH = +870 kJ mol^{-1}, corresponding to a large absorption of energy. If, however, we wish to compare the stabilities of the two ions in the crystalline or dissolved states this ΔH will be altered by a (large) amount corresponding to the difference between the interactions of Cu^+ and Cu^{2+} with their surroundings, as represented by lattice energies or solvation energies. These will be much greater for Cu^{2+} than for Cu^+, so that $\Delta H'$ may become negative, that is, ionic cuprous salts are less stable than cupric. With increasing covalent character of the $Cu-X$ bonds, $\Delta H'$ again becomes positive, and in the case of the iodide and the cyanide it is the cuprous compound which is more stable. The actual configuration of a coordinating molecule may be important in determining the relative stabilities of cuprous and cupric compounds, as shown by the following figures for $(Cu^{II})/(Cu^{I})^2$ in the presence of various diamines.

$$(Cu^{II})/(Cu^{I})^2$$

Ethylene diamine	$\sim 10^5$
Trimethylene diamine	$\sim 10^4$
Pentamethylene diamine	3×10^{-2}
(cf. Ammonia	2×10^{-2})

Whereas ammonia stabilizes Cu^I in the reaction

$$2Cu(NH_3)_2^+ \rightleftharpoons Cu(NH_3)_4^{2+} + Cu,$$

the first two diamines stabilize the cupric state. These compounds can form chelate complexes with Cu^{II} but apparently not with Cu^I, while pentamethylene diamine presumably can be attached by only one NH_2 to either Cu^{II} or Cu^I and therefore behaves like ammonia.

Some features of the crystal chemistry of silver are worthy of special notice.

(i) The c.n. of $Ag(I)$ ranges from 2, 3, or 4 in compounds in which the bonds are essentially covalent in character to 6, 7, or 8 in oxy compounds where the bonds are ionic. A similar range is found for $Hg(II)$, another B subgroup element (Chapter 26). The lower c.n.s are not exhibited by the more electropositive A subgroup elements such as Ca in normal crystalline compounds, though there is necessarily only 2-coordination in vapour molecules of the dihalides.

(ii) The distances between Ag atoms suggest metal–metal interactions in the stable orthorhombic form of $AgNO_3$[5] (3.22 Å) and in Ag_2HPO_4[6] (3.08 Å) and in chains of Ag atoms in the trisulphimide, $Ag_3(NSO_2)_3 . 3H_2O$[7] (3.03 Å) and the fulminate (2.8-2.9 Å), p. 934; compare the Ag–Ag distance in the metal (2.89 Å). The limit of such interactions is the formation of a cluster of Ag atoms, as in $Ag(imidazole)_2ClO_4$.[8] In the nitrate there are isolated linear cations and no Ag–Ag distances less than 5.0 Å, but in the perchlorate there are groups of 6 Ag atoms, (a), in which the outer Ag–Ag bonds (3.05 Å) are appreciably shorter than

(a)

the central ones (3.49 Å). The coordination group of Ag_1 is completed by 2 N atoms of imidazole molecules (Ag–2N, 2.08 Å) and that of Ag_2 by 2 N at a similar distance and 2 O of ClO_4^- at around 3 Å. Octahedral Ag_6 clusters have been found in dehydrated Ag-exchanged zeolites.[9] See also p. 1110 for bridged tetrahedral Ag complexes.

(iii) Of special chemical interest are compounds with such unusual formulae as Ag_2ClNO_3[10] and the isostructural bromide, Ag_2INO_3,[11] $Ag_3I(NO_3)_2$,[12] Ag_3SNO_3,[13] $Ag_4Te(NO_3)_2$,[14] and $Ag_7O_8NO_3$.[15] The structural feature common to a number of these compounds is the formation of a 3D framework incorporating the Ag and halogen or S(Te) atoms and accommodating the NO_3^- ions in cavities in the framework. For example, in Ag_2BrNO_3 there is a 3D framework formed from trigonal bipyramidal $BrAg_5$ groups sharing 2 vertices and 2 edges in which one half of the Ag atoms are bonded to 2 Br and the remainder to 3 Br. The topological diagram (b) shows how the composition Ag_2Br is achieved with these c.n.s. The

(b)

NO_3^- ions are situated in cavities surrounded by 7 Ag, and the coordination groups of Ag atoms are completed by O atoms of NO_3^- ions: $Ag-(2Br + 6O)$ or $Ag-(3Br + 4O)$. In both Ag_3SNO_3 and $Ag_7O_8NO_3$ there are highly symmetrical 3D frameworks with cubic symmetry. In the former the S atoms are at the nodes of a 6-connected net and joined through 2-connected Ag atoms (S–Ag–S, 157°); the NO_3^- ions occupy interstices in the framework. Alternatively, the framework of this remarkable compound, which is formed from CS_2 and concentrated $AgNO_3$ solution, may be described as built of SAg_6 groups, with a configuration intermediate between a regular octahedron and a trigonal prism; compare the structure of Ag_2BrNO_3 built from $BrAg_5$ groups. In $Ag_7O_8NO_3$ the 8 O atoms are at the nodes of a (different) 6-connected 3D net, the links of which form a space-filling assembly of rhombi-cuboctahedra, cubes, and tetrahedra (Fig. 25.1) which may be compared with the closely related space-filling of Fig. 3.47(e) on p. 138. We refer again to this compound on p. 1103. Another remarkable 3D framework is found in β-$Ag_4Te(NO_3)_2$ and the isostructural perchlorate. Here it is built of $TeAg_5$ groups, again joined through Ag atoms, and with the anions in the interstices. The α form of this compound has a quite different structure to which we refer on p. 1100.

(1) JCS 1952 3686
(2) JACS 1965 87 3534
(3) JACS 1965 87 2496
(4) JINC 1964 26 1122
(5) AC 1978 B34 1457
(6) AC 1978 B34 3723
(7) AC 1974 B30 2721
(8) JCS D 1980 1682
(9) JACS 1977 99 7055; 1978 100 175
(10) AC 1979 B35 1432
(11) AC 1979 B35 302
(12) ZK 1970 132 87
(13) ZaC 1959 299 328
(14) AC 1969 B25 2645
(15) AC 1965 19 180

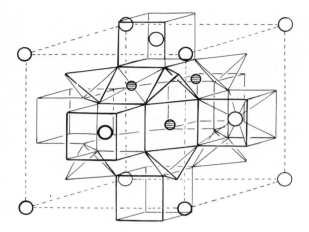

FIG. 25.1. The unit cell of $Ag_7O_8NO_3$. The 4- and 8-coordinated Ag atoms, of which only a few are shown, are distinguished as shaded and open circles. The nitrate ions (not shown) are situated in the large rhombicuboctahedral cavities.

Superionic conductors

We noted on p. 410 the high ionic conductivity of α-AgI. Other relatively simple compounds of Ag with this property include α-Ag_2Te, in which one-half of the Ag atoms and the Te atoms form a sphalerite structure in which the remaining Ag atoms are statistically distributed, and α-$Ag_4Te(NO_3)_2$, in which one-quarter of the Ag atoms together with the Te atoms form a sodium chloride framework within which the remaining Ag and NO_3 ions are distributed at random.[1] The ionic conductivity of α-AgI arises because the Ag^+ ions can move through the rather loosely packed b.c.c. arrangement of the easily deformable I atoms. In Chapter 5 we were concerned with structures in which tetrahedral AX_4 groups share vertices and/or edges; faces are not shared owing to the close approach of A atoms which would result. In recent years numerous compounds have been made which exhibit ionic conductivity to varying degrees in which AgI is combined with amines or other organic molecules. They contain regions of face-sharing I_4 tetrahedra which extend throughout the crystal and are only partially occupied. By passing through the centres of faces the Ag ions can travel through the crystal via the empty tetrahedra. It is necessary that there are fewer Ag ions than tetrahedral interstices and that there are reasonable paths through the crystal. Examples of these extraordinary compounds, made by solid state reactions between AgI and, for example, an amine salt, include Ag_2I_4A, Ag_6I_8A, $Ag_{31}I_{39}A_4$, in which A is $(CH_3)_3N(CH_2)_2N(CH_3)_3^{2+}$;[2] for other references see ref. (3).

Other compounds exhibiting high ionic conductivity include β-alumina (p. 598), β-eucryptite, $LiAlSiO_4$,[4] Li_5AlO_4,[5] Li_3N (p. 833), $Ag_{13}I_9(WO_4)_2$,[6] intercalation compounds (p. 758), and a group of 'icosahedral' structures. A feature of the latter is a framework consisting of interpenetrating icosahedra. An atom at the body-centre of an icosahedron produces, with the 20 triangular faces, a set of tetrahedra, some of which are regular while others are less regular and only partially occupied. For example, in Cu_6PS_5Br the P atoms occupy the regular tetrahedra in the S_5Br framework and Cu atoms occupy a proportion of the less regular ones. The Cu atoms can move through the structure by passing through tetrahedron faces. A more complex example is $Cd_{13}P_4S_{22}I_2$.[7] For references to similar and related structures (e.g. Ag_8SiS_6) see refs. (3) and (8).

(1) AC 1975 **B31** 2837
(2) AC 1976 **B32** 1248
(3) AC 1978 **B34** 71
(4) JSSC 1980 **34** 199

(5) JSSC 1979 **29** 393
(6) JSSC 1977 **21** 331
(7) AC 1976 **B32** 2825
(8) AC 1978 **B34** 64

Compounds of Cu(III) and Cu(IV)

Very few purely inorganic compounds of Cu(III) are known. Salts such as Cs_3CuF_6, made by high-pressure fluorination of a mixture of CsCl and $CuCl_2$, contain the high-spin Cu^{3+} ion, and are paramagnetic with $\mu_{eff.}$ corrresponding to 2 unpaired electrons. Other compounds of Cu(III) are diamagnetic, and the metal forms either 4 coplanar or 5 square pyramidal bonds. Complex oxides formed with alkali metals include $KCuO_2$ and $Li_6Cu_2O_6$,[1] the latter being isostructural with the corresponding Au compound (p. 1145). The diperiodato and ditellurato complexes of Cu(III) and

(a)   (b)

Ag(III) have structures of type (a) consisting of two octahedral groups linked by a Cu atom (Cu–O, 1.9 Å) which forms a fifth bond to a water molecule.[2a] These ions were originally assigned charges of –7 and –9, but chemical analysis, i.r., and conductimetric measurements[2b] show that there are respectively 2 and 4 protons incorporated in these ions, which are therefore $[M(HIO_6)_2]^{5-}$ and $[M(H_2TeO_6)_2]^{5-}$. Numerous complexes with organic ligands have been prepared. In the violet crystals of $Br_2CuS_2CN(C_4H_9)_2$,[3] (b), which is isostructural with the Au(III) compound, the metal forms 4 coplanar bonds; both Cu–Br (2.31 Å) and Cu–S (2.19 Å) are shorter than the corresponding bonds to Cu(II). Another example of the formation of square coplanar bonds is the ion formed with *o*-phenylene bisbiuret, in which Cu(III) forms 4 bonds to N (1.82–1.89 Å).[4]

The simplest example of a compound of Cu(IV) is the brick-red Cs_2CuF_6, apparently with a distorted K_2PtCl_6 type of structure,[5] which is made, like Cs_3CuF_6, by h.p. fluorination of a mixture of CsCl and $CuCl_2$. Oxides such as $BaCuO_{2.63}$ presumably contain Cu(IV) and also Cu in a lower oxidation state.[6]

(1)   NW 1976 **63** 387   (4)   IC 1977 **16** 2478
(2a) NW 1960 **47** 377   (5)   AnCIE 1973 **12** 582
(2b) ICA 1977 **22** 7   (6)   JCS D 1975 1061
(3)   IC 1968 7 810

Higher oxidation states of Ag

The simple Ag^{2+} ion has been produced in concentrated nitric acid solution, but apart from the fluoride AgF_2 compounds of Ag(II) can be prepared only in the presence of molecules which will coordinate to the metal and form complex ions. The structure of AgF_2,[1] which is quite different from that of CuF_2, is described on p. 275. The nitrate is formed by anodic oxidation of $AgNO_3$ solution in the presence of pyridine and isolated as $[Ag(py)_4](NO_3)_2$ and the persulphate is formed by double decomposition and isolated as $[Ag(py)_4]S_2O_8$. Many other coordination compounds of Ag(II) have been prepared, and their magnetic moments (around

(a)

2 BM) correspond to 1 unpaired electron. An incomplete X-ray study[2] of the iso-structural cupric and argentic salts of picolinic acid, (a), provides the only evidence for the coplanar arrangement of four bonds formed by Ag(II) in a finite complex. A preliminary study of $AgL_2.H_2O$ (L = pyridine-2,6-dicarboxylate) indicates a distorted octahedral structure in which the two ligands coordinate to the metal with different Ag–O and Ag–N bond lengths (Ag–O, 2.20 and 2.54 Å; Ag–N, 2.08 and 2.20 Å).[3]

Diamagnetic compounds containing exclusively Ag(III) presumably include the yellow salts $KAgF_4$ and $CsAgF_4$ (which are readily decomposed by moisture), salts such as $K_5[Ag(HIO_6)_2].10H_2O$ and $Na_5[Ag(H_2TeO_6)_2].18H_2O$ which are probably structurally similar to the Cu(III) compounds, and the red salts (sulphate, nitrate, etc.) of the very stable ethylenebisbiguanide complex (b). In $(Ag^{III}ebs)SO_4(SO_4H).H_2O$[4] there are four strong Ag–N bonds (1.98 Å) and two very much weaker bonds to O atoms of anions (Ag–O, 2.94 Å).

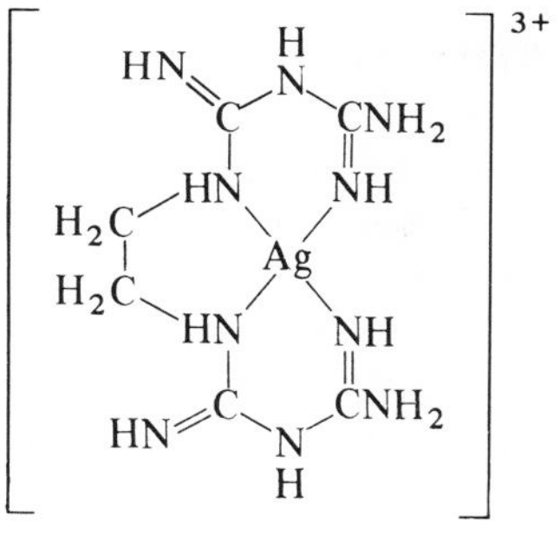

(b)

The oxide AgO is prepared by slow addition of $AgNO_3$ solution to an alkali persulphate solution, and is commercially available. It is a black crystalline powder which is a semiconductor and is diamagnetic. It is $Ag^I Ag^{III} O_2$, and its structure is a distorted form of the PtS structure, another variant of which is the structure of CuO (tenorite). In CuO all Cu atoms have 4 coplanar neighbours, but in AgO Ag(I) has two collinear O neighbours (Ag^I–O, 2.18 Å), the other two atoms of the original square planar coordination group being at 2.66 Å, while Ag^{III} has 4 coplanar O neighbours at 2.05 Å.[5] Electrolysis of an aqueous solution of $AgNO_3$ with a Pt anode gives a compound with the empirical composition Ag_7NO_{11} as black cubic crystals with a metallic lustre. This is an oxynitrate, $Ag(Ag_6O_8)NO_3$[6] and the corresponding fluoride, $Ag(Ag_6O_8)F$ apparently has a closely related structure which has been illustrated in Fig. 25.1. One-seventh of the Ag atoms (presumably Ag^+) are situated within the cubes and have 8 O neighbours at 2.52 Å, as in $AgClO_3$. The remainder are all equivalent and are situated at the centres of those square faces of rhombicuboctahedra which are not shared with cubes, and therefore have a planar square coordination group (Ag O, 2.05 Å). There is clearly interchange of electrons between Ag atoms in higher oxidation states, to give a neutral framework of composition Ag_6O_8, leading to the colour and semiconductivity (compare the bronzes, p. 612).

(1) JPCS 1971 **32** 543
(2) JCS 1936 775
(3) JACS 1969 **91** 7769
(4) AC 1975 **B31** 131
(5) JES 1961 **108** 819
(6) AC 1965 **19** 180

The structural chemistry of Cu(I), Ag(I), and Au(I)

The simplest possibilities are the formation of two collinear (sp) or four tetrahedral (sp³) bonds. In addition, Cu(I) and Ag(I) form three bonds in a number of crystals, and we shall give examples of this bond arrangement after dealing with the two simpler ones.

The formation of two collinear bonds by Cu(I), Ag(I), *and* Au(I)

The formation of only two collinear bonds by Cu(I) is very rare, and is observed only with the electronegative O atom, in Cu_2O (with which Ag_2O is isostructural), $CuFeO_2$ and $CuCrO_2$ (p. 577) and also apparently in KCuO, with which KAgO, CsAgO, and CsAuO are isostructural.[1a] Although two collinear bonds are formed to N atoms in the diazoaminobenzene compound (a), (of length 1.92 Å) the Cu–Cu distance (2.45 Å) is less than in metallic Cu (2.56 Å) and presumably indicates some interaction between the metal atoms.[1b] The formation of metal–metal bonds in addition to 2, 3, or 4 bonds to non-metals seems to be a feature of the structural

$$C_6H_5N-Cu-NC_6H_5$$

(a)

chemistry of Cu(I) and Ag(I). The reluctance of Cu(I) to form only two collinear bonds leads to many differences between the chemistry of Cu(I) on the one hand and Ag(I) and Au(I) on the other. For example, whereas Ag(I) and Au(I) form two collinear bonds in AgCN and AuCN and also in the $M(CN)_2^-$ ions, CuCN has a much more complex (unknown) structure with 36 CuCN in the unit cell,[2] and in $KCu(CN)_2$ Cu(I) forms three bonds (p. 1112). With long-chain amines cuprous halides form 1:1 and 1:2 complexes. The former are tetrameric and presumably similar structurally to $[CuI.As(C_2H_5)_3]_4$ (p. 1110), while the 1:2 complexes are dimeric in benzene solution and may be bridged molecules, $(RNH_2)_2Cu.X.Cu(NH_2R)_2$. The compounds formed by Cu(I) and Au(I) of the type P(N)–M–C, where P(N) is a phosphine (amine) and C an acetylene, have entirely different structures. The gold compounds form linear molecules, for example,

$$(i\text{-}C_3H_7)H_2N\underset{2.03\,\text{Å}}{\underline{\hspace{1cm}}}Au\underset{1.94\,\text{Å}}{\underline{\hspace{1cm}}}C\equiv C(C_6H_5)^{(3)}$$

whereas the phosphine $(CH_3)_3P-Cu-C\equiv C(C_6H_5)$ is tetrameric (p. 1111), though with a quite different structure from $(R_3PCuI)_4$.

There are numerous examples of Ag(I) forming 2 collinear (or approximately collinear) bonds, in finite groups, chains, and certain 3D frameworks. Examples of finite groups include the linear anion in $K[Ag(CN)_2]$ and the linear cation in $[Ag(NH_3)_2]_2SO_4$. In AgCN there are infinite linear chains of metal atoms linked through CN groups (Ag–Ag, 5.26 Å) and there are similar chains in $Ag_3CN(NO_3)_2$,[4] where the NO_3^- ions and the other Ag^+ ions lie between the chains. Approximately collinear bonds (165-170°) are formed by Ag(I) in $KAgCO_3$,[5] (b), in $Na(AgSO_3)$.$2H_2O$,[6] (c), where Ag is bonded to O and S (Ag–O, 2.19 Å; Ag–S, 2.47 Å), in the dicyanamide $Ag[N(CN)_2]$,[7] (d), in AgSCN (p. 936), and in the complex sulphides proustite, Ag_3AsS_3, and pyrargyrite, Ag_3SbS_3. In the compound formed by $AgNO_3$

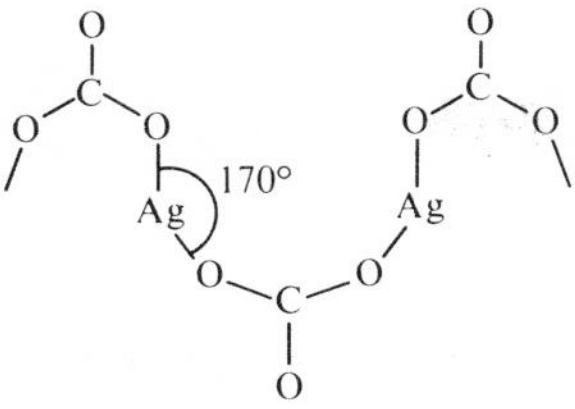

(b)

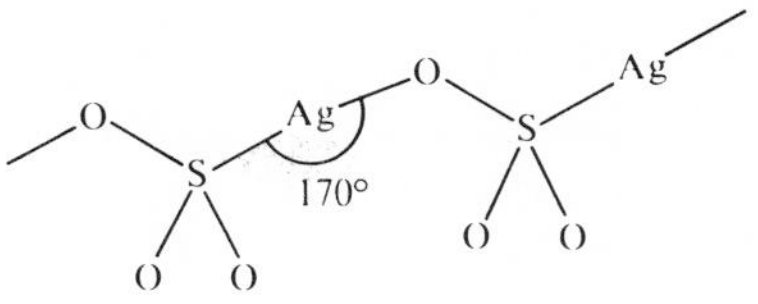

(c)

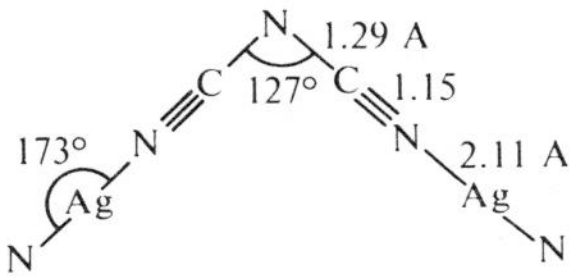

(d)

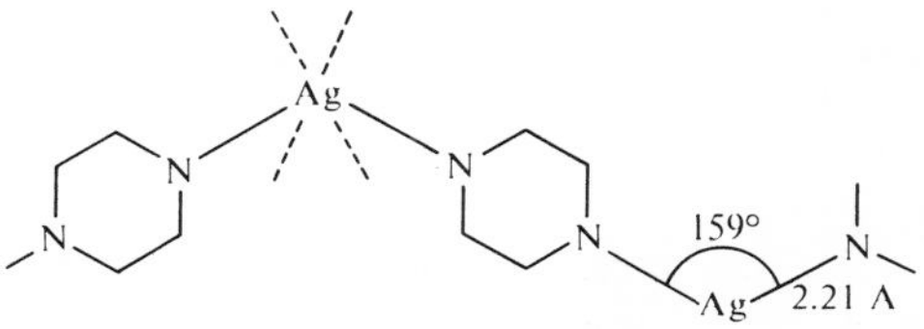

(e)

with pyrazine, $AgNO_3.N_2C_4H_4$[8] Ag forms two bonds to N but also four weak bonds to O (two of length 2.72 Å, two of 2.94 Å), (e). We have already noted that Ag acts as a (2-connected) link between S atoms in the 3D framework of Ag_3SNO_3. In the (colourless) crystals of $Ag_2SO_2(NH)_2$[9] Ag is bonded to two N atoms of different $SO_2(NH)_2^{2-}$ ions to form double chains but the deep-red $Ag_4SO_2N_2$[10] has a much more complex structure. There is a complex 3D network containing four sets of non-equivalent Ag atoms each of which has two approximately collinear ligands (2 N or O and N) but there are also many short Ag–Ag distances (2.9-3.0 Å) similar to that in metallic Ag (2.88 Å).

For Au(I) two is the preferred coordinaton number. The most stable amino derivatives of AuCl are $H_3N.AuCl$ and $(H_3N-Au-NH_3)Cl$, and whereas with ligands such as thioacetamide Cu and Ag form salts of type (f), Au forms only the rather unstable salt (g). With trialkyl phosphines and arsines cuprous and argentous

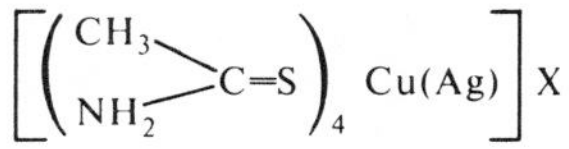

(f)

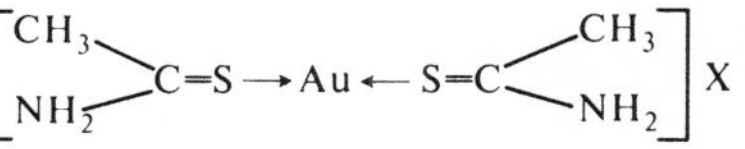

(g)

halides form the tetrameric molecules $[R_3P(As).Cu(Ag)X]_4$, but Au forms only $R_3P(As).AuX$. Examples of finite complexes in which Au(I) forms 2 collinear bonds include the anion in $K[Au(CN)_2]$,[11] isostructural with the Ag compound, the $AuCl_2^-$ ion in $Cs_2(Au^ICl_2)(Au^{III}Cl_4)$, p. 466, $Cl-Au-PCl_3$,[12] various molecules containing phosphines or arsines such as $Cl-Au-P\phi_3$[13] (Au–Cl, 2.28; Au–P, 2.24 Å), $H_3C-Au-P\phi_3$[14] (Au–C, 2.12; Au–P, 2.28 Å), and $Br-Au-As\phi_3$[15] (Au–Br, 2.38; Au–As, 2.34 Å), and bridged complexes such as (h):[16]

$$\phi_3 P \diagdown \qquad\qquad\qquad \diagup P\phi_3 \; \oplus$$
$$Au \quad 82° \quad Au$$
$$Cl \quad 2.34 \text{ A}$$

(h)

There are linear chains in AuCN, isostructural with AgCN, and zigzag chains in AuCl[17] and AuI:[18]

Cl 2.36 A
Au 92° Au
Cl   Cl

I
Au 72° Au Au
I 2.6 A

We have seen that there is evidence for appreciable metal–metal interactions in some compounds of Ag, and we shall note later that there are Cu–Cu distances in certain Cu(I) compounds not much greater than those in the metal. There are relatively short Au–Au distances, in the range 3.1–3.4 Å, in molecules such as (h) and in the AuCl chain but these are a consequence of the Cl bond angle and the Au–Cl bond length. However, there are examples of short distances between pairs of Au atoms which are not necessitated by the geometry of the molecule and appear to be a feature of many Au(I) compounds. In salts containing the ion (i)[19] the cations are associated in pairs, and in (j)[20] there would not appear to be other constraints to explain the short Au–Au distance. In the bridged molecule (k)[21] the planes of the $S_2CN(C_3H_7)_2$ portions of the molecule are inclined at an angle of 50°, and the distance between the Au atoms is shorter than in the metal (2.88 Å).

$$\phi_3 \qquad\qquad \phi_3$$
$$P \qquad\qquad P$$
$$\oplus \qquad\qquad\qquad\qquad \oplus$$
$$Au\text{---}3.24 \text{ Å}\text{---}Au$$
$$\phi_3 P\text{---}Au\text{---}S \qquad S\text{---}Au\text{---}P\phi_3$$
$$\approx 90°$$
$$Au\text{---}3.36 \text{ Å}\text{---}Au$$
$$P \qquad\qquad P$$
$$\phi_3 \qquad (i) \qquad \phi_3$$

$$CH_2\text{---}CH_2$$
$$\phi\diagdown \qquad\qquad\qquad \diagup\phi$$
$$\phi\diagup P \qquad\qquad P\diagdown\phi$$
$$Au\text{----}3.05 \text{ Å}\text{----}Au$$
$$Cl \qquad\qquad Cl$$

(j)

$$2.28 \text{ Å}$$
$$S\text{---}Au\text{---}S$$
$$H_7C_3\diagdown \qquad\qquad\qquad\qquad\qquad \diagup C_3H_7$$
$$N\text{---}C \qquad 2.76 \qquad C\text{---}N$$
$$H_7C_3\diagup \qquad\qquad\qquad\qquad\qquad \diagdown C_3H_7$$
$$S\text{---}Au\text{---}S$$

(k)

In the limit Au—Au interactions lead to the formation of metal clusters. In $[Au_9(PR_3)_8]^{3+}$ there is a central Au atom surrounded by 8 Au at the vertices of an incomplete icosahedron.[22] In $Au_{11}X_3L_7$,[23] in which L is a phosphine and X is I or SCN, there is a central metal atom bonded to 10 Au at the vertices of a polyhedron related to an icosahedron, one triangular face being replaced by a single atom. The 3 I and 7 P ligands are bonded to the outer 10 Au atoms.

(1a)	ZaC 1968 **360** 113	(12)	RTC 1962 **81** 307
(1b)	AC 1961 **14** 480	(13)	AC 1976 **B32** 962
(2)	AlC 1957 **28** 316	(14)	AC 1977 **B33** 137
(3)	AC 1967 **23** 156	(15)	AC 1975 **B31** 624
(4)	AC 1965 **19** 815	(16)	AC 1980 **B36** 1486
(5)	JCS 1963 2807	(17)	JLCM 1974 **38** 71
(6)	AC 1973 **B29** 623	(18)	ZK 1959 **112** 80
(7)	AC 1977 **B33** 697	(19)	AC 1980 **B36** 2777
(8)	IC 1966 **5** 1020	(20)	AC 1980 **B36** 2775
(9)	AC 1977 **B33** 3595	(21)	ACSc 1972 **26** 3855
(10)	AC 1980 **B36** 1044	(22)	CC 1971 1423
(11)	AC 1959 **12** 709	(23)	JCS D 1972 1481

The formation of four tetrahedral bonds by Cu(I) *and* Ag(I)

No example of Au(I) forming four tetrahedral bonds has yet been established by a structural study, but tetrahedral bonds are formed in numerous compounds of Cu(I) and Ag(I), the simplest examples being the cuprous halides and AgI with the zinc-blende structure (p. 410).

Finite complexes of Cu(I) include the $Cu(CN)_4^{3-}$ ion (mean Cu—C, 2.00 Å),[1] the thioacetamide complex $[Cu\{SC(NH_2)(CH_3)\}_4]Cl^{[2]}$ (with which the Ag compound is isostructural), and the assortment of molecules shown in Fig. 25.2. In the molecule $CH_3COOCu(P\phi_3)_2^{[2a]}$ the bidentate acetato group replaces the NO_3 group in Fig. 25.2(b).

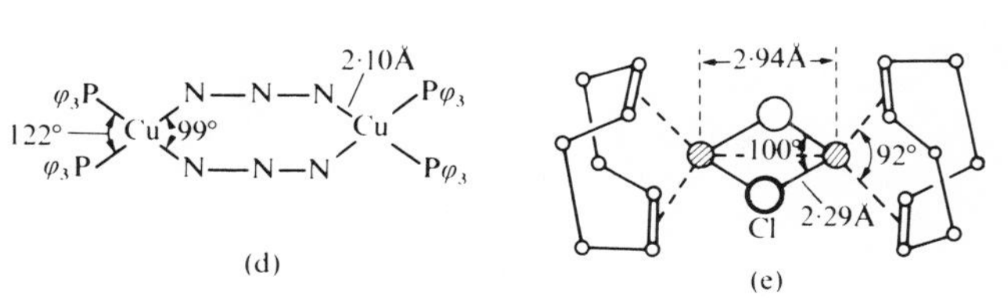

FIG. 25.2. Some molecules formed by Cu(I). (a) JACS 1967 **89** 3929; (b) IC 1969 **8** 2750; (c) AC 1970 **B26** 515; (d) JACS 1970 **92** 738; (e) JACS 1963 **85** 1009.

There is considerable distortion of the tetrahedral bond angles in the dimeric azido complex, in which the N_3 group is apparently symmetrical (mean N–N, 1.18 Å), as compared with the normal unsymmetrical form (N–N, 1.15 and 1.21 Å) in transition-metal complexes in which it is attached to a metal through one N atom only. We include the cyclopentadienyl compound (c) since here C_5H_5 contributes 5 electrons and Cu acquires the Kr configuration as in the tetrahedral complexes. In $C_6H_6 . CuAlCl_4$[3] Cu has 3 Cl neighbours of different $AlCl_4^-$ ions and the C_6H_6 ring, which is distorted towards a cyclohexatriene structure.

An unusually interesting finite complex ion is found in $[Co(NH_3)_6]_4Cu_5^ICl_{17}$.[3a] Four very flat tetrahedral $CuCl_4$ groups share Cl atoms with a central $CuCl_4$ nucleus forming a complex ion $Cu_5Cl_{16}^{11-}$, (a), of the same general type as the $PCr_4O_{16}^{3-}$ ion

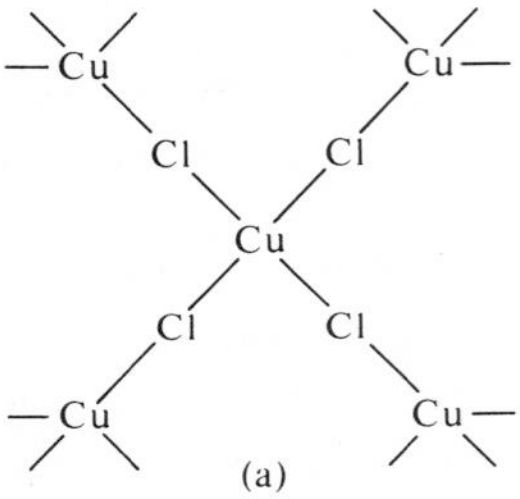

(a)

(p. 861) and the $Si_5O_{16}^{12-}$ ion in zunyite (p. 1013). These pack together with the cations and additional Cl^- ions which occupy holes in the structure. A point of interest is that this complex is formed in preference to simple $CuCl_4^{2-}$ ions, which in fact exist, together with trigonal bipyramidal $Cu^{II}Cl_5^{3-}$ ions, in the dark-brown to black compounds[3b] intermediate between the orange end-members, $[Co(NH_3)_6]_4$-Cu_5Cl_{17} and $[Co(NH_3)_6]Cu^{II}Cl_5$. Both of these compounds are cubic with a very large unit cell ($a = 21.8$ Å) containing 8 formula-weights of the Cu(I) compound and 32 of the Cu(II) compound. We mention shortly some finite polymeric molecules in which there are metal–metal bonds.

In tris-thiourea Cu(I) chloride, $[Cu\{SC(NH_2)_2\}_3]Cl$[4] there are infinite chain ions, (b), in which one of the three thiourea molecules forms a bridge between two Cu atoms:

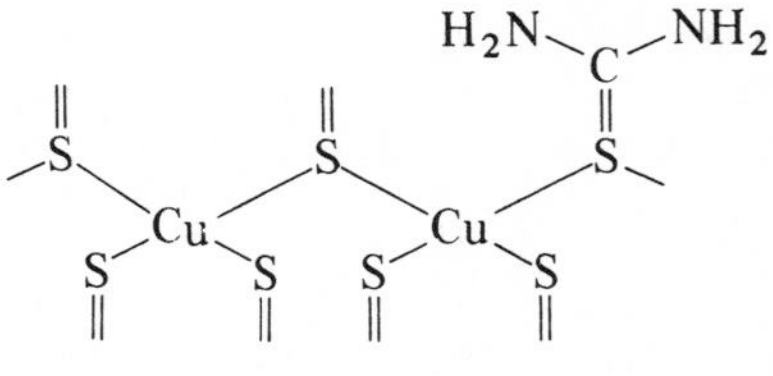

(b)

TABLE 25.1

Complex halides of Cu(I) *and* Ag(I)

Nature of M–X complex	Fig. 25.3	Examples	Reference
MX$_3$ chain		K$_2$CuCl$_3$, Cs$_2$AgCl$_3$, Cs$_2$AgI$_3$, (NH)$_4$)$_2$CuCl$_3$, (NH$_4$)$_2$CuBr$_3$, K$_2$AgI$_3$, Rb$_2$AgI$_3$, (NH$_4$)$_2$AgI$_3$	ZK 1973 **137** 225 AC 1975 **B31** 2339
MX$_2$ chain		[Ni(en)$_2$](AgBr$_2$)$_2$ (pq)(CuCl$_2$)$_2$ [(CH$_3$)$_3$N(CH$_2$)$_2$N(CH$_3$)$_3$](AgI$_2$)$_2$	ACSc 1969 **23** 3498 JCS A 1969 1520 AC 1975 **B31** 2341
M$_2$X$_3$ chain	(a)	CsCu$_2$Cl$_3$, CsAg$_2$I$_3$ (C$_6$H$_5$N$_2$)(Cu$_2$Br$_3$)	AC 1954 7 176 CC 1965 299
M$_3$X$_5$ chain	(b)	(Cu$_3$Cl$_5$)(L$_2$Cu^{II})	IC 1971 **10** 138
MXL′ chain	(c)	CuCl.C$_6$H$_5$CN	AC 1976 **B32** 1586
	(d)	CuI.CHNC	JCS 1960 2303
MXL″ chain	(e)	CuBr[(C$_2$H$_5$)$_2$P–P(C$_2$H$_5$)$_2$]	ZaC 1970 **372** 150
M$_2$X$_2$L″ layer	(f)	Cu$_2$Cl$_2$.N$_2$(CH$_3$)$_2$	AC 1960 **13** 28

Chains built from CuX$_4$ tetrahedra form the infinite anions in a number of complex halides of Cu(I) and Ag(I)—Table 25.1. The two simplest are the vertex-sharing MX$_3$ (pyroxene) chain and the MX$_2$ chain in which the tetrahedral groups share opposite edges (as in BeCl$_2$ and SiS$_2$). There is some angular distortion of the CuCl$_4$ tetrahedra in the chain anion in the 'paraquat' salt, pq^{2+}(CuCl$_2$)$_2$, where pq^{2+} represents N,N′-dimethyl-4,4′-bipyridylium ion. The cation is an electron-acceptor, and the black colour, semiconductivity, and paramagnetism of this salt suggest that there is some interchange of electrons between Cu(I) and Cu(II) in the chains. The M$_2$X$_3$ chain found in CsCu$_2$Cl$_3$ and CsAg$_2$I$_3$ consists of double rows of tetrahedra in which each MX$_4$ groups shares three edges and two vertices with other similar groups (Fig. 25.3(a)). Chains of a similar type form the anion in the diazonium salt (C$_6$H$_5$N$_2$)(Cu$_2$Br$_3$), an intermediate in the Sandmeyer reaction. The more complex M$_3$X$_5$ chain of Fig. 25.3(b) is found in (Cu$_3^I$Cl$_5$)(Cu^{II}L$_2$), where L is N-benzoyl-hydrazine. Within the chain distorted CuCl$_4$ tetrahedra share a vertex and either two or three edges. Each Cu(I) is bonded (through Cl) to 3 or 4 other Cu(I) atoms in the chain, and certain of the Cl atoms (shaded) complete the (4+2)-coordination around the Cu(II) atoms.

The simplest chain which includes a ligand other than halogen is the 'ladder' of composition CuXL′, Fig. 25.3(c), where L′ is a ligand forming a bond to one metal atom only. This is the type of chain in CuCl.C$_6$H$_5$CN and CuBr.C$_6$H$_5$CN, but CuI.CH$_3$NC consists of double chains, Fig. 25.3(d), in which the centrally situated Cu atoms are bonded to 4 I as in CuI and the outer ones to 2 I and 2 CH$_3$NC molecules. A ligand L″ capable of bonding at both ends can form chains of the kind shown in Fig. 25.3(e) or layers as in Cu$_2$Cl$_2$.N$_2$(CH$_3$)$_2$, Fig. 25.3(f).

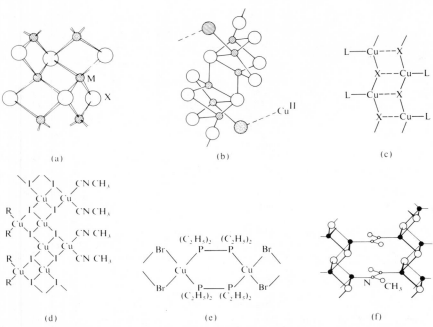

FIG. 25.3 Some infinite complexes containing Cu(I): (a) the M_2X_3 chain in $CsCu_2Cl_3$ and $C_6H_5N_2(Cu_2Br_3)$; (b) the $(Cu_3Cl_5)_n^{2n-}$ ion; (c) the MXL' chain; (d) the linear molecule $(CuI.CH_3NC)_n$; (e) the chain in $CuBr(Et_2P.PEt_2)$; (f) part of a layer in $Cu_2Cl_2.N_2(CH_3)_2$.

We noted earlier the formation of one metal–metal bond by Cu(I), in addition to two bonds to N, in $[Cu(\phi NNN\phi)]_2$. There are other polymeric complexes in which Cu(I) or Ag(I) has close metal neighbours in addition to 3 or 4 non-metal atoms at normal distances. The tetrameric molecule $Cu_4I_4(AsEt_3)_4$ consists of a central tetrahedron of Cu atoms surrounded by tetrahedra of I atoms and triethylarsine molecules (Fig. 25.4(a)). In this molecule Cu has, in addition to 3 I and 1 As at normal single bond distances, 3 Cu at 2.78 Å; compare the interatomic distance in metallic Cu (2.56 Å).[5] Many molecules $M_4X_4L_4$ are formed by Cu and Ag with

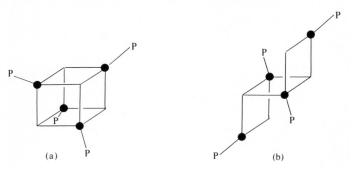

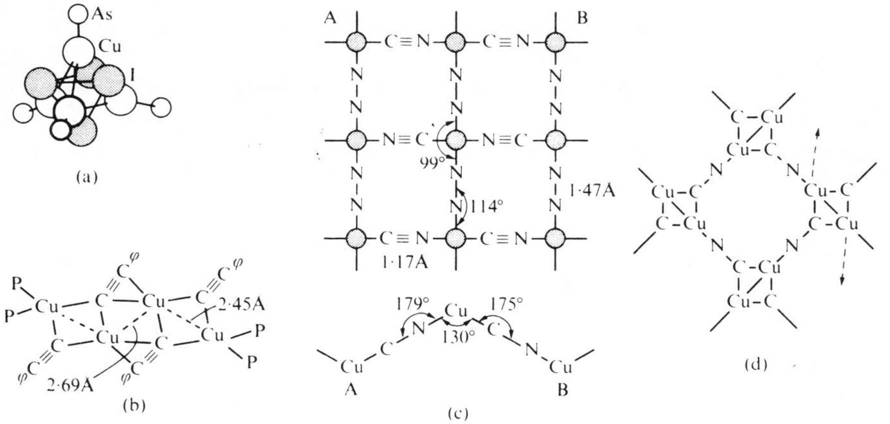

FIG. 25.4. (a) The molecule $Cu_4I_4[As(CH_3)_3]_4$; (b) $[(CH_3)_3PCuC{\equiv}C(C_6H_5)]_4$; (c) layer in $CuCN.N_2H_4$; (d) layer in $CuCN.NH_3$ (NH_3 omitted).

phosphines and arsines, but with large halogen and very bulky groups on P or As the chair-like configuration (b) is adopted; for example, $Cu_4Cl_4(P\phi_3)_4$ (a), but $Cu_4Br_4(P\phi_3)_4$, $Cu_4I_4(P\phi_3)_4$, and $Ag_4I_4(P\phi_3)_4$, (b).[5a] The compound $(CH_3)_3P.Cu.C{\equiv}C(C_6H_5)$ forms tetrameric molecules of a quite different kind. The bonding system is obviously complex, involving Cu–C bonds (1.96–2.22 Å) and Cu–Cu bonds of two different lengths (Fig. 25.4(b)).[6]

Several cuprous compounds form layer structures based on the simplest plane 4-connected net. In the puckered layers in $CuCN.N_2H_4$,[7] Fig. 25.4(c), the bond lengths are: Cu–C, 1.93 Å, Cu–N, 1.95 Å, and Cu–2N (of N_2H_4), 2.17 Å; the Cu(I) bond angles being 99° (three), 113° (two), and 130° (one). Nitriles of aliphatic dibasic acids can function as bridging ligands, $-NC-(CH_2)_n-CN-$ giving compounds of the type $(CuR_2)X$ where R is the nitrile and X, for example, NO_3^- or ClO_4^-. All the salts $Cu(succinonitrile)_2ClO_4$,[8] $Cu(glutaronitrile)_2NO_3$,[9] and $Ag(adiponitrile)_2$-ClO_4[10] form layers based on the plane 4-gon net. The structure of $Cu(adiponitrile)_2$-NO_3,[11] which consists of six interpenetrating identical 3D nets of the cristobalite type, was mentioned in Chapter 3.

The ammine $CuCN.NH_3$ is built of a unique kind of CuCN layer, Fig. 25.4(d), to each Cu of which one NH_3 is attached, the Cu–NH_3 bond being approximately perpendicular to the plane of the layer.[12] The distinction between C and N atoms is not certain, but the arrangement shown appears most likely. Within the planar C_2Cu_2 units (which are inclined to the plane of the figure) the Cu–Cu distance (2.42 Å) is shorter than in the metal, and in addition to this bond Cu has four neighbours in a somewhat distorted tetrahedral arrangement (Cu–N, 1.98 Å; Cu–NH_3, 2.07 Å; Cu–C, 2.09 and 2.13 Å). The Cu–C bonds are abnormally long, and the nature of the bonding is not clear.

Some further examples of tetrahedral Cu(I) are included in the section on compounds of Cu(I) and Cu(II).

We may include here some examples of Ag(I) forming tetrahedral bonds in finite complexes. In the complex anion in $(NH_4)_9Cl_2[Ag(S_2O_3)_4]^{(13)}$ Ag(I) forms bonds to S atoms (Ag–S, 2.58 Å) of four $S_2O_3^{2-}$ ligands. In the two related molecules:

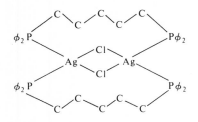 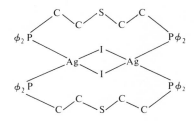

the Ag atoms are bridged by two halogen atoms and also by 7-membered chain molecules terminated by P atoms.[14, 15] For clarity all H atoms are omitted from CH_2 groups.

(1)	AC 1968 **B24** 269	(7)	AC 1966 **20** 279
(2)	JCS 1962 1748	(8)	AC 1969 **B25** 1518
(2a)	JACS 1963 **85** 4046	(9)	BCSJ 1959 **32** 1216
(3a)	AC 1973 **B29** 2559	(10)	IC 1969 **8** 2768
(3b)	AC 1975 **B31** 978	(11)	BCSJ 1959 **32** 1221
(4)	AC 1964 **17** 928	(12)	AC 1965 **19** 192
(5)	IC 1974 **13** 1899	(13)	AC 1972 **B28** 2079
(5a)	IC 1977 **16** 3267	(14)	AC 1976 **B32** 2521
(6)	AC 1966 **21** 957	(15)	AC 1975 **B31** 1194

The formation of three bonds by Cu(I), Ag(I), *and* Au(I)

We have quoted a number of examples of the formation by Cu(I) of 4 tetrahedral bonds to CN or to S atoms of ligands such as thioacetamide or thiourea. In other compounds Cu(I) forms 3 coplanar bonds to C(N) or S. As already noted, the structure of $KCu(CN)_2^{(1)}$ is quite different from that of the isostructural $KAg(CN)_2$ and $KAu(CN)_2$, which contain linear $(NC–M–CN)^-$ ions. The complex ions consist of helical chains (Fig. 25.5(a)) bonded by K^+ ions lying between them. In these chains the Cu bonds are definitely not coplanar, whereas in the chains in $NaCu(CN)_2.2H_2O^{(2a)}$ which have essentially the same type of structure, the Cu bonds are coplanar. In $Na_2Cu(CN)_3.3H_2O^{(2b)}$ there are discrete planar $[Cu(CN)_3]^{2-}$ ions (Cu–C, 1.94 Å). In addition to the anhydrous $KCu(CN)_2$ and $K_3Cu(CN)_4$ (which contains tetrahedral $[Cu(CN)_4]^{3-}$ ions, p. 1107), potassium forms $KCu_2(CN)_3.H_2O.^{(2c)}$ In this crystal each Cu(I) forms 3 nearly coplanar bonds, as in $KCu(CN)_2$, (interbond angles, two of 112° and one of 134°) in puckered layers of the simplest possible type (Fig. 25.5(b)). The lengths of the Cu–C and Cu–N bonds are very similar to those in $KCu(CN)_2$, and the shortest Cu–Cu distance is 2.95 Å, as compared with 2.84 Å in $KCu(CN)_2$. The H_2O molecules are not bonded

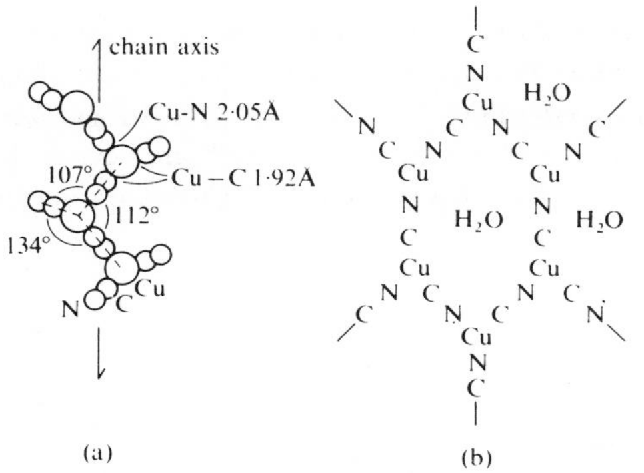

FIG. 25.5. (a) The helical chain ion in $KCu(CN)_2$; (b) the layer anion in $KCu_2(CN)_3.H_2O$.

to the metal atoms, but occupy holes in the layers, the latter being held together by the K^+ ions. We have noted the quite different behaviour of $Cu(I)$ in $CuCN.N_2H_4$ and $CuCN.NH_3$ in the previous section.

Cuprous chloride forms crystalline complexes with numerous unsaturated hydrocarbons, of which one has already been mentioned. In several of these $Cu(I)$ forms two bonds to Cl and a third to a multiple bond of the hydrocarbon, the three bonds from Cu being coplanar. In the 1,5-cyclooctatetraene complex, $CuCl.C_8H_8$,[3a] the Cu and Cl atoms form an infinite chain, (a), while in the complexes with norborna-

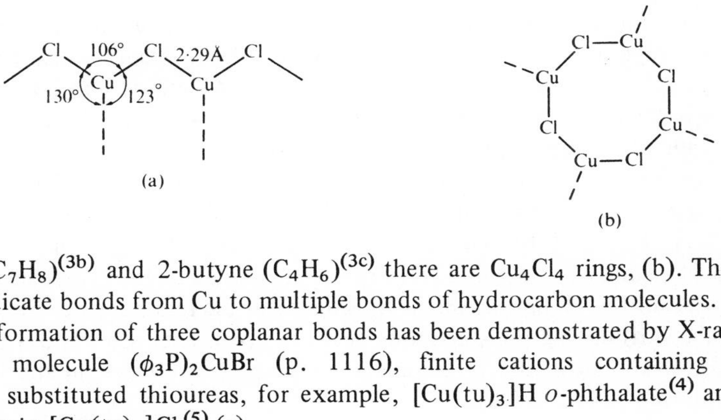

diene (C_7H_8)[3b] and 2-butyne (C_4H_6)[3c] there are Cu_4Cl_4 rings, (b). The broken lines indicate bonds from Cu to multiple bonds of hydrocarbon molecules.

The formation of three coplanar bonds has been demonstrated by X-ray studies of the molecule $(\phi_3P)_2CuBr$ (p. 1116), finite cations containing thiourea (tu), or substituted thioureas, for example, $[Cu(tu)_3]H$ *o*-phthalate[4] and of the chain ion in $[Cu(tu)_2]Cl$,[5] (c).

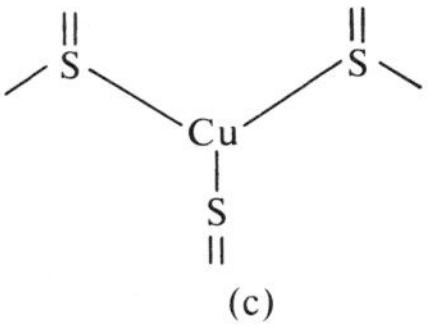

In polymeric complexes with certain S-containing ligands Cu(I) and Ag(I) form bonds to 3 S atoms and there are also 3 or 4 metal–metal bonds. The nucleus of the molecule of $Cu_4[(i\text{-}C_3H_7O)_2PS_2]_4$ [6] is a tetrahedral Cu_4 group having four shorter (2.74 Å) and two longer (2.94 Å) edges to which are bonded four ligand molecules as shown in Fig. 25.6(a). Each Cu has 3 S and 3 Cu nearest neighbours. There is also a

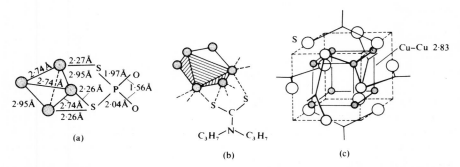

(a)   (b)   (c)

FIG. 25.6.   Details of some polymeric Cu(I) and Ag(I) complexes (see text). In (b) the long Ag–Ag bonds are the edges of the unshaded face and of the opposite face of the Ag_6 octahedron.

tetrahedral nucleus of metal atoms in $[(C_2H_5)_2NCS_2Cu]_4$.[7a] The type of polymer appears to be determined by the geometry of the ligand, for both $[(C_3H_7)_2NCOSCu]_6$ (ref. 7b) and $[(C_3H_7)_2NCS_2Ag]_6$ [7c] are hexameric. In the Cu_6 octahedron in the former Cu–Cu ranges from 2.70 to 3.06 Å (mean 2.88 Å). The hexameric molecule of silver dipropyldithiocarbamate, $[(C_3H_7)_2NCS_2Ag]_6$, contains a distorted octahedral Ag_6 nucleus which has six short edges (2.9–3.2 Å) and six longer ones (3.45–4.0 Å), the latter being the edges of a pair of opposite faces. The ligands behave in very much the same way as in the tetramer (a), one S being bonded to one Ag and the other to two Ag atoms. As a result each Ag is bonded to 3 S (Ag–S, 2.43–2.56 Å) and also has 2 Ag neighbours (Ag–Ag, 2.9, 3.0 Å) and 2 more distant Ag (3.45, 4.0 Å)–Fig. 25.6(b). The anion in salts $(Cu_8L_6)(As\phi_4)_4$,[8] in which L is the ligand (d), consists of a cubic cluster of 8 Cu atoms of edge 2.83 Å surrounded by 12 S atoms of the six ligands which are situated at the mid-points of the edges of a larger cube surrounding the Cu_8 group (Fig. 25.6(c)). Each Cu has 3 S and 3 Cu atoms as nearest neighbours at distances very similar to those in the tetramer described above (Cu–S, 2.25 Å).

The unique structure adopted by silver tricyanomethanide, $Ag[C(CN)_3]$,[9a] is

$$\left[\begin{array}{c} S \\ \diagdown \\ C - C \diagup \begin{array}{c} CN \\ \diagdown \\ CN \end{array} \\ \diagup \\ S \end{array} \right]^{2-}$$

(d)

illustrated in Chapter 3 (p. 105). It consists of pairs of interwoven hexagonal layers in which Ag is approximately 0.5 Å out of the plane of its three N neighbours. Presumably single plane layers would have too open a structure — compare $KCu_2(CN)_3.H_2O$, in which there are K^+ ions and H_2O molecules to occupy the large holes in the sheets. Modified forms of this structure are adopted by $Ag[C(CN)_2NO]$[9b] and $Ag[C(CN)_2NO_2]$[9c] but both structures are more complex than that of the tricyanomethanide. In the network of the former Ag is bonded to 2 N and O of the ligands but also forms a bond to the N atom of the NO group, resulting in a (distorted) tetrahedral coordination of the metal atom. In the latter the NO_2 group bonds to Ag not only through the N atom but also through both O atoms (to the same Ag atom).

Three bonds are formed by Cu(I) in certain sulphides, the bonding in which is discussed in a later section of this chapter, and also by Ag(I) in a number of S-containing compounds, but here also there is often no simple interpretation of the bonding. In Ag_3AsS_3 there is both planar and pyramidal 3-coordination of Ag,[10] while in $Ag(thiourea)_2Cl$[11] Ag has three approximately coplanar S neighbours but also one Cl at a rather large distance (2.95 Å).

Although the stereochemistry of Ag(I) is simpler than that of Cu(I) in some compounds the reverse is true in many complexes with organic compounds. For example, the structure of $C_6H_6.AgAlCl_4$[12] is much less simple than that of the Cu(I) compound (p. 978), Ag forming four bonds to Cl (2.59, 2.77, 2.80, and 3.03 Å) and a fifth to one bond of the benzene molecule. In $(C_6H_5C{\equiv}C)Ag[P(CH_3)_3]$[13] there are two types of Ag atom with entirely different environments. One type forms 2 collinear bonds to terminal C atoms of $C{\equiv}C.C_6H_5$ and the other two bonds to $P(CH_3)_3$ molecules (P–Ag–P, 118°) and two bonds to triple bonds of the hydrocarbon.

Certain silver oxy-salts, in particular the perchlorate, are notable for forming crystalline complexes with organic compounds. For $AgClO_4.C_6H_6$ and $AgNO_3.C_8H_8$ see p. 978. In the compound with dioxane, $AgClO_4.3C_4H_8O_2$,[14] the Ag atoms are at the points of a simple cubic lattice, the ClO_4^- at the body-centre, and the dioxane molecules along the edges. The metal ions are therefore surrounded octahedrally by six O atoms (at about 2.46 Å), presumably bonded by some sort of ion-dipole bonds. Both the perchlorate ions and the dioxane molecules are rotating, probably quite freely.

Three coplanar bonds of normal lengths are formed by Au(I) in ions such as

(e)   (f)   (g)

$[(\phi_3P)_3Au]^{+}$[15] and in the molecule $(\phi_3P)_2AuCl$[16] with bond angles close to $120°$; the molecule $(\phi_3P)_2CuBr$[17] has a similar structure, the angle P–Cu–P being $126°$. With one bidentate ligand there is one small angle and also unsymmetrical attachment of the bidentate ligand, one bond being much stronger than the other. In an extreme case, such as the diethyldithiocarbamate complex, (e),[18] the two stronger bonds to Au are almost collinear, and there are also intermediate cases such as $\phi_3PAu(2:2'$-bipyridyl),[19] (f), and $[(CH_3)_2P\phi]_2Au(SnCl_3)$,[20] (g). The structures of 3-covalent Au(I) complexes have been discussed in ref. (19).

(1)	JPC 1957 **61** 1388	(9a)	IC 1966 **5** 1193
(2a)	IC 1977 **16** 250	(9b)	AC 1974 **B30** 1117
(2b)	IC 1978 **17** 1945	(9c)	AC 1974 **B30** 147
(2c)	AC 1962 **15** 397	(10)	AC 1968 **B24** 77
(3a)	IC 1964 **3** 1529	(11)	IC 1968 **7** 1351
(3b)	IC 1964 **3** 1535	(12)	JACS 1966 **88** 3243
(3c)	AC 1957 **10** 801	(13)	AC 1966 **20** 502
(4)	AC 1977 **B33** 3772	(14)	AC 1956 **9** 741
(5)	AC 1970 **B26** 1474	(15)	AC 1980 **B36** 3105
(6)	IC 1972 **11** 612	(16)	IC 1974 **13** 805
(7a)	AK 1963 **20** 481	(17)	IC 1973 **12** 213
(7b)	ACSc 1970 **24** 1355	(18)	JCMS 1972 **2** 7
(7c)	ACSc 1969 **23** 825	(19)	AC 1976 **B32** 2712
(8)	JACS 1968 **90** 7357	(20)	AC 1978 **B34** 278

Salts containing Cu(I) and Cu(II)

We comment later in this chapter on the difficulty of assigning oxidation numbers to Cu atoms in the sulphides, and in Chapter 17 we noted the structure of KCu_4S_3. Apart from these semi-metallic compounds there are normal coloured salts containing Cu(I) and Cu(II) each with an entirely different stereochemistry. In the red salt $Cu_2^ISO_3 . Cu^{II}SO_3 . 2H_2O$[1] Cu(I) has a tetrahedral environment (though Cu–S, $2.14\,\text{Å}$, is remarkably similar to Cu–O, $2.11–2.14\,\text{Å}$), and Cu(II) has a typical (4+2)-octahedral coordination group. In the pale-violet $Na_4[Cu^{II}(NH_3)_4][Cu^I(S_2O_3)_2]_2$[2] the $Cu(NH_3)_4^{2+}$ ions are planar (with no other close neighbours of Cu(II) and the anion consists of chains (a), in which Cu(I) has four tetrahedral S neighbours at $2.36\,\text{Å}$. In contrast to the pale-yellow α and β CuNCS and the light-blue $Cu(NCS)_2$-$(NH_3)_2$ the compound $Cu_2(NCS)_3(NH_3)_3$ is greenish-black.[3] There is tetrahedral

(a)   (b)   (c)

coordination of Cu(I), by 2 N and 2 S of NCS groups, and tetragonal pyramidal coordination of Cu(II), (b). The Cu(I) and Cu(II) atoms presumably interact through the S atoms, (c). For (blue) $(Cu^{I}Cl_2)[Cu^{II}(OAs\phi_3)_4]$ see p. 1122.

The hydrated complex cyanide $Cu_3en_2(CN)_4 . H_2O^{(4)}$ has an interesting structure. The Cu(I) atoms linked through CN groups form a diamond-like framework, and square pyramidal $(Cu^{II}en_2H_2O)^{2+}$ groups occupy interstices. The structural formula is accordingly $[Cu_2^{I}(CN)_4]^{2-}[Cu^{II}en_2H_2O]^{2+}$.

(1) ACSc 1965 **19** 2189
(2) AC 1966 **21** 605
(3) IC 1969 **8** 304
(4) AC 1972 **B28** 858

The structural chemistry of cupric compounds

This subject is of quite outstanding interest, for in one oxidation state this element shows a greater diversity in its stereochemical behaviour than any other element. Before commencing this survey it is interesting to note that a number of simple cupric compounds are either difficult to prepare or are not known. Although crystalline $Cu(OH)_2$ can be prepared by special methods (for example, by treating $Cu_4(OH)_6SO_4$ with aqueous NaOH, dissolving the hydroxide in concentrated NH_4OH and removing the NH_3 slowly *in vacuo* over H_2SO_4), the ready decomposition of the freshly precipitated compound is well known. (For the structure of $Cu(OH)_2$ see p. 633.) Two hydroxycarbonates, azurite, $Cu_3(OH)_2(CO_3)_2$, and malachite, $Cu_2(OH)_2CO_3$, are well-known minerals, but the normal carbonate, $CuCO_3$, is not stable at ordinary temperatures under atmospheric pressure. It has, however, been prepared from CuO or from the hydroxycarbonates by reaction with CO_2 under 20 kbar pressure at 500°C. Its structure is described on p. 1127. A number of oxy-salts (e.g. nitrate and perchlorate) crystallize from aqueous solution as hydrates which cannot be dehydrated to the anhydrous salts, but both $Cu(NO_3)_2$ and $Cu(ClO_4)_2$ have been prepared by other methods (p. 823). They can both be volatilized without decomposition and dissolve in a number of polar organic solvents. The structures of some of the numerous hydroxy-salts are described in other chapters ($Cu(OH)Cl$ and $Cu_2(OH)_3Cl$ in Chapter 10, and $Cu(OH)IO_3$ and $Cu_2(OH)_3NO_3$ in Chapter 14). (See also p. 169 for the relationship of atacamite, $Cu_2(OH)_3Cl$, to the NaCl structure.) The compounds CuI_2, $Cu(CN)_2$, and $Cu(SCN)_2$ are extremely unstable, and CuS is not a simple cupric compound (see later). Because of the instability or non-existence of Cu(II) compounds containing the less electronegative ligands most of our structural information concerns compounds in which the metal is bonded to F, Cl, N, or O, the bonds having considerable ionic character.

The stereochemistry of Cu(II) is summarized in Fig. 25.7. Four or five bonds are formed of length corresponding to normal single bonds. Four bonds are coplanar or directed towards the vertices of a flattened tetrahedron; regular tetrahedral coordination is not observed. Five bonds of approximately single bond length are directed towards the vertices of a trigonal bipyramid; this bond arrangement is rare. The most frequently observed coordination group of Cu(II) is a group of 4 coplanar

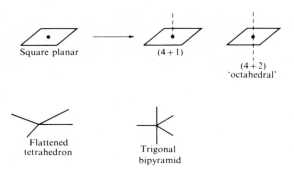

FIG. 25.7. Stereochemistry of Cu(II).

neighbours with one or two more distant neighbours completing a distorted tetragonal pyramid or distorted octahedron. In many crystals in which the coordination group of Cu(II) consists of 6 F (Cl, Br) or 6 O there is Jahn–Teller distortion of the octahedral group, almost always to (4+2)-coordination, the axial bonds being about 12 per cent longer than the equatorial ones.

From the geometrical standpoint the square coplanar arrangement may be regarded as the limit of the extension of an octahedron along a 4-fold axis; it may alternatively be regarded as the limit of compression of a tetrahedron along a 2-fold axis (normal to an edge). In the former case the six equal *bonds* become two long and four short, and in the latter case the six equal *angles* of a regular tetrahedral bond arrangement (109½°) become two larger (in the limit 180°) and four smaller (in the limit 90°):

$$\text{six of } 109\tfrac{1}{2}° \longrightarrow \begin{cases} \text{two equal angles} \longrightarrow 180° \\ \quad (> 109\tfrac{1}{2}°) \\ \text{four equal angles} \longrightarrow 90° \\ \quad (< 109\tfrac{1}{2}°) \end{cases}$$

regular tetrahedron : (flattened tetrahedron) : square planar

We show in Table 25.2 examples of these intermediate configurations which are described later in the text; for $CuCr_2O_4$ see p. 596.

In this chapter we shall not adopt a strictly systematic treatment, dealing in turn with the bond arrangements of Fig. 25.7, for a number of reasons. There are few examples of the formation of only 4 bonds, whether coplanar or tetrahedral; examples of the latter will be referred to in various places. Also the distinction between 4-, (4+1)-, and (4+2)-coordination is not always clear-cut. It seems more appropriate therefore to deal separately with certain bond arrangements but also to keep together chemically related groups of compounds.

TABLE 25.2

Configurations intermediate between square coplanar and regular tetrahedral in Cu(II) *compounds*

Compound	Interbond angles	
	4 of	2 of
(Square coplanar)	90°	180°
Chelates	93°	154°
	96°	141°
Cu(imidazole)$_2$	97°	140°
(CuCl$_4$)$^{2-}$	100°	130°
CuCr$_2$O$_4$	103°	123°
(Regular tetrahedron)	6 of 109$\frac{1}{2}$°	

The formation of 4, (4+1), *or* (4+2) *bonds*

The available evidence appears to indicate that having formed four strong coplanar bonds Cu(II) forms if possible one or two additional weaker bonds perpendicular to the plane of the four coplanar bonds. In some cases (for example, chelate molecules) steric factors limit the c.n. to 4 or (4+1), and therefore we deal separately with these compounds. Naturally there must always be neighbours of some kind additional to 4 coplanar neighbours, and it is necessary to consider their distances from the metal atom in order to decide whether they should be regarded as bonded to Cu(II). An alternative to (4+1)- or (4+2)-coordination is to rearrange the four close neigbours to form a flattened tetrahedral group.

The fact that Ni(II), Pd(II), and Pt(II) form four strong coplanar bonds suggests a comparison of the structures of their compounds with those of Cu(II). However, no cupric compounds other than the phthalocyanine and similar complexes are isostructural with compounds of these Group VIII metals; the closest resemblances are those between CuO and PdO and between CuCl$_2$ and the linear molecules of PdCl$_2$ and PtCl$_2$. We refer to these structures shortly.

In fact, the only simple salts known to be isostructural with those of Cu(II), d^9, are certain compounds of Cr(II), d^4, as might be expected from ligand field considerations. This similarity does not, however, extend to all compounds of these elements in this oxidation state. For example, while CrBr$_2$ and CrI$_2$ have structures similar to CuCl$_2$ and CuBr$_2$, CrCl$_2$ has a distorted rutile-type structure (but distorted in a different way from CrF$_2$ and the isostructural CuF$_2$) in contrast to the CdI$_2$-like structure of CuCl$_2$. In contrast to the complex structure of CuS, CrS has the type of structure that might have been expected for CuS (and CuO); this is a structure in which Cr has four approximately equidistant coplanar neighbours and two more completing a distorted octahedron, that is, the (4+2)-coordination characteristic of many Cu(II) compounds. (The oxide CrO is not known.) Examples of pairs of isostructural Cu(II) and Cr(II) compounds include the difluorides and dibromides, $KCuF_3$ and $KCrF_3$, and the dimeric acetates $[M(CH_3COO)_2.H_2O]_2$ (p. 1130).

Crystalline $CuO^{(1)}$ (the mineral tenorite) provides the simplest example of Cu(II) forming four coplanar bonds. The structure is a distorted version of the PdO (PtS) structure in which the O–Cu–O angles are two of $84\frac{1}{2}°$ and two of $95\frac{1}{2}°$; Cu has 4 O' neighbours at 1.96 Å, the next nearest neighbours being two O" at 2.78 Å. The ratio of these distances is much greater than for the usual distorted octahedral coordination of Cu(II), and the line O"–Cu–O" is inclined at 17° to the normal to the $Cu(O')_4$ plane. The shortest distance Cu–Cu is 2.90 Å. Presumably, alternative structures such as a distorted NaCl structure with (4+2)-coordination for Cu(II), implying also this type of coordination of O, would be less stable than the observed structure with planar coordination of Cu(II) and tetrahedral coordination of O, but the relevant lattice energy calculations have not been made.

A structure has been suggested for the mineral paramelaconite, which has the strange formula $Cu_{16}O_{14}.^{(2a)}$ The structure is related to that of tenorite but has vacant O sites resulting in a rearrangement of the monoclinic tenorite structure to a more symmetrical (tetragonal) one. The proposed structure does not distinguish between Cu(I) and Cu(II) and gives all Cu atoms a distorted square planar coordination. There is also apparently square planar coordination of Cu(II) in $CaCu_2O_3$ and Sr_2CuO_3 the structures of which have not been determined in detail.$^{(2b)}$ The crystalline derivative of imidazole, $Cu(C_3N_2H_3)_2,^{(3)}$ is a 3D network of Cu atoms linked by imidazole ligands (a), each N forming a bond to a metal atom. The Cu

(a)

atoms are of two kinds, one-half having 4 coplanar N neigbbours (interbond angles close to 90°) and no other close neighbours, and the remainder a flattened tetrahedral arrangement of 4 neighbours (Table 25.2). It is evidently not possible to bring additional N atoms closer to the Cu(II) atoms owing to the way in which the imidazole rings are packed.

In contrast to the normal rutile structure of PdF_2, CuF_2 has a distorted variant of that structure with (4+2)-coordination of Cu^{2+}, the four stronger bonds linking Cu and F into layers (Fig. 6.7(a), p. 249). The bond lengths Pd^{II}–Cl and Cu^{II}–Cl are the same (2.30 Å) and planar chains of exactly the same type are found in $CuCl_2$ and in one form of $PdCl_2$:

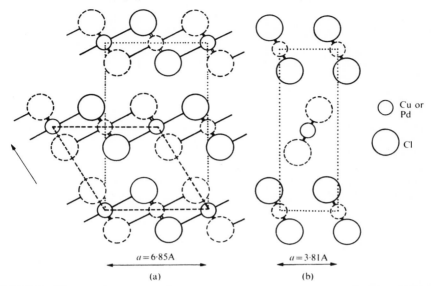

FIG. 25.8. The crystal structures of (a) $CuCl_2$, and (b) $PdCl_2$ viewed along the direction of the chains showing how these are packed differently in the two crystals. Atoms at $y = 0$ and $y = \frac{1}{2}$ are represented by full and dotted circles respectively. In (a) the broken lines enclose the monoclinic cell, and the dotted lines indicate the pseudorhombic unit cell.

However, the chains pack together quite differently in the two halides (Fig. 25.8). In $PdCl_2$ the next nearest neigbours of Pd are 4 more Cl at 3.85 Å arranged at the corners of a rectangle, the plane of which is perpendicular to that of the chain; in $CuCl_2$ the chains are packed side-by-side to form a CdI_2-like layer in which $Cu(II)$ has 2 additional Cl neighbours at 2.95 Å completing a very distorted octahedral coordination group (Fig. 25.9).

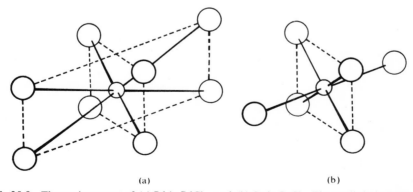

FIG. 25.9. The environment of (a) Pd in $PdCl_2$, and (b) Cu in $CuCl_2$. The small circles represent metal atoms.

Finite complexes containing only monodentate ligands range from cations through neutral molecules to anions:

$$\left[\begin{array}{c} L \diagdown \diagup L \\ Cu \\ L \diagup \diagdown L \end{array}\right]^{2+} \quad \begin{array}{c} L \diagdown \diagup X \\ Cu \\ X \diagup \diagdown L \end{array} \quad \left[\begin{array}{c} X \diagdown \diagup L \diagdown \diagup X \\ Cu \qquad Cu \\ X \diagup \diagdown L \diagup \diagdown X \end{array}\right]^{2-} \quad \left[\begin{array}{c} X \diagdown \diagup X \\ Cu \\ X \diagup \diagdown X \end{array}\right]^{2-}$$

$$\left[\begin{array}{c} X \diagdown \diagup X \diagdown \diagup X \\ Cu \qquad Cu \\ X \diagup \diagdown X \diagup \diagdown X \end{array}\right]^{2-}$$

and others noted in a later section on halogen compounds. Apparently *cis* and *trans* isomers of strictly planar complexes CuL_2X_2 have not been characterized. The dichloro dilutidine complex exists in two forms, a green *trans* planar form and a yellow *cis* isomer with a somewhat distorted configuration.[4] The state of Cu(II) in

(lutidine)

complexes of the above types in solution is not generally known—completion of the octahedron by solvation is likely where this is sterically possible—certainly in solids the formation of *only* four bonds is rare and then it is due to the geometry of the complex. Thus in $[Cu(pyridine oxide)_4](BF_4)_2$[5] and the perchlorate[6] Cu has 4 coplanar O neighbours at approximately 1.95 Å but no more neighbours closer than 3.35 Å. Similarly, in $(CuCl_2)[Cu(OAs\phi_3)_4]$[6a] the formation of more than 4 coplanar bonds is prevented by the very bulky ligands; the blue colour shows also that interaction between Cu(I) and Cu(II) is prevented. In the bridged cation in the compound (b)[7] Cu is bonded to only 4 coplanar neighbours, the next nearest neighbours being at 4.78 Å. On the other hand, Cu(II) forms 5 bonds in the molecule (c),[7a] in which the bond arrangement is not clear-cut square pyramidal or trigonal bipyramidal, and in (d),[7b] where the metal forms five square pyramidal bonds. Both $Cu(pyridine oxide)_2(NO_3)_2$[8] and $Cu(pyridine oxide)_2Cl_2$[9] form dimers in their crystals, each Cu forming a weaker fifth bond to the O atom of a pyridine

(b)

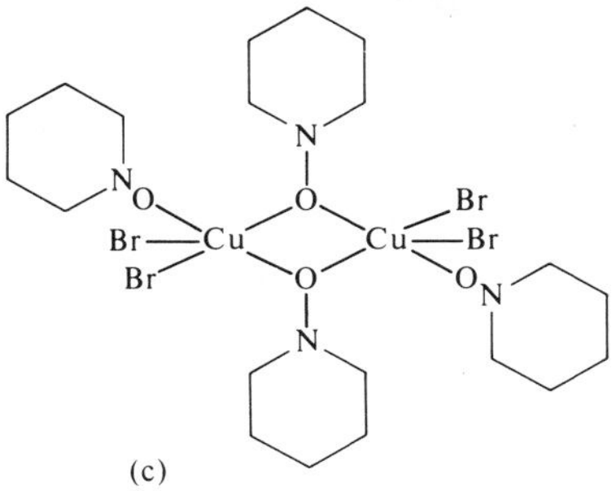

(c)

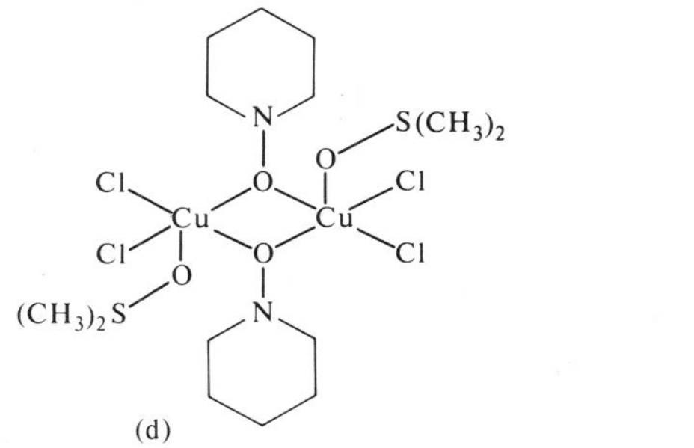

(d)

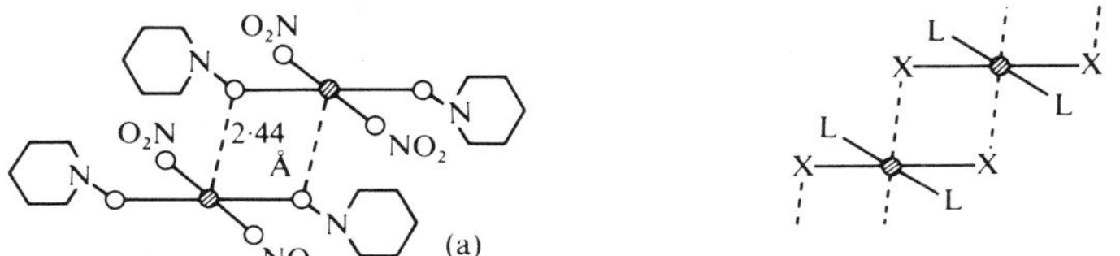

(a)

(b)

FIG. 25.10. (a) Dimers in Cu(pyridine oxide)$_2$(NO$_3$)$_2$; (b) chains CuL$_2$X$_2$.

oxide molecule (Fig. 25.10(a)). For suitable ligands X and suitable geometry of the monomer this process leads to the formation of the infinite chain of Fig. 25.10(b). In some cases the bridging 'bonds' are too long to be regarded as much more than van der Waals bonds (e.g. 3.05 Å in Cu(pyridine)$_2$Cl$_2$).[10] A review of complexes of Cu(II) with aromatic N-oxides is available.[11]

(1) AC 1970 **B26**. 8	(5) AC 1969 **B25** 1595	(7b) AC 1975 **B31** 2362
(2a) AC 1978 **B34** 22	(6) AC 1969 **B25** 1378	(8) AC 1969 **B25** 2046
(2b) ZaC 1969 **370** 134	(6a) AC 1981 **B37** 232	(9) IC 1967 **6** 951
(3) AC 1960 **13** 1027	(7) AC 1970 **B26** 2096	(10) AC 1957 **10** 307
(4) IC 1969 **8** 308	(7a) AC 1972 **B28** 126	(11) IC 1969 **8** 1879

Structures of chelate cupric compounds

Many chelate cupric compounds have been made, particularly with bidentate and tetradentate ligands. In a simple compound such as Cu(en)$_2$(BF$_4$)$_2$[1] there is no difficulty in forming two additional weaker Cu–F bonds (2.56 Å) to F atoms of the anions. If, however, the Cu(II) atom is at the centre of a large tetradentate molecule such as phthalocyanin and forms four coplanar bonds to atoms of that molecule, or if the four atoms bonded to Cu(II) belong to two large coplanar molecules as in many chelates CuR$_2$, then no other bonds from Cu shorter than about 3.4 Å are possible since this is the minimum van der Waals separation of adjacent parallel molecules (Fig. 25.11(a)). Only four coplanar bonds are formed by Cu(II) in the bis-salicylaldiminato complex (a),[2] but such cases are rare, and a number of interesting possibilities are realized in crystals consisting of molecules of this general type.

(i) If parts of the ligands can be bent out of the plane of the central Cu atom and its four coplanar neighbours it may be possible to have two additional Cu–N or Cu–O bonds of, say 2½ Å while maintaining the necessary 3½ Å between successive

(a)   (b)

molecules (Fig. 25.11(b)). This occurs in the cupric derivative of benzene azo-β-naphthol, (b).[3]

(ii) Alternatively there may be other atoms in the chelating molecules to which Cu can form the two additional bonds, when plane molecules may be stacked as in Fig. 25.11(c) with a sideways shift of each successive molecule. The salicylaldoxime compound[4] (c), has this type of structure. Such additional O atoms are present in

(c)   (d)

the molecule of dimethylglyoxime (DMG), but here the presence of the bulky methyl groups leads, for the Ni compound, to stacking of the molecules directly above one another, alternate molecules being rotated through 90° to facilitate packing of the CH$_3$ groups (Fig. 25.11(d)). The only intermolecular bonds formed by Ni are those to Ni atoms of neighbouring molecules at 3.25 Å. This type of packing is adopted in one crystalline form of bis-(N-methylsalicylaldiminato)Cu(II), (d),[5] in which Cu has, in addition to its four coplanar neighbours within the molecule, no other neighbours nearer than 2 Cu at 3.33 Å.

(iii) Cu(DMG)$_2$[6] is not isostructural with the Ni compound although the structure of Fig. 25.11(d) would be geometrically possible. Stacking of the molecules to form infinite columns as in Fig. 25.11(c) is not possible because of the CH$_3$ groups, but this type of relation is possible between a *pair* of molecules which are bent to increase the clearance between them. Crystalline Cu(DMG)$_2$ consists of

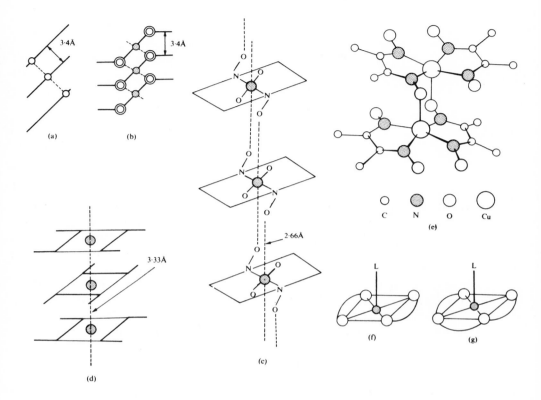

FIG. 25.11. Structures of chelated cupric compounds (see text).

dimers (Fig. 25.11(e)). The two molecules of DMG bonded to a given Cu atom are inclined at an angle of 23° to one another, and the metal atom is displaced slightly (by 0.3 Å) out of the plane of its four N atoms towards an O atom of the other half of the dimer. The Cu–N bonds are of normal length (1.95 Å) but the fifth bond (Cu–O, 2.30 Å) is much longer than a normal one (1.9–2.0 Å).

(iv) Dimers of the same type are formed in the chelates formed with 8-hydroxy-quinoline[7] and NN′-ethylene salicylideneimine, but the corresponding propylene compound forms a monohydrate (Fig. 25.11(f)) in which the fifth bond is formed to a water molecule (Cu–OH$_2$, 2.53 Å).[8] Other square pyramidal monomers of this type include the chelates with *o*-hydroxyacetophenone (L = 4-methyl pyridine)[9] and triene (L = SCN) in [Cu(trien)SCN]NCS.[10] (Fig. 25.11(g)). In all these molecules the fifth bond is appreciably longer than a normal single bond.

(v) If large alkyl groups are substituted for H on N in the salicylaldiminato ligand in chelates CuL$_2$, (e), there is slight buckling of the molecule and the bond arrangement becomes 'flattened tetrahedral':

(e)   (f)

$R = C_2H_5^{(11)}$ bond angles four of 93° and two of 154°; $R = t\text{-}C_4H_9,^{(12)}$ bond angles four of 96° and two of 141°. In the dipyrromethene complex (f) interference between the methyl groups of the two ligands leads to a very non-planar Cu coordination. The ligands remain planar, but the angle between their planes is 66°.[13]

(vi) We noted on p. 1110 the 'cubane-like' molecules such as $Cu_4X_4(PR_3)_4$ formed by Cu(I) in which X is Cl, Br, or I and PR_3 is a substituted phosphine. The molecules $Cu_4X_4L_4$ formed by Cu(II), in which L is an alkyl substituted amino-ethanolato ligand $R_2N.CH_2.CH_2.O-$ and X is Cl, Br, NCO, or NCS, form a variety of polymeric structures.[14] In the tetramers Cu and O atoms are situated at alternate vertices of a cube, the ligands spanning the four edges marked in the sketch, in which only one of the four organic ligands is shown. Each Cu is bonded to 5 ligands, 3 O, N, and X, but the coordination polyhedron is not usually of a very well-defined

shape. Values quoted for Cu—Cu (around 3.3 Å) are much higher than in the $Cu_4X_4(PR_3)_4$ molecules (2.8–2.9 Å).

(1)	AC 1968 **B24** 730	(8)	JCS 1960 2639
(2)	JCS A 1966 680	(9)	AC 1969 **B25** 2245
(3)	AC 1961 14 961	(10)	IC 1969 8 2763
(4)	AC 1964 17 1109, 1113	(11)	AC 1967 23 537
(5)	AC 1961 **14** 1222	(12)	JCS A 1966 685
(6)	JCS A 1970 218	(13)	JCS A 1969 2556
(7)	AC 1967 **22** 476	(14)	AC 1977 **B33** 1877

The structures of oxy-salts

Crystalline compounds containing ions CO_3^{2-}, NO_3^-, or $R.COO^-$ provide many examples of ligands bridging metal atoms to form finite complexes, chains, layers, or 3D framework structures, in which there is usually (4+1)- or (4+2)-coordination of Cu(II).

Three of these possibilities are realized in $Na_2Cu(CO_3)_2.3H_2O$,[1a] $Na_2Cu(CO_3)_2$, (ref. 1b), $Cu(NH_3)_2CO_3$,[1c] and $CuCO_3$.[1d] In $Cu(NH_3)_2CO_3$ there are simple chains (a), in which there is square pyramidal coordination of the metal. In the more complex chain of $Na_2Cu(CO_3)_2.3H_2O$, (b), the coordination may be described as (4+1) or (4+1+1) depending on the interpretation of the sixth weak bond. The anhydrous salt $Na_2Cu(CO_3)_2$ consists of puckered layers similar to those of Fig. 25.12(a) based on the planar 4^4 net. The plane of the CO_3 group is inclined to that of the layer, and in addition to the 4 coplanar O atoms (at 1.92 Å) Cu has a fifth O neighbour at 2.77 Å which is the non-bridging atom of the CO_3 group; contrast (b), where the *sixth* bond is to an O atom bonded to Cu. In the structure of $CuCO_3$, determined from a powder photograph, square pyramidal CuO_5 groups form a 3D structure in which Cu is bonded to O atoms of five different CO_3 groups. Layers of type (d) are displaced relative to one another so that each Cu forms a fifth bond to an O atom (shaded circle) of the layer below.

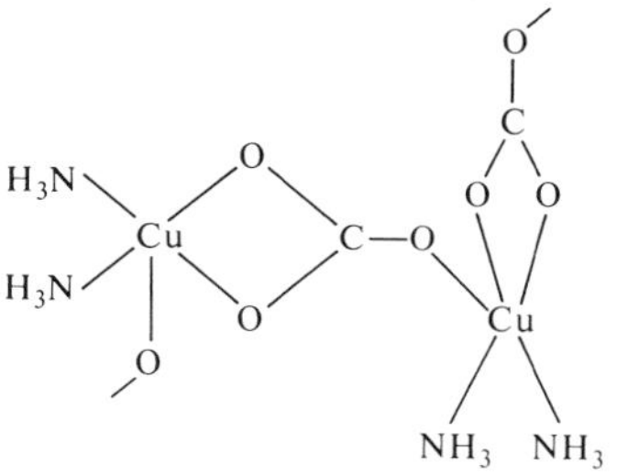

(a)

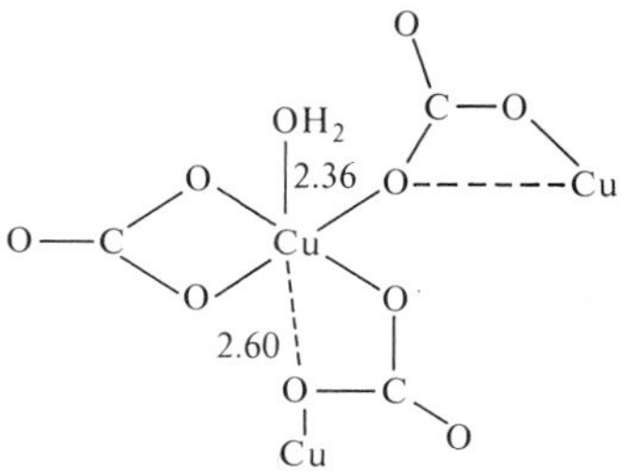

(b)

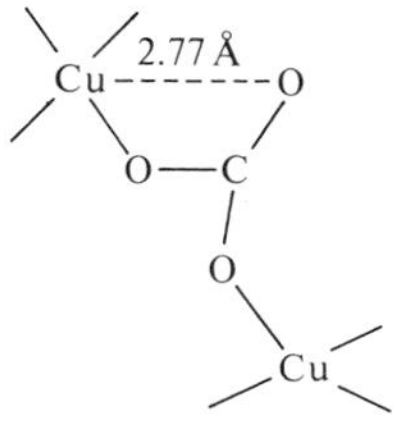

(c)

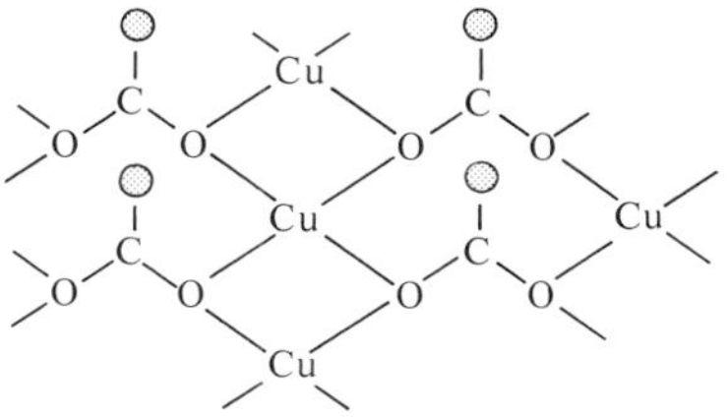

(d)

The structures of one of the two crystalline forms of anhydrous $Cu(NO_3)_2$ and those of a number of complexes formed with organic molecules are of interest not only in connection with the stereochemistry of Cu(II) but also as examples of the behaviour of the NO_3^- ligand. (The structure of the vapour molecule $Cu(NO_3)_2$ is described on p. 823.) The molecule $Cu(en)_2(NO_3)_2$ (p. 1134) is an example of NO_3^- behaving as a monodentate ligand, it being bonded to Cu(II) through two long bonds (2.59 Å) in a (4+2)-coordination group. The basic framework of α-$Cu(NO_3)_2$[2a] is a corrugated layer (4-gon net) in which each NO_3 bridges two Cu atoms as shown diagrammatically in Fig. 25.12(a). One-half of the anions (N*) also form two weak bonds, both from the same O atom. The longer bond (2.68 Å) is to one of the Cu atoms in its own layer, the other (2.43 Å) to Cu of an adjacent layer. Each Cu thus forms 4 stronger coplanar bonds (Cu–O, 1.98 Å) and 2 weaker ones (2.43 and 2.68 Å) (Fig. 25.12(b)). The same type of corrugated layer occurs in $Cu(NO_3)_2 . CH_3NO_2$.[2b] As in α-$Cu(NO_3)_2$ all NO_3^- are bridging ions, and to each Cu is bonded a CH_3NO_2 molecule through an O atom (Cu–O, 2.31 Å), giving (4+1)-coordination (Fig.

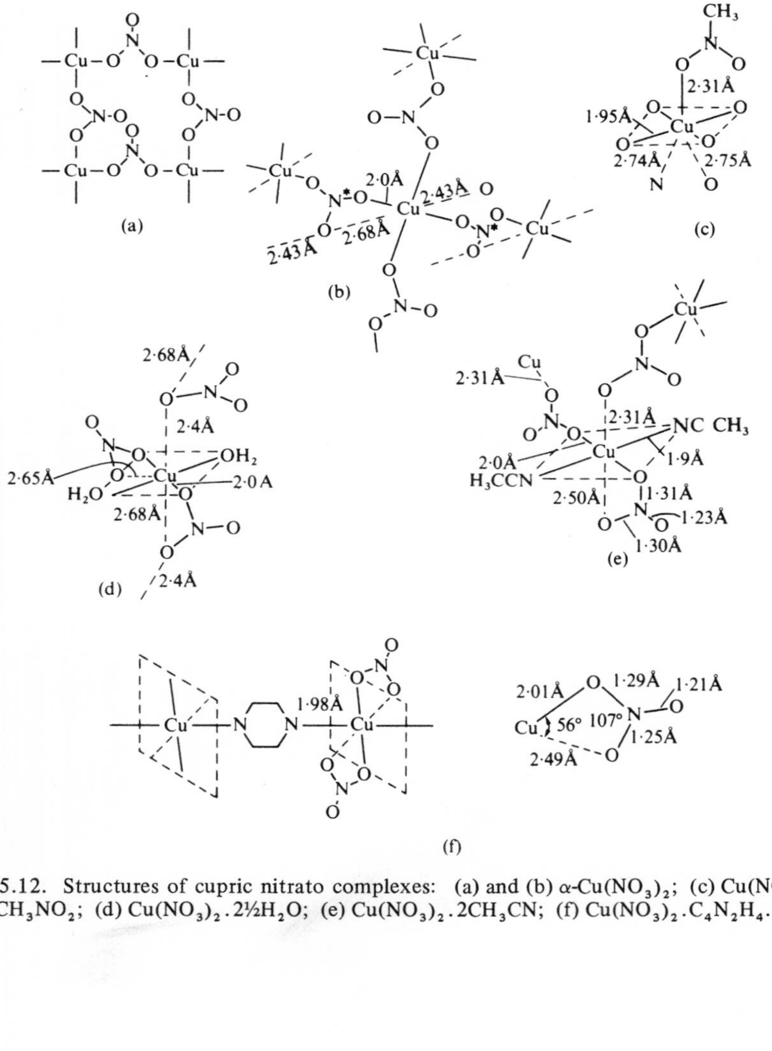

FIG. 25.12. Structures of cupric nitrato complexes: (a) and (b) α-$Cu(NO_3)_2$; (c) $Cu(NO_3)_2 .$ CH_3NO_2; (d) $Cu(NO_3)_2 . 2\frac{1}{2}H_2O$; (e) $Cu(NO_3)_2 . 2CH_3CN$; (f) $Cu(NO_3)_2 . C_4N_2H_4$.

25.12(c)). (There are also some still weaker interactions, Cu–N, 2.74 Å, and Cu–O, 2.75 Å.)

In $Cu(NO_3)_2 \cdot 2\frac{1}{2}H_2O^{(3)}$ the four strong bonds from Cu define groups $Cu(NO_3)_2$-$(H_2O)_2$ (Fig. 25.12(d)) and as in the anhydrous nitrate there are two kinds of non-equivalent NO_3^- ions. Here both types of NO_3^- ion are behaving as 'pseudo-bidentate' ligands, and a second bond from O of one NO_3^- links the square coplanar groups into infinite chains. The environment of Cu(II) is very similar to that in the nitro-methane complex, namely: Cu–4O, 1.97 Å; 1O, 2.39 Å; and 2O, 2.65 and 2.68 Å. One-fifth of the H_2O molecules are not bonded to Cu but are hydrogen bonded to O atoms of NO_3^- and H_2O.

In $Cu(NO_3)_2 \cdot 2CH_3CN^{(4)}$ also there are two types of NO_3^- ion, pseudo-bidentate and bridging. The bridging is here unsymmetrical (Fig. 25.12(e)) and links the tetragonal pyramidal groups into chains. A simpler behaviour of the NO_3^- ligand is found in the chain structure of the complex with pyrazine, $Cu(NO_3)_2 \cdot C_4N_2H_4^{(5)}$ (Fig. 25.12(f)). Here all these groups are behaving as unsymmetrical bidentate ligands. Owing to the dimensions of the NO_3^- ion the two weaker Cu–O bonds are inclined at 34° to the normal to the plane of the four strong bonds.

We noted in Chapter 18 the behaviour of NO_3^- as a monodentate, symmetrical bidentate, and bridging ligand. We summarize in Fig. 25.13 the types of NO_3^- ligand

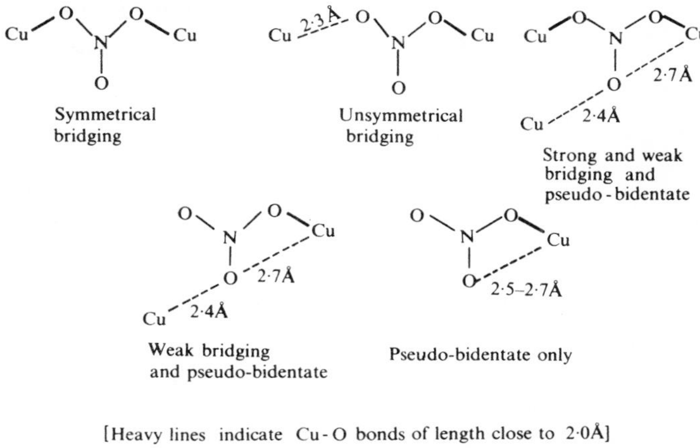

[Heavy lines indicate Cu-O bonds of length close to 2·0Å]

FIG. 25.13. Behaviour of NO_3^- ligand in cupric nitrato complexes.

observed in the compounds of Fig. 25.12. There appears to be a correlation between the dimensions of the ion and its behaviour as a ligand, though it is not certain that the accuracy of location of the light atoms in all these structures justifies detailed discussion at this time.

Cupric salts of carboxylic acids illustrate some of the many types of complex that can be formed by bridging –OC(R)O– groups. In anhydrous cupric formate[6] there is a 3D network of Cu atoms linked by $-O.CH.O-$ groups to give square coordination around Cu(II), but these planar CuO_4 groups are drawn together in pairs so that each Cu forms a fifth bond (2.40 Å) as in Cu(DMG)$_2$. Although the formula [$NH_2(CH_3)_2$] [$Cu(OOCH)_3$][7] is more complex than Cu(OOCH)$_2$ the structure of this compound is based on a simpler 3D framework. The idealized structure is a simple P lattice consisting of 6-connected Cu atoms (4 O, 1.97; 2 O, 2.49 Å) joined through bridging (2-connected) OOCH groups, that is, the simplest type of 3D AX$_3$ structure. A cation occupies each 'cubic' unit cell. In the tetrahydrate of cupric formate[8] there are infinite chains of composition Cu(HCOO)$_2$-(H$_2$O)$_2$, in which Cu has 4 O at 2.00 Å and 2 H$_2$O at 2.36 Å, joined together by hydrogen bonding through the other two water molecules as shown in Fig. 25.14(a). In the monohydrate of the acetate, on the other hand, there are dimers

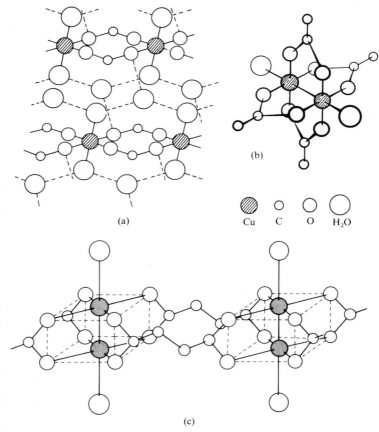

FIG. 25.14.   (a) Crystal structure of Cu(HCOO)$_2$.4H$_2$O;   (b) dimeric units in cupric acetate monohydrate;   (c) chain structure of cupric succinate dihydrate.

$Cu_2(CH_3COO)_4 . 2H_2O$ of the kind shown in Fig. 25.14(b). The distorted octahedral coordination group around Cu is composed of 4 coplanar O atoms (of acetate groups) at 1.97 Å, one H_2O at 2.20 Å, and the other Cu atom of the dimer at 2.65 Å.[9] This last distance is close to Cu–Cu in metallic copper. The corresponding Cr(II)[10] and Rh(II)[11] salts have the same structure, and the same type of dimer is formed in $Cu_2(HCOO)_4[OC(NH_2)_2]_2$[12] and in both crystalline forms of $Cu_2(CH_3COO)_4$-$(C_5H_5N)_2$,[13, 14] urea or pyridine molecules replacing water in Fig. 25.14(b). As in the case of the Cu–H_2O bond, the Cu–N bond (2.19 Å) is appreciably longer than a normal single bond. In these molecules Cu is displaced slightly (about 0.2 Å) out of the plane of the 4 O atoms and away from the other Cu atom.

In the dihydrate of cupric succinate[15] one-half of the water is present as isolated molecules of water of crystallization and the remainder is present in dimeric complexes like those of $Cu_2(ac)_4(H_2O)_2$. The succinate ions form not only the bridges in the dimers but also link these units into infinite chains (Fig. 25.14(c)).

Somewhat unexpectedly, the same dimeric units are found in *anhydrous* salts of long-chain acids, for example, the decanoate, $Cu[CH_3(CH_2)_8COO]_2$.[16] The molecules are packed end-to-end and the COO groups of four molecules bridge two Cu atoms as in $Cu_2(ac)_4(H_2O)_2$. The sixth coordination position, an H_2O molecule in Fig. 25.14(b), is here occupied by an O atom of a neighbouring dimer, so that Cu is bonded to 5 O and 1 Cu.

Finally we mention two examples of chain structures which are also adopted by compounds other than oxy-salts. In $Na_2Cu(C_2O_4)_2 . 2H_2O$[17] planar $Cu(C_2O_4)_2$ groups stack into columns of the type of Fig. 25.10(b) in which Cu is bonded to 4 O at 1.93 Å and 2 O at 2.79 Å. In the trihydrate of cupric benzoate[18] there are chains CuX_2B of the type illustrated in Fig. 25.16(c), p. 1138, in which X is H_2O and B is the benzoate group, $-OC(C_6H_5)O-$; the remaining H_2O and $C_6H_5COO^-$ ions are accommodated between the chains.

It is interesting to note that (4+2)-coordination of Cu(II) is found in α-$Cu(NO_3)_2$, $CuSO_4$[19] (Cu–4O, 1.9–2.0; 2O, 2.4 Å), and $CuSO_4 . 5H_2O$ (p. 678), but (4+1)-coordination in $Cu(NH_3)_4SO_4 . H_2O$[20] (Cu–4N, 2.05 Å; Cu–OH_2, 2.59 Å), $CuCO_3$, and azurite, $Cu_3(OH)_2(CO_3)_2$.[21]

(1a)	AC 1975 **B31** 1313	(12)	IC 1970 9 1626
(1b)	JCS D 1972 1913	(13)	JCS 1961 5 244
(1c)	AC 1972 **B28** 1607	(14)	AC 1964 17 633
(1d)	ZaC 1974 **410** 138	(15)	AC 1966 **20** 824
(2a)	JCS 1965 2925	(16)	AC 1974 **B30** 2912
(2b)	AC 1966 **20** 210	(17)	AC 1980 **B36** 2145
(3)	AC 1970 **B26** 1203	(18)	IC 1965 4 626
(4)	AC 1968 **B24** 396	(19)	AC 1961 14 321
(5)	AC 1970 **B26** 979	(20)	AC 1955 8 137
(6)	JCS 1961 3289	(21)	AC 1958 11 866
(7)	AC 1973 **B29** 1752		
(8)	AC 1954 7 482		
(9)	AC 1953 6 227		
(10)	AC 1953 6 501		
(11)	DAN 1962 **146** 1102		

The formation of trigonal bipyramidal bonds by Cu(II)

The simplest example is the anion in $[Cr(NH_3)_6](CuCl_5)$.[1] The fact that the equatorial bonds (2.39 Å) are longer than the axial ones (2.30 Å) indicates that a simple electrostatic treatment of ligand electron-pair repulsions is inadequate and that account must also be taken of the interaction of the asymmetric d electrons with the ligand electron pairs. Another simple example of this arrangement of five monodentate ligands is the anion in $Ag[Cu(NH_3)_2(NCS)_3]$[2] (Fig. 25.15(a)). Ions or molecules containing polydentate ligands include iodo-bis-(2,2'-dipyridyl)Cu(II) iodide,[3] (b), $[Cu(tren)NCS]SCN$,[4] (c), and $[Cu(Me_6tren)Br]Br$.[5] In the latter case this stereochemistry is not peculiar to Cu(II) for there are similar compounds of Ni and Co. Note the square pyramidal configuration of $[Cu(trien)SCN]^+$ (p. 1125). In contrast to $[Cr(NH_3)_6]CuCl_5$ there are bridged $Cu_2Cl_8^{4-}$ ions, (d) in $Co(en)_3CuCl_5\cdot H_2O$,[6] and, of course, additional Cl^- ions. This compound should be formulated $[Co(en)_3]_2(Cu_2Cl_8)Cl_2\cdot 2H_2O$.

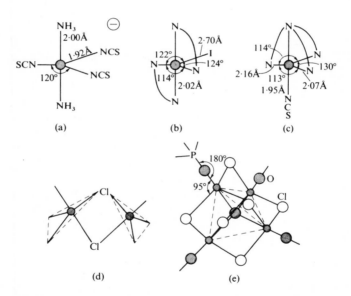

FIG. 25.15. Trigonal bipyramidal coordination of Cu(II) (see text).

There is trigonal bipyramidal arrangement of bonds from Cu(II) in the tetrahedral molecules $Cu_4OCl_6L_4$ (p. 95) in which L is, for example, pyridine, 2-methyl pyridine,[7] or $(C_6H_5)_3PO$ (Fig. 25.15(e)).[8] The central O is bonded tetrahedrally to 4 Cu (Cu–O, 1.90 Å) and each pair of Cu atoms is bridged by Cl (Cu–Cl, 2.38 Å). The $Cu_4OCl_6(\phi_3PO)_4$ molecule is notable as one of the comparatively rare examples of O forming two collinear bonds. The Cu–Cu distances in these molecules (around 3.1 Å) are too large for metal–metal interactions, and the magnetic moments (2.2 BM) are larger than the usual range (1.8–1.9 BM) observed for trigonal bipyramidal $Cu(II)$.[9] A similar moment is observed for the ion $(Cu_4OCl_{10})^{4-}$ in its $[N(CH_3)_4]^+$ salt,[10] which has the same type of structure as Fig. 25.15(e) with $L = Cl$, but the moment is lower (1.84 BM) for the cubane-like complex Cu_4L_4[11] formed with the tridentate Schiff base (a) in which also there is trigonal bipyramidal coordination of $Cu(II)$. These magnetic moments are discussed in ref. 10.

$$\begin{array}{c} H_3C \\ \quad\diagdown \\ \quad\; C\!\!-\!\!O\cdots M \\ HC \\ \quad\; C\!\!-\!\!N \\ H_3C \quad\quad C\!\!-\!\!C \\ \quad\quad\quad H_2\;H_2 \end{array}$$

(a)

In crystalline $Cu_2O(SO_4)$,[12] which is formed as brown crystals by heating $CuSO_4 \cdot 5H_2O$ to $650°C$, one half of the metal atoms have octahedral (4+2)-coordination and the remainder trigonal bipyramidal coordination. Both types of coordination are found also in olivenite, $Cu_2(AsO_4)OH$[13] (Cu–4O, 1.99, and 2O, 2.37, or 2O, 1.95, and 3O, 2.07 Å). There is also trigonal bipyramidal coordination of $Cu(II)$ in Cu_3WO_6[14] and very distorted octahedral coordination of W (3O, 1.79, 3O, 2.09 Å).

(1)	IC 1968 7 1111	(8)	IC 1967 6 495
(2)	ACSi 1966 **32** 162	(9)	TFS 1963 **59** 1055
(3)	JCS 1963 5691	(10)	IC 1969 **8** 1982
(4)	JACS 1967 **89** 6131	(11)	CC 1968 1329
(5)	AC 1968 **B24** 595	(12)	AC 1963 **16** 1009
(6)	IC 1971 **10** 1061	(13)	AC 1977 **B33** 2628
(7)	IC 1970 **9** 1619	(14)	ACSc 1969 **23** 221

Octahedral complexes

Examples of the simplest type of finite octahedral ion or molecule CuL_6, (a), include $Cu(OH)_6^{4-}$, $Cu(NH_3)_6^{2+}$, $Cu(NO_2)_6^{4-}$, and $Cu(H_2O)_6^{2+}$. In the distorted K_2PtCl_6 structure of $Ba_2[Cu(OH)_6]$[1] Cu has 4 OH at 1.97 Å and 2 OH at 2.81 Å forming the usual elongated octahedral group, and in the tetragonal forms of $Cu(NH_3)_6X_2$[2a]

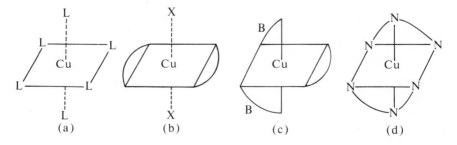

(a)   (b)   (c)   (d)

(X = Cl, Br, or I) the corresponding bond lengths are 2.1 and 2.6 Å. Similar (4+2)-coordination is found in $Cu(en)_2X_2$,[2b] (b), (X = SCN, NO_3, etc.), and in $CuCl_2(pyrazole)_4$[2c] (Cu—4N, 2.02; Cu—2Cl, 2.84 Å). In the cubic forms of $Cu(NH_3)_6Br_2$ and $Cu(NH_3)_6I_2$ there is presumably statistical orientation of statically deformed groups or some kind of dynamic Jahn–Teller effect.

Salts $A_2B[Cu(NO_2)_6]$ in which A is K, Rb, or Tl and B is Pb, Ca, Sr, or Ba present a similar problem. Diffraction studies[3] show that in the cubic high-temperature forms of all these compounds Cu has 6 equally distant N neighbours (at 2.12 Å). In the low-temperature (orthorhombic) forms of the alkaline-earth compounds there

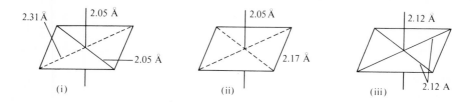

is the normal (4+2)-coordination, (i), but the Pb compounds show 'compressed octahedral' (2+4)-coordination, (ii). It seems probable that the latter configuration is the result of a 2D dynamic Jahn–Teller distortion, interchange of long and short bonds in the equatorial plane, which results in four equatorial bonds with a time-average mean length greater than that of the axial bonds. Interchange between the long and short bonds along all three axes results in 6 bonds of the same length, (iii). The difference between the behaviours of the Pb and alkaline-earth compounds is attributed to the greater polarizability of Pb^{2+}, which tolerates more distortion of its (oxygen) coordination group than the more rigid alkaline-earth ions, so that a transition to the statically deformed (4+2)-coordination does not take place. A brief review[4] emphasizes the importance of spectroscopic (polarized electronic and e.s.r.) and magnetic data in studies of the Jahn–Teller effect.

A rhombic distortion of $Cu(H_2O)_6^{2+}$ groups has been found in two salts. The structure of the cupric Tutton's salt, $Cu(NH_4)_2(SO_4)_2.6H_2O$,[5] is generally similar to those of the Mg and Zn compounds (in which there is little, if any, distortion of the $M(H_2O)_6$ octahedra), but there is definitely a non-tetragonal distortion of $Cu(H_2O)_6^{2+}$, and a similar effect is found in $Cu(ClO_4)_2.6H_2O$:[6]

	2 O at	2 O at	2 O at
$Cu(NH_4)_2(SO_4)_2.6H_2O$	1.96	2.10	2.22 Å
$Cu(ClO_4)_2.6H_2O$	2.09	2.16	2.28 Å

Complexes of type (c) are found at positions with site symmetry 32 (D_3) in $Cu(en)_3SO_4$[7] (Cu—N, 2.15 Å) and in $CuB_3(ClO_4)_2$[8] where B represents the ligand (e). The neutral molecule $CuB_2(ClO_4)_2$[9] has the expected structure of type (a) with long bonds (2.55 Å) to O atoms of the ClO_4 ligands, but in $CuB_3(ClO_4)_2$, which

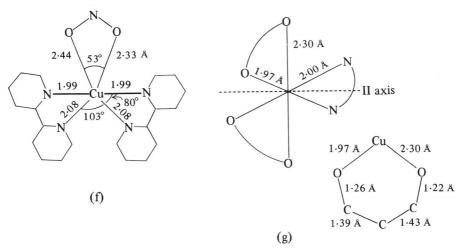

(e)

is isostructural with the Co and Mg compounds, Cu has six equidistant O neighbours (Cu–O, 2.07 Å). In an unsymmetrical complex such as [Cu(bipyr)$_2$NO$_2$]NO$_3$[10] there is necessarily unsymmetrical octahedral coordination because of the different dimensions of the ligands, (f); but note the different Cu–N bond lengths and the two long bonds to O atoms of NO$_2$. There would appear to be a definite Jahn–Teller effect in Cu(bipyridine)(F$_3$C.CO.CH.CO.CF$_3$)$_2$,[11] in which there are unequal Cu–O bond lengths and moreover distortion of the normally symmetrical ligand, (g).

In Cu(diethylenetriamine)$_2$(NO$_3$)$_2$ of type (d), there is tetragonal distortion, with two shorter *trans* Cu–N bonds (2.01 Å) and four longer bonds (mean 2.22 Å).[12]

(1)   AC 1973 **B29** 1929
(2a) IC 1967 **6** 126
(2b) AC 1964 **17** 254, 1145
(2c) AC 1975 **B31** 2479
(3)   AC 1976 **B32** 326, 668, 1278, 2524
(4)   JSSC 1979 **27** 71
(5)   AC 1966 **20** 659
(6)   ZK 1961 **115** 97
(7)   IC 1970 **9** 1858
(8)   JACS 1968 **90** 5623
(9)   IC 1970 **9** 162
(10) JCS A 1969 1248
(11) JACS 1969 **91** 1859
(12) JCS A 1969 883

Cupric halides – simple and complex

Some details of the structures of these compounds are given in Table 25.3. The structure types (second column) are the 'ideal' structures with approximately regular

TABLE 25.3

Crystal structures of cupric halogen compounds

Compound	Octahedral structure type	Coordination group of Cu (or Cr)		Reference
		4X (Å)	2X (Å)	
CuF_2 ⎫	Rutile	4 F 1.93	2 F 2.27	JACS 1957 **79** 1049
CrF_2 ⎭		4 F 2.00	2 F 2.43	PCS 1957 232
Na_2CuF_4	Na_2MnCl_4	4 F 1.91	2 F 2.37	ZaC 1965 **336** 200
$CuF_2.2H_2O$		2 F 1.90	2 F 2.47	JCP 1962 **36** 50 (n.d.)
		2 O 1.94		
$KCuF_3$	Perovskite	2 F 1.89	2 F 2.25	AC 1979 **B35** 1303;
		2 F 1.96		AC 1980 **B36** 1264
$NaCuF_3$		2 F 1.88	2 F 2.26 ⎫	ZaC 1974 **409** 11;
		2 F 1.97	⎬	SSC 1979 **31** 393
K_2CuF_4	K_2NiF_4	4 F 1.92	2 F 2.22 ⎭	
Ba_2CuF_6		2 F 1.85	2 F 2.32	ZN 1977 **32b** 476
		2 F 1.93		
$CsCuCl_3$ (l.t.)	$CsNiCl_3$	2 Cl 2.28	2 Cl 2.78	IC 1966 **5** 277
		2 Cl 2.35		
(h.t.)	$CsNiCl_3$	3 Cl 2.39		AC 1974 **B30** 1053
		3 Cl 2.51		
$[(CH_3)_2NH_2]CuCl_3$		2.25–2.35	2.73 (one)	JCP 1966 **44** 39
$(NH_4)_2CuCl_4$	K_2NiF_4	2.31	2.79	JCP 1964 **41** 2243
$LiCuCl_3.2H_2O$		2.33	O 2.60,	AC 1963 **16** 1037;
			Cl 2.90	JCP 1963 **39** 2923 (n.d.)
$CuCl_2(en)_2.H_2O$		4 N 2.00	O 2.62,	JCS A 1967 1435
			Cl 2.81	
$CuCl_2.2H_2O$	$CoCl_2.2H_2O$	2 Cl 2.29	2.94	ACSc 1970 **24** 3510
		2 O 1.96		
$K_2CuCl_4.2H_2O$		2 Cl 2.29	2.90	AC 1970 **B26** 827 (n.d.)
		2 O 1.97		
Cu_2OCl_2		2 Cl 2.29	2 Cl 3.13	ZN 1977 **B32** 380
		2 O 1.94		
$CuCl_2$	CdI_2	2.30	2.95	JCS 1947 1670
$(C_2H_5NH_3)_2CuCl_4$		2.28	2.98	ICA 1970 **4** 367
$KCuCl_3$	NH_4CdCl_3	2.29	3.03	JCP 1963 **38** 2429
$CuCl_2(pyr)_2$	$\alpha\text{-}CoCl_2(pyr)_2$	2 N 2.02	3.05	AC 1957 **10** 307
		2 Cl 2.28		
$CuBr_2$	CdI_2	2.40	3.18	JACS 1947 **69** 886
Cs_2CuCl_4	⎫			JPC 1961 **65** 50
$[(C_2H_5)_3NH]_2CuCl_4$	⎮			AC 1973 **B29** 241
$[(CH_3)_4N]_2CuCl_4$	⎮	2.19–2.25		AC 1975 **B31** 289
$[C_6H_5(CH_2)_2NH_2CH_3]_2CuCl_4$	⎬	See text		IC 1974 **13** 2106
$(DMBP)CuCl_4$	⎮			AC 1969 **B25** 1691
$(\phi_4As)_2Cu_2Cl_6$	⎭	2.21, 2.32		AC 1974 **B30** 207
Cs_2CuBr_4		2.39		AC 1960 **13** 807
$Cs_3Cu_2Cl_7.2H_2O$		See text		AC 1971 **B27** 1538
$Cu_3Cl_6(H_2O)_2.2C_4H_8SO_2$				IC 1974 **8** 143
$Cu_3Cl_6(C_6H_7NO)_2.2H_2O$				IC 1968 **7** 2035
$Cu_3Cl_6(CH_3CN)_2$	⎫			JCP 1964 **40** 838
$Cu_5Cl_{10}(C_3H_7OH)_2$	⎬	See text		
$Cu_4Cl_{10}[(CH_3)_3NH]_2$	⎭			AC 1976 **B32** 2516

octahedral coordination; they have all been described in earlier chapters, and earlier in this chapter we have referred to the cupric halides. The following points amplify the Table.

(i) The structure of K_2CuF_4 is not the simple K_2NiF_4 structure but has a doubled c axis. In each layer there is alternation of long and short bonds as shown at (h), but in successive layers the positions of the two kinds of bonds are interchanged.

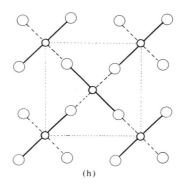

(h)

The description of the coordination in these compounds as a tetragonal $(4+2)$-coordination is apparently an approximation, for a significant orthorhombic distortion has been found in $KCuF_3$, $NaCuF_3$, Ba_2CuF_6, and $CsCuCl_3$ (l.t.).

(ii) Different kinds of distortion of the $CsNiCl_3$ structure are found in the low- and high-temperature forms of $CsCuCl_3$. In the former there is approximately $(4+2)$-coordination, but in the latter $(3+3)$-coordination.

(iii) The structure of the anion in salts $M_2(CuCl_4)$ is very sensitive to the nature of the cation. If the cation is small there is octahedral $(4+2)$-coordination, as in the $(NH_4)^+$, $(CH_3NH_3)^+$, and $(C_2H_5NH_3)^+$ salts, but when the cation is large, as in the Cs^+, $(CH_3)_4N^+$, $(C_2H_5)_3NH^+$, and N,N'-dimethyl-4,4'-bipyridinium (DMBP) salts, a flattened tetrahedral ion is found. The bond angles are close to $130°$ (two) and $100°$ (four); quite similar bond angles are found in the distorted spinel structure of $CuCr_2O_4$. In these ions the tetrahedron is not only flattened but also twisted. The dihedral angle Cl_1CuCl_2/Cl_3CuCl_4 for a regular tetrahedron is $90°$, but it ranges from approximately $74°$ in Cs_2CuCl_4 to $64°$ in $[(C_2H_5)_3NH]_2CuCl_4$. The same type of distortion is found in the bridged anion in $(\phi_4As)CuCl_3$. The bond arrangement is intermediate between square planar and regular tetrahedral (large angles $145°$) and the dihedral angle Cl_tCuCl_t/Cl_bCuCl_b is $66°$; see also p. 1139.

In combination with a very non-spherical cation such as $(C_6H_5CH_2CH_2NH_2CH_3)^+$ a more complex behaviour is observed. In the yellow h.t. form there are irregular tetrahedral $CuCl_4^{2-}$ ions (large angles $123°$ and $138°$, Cu--Cl, 2.23 and 2.19 Å), but in the green l.t. form there are planar centrosymmetrical $CuCl_4^{2-}$ ions. In neither form does Cu(II) have other neighbours closer than 3.5 Å. The l.t. form is apparently the only example of a square planar $CuCl_4^{2-}$ ion, Cu having no additional neighbours in the usual range (2.8–3.05 Å). See also $[Pt(NH_3)_4][CuCl_4]$, p. 1238.

The description of the anions in the chloro compounds depends on the interpretation placed on the two Cu–Cl bond lengths. The description in terms of distorted octahedral coordination groups brings together a number of structures based on octahedral chains of the types described in Chapter 5. The simple edge-sharing chain has the composition $(AX_4)_n^{2n-}$ if all the larger circles in Fig. 25.16(a) represent

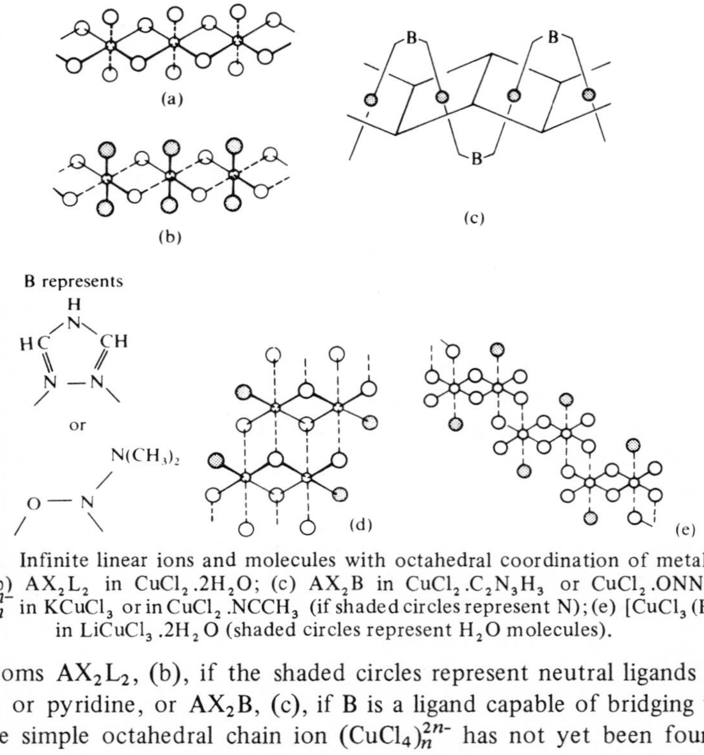

FIG. 25.16. Infinite linear ions and molecules with octahedral coordination of metal atoms: (a) AX_4; (b) AX_2L_2 in $CuCl_2.2H_2O$; (c) AX_2B in $CuCl_2.C_2N_3H_3$ or $CuCl_2.ONN(CH_3)_2$; (d) $(CuCl_3)_n^{n-}$ in $KCuCl_3$ or in $CuCl_2.NCCH_3$ (if shaded circles represent N); (e) $[CuCl_3(H_2O)]_n^{n-}$ in $LiCuCl_3.2H_2O$ (shaded circles represent H_2O molecules).

halogen atoms AX_2L_2, (b), if the shaded circles represent neutral ligands such as H_2O, NH_3 or pyridine, or AX_2B, (c), if B is a ligand capable of bridging two Cu atoms. The simple octahedral chain ion $(CuCl_4)_n^{2n-}$ has not yet been found in a complex cupric chloride, but the $CuCl_2.L_2$ chain is the structural unit in $CuCl_2.2H_2O$ and other compounds listed in Table 25.4. Examples of the $CuCl_2B$ chain of Fig. 25.16(c) include the complexes with 1:2:4-triazole and $(CH_3)_2NNO$. In the latter compound the bond lengths found were: Cu–4Cl, 2.29 Å, Cu–O, 2.29 Å, and Cu–N, 3.00 Å, suggesting (4+1)- rather than (4+2)-coordination.

The double chain ion $(CuCl_3)_n^{n-}$ of Fig. 25.16(d) is the anion in $KCuCl_3$, with which $KCuBr_3$ is isostructural; it also represents the linear molecule $CuCl_2.NC(CH_3)$. In the AX_4 chain of Fig. 25.16(a) the AX_6 octahedra share two opposite edges. In the chain of Fig. 25.16(e), necessarily of the same composition, a different pair of edges is shared, giving a staggered arrangement of pairs of octahedra. This is the

TABLE 25.4

A family of octahedral chain structures

Type of complex	Regular octahedral coordination	Distorted octahedral coordination	Reference to Cu compound
Single chain Fig. 25.16			
(a) AX_4^{2-}	Na_2MnCl_4	Na_2CuF_4	ZaC 1965 **336** 200
(b) AX_2L_2	$CoCl_2 . 2H_2O$	$CuCl_2 . 2H_2O$ $CuF_2 . 2H_2O$	See Table 25.3
	$CdCl_2(NH_3)_2$	α-$CuBr_2(NH_3)_2$	AC 1959 **12** 739
	α-$CoCl_2(pyr)_2$	$CuCl_2(pyr)_2$	AC 1975 **B31** 632
(c) AX_2B		$CuCl_2$ (triazole)	AC 1962 **15** 964
		$CuCl_2(Me_2NNO)$	AC 1969 **B25** 2460
Double chain			
(d) AX_3^-	NH_4CdCl_3	$KCuCl_3$	See Table 25.3
AX_2L		$CuCl_2(NCCH_3)$	JCP 1964 **40** 838
(e) AX_4^-	$TcCl_4$		
AX_3L^-		$Li[CuCl_3(H_2O)] . H_2O$	See Table 25.3

anion in $LiCuCl_3 . 2H_2O$, the shaded circles representing H_2O molecules. The composition of the chain is $CuCl_3(H_2O)$; the second H_2O and the Li^+ ions are situated between the chains. These structures are summarized in Table 25.4.

By describing the structures of these compounds in this way we have emphasized the octahedral (4+2)-coordination of Cu(II). Alternatively we may describe the systems formed by the four stronger Cu—X bonds. These are layers in CuF_2, chains in $CsCuCl_3$ and $CuCl_2$, ions $Cu_2Cl_6^{2-}$ in $KCuCl_3$ and $LiCuCl_3 . 2H_2O$, and $CuCl_2(H_2O)_2$ molecules in $CuCl_2 . 2H_2O$ and $K_2CuCl_4 . 2H_2O$. The structure of this last compound may therefore be described as an aggregate of $CuCl_2(H_2O)_2$ molecules, K^+, and Cl^- ions; its does not contain $CuCl_4^{2-}$ ions. The infinite helical chains $(CuCl_3)_n^{n-}$ in $CsCuCl_3$ are of the type (a) while those in $CuCl_2$ are formed from planar $CuCl_4$ groups sharing opposite edges, (b).

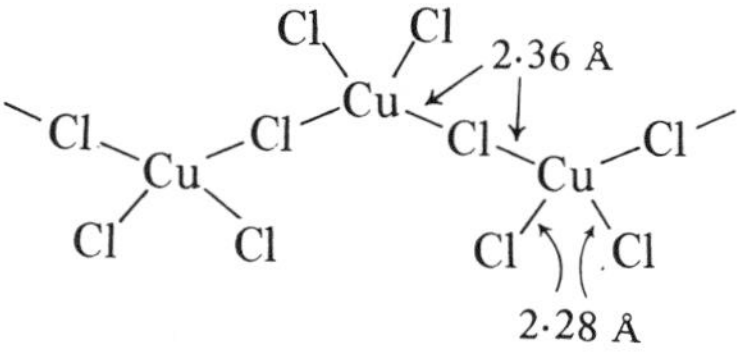

(a) 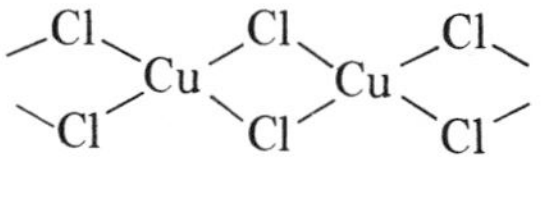   (b)

As in the case of $CuCl_4^{2-}$ the structure of $Cu_2Cl_6^{2-}$ is very sensitive to its environment. In $[(C_6H_5)_4As]_2Cu_2Cl_6$ there is no interaction between the anions owing to the presence of the bulky cations, the closest contacts being Cl—Cl of 6.76 Å. The following figures suggest that the extent of distortion of the Cu coordination from planar to flattened tetrahedral *increases* as the extent of interaction with its environment *decreases*; compare with Table 25.2 (p. 1119):

	Additional Cl *neighbours*	*Trans* Cl–Cu–Cl *angle*
$LiCuCl_3 . 2H_2O$	2.92 Å	180°
$[(CH_3)_2NH_2]CuCl_3$	2.73	166°
$[(C_6H_5)_4As]CuCl_3$	None	145°

Two structures are of some interest in this connection. In $[(CH_3)_2NH_2]CuCl_3$ planar $Cu_2Cl_6^{2-}$ ions are linked into chains by a fifth 'bond' (2.73 Å) from each Cu atom. The sixth octahedral position is occupied by a methyl group (Cu–CH_3, 3.78 Å), so that Cu(II) is there exhibiting 5-coordination (tetragonal pyramidal) (c). In a 2-picoline-N oxide complex, $Cu_3Cl_6(C_6H_7NO)_2 . 2H_2O$ dimers $Cu_2Cl_4L_2$ are bridged by $CuCl_2(H_2O)_2$ molecules through Cu–Cl bonds of two lengths, 2.65 Å and 2.96 Å, to form chains in which there is both 5- and 6-coordination of Cu(II) atoms (d).

(c)   (d)

The $Cu_2Cl_6^{2-}$ ion is simply a portion of the $CuCl_2$ chain, and there are some compounds of $CuCl_2$ with organic molecules which have structures very closely related to these two systems; they are prepared by dissolving anhydrous $CuCl_2$ in hot solvents such as acetonitrile or n-propanol and crystallizing the resultant solutions. The yellow-red $CuCl_2 . CH_3CN$ consists of dimeric molecules (e) which are packed in the same way as the $Cu_2Cl_6^{2-}$ ions in $KCuCl_3$. The longer Cu–Cl bonds link each molecule to *two* others to form chains (Table 25.4), the CH_3CN molecules

(e)   (f)

projecting from both sides of the double-octahedral chain as indicated by the shaded circles (N atoms) in Fig. 25.16(d). The packing of the longer molecules $Cu_3Cl_6(CH_3CN)_2$ (f), and $Cu_5Cl_{10}(C_3H_7OH)_2$ is different, the longer Cu–Cl bonds linking atoms of one molecule to atoms of *four* adjacent ones, so that layer structures are formed. The structures of the layers are different in these two compounds; the layer in the former is shown in Fig. 25.17.

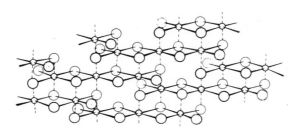

FIG.25.17 Portion of layer of $Cu_3Cl_6(CH_3CN)_2$ showing how each molecule is bonded to four others by long Cu–Cl bonds (dotted). Only N atoms (shaded) of CH_3CN molecules are shown.

The salt $[(CH_3)_3NH]_2Cu_4Cl_{10}$ contains planar ions of type (f), with all ligands Cl, which are stacked so that the two inner Cu atoms have $(4+1+1)$-coordination (4 Cl, 2.29; 1 Cl, 2.75; and 1 Cl, 3.23 Å) while the outer ones have only 1 Cl neighbour (at 2.94 Å) in addition to $2 Cl_t$, 2.24, and $2 Cl_b$, 2.29 Å.

We note the bridged $Cu_2Cl_8^{4-}$ ion (p. 1132) as an example of Cu(II) forming trigonal bipyramidal bonds. There are apparently bridged ions $[Cu_2Cl_5(H_2O)_2]^-$ in the salt $Cs_3Cu_2Cl_7.2H_2O$. The $(4+2)$-coordination is unusually unsymmetrical (g),

$$
\begin{array}{ccccc}
Cl\diagdown & & \overset{3.03\ \text{Å}}{\cdots}Cl & & \overset{2.74\ \text{Å}}{\cdots}Cl^{-} \\
H_2O\!\!-\!\!\!-\!\!\!-\!\!\!-Cu\!\!\diagdown & \!\!\!-\!\!\!-Cl\!\!-\!\!\!-\!\!\!-Cu\!\!\diagdown & \!\!\!-\!\!\!-OH_2 \\
Cl^{-}\cdots\overset{2.71\ \text{Å}}{} & & Cl\cdots\overset{3.16\ \text{Å}}{} & & Cl
\end{array}
$$

(g)

but the range of Cu–Cl bond lengths (2.09–2.41 Å) suggests that further study of this structure may be desirable.

Cupric hydroxy-salts

Copper(II) forms an extraordinary variety of hydroxy-oxysalts,[1] many of which occur as (secondary) minerals. Examples include $Cu(OH)IO_3$, $Cu_2(OH)_3NO_3$, $Cu_2(OH)_2CO_3$ (malachite),[2] $Cu_2(OH)PO_4$ (libethenite), $Cu_3(OH)_3PO_4$, $Cu_5(OH)_4$-$(PO_4)_2$, and $Cu_4(OH)_6SO_4$ (brochantite). The structures of $Cu(OH)IO_3$ and $Cu_2(OH)_3NO_3$ have been described in Chapter 14, where the general features of the structures of some hydroxy-salts are noted, $CuB(OH)_4Cl$ (bandylite) in Chapter 24, and the (layer) structure of $CuHg(OH)_2(NO_3)_2(H_2O)_2$[3] on p. 120. In most of these compounds Cu(II) has 4 coplanar O and 2 more distant ones completing a distorted octahedral coordination group, but in some cases the sixth O is so far away from Cu that the coordination is preferably described as tetragonal pyramidal (for example, in $Cu_3(OH)_3AsO_4$, clinoclase).[4]

The structures of two forms of $Cu_2(OH)_3Cl$, atacamite and botallackite, and of $Cu_2(OH)_3Br$ have been described under 'Hydroxyhalides' (p. 489), where references are given. Atacamite has a framework structure which, in its idealized form, is derivable from the NaCl structure, as shown in Fig. 4.22 (p. 170). Botallackite and $Cu_2(OH)_3Br$ have layer structures of the CdI_2 type. In all these structures there is the characteristic distorted (4+2)-coordination.

$$Cu_2(OH)_3Cl \qquad\qquad Cu_2(OH)_3Br$$

$$Cu \begin{cases} \dfrac{4\,OH\ 2{\cdot}02\,\text{Å}}{2\,Cl\ 2{\cdot}76} \end{cases} \text{ or } \dfrac{4\,OH\ 2{\cdot}00\,\text{Å}}{\begin{array}{l}OH\ 2{\cdot}36 \\ Cl\ 2{\cdot}76\end{array}} \qquad Cu \begin{cases} \dfrac{4\,OH\ 2{\cdot}0\,\text{Å}}{2\,Br\ 3{\cdot}0} \end{cases} \text{ or } \dfrac{4\,OH\ 2{\cdot}0\,\text{Å}}{\begin{array}{l}OH\ 2{\cdot}3 \\ Br\ 2{\cdot}8\end{array}}$$

The metal atoms do not all have the same environment in these hydroxyhalides, one half being surrounded by $4\,OH + 2\,X$ and the remainder by $5\,OH + X$. These coordination groups in a compound M_2A_3B are a consequence of the relative numbers of A and B atoms and of the coordination numbers of M (6), A (3), and B (3). It is not possible for all the M atoms in such a structure to have the same environment, and the fact that the M atoms are (crystallographically) of two kinds has no chemical significance.

In Fig. 25.18 are shown elevations and plans of portions of the structures of $CuCl_2$, $Cu_2(OH)_3Br$, and $Cu_2(OH)_3Cl$. Although atacamite does not have a layer structure, it can be dissected into approximately close-packed layers of the same type as occur in the hydroxybromide, as described on p. 490 and illustrated in Fig. 25.18(e) and (f).

(1)　For earlier references see:　FM 1961 **39** 59;　AC 1963 **16** 124　　(3)　AC 1969 **B25** 800
(2)　AC 1967 **22**, 146, 359　　　　　　　　　　　　　　　　　　　　(4)　AC 1965 **18** 777

The sulphides of copper

The Cu–S system is remarkably complex, and the structures of CuS and Cu_2S are not consistent with their formulation as cupric and cuprous sulphides. The phases stable under atmospheric pressure are set out in Table 25.5. For $CuS_{1.9}$ see p. 759.

Covellite, which can be ground to a blue powder, is a moderate conductor of electricity and has an extraordinary and unique structure. One-third of the metal atoms have three S neighbours (at 2.19 Å) at the corners of a triangle, and the remainder have four S neighbours arranged tetrahedrally (at 2.32 Å). Moreover,

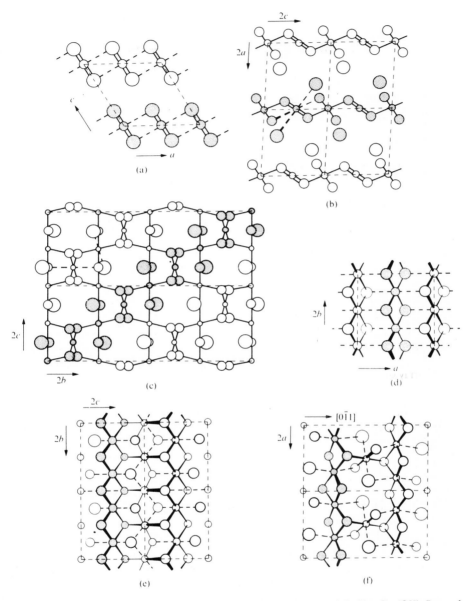

FIG. 25.18.   (a), (b), (c) Elevations of the crystal structures of CuCl$_2$, Cu$_2$(OH)$_3$Br, and Cu$_2$(OH)$_3$Cl (atacamite).   (d), (e), (f) Projections of portions of these structures (the atoms shaded in (a)–(c)) showing a layer of metal atoms (smallest circles) in the plane of the paper, and close-packed layers of halogen atoms (or halogen atoms and OH groups) above and below the plane of the paper. The largest circles represent halogen atoms. Full lines indicate short Cu–OH or Cu–X bonds, and the heavier broken lines long Cu–OH or Cu–X bonds. (See text).

TABLE 25.5
The sulphides of copper

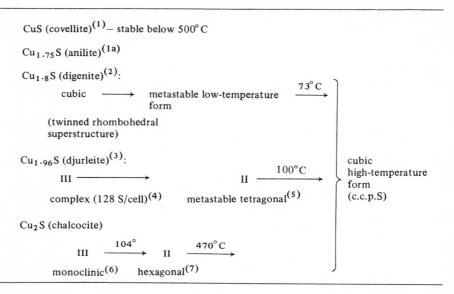

CuS (covellite)[1] – stable below 500°C

Cu$_{1.75}$S (anilite)[1a]

Cu$_{1.8}$S (digenite)[2]:

cubic $\longrightarrow$ metastable low-temperature form $\xrightarrow{73°C}$

(twinned rhombohedral superstructure)

Cu$_{1.96}$S (djurleite)[3]:

III $\longrightarrow$ complex (128 S/cell)[4]   II $\xrightarrow{100°C}$ metastable tetragonal[5]

cubic high-temperature form (c.c.p.S)

Cu$_2$S (chalcocite)

III $\xrightarrow{104°}$ II $\xrightarrow{470°C}$

monoclinic[6]   hexagonal[7]

two-thirds of the S atoms are present as S_2 groups like those in pyrites, so that if we regard this as a normal covalent structure it would be represented $Cu_4^I Cu_2^{II}(S_2)_2 S_2$.

In anilite there is an ordered arrangement of Cu atoms in the interstices of a c.c.p. array of S atoms, and the other three distinct species with formulae near to or equal to Cu_2S all have a high-temperature form based on c.c.p. S atoms in which there is statistical arrangement of the Cu atoms, probably in tetrahedral holes, this $Cu_{2-x}S$ phase being stable over the composition range $x = 0.0$–0.2. In high-digenite there is on the average 9/10 of a Cu atom within each S_4 tetrahedron but not centrally situated. In the intermediate metastable form some of the Cu atoms have moved so that now some S_4 tetrahedra are unoccupied and in the occupied tetrahedra Cu is ststistically distributed over 4 equivalent (off-centre) positions.

The extremely complex structure of Cu$_{1.96}$S-III is not known, but in the metastable tetragonal form there is almost perfect cubic closest packing of the S atoms in which Cu occupies trigonal holes (Cu–3S, 2.31 Å); S is surrounded by 6 Cu at the vertices of a trigonal prism. There is very slight displacement of the Cu atom towards a tetrahedral hole, and in addition to its three S neighbours Cu has 2 Cu at 2.64 Å and 2 more at 2.97 Å. In hexagonal chalcocite-II the Cu atoms are disordered in a h.c.p. array of S atoms, being statistically distributed in three types of site. In the unit cell containing 2 S atoms there is the following average occupancy of sites: 1.74 Cu in sites of 3-coordination (Cu–S, 2.28 Å), 1.42 Cu in sites of tetrahedral coordination with Cu non-central (Cu–3S, 2.59 Å; Cu–1S, 2.15 Å), and 0.84 Cu in sites of 2-coordination, on edges of S_4 tetrahedra (Cu–2S, 2.06 Å). In addition, there are separations of only 2.59 Å between the 3- and 4-coordinated Cu atoms.

A very careful study of low-chalcocite, which was originally referred to an ortho-rhombic cell containing 96 Cu_2S, has shown that the true symmetry is certainly not higher than monoclinic. In the proposed structure, which has 48 Cu_2S in the unit cell, all Cu atoms are 3-coordinated, mostly in or near positions of planar coordination in a h.c.p. array of S atoms. Certain features of the structure, notably two very long Cu–S distances (2.88 Å, compared with 2.33 Å, mean), suggest that although this structure must be very near to the true one, there may still be complications due to submicroscopic twinning.

The physical properties of these compounds, which include metallic lustre, intrinsic semiconductivity, and in some case ductility, the close contacts between Cu atoms in some of the phases, and the variety of arrangements of the bonds from the Cu atoms, indicate that these are semi-metallic phases, as also are some of the numerous complex sulphides. On the grounds that the chalcopyrite (zinc-blende superstructure) structure of $CuFeS_2$[8] is adopted by $CuAlS_2$ and $CuGaS_2$ we could formulate this compound as $Cu^IFe^{III}S_2$, with tetrahedral coordination of Cu^I, but in stromeyerite, $AgCuS$,[9] there is trigonal coordination of Cu^I. It does not seem possible to assign oxidation numbers to Cu in the numerous complex sulphides (e.g. $CuFe_2S_3$, cubanite (p. 781), $Cu_3Fe_4S_6$, Cu_5FeS_4, bornite,[10] and $Cu_9Fe_9S_{16}$[11]) in such a way as to give a consistent stereochemistry of Cu(I) and Cu(II).

(1)	AM 1954 **39** 504	(6)	NPS 1971 **232** 69
(1a)	AC 1970 **B26** 915	(7)	Min. Soc. America Special Paper 1 (1963) 164
(2)	AM 1963 **48** 110	(8)	AC 1973 **B29** 579
(3)	ACSc 1958 **12** 1415	(9)	ZK 1955 **106** 299
(4)	ZK 1967 **125** 404	(10)	AC 1975 **B31** 2268
(5)	AC 1964 **17** 311	(11)	AC 1973 **B29** 2365

The structural chemistry of Au(III)

In this oxidation state gold forms four coplanar (dsp^2) bonds, and information about the structural chemistry of Au(III) comes from (a) auric compounds and (b) compounds containing both Au(I) and Au(III) or both Ag(I) and Au(III).

Gold has comparatively little tendency to form oxy or hydroxy compounds in either of its oxidation states; Au_2O, $AuOH$, and $Au(OH)_3$ are unknown, but Au_2O_3 is well characterized and a number of complex oxides have been prepared. The structures of Au_2O_3,[1a] $AuOCl$,[1b] and $Na_6Au_2O_6$[1c] have been determined, and for other compounds there are X-ray powder photographic data.[1d] In Au_2O_3 three of the four coplanar O neighbours of each Au are bonded to 3 Au and the fourth to only 1 Au so that a 3D (3,4)-connected net is formed. A somewhat similar bond arrangement is found in $AuOCl$, except that Cl is bonded to 1 Au only. The O and Au atoms together form a 3D 3-connected net which is a system of 4- and 12-membered rings.[1e] In $Na_6Au_2O_6$ there are $(Au_2O_6)^{6-}$ ions with a planar structure

$$Au_2O_3 \qquad\qquad AuOCl$$

similar to that of the Au_2Cl_6 molecule described below (Au–O_t, 2.12; Au–O_b, 2.17 Å).

The square coplanar structure has been confirmed for the anions in $K(AuF_4)$[2a] (Au–F, 1.95 Å); $K[Au(NO_3)_4]$ (p. 824); $(As\phi_4)(AuCl_4)$[2b] (Au–Cl, 2.27 Å); $K[Au(CN)_4].H_2O$[2c] (Au–C, 1.98 Å); and $HAu(CN)_4.2H_2O$[2d] The last compound is a strong acid, made by passing a solution of $KAu(CN)_4$ through a H⁺-exchange resin, and the two H_2O molecules are present in the crystal as hydrogen-bonded $H_5O_2^+$ ions. Planar ions containing Au^{III} are present also in $Cs_2(Au^ICl_2)$-$(Au^{III}Cl_4)$, p. 466, in $K_5[Au^I(CN)_2]_4[Au^{III}(CN)_2I_2].2H_2O$,[2e] a compound that is jet black at ordinary temperatures but becomes deep red at 77 K, and also in $(NH_4)_6(AuCl_4)_3(Ag_2Cl_5)$.[2f] This last compound is deep-red and contains planar $AuCl_4^-$ and planar bridged $Ag_2Cl_5^{3-}$ ions.

The structures of two auric halides are known. Crystalline AuF_3 has a unique helical (chain) structure formed from planar AuF_4 groups sharing *cis* F atoms (Fig. 25.19).[3a] The lengths of the Au–F bonds in the chains are 1.91 Å and 2.04 Å (bridging), and the F bond angle is 116°. Weaker bonds of length 2.69 Å complete octahedral groups around the Au(III) atoms. The vapour density of the chloride

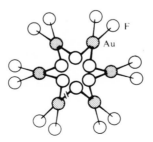

FIG. 25.19. The helical AuF_3 molecule viewed along its length.

between 150 and 260 °C indicates Au_2Cl_6 molecules, and the elevation of the b.p. of bromine by the bromide shows that this compound exists in solution as Au_2Br_6 molecules. An X-ray study[3b] showed that crystalline $AuCl_3$ consists of planar dimers:

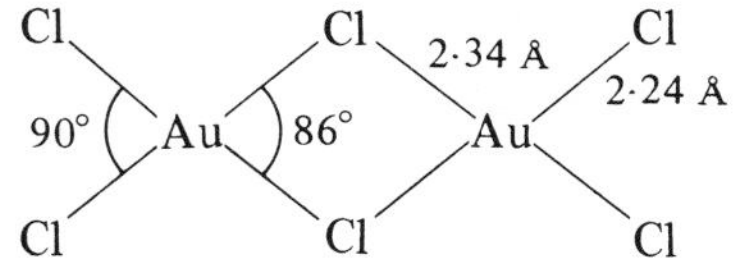

The compound $Au_2Br_2(C_2H_5)_4$[4] consists of bridged dimeric molecules, and the coplanar configuration of four bonds from Au(III) has also been confirmed in $AuBr_3P(CH_3)_3$[5] and in the compounds (a)-(c):

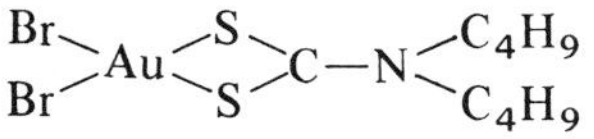

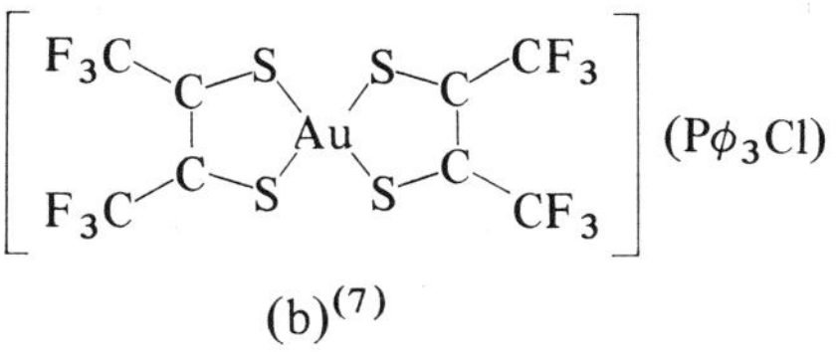

There appears to be weak additional bonding in (c) (as in the Ni and Pt compounds, where M–I distances are 3.21 Å and 3.50 Å). The sum of the 'covalent radii' gives Au–I = 2.73 Å for a single bond.

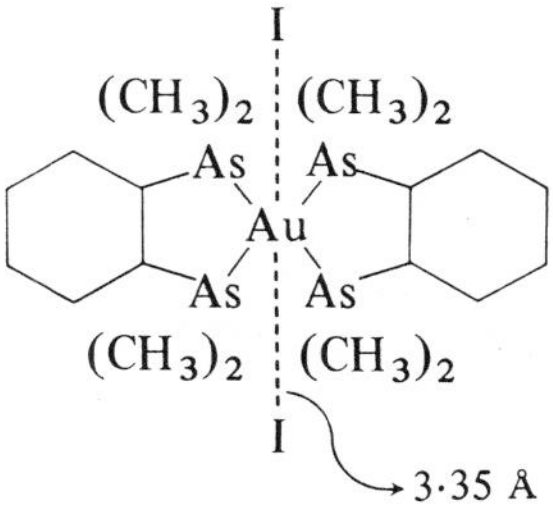

The action of AgCN on $Au_2Br_2(C_2H_5)_4$ gives the cyanide, which has a 4-fold molecular weight. The bridged structure noted above for the bromo compound is possible bcause the angle between two bonds from a halogen atom can be approximately 90°, but a similar structure is not possible for the cyano compound since

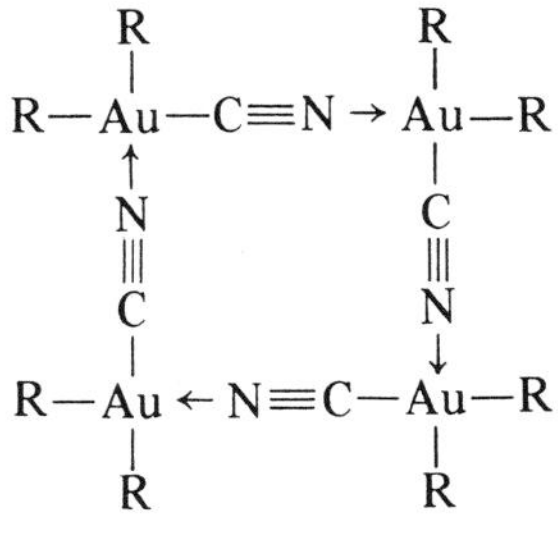

(d)

the atoms linked by the CN group must be collinear with the C and N atoms. Accordingly this compound forms a square molecule of type (d), the structure having been established for the case $R = n\text{-}C_3H_7$.[9]

Dimethyl auric hydroxide, $[(CH_3)_2 AuOH]_4$ is also a cyclic tetramer, (e), but there is apparently a strange asymmetry and the interbond angles at Au(III) range from $81°$ to $106°$; the range of Au–O bond lengths, 1.87–2.40 Å, is inexplicably large.[10]

Examples of compounds containing Au(I) and Au(III) include $Cs_2(AuCl_2)(AuCl_4)$ and $(C_6H_5CH_2)_2S \cdot AuBr_2$, which have already been noted, and the compound (f).[11] All these compounds contain Au(I) forming 2 collinear and Au(III) forming 4

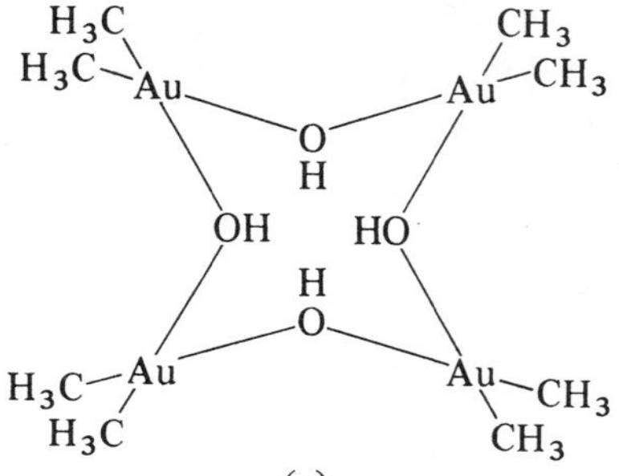

(e)

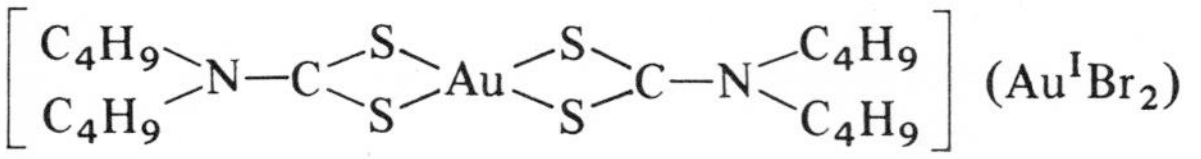

(f)

coplanar bonds. In contrast to the black $Cs_2(AuCl_2)(AuCl_4)$ there is no interaction between Au(I) and Au(III) in the compound (f), which forms yellow needles. In crystalline $[Au^{III}(\text{dimethylglyoxime})_2](Au^ICl_2)$ Au(III) forms two weak bonds in addition to the four coplanar bonds. The crystal consists of planar $[Au^{III}(DMG)_2]^+$ and linear $(Au^ICl_2)^-$ ions so arranged that there are chains of alternate Au(I) and Au(III) atoms in which the metal–metal distance is 3.26 Å.[12]

(1a) AC 1979 **B35** 1435
(1b) AC 1979 **B35** 2380
(1c) NW 1976 **63** 387
(1d) ZaC 1968 **359** 36
(1e) ACA Monograph No. 8 (1979), p. 16
(2a) JCS A 1969 1936
(2b) AC 1975 **B31** 2687
(2c) AC 1970 **B26** 422
(2d) AC 1972 **B28** 1629
(2e) AC 1972 **B28** 2635
(2f) AC 1975 **B31** 2149

(3a) JCS A 1967 478
(3b) AC 1958 **11** 284
(4)   JCS 1937 1690
(5)   JCS 1946 428
(6)   IC 1968 **7** 810
(7)   IC 1968 **7** 2636
(8)   IC 1969 **8** 1661
(9)   PRS 1939 A **173** 147
(10) JACS 1968 **90** 1131
(11) IC 1968 **7** 805
(12) JACS 1954 **76** 3101

26

The elements of subgroups IIB, IIIB, and IVB

Introduction

Before the structural chemistry of these elements is considered in more detail attention should be drawn to one feature common to a number of them. We have seen that the elements Cu, Ag, and Au can make use of d electrons of the penultimate quantum group and in the case of Cu can lose one or two of the 3d electrons to form ions Cu^{2+} and Cu^{3+}. Some elements of the later B subgroups show a quite different behaviour. In addition to forming the normal ion M^{N+} by loss of all the N electrons of the outermost shell (N being the number of the Periodic Group) there is loss of only the p electron(s), the pair of s electrons remaining associated with the core – the so-called 'inert pair'. For a monatomic ion this implies that M must have at least 3 electrons in the valence shell, and we have therefore to examine the evidence for the existence of ions M^+ in Group IIIB and ions M^{2+} in Group IVB. Mercury retains the structure 78(2) in its monatomic vapour and would have the same effective atomic number in $(Hg–Hg)^{2+}$ if the free ion exists; this point is mentioned later. The very small degree of ionization of mercuric halides was regarded by Sidgwick as evidence for the inertness of the pair of 6s electrons of Hg, but the Hg^{2+} ion obviously exists in crystalline HgF_2 (fluorite structure). Evidence for the existence of ions may be adduced from the properties of compounds in solution or in the fused state, or from the nature of their crystal structures.

Ga^+, In^+, and Tl^+

All the monohalides of Ga are known as vapour species, being prepared by methods such as heating $Ga+CaF_2$ in a Knudsen cell or by vaporizing $GaCl_2$, but only GaI has been prepared as a solid (actually $GaI_{1.06}$);[1] its structure is not known. Reduction of the dibromide by metal proceeds only as far as $GaBr_{1.30}$.[2] Both Ga_2O and Ga_2S have been prepared, but their structures also await examination. However, the Ga^+ ion exists in the crystalline 'dichloride', $Ga^+(GaCl_4)^-$,[3] and the fused salt (at 190 °C) also has this ionic structure for its Raman spectrum is very similar to that of the $GaCl_4^-$ ion formed in a solution of $GaCl_3$ containing excess Cl^- ions. The fused dibromide is similar.[4] The Ga^+ ion can be introduced into the β-alumina structure (q.v.) and then replaced by Na^+ in molten NaCl (810 °C) to form the original compound.[5]

All the crystalline monohalides of In exist except InF, which is known only as a vapour species, though these compounds are much less stable than the thallous compounds. The In^+ ion is present in crystalline InBr and InI (see Chapter 9), but it is not stable in water, in which both InCl and $InCl_2$ form In^{3+} + In. Solid $InCl_2$ is diamagnetic and is presumably $In(InCl_4)$; In also forms the chlorides In_4Cl_5, In_4Cl_7, and In_2Cl_3, the latter presumably being $In_3^I(In^{III}Cl_6)$. The Tl^+ ion is certainly stable, for the soluble TlOH is a strong base like KOH, and the halides have ionic crystal structures; note, however, the unsymmetrical structures of TlF and yellow TlI (Chapter 9).

For compounds apparently containing divalent Ga, In, or Tl, there are two simple possibilities: equal numbers of M^+ and M^{3+} (or M^I and M^{III} in a covalent compound) or ions $(M-M)^{4+}$ analogous to Hg_2^{2+}. The former alternative has been confirmed in a number of compounds; there is no evidence for binuclear ions. The diamagnetic GaS[6] is not an ionic crystal but contains Ga bonded to four tetrahedral neighbours (1 Ga + 3 S) to form a layer structure (Fig. 26.1). With the equivalence of all the Ga atoms in GaS compare the presence of equal numbers of Tl(I) and Tl(III) in TlS (p. 1171). The bond lengths in GaS are: Ga–1 Ga, 2.45 Å; Ga–3S, 2.33 Å. One of the four forms of GaSe has the structure of Fig. 26.1, as also does InSe[7] (In–In, 2.82; In–3Se, 2.64 Å).

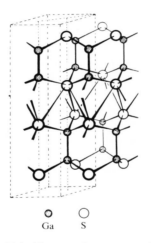

Ga   S

FIG. 26.1. The crystal structure of GaS.

InS[8] may be described as built of puckered 3-connected layers similar to those in GeS and SnS), but the layers are translated relative to their positions in GeS so that there are close In–In contacts between the layers (2.80 Å). The structures are compared in Fig. 26.2. As a result In has a distorted tetrahedral arrangement of 4 nearest neighbours (3 S, 2.57 Å; 1 In, 2.80 Å). A S atom has 3 In neighbours and also 1 S at the surprisingly short distance of 3.09 Å (in addition to 4 S at 3.71 Å and 2 S at 3.94 Å); confirmation of this distance would seem desirable.

In InTe[9] there are two kinds of non-equivalent In atom, the environments of which suggest the formulation $In^+In^{3+}(Te^{2-})_2$. The atoms of one kind have 8 Te at

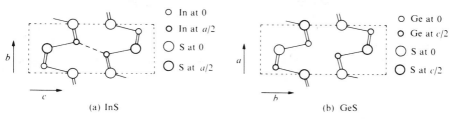

FIG. 26.2. Projections of the structures of (a) InS, (b) GeS (layers perpendicular to plane of paper), showing one In--In bond in (a) as a broken line.

3.58 Å forming a distorted square antiprismatic coordination group while the others have 4 tetrahedral neighbours at 2.82 Å.

(1) JACS 1955 **77** 4217
(2) JACS 1958 **80** 1530
(3) JINC 1957 **4** 84
(4) JCS 1958 1505
(5) IC 1969 **8** 994

(6) AC 1976 **B32** 983
(7) AC 1975 **B31** 1252
(8) NW 1954 **41** 448
(9) AC 1976 **B32** 2689

Ge^{2+}, Sn^{2+}, and Pb^{2+}

Compounds of divalent Ge are well known. There is no evidence that GeO is a stable phase at temperatures below 1000 K (compare SiO), but compounds stable at ordinary temperatures include GeS, all four dihalides, and complex halides such as $MGeCl_3$. The Ge^{2+} ion is not stable in water, and its crystal structure shows that GeF_2 is not a simple ionic crystal (p. 1174). There seem to be no simple ionic crystalline stannous compounds. In solution Sn^{2+} presumably exists as complexes; its easy conversion into Sn^{4+} gives stannous compounds their reducing properties. Lead presents a quite different picture. The stable ion is Pb^{2+}. This has no reducing properties, and there is no evidence that Pb^{4+} can exist in aqueous solution, but it certainly exists in crystalline compounds such as PbO_2 (rutile structure).

The structural chemistry of zinc

The structures of many of the simple compounds of the IIB elements have been described in earlier chapters. Cadmium, like zinc, has only one valence state in its normal chemistry and its structural chemistry presents no points of special interest. We therefore confine our attention here to zinc and mercury.

From the geometrical standpoint the structural chemistry of zinc is comparatively simple. There is only one valence state to consider (Zn^{II}) and in most molecules and crystals the metal forms 4 tetrahedral or 6 octahedral bonds; two collinear bonds are formed in the gaseous ZnX_2 molecules and presumably in $Zn(CH_3)_2$, and some examples of 5-coordination are noted later.

A comparison of the structures of compounds of Be, Mg, Zn, and Cd reveals an interesting point; we exclude Hg from this series because there is practically no resemblance between the structural chemistries of Zn and Hg apart from the facts that one form of HgS has the zinc-blende structure and that both ZnI_2 and one form of HgI_2 are built of 'super-tetrahedra' M_4I_{10} (p. 195). In the following group of compounds italic type indicates tetrahedral coordination of the metal atom in the crystal; in other cases the metal has 6 octahedral neighbours.

BeO	MgO	*ZnO*	CdO
BeS	MgS	*ZnS*	*CdS*
BeF$_2$	MgF$_2$	ZnF$_2$	CdF$_2$
Be Cl$_2$	MgCl$_2$	*ZnCl$_2$*	CdCl$_2$
β-Be(OH)$_2$	Mg(OH)$_2$	*Zn(OH)$_2$*	Cd(OH)$_2$

Apart from the halides (for which see Chapter 9) the Be and Zn compounds are isostructural. Octahedral coordination of Zn is found in ZnF_2 and also in the following compounds, all of which are isostructural with the corresponding Mg compounds: $ZnCO_3$, $ZnWO_4$, $ZnSb_2O_6$, $Zn(ClO_4)_2 .6H_2O$, and $ZnSO_4 .7H_2O$. Comparison with the Be and Cd compounds is not possible for all these salts either because Be does not form the analogous compound or because their structures are not known. Compounds in which Zn, like Be, has 4 tetrahedral oxygen neighbours include ZnO, $Zn(OH)_2$, Zn_2SiO_4 (contrast the 6-coordination of Mg in Mg_2SiO_4), and complex oxides such as K_2ZnO_2 (chains of edge-sharing ZnO_4 tetrahedra)[1] and $SrZnO_2$ (layers of vertex-sharing tetrahedra).[2] These differences in oxygen coordination number are not due to the relative sizes of the ions Be^{2+}, Mg^{2+}, and Zn^{2+}, for the last two have similar radii; moreover, there is 6-coordination of Mg^{2+} by Cl^- in $MgCl_2$ but 4-coordination of the metal in $ZnCl_2$. The tetrahedral coordination by Cl persists in $ZnCl_2 .1\frac{1}{3}H_2O$[3] and in $ZnCl_2 .\frac{1}{2}HCl.H_2O$.[4] In the former all the Cl is associated with two-thirds of the Zn atoms in tetrahedral $ZnCl_4$ groups which share two vertices to form chain $(ZnCl_3)_n^{n-}$. The remaining Zn atoms lie between the chains surrounded octahedrally by 2 Cl and 4 H_2O; note that the more electronegative O (of H_2O) belongs to the octahedral coordination groups. The structural formula could be written $(Zn^{tetr.}Cl_3)_2[Zn^{oct.}(H_2O)_4]$ though this does not show the actual coordination numbers of the two kinds of Zn atom (ion). In the second compound $ZnCl_4$ tetrahedra each share 3 vertices to form a 3D framework of composition $(Zn_2Cl_5)_n^{n-}$ which forms around $(H_5O_2)^+$ ions, i.e. $(Zn_2Cl_5)^-(H_5O_2)^+$—see Chapter 15.

It would seem that the tetrahedral Zn—O bonds have less ionic character than octahedral ones. Evidently the character of a Zn—O bond will depend on the environment of the *oxygen* atom, though a satisfactory discussion of this question is not yet possible. In ZnO the O atom is forming four equivalent bonds (a), in Zn_2SiO_4 bonds to 2 Zn and 1 Si (b), and in $ZnCO_3$ bonds to 2 Zn and 1 C of a CO_3^{2-} ion, within which the C—O bonds are covalent in character, (c).

Both tetrahedrally and octahedrally coordinated Zn occur in a number of crystalline oxy-compounds, including $Zn_2Mo_3O_8$, $Zn_2(OH)_2SO_4$, γ-$Zn_3(PO_4)_2$, $ZnMn_3O_7 .3H_2O$, $Zn_5(OH)_8Cl_2 .H_2O$, and $Zn_5(OH)_6(CO_3)_2$. The structures of the

(a)   (b)   (c)

hydroxy-salts are described in other chapters. In some crystals there appears to be a clear-cut difference in length between the tetrahedral and octahedral Zn–O bonds, for example 2.02 and 2.16 Å respectively in $Zn_5(OH)_8Cl_2.H_2O$, 1.95 and 2.10 Å in the hydroxycarbonate, and 1.96 and 2.10 Å in $Zn_3(PO_4)_2.4H_2O$.[5] In $ZnMn_3O_7.3H_2O$ the octahedral coordination group is made up of 3 O of H_2O molecules (Zn–O, ≈2.15 Å) and 3 O of the Mn_3O_7 layer (Zn–O, ≈1.95 Å), and in the inverse spinel $Zn(Sb_{0.67}Zn_{1.33})O_4$ there is apparently little difference in length between the two types of Zn–O bond (all close to 2.05 Å).[6] It is probably not profitable to discuss this subject in more detail until more precise information on bond lengths is available.

A similar difference in M–O bond character presumably exists in oxy-compounds of divalent lead. Many anhydrous oxy-salts of Pb^{II} are isostructural with those of Ba (and sometimes Sr and Ca):

	C.n. of M
$PbSO_4$, $BaSO_4$	12
$PbCO_3$, $BaCO_3$	9
$PbWO_4$, $BaWO_4$	8

in which it is reasonable to conclude that the Pb is present as (colourless) Pb^{2+} ions with radius close to those of Ba^{2+} and Sr^{2+}. With these compounds compare the coloured PbO (two forms) in which Pb has only 4 nearest neighbours, in contrast to the NaCl structure of the colourless BaO and SrO.

(1) ZaC 1968 **360** 7
(2) ZaC 1961 **312** 87
(3) AC 1970 **B26** 1679
(4) AC 1970 **B26** 1544
(5) AC 1975 **B31** 2026
(6) AC 1963 **16** 836

5-*covalent zinc*

Monomeric chelate complexes with a trigonal bipyramidal bond arrangement include $[Zn(tren)NCS]SCN^{(1)}$ and $ZnCl_2(terpyridyl)$,[2] Fig. 26.3(a). The stereochemistry of such complexes is strongly influenced by the structure of the polydentate ligand, but there are other compounds which, like the cupric compounds discussed earlier, emphasize the importance of considering the structure as a whole. Some chelates such as the 8-hydroxyquinolinate[3] form dihydrates which consist of octahedral molecules (Fig. 26.3(b)), while others form monohydrates in which the configur-

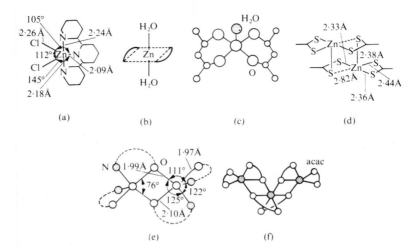

FIG. 26.3. Zn forming 5 or 6 bonds in molecules.

ation of the molecule is intermediate between tetragonal pyramidal and trigonal bipyramidal. In the monoaquo bis-(acetylacetonate) molecule,[4] Fig. 26.3(c), the configuration is nearer the former, with all Zn–O = 2.02 Å and the Zn atom 0.4 Å above the base of the pyramid, but we saw in Chapter 3 that the choice of description for this type of configuration is somewhat arbitrary. The configuration of the monohydrated complex with NN′-disalicylidene ethylene diamine is of the same general type, with Zn 0.34 Å above the plane of the O_2N_2 square.[5] The reason for the formation by these chelates of a monohydrate rather than a dihydrate is presumably associated with the packing of the molecules in the crystal. In crystals of the last compound there is strong hydrogen bonding of H_2O to two O atoms of an adjacent molecule (O–H---O ≈2.5 Å), and this may be a decisive factor in determining the composition and structure of the crystalline compound.

In the foregoing examples Zn achieves 5-coordination in a monomeric complex. A fifth bond is formed in some compounds by dimerization, as in Fig. 26.3(d), a type of structure already noted for a number of Cu^{II} complexes. In the diethyl-dithiocarbamate[6] apparently one of the four short bonds is formed to a S atom of the second molecule of the dimer. A further possibility is realized in bis-(N-methylsalicylaldiminato) zinc, Fig. 26.3(e), in which the five bonds from each Zn are formed within a simple bridged dimer;[7] the Mn and Co compounds are isostructural. A rather surprising structure is adopted by anhydrous $Zn(acac)_2$. We note elsewhere the polymeric structures of $[Co(acac)_2]_4$ and $[Ni(acac)_2]_3$ formed from respectively 4 and 3 octahedral coordination groups. The Zn compound is trimeric but structurally quite unlike the linear Ni compound. The central Zn is octahedrally coordinated, but the terminal Zn atoms are 5-coordinated (Fig. 26.3(f)), the bond arrangement being described as approximately trigonal bipyramidal.[8]

(1) JACS 1968 **90** 519
(2) AC 1966 **20** 924
(3) AC 1964 **17** 696
(4) AC 1963 **16** 748

(5) JCS A 1966 1822
(6) AC 1965 **19** 898
(7) AC 1966 **5** 400
(8) AC 1968 **B24** 904

Coordination compounds with tetrahedral or octahedral Zn bonds

Ligands such as O_2PR_2, S_2CR_2, and S_2PR_2 can behave as simple bridging ligands or as bidentate ligands, and so form a variety of finite and infinite complexes. Zinc dimethyldithiocarbamate[1] forms dimeric molecules, Fig. 26.4(a), in which the ligand acts in both these ways, Zn forming somewhat irregular tetrahedral bonds. The diethylthiophosphinate forms dimers of the same kind, $S_2P(C_2H_5)_2$ replacing $S_2CN(CH_3)_2$.[2] Larger rings are, of course, possible, as in the tetrameric Zn isopropylxanthate.[3] Two different types of chain in Zn compounds are illustrated on p. 866. If all the ligands behave as simple 2-connected ligands a 4-connected layer can be formed, of which the simplest is the planar 4^4 net (Fig. 26.4(b)). This is the structure of the ethylxanthates of Zn, Cd, and Hg.[4] The S_4 tetrahedra around the metal atom become less regular in the series Zn, Cd, Hg, as may be seen from the values of the largest and smallest tetrahedral bond angles, namely, 115°, 121°, 148°. and 103°, 94°, and 84°, respectively. In the Hg compound the bond lengths adjacent to the largest angle are 2.42 Å and those adjacent to the smallest angle have a mean length of 2.82 Å.

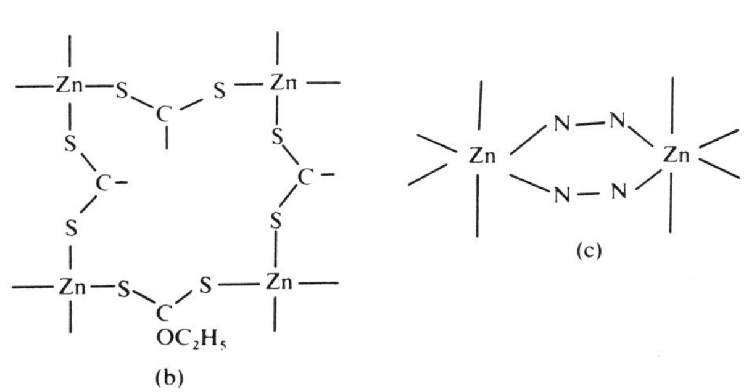

(a)

(b)

(c)

FIG. 26.4.   Zn forming tetrahedral bonds in (a) $Zn[S_2CN(CH_3)_2]_2$ and (b) $Zn(S_2COC_2H_5)_2$, and octahedral bonds in (c) $Zn(N_2H_4)_2Cl_2$.

In salts $Zn(N_2H_4)_2X_2$ the hydrazine molecules link the Zn atoms into infinite chains, Fig. 26.4(c), and the two X ligands complete the octahedral coordination group of Zn. In the chloride,[5] the Zn–Cl distances are surprisingly large (2.58 Å) compared with values around 2.25 Å for tetrahedral $ZnCl_4^{2-}$ in $(NH_4)_3ZnCl_4.Cl$ etc. A similarly long Zn–Cl bond (2.53 Å) is found in the *trans* octahedral molecules of the biuret complex,[6] in which $H_2N.CO.NH.CO.NH_2$ behaves as a bidentate ligand like acac in $Zn(acac)_2(H_2O)_2$, Fig. 26.3(b). In the dihydrazine isothiocyanate[7] all six Zn–N bonds are said to be equal in length (2.17 Å).

(1) AC 1966 **21** 536
(2) JCS A 1970 714
(3) AC 1972 **B28** 1697
(4) AC 1980 **B36** 1367
(5) AC 1963 **16** 498
(6) AC 1963 **16** 343
(7) AC 1965 **18** 367

The structural chemistry of mercury

Mercury forms two series of compounds, mercurous and mercuric, but the former are not compounds of monovalent Hg in the sense that cuprous compounds, for example, are derivatives of monovalent Cu. A number of elements form compounds in which there are metal–metal bonds but mercury is unique in forming, in addition to Hg^{2+} and the normal mercuric compounds, a series of compounds based on the grouping $(-Hg-Hg-)$. The Zn_2^{2+} ion has been identified in $Zn/ZnCl_2$ melts[1] and Cd_2^{2+} presumably exists in $Cd_2(AlCl_4)_2$,[2] but only the mercurous ion, Hg_2^{2+}, is stable in (acid) aqueous solution and in simple salts.

Mercury cations

Mercury is unique in forming salts containing linear systems of directly bonded metal atoms. These salts are formed, like those containing the S, Se, and Te cations, in highly acidic non-aqueous solvents and are stable crystalline solids containing ions such as $(AlCl_4)^-$ and $(AsF_6)^-$. Mercury dissolves in HSO_3F to form a yellow solution containing the Hg_3^{2+} cation which has been studied in $Hg_3(AlCl_4)_2$,[3] prepared by reacting Hg and $HgCl_2$ in molten $AlCl_3$. This compound consists of molecules (a), and the same linear Hg_3 group is found in $Hg_3(AsF_6)_2$.[4] The dark red crystals of $Hg_4(AsF_6)_2$[5] contain the analogous Hg_4 group, with a somewhat

(a)

longer central bond (2.70 Å). The most remarkable compound in this group has the analytical composition Hg_3AsF_6 and is prepared from Hg and AsF_5 in liquid SO_2. It forms golden yellow crystals with metallic appearance and is an anisotropic superconductor. The crystals contain infinite linear, mutually perpendicular and non-intersecting, chains of Hg atoms embedded in a 3D array of $(AsF_6)^-$ ions. The chains run parallel to the tetragonal a axes, and since the Hg–Hg bond length (2.64 Å) is not commensurate with a (7.54 Å) the formula deduced from the structure is $Hg_{2.86}AsF_6$ (7.54/2.64 = 2.86).[6] Although it is customary to describe these Hg_n groups as cations Hg_3^{2+} and Hg_4^{2+} the 'salts' are covalent molecules like mercurous compounds; the only structure which does not contain finite molecules is that of Hg_3AsF_6, in which each Hg carries a formal charge of $+\frac{1}{3}$.

(1) JCS A 1967 1122
(2) JACS 1961 83 76
(3) IC 1972 11 833

(4) IC 1973 12 1343
(5) CC 1973 723
(6) IC 1978 17 646

Mercurous compounds

The 'mercurous' ion in aqueous solution is not Hg^+ but Hg_2^{2+} (presumably, $(H_2O-Hg-Hg-OH_2)^{2+}$), and as far as is known all mercurous compounds contain pairs of Hg atoms in linear groups X–Hg–Hg–X in which the bonds formed by Hg are two collinear sp bonds as in molecules of mercuric compounds X–Hg–X. Apparently the bonds from Hg are not always strictly collinear in these compounds. The angle Hg–Hg–OH$_2$ in the nitrate is 167.5°, and bond angles in the range 146-157° are recorded in the anhydrous arsenate. Structural data for some simple compounds are listed in Table 26.1; the oxide and hydroxide are not known. Even Hg_2F_2 has the same linear molecular structure as the other halides, and the (hydrated) nitrate and perchlorate do not contain $Hg-Hg^{2+}$ ions but linear groups $H_2O-Hg-Hg-OH_2^{2+}$. Since Hg in these compounds has only two nearest neighbours its coordination group is necessarily completed by a number of more distant neighbours. In the halides Hg_2X_2 these complete a distorted octahedral group, but the number and arrangement of next nearest neighbours of Hg in crystalline mercurous compounds is variable. In the bromate, which consists of linear molecules $O_3Br-Hg-Hg-BrO_3$, two more O atoms complete a very distorted tetrahedral

TABLE 26.1
Interatomic distances in mercurous compounds

	Hg–Hg	Hg–X	Hg–4X	Reference
Hg_2F_2	2.51 Å	2.14 Å	2.72 Å $\Big\}$	CC 1971 466
Hg_2Cl_2	2.53	2.43	3.21	
Hg_2Br_2	2.49	2.71	3.32	
Hg_2I_2	[2.72	2.68	3.51]	[JACS 1926 48 2113]
$Hg_2(NO_3)_2 \cdot 2H_2O$				AC 1975 B31 2174
$Hg_2(ClO_4)_2 \cdot 4H_2O$				ACSc 1966 20 553
$Hg_2(BrO_3)_2$	2.51 (mean)	Hg–2O, 2.14 (mean)		ACSc 1967 21 2834
Hg_2SO_4				ACSc 1969 23 1607
$Hg_2(OOC.CF_3)_2$				AC 1974 B30 144
$(Hg_2)_3(AsO_4)_2$				AC 1973 B29 1666

(a)   (b)

group. In the arsenate the stronger Hg–O bonds form chains, (a), and two more bonds (2.4–2.7 Å) to O atoms of other chains complete the tetrahedral coordination of Hg. The sulphate consists of infinite linear molecules (b), and the next nearest neighbours of Hg are 3 O atoms of adjacent chains at 2.50, 2.72, and 2.93 Å.

A most interesting structure has been found for $[(Hg_2)_3O_2H]Cl_3$[1] which is found as the mineral eglestonite. The cation consists of four interpenetrating cubic (10,3) nets (p. 112) in which O forms 3 bonds, these atoms being connected through pairs of Hg atoms ($\supset$O–Hg–Hg–O$\subset$) situated along each link of the net (Hg–Hg, 2.52; Hg–O, 2.17 Å). The nets are held together by O–H–O bonds (2.6 Å) and by the Cl$^-$ ions situated in the interstices (Hg–3 Cl, 3.05 Å).

(1)   Tschermaks Min. Petr. Mitt. 1976 **23** 105

Mercuric compounds

In most of its compounds Hg(II) forms either two collinear (sp) or four tetrahedral (sp^3) bonds. There are no examples as yet of the formation of six equivalent octahedral bonds, though this may possibly occur in compounds such as $[Hg(C_5H_5NO)_6]$-$(ClO_4)_2$.[1a] The planar symmetrical ion $Hg(SiMe_3)_3^-$[1b] has been found in combination with the very complex polynuclear cation $[Mg_4(OCH_2CH_2OCH_3)_6$-$(CH_3OCH_2CH_2OCH_3)_2]^+$, and a few examples of 3- and 5-covalent Hg(II) are noted on p. 1162. The only simple ionic compound is HgF_2 (fluorite structure), in which Hg^{2+} has 8 equidistant F$^-$ neighbours. Even oxygen is not sufficiently electronegative to ionize Hg, and we find Hg forming two short collinear bonds to O in both forms of HgO and also in crystals such as $HgSO_4 . H_2O$,[2a] where Hg has a distorted octahedral coordination group of 5 O (of anions) and 1 H_2O, (a). A feature of many

(a)   (b)

crystalline hydroxy-salts Hg(OH)X in which X is F,[2b] ClO₃, BrO₃, or NO₃, is the planar zigzag chain (b), in which Hg forms two collinear bonds (Hg–O around 2.10–2.15 Å) and has four more distant neighbours completing an octahedral group as in HgSO₄.H₂O. (For an alternative description of Hg(OH)F as a rutile-like structure see p. 250.)

In HgCN(NO₃)[2c] linear chains —Hg–CN–Hg–CN– are stacked with their axes parallel, and the coordination group of Hg(II) is completed by three pairs of O atoms of NO₃⁻ ions. Here the metal atom forms two short collinear bonds (2.06 Å) to C or N (which were not distinguishable) and six Hg–O bonds (2.73 Å) in the equatorial plane (compare uranyl compounds, Chapter 28).

Molecules (other than the dihalides, for which see later) which have been shown to be linear include H₃C–Hg–CH₃; F₃C–Hg–CF₃; mercaptides RS–Hg–SR;[3a] C₂H₅S–Hg–Cl; Cl–Hg–SCN, Br–Hg–SCN; φ–Hg–CN[3b] (Hg–C₆H₅, 2.05; Hg–CN, 2.09 Å); and φ–Hg–φ[3c] (Hg–C, 2.085 Å). From the bond lengths in H₃CHgCl and H₃CHgBr[4] (Hg–C, 2.07 Å; Hg–Cl, 2.28 Å; and Hg–Br, 2.41 Å) the radius of 2-covalent Hg has been deduced as 1.30 Å. The value 1.48 Å, on the Pauling scale, has been suggested for tetrahedral Hg(II).

In some crystals in which Hg(II) forms only two strong bonds there are only two weakly bonded next nearest neighbours completing a very distorted tetrahedral group as opposed to the distorted (2+4) octahedral coordination noted above. These cases are intermediate between the collinear molecules and normal tetrahedral molecules (ions). A combined X-ray and neutron diffraction study of Hg(CN)₂[5] showed that there are two additional weak Hg–N bonds in a plane perpendicular to that of the two Hg–C bonds, (c), the angle between the latter being 171(±2)°. Other examples of very distorted tetrahedral coordination with one large angle between two bonds of normal length and a small angle (80–100°) between two weaker bonds include the dithizone complex, Hg(C₁₃H₁₁N₄S)₂.2C₅H₅N,[6] (S–Hg–S, 155°; N---Hg---N, 102°) and HgCl₂[OAs(C₆H₅)₃]₂,[7] (Cl–Hg–Cl, 147°; O---Hg---O, 93°).

(c)

Normal tetrahedral coordination of Hg(II) appears to be more characteristic of ligands such as S and I rather than the more electronegative O and the lighter halogens. Examples of 3D structures mentioned later include HgI₂ and HgS. Finite ions or molecules include (HgI₄)²⁻, (HgS₄)⁶⁻ (in K₆HgS₄[8a]), the anion in [Hg(SCN)₄][Cu(en)₂],[8b] and the thioxan complex (d).[9] In the infinite chain (e) the trithiane molecules act as bridges between the HgCl₂ groups.[10] The layer in CoHg₂(SCN)₆.C₆H₆[11] is a rather complex example of a layer based on the planar

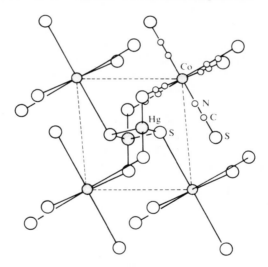

(d)

(e)

4-gon net. Octahedral $Co(SCN)_6$ groups are linked through their terminal S atoms by pairs of Hg atoms which are bridged by two S atoms (Fig. 26.5); the C_6H_6 molecules are accommodated between the layers. In the diamond-like structure of $CoHg(SCN)_4^{(12)}$ the —SCN— ligands link Co and Hg into a 3D framework in which Co has four tetrahedral N and Hg four tetrahedral S neighbours.

FIG. 26.5. Portion of layer in $CoHg_2(SCN)_6 \cdot C_6H_6$.

(1a) IC 1962 **1** 182
(1b) JACS 1978 **100** 7761
(2a) AC 1980 **B36** 23
(2b) AC 1979 **B35** 949
(2c) IC 1971 **10** 2331
(3a) JCP 1964 **40** 2258
(3b) AC 1976 **B32** 2680
(3c) AC 1977 **B33** 587
(4)   JCP 1954 **22** 92

(5)   ACSc 1958 **12** 1568
(6)   JCS 1958 4136
(7)   ACSc 1963 **17** 1363
(8a) ZaC 1978 **443** 201
(8b) AC 1953 **6** 651
(9)   JCS A 1967 271
(10) JCS A 1966 1190
(11) HCA 1964 **47** 1889
(12) AC 1968 **B24** 653

Mercuric halides – simple and complex. In contrast to HgF_2, which crystallizes with the fluorite structure, $HgCl_2$, $HgBr_2$, and the yellow form of HgI_2 crystallize as linear molecules X–Hg–X; four halogen atoms of other molecules complete very distorted octahedral coordination groups around Hg(II). (Alternatively, the $HgBr_2$ structure may be described as built of very distorted CdI_2-type layers.) The red form of HgI_2 has a quite different layer structure in which each Hg atom has four equidistant I neighbours arranged tetrahedrally. In the remarkable structure assigned to an orange form of HgI_2 multiple groups of four tetrahedra, similar to the P_4O_{10} molecule, replace the single tetrahedra in the layer of red HgI_2. The molecules $HgCl_2$, $HgBr_2$, and HgI_2 have been shown by electron diffraction to form linear molecules in the vapour state. Interatomic distances in these halides are included in Table 26.2.

TABLE 26.2

Interatomic distances in mercuric compounds

Compound	Numbers of X atoms around Hg in crystal			Reference
	2 at	2 at		
HgF_2		8 at 2.40 Å		
$HgCl_2$	2.28	2 at 3.38	3.46	AC 1980 **B36** 2132
$HgBr_2$	2.48	4 at 3.23		ZK 1931 77 122
HgI_2 (yellow)	2.62	4 at 3.51⎫		IC 1967 6 396
HgI_2 (red)		4 at 2.78⎭		
HgI_2 (orange)		4 at 2.68		ZK 1968 **128** 97
NH_4HgCl_3	2.34	4 at 2.96		ZK 1938 **100** 208
$NH_4HgCl_3.H_2O$	2.37	2 at 2.75	3.22	AC 1974 **B30** 1603
$K_2HgCl_4.H_2O$	2.29	2 at 2.80	3.15 ⎫	DAN 1955 **102** 1115
$CsHgCl_3$	2.29	4 at 2.70	⎭	

In the linear HgX_2 molecules in the vapour the following bond lengths have been found: Hg–Cl, 2.25 Å (BCSJ 1973 **46** 410); Hg–Br, 2.44 Å and Hg–I, 2.61 Å (TFS 1937 **33** 852).

Many complex halides are known, though less is known of fluorides than of compounds containing the other halogens. The coordination of Hg is very dependent on the nature of the cation, and salts of small and very large cations often have structures of quite different kinds. Finite ions of all the following types have been found: HgX_3^-, HgX_4^{2-}, HgX_5^{3-}, and HgX_6^{3-} and also the bridged ion $Hg_2I_6^{2-}$, usually associated with large cations. An example is the trigonal bipyramidal anion in $[Cr(NH_3)_6](HgCl_5)$;[1] this unusual type of coordination is also found for the $CuCl_5^{3-}$ and $CdCl_5^{3-}$ ions in similar salts.[2] Iodides stand rather apart from the other halides, showing a preference for tetrahedral coordination, just as HgI_2 is the only binary halide in which there is tetrahedral coordination of Hg

There are discrete HgX_6^{4-} ions in salts Tl_4HgX_6[3] with the characteristic (2+4)-distortion of the octahedral group found in $HgCl_2$, and similar groups form more

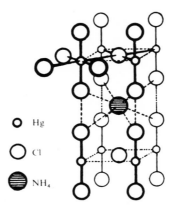

(a)

complex aggregates in a number of complex chlorides. They include $K_2HgCl_4.H_2O$ built of edge-sharing chains (a) and $NH_4HgCl_3.H_2O$ which contains double octahedral chains (details in Table 26.2). In NH_4HgCl_3 the octahedra share four equatorial edges to form layers in which the Hg—Cl bonds are so much longer (2.96 Å) than those perpendicular to the layers (2.34 Å), that this crystal is an aggregate of $HgCl_2$ molecules, Cl^-, and NH_4^+ ions (Fig. 26.6). The salt $CsHgCl_3$ has a deformed perovskite structure in which Hg has 2 Cl neighbours at 2.29 Å and 4 Cl at 2.70 Å.

O Hg

◯ Cl

⊜ NH₄

FIG. 26.6. The crystal structure of NH_4HgCl_3.

On the other hand, in certain salts containing very large cations there is trigonal coordination of Hg(II) as in, for example, $(CH_3)_4N(HgCl_3)$ and the isostructural bromide[4] and iodide. In the bromide, which was studied in detail, approximately planar $HgBr_3^-$ groups (Hg–Br, 2.52 Å) are linked into chains by a fourth bond from each Hg (Hg–Br, 2.90 Å), a system intermediate between discrete $HgBr_3^-$ ions and chains formed from $HgBr_4$ tetrahedra sharing two vertices as in a metasilicate chain. In $(CH_3)_3SHgI_3$[5] there is also approximately planar coordination of Hg (Hg–3I, 2.71 Å) but in this case the HgI_3 groups are linked into chains by very much weaker bonds (Hg–I, 3.52 and 3.69 Å), the five bonds forming very distorted trigonal bipyramidal coordination groups around the metal atoms.

Tetrahedral coordination of Hg(II) by halogen atoms occurs in only one of the dihalides (red HgI_2) but it is a feature also of complex halides such as Ag_2HgI_4 and

Cu_2HgI_4 (p. 780). However, the tetrahedral ions HgX_4^{2-} (including HgF_4^{2-}) are stable in combination with large cations, as in the mercurichloride of the alkaloid perloline (Hg–Cl, 2.50 Å),[6] in $[N(CH_3)_4]_2HgBr_4$,[7] and $[(CH_3)_3S]_2HgI_4$. Note the large difference between Hg -Cl when two collinear bonds are formed (2.3 Å) and the value 2.5 Å for tetrahedral bonds. Both types of bond occur in $C_8H_{12}S_2 . Hg_2Cl_4$,[8] in which some of the Hg atoms form two approximately collinear bonds (Hg–Cl, 2.30 Å; Cl–Hg–Cl, 168°) and the remainder form four tetrahedral bonds (Hg–2Cl, 2.51 Å; Hg–2S, 2.53 Å). The bridged ion $(Hg_2I_6)^{2-}$[9] is structurally similar to the Al_2X_6 molecules.

Mercuric halides combine with tertiary phosphines and arsines (e.g. $As(C_4H_9)_3$) to form crystalline products $(R_3As)_m(MX_2)_n$. The compound $[(C_4H_9)_3As]_2[HgBr_2]_2$ consists of molecules in which the two Hg atoms are joined by halogen atoms:

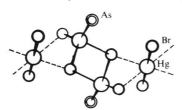

This compound is therefore similar in general type to the gold compounds already described (p. 1147) and to the bridged palladium compounds mentioned later (p. 1236), except that the bonds from the metal atoms are disposed tetrahedrally instead of being coplanar. The product richer in mercuric bromide and having the empirical formula $[(C_4H_9)_3As]_2[HgBr_2]_3$ is found to be a mixed crystal of $[(C_4H_9)_3As]_2-Hg_2Br_4$, the above bridged molecules, and $HgBr_2$. In this crystal the environment of a Hg atom of a $HgBr_2$ molecule is very similar to that in $HgBr_2$ itself. There are two

FIG. 26.7. The environment of a mercury atom of a $HgBr_2$ molecule in $[(C_4H_9)_3As]_2Hg_2Br_4 .-HgBr_2$.

near Br neighbours and four more distant ones which belong to bridged molecules, as shown in Fig. 26.7. An unsymmetrical bridge is found in $Hg_2Cl_4(Se . P\phi_3)_2$:[10]

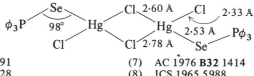

(1) JCS D 1975 2591
(2) JCS A 1971 3628
(3) ZaC 1973 **401** 217
(4) AC 1963 **16** 397
(5) AC 1966 **20** 20
(6) PCS 1963 171

(7) AC 1976 **B32** 1414
(8) JCS 1965 5988
(9) JCMS 1972 **2** 183
(10) JCS A 1969 2501
For other references see Table 26.2

Oxychlorides and related compounds.   Like many other B sub-group elements mercury forms numerous oxy- and thio-salts. In some of these compounds the oxidation state of individual Hg atoms is not obvious as, for example, in the complex structure of $Hg_4O_2Cl_2$.[1] Various oxychlorides of Hg(II) may be prepared by the slow hydrolysis of $HgCl_2$ solution containing marble chips. At least four compounds are well characterized in the family $nHgO.HgCl_2$: $n = ½$, Hg_3OCl_4 (colourless);[2] $n = 2$, two compounds, black $Hg_3O_2Cl_2$[3] and red $Hg_6O_4Cl_4$;[4] and $n = 3$, the yellow mineral kleinite. The oxybromide $Hg_5O_4Br_2$[5] is a member of the Br family, with $n = 4$. The structure of Hg_3OCl_4 shows it to be an ionic compound, trichloro-mercury oxonium chloride, for the crystal is an aggregate of Cl^- and $OHg_3Cl_3^+$ ions.

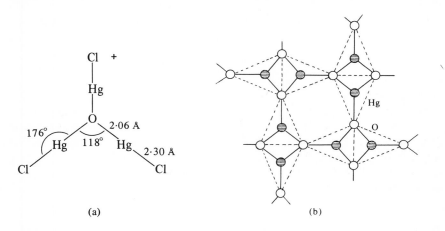

(a)   (b)

The latter have the form of flat pyramids with the configuration (a). In the black $Hg_3O_2Cl_2$ there are Hg atoms with quite different environments. Hg' atoms forming 3 bonds (in flat pyramidal HgO_3 groups) and O atoms bonded to 3 Hg atoms lie at alternate points of the planar 4.8^2 net, (b). These layers are joined into a 3D framework through Hg'' atoms forming 2 collinear bonds to O atoms of the layers. The Cl^- ions occupy interstices in the 3D cationic framework. There is an obvious topological resemblance to Si_2N_2O, in which the Si and N atoms occupy alternate points of the planar 6^3 net and the layers are joined through 2-connected O atoms:

coordination number	4	3	2
	Si_2	N_2	O
	O_2	Hg'_2 •	Hg''

The more complex structure of $Hg_6O_4Cl_2$ contains six kinds of non-equivalent Hg atoms, and bears no obvious relation to that of $Hg_3O_2Cl_2$.

The structures of two of the three polymorphs of $Hg_3S_2Cl_2$ are accurately known. In both there are continuous frameworks $(Hg_3S_2)_n^{2n-}$ built from pyramidal SHg_3 groups sharing Hg atoms (S–Hg–S, 166°; Hg–S–Hg, 92° (α form), 95° (γ form)). In the α form this is a 3D network (see p. 112) and in the γ form a 2D system; Cl^- ions complete distorted octahedral groups around Hg (Hg–2S, $\approx 2.4\,Å$; Hg–4Cl at distances from 2.7 to 3.5 Å).[6]

The structures of crystalline compounds of B subgroup elements in which the metal forms a small number of stronger bonds can be described in terms of the molecules or ions delineated by the stronger bonds. With the description of NH_4HgCl_3 as $HgCl_2$ molecules, NH_4^+, and Cl^- ions, compare the structure of $2HgO.NaI$,[7] which contain infinite zigzag chains like those in orthorhombic HgO embedded in a mixture of Na^+ and I^- ions. Three I^- neighbours complete a

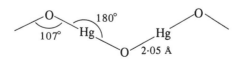

tetragonal pyramidal group around Hg(II). The structure of $NH_4Pb_2Br_5$ will be noted later in this chapter as a further example of a compound of this type — $NH_4^+(PbBr_2)_2Br^-$.

(1) ACSc 1968 **22** 2529
(2) AK 1964 **22** 517, 537
(3) AC 1974 **B30** 1907
(4) AC 1978 **B34** 79

(5) AK 1968 **28** 279
(6) AC 1968 **B24** 156, 1661
(7) ACSc 1964 **18** 1305

Mercuric oxide and sulphide. The two well-known forms of HgO are the stable orthorhombic and the metastable hexagonal forms. The former is made, for example, by heating the nitrate in air and the latter by the prolonged action of NaOH on K_2HgI_4 solution at 50°C. Both form red crystals which grind to a yellow powder like the precipitated form. The former is built of infinite planar zigzag chains:[1]

and the next nearest neighbours of Hg are 4 O atoms of different chains at 2.81–2.85 Å. The hexagonal form[2] is isostructural with cinnabar (HgS) and is built of helical chains in which the bond angles and bond lengths are the same as in the planar chains of the orthorhombic form. Two further O atoms at 2.79 Å and two at 2.90 Å complete a very distorted octahedral coordination group around Hg. The determination of these structures provides a good example of the use of both X-ray and neutron diffraction. It is not possible to locate the O atoms by X-ray diffraction but it is possible using neutrons, for which the relative scattering amplitudes are much more closely similar and moreover independent of the angle of scattering. Apparently at least two other forms of HgO can be prepared hydrothermally.

Mercuric sulphide, HgS, is dimorphic. Metacinnabarite, a rare mineral, crystallizes with the zinc-blende structure, in which Hg(II) forms tetrahedral bonds (Hg–S, 2.53 Å).[3] The more common form, cinnabar, has a structure[4] which is unique among those of monosulphides, for the crystal is built of helical chains in which Hg has two nearest neighbours (at 2.36 Å), two more at 3.10 Å, and two at 3.30 Å. Interbond angles are S–Hg–S, 172° and Hg–S–Hg, 105°. Cinnabar is notable for its extraordinarily large optical rotatory power which is, of course, a property of the solid only.

Two other oxy-compounds may be mentioned here, although the details of their structures are not established. Alkali-metal compounds of the type Na_2HgO_2[5] apparently contain linear $(O–Hg–O)^{2-}$ ions (Hg–O ≈ 1.95 Å), and one of the two forms of HgO_2[6] (prepared from HgO and H_2O_2 at $-15\,°C$) consists of chains

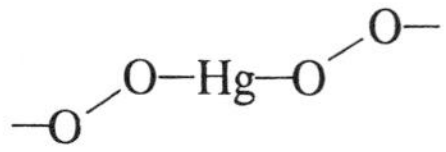

$$-O \diagup ^{O-Hg-O} \diagdown ^{O-}_{O-}$$

The relation of the structure of β-HgO_2 to PdS_2 and AgF_2 is described on p. 275.

(1) ACSc 1964 **18** 1305   (4) ACSc 1950 **4** 1413
(2) ACSc 1958 **12** 1297   (5) ZaC 1964 **329** 110
(3) ACSc 1965 **19** 522   (6) AK 1959 **13** 515

Mercury–nitrogen compounds. By the interaction of mercuric salts with ammonia in aqueous solution a number of compounds can be obtained, depending on the conditions. They include:

$Hg(NH_3)_2X_2$, diamminohalides (X = Cl, Br)
$HgNH_2X$, aminohalides (X = F,* Cl, Br)
Hg_2NHX_2, iminodihalides (X = Cl, Br)

and Hg_2NX, Millon's base $(Hg_2N.OH.xH_2O)$ and salts (X = Cl, Br, I, NO_3, and ClO_4). The chlorides $Hg(NH_3)_2Cl_2$ and $HgNH_2Cl$ have long been known as fusible and infusible white precipitates respectively.

The structural feature common to all these compounds is the presence of complexes containing Hg^{II} forming two collinear (sp) bonds and N forming four tetrahedral bonds, though in the iminodihalide Hg_2NHBr_2 there is also an unusual (3+2)-coordination of one-quarter of the Hg atoms.

The diamminohalide $Hg(NH_3)_2Cl_2$ (with which the dibromide is isostructural) has a very simple cubic structure[2] (Fig. 26.8(a)) in which the unit cell contains

*This compound is apparently not structurally similar to the chloride and bromide.[1] It may be a 3-dimensional framework of the Millon's base type with NH_4^+ and F^- ions in the interstices, that is $(Hg_2N)F.NH_4F$.

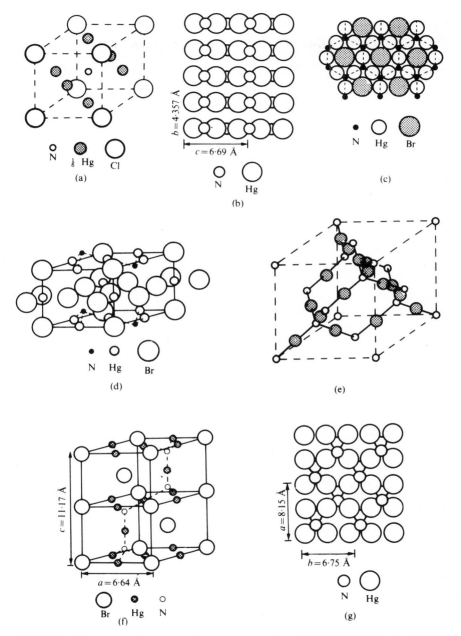

FIG. 26.8. Structures of mercury–nitrogen compounds: (a) $Hg(NH_3)_2Cl_2$; (b) $-Hg-NH_2-$ chains in orthorhombic $HgNH_2Cl$; (c) $[Hg_3(NH)_2]$ layer in Hg_2NHBr_2; (d) unit cell of Hg_2NHBr_2; (e) NHg_2NO_3; (f) NHg_2Br; (g) layer in $Hg_2N_2H_2Cl_2$.

½[Hg(NH$_3$)$_2$Cl$_2$], so that only one-sixth of the Hg positions are occupied (statistically). Each Hg position has two equidistant NH$_3$ neighbours and every NH$_3$ will have, on the average, one Hg neighbour. Accordingly there are isolated (H$_3$N.Hg.NH$_3$)$^{2+}$ ions randomly arranged. In HgNH$_2$Br[3] the positions available for Hg, N, and Br are the same as in Hg(NH$_3$)$_2$Br$_2$ but there are twice as many Hg atoms, so that each shaded circle in Fig. 26.8(a) now represents Hg/3. Since any particular Hg atom has two equidistant N atoms as nearest neighbours and since the number of Hg and N atoms is the same, each N atom will have an average of two Hg neighbours. There must therefore be chains

but these are not straight and parallel over long distances but bent at each N atom and irregularly arranged in the crystal.

Independent X-ray studies of HgNH$_2$Cl[4] and HgNH$_2$Br[5] led to the assignment of an orthorhombic structure (Fig. 26.8(b)) also containing infinite chain ions bound together by halide ions. In this (regular) structure the zigzag chains are arranged in parallel array, but it appears[6] that this structure is not characteristic of pure HgNH$_2$X, being formed only when there is a small amount of halide HgX$_2$ in solid solution. (The nuclear resonance spectrum of 'infusible white precipitate' is consistent with the presence of NH$_2$ groups[7] and eliminates older formulae involving NH$_4^+$ or NH$_3$.)

With increasing Hg:N ratio more extended complexes become possible. In Hg$_2$NHBr$_2$[8] three-quarters of the Hg atoms form hexagonal layers of composition (NH)$_2$Hg$_3$ in which N is attached to three Hg and one H (Fig. 26.8(c)), the remainder of the Hg atoms being situated between the layers surrounded by three coplanar Br atoms at 2.6 ± 0.1 Å and two more at 3.08 Å completing a trigonal bipyramidal coordination group. A unit cell of the structure is shown in Fig. 26.8(d).

Finally, 3-dimensional NHg$_2$ frameworks similar to those in the forms of SiO$_2$ occur in Millon's base, NHg$_2$OH.xH$_2$O (x = 1 or 2), and its salts.[9] The framework in the nitrate, NHg$_2$NO$_3$, is of the cristobalite type, and other salts also have this cubic structure if prepared from the nitrate (Fig. 26.8(e)), but it seems likely that the hexagonal (tridymite-like) structure of Fig. 26.8(f) is more stable for the halides. The base itself and the bromide and iodide can be prepared initially in this hexagonal form or by heating the cubic modifications. In these frameworks N forms four tetrahedral bonds and Hg two collinear bonds (Hg–N ≈ 2.07 Å). The OH$^-$, X$^-$ or other ions occupy the interstices in the frameworks.

Many heavy-metal salts are reduced to the metal by hydrazine, and this is true of mercuric oxy-salts. From the chloride and bromide, however, compounds of the types [Hg(N$_2$H$_4$)]X$_2$, [Hg(N$_2$H$_4$)$_2$]X$_2$ and (Hg$_2$N$_2$H$_2$)X$_2$ may be prepared. The chloride $(Hg_2N_2H_2)Cl_2$, is closely related structurally to $HgNH_2Cl$, but whereas $H_2N{<}$ can form only chains the unit ${>}HN-NH{<}$ forms layers[10] (Fig. 26.8(g)), between which are situated the Cl^- ions.

Summarizing, these mercury–nitrogen compounds provide examples of all four types of Hg–N complex, namely:

finite $(H_3N-Hg-NH_3)^{2+}$ ions in $Hg(NH_3)_2X_2$;
chains $-NH_2-Hg-NH_2-Hg-$ in $HgNH_2X$;
layers $Hg_3(NH)_2$ in Hg_2NHBr_2;
$Hg_2N_2H_2$ in $(Hg_2N_2H_2)Cl_2$;

and 3-dimensional frameworks in Hg_2NX.

(1)	ZaC 1956 **287** 24	(6)	ZaC 1954 **275** 141
(2)	ZK 1936 **94** 231	(7)	JCS 1954 3697
(3)	ZaC 1952 **270** 145	(8)	AC 1955 8 723
(4)	AC 1951 4 266	(9)	ZaC 1953 **274** 323; AC 1951 4 156; AC 1954 7 103
(5)	AC 1952 5 604	(10)	ZaC 1956 **285** 5; ZaC 1957 **290** 24

The structural chemistry of gallium and indium

To our earlier remarks on the lower valence states of these elements we add here a summary of their structural chemistry in the trivalent state. Many Ga(III) and In(III) (and a few Tl(III)) compounds have similar structures, and there is a strong resemblance to Al, which forms compounds structurally similar to all those of Table 26.3. (The structures of most of these compounds are described in other chapters.) Like Al, Ga exhibits both tetrahedral and ocrahedral coordination by Cl and O; in β-Ga_2O_3 there are both kinds of coordination in the same crystal. For $Ga(CH_3)_3$ and $In(CH_3)_3$ see p. 979.

More complex examples of tetrahedrally bonded Ga(III) include $Na\{Ga[OSi(CH_3)_3]_4\}$[1] and the cyclic molecule $Ga_4(OH)_4(CH_3)_8$,[2] (a), which has a structure similar to that of the corresponding gold compound but with tetrahedral instead of coplanar bonds from the metal atoms.

TABLE 26.3
Environment of Ga(III) *and* In(III) *in simple compounds*

Tetrahedral coordination	Octahedral coordination
Li(GaH$_4$)	GaF$_3$
Ga$_2$Cl$_6$ and In$_2$Cl$_6$ (vapour molecules)	InCl$_3$ (and TlCl$_3$)
GaPO$_4$ (cristobalite structure)	GaSbO$_4$ (statistical rutile structure)
GaN, InN (wurtzite structure)	α-Ga$_2$O$_3$ (corundum structure but
GaP, InP (zinc-blende structure)	also C-M$_2$O$_3$ structure like In$_2$O$_3$)
	In(OH)$_3$
Ga$_2$S$_3$	
Both tetrahedral and octahedral coordination in β-Ga$_2$O$_3$, In$_2$S$_3$	

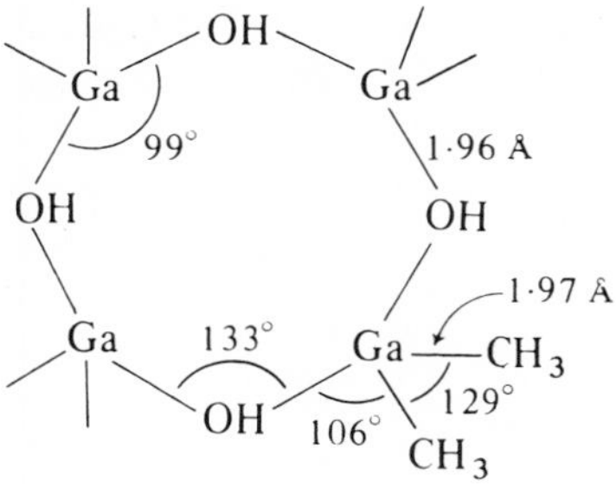

(a)

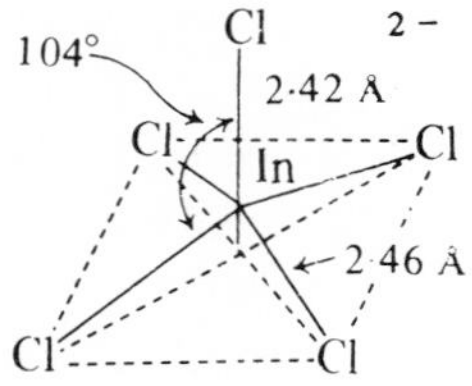

(b)

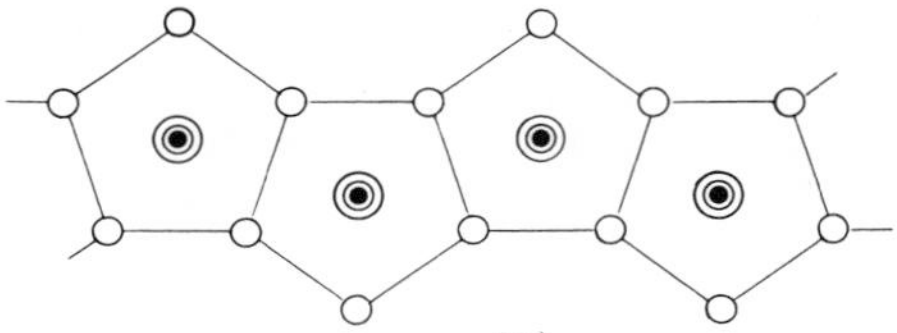

(c)

In $[N(C_2H_5)_4]_2InCl_5$[3] In forms five bonds directed towards the vertices of a square pyramid, the metal atom being 0.6 Å above the base; compare the similar structure of $Sb(C_6H_5)_5$. The configuration of the $InCl_5^{2-}$ ion, (b), is close to that calculated for minimum Cl–Cl repulsions using a simple inverse square law. Contrast the trigonal bipyramidal structure of the isoelectronic $SnCl_5^-$ ion.

To complete this very brief account of the structural chemistry of In(III) we note that in addition to 4-, 5-, and 6-coordination In also exhibits 7- and 8-coordination. In $N_2H_5[InF_4(H_2O)]$[4] the anion consists of edge-sharing pentagonal bipyramidal groups; the 'PaCl$_5$ chain', (c), is a prominent feature of the structural chemistry of 5f elements. The single circles represent H_2O and the axial bonds F–In–F are normal to the plane of the paper. There is presumably 8-coordination of In in the anion in $[N(C_2H_5)_4][In(NO_3)_4]$[5] which is isoelectronic with $Sn(NO_3)_4$.

(1) JCS 1963 3200
(2) JACS 1959 81 3907
(3) IC 1969 8 14
(4) AC 1976 **B32** 948
(5) JCS A 1966 1081

The structural chemistry of thallium

The structures of some simple thallous and thallic compounds have been noted in earlier chapters. The thallous halides show a remarkable resemblance to those of Ag(I) in their colours and solubilities, though their crystal structures are different owing to the different sizes and polarizabilities of the cations. TlF, like AgF, is soluble in water and the other halides are very insoluble. TlOH is a strong base, soluble in water like the alkali hydroxides, and Tl_2SO_4 is isostructural with K_2SO_4.

Compounds which apparently contain Tl(II) in fact contain Tl(I) and Tl(III). For example, both $TlCl_2$ and Tl_2Cl_3 are diamagnetic, whereas $Tl(II)$ would have one unpaired electron. The former is presumably similar to $TlBr_2^{(1)}$ which is isostructural with Ga_2Cl_4, that is, it is $Tl^I(Tl^{III}Br_4)$ and contains the same tetrahedral ion as $CsTlBr_4$ (Tl—Br, 2.51 Å). The complex structure of Tl_2Cl_3 has not been determined; the compound is presumably $Tl_3^I(Tl^{III}Cl_6)$ containing the same octahedral ions as $K_3TlCl_6 . 2H_2O$ and similar salts. The compound TlI_3 is not thallic iodide but is isostructural with NH_4I_3 and is therefore $Tl^+I_3^-$.[2] The halogen complexes of $Tl(III)$ have been reviewed.[3]

Sulphides of thallium include Tl_2S,[4a] TlS,[4b] TlS_2, Tl_4S_3,[4c] and Tl_2S_5,[4d] the last being the pentasulphide $Tl_2^+(S_5)^{2-}$. Like Tl_2O (see below) Tl_2S has a layer structure of the anti-CdI_2 type; compare Cs_3O with the anti-ZrI_3 structure. The sulphides TlS and Tl_4S_3 both contain $Tl(I)$ and $Tl(III)$, the latter in tetrahedral coordination. In Tl_4S_3 there are simple vertex-sharing chains $Tl^{III}S_3$ (Tl—4S, 2.55 Å) between which lie the $Tl(I)$ atoms with no S neighbours nearer than 2.9 Å. The structural formula is therefore $Tl_3^I(Tl^{III}S_3)$. In crystalline TlS also the metal atoms are of two kinds, one set having 4 (tetrahedral) and the other 8 nearest neighbours. The former are in chains of edge-sharing tetrahedra (as in the normal form of SiS_2) whereas the latter lie between the chains in position of 8-coordination. The large distance between these latter Tl atoms and their eight S neighbours (3.32 Å) compared with the Tl—S distance of only 2.60 Å in the chains suggests the formulation of the compound as an ionic thallous thallic sulphide, $Tl^I(Tl^{III}S_2)$.

In addition to the tetrahedral and octahedral coordination of $Tl(III)$ or Tl^{3+} in halide complexes and other compounds there may be 5-coordination in compounds such as $(C_6H_5)_2L_2Tl(NO_3)$, where L is $OP(C_6H_5)_3$, but their structures are not yet known.[5]

In its covalent compounds thallium shows no reluctance to utilizing the two 6s electrons for bond formation. Indeed monoalkyl derivatives Tl(Alk) in which the valence group would be (2,2) are not known, whereas trialkyls (valence group 6) are known and the most stable alkyl derivatives are the dialkyl halides such as $[Tl(CH_3)_2]I$. These are ionic compounds—$[Tl(CH_3)_2]OH$ being a strong base—and in the $Tl(Alk)_2^+$ ions the thallium atom has the same outer electronic structure as mercury in CH_3—Hg—CH_3, viz. (4). Accordingly the $(CH_3$—Tl—$CH_3)^+$ ion is linear, as has been demonstrated in many salts $Tl(CH_3)_2X$, where X is I, CN, N_3, etc.[6] The $Tl(CH_3)_2$ group is slightly bent (around 170°) in chelates (X = OAc, acac)[7] in which Tl is bonded to 2 O of the organic radical and by two more bonds to O

(CH₃)₂Tl(OAc)   (CH₃)₂Tl(acac)

atoms of neighbouring molecules. In the acetate the four bonds are of almost the same length but in the acetylacetonate there is a marked difference between the lengths of the two pairs of Tl–O bonds, as shown in the accompanying sketches, where the molecules are viewed along the H_3C–Tl–CH_3 axis.

Thallous alkoxides exist as tetramers in benzene solution, and an X-ray study of $(TlOCH_3)_4$[8] shows a tetrahedral arrangement of the metal atoms as in the idealized structure shown above, but the alkoxyl groups were not located. In such molecules Tl forms pyramidal bonds, with valence group $(2,\underline{6})$.

Much of the interest in the structures of compounds of Tl(ɪ) lies in the stereo-activity or otherwise of the lone pair of electrons. In various oxy-compounds c.n.s range from 3 to much higher values characteristic of normal large cations (e.g. 10 in one form of $TlNO_3$). Low c.n.s are invariably associated with stereoactivity of the lone pair, there being typically 3 or 4 short bonds lying all to one side of Tl(ɪ) with very long secondary bonds (>3.1 Å) on the other side. In other structures there are 6 or more bonds distributed uniformly around the Tl atom with lengths in the range 2.9–3.2 Å. An attempt has been made to relate these facts to the Lewis base strength of the anion. For the calculations of these 'base strengths', which are not straightforward, and for references to the structures of numerous Tl(ɪ) compounds, the reader is referred to the original paper.[9]

Three structures which illustrate the stereoactivity of the lone pair are those of Tl_2O,[10] Tl_3BO_3,[11] and $Tl_3F(CO_3)$.[12] The structure of Tl_2O is a polytype of the anti-CdI_2 structure, with the 12-layer sequence *cchh* (p. 157). In Tl_3BO_3 (Fig. 24.11(b), p. 1067) columns of Tl atoms enclose empty tunnels into which the lone pairs are directed. In both these structures Tl has a pyramidal arrangement of 3 O (at distances close to 2.50 Å) and in Tl_3BO_3 there are 3 metal atoms at 3.60 Å. In Tl_2O the metal atoms are nearly close-packed, and in addition to the 3 O atoms Tl has 12 Tl neighbours at distances in the range 3.51–3.86 Å; compare 3.36 and

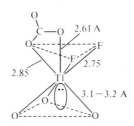

3.46 Å in the b.c.c. and h.c.p. forms of the metal. In $Tl_3F(CO_3)$ Tl has 4 close neighbours to one side and 3 more distant on the other which complete a distorted monocapped octahedral group.

(1)   JCS 1963 3459
(2)   AC 1963 **16** 71
(3)   IC 1965 **4** 502
(4a) ZK 1939 **101** 367; K 1970 **15** 471
(4b) ZaC 1949 **260** 110
(4c) AC 1973 **B29** 2334
(4d) AC 1975 **B31** 1675
(5)   JCS 1965 6107

(6) AC 1975 **B31.** 1922
(7) AC 1975 **B31** 1929
(8) JINC 1962 **24** 357
(9) AC 1980 **B36** 1802

(10) ZaC 1971 **381** 266
(11) CR 1973 **276C** 177
(12) AC 1973 **B29** 498

The structural chemistry of germanium

The most important characteristics of this element are perhaps most easily seen by comparing it with silicon. The resemblance of Ge(IV) to silicon is very marked. Not only are the ordinary forms of elementary Ge and Si isostructural but so also are GeI_4 and SiI_4, hexagonal GeO_2[1] and SiO_2 (high-quartz), tetragonal GeO_2 and the rutile (high-pressure) form of SiO_2, and many oxy-compounds of the two elements. The Ge analogues of all the major types of silicates and aluminosilicates have been prepared, ranging from those containing finite ions to chain, layer, and 3D framework structures.[2] Examples of germanates containing various types of complex ion which are isostructural with the corresponding silicates include: Be_2GeO_4 and Zn_2GeO_4 with the phenacite and willemite structures respectively; $Sc_2Ge_2O_7$ and $Sc_2Si_2O_7$; $BaTiGe_3O_9$ with the same type of cyclic ion as in benitoite; and $CaMg(GeO_3)_2$ with a chain ion similar to that in diopside. The extent of this structural resemblance to Si is seen from the facts that two crystalline forms of Ca_2GeO_4 are isostructural with two forms of Ca_2SiO_4 while Ca_3GeO_5 crystallizes with no fewer than four of the structures of Ca_3SiO_5.

There is rather less resemblance between the structures of Si–S and Ge–S compounds. The structures of the normal forms of SiS_2 and GeS_2 are different, the former consisting of edge-sharing chains while the latter has a unique 3D structure containing Ge_3S_3 rings (and, of course, larger, $Ge_{10}S_{10}$ and $Ge_{11}S_{11}$, rings) in which only vertices are shared.[3] However, when subjected to high pressure and high temperature both of these sulphides form a cristobalite-like structure.[4] In this form of GeS_2 the Ge–S–Ge angles are close to the regular tetrahedral value, and the GeS_4 coordination groups are compressed along the $\bar{4}$ axis, resulting in two S–Ge–S angles of $118°$ and four of $105°$. The ambient pressure h.t. form of GeS_2 consists of layers in which equal numbers of GeS_4 tetrahedra share 4 vertices or 2 vertices and 1 edge.[5] Like the normal form of GeS_2 this layer, which is described on p. 199, contains rings of 3 GeS_4 tetrahedra (in addition to larger rings). Complex ions formed from vertex-sharing GeS_4 tetrahedra include the finite A_2X_7 ion in $Na_6(Ge_2S_7)$,[6] the infinite AX_3 chain ion in, for example, Na_2GeS_3 and $PbGeS_3$,[7] and the finite $Ge_4S_{10}^{4-}$ anion in $Cs_4Ge_4S_{10}.4H_2O$, $Ba_2Ge_4S_{10}$, and in the isostructural $Na_4Ge_4S_{10}$ and $Na_4Si_4S_{10}$.[8] This ion has the 'super-tetrahedral' A_4X_{10} structure of which examples are summarized on p. 196. One edge is shared between two GeS_4 groups to form the A_2X_6 ion in salts such as $Na_4Ge_2S_6.14H_2O$ and $Tl_4Ge_2S_6$.[9]

Silanes up to Si_7H_{16} have been prepared, and germanes to Ge_9H_{20} have been characterized. Reduction of an aqueous germanate solution by KBH_4 gives a good yield of GeH_4, from which the higher members are produced by the action of a spark discharge. Some simple molecules GeX_4 are included in Table 21.1, p. 914, and the molecules $H_3Ge-O-GeH_3$ and $\phi_3Ge-O-Ge\phi_3$ in Table 11.3 on p. 500.

The major differences between these elements may be summarized: (i) the much greater stability of Ge(II) than Si(II); (ii) the greater tendency of Ge(IV) to form 6 bonds; and (iii) the formation of salts by Ge(IV).

(i) The divalent state does not enter into normal silicon chemistry, though it may be important at higher temperatures (see SiO, p. 982), but as noted on p. 1151 a number of compounds of Ge(II) are stable at ordinary temperatures. As in the case of Tl(I) compounds the structures of Ge(II) compounds are of special interest because of the potential stereo-activity of the lone pair. The suggestion that a lone pair behaves as if it occupies a volume similar to that of O or F ligands seems to be supported by the fact that it has much more effect (and of a different kind) in oxides and fluorides than in, for example, sulphides and iodides. In the structure of GeF_2[10] trigonal pyramidal GeF_3 groups share two F atoms (F bond angle 157°) to form infinite chains (Fig. 26.9). The bond to the unshared F atom is appreciably

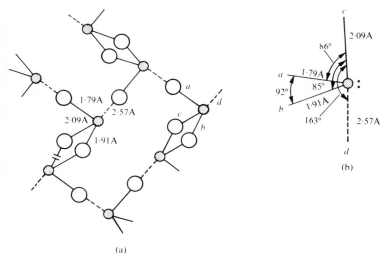

FIG. 26.9. The crystal structure of GeF_2: (a) projection along the chains, (b) environment of Ge.

shorter (1.79 Å) than those in the chain (1.91, 2.09 Å), and weaker bonds (2.57 Å) link the chains into a 3D structure; compare the arrangement of bonds from Sn(II) in $NaSn_2F_5$ (p. 1184):

		GeF_2	$(Sn_2F_5)^-$		GeF_2	$(Sn_2F_5)^-$
(For the labelling	a	1·79 Å	2·08 Å	ab	92°	89°
of the bonds see	b	1·91 Å	2·07 Å	cd	163°	142°
Fig. 26.9).	c	2·09 Å	2·22 Å	ac	86°	81°
	d	2·57 Å	2·53 Å	bc	85°	84°

Regarding Ge as forming three pyramidal bonds the valence group is $(2,\underline{6})$; if the much more distant fourth F is included the valence group would be $(2,\underline{8})$, derived from a trigonal bipyramid, as shown at (b) in Fig. 26.9. (Gaseous GeF_2 exists only at high temperatures in the presence of excess metal, and the vapour contains polymeric species up to $Ge_3F_5^+$; SnF_2 behaves similarly.)[11]

The structure of $GeCl_2$ does not appear to be known, but in the distorted perovskite structure of the low-temperature form of $CsGeCl_3$[12] there is $(3+3)$-coordination, and in $GeBr_2$ also there is very distorted octahedral coordination, with 3 Br at a mean distance of 2.60 Å and 3 more at distances from 3.0 to 3.74 Å. The CdI_2 structure assigned to GeI_2 in an early X-ray study implies octahedral coordination of Ge(II), but it may be necessary to check whether or not Ge has 6 equidistant S neighbours. There is distorted octahedral $(3+3)$ coordination in GeS,[13] the structure of which is compared with that of SnS[13a] on p. 1186.

(ii) We noted at the beginning of Chapter 23 that the number of molecules, ions, or crystals in which Si forms 6 bonds is small. Although the structural chemistry of Ge(IV) is very largely based on 4-coordinated (or 4-covalent) Ge examples of octahedral coordination are more numerous than in the case of Si. There is 4-coordination of Ge in the soluble quartz-like form of GeO_2, but in the stable (tetragonal) form there is 6-coordination (Ge–4 O, 1.87 Å; Ge–2 O, 1.90 Å).[14]

Examples of 6-coordinated Ge(IV) include molecules such as *trans*-$GeCl_4(pyr)_2$[15] (Ge–Cl, 2.27 Å), the GeF_6^{2-} and $GeCl_6^{2-}$ ions, and $FeGe(OH)_6$,[16] which is isostructural with $FeSn(OH)_6$ and $NaSb(OH)_6$ (Ge–6OH, 1.96 Å). There is both tetrahedral and octahedral coordination of Ge in $K_3HGe_7O_{16}.4H_2O$,[2] which has an elegant framework structure built of GeO_4 and GeO_6 groups (Fig. 5.47, p. 237) and possesses zeolitic properties, and in $Na_4Ge_9O_{20}$.[17]

(iii) The more metallic character of Ge is shown in the formation of salts such as $Ge(SO_4)_2$, $Ge(ClO_4)_4$, and $GeH_2(C_2O_4)_3$, which have no silicon analogues.[18]

(1)	AC 1964 **17** 842	(11)	IC 1968 **7** 608
(2)	FM 1966 **43** 230	(12)	ACSc 1965 **19** 421
(3)	AC 1976 **B32** 1188	(13)	AC 1978 **B34** 1322
(4)	Sci 1965 **149** 535	(13a)	IC 1965 **4** 1363
(5)	AC 1975 **B31** 2060	(14)	AC 1971 **B27** 2133
(6)	RCM 1974 **11** 13	(15)	JCS 1960 366
(7)	RCM 1972 **9** 757; AC 1974 **B30** 1391	(16)	AC 1961 **14** 205
(8)	JSSC 1973 **8** 195	(17)	ACSc 1963 **17** 617
(9)	AC 1978 **B34** 2614	(18)	ACSc 1963 **17** 597
(10)	JCS A 1966 30		

The structural chemistry of tin and lead

A comparison of the structural chemistry of these two elements reveals some interesting resemblances and also some remarkable differences. The atoms of each element have the same outer electronic structure, two s and two p electrons, and each has oxidation states of 2 and 4. The more metallic nature of lead is shown by the difference between the structures of the elements (Chapter 29) and by many differences between stannous and plumbous compounds. It will be convenient to deal first with Sn(IV) and Pb(IV) since the structural chemistry is more straightforward for the high oxidation state. Few compounds of Sn(II) and Pb(II) are isostructural and compounds containing these metals in both oxidation states are different for the two elements (e.g. Sn_2S_3 and Pb_3O_4).

Stannic and plumbic compounds

Bond arrangements found are tetrahedral, trigonal bipyramidal, and octahedral for completely shared valence groups of 8, 10, and 12 electrons respectively. Higher coordination numbers are exhibited by Sn(IV), 7, and 8, and by Pb(IV), 8, in certain complexes formed by chelating ligands, most or all of the bonds being formed to O atoms.

Tetrahedral coordination. The simplest examples are molecules MX_4 and $MX_{4-x}Y_x$, a number of which have been shown to be tetrahedral either in the vapour or crystalline states:

$SnCl_4$, $SnBr_4$, SnI_4, $Sn(C_6H_5)_4$[1a] (crystal); $(CH_3)_3SnX$ etc., $PbCl_4$, $Pb(CH_3)_4$ and $Pb_2(CH_3)_6$ in the vapour state (references in Table 21.1, p. 914). Like Si in $(SiR)_4S_6$ Sn(IV) forms molecules of the adamantane type based on the tetrahedral skeleton A_4X_6 of Fig. 3.18(c), p. 94. An example is $[Sn(CH_3)]_4S_6$,[1b] in which Sn is bonded to CH_3 at 2.15 Å and to 3 S at 2.39 Å. In contrast to the tetrahedral structure of $(CH_3)_3SnCl$ in the vapour state there is considerable distortion of the molecule in the crystal, weak bridging Sn—Cl bonds suggesting a tendency to polymerize to an octahedral chain structure:[1c]

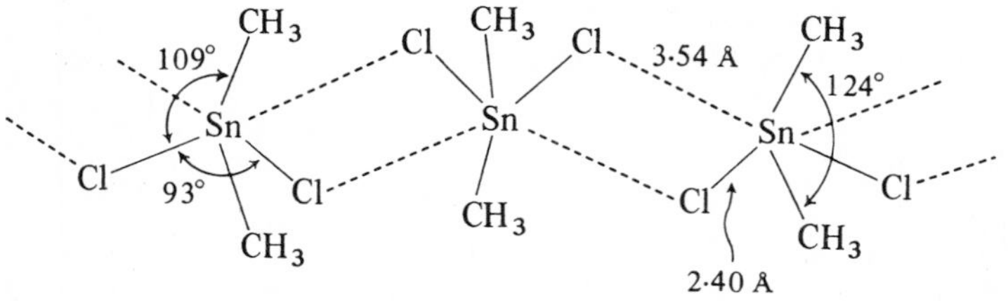

Some compounds $(CH_3)_3SnX$ and $(CH_3)_2SnX_2$ show a clear-cut change from a tetrahedral monomer to a polymeric structure with 5- or 6-coordination of Sn (see later).

Monomeric $Sn^{II}R_2$ compounds generally polymerize to Sn^{IV} compounds, either linear (for example, $H[Sn(C_6H_5)_2]_6H$) or cyclic molecules. An example of the latter is $[Sn(C_6H_5)_2]_6$, Fig. 26.10(a), which has been studied in the crystalline complex with *m*-xylene.[2] The Sn—Sn bond length in the ring (2.78 Å) is the same as in grey tin (2.80 Å). There is also tetrahedral coordination of Sn in complexes containing the ligand —$SnCl_3$ such as $(C_8H_{12})_3Pt_3(SnCl_3)_2$ (see p. 437). In these complexes Sn is bonded to 3 Cl and 1 Pt atom.

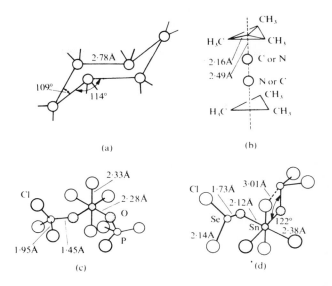

FIG. 26.10. Stereochemistry of Sn(IV):   (a) cyclic $[Sn(C_6H_5)_2]_6$ molecule (phenyl groups omitted); (b) infinite chains in $(CH_3)_3SnCN$; (c) $SnCl_4 . 2POCl_3$; (d) $SnCl_4 . 2SeOCl_2$.

There are numerous molecules in which Sn is bonded to other metal atoms to which are attached ligands such as CO, C_5H_5, and phosphines. Four bonds from Sn in such cases are arranged tetrahedrally with various degrees of distortion from regular tetrahedral bond angles. For example, in $Sn[Fe(CO)_4]_4$[3] there is appreciable distortion to bring together one pair of Fe atoms (Fe—Fe, 2.87 Å). The special interest of these compounds lies in the metal–metal bonding.

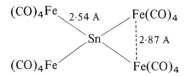

There is tetrahedral coordination of Sn(IV) in numerous thiostannates, but there is also 5- and 6-coordination of the metal atom in some of these compounds. For this reason we devote a short section later to this group of compounds.

Trigonal bipyramidal coordination.   The simplest example is the $(SnCl_5)^-$ ion[4] which has been studied in the salt of the substituted cyclobutenium ion (a). Both 5- and 6-coordination of Sn(IV) occur in crystalline $[(CH_3)_2SnCl(terpyridyl)]^+$-$[(CH_3)_2SnCl_3]^-$, which is an assembly of the ions (b) and (c).[5] Crystalline compounds $(CH_3)_3SnX$ do not consist of tetrahedral molecules like, for example, $(CH_3)_3GeCN$[6] or of ions $Sn(CH_3)_3^+$ and X^-. They consist of infinite

(a)   (b)   and   (c)

linear molecules in which planar $Sn(CH_3)_3$ groups are bonded through X atoms. The CH_3 groups occupy the equatorial positions of the trigonal bipyramidal coordination group of $Sn(IV)$. Numerous compounds of this type have been studied in which X is F,[7] OH,[8] CN,[9] $-O.C(CH_3).O-$,[10] and dicyanamide.[10a] In the zigzag chain of $(CH_3)_3SnOH$ the O bond angle is 138° and the Sn—O bond lengths are 2.20 and 2.26 Å in the slightly unsymmetrical bridges. In $(CH_3)_3Sn(dicyanamide)$ the angular bridging ligand has the dimensions shown. (Contrast $(CH_3)_2Sn(NCNCN)_2$, in which there is *octahedral* coordination of the metal, as in $Sn(CH_3)_2F_2$; there is a square net of Sn atoms which are bridged by —NCNCN— groups, the octahedron being completed by two axial CH_3 groups.) The compounds $(CH_3)_3PbOH$ and $(CH_3)_3Pb(CH_3COO)$ are isostructural with the Sn compounds, and are of special interest as rare examples of trigonal bipyramidal coordination of $Pb(IV)$.

Octahedral coordination. The structures of crystalline SnF_4[11] and $SnF_2(CH_3)_2$[12] have been described in Chapter 5. Octahedral SnF_6 and $SnF_4(CH_3)_2$ groups respectively share four equatorial vertices (F atoms) to form infinite layers:

	Sn—F (bridging)	Sn—F (terminal)	Sn—C
SnF_4	2.02 Å	1.88 Å	
$SnF_2(CH_3)_2$	2.12 Å		2.08 Å

Numerous octahedral ions MX_6^{2-} are formed by these elements. For example, Rb_2SnCl_6 and Rb_2PbCl_6 both crystallize with the K_2PtCl_6 structure; ions intermediate between SnF_6^{2-} and $Sn(OH)_6^{2-}$ have been studied in solution[13] and isolated in salts such as $M_2SnF_5(OH)$ and $M_2SnF_4(OH)_2$.[14]

In the crystalline compounds $SnCl_4.A_2$, where A is, for example, $POCl_3$[15] or $SeOCl_2$,[16] two O atoms of the A molecules complete octahedral coordination groups around $Sn(IV)$ with the *cis* configuration (Fig. 26.10(c) and (d)). The compound with C_4H_8S (tetrahydrothiophene)[17] is exceptional not only in having the *trans* configuration, attributable to the bulky ligand, but also in having a high dipole moment in solution (5 D) for which there is no obvious explanation. Other examples of octahedral Sn(IV) complexes include the molecule of dimethyl Sn bis-(8-hydroxyquinolinate), (a),[18] and $SnCl_4[NC(CH_2)_3CN]$,[19] (b), in which the

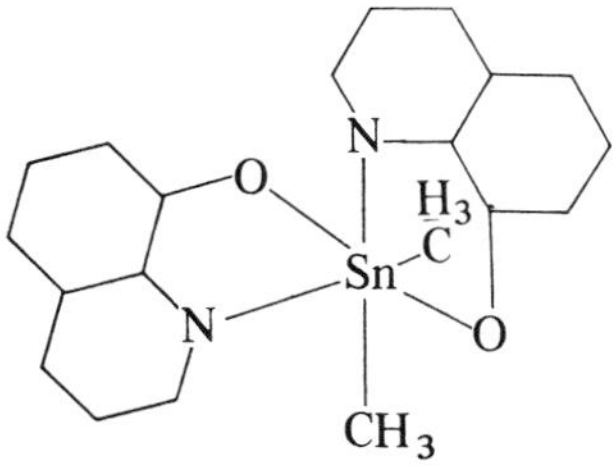

(a)

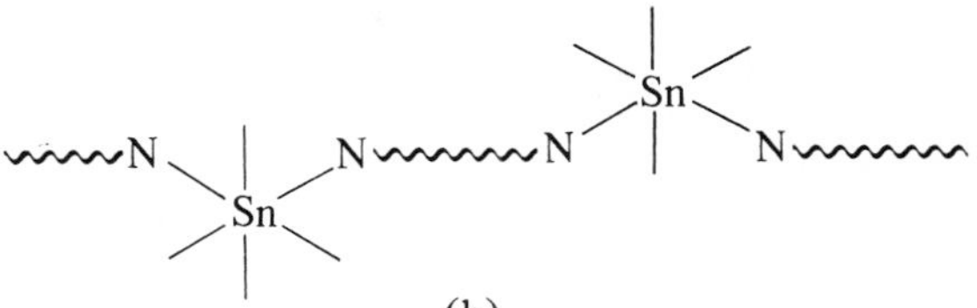

(b)

glutaronitrile molecules link the Sn atoms into infinite chains (note the *cis* N atoms in the coordination group of Sn).

There is octahedral coordination of the metal atoms in the dioxides of Sn and Pb; these are essentially ionic crystals. The structure of $Sn(OH)_4$ is not known. The gel-like $Sn(OH)_4$, precipitated from $SnCl_4$ solution by NH_4OH, dries to constant weight at 110 °C with the composition SnO_3H_2. Dehydration to SnO_2 above 600 °C apparently takes place through a number of definite crystalline phases with the compositions $Sn_2O_5H_2$, $Sn_4O_9H_3$, and $Sn_8O_{16}H_2$.[20] An example of a binuclear molecule with a double hydroxyl bridge is the compound with the empirical formula $SnCl_2(OH)(C_2H_5)(H_2O)$,[21] (c).

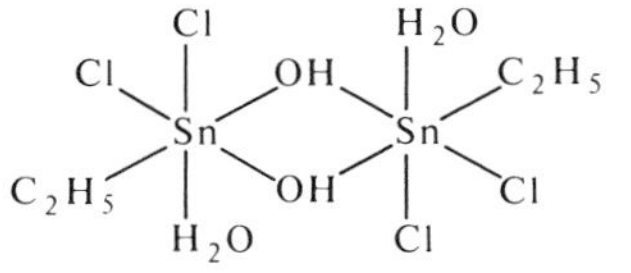

(c)

In SnS_2 (CdI_2 structure) Sn has 6 octahedral neighbours, and S forms 3 pyramidal bonds. Lead forms no disulphide under atmospheric pressure, but two high-pressure forms of PbS_2 have been made,[22] one of which probably has the CdI_2 structure, like SnS_2. For references to octahedral Sn(IV) complexes see ref. (23).

7- and 8-coordinated Sn(IV); *8-coordinated* Pb(IV). Examples of structural studies which definitely establish c.n.s greater than 6 for Sn(IV) or Pb(IV) are few in number, and all involve coordination wholly or largely by pairs of O atoms in special chelating ligands.

In the monohydrate $SnY(H_2O)$,[24] where H_4Y = edta, Sn(IV) is bonded to three pairs of O atoms of the ligand and to one H_2O molecule. There is no simple description of the geometry of the 7-coordination polyhedron, which is very similar to that in $Mn^{II}HY(H_2O)$ — illustrated in Fig. 27.3(d). With the tropolonato (T) ligand, (a), the metal forms chelate complexes SnT_3Cl and $SnT_3(OH)$ the configuration of which approximates to pentagonal bipyramidal, (b).[25] In the ligand T the distance

(a)   (b)

between the O atoms is 2.56 Å. In the bidentate NO_3^- ion it is only 2.14 Å, and in the $Sn(NO_3)_4$ molecule[26] there is a tetrahedral arrangement of four NO_3^- groups around the 8-coordinated $Sn(IV)$ atom. This special type of 8-coordination is also possible with the acetato ligand, in which O–O is similar to that in NO_3^-; 8-coordination of $Pb(IV)$ is found in $Pb(ac)_4$, the coordination polyhedron having the form of a flattened triangulated dodecahedron.[27]

(1a)	JCS A 1970 911	(13)	PCS 1964 407
(1b)	AC 1972 **B28** 2323	(14)	ZaC 1963 **325** 442
(1c)	JCS A 1970 2862	(15)	ACSc 1963 **17** 759
(2)	IC 1963 **2** 1310	(16)	AC 1960 **13** 656
(3)	JCS A 1967 382	(17)	JCS 1965 1581
(4)	JACS 1964 **86** 733	(18)	IC 1967 **6** 2012
(5)	JCS A 1968 3019	(19)	IC 1968 **7** 1135
(6)	IC 1966 **5** 511	(20)	IC 1967 **6** 1294
(7)	JCS 1964 2332	(21)	AC 1976 **B32** 923
(8)	AC 1978 **B34** 129	(22)	IC 1966 **5** 2067
(9)	IC 1966 **5** 507	(23)	JCS A 1970 1257
(10)	AC 1975 **B31** 2740	(24)	IC 1971 **10** 2313
(10a)	IC 1971 **10** 1938	(25)	JACS 1970 **92** 3636
(11)	NW 1962 **49** 254	(26)	JCS A 1967 1949
(12)	IC 1966 **5** 995	(27)	AC 1963 **16** A34

Structures of some thiostannates

We shall not attempt here a survey of $Sn(IV)$–S compounds but merely describe the structures of a few thiostannates as an introduction to a very interesting field of crystal chemistry. In these compounds $Sn(IV)$ exhibits c.n.s of 4 (tetrahedral), 5 (trigonal bipyramidal), and 6 (octahedral); their structures therefore have features in common with those of compounds of Ge and Pb. Thiostannates of Na include the following (in order of decreasing S:Sn ratio): Na_4SnS_4, $Na_6Sn_2S_7$, $Na_4Sn_2S_6$, Na_2SnS_3,[1] and $Na_4Sn_3S_8$.[2] In the first three there is tetrahedral coordination of Sn in finite ions which have also been studied in anhydrous Ba salts and in hydrated salts such as $Na_4SnS_4 \cdot 14H_2O$ and $Na_4Sn_2S_6 \cdot 14H_2O$. In Na_2SnS_3, as in SnS_2, there is octahedral coordination of Sn. The structure is a NaCl superstructure in which the octahedral interstices between c.p. S layers are occupied alternately by ($\frac{1}{3}Na + \frac{2}{3}Sn$) and (all Na). The pattern of sites occupied by Sn in the first type of layer is that of Fig. 4.24(e), p. 173, and an isolated layer would have the composition $NaSn_2S_6$. Such layers alternating with Na_3 layers give the composition Na_2SnS_3. In this structure each SnS_6 octahedron shares 3 edges with other SnS_6 octahedra. In $PbSnS_3$ (NH_4CdCl_3 structure) each SnS_6 group shares 4 edges and in SnS_2 (CdI_2 structure) 6 edges. We might add to this series the structure of La_2SnS_5,[3] which consists of rutile-type chains, in which each SnS_6 group shares 2 edges, together with additional S atoms as noted on p. 214.

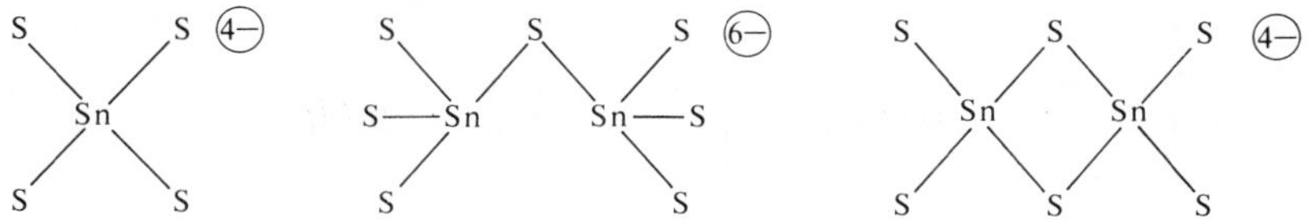

From the structural standpoint the most interesting of the Na thiostannates is $Na_4Sn_3S_8$, in which there are Sn atoms of two kinds, one-third (Sn_I) having tetrahedral and two-thirds (Sn_{II}) trigonal bipyramidal coordination. The latter groups share 2 edges (dotted lines in (a)) with other similar groups and 1 vertex

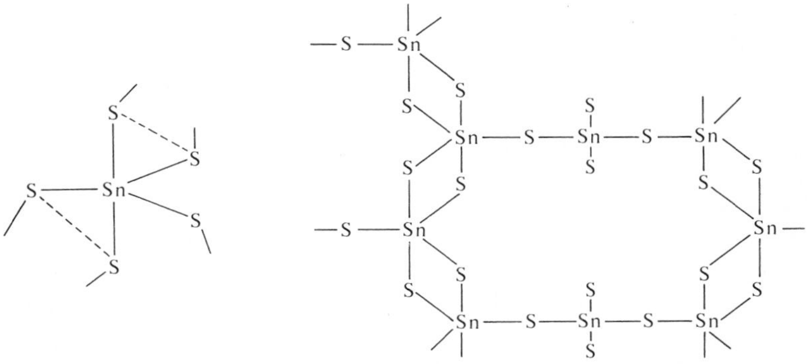

(a)   (b)

with a tetrahedral SnS_4 group. The SnS_4 groups merely act as (2-connected) links between the SnS_5 groups, the latter forming a 3D 3-connected system which is simply an elaboration of the (10,3)-b net of Fig. 3.29(a), p. 113. The rings of $6 Sn_{II}$ (and $2 Sn_I$) atoms shown in the diagrammatic projection (b) are actually helices in the 3D structure.

Mean bond lengths in these compounds are: Sn—4S (tetrahedral), 2.40 Å; Sn—5S (trigonal bipyramidal), 2.59 Å (axial), and 2.42 Å (equatorial); Sn—6S (octahedral), 2.55 Å. References to the other compounds mentioned will be found in refs. (1) and (2).

(1) AC 1974 **B30** 2620   (3) AC 1974 **B30** 2283
(2) JSSC 1975 **14** 152

Stannous and plumbous compounds

Examples of isostructural compounds of Sn(II) and Pb(II) are few in number; they include the tetragonal forms of the monoxides, the oxyhydroxides $M_6O_4(OH)_4$, $CsSnI_3$ and $CsPbI_3$ (see below). Both Sn(II) and Pb(II) form covalent compounds in which they form a small number of bonds the arrangement of which shows that the lone pair is stereochemically active, but the effect of the lone pair is much more evident in the Sn than in the Pb compounds and more evident the more electronegative the ligands; in this respect Sn(II) shows more resemblance to Ge(II) than to Pb(II). Regular octahedral coordination of Pb(II) is found in PbS (NaCl structure) and PbI_2 (CdI₂ structure) but is rare in Sn(II) compounds—see SnS and SnI_2 (later). There is no characteristic 'lone pair effect' in the l.t. forms of $CsPbI_3$ or $CsSnI_3$ or in $KPbI_3 . 2H_2O$, though some distortion of the octahedral coordination groups is to be expected in the double octahedral chain present in all these compounds.[1] In PbF_2, $PbCl_2$, and $PbBr_2$ Pb forms the colourless Pb^{2+} ion in structures of high coordination, 8 or (7+2), p. 416—contrast SnF_2 and $SnCl_2$ (below)—and high c.n.s, 8–12, are also found in plumbous oxy salts, as noted on p. 1153. We refer to the structures of $SnWO_4$ and $SnSO_4$ later; $SnCO_3$ and $Sn(NO_3)_2$ do not exist. Details of the structures of the Pb(II) halides and oxy salts are included in other chapters. We shall deal here in some detail with stannous compounds and conclude with a few examples of covalent Pb(II) compounds.

Whereas the structures of $SnCl_2$ and $SnBr_2$ are related to those of $PbCl_2$ and $PbBr_2$, modified to give Sn(II) a small number of closest neighbours consistent with the stereo-activity of the lone pair, the structures of SnF_2 and SnI_2 are quite different from those of the corresponding Pb halides. Stannous fluoride is apparently trimorphic,[1a] having in addition to the normal (α) form two metastable forms with structures similar to those of GeF_2 and TeO_2 (paratellurite). The normal form consists of tetrameric molecules (a) in which the mean Sn—F bond length in the Sn_4F_4 ring is 2.18 Å and the length of the extra-cyclic bonds is 2.05 Å.[1b] There are Sn(II) atoms with two quite different arrangements of ligands, (b) and (c), where broken lines represent intermolecular bonds, but the three shortest bonds from each are those within the Sn_4F_8 molecule and are pyramidal with mean bond angle 84°. However, having regard to all the ligands the bond arragements may be described as tetrahedral SnF_3E and octahedral SnF_5E, where E stands for the lone pair occupying an orbital.

In the 1:1 compound $SnF_2 . AsF_5$ there are cyclic cations $Sn_3F_3^{3+}$ (d), and the structural formula is therefore $(Sn_3F_3)(AsF_6)_3$.[1c] Here there is the characteristic

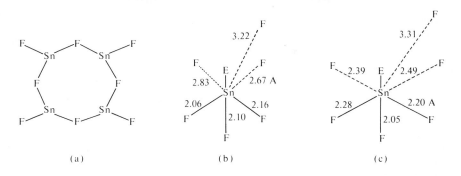

(a)   (b)   (c)

'one-sided' 4-coordination (e), variously described as square pyramidal or as derived from a trigonal bipyramid, SnX_4E. There are two further sets of neighbours, 2 F at 2.85 Å and 2 F at 3.05 Å.

(d)   (e)   (f)

There are many halides of Sn(II) containing different halogens. The structure of $SnClF^{(2)}$ could be described as a very distorted $PbCl_2$ structure, but the four close neighbours (1 F at 2.18; 2 F at 2.39; and 1 Cl at 2.52 Å) delineate chains, (f), very similar to those in $L-SbOF$ (p. 894). In $Sn_2ClF_3^{(3)}$ the Sn and F atoms form a 3D 3-connected net of Sn atoms joined through F atoms (the cubic net of Fig. 3.28(a), p. 111) in which Sn forms three pyramidal bonds ($Sn-F$, 2.15 Å, interbond angles close to 80°). The Cl^- ions occupy the interstices in the 3D $Sn_2F_3^+$ cationic framework, the shortest $Sn-Cl$ distances being 3.14 Å. In $Sn_2IF_3^{(3a)}$ there are $Sn_2F_3^+$ layers in which Sn is 4-coordinated ($Sn-F$, 2.20, 2.23 and 2.38 Å (two)). The layers are built from chains similar to those in SnClF; Cl is replaced by F and the chains joined by sharing these F atoms. In $Sn_3BrF_5^{(3)}$ there is a 2D cation $Sn_3F_5^+$, the layer being the 3-connected 4.8^2 net (p. 82) with additional '2-connected' Sn atoms along the links common to the 8-gons; there is, of course, a F atom between each pair of Sn atoms forming the net and a third (unshared) F atom attached to each '2-connected' Sn atom (see p. 110). This layer could alternatively be described as formed from the Sn_4F_4 rings of SnF_2 joined through additional SnF_3 groups, the latter sharing only two F atoms.

Mixed valence fluorides are formed by Ge, Sn, and Pb. In $Sn_3F_8^{(3b)}$ pyramidal SnF_3 groups form chains which are cross-linked by octahedral $Sn^{IV}F_6$ groups.

In the system $NaF-SnF_2-H_2O$ the only compounds formed are $NaSn_2F_5$ and $Na_4Sn_3F_{10}$, while KF forms $KSnF_3.\frac{1}{2}H_2O$; note the absence of M_2SnF_4. In $KSnF_3.\frac{1}{2}H_2O^{(4)}$ the infinite anion consists of square pyramidal SnF_4 groups sharing opposite vertices (Fig. 26.11(a)), and the water molecules are not bonded to the Sn

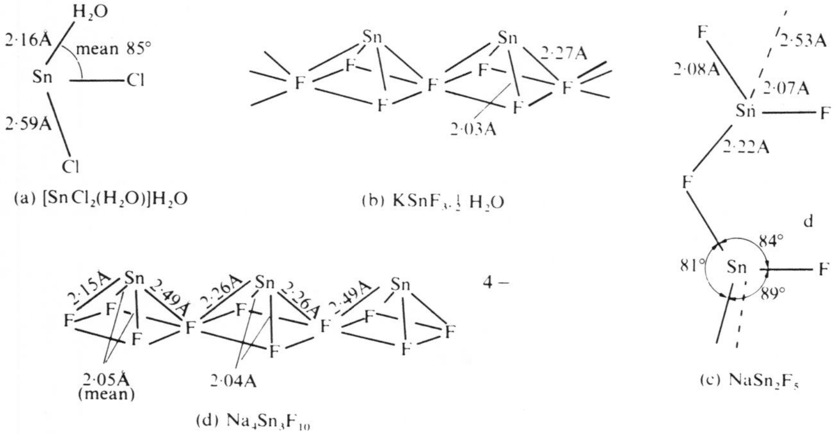

(a) $[SnCl_2(H_2O)]H_2O$ (b) $KSnF_3.\frac{1}{2}H_2O$

(d) $Na_4Sn_3F_{10}$

(c) $NaSn_2F_5$

FIG. 26.11. Halogen–metal complexes in stannous compounds: (a) $KSnF_3.\frac{1}{2}H_2O$; (b) $NaSn_2F_5$; (c) $Na_4Sn_3F_{10}$.

atoms. In $NaSn_2F_5^{(5)}$ pairs of pyramidal SnF_3 groups share a F atom to form $Sn_2F_5^-$ ions which are then linked through weaker bonds (2.53 Å) to form chains (Fig. 26.11(b)). The terminal and bridging bonds are very similar in length to those in $KSnF_3.\frac{1}{2}H_2O$. In $Na_4Sn_3F_{10}^{(6)}$ there are complex anions formed from three square pyramidal SnF_4 groups, but there is a large range of bond lengths (Fig. 26.11(c)). In particular the outer bridging bonds are very long, suggesting a tendency to split into a central SnF_4^{2-} ion and two SnF_3^- ions.

The SnX_4E arrangement of 4 bonds from $Sn(\textsc{ii})$ is found, together with trigonal pyramidal coordination, in Sn_2OF_2.$^{(7)}$ Sub-units of type (g), in which the lone

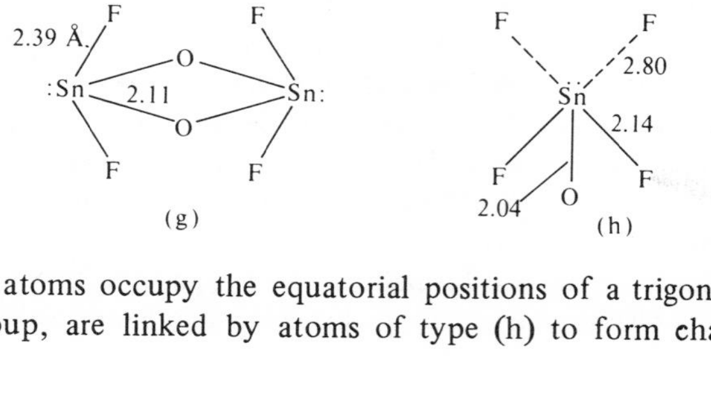

(g)

(h)

pair and two O atoms occupy the equatorial positions of a trigonal bipyramidal coordination group, are linked by atoms of type (h) to form chains which are

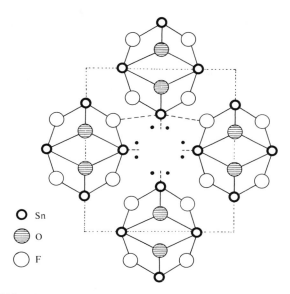

O Sn

◎ O

◯ F

FIG. 26.12. The structure of Sn_2OF_2 viewed along the direction of the chains.

viewed end-on in Fig. 26.12. The packing of the chains shows the importance of the lone pairs, for whereas the Sn atoms of type (h) have two additional F neighbours in adjacent chains those of type (g) have no next nearest F or O neighbours but only Sn atoms (at 3.29 Å)—compare the Sn (Pb) atoms on the outer surfaces of the layers in the monoxides and the Tl(I) atoms around the tunnels in Tl_3BO_3 (p. 1067).

Although the structure of $SnCl_2$[8a] could be described as a very distorted $PbCl_2$ structure the shortest bonds define pyramidal $SnCl_3$ groups, in which the interbond angles are 80° (two) and 105° (one) and the bond lengths 2.66 Å (one) and 2.78 Å (two). The more distant neighbours are at 3.06 Å (2), 3.22 Å (1), 3.30 Å (1), and 3.86 Å (2). The Raman spectrum of molten $SnCl_2$[8b] is very similar to that of the solid, suggesting that the liquid also consists of chains of linked $SnCl_3$ groups. The compound originally formulated as $SnCl_2.2H_2O$ is in fact $[SnCl_2(H_2O].H_2O$,[9] consisting of pyramidal $SnCl_2(H_2O)$ groups (Sn—Cl, 2.59; Sn—O, 2.16 Å; mean bond angle 85°) and separate H_2O molecules. There are pyramidal $SnCl_3^-$ ions in $K_2(SnCl_3)Cl.H_2O$,[10] previously written $K_2SnCl_4.H_2O$, and in $CsSnCl_3$.[11] (Other pyramidal anions include those in $K[Sn(O_2CH)_3]$[12a] and $Sr[Sn(CH_2ClCOO)_3]_2$[12b] in which Sn—O is close to 2.15 Å and the O—Sn—O bond angles lie in the range 80–90°.

The structure of $SnBr_2$[13] is of the same general type as those of $PbBr_2$ and $SnCl_2$ with very irregular 8-coordination, the nearest neighbours being at 2.81 (one), 2.90 (two), and 3.11 Å (two). That of SnI_2[14] is a unique AX_2 structure of unexpected complexity which is shown in projection in Fig. 26.13. One-third of the Sn atoms (Sn′) are in positions of nearly regular octahedral coordination in rutile-like chains which are cross-linked by double chains containing the remaining metal atoms. The latter (Sn") form 5 shorter and 2 longer bonds to I atoms:

$$
\text{Sn''–I} \quad \left\{ \begin{array}{ll} a & 3.00\,\text{Å} \\ 2b & 3.20 \\ 2c & 3.25 \\ 2d & 3.72 \end{array} \right\} \quad \text{mean } 3.18\,\text{Å} \qquad \text{Sn'–I} \quad 6 \text{ at } 3.16\,\text{Å}
$$

The bonds b link the double chains to the rutile-like chains to form layers in the direction of the arrows in Fig. 26.13. These layers are linked by the weaker d bonds into a 3D structure. All the five shorter bonds lie to one side of Sn", the coordination of which may be described as monocapped trigonal prismatic. The two longer bonds go to atoms of one edge; with this (5+2)-coordination compare the (7+2)-coordination of Pb in $PbCl_2$. The two longer bonds, with electrostatic bond strength 1/6 as compared with 1/3 for the other five, can be regarded as replacing the sixth octahedral bond formed by Sn(II).

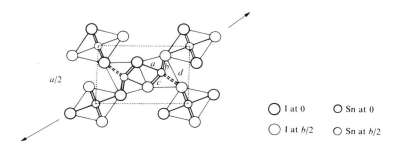

O I at 0   O Sn at 0

O I at $b/2$ O Sn at $b/2$

FIG. 26.13.   Projection of one-half of a unit cell of the crystal structure of SnI_2. The chains containing the two kinds of Sn atom are perpendicular to the paper.

Simple dihydroxides $M(OH)_2$ are not formed by Sn (or Pb); instead they form $M_6O_4(OH)_4$[15] similar in structure to $U_6O_4(OH)_4^{12-}$ (Fig. 14.15, p. 646). By careful heating the Sn compound may be converted into a metastable red form of SnO. The usual blue-black form is isostructural with tetragonal PbO, in which the metal forms 4 square pyramidal bonds (Sn–O, 2.22 Å).[15a]

The structure of GeS and isostructural SnS is a layer structure similar to that of black P (see pp. 103 and 839) which may also be described as a very deformed version of the NaCl structure in which the metal atom has 3 pyramidal neighbours; compare the environments of a metal atom in GeS, SnS, and PbS:

$$
\text{Ge} \left\{ \begin{array}{ll} 3\,\text{S} & 2.44\,\text{Å} \\ 2\,\text{S} & 3.27 \\ 1\,\text{S} & 3.28 \end{array} \right. \qquad \text{Sn} \left\{ \begin{array}{ll} 1\,\text{S} & 2.62\,\text{Å} \\ 2\,\text{S} & 2.68 \\ 2\,\text{S} & 3.27 \\ 1\,\text{S} & 3.39 \end{array} \right. \qquad \begin{array}{ll} \text{Pb} & 6 \text{ equidistant} \\ & \text{S at } 2.97\,\text{Å}. \end{array}
$$

There is a similar bond arrangement around Sn in $BaSnS_2$,[16a] namely, metal atom off-centre in an octahedral group of S atoms (3 S at 2.57, 3 S at 3.59–3.79 Å). The structure may be regarded as a distorted NaCl superstructure in the same way that SnS is a distorted form of the simple NaCl structure.

The structure of SnP_3[16b] is closely related to that of SnS. It consists of As-like layers (Fig. 3.2, p. 67) which have a less buckled configuration of the plane 6^3 net than that in black P. One-quarter of the P atoms are replaced by Sn as noted on p. 843. $Sn(II)$ is bonded to 3 P at 2.66 Å and has 3 P at 2.93 Å in the adjacent layer.

The structures of $SnSO_4$ and $SnWO_4$ are distorted forms of the high-coordination structures of $PbSO_4$ and $PbWO_4$. In the Pb compounds there are symmetrical arrangements of 12 O and 8 O respectively around the metal ions, the lone pair presumably occupying the s orbital; in the Sn compounds the lone pair is stereo-active and the coordination groups are distorted. In $SnSO_4$[17] Sn has 3 pyramidal neighbours at 2.26 Å and 9 more at distances of 2.95–3.34 Å, a total of 12. The high-temperature cubic form of $SnWO_4$[18a] contains WO_4^{2-} tetrahedra and there is distorted octahedral coordination of Sn with 3 O at 2.21 and 3 O at 2.81 Å, the lone pair projecting through the enlarged octahedron face (monocapped octahedral coordination). In low-$SnWO_4$ (orthorhombic)[18b] there is distorted octahedral coordination of both Sn and W (W—O, 1.80–2.14 Å), but the coordination group of $Sn(II)$ is different from that in the h.t. form. The 8-coordination group of the scheelite structure breaks down into sets of 2 O at 2.18 Å, 2 O at 2.39 Å (the typical 'one-sided' 4-coordination), with 2 O at 2.83 Å and 2 much more distant at 3.44 Å.

We may summarize as follows. (i) If the ligands are F or O the coordination group is an unsymmetrical arrangement of a small number of ligands, usually 3 or 4 but sometimes 5, all lying to one side of the metal atom. The effect of the lone pair is less marked with the larger (and less electronegative) halogens and chalcogens and is not observed at all in some SnI_6 or $SnTe_6$ coordination groups; SnTe has, like PbS, PbSe, and PbTe, the NaCl structure. (ii) There are many structural resem-blances between compounds of $Sn(II)$ and those of other 'lone pair elements', notably $Ge(II)$, $Sb(III)$, $Se(IV)$, and $I(V)$. (iii) The type of distortion of the bond arrangement due to the presence of the lone pair is very sensitive to the geometry and topology of the structure as a whole, as evidenced by the fact that different bond arrangements may be found in the same crystal; two in α-SnF_2, Sn_2OF_2, and $Sn_{10}W_{16}O_{46}$,[19] and three in $Sn_3(PO_4)_2$.[20] In this orthophosphate, as in Sn_2OF_2 and $TlBO_3$, the lone pairs point into tunnels in the structure. The gross geometry of such structures, together with the immediate bond arrangement around the metal atom and calculations of the molecular volume per anion in a number of oxides, oxy salts and fluorides, show that the lone pairs may be regarded as occupying a volume similar to that of O^{2-} or F^-. Two ways of systematizing the crystal chemistry of $Sn(II)$ and other lone pair elements have been suggested.

In the electrostatic approach[21a, b] it is assumed that the coordination group of a 'normal' ion of the size of Sn^{2+} (i.e. without a lone pair) would be a regular octahedron, but that extra stabilization energy can be gained by mixing of the ground state $5s^2$ with the first excited state $5s^1p^1$ resulting in distortion of the octahedral group; compare the crystal field treatment of transition metal ions, p. 322). Alternatively[22] the ligands together with the lone pair are regarded as occupying a number of hybrid orbitals; for example, a trigonal pyramidal arrangement of ligands is described as tetrahedral sp^3, (SnX_3E), and an arrangement of four ligands as derived from an sp^3d trigonal bipyramidal arrangement of five orbitals (SnX_4E). In this connection we should note the difficulty of deciding how to describe less regular coordination groups; see, for example, p. 76 and ref. (23).

In essentially covalent molecules Pb(II) forms 2, 3, or 4 bonds. The molecules $PbCl_2$, $PbBr_2$, and PbI_2 (and also the Sn halides) are non-linear (p. 444), but the interbond angles do not appear to have been determined. There is irregular 7- or 8-coordination of Pb^{2+} in the monoclinic form of $NH_4Pb_2Cl_5$, one of a family of isostructural compounds AB_2X_5 in which A is K, Rb, Tl, or NH_4, B is Pb or Sr, and X is Cl or Br.[24] In the tetragonal form of $NH_4Pb_2Br_5$[25] there is irregular 8-coordination of Pb^{2+}, with 2 Br appreciably closer (2.89 Å) than the others (3.16–3.35 Å), and in $[Co(NH_3)_6](Pb_4Cl_{11})$[26] a 3D framework anion is built from monocapped trigonal prismatic $PbCl_7$ coordination groups.

(a)   (b)

A rather irregular pyramidal arrangement of three bonds is formed by Pb in $Pb(N_2S_2)NH_3$,[27] (a), in which the system PbN_2S_2 is planar, Pb–S, 2.73, Pb–NH_3, 2.24, and Pb–N_{ring}, 2.29 Å. The square pyramidal bond arrangement is found in molecules of the diethyl thiocarbamate,[28] (b), and the closely related ethyl-xanthate[29] and diethoxydithiophosphate.[30] Non-planar molecules of Pb phthalo-cyanin[31] stack in columns with the Pb atoms vertically above one another separated by only 3.73 Å, not much greater than the Pb–Pb distance in the metal (3.48 Å). Presumably an appreciable degree of ionic character is responsible for the much less regular arrangements of 7 or 8 neighbours found in compounds such as $PbCl_2$-(thiourea)$_2$[32] and Pb(thiourea) acetate), $Pb[SC(NH_2)_2](C_2H_3O_2)_2$.[33]

In contrast to the colourless PbF_2, $PbCl_2$, and oxy-salts such as the nitrate, sulphate, etc. containing Pb^{2+} ions, the oxides are highly coloured. (Note also that whereas $PbSO_4$ and $PbMoO_4$ are colourless and $PbCrO_4$ pale-yellow, coprecipitated solid solutions of these compounds with appropriate $CrO_4:SO_4:MoO_4$ ratios have intense scarlet colour, the scarlet chrome pigments.) In tetragonal PbO and Ag_2PbO_2[34] a Pb atom is bonded to 4 O atoms at the basal vertices of a square pyramid, while in Pb_3O_4, Pb_2TiO_4, and $PbCu(OH)_2SO_4$[35] Pb^{II} is 3-coordinated, and an analogy may be drawn witn $Sb(III)$. Instead of emphasizing the $Pb^{IV}O_6$ octahedra in Fig. 12.15, p. 559, we could distinguish $(Pb_2^{II}O_4)_n^{4n-}$ chains in which Pb^{II} forms three pyramidal bonds, like Sb^{III} in $MgSb_2O_4$. We have already noted the NaCl structure of PbS, but the physical properties of galena, for example, its opacity and brilliant metallic lustre, are not those of simple ionic crystals.

For $Pb(C_5H_5)_2$ see p. 974; for $Pb_4(OH)_4^{4+}$ and $OPb_6(OH)_6^{4+}$, p. 627, and for $Pb_6O_4(OH)_4$ and $Sn_6O_4(OH)_4$, pp. 376 and 1186.

(1)   AC 1980 **B36** 683, 782
(1a) JSSC 1980 **33** 1
(1b) JSSC 1979 **30** 335
(1c) AC 1977 **B33** 232
(2)   AC 1976 **B32** 3199
(3)   JCS D 1977 865
(3a) AC 1978 **B34** 3308
(3b) CC 1973 944
(4)   AC 1968 **B24** 803
(5)   AC 1964 **17** 1104
(6)   AC 1970 **B26** 19
(7)   AC 1977 **B33** 1489
(8a) AC 1962 **15** 1051
(8b) JCP 1967 **47** 1923
(9)   JCS 1961 3954
(10) JINC 1962 **24** 1039
(11) ACSc 1970 **24** 150
(12a) ACSc 1969 **23** 3071
(12b) AC 1980 **B36** 1935
(13) ACSc 1975 **A29** 956
(14) AC 1972 **B28** 2965
(15) N 1968 **219** 372
(15a) AC 1980 **B36** 2763

(16a) AC 1973 **B29** 1480
(16b) JSSC 1972 **5** 441
(17) AC 1972 **B28** 864
(18a) AC 1972 **B28** 3174
(18b) AC 1974 **B30** 2088
(19) AC 1980 **B36** 15
(20) AC 1977 **B33** 1812
(21a) JCS 1959 3815
(21b) JSSC 1974 **11** 214
(22) JSSC 1975 **13** 142
(23) JSSC 1979 **30** 23
(24) AC 1977 **B33** 259
(25) JCS 1937 119
(26) AC 1979 **B35** 295
(27) ZaC 1966 **343** 315
(28) K 1956 **1** 514
(29) AC 1966 **21** 350
(30) AC 1972 **B28** 1034
(31) AC 1973 **B29** 2290
(32) AC 1959 **12** 727
(33) AC 1960 **13** 898
(34) ACSc 1950 **4** 613
(35) AC 1961 **14** 747

27

Group VIII and other transition metals

Introduction

This chapter is largely concerned with the structural chemistry of Fe, Co, Ni, Pd, and Pt, for the most part in finite complexes. The structures of the simpler compounds of these and other transition metals have been described under Halides, Oxides, etc. in the appropriate chapters. Other groups of compounds described in earlier chapters include hydrido compounds, oxo-, peroxo-, and superoxo-compounds, carbonyls, and nitrosyls, in Chapters 11, 12, and 18, respectively. We note here a few general points.

The formation of complexes with certain ligands or with particular types or arrangements of bonds seems to be characteristic of small numbers of transition metals. For example, there are metal–metal bonds in compounds of many elements (Chapter 7) but Nb, Ta, Mo, W, Re (and Tc) are notable for forming metal-cluster compounds with the halogens (p. 432); some other groups of complexes containing metal–metal bonds are summarized in Table 7.9 (p. 310). The formation of complexes containing, in addition to other ligands, a small number of O atoms strongly bonded to the metal is an important feature of the chemistry of V(IV) (see pp. 689 and 1194 for vanadyl compounds), Mo(v) and Mo(vI) (p. 507). Complexes containing neutral molecules such as CO, NO, NH_3, etc. have been known for a long time, and more recently complexes have been prepared containing N_2 (Co, Ru) and SO_2 (Ir, Ru). It appears that certain ligands bond to transition metals in two quite different ways, for example, SO_2 (p. 714), NO (p. 812), and SCN^- (p. 935). There are interesting differences between compounds formed with molecular N_2 and O_2. In the N_2 compounds (p. 790) there is end-on coordination of the N_2 molecule, which retains the same structure as the free molecule, but there is a radical rearrangement of O_2 when it bonds to Ir. Morevoer, this rearrangement differs in $IrCl(CO)O_2(P\phi_3)_2$, which is formed reversibly from $IrCl(CO)(P\phi_3)_2$ in benzene solution, and the analogous I compound, which is formed irreversibly (p. 505).

Of special interest are the multiple bonds to N studied in some compounds of Os and Re. In the anion in $K_2[Os^{VI}NCl_5]$,[1] (a), the bond to N is a formal triple bond, as in the 5-coordinated complex $Re^VNCl_2(P\phi_3)_2$,[2] which has a configuration intermediate between tetragonal pyramidal and trigonal bipyramidal. In the molecule (b)[3] the Re^V—N bond is longer (double?) and in (c)[4] longer still; this unexpected result suggests that steric factors may affect the metal–nitrogen bond length. Note the lengthening of the M—Cl bond *trans* to M—N in both (a) and (c).

The anions in $(\phi_4As)(RuNCl_4)$,[4a] (d), and $(\phi_4As)(OsNCl_4)$[4b] are examples of square pyramidal complexes with triple bonds to N (1.57 and 1.60 Å respectively).

Certain bond arrangements are favoured by particular ligands in finite complexes. The unusual trigonal prismatic coordination of the metal by 6 S in the disulphides

N
‖ 1·61 A
Cl⟍   ⟋Cl
  Os⟍2·36 A
Cl⟋   ⟍Cl
2·61 A |
  Cl

(a)

   CH_3
   N
  ‖ 1·685 A
Cl⟍   ⟋Pϕ_2Et
  Re
Etϕ_2P⟋ ⟍Cl
2·41 A |
  Cl

(b)

   N
  | 1·79 A
Et$_2\phi$P⟍   ⟋Cl
  Re 2·45 A
Et$_2\phi$P⟋   ⟍PϕEt$_2$
2·56 A |
  Cl

(c)

  N   ⊖
  ‖
  Ru
Cl⟋   ⟍Cl
⟋   ⟍
Cl   Cl

(d)

of Nb, Ta, Mo, W, and Re is also found for a number of these metals in chelate complexes with certain S-containing ligands. Examples include $Mo(SCHCHS)_3$ and the isostructural W compound,[5] $Re(S_2C_2\phi_2)_3$,[6a] $V(S_2C_2\phi_2)_3$, and the isostructural Cr compound.[6b] The dodecahedral bond arrangement has been found so far for very few metals in covalent complexes; see, for example, the octacyano complexes of Mo and W (p. 941). The trigonal bipyramidal arrangement of five bonds has been found in a number of transition-metal complexes, including the following d^8 complexes of the heavier Group VIII elements: $Os^0(CO)_3(P\phi_3)_2$,[7] $Rh^IH(CO)$-$(P\phi_3)_3$,[8] and $Ir^ICl(CO)_2(P\phi_3)_2$.[9] The ligands in these complexes are necessarily those which stabilize low oxidation states, as in the case of the tetrahedral $Ni(CO)_4$ and $Pt^0CO(P\phi_3)_3$;[10] the last molecule is of special interest since Pt does not form a simple carbonyl.

The formal oxidation state may be defined as the charge left on the metal atom when the attached ligands are removed in their closed-shell configurations, for example, NH_3 molecule, Cl^-, NO^+, or H^- (if directly bonded to the metal). The oxidation states recognized for Mn then cover the whole range from +7 (d^0), in Mn_2O_7, to -3 (d^{10}) in $Mn(NO)_3CO$; they are nearly as numerous for Cr and Fe and fall off rapidly on each side of these elements. All the 3d elements from Ti to Ni exhibit 'zero valence', but the low oxidation states are observed only with special ligands which are of two types. The π-acceptor (π-acid) ligands (e.g. CO, CN^-, RNC, PR_3, AsR_3) have lone pairs which form a σ bond to the metal and they also have vacant orbitals of low energy to which the metal can 'back donate' some of the negative charge acquired in the formation of the σ bonds. Less important numerically are the π-complexes formed by unsaturated organic molecules and ions (e.g. C_5H_5, C_6H_6) in which all bonding between metal and ligand involves ligand π orbitals;

their structures are described in Chapter 22 and later in this chapter (Pd and Pt compounds). The highest oxidation states are exhibited in combination with the most electronegative elements O and F, but two points are worth noting in this connection. First, the highest oxidation state is not in all cases exhibited in combination with F, the most electronegative element; for example, Mn and Tc exhibit their 'group valence' in Mn_2O_7 and Tc_2O_7 but they do not form heptafluorides, and Ru and Os form tetroxides but not octafluorides (Table 27.1). Second, several

TABLE 27.1
Highest oxidation states in binary fluorides and oxides

Fluorides					Oxides				
Groups	Mn	Fe	Co	Ni	*Groups*	Mn	Fe	Co	Ni
III–VI	4	3	3	2	III–VI	7	3	2	2
3–6	Tc	Ru	Rh	Pd	3–6	Tc	Ru	Rh	Pd
	6	6	6	4		7	8	3	2
	Re	Os	Ir	Pt		Re	Os	Ir	Pt
	7	6	6	6		7	8	4	4

elements exhibit higher oxidation states in complex fluoro- or oxy-ions than in simple fluorides or oxides. Compare $Cs_2Co^{IV}F_6$ and $Cs_2Ni^{IV}F_6$ with CoF_3 and NiF_2, the highest fluorides at present known. a somewhat similar difference is found in the oxy-compounds. Iron rises to Fe(VI) in $(FeO_4)^{2-}$ as compared with Fe(III) in the highest oxide. The highest simple oxide of Co is CoO (there is no evidence for the existence of pure Co_2O_3), but Co^{3+} exists in Co_3O_4 and Co(IV) may exist in Ba_2CoO_4, though this is not certain. The only Co^{3+} salts known are CoF_3 (anhydrous and hydrated), $Co_2(SO_4)_3 . 18H_2O$ and alums, and $Co(NO_3)_3$. The last salt has been prepared as dark green hygroscopic crystals by the action of N_2O_5 on CoF_3. It is reduced by water.[11] For its structure see p. 823. Similarly there is no evidence for the existence of pure anhydrous Ni_2O_3 but Ni^{3+} is present in $LiNiO_2$. (In Table 27.1 we have disregarded the compound $IrO_{2.7}$ described as formed by fusing Ir with Na_2O_2.)

Following brief sections on finite complexes of Ti and V and on complexes in which certain metals exhibit unusually high coordination numbers, we deal in some detail with the structural chemistry of Fe, Co, Ni, Pd, and Pt.

(1)   IC 1969 8 709
(2)   IC 1967 6 204
(3)   IC 1969 8 703
(4)   IC 1967 6 197
(4a) AC 1975 **B31** 2667
(4b) JCMS 1975 5 83
(5)   JACS 1965 87 5798
(6a) IC 1966 5 411
(6b) IC 1967 6 1844
(7)   IC 1969 8 419
(8)   AC 1965 18 511
(9)   IC 1969 8 2714
(10) IC 1969 8 2109
(11) JCS A 1969 2699

The stereochemistry of Ti(IV) in some finite complexes

In finite complexes Ti(IV) exhibits all the c.n.s 4, 5, 6, 7, and 8, though the last two require special bidentate ligands. There is regular tetrahedral coordination in $TiBr_4$ and TiI_4 and, if C_5H_5 is regarded as one ligand, less regular tetrahedral coordination in $Ti(C_5H_5)_2(N_3)_2$,[1a] $Ti_2Cl_4O(C_5H_5)_2$, Fig. 27.1(a),[1b] and the cyclic [TiOCl-$(C_5H_5)]_4$.[1c] The preferred c.n., particularly with O or oxygen-containing ligands is 6, and in crystalline oxides 4-coordination is rare; an example is β-Ba_2TiO_4.[1d] Thus, the tetrahalides readily form adducts with molecules such as $POCl_3$ and ethyl acetate in which octahedral coordination is attained, and the alkoxides form tetramers in which the metal is octahedrally coordinated. With acetylacetone Ti forms the octahedral molecule $Ti(acac)_2Cl_2$ and, like Si, the ion $[Ti(acac)_3]^+$, but does not, like Zr, form $M(acac)_4$. With special ligands such as *diarsine*, $S_2CN(C_2H_5)_2$, and the bidentate nitrate ion Ti forms 8-coordinated complexes. (For $Ti(NO_3)_4$ see p. 823.) With diarsine the presumably octahedral complex $TiCl_4$(diarsine) is formed and also $TiCl_4$(diarsine)$_2$,[2a] which has a dodecahedral structure, Fig. 27.1(b); similar complexes are formed by Zr, Hf, V, and Nb; The As atoms occupy

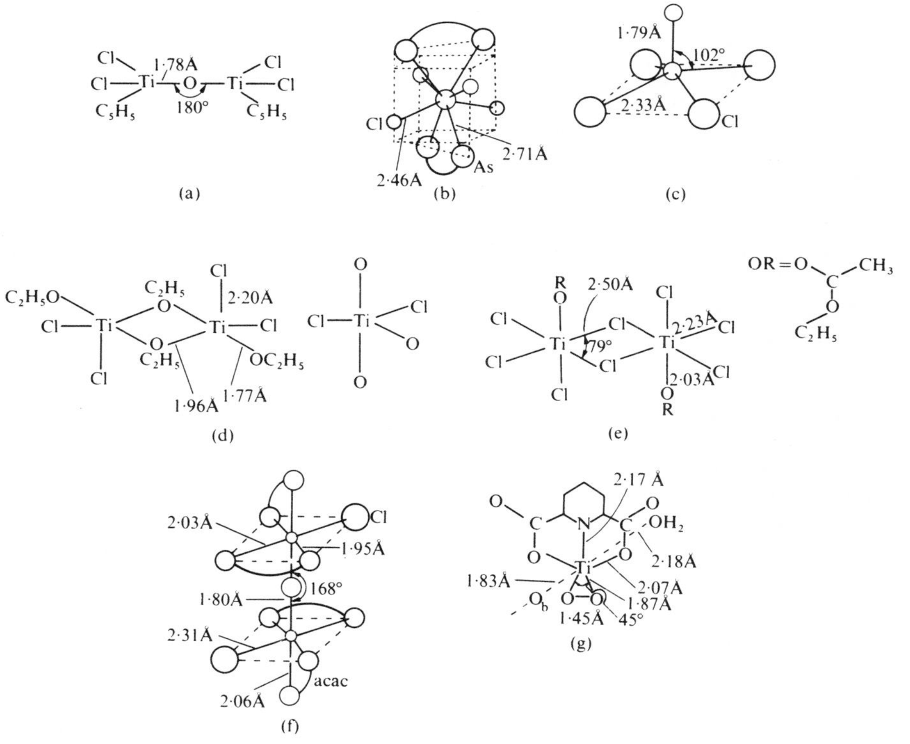

FIG. 27.1. Structures of ions and molecules containing Ti (see text).

vertices of type A in Fig. 3.7(a) and the Cl atoms the vertices of type B, with Ti–Cl, 2.46 Å, and Ti–As, 2.71 Å. The molecule $Ti(S_2CNEt_2)_4$ also has a dodecahedral configuration, as also has the analogous V compound[2b] and the closely related $V(CH_3CS_2)_4$.[2c]

Examples of 5-coordination include the square pyramidal anion (c) in $(TiCl_4O)$-$(NEt_4)_2$,[3] which is isostructural with the corresponding V compound, and the bridged molecule (d),[4] in which the coordination is described as close to trigonal bipyramidal. In the related (red) phenoxy-compound the (planar) bridge was found to be unsymmetrical, with Ti–O, 1.91 Å and 2.12 Å.[5]

The tetrameric molecule of an alkoxide such as $[Ti(OCH_3)_4]_4$[6] has been illustrated in Fig. 5.8(h), p. 200, as an assembly of four octahedra. In this tetramer there are two pairs of non-equivalent Ti atoms in coordination octahedra which share 2 or 3 edges, and the Ti–O bond lengths range from 1.8 to 2.2 Å; some O atoms are bonded to one Ti only, others to two or three. The alkoxides are colourless, in contrast to the yellow or red compounds of type (d). Another type of bridged molecule is that of $Ti_2Cl_8(CH_3COOC_2H_5)_2$,[7] (e), in which an ethyl acetate ion is bonded to each Ti atom and two Cl atoms form the bridge between the two edge-sharing octahedral coordination groups.

We noted above the acetylacetone complexes $Ti(acac)_2Cl_2$ and $[Ti(acac)_3]^+$. In the orange-yellow $Ti_2Cl_2O(acac)_4$,[8] which crystallizes with one molecule of $CHCl_3$ as solvent of crystallization, there is an approximately linear Ti–O–Ti bridge and short bridging Ti–O bonds (Fig. 27.1(f)). The Ti–O–Ti bridge in the red dipicolinic acid salt $K_2[Ti_2O_5(C_7H_3O_4N)_2].5H_2O$[9] is strictly linear ($O_b$ lying on a 2-fold axis), and the coordination group around Ti, (g), is similar to that in $K_2W_2O_{11}.4H_2O$ (p. 505) which also contains peroxo-groups bonded to the metal atoms. In Fig. 27.1(g) the broken lines represent bonds perpendicular to the plane of the chelate ligand and the peroxo-group, so that the coordination around Ti may be described as either pentagonal bipyramidal or as octahedral if the O_2 groups is counted as one ligand.

(1a) AC 1977 **B33** 578
(1b) JACS 1959 **81** 5510
(1c) AC 1970 **B26** 716
(1d) AC 1973 **B29** 2009
(2a) JCS 1962 2462
(2b) CC 1971 365
(2c) AC 1972 **B28** 1298

(3) AC 1968 **B24** 282
(4) AC 1968 **B24** 281
(5) IC 1966 5 1782
(6) AC 1968 **B24** 1107
(7) AC 1966 20 739
(8) IC 1967 6 963
(9) IC 1970 9 2391

The stereochemistry of V in some coordination complexes

The structural chemistry of V in various oxidation states (II–V) in binary compounds has been noted in previous chapters, in particular the unusual coordination of V(v) by oxygen in oxides and oxy-salts, where there is square pyramidal or irregular octahedral coordination, in addition to tetrahedral coordination; see pp. 565 and

608. In the 5- or 6-coordinated compounds there are 4 or 5 bonds in the range 1.6–2.0 Å and usually, but not always, one longer bond which is sometimes so long (e.g. 2.8 Å in V_2O_5) as to justify the description of the stereochemistry as square pyramidal rather than distorted octahedral. This irregular 6-coordination is also found in methyl vanadate, $VO(OCH_3)_3$,[1] which consists of chains of octahedral coordination groups each sharing two (non-opposite) edges. These groups are very distorted, there being five different V–O bond lengths, the two shortest (1.54 Å and 1.74 Å) to unshared O atoms (Fig. 27.2(a)). The V$\overset{O}{\underset{O}{<>}}$V bridges are alternately symmetrical (all four V–O approximately 2.02 Å) and unsymmetrical (V–O, 1.85 Å and 2.27 Å), and the six V–O bonds from any V(v) atom have the lengths: 1.54, 1.74, 1.85, 2.02 (two), and 2.27 Å.

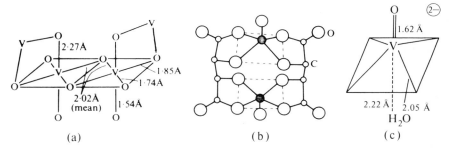

FIG. 27.2. (a) Portion of chain in $VO(OCH_3)_3$; (b) the divanadyl-*dl*-tartrate ion; (c) the $[VO(NCS)_4]^{2-}$ ion.

The arrangement of the five nearest neighbours in many vanadyl complexes is square pyramidal with a very short apical V=O bond of length close to 1.6 Å. Some examples are included in Table 27.2, which includes the interesting divanadyl-*dl*-tartrate ion of Fig. 27.2(b). The distance to the 4 more distant neighbours is similar to V–O (1.98 Å) in $V^{III}(acac)_3$,[2] a compound that is easily oxidized to $VO(acac)_2$.

TABLE 27.2
Bond lengths in vanadyl complexes

	V–O bond lengths			*Reference*
	one	two	two	
$VOCl_2(tmu)_2$	1.61	2.00	2.34 (V–Cl)	AC 1970 **B26** 872
$VOCl_2(OP\phi_3)_2$	1.58	1.99	2.31 (V–Cl)	AC 1980 **B36** 1198
$VO(acac)_2$	1.57	1.97 (four)		JCP 1965 **43** 3111
$Na_4[VO\text{-}dl\text{-tartrate}].12H_2O$	1.62	1.91	2.00	IC 1968 7 356
$(NH_4)_4[VO\text{-}dl\text{-tartrate}].2H_2O$	1.60	1.86	2.02	JCS A 1967 1312

In a number of hydrated vanadyl salts, however, there is a sixth O atom close enough (2.2 Å) to be regarded as the sixth octahedral neighbour as, for example, in $(NH_4)_2[VO(C_2O_4)_2H_2O]$,[3] $(NH_4)_2[VO(SCN)_4].5H_2O$,[4] Fig. 27.2(c), and in certain hydrated vanadyl sulphates which have structures strikingly similar to those of hydrated Mg sulphates, as described on p. 689.

In contrast to the square pyramidal structures of $VOCl_2(OP\phi_3)_2$ and $VOCl_2$-(tetramethylurea)$_2$,[5] the molecule $VOCl_2(NMe_3)_2$[6] has the trigonal bipyramidal structure (a) in which the equatorial bond angles are 120° to within the experimental

error. It seems that this configuration is favoured by the shape of the tertiary amine, for the methyl groups are in the staggered positions relative to the equatorial ligands, a view supported by the fact that the V(III) molecule $VCl_3(NMe_3)_2$,[7] (b), has almost exactly the same structure.

We comment elsewhere on the rarity of trigonal prism coordination, which has been observed in complexes $M(S_2C_2\phi_2)_3$ in which M is V, Cr, Mo, or Re. In the V complex[8] V–S = 2.34 Å and S–V–S = 82° (mean), and the shortest distances between S atoms of different ligands are close to 3.1 Å; the dimensions of the Re and Mo complexes are almost identical to those of the V complex. It has been suggested that some type of weak interaction between the S atoms stabilizes this configuration, which is not found for all chelate sulphur-containing ligands. In the maleonitrile dithiolate ion in $[V(mnt)_3](NMe_4)_2$[9] the 6-coordination around V is irregular (V–S, 2.36 Å), with 2.93 Å as the shortest *inter*-ligand S–S distance. The coordination group is intermediate between octahedral and trigonal prismatic, as shown by the angle S–V–S for pairs of S atoms that are farthest apart. For a regular octahedron this angle would be 180° or 173° for the nearest approach to a regular octahedron taking into account the constraints due to the rigid bidentate ligand. In $V(S_2C_2\phi_2)_3$ this is 136° (trigonal prismatic coordination), and in the dithiolate ion it is 159°. We have included these complexes with compounds of V(IV) though the formal oxidation state of the metal could be regarded as zero, depending on the degree of delocalization of the four electrons on the three chelate ring systems.

The anion in $K_4[V(CN)_7].2H_2O$[10] may be noted here as an example of V(III) in pentagonal bipyramidal coordination.

(1)   IC 1966 5 2131
(2)   AC 1969 **B25** 1354
(3)   AC 1976 **B32** 82
(4)   JCS 1963 5745
(5)   AC 1970 **B26** 872
(6)   JCS A 1968 1000
(7)   JCS A 1969 1621
(8)   IC 1967 6 1844
(9)   JACS 1967 89 3353
(10)   JACS 1972 94 4345

The structural chemisty of Cr(IV), Cr(V), and Cr(VI)

Compounds of Cr(IV)

In addition to the well known ferromagnetic CrO_2 (rutile structure) CrF_4 and $CrBr_4$ have been synthesized; their structures are at present unknown. The complex fluoride K_2CrF_6 contains octahedral CrF_6^{2-} ions.

TABLE 27.3

Compounds of Cr *in higher oxidation states*

Cr(IV)	Cr—O (Å)			Reference
CrO_2	2 at 1.88, 4 at 1.92			JAP 1962 **33** 1193
Sr_2CrO_4	4 at 1.66–1.95, mean 1.82			AK 1966 **26** 157
K_2CrF_6				ZaC 1956 **286** 136
Cr(V)				
$Ca_5OH(CrO_4)_3$	4 at 1.66			ACSc 1965 **19** 177
$Ca_2Cl(CrO_4)$	4 at 1.70 (mean)			
	Angles 105° (4), 119° (2)			AC 1967 **23** 166
Cr(VI)	Cr—O (bridge)	Cr—O (term.)	Cr—O—Cr	
CrO_3	1.75	1.60	143°	AC 1970 **B26** 222
$K_2Cr_2O_7$	1.79	1.63	126°	CJC 1968 **46** 933
$(NH_4)_2CrO_4$	—	1.66	–	AC 1970 **B26** 437
Cr(III) and Cr(VI)	Cr^{III}—6 O	Cr^{VI}—4 O		
Cr_5O_{12}	1.97	1.65		ACSc 1965 **19** 165
KCr_3O_8	1.97	1.60		ACSc 1958 **12** 1965
$LiCr_3O_8$	2.05	1.66		AK 1966 **26** 131
$CsCr_3O_8$	1.96	1.63		AK 1966 **26** 141
Cr(IV) and Cr(VI)	Cr^{IV}—6 O	Cr^{VI}—4 O		
$K_2Cr_3O_9$	1.97	1.65		ACSc 1969 **23** 1074

For the preparation of high-pressure oxides see ACSc 1968 **22** 2565.

The olive-green Ba_2CrO_4, with a magnetic moment of 2.82 BM, corresponding to Cr(IV), is isostructural with Ba_2FeO_4 and Ba_2TiO_4 (notable as one of the few examples of Ti^{4+} tetrahedrally coordinated by oxygen) and like the green Ba_3CrO_5 is a compound of Cr(IV). In Sr_2CrO_4 rather distorted CrO_4^{4-} groups were found, which may be compared with the nearly regular octahedral coordination in CrO_2 (Table 27.3). Under high pressure Sr_2CrO_4 changes to the K_2NiF_4 structure (octahedral coordination of Cr(IV)). Oxides $MCrO_3$ (M = Ca, Sr, Ba, Pb) have been prepared under pressure.

Compounds of Cr(v)

These include a number of compounds isostructural with silicates or phosphates, for example:

$NdCrO_4$, $YCrO_4$: (zircon structure);
$Ca_5OH(CrO_4)_3$: apatite structure, Cr^V–O, 1.66 Å;
$Ca_2Cl(CrO_4)$: isostructural with $Ca_2Cl(PO_4)$.

Cr(v) is tetrahedrally coordinated by oxygen and has a similar radius to Cr(vi). The complete structure determination of $Ca_2Cl(CrO_4)$ shows that all the CrO_4^{3-} ions are equivalent, i.e. this is not a compound containing Cr(iv) and Cr(vi). The magnetic moment has the expected value, 1.7 BM. The CrO_4^{3-} tetrahedra are slightly distorted.

Compounds of Cr(vi)

The structural chemistry of Cr(vi) is very similar to that of S(vi), and accurate structural data are available for several simple compounds.

The crystalline trioxide is built of infinite chains formed by the linking of CrO_4 tetrahedra by two vertices:

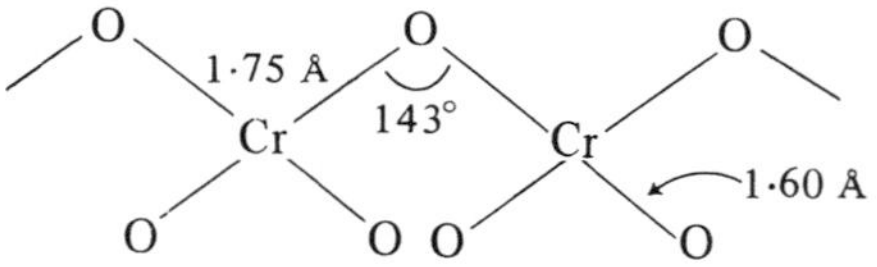

There are only van der Waals forces between the chains, consistent with the comparatively low melting point (197 °C).

In simple crystalline chromates there are tetrahedral CrO_4^{2-} ions in which Cr–O is 1.66 Å (in the ammonium salt). The tetrahedral CrO_3F^- ion (Cr–O, 1.58 Å) is the analogue of SO_3F^-. The sizes of O and F being very similar, $KCrO_3F$ crystallizes with the scheelite ($CaWO_4$) structure with random arrangement of O and F.

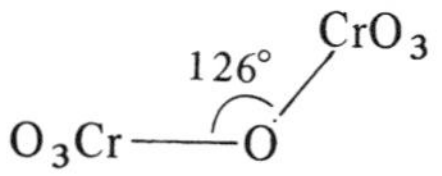

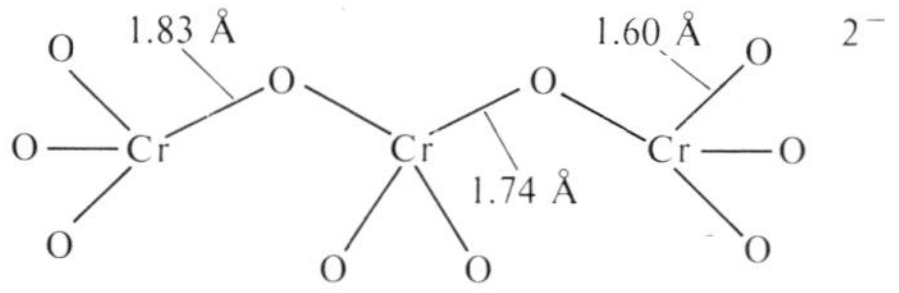

$KCrO_3Cl$ has a distorted form of the same structure due to the much greater difference between Cr–O and Cr–Cl (2.12 Å). Polychromate ions result from the sharing of vertices between limited numbers of CrO_4 tetrahedra, the simplest being the dichromate ion, $Cr_2O_7^{2-}$ (Table 27.3). Studies have been made of the trichromate ion, in which there are three different Cr—O bond lengths with the mean values shown,[1] and of the tetrachromate ion in $Rb_2Cr_4O_{13}$.[2] There are also ions of this type containing one P or As atom; for the ions $PCrO_7^{3-}$ to $PCr_4O_{16}^{3-}$ see p. 861 and for $AsCr_2O_{10}^{3-}$ see p. 902.

The structures of the tetrahedral molecules CrO_2F_2 and CrO_2Cl_2 are included in Table 10.12 (p. 483).

(1)   AC 1980 **B36** 135   (2)   AC 1973 **B29** 2141

Compounds containing Cr *in two oxidation states*

The structure of Cr_2F_5, a compound of Cr(II) and Cr(III), has been described in Chapter 5. We note here some compounds which contain Cr(III) and Cr(VI) or Cr(IV) and Cr(VI).

A number of oxides have been prepared which are intermediate between CrO_2 and CrO_3. They include Cr_2O_5 and Cr_3O_8, resulting from the thermal decomposition of CrO_3 under oxygen pressures up to 1 kbar in the temperature range 200–400°C, and an oxide Cr_5O_{12} which forms under oxygen pressures up to 3 kbar. This last oxide is a compound of Cr(III) and Cr(VI), $Cr_2^{III}(Cr^{VI}O_4)_3$, in which Cr(III) atoms occupy positions of octahedral coordination (Cr—O, 1.97 Å) and Cr(VI) atoms positions of tetrahedral coordination (Cr—O, 1.65 Å) in an approximately c.c.p. assembly of oxygen atoms. The black metallic-looking KCr_3O_8 prepared by heating $K_2Cr_2O_7$ with CrO_3 in oxygen under pressure is also a compound of Cr(III) and Cr(VI), i.e. $KCr^{III}(Cr^{VI}O_4)_2$. Octahedral $Cr^{III}O_6$ and tetrahedral $Cr^{VI}O_4$ groups are linked to form layers of composition Cr_3O_8 which are held together by the potassium ions. The corresponding lithium salt has a quite different structure which is essentially the same as that of a number of chromates $MCr^{VI}O_4$ (M = Co, Cu, Zn, Cd), phosphates (Cr, In, Tl), and sulphates (Mn, Ni, Mg). The octahedral holes between O atoms of CrO_4 groups are occupied *at random* by Li^+ and Cr^{3+}. (This structure may also be described as an approximately cubic closest packing of O atoms with 1/8 of the tetrahedral and 1/4 of the octahedral holes occupied.

The compounds $M_2Cr_3O_9$ (M = Na, K, Rb) have been prepared under pressure as red crystals insoluble in water. The structure consists of rutile-type chains of $Cr^{IV}O_6$ octahedra which are cross-linked by $Cr^{VI}O_4$ tetrahedra.

Higher coordination numbers of metals in finite complexes

We have collected together in this section some examples of finite complexes formed with polydentate ligands in which transition metals exhibit coordination numbers from 7 to 10. For chelates of 4f metals see p. 77.

We noted in Chapter 10 two rather symmetrical arrangements of 7 nearest neighbours found in oxy- and fluoro-compounds of transition metals, namely, pentagonal bipyramidal (ZrF_7^{3-}), and monocapped trigonal prism (NbF_7^{2-}). The ions

$$\begin{array}{l} H_2C \stackrel{\displaystyle N \diagdown CH_2COOH}{\diagup CH_2COOH} \\ H_2C \diagdown N \stackrel{\displaystyle CH_2COOH}{\diagdown CH_2COOH} \end{array}$$

$$\begin{array}{l} H_2C \stackrel{\displaystyle C}{\diagup} \stackrel{\displaystyle H_2}{\diagdown} CH \stackrel{\displaystyle N \diagdown CH_2COOH}{\diagup CH_2COOH} \\ H_2C \diagdown C \diagup CH \diagdown N \stackrel{\displaystyle CH_2COOH}{\diagdown CH_2COOH} \\ H_2 \end{array}$$

formed from ethylenediamine tetracetic acid and the closely related diamino cyclo-hexane acid are potentially sexadentate ligands, for there are 4 O atoms and 2 N atoms which can coordinate to a metal atom. In the formulae in this section and in Fig. 27.3 we use the abbreviations edta and dcta to mean the ions (4−) formed

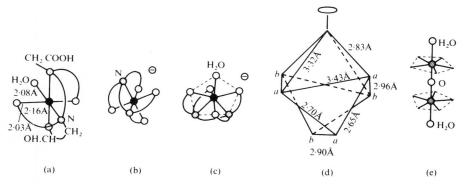

FIG. 27.3. Structures of (a) [Ni(edta)H$_2$(H$_2$O)], (b) [Co(edta)]$^-$, (c) [Fe(edta)(H$_2$O)]$^-$, (d) coordination polyhedron of [MnH(edta)(H$_2$O)]$^-$, (e) [(H$_2$O)LFe—O—FeL(H$_2$O)]$^{4+}$, where the pentagon represents the ligand L shown on p. 1201.

from these acids. Three different types of complex formed with Ni, Co, and Fe are illustrated in Fig. 27.3(a)–(c). The Ni(II) and Co(III) complexes are octahedral, but in the former, (a), the ligand forms only five bonds to Ni, the metal atom completing its coordination group with a water molecule.[1] (Although only 2 H of edta are replaced by Ni only one acetate group is unattached and presumably carries one H as OH. The location of the other H atom is not clear, but it has been provisionally allocated as shown, on the basis of the Ni—O bond lengths.) In the (diamagnetic) Co(III) complex,[2] (b), edta is behaving as a sexadentate ligand. On the other hand, in Rb[Fe^{III}(H$_2$O)edta].H$_2$O[3] the anion is a 7-coordinated complex, Fig. 27.3(c), with one H$_2$O in addition to the 6 bonds from edta. The bond arrangement is very approximately pentagonal bipyramidal, two O atoms and two N atoms of edta and one H$_2$O molecule forming the equatorial group of five ligands. In contrast to Ni(II), Mn(II) forms a salt of a hydrated aquo-complex, Mn^{2+}[Mn^{II}(H$_2$O)H(edta)]$_2$-8H$_2$O,[4] with replacement of only 3 H atoms of edta. The Mn^{2+} ions in the crystal have normal octahedral coordination (by 4 H$_2$O and 2 O atoms of anions), but within the complex Mn(II) is 7-coordinated. The bond arrangement is, however, different from that in the iron complex (c) and may be regarded as a skew form of the NbF_7^{2-} coordination group. A small clockwise rotation of the face *aaa* and an equivalent anticlockwise rotation of *bbb*, Fig. 27.3(d), would convert this coordinaation group into a monocapped trigonal prism. The Fe(III) complex with dcta has a configuration very similar to that of the Mn complex (d); it was studied in $Ca[Fe^{III}(H_2O)dcta]_2 . 8H_2O$.[5]

It is evident that the geometries of these 7-coordinated complexes are very much dependent on the constraints due to the shapes of the polydentate ligands. A more extreme example is the pentagonal bipyramidal coordination in complexes formed by the pentadentate ligand L, such as the cations in the salts $[FeL(NCS)_2]ClO_4$,

(L)

(where the NCS groups occupy the apical positions with Fe–N, 2.01 Å as compared with 2.23 Å for the equatorial bonds), and $[(H_2O)LFe-O-FeL(H_2O)](ICO_4)_4$.[6] The cation in this salt, Fig. 27.3(e), is also of interest as an example of a bridged oxo-compound of Fe(III) with a linear Fe–O–Fe bridge.

With bidentate ligands of the type $R.CO.CH.CO.R$ yttrium and the smaller 4f ions form a number of 7-coordinated complexes in which the seventh ligand is a water molecule. These are noted in Chapter 3 in our discussion of 7-coordination.

Although Zr is 6-coordinated by F or O in a few complex fluorides or oxides there is a marked preference for higher coordination numbers. In a number of cases this is 7-coordination as in $(NH_4)_3ZrF_7$, in monoclinic ZrO_2, and in ZrOS. The Zr^{4+} ion is obviously a borderline case, for in contrast to the above examples there is 8-coordination (dodecahedral) in K_2ZrF_6, in the high-temperature forms of ZrO_2, and in numerous hydrated and hydroxy-salts (p. 645). From the energy standpoint there is clearly very little to choose between the two predominant types of 8-coordination, namely antiprismatic and dodecahedral. In a 3D structure the crystal energy can be minimized by adjusting the shapes of all the various coordination groups, and in extreme cases these geometrical requirements actually determine the *composition* of the salt that crystallizes from solution, as in the salts Na_3TaF_8, K_2TaF_7, and $CsTaF_6$, or $(NH_4)_3ZrF_7$ and K_2ZrF_6. In *finite* complexes containing chelate ligands the choice between the two coordination polyhedra is probably determined by 'non-bonded' interactions between atoms of the ligands, and we find

dodecahedral: $Na_4[Zr(C_2O_4)_4] . 3H_2O$ IC 1963 2 250
$K_2[Zr(NTA)_2] . H_2O$[a]   JACS 1965 87 1610

antiprismatic: Zr(acac)$_4$ IC 1963 2 243
                  [Zr$_4$(OH)$_8$(H$_2$O)$_{16}$].Cl$_8$.12H$_2$O   See p. 646

(a) NTA is the nitrilo triacetate ion N(CH$_2$COO)$_3^{3-}$.

The 9-coordination of La^{3+} and the larger 4f ions in both finite complexes and 3D structures has been noted in previous chapters. We have seen that 6-fold coordination by the ligand edta is possible for small ions M^{2+} and M^{3+}. Even the smallest 4f ion is too large for 'octahedral' coordination by edta, and with the larger ions the ligand occupies only part of the coordination sphere, the remainder of which can be filled with, for example, H$_2$O molecules. The very symmetrical 9-coordination group found in a number of 4f compounds is not consistent with the geometry of some polydentate ligands, and we find coordination groups more closely related to the dodecahedral 8-coordination group but with additional ligands giving 9-coordination or the much less common 10-coordination. Thus in the 9-coordinated anion in K[La(H$_2$O)$_3$edta].5H$_2$O the 4 O and 2 N atoms of edta and one H$_2$O molecule occupy seven of the vertices of the triangulated dodecahedron characteristic of 8-coordination, and there are isostructural salts in which the 4f metal is Nd, Gd, Tb, and Er.[7] The arrangement of seven of the ligand atoms around La is similar in [LaH(edta)(H$_2$O)$_4$].3H$_2$O,[8] where there are 4 H$_2$O molecules completing a 10-coordination group. It should, however, be remarked that this polyhedron approximates closely to a bicapped square antiprism.

High coordination numbers are also possible if a group such as NO$_3^-$ or CH$_3$COO$^-$ behaves as a bidentate ligand. In Chapter 18 we noted that 8 O atoms (four NO$_3^-$ ions) surround the metal in [Co(NO$_3$)$_4$]$^{2-}$ and Ti(NO$_3$)$_4$ and 12 O (six bidentate NO$_3^-$ ions) in nitrato ions of Ce(III), Ce(IV), and Th(IV). Three bidentate nitrate ions and two bipyridyl molecules give La 10-coordination in La(NO$_3$)$_3$.(bipyridyl)$_2$ in an arrangement that has been described as a bicapped dodecahedron. A similar description could be given of the 'pseudo-octahedral' 10-coordination group in Th(NO$_3$)$_4$(OPϕ_3)$_2$ (p. 827); for references to these compounds see the section on metal nitrates and nitrato complexes in Chapter 18.

(1) JACS 1959 81 556   (5) JACS 1966 88 3228
(2) JACS 1959 81 549   (6) JACS 1967 89 720
(3) IC 1964 3 34   (7) JACS 1965 87 1612
(4) IC 1964 3 27   (8) JACS 1965 87 1611

The structural chemistry of iron

Although the formal oxidation states from and including –2 to +6 are recognized, only Fe(II) and Fe(III) are of practical importance. Oxidation states lower than +2 involve π-type ligands and the states –2, –1, and +1 apply to compounds such as Fe(CO)$_2$(NO)$_2$ and Fe(NO)$_2$X and the ion [Fe(H$_2$O)$_5$NO]$^{2+}$ where the meaning of the term oxidation state is somewhat dubious. The diamagnetic compound

$Fe(NO)_2I$ exists as dimers, (a),[1] very similar in structure to (b)[2] except for the larger Fe—Fe distance. Unless there is some sort of superexchange interaction involving the I atoms in (a) it would seem necessary to account for the diamagnetism by postulating metal-metal interaction of some kind, and to suppose that the large Fe—Fe distance, as compared with that in (b), is due to the size of I. The bond arrangement around Fe is tetrahedral. Crystalline $Co(NO)_2I$ has a tetrahedral chain

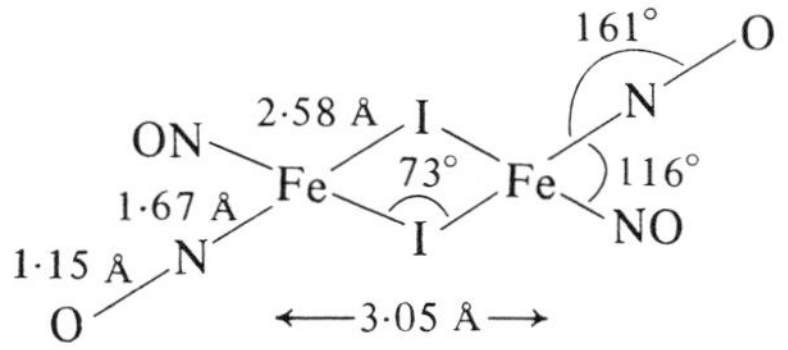

(a)

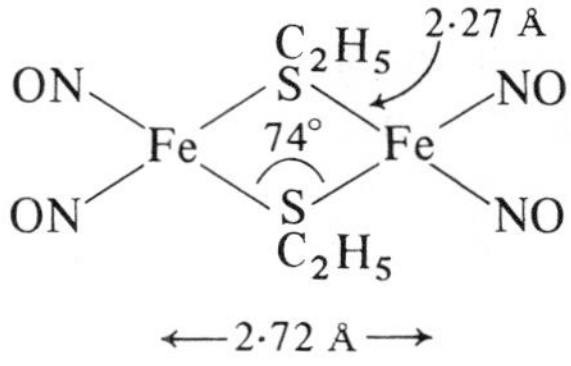

(b)

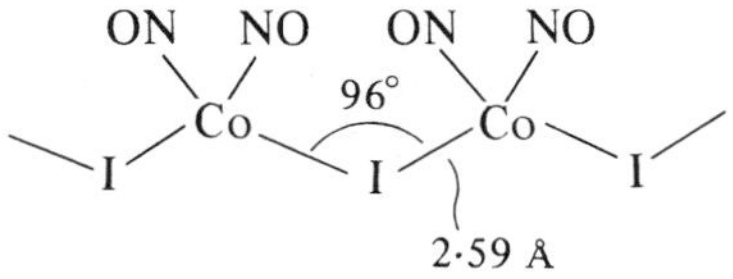

(c)

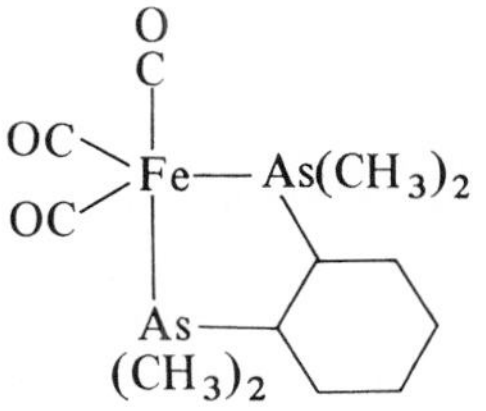

(d)

structure (c).[1] The small bond angle at I in (a) seems to be a feature of metal-metal bonded systems; compare

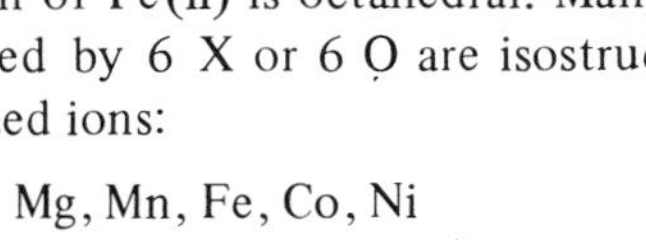

The only bond arrangement established for Fe(0) is the trigonal bipyramidal arrangement of five bonds, as in $Fe(CO)_5$, and $Fe(CO)_3$(diarsine)[3] (d).

The normal coordination of Fe(II) is octahedral. Many of the ionic compounds in which Fe^{2+} is surrounded by 6 X or 6 O are isostructural with those of other metals forming similarly sized ions:

MO (NaCl structure): Mg, Mn, Fe, Co, Ni
MF_2 (rutile structure)
$MSO_4 \cdot 7H_2O$
$K_2M(SO_4)_2 \cdot 6H_2O$ $\qquad$ M = Mg, Mn, Fe, Co, Ni, Zn
$M_3Bi_2(NO_3)_{12} \cdot 24H_2O$

The structural similarity between these $M(II)$ compounds does not, however, extend to the more covalent compounds. For example, the complex cyanides formed by Fe, Co, and Ni are of quite different kinds:

$$K_4[Fe(CN)_6], \quad K_3[Co(CN)_5], \quad K_2[Ni(CN)_4], \quad \text{and} \quad K_4Ni_2(CN)_6 \text{ (p. 942).}$$

(The Co compound was incorrectly described in the older literature as $K_4[Co(CN)_6]$.) The complex thiocyanates are:

$$Na_4[Fe(NCS)_6].12H_2O, \quad Na_2[Co(NCS)_4].4H_2O, \quad \text{and} \quad Na_4[Ni(NCS)_6].12H_2O,$$

of which the Ni compound was earlier assigned the formula $Na_2[Ni(NCS)_4].8H_2O$.

The octahedral complexes of $Fe(II)$ are high-spin ($\mu \approx 5$ BM) except for those containing strong-field ligands such as $Fe(CN)_6^{4-}$ and $Fe(CNR)_6^{2+}$, which are diamagnetic. Neutral molecules include *trans*-$Fe(NCS)_2(pyr)_4$,[4] (e).

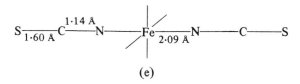

(e)

Tetrahedral coordination of $Fe(II)$ has been demonstrated in the anion in crystalline $(FeCl_4)(Me_4N)_2$,[4a] ($Fe-Cl$, 2.29 Å), and deduced for $(FeBr_4)^{2-}$ and $(FeI_4)^{2-}$ from the existence of salts isostructural with the corresponding compounds of Zn, Mn, Co, and Ni. High spin $Fe(II)$, d^6, has the expected moment of approximately 5 BM corresponding to 4 unpaired electrons. There are also molecules $FeCl_2L_2$ and ions FeL_4^{2+} formed with ligands such as ϕ_3PO. Like $(NiCl_4)^{2-}$ the $(FeCl_4)^{2-}$ ion is stable only in the presence of large cations.

Square coplanar coordination of $Fe(II)$ is observed in the phthalocyanin ($\mu = 3.96$ BM) and also in gillespite (p. 1025). In this crystal Fe^{2+} has 4 O at 1.97 Å (compare $Fe-6O$ (octahedral), 2.14 Å) and no other neighbours closer than 3.98 Å. The visible and i.r. absorption spectra have been discussed in terms of high-spin Fe^{2+}.[5]

Other coordination numbers observed with special types of ligand include 5 in complexes $FeLX^+$, where L is a tetradentate ligand such as $N(CH_2CH_2P\phi_2)_3$,[5a] or $N(CH_2CH_2NH_2)_3$, *tren*, and 8 (dodecahedral) in $[Fe(1,8\text{-naphthyridine})_4](ClO_4)_2$.

The structural chemistry of $Fe(III)$ is largely that of octahedral coordination, in crystalline halides, Fe_2O_3, and numerous finite complexes ($Fe(en)_3^{3+}$, $Fe(acac)_3$, $Fe(C_2O_4)_3^{3-}$, etc.), but tetrahedral coordination is less rare than for $Fe(II)$. Thus there is octahedral coordination of the metal in crystalline $FeCl_3$ but tetrahedral coordination in the vapour dimer Fe_2Cl_6 and in the $FeCl_4^-$ ion, and there is both tetrahedral and octahedral coordination of Fe^{3+} in the complex of $FeCl_3$ with dimethyl sulphoxide (L), $FeCl_3L_2$. This consists of equal numbers of tetrahedral $(FeCl_4)^-$ ions ($Fe-Cl$, 2.16 Å) and octahedral *trans* $(FeCl_2L_4)^+$ ions ($Fe-Cl$, 2.34 Å; $Fe-O$, 2.01 Å).[6] The simplest salts containing the $(FeCl_4)^-$ ion are the alkali salts $MFeCl_4$; others contain cations such as PCl_4^+ and $As\phi_4^+$.

There is both tetrahedral and octahedral coordination of Fe^{3+} in Fe_3O_4 (inverse spinel structure), and numerous alkali metal salts have been prepared containing ions similar to those in silicates, such as FeO_4^{5-}, $Fe_2O_7^{8-}$, and others noted on p. 514.

The complex (f), containing the ligands thio-*p*-toluoyl disulphide and dithio-*p*-toluate, is of interest as containing 4-membered rings (FeS_2C) and 5-membered rings (FeS_3C).[7] As in the case of Fe(II) compounds the magnetic moments of Fe(III) compounds correspond to high spin (d^5, 5.9 BM) except for complexes such as $[Fe(CN)_6]^{3-}$ and $[Fe(dipyr)_3]^{3+}$ containing strong-field ligands.

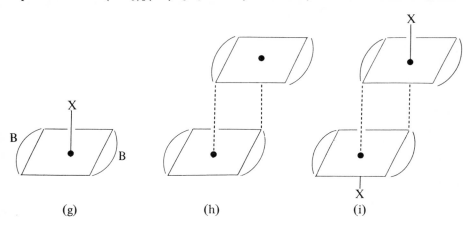

(f)

The simplest type of 5-coordinated complex is an ion such as $(FeCl_5)^{2-}$; spectroscopic data (Raman and i.r.)[8a] for $(C_8H_{18}N_2)^{2+}(FeCl_5)^{2-}$ indicate a trigonal bipyramidal configuration.

The square pyramidal bond arrangement is found in a number of finite complexes. Apart from $FeH(SiCl_3)_2(CO)C_5H_5$[8b] they are mostly monomers containing two

(g) (h) (i)

bidentate ligands, (g), or dimers in which a fifth weaker bond is formed as shown at (h). The metal atom is situated about ½ Å above the centre of the square base of the pyramid. Dimerization of a molecule of type (g) gives a complex of type (i) in which the metal atom has distorted octahedral coordination.

As in the case of Fe(II) other c.n.s are found with special ligands. For example, there is planar 3-coordination in $Fe[N(SiMe_3)_2]_3$, which is isostructural with the Al compound (p. 796), and dodecahedral 8-coordination in the $[Fe(NO_3)_4]^-$ ion (p. 824). Examples include the following:

	Bond lengths (Å)	Reference
(g) $Fe[S_2CN(C_2H_5)_2]_2Cl$	4 S 2.32, Cl 2.27	IC 1967 6 712
Fe(salen)Cl	2 O 1.88, 2 N 2.08	JCS A 1967 1598
	Cl 2.24	
(h) $\{Fe[S_2C_2(CN)_2]_2\}_2$ $(n\text{-}Bu_4N)_2$	4 S 2.23, S 2.46	IC 1967 6 2003
(i) $[Fe(salen)Cl]_2$	2 N 2.10, Cl 2.29	JCS A 1967 1900
	3 O 1.90, 1.98, 2.18	

Salen = NN'-*bis*-salicylideneiminato.

Structural studies of the (very few) compounds containing Fe in higher oxidation states appear to be limited to the following.

Fe(IV): Ba_2FeO_4,[9] isostructural with Ba_2TiO_4 and β-Ca_2SiO_4 (tetrahedral coordination of Fe(IV)). In oxides $BaFeO_{3-x}$ x tends to zero, and the proportion of Fe(IV) increases, with increasing O_2 pressure, which favours octahedral coordination of Fe(IV).[10] The structures are of the c.p. ABO_3 types described on p. 583.

The most stable finite complexes containing Fe(IV) are the trisdithiocarbamate cations such as the pyrrolidyldithiocarbamate ion (j); they are prepared by oxidation of the corresponding Fe(III) molecule and isolated as BF_4^- or ClO_4^- salts. The structure of (j)[11] is very similar to that of $Fe(S_2CNR)_3$, in which Fe—S is 2.41 Å.

(j)

The arrangement of the 6 bonds around Fe in (j) is intermediate between regular octahedral and trigonal prismatic.

Fe(VI): K_2FeO_4[12]—isostructural with K_2SO_4; $BaFeO_4$—isostructural with $BaSO_4$. The magnetic moments of these compounds are close to the value (2.83 BM) expected for 2 unpaired electrons (d^2). For the reduction of the oxidation state of Fe by pressure, for example, $SrFeO_3 \rightarrow SrFeO_{2.86}$, see reference (13).

(1)   JACS 1969 **91** 1653
(2)   AC 1958 **11** 599
(3)   AC 1967 **22** 296
(4)   ACSc 1967 **21** 2028
(4a) IC 1975 **14** 348
(5)   IC 1966 **5** 1268
(5a) IC 1978 **17** 810
(6)   AC 1967 **23** 581

(7)   JACS 1968 **90** 3281
(8a) JINC 1979 **41** 469
(8b) IC 1970 **9** 447
(9)   ZaC 1956 **383** 330; NBS 1969 **73A** 425
(10) JPC 1964 **68** 3786
(11) JACS 1974 **96** 3647
(12) ZaC 1950 **263** 175
(13) JCP 1969 **51** 3305, 4353

The structural chemistry of cobalt

We shall be concerned here chiefly with two topics, the stereochemistry of Co(II) in finite complexes and the extensive chemistry of octahedral coordination complexes of Co(III). The formal oxidation states –1, 0, and +1 are recognized in complexes formed with π-bonding ligands as in $[Co^{-1}(CO)_4]^-$, $Co^{-1}(CO)_3NO$ (p. 967), $Co^0(PMe_3)_4$,[1a] and $Co^{+1}X(PR_3)_3$, all expected to be tetrahedral, and the presumably octahedral $[Co^{+1}(bipyr)_3]^+$.

Co(I). Tetrahedral coordination has been demonstrated in crystals of (a),[1b] and several structural studies have established the arrangement of 5 bonds. The tetragonal pyramidal structure of the molecule $Co[S_2CN(CH_3)_2]_2NO$ is described on

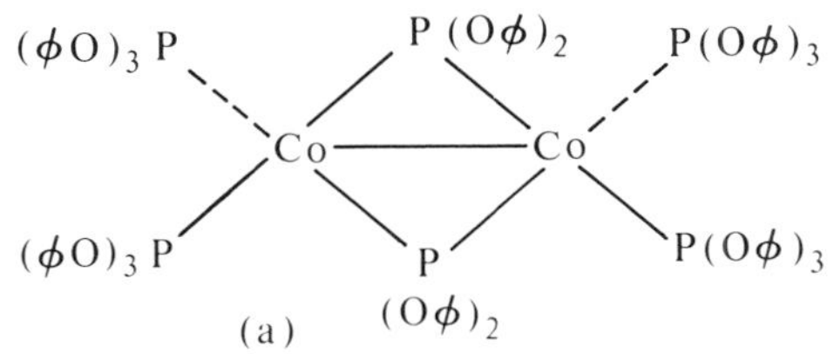

(a)

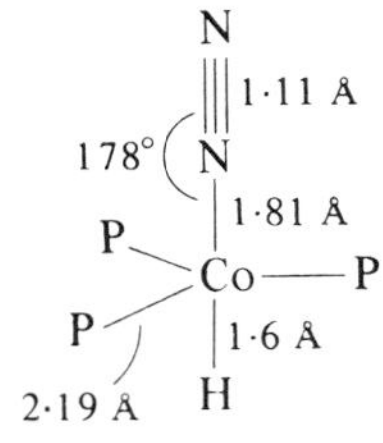

(b)

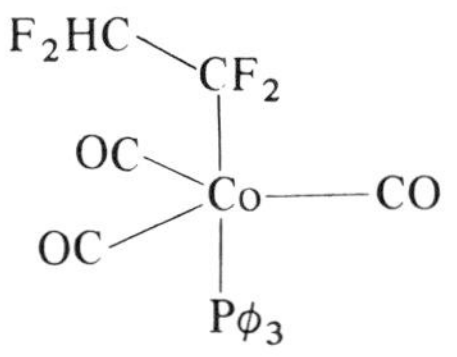

(c)

p. 813, and X-ray studies have established the trigonal bipyramidal structure of the following complexes. The molecule $CoH(P\phi_3)_3N_2^{(1c)}$ is of special interest as an example of the coordination of molecular nitrogen to a metal atom, (b). The bond length in the N_2 ligand is similar to that in the free molecule, and the length of the Co—N bond suggests some multiple bond character (single Co—N, 1.95 Å). There is apparently similar coordination of N_2 to the metal atom in the $[Ru^{II}(NH_3)_5N_2]^{2+}$ ion,[1d] but owing to disorder in the crystal less accurate data were obtained. In our third example, $Co(P\phi_3)(CO)_3(CF_2.CHF_2)$,[2] (c), the three CO ligands occupy the equatorial positions. The trigonal bipyramidal ion $[Co(CNCH_3)_5]^+$, studied in the perchlorate,[3] was mentioned on p. 946.

Co(III). With very few exceptions, such as the heteropolyacid ion $(CoW_{12}O_{40})^{5-}$ (p.523), the bond arrangement is octahedral. We deal separately with examples of the extremely numerous cobaltammines. High-spin Co(III), d^6, is rare; it occurs in K_3CoF_6 but not in $Co(NH_3)_3F_3$ or most cobaltammines.

Co(IV). The pale-blue compound originally thought to be K_3CoF_7 was later shown to be K_3CoF_6. However, the brownish-yellow Cs_2CoF_6 has the K_2PtCl_6 structure,[4] and $BaCoO_3$ has the $BaNiO_3$ structure, with distorted h.c.p. Ba and O and Co—Co, 2.38 Å, in the columns of face-sharing CoO_6 octahedra.[5] In contrast to the octahedral coordination in these compounds there is tetrahedral coordination of Co(IV) in a number of alkali oxy compounds which contain ions similar to those formed by Si: discrete CoO_4^{4-} in Na_4CoO_4[6] and Li_8CoO_6,[7] isostructural with Li_8SiO_6,[8] $Co_2O_7^{6-}$ ions in the red-violet $K_6Co_2O_7$,[9] and chains in the black K_2CoO_3,[10] which has a structure similar to that of Rb_2TiO_3.[11]

(1a)	JACS 1977 **99** 739	(5)	AC 1977 **B33** 1299
(1b)	IC 1977 **16** 430	(6)	ZaC 1975 **417** 35
(1c)	IC 1969 8 2719	(7)	ZaC 1973 **398** 54
(1d)	AC 1968 **B24** 1289	(8)	NW 1973 **60** 256
(2)	JCS A 1967 2092	(9)	ZaC 1974 **409** 152
(3)	IC 1965 4 318	(10)	ZaC 1974 **408** 75
(4)	ZaC 1961 **308** 179	(11)	ZaC 1974 **408** 60

The stereochemistry of cobalt (II) $-d^7$

We have to discuss the spatial arrangement of 4, 5, or 6 bonds. At one time many deductions of bond arrangements were made from magnetic moments, which would be expected to have the following (spin-only) values:

		μ_{eff}
low spin	planar	
	octahedral	} 1.73 BM
high spin	tetrahedral	
	octahedral	} 3.87

However, the magnetochemistry of Co(II) is complicated by spin–orbital interactions. Observed moments are almost invariably higher than the above values, and for some time it was supposed that typical ranges for the two kinds of complex were 1.8–2.1 and 4.3–5.6 BM. More recently many intermediate values have been recorded for octahedral complexes (e.g. 2.63 BM for Co(terpyr)$_2$Br$_2$.H$_2$O) and in fact these now cover the entire range between 2 and 4 BM. The moment is evidently very sensitive to the environment of the Co(II) atom and can no longer be regarded as a reliable indication of the stereochemistry. With the discovery of high-spin square planar complexes with moments around 4 BM magnetic moments do not even distinguish between planar and tetrahedral bonds. Since it is necessary also to be sure of the coordination number of the metal, which cannot always be correctly deduced from the chemical formula, we shall confine our examples to those for which diffraction studies have been made.

Co(II) *forming* 4 *bonds*. The preferred arrangement of four bonds is the tetrahedral one, which is found in the CoX$_4^{2-}$ ions (X = Cl, Br, I), which are stable in the salts of large cations (Cs$_2$CoCl$_4$, (NBu$_4$)$_2$CoI$_4$, and Cs$_3$(CoCl$_4$)Cl), in the [Co(NCS)$_4$]$^{2-}$ ion in, for example, K$_2$Co(NCS)$_4$.3H$_2$O[1] and in the blue forms of compounds with organic amines such as CoCl$_2$(pyridine)$_2$.[2] In the pink or violet forms of these compounds there are infinite chains of octahedral coordination groups which share a pair of opposite edges. The imidazole compound Co(N$_2$C$_3$H$_3$)$_2$[3] is an example of a 3D complex, in which each Co is bonded to 4 others through the organic ligands (compare Cu(N$_2$C$_3$H$_3$)$_2$, p. 1120). There is tetrahedral coordination of Co^{2+} and octahedral coordination of Co^{3+} in Co$_3$O$_4$ (Co–O, 1.99 and 1.89 Å respectively);[3a] compare Co^{2+}–O in CoO, 2.13 Å.

Square coplanar bonds are formed only in complexes with certain polydentate ligands or combinations of ligands such that steric factors prevent a tetrahedral configuration. They include the phthalocyanin (low-spin), the bis-salicylaldehyde diimine complex (a)[4] (in which the maximum deviation of any atom from the mean plane of the complex is 0.6 Å and Co–O(N) is 1.85 Å), and the ion (b) in the salt [Co(mnt)$_2$][N(C$_4$H$_9$)$_4$]$_2$.[5] The magnetic moment of this salt (3.92 BM)

(a)   (b)

(c)

corresponds to 3 unpaired electrons (d^7). The Ni(II) compound is isostructural. Ions [M(mnt)$_2$]$^-$ are formed by Ni(III) and Cu(III).[6] The similar ion formed by Co(III) and the ligand (c) has a magnetic moment of 3.18 BM, slightly higher than the spin-only value for 2 unpaired electrons (d^6); it also is planar.[7]

The planar molecule *trans*-Co(PEt$_2\phi$]$_2$(mesityl)$_2$[8] is an example of a spin-paired d^7 complex with monodentate ligands.

(1)   AC 1975 **B31** 613
(2)   AC 1975 **B31** 1543
(3)   AC 1975 **B31** 2369
(3a) AC 1973 **B29** 362
(4)   AC 1969 **B25** 1675

(5)   IC 1964 3 1500
(6)   IC 1964 3 1507
(7)   IC 1968 7 741
(8)   JCS 1963 3411

Co(II) *forming 5 bonds.* We shall, somewhat arbitrarily, divide the examples into three groups: (i) those approximating to trigonal bipyramidal complexes, with monodentate or 'tripod-like' tetradentate ligands; (ii) those with a fairly characteristic square pyramidal structure, with the metal atom 0.3–0.6 Å above the centre of the base; and (iii) dimeric complexes formed from square planar complexes ML$_2$ in which each M atom forms a fifth (longer) bond to an atom in the other half of the dimer. Many complexes have configurations intermediate between trigonal bipyramidal and square pyramidal, and in some cases the 5-coordination is to be regarded as indicating a failure to attain 6-coordination owing to the bulky nature of the ligand. We shall include these intermediate structures in classes (i) or (ii).

(i) The two simple types to which we have referred are illustrated in Fig. 27.4(a) and (b). Examples of (a) include Co(2-picoline N oxide)$_5$(ClO$_4$)$_2$,[1] with axial bonds (2.10 Å) longer than the equatorial ones (1.98 Å), and the less symmetrical Co(CO)$_4$SiCl$_3$.[2] In [CoBr(Me$_6$tren)]Br,[3] Fig. 27.4(b), Co lies 0.3 Å below the equatorial plane, giving N^1–Co–N^2 81° and N^2–Co–N^3 118°, and the axial Co–N^1, 2.15 Å, slightly longer than the equatorial Co–N^2 (2.08 Å). These distortions are presumably associated with the geometry of the ligand. There is

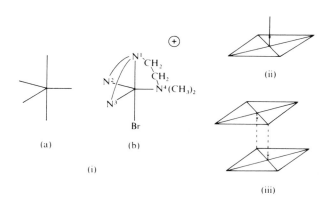

FIG. 27.4. Co(II) forming five bonds (see text).

more distortion in $Co(QP)Cl$,[4] where the equatorial bond angles are 109°, 113°, and 138° (QP is the P analogue of QAS, p. 1240), and still more in $CoBr_2(HP\phi_2)_3$,[5] where the equatorial bond angles are 98°, 126°, and 136° and there is a marked difference between the lengths of the two Co–Br bonds (both equatorial, 2.33 Å and 2.54 Å). (The H atom is omitted from the formula throughout ref. (5).) Intermediate configurations are also found with bulky amine ligands, as in the high-spin complexes $Co(Et_4dien)Cl_2$[6] and $Co(Me_5dien)Cl_2$,[7] in which the ligands are:

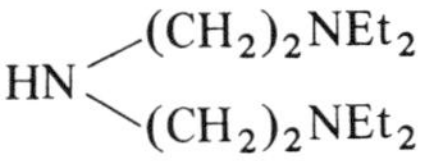

(ii) Square pyramidal monomeric complexes, Fig. 27.4 (ii), include $CoCl_2$-(L_3),[8] where L_3 is the tridentate ligand (a), the structure of which is very similar to that of $ZnCl_2(terpyr)$, p. 1153. The metal atom lies 0.4 Å above the base of the pyramid.

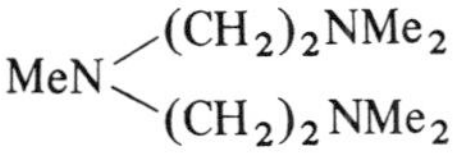

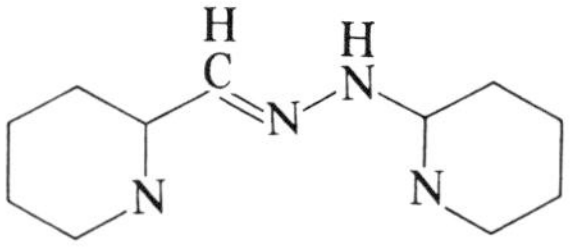

(a) 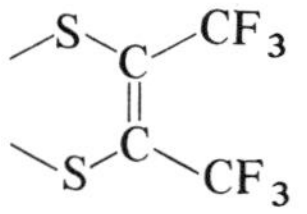 (b) 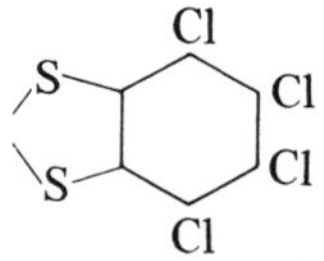 (c)

(iii) Dimers of the type shown in Fig. 27.4 (iii) exist in $[Co(b)_2]_2$[9] and $[Co(c)_2]_2(NBu_4)_2$,[10] in which (b) and (c) are the ligands shown above. In both complexes the metal atom lies slightly above the plane of the four close S ligands (0.37 Å and 0.26 Å respectively), and the vertical bonds to the fifth S are longer than the four basal bonds (approximately 2.40 Å and 2.18 Å respectively).

The arrangement of P atoms around Co in $CoH(PF_3)_4$ is distorted tetrahedral,[11] with bond angles ranging from 102° to 118°, but since H is presumably bonded to Co this is an example of 5-covalent Co. (The H atom was not located in the X-ray study.)

(1)   IC 1970 9 767
(2)   IC 1967 6 1208
(3)   IC 1967 6 955
(4)   CC 1967 763
(5)   IC 1966 5 879
(6)   IC 1967 6 483
(7)   IC 1969 8 2729
(8)   JCS A 1966 1317
(9)   IC 1965 4 1729
(10) JACS 1968 90 4253
(11) IC 1970 9 2403

$Co(II)$ *forming 6 bonds.* With one exception, noted below, the bond arrangement is octahedral, and this is found in 3D complexes in crystalline binary compounds $(CoO, CoX_2,$ etc.), in layer structures such as K_2CoF_4 and $BaCoF_4$, in chain structures of the type $CoCl_2(pyr)_2$, and in numerous finite complexes which include *trans*-$Co(NC_5H_5)_4Cl_2$, $Co(dimethylglyoxime)_2Cl_2$, and $Co(acac)_2(pyr)_2$.[1] In its complexes with acetylacetone Co(II) consistently achieves 6-coordination. The acetylacetonate sublimes as the tetramer, $Co_4(acac)_8$,[2] but from solution in inert solvents containing a little water various polymeric species can be obtained: $Co_3(acac)_6 \cdot H_2O$,[3] $Co_2(acac)_4(H_2O)_2$,[4] and finally the simple monomeric $Co(acac)_2(H_2O)_2$.[5] These correspond to splitting the tetrameric $[Co(acac)_2]_4$, Fig. 27.5(a), at a shared face (*A*), giving (b), or at both (*A*) and (*B*), leaving $Co_2(acac)_2(H_2O)_2$, (c); an unoccupied bond position is filled with a H_2O molecule.

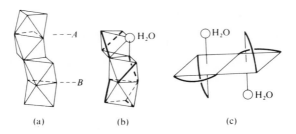

FIG. 27.5. The molecular structures of (a) $Co_4(acac)_8$, (b) $Co_3(acac)_6 \cdot H_2O$, (c) $Co_2(acac)_4(H_2O)_2$.

In the red diamagnetic form of $[Co_2(CNCH_3)_{10}](ClO_4)_4$,[6] one of the octahedral bonds from each Co is to a second metal atom. In this ion Co(II) is isoelectronic with Mn(0) and forms the same type of structure as $Mn_2(CO)_{10}$ (staggered configuration. Co—Co, 2.74 Å). The blue solution of this compound and the blue crystalline form presumably contain $[Co(CNCH_3)_5]^{2+}$, the paramagnetism corresponding to one unpaired electron.

The very irregular 6-coordination of Co(II) in $Co(NO_3)_2[OP(CH_3)_3]_2$[7] is noted in Chapter 18.

An unusual kind of isomerism is exhibited by $Co(CNR)_4I_2$,[8] in which R is, for example, C_6H_5. The paramagnetic isomer consists of finite octahedral molecules with the *trans* configuration (and approximately collinear Co—C—N—R bonds) containing low-spin Co(II). The other isomer is diamagnetic and contains dimeric ions (a) in which there is presumably spin exchange in the Co—I—Co portion of the ion.

$$\left[I-\overset{\diagdown\diagup}{\underset{\diagup\diagdown}{Co}} -I-\overset{\diagdown\diagup}{\underset{\diagup\diagdown}{Co}} -I \right]^+ \quad I^-$$

(a)

The ion $[Co_3(OCH_2CH_2NH_2)_6]^{2+}$ is the sole example of trigonal prism coordination of Co(II) by oxygen. In this ion (Fig. 27.6) there is octahedral coordination of the terminal Co(III) atoms but trigonal prism coordination of the central Co(II).[9] It is possible that this unusual coordination is connected with the geometry of the ligands, there being apparently better packing of the CH_2 groups (which are not

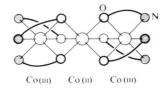

$Co_{(III)}$ $Co_{(II)}$ $Co_{(III)}$

FIG. 27.6. The structure of the ion $[Co_3(OCH_2CH_2NH_2)_6]^{2+}$ (diagrammatic).

shown in Fig. 27.6) with trigonal prismatic than with octahedral coordination around the central Co atom.

(1)   IC 1968 7 1117
(2)   IC 1965 **4** 1145
(3)   JACS 1968 **90** 38
(4)   IC 1966 **5** 423
(5)   AC 1959 **12** 703

(6)   IC 1964 **3** 1495
(7)   JACS 1963 **85** 2402
(8)   AC 1975 **B31** 40
(9)   JACS 1969 **91** 2394

$Co(II)$ *forming* 8 *bonds*. There is coordination by 4 bidentate ligands in a number of complexes, in which the arrangement of the 4 *pairs* of O atoms may be described as tetrahedral. Examples include $[Co(O_2C.CF_3)_4](As\phi_4)_2$[1] and $[Co(NO_3)_4]$-$(As\phi_4)_2$.[2] In complexes of this general type there are 8 O atoms (of four bidentate groups) around the metal atom and these form two groups, 4 A + 4 B, which separately define two interpenetrating (non-regular) tetrahedra (Fig. 27.7(a)). The 8 atoms lie at the vertices of a dodecahedron (Fig. 27.7(b)), and in $Ti(NO_3)_4$ the

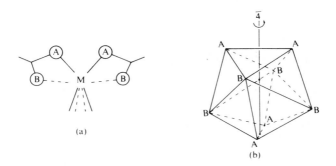

(a)   (b)

FIG. 27.7. Coordination of metal by four bidentate ligands (see text).

two sets of M—O distances (to A and B atoms) are equal. In $[Co(NO_3)_4]^{2-}$ the distances to A and B atoms are appreciably different (2.07 Å and 2.45 Å) so that Co is forming 4 stronger tetrahedral bonds (i.e. (4+4) distorted 8-coordination). In the trifluoroacetato complex the distortion from dodecahedral coordination is much greater, the two Co—O distances being 2.00 Å and 3.11 Å. The four B atoms are coplanar with Co, and the angle O_A—Co—O_A is 97°, so that the coordination is much closer to regular tetrahedral. The magnetic moments of both complexes are close to 4.6 BM, typical of Co(II) high-spin complexes.

(1)   IC 1966 5 1420   (2)   IC 1966 5 1208

Co(III) *forming* 6 *octahedral bonds: cobaltammines.* Finite complexes in which Co(III) forms octahedral bonds are numerous. The ligands may be monodentate, as in $[Co(NO_2)_6]^{3-}$, $[Co(CN)_6][Co(NH_3)_6]$,[1] or $[Co(CO_3)_2(H_2O)_4]^{2-}$,[2] bidentate, as in $Co(acac)_2$[3] or CoL_3[4] where $L = S_2CS(C_2H_5)$, $S_2CO(C_2H_5)$, or $S_2CN(C_2H_5)_2$, or tridentate, as in $Co(N_3\phi_2)_3$.[5] The structure of the diazoaminobenzene (1,3-diphenyltriazene) complex, illustrated in Fig. 27.8(a), (p. 1219), is of interest and may be compared with the dimeric structure of the cuprous derivative (p. 1104). The largest group of octahedral cobaltic complexes comprises the cobaltammines, which include many examples of the types of complex noted above. We shall apply the name cobaltammine to any octahedral Co(III) complex in which some or all of the ligands are NH_3 or amine molecules.

The discovery that $CoCl_3$ could form stable compounds with 3, 4, 5, or 6 molecules of NH_3 led to Werner's suggestion that in 'molecular compounds' such as $CoCl_3 . 6NH_3$ the six NH_3 molecules are arranged in the first 'coordination sphere' around the metal atom and the three Cl atoms in an outer sphere. This was consistent with the chemical behaviour of the compounds, with the numbers of ions formed from one 'molecule', and with the numbers of isomers of mixed coordination groups such as $[Co(NH_3)_4Cl_2]^+$. Some or all of the NH_3 molecules in $[Co(NH_3)_6]^{3+}$ may be replaced by other ligands such as Cl^-, H_2O, NO_2^-, CO_3^{2-}, $C_2O_4^{2-}$, $NH_2 . CH_2 . CH_2 . NH_2$(en), etc. Of the ammines of $CoCl_3$ containing 3, 4, 5, and 6 NH_3 the first is a non-electrolyte while the others ionize to give respectively 2, 3, and 4 ions from one formula-weight, so that they are $Co(NH_3)_3Cl_3$, $[Co(NH_3)_4Cl_2]Cl$, $[Co(NH_3)_5Cl]Cl_2$, and $[Co(NH_3)_6]Cl_3$. Owing to the large number of combinations of atoms and groups which may be attached to Co these compounds are very numerous. The only comparable group of compounds is that formed by Cr(III), but these have been much less studied in recent years.

Many cobaltammines are easily prepared by oxidizing ammoniacal solutions of Co(II) salts by a current of air. For example, if the solution contains $CoCl_2$ and NO_2^- or CO_3^{2-} ions the brown Erdmann's salt, $[Co(NH_3)_2(NO_2)_4]NH_4$, or the red $[Co(NH_3)_4CO_3]Cl$ may be crystallized out. Crystals of the bridged salt $[(NH_3)_5Co.O_2.Co(NH_3)_5](SCN)_4$ are formed in a strongly ammoniacal solution of $Co(SCN)_2$ which is left exposed to the atmosphere. In such preparations mixtures of cobaltammine ions are formed, both mononuclear and polynuclear, solubility differences determining the nature of the product obtained in a particular preparation. As an illustration of the stability of certain cobaltammines we may quote the action of concentrated H_2SO_4 on $[Co(NH_3)_6]Cl_3$ to form the sulphate of the complex, which is not disrupted in the process. The aquo-pentammine, $[Co(NH_3)_5H_2O]Cl_3$, retains its water up to a temperature of $100°C$, and even then the complex does not break up but a rearrangement takes place to form $[Co(NH_3)_5Cl]Cl_2$.

Cobaltammines, including $Co(NH_3)_3F_3$ (but not K_3CoF_6, which has $\mu_{eff.} = 5.3$ BM), are diamagnetic except in special cases where an unpaired electron is introduced with a ligand such as O_2^- (see later). Their stability is presumably associated with the fact that with 12 bonding electrons added to the d^6 configuration of Co^{III} there are 18 electrons to fill the 9 available orbitals; for Cr^{III} there is a

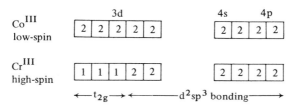

single electron in each of the t_{2g} orbitals. (For Cr^{III} there is no difference between the magnetic moment for high-spin and low-spin complexes.)

(1) AC 1973 **B29** 822
(2) AC 1976 **B32** 466
(3) AC 1974 **B30** 822

(4) AC 1972 **B28** 2231
(5) JACS 1967 **89** 1530

The isomerism of cobaltammines. The octahedral arrangement of the six ligands has now been established by X-ray studies of many crystalline cobaltammines, but when it was first suggested it was not by any means generally accepted. However, it was consistent with the existence and numbers of isomers of many complexes. For example, there should be only two isomers (*cis* and *trans*) of an octahedral complex CoA_4B_2 but three for a plane hexagonal or trigonal prism model, and the fact that only two isomers of the numerous complexes of this type have ever been isolated is strong presumptive evidence in favour of the octahedral model. The demonstration of the optical activity of $K_3Co(C_2O_4)_3$ provided striking confirmation of the octahedral arrangement of the bonds from the Co atom. It is now well known that any octahedral complex MA_3 in which A represents a bidentate group such as $C_2O_4^{2-}$ or 'en' is enantiomorphic, but it will be appreciated that the resolution of such a compound appeared surprising, for it was previously formulated as a double salt, $3K_2C_2O_4.Co_2(C_2O_4)_3$. Nevertheless, it is a curious fact that the resolution of salts like $K_3[Co(C_2O_4)_3]$ and $[Co(NH_2.CH_2.CH_2.NH_2)_3]Br_3$ into their optical antimers was not considered by some chemists to prove Werner's hypothesis. There still lingered a belief that the optical activity was in some way dependent on the presence of an organic radical in the complex. The last doubts were removed when Werner resolved a purely inorganic cobaltammine, $[Co_4(OH)_6(NH_3)_{12}]Cl_6$. The type of structure suggested for the cation is compounds of this kind (Fig. 27.8(e)) (p. 1219) has been confirmed in a number of salts of $[Co_4(OH)_6(en)_6]^{6+}$.

Some idea of the complexity of the chemistry of the cobaltammines may be gathered from the fact that no fewer than nine compounds with the empirical formula $Co(NH_3)_3(NO_2)_3$ are said to have been prepared. There are many possible types of isomerism and polymerism, as the following examples will show.

Isomers have the same molecular weight but differ in physical and/or chemical properties. The following kinds of isomerism may be distinguished.

(i) *Stereoisomerism.* The isomers contain exactly the same set of bonds, that is, the same pairs of atoms are linked together in the isomers.

(a) The positions of all the atoms relative to one another are the same. The only difference between the isomers is that they are related as object and mirror image (*optical isomerism*).

(b) The constituent atoms are arranged differently in space relative to one another (*geometrical isomerism*).

Optically active complexes which have been resolved include $[Co(C_2O_4)_3]^{3-}$, $[Co(en)_3]^{3+}$, and $[Co(en)_2Cl_2]^+$. A more complex type of isomerism can arise when the chelate group is unsymmetrical For a symmetrical group occupying two coordination positions (e.g. C_2O_4), which we may represent $A-A$, there is only the one stereoisomer (I) which, of course, exists in right- and left-handed modifications. If, however, the group is unsymmetrical there are two possible forms (II and III) exactly analogous to the two stereoisomers of a complex CoA_3B_3, and each form is

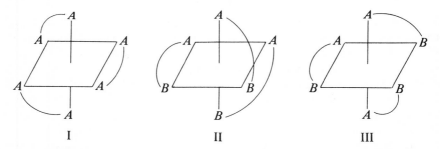

enantiomorphic. Two such forms of the tri-glycine complex $Co(NH_2CH_2COO)_3$ have been prepared.

All three isomers have been made of the $Co(dien)_2^{3+}$ ion (dien = $H_2N.CH_2.CH_2.-NH.CH_2.CH_2.NH_2$) and the structure of the s-facial isomer determined in the tribromide:

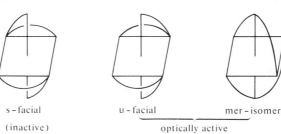

s-facial   u-facial   mer-isomer

(inactive)   optically active

Some bridged complexes provide interesting examples of enantiomorphism. For example, $[(en)_2Co(NH_2)(NO_2)Co(en)_2]Br_4$ exists in *d*- and *l*-forms and also in a meso form:

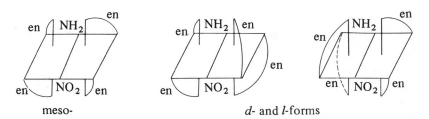

meso-   *d*- and *l*-forms

The active forms of this compound were obtained by Werner by resolution with *d*-bromcamphorsulphonic acid.

Many pairs of geometrical isomers have been made, such as the 'violeo' and 'praseo' series, *cis* and *trans* $[Co(NH_3)_4Cl_2]$ Cl.

(ii) *Structural isomerism*. This includes all cases of isomerism where different pairs of atoms are linked together in the various isomers. These may therefore possess very different chemical properties. We may distinguish three simple kinds of structural isomerism.

(a) A number of cobaltammines containing groups such as NO_2 or SCN exist in two forms, the isomerism being due to the possibility of linking the group to the metal atom in the two ways, $M-NO_2$ or $M-ONO$ and $M-SCN$ or $M-NCS$ respectively. When aquo-pentammine cobaltic chloride $[Co(NH_3)_5H_2O]Cl_3$ is treated with nitrous acid under the correct conditions a red compound is formed which evolves nitrous fumes if treated with dilute mineral acids and which at ordinary temperatures is slowly (and at $60\,^\circ C$ rapidly) converted into a brown isomer. The latter is quite stable towards dilute acids. The isomers are formulated $[Co(NH_3)_5ONO]Cl_2$ and $[Co(NH_3)_5NO_2]Cl_2$. Pairs of 'linkage' isomers of which structural studies have been made include $[Co(NH_3)_5SCN]Cl_2.H_2O$ and $[Co(NH_3)_5NCS]Cl_2$, $[Co(NH_3)_5NCCo(CN)_5].H_2O$ and $[Co(NH_3)_5CNCo(CN)_5].$- H_2O (see p. 943 and Table 27.4).

(b) The term 'ionization isomerism' has been applied to the isomerism of pairs of compounds like $[Co(NH_3)_5Br]SO_4$ and $[Co(NH_3)_5SO_4]Br$. There are also more interesting cases such as $[Co(NH_3)_5SO_3]NO_3$ and $[Co(NH_3)_5NO_2]SO_4$. In these pairs, as in those in (c), the isomerism applies only to the solid state, for the actual coordination complexes have different compositions.

(c) Since many of the octahedral complexes are ions and since there are positively and negatively charged complexes, as in the series

$$Co(NH_3)_4(NO_2)_2^+, \ Co(NH_3)_3(NO_2)_3, \ Co(NH_3)_2(NO_2)_4^-,$$

it should obviously be possible to prepare crystalline salts in which both the cation and the anion are octahedral complexes. We could then have pairs of salts with the

TABLE 27.4
References to the structures of cobaltammines

Compound	Reference
$[Co(NH_3)_6]Cl_3$	AC 1978 **B34** 915
$[Co(NH_3)_6]I_3$	AC 1969 **B25** 168
$[Co(CN)_6][Cr(en)_3] \cdot 6 H_2O$	IC 1968 7 2333
$Co(N_3\phi_2)_3$	JACS 1967 **89** 1530
$[Co(NH_3)_5NO_2]Br_2$	AC 1968 **B24** 474
$[Co(NH_3)_5NO_2]Cl_2$	ACSc 1968 **22** 2890
$[Co(NH_3)_5CO_3]Br \cdot H_2O$	JCS 1965 3194
$[Co(NH_3)_5N_3](N_3)_2$	AC 1964 **17** 360
$[Co(NH_3)_5SCN]Cl_2 \cdot H_2O$ and $[Co(NH_3)_5NCS]Cl_2$	AC 1972 **B28** 1908
$[Co(en)_2(N_3)_2]NO_3$	AC 1968 **B24** 1638
$[Co(en)_2Cl_2]Cl \cdot HCl \cdot 2 H_2O$	BCSJ 1952 **25** 331
$[Co(NH_3)_4CO_3]Br$	JCS 1962 586; ACSc 1963 **17** 1630
$Co(NH_3)_3(NO_2)_3$-*mer*	IC 1971 **10** 1057
$Co(NH_3)_3(NO_2)_3$-*fac*	AC 1979 **B35** 1020
$Co(NH_3)_3(NO_2)_2Cl$	BCSJ 1953 **26** 420
$[Co(NH_3)_3(H_2O)Cl_2]Cl$	BCSJ 1952 **25** 328
$[Co(NH_3)_2(NO_2)_4]Ag$	ZK 1936 **95** 74
$[Co(dien)_2]Br_3$	AC 1972 **B28** 470
$[Co(NH_3)_5CoNCCo(CN)_5] \cdot H_2O$	IC 1971 **10** 1492
$[(NH_3)_5NCCo(CN)_5] \cdot H_2O$	IC 1971 **10** 1492
$[(NH_3)_5Co \cdot NH_2 \cdot Co(NH_3)_5](NO_3)_5$	AC 1968 **B24** 283
$\left[(NH_3)_4Co \underset{OH}{\overset{OH}{<}} Co(NH_3)_4 \right] Cl_4 \cdot 4 H_2O$	JCS 1962 4429; ACSc 1963 17 85
$\left[(NH_3)_4Co \underset{NH_2}{\overset{Cl}{<}} Co(NH_3)_4 \right] Cl_4 \cdot 4 H_2O$	IC 1970 9 2131
$\left[(NH_3)_3Co \underset{OH}{\overset{OH}{-}} OH - Co(NH_3)_3 \right] Br_3$	AC 1977 **B33** 700
$[(NH_3)_3Co(OH)_2(CH_3COO)Co(NH_3)_3]Br_3 \cdot 3H_2O$	AC 1977 **B33** 3185
$[(en)_2Co(NH_2)(SO_4)Co(en)_2]Br_3$	AC 1971 **B27** 1744
$[Co_3(NH_3)_8(OH)_2(NO_2)_2(CN)_2](ClO_4)_3 \cdot NaClO_4 \cdot 2 H_2O$	AC 1970 **B26** 1709
$[Co_4(OH)_6(en)_6]^{6+}$	AC 1978 **B34** 807

same composition of which the following are the simplest types: $(M^1A_6)(M^2B_6)$ and $(M^1B_6)(M^2A_6)$ containing two different metals. The pairs of salts $[Co(NH_3)_6]$-$[Cr(CN)_6]$, $[Cr(NH_3)_6][Co(CN)_6]$ and $[Co(NH_3)_4(H_2O)_2][Cr(CN)_6]$, $[Cr(NH_3)_4$-$(H_2O)_2][Co(CN)_6]$ have been prepared.

(iii) *Polymerism*. Consider a salt of the type $(MA_nB_{6-n})(MA_{6-n}B_n)$ in which the central metal atom is the same in both complexes. Crystals of the compound consist of equal numbers of octahedral complexes of two kinds, so that the simplest structural formula is that given above. The empirical formula is, however, MA_3B_3, which corresponds to an entirely different compound, in which all the structural units are identical. The compounds $Co(NH_3)_3(NO_2)_3$ and $[Co(NH_3)_6][Co(NO_2)_6]$

are related in this way, and $[(NH_3)_3Co(OH)_3Co(NH_3)_3]Cl_3$ has the same composition as $[Co\{(OH)_2Co(NH_3)_4\}_3]Cl_6$.

The structures of cobaltammines. The structures of many cobaltammines have been established by X-ray crystallographic studies. The special interest attaching to a particular compound may lie in the bond lengths, in the mode of attachment of the ligands to the metal, in their relative arrangement (geometrical isomerism), or in the actual gross stereochemistry in the case of more complex (bridged) complexes. For example, in $[Co(NH_3)_6]I_3$ the length of the bond Co^{III}–N is 1.94 Å, as compared with Co^{II}–N, 2.11 Å in $[Co(NH_3)_6]Cl_2$ (and planar Co^{II}–N, 1.85 Å).

In $[Co(NH_3)_4CO_3]Br$ there is a bidentate CO_3^{2-} ligand, Fig. 27.8(b); the lengthening of the Co–N bonds *trans* to the bidentate ligand is probably real. Contrast the monodentate CO_3^{2-} group in $[Co(NH_3)_5CO_3]Br.H_2O$, where there is apparently hydrogen bonding between one O and one NH_3 (c).

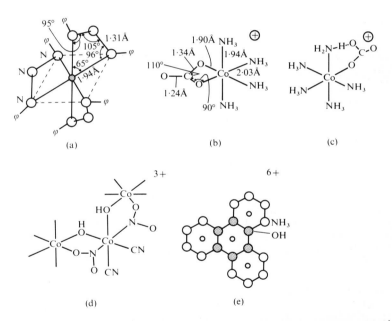

(a)   (b)   (c)

(d)   (e)

FIG. 27.8 The structures of Co(III) complexes: (a) $Co(N_3\phi_2)_3$; (b) $[Co(NH_3)_4CO_3]^+$; (c) $[Co(NH_3)_5CO_3]^+$; (d) $[Co_3(NH_3)_8(OH)_2(NO_2)_2(CN)_2]^{3+}$; (e) probable structure of $[Co_4(OH)_6(NH_3)_{12}]^{6+}$ (see p. 201).

In complexes CoA_4B_2 (and more complex types) the structural study is necessary to establish unambiguously the relative arrangement of the ligands, e.g. the *trans* arrangement of the Cl atoms in $[Co(NH_3)_3(H_2O)Cl_2]^+$ and of the two NH_3 molecules in $[Co(NH_3)_2(NO_2)_4]^-$. The latter ion, of Erdmann's salt, has been assigned the *cis* configuration on the basis of self-consistent chemical evidence.

The three types of binuclear cobaltammine involve a single, double, or triple bridge between the metal atoms, the bridging ligands being Cl, OH, NH_2, or O_2 in most cases. For a bridged cyano compound see p. 943 and Table 27.4. The structures of the red diamagnetic peroxo- and the green paramagnetic superoxo- compounds are described in Chapter 11. Examples of bridged complexes are included in Table 27.4; some details are:

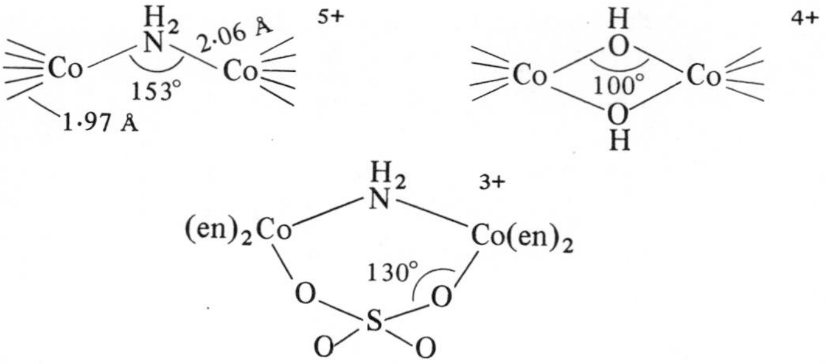

The ion $[Co_3(NH_3)_8(OH)_2(NO_2)_2(CN)_2]^{3+}$ is of special interest as containing not only bridging OH groups but also NO_2^- ions bridging through O and N atoms (Fig. 27.8(d)); in $[(en)_2Co(NH_2)(SO_4)Co(en_2)]^{3+}$ there is a bridging SO_4 group (see above).

Bridging can proceed further as in complexes such as

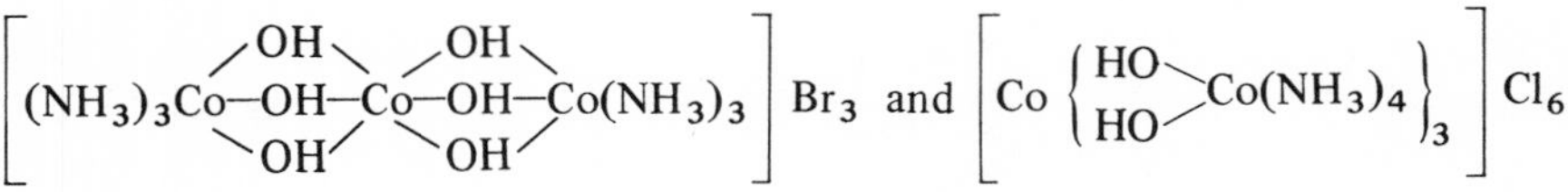

the second of which, Fig. 27.8(e), has already been mentioned in connection with the optical activity of cobaltammines. It is noteworthy that the closely related ion $[Cr_4(OH)_6(en)_6]^{6+}$ has the different structure illustrated in Fig. 5.10 (p. 202).

The structural chemistry of nickel

The stereochemistry of $Ni(II)–d^8$

The following bond arrangements have been established by structural studies of finite ions or molecules:

C.N.	Low spin (diamagnetic)	High spin (paramagnetic)
4	Coplanar	Tetrahedral
5	Trigonal bipyramidal	Trigonal bipyramidal
	Tetragonal pyramidal	Tetragonal pyramidal
6	(Distorted octahedral)	Octahedral

Much confusion in the literature on this subject has been due to the placing of too much reliance on magnetic data of dubious quality and to incorrect deductions of coordination numbers from chemical formulae. There are regrettably few examples of compounds of which both thorough magnetic and structural studies have been made. The following facts have been established:

(a) diamagnetism is indicative of planar as opposed to tetrahedral coordination if the c.n. is known to be 4;

(b) 5-coordinated complexes may be either diamagnetic or paramagnetic;

(c) paramagnetism is exhibited by 4-, 5-, and 6-coordinated Ni(II). The spin-only moment for d^8 is 2.83 BM, but spin-orbital interactions in *magnetically dilute* systems always lead to higher values. It is probably safe to interpret values of $\mu > 3.4$ BM as indicative of tetrahedral coordination and values < 3.2 BM as associated with octahedral coordination, the interpretation of values in the intermediate range is not always clear. Examples of magnetic moments of compounds whose structures have been established include:

		μ
Tetrahedral:	$(NiCl_4)[N(C_2H_5)_4]_2$	3.90 BM
Trigonal bipyramidal:	$NiCl_3(H_2O)[N(C_2H_4)_3NCH_3]$	3.7
Octahedral:	$Ni(pyr)_4(ClO_4)_2$	3.24

The coordination number of the metal atom in a complex cannot in general be deduced simply from the chemical formula because a ligand may be shared between two (or more) metal atoms in a polymeric (finite or infinite) grouping, or a particular ligand may behave in different ways, as illustrated by the behaviour of acac in Pt compounds (p. 1242). For example, the (paramagnetic) nickel acetylacetonate is not a tetrahedral molecule but a trimer, $[Ni(acac)_2]_3$, containing octahedrally coordinated Ni, and compounds MX_2L_2 may be either finite (that is, planar or tetrahedral) or an infinite chain of edge-sharing octahedral groups. The stereochemistry of Ni is further complicated by the fact that there is evidently little difference in stability between the following pairs of bond arrangements for certain types of complex:

(i) square planar (4) and tetrahedral (4);

(ii) trigonal bipyramidal (5) and square pyramidal (5);

(iii) square planar (4) and octahedral (6).

(i) Many compounds NiX_2L_2 exist in two forms. For example, $NiBr_2(PEt\phi_2)_2$ crystallizes as a dark-green paramagnetic form from polar solvents and as a brown diamagnetic form from CS_2. Both forms are monomeric in solution and presumably tetrahedral and planar complexes respectively. (The blue and yellow forms of $Ni(quinoline)_2Cl_2$ are both paramagnetic and may be examples of finite tetrahedral and octahedral chain structures.) Square planar–tetrahedral isomerism seems to have been observed only with certain series of Ni complexes, and pure specimens of both isomers have been isolated only with diphenyl alkyl ligands. In series of compounds $NiX_2(PR_3)_2$, with X changing from Cl to Br to I, and R_3 from $(alkyl)_3$

to (phenyl)$_3$, the change in configuration usually occurs around $NiBr_2(PR\phi_2)_2$, that is, the Cl compounds are mostly planar, the I compounds mostly tetrahedral, and the Br compounds often exist in both forms.[1] In one form of $NiBr_2[P\phi_2$-(benzyl)]$_2$ the crystal contains both planar and tetrahedral isomers. This compound crystallizes in a red diamagnetic form and also in a green paramagnetic form (moment 2.7 BM). In the latter, one-third of the molecules are square planar and two-thirds tetrahedral; the name *interallogon* compound has been suggested for a crystal of this type containing two geometrical isomers with different geometries.[2] Examples of complexes with bond arrangements intermediate between square planar and tetrahedral are noted later.

(ii) Numerous 5-covalent diamagnetic complexes have been studied, containing both mono- and poly-dentate ligands (examples are given later) and here again two configurations of a complex have been found in the same crystal. The $Ni(CN)_5^{3-}$ ion is stable only in the presence of large cations—at low temperatures K^+ is sufficiently large—and its configuration is dependent on the nature of the cation. In $[Cr(tn)_3][Ni(CN)_5].2H_2O$ (tn = 1,3-propane diamine) all the anions are square pyramidal (a), but in $[Cr(en)_3][Ni(CN)_5].1\frac{1}{2}H_2O$ there are ions of both types (a) and (b).[3]

(a)   (b)

(iii) Compounds $Ni(pyr)_4(ClO_4)_2$, in which 'pyr' represents pyridine or a substituted pyridine, exist in yellow diamagnetic and blue paramagnetic forms. The former contain planar ions, (c), the next nearest neighbours of Ni being O atoms at

(c)   (d)

3.34 Å, while the latter consist of octahedral molecules, (d).[4] The compounds studied were the 3,5-dimethylpyridine compound, (c), and the 3,4-dimethylpyridine compound, (d). The compounds containing the bidentate ligand meso-stilbene diamine, $C_6H_5.CH(NH_2)-CH(NH_2).C_6H_5$, are still more interesting.[5] The blue and yellow forms are interconvertible in solution, and while the hydrated blue form ($\mu = 3.16$ BM) of the dichloroacetate contains octahedral ions, (e), the yellow form ($\mu = 2.58$ BM) contains planar ions of type (f) and octahedral molecules (g) in the

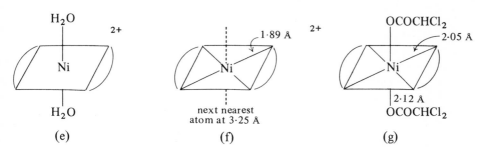

(e)	(f)	(g)

ratio 1:2. This is the same ratio of diamagnetic to paramagnetic complexes as in $NiBr_2(P\phi_2Bz)_2$ mentioned above, and results in a similar intermediate value of μ_{eff}.

We referred in Chapter 3 to the clathrate compounds based on layers of the composition $Ni(CN)_2.NH_3$. In the clathrates containing C_6H_6 or $C_6H_5NH_2$ the layers are directly superposed, with NH_3 molecules of different layers directed towards one another, so that the interlayer spacing is 8.3 Å and there are large cavities between the layers (Fig. 1.9(a), p. 31).[6] In the biphenyl clathrate, $Ni(NH_3)_2.Ni(CN)_4.2C_{12}H_{10}$,[7] the inter-layer distance increases to 12.65 Å. In the hydrate and in the anhydrous compound the layers are packed much more closely (inter-layer spacing 4.4 Å), the NH_3 molecules of one layer projecting towards the centres of the rings of adjacent layers (pseudo body-centred).[8] Similar compounds may be made in which the octahedrally coordinated metal atoms are Cu, Cd, or Mn[9] as, for example, in the dioxane clathrate, $Cd(NH_3)_2.Ni(CN)_4.2C_4H_8O_2$.[10]

(1) IC 1965 **4** 1701	(6) JCS 1958 3412
(2) JCS A 1970 1688	(7) AC 1974 **B30** 292
(3) IC 1970 **9** 2415	(8) ZK 1966 **123** 391
(4) AC 1968 **B24** 745, 754	(9) AC 1974 **B30** 687
(5) IC 1964 **3** 468	(10) ZK 1976 **144** 91

Ni(II) *forming 4 coplanar bonds.* Planar diamagnetic complexes include the $Ni(CN)_4^{2-}$ ion, which has been studied in the K[1] and other salts[2] (Ni–C, 1.85 Å), and molecules such as (a)[3] and (b).[4] The numerous complexes with bidentate ligands include the thio-oxalate ion, (c),[5] molecules of the types (d)[6] and (e),[7] and the glyoximes, (f),[8] notable for the short intramolecular hydrogen bonds. Many divalent metals form phthalocyanins,[9] M(II) replacing 2 H in $C_{32}H_{18}N_8$ (Fig. 27.9).

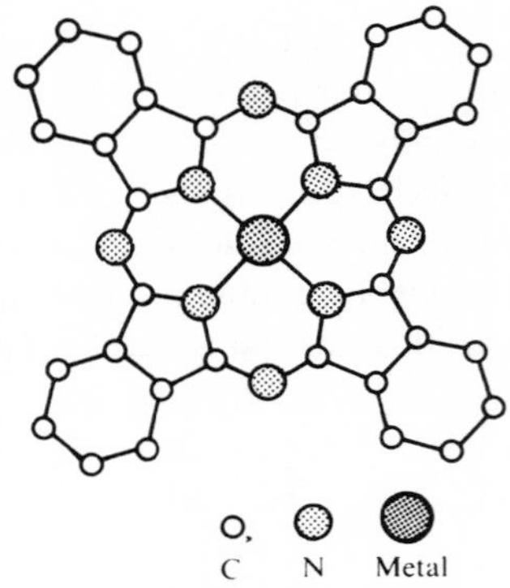

FIG. 27.9. Molecule of a metal phthalocyanin, $C_{32}H_{16}N_8M$.

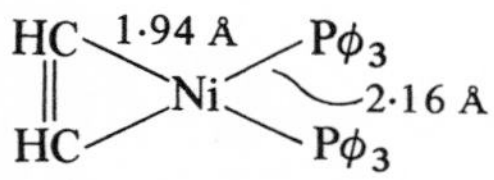

(a)

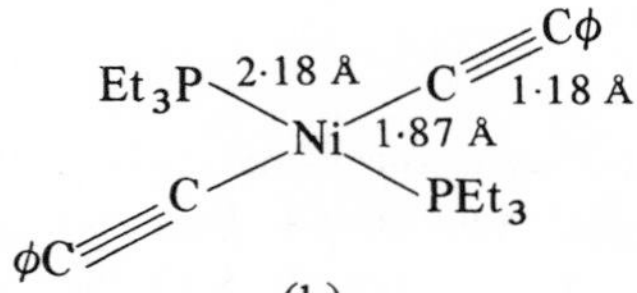

(b)

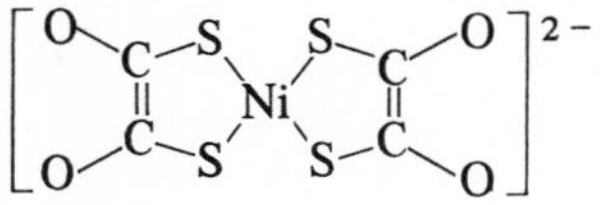

(c)

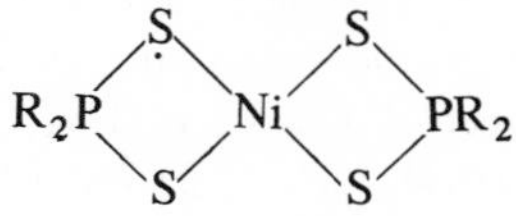

(d)

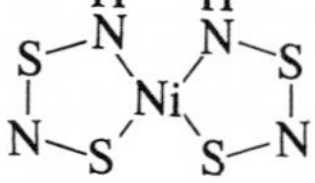

(e)

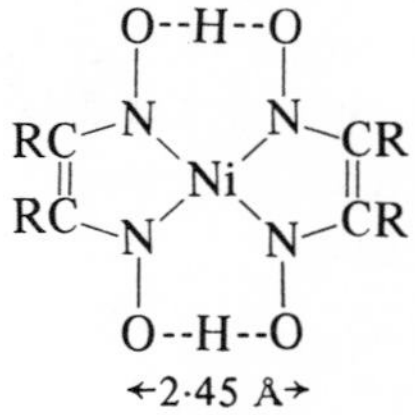

(f)

Polymeric species include finite molecules and chains, In Ni_3L_4 (L = $H_2N.CH_2.-CH_2.S$),[10] (g), the three Ni atoms are collinear with a planar arrangement of bonds around each. The whole molecule is twisted owing to the symmetrical arrangement of the chelates around the central Ni atom and the *cis* configuration around the terminal Ni atoms. The Ni—Ni distance in this complex is 2.73 Å.

(g)

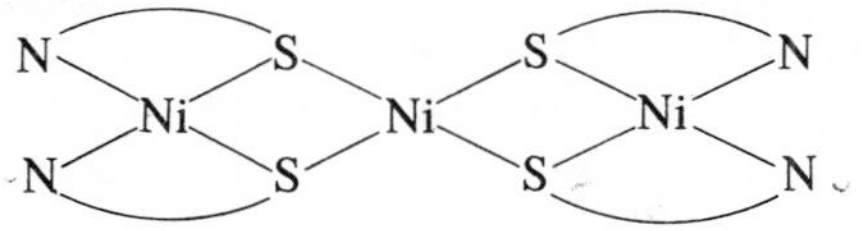

Nickel mercaptides are mixtures of insoluble polymeric materials and hexamers. The molecule of $[Ni(SC_2H_5)_2]_6$[11] is shown diagrammatically in Fig. 27.10. The six Ni atoms are coplanar and are bridged by 12 mercaptan groups. The bond arrangement around Ni is square planar, with all angles in the Ni_2S_2 rings equal to $83°$, Ni–S, 2.20 Å, and Ni--Ni, 2.92 Å (mean). The dimethylpyrazine complex, $NiBr_2L$,[12] (h), is an example of a chain in which Ni is 4-coordinated by 2 Br and by 2 N atoms of the bridging ligands:

(h)

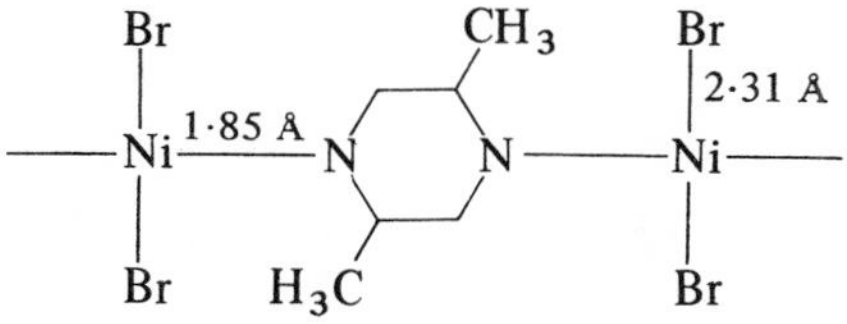

Departures from exact coplanarity of Ni and its four bonded atoms may occur in unsymmetrical complexes, for example, those containing polydentate ligands. In the (diamagnetic) complex (i),[13] there is slight distortion (tetrahedral rather than square pyramidal), the angles S_1NiN_1 and S_2NiN_2 being $173°$. Large deviations from planarity, leading to paramagnetic tetrahedral complexes, can result if there is steric interference between atoms forming parts of independent ligands. In the tetramethyl dipyrromethenato complex (j)[14] there is interference between the

(i)

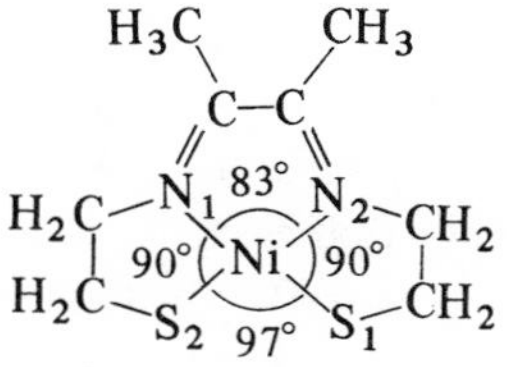

(j)

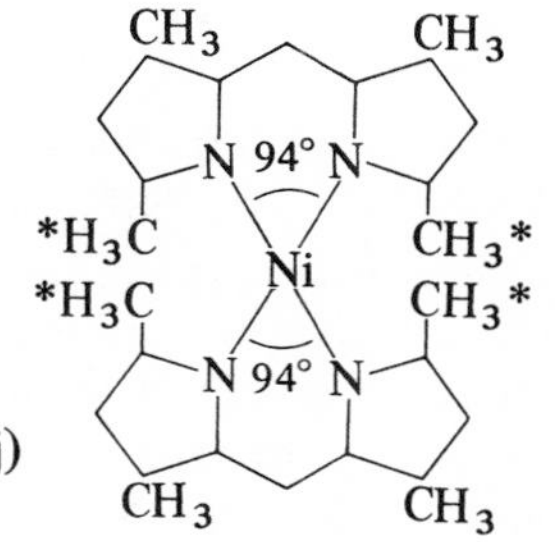

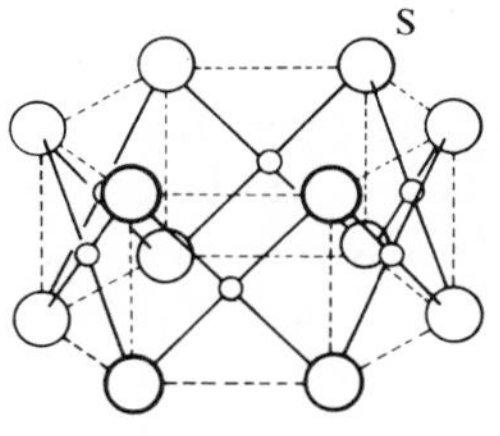

FIG. 27.10. The molecule $[Ni(SC_2H_5)_2]_6$ (diagrammatic).

CH_3* groups, and the angle between the planes of the two ligands is 76°; it is not obvious why this angle is not 90°. The 3-substituted bis(N-isopropylsalicylaldi-minato) Ni complexes (k),[15] are of interest in this connection, for they fall into

(k)

two groups; tetrahedral ($\mu \approx 3.3$ BM) or planar diamagnetic according to the nature of the substituent in the 3 position. The detailed stereochemistry of these compounds is not readily understandable, for the 3-CH_3 compound is planar while the 3-H and 3-C_2H_5 compounds are tetrahedral.

(1)   ACSc 1964 **18** 2385
(2)   ACSc 1969 **23** 14, 61; AC 1970 **B26** 361
(3)   AC 1968 **B24** 108
(4)   JCS A 1967 1750
(5)   JCS 1935 1475
(6)   IC 1968 **7** 2140, 2625; AC 1969 **B25** 909, 1294, 1939
(7)   ZaC 1968 **363** 159
(8)   AC 1967 **22** 468; AC 1972 **B28** 2318

(9)   JCS 1937 219
(10)   IC 1970 **9** 1878
(11)   JACS 1965 **87** 5251
(12)   IC 1964 **3** 1303
(13)   IC 1965 **4** 1726
(14)   IC 1970 **9** 783
(15)   AC 1967 **22** 780

Ni(II) *forming* 4 *tetrahedral bonds.* The simplest examples are the complexes NiX_4^{2-}, NiX_3L^-, NiX_2L_2, and NiL_2', where L is a monodentate and L' is a bidentate ligand. The failure to recognize, until fairly recently, the existence of halide ions NiX_4^{2-} was due, not to their intrinsic instability, but to their instability in aqueous solution, where X is displaced by the more strongly coordinating H_2O molecule. The $NiCl_4^{2-}$ ion can be formed in melts and studied in solid solutions in, for example, Cs_2ZnCl_4, but Cs_2NiCl_4 dissociates on cooling, On the other hand, Cs_3NiCl_5 (analogous to $Cs_3(CoCl_4)Cl$) can be prepared from the molten salts and quenched to room temperature, but on slow cooling it breaks down to a mixture of CsCl and $CsNiCl_3$ (in which there is octahedral coordination of Ni). By working in alcoholic solution salts of large organic cations can be prepared, for example, $(NEt_4)_2NiCl_4$ (blue, $\mu_{eff.}$, 3.9 BM) and $(As\phi_3Me)_2NiI_4$ (red, $\mu_{eff.}$, 3.5 BM). There is no Jahn–Teller distortion in the $NiCl_4^{2-}$ ion, which has a regular tetrahedral shape in $(As\phi_3Me)_2$-$NiCl_4^{(1)}$ but is slightly flattened (two angles of 107° and four of 111°) in the (NEt_4) salt,[2] presumably the result of crystal packing forces. Ions NiX_4^{2-} have been prepared in which X is Cl, Br, I, NCS, or NCO.[3] Ions NiX_3L^- which have been studied include those in the salts [NiBr$_3$(quinoline)](Asϕ_4)[4] and [NiI$_3$(Pϕ_3)]-[N(n-C$_4$H$_9$)$_4$].[5] In molecules NiX$_2$L$_2$ there may be considerable distortion from regular tetrahedral bond angles as, for example, Br—Ni—Br, 126° in NiBr$_2$(Pϕ_3)$_2$,[6] and Cl—Ni—Cl, 123° and P—Ni—P, 117°, in NiCl$_2$(Pϕ_3)$_2$.[7] We have already commented on the planar-tetrahedral isomerism of molecules NiX$_2$L$_2$ and of molecules NiL$_2'$ containing certain bidentate ligands.

Some more exotic examples of molecules in which Ni forms tetrahedral bonds are shown at (a)–(d).

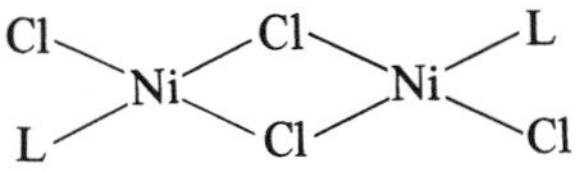

L = cyclobutadiene[8]

(a)

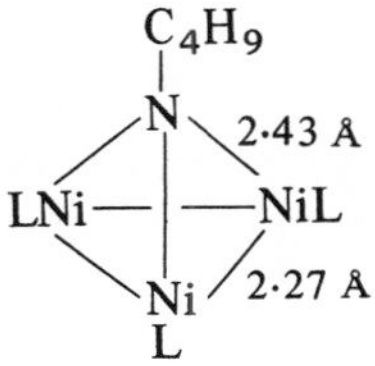

L = cyclopentadiene[9]

(b)

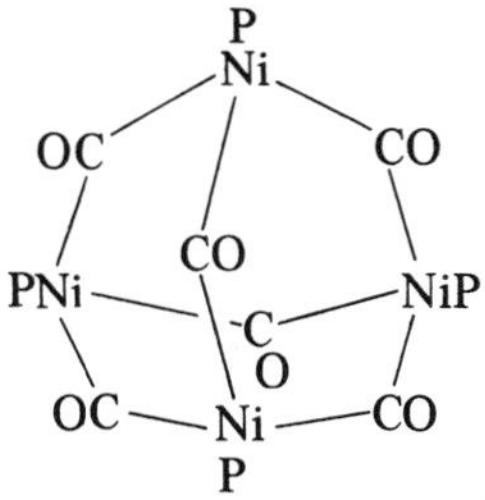

P = P(C$_2$H$_4$CN)$_3$[10]

(c)

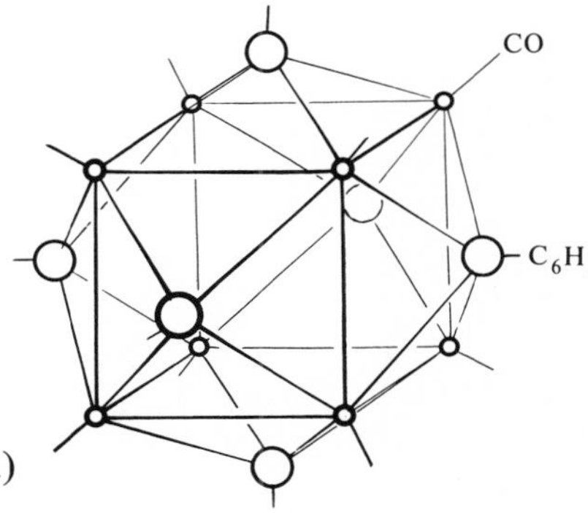

(d)

In Ni$_4$(CO)$_6$[P(C$_2$H$_4$CN)$_3$]$_4$ Ni acquires a closed shell configuration if the three metal–metal bonds (2.57 Å, compare 2.49 Å in the metal) are included with those to the four nearest neighbours (Ni–3C, 1.89 Å; Ni–P, 2.16 Å), and for this reason Ni should perhaps be regarded as 7- rather than 4-coordinated in this compound.

The bonding situation is similar in the cubic molecule Ni$_8$(CO)$_8$(Pϕ)$_6$[11] if the Ni–Ni bonds (2.65 Å) are counted as electron-pair bonds; not all ligands are shown in (d).

(1)   IC 1966 **5** 1498
(2)   AC 1967 **23** 1064
(3)   JINC 1964 **26** 2035
(4)   IC 1968 **7** 2303
(5)   IC 1968 **7** 2629
(6)   JCS A 1968 1473
(7)   JCS 1963 3625
(8)   HCA 1962 **45** 647
(9)   IC 1968 **7** 261
(10) JACS 1967 **89** 5366
(11) JACS 1976 **98** 5046

Ni(II) *forming* 5 *bonds.* The two most symmetrical configurations, trigonal bipyramidal and square (tetragonal) pyramidal, are closely related geometrically, so that descriptions of configurations intermediate between the two extremes are sometimes rather arbitrary. However, complexes with geometries very near to both configurations are found for both, diamagnetic and paramagnetic compounds of Ni(II), and in one case (p. 1222) two distinct configurations of the $Ni(CN)_5^{3-}$ ion are found in the same crystal. The choice of configuration is influenced by the geometry of the ligand if this is polydentate. Apart from the $Ni(CN)_5^{3-}$ ion nothing is known of the structures of ions NiX_5^{3-}. In salts such as $Rb_3Ni(NO_2)_5$ there may be bridging NO_2 groups in an octahedral chain, as in $[Ni(en)_2(NO_2)]^+$. Complexes with mono-dentate ligands include the diamagnetic molecule (a) with a nearly ideal trigonal bipyramidal configuration (though with $P\phi(OEt)_2$ ligands an intermediate configuration is adopted[1]) and the paramagnetic complex (b) (μ = 3.7 BM).[2]

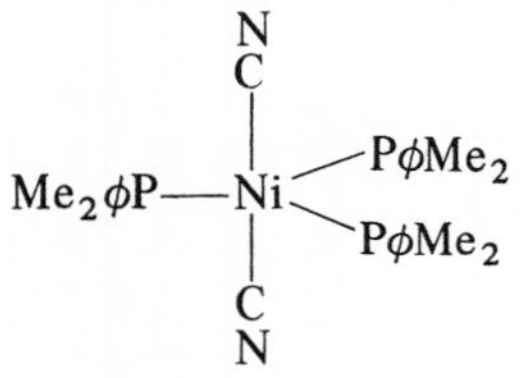

(a) 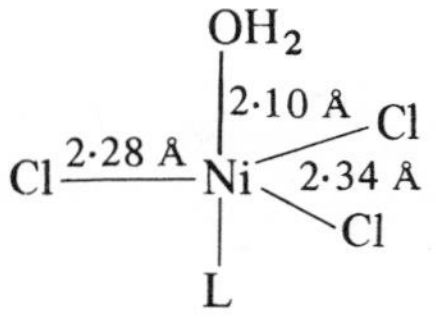   (b) 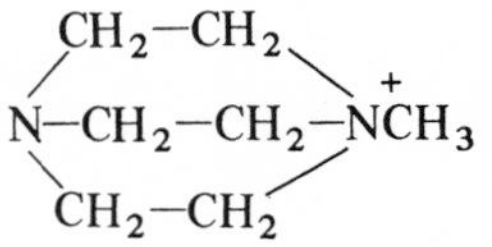   (L)

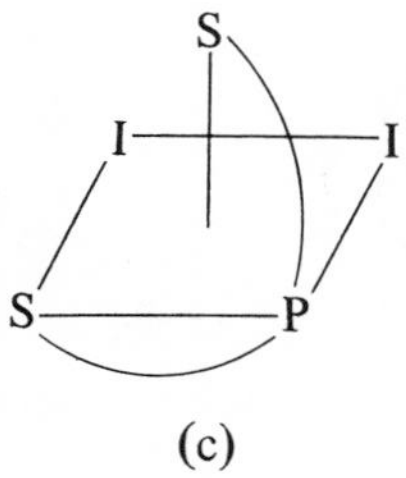

(c) 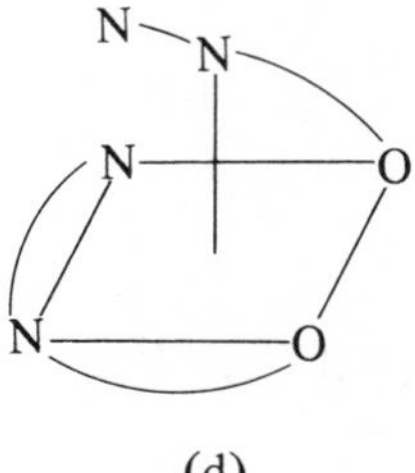   (d)

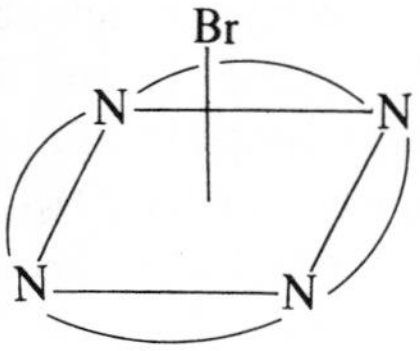

(e) 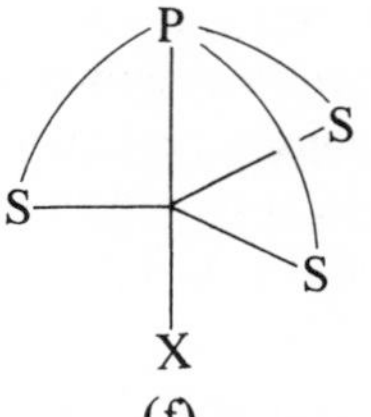   (f)

Complexes with polydentate ligands include the types (c)–(f). We shall not discuss all these cases individually since the configuration often depends on the detailed structure of the (organic) ligand. For example, in derivatives of Schiff bases

R.C_6H_3(OH)CH=N.$CH_2CH_2NEt_2$ (type (d)) the compound NiL_2 may be diamagnetic planar, paramagnetic 5-coordinated, or paramagnetic octahedral, depending on the nature of the substituent R.[3] The 'tripod-like' tetradentate ligands such as TSP and TAP tend to favour the trigonal bipyramidal configuration, as in the

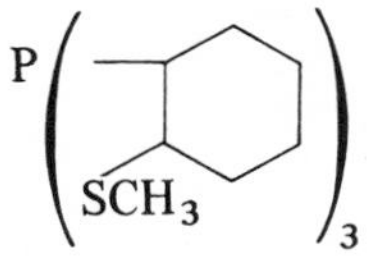

$$P\left(\underset{\overset{|}{SCH_3}}{-\bigcirc}\right)_3 \qquad \text{and} \qquad P[-(CH_2)_3As(CH_3)_2]_3$$

(TSP)   (TAP)

diamagnetic compounds [Ni(TSP)Cl]ClO_4[4] and [Ni(TAP)CN]ClO_4.[5] References to a number of square pyramidal complexes are given in reference (6).

The structure of $NiBr_2$(TAS),[7] in which TAS is the tridentate ligand

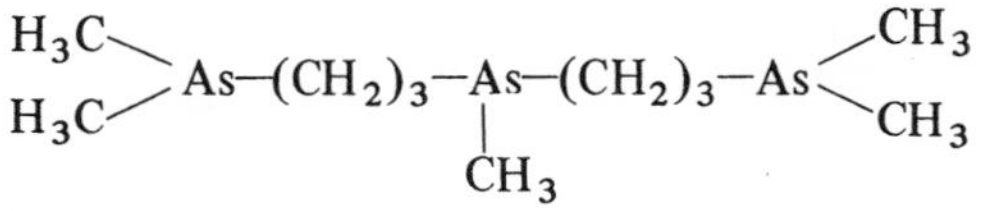

$$\underset{H_3C}{\overset{H_3C}{>}}As-(CH_2)_3-\underset{\overset{|}{CH_3}}{As}-(CH_2)_3-As\underset{CH_3}{\overset{CH_3}{<}} \qquad \text{(TAS)}$$

is described on p. 1239, where it is compared with a 5-coordinated Pd complex with somewhat similar geometry.

(1)   IC 1969 8 1084, 1090   (5)   JACS 1967 89 3424
(2)   IC 1969 8 2734   (6)   IC 1969 8 1915
(3)   JACS 1965 87 2059   (7)   PCS 1960 415
(4)   IC 1969 8 1072

Ni(II) *forming* 6 *octahedral bonds.* The simplest examples of octahedrally coordinated Ni^{2+} are the crystalline monoxide and the dihalides. Octahedral complexes mentioned elsewhere include the cation in [Ni(H_2O)$_6$]$SnCl_6$, the trimeric [Ni(acac)$_2$]$_3$[1a] and Ni_6(CF_3COCHCOCH$_3$)$_{10}$(OH)$_2$(H_2O)$_2$[1b] both of which are illustrated in Chapter 5 (Fig. 5.9 (p. 202) and Fig. 5.11 (p. 203)).

Many octahedral paramagnetic complexes of Ni(II) have been studied, and examples of the general types (a)–(d) include:

(a)   Ni(pyrazole)$_4Cl_2$ AC 1969 **B25** 595
(b)   Ni(acac)$_2$(pyr)$_2$ IC 1968 7 2316
      Ni(en)$_2$(NCS)$_2$ AC 1963 **16** 753
(c)   [Ni(en)$_3$](NO$_3$)$_2$ AC 1960 **13** 639
(d)   Ni(tren)(NCS)$_2$ ACSc 1959 **13** 2009

In contrast to Ni(tu)$_4Cl_2$, which forms finite molecules in which there is an (unexplained) asymmetry (the lengths of the two *trans* Ni–Cl bonds being 2.40 and 2.52 Å),[2] the thiocyanate Ni(tu)$_2$(NCS)$_2$ consists of infinite chains of edge-sharing

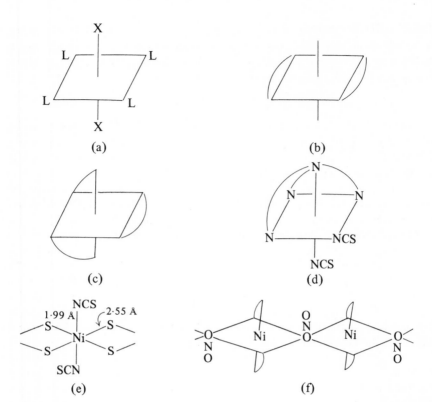

(a)

(b)

(c)

(d)

(e)

(f)

octahedra, (e);[3] in contrast to $Ni(en)_2(NCS)_2$, type (b), the nitrito compound $Ni(en)_2ONO$ is built of chains of vertex-sharing octahedra, (f).[4]

The situation with regard to diamagnetic octahedral complexes is still unsatisfactory. Halogen compounds $Ni(diarsine)_2X_2$ behave in solution in CH_3NO_2 as uni-univalent electrolytes, but in crystals of the brown form of $Ni(diarsine)_2I_2$ there are octahedral molecules of type (b) in which Ni–As is 2.29 Å but Ni–I, 3.21 Å.[5] This Ni–I bond length may be compared with 2.55 Å in $[NiI_3(P\phi_3)]^-$ (tetrahedral) and 2.54 Å in $NiI_2[(C_6H_5)P(C_6H_4.SCH_3)_2]$ (square pyramidal, ref. (6) of previous section). The nature of the Ni–I bonds is uncertain, for the close approach of I to Ni is prevented by the methyl groups which project above and below the equatorial plane. Structural studies have apparently not been made of compounds such as $[Ni(diarsine)_3](ClO_4)_2$ which are often quoted as examples of octahedral spin-paired Ni(II). (The electronic structure of the paramagnetic $[Ni(diarsine)_2Cl_2]Cl$, containing one unpaired electron, is uncertain.[6])

(1a) IC 1965 4 456
(1b) IC 1969 8 1304
(2)   JCS 1963 1309
(3)   AC 1966 20 349

(4)   JCS 1962 3845
(5)   AC 1964 17 592
(6)   JACS 1968 90 1067

Other oxidation states of Ni

Ni(o). Only tetrahedral complexes are known; they include $Ni(CO)_4$, $Ni(PF_3)_4$, $Ni(CNR)_4$, and the $Ni(CN)_4^{4-}$ ion in the yellow potassium salt. Structural information regarding these complexes is summarized in Chapter 22.

Ni(I). The structure of the ion $[Ni_2(CN)_6]^{4-}$ is described on p. 942.

Ni(III). Compounds of Ni(III) may include $NiBr_3(PEt_3)_2$, formed by oxidation of *trans*-$NiBr_2(PEt_3)_2$, and $Ni(diarsine)_2Cl_3$, which results from the oxidation of the dichloro compound by O_2 in the presence of excess Cl^- ions. Both these compounds have magnetic moments corresponding to 1 unpaired electron, but their structures are not yet known. For the former a trigonal bipyramidal configuration would be consistent with its dipole moment,[1] and this structure is likely in view of the structure of the compound $Ni^{III}Br_3.P_2'.0.5(Ni^{II}Br_2P_2').C_6H_6$,[2] in which $P' = P(C_6H_5)(CH_3)_2$. This compound consists of *trans* planar molecules $NiBr_2P_2'$ with twice as many trigonal bipyramidal molecules $NiBr_3P_2'$ in which P occupies axial positions (Ni–P, 2.27 Å; Ni–Br, 2.35 Å). (There is a slight distortion of the latter molecules, with one equatorial bond about 0.03 Å longer than the other two and an angle of 133° opposite this longer bond. The Ni–P bonds are not significantly different in length from those in the square Ni(II) complex; the Ni–Br bonds are approximately 0.05 Å longer.) The diarsine complex is presumably $[Ni(diarsine)_2Cl_2]Cl$ containing Ni(III) forming octahedral bonds, though there is some doubt as to whether such octahedral complexes should be regarded as compounds of Ni(III) since e.s.r. measurements suggest that they should perhaps be regarded as Ni(II) compounds in which the unpaired electron spends a large part of its time on the As atom.[3]

Structural information about simple Ni(III) compounds is limited to those containing F or O. K_3NiF_6 has been assigned the K_3FeF_6 structure and has a moment of 2.5 BM, intermediate between the values for low- and high-spin.[4] It contains octahedrally coordinated Ni(III). Oxides $LnNiO_3$ formed with the 4f elements have the perovskite structure,[5] while $NiCrO_3$ has a statistical corundum structure; it is concluded from the magnetic properties that this compound contains high-spin Ni(III).[6]

Ni(IV). The only definite structural information relates to the red salts K_2NiF_6[7] and Cs_2NiF_6.[8] They are diamagnetic and have the K_2SiF_6 structure, with octahedral coordination of Ni(IV). The oxide $BaNiO_3$ has been quoted for many years as an example of a Ni(IV) compound,[9] but there is still doubt about its composition ($BaNiO_{2.5}$?) and the interpretation of its magnetic properties.

The diarsine cation mentioned above can be further oxidized to $[Ni(diarsine)_2Cl_2]^{2+}$ which is isolated as the deep-blue perchlorate, possibly containing Ni(IV).

(1) ACSc 1963 17 1126
(2) IC 1970 9 453
(3) JACS 1968 90 1067
(4) ZaC 1961 308 179
(5) JSSC 1971 3 582
(6) JAP 1969 40 434
(7) ZaC 1949 258 221
(8) ZaC 1956 286 136
(9) AC 1951 4 148

The structural chemistry of Pd and Pt

We shall discuss first the compounds of Pd(II) and Pt(II). The zero-valent state is confined to compounds containing CO, PF_3, or PR_3 ligands, such as the tetrahedral $Pt(PF_3)_4$ molecule[1] and molecules such as $Pt(CO)(P\phi_3)_3$ but *not* $M(CO)_4$. In the metal cluster carbonyls (p. 964) the formal oxidation state is +1.

(1)   JMSt 1976 **31** 73

Planar complexes of Pd(II) *and* Pt(II)

In 1893 Werner suggested that the isomerism of certain compounds of divalent Pd or Pt could be explained if the four atoms attached to the metal atom were coplanar with that atom. A molecule Pta_2b_2 could have *cis* and *trans* isomers, whereas if the bonds were arranged tetrahedrally there would be only one isomer. Similarly, three isomers of a planar molecule *Mabcd* are possible as compared with only one configuration of a tetrahedral complex. In pre-structural days the only approach to this problem was to study geometrical and optical isomerism, and much ingenuity was exercised in making appropriate molecules. It is not now necessary to relate this interesting chapter of the history of chemistry, since the coplanar arrangement of four bonds from Pd(II) or Pt(II) has been demonstrated in many molecules and ions, following the determination, in 1922, of the planar structure of the $PdCl_4^{2-}$ and $PtCl_4^{2-}$ ions in their K salts by X-ray diffraction. The structures of binary compounds, halides, oxides, sulphides, etc., are described in other chapters; here we shall give examples of finite molecules and ions which present points of special interest. Because of the virtual identity of bond lengths Pd—X and Pt—X many compounds of Pd(II) and Pt(II) are isostructural, but since there are appreciable differences between the chemical properties of these two metals we shall deal with them separately. It should perhaps be emphasized that our examples are necessarily limited to those of which structural studies have been made; not all the types of compound formed by Pd are formed by Pt, and vice versa.

Pd(II) *compounds*

The structures of square planar mononuclear complexes have been established by diffraction studies of compounds such as K_2PdCl_4,[1] $[Pd(NH_3)_4]Cl_2.H_2O$,[2] numerous complex cyanides (many of which are isostructural with the Ni and Pt analogues), for example, $Ca[Pd(CN)_4].5H_2O$, and $Na_2[Pd(CN)_4].3H_2O$, molecules containing chelate groups (e.g. $Pd(en)Cl_2$,[3] and complexes such as the dithiooxalate ion, glyoximes, phthalocyanin, and related molecules. It was noted on p. 452 that the structure of $[Pd(NH_3)_4]Cl_2.H_2O$ is closely related to that of $K_2(PdCl_4)$, Fig. 10.1, the positons of the anions and cations being interchanged and H_2O molecules occupying interstices in the structure. It appears that the holes in the structure are rather too large for the H_2O molecules, which are apparently disordered over positions close to the centre of each hole. In $[Pd(en)_2]$- $[Pd(en)(S_2O_3)_2]$[4] both ions are square planar, and the S_2O_3 ligand is bonded through S, (a). The dissolution of Pd in HNO_3 followed by treatment with NH_3 does not give the expected $[Pd(NH_3)_4](NO_3)_2$ but the bright-yellow compound $[Pd(NH_3)_3NO_2]_2[Pd(NH_3)_4](NO_3)_4$ which contains two kinds of planar ion in the proportion of 2:1.[5] For $K_2[Pd(SO_3)_2].H_2O$ and $Na_6[Pd(SO_3)_4].2H_2O$ see p. 719.

In the 2,2'-dipyridyl imine complex (b),[6] in which there is interference between the H* atoms, the bond arrangement around Pd remains square planar and the ligands distort, in contrast to the Ni methanato complex mentioned on p. 1225.

There are numerous binuclear (bridged) compounds of Pd(II). Early X-ray studies established the planar *trans* configuration of (c) and (d):

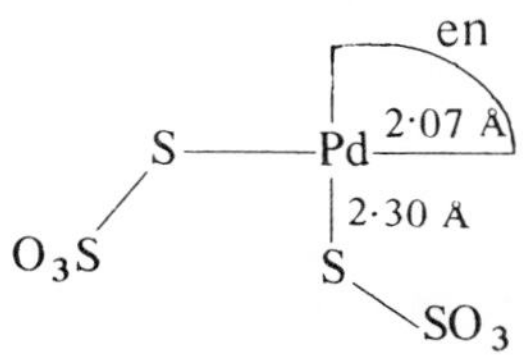

(a)

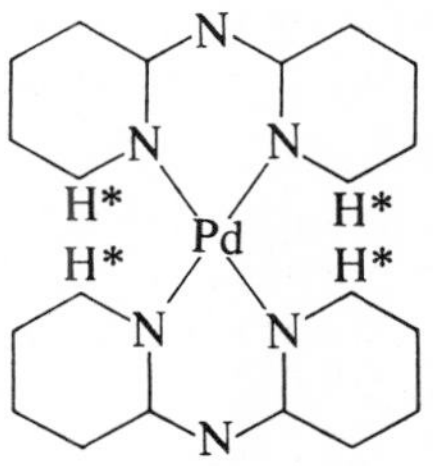

(b)

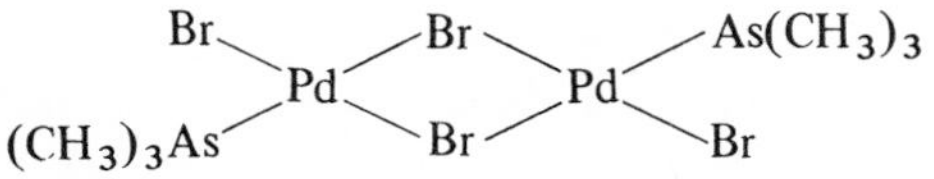

(c)

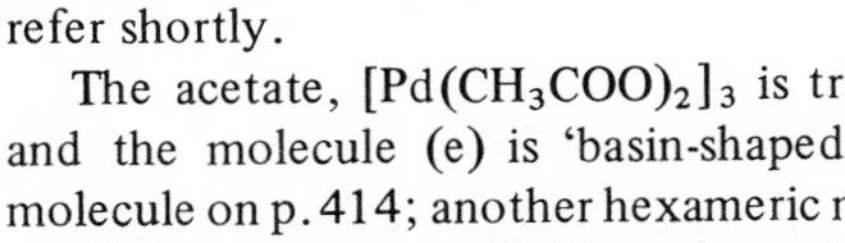

(d)

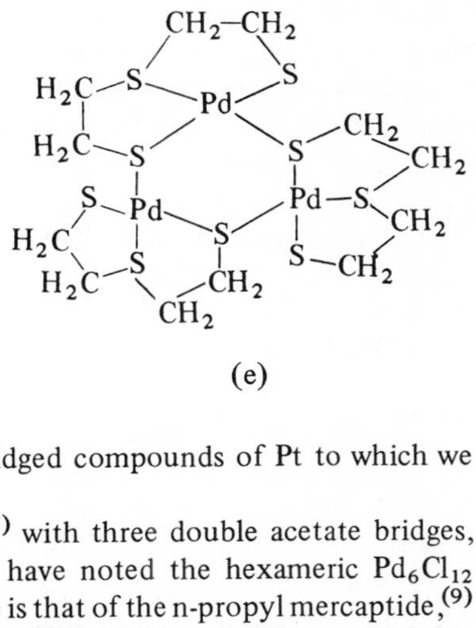

(e)

but there have been more recent studies of bridged compounds of Pt to which we refer shortly.

The acetate, $[Pd(CH_3COO)_2]_3$ is trimeric,[7] with three double acetate bridges, and the molecule (e) is 'basin-shaped'.[8] We have noted the hexameric Pd_6Cl_{12} molecule on p. 414; another hexameric molecule is that of the n-propyl mercaptide,[9] which has a structure similar to that of the Ni compound (Fig. 27.10).

The dithioacetate is of unusual interest for there are two crystalline forms, one consisting of both planar monomers and dimers, and the other of dimers only.[10] The dimer is a tetrakis complex of the type (b) of Table 7.9, p. 310, in which Pd forms a fifth bond to the other metal atom (Pd–Pd, 2.75 Å).

Examples of the rare compounds in which Pd(II) does *not* form square coplanar bonds include $[PdAlCl_4(C_6H_6)]_2$ and $[PdAl_2Cl_7(C_6H_6)]_2$ which are formed when

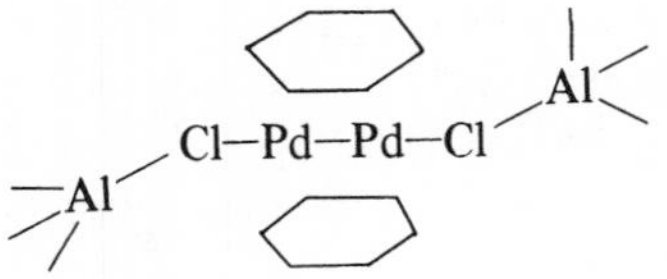

(f) 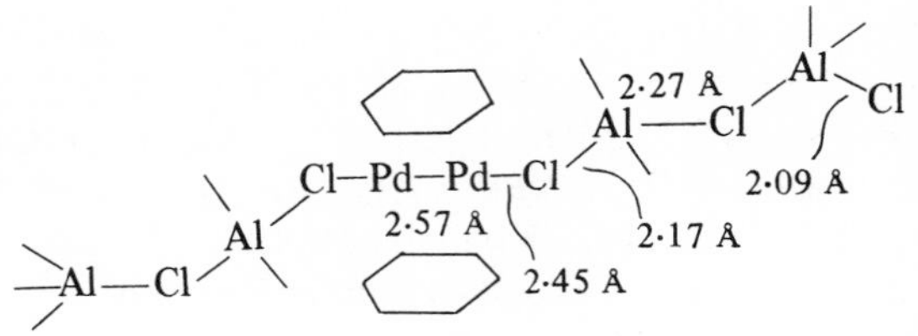   (g)

AlCl$_3$, Al metal, and PdCl$_2$ are reacted together in boiling benzene.[11] The molecules have the structures sketched at (f) and (g). These molecules are notable for the short Pd—Pd bonds, which are shorter than in the metal (2.75 Å), and long Pd—Cl bonds.

(1) AC 1972 **B28** 393
(2) AC 1976 **B32** 634
(3) AC 1975 **B31** 1672
(4) AC 1970 **B26** 1698
(5) IC 1971 **10** 651
(6) AC 1965 **18** 845
(7)   CC 1970 658
(8)   AC 1969 **B25** 1659
(9)   AC 1971 **B27** 2292
(10) IC 1979 **18** 2258
(11) JACS 1970 **92** 289

Compounds of Pt(II)

There is no evidence for the formation of more than four bonds (which are coplanar) except in the cases noted on pp. 1239 and 1240. As in the case of Pt(IV) (p. 1243) the formulae of certain compounds suggest other coordination numbers, but structural studies have confirmed 4-coordination. In [Pt(acac)$_2$Cl]K,[1] (a), 5-coordination is avoided by coordination to C, and an X-ray study of the compound originally formulated as [Pt(NH$_3$)$_4$(CH$_3$CN)$_2$]Cl$_2$.H$_2$O shows that this is not an

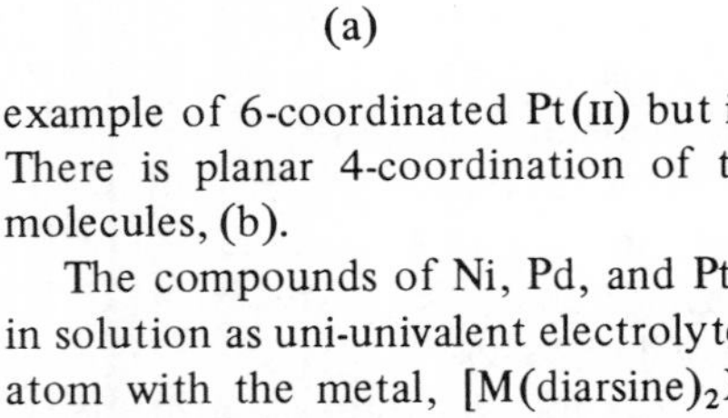

(a) 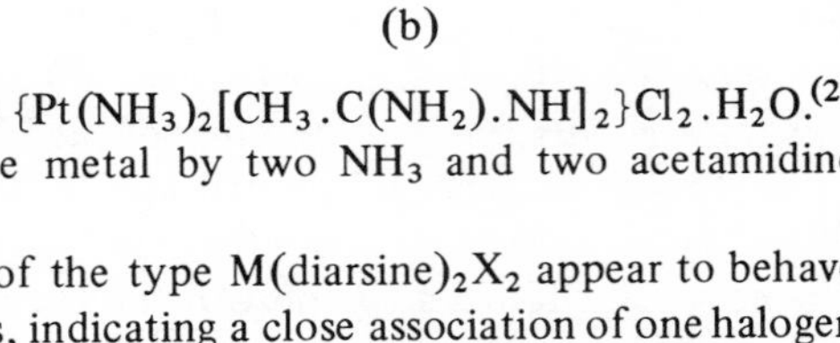   (b)

example of 6-coordinated Pt(II) but is {Pt(NH$_3$)$_2$[CH$_3$.C(NH$_2$).NH]$_2$}Cl$_2$.H$_2$O.[2] There is planar 4-coordination of the metal by two NH$_3$ and two acetamidine molecules, (b).

The compounds of Ni, Pd, and Pt of the type M(diarsine)$_2$X$_2$ appear to behave in solution as uni-univalent electrolytes, indicating a close association of one halogen atom with the metal, [M(diarsine)$_2$X]$^+$. In the crystalline state[3] they form very distorted octahedral units with Ni—As, 2.3 Å, and Pd(Pt)—As, 2.4 Å, but M—X very

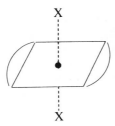

much longer than a normal covalent bond. In Pt(diarsine)$_2$Cl$_2$ the length of the Pt—Cl bond is 4.16 Å, suggesting Cl$^-$ ions resting on the four CH$_3$ groups projecting out of the equatorial plane of the two diarsine molecules. The *shorter* M—I distances (3.2, 3.4, and 3.5 Å for the Ni, Pd, and Pt compounds) may be due to considerable polarization of the I$^-$ ions; they are intermediate between the covalent radius sums and the sums of the van der Waals (or ionic) radii. The Ni—I distance has been compared with the much smaller Ni—I distances in paramagnetic tetrahedral and square pyramidal molecules on p. 1230.

Like Pd(II), Pt(II) forms numerous square planar complexes with monodentate ligands, and examples of structural studies include those of PtHBr(PEt$_3$)$_2$ (p. 354), *cis*-PtCl$_2$(PMe$_3$)$_2$,[4] *cis*- and *trans*-PtCl$_2$(NH$_3$)$_2$,[5] [PtCl(CO)(PEt$_3$)$_2$]BF$_4$,[6] and (PtCl$_3$NH$_3$)K.H$_2$O.[7] Complexes containing bidentate ligands include Pt(S$_2$N$_2$H)$_2$ (ref. 8), the dimethyl glyoxime,[9] and diglycine.[10] For K$_3$[Pt{(SO$_3$)$_2$H}Cl$_2$] see p. 360.

In certain complexes there is observed a lengthening of the Pt—X bonds which are *trans* to a π-bonded ligand ('*trans*-effect'). This is seen in PtCl$_2$(PMe$_3$)$_2$, (c), and also in the bridged molecule (d),[11] where it leads to an unsymmetrical bridge (and three different Pt—Cl bond lengths). A similar effect is observed in Pt$_2$Cl$_4$(PPr$_3$)$_2$.[12]

$$
\begin{array}{c}
\text{Cl} \diagdown \overset{2\cdot38\ \text{Å}}{} \diagup \text{PMe}_3 \\
\qquad \text{Pt} \diagdown \overset{}{=2\cdot25\ \text{Å}} \qquad \text{contrast} \\
\text{Cl} \diagup \diagdown \text{PMe}_3
\end{array}
$$

(c)

$$
\begin{array}{c}
\text{Cl} \diagdown \overset{2\cdot29\ \text{Å}}{} \diagup \text{PEt}_3 \\
\qquad \text{Pt} \overset{2\cdot31\ \text{Å}}{} \\
\text{Et}_3\text{P} \diagup \diagdown \text{Cl}
\end{array}
\qquad
\begin{array}{c}
\overset{2\cdot39\ \text{Å}}{} \qquad \overset{2\cdot31\ \text{Å}}{} \\
\text{Cl} \diagdown \diagup \text{Cl} \diagdown \diagup \text{AsMe}_3 \\
\overset{2\cdot27\ \text{Å}}{} \text{Pt} \text{Pt} \\
\text{Me}_3\text{As} \diagup \diagdown \text{Cl} \diagdown \text{Cl} \\
\overset{}{2\cdot31\ \text{Å}}
\end{array}
$$

(d)

More complex molecules include the dimer Pt$_2$(S$_2$C.CH$_3$)$_4$ of the dithioacetate, which has the same structure as the Pd compound (see Table 7.9, p. 310) and in which Pt forms 4 coplanar bonds to S (Pt—S, 2.32 Å) and a fifth to the other Pt atom of length 2.77 Å, the tetrameric [Pt(CH₃COO)₂]₄ molecules in both crystalline forms of the diacetate,[13] and the molecule $Pt_4(CH_3COO)_6(NO)_2$.[14] In the former the Pt atoms form a nearly square group. Of the eight bridging acetato groups four lie in the plane of the Pt atoms and the other four are perpendicular to, and alternately above and below, this plane. The metal atom is therefore bonded octa-hedrally to 4 O (2 at 2.00 and 2 at 2.16 Å) and to 2 Pt (at 2.495 Å), as shown diagrammitically at (e).

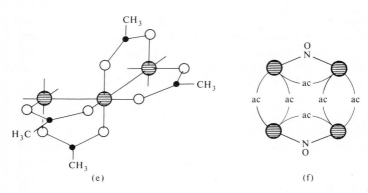

(e)   (f)

In $Pt_4(ac)_6(NO)_2$, on the other hand, there are no metal–metal bonds. The Pt atoms form a rectangle with sides 2.94 and 3.31 Å, and there are 6 bridging CH_3COO and 2 bridging NO groups, (f). The dichloride forms a hexameric molecule Pt_6Cl_{12} similar to that in one form of crystalline $PdCl_2$.

(1)   JCS A 1969 485
(2)   JINC 1962 **24** 801
(3)   JINC 1962 **24** 791, 797; AC 1964 **17** 1517
(4)   IC 1967 **6** 725
(5)   JCS A 1966 1609
(6)   JACS 1967 **89** 3360
(7)   IC 1970 **9** 778

(8)   JINC 1958 **7** 421
(9)   AC 1959 **12** 1027
(10) AC 1969 **B25** 1203
(11) JCS A 1970 168
(12) AC 1969 **B25** 1760
(13) AC 1978 **B34** 1857, 3576
(14) JCS D 1973 1194

Bridged compounds of Pd *and* Pt

Solutions containing PdX_4^{2-} or PtX_4^{2-} ions react with tetramethyl ammonium salts to give compounds with empirical formulae of the type $(NR_4)PdX_3$ or $(NR_4)PtX_3$ in which X is Br or Cl. An X-ray study of $[N(C_2H_5)_4]PtBr_3$[1] showed that it contains a bridged ion and accordingly salts of this kind should be formulated $(NR_4)_2(Pt_2X_6)$. By the reaction of $PtCl_2$ with melted L_2PtCl_2 or of *cis* L_2PtCl_2 with K_2PtCl_4 in solution bridged compounds $Pt_2Cl_4L_2$ are obtained, L being one of the neutral ligands CO, C_2H_4, NH_3, $P(OR)_3$, PR_3, AsR_3, SR_2, etc. We have already noted that Pd forms similar compounds, and that the molecule $Pd_2Br_4[As(CH_3)_3]_2$ is planar, apart from the methyl groups, with the *trans* configuration I:

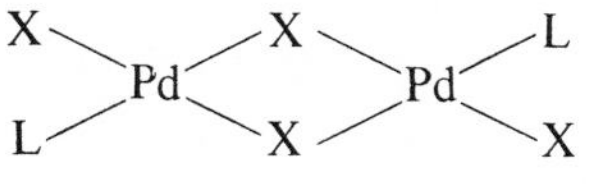

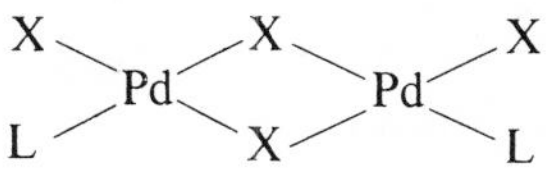

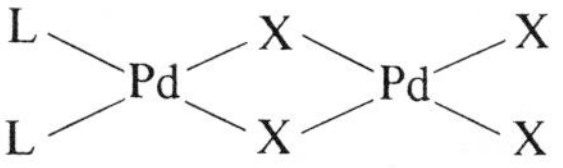

Although three isomers are theoretically possible, all the compounds in which L is one of the ligands listed above and X is a halogen are known in one form only, the *trans* isomer I, as shown by dipole moments.

The bridge between the metal atoms may also be formed by

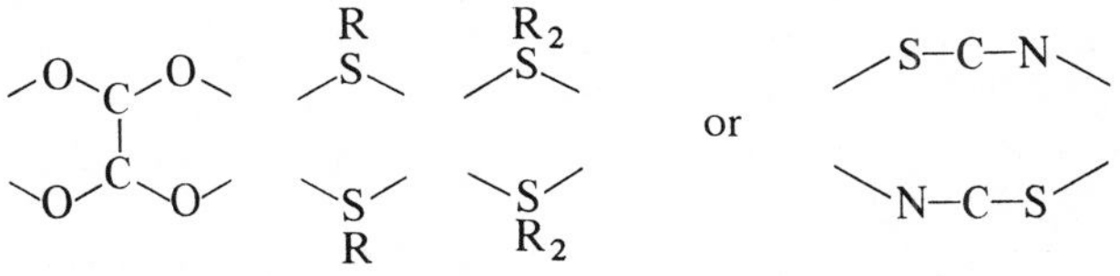

Two isomers of the ethyl-thiol compound $(Pr_3P)_2Pt_2(SR)_2Cl_2$ have been isolated, with dipole moments 10.3 D (stable form) and zero, from which it follows that the stable form is the *cis* isomer IV and the labile isomer V:[2]

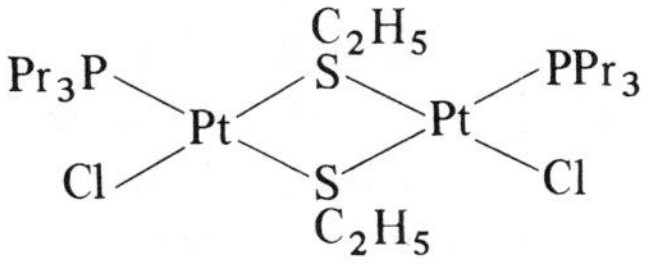

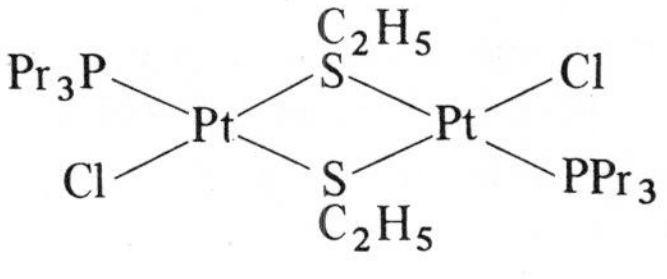

It is interesting that in $Pt_2Br_4(SEt_2)_2$ the bridge consists of two $S(C_2H_5)_2$ groups whereas in the closely related $Pd_2Br_4(SMe_2)_2$ the two Br atoms form the bridge:[3]

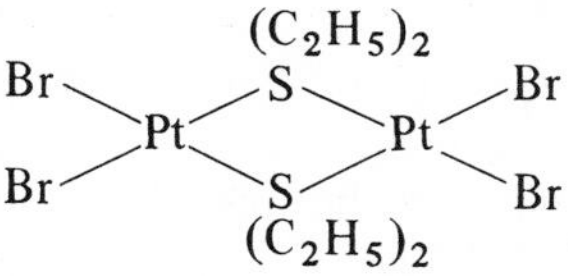

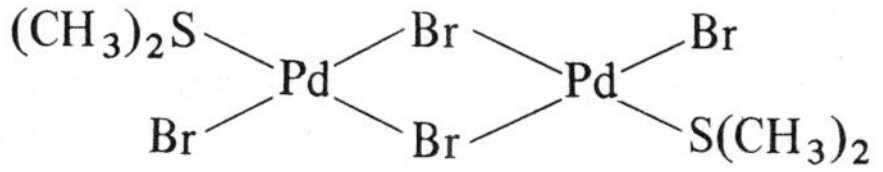

X-ray examination of the two isomers of $Pt_2Cl_2(SCN)_2(Pr_3P)_2^{(4)}$ shows that they have the structures VI and VII:

VI

VII

(1)   AC 1975 **B31** 2530
(2)   JCS 1953 2363

(3)   JCS A 1968 1852
(4)   JCS A 1970 2770

Some highly-coloured compounds of Pt

A number of crystalline salts consist of alternate planar ions of two kinds stacked vertically above one another. The anions and cations of 'Magnus's green salt', $[Pt(NH_3)_4]PtCl_4$, are respectively red and colourless in solution, and the green colour and abnormal dichroism of the crystals are associated with the metal–metal bonds, which are of the same length (3.25 Å) as Ni–Ni in Ni dimethylgloxime. Many salts with the same structure are pink, for example, $[Pd(NH_3)_4]PdCl_4$ and $[Pd(NH_3)_4]PtCl_4$; the unusual optical properties arise only when both ions contain Pt(II) and when Pt–Pt is short. Thus $[Pt(NH_3)_4]PtCl_4$ and $[Pt(CH_3NH_2)_4]PtCl_4$ are green but $[Pt(NH_2C_2H_5)_4]PtCl_4$ is pink. In the last compound all atoms of the cation are not coplanar, the planes PtN_4 and $PtCl_4$ being inclined at 29° and the Pt–Pt distance is 3.62 Å.[1]

The deep purple Millon's salt, $PtCl_4.Cu(NH_3)_4$[1a] consists of the two kinds of square planar ions stacked as in Magnus's green salt (Cu–Pt, 3.22 Å), but the isomeric green (Becton's) salt, $Pt(NH_3)_4.CuCl_4$ consists of layers of the same type as in $(NH_4)_2CuCl_4$ with planar $Pt(NH_3)_4^{2+}$ ions situated between them. The distortion of the octahedral coordination group of Cu(II) is greater than usual (4 Cl at 2.29 Å; 2 Cl at 3.26 Å); it may be compared with values given in Table 25.3, p. 1136.

Certain hydrated platinocyanides have interesting optical properties. Alkali salts such as $K_2Pt(CN)_4.3H_2O$ are yellow, but $MgPt(CN)_4.7H_2O$ is red-purple with green surface reflection from certain faces; both crystallize from colourless solutions. The mode of stacking of the planar $Pt(CN)_4^{2-}$ ions in these normal-valence compounds leads to Pt–Pt separations appreciably greater than that (2.78 Å) in the metal; for example, 3.3–3.7 Å in the hydrated Ba, K, Na, and Cs salts.[2] There has been great interest in recent years in the coloured compounds formed by partial oxidation of, for example, $K_2Pt(CN)_4$ or partial reduction of $K_2[Pt(CN)_4X_2]$, in which the formal oxidation number of Pt is non-integral. The Pt atoms are all equivalent and the complexes are stacked in columns with short Pt–Pt distances, less than 3 Å. Examples include $K_{1.75}Pt(CN)_4.1\frac{1}{2}H_2O$ (Pt–Pt, 2.96 Å) and $K_2Pt(CN)_4.Br_{0.3}.3H_2O$,[3] much studied as a 1-dimensional conductor.

In another type of salt intense colour results from interaction between Pt(II) and Pt(IV). The deeply-coloured salt with the composition $Pt(NH_3)_2Br_3$ contains planar

$Pt^{II}(NH_3)_2Br_2$ and octahedral $Pt^{IV}(NH_3)_2Br_4$ molecules.[4] These are arranged in columns in the crystal (Fig. 27.11), the planar and octahedral molecules alternating. The Pt(II) atom appears to be forming two additional weak bonds (3.03 Å) to Br atoms of $Pt^{IV}(NH_3)_2Br_4$ molecules; compare Pt−Br within the complexes, 2.50 Å. Other salts of this kind include $[Pt^{II}(en)Br_2][Pt^{IV}(en)Br_4]$[5] and Wolfram's red salt, $[Pt^{II}(C_2H_5NH_2)_4]^{2+}[Pt^{IV}(C_2H_5NH_2)_4Cl_2]^{2+}.Cl_4.4H_2O$.[6] In the latter the Cl⁻ ions (and H_2O molecules) lie between the chains, which are of the same general type as those of Fig. 27.11. The structure of the (anhydrous) Br analogue of Wolfram's salt has also been determined.[7]

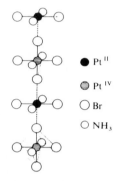

FIG. 27.11. Portion of the infinite chain in $Pt(NH_3)_2Br_2.Pt(NH_3)_2Br_4$.

(1)   AC 1971 **B27** 480

(1a) AC 1975 **B31** 2220

(2)   IC 1976 **15** 74; AC 1977 **B33** 884, 1293, 1976

(3)   IC 1976 **15** 2446, 2455

(4)   AC 1958 **11** 624

(5)   JACS 1961 **83** 2814

(6)   AC 1961 **14** 475

(7)   AC 1966 **21** 177

Pd(II) and Pt(II) forming 5 bonds

Very few structural studies have been made of complexes in which these elements form five bonds. The configuration of the Pd complexes appears to approximate to tetragonal pyramidal but with one abnormally long bond. The molecule $PdBr_2L_3$,[1] in which L is the phosphine shown in Fig. 27.12(a), has a configuration close to tetragonal pyramidal but with the axial Pd−Br bond much longer than the equatorial ones. In addition, Br^2 lies below the equatorial plane, and in this respect the structure resembles that of the Ni compound $NiBr_2(TAS)$[2] (b), where TAS is the tridentate ligand shown on p. 1229. The distortion from tetragonal pyramidal configuration is much greater in the Ni compound and is tending towards trigonal bipyramidal (when the angles 95°, 111°, and 154° would all be 120°). There is apparently no distortion of this type in the ion $[Pd(TPAS)Cl]^+$,[3] (c), containing the tetradentate ligand TPAS, but here also one of the five bonds (Pd−As^2, 2.86 Å) is much longer than normal (mean of other three, 2.37 Å).

Another example of the formation by Pd(II) of four stronger and one weaker

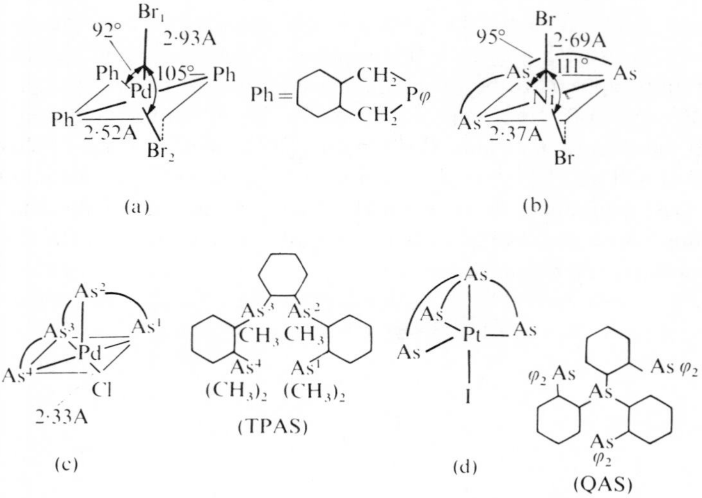

FIG. 27.12. 5-coordinated complexes of Pd, Ni, and Pt (see text).

bond is provided by crystalline *trans* PdI$_2$(PMe$_2\phi$)$_2$.[4] Here the complex is a square planar one (Pd–P, 2.34 Å; Pd–I, 2.63 Å), and the molecules are arranged so that on one side Pd has an I atom of another molecule as its next nearest neighbour (at 3.28 Å); on the other side of the square planar coordination group Pd has two H atoms of different C$_6$H$_5$ groups as next nearest neighbours. This case is, of course, not strictly comparable with (a) and (c) where Pd forms five bonds within each finite molecule or ion.

Pt(II) shows as much reluctance as Pd(II) to form five bonds, and the only examples of 5-covalent Pt(II) are both special cases. One is the ion (d) of Fig. 27.12 in [Pt(QAS)I](Bϕ_4),[5] of which no details are available; it is described as trigonal bipyramidal, with normal single covalent bonds. Here the 3-fold symmetry of the tetradentate ligand might be expected to favour the trigonal bipyramidal structure. The other compound of which a preliminary study has been made is (ϕ_3PMe)$_3$-[Pt(SnCl$_3$)$_5$],[6] in which the five ligands are attached to Pt by Pt–Sn bonds. Here also the configuration is described as trigonal bipyramidal, but the details of the structure could not be determined owing to disorder in the crystal.

(1) JCS 1964 1803
(2) PCS 1960 415
(3) JCS 1967 1650
(4) CC 1965 237
(5) PCS 1961 170
(6) JACS 1965 87 658

Pd(II) *and* Pt(II) *forming* 6 *octahedral bonds*

The reluctance of Ni(II) to form low-spin octahedral complexes, which has already been noted, is also shown by Pd(II) and Pt(II). There is octahedral coordination of Pd(II) in PdF$_2$ (rutile structure) and in Pd^{II}(Pd^{IV}F$_6$), but the formation of six covalent bonds would require the use of a d orbital of the outermost shell in addition to one $(n-1)$d, one ns, and three np, possibly as four coplanar dsp^2 and two pd hybrids. Certainly the two bonds completing octahedral coordination of low-spin complexes of these elements are much longer than the normal single bonds. Known examples have 4 As coordinated to the metal, and it has been suggested that with a planar arrangement of 4 As around M, back donation of electrons from the metal (increasing its positive charge) enables the metal atom to bond to more highly polarizable anions such as I$^-$. With only 2 As bonded to the metal there is not sufficient back donation, and we find in (a)[1] only four bonds of normal length. In (b),[2] the Pd–I bonds are very weak, though this may be partly due to the fact that I is resting on the four methyl groups. In nitrobenzene solution Pt(diarsine)$_2$I$_2$ apparently behaves as a uni-univalent electrolyte, [Pt(diarsine)$_2$I]I. Its structure is

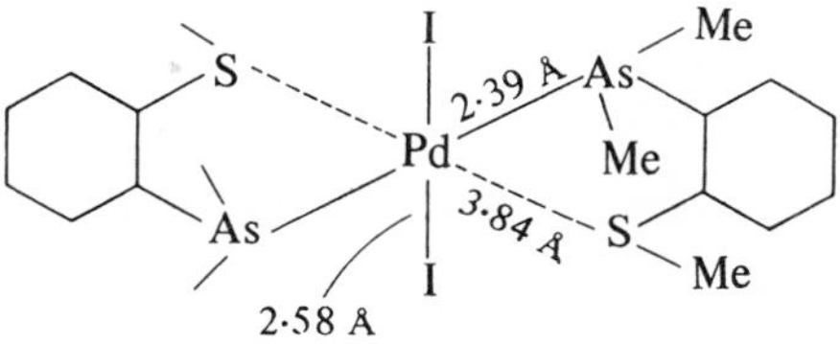

(a)

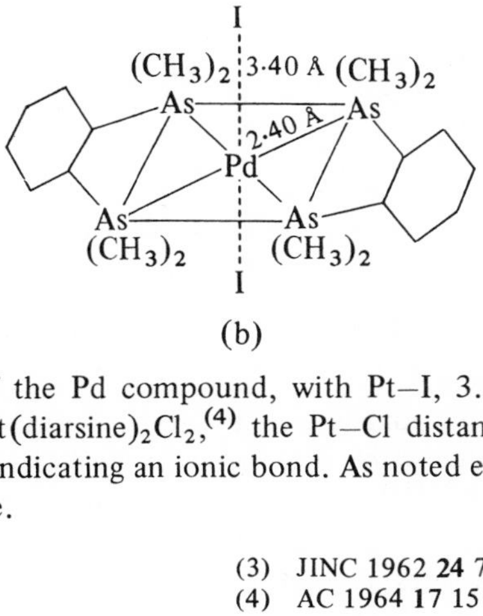

(b)

very similar to that of the Pd compound, with Pt–I, 3.50 Å.[3] However, in the dichloro compound, Pt(diarsine)$_2$Cl$_2$,[4] the Pt–Cl distance is 4.16 Å, which can only be interpreted as indicating an ionic bond. As noted earlier, the Ni compounds present a similar picture.

(1)   AC 1970 **B26** 1655   (3)   JINC 1962 **24** 791
(2)   JINC 1962 **24** 797   (4)   AC 1964 **17** 1517

Octahedral coordination of Pt(IV): *trimethyl platinum chloride and related compounds*

We have referred in earlier chapters to the octahedral coordination of Pt(IV) in the finite PtF$_6^{2-}$, PtCl$_6^{2-}$, and Pt(OH)$_6^{2-}$ ions, in chain structures (PtCl$_4$ and PtI$_4$), and in 3D structures (PtF$_4$ and PtO$_2$). The ion [Pt(CN)$_4$Br$_2$]$^{2-}$ has been shown to have an octahedral configuration in the Na and Rb salts.[1]

Trimethyl Pt(IV) chloride and certain of its derivatives are of quite unusual interest because their empirical formulae suggest coordination numbers 4, 5, and 7 for Pt^{IV}. Platinum forms a number of compounds $Pt(CH_3)_3X$, in which X is Cl, I, OH or SH, which exist as tetrameric molecules (Fig. 27.13(a)) in which Pt and X atoms occupy alternate vertices of a (distorted) cube. These compounds are notable

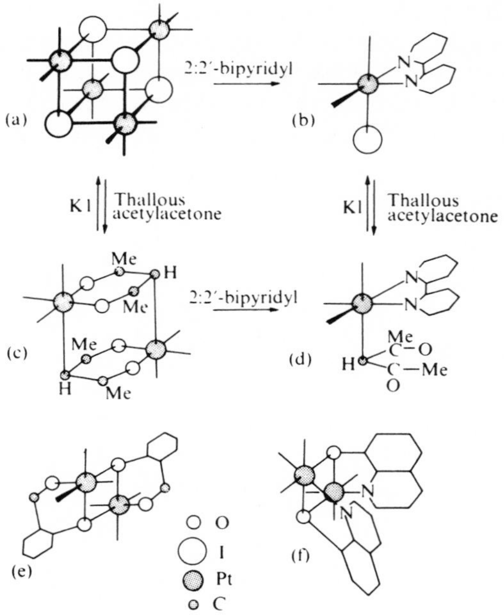

FIG. 27.13. Trimethylplatinum halides and related compounds.

for the facts that the hydroxide was formerly described as the tetramethyl compound[2a] and the iodide as $Pt_2(CH_3)_6$.[2b] In the early study of $[Pt(CH_3)_3Cl]_4$ Pt–Cl was found to be 2.48 Å;[3a] the value, 2.58 Å, determined in $[Pt(C_2H_5)_3Cl]_4$[3b] should be more reliable. The following figures for the hydroxide show that there is some distortion from the highest possible cubic symmetry:[4]

O–Pt–O	78°	Pt–C	2.04 Å
C–Pt–C	87°	Pt–O	2.22
Pt–O–Pt	101°	Pt–Pt	3.43

A neutron diffraction study of this compound has also been made.[5] Since OH or Cl together provide 5 electrons (one normal and two dative covalent bonds) Pt acquires the same electronic structure as in $PtCl_6^{2-}$.

The bipyridyl derivative (Fig. 27.13(b)) is a normal monomeric octahedral Pt^{IV} molecule, but the empirical formulae of the β-diketone compounds $Pt(CH_3)_3$-$(R.CO.CH.CO.R)$, (c),[6] and $Pt(CH_3)_3(bipyr)(R.CO.CH.CO.R)$, (d),[7] would suggest 5- and 7-coordination of the metal respectively, since β-diketones normally chelate through both O atoms. In fact (c) is a dimer in which each diketone molecule chelates to a Pt atom and is also attached to the other Pt atom through the active methylene carbon atom. This Pt–CH bond is very much longer (2.4 Å) than the Pt–CH_3 bonds (2.02 Å). (The 6-rings are actually boat-shaped.) In the molecule (d) use of the O atoms of the diketone would give 7-coordination of Pt^{IV}; instead, the metal is bonded only to the CH group, so that there is octahedral bonding by Pt as in (a), (b), and (c). The bond to the methylene carbon is similar to that in (c), namely, 2.36 Å. In $[(CH_3)_3Pt(acac)]_2(en)$[8] the ethylenediamine bridge completes the octahedral coordination group of each Pt atom, and acac coordinates to each Pt through its O atoms in the normal way:

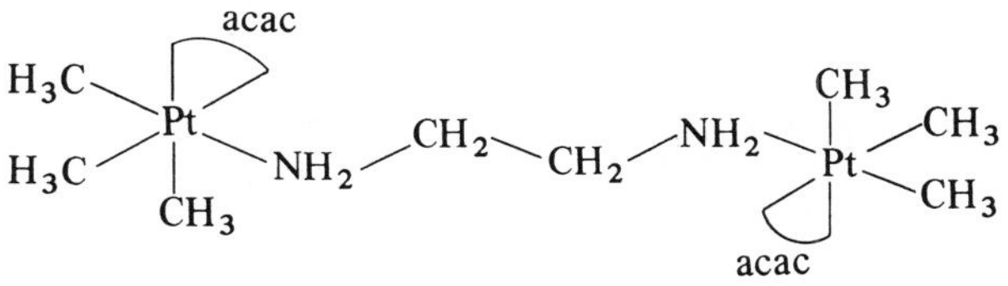

The molecules of salicylaldehyde, I, and 8-hydroxyquinoline, II, normally form chelate compounds in which two O atoms or O and N are bonded to the metal atom:

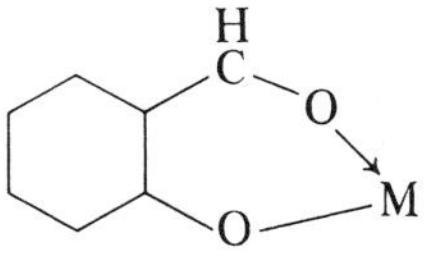

(I) 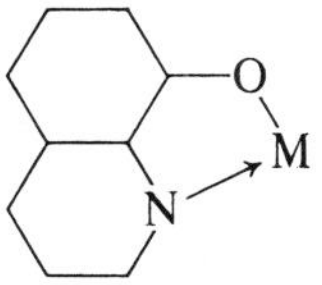   (II)

so that the compounds with the empirical formulae $(C_7H_5O_2)Pt(CH_3)_3$ and $C_9H_6NO)Pt(CH_3)_3$ would appear to contain 5-coordinated Pt^{IV}. The crystalline compounds are composed of dimers, (Fig. 27.13(e)),[9] and (f),[10] both containing

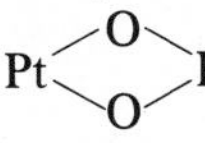

Pt bridges in which an O atom of the ligand forms bonds to both Pt atoms.

In (f) the bond lengths are: Pt–O, 2.24 Å; Pt–N, 2.13 Å; and Pt–C, 2.06 Å.

In all these compounds there is octahedral coordination of Pt^{IV} which is achieved in a variety of ways in compounds which present the possibility of coordination numbers of 4, 5, and 7.

(1) AC 1977 **B33** 558, 887
(2a) AC 1968 **B24** 287
(2b) AC 1968 **B24** 157
(3a) JACS 1947 **69** 1561

(3b) JCS A 1971 90
(4) IC 1968 7 2165
(5) JOC 1968 **14** 447
(6) PRS A 1969 **254** 205, 218; JCS A 1969 2282 (n.d.)

(7)   PRS A 1962 **266** 527   (9)   JCS A 1967 1955
(8)   JCS 1965 630   (10)   JCS 1965 6899

Olefine compounds

It was originally thought that in coordination compounds a bond is formed between an atom with one or more lone pairs of electrons and a metal atom requiring electrons to complete a stable group of valence electrons. This does not, however, account for the formation of compounds such as, for example, $(C_2H_4PtCl_3)K$, by olefines (which have no lone pairs) or for the fact that basicity does not run parallel with donor properties in series of ligands, as it might be expected to do. For example, basicity falls rather rapidly in the series NH_3, PH_3, AsH_3, SbH_3, but PR_3 and AsR_3 form complexes more readily than NR_3 (R is an organic radical).

The elements forming reasonably stable olefine compounds are those with filled d orbitals having energies close to those of the s and p orbitals of the valence shell:

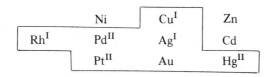

Examples include: $C_2H_4Pt(NH_3)Cl_2$, C_2H_4CuCl, $C_6H_{10}Ag^+$.aq., $C_2H_4Hg^{2+}$.aq., and bridged compounds $(diene)_2Rh_2^ICl_2$ where diene indicates, for example, cycloocta-1:5-diene. It is unlikely that the bond between C_2H_4 and Pt in $(C_2H_4PtCl_3)K$ is formed by donation of π electrons to Pt, for C_2H_4 does not combine even at low temperatures with an acceptor molecule such as $B(CH_3)_3$, whereas amines readily do so. Similarly, the P atom in PF_3 forms no compound with the very strong acceptor BF_3, showing that its lone pair is very inert, yet it combines with Pt to form the stable volatile carbonyl-like compounds $(PF_3)_2PtCl_2$ and $(PF_3PtCl_2)_2$. Also, $Ni(PF_3)_4$ has been prepared as a volatile liquid resembling $Ni(CO)_4$. These facts suggest the possibility of a similar type of bonding in compounds in which C_2H_4, CO, PF_3, and possibly other ligands are attached to Pt and similar metals, and suggest that an essential part is played by d orbitals.

In the free PF_3 molecule the inertness of the lone pair is due to the withdrawal of electrons by the very electronegative F. It is supposed that there is interaction between the filled d orbitals of Pt and the d orbitals of P (Fig. 27.14(a)), as suggested earlier by Pauling for the interaction of CO with Pt (Fig. 27.14(b)), this d_π bonding being assisted by the electronegative F atoms attached to P. At the same time this process would tend to make the lone pair of electrons on the P atom more readily available for σ bond formation. The Pt—CO bond may be described as a combination of the ordinary dative (σ) bond, due to overlap of the carbon orbital containing the lone pair and an empty $5d6s6p^2$ orbital, with a π bond resulting from overlap of a *filled* orbital of Pt and the empty p-molecular orbital of CO. As regards the nature

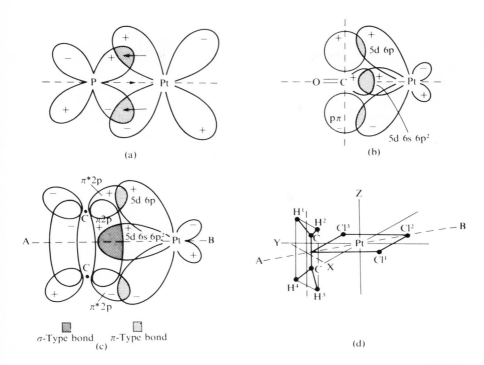

FIG. 27.14. Bonding between Pt and (a) PF$_3$, (b) CO, (c) C$_2$H$_4$. (d) The ion [Pt(C$_2$H$_4$)Cl$_3$]$^-$.

of the filled orbital used for the π bonding, it is thought likely that some sort of hybrid 5d6p orbital will be preferred to a simple 5d orbital.

A rather similar combination of σ and π bonds may be postulated for the C$_2$H$_4$–Pt bond, except that here both the orbitals of C$_2$H$_4$ are π orbitals, namely, the 2pπ 'bonding' orbital containing the two π electrons and the (empty) 2pπ* 'antibonding' orbital (Fig. 27.14(c)). A feature of this bond is that the bonding molecular orbital of the olefine is still present. The infra-red spectra of K[Pt(C$_3$H$_6$)Cl$_3$] and (C$_3$H$_6$)$_2$Pt$_2$Cl$_4$ show the C=C bond is still present, though its stretching frequency is lowered by 140 cm^{-1} as compared with the free olefine, the bond being weakened by use of some of its electrons in the bond to the Pt atom. In the spectra of K[Pt(C$_2$H$_4$)Cl$_3$] and (C$_2$H$_4$)$_2$Pt$_2$Cl$_4$ the C=C absorption band is very weak, indicating symmetrical bonding of C$_2$H$_4$ to Pt as indicated in Fig. 27.14(d). From the dipole moments of molecules such as *trans* C$_2$H$_4$PtCl$_2$.NH$_2$C$_6$H$_4$.CH$_3$ and *trans* C$_2$H$_4$PtCl$_2$.NH$_2$C$_6$H$_4$Cl it is estimated that the Pt–C$_2$H$_4$ bond has about one-third double-bond character.

The existence of very stable complexes of PtX$_2$ with cyclooctatetraene (a) and CH$_2$=CH.CH$_2$CH$_2$.CH=CH$_2$ (b) is consistent with this view of the metal–olefine bonds because the π bonds in these *cis* molecules must use two *different* d orbitals

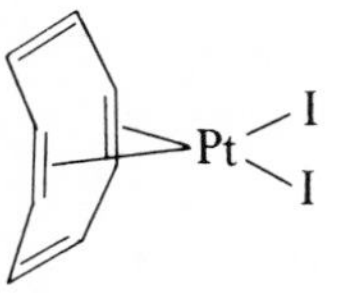

(a) 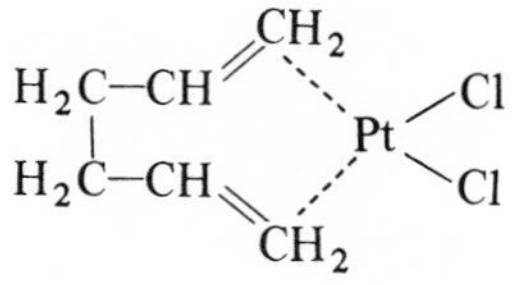   (b)

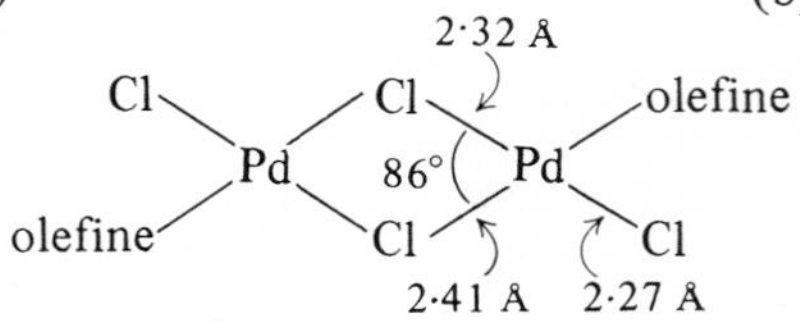

(c)

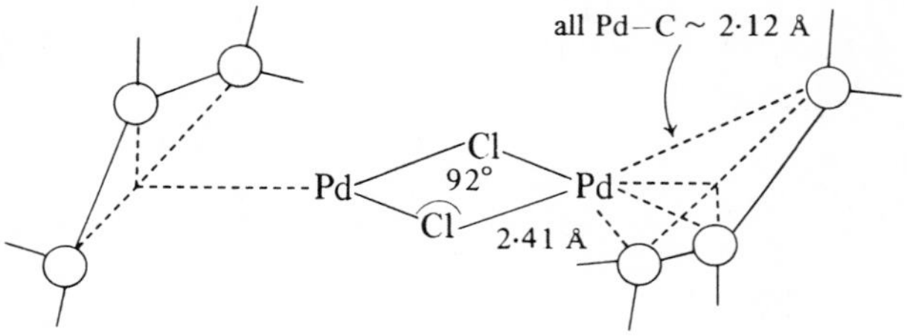

(d)

of the Pt atom. On the other hand, the *trans* compound $(C_2H_4)_2PtCl_2$, in which the same orbitals of Pt would be used to bind both ethylene molecules, is apparently much less stable than the *cis* isomer. The structure of a *cis* dichloro compound of type (b), $PtCl_2 . C_{10}H_{16}$, has been determined.[1]

The hypothesis that olefines are linked to Pt or Pd by interaction between the π orbital of the double bond and a dsp^2 orbital of the metal is supported by the crystal structure of the (dimeric) styrene–$PdCl_2$[2] and ethylene–$PdCl_2$[3] complexes. Both these bridged molecules have the *trans* configuration (c), and the plane of the ethylene molecule or of $H_2C=CH–C$ in the case of the styrene compound, $[(C_6H_5 . CH=CH_2)PdCl_2]_2$, is perpendicular (or approximately so) to the plane containing Pd, the two ethylene C atoms and Cl. An interesting point is that the bridge system is unsymmetrical in both the ethylene and styrene compounds, the Pd–Cl bonds opposite to the Pd–olefine bonds being longer than the other two. In the symmetrical molecule $Pd_2Cl_2(C_3H_5)_2$[4] (d), all the Pd–Cl (bridge) bonds are long (2.41 Å).

The structures of the molecule $PtCl_2(C_2H_4)NH(CH_3)_2$[5] and the ion in Zeise's salt, $K(PtCl_3 . C_2H_4).H_2O$,[6] (see also Fig. 27.14(d)), are shown at (e) and (f). The structure of Zeise's salt is very similar geometrically to that of $K(PtCl_3NH_3).H_2O$.

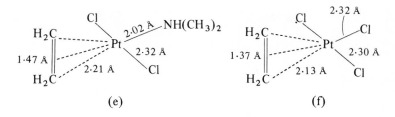

(e) (f)

Within the double layer there are ionic bonds between anions and cations and also $H_2O–Cl$ bonds (3.28 Å), and between the layers van der Waals bonds between C_2H_4 molecules; in the ammine there are N–H---Cl bonds between the layers. In the ion (f) C=C is very nearly perpendicular to the $PtCl_3$ plane, its centre being approximately 0.2 Å above that plane. There is very little difference between the lengths of the Pt–Cl bonds *cis* and *trans* to C_2H_4; the latter may be slightly longer, but the difference is certainly very small and possibly within the limits of accuracy of the structure determination.

(1)   AC 1965 **18** 237
(2)   JACS 1955 **77** 4987
(3)   JACS 1955 **77** 4984
(4)   AC 1965 **18** 331
(5)   AC 1960 **13** 149
(6)   AC 1971 **B27** 366

28

The lanthanides and actinides

The crystal chemistry of the lanthanides (rare-earths)

In the fourteen elements (Ce to Lu) which follow La the 4f shell is being filled while the outer structure of the atoms is either $5d^1 6s^2$ or $5d^0 6s^2$. Since many of the compounds of La are structurally similar to those of the rare-earths we shall include this element in our discussion.

The characteristic valence of the lanthanides is three, but there are some interesting differences between the rare-earths. Ce, Pr, and Tb exhibit tetravalence, and Sm, Eu, Yb, and possibly Tm, divalence, that is, from the elements following La and Gd it is possible to remove two 4f electrons (in addition to two 6s electrons), and the

		4	4	?						4	?					
(La)	Ce	Pr	Nd	Pm	Sm	Eu	(Gd)	Tb	Dy	Ho	Er	Tm	Yb	(Lu)		
				?	2	2						?	2			

elements immediately preceding Gd and Lu can form ions M^{2+} in addition to M^{3+}. The radii of the 4f ions M^{3+} fall steadily throughout the series (p. 314), but this *lanthanide contraction* is not shown by Eu and Yb in the metals or in compounds such as the hexaborides, where the lower valence of these two elements is evident from the larger radii compared with those of neighbouring elements. The rare-earths therefore fall into two series, with a break at Gd, showing the same sequence of valences in each. In Gd there is one electron in each of the seven 4f orbitals, so that further electrons have to be paired with those already present, and we find a marked resemblance between atoms having the same number of doubly occupied orbitals and those with singly occupied orbitals—with Gd compare Mn, the middle member of the 3d series of elements.

Trivalent lanthanides

A feature of the crystal chemistry of 4f compounds is that numerous structures are found for each group of compounds (for example, halides) owing to the variation in ionic (or atomic) radius throughout the series. In the compounds of the earlier 4f elements there is often a high c.n. of the metal ion which falls to lower values as the ionic radius decreases. For example, there are at least seven structures for trihalides, with (somewhat irregular) 9-coordination of M in the LaF_3 and YF_3 structures, 9- or (8+1)-coordination in other halides of the earlier 4f elements, and 6-coordination (YCl_3 or BiI_3 structures) in the compounds of the later elements. (For a summary see Table 9.19, p. 421.) A number of 4f trihydroxides have the same 9-coordinated (UCl_3) structure as the trichlorides of the earlier rare-earths (p. 422). Oxyhalides are described in Chapter 10.

In oxides M_2O_3 there is 6- and/or 7-coordination of M, the La_2O_3 structure being notable for the unusual (monocapped octahedral) coordination of M (Fig. 12.7, p. 546). This structure is also described in a different way on p. 1271 in connection with the structures of compounds such as U_2N_2S and U_2N_2Te; it is also the structure of Ce_2O_2S and other 4f oxysulphides. Sesquisulphides adopt one or other of three structures, in which the c.n. of M falls from 8 to 6:

	Ce_2S_3	α-Gd_2S_3	Ho_2S_3	Yb_2S_3
c.n. of M	8	8 and 7	7 and 6	6

Other compounds are described in other chapters:

borides: p. 1052;	hydrides: p. 346;
carbides: p. 947;	nitrides: p. 832.

Tetravalent lanthanides

The oxidation state IV is firmly established only for Ce, Pr, and Tb, though the preparation of complex fluorides containing Nd(IV) and Dy(IV) has been claimed. The only lanthanide ion M^{4+} stable in aqueous solution is Ce^{4+}, and this is probably always present in complexes, as in the nitrato complex in $(NH_4)_2[Ce(NO_3)_6]$ (p. 825). The only known binary solid compounds of these elements in this oxidation state are:

CeO_2	PrO_2	TbO_2	(fluorite structure)
CeF_4	—	TbF_4	(UF$_4$ structure).

Some complex fluorides and complex oxides have been prepared, for example, alkaline-earth compounds $MCeO_3$ with the perovskite structure (p. 584). The metal-oxygen systems are complex, there being various ordered phases intermediate between the sesquioxide (p. 543) and the dioxide. In the Tb—O system[1] there are $TbO_{1.715}$ (Tb_7O_{12}), $TbO_{1.81}$, and $TbO_{1.83}$, and one or more of these phases occur in the Ce—O[2] and Pr—O[3] systems. For Tb_7O_{12} and Pr_6O_{11} see p. 607.

(1) JACS 1961 83 2219   (3) JACS 1954 76 5239
(2) JINC 1955 1 49

Divalent lanthanides

In the lanthanides the oxidation state II is most stable for Eu, appreciably less so for Yb and Sm (in that order), and extremely unstable for Tm and Nd. It appears that Eu is probably the only lanthanide which forms a monoxide, the compounds previously described as monoxides of Sm and Yb being Sm_2ON and Yb_2OC. Of the many 4f compounds MX which crystallize with the NaCl structure only EuO, SmS, EuS, and YbS contain M(II); all other compounds MS, for example, the bright yellow CeS (and also LaS) are apparently of the type $M^{3+}S^{2-}$ (e). The cell dimensions of the 4f compounds MS decrease steadily with increasing atomic number except those of SmS, EuS, and YbS, the points for which fall far above the curve;[1] all compounds MN, including those of Sm, Eu, and Yb, are normal compounds of M(III). Europium forms an oxide Eu_3O_4 ($Eu^{II}Eu_2^{III}O_4$) which is isostructural with $SrEu_2O_4$ (p. 601).

All dihalides of Eu, Sm, and Yb are known, but less is known of the dihalides of Nd and Tm. Of the structures of the Yb dihalides only that of YbI_2 (CdI_2 structure) appears to be known. The structures of Eu and Sm dihalides are summarized in Table 9.15 (p. 415) and the accompanying text, where the similarity to the alkaline-earth compounds is stressed. Compounds MI_2 formed from the metal and MI_3 by La, Ce, Pr, and Gd[2] are not compounds of M(II), but have metallic properties and may be formulated $M^{III}I_2$ (e).

The ionic radius of Yb^{2+} is close to that of Ca^{2+} and those of Sm^{2+} and Eu^{2+} are practically identical to that of Sr^{2+}. Accordingly many compounds of these three 4f elements are isostructural with the corresponding alkaline-earth compounds, for example, dihalides, EuO and MS with the alkaline-earth sulphides (NaCl structure), $EuSO_4$ and $SmSO_4$ with $SrSO_4$ (and $BaSO_4$), three polymorphs of EuB_2O_4 are isostructural with those of SrB_2O_4 (p. 1070), and EuB_4O_7 with SrB_4O_7. The structures of Eu borates illustrate the effect of crystal structure on chemical properties. The temperatures at which rapid oxidation sets in can be determined by thermogravimetric analysis in an oxidizing atmosphere:

	Type of borate ion	*Oxidation temperature*
$Eu_3B_2O_6$	BO_3^{3-} ion	673 °C
$Eu_2B_2O_5$	$B_2O_5^{4-}$ ion	683
EuB_2O_4	$(BO_2)_n^{n-}$ chain	753
EuB_4O_7	3D network	1033

In EuB_4O_7 the Eu^{2+} ions occupy cavities in a 3D B_4O_7 framework, which is a more rigid, and thermally stable, structure than those of borates containing finite or chain ions.

(1)   AC 1965 **19** 214
(2)   IC 1965 **4** 88

The actinides

Introduction

Just as the lanthanides form a series of closely related elements following La in which the characteristic ions M^{3+} have from 1 to 14 4f electrons, so the actinide series might be expected to include the 14 elements following the prototype Ac (which, like La, is a true member of Group III), with from 1 to 14 electrons entering the 5f in preference to the 6d shell. In fact there are probably no 5f electrons in Th and the number in Pa is uncertain, and these elements are much more characteristically members of Groups IV and V respectively than are the corresponding lanthanides Ce and Pr. Thus the chemistry of Th is essentially that of Th(IV), whereas there is an extensive chemistry of Ce(III) but only two solid binary compounds of Ce(IV), namely, the oxide and fluoride. In contrast to Pa, the most stable oxidation state of which is V, there are no compounds of Pr(V).

Owing to the comparable energies of the 5f, 6d, 7s, and 7p levels, the chemistry of the actinides is much more complex than the predominantly ionic chemistry of the trivalent lanthanides. The oxidation state III is common to all actinides, though it is observed only in the solid state for Th and it is unimportant for Pa. Except for these two elements the behaviour of 4f and 5f elements in this oxidation state is similar, both as regards their solution chemistry and the structures of compounds in the solid state—witness the trifluorides AcF_3 and UF_3–CmF_3 with the LaF_3 structure and the corresponding trichlorides with the same structure as $LaCl_3$. (Compounds MO and MS formed by U, Np, and Pu are not regarded as compounds of M(II), for compounds MC and MN have the same (NaCl) structure—see p. 1252.) In contrast to the exceptional II and IV oxidation state of certain 4f elements the actinides as a group exhibit all oxidaton states from III to VI, and there is no parallel in the 4f series either to the elements Th and Pa or to the closely related group consisting of U, Np, Pu, and Am, all of which exhibit oxidation numbers from III to VI inclusive (Table 28.1). Other differences between the lanthanides and actinides include the formation of ions MO_2^{2+} (by U, Np, Pu, and Am) and MO_2^+ (by Np and Am; the U and Pu ions are very unstable), the greater tendency of the actinides to form complexes, and the more complex structures of the metals themselves; Am is the first actinide to crystallize with a close-packed structure like the majority of the lanthanides.

TABLE 28.1

The oxidation states of the actinide elements

Ac	Th	Pa	U	Np	Pu	Am	Cm	Bk	Cf
3	**3**	3	3	3	3	**3**	**3**	**3**	3
	4	4	4	4	**4**	4	4	4	
		5	5	**5**	5	5			
			6	6	6	6			

(Heavy type indicates the most stable oxidation state.)

The radii of actinide ions M^{3+} and M^{4+} decrease with increasing atomic number in much the same way as do those of the lanthanides, the radius of M^{3+} being about 0.10 Å greater than that of M^{4+} (e.g. U^{3+}, 1.03 Å; U^{4+}, 0.93 Å). The screening effect of the f electrons does not entirely compensate for the increased nuclear charge, and since the outermost electronic structure remains the same the radii decrease ('lanthanide' and 'actinide' contractions). The effect is clearly seen in series of isostructural compounds such as the dioxides (fluorite structure) or trifluorides (LaF_3 structure).

The crystal chemistry of thorium

Chemically thorium is essentially a member of Group IV of the Periodic Table. It is tetravalent in most of its compounds, it is the most electropositive of the tetravalent elements, and its chemistry strongly resembles that of Hf. Its ionic radius (Th^{4+}) is, however, closer to that of Ce^{4+} than to that of Hf^{4+}, and in its crystal chemistry it has ·much in common not only with U^{IV} but also with Ce^{IV}. For example, like Ce^{IV} Th forms no normal carbonate but complex carbonates of the type $(NH_4)_2M(CO_3)_3 \cdot 6H_2O$, and both elements form complex sulphates and nitrates, respectively $K_4M(SO_4)_4$ and $K_2M(NO_3)_6$. Again, many compounds of Th are isostructural with compounds of U^{IV} and the transuranium elements, as shown in Table 28.2 for some

TABLE 28.2

Structures of compounds of actinide elements

	Ac	Th	Pa	U	Np[a]	Pu	Am[b],[c]	Cm	Structure
MC		N		N		N			N = NaCl
MN		N		N	N	N			
MO		N	N	N	N	N	N	N	
MS[d]		N		N		N			
M$_2$O$_3$	A					A/C	A/C	A/C	A or C—M$_2$O$_3$
M$_2$S$_3$	D[e]	S[d]		S[d]	S[d]	D[e]	D[e]		D = Ce$_2$S$_3$ (p. 621)
									S = Sb$_2$S$_3$
MO$_2$		F	F	F	F	F	F	F	F = CaF$_2$
MS$_2$		P[d]							P = PbCl$_2$
MC$_2$				C	C				C = CaC$_2$ (p. 757)
MSi$_2$		Si		Si	Si				Si = ThSi$_2$ (p. 792)
MOS		B[d]	B	B[d]	B[d]				
MOCl	B					B	B		B = BiOCl(PbFCl)
MF$_3$	L			L	L	L	L	L	L = LaF$_3$
MF$_4$		U	U	U	U	U	U[b]	U	U = UF$_4$
MCl$_3$	V[f]			V[f]	V[f]	V[f]	V[f]	V	V = UCl$_3$
MCl$_4$		W	W	W	W				W = UCl$_4$

(a) JACS 1953 75 1236.   (b) JACS 1953 75 4560.   (c) JACS 1954 76 2019.   (d) AC 1949 2 291.   (e) AC 1949 2 57.   (f) AC 1948 1 265.

of the simpler compounds of these elements. On the one hand, ThF_4 is isostructural with CeF_4, ZrF_4, and HfF_4, and on the other with PaF_4, UF_4, NpF_4, PuF_4, and AmF_4. Some complex fluorides M_2ThF_6 are structurally similar to U(IV) compounds, but other complex fluorides M_nThF_{4+n} have distinctive structures, some of which are noted in a later section. The remarkable framework in compounds $M_7^I M_6^{IV} F_{31}$ (p. 472) is formed not only by Zr and Th but also by Pr, Pa, U, Np. Pu, Am, and Cm. Compounds of Th mentioned in other chapters include ThC_2 (p. 949) and Th_3N_4 (p. 835). There is a note on the sulphides on p. 1272.

As noted earlier, Th(III) is known only in the solid state, in $ThCl_3$, $ThBr_3$ and ThI_3,[1, 2] Th_2S_3 (Table 28.2), and oxyfluorides. The latter include ThOF (statistical fluorite structure)[3] and $ThO_{0.5}F_{2.5}$[4] (superstructure of LaF_3 type), this compound presumably containing equal numbers of Th^{3+} and Th^{4+} ions. The properties of ThI_2, the structure of which is described on p. 415, suggest formulation as $Th^{IV}I_2(e)_2$.

(1) JCS 1964 5450   (3) CR 1968 **266** 1056
(2) RJIC 1966 **11** 1318   (4) MRB 1969 **4** 443

The crystal chemistry of protoactinium

Although the most stable oxidation state of this element is v few examples are yet known of crystalline compounds of Pa(v) which are isostructural with those of Nb or Ta. Examples probably include PaO_2Br, NbO_2Br, NbO_2I, and TaO_2I, which apparently have the same structure as UO_2Br (see later). Usually Pa compounds are isostructural with those of U(v) or they have structures peculiar to this element, as shown by the examples of Table 28.3. The c.n. of Pa^{5+} is generally higher than that

TABLE 28.3

Comparison of halides of Nb (and Ta), Pa(v), and U(v)

C.n. of M		*C.n. of* M		*C.n. of* M	
NbF_5 TaF_5	6	PaF_5	7	$UF_5\ \alpha$	6 (octahedral)
Tetramer M_4F_{20} (octahedral)		Isostructural with β-UF_5		$UF_5\ \beta$	7 (Fig. 28.2)
$NbCl_5$ $TaCl_5$	6	$PaCl_5$	7	UCl_5	6
Dimer M_2Cl_{10} (octahedral)		Pentagonal bipyramidal chain		U_2Cl_{10} dimer	
$RbNbF_6$ $RbTaF_6$	6	$RbPaF_6$	8 and	$RbUF_6$	8
Octahedral MF_6^-		Dodecahedral chain ion			
K_2NbF_7 K_2TaF_7	7	K_2PaF_7	9	K_2UF_7	?
Monocapped trigonal prism MF_7^{2-}		Tricapped trigonal prism chain			?
Na_3TaF_8	8	Na_3PaF_8	8 and	Na_3UF_8	8
Antiprism MF_8^{3-}		Cubic coordination (fluorite superstructure)			

of Nb^{5+} (or Ta^{5+}), consistent with the larger radius of that ion (about 0.9 Å) as compared with about 0.7 Å for the group V ions. In view of the fact that recent work on Nb_2O_5 and Ta_2O_5 has shown that the structural chemistry of these oxides is extremely complex, it is likely that earlier statements on the similarity of Pa_2O_5 to these oxides should be checked. Certainly there is no resemblance between the complex oxides formed by Pa(v) with 4f elements and those formed by Nb and Ta.

For example, Pa does not form the analogues of MNb_3O_9 and MTa_3O_9 but instead forms extensive solid solutions M_2O_3–Pa_2O_5 with fluorite-like structures.[1]

Compounds of Pa(IV) are structurally similar to those of other actinide (IV) compounds. For example, all the compounds PaF_4, $PaCl_4$, $LiPaF_5$, and $(NH_4)_4PaF_8$ are isostructural with the corresponding compounds of U(IV);[2] see also Table 28.2 for PaO_2 and $PaOS$.

The structure of $PaOCl_2$[3] (and the isostructural Th, U, and Np compounds) is unexpectedly complex. There are tightly-knit chains of composition $Pa_3O_3Cl_4(Cl_4^*)$ of which the Cl* atoms are shared with other similar chains, forming a 3D structure with the composition $PaOCl_2$. There is 2- and 3-coordination of Cl, 3- and 4-coordination of O, and 7-, 8-, and 9-coordination of the Pa atoms. This structure has been confirmed by a later study[4] of the isostructural $UOCl_2$.

(1)   AC 1967 **23** 740   (3)   AC 1968 **B24** 304
(2)   IC 1967 **6** 544   (4)   AC 1974 **B30** 175
For references to complex halides pf Pa(V) see Table 28.6 (p. 1263).

The crystal chemistry of uranium

The structures of the metal, hydrides, and carbides are described in other chapters, as also are certain halides MX_3, MX_4, and MX_5. Here we devote sections to certain halide structures peculiar to U, complex fluorides of Th and U, oxides of U, uranyl compounds and uranates, nitrides and related compounds, and conclude with a note on the sulphides of U, Th, and Ce.

Halides of uranium. Our present knowledge of the structures of the majority of the 5f halides is summarized (with references) under Trihalides, Tetrahalides, etc. in Chapter 9. The halides of U include:

	C.n. of U		C.n. of U
UF_3 : LaF_3 structure	9 + 2	UF_4 : ZrF_4 structure	8
UCl_3 ⎫		UCl_4 : $ThCl_4$ structure	8
UCl_3 ⎬ UCl_3 structure	9	UBr_4 (see below)	7
UBr_3 ⎭			
		$UF_5\ \alpha$ ⎫ see text	6
		$UF_5\ \beta$ ⎭	8
UI_3 : $PuBr_3$ structure	8	UCl_5 : U_2Cl_{10} dimers	6
		UF_6 ⎫ molecular	6
		UCl_6 ⎭	

We comment here on the two forms of UF_5[1] and on U_2F_9,[2] and include the structure of UBr_4 in the next section, where we describe a number of oxyhalides in which there is pentagonal bipyramidal coordinaton of U.

The two crystalline forms of the pentafluoride illustrate two ways in which a ratio of $5F:1U$ can be attained with different c.n.s of U. In α-UF_5, Fig. 28.1(a),

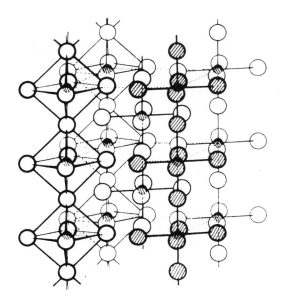

FIG. 28.1. The crystal structure of α-UF₅. (Small circles represent U atoms).

there are infinite chains formed from octahedral UF_6 groups sharing a pair of opposite vertices. In β-UF_5 there is distorted square antiprismatic coordination of U, 6 F being shared with other coordination groups; the bond lengths are: U–2F, 1.96 Å; U–6F, 2.27 Å (mean). The mean U–F distance in both structures is close to 2.20 Å.

In the black U_2F_9 all the U atoms have nine F neighbours (at 2.31 Å), and since all the U atoms are crystallographically equivalent there is presumably interchange of electrons between ions with different charges (e.g. U^{4+} and U^{5+}), accounting for the colour. The same metal and fluorine positions are occupied by Th^{4+} and F^- in $NaTh_2F_9$[2] but in addition Na^+ ions occupy two-thirds of the available octahedral holes between six F^- ions (Na–6F, 2.34 Å; ·Th–9F, 2.40 Å: very similar to the corresponding distances in β_2-Na_2ThF_6).

(1) (α) AC 1949 **2** 296; (β) AC 1976 **B32** 3311   (2) AC 1949 **2** 390

Pentagonal bipyramidal coordination of U *in oxyhalides and oxyhalide ions.* Oxyhalides of U are numerous; for example, oxychlorides include UOCl (3), $UOCl_2$ (4), $U_2O_2Cl_5$ (4.5),[1] UO_2Cl (5), $UOCl_3$ (5), $U_2O_4Cl_3$ (5.5), and UO_2Cl_2 (6), the numbers in parentheses being the (mean) oxidation numbers of U. Oxyhalide ions also are numerous. The U^{VI} ions $UO_2Cl_4^{2-}$[2] and $UO_2Br_4^{2-}$[3] have an octahedral (*trans*) structure, but in the three oxyfluoride ions we shall describe there is pentagonal bipyramidal coordination of U, as in all the compounds of Table 28.4.

TABLE 28.4
Pentagonal bipyramidal coordination of U *in oxyhalides and oxyhalide ions*

X:M	Formula of oxyhalide or oxyhalide ion	Type of structure	Reference
7	$UO_2F_5^{3-}$	Finite	AC 1954 **7** 783
6.5	$U_2O_4F_9^{5-}$	Finite	AC 1974 **B30** 768
6	$U_2O_4F_8^{4-}$	Finite	AC 1972 **B28** 2011, 2617
6	$U_2O_4Cl_2(OH)_2(H_2O)_4$	Finite	ACSc 1969 **23** 791
5	$(U_4O_{10}Cl_4(OH)_2(H_2O)_4]^{2-}$	Finite	AC 1977 **B33** 2477
5	β-UOF_4	3D	AC 1974 **B30** 1701
5	$UO_2Cl_2.H_2O$	Chain	AC 1974 **B30** 169
4	$UO_3Cl_{0:9}^{0:9-}$	Double chain	AC 1964 **17** 41
	$(PaOBr_3$		AC 1975 **B31** 1382)
4	UBr_4	Layer	AC 1974 **B30** 2664
4	UO_2Cl_2	3D	AC 1973 **B29** 1073
3.5	$U_2O_2Cl_5$	3D	JSSC 1980 **32** 297
3	UO_2Br	Layer	AC 1977 **B33** 2542

Another feature common to several of the structures of this section is the presence of the distinctive $PaCl_5$-type chain formed from pentagonal bipyramidal groups sharing two edges. This is found as a simple chain or as a sub-unit from which more complex structures are built.

(1) JSSC 1980 **32** 297   (3) AC 1974 **B30** 840
(2) AC 1976 **B32** 1541

In addition to the simple $UO_2F_5^{3-}$ ion, (a), there are binuclear ions $U_2O_4F_9^{5-}$, (b), and $U_2O_4F_8^{4-}$, (c), formed from bipyramids sharing a vertex or edge respectively. In the latter the two O–U–O axes are parallel, but in $U_2O_4F_9^{5-}$ they are tilted towards one another, the angle between these axes being 120°. Note that salts containing the ion (c) are preferably formulated as, for example, $Cs_4U_2O_4F_8.2H_2O$ rather than $Cs_2UO_2F_4.H_2O$; contrast $Cs_2UO_2Cl_4$, which contains the octahedral ion $UO_2Cl_4^{2-}$. Pairs of pentagonal bipyramidal groups sharing an edge form the molecules in the dihydrate of $UO_2Cl(OH)$, (d).

A more complex example of a finite complex containing the same ligands is the tetranuclear anion in $K_2[(UO_2)_4O_2(OH)_2Cl_4(H_2O)_4].2H_2O$, (e). In this ion there

(a)   (b)   (c)

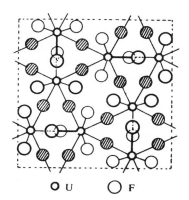

(d)   (e)

are two different kinds of U coordination group, namely, (3 O, 2 Cl, OH, and H_2O) and (4 O, Cl, OH, and H_2O).

The structure of β-UOF_4 (Fig. 28.2) is apparently similar to that originally assigned to β-UF_5. In the 7-coordination groups O occupies an equatorial position and each of the four equatorial F atoms bridges to a different U atom.

○ U   ○ F

FIG. 28.2. Projection of the crystal structure of β-UOF_4. The bridging F atoms are shown as shaded circles and the pairs of overlapping open circles represent O atoms.

The structure of $UO_2Cl_2 \cdot H_2O$ is quite different from, and much simpler than, the structure of the anhydrous compound, which is described shortly. It consists of $PaCl_5$-type chains

in which the bond lengths are: U–O, 1.72; U–OH_2, 2.46; and U–Cl, 2.78 Å. The anion in $Cs_xUO_3Cl_x$ ($x \approx 0.9$) consists of the double chain of Fig. 28.3. There is a similar chain in $PaOBr_3$, in which the O–U–O groups are replaced by Br–Pa–Br

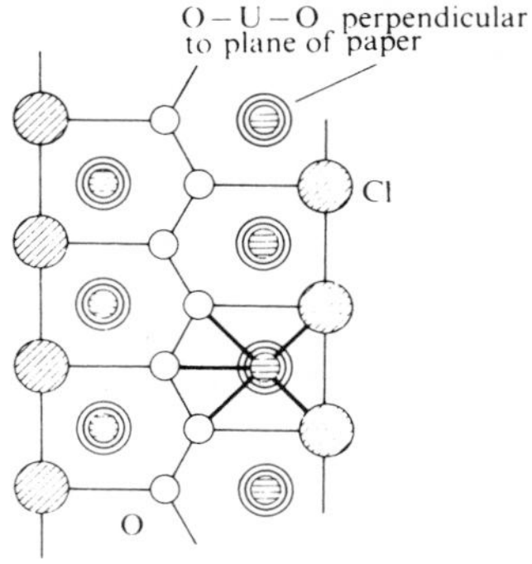

FIG. 28.3. The infinite chain ion in $Cs_x(UO_2)OCl_x$.

and there are 3 O atoms in the equator of each PaO_3Br_4 coordination group. ($Pa-O$, 2.18; $Pa-Br_{equat.}$, 2.91; and $Pa-Br_{axial}$, 2.61 Å).

The structure of UBr_4 may be represented as constructed from $PaCl_5$-type chains which are linked by sharing one (slanting) edge of each UBr_7 group to form a layer based on the planar 6^3 net. This is shown diagrammatically in Fig. 28.4(a) and (b). The heavier lines in (a) indicate the shared edges and (b) is an end-on view of the layer, viewed in the direction of the arrow in (a). In the closely related structure of UO_2Cl_2 chains built of UO_3Cl_4 groups, Fig. 28.4(c), are joined so that each equatorial O is also an apical O of another chain. The chains are tilted as shown diagrammatically at (d), which shows how a second series of chains are formed perpendicular to the plane of the paper, only one of the uranyl O atoms being shared in this way. The result is a diamond-like 3D framework built from pentagonal bipyramidal groups sharing 2 edges and 2 vertices. With this structure compare the very simple layer structure of UO_2F_2 (Fig. 28.8(c)), with 8-coordinated U, and the simple chain structure of $UO_2Cl_2 . H_2O$ already described.

Lateral joining of two $PaCl_5$-type chains gives the layer of Fig. 28.4(e), which is then joined to similar layers through the axial O atoms in a direction normal to the plane of the paper, to form a 3D structure of composition $U_2O_2Cl_5$. If all non-metal atoms are of the same kind the structure has the composition A_2X_7 as in $RbIn_2F_7$, where Rb ions occupy positions (x) in the tunnels. If additional M and O atoms are placed in the tunnels as shown at (f), so that M is octahedrally coordinated, this represents the structure of β-U_3O_8. In UO_2Br pentagonal bipyramidal UO_5Br_2 groups form layers which are normal to the paper in Fig. 28.4(g), there being $-O-U-O-U-O-$ chains ($U-O$, 2.05 Å) perpendicular to the plane of the paper. (For $UOCl_2$ see p. 1254.)

Complex fluorides of the 5f elements. As already pointed out in Chapter 10 the closest relation between the crystal structures of simple and complex halides is found for certain of the latter which have the structures of simple halides but with

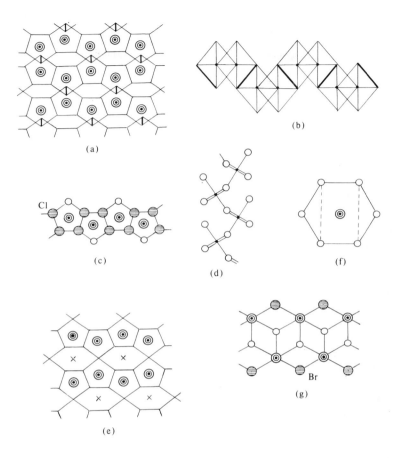

FIG. 28.4. Pentagonal bipyramidal coordination in halogen compounds of U: (a) and (b) UBr₄, (c) and (d) UO₂Cl₂, (e) U₂O₂Cl₅, (f) β-U₃O₈, (g) UO₂Br.

statistical distribution of two or more kinds of metal ions in the cation positions. Examples include:

$$\left.\begin{array}{c} \text{KLaF}_4 \\ \alpha\text{-K}_2\text{UF}_6 \end{array}\right\} \text{ disordered CaF}_2 \text{ structure;} \qquad \left.\begin{array}{c} \text{BaThF}_6 \\ \text{BaUF}_6 \end{array}\right\} \text{ disordered LaF}_3 \text{ structure.}$$

The statistical fluorite structure is to be expected for a compound A_mBF_n when both A and B are large enough for 8-coordination by F. For smaller A ions other structures are found, as for example, NaLaF₄ and Na₂ThF₆ with the β₂-Na₂ThF₆ structure, which is described later, in which there is 6-coordination of A and 9-coordination of B ions. Other ways in which a more complex structure may be related to an A_mX_n structure are exemplified by γ-Na₂UF₆ and Na₃UF₈ (fluorite

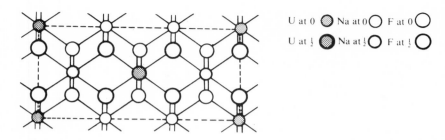

FIG. 28.5. Projection of the crystal structure of γ-Na$_2$UF$_6$. The heights of the ions above the plane of the paper are given as fractions of the length of the b axis.

superstructures), Na$_3$UF$_7$ (anion defective fluorite structure), and Na$_2$UF$_8$ (cation defective fluorite structure.

The γ-Na$_2$UF$_6$ *structure* is a slightly deformed fluorite structure in which both types of positive ion are 8-coordinated. It is illustrated in Fig. 28.5.

The Na$_3$UF$_7$ *structure* also is closely related to the fluorite structure. It is tetragonal, with a = 5.45 Å and c = 10.896 Å. A projection of the structure (Fig. 28.6) shows that the doubling of one axis is due to the regular arrangement of U in one-quarter of the cation positions. One-eigth of the F positions of the fluorite structure are unoccupied, as shown at the left in Fig. 28.6, with the result that both

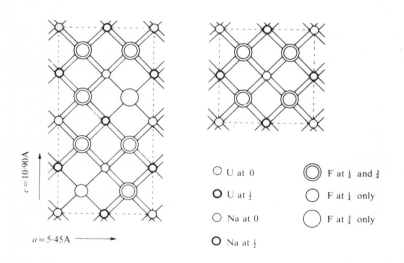

FIG. 28.6. Projection on (010) of the (tetragonal) crystal structure of Na$_3$UF$_7$ (left), to which the key refers. The heights are fractions of b (5.45 Å) above the plane of the paper. At the right is shown a projection of the (cubic) fluorite structure (compare Fig. 6.9, p. 252), in which the heights of the atoms are indicated in the same way as for Na$_3$UF$_7$.

Na^+ and U^{4+} are 7-coordinated. Summarizing, all the following complex fluorides of U have structures related to the fluorite structure:

Statistical	Superstructures	Anion defective	Cation defective
High-K_2UF_6	Na_2UF_6 Na_3UF_8	Na_3UF_7	Na_2UF_8

In the complex fluorides we are describing there are high c.n.s of the 4f and 5f elements, and of special interest are (i) the persistence of these high c.n.s down to low F:M ratios, and (ii) the unexpected variations in c.n. in certain series of compounds, notably some complex fluorides of thorium. The first point is illustrated by the structures of compounds of U(IV), some of which are set out in Table 28.5. The same type of 9-coordination (tricapped trigonal prism) is found in $LiUF_5$ and in two forms of K_2UF_6.

TABLE 28.5

Coordination of U(IV) *in some complex fluorides*

Compound	C.N. of U(IV)	Coordination polyhedron	Reference
$LiUF_5$	9	t.c.t.p.	AC 1966 **21** 814
Rb_2UF_6	8	Dodecahedron	IC 1969 **8** 33
β_2-Na_2UF_6 β_1-K_2UF_6 δ-Na_2UF_6	9	t.c.t.p.	AC 1948 **1** 265; AC 1969 **B25** 2163 AC 1979 **B35** 1198
Na_3UF_7	7	Fig. 28.4	AC 1948 **1** 265
K_3UF_7	7	Pentagonal bipyramid	AC 1954 **7** 792
Li_4UF_8	8 (+1)	See text	JINC 1967 **29** 1631
$(NH_4)_4UF_8$	8	Distorted antiprism	AC 1970 **B26** 38

The β_2-Na_2ThF_6 *and* β_1-K_2UF_6 *structures*. These two structures are very closely related. Projections on the base of the hexagonal unit cell are given in Fig. 28.7. In both there is 9-coordination of the heavy metal ions, with fluorine ions at the apices of a trigonal prism and three beyond the centres of the vertical prism faces. The arrangement of Th (or U) and F is the same as that of Sr and H_2O in $SrCl_2.6H_2O$ (Fig. 15.14, p. 679). The alkali-metal ions are situated between these

β_2-Na_2ThF_6 *structure*	β_1-K_2UF_6 *structure*
β_2-Na_2ThF_6	β_1-K_2ThF_6
β_2-Na_2UF_6	β_1-K_2UF_6
β_2-K_2UF_6	
	β_1-$KLaF_4$
$NaPuF_4$	β_1-$KCeF_4$

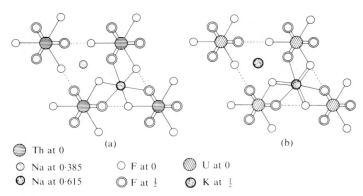

Th at 0

Na at 0·385   ○ F at 0   U at 0

Na at 0·615   ○ F at ½   K at ½

FIG. 28.7. The crystal structures of (a) β_2-Na$_2$ThF$_6$, and (b) β_1-K$_2$UF$_6$. The heights of atoms are given as fractions of the lengths of the hexagonal c axes, which are normal to the plane of the paper.

chains in positions of 6- or 9-coordination in β_2-Na$_2$ThF$_6$ and β_1-K$_2$UF$_6$ respectively. The former structure is therefore generally preferred by sodium salts and the latter by potassium salts. In the isostructural compounds ABX$_4$ there is A$_{\frac{3}{2}}$B$_{\frac{3}{2}}$X$_6$ in the unit cell instead of, for example, Na$_2$ThF$_6$ or K$_2$UF$_6$, that is, the same total number of cations. Of these one B ion occupies the Th (or U) position and $\frac{3}{2}$A$+\frac{1}{2}$B the Na (or K) positions, statistically. (Compare, for example, certain of the spinels, p. 593.) The structure of NaNdF$_4$ (and isostructural Ce and La salts) is similar to but different from the β_2-Na$_2$ThF$_6$ structure originally assigned; there is 9-coordination (t.c.t.p.) of the 4f ion.[1]

The structures of compounds below the line in Table 28.5 differ from those above the line in containing discrete UF$_7$ or UF$_8$ groups. The structure of Na$_3$UF$_7$ has already been noted. One form of K$_3$UF$_7$ is isostructural with K$_3$UO$_2$F$_5$; the other is similar structurally to K$_3$ZrF$_7$ and (NH$_4$)$_3$ZrF$_7$ (p. 474). In both polymorphs the 7 F atoms are arranged at the vertices of a pentagonal bipyramid (U—F, 2.26 Å).

A number of 8-coordination polyhedra are found for U(IV)—contrast the cubic coordination of U(V) and U(VI) in Na$_3$UF$_8$ and Na$_2$UF$_8$ respectively. It should be remarked that whereas in LiUF$_5$ there is t.c.t.p. coordination of U with all U—F distances between 2.26 and 2.59 Å, the coordination polyhedron of U in Li$_4$UF$_8$ is strictly a bicapped trigonal prism (8 F at 2.21–2.39 Å). There is a ninth neighbour (at 3.3 Å) beyond the third prism face, and if this is included in the coordination group the structure no longer contains discrete UF$_8^{4-}$ ions, since this more distant F$^-$ is necessarily shared between UF$_9$ groups.

Our second point concerns the complex fluorides of Th, which are remarkable in several respects (Table 28.6). Both (NH$_4$)$_3$ThF$_7$ and (NH$_4$)$_4$ThF$_8$ contain infinite linear ions of composition (ThF$_7$)$_n^{3n-}$ formed from tricapped trigonal prism groups sharing two edges, though the details of the chains are different in the two crystals. The eighth F in (NH$_4$)$_4$ThF$_8$ is a discrete F$^-$ ion, not bonded to Th. For (NH$_4$)$_4$ThF$_8$

TABLE 28.6

Coordination of metal in some complex fluorides of Th^{4+} *and* Pa^{5+}

Compound	C.N. of M	Coordination polyhedron	Reference
$RbTh_3F_{13}$	9	t.c.t.p. sharing edges and vertices	AC 1971 **B27** 1823
$\beta_2\text{-}Na_2ThF_6$ $\beta_1\text{-}K_2ThF_6$	9	t.c.t.p. sharing two faces	See p. 1261
$(NH_4)_3ThF_7$	9	t.c.t.p. sharing two edges	AC 1971 **B27** 2279
$(NH_4)_4ThF_8$	9	t.c.t.p. sharing two edges	AC 1969 **B25** 1958
$(NH_4)_5ThF_9$	8	Dodecahedron	AC 1971 **B27** 829
$K_7Th_6F_{31}$	8	Antiprism	AC 1971 **B27** 2290
$RbPaF_6$ $KPaF_6$ NH_4PaF_6	8	Dodecahedra sharing 2 edges	AC 1968 **B24** 1675 IC 1966 5 659
K_2PaF_7	9	t.c.t.p. sharing 2 edges	JCS A 1967 1429
Na_3PaF_8	8	Cube	JCS A 1969 1161

there are two types of structure more obvious than the one adopted, namely, discrete ThF_8^{4-} ions (which actually occur in $(NH_4)_5ThF_9$) or a chain of tricapped trigonal prisms sharing two *vertices*. Instead of either of these structures there are chains of composition $(ThF_7)_n^{3n-}$ similar to those in K_2PaF_7 (tricapped trigonal prisms sharing two *edges*) plus additional F^- ions. More remarkable still is $(NH_4)_5ThF_9$, which might have been expected to contain ThF_9^{5-} ions or F^- ions plus $(ThF_8)_n^{4n-}$ chains built from ThF_9 groups sharing two vertices (thus retaining 9-coordination). In fact this crystal contains discrete dodecahedral ThF_8^{4-} plus F^- ions, that is, the c.n. of Th is lower in $(NH_4)_5ThF_9$ than in some compounds with lower F:Th ratios (for example, $(NH_4)_4ThF_8$, $(NH_4)_3ThF_7$, and $RbTh_3F_{13}$, in all of which Th is 9-coordinated) but the same (8) as in $K_7Th_6F_{31}$ with an intermediate F:Th ratio. Note that the 8-coordination of Th is dodecahedral in $(NH_4)_5ThF_9$ but antiprismatic in $K_7Th_6F_{31}$, which is isostructural with $Na_7Zr_6F_{31}$ (p. 472).

(1) IC 1965 4 881

Oxides of uranium. The oxygen chemistry of this element is extraordinarily complex, for although there are oxides with simple formulae such as UO (NaCl structure), UO_2, and UO_3, the last two have ranges of composition which depend on the temperature and there are numerous intermediate phases with ordered structures. Moreover, UO_3 is polymorphic, as also is U_3O_8. We summarize here what is known of oxides in the range UO_2–UO_3; references to structural studies are given in Table 28.7.

At ordinary temperatures UO_2 has the normal fluorite structure. At higher temperatures some of the O atoms move into the large interstices at $(\frac{1}{2}\frac{1}{2}\frac{1}{2})$ etc. and UO_2 takes up oxygen to the composition $UO_{2.25}$. A n.d. study of the disordered solid solution at the intermediate composition $UO_{2.13}$ shows three types of O atom,

TABLE 28.7
The higher oxides of uranium

Oxide	Polymorph	Coordination of U	Reference
$UO_{2.13}$	–		N 1963 **197** 755
U_4O_9	–	See text	AC 1972 **B28** 785
$UO_{2.6}$	–		JINC 1964 **26** 1829
U_3O_8	α	All 7 (pentagonal bipyramidal)	JACr 1970 **3** 94
	β	2 U, 7 (pentagonal bipyramidal)	AC 1969 **B25** 2505;
		1 U, 6 (distorted octahedral)	AC 1970 **B26** 656
UO_3	α	Not known	RTC 1966 **85** 135;
			JINC 1964 **26** 1829
	β	All 6 or 7 (three kinds of U atom)	AC 1966 **21** 589
	γ	All 6 (distorted octahedral)	AC 1963 **16** 993
	δ ($UO_{2.8}$)	6 (regular octahedral, 2·07 Å, ReO_3 structure)	JINC 1955 **1** 309
	High pressure	All 7 (pentagonal bipyramidal)	AC 1966 **20** 292

O_I close to the original fluorite positions (but moved slightly towards the large interstitial holes), and O_{II} and O_{III} in positions about 1 Å from these large holes. At the composition $UO_{2.25}$ (U_4O_9) these three types of O atom are arranged in an orderly way forming a superstructure referable to a much larger unit cell (64 times the volume of the UO_2 cell).

The phase $UO_{2.6}$, both forms of $UO_{2.67}$ (U_3O_8), and the orthorhombic α-UO_3 (see later) all appear to have related structures, but not all of these are known in detail.

Two forms of U_3O_8 are known, the α form being usually encountered. Their structures are closely related, but differ as follows. In α-U_3O_8 all U atoms are 7-coordinated, with 6 O in the range 2.07–2.23 Å but with the seventh at 2.44 Å for one-third of the U atoms and at 2.71 Å for the remainder. For a projection of this structure see Fig. 12.8(c), p. 548. In the β form (prepared by heating α to 1350 °C in air and cooling slowly to room temperature) there is both distorted octahedral and 7-coordination (pentagonal bipyramidal). In contrast to the α form one-third of the U atoms (U^{VI}?) have two close O neighbours (1.89 Å), but the remaining U atoms have quite different environments:

$$U_I \ : \ 1.89\,(2) \quad 2.11\,(1) \quad 2.30\,(2) \quad 2.37\,(2) \ \text{Å}$$
$$U_{II} \ : \ 2.09\,(4) \quad 2.28\,(2)$$
$$U_{III} \ : \ 2.02\,(1) \quad 2.08\,(2) \quad 2.29\,(2) \quad 2.40\,(2)$$

(The numbers in parentheses are the numbers of neighbours at the particular distance.)

At atmospheric pressure UO_3 appears to have six polymorphs, and a seventh has been prepared under high pressure. The original 'hexagonal α-UO_3' studied by Zachariasen is apparently a highly twinned form of the orthorhombic phase. The observed density of α-UO_3 never exceeds about 7.3 g/cm^{-3} as compared with the value $8.4 \, g/cm^{-3}$ calculated from the unit cell dimensions. This difference is due to vacancies in both the cation and anion sites. The absence of certain U atoms in the vertex-sharing pentagonal bipyramids, resulting in the formation of short U—O bonds would account for the strong i.r. absorption around 930 cm^{-1} which is characteristic of the UO_2^{2+} ion. The δ form has the cubic ReO_3 structure, with 6 equidistant octahedral neighbours at 2.07 Å; it was determined for a crystal with the composition $UO_{2.8}$. In the β and γ forms U has 6 and/or 7 nearest O neighbours (at distances in the approximate range 1.8–2.4 Å), the coordination being irregular; only in the high-pressure form are there well-defined UO_2 groups with two short collinear bonds (1.83 Å) and five U—O 2.20–2.56 Å.

Uranyl compounds. An interesting feature of the chemistry of U and certain other 5f elements is the formation of ions MO_2^+ and MO_2^{2+}, the former by Pu(v) and Am(v) and the latter by U(vi), Np(vi), Pu(vi), and Am(vi). Uranyl compounds contain a linear group O—U—O with short U—O bonds (1.7–1.8 Å) and an equatorial group of 4, 5, or 6 O atoms completing 6-, 7-, or 8-coordination of U (Table 28.8). Different total coordination numbers are found even in different polymorphs of a compound (e.g. $UO_2(OH)_2$).

'Hydrates' of UO_3 include $UO_3.\frac{1}{2}H_2O$, $UO_3.H_2O$, and $UO_3.2H_2O$. The structure of the 'hemihydrate' is not known. There appear to be three forms of the 'mono-hydrate', which is actually $UO_2(OH)_2$, and possibly two of the 'dihydrate' $(UO_2(OH)_2.H_2O)$. The latter has been studied as the mineral schoepite; it consists of $UO_2(OH)_2$ layers with H_2O molecules interleaved between them. The type of layer is similar to that in α-$UO_2(OH)_2$. The structures of all three polymorphs of $UO_2(OH)_2$ have been determined. The β form consists of layers of octahedral $UO_2(OH)_4$ groups which share all four equatorial OH groups to form a layer based on the simplest 4-connected planar net (Fig. 28.8(a)). The H atoms have been located in both α and β forms of $UO_2(OH)_2$ by neutron diffraction, confirming the formation of O—H—O bonds (2.76 and 2.80 Å in α and β respectively) between the layers. The structure of γ-$UO_2(OH)_2$ is very similar to that of the β form. The β form of $UO_2(OH)_2$ is very easily converted by slight pressure to the α form (with increase in density from 5.73 to 6.73 g/cm^{-3}), in which U has 6 OH neighbours in a plane approximately perpendicular to the UO_2 groups, the layer having the form of Fig. 28.8(b).

In $U_3O_8(OH)_2$ there is both octahedral and pentagonal bipyramidal coordination of U, in each case with two short U—O bonds (1.83 and 1.76 Å respectively). The positions of the H atoms in the 7-coordination groups have been confirmed by a n.d. study of $U_3O_8(OD)_2$.

The 'tetragonal' layer of Fig. 28.8(a) and the 'hexagonal' layer of Fig. 28.8(b), or its more symmetrical variant (c), form the bases of the structures of a number of uranyl compounds and uranates.

Crystalline UO_2F_2 consists of layers of the type shown in Fig. 28.8(c), in which linear UO_2 groups are perpendicular to the plane of the paper and the U atoms also

TABLE 28.8
Coordination of U *in uranyl compounds*

Compound	U$-$2O in UO$_2$ group	Additional neighbours	Reference
α-UO$_2$(OH)$_2$	1.71 Å	6 OH, 2.48 Å	AC 1971 **B27** 1088
β-UO$_2$(OH)$_2$	1.81	4 OH, 2.30	AC 1970 **B26** 1775;
			AC 1971 **B27** 2018
γ-UO$_2$(OH)$_2$	1.81	4 OH, 2.30	AC 1972 **B28** 3469
UO$_2$(OH)$_2$.H$_2$O			AM 1960 **45** 1026
U$_3$O$_8$(OH)$_2$	1.83	4 O	AC 1972 **B28** 117 (X-ray)
	1.76	3 O, 2 OH	AC 1974 **B30** 151 (n.d.)
UO$_2$F$_2$	1.74	6 F, 2.43	AC 1970 **B26** 1540 (n.d.)
UO$_2$Cl$_2$	See Table 28.4		
β-UO$_2$SO$_4$	1.77	5 O, 2.38	AC 1978 **B34** 3734
UO$_2$CO$_3$	(1.7)	6 O of CO$_3^{2-}$, 2.49 Å	AC 1955 8 847
UO$_2$(NO$_3$)$_2$.6H$_2$O	1.76	2 H$_2$O, 4 O of NO$_3^-$	AC 1965 **19** 536 (n.d.)
UO$_2$(NO$_3$)$_2$.2H$_2$O	1.76	2 H$_2$O, 4 O of NO$_3^-$	IC 1971 **10** 323 (n.d.)
Rb(UO$_2$)(NO$_3$)$_3$	1.78	6 O of NO$_3^-$, 2.48 Å	AC 1965 **19** 205
UO$_2$(HCOO)$_2$.H$_2$O	1.76	5 O, 2.41	AC 1977 **B33** 1848
UO$_2$(C$_2$O$_4$).3H$_2$O	1.63	5 O, 2.49	AC 1972 **B28** 3178
UO$_2$(C$_2$O$_4$)$_3$(NH$_4$)$_4$	1.69	6 O, 2.43, 2.57	JCS D 1973 1610
UO$_2$(C$_2$O$_4$)$_2$(NH$_4$)$_2$	1.77	5 O, 2.37	JCS D 1973 1614
(UO$_2$)$_2$(C$_2$O$_4$)$_3$(NH$_4$)$_2$	1.76	5 O, 2.38	JCS D 1973 1616
(UO$_2$)$_2$(C$_2$O$_4$)$_3$K$_2$.4H$_2$O	1.76	6 O, 2.45	AC 1975 **B31** 2277
(UO$_2$)$_2$(C$_2$O$_4$)$_5$K$_6$.10H$_2$O	1.82	5 O, 2.38	AC 1976 **B32** 2497
Na(UO$_2$)(OOC.CH$_3$)$_3$	1.71	6 O, 2.49	AC 1959 **12** 526
Cs$_2$(UO$_2$)(SO$_4$)$_3$	(1.74)	5 O, 2.37–2.47	JINC 1960 **15** 338
[UO$_2$(CH$_3$COO)$_2$(ϕ_3PO)]$_2$	(1.78)	5 O, 2.36	IC 1969 8 320
NpO$_2$F$_2$			AC 1949 **2** 388
KAmO$_2$F$_2$			JACS 1954 76 5235
KAmO$_2$CO$_3$			IC 1964 **3** 1231

have 6 rather more distant neighbours forming a puckered hexagon (or flattened octahedron). There are no primary bonds between the layers, and a characteristic feature of this structure is the presence of stacking faults, which have been studied in a later n.d. investigation (see Table 28.8). NpO$_2$F$_2$ has a similar structure. If U(vi) in this layer is replaced by Am(v) we have the 2D ion in K(AmO$_2$F$_2$), in which K$^+$ ions are accommodated between the layers. The structures of UO$_2$Cl$_2$ and UO$_2$Cl$_2$.H$_2$O have been described on p. 1257 (references in Table 28.4).

Two studies of UO$_2$(NO$_3$)$_2$.6H$_2$O give essentially similar structures showing that this is a tetrahydrate of the complex (a), in which two H$_2$O molecules and two bidentate NO$_3^-$ ions form a planar equatorial group of 6 O around U. This is the structural unit in the dihydrate, in which very similar U$-$O distances were found (U$-$2O, 1.76 Å; U$-$OH$_2$, 2.45 Å; U$-$ONO$_2$, 2.50 Å). A somewhat similar coordi-

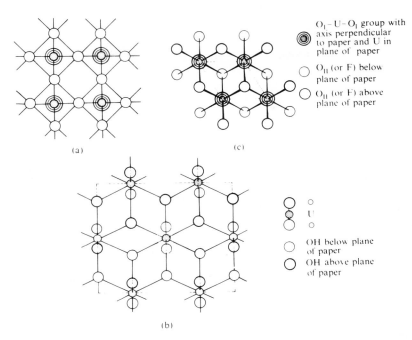

FIG. 28.8. Forms of $UO_2(O_2)$ layers: (a) tetragonal layer in β-$UO_2(OH)_2$ and $BaUO_4$, (b) hexagonal layer in α-$UO_2(OH)_2$, (c) hexagonal layer in UO_2F_2 and $CaUO_4$. In (a) and (c) O—U—O is perpendicular to the plane containing the U atoms; in (b) it is inclined at 70° to the plane of the U atoms.

nation of U, but with three bidentate NO_3^- ions, is found in $Rb(UO_2)(NO_3)_3$ (b). A n.d. redetermination of the structure of $UO_2(NO_3)_2 \cdot 2H_2O$ gives the same U—O bond length (1.76 Å) in the uranyl group as the n.d. study of the hexahydrate.

The acetate ion, $H_3C.COO^-$ (ac), can behave like NO_3^- as a bidentate or bridging ligand. Thus we find $UO_2(ac)_2L_2$ with the octahedral structure (a), in which H_2O has been replaced by L (ϕ_3PO or ϕ_3AsO), and $[UO_2(ac)_2L]_2$ with the bridged structure of type (c), containing both bridging and bidentate acetate groups. The

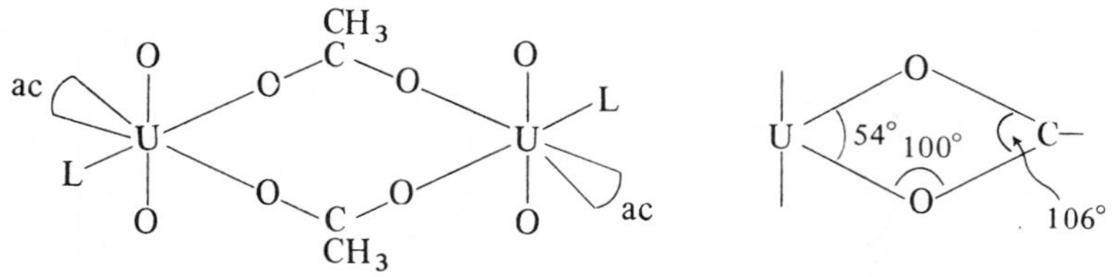

(c)

geometry of the bidentate CH_3COO^- group is very similar to that of the bidentate nitrate ion (p. 822).

There are numerous complex oxalates containing the UO_2 group, some of which are notable for their low solubility in water and their relatively high stability. These compounds are numerous because the $C_2O_4^{2-}$ ligand can behave in a variety of ways, sometimes in more than one way in the same compound:

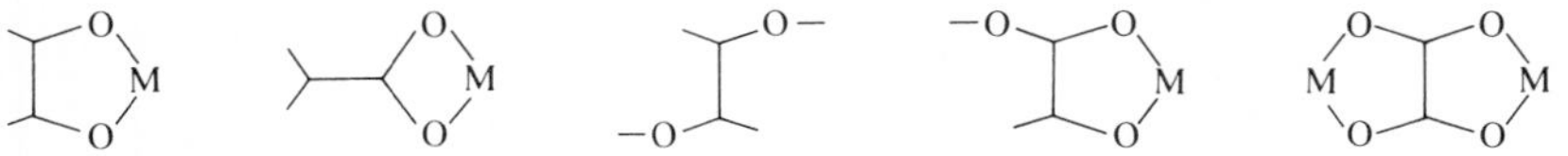

and also the coordination group of U may be pentagonal or hexagonal bipyramidal. Assuming that U—O for bonds additional to those in the uranyl group is close to 2.4 Å, and assuming a minimum non-bonded O—O separation of 2.7-2.8 Å, the UO_2 group can accommodate 5 additional coplanar O atoms around the equator to form a pentagonal bipyramidal group or 6 O atoms if some or all belong to bidentate ligands such as NO_3^- or CH_3COO^-, in which O—O is smaller (2.1-2.2 Å). In the oxalato group there is, of course, a greater distance (2.7 Å) between the O atoms attached to different C atoms, and in the anion in $(NH_4)_4[UO_2(C_2O_4)_3]$ there are two different kinds of bidentate oxalato groups, (d). (Alternatively, 6 O atoms can be accommodated if they are alternately above and below the equatorial plane, as in $CaUO_4$, Fig. 28.8(c).) Simple examples of uranyl oxalato complexes include the finite mono- and binuclear ions (d) and (e) and the chains (f) and (g). For a $C_2O_4:UO_2$ ratio of 3:2 3-connected nets can be formed, as in the double chain (ladder) ion (h) and the planar 6^3 layer (i) formed from UO_2 groups each bonded to 3 tetradentate oxalato ligands.

Uranates and complex oxides of uranium. From their empirical formulae uranates would appear to be normal ortho- and pyro-salts containing UO_4^{2-} or $U_2O_7^{2-}$ ions. However, all the known uranates (alkali, alkaline-earth, Ag) are insoluble in water, including $Na_2U_2O_7.6H_2O$—contrast the soluble alkali chromates, molybdates, and tungstates. In fact the alkali and alkaline-earth uranates $M_2^IUO_4$ and $M^{II}UO_4$ contain infinite 2D ions of the same general types as the tetragonal or hexagonal layers of Fig. 28.8:

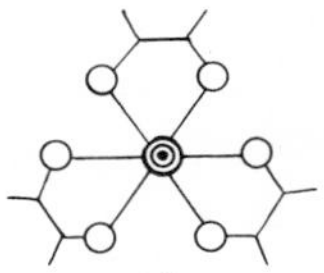

(d)

$[UO_2(C_2O_4)_3]^{4-}$ $(NH_4)_4$

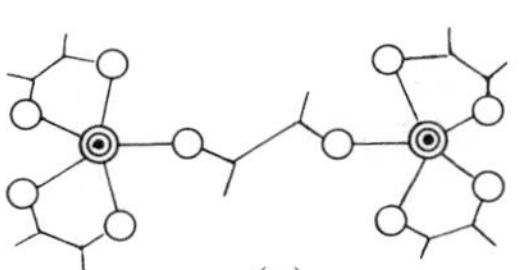

(e)

$[(UO_2)_2(C_2O_4)_5]^{6-}K_6.10H_2O$

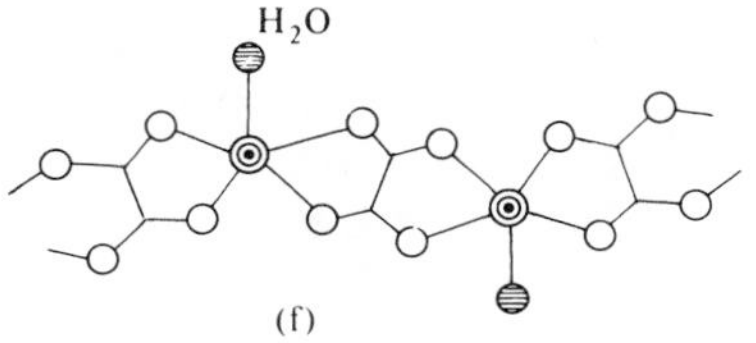

H_2O

(f)

$UO_2C_2O_4.3H_2O$

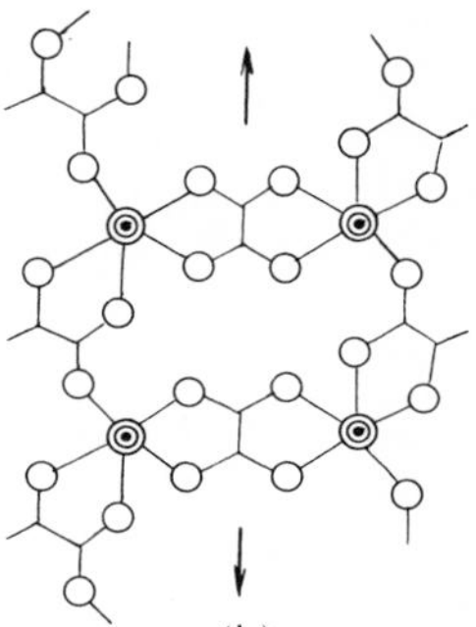

(h)

$[(UO_2)_2(C_2O_4)_3]_n^{2n-}$ $(NH_4)_2$

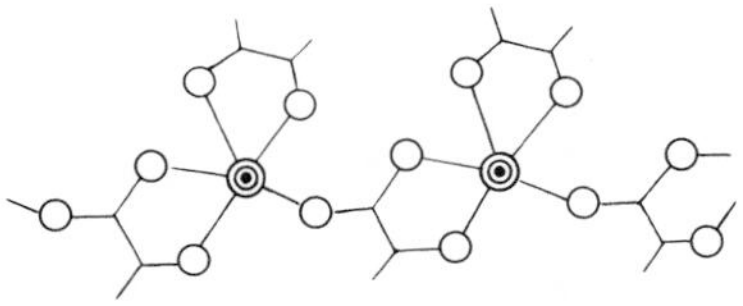

(g)

$[UO_2(C_2O_4)_2]_n^{2n-}$ $(NH_4)_2$

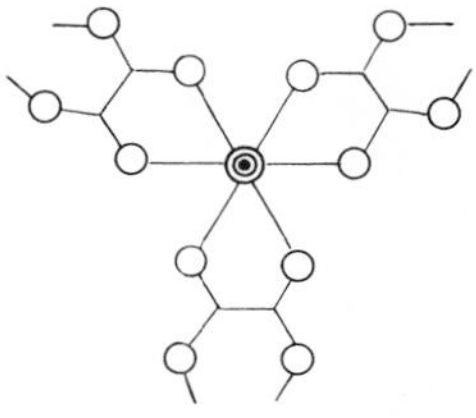

(i)

$[(UO_2)_2(C_2O_4)_3]_n^{2n-}$ $K_2.4H_2O$

hexagonal layer: Na_2UO_4, K_2UO_4,[1] $CaUO_4$[2] (and rhombohedral $SrUO_4$)

tetragonal layer: $SrUO_4$ (rhombic), $BaUO_4$.[2]

In these 'uranates' there are well defined UO_2 groups, but this is not a feature of all complex oxides of U^{VI}, as may be seen from the following figures which show (in parentheses) the number of O neighbours of U at the particular distance:

$CaUO_4$: 1·96 (2)	2·30 (6)	Ca_2UO_5: U_I 2·02 (4), 2·25 (2) U_{II} 1·95 (2), 2·13 (2), 2·21 (2)
$SrUO_4$: 1·87 (2) $BaUO_4$: 1·89 (2)	2·20 (4) 2·20 (4)	Ca_3UO_6: 2·03 (3), 2·11 (2), 2·18 (1)

The distinction between two close and the remaining neighbours can still be detected for one-half of the U atoms in Ca_2UO_5 but it has disappeared entirely in Ca_3UO_6.

There are equally striking differences between the environments of U in other complex oxides of U^{VI}. In $UTeO_5$ there are $PaCl_5$-like chains of composition UO_5 bonded through Te^{IV} atoms in characteristic 'one-sided' 4-coordination.[2a] The length of the axial U—O bonds in the pentagonal bipyramidal coordination groups is 1.83 Å and the mean length of the five equatorial bonds is 2.39 Å. On the other hand, the high-temperature form of Bi_2UO_6,[2b] which is a catalyst for the conversion of toluene to benzene, has a more typical complex oxide structure which may be described as a distorted hexagonal superstructure of fluorite. The length of the axial bonds is 2.1 Å and the six nearly coplanar equatorial bonds have a length of 2.4 Å.

In contrast to the 2D ions in $SrUO_4$ and $BaUO_4$ there is linking of quite similar distorted UO_6 octahedra into infinite chains in the rutile-type superstructure of $MgUO_4$: U—2O, 1.92 Å; U—4O, 2.18 Å.[3a] This formation of two stronger bonds by U leads to a similar, and unusual, unsymmetrical environment of Mg: Mg—2O, 1.98 Å; Mg—4O, 2.19 Å.

The 3D framework of $UO_2(O_4)$ octahedra which forms the anion in BaU_2O_7[3b] has been described and illustrated in Chapter 5 and Fig. 5.28.

In all the uranyl compounds and in some of the uranates U has two nearest O neighbours and from four to six next nearest neighbours:

$$4 \text{ in } BaUO_4, MgUO_4$$
$$5 \text{ in } K_3UO_2F_5, Cs_2(UO_2)(SO_4)_3$$
$$6 \text{ in } CaUO_4, UO_2F_2.$$

Values of the short U—O distance range from around 1.7 Å to 1.96 Å, and the longer U—O distances are usually in the range 2.2–2.5 Å. A bond strength (number) s may be assigned to each U—O bond such that Σs = valence (i.e. 2 for Ba, 6 for U, and so on), the bond strength being derived from an equation of the type suggested by Pauling for metals: $D_1 - D_s = 2k \log s$, where s is the bond strength, D_1 the single bond length, D_s the bond length, and k a constant (here 0.45).[4] Bond strengths estimated in this way are, for example: 1.75 Å ($s = 2$); 2.05 Å ($s = 1$); and 2.45 Å ($s = 0.33$). Although these bonds strengths correspond to electrostatic bond strengths in ionic crystals it is not implied that the bonds are simple ionic bonds. Where U has two close O neighbours and a group of 4, 5, or 6 additional equatorial neighbours the geometry of the coordination groups suggests that the bonds have some covalent character. The six next nearest neighbours of U in UO_2F_2 and $CaUO_4$ are only 0.5 Å above or below the equatorial plane (perpendicular to the axis of the UO_2 group); in $Rb(UO_2)(NO_3)_3$ they are actually coplanar, being three pairs of O atoms of bidentate NO_3^- ions. A satisfactory description of the bonds in these crystals cannot yet be given, but the possibility of f hybridization in the formation of 2+6 bonds has been discussed.[5]

(1)   AC 1948 **1** 281   (2b) AC 1975 **B31** 127   (4)   AC 1954 7 795
(2)   AC 1969 **B25** 787   (3a) AC 1954 7 788   (5)   JCS 1956 3650
(2a) AC 1973 **B29** 1251   (3b) JINC 1965 27 1521

Nitrides and related compounds of Th *and* U; *the* La_2O_3 *and* Ce_2O_2S *structures*

Both Th and U form 'interstitial' nitrides MN (and also MC and MO, all with the NaCl structure). In addition Th forms the typical Group IV nitride Th_3N_4, while U forms UN_2 (fluorite structure) and U_2N_3 (Mn_2O_3 structure).[1] Closely related to the nitrides are compounds such as Th_2N_2O and U_2N_2S which are members of a large group of compounds M_2Y_2X formed by 4f and 5f metals; some examples are set out in Table 28.9.

TABLE 28.9

Compounds with the La_2O_3 (Ce_2O_2S) *and* U_2N_2X *structures*

		La_2O_2O	$Th_2N_2(NH)^{(6)}$	$Th_2N_2O^{(5)}$		
	$Ce_2O_2S^{(1)}$	S	$Th_2(N, O)_2P^{(3)}$	S	$U_2N_2P^{(3)}$	U_2N_2S
	Se		As	Se	As	Se
$Ce_2O_2Sb^{(2)}$	Ce_2O_2Te	La_2O_2Te	$Th_2N_2Sb^{(4)}$	Th_2N_2Te	$U_2N_2Sb^{(4)}$	$U_2N_2T_e$
Bi		Nd_2O_2Te	Bi		Bi	

(1) AC 1949 **2** 60. (2) AC 1971 **B27** 853. (3) AC 1969 **B25** 294. (4) AC 1970 **B26** 823. (5) AC 1966 **21** 838. (6) ZaC 1968 **363** 245.

The compounds in which X = O, S, Se, P, or As have the A-M_2O_3 or La_2O_3 structure, while those containing larger X atoms (Te, Sb, or Bi) have a different structure, which is also that of $(Na_{1/4}Bi_{3/4})_2O_2Cl$. In both structures there is a rigid layer of composition MY (M_2Y_2) built of YM_4 tetrahedra, which in the La_2O_3 structure share 3 edges (Fig. 28.9(a)) and in the U_2N_2X structure, (b), share 4 edges. The latter is the same layer as in tetragonal PbO or LiOH. In the La_2O_3 structure the third O (X) is situated between the layers surrounded octahedrally by 6 M, while the M atom (on the surface of the 'layer') has 4 O (Y) neighbours within the layer and 3 O (X) neighbours between the layers, as shown at (c). The O (Y) atoms within the layers have 4 tetrahedral M (La) neighbours. This La_2O_3 (Ce_2O_2S) structure of $7:\frac{3}{4}$ coordination is adopted by a number of 4f sesquioxides M_2O_3 and oxysulphides M_2O_2S, by Ac_2O_3, Pu_2O_3, β-Am_2O_3, Pu_2O_2S, and also by β-Al_2S_3. In La_2O_3 there is a difference of about 0.3 Å between the distance La–4O within the tetrahedral layer and La–3O between the layers; the corresponding difference is, of course, much greater in the U compounds, where Y = N and X = P, As, S, or Se, and Ce_2O_2S, where Y = O and X = S:

M_2Y_2X	M—4 Y	M—3 X
La_2O_2O	2.38–2.45 Å	2.72 Å
U_2N_2S	2.28	2.87
U_2N_2As	2.27	2.97
Ce_2O_2S	2.36	3.04

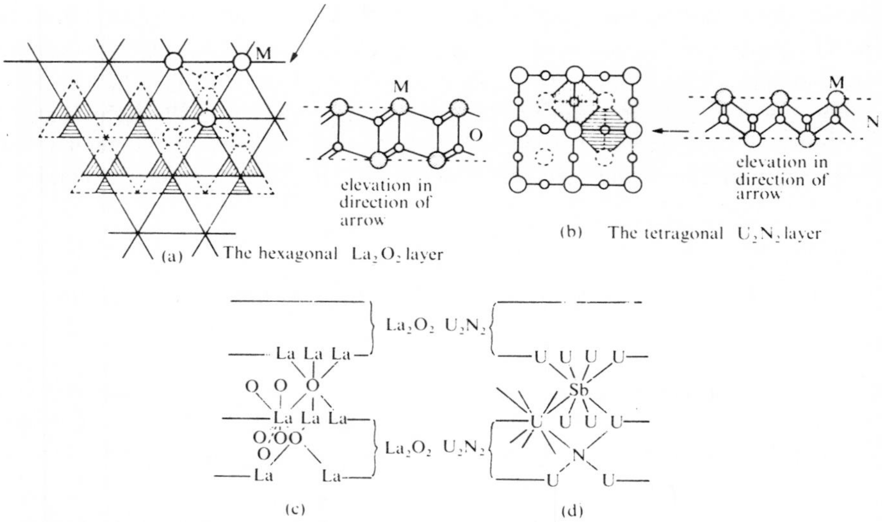

FIG. 28.9. Layer of tetrahedra in (a) $(La_2O_2)O$, (b) U_2N_2X. The diagrammatic elevations (c) and (d) show the coordination of the two types of O atom in La_2O_3 and of N and Sb in U_2N_2Sb.

In the U_2N_2X structure the tetragonal layers (Fig. 28.9(b)) stack so as to provide positions of 8-coordination (cubic) for the larger X atoms (Sb, Bi, Te) between them, and the coordination groups of the atoms are:

$$U \quad : \quad 4\,N + 4\,Sb \text{ (cubic)}$$
$$N \quad : \quad 4\,U \text{ (tetrahedral)}$$
$$Sb \quad : \quad 8\,U \text{ (cubic)}$$

as shown diagrammatically in Fig. 28.9(d).

(1)   IC 1965 **4** 115

Sulphides of U, Th, *and* Ce

Most of these compounds are very refractory materials melting at temperatures of 1800 °C or higher (CeS melts at 2450 ± 100 °C), and some are highly coloured: CeS, brass-yellow; Ce_2S_3, red; but Ce_3S_4, black; the uranium compounds are all metallic-grey to black in colour. The following compounds are known:

CeS	ThS	US
Ce_2S_3	Th_2S_3	U_2S_3
	ThS_2	US_2
Ce_3S_4	Th_7S_{12}	
	$(ThS_{1.71-1.76})$	

Although the monosulphides all have the NaCl structure they are evidently not simple ionic crystals. ThS is, like all the thorium sulphides, diamagnetic, but the susceptibilities of the paramagnetic CeS and US indicate respectively 1 and 2 unpaired electrons. Assuming that electrons not used for bonding would remain unpaired in the orbitals available and that, judging from the magnetic properties of the transition metals, d electrons will pair to form metallic bonds, the paramagnetism may be attributed to unpaired f electrons. We then have the following picture:

CeS	ThS	US	Ce_2S_3	Th_2S_3	U_2S_3	
1	2	2	–	1	2	d electrons used in metallic bonding
1	–	2	1	–	1	unpaired f electrons

Calculated lattice energies (using estimated ionization potentials where necessary) increase in the order BaS, CeS, ThS, that is, as the number of d electrons available for metallic bonding increases from 0 to 1 to 2. The bond lengths M–S show a decrease in the same order in both MS and M_2S_3:

	BaS	CeS	ThS	(US)
M–S	3.17	2.88	2.84	2.74 Å

	(La_2S_3)	Ce_2S_3	Th_2S_3	U_2S_3
	(3.01)	2.98	2.90	2.82 Å

whereas Th is normally larger than Ce. The comparison is perhaps less justifiable for the sesquisulphides because they do not all have the same structure.[1] Those of Th, U, and Np have the Sb_2S_3 structure, but the sesquisulphides of La, Ce, Ac, Pu, and Am have a defect Th_3P_4 structure. The Th_3P_4 structure is a 3D array of Th and P atoms in which Th is bonded to 8 equidistant P and P to 6 Th. In these sesqui-sulphides $10\frac{2}{3}$ metal atoms are distributed statistically over the 12 metal sites occupied in Th_3P_4; in Ce_3S_4 all the metal sites are occupied.

The sulphide Th_7S_{12} has a disordered structure[2] in which Th is surrounded by either 8 or 9 S, and ThS_2, like β-US_2, has the $PbCl_2$ structure.

(1) For references see Table 28.2 (p. 1252) (2) AC 1949 **2** 288

29

Metals and alloys

The structures of the elements

We have already mentioned the structures of a number of non-metals in earlier chapters. In this concluding chapter we shall be concerned chiefly with the structures of the metallic elements and intermetallic compounds. We shall confine our attention to metals in the solid state and generally to the forms stable under atmospheric pressure. High-pressure polymorphs, which are numerous, will be mentioned only if they are of special interest. Not very much is known of the structures of liquid metals; in the few cases in which diffraction studies have been made the structural information is limited to the average numbers of neighbours within particular ranges of distance, and typical numbers of nearest neighbours are: liquid K (70 °C), 8; Li (200 °C), 9.8; Hg and Al, 8–9,[1] The existence of diatomic molecules in the vapours of a number of metals has been demonstrated and their dissociation energies determined; they are noted in the discussion of metal–metal bonds in Chapter 7 (p. 305). Diatomic molecules have also been detected in the vapours above certain metal solutions, for example, GeCr and GeNi.[2] Before proceeding to the inter-metallic systems we shall review briefly the structures of all the elements in the solid state as far as they are known. For this purpose it is convenient to consider them in groups as indicated below.

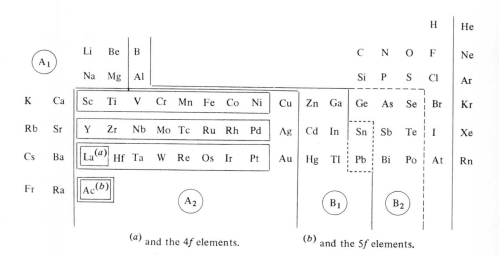

(a) and the 4f elements. (b) and the 5f elements.

The groups are:

(1) The noble gases.
(2) Hydrogen, the non-metals of the first two short Periods (excluding B) and the later B subgroup elements.
(3) Boron, aluminium, the elements of subgroups IIB and IIIB, Sn, and Pb.
(4) The transition elements and those of IB.
(5) The typical and A subgroup elements of Groups I and II.

We shall discuss the structures of intermetallic compounds in terms of the groups A_1, A_2, B_1, and B_2 (see above), of which A_1 and A_2 comprise the A subgroup elements together with the Group VIII triads and Cu, Ag, and Au, while B_1 and B_2 include the more metallic B subgroup elements apart from those of IB. The reason for placing these latter elements with the transition metals rather than with the B subgroup elements will be apparent later.

(1)   AC 1965 **19** 807   (2)   JCP 1968 **49** 3579

(1) *The noble gases*

All these elements except radon have been studied in the crystalline state; they are monatomic in all states of aggregation. Details are given in Table 29.1. Helium is unique among the elements in that it forms a true solid only under pressure, a minimum of 25 atm being necessary. The predominance of the c.c.p. structure is

TABLE 29.1
Crystal structures of the noble gases, $H_2, N_2, O_2,$ *and* F_2

Element	Crystalline forms	Reference
He	Three polymorphs of each isotope (^{3}He and ^{4}He): b.c.c. → h.c.p. → c.c.p. with increasing pressure.	PRL 1962 **8** 469
Ne	c.c.p.	
Ar	c.c.p. and h.c.p. (metastable)	JCP 1964 **41** 1078
Kr	c.c.p.	
Xe	c.c.p.	
For calculations of lattice energies see: JCP 1964 **40** 2744		
H_2	Normally h.c.p. but also f.c.c. under special conditions of crystallization	JCP 1966 **45** 834
N_2	α cubic	AC 1974 **B30** 929
	β hexagonal	JCP 1964 **41** 756
	γ tetragonal	JCP 1970 **52** 6000
O_2	α monoclinic	JCP 1967 **47** 592
	β rhombohedral	AC 1962 **15** 845
	γ cubic	AC 1964 **17** 777
	Also α' and amorphous	AC 1969 **B25** 2515
F_2	α monoclinic (below 45.6 K)	JSSC 1970 **2** 225
	β cubic	JCP 1964 **41** 760

not reconcilable with calculations of lattice energies, which suggest that h.c.p. should be preferred.

(2) *Non-metals and the later* B *subgroup elements*

Crystalline hydrogen, oxygen, and nitrogen consists of diatomic molecules. Solid H_2 has a h.c.p. structure in which the rotation of the molecules probably persists at temperatures down to the absolute zero. In the low-temperature (α) form of N_2 the centres of the molecules are situated at the points of a f.c.c. lattice, but each molecule points in a different [111] direction. The γ form is a high-pressure polymorph which has been studied under a pressure of 4015 atmospheres at 20.5 K. The (isostructural) high-temperature forms of O_2 and F_2 have an interesting cubic structure with 8 molecules in the unit cell, of which 2 are spherically disordered while the remaining 6 behave as oblate spheroids. A spherically disordered molecule (A in Fig. 29.1) has 12 F_2 neighbours at 3.7 Å, while a molecule of type B has 2 neighbours at 3.3 Å, 4 at 3.7 Å, and 8 at 4.1 Å. Van der Waals distances between (stationary) F_2 molecules would be expected to be 2.7 Å for lateral contacts and approximately 4.1 Å for end-to-end contacts. The shortest intermolecular contacts are within 'chains' of molecules running parallel to the cubic axes. This structure is clearly not consistent with any form of dimer, as at one time suggested for γ-O_2.

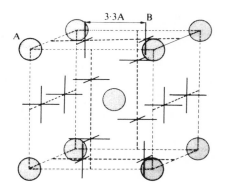

FIG. 29.1.   The crystal structure of the high-temperature form of F_2. The two molecules at (000) and (½½½) are spherically disordered. The other six, at (¼½0) etc., show an oblate spheroidal distribution of electron density. These molecules are represented by their major axes, the minor axes lying along the broken lines.

The low-temperature (α) form of F_2 apparently consists of a distorted closest packing of molecules in which each F *atom* has 12 nearest neighbours in addition to the other atom of its own molecule (at 1.49 Å). The molecular axes are nearly perpendicular to the plane of a layer and the interatomic distances are greater within a molecular layer (F–9F, 3.24 Å) than between the layers (F–3F, 2.84 Å). Contrast the isostructural Cl_2, Br_2, and I_2 (p. 388).

As noted in Chapter 3 many of the elements of this group crystallize at atmospheric pressure with structures in which an atom has 8-N nearest neighbours, N being the ordinal number of the Periodic Group:

VII diatomic molecules F_2, etc.;

VI rings (S_6, S_8, S_{12}, Se_8) or chains (fibrous S, 'metallic' Se and Te);

V tetrahedral molecules in white P and presumably in the metastable yellow forms of As and Sb, or layer structures (P, red and black, As, Sb, and Bi);

IV diamond structure (C, Si, Ge, grey Sn).

The structures of these elements are described in earlier chapters. The graphite structure is obviously exceptional, and the heavier elements in the subgroups IVB–VIB become increasingly metallic. In IVB Sn also forms the more metallic white polymorph, and Pb is a metal. In VB there is a progressive decrease in the difference between the distances to the 3 nearest and 3 next-nearest neighbours in the layer structures of As, Sb, and Bi, while in VIB the interatomic distances between the chains in Se and Te indicate interactions between the chains stronger than normal van der Waals bonding (p. 285). The last member of the VIB subgroup is a metal. An X-ray study of metallic Po (in which a powder photograph was obtained from 100 μg) showed this metal to be dimorphic, one form being simple cubic and the other rhombohedral.[1] In each form Po has only 6 (equidistant) nearest neighbours (at 3.37 Å).

(1) JCP 1957 **27** 985

(3) *Aluminium, the elements of subgroups* IIB *and* IIIB, *tin, and lead*

These elements are grouped together because: (i) certain of them have peculiar structures which do not conform with the principles governing those of the elements in the adjacent groups in our classification; and (ii) those which have typical metallic structures show certain abnormalities which make it convenient to distinguish them from the 'true' metals of our fourth and fifth groups. Of the IIB subgroup elements zinc and cadmium crystallize with a distorted form of hexagonal closest packing, the axial ratio being nearly 1.9 instead of 1.63 as in perfect hexagonal closest packing. Each atom has six nearest neighbours in its own plane, the other six (three above and three below) being at a rather greater distance. The distances to these two sets of neighbours are 2.659 and 2.906 Å in Zn and 2.973 and 3.286 Å in Cd. It is not satisfactory simply to regard the atoms in these structures as packing like prolate ellipsoids, for in the close-packed ϵ phase of Cu–Zn (p. 1311) the axial ratio is *less* than 1.63, that is, we should have to assume that the atoms are behaving at some compositions as prolate and at others as oblate spheroids. It has been suggested that there is some tendency to form covalent bonds (dp^2 hybrids) in the basal planes, so accounting for the smaller interatomic distances in the close-packed planes. The striking decrease in the axial ratio of $MgCd_3$ from 1.72 to 1.63 on forming the ordered superstructure at 80 °C (this superstructure consists of identical close-packed layers of type (d), Fig. 4.15) has been attributed to the decrease in the number of Cd—Cd contacts on formation of the ordered structure. (There is presumably also some kind of directional bonding in Sc, Y, and the hexagonal rare-earths with axial ratio *less* than 1.63, for example, $c:a \approx 1.57$ for Y, Ho, Er, and Tm.)

The normal (α) form of crystalline Hg has a quite different (rhombohedral) structure which may be derived from a simple cubic packing by distorting it so that the interaxial angle changes from $90°$ to $70\frac{1}{2}°$. Each atom has 6 nearest neighbours at 2.99 Å and 6 more at 3.47 Å. Alternatively the structure may be described as resulting from compression of c.c.p. along a body-diagonal, so that it is referable to either of the rhombohedral cells (compare Fig. 6.3, p. 243):

$$Z = 1: \quad a = 2.99 \text{ Å}, \quad \alpha = 70\frac{1}{2}°,$$
or
$$Z = 4: \quad a = 4.58 \text{ Å}, \quad \alpha = 98°.$$

A second (β) form is produced under pressure and after formation is the stable form below 79 K. It is b.c. tetragonal and results from compressing c.c.p. along a cube edge, giving each atom 2 nearest neighbours (2.83 Å) and 8 more distant ones (3.16 Å).[1] The structure is also adopted by Cd—Hg alloys containing 37–74 atomic per cent Hg.

Aluminium is cubic close packed at all temperatures under atmospheric pressure; partial conversion under pressure to a h.c.p. polymorph has been reported. It is notable for its low melting point, which is only 8 °C higher than that of Mg, in contrast to the rise in melting point with increase in valence in the first row and A subgroup elements:

Li	Be	B	C
178	1285	2300	3500 °C
Na	Mg	Al	Si
97	650	658	1410
K	Ca	Sc	Ti
64	845	1540	1675

The structures of Ga, In, and Tl are all different. In the normal form of Ga an atom has the following set of neighbours: 1 at 2.47; 2 at 2.70; 2 at 2.74; and 2 at 2.79 Å; three high-pressure polymorphs have been reported, but further work is required on some of their structures.[2] Gallium is remarkable not only for its low melting point (around 30 °C) but also for the association of the atoms in pairs in the normal form of the metal. It is reported that the X-ray diffraction pattern of the liquid is different from that of a simple c.p. liquid metal such as mercury, suggesting that the Ga_2 units may persist in the liquid. Indium crystallizes with a tetragonal structure which is only a slightly distorted form of cubic closest packing, each atom having 4 neighbours at 3.25 Å and 8 more at 3.38 Å. Thallium is h.c.p. (and possibly b.c.c. at high temperatures), while lead is c.c.p. with the h.c.p. structure under high pressure.

Indium, thallium, and lead form a group of close-packed metals isolated from

TABLE 29.2

Interatomic distances in the elements (to nearest neighbours)

Cu	Zn	Ga	Ge	As
2·55	2·66	2·70(a)	2·44	2·51 A
Ag	Cd	In	Sn	Sb
2·88	2·97	3.34(a)	2·80(b) 3.18(a)	2·90
Au	Hg	Tl	Pb	Bi
2·88	2·99	3·40	3·49	3·10

(a) Weighted means for 7 (Ga), 12 (In), and 6 (white Sn) nearest neighbours. (b) Grey tin.

the rest of the close-packed metals. As may be seen from Table 29.2, the interatomic distances in these metals are very large compared with those in the neighbouring elements. Larger distances are to be expected on account of the much higher coordination number, but the distances are much greater than in the close-packed silver and gold, whereas there is very little variation in interatomic distance throughout the whole series of elements from Cu to As or from Ag to Sb, excluding In and white Sn. The figure 2.80 Å given for tin in Table 29.2 is the interatomic distance in grey tin (diamond structure). The normal form of this element is white tin, which has a less symmetrical structure than the grey variety. In white tin (Fig. 29.2) each atom has 4 nearest neighbours at 3.02 Å, forming a very flattened

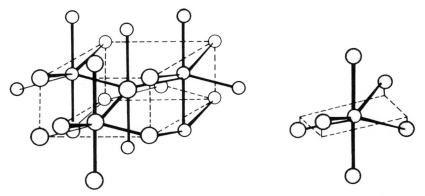

FIG. 29.2. The crystal structure of white tin, showing the deformed octahedral group of neighbours around each atom.

tetrahedron around it, and 2 more at 3.18 Å. Although the four nearest neighbours are at a greater distance than the four nearest neighbours in grey Sn (at 2.80 Å) there are also 4 Sn at 3.77 Å and 8 Sn at 4.41 Å, as compared with the 12 next nearest neighbours at 4.59 Å in grey Sn. White Sn is accordingly much more dense

$(7.31\,\text{g}\,\text{cm}^{-3})$ than grey Sn $(5.75\,\text{g}\,\text{cm}^{-3})$, in spite of the fact that it is the high-temperature form. For the relation of white to grey Sn[3] see also p. 121. The suggestion that Sn is in different valence states in its two polymorphs is supported by the fact that if white and grey Sn are dissolved in concentrated HCl the compounds isolated from the resulting solutions are $SnCl_2 . 2H_2O$ and $SnCl_4 . 5H_2O$ respectively.[4]

Under high pressure (120 kbar) Ge, which is a semiconductor, is converted into a highly conducting form with the white tin structure.[5] On decompression at room temperature this changes into the high-pressure (tetragonal) form illustrated in Fig. 3.41(d), p. 130, which is some 11 per cent more dense that the normal cubic form (5.91 as compared with $5.33\,\text{g}\,\text{cm}^{-3}$). The bond length (2.48 Å) in this form is very similar to that in cubic Ge, but there is a very considerable range of bond angles $(88-135°)$.[6]

It is considered probable that in Tl and Pb, and possibly also in In, the atoms are only partly ionized, for example, to the stages Tl^+ and Pb^{2+} in the former elements. The peculiar stability of the pair of s electrons in the outer shells of these elements has already been noted in Chapter 26.

(1)   JCP 1959 **31** 1628
(2)   AC 1969 **B24** 995; AC 1972 **B28** 1974; AC 1973 **B29** 367
(3)   PRS 1963 A **272** 503
(4)   JCP 1956 **24** 1009
(5)   Sc 1963 **139** 762
(6)   AC 1964 **17** 752

(4)   *The transition elements and those of subgroup* IB

This group of more than 50 elements, including the 4f and 5f elements, comprises most of the metals. Although a few metals, notably Mg, Zn, Cd, Sc, Y, Ru, Rh, Pd, Os, Ir, and Pt, are known in only one crystalline from, most metals undergo structural changes when temperature and/or pressure is changed. Moreover, the structures of special forms of some metals, particularly sputtered or electrodeposited thin films, are still open to doubt. For these reasons Table 29.3 shows only the structures at normal temperature and pressure; the polymorphism of certain elements is noted in the text. With few exceptions metals crystallize with one or more of three structures, the hexagonal and cubic close-packed structures and the body-centred cubic structure.

The two simplest forms of closest packing, hexagonal and cubic, have already been described in Chapter 4. In Fig. 29.3 we show these in their conventional orientations together with the body-centred cubic structure. Sufficient atoms of adjacent unit cells are depicted to show the full set of nearest neighbours of one atom. The body-centred cubic structure is slightly less closely packed than the others, the effective volume per atom (of radius a) being $6.16a^3$ as compared with $5.66a^3$ in close-packed structures. In the latter an atom has twelve equidistant nearest neighbours, the next set of neighbours being 41 per cent farther away.

TABLE 29.3
The crystal structures of metals at room temperature

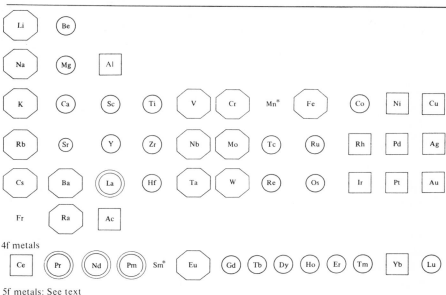

4f metals

5f metals: See text

The primary object of this Table is to emphasize the widespread occurrence of the b.c.c. and the c.p. structures:

*Structure at room temperature is complex and is described in the text.

Certain metals show a more complex behaviour and are further discussed in the text, for example, Mn, W, 4f and 5f metals (Table 29.4, p. 1019).

(a) The electrodeposited h.c.p. structure is now known to be that of the hydride CrH (p. 298).

(b) Normally close-packed *chh* type.

(c) A c.c.p. form of La has been described and an intermediate h.c.p. form of Yb (260°–720°C) is apparently stabilized by impurities. For references to b.c.c. and c.p. metals see: JACS 1963 85 1238.

The body-centred cubic structure is commonly the high-temperature form of metals which are close-packed at lower temperatures. The transition temperature is above room temperature for many metals (Ca, Sr, La, Ce, etc., Tl, Ti, Zr, Th, Mn) but below room temperature for Li and Na (78 K and 36 K respectively). In this structure an atom has 8 equidistant nearest neighbours, but the six next nearest neighbours are only 15 per cent farther away. Neglecting this comparatively small

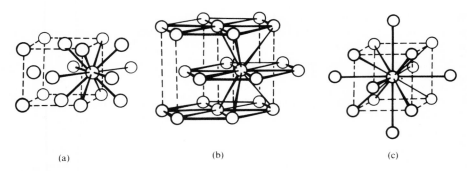

FIG. 29.3. The three commonest metallic structures: (a) cubic close-packing, (b) hexagonal close-packing, and (c) body-centred cubic.

difference, or alternatively, adopting the more logical definition of coordination number based on the *polyhedral domain* of an atom (p. 68), the c.n. in this structure is 14. Note that if this structure is compressed or extended along one of the 4-fold axes to become body-centred tetragonal with axial ratio $\sqrt{\frac{2}{3}}$ or $\sqrt{2}$ respectively, the number of nearest neighbours becomes ten or twelve. (Compare the structure of Pa, p. 143.) The latter case corresponds to the face-centred cubic structure.

Two features of Table 29.3 may be noted. One is the presence of the central group of b.c.c. structures of V, Nb, Ta, Cr, Mo, and W, and, if we include high-temperature forms, of Ti, Zr, and Hf (above 882, 865, and 1995 °C respectively). Ti and Zr also adopt a hexagonal structure at room temperature under high pressure. In this structure one-third of the atoms are 14-coordinated (hexagonal prism capped on the hexagonal faces) and two-thirds are 11-coordinated (trigonal prism capped on all faces). The other feature is that there are similar sequences of structures from Zr to Ag and from Hf to Au but a more complex situation in the 3d series. The complex polymorphism of Mn is described shortly, and Fe is also abnormal in one respect. Normally a particular polymorph is, under a given pressure, stable over one temperature range, but in the case of Fe the f.c.c. structure is stable over a certain temperature range, on both sides of which the b.c.c. structure is stable (p. 1323). Cobalt exhibits a more complex behaviour. At temperatures above 500 °C the structure is cubic close-packed, but on cooling it becomes partly hexagonal The sequence of layers is not, however, *ABAB* . . . or *ABCABC* . . . throughout but a random sequence. Thus in a typical specimen of 'hexagonal close-packed' cobalt at room temperature the probability that alternate planes are similar is less than one; a value 0.90 was found in one specimen.

Manganese. The normal (α) form has a complex cubic structure ($Z = 58$) which is further discussed on p. 1306. In β-Mn (stable between 727 and 1095 °C, but can be quenched to room temperature) there are two kinds of atom with different environments, each set having 12 nearest neighbours at distances from 2.36 Å to 2.67 Å.

At 1095 °C the γ form appears. Quenched material has tetragonal symmetry corresponding to a distorted form of the c.c.p. structure which it possesses at the high temperature. At 1134 °C there is a further transformation to a b.c.c. structure, this form persisting up to the melting point (1245 °C).

Tungsten. This element is normally encountered in the b.c. cubic (α) form, but a second (β) form, described as being stable at temperatures below about 650 °C, has long been recognized. This form has to be prepared at temperatures below 650° C, above which it is irreversibly, though rather sluggishly, converted into α-W, by chemical methods such as reduction of an oxide. The claim that 'β-W' is in fact an oxide W_3O and not a polymorph of the metal (see p. 570) is not universally accepted,[1] for it has been stated[2] that β-W can be prepared with less than 0.01 O atoms per atom of W. In any case it is convenient to refer to the structure of Fig. 29.4(a) as the β-W structure. It is adopted by a number of intermetallic compounds and silicides, including NiV_3, GeV_3, SiV_3, and $SiCr_3$.

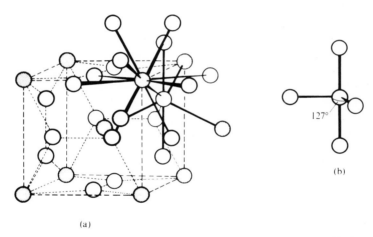

FIG. 29.4. (a) The 'β-tungsten' structure. As regards environment the atoms are of two types. The twelve equidistant neighbours of an atom of the first type (shaded circles) and the six nearest neighbours of an atom of the second type (open circles) are shown. (b) The arrangement of the four nearest neighbours of an atom in metallic uranium (α form).

The 4f (lanthanide or rare-earth) metals. All of these metals except Pm are known to be polymorphic, some with as many as four different structures at various temperatures and pressures. The abnormal chemical properties of the elements immediately preceding Gd and Lu have already been noted in our account of the 4f elements in Chapter 28. In the elementary state also Eu and Yb are abnormal. The majority of the 4f metals crystallize at ordinary temperatures with the *h* or *hc* type of closest packing, though the 9-layer sequence *chh* is the normal structure of Sm.[3] On the other hand Eu is b.c.c. and Yb c.c.p. The metallic radii of the other

4f elements fall steadily from about 1.82 Å (Pr) to 1.72 Å (Lu), but Eu (2.06 Å) and Yb (1.94 Å) have much larger radii, presumably due to the withdrawal of one electron into an inner orbital. The behaviour of Yb under pressure is extremely interesting.[4] At 20 kbar the metal becomes semiconducting, and at 40 kbar it again becomes a metallic conductor. This latter change is accompanied by a structural change (f.c.c. → b.c.c.), that is, the c.n. has altered from 12 to 8+6 with a decrease in volume of 3.2 per cent at the transition point. The radius of Yb in the b.c.c. structure (1.75 Å, corresponding to 1.82 Å for 12-coordination) is considerably less than that of Yb(II) in the normal form (1.94 Å) and lies on the curve for other 4f metals M(III), showing that the transition at the higher pressure corresponds to a change from Yb(II) to Yb(III). This transformation requires higher pressure as the temperature is reduced. With this behaviour contrast that of Ce, which under high pressure changes to another f.c.c. phase, but this change takes place at *lower* pressures as the temperature is reduced—at atmospheric pressure at 109 K. The volume decrease of 16.5 per cent accompanying this change is attributed to the promotion of a 4f electron to a bonding orbital. Both f.c.c. phases can coexist at low temperatures, the cell dimensions at 90 K being 5.14 Å and 4.82 Å.[5]

The 5f (or actinide) metals. The elements between Th (c.c.p.) and Am (close-packed *ABAC*...) show a very complex behaviour in the elementary state. Not only is polymorphism common, Pu having as many as six forms, but a number of the structures are peculiar to the one element. This is true of the body-centred tetragonal structure of Pa (see p. 143), in which an atom has ten practically equidistant neighbours, and of the α-U, β-U, α-Np, β-Np, and γ-Pu structures.

Uranium is trimorphic:

α-U   (660 °C) β-U   (760 °C) γ-U
orthorhombic   tetragonal   b.c.c.

The structure of α-U could be regarded as a very deformed version of hexagonal closest packing in which each atom has only four nearest neighbours, two at 2.76 Å and two at 2.85 Å. The arrangement of these neighbours is very unusual, for they lie in two perpendicular planes at four of the five vertices of a trigonal bipyramid (Fig. 29.4(b)); for further remarks on this structure see p. 1292. β-U apparently has a typical metallic structure with high c.n. (12 or more), but there is not yet agreement as to the details of this structure.

Neptunium shows some resemblance to uranium:

α-Np   (278 °C) β-Np   (540 °C) γ-Np
orthorhombic   tetragonal   b.c.c.(?)

In the α form each atom has only four nearest neighbours. For one-half of the atoms these are arranged as in α-U, but for the others they are arranged at one polar and three equatorial vertices of a trigonal bipyramid. In the β form also each atom forms four strong bonds, but here one half of the atoms have their nearest neighbours at the vertices of a deformed tetrahedron while for the others they lie at the basal vertices of a very flat tetragonal pyramid.

As already mentioned, Pu is notable for having six polymorphs. The α-form has a complex monoclinic structure in which there are 8 kinds of non-equivalent atom, each of which forms a number of short bonds (2.57–2.78 Å) with the remainder in the range 3.19–3.71 Å (compare α-U, α-Np, and β-Np), giving total c.n.s of 12–16. Three-quarters of the atoms form 4 short and 10 long, one-eighth form 3 short and 13 long, and the remainder 5 short and 7 long bonds. The β form also has a complex monoclinic structure (Z = 34, 7 kinds of non-equivalent atom). Coordination numbers range from 12 to 14, and there is no obvious relation to the α or γ structures. In the γ form (Fig. 29.5) Pu has 10 nearest neighbours: 4 at 3.026 Å, 4 at 3.288 Å, and 2

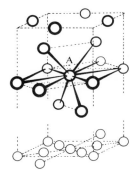

FIG. 29.5. The crystal structure of γ-plutonium showing the atom A and its ten nearest neighbours projected on to the base of the unit cell.

at 3.159 Å. The three high-temperature forms have much simpler structures, (Table 29.4), the δ′ structure being a slightly distorted version of δ.

In two studies of Am metal the *hc* form was found but with rather different densities and cell dimensions (Table 29.4). Although the less dense form was not reproduced in the later study it is conceivable that there are two forms of the metal with the same structure, as in the case of Ce. In the case of Cm also the *hc* structure reported earlier was not found in a later study; here the difference between the densities of two (close-packed) forms is more than a little surprising.

(1) JPC 1956 **60** 1148
(2) ZaC 1957 **293** 241
(3) AC 1970 **B26** 1043
(4) IC 1963 **2** 618
(5) JCP 1950 **18** 145

(5) *The typical and* A *subgroup elements of groups* I *and* II

The metals of this group call for little discussion. The alkali metals adopt the b.c.c. structure under ordinary conditions, but Li can be converted to c.c.p. by cold-working at low temperatures, and both Li and Na convert spontaneously to h.c.p. below 78 K and 36 K respectively. High-pressure forms of all except Na have been reported. The normal form of Be, Mg, Ca, and Sr is h.c.p., but Ba and Ra are b.c.c.

TABLE 29.4
Crystal structures of 5f *metals*

Element		Crystal structure	Reference
Pa		Body-centred tetragonal: c.n. 10, 8 at 3.21 Å, 2 at 3.24 Å	AC 1965 **18** 815
		F.c.c. (c.n. 12 at 3.55 Å)	INCL 1971 **7** 977
U	α	Orthorhombic (A 20 structure) ⎱ see text	AC 1970 **B26** 129
	β	Tetragonal (Z = 30) ⎰	AC 1971 **B27** 1740
	γ	Body-centred cubic (U—U, 3.01 Å)	
Np	α	Orthorhombic ⎫	
	β	Tetragonal ⎬ See text	AC 1952 **5** 660, 664
	γ	Cubic (b.c.?) ⎭	
Pu	α	Monoclinic (c.n. 12–16)	AC 1963 **16** 777
	β	Monoclinic (c.n. 12–14)	AC 1963 **16** 369
	γ	F.c. orthorhombic (c.n. 10)	AC 1955 **8** 431
	δ	F.c. cubic (Pu–Pu, 3.28 Å) ⎱	JCP 1955 **23** 365
	ϵ	B.c. cubic (Pu–Pu, 3.15 Å) ⎰	
	δ'	B.c. tetragonal	TAIME 1956 **206** 1256
Am		F.c.c. and *hc* (density 13.67 g cm⁻³, *r* = 1.73 Å)	JINC 1962 **24** 1025
		hc (density 11.87 g cm⁻³, *r* = 1.82 Å)	JACS 1956 **78** 2340
Cm		F.c.c. (density 19.2 g cm⁻³, *r* = 1.55 Å)	JCP 1969 **50** 5066
		hc (density 13.5 g cm⁻³, *r* = 1.74 Å)	JINC 1964 **26** 271
Bk		F.c.c. (c.n. 12 at 3.53 Å) ⎱	JINC 1971 **33** 3345
		hc (c.n. 12, 6 at 3.416, 6 at 3.398 Å) ⎰	

No polymorphs of Mg are known, but Be, Ca, and Sr transform to b.c.c. at higher temperatures, with c.c.p. stable over an intermediate temperature range (213–621 °C) in the case of Sr. High-pressure transformations have been reported for Be, Ca, and Ba.

Interatomic distances in metals: metallic radii

Apart from the intrinsic interest of the interatomic distances in metals, it is useful to have a set of radii to refer to when discussing the structures of alloys. Since the c.n. 12 is the most common in metals, it is usual to draw up a standard set of radii for this coordination number. For the metals with ideal close-packed structures the radii are simply one-half the distances between an atom and its twelve equidistant nearest neighbours. Many structures, however, deviate slightly from ideal hexagonal closest packing in such a way that six of the neighbours are slightly farther away than the other six, for example:

	Axial ratio	6 at	6 at	Mean
Be	1.5848	2.2235	2.2679	2.25 Å
Y	1.588	3.595	3.663	3.63
α-Zr	1.589	3.166	3.223	3.19

In such cases the mean of the two distances is taken. (We are not here referring to zinc and cadmium which show a very much larger deviation from closest packing, with axial ratios 1.856 and 1.885 respectively.) For metals which crystallize with structures of lower coordination the radii for 12-coordination have to be derived in other ways. From a study of the interatomic distances in many metals and alloys Goldschmidt found that the apparent radius of a metal atom varies with the coordination numbers in the following way. The relative radii for different coordination numbers are:

C.n.	12	1.00
	8	0.97
	6	0.96
	(4	0.88)

For the alkali metals, for example, the radius for c.n. 12 is derived from one-half the observed interatomic distance in the b.c.c. structure by multiplying by 1.03. This ratio, which is equal to $2^{5/6}/3^{1/2}$, results from assuming that no change of volume accompanies the transition from 8- to 12-coordination (see also p. 1284). For some of the B subgroup elements which crystallize with structures of very low coordination (for example, Ge, Sn) the problem is more difficult. The use of a ratio such as that given above is difficult to justify for such a large change in c.n., since the bonds in structures such as the diamond structure are probably nearer to covalent than to purely metallic bonds. There are alternative ways of deriving radii for c.n. 12 for such elements. Many of these elements form alloys with true metals in which both elements exhibit high coordination numbers. The radius of the B subgroup element can be found from the interatomic distance in such an alloy, for example, that of Sb from the Ag–Sb distance in the hexagonal close-packed Ag_3Sb or that of Ge from the Cu–Ge distance in the hexagonal close-packed Cu_3Ge. This method is open to objection since there is probably not true metallic binding in such compounds and it is necessary, of course, to assume that the radius of Ag, for example, in Ag_3Sb is the same as in pure silver. The second method is to derive the radius of the B subgroup element from the variation in cell dimensions of the solid solution it forms in a true metal. By extrapolation, the cell dimension of the hypothetical form of the B subgroup element with c.n. 12, and hence the interatomic distance therein, may be obtained, assuming a linear relation between cell dimension and concentration of solute atoms. Unfortunately this relation is not always linear, thus necessitating the use of empirical relations. Also, it is not certain that the radius of the solute element is not affected by the nature of the solvent metal, for there may be alterations in the state of ionization of an atom in different alloys. Nevertheless it is possible to derive a set of radii for many elements for c.n. 12 which are valuable when considering the structures of alloys. Such a set of radii is given in Table 29.5.

The actual interatomic distances in the elements are plotted against atomic number in Fig. 29.6, which shows a number of points of interest. The general

TABLE 29.5
Metallic radii for 12-coordination (Å)

Li	Be													
1.57	1.12													

Na	Mg	Al												
1.91	1.60	1.43												

K	Ca	Sc	Ti	V	Cr	Mn	Fe	Co	Ni	Cu	Zn	Ga	Ge	
2.35	1.97	1.64	1.47	1.35	1.29	1.37	1.26	1.25	1.25	1.28	1.37	1.53	1.39	

Rb	Sr	Y	Zr	Nb	Mo	Tc	Ru	Rh	Pd	Ag	Cd	In	Sn	Sb
2.50	2.15	1.82	1.60	1.47	1.40	1.35	1.34	1.34	1.37	1.44	1.52	1.67	1.58	1.61

Cs	Ba	La	Hf	Ta	W	Re	Os	Ir	Pt	Au	Hg	Tl	Pb	Bi
2.72	2.24	1.88	1.59	1.47	1.41	1.37	1.35	1.36	1.39	1.44	1.55	1.71	1.75	1.82

4f elements: Ce (1·82)– Lu (1·72) but Eu, 2·06, Yb, 1·94

5f elements: Th   Pa   U   Np   Pu   Am   Cm
                     1.80 1.63 1.56 1.56 1.64 See Table 29.4

periodic variation is apparent, as also is the fact that there is no general increase in the size of atoms with increasing atomic number. For example, Th (a.n. 90) is not much larger than Li (a.n. 3) and actually smaller than Na (a.n. 11), owing to the tighter binding of the electrons in the heavier elements with increased nuclear charge. A second point is that although the transition and B subgroup elements of the second long period are larger than those of the first, there is no similar increase in size from the second to the third long period. Compare, for example, the atomic diameters of Cu, 2.551 Å, Ag, 2.883 Å, and Au, 2.877 Å. We have already referred to the lanthanide contraction and the very close resemblance in chemical properties between such pairs of elements as Zr and Hf or Nb and Ta. The radii of the intervening 4f elements fall from 1.82 Å (Ce) to 1.72 Å (Lu) while the 4f sub-shell is being filled, except that Eu and Yb have much larger radii, as mentioned on p. 1284. The variation of atomic radii in the long periods, with minima in the neighbourhood of the Group VIII triads, corresponds to maxima or minima in a number of physical properties—compressibility, tensile strength, hardness, and melting point. Pauling has suggested that if an element such as iron can utilize its 3d orbitals as well as those of the valence shell in its carbonyls and in the ferrocyanide and other complex ions, this may also be the case in the metal. Assuming that for the elements of the first long period 3d, 4s, and 4p orbitals (nine in number) are available for bond formation, then the maximum number of bonds could be formed when there are nine electrons available per atom. This occurs at Co, Rh, and Ir, for the first, second, and third long periods respectively. In the preceding elements there are fewer available electrons and in the later ones more electrons than orbitals available. Pauling has considered in some detail the nature of the bonds in many metals

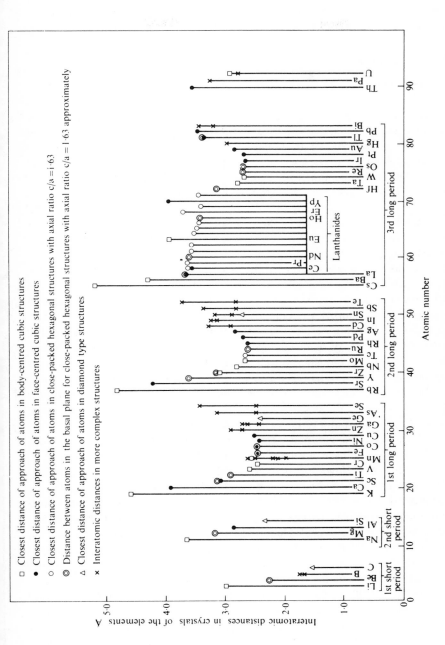

FIG. 29.6. Interatomic distances in metals.

and some alloys in relation to the bond lengths in the crystals, and his theory of resonance in metallic structures will be briefly outlined.

Theories of metallic bonding

In the conventional theory of the metallic state, which has developed from the early electron-gas theories of Drude and Lorentz, the methods of wave mechanics are applied to the behaviour of electrons in a 3-dimensional periodic field, the periodicity of which is that of the crystal lattice. The possible states of the electrons are described in terms of permitted energy bands (zones) which are separated by ranges of forbidden energies, and such a theory gives a satisfactory picture of normal conductors, semiconductors, and insulators. It has had considerable success in calculating properties such as lattice spacings and energies and the compressibilities of a number of metals, assuming only the type of crystal structure (for example, f.c.c.), but it does not offer an explanation of mechanical properties outside the elastic range, for these are dependent on the secondary structure (mosaic structure, dislocations, etc.). We propose to say no more about the wave-mechanical theory of metals here but only emphasize that in the band theory the assembly of electrons is treated as a whole, first in a simple periodic 3-dimensional field, and then later in a field which is modified by supposing that in the vicinity of atoms the wave functions have symmetry characteristics resembling those (the s, p, and d functions) of free atoms. The wave function of each electron extends over the whole crystal, and in this sense the band theory of metals may be compared with the molecular orbital treatment of molecules, though as is well known it is not in practice possible to deal with genuine molecular orbitals. The analogue of the valence-bond treatment would be a theory of metals in which the bonds from a metal atom to its neighbours are considered in terms of the electronic structure of the atom, regarding the whole crystal as an assembly of atoms linked by localized bonds.

Pauling[1] adopts the viewpoint that there is no essential difference between metallic bonds and ordinary covalent bonds, a view originally put forward by V. M. Goldschmidt in 1928. However, there are usually very high coordination numbers in metallic crystals as compared with normal covalent crystals, and, moreover, in a metal such as sodium only four orbitals (one s and three p) are available for forming the 8+6 bonds of the b.c.c. structure. Pauling therefore assumes that all or most of the outer electrons of the atom, including the d electrons in the case of the transition metals, take part in bond formation, and that there is resonance of a new type, to which we refer shortly. These concepts imply fractional bond orders and valences. The drop in atomic diameter from K through Ca, Sc, and Ti to V (and similarly from Rb to Nb and from Cs to Ta) and the approximate constancy of size from the element of Group V through the Group VIII triad in each series of transition metals is interpreted to mean that on passing from K to V the numbers of bonding electrons increase from 1 to 5 per atom, with a regular increase in the number of covalent bonds between which resonance can occur and hence a steady drop in interatomic distance. It is then supposed that in the elements from Cr to Ni not all of the nine available orbitals (s, three p, and five d) are used in bonding, but that only 5.78 are stable spd *bonding* orbitals, 2.44 being *atomic non-bonding* d orbitals and the remaining 0.78 (the *metallic* orbital) being necessary to allow non-synchronized resonance between the individual valence bonds. These numbers are derived from the saturation moments of (ferromagnetic) iron, cobalt, and nickel. Pauling's electronic structures for a number of metals are shown in Table 29.6.

TABLE 29.6
Electronic structures in the metallic state (Pauling)

Metal	Atomic d orbitals		Saturation moment (BM)		Total number of electrons in bonding 3d orbitals	Total number of 3d and 4s electrons
	+	−	Assumed	Observed		
Cr	0.22	0	0.22	−	5.78	6
Mn	1.22	0	1.22	−	5.78	7
Fe	2.22	0	2.22	2.22	5.78	8
	3.22					
Co	2.44	0.78	1.66	1.61	5.78	9
	4.22					
Ni	2.44	1.78	0.66	0.61	5.78	10

For Cr, Mn, and Fe the number of electrons assigned to atomic d orbitals is less than the number of orbitals so that there is no pairing of spins, but in Co, for example, there are 3.22 electrons to be accommodated in 2.44 orbitals, and therefore 0.78 are paired and the moment drops to 2.44−0.78 = 1.66 BM.

This picture of a gradual building up of a core of 3d electrons on passing from Cr to Ni might account for the difference (leading, for example, to superstructure formation in solid solutions) between elements apparently so similar as Fe and Co or Co and Ni. With this may be compared the conventional explanation of the magnetic properties of, for example, Ni, that on the average 9.39 of the ten outer electrons occupy the 3d shell and that the remaining 0.61 electron occupies the 4s shell and is responsible for the cohesive forces, the 'hole' of 0.61 electron in the 3d level giving rise to the magnetic moment of this value (in Bohr magnetons). For Co the cohesion would be due to 0.61 s electron compared with 0.22 for Fe and 1.00 for Cu, corresponding to magnetic moments: Fe, 2.22; Co, 1.61; and Cu, diamagnetic. The difference between normal and metallic resonance may be illustrated by considering grey and white tin. In grey tin the atoms are forming normal tetrahedral sp^3 bonds, using only four of the nine (d^5sp^3) orbitals. There is no 'metallic orbital' available so that the crystals are essentially non-metallic. In white tin it is supposed that the fourteen outer electrons of Sn are in eight of the nine orbitals—six pairs in six (d^5s) hybrids and two electrons in two other orbitals, leaving one 'spare' orbital. Two bonds are then supposed to resonate among six bond positions but not necessarily synchronously because of the existence of the metallic orbital. The resonance involves neutral 2-covalent Sn, 3-covalent Sn^-, and 1-covalent Sn^+, and it is supposed that the crystal is stabilized by this resonance and also by the energy resulting from interchange of one electron between p and s orbitals.

The figures in Table 29.6 imply valences of 5.78 for the transition metals Cr, Mn, Fe, Co, and Ni, and the same is assumed for the series Mo–Pd, W–Pt, and U–Cm. It is less easy to justify the valences of 5.44, 4.44, and 3.44 resepctively for the elements of Groups IB, IIB, and IIIB, which follow from the interpretation of interatomic distances in terms of an essentially empirical equation. This relates the metallic radius $R(n)$ for a bond of order n (i.e. one involving n electron pairs) to the single-bond radius $R(1)$:

$$R(1) - R(n) = 0.300 \log_{10} n \tag{1}$$

and values of $R(1)$ for many metals have been tabulated (Pauling, ref. (1)). The values of the bond-order n are in general non-integral. For a close-packed structure, in which each atom has twelve equidistant neighbours, $n = v/12$, where v is the valence. For example, for the hexagonal close-packed form of Zr, $n = \frac{1}{3}$, and for the body-centred cubic form, $n = \frac{1}{2}$ (neglecting the six more distant neighbours). For more complex structures in which there are various numbers of bonds of different lengths the bonding powers of the atoms are divided up among the different bonds in accordance with the interatomic distances, so that $\Sigma Nn = v$. The application of these ideas to uranium will illustrate the method.

One form of metallic uranium has the body-centred cubic structure, from which $R(1)$ is found to be 1.421 Å, assuming $v = 5.78$. The interatomic distances in the orthorhombic α form then correspond to the following bond orders:

$$
\text{U}
\begin{cases}
N & & n \\
2 \text{ at } 2.76\,\text{Å} & & 1.36 \\
2 \quad\;\; 2.85 & & 0.96 \\
4 \quad\;\; 3.27 & & 0.19 \\
4 \quad\;\; 3.36 & & 0.14
\end{cases}
\Sigma Nn = v = 5.96
$$

This treatment has also been applied to FeSi,[2] with which CrSi, MnSi, ReSi, and CoSi are isostructural. In the FeSi structure each atom has 7 neighbours of the other kind at the distances 2.29 (1), 2.34 (3), and 2.52 (3), and 6 further neighbours of its own kind at 2.75 Å (for Fe) or 2.78 Å (for Si). Application of equation (1) using the $R(1)$ values from elementary iron and silicon gives, however, a calculated valence of 6.85 (ΣNn) for Si, whereas the value 4 is to be expected. In order to retain a valence of 4 for Si it is necessary to neglect the bonding of the Si to 6 Si at 2.78 Å (to which bond equation (1) assigns a bond-order of 0.19), to assume for Fe a valence of 6 instead of the 5.78 previously assumed, and to use different single-bond radii for Fe^{VI} for the Fe–Si and Fe–Fe bonds on the grounds that these (hybrid) bonds have different amounts of d character. More recently it has been pointed out[3] that the Pauling equation is not consistent with the interatomic distances for 12- and 8-coordination in most of the metals which show a transition from a c.p. to the b.c.c. structure, the distances being in the ratio 1.03 : 1, and that it is questionable whether the equation should be used to discuss interatomic distances in intermetallic compounds.

In later developments of the Pauling theory attempts are made to interpret the fractional valences as averages of integral valences, and in addition the hypothesis was advanced that a special stability is to be associated with bond numbers that are simple fractions ($\frac{1}{2}$, $\frac{1}{3}$, etc.). The criticism has been made that the Pauling theory is essentially a discussion of known experimental data in terms of empirical assumptions for which no independent evidence is available, that a large number of arbitrary assumptions have been made, and that no useful generalizations have yet resulted. The need is not for further discussions of structures but for the calculation of physical properties by fundamental methods which do not assume the values of the properties concerned, or, alternatively, the generalization of facts to a number substantially greater than that of the assumptions involved.

Other authors have attempted to develop hybrid bond theories to account for the apparent directional characteristics of metallic bonding. Not only has the concept of varying amounts of d orbitals been used but also the concept of partial occupation of hybrid orbitals.[4] Four tetragonal pyramidal d^4 bonds might be regarded as forming eight cubic 'delocalized' bonds in an electron-deficient system comparable with the π-electron systems of conjugated molecules. Alternatively eight equivalent bonds directed towards the vertices of a cube may arise from tetrahedral sd^3 and tetragonal pyramidal d^4 hybrids, and six longer bonds in the b.c.c. structure from delocalized d^3 trigonal prism hybrids.

(1) JACS 1947 **69** 542; PRS A 1949 **196** 343   (3) JACS 1963 **85** 1238
(2) AC 1948 **1** 212   (4) PRS A 1957 **240** 145

Solid solutions

A characteristic property of metals is that if two (or more) are melted together in suitable proportions a homogeneous solution often results. When cooled this is called a solid solution because, as in the case of a liquid solution, the solute and solvent atoms (applying the term solvent to the metal which is in excess) are arranged at random. Random arrangement of the two kinds of metal atom is always found if the alloy is cooled rapidly (quenched). In certain solid solutions with particular concentrations of solute a regular atomic arrangement develops on slow cooling or appropriate subsequent heat treatment, and we shall describe shortly some of the features of this *superstructure* formation. It was necessary to qualify the first statement in this paragraph because (i) solid solutions are not formed by all pairs of metals, and (ii) when they are, the range of composition over which solid solutions are formed varies from the one extreme, complete miscibility, to the other, immiscibility, depending on the metals concerned. In what follows we are referring to substitutional solid solutions, that is, those in which the atoms of the solute replace some of those of the solvent in the structure of the latter. There is another kind of solid phase, the interstitial solid solution, in which the small atoms of some of the lighter non-metals occupy the interstices between the atoms in metal structures. We deal with these interstitial solid solutions in a later section. It is possible to have a combination of substitutional and interstitial solid solution in ternary or higher systems. For example, an austenitic manganese steel is a substitutional solid solution of manganese in iron and also an interstitial solution of carbon in the (Fe, Mn) structure.

The conditions determining the formation of solid solutions are as follows:

(1) The tendency to form solid solutions is small if the metals are chemically dissimilar, In general we may say that extended solid solution formation is common between metals of our classes A_1 and A_2, subject to the further restrictions discussed below, and between metals of the same subgroup of the Periodic Table. Thus the following pairs of metals form continuous series of solid solutions: K–Rb; Ag–Au; Cu–Au; As–Sb; Mo–W; and Ni–Pd. For elements of greatly differing electronegativity, for example, an A subgroup metal and a member of one of the later B subgroups, not only are the structures of the pure elements quite different but, owing to the difference in electronegativity, compound formation is likely to occur rather than solid solution. For this reason we divided the elements into groups A_1, A_2, B_1, and B_2 and we shall later consider in turn the types of intermetallic compounds formed between pairs of elements from different groups.

(2) For two elements in the same group in our classification the range of composition over which solid solutions are formed depends on the relative sizes of the two atoms. This is to be expected, since if some of the atoms in a structure are replaced (at random) by others of a different size, distortion of the structure must occur and the cell dimensions alter as the concentration of the solute increases. To a first approximation they vary linearly with the atomic percentage of the solute (Vegard's law), though in many cases this law is not exactly obeyed. If the difference between the radii of the metals is greater than about 15 per cent (of the radius of the solvent atom) there is no extensive formation of solid solutions. There is rather more tolerance at high temperatures, but then precipitation-hardening generally occurs on quenching. For metals crystallizing with very different types of structure the application of this 'relative size' criterion is complicated by the difference in coordination number in the structures of the two metals. We need not, however, go into this point as we shall only be considering solid solutions of metals with typical metallic structures of high coordination.

(3) The mutual solubilities of metals are not reciprocal. Hume-Rothery observed that, other things being equal, a metal of lower valence is likely to dissolve more of one of higher valence than vice versa—the 'relative valence effect'. Though there may be exceptions to this generalization, it is nevertheless a useful one. The following figures, taken from the much more extensive data quoted by Hume-Rothery, give some idea of the striking differences in solubilities in cases where the size factor

is favourable.

Solubility of:	Zn in Ag	37.8% atomic (Zn)
	Ag in Zn	6.3% atomic (Ag)
	Zn in Cu	38.4% atomic (Zn)
	Cu in Zn	2.3% atomic (Cu)

We shall now describe some of the changes in structure which take place in certain substitutional solid-solutions on cooling slowly or annealing.

Order–disorder phenomena and superstructures

Some alloys which are random solutions when quenched from the molten state undergo rearrangement when given a suitable heat treatment or in some cases simply when cooled sufficiently slowly. Such a structural change, which begins at a temperature often hundreds of degrees below the melting point, results in a change from the random arrangement to one in which there is regular alternation of atoms of different kinds throughout the structure. We may illustrate such a change by showing the structure of a portion of the alloy CuAu (a) as a random solid solution and (b) as the ordered superstructure (Fig. 29.7).

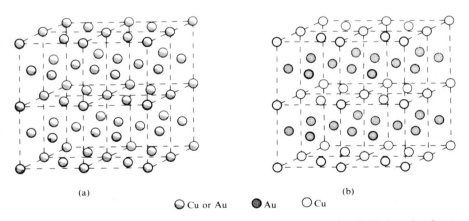

(a)   (b)

◑ Cu or Au   ● Au   ○ Cu

FIG. 29.7. The crystal structure of the alloy CuAu in (a) the disordered, and (b) the ordered state.

The ordered structure has lower (tetragonal) symmetry than the disordered, being a slightly deformed cubic closest packing of the atoms. The structures of Fig. 29.7 correspond to temperatures (a) above 420 °C and (b) below 380 °C. At intermediate temperatures the structure is more complex, being a superstructure with a 10-fold unit cell. Moreover the period of the superstructure can be altered by adding a third element of different valence, that is, it is affected by the electron:atom ratio. The possibility of rearrangement to form a superstructure is determined by the thermal energy of the atoms, the difference in potential energies of the ordered and disordered states, and the magnitude of the energy barrier that has to be surmounted before two atoms can change places. The formation of a superstructure is a cooperative phenomenon like the loss of ferromagnetism, which occurs at the Curie point, or the onset of rotation of ions or molecules in crystals. A characteristic of such processes is that the behaviour of a particular atom (or molecule) is affected by that of its neighbours. Once the process has started it speeds up, giving a curve like that shown in Fig. 29.8(a) for the specific heat of β-brass (see below). The rotation of one methane molecule, for example, in a crystal of that compound loosens the attachments of the others so that not only is the process facilitated for each succeeding molecule but also the thermal energy of the molecules is increasing all the time since the temperature is rising. Although order–disorder transformations may sometimes be loosely compared with melting, since a regular arrangement of atoms gives place to a random one and movement of the atoms relative to one another takes place, it is clear that they differ from the process of melting in certain important respects. There is no general collapse of the whole structure and the transformation takes place over a wide range of temperature, though in this latter respect the melting of a glass is similar. In a liquid, however, the arrangement of the atomic centres is less regular than in a solid and, moreover, is constantly changing, whereas in both the disordered and the ordered phases of an alloy the positions of the atomic centres are the same, or the same to within very close limits.

Starting with an ordered alloy at a low temperature two types of order–disorder transformation may be recognized:

(a) Order decreases continuously, passing through all intermediate states, and finally disappears completely at the critical temperature. The specific heat shows an

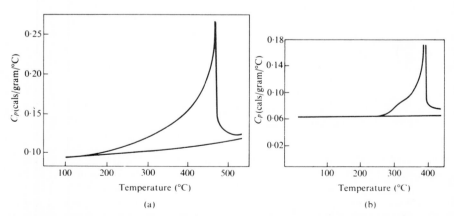

FIG. 29.8. The variation with temperature, on heating, of the specific heat of (a) β-brass, and (b) Cu_3Au (after cooling at 30 °C per hour). In each case the upper curve is drawn through the experimental values and the lower curve is that to be expected in the absence of a transformation.

abnormally great increase with rising temperature and at the critical temperature falls suddenly to a value only slightly above the theoretical value.

(b) The behaviour is similar to that described in (a), except that there is still some order left when the critical temperature is reached so that the last stage of the rearrangement takes place suddenly. In this case there is also a latent heat in addition to the specific heat anomaly.

We shall now give some examples of order–disorder transformations observed in alloys of two types, XY and X_3Y.

β-Brass

The structures of the ordered and disordered phases are illustrated in Fig. 29.9. The atoms are arranged on a body-centred cubic lattice, in one case in a regular way and in the other at random. Instead of drawing a large portion of the structure to show the random arrangement it is more convenient to use a circle with intermediate shading to represent an atom of either kind. The probability of such an atom being either Cu or Zn is one-half. In this alloy the transformation takes place over a temperature range of some $300\,^\circ C$. Below $470\,^\circ C$ it takes place continuously and reversibly, there being no sudden change in the degree of order at the transition point and therefore no latent heat. The specific heat/temperature curve for a change of this sort is shown (for CuZn) in Fig. 29.8(a). To account for the fact that the specific heat is still abnormal above the transition point it has been suggested that although the long-distance order has disappeared there is still a tendency for atoms to prefer unlike atoms as nearest neighbours.

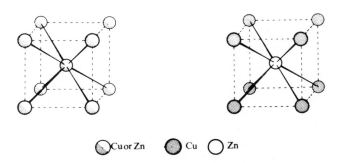

Cu or Zn   ● Cu   ○ Zn

FIG. 29.9. The structures of the disordered and ordered forms of β-brass.

Alloys X_3Y

Order–disorder transformations have been observed in the following alloys: Cu_3Au, Cu_3Pd, Cu_3Pt, Ni_3Fe, and Fe_3Al. In the first four cases the atoms occupy the positions of a face-centred cubic structure, and the relation of the ordered to the disordered structure is that shown in Fig. 29.10(a). These transformations are

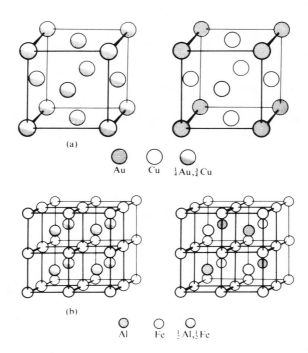

FIG. 29.10. The crystal structures of (a) Cu₃Au, and (b) Fe₃Al in the disordered and ordered states.

sluggish and inhibited by quenching from high temperatures, and annealing for several months is necessary to produce a high degree of order in Ni₃Fe. The specific heat/temperature curve for Cu₃Au is shown in Fig. 29.8(b). Investigation of the variation with temperature of the electrical resistance of Cu₃Au shows that a marked degree of order sets in at the critical temperature, for on cooling through this temperature a sudden drop of 20 per cent occurs. Accordingly there is a latent heat associated with the order–disorder transformation in Cu₃Au. The absence of the expected latent heat in the case of Ni₃Fe is probably due to the extreme slowness of the rearrangement. The transformation in Fe₃Al is illustrated in Fig. 29.10(b).

Certain other alloys X₃Y provide interesting examples of superstructures and also of close-packed structures containing atoms of two types. The structures of Al₃Ti and Al₃Zr are superstructures derived from the cubic close-packed structure of Al. A unit cell of the (tetragonal) structure of Al₃Ti is shown in Fig. 29.11 together with an equivalent portion of the structure of Al. One-quarter of the Al atoms are replaced by Ti in a regular manner, and it will be seen that the plane corresponding to (111) of Al is now a close-packed X₃Y layer of the type shown in Fig. 4.15(c). The more complex structure of Al₃Zr is closely related to that of Al₃Ti but contains both types of X₃Y layer (Fig. 4.15(c) and (d)). It will be appreciated

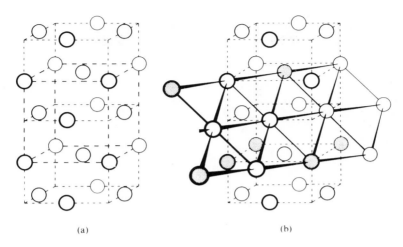

FIG. 29.11. The crystal structures of (a) Al (two unit cells), and (b) Al_3Ti (one unit cell). A portion of one close-packed Al_3Ti layer is emphasized in (b) where the shaded circles represent Ti atoms.

that the relation of Al_3Ti to Al is similar to that of Fe_3Al (and the isostructural Fe_3Si) to Fe. Al_3Ti arises by the regular replacement (by Ti) of one-quarter of the Al in a face-centred cubic structure and Fe_3Al by the regular replacement of one-quarter of the Fe by Al in the body-centred cubic structure (Fig. 29.10(b)). Another group of alloys have structures which are superstructures derived from *hexagonal* or more complex types of closest packing. In Ni_3Sn (and the isostructural Mg_3Cd and Cd_3Mg) and Ni_3Ti the layers are of the type of Fig. 4.15(d) but the structures differ in the sequence of layers (h in Ni_3Sn, hc in Ni_3Ti).

Examples of intermetallic phases X_3Y based on more complex c.p. layer sequences are included in Table 4.4 (p. 159). It is tempting to speculate on the reasons for the choice of different layer sequences in these compounds. The cubic Cu_3Au structure implies that all interatomic distances are the same, and this structure is adopted by, for example, the compounds Al_3M formed by the smaller 4f metals (e.g. Er). In the hexagonal or rhombohedral structures adjustment is possible to give two different interatomic distances as required by the larger 4f metals and Y:[1]

Layer sequence	Al_3Er c	Low-Al_3Y h	High-Al_3Y chh
	Er – 12 Al 2.98 Å	Y – 12 Al 3.11 Å	Y – 12 Al 3.07 Å
	Al $\begin{cases} 4\ Er \\ 8\ Al \end{cases}$ 2.98	Al $\begin{cases} 4\ Y\ \ \ 3.11 \\ 8\ Al\ \ 2.96 \end{cases}$	Al $\begin{cases} 4\ Y\ \ \ 3.07 \\ 8\ Al\ \ 2.98 \end{cases}$

The polymorphism of Al_3Pu[2] suggests that the relative sizes of the atoms are more important at lower temperatures. This compound has at least three polymorphs with different layer sequences:

————1027 °C————1210 °C————1400 °C

9-layer	6-layer	3-layer
chh	*cch*	*c*

The compound WAl_5 is an interesting example of a more complex formula arising in a close-packed structure by the alternation of layers of composition WAl_2 and Al_3 in the sequence $ABAC$. . . .[3]

(1)   AC 1967 **23** 729   (3)   AC 1955 **8** 349
(2)   JNM 1965 **15** 1, 57; AC 1965 **19** 184

The structures of alloys

We shall now describe briefly some of the structures adopted by alloys. We shall limit our survey to binary systems. Adopting the classification of the elements indicated on p. 1274, we have to consider three main classes of alloys:

I. Alloys of two A metals, AA

II. Alloys of an A and a B subgroup metal, AB
$$\begin{cases} A_1B_1 \\ A_2B_1 \\ A_1B_2 \\ A_2B_2 \end{cases}$$

III. Alloys of two B subgroup metals, BB.

Since we subdivide the A metals into two groups A_1 and A_2 and also distinguish the earlier from the later B subgroup metals as B_1 and B_2 respectively we make further subdivisions in class II as shown; we shall not deal systematically with the three subdivisions in each of the classes I and III. We have dealt with several systems of type AA in our discussion of solid solutions and superstructures, and the only other examples we shall mention of class I alloys are some phases formed by transition metals (group (c), p. 1304).

It is not easy to draw a hard-and-fast line between truly metallic alloys and homopolar compounds in some cases, particularly those involving elements of the later B subgroups (As, Sb, Se, Te), and it is also not convenient to adhere rigidly to the classification of elements into the groups A_1, A_2, B_1, and B_2. We shall make our classification somewhat flexible so as to allow us to bring together families of structures with a common structural theme. Size factors play an important part in determining the appearance of the so-called Laves phases with the closely related $MgZn_2$, $MgCu_2$, and $MgNi_2$ structures. From the structural standpoint these phases are probably most closely related to the σ phases formed by transition elements, so that although the Laves phases are not typically combinations of transition (A_2) metals they are conveniently mentioned in connection with the σ phases.

(a) *The* NaTl *and related structures*

Two structures commonly found for phases of composition A_1B_1 are the caesium chloride and 'sodium thallide' structures. Among the alloys crystallizing with the former structure are LiHg, LiTl, MgTl, CaTl, and SrTl. The structure needs no description here, but we shall refer to these alloys in section (d). In the NaTl structure (which is also that of LiZn, LiCd, LiAl, LiGa, LiIn, and NaIn) the coordinates of the atoms are the same as in the CsCl structure (that is, those of a body-centred cubic lattice), but the arrangement of the two kinds of atoms is such as to give each atom four nearest neighbours of its own kind and four of the other kind (Fig. 29.12). The interatomic distances in alloys with the NaTl structure are very much shorter (some 12 per cent for NaTl itself) than the sums of the radii derived from the structures of the pure metals.

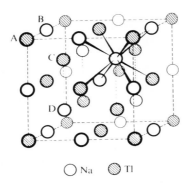

FIG. 29.12. The crystal structure of NaTl.

It is convenient to include here compounds formed by those B subgroup metals which lie on the borderline between B_1 and B_2 elements in our classification on p. 1274. The same atomic positions as those of the NaTl structure (that is, those of the b.c.c. lattice) are occupied in some phases X_3Y, but the arrangements of the X and Y atoms lead to quite different structures. Li_3Bi, Li_3Sb, and Li_3Pb have the same structure as the ordered form of Fe_3Al (Fig. 29.10(b)). In Cs_3Sb, which is of interest on account of its photoemissivity, Cs atoms occupy the sites A and C of Fig. 29.12 while equal numbers of Cs and Sb atoms occupy at random the sites B and D. Since the combination of B and D sites corresponds to a diamond-like arrangement it has been suggested that ionization takes place: $3Cs + Sb \rightarrow 2Cs^+ + Cs^{3-} + Sb^+$, and that Cs^{3-} and Sb^+ form tetrahedral $6s6p^3$ and $5s5p^3$ bonds respectively, the Cs^+ ions occupying the interstices in the diamond network.[1] These structures are summarized in Table 29.7, and for completeness we add two other compounds already mentioned in earlier chapters (Bi_2OF_4, p. 422, and NaY_3F_{10}, p. 421).

A feature of many compounds of the type A_1B_1 is the tendency of the B_1 atoms to group together, the type of group depending on the relative numbers of A_1 and B_1 atoms. For example, in the Na–Hg system, which is notable for the number of different phases formed, three of the phases are (in order of increasing Hg content) Na_3Hg_2, NaHg, and $NaHg_2$.[2] In the first there are nearly square Hg_4 groups which

TABLE 29.7

Structures of the CsCl–NaTl *family*

| | 4-fold positions of Fig. 29.12 | | | |
	A	B	C	D
CsCl	Cs	Cs	Cl	Cl
NaTl	Na	Tl	Na	Tl
Cs_3Sb	Cs	Cs, Sb	Cs	Cs, Sb
Li_3Bi	Bi	Li	Li	Li
Fe_3Al (ordered)	Al	Fe	Fe	Fe
Fe_3Al (disordered)	Fe	Fe	Fe, Al	Fe, Al
Bi_2OF_4	4 Bi	←	2 O + 8 F	→
NaY_3F_{10}	Na, 3 Y (Fig. 9.7)	←	10 F	→

are isolated units entirely surrounded by Na atoms. In NaHg pairs of Hg atoms (Hg–Hg, 3.05 Å) are linked into infinite chains by slightly longer bonds (3.22 Å), while $NaHg_2$ has the AlB_2 structure (p. 1055) with layers of linked Hg atoms. In Na_2Tl there are tetrahedral Tl_4 groups which form part of the icosahedral coordination group of 9 Na + 3 Tl around Tl.[3] There are also tetrahedral groups of atoms of the Group IV element in NaSi and $BaSi_2$, KGe, KSn, and NaPb.[4] With NaHg and the complex NaPb structure contrast the simple NaTl structure illustrated in Fig. 29.12.

(1)   PRS A 1957 **239** 46
(2)   AC 1964 7 277
(3)   AC 1967 **22** 836
(4)   AC 1953 **6** 197

(b) *The* BaCu, $CaCu_5$, XY_{11}, *and* XY_{13} *structures*

Here we describe a number of structures formed typically by larger A_1 atoms, including in this case such atoms as La, Ce, and Th, and the much smaller B_1 atoms, including the very small Be and in some cases 3d metals such as Fe, Co, Ni, and Cu. Features of these structures are the high c.n. of the larger atom and the formation by the smaller B_1 atoms of networks between, or enclosing, the large A_1 atoms. The BaCu structure, which is also that of SrCu,[1] consists of c.p. layers of Ba atoms, alternate pairs of which are directly superposed, with Cu atoms in trigonal prismatic holes between the latter, as shown diagrammatically in Fig. 4.10, p. 153. The Cu atoms form planar 6^3 nets, and therefore have a trigonal prismatic set of 6 Ba neighbours and 3 Cu at 2.60 Å completing a t.c.t.p. coordination group. The neighbours of Ba are 6 Cu and a total of 10 Ba, at distances from 4.17 to 4.73 Å. The $CaCu_5$ structure[2] is built from alternate layers of the types shown in Fig. 29.13(a) and (b), so that each Ca is surrounded by 6 Cu in one plane (at 2.94 Å) and by 2 more sets of 6 in adjacent planes (at 3.27 Å), making a total c.n. of 18. For this structure R_A should be less than 1.6 R_B; it is in fact found for values of R_A between 1.37 and 1.58 R_B. Examples of phases with this structure include

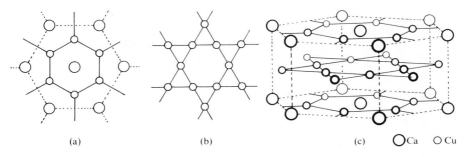

(a)   (b)   (c)   ⬤Ca ◯Cu

FIG. 29.13. The crystal structure of $CaCu_5$ (see text).

$CaNi_5$, $CaZn_5$, $LaZn_5$, $ThZn_5$, $CeCo_5$, and $ThCo_5$. Sr and Ba are near the upper size limit, and variants of this structure are found for $SrZn_5$ and $BaZn_5$.[3] By replacing in a regular manner certain of the X atoms in XY_5 by pairs of Y atoms there arise the related structures of Th_2Fe_{17},[4] $ThMn_{12}$, and $TiBe_{12}$.[5]

In the XY_{11}[6] and XY_{13} structures the X atoms occupy holes in which they are surrounded by 22 and 24 Y atoms respectively. For a given Y atom the smaller X metals form the XY_{11} phase and the larger ones the XY_{13} phase, while if X is too large even the XY_{13} phase is not formed (Table 29.8). This XY_{13} structure is of

TABLE 29.8

Phases with the XY_{11} *and* XY_{13} *structures*

Radius (Å)		Phases formed		
Cs	2.72			$CsCd_{13}$
Rb	2.50		$RbZn_{13}$	$RbCd_{13}$
K	2.35		KZn_{13}	KCd_{13}
Ba	2.24		$BaZn_{13}$	$BaCd_{11}$
Sr	2.15	$SrBe_{13}$	$SrZn_{13}$	$SrCd_{11}$
Ca	1.97	$CaBe_{13}$	$CaZn_{13}$	
Na	1.91	—	$NaZn_{13}$	
La	1.88	$LaBe_{13}$	$LaZn_{11}$	
U	1.56	UBe_{13}		

particular interest on account of the coordination polyhedra. The X atom has 24 Y neighbours at the vertices of a nearly regular snub cube, which is one of the less familiar Archimedean solids and has 6 square faces parallel to those of a cube and 32 equilateral triangular faces. The unique Y atom at the origin has a nearly regular icosahedral coordination group of 12 Y atoms, while the coordination group of the other twelve Y atoms is much less regular, as shown by the figures for $RbZn_{13}$:[7]

Rb –   24 Zn$_{II}$ 3.62 Å
Zn$_I$ –   12 Zn$_{II}$ 2.68   (compare the shortest Zn–Zn
Zn$_{II}$ – $\begin{cases} 10\ Zn_{I,\,II} & 2.61\text{–}2.96 \\ 2\ Rb & 3.62 \end{cases}$ in Zn metal, 2.66 Å)

(1) AC 1980 **B36** 1288   (5) AC 1952 **5** 85
(2) ZaC 1940 **244** 17   (6) AC 1953 **6** 627
(3) AC 1956 **9** 361   (7) AC 1971 **B27** 862
(4) AC 1969 **B25** 464

(c) *Transition metal σ phases and Laves phases*

Between electron compounds and 'normal valence' compounds, which are character-ized by relatively small numbers of structures, there lie large groups of intermetallic compounds in which the structural principles are less clear. In the structures to be described in this section the importance of geometrical factors is becoming evident, a special feature of these structures being the high coordination numbers which range from 12 to 16.

In a sphere packing the smallest hole enclosed within a polyhedral group of spheres is obviously a tetrahedral hole, and therefore the closest packing is achieved if the number of such holes is the maximum possible. In a 3-dimensional closest-packing of N equal spheres there are $2N$ tetrahedral holes but also N octahedral holes; more dense packings with only tetrahedral holes are possible, but these extend indefinitely in only one or two dimensions.[1] A local density higher than that of the ordinary closest packings is achieved, for example, by packing twenty spheres above the faces of an icosahedron, and this grouping of atoms is in fact found in some of the structures we shall describe. However, this type of packing of equal spheres cannot be extended to fill space, for holes appear which reduce the mean density of the packing to a value below that of closest packing. If, on the other hand, there are larger spheres to fill the holes (surrounded by 14–16 spheres) between the icosahedral groups a very efficient packing results, and this suggests at least a partial explanation of the occurrence of the structures we shall describe.

From the fact that the smallest kind of interstice between spheres in contact is a tetrahedral hole it follows that we should expect to find coordination polyhedra with only triangular faces, in contrast to those in, for example, cubic closest packing which have square in addition to triangular faces. Moreover, it seems likely that the preferred coordination polyhedra will be those in which five or six triangular faces (and hence five or six edges) meet at each vertex, since the faces are then most nearly equilateral. It follows from Euler's relation (p. 69) that for such a polyhedron, $v_5 + 0v_6 = 12$, where v_5 and v_6 are the numbers of vertices at which five or six edges meet, so that starting from the icosahedron ($v_5 = 12$) we may add 6-fold vertices to form polyhedra with more than twelve vertices.

It can readily be shown that there is no polyhedron of this family with $Z = 13$, only one for $Z = 14$, one for $Z = 15$, and two for $Z = 16$ (one of which has a pair of adjacent 6-fold vertices). Polyhedra with $Z > 16$ have at least one pair of adjacent

6-fold vertices. It is found that the three coordination polyhedra for c.n. > 12 in a number of alloy structures are in fact those for Z = 14, 15, and 16 which have no adjacent 6-fold vertices; they are illustrated in Fig. 29.14. The icosahedron (a) may be derived from a pentagonal antiprism by adding atoms above the mid-points of the opposite pentagonal faces; the 14-hedron (b) is similarly derived from a hexagonal antiprism. Since the distance from the centre to a vertex of a regular icosahedron is about 0.95 of the edge length this group may be regarded as a coordination polyhedron for radius ratio 0.90. In the 14-hedron (b) the twelve spheres of the antiprism are equal and different in radius from the central sphere, but also different in size from the remaining two which cap the hexagonal faces. The 15-hedron (c) may be described as a hexagonal antiprism with one atom beyond one hexagonal face and two above the other. If the polyhedron (d) for c.n. 16 consists of a truncated tetrahedral group of 12 B atoms around a central A atom plus a tetrahedral group of 4 A atoms which lie above the centres of the hexagonal faces it is sometimes referred to as the Friauf polyhedron, after the 'Friauf phases' $MgCu_2$ and $MgZn_2$, though these together with $MgNi_2$ are also called the 'Laves phases'. We discuss later the description of this coodination group as one corresponding to the radius ratio 1.23.

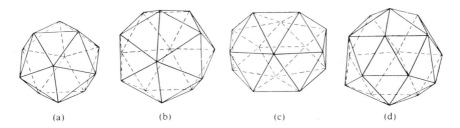

(a)   (b)   (c)   (d)

FIG. 29.14. Coordination polyhedra in transition metal structures.

In the 14-hedron the lines joining the centre of the polyhedron to the 6-fold vertices are collinear, in the 15-hedron they are coplanar (at $120°$), and in the 16-hedron they are directed towards the vertices of a tetrahedron; compare the stereochemistry of carbon. Frank and Kasper have pointed out that the basic geometry of certain structures in which the coordination numbers of all the atoms are 12, 14, 15 and/or 16 can be described in a very elegant way.[2] They describe sites with c.n. > 12 as *major sites*, the lines joining major sites which have six neighbours in common as *major ligands*, and the system of major ligands as the *major skeleton*. Since a major ligand connects two points of c.n. 14, 15, or 16 and passes through a 6-fold vertex there is a ring of six atoms around every major ligand, and the major skeleton generally defines the complete structure uniquely, apart from small displacements and distortions. If the only sites in a structure with c.n. > 12 are 14-coordinated then the major skeleton is simply a set of non-

TABLE 29.9

Coordination numbers in some intermetallic phases

Compound	Z	No. of atoms with c.n.					Major ligand network	Reference
		12	13	14	15	16		
Cu_4Cd_3	1124	736	–	120	144	124		AC 1967 **23** 586
$Mg_2Al_3(\beta)^{(a)}$	1168	672				252·	For description in terms	AC 1965 **19** 401
$\epsilon\text{-}Mg_{23}Al_{30}{}^{(b)}$	53	24	2	13	–	8	of Friauf polyhedra see	AC 1968 **B24** 1004
R-phase	53	27	–	12	6	8	refs. to $Mg_{23}Al_{30}$ and	AC 1960 **13** 575
$\gamma\text{-}Mg_{17}Al_{12}$							Cu_4Cd_3	
χ-phase	58	24	24	–	--	10		JM 1956 **8** 265
α-Mn								AC 1970 **B26** 1499
δ-phase (Mo—Ni)	56	24	–	20	8	4	4 interpenetrating	AC 1963 **16** 997
P-phase							(3,4)-connected nets	AC 1957 **10** 1
σ-phase (Ni–V)	30	10	–	16	4	–	2 D + 1 D	AC 1956 **9** 289
Zr_4Al_3	7	3	–	2	2	–		AC 1960 **13** 56
$MgZn_2$ (C 14)	12	8	–	–	–	4		
$MgCu_2$ (C 15)	24	16	–	–	–	8	3 D 4-connected nets	AC 1968 **B24** 7.1415
$MgNi_2$ (C 36)	24	16	–	–	–	8		
$\mu\text{-}Fe_7W_6$	13	7	–	2	2	2	3 D + 2 D	AC 1962 **15** 543
$Mg_{32}(Al, Zn)_{49}$	162	98	--	12	12	40	See text	AC 1957 **10** 254

(a) Also 244 atoms with miscellaneous types of coordination (c.n.'s 10–16).
(b) Also 6 atoms with c.n. 11.

intersecting lines, because there are only two major ligands from any point, but in structures with sites of 15- or 16-coordination the major skeleton may be a planar or 3-dimensional net. In the σ phase to be described shortly the major skeleton formed by joining the 15-coordinated sites is a planar net, and in the Laves phases the skeletons formed by linking the 16-coordinated sites correspond to the diamond, wurtzite, and carborundum-III structures.

In α-Mn there are four kinds of crystallographically non-equivalent atoms with the c.n.s shown in Table 29.9 and Mn—Mn distances ranging from 2.26 to 2.93 Å. It has been supposed for a long time that the complexity of this structure is due to the presence of atoms in different valence states, but there has been no generally accepted interpretation of the structure. In the complex σ phases formed by a number of transition metals there are no fewer than five crystallographically different kinds of atom with c.n.s ranging from 12 to 15. Of the 30 atoms in the unit cell 10 have the icosahedral coordination group of Fig. 29.14(a), 16 the 14-group of (b), and 4 the c.n. 15 (c). Neutron diffraction studies of Ni—V σ phases (Ni_9V_{21}, $Ni_{11}V_{19}$, $Ni_{13}V_{17}$) and σ-FeV, and X-ray studies of σ-MnMo and σ-FeMo show that there is a definite segregation of certain atoms into sites of highest c.n. and of others into sites of low c.n. If we distinguish between atoms to the left and right of

$$V \quad Cr \quad Mn \quad Fe \quad Co \quad Ni$$
$$Mo$$
$$(a) \qquad \qquad (b)$$

Mn as (a) and (b), we find (a) atoms with c.n. 15, both (a) and (b) in positions of c.n. 14, and only (b) in (icosahedral) 12-coordination.

In the μ phases Fe_7W_6, Fe_7Mo_6, and the corresponding Co compounds, Fe or Co is found in positions of 12-coordination (icosahedral) while Mo or W is found with higher c.n.s (14–16) for which the coordination polyhedra are also those of Fig. 29.14. Further examples of structures of high c.n. are included in Table 29.9, which includes references to some of the more recent work. It is interesting that of the three phases in the Mg–Al system the one with the simplest formula (Mg_2Al_3) has the most complex structure, with which may be compared the equally complex structure of Cu_4Cd_3. We have noted the description of these structures in terms of the 'major skeleton'. An alternative, for structures in which there is a reasonable proportion of 16-coordinated atoms, is to describe them in terms of the linking of the Friauf polyhedra, since by sharing some or all of their hexagonal faces these form a rigid framework of centred truncated tetrahedra. For references to such descriptions see Table 29.9, in which Z is the number of atoms in the unit cell.

The structure of $Mg_{32}(Al, Zn)_{49}$ provides a beautiful illustration of the importance of geometrical factors in determining a crystal structure, for no well-defined Brillouin zone could be found to account for its stability. Starting from one atom surrounded by an icosahedral group of 12 others, Fig. 29.15(a), 20 more are placed beyond the mid-points of the icosahedron faces at the vertices of a pentagonal dodechaedron, (b). Beyond the faces of the latter 12 more atoms are placed forming a larger

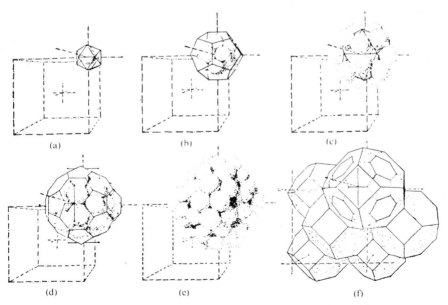

FIG. 29.15. The crystal structure of $Mg_{32}(Al, Zn)_{49}$ (see text).

icosahedron (c). With the previous 20 these 32 atoms lie at the vertices of a rhombic triacontahedron. This figure has 30 rhombus-shaped faces. Atoms are now placed beyond the centres of the 60 triangular half-rhombs of the triacontahedron; these lie at the vertices of the truncated icosahedron (d). The addition of 12 more atoms gives an outer group of 72 atoms which lie on the faces of a truncated octahedraon (e). Such groups can now be packed to fill space by sharing the atoms on their surfaces (f). Within each truncated octahedron there is a nucleus of 45 atoms, and since each atom of the outer group of 72 is shared with another similar group the number of atoms per unit cell is $2[45+\tfrac{1}{2}(72)] = 162$. In the resulting structure

$$
\begin{array}{lll}
\text{98 atoms have c.n. 12} & \text{Zn, Al} \left.\right\} & \\
\text{12 atoms have c.n. 14} \left.\right\} & & \left.\right\} \ \mathrm{Mg_{32}(Al, Zn)_{49}} \\
\text{12 atoms have c.n. 15} \left.\right\} \mathrm{Mg} & & \\
\text{and} \quad \text{40 atoms have c.n. 16} \left.\right\} & &
\end{array}
$$

In the Laves phases AB_2 typified by $MgZn_2$, $MgCu_2$, and $MgNi_2$, with three closely related structures, the coordination of A by 12 B + 4 A is consistent with the view that the structures are determined *primarily* by size factors. These phases are formed by a great variety of elements from many Periodic Groups, and the same element may be A or B in different compounds. In some ternary systems two or all of the structures occur at different compositions, the change from one structure to another being connected with electron–atom ratios, suggesting that factors of more than one kind may be important in determining the choice of one of a number of closely related structures.

The structures of certain crystals are derived from close-packed assemblies from which a proportion of the atoms have been removed in a regular way, as in the ReO_3 structure in which O atoms occupy three-quarters of the positions of cubic closest packing. It is of interest to note here some ways of removing one-half of the atoms from different kinds of closest packings. From a close-packed layer A we remove one-quarter of the atoms as shown in Fig. 29.16(a), and from the layer B

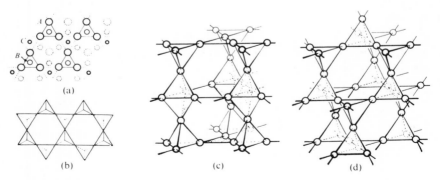

FIG. 29.16. The structures of the Laves phases (see text).

above we remove three-quarters of the atoms. This leaves a set of isolated tetrahedra. Below the *A* layer we could place either another *B* layer (leaving pairs of tetrahedra base to base) or a layer in position *C*, in which case we are left with a layer of tetrahedra with vertices pointing alternately up and down (Fig. 29.16(b)).

Certain sequences of layers of this kind form space-filling assemblies of tetrahedra and truncated tetrahedra, as shown in Fig. 29.16(c) and (d). That of Fig. 29.16(d) is of particular interest as a system of tetrahedra linked entirely through vertices, and it is alternatively derived by placing atoms at the mid-points of the bonds in the diamond structure and joining them to form tetrahedra around the points of that structure. The centres of the truncated tetrahedral holes in (d) also correspond to the positions of the atoms of the diamond structure; in (c) they correspond to the wurtzite structure. The number of these large holes is equal to one-half the number of atoms forming the tetrahedral network, so that if we place an atom A in each hole (that is, replacing a tetrahedral group of four B atoms in the original close-packed assembly) we have a compound of formula AB_2 in which A is surrounded by twelve B at the vertices of a truncated tetrahedron and is also linked tetrahedrally to four A neighbours lying beyond the hexagonal faces of that polyhedron. These are the structures of the Laves phases, $MgCu_2$, $MgZn_2$, and $MgNi_2$, which are related as follows:

	Sequence of close-packed layers		*Network formed by* A *atoms*
$MgCu_2$	*ABC* ...	(*c*)	Diamond
$MgZn_2$	*ABAC* ...	(*hc*)	Wurtzite
$MgNi_2$	*ABC BAC BC* ...	(*cchc*)	Carborundum-III

These structures provide an elegant example of the interrelations of nets, open packings of polyhedra, space-filling arrangements of polyhedra, and the closest packing of equal spheres.

It has been remarked[3] that the description of the Laves phases as suitable for atoms with radius ratio $r_A : r_B = 1.225$ is an over-simplification, for the observed range of the ratio is 1.05–1.67. In the simplest of these structures, the cubic $MgCu_2$ structure, the relative interatomic distances are the following fractions of the cell edge *a*:

$$A-A \ 0.433; \quad B-B \ 0.354; \quad and \quad A-B \ 0.414$$

from which it is evident that all three types of contact are not simultaneously possible for rigid spheres. Radii r_A and r_B may be chosen consistent with any two of the three types of contact:

$r_A : r_B$ 1·09	1·225	1·34
r_A 0·216	0·216	0·237
r_B 0·198	0·177	0·177
Contacts: A—A	A—A	A—B
A—B	B—B	B—B
B compressed	No A—B contacts	A compressed

The majority of compounds with the $MgCu_2$ structure have $r_A : r_B > 1.225$, suggesting that it is important to ensure A–B contacts even at the expense of compressing the A atoms, and the largest radius ratios are found for the largest electronegativity difference between A and B.

A similar difficulty arises in the β-W (A 15) structure (Fig. 29.4(a)) for a compound A_3B (i.e. atom B at the origin), since the environments of the atoms are:

$$A \begin{cases} 2\,A & 0.500a \\ 4\,B & 0.559 \\ 8\,A & 0.613 \end{cases} \quad \text{and } B-12\,A \ 0.559a$$

where a is the length of the edge of the cubic unit cell. There is evidently compression of the A atoms in chains parallel to the cubic axes, and in fact the stability and widespread adoption of this structure by binary intermetallic phases may be due to special interactions between pairs of A atoms, leading to interatomic distances which are not reconcilable with rigid spherical atoms.

(1) PRR 1952 7 303   (3) AC 1968 **B24** 7, 1415
(2) AC 1958 **11** 184; AC 1959 **12** 483

(d) *Electron compounds*

In systems A_2B_1, containing a transition metal or Cu, Ag, or Au and a metal of one of the earlier B subgroups, one or more of three characteristic structures are generally found. These are termed the β, γ, and ϵ structures, the solid solution at one end of the phase diagram being designated the α phase. These β, γ, and ϵ phases are not necessarily stable down to room temperature. In the Cu–Al system, for example, the β phase is not stable at temperatures below about 540 °C. The structures of these phases are:

  β 'body-centred cubic'*
  γ complex cubic structure containing 52 atoms in the unit cell
  ϵ hexagonal close-packed.

Although we shall assign formulae to these phases, which are often termed *electron compounds* for reasons that will be apparent later, they may appear over considerable ranges of composition. Moreover, although we might expect the β (body-centred) structure to contain at least approximately equal numbers of atoms of the two kinds, it sometimes appears with a composition very different from this. Thus, although the coordinates of the positions occupied in these phases are always the same, the distribution of a particular kind of atom over these positions is variable. In the Ag–Cd system the β phase is homogeneous at 50 per cent Cd and has the CsCl structure, but in the Cu–Sn and Cu–Al systems it appears with the approximate compositions Cu_5Sn and Cu_3Al respectively. In these cases there is random arrangement of the two types of atom in the body-centred structure. It is interesting that

*In this section the term 'body-centred cubic' implies only that all atoms occupy positions (000) or (½½½) in any unit cell.

whereas Cu_3Al has the β structure, Ag_3Al and Au_3Al crystallize with the β-Mn structure (p. 1282), and the same difference is found between Cu_5Sn (β structure) and Cu_5Si (β-Mn structure). In the case of alloys with the γ structure, the rather complex formulae such as Ag_5Cd_8, Cu_9Al_4, Fe_5Zn_{21}, and $Cu_{31}Sn_8$ are based on the number of atoms (52) in the unit cell. It will be seen that the total numbers of atoms in the formulae quoted are respectively 13, 13, 26, and 39. In the Ag–Cd system the γ phase is stable over the range 57–65 atomic per cent Cd, allowing a considerable choice of formulae, that chosen (Ag_5Cd_8) being consistent with the crystal structure. A selection of phases with the β, γ, and ϵ structures is given in Table 29.10.

TABLE 29.10

Relation between electron : atom ratio and crystal structure

Electron : atom ratio 3 : 2		*Electron : atom ratio 21 : 13*	*Electron : atom ratio 7 : 4*
β b-c structure	β Mn cubic structure	'γ brass' structure	ϵ close-packed hexagonal structure
CuBe	Ag_3Al	Cu_5Zn_8	$CuZn_3$
CuZn	Au_3Al	Cu_9Al_4	Cu_3Sn
Cu_3Al	Cu_5Si	Fe_5Zn_{21}	$AgZn_3$
Cu_5Sn	$CoZn_3$	Ni_5Cd_{21}	Ag_5Al_3
CoAl		$Cu_{31}Sn_8$	Au_3Sn
(for MgTl, etc., see later)		$Na_{31}Pb_8$	

A striking feature of this table is the variety of formulae of alloys with a particular structure. Hume-Rothery first pointed out that these formulae could be accounted for if we assume that the appearance of a particular structure is determined by the ratio of valence electrons to atoms. Thus for all the formulae in the first two columns we have an electron : atom ratio of $3:2$, for the third column $21:13$, and for the fourth $7:4$, if we assume the normal numbers of valence electrons for all the atoms except the triads in Group VIII of the Periodic Table. These fit into the scheme only if we assume that they contribute no valence electrons, as may be seen from the following examples:

CuBe	$(1 + 2)/2$		Cu_5Zn_8	$(5 + 16)/13$		$CuZn_3$	$(1 + 6)/4$	
Cu_3Al	$(3 + 3)/4$	$\frac{3}{2}$	Cu_9Al_4	$(9 + 12)/13$	$\frac{21}{13}$	Cu_3Sn	$(3 + 4)/4$	$\frac{7}{4}$
Cu_5Sn	$(5 + 4)/6$		Fe_5Zn_{21}	$(0 + 42)/26$		Ag_5Al_3	$(5 + 9)/8$	
CoAl	$(0 + 3)/2$		$Na_{31}Pb_8$	$(31 + 32)/39$				

Certain other alloys crystallizing with the CsCl structure, MgTl, CaTl, etc., which have already been mentioned, have sometimes been regarded as exceptions to the Hume-Rothery rules. They fall into line with the other β structures only if we assume that Tl provides one valence electron. Since, however, the radii of the metal atoms in these alloys, as in those with the NaTl structure, are smaller than the normal values, it is probably preferable not to regard these as β electron compounds.

Although the alloys with compositions giving the above comparatively simple electron:atom ratios generally fall within the range of homogeneity of the particular phases, it now appears that the precise values of those ratios, 3/2, 21/13, and 7/4, have no special significance. By applying wave mechanics to determine the possible energy states of electrons in metals it has been found possible to derive theoretical values for the electron:atom ratios at the boundaries of the α, β, and γ phases, the α phase being the solid solution with the close-packed structure of one of the pure metals. In comparing these values with the electron:atom ratios found experimentally we have to remember that phase boundaries may change with temperature, that is, the tie-lines separating the regions of stability of different phases on the phase diagram are not necessarily parallel to the temperature axis. This is invariably the case with β phases, the range of composition over which the phase is stable decreasing at lower temperatures. In Table 29.11 are given the experimental values of the electron:atom ratios for four systems in which both β and γ phases occur. The second column gives the electron:atom ratio for maximum solubility in the α phase, the third for the β phase boundary with smallest electron concentration, and the fourth for the boundaries of the γ phase. It will be seen that there is general agreement with the theoretical values, particularly in the second and third columns, though all the values for the γ phase boundaries exceed the theoretical electron:atom ratio.

TABLE 29.11

Experimental electron : atom ratios

System	Electron : atom ratios for		
	Maximum solubility in α phase	β phase boundary with smallest electron concentration	γ phase boundaries
Cu—Al	1.408	1.48	1.63–1.77
Cu—Zn	1.384	1.48	1.58–1.66
Cu—Sn	1.270	1.49	1.67–1.67
Ag—Cd	1.425	1.50	1.59–1.63
Theoretical	1.362	1.480	1.538

(e) *Some aluminium-rich alloys* A_2B_1

We have already noted that we have yet little understanding of the principles determining the structures of many phases formed by transition metals. Here we shall simply indicate some of the features of a number of structures which have been discussed in more detail elsewhere.[1]

Some of these compounds have simple structures with 8-coordination of the transition metal, for example,

CsCl structure	FeAl, CoAl, NiAl
Ni_2Al_3 structure	Pd_2Al_3 (also Ni_2In_3, Pt_2Ga_3, etc.)
CaF_2 structure	$PtAl_2$, $AuAl_2$, etc.

[The Ni_2Al_3 structure is a distorted CsCl structure in which one-third of the body-centring (Ni) positions are unoccupied.]

Certain other structures can be dissected into the 5-connected net of Fig. 29.17(a) and/or simple square nets. Depending on the relative orientations and translations of successive nets of these kinds (composed of Al atoms) there are formed polyhedral holes between eight or nine Al atoms in which the transition-metal atoms are found. A simple example is the $CuAl_2$-θ structure, in which alternate layers of type (a) are related by the translation *ac*; the Cu atoms are situated between the layers surrounded by eight Al at the vertices of the square antiprism of Fig. 29.17(b). In Co_2Al_9 (Fig. 29.17(c)) the sequence: square net, type (a) net, square net rotated and translated relative to the first, provides positions of 9-fold coordination for the Co atoms.

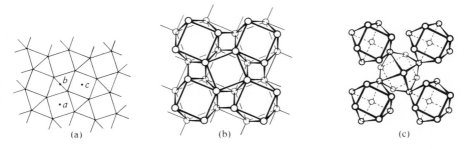

FIG. 29.17. The $CuAl_2$ (θ) and Co_2Al_9 structures: (a) Al layer of the $CuAl_2$ structure, (b) 8-coordination of Cu in $CuAl_2$ between two layers of type (a), (c) 9-coordination of Co in Co_2Al_9.

The structures of Co_2Al_5 and $FeAl_3$ may be interpreted as close-packed structures which have been modified so as to permit 9- and 10- (instead of 12-) coordination of the transition metal. In Co_2Al_5 (Fig. 29.18) close-packed layers at heights $c = \frac{1}{4}$ and $\frac{3}{4}$ are split apart by Co atoms X and Y (which have nine Al neighbours) with the result that layers are formed at $c \approx 0$ and $c \approx \frac{1}{2}$ (the broken lines in Fig. 29.18(b)) which are of the type of Fig. 29.18(c). The Co atoms in the original close-packed layers are 10-coordinated. The very complex structure of $FeAl_3$ may be dissected into layers of two kinds, flat layers as in Fig. 29.19(a) alternating with puckered layers of type (b). The flat layers are made up of close-packed regions and 'misfit' regions, and the triangles of the (b) layers lie

FIG. 29.18.   The structure of Co_2Al_5:   (a) projection along *c*-axis of atoms at height ≈ ¼, (b) elevation of one unit cell viewed in direction of arrow in (a)—X and Y are atoms which split such planes into two,   (c) projection as in (a) showing atoms at height ≈ 0.

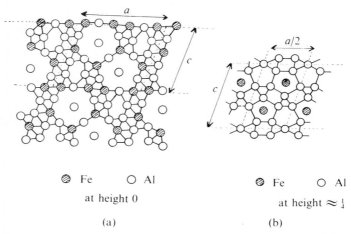

⊘ Fe	○ Al		⊘ Fe	○ Al
at height 0			at height ≈ ¼	
(a)			(b)	

FIG. 29.19.   The crystal structure of $FeAl_3$ showing the two kinds of layer into which the structure may be dissected.

approximately above and below the Fe atoms of an (a) layer. These Fe atoms have ten Al neighbours, whereas those in a (b) layer have nine Al neighbours.

(1)   See, for example: AC 1955 **8** 175;   AcM 1954 **2** 684;   AcM 1956 **4** 172

(f)   *Systems* A_1B_2

The systems A_1B_2 and A_2B_2, containing elements of the later B subgroups, call for little discussion here since arsenides have been dealt with to some extent in Chapter 20, and sulphides, selenides, and tellurides in Chapter 17. The phase diagrams of systems A_1B_2 are generally very simple, showing very restricted solid solution and only one, usually very stable, compound with a formula conforming to the ordinary valences of the elements (Mg_2Ge, Mg_3As_2, $MgSe$, etc.). These intermetallic compounds have simple structures which are similar to those of simple salts and they are electrical insulators. The structures of some of these compounds are set out below.

$$\left.\begin{array}{l} Mg_2Si \\ Mg_2Ge \\ Mg_2Sn \\ Mg_2Pb \end{array}\right\} anti\text{-}CaF_2 \qquad \left.\begin{array}{l} Mg_3P_2 \\ Mg_3As_2 \\ Mg_3Sb_2 \\ Mg_3Bi_2 \end{array}\right\}\begin{array}{l} anti\text{-}Mn_2O_3 \\ \\ anti\text{-}La_2O_3 \end{array} \qquad \left.\begin{array}{l} MgS \\ MgSe \end{array}\right\} NaCl \\ MgTe\ wurtzite$$

The fact that compounds such as Mg_2Si to Mg_2Pb have such high resistances and crystallize with the antifluorite structure does not mean that they are ionic crystals. Wave-mechanical calculations show that in these crystals the number of energy states of an electron is equal to the ratio of valence electrons: atoms (8/3) so that, as in other insulators, the electrons cannot become free (that is, reach the conduction band) and so conduct electricity. That the high resistance is characteristic only of the crystalline material and is not due to ionic bonds between the atoms is confirmed by the fact that the conductivity of molten Mg_2Sn, for example, is about the same as that of molten tin.

As indicated earlier, the dividing line between B_1 and B_2 metals is somewhat uncertain. We have included here some compounds of Pb, Sb, and Bi; certain other phases containing these metals were included with the A_1B_1 structures on p. 1301.

As examples of structures found in systems A_2B_2 we shall mention first the NiAs structure and then the structures of some Bi-rich phases formed by certain transition elements.

(g) Phases A_2B_2 with the nickel arsenide structure

The A_2 metals and the elements of the earlier B subgroups (B_1 metals) form the electron compounds already discussed. With the metals of the later B subgroups the A_2 metals, like the A_1, tend to form intermetallic phases more akin to simple homopolar compounds, with structures quite different from those of the pure metals. The nickel arsenide structure has, like typical alloys, the property of taking up in solid solution a considerable excess of the transition metal. From Table 29.12 it will be seen that many A_2B_2 compounds crystallize with this structure, which has been illustrated on p. 753, where it is discussed in more detail. The following

TABLE 29.12

Compounds crystallizing with the NiAs structure

	Cu	Au	Cr	Mn	Fe	Co	Ni	Pd	Pt
Sn	CuSn	AuSn			FeSn		NiSn	PdSn	PtSn
Pb									PtPb
As				MnAs			NiAs		
Sb			CrSb	MnSb	FeSb	CoSb	NiSb	PdSb	PtSb
Bi				MnBi			NiBi		PtBi
Se			CrSe		FeSe	CoSe	NiSe		
Te			CrTe	MnTe	FeTe	CoTe	NiTe	PdTe	PtTe

compounds crystallize with the pyrites structure:

$$PtP_2$$
$$PdAs_2 \qquad PtAs_2$$
$$AuSb_2 \qquad PdSb_2 \qquad PtSb_2$$
$$\alpha\text{-}PtBi_2$$

(h) *Some bismuth-rich alloys* A_2B_2

The structural principles in a group of alloys of Bi with Ni, Pd, and Rh are not simple, for although there is in some cases an obvious attempt by Bi to form three strongest pyramidal bonds (as in metallic bismuth itself) there is also a tendency to attain the much higher coordination numbers characteristic of the metallic state. For example, RhBi has a deformed NiAs structure in which the c.n. of Rh is raised to 8 and that of Bi to 12. An interesting feature of a number of these alloys is that they may be described in terms of the packing of monocapped trigonal prisms (Fig. 29.20(a)). The transition-metal atoms occupy the centres of these polyhedra but

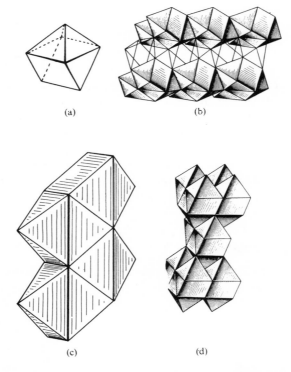

(a)   (b)

(c)   (d)

FIG. 29.20. Structures of some bismuth-rich alloys: (a) coordination polyhedron around transition metal atom, (b)–(d) packing of these coordination polyhedra in $\alpha\text{-}Bi_2Pd$, Bi_3Ni, and BiPd.

are also bonded to similar atoms in neighbouring polyhedra. For example, in Bi_3Ni (Fig. 29.20(c)) these polyhedra are linked into columns and a Ni atom is bonded to two Ni in neighbouring polyhedra (at 2.53 Å) as well as to seven Bi, making a total c.n. of 9. Figures 29.20(b) and (d) show how these Bi_7 polyhedra are packed in α-Bi_2Pd and BiPd respectively; in all these packings there are (empty) tetrahedral and octahedral holes between the Bi_7 polyhedra, which do not, of course, form space-filling assemblies. In α-Bi_4Rh there is 8-coordination of Rh by Bi (square antiprism), so that the characteristic c.n.s of the transition-metal atoms in a number of these alloys are 8 or 9 (7 Bi + 2 transition metal) while those of Bi are high, 11–13.

In contrast to α-Bi_2Pd the high-temperature β form has a more symmetrical (tetragonal) structure which is a superstructure of the f.c.c. structure, of the same general type as Cr_2Al (Fig. 29.21), which is the analogous b.c.c. superstructure. Note that although both Cr_2Al and β-Bi_2Pd have the same space group ($F/4mmm$) and the same positions are occupied, namely Al(Pd) in (000, ½½½) and Cr(Bi) in $\pm$(00z, ½½½+z), for Cr_2Al $c:a \approx 3$ (z = 0.32), while β-Bi_2Pd has $c:a \approx 3\sqrt{2}$ (z = 0.36). These compounds are not isostructural; see the discussion of this topic in Chapter 6.

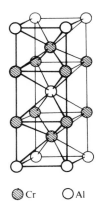

⊘ Cr   ◯ Al

FIG. 29.21. The crystal structure of Cr_2Al.

(i) *Systems* BB

Alloys containing elements of only the earlier B subgroups have typically metallic properties. Solid solutions are formed to an appreciable extent only by elements of the same subgroup and, of course, the relative size criterion applies as in other systems. Thus Cd and Hg form solid solutions over quite large ranges of composition and Cd and Zn over smaller ranges; compare the radii: Zn, 1.37; Cd, 1.52; and Hg, 1.55 Å. Cadmium and tin, on the other hand, are practically immiscible. When both the metals belong to later B subgroups a 1:1 compound with the NaCl structure occurs in a number of cases, for example, SnSe, SnTe, PbSe, and PbTe. Finally the zinc-blende or wurtzite structure is generally found for 1:1 compounds in which the average number of valence electrons per atom is four, that is, when the atoms belong to the Nth and $(8-N)$th subgroups. Examples of compounds with these structures are the sulphides, selenides, and tellurides of Zn, Cd, and Hg, and GaAs, GaSb, and InSb. (Compounds such as BeS and AlP also crystallize with the zinc-blende structure, which is not restricted to elements of the B subgroups.)

In contrast to InSb, TlSb (and also TlBi) has the CsCl structure; InBi is also exceptional in having the B 10 structure (p. 270) in which In has four Bi neighbours arranged tetrahedrally, but Bi has four In neighbours on one side at the basal vertices of a square pyramid.

The formulae of alloys

Substitutional solid solutions can have any composition within the range of miscibility of the metals concerned, and there is random arrangement of the atoms over the sites of the structure of the solvent metal. At particular ratios of the numbers of atoms superstructures may be formed, and an alloy with either of the two extreme structures, the ordered and disordered, but with the same composition in each case, can possess markedly different physical properties. Composition therefore does not completely specify such an alloy. Interstitial solid solutions also have compositions variable within certain ranges. The upper limit to the number of interstitial atoms is set by the number of holes of suitable size, but this limit is not necessarily reached, as we shall see later. When a symmetrical arrangement is possible for a particular ratio of interstitial to parent lattice atoms this is adopted. In intermediate cases the arrangement of the interstitial atoms is random.

When we come to alloys which are described as intermetallic compounds as opposed to solid solutions, we find that in some cases the ratios of the numbers of atoms of different kinds which we should expect to find after examination of the crystal structure are never attained in practice. When dealing with electron compounds we noted many cases of alloys with quite different types of formulae which crystallize with the same structure (e.g. CuZn, Cu_3Al, and Cu_5Sn all with the β structure). We know that in cases of this kind the compositions are determined by the electron:atom ratios. At an appropriate temperature an electron compound, like a solid solution, is stable over a range of composition, and the particular formulae adopted were selected to conform with the numbers of equivalent positions in the crystal structure (e.g. Cu_9Al_4 rather than, say, Cu_9Al_5 which also lies within the range of stability of this phase) and/or Hume-Rothery's simple electron:atom ratios. We have seen that the precise values of these ratios, 3/2, 21/13, and 7/4, have no theoretical significance. An alloy such as Ag_3Al is completely disordered, there being a total of 20 atoms in the unit cell in two sets of 12-fold and 8-fold positions. There are, however, alloys other than those with the β, γ, and ϵ structures, the compositions of which are never those of the ideal structures. In the Cu–Al system, for example, there is a θ phase with ideal formula $CuAl_2$, the structure of which has been illustrated in Fig. 29.17. In this (tetragonal) structure there are, in one unit cell, four Cu and eight Al atoms, the symmetry requiring these numbers of the two kinds of atom. If, however, an alloy is made up with the composition $CuAl_2$ it is not homogeneous but consists of a mixture of alloys with the $CuAl_2$ and $CuAl$ structures. In other words, the $CuAl_2$ structure is not stable with the Cu:Al ratio equal to 1:2 but prefers a rather greater proportion of aluminium. Thus, although this θ phase is stable over a certain small range of composition, the alloy $CuAl_2$ lies outside this range. The laws applicable to conventional chemical compounds (see, however, FeS, p. 754) do not hold in metal systems, and these facts concerning $CuAl_2$ are not surprising when we remember that the approximate compositions of the phases in this system with the β and γ structures are Cu_3Al and Cu_9Al_4 respectively. The β phase in the Cr—Al system provides another example of the same phenomenon. The body-centred solid solution of Al in Cr is stable at high temperatures up to some 30 per cent Al, that is, beyond the composition Cr_2Al. If, however, an alloy containing about 25 per cent Al is cooled slowly, the body-centred cubic structure changes to the tetragonal β structure illustrated in Fig. 29.21. As may be seen from the diagram, this structure is closely related to the body-centred cubic structure and its ideal composition is clearly Cr_2Al. Although this β structure is stable over a considerable range of composition, the alloy Cr_2Al lies outside this range. The slow cooling of an alloy with the exact composition Cr_2Al therefore gives an inhomogeneous mixture which actually consists of an Al-rich Cr_2Al component and some body-centred cubic solid solution.

Interstitial carbides and nitrides

We saw in Chapter 4 that from the geometrical standpoint the structures of many inorganic compounds, particularly halides and chalconides, may be regarded as assemblies of close-packed non-metal atoms (ions) in which the metal atoms occupy tetrahedral or octahedral interstices between 4 or 6 c.p. non-metal atoms. The numbers of such interstices are respectively $2N$ and N in an assembly of N c.p. spheres, and occupation of some or all of them in hexagonal or cubic closest packing gives rise to the following simple structures:

Interstices occupied	h.c.p.	c.c.p.	Formula
All tetrahedral	—	antifluorite	A_2X
$\frac{1}{2}$ tetrahedral	wurtzite	zinc-blende	AX
All octahedral	NiAs	NaCl	AX
$\frac{1}{2}$ octahedral	$\left\{\begin{array}{l}CaCl_2\ (rutile)\\CdI_2\end{array}\right.$	$\left\{\begin{array}{l}atacamite\\CdCl_2\end{array}\right.$	AX_2

The description of these structures in terms of the closest packing of the halide or chalconide ions is both convenient and, certainly for the octahedral structures, realistic (see p. 161), because in most cases these ions are appreciably larger than the metal ions. At the other extreme there are many compounds of metals with the smaller non-metals in which the non-metal atoms occupy interstices between c.p. metal atoms. For structural reasons it is preferable to deal separately with hydrides (p. 344) and borides (p. 1052). In the latter compounds B—B bonds are an important feature of many of the structures, and their formulae and structures are generally quite different from those of carbides and nitrides. The structures of carbides of the type MC_2 have been described in Chapter 22. In the LaC_2 and ThC_2 structures the carbon atoms are in pairs as C_2^{2-} ions. Although these structures may be regarded as derived from c.c.p. metal structures with the C_2^{2-} ions in octahedral interstices, there is considerable distortion from cubic symmetry owing to the large size and non-spherical shape of these ions, and these carbides are therefore not to be classed with the interstitial compounds. Certain oxides are sometimes included with the interstititial carbides and nitrides; these will be mentioned later. It is convenient to deal separately with the carbides and nitrides of Fe, which are much more chemically reactive than, and structurally different from, the compounds to be considered here.

Interstitial carbides and nitrides have many of the properties characteristic of intermetallic compounds, opacity (contrast the transparent salt-like carbides of Ca, etc.), electrical conductivity, and metallic lustre. In contrast to pure metals, however, these compounds are mostly very hard and they melt at very high temperatures. Compounds of the type MX are generally derived from cubic close packing; those of the type M_2X from hexagonal close packing. The melting points and hardnesses of some interstitial compounds are set out below. These compounds may be prepared by heating the finely divided metal with carbon, or in a stream of ammonia, to temperatures of the order of 2200 °C for carbides, or 1100–1200 °C for nitrides. Alternatively, the metal, in the form of a wire, may be heated in an atmosphere of a hydrocarbon or of nitrogen. The solid solution of composition 4 TaC + ZrC melts at the extraordinarily high temperature 4215 K. These compounds are very inert chemically except towards oxidizing agents. Their electrical conductivities are high and decrease with rising temperature as in the case of metals, and some exhibit supraconductivity. (The hardness is according to Mohs' scale, on which that of diamond is 10.)

	M.P. (°K)	Hardness		M.P. (°K)	Hardness
TiC	3410	8–9	TiN	3220	8–9
HfC	4160		ZrN	3255	8
W_2C	3130	9–10	TaN	3360	
NbC	3770				

The term 'interstitial compound' or 'interstitial solid solution' was originally given to these compounds because it was thought that they were formed by the interpenetration of the non-metal atoms into the interstices of the metal structure, implying that no gross rearrangement of the metal atoms accompanied the formation of the interstitial phase. This view of their structures was apparently supported by their metallic conductivity, by the variable composition of many of these phases, and by the fact that there is an upper limit to the size of the 'interstitial' atom compared with that of the metal atom. While it is certainly true that the C or N atoms occupy interstices (usually octahedral) in an array of (usually) c.p. metal atoms, it is now known that the arrangement of the metal atoms in the interstitial compound is generally different from that in the metal from which it is formed, although initially the metal structure may be retained if a solid solution is formed. Thus Ti dissolves nitrogen to the stage $TiN_{0.20}$, this phase being a solid solution of N in the h.c.p. structure of α-Ti. At the composition $TiN_{0.50}$ the ϵ phase has the anti-rutile structure, with distorted h.c.p. Ti, , but at $TiN_{0.60}$ (at $900\,^\circ C$) the arrangement of metal atoms becomes c.c.p. (defect NaCl structure).[1] The high-temperature form of the metal (above $880\,^\circ C$) is b.c.c.

In Table 29.13 are listed the structures of compounds MC and MN with metallic character and known crystal structure. Of all the metals in Table 29.13 forming a carbide MC or a nitride MN with the NaCl structure only four have the cubic close-packed structure. For all the other compounds the arrangement of metal atoms in the compound MX is *different* from that in the metal itself. (This is also true of many hydrides—see p. 346). Also, although many of these compounds exhibit variable composition, some do not, for example, UC, UN, and UO. In any case, variable composition is not confined to interstitial compounds—see, for example, the note on non-stoichiometric compounds, p. 5.

TABLE 29.13

The structures of metals and of interstitial compounds MX

Metal	Structure	Carbide	Nitride	Oxide
Sc	A 1, A 3	–	B 1	–
La	A 1, A 3	–	B 1	–
Ce	A 1, A 3	–	B 1	–
Pr	A 3	–	B 1	–
Nd	A 3	–	B 1	–
Ti	A 2, A 3	B 1	B 1	B 1 (p. 562)
Zr	A 2, A 3	B 1	B 1	B 1
Hf	A 2, A 3	B 1	B 1 (?)	–
Th	A 1	B 1	B 1	B 1
V	A 2	B 1	B 1	B 1
Nb	A 2	B 1	B 1[b]	See text
Ta	A 2	B 1	See text	–
Cr	A 2	Hex. ?	B 1	–
Mo	A 2, A 3	WC	WC	–
W	A 2	WC	WC	–
U(γ)	A 2 [a]	B 1	B 1	B 1

A 1 indicates cubic close-packed, A 2, body-centred cubic, A 3, hexagonal close-packed, B 1, NaCl structure.

(a) Also α and β forms of lower symmetry.

(b) B 1 structure for composition $NbN_{0.9}O_{0.1}$.

Of the MX structures listed in Table 29.13 other than the NaCl structure, the WC structure is illustrated diagrammatically in Fig. 4.8, p. 152 and the ϵ-TaN structure in Fig. 29.22. In addition to the β-phase Ta_2N (isostructural with β-Nb_2N, β-V_2N, ϵ-V_2C, and ϵ-Fe_2N, p. 171) and δ-TaN, a high-temperature phase with the NaCl structure, tantalum forms ϵ-TaN.[2] This structure is notable for the square pyramidal coordination of N by 5 Ta (4 at 2.16 and 1 at 2.04 Å). Ta_I has a coplanar triangular group of 3N neighbours (at 2.04 Å) and in addition 2 Ta at 2.91 and 12 Ta at 3.34 Å, while Ta_{II} is at the centre of a trigonal prismatic group of 6 N (at 2.16 Å) and has the following sets of Ta neighbours, 2 at 2.91, 3 at 3.00, and 6 at 3.34 Å).

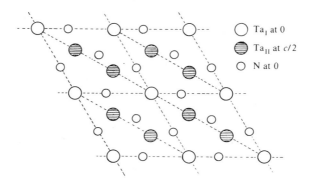

FIG. 29.22. Projection of four unit cells of the TaN structure

It would seem that the salient characteristics of these compounds are: (a) adoption in most cases of the NaCl structure irrespective of the structure of the metal; (b) high melting point and hardness; and (c) electrical conductivity. Rundle therefore suggested[3] that these properties indicate metal–non-metal bonds of considerable strength, the bonds from the non-metal atoms being directed octahedrally but not localized (to account for the conductivity). The compounds are regarded as electron-deficient compounds in which the non-metal atoms form six octahedral bonds either (1) by using three 2p orbitals (for three electron pairs), the 2s orbital being occupied by an electron pair, or (2) by using two hybrid sp orbitals (bond angle 180°) and two p orbitals, which are perpendicular to the hybrid orbitals and to each other. The six (octahedral) bonds would then become equivalent by resonance. In (1) the bonds would be '$\frac{1}{2}$-bonds'; in (2) they would be '$\frac{2}{3}$-bonds', using Pauling's nomenclature (p. 1292). Use of the 2s orbital by an unshared pair or electrons would be expected if the non-metal is sufficiently electronegative compared with the metal (as in 'suboxides', MO), or possibly in the nitrides of the more electropositive metals. On this view the total number of valence electrons is used for the metal–non-metal bonds in the Group IIIA nitrides and the Group IVA carbides, so that the metal–metal distances would be the result of the M–X bonding. For Group IV nitrides and Group V carbides there is one electron per metal atom available for metal–metal bonds, which would therefore have bond-number 1/12 in the NaCl structure. In the Group V nitrides and Group VI carbides there are two electrons per metal atom for M–M bonds which would accordingly be $\frac{1}{6}$-bonds. These bonds are now sufficiently strong to influence the stability of the NaCl structure. Accordingly the Nb–N and Ta–N systems show a complex behaviour, and for the carbides and nitrides of Cr, Mo, and W we also find structures other than *B* 1 for the compounds MX. (UC and UN are exceptional, possibly because U is not hexavalent in these compounds.)

An interesting feature of this treatment of these compounds is that it offers some explanation of the fact that they are limited to compounds MC and MN (and sometimes MO) of the elements of Groups IIIA, IVA, VA, and VIA, owing to the requirements that: (1) one element (M) must have more stable bond orbitals than valence electrons, and must therefore generally be a metal; (2) the second element must have relatively few bond orbitals, and is therefore generally a non-metal; and (3) the electronegativities of the two elements must not differ so much that the bond is essentially ionic (hence all fluorides and some oxides are excluded). Boron, for example, is classified with the metals in this scheme, so that borides are structurally different from the carbides and nitrides. Only metals having more than four stable bond orbitals will require C, N, and O to use a single orbital for more than one bond; these are the A subgroup metals. In the B subgroup metals the d levels below the valence group are filled, as for example in Ga, In, and Tl, which have tetrahedral (sp^3) orbitals and therefore form normal, as opposed to interstitial, mononitrides. Of the A subgroup metals, the alkalis and alkaline-earths are too electropositive to form essentially covalent bonds with C, N, and O; hence the interstitial compounds begin in Group III.

(1) ACSc 1962 **16** 1255
(2) AC 1978 **B34** 261
(3) AC 1948 **1** 180

Iron and steel

A simple steel consists of iron containing a small amount of carbon. Many steels now used for particular purposes also contain one or more of a number of metals (Cr, Mo, W, V, Ni, Mn) which modify the properties of the steel. We shall restrict our remarks to the simple Fe–C system. The properties of the various kinds of 'iron' and of steels which make these materials so valuable are dependent on the amount of carbon and the way in which it is distributed throughout the metal. The Fe–C system is, for two reasons, more complicated than the metal–non-metal systems giving the interstitial compounds just described. Firstly, iron is dimorphic. The form stable at ordinary temperatures is called α-iron. It has the body-centred cubic structure and is ferromagnetic. The body-centred structure is stable up to 906 °C and again from 1401 to 1530 °C. the melting point. Over the intermediate range of temperature, 906–1401 °C, the structure is face-centred cubic (γ-iron), in which form the metal is non-magnetic. The Curie point (766 °C) is lower than the α-γ transition point, and the term β-iron is applied to iron in the temperature range 766–906 °C. Since there is no change in atomic arrangement at the Curie point we shall not refer to β-iron in what follows. The second complicating factor is that the C : Fe radius ratio (0.60) lies near the critical limit for the formation of interstitial solid solutions. Accordingly, in addition to the latter a carbide Fe_3C is formed. Thus according to the carbon content and heat treatment the carbon may be present either in the free state as graphite, in solid solution, or as cementite (Fe_3C).

Iron is obtained by smelting oxy-ores with coke, so that in the melt there is an excess of carbon. Molten iron dissolves up to 4.3 per cent of carbon, the eutectic mixture solidifying at 1150 °C. Pig-iron (cast-iron) contains therefore about 4 per cent C. There is also up to 2 per cent Si from the clays associated with the ores. The carbon in cast-iron may be in the form of cementite (white cast-iron) or graphite (grey cast-iron), depending on the silicon content. The presence of silicon favours the decomposition of cementite into graphite, and in general some of the carbon in cast-iron is present in both forms. The material which solidifies at 1150 °C containing 4.3 per cent C is a mixture: the maximum solubility of C in γ-Fe at that temperature, that is, in a homogeneous solid solution, is 1.9 per cent. This solubility falls to 0.9 per cent at 690 °C, the lowest temperature at which γ-Fe is rendered stable by the presence of carbon. (Iron can be retained in the non-magnetic γ form at ordinary temperatures by adding elements such as Mn and Ni which form solid solutions with γ- but not with α-iron.) Removal of all but the last traces of impurity from pig-iron gives wrought iron, the purest commercial iron. Steels contain up to 1.5 per cent C; mild steels from about 0.1 to 0.5 per cent. The production of steels therefore involves either the controlled reduction of the amount of carbon if they are made from pig-iron or the controlled addition of carbon if they are made from wrought iron (the process of cementation). We may summarize the processes which take place in the production of steels as follows.

Above 906 °C the steel is in the form of a (non-magnetic) solid solution of carbon in γ-iron (austenite). This is a simple interstitial solid solution in which the carbon atoms are arranged at random since there are not sufficient to form a regular structure. It is fairly certain that in austenite the carbon atoms occupy octahedral holes in the γ-Fe lattice. When austenite is cooled slowly, the first process which takes place is the separation of the excess carbon as cementite, since the solubility of carbon falls to 0.9 per cent at 690 °C. Below this temperature γ-Fe is no longer stable, and the solid solution of C in γ-Fe changes at 690 °C into a eutectoid mixture of ferrite and cementite. Ferrite is nearly pure α-Fe; it contains some 0.06 per cent of C in interstitial solid solution. The remaining carbon goes into the cementite. Pearlite, which is the name given to this eutectoid mixture, has a fine-grained banded structure with a pearly lustre and is very soft. The other extreme form of heat treatment is to quench the austenite to a temperature below 150 °C, when martensite is formed. This is a super-saturated solid solution of C in α-Fe and may contain up to 1.6 per cent C. It is very hard, extreme hardness being a characteristic of quenched steels. (The original γ solid solution, austenite, can be preserved after quenching only if other metals are present, as mentioned above.) The hard, brittle quenched steels are converted into more useful steels by the process of tempering. This consists in reheating the martensite to temperatures from 200 to 300 °C. The object of tempering is the controlled conversion of the quenched solid solution into ferrite and cementite. The mixture produced by tempering has a coarser texture than pearlite and is termed sorbite. The tempering reduces the hardness but increases the toughness of the steel. Sorbite is thus an intermediate product in the sequence: austenite–martensite–sorbite–pearlite. These forms of heat treatment are summarized below.

It is interesting that the structure of cementite is not related in any simple way to those of α- or γ-Fe. In the γ structure each Fe atom has twelve equidistant nearest neighbours (at 2.52 Å, the value obtained by extrapolation to room temperature), and in the α structure eight, at 2.48 Å. In cementite some of the Fe atoms have twelve neighbours at distances ranging from 2.52 to 2.68 Å and the others eleven at distances from 2.49 to 2.68 Å. The most interesting feature of the structure is the environment of the carbon atoms. The six nearest neighbours lie at the apices of a distorted trigonal prism, but the range of Fe–C distances is considerable, 1.89–2.15 Å, and two further neighbours at 2.31 Å might also be included in the coordination group of the carbon atoms. The reason for this unsymmetrical environment—contrast the octahedral environment in austenite—is not known.

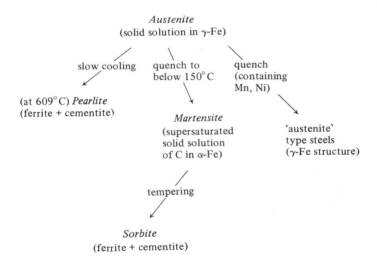

The process of case-hardening of steel provides an interesting example of the formation of interstitial compounds. In one method both C and N are introduced into the surface of steel either by immersion in a molten mixture of $NaCN, Na_2CO_3$, and NaCl at about 870 °C or by heating in an atmosphere of H_2, CO, and N_2 to which controlled amount of NH_3 and CH_4 are added. By these means both C and N are introduced. Although Fe does not react with molecular N_2 certain steels can be case-hardened by the action of ammonia at temperatures around $500\,^\circ$C. The Fe–N phase diagram is complex, and phases formed include:

  α solid solution (about 1.1 atomic per cent N at $500\,^\circ$C)
  γ phase (stable only above $600\,^\circ$C, at which temperature composition is approximately $Fe_{10}N$)
  γ' also f.c.c., Fe_4N
  ϵ h.c.p. (composition range extends from about Fe_3N to Fe_2N at low temperatures, but composition varies considerably with temperature)
  ζ Fe_2N (orthorhombic, slightly deformed variant of ϵ).

We mention these phases because they illustrate once again how the arrangement of metal atoms changes with increasing concentration of interstitial atoms and also because an outstanding feature of these nitrides is that with the exception of the γ phase (stable only at high temperatures and with rather low concentrations of N atoms) they have ordered arrangements of the N atoms in the octahedral interstices.[1]

(1) AC 1952 5 404

Formula index

Al(CH$_3$)$_3$ 979
Al(CH$_3$)$_2$(C$_5$H$_5$)   974
Al(C$_2$H$_5$)$_2$(C$_5$H$_5$)$_2$Cl$_2$Ti   976
[Al(CH$_3$)$_2$Cl]$_2$ 981
[Al(C$_2$H$_5$)$_2$F]$_4$ 981
[Al(C$_2$H$_5$)$_6$F]K   384
Al$_2$(CH$_3$)$_8$Mg   980
Al$_2$(CH$_3$)$_5$N(C$_6$H$_5$)$_2$ 980
Al$_2$(CH$_3$)$_{10}$O$_2$Si$_2$ 496
Al$_2$CdS$_4$ 779
AlCl   441
AlCl$_3$ 419, 444
AlCl$_3$·6H$_2$O   674
(AlCl$_4$)$_2$Cd$_2$ 1156
(AlCl$_4$)$_2$Co   176
(AlCl$_4$)SeCl$_3$ 739
(AlCl$_4$)$_2$Se$_8$ 703
AlCl$_4$Cu(C$_6$H$_6$)   978
Al$_2$Cl$_7^-$ 464
AlCo   1313
Al$_5$Co$_2$ 1313
Al$_9$Co$_2$ 1313
AlCr$_2$ 1317
Al$_2$Cu   1313
AlCu$_3$ 1310, 1311
Al$_4$Cu$_9$ 1311, 1318
Al$_3$Er   1299
AlF   441
AlF$_3$ 417, 444
Al(F, OH)$_{3.\frac{3}{8}}$H$_2$O   604
(AlF$_4$)Tl   469
(AlF$_5$)Sr   453
(AlF$_5$)Tl$_2$ 453, 470
(AlF$_5$·H$_2$O)K$_2$ 454
(AlF$_6$)M$_3$ 460, 462
AlF$_7$MgNa$_2$ 906
Al$_2$F$_{12}$Li$_3$Na$_3$ 468
Al$_3$F$_{14}$Na$_5$ 468
AlFe   1313
AlFe$_3$ 1297
Al$_3$Fe   1314
AlH$_3$ 344
AlH$_3$.2N(CH$_3$)$_3$ 345
AlH$_3$.N$_2$[CH$_2$]$_2$(CH$_3$)$_4$ 345
AlH$_4$Li   351
AlH$_4$Na   351
AlI$_3$ 419, 444
AlLi   1301
AlLi(C$_2$H$_5$)$_4$ 980
AlLiO$_2$ 577
Al$_3$Mg$_2$ 1306
Al$_{12}$Mg$_{17}$ 1306
Al$_{30}$Mg$_{23}$ 1306
AlN   835
Al(NH$_2$)$_4$Na   797
AlN(CH$_3$)$_3$Cl$_3$ 792
Al$_3$N$_3$(CH$_3$)$_6$(C$_2$H$_4$)$_3$ 792

Al$_4$N$_4$(C$_6$H$_5$)$_8$ 792
Al(NO$_3$)$_3$ 822
[Al(NO$_3$)$_4$]$^-$ 822
Al(NO$_3$)$_5^{2-}$ 822
Al[N(SiMe$_3$)$_2$]$_3$ 793, 795
AlNbO$_4$ 610
AlNi   1313
Al$_3$Ni$_2$ 1313
Al$_2$O$_3$, $-\alpha$ 544, 552
Al$_2$O$_3$, $-\beta$ 598
Al$_2$O$_3$, $-\gamma$ 552
Al$_2$O$_4$M   594
AlOBr   485
AlOCl   485
AlOI   485
Al(OH)$_3$ 552, 635
[Al(OH)$_6$]$_2$Ca$_3$ 637
AlO.OH, $-\alpha$ 552, 639
AlO.OH, $-\gamma$ 552, 641
[AlO(OH)$_2$]$_2$Ba   1018
Al$_2$O$_3$(CH$_3$)$_9$ 496
Al$_2$O$_2$(CH$_3$)$_4$[Si(CH$_3$)$_3$]$_2$ 496
[Al$_2$(OH)$_{10}$]Ba$_2$ 637
[Al$_2$(OH)$_6$O]K$_2$ 626
[Al$_2$(OH)$_2$(H$_2$O)$_8$](SO$_4$)$_2$.2H$_2$O   645
[Al$_{13}$O$_4$(OH)$_{24}$(H$_2$O)$_{12}$]$^{7-}$ 524
[Al$_{13}$O$_4$(OH)$_{25}$(H$_2$O)$_{11}$]$^{6-}$ 524
Al$_2$O(2-methyl-8-quinolinol)$_4$ 509
Al$_{32}$Ca$_3$O$_{51}$ 598
AlP   835
AlPO$_4$ 859
Al$_3$Pd$_2$ 1313
Al$_2$Pt   1313
Al$_3$Pu   1299
Al$_2$S$_3$ 764, 1271
Al$_2$S$_4$Zn   774, 780
AlSb   835
AlSbO$_4$ 966
Al$_2$Se$_3$ 780
(AlSi$_3$O$_8$)K   1034
(AlSi$_3$O$_8$)Na   1034
Al$_2$SiO$_5$ 1017, 1023
(Al$_2$Si$_2$O$_8$)Ba   1034
(Al$_2$Si$_2$O$_8$)Ca   1034
AlTh$_2$H$_4$ 351
Al$_3$Ti   1298
AlTiCl$_2$(C$_2$H$_5$)$_2$(C$_5$H$_5$)$_2$ 976
Al$_2$TiO$_5$ 603
Al$_5$W   1300
AlWO$_4$ 591
Al$_2$(WO$_4$)$_3$ 606
Al$_3$Y   1299
Al$_2$ZnS$_4$ 780
(Al, Zn)$_{49}$Mg$_{32}$ 1306
Al$_3$Zr   1298
Al$_3$Zr$_4$ 1306

Subject index